s Block elements	d Block elements
p Block elements	f Block elements

18

					4.002602
13	**14**	**15**	**16**	**17**	2 **He**

10.811	12.0107	14.00674	15.9994	18.9984032	20.1797
3	4, 2, −4	5, 4, 3, 2, −3	−2, −1	−1	
2.0	2.6	3.0	3.4	4.0	
5 **B** ●	6 **C** ●	7 **N** ●	8 **O** ●	9 **F** ●	10 **Ne**

26.981538	28.0855	30.973761	32.066	35.4527	39.948
3	4, −4	5, 3, −3	6, 4, 2, −2	7, 5, 3, 1, −1	
1.6	1.9	2.2	2.6	3.2	
13 **Al**	14 **Si** ●	15 **P** ●	16 **S** ●	17 **Cl** ●	18 **Ar**

10 | **11** | **12**

58.6934	63.546	65.39	69.723	72.61	74.92160	78.96	79.904	83.80
3, 2, 0	2, 1	2	3	4	5, 3, −3	6, 4, −2	7, 5, 3, 1, −1	2
1.9	1.9	1.7	1.8	2.0	2.2	2.6	3.0	3.0
28 **Ni** ●	29 **Cu** ●	30 **Zn** ●	31 **Ga**	32 **Ge**	33 **As** ●	34 **Se** ●	35 **Br**	36 **Kr**

106.42	107.8682	112.411	114.818	118.710	121.760	127.60	126.90447	131.29
4, 2, 0	2, 1	2	3	4, 2	5, 3, −3	6, 4, −2	7, 5, 1, −1	8, 6, 4, 2
2.2	1.9	1.7	1.8	2.0	2.1	2.1	2.7	2.6
46 **Pd**	47 **Ag**	48 **Cd**	49 **In**	50 **Sn**	51 **Sb**	52 **Te**	53 **I** ●	54 **Xe**

195.078	196.96655	200.59	204.3833	207.2	208.98038	208.9824*	209.9871*	222.0176*
4, 2, 0	3, 1	2, 1	3, 1	4, 2	5, 3	6, 4, 2	7, 5, 3, 1, −1	2
2.3	2.5	2.0	2.0	2.3	2.0	2.0		
78 **Pt**	79 **Au**	80 **Hg**	81 **Tl**	82 **Pb** ●	83 **Bi**	84 **Po**	85 **At**	86 **Rn**

269	272	277		289		289		293
110	111	112		114		116		118

157.25	158.92534	162.50	164.93032	167.26	168.93421	173.04	174.967
3	4, 3	3	3	3	3, 2	3, 2	3
1.2	1.2	1.2	1.2	1.2	1.3		1.0
64 **Gd**	65 **Tb**	66 **Dy**	67 **Ho**	68 **Er**	69 **Tm**	70 **Yb**	71 **Lu**

247.0703*	247.0703*	251.0796*	252.0829*	257.0951*	258.0986*	259.1009*	260.1053*
4, 3	4, 3	4, 3	3	3	3	3, 2	3
96 **Cm**	97 **Bk**	98 **Cf**	99 **Es**	100 **Fm**	101 **Md**	102 **No**	103 **Lr**

ORGANIC CHEMISTRY

Structure and Function

Third Edition

ORGANIC CHEMISTRY

Structure and Function

K. Peter C. Vollhardt
University of California at Berkeley

Neil E. Schore
University of California at Davis

W. H. Freeman and Company
NEW YORK

About the Cover: *Brevetoxin B (shown as a ball-and-stick model) is a potent marine neurotoxin responsible for massive fish kills, mollusk poisoning, and human food poisoning along the coast of Florida, the Gulf of Mexico, and many other parts of the world. It is associated with the explosive growth, or bloom, of the algae dinoflagellate* Ptychodiscus brevis *under certain favorable conditions of temperature, salinity, and sunlight, causing a phenomenon called "red tide." The front cover depicts a red tide on May 6, 1976, approaching the coast of Oshima Island, Japan. The back cover shows space-filling and bond-line formulas of the unique structure of brevetoxin B consisting of a single carbon chain arranged in a rigid ladder-like framework and composed of 11 contiguous* trans *fused ether rings. Brevetoxin B was made in the laboratory by total synthesis in 1994, requiring 83 steps from 2-deoxyribose and 12 years of effort, spearheaded by Professor Kyriacos Costa Nicolaou (who also supplied the original photograph) at the Scripps Research Institute and the University of California at San Diego. Synthetic strategies are discussed in Chapter 8 and subsequently throughout the text.*

Acquisitions Editor: Michelle Russel Julet

Development Editor: Randi Rossignol

Project Editor: Mary Louise Byrd

Marketing Manager: Kimberly Manzi

Cover Designer: Cambraia Magalhães

Text Designer: Circa 86, Inc.

Design Coordinator: Cambraia Magalhães

Illustration Coordinator: Bill Page

Illustrations: Network Graphics

Production Coordinator: Ellen Cash

Associate Media and Supplements Editor: Matthew P. Fitzpatrick

Composition: York Graphic Services

Manufacturing: RR Donnelly & Sons Company

Vollhardt, K. Peter C.
    Organic chemistry: structure and function/K. Peter C.
Vollhardt, Neil E. Schore. — 3rd ed.
      p. cm.
    Includes index.
    ISBN 0-7167-2721-8
    1. Chemistry, Organic.   I. Schore, Neil Eric, 1948–   .
  II. Title.
QD251.2.V65   1998   97–22455
547 — dc21   CIP

Printed in the United States of America

Second printing 1999

ORGANIC CHEMISTRY
Structure and Function

CONTENTS OVERVIEW

v

ORGANIC CHEMISTRY
Structure and Function

CONTENTS

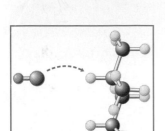

③   Reactions of Alkanes: Bond-Dissociation Energies, Radical Halogenation, and Relative Reactivity   93

④   Cyclic Alkanes   129

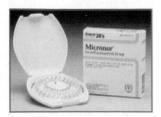

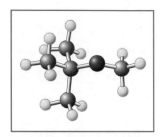

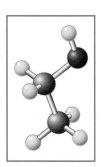

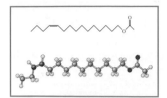

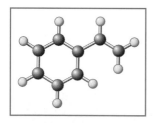

15 Unusual Stability of the Cyclic Electron Sextet: Benzene, Other Cyclic Polyenes, and Electrophilic Aromatic Substitution

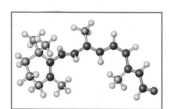

18 **Enols and Enones: α,β-Unsaturated Alcohols, Aldehydes, and Ketones** 780

19 **Carboxylic Acids** 825

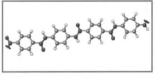

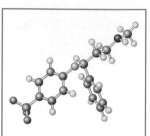

21   Amines and Their Derivatives: Functional Groups Containing Nitrogen   936

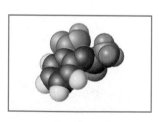

22   Chemistry of Benzene Substituents: Alkylbenzenes, Phenols, and Benzenamines   982

23 Ester Enolates and Acyl Anion Equivalents: Syntheses of β-Dicarbonyl and α-Hydroxycarbonyl Compounds

24 Carbohydrates: Polyfunctional Compounds in Nature

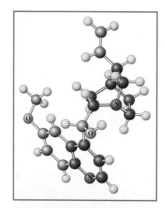

25 Heterocycles: Heteroatoms in Cyclic Organic Compounds   1116

26 Amino Acids, Peptides, Proteins, and Nucleic Acids: Nitrogen-Containing Polymers in Nature   1156

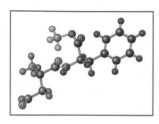

PREFACE

Structure and Function Motif

Too many students find that organic chemistry is an overwhelming parade of facts. Our goals are to dispel this notion and, more important, to help students learn and understand organic chemistry. The best way to do this is to provide a framework, or scaffolding, around which students can organize their thoughts. The framework that we provide is the accessible notion that an understanding of structure will lead to an understanding of function.

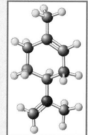 

S-(−)-Limonene   R-(+)-Limonene

A Uniform Organization Emphasizes the Relation Between Structure and Function

Much like a language, in which grammar would be dangling without the "meat" of vocabulary, the text develops material as a juxtaposition of structure and function. Thus, Chapter 1 provides the fundamentals of structure and bonding, specifically as they will become useful in understanding organic chemistry. Chapter 2 then follows with an introduction to the structural features of the alkanes and how they "function" in the simplest sense—namely, conformational mobility. Chapter 3 relates bond-dissociation energies to the lead reaction: radical halogenation (functionalization) of alkanes. Chapter 4 repeats the motif of Chapter 2 but with cycloalkanes as its center of focus.

The structure of the haloalkanes and how it determines their fate in its nucleophilic substitution and elimination reactions are the topics of Chapters 6 and 7. Each subsequent functional group is covered according to the same scheme: naming, structure, spectroscopy, preparations, reactions, and biological and other applications.

Silver fir tree cone Orange peel

Students Are Given the Structural "Tools" They Need to Understand Function

The interplay between structure and function that gives a hierarchy to individual chapters also confers a hierarchy on the text as a whole. This is why we introduce stereochemistry in Chapter 5. Students learn about stereochemical principles so that they are prepared to understand the substitution and elimination reactions of haloalkanes (Chapters 6 and 7) and the addition reactions of alkenes (Chapter 12). Moreover, this hierarchy allows the mechanistic discussion of all new important reactions to take place concurrently rather than being scattered in different places throughout the text. Such a unified presentation of mechanisms benefits the student enormously.

Alcohols (Chapters 8 and 9) with the simplest oxygen-containing function are treated early because their chemistry sets the stage for understanding their central role in synthesis. Similarly, carbocations (and their rearrangements; see Section 9-3) appear before a discussion of the Markovnikov rule, alkenes (Chapter 12) before conjugated polyenes (Chapter 14), and conjugated polyenes before aromatic systems. Early coverage of spectroscopy reinforces the interplay between structure and function. This organization allows students to apply spectroscopic techniques in the context of the functional group. Infrared spectroscopy, for example, is introduced in Chapter 11 on alkenes.

The Lead Reaction: Radical Halogenation of Methane

The first, more detailed discussion of a reaction and its mechanism, the "lead reaction," is presented very early in Chapter 3. For several reasons, the best (most logical) choice is the chlorination of methane. First, all chemical reactions require bond making and bond breaking. The radical halogenation of methane allows the introduction of the concepts of bond-dissociation energies and the stability and structure of the ensuing radicals. This leads to an understanding of the thermal stability of the simplest organic bonds, C—H and C—C, and, hence, of why organic materials are capable of existence. The students learn that to "activate" a C—H bond, a reactive agent is required. Second, the lead reaction, because it purposely does not include ionic species, can be analyzed thermodynamically by calculating enthalpies of the overall process, as well as individual steps. This exercise is fundamental and gives the student the basic tool of "eyeballing" the relative feasibility of all future transformations. It also serves as a first application of potential energy diagrams to a chemical process. Third, the generalization of the chlorination of methane to the halogenation of other alkanes permits the simple introduction of the concepts of reactivity and selectivity, a feeling for the statistics needed to deal with molecules endowed with several equally reactive sites, and practical applications of these principles.

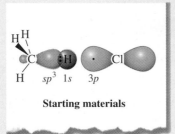

Starting materials

A Book for Real Students

We are aware of the challenges this course presents to students. Our students have taught us what these challenges are and we have applied this understanding in the following ways:

Early Coverage of Acids and Bases

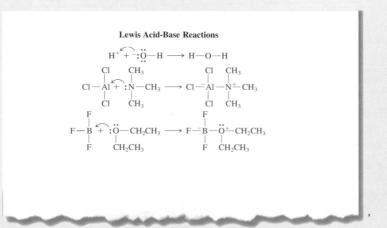

Lewis Acid-Base Reactions

We have moved the discussion of acids and bases to Section 2-9 and expanded it to provide the student with an early review of this aspect of general chemistry as it applies to organic systems. This treatment now includes explicitly Lewis acids and bases and sets the student up for a general understanding of the similarity between such diverse processes as nucleophilic trapping of carbocations and solvation (e.g., of Grignard reagents or by crown ethers or ionophores) and for the role of metal halides in Friedel-Crafts alkanoylation.

Improved Presentation of Reaction Mechanisms

The presentation of reaction mechanisms has been improved by the increased use of arrows to better show electron flow. The introduction of icons for a "reaction" and its "mechanism" serves to emphasize the "vocabulary-grammar" duality of the two types of schemes. In addition, we have added a new section on electron-pushing arrows (Section 6-4) to familiarize the student explicitly with this technique.

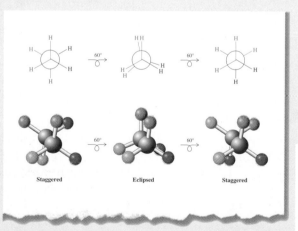

New Ways to Visualize Organic Chemistry

We have now included computer-generated pictures of ball-and-stick and space-filling models. These pictures encourage students to build actual models. They also provide students with lowest-energy conformations, guiding them in the construction of realistic assemblies. Finally, space-filling renditions create a more accurate impression of size, shape, and the extent of orbitals.

The first and second editions emphasized the importance of building molecular models as an aid in visualizing three-dimensional structure and dynamics. We have highlighted this emphasis by a new icon at numerous locations. Ball-and-stick model kits are available for purchase through the publisher.

Innovative Approaches to Problem Solving

SOLUTION

Before we start a random trial and error approach to solving this problem, it is better to take an inventory of what is given. First, we are given cyclohexane, and we note that this unit shows up as a substituent in tertiary alcohol A. Second, a total of seven additional carbons appears in the product, so our synthesis will require some additional stitching together of smaller fragments because we cannot use compounds containing more than four carbons. Third, target A is a tertiary alcohol, which should be amenable to the retrosynthetic analysis introduced in Section 8-9 (M = metal):

• *New Chapter Integration Problems* are solved problems that emphasize concept integration both within and between chapters. The solution is worked out step-by-step, teaching the art of problem solving in general and specifically demonstrating how one set of learned skills builds on and interacts with preceding ones. Particular emphasis is placed on problem analysis, deductive reasoning, and logical conclusions.

Team Problem

47. Consider the general substitution-elimination reactions of the bromoalkanes.

$$R-Br \xrightarrow{\text{Nu/Base}} R-Nu + \text{alkene}$$

How do the reaction mechanisms and product formation differ when the structure of the substrate and reaction conditions change? To begin to unravel the nuances of bimolecular and unimolecular substitution and elimination reactions, focus on the treatment of bromoalkanes A through D under conditions (a) through (e). Divide the problem evenly among yourselves so that each of you tackles the questions of reaction mechanism(s) and qualitative distribution of product(s), if any. Reconvene to discuss your conclusions and come to a consensus. When you are explaining a reaction mechanism to the rest of the team, use curved arrows to show the flow of electrons. Label the stereochemistry of starting materials and products as R or S, as appropriate.

• *The Team Problem* is also new to each chapter. Team Problems encourage discussion and collaborative learning among students. Although these problems could be assigned in a classroom setting, they are written so as to be perfectly workable in an unstructured, casual setting, such as a library, coffee shop, study hall, or home. The idea is to stimulate "cross-talk," an exchange of information and ideas, and support among students.

Preprofessional Problems

55. The enantiomer of

(a) is $CH_3CH_2-\overset{\overset{\displaystyle Cl}{|}}{\underset{\underset{\displaystyle CH_3}{|}}{C}}-H$ **(b)** can exist only at low temperatures

(c) is nonisomeric   **(d)** is incapable of existence

• *Preprofessional Problems* will be appreciated by students who are planning careers in medicine or related fields. In a new multiple-choice format, they are typical of those that appear on the MCAT, GRE, and DAT.

New Reaction Summary Road Maps

From Chapter 8 onward, the chemistry of each functional group is shown in condensed form through two types of **reaction summary road maps**, providing "the functional group at a glance." The first type depicts the function as the origin of multiple reaction arrows, each labeled with a particular reagent, ending in a specific product. Section numbers indicate where the transformation is discussed in the text,

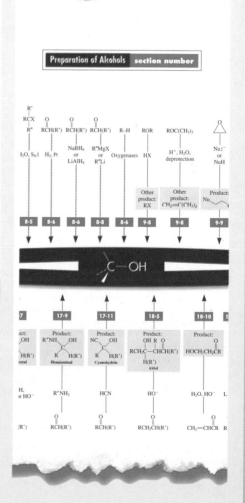

and color distinguishes past from future chemistry. This map provides information about the reactions of the functional group—that is, what it does. The second type of map is very similar, but the reaction arrows are reversed—that is, pointing toward the functionality. This map provides information about the function's possible origins—that is, its precursor functional groups. Thus, a specific reaction A→B may appear in two separate schemes, one with A and the other with B as the center. These maps are an important aid in synthetic-retrosynthetic analysis and a check on the student's "vocabulary" of synthetic methodology.

A Book for the Real World

Modern Biological and Industrial Applications

In every chapter of this textbook you will find many references to biological, medical, and industrial applications of organic chemistry. Much of this material is new to this edition. For instance, in Chapter 3 there is a new section on the effects of chlorine compounds on the atmosphere. In Chapter 5 you will find new Chemical Highlights on chiral drugs and the handedness of nature. New material on polyethers and their medical uses is found in Chapter 9. We have expanded our coverage of polymers, including polymer synthesis using dienes and nitriles (Chapter 14), and biodegradable polyester plastics (Chapter 19). Chapter 26 discusses the mechanism of chymotrypsin activity. Finally, in keeping with our emphasis on synthesis,

---

CHEMICAL HIGHLIGHT 5-5　　**Why Is Nature "Handed"?**

In this chapter, we have seen that many of the organic molecules in nature are chiral. More importantly, most natural compounds in living organisms not only are chiral, but also are present in only one enantiomeric form. An example of an entire class of such compounds consists of the *amino acids,* which are the component units of *polypeptides*. The large polypeptides in nature are called *proteins* or, when they catalyze biotransformations, *enzymes*.

Absolute Configuration of Natural Amino Acids and Polypeptides

Amino acid
(R variable)

Amino acid 1　Amino acid 2　Amino acid 3
Polypeptide

Being made up of smaller chiral pieces, enzymes arrange themselves into bigger conglomerates that also are chiral and show handedness. Thus, much as a right hand will readily distinguish another right hand from a left hand, enzymes (and other biomolecules) have "pockets" that, by virtue of their stereochemically defined features, are capable of recognizing and processing only one of the enantiomers in a racemate. The differences in physiological activity of the two enantiomers of a

---

throughout the text students are guided through the syntheses of many biologically important compounds, including norethynodrel (Chapter 17), steroids (Chapter 18), and Prozac (Chapter 21).

New Chapter Introductions

When you hear or read the word *steroids,* two things probably come to mind immediately: athletes who illegally "take steroids" to develop their muscles and "the pill" used for birth control. But what do you know about steroids aside from this general association? What is their structure? How does one steroid differ from another? Where are they found in nature?

An example of a naturally occurring steroid is diosgenin, obtained from the Mexican yam and used as a starting material for the synthesis of several commercial steroids. Most striking is the number of *rings* in the compound.

The introduction to each chapter has been "spiced up" with thought-provoking questions relating the relevance of the chapter's material to everyday experience. These general questions find answers on further reading, thus prompting the student's interest and participation.

A Book About Real Chemistry

Emphasis on Synthetic Strategy

The importance of synthesis is stressed starting on page 3, and the considerations entering into the development of a good synthetic strategy and the avoidance of pitfalls are developed throughout the text. Since the innovative introduction of Section 8-9 on retrosynthetic analysis in the first edition, this aspect of the text has received much positive feedback from teachers and students. The present edition has added a slightly more explicit treatment of linear versus convergent synthesis to this section. Multistep partial and total syntheses are pointed out where appropriate in the various functional-group treatments. Particular emphasis is placed on stereo- and regioselectivity (Chapter 12 and Section 16-5), biological and medicinal relevance (e.g., Sections 9-11, 12-16, 18-12, and 19-4 and Chapters 24 through 26), and the importance of materials synthesis (e.g., Sections 12-13 through 12-15, 13-11, 14-10, and 21-12). These discussions are then extensively reinforced in the In-Chapter Exercises and End-of-Chapter Problems and highlighted in numerous Chemical Highlights.

Retrosynthetic analysis simplifies synthesis problems

Many compounds that are commercially available and inexpensive are also small, containing six or fewer carbon atoms. Therefore, the most frequent task facing the synthetic planner is that of building up a larger, complicated molecule from smaller, simple fragments. The best approach to the preparation of the target is to work its synthesis *backward* on paper, an approach called **retrosynthetic analysis*** (*retro*, Latin, backward). In this analysis, strategic carbon–carbon

Early and More Unified Presentation of Spectroscopy

Our first edition broke ground by introducing spectroscopy right after alcohol chemistry. Early coverage, beginning with NMR in Chapter 10, offers opportunities to practice the application of spectroscopic methods to many kinds of compounds. After NMR, we cover IR- and UV-visible spectroscopy in Chapters 11 and 14 in the context of functional groups. Courses can include each of the principal types of spectroscopy in the first half of the text.

Early coverage of spectroscopy reinforces the interplay between structure and function. This organization allows students to apply spectroscopic techniques in the context of the functional group. Infrared spectroscopy, for example, is introduced in Chapter 11 on alkenes.

We have also added new discussions (e.g., Section 10-9) and problems that unify the application of spectroscopic techniques in structure determination.

An Innovative Package:

SUPPLEMENTS AND LABORATORY MANUALS

The CD-ROM found in the back of this text is a multimedia learning tool developed by W. H. Freeman and Company in conjunction with Sumanas, Inc. All the features of the CD function within the context of the book's coverage. Many of the structures mentioned in the book are depicted as three-dimensional animations with multiple-display options through a molecular-modeling program. These animations and many other molecular-level simulations bring the concepts of the book to life. Practice tools, such as interactive quizzes in every chapter and a preprofessional examination, help students review for exams. Spectroscopy and NMR exercises along with other interactive exercises help students master difficult concepts. Presentation software for instructors allows them to prepare a series of illustrations and animations for lecture.

• *The Study Guide* is written by Neil Schore, providing a direct link from the text to the supplement. Sample problems are worked out, and the solutions to the End-of-Chapter Problems are given. "Keys to the Chapter" sections point out pitfalls

of faulty logic and help students visualize the solution steps for various exercises. Tables summarize the spectral features associated with each functional group. A glossary of key terms is also provided.

Available in ground glass and flexible connector versions.

• *The Test Bank*, by Charles M. Garner and Kevin G. Pinney of Baylor University, is new to this edition. With the Windows and Macintosh software of the computerized versions, instructors can easily change and add questions as well as import their own electronic drawings.

• *The Maruzen Molecular Structure Model Set* and *Space-Filling Model Set* are also available for student purchase. These essential tools can be used to present orbitals; single, double, and triple bonds; and locations of atoms.

• *Experimental Organic Chemistry: Macroscale* and *Microscale*. With these texts by Jerry R. Mohrig, Christina Noring Hammond, Terence C. Morrill, and Douglas C. Neckers, the laboratory becomes a place of discovery and critical thinking. Instead of simply following directions, students immerse themselves in the experimental process. Instructors will appreciate the versatility of the manuals' balanced approach, with enough experiments to use macroscale glassware, microscale glassware, or a combination of both. Innovative discovery-based experiments and multiweek projects encourage students in scientific investigation. A CD-ROM of techniques accompanies both texts.

A KEY TO THE FUNCTIONAL USE OF COLOR

We use color consistently and functionally to help students master basic principles, including nomenclature, orbitals, sequence rules in stereochemistry, the relation of spectral lines to functional groups, topological changes in molecular transformations, and the reactivity of functional groups. Color is suspended in exercises, chapter reviews, and problems, however, because it is important to learn how not to rely on it. In this edition, we have carefully reevaluated the application of color in reaction schemes and simplified its use.

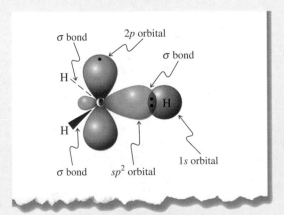

For example, wherever possible, s orbitals are shown in red, $2p$ orbitals in blue, sp^n hybrids in purple, and $3p$ orbitals in green.

Color shows the relation of the names of organic molecules to their structures. In the illustration shown at the top of the next page, which is from Chapter 11, the functional group that gives the molecule its unique chemical behavior and other substituents are clearly differentiated from the stem.

$$\underset{1}{CH_3}\underset{2}{\overset{OH}{\underset{|}{CH}}}$$

$$\underset{3}{\overset{|}{CH}}CH_2CH_3$$

$$\underset{H_3C}{\overset{Cl}{\underset{5}{\diagdown}}}\underset{4}{\overset{}{C}}{=}C\underset{H}{\diagup}$$

(Z)-5-Chloro-3-ethyl-4-hexen-2-ol
(The two stereocenters are unspecified)

Color is used to associate spectral features with certain molecular units. For example, in the adjoining spectrum, the three colors show how the three nonequivalent hydrogens give rise to three distinct "peaks"—an observation that will help the students identify a molecule when they know its spectrum.

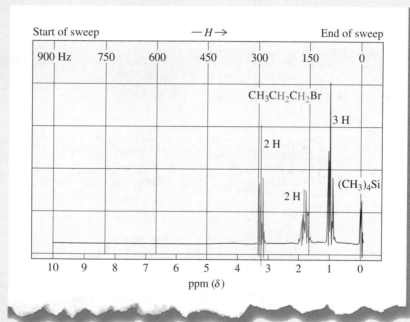

Color offers clues to a molecule's stereochemistry, or the arrangement of its atoms in space. The student will see in Chapter 5 that substituents in three dimensions can be assigned a priority according to certain "sequence rules," and this assignment has been indicated, in diminishing order of priority, by red, blue, green, and black.

Remember the use of color to denote group priorities:
Highest—red
Second highest—blue
Third highest—green
Lowest—black

$$\underset{ClH_2C}{\overset{H}{\underset{4}{\diagup}}}\underset{3}{\overset{|}{\underset{CH_3CH_2}{C^2}}}\diagdown Br$$

Optically active
2R

STEP 4. Trapping by bromide

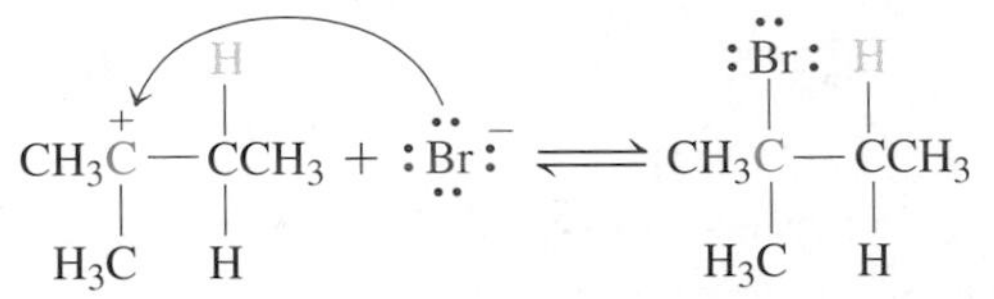

$$CH_3\overset{+}{\underset{\underset{H_3C}{|}}{C}}\!-\!\underset{\underset{H}{|}}{C}CH_3 \; + \; :\!\overset{..}{\underset{..}{Br}}\!:^- \;\; \Longleftrightarrow \;\; CH_3\overset{:\overset{..}{Br}:}{\underset{\underset{H_3C}{|}}{C}}\!-\!\underset{\underset{H}{|}}{C}CH_3$$

Most important, color frequently shows how the functional groups transform in the reaction mechanism. Electron-rich, or "nucleophilic," parts are shown in red; electron-deficient, or "electrophilic," fragments are blue; and radicals and leaving groups are green. Red arrows in these transformations indicate the movements of electrons.

ACKNOWLEDGMENTS

We are grateful to the following professors who reviewed the manuscript for the third edition:

Steven Angle, University of California at Riverside
Jeffrey Arterburn, New Mexico State University
Ronald Blankespoor, Calvin College
Frances Blase, Haverford College
Richard Broene, Bowdoin College
Patrick Buick, University of Toledo
Dee Ann Casteel, Bucknell University
Dana Chatellier, University of Delaware
James Deyrup, University of Florida
Morris Fishman, New York University
Thomas Flechtner, Cleveland State University
Francis Flores, California State Polytechnic University
Marcia France, Washington & Lee University
Andrew French, Radford University
Charles Garner, Baylor University
Rainer Glaser, University of Missouri at Columbia
Frank Guziec, Southwestern University
William Hagan, College of St. Rose
Michael Heagy, New Mexico Tech
Eamonn Healy, St. Edward's University
Steven Kass, University of Minnesota at Minneapolis
Ross Kelly, Boston College
Robert Kulawiec, Georgetown University
Mark Kurth, University of California at Davis
David Lemal, Dartmouth College
Todd Lowary, Ohio State University
Ronald Magid, University of Tennessee at Knoxville
Roger Murray, University of Delaware
Thomas Newton, University of Southern Maine

Patrick O'Bannon, Kenyon College
Daniel O'Leary, Pomona College
Kenneth Piers, Calvin College
Michael Rathke, Michigan State University
Gretchen Rehburg, Bucknell University
John Richard, State University of New York at Buffalo
Carl Rhodes, Stanford University
Adrian Schwan, University of Guelph
Larry Scott, Boston College
Raymond Shelden, La Sierra University
Jan Shepard, Millersville University of Pennsylvania
Sam Stevenson, Northeast State Technical Community College
Chaim Sukenik, Case Western Reserve University
Julie Tan, Cumberland College
Emmanuel Theodorakis, University of California, San Diego
Peter Trumper, University of Northern Maine at Orono
Kent Vrana, Wake Forest University
Jeffrey Ward, Georgetown University
Kraig Wheeler, Delaware State University
James White, Pepperdine University
John Williams, Temple University
John Wood, Indiana University of Pennsylvania

Peter Vollhardt thanks Administrative Assistants Kim Steele and Bonnie Kirk, for typing, photocopying, and coordinating manuscripts; graduate students Adam Matzger and Dan Holmes, for Spartan and artistic ideas; Kevin Cammack and Michael Eichberg, for running 300 MHz NMR and other spectra; and all four plus Dr. Christoph Erben, Sriram Kumaraswamy, Jennifer Moore, and Ian Wasser, for their assistance in checking page proofs.

We are indebted to Professor Richard Bozak for contributing the Preprofessional Problems and to Nancy Cox-Konopelski for contributing the Team Problems.

We express special gratitude to Professor Ronald Magid, his colleagues, and his students at the University of Tennessee at Knoxville, for the time and effort they have devoted to uncovering numerous typographical and factual errors. They deserve a great deal of credit for whatever improvements in accuracy and clarity we have achieved in this revision.

K. PETER C. VOLLHARDT was born in Madrid in 1946, raised in Buenos Aires and Munich, studied at the University of Munich, received his Ph.D. with Professor Peter Garratt at the University College, London, and was a postdoctoral fellow with Professor Bob Bergman (then) at the California Institute of Technology. He moved to Berkeley in 1974, when he began his efforts toward the development of organocobalt reagents in organic synthesis, the preparation of theoretically interesting hydrocarbons, the assembly of novel transition metal arrays with potential in catalysis, and the discovery of a parking space. Among other pleasant experiences, he was a Studienstiftler, Adolf Windaus medalist,

Humboldt Senior Scientist, ACS Organometallic Awardee, Otto Bayer Prize Awardee, and A. C. Cope Scholar. He is the current Chief Editor of SYNLETT. Among his more than 240 publications, he especially treasures this textbook in organic chemistry, translated into seven languages. Peter is married to Marie-José Sat, a French artist, and they have two children, Paloma (b. 1994), whose picture you can admire in Chapter 5, and Julien (b. 1997), who refused to pose.

NEIL SCHORE was born in Newark, New Jersey, in 1948. His education took him through the public schools of the Bronx, New York, and Ridgefield, New Jersey, after which he completed a B.A. with honors in chemistry at the University of Pennsylvania in 1969. Moving back to New York, he worked with Professor Nicholas Turro at Columbia University, studying photochemical and photophysical processes of organic compounds for his Ph.D. thesis. He first met Peter Vollhardt when he and Peter were doing postdoctoral work in Professor Robert Bergman's laboratory at Cal Tech in the 1970s. Since joining the U. C. Davis faculty in 1976, he has taught organic chemistry to over 10,000 nonchemistry majors, winning three teaching awards, and published over 70 papers in various areas related to organic synthesis. Neil is married to Carrie Erickson, a microbiologist at the U. C. Davis School of Veterinary Medicine. They have two children, Michael (b. 1981) and Stefanie (b. 1983), both of whom carried out experiments for this book.

ORGANIC CHEMISTRY

Structure and Function

Structure and Bonding in Organic Molecules

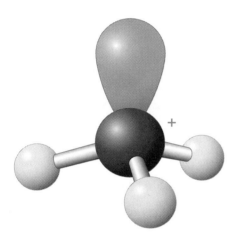

The hydronium ion, derived by protonation of water, is responsible for the acidity of aqueous solutions below pH 7. Similar protonation of an oxygen atom of the dye molecule responsible for the blue color of cornflowers gives rise to the color of the red poppy.

How does your body function? Why did your muscles ache this morning after last night's long jog? What is in the pill you took to get rid of that headache you got after studying all night? What happens to the gasoline you pour into the gas tank of your car? What is the molecular composition of the things you wear? What is the difference between a cotton shirt and one made of silk? What is the origin of the odor of garlic? You will find the answers to these questions, and many others that you may have asked yourself, in this book on organic chemistry.

Chemistry is the study of the structure of molecules and the rules that govern their interactions. As such, it interfaces closely with the fields of biology, physics, and mathematics. What, then, is organic chemistry? What distinguishes it from other chemical disciplines, such as physical, inorganic, or nuclear chemistry? A common definition provides a partial answer: *Organic chemistry is the chemistry of carbon and its compounds.* These compounds are called **organic molecules.**

Toothbrushes and toothpaste consist of a mixture of many organic molecules.

Organic molecules constitute the chemical bricks of life. Fats, sugars, proteins, and the nucleic acids are compounds in which the principal component is carbon. So are countless substances that we take for granted in everyday use. Virtually all the clothes that we wear are made of organic molecules—some of natural fibers, such as cotton and silk; others artificial, such as polyester. Toothbrushes, toothpaste, soaps, shampoos, deodorants, perfumes—all contain organic compounds, as do furniture, carpets, the plastic in light fixtures and cooking utensils, paintings, food, and countless other items.

Organic substances such as gasoline, medicines, pesticides, and polymers have improved the quality of our lives. Yet the uncontrolled disposal of organic chemicals has polluted the environment, causing deterioration of animal and plant life, as well as injury and disease to humans. If we are to create useful molecules—and learn to control their effects—we need a knowledge of their properties and an understanding of their behavior. We must be able to apply the principles of organic chemistry. This chapter explains how the basic ideas of chemical structure and bonding apply to organic molecules.

1-1 The Scope of Organic Chemistry: An Overview

A goal of organic chemistry is to relate the structure of a molecule to the reactions that it can undergo. We can then study the steps by which each type of reaction takes place, and we can learn to create new molecules by applying those processes.

Thus, it makes sense to classify organic molecules according to the subunits and bonds that determine their chemical reactivity: These determinants are groups of atoms called **functional groups.** The study of the various functional groups and their respective reactions provides the structure of this book.

Functional groups determine the reactivity of organic molecules

$$H_3C—CH_3$$
Ethane

We begin with the **alkanes,** which contain the basic carbon framework of organic molecules. The alkanes are simple **hydrocarbons,** organic compounds composed of only hydrogen and carbon and lacking functional groups. As with other classes of molecules, we discuss the systematic rules for naming them, their structures, and their physical properties. An example of an alkane is ethane. Its structural mobility will be the starting point for a review of thermodynamics and kinetics. This review is then followed by a discussion of the strength of alkane bonds, which can be broken by heat, light, or chemical reagents. We shall illustrate these processes with the chlorination of alkanes (Chapter 3).

A Chlorination Reaction

$$CH_4 + Cl_2 \xrightarrow{\text{Energy}} CH_3—Cl + HCl$$

Next we shall look at cyclic alkanes (Chapter 4), which contain carbon atoms in a ring. This arrangement can lead to new properties and changes in reactivity. The recognition of a new type of isomerism in cycloalkanes bearing two or more substituents—either on the same side or on opposite sides of the ring plane—sets the stage for a general discussion of **stereoisomerism,** exhibited by compounds with the

Cyclohexane

same connectivity but differing in the relative positioning of their component atoms in space (Chapter 5).

We shall then study the haloalkanes, our first example of compounds containing a functional group—the carbon–halogen bond. The haloalkanes participate in two types of organic reactions: substitution and elimination (Chapters 6 and 7). In a **substitution** reaction, one halogen atom may be replaced by another; in an **elimination** process, adjacent atoms may be removed from a molecule to generate a double bond.

A Substitution Reaction

$$CH_3-Cl + K^+I^- \longrightarrow CH_3-I + K^+Cl^-$$

An Elimination Reaction

$$\underset{\underset{H}{|} \quad \underset{I}{|}}{CH_2-CH_2} + K^+\ ^-OH \longrightarrow H_2C=CH_2 + HOH + K^+I^-$$

Like the haloalkanes, each of the major classes of organic compounds is characterized by a particular functional group. For example, the carbon–carbon triple bond is the functional group of alkynes; ethyne, a well-known alkyne, is the chemical burned in a welder's torch (Chapter 13). A carbon–oxygen double bond fulfills this role for aldehydes and ketones, the starting materials in many industrial processes (Chapters 16 and 17); and the amines, which include drugs such as nasal decongestants and amphetamines, contain nitrogen in their functional group (Chapter 21). We shall study a number of tools for identifying these molecular subunits, including various forms of spectroscopy (Chapters 10, 11, 14, and 20).

Subsequently, we shall encounter several important classes of organic molecules that are especially crucial in biology and industry. Many of these classes, such as the carbohydrates (Chapter 24) and amino acids (Chapter 26), contain multiple functional groups. However, in *every* class of organic compounds, the principle remains the same: *The structure of the molecule is related to the reactions that it can undergo.*

$$HC\equiv CH$$
An alkyne

$$H_2C=O$$
An aldehyde

$$\underset{\text{A ketone}}{\overset{\overset{\displaystyle O}{\|}}{H_3C-C-CH_3}}$$

$$H_3C-NH_2$$
An amine

Synthesis is the making of new molecules

Carbon compounds are called "organic" because it was originally thought that they could be produced only from living organisms. In 1828, Friedrich Wöhler* proved this idea to be false when he converted the inorganic salt lead cyanate into urea, an organic product of protein metabolism in mammals. [The average human excretes 30 g (grams) of urea each day.]

Wöhler's Synthesis of Urea

$$\underset{\textbf{Lead cyanate}}{Pb(OCN)_2} + \underset{\textbf{Water}}{2\ H_2O} + \underset{\textbf{Ammonia}}{2\ NH_3} \longrightarrow \underset{\textbf{Urea}}{2\ H_2N\overset{\overset{\displaystyle O}{\|}}{C}NH_2} + \underset{\textbf{Lead hydroxide}}{Pb(OH)_2}$$

*Professor Friedrich Wöhler (1800–1882), University of Göttingen, Germany. In this and subsequent biographical notes, only the scientist's last known location of activity will be mentioned, even though much of his or her career may have been spent elsewhere.

CHEMICAL HIGHLIGHT 1-1 Saccharin: One of the Oldest Synthetic Organic Compounds in Commercial Use

Saccharin was synthesized in the course of a study of the oxidation of organic chemicals containing sulfur and nitrogen. Its sweetness was discovered by Ira Remsen[*] in 1879, a time when chemists routinely *tasted* every new compound they made. This was an extremely dangerous practice, one that you should not

[*]Professor Ira Remsen (1846–1927), Johns Hopkins University, Baltimore.

Familiar saccharin-containing packets.

observe under any circumstances, even with supposedly "safe" compounds that you may encounter in your laboratory. For example, had Remsen tasted brevetoxin B (illustrated on the cover of this book), he would have immediately felt a prickly sensation in his mouth and fingers, rapidly followed by hot and cold sensations, breathing problems, paralysis, and death.

Saccharin is 300 times as sweet as sugar and virtually nontoxic. It has proved to be a lifesaver for countless diabetics and of great value to people who need to control their caloric intake. The possibility that saccharin may be *carcinogenic*—that is, capable of causing cancer—was raised in the 1960s. In the 1970s, a connection was found between high doses of saccharin and bladder tumors in rats. Experiments completed in 1990 demonstrated that saccharin does not cause cancer directly, but at very high doses it promotes accelerated cell division, which may increase the likelihood of cell mutation and tumor formation. Warning labels are required on saccharin-containing products sold in the United States. These studies illustrate how society must balance the benefits that synthetic substances bring to our daily lives with the possible risks associated with their use.

Synthesis, or the making of molecules, is a very important part of organic chemistry (Chapter 8). Since Wöhler's time, more than 10 million organic substances have been synthesized from simpler materials, both organic and inorganic. These substances include many that also occur in nature, such as the penicillin antibiotics, as well as entirely new compounds. Some, like cubane, which gave chemists the opportunity to study special kinds of bonding and reactivity, are of largely theoretical interest. Others, like the artificial sweetener saccharin, have become a part of everyday life.

Typically, the goal of synthesis is to construct complex organic chemicals from simpler, more readily available ones. To be able to convert one molecule into another, chemists must know organic reactions. They must also know the physical conditions that govern such processes, such as temperature, pressure, solvent, and molecular structure. This knowledge is equally valuable in analyzing biological transformations.

As we study the chemistry of each functional group, we shall develop the tools both for planning effective syntheses and for predicting the processes that take place in nature. But how? The answer lies in looking at reactions step by step.

Benzylpenicillin

Cubane

Saccharin

Reactions are the vocabulary and mechanisms are the grammar of organic chemistry

When we introduce a chemical reaction, we will first show just the starting compounds, or **reactants** (also called **substrates**), and the **products.** In the chlorination process mentioned earlier, the substrates—methane, CH_4, and chlorine, Cl_2—may undergo a reaction to give chloromethane, CH_3Cl, and hydrogen chloride, HCl. The overall transformation was described as $CH_4 + Cl_2 \rightarrow CH_3Cl + HCl$. However, even a simple reaction like this one may proceed through a complex sequence of steps. The reactants could have first formed one or more *unobserved* substances—call these X—that rapidly changed into the observed products. These underlying details of the reaction constitute the **reaction mechanism.** In our example, the mechanism consists of a two-step sequence: $CH_4 + Cl_2 \rightarrow X$ followed by $X \rightarrow CH_3Cl + HCl$. Each step may have a part in determining whether the overall reaction will proceed.

Substance X in our chlorination reaction is an example of a **reaction intermediate,** a species formed on the pathway between reactants and products. We shall learn the mechanism of this chlorination process and the true nature of the reaction intermediates in Chapter 3.

How can we determine reaction mechanisms? The strict answer to this question is, We cannot. All we can do is amass circumstantial evidence that is consistent with (or points to) a certain sequence of molecular events that connect starting materials and products ("the postulated mechanism"). To do so, we exploit the fact that organic molecules are no more than collections of bonded atoms. We can, therefore, study how, when, and how fast bonds break and form, in which way they do so in three dimensions, and how changes in substrate structure affect the outcome of reactions. Thus, although we cannot strictly prove a mechanism, we can certainly rule out many (or even all) reasonable alternatives and propose a most likely pathway.

In a way, the "learning" and "using" of organic chemistry is much like learning and using a language. You need the vocabulary (i.e., the reactions) to be able to use the right words, but you also need the grammar (i.e., the mechanisms) to be able to converse intelligently. Neither one on its own gives complete knowledge and understanding, but together they form a powerful means of communication, rationalization, and predictive analysis. To highlight the interplay between reaction and mechanism, icons are displayed in the margin at appropriate places throughout the text.

Before we begin our study of the principles of organic chemistry, let us review some of the elementary principles of bonding. We shall find these concepts useful in understanding and predicting the chemical reactivity and the physical properties of organic molecules.

1-2 Coulomb Forces: A Simplified View of Bonding

The bonds between atoms hold a molecule together. But why are there bonds? Two atoms form a bond only if their interaction is energetically favorable; that is, if energy—heat, for example—is released when the bond is formed. Conversely, breaking that bond requires the input of the same amount of energy.

The two main causes of the energy release associated with bonding are based on fundamental laws of physics:

1. Opposite charges attract each other.
2. Electrons spread out in space.

Bonds are made by simultaneous Coulombic attraction and electron exchange

Each atom consists of a nucleus, containing electrically neutral particles, or neutrons, and positively charged protons. Surrounding the nucleus are negatively charged electrons, equal in number to the protons so that the net charge is zero. As two atoms approach each other, the positively charged nucleus of the first attracts the electrons of the second; similarly, the nucleus of the second attracts the electrons of the first. This sort of bonding is described by **Coulomb's* law:** Opposite charges attract each other with a force inversely proportional to the square of the distance between the centers of the charges.

Coulomb's Law

$$\text{Attracting force} = \text{constant} \times \frac{(+)\,\text{charge} \times (-)\,\text{charge}}{\text{distance}^2}$$

This attractive force causes energy to be released as the atoms are brought together. This energy is called the **bond strength.**

When the atoms reach a certain closeness, no more energy is released. The distance between the two nuclei at this point is called the **bond length** (Figure 1-1).

*Lieutenant-Colonel Charles Augustin de Coulomb (1736–1806), Inspecteur Général of the University of Paris, France.

FIGURE 1-1

The changes in energy, *E*, that result when two atoms are brought into close proximity. At the separation defined as bond length, maximum bonding is achieved.

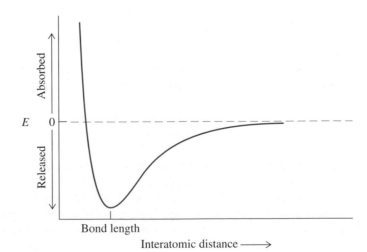

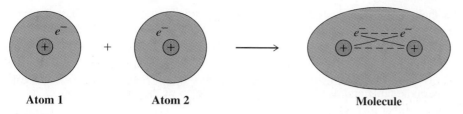

FIGURE 1-2 ─────────────────────

Covalent bonding. Attractive (solid line) and repulsive (dashed line) forces in the bonding between two atoms. The large circles represent areas in space in which the electrons are found around the nucleus. The small circle around the plus sign stands for the nucleus.

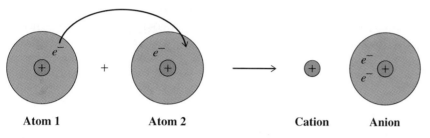

FIGURE 1-3 ─────────────────────

Ionic bonding. An alternative mode of bonding results from the complete transfer of an electron from atom 1 to atom 2, thereby generating two ions whose opposite charges attract each other.

Bringing the atoms closer together than this distance results in a sharp *increase* in energy. Why? Just as opposite charges attract, like charges repel. If the atoms are too close, the electron–electron and nuclear–nuclear repulsions become stronger than the attractive forces. When the nuclei are the appropriate bond length apart, the electrons are spread out around both nuclei, and attractive and repulsive forces balance for maximum bonding. The energy content of the two-atom system is then at a minimum, the most stable situation (Figure 1-2).

An alternative to this type of bonding results from the complete transfer of an electron from one atom to the other. The result is two charged *ions:* one positively charged, a *cation,* and one negatively charged, an *anion* (Figure 1-3). Again, the bonding is based on Coulombic attraction, this time between two ions.

The Coulombic bonding models of attracting and repelling charges shown in Figures 1-2 and 1-3 are highly simplified views of the interactions that take place in the bonding of atoms. Nevertheless, these models explain many of the properties of organic molecules.

We have seen that attraction between negatively and positively charged particles is a basis for bonding. How does this concept work in real molecules?

1-3 Ionic and Covalent Bonds: The Octet Rule

Two extreme types of bonding explain the interactions between atoms in organic molecules:

1. A **covalent bond** is formed by the sharing of electrons (as shown in Figure 1-2).
2. An **ionic bond** is based on the electrostatic attraction of two ions with opposite charges (as shown in Figure 1-3).

We shall see that many atoms bind to carbon in a way that is intermediate between these extremes: Some ionic bonds have covalent character, and some covalent bonds are partly ionic.

What are the factors that account for the two types of bonds? To answer this question, let us return to the atoms and their compositions. We shall start by looking at the periodic table and at how the electronic makeup of the elements changes as the atomic number increases.

The periodic table underlies the octet rule

The partial periodic table depicted in Table 1-1 includes those elements most widely found in organic molecules: carbon (C), hydrogen (H), oxygen (O), nitrogen (N), sulfur (S), chlorine (Cl), bromine (Br), and iodine (I). Certain reagents, indispensable for synthesis and commonly used, contain elements such as lithium (Li), magnesium (Mg), boron (B), and phosphorus (P). (If you are not familiar with these elements, you should learn Table 1-1.)

EXERCISE 1-1

(a) Redraw Figure 1-1 for a weaker bond than the one depicted. (b) Write Table 1-1 from memory.

The elements in the periodic table are listed according to their atomic number, or nuclear charge (number of protons), which also equals their number of electrons. This number increases by one with each element listed. The electrons occupy energy levels, or "shells," each with a fixed capacity. For example, the first shell has room for two electrons; the second, eight; and the third, eighteen. Helium, with two electrons in its shell, and the other noble gases, with eight electrons (called **octets**) in their outermost shells, are especially stable. These elements show very little chemical reactivity. All other elements lack octets in their outermost electron shells. *They will tend to form molecules in such a way as to reach an octet in the outer electron shell and attain a noble-gas configuration.* In the next two sections, we describe two extreme ways in which this goal may be accomplished: by the formation of pure ionic or pure covalent bonds.

TABLE 1-1	Partial Periodic Table							
Period							**Halogens**	**Noble gases**
First	H^1							He^2
Second	$Li^{2,1}$	$Be^{2,2}$	$B^{2,3}$	$C^{2,4}$	$N^{2,5}$	$O^{2,6}$	$F^{2,7}$	$Ne^{2,8}$
Third	$Na^{2,8,1}$	$Mg^{2,8,2}$	$Al^{2,8,3}$	$Si^{2,8,4}$	$P^{2,8,5}$	$S^{2,8,6}$	$Cl^{2,8,7}$	$Ar^{2,8,8}$
Fourth	$K^{2,8,8,1}$						$Br^{2,8,18,7}$	$Kr^{2,8,18,8}$
Fifth							$I^{2,8,18,18,7}$	$Xe^{2,8,18,18,8}$

Note: The superscripts indicate the number of electrons in each principal shell of the atom.

In pure ionic bonds, electron octets are formed by transfer of electrons

Sodium (Na), a reactive metal, interacts with chlorine, a reactive gas, in a violent manner to produce a stable substance: sodium chloride. Similarly, sodium reacts with fluorine (F), bromine, or iodine to give the respective salts. Other alkali metals, such as lithium and potassium (K), undergo the same reactions. These transformations succeed because both reaction partners attain noble-gas character by the *transfer of outer-shell electrons,* called **valence electrons,** from the alkali metals on the left side of the periodic table to the halogens on the right.

Let us see how this works for the ionic bond in sodium chloride. Why is the interaction energetically favorable? First, it takes energy to remove an electron from an atom. This energy is the **ionization potential (IP)** of the atom. For sodium gas, the ionization energy amounts to 119 kcal mol^{-1}.* Conversely, energy may be released when an electron attaches itself to an atom. For chlorine, this energy, called its **electron affinity (EA),** is -83 kcal mol^{-1}. These two processes result in the transfer of an electron from sodium to chlorine. Together, they require a net energy *input* of $119 - 83 = 36$ kcal mol^{-1}.

$$Na^{2,8,1} \xrightarrow{-1\,e} [Na^{2,8}]^+ \qquad IP = 119 \text{ kcal mol}^{-1}$$

Sodium cation
(Neon configuration) · Energy input required

$$Cl^{2,8,7} \xrightarrow{+1\,e} [Cl^{2,8,8}]^- \qquad EA = -83 \text{ kcal mol}^{-1}$$

Chloride anion
(Argon configuration) · Energy released

$$Na + Cl \longrightarrow Na^+Cl^- \qquad \text{Total} = 36 \text{ kcal mol}^{-1}$$

Why, then, do the atoms readily form NaCl? The reason is their electrostatic attraction, which pulls them together in an ionic bond. At the most favorable interatomic distance [about 2.8 Å (angstroms) in the gas phase], this attraction releases about 120 kcal mol^{-1} (see Figure 1-1). This energy release is enough to make the reaction of sodium with chlorine energetically highly favorable ($+36 - 120 = -84$ kcal mol^{-1}).

More than one electron may be donated (or accepted) to achieve favorable electronic configurations. Magnesium, for example, has two valence electrons. Donation to an appropriate acceptor produces the corresponding doubly charged cation with the electronic structure of neon. In this way, the ionic bonds of typical salts are formed.

Formation of Ionic Bonds by Electron Transfer

$$Na^{2,8,1} + Cl^{2,8,7} \longrightarrow [Na^{2,8}]^+ \, [Cl^{2,8,8}]^-, \text{ or NaCl}$$

A more convenient way of depicting valence electrons is by means of dots around the symbol for the element. In this case, the letters represent the nucleus and all the electrons in the inner shells, together called the **core configuration.**

---

*This book will cite energy values in the traditional units of kcal mol^{-1}, in which mol is the abbreviation for mole and a kilocalorie (kcal) is the energy required to raise the temperature of 1 kg (kilogram) of water by 1°C. In SI units, energy is expressed in joules (kg m^{2}s^{-2}, or kilogram-meter^2 per second^2). A joule (J) is the energy required to raise a mass of 1 kg every second by 1 m s^{-1}. The conversion factor is 1 kcal = 4184 J = 4.184 kJ (kilojoule).

Valence Electrons as Electron Dots

$$\text{Li} \cdot \quad \cdot \text{Be} \quad \cdot \text{B} \cdot \quad \cdot \overset{\cdot}{\text{C}} \cdot \quad \cdot \overset{\cdot}{\text{N}} \cdot \quad : \overset{\cdot}{\text{O}} \cdot \quad : \overset{\cdot\cdot}{\text{F}} \cdot$$

$$\text{Na} \cdot \quad \cdot \text{Mg} \quad \cdot \text{Al} \cdot \quad \cdot \overset{\cdot}{\text{Si}} \cdot \quad \cdot \overset{\cdot}{\text{P}} \cdot \quad : \overset{\cdot}{\text{S}} \cdot \quad : \overset{\cdot\cdot}{\text{Cl}} \cdot$$

Electron-Dot Picture of Salts

$$\text{Na} \cdot + \cdot \overset{\cdot\cdot}{\underset{\cdot\cdot}{\text{Cl}}} : \xrightarrow{1\,e \text{ transfer}} \text{Na}^+ : \overset{\cdot\cdot}{\underset{\cdot\cdot}{\text{Cl}}} :^-$$

$$\cdot \overset{\cdot}{\text{Mg}} + 2 \cdot \overset{\cdot\cdot}{\underset{\cdot\cdot}{\text{Cl}}} : \xrightarrow{2\,e \text{ transfer}} \text{Mg}^{2+} [: \overset{\cdot\cdot}{\underset{\cdot\cdot}{\text{Cl}}} :]_2^-$$

The hydrogen atom may either lose an electron to become a bare nucleus, the **proton,** or accept an electron to form the **hydride ion,** $[\text{H}:]^-$, which possesses the helium configuration. Indeed, the hydrides of lithium, sodium, and potassium (Li^+H^-, Na^+H^-, and K^+H^-) are commonly used reagents.

$$\text{H} \cdot \xrightarrow{-1\,e} [\text{H}]^+ \qquad \text{Bare nucleus} \qquad \text{IP} = 314 \text{ kcal mol}^{-1}$$
Proton

$$\text{H} \cdot \xrightarrow{+1\,e} [\text{H}:]^- \qquad \text{Helium configuration} \qquad \text{EA} = -18 \text{ kcal mol}^{-1}$$
Hydride ion

EXERCISE 1-2

Draw electron-dot pictures for ionic LiBr, Na_2O, BeF_2, AlCl_3, and MgS.

In covalent bonds, electron octets are formed by sharing electrons

Formation of ionic bonds between two identical elements is difficult because the electron transfer is usually very unfavorable. For example, in H_2, formation of H^+H^- would require an energy input of nearly 300 kcal mol^{-1}. For the same reason, none of the halogens, F_2, Cl_2, Br_2, and I_2, has an ionic bond. The high IP of hydrogen also prevents the bonds in the hydrogen halides from being ionic. For elements nearer the center of the periodic table, the formation of ionic bonds is unfeasible, because it becomes more and more difficult to donate or accept enough electrons to attain the noble-gas configuration. Such is the case for carbon. This element would have to shed four electrons to reach the helium electronic structure or add four electrons for a neon-like arrangement. The large amount of charge that would develop makes these processes very energetically unfavorable.

$$\text{C}^{4+} \quad \xleftarrow{-4\,e} \quad \cdot \overset{\cdot}{\underset{\cdot}{\text{C}}} \cdot \quad \xrightarrow{+4\,e} \quad : \overset{\cdot\cdot}{\underset{\cdot\cdot}{\text{C}}} :^{4-}$$
Helium   **Neon**
configuration   **configuration**

Instead, **covalent bonding** takes place: The elements *share* electrons so that each attains a noble-gas configuration. Typical products of such sharing are H_2 and HCl. In HCl, the chlorine atom assumes an octet structure by sharing one of its valence electrons with that of hydrogen. Similarly, the chlorine molecule, Cl_2, is diatomic because both component atoms gain octets by sharing two electrons. Such bonds are called **covalent single bonds.**

Electron-Dot Picture of Covalent Single Bonds

$$\text{H} \cdot + \cdot \text{H} \longrightarrow \text{H} : \text{H}$$

$$\text{H} \cdot + \cdot \overset{\cdot\cdot}{\underset{\cdot\cdot}{\text{Cl}}} : \longrightarrow \text{H} : \overset{\cdot\cdot}{\underset{\cdot\cdot}{\text{Cl}}} :$$

$$: \overset{\cdot\cdot}{\underset{\cdot\cdot}{\text{Cl}}} \cdot + \cdot \overset{\cdot\cdot}{\underset{\cdot\cdot}{\text{Cl}}} : \longrightarrow : \overset{\cdot\cdot}{\underset{\cdot\cdot}{\text{Cl}}} : \overset{\cdot\cdot}{\underset{\cdot\cdot}{\text{Cl}}} :$$

Because carbon has four valence electrons, it must acquire a share of four electrons to gain the neon configuration, as in methane. Nitrogen has five valence electrons and needs three to share, as in ammonia; and oxygen, with six valence electrons, requires only two to share, as in water.

$$\begin{array}{ccc} \text{H} & & \\ \text{H} : \overset{\cdot\cdot}{\text{C}} : \text{H} & \text{H} : \overset{\cdot\cdot}{\text{N}} : \text{H} & \text{H} : \overset{\cdot\cdot}{\underset{\cdot\cdot}{\text{O}}} : \text{H} \\ \text{H} & \text{H} & \end{array}$$

$$\quad\;\text{Methane} \qquad \text{Ammonia} \qquad \text{Water}$$

It is possible for one atom to supply both of the electrons required for covalent bonding. This occurs upon addition of a proton to ammonia, thereby forming $NH_4{}^+$, or to water, thereby forming H_3O^+.

$$\begin{array}{c} \text{H} \\ \text{H} : \overset{\cdot\cdot}{\text{N}} : + \text{H}^+ \\ \text{H} \end{array} \longrightarrow \left[\begin{array}{c} \text{H} \\ \text{H} : \overset{\cdot\cdot}{\text{N}} : \text{H} \\ \text{H} \end{array} \right]^+ \qquad \begin{array}{c} \overset{\cdot\cdot}{} \\ \text{H} : \overset{\cdot\cdot}{\underset{\cdot\cdot}{\text{O}}} : + \text{H}^+ \\ \text{H} \end{array} \longrightarrow \left[\begin{array}{c} \overset{\cdot\cdot}{} \\ \text{H} : \overset{\cdot\cdot}{\underset{\cdot\cdot}{\text{O}}} : \text{H} \\ \text{H} \end{array} \right]^+$$

$$\qquad\qquad\quad \text{Ammonium} \qquad\qquad\qquad\qquad\qquad \text{Hydronium}$$
$$\qquad\qquad\qquad \text{ion} \qquad\qquad\qquad\qquad\qquad\qquad\quad \text{ion}$$

Besides two-electron (**single**) bonds, atoms may form four-electron (**double**) and six-electron (**triple**) bonds to gain noble-gas configurations. Atoms that share more than one electron pair are found in ethene and ethyne.

$$\begin{array}{cc} \text{H} & \text{H} \\ \quad \overset{\cdot}{} \;\; \overset{\cdot}{} \\ \;\;\; \text{C} :: \text{C} \\ \;\; \overset{\cdot}{} \;\; \overset{\cdot}{} \\ \text{H} & \text{H} \end{array} \qquad \text{H} : \text{C} ::: \text{C} : \text{H}$$

$$\quad\;\; \textbf{Ethene} \qquad\qquad\qquad \textbf{Ethyne}$$
$$\textbf{(Ethylene)*} \qquad\qquad \textbf{(Acetylene)*}$$

EXERCISE 1-3

Draw electron-dot structures for F_2, CF_4, CH_2Cl_2, PH_3, BrI, OH^-, NH_2^-, and CH_3^-. (Where applicable, the first element is at the center of the molecule.) Make sure that all atoms have noble-gas electron configurations.

---

*In labels of molecules, systematic names (introduced in Section 2-3) will be given first, followed in parentheses by so-called common names that are still in frequent usage.

In most organic bonds, the electrons are not shared equally: polar covalent bonds

The preceding two sections presented two extreme ways in which atoms attain noble-gas configurations by entering into bonding: pure ionic and pure covalent. In reality, most bonds are of a nature that lies between these two extremes: **polar co-valent.** Thus, the ionic bonds in most salts have some covalent character; conversely, the covalent bonds to carbon have some ionic or polar character. Recall (Section 1-2) that both sharing of electrons *and* Coulombic attraction contribute to the stability of a bond. How polar are polar covalent bonds and what is the direction of the polarity? We can answer these questions by considering the periodic table and keeping in mind that the positive nuclear charge of the elements increases from left to right. Thus, the elements on the left of the periodic table are often called **electropositive,** electron donating, or "electron pushing," because their electrons are held by the nucleus less tightly than are those of elements to the right. The latter are therefore described as being **electronegative,** electron accepting, or "electron pulling." Table 1-2 lists the relative electronegativity of some elements. On this scale, fluorine, the most electronegative of them all, is assigned the value 4.

Consideration of Table 1-2 readily explains why the most ionic (least covalent) bonds occur between elements at the two extremes (e.g., the alkali metal salts, such as sodium chloride). On the other hand, the purest covalent bonds are formed between atoms of equal electronegativity (i.e., identical elements, as in H_2, N_2, O_2, F_2, and so on) or in carbon–carbon bonds. However, most covalent bonds are between atoms of differing electronegativity, resulting in their **polarization.** The polarization of a bond is the consequence of a shift of the center of electron density in the bond toward the more electronegative atom. It is indicated in a very qualitative manner by designating a partial positive charge, δ^+, and partial negative charge, δ^-, to the respective less or more electronegative atom. The larger the difference in electronegativity, the bigger the charge separation.

Polar Bonds

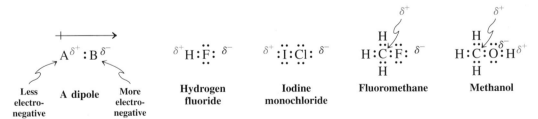

Molecules Can Have Polar Bonds but No Net Polarization

The separation of opposite charges is called an electric **dipole,** symbolized by an arrow crossed at its tail and pointing from positive to negative. A polarized bond can impart polarity to a molecule as a whole, as in HF, HCl, and CH_3F. In symmetrical structures, however, the polarizations of the individual bonds may cancel, thus leading to molecules with no net polarization, such as CO_2 and CCl_4. To know whether a molecule is polar, we have to know its shape, because the net polarity is the vector sum of the bond dipoles.

TABLE 1-2	Electronegativities of Selected Elements					
H 2.2						
Li 1.0	Be 1.6	B 2.0	C 2.6	N 3.0	O 3.4	F 4.0
Na 0.9	Mg 1.3	Al 1.6	Si 1.9	P 2.2	S 2.6	Cl 3.2
K 0.8						Br 3.0
						I 2.7

Note: Values established by L. Pauling and updated by A. L. Allred (see *Journal of Inorganic and Nuclear Chemistry,* 1961, *17,* 215).

Electron repulsion controls the shapes of molecules

Molecules adopt shapes in which electron repulsion is minimized. In diatomic species such as H_2 or LiH, there is only one bonding electron pair and one possible arrangement of the two atoms. However, beryllium fluoride, BeF_2, is a triatomic species. Will it be bent or linear? Electron repulsion is at a minimum in a **linear** structure, because the bonding and nonbonding electrons are placed as far from each other as possible, at 180°.* Linearity is also expected for other derivatives of beryllium, as well as of other elements in the same column of the periodic table.

BeF₂ Is Linear   **BCl₃ Is Trigonal**

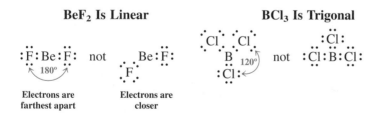

In boron trichloride, the three valence electrons of boron allow it to form covalent bonds with three chlorine atoms. Electron repulsion enforces a regular **trigonal** arrangement—that is, the three halogens are at the corners of an equilateral triangle, the center of which is occupied by boron, and the bonding (and nonbonding) electron pairs of the respective chlorine atoms are at maximum distance from each other, that is, 120°. Other derivatives of boron, and the analogous compounds with other elements in the same column of the periodic table, are again expected to adopt trigonal structures.

*This is true only in the gas phase. At room temperature, BeF_2 is a solid (it is used in nuclear reactors) that exists as a complex network of linked Be and F atoms, not as a distinct linear triatomic structure.

Applying this principle to carbon, we can see that methane, CH_4, has to be **tetrahedral.** Pointing its four valences toward the vertices of a tetrahedron is the best arrangement for minimizing electron repulsion.

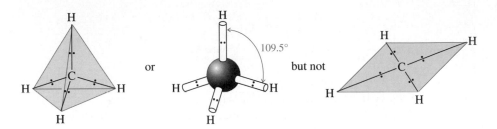

This method for determining molecular shape by minimizing electron repulsion is called the *valence shell electron pair repulsion* (*VSEPR*) method. Note that we often draw molecules such as BCl_3 and CH_4 as if they were flat and had 90° angles. *This depiction is for ease of drawing only.* Do *not* confuse such drawings with the true molecular shapes (trigonal for BCl_3 and tetrahedral for CH_4).

EXERCISE 1-4

Show the bond polarization in H_2O, SCO, SO, IBr, CH_4, $CHCl_3$, CH_2Cl_2, and CH_3Cl by using dipole arrows to indicate separation of charge. (In the last four examples, place the carbon in the center of the molecule.)

EXERCISE 1-5

Ammonia, $:NH_3$, is not trigonal but pyramidal, with bond angles of 107.3°. Water, $H_2\overset{..}{\underset{..}{O}}$, is not linear but bent (104.5°). Why? (**Hint:** Consider the effect of the nonbonding electron pairs.)

To summarize, there are two extreme types of bonding, ionic and covalent. Both derive favorable energetics from Coulomb forces and the attainment of noble-gas electronic structures. Most bonds are better described as something between the two types: the polar covalent (or covalent ionic) bonds. Polarity in bonds may give rise to polar molecules. The outcome depends on the shape of the molecule, which is determined in a simple manner by arrangement of its bonds and nonbonding electrons to minimize electron repulsion.

1-4 Electron-Dot Model of Bonding: Lewis Structures

The drawings in the preceding section, with pairs of electron dots representing bonds, are also called **Lewis* structures.** In this section, rules are given for writing such structures correctly and for keeping track of valence electrons.

Lewis structures are drawn by following simple rules

The procedure for drawing correct electron-dot structures is straightforward, as long as the following rules are observed.

*Professor Gilbert N. Lewis (1875–1946), University of California at Berkeley.

RULE 1. *Draw the molecular skeleton.* As an example, consider methane. The molecule has four hydrogen atoms bonded to one central carbon atom.

$$
\begin{array}{c}
\text{H} \\
\text{H C H} \\
\text{H}
\end{array}
\qquad
\text{H H C H H}
$$

Correct   Incorrect

RULE 2. *Count the number of available valence electrons.* Add up all the valence electrons of the component atoms. Special care has to be taken with charged structures (anions or cations), in which case the appropriate number of electrons has to be added or subtracted to account for extra charges.

CH_4	4 H	4×1 electron	=	4 electrons
	1 C	1×4 electrons	=	4 electrons
		Total		8 electrons

H_3O^+	3 H	3×1 electron	=	3 electrons
	1 O	1×6 electrons	=	6 electrons
	Charge	$+1$	=	-1 electron
		Total		8 electrons

HBr	1 H	1×1 electron	=	1 electron
	1 Br	1×7 electrons	=	7 electrons
		Total		8 electrons

NH_2^-	2 H	2×1 electron	=	2 electrons
	1 N	1×5 electrons	=	5 electrons
	Charge	-1	=	$+1$ electron
		Total		8 electrons

RULE 3. (The **octet rule**) *Depict all covalent bonds by two shared electrons, giving as many atoms as possible a surrounding electron octet, except for H, which requires a duet.* Make sure that the number of electrons used is *exactly* the number counted according to rule 2. Elements at the right in the periodic table may contain pairs of valence electrons not used for bonding, called **lone electron pairs** or just **lone pairs.**

Consider, for example, hydrogen bromide. The shared electron pair supplies the hydrogen atom with a duet, the bromine with an octet, because the bromine carries three lone electron pairs. Conversely, in methane, the four C–H bonds satisfy the requirement of the hydrogens and, at the same time, furnish the octet for carbon. Examples of correct and incorrect Lewis structures for HBr are shown below.

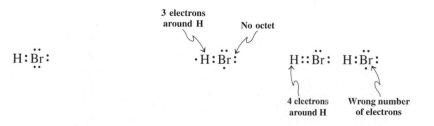

Correct Lewis Structure   **Incorrect Lewis Structures**

Frequently, the number of valence electrons is not sufficient to satisfy the octet rule only with single bonds. In this event, double bonds (two shared electron pairs) and even triple bonds (three shared pairs) are necessary to obtain octets. An example is the nitrogen molecule, N_2, which has ten valence electrons. An N–N single bond would leave both atoms with electron sextets, and a double bond provides only one nitrogen atom with an octet. It is the molecule with a triple bond that satisfies both atoms.

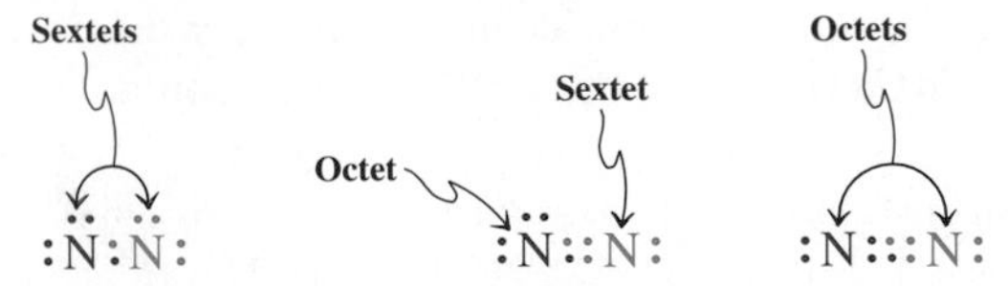

Sextets

Sextet

Octets

Octet

:N:N: :N::N: :N:::N:

Single bond   **Double bond**   **Triple bond**

Further examples of molecules with double and triple bonds are shown below.

Correct Lewis Structures

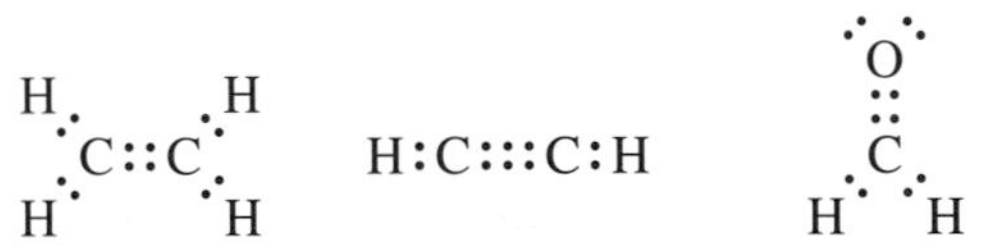

H . . H
   C::C
H ·   · H

Ethene
(Ethylene)

H:C:::C:H

Ethyne
(Acetylene)

. .
O
. .
C
H   H

Formaldehyde

In practice, you may find a simple sequence useful: First, connect all mutually bonded atoms in your structure by single bonds (i.e., shared electron pairs); second, if there are any electrons left, distribute them as lone electron pairs to maximize the number of octets; and finally, if some of the atoms lack octet structures, change as many lone electron pairs into shared electron pairs as required to complete the octet shells (see also the Chapter Integration Problem on p. 40).

EXERCISE 1-6

Draw Lewis structures for the following molecules: HI, $CH_3CH_2CH_3$, CH_3OH, HSSH, SiO_2 (OSiO), O_2, CS_2 (SCS).

RULE 4. *Assign charges to atoms in the molecule.* Each lone pair contributes two electrons to the valence electron count of an atom in a molecule, and each bonding (shared) pair contributes one. An atom is charged if this total is different from the outer-shell electron count in the free, nonbonded atom. Thus we have the formula

$$\text{Charge} = \begin{pmatrix} \text{number of outer-shell} \\ \text{electrons on the} \\ \text{free, neutral atom} \end{pmatrix} - \begin{pmatrix} \text{number of unshared} \\ \text{electrons on the atom} \\ \text{in the molecule} \end{pmatrix} - \frac{1}{2} \begin{pmatrix} \text{number of bonding} \\ \text{electrons surrounding the} \\ \text{atom in the molecule} \end{pmatrix}$$

+
· ·
H:O:H
· ·
H

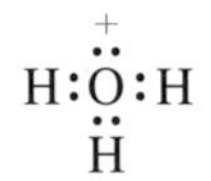

Hydronium ion

As an example, which atom bears the positive charge in the hydronium ion? Each hydrogen has a valence electron count of 1 from the shared pair in its bond to oxygen. Because this value is the same as the electron count in the free atom, the charge on each hydrogen is zero. The electron count on the oxygen in the hydronium ion is 2 (the lone pair) + 3 (half of 6 bonding electrons) = 5. This value is one short of the number of outer-shell electrons in the free atom, thus giving the oxygen a charge of +1. Hence the positive charge is assigned to oxygen.

+
:N:::O:

Nitrosyl cation

Another example is the nitrosyl cation, NO^+. The molecule bears a lone pair on nitrogen, in addition to the triple bond connecting the nitrogen to the oxygen atom. This gives nitrogen five valence electrons, a value that matches the count in the free atom; therefore the nitrogen atom has no charge. The same number of valence electrons (5) is found on oxygen. Because the free oxygen atom requires six valence electrons to be neutral, the oxygen in NO^+ possesses the +1 charge. Other examples follow on the next page.

$$H:\overset{\bar{\cdot}}{\underset{\cdot\cdot}{\underset{H}{C}}}:H$$

Methyl anion

$$H:\overset{\cdot\cdot}{\underset{\cdot\cdot}{\underset{H}{C}}}:\overset{\cdot\cdot}{\underset{\cdot\cdot}{S}}:^-$$

**Methanethiolate
ion**

$$\overset{\overset{\displaystyle H}{\overset{\cdot\cdot}{\overset{+}{O}}}}{\underset{H\quad\quad H}{\underset{\cdot}{C}}}$$

**Protonated
formaldehyde**

Sometimes the octet rule leads to charges on atoms even in neutral molecules. The Lewis structure is then said to be **charge separated.** An example is carbon monoxide, CO. Some compounds containing nitrogen–oxygen bonds, such as nitric acid, HNO_3, also exhibit this behavior.

$$:\overset{-}{C}\equiv\overset{+}{O}:$$

Carbon monoxide

The octet rule does not always hold

The octet rule strictly holds only for the elements of the second row and then only if there is a sufficient number of valence electrons to satisfy it. Thus, there are three exceptions to be considered.

Nitric acid

EXCEPTION 1. You will have noticed that all our examples of "correct" Lewis structures contain an even number of electrons; that is, all are distributed as bonding or lone pairs. This distribution is not possible in species having an odd number of electrons, such as nitrogen oxide (NO) and neutral methyl (methyl radical, ·CH_3; see Section 3-1).

$$:\overset{\cdot}{N}::\overset{\cdot}{\underset{\cdot\cdot}{O}}$$

Nitrogen oxide

$$H:\overset{\cdot}{\underset{H}{C}}:H$$

Methyl radical

$$H:\overset{\cdot\cdot}{Be}:H$$

Beryllium hydride

$$\overset{H\quad\cdot\cdot\quad H}{\underset{H}{B}}$$

Borane

EXCEPTION 2. Some compounds of the early second-row elements, such as BeH_2 and BH_3, have a deficiency of valence electrons.

Compounds falling under exceptions 1 and 2 reveal the consequences of being denied octet configurations: they are unusually reactive and transform readily in reactions that lead to octet structures. For example, ·CH_3 dimerizes spontaneously to ethane, CH_3–CH_3, and BH_3 reacts with hydride, H^-, to give borohydride, BH_4^-.

$$H:\overset{\overset{\displaystyle H}{\cdot\cdot}}{\underset{\underset{\displaystyle H}{\cdot\cdot}}{C}}\cdot \;+\; \cdot\overset{\overset{\displaystyle H}{\cdot\cdot}}{\underset{\underset{\displaystyle H}{\cdot\cdot}}{C}}:H \;\longrightarrow\; H:\overset{\overset{\displaystyle H}{\cdot\cdot}}{\underset{\underset{\displaystyle H}{\cdot\cdot}}{C}}:\overset{\overset{\displaystyle H}{\cdot\cdot}}{\underset{\underset{\displaystyle H}{\cdot\cdot}}{C}}:H$$

Ethane

$$\overset{H\quad\cdot\cdot\quad H}{\underset{\underset{\displaystyle H}{B}}{}}\cdot \;+\; :H^- \;\longrightarrow\; H:\overset{\overset{\displaystyle H}{\cdot\cdot}}{\underset{\underset{\displaystyle H}{\cdot\cdot}}{B}}:H^-$$

Borohydride

EXCEPTION 3. Beyond the second row, the simple Lewis model is not strictly applied, and elements may be surrounded by more than eight valence electrons, a feature referred to as **valence shell expansion.** For example, not only are phosphorus and sulfur (as relatives of nitrogen and oxygen) trivalent and divalent, respectively, and Lewis octet structures readily formulated for their derivatives, but they form stable compounds of higher valency, among them the familiar phosphoric and sulfuric acids. Some examples of octet and expanded-octet molecules containing these elements are shown below.

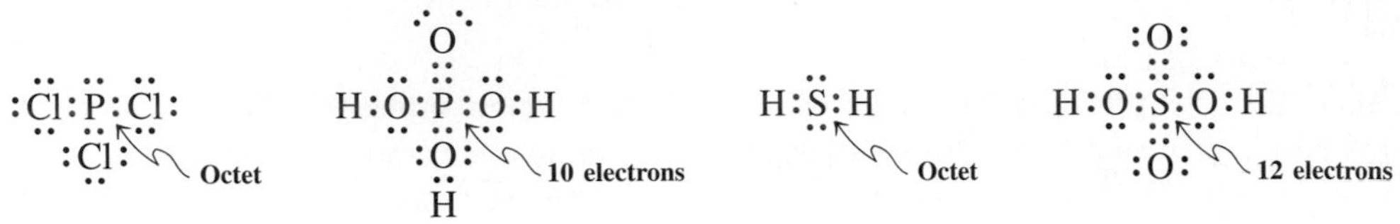

Phosphorous trichloride	Phosphoric acid	Hydrogen sulfide	Sulfuric acid

An explanation for this apparent violation of the octet rule is found in a more sophisticated description of atomic structure by quantum mechanics (Section 1-6). However, you will notice that, even in these cases, you can construct dipolar forms in which the Lewis octet rule is preserved (see Section 1-5).

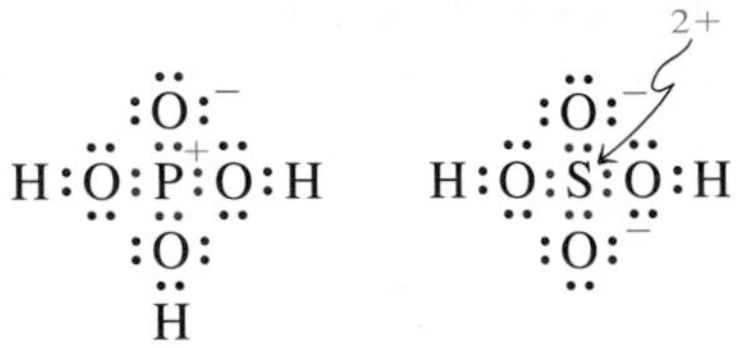

Covalent bonds can be depicted as straight lines

Electron-dot structures can be cumbersome, particularly for larger molecules. It is simpler to represent covalent single bonds by single straight lines; double bonds are represented by two lines and triple bonds by three. Lone electron pairs can either be shown as dots or simply omitted. The use of such notation was first suggested by the German chemist August Kekulé,* long before electrons were discovered; structures of this type are often called **Kekulé structures.**

Straight-Line Notation for the Covalent Bond

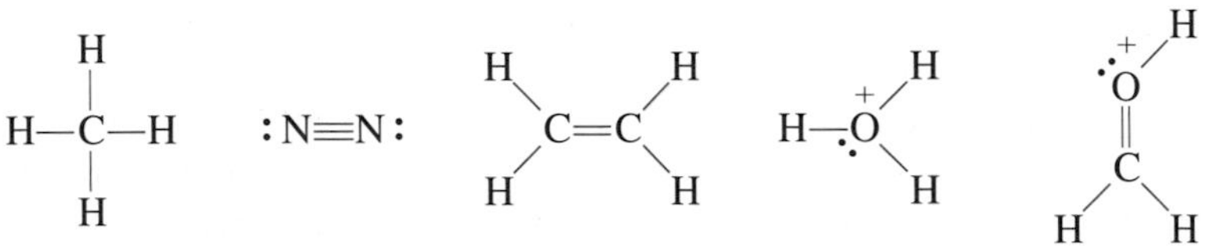

Methane	Diatomic nitrogen	Ethene	Hydronium ion	Protonated formaldehyde

---

EXERCISE 1-7

Draw Lewis structures of the following molecules, including the assignment of any charges to atoms (the order in which the atoms are attached is given in parentheses when it may not be obvious from the formula as it is commonly written): SO, F_2O (FOF), $HClO_2$ (HOClO), BF_3NH_3 (F_3BNH_3), $CH_3OH_2^+$ ($H_3COH_2^+$), $Cl_2C{=}O$, CN^-, C_2^{2-}.

*Professor F. August Kekulé von Stradonitz (1829–1896), University of Bonn, Germany.

In summary, Lewis structures describe bonding by the use of electron dots or straight lines. Whenever possible, they are drawn so as to give hydrogen an electron duet and other atoms an electron octet. Charges are assigned to each atom by evaluating its electron count.

1-5 Resonance Forms

In organic chemistry, we also encounter molecules for which there are *several* correct Lewis structures.

The carbonate ion has several correct Lewis structures

Let us consider the carbonate ion, CO_3^{2-}. Following our rules, we can easily draw a Lewis structure (A) in which every atom is surrounded by an octet. The two negative charges are located on the bottom two oxygen atoms; the third oxygen is neutral, connected to the central carbon by a double bond and bearing two lone pairs. But why choose the bottom two oxygen atoms as the charge carriers? There is no reason at all—it is a completely arbitrary choice. We could equally well have drawn structures B or C to describe the carbonate ion. The three correct Lewis pictures are called **resonance forms.**

Resonance Forms of the Carbonate Ion

The individual resonance forms are connected by double-headed arrows and all placed within one set of square brackets. They have the characteristic property of being interconvertible by *electron-pair movement only,* the nuclear positions in the molecule remaining *unchanged.* Note that, to turn A into B and then into C, we have to shift two electron pairs in each case. Such movement of electrons can be depicted by curved arrows, a procedure informally called "electron pushing."

The use of curved arrows to depict electron-pair movement is a useful technique that will prevent us from making the common mistake of changing the total number of electrons when we draw resonance forms. It is also advantageous in keeping track of electrons when formulating mechanisms (Section 6-4).

But what is its true structure?

Does the carbonate ion have one uncharged oxygen atom bound to carbon through a double bond and two other oxygen atoms bound through a single bond each, both bearing a negative charge, as suggested by the Lewis structures? *The answer is no.* If that were true, the carbon–oxygen bonds would be of different lengths, because double bonds are normally shorter than single bonds. But the carbonate ion is *perfectly symmetrical* and contains a trigonal central carbon, all C–O bonds being of equal length—between the length of a double and that of a single bond. The negative charge is evenly distributed over all three oxygens: It is said to be **delocalized.**

In other words, none of the individual Lewis representations of this molecule is correct on its own. Rather, *the true structure is a composite of A, B, and C.* The resulting picture is called a **resonance hybrid.** Because A, B, and C are equivalent (i.e., each is composed of the same number of atoms, bonds, and electron pairs), they contribute equally to the true structure of the molecule, but none of them by itself accurately represents it.

The word *resonance* may imply to you that the molecule vibrates or equilibrates from one form to another. This inference is incorrect. The molecule *never* looks like any of the individual resonance forms; it has only one structure, the resonance hybrid. Unlike substances in ordinary chemical equilibria, resonance forms are *not* real, although each makes a partial contribution to reality.

Dotted-Line Notation of Carbonate as a Resonance Hybrid

An alternative convention used to describe resonance hybrids such as carbonate is to represent the bonds as a combination of solid and dotted lines. The $\frac{2}{3}-$ sign here indicates that a partial charge ($\frac{2}{3}$ of a negative charge) resides on each oxygen atom. The equivalence of all three carbon–oxygen bonds and all three oxygens is clearly indicated by this convention. Other examples of resonance hybrids are the acetate anion and the 2-propenyl (allyl) cation.

Acetate anion

2-Propenyl (allyl) cation

When drawing resonance forms, keep in mind that (1) pushing one electron pair toward one atom and away from another results in a movement of charge; (2) the relative positions of all the atoms stay unchanged—only electrons are moved; (3) equivalent resonance forms contribute equally to the resonance hybrid; and (4) the arrows connecting resonance forms are double headed ($\leftrightarrow$).

EXERCISE 1-8

Draw two resonance forms for nitrite ion, NO_2^-. What can you say about the geometry of this molecule (linear or bent)? (**Hint:** Consider the effect of electron repulsion exerted by the lone pair on nitrogen.)

Not all resonance forms are equivalent

The carbonate and acetate anions and the 2-propenyl cation all have equivalent octet resonance forms. However, many molecules are described by resonance forms that are not equivalent. An example is the enolate anion. The two resonance forms differ in the locations of both the double bond and the charge.

The Two Nonequivalent Resonance Forms of the Enolate Ion

Although both forms are contributors to the true structure of the anion, we shall see that one contributes more than the other. The question is, which one? If we (greatly) extend our consideration of nonequivalent resonance forms to those devoid of octets, the question becomes more general.

[Octet ⟷ Nonoctet] Resonance Forms

Formaldehyde   Sulfuric acid

Such an extension requires that we relax our definitions of "correct" and "incorrect" Lewis structures and broadly regard *all* resonance forms as potential contributors to the true picture of a molecule. The task is then to recognize which resonance form is the most important one. In other words, which one is the **major resonance contributor?** Here are some guidelines.

GUIDELINE 1. *Structures with a maximum of octets are most important.* In the enolate ion, all component atoms in either structure are surrounded by octets. Consider, however, the nitrosyl cation, NO^+: The better resonance form has a positive charge on oxygen with electron octets around both atoms; the other form places the positive charge on nitrogen, thereby resulting in an electron sextet on this atom. Because of the octet rule, the second structure contributes less to the hybrid. Thus, the N–O linkage is closer to being a triple than a double bond, and more of the positive charge is on oxygen than on nitrogen. Similarly, the dipolar resonance form for formaldehyde (shown earlier) generates an electron sextet around carbon, rendering it a minor resonance contributor. The possibility of valence shell expansion for third-row elements (Section 1-4) makes the non-charge-separated picture of sulfuric acid with 12 electrons around sulfur a feasible resonance form, but the dipolar octet structure is better.

Major resonance contributor   Minor resonance contributor

Nitrosyl cation

GUIDELINE 2. *Charges should be preferentially located on atoms with compatible electronegativity.* Consider again the enolate ion. Which is the major contributing resonance form? Guideline 2 requires it to be the first, in which the negative charge resides on the more electronegative oxygen atom.

Looking again at NO^+, you might find guideline 2 confusing. The major resonance contributor to NO^+ has the positive charge on the more electronegative oxygen. In cases such as this, *the octet rule overrides the electronegativity criterion;* that is, guideline 1 takes precedence over guideline 2.

GUIDELINE 3. *Structures with less (opposite) charge separation are greater resonance contributors than those with more charge separation.* This rule is a simple consequence of Coulomb's law: Separating opposite charges requires energy; hence neutral structures are better than dipolar ones.

Formic acid

Carbon monoxide

In some cases, to ensure octet Lewis structures, charge separation is necessary; that is, guideline 1 takes precedence over guideline 3. An example is carbon monoxide. Other examples are phosphoric and sulfuric acids, although valence shell expansion allows the formulation of expanded octet structures (see also Section 1-4 and guideline 1).

When there are several charge-separated resonance forms that comply with the octet rule, the most favorable is the one in which the charge distribution best accommodates the relative electronegativities of the component atoms (guideline 2). In diazomethane, for example, nitrogen is more electronegative than carbon, thus allowing a clear choice between the two resonance contributors.

Diazomethane

EXERCISE 1-9

Draw resonance forms for the following two molecules. Indicate the more favorable resonance contributor in each case. **(a)** CNO^-; **(b)** NO^-.

In summary, there are molecules that cannot be described accurately by one Lewis structure but exist as hybrids of several extreme resonance forms. To find the most important resonance contributor, consider the octet rule, make sure that there is a minimum of charge separation, and place on the relatively more electronegative atoms as much negative and as little positive charge as possible.

1-6 Atomic Orbitals: A Quantum Mechanical Description of Electrons Around the Nucleus

So far, we have considered bonds in terms of electron pairs arranged around the component atoms in such a way as to maximize noble-gas configurations (e.g., Lewis octets) and minimize electron repulsion. This approach is useful as a descriptive and predictive tool with regard to the number and location of electrons in molecules. However, it does not answer some simple questions that you may have asked yourself while dealing with this material. For example, why are some Lewis structures "incorrect" or, ultimately, why are noble gases relatively stable? Why are some bonds stronger than others, and how can we tell? What is so good about the two-electron bond, and what do multiple bonds look like? To get some answers, we will start by learning more about the way in which the electrons are distributed around the nucleus, both spatially and energetically. The simplified treatment presented here has as its basis the theory of quantum mechanics developed independently in the 1920s by Heisenberg, Schrödinger, and Dirac.* In this theory, the movement of an electron around a nucleus is expressed in the form of equations that are very similar to those characteristic of waves. The solutions to these equations, called **atomic orbitals,** allow us to describe the probability of finding the electron in a certain region in space. The shape of these domains depends on the energy of the electron.

The electron is described by wave equations

The classical description of the atom (Bohr† theory) assumed that electrons move on more or less defined trajectories around the nucleus. Their energy was thought to relate to their distance from the nucleus. This view is intuitively appealing because it coincides with our physical understanding of classical mechanics. Yet it is incorrect for several reasons.

First, the classical picture of an electron moving in an orbit requires (as does any moving charge) the emission of electromagnetic radiation. The resulting energy loss from the system would cause the electron to spiral toward the nucleus, a prediction that is completely at odds with reality.

Second, in the classical picture, an electron can have any energy, so it can have any of an infinite number of orbits of differing radii. This, again, is not what is observed. Rather, only certain defined energies, called **energy states,** are possible for an electron around a nucleus. Thus, classical mechanics does not satisfactorily explain atomic structure and, ultimately, bonding.

A better model is afforded by considering the wave nature of moving particles. Matter of mass m that moves with velocity v has a wavelength λ.

de Broglie‡ Wavelength

$$\lambda = \frac{h}{mv}$$

in which h is Planck's§ constant. As a result, an orbiting electron can be described by equations that are the same as those used in classical mechanics to describe waves (Figure 1-4; see page 24). The latter have amplitudes with alternating positive and negative signs. Points at which the sign changes are called **nodes.** Waves that interact in phase reinforce each other, as shown in Figure 1-4B. Those out of phase interfere with each other to make smaller waves (and possibly even cancel each other), as shown in Figure 1-4C.

*Professor Werner Heisenberg (1901–1976), University of Munich, Germany, Nobel Prize 1932 (physics); Professor Erwin Schrödinger (1887–1961), University of Dublin, Ireland, Nobel Prize 1933 (physics); Professor Paul Dirac (1902–1984), Florida State University, Tallahassee, Nobel Prize 1933 (physics).
†Professor Niels Bohr (1885–1962), University of Copenhagen, Denmark, Nobel Prize 1922 (physics).
‡Prince Louis-Victor de Broglie (1892–1987), Nobel Prize 1929 (physics).
§Professor Max K. E. L. Planck (1858–1947), University of Berlin, Germany, Nobel Prize 1918 (physics).

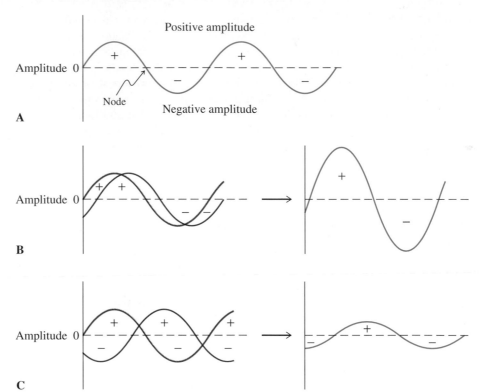

FIGURE 1-4

(A) A wave. The signs of the amplitude are assigned arbitrarily. At points of zero amplitude, called nodes, the wave changes sign.
(B) Waves with amplitudes of like sign (in phase) reinforce each other to make a larger wave. (C) Waves out of phase subtract from each other to make a smaller wave.

Note: The + and − signs in Figure 1-4 refer to signs of the mathematical functions describing the wave amplitudes and have nothing to do with electrical charges.

This theory of electron motion is called **quantum mechanics.** The equations developed in this theory, the **wave equations,** have a series of solutions called **wave functions,** usually described by the Greek letter psi, ψ. Their values around the nucleus are not directly identifiable with any observable property of the atom. However, *the squares (ψ^2) of their values at each point in space describe the probability of finding an electron at that point.* The physical realities of the atom make solutions attainable only for certain *specific energies.* The system is said to be **quantized.**

EXERCISE 1-10

Draw a picture similar to Figure 1-4 of two waves overlapping such that their amplitudes cancel each other.

Atomic orbitals have characteristic shapes

Plots of wave functions in three dimensions typically have the appearance of spheres or dumbbells with flattened or teardrop-shaped lobes. For simplicity, we may regard artistic renditions of **atomic orbitals** as indicating the regions in space in which the electron is likely to be found. Nodes separate portions of the wave function with opposite mathematical signs. The value of the wave function at a node is zero; therefore the probability of finding electron density there is zero. Higher energy wave functions have more nodes than do those of low energy.

Let us consider the shapes of the atomic orbitals for the simplest case, that of the hydrogen atom, consisting of a proton surrounded by an electron. The single lowest energy solution of the wave equation is called the 1*s* orbital, the number one referring to the first (lowest) energy level. An orbital label also denotes the shape and num-

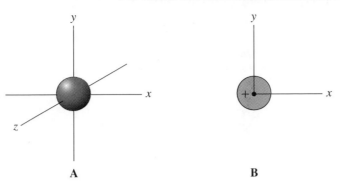

FIGURE 1-5

Representations of a 1s orbital. (A) The orbital is spherically symmetric in three dimensions. (B) A simplified two-dimensional view. The plus sign denotes the mathematical sign of the wave function and is *not a charge.*

ber of nodes of the orbital. The 1s orbital is *spherically symmetric* (Figure 1-5) and has no nodes. This orbital can be represented pictorially as a sphere (Figure 1-5A) or simply as a circle (Figure 1-5B).

The next higher energy wave function, the 2s orbital, also is unique and, again, spherical. The 2s orbital is larger than the 1s orbital; the higher energy 2s electron is on the average farther from the positive nucleus. In addition, the 2s orbital has one node, a spherical surface of zero electron density separating regions of the wave function of opposite sign (Figure 1-6). Like that of classical waves, the sign of the wave function on either side of the node is arbitrary, as long as it changes at the node. Remember that the sign of the wave function is not related to "where the electron is." As mentioned earlier, the probability of electron occupancy at any point of the orbital is given by the square of the value of the wave function. Moreover, the node does not constitute a barrier of any sort to the electron, which, in this description, is regarded not as a particle but as a wave.

After the 2s orbital, the wave equations for the electron around a hydrogen atom have three energetically equivalent solutions, the $2p_x$, $2p_y$, and $2p_z$ orbitals. Solutions of equal energy of this type are called **degenerate** (*degenus,* Latin, without genus or

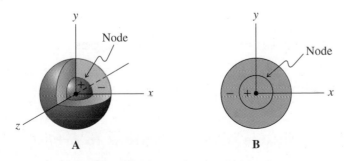

FIGURE 1-6

Representations of a 2s orbital. Notice that it is larger than the 1s orbital and that a node is present. The + and − denote the sign of the wave function. (A) The orbital in three dimensions, with a section removed to allow the visualization of the node. (B) The more conventional two-dimensional representation of the orbital.

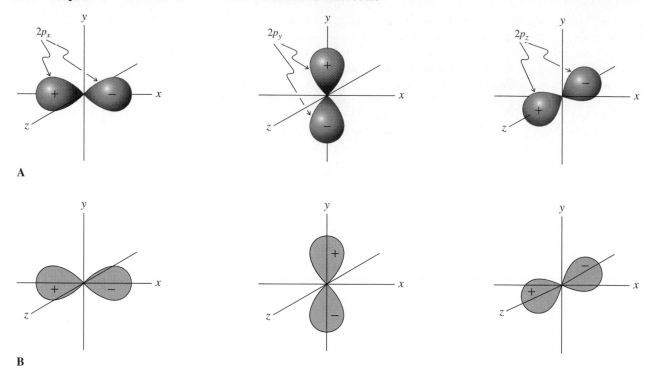

FIGURE 1-7

Representations of 2p orbitals (A) in three dimensions and (B) in two dimensions. Remember that the + and − signs refer to the wave functions and *not* to electrical charges. Lobes of opposite sign are separated by a nodal plane that is perpendicular to the axis of the orbital. For example, the p_x orbital is divided by a node in the *yz* plane.

kind). As shown in Figure 1-7, *p* orbitals consist of two lobes that resemble a solid figure eight. A *p* orbital is characterized by its directionality in space. The orbital axis can be aligned with any one of the *x, y,* and *z* axes, hence the labels p_x, p_y, and p_z. The two lobes of opposite sign of each orbital are separated by a nodal plane through the atom's nucleus and perpendicular to the orbital axis.

The third set of solutions furnishes the 3*s* and 3*p* atomic orbitals. They are similar in shape to, but more diffuse than, their lower energy counterparts and have two nodes. Still higher energy orbitals (3*d*, 4*s*, 4*p*, etc.) are characterized by an increasing number of nodes and a variety of shapes. They are of much less importance in organic chemistry than are the lower orbitals. To a first approximation, the shapes and nodal properties of the atomic orbitals of other elements are very similar to those of hydrogen. Therefore, we may use *s* and *p* orbitals in a description of the electronic configurations of helium, lithium, and so forth.

The Aufbau principle assigns electrons to orbitals

Approximate relative energies of the atomic orbitals up to the 5*s* level are shown in Figure 1-8. With its help, we can give an electronic configuration to every atom in the periodic table. To do so, we follow three rules for assigning electrons to atomic orbitals:

1. Lower energy orbitals are filled before those with higher energy.
2. No orbital may be occupied by more than two electrons, according to the **Pauli*** **exclusion principle.** Furthermore, these two electrons must differ in the orientation of their intrinsic angular momentum, their **spin.** There are two possible directions of the electron spin, usually depicted by vertical arrows pointing in opposite directions. An orbital is filled when it is occupied by two electrons of opposing spin, frequently referred to as **paired electrons.**
3. Degenerate orbitals, such as the p orbitals, are first occupied by one electron each, all of these electrons having the same spin. Subsequently, three more, each of opposite spin, are added to the first set. This assignment is based on **Hund's**[†] **rule.**

With these rules in hand, the determination of electronic configuration becomes simple. Helium has two electrons in the $1s$ orbital and its electronic structure is abbreviated $(1s)^2$. Lithium $[(1s)^2(2s)^1)]$ has one and beryllium $[(1s)^2(2s)^2]$ two additional electrons in the $2s$ orbital. In boron $[(1s)^2(2s)^2(2p)^1]$, we begin the filling of

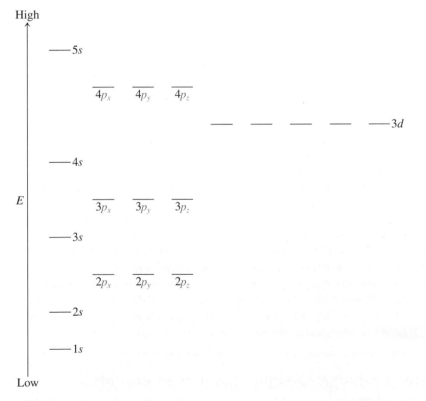

FIGURE 1-8
Approximate relative energies of atomic orbitals, corresponding roughly to the order in which they are filled in atoms. Orbitals of lowest energy are filled first; degenerate orbitals are filled according to Hund's rule.

That the description of electrons as waves is not simply a mathematical construct but is "visibly real" was demonstrated by researchers at IBM in 1993. Using a device called a scanning tunneling microscope, which allows pictures to be taken at the atomic level, they generated this computer-enhanced view of a circle of iron atoms deposited on a copper surface. The image, which they called a "quantum corral," reveals the electrons moving in waves over the surface, the maxima defining the "corral" hovering over the individual iron atoms. [*Photograph courtesy of Dr. Donald Eigler, IBM, San Jose, California.*]

the three degenerate $2p$ orbitals. This pattern continues with carbon and nitrogen, and then the addition of electrons of opposite spin for oxygen, fluorine, and neon fills all p levels. The electronic configurations of four of the elements are depicted in Figure 1-9. Atoms with completely filled sets of atomic orbitals are said to have a **closed-shell configuration.** For example, helium, neon, and argon have this attribute (Figure 1-10). Carbon, in contrast, has an **open-shell configuration.**

The process of adding electrons one by one to the orbital sequence shown in Figure 1-8 is called the **Aufbau principle** (*Aufbau,* German, build up). It is easy to see that the Aufbau principle affords a rationale for the stability of the electron octet and duet. These numbers are required for closed-shell configurations. For helium, the closed-shell configuration is a $1s$ orbital filled with two electrons of opposite spin. In neon, the $2s$ and $2p$ orbitals are occupied by an additional eight electrons; in argon, the $3s$ and $3p$ levels accommodate eight more (Figure 1-10). The availability of $3d$ orbitals for the third-row elements provides an explanation for the phenomenon of valence shell expansion (Section 1-4) and the loosening up of the strict application of the octet rule beyond neon.

EXERCISE 1-11

Using Figure 1-8, draw the electronic configurations of sulfur and phosphorus.

To summarize, the motion of electrons around the nucleus is described by wave equations. Their solutions, atomic orbitals, can be symbolically represented as regions in space, with each point given a positive, negative, or zero (at the node) numerical value, the square of which represents the probability of finding the electron there. The Aufbau principle allows us to assign electronic configurations to all atoms.

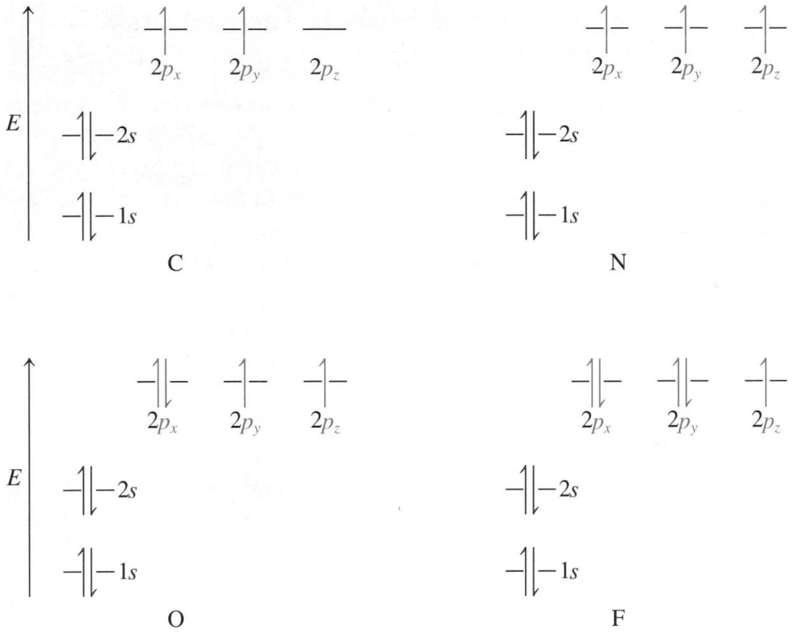

FIGURE 1-9
The most stable electronic configurations of carbon, $(1s)^2(2s)^2(2p)^2$; nitrogen, $(1s)^2(2s)^2(2p)^3$; oxygen, $(1s)^2(2s)^2(2p)^4$; and fluorine $(1s)^2(2s)^2(2p)^5$. Notice that the unpaired electron spins in the p orbitals are in accord with Hund's rule, and the paired electron spins in the filled $1s$ and $2s$ orbitals are in accord with the Pauli principle and Hund's rule. The order of filling the p orbitals has been arbitrarily chosen as p_x, p_y, and then p_z. Any other order would have been equally good.

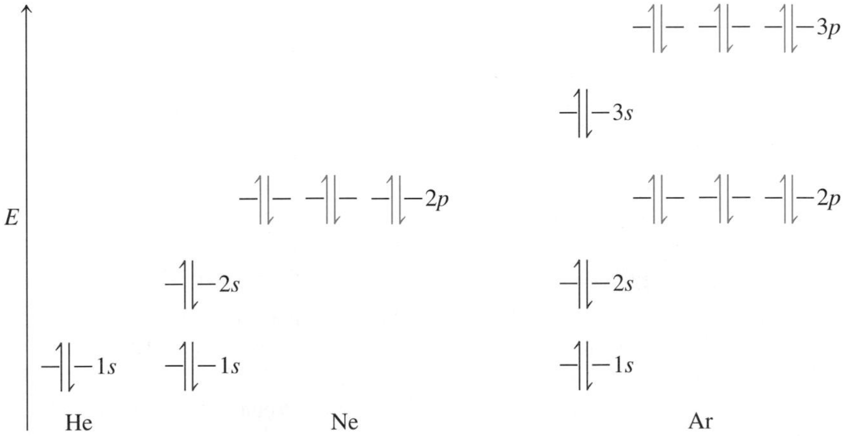

FIGURE 1-10
Closed-shell configurations of the noble gases helium, neon, and argon.

1-7 Molecular Orbitals and Covalent Bonding

We shall now see how covalent bonds are constructed from the overlap of atomic orbitals.

The bond in the hydrogen molecule is formed by the overlap of 1s atomic orbitals

Let us begin by looking at the simplest case: the bond between the two hydrogen atoms in H_2. In a Lewis structure of the hydrogen molecule, we would write the bond as an electron pair shared by both atoms to give each a helium configuration. How do we construct H_2 by using atomic orbitals? An answer to this question was developed by Pauling*: *Bonds are made by the in-phase overlap of atomic orbitals.* What is meant by that? Recall that atomic orbitals are solutions of wave equations. Like waves, they may interact in a reinforcing way (Figure 1-4B) if the overlap is between areas of the wave function of the same sign, or *in phase*. They may also interact in a destructive way if the overlap is between areas of opposite sign, or *out of phase* (Figure 1-4C).

The in-phase overlap of the two 1s orbitals results in a new orbital of lower energy called a **bonding molecular orbital** (Figure 1-11). In the bonding combination, the wave function in the space between the nuclei is strongly reinforced. Thus, the probability of finding the electrons occupying this molecular orbital in that region is very high: a condition for bonding between the two atoms. This picture is strongly reminiscent of that shown in Figure 1-2. The use of two wave functions with *positive* signs for representing the in-phase combination of the two 1s orbitals in Figure 1-11 is arbitrary. Overlap between two *negative* orbitals would give identical results. In other words, it is overlap between *like* lobes that makes a bond, regardless of the sign of the wave function.

On the other hand, out-of-phase overlap between the same two atomic orbitals results in a destabilizing interaction and formation of an **antibonding molecular orbital.** In the antibonding molecular orbital, the amplitude of the wave function is canceled in the space between the two atoms, thereby giving rise to a node.

Thus, the net result of the interaction of the two 1s atomic orbitals of hydrogen is the generation of two molecular orbitals. One is bonding and lower in energy; the

*Professor Linus Pauling (1901–1994), Stanford University, Nobel Prizes 1954 (chemistry) and 1963 (peace).

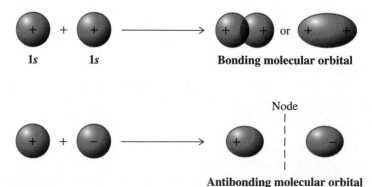

FIGURE 1-11

In-phase (bonding) and out-of-phase (antibonding) combinations of 1s atomic orbitals. The + and − signs denote the *sign* of the wave function, not charges. Electrons in bonding molecular orbitals have a high probability of occupying the space *between* the atomic nuclei, as required for good bonding (compare Figure 1-2). The antibonding molecular orbital has a nodal plane, where the probability of finding electrons is zero. Electrons in antibonding molecular orbitals are most likely to be found *outside* the space between the nuclei and therefore do not contribute to bonding.

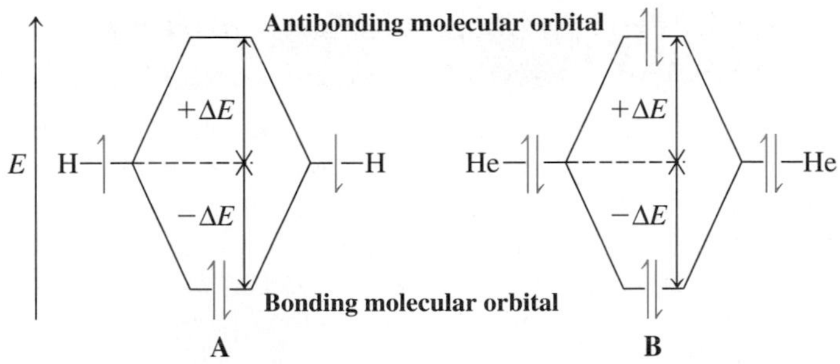

FIGURE 1-12

Schematic representation of the interaction of two (A) singly (as in H_2) and (B) doubly (as in He_2) occupied atomic orbitals to give two molecular orbitals (MO). (Not drawn to scale.) Formation of an H–H bond is favorable because it stabilizes two electrons. Formation of an He–He bond stabilizes two electrons (in the bonding MO) but destabilizes two others (in the antibonding MO). Bonding between He and He thus results in no net stabilization. Therefore, helium is monatomic.

other is antibonding and higher in energy. Because the total number of electrons available to the system is only two, they are placed in the lower energy molecular orbital: the two-electron bond. The result is a decrease in total energy, thereby making H_2 more stable than two free hydrogen atoms. This difference in energy levels corresponds to the strength of the H–H bond. The interaction can be depicted schematically in an energy diagram (Figure 1-12A).

It is now readily understandable why hydrogen exists as H_2, whereas helium is monatomic. The overlap of two filled atomic orbitals, as in helium, leads to bonding and antibonding orbitals, *both of which are filled* (Figure 1-12B). Therefore, making a He–He bond does not decrease the total energy.

The overlap of atomic orbitals gives rise to sigma and pi bonds

The formation of molecular orbitals by an overlap of atomic orbitals applies not only to the $1s$ orbitals of hydrogen, but also to other atomic orbitals. The amount of energy by which the bonding level drops and the antibonding level is raised is called the **energy splitting.** It indicates the strength of the bond being made and depends on a variety of factors. For example, overlap is best between orbitals of similar size and energy. Therefore, two $1s$ orbitals will interact with each other more effectively than a $1s$ and a $3s$.

Geometric factors also affect the degree of overlap. This consideration is important for orbitals with directionality in space, such as p orbitals. Such orbitals give rise to two types of bonds: one in which the atomic orbitals are aligned along the internuclear axis (parts A, B, C, and D in Figure 1-13) and the other in which they are perpendicular to it (part E). The first type is called a **sigma (σ) bond,** the second a **pi (π) bond.** All carbon–carbon single bonds are of the σ type; however, we shall find that double and triple bonds also have π components (Section 1-9).

EXERCISE 1-12

Construct a molecular-orbital and energy-splitting diagram of the bonding in He_2^+. Is it favorable?

FIGURE 1-13

Bonding between atomic orbitals. (A) 1s and 1s (e.g., H_2), (B) 1s and 2p (e.g., HF), (C) 2p and 2p (e.g., F_2), (D) 2p and 3p (e.g., FCl) aligned along internuclear axes, σ bonds; (E) 2p and 2p perpendicular to internuclear axis, a π bond. Note the arbitrary use of + and − signs to indicate in-phase interactions of the wave functions.

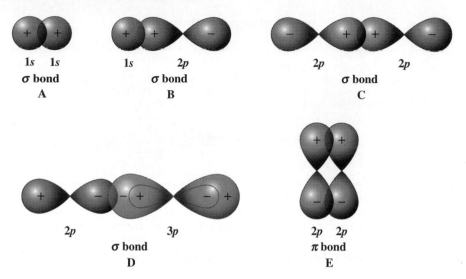

We have come a long way in our description of bonding. First, we thought of bonds in terms of Coulomb forces, then in terms of covalency and shared electron pairs, and now we have a quantum mechanical picture. Bonds are a result of the overlap of atomic orbitals. The two bonding electrons are placed in the bonding molecular orbital. Because it is stabilized relative to the two initial atomic orbitals, energy is given off during bond formation. This decrease in energy represents the bond strength.

1-8 Hybrid Orbitals: Bonding in Complex Molecules

Let us now construct bonding schemes for more complex molecules by using quantum mechanics. How can we use atomic orbitals to build linear (as in BeH_2), trigonal (as in BH_3), and tetrahedral molecules (as in CH_4)?

Mixing orbitals in a single atom gives hybrid orbitals

Consider the molecule beryllium hydride, BeH_2. Beryllium has two electrons in the 1s orbital and two electrons in the 2s orbital. Without unpaired electrons, this arrangement does not appear to allow for bonding.

However, it takes a relatively small amount of energy to promote one electron from the 2s orbital to one of the 2p levels (Figure 1-14), energy to be readily regained by bond formation. Thus, in the $1s^2 2s^1 2p^1$ configuration, there are now two singly filled atomic orbitals available for bonding overlap. One could propose bond formation by overlap of the Be 2s orbital with the 1s orbital of one H, on the one hand, and the Be 2p orbital with the second H, on the other (Figure 1-15). This scheme predicts two different bonds of unequal length, probably at an angle. However, the theory of electron repulsion predicts that compounds such as BeH_2 should have *linear* structures (Section 1-3). Experiments on related compounds confirm this prediction and also show that the bonds to beryllium are of *equal* length.*

*These predictions cannot be tested for BeH_2 itself, which exists as a complex network of Be and H atoms. However, both BeF_2 and $Be(CH_3)_2$ exist as individual molecules in the gas phase and possess the predicted structures.

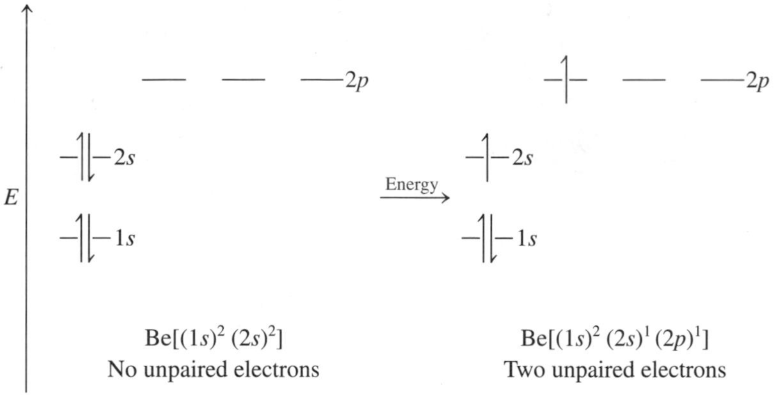

$$Be[(1s)^2 (2s)^2]$$
No unpaired electrons

$$Be[(1s)^2 (2s)^1 (2p)^1]$$
Two unpaired electrons

FIGURE 1-14

Promotion of an electron in beryllium to allow the use of both valence electrons in bonding.

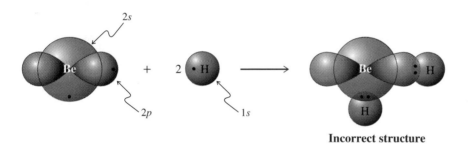

Incorrect structure

FIGURE 1-15

Possible but incorrect bonding in BeH$_2$ by separate use of a 2s and a 2p orbital on beryllium. The node in the former is not shown. Moreover, the other two empty p orbitals and the lower energy filled 1s orbital are omitted for clarity. The dots indicate the valence electrons.

sp Hybrids produce linear structures

How can we explain this geometry in orbital terms? To answer this question, we use a quantum mechanical approach called **orbital hybridization.** Like the mixing of atomic orbitals on different atoms to form molecular orbitals, the mixing of atomic orbitals on the same atom forms new **hybrid orbitals.**

When we mix the 2s and one of the 2p wave functions on beryllium, we obtain two new hybrids, called *sp* orbitals, made up of 50% s and 50% p character. This treatment rearranges the orbital lobes in space, as shown in Figure 1-16 (see page 34). The major parts of the orbitals, also called front lobes, point away from each other at an angle of 180°. There are two additional minor back lobes (one for each *sp* hybrid) with opposite sign. The remaining two p orbitals are left unchanged.

Overlap of the *sp* front lobes with two hydrogen 1s orbitals yields the bonds in BeH$_2$. The 180° angle that results from this hybridization scheme minimizes electron repulsion. The oversized front lobes of the hybrid orbitals also overlap better than do lobes of unhybridized orbitals; the result is improved bonding.

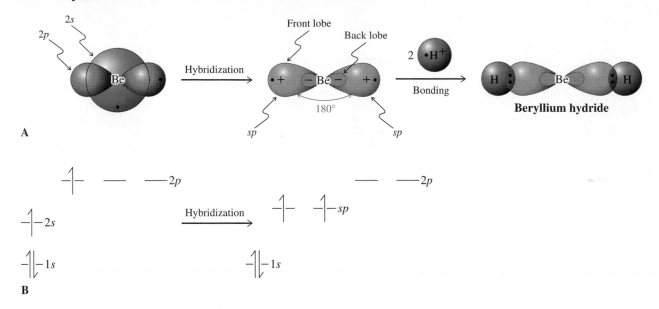

FIGURE 1-16

Hybridization in beryllium to create two sp hybrids. (A) The resulting bonding gives BeH_2 a linear structure. Again, both remaining p orbitals and the $1s$ orbital have been omitted for clarity. The sign of the wave function for the large sp lobes is opposite that for the small lobes. (B) The energy changes occurring on hybridization. The $2s$ orbital and one $2p$ orbital combine into two sp hybrids of intermediate energy. The $1s$ and remaining $2p$ energies remain the same.

Note that hybridization does not change the overall number of orbitals available for bonding. Hybridization of the four orbitals in beryllium gives a new set of four: two sp hybrids and two essentially unchanged $2p$ orbitals. We shall see shortly that carbon uses sp hybrids when it forms triple bonds.

sp^2 Hybrids create trigonal structures

Now let us consider the group of elements in the periodic table with three valence electrons. What bonding scheme can be derived for borane, BH_3? Promotion of a $2s$ electron in boron to one of the $2p$ levels gives the three singly filled atomic orbitals (one $2s$, two $2p$) needed for forming three bonds. Mixing these atomic orbitals creates *three* new hybrid orbitals, which are designated sp^2 to indicate the component atomic orbitals (Figure 1-17). The third p orbital is left unchanged, so the total number of orbitals stays the same—namely, four.

The front lobes of the three sp^2 orbitals of boron overlap the respective $1s$ orbitals of the hydrogen atoms to give trigonal planar BH_3. Again, hybridization minimizes electron repulsion and improves overlap, conditions giving stronger bonds. The remaining unchanged p orbital is perpendicular to the plane incorporating the sp^2 hybrids. It is empty and does not enter significantly into bonding.

The molecule BH_3 is **isoelectronic** with the methyl cation, CH_3^+; that is, they have the same number of electrons. Bonding in CH_3^+ requires three sp^2 hybrid orbitals, and we shall see shortly that carbon uses sp^2 hybrids in double-bond formation.

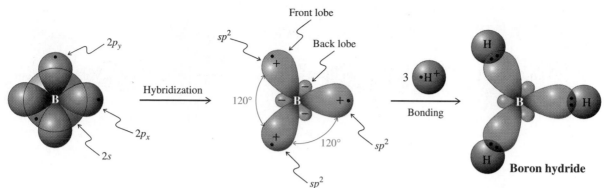

FIGURE 1-17
Hybridization in boron to create three sp^2 hybrids. The resulting bonding gives BH_3 a trigonal planar structure. There are three front lobes of one sign and three back lobes of opposite sign. The remaining p orbital (p_z) is perpendicular to the molecular plane (the plane of the page; one p_z lobe is above, the other below that plane) and has been omitted. In analogy to Figure 1-16B, the energy diagram for the hybridized boron features three singly occupied, equal-energy sp^2 levels and one remaining empty $2p$ level, in addition to the filled $1s$ orbital.

sp^3 Hybridization explains the shape of tetrahedral carbon compounds

Consider the element whose bonding is of most interest to us: carbon. Its electronic configuration is $(1s)^2(2s)^2(2p)^2$, with two unpaired electrons residing in two $2p$ orbitals. Promotion of one electron from $2s$ to $2p$ results in four singly filled orbitals for bonding. We have learned that the arrangement of the four C–H bonds of methane in space that would minimize electron repulsion is tetrahedral (Section 1-3). To be able to achieve this geometry, the $2s$ orbital on carbon is hybridized with *all three* $2p$ orbitals to make *four* equivalent sp^3 orbitals with tetrahedral symmetry, each occupied by one electron. Overlap with four hydrogen $1s$ orbitals furnishes methane with four equal C–H bonds. The HCH bond angles are typical of a tetrahedron: 109.5° (Figure 1-18).

Any combination of atomic and hybrid orbitals may overlap to form bonds. For example, the four sp^3 orbitals of carbon combine with four chlorine $2p$ orbitals to result in tetrachloromethane, CCl_4. Carbon–carbon bonds are generated by overlap of

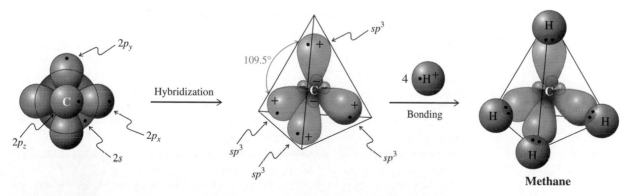

FIGURE 1-18
Hybridization in carbon to create four sp^3 hybrids. The resulting bonding gives CH_4 and other carbon compounds tetrahedral structures. The sp^3 hybrids contain small back lobes of sign opposite that of the front lobes. In analogy to Figure 1-16B, the energy diagram of sp^3-hybridized carbon contains four singly occupied, equal-energy sp^3 levels, in addition to the filled $1s$ orbital.

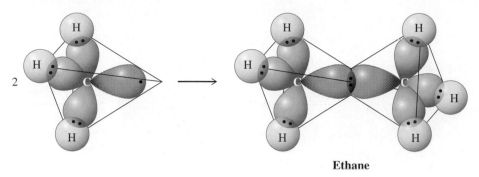

Ethane

FIGURE 1-19 _____
Overlap of two sp^3 orbitals to form the carbon–carbon bond in ethane.

hybrid orbitals. In ethane, CH_3–CH_3 (Figure 1-19), this bond consists of two sp^3 hybrids, one from each of two CH_3 units. Any hydrogen atom in methane and ethane may be replaced by CH_3 or other groups to give new combinations.

In all of these molecules, and countless more, *carbon is approximately tetrahedral.* It is this ability of carbon to form chains of atoms bearing a variety of additional substituents that gives rise to the extraordinary diversity of organic chemistry.

Hybrid orbitals may contain lone electron pairs: ammonia and water

What sort of orbitals describe the bonding in ammonia and water (see Exercise 1-5)? Let us begin with ammonia. The electronic configuration of nitrogen, $(1s)^2(2s)^2(2p)^3$, explains why nitrogen is trivalent, three covalent bonds being needed for octet formation. We could use p orbitals for overlap, leaving the nonbonding electron pair in the $2s$ level. However, this arrangement does not minimize electron repulsion. The best solution is again sp^3 hybridization. Three of the sp^3 orbitals are used to bond to the hydrogen atoms, and the fourth contains the lone electron pair. The HNH bond angles (107.3°) in ammonia are almost tetrahedral (Figure 1-20).

Similarly, the bonding in water is best described by sp^3 hybridization on oxygen. The HOH bond angle is 104.5°, again close to tetrahedral.

The effect of the lone electron pairs explains why the bond angles in NH_3 and H_2O are reduced below the tetrahedral value of 109.5°. Because they are not shared, the lone pairs are relatively close to the nitrogen or oxygen. As a result, they exert increased repulsion on the electrons in the bonds to hydrogen, thereby leading to the observed bond-angle compression.

FIGURE 1-20 _____
Bonding and electron repulsion in ammonia and water. The arcs indicate increased electron repulsion by the lone pairs located close to the central nucleus.

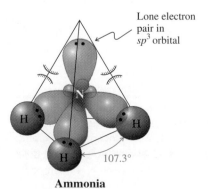

Lone electron pair in sp^3 orbital

107.3°

Ammonia

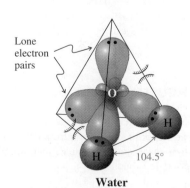

Lone electron pairs

104.5°

Water

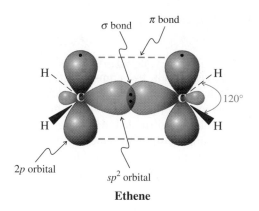

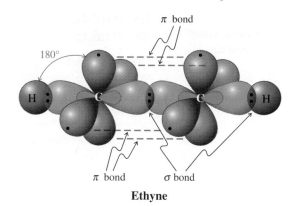

Ethene   **Ethyne**

FIGURE 1-21

The double bond in ethene (ethylene) and the triple bond in ethyne (acetylene).

Pi bonds are present in ethene (ethylene) and ethyne (acetylene)

The double bond in alkenes, such as ethene (ethylene), and the triple bond in alkynes, such as ethyne (acetylene), are the result of the ability of the atomic orbitals of carbon to adopt sp^2 and sp hybridization, respectively. Thus, the σ bonds in ethene are derived entirely from carbon-based sp^2 hybrid orbitals: Csp^2–Csp^2 for the C–C bond, and Csp^2–H1s for holding the four hydrogens (Figure 1-21). In contrast with BH_3, with an empty p orbital on boron, the leftover unhybridized p orbitals on the ethene carbons are occupied by one electron each, overlapping to form a π bond (recall Figure 1-13E). In ethyne, the σ frame is made up of bonds consisting of Csp hybrid orbitals. The arrangement leaves *two* singly occupied p orbitals on each carbon and allows the formation of two π bonds (Figure 1-21).

EXERCISE 1-13

Draw a scheme for the hybridization and bonding in methyl cation, CH_3^+, and methyl anion, CH_3^-.

In summary, to minimize electron repulsion and maximize bonding in triatomic and larger molecules, we apply the concept of atomic-orbital hybridization to construct orbitals of appropriate shape. Combinations of s and p atomic orbitals create hybrids. Thus, a 2s and a 2p orbital mix to furnish two linear sp hybrids, the remaining two p orbitals being unchanged. Combination of the 2s with two p orbitals gives three sp^2 hybrids used in trigonal molecules. Finally, mixing the 2s with all three p levels results in the four sp^3 hybrids that produce the geometry around tetrahedral carbon.

1-9 Structures and Formulas of Organic Molecules

A good understanding of the nature of bonding allows us to learn how chemists determine the identity of organic molecules and depict their structures. Do not underestimate the importance of the latter task. Sloppiness in drawing molecules had been the source of many errors in the literature and, perhaps of more immediate concern, in organic chemistry examinations.

Ethanol and Methoxymethane: Two Isomers

H—C—C—O—H (with H H above and H H below the carbons)

Ethanol
(b.p. 78.5°C)

H—C—O—C—H (with H above and H below each carbon)

Methoxymethane
(Dimethyl ether)
(b.p. −23°C)

To establish the identity of a molecule, we determine its structure

Organic chemists have many diverse techniques at their disposal with which to determine molecular structure. **Elemental analysis** reveals the **empirical formula,** which summarizes the kinds and ratios of the elements present. However, other procedures are usually needed to determine the molecular formula and to distinguish between structural alternatives. For example, the molecular formula C_2H_6O corresponds to *two* known substances: ethanol and methoxymethane (dimethyl ether). But we can tell them apart on the basis of their physical properties—for example, their melting points, boiling points (b.p.), refractive indices, specific gravities, and so forth. Thus ethanol is a liquid (b.p. 78.5°C) commonly used as a laboratory and industrial solvent and present in alcoholic beverages. In contrast, methoxymethane is a gas (b.p. −23°C) used as a refrigerant in place of Freon. Their other physical and chemical properties differ as well. Molecules such as these, which have the same molecular formula but differ in the sequence (**connectivity**) in which the atoms are held together, are called **constitutional** or **structural isomers.**

EXERCISE 1-14

Construct as many isomers with the molecular formula C_4H_{10} as you can.

Two naturally occurring substances illustrate the biological consequences of such structural differences. Prostacyclin I_2 prevents blood inside the circulatory system from clotting. Thromboxane A_2, which is released when bleeding occurs, *induces* platelet aggregation, causing clots to form over wounds. Incredibly, these compounds are constitutional isomers (both have the molecular formula $C_{20}H_{32}O_5$) with only relatively minor connectivity differences. Indeed, they are so closely related that they are synthesized in the body from a common starting material (see Section 19-13 for details).

When a compound is isolated in nature or from a reaction, a chemist may attempt to identify it by matching its properties with those of known materials. Suppose, however, that the chemical under investigation is new. In this case, structural elucidation requires the use of other methods, most of which are various forms of spectroscopy. These methods will be dealt with and applied often in later chapters.

The most complete methods for structure determination are X-ray diffraction of single crystals and electron diffraction or microwave spectroscopy of gases. These techniques reveal the exact position of every atom, as if viewed under very powerful magnification. The structural details that emerge in this way for the two isomers ethanol and methoxymethane are depicted in the form of ball-and-stick models in Figure 1-22A and B. Note the tetrahedral bonding around the carbon atoms and the bent arrangement of the bonds to oxygen, which is hybridized as in water. A more

FIGURE 1-22 —————

Three-dimensional representations of (A) ethanol and (B) methoxymethane, depicted by ball-and-stick molecular models. Bond lengths are given in angstrom units, bond angles in degrees. (C) Space-filling rendition of methoxymethane, taking into account the effective size of the electron "clouds" around the component nuclei.

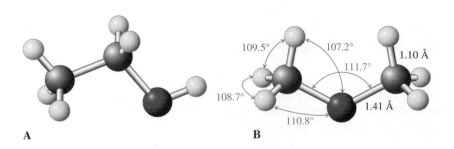

A   B

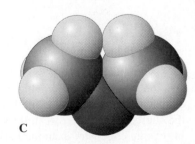

C

accurate picture of the actual size of methoxymethane and its component units is given in Figure 1-22C, a space-filling model.

The perception of organic molecules in three dimensions is essential for understanding their structures and, frequently, their reactivities. You may find it difficult to visualize the spatial arrangements of the atoms in even very simple systems. A good aid is a molecular model kit. You should acquire one and practice the assembly of organic structures. To encourage you in this practice and to indicate particularly good examples where building a molecular model can help you, the icon displayed in the margin will appear at the appropriate places in the text.

EXERCISE 1-15

Repeat Exercise 1-14, using your molecular model kit to construct as many isomers with the molecular formula C_4H_{10} as you can. Draw each isomer.

Several types of drawings are used to represent molecular structures

The representation of molecular structures is not new to us. It was first addressed in Section 1-4, which outlined rules for drawing Lewis structures. We learned that bonding and nonbonding electrons are depicted as dots. A simplification is the straight-line notation (Kekulé structure), with lone pairs (if present) added again as dots. To simplify even further, chemists use **condensed formulas** in which most single bonds and lone pairs have been omitted. The main carbon chain is written horizontally, the attached hydrogens usually to the right of the associated carbon atom. Other groups (the **substituents** on the main stem) are added through connecting vertical lines.

The most economical notation of all is the **bond-line formula.** It portrays the carbon frame by zigzag straight lines, omitting all hydrogen atoms. Each terminus represents a methyl group and each apex a carbon atom.

Kekulé	Condensed	Bond-Line Formulas

Kekulé structures, condensed formulas, and bond-line formulas:

$CH_3CH_2CH_3$

$CH_3CHCH_2CH_2Br$ (with Br substituent)

$CH_3CCH{=}CH_2$ (with O)

$HC{\equiv}CCH_2OH$

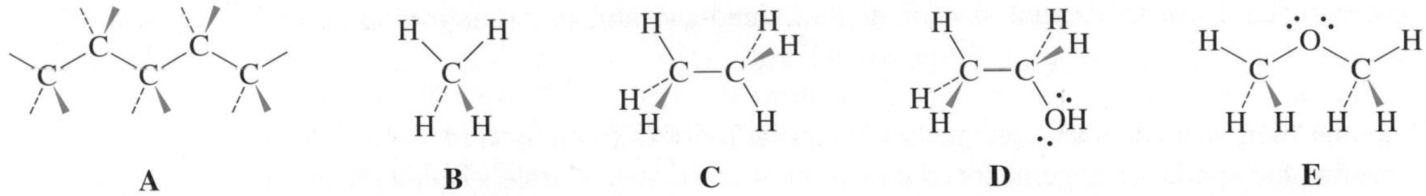

A	B	C	D	E

FIGURE 1-23 _____

Dashed (red) and wedged (blue) line notation for (A) a carbon chain; (B) methane; (C) ethane; (D) ethanol; and (E) methoxymethane. Atoms attached by ordinary straight lines lie in the plane of the page. Groups at the ends of dashed lines lie below that plane; groups at the ends of wedges lie above it.

EXERCISE 1-16

Draw condensed and bond-line formulas for each C_4H_{10} isomer.

Figure 1-22 calls attention to a problem: How can we draw the three-dimensional structures of organic molecules accurately, efficiently, and in accord with generally accepted conventions? For tetrahedral carbon, this problem is solved by the **dashed-wedged line notation.** It uses a zigzag convention to depict the main carbon chain, now defined to lie *in the plane* of the page. Each apex (carbon atom) is then connected to two additional lines, one dashed and one wedged, both pointing away from the chain. These represent the remaining two bonds to carbon; the dashed line corresponds to the bond that lies *below the plane* of the page and the wedged line to that lying *above that plane* (Figure 1-23). Substituents are placed at the appropriate termini. This convention is applied to molecules of all sizes, even methane (see Figure 1-23B–E).

EXERCISE 1-17

Draw dashed-wedged line formulas for each C_4H_{10} isomer.

In summary, determination of organic structures relies on the use of several experimental techniques, including elemental analysis and various forms of spectroscopy. Molecular models are useful aids for the visualization of the spatial arrangements of the atoms in structures. Condensed and bond-line notations are useful shorthand approaches to drawing two-dimensional representations of molecules, whereas dashed-wedged line formulas provide a means of depicting the atoms and bonds in three dimensions.

CHAPTER INTEGRATION PROBLEM

Propyne can be deprotonated twice with very strong base (i.e., the base removes two protons) to give a dianion.

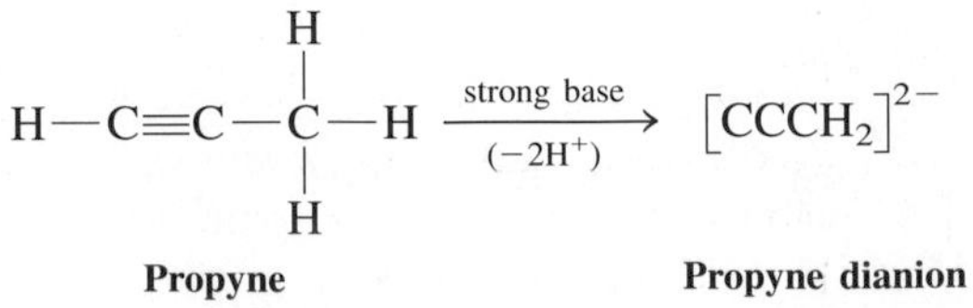

Propyne   **Propyne dianion**

Two resonance forms can be constructed in which all three carbons have Lewis octets.

a. Draw both structures and indicate which is the more important one.

SOLUTION

Let us analyze the problem one step at a time:

Step 1. What structural information is embedded in the picture given for propyne dianion? *Answer (Section 1-4, Rule 1):* The picture shows the connectivity of the atoms: a chain of three carbons, one of the terminal atoms bearing two hydrogens.

Step 2. How many valence electrons are available? *Answer (Section 1-4, Rule 2):*

$$
\begin{array}{llll}
2\,H & = & 2 \times 1 \text{ electron} & = \ \ 2 \text{ electrons} \\
3\,C & = & 3 \times 4 \text{ electrons} & = 12 \text{ electrons} \\
\underline{\text{Charge} = -2} & & & = \ \ 2 \text{ electrons} \\
\text{Total} & & & \ \ 16 \text{ electrons}
\end{array}
$$

Step 3. How do we get a Lewis octet structure for this ion? *Answer (Section 1-4, Rule 3):* Using the connectivity given in the structure for propyne dianion, we can immediately dispose of eight of the available electrons:

$$
\begin{matrix}
& & \text{H} \\
& & \overset{\cdot\cdot}{} \\
\text{C} : \text{C} : & \text{C} : \text{H}
\end{matrix}
$$

   Now, let us use the remaining eight electrons in the form of lone electron pairs to give as many carbons as possible octet surroundings. A good place to start is at the right, because that carbon requires only two electrons for this purpose, the center carbon needs two lone pairs, and, finally, the carbon at the left has to make do (for the time being) with one additional pair of electrons:

$$
: \text{C} : \overset{\cdot\cdot}{\underset{\cdot\cdot}{\text{C}}} : \overset{\overset{\text{H}}{\cdot\cdot}}{\text{C}} : \text{H}
$$

This structure leaves the carbon at the left with only four electrons. Thus, we have to change the two lone pairs at the center into two shared pairs, furnishing the following dot structure:

$$
: \text{C} ::: \text{C} : \overset{\overset{\text{H}}{\cdot\cdot}}{\underset{\cdot\cdot}{\text{C}}} : \text{H}
$$

Step 4. Now every atom has its duet or octet satisfied, but we still have to concern ourselves with charges. What are the charges on each atom? *Answer (Section 1-4, Rule 4):* Starting again at the right, we can quickly ascertain that the hydrogens are charge neutral. Each is attached to carbon through a shared electron pair, giving it an effective electron count of one, the same as in a free, neutral hydrogen atom. On the other hand, the carbon atom bears three shared electron pairs and one lone pair, thus having an effective electron count of five, one more electron than the number associated with the neutral nucleus. Hence one of the negative charges is located on the carbon at the right. The central carbon nucleus is surrounded by four shared electron pairs and is therefore neutral. Finally, the carbon at the left is attached to its neighbor by three shared pairs and has, in addition, two unbound electrons, giving it the other negative charge. The result is

$$\overset{\displaystyle H}{\underset{}{:\overset{..}{\underset{..}{C}}:::C:\overset{..}{C}:H}}$$

Step 5. We can now address the question of resonance forms. Is it possible to move pairs of electrons in such a way as to generate another Lewis octet form? *Answer (Section 1-5):* Yes, by shifting the lone pair on the carbon at the right into a sharing position and at the same time moving one of the three shared pairs to the left into an unshared location:

$$\left[\; :\bar{C}:::C:\overset{\overset{\displaystyle H}{..}}{C}:H \longleftrightarrow \; :\overset{2-}{C}::C::\overset{\overset{\displaystyle H}{..}}{\underset{..}{C}}:H \; \right]$$

The consequence of this movement is the transfer of the negative charge from the carbon at the right to the carbon at the left, the latter therefore becoming doubly negative.

Step 6. Which one of the two resonance pictures is more important? *Answer (Section 1-5):* Electron repulsion makes the structure at the left a more important resonance contributor.

A final point: You could have derived the first resonance structure much more quickly by considering the information given in the reaction scheme leading to the dianion. Thus, the bond-line formula of propyne represents its Lewis structure, and the process of removing a proton each from the respective terminal carbons leaves these carbons with two lone electron pairs and the associated charges:

$$H-C\equiv C\underset{\underset{\displaystyle H}{|}}{\overset{\overset{\displaystyle H}{|}}{C}}-H \xrightarrow{-2H^+} :\bar{C}\equiv C-\overset{\overset{\displaystyle H}{|}}{\underset{..}{C}}-H$$

The important lesson to be learned from this final point is that, whenever you are confronted with a problem, you should take the time to complete an inventory (write it down) of all the information given explicitly or implicitly in stating the problem.

b. Provide an orbital drawing of propyne dianion, given the following hybridization: $[CspCspCsp^2H_2]^{2-}$.

SOLUTION
You can construct an orbital picture simply by attaching one half (the CH_2 group) of the rendition of ethene in Figure 1-21 to that of ethyne without its hydrogens.

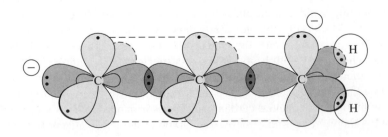

You can clearly see how the doubly filled p orbital of the CH_2 group enters into overlap with one of the π bonds of the alkyne unit, allowing the charge to delocalize as represented by the two resonance structures.

IMPORTANT CONCEPTS

1. Organic chemistry is the chemistry of **carbon** and its compounds.
2. **Coulomb's law** relates the attractive force between particles of opposite electrical charge to the distance between them.
3. **Ionic bonds** result from Coulombic attraction of oppositely charged ions. These ions are formed by the complete transfer of electrons from one atom to another, typically to achieve a noble-gas configuration.
4. **Covalent bonds** result from electron sharing between two atoms. Electrons are shared to allow the atoms to attain noble-gas configurations.
5. **Bond length** is the average distance between two covalently bonded atoms. Bond formation releases energy; bond breaking requires energy.
6. **Polar bonds** are formed between atoms of differing electronegativity (a measure of an atom's ability to attract electrons).
7. The **shape of molecules** is strongly influenced by electron repulsion.
8. **Lewis structures** describe bonding by the use of valence electron dots. They are drawn so as to give hydrogen an electron duet, the other atoms electron octets (octet rule). Charge separation should be minimized but may be enforced by the **octet rule.**
9. When two or more Lewis structures differing only in the positions of the electrons are needed to describe a molecule, they are called **resonance forms.** None correctly describes the molecule, its true representation being an average (**hybrid**) of all its Lewis structures. If the resonance forms of a molecule are unequal, those which best satisfy the rules for writing Lewis structures and the electronegativity requirements of the atoms are more important.
10. The **de Broglie relation** describes the wavelength of a moving particle in terms of its mass and velocity.
11. The motion of electrons around the nucleus can be described by **wave equations.** The solutions to these equations are **atomic orbitals,** which roughly delineate regions in space in which there is a high probability of finding electrons.
12. An *s* **orbital** is spherical; a *p* **orbital** looks like two touching teardrops or a "spherical figure eight." The

mathematical sign of the orbital at any point can be positive, negative, or zero (node). With increasing energy, the number of nodes increases. Each orbital can be occupied by a maximum of two electrons of opposite spin (**Pauli exclusion principle, Hund's rule).**
13. The process of adding electrons one by one to the atomic orbitals, starting with those of lowest energy, is called the **Aufbau principle.**
14. A **molecular orbital** is formed when two atomic orbitals overlap to generate a bond. Atomic orbitals of the same sign overlap to give a **bonding molecular orbital** of lower energy. Atomic orbitals of opposite sign give rise to an **antibonding molecular orbital** of higher energy and containing a node. The number of molecular orbitals equals the number of atomic orbitals from which they derive.
15. Bonds made by overlap along the internuclear axis are called σ **bonds**; those made by overlap of p orbitals perpendicular to the internuclear axis are called π **bonds.**
16. The mixing of orbitals on the same atom results in new **hybrid orbitals** of different shape. One s and one p orbital mix to give two **linear sp hybrids,** used, for example, in the bonding of BeH_2. One s and two p orbitals result in three **trigonal sp^2 hybrids,** used, for example, in BH_3. One s and three p orbitals furnish four **tetrahedral sp^3 hybrids,** used, for example, in CH_4. The orbitals that are not hybridized stay unchanged. Hybrid orbitals may overlap with each other. Overlapping sp^3 hybrid orbitals on different carbon atoms form the carbon–carbon bonds in ethane and other organic molecules. Hybrid orbitals may also be occupied by lone electron pairs, as in NH_3 and H_2O.
17. The composition (i.e., ratios of types of atoms) of organic molecules is revealed by **elemental analysis. The molecular formula** gives the number of atoms of each kind.
18. Molecules that have the same molecular formula but different connectivity order of their atoms are called **constitutional** or **structural isomers.** They have different properties.
19. **Condensed** and **bond-line formulas** are abbreviated representations of molecules. **Dashed-wedged line drawings** illustrate molecular structures in three dimensions.

PROBLEMS

18. Draw a Lewis structure for each of the following molecules and assign charges where appropriate. The order in which the atoms are connected is given in parentheses.

$$O$$

(a) ClF
(b) BrCN
(c) $SOCl_2$ (ClSCl)
(d) CH_3NH_2
(e) CH_3OCH_3
(f) N_2H_2 (HNNH)
(g) CH_2CO
(h) HN_3 (HNNN)
(i) N_2O (NNO)

19. Using electronegativity values from Table 1-2 (in Section 1-3), identify polar covalent bonds in several of the structures in Problem 18 and label the atoms δ^+ and δ^-, as appropriate.

20. Draw a Lewis structure for each of the following species. Again, assign charges where appropriate.

(a) H^-
(b) CH_3^-
(c) CH_3^+
(d) CH_3
(e) $CH_3NH_3^+$
(f) CH_3O^-
(g) CH_2
(h) HC_2^- (HCC)
(i) H_2O_2 (HOOH)

21. Each of the following structures contains at least one error, such as an incorrect number of electrons or bonds on one or more atoms or a violation of the octet rule. Identify the errors and suggest corrections for them.

(a) H_2OCH_3
(b) H_2COH
(c) $:CH_4$
(d) NH_3OH

$$\begin{array}{c} CH_3 \\ | \\ \text{(e) } CH_3-CH_2-CH_3 \end{array}$$

(f) $CH_3-CH_2=CH$

$$\begin{array}{c} H \\ | \\ \text{(g) } O=N=O \end{array}$$

(h) $H_2C\equiv CH_2$

$$\text{(i)} \quad \begin{array}{c} H \quad\quad H \\ | \quad\quad | \\ H-C-H-C-H \\ | \quad\quad | \\ H \quad\quad H \end{array}$$

(j) $C=O=O$

22. **(a)** The structure of the bicarbonate (hydrogen carbonate) ion, HCO_3^-, is best described as a hybrid of several contributing resonance forms, two of which are shown here.

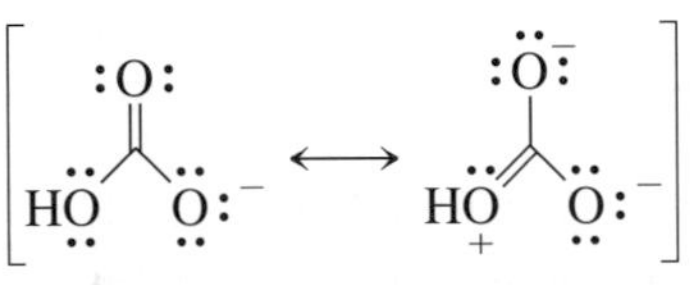

(i) Draw at least one additional resonance form. (ii) Using curved "electron pushing" arrows, show how these Lewis structures may be interconverted by movement of electron pairs. (iii) Determine which form or forms will be the major contributor(s) to the real structure of bicarbonate, explaining your answer on the basis of the criteria in Section 1-5.

(b) Draw two resonance forms for formaldehyde oxime, H_2CNOH. As in parts (ii) and (iii) of (a), use curved arrows to interconvert the resonance forms and determine which form is the major contributor.

(c) Repeat the exercises in (b) for formaldehyde oximate anion, $[H_2CNO]^-$.

23. Several of the compounds in Problem 18 can have resonance forms. Identify these molecules and write an additional resonance Lewis structure for each. In each case, indicate the major contributor to the resonance hybrid.

24. Draw two or three resonance forms for each of the following species. Indicate the major contributor or contributors to the hybrid in each case.

(a) OCN^-
(b) CH_2CHNH^-
(c) $HCONH_2$ ($HCNH_2$)
$\overset{O}{}$
(d) O_3 (OOO)
(e) $CH_2CHCH_2^-$
(f) SO_2 (OSO)
(g) $HOCHNH_2^+$
(h) CH_3CNO

25. Use a molecular-orbital analysis to predict which species in each of the following pairs has the stronger bonding between atoms. (**Hint:** Refer to Figure 1-12).
(a) H_2 or H_2^+
(b) He_2 or He_2^+
(c) O_2 or O_2^+
(d) N_2 or N_2^+

26. Describe the hybridization of each carbon atom in each of the following structures. Base your answer on the geometry about the carbon atom.
(a) CH_3Cl (b) CH_3OH (c) $CH_3CH_2CH_3$ (d) $CH_2{=}CH_2$ (trigonal carbons)

(e) $HC{\equiv}CH$ (linear structure)   (f) 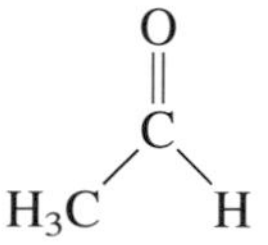   (g)

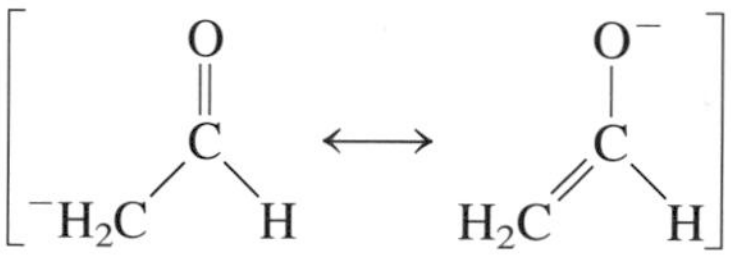

27. Depict the following condensed formulas in Kekulé (straight-line) notation.

(a) CH_3CN (b) 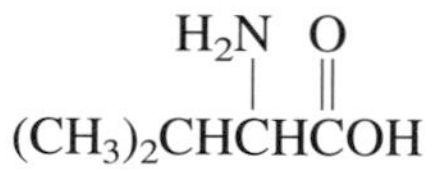 $(CH_3)_2CHCHCOH$ (c) 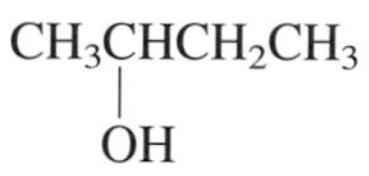 $CH_3CHCH_2CH_3$

(d) $CH_2BrCHBr_2$ (e) 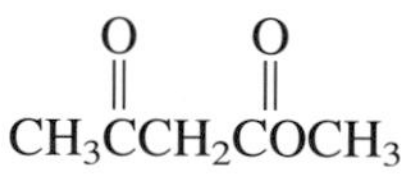 $CH_3CCH_2COCH_3$ (f) $HOCH_2CH_2OCH_2CH_2OH$

28. Convert the following bond-line formulas into Kekulé (straight-line) structures.

(a) 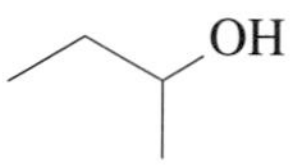   (b) 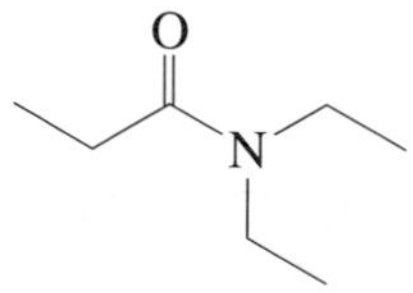   (c)

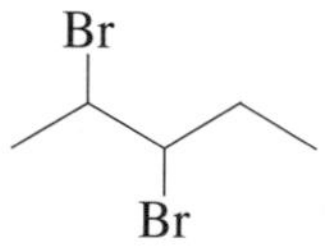

(d) 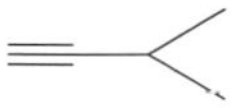   (e)    (f)

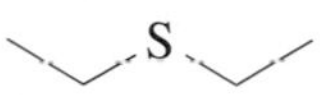

29. Convert the following dashed-wedged line formulas into condensed formulas.

(a) 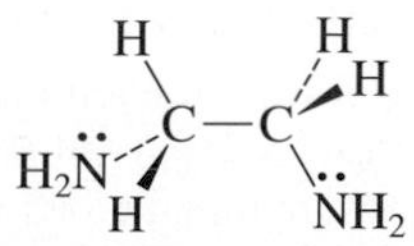   (b) 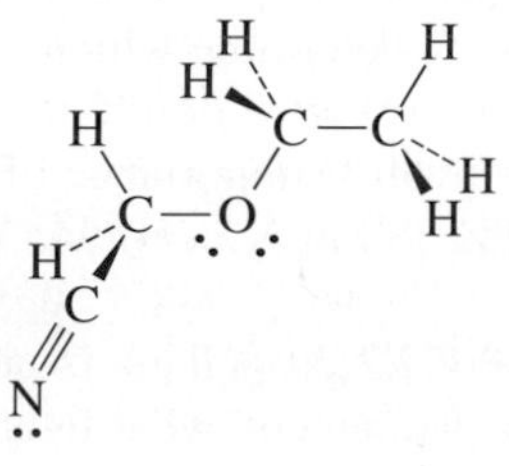   (c)

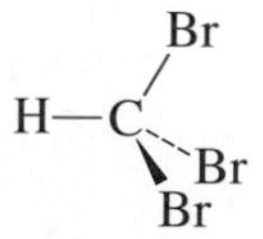

30. Depict the following straight-line formulas in their condensed form.

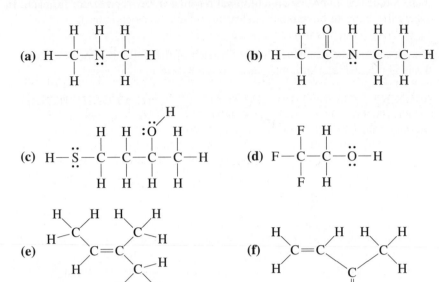

31. Redraw the structures depicted in Problems 27 and 30 using bond-line formulas.

32. Convert the following condensed formulas into dashed-wedged line structures.

(a) CH_3CHOCH_3 with CN above
(b) $CHCl_3$
(c) $(CH_3)_2NH$
(d) $CH_3CHCH_2CH_3$ with SH above

33. Construct as many isomers of each molecular formula as you can for
(a) C_5H_{12}; (b) C_3H_8O. Draw both condensed and bond-line formulas for each isomer.

34. Draw condensed formulas showing the multiple bonds, charges, and lone electron pairs (if any) for each molecule in the following pairs of constitutional isomers. (**Hint:** First make sure that you can draw a proper Lewis structure for each molecule.) Do any of these pairs consist of resonance forms?

(a) $HCCCH_3$ and H_2CCCH_2
(b) CH_3CN and CH_3NC
(c) CH_3CH (with O above) and H_2CCHOH

35. Two resonance forms can be written for a bond between trivalent boron and an atom with a lone pair of electrons. (a) Formulate them for (i) $(CH_3)_2BN(CH_3)_2$; (ii) $(CH_3)_2BOCH_3$; (iii) $(CH_3)_2BF$. (b) Using the guidelines in Section 1-5, determine which form in each pair of resonance forms is more important. (c) How do the electronegativity differences between N, O, and F affect the relative importance of the resonance forms in each case? (d) Predict the hybridization of N in (i) and O in (ii).

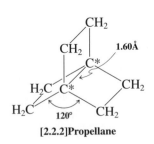

[2.2.2]Propellane

36. The unusual molecule [2.2.2]propellane is pictured in the margin. On the basis of the given structural parameters, what hybridization scheme best describes

the carbons marked by asterisks? (Make a model to help you visualize its shape.) What types of orbitals are used in the bond between them? Would you expect this bond to be stronger or weaker than an ordinary carbon–carbon single bond (which is usually 1.54 Å long)?

37. (a) On the basis of the information in Problem 26, give the likely hybridization of the orbital that contains the unshared pair of electrons (responsible for the negative charge) in each of the following species: $CH_3CH_2^-$; $CH_2=CH^-$; $HC\equiv C^-$. **(b)** Electrons in sp, sp^2, and sp^3 orbitals do not have identical energies. Because the $2s$ orbital is lower in energy than a $2p$, the more s character a hybrid orbital has, the lower its energy will be. Therefore the sp^3 ($\frac{1}{4}s$ and $\frac{3}{4}p$ in character) is highest in energy, and the sp ($\frac{1}{2}s$, $\frac{1}{2}p$) lowest. Use this information to determine the relative abilities of the three anions in (a) to accommodate the negative charge. **(c)** The strength of an acid HA is related to the ability of its conjugate base A^- to accommodate negative charge. In other words, the ionization $HA \rightleftharpoons H^+ + A^-$ is favored for a more stable A^-. Although CH_3CH_3, $CH_2=CH_2$, and $HC\equiv CH$ are all weak acids, they are not equally so. On the basis of your answer to (b), rank them in order of acid strength.

38. A number of substances containing positively polarized carbon atoms have been labeled as "cancer suspect agents" (i.e., suspected carcinogens or cancer-inducing compounds). It has been suggested that the presence of such carbon atoms is responsible for the carcinogenic properties of these molecules. Assuming that the extent of polarization is proportional to carcinogenic potential, how would you rank the following compounds with regard to cancer-causing potency?

(a) CH_3Cl **(b)** $(CH_3)_4Si$ **(c)** $ClCH_2OCH_2Cl$
(d) CH_3OCH_2Cl **(e)** $(CH_3)_3C^+$

(Note: Polarization is only one of the many factors known to be related to carcinogenicity. Moreover, none of them shows the type of straightforward correlation implied in this question.)

39. The structure of the substance lynestrenol, a component of certain oral contraceptives, is presented below. Locate an example of each of the following types of bonds or atoms: **(a)** a highly polarized covalent bond; **(b)** a nearly unpolarized covalent bond; **(c)** an sp-hybridized carbon atom; **(d)** an sp^2-hybridized carbon atom; **(e)** an sp^3-hybridized carbon atom; **(f)** a bond between atoms of different hybridization.

Lynestrenol

Team Problems

Team problems are meant to encourage discussion and collaborative learning. Try to solve team problems with a partner or small study group. Notice that the problems are divided into parts. Rather than tackling each part individually, discuss each section of the problem together. Try out the vocabulary that you learned in the chapter to question and convince yourselves that you are on the right track, before you move to the next part. In general, the more you use the terms and apply the concepts presented in the text, the better you will become at correlating molecular structure and reactivity and thus visualizing bond breaking and bond making. You will begin to see the elegant patterns of organic chemistry and will not be a slave to memorization. The collaborative process used in partner or group study will force you to articulate your ideas. Talking out a solution with an "audience" instead of to yourself builds in checks and balances. Your teammates will not let you get away with, "Well, you know what I mean," because they probably do not. You become responsible to others as well as yourself. By learning from and teaching others, you solidify your own understanding.

40. Consider the following reaction:

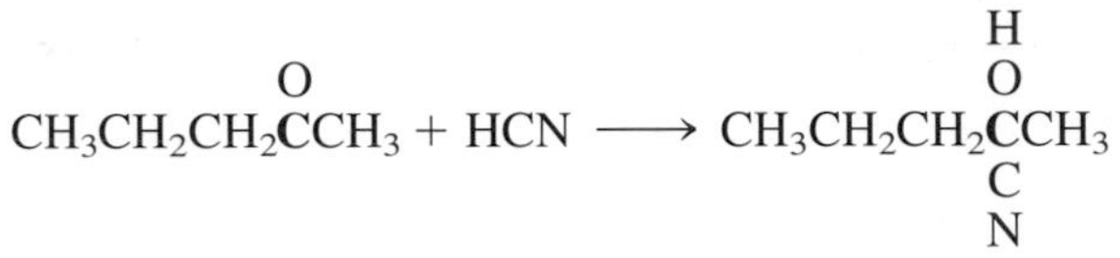

$$CH_3CH_2CH_2CCH_3 + HCN \longrightarrow CH_3CH_2CH_2\underset{\underset{N}{\overset{\overset{\displaystyle H}{\overset{\displaystyle O}{|}}}{C}}{C}CH_3$$

A   B

(a) Draw these condensed formulas as Lewis dot structures. Label the geometry and hybridization of the bold carbons in compounds A and B. Did the hybridization change in the course of the reaction?
(b) Draw the condensed formulas as bond-line structures.
(c) Examine the components of the reaction in light of bond polarity. Using the notation for partial charge separation, δ^+ and δ^-, indicate, on the bond-line structures, any polar bonds.
(d) This reaction is actually a two-step process: cyanide attack followed by protonation. Depict these processes by using the same "electron pair pushing" technique that we employed for resonance structures in Section 1-5, but now show the flow of electrons for the two steps. Clearly position the beginning (an electron pair) and the end (a positively polarized or charged nucleus) of your arrows.

Preprofessional Problems

Preprofessional problems are included to give you practice solving the type of problems found on exams required for entry into professional schools, such as the MCAT, DAT, chemistry GRE, and ACS exams, as well as on many undergraduate tests. Do these multiple-choice questions as you go through this course and then return to them before you take a professional school exam. These questions are to be answered "closed-book"—that is, no periodic table, calculator, or the like.

41. A certain organic compound was found on combustion analysis to contain 84% carbon and 16% hydrogen (C = 12.0, H = 1.00). A molecular formula for the compound could be
(a) CH_4O (b) $C_6H_{14}O_2$ (c) C_7H_{16}
(d) C_6H_{10} (e) $C_{14}H_{22}$

42. The compound

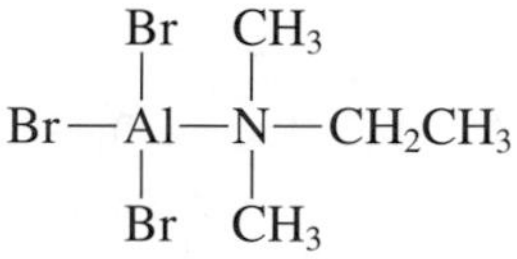

$$\begin{array}{ccc} & Br & CH_3 \\ & | & | \\ Br-&Al-N&-CH_2CH_3 \\ & | & | \\ & Br & CH_3 \end{array}$$

has a formal charge of

(a) −1 on N   (b) +2 on N   (c) −1 on Al

(d) +1 on Br   (e) none of the above

43. The arrow in the structure points to a bond that is formed by 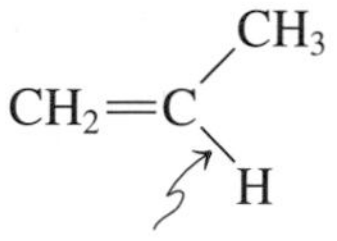
    (a) overlap of an s orbital on H and an sp^2 orbital on C
    (b) overlap of an s orbital on H and an sp orbital on C
    (c) overlap of an s orbital on H and an sp^3 orbital on C
    (d) none of the above

$$CH_2{=}C\begin{array}{c} CH_3 \\ \\ H \end{array}$$

44. Which compound has bond angles nearest to 120°?
    (a) $O{=}C{=}S$ (b) CHI_3 (c) $H_2C{=}O$
    (d) $H-C{\equiv}C-H$ (e) CH_4

45. The pair of structures that are resonance hybrids is
    (a) $H\overset{\cdot\cdot}{\underset{\cdot\cdot}{O}}-\overset{+}{C}HCH_3$ and $H\overset{\cdot\cdot}{\underset{\cdot\cdot}{O}}{=}\overset{+}{C}HCH_3$

    (b) 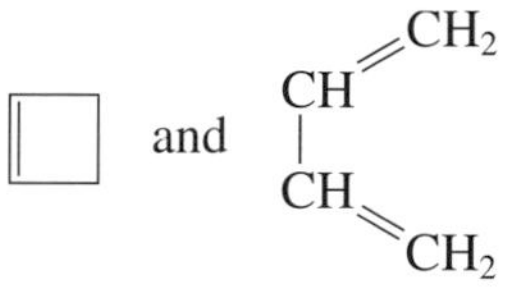 □ and $\begin{array}{c} CH{=}CH_2 \\ | \\ CH{=}CH_2 \end{array}$

    (c) $CH_3\overset{:O:}{\overset{\|}{C}}H$ and $CH_2{=}\overset{:\overset{\cdot\cdot}{O}-H}{C}H$

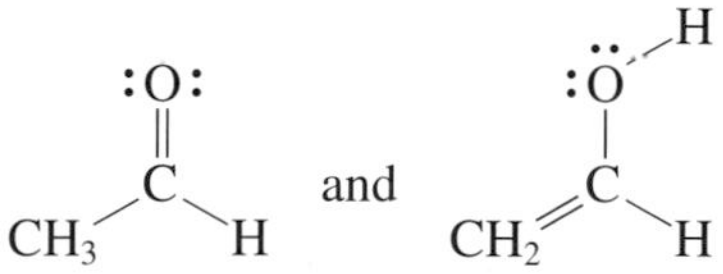

    (d) $CH_3\overset{+}{C}H_2$ and $\overset{+}{C}H_2CH_3$

Alkanes

Molecules Lacking Functional Groups

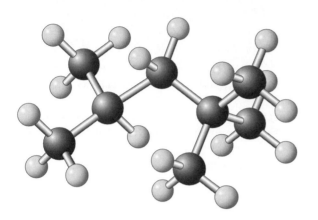

Different blends of alkanes and other additives give rise to gasolines with different octane number ratings.

Turn to page 89 of this book and look at the structures of the molecules illustrated in Problem 29. Each one contains a variety of types of bonds between various elements. Can we predict what kinds of chemical reactivity will be displayed by these substances? This chapter will begin to answer this question with a brief description of functional groups: the places in molecules where reactions tend to occur. Next we shall examine in depth the simplest class of organic molecules, the alkanes. If you have an appropriate kit, make a model of the structure shown at the top of this page. Does your model look exactly like the picture? Can it adopt other shapes by rotation of the atoms about bonds? This molecule is called 2,2,4-trimethylpentane, an alkane used in gasoline. As we proceed through this chapter, we shall explore the names, physical properties, and structural mobility of the members of the alkane family.

2-1 Functional Groups: Centers of Reactivity

Many organic molecules consist predominantly of a backbone of carbons linked by single bonds, with only hydrogen atoms attached. However, they may also contain doubly or triply bonded carbons, as well as other elements. These atoms or groups of atoms tend to be sites of comparatively high chemical reactivity and are referred to as **functional groups** or **functionalities.** Such groups have characteristic properties, and *they control the reactivity of the molecule as a whole.*

Hydrocarbons are molecules that contain only hydrogen and carbon

We begin our study with hydrocarbons, which have the general empirical formula $C_x H_y$. Those containing only single bonds, such as methane, ethane, and propane, are called **alkanes.** Molecules such as cyclohexane, whose carbons form a ring, are called **cycloalkanes.** *Alkanes lack functional groups;* as a result, they are relatively nonpolar and unreactive. The properties and chemistry of the alkanes are described in this chapter and in Chapters 3 and 4.

Alkanes

$$CH_4 \qquad CH_3{-}CH_3 \qquad CH_3{-}CH_2{-}CH_3$$

Methane   Ethane   Propane   Cyclohexane

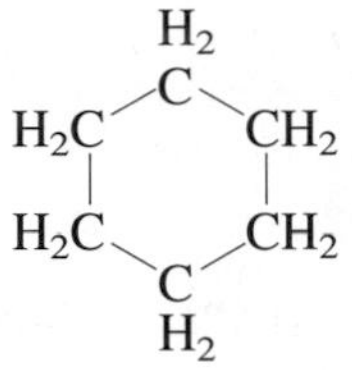

Double and triple bonds are the functional groups of **alkenes** and **alkynes,** respectively. Their properties and chemistry are the topics of Chapters 11 through 13.

Alkenes and Alkynes

$$CH_2{=}CH_2 \qquad \qquad C{=}CH_2 \qquad HC{\equiv}CH \qquad CH_3{-}C{\equiv}CH$$

Ethene   Propene   Ethyne   Propyne
(Ethylene)   (Acetylene)

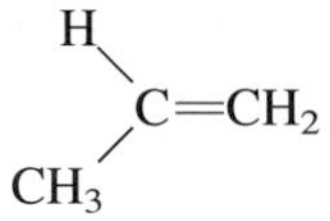

A special hydrocarbon is **benzene,** C_6H_6, in which three double bonds are incorporated into a six-membered ring. Benzene and its derivatives are traditionally called **aromatic,** because some substituted benzenes do have a strong fragrance. Aromatic compounds are discussed in Chapters 15, 16, 22, and 25.

Aromatic Compounds

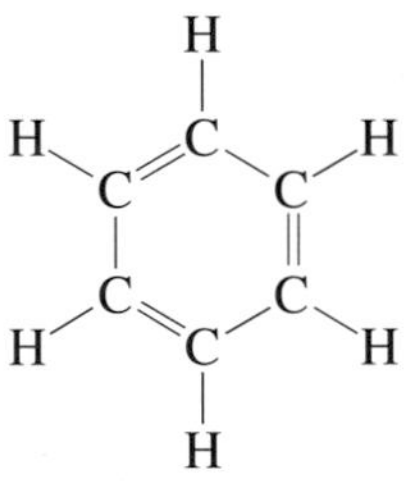

Benzene 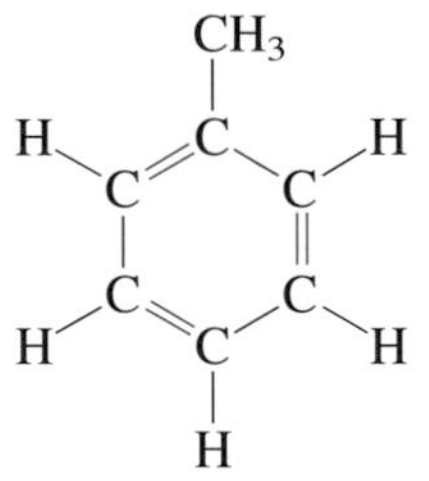   Methylbenzene
(Toluene)

Many functional groups contain polar bonds

Polar bonds determine the behavior of many classes of molecules. Recall that polarity is due to a difference in the electronegativity of two atoms bound to each other (Section 1-3). Chapters 6 and 7 will introduce the **haloalkanes,** which contain polar carbon–halogen bonds as their functional groups. Another example is the **hydroxy** group, –O–H, characteristic of **alcohols.** The characteristic functional unit of **ethers** is an oxygen atom bonded to two carbon atoms. The functional group in alcohols and those in some ethers can be converted into a large variety of other functionalities and are therefore important in synthetic transformations. This chemistry is the subject of Chapters 8 and 9.

Haloalkanes		**Alcohols**		**Ethers**	

Haloalkanes

CH_3Cl CH_3CH_2Cl
Chloromethane   **Chloroethane**
(Methyl chloride)   **(Ethyl chloride)**
(Topical anesthetics)

Alcohols

CH_3OH CH_3CH_2OH
Methanol   **Ethanol**
(Wood alcohol)   **(Grain alcohol)**

Ethers

CH_3OCH_3 $CH_3CH_2OCH_2CH_3$
Methoxymethane   **Ethoxyethane**
(Dimethyl ether)   **(Diethyl ether)**
(A refrigerant)   **(An inhalation anesthetic)**

The **carbonyl** function, C=O, is found in **aldehydes,** in **ketones,** and, in conjunction with an attached –OH, in the **carboxylic acids.** Aldehydes and ketones are discussed in Chapters 17 and 18, the carboxylic acids and their derivatives in Chapters 19 and 20.

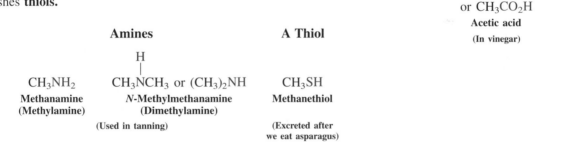

Aldehydes

$$\overset{\displaystyle O}{\overset{\|}{HCH}}$$
Formaldehyde
(A disinfectant)

$$\overset{\displaystyle O}{\overset{\|}{CH_3CH}} \text{ or } CH_3CHO$$
Acetaldehyde
(A hypnotic)

Ketones

$$\overset{\displaystyle O}{\overset{\|}{CH_3CCH_3}}$$
Propanone
(Acetone)

$$\overset{\displaystyle O}{\overset{\|}{CH_3CH_2CCH_3}}$$
Butanone
(Methyl ethyl ketone)
(Common solvents)

Carboxylic Acids

$$\overset{\displaystyle O}{\overset{\|}{HCOH}} \text{ or } HCOOH \text{ or } HCO_2H$$
Formic acid
(Strong irritant)

Other elements give rise to further characteristic functional groups. For example, alkyl nitrogen compounds are **amines.** The replacement of oxygen in alcohols by sulfur furnishes **thiols.**

$$\overset{\displaystyle O}{\overset{\|}{CH_3COH}} \text{ or } CH_3COOH \text{ or } CH_3CO_2H$$
Acetic acid
(In vinegar)

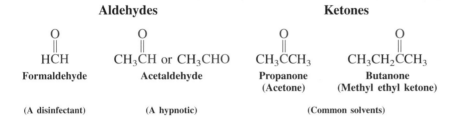

Amines

CH_3NH_2
Methanamine
(Methylamine)

$$CH_3\overset{\displaystyle H}{\overset{|}{N}}CH_3 \text{ or } (CH_3)_2NH$$
N-Methylmethanamine
(Dimethylamine)
(Used in tanning)

A Thiol

CH_3SH
Methanethiol
(Excreted after we eat asparagus)

R represents a part of an alkane molecule

Table 2-1 depicts a selection of common functional groups, the class of compounds to which they give rise, a general structure, and an example. In the general structures, we commonly use the symbol **R** (for *radical* or *residue*) to represent an **alkyl group,** a molecular fragment derived by removal of one hydrogen atom from an alkane (Section 2-3). Therefore, a general formula for a haloalkane is R–X, in which R stands for any alkyl group and X for any halogen. Alcohols are similarly represented as R–O–H. In structures that contain multiple alkyl groups, we add a prime (′) or double prime (″) to R to distinguish groups that differ in structure from one another. Thus a general formula for an ether in which both alkyl groups are the same (a **symmetrical ether**) is R–O–R, whereas an ether with two dissimilar groups (an **unsymmetrical ether**) is represented by R–O–R′.

The alkanes $C_{29}H_{60}$ and $C_{31}H_{64}$ constitute the waxy, water-repellent coatings on these wild lupine leaves.

TABLE 2-1	Common Functional Groups		
Compound class	**General structure**[a]	**Functional group**	**Example**
Alkanes	R—H	None	$CH_3CH_2CH_2CH_3$ **Butane**
Haloalkanes	R—X (X = F, Cl, Br, I)	—X	CH_3CH_2—Br **Bromoethane**
Alcohols	R—OH	—OH	$(CH_3)_2C$—OH (with H) **2-Propanol (Isopropyl alcohol)**
Ethers	R—O—R′	—O—	CH_3CH_2—O—CH_3 **Methoxyethane (Ethyl methyl ether)**
Thiols	R—SH	—SH	CH_3CH_2—SH **Ethanethiol**
Alkenes	(H)R\(H)R C=C R(H)/R(H)	C=C	CH_3/CH_3 C=CH_2 **2-Methylpropene**
Alkynes	(H)R—C≡C—R(H)	—C≡C—	$CH_3C≡CCH_3$ **2-Butyne**
Aromatic compounds	(benzene ring with R(H))	(benzene ring)	(toluene ring) **Methylbenzene (Toluene)**
Aldehydes	R—C(=O)—H	—C(=O)—H	CH_3CH_2CH(=O) **Propanal**
Ketones	R—C(=O)—R′	—C(=O)—	$CH_3CH_2CCH_2CH_2CH_3$ **3-Hexanone**
Carboxylic acids	R—C(=O)—O—H	—C(=O)—OH	CH_3CH_2COH **Propanoic acid**
Anhydrides	R—C(=O)—O—C(=O)—R′(H)	—C(=O)—O—C(=O)—	$CH_3CH_2COCCH_2CH_3$ **Propanoic anhydride**

[a]The letter R denotes an alkyl group. Different alkyl groups can be distinguished by adding primes to the letter R: R′, R″, and so forth.

Compound class	General structure^a	Functional group	Example
Esters	$\underset{\displaystyle}{(H)R—\overset{\displaystyle\overset{O}{\|\|}}{C}—O—R'}$	$\overset{\displaystyle\overset{O}{\|\|}}{—C—O—}$	$\underset{\substack{\textbf{Methyl propanoate}\\\textbf{(Methyl propionate)}}}{CH_3CH_2\overset{\displaystyle\overset{O}{\|\|}}{C}OCH_3}$
Amides	$\underset{\displaystyle\underset{R''(H)}{\|}}{R—\overset{\displaystyle\overset{O}{\|\|}}{C}—N—R'(H)}$	$\underset{\displaystyle\|}{—\overset{\displaystyle\overset{O}{\|\|}}{C}—N—}$	$\underset{\textbf{Butanamide}}{CH_3CH_2CH_2\overset{\displaystyle\overset{O}{\|\|}}{C}NH_2}$
Nitriles	$R—C\equiv N$	$—C\equiv N$	$\underset{\substack{\textbf{Ethanenitrile}\\\textbf{(Acetonitrile)}}}{CH_3C\equiv N}$
Amines	$\underset{\displaystyle\underset{R''(H)}{\|}}{R—N—R'(H)}$	$—N\overset{\diagup}{\diagdown}$	$\underset{\substack{\textbf{N,N-Dimethylmethanamine}\\\textbf{(Trimethylamine)}}}{(CH_3)_3N}$

2-2 Straight-Chain and Branched Alkanes

We begin with the alkanes, hydrocarbons that contain only single bonds. They are classified into several types according to structure: the linear **straight-chain alkanes;** the **branched alkanes,** in which the carbon chain contains one or several branching points; and the cyclic alkanes, or **cycloalkanes,** which will be covered in Chapter 4.

A Straight-Chain Alkane   **A Branched Alkane**   **A Cycloalkane**

$$CH_3—CH_2—CH_2—CH_3 \qquad CH_3—\underset{\displaystyle\underset{CH_3}{\|}}{\overset{\displaystyle\overset{CH_3}{\|}}{C}}—H \qquad \underset{\displaystyle CH_2—CH_2}{\overset{\displaystyle CH_2—CH_2}{\big|\qquad\big|}}$$

Butane, C_4H_{10} 2-Methylpropane, C_4H_{10} Cyclobutane, C_4H_8
                           (Isobutane)

The alkanes form a homologous series

In the straight-chain alkanes, each carbon is bound to its two neighbors and to two hydrogen atoms. Exceptions are the two terminal carbon nuclei, which are bound to only one carbon atom and three hydrogen atoms. Several general formulas may be written for the straight-chain alkane series:

$$H—(CH_2)_n—H \quad CH_3—(CH_2)_{n-1}—H \quad CH_3—(CH_2)_{n-2}—CH_3$$

Each member of this series differs from the next lower one by the addition of a methylene group, $—CH_2—$. Molecules that are related in this way are **homologs** of each other (*homos,* Greek, same as), and the series is a **homologous series.** Methane ($n = 1$) is the first member of the homologous series of the alkanes, ethane ($n = 2$) the second, and so forth.

Branched alkanes are constitutional isomers of straight-chain alkanes

Branched alkanes are derived from the straight-chain systems by removal of a hydrogen from a methylene (CH_2) group and replacement with an alkyl group. Both branched and straight-chain alkanes have the same general formula, C_nH_{2n+2}. The smallest branched alkane is 2-methylpropane. It has the same molecular formula as that of butane (C_4H_{10}) but different connectivity; the two compounds therefore form a pair of constitutional isomers (Section 1-9).

For the higher alkane homologs ($n > 4$), more than two isomers are possible. There are three pentanes, C_5H_{12}, as shown below. There are five hexanes, C_6H_{14}; nine heptanes, C_7H_{16}; and eighteen octanes, C_8H_{18}.

The Isomeric Pentanes

$$CH_3-CH_2-CH_2-CH_2-CH_3$$

Pentane

2-Methylbutane
(Isopentane)

2,2-Dimethylpropane
(Neopentane)

The number of possibilities in connecting n carbon atoms to each other and to $2n + 2$ surrounding hydrogen atoms increases dramatically with the size of n (Table 2-2).

EXERCISE 2-1

(**a**) Draw the structures of the five isomeric hexanes. (**b**) Draw the structures of all the possible next higher and lower homologs of 2-methylbutane.

TABLE 2-2 Number of Possible Isomeric Alkanes, C_nH_{2n+2}

n	Isomers
1	1
2	1
3	1
4	2
5	3
6	5
7	9
8	18
9	35
10	75
15	4,347
20	366,319

2-3 Naming the Alkanes

The multiplicity of ways of assembling carbon atoms and attaching various substituents accounts for the existence of the very large number of organic molecules. This diversity poses a problem: How can we systematically differentiate all these compounds by name? Is it possible, for example, to name all the C_6H_{14} isomers so that information on any of them (such as boiling points, melting points, reactions) might easily be found in the index of a handbook? And is there a way to name a compound that we have never seen in such a way as to be able to draw its structure?

This problem of naming organic molecules has been with organic chemistry from its very beginning, but the initial method was far from systematic. Compounds have been named after their discoverers ("Nenitzescu's hydrocarbon"), after localities ("sydnones"), after their shapes ("cubane," "basketane"), and after their natural sources ("vanillin"). Many of these **common** or **trivial names** are still widely used. However, there now exists a precise system for naming the alkanes. **Systematic nomenclature,** in which the name of a compound describes its structure, was first introduced by a chemical congress in Geneva, Switzerland, in 1892. It has continually been revised since then, mostly by the International Union of Pure and Applied Chemistry (IUPAC). Table 2-3 gives the systematic names of the first twenty straight-chain alkanes. Their stems, mainly of Latin or Greek origin, reveal the number of carbon atoms in the chain. For example, the name heptadecane is composed of the Greek

TABLE 2-3	Names and Physical Properties of Straight-Chain Alkanes, C_nH_{2n+2}				
n	Name	Formula	Boiling point (°C)	Melting point (°C)	Density at 20°C (g ml^{-1})
1	Methane	CH_4	−161.7	−182.5	0.466 (at −164°C)
2	Ethane	CH_3CH_3	−88.6	−183.3	0.572 (at −100°C)
3	Propane	$CH_3CH_2CH_3$	−42.1	−187.7	0.5853 (at −45°C)
4	Butane	$CH_3CH_2CH_2CH_3$	−0.5	−138.3	0.5787
5	Pentane	$CH_3(CH_2)_3CH_3$	36.1	−129.8	0.6262
6	Hexane	$CH_3(CH_2)_4CH_3$	68.7	−95.3	0.6603
7	Heptane	$CH_3(CH_2)_5CH_3$	98.4	−90.6	0.6837
8	Octane	$CH_3(CH_2)_6CH_3$	125.7	−56.8	0.7026
9	Nonane	$CH_3(CH_2)_7CH_3$	150.8	−53.5	0.7177
10	Decane	$CH_3(CH_2)_8CH_3$	174.0	−29.7	0.7299
11	Undecane	$CH_3(CH_2)_9CH_3$	195.8	−25.6	0.7402
12	Dodecane	$CH_3(CH_2)_{10}CH_3$	216.3	−9.6	0.7487
13	Tridecane	$CH_3(CH_2)_{11}CH_3$	235.4	−5.5	0.7564
14	Tetradecane	$CH_3(CH_2)_{12}CH_3$	253.7	5.9	0.7628
15	Pentadecane	$CH_3(CH_2)_{13}CH_3$	270.6	10	0.7685
16	Hexadecane	$CH_3(CH_2)_{14}CH_3$	287	18.2	0.7733
17	Heptadecane	$CH_3(CH_2)_{15}CH_3$	301.8	22	0.7780
18	Octadecane	$CH_3(CH_2)_{16}CH_3$	316.1	28.2	0.7768
19	Nonadecane	$CH_3(CH_2)_{17}CH_3$	329.7	32.1	0.7855
20	Icosane	$CH_3(CH_2)_{18}CH_3$	343	36.8	0.7886

Propane, stored under pressure in liquefied form in canisters such as these, is a common fuel for torches, lanterns, and outdoor cooking stoves.

word *hepta,* seven, and the Latin word *decem,* ten. The first four alkanes have special names that have been accepted as part of the systematic nomenclature but also all end in **-ane.** It is important to know these names, because they serve as the basis for naming a large fraction of all organic molecules. A few smaller branched alkanes have common names that still have widespread use. They make use of the prefixes **iso-** and **neo-,** as in isobutane, isopentane, and neohexane.

$$CH_3-\overset{\displaystyle CH_3}{\underset{\displaystyle II}{C}}-(CH_2)_n-CH_3$$

An isoalkane
(e.g., $n = 1$, Isopentane)

$$CH_3-\overset{\displaystyle CH_3}{\underset{\displaystyle CH_3}{C}}-(CH_2)_n-H$$

A neoalkane
(e.g., $n = 2$, Neohexane)

EXERCISE 2-2

Draw the structures of isohexane and neopentane.

TABLE 2-4	Branched Alkyl Groups			
Structure	Common name	Example of common name in use	Systematic name	Designation
$\text{CH}_3-\overset{\overset{\displaystyle CH_3}{\mid}}{\underset{\underset{\displaystyle H}{\mid}}{C}}-$	Isopropyl	$\text{CH}_3-\overset{\overset{\displaystyle CH_3}{\mid}}{\underset{\underset{\displaystyle H}{\mid}}{C}}-\text{Cl}$ (Isopropyl chloride)	1-Methylethyl	Secondary
$\text{CH}_3-\overset{\overset{\displaystyle CH_3}{\mid}}{\underset{\underset{\displaystyle H}{\mid}}{C}}-\text{CH}_2-$	Isobutyl	$\text{CH}_3-\overset{\overset{\displaystyle CH_3}{\mid}}{\underset{\underset{\displaystyle H}{\mid}}{C}}-\text{CH}_3$ (Isobutane)	2-Methylpropyl	Primary
$\text{CH}_3-\text{CH}_2-\overset{\overset{\displaystyle CH_3}{\mid}}{\underset{\underset{\displaystyle H}{\mid}}{C}}-$	*sec*-Butyl	$\text{CH}_3-\text{CH}_2-\overset{\overset{\displaystyle CH_3}{\mid}}{\underset{\underset{\displaystyle H}{\mid}}{C}}-\text{NH}_2$ (*sec*-Butyl amine)	1-Methylpropyl	Secondary
$\text{CH}_3-\overset{\overset{\displaystyle CH_3}{\mid}}{\underset{\underset{\displaystyle CH_3}{\mid}}{C}}-$	*tert*-Butyl	$\text{CH}_3-\overset{\overset{\displaystyle CH_3}{\mid}}{\underset{\underset{\displaystyle CH_3}{\mid}}{C}}-\text{Br}$ (*tert*-Butyl bromide)	1,1-Dimethylethyl	Tertiary
$\text{CH}_3-\overset{\overset{\displaystyle CH_3}{\mid}}{\underset{\underset{\displaystyle CH_3}{\mid}}{C}}-\text{CH}_2-$	Neopentyl	$\text{CH}_3-\overset{\overset{\displaystyle CH_3}{\mid}}{\underset{\underset{\displaystyle CH_3}{\mid}}{C}}-\text{CH}_2-\text{OH}$ (Neopentyl alcohol)	2,2-Dimethylpropyl	Primary

Alkyl groups

CH_3-

Methyl

CH_3-CH_2-

Ethyl

$\text{CH}_3-\text{CH}_2-\text{CH}_2-$

Propyl

As mentioned in Section 2-2, an **alkyl** group is formed by the removal of a hydrogen from an alkane. It is named by replacing the ending -ane in the corresponding alkane by **-yl,** as in methyl, ethyl, and propyl. Table 2-4 shows a few branched alkyl groups having common names. Note that some use the prefixes *sec-* (or *s-*), which stands for secondary, and *tert-* (or *t-*), for tertiary. To apply these prefixes, we must first see how to classify carbon atoms in organic molecules. A **primary** carbon is one attached to only one other carbon atom. For example, all carbon atoms at the ends of alkane chains are primary. The hydrogens attached to such carbons are designated primary hydrogens, and an alkyl group created by removing a primary hydrogen also is called primary. A **secondary** carbon is attached to two other carbon atoms, and a **tertiary** carbon to three others. Their hydrogens are labeled similarly. As shown in Table 2-4, removal of a secondary hydrogen results in a secondary alkyl group, and removal of a tertiary hydrogen in a tertiary alkyl group. Finally, a carbon bearing four alkyl groups is called **quaternary.**

Primary, Secondary, and Tertiary Carbons and Hydrogens

Primary C

Secondary C

CH_3

Tertiary C

$\text{CH}_3\text{CH}_2\text{CCH}_2\text{CH}_3$

Primary H

H

Secondary H

Tertiary H

3-Methylpentane

Label the primary, secondary, and tertiary hydrogens in 2-methylpentane (isohexane).

The information in Table 2-3 enables us to name the first twenty straight-chain alkanes. How do we go about naming branched systems? A set of IUPAC rules makes this a relatively simple task, as long as they are followed carefully and in sequence.

IUPAC RULE 1. *Find the longest chain in the molecule and name it.* This task is not as easy as it seems. The problem is that, in the condensed formula, complex alkanes may be written in ways that mask the identity of the longest chain. In the following examples, the longest chain, or **stem chain,** is clearly marked; the alkane stem gives the molecule its name. Groups other than hydrogen attached to the stem chain are called **substituents.**

> The stem chain is shown in black in the examples in this section.

Methyl ⟶ CH₃

CH₃CHCH₂CH₃

A methyl-substituted butane
(A methylbutane)

CH₃
|
CH₃CH   CH₂CH₂CH₂CH₃
| | ⟵ Ethyl
CH₃CHCH₂CH₂CHCH₂CH₃

An ethyl- and methyl-substituted decane
(An ethylmethyldecane)

If a molecule has two or more chains of equal length, the chain with the largest number of substituents is the base stem chain.

CH₃   CH₃
| |
CH₃CHCHCHCHCH₂CH₃   not   CH₃CHCHCHCHCH₂CH₃
| | | |
CH₃ CH₂   CH₃ CH₂
    | |
    CH₂   CH₂
    | |
    CH₃   CH₃

4 substituents   **3 substituents**
A heptane   **A heptane**
Correct stem chain   **Incorrect stem chain**

Here are two more examples, drawn with the use of bond-line notation:

Methyl ⟶   ⟵ Ethyl

A methylbutane   **An ethylmethyldecane**

IUPAC RULE 2. *Name all groups attached to the longest chain as alkyl substituents.* For straight-chain substituents, Table 2-3 can be used to derive the alkyl name. However, what if the substituent chain is branched? In this case, the same IUPAC rules apply to such complex substituents: First, find the longest chain in the substituent; next, name all *its* substituents.

IUPAC RULE 3. *Number the carbons of the longest chain beginning with the end that is closest to a substituent.*

$$\underset{\underset{\text{not 4}\quad 3\quad 2\quad 1}{1\quad 2\quad 3\quad 4}}{\text{CH}_3\overset{\overset{\text{CH}_3}{|}}{\text{CH}}\text{CH}_2\text{CH}_3}$$

If there are two substituents at *equal* distance from the two ends of the chain, use the alphabet to decide how to number. The substituent to come first in alphabetical order is attached to the carbon with the lower number.

$$\underset{\underset{\text{Ethyl before methyl}}{1\quad 2\quad 3\quad 4\quad 5\quad 6\quad 7\quad 8}}{\text{CH}_3\text{CH}_2\overset{\overset{\text{CH}_3\text{CH}_2}{|}}{\text{CH}}\text{CH}_2\text{CH}_2\overset{\overset{\text{CH}_3}{|}}{\text{CH}}\text{CH}_2\text{CH}_3}$$

Butyl before propyl

What if there are three or more substituents? Then number the chain in the direction that gives the lower number at the *first difference* between the two possible numbering schemes. This procedure follows the **first point of difference principle.**

$$\text{CH}_3\text{CH}_2\overset{\overset{\text{CH}_3}{|}}{\text{CH}}\text{CH}_2\text{CH}_2\text{CH}_2\text{CH}_2\overset{\overset{\text{CH}_3}{|}}{\text{CH}}\text{CH}_2\overset{\overset{\text{CH}_3}{|}}{\text{CH}}\text{CH}_2\text{CH}_3$$

1 2 3 4 5 6 7 8 9 10 11 12
12 11 10 9 8 7 6 5 4 3 2 1

Numbers for substituted carbons:
← 3, **8**, and 10 (incorrect)
← 3, **5**, and 10 (correct; 5 lower than 8)

3,5,10-Trimethyldodecane

Substituent groups are numbered outward from the main chain, with C1 of the group being the carbon attached to the main stem.

IUPAC RULE 4. *Write the name of the alkane by first arranging all the substituents in alphabetical order (each preceded by the carbon number to which it is attached and a hyphen) and then add the name of the stem.* Should a molecule contain more than one of a particular substituent, its name is preceded by the prefix di, tri, tetra, penta, and so forth. The positions of attachment to the stem are given collectively before the substituent name and are separated by commas. These prefixes, as well as *sec-* and *tert-*, are not considered in the alphabetical ordering, except when they are part of a complex substituent name.

5-Ethyl-2,2-dimethyloctane

("di" not counted in alphabetical ordering) but

5-(1,1-Dimethylethyl)-3-ethyloctane

("di" counted: part of substituent name)

$$\underset{\text{2-Methylbutane}}{\text{CH}_3\overset{\overset{\text{CH}_3}{|}}{\text{CH}}\text{CH}_2\text{CH}_3}$$

$$\underset{\text{2,3-Dimethylbutane}}{\text{CH}_3\overset{\overset{\text{CH}_3}{|}}{\text{CH}}\underset{\underset{\text{CH}_3}{|}}{\text{CH}}\text{CH}_3}$$

$$\underset{\text{4-Ethyl-2,2,7-trimethyloctane}}{\text{CH}_3\overset{\overset{\text{CH}_3}{|}}{\text{CH}}\text{CH}_2\text{CH}_2\underset{\underset{\text{CH}_3\text{CH}_2}{|}}{\text{CH}}\text{CH}_2\overset{\overset{\text{CH}_3}{|}}{\underset{\underset{\text{CH}_3}{|}}{\text{C}}}\text{CH}_3}$$

4,5-Diethyl-3,6-dimethyldecane

$$\underset{\text{3-Ethyl-2-methylpentane}}{\text{CH}_3\text{CH}_2\overset{\overset{\text{CH}_2\text{CH}_3}{|}}{\text{CH}}\underset{\underset{\text{CH}_3}{|}}{\text{CH}}\text{CH}_3}$$

Although the common group names in Table 2-4 are permitted by IUPAC, it is preferable to use systematic names. Such complex names are usually enclosed in parentheses, to avoid possible ambiguities.

If a particular complex substituent is present more than once, its name is preceded by the prefix bis, tris, tetrakis, pentakis, and so on. In a substituent chain, the carbon numbered one (C1) is *always* the carbon atom bound to the principal chain.

Complex alkyl group has carbon-1 attached to the base stem

First substituent at position 2 determines numbering

Longest chain chosen has highest number of substituents

4-(1-Ethylpropyl)-2,3,5-trimethylnonane

CH_3
|
CH_3CH
|
$CH_3CH_2CH_2CHCH_2CH_2CH_3$

4-(1-Methylethyl)heptane
(4-Isopropylheptane)

5,8-Bis(1-methylethyl)-dodecane

Further instructions on nomenclature will be presented when new classes of compounds, such as the cycloalkanes and haloalkanes, are introduced.

EXERCISE 2-4

Write down the names of the preceding eight branched alkanes, close the book, and reconstruct their structures from those names.

To summarize, four rules should be applied in sequence when naming a branched alkane: (1) find the longest chain; (2) find the names of all the alkyl groups attached to the stem; (3) number the chain; (4) name the alkane, with substituent names in alphabetical order and preceded by numbers to indicate their locations.

2-4 Structural and Physical Properties of Alkanes

What do the structures of alkanes look like in three dimensions? What are their physical appearances, and what are their physical properties? These questions will be addressed next.

Alkanes exhibit regular molecular structures and properties

The structural features of the alkanes are remarkably regular. The carbon atoms are tetrahedral, with bond angles close to 109° and with regular C–H (≈ 1.10 Å) and

C–C ($\approx$ 1.54 Å) bond lengths. Alkane chains often adopt the zigzag patterns used in bond-line notation (Figure 2-1). To depict three-dimensional structures, we shall make use of the dashed-wedged line notation (see Figure 1-23). The main chain and a hydrogen at each end are drawn in the plane of the page (Figure 2-2).

EXERCISE 2-5

Draw zigzag dashed-wedged line structures for 2-methylbutane and 2,3-dimethylbutane.

The regularity in alkane structures suggests that their physical constants would follow predictable trends. Indeed, inspection of the data presented in Table 2-3 reveals regular incremental increases along the homologous series. For example, at room temperature (25°C), the lower homologs of the alkanes are gases or colorless liquids, the higher homologs waxy solids. From pentane to pentadecane, each additional CH_2 group causes a 20°–30°C increase in boiling point (Figure 2-3).

FIGURE 2-2
Dashed-wedged line structures of methane through pentane. Note the zigzag arrangement of the principal chain and two terminal hydrogens.

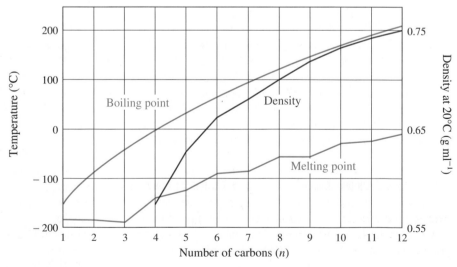

FIGURE 2-3
The physical constants of straight-chain alkanes. Their values increase with increasing size because molecular weights and London forces increase. Note that even-numbered systems have somewhat higher melting points than expected; these systems are more tightly packed in the solid state (notice their higher densities), thus allowing for stronger attractions between molecules.

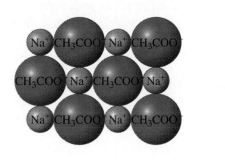

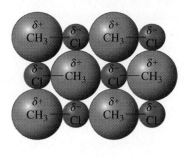

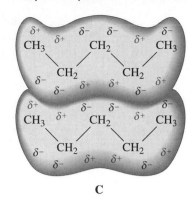

A

B

C

FIGURE 2-4

(A) Coulombic attraction in an ionic compound: crystalline sodium acetate, the sodium salt of acetic acid (in vinegar). (B) Dipole–dipole interactions in solid chloromethane. The polar molecules arrange to allow for favorable Coulombic attraction. (C) London forces in crystalline pentane. In this simplified picture, the electron clouds as a whole mutually interact to produce partial charges of opposite sign. The charge distributions in the two molecules change continually as the electrons continue to correlate their movements.

Attractive forces between molecules govern the physical properties of alkanes

Why are the physical properties of alkanes predictable? Such trends exist because of **intermolecular** or **van der Waals* forces.** Molecules exert several types of attractive forces on each other, causing them to aggregate into organized arrangements as solids and liquids. Most solid substances exist as highly ordered crystals. *Ionic* compounds, such as salts, are rigidly held in a crystal lattice, mainly by strong Coulomb forces. Nonionic but *polar* molecules, such as chloromethane (CH_3Cl), are attracted by weaker dipole–dipole interactions, also of Coulombic origin (Sections 1-2 and 6-2). Finally, the *nonpolar* alkanes attract each other by **London† forces,** which are due to **electron correlation.** When one alkane molecule approaches another, repulsion of the electrons in one molecule by those in the other results in correlation of their movement. Electron motion causes temporary bond polarization in one molecule; correlated electron motion in the bonds of the other induces polarization in the opposite direction, resulting in attraction between the molecules. Figure 2-4 is a simple picture comparing ionic, dipolar, and London attractions.

London forces are very weak. In contrast with Coulomb forces, which change with the square of the distance between charges, London forces fall off as the sixth power of the distance between molecules. There is also a limit to how close these forces can bring molecules together. At small distances, nucleus–nucleus and electron–electron repulsions outweigh these attractions.

How do these forces account for the physical constants of elements and compounds? The answer is that it takes energy, usually in the form of heat, to melt solids

---

*Professor Johannes D. van der Waals (1837–1923), University of Amsterdam, Netherlands, Nobel Prize 1910 (physics).

†Professor Fritz London (1900–1954), Duke University, North Carolina. *Note:* In older references the term "van der Waals forces" referred exclusively to what we now call *London forces; van der Waals forces* now refers collectively to *all* intermolecular attractions.

and boil liquids. For example, to cause melting, the attractive forces responsible for the crystalline state must be overcome. In an ionic compound, such as sodium acetate (Figure 2-4A), the strong interionic forces require a rather high temperature (324°C) for the compound to melt. In alkanes, melting points rise with increasing molecular size: Molecules with relatively large surface areas are subject to greater London attractions. However, these forces are still relatively weak, and even high molecular weight alkanes have rather low melting points. For example, the straight-chain alkanes $C_{29}H_{60}$ and $C_{31}H_{64}$, waxy solids present in the protective coatings of plant leaves, have melting points below 70°C.

For a molecule to escape these same attractive forces in the liquid state and enter the gas phase, more heat has to be applied. When the vapor pressure of a liquid equals atmospheric pressure, boiling occurs. Like melting points, boiling points rise with increasing molecular weight: Heavy compounds require more kinetic energy to leave the liquid state. Boiling points of compounds are also relatively high if the intermolecular forces are relatively large. These effects lead to the smooth increase in boiling points seen in Figure 2-3.

Branched alkanes have smaller surface areas than do their straight-chain isomers. As a result, they are generally subject to smaller London attractions and are unable to pack as well in the crystalline state. The weaker attractions result in lower melting and boiling points. Branched molecules with highly compact shapes are exceptions. For example, 2,2,3,3-tetramethylbutane melts at +101°C because of highly favorable crystal packing (compare octane, m.p. –57°C). On the other hand, the greater surface area of octane, compared with that of the more spherical 2,2,3,3-tetramethylbutane, is clearly demonstrated in their boiling points (126°C and 106°C, respectively). Crystal packing differences also account for the slightly lower than expected melting points of odd-membered straight-chain alkanes relative to those of even-membered systems (Figure 2-3).

In summary, straight-chain alkanes have regular structures. Their melting points, boiling points, and densities increase with molecular weight because of increasing attraction between molecules.

2,2,3,3-Tetramethylbutane

2-5 Rotation About Single Bonds: Conformations

We have considered how intermolecular forces can affect the physical properties of molecules. These forces act *between* molecules. In this section, we shall examine how the forces present *within* molecules (i.e., intramolecular forces) make some arrangements in space energetically more favorable than others.

Rotation interconverts the conformations of ethane

If we build a molecular model of ethane, we can see that the two methyl groups are readily rotated with respect to each other. The energy required to move the hydrogen atoms past each other, the *barrier to rotation,* is only 3 kcal mol^{-1}. This value turns out to be so low that chemists speak of "free rotation" of the methyl groups. In general, *there is free rotation about all single bonds.*

Figure 2-5 depicts the rotational movement in ethane by the use of dashed-wedged line structures (Section 1-9). There are two extreme ways of drawing ethane: the staggered conformation and the eclipsed one. If the **staggered conformation** is viewed along the C–C axis, each hydrogen atom on the first carbon is seen to be positioned perfectly between two hydrogen atoms on the second. The second extreme is derived

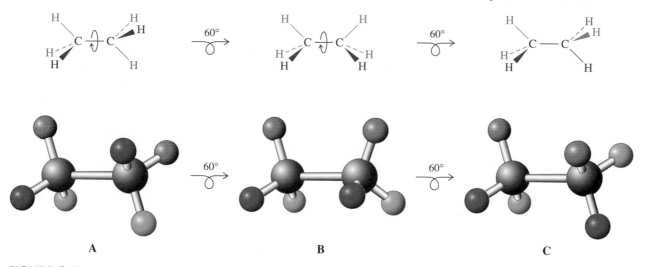

FIGURE 2-5

Rotation in ethane: (A and C) staggered conformations; (B) eclipsed. There is virtually "free rotation" between conformers.

from the first by a 60° turn of one of the methyl groups about the C–C bond. Now, if this **eclipsed conformation** is viewed along the C–C axis, all hydrogen atoms on the first carbon are directly opposite those on the second—that is, those on the first eclipse those on the second. A further 60° turn converts the eclipsed form into a new but equivalent staggered arrangement. Between these two extremes, rotation of the methyl group results in numerous additional positions, referred to collectively as **skew conformations.**

The many forms of ethane (and, as we shall see, substituted analogs) created by such rotations are **conformations** (also called **conformers** and **rotamers**). All of them rapidly interconvert at room temperature. The study of their thermodynamic and kinetic behavior is **conformational analysis.**

Newman projections depict the conformations of ethane

A simple alternative to the dashed-wedged line structures for illustrating the conformers of ethane is the **Newman* projection.** We can arrive at a Newman projection from the dashed-wedged line picture by turning the molecule out of the plane of the page toward us and viewing it along the C–C axis (Figure 2-6A and B). In this notation, the front carbon obscures the back carbon, but the bonds emerging from both are clearly seen. The front carbon is depicted as the point of juncture of the three bonds attached to it, one of them usually drawn vertically and pointing up. The back carbon is a circle (Figure 2-6C). The bonds to this carbon project from the outer edge of the circle. The extreme conformational shapes of ethane are readily drawn in this way (Figure 2-7). To make the three rear hydrogen atoms more visible in eclipsed conformations, they are drawn somewhat rotated out of the perfectly eclipsing position.

*Professor Melvin S. Newman (1908–1993), Ohio State University.

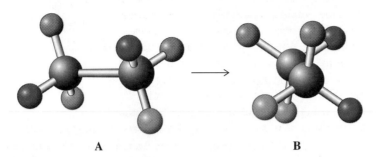

FIGURE 2-6

Representations of ethane. (A) Side-on views of the molecule. (B) End-on views of ethane, showing the carbon atoms directly in front of each other and the staggered positions of the hydrogens. (C) Newman projection of ethane derived from the view shown in (B). The "front" carbon is represented by the intersection of the bonds to its three attached hydrogens. The bonds from the remaining three hydrogens connect to the large circle, which represents the "back" carbon.

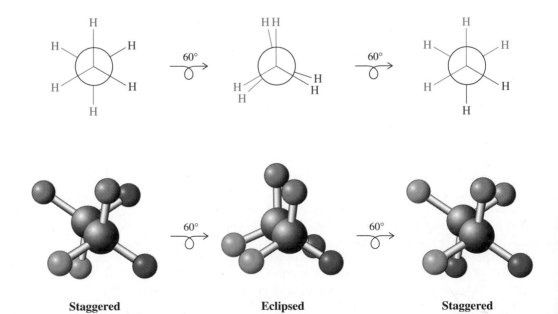

FIGURE 2-7

Newman projections and ball-and-stick models of staggered and eclipsed rotamers of ethane. In these representations, the back carbon is rotated clockwise in increments of 60°.

2-6 Potential-Energy Diagrams

As mentioned earlier, about 3 kcal mol^{-1} of heat is required to rotate the methyl groups in ethane. What is the reason for this requirement?

The rotamers of ethane have different potential energies

The various rotamers of ethane do not all have the same potential energies. A simple explanation is based on electron repulsion. As one methyl group turns about the C–C axis, starting from a staggered conformation, the distance between the hydrogen atoms of the respective methyl groups begins to diminish, resulting in increasing repulsion between the bonding pairs of electrons in the C–H bonds. Thus, the potential energy of the system rises steadily as the methyl group rotates from staggered through skew to eclipsed conformations. At the point of eclipsing, the molecule has its highest energy content, because at this stage the two sets of six bonding electrons are closest. This point is 3 kcal mol^{-1} above the lowest energy state of the molecule, the staggered rotamer. The change in energy resulting from bond rotation is called **rotational** or **torsional energy.**

Potential-energy diagrams are a convenient way to depict energy changes

The differences in potential energy between rotamers can be pictured by plotting the energy changes against the degree of rotation (Figure 2-8). Such a plot is called a **potential-energy diagram.** Potential-energy diagrams are useful in the description

FIGURE 2-8 ————
Potential-energy diagram of the rotational isomerism in ethane. Because the eclipsed conformations have the highest energy, they correspond to peaks in the diagram. These maxima may be viewed as transition states (TS) between the more stable staggered rotamers. The activation energy (E_a) corresponds to the barrier to rotation.

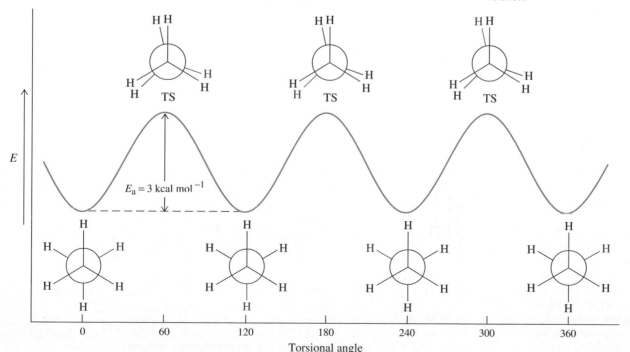

of other chemical processes as well. Changes in potential energy are plotted against a **reaction coordinate,** which describes the progress of the process or reaction. In the diagram for rotation of ethane, the reaction coordinate is degrees of rotation (Figure 2-8), usually called the **torsional angle.** Ethane is best described in its staggered conformation. In fact, the eclipsed rotamer has only a fleeting lifetime (of the order of 10^{-12} s) as the hydrogens rapidly move past each other, equilibrating one staggered arrangement with another. Because eclipsed conformations have the highest energy, they are maxima in potential-energy diagrams. Such points are called **transition states** (TS), marking the transition from one staggered rotamer to another. The energy of the transition state can be viewed as the barrier to be overcome when the molecule goes from one staggered arrangement to the next. This energy is called the **activation energy,** E_a, for the rotational process. The lower its value, the faster the rotation.

Collisions supply the energy to get past the activation-energy barrier

Where do organic molecules get the energy to overcome the barrier to rotation? Molecules have *kinetic energy* as a result of their motion, but at room temperature the average kinetic energy is only about 0.6 kcal mol^{-1}, far below the activation-energy barrier. To pick up enough energy, molecules must collide with each other or with the walls of the container. Each collision transfers energy from one molecule to another.

A graph called a **Boltzmann* distribution curve** depicts the distribution of kinetic energy. Figure 2-9 shows that, although most molecules have only average speed at any given temperature, some molecules have kinetic energies that are much higher. In ethane, part of this energy may be used to overcome the activation-energy barrier. Because continual collisions rapidly redistribute the kinetic energy, all molecules eventually get past this barrier. That is why we can speak of "free rotation."

The shape of the Boltzmann curve depends on the temperature. We can see that at higher temperatures, as the average kinetic energy increases, the curve flattens and shifts toward higher energies. More molecules now have energy higher than is required by the transition state, so the speed of rotation increases. Conversely, at lower temperatures, the rate of rotation decreases.

*Professor Ludwig Boltzmann (1844–1906), University of Vienna, Austria.

FIGURE 2-9

Boltzmann curves at two temperatures. At the higher temperature (green curve), there are more molecules of kinetic energy E than at the lower temperature (blue curve). Molecules with higher kinetic energy can more easily overcome the activation-energy barrier.

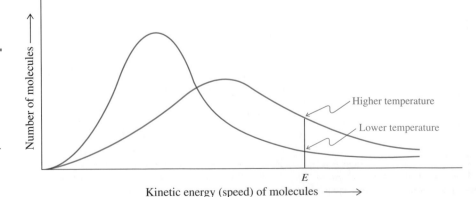

In summary, intramolecular forces control the arrangement of substituents on neighboring and bonded carbon atoms. In ethane, the relatively stable staggered conformations are interconverted through higher energy transition states in which substituents are eclipsed. To reach these transition states, molecules have to absorb the kinetic energy of others through collisions. The energy distribution of a collection of molecules at any given temperature is depicted by a Boltzmann curve. The energetics of rotation about the C–C bond is conveniently pictured in a potential-energy diagram.

2-7 Rotation in Substituted Ethanes

How does the potential-energy diagram change when a substituent is added to ethane? Consider, for example, propane, whose structure is similar to that of ethane, except that a methyl group replaces one of ethane's hydrogen atoms.

Steric hindrance raises the energy barrier to rotation

A potential-energy diagram for the rotation about a C–C bond in propane is shown in Figure 2-10. The Newman projections of propane differ from those of ethane only by the substituted methyl group. Again, the extreme conformations are staggered

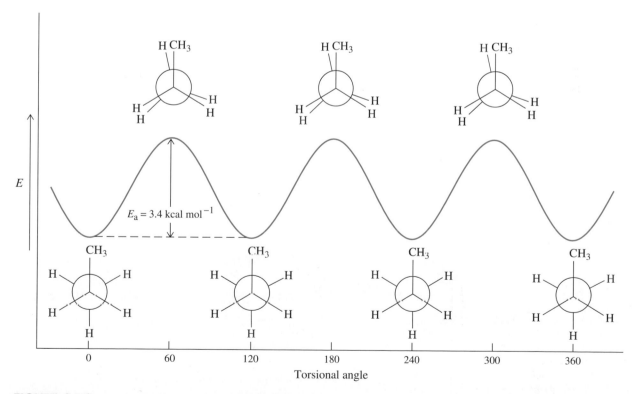

FIGURE 2-10

Potential-energy diagram of the rotational isomerism in propane. Steric hindrance increases the relative energy of the eclipsed form.

and eclipsed. However, the activation barrier separating the two is 3.4 kcal mol^{-1}, slightly higher than that for ethane. This energy difference is due to unfavorable interference between the methyl substituent and the eclipsing hydrogen in the transition state, a phenomenon called **steric hindrance.** This effect arises from the fact that two molecular fragments cannot occupy the same region in space.

Steric hindrance in propane is actually worse than the E_a value for rotation indicates. Methyl substitution raises the energy not only of the eclipsed conformation, but also of the staggered (lowest energy, or *ground* state) one, the latter to a lesser extent because of less hindrance. Because the activation energy is equal to the *difference* in energy between ground and transition states, the net result is a small increase in E_a.

There can be more than one staggered and one eclipsed conformation: conformational analysis of butane

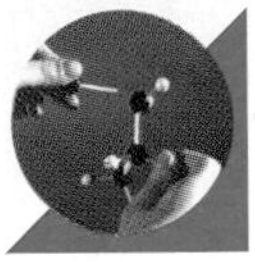

If we build a model and look at the rotation about the central C–C bond of butane, we find that there are more conformations than one staggered and one eclipsed (Figure 2-11). Consider the staggered conformer in which the two methyl groups are as far away from each other as possible. This arrangement, called *anti* (i.e., opposed), is the most stable because steric hindrance is minimized. Rotation of the rear carbon in the Newman projections in either direction (in Figure 2-11, the direction is clockwise) produces an eclipsed conformation with two CH$_3$–H interactions. This rotamer is 3.8 kcal mol^{-1} higher in energy than the *anti* precursor. Further rotation furnishes a *new* staggered structure in which the two methyl groups are closer than they are in the *anti* conformation. To distinguish this conformer from the others, it is named *gauche* (*gauche,* French, in the sense of awkward, clumsy). As a consequence of

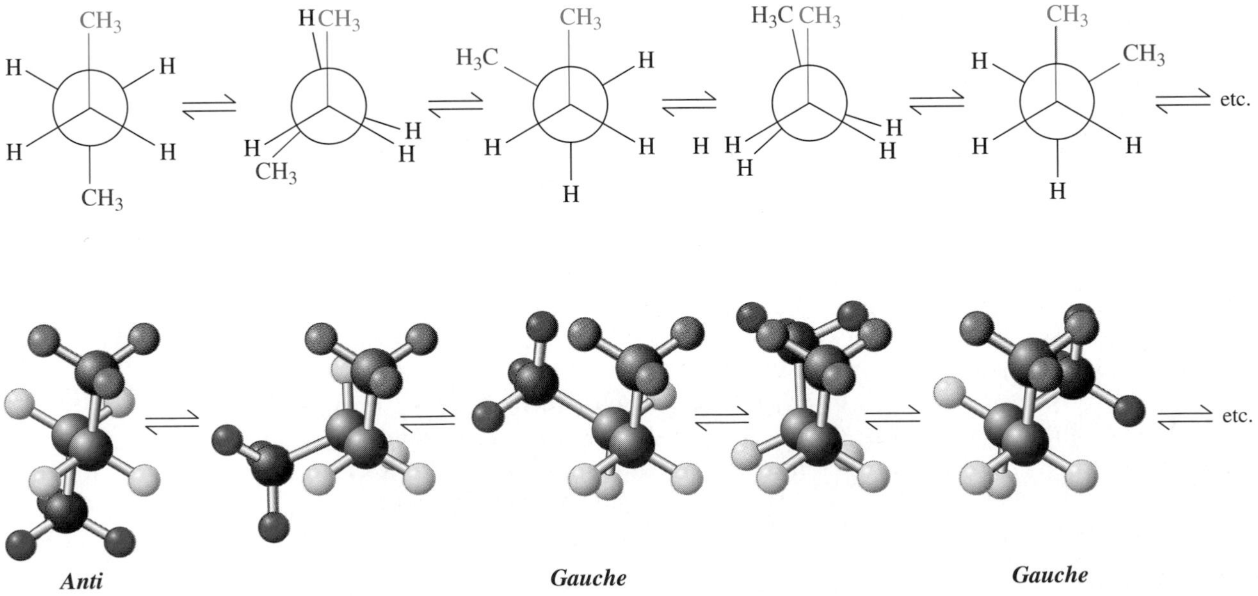

FIGURE 2-11

Clockwise rotation of the rear carbon along the C2–C3 bond in a Newman projection (top) and a ball-and-stick model (bottom) of butane.

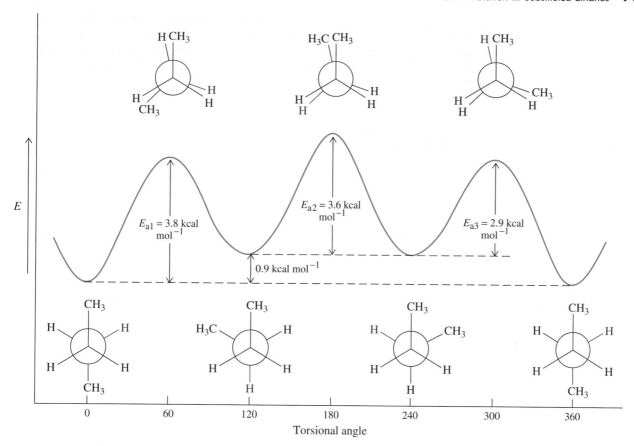

FIGURE 2-12

Potential-energy diagram of the rotation about the C2–C3 bond in butane. There are three processes: *anti* → *gauche* conversion with $E_{a1} = 3.8$ kcal mol^{-1}; *gauche* → *gauche* rotation with $E_{a2} = 3.6$ kcal mol^{-1}; and *gauche* → *anti* transformation with $E_{a3} = 2.9$ kcal mol^{-1}.

steric hindrance, the *gauche* conformer is higher in energy than the *anti* conformer by about 0.9 kcal mol^{-1}.

Further rotation (Figure 2-11) results in a *new* eclipsed arrangement in which the two methyl groups are superposed. Because the two bulkiest substituents eclipse in this rotamer, it is energetically highest, 4.5 kcal mol^{-1} higher than the most stable *anti* structure. Further rotation produces another *gauche* conformer. The activation energy for *gauche* ⇌ *gauche* interconversion is 3.6 kcal mol^{-1}. A potential-energy diagram summarizes the energetics of the rotation (Figure 2-12). The most stable *anti* conformer is the most abundant in solution (about 72% at 25°C). Its less stable *gauche* counterpart is present in lower concentration (28%).

We can see from Figure 2-12 that knowing the difference in thermodynamic stability of two conformers (e.g., 0.9 kcal mol^{-1} between the *anti* and *gauche* isomers) and the activation energy for proceeding from the first to the second (e.g., 3.8 kcal mol^{-1}) allows us to estimate the activation barrier of the reverse reaction. In this case, E_a for the *gauche*-to-*anti* conversion is $3.8 - 0.9 = 2.9$ kcal mol^{-1}.

EXERCISE 2-6

Draw the expected potential-energy diagram for the rotation about the C2–C3 bond in 2,3-dimethylbutane. Include the Newman projections of each staggered and eclipsed conformation.

In summary, if two substituents (one on each carbon atom) are 180° apart in a staggered Newman projection, they belong to an *anti* conformer. If they are 60° apart, the conformer is *gauche. Anti* conformations are usually more stable than their *gauche* counterparts. Conformational analysis is the study of the changes in potential energy that take place on rotation about single bonds.

2-8 Kinetics and Thermodynamics of Conformational Isomerism and of Simple Reactions

The *anti* $\rightleftharpoons$ *gauche* conformational isomerism is a typical example of an equilibrium between two distinct species. Although no bonds are broken or made, as in the usual chemical reaction, the process is controlled by the same physical criteria as are ordinary reactions.

We shall review two basic principles governing chemical reactions:

1. **Chemical thermodynamics** deals with the changes in energy that take place when processes such as conformational changes or chemical reactions occur, a feature that controls the *extent* to which a reaction will go to completion.
2. **Chemical kinetics** concerns the velocity or rate at which the concentrations of reactants and products change, in other words the *speed* at which a reaction will go to completion.

The two phenomena are frequently related. Reactions that are thermodynamically very favorable often proceed faster than do less favorable ones. Conversely, some reactions are faster than others even though they result in a comparatively less stable product. A transformation that yields the most stable products is said to be under **thermodynamic control.** Its outcome is determined by the net favorable change in energy in going from starting materials to products. A reaction in which the product obtained is the one formed fastest is defined as being under **kinetic control.** Let us put these statements on a more quantitative footing.

Equilibria are governed by the thermodynamics of chemical change

All chemical reactions are reversible, and reactants and products interconvert to various degrees. When the concentrations of reactants and products no longer change, the reaction is in a **state of equilibrium.** In many cases, equilibrium lies extensively (say, more than 99.9%) on the side of the products. When this occurs, the reaction is said to have *gone to completion.* (In such cases, the arrow indicating the reverse reaction is usually omitted.) Equilibria are described by equilibrium constants, K. To find an equilibrium constant, divide the arithmetic product of the concentrations of the components on the right side of the reaction by that of the components on the left, all given in units of moles per liter (mol L^{-1}). A large value for K indicates that a reaction will go to completion; it is said to have a large **driving force.**

Typical Chemical Equilibria

$$A \xrightleftharpoons{K} B \qquad K = \frac{[B]}{[A]}$$

$$A + B \xrightleftharpoons{K} C + D \qquad K = \frac{[C][D]}{[A][B]}$$

TABLE 2-5	Equilibria and Free Energy for $A \rightleftharpoons B$; $K = [B]/[A]$		
	Percentage		
K	**B**	**A**	**$\Delta G°$** (kcal mol^{-1} at 25°C)
0.01	0.99	99.0	+2.73
0.1	9.1	90.9	+1.36
0.33	25	75	+0.65
1	50	50	0
2	67	33	−0.41
3	75	25	−0.65
4	80	20	−0.82
5	83	17	−0.95
10	90.9	9.1	−1.36
100	99.0	0.99	−2.73
1,000	99.9	0.1	−4.09
10,000	99.99	0.01	−5.46

If a reaction has gone to completion, a certain amount of energy has been released. The equilibrium constant can be related directly to the thermodynamic function of the **Gibbs* standard free energy change, $\Delta G°$,†** at equilibrium:

$$\Delta G° = -RT \ln K = -2.303 \, RT \log K \text{ (in kcal mol}^{-1})$$

in which R is the gas constant (1.986 cal deg^{-1} mol^{-1}) and T is the absolute temperature in kelvins‡ (K). A negative value for $\Delta G°$ signifies a release of energy. The equation shows that a large value for K indicates a large favorable free energy change. At room temperature (25°C, 298 K), the preceding equation becomes

$$\Delta G° = -1.36 \log K \text{ (in kcal mol}^{-1})$$

This expression tells us that an equilibrium constant of 10 would have a $\Delta G°$ of −1.36 kcal mol^{-1}, and, conversely, a K of 0.1 would have a $\Delta G° = +1.36$ kcal mol^{-1}. Because the relation is logarithmic, changing the $\Delta G°$ value affects the K value exponentially. When $K = 1$, starting materials and products are present in equal concentrations and $\Delta G°$ is zero (Table 2-5).

*Professor Josiah Willard Gibbs (1839–1903), Yale University, Connecticut.
†The descriptor $\Delta G°$ refers to the free energy of a reaction with the molecules in their standard states (e.g., ideal molar solutions) after it has reached equilibrium.
‡Temperature intervals in kelvins and degrees Celsius are identical. Temperature units are named after Lord Kelvin, Sir William Thomson (1824–1907), University of Glasgow, Scotland, and Anders Celsius (1701–1744), University of Uppsala, Sweden.

EXERCISE 2-7

Calculate the equilibrium concentration of *gauche* butane at 25°C and at 100°C. Use data from Figure 2-12. (**Hint:** Recall that $\Delta G°$ for a process in the thermodynamically favored direction—less stable to more stable—is negative.)

The free energy change is related to changes in bond strengths and the degree of order in the system

The Gibbs standard free energy change is related to two other thermodynamic quantities: the change in **enthalpy,** $\Delta H°$, and the change in **entropy,** $\Delta S°$.

Gibbs Standard Free Energy Change

$$\Delta G° = \Delta H° - T\Delta S°$$

In this equation, T is again in kelvins and $\Delta H°$ in kcal mol^{-1}, whereas $\Delta S°$ is in cal deg^{-1} mol^{-1}, also called entropy units (e.u.).

The **enthalpy change,** $\Delta H°$, is the heat of a reaction at constant pressure. Enthalpy changes in an organic chemical reaction relate mainly to changes in bond strengths in the course of the reaction. Thus, the value of $\Delta H°$ can be estimated by subtracting the sum of the strengths of the bonds formed from that of the bonds broken.

Enthalpy Change in a Reaction

$$\left(\begin{matrix}\text{Sum of strengths} \\ \text{of bonds broken}\end{matrix}\right) - \left(\begin{matrix}\text{sum of strengths} \\ \text{of bonds formed}\end{matrix}\right) = \Delta H°$$

If the bonds formed are stronger than those broken, the value of $\Delta H°$ is negative and the reaction is defined as **exothermic** (releasing heat). In contrast, a positive $\Delta H°$ is characteristic of an **endothermic** (heat-absorbing) process. An example of an exothermic reaction is the combustion of methane, the main component of natural gas, to carbon dioxide and liquid water. This process has a $\Delta H°$ value of -213 kcal mol^{-1}.

$$CH_4 + 2\ O_2 \rightarrow CO_2 + 2\ H_2O_{liq} \qquad \Delta H° = -213 \text{ kcal mol}^{-1}$$

The exothermic nature of this reaction is due to the very strong bonds formed in the products. Many hydrocarbons release a lot of energy on combustion and are therefore valuable fuels.

If the enthalpy of a reaction strongly depends on changes in bond strength, what is the significance of $\Delta S°$? The **entropy change,** $\Delta S°$, is a measure of the changes in the order of a system. The value of $S°$ increases with increasing disorder. Because of the negative sign in front of the $T\Delta S°$ term in the equation for $\Delta G°$, a positive value for $\Delta S°$ makes a negative contribution to the free energy of the system. In other words, going from order to disorder is thermodynamically favorable.

What is meant by disorder in a chemical reaction? Consider a transformation in which the number of reacting molecules differs from the number of product molecules formed. For example, upon strong heating, 1-pentene undergoes cleavage into ethene and propene. This process, in which two molecules are made from one, has a relatively large positive $\Delta S°$. The increased number of particles present after bond cleavage means greater freedom of motion, thus representing an increase in disorder for the system.

$$CH_3CH_2CH_2CH{=}CH_2 \longrightarrow CH_2{=}CH_2 + CH_3CH{=}CH_2 \qquad \begin{array}{l} \Delta H° = +22.4 \text{ kcal mol}^{-1} \\ \Delta S° = +33.3 \text{ e.u.} \end{array}$$

        1-Pentene   **Ethene**   **Propene**
                                  (Ethylene)

EXERCISE 2-8

Calculate the $\Delta G°$ at 25°C for the preceding reaction. Is it thermodynamically feasible at 25°C? What is the effect of increasing T on $\Delta G°$? What is the temperature at which the reaction becomes favorable?

In contrast, disorder and entropy decrease when the number of product molecules is less than the number of molecules of starting materials. For example, the reaction of ethene (ethylene) with hydrogen chloride to give chloroethane is exothermic by -15.5 kcal mol^{-1}, but the entropy makes an unfavorable contribution to the $\Delta G°$; $\Delta S° = -31.3$ e.u.

$$CH_2{=}CH_2 + HCl \longrightarrow CH_3CH_2Cl \qquad \begin{array}{l} \Delta H° = -15.5 \text{ kcal mol}^{-1} \\ \Delta S° = -31.3 \text{ e.u.} \end{array}$$

EXERCISE 2-9

Calculate the $\Delta G°$ at 25°C for the preceding reaction. In your own words, explain why a reaction that combines two molecules into one should have a large negative entropy change.

The rate of a chemical reaction depends on the activation energy

How fast is equilibrium established? The thermodynamic features of chemical reactions do not by themselves tell us anything about their rates. Consider the conversion of *gauche* butane into the *anti* rotamer (Figure 2-12). This process is thermodynamically favorable by only a small amount, and yet equilibrium is established exceedingly rapidly, even at very low temperatures. Now compare that with the combustion of methane considered earlier. This process releases 213 kcal mol^{-1}, a huge amount of energy, but we know that methane does not spontaneously ignite in air at room temperature. Why is the much more favorable combustion process so much slower? The answer is that rates of chemical processes are controlled by activation energies. We have already seen that E_a for bond rotation in butane is very low, which corresponds to a low-energy transition state, through which the molecule may pass very rapidly. Conversely, the transition state for methane combustion is very high in energy, corresponding to a high E_a and a very low rate (Figure 2-13).

How can there be such high activation energies for exothermic reactions? A simple answer is that partial bond-breaking usually precedes partial bond formation. Thus, before energy is released through bonding, energy must be expended to break bonds. The transition state is the point at which the initial energy input is compensated by a corresponding release of energy.

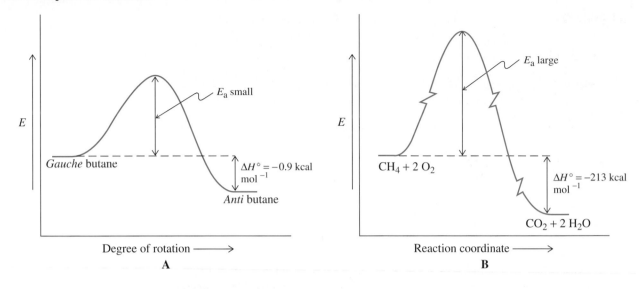

FIGURE 2-13
Comparison of potential-energy diagrams for (A) *gauche–anti* conversion in butane and (B) the combustion of methane. Comparison of the activation energies explains why bond rotation in butane is so much faster, even though the combustion reaction of methane is so much more thermodynamically favorable, as shown by its large negative $\Delta H°$. The diagrams are not drawn to scale.

The concentration of reactants can affect reaction rates

Consider the addition of reagent A to reagent B to give C:

$$A + B \longrightarrow C$$

In many transformations of this type, increasing the concentration of either reactant increases the rate of the reaction. In such cases, the transition-state structure incorporates both molecules A and B. The rate is expressed by

$$\text{Rate} = k[\text{A}][\text{B}] \text{ in units of mol L}^{-1} \text{ s}^{-1}$$

in which the proportionality constant, *k,* is also called the **rate constant** of the reaction. A reaction for which the rate depends on the concentrations of two molecules in this way is said to be of **second order.**

There are processes whose rate depends on the concentration of only one reactant, such as the hypothetical reaction

$$A \longrightarrow B$$
$$\text{Rate} = k[\text{A}] \text{ in units of mol L}^{-1} \text{ s}^{-1}$$

A reaction of this type is said to be of **first order.** Rotation about a carbon–carbon bond follows such a rate law.

EXERCISE 2-10

The dependence of reaction rates on reactant concentrations means that reactions slow down as starting materials are used up. For example, for a process following the first-order rate law, rate = $k[A]$, when half of A is consumed (i.e., after 50% conversion of the starting material), the reaction rate is reduced to half of its initial value. What would be the reduction in rate of a second-order reaction (in which rate = $k[A][B]$) after 50% conversion of the starting material?

The Arrhenius equation describes how temperature affects reaction rates

Temperature also greatly affects reaction rates. An increase in temperature leads to faster reactions. The kinetic energy of molecules increases when they are heated, which means that a larger fraction of them have sufficient energy to overcome the activation barrier (Figure 2-9). A useful rule of thumb applies to many reactions: Raising the reaction temperature by 10 degrees (Celsius) causes the rate to increase by a factor of 2 or 3. The Swedish chemist Arrhenius* noticed the dependence of reaction rate on temperature. He found that his measured data conformed to the equation

> **Arrhenius Equation**
>
> $$k = Ae^{-E_a/RT}$$

The A term is the maximum rate constant that the reaction would have if every molecule had sufficient collisional energy to overcome the activation barrier: At very high temperature, E_a/RT is small, $e^{-E_a/RT}$ approaches 1, and k nearly equals A. Each reaction has its own characteristic value for A. The Arrhenius equation describes how rates of reactions with different activation energies vary with temperature.

EXERCISE 2-11

(a) Calculate $\Delta G°$ at 25°C for the reaction $CH_3CH_2Cl \rightarrow CH_2{=}CH_2 + HCl$ (the reverse of the reaction in Exercise 2-9). **(b)** Calculate $\Delta G°$ at 500°C for the same reaction. (**Hint:** Apply $\Delta G° = \Delta H° - T\Delta S°$ and do not forget to first convert degrees Celsius into kelvins.)

EXERCISE 2-12

For the reaction in Exercise 2-11, $A = 10^{14}$ and $E_a = 58.4$ kcal mol^{-1}. Using the Arrhenius equation, calculate k at 500°C for this reaction. $R = 1.986$ cal deg^{-1} mol^{-1}.

This completes our review of the thermodynamic and kinetic relations governing many organic transformations. All reactions are described by equilibrating starting materials and products. On which side the equilibrium lies depends on the size of the equilibrium constant, in turn related to the Gibbs free energy changes, $\Delta G°$. An increase in the equilibrium constant by a factor of 10 is associated with a change in $\Delta G°$ of about -1.36 kcal mol^{-1} at 25°C. The free energy change of a reaction is

---

*Professor Svante Arrhenius (1859–1927), Technical Institute of Stockholm, Sweden, Nobel Prize 1903 (chemistry), director of the Nobel Institute from 1905 until shortly before his death.

composed of changes in enthalpy, $\Delta H°$, and entropy, $\Delta S°$. Contributions to the former stem mainly from variations in bond strengths, to the latter from the relative disorder in starting materials and products. Whereas these terms define the position of an equilibrium, the rate at which it is established depends on the concentration of starting materials, the activation barrier separating reactants and products, and the temperature. The relation between rate, E_a, and T is expressed by the Arrhenius equation.

2-9 Acids and Bases: A Review

In this chapter we have begun our study of organic compounds with the nonfunctional alkanes, and we have briefly examined the variety of functionalized compounds that may be derived from them. We have also looked into the energetic aspects of one of the physical processes common to virtually all organic molecules, rotation about single bonds. Now we turn to a fundamental application of thermodynamics, the chemistry of acids and bases. We shall see that acid-base processes provide models for the reactivity of polar organic functional groups. This section reviews the way in which acids and bases interact and how this process is quantified.

Acid and base strengths are measured by equilibrium constants

Brønsted and Lowry have given us a simple definition of acids and bases: An **acid** is a proton donor and a **base** is a proton acceptor. Acidity and basicity are commonly measured in water. An acid donates protons to water to give the hydronium ion, whereas a base removes them to give the hydroxide ion. Examples are hydrogen chloride for the former and sodium methoxide for the latter.

$$H—Cl: + HOH \rightleftharpoons H—O:^+ + :Cl:^-$$

Hydronium ion

$$CH_3O:^- Na^+ + HOH \rightleftharpoons CH_3OH + Na^+ + {}^-:OH$$

Hydroxide ion

Water itself is neutral. It forms an equal number of hydronium and hydroxide ions by self-dissociation. The process is described by the equilibrium constant K_w, the self-ionization constant of water. At 25°C,

$$H_2O + H_2O \underset{}{\overset{K_w}{\rightleftharpoons}} H_3O^+ + OH^- \qquad K_w = [H_3O^+][OH^-] = 10^{-14}\ mol^2\ L^{-2}$$

From the value for K_w, it follows that the concentration of H_3O^+ in pure water is 10^{-7} mol L^{-1}.

The pH is defined as the negative logarithm of the value for $[H_3O^+]$.

$$pH = -\log [H_3O^+]$$

Thus, for pure water, the pH is +7. An aqueous solution with a pH lower than 7 is acidic; that with a pH higher than 7 is basic.

The acidity of a general acid, HA, is expressed by the following general equation, together with its associated equilibrium constant.

$$HA + H_2O \xrightleftharpoons{K} H_3O^+ + A^- \qquad K = \frac{[H_3O^+][A^-]}{[HA][H_2O]}$$

Because, in aqueous solution, $[H_2O]$ is constant at 55 mol L^{-1}, that number may be incorporated into a new constant, the **acidity constant,** K_a.

$$K_a = K[H_2O] = \frac{[H_3O^+][A^-]}{[HA]} \text{ mol } L^{-1}$$

Like the concentration of H_3O^+ and its relation to pH, this measurement may be put on a logarithmic scale by the corresponding definition of pK_a.

$$pK_a = -\log K_a$$

The pK_a is the pH at which the acid is 50% dissociated. An acid with a pK_a lower than 1 is defined as strong, one with a pK_a higher than 4 as weak. The acidities of several common acids are compiled in Table 2-6 and compared with those of com-

TABLE 2-6	Relative Strengths of Common Acids (25°C)	
Acid	K_a	pK_a
Hydrogen iodide, HI (strongest acid)	1.6×10^5	−5.2
Sulfuric acid, H_2SO_4	1.0×10^5	−5.0[a]
Hydrogen bromide, HBr	5.0×10^4	−4.7
Hydrogen chloride, HCl	160	−2.2
Hydronium ion, H_3O^+	50	−1.7
Methanesulfonic acid, CH_3SO_3H	16	−1.2
Hydrogen fluoride, HF	6.3×10^{-4}	3.2
Acetic acid, CH_3COOH	2.0×10^{-5}	4.7
Hydrogen cyanide, HCN	6.3×10^{-10}	9.2
Methanethiol, CH_3SH	1.0×10^{-10}	10.0
Methanol, CH_3OH	3.2×10^{-16}	15.5
Water, H_2O	2.0×10^{-16}	15.7
Ammonia, NH_3	1.0×10^{-35}	35
Methane, CH_4 (weakest acid)	$\sim 1.0 \times 10^{-50}$	~50

Note: $K_a = [H_3O^+][A^-]/[HA]$ mol L^{-1}.
[a]First dissociation equilibrium.

pounds with higher pK_a values. Sulfuric acid and, with the exception of HF, the hydrogen halides, are very strong acids. Hydrogen cyanide, water, methanol, ammonia, and methane are decreasingly acidic, the last two being exceedingly weak.

Like acid dissociation, the protonation of bases and their basicity can be described by a corresponding set of equations. The basicity of a base, A^-, is governed by the equilibrium constant K'.

$$A^- + H_2O \underset{}{\overset{K'}{\rightleftharpoons}} HO^- + HA \qquad K' = \frac{[HO^-][HA]}{[A^-][H_2O]}$$

By incorporation of the constant value for $[H_2O]$, this equilibrium constant transforms into the basicity constant, K_b, and gives rise to a set of pK_b values.

$$K_b = K'[H_2O] = \frac{[HO^-][HA]}{[A^-]} \text{ mol } L^{-1}$$

For an acid HA and its derived base A^-, K_a and K_b are related by simple multiplication.

$$K_a \times K_b = \frac{[H_3O^+][A^-]}{[HA]} \times \frac{[HO^-][HA]}{[A^-]} = [H_3O^+][HO^-] = K_w = 10^{-14}$$

We see that the product of the two is equal to the self-ionization constant of water. Hence,

$$pK_a + pK_b = 14$$

Therefore, if we know the pK_a of an acid HA, we automatically know the pK_b of A^-. Because of this relation, the species A^- derived from HA is frequently referred to as its **conjugate** base (*conjugatus,* Latin, joined). Conversely, a species HA would be the conjugate acid of base A^-. For example, Cl^- is the conjugate base of HCl, and CH_3OH is the conjugate acid of CH_3O^-. Or, HCl may be viewed as the conjugate acid of Cl^-, and CH_3O^- as the conjugate base of CH_3OH. It follows from this discussion that the conjugate base of a strong acid is weak, as is the conjugate acid of a strong base.

EXERCISE 2-13

Write the formula for the conjugate base of each of the following acids. **(a)** Sulfurous acid, H_2SO_3; **(b)** chloric acid, $HClO_3$; **(c)** hydrogen sulfide, H_2S; **(d)** dimethyloxonium, $(CH_3)_2OH^+$; **(e)** hydrogen sulfate, HSO_4^-.

EXERCISE 2-14

Write the formula for the conjugate acid of each of the following bases. **(a)** Dimethylamide, $(CH_3)_2N^-$; **(b)** sulfide, S^{2-}; **(c)** ammonia, NH_3; **(d)** propanone (acetone), $(CH_3)_2C{=}O$; **(e)** 2,2,2-trifluoroethoxide, $CF_3CH_2O^-$.

EXERCISE 2-15

Which is the stronger acid, nitrous (HNO_2, $pK_a = 3.3$) or phosphorous acid (H_3PO_3, $pK_a = 1.3$)? Calculate K_a for each.

We can predict relative acid and base strengths

Are there structural features that allow us to predict, at least qualitatively, the strength of an acid HA (and hence the weakness of its conjugate base)? Yes, there are several. Prominent among them are two:

1. The increasing *size* of A as we proceed down a column in the periodic table. This trend is seen in the ordering of the acid strengths of the hydrogen halides: HI > HBr > HCl > HF. Electron repulsion in the anionic conjugate base diminishes as the volume of space available for the charge to "spread out" increases.

2. The ability of the conjugate base A$^-$ to accommodate the negative charge in either or both of two ways:

   (a) The increasing *electronegativity* of A as we proceed from left to right across a row in the periodic table. The more electronegative the atom to which the acidic proton is attached, the more acidic the latter will be. For example, the decreasing order of acidity in the series HF > H_2O > H_3N > H_4C parallels the decreasing electronegativity of A (Table 1-2). In the hydrogen halides, this trend is outweighed by the size of A.

   (b) The *resonance* in A$^-$ that allows delocalization of charge over several atoms. For example, acetic acid is more acidic than methanol. In both cases, an O–H bond dissociates into ions (heterolytic cleavage, Section 3-1). However, unlike methoxide, the acetate ion has two resonance structures to better accommodate the charge (Section 1-5) and is the weaker base.

Acetic Acid Is More Acidic Than Methanol Because of Resonance

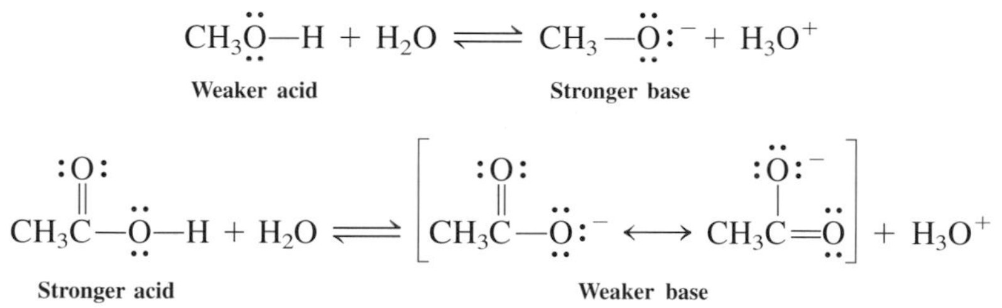

$$CH_3\overset{..}{\underset{..}{O}}\!\!-\!\!H + H_2O \rightleftharpoons CH_3\!-\!\overset{..}{\underset{..}{O}}\!:^- + H_3O^+$$

Weaker acid   Stronger base

$$CH_3\overset{\displaystyle :O:}{\overset{\|}{C}}\!\!-\!\!\overset{..}{\underset{..}{O}}\!\!-\!\!H + H_2O \rightleftharpoons \left[CH_3\overset{\displaystyle :O:}{\overset{\|}{C}}\!\!-\!\!\overset{..}{\underset{..}{O}}\!:^- \longleftrightarrow CH_3\overset{\displaystyle :\overset{..}{O}:^-}{\overset{|}{C}}\!\!=\!\!\overset{..}{\underset{..}{O}} \right] + H_3O^+$$

Stronger acid   Weaker base

The effect of resonance is even more pronounced in sulfuric acid. The availability of *d* orbitals on sulfur enables us to write valence-shell-expanded Lewis structures containing as many as 12 electrons (Sections 1-4 and 1-5). Alternatively, charge-separated structures with one or two positive charges on sulfur can be used. Both representations indicate that the pK_a of H_2SO_4 should be low.

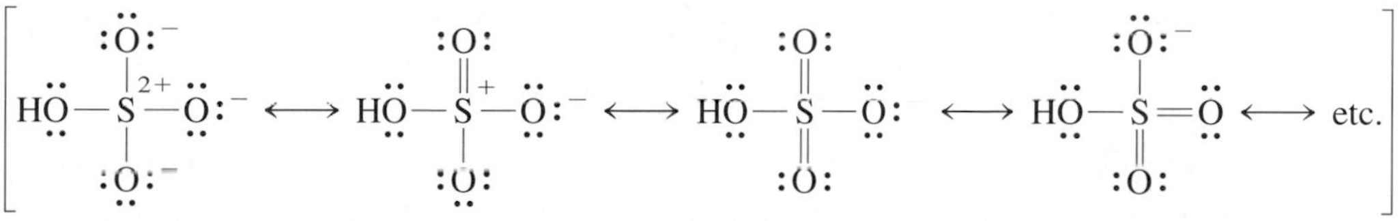

$$\left[\begin{array}{cccc} :\overset{..}{O}:^- & :O: & :O: & :\overset{..}{O}:^- \\ | & \| & \| & | \\ H\overset{..}{O}\!-\!\overset{2+}{S}\!-\!\overset{..}{O}:^- & H\overset{..}{O}\!-\!\overset{+}{S}\!-\!\overset{..}{O}:^- & H\overset{..}{O}\!-\!S\!-\!\overset{..}{O}: & H\overset{..}{O}\!-\!S\!=\!\overset{..}{O} \\ | & | & \| & \| \\ :\overset{..}{O}:^- & :\overset{..}{O}: & :O: & :O: \end{array} \longleftrightarrow \text{etc.} \right]$$

Hydrogen sulfate ion

The sulfonic acids also are quite strong, for the same reason. Consequently their conjugate bases, the sulfonates, are weak bases. As a rule, the acidity of HA increases to the right and down in the periodic table. Therefore, the basicity of A$^-$ *decreases* in the same fashion.

The same molecule may act as an acid under one set of conditions and as a base under another. Water is the most familiar example of this behavior, but many other substances possess this capability as well. For instance, nitric acid acts as an acid in the presence of water but behaves as a base toward the more powerfully acidic H_2SO_4:

Nitric Acid Acting as an Acid

$$HNO_3 + H_2O \rightleftharpoons NO_3^- + H_3O^+$$

Nitric Acid Acting as a Base

$$H_2SO_4 + HNO_3 \rightleftharpoons HSO_4^- + H_2NO_3^+$$

Similarly, acetic acid protonates water, as shown earlier in this section, but is protonated by stronger acids such as HBr:

$$HBr + CH_3CO_2H \rightleftharpoons Br^- + CH_3CO_2H_2^+$$

EXERCISE 2-16

Suggest a structure for $CH_3CO_2H_2^+$. [**Hint:** Try placing the proton first on one, and then on the other of the two oxygen atoms in the molecule ($CH_3\overset{\displaystyle O}{\overset{\|}{C}}-OH$), and consider which of the two resulting structures is better stabilized by resonance.]

Lewis acids and bases interact by sharing an electron pair

A more generalized description of acid-base interaction in terms of electron sharing was introduced by Lewis. A **Lewis acid** is a species that is at least two electrons short of a closed shell, whereas a **Lewis base** contains at least one lone pair of electrons.

Lewis Acids

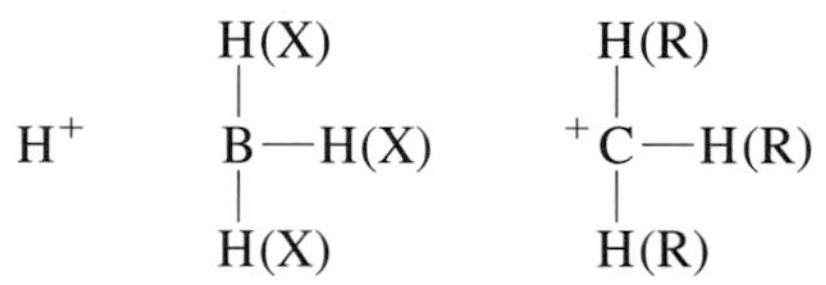

$$H^+ \qquad \overset{\displaystyle H(X)}{\underset{\displaystyle H(X)}{B}}-H(X) \qquad \overset{\displaystyle H(R)}{\underset{\displaystyle H(R)}{\overset{+}{C}}}-H(R) \qquad \begin{array}{c} MgX_2,\ AlX_3,\ \text{many} \\ \text{transition metal} \\ \text{halide salts} \end{array}$$

Lewis Bases

$$^-:\ddot{O}H(R) \qquad :\ddot{O}\underset{\displaystyle H(R)}{-}H(R) \qquad :\ddot{S}\underset{\displaystyle H(R)}{-}H(R) \qquad :N\underset{\displaystyle H(R)}{\overset{\displaystyle H(R)}{-}}H(R) \qquad :P\underset{\displaystyle H(R)}{\overset{\displaystyle H(R)}{-}}H(R) \qquad :\ddot{X}:$$

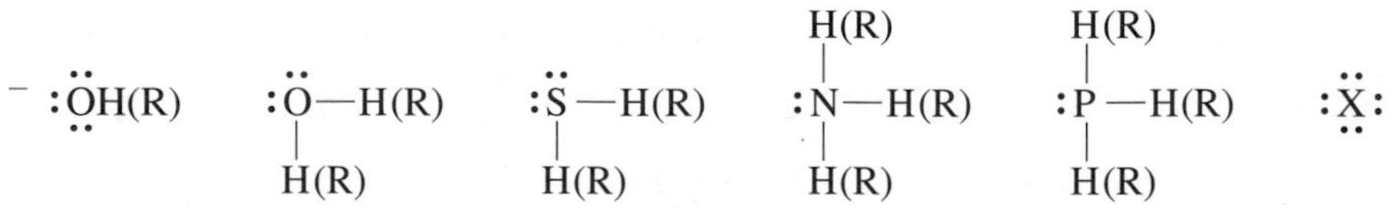

A Lewis base shares its lone pair with a Lewis acid to form a new covalent bond. From Section 1-5, we know that organic chemists routinely depict the movement of electron pairs through the use of curved arrows. A Lewis base–Lewis acid interaction may therefore be pictured by means of an arrow pointing in the direction that the electron pair moves—from the base to the acid. The Brønsted acid-base reaction between hydroxide ion and a proton is an example of a Lewis acid-base process as well.

Lewis Acid-Base Reactions

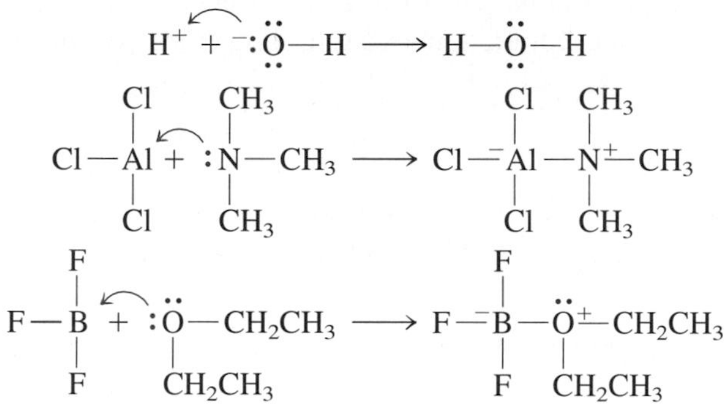

$$H^+ + \ ^-\!:\!\ddot{\underset{..}{O}}\!-\!H \longrightarrow H\!-\!\ddot{\underset{..}{O}}\!-\!H$$

$$Cl\!-\!\underset{\underset{Cl}{|}}{\overset{\overset{Cl}{|}}{Al}} + \ :\!\underset{\underset{CH_3}{|}}{\overset{\overset{CH_3}{|}}{N}}\!-\!CH_3 \longrightarrow Cl\!-\!\underset{\underset{Cl}{|}}{\overset{\overset{Cl}{|}}{Al}}\!-\!\underset{\underset{CH_3}{|}}{\overset{\overset{CH_3}{|}}{N^\pm}}\!-\!CH_3$$

$$F\!-\!\underset{\underset{F}{|}}{\overset{\overset{F}{|}}{B}} + \ :\!\ddot{\underset{..}{O}}\!-\!CH_2CH_3 \longrightarrow F\!-\!\underset{\underset{F}{|}}{\overset{\overset{F}{|}}{B}}\!-\!\ddot{O^\pm}\!-\!CH_2CH_3$$

with the lower CH_2CH_3 and CH_2CH_3 groups.

The dissociation of a Brønsted acid HA is just the reverse of the combination of the Lewis acid H^+ and the Lewis base A^-. We write it as follows:

Dissociation of a Brønsted Acid

$$H\!-\!A \longrightarrow H^+ + \ :\!A^-$$

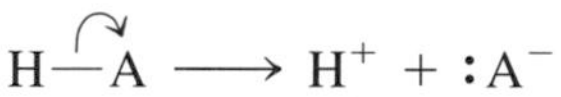

Notice that the curved arrow is from the bond *to A,* the direction in which the *pair of electrons* moves. *The curved arrow never points to the hydrogen atom in the dissociation of a Brønsted acid.*

Many processes in organic chemistry include either Lewis acid-base reactions or their reverse, dissociation of a covalent bond into ions. For example, certain haloalkanes are capable of ionization to give halide ions and alkyl cations, a process similar to dissociation of a Brønsted acid. As the curved arrow indicates, the electron pair originally constituting the C–Cl covalent bond shifts onto Cl, giving it a negative charge and leaving the carbon atom positive.

Dissociation of a Haloalkane into a Halide Ion and an Alkyl Cation

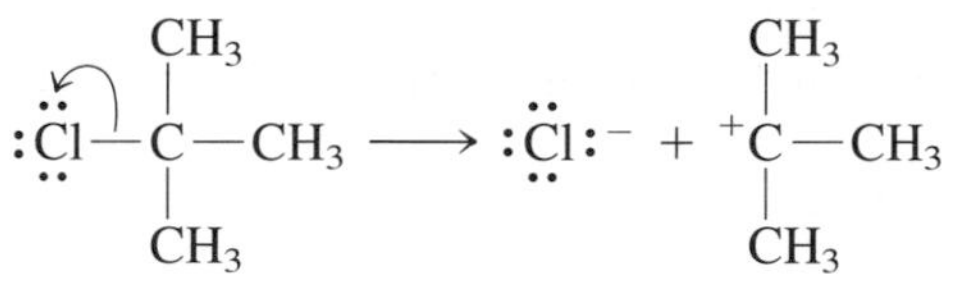

$$:\!\ddot{\underset{..}{Cl}}\!-\!\underset{\underset{CH_3}{|}}{\overset{\overset{CH_3}{|}}{C}}\!-\!CH_3 \longrightarrow \ :\!\ddot{\underset{..}{Cl}}\!:^- + \ ^+\!\underset{\underset{CH_3}{|}}{\overset{\overset{CH_3}{|}}{C}}\!-\!CH_3$$

When the alkyl cation has formed, it may react with a Lewis base such as water. These processes are two of the steps in the conversion of haloalkanes into alcohols (Section 7-1).

Lewis Acid-Base Reaction of Water and an Alkyl Cation

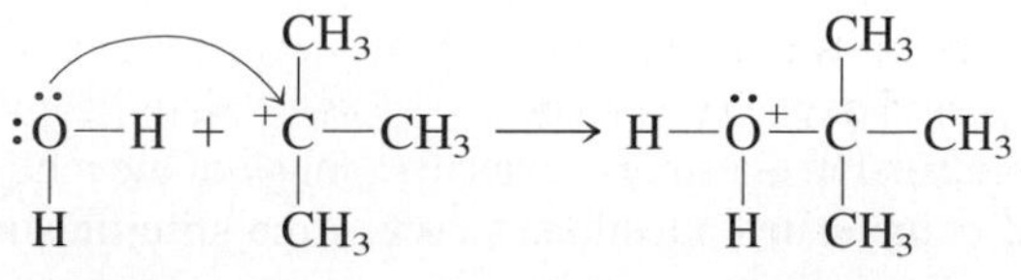

$$:\!\ddot{\underset{\underset{H}{|}}{O}}\!-\!H + \ ^+\!\underset{\underset{CH_3}{|}}{\overset{\overset{CH_3}{|}}{C}}\!-\!CH_3 \longrightarrow H\!-\!\ddot{\underset{\underset{H}{|}}{O^+}}\!-\!\underset{\underset{CH_3}{|}}{\overset{\overset{CH_3}{|}}{C}}\!-\!CH_3$$

In summary, in Brønsted-Lowry terms, acids are proton donors and bases are proton acceptors. Acid-base interactions are governed by equilibria, which are quantitatively described by an acidity constant K_a. Removal of a proton from an acid generates its conjugate base; attachment of a proton to a base forms its conjugate acid. Lewis bases donate an electron pair to form a covalent bond with Lewis acids, a process depicted by a curved arrow pointing from the lone pair of the base toward the acid.

CHAPTER INTEGRATION PROBLEM

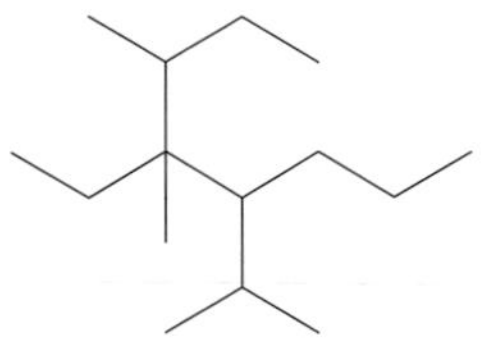

Consider the alkane shown in the margin.

a. Name this molecule according to the IUPAC system.

SOLUTION

Step 1. Locate the main, or stem, chain, the longest one in the molecule (shown in black in the illustration below). Do not be misled: The drawing of the stem chain can have almost any shape. The stem has eight carbons, so the base name is **octane.**

Step 2. Identify and name all substituents (shown in color): two **methyl** groups, an **ethyl** group, and a fourth, branched substituent. The branched substituent is named by first giving the number 1 (italicized in the illustration below) to the carbon that connects it to the main stem. By numbering away from the stem, we reach the number 2; therefore, the substituent is a derivative of the ethyl group (in green), onto which is attached a methyl group (red) at carbon 1. Thus, this substituent is called a **1-methylethyl** group.

Step 3. Number the stem chain, starting at the end closest to a carbon bearing a substituent. The numbering shown below gives a methyl-substituted carbon the number 3. Numbering the opposite way would have C4 as the lowest-numbered substituted carbon.

Step 4. Arrange the names of substituents alphabetically in the final name: *ethyl* comes first; then *methyl* comes before *methylethyl* (the "di" in *dimethyl,* denoting two methyl groups, is not considered in alphabetization because it is a multiplier of a substituent name and is therefore not considered part of the name).

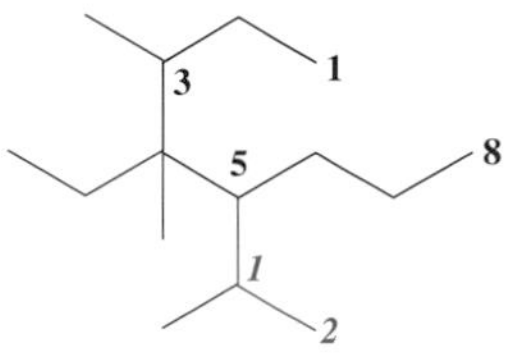

4-Ethyl-3,4-dimethyl-5-(1-methylethyl)octane

b. Draw structures to represent rotation about the C6–C7 bond. Correlate the structures that you draw with a qualitative potential energy diagram.

SOLUTION

Step 1. Identify the bond in question. Notice that much of the molecule can be treated simply as a large, complicated substituent on C6, the specific structure of which is unimportant. For the purpose of this question, this large substituent may be replaced by "R." The "action" in this problem takes place between C6 and C7:

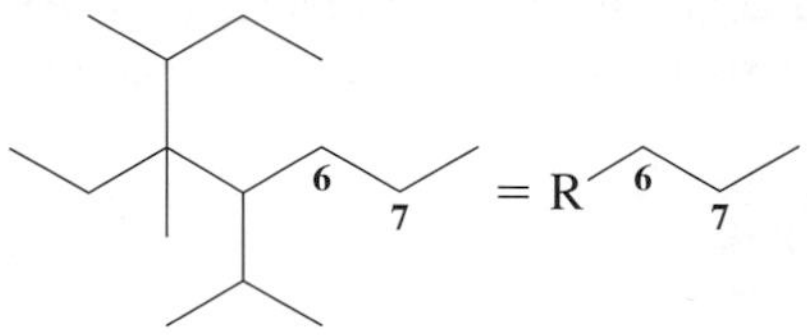

Step 2. Recognize that step 1 simplified the problem: rotation about the C6–C7 bond will give results very similar to rotation about the C2–C3 bond in butane. The only difference is that a large "R" group has replaced one of the smaller methyl groups of butane.

Step 3. Draw conformations modeled after those of butane (Section 2-7) and superimpose them on an energy diagram similar to that in Figure 2-12. The only difference between this diagram and that for butane is that we do not know the exact heights of the energy maxima relative to the energy minima. However, we can expect them to be higher, qualitatively, because our "R" group is larger than a methyl group and thus can be expected to cause greater steric hindrance.

c. Two alcohols derived from this alkane are illustrated in the margin. Alcohols are categorized on the basis of the type of carbon atom that contains the –OH group (primary, secondary, or tertiary). Characterize the alcohols shown in the margin.

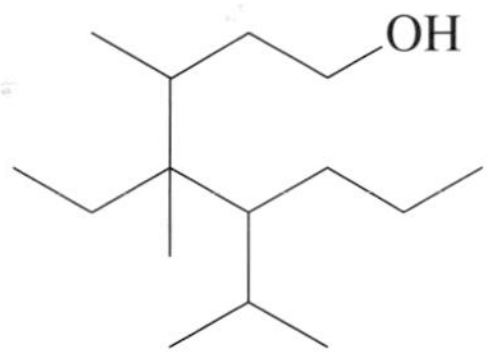

Alcohol 1

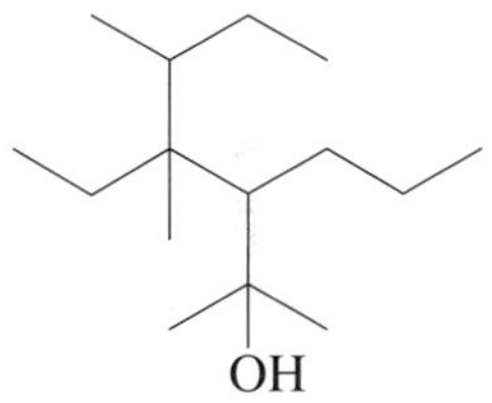

SOLUTION

In alcohol 1, the –OH group is located on a carbon atom that is directly attached to one other carbon, a primary carbon. Therefore, alcohol 1 is a primary alcohol. Similarly, the –OH group in alcohol 2 resides on a tertiary carbon (one attached to three other carbon atoms). It is a tertiary alcohol.

Alcohol 2

d. The –O–H bond in an alcohol is acidic to a similar degree to that in water. Primary alcohols have $K_a \approx 10^{-16}$; tertiary alcohols $K_a \approx 10^{-18}$. What are the approximate pK_a values for alcohols 1 and 2? Which is the stronger acid?

SOLUTION

The pK_a for alcohol 1 is approximately 16 ($-\log K_a$); that for alcohol 2 is about 18. Alcohol 1, with the lower pK_a value, is the stronger acid.

e. In which direction does the following equilibrium lie? Calculate K, the equilibrium constant, and $\Delta G°$, the free energy change, associated with the reaction as written in the left-to-right direction.

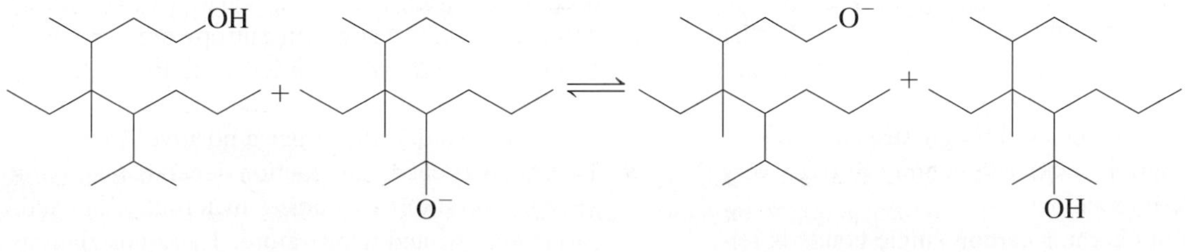

SOLUTION

The stronger acid (alcohol 1) is on the left; the weaker (alcohol 2) is on the right. Recall the relation between conjugate acids and bases: stronger acids have weaker conjugate bases, and vice versa. Relatively speaking, therefore, we have

$$\begin{array}{ccccccc}
\text{Alcohol 1} & + & \begin{array}{c}\text{Conjugate base} \\ \text{of alcohol 2}\end{array} & \rightleftharpoons & \begin{array}{c}\text{Conjugate base} \\ \text{of alcohol 1}\end{array} & + & \text{Alcohol 2} \\
\textbf{(Stronger acid)} & & \textbf{(Stronger base)} & & \textbf{(Weaker base)} & & \textbf{(Weaker acid)}
\end{array}$$

The equilibrium lies to the right, on the side of the *weaker* acid-base pair. Recall that $K > 1$ and $\Delta G° < 0$ for a reaction that is thermodynamically favorable as written from left to right; *use* this information to be sure to get the magnitude of K and the sign of $\Delta G°$ correct. The equilibrium constant, K, for the process is the ratio of the K_a values, $(10^{-16}/10^{-18}) = 10^2$ (not 10^{-2}). With reference to Table 2-5, a K value of 100 corresponds to a $\Delta G°$ of -2.73 kcal mol^{-1} (not $+2.73$). If the reaction were written in the opposite direction, with the equilibrium lying to the left, the correct values would be those in parentheses.

IMPORTANT CONCEPTS

1. An organic molecule may be viewed as being composed of a carbon skeleton with attached **functional groups.**
2. **Hydrocarbons** are made up of carbon and hydrogen only. Hydrocarbons possessing only single bonds are also called **alkanes.** They do not contain functional groups. An alkane may exist as a single continuous chain or it may be branched or cyclic. The empirical formula for the **straight-chain** and **branched alkanes** is C_nH_{2n+2}.
3. Molecules that differ only in the number of methylene groups, CH_2, in the chain are called **homologs** and are said to belong to a homologous series.
4. A **primary carbon** is attached to only one other carbon. A **secondary carbon** is attached to two and a **tertiary** to three other carbon atoms. The hydrogen atoms bound to such carbon atoms are likewise designated primary, secondary, or tertiary.
5. The **IUPAC rules** for naming saturated hydrocarbons are (a) find the longest continuous chain in the molecule and name it; (b) name all groups attached to the longest chain as alkyl substituents; (c) number the carbon atoms of the longest chain; (d) write the name of the alkane, citing all substituents as prefixes arranged in alphabetical order and preceded by numbers designating their positions.
6. Alkanes attract each other through weak **London forces,** polar molecules through stronger dipole–dipole interactions, and salts mainly through very strong ionic interactions.
7. Rotation about carbon–carbon single bonds is rel-

atively easy and gives rise to **conformations** (conformers, rotamers). Substituents on adjacent carbon atoms may be **staggered** or **eclipsed.** The eclipsed conformation is a transition state between staggered conformers. The energy required to reach the eclipsed state is called the activation energy for rotation. When both carbons bear alkyl or other groups, there may be additional conformers: Those in which the groups are in close proximity (60°) are *gauche;* those in which the groups are directly opposite (180°) each other are *anti.* Molecules tend to adopt conformations in which steric hindrance, as in *gauche* conformations, is minimized.

8. Chemical reactions can be described as equilibria controlled by **thermodynamic** and **kinetic** parameters. The change in the **Gibbs free energy, $\Delta G°$,** is related to the **equilibrium constant** by $\Delta G° = -RT \ln K = -1.36 \log K$ (at 25°C). The free energy has contributions from changes in **enthalpy, $\Delta H°$,** and **entropy, $\Delta S°$:** $\Delta G° = \Delta H° - T\Delta S°$. Changes in enthalpy are mainly due to differences between the strengths of the bonds made and those of the bonds broken. A reaction is **exothermic** when the former is larger than the latter. It is **endothermic** when there is a net loss in combined bond strengths. Changes in entropy are controlled by the relative degree of order in starting materials compared with that in products. The greater the increase in disorder, the larger a positive $\Delta S°$.
9. The rate of a chemical reaction depends mainly on the concentrations of starting material(s), the activation energy, and temperature. These correlations

are expressed in the **Arrhenius equation:** rate constant $k = Ae^{-E_a/RT}$.

10. If the rate depends on the concentration of only one starting material, the reaction is said to be of **first order.** If the rate depends on the concentrations of two reagents, the reaction is of **second order.**

11. **Brønsted acids** are proton donors; **bases** are proton acceptors. Acid strength is measured by the **acidity constant** K_a; $pK_a = -\log K_a$. Acids and their deprotonated forms have a **conjugate** relation. **Lewis acids** and **bases** are electron pair acceptors and donors, respectively.

PROBLEMS

17. For each example in Table 2-1, identify all polarized covalent bonds and label the appropriate atoms with partial positive or negative charges. (Do not consider carbon–hydrogen bonds.)

18. Circle and identify by name each functional group in the compounds pictured.

(a) 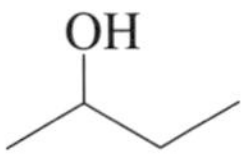 (b) 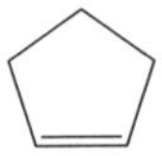 (c)  (d)

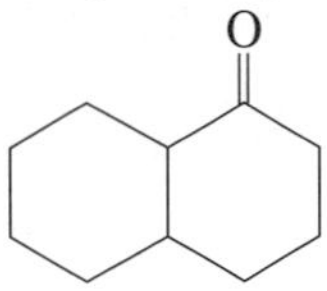

(e) 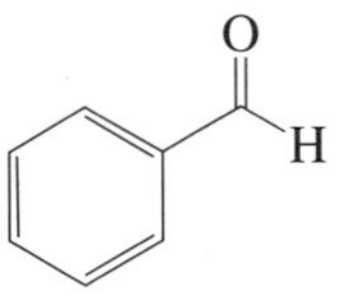 (f) 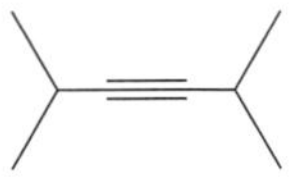 (g)

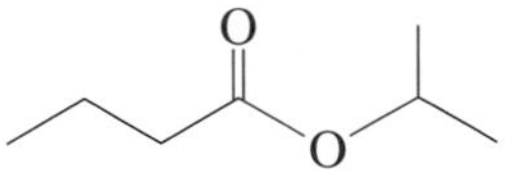

(h) 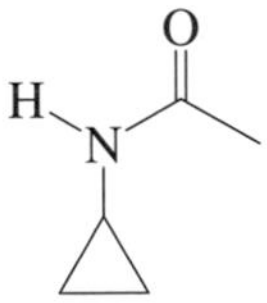 (i) 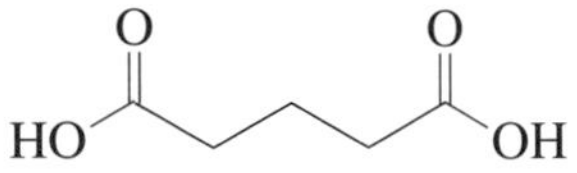 (j)

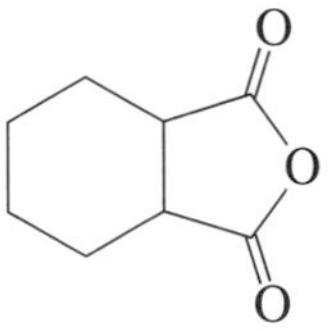

19. On the basis of electrostatics (Coulomb attraction), predict which atom in each of the following organic molecules is likely to react with the indicated reagent. Write "no reaction" if none seems likely. (See Table 2-1 for the structures of the organic molecules.) **(a)** Bromoethane, with the oxygen of HO^-; **(b)** propanal, with the nitrogen of NH_3; **(c)** methoxyethane, with H^+; **(d)** 3-pentanone, with the carbon of CH_3^-; **(e)** ethanenitrile (acetonitrile), with the carbon of CH_3^+; **(f)** butane, with HO^-.

20. Name the following molecules according to the IUPAC system of nomenclature.

(a) 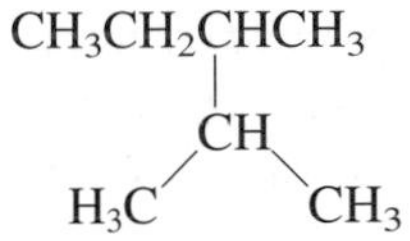 (b) 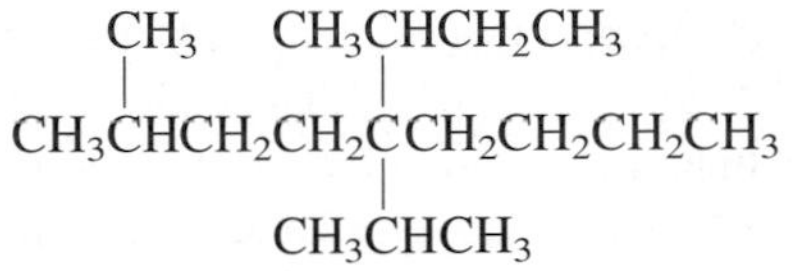 (c) 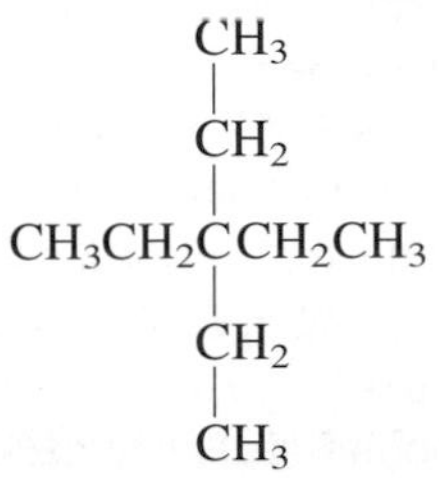 (d)

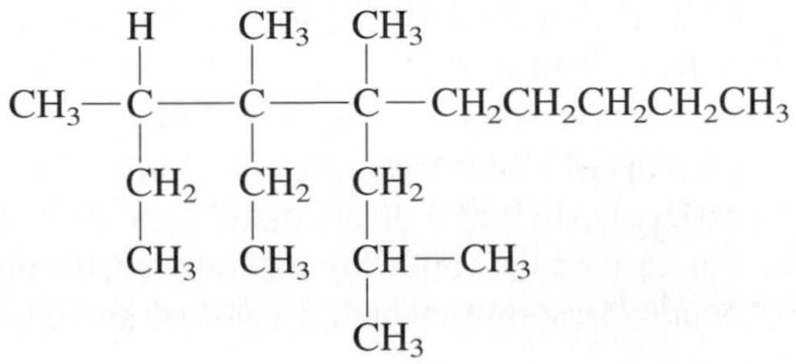

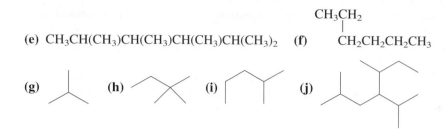

(e) $CH_3CH(CH_3)CH(CH_3)CH(CH_3)CH(CH_3)_2$ (f)

21. Convert the following names into the corresponding molecular structures. After doing so, check to see if the name of each molecule as given here is in accord with the IUPAC system of nomenclature. If not, name the molecule correctly. (a) 2-methyl-3-propylpentane; (b) 5-(1,1-dimethylpropyl)nonane; (c) 2,3,4-trimethyl-4-butylheptane; (d) 4-*tert*-butyl-5-isopropylhexane; (e) 4-(2-ethyl-butyl)decane; (f) 2,4,4-trimethylpentane; (g) 4-*sec*-butylheptane; (h) isoheptane; (i) neoheptane.

22. Draw and name all possible isomers of C_7H_{16} (isomeric heptanes).

23. Identify the primary, secondary, and tertiary carbon atoms and the hydrogen atoms in each of the following molecules: (a) ethane; (b) pentane; (c) 2-methylbutane; (d) 3-ethyl-2,2,3,4-tetramethylpentane.

24. Identify each of the following alkyl groups as being primary, secondary, or tertiary, and give it a systematic IUPAC name.

(a)
$$\overset{\displaystyle CH_3}{\underset{\displaystyle |}{-CH_2-CH-CH_2-CH_3}}$$
(b)
$$\overset{\displaystyle CH_3}{\underset{\displaystyle |}{CH_3-CH-CH_2-CH_2-}}$$
(c)
$$\overset{\displaystyle CH_3 \quad CH_3}{\underset{\displaystyle | \quad\quad |}{CH_3-CH-\!\!-CH-}}$$

(d)
$$\overset{\displaystyle CH_3-CH_2}{\underset{\displaystyle |}{CH_3-CH_2-CH-CH_2-}}$$
(e)
$$\overset{\displaystyle CH_3-CH-}{\underset{\displaystyle |}{CH_3-CH_2-CH-CH_3}}$$
(f)
$$\overset{\displaystyle CH_3-CH_2}{\underset{\displaystyle |}{CH_3-CH_2-C-CH_3}}$$

25. Rank the following molecules in order of increasing boiling point (*without* looking up the real values): (a) 2-methylhexane; (b) heptane; (c) 2,2-dimethylpentane; (d) 2,2,3-trimethylbutane.

26. Draw dashed-wedged line structures for the following molecules in the conformations indicated: (a) staggered propane; (b) eclipsed propane; (c) *anti* butane; (d) *gauche* butane. (**Hint:** Refer to Figure 2-50.)

27. Using Newman projections, draw each of the following molecules in its most stable conformation with respect to the bond indicated: (a) 2-methylbutane, C2–C3 bond; (b) 2,2-dimethylbutane, C2–C3 bond; (c) 2,2-dimethylpentane, C3–C4 bond; (d) 2,2,4-trimethylpentane, C3–C4 bond.

28. At room temperature, 2-methylbutane exists primarily as two alternating conformations of rotation about the C2–C3 bond. About 90% of the molecules exist in the more favorable conformation and 10% in the less favorable one. (a) Calculate the free energy change ($\Delta G°$, more favorable conformation − less favorable conformation) between these conformations. (b) Draw a potential-energy diagram for rotation about the C2–C3 bond in 2-methylbutane. To the best of your ability, assign relative energy values to all the conformations on your diagram. (c) Draw Newman projections for all staggered and eclipsed rotamers in (b) and indicate the two most favorable ones.

29. For each of the following naturally occurring compounds, identify the compound class(es) to which it belongs, and circle all functional groups.

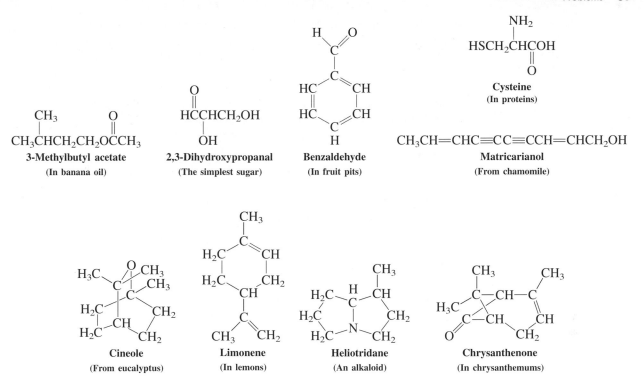

3-Methylbutyl acetate
(In banana oil)

2,3-Dihydroxypropanal
(The simplest sugar)

Benzaldehyde
(In fruit pits)

Cysteine
(In proteins)

Matricarianol
(From chamomile)

Cineole
(From eucalyptus)

Limonene
(In lemons)

Heliotridane
(An alkaloid)

Chrysanthenone
(In chrysanthemums)

30. Give IUPAC names for all alkyl groups marked by dashed lines in each of the following biologically important compounds. Identify each group as a primary, secondary, or tertiary alkyl substituent.

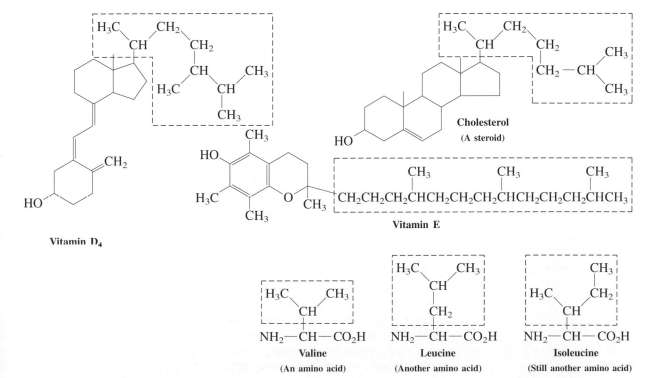

Vitamin D₄

Vitamin E

Cholesterol
(A steroid)

Valine
(An amino acid)

Leucine
(Another amino acid)

Isoleucine
(Still another amino acid)

31. The equation relating $\Delta G°$ to K contains a temperature term. Refer to your answer to Problem 28(a) to calculate the answers to the questions that follow. You will need to know that $\Delta S°$ for the formation of the more stable conformer of 2-methylbutane from the next most stable conformer is $+1.4$ cal deg^{-1} mol^{-1}. **(a)** Calculate the enthalpy difference ($\Delta H°$) between these two conformers from the equation $\Delta G° = \Delta H° - T\Delta S°$. How well does this agree with the $\Delta H°$ calculated from the number of *gauche* interactions in each conformer? **(b)** Assuming that $\Delta H°$ and $\Delta S°$ do not change with temperature, calculate $\Delta G°$ between these two conformations at the following three temperatures: $-250°C$; $-100°C$; $+500°C$. **(c)** Calculate K for these two conformations at the same three temperatures.

32. The hydrocarbon propene ($CH_3–CH=CH_2$) can react in two different ways with bromine (Chapters 12 and 14).

(i) $CH_3—CH=CH_2 + Br_2 \longrightarrow CH_3—\overset{\displaystyle Br}{\underset{\displaystyle |}{C}}H—\overset{\displaystyle Br}{\underset{\displaystyle |}{C}}H_2$

(ii) $CH_3—CH=CH_2 + Br_2 \longrightarrow \underset{\displaystyle |}{\underset{\displaystyle Br}{C}}H_2—CH=CH_2 + HBr$

Bond	Average Strength
C—C	83
C=C	146
C—H	99
Br—Br	46
H—Br	87
C—Br	68

(a) Using the bond strengths (kcal mol^{-1}) given in the margin, calculate $\Delta H°$ for each of these reactions. **(b)** $\Delta S° \approx 0$ cal deg^{-1} mol^{-1} for one of these reactions and -35 cal deg^{-1} mol^{-1} for the other. Which reaction has which $\Delta S°$? Briefly explain your answer. **(c)** Calculate $\Delta G°$ for each reaction at 25°C and at 600°C. Are both of these reactions thermodynamically favorable at 25°C? At 600°C?

33. Using the Arrhenius equation, calculate the effect on k of increases in temperature of 10, 30, and 50 degrees (Celsius) for the following activation energies. Use 300 K (approximately room temperature) as your initial T value, and assume that A is a constant. **(a)** $E_a = 15$ kcal mol^{-1}; **(b)** $E_a = 30$ kcal mol^{-1}; **(c)** $E_a = 45$ kcal mol^{-1}.

34. The Arrhenius equation can be reformulated in a way that permits the experimental determination of activation energies. For this purpose, we take the natural logarithm of both sides and convert into the base 10 logarithm.

$$\ln k = \ln (Ae^{-E_a/RT}) = \ln A - E_a/RT \quad \text{becomes} \quad \log k = \log A - \frac{E_a}{2.3RT}$$

The rate constant k is measured at several temperatures T and a plot of log k versus $1/T$ is prepared, a straight line. What is the slope of this line? What is its intercept (i.e., the value of log k at $1/T = 0$)? How is E_a calculated?

35. (i) Determine whether each species in the following equations is acting as a Brønsted acid or base, and label it. (ii) Indicate whether the equilibrium lies to the left or to the right. (iii) Estimate K for each equation if possible. (**Hint:** Use the data in Table 2-6.)
 (a) $H_2O + HCN \rightleftharpoons H_3O^+ + CN^-$
 (b) $CH_3O^- + NH_3 \rightleftharpoons CH_3OH + NH_2^-$
 (c) $HF + CH_3COO^- \rightleftharpoons F^- + CH_3COOH$
 (d) $CH_3^- + NH_3 \rightleftharpoons CH_4 + NH_2^-$
 (e) $H_3O^+ + Cl^- \rightleftharpoons H_2O + HCl$
 (f) $CH_3COOH + CH_3S^- \rightleftharpoons CH_3COO^- + CH_3SH$

36. Identify each of the following species as either a Lewis acid or a Lewis base, and write an equation illustrating a Lewis acid-base reaction for each one. Use curved arrows to depict electron-pair movement. Be sure that the product of each reaction is depicted by a complete, correct Lewis structure.

(a) CN^- (b) CH_3OH

(c) $(CH_3)_2CH^+$ (d) $MgBr_2$

(e) CH_3BH_2 (f) CH_3S^-

37. Reexamine your answers to Problem 19. Rewrite each one in the form of a complete equation describing a Lewis acid-base process, showing the product and using curved arrows to depict electron-pair movement. [**Hint:** For (b) and (d), start with a Lewis structure that represents a second resonance form of the starting organic molecule.]

Team Problem

38. Consider the difference in the rate between the following two second-order substitution reactions.

Reaction 1: The reaction of bromoethane and iodide ion to produce iodoethane and bromide ion is second order; that is, the rate of the reaction depends on the concentrations of both bromoethane and iodide ion:

$$\text{Rate} = k[CH_3CH_2Br][I^-] \text{ mol } L^{-1} \text{ s}^{-1}$$

Reaction 2: The reaction of 1-bromo-2,2-dimethylpropane (neopentyl bromide) with iodide ion to produce neopentyl iodide and bromide ion is more than 10,000 times slower than the reaction of bromoethane with iodide ion.

$$\text{Rate} = k[\text{neopentyl bromide}][I^-] \text{ mol } L^{-1} \text{ s}^{-1}$$

(a) Formulate each reaction by using bond-line structural drawings in your reaction scheme.

(b) Identify the reactive site of the starting haloalkane as primary, secondary, or tertiary.

(c) Discuss how the reaction might take place; that is, how would the species have to interact in order for the reaction to proceed. Remember that, because the reaction is second order, *both* reagents must be present in the transition state. Use your model kits to help you visualize the trajectory of approach of the iodide ion toward the bromoalkane that enables the simultaneous iodide bond-making and bromide bond-breaking required by the second-order kinetics of these two reactions. Of all the possibilities, which one best explains the experimentally determined difference in rate between the reactions?

(d) Use dashed-wedged line structures to make a three-dimensional drawing of the trajectory upon which you agree.

Preprofessional Problems

39. The compound 2-methylbutane has
  (a) no secondary H's
  (b) no tertiary H's
  (c) no primary H's
  (d) twice as many secondary H's as tertiary H's
  (e) twice as many primary H's as secondary H's

40.

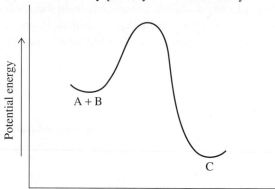

This energy profile diagram represents
  (a) an endothermic reaction
  (b) an exothermic reaction
  (c) a fast reaction
  (d) a thermolecular reaction

41. In 4-(1-methylethyl)heptane, any H–C–C angle has the value
  (a) 120°   **(b)** 109.5°   **(c)** 180°
  (d) 90°   **(e)** 360°

42. The structural representation shown in the margin is a Newman projection of the conformer of butane that is

  (a) *gauche* eclipsed **(b)** *anti* gauche **(c)** *anti* staggered **(d)** *anti* eclipsed

Reactions of Alkanes

Bond-Dissociation Energies, Radical Halogenation, and Relative Reactivity

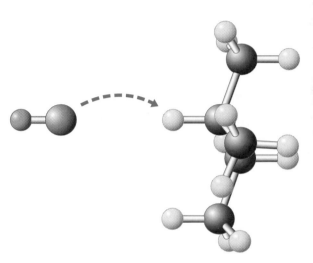

Combustion of alkanes releases most of the energy that powers modern industrialized society. How does this process occur? Is it related to the chemistry that converts alkanes into other organic molecules? We will find that both transformations rely on the same essential step: the breaking of a bond, or **bond dissociation.**

Many liquid and solid alkanes are obtained cheaply from petroleum (Section 3-3). Thus, nature has given us large quantities of hydrocarbons that can be used as "chemical feedstocks," or starting materials, for the synthesis of other organic molecules. Alkanes are also produced naturally by the slow decomposition of animal and vegetable matter in the presence of water but in the absence of oxygen, a process lasting millions of years. The smaller alkanes—methane, ethane, propane, and butane—are present in natural gas, methane being by far its major component. In the United States, natural gas is a major source of energy, with annual production in the hundreds of millions of tons.

As stated in Chapter 2, alkanes are organic chemicals that lack functional groups. Chapter 3 begins by explaining how bond dissociation in alkanes can be made to occur. Next, we learn how to introduce functional groups, to turn alkanes into compounds useful for synthesis, a process called **functionalization.** An important functionalization reaction of alkanes is **halogenation,** in which a hydrogen atom is replaced by a halogen. For each of these processes, we shall use a description of its **mechanism** to explain the conditions under which each takes place. You will see that these mechanistic concepts also explain the effects of halogen-containing chemicals on the stratospheric ozone layer. Finally, a discussion of alkane combustion leads to a description of the methods used to establish the heat contents and relative stabilities of molecules.

Reactive odd-electron species such as halogen atoms and hydroxy radicals (shown here) are involved in the oxidative degradation of organic materials and contribute to weathering.

3-1 Strength of Alkane Bonds: Radicals

Chapter 1 explained how bonds are formed and that energy is released on bond formation. For example, bringing two hydrogen atoms into bonding distance produces 104 kcal mol^{-1} of heat (refer to Figures 1-1 and 1-12).

$$H\cdot + H\cdot \longrightarrow H-H \qquad \Delta H° = -104 \text{ kcal mol}^{-1}$$

Consequently, breaking such a bond *requires* heat, in fact the same amount of heat that was released when the bond was made. This energy is called **bond-dissociation energy,** $DH°$, or **bond strength.**

$$H-H \longrightarrow H\cdot + H\cdot \qquad \Delta H° = DH° = 104 \text{ kcal mol}^{-1}$$

Radicals are formed by homolytic cleavage

In our example, the bond breaks in such a way that the two bonding electrons divide equally between the two participating atoms or fragments. This process is called **homolytic cleavage** or **bond homolysis.**

> **Homolytic Cleavage**
>
> $$A-B \longrightarrow A\cdot + \cdot B$$
> **Radicals**

Chlorine atom

Methyl radical

Ethyl radical

The fragments that form have an unpaired electron, for example, $H\cdot$, $Cl\cdot$, $CH_3\cdot$, and $CH_3CH_2\cdot$. When these species are composed of more than one atom, they are called **radicals.** Because of the unpaired electron, radicals and free atoms are very reactive and usually cannot be isolated. However, radicals and free atoms are present in low concentration as unobserved *intermediates* in many reactions, such as the oxidation of fats that leads to the spoilage of perishable foods (Chapter 22).

In Section 2-9 we were introduced to an alternative way of breaking a bond, in which the entire bonding electron pair is donated to one of the atoms. This process is **heterolytic cleavage** and results in the formation of **ions.**

> **Heterolytic Cleavage**
>
> $$A-B \longrightarrow A^+ + :B^-$$
> **Ions**

Homolytic cleavage may be observed in nonpolar solvents or even in the gas phase. In contrast, heterolytic cleavage normally occurs in polar solvents, which are capable of stabilizing ions, and is restricted to situations in which the electronegativities of atoms A and B and the groups attached to them stabilize positive and negative charges, respectively.

Dissociation energies, $DH°$, refer only to homolytic cleavages. They have characteristic values for the various bonds that can be formed between the elements. Table 3-1 lists dissociation energies of some common bonds. Note the relatively strong bonds to hydrogen, as in H–F and H–OH. Even though these bonds have high $DH°$ values, they readily undergo *heterolytic* cleavage in water to H^+ and F^- or HO^-; *do not confuse homolytic with heterolytic processes.*

Bonds are strongest when made by overlapping orbitals that are closely matched in energy and size. For example, the strength of the bonds between hydrogen and the

TABLE 3-1	Bond-Dissociation Energies of Various A–B Bonds ($DH°$ in kcal mol^{-1})						
	B in A–B						
A in A–B	**–H**	**–F**	**–Cl**	**–Br**	**–I**	**–OH**	**–NH$_2$**
H—	104	135	103	87	71	119	107
CH$_3$—	105	110	85	71	57	93	80
CH$_3$CH$_2$—	98	107	80	68	53	92	77
CH$_3$CH$_2$CH$_2$—	98	107	81	68	53	91	78
(CH$_3$)$_2$CH—	94.5	106	81	68	53	92	93
(CH$_3$)$_3$C—	93	110	81	67	52	93	93

Note: These numbers are being revised continually because of improved methods for their measurement. Some of the values given here may be in (small) error.

halogens decreases in the order F > Cl > Br > I, because the *p* orbital of the halogen contributing to the bonding becomes larger and more diffuse along the series. Thus, the efficiency of its overlap with the relatively small 1*s* orbital on hydrogen diminishes. A similar trend holds for bonding between the halogens and carbon.

EXERCISE 3-1

Compare the bond-dissociation energies of CH$_3$–F, CH$_3$–OH, and CH$_3$–NH$_2$. Why do the bonds get weaker along this series even though the orbitals participating in overlap become better matched in size and energy? (**Hint:** Consider Figure 1-2 and Table 1-2 for a simple explanation.)

The stability of radicals determines the C–H bond strengths

How strong are the C–H and C–C bonds in alkanes? The bond-dissociation energies of various alkane bonds are given in Table 3-2. Note that bond energies generally

TABLE 3-2	Bond-Dissociation Energies for Some Alkanes		
Compound	$DH°$ (kcal mol^{-1})	**Compound**	$DH°$ (kcal mol^{-1})
CH$_3$⊢H	105	CH$_3$⊢CH$_3$	90
C$_2$H$_5$⊢H	98	C$_2$H$_5$⊢CH$_3$	86
C$_3$H$_7$⊢H	98	C$_3$H$_7$⊢CH$_3$	87
(CH$_3$)$_2$CHCH$_2$⊢H	98	C$_2$H$_5$⊢C$_2$H$_5$	82
(CH$_3$)$_2$CH⊢H	94.5	(CH$_3$)$_2$CH⊢CH$_3$	86
(CH$_3$)$_3$C⊢H	93	(CH$_3$)$_3$C⊢CH$_3$	84
		(CH$_3$)$_3$C⊢C(CH$_3$)$_3$	72

Note: See footnote for Table 3-1.

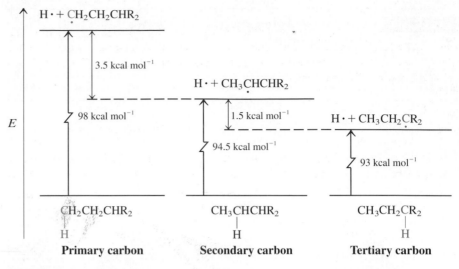

FIGURE 3-1

The different energies needed to form radicals from an alkane $CH_3CH_2CHR_2$. Radical stability increases from primary to secondary to tertiary.

decrease with the progression from methane to primary, secondary, and tertiary carbon. For example, the C–H bond in methane has a high $DH°$ value of 105 kcal mol^{-1}. In ethane, this bond energy is less: $DH° = 98$ kcal mol^{-1}. The latter number is typical for primary C–H bonds, as can be seen for the bond in propane. The secondary C–H bond is even weaker, with a $DH°$ of 94.5 kcal mol^{-1}, and a tertiary carbon atom bound to hydrogen has a $DH°$ of only 93 kcal mol^{-1}.

$$CH_3\text{—}H \longrightarrow CH_3\cdot + H\cdot \qquad DH° = 105 \text{ kcal mol}^{-1}$$
$$R\text{—}H \longrightarrow R\cdot + H\cdot$$

C–H bond is weaker and easier to break ↓	R primary	$DH° = 98$ kcal mol^{-1}	**Radical formed is more stable** ↓
	secondary	$DH° = 94.5$ kcal mol^{-1}	
	tertiary	$DH° = 93$ kcal mol^{-1}	

A similar trend is seen for C–C bonds. The extremes are the central linkages in ethane ($DH° = 90$ kcal mol^{-1}) and 2,2,3,3-tetramethylbutane ($DH° = 72$ kcal mol^{-1}).

Why do all of these dissociations exhibit different $DH°$ values? One explanation is that the radicals formed have different energies. Radical stability *increases* along the series from primary to secondary to tertiary; consequently, the energy required to create them *decreases* (Figure 3-1).

Stability of Alkyl Radicals

$CH_3\cdot$ < primary < secondary < tertiary

EXERCISE 3-2

Which C–C bond would break first, the bond in ethane or that in 2,2-dimethylpropane?

In summary, bond homolysis in alkanes yields radicals and free atoms. The heat required to do so is called the bond-dissociation energy, $DH°$. Its value is characteristic only for the bond between the two participating elements. Bond-breaking that results in tertiary radicals demands less energy than that furnishing secondary radicals; the latter are in turn formed more readily than primary radicals. The methyl radical is the most difficult to obtain in this way.

3-2 Structure of Alkyl Radicals: Hyperconjugation

What is the reason for the ordering in stability of alkyl radicals? To answer this question, we need to inspect the alkyl radical structure more closely. Radicals are stabilized by electron delocalization in which the p orbital at the radical center overlaps with a neighboring C–H σ bond.

Consider the structure of the methyl radical, formed by removal of a hydrogen atom from methane. It could be described as an sp^3-hybridized carbon with three sp^3 C–H bonds, the odd electron occupying the fourth sp^3 molecular orbital. Spectral measurements, however, have shown that the methyl radical, and probably other alkyl radicals, adopt a *nearly planar* configuration, more accurately described by sp^2 hybridization (Figure 3-2). The unpaired electron occupies the remaining p orbital perpendicular to the molecular plane.

Let us see how the planar structures of alkyl radicals help explain their relative stabilities. Figure 3-3A (see page 98) shows that there is a conformer in the ethyl radical in which a C–H bond of the CH_3 group is aligned with and overlaps one of the lobes of the singly occupied p orbital on the radical center. This arrangement allows the bonding pair of electrons in the σ orbital to delocalize into the partly empty p lobe, a phenomenon called **hyperconjugation.** The interaction between a filled orbital and a singly occupied orbital has a net stabilizing effect (recall Exercise 1-12). Both hyperconjugation and resonance (Section 1-5) are forms of electron delocalization. They are distinguished by type of orbital: resonance normally refers to π-type overlap of p orbitals, whereas hyperconjugation incorporates overlap with the orbitals of σ bonds.

As further hydrogen atoms on the radical carbon are replaced successively by alkyl groups, the number of hyperconjugation interactions increases (Figure 3-3B). The order of stability of the radicals is a consequence of this effect. Notice in Figure 3-1 that the degree of stabilization arising from each hyperconjugation interaction is relatively small (1.5–3.5 kcal mol^{-1}); we shall see later that stabilization of radicals by resonance is considerably greater (Chapter 14). Another contribution to the relative

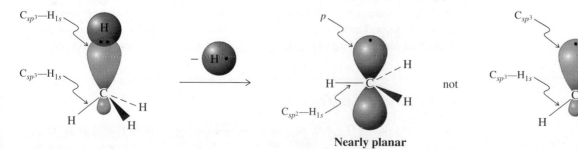

FIGURE 3-2 _____

The hybridization change upon formation of a methyl radical from methane. The nearly planar arrangement is reminiscent of the hybridization in BH_3 (Figure 1-17).

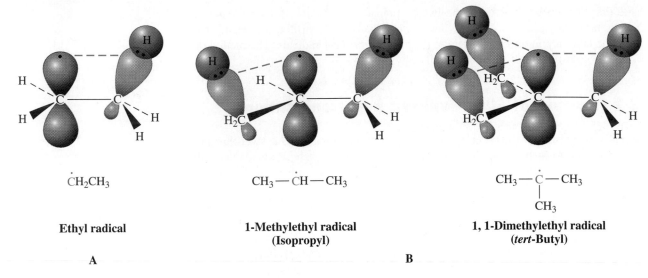

$\overset{.}{C}H_2CH_3$

Ethyl radical

$CH_3 - \overset{.}{C}H - CH_3$

**1-Methylethyl radical
(Isopropyl)**

$CH_3 - \overset{.}{C} - CH_3$
$\quad\quad\quad |$
$\quad\quad\quad CH_3$

**1, 1-Dimethylethyl radical
(*tert*-Butyl)**

A

B

FIGURE 3-3

Hyperconjugation (green dashed lines) resulting from the donation of electrons in filled sp^3 hybrids to the partly filled p orbital in (A) ethyl and (B) 1-methylethyl and 1,1-dimethylethyl radicals. The resulting delocalization of electron density has a net stabilizing effect.

stability of secondary and tertiary radicals is the greater relief of steric crowding between the substituent groups as the geometry changes from tetrahedral in the alkane to planar in the radical.

3-3 Conversion of Petroleum: Pyrolysis

A knowledge of bond-dissociation energies helps us understand the high-temperature reactivity of hydrocarbons. Consider, for example, the conversion of crude petroleum into gasoline and other volatile materials. Distillation alone does not meet the demand for these desired lower molecular weight hydrocarbons. Additional heating is required to break up longer carbon chains into smaller fragments. How does this occur? Let us look first at the behavior of simple alkanes under these conditions and then move on to petroleum.

High temperatures cause bond homolysis

When alkanes are heated to a high temperature, both C–H bonds and C–C bonds rupture, a process called **pyrolysis.** In the absence of oxygen, the resulting radicals can combine to form new higher or lower alkanes. They can also remove hydrogen atoms from the carbon atom adjacent to another radical center to give alkenes, a process called *hydrogen abstraction*. Indeed, very complicated mixtures of alkanes and alkenes form in pyrolyses. Under special conditions, however, these transformations can be controlled to obtain a large proportion of hydrocarbons of a defined chain length.

Pyrolysis of Hexane

Examples of cleavage into radicals

$$\underset{\textbf{Hexane}}{\overset{1\quad 2\quad 3\quad 4}{CH_3CH_2CH_2CH_2CH_2CH_3}}$$

$\xrightarrow{\text{C1, C2 cleavage}} CH_3\cdot \ + \ \cdot CH_2CH_2CH_2CH_2CH_3$

$\xrightarrow{\text{C2, C3 cleavage}} CH_3CH_2\cdot \ + \ \cdot CH_2CH_2CH_2CH_3$

$\xrightarrow{\text{C3, C4 cleavage}} CH_3CH_2CH_2\cdot \ + \ \cdot CH_2CH_2CH_3$

Examples of radical combination reactions

$$CH_3\cdot + \cdot CH_2CH_3 \longrightarrow CH_3CH_2CH_3$$
Propane

$$CH_3CH_2CH_2CH_2CH_2\cdot + \cdot CH_2CH_2CH_3 \longrightarrow CH_3CH_2CH_2CH_2CH_2CH_2CH_2CH_3$$
Octane

Examples of hydrogen abstraction reactions

$$\begin{array}{cc} & H \\ & | \\ CH_3CH_2\cdot + CH_3CH-CH_2\cdot \longrightarrow & CH_3CH_2 + CH_3CH=CH_2 \end{array}$$
Ethane   **Propene**

$$\begin{array}{cc} H & H \\ | & | \\ CH_2CH_2\cdot + CH_3CH_2CH_2\cdot \longrightarrow & CH_2=CH_2 + CH_3CH_2CH_2 \end{array}$$
Ethene   **Propane**

Such control frequently requires the use of special catalysts, such as crystalline sodium aluminosilicates, also called zeolites. For example, zeolite-catalyzed pyrolysis of dodecane yields a mixture in which hydrocarbons containing from three to six carbons predominate.

$$\text{Dodecane} \xrightarrow{\text{Zeolite, 482°C, 2 min}} C_3 + C_4 + C_5 + C_6 + \text{other products}$$
            17%   31%   23%   18%   11%

Petroleum is an important source of alkanes

Breaking an alkane down into smaller fragments is also known as **cracking.** Such processes are important in the oil-refining industry for the production of gasoline and other liquid fuels from petroleum.

CHEMICAL HIGHLIGHT **3-1**   **The Function of a Catalyst**

What is the function of the zeolite catalyst? As shown in the illustration, a *catalyst* is a substance that speeds up a reaction; that is, it increases the rate at which equilibrium is established. It does so by allowing reactants and products to be interconverted by a new pathway that has a lower activation energy (E_{cat}) than that of the reaction in the absence (E_a) of the catalyst. Apart from zeolites and other mineral-derived surfaces, many metals act as catalysts. In nature, *enzymes* usually fulfill this function (Chapter 26). The presence of catalysts allows many transformations to take place at lower temperatures and under generally milder conditions.

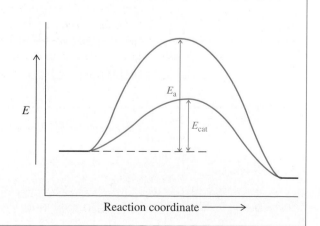

The Alyeska Pipeline Marine Terminal, Valdez, Alaska. Alaska is second only to Texas in oil production in the United States.

$CH_3CH_2CH_2CH_2CH_2CH_2CH_3$
Heptane

$\xrightarrow{\substack{\text{Pt-SiO}_2\text{-Al}_2\text{O}_3, \\ 500°C, \\ 20 \text{ atm } H_2}}$

$+ \; 4 \; H_2$

**Methylbenzene
(Toluene)**

As mentioned in the introduction to this chapter, petroleum, or crude oil, is believed to be the product of microbial degradation of living organic matter that existed several hundred million years ago. Crude oil, a dark viscous liquid, is primarily a mixture of several hundred different hydrocarbons, particularly straight-chain alkanes, some branched alkanes, and varying quantities of aromatic hydrocarbons. Distillation yields several fractions with a typical product distribution, as shown in Table 3-3. The composition of petroleum varies widely, depending on the origin of the oil.

To increase the quantity of the much-needed gasoline fraction, the oils with higher boiling points are cracked by pyrolysis. Originally (in the 1920s), this process required high temperatures (800°–1000°C), but modern cracking processes use catalysts, such as zeolites, at relatively low temperatures (500°C). Cracking the residual oil from crude petroleum distillation gives approximately 30% gas, 50% gasoline, and 20% higher molecular weight oils and a residue called coke.

Another process converts alkanes into aromatic hydrocarbons with approximately the same number of carbon atoms. The aromatics are highly efficient fuels and are used as feedstocks for the chemical industry. Because the process reforms a new hydrocarbon from an old one, it is referred to as **reforming.** An example of reforming is the conversion of heptane into methylbenzene (toluene). Hundreds of millions of liters of reformate gasoline are produced in the United States alone.

TABLE 3-3	Product Distribution in a Typical Distillation of Crude Petroleum		
Amount (% of volume)	**Boiling point (°C)**	**Carbon atoms**	**Products**
1–2	<30	C_1–C_4	Natural gas, methane, propane, butane, liquefied petroleum gas (LPG)
15–30	30–200	C_4–C_{12}	Petroleum ether ($C_{5,6}$), ligroin (C_7), naphtha, straight-run gasoline[a]
5–20	200–300	C_{12}–C_{15}	Kerosene, heater oil
10–40	300–400	C_{15}–C_{25}	Gas oil, diesel fuel, lubricating oil, waxes, asphalt
8–69	>400 (Nonvolatiles)	>C_{25}	Residual oil, paraffin waxes, asphalt (tar)

[a]This refers to gasoline straight from petroleum, without having been treated in any way.

CHEMICAL HIGHLIGHT 3-2

Petroleum and Gasoline: Our Main Energy Sources

Oil and natural gas supply most of the U.S. energy requirement. Other industrialized nations have similar dependence on petroleum as a source of energy. Yearly production of natural gas in the United States approximates 500 million liters. In 1995, U.S. energy sources apart from gas (24%) and oil (38%) were coal (23%), nuclear power (8%), and hydroelectric power (4%). Domestic annual oil production peaked in 1971 at 4.2 billion barrels. In 1995, it had dropped below 3 billion barrels, with some 2 billion barrels imported. (One barrel equals 42 gallons, about 158 liters.)

The dependence of many countries on imported oil has had important economic and political consequences, as demonstrated by the wild swings in oil prices that accompanied the Iraqi invasion of Kuwait in 1990 and the subsequent Persian Gulf War. Renewed efforts are being undertaken to decrease economic dependence on imported oil and to develop new energy sources to satisfy demand when oil and gas reserves are depleted.

The United States, with more than 200 million motor vehicles, has the highest annual energy consumption of any nation in the world. More than 70% of that need is met by fossil fuels.

To summarize, the pyrolysis of alkanes often leads to complex mixtures of hydrocarbons, control being attained in the presence of special catalysts. Are there other reactions of alkanes? If so, can they be carried out with even greater control? Moreover, can they introduce functional groups into the chain? The sections that follow will answer these questions.

3-4 Chlorination of Methane: The Radical Chain Mechanism

We have seen that alkanes undergo chemical transformations when subjected to pyrolysis, and that these processes include the formation of radical intermediates. Do alkanes participate in other reactions? In this section, we shall consider the effect of exposing an alkane, methane, to a halogen, chlorine. A **chlorination** reaction, in which radicals again play a key role, takes place to produce chloromethane and hydrogen chloride. We shall analyze each step in this transformation to establish the *mechanism* of the reaction.

Chlorine converts methane into chloromethane

When methane and chlorine gas are mixed in the dark at room temperature, there is no reaction. The mixture must be heated to a temperature above 300°C (Δ) *or* irradiated with ultraviolet light ($h\nu$) before reaction occurs. One of the two initial products is chloromethane, derived from methane in which a hydrogen atom is removed and replaced by chlorine. The other product of this transformation is hydrogen chloride. Further substitution leads to dichloromethane (methylene chloride), CH_2Cl_2; trichloromethane (chloroform), $CHCl_3$; and tetrachloromethane (carbon tetrachloride), CCl_4.

Why should this reaction proceed? Consider its $\Delta H°$. Note that a C–H bond in methane ($DH° = 105$ kcal mol^{-1}) and a Cl–Cl bond ($DH° = 58$ kcal mol^{-1}) are broken, whereas the C–Cl bond of chloromethane ($DH° = 85$ kcal mol^{-1}) and an H–Cl linkage ($DH° = 103$ kcal mol^{-1}) are formed. The net result is the release of 25 kcal mol^{-1} in forming stronger bonds: The reaction is substantially *exothermic*.

$$CH_3 \!-\! H + \ddot{\underset{..}{Cl}} \!-\! \ddot{\underset{..}{Cl}}: \xrightarrow{\Delta \text{ or } h\nu} CH_3 \!-\! \ddot{\underset{..}{Cl}}: + H \!-\! \ddot{\underset{..}{Cl}}:$$

$$\quad\quad 105 \quad\quad 58 \quad\quad\quad\quad\quad\quad\quad\quad 85 \quad\quad 103$$

$$DH° \text{ (kcal mol}^{-1})\quad\quad\quad\quad\quad \textbf{Chloromethane}$$

$\Delta H°$ = energy input − energy output
    = $\Sigma DH°$ (bonds broken) − $\Sigma DH°$ (bonds formed)
    = $(105 + 58) - (85 + 103)$
    = -25 kcal mol^{-1}

Why then does the thermal chlorination of methane not occur at room temperature? The fact that a reaction is exothermic does not necessarily mean that it should proceed rapidly and spontaneously. Remember (Section 2-8) that the rate of a chemical transformation depends on its activation energy, which in this case is evidently high. Why is this so? What is the function of irradiation when the reaction does proceed at room temperature? Answering these questions requires an investigation of the mechanism of the reaction.

The mechanism explains the experimental conditions required for reaction

A **mechanism** is a detailed, step-by-step description of all the changes in bonding that occur in a chemical reaction (Section 1-1). Even simple reactions may consist of several separate steps. The mechanism shows the sequence in which bonds are broken and formed, as well as the energy changes associated with each step. This information is of great value in both analyzing possible transformations of complex molecules and understanding the experimental conditions required for reactions to occur.

The mechanism for the chlorination of methane consists of three stages: initiation, propagation, and termination. Let us look at these stages and the experimental evidence for each of them in more detail.

The chlorination of methane can be studied step by step

Experimental observation. Clorination occurs when a mixture of CH_4 and Cl_2 is either heated to 300°C or irradiated with light, as mentioned earlier. Under such conditions, methane by itself is completely stable, but Cl_2 undergoes homolysis to two atoms of chlorine.

Interpretation. The first step in the mechanism of chlorination of methane is the heat- or light-induced homolytic cleavage of the Cl–Cl bond (which happens to be the weakest bond in the starting mixture, with $DH° = 58$ kcal mol^{-1}). This event is required to start the chlorination process and is therefore called the **initiation** step.

As implied by its name, the initiation step generates reactive species (in this case, chlorine atoms) that permit the subsequent steps in the overall reaction to take place.

INITIATION

$$:\overset{\cdot\cdot}{\underset{\cdot\cdot}{Cl}}-\overset{\cdot\cdot}{\underset{\cdot\cdot}{Cl}}: \xrightarrow{\Delta \text{ or } h\nu} \quad 2 :\overset{\cdot\cdot}{\underset{\cdot\cdot}{Cl}}\cdot \qquad \begin{aligned}\Delta H^\circ &= DH^\circ(Cl_2) \\ &= +58 \text{ kcal mol}^{-1}\end{aligned}$$

Chlorine atom

> **NOTE: In this scheme and in those that follow, all radicals and free atoms are in green.**

Experimental observation. Only a relatively small number of initiation events are necessary to enable a great many methane and chlorine molecules to undergo conversion into products.

Interpretation. After initiation has taken place, the subsequent steps in the mechanism are self-sustaining, or self-propagating; that is, they can occur many times without the addition of further chlorine atoms from the homolysis of Cl_2. Two **propagation** steps fulfill this requirement. In the first step, a chlorine atom attacks methane by abstracting a hydrogen atom. The resulting products are hydrogen chloride and a methyl radical.

PROPAGATION STEP 1

$$:\overset{\cdot\cdot}{\underset{\cdot\cdot}{Cl}}\cdot + \text{H}\underset{\underset{\text{H}}{|}}{\overset{\overset{\text{H}}{|}}{-}}\text{C}-\text{H} \longrightarrow :\overset{\cdot\cdot}{\underset{\cdot\cdot}{Cl}}-\text{H} + \overset{\text{H}}{\underset{\text{H}}{\diagdown}}\text{C}-\text{H} \qquad \begin{aligned}\Delta H^\circ &= DH^\circ(CH_3-H) \\ &\quad - DH^\circ(H-Cl) \\ &= +2 \text{ kcal mol}^{-1}\end{aligned}$$

$$\underset{105}{} \qquad \underset{103}{}$$

$$DH^\circ \text{ (kcal mol}^{-1}) \qquad\qquad \textbf{Methyl radical}$$

FIGURE 3-4

Approximate molecular-orbital description of the abstraction of a hydrogen atom by a chlorine atom to give a methyl radical and hydrogen chloride. Notice the rehybridization at carbon in the planar methyl radical. The additional three nonbonded electron pairs on chlorine have been omitted. The orbitals are not drawn to scale. The symbol ‡ identifies the transition state.

The ΔH° for this transformation is positive; its equilibrium is slightly unfavorable. What is its activation energy, E_a? Is there enough heat to overcome this barrier? In this case, the answer is yes. A molecular-orbital description of the transition state (Section 2-6) of hydrogen removal from methane (Figure 3-4) reveals the details of

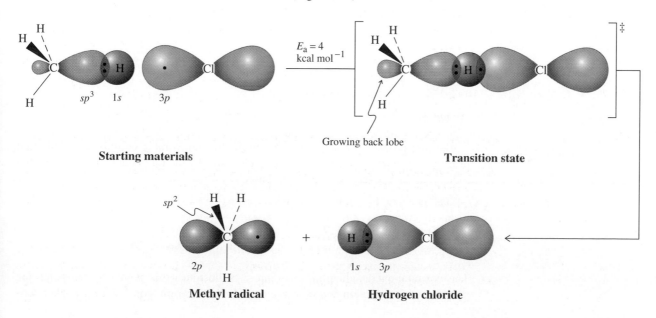

Starting materials

sp^3 $1s$ $3p$

$E_a = 4$ kcal mol^{-1}

Growing back lobe

Transition state

sp^2

$2p$

Methyl radical

$1s$ $3p$

Hydrogen chloride

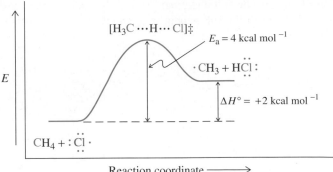

FIGURE 3-5 ──────────────────────
Potential-energy diagram of the reaction of methane with a chlorine atom. Partial bonds in the transition state are depicted by dotted lines. This process, propagation step 1 in the radical chain chlorination of methane, is slightly endothermic.

the process. The reacting hydrogen is positioned between the carbon and the chlorine, partly bound to both: H–Cl bond formation has occurred to about the same extent as C–H bond-breaking. The transition state, which is labeled by the symbol ‡, is located only about 4 kcal mol^{-1} above the starting materials. A potential-energy diagram describing this step is shown in Figure 3-5.

Propagation step 1 gives one of the products of the chlorination reaction: HCl. What about the organic product, CH_3Cl? Chloromethane is formed in the *second* propagation step. Here the methyl radical abstracts a chlorine atom from one of the starting Cl_2 molecules, thereby furnishing chloromethane *and a new chlorine atom*. The latter reenters propagation step 1 to react with a new molecule of methane. Thus one propagation cycle is closed, and a new one begins, *without the need for another initiation step to take place*. Note how exothermic propagation step 2 is, -27 kcal mol^{-1}. It supplies the overall driving force for the reaction of methane with chlorine.

PROPAGATION STEP 2

$$\text{H}-\overset{\overset{\textstyle H}{|}}{\underset{\underset{\textstyle H}{|}}{\text{C}}}\cdot \; + \; :\overset{..}{\underset{..}{\text{Cl}}}-\overset{..}{\underset{..}{\text{Cl}}}: \;\longrightarrow\; \text{H}-\overset{\overset{\textstyle H}{|}}{\underset{\underset{\textstyle H}{|}}{\text{C}}}-\text{Cl} \; + \; \cdot\overset{..}{\underset{..}{\text{Cl}}}: \qquad \begin{aligned}\Delta H^\circ &= DH^\circ(Cl_2) - DH^\circ(CH_3\text{-}Cl)\\ &= -27 \text{ kcal mol}^{-1}\end{aligned}$$

58   85

DH° (kcal mol^{-1})

Because propagation step 2 is exothermic, the unfavorable equilibrium in the first propagation step is pushed toward the product side by the rapid depletion of its methyl radical product in the subsequent reaction.

$$CH_4 + Cl\cdot \; \rightleftharpoons \; CH_3\cdot + HCl \overset{Cl_2}{\rightleftharpoons} CH_3Cl + Cl\cdot + HCl$$

Slightly   Very favorable;
unfavorable   "drives" first equilibrium

The potential-energy diagram in Figure 3-6 illustrates this point by continuing the progress of the reaction begun in Figure 3-5. Propagation step 1 has the higher activation energy and is therefore slower than step 2. The diagram also shows that the overall $\Delta H°$ of the reaction is made up of the $\Delta H°$ values of the propagation steps: $+2 - 27 = -25$ kcal mol^{-1}. You can see that this should be so by adding the equations for the two.

$$
\begin{array}{lc}
 & \Delta H° \text{ (kcal mol}^{-1}) \\
:\ddot{C}l\cdot + CH_4 \longrightarrow CH_3\cdot + H\ddot{C}l: & +2 \\
CH_3\cdot + Cl_2 \longrightarrow CH_3\ddot{C}l: + :\ddot{C}l\cdot & -27 \\
\hline
CH_4 + Cl_2 \longrightarrow CH_3\ddot{C}l: + H\ddot{C}l: & -25
\end{array}
$$

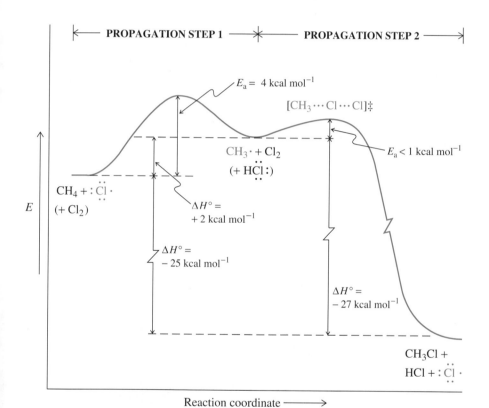

FIGURE 3-6

Complete potential-energy diagram for the formation of CH_3Cl from methane and chlorine. Propagation step 1 has the higher transition-state energy and is therefore slower. The $\Delta H°$ of the overall reaction $CH_4 + Cl_2 \rightarrow CH_3Cl + HCl$ amounts to -25 kcal mol^{-1}, the sum of the $\Delta H°$ values of the two propagation steps.

Experimental observation. Small amounts of *ethane* are identified among the products of chlorination of methane.

Interpretation. Radicals and free atoms are capable of undergoing direct covalent bonding with one another. In the methane chlorination process, three such combination processes are possible, one of which, the reaction of two methyl groups, furnishes ethane. The concentrations of radicals and free atoms in the reaction mixture are very low, however, and hence the chance of one radical or free atom finding another is small. Such combinations are therefore relatively infrequent. When such an event does take place, the propagation of the chains giving rise to the radicals or atoms is terminated. We thus describe these combination processes as **termination** steps.

CHAIN TERMINATION

$$:\ddot{\text{C}}\text{l}\cdot \ + \ :\ddot{\text{C}}\text{l}\cdot \ \longrightarrow \ \text{Cl}_2$$

$$:\ddot{\text{C}}\text{l}\cdot \ + \ \text{CH}_3\cdot \ \longrightarrow \ \text{CH}_3\ddot{\text{C}}\text{l}:$$

$$\text{CH}_3\cdot \ + \ \text{CH}_3\cdot \ \longrightarrow \ \text{CH}_3{-}\text{CH}_3$$

The mechanism for the chlorination of methane is an example of a **radical chain mechanism.**

A Radical Chain Mechanism

Initiation	Propagation steps	Chain termination
$X_2 \longrightarrow 2\ :\ddot{X}\cdot$	$:\ddot{X}\cdot + RH \longrightarrow R\cdot + H\ddot{X}:$	$:\ddot{X}\cdot + :\ddot{X}\cdot \longrightarrow X_2$
	$X_2 + R\cdot \longrightarrow R\ddot{X}: + :\ddot{X}\cdot$	$R\cdot + :\ddot{X}\cdot \longrightarrow RX$
		$R\cdot + R\cdot \longrightarrow R_2$

Only a few halogen atoms are necessary for initiating the reaction, because the propagation steps are self-sufficient in $:\ddot{X}\cdot$. The first propagation step consumes a halogen atom, the second produces one. The newly generated halogen atom then reenters the propagation cycle in the first propagation step. In this way, a *radical chain* is set in motion that can drive the reaction for many thousands of cycles.

EXERCISE 3-3

Chlorination of ethane furnishes chloroethane. Write a mechanism for this transformation and calculate $\Delta H°$ for each step (see Tables 3-1 and 3-2).

One of the practical problems in chlorinating methane is the control of product selectivity. As mentioned earlier, the reaction does not stop at the formation of chloromethane but continues to form di-, tri-, and tetrachloromethane by further substitution. A practical solution to this problem is the use of a large excess of methane in the reaction. Under such conditions, the reactive intermediate chlorine atom is at any given moment surrounded by many more methane molecules than product CH_3Cl. Thus, the chance of $Cl\cdot$ finding CH_3Cl to eventually make CH_2Cl_2 is greatly diminished, and product selectivity is achieved.

In summary, chlorine transforms methane into chloromethane. The reaction proceeds through a mechanism in which heat or light causes a small number of Cl_2 molecules to undergo homolysis to chlorine atoms (initiation). The latter induce and maintain a radical chain sequence consisting of two (propagation) steps: (1) hydrogen abstraction to generate the methyl radical and HCl and (2) conversion of $CH_3\cdot$ by Cl_2 into CH_3Cl and regenerated $Cl\cdot$. The chain is terminated by various combinations of radicals and free atoms. The heats of the individual steps are calculated by comparing the strengths of the bonds that are being broken with those of the bonds being formed.

3-5 Other Radical Halogenations of Methane

Fluorine and bromine, but not iodine, also react with methane by similar radical mechanisms to furnish the corresponding halomethanes. The dissociation energies of X_2

(X = F, Br, I) are lower than that of Cl_2, thus ensuring ready initiation of the radical chain (Table 3-4).

Fluorine is most reactive, iodine least reactive

Let us compare the enthalpies of the two propagation steps in the different halogenations of methane (Table 3-5). It is apparent that there are quite striking differences in the driving force for hydrogen abstraction. For fluorine, this step is exothermic by -30 kcal mol^{-1}. We have already seen that, for chlorine, the same step is slightly endothermic; for bromine, it is substantially so, and for iodine even more so. This trend has its origin in the decreasing bond strengths of the hydrogen halides in the procession from fluorine to iodine (Table 3-1). The strong hydrogen–fluorine bond is the cause of the high reactivity of fluorine atoms in hydrogen abstraction reactions. Fluorine is more reactive than chlorine, chlorine is more reactive than bromine, and the least reactive halogen atom is iodine.

The contrast between fluorine and iodine is illustrated by comparing potential-energy diagrams for their respective hydrogen abstractions from methane (Figure 3-7) (see page 108). The highly exothermic reaction of fluorine has a negligible activation barrier. Moreover, in its transition state, the fluorine atom is relatively far from the hydrogen that is being transferred, and the H–CH_3 distance is only slightly greater than that in CH_4 itself. Why should this be so? At the transition state the energy needed for (partial) bond-breaking exactly equals that gained by (partial) bond-making. In the present case, the full H–CH_3 bond is 30 kcal mol^{-1} weaker than that of H–F (Table 3-1). Only a small shift of the H toward the F· is necessary for bonding between the two to overcome that between hydrogen and carbon. If we view the reaction coordinate as a measure of the degree of hydrogen shift from C to F, the transition state is reached *early* and is much closer in appearance to the starting materials than to the products. *Early transition states are frequently characteristic of fast, exothermic processes.*

On the other hand, reaction of I· with CH_4 has a very high E_a (at least as large as its endothermicity, $+34$ kcal mol^{-1}; Table 3-1). Thus, the transition state is not reached until the H–C bond is nearly completely broken and the H–I bond is almost fully formed. The transition state is said to be *late:* It is substantially further along the reaction coordinate and is much closer in structure to the products of this process, CH_3· and HI. *Late transition states are frequently typical of relatively slow, endothermic transformations.* Together these rules are known as the **Hammond* postulate.**

*Professor George S. Hammond (b. 1921), Georgetown University, Washington, D.C.

TABLE 3-4 DH° Values for the Elemental Halogens	
Halogen	**DH° (kcal mol^{-1})**
F_2	37
Cl_2	58
Br_2	46
I_2	36

Relative Reactivities of X· in Hydrogen Abstractions
F· > Cl· > Br· > I·

TABLE 3-5	Enthalpies of the Propagation Steps in the Halogenation of Methane (kcal mol^{-1})			
Reaction	**F**	**Cl**	**Br**	**I**
:X· + CH_4 $\longrightarrow$ ·CH_3 + HX:	-30	$+2$	$+18$	$+34$
·CH_3 + X_2 $\longrightarrow$ CH_3X: + :X·	-73	-27	-25	-21
CH_4 + X_2 $\longrightarrow$ CH_3X: + HX:	-103	-25	-7	$+13$

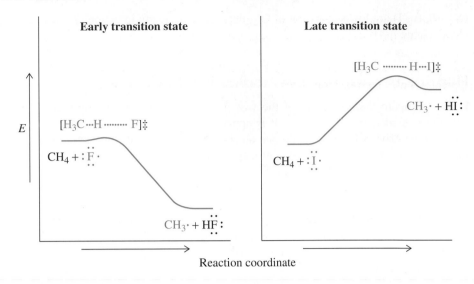

FIGURE 3-7

Potential-energy diagrams: *(left)* the reaction of a fluorine atom with CH_4, an exothermic process with an early transition state; and *(right)* the reaction of an iodine atom with CH_4, an endothermic transformation with a late transition state. Both are thus in accord with the Hammond postulate.

The second propagation step is exothermic

Let us now consider the second propagation step for each halogenation in Table 3-5. This process is exothermic for all the halogens. Again, the reaction is fastest and most exothermic for fluorine. The combined enthalpies of the two steps for the fluorination of methane result in a $\Delta H°$ of -103 kcal mol^{-1}. Indeed, this value is so large that, at sufficiently high concentrations of methane and fluorine gas, a violent reaction occurs. The formation of chloromethane is less exothermic, and that of bromomethane even less so. In the latter case, the appreciably endothermic nature of the first step ($\Delta H° = +18$ kcal mol^{-1}) is barely overcome by the enthalpy of the second ($\Delta H° = -25$ kcal mol^{-1}), resulting in an energy change of only -7 kcal mol^{-1} for the overall substitution. Finally, inspection of the thermodynamics of iodination reveals why iodine does not react with methane to furnish methyl iodide and hydrogen iodide. The first step costs so much energy that the second step, although exothermic, cannot drive the reaction.

EXERCISE 3-4

Predict the product distribution of the reaction of methane with an equimolar mixture of chlorine and bromine at low conversion.

In summary, fluorine, chlorine, and bromine react with methane to give halomethanes. All three reactions follow the radical chain mechanism described for chlorination. In these processes, the first propagation step is always the slower of the two. It becomes more exothermic and its activation energy decreases in the procession from bromine to chlorine to fluorine. This trend explains the relative reactivity of the halogens, fluorine being the most reactive. Iodination of methane is endothermic and does not occur. Strongly exothermic reaction steps are often characterized by early transition states. Conversely, endothermic or relatively less exothermic steps typically have late transition states.

3-6 Chlorination of Higher Alkanes: Relative Reactivity and Selectivity

What happens in the radical halogenation of other alkanes? Will the different types of R–H bonds—namely, primary, secondary, and tertiary—react in the same way as those in methane? Let us consider the chlorination of ethane, then propane, and finally 2-methylpropane.

The monochlorination of ethane gives chloroethane as the product.

Chlorination of Ethane

$$CH_3CH_3 + Cl_2 \xrightarrow{\Delta \text{ or } h\nu} CH_3CH_2Cl + HCl \qquad \Delta H^\circ = -27 \text{ kcal mol}^{-1}$$

Chloroethane

This reaction proceeds by a radical chain mechanism analogous to the one observed for methane. The propagation steps include formation of the ethyl radical from ethane and the chlorine atom, followed by generation of chloroethane and the release of another chlorine atom (Exercise 3-3). The greatest difference between the two mechanisms lies in the change in values for ΔH°. Thus, the abstraction of a hydrogen from ethane is no longer endothermic, as in methane, but favorable by -5 kcal mol^{-1}. The reason is the weaker C–H bond of ethane ($DH^\circ = 98$ kcal mol^{-1}).

Propagation Steps in the Mechanism of the Chlorination of Ethane

$$CH_3CH_3 + :\ddot{C}l\cdot \longrightarrow CH_3CH_2\cdot + H\ddot{C}l: \qquad \Delta H^\circ = -5 \text{ kcal mol}^{-1}$$

$$CH_3CH_2\cdot + Cl_2 \longrightarrow CH_3CH_2\ddot{C}l: + :\ddot{C}l\cdot \qquad \Delta H^\circ = -22 \text{ kcal mol}^{-1}$$

What can be expected for the next homolog, propane?

Secondary C–H bonds are more reactive than primary ones

The eight hydrogen atoms in propane fall into two groups: six primary and two secondary hydrogens. If chlorine atoms were to abstract and replace primary and secondary hydrogens at equal rates, we should expect to find a product mixture containing three times as much 1-chloropropane as 2-chloropropane. We call this outcome a **statistical product ratio,** because it derives from the statistical fact that in propane there are three times as many primary sites for reaction as secondary sites. In other words, it is three times as likely for a chlorine atom to collide with a primary

$$CH_3CH_2CH_3$$

Propane

Six primary hydrogens (blue)
Two secondary hydrogens (red)

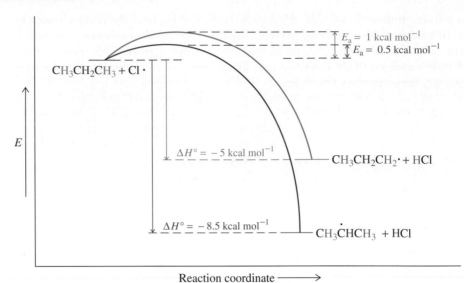

FIGURE 3-8
Hydrogen abstraction by a chlorine atom from the secondary carbon in propane is more exothermic and faster than that from the primary carbon.

hydrogen in propane, of which there are six, as with a secondary hydrogen atom, of which there are only two.

However, secondary C–H bonds are weaker than primary ones ($DH° = 94.5$ versus 98 kcal mol^{-1}). Abstraction of a secondary hydrogen is therefore more exothermic and proceeds with a smaller activation barrier (Figure 3-8). We might thus expect secondary hydrogens to react faster, leading to more 2-chloro- than 1-chloropropane. What is actually observed? At 25°C, the experimental ratio of 1-chloropropane:2-chloropropane is found to be 43:57. This result indicates that both statistical and bond-energy factors determine the product formed.

Chlorination of Propane

$$Cl_2 + CH_3CH_2CH_3 \xrightarrow{h\nu} CH_3CH_2CH_2Cl + CH_3\overset{\overset{\displaystyle Cl}{|}}{C}HCH_3 + HCl$$

$\qquad\qquad\qquad\qquad\qquad$ **1-Chloropropane** $\qquad$ **2-Chloropropane**

Expected statistical ratio	3	:	1
Expected C–H bond reactivity ratio	Less	:	More
Experimental ratio (25°C)	43	:	57

We can calculate the *relative reactivity of secondary and primary hydrogens* in chlorination by factoring out the statistical contribution to the product ratio.

$$\frac{\text{Relative reactivity of a secondary hydrogen}}{\text{Relative reactivity of a primary hydrogen}} = \frac{\left(\begin{array}{c}\text{yield of product from}\\\text{secondary hydrogen abstraction}\end{array}\right)\Big/\left(\begin{array}{c}\text{number of}\\\text{secondary hydrogens}\end{array}\right)}{\left(\begin{array}{c}\text{yield of product from}\\\text{primary hydrogen abstraction}\end{array}\right)\Big/\left(\begin{array}{c}\text{number of}\\\text{primary hydrogens}\end{array}\right)} = \frac{57/2}{43/6} \approx 4$$

In other words, each secondary hydrogen in the chlorination of propane at 25°C is about four times as reactive as each primary one. We say that chlorine exhibits a **selectivity** of 4:1 in the removal of secondary versus primary hydrogens at 25°C.

We could predict from this analysis that *all* secondary hydrogens are four times as reactive as all primary hydrogens in *all* radical chain reactions. Is this prediction true? Unfortunately, no. Although secondary C–H bonds generally undergo dissociation faster than do their primary counterparts, their relative reactivity very much depends on the nature of the attacking species, X·, the strength of the resulting H–X bond, and the temperature. For example, at 600°C, the chlorination of propane results in the statistical distribution of products: roughly three times as much 1-chloropropane as 2-chloropropane. At such a high temperature, virtually every collision between a chlorine atom and any hydrogen in the propane molecule takes place with sufficient energy to lead to successful reaction. Chlorination is said to be **unselective** at this temperature, and the product ratio is governed by statistical factors.

EXERCISE 3-5

What do you expect the products of monochlorination of butane to be? In what ratio will they be formed at 25°C?

Tertiary C–H bonds are more reactive than secondary ones

Let us now find the relative reactivity of a tertiary hydrogen in the chlorination of alkanes. For this purpose, we expose 2-methylpropane, a molecule containing one tertiary and nine primary hydrogens, to chlorination conditions at 25°C. The resulting two products, 2-chloro-2-methylpropane (*tert*-butyl chloride) and 1-chloro-2-methylpropane (isobutyl chloride), are formed in the relative yields of 36 and 64%, respectively.

Chlorination of 2-Methylpropane

$$Cl_2 + \underset{\underset{CH_3}{|}}{\overset{\overset{CH_3}{|}}{CH_3-C-H}} \xrightarrow{h\nu} \underset{\underset{CH_3}{|}}{\overset{\overset{CH_3}{|}}{ClCH_2-C-H}} + \underset{\underset{CH_3}{|}}{\overset{\overset{CH_3}{|}}{CH_3-C-Cl}} + HCl$$

	64%	36%
	1-Chloro-2-methylpropane	**2-Chloro-2-methylpropane**
	(Isobutyl chloride)	**(*tert*-Butyl chloride)**

Expected statistical ratio	9 : 1
Expected C–H bond reactivity ratio	Less : More
Experimental ratio (25°C)	64 : 36

We can determine the reactivity of tertiary relative to primary hydrogens as follows: we combine the experimental result of 64% primary chlorination and 36% tertiary chlorination in 2-methylpropane with the statistical presence of nine primary hydrogens and one tertiary hydrogen atom in the starting alkane. Dividing the proportionate amount observed of each product by the number of hydrogens that contribute to its formation gives a measure of reactivity per hydrogen atom:

$$\frac{\text{Relative reactivity of a tertiary hydrogen}}{\text{Relative reactivity of a primary hydrogen}} = \frac{\left(\begin{array}{c}36\% \text{ tertiary} \\ \text{chlorination}\end{array}\right) \Big/ \left(\begin{array}{c}1 \text{ tertiary} \\ \text{hydrogen}\end{array}\right)}{\left(\begin{array}{c}64\% \text{ primary} \\ \text{chlorination}\end{array}\right) \Big/ \left(\begin{array}{c}9 \text{ primary} \\ \text{hydrogens}\end{array}\right)} = \frac{36/1}{64/9} \approx 5$$

$$\underset{\underset{CH_3}{|}}{\overset{\overset{CH_3}{|}}{CH_3-C-H}}$$

2-Methylpropane

Nine primary hydrogens (blue)
One tertiary hydrogen (red)

Thus, tertiary hydrogen atoms are about five times as reactive as primary ones. This selectivity, again, decreases at higher temperatures. However, we can say that, at 25°C, the relative reactivities of the various C–H bonds in chlorinations are roughly

$$\text{Tertiary : secondary : primary} = 5:4:1$$

The result agrees well with the relative reactivity expected from consideration of bond strength: The tertiary C–H bond is weaker than the secondary, and the latter in turn is weaker than the primary.

We can verify this ordering by looking at the competition among all three types of hydrogens within a single substrate. 2-Methylbutane, which contains nine primary hydrogens, two secondary hydrogens, and one tertiary hydrogen, is an example. Because this molecule has two types of primary hydrogens, one set of six and one set of three, reaction with chlorine yields a total of four different monochlorination products.

Chlorination of 2-Methylbutane

27%	14%	36%	23%
1-Chloro-2-methylbutane	**1-Chloro-3-methylbutane**	**2-Chloro-3-methylbutane**	**2-Chloro-2-methylbutane**
(Chlorination at A)	(Chlorination at B)	(Chlorination at C)	(Chlorination at D)

Substitution at primary carbons | Substitution at the secondary carbon | Substitution at the tertiary carbon

The combined yield of the two primary halide products is 41% (1-chloro-2-methylbutane plus 1-chloro-3-methylbutane), the secondary halide 2-chloro-3-methylbutane is formed in 36% yield, and the tertiary halide in 23%. Therefore,

$$\text{Primary : secondary : tertiary halide} = 41:36:23$$
$$\textit{Relative reactivity}\quad \text{primary : secondary : tertiary} = 41/9:36/2:23/1$$
$$= 1:4:5$$

as expected.

EXERCISE 3-6

Give products and the ratio in which they are expected to form for the monochlorination of 3-methylpentane at 25°C. Be careful to take into account the number of hydrogens in each distinct group in the starting alkane.

To summarize, the relative reactivity of primary, secondary, and tertiary hydrogens follows the trend expected on the basis of their relative C–H bond strengths. Relative reactivity ratios can be calculated by factoring out statistical considerations. These ratios are temperature dependent, with greater selectivity at lower temperatures.

3-7 Selectivity in Radical Halogenation with Fluorine and Bromine

How selectively do halogens other than chlorine halogenate the alkanes? Table 3-5 and Figure 3-7 show that fluorine is the most reactive halogen: Hydrogen abstraction is highly exothermic and has negligible activation energy. Conversely, bromine is much less reactive, because the same step has a large positive $\Delta H°$ and a high activation barrier. Does this difference affect their selectivity in halogenation of alkanes?

To answer this question, consider the reactions of fluorine and bromine with 2-methylpropane. Single fluorination at 25°C furnishes two possible products, in a ratio very close to that expected statistically.

	2-Fluoro-2-methylpropane	:	1-Fluoro-2-methylpropane
	(*tert*-Butyl fluoride)		(Isobutyl fluoride)
Observed	14	:	86
Expected	1	:	9

Fluorine thus displays very little selectivity. Why? Because the transition states for the two competing processes are reached very early, their energies and structures are similar to each other, as well as similar to those of the starting material (Figure 3-9).

Fluorination of 2-Methylpropane

$$F_2 + (CH_3)_3CH \xrightarrow{h\nu} (CH_3)_3CF + FCH_2-\underset{\underset{CH_3}{|}}{\overset{\overset{CH_3}{|}}{C}}-H + HF$$

14%   **86%**

2-Fluoro-2-methylpropane   **1-Fluoro-2-methylpropane**
(*tert*-Butyl fluoride)   (Isobutyl fluoride)

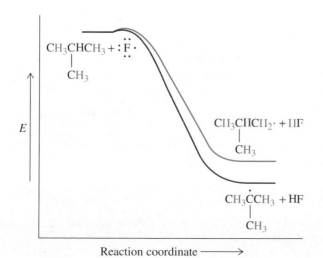

FIGURE 3-9

Potential-energy diagram for the abstraction of a primary or a tertiary hydrogen by a fluorine atom from 2-methylpropane. The energies of the respective early transition states are almost the same and barely higher than that of starting material (i.e., both E_a values are close to zero), resulting in little selectivity.

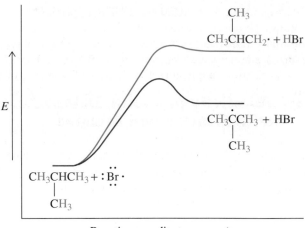

FIGURE 3-10

Potential-energy diagram for the abstraction of a primary or a tertiary hydrogen by a bromine atom from 2-methylpropane. The two late transition states are dissimilar in energy, indicative of the energy difference between the resulting primary and tertiary radicals, respectively, leading to the products with greater selectivity.

Conversely, *bromination of the same compound is highly selective,* giving the tertiary bromide almost exclusively. Hydrogen abstractions by bromine have *late* transition states in which extensive C–H bond-breaking and H–Br bond-making have occurred. Thus, their respective structures and energies resemble those of the corresponding radical products. As a result, the activation barriers for the reaction of bromine with primary and tertiary hydrogens, respectively, will differ by almost as much as the difference in stability between primary and tertiary radicals (Figure 3-10), a difference leading to the observed high selectivity (more than 1700:1).

Bromination of 2-Methylpropane

$$Br_2 + (CH_3)_3CH \xrightarrow{h\nu} (CH_3)_3CBr + BrCH_2\!-\!\overset{\displaystyle CH_3}{\underset{\displaystyle CH_3}{\overset{|}{\underset{|}{C}}}}\!-\!H + HBr$$

>99%   <1%

2-Bromo-2-methylpropane   **1-Bromo-2-methylpropane**
(***tert*-Butyl bromide**)   (**Isobutyl bromide**)

In summary, increased reactivity goes hand in hand with reduced selectivity in radical substitution reactions. Fluorine and chlorine, the more reactive halogens, discriminate between the various types of C–H bonds much less than does the less reactive bromine (Table 3-6).

TABLE 3-6	Relative Reactivities of the Four Types of Alkane C–H Bonds in Halogenations		
C–H bond	**F·** (25°C, gas)	**Cl·** (25°C, gas)	**Br·** (150°C, gas)
CH_3—H	0.5	0.004	0.002
RCH_2—H[a]	1	1	1
R_2CH—H	1.2	4	80
R_3C—H	1.4	5	1700

[a]For each halogen, reactivities with four types of alkane C–H bonds are normalized to the reactivity of the primary C–H bond.

3-8 Synthetic Radical Halogenation

How can we devise a successful and cost-effective alkane halogenation? We must take into account selectivity, convenience, efficiency, and price.

Fluorinations are unattractive, because fluorine is relatively expensive and corrosive; and, even worse, its reactions are often violently uncontrollable. Radical iodinations, at the other extreme, fail because of unfavorable thermodynamics.

Chlorinations are important, particularly in industry, simply because chlorine is cheap. (It is prepared by electrolysis of sodium chloride, ordinary table salt.) The drawback to chlorination is low selectivity, so the process results in mixtures of isomers that are difficult to separate. To circumvent the problem, an alkane that contains a single type of hydrogen can be used as a substrate, thus giving (at least initially) a single product. Cyclopentane is one such alkane.

Chlorination of a Molecule with Only One Type of Hydrogen

Cyclopentane
(Large excess)

Chlorocyclopentane

To minimize overhalogenation, chlorine is used as the limiting reagent (Section 3-4). Even then, multiple substitution can complicate the reaction. Conveniently, the more highly chlorinated products have higher boiling points and can be separated by distillation.

On an industrial scale, alkanes are chlorinated in large vessels fitted with elaborate controls to ensure smooth, safe operation. In the research laboratory, the use of chlorine gas is often avoided, because it is toxic, corrosive, and difficult to measure accurately. Other chlorinating agents have been developed that can be handled more safely and accurately. These agents are usually liquids or solids, such as sulfuryl chloride (SO_2Cl_2) and *N*-chlorobutanimide (*N*-chlorosuccinimide, NCS).

Sulfuryl chloride
(b.p. 69°C)

N-**Chlorobutanimide**
(*N*-**Chlorosuccinimide**)
(m.p. 148°C)

EXERCISE 3-7

Which of the following compounds will give a monochlorination product with reasonable selectivity: propane, 2,2-dimethylpropane, cyclohexane, methylcyclohexane?

Because bromination is selective (and bromine is a liquid), it is frequently the method of choice for halogenating an alkane on a relatively small scale in the research laboratory. Reaction occurs at the more substituted carbon, even in statistically unfavorable situations. Typical solvents are chlorinated methanes (CCl_4, $CHCl_3$, CH_2Cl_2), which are comparatively unreactive with bromine.

Bromine is obtained from aqueous sodium bromide solutions (found in natural brines) by treatment with chlorine. Bromine is used less in industry because of its relatively greater cost per mole compared with chlorine. A popular solid brominating agent is *N*-bromobutanimide (*N*-bromosuccinimide, NBS; see Section 14-2), an analog of NCS.

In summary, even though more expensive, bromine is the reagent of choice for selective radical halogenations. Chlorinations furnish product mixtures, a problem that can be minimized by choosing alkanes with only one type of hydrogen and treating them with a deficiency of chlorine. For ease in handling, research chemists often

CHEMICAL HIGHLIGHT **3-3**

Chlorination, Chloral, and DDT

**1,1,1-Trichloro-2,2-bis(4-chlorophenyl)ethane
(DDT)**

Chlorination of ethanol in the production of trichloroacetaldehyde, CCl_3CHO, was first described in 1832. The hydrated form is commonly called *chloral* and is a powerful hypnotic with the nickname "knockout drops." Chloral is also a key reagent in the synthesis of the powerful insecticide DDT (an abbreviation derived from the nonsystematic name *d*ichloro*d*iphenyl*t*richloroethane).

The use of DDT in the control of insect-borne diseases has saved many millions of lives in the past half-century, chiefly through the decimation of the *Anopheles* mosquito, the main carrier of the parasite that causes malaria. Although its toxicity toward mammals is low (the fatal human dose is about 500 mg kg^{-1} of body weight), DDT is very resistant

to biodegradation. Its accumulation in the food chain makes it a hazard to birds and fish, and consequently it has been banned by the U.S. Environmental Protection Agency since 1972.

Eggshells damaged by high concentrations of pesticide residues.

---

prefer solid reagents, such as *N*-bromobutanimide, or liquids, such as sulfuryl chloride, to the respective liquid and gaseous reagents.

3-9 Synthetic Chlorine Compounds and the Stratospheric Ozone Layer

We have seen how bond homolysis can be caused by both heat and light. Such chemical events can occur on a grand scale in nature and may have substantial environmental consequences. This section will explore an example of radical chemistry that has had a significant effect on our lives and will continue to do so well into the next century.

The ozone layer shields Earth's surface from high-energy ultraviolet light

Earth's atmosphere consists of several distinct layers. The lowest layer, extending to about 15 km in altitude, is the troposphere, the region where weather occurs. Above the troposphere, the stratosphere extends upward to an altitude of some 50 km. Although the density of the stratosphere is too low to sustain terrestrial life, the stratosphere is the home of the **ozone layer,** which plays a critical role in the ability of life to survive on Earth. Ozone (O_3) and ordinary molecular oxygen (O_2) equilibrate in the stratosphere by the action of ultraviolet light from the sun:

Interconversion of Ozone and Molecular Oxygen in the Stratosphere

$$O_2 \xrightarrow{h\nu} 2O$$

$$O_2 + O \longrightarrow O_3$$

$$O_3 \xrightarrow{h\nu} O_2 + O$$

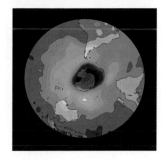

A color-enhanced view of the upper atmosphere above Antarctica during October 1996, showing a region in which the concentration of ozone has dropped below 35% of normal (dark gray).

In the first two reactions high-energy solar radiation splits O_2 into oxygen atoms, which may combine with other molecules of O_2 to produce ozone, a bluish gas with a characteristically sharp and penetrating odor. Occasionally, ozone may be detected in the vicinity of high-voltage equipment, in which electrical discharges cause the conversion of O_2 into O_3. In urban areas, nitrogen dioxide (NO_2) is produced through oxidation of nitric oxide (NO), a product of high temperature combustion. Nitrogen dioxide, in turn, is split by sunlight to release oxygen atoms, which produce O_3 in the lower atmosphere. The presence of ozone as an air pollutant near Earth's surface causes severe irritation of the respiratory membranes and the eyes. However, in the upper atmosphere, the third reaction in the interconversion of ozone and molecular oxygen shown above occurs, in which ozone absorbs UV light in the 200 to 300 nm wavelength range. Irradiation at these wavelengths is capable of destroying the complex molecules that make up biochemical systems. Ozone serves as a natural atmospheric filter to prevent this light from reaching the surface, thereby protecting Earth's life from damage.

CFCs release chlorine atoms upon ultraviolet irradiation

Chlorofluorocarbons (CFCs), or **Freons,** are alkanes in which all of the hydrogens have been replaced by fluorine and chlorine. In general, chlorofluorocarbons are thermally stable, essentially odorless, and nontoxic gases. Among their many commercial applications, their use as refrigerants predominates, because they absorb large quantities of heat upon vaporization. Compression liquefies gaseous CFC, which flows through the cooling coils in refrigerators, freezers, and air conditioners. The liquid absorbs heat from the environment outside the coils and evaporates. The gas then reenters the compressor to be liquefied again, and the cycle continues.

CFCs are among the most effective and widely used synthetic organic compounds in modern society. Why, then, have the nations of the world, with almost unprecedented unanimity, agreed to phase them completely out of use? This remarkable event dates its origins to the late 1960s and early 1970s, when chemists Johnston, Crutzen, Rowland, and Molina* pointed to the existence of radical mechanisms that could convert several kinds of compounds, including CFCs, into reactive species capable of destroying ozone in the Earth's stratosphere.

Upon irradiation by UV light, the weaker C–Cl bonds in the CFC molecules undergo homolysis, giving rise to atomic chlorine.

No CFCs!

Common CFCs

CCl_3F

CFC-11

CCl_2F_2

CFC-12

CCl_2FCClF_2

CFC-113

---

*Professor Harold S. Johnston (b. 1920), University of California at Berkeley; Professor Paul Crutzen (b. 1933), Max Planck Institute, Mainz, Germany, Nobel Prize 1995 (chemistry); Professor F. Sherwood Rowland (b. 1927), University of California at Irvine, Nobel Prize 1995 (chemistry); Professor Mario Molina (b. 1943), Massachusetts Institute of Technology, Nobel Prize 1995 (chemistry).

INITIATION STEP

$$F_3C - \overset{..}{\underset{..}{Cl}}: \overset{hv}{\longrightarrow} F_3C\cdot + :\overset{..}{\underset{..}{Cl}}\cdot$$

Chlorine atoms, in turn, react efficiently with ozone in a radical chain sequence.

PROPAGATION STEPS

$$:\overset{..}{\underset{..}{Cl}}\cdot + O_3 \longrightarrow \cdot ClO + O_2$$

$$\cdot ClO + O \longrightarrow O_2 + :\overset{..}{\underset{..}{Cl}}\cdot$$

The net result of these two steps is the conversion of a molecule of ozone and an oxygen atom into two molecules of ordinary oxygen. As in the other radical chain processes that we have examined in this chapter, however, the reactive species consumed in one propagation step ($:\overset{..}{\underset{..}{Cl}}\cdot$) is regenerated in the other. As a consequence, a small concentration of chlorine atoms is capable of destroying many molecules of ozone. Does such a process actually occur in the atmosphere?

Stratospheric ozone has decreased by about 3% since 1978

Since measurements of atmospheric composition were first made, measurable decreases in stratospheric ozone have been recorded. These changes are seasonal, being most severe in the winter. They show extreme variations with latitude: Large reductions in the ozone layer over the Antarctic were noticeable as early as 1978. Satellite measurements confirmed that total ozone content in this part of the atmosphere in 1987 was less than half of its usual value, and by 1994 it had dropped further, to less than one-third of normal. Localized regions of the Antarctic had no ozone layer above them at all: An "ozone hole" was observed. Ozone amounts north of the Arctic Circle sank to 45% of normal, the lowest readings ever, in the winter of 1996. Stratospheric clouds, which can form only in the extreme cold of the polar regions, appeared to correlate with ozone-hole formation.

Reduction in total ozone above the temperate regions of the Northern Hemisphere currently averages 3%, approximately the worldwide mean, but 6% depletion is measured at 40°N latitude during the winter months. Epidemiological studies suggest that a reduction in stratospheric ozone density of 1% might be expected to give rise to a 1–3% increase in skin cancers. As a consequence, considerable effort has been made to identify the causes of ozone depletion. Are CFCs responsible? Or could natural sources of atmospheric chlorine or other substances be contributing significantly?

The answers to these questions were obtained in systematic studies of satellite observations made from 1987 to 1994: Chlorine monoxide (ClO), a critical component in the ozone-destroying chain reaction illustrated earlier, rises to more than 500 times normal levels in regions of the Antarctic ozone hole, where O_3 concentration is the lowest. Furthermore, this chlorine monoxide and at least 75% of stratospheric chlorine come from CFCs. This connection has been proved by the observation of corresponding concentrations of gaseous hydrogen fluoride (HF) in the stratosphere as well. Neither HF nor any other gaseous fluorine compound occurs naturally or is produced by natural chemical processes anywhere on (or in) this planet. However, CFC decomposition in the presence of hydrocarbons in the atmosphere is known to produce HF. The observations demonstrate that natural sources such as volcanic eruptions or sea spray are not major contributors to stratospheric chlorine, relative to CFCs.

Volcanic aerosols do, however, contribute to depletion of the ozone layer indirectly, by interfering with chemical processes that remove stratospheric chlorine.

The world is searching for CFC substitutes

The Montreal Protocol on Substances That Deplete the Ozone Layer was signed in 1987 and called for a 50% reduction in output of CFCs by 1998. Increasingly alarming evidence regarding ozone depletion led to amendments in 1990 and again in 1992, finally setting an advanced deadline for complete rather than partial phaseout of CFC production: December 31, 1995, marked the end of production of virtually all CFCs in the industrialized world. Meanwhile, hydrochlorofluorocarbons (HCFCs) and hydrofluorocarbons (HFCs) have been developed as replacements for CFCs in commercial applications. HCFCs are more chemically reactive than CFCs and are much more prone to decomposition at lower atmospheric altitudes. Their threat to stratospheric ozone is therefore less, because a smaller proportion survives the time necessary to diffuse to the upper atmosphere. Currently, HFC-134a is replacing CFC-12 in refrigerator and motor vehicle air conditioner compressors. HCFCs-22 and -141b have replaced CFC-11 in the manufacture of foam insulation.

Hydrofluorocarbons have been demonstrated to be safe for the ozone layer. However, hydrochlorofluorocarbons are potential ozone-destroying agents and are themselves scheduled for total phaseout by no later than 2030. Efforts to replace all HCFCs with HFCs are actively underway. It is hoped that this worldwide effort will finally bring the depletion of the ozone layer to a halt by the year 2000. Recovery to nearly normal levels is expected over the next century.

CFC Substitutes

CH_2FCF_3

HFC-134a

$CHClF_2$

HCFC-22

$CHCl_2CF_3$

HCFC-123

CH_3CCl_2F

HCFC-141b

CH_3CClF_2

HCFC-142b

3-10 Combustion and the Relative Stabilities of Alkanes

Let us review what we have learned in this chapter so far. We started by defining bond strength as the energy required to cleave a molecule homolytically. Some typical values were then presented in Tables 3-1 and 3-2 and explained through a discussion of relative radical stabilities, a major factor being the varying extent of hyperconjugation. We then used this information to calculate the $\Delta H°$ values of the steps making up the mechanism of radical halogenation, a discussion leading to an understanding of reactivity and selectivity. It is clear that knowing bond-dissociation energies is a great aid in the thermochemical analysis of organic transformations, an idea that we shall explore on numerous occasions later on. How are these numbers found experimentally?

Chemists determine bond strengths by first establishing the relative heat contents of entire molecules, or their relative positions along the energy axis in our potential-energy diagrams. The reaction chosen for this purpose is complete oxidation (literally, "burning"), or **combustion,** a process common to almost all organic structures, in which all carbon atoms are converted into CO_2 (gas) and all of the hydrogens into H_2O (liquid).

Both products in the combustion of alkanes have a very low energy content, and hence their formation is associated with a large negative $\Delta H°$, released as heat.

$$2 \, C_nH_{2n+2} + (3n + 1) \, O_2 \longrightarrow 2n \, CO_2 + (2n + 2) \, H_2O + \text{heat}$$

The heat released in the burning of a molecule is called its **heat of combustion,** $\Delta H°_{comb}$, many of which have been measured with high accuracy, thus allowing com-

TABLE 3-7	Heats of Combustion (kcal mol^{-1}, normalized to 25°C) of Various Organic Compounds	
Compound (state)	**Name**	ΔH°_{comb}
CH_4 (gas)	Methane	−212.8
C_2H_6 (gas)	Ethane	−372.8
$CH_3CH_2CH_3$ (gas)	Propane	−530.6
$CH_3(CH_2)_2CH_3$ (gas)	Butane	−687.4
$(CH_3)_3CH$ (gas)	2-Methylpropane	−685.4
$CH_3(CH_2)_3CH_3$ (gas)	Pentane	−845.2
$CH_3(CH_2)_3CH_3$ (liquid)	Pentane	−838.8
$CH_3(CH_2)_4CH_3$ (liquid)	Hexane	−995.0
⬡ (liquid)	Cyclohexane	−936.9
CH_3CH_2OH (gas)	Ethanol	−336.4
CH_3CH_2OH (liquid)	Ethanol	−326.7
$C_{12}H_{22}O_{11}$ (solid)	Cane sugar (sucrose)	−1348.2

Note: Combustion products are CO_2 (gas) and H_2O (liquid).

parisons of the relative energy content of the alkanes (Table 3-7) and other compounds. Such comparisons have to take into account the physical state of the compound undergoing combustion (gas, liquid, solid). For example, the difference between the heats of combustion of liquid and gaseous ethanol corresponds to its heat of vaporization, $\Delta H^\circ_{vap} = 9.7$ kcal mol^{-1}.

It is not surprising that the ΔH°_{comb} of alkanes increases with chain length, simply because there is more carbon and hydrogen to burn along the homologous series. Con-

CHEMICAL HIGHLIGHT 3-4   Enzymatic Oxidation of the Alkanes

Combustion is one way to activate the ordinarily quite unreactive alkane. Unfortunately, at high temperatures, oxidation is relatively unselective and destroys much of the starting molecule. A milder approach, called *enzymatic activation,* uses enzymes, the catalysts in living systems (see Chemical Highlight 3-1 and Chapter 26).

Enzymatic Alkane Activation

$$R—H \xrightarrow{\text{Enzyme, } O_2} R—OH$$
Alkanes   **Alcohols**
(C_1–C_8)

The monooxygenases, found in mammalian tissue, catalyze the oxidation of drugs, steroids (Chapter 4), and fatty acids (Chapter 19). In microbial systems, these same enzymes catalyze the oxidation of alkanes. Thus, an enzyme from *Methylococcus capsulatus* inserts oxygen into a number of hydrocarbons, to give alcohols.

Environmental pollutants can promote *non*enzymatic oxidations of molecules of biological importance, such as lipids, the building blocks of fatty tissue (Chapter 20). This chemistry will be discussed in Chapter 22.

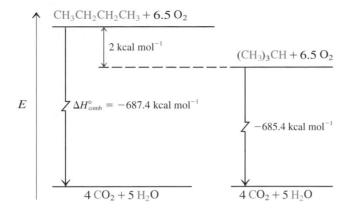

FIGURE 3-11 ————

Butane has a higher energy content than does 2-methyl-propane, as measured by the release of energy on combustion. Butane is therefore thermodynamically less stable than its isomer.

versely, isomeric alkanes contain the same number of carbons and hydrogens, and one might expect that their respective combustions would be equally exothermic. However, that is not the case.

A comparison of the heats of combustion of isomeric alkanes reveals that their values are usually *not* the same. Consider butane and 2-methylpropane. The combustion of butane has a ΔH°_{comb} of -687.4 kcal mol^{-1}, whereas its isomer releases $\Delta H^\circ_{comb} = -685.4$ kcal mol^{-1}, 2 kcal mol^{-1} less (Table 3-7). This finding shows that 2-methylpropane has a *smaller* energy content than does butane, because combustion yielding the identical kind and number of products produces less energy (Figure 3-11). Butane is said to be *thermodynamically less stable* than its isomer.

<hr>

EXERCISE 3-8

The hypothetical thermal conversion of butane into 2-methylpropane should have a $\Delta H^\circ = -2.0$ kcal mol^{-1}. What value do you obtain by using the bond-dissociation data in Table 3-2?

To summarize, the heats of combustion values of alkanes and other organic molecules give quantitative estimates of their energy content and, therefore, their relative stabilities.

CHAPTER INTEGRATION PROBLEM

Iodomethane reacts with hydrogen iodide under free radical conditions ($h\nu$) to give methane and iodine. The overall equation of the reaction is

$$CH_3I + HI \xrightarrow{h\nu} CH_4 + I_2$$

a. Write a mechanism for this process including initiation, propagation, and at least one termination step. Use bond strength data from Tables 3-1 and 3-4.

SOLUTION

Step 1. Begin by proposing a likely **initiation** step. Recall—from Section 3-4, for example—that initiation steps of radical reactions include cleavage of the *weakest* bond in the starting compounds. According to Tables 3-1 and 3-4, that is the carbon–

iodine bond in CH_3I, with $DH° = 57$ kcal mole^{-1}. Therefore

Initiation Step

$$H_3C-I \xrightarrow{h\nu} H_3C\cdot + :\overset{\cdot\cdot}{\underset{\cdot\cdot}{I}}\cdot$$

Step 2. Again following the model in Section 3-4, propose a **propagation** step in which one of the species produced in the initiation step reacts with one of the molecules shown in the overall equation of the reaction. Try to design the step so that one of its products corresponds to a molecule formed in the overall reaction and the other is a species that can give rise to a second propagation step. The possibilities are

(i) $H_3C\cdot + HI \longrightarrow CH_4 + :\overset{\cdot\cdot}{\underset{\cdot\cdot}{I}}\cdot$

(ii) $H_3C\cdot + CH_3I \longrightarrow CH_4 + \cdot CH_2I$

(iii) $:\overset{\cdot\cdot}{\underset{\cdot\cdot}{I}}\cdot + HI \longrightarrow I_2 + H\cdot$

(iv) $:\overset{\cdot\cdot}{\underset{\cdot\cdot}{I}}\cdot + CH_3I \longrightarrow I_2 + \cdot CH_3$

Propagation steps (i) and (ii) both convert methyl radical into methane by removing a hydrogen atom from HI and CH_3I, respectively. Processes (iii) and (iv) show the removal of an iodine atom by another iodine atom from either HI or CH_3I, giving I_2. All four propagation steps convert a molecule of starting material in the overall equation of the reaction into a molecule of product. How do we choose the correct steps? *Look at the radical products of each hypothetical propagation step. The two correct steps are those whose product radicals are each other's reactants.* Propagation step (i) consumes a methyl radical and produces an iodine atom. Step (iv) consumes iodine, and produces methyl. Therefore, steps (i) and (iv) are the correct steps of a propagation cycle.

Propagation Steps

$$H_3C\cdot + HI \longrightarrow CH_4 + :\overset{\cdot\cdot}{\underset{\cdot\cdot}{I}}\cdot$$

$$:\overset{\cdot\cdot}{\underset{\cdot\cdot}{I}}\cdot + CH_3I \longrightarrow I_2 + \cdot CH_3$$

You can check this answer by adding up all species on the left and right sides of these equations, respectively, to see if they correspond to the reactants and products of the overall reaction. They do: the radicals cancel on both sides of the equations, leaving only the correct molecules.

Step 3. Finally, combination of *any pair of radicals* to give a single molecule constitutes a legitimate **termination** step. There are three:

$$2 :\overset{\cdot\cdot}{\underset{\cdot\cdot}{I}}\cdot \longrightarrow I_2$$

$$2 H_3C\cdot \longrightarrow H_3C-CH_3 \quad \text{(ethane)}$$

$$H_3C\cdot + :\overset{\cdot\cdot}{\underset{\cdot\cdot}{I}}\cdot \longrightarrow CH_3I$$

b. Calculate the enthalpy changes, $\Delta H°$, associated with the overall reaction and all of the mechanistic steps. Use Tables 3-1, 3-2, and 3-4, as appropriate.

Solution

Breaking a bond requires energy *input,* forming a bond gives rise to energy *output,* and $\Delta H° = $ (energy in) $-$ (energy out). For the overall reaction, we have the following bond strength values to consider:

$$CH_3-I + H-I \xrightarrow{h\nu} CH_3-H + I-I$$

$DH°$:   57   71   105   36

The answer is $\Delta H° = (57 + 71) - (105 + 36) = -13$ kcal mole^{-1} (see also Table 3-5).

For the mechanistic steps, the same principle applies. With one exception, the same four $DH°$ values just shown are all that you need, because they correspond to the only four bonds that are either made or broken in any of the steps in the mechanism.

Initiation step: $\Delta H° = DH°$ (CH$_3$–I) $= +57$ kcal mole^{-1}

Propagation step (i): $\Delta H° = DH°$ (H–I) $- DH°$ (CH$_3$–H) $= -34$ kcal mole^{-1}

Propagation step (iv): $\Delta H° = DH°$ (CH$_3$–I) $- DH°$ (I–I) $= +21$ kcal mole^{-1}

Notice that the sum of the $\Delta H°$ values for the two propagation steps equals $\Delta H°$ for the overall reaction. *This is always true.*

Termination steps: $\Delta H° = -DH°$ for the bond formed; -36 kcal mole^{-1} for I$_2$, -57 kcal mole^{-1} for CH$_3$I, and -90 kcal mole^{-1} for the C–C bond in ethane.

IMPORTANT CONCEPTS

1. The $\Delta H°$ of **bond homolysis** is defined as the **bond-dissociation energy, $DH°$**. Bond homolysis gives radicals or free atoms.

2. The C–H bond strengths in the alkanes decrease in the order

CH$_3$—H > RCH$_2$—H > R—CH > R—C—H

Methyl (Strongest)   **Primary**   **Secondary**   **Tertiary** (Weakest)

because the order of stability of the corresponding alkyl radicals is

CH$_3$· < RCH$_2$· < R—CH· < R—C·

Methyl (Least stable)   **Primary**   **Secondary**   **Tertiary** (Most stable)

This is the order of increasing **hyperconjugative stabilization.**

3. **Catalysts** speed up the establishment of an equilibrium between starting materials and products.

4. Alkanes react with halogens (except iodine) by a **radical chain mechanism** to give haloalkanes. The mechanism consists of **initiation** to create a halogen atom, two **propagation steps,** and various **termination steps.**

5. In the first propagation step, the slower of the two, a hydrogen atom is abstracted from the alkane chain, a reaction resulting in an alkyl radical and

HX. Hence, **reactivity** increases from I$_2$ to F$_2$. **Selectivity** decreases along the same series, as well as with increasing temperature.

6. The **Hammond postulate** states that fast, exothermic reactions are typically characterized by **early transition states,** which are similar in structure to the starting materials. On the other hand, slow, endothermic processes usually have **late** (productlike) **transition states.**

7. The $\Delta H°$ for a reaction may be calculated from the $DH°$ values of the bonds affected in the process as follows: $\Delta H° = \Sigma\, DH°_{\text{bonds broken}} - \Sigma\, DH°_{\text{bonds formed}}$.

8. The $\Delta H°$ for a radical halogenation process equals the sum of the $\Delta H°$ values for the propagation steps.

9. The relative reactivities of the various types of alkane C–H bonds in halogenations can be estimated by factoring out statistical contributions. They are roughly constant under identical conditions and follow the order

$$CH_4 \;<\; \begin{matrix}\text{primary}\\ \text{CH}\end{matrix} \;<\; \begin{matrix}\text{secondary}\\ \text{CH}\end{matrix} \;<\; \begin{matrix}\text{tertiary}\\ \text{CH}\end{matrix}$$

In radical chlorinations of alkanes at 25°C, the approximate relative reactivities of the tertiary : secondary : primary positions are 5 : 4 : 1. In fluorinations, these ratios are about 1.4 : 1.2 : 1, whereas, in brominations (150°), they are 1700 : 80 : 1.

10. Chemists often use halogenating agents other than the halogens. Examples are sulfuryl chloride, SO$_2$Cl$_2$, and *N*-chloro- and *N*-bromobutanimide.

11. The $\Delta H°$ of the combustion of an alkane is called the **heat of combustion, $\Delta H°_{\text{comb}}$**. The heats of combustion of isomeric compounds provide an experimental measure of their relative stabilities.

PROBLEMS

9. Label the primary, secondary, and tertiary hydrogens in each of the following compounds.

   (a) $CH_3CH_2CH_2CH_3$ (b) $CH_3CH_2CH_2CH_2CH_3$

   (c)

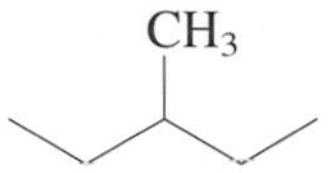

   (d)

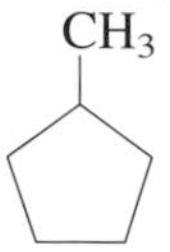

10. Within each of the following sets of alkyl radicals, name each radical; identify each as either primary, secondary, or tertiary; rank in order of decreasing stability; and sketch an orbital picture of the most stable radical, showing the hyperconjugative interaction(s).
    (a) $CH_3CH_2\dot{C}HCH_3$ and $CH_3CH_2CH_2CH_2\cdot$
    (b) $(CH_3CH_2)_2CHCH_2\cdot$ and $(CH_3CH_2)_2\dot{C}CH_3$
    (c) $(CH_3)_2CH\dot{C}HCH_3$, $(CH_3)_2\dot{C}CH_2CH_3$, and $(CH_3)_2CHCH_2CH_2\cdot$

11. Write as many products as you can think of that might result from the pyrolytic cracking of propane. Assume that the only initial process is C–C bond cleavage.

12. Answer the question posed in Problem 11 for (a) butane and (b) 2-methylpropane. Use the data in Table 3-2 to determine the bond most likely to cleave homolytically, and use that bond cleavage as your first step.

13. Calculate $\Delta H°$ values for the following reactions. (a) $H_2 + F_2 \rightarrow 2\ HF$; (b) $H_2 + Cl_2 \rightarrow 2\ HCl$; (c) $H_2 + Br_2 \rightarrow 2\ HBr$; (d) $H_2 + I_2 \rightarrow 2\ HI$; (e) $(CH_3)_3CH + F_2 \rightarrow (CH_3)_3CF + HF$; (f) $(CH_3)_3CH + Cl_2 \rightarrow (CH_3)_3CCl + HCl$; (g) $(CH_3)_3CH + Br_2 \rightarrow (CH_3)_3CBr + HBr$; (h) $(CH_3)_3CH + I_2 \rightarrow (CH_3)_3CI + HI$.

14. For each compound in Problem 9, determine how many constitutional isomers can form upon monohalogenation. (**Hint:** Identify all groups of hydrogens that reside in distinct structural environments in each molecule.)

15. (a) Using the information given in Sections 3-6 and 3-7, write the products of the radical monochlorination of (i) pentane and (ii) 3-methylpentane. (b) For each, estimate the ratio of the isomeric monochlorination products that would form at 25°C. (c) Using the bond-strength data from Table 3-1, determine the $\Delta H°$ values of the propagation steps for the chlorination of 3-methylpentane at C3. What is the overall $\Delta H°$ value for this reaction?

16. Write in full the mechanism for monobromination of methane. Be sure to include initiation, propagation, and termination steps.

17. Write a mechanism for the radical bromination of the hydrocarbon benzene, C_6H_6 (for structure, see Section 2-1). Use propagation steps similar to those in the halogenation of alkanes, as presented in Sections 3-4 through 3-6. Calculate $\Delta H°$ values for each step and for the reaction as a whole. How does this reaction compare thermodynamically with the bromination of other hydrocarbons? Data: $DH°_{C_6H_5-H} = 112$ kcal mol^{-1}; $DH°_{C_6H_5-Br} = 81$ kcal mol^{-1}.

18. Write the major organic product(s), if any, of each of the following reactions.

    (a) $CH_3CH_3 + I_2 \xrightarrow{\Delta}$ (b) $CH_3CH_2CH_3 + F_2 \longrightarrow$

(c) [structure: methylcyclopentane] $+ Br_2 \xrightarrow{\Delta}$

(d) $CH_3\overset{\overset{\displaystyle CH_3}{|}}{CH}-CH_2-\overset{\overset{\displaystyle CH_3}{|}}{\underset{\underset{\displaystyle CH_3}{|}}{C}}CH_3 + Cl_2 \xrightarrow{h\nu}$

(e) $CH_3\overset{\overset{\displaystyle CH_3}{|}}{CH}-CH_2-\overset{\overset{\displaystyle CH_3}{|}}{\underset{\underset{\displaystyle CH_3}{|}}{C}}CH_3 + Br_2 \xrightarrow{h\nu}$

19. Calculate product ratios in each of the reactions in Problem 18. Use relative re-activity data for F_2 and Cl_2 at 25°C and for Br_2 at 150°C (Table 3-6).

20. Which, if any, of the reactions in Problem 18 give the major product with reasonable selectivity (i.e., are useful "synthetic methods")?

21. Predict the major product(s) of radical monobromination of each of the following compounds (identified by their common names). Point out any reaction that gives the major product with reasonable selectivity. Except for twistane, all the hydrocarbons shown are derived from molecules representative of the class of naturally occurring compounds called terpenes (see Section 4-7).

(a) H_3C- [cyclohexane structure] $-CH(CH_3)_2$

Menthane

(b) [bicyclic structure]

Pseudoguaiane

(c) [twistane structure]

Twistane

(d) [decalin structure] $(CH_3)_2CH$

Eudesmane

22. Write balanced equations for the combustion of each of the following substances (molecular formulas may be obtained from Table 3-7): **(a)** methane; **(b)** propane; **(c)** cyclohexane; **(d)** ethanol; **(e)** sucrose.

23. Propanal ($CH_3CH_2\overset{\overset{\displaystyle O}{||}}{C}H$) and propanone (acetone; $CH_3\overset{\overset{\displaystyle O}{||}}{C}CH_3$) are isomers with the formula C_3H_6O. The heat of combustion of propanal is -434.1 kcal mol^{-1}, that of propanone -427.9 kcal mol^{-1}. **(a)** Write a balanced equation for the combustion of either compound. **(b)** What is the energy difference between propanal and propanone? Which has the lower energy content? **(c)** Which substance is more thermodynamically stable, propanal or propanone? (**Hint:** Draw a diagram similar to that in Figure 3-11.)

24. Propose a mechanism for chlorination of CH_4, using sulfuryl chloride, SO_2Cl_2. (**Hint:** Follow the usual model for a radical chain process, substituting SO_2Cl_2 for Cl_2 where appropriate.)

25. Use the Arrhenius equation (Section 2-8) to estimate the ratio of the rate constants k for the reactions of a C–H bond in methane with a chlorine atom and with a bromine atom at 25°C. Assume that the A values for the two processes are equal, and use $E_a = 19$ kcal mol^{-1} for the reaction between Br· and CH_4.

26. Reexamine Exercise 3-4 regarding the reaction between methane and an equimolar mixture of Br_2 and Cl_2. As this reaction proceeds, CH_3Br is observed to eventually form in significantly greater quantities than would be expected, considering the very large difference in reactivity between Cl· and Br· toward C–H bonds (Problem 25). Suggest an explanation. (**Hint:** The mechanism for formation of CH_3Br in this situation differs in one important way from that of the reaction between CH_4 and Br_2 alone. In analyzing this problem, consider which step in the radical chain mechanism determines the structure of the *organic* product.)

27. When an alkane with different types of C–H bonds, such as propane, reacts with an equimolar mixture of Br_2 and Cl_2, the selectivity in the formation of the brominated products is much worse than that observed when reaction is carried out with Br_2 alone. (In fact, it is very similar to the selectivity for *chlorination*.) Explain. (**Hint:** Recall Problem 26, and consider which step in the radical chain mechanism is responsible for the *selectivity* of halogenation.)

28. A hypothetical alternative mechanism for the halogenation of methane has the following propagation steps.

$$\textbf{(i)} \ \ X· + CH_4 \longrightarrow CH_3X + H·$$
$$\textbf{(ii)} \ \ H· + X_2 \longrightarrow HX \ \ + X·$$

(**a**) Using $DH°$ values from appropriate tables, calculate $\Delta H°$ for these steps for any one of the halogens. (**b**) Compare your $\Delta H°$ values with those for the accepted mechanism (Table 3-5). Do you expect this alternative mechanism to compete successfully with the accepted one? (**Hint:** Consider activation energies.)

29. The addition of certain materials called radical inhibitors to halogenation reactions causes the reactions to come to a virtually complete stop. An example is the inhibition by I_2 of the chlorination of methane. Explain how this inhibition might come about. (**Hint:** Calculate $\Delta H°$ values for possible reactions of the various species present in the system with I_2, and evaluate the possible further reactivity of the products of these I_2 reactions.)

30. One additional piece of experimental evidence in support of the radical chain mechanism is the observation that traces of oxygen, O_2, strongly inhibit halogenation reactions of alkanes. (**a**) What is unusual about the electronic structure of O_2 that may be relevant in this context? (**Hint:** Refer to Problem 25 of Chapter 1.) (**b**) Suggest a process by which O_2 interferes with a key step in alkane halogenation.

31. Typical hydrocarbon fuels (e.g., 2,2,4-trimethylpentane, a common component of gasoline) have very similar heats of combustion when calculated in kilocalories *per gram*. (**a**) Calculate heats of combustion per gram for several representative hydrocarbons in Table 3-7. (**b**) Make the same calculation for ethanol (Table 3-7). (**c**) In evaluating the feasibility of "gasohol" (90% gasoline and 10% ethanol) as a motor fuel, it has been estimated that an automobile running on pure ethanol would get approximately 40% fewer miles per gallon than would an identical automobile running on standard gasoline. Is this estimate consistent with the results in (a) and (b)? What can you say in general about the fuel capabilities of oxygen-containing molecules relative to hydrocarbons?

32. Two simple organic molecules that are in use as fuel additives are methanol (CH_3OH) and 2-methoxy-2-methylpropane [*tert*-butyl methyl ether, $(CH_3)_3COCH_3$]. The $\Delta H°_{comb}$ values for these compounds in the gas phase are

-182.6 kcal mol^{-1} for methanol and -809.7 kcal mol^{-1} for 2-methoxy-2-methylpropane. **(a)** Write balanced equations for the complete combustion of each of these molecules to CO_2 and H_2O. **(b)** Using Table 3-7, compare the ΔH°_{comb} values for these compounds with those for alkanes with similar molecular weights.

33. Ordinary glassblowing torches are fueled by natural gas. However, welders require much hotter temperatures for their work and often use torches fueled by ethyne (acetylene, HC≡CH). **(a)** Write a balanced equation for the combustion of ethyne. **(b)** The heat of combustion of ethyne is -310.7 kcal mol^{-1}. Compare this value with that for propane, an important component of natural gas, both per mole and per gram. Does this explain the hotter flame of ethyne?

34. Figure 3-8 compares the reactions of Cl· with the primary and secondary hydrogens of propane. **(a)** Draw a similar diagram comparing the reactions of Br· with the primary and secondary hydrogens of propane. (**Hint:** First obtain the necessary DH° values from Table 3-1 and calculate ΔH° for both the primary and the secondary hydrogen abstraction reactions. Other data: $E_a = $ 13 kcal mol^{-1} for Br· reacting with a primary C–H bond and $E_a = $ 10 kcal mol^{-1} for Br· reacting with a secondary C–H bond. **(b)** Which among the transition states of these reactions would you call "early," and which "late"? **(c)** Judging from the locations of the transition states of these reactions along the reaction coordinate, should they show greater or lesser radical character than do the corresponding transition states for chlorination (Figure 3-8)? **(d)** Is your answer to (c) consistent with the selectivity differences between Cl· reacting with propane and Br· reacting with propane? Explain.

35. Two of the propagation steps in the Cl·/O_3 system consume ozone and oxygen atoms (which are necessary for the production of ozone), respectively (Section 3-9).

$$Cl + O_3 \longrightarrow ClO + O_2$$
$$ClO + O \longrightarrow Cl + O_2$$

Calculate ΔH° for each of these propagation steps. Use the following data: DH° for ClO $= 56$ kcal mol^{-1}; DH° for $O_2 = 120$ kcal mol^{-1}; DH° for an O–O_2 bond in $O_3 = 26$ kcal mol^{-1}. Write the overall equation described by the combination of these steps and calculate its ΔH°. Comment on the thermodynamic favorability of the process.

Team Problem

36. **(a)** Provide an IUPAC name for each of the isomers that you drew in Exercise 2-1(a). **(b)** For each isomer that you drew and named here, give all the free radical monochlorination and monobromination products that are structurally isomeric. **(c)** Referring to Table 3-6, discuss which starting alkane and which halogen will yield the least number of isomeric products.

Preprofessional Problems

37. The reaction $CH_4 + Cl_2 \rightarrow CH_3Cl + HCl$ is an example of
    (a) neutralization   **(b)** an acidic reaction
    (c) an isomerization   **(d)** an ionic reaction
    (e) a radical chain reaction

38.

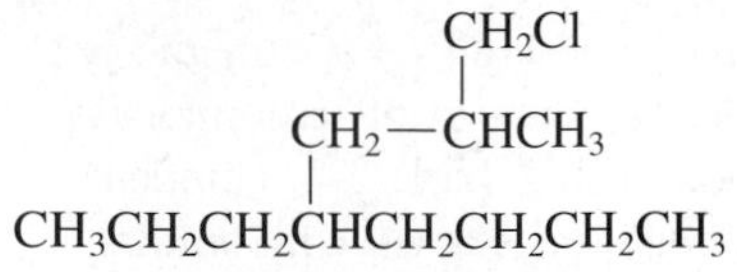

$$CH_2Cl$$
$$|$$
$$CH_2—CHCH_3$$
$$|$$
$$CH_3CH_2CH_2CHCH_2CH_2CH_2CH_3$$

The sum of all the digits that appear in the (IUPAC) name for this compound is which of the following?

(**a**) five   (**b**) six   (**c**) seven   (**d**) eight   (**e**) nine

39. In a competition reaction, equimolar amounts of the four alkanes shown were allowed to react with a limited amount of Cl_2 at 300°C. Which one of these alkanes would be depleted most from the mixture?

(**a**) pentane   (**b**) 2-methylpropane   (**c**) butane   (**d**) propane

40. The reaction of CH_4 with Cl_2 to yield CH_3Cl and HCl is well known. On the basis of the values in the short table below, the enthalpy $\Delta H°$ (kcal mol^{-1}) of this reaction is

(**a**) $+135$ (**b**) -135 (**c**) $+25$ (**d**) -25

Bond Dissociation Energies $DH°$ (kcal mol^{-1})			
H–Cl	103	Cl–Cl	58
H$_3$C–Cl	85	H$_3$C–H	105

Cyclic Alkanes

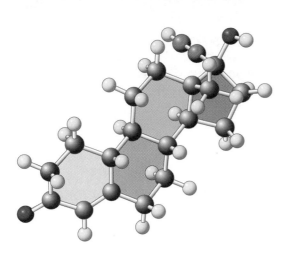

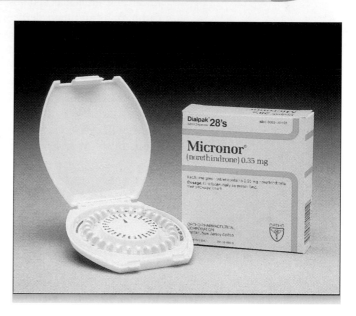

The polycyclic alkane framework of the steroids is exemplified by norethindrone, one of the active ingredients in the birth control pill.

When you hear or read the word *steroids,* two things probably come to mind immediately: athletes who illegally "take steroids" to develop their muscles and "the pill" used for birth control. But what do you know about steroids aside from this general association? What is their structure? How does one steroid differ from another? Where are they found in nature?

An example of a naturally occurring steroid is diosgenin, obtained from the Mexican yam and used as a starting material for the synthesis of several commercial steroids. Most striking is the number of *rings* in the compound.

Diosgenin

The root of the Mexican yam.

Hydrocarbons containing single-bonded carbon atoms arranged in rings are known as **cyclic alkanes, carbocycles** (in contrast with heterocycles, Chapter 25), or **cyclo-alkanes.** The majority of organic compounds occurring in nature contain rings. Indeed, so many fundamental biological functions depend on the chemistry of ring-containing compounds that life as we know it could not exist in their absence. This chapter deals with the names, physical properties, structural features, and conformational characteristics of the cycloalkanes. Members of this class of compounds serve to review and amplify some of the principles presented in Chapter 2 regarding the linear and branched alkanes. We end with the biochemical significance of selected carbocycles and their derivatives, including some common flavoring agents, cholesterol, and other biological regulators.

4-1 Names and Physical Properties of Cycloalkanes

How do the cycloalkanes differ in their names and physical properties from their noncyclic (also called *acyclic*) analogs containing the same number of carbons?

The names of the cycloalkanes follow IUPAC rules

We can construct a molecular model of a cycloalkane by removing two terminal hydrogen atoms from a model of a straight-chain alkane and allowing the terminal carbons to form a bond. The general formula of a cycloalkane is C_nH_{2n}. The system for naming members of this class of compounds is straightforward: Alkane names are preceded by the prefix **cyclo-**. Three members in the homologous series—starting with the smallest, cyclopropane—are shown in the margin, written both in condensed form and in bond-line notation.

$$\begin{array}{c} H_2 \\ C \\ H_2C{-\!-}CH_2 \end{array} \qquad \triangle$$
Cyclopropane

$$\begin{array}{c} H_2C{-\!-}CH_2 \\ | \qquad | \\ H_2C{-\!-}CH_2 \end{array} \qquad \square$$
Cyclobutane

$$\begin{array}{c} H_2 \\ C \\ H_2C \qquad CH_2 \\ H_2C \qquad CH_2 \\ C \\ H_2 \end{array} \qquad \hexagon$$
Cyclohexane

EXERCISE 4-1

Make molecular models of cyclopropane through cyclododecane. Compare the relative conformational flexibility of each ring with that of others within the series and with that of the corresponding straight-chain alkanes.

Naming a substituted cyclic alkane requires the numbering of the individual ring carbons only if more than one substituent is attached to the ring. In monosubstituted systems, the carbon of attachment is defined as carbon 1 of the ring. For polysubstituted compounds, take care to provide the lowest possible numbering sequence. When two such sequences are possible, the alphabetical order of the substituent names takes precedence. Radicals derived from cycloalkanes by abstraction of a hydrogen atom are **cycloalkyl radicals.** Substituted cycloalkanes are therefore sometimes named as cycloalkyl derivatives. In general, the smaller unit is treated as a substituent to the larger one—for example, propylcyclopentane (not cyclopentylpropane) and cyclohexyloctane (not octylcyclohexane).

$$\triangle{-}CH_3$$
Methylcyclopropane

$$\square\begin{array}{l}{-}CH_3 \\ CH_2CH_3\end{array}$$
1-Ethyl-1-methylcyclobutane

1-**Chloro**-2-methyl-4-**propyl**cyclopentane

(Not 2-chloro-1-methyl-4-propylcyclopentane)

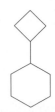
Cyclobutylcyclohexane

Disubstituted cycloalkanes possess isomers

Inspection of molecular models of disubstituted cycloalkanes in which the two substituents are located on different carbons shows that *two isomers are possible* in each case. In one, the two substituents are positioned on the *same* face, or side, of the ring; in the other, on *opposite* faces. Substituents on the same face are called **cis** (*cis,* Latin, on the same side); those on opposite faces, **trans** (*trans,* Latin, across).

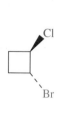
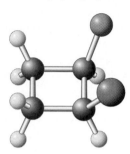

cis-**1,2-Dimethylcyclopropane**   *trans*-**1,2-Dimethylcyclopropane**

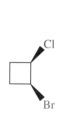
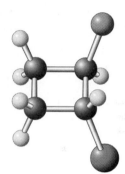

cis-**1-Bromo-2-chlorocyclobutane**   *trans*-**1-Bromo-2-chlorocyclobutane**

Cis and trans isomers are **stereoisomers**—compounds that have identical connectivities (i.e., their atoms are attached in the same sequence) but differ in the arrangement of their atoms in space. They are distinct from constitutional or structural isomers (Sections 1-9 and 2-2), which are compounds with differing sequences of atoms. Conformations (Sections 2-5 through 2-7) also are stereoisomers by the preceding definition. However, unlike cis and trans isomers, which can be interconverted only by *breaking* bonds (try it on your models), conformers are readily equilibrated by *rotation* about bonds. The subject of stereoisomerism will be discussed in more detail in Chapter 5.

Dashed-wedged line structures can be used to depict the three-dimensional arrangement of substituted cycloalkanes. The positions of any remaining hydrogens are not always shown. Structural and cis-trans isomerisms give rise to a variety of structural possibilities in substituted cycloalkanes. For example, there are eight isomeric bromomethylcyclohexanes (three of which are shown below), all with different and distinct physical and chemical properties.

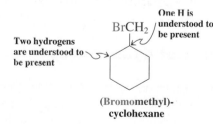

(**Bromomethyl**)-   **1-Bromo-1-methyl**-   *cis*-**1-Bromo-2-**
cyclohexane   **cyclohexane**   **methylcyclohexane**

Give the structures and names of the other five isomeric bromomethylcyclohexanes.

TABLE 4-1	Physical Properties of Various Cycloalkanes		
Cycloalkane	Boiling point (°C)	Melting point (°C)	Density at 20°C (g mL^{-1})
Cyclopropane	−32.7	−127.6	0.617^b
Cyclobutane	−12.5	−50.0	0.720
Cyclopentane	49.3	−93.9	0.7457
Cyclohexane	80.7	6.6	0.7785
Cycloheptane	118.5	−12.0	0.8098
Cyclooctane	148.5	14.3	0.8349
Cyclododecane	160 (100 torr)	64	0.861
Cyclopentadecane	110 (0.1 torr^a)	66	0.860

^{a}Sublimation point.
^{b}At 25°C.

The properties of the cycloalkanes differ from those of their straight-chain analogs

The physical properties of a few cycloalkanes are recorded in Table 4-1. Note that, compared with the corresponding straight-chain alkanes (Table 2-3), the cycloalkanes have higher boiling and melting points, as well as higher densities. These differences are due in large part to increased London interactions of the relatively more rigid and more symmetric cyclic systems. In comparing lower cycloalkanes possessing an odd number of carbons with those having an even number, we find a pronounced alternation in their melting points. This phenomenon has been ascribed to differences in crystal packing forces between the two series.

In summary, names of the cycloalkanes are derived in a straightforward manner from those of the straight-chain alkanes. In addition, the position of a single substituent is defined to be C1. Disubstituted cycloalkanes can give rise to cis and trans isomers, depending on the location of the substituents. Physical properties parallel those of the straight-chain alkanes, except that the individual values for boiling and melting points and for densities are higher for the cyclic compounds of equal carbon number.

4-2 Ring Strain and the Structure of Cycloalkanes

The molecular models made for Exercise 4-1 reveal obvious differences between cyclopropane, cyclobutane, cyclopentane, and so forth, and the corresponding straight-chain alkanes. One notable feature of the first two members in the series is how difficult it is to close the ring without breaking the plastic tubes used to represent bonds. This problem is called **ring strain.** The reason for it lies in the tetrahedral carbon model. The C–C–C bond angles in, for example, cyclopropane (60°) and cyclobutane (90°) differ considerably from the tetrahedral value. As the ring size increases, strain diminishes. Thus, cyclohexane can be assembled without distortion or strain.

Does this observation tell us anything about the relative stability of the cyclo-alkanes—for example, as measured by their heats of combustion, $\Delta H°_{comb}$? How does strain affect structure? This section and Section 4-3 address these questions.

The heats of combustion of the cycloalkanes reveal the presence of ring strain

Section 3-10 introduced one measure of the stability of a molecule: its heat content. We also learned that the heat content of an alkane could be estimated by measuring its heat of combustion, $\Delta H°_{comb}$ (Table 3-7). To determine the stability of each cy-cloalkane, we can compare its heat of combustion with the value measured for the analogous straight-chain molecule. The (negative) $\Delta H°_{comb}$ values for the straight-chain alkanes increase by about the same amount with each successive member of the series.

$\Delta H°_{comb}$ **Values for the Series of Straight-Chain Alkanes**

$$CH_3CH_2CH_3 \text{ (gas)} \qquad -530.6$$
$$CH_3CH_2CH_2CH_3 \text{ (gas)} \qquad -687.4 \qquad \left.\begin{array}{r} -156.8 \\ -157.8 \end{array}\right\} \text{ kcal mol}^{-1}$$
$$CH_3(CH_2)_3CH_3 \text{ (gas)} \qquad -845.2$$

There appears to be a regular increment of about 157 kcal mol^{-1} for each additional CH_2 group. When averaged over a large number of alkanes, this value approaches 157.4 kcal mol^{-1}.

What does this tell us about cycloalkanes? Because these molecules have the general formula $(CH_2)_n$, we might expect their approximate $\Delta H°_{comb}$ to be $-(n \times 157.4)$ kcal mol^{-1} (Table 4-2, column 2). However, the measured heats of combustion turn

TABLE 4-2	Calculated and Experimental Heats of Combustion (kcal mol^{-1}) of Various Cycloalkanes			
Ring size (C_n)	$\Delta H°_{comb}$ (calculated)	$\Delta H°_{comb}$ (experimental)	Total strain	Strain per CH_2 group
3	−472.2	−499.8	27.6	9.2
4	−629.6	−655.9	26.3	6.6
5	−787.0	−793.5	6.5	1.3
6	−944.4	−944.5	0.1	0.0
7	−1101.8	−1108.2	6.4	0.9
8	−1259.2	−1269.2	10.0	1.3
9	−1416.6	−1429.5	12.9	1.4
10	−1574.0	−1586.0	14.0	1.4
11	−1731.4	−1742.4	11.0	1.1
12	−1888.8	−1891.2	2.4	0.2
14	−2203.6	−2203.6	0.0	0.0

Note: The calculated numbers are based on the value of −157.4 kcal mol^{-1} for a CH_2 group.

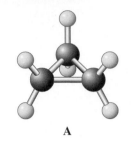

A

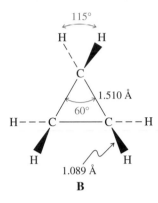

B

FIGURE 4-1
Cyclopropane: (A) molecular
model; (B) bond lengths and
angles.

out to be *larger in magnitude* (Table 4-2, column 3). For example, cyclopropane should have a $\Delta H°_{comb}$ of -472.2 kcal mol^{-1}, but the experimental value is -499.8 kcal mol^{-1}. The difference between expected and observed values is 27.6 kcal mol^{-1}. It is attributed to a property of cyclopropane of which we are already aware because of the model we built: *ring strain*. The strain per CH_2 group in this molecule is 9.2 kcal mol^{-1}. A similar calculation for cyclobutane (Table 4-2) reveals a ring strain of 26.3 kcal mol^{-1}, or about 6.6 kcal mol^{-1} per CH_2 group. In cyclopentane, this effect is much smaller, the total strain amounting to only 6.5 kcal mol^{-1}, and cyclohexane is virtually strain free. However, succeeding members of the series again show considerable strain until we reach very large rings. Because of these trends, organic chemists have loosely defined four groups of cycloalkanes.

1. *Small rings* (cyclopropane, cyclobutane)
2. *Common rings* (cyclopentane, cyclohexane, cycloheptane)
3. *Medium rings* (from eight- to twelve-membered)
4. *Large rings* (thirteen-membered and larger)

What kinds of effects contribute to the ring strain in cycloalkanes? We shall answer this question by exploring the detailed structures of several of these compounds.

Strain affects the structures and conformations of the smaller cycloalkanes

As we have just seen, the smallest cycloalkane, *cyclopropane,* is much less stable than expected for three methylene groups. Why should this be? The reason is twofold: torsional strain and bond-angle strain.

The structure of the cyclopropane molecule is represented in Figure 4-1. We notice first that all methylene hydrogens are eclipsed, much like the hydrogens in the eclipsed conformation of ethane (Section 2-5). We know that the energy of the eclipsed form of ethane is raised above that of the more stable staggered conformation because of **eclipsing (torsional) strain.** This effect is also present in cyclopropane. Moreover, the carbon skeleton in cyclopropane is by necessity flat and quite rigid, and bond rotation that might relieve eclipsing strain is very difficult.

Second, we notice that cyclopropane has C–C–C bond angles of 60°, a significant deviation from the "natural" tetrahedral bond angle of 109.5°. How is it possible for three supposedly tetrahedral carbon atoms to maintain a bonding relation at such highly distorted angles? The problem is perhaps best illustrated in Figure 4-2, in which the bonding in the strain-free "open cyclopropane," the trimethylene diradical ·$CH_2CH_2CH_2$·, is compared with that in the closed form. We can see that the two ends of the trimethylene diradical cannot "reach" far enough to close the ring without "bending" the two C–C bonds already present. However, if all three C–C bonds in cyclopropane adopt a bent configuration (interorbital angle 104°, see Figure 4-2B), overlap is sufficient for bond formation. The energy needed to distort the tetrahedral carbons enough to close the ring is called **bond-angle strain.** The ring strain in cyclopropane is derived from a combination of eclipsing and bond-angle contributions.

As a consequence of its structure, cyclopropane has relatively weak C–C bonds ($DH° = 65$ kcal mol^{-1}). This value is low (recall that the C–C strength in ethane is 90 kcal mol^{-1}) because breaking the bond opens the ring and relieves ring strain. Therefore, cyclopropane undergoes several unusual reactions. For example, reaction with hydrogen in the presence of a palladium catalyst opens the ring to give propane.

$$\triangle \; + \; H_2 \; \xrightarrow{\text{Pd catalyst}} \; CH_3CH_2CH_3 \qquad \Delta H° = -37.6 \text{ kcal mol}^{-1}$$
Propane

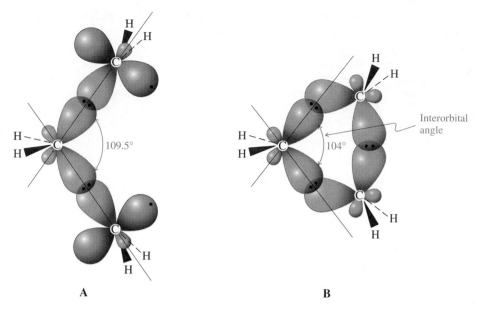

A

B

FIGURE 4-2

Molecular-orbital picture of (A) the trimethylene diradical and (B) the bent bonds in cyclopropane. Only the hybrid orbitals forming C–C bonds are shown. Note the interorbital angle of 104° in cyclopropane.

Trans-1,2-dimethylcyclopropane is more stable than *cis*-1,2-dimethylcyclopropane. Why? Draw a picture to illustrate your answer. Which isomer liberates more heat on combustion?

What about higher cycloalkanes? The structure of *cyclobutane* (Figure 4-3) reveals that this molecule is not planar but puckered, with an approximate bending angle of 26°. The nonplanar structure of the ring, however, is not very rigid. The molecule "flips" rapidly from one puckered conformation to the other. Construction of a molecular model shows why distorting the four-membered ring from planarity is favorable: It partly relieves the strain introduced by the eight eclipsing hydrogens. Moreover, bond-angle strain is considerably reduced relative to that in cyclopropane,

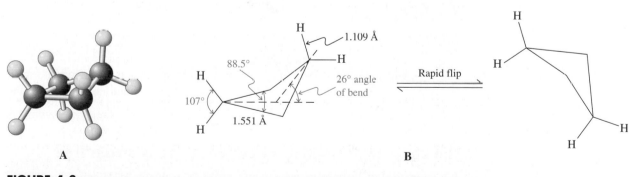

A

Rapid flip

B

FIGURE 4-3

Cyclobutane: (A) molecular model; (B) bond lengths and angles. The nonplanar molecule "flips" rapidly from one conformation to another.

FIGURE 4-4 _____
Cyclopentane: (A) molecular
model of the half-chair
conformation; (B) bond
lengths and angles. The
molecule is flexible, with little
strain.

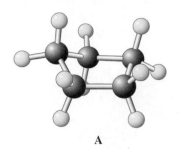

A

B

although maximum overlap is, again, only possible with the use of bent bonds. The C–C bond strength in cyclobutane also is low (about 63 kcal mol^{-1}) because of the release of ring strain on ring opening and the consequences of relatively poor overlap in bent bonds. Cyclobutane is less reactive than cyclopropane but undergoes similar ring-opening processes.

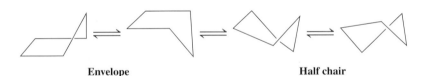

Butane

Cyclopentane might be expected to be planar because the angles in a regular pentagon are 108°, close to tetrahedral. However, such a planar arrangement would have *ten* H–H eclipsing interactions. The puckering of the ring reduces this effect, as indicated in the structure of the molecule (Figure 4-4). Although puckering relieves eclipsing, it also increases bond-angle strain. The conformation of lowest energy is a compromise in which the energy of the system is minimized. There are two puckered conformations possible for cyclopentane: the **envelope** and the **half chair.**

Envelope   **Half chair**

There is little difference in energy between them, and the activation barriers for rapid interconversion are extremely low. Overall, cyclopentane has relatively little ring strain and hence does not show the unusual reactivity of three- or four-membered rings.

4-3 Cyclohexane: A Strain-Free Cycloalkane

The cyclohexane ring is one of the most abundant and important structural units in organic chemistry. Its substituted derivatives exist in many natural products (see Section 4-7), and an understanding of its conformational mobility is an important aspect of organic chemistry. Table 4-2 reveals that, within experimental error, cyclohexane is unusual in that it is free of bond-angle or eclipsing strain. Why?

The chair conformation of cyclohexane is strain free

A hypothetical planar cyclohexane would suffer from twelve H–H eclipsing interactions and sixfold bond-angle strain (a regular hexagon requires 120° bond angles). However, one conformation of cyclohexane, obtained by moving carbons 1 and 4 out

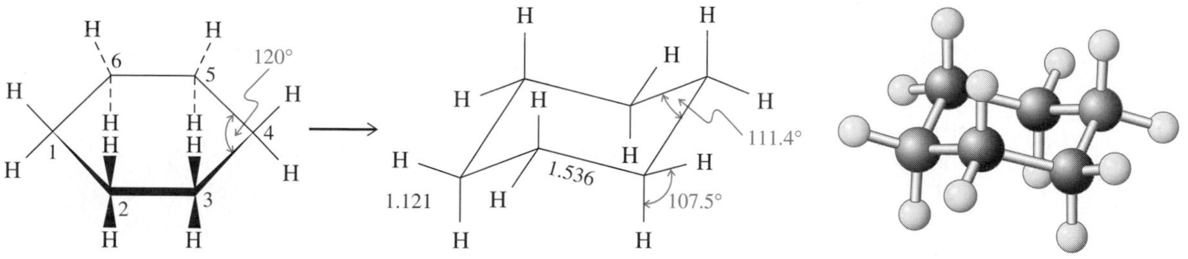

A
Planar cyclohexane
(120° bond angles;
12 eclipsing hydrogens)

B
Chair cyclohexane
(Nearly tetrahedral bond angles;
no eclipsing hydrogens)

C

FIGURE 4-5

Conversion of the (A) hypothetical planar cyclohexane into the (B) chair conformation, showing bond lengths and angles; (C) molecular model. The chair conformation is strain free.

of planarity in opposite directions, is in fact strain free (Figure 4-5). This structure is called the **chair conformation** of cyclohexane (because it resembles a chair), in which eclipsing is completely prevented and the bond angles are very nearly tetrahedral. As seen in Table 4-2, the calculated $\Delta H°_{comb}$ of cyclohexane (-944.4 kcal mol^{-1}) based on a strain-free $(CH_2)_6$ model is very close to the experimentally determined value (-944.5 kcal mol^{-1}).

Looking at the molecular model of cyclohexane enables us to recognize the conformational stability of the molecule. If we view it along (any) one C–C bond, we can see the staggered arrangement of all substituent groups along it. We can visualize this arrangement by drawing a Newman projection of that view (Figure 4-6). Because of its lack of strain, cyclohexane is as inert as a straight-chain alkane.

EXERCISE 4-4

Draw Newman projections of the carbon–carbon bonds in cyclopropane, cyclobutane, and cyclopentane in their most stable conformations. Use the models that you prepared for Exercise 4-1 to assist you and refer to Figure 4-6. What are the approximate torsional angles between the C–H bonds in each?

Cyclohexane also has several less stable conformations

Other, less stable conformations of cyclohexane are nevertheless readily accessible to the molecule. One is the **boat form,** in which carbons 1 and 4 are out of the plane

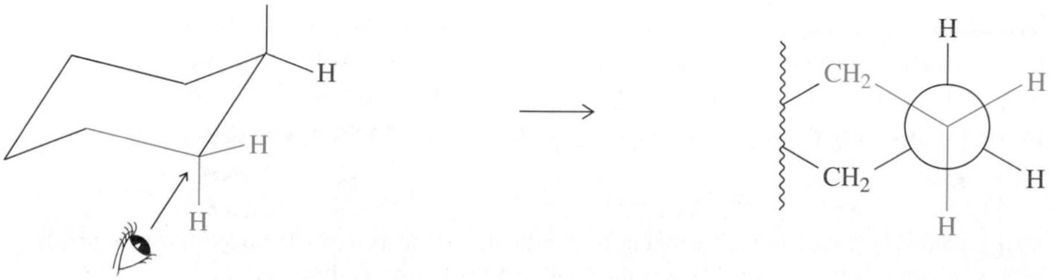

FIGURE 4-6

View along one of the C–C bonds in the chair conformation of cyclohexane. Note the staggered arrangement of all substituents.

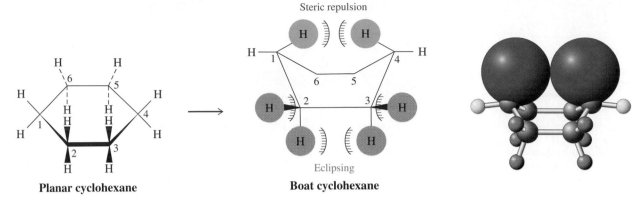

Planar cyclohexane

Boat cyclohexane

FIGURE 4-7

Conversion of the hypothetical planar cyclohexane into the boat form. In the latter form, the hydrogens on carbons 2, 3, 5, and 6 are eclipsed, thereby giving rise to torsional strain. The "inside" hydrogens on carbons 1 and 4 interfere with each other sterically in a transannular interaction. The space-filling size of these two hydrogens is depicted in the ball-and-stick model on the right.

in the *same* direction (Figure 4-7). The boat is less stable than the chair form by 6.9 kcal mol^{-1}. One reason for this difference is the eclipsing of eight hydrogen atoms at the base of the boat. Another is steric hindrance (Section 2-7) due to the close proximity of the two inside hydrogens in the boat framework. The distance between these two hydrogens is only 1.83 Å, small enough to create an energy of repulsion of about 3 kcal mol^{-1}. This effect is an example of **transannular strain,** that is, strain resulting from steric crowding of two groups across a ring (*trans,* Latin, across; *anulus,* Latin, ring).

Boat cyclohexane is fairly flexible. If one of the C–C bonds is twisted relative to another, this form can be somewhat stabilized by partial removal of the transannular interaction. The new conformation obtained is called the **twist-boat** (or **skew-boat**) **conformation** of cyclohexane (Figure 4-8). The stabilization relative to the boat form amounts to about 1.4 kcal mol^{-1}. As shown in Figure 4-8, two twist-boat forms are possible. They interconvert rapidly, with the boat conformer acting as a *transition state* (verify this with your model). Thus, the boat cyclohexane is not a normally isolable species, the twist-boat form is present in very small amounts, and the chair form is the major conformer (Figure 4-9). The activation barrier separating the most stable chair from the boat manifold is 10.8 kcal mol^{-1}. We shall see that the equilibration depicted in Figure 4-9 has important structural consequences with respect to the positions of substituents on the cyclohexane ring.

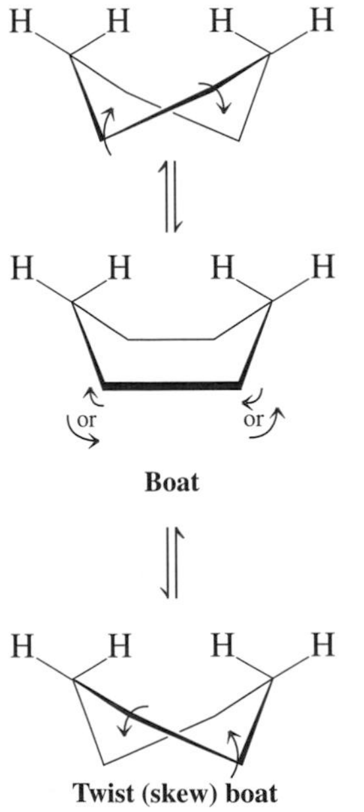

Boat

Twist (skew) boat

FIGURE 4-8

Twist-boat to twist-boat flipping of cyclohexane proceeds through the boat conformation.

Cyclohexane has axial and equatorial hydrogen atoms

The chair-conformation model of cyclohexane reveals that the molecule has two types of hydrogens. Six carbon–hydrogen bonds are nearly parallel to the principal molecular axis (Figure 4-10) and hence are referred to as **axial;** the other six are nearly perpendicular to the axis and close to the equatorial plane and are therefore called **equatorial.***

*An equatorial plane is defined as being perpendicular to the axis of rotation of a rotating body and equidistant from its poles, such as the equator of the planet Earth.

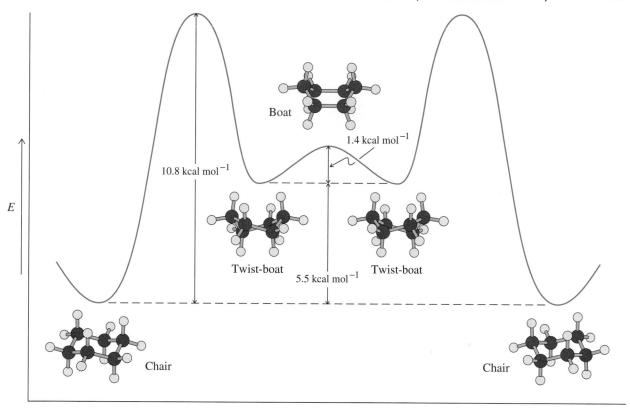

E

10.8 kcal mol^{-1}

Boat

1.4 kcal mol^{-1}

Twist-boat

Twist-boat

5.5 kcal mol^{-1}

Chair

Chair

Reaction coordinate to conformational interconversion ⟶

FIGURE 4-9

Potential energy diagram for the chair–chair interconversion of cyclohexane through the twist-boat and boat forms. In the procession from left to right, the chair is converted into a twist boat (by the twisting of one of the C–C bonds) with an activation barrier of 10.8 kcal mol^{-1}. The twist-boat form flips (as shown in Figure 4-8) through the boat conformer as a transition state (1.4 kcal mol^{-1} higher in energy) into another twist-boat structure, which relaxes back into the (ring-flipped) chair cyclohexane. Use your molecular models to visualize these changes.

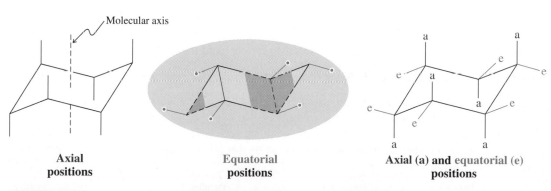

Molecular axis

Axial positions

Equatorial positions

Axial (a) and equatorial (e) positions

FIGURE 4-10

The axial and equatorial positions of hydrogens in the chair form of cyclohexane.

Being able to draw cyclohexane chair conformations will help you learn the chemistry of six-membered rings. Several rules are useful.

How to Draw Chair Cyclohexanes

1. Draw the chair so as to place the C2 and C3 atoms below and slightly to the right of C5 and C6, with apex 1 pointing downward on the left and apex 4 pointing upward on the right.

> The bond between C1 and C6 is also parallel to that between C3 and C4.

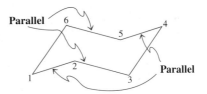

2. Add all the axial bonds as vertical lines, pointing downward at C1, C3, and C5 and upward at C2, C4, and C6.

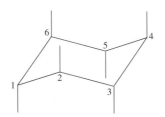

3. Draw the two equatorial bonds at C1 and C4 at a slight angle from horizontal, pointing upward at C1 and downward at C4, parallel to the bond between C2 and C3 (or between C5 and C6).

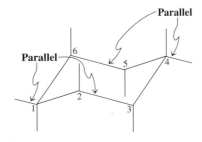

4. This rule is the most difficult to follow: Add the remaining equatorial bonds at C2, C3, C5, and C6 by aligning them *parallel* to the C–C bond "once removed," as shown below.

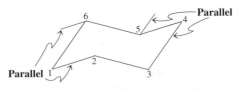
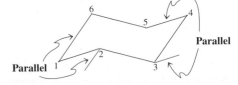

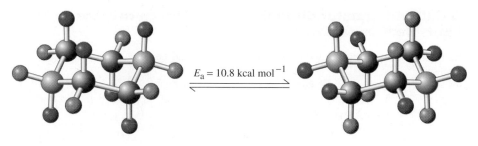

$E_a = 10.8 \text{ kcal mol}^{-1}$

FIGURE 4-11

Chair–chair interconversion ("ring flipping") in cyclohexane. In the process, which is rapid at room temperature, a (green) carbon at one end of the molecule moves up, while its counterpart (also green) at the other end moves down. All groups originally in axial positions (red in the structure at the left) become equatorial, and those that start in equatorial positions (blue) become axial.

Conformational flipping interconverts axial and equatorial hydrogens

What happens to the identity of the equatorial and axial hydrogens when we let chair cyclohexane equilibrate with its boat forms? You can follow the progess of conformational interconversion in Figure 4-9 with the help of molecular models. Starting with the chair structure on the left, you can simply "flip" the CH_2 group farthest to the left (C1 in the preceding section) upward through the equatorial plane to generate the boat conformers. If you now return the molecule to the chair form not by a reversal of the movement but by the equally probable alternative—namely, the flipping downward of the opposite CH_2 group (C4)—you will recognize that the original sets of axial and equatorial positions have traded places. In other words, cyclohexane undergoes chair–chair interconversions ("flipping") in which all axial hydrogens in one chair become equatorial in the other and vice versa (Figure 4-11). The activation energy for this process is 10.8 kcal mol^{-1} (Figure 4-9). As suggested in Sections 2-5 through 2-7, this value is so low that, at room temperature, two equivalent chair forms interconvert rapidly (approximately 100,000 times per second).

To summarize, the discrepancy between calculated and measured heats of combustion in the cycloalkanes can be largely attributed to three forms of strain: bond angle (deformation of tetrahedral carbon), eclipsing (torsional), and transannular (across the ring). Because of strain, the small cycloalkanes are chemically reactive, undergoing ring-opening reactions. Cyclohexane is strain free. It has a lowest-energy chair, as well as additional higher-energy conformations, particularly the boat and twist-boat structures. Chair–chair interconversion is rapid at room temperature; it is a process in which equatorial and axial hydrogen atoms interchange their positions.

4-4 Substituted Cyclohexanes

We can now apply our knowledge of conformational analysis to substituted cyclohexanes. Let us look at the simplest alkylcyclohexane, methylcyclohexane.

Axial and equatorial methylcyclohexanes are not equivalent in energy

In methylcyclohexane, the methyl group occupies either an equatorial or an axial position.

No 1,3-diaxial interactions
More stable

1,3-Diaxial interactions
Less stable

Ratio = 95:5

Are the two forms equivalent? Clearly not. In the equatorial conformer, the methyl group extends into space away from the remainder of the molecule. In contrast, in the axial conformer, the methyl substituent is close to the other two axial hydrogens on the same side of the molecule. The distance to these hydrogens is small enough (about 2.7 Å) to result in steric repulsion, another example of transannular strain. Because this effect is due to axial substituents on carbon atoms that have a 1,3-relation (in the drawing, 1,3 and 1,3′), it is called a **1,3-diaxial interaction.** This interaction is the same as that resulting in the *gauche* conformation of butane (Section 2-7). Thus, the axial methyl group is *gauche* to two of the ring carbons (C3 and C3′); when it is in the equatorial position, it is *anti* to the same nuclei.

The two forms of chair methylcyclohexane are in equilibrium. *The equatorial conformer is more stable* by 1.7 kcal mol^{-1} and is favored by a ratio of 95:5 at 25°C (Section 2-8). The activation energy for chair–chair interconversion is similar to that in cyclohexane itself (about 11 kcal mol^{-1}), and equilibrium between the two conformers is established rapidly at room temperature.

The unfavorable 1,3-diaxial interactions to which an axial substituent is exposed are readily seen in Newman projections of the ring C–C bond bearing that substituent. In contrast with that in the axial form (*gauche* to two ring bonds), the substituent in the equatorial conformer (*anti* to the two ring bonds) is away from the axial hydrogens (Figure 4-12).

EXERCISE 4-5

Calculate K for equatorial versus axial methylcyclohexane from the $\Delta G°$ value of 1.7 kcal mol^{-1}. Use the expression $\Delta G°$ (in kcal mol^{-1}) $= -1.36 \log K$. (**Hint:** If $\log K = x$, then $K = 10^x$.) How well does your result agree with the 95:5 conformer ratio stated in the text?

The energy difference between the axial and the equatorial forms of many mono-substituted cyclohexanes has been measured; several are given in Table 4-3. In many cases (but not all), particularly for alkyl substituents, the energy difference between the two forms increases with the size of the substituent, a direct consequence of in-

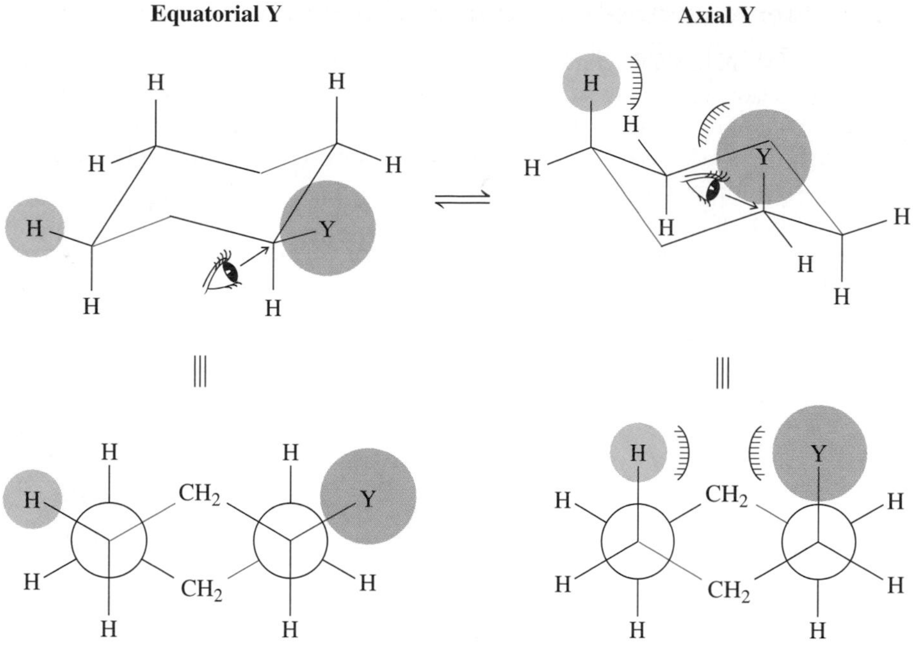

Equatorial Y   **Axial Y**

FIGURE 4-12

Newman projections of a substituted cyclohexane. The conformation with axial Y is less stable because of 1,3-diaxial interactions. Axial Y is *gauche* to the ring bonds shown in green; equatorial Y is *anti*.

creasing unfavorable 1,3-diaxial interactions. This effect is particularly pronounced in (1,1-dimethylethyl)cyclohexane (*tert*-butylcyclohexane). The energy difference here is so large (about 5 kcal mol^{-1}) that very little (about 0.01%) of the axial conformer is present at equilibrium.

TABLE 4-3	Change in Free Energy on Flipping from the Cyclohexane Conformer with the Indicated Substituent Equatorial to the Conformer with the Substituent Axial		
Substituent	**$\Delta G°$ (kcal mol^{-1})**	**Substituent**	**$\Delta G°$ (kcal mol^{-1})**
H	0	F	0.25
CH_3	1.70	Cl	0.52
CH_3CH_2	1.75	Br	0.55
$(CH_3)_2CH$	2.20	I	0.46
$(CH_3)_3C$	≈ 5	HO	0.94
HO—C(=O)	1.41	CH_3O	0.75
CH_3O—C(=O)	1.29	H_2N	1.4

Note: In all examples, the equatorial form is more stable.

Substituents compete for equatorial positions

To predict the more stable conformer of a more highly substituted cyclohexane, the cumulative effect of placing substituents either axially or equatorially must be considered, in addition to their potential mutual 1,3-diaxial or 1,2-*gauche* (Section 2-7) interactions. For many cases, we can ignore the last two and simply apply the values of Table 4-3 for a prediction.

Let us look at some isomers of dimethylcyclohexane to illustrate this point. In 1,1-dimethylcyclohexane, one methyl group is always equatorial and the other axial. The two chair forms are identical, and hence their energies are equal.

One CH₃ axial
One CH₃ equatorial

One CH₃ axial
One CH₃ equatorial

1,1-Dimethylcyclohexane

(Conformations equal in energy, equally stable)

Similarly, in *cis*-1,4-dimethylcyclohexane, both chairs have one axial and one equatorial substituent and are of equal energy.

The bonds to both methyl groups point downward; they are cis (i.e., on the same side of the ring) *regardless* of conformation.

One axial, one equatorial

One axial, one equatorial

cis-**1,4-Dimethylcyclohexane**

On the other hand, the trans isomer can exist in two different chair conformations: one having two axial methyl groups (diaxial) and the other having two equatorial groups (diequatorial).

The bond to one methyl group points downward, the other upward. They are trans (i.e., on opposite sides of the ring) *regardless* of conformation.

Diequatorial methyls
More stable

Diaxial methyls
Less stable: +3.4 kcal mol⁻¹

trans-**1,4-Dimethylcyclohexane**

Experimentally, the latter is preferred over the former by 3.4 kcal mol^{-1}, exactly twice the $\Delta G°$ value for monomethylcyclohexane. Indeed, this additive behavior of the data given in Table 4-3 applies to many other substituted cyclohexanes. For example, the $\Delta G°$ (diaxial $\rightleftharpoons$ diequatorial) for *trans*-1-fluoro-4-methylcyclohexane is −1.95 kcal mol^{-1} [−(1.70 kcal mol^{-1} for CH$_3$ plus 0.25 kcal mol^{-1} for F)]. Conversely, in *cis*-1-fluoro-4-methylcyclohexane, the two groups compete for the equatorial positions and the corresponding $\Delta G° = -1.45$ kcal mol^{-1} [−(1.70 kcal mol^{-1} minus 0.25 kcal mol^{-1})], with the larger methyl winning out over the smaller fluorine.

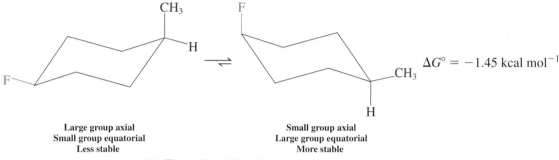

$\Delta G° = -1.45$ kcal mol^{-1}

Large group axial　　　　　**Small group axial**
Small group equatorial　　**Large group equatorial**
Less stable　　　　　　　　**More stable**

cis-**1-Fluoro-4-methylcyclohexane**

EXERCISE 4-6

Calculate $\Delta G°$ for the equilibrium between the two chair conformers of **(a)** 1-ethyl-1-methylcyclohexane; **(b)** *cis*-1-ethyl-4-methylcyclohexane; **(c)** *trans*-1-ethyl-4-methyl-cyclohexane.

EXERCISE 4-7

Draw both chair conformations for each of the following isomers: **(a)** *cis*-1,2-dimethyl-cyclohexane; **(b)** *trans*-1,2-dimethylcyclohexane; **(c)** *cis*-1,3-dimethylcyclohexane; **(d)** *trans*-1,3-dimethylcyclohexane. Which of these isomers always have equal numbers of axial and equatorial substituents? Which exist as equilibrium mixtures of diaxial and diequatorial forms?

EXERCISE 4-8

Although the substituent values in Table 4-3 are additive and may be used to indicate the position of the equilibrium between two substituted cyclohexane conformers, the observed $\Delta G°$ values can be perturbed by additional 1,3-diaxial or 1,2-*gauche* interactions between groups. For example, like *trans*-1,4-dimethylcyclohexane, its isomers *cis*-1,3- and *trans*-1,2-dimethylcyclohexane exist in a diequatorial–diaxial equilibrium and hence should exhibit the same $\Delta G°$ value of 3.4 kcal mol^{-1}. However, the measured values are larger (3.7 kcal mol^{-1}) for the former and smaller (2.5 kcal mol^{-1}) for the latter. Explain. (**Hints:** For *cis*-1,3-dimethylcyclohexane, look closely at all 1,3-diaxial interactions and compare them with those of diaxial *trans*-1,4-dimethylcyclohexane. For the *trans*-1,2-isomer, take into consideration the proximity of the two methyl groups [see *gauche–anti* butane, Section 2-7].)

In summary, the conformational analysis of cyclohexane enables us to predict the relative stability of its various conformers and even to approximate the energy differences between two chair conformations. Bulky substituents, particularly a 1,1-dimethylethyl group, tend to shift the chair–chair equilibrium toward the side in which the large substituent is equatorial.

4-5 Larger Cycloalkanes

Do similar relations hold for the larger cycloalkanes? Table 4-2 shows that cycloalkanes with rings larger than that of cyclohexane also have more strain. This strain is due to a combination of bond-angle distortion, partial eclipsing of hydrogens, and transannular steric repulsions. It is not possible for medium-sized rings to relieve all of these strain-producing interactions in a single conformation. Instead, a compromise solution is found in which the molecule equilibrates among several geometries that are very close in energy.

Essentially strain-free conformations are attainable only for large-sized cycloalkanes, such as cyclotetradecane (Table 4-2). In such rings, the carbon chain adopts a structure very similar to that of the straight-chain alkanes (Section 2-4), having staggered hydrogens and an all-*anti* configuration. However, even in these systems, the attachment of substituents usually introduces various amounts of strain. Most cyclic molecules described in this book are not strain free.

4-6 Polycyclic Alkanes

The cycloalkanes discussed so far contain only one ring and therefore may be referred to as monocyclic alkanes. In more complex structures—the bi-, tri-, tetra-, and higher polycyclic hydrocarbons—two or more rings share carbon atoms. We shall see the structural variety possible in these compounds, many of which, when bearing alkyl and functional groups, exist in nature.

Polycyclic alkanes may contain fused or bridged rings

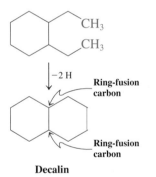

Decalin

Molecular models of polycyclic alkanes can be readily constructed by linking the carbon atoms of two alkyl substituents in a monocyclic alkane. For example, if two hydrogen atoms are removed from the methyl groups in 1,2-diethylcyclohexane, thereby allowing a new C–C bond to form, the result is a new molecule with the common name decalin. In decalin, two cyclohexanes share two adjacent carbon atoms, and the two rings are said to be **fused.** Compounds constructed in this way are called **fused bicyclic** ring systems, and the shared carbon atoms are called the **ring-fusion carbons.** Groups attached to ring-fusion carbons are called **ring-fusion substituents.**

If we treat a molecular model of *cis*-1,3-dimethylcyclopentane in the same way, we obtain another carbon skeleton, that of norbornane. Norbornane is an example of a **bridged bicyclic** ring system. In bridged bicyclic systems, two nonadjacent carbon atoms, the **bridgehead** carbons, belong to both rings.

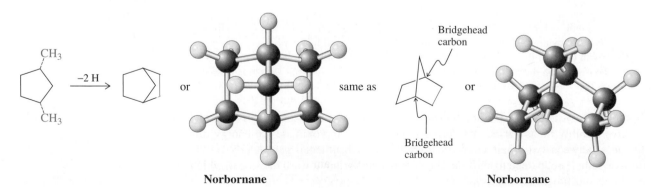

Norbornane   **Norbornane**

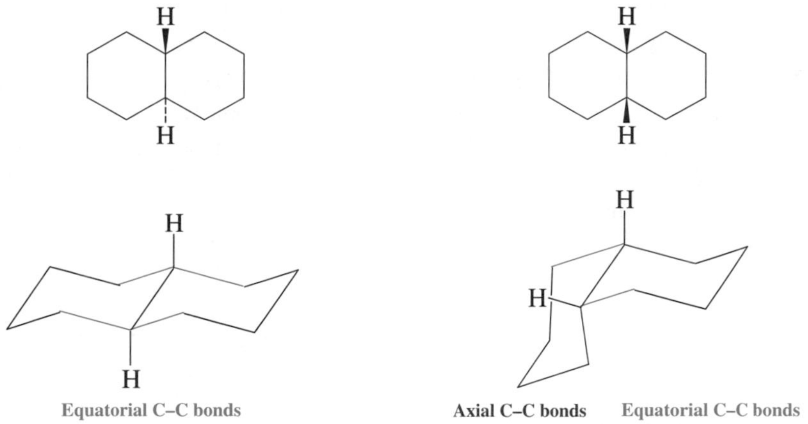

Equatorial C–C bonds

trans-Decalin

Axial C–C bonds   Equatorial C–C bonds

cis-Decalin

FIGURE 4-13

Conventional drawings and chair conformations of *trans*- and *cis*-decalin. The trans isomer contains only equatorial carbon–carbon bonds at the ring fusion, whereas the cis isomer possesses two equatorial C–C bonds (green) and two axial C–C linkages (red), one with respect to each ring.

If we think of one of the rings as a substituent on the other, we can identify stereochemical relations at ring fusions. In particular, bicyclic ring systems can be cis or trans fused. The stereochemistry of the ring fusion is most easily determined by inspecting the ring-fusion substituents. For example, the ring-fusion hydrogens of *trans*-decalin are trans with respect to each other, whereas those of *cis*-decalin have a cis relation (Figure 4-13).

EXERCISE 4-9

Construct molecular models of both *cis*- and *trans*-decalin. What can you say about their conformational mobility?

Do hydrocarbons have strain limits?

Seeking the limits of strain in hydrocarbon bonds is a fascinating area of research that has resulted in the synthesis of many exotic molecules. What is surprising is how much bond-angle distortion a carbon atom is able to tolerate. A case in point in the bicyclic series is bicyclobutane, whose strain energy is estimated to be 66.5 kcal mol^{-1}, making it remarkable that the molecule exists at all.

A series of strained compounds attracting the attention of synthetic chemists possess a carbon framework geometrically equivalent to the Platonic solids: the *tetrahedron* (tetrahedrane), the *hexahedron* (cubane), and the pentagonal *dodecahedron* (dodecahedrane). (See their structures in the margin on page 148.) In these polyhedra, all faces are composed of equally sized rings—namely, cyclopropane, cyclobutane, and cyclopentane, respectively. The hexahedron was synthesized first in 1964, a C_8H_8 hydrocarbon shaped like a cube and accordingly named cubane. The experimental strain energy (157 kcal mol^{-1}) is approximately equal to the total strain of six cyclobutanes. Although tetrahedrane itself is unknown, a tetra(1,1-dimethylethyl) derivative was synthesized in 1978. Despite the measured strain (from ΔH°_{comb}) of 129 kcal mol^{-1}, the compound is stable and has a melting point of 135°C. The synthesis

Bicyclobutane

**Tetrakis(1,1-dimethylethyl)-
tetrahedrane**

Tetrahedrane (C_4H_4)

Cubane (C_8H_8)

Dodecahedrane ($C_{20}H_{20}$)

of dodecahedrane was achieved in 1982. It required 23 synthetic operations, starting from a simple cyclopentane derivative. The last step gave 1.5 mg of pure compound. Although small, this amount was sufficient to permit complete characterization of the molecule. Its melting point at 430°C is extraordinarily high for a C_{20} hydrocarbon and is indicative of the symmetry of the compound. For comparison, icosane, also with 20 carbons, melts at 36.8°C (Table 2-3).

In summary, carbon atoms in bicyclic compounds are shared by rings in either fused or bridged arrangements. A great deal of strain may be tolerated by carbon in its bonds, particularly to other carbon atoms. This capability has allowed the preparation of molecules in which carbon is severely deformed from its tetrahedral shape.

4-7 Carbocyclic Products in Nature

Let us now take a brief look at the variety of cyclic molecules created in nature. **Natural products** are organic compounds produced by living organisms. Some of these compounds, such as methane, are extremely simple; others have great structural complexity. Scientists have attempted to classify the multitude of natural products in various ways. Generally, four schemes are followed, in which these products are classified according to (1) chemical structure, (2) physiological activity, (3) organism or plant specificity (taxonomy), and (4) biochemical origin.

There are many reasons why organic chemists are interested in natural products. Many of these compounds are powerful drugs, others function as coloring or flavoring agents, and yet others are important raw materials. A study of animal secretions furnishes information concerning the ways in which animals use chemicals to mark trails, harm their predators, and attract the opposite sex. Investigations of the biochemical pathways by which an organism metabolizes and otherwise transforms a compound are sources of insight into the detailed workings of the organism's bodily functions. Two classes of natural products, terpenes and steroids, have received particularly close attention from organic chemists.

Terpenes are constructed in plants from isoprene units

Most of you have smelled the strong odor emanating from freshly crushed plant leaves or orange peels. This odor is due to the liberation of a mixture of volatile compounds called **terpenes,** usually containing 10, 15, or 20 carbon atoms. Terpenes are used as food flavorings (the extracts from cloves and peppermint), as perfumes (roses, lavender, sandalwood), and as solvents (turpentine).

Lavender field in the Côte du Rhone region, France.

CHEMICAL HIGHLIGHT   4-1

Cubane Derivatives Have Potential as Explosives

The synthesis of cubane and its derivatives has been scaled up to the kilogram level because of potential applications of such compounds as explosive materials and highly energetic fuels. A key functionalization step of cubane is the radical chlorocarbonylation of commercial cubanecarboxylic acid to a "tetrahedral" tetrasubstituted system, which can be converted into the powerful explosive tetranitrocubane.

produce a (usually destructive) shock wave in the surrounding medium. The reaction can be initiated by the mechanical impact of friction, by heat (such as sparks), or by detonating shock (blasting caps). The ring strain in cubane clearly contributes to the thermal lability of tetranitrocubane, and the molecular formula, $C_8H_4N_4O_8$, indicates a composition conducive to gaseous product formation (e.g., 8 CO + 2 N_2 + 2 H_2, to mention just one hypothetical

Cubanecarboxylic acid —COOH

$\xrightarrow[\text{– HCl, – CO}_2]{\text{(COCl)}_2, \, h\nu}$

ClOC, COCl, COCl, ClOC "Tetrahedral" cubane derivative

O_2N, NO_2, O_2N, NO_2 **Tetranitrocubane**

O_2N, NO_2, O_2N, NO_2, O_2N, NO_2, O_2N, NO_2 **Octanitrocubane**

Explosives are generally compounds capable of extremely rapid decomposition with generation of high heat and a large quantity of gaseous products to

outcome). Octanitrocubane is predicted to be the most explosive nitrocubane derivative but is as yet unknown.

Terpenes are synthesized in the plant by the linkage of at least two molecular units containing five carbon atoms. The structure of these units is like that of 2-methyl-1,3-butadiene (isoprene), and so they are referred to as **isoprene units.** Depending on how many isoprene units are incorporated into the structure, terpenes are classified as mono- (C_{10}), sesqui- (C_{15}), and diterpenes (C_{20}). (The isoprene building units are shown in color in the examples given here.)

CH_3

2-Methyl-1,3-butadiene
(Isoprene)

Isoprene unit in terpenes
(Some contain double bonds)

Chrysanthemic acid is a monocyclic terpene containing a three-membered ring. Its esters are found in the flower heads of pyrethrum (*Chrysanthemum cinerariae-folium*) and are naturally occurring insecticides. A cyclobutane is present in grandisol, the sex-attracting chemical used by the male boll weevil (*Anthonomus grandis*). Menthol (peppermint oil) is an example of a substituted cyclohexane natural product, whereas camphor (from the camphor tree) and β-cadinene (from juniper and cedar) are simple bicyclic terpenes, the first a norbornane system, the second a decalin derivative. **Taxol** (paclitaxel) is a complex, functionalized diterpene isolated from the bark of the Pacific yew tree, *Taxus brevifolia,* in 1962 as part of a National Cancer Institute program in search of natural products exhibiting anticancer activity. Taxol proved to be perhaps the most interesting of more than 100,000 compounds extracted from more

The flavor of peppermint in the products shown is due mainly to menthol in the extracts of the peppermint plant (*Mentha piperita*).

than 35,000 plant species and constitutes a clinically approved, powerful new weapon in the arsenal against human cancerous tumor growth. Because roughly six trees must be sacrificed to treat one patient, many efforts are underway to improve efficacy and availability and to increase yields. Many of these efforts have been undertaken by synthetic organic chemists, leading to the first two total syntheses of taxol in 1994.

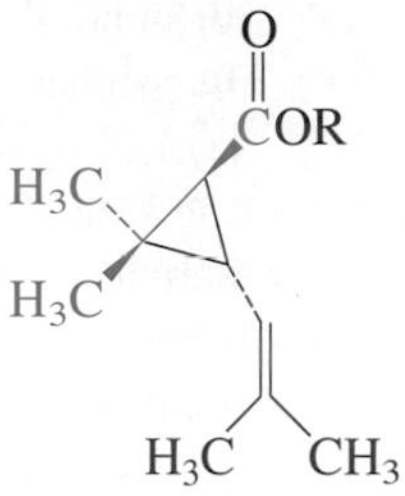

trans-Chrysanthemic acid (R = H)
trans-Chrysanthemic esters (R ≠ H)

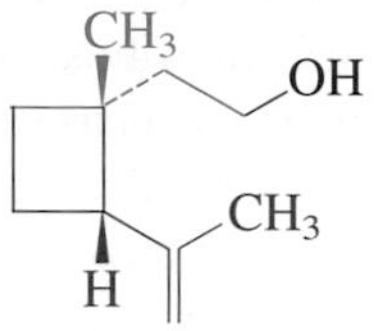

Grandisol

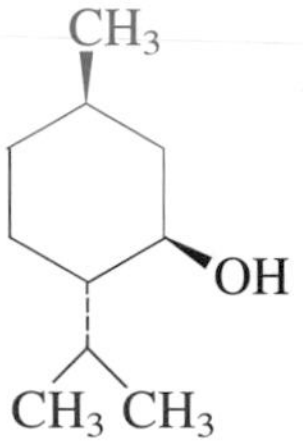

Menthol

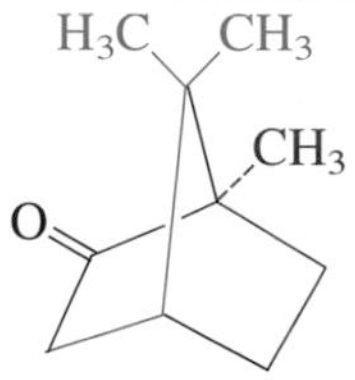

Camphor

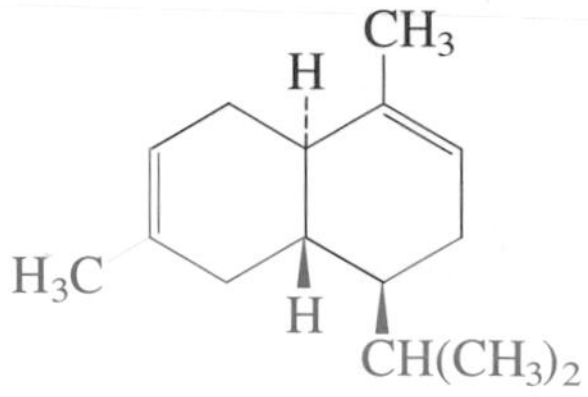

β-Cadinene

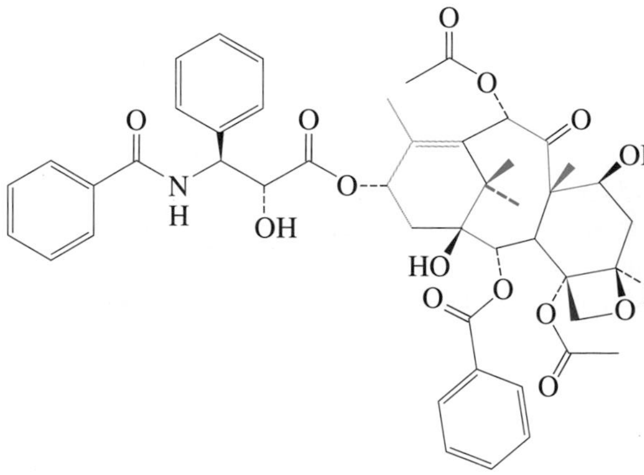

Taxol

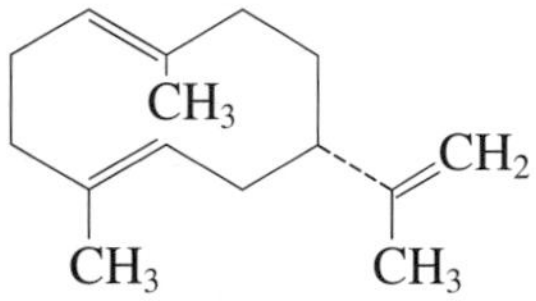

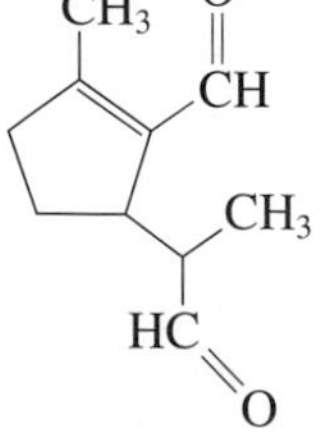

EXERCISE 4-10

Draw the preferred chair conformation of menthol.

EXERCISE 4-11

The structures of two terpenes utilized by insects in defense are shown in the margin. Classify them as mono-, sesqui-, or diterpenes. Identify the isoprene units in each.

EXERCISE 4-12

After reviewing Section 2-1, specify the functional groups present in the terpenes shown in Section 4-7.

Steroids are tetracyclic natural products with powerful physiological activities

Steroids are abundant in nature, and many derivatives have physiological activity. Steroids frequently function as **hormones,** which are regulators of biochemical activity. In the human body, for example, they control sexual development and fertility, in addition to other functions. Because of this feature, many steroids, often the products of laboratory synthesis, are used in medicines in, for example, the treatment of cancer, arthritis, or allergies, and in birth control.

In the steroids, three cyclohexane rings are fused in such a way as to form an angle. The ring junctions are usually trans, as in *trans*-decalin. The fourth ring is a cyclopentane; its addition gives the typical tetracyclic structure. The four rings are labeled A, B, C, D, and the carbons are numbered according to a scheme specific to steroids. Many steroids have methyl groups attached to C10 and C13 and oxygen at C3 and C17. In addition, longer side chains may be found at C17. The trans fusion of the rings allows for a least-strained all-chair configuration in which the methyl groups and hydrogen atoms at the ring junctions occupy axial positions.

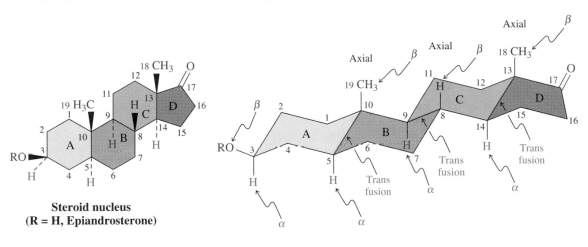

Steroid nucleus
(R = H, Epiandrosterone)

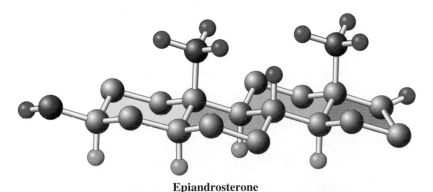

Epiandrosterone

Groups attached above the plane of the steroid molecule as written are β substituents, whereas those below are referred to as α. Thus, the structure of the steroid nucleus shown here has a 3β-OR, 5α-H, 10β-CH$_3$, and so forth. The axial methyl groups are also called **angular** methyls because they protrude sharply from the general framework (*angulus,* Latin, at an angle; being at a sharp corner).

Among the most abundant steroids, cholesterol is present in almost all human and animal tissue, particularly in the brain and the spinal cord.

Cholesterol

Cholic acid

Cortisone

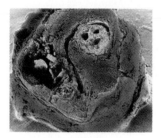

Cross section of a human coronary artery of the heart blocked by cholesterol plaques. The twisted artery wall is shown in brown, the inner open space in blue, and the deposits in yellow.

In fact, it is so concentrated in the spinal cord of cattle that it is isolated from that tissue by simple extraction. The adult human body contains from 200 to 300 g of cholesterol; gallstones may consist entirely of it. This steroid has been implicated in several circulatory diseases because it precipitates in the arteries, thereby causing arteriosclerosis and heart disease. Although its biological function in the body is not completely understood, it is a precursor of steroid hormones and bile acids. Bile acids are produced in the liver as part of a fluid delivered to the duodenum to aid in the emulsification, digestion, and absorption of fats. An example is cholic acid.

Cortisone, used extensively in the treatment of rheumatoid inflammations, is one of the adrenocortical hormones produced by the outer part (cortex) of the adrenal glands. These hormones participate in regulating the electrolyte and water balance in the body, as well as protein and carbohydrate metabolism.

The sex hormones are divided into three groups: (1) the male sex hormones, or *androgens;* (2) the female sex hormones, or *estrogens;* and (3) the pregnancy hormones, or *progestins.* Testosterone is the principal male sex hormone. Produced by the testes, it is responsible for male (masculine) characteristics (deep voice, facial hair, general physical constitution). Synthetic testosterone analogs are used in medicine to promote muscle and tissue growth (anabolic steroids; *ana-,* Greek, up—i.e., "anabolic," the opposite of "metabolic"), for example, in patients with muscular atrophy. Unfortunately, such steroids are also abused and consumed illegally, most commonly by "body builders" and athletes, even though the health risks are numerous, including liver cancer, coronary heart disease, and sterility. Estradiol is the principal female sex hormone. It was first isolated by extraction of four tons of sow ovaries, yielding only a few milligrams of pure steroid. Estradiol is responsible for the development of the secondary female characteristics and participates in the control of the menstrual cycle. An example of a progestin is progesterone, responsible for preparing the uterus for implantation of the fertilized egg.

Testosterone

Estradiol

Progesterone

CHEMICAL HIGHLIGHT 4-2

Controlling Fertility: From "the Pill" to RU-486

The menstrual cycle is controlled by three protein hormones from the pituitary gland. The follicle-stimulating hormone (FSH) induces the growth of the egg, and the luteinizing hormone (LH) induces its release from the ovaries and the formation of an ovarian tissue called the *corpus luteum*. The third pituitary hormone (luteotropic hormone, also called luteotropin or prolactin), stimulates the corpus luteum and maintains its function.

As the cycle begins and egg growth is initiated, the tissue around the egg secretes increasing quantities of estrogens. When a certain concentration of estrogen in the bloodstream has been reached, the production of FSH is turned off. The egg is released at this stage in response to LH. At the time of ovulation, LH also triggers the formation of the corpus luteum, which in turn begins to secrete increasing amounts of progesterone. This last hormone suppresses any further ovulation by turning off the production of LH. If the egg is not fertilized, the corpus luteum regresses and the ovum and the *endometrium* (uterine lining) are expelled (menstruation). Pregnancy, on the other hand, leads to increased production of estrogens and progesterone to prevent pituitary hormone secretion and thus renewed ovulation.

The birth control pill consists of a mixture of synthetic potent estrogen and progesterone derivatives

(more potent than the natural hormones), which, when taken throughout most of the menstrual cycle, prevent both development of the ovum and ovulation by turning off production of both FSH and LH. The female body is essentially being tricked into believing that it is pregnant. Some of the commercial pills contain a combination of norethindrone and

Norethindrone

Ethynylestradiol

ethynylestradiol. Other preparations consist of similar analogs with minor structural variations.

RU-486 (mifepristone) is a synthetic steroid that blocks the effects of progesterone. The fertilized egg is not implanted, because the necessary preparation of the endometrium has been prevented. RU-486 has been used in France since 1988 as a "morning after" pill. After much discussion and testing, the Food and Drug Administration approved the drug for the U.S. market in 1997.

RU-486 (mifepristone)

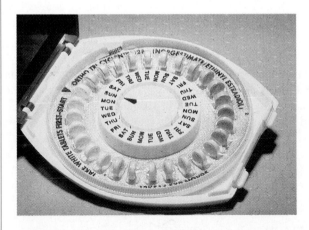

"Pill" dispenser.

The structural similarity of the steroid hormones is remarkable, considering their widely divergent activity. Steroids are the active ingredients of "the pill," functioning as an antifertility agent for the control of the female menstrual cycle and ovulation. It is estimated that 50 to 60 million women throughout the world take "the pill" as the primary form of contraception.

In summary, there is great variety in the structure and function of naturally occurring organic products, as manifested in the terpenes and the steroids. Natural products will be frequently introduced in subsequent chapters to illustrate the presence and chemistry of a functional group, to demonstrate synthetic strategy or the use of reagents, to picture three-dimensional relations, and to exemplify medicinal applications. Several classes of natural products will be discussed more extensively: fats (Sections 19-12 and 20-4), carbohydrates (Chapter 24), alkaloids (Section 25-7), and amino and nucleic acids (Chapter 26).

CHAPTER INTEGRATION PROBLEM

a. 1, 2, 3, 4, 5, 6-Hexachlorocyclohexane exists as a number of cis-trans isomers. Using the flat cyclohexane stencil and dashed-wedged lines, draw all of them.

SOLUTION

Before starting to solve this part by random trial and error, consider a more systematic approach. In this approach, we start with the simplest case of all chlorines positioned cis to one another (using all wedged lines) and then look at the various permutations obtained by placing a progressively increasing number of the substituents trans (dashed lines). We can stop at the stage of three chlorines "up" and three chlorines "down" because "two up and four down" is the same as "four up and two down," and so on. The first two cases, "six up" (A) and "five up and one down" (B), are unique because only one structural possibility exists for each:

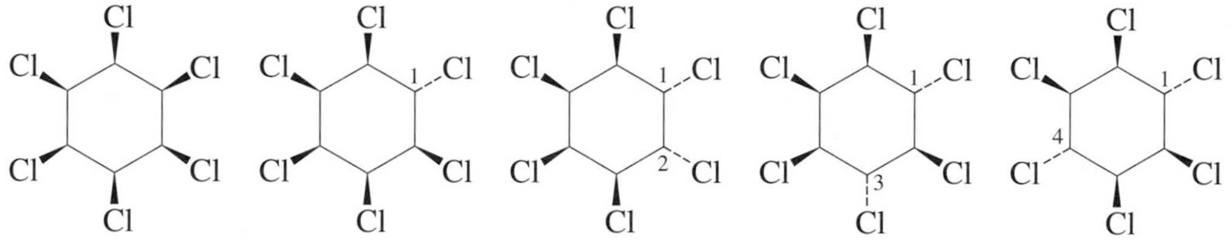

Now we look at the "four up and two down" isomers. There are three differing ways of placing two chlorines "down": their positioning along the six-membered ring can be only 1, 2 (C), 1, 3 (D), or 1, 4 (E), because the options 1, 5 and 1, 6 are the same as 1, 3 and 1, 2, respectively. Finally, a similar line of thought leads to the realization that placing three chlorines "down" can be done only in three unique ways: 1, 2, 3 (F), 1, 2, 4 (G), and 1, 3, 5 (H):

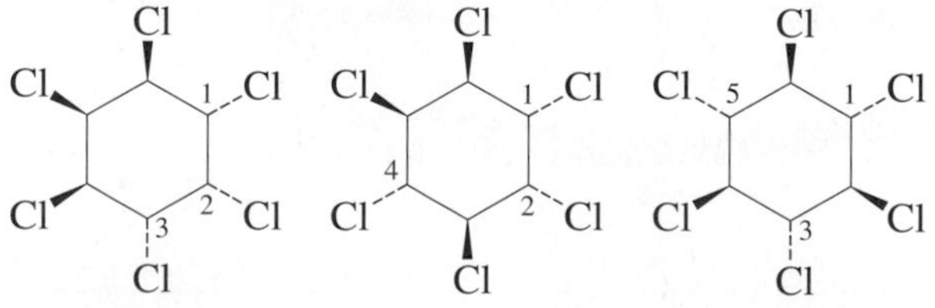

b. The so-called γ-isomer is an insecticide (lindane, gammexane, kwell) with the following structure:

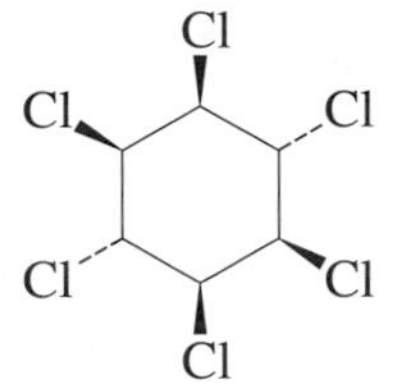

γ-Hexachlorocyclohexane

Draw the two chair conformations of this compound. Which one is more stable?

SOLUTION

Note that the γ-isomer corresponds to our structure E in (a). To help in converting the flat cyclohexane stencil structures into their chair renditions, look at Figure 4-11 and note the alternating relation of the members of the two sets of hydrogens. As we proceed around the ring, we can see that neighboring members (i.e., those in a "1,2"-relation) of either set (i.e., axial or equatorial neighbors) always have a trans disposition. On the other hand, the 1,3-relation is always cis, whereas that describable as 1,4 is again trans. Conversely, in a dashed-wedged line structure, when two neighboring substituents (i.e., 1,2-relation) are cis, one of the substituents will be axial and the other equatorial in either of the two chair conformations of the molecule. When they are trans, one chair picture will show them diequatorial, the other diaxial. This assignment alternates in 1,3-related substituents: cis results in a diaxial-diequatorial pair of chair conformers, trans in an axial-equatorial or equatorial-axial pair, and so on. These relations are summarized in Table 4-4.

The two chair forms of γ-hexachlorocyclohexane thus look as shown here:

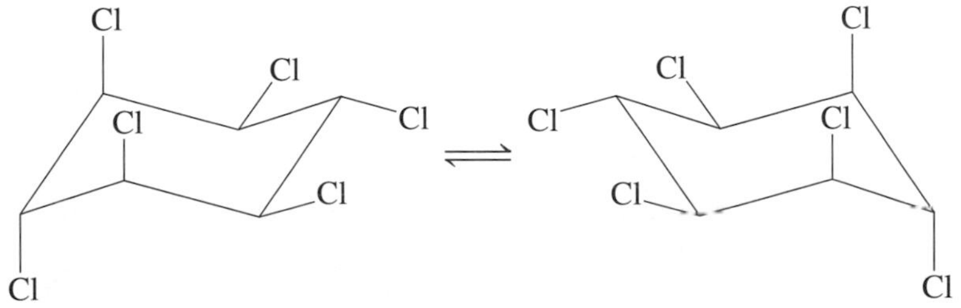

$$\Delta G° = 0 \text{ kcal mol}^{-1}$$

Either structure has three equatorial and three axial chlorine substituents; hence they are equal in energy.

TABLE 4-4	Relation of *Cis-Trans* Stereochemistry in Substituted Cyclohexanes to Equatorial-Axial Positions in the Two-Chair Forms	
cis-1,2	Axial-equatorial	Equatorial-axial
trans-1,2	Axial-axial	Equatorial-equatorial
cis-1,3	Axial-axial	Equatorial-equatorial
trans-1,3	Axial-equatorial	Equatorial-axial
cis-1,4	Axial-equatorial	Equatorial-axial
trans-1,4	Axial-axial	Equatorial-equatorial

c. For which isomer do you expect the energy difference between the two cyclohexane chair forms to be largest? Estimate the $\Delta G°$ value.

SOLUTION

The biggest $\Delta G°$ for chair–chair flip is that between an all-equatorial and all-axial hexachlorocyclohexane. Application of Table 4-4 and inspection of Figure 4-11 reveal that this relation holds only for the all-trans isomer. Table 4-3 gives the $\Delta G°$ (equatorial-axial) for Cl as 0.52 kcal mol^{-1}; hence, for our example, $\Delta G° = 6 \times 0.52 = 3.12$ kcal mol^{-1}.

$\Delta G° = +3.12$ kcal mol^{-1}

all-*trans*-**Hexachlorohexane**

Note that this value is only an estimate. For example, it ignores the six Cl–Cl *gauche* interactions in the all-equatorial form, which would reduce the energy difference between the two conformers. However, it also disregards the six 1,3-diaxial interactions in the all-axial chair, which would counteract this effect.

IMPORTANT CONCEPTS

1. **Cycloalkane** nomenclature is derived from that of the straight-chain alkanes.

2. All but the 1,1-disubstituted cycloalkanes exist as two isomers: If both substituents are on the same face of the molecule, they are **cis;** if they are on opposite faces, they are **trans.** Cis and trans isomers are **stereoisomers**—compounds that have identical connectivities but differ in the arrangement of their atoms in space.

3. Some cycloalkanes are **strained.** Distortion of the bonds about tetrahedral carbon introduces **bond-angle strain. Eclipsing (torsional) strain** results from the inability of a structure to adopt staggered conformations about C–C bonds. Steric repulsion between atoms across a ring leads to **transannular strain.**

4. Bond-angle strain in the small cycloalkanes is largely accommodated by the formation of **bent bonds.**

5. Bond-angle, eclipsing, and other strain in the cycloalkanes larger than cyclopropane (which is by necessity flat) can be accommodated by deviations from planarity.

6. Ring strain in the small cycloalkanes gives rise to reactions that result in opening of the ring.

7. Deviations from planarity lead to conformationally mobile structures, such as **chair, boat,** and **twist-boat** cyclohexane. Chair cyclohexane is almost strain free.

8. Chair cyclohexane contains two types of hydrogens: **axial** and **equatorial.** These interconvert rapidly at room temperature by a conformational **chair–chair ("flip") interconversion,** with an activation energy of 10.8 kcal mol^{-1}.

9. In monosubstituted cyclohexanes, the $\Delta G°$ of equilibration between the two chair conformations is substituent dependent. Axial substituents are exposed to **1,3-diaxial interactions.**

10. In more highly substituted cyclohexanes, substituent effects are often **additive,** the bulkiest substituents being the most likely to be equatorial.

11. Completely strain-free cycloalkanes are those that can readily adopt an all-*anti* conformation and lack transannular interactions.

12. **Bicyclic** ring systems may be **fused** or **bridged.** Fusion can be cis or trans.

13. Natural products are generally classified according to structure, physiological activity, taxonomy, and biochemical origin. Examples of the last class are the **terpenes,** of the first the **steroids.**

14. Terpenes are made up of **isoprene** units of five carbons.

15. Steroids contain three angularly fused cyclohexanes (A, B, C rings) attached to the cyclopentane D ring. Beta substituents are above the molecular plane, alpha substituents below.

16. An important class of steroids are the **sex hormones,** which have a number of physiological functions, including the control of fertility.

PROBLEMS

13. Write as many structures as you can that have the formula C_5H_{10} and contain one ring. Name them.

14. Name the following molecules according to the IUPAC nomenclature system.

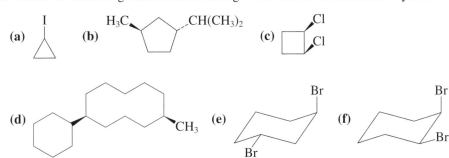

15. Draw structural representations of each of the following molecules: **(a)** *trans*-1-chloro-2-ethylcyclopropane; **(b)** *cis*-1-bromo-2-chlorocyclopentane; **(c)** 2-chloro-1,1-diethylcyclopropane; **(d)** *trans*-2-bromo-3-chloro-1,1-diethylcyclopropane; **(e)** *cis*-1,3-dichloro-2,2-dimethylcyclobutane; **(f)** *cis*-2-chloro-1,1-difluoro-3-methylcyclopentane.

16. The kinetic data for the radical chain chlorination of several cycloalkanes (see the adjoining table) illustrate that the C–H bonds of cyclopropane and, to a lesser extent, cyclobutane are somewhat abnormal. **(a)** What do these data tell you about the strength of the cyclopropane C–H bond and the stability of the cyclopropyl radical? **(b)** Suggest a reason for the stability characteristics of the cyclopropyl radical. (**Hint:** Consider bond-angle strain in the radical relative to cyclopropane itself.)

17. Use the data in Tables 3-2 and 4-2 to estimate the $DH°$ value for a C–C bond in **(a)** cyclopropane; **(b)** cyclobutane; **(c)** cyclopentane; and **(d)** cyclohexane.

18. Draw each of the following substituted cyclobutanes in its two interconverting "puckered" conformations (Figure 4-3). When the two conformations differ in energy, identify the more stable shape and indicate the form(s) of strain that raise the relative energy of the less stable one. (**Hint:** Puckered cyclobutane has axial and equatorial positions similar to those in chair cyclohexane.)
    (a) Methylcyclobutane
    (b) *cis*-1,2-Dimethylcyclobutane
    (c) *trans*-1,2-Dimethylcyclobutane
    (d) *cis*-1,3-Dimethylcyclobutane
    (e) *trans*-1,3-Dimethylcyclobutane

Reactivity per Hydrogen Toward Cl·	
Cycloalkane	Reactivity
Cyclopentane	0.9
Cyclobutane	0.7
Cyclopropane	0.1

Note: Relative to cyclohexane = 1.0; at 68°C, hv, CCl_4 solvent.

Which is more stable: *cis-* or *trans*-1,2-dimethylcyclobutane; *cis-* or *trans*-1,3-dimethylcyclobutane?

19. For each of the following cyclohexane derivatives, indicate (i) whether the molecule is a cis or trans isomer and (ii) whether it is in its most stable conformation. If your answer to (ii) is no, flip the ring and draw its most stable conformation.

(a) 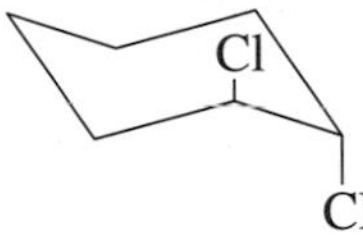   (b)

(c) 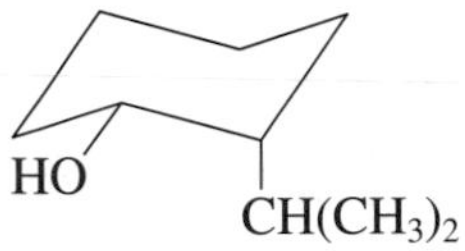   (d)

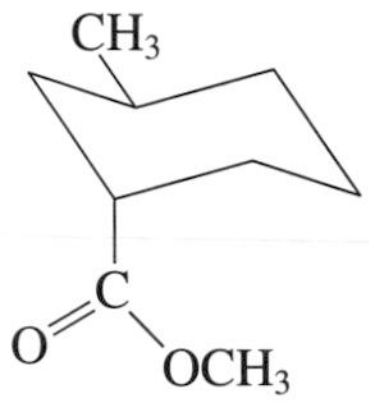

(e) 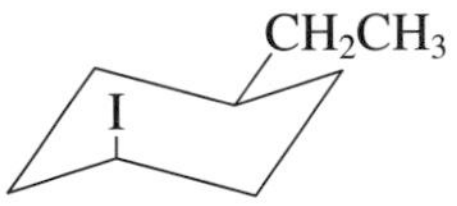   (f) 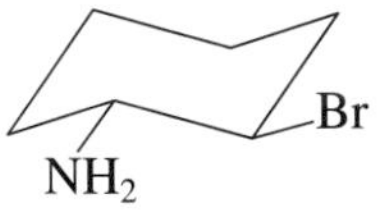   (g)

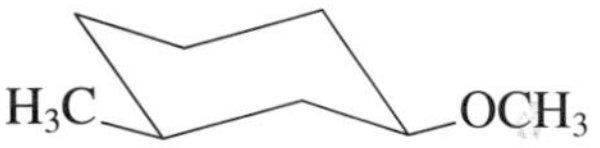

(h) 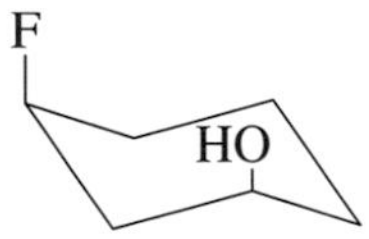   (i) 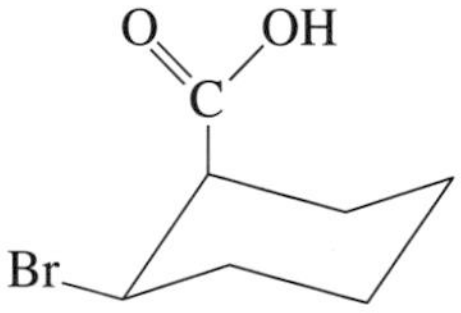   (j)

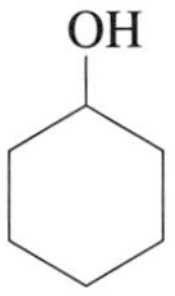

20. Using the data in Table 4-3, calculate the $\Delta G°$ for ring flip to the other conformation of the molecules depicted in Problem 19. Make sure that the sign (i.e., positive or negative) of your values is correct.

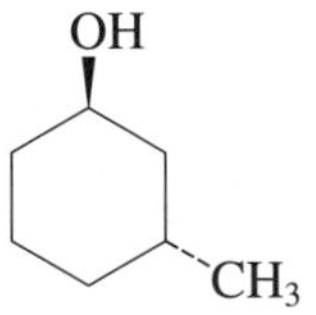

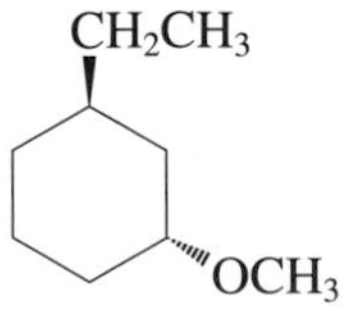

21. Draw the most stable conformation for each of the following substituted cyclohexanes; then, in each case, flip the ring and redraw the molecule in the higher energy chair conformation: **(a)** cyclohexanol; **(b)** *trans*-3-methylcyclohexanol (see structures in the margin); **(c)** *cis*-1-(1-methylethyl)-3-methylcyclohexane; **(d)** *trans*-1-ethyl-3-methoxycyclohexane (see structures in the margin); **(e)** *trans*-1-chloro-4-(1,1-dimethylethyl)cyclohexane.

22. For each molecule in Problem 21, estimate the energy difference between the most stable and next best conformation. Calculate the approximate ratio of the two at 300 K.

23. Sketch a potential-energy diagram (similar to that in Figure 4-9) for methylcyclohexane showing the two possible chair conformations at the left and right ends of the reaction coordinate for conformational interconversion.

24. Draw all the possible all-chair conformers of cyclohexylcyclohexane.

Cyclohexanol

***trans*-3-Methylcyclohexanol**

***trans*-1-Ethyl-3-methoxycyclohexane**

25. What is the most stable of the four *boat* conformations of methylcyclohexane, and why?

26. The most stable conformation of *trans*-1,3-bis(1,1-dimethylethyl)cyclohexane is not a chair. What conformation would you predict for this molecule? Explain.

27. The bicyclic hydrocarbon formed by the fusion of a cyclohexane ring with a cyclopentane ring is known as hexahydroindane. Using the drawings of *trans*- and *cis*-decalin for reference (Figure 4-13), draw the structures of *trans*- and *cis*-hexahydroindane, showing each ring in its most stable conformation.

Hexahydroindane

28. On viewing the drawings of *cis*- and *trans*-decalin in Figure 4-13, which do you think is the more stable isomer? Estimate the energy difference between the two isomers. Answer the same question for cis and trans dimethyl-substituted decalins.

cis-9,10-Dimethyldecalin   **trans-9,10-Dimethyldecalin**

29. Identify each of the following molecules as a monoterpene, a sesquiterpene, or a diterpene (all names are common).

(a) **Geraniol**

(b) **Eremanthin**

(c) **Eudesmol**

(d) **Ipomeamarone**

(e) **Genipin**

(f) **Castoramine**

(g) **Cantharidin**

(h) **Vitamin A**

30. Find the 2-methyl-1,3-butadiene (isoprene) units in each of the naturally occurring organic molecules pictured in Problem 29.

31. Circle and identify by name all the functional groups in any three of the steroids illustrated in Section 4-7. Label any polarized bonds with partial positive and negative charges (δ^+ and δ^-).

32. Several additional examples of naturally occurring molecules with strained ring structures are shown here.

H₂N   COOH

**1-Aminocyclopropane-
carboxylic acid**

(Present in plants, this molecule
plays a role in the ripening of fruits
and the dropping of autumn leaves)

H₃C   CH₃

CH₃

α-Pinene

(Present in cedar-wood oil)

H₃C

--H

CH₃

O   CH₃
CH₃

Africanone

(Also a plant-leaf oil)

O   CH₃ H₃C   O

HN   NH

O   N   H   H   N   O

HOCH₂   O   O   CH₂OH

HO   OH

Thymidine dimer

(A component of DNA that has been
exposed to ultraviolet light)

Identify the terpenes (if any) in the preceding group of structures. Find the 2-methyl-1,3-butadiene units in each structure and classify the latter as a mono-, sesqui-, or diterpene.

33. If cyclobutane were flat, it would have exactly 90° C–C–C bond angles and could conceivably use pure *p* orbitals in its C–C bonds. What would be a possible hybridization for the carbon atoms of the molecule that would allow all the C–H bonds to be equivalent? Exactly where would the hydrogens on each carbon be located? What are the real structural features of the cyclobutane molecule that contradict this hypothesis?

34. Compare the structure of cyclodecane in an all-chair conformation with that of *trans*-decalin. Explain why all-chair cyclodecane is highly strained, and yet *trans*-decalin is nearly strain free. Make models.

All-chair cyclodecane   *trans*-**Decalin**

35. Fusidic acid is a steroidlike microbial product that is an extremely potent antibiotic with a broad spectrum of biological activity. Its molecular shape is most unusual and has supplied important clues to researchers investigating the methods by which steroids are synthesized in nature.

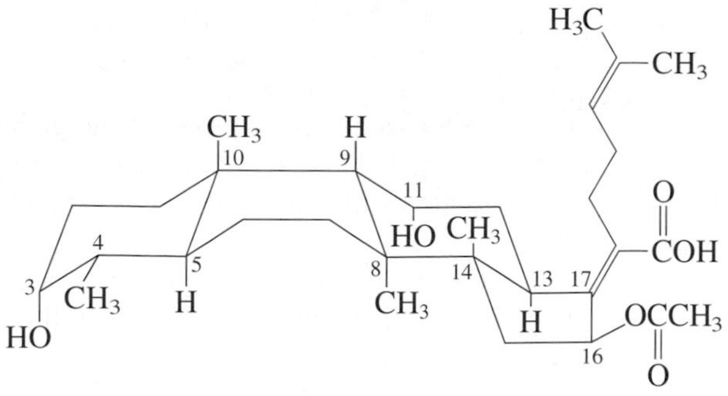

Fusidic acid

(a) Locate all the rings in fusidic acid and describe their conformations. **(b)** Identify all ring fusions in the molecule as having either cis or trans geometry. **(c)** Identify all groups attached to the rings as being either α- or β-substituents. **(d)** Describe in detail how this molecule differs from the typical steroid in structure and stereochemistry. (As an aid to answering these questions, the carbon atoms of the framework of the molecule have been numbered.)

36. The enzymatic oxidation of alkanes to produce alcohols (see Chemical Highlight 3-4) is a simplified version of the reactions that produce the adrenocortical steroid hormones. In the biosynthesis of corticosterone from progesterone (Section 4-7), two such oxidations take place successively. It is thought that the monooxygenase enzymes act as complex oxygen-atom donors in these reactions. A suggested mechanism, as applied to cyclohexane, consists of the two steps shown below the biosynthesis.

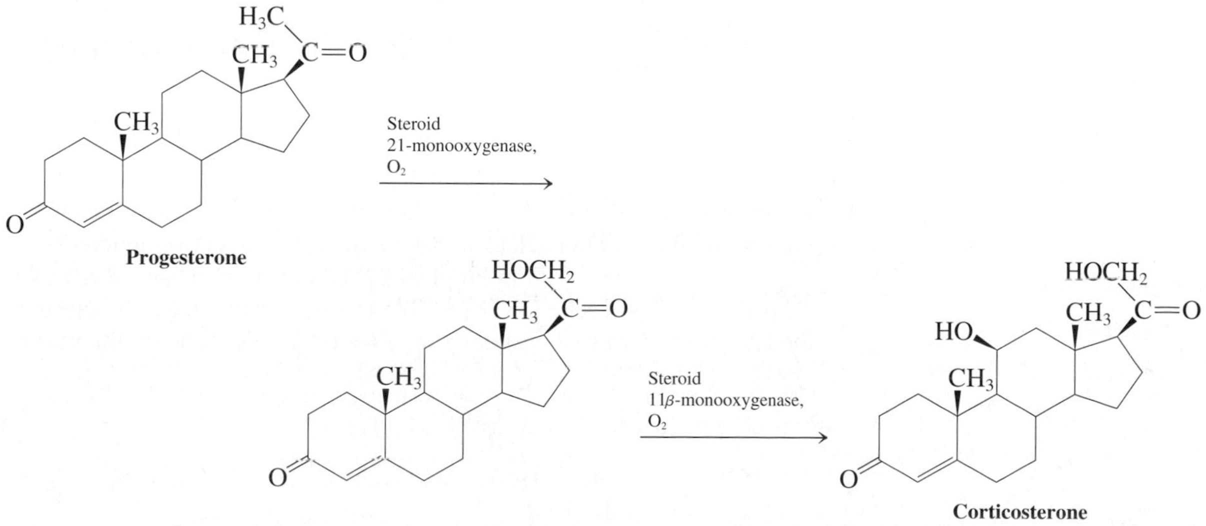

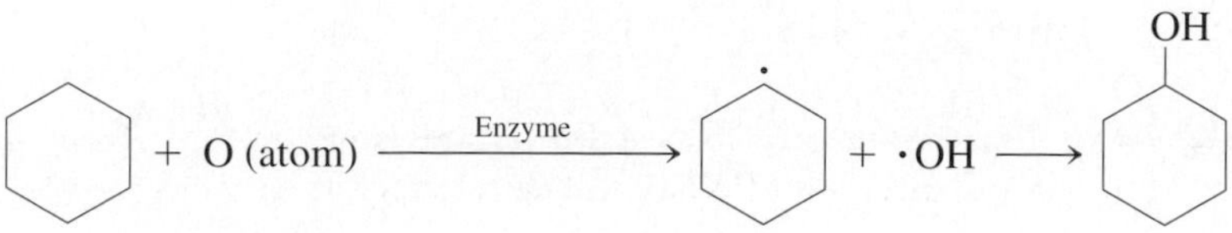

Progesterone

Corticosterone

Calculate $\Delta H°$ for each step and for the overall oxidation reaction of cyclohexane. Use the following $DH°$ values: cyclohexane C–H bond, 96.5 kcal mol^{-1}; bond in O–H radical, 102.5 kcal mol^{-1}; cyclohexanol C–O bond, 93 kcal mol^{-1}.

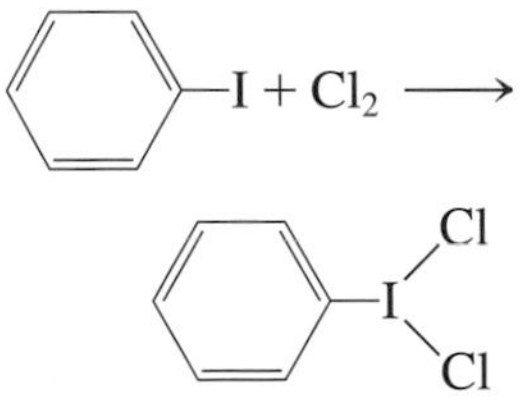

37. Like sulfuryl chloride and NCS (Section 3-8), iodobenzene dichloride, formed by the reaction of iodobenzene and chlorine, is a reagent for the chlorination of alkane C–H bonds. Chlorinations in which iodobenzene dichloride is used are initiated by light.

Iodobenzene dichloride

(a) Propose a radical chain mechanism for the chlorination of a typical alkane RH by iodobenzene dichloride. To get you started, the overall equation for the reaction is given below, as is the initiation step.

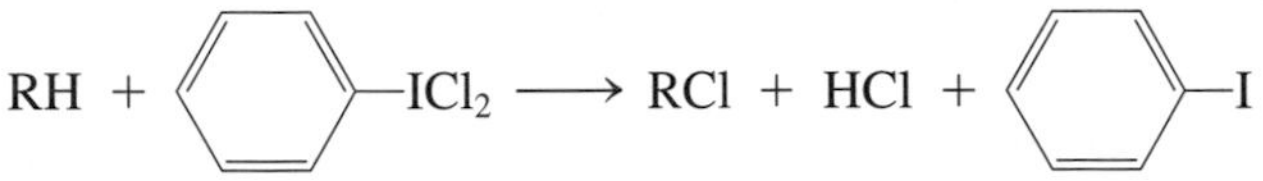

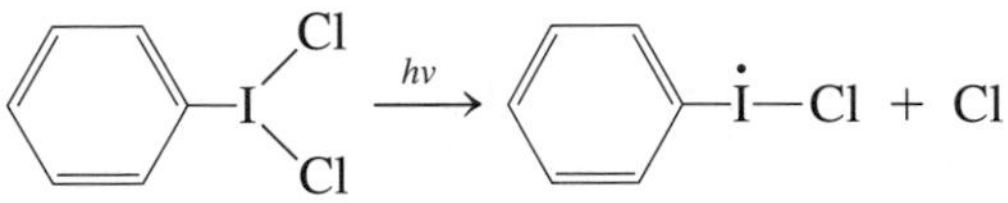

(b) Radical chlorination of typical steroids by iodobenzene dichloride gives, predominantly, three isomeric monochlorination products. On the basis of both reactivity (tertiary, secondary, primary) considerations and steric effects (which might hinder the approach of a reagent toward a C–H bond that might otherwise be reactive), predict the three major sites of chlorination in the steroid molecule. Either make a model or carefully analyze the drawings of the steroid nucleus in Section 4-7.

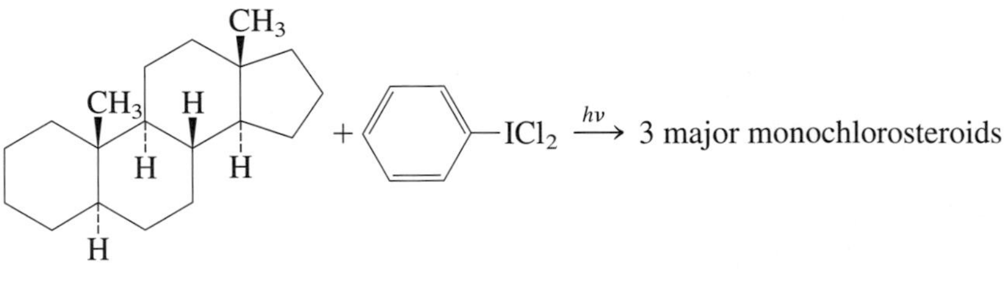

38. As Problem 36 indicates, the enzymatic reactions that introduce functional groups into the steroid nucleus in nature are highly selective, unlike the laboratory chlorination described in Problem 37. However, by means of a clever adaptation of this reaction, it is possible to partly mimic nature's selectivity in the laboratory. Two such examples are illustrated below and on the next page.

(a)

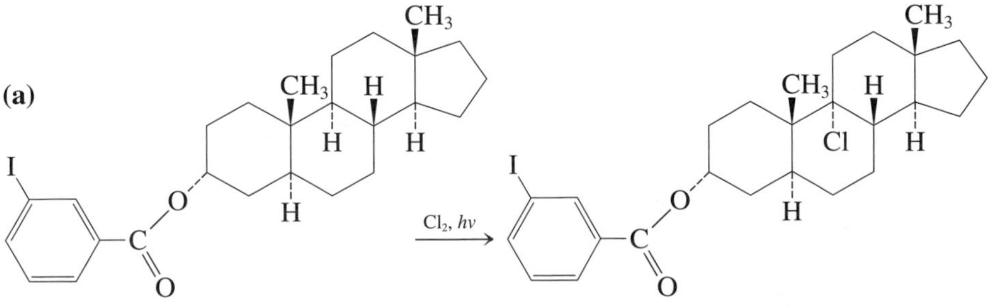

(b)

Propose reasonable explanations for the results of these two reactions. Make a model of the product of the addition of Cl_2 to each iodocompound (compare Problem 37) to help in analyzing each system.

Team Problem

39. Consider the following compounds:

Conformational analysis reveals that, though compound A exists in a chair conformation, compound B does not.

(a) Make a model of A. Draw chair conformations and label the substituents as equatorial or axial. Circle the most stable conformation. (Note that the carbonyl carbon is sp^2 hybridized and therefore the attached oxygen is neither equatorial nor axial. Do not let that lead you astray.)

(b) Make a model of B. Consider both transannular and *gauche* interactions in your analysis of its two chair forms. Discuss the steric problems of these conformations in comparison with those of A. Illustrate the key points of your discussion with Newman projections. Suggest a less sterically encumbered conformation for B.

Preprofessional Problems

40. Which of the following cycloalkanes has the greatest ring strain?

 (a) Cyclopropane **(b)** Cyclobutane **(c)** Cyclohexane **(d)** Cycloheptane

41. The following molecule has

(a) one axial chlorine and one sp^2 carbon (b) one axial chlorine and two sp^2 carbons (c) one equatorial chlorine and one sp^2 carbon (d) one equatorial chlorine and two sp^2 carbons.

42. In this compound

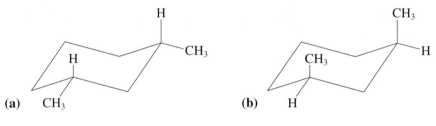

(a) the D is equatorial   (b) the methyls are both equatorial
(c) the Cl is axial   (d) the deuterium is axial

43. Which of the following isomers has the smallest heat of combustion?

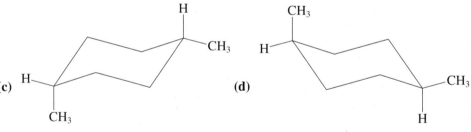

Stereoisomers

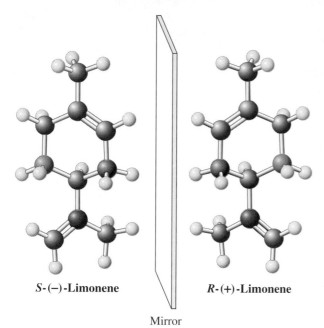

S-(−)-Limonene   **R-(+)-Limonene**

Mirror

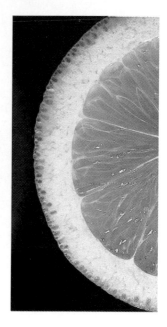

Have you ever looked at yourself in the mirror in the morning and exclaimed: "That can't be me!" Well, you were right. What you see, your mirror image, is not identical with your image: Your image and mirror image are *nonsuperimposable*. You can demonstrate this fact by trying to shake hands with your mirror counterpart: As you reach out with your right hand, your mirror image will offer you its left hand! We shall see that many molecules have this property—namely, that image and mirror image are nonsuperimposable and therefore not identical. How do we classify such structures?

Because they have the same molecular formula, these molecules are isomers but of a different kind from those encountered so far. The preceding chapters dealt with two kinds of isomerism: constitutional (also called structural) and stereo (Figure 5-1). **Constitutional isomerism** describes compounds that have identical molecular formulas but differ in the order in which the individual atoms are connected (Section 1-9).

Image and mirror image of the hydrocarbon limonene smell quite different. The *S*-image is present in the cones of fir trees and has a turpentine-like odor; the *R*-mirror image gives oranges their characteristic fragrance.

Constitutional Isomers

C_4H_{10} $CH_3\,CH_2\,CH_2\,CH_3$ $H_3\,C\!\!-\!\!\underset{\displaystyle CH_3}{\overset{\displaystyle CH_3}{CH}}$

Butane   **2-Methylpropane**

C_2H_6O $CH_3\,CH_2OH$ $CH_3O\,CH_3$

Ethanol   **Methoxymethane (Dimethyl ether)**

Paloma Sat-Vollhardt and her mirror image.

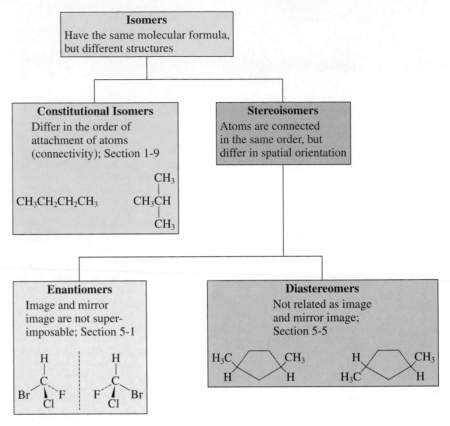

FIGURE 5-1
Relations among isomers of various types.

EXERCISE 5-1

Are cyclopropylcyclopentane and cyclobutylcyclobutane isomers?

Stereoisomerism describes isomers whose atoms are connected in the same order but differ in their spatial arrangement. Examples of stereoisomers include the relatively stable cis-trans isomers and the rapidly equilibrating conformational ones (Section 4-1).

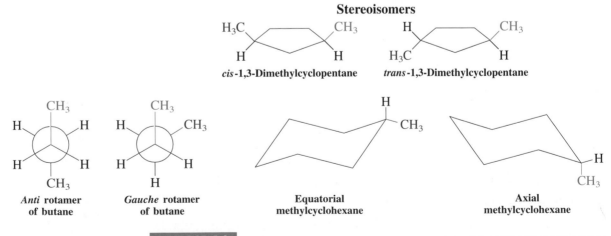

EXERCISE 5-2

Draw additional stereoisomers of methylcyclohexane. (**Hint:** Use molecular models in conjunction with Figure 4-8.)

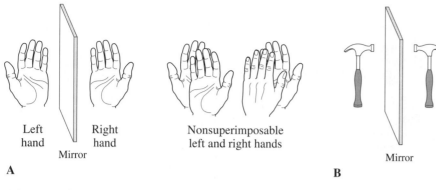

A
B

FIGURE 5-2

(A) Left and right hands as models for enantiomeric relations. Like these mirror images, chiral molecules cannot be superimposed on their corresponding enantiomers. (B) Image and mirror image of an achiral hammer are superimposable.

This chapter introduces another type of stereoisomerism, **mirror-image stereoisomerism.** Molecules in this class are said to possess "handedness," referring to the fact that your left hand is not superimposable on your right hand, yet one hand can be viewed as the mirror image of the other (Figure 5-2A). The property of handedness in molecules is very important in nature, because most biologically relevant compounds are either "left-" or "right-handed." As such, they will react differently with each other, much as shaking your friend's right hand is very different from shaking his or her left hand. A summary of isomeric relations is depicted in Figure 5-1.

5-1 Chiral Molecules

How can a molecule exist as two nonsuperimposable mirror images? Consider the radical bromination of butane. This reaction proceeds mainly at one of the secondary carbons to furnish 2-bromobutane. A molecular model of the starting material *seems* to show that either of the two hydrogens on that carbon may be replaced to give only one form of 2-bromobutane (Figure 5-3). Is this really true, however?

FIGURE 5-3

Replacement of one of the secondary hydrogens in butane results in two stereoisomeric forms of 2-bromobutane.

Chiral molecules cannot be superimposed on their mirror images

Look more closely at the 2-bromobutanes obtained by replacing either of the methylene hydrogens with bromine. In fact, the two structures are nonsuperimposable and therefore *not identical* (see page 168). The two molecules are related as object and mirror image, and to convert one into the other would require the breaking of bonds. A molecule

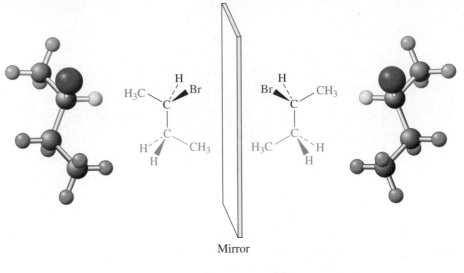

Mirror

**The two enantiomers of 2-bromobutane
are nonsuperimposable**

that is not superimposable on its mirror image is said to be **chiral.** Each isomer of the image–mirror image pair is called an **enantiomer** (*enantios,* Greek, opposite). In our example of the bromination of butane, a 1 : 1 mixture of enantiomers is formed.

In contrast with chiral molecules, such as 2-bromobutane, compounds having structures that *are* superimposable on their mirror images are **achiral.** Examples of chiral

Enantiomers

and achiral molecules are shown above. The first two chiral structures depicted are enantiomers of each other.

All the chiral examples contain an atom that is connected to four *different* substituent groups. Such a nucleus is called an **asymmetric atom** (e.g., asymmetric carbon) or a **stereocenter.** Centers of this type are sometimes denoted by an asterisk. *Molecules with one stereocenter are always chiral.* (We shall see in Section 5-5 that structures incorporating more than one such center need *not* be chiral.)

Mirror plane
(C* = a stereocenter based on
asymmetric carbon)

EXERCISE 5-3

Among the natural products shown in Section 4-7, which are chiral and which are achiral? Give the number of stereocenters in each case.

The symmetry in molecules helps to distinguish chiral structures from achiral ones

The word *chiral* is derived from the Greek *cheir,* meaning "hand" or "handedness." Human hands have the mirror-image relation that is typical of enantiomers (see Figure 5-2A). Among the many other objects that are chiral are shoes, ears, screws, and spiral staircases. On the other hand, there are many achiral objects, such as balls, ordinary water glasses, hammers (Figure 5-2B), and nails.

Many chiral objects, such as spiral staircases, do not have stereocenters. This statement is true for many chiral molecules. *Remember that the only criterion for chirality is the nonsuperimposable nature of object and mirror image.* In this chapter, we shall confine our discussion to molecules that are chiral as a result of the presence of stereocenters. But how do we determine whether a molecule is chiral or not? As you have undoubtedly already noticed, it is not always easy to tell. A foolproof way is to construct molecular models of the molecule and its mirror image and look for superimposability. However, this procedure is very time consuming. A simpler method is to look for symmetry in the molecule under investigation.

For most organic molecules, we have to consider only one test for chirality: the presence of a plane of symmetry. A **plane of symmetry (mirror plane)** is one that

---

CHEMICAL HIGHLIGHT 5-1 — Chiral Substances in Nature

Many organic compounds exist in nature as only one enantiomer, some as both. For example, natural *alanine* (systematic name: 2-aminopropanoic acid) is an abundant amino acid that is found in only one form. *Lactic acid* (2-hydroxypropanoic acid), however, is present in blood and muscle fluid as one enantiomer but in sour milk and some fruits and plants as a mixture of the two.

may be thought of as bearing four different groups, if we consider the ring itself to be two separate and different substituents. They are different because, starting from the stereocenter, the clockwise sequence of atoms differs from the counterclockwise sequence. Carvone is found in nature in both enantiomeric forms. The characteristic odor of caraway and dill seed is due to the enantiomer shown, whereas the flavor of spearmint is due to the other enantiomer.

2-Aminopropanoic acid
(Alanine)

2-Hydroxypropanoic acid
(Lactic acid)

Another case is *carvone* [2-methyl-5-(1-methylethenyl)-2-cyclohexenone], which contains a stereocenter in a six-membered ring. This carbon atom

2-Methyl-5-(1-methylethenyl)-2-cyclohexenone
(Carvone)

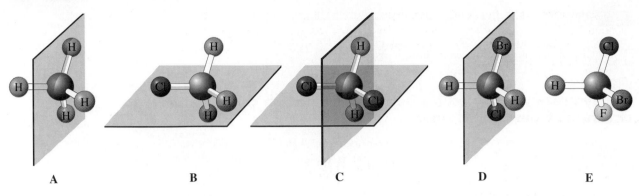

FIGURE 5-4

Examples of planes of symmetry: (A) methane has four planes of symmetry, only one of which is shown; (B) chloromethane has three such planes, only one of which is shown; (C) dichloromethane has only two; (D) bromochloromethane only one; and (E) bromochlorofluoromethane has none. Chiral molecules cannot have a plane of symmetry.

bisects the molecule so that the part of the structure lying on one side of the plane mirrors the part on the other side. For example, methane has six planes of symmetry, chloromethane has three, dichloromethane two, bromochloromethane one, and bromochlorofluoromethane none (Figure 5-4).

How do we use this idea to distinguish a chiral molecule from an achiral one? *Chiral molecules cannot have a plane of symmetry.* For example, the first four methanes in Figure 5-4 are clearly achiral because of the presence of a mirror plane. You will be able to classify most molecules in this book as chiral or achiral simply by identifying the presence or absence of a plane of symmetry.

Implicit in our definition of achirality and the associated mirror plane is a practical understanding of the "structure" of a molecule. Typically, for the practicing organic chemist, this understanding translates into what is distinctly observable or, better, isolable at room temperature as a stable entity. As we learned in Chapters 2 and 4, such an entity may exist as a mixture of rapidly equilibrating rotamers or conformational isomers. These processes may render a compound achiral if they include rapid equilibration between enantiomeric forms (e.g., through the intermediacy of an achiral isomer or the intervention of an achiral transition state). For example, although we recognize that *gauche* butane is chiral, butane is defined as achiral because rotation about the C2–C3 bond leads to enantiomerization. Similarly, *cis*-1,2-dimethylcyclohexane (Section 4-4) is defined as achiral because of rapid chair–chair interconversion between the two enantiomers through an achiral form (see also Section 5-6). Such molecules are said to have "average symmetry." Most chiral molecules we will encounter in this book owe their asymmetry to a rigid stereocenter, such as a carbon atom bearing four different substituents.

EXERCISE 5-4

Draw pictures of the following common achiral objects, indicating the plane of symmetry in each: a ball, an ordinary water glass, a hammer, a chair, a suitcase, a toothbrush.

EXERCISE 5-5

Write the structures of all dimethylcyclobutanes. Specify those that are chiral. Show the mirror planes in those that are not.

To summarize, a chiral molecule exists in either of two stereoisomeric forms called enantiomers. These enantiomers are related as object and nonsuperimposable mirror image. Most chiral organic molecules contain stereocenters, although chiral structures that lack such centers do exist. A molecule that contains a plane of symmetry is achiral.

5-2 Optical Activity

Considering their close similarity, we may wonder how it is possible to distinguish one enantiomer from another. This task is indeed a very difficult one, because most *physical* properties of enantiomers are identical. A notable exception is the interaction with a special type of light.

Our first example of a chiral molecule was the two enantiomers of 2-bromobutane. If we were to isolate each enantiomer in pure form, we would find that we cannot distinguish between them on the basis of their physical properties, such as boiling points, melting points, and densities. This result should not surprise us: Their bonds are identical and so are their energy contents. However, when plane-polarized light (which will be defined shortly) is passed through a sample of one of the enantiomers, the plane of polarization of the incoming light is *rotated* in one direction (either clockwise or counterclockwise). When the same experiment is repeated with the other enantiomer, the plane of the polarized light is rotated by exactly the same amount *but in the opposite direction.*

An enantiomer that rotates the plane of light in a clockwise sense as the viewer faces the light source is **dextrorotatory** (*dexter,* Latin, right), and the compound is (arbitrarily) referred to the (+) enantiomer. Consequently, the other enantiomer, which will effect counterclockwise rotation, is **levorotatory** (*laevus,* Latin, left) and called the (−) enantiomer. This special interaction with light is called **optical activity,** and enantiomers are frequently called **optical isomers.**

Optical rotation is measured with a polarimeter

What is plane-polarized light, and how is its rotation measured? Ordinary light can be thought of as bundles of electromagnetic-field waves that oscillate simultaneously in all planes perpendicular to the direction of the light beam. When such light is passed through a material called a polarizer, all but one of these light waves are "filtered" away, and the resulting beam oscillates in only one plane: **plane-polarized light** (Figure 5-5).

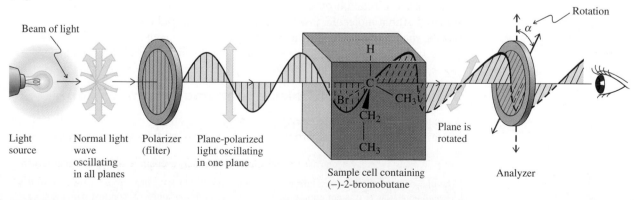

FIGURE 5-5

Measuring the optical rotation of the (−) enantiomer of 2-bromobutane with a polarimeter.

When light travels through a molecule, the electrons around the nuclei and in the various bonds interact with the electric field of the light beam. If a beam of plane-polarized light is passed through a chiral substance, the electric field interacts differently with, say, the "left" and "right" halves of the molecule. This interaction results in a rotation of the plane of polarization, called **optical rotation;** the sample giving rise to it is referred to as **optically active.**

Optical rotations are measured by using a **polarimeter** (Figure 5-5). In this instrument, light is first plane polarized and subsequently traverses a cell containing the sample. Rotation of the plane is detected by another polarizer—the analyzer—to maximize transmittance of the light beam to the eye of the observer. The measured rotation (in degrees) is the **observed optical rotation,** α, of the sample. Its value depends on the concentration and structure of the optically active molecule, the length of the sample cell, the wavelength of the light, the solvent, and the temperature. To avoid ambiguities, chemists have agreed on a standard value of α, the **specific rotation,** $[\alpha]$, for each compound. This quantity (which is solvent dependent) is defined as

Specific Rotation*

$$[\alpha]_\lambda^{t^\circ} = \frac{\alpha}{l \cdot c}$$

where $[\alpha]$ = specific rotation

       t = temperature in degrees Celsius

       λ = wavelength of incident light; for the sodium D lamp, indicated simply by "D"; it is 589 mm, the yellow emission line of hot sodium vapor

       α = observed optical rotation in degrees

       l = length of sample container in decimeters; its value is frequently 1 (i.e., 10 cm)

       c = concentration (grams per milliliter of solution)

EXERCISE 5-6

A solution of 0.1 g mL^{-1} of common table sugar (the naturally occurring form of sucrose) in water in a 10-cm cell exhibits a clockwise optical rotation of 6.65°. Calculate $[\alpha]$. Does this information tell you $[\alpha]$ for the enantiomer of natural sucrose?

The specific rotation of an optically active molecule is a physical constant characteristic of that molecule, just like its melting point, boiling point, and density. Four specific rotations are recorded in Table 5-1.

Optical rotation indicates enantiomeric composition

As mentioned, enantiomers rotate plane-polarized light by equal amounts but in opposite directions. Thus, in 2-bromobutane the ($-$) enantiomer rotates this plane coun-

---

*The dimensions of $[\alpha]$ are deg cm^2 g^{-1}, the units (for $l = 1$) 10^{-1} deg cm^2 g^{-1}. Because of their awkward appearance, it is common practice to give $[\alpha]$ without units, in contrast with the observed rotation α (degrees).

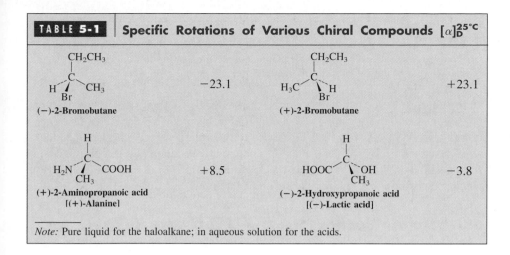

TABLE 5-1	Specific Rotations of Various Chiral Compounds $[\alpha]_D^{25°C}$

(−)-2-Bromobutane −23.1

(+)-2-Bromobutane +23.1

(+)-2-Aminopropanoic acid
[(+)-Alanine] +8.5

(−)-2-Hydroxypropanoic acid
[(−)-Lactic acid] −3.8

Note: Pure liquid for the haloalkane; in aqueous solution for the acids.

terclockwise by 23.1°, its mirror image (+)-2-bromobutane clockwise by 23.1°. It follows that a 1:1 mixture of (+) and (−) enantiomers shows no rotation and is therefore optically inactive. Such a mixture is called a **racemic mixture.** If one enantiomer equilibrates with its mirror image it is said to undergo **racemization.** For example, optically active acids such as (+)-alanine (Table 5-1) have been found to undergo very slow racemization in fossil deposits by tertiary C–H bond-breaking, a process resulting in reduced optical activity.

Optical activity can be measured in a mixture of enantiomers, but only if the enantiomers are present in unequal amounts. Using the value of the measured rotation, we can calculate the composition of such a mixture. For example, if a solution of (+)-alanine from a fossil exhibits an $[\alpha]$ of only +4.25 (i.e., one-half the value for the pure enantiomer), we can deduce that 50% of the sample is pure (+) isomer and the other 50% is racemic. It is said to have 50% **enantiomer excess.** Because the racemic portion consists of equal amounts of (+) and (−), the actual composition of the sample is 75% (+) and 25% (−), as shown here.

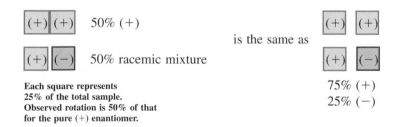

(+) (+) 50% (+)

(+) (−) 50% racemic mixture

**Each square represents
25% of the total sample.
Observed rotation is 50% of that
for the pure (+) enantiomer.**

is the same as

(+) (+)

(+) (−)

75% (+)
25% (−)

The 25% (−) enantiomer cancels the rotation of a corresponding amount of (+). This mixture is called 50% (i.e., 75% – 25%) *optically pure:* The observed optical rotation is one-half that of the pure dextrorotatory enantiomer.

Optical Purity

$$\% \text{ optical purity} = \left(\frac{[\alpha]_{\text{observed}}}{[\alpha]} \cdot 100 \right) = \text{enantiomer excess}$$

EXERCISE 5-7

What is the optical rotation of a sample of (+)-2-bromobutane that is 75% optically pure? What percentages of (+) and (−) enantiomers are present in this sample? Answer the same questions for samples of 50% and 25% optical purity.

In summary, two enantiomers can be distinguished by their optical activity; that is, their interaction with plane-polarized light as measured in a polarimeter. One enantiomer always rotates such light clockwise (dextrorotatory), the other counterclockwise (levorotatory) by the same amount. The specific rotation, $[\alpha]$, is a physical constant possible only for chiral molecules. The interconversion of enantiomers leads to racemization and the disappearance of optical activity.

5-3 Absolute Configuration: *R–S* Sequence Rules

How do we establish the structure of one pure enantiomer of a chiral compound? And, once we know the answer, is there a way to name it unambiguously and distinguish it from its mirror image?

X-ray diffraction can establish the absolute configuration

Virtually all the physical characteristics of one enantiomer are identical with those of its mirror image, except for the sign of optical rotation. Is there a correlation between the sign of optical rotation and the actual spatial arrangement of the substituent groups, the **absolute configuration?** Is it possible to determine the structure of an enantiomer by measuring its $[\alpha]$ value? The answer to both questions is, unfortunately, no. *There is no straightforward correlation between the sign of rotation and the structure of the particular enantiomer.* For example, conversion of lactic acid (Table 5-1) into its sodium salt changes the sign (and degree) of rotation, even though the absolute configuration at the stereocenter is unchanged.

If the sign of rotation does not tell us anything about structure, how do we know which enantiomer of a chiral molecule is which? Or, to put it differently, how do we know that the levorotatory enantiomer of 2-bromobutane has the structure indicated in Table 5-1 (and therefore the dextrorotatory enantiomer the mirror-image configuration)? The answer is that such information can be obtained only through single-crystal X-ray diffraction analysis (Section 1-9). This does not mean that every chiral compound must be submitted to X-ray analysis to ascertain its structure. Absolute configuration can also be established by chemical correlation with a molecule whose own structure has been proved by this method. For example, knowing the stereocenter in (−)-lactic acid by X-ray also provides the absolute configuration of the (+)-sodium salt (i.e., the same).

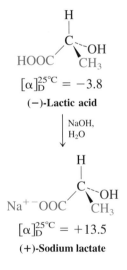

$[\alpha]_\text{D}^{25°C} = -3.8$

(−)-Lactic acid

NaOH, H_2O

$[\alpha]_\text{D}^{25°C} = +13.5$

(+)-Sodium lactate

Stereocenters are labeled *R* or *S*

To name enantiomers unambiguously, we need a system that allows us to indicate the handedness in the molecule, a sort of "left hand" versus "right hand" nomenclature. Such a system was developed by three chemists, R. S. Cahn, C. Ingold, and V. Prelog.*

---

*Dr. Robert S. Cahn (1899–1981), Fellow of the Royal Institute of Chemistry, London; Professor Christopher Ingold (1893–1970), University College, London; Professor Vladimir Prelog (1906–1998), Swiss Federal Institute of Technology (ETH), Zürich, Nobel Prize 1975 (chemistry).

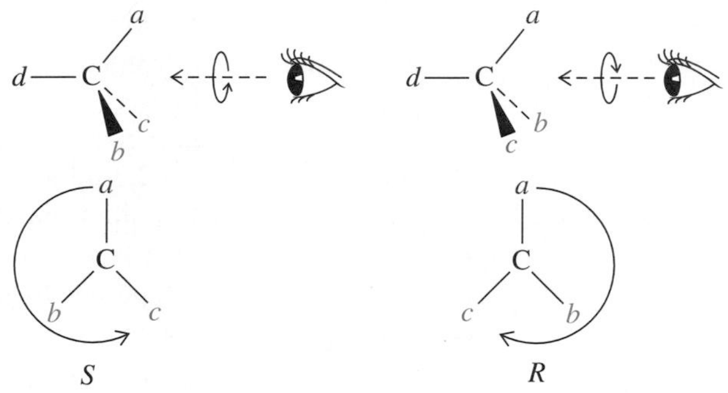

FIGURE 5-6

Assignment of *R* or *S* configuration at a tetrahedral stereocenter. The group of lowest priority is placed as far away from the observer as possible. In many of the structural drawings in this chapter, the color scheme shown here is used to indicate the priority of substituents—in decreasing order, red > blue > green > black.

Let us see how the handedness around an asymmetric carbon atom is labeled. The first step is to rank all four substituents in the order of decreasing priority, the rules of which will be described shortly. Substituent *a* has the highest priority, *b* the second highest, *c* the third, and *d* the lowest. Next, we position the molecule (mentally, on paper, or by using a molecular model set) so that the lowest priority substituent is placed as far away from us as possible (Figure 5-6). This process results in two (and

---

CHEMICAL HIGHLIGHT 5-2 | **Absolute Configuration: A Historical Note**

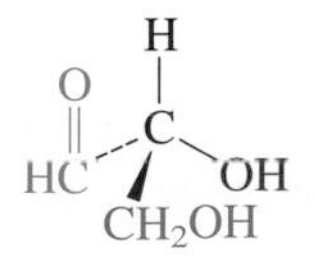

$[\alpha]_D^{25°C} = +8.7$

D-(+)-2,3-Dihydroxypropanal
[D-(+)-Glyceraldehyde]

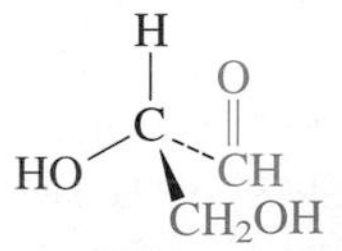

$[\alpha]_D^{25°C} = -8.7$

L-(−)-2,3-Dihydroxypropanal
[L-(−)-Glyceraldehyde]

Before the X-ray diffraction technique was developed, the absolute configurations of chiral molecules were unknown. Amusingly, the first assignment of a three-dimensional structure to a chiral molecule was a

guess made more than a century ago. The naturally occurring dextrorotatory enantiomer of 2,3-dihydroxypropanal (glyceraldehyde) was arbitrarily assigned the structure shown above and labeled "D-glyceraldehyde." The label "D" was not used to refer to the sign of rotation of plane-polarized light but to the relative arrangement of the substituent groups.

The other isomer was called L-glyceraldehyde. All chiral compounds that could be converted into D-(+)-glyceraldehyde by reactions that did not affect the configuration at the stereocenter were assigned the D configuration, and their mirror images the L. In 1951, the absolute configurations of these compounds became known and the original guess was found to be correct. D,L nomenclature is still used for sugars (Chapter 24) and amino acids (Chapter 26).

D-Configurations

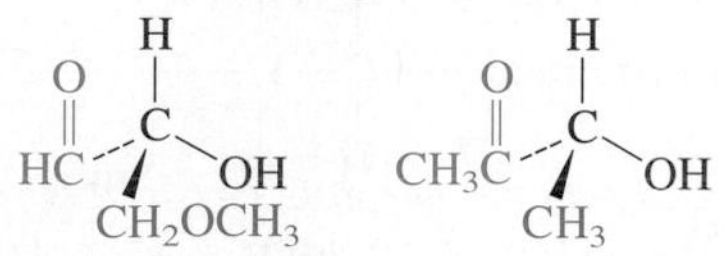

L-Configurations

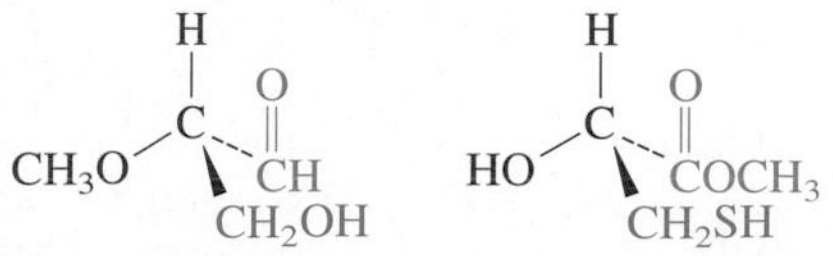

only two) possible arrangements of the remaining substituents. If the progression from *a* to *b* to *c* is counterclockwise, the configuration at the stereocenter is named *S* (*sinister,* Latin, left). Conversely, if the progression is clockwise, the center is *R* (*rectus,* Latin, right). The symbol *R* or *S* is added as a prefix in parentheses to the name of the chiral compound, as in (*R*)-2-bromobutane and (*S*)-2,3-dihydroxypropanal. A racemic mixture is designated *R,S*, as in (*R,S*)-bromochlorofluoromethane. The sign of the rotation of plane-polarized light may be added if it is known, as in (*S*)-(+)-2-bromobutane and (*R*)-(+)-2,3-dihydroxypropanal. It is important to remember, however, that the symbols *R* and *S* are *not* necessarily correlated with either sign of α.

Sequence rules assign priorities to substituents

Before applying the *R,S* nomenclature to a stereocenter, we must first assign priorities by using sequence rules.

RULE 1. We look first at the atoms attached directly to the stereocenter. A substituent atom of higher atomic number takes precedence over one of lower atomic number. Consequently, the substituent of lowest priority is hydrogen. In regard to isotopes, the atom of higher atomic mass receives higher priority.

AN = atomic number

(*R*)-1-Bromo-1-iodoethane

RULE 2. What if two substituents have the same rank when we consider the atoms directly attached to the stereocenter? In such a case, we proceed along the two respective substituent chains until we reach a point of difference.

For example, an ethyl substituent takes priority over methyl. Why? At the point of attachment to the stereocenter, each substituent has a carbon nucleus, equal in priority. Farther from that center, however, methyl has only hydrogen atoms, but ethyl has a carbon atom (higher in priority).

—CH₂CH₂CH₂CH₃
ranks lower in priority than

ranks lower in priority than

However, 1-methylethyl takes precedence over ethyl because, at the first carbon, ethyl bears only one other carbon substituent, but 1-methylethyl bears two. Similarly, 2-methylpropyl takes priority over butyl but ranks lower than 1,1-dimethylethyl.

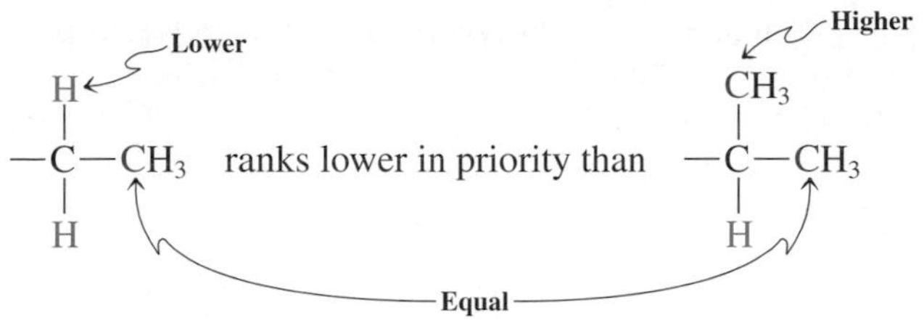

Ethyl **1-Methylethyl (Isopropyl)**

We must remember that the decision on priority is made at the *first* point of difference along otherwise similar substituent chains. When that point has been reached, the constitution of the remainder of the chain is irrelevant.

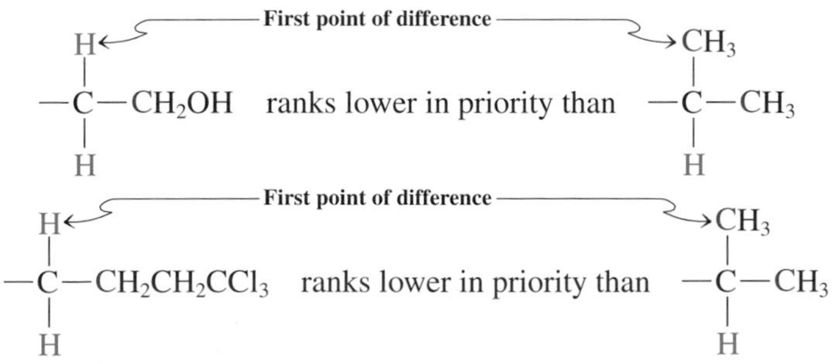

When we reach a point along a substituent chain at which it branches, we choose the branch that is higher in priority. When two substituents have similar branches, we rank the elements in those branches until we reach a point of difference.

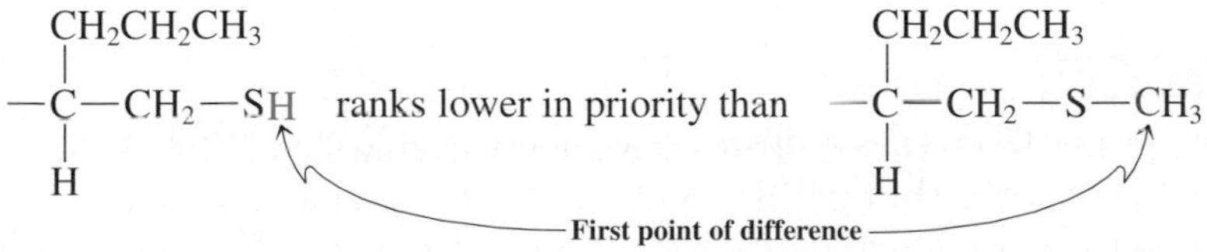

Some examples are shown below.

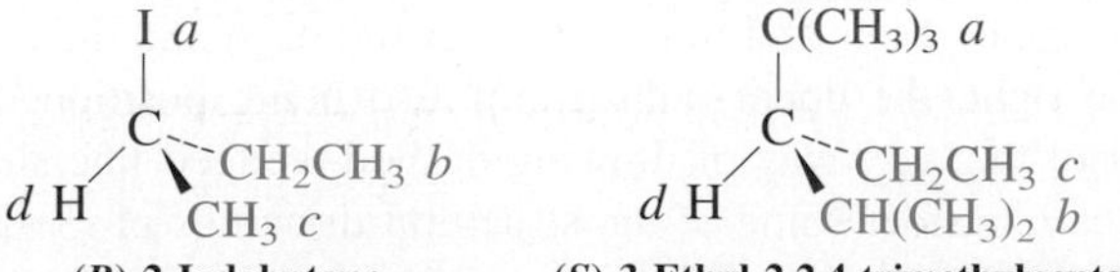

(*R*)-2-Iodobutane **(*S*)-3-Ethyl-2,2,4-trimethylpentane**

RULE 3. Double and triple bonds are treated as if they were single, and the atoms in them are duplicated or triplicated at each end by the respective atoms at the other end of the multiple bond.

> The red atoms shown in the groups on the right side of the display are not really there. They are added only for the purpose of assigning a relative priority to each of the corresponding groups to their left.

Examples are shown in the margin.

EXERCISE 5-8

Draw the structures of the following substituents and within each group rank them in order of decreasing priority. **(a)** Methyl, bromomethyl, trichloromethyl, ethyl; **(b)** 2-methylpropyl (isobutyl), 1-methylethyl (isopropyl), cyclohexyl; **(c)** butyl, 1-methylpropyl (*sec*-butyl), 2-methylpropyl (isobutyl), 1,1-dimethylethyl (*tert*-butyl); **(d)** ethyl, 1-chloroethyl, 1-bromoethyl, 2-bromoethyl.

EXERCISE 5-9

Assign the absolute configuration of the molecules depicted in Table 5-1.

EXERCISE 5-10

Draw one enantiomer of your choice (specify which, *R* or *S*) of 2-chlorobutane, 2-chloro-2-fluorobutane, and (HC≡C)(CH₂=CH)C(Br)(CH₃).

To correctly assign the stereostructure of stereoisomers, we must develop a fair amount of three-dimensional "vision," or "stereoperception." In the structures that have been used to illustrate the priority rules, the lowest priority substituent is located at the left of the carbon center and in the plane of the page and the remainder of the substituents at the right, the upper-right group also being positioned in this plane. However, this is not the only way of drawing dashed-wedged line structures; others are equally correct. Consider some of the structural drawings of (*S*)-2-bromobutane. These are simply different views of the same molecule.

Six Ways of Depicting (S)-2-Bromobutane

To summarize, the sign of optical rotation cannot be used to establish the absolute configuration of a stereoisomer. Instead, X-ray diffraction (or chemical correlations) must be used. We can express the absolute configuration of the chiral molecule as *R* or *S* by applying the sequence rules, which allow us to rank all substituents in order of decreasing priority. Turning the structures so as to place the lowest priority group at the back causes the remaining substituents to be arranged in clockwise (*R*) or counterclockwise (*S*) fashion.

5-4 Fischer Projections

A **Fischer* projection** is a simplified way of depicting tetrahedral carbon atoms and their substituents in two dimensions. With this method, the molecule is drawn in the form of a cross, the central carbon being at the point of intersection. The horizontal

*Professor Emil Fischer (1852–1919), University of Berlin, Nobel Prize 1902 (chemistry).

lines signify bonds directed *toward* the viewer; the vertical lines are pointing *away*. Dashed-wedged line structures have to be arranged in this way to facilitate their conversion into Fischer projections.

Conversion of the Dashed-Wedged Line Structures of 2-Bromobutane into Fischer Projections (of the Stereocenter)

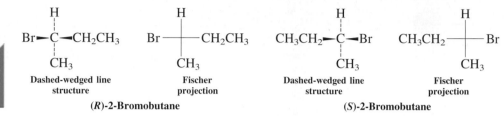

Dashed-wedged line structure   Fischer projection

(*R*)-2-Bromobutane

Dashed-wedged line structure   Fischer projection

(*S*)-2-Bromobutane

You will notice that just as there are several ways of depicting a molecule in the dashed-wedged line notation, there are several correct Fischer projections of the same stereocenter.

Two Additional Projections of (*R*)-2-Bromobutane

A simple mental procedure that will allow you to safely convert any dashed-wedged line structure into a Fischer projection is to picture yourself at the molecular level and grasp any two substituents with your hands while facing the central carbon. If you then imagine descending on the page while holding the molecule, the two remaining substituents positioned behind that center will dock on the surface and then submerge below it, and your left and right hands will position the two horizontal, wedged-line groups in the proper orientation. This procedure places the two remaining dashed-line groups (juxtaposing your head and feet, respectively) vertically.

A Simple Mental Exercise: Conversion of Dashed-Wedge Line Structures into Fischer Projections

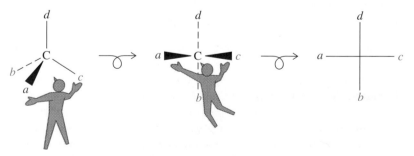

Having achieved this conversion, you can change one Fischer projection into another of the same molecule by using certain manipulations: rotations and substituent switches. However, we shall see next that care has to be taken not to inadvertently convert *R* into *S* configurations.

Rotating a Fischer projection may or may not change the absolute configuration

What happens when we rotate a Fischer projection by 90°? Does the result depict the spatial arrangement of the original molecule? The definition of a Fischer projection—horizontal bonds are pointed above, vertical ones below the plane of the page—tells us that the answer is clearly no, because this rotation has *switched* the relative spatial disposition of the two sets: the result is a picture of the enantiomer. On the other hand, rotation by 180° is fine, because horizontal and vertical lines have not been interchanged: the resulting drawing represents the same enantiomer.

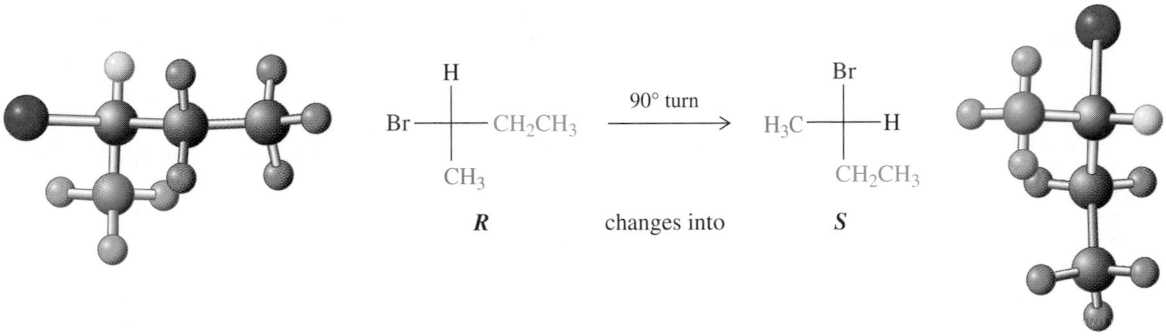

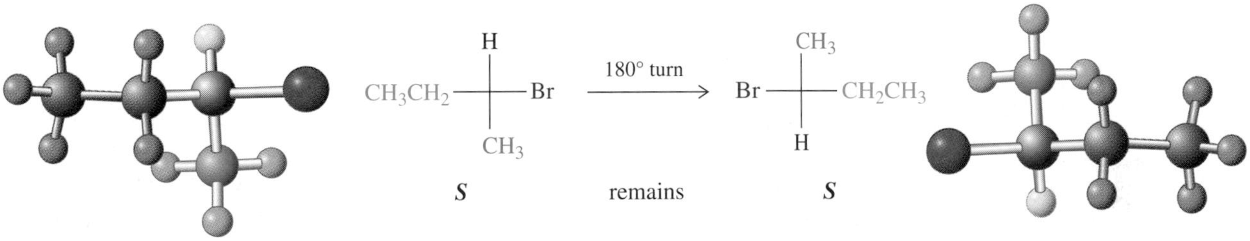

EXERCISE 5-11

Draw Fischer projections for all the molecules in Exercises 5-9 and 5-10.

Exchanging substituents in a Fischer projection also changes the absolute configuration

As is the case for dashed-wedged line structures, there are several Fischer projections of the same enantiomer, a situation that may lead to confusion. How can we quickly ascertain whether two Fischer projections are depicting the same enantiomer or two mirror images? We have to find a sure way to convert one Fischer projection into another in a manner that either leaves the absolute configuration unchanged or converts it into its opposite. It turns out that this task can be achieved by simply making substituent groups trade places. As we can readily verify by using molecular models, any *single* such exchange turns one enantiomer into its mirror image. Two such exchanges (we may select different substituents ever time) produce the original absolute configuration. As shown by the dashed-wedged line structures below, this operation merely results in a different view of the same molecule, rotated 120° about the C–Cl bond.

$$\underset{\mathbf{S}}{\overset{\text{CH}_3}{\underset{\text{CH}_2\text{CH}_3}{\text{Cl}\!-\!\!\!\!\!-\!\!\!\!\!-\text{Br}}}} \xrightarrow{\text{CH}_3 \leftarrow_{\!\!\!\!\!\sigma\!\!\!\!\!} \rightarrow \text{Br}} \underset{\mathbf{R}}{\overset{\text{Br}}{\underset{\text{CH}_2\text{CH}_3}{\text{Cl}\!-\!\!\!\!\!-\!\!\!\!\!-\text{CH}_3}}} \xrightarrow{\text{CH}_3\text{CH}_2 \leftarrow_{\!\!\!\!\!\sigma\!\!\!\!\!} \rightarrow \text{Br}} \underset{\mathbf{S}}{\overset{\text{CH}_2\text{CH}_3}{\underset{\text{Br}}{\text{Cl}\!-\!\!\!\!\!-\!\!\!\!\!-\text{CH}_3}}}$$

$$\mathbf{S} \qquad \text{changes into} \qquad \mathbf{R} \qquad \text{changes back into} \qquad \mathbf{S}$$

(The double arrow denotes two groups trading places)

We now have a simple way of establishing whether two different Fischer projections depict the same or opposite configurations. If the conversion of one structure into another takes an even number of exchanges, the structures are identical. If it requires an odd number of such moves, the structures are mirror images of each other.

Consider, for example, the two Fischer projections A and B. Do they represent molecules having the same configuration? The answer is found quickly. We convert A into B by two exchanges; so A equals B.

$$\underset{\mathbf{A}}{\overset{\text{Cl}}{\underset{\text{CH}_2\text{CH}_3}{\text{H}\!-\!\!\!\!\!-\!\!\!\!\!-\text{CH}_3}}} \xrightarrow{\text{H} \leftarrow_{\!\!\!\!\!\sigma\!\!\!\!\!} \text{Cl}} \overset{\text{H}}{\underset{\text{CH}_2\text{CH}_3}{\text{Cl}\!-\!\!\!\!\!-\!\!\!\!\!-\text{CH}_3}} \xrightarrow{\text{Cl} \leftarrow_{\!\!\!\!\!\sigma\!\!\!\!\!} \rightarrow \text{CH}_3} \underset{\mathbf{B}}{\overset{\text{H}}{\underset{\text{CH}_2\text{CH}_3}{\text{H}_3\text{C}\!-\!\!\!\!\!-\!\!\!\!\!-\text{Cl}}}}$$

EXERCISE 5-12

Draw the dashed-wedged line structures corresponding to Fischer projections A and B, above. Is it possible to transform A into B by means of a rotation about a single bond? If so, identify the bond and the degree of rotation required. Use models if necessary.

Fischer projections tell us the absolute configuration

When we deal with stereochemical problems, an accurate perception of space is very useful. However, Fischer projections allow us to assign absolute configurations without having to visualize the three-dimensional arrangement of the atoms. For this purpose, we first draw the molecule as a (any) Fischer projection. Next, we rank all the substituents in accord with the sequence rules. Finally, we exchange two groups so that the lowest priority substituent is at the top, and then we exchange any other pair (to make sure that the absolute configuration stays unchanged from the original). On completion of these manipulations, we find that the three groups of priority, a, b, and c, are arranged in either clockwise or counterclockwise fashion, in turn corresponding to either the R or the S configuration.

Although this procedure provides you with a fail-safe mechanism with which to assign the absolute configuration of a stereocenter, you must not forget its three-dimensional origin. It is important that you continue to work with molecular models or other three-dimensional representations of molecules to improve your ability to think of them in space.

EXERCISE 5-13

What is the absolute configuration of the following molecules?

$$H-\overset{\displaystyle Br}{\underset{\displaystyle CH_3}{|}}-D \qquad F-\overset{\displaystyle Cl}{\underset{\displaystyle I}{|}}-Br \qquad H_2N-\overset{\displaystyle CH_3}{\underset{\displaystyle H}{|}}-\overset{\displaystyle COH}{\underset{\displaystyle O}{\|}}$$

EXERCISE 5-14

Convert the Fischer projections in Exercise 5-13 into dashed-wedged line formulas and determine their absolute configurations by using the procedure described in Section 5-3. When the lowest priority group is at the top in a Fischer projection, is it in front of the plane of the page or behind it? Does this explain why the procedure outlined above for determination of configuration from Fischer projections succeeds?

In summary, a Fischer projection is a convenient way of drawing chiral molecules. We can rotate such projections in the plane by 180° but not by 90°. Switching substituents reverses absolute configuration, if done an odd number of times, but leaves it intact when the number of such exchanges is even. By placing the substituent of lowest priority on top, we can readily assign the absolute configuration.

5-5 Molecules Incorporating Several Stereocenters: Diastereomers

Many molecules contain several stereocenters. Because the configuration about each center can be *R* or *S*, several possible structures emerge, all of which are isomeric.

Two stereocenters can give four stereoisomers: chlorination of 2-bromobutane at C3

Section 5-1 described how a carbon-based stereocenter can be created by the radical halogenation of butane. Let us now consider the chlorination of racemic 2-bromobutane to give (among other products) 2-bromo-3-chlorobutane. The introduction of a chlorine atom at C3 produces a new stereocenter in the molecule. This center may have either the *R* or the *S* configuration, assignable by using the sequence rules that apply to molecules with only one such center.

How many stereoisomers are possible for 2-bromo-3-chlorobutane? There are four, as can be seen by completing a simple exercise in permutation. Each stereocenter can be either *R* or *S*, and, hence, the possible combinations are *RR*, *RS*, *SR*, and *SS* (Figure 5-7).

Because all horizontal lines in Fischer projections signify bonds directed toward the viewer, the result is a representation of a molecule in an *eclipsed* conformation. Therefore, the first step in converting a staggered Newman or dashed-wedged line representation of a molecule into a Fischer projection is to rotate the molecule to form an eclipsed rotamer. To make stereochemical assignments, one treats each stereocenter

$$CH_3\overset{*}{\underset{\displaystyle Br}{\overset{\displaystyle H}{C}}}CH_2CH_3$$

One stereocenter

$$Cl_2, hv \quad \Big\downarrow \quad -HCl$$

$$CH_3\overset{**}{\underset{\displaystyle Br}{\overset{\displaystyle H}{C}}\underset{\displaystyle H}{\overset{\displaystyle Cl}{C}}}CH_3$$

Two stereocenters

2-Bromo-3-chlorobutane

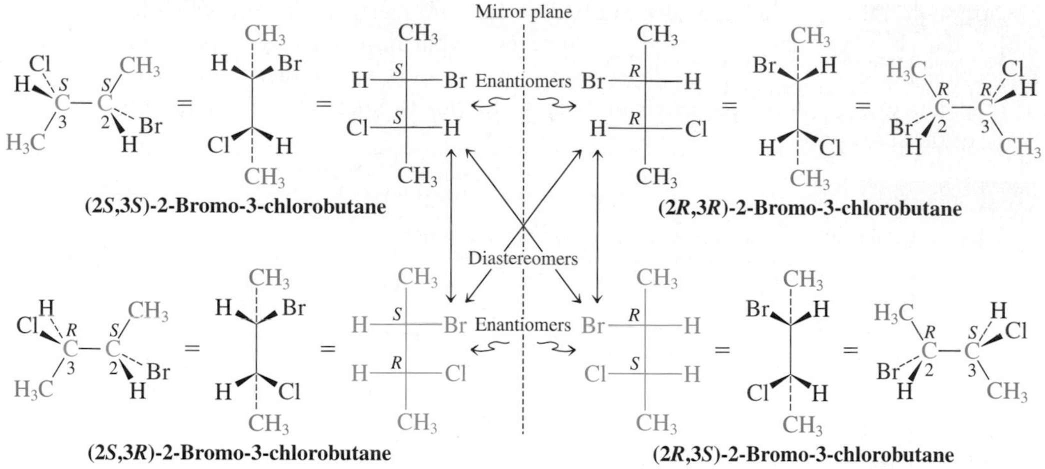

(2S,3S)-2-Bromo-3-chlorobutane   (2R,3R)-2-Bromo-3-chlorobutane

(2S,3R)-2-Bromo-3-chlorobutane   (2R,3S)-2-Bromo-3-chlorobutane

FIGURE 5-7

The four stereoisomers of 2-bromo-3-chlorobutane. Each molecule is the enantiomer of one of the other three (its mirror image) and is at the same time a diastereomer of each of the remaining two. For example, the 2R,3R isomer is the enantiomer of the 2S,3S compound and is diastereomerically related to both the 2S,3R and the 2R,3S structures. Notice that two structures are enantiomers only when they possess the opposite configuration at *every* stereocenter.

separately, and the group containing the other stereocenter is regarded as a simple substituent (Figure 5-8).

By looking closely at the structures of the four stereoisomers (Figure 5-7), we see that there are two related pairs of compounds: an *R,R/S,S* pair and an *R,S/S,R* pair. The members of each individual pair are mirror images of each other and therefore enantiomers. Conversely, each member of one pair is not a mirror image of either member of the other pair; hence, they are not enantiomeric with respect to each other. Stereoisomers that are not related as object and mirror image are called **diastereomers** (*dia,* Greek, across).

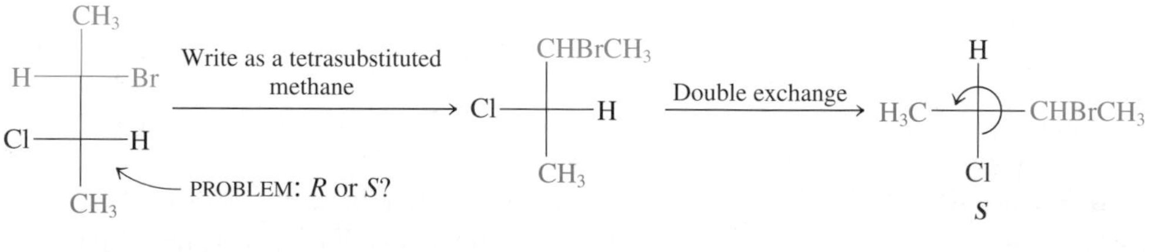

SOLUTION: The center under scrutiny is *S*.

FIGURE 5-8

Assigning the absolute configuration at C3 in 2-bromo-3-chlorobutane. We consider the group containing the stereocenter C2 merely as one of the four substituents. Priorities (also noted in color) are assigned in the usual way ($Cl > CHBrCH_3 > CH_3 > H$), giving rise to the representation shown in the center. Two exchanges place the substituent of lowest priority (hydrogen) at the top of the Fischer projection to facilitate assignment.

The two amino acids isoleucine and alloisoleucine are depicted below in staggered conformations. Convert both into Fischer projections. (Keep in mind that Fischer projections are views of molecules *in eclipsed conformations.*) Are these two compounds enantiomers or diastereomers?

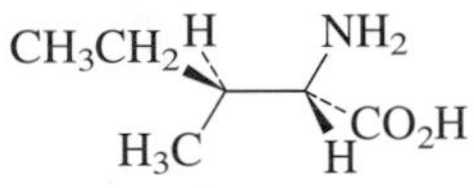

CH₃CH₂ H   NH₂

H₃C   H CO₂H

Isoleucine

H₃C H   NH₂

CH₃CH₂   H CO₂H

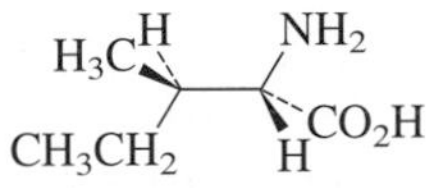

Alloisoleucine

In contrast with enantiomers, diastereomers, because they are *not* mirror images of each other, are distinct molecules with *different physical and chemical properties* (see, e.g., Chemical Highlight 5-3). Their steric interactions and energies differ. They can be separated by fractional distillation or crystallization or by chromatography. They have different melting and boiling points and different densities, just as constitutional isomers do. In addition, they have different specific rotations.

What are the stereochemical relations (identical, enantiomers, diastereomers) of the following four molecules? Assign absolute configurations at each stereocenter.

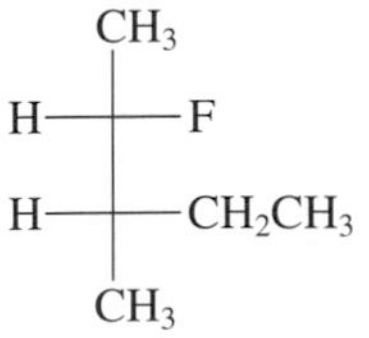

CH₃
H——F
H——CH₂CH₃
CH₃
1

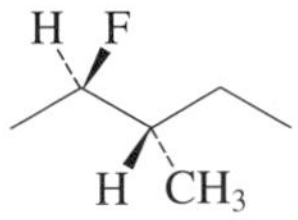

H F
H CH₃
2

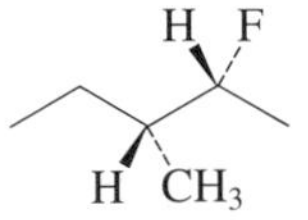

H F
H CH₃
3

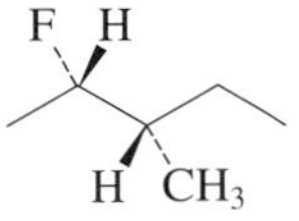

F H
H CH₃
4

Cis and trans isomers are cyclic diastereomers

It is instructive to compare the stereoisomers of 2-bromo-3-chlorobutane with those of a cyclic analog, 1-bromo-2-chlorocyclobutane (Figure 5-9). In both cases, there are four stereoisomers: *R,R*, *S,S*, *R,S*, and *S,R*. In the cyclic compound, however, the stereoisomeric relation of the first pair to the second is easily recognized: One pair has cis stereochemistry, the other trans. Cis and trans isomers (Section 4-1) in cycloalkanes are in fact diastereomers.

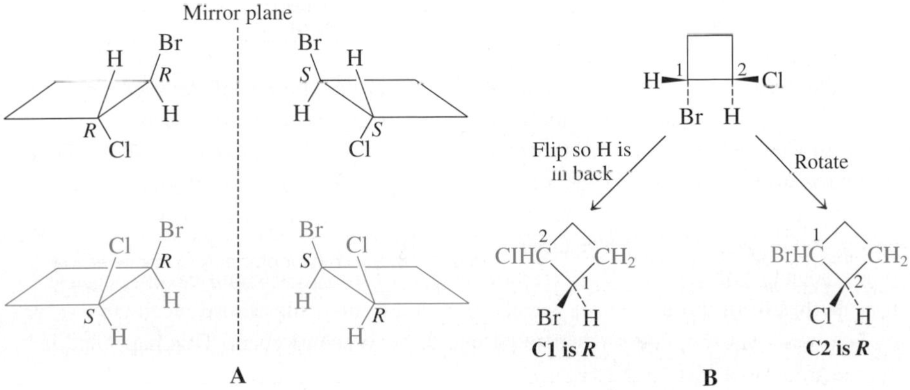

Mirror plane

H Br   Br H
  R   S
H   H
R   S
Cl   Cl

Cl Br   Br Cl
  R   S
H   H
S   R
H   H

A

H—1 2—Cl
   Br H

Flip so H is in back   Rotate

ClHC CH₂   BrHC CH₂
   Br H   Cl H
C1 is R   **C2 is R**

B

FIGURE 5-9
(A) The diastereomeric relation of *cis*- and *trans*-bromo-2-chlorobutane. (B) Stereochemical assignment of the *R,R* stereoisomer. Recall that the color scheme indicates the priority order of the groups around each stereocenter: red > blue > green > black.

More than two stereocenters means still more stereoisomers

What structural variety do we expect for a compound having three stereocenters? We may again approach this problem by permuting the various possibilities. If we label the three centers consecutively as either *R* or *S*, the following sequence emerges:

| *RRR* | *RRS* | *RSR* | *SRR* | *RSS* | *SRS* | *SSR* | *SSS* |

a total of eight stereoisomers. They can be arranged to reveal a division into four enantiomer pairs of diastereomers.

Image	*RRR*	*RRS*	*RSS*	*SRS*
Mirror image	*SSS*	*SSR*	*SRR*	*RSR*

Generally, *a compound with n stereocenters can have a maximum of 2^n stereoisomers.* Therefore, a compound having three such centers gives rise to a maximum of eight stereoisomers; one having four produces sixteen; one having five, thirty-two; and so forth. The structural possibilities are quite staggering for larger systems.

EXERCISE 5-17

Draw all the stereoisomers of 2-bromo-3-chloro-4-fluoropentane.

In summary, the presence of more than one stereocenter in a molecule gives rise to diastereomers. These are stereoisomers that are not related to each other as object and mirror image. Whereas enantiomers have opposite configurations at every respective stereocenter, two diastereomers do not. A molecule with *n* stereocenters may

---

CHEMICAL HIGHLIGHT 5-3 — Stereoisomers of Tartaric Acid

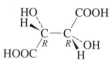

(+)-Tartaric acid

$[\alpha]_D^{20°C} = +12.0$
m.p. 168–170°C
Density (g mL^{-1}) $d = 1.7598$

(−)-Tartaric acid

$[\alpha]_D^{20°C} = -12.0$
m.p. 168–170°C
$d = 1.7598$

meso-Tartaric acid

$[\alpha]_D^{20°C} = 0$
m.p. 146–148°C
$d = 1.666$

Tartaric acid (systematic name: 2,3-dihydroxy-butanedioic acid) is a naturally occurring dicarboxylic acid containing two stereocenters with identical substitution patterns. Therefore it exists as a pair of enantiomers (which have identical physical properties but which rotate plane-polarized light in opposite directions) and an achiral meso compound (with different physical and chemical properties from those of the chiral diastereomers).

The dextrorotatory enantiomer of tartaric acid is widely distributed in nature. It is present in many fruits (fruit acid), and its monopotassium salt is found as a deposit during the fermentation of grape juice. Pure levorotatory tartaric acid is rare, as is the meso isomer.

Tartaric acid is of historical significance, because it was the first chiral molecule whose racemate was separated into the two enantiomers. This happened in

exist in as many as 2^n stereoisomers. In cyclic compounds, cis and trans isomers are diastereomers.

5-6 Meso Compounds

We saw that the molecule 2-bromo-3-chlorobutane contains two distinct stereocenters, each with a *different* halogen substituent. How many stereoisomers are to be expected if both centers are identically substituted?

Two identically substituted stereocenters give rise to only three stereoisomers

Consider, for example, 2,3-dibromobutane, which can be obtained by the radical bromination of 2-bromobutane. As we did for 2-bromo-3-chlorobutane, we have to consider four structures, resulting from the various permutations in *R* and *S* configurations (Figure 5-10; see page 188).

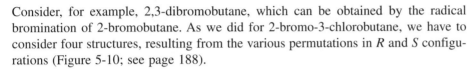

One stereocenter   Two stereocenters
2,3-Dibromobutane

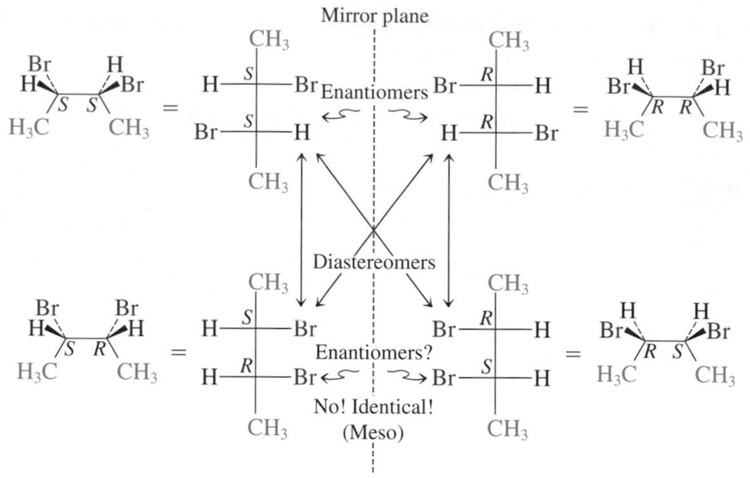

FIGURE 5-10

The stereochemical relations of the stereoisomers of 2,3-dibromobutane. The lower pair consists of identical structures. (Make a model.)

The first pair of stereoisomers, with *R,R* and *S,S* configurations, is clearly recognizable as a pair of enantiomers. However, a close look at the second pair reveals that (*S,R*) and mirror image (*R,S*) are superimposable and therefore identical. Thus, the *S,R* diastereomer of 2,3-dibromobutane is achiral and not optically active, even though it contains two stereocenters. The identity of the two structures can be readily confirmed by using molecular models.

A compound that contains two (or, as we shall see, even more than two) stereocenters but is superimposable with its mirror image is a **meso compound** (*mesos,* Greek, middle). A characteristic feature of a meso compound is the *presence of a mirror plane,* which divides the molecule such that one half is the mirror image of the other half. For example, in 2,3-dibromobutane, the 2*R* center is the reflection of the 3*S* center. This arrangement is best seen in an eclipsed dashed-wedged line structure (Figure 5-11). The presence of a mirror plane in *any* energetically accessible conformation of a molecule (Sections 2-5 and 2-7) is sufficient to make it achiral (Section 5-1). As a consequence, 2,3-dibromobutane exists in the form of three stereoisomers only: a pair of (necessarily chiral) enantiomers and an achiral meso diastereomer.

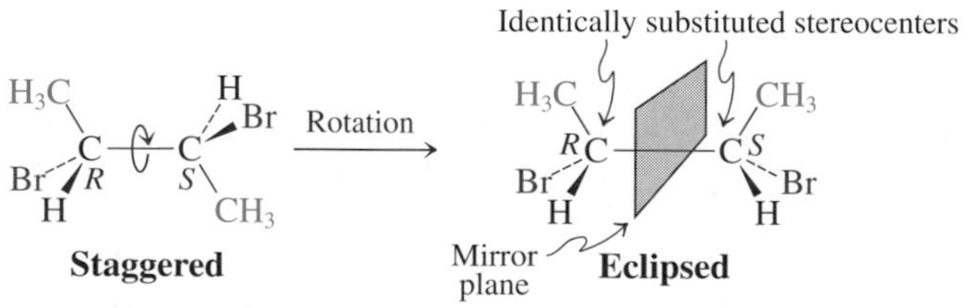

FIGURE 5-11

meso-2,3-Dibromobutane contains a mirror plane when rotated into the eclipsed conformation shown. A molecule with more than one stereocenter is meso and achiral as long as it contains a mirror plane in any readily accessible conformation. Meso compounds possess identically substituted stereocenters.

Meso diastereomers can exist in molecules with more than two stereocenters. Examples are 2,3,4-tribromopentane and 2,3,4,5-tetrabromohexane.

Meso Compounds with Multiple Stereocenters

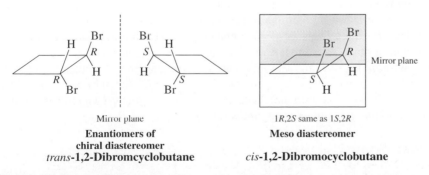

EXERCISE 5-18

Draw all the stereoisomers of 2,4-dibromo-3-chloropentane.

Cyclic compounds may also be meso

It is again instructive to compare the stereochemical situation in 2,3-dibromobutane with that in an analogous cyclic molecule: 1,2-dibromocyclobutane. We can see that *trans*-1,2-dibromocyclobutane exists as two enantiomers (*R,R* and *S,S*) and may therefore be optically active. The cis isomer, however, has a mirror plane and is meso, achiral, and optically inactive (Figure 5-12).

Notice that we have drawn the ring in a planar shape in order to illustrate the mirror symmetry, although we know from Chapter 4 that cycloalkanes with four or more carbons in the ring are not flat. Is this justifiable? Generally yes, because such compounds, like their acyclic analogs, possess a variety of conformations that are readily accessible at room temperature (Sections 4-2 through 4-4 and Section 5-1). At least one of these conformations will contain the necessary mirror plane to render achiral any cis-disubstituted cycloalkane with identically constituted stereocenters. For simplicity, cyclic compounds may usually be treated *as if they were planar* for the purpose of identifying a mirror plane.

Enantiomers of chiral diastereomer
trans-**1,2-Dibromcyclobutane**

Meso diastereomer
cis-**1,2-Dibromocyclobutane**

FIGURE 5-12

The trans isomer of 1,2-dibromocyclobutane is chiral; the cis isomer is a meso compound and optically inactive.

In summary, molecules with two or more identically substituted stereocenters may exist as meso stereoisomers. Meso compounds are superimposable on their mirror images and therefore achiral.

5-7 Stereochemistry in Chemical Reactions

We have seen that a chemical reaction can introduce chirality into a molecule. Let us examine more closely the conversion of achiral butane into chiral-2-bromobutane, which gives racemic material. We shall also see that the chiral environment of a stereocenter already present in a molecule exerts control on the stereochemistry of a reaction that introduces a second stereocenter. We begin with another look at the radical bromination of butane.

The radical mechanism explains why the bromination of butane results in a racemate

The radical bromination of butane at C2 creates a chiral molecule (Figure 5-3). This happens because one of the methylene hydrogens is replaced by a new group, furnishing a stereocenter—a carbon atom with four different substituents.

In the first step of the mechanism for radical halogenation (Sections 3-6 and 3-7), one of these two hydrogens is abstracted by the attacking bromine atom. It does not matter which of the two is removed: This step does not generate a stereocenter. It furnishes a planar, sp^2-hybridized, and therefore achiral radical. The radical center has two equivalent reaction sites—the two lobes of the p orbital (Figure 5-13)—equally susceptible to attack by bromine in the second step. We can see that the two transition states resulting in the respective enantiomers of 2-bromobutane are mirror images of each other. They are enantiomeric and therefore energetically equivalent. The rates of formation of *R* and *S* products are hence equal, and a racemate is formed. In general, *the formation of chiral compounds* (e.g., 2-bromobutane) *from achiral reactants* (e.g., butane and bromine) *yields racemates.* Or, *optically inactive starting materials furnish optically inactive products.**

*We shall see later that it *is* possible to generate optically active products from optically inactive starting materials if we use an optically active reagent.

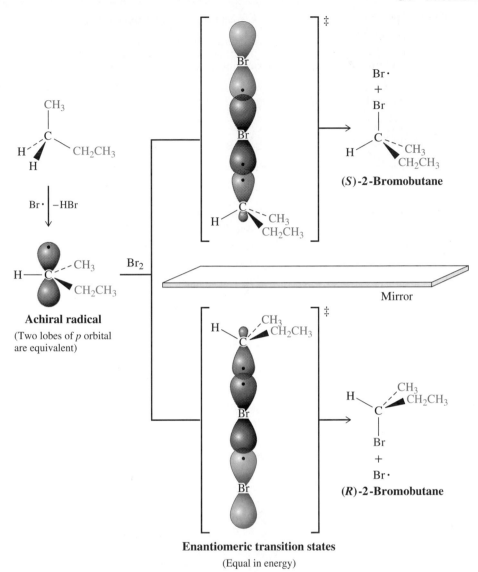

FIGURE 5-13

The creation of racemic 2-bromobutane from butane by radical bromination at C2. Abstraction of either methylene hydrogen by bromine gives an achiral radical. Reaction of Br_2 with this radical is equally likely at either the top or the bottom face, a condition leading to a racemic mixture of products.

The presence of a stereocenter affects the outcome of the reaction: chlorination of (S)-2-bromobutane

Now we understand why the halogenation of an achiral molecule gives a racemic halide. What products can we expect from the halogenation of a chiral, enantiomerically pure molecule?

For example, consider the radical chlorination of the S enantiomer of 2-bromobutane. In this case, the chlorine atom has several options for attack: the two terminal methyl groups, the single hydrogen at C2, and the two hydrogens on C3. Let us examine each of these reaction paths.

Chlorination of (S)-2-Bromobutane at Either C1 or C4

$$\underset{\substack{\text{Optically active}\\ \mathbf{2R}}}{\overset{\text{H}}{\underset{\underset{4\quad 3}{\text{CH}_3\text{CH}_2}}{\underset{\text{ClH}_2\text{C}}{\overset{1}{\underset{}{\overset{2}{\text{C}}}}}}\text{Br}}} \xleftarrow[-\text{HCl}]{\underset{\text{at C1}}{\text{Cl}_2, \, hv}} \underset{\substack{\text{Optically active}\\ \mathbf{2S}}}{\overset{\text{H}}{\underset{\underset{4\quad 3}{\text{CH}_3\text{CH}_2}}{\underset{\text{H}_3\text{C}}{\overset{1}{\underset{}{\overset{2}{\text{C}}}}}}\text{Br}}} \xrightarrow[-\text{HCl}]{\underset{\text{at C4}}{\text{Cl}_2, \, hv}} \underset{\substack{\text{Optically active}\\ \mathbf{3S}}}{\overset{\text{H}}{\underset{\underset{1\quad 2}{\text{ClCH}_2\text{CH}_2}}{\underset{\text{H}_3\text{C}}{\overset{4}{\underset{}{\overset{3}{\text{C}}}}}}\text{Br}}}$$

Chlorination of either terminal methyl group is straightforward, proceeding at C1 to give 2-bromo-1-chlorobutane or at C4 to give 3-bromo-1-chlorobutane. In the latter, the original C4 has now become C1, to maintain the lowest possible substituent numbering. *Both of these chlorination products are optically active because the original stereocenter is left intact.* Note, however, that conversion of the C1 methyl into a chloromethyl unit changes the sequence of priorities around C2. Thus, although the stereocenter itself does not participate in the reaction, its designated configuration changes from *S* to *R*.

What about halogenation at C2, the stereocenter? The product from chlorination at C2 of (*S*)-2-bromobutane is 2-bromo-2-chlorobutane. Even though the substitution pattern at the stereocenter has changed, the molecule remains chiral. However, an attempt to measure the [α] value for the product would reveal the absence of optical activity: *Halogenation at the stereocenter leads to a racemic mixture.* How can this be explained? For the answer, we must look again at the structure of the radical formed in the course of the reaction mechanism.

A racemate forms in this case because hydrogen abstraction from C2 furnishes a planar, sp^2-hybridized, achiral radical.

$$\underset{\substack{\text{Optically active}\\ \mathbf{2S}}}{\overset{\text{H}}{\underset{\underset{}{\text{CH}_3\text{CH}_2}}{\underset{\text{H}_3\text{C}}{\text{C}}}}\text{Br}} \xrightarrow[-\text{HCl}]{\text{Cl}\cdot} \underset{\text{Achiral}}{\overset{\text{H}_3\text{C}}{\underset{\text{CH}_3\text{CH}_2}{\text{C}}}\text{Br}} \xrightarrow[-\text{Cl}\cdot]{\text{Cl}_2} \underset{\substack{\text{Optically inactive}\\ \text{(A racemate)}\\ \mathbf{50\%\ 2S}}}{\overset{\text{Cl}}{\underset{\underset{}{\text{Br}}}{\underset{\text{CH}_3\text{CH}_2}{\text{C}}}}} + \underset{\mathbf{50\%\ 2R}}{\overset{\text{H}_3\text{C}}{\underset{\text{Cl}}{\text{C}}}\overset{\text{Br}}{}}$$

Chlorination can occur from either side through enantiomeric transition states of equal energy, as in the bromination of butane (Figure 5-13), producing (*S*)- and (*R*)-2-bromo-2-chlorobutane at equal rates and in equal amounts. The reaction is an example of a transformation in which an optically active compound leads to an optically inactive product (a racemate).

EXERCISE 5-21

What other halogenations of (*S*)-2-bromobutane would furnish optically inactive products?

The chlorination of (*S*)-2-bromobutane at C3 does not affect the existing chiral center. However, *the formation of a second stereocenter gives rise to diastereomers.* Specifically, attachment of chlorine to the left side of C3 in the drawing on the next page gives (2*S*,3*S*)-2-bromo-3-chlorobutane, whereas attachment to the right side gives its 2*S*,3*R* diastereomer.

Chlorination of (S)-2-Bromobutane at C3

The chlorination at C2 gives a 1 : 1 mixture of enantiomers. Does the reaction at C3 also give an equimolar mixture of diastereomers? The answer is no. This finding is readily explained on inspection of the two transition states leading to the product (Figure 5-14). Abstraction of either one of the hydrogens results in a radical center at C3. In contrast with the radical formed in the chlorination at C2, however, the two

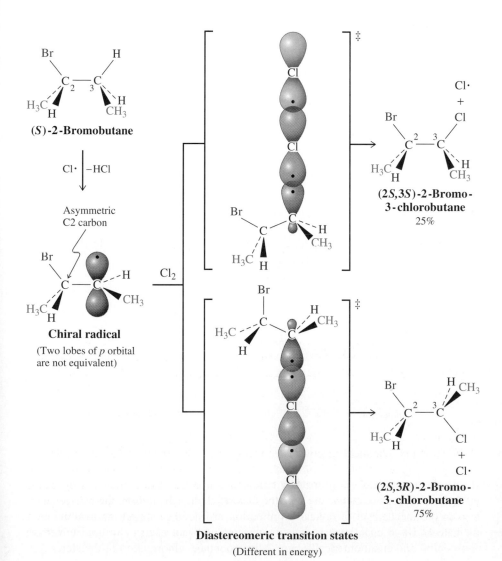

FIGURE 5-14

The chlorination of (S)-2-bromobutane at C3 produces the two diastereomers of 2-bromo-3-chlorobutane in unequal amounts as a result of the chirality at C2.

faces of this radical are *not* mirror images of each other, because the radical retains the asymmetry of the original molecule as a result of the presence of the stereocenter at C2. Thus, the two sides of the *p* orbital are not equivalent.

What are the consequences of this nonequivalency? If the rate of attack at the two faces of the radical differ, as one would predict on steric grounds, then the rates of formation of the two diastereomers should be different, as is indeed found: (2*S*,3*R*)-2-Bromo-3-chlorobutane is preferred over the 2*S*,3*S* isomer by a factor of 3 (see Figure 5-14). The two transition states leading to products are not mirror images of each other and are not superimposable: They are diastereomeric. They therefore have different energies and represent different pathways.

EXERCISE 5-22

Write the structures of the products of monobromination of (*S*)-2-bromopentane at each carbon atom. Name the products and specify whether they are chiral or achiral, whether they will be formed in equal or unequal amounts, and which will be in optically active form.

Stereoselectivity is the preference for one stereoisomer

A reaction that leads to the predominant (or exclusive) formation of one of several possible stereoisomeric products is **stereoselective.** For example, the chlorination of (*S*)-2-bromobutane at C3 is stereoselective, as a result of the chirality of the radical intermediate. The corresponding chlorination at C2, however, is not stereoselective: The intermediate is achiral and a racemate is formed.

How much stereoselectivity is possible? The answer depends very much on substrate, reagents, the particular reaction in question, and conditions. Enzymes in nature manage to convert achiral compounds into chiral molecules with very high stereoselectivity. They are capable of this task because enzymes themselves have handedness and therefore convert achiral materials into those that are compatible with their own chirality. An example is the enzyme-catalyzed oxidation of dopamine to (−)-norepinephrine discussed in detail in Problem 52 at the end of the chapter. The chiral reaction environment created by the enzyme gives rise to 100% stereoselectivity in favor of the enantiomer shown. The situation is very similar to shaping flexible achiral objects with your hands. For example, clasping a piece of modeling clay with your left hand furnishes a shape that is the mirror image of that made with your right hand.

HO—⟨benzene ring⟩—CH$_2$CH$_2$NH$_2$ Dopamine →$\xrightarrow{\beta\text{-monooxygenase, O}_2}$ HO—⟨benzene ring⟩—$\overset{\text{HO}\quad\text{H}}{\underset{}{C}}$—CH$_2NH_2$

Dopamine   **(−)-Norepinephrine**

In summary, chemical reactions, as exemplified by radical halogenation, can be stereoselective or not. Starting from achiral materials, such as butane, a racemic (non-stereoselective) product is formed by halogenation at C2. The two hydrogens at the methylene carbons of butane are equally susceptible to substitution, the halogenation step in the mechanism of radical bromination proceeding through an achiral intermediate and two enantiomeric transition states of equal energy. Similarly, starting from chiral and enantiomerically pure 2-bromobutane, chlorination of the stereocen-

CHEMICAL HIGHLIGHT 5-4 Chiral Drugs: Racemic or Enantiomerically Pure?

Until recently, most synthetic chiral medicines were prepared as racemic mixtures and sold as such. The reasons were mainly of a practical nature. Reactions that convert an achiral into a chiral molecule ordinarily produce racemates (Section 5-7). In addition, often both enantiomers have comparable physiological activity or one of them (the "wrong" one) is inactive; therefore resolution was deemed unnecessary. Finally, resolution of racemates on a large scale is expensive and substantially adds to the costs of drug development.

However, in several cases, one of the enantiomers of a drug has been found to act as a blocker of the biological receptor site, thus diminishing the activity of the other enantiomer. Worse, one of the enantiomers may have a completely different, and sometimes toxic, spectrum of activity. A tragic example of the latter is the sedative *thalidomide,* which was marketed in 1960

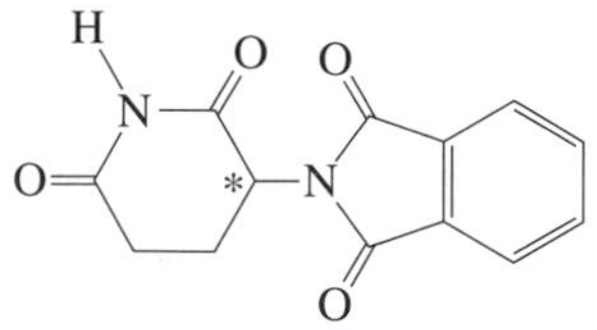

Thalidomide

in Europe as a racemic mixture. Consumption of this drug by pregnant women led to serious birth defects in hundreds of babies. Subsequent studies showed that the *S* enantiomer is teratogenic in some laboratory animals, whereas the *R* form is not. The problem is further complicated by the finding that each enantiomer can racemize at physiological pH.

Because of these (and other) developments, the U.S. Food and Drug Administration revised its guidelines for the commercialization of chiral drugs, making it more advantageous for companies to produce single enantiomers of medicinal products. The result has been a flurry of research activities designed to improve resolution of racemates or, even better, to develop methods of *enantioselective synthesis.* The essence of this approach is that used by nature in enzyme-catalyzed reactions (see the oxidation of dopamine in Section 5-7): An achiral starting material is converted into the chiral product in the presence of an enantiopure environment, often a chiral catalyst. Because in such an environment enantiomeric transition states (Figure 5-13) become diastereomeric (Figure 5-14; note that, in this case, the chiral "environment" of the reacting carbon is provided by the neighboring stereocenter), high stereoselectivity can be achieved. As shown below, such methods have been applied to the syntheses of drugs such as the antiarthritic *naproxen* and the antihypertensive *propanolol* in high enantiomeric purity.

To give you an idea of the importance of this emerging technology, the worldwide market for chiral drugs is estimated to be in excess of \$40 billion in 1997.

(C* = a new stereocenter)

(*R*)-Naproxen

(*S*)-Propanolol

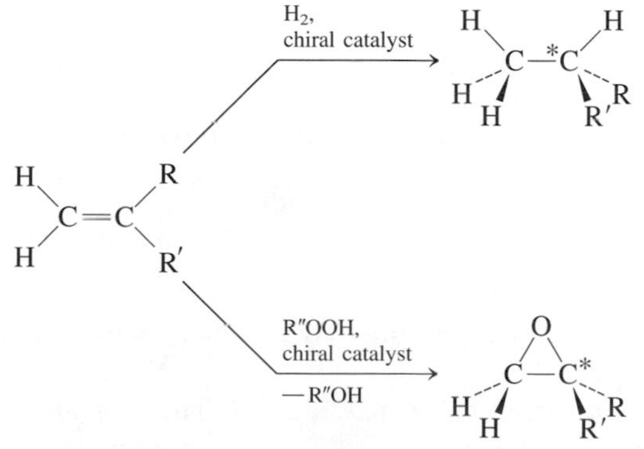

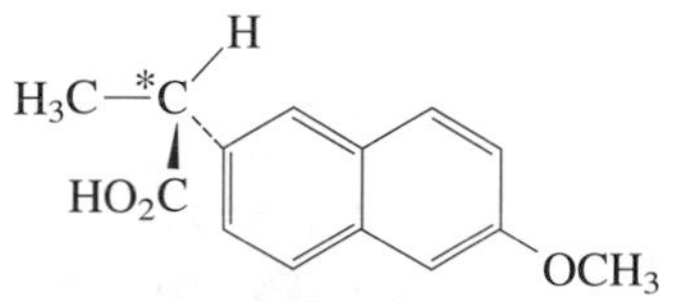

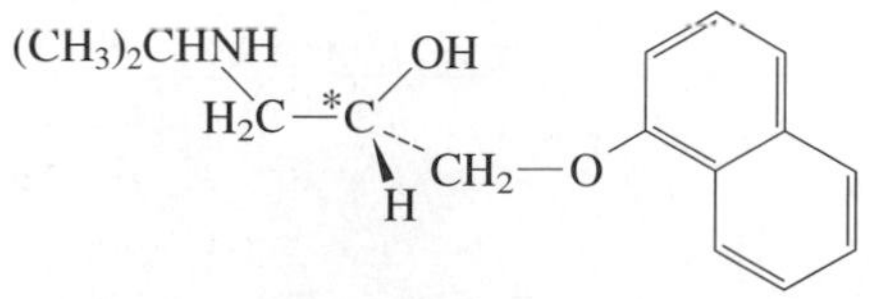

ter also gives a racemic product. However, stereoselectivity is possible in the formation of a new stereocenter, because the chiral environment retained by the molecule results in two unequal modes of attack on the intermediate radical. The two transition states have a diastereomeric relation, a condition that leads to the formation of products at unequal rates.

5-8 Resolution: Separation of Enantiomers

As we know, the generation of a chiral structure from an achiral starting material furnishes a racemic mixture. How, then, can pure enantiomers of a chiral compound be obtained?

One possible approach is to start with the racemate and separate one enantiomer from the other. This process is called the **resolution** of enantiomers. Some enan-

---

CHEMICAL HIGHLIGHT 5-5 | Why Is Nature "Handed"?

In this chapter, we have seen that many of the organic molecules in nature are chiral. More importantly, most natural compounds in living organisms not only are chiral, but also are present in only one enantiomeric form. An example of an entire class of such compounds consists of the *amino acids*, which are the component units of *polypeptides*. The large polypeptides in nature are called *proteins* or, when they catalyze biotransformations, *enzymes*.

Absolute Configuration of Natural Amino Acids and Polypeptides

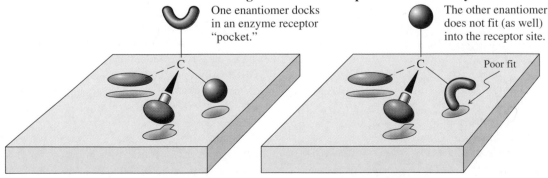

Being made up of smaller chiral pieces, enzymes arrange themselves into bigger conglomerates that also are chiral and show handedness. Thus, much as a right hand will readily distinguish another right hand from a left hand, enzymes (and other biomolecules) have "pockets" that, by virtue of their stereochemically defined features, are capable of recognizing and processing only one of the enantiomers in a racemate. The differences in physiological activity of the two enantiomers of a

Schematized Enantiomer Recognition in the Receptor Site of an Enzyme

One enantiomer docks in an enzyme receptor "pocket."

The other enantiomer does not fit (as well) into the receptor site.

Poor fit

tiomers, such as those of tartaric acid, crystallize into mirror-image shapes, which can be manually separated (as done by Pasteur; see Chemical Highlight 5-3). However, this process is time consuming, uneconomic for anything but minute-scale separations, and applicable only in rare cases.

A better strategy for resolution is based on the difference in the physical properties of diastereomers. Suppose we can find a reaction that converts a racemate into a mixture of diastereomers. All the R forms of the original enantiomer mixture should then be separable from the corresponding S forms by fractional crystallization, distillation, or chromatography of the diastereomers. How can such a process be developed? The trick is to add an enantiomerically pure reagent that will attach itself to the components of the racemic mixture. For example, we can imagine reaction of a racemate, $X_{R,S}$ (in which X_R and X_S are the two enantiomers), with an optically pure compound Y_S (the choice of the S configuration is arbitrary; the pure R mirror image would work just as well). The reaction produces two optically active diastereomers,

Pasteur's polarimeter and crystals of (+)- and (−)-tartaric acid.

chiral drug are based on this recognition (Chemical Highlight 5-4). A good analogy is that of a chiral key fitting only its image (not mirror image) lock. The chiral environment provided by these structures is also able to effect highly enantioselective conversions of achiral starting materials into enantiopure, chiral

The 4.5-billion-year-old meteorite from Mars which has led scientists to believe that Mars may have once harbored life, on exhibit at the Smithsonian's National Museum of Natural History, Washington, D.C.

products (see Chemical Highlights 5-4 and 23-2). In this way, how nature preserves and proliferates its own built-in chirality can be readily understood (at least in principle).

What is more difficult to understand is how the enantiomeric homogeneity of nature arose in the first place; in other words, why was only one stereochemical configuration of the amino acids chosen but not the other? Trying to understand this mystery has fascinated many scientists, because it is very likely linked to the evolution of life as we know

it. Speculation ranges from the invocation of a chance separation of enantiomers ("spontaneous resolution") to the postulate of the operation of a chiral physical force, such as handed radiation (as observed during the decay of radioactive elements). Another hypothesis suggests that enantiomeric excess (and perhaps life itself) was simply imported from another planet, with meteorites as carriers (thus really begging the question). A lot of effort has been expended in trying to detect nonracemic amino acids in meteor (and other planetary) samples, so far without success.

X_RY_S and X_SY_S, separable by standard techniques (Figure 5-15). Now the bond between X and Y in each of the separated and purified diastereomers is broken, liberating X_R and X_S in their enantiomerically pure states. In addition, the optically active agent Y_S may be recovered and reused in further resolutions.

What we need, then, is a readily available, enantiomerically pure compound, Y, that can be attached to the molecule to be resolved in an easily reversible chemical reaction. In fact, nature has provided us with a large number of pure optically active molecules that can be used. An example is (+)-2,3-dihydroxybutanedioic acid [(+)-(*R,R*)-tartaric acid]. A popular reaction employed in the resolution of enantiomers is salt formation between acids and bases. For example, (+)-tartaric acid functions as an effective resolving agent of racemic amines. Figure 5-16 shows how this works for 3-butyn-2-amine. The racemate is first treated with (+)-tartaric acid to form two diastereomeric tartrate salts. The salt incorporating the *R*-amine crystallizes on standing and can be filtered away from the solution, which contains the more soluble salt of the *S*-amine. Treatment of the (+)-salt with aqueous base liberates the free amine, (+)-(*R*)-3-butyn-2-amine. Similar treatment of the solution gives the (−)-*S* enantiomer (evidently slightly less pure: Note the slightly lower optical rotation). This process is just one of many ways in which the formation of diastereomers can be used in the resolution of racemates.

A very convenient way of separating enantiomers without the necessity of isolating diastereomers is by so-called **chiral chromatography.** The principle is the same

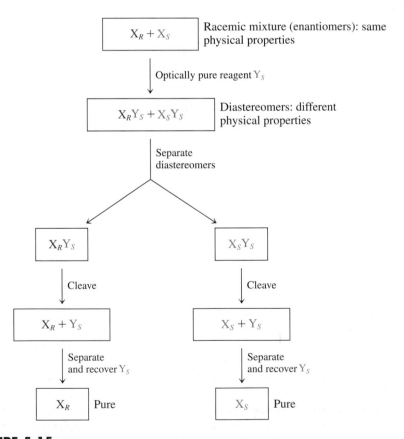

FIGURE 5-15

Flowchart for the separation (resolution) of two enantiomers. The procedure is based on conversion into separable diastereomers by means of reaction with an optically pure reagent.

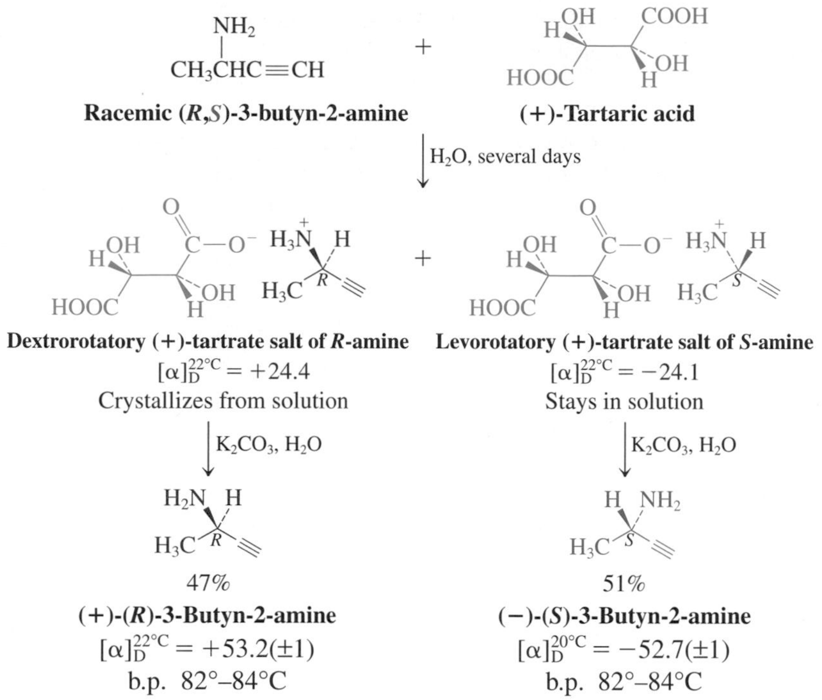

FIGURE 5-16

Resolution of 3-butyn-2-amine with (+)-2,3-dihydroxybutanedioic [(+)-tartaric] acid. It is purely accidental that the [α] values for the two diastereomeric tartrate salts are similar in magnitude and of opposite sign.

as that illustrated in Figure 5-16, except that the optically active auxiliary [such as (+)-tartaric acid or any other suitable cheap optically active compound] is immobilized on a solid support (such as silica gel, SiO_2, or aluminum oxide, Al_2O_3). This material is then used to fill a tube ("column") of varying length, and a solution of the racemate is allowed to pass through it. The individual enantiomers will reversibly bind to the chiral support to different extents (because this interaction is diastereomeric) and therefore be held on the column for different lengths of time (retention time). Therefore, one enantiomer will elute from the column before the other, enabling separation.

CHAPTER INTEGRATION PROBLEM

Selectivity in chemical reactions is a primary goal of the synthetic chemist. We have learned how such selectivity may be achieved, at least to some extent, in radical halogenations: in Sections 3-6 and 3-7, with respect to the type of hydrogen to be replaced (e.g., primary versus secondary versus tertiary) and, in Section 5-7, with respect to stereochemistry. You will have recognized that, because of the reactivity of radicals and the planarity of the carbon-centered radical intermediates, radical halogenations often lack selectivity and any synthetic plan considering them must take into account all possible outcomes of a proposed conversion. For example, looking again at our picture of the generalized steroid nucleus (Section 4-7), you can see that there are many types of hydrogens, all of which would, in principle, be susceptible to abstraction by a halogen atom.

Because the steroids are important biological molecules, their selective functionalization has been the focus of the attention of many researchers and, by developing carefully controlled conditions with special halogenating agents, they have been able to restrict attack not only to the tertiary centers, but also selectively to either C5, C9, or C14 (see also Problems 36–38 in Chapter 4). The following problem illustrates the kind of analyses that they undertook with a less complex cyclohexane fragment of the steroid nucleus.

How many products are there of the radical monobromination of (S)-1-bromo-2,2-dimethylcyclohexane at C1 and C3? Draw the structure of the starting material, name the resulting dibromodimethylcyclohexanes, label them as chiral or achiral, specify whether they will be formed in equal or unequal amounts, and state whether they will be optically active or not.

SOLUTION

We begin by drawing the structure of our starting material, first ignoring stereochemistry (A).

In order to avoid the complication of carbon atoms changing their numbers during the reactions of molecule A, we have used names that do not conform to IUPAC numbering rules for several of the structures in this problem.

We then designate the priority sequence (B) according to the rules in Section 5-3. We now have a choice of two enantiomeric arrangements (C and D), and the task is to orient the molecule in our minds in such a way as to place the substituent of lowest priority (H atom) as far away as possible. To assist in this mental exercise, picture yourself at the molecular scale (shrunk by a factor of 10^{10}) and stand on the stereocenter in question with the C–H bond pointing away from you. The three remaining substituents will now surround you either in clockwise (R) or counterclockwise (S) fashion: D is the correct structure of the S enantiomer.

Now we are ready to introduce bromine at either C1 or C3. It is important here to remember the mechanism of the free-radical halogenation: The crucial intermediate is a radical center—in our case, at either C1 (E) or C3 (F)—that can be attacked by halogen from either side of the p orbital (Section 3-4).

In E, the molecule is symmetrical, and the rate of attack from the top will be equal to that from the bottom. If the halogenation were to be executed by using F_2 or Cl_2, C1 would remain a stereocenter, the R and S enantiomers having been formed in equal amounts (racemate; Section 5-7, Figure 5-13). However, in our case, bromination at C1 removes the asymmetry of this carbon: Compound G, 1,1-dibromo-2-2-dimethylcyclohexane, is achiral and hence not optically active.

Turning to F, the situation is different. Here, the presence of the unchanged original stereocenter (C1 in D) makes the two faces of the intermediate radical center unequal: Two diastereomers (H and I) are formed at unequal rates and therefore in unequal amounts (Section 5-7, Figure 5-14). In H, *cis*-1,3-dibromo-2,2-dimethylcyclohexane, the second bromine is attached in such a way as to introduce a mirror plane into the molecule: H is a meso compound, achiral, and hence not optically active (Section 5-6). Another way of describing what has happened is that the chirality of C1 in D—namely, S—is canceled out by the introduction of its "mirror image" at C3—namely, R. The two stereoisomers are indistinguishable because $(1S,3R)$-H is the same as $(1R,3S)$-H. (You can verify that statement by simply rotating compound H about the dashed line representing the mirror plane.)

On the other hand, compound I, $(1S,3S)$-1,3-dibromo-2,2-dimethylcyclohexane, contains no mirror plane: The molecule is chiral, enantiomerically pure, and, hence, optically active. In other words, the reaction leaves the stereochemical integrity and identity of C1 intact, generating only one enantiomer of the product, which is non-superimposable with its mirror image, the $(1R,3R)$ diastereomer (Section 5-5).

IMPORTANT CONCEPTS

1. Isomers have the same molecular formula but are different compounds. Constitutional (structural) isomers differ in the order in which the individual atoms are connected. Stereoisomers have the same connectivity but differ in the three-dimensional arrangement of the atoms. **Mirror-image stereoisomers** are related to each other as image and mirror image.

2. An object that is not superimposable on its mirror image is **chiral.**

3. A carbon atom bearing four different substituents **(asymmetric carbon)** is an example of a **stereocenter.**

4. Two stereoisomers that are related to each other as image–nonsuperimposable mirror image are called **enantiomers.**

5. A compound containing one stereocenter is chiral and exists as a pair of enantiomers. A 1:1 mixture of enantiomers is a **racemate (racemic mixture).**

6. Chiral molecules cannot have a plane of symmetry (mirror plane). If a molecule has a **mirror plane,** then it is **achiral.**

7. **Diastereomers** are stereoisomers that are not related to each other as object to mirror image. Cis and trans isomers of cyclic compounds are examples of diastereomers.

8. Two stereocenters in a molecule result in as many as four stereoisomers—two diastereomerically related pairs of enantiomers. The maximum number of stereoisomers that a compound with n stereocenters can have is 2^n. This number is reduced when equivalently substituted stereocenters give rise to a plane of symmetry. A molecule containing stereocenters *and* a mirror plane is identical with its mirror image (achiral) and is called a **meso compound.** The presence of a mirror plane in any energetically accessible conformation of a molecule is sufficient to make it achiral.

9. Most of the physical properties of enantiomers are the same. A major exception is their interaction with **plane-polarized light:** One enantiomer will rotate the polarization plane clockwise **(dextrorotatory),** the other counterclockwise **(levorotatory).** This phenomenon is called **optical activity.** The extent of the rotation is measured in degrees and is expressed by the **specific rotation, $[\alpha]$.** Racemates and meso compounds show zero rotation. The **optical purity** of an unequal mixture of enantiomers is given by

$$\% \text{ optical purity} = \left(\frac{[\alpha]_{observed}}{[\alpha]} \right) \cdot 100$$

10. The "handedness" of a stereocenter (its absolute configuration) is revealed by X-ray diffraction and can be assigned as **R** or **S** by using the **sequence rules** of Cahn, Ingold, and Prelog.

11. **Fischer projections** provide stencils for the quick drawing of molecules with stereocenters.

12. Chirality can be introduced into an achiral compound by radical halogenation. When the transition states are enantiomeric (related as object and mirror image), the result is a racemate because the faces of the planar radical react at equal rates.

13. Radical halogenation of a chiral molecule containing one stereocenter will give a racemate if the reaction takes place at the stereocenter. When reac-tion elsewhere leads to two diastereomers, they will be formed in unequal amounts.

14. The preference for the formation of one stereoisomer, when several are possible, is called **stereoselectivity.**

15. The separation of enantiomers is called **resolution.** It is achieved by the reaction of the racemate with the pure enantiomer of a chiral compound to yield separable diastereomers. Chemical removal of the chiral reagent frees both enantiomers of the original racemate. Another way of separating enantiomers is by **chiral chromatography** on an optically active support

PROBLEMS

23. Classify each of the following common objects as being either chiral or achiral. Assume in each case that the object is in its simplest form, without decoration or printed labels. **(a)** A ladder; **(b)** a door; **(c)** an electric fan; **(d)** a refrigerator; **(e)** Earth; **(f)** a baseball; **(g)** a baseball bat; **(h)** a baseball glove; **(i)** a flat sheet of paper; **(j)** a fork; **(k)** a spoon; **(l)** a knife.

24. Each part of this problem lists two objects or sets of objects. As precisely as you can, describe the relation between the two sets, using the terminology of this chapter; that is, specify whether they are identical, enantiomeric, or diastereomeric. **(a)** An American toy car compared with a British toy car (same color and design but steering wheels on opposite sides); **(b)** two left shoes compared with two right shoes (same color, size, and style); **(c)** a pair of skates compared with two left skates (same color, size, and style); **(d)** a right glove on top of a left glove (palm to palm) compared with a left glove on top of a right glove (palm to palm; same color, size, and style).

25. For each pair of the following molecules, indicate whether its members are identical, structural isomers, conformers, or stereoisomers. How would you describe the relation between conformations when they are maintained at a temperature too low to permit them to interconvert?

(g)

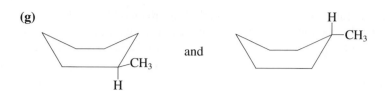

and

(h) Cl, H, H, CH₃   and   H, CH₃, Cl, H

26. Which of the following compounds are chiral? (**Hint:** Look for stereocenters.)
(a) 2-Methylheptane   (b) 3-Methylheptane   (c) 4-Methylheptane
(d) 1,1-Dibromopropane   (e) 1,2-Dibromopropane   (f) 1,3-Dibromopropane
(g) Ethene, $H_2C\!=\!CH_2$ (h) Ethyne, $HC\!\equiv\!CH$

(i) Benzene,   (Note: Like ethene, benzene contains all sp^2-
hybridized carbons and is therefore planar.)

(j) Epinephrine,   HO—⟨⟩—$\overset{\text{OH}}{\underset{}{\text{CHCH}_2\text{NHCH}_3}}$ HO

(k) Vanillin, HO—⟨⟩—$\overset{\text{O}}{\underset{}{\text{CH}}}$ CH_3O

(l) Citric acid, $HOCCH_2CCH_2COH$ with O, OH, O, and C/O OH groups

(m) Ascorbic acid, $HOCH_2$, $HOCH$, H, HO, OH, O

(n) *p*-Menthane-1,8-diol (terpin hydrate),   HO CH₃ ... $H_3C\!-\!\overset{}{\underset{\text{CH}_3}{\text{C}}}\!-\!OH$

(o) Meperidine (demerol),   CH₃, N, OCH_2CH_3, O

27. Each of the following molecules has the molecular formula $C_5H_{12}O$ (check for
yourself). Which ones are chiral?

(a) OH   (b)   OH   (c)   OH

(d) OH   (e)   OH   (f)   OH

28. Which of the following cyclohexane derivatives are chiral? For the purpose of determining the chirality of a cyclic compound, the ring may generally be treated as if it were planar.

(a) 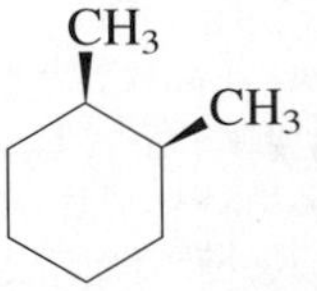

(b) 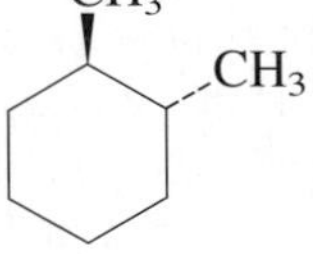

(c) 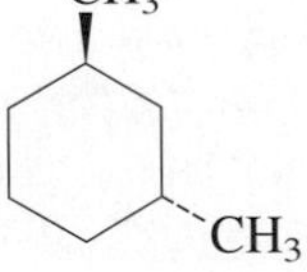

(d) 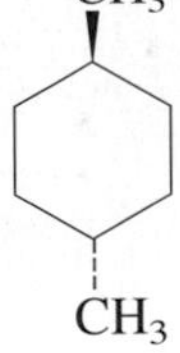

29. For each pair of structures shown below, indicate whether the two species are constitutional isomers, enantiomers, diastereomers of one another, or identical molecules.

(a) 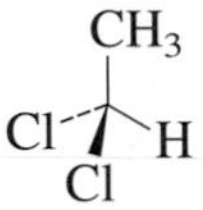

and 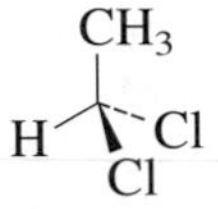

(b) 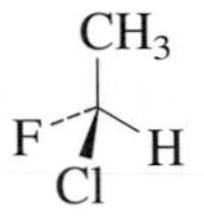

and 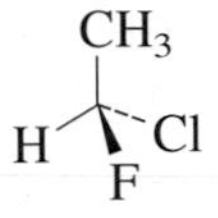

(c) 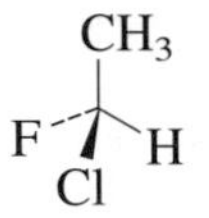

and 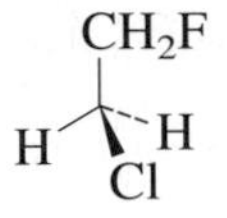

(d) 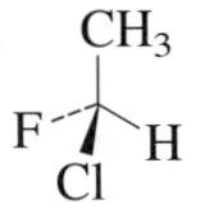

and 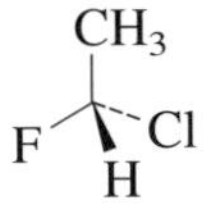

(e) 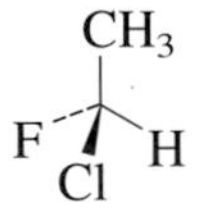

and 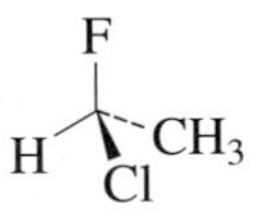

(f) 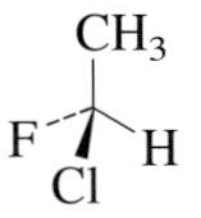

and 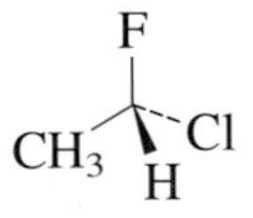

(g) 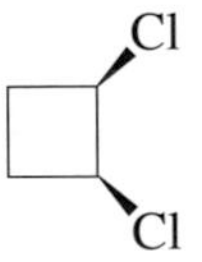

and 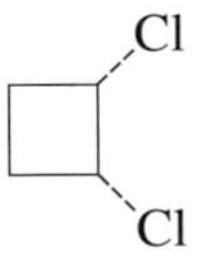

(h) 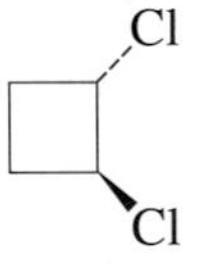

and 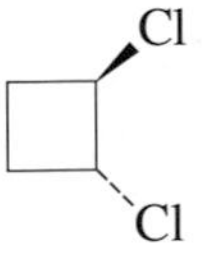

(i) 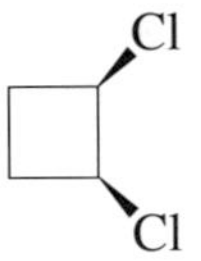

and 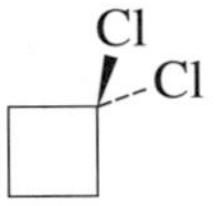

(j) 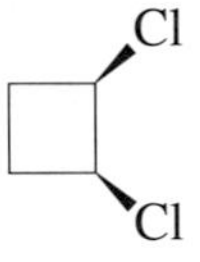

and 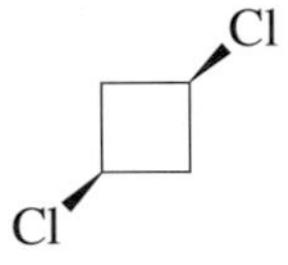

(k) 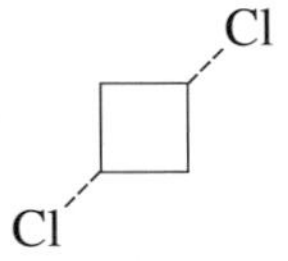

and 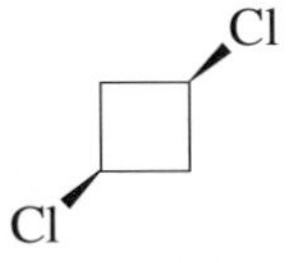

(l) 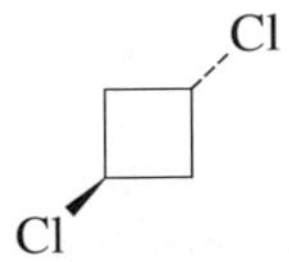

and 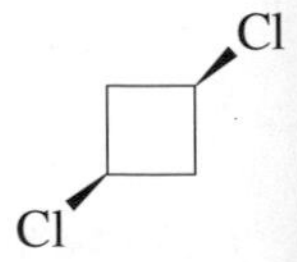

(m) 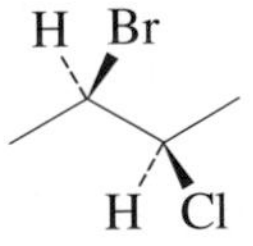

and 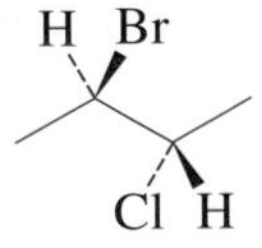

(n) 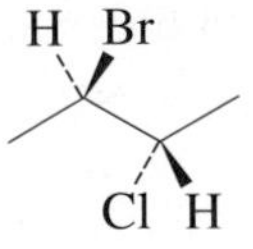

and 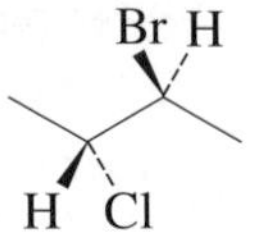

(o) 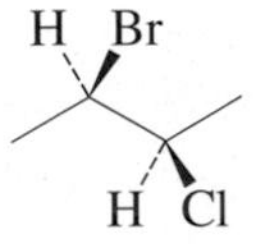

and 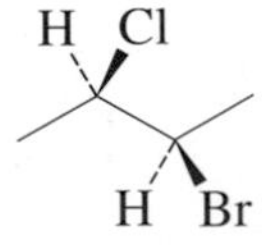

(p) 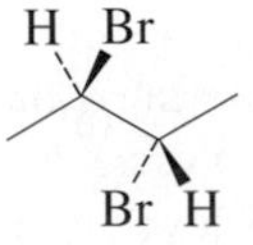

and 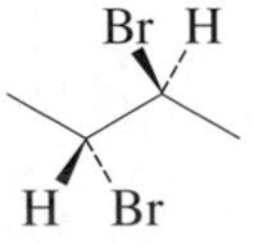

30. For each of the following formulas, identify every structural isomer containing one or more stereocenters, give the number of stereoisomers for each, and draw and fully name at least one of the stereoisomers in each case.
  (a) C_7H_{16} **(b)** C_8H_{18} **(c)** C_5H_{10}, with one ring

31. Assign the appropriate designation of configuration (R or S) to the stereocenter in each of the following molecules. (**Hint:** Regarding cyclic structures containing stereocenters, treat the ring as if it were two separate substituents that happen to be attached to each other at the far end of the molecule—look for the first point of difference, just as you would for acyclic structures.)

(a) 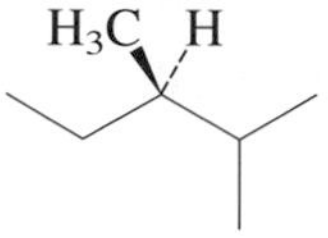 (b) 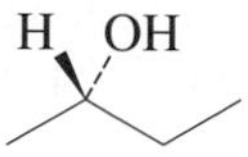 (c) 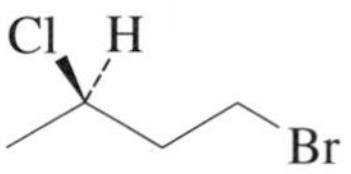 (d)

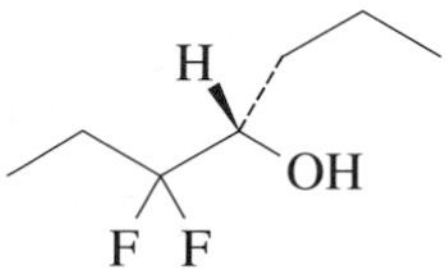

(e) 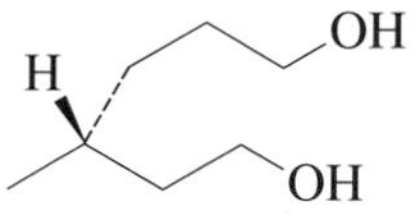 (f) 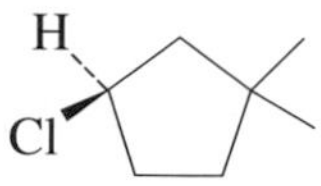 (g) 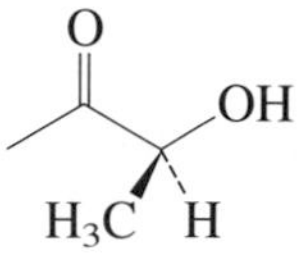 (h)

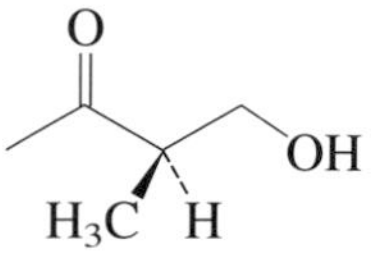

(i) 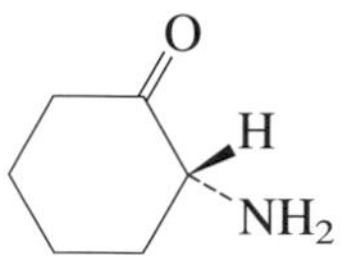 (j)

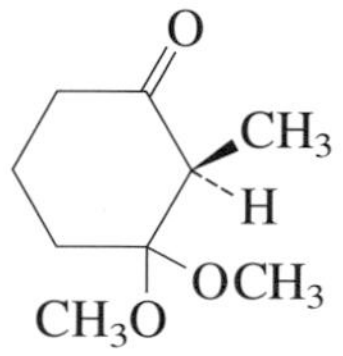

32. Mark the stereocenters in each of the chiral molecules in Problem 27. Draw any single stereoisomer of each of these molecules, and assign the appropriate designation (R or S) to each stereocenter.

33. The two isomers of carvone [systematic name: 2-methyl-5-(1-methylethenyl)-2-cyclohexenone] are drawn here. Which is R and which is S?

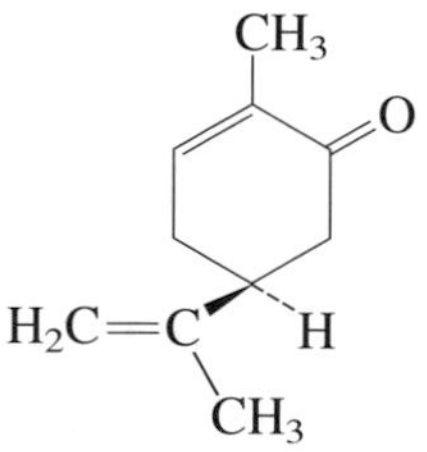

(+)-Carvone
(In caraway seeds)

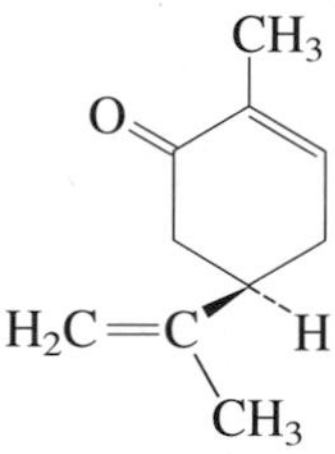

(−)-Carvone
(In spearmint)

34. Draw structural representations of each of the following molecules. Be sure that your structure clearly shows the configuration at the stereocenter. (**Hint:** You may find it useful to first draw the enantiomer whose configuration is easiest for you to determine and then, if necessary, modify your structure to fit the one requested in the problem.) **(a)** (R)-2-chloropentane; **(b)** (S)-2-methyl-3-bromohexane; **(c)** (S)-1,3-dichlorobutane; **(d)** (R)-2-chloro-1,1,1-trifluoro-3-methylbutane.

35. Draw structural representations of each of the following molecules. Be sure that your structure clearly shows the configuration at each stereocenter.
  (a) (R)-3-bromo-3-methylhexane; **(b)** ($3R,5S$)-3,5-dimethylheptane;
  (c) ($2R,3S$)-2-bromo-3-methylpentane; **(d)** (S)-1,1,2-trimethylcyclopropane;
  (e) ($1S,2S$)-1-chloro-1-trifluoromethyl-2-methylcyclobutane; **(f)** ($1R,2R,3S$)-1,2-dichloro-3-ethylcyclohexane.

36. For each of the following questions, assume that all measurements are made in 10-cm polarimeter sample containers. **(a)** A 10-mL solution of 0.4 g of optically active 2-butanol in water displays an optical rotation of $-0.56°$. What is its specific rotation? **(b)** The specific rotation of sucrose (common sugar) is $+66.4$. What would be the observed optical rotation of such a solution containing 3 g of sucrose? **(c)** A solution of pure (S)-2-bromobutane in ethanol is found to have an observed $\alpha = 57.3°$. If $[\alpha]$ for (S)-2-bromobutane is 23.1, what is the concentration of the solution?

37. Natural epinephrine, $[\alpha]_D^{25°C} = -50$, is used medicinally. Its enantiomer is medically worthless and is, in fact, toxic. You, a pharmacist, are given a solution said to contain 1 g of epinephrine in 20 mL of liquid, but the optical purity is not specified. You place it in a polarimeter (10-cm tube) and get a reading of $-2.5°$. What is the optical purity of the sample? Is it safe to use medicinally?

38. Sodium hydrogen (S)-glutamate [(S)-monosodium glutamate], $[\alpha]_D^{25°C} = +24$, is the active flavor enhancer known as MSG. The condensed formula of MSG is shown in the margin. **(a)** Draw the structure of the S enantiomer of MSG. **(b)** If a commercial sample of MSG were found to have a $[\alpha]_D^{25°C} = +8$, what would be its optical purity? What would be the percentages of the S and R enantiomers in the mixture? **(c)** Answer the same questions for a sample with $[\alpha]_D^{25°C} = +16$.

$$\underset{\underset{O}{\overset{\overset{NH_2}{|}}{}}}{HOCCHCH_2CH_2CO^-Na^+}\overset{O}{\overset{\|}{}}$$

39. For each of the following compounds, mark each stereocenter, assign an R or S designation, and draw a clear picture of the molecule's enantiomer.

(a)

(b)

(c)

(d)

(e)

(f)

(g) Chlorpheniramine
(As in Coricidin decongestant)

(Note: The carbons in benzene or benzenelike rings are treated in the same way as those in alkenes. Use sequence rule 3 from Section 5-3.)

(h) Limonene
(From trees, fruits, etc.)

40. For each of the following pairs of structures, indicate whether the two compounds are identical or enantiomers of each other.

(a)

(b)

(c) Cl—CF$_3$ (CH$_3$ top, OCH$_3$ bottom) and F$_3$C—CH$_3$ (OCH$_3$ top, Cl bottom)

(d) H$_2$N—C—CO$_2$H (H top, CH(CH$_3$)$_2$ bottom) and H—CH(CH$_3$)$_2$ (NH$_2$ top, CO$_2$H bottom)

41. Determine the *R* or *S* designation for each stereocenter in the structures in Problem 40.

42. Redraw each of the following molecules as a Fischer projection; then assign *R* or *S* designations to each stereocenter.

(a) H$_3$C—Cl, H, Cl, CH$_3$ (C—C with substituents)

(b) CO$_2$H, OHC, CH$_3$, HO, CH$_3$, OH

(c) H$_2$N, OH, H, CH$_3$, COOH (C—C)

(d) H$_3$C, Br, H, H, Cl, CH$_3$ (C—C)

43. The compound pictured in the margin is a sugar called (−)-arabinose. Its specific rotation is −105. **(a)** Draw the enantiomer of (−)-arabinose. **(b)** Does (−)-arabinose have any other enantiomers? **(c)** Draw a diastereomer of (−)-arabinose. **(d)** Does (−)-arabinose have any other diastereomers? **(e)** If possible, predict the specific rotation of the structure that you drew for (a). **(f)** If possible, predict the specific rotation of the structure that you drew for (c). **(g)** Does (−)-arabinose have any optically inactive diastereomers? If it does, draw one.

44. Write the complete IUPAC name of the following compound (do not forget stereochemical designations).

CH$_2$CH$_3$ / H—C—CH$_2$CH$_2$Cl (Cl below)   C$_5$H$_{10}$Cl$_2$

Reaction of this compound with 1 mol of Cl$_2$ in the presence of light produces several isomers of the formula C$_5$H$_9$Cl$_3$. For each part of this problem, give the following information: How many stereoisomers are formed? If more than one is formed, are they generated in equal or unequal amounts? Designate every stereocenter in each stereoisomer as *R* or *S*.
(a) Chlorination at C3   **(b)** Chlorination at C4   **(c)** Chlorination at C5

45. Monochlorination of methylcyclopentane can result in several products. Give the same information as that requested in Problem 44 for the monochlorination of methylcyclopentane at C1, C2, and C3.

46. Draw all possible products of the chlorination of (*S*)-2-bromo-1,1-dimethylcyclobutane. Specify whether they are chiral or achiral, whether they are formed in equal or unequal amounts, and which are optically active when formed.

47. Illustrate how to resolve racemic 1-phenylethanamine (shown in the margin), using the method of reversible conversion into diastereomers.

48. Draw a flowchart that diagrams a method for the resolution of racemic 2-hydroxypropanoic acid (lactic acid, Table 5-1), using (*S*)-1-phenylethanamine.

H / C=O
HO——H
H——OH
H——OH
CH$_2$OH
(−)-Arabinose

NH$_2$
C$_6$H$_5$CHCH$_3$
1-Phenylethanamine

49. How many different stereoisomeric products are formed in the monobromination of **(a)** racemic *trans*-1,2-dimethylcyclohexane and **(b)** pure (*R*,*R*)-1,2-dimethylcyclohexane? **(c)** For your answers to (a) and (b), indicate whether you expect equal or unequal amounts of the various products to be formed. Indicate to what extent products can be separated on the basis of having different physical properties (e.g., solubility, boiling point).

50. Make a model of *cis*-1,2-dimethylcyclohexane in its most stable conformation. If the molecule were rigidly locked into this conformation, would it be chiral? (Test your answer by making a model of the mirror image and checking for superimposability.)

Flip the ring of the model. What is the stereoisomeric relation between the original conformation and the conformation after flipping the ring? How do the results that you have obtained in this problem relate to your answer to Problem 28(a).

51. Morphinane is the parent substance of the broad class of chiral molecules known as the morphine alkaloids. Interestingly, the (+) and (−) enantiomers of the compounds in this family have rather different physiological properties. The (−) compounds, such as morphine, are "narcotic analgesics" (painkillers), whereas the (+) compounds are "antitussives" (ingredients in cough syrup). Dextromethorphan is one of the simplest and most common of the latter.

Morphinane   **Dextromethorphan**

(a) Locate and identify all the stereocenters in dextromethorphan. **(b)** Draw the enantiomer of dextromethorphan. **(c)** As best you can (it is not easy), assign *R* and *S* configurations to all the stereocenters in dextromethorphan.

52. The enzymatic introduction of a functional group into a biologically important molecule is not only specific with regard to the location at which the reaction occurs in the molecule (see Chapter 4, Problem 36), but also usually specific in the stereochemistry obtained. The biosynthesis of epinephrine first requires that a hydroxy group be introduced specifically to produce (−)-norepinephrine from the achiral substrate dopamine. (The completion of the synthesis of epinephrine will be presented in Problem 59 of Chapter 9.) Only the (−) enantiomer is functional in the appropriate physiological manner, so the synthesis must be highly stereoselective.

Dopamine   **(−)-Norepinephrine**

(a) Is the configuration of (−)-norepinephrine *R* or *S*? **(b)** In the absence of an enzyme, would the transition states of a radical oxidation leading to (−)- and

(+)-norepinephrine be of equal or unequal energy? What term describes the relation between these transition states? **(c)** In your own words, describe how the enzyme must affect the energy of these transition states to favor production of the (−) enantiomer. Does the enzyme have to be chiral or can it be achiral?

Team Problem

53. Studies have shown that one isomeric form of compound A is an effective agent against certain types of neurodegenerative disorders. Recognize that structure A contains a decalin-type system, as illustrated in structure B, and that the nitrogen can be treated just like a carbon.

(a) Use your model kits to analyze the ring juncture. Make models of the cis as well as the trans ring juncture of structure B. You should have four different models. Identify the stereochemical relation between them as diastereomeric or enantiomeric. Draw the isomers and assign the *R* or *S* configuration to the stereocenters at the ring fusion.
(b) Although the trans ring juncture is the energetically more favorable one, the compound with cis ring juncture is the stereoisomer of structure A that shows biological activity. Make models of structure A that have the cis ring juncture exclusively. Set the stereochemistry of C3 as shown in structure A and vary the center at C6 in relation to that at C3. Again, there are four different models. Draw them and convince yourselves that none of them are enantiomers by assigning the *R* or *S* configuration to all four of the stereocenters in each of the compounds.
(c) The stereoisomer of compound A that shows the greatest biological activity has a cis ring fusion with substituents at C3 and C6 that are both equatorial. Which of the stereoisomers that you drew encompasses these constraints? Identify it by recording the absolute configuration at C3, C4a, C6, and C8a.

Preprofessional Problems

54. Which compound will *not* exhibit optical activity? (Note that these are all Fischer projections.)

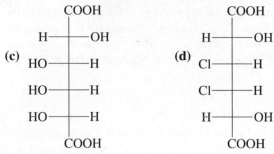

55. The enantiomer of

$$\underset{\underset{CH_3}{|}}{H-\underset{|S}{C}-CH_2CH_3}$$
$$\overset{|}{Cl}$$

(a) is $CH_3CH_2-\underset{\underset{CH_3}{|}}{\overset{\overset{Cl}{|}}{\underset{R}{C}}}-H$

(b) can exist only at low temperatures

(c) is nonisomeric

(d) is incapable of existence

56. The molecule that is of the *R* configuration according to the Cahn-Ingold-Prelog convention is (remember these are Fischer projections):

(a) $H_3C\overset{\overset{H}{|}}{\underset{\underset{CH_3}{|}}{——}}CH_2Cl$ **(b)** $H_3C\overset{\overset{H}{|}}{\underset{\underset{CH_2Br}{|}}{——}}CH_2Cl$ **(c)** $H_3C\overset{\overset{CH_2Br}{|}}{\underset{\underset{H}{|}}{——}}CH_2Cl$

(d) $H_3C\overset{\overset{H}{|}}{\underset{\underset{CH_2Br}{|}}{——}}CH_2F$ **(e)** $H_3C\overset{\overset{CH_2Br}{|}}{\underset{\underset{CH_2Cl}{|}}{——}}CH_2Br$

57. Which compound below is *not* a meso compound?

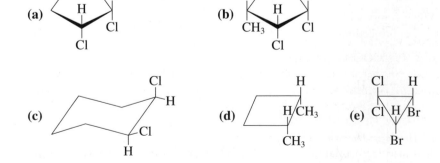

Properties and Reactions of Haloalkanes

Bimolecular Nucleophilic Substitution

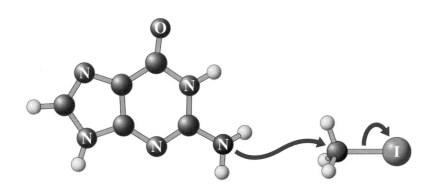

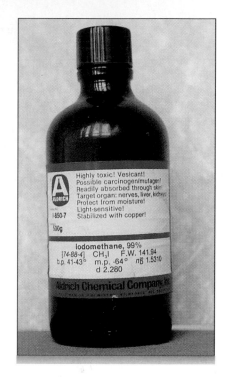

The nucleophilic substitution reaction between iodomethane and a nitrogen atom in the molecule guanine, a constituent of DNA, may cause abnormalities in cell function serious enough to cause cancer.

Organic chemistry provides us with myriad ways to convert one substance into another. The products of these transformations are literally all around us. Recall from Chapter 2, however, that functional groups are the centers of reactivity in organic molecules; before we can make practical use of organic chemistry, we must develop our ability to work with these functional groups. In Chapter 3 we examined halogenation of alkanes, a process by which the carbon–halogen group is introduced into an initially unfunctionalized structure. Where do we go from here?

In this chapter we turn to the chemistry of the products of halogenation, the haloalkanes. We shall see how the polarized carbon–halogen bond governs the reactivity of these substances and how it can be converted into other functional groups. On the basis of the kinetics observed for a common reaction of haloalkanes, we introduce a new mechanism and learn the effects of different solvents on its progress. We shall also learn principles that apply in general to the mechanistic behavior of molecules with polar functional groups. Finally, we shall begin to practice the application of these principles and see the role that they play in many conversions of halogenated organic compounds into other substances, such as amino acids, the building blocks of proteins.

Laboratory Preparation of the Amino Acid Alanine (from Chapter 26)

$$\underset{\underset{\displaystyle CH_3-CH-COOH}{|}}{Br} + NH_3 \longrightarrow \underset{\underset{\displaystyle \underset{\textbf{Alanine}}{CH_3-CH-COOH}}{|}}{NH_2} + HBr$$

We start with the rules for naming haloalkanes.

6-1 Naming the Haloalkanes

We learned in Chapter 2 that alkanes are depicted by the general formula R–H, where R denotes an alkyl group. In a similar way, the **haloalkanes** are represented as R–X, in which X corresponds to a halogen atom.

In the systematic (IUPAC) nomenclature, the halogen is treated as a substituent to the alkane framework.

CH₃I
Iodomethane   **Fluorocyclohexane**   **2-Bromo-2-methyl**propane

The longest alkane chain is numbered so that the first substituent from either end receives the lowest number. As usual, substituents are ordered according to the alphabet. Complex appendages are named according to the rules used for complex alkyl substituents (Section 2-3).

$$ICH_2\underset{\underset{\displaystyle H}{|}}{\overset{\overset{\displaystyle CH_3}{|}}{C}}CH_3$$

1-Iodo-2-methylpropane

(**1-Iodoethyl**)cyclooctane

6-(2-Chloro-2,3,3-trimethylbutyl)undecane

$$CH_3CH_2CH_2CH_2CH_2CHCH_2CH_2CH_2CH_2CH_3$$

Common names are based on the older term *alkyl halide.* For example, the three structures at the beginning of this section have the common names methyl iodide, cyclohexyl fluoride, and *tert*-butyl bromide, respectively. Some chlorinated solvents have common names: for example, carbon tetrachloride, CCl_4; chloroform, $CHCl_3$; and methylene chloride, CH_2Cl_2.

EXERCISE 6-1

Draw the structures of (2-iodoethyl)cyclooctane and 5-butyl-3-chloro-2,2,3-trimethyl-decane.

In summary, haloalkanes are named in accord with the rules that apply to naming the alkanes (Section 2-3), the halo substituent being treated the same as alkyl groups.

6-2 Physical Properties of Haloalkanes

The physical properties of the haloalkanes are quite distinct from those of the corresponding alkanes. To understand these differences, we must consider the size of the halogen substituent and the polarity of the carbon–halogen bond. Let us see how these factors affect bond strength, bond length, molecular polarity, and boiling point.

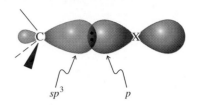

FIGURE 6-1 ——————
Bond between an alkyl carbon and a halogen. The size of the *p* orbital is substantially larger than that shown for X = Cl, Br, or I.

The bond strength of C–X decreases as the size of X increases

The bond between carbon and a halogen is made up mainly by the overlapping of an *sp*³ hybrid orbital on carbon with a *p* orbital on the halogen (Figure 6-1). In the progression from fluorine to iodine in the periodic table, the size of the halogen *p* orbital increases, and the electron cloud around the halogen atom becomes more diffuse. Consequently, its overlap with the carbon orbital and hence the C–X bond strength diminish. For example, the C–X bond dissociation energies in the halomethanes, CH₃X, decrease along the series; at the same time, the C–X bond lengths increase (Table 6-1).

The C–X bond is polarized

A characteristic of the haloalkanes is their polar C–X bond. Recall from Section 1-3 that the halogens are more electronegative than is carbon. Thus, the electron density along the C–X bond is displaced in the direction of X, thereby giving the halogen a partial negative charge (δ^-) and the carbon a partial positive charge (δ^+). How does this bond polarization govern the chemical behavior of the haloalkanes? We shall see, for example, that anions and other electron-rich species can attack the positively polarized carbon atom. Cations and other electron-deficient species, however, attack the halogen.

TABLE 6-1	C–X Bond Lengths and Bond Strengths in CH₃X	
Halo-methane	Bond length (Å)	Bond strength (kcal mol⁻¹)
CH₃F	1.385	110
CH₃Cl	1.784	85
CH₃Br	1.929	71
CH₃I	2.139	57

| The Polar Character of the C–X Bond

Haloalkanes have higher boiling points than the corresponding alkanes

Does the polarity of the C–X bond affect the physical properties of the haloalkanes? Yes, their boiling points are generally higher than those of the corresponding alkanes (Table 6-2). The most important contributor to this effect is Coulombic attraction between the δ^+ and δ^- ends of C–X bond dipoles in the liquid state *(dipole–dipole interaction).*

Boiling points also rise with increasing size of X, the result of increased molecular weight and greater London interactions (Section 2-4). Recall that London forces arise from mutual correlation of electrons among molecules. This effect is strongest when the outer electrons are not held very tightly around the nucleus, as in the heavier atoms. To measure it, we define the **polarizability** of an atom as the degree to

Dipole–Dipole Attraction

TABLE 6-2	Boiling Points of Haloalkanes (R–X)				
	Boiling point (°C)				
R	X = H	F	Cl	Br	I
CH_3	−161.7	−78.4	−24.2	3.6	42.4
CH_3CH_2	−88.6	−37.7	12.3	38.4	72.3
$CH_3(CH_2)_2$	−42.1	−2.5	46.6	71.0	102.5
$CH_3(CH_2)_3$	−0.5	32.5	78.4	101.6	130.5
$CH_3(CH_2)_4$	36.1	62.8	107.8	129.6	157.0
$CH_3(CH_2)_7$	125.7	142.0	182.0	200.3	225.5

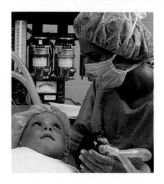

Inhalation anesthetics such as Halothane, $CF_3CHBrCl$, derive their biological activity from the polar nature of their C–X bonds.

which its electron cloud is deformed under the influence of an external electric field. The more polarizable an atom, the more effectively it will enter into London interactions, and the higher will be the boiling point.

To summarize, the halogen orbitals become increasingly diffuse along the series F, Cl, Br, I. Hence, (1) the C–X bond strength decreases; (2) the C–X bond becomes longer; (3) for the same R, the boiling points increase; (4) the polarizability of X becomes greater; and (5) London interactions improve. We shall see next that these interrelated effects also play an important role in the reactions of haloalkanes.

6-3 Nucleophilic Substitution

Haloalkanes often react with substances containing an unshared electron pair. This reagent can be an anion, such as iodide ($:\ddot{I}:^-$), or a neutral species, such as ammonia ($:NH_3$). Each reagent can attack the haloalkane and replace the halide, a process called **nucleophilic substitution.** A great many species are transformed in this way, particularly in solution. The process occurs widely in nature and can be controlled effectively even on an industrial scale. Let us see how it works.

Nucleophiles attack electrophilic centers

We noted that in the polarization of the carbon–halogen bond, the carbon atom acquires a partial positive charge. As a result, this center exhibits a tendency to react with species possessing unshared pairs of electrons: The carbon is said to be **electrophilic** (literally, "electron loving"; *philos*, Greek, loving). In turn, atoms bearing lone pairs are described as **nucleophilic** ("nucleus loving"). By definition, the terms *nucleophile* and *Lewis base* (Section 2-9) are synonymous: *All nucleophiles are Lewis bases.* In common usage, we employ the term *nucleophile* to describe a Lewis basic species that is attacking a *nonhydrogen* electrophilic atom. Nucleophiles, often denoted by the abbreviation Nu, may be negatively charged or neutral, but every nucleophile contains at least one unshared pair of electrons.

CHEMICAL HIGHLIGHT **6-1** **Halogenated Organic Compounds and the Environment**

Today, more than 15,000 halogenated organic compounds are manufactured for commercial use. The largest single industrial use of chlorine is in the preparation of polychloroethene-containing plastics [polyvinyl chloride (PVC), Section 12-14]. More than 6 million tons of PVC-based materials are produced annually in the United States. Other significant applications of chlorinated organics include solvents, industrial lubricants and insulators, herbicides, and insecticides. Some of these substances persist in the environment after disposal and are suspected of causing a variety of adverse health effects in both humans and wildlife. Insecticides such as DDT (Chemical Highlight 3-3), chlordane, and the hexachlorocyclohexanes (e.g., lindane) are known endocrine disruptors, causing reproductive abnormalities in animals. These environmentally very persistent substances are still present in significant amounts, despite their use in the United States having been restricted or banned since the early 1970s. In the same category are the PCBs (polychlorinated biphenyls), widely used as insulating fluids in electrical transmission equipment since the 1920s

PCBs, formerly used as insulating fluids in power transformers, have been banned since the 1970s.

Chlordane

Lindane

Decachlorobiphenyl (a PCB)

until their restriction in 1976, and a major contaminant in the waters of the Great Lakes basin.

Several solutions to the problems of use and disposal of chlorinated organics have been developed. For example, carbon dioxide at moderately elevated pressure and temperature converts into a fluid that can extract the caffeine from coffee beans, replacing dichloromethane for this purpose. Properly controlled incineration can destroy halocarbon wastes with minimal environmental effect. The difficult problem of decontamination of polluted sites is being addressed by several innovative technologies. One of them, bioremediation, employs microorganisms that feed on chlorinated organics. Indeed, by the early 1990s, exotic anaerobic organisms naturally present in the upper Hudson River had removed most of the chlorine from the PCBs in the river sediments, converting these molecules into substances more readily biodegradable by conventional aerobes. Whether human efforts to develop bioremediation into a practical technology will ultimately succeed on a large scale is very much an open question at the present time. For discussions on related topics, see Chemical Highlights 6-2 and 22-1.

The nucleophilic substitution of a haloalkane is described by either of two general equations.

Nucleophilic Substitutions

$$Nu{:}^- \quad + \quad \overset{\delta^+}{R}{-}\overset{\delta^-}{\ddot{X}}{:} \quad \longrightarrow \quad R{-}Nu \quad + \quad {:}\ddot{X}{:}^-$$

Nucleophile   Leaving group

Electrophile

$$Nu{:} \quad + \quad \overset{\delta^+}{R}{-}\overset{\delta^-}{\ddot{X}}{:} \quad \longrightarrow \quad [R{-}Nu]^+ \quad + \quad {:}\ddot{X}{:}^-$$

Nucleophile   Leaving group

Electrophile

In the first example, a negatively charged nucleophile reacts with a haloalkane to yield a neutral substitution product. In the second example, an uncharged Nu produces a positively charged product. In both cases, the group displaced is the halide ion, $:\ddot{X}:^-$, which is called the **leaving group.** Specific examples of these two types of nucleophilic substitution are shown in Table 6.3. As will be the case in many equations and mechanisms that follow, nucleophiles, electrophiles, and leaving groups are shown here in red, blue, and green, respectively. The general term **substrate** (*substratus,* Latin, to have been subjected) is applied to the organic starting material—in this case, the haloalkane—which is the target of attack by a nucleophile.

Nucleophilic substitution exhibits considerable diversity

Note that Table 6-3 depicts only primary and secondary halides. Tertiary substrates behave differently toward these nucleophiles, and even secondary halides may give other products in addition to those of substitution. These reactions will be addressed in Chapter 7. The "cleanest" nucleophilic substitutions are obtained with methyl and primary haloalkanes.

Let us inspect these transformations in greater detail. In reaction 1, a hydroxide ion, typically derived from sodium or potassium hydroxide, displaces chloride from chloromethane to give methanol. This substitution is a general synthetic method for converting a primary haloalkane into an alcohol.

A variation of this transformation is reaction 2. Methoxide ion reacts with iodoethane to give methoxyethane, an example of the synthesis of an ether (Section 9-6).

In reactions 1 and 2, the species attacking the haloalkane is an anionic oxygen nucleophile. Reaction 3 shows that a halide ion may function not only as a leaving group, but also as a nucleophile.

Reaction 4 depicts a carbon nucleophile, cyanide (often supplied as sodium cyanide, $Na^{+-}CN$), and leads to the formation of a new carbon–carbon bond, an important means of lengthening the carbon chain.

Reaction 5 shows the sulfur analog of reaction 2, demonstrating that nucleophiles in the same column of the periodic table react similarly to give analogous products. This conclusion is also borne out by reactions 6 and 7. However, the nucleophiles in these two reactions are *neutral,* and the expulsion of the negatively charged leaving group results in a cationic species, an ammonium or phosphonium salt.

All of the nucleophiles shown in Table 6-3 are quite reactive, but not all for the same reasons. Some are reactive because they are strongly basic (HO^-, CH_3O^-). Oth-

TABLE 6-3	The Diversity of Nucleophilic Substitution			
Reaction number	Substrate	Nucleophile	Product	Leaving group
1.	CH_3Cl Chloromethane	+ HO^-	⟶ CH_3OH Methanol	+ Cl^-
2.	CH_3CH_2I Iodoethane	+ CH_3O^-	⟶ $CH_3CH_2OCH_3$ Methoxyethane	+ I^-
3.	$CH_3\overset{H}{\underset{Br}{C}}CH_2CH_3$ 2-Bromobutane	+ I^-	⟶ $CH_3\overset{H}{\underset{I}{C}}CH_2CH_3$ 2-Iodobutane	+ Br^-
4.	$CH_3\overset{H}{\underset{CH_3}{C}}CH_2I$ 1-Iodo-2-methyl-propane	+ $N{\equiv}C^-$	⟶ $CH_3\overset{H}{\underset{CH_3}{C}}CH_2C{\equiv}N$ 3-Methylbutane-nitrile	+ I^-
5.	Br (Bromocyclohexane)	+ CH_3S^-	⟶ SCH_3 (Methylthiocyclohexane)	+ Br^-
6.	CH_3CH_2I Iodoethane	+ $:NH_3$	⟶ $CH_3CH_2\overset{H}{\underset{H}{\overset{+}{N}}}$ Ethylammonium iodide	+ I^-
7.	CH_3Br Bromomethane	+ $:P(CH_3)_3$	⟶ $CH_3\overset{CH_3}{\underset{CH_3}{\overset{+}{P}}}CH_3$ Tetramethylphosphonium bromide	+ Br^-

Note: Remember that nucleophiles are red, electrophiles are blue, and leaving groups are green.

ers are weak bases (I^-) whose nucleophilicity derives from other characteristics. Notice that, in each example, the leaving group is a halide ion. Halides are unusual in that they may serve as leaving groups as well as nucleophiles (therefore making reaction 3 reversible). However, the same is *not* true of some of the other nucleophiles in Table 6-3 (in particular, the strong bases); the equilibria of their reactions lie strongly in the direction shown. These topics are addressed in Sections 6-8 and 6-9, as are factors that affect the reversibility of displacement reactions. First, however, we shall examine the mechanism of nucleophilic substitution.

EXERCISE 6-2

What are the substitution products of the reaction of 1-bromobutane with (a) $:\ddot{I}:^-$; (b) $CH_3CH_2:\ddot{O}:^-$; (c) N_3^-; (d) $:As(CH_3)_3$; (e) $(CH_3)_2\ddot{S}e$?

EXERCISE 6-3

Suggest starting materials for the preparation of **(a)** $(CH_3)_4N^+I^-$; **(b)** $CH_3SCH_2CH_3$.

In summary, nucleophilic substitution is a fairly general reaction for primary and secondary haloalkanes. The halide functions as the leaving group, and several types of nucleophilic atoms enter into the process.

6-4 Reaction Mechanisms Involving Polar Functional Groups: Using "Electron-Pushing" Arrows

In our consideration of radical halogenation in Chapter 3, we found that a knowledge of its mechanism was helpful in explaining the experimental characteristics of the process. The same will be the case for nucleophilic substitution, and, indeed, virtually every chemical process that we encounter. Nucleophilic substitution is an example of a polar reaction: it includes charged species and polarized bonds. Recall (Chapter 2) that an understanding of electrostatics is essential if we are to comprehend how such processes take place. Opposite charges attract—nucleophiles are attracted to electrophiles—and this principle provides us with a basis for understanding the mechanisms of polar organic reactions. In this section, we shall expand the concept of *electron flow* first introduced in the context of acid-base reactions (Section 2-9) and learn the conventional methods for illustrating polar reaction mechanisms by *moving electrons* from electron-rich to electron-poor sites. In subsequent sections, we shall apply these ideas specifically to nucleophilic substitution.

Curved arrows depict the movement of electrons

As we learned in Section 2-9, acid-base processes require electron movement. Let us briefly examine the Brønsted-Lowry process in which the acid HCl donates a proton to a molecule of water in aqueous solution:

$$HCl + H_2O \rightleftharpoons H_3O^+ + Cl^-$$

In this process, a lone pair originally on the oxygen atom of water has become part of a new bond in the hydronium ion. Likewise, the pair of electrons that constituted the H–Cl bond has shifted to the chlorine atom, converting it into a negatively charged chloride ion. We employ two curved arrows to denote the movement of these two pairs of electrons:

Depiction of a Brønsted-Lowry Acid-Base Reaction by Using Curved Arrows

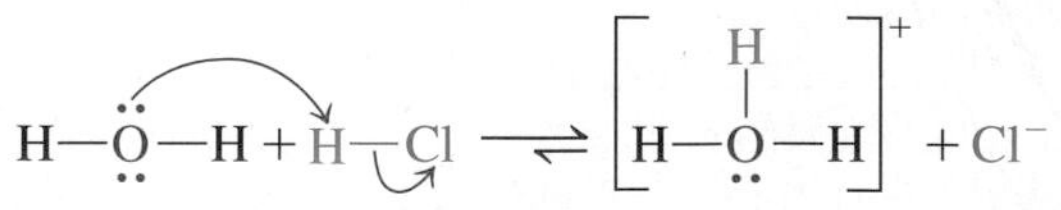

Notice that the arrow starting at the lone pair on oxygen and ending at the hydrogen of HCl does *not* imply that the lone electron pair departs from oxygen completely; it just becomes a *shared* pair between that oxygen atom and the atom to which the arrow points. In contrast, however, the arrow beginning at the H–Cl bond and pointing toward the chlorine atom *does* signify cleavage of the bond; that electron pair becomes separated from hydrogen and ends up entirely on the chloride ion.

Use curved arrows to depict the flow of electrons in each of the following acid-base reactions. **(a)** Hydrogen ion + hydroxide ion; **(b)** fluoride ion + boron trifluoride, BF_3; **(c)** ammonia + hydrogen chloride; **(d)** hydrogen sulfide, H_2S, + sodium methoxide, $NaOCH_3$; **(e)** dimethyloxonium ion, $(CH_3)_2OH^+$, + water; **(f)** the self-ionization of water to give hydronium ion and hydroxide ion.

What about mechanisms in organic chemistry? We shall find that they share many of the features of acid-base reactions. As stated in Section 6-3, a nucleophile is nothing more than a Lewis base that attacks an electrophilic atom other than hydrogen. Electrophiles and Lewis acids are related but not in quite the same way: Lewis acids constitute the *subset* of electrophiles that are at least two electrons short of a closed shell. As we have seen, the carbon atom in a haloalkane possesses a filled outer shell; it is electrophilic by virtue of its partial positive charge, which renders it susceptible to attack by nucleophiles. Nucleophilic substitution is just one of many kinds of processes in which electrophiles and nucleophiles interact. Several examples are shown here, with curved arrows representing electron-pair movement.

Curved-Arrow Representations of Several Common Types of Mechanisms

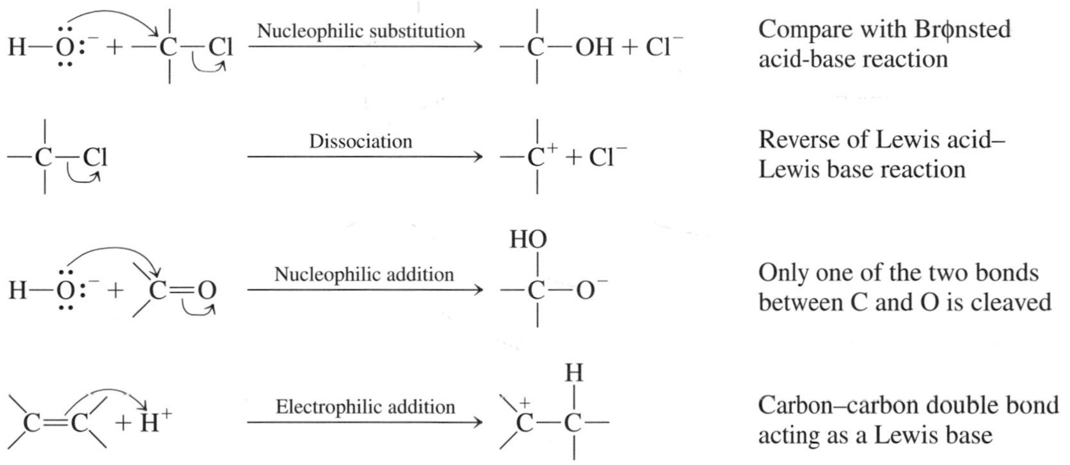

The first and third examples illustrate a characteristic property of electron movement: If an electron pair moves toward an atom, that atom must have a "place to put that electron pair," so to speak. In nucleophilic substitution, the carbon atom in a haloalkane begins with a filled outer shell; another electron pair cannot be added without displacement of the electron pair bonding carbon to halogen. The two electron pairs can be viewed as "flowing" in a synchronous manner: As one pair arrives at the closed-shell atom, the other departs, thereby preventing violation of the octet rule. When you use the curved-arrow method to depict electron movement, *it is absolutely essential to keep in mind the rules for drawing Lewis structures.* Proper use of electron-pushing arrows, however, helps in drawing correct structures, because all electrons are moved to their proper destinations.

There are other types of processes, but, surprisingly, *not that many.* One of the most powerful consequences of studying organic chemistry from a mechanistic point of view is the way in which this approach highlights similarities between types of polar reactions even if the specific atoms and bonds are not the same.

EXERCISE 6-5

Identify the electrophilic and nucleophilic sites in the four mechanisms shown earlier as curved-arrow representations.

EXERCISE 6-6

Propose a curved-arrow depiction of the flow of electrons in the following processes, which will be considered in detail in this chapter and in Chapter 7.

(a) $-\overset{|}{\underset{|}{C}}{}^+ + Cl^- \longrightarrow -\overset{|}{\underset{|}{C}}-Cl$ 　(b) $HO^- + \overset{H}{\underset{|}{\overset{|}{C}}-\overset{|}{\underset{|}{C}}-} \longrightarrow H_2O + \overset{}{\underset{}{C}}{=}\overset{}{\underset{}{C}}$

In summary, curved arrows depict movement of electron pairs in reaction mechanisms. Electrons move from nucleophilic, or Lewis basic, atoms toward electrophilic, or Lewis acidic, sites. If a pair of electrons approaches an atom already containing a closed shell, a pair of electrons must depart from that atom so as not to exceed the maximum capacity of its valence orbitals.

6-5 A First Look at the Nucleophilic Substitution Mechanism: Kinetics

$CH_3Cl + NaOH$

$\downarrow$ H_2O, Δ

$CH_3OH + NaCl$

Many questions can be raised at this stage. What are the kinetics of nucleophilic substitution, and how does this information help us determine the underlying mechanism? What happens with optically active haloalkanes? Can we predict relative rates of substitution? These questions will be addressed in the remainder of this chapter.

When a mixture of chloromethane and sodium hydroxide in water is heated (denoted by the uppercase Greek letter *delta,* Δ, at the right of the arrow in the equation in the margin), a high yield of two compounds—methanol and sodium chloride—is the result. This outcome, however, does not tell us anything about *how* starting materials are converted into products. What experimental methods are available for answering this question?

One of the most powerful techniques employed by chemists is the measurement of the *kinetics* of the reaction (Section 2-8). By comparing the rate of product formation beginning with several different concentrations of the starting materials, we can establish the **rate law** for a chemical process. Let us see what this experiment tells us about the reaction of chloromethane with sodium hydroxide.

The reaction of chloromethane with sodium hydroxide is bimolecular

We can monitor rates by measuring either the disappearance of one of the reactants or the appearance of one of the products. When we apply this method to the reaction of chloromethane with sodium hydroxide, we find that the rate depends on the initial concentrations of *both* of the reagents. For example, doubling the concentration of hydroxide doubles the rate at which the reaction proceeds. Likewise, at a fixed hydroxide concentration, doubling the concentration of chloromethane has the same effect. Doubling the concentrations of both increases the rate by a factor of 4. These results are consistent with a *second-order* process (Section 2-8), which is governed by the following rate equation.

$$Rate = k[CH_3Cl][HO^-] \text{ mol } L^{-1} s^{-1}$$

All the examples given in Table 6-3 exhibit such second-order kinetics: Their rates are directly proportional to the concentration of both substrate and nucleophile.

EXERCISE 6-7

When a solution containing 0.01 M sodium azide ($Na^+N_3^-$) and 0.01 M iodomethane in methanol at 0°C is monitored kinetically, the results reveal that iodide ion is produced at a rate of 3.0×10^{-10} mol L^{-1} s^{-1}. Write the formula of the organic product of this reaction and calculate its rate constant k. What would be the rate of appearance of I^- for each of the following initial concentrations of reactants? **(a)** $[NaN_3] = 0.02$ M; $[CH_3I] = 0.01$ M. **(b)** $[NaN_3] = 0.02$ M; $[CH_3I] = 0.02$ M. **(c)** $[NaN_3] = 0.03$ M; $[CH_3I] = 0.03$ M.

What kind of mechanism is consistent with a second-order rate law? The simplest is one in which the two reactants interact in a single step. We call such a process **bimolecular,** and the general term applied to substitution reactions of this type is **bimolecular nucleophilic substitution,** abbreviated as S_N2 (S stands for substitution, N for nucleophilic, and 2 for bimolecular).

Bimolecular nucleophilic substitution is a concerted, one-step process

Bimolecular nucleophilic substitution is a one-step process: The nucleophile attacks the haloalkane, with simultaneous expulsion of the leaving group. Bond-making takes place *at the same time* as bond-breaking. Because the two events occur "in concert," we call this process a **concerted** reaction.

We can envisage two stereochemically distinct alternatives for such concerted displacements. The nucleophile could approach the substrate from the same side as the leaving group, one group exchanging for the other. This pathway is called **frontside displacement** (Figure 6-2). The second possibility is a **backside displacement,** in which the nucleophile approaches carbon from the side opposite the leaving group

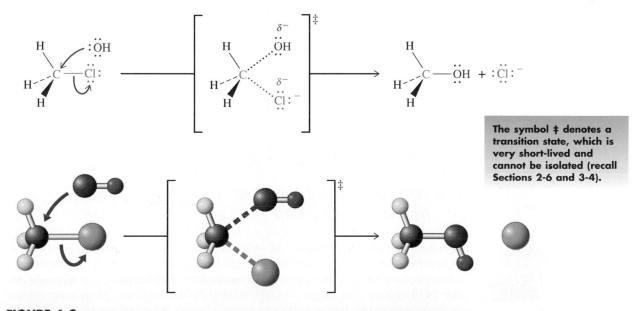

> The symbol ‡ denotes a transition state, which is very short-lived and cannot be isolated (recall Sections 2-6 and 3-4).

FIGURE 6-2

Frontside nucleophilic substitution. The concerted nature of bond-making (to OH) and bond-breaking (from Cl) is indicated by the dotted lines.

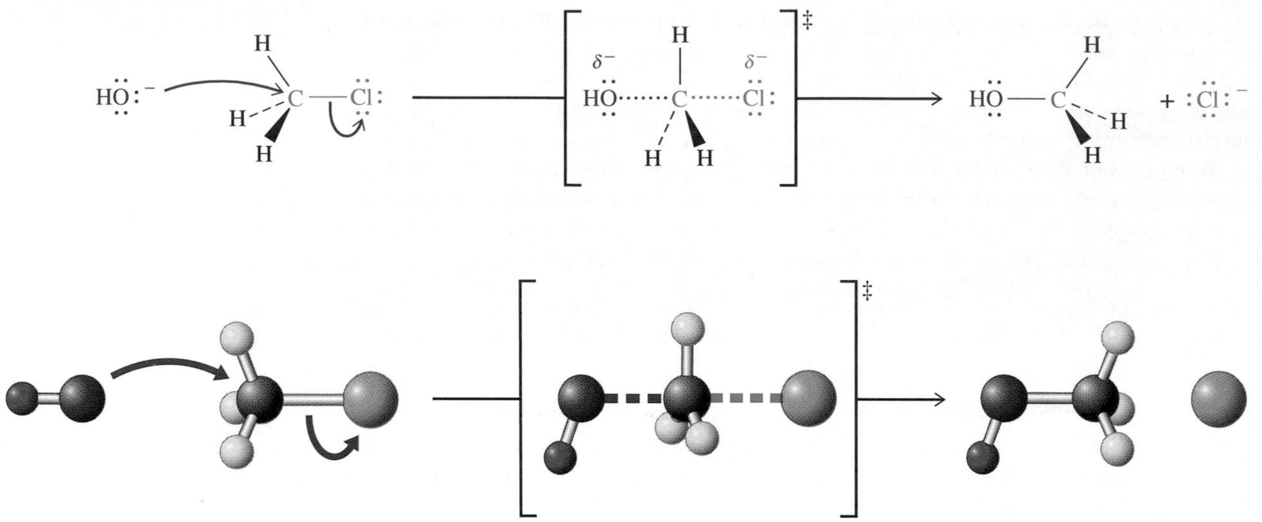

FIGURE 6-3

Backside nucleophilic substitution. Attack is from the side *opposite* the leaving group.

(Figure 6-3). In both equations, we use curved arrows to denote the *movement of electron pairs*. An electron pair from the negatively charged hydroxide oxygen moves toward carbon, creating the C–O bond, while that of the C–Cl linkage shifts onto chlorine, thereby expelling the latter as $:\ddot{C}l:^-$. In either of the two respective transition states, the negative charge is distributed over both the oxygen and the chlorine atoms.

EXERCISE 6-8

Draw representations of the hypothetical frontside and backside displacement mechanisms for the S_N2 reaction of sodium iodide with 2-bromobutane (Table 6-3). Use arrows like those shown in Figures 6-2 and 6-3 to represent electron-pair movement.

In summary, the reaction of chloromethane with hydroxide to give methanol and chloride, as well as the related transformations of a variety of nucleophiles with haloalkanes, are examples of the bimolecular process known as the S_N2 reaction. Two single-step mechanisms—frontside attack and backside attack—may be envisioned for the reaction. Both are concerted processes, consistent with the second-order kinetics obtained experimentally. Can we distinguish between the two? To answer this question, we return to a topic that we have considered in detail: stereochemistry.

6-6 Frontside or Backside Attack? Stereochemistry of the S_N2 Reaction

When we compare the structural drawings in Figures 6-2 and 6-3 with respect to the arrangement of their component atoms in space, we note immediately that, in the first conversion, the three hydrogens stay put and to the left of the carbon, whereas, in the second, they have "moved" to the right. In fact, the two methanol pictures are related as object and mirror image. In this example, the two are superimposable and therefore indistinguishable—properties of an achiral molecule. The situation is entirely different for a chiral haloalkane in which the electrophilic carbon is a stereocenter.

The S$_N$2 reaction is stereospecific

Consider the reaction of (*S*)-2-bromobutane with iodide ion. Frontside displacement should give rise to 2-iodobutane with the *same* configuration as that of the starting material; backside displacement should furnish a product with the *opposite* configuration.

What is actually observed? It is found that (*S*)-2-bromobutane gives (*R*)-2-iodobutane on treatment with iodide: *This and all other S$_N$2 reactions proceed with* **inversion of configuration.** A process whose mechanism requires that each stereoisomer of the starting material transform into a specific stereoisomer of product is described as **stereospecific.** The S$_N$2 reaction is therefore stereospecific, proceeding by a backside displacement mechanism to give inversion of configuration at the site of the reaction.

Stereochemistry of the Backside Displacement Mechanism for S$_N$2 Reactions

I$^-$ H$_3$C $\cdots$ C $-$ Br $\longrightarrow$ $\left[\delta^- \text{I} \cdots\cdots \text{C} \cdots\cdots \text{Br}^{\delta^-} \right]^{\ddagger}$ $\longrightarrow$ I $-$ C $\cdots$ CH$_3$ + Br$^-$

S	Backside displacement	**R**
(Chiral and optically active)		(Chiral and optically active; configuration inverted)

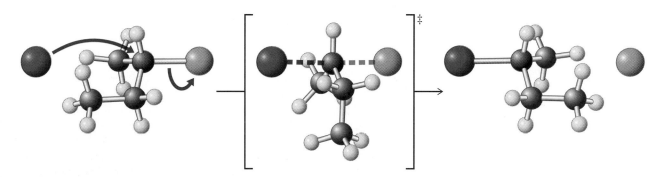

EXERCISE 6-9

Write the products of the following S$_N$2 reactions. **(a)** (*R*)-3-chloroheptane + Na$^+$ $^-$SH; **(b)** (*S*)-2-bromooctane + N(CH$_3$)$_3$; **(c)** (3*R*,4*R*)-4-iodo-3-methyloctane + K$^+$ $^-$SeCH$_3$.

EXERCISE 6-10

Write the structures of the products of the S$_N$2 reactions of cyanide ion with **(a)** *meso*-2,4-dibromopentane (double S$_N$2 reaction); **(b)** *trans*-1-iodo-4-methylcyclohexane.

The transition state of the S$_N$2 reaction can be described in an orbital picture

The transition state for the S$_N$2 reaction can be described in orbital terms, as shown in Figure 6-4. As the nucleophile approaches the back lobe of the *sp*^{3} hybrid orbital used by carbon to bind the halogen atom, the rest of the molecule becomes planar at the transition state by changing the hybridization at carbon to *sp*^{2}. The negative charge is no longer located entirely on the nucleophile; it is also partly on the leaving group.

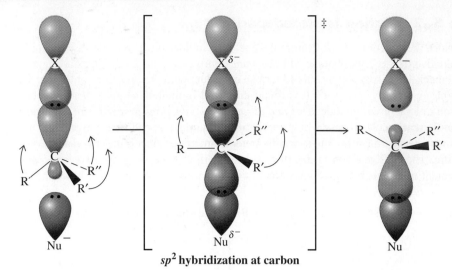

sp² **hybridization at carbon**

FIGURE 6-4
Molecular-orbital description of the S$_N$2 reaction. The process is reminiscent of the inversion of an umbrella exposed to gusty winds.

As the reaction proceeds to products, the inversion motion is completed, the carbon returns to the tetrahedral *sp³* configuration, and the leaving group becomes a fully charged anion. A depiction of the course of the reaction using a potential energy–reaction coordinate diagram is shown in Figure 6-5.

6-7 Consequences of Inversion in S$_N$2 Reactions

What are the consequences of the inversion of stereochemistry in the S$_N$2 reaction? Because the reaction is stereospecific, we can design ways to use displacement reactions to synthesize a desired stereoisomer.

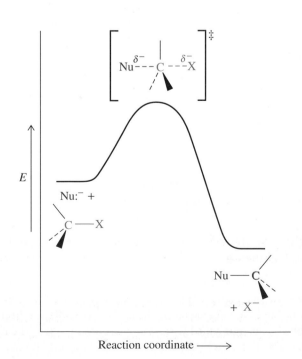

FIGURE 6-5
Potential energy diagram for an S$_N$2 reaction. The process takes place in a single step, with a single transition state.

We can synthesize a specific enantiomer by using S$_N$2 reactions

Consider the conversion of 2-bromooctane into 2-octanethiol in its reaction with hydrogen sulfide ion, HS⁻. If we were to start with optically pure *R* bromide, we would obtain only *S* thiol and none of its *R* enantiomer.

Inversion of Configuration of an Optically Pure Compound by S$_N$2 Reaction

HS⁻ +
      H
      C—Br
CH$_3$(CH$_2$)$_4$CH$_2$ CH$_3$

⟶   HS—C   + Br⁻
      H
      CH$_2$(CH$_2$)$_4$CH$_3$
      CH$_3$

(*R*)-2-Bromooctane
([α] = −34.6)

(*S*)-2-Octanethiol
([α] = +36.4)

> **Color code for priorities** (see Section 5-3)
> Highest: red
> Second highest: blue
> Third highest: green
> Lowest: black

But what if we wanted to convert (*R*)-2-bromooctane into the *R* thiol? One technique uses a sequence of *two* S$_N$2 reactions, each resulting in inversion of configuration at the stereocenter. For example, an S$_N$2 reaction with iodide would first generate (*S*)-2-iodooctane. We would then use this halide with an inverted configuration as the substrate in a second displacement, now with HS⁻ ion, to furnish the *R* thiol. This double inversion sequence of two S$_N$2 processes gives us the result we desire, a net **retention of configuration.**

Using Double Inversion to Give Net Retention of Configuration

      H
      C—Br
CH$_3$(CH$_2$)$_4$CH$_2$ CH$_3$
  $\xrightarrow[{-Br}]{I^-}$
      H
I—C
      CH$_2$(CH$_2$)$_4$CH$_3$
      CH$_3$
  $\xrightarrow[{-I}]{HS^-}$
      H
      C—SH
CH$_3$(CH$_2$)$_4$CH$_2$ CH$_3$

(*R*)-2-Bromooctane
([α] = −34.6)

(*S*)-2-Iodooctane
([α] = +46.3)

(*R*)-2-Octanethiol
([α] = −36.4)

EXERCISE 6-11

As we saw for carvone (Chapter 5, Problem 33), enantiomers can sometimes be distinguished by odor and flavor. 3-Octanol and some of its derivatives are examples: The dextrorotatory compounds are found in natural peppermint oil, whereas their (−) counterparts contribute to the essence of lavender. Show how you would synthesize optically pure samples of each enantiomer of 3-octyl acetate, starting with (*S*)-3-iodooctane.

          O
          ‖
         OCCH$_3$
         |
CH$_3$CH$_2$CHCH$_2$CH$_2$CH$_2$CH$_2$CH$_3$
3-Octyl acetate

EXERCISE 6-12

Treatment of (*S*)-2-iodooctane with NaI causes the optical activity of the starting material to disappear. Explain.

In substrates bearing more than one stereocenter, inversion will take place *only* at the carbons that undergo reaction with the incoming nucleophile. Note that the reaction of (2*S*,4*R*)-2-bromo-4-chloropentane with excess cyanide ion results in a meso product.

S$_N$2 Reactions of Molecules with Two Stereocenters

(structure 2S,4R) + $^-$CN (Excess) →(Ethanol)→ (structure 2R,4S: Meso) + Br$^-$ + Cl$^-$

(structure 2S,3R) + I$^-$ →(Propanone (acetone))→ (structure 2R,3R) + Br$^-$

EXERCISE 6-13

As an aid in the prediction of stereochemistry, organic chemists often use the guideline "diastereomers produce diastereomers." Replace the starting compound in each of the two preceding examples with one of its diastereomers, and write the product of S$_N$2 displacement with the nucleophile shown. Are the resulting structures in accord with this "rule"?

Similarly, nucleophilic substitution of a substituted halocycloalkane may change the stereochemical relation between the substituents.

(cyclohexane structure with Br) →(NaI, propanone (acetone))→ (cyclohexane structure with I) + NaBr

Cis → **Trans**

In summary, inversion of configuration in the S$_N$2 reaction has distinct stereochemical consequences. Optically active substrates give optically active products, unless the nucleophile and the leaving group are the same or meso compounds are formed. In cyclic systems, cis and trans stereochemical relations may be interconverted.

6-8 S$_N$2 Reactivity and Leaving-Group Ability

The relative facility of S$_N$2 displacements depends on a variety of factors, including the nature of the leaving group, the relative reactivity of the nucleophile, and the structure of the alkyl group in the substrate. We can enhance our understanding of the mechanism of the process by measuring relative reaction rates as we systematically vary each component. We shall begin with leaving groups, first considering the halides and then generalizing to other groups that can function in this capacity as well. Subsequent sections will address the nucleophile and the alkyl part of the substrate.

Leaving-group ability is a measure of the ease of its displacement

As a general rule, nucleophilic substitution will occur only when the group being displaced, X, is readily able to depart, taking with it the electron pair of the C–X bond. Are there structural features that might allow us to predict, at least qualitatively, whether a leaving group is "good" or "bad"? Not surprisingly, the relative ease with which it can be displaced, its **leaving-group ability,** can be correlated with its capacity to accommodate a negative charge. Remember that a certain amount of negative charge is transferred to the leaving group in the transition state of the reaction (Figure 6-4).

For the halogens, leaving-group ability increases along the series from fluorine to iodine. Thus, iodide is regarded as a "good" leaving group; fluoride, however, is so "poor" that S_N2 reactions of fluoroalkanes are rarely observed.

Leaving-Group Ability

$$I^- > Br^- > Cl^- > F^-$$
Best   Worst

EXERCISE 6-14

Predict the product of the reaction of 1-chloro-6-iodohexane with one equivalent of sodium methylselenide ($Na^{+}{}^{-}SeCH_3$).

Halides are not the only groups that can be displaced by nucleophiles in S_N2 reactions. Other examples of good leaving groups are sulfur derivatives of the type $ROSO_3^-$ and RSO_3^-, such as methyl sulfate ion, $CH_3OSO_3^-$, and various sulfonate ions. Alkyl sulfate and sulfonate leaving groups are used so often that trivial names, such as mesylate, triflate, and tosylate, have found their way into the chemical literature.

Sulfate and Sulfonate Leaving Groups

| Methyl sulfate ion | Methanesulfonate ion (Mesylate ion) | Trifluoromethanesulfonate ion (Triflate ion) | 4-Methylbenzenesulfonate ion (*p*-Toluenesulfonate ion, tosylate ion) |

Weak bases are good leaving groups

Is there some characteristic property that distinguishes good leaving groups from poor ones? Yes: *Leaving-group ability is inversely related to base strength.* Weak bases are best able to accommodate negative charge and are the best leaving groups. Among the halides, iodide is the weakest base and therefore the best leaving group in the series. Sulfates and sulfonates are weak bases as well. Table 6-4 (see page 228) lists a variety of species in ascending order of base strength, which is quantified by a **basicity constant,** K_b.

Is there a way to recognize weak bases readily? The weaker X^- is as a base, the stronger is its conjugate acid HX. Therefore, *good leaving groups are the conjugate bases of strong acids.* This rule applies to the four halides: HF is the weakest of the conjugate acids, HCl is stronger, and HBr and HI are stronger still. In our review of acids and bases (Section 2-9), we outlined some guidelines that may be used to evaluate relative strengths of acids. It is perhaps worthwhile to repeat them: The strength of an acid, HA, increases with the following structural features.

1. The increasing *size* of A in the procession down a column in the periodic table. Therefore, the acid strengths of the hydrogen halides follow the order HI >

TABLE 6-4	Base Strengths and Leaving Groups	
Leaving group	K_b
Good leaving groups (weaker bases)	
I^-	6.3×10^{-20}
HSO_4^-	1.0×10^{-19}
Br^-	2.0×10^{-19}
Cl^-	6.3×10^{-17}
H_2O	2.0×10^{-16}
$CH_3SO_3^-$	6.3×10^{-16}
Poor leaving groups (stronger bases)	
F^-	1.6×10^{-11}
$CH_3CO_2^-$	5.0×10^{-10}
NC^-	1.6×10^{-5}
CH_3S^-	1.0×10^{-4}
CH_3O^-	32
HO^-	50
H_2N^-	1.0×10^{21}
H_3C^-	$\sim 1.0 \times 10^{36}$

$HBr > HCl > HF$. Consequently, I^- is the weakest conjugate base and, as noted earlier, the best leaving group, whereas F^-, the strongest base in the series, is difficult to displace in S_N2 processes.

2. The ability of the conjugate base, A^-, to accommodate the negative charge in either or both of two ways:

(a) The increasing *electronegativity* of A in the procession from left to right across a row in the periodic table. For example, the decreasing order of acidity in the series $HF > H_2O > NH_3 > CH_4$ parallels the decreasing electronegativity of A. The leaving-group ability of the corresponding conjugate bases follows suit: F^- is a poor leaving group, but HO^- is worse—much too strong a base to depart in ordinary S_N2 processes. The even stronger bases NH_2^- and CH_3^- are similarly incapable of leaving in displacement reactions.

(b) The resonance in A^- that allows delocalization of charge over several atoms. The sulfate and sulfonate anions illustrated earlier in this section possess significant resonance stabilization, thus explaining the considerable strength of their conjugate acids (Table 2-6) and their own value as excellent leaving groups.

EXERCISE 6-15

Predict the relative acidities within each of the following groups. **(a)** H_2S, H_2Se; **(b)** PH_3, H_2S; **(c)** $HClO_3$, $HClO_2$; **(d)** HBr, H_2Se; **(e)** NH_4^+, H_3O^+. Within each of the groups, identify the conjugate bases and predict their relative leaving-group abilities.

EXERCISE 6-16

Predict the relative basicities within each of the following groups. **(a)** ^-OH, ^-SH; **(b)** $^-PH_2$, ^-SH; **(c)** I^-, ^-SeH; **(d)** $HOSO_2^-$, $HOSO_3^-$. Predict the relative acidities of the conjugate acids within each group.

In summary, the leaving-group ability of a substituent is roughly proportional to the strength of its conjugate acid. Both depend on the ability of the leaving group to accommodate negative charge. In addition to the halides Cl^-, Br^-, and I^-, sulfates and sulfonates (such as methane- and 4-methylbenzenesulfonates) are good leaving groups. Good leaving groups are weak bases, the conjugate bases of strong acids. We shall return to uses of sulfates and sulfonates as leaving groups in synthesis in Section 9-4.

6-9 Effect of Nucleophilicity on the S_N2 Reaction

Now that we have looked at the effect of the leaving group, let us turn to a consideration of nucleophiles. How can we predict their relative nucleophilic strength, their **nucleophilicity?** We shall see that nucleophilicity depends on a variety of factors: charge, basicity, solvent, polarizability, and the nature of substituents. To grasp the relative importance of these effects, let us analyze the outcome of a series of comparative experiments.

Increasing negative charge increases nucleophilicity

If the same nucleophilic atom is used, does charge play a role in the reactivity of a given nucleophile as determined by the rate of its S_N2 reaction? The following experiments answer this question.

EXPERIMENT 1

$$CH_3Cl + HO^- \longrightarrow CH_3OH + Cl^- \qquad \text{Fast}$$
$$CH_3Cl + H_2O \longrightarrow CH_3OH_2^+ + Cl^- \qquad \text{Very slow}$$

EXPERIMENT 2

$$CH_3Cl + H_2N^- \longrightarrow CH_3NH_2 + Cl^- \qquad \text{Very fast}$$
$$CH_3Cl + H_3N \longrightarrow CH_3NH_3^+ + Cl^- \qquad \text{Slower}$$

Conclusion. Of a pair of nucleophiles containing the same reactive atom, the species with a negative charge is the more powerful nucleophile. Or, of a base and its conjugate acid, the base is always more nucleophilic. This finding is intuitively very reasonable. Because nucleophilic attack is characterized by the formation of a bond with an electrophilic carbon center, the more negative the attacking species, the faster the reaction should be.

EXERCISE 6-17

Predict which member in each of the following pairs is a better nucleophile. **(a)** HS^- or H_2S; **(b)** CH_3SH or CH_3S^-; **(c)** CH_3NH^- or CH_3NH_2; **(d)** HSe^- or H_2Se.

Nucleophilicity decreases to the right in the periodic table

Experiments 1 and 2 compared pairs of nucleophiles containing the same nucleophilic element (e.g., oxygen in H_2O versus HO^- and nitrogen in H_3N versus H_2N^-). What about nucleophiles of similar structure but with different nucleophilic atoms? Let us examine the elements along one row of the periodic table.

EXPERIMENT 3

$$CH_3CH_2Br + H_3N \longrightarrow CH_3CH_2NH_3^+ + Br^- \qquad \text{Fast}$$
$$CH_3CH_2Br + H_2O \longrightarrow CH_3CH_2OH_2^+ + Br^- \qquad \text{Very slow}$$

EXPERIMENT 4

$$CH_3CH_2Br + H_2N^- \longrightarrow CH_3CH_2NH_2 + Br^- \qquad \text{Very fast}$$
$$CH_3CH_2Br + HO^- \longrightarrow CH_3CH_2OH + Br^- \qquad \text{Slower}$$

Conclusion. Nucleophilicity again appears to correlate with basicity: The more basic species is the more reactive nucleophile. Therefore, in the procession from the left to the right of the periodic table, nucleophilicity decreases. The approximate order of reactivity for nucleophiles in the first row is

$$H_2N^- > HO^- > NH_3 > F^- > H_2O$$

EXERCISE 6-18

In each of the following pairs of molecules, predict which is the more nucleophilic. **(a)** Cl^- or CH_3S^-; **(b)** $P(CH_3)_3$ or $S(CH_3)_2$; **(c)** $CH_3CH_2Se^-$ or Br^-; **(d)** H_2O or HF.

Should basicity and nucleophilicity be correlated?

The parallels between nucleophilicity and basicity first described in Section 6-3 are intuitively reasonable: Strong bases typically make good nucleophiles. However, a fundamental difference between the two properties is based on how they are measured. Basicity is a *thermodynamic* property, measured by an equilibrium constant:

$$A^- + H_2O \overset{K}{\rightleftharpoons} AH + HO^- \qquad K = \text{equilibrium constant}$$

In contrast, nucleophilicity is a *kinetic* phenomenon, quantified by comparing rates of reactions:

$$Nu^- + R—X \xrightarrow{k} Nu—R + X^- \qquad k = \text{rate constant}$$

Despite these inherent differences, we have observed good correlation between basicity and nucleophilicity in the cases of charged versus neutral nucleophiles along a row of the periodic table. What happens if we look at nucleophiles in a column of the periodic table?

Solvation impedes nucleophilicity

If it is a general rule that nucleophilicity correlates with basicity, then the elements considered from top to bottom of a column of the periodic table should show decreasing nucleophilic power. Recall (Section 6-8) that basicity decreases in an analogous fashion. To test this prediction, let us consider another series of experiments.

EXPERIMENT 5

$$CH_3CH_2CH_2O\overset{\displaystyle O}{\underset{\displaystyle O}{\overset{\|}{\underset{\|}{S}}}}CH_3 + Cl^- \xrightarrow{CH_3OH} CH_3CH_2CH_2Cl + {}^-O_3SCH_3 \qquad \textbf{Slow}$$

$$CH_3CH_2CH_2O\overset{\displaystyle O}{\underset{\displaystyle O}{\overset{\|}{\underset{\|}{S}}}}CH_3 + Br^- \xrightarrow{CH_3OH} CH_3CH_2CH_2Br + {}^-O_3SCH_3 \qquad \textbf{Faster}$$

$$CH_3CH_2CH_2O\overset{\displaystyle O}{\underset{\displaystyle O}{\overset{\|}{\underset{\|}{S}}}}CH_3 + I^- \xrightarrow{CH_3OH} CH_3CH_2CH_2I + {}^-O_3SCH_3 \qquad \textbf{Fastest}$$

EXPERIMENT 6

$$CH_3CH_2CH_2Br + CH_3O^- \xrightarrow{CH_3OH} CH_3CH_2CH_2OCH_3 + Br^- \qquad \textbf{Not very fast}$$

$$CH_3CH_2CH_2Br + CH_3S^- \xrightarrow{CH_3OH} CH_3CH_2CH_2SCH_3 + Br^- \qquad \textbf{Very fast}$$

Conclusion. Nucleophilicity *increases* in the procession down the periodic table, a trend *directly opposing* that expected from the basicity of the nucleophiles tested. Sulfur nucleophiles are more reactive than the analogous oxygen systems but less so than their selenium counterparts. Similarly, among the halides, iodide is the fastest, although it is the weakest base.

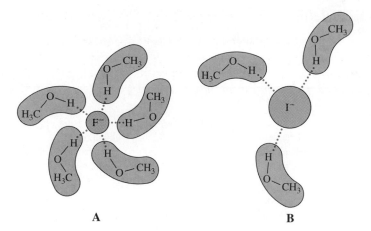

A   B

FIGURE 6-6

Approximate representation of the difference between solvation of (A) a small anion (F$^-$) and (B) a large anion (I$^-$). The tighter solvent shell around the smaller F$^-$ impedes its ability to participate in nucleophilic substitution reactions.

How can these trends be explained? What accounts for the increasing nucleophilicity of *negatively charged nucleophiles* from the top to the bottom of a column of the periodic table? An important explanation is the interaction of solvent with the different anions.

When a solid dissolves, the intermolecular forces that held it together (Section 2-4) are replaced by interactions between molecules and solvent. In solution, molecules or ions are surrounded by solvent molecules; they are said to be **solvated.** Salts dissolve only in polar solvents, because only in these solvents is there enough solvation to separate the oppositely charged ions.

How does the solvation affect the strength of a nucleophile? Generally, solvation weakens the nucleophile by forming a shell of solvent molecules around the nucleophile and impeding its ability to attack an electrophile. Moreover, *smaller anions are more tightly solvated than are larger ones* because their charge is more concentrated. Figure 6-6 depicts this effect for methanol. The smaller fluoride ion is much more heavily solvated than is the larger iodide. Is this true in other solvents as well?

Protic and aprotic solvents: the effect of hydrogen bonding

In some polar solvents, such as methanol, ethanol, and water, a hydrogen is attached to an electronegative atom, Y. These solvents contain highly polarized $^{\delta+}$H–Y$^{\delta-}$ bonds, in which the hydrogen has protonlike character and can interact particularly strongly with anionic nucleophiles (Figure 6-6). (We shall study these interactions, called **hydrogen bonds,** more closely in Chapter 8.) Such solvents, called **protic,** are commonly used in nucleophilic substitutions.

Other solvents useful in S$_N$2 reactions are highly polar but **aprotic:** They lack positively polarized hydrogens. Polar, aprotic solvents such as propanone (acetone) are capable of dissolving salts. However, because they do not form hydrogen bonds, these solvent molecules solvate anionic nucleophiles relatively weakly. The result is that the reactivity of the nucleophile is raised, sometimes dramatically. For example, bromomethane reacts with potassium iodide 500 times as fast in propanone (acetone) as in methanol. Several other polar aprotic solvents are shown in Table 6-5; all lack protons capable of hydrogen bonding.

Table 6-6 (see page 232) compares the rates of S$_N$2 reactions of iodomethane with chloride in three protic solvents—methanol, formamide, and *N*-methylformamide—and one *aprotic* solvent, *N,N*-dimethylformamide (DMF). The rate of reaction in DMF is more than a million times as great as it is in methanol. Recall that small anionic nucleophiles are most heavily solvated by protic solvents, explaining why the nucleophilicity of the halides increases from top to bottom of the periodic table *in protic*

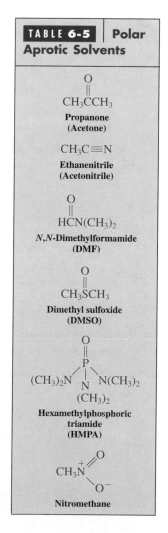

TABLE 6-5 | Polar Aprotic Solvents

$$CH_3\overset{\overset{\displaystyle O}{\|}}{C}CH_3$$

Propanone (Acetone)

$$CH_3C{\equiv}N$$

Ethanenitrile (Acetonitrile)

$$H\overset{\overset{\displaystyle O}{\|}}{C}N(CH_3)_2$$

***N,N*-Dimethylformamide (DMF)**

$$CH_3\overset{\overset{\displaystyle O}{\|}}{S}CH_3$$

Dimethyl sulfoxide (DMSO)

$$(CH_3)_2N\overset{\overset{\displaystyle O}{\|}}{\underset{\underset{\displaystyle (CH_3)_2}{N}}{P}}N(CH_3)_2$$

Hexamethylphosphoric triamide (HMPA)

$$CH_3\overset{+}{N}\overset{\nearrow O}{\underset{\searrow O^-}{}}$$

Nitromethane

TABLE 6-6	**Relative Rates of S$_N$2 Reactions of Iodomethane with Chloride Ion in Various Solvents** $$CH_3I + Cl^- \xrightarrow[k_{rel}]{\text{Solvent}} CH_3Cl + I^-$$			
	Solvent			
Formula	**Name**		**Classification**	**Relative rate** (k_{rel})
CH$_3$OH	Methanol		Protic	1
HCONH$_2$	Formamide		Protic	12.5
HCONHCH$_3$	*N*-Methylformamide		Protic	45.3
HCON(CH$_3$)$_2$	*N,N*-Dimethylformamide		Aprotic	1,200,000

solvents. Switching to an *aprotic* solvent reduces the solvation and increases the re-activity of all anions, but the effect is greatest for small anionic nucleophiles. Indeed, in polar aprotic solvents, the opposing factors of base strength and polarizability sub-stantially reduce the differences in nucleophilicity among the halides and may even lead to inversion of the reactivity order under some conditions.

Increasing polarizability improves nucleophilic power

FIGURE 6-7

Comparison of I$^-$ and F$^-$ in the S$_N$2 reaction. (A) The larger iodide is a better nucleophile, because its polarizable 5p orbital is distorted toward the electrophilic carbon atom. (B) The tight, less polarizable 2p orbital on fluoride does not interact as effectively with the electrophilic carbon at a point along the reaction coordinate comparable to the one for (A).

The solvation effects just described should be very pronounced only for charged nu-cleophiles. Nevertheless, the degree of nucleophilicity increases down the periodic table, even for *uncharged nucleophiles,* for which solvent effects should be much less strong: For example, H$_2$Se > H$_2$S > H$_2$O, and PH$_3$ > NH$_3$. Therefore, there must be an additional explanation for the observed trend in nucleophilicity.

This explanation lies in the polarizability of the nucleophile. Larger elements have larger, more diffuse, and more polarizable electron clouds. These electron clouds al-low for more effective orbital overlap in the S$_N$2 transition state (Figure 6-7). The re-sult is a lower transition-state energy and faster nucleophilic substitution.

EXERCISE 6-19

Which species is more nucleophilic: **(a)** CH$_3$SH or CH$_3$SeH; **(b)** (CH$_3$)$_2$NH or (CH$_3$)$_2$PH?

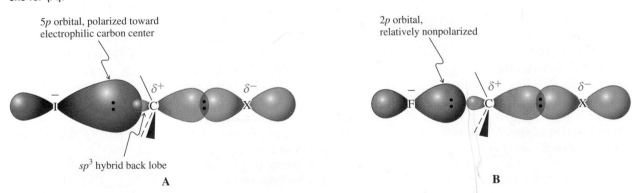

5p orbital, polarized toward electrophilic carbon center

sp^3 hybrid back lobe

A

2p orbital, relatively nonpolarized

B

Sterically hindered nucleophiles are poorer reagents

We have seen that the bulk of the surrounding solvent may adversely affect the power of a nucleophile, another example of steric hindrance (Section 2-7). Such hindrance may also be built into the nucleophile itself in the form of bulky substituents. The effect on the rate of reaction can be seen in Experiment 7.

EXPERIMENT 7

$$CH_3I + CH_3O^- \longrightarrow CH_3OCH_3 + I^- \quad \text{Fast}$$

$$CH_3I + CH_3\overset{\overset{\displaystyle CH_3}{|}}{\underset{\underset{\displaystyle CH_3}{|}}{C}}O^- \longrightarrow CH_3O\overset{\overset{\displaystyle CH_3}{|}}{\underset{\underset{\displaystyle CH_3}{|}}{C}}CH_3 + I^- \quad \text{Slower}$$

Conclusion. Sterically bulky nucleophiles react more slowly.

EXERCISE 6-20

Which of the two nucleophiles in the following pairs will react more rapidly with bromomethane?

(a) CH_3S^- or $CH_3\overset{\overset{\displaystyle CH_3}{|}}{C}HS^-$ (b) $(CH_3)_2NH$ or $(CH_3\overset{\overset{\displaystyle CH_3}{|}}{C}H)_2NH$

Nucleophilic substitutions may be reversible

The halide ions Cl^-, Br^-, and I^- are both good nucleophiles and good leaving groups. Therefore, their S_N2 reactions are reversible. In such situations, the directions in which the equilibria lie must be determined experimentally. Bond-strength calculations alone cannot be used to determine the thermodynamics, because these calculations do not take into account the energetics of ion solvation, a complex problem. For example, in propanone (acetone), the reactions between lithium chloride and primary bromo- and iodoalkanes are reversible, with the equilibrium on the side of the chloroalkane products:

$$CH_3CH_2CH_2CH_2I + LiCl \underset{}{\overset{\text{Propanone}}{\underset{\text{(acetone)}}{\rightleftharpoons}}} CH_3CH_2CH_2CH_2Cl + LiI$$

This result correlates with the relative bond strengths in the product and starting material ($DH°_{C-Cl} = 80$ kcal mol^{-1} versus $DH°_{C-I} = 53$ kcal mol^{-1}). However, this equilibrium may be driven in the reverse direction by a simple "trick": Whereas all of the lithium halides are soluble in acetone, solubility of the sodium halides decreases dramatically in the order $NaI > NaBr > NaCl$, the last being virtually insoluble in this solvent. Indeed, the reaction between NaI and a primary or secondary chloroalkane in acetone is *completely* driven to the side of the iodoalkane (the reverse of the reaction just shown) by the precipitation of NaCl:

$$CH_3CH_2CH_2CH_2Cl + NaI \underset{}{\overset{\text{Propanone}}{\underset{\text{(acetone)}}{\rightleftharpoons}}} CH_3CH_2CH_2CH_2I + NaCl\downarrow$$

**Insoluble
in acetone**

TABLE 6-7	Relative Rates of Reaction of Various Nucleophiles with Iodomethane in Methanol	
Nucleophile	**Relative rate**
CH_3OH	1
NO_3^-	~ 32
F^-	500
$CH_3\overset{\overset{\displaystyle O}{\|}}{C}O^-$	20,000
Cl^-	23,500
$(CH_3CH_2)_2S$	219,000
NH_3	316,000
CH_3SCH_3	347,000
N_3^-	603,000
Br^-	617,000
CH_3O^-	1,950,000
CH_3SeCH_3	2,090,000
CN^-	5,010,000
$(CH_3CH_2)_3As$	7,940,000
I^-	26,300,000
HS^-	100,000,000

The direction of the equilibrium in reaction 3 of Table 6-3 may be manipulated in exactly the same way. However, when the nucleophile in an S_N2 reaction is a strong base (e.g., HO^- or CH_3O^-; see Table 6-4), it will be incapable of acting as a leaving group. In such cases, K_{eq} will be very large, and displacement will essentially be an irreversible process (Table 6-3, reactions 1 and 2).

To summarize, nucleophilicity is controlled by a number of factors. Increased negative charge and progression from right to left and down the periodic table generally increase nucleophilic power. Table 6-7 compares the reactivity of a range of nucleophiles relative to that of methanol (arbitrarily set at 1). We can confirm the validity of the conclusions of this section by inspecting the various entries. The use of aprotic solvents improves nucleophilicity, especially of smaller anions, by eliminating hydrogen bonding.

6-10 Effect of the Alkyl Group on the S_N2 Reaction

Finally, does the structure of the substrate, particularly in the vicinity of the center bearing the leaving group, affect the rate of nucleophilic attack? Once again, we can get a sense of comparative reactivities by looking at relative rates of reaction. Let us examine the kinetic data that have been obtained.

Branching at the reacting carbon decreases the rate of the S_N2 reaction

What happens if we successively replace each of the hydrogens in a halomethane with a methyl group? Will this affect the rate of its S_N2 reactions? In other words, what are the relative bimolecular nucleophilic reactivities of methyl, primary, secondary, and tertiary halides? Kinetic experiments show that reactivities rapidly decrease in the order shown in Table 6-8.

We can find an explanation by comparing the transition states for these three substitutions. Figure 6-8A shows this structure for the reaction of chloromethane with hydroxide ion. The carbon is surrounded by the incoming nucleophile, the outgoing leaving group, and three substituents (all hydrogen in this case). Although the presence of these five groups increases the crowding about the carbon relative to that in the starting halomethane, the hydrogens do not give rise to serious steric interactions with the nucleophile, because of their small size. However, replacement of one hydrogen by a methyl group, as in a haloethane, creates substantial steric repulsion with the incoming nucleophile, thereby raising the transition-state energy (Figure 6-8B). This effect significantly retards nucleophilic attack. If we continue to replace hydrogen atoms with methyl groups, we find that steric hindrance to nucleophilic attack increases dramatically. The two methyl groups in the secondary substrate severely shield the backside of the carbon attached to the leaving group; the rate of reaction diminishes considerably (Figure 6-8C and Table 6-8). Finally, in the tertiary substrate, in which a third methyl group is present, access to the backside of the halide-bearing carbon is entirely blocked (Figure 6-8D); the transition state for S_N2 substitution is energetically inaccessible, and displacement of a tertiary halide by this mechanism is not observed. To summarize, as we successively replace the hydrogens of a halomethane by methyl groups (or alkyl groups in general), S_N2 reactivity decreases in the following order:

Relative S$_N$2 Displacement Reactivity of Haloalkanes

Methyl	>	primary	>	secondary	>	tertiary
Fast		**Slower**		**Very slow**		**Not at all**

EXERCISE 6-21

Predict the relative rates of the S$_N$2 reaction of cyanide with these pairs of substrates.

(a) [structure: bromocyclohexane] and [structure: CH$_3$, Br on cyclohexane]

(b) CH$_3$CH$_2$$\overset{\displaystyle CH_3}{\underset{\displaystyle CH_3}{C}}$Br   and   CH$_3CH_2CH_2$Br

Now that we have seen the effect of major structural changes on substrate reactivity in the S$_N$2 process, we are in a position to evaluate the effects of more subtle structural modifications. In all cases, we shall find that steric hindrance to attack at the backside of the reacting carbon is the most important consideration.

Lengthening the chain by one or two carbons reduces S$_N$2 reactivity

As we have seen, the replacement of one hydrogen atom in a halomethane by a methyl group (Figure 6-8B) causes significant steric hindrance and reduction of the rate of S$_N$2 reaction. Chloroethane is about two orders of magnitude less reactive than chloromethane in S$_N$2 displacements. Will elongation of the chain of the primary alkyl substrate by the addition of methylene (CH$_2$) groups further reduce S$_N$2 reactivity? Kinetic experiments reveal that 1-chloropropane reacts about half as fast as chloroethane with nucleophiles such as I$^-$.

Does this trend continue as the chain gets longer? The answer is *no:* Higher haloalkanes, such as 1-chlorobutane and 1-chloropentane, react at about the same rate as does 1-chloropropane.

TABLE 6-8 Relative Rates of S$_N$2 Reaction of Branched Bromoalkanes with Iodide

Bromoalkane	Rate
CH$_3$Br	145
CH$_3$CH$_2$Br	1
CH$_3$$\overset{\displaystyle CH_3}{C}$HBr	0.0078
CH$_3$$\overset{\displaystyle CH_3}{\underset{\displaystyle CH_3}{C}}$Br	Negligible

FIGURE 6-8
Transition states for S$_N$2 reactions of hydroxide ion with (A) chloromethane, (B) chloroethane, (C) 2-chloropropane, and (D) 2-chloro-2-methylpropane.

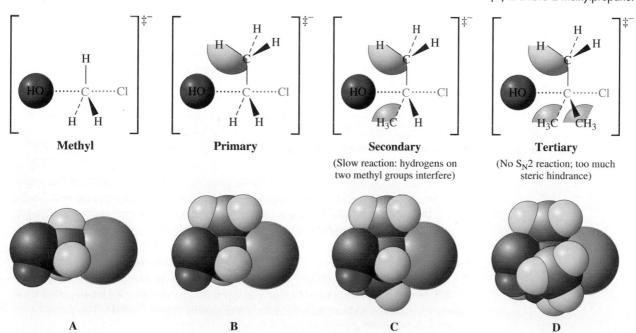

Methyl

Primary

Secondary
(Slow reaction: hydrogens on two methyl groups interfere)

Tertiary
(No S$_N$2 reaction; too much steric hindrance)

A

B

C

D

CHEMICAL HIGHLIGHT **6-2** | **The Dilemma of Bromomethane: Highly Useful but Also Highly Toxic**

Bromomethane, CH_3Br, is a substance with numerous uses. Easy and inexpensive to prepare, it is employed as an insect fumigant for large storage spaces such as warehouses and railroad boxcars. It is also effective in eradicating insect infestations in soil and around a number of major crops, including potatoes and tomatoes. Not surprisingly, it owes part of its value to its high toxicity, which can be attributed largely to its S_N2 reactivity. The chemistry of life is highly dependent on several classes of molecules containing nucleophilic groups such as amines ($-NH_2$ and related functions) and thiols ($-SH$). The biochemical roles of these substituents are many and varied, as well as being critical to the survival of living organisms. Highly reactive electrophiles such as bromomethane wreak havoc on this biochemistry by indiscriminately *alkylating* such nucleophilic atoms— by reacting through the S_N2 mechanism to attach alkyl groups (in this case, a methyl group) to them (see, for example, the reaction below). Some of these processes can generate HBr as a by-product, which amplifies the danger posed by this material to living systems.

The toxicity of bromomethane is not limited to insects. Human exposure is known to cause numerous health problems: Direct contact causes burns to the skin, chronic exposure leads to kidney, liver, and central nervous system damage, and inhalation of high concentrations can lead to the destruction of lung tissue, to pulmonary edema, and to death. The limit set for bromomethane exposure in the workplace

is a concentration of 20 parts per million of bromomethane vapor in ambient air. As is the case for so many substances that have been found to be useful in large-scale applications in our society, bromomethane's toxicity poses a dilemma that requires the most responsible control of its use. The resolution between the issues of utility and safety does not always come easily, and the costs—human, environmental, and economic—must be assessed most carefully.

Plant pathologist Frank Westerlund demonstrates the "bromomethane difference" between strawberries grown in a fumigated plot (*right*) and those not (*left*). The latter are withered by verticillium wilt, a fungus.

$$R\overset{..}{\underset{..}{S}}H + CH_3-Br \longrightarrow R-\overset{CH_3}{\underset{..}{S^+}}-H + Br^- \longrightarrow R-\overset{..}{\underset{..}{S}}-CH_3 + HBr$$

Again, an examination of the transition states to backside displacement provides an explanation for these observations. In Figures 6-9A and 6-9B, one of the hydrogens on the methyl carbon of chloroethane is partially obstructing the path of attack of the incoming nucleophile. The 1-halopropanes have an additional methyl group near the reacting carbon center. If reaction occurs from the most stable *anti* rotamer of the substrate, the incoming nucleophile faces severe steric hindrance (Figure 6-9C). However, rotation to a *gauche* conformation before attack gives an S_N2 transition state similar to that derived from a haloethane (Figure 6-9D). The propyl substrate exhibits only a small decrease in reactivity relative to the ethyl, the decrease

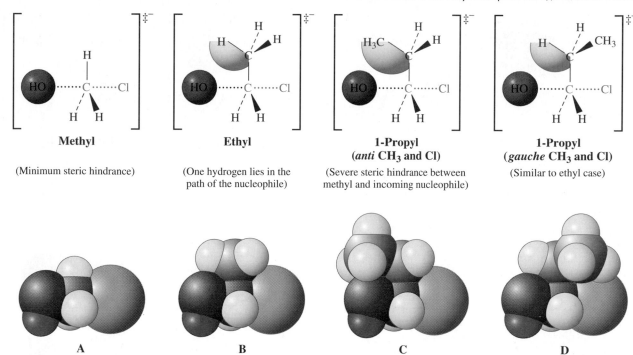

Methyl

(Minimum steric hindrance)

Ethyl

(One hydrogen lies in the path of the nucleophile)

1-Propyl
(*anti* CH$_3$ and Cl)

(Severe steric hindrance between methyl and incoming nucleophile)

1-Propyl
(*gauche* CH$_3$ and Cl)

(Similar to ethyl case)

A   B   C   D

FIGURE 6-9

Dashed-wedged line and space-filling drawings of the transition states for S$_N$2 reactions of hydroxide ion with (A) chloromethane; (B) chloroethane; and (C and D) two rotamers of 1-chloropropane: (C) *anti* and (D) *gauche*. The shaded areas in the dashed-wedged line drawings highlight steric interference with the incoming nucleophile. This interference is illustrated strikingly in the space-filling drawing. Partial charges have been omitted for clarity. (See Figure 6-3.)

resulting from the energy input needed to attain a *gauche* conformation. Further chain elongation has no effect, because the added carbon atoms do not increase steric hindrance around the reacting carbon in the transition state.

Branching next to the reacting carbon also retards substitution

What about multiple substitution at the position *next* to the electrophilic carbon? Let us compare the reactivities of bromoethane and its derivatives (Table 6-9). A dramatic decrease in rate is seen on further substitution: 1-Bromo-2-methylpropane is two orders of magnitude less reactive toward iodide than is 1-bromopropane, and 1-bromo-2,2-dimethylpropane is virtually inert. Branching at positions farther from the site of reaction has a much smaller effect.

Recalling Figure 6-9, we know that rotation into a *gauche* conformation is necessary to permit nucleophilic attack on a 1-halopropane. We can use the same picture to understand the data in Table 6-9. For a 1-halo-2-methylpropane, the only conformation that permits the nucleophile to approach the backside of the reacting carbon experiences *two gauche* methyl–halide interactions, a considerably worse situation (Figure 6-10B). With the addition of a third methyl group, as in a 1-halo-2,2-dimethylpropane, backside attack is blocked almost completely (Figure 6-10C).

TABLE 6-9 Relative Reactivities of Branched Bromoalkanes with Iodide	
Bromoalkane	**Relative rate**
H—CCH$_2$Br (with H above and below)	1
CH$_3$CCH$_2$Br (with H above and below)	0.8
CH$_3$CCH$_2$Br (with CH$_3$ above and H below)	0.03
CH$_3$CCH$_2$Br (with CH$_3$ above and CH$_3$ below)	1.3×10^{-5}

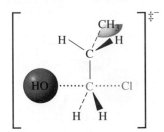

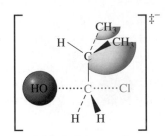

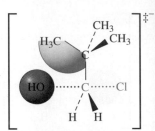

1-Propyl
(*gauche* CH₃ and Cl)

2-Methyl-1-propyl
(two *gauche* CH₃ and Cl)

(High energy transition state: reaction is slower)

2,2-Dimethyl-1-propyl

(All conformations experience severe steric hindrance)

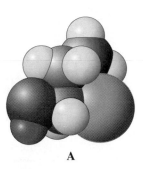

A   B   C

FIGURE 6-10

Dashed-wedged line and space-filling renditions of the transition states for S$_N$2 reactions of hydroxide ion with (A) 1-chloropropane, (B) 1-chloro-2-methylpropane, and (C) 1-chloro-2,2-dimethylpropane. Increasing steric hindrance from a second *gauche* interaction reduces the rate of reaction in (B). S$_N$2 reactivity in (C) is eliminated almost entirely because a methyl group prevents backside attack by the nucleophile in all accessible conformations of the substrate. (See also Figures 6-8 and 6-9.)

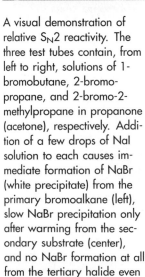

A visual demonstration of relative S$_N$2 reactivity. The three test tubes contain, from left to right, solutions of 1-bromobutane, 2-bromo-propane, and 2-bromo-2-methylpropane in propanone (acetone), respectively. Addition of a few drops of NaI solution to each causes immediate formation of NaBr (white precipitate) from the primary bromoalkane (left), slow NaBr precipitation only after warming from the secondary substrate (center), and no NaBr formation at all from the tertiary halide even after extended heating (right).

EXERCISE 6-22

Predict the order of reactivity in the S$_N$2 reaction of

versus

In summary, the structure of the alkyl part of a haloalkane can have a pronounced effect on nucleophilic attack. Simple chain elongation beyond three carbons has little effect on the rate of the S$_N$2 reaction. However, increased branching leads to strong steric hindrance and rate retardation.

CHAPTER INTEGRATION PROBLEM

a. Write a mechanism and final product for the reaction between sodium ethoxide, NaOCH$_2$CH$_3$, and bromoethane, CH$_3$CH$_2$Br, in ethanol solvent, CH$_3$CH$_2$OH.

SOLUTION

The mechanism is backside attack in which the nucleophilic atom of the reagent attacks the atom of the substrate that contains the leaving group (Section 6-6). We begin by identifying each of these components. The nucleophilic atom is the negatively charged oxygen atom in ethoxide ion, $CH_3CH_2O^-$. Attack occurs at the carbon attached to bromine in the substrate molecule, CH_3CH_2Br:

$$CH_3CH_2O^- + \overset{H_3C}{\underset{H}{\overset{|}{C}}}-Br \longrightarrow CH_3CH_2O-CH_2CH_3 + Br^-$$

The products are bromide ion and ethoxyethane, $CH_3CH_2OCH_2CH_3$, an ether.

b. How would the preceding reaction be affected by each of the following changes?
   1. Replace bromoethane with fluoroethane.
   2. Replace bromoethane with bromomethane.
   3. Replace sodium ethoxide with sodium ethanethiolate, $NaSCH_2CH_3$.
   4. Replace ethanol with dimethylformamide (DMF).

SOLUTION

1. Table 6-4 tells us that fluoride is a stronger base than bromide and, therefore, a poorer leaving group. The reaction would still take place but would be very much slower. (The actual rate decrease is on the order of 10^{-4}.)
2. The carbon containing the leaving group in bromomethane is less sterically hindered than that in bromoethane, so the rate of reaction would increase (Section 6-10). The product of the reaction would be $CH_3OCH_2CH_3$, methoxyethane.
3. Both ethoxide and ethanethiolate are negatively charged. Oxygen in ethoxide is more basic than sulfur in ethanethiolate (Table 6-4), but the sulfur atom in the latter is larger, more polarizable, and less tightly solvated in the hydrogen-bonding ethanol solvent (compare Figure 6-6). We know that strong bases are good nucleophiles, but base strength is outweighed by the increased polarizability and reduced solvation of the larger atoms within the same column of the periodic table (Section 6-9). Ethanethiolate reacts hundreds of times as fast, giving as a product $CH_3CH_2SCH_2CH_3$, an example of a sulfide (Section 9-10).
4. Conversion from a protic, hydrogen-bonding solvent into a polar, aprotic one accelerates the reaction enormously by reducing solvation of the negatively charged oxygen atom (compare Table 6-6).

c. Which of the following compounds would be expected to react in an S_N2 manner at a reasonable rate with sodium azide, NaN_3, in ethanol? Which will not? Why not?

SOLUTION

What are the structural requirements for S_N2 reactions to occur? We need a strong nucleophile, a suitable substrate, and a good leaving group. The nucleophile is already specified, so we turn to the other two ingredients, ruling out systems that lack one or the other. Substrates (i), (iv), and (vi) lack good leaving groups: $^-NH_2$, ^-OH, and ^-CN are too strongly basic for this purpose (Table 6-4). These compounds will not undergo S_N2 displacement. Substrate (ii) contains a good leaving group, but the reaction site is a tertiary carbon and incapable of following the S_N2 mechanism. That leaves substrates (iii) and (v), both of which are primary haloalkanes with good leaving groups and no branching at the position adjacent to the site of displacement (Section 6-10). They will transform readily by the S_N2 mechanism.

IMPORTANT CONCEPTS

1. A **haloalkane,** commonly termed an alkyl halide, consists of an alkyl group and a halogen.
2. The physical properties of the haloalkanes are strongly affected by the polarization of the C–X bond and the polarizability of X.
3. Reagents bearing lone electron pairs are called **nucleophilic** when they attack positively polarized centers (other than protons). The latter are called **electrophilic.** When such a reaction leads to displacement of a substituent, it is a **nucleophilic substitution.** The group being displaced by the nucleophile is the **leaving group.**
4. The kinetics of the reaction of nucleophiles with primary (and most secondary) haloalkanes are second order, indicative of a **bimolecular** mechanism. This process is called **bimolecular nucleophilic substitution (S_N2 reaction).** It is a **concerted reaction,** one in which bonds are simultaneously broken and formed. Curved arrows are typically used to depict the flow of electrons as the reaction proceeds.
5. The S_N2 reaction is **stereospecific** and proceeds by **backside displacement,** thereby producing **inversion of configuration** at the reacting center.

6. An orbital description of the S_N2 transition state includes an sp^2-hybridized carbon center, partial bond-making between the nucleophile and the electrophilic carbon, and simultaneous partial bond-breaking between that carbon and the leaving group. Both the nucleophile and the leaving group bear partial charges.
7. **Leaving group ability,** a measure of the ease of displacement, is roughly proportional to the strength of the conjugate acid. Especially good leaving groups are weak bases such as chloride, bromide, iodide, and the sulfonates.
8. **Nucleophilicity** increases (a) with negative charge, (b) for elements farther to the left and down the periodic table, and (c) in polar aprotic solvents.
9. **Polar aprotic solvents** accelerate S_N2 reactions because the nucleophiles are well separated from their counterions but are not tightly solvated.
10. **Branching** at the reacting carbon or at the carbon next to it in the substrate leads to steric hindrance in the S_N2 transition state and decreases the rate of bimolecular substitution.

PROBLEMS

23. Name the following molecules according to the IUPAC system.

    (a) CH_3CH_2Cl (b) $BrCH_2CH_2Br$ (c) $CH_3CH_2CHCH_2F$
    $\qquad\qquad\qquad\qquad\qquad\qquad\qquad\qquad\quad |$
    $\qquad\qquad\qquad\qquad\qquad\qquad\qquad\qquad CH_2CH_3$

    (d) $(CH_3)_3CCH_2I$ (e) ⬡—CCl_3 (f) $CHBr_3$

24. Draw structures for each of the following molecules. **(a)** 3-Ethyl-2-iodopentane; **(b)** 3-bromo-1,1-dichlorobutane; **(c)** *cis*-1-(bromomethyl)-2-(2-chloroethyl)cyclobutane; **(d)** (trichloromethyl)cyclopropane; **(e)** 1,2,3-trichloro-2-methylpropane.

25. Draw and name all possible structural isomers having the formula C_3H_6BrCl.

26. Draw and name all structurally isomeric compounds having the formula $C_5H_{11}Br$.

27. For each structural isomer in Problems 25 and 26, identify all stereocenters and give the total number of stereoisomers that can exist for the structure.

28. For each reaction in Table 6-3, identify the nucleophile, its nucleophilic atom (draw its Lewis structure first), the electrophilic atom in the organic substrate, and the leaving group.

29. A second Lewis structure can be drawn for one of the nucleophiles in Problem 28. **(a)** Identify it and draw its alternate structure (which is simply a second resonance form). **(b)** Does this second resonance form predict the presence of another nucleophilic atom in the nucleophile? If so, rewrite the reaction of Problem 28, using the new nucleophilic atom, and write a correct Lewis structure for the product.

30. For each reaction shown here, identify the nucleophile, its nucleophilic atom, the electrophilic atom in the substrate molecule, and the leaving group. Write the organic product of the reaction.

(a) $CH_3I + NaNH_2 \rightarrow$

(b) 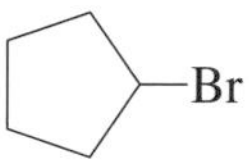 $-Br + NaSH \rightarrow$

(c) 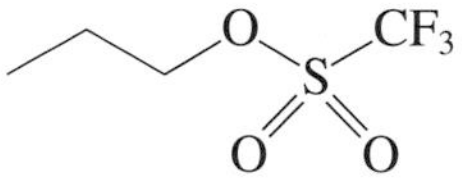 $+ NaI \rightarrow$

(d) 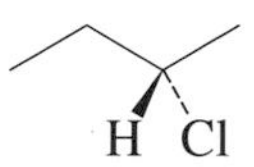 $+ NaN_3 \rightarrow$

(e) $CH_3Cl +$ $\rightarrow$

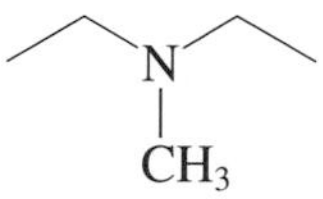

(f) 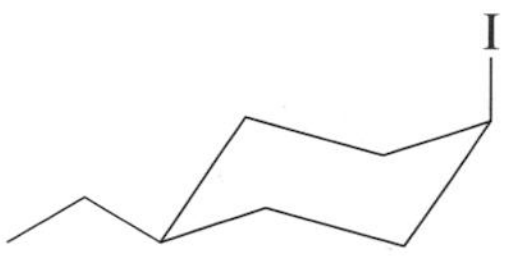 $+ KSeCN \rightarrow$

31. A solution containing 0.1 M CH_3Cl and 0.1 M KSCN in DMF reacts to give CH_3SCN and KCl with an initial rate of 2×10^{-8} mol L^{-1} s^{-1}. **(a)** What is the rate constant for this reaction? **(b)** Calculate the initial reaction rate for each of the following sets of reactant concentrations: (i) $[CH_3Cl]$ = 0.2 M, $[KSCN]$ = 0.1 M; (ii) $[CH_3Cl]$ = 0.2 M, $[KSCN]$ = 0.3 M; (iii) $[CH_3Cl]$ = 0.4 M, $[KSCN]$ = 0.4 M.

32. Write the product of each of the following bimolecular substitutions. The solvent is indicated above the reaction arrow.

(a) $CH_3CH_2CH_2Br + Na^+I^- \xrightarrow{\text{Propanone (acetone)}}$

(b) $(CH_3)_2CHCH_2I + Na^+{}^-CN \xrightarrow{\text{DMSO}}$

(c) $CH_3I + Na^+{}^-OCH(CH_3)_2 \xrightarrow{\text{(CH}_3)_2\text{CHOH}}$

(d) $CH_3CH_2Br + Na^+{}^-SCH_2CH_3 \xrightarrow{\text{CH}_3\text{OH}}$

(e) 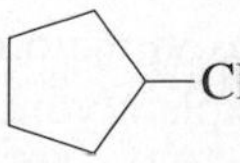 $-CH_2Cl + CH_3CH_2SeCH_2CH_3 \xrightarrow{\text{Propanone (acetone)}}$

(f) $(CH_3)_2CHOSO_2CH_3 + N(CH_3)_3 \xrightarrow{\text{(CH}_3\text{CH}_2)_2\text{O}}$

33. Determine the *R/S* designations for both starting materials and products in the following S$_N$2 reactions. Which of the products are optically active?

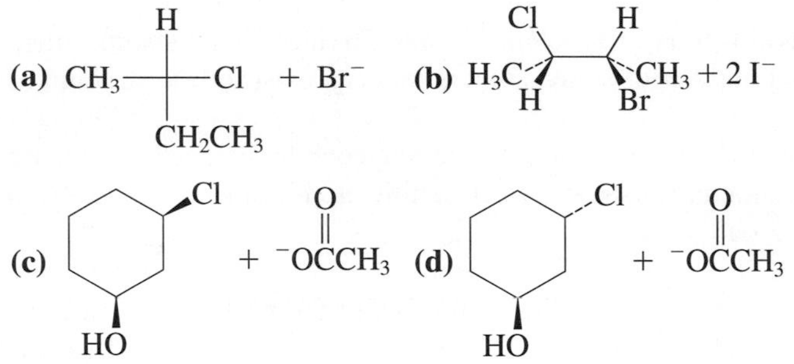

(a) CH$_3$—|—Cl + Br$^-$ (b) H$_3$C—CH—CH—CH$_3$ + 2 I$^-$
   (with CH$_2$CH$_3$)

(c) [cyclohexane with Cl and HO substituents] + $^-$OCCH$_3$ (O)   (d) [cyclohexane with Cl and HO substituents] + $^-$OCCH$_3$ (O)

34. List the product(s) of the reaction of 1-bromopropane with each of the following reagents. Write "no reaction" where appropriate. (**Hint:** Carefully evaluate the nucleophilic potential of each reagent.)
   (a) H$_2$O   (b) H$_2$SO$_4$ (c) KOH   (d) CsI   (e) NaCN
   (f) HCl   (g) (CH$_3$)$_2$S   (h) NH$_3$ (i) Cl$_2$ (j) KF

35. Formulate the potential product of each of the following reactions. As you did in Problem 34, write "no reaction" where appropriate. (**Hint:** Identify the expected leaving group in each of the substrates and evaluate its ability to undergo displacement.)

(a) CH$_3$CH$_2$CH$_2$CH$_2$Br + K$^+$ $^-$OH $\xrightarrow{\text{CH}_3\text{CH}_2\text{OH}}$ (b) CH$_3$CH$_2$I + K$^+$Cl$^-$ $\xrightarrow{\text{DMF}}$

(c) [phenyl]—CH$_2$Cl + Li$^+$ $^-$OCH$_2$CH$_3$ $\xrightarrow{\text{CH}_3\text{CH}_2\text{OH}}$ (d) (CH$_3$)$_2$CHCH$_2$Br + Cs$^+$I$^-$ $\xrightarrow{\text{CH}_3\text{OH}}$

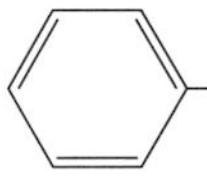

(e) CH$_3$CH$_2$CH$_2$Cl + K$^+$ $^-$SCN $\xrightarrow{\text{CH}_3\text{CH}_2\text{OH}}$ (f) CH$_3$CH$_2$F + Li$^+$Cl$^-$ $\xrightarrow{\text{CH}_3\text{OH}}$

(g) CH$_3$CH$_2$CH$_2$OH + K$^+$I$^-$ $\xrightarrow{\text{DMSO}}$ (h) CH$_3$I + Na$^+$ $^-$SCH$_3$ $\xrightarrow{\text{CH}_3\text{OH}}$

(i) CH$_3$CH$_2$OCH$_2$CH$_3$ + Na$^+$ $^-$OH $\xrightarrow{\text{H}_2\text{O}}$ (j) CH$_3$CH$_2$I + K$^+$ $^-$OCCH$_3$ (O) $\xrightarrow{\text{DMSO}}$

36. Show how each of the following transformations might be achieved.

(a) (*R*)-CH$_3$CHCH$_2$CH$_3$ (OSO$_2$CH$_3$) $\longrightarrow$ (*S*)-CH$_3$CHCH$_2$CH$_3$ (N$_3$)   (b) [Fischer projection CH$_3$, H—Br, CH$_3$O—H, CH$_3$] $\longrightarrow$ [Fischer projection CH$_3$, H—CN, CH$_3$O—H, CH$_3$]

(c) [bicyclic ring with H, H and Br] $\longrightarrow$ [bicyclic ring with H, H and SCH$_3$]   (d) [piperidine with N-CH$_3$] $\longrightarrow$ [piperidinium with N$^+$ CH$_3$ CH$_3$]

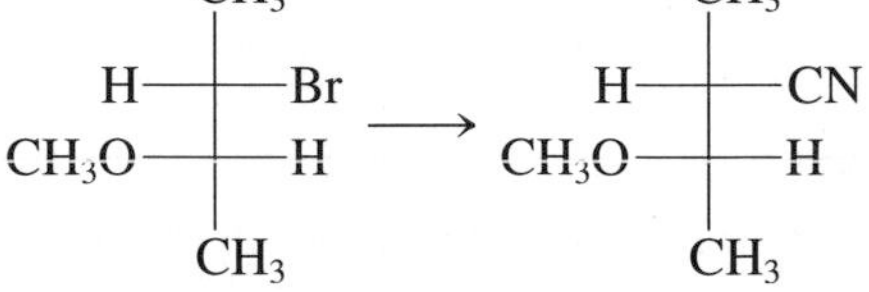

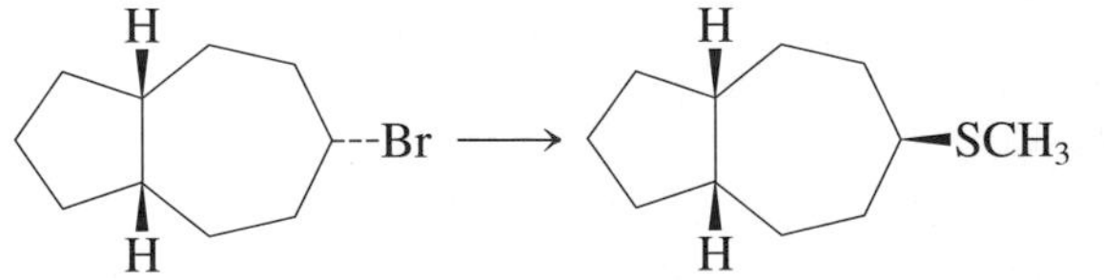

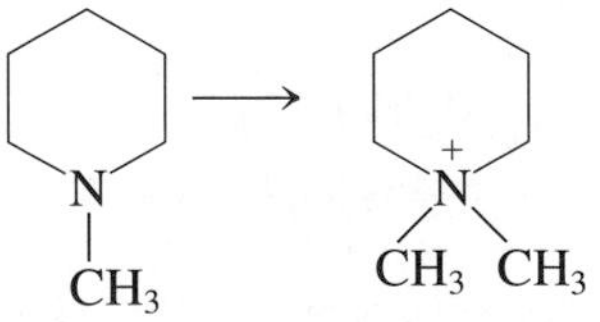

37. Rank the members of each of the following groups of species in the order of basicity, nucleophilicity, and leaving-group ability. Briefly explain your answers. **(a)** H$_2$O, HO$^-$, CH$_3$CO$_2$$^-$; **(b)** Br$^-$, Cl$^-$, F$^-$, I$^-$; **(c)** $^-$NH$_2$, NH$_3$, $^-$PH$_2$; **(d)** $^-$OCN, $^-$SCN; **(e)** F$^-$, HO$^-$, $^-$SCH$_3$; **(f)** H$_2$O, H$_2$S, NH$_3$.

38. Write the product(s) of each of the following reactions. Write "no reaction" as your answer, if appropriate.

(a) $CH_3CH_2CH_2CH_3 + Na^+Cl^- \xrightarrow{CH_3OH}$

(b) $CH_3CH_2Cl + Na^{+-}OCH_3 \xrightarrow{CH_3OH}$

(c) $+ Na^+I^- \xrightarrow{\text{Propanone (acetone)}}$

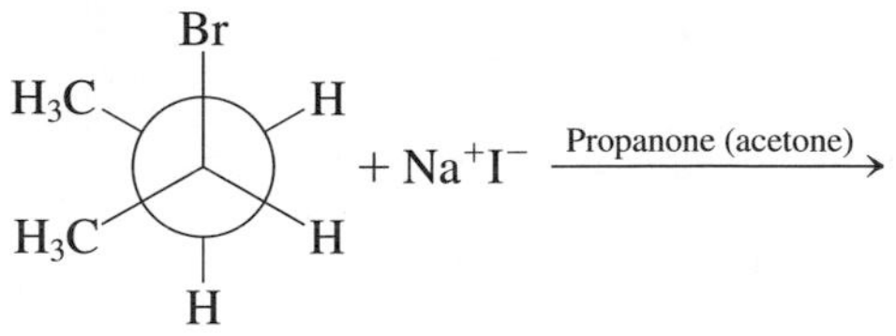

(d) $+ Na^{+-}SCH_3 \xrightarrow{\text{Propanone (acetone)}}$

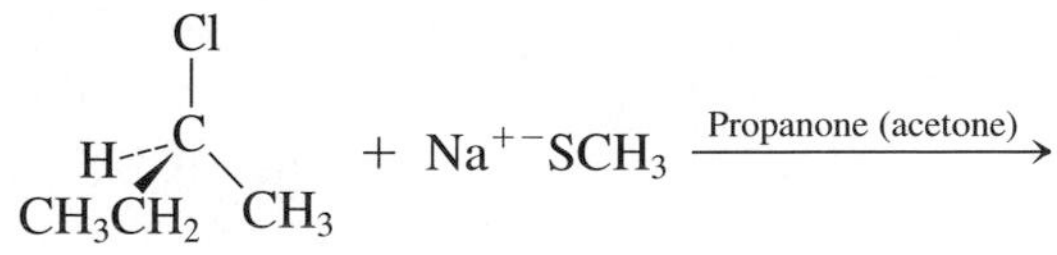

(e) $CH_3CHCH_3 + Na^{+-}CN \longrightarrow$

(f) $CH_3CHCH_3 + HCN \xrightarrow{CH_3CH_2OH}$

(g) $CH_3CHCH_3 + Na^{+-}CN \xrightarrow{CH_3CH_2OH}$

(h) $+ K^{+-}SCN \xrightarrow{CH_3OH}$

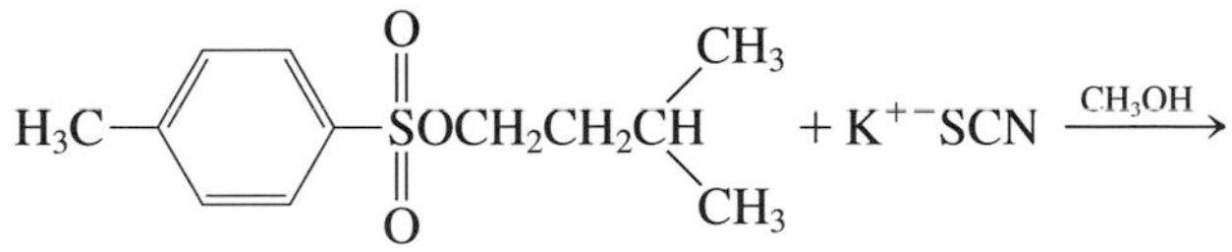

(i) $CH_3CH_2NH_2 + Na^+Br^- \xrightarrow{DMSO}$

(j) $CH_3I + Na^{+-}NH_2 \xrightarrow{NH_3}$

(k) Product of (j) + more $CH_3I \longrightarrow$

(l) $+ Na^{+-}SH \xrightarrow{CH_3OH}$

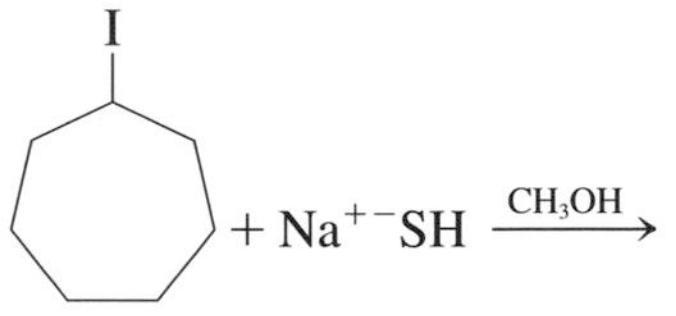

(m) $+ Na^{+-}SH \xrightarrow{CH_3OH}$

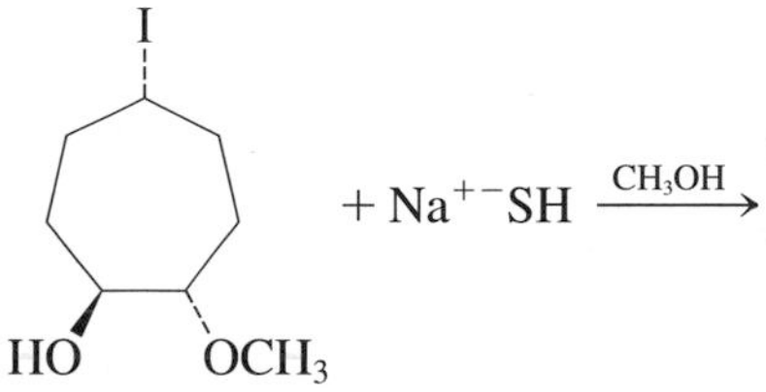

(n)

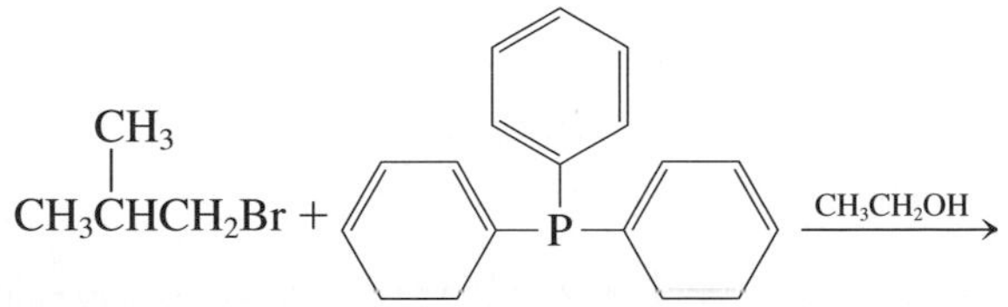

39. Using the information in Chapters 3 and 6, propose the best possible synthesis of each of the following compounds with propane as your organic starting material and any other reagents needed. [**Hint:** On the basis of the information in Section 3-7, you should not expect to find very good answers for (a), (c), and (e). One general approach is best, however.]
(a) 1-Chloropropane **(b)** 2-Chloropropane **(c)** 1-Bromopropane
(d) 2-Bromopropane **(e)** 1-Iodopropane **(f)** 2-Iodopropane

40. Propose two syntheses of *trans*-1-methyl-2-(methylthio)cyclohexane (shown in the margin), beginning with the starting compound **(a)** *cis*-1-chloro-2-methylcyclohexane; **(b)** *trans*-1-chloro-2-methylcyclohexane.

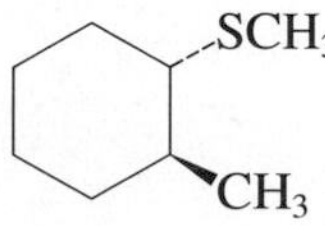

41. Rank each of the following sets of molecules in order of increasing S_N2 reactivity.

(a) CH_3CH_2Br, CH_3Br, $(CH_3)_2CHBr$

(b) $(CH_3)_2CHCH_2CH_2Cl$, $(CH_3)_2CHCH_2Cl$, $(CH_3)_2CHCl$

(c) CH_3CH_2Cl, CH_3CH_2I, ⟨cyclohexyl⟩$-Cl$

(d) $(CH_3CH_2)_2CHCH_2Br$, $CH_3CH_2CH_2CHBr$, $(CH_3)_2CHCH_2Br$
                                                 CH_3

42. Predict the effect of the changes given below on the rate of the reaction

$CH_3Cl + {}^-OCH_3 \xrightarrow{CH_3OH} CH_3OCH_3 + Cl^-$. **(a)** Change substrate from CH_3Cl to CH_3I; **(b)** change nucleophile from CH_3O^- to CH_3S^-; **(c)** change substrate from CH_3Cl to $(CH_3)_2CHCl$; **(d)** change solvent from CH_3OH to $(CH_3)_2SO$.

43. The following table presents rate data for the reactions of CH_3I with three different nucleophiles in two different solvents. What is the significance of these results regarding relative reactivity of nucleophiles under different conditions?

Nucleophile	k_{rel}, CH_3OH	k_{rel}, DMF
Cl^-	1	1.2×10^6
Br^-	20	6×10^5
$NCSe^-$	4000	6×10^5

44. Rings are readily prepared by means of intramolecular S_N2 reactions. An example and its mechanism are:

$$ClCH_2CH_2CH_2CH_2O{-}H + {}^-OH \underset{\text{reaction}}{\overset{\text{Acid-base}}{\rightleftharpoons}} H_2O + \overset{\displaystyle CH_2CH_2CH_2CH_2O^-}{\underset{\displaystyle Cl}{|}} \longrightarrow ⟨ring O⟩ + Cl^-$$

**Intramolecular
displacement reaction**

Explain the outcome of the following transformations mechanistically. (**Hint:** Notice in the example how an acid-base reaction leads to a stronger nucleophile at one end of the molecule. Use this in your answers.)

(a) $HSCH_2CH_2Br + NaOH \xrightarrow{CH_3CH_2OH}$ ⟨thiirane S⟩

(b) $BrCH_2CH_2CH_2CH_2CH_2Br + NaOH \xrightarrow{CH_3OH}$ ⟨tetrahydropyran O⟩

(c) $BrCH_2CH_2CH_2CH_2CH_2Br + NH_3 \xrightarrow{CH_3CH_2OH}$ ⟨piperidine N-H⟩

45. S_N2 reactions of halocyclopropane and halocyclobutane substrates are very much slower than those of analogous acyclic secondary haloalkanes. Suggest an explanation for this finding. (**Hint:** Consider the effect of bond-angle strain on the energy of the transition state; see Figure 6-4.)

46. Nucleophilic attack on halocyclohexanes is also somewhat retarded compared with that on acyclic secondary haloalkanes, even though in this case bond-angle strain is *not* an important factor. Explain. (**Hint:** Make a model, and refer to Chapter 4 and Section 6-10.)

Team Problem

47. Compounds A through H are isomeric bromoalkanes with the molecular formula $C_5H_{11}Br$. With your team, draw all eight constitutional isomers. Indicate any stereocenter(s), but do not label it (them) as *R* or *S* until you have completed your analysis. Using the data below, assign structures to A through H. Divide the problem into equal parts to share the effort of finding a solution. Reconvene and discuss your analysis. At this point, you should indicate the stereochemistry with wedged and dashed lines as appropriate.

- Treatment of compounds A through G with NaCN in DMF followed second-order kinetics and showed the following relative rates:

$$A \cong B > C > D \cong E > F >>> G$$

- Compound H does not undergo the S_N2 reaction under the preceding conditions.
- Compounds C, D, and F were found to be optically active, each having *S* absolute configuration at the stereocenter. Substitution reactions of D and F with NaCN in DMF proceeded with inversion of configuration, while treatment of C in the same way proceeded with retention of configuration.

Preprofessional Problems

48. The S_N2 reaction mechanism best applies to
   (**a**) cyclopropane and H_2 (**b**) 1-chlorobutane and aqueous NaOH
   (**c**) KOH and NaOH   (**d**) ethane and H_2O

49. The reaction $CH_3Cl + OH^- \longrightarrow CH_3OH + Cl^-$ is first order in both chloromethane and hydroxide. Given the rate constant $k = 3.5 \times 10^{-3}$ mol $L^{-1}s^{-1}$, what is the observed rate at the following concentrations?

$$[CH_3Cl] = 0.50 \text{ mol } L^{-1}; [OH^-] = 0.015 \text{ mol } L^{-1}$$

   (**a**) 2.6×10^{-5} mol L^{-1} s^{-1} (**b**) 2.6×10^{-6} mol L^{-1} s^{-1} (**c**) 2.6×10^{-3} mol L^{-1} s^{-1}
   (**d**) 1.75×10^{-3} mol L^{-1} s^{-1} (**e**) 1.75×10^{-5} mol L^{-1} s^{-1}

50. Which ion is the strongest nucleophile in aqueous solution?
   (**a**) F^- (**b**) Cl^- (**c**) Br^- (**d**) I^- (**e**) all of these are equally strong

51. Only one of the following processes will occur measurably at room temperature. Which one?

(a) $:\overset{..}{\underset{..}{F}}-\overset{..}{\underset{..}{Cl}}:$

(b) $:N\equiv C:^{-} \curvearrowright CH_3-I$

(c) $:N\equiv N: \curvearrowright CH_3-I$

(d) $:\overset{..}{\underset{..}{O}}=\overset{..}{\underset{..}{O}}: \curvearrowright CH_2=CH_2$

Further Reactions of Haloalkanes

Unimolecular Substitution and Pathways of Elimination

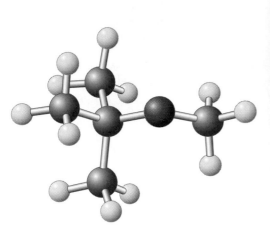

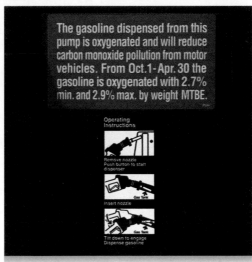

2-Methoxy-2-methylpropane, also known as methyl *tert*-butyl ether (MTBE), may be prepared by solvolysis; it replaces lead as an octane-rating enhancer in gasoline.

Is the S_N2 displacement process the only reaction available to haloalkanes? Are other mechanisms for substitution possible? Finally, are there other, fundamentally different types of transformations that haloalkanes undergo? In this chapter we shall see that haloalkanes can indeed follow reaction pathways other than S_N2 displacement, especially if the haloalkanes are tertiary or secondary. We shall see that bimolecular substitution is only one of *four* possible modes of reaction. The other three modes are unimolecular substitution and two types of elimination processes. The elimination processes give rise to double bonds through loss of HX and serve as our first method of entry into multiply bonded organic compounds.

7-1 Solvolysis of Tertiary and Secondary Haloalkanes

We have seen that the rate of the S_N2 reaction diminishes drastically when the reacting center changes from primary to secondary to tertiary. For example, although the S_N2 reactivity of bromomethane and bromoethane with iodide ion in propanone (acetone) is high, 2-bromopropane is much less reactive, and 2-bromo-2-methylpropane is essentially inert. However, these observations pertain only to *bimolecular* substitution. Secondary and tertiary halides do undergo substitution, but by another mechanism. This section will show that, in fact, these substrates transform readily, even in the presence of weak nucleophiles, to give substitution products.

For example, when 2-bromo-2-methylpropane (*tert*-butyl bromide) is mixed with water, it is rapidly converted into 2-methyl-2-propanol (*tert*-butyl alcohol) and hydrogen bromide. Water is the nucleophile here, even though it is poor in this capacity. Such a transformation, in which a substrate undergoes substitution by *solvent* molecules, is called **solvolysis.** When the solvent is water, the term **hydrolysis** is applied.

Reminder
Nucleophile: red
Electrophile: blue
Leaving group: green

An Example of Solvolysis: Hydrolysis

$$\underset{\substack{\text{2-Bromo-2-methylpropane} \\ (\textit{tert}\text{-Butyl bromide})}}{CH_3\overset{\overset{\displaystyle CH_3}{|}}{\underset{\underset{\displaystyle CH_3}{|}}{C}}Br} + H-OH \underset{\text{Relatively fast}}{\rightleftharpoons} \underset{\substack{\text{2-Methyl-2-propanol} \\ (\textit{tert}\text{-Butyl alcohol})}}{CH_3\overset{\overset{\displaystyle CH_3}{|}}{\underset{\underset{\displaystyle CH_3}{|}}{C}}OH} + HBr$$

Methyl and Primary Haloalkanes: Unreactive in Solvolysis

2-Bromopropane is hydrolyzed similarly, albeit much more slowly, whereas 1-bromopropane, bromoethane, and bromomethane are relatively unaffected by these conditions.

CH_3Br
CH_3CH_2Br
$CH_3CH_2CH_2Br$
Essentially no reaction with H_2O at room temperature

Hydrolysis of a Secondary Haloalkane

$$\underset{\substack{\text{2-Bromopropane} \\ (\text{Isopropyl bromide})}}{CH_3\overset{\overset{\displaystyle CH_3}{|}}{\underset{\underset{\displaystyle H}{|}}{C}}Br} + H-OH \underset{\text{Relatively slow}}{\rightleftharpoons} \underset{\substack{\text{2-Propanol} \\ (\text{Isopropyl alcohol})}}{CH_3\overset{\overset{\displaystyle CH_3}{|}}{\underset{\underset{\displaystyle H}{|}}{C}}OH} + HBr$$

Solvolysis also takes place in alcohol solvents.

Solvolysis of 2-Chloro-2-methylpropane in Methanol

$$\underset{\substack{\text{2-Chloro-} \\ \text{2-methylpropane}}}{CH_3\overset{\overset{\displaystyle CH_3}{|}}{\underset{\underset{\displaystyle CH_3}{|}}{C}}Cl} + \underset{\text{Solvent}}{CH_3OH} \rightleftharpoons \underset{\substack{\text{2-Methoxy-} \\ \text{2-methylpropane}}}{CH_3\overset{\overset{\displaystyle CH_3}{|}}{\underset{\underset{\displaystyle CH_3}{|}}{C}}OCH_3} + HCl$$

TABLE 7-1

Relative Reactivities of Various Bromoalkanes with Water

Bromoalkane	Relative rate
CH_3Br	1
CH_3CH_2Br	1
$(CH_3)_2CHBr$	12
$(CH_3)_3CBr$	1.2×10^6

The relative rates of reaction of 2-bromopropane and 2-bromo-2-methylpropane with water to give the corresponding alcohols are shown in Table 7-1 and are compared with the corresponding rates of hydrolysis of their unbranched counterparts. Although the process gives the products expected from an S_N2 reaction, the order of reactivity is *reversed* from that found under typical S_N2 conditions. Thus, primary halides are very slow in their reactions with water, secondary halides are more reactive, and tertiary systems are about *1 million times* as fast as primary ones.

These observations suggest that the mechanism of solvolysis of secondary and, especially, tertiary haloalkanes must be different from that of bimolecular substitution. To understand the details of this transformation, we will use the same methods that we used to study the S_N2 process: kinetics, stereochemistry, and the effect of substrate structure and solvent on reaction rates.

H₃C CH₂Br   H₃C Br

A   B

EXERCISE 7-1

Whereas compound A (shown in the margin) is completely stable in ethanol, B is rapidly converted into another compound. Explain.

7-2 Unimolecular Nucleophilic Substitution

In this section we shall learn about a new pathway for nucleophilic substitution. Recall that the S_N2 reaction has second-order kinetics, generates products stereospecifically, and is fastest with halomethanes, successively slower with primary and secondary halides, and does not take place with tertiary substrates at all. In contrast, solvolyses follow a *first-order* rate law, are *not* stereospecific, and are characterized by the *opposite* order of reactivity. Let us see how these findings can be accommodated mechanistically.

Solvolysis follows first-order kinetics

In Chapter 6 the reaction's kinetics revealed a bimolecular transition state: The rate of the S_N2 reaction is proportional to the concentration of both the haloalkane and the nucleophile. Similar studies have been carried out by varying the concentrations of 2-bromo-2-methylpropane and water in formic acid (a polar solvent of very low nucleophilicity) and measuring the rates of solvolysis. The results of these experiments show that *the rate of hydrolysis of the bromide is proportional to the concentration of only the starting halide,* **not** *the water.*

$$\text{Rate} = k[(CH_3)_3CBr] \text{ mol } L^{-1} s^{-1}$$

What does this observation mean? First, it is clear that the haloalkane has to undergo some transformation on its own before anything else takes place. Second, because the final product contains a hydroxy group, water (or, in general, any nucleophile) must enter the reaction, but at a later stage and not in a way that will affect the rate law. The only way to explain this behavior is to postulate that any steps that follow the initial reaction of the halide are relatively fast. In other words, *the observed rate is that of the slowest step in the sequence:* the **rate-determining step.** It follows that only those species taking part in the transition state of this step enter into the rate expression: in this case, only the starting haloalkane.

In analogy, think of the rate-determining step as a bottleneck. Imagine a water hose with several attached clamps restricting the flow (Figure 7-1). We can see that the rate at which the water will spew out of the end is controlled by the narrowest constriction. If we were to reverse the direction of flow (to model the reversibility of a reaction), again the rate of flow would be controlled by this point. Such is the case in transformations consisting of more than one step—for example, solvolysis. What, then, are the steps in our example?

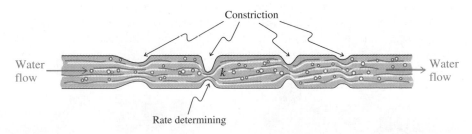

FIGURE 7-1

The rate, k, at which water flows through a hose is controlled by the narrowest constriction.

The mechanism of solvolysis includes carbocation formation

The hydrolysis of 2-bromo-2-methylpropane is said to proceed by **unimolecular nucleophilic substitution,** abbreviated S_N1. The number 1 indicates that only one molecule, the haloalkane, participates in the rate-determining step: The rate of the reaction does *not* depend on the concentration of the nucleophile. The mechanism consists of three steps.

STEP 1. The rate-determining step is the dissociation of the haloalkane to an alkyl cation and bromide.

Dissociation of Halide to Form a Carbocation

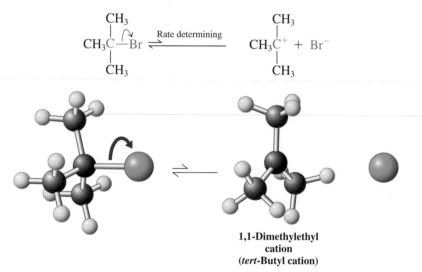

1,1-Dimethylethyl cation
(*tert*-**Butyl cation**)

This conversion is an example of heterolytic cleavage. The hydrocarbon product contains a positively charged central carbon atom attached to three other groups and bearing only an electron sextet. Such a structure is called a **carbocation.**

STEP 2. The 1,1-dimethylethyl (*tert*-butyl) cation formed in step 1 is a powerful electrophile that is immediately trapped by the surrounding water. This process can be viewed as a nucleophilic attack by the solvent on the electron-deficient carbon.

Nucleophilic Attack by Water

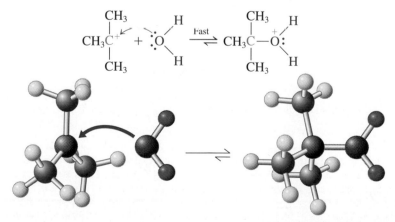

An alkyloxonium ion

The resulting species is an example of an **alkyloxonium ion,** the conjugate acid of an alcohol—in this case 2-methyl-2-propanol, the eventual product of the sequence.

STEP 3. Like the hydronium ion, H_3O^+, the first member of the series of oxonium ions, all alkyloxonium ions are strong acids. They are therefore readily deprotonated by the water in the reaction medium to furnish the final alcohol.

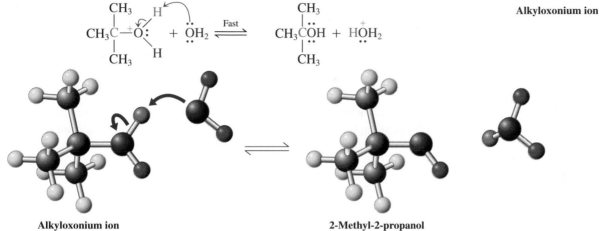

Deprotonation

Alkyloxonium ion

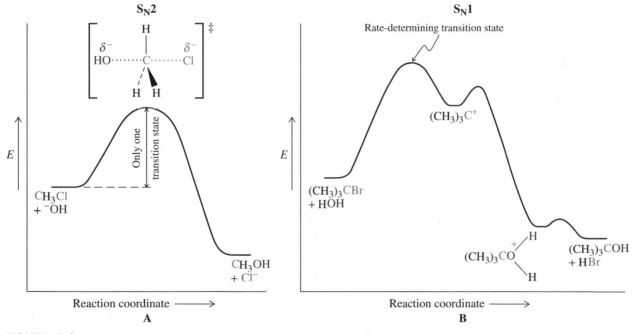

Alkyloxonium ion
(Strongly acidic)

2-Methyl-2-propanol

Figure 7-2 compares the potential-energy diagrams for the S_N2 reaction of chloromethane with hydroxide ion and the S_N1 reaction of 2-bromo-2-methylpropane

FIGURE 7-2

Potential-energy diagrams for (A) S_N2 reaction of chloromethane with hydroxide and (B) S_N1 hydrolysis of 2-bromo-2-methylpropane. Whereas the S_N2 process takes place in a single step, the S_N1 mechanism consists of three distinct events: rate-determining dissociation of the haloalkane into a halide ion and a carbocation, nucleophilic attack by water on the carbocation to give an alkyloxonium ion, and proton loss to furnish the final product. Note: For clarity, inorganic species have been omitted from the intermediate stages of (B).

with water. The latter exhibits three transition states, one for each step in the mechanism. The first has the highest energy—and thus is rate determining—because it requires the separation of opposite charges.

EXERCISE 7-2

Using the bond-strength data in Table 3-1, calculate the $\Delta H°$ for the hydrolysis of 2-bromo-2-methylpropane to 2-methyl-2-propanol and hydrogen bromide.

All three steps of the mechanism of solvolysis are reversible. The overall equilibrium can be driven in either direction by the suitable choice of reaction conditions. Thus, a large excess of nucleophilic solvent ensures complete solvolysis. In Chapter 9 we shall see how this reaction can be reversed to permit the synthesis of tertiary haloalkanes from alcohols.

In summary, the kinetics of haloalkane solvolysis leads us to a mechanism in which initial dissociation to form a carbocation is the crucial, rate-determining step. Can we back up our mechanistic hypothesis with other experimental observations?

7-3 Stereochemical Consequences of S$_N$1 Reactions

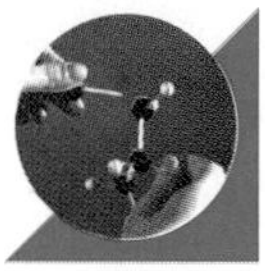

The proposed mechanism of unimolecular nucleophilic substitution has predictable stereochemical consequences because of the structure of the intermediate carbocation. To minimize electron repulsion, the positively charged carbon assumes trigonal planar geometry, the result of sp^2 hybridization (Sections 1-3 and 1-8). Such an intermediate is therefore achiral (make a model). Hence, starting with an optically active tertiary (or secondary) haloalkane in which the stereocenter bears the departing halogen, we should obtain racemic S$_N$1 products (Figure 7-3). This result is, in fact, observed in many solvolyses. In general, the formation of racemic products from optically active substrates is strong evidence for the intermediacy of a symmetrical, achiral species, such as a carbocation, in the course of a reaction.

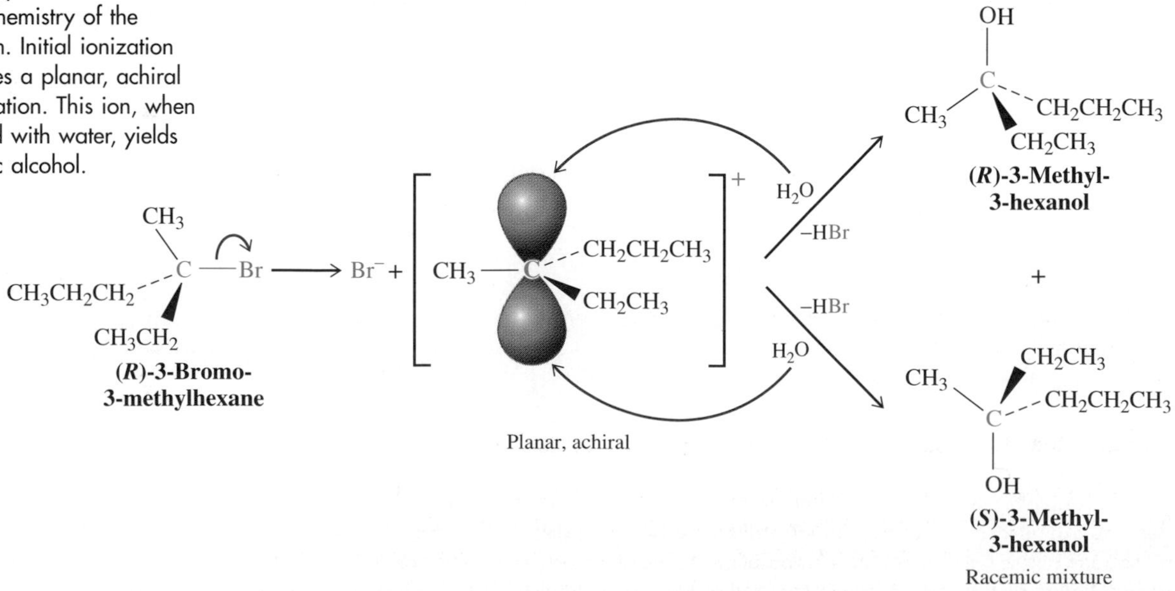

FIGURE 7-3

The mechanism of hydrolysis of (R)-3-bromo-3-methylhexane predicts the stereochemistry of the reaction. Initial ionization furnishes a planar, achiral carbocation. This ion, when trapped with water, yields racemic alcohol.

(R)-3-Bromo-3-methylhexane

Planar, achiral

(R)-3-Methyl-3-hexanol

+

(S)-3-Methyl-3-hexanol

Racemic mixture

CHEMICAL HIGHLIGHT 7-1 **Incomplete Racemization in S_N1 Reactions**

Many solvolyses of optically pure compounds in which the leaving group is attached to a stereocenter give racemic mixtures, but many others do not. It is often found that unequal amounts of *R* and *S* products are obtained from S_N1 reactions, with the major enantiomer being the one in which inversion of the reacting center has taken place. The hydrolyses of optically pure samples of (1-chloro-) and (1-bromoethyl)benzene illustrate this phenomenon. The bromo compound reacts with water to give racemic 1-phenylethanol. In contrast, the chloro analog gives a 15–20% excess of the alcohol enantiomer with inverted configuration at the site of displacement.

Why does this result occur? Chloride ion is a good, but not particularly outstanding, leaving group. Its departure to give the secondary carbocation is sluggish. As a consequence, it has a tendency to

hover in the vicinity of the cationic carbon, partially shielding one face of the latter from attack by solvent. Attack by water on the opposite face of the carbocation is not inhibited, leading to a net excess of inverted product. In contrast, the halogen in the bromo compound is an excellent leaving group; in its departure, it does not shield the nearer face of the carbocation. Formation of a completely racemized hydrolysis product is observed.

Interestingly, this result is noticeable, at least qualitatively, without sophisticated equipment: The two enantiomeric 1-phenylethanols have noticeably different odors. If we start with (*R*)-chloride, the odor of the hydrolysis product mixture, containing excess (*S*)-alcohol, differs from that of the product mixture from hydrolysis of the (*S*)-chloride, in which (*R*)-alcohol is present in excess.

	S	*R*	
When X is Br:	50%	50%	Racemic
When X is Cl:	41%	59%	Excess of inversion

EXERCISE 7-3

(*R*)-3-Bromo-3-methylhexane loses its optical activity when dissolved in nitromethane, a highly polar but nonnucleophilic solvent. Explain.

EXERCISE 7-4

Hydrolysis of molecule A (shown in the margin) gives two alcohols. Explain.

A

7-4 Effects of Solvent, Leaving Group, and Nucleophile on Unimolecular Substitution

As in S_N2 reactions, varying the solvent, the leaving group, and the nucleophile greatly affects unimolecular substitution.

Polar solvents accelerate the S_N1 reaction

Heterolytic cleavage of the C–X bond in the rate-determining step of the S_N1 reaction entails a transition-state structure that is highly polarized (Figure 7-4; see page 254), eventually leading to two fully charged ions. In contrast, in a typical S_N2 transition state, charges are not created; rather, they are dispersed (see Figure 6-4).

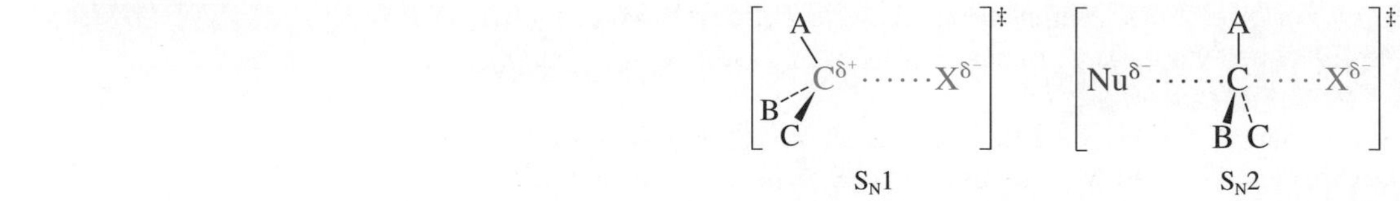

$$\left[\begin{array}{c} A \\ \overset{\displaystyle |}{\underset{\displaystyle B}{\diagdown}}C^{\delta+}\cdots\cdots X^{\delta-} \\ C \end{array} \right]^{\ddagger} \qquad \left[Nu^{\delta-}\cdots\cdots \overset{\displaystyle A}{\underset{\displaystyle B\ C}{\diagdown\!\!|\,}}C\cdots\cdots X^{\delta-} \right]^{\ddagger}$$

$$S_N1 \qquad\qquad\qquad S_N2$$

FIGURE 7-4

The respective transition states for the S_N1 and S_N2 reactions explain why the former is strongly accelerated by polar solvents. Heterolytic cleavage entails charge separation, a process aided by polar solvation.

Because of this polar transition state, the rate of an S_N1 reaction increases as solvent polarity is increased. The effect is particularly striking when the solvent is changed from aprotic to protic. For example, hydrolysis of 2-bromo-2-methylpropane is much faster in pure water than in a $9:1$ mixture of propanone (acetone) and water. The protic solvent accelerates the S_N1 reaction, because it stabilizes the transition state shown in Figure 7-4 by hydrogen bonding with the leaving group. Remember that, in contrast, the S_N2 reaction is accelerated in polar *aprotic* solvents, mainly because of a solvent effect on the reactivity of the nucleophile and *not* of the substrate.

Effect of Solvent on the Rate of an S_N1 Reaction

		Relative rate
$(CH_3)_3CBr \xrightarrow{\text{100\% H}_2\text{O}} (CH_3)_3COH + HBr$		400,000
$(CH_3)_3CBr \xrightarrow{\text{90\% propanone (acetone), 10\% H}_2\text{O}} (CH_3)_3COH + HBr$		1

The solvent nitromethane, CH_3NO_2 (see Table 6-5), is exceptional in being both highly polar and essentially nonnucleophilic. It therefore is useful in studies of S_N1 reactions with nucleophiles other than solvent molecules.

The S_N1 reaction speeds up with better leaving groups

Because the leaving group departs in the rate-determining step of the S_N1 reaction, it is not surprising that the rate of the reaction increases as the leaving-group ability of the departing group improves. Thus, tertiary iodoalkanes are more readily solvolyzed than are the corresponding bromides, and the latter are in turn more reactive than chlorides. Sulfonates are particularly prone to departure.

Relative Rate of Solvolysis of RX (R = Tertiary Alkyl)

$$X = -OSO_2R' > -I > -Br > -Cl$$

The strength of the nucleophile affects the product distribution but not the reaction rate

Does changing the nucleophile affect the rate of S_N1 reaction? The answer is no. Recall that, in the S_N2 process, the rate of reaction increases significantly as the nucleophilicity of the attacking species improves. However, because the rate-determining step of unimolecular substitution does *not* include the nucleophile, changing its structure (or concentration) should *not* alter the rate of disappearance of the haloalkane.

Nevertheless, when two or more nucleophiles compete for capture of the intermediate carbocation, their relative strengths and concentrations may greatly affect the *product distribution.*

For example, solvolysis of a 0.1 M solution of 2-chloro-2-methylpropane in methanol gives the expected 2-methoxy-2-methylpropane, with a rate constant k_1. Quite a different result is obtained when the same experiment is carried out in the presence of an equivalent amount of sodium azide: The product is 1,1-dimethylethyl (*tert*-butyl) azide, still formed at the *same* rate. In this case, the much more powerful nucleophile N$_3^-$ (see Table 6-7) wins out in competition with methanol. The rate of disappearance of 2-chloro-2-methylpropane is determined by k_1 (regardless of the product eventually formed), but the relative yields of the *products* depend on the relative reactivities of the competing nucleophiles (k_{CH_3OH} is much smaller than $k_{N_3^-}$).

Competing Nucleophiles in the S$_N$1 Reaction

$$(CH_3)_3CCl$$
$$+$$
$$CH_3OH \xrightarrow[\text{Rate determining}]{k_1} (CH_3)_3C^+ + Cl^-$$
$$+$$
$$NaN_3$$

$$\xrightarrow{k_{CH_3OH}} (CH_3)_3COCH_3 + HCl$$

2-Methoxy-2-methylpropane

$$\xrightarrow{k_{N_3^-}} (CH_3)_3CN_3 + NaCl$$

1,1-Dimethylethyl azide (*tert*-Butyl azide)

EXERCISE 7-5

A solution of 1,1-dimethylethyl (*tert*-butyl) methanesulfonate in polar aprotic solvent containing equal amounts of sodium fluoride and sodium bromide produces 75% 2-fluoro-2-methylpropane and only 25% 2-bromo-2-methylpropane. Explain. (**Hint:** Refer to Section 6-9 and Problem 43 in Chapter 6 for information regarding relative nucleophilic strengths of the halide ions in aprotic solvents.)

To summarize, we have seen further evidence supporting the S$_N$1 mechanism for the reaction of tertiary (and secondary) haloalkanes with certain nucleophiles. The stereochemistry of the process, the effects of the solvent and the leaving-group ability on the rate, and the absence of such effects when the strength of the nucleophile is varied are consistent with the unimolecular route. The next question to be answered is, Why? What is so special about tertiary haloalkanes that they undergo conversion by the S$_N$1 pathway, whereas primary systems follow S$_N$2? How do secondary haloalkanes fit into this scheme?

7-5 Effect of the Alkyl Group on the S$_N$1 Reaction: Carbocation Stability

Somehow, the degree of substitution at the reacting carbon must control the pathway followed in the reaction of haloalkanes (and related derivatives) with nucleophiles. We shall see that only secondary and tertiary systems can form carbocations. For this reason, tertiary halides, *whose steric bulk prevents them from undergoing S$_N$2 reactions,* transform solely by the S$_N$1 mechanism, primary haloalkanes only by S$_N$2, and secondary haloalkanes by either route, depending on conditions.

Carbocation stability increases from primary to secondary to tertiary

We have learned that primary haloalkanes undergo *only* direct nucleophilic substitution. In contrast, secondary systems often transform through carbocation intermediates, and tertiary systems always do. The reasons for this difference are twofold: First, steric hindrance increases along the series, thereby slowing down S$_N$2; and, second, increasing alkyl substitution stabilizes carbocationic centers. Only secondary and tertiary cations are energetically feasible under the conditions of the S$_N$1 reaction.

Relative Stability of Carbocations

$$CH_3CH_2CH_2\overset{+}{C}H_2 \quad < \quad CH_3CH_2\overset{+}{C}HCH_3 \quad < \quad (CH_3)_3\overset{+}{C}$$

Primary   <   **Secondary**   <   **Tertiary**

Now we can see why tertiary haloalkanes solvolyze so readily. Because tertiary carbocations are more stable than their less substituted relatives, they form more easily. But what is the reason for this order of stability?

Hyperconjugation stabilizes positive charge

Note that the order of carbocation stability parallels that of the corresponding radicals. Both trends have their roots in the same phenomenon: *hyperconjugation.* Recall from Section 3-2 that hyperconjugation is the result of overlap of a *p* orbital with a neighboring bonding molecular orbital, such as that of a C–H or a C–C bond. In a radical, the *p* orbital is singly filled; in a carbocation, it is empty. In both cases, the alkyl group donates electron density to the electron-deficient center and thus stabilizes it. Figure 7-5 compares the methyl cation, devoid of hyperconjugation, with the much more stable 1,1-dimethylethyl (*tert*-butyl) cation. Figure 7-6 shows the structure of the latter as derived from X-ray diffraction measurements.

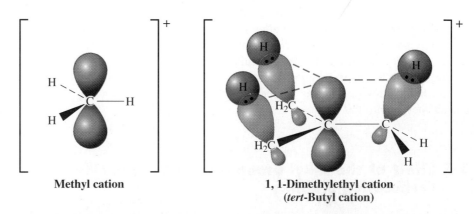

Methyl cation   **1, 1-Dimethylethyl cation** (*tert*-**Butyl cation**)

FIGURE 7-5

The methyl cation is not stabilized by hyperconjugation (left), whereas the 1,1-dimethylethyl (*tert*-butyl) cation benefits from three hyperconjugative interactions (right).

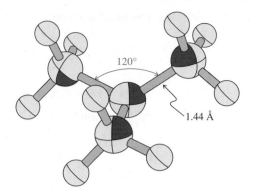

FIGURE 7-6

X-ray crystal structure determination for the 1,1-dimethylethyl (*tert*-butyl) cation. The four carbons lie in a plane with 120° C–C–C bond angles, consistent with sp^2 hybridization at the central carbon. The C–C bond length is 1.44 Å, shorter than normal single bonds, a consequence of hyperconjugative overlap.

Secondary systems undergo both S$_N$1 and S$_N$2 reactions

As you will have gathered from the preceding discussion, secondary haloalkanes exhibit the most varied substitution behavior. Do they prefer a bimolecular substitution pathway or are they more likely to enter into carbocation formation? *Both* are possible: Steric hindrance slows but does not preclude direct nucleophilic attack. At the same time, dissociation becomes competitive because of the relative stability of secondary carbocations. The pathway chosen depends on the reaction conditions: the solvent, the leaving group, and the nucleophile.

If we use a substrate carrying a very good leaving group, a nucleophile that is poor, and a polar protic solvent (S$_N$1 conditions), *unimolecular* substitution is favored. If we employ a high concentration of a good nucleophile, a polar aprotic solvent, and a haloalkane bearing a reasonable leaving group (S$_N$2 conditions), *bimolecular* substitution predominates. Table 7-2 summarizes our observations regarding the reactivity of haloalkanes toward nucleophiles.

TABLE 7-2	Reactivity of R–X in Nucleophilic Substitutions: R–X + Nu⁻ ⟶ R–Nu + X⁻	
R	**S$_N$1**	**S$_N$2**
CH$_3$	Not observed in solution (methyl cation too high in energy)	Frequent; fast with good nucleophiles and good leaving groups
Primary	Not observed in solution (primary carbocations too high in energy)^a	Frequent; fast with good nucleophiles and good leaving groups, slow when branching at C2 is present in R
Secondary	Relatively slow; best with good leaving groups in polar protic solvents	Relatively slow; best with high concentrations of good nucleophiles in polar aprotic solvents
Tertiary	Frequent; particularly fast in polar, protic solvents and with good leaving groups	Extremely slow

^{a}Exceptions are resonance-stabilized carbocations; see Chapter 14.

Substitution of a Secondary Haloalkane Under S_N2 Conditions

$$H_3C\underset{\underset{H}{|}}{\overset{CH_3}{C}}-Br + CH_3S^- \xrightarrow{\text{Propanone (acetone)}} CH_3S-\underset{\underset{H}{|}}{\overset{CH_3}{C}}CH_3 + Br^-$$

Substitution of a Secondary Substrate Under S_N1 Conditions

$$H_3C\overset{H_3C}{\underset{H}{C}}-OSCF_3 \xrightarrow{H_2O} H_3C\overset{H_3C}{\underset{H}{C}}-OH + CF_3SO_3H$$

EXERCISE 7-6

Explain the following results.

(a) [structure] $Cl\ H$ (R) $+ \ ^-CN \xrightarrow{\text{Propanone (acetone)}}$ [structure] $H\ CN$ (S)

(b) [structure] $I\ H$ (R) $+ CH_3OH \longrightarrow$ [structure] OCH_3 ($R + S$)

A visual demonstration of relative S_N1 reactivity. The three test tubes contain, from left to right, solutions of 1-bromobutane, 2-bromopropane, and 2-bromo-2-methylpropane in ethanol, respectively. Addition of a few drops of AgNO$_3$ solution to each causes immediate formation of a heavy AgBr precipitate from the *tert*-bromoalkane (right), less AgBr precipitation from the secondary substrate (center), and very little AgBr formation from the primary halide (left).

In contrast with S_N2 processes, S_N1 reactions are of limited use in synthesis because the chemistry of carbocations is complex. As we shall see in Chapter 9, these species are prone to rearrangements, frequently resulting in complicated mixtures of products. In addition, carbocations undergo another important reaction, as we shall see next: *loss of a proton* to furnish a double bond.

To summarize, tertiary haloalkanes are reactive in the presence of nucleophiles even though they are too sterically hindered to undergo S_N2 reactions: The tertiary carbocation is readily formed because it is stabilized by hyperconjugation. Subsequent trapping by a nucleophile, such as a solvent (solvolysis), results in the product of nucleophilic substitution. Primary haloalkanes do not react in this manner: The primary cation is too highly energetic (unstable) to be formed in solution. The primary substrate follows the S_N2 route. Secondary systems are converted into substitution products through either pathway, depending on the nature of the leaving group, the solvent, and the nucleophile.

7-6 Unimolecular Elimination: E1

We know that carbocations are readily trapped by nucleophiles through attack at the positively charged carbon. This is not their only mode of reaction, however. An alternative is deprotonation, furnishing a new class of compounds, the alkenes. Starting from a branched haloalkane, the overall transformation constitutes the removal of HX with the simultaneous generation of a double bond. The general term for such a process is **elimination,** abbreviated **E.**

Elimination

$$\underset{X}{\overset{H}{\underset{\diagup}{\diagdown}}}C-C\diagup \xrightarrow{\text{Base: :B}^-} \diagdown C=C\diagup + \text{H—B} + \text{X}^-$$

Eliminations can take place by several mechanisms. Let us establish the one that is followed in solvolysis.

When 2-bromo-2-methylpropane is dissolved in methanol, it rapidly disappears. As expected, the major product, 2-methoxy-2-methylpropane, arises by solvolysis. However, there is also a significant amount of another compound, 2-methylpropene, the product of *elimination* of HBr from the original substrate. Thus, in competition with the S_N1 process, which leads to displacement of the leaving group, another mechanism transforms the tertiary halide, giving rise to the alkene. What is it? Is it related to the S_N1 reaction? Once again we turn to a kinetic analysis and find that the rate of alkene formation depends on the concentration of *only* the starting halide; the reaction is first order. Because they are unimolecular, eliminations of this type are labeled **E1.** *The rate-determining step in the E1 process is the same as that in S_N1 reactions: dissociation to a carbocation. This intermediate then has a second pathway at its disposal along with nucleophilic trapping: loss of a proton from a carbon adjacent to the one bearing the positive charge.*

$$(\text{CH}_3)_3\text{CBr} \underset{}{\overset{\text{CH}_3\text{OH}}{\rightleftharpoons}} \text{H}_3\text{C}-\overset{+}{\underset{\text{CH}_3}{\overset{\text{CH}_3}{\text{C}}}} + \text{Br}^-$$

2-Bromo-2-methyl-propane

E1 ↙   ↘ S_N1 CH₃OH

$$\underset{\text{CH}_3}{\overset{\text{CH}_3}{\text{H}_2\text{C}=\text{C}}} + \text{H}^+ + \text{Br}^- \qquad\qquad (\text{CH}_3)_3\text{COCH}_3 + \text{H}^+ + \text{Br}^-$$

20%   80%

2-Methylpropene   **2-Methoxy-2-methylpropane**

How exactly is the proton lost? Figure 7-7 (see page 260) depicts this process with orbitals. Although we often show protons that evolve in chemical processes by using the notation H^+, "free" protons do not participate under the conditions of ordinary organic reactions. A Lewis base (Section 2-9) typically removes the proton. In aqueous solution, water plays this role, giving H_3O^+; here, the proton is carried off by CH_3OH as $CH_3OH_2^+$, an alkyloxonium ion. The carbon left behind rehybridizes from sp^3 to sp^2. As the C–H bond breaks, its electrons shift to overlap in a π fashion with the vacant p orbital at the neighboring cationic center. The result is a hydrocarbon containing a double bond: an alkene. The complete mechanism is as follows.

The E1 Reaction Mechanism

$$\underset{\text{CH}_3}{\overset{\text{CH}_3}{\text{CH}_3\text{C}-\text{Br}}} \underset{}{\overset{\text{CH}_3\text{OH}}{\rightleftharpoons}} \text{Br}^- + \underset{\text{H}_3\text{C}}{\overset{\text{H}_3\text{C}}{\overset{+}{\text{C}}-\text{C}}}\overset{\text{H}}{\underset{\text{H}}{}} \longrightarrow \underset{\text{H}_3\text{C}}{\overset{\text{H}_3\text{C}}{\text{C}=\text{C}}}\overset{\text{H}}{\underset{\text{H}}{}} + \underset{\text{H}}{\overset{\text{H}}{}}\overset{+}{\text{O}}\text{CH}_3$$

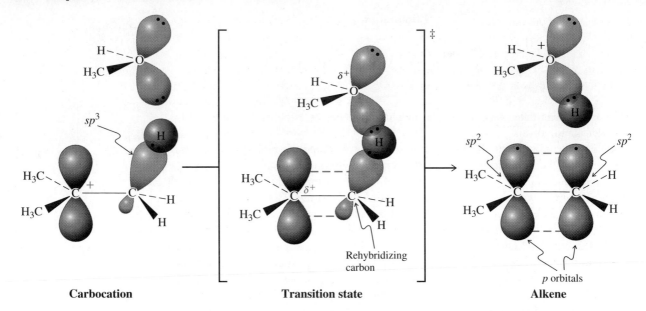

Carbocation   **Transition state**   **Alkene**

FIGURE 7-7

The alkene-forming step in unimolecular elimination (E1): deprotonation of a 1,1-dimethylethyl (*tert*-butyl) cation by the solvent methanol. In an orbital description of proton abstraction, an electron pair on the oxygen atom in the solvent attacks a hydrogen on a carbon adjacent to that bearing the positive charge. The proton is transferred, leaving an electron pair behind. As the carbon rehybridizes from sp^3 to sp^2, these electrons redistribute over the two p orbitals of the new double bond.

Any hydrogen positioned on *any carbon next to the center bearing the leaving group* can participate in the E1 reaction. The 1,1-dimethylethyl (*tert*-butyl) cation has nine such hydrogens, each of which is equally reactive. In this case, the product is the same regardless of the identity of the proton lost. In other cases, more than one product may be obtained. These pathways will be discussed in more detail in Chapter 11.

The E1 Reaction Can Give Product Mixtures

$$(CH_3CH_2)_2CH-\underset{\underset{Cl}{|}}{\overset{\overset{CH_3}{|}}{C}}-CH(CH_3)_2 \xrightarrow[-HCl^*]{CH_3OH,\ \Delta} (CH_3CH_2)_2CH-\underset{\underset{OCH_3}{|}}{\overset{\overset{CH_3}{|}}{C}}-CH(CH_3)_2$$

$$\mathbf{S_N1\ product}$$

$+$

$$\underset{(CH_3CH_2)_2CH}{\overset{\overset{CH_2}{\|}}{C}}\underset{CH(CH_3)_2}{\qquad} + \underset{(CH_3CH_2)_2CH}{\overset{CH_3}{C}}=\underset{CH_3}{\overset{CH_3}{C}} + \underset{CH_3CH_2}{\overset{CH_3CH_2}{C}}=\underset{CH(CH_3)_2}{\overset{CH_3}{C}}$$

$$\mathbf{E1\ products}$$

The nature of the leaving group should have no effect on the ratio of substitution to elimination, because the carbocation formed is the same in either case. This is in-

*This notation indicates that the elements of the acid have been removed from the starting material. In reality, the proton ends up protonating the base. This system will be used occasionally in other elimination reactions in this book.

deed observed qualitatively (Table 7-3). The product ratio may be affected by the addition of base, but at low base concentration this effect is usually small. Recall that strong bases are usually strong nucleophiles as well (Section 6-9), so addition of a base will generally not greatly favor deprotonation of the carbocation at the expense of nucleophilic attack, and the ratio of E1 to S_N1 products remains approximately constant. However, at high concentrations of strong base, the proportion of elimination rises dramatically. This effect is not the consequence of a change in the E1 : S_N1 ratio, however. Instead, a new pathway for elimination becomes important. This reaction is the subject of the next section.

TABLE 7-3	Ratio of S_N1 to E1 Products in the Hydrolyses of 2-Halo-2-methyl-propanes at 25°C	
X in (CH$_3$)$_3$CX		Ratio S_N1 : E1
Cl		95 : 5
Br		95 : 5
I		96 : 4

EXERCISE 7-7

When 2-bromo-2-methylpropane is dissolved in aqueous ethanol at 25°C, a mixture of $(CH_3)_3COCH_2CH_3$ (30%), $(CH_3)_3COH$ (60%), and $(CH_3)_2C{=}CH_2$ (10%) is obtained. Explain.

To summarize, carbocations formed in solvolysis reactions are not only trapped by nucleophiles to give S_N1 products but also deprotonated in an elimination (E1) reaction. In this process, the nucleophile (usually the solvent) acts as a base.

7-7 Bimolecular Elimination: E2

In addition to S_N2, S_N1, and E1 reactions, there is a fourth pathway by which haloalkanes may react with nucleophiles *that are also strong bases:* elimination by a *bimolecular* mechanism.

Strong bases effect bimolecular elimination

The preceding section taught us that unimolecular elimination may compete with substitution. A dramatic change of the kinetics is observed at higher concentrations of strong base, however. The rate of alkene formation becomes proportional to the concentrations of both the starting halide *and* the base: The kinetics of elimination are now second order, and the process is called **bimolecular elimination,** abbreviated **E2.**

Kinetics of the E2 Reaction of 2-Chloro-2-methylpropane

$$(CH_3)_3CCl \ + \ Na^+{}^-OH \ \xrightarrow{\ k\ } \ CH_2{=}C(CH_3)_2 \ + \ NaCl \ + \ H_2O$$
$$Rate \ = \ k[(CH_3)_3CCl][^-OH] \ mol \ L^{-1} \ s^{-1}$$

What causes this change in mechanism? Strong bases (such as hydroxide, HO$^-$, and alkoxides, RO$^-$) can attack haloalkanes before carbocation formation. The target is a hydrogen on a carbon atom *next to* the one carrying the leaving group. This reaction pathway is not restricted to tertiary halides, although, in secondary and primary systems, it must compete with the S_N2 process.

Competition Between E2 and S_N2 Reactions

$$CH_3CH_2CH_2Br \ \xrightarrow{CH_3O^-Na^+,\ CH_3OH} \ CH_3CH_2CH_2OCH_3 \ +$$

H₂C=CH₂ structure with H, H, H₃C, H substituents

92%   8%

1-Methoxypropane   **Propene**

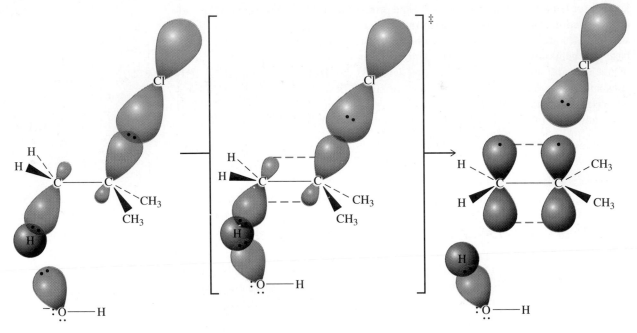

FIGURE 7-8
Orbital description of the E2 reaction of 2-chloro-2-methylpropane with hydroxide ion.

EXERCISE 7-8

What products do you expect from the reaction of bromocyclohexane with hydroxide ion?

EXERCISE 7-9

Give the products (if any) of the E2 reaction of the following substrates: CH_3CH_2I; CH_3I; $(CH_3)_3CCl$; $(CH_3)_3CCH_2I$.

E2 reactions proceed in one step

The bimolecular elimination mechanism consists of a *single step*. The bonding changes that occur in its transition state are shown here with electron pushing arrows; in Figure 7-8, they are shown with orbitals. Three changes take place:

1. Deprotonation by the base
2. Departure of the leaving group

The E2 Reaction Mechanism

3. Rehybridization of the reacting carbon centers from sp^3 to sp^2 to furnish the two p orbitals of the emerging double bond

All three take place *simultaneously*: The E2 reaction is a one-step, *concerted* process.

Notice that the E1 (Figure 7-7) and E2 mechanisms are very similar, differing only in the sequence of events. In the bimolecular reaction, proton abstraction and leaving-group departure are simultaneous. In the E1 process, the halide leaves first, to be followed by an attack by the base. A good way of thinking about the difference is to imagine that the strong base participating in the E2 reaction is more aggressive. It does not wait for the tertiary or secondary halide to dissociate but attacks the substrate directly.

Experiments elucidate the detailed structure of the E2 transition state

What is the experimental evidence in support of a one-step process with a transition state like that depicted in Figure 7-8? There are three pieces of relevant information. First, the second-order rate law requires that both the haloalkane and the base take part in the rate-determining step. Second, better leaving groups result in faster eliminations. This observation implies that the bond to the leaving group is partially broken in the transition state.

Relative Reactivity in the E2 Reaction

$$RCl < RBr < RI$$

Explain the result in the reaction shown below.

$$Cl{-}\hspace{-2pt}\bigcirc\hspace{-2pt}\bigcirc\hspace{-2pt}{-}I \xrightarrow{\text{CH}_3\text{O}^-} Cl{-}\hspace{-2pt}\bigcirc\hspace{-2pt}\bigcirc$$

The third observation is one that not only strongly suggests that both the C–H and the C–X bonds are broken in the transition state, but also describes their relative orientation in space when this event takes place. Figure 7-8 illustrates a characteristic feature of the E2 reaction: its stereochemistry. The substrate is pictured as reacting in a conformation that places the breaking C–H and C–X bonds in an *anti* relation. How can we establish the structure of the transition state with such precision? For this purpose, we can use the principles of conformation and stereochemistry. Treatment of *cis*-1-bromo-4-(1,1-dimethylethyl)cyclohexane with strong base leads to rapid bimolecular elimination to the corresponding alkene. In contrast, under the same conditions the *trans* isomer reacts only very slowly. Why? When we examine the most stable chair conformation of the *cis* compound, we find that two hydrogens are located *anti* to the axial bromine substituent. This geometry is very similar to that required by the E2 transition state, and consequently elimination is easy. Conversely, the *trans* system has no C–H bonds aligned *anti* to the equatorial leaving group (make a model). E2 elimination in this case would require either ring-flip to a diaxial conformer (see Section 4-4) or removal of a hydrogen *gauche* to the bromine, both energetically costly. The latter would be an example of an elimination proceeding through an unfavorable *syn* transition state (*syn*, Greek, together). We will return to E2 elimination in further detail in Chapter 11.

Anti Elimination Occurs Readily for _cis_- but Not for
trans-1-Bromo-4-(1,1-dimethylethyl)cyclohexane

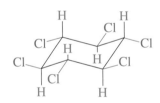

cis-1-Bromo-4-
(1,1-dimethylethyl)-
cyclohexane

(**Two _anti_** hydrogens)

trans-1-Bromo-4-
(1,1-dimethylethyl)-
cyclohexane

(**No _anti_ hydrogens; only**
anti ring carbons)

EXERCISE 7-11

The rate of elimination of _cis_-1-bromo-4-(1,1-dimethylethyl)cyclohexane is proportional to the concentration of both substrate and base, but that of the _trans_ isomer is proportional _only_ to the concentration of the substrate. Explain.

EXERCISE 7-12

The isomer of 1,2,3,4,5,6-hexachlorocyclohexane shown in the margin undergoes E2 elimination 7000 times _more slowly_ than any of its stereoisomers. Explain.

In summary, strong bases react with haloalkanes not only by substitution, but also by elimination. The kinetics of these reactions are second order, an observation pointing to a bimolecular mechanism. An _anti_ transition state is preferred, in which the base abstracts a proton at the same time as the leaving group departs.

7-8 Competition Between Substitution and Elimination

The multiple reaction pathways, S_N2, S_N1, E2, and E1, that haloalkanes may follow in the presence of nucleophiles may seem confusing. Given the many parameters that affect the relative importance of these transformations, are there some simple guidelines that might allow us to predict, at least roughly, what the outcome of any particular reaction will be? The answer is a cautious yes. This section will explain how consideration of _base strength_ and _steric bulk_ of the reacting species can help us decide whether substitution or elimination will predominate.

Weakly basic nucleophiles give substitution

Good nucleophiles that are weaker bases than hydroxide give good yields of S_N2 products with primary and secondary halides and of S_N1 products with tertiary substrates. Examples include I^-, Br^-, RS^-, N_3^-, $RCOO^-$, and PR_3. Thus, 2-bromopropane reacts with both iodide and acetate ions cleanly through the S_N2 pathway, with virtually no competing elimination.

$$\underset{\overset{|}{H}}{\overset{\overset{CH_3}{|}}{CH_3CBr}} + Na^+ I^- \xrightarrow{\text{Propanone (acetone)}} \underset{\overset{|}{H}}{\overset{\overset{CH_3}{|}}{CH_3CI}} + Na^+ Br^-$$

$$\underset{\overset{|}{H}}{\overset{\overset{CH_3}{|}}{CH_3CBr}} + \underset{}{\overset{\overset{O}{\|}}{CH_3CO^-}} Na^+ \xrightarrow{\text{Propanone (acetone)}} \underset{\underset{100\%}{\overset{|}{H}}}{\overset{\overset{CH_3}{|}}{CH_3COCCH_3}} + Na^+ Br^-$$

2-Chloro-2-methylpropane transforms with sodium azide in methanol to 1,1-dimethylethyl (*tert*-butyl) azide (Section 7-4) through the S_N1 mechanism.

$$\underset{\overset{|}{CH_3}}{\overset{\overset{CH_3}{|}}{CH_3CCl}} \xrightarrow{NaN_3, CH_3OH} \underset{\underset{\textbf{Major}}{\overset{|}{CH_3}}}{\overset{\overset{CH_3}{|}}{CH_3CN_3}} + \underset{\underset{\textbf{Minor}}{\overset{|}{CH_3}}}{\overset{\overset{CH_3}{|}}{CH_3COCH_3}} + \underset{CH_3 \quad CH_3}{\overset{\overset{CH_2}{\|}}{C}}$$

Weak nucleophiles such as water and alcohols react at appreciable rates only with secondary and tertiary halides, substrates capable of following the S_N1 pathway. Unimolecular elimination is usually only a minor side reaction.

$$\underset{}{\overset{\overset{Br}{|}}{CH_3CH_2CHCH_2CH_3}} \xrightarrow{H_2O, CH_3OH, 80°C} \underset{85\%}{\overset{\overset{OH}{|}}{CH_3CH_2CHCH_2CH_3}} + \underset{15\%}{CH_3CH=CHCH_2CH_3}$$

Strongly basic nucleophiles give more elimination as steric bulk increases

We have seen (Section 7-7) that strong bases may give rise to elimination through the E2 pathway. Is there some straightforward way to predict how much elimination will occur in competition with substitution in any particular situation? Yes, but other factors need to be considered. Let us examine the reactions of sodium ethoxide, a strong base, with several halides, measuring the relative amounts of ether and alkene produced in each case.

	Ether (Substitution product)	**Alkene** (Elimination product)

$$CH_3CH_2CH_2Br \xrightarrow[-\text{HBr}]{CH_3CH_2O^- Na^+, CH_3CH_2OH} \underset{91\%}{CH_3CH_2CH_2OCH_2CH_3} + \underset{9\%}{\overset{H_3C \quad\quad H}{\underset{H \quad\quad\quad H}{C=C}}}$$

$$\underset{\overset{|}{H}}{\overset{\overset{CH_3}{|}}{CH_3CCH_2Br}} \xrightarrow[-\text{HBr}]{CH_3CH_2O^- Na^+, CH_3CH_2OH} \underset{\underset{40\%}{\overset{|}{H}}}{\overset{\overset{CH_3}{|}}{CH_3CCH_2OCH_2CH_3}} + \underset{60\%}{\overset{H_3C \quad\quad H}{\underset{H_3C \quad\quad\quad H}{C=C}}}$$

$$CH_3\overset{\displaystyle CH_3}{\underset{\displaystyle H}{CBr}} \xrightarrow[-\,HBr]{CH_3CH_2O^-Na^+,\ CH_3CH_2OH} CH_3\overset{\displaystyle CH_3}{\underset{\displaystyle H}{COCH_2CH_3}} + \overset{\displaystyle H_3C}{\underset{\displaystyle H}{}}C{=}C\overset{\displaystyle H}{\underset{\displaystyle H}{}}$$

$$13\% \qquad\qquad 87\%$$

Reactions of simple primary halides with strongly basic nucleophiles give mostly S_N2 products. As steric bulk is increased around the carbon bearing the leaving group, substitution is retarded relative to elimination because an attack at carbon is subject to more steric hindrance than is an attack on hydrogen. Thus, branched primary substrates give about equal amounts of S_N2 and E2 reaction, whereas E2 is the major outcome with secondary substrates.

The S_N2 mechanism is not an option for tertiary halides. S_N1 and E1 pathways compete under neutral or weakly basic conditions. However, high concentrations of strong base give exclusive E2 reaction.

Sterically hindered basic nucleophiles favor elimination

We have seen that primary haloalkanes react by substitution with good nucleophiles, including strong bases. The situation changes when the steric bulk of the nucleophile hinders attack at the electrophilic carbon. In this case, elimination may predominate, even with primary systems, through deprotonation at the less hindered periphery of the molecule.

$$CH_3CH_2CH_2CH_2Br \xrightarrow[-\,HBr]{(CH_3)_3CO^-K^+,\ (CH_3)_3COH} CH_3CH_2CH{=}CH_2 + CH_3CH_2CH_2CH_2OC(CH_3)_3$$

$$85\% \qquad\qquad 15\%$$

Two examples of sterically hindered bases are potassium *tert*-butoxide and lithium diisopropylamide (LDA). When used in elimination reactions, they are frequently dissolved in their conjugate acids, 2-methyl-2-propanol and *N*-(1-methylethyl)-1-methylethanamine (diisopropylamine), respectively.

Sterically Hindered Bases

Potassium *tert*-butoxide

Lithium diisopropylamide
(LDA)

In summary, we have identified three principal factors that affect the competition between substitution and elimination: basicity of the nucleophile, steric hindrance in the haloalkane, and steric bulk around the nucleophilic (basic) atom.

FACTOR 1. Base strength of the nucleophile

Weak Bases	**Strong Bases**
H_2O^*, ROH^*, PR_3, halides, RS^-, N_3^-, NC^-, $RCOO^-$	HO^-, RO^-, H_2N^-, R_2N^-
Substitution more likely	Likelihood of elimination increased

FACTOR 2. Steric hindrance around the reacting carbon

Sterically Unhindered	**Sterically Hindered**
Primary haloalkanes	Branched primary, secondary, tertiary haloalkanes
Substitution more likely	Likelihood of elimination increased

FACTOR 3. Steric hindrance in the nucleophile (strong base)

Sterically Unhindered	**Sterically Hindered**
HO^-, CH_3O^-, $CH_3CH_2O^-$, H_2N^-	$(CH_3)_3CO^-$, $[(CH_3)_2CH]_2N^-$
Substitution may occur	Elimination strongly favored

For simple predictive purposes, we assume that their relative importance is equal in determining the ratio of elimination to substitution. Thus, the "majority rules." This method of analysis is quite reliable. Verify that it applies to the examples of this section and the summary section that follows.

EXERCISE 7-13

Which nucleophile in each of the following pairs will give a higher elimination:substitution product ratio in reaction with 1-bromo-2-methylpropane?

$$CH_3$$
$$|$$

(a) $N(CH_3)_3$, $P(CH_3)_3$ **(b)** H_2N^-, $(CH_3CH)_2N^-$ **(c)** I^-, Cl^-

---

EXERCISE 7-14

In all cases where substitution and elimination compete, higher reaction temperatures lead to greater proportions of elimination products. Thus, the amount of elimination accompanying hydrolysis of 2-bromo-2-methylpropane doubles as the temperature is raised from 25° to 65°C, and that from reaction of 2-bromopropane with ethoxide rises from 80% at 25°C to nearly 100% at 55°C. Explain.

7-9 Summary of Reactivity of Haloalkanes

Primary, secondary, and tertiary haloalkanes may react with nucleophiles through different pathways.

PRIMARY HALOALKANES. Unhindered primary alkyl substrates always react in a bimolecular way and almost always give predominantly substitution products, except when sterically hindered strong bases, such as potassium *tert*-butoxide, are employed. In these cases, the S_N2 pathway is slowed down sufficiently for steric reasons to allow the E2 mechanism to take over. Another way of reducing substitution is to in-

*Reacts only with S_N1 substrates; no reaction with simple primary halides.

troduce branching. However, even in these cases, good nucleophiles still furnish predominantly substitution products. Only strong bases, such as alkoxides, RO^-, or amides, R_2N^-, tend to react by elimination.

Reactivity of Primary Haloalkanes R–X with Nucleophiles (Bases)

For unhindered primary R–X:

S_N2 with good nucleophiles that are not strongly basic

$$CH_3CH_2CH_2Br + {}^-CN \xrightarrow{\text{Propanone (acetone)}} CH_3CH_2CH_2CN + Br^-$$

S_N2 with good nucleophiles that are also strong bases

$$CH_3CH_2CH_2Br + CH_3O^- \xrightarrow{CH_3OH} CH_3CH_2CH_2OCH_3 + Br^-$$

But E2 with strong, hindered base

$$CH_3CH_2CH_2Br + CH_3\overset{\overset{\displaystyle CH_3}{|}}{\underset{\underset{\displaystyle CH_3}{|}}{C}}O^- \xrightarrow[-HBr]{(CH_3)_3COH} CH_3CH=CH_2$$

No (or exceedingly slow) reaction with poor nucleophiles (CH_3OH)

For branched primary R–X:

S_N2 with good nucleophiles (although slow compared with unhindered R–X)

$$CH_3\overset{\overset{\displaystyle CH_3}{|}}{\underset{\underset{\displaystyle H}{|}}{C}}CH_2Br + I^- \xrightarrow{\text{Propanone (acetone)}} CH_3\overset{\overset{\displaystyle CH_3}{|}}{\underset{\underset{\displaystyle H}{|}}{C}}CH_2I + Br^-$$

E2 with strong base (not necessarily hindered)

$$CH_3\overset{\overset{\displaystyle CH_3}{|}}{\underset{\underset{\displaystyle H}{|}}{C}}CH_2Br + CH_3CH_2O^- \xrightarrow[-HBr]{CH_3CH_2OH} CH_3\overset{\overset{\displaystyle CH_3}{|}}{C}=CH_2$$

No (or exceedingly slow) reaction with poor nucleophiles

SECONDARY HALOALKANES. Secondary alkyl systems undergo, depending on conditions, both eliminations and substitutions by either possible pathway: uni- or bimolecular. Good nucleophiles favor S_N2, strong bases result in E2, and weakly nucleophilic polar media give mainly S_N1 and E1.

Reactivity of Secondary Haloalkanes R–X with Nucleophiles (Bases)

S_N1 and E1 when X is a good leaving group in a highly polar medium with weak nucleophiles

$$CH_3\overset{\overset{\displaystyle CH_3}{|}}{\underset{\underset{\displaystyle H}{|}}{C}}Br \xrightarrow[-HBr]{CH_3CH_2OH} CH_3\overset{\overset{\displaystyle CH_3}{|}}{\underset{\underset{\displaystyle H}{|}}{C}}OCH_2CH_3 + CH_3CH=CH_2$$

Major   **Minor**

S_N2 with high concentrations of good, weakly basic nucleophiles

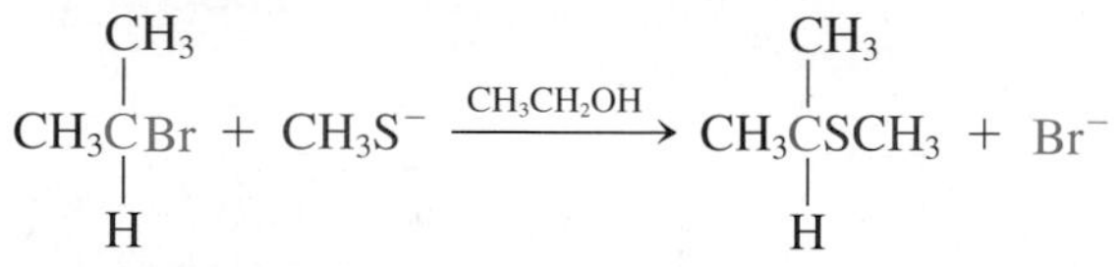

$$CH_3\overset{\displaystyle CH_3}{\underset{\displaystyle H}{C}}Br + CH_3S^- \xrightarrow{CH_3CH_2OH} CH_3\overset{\displaystyle CH_3}{\underset{\displaystyle H}{C}}SCH_3 + Br^-$$

E2 with high concentrations of strong base (for example, HO^- or RO^- in alcohol solvent)

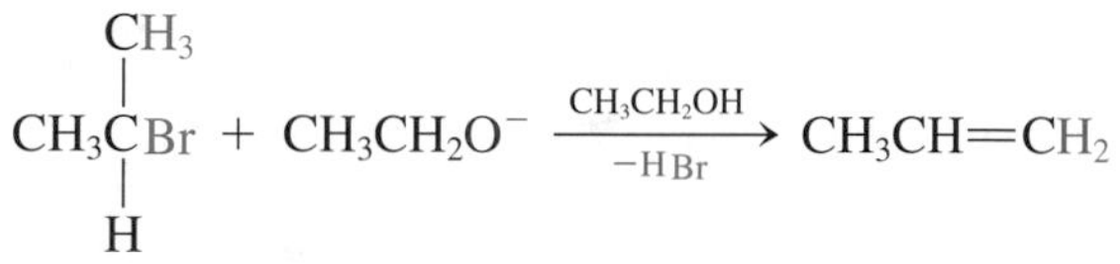

$$CH_3\overset{\displaystyle CH_3}{\underset{\displaystyle H}{C}}Br + CH_3CH_2O^- \xrightarrow[-HBr]{CH_3CH_2OH} CH_3CH{=}CH_2$$

TERTIARY HALOALKANES. Tertiary systems eliminate (E2) with concentrated strong base and are substituted in nonbasic media (S_N1). Bimolecular substitution is not observed, but elimination by E1 accompanies S_N1.

Reactivity of Tertiary Haloalkanes R–X with Nucleophiles (Bases)

S_N1 and E1 in polar solvents when X is a good leaving group and dilute or no base is present

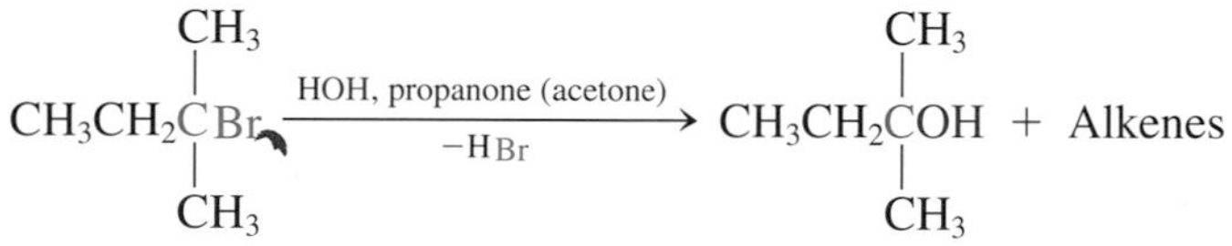

$$CH_3CH_2\overset{\displaystyle CH_3}{\underset{\displaystyle CH_3}{C}}Br \xrightarrow[-HBr]{HOH,\ propanone\ (acetone)} CH_3CH_2\overset{\displaystyle CH_3}{\underset{\displaystyle CH_3}{C}}OH + Alkenes$$

E2 with high concentrations of strong base

$$CH_3CH_2\overset{\displaystyle \overset{CH_3}{|}\underset{|}{CH_2}}{\underset{\displaystyle \underset{|}{CH_2}\underset{|}{CH_3}}{C}}Cl \xrightarrow[-HCl]{CH_3O^-,\ CH_3OH} CH_3CH_2\overset{\displaystyle \overset{CH_3}{|}\underset{}{CH_2}}{C}{=}CHCH_3$$

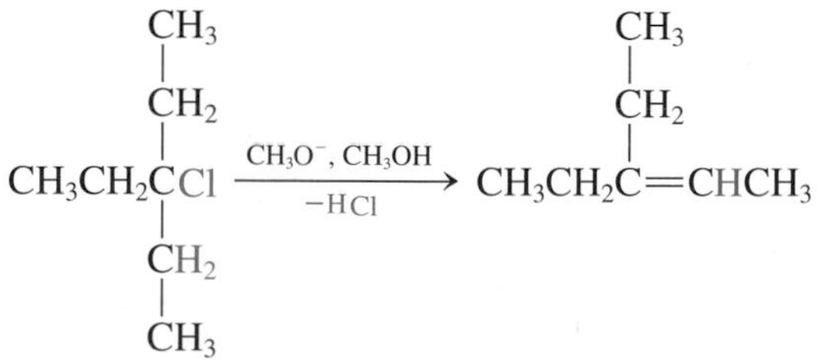

Table 7-4 (see page 270) summarizes the most likely mechanisms by which the haloalkanes undergo substitution and elimination.

EXERCISE 7-15

Predict which reaction in each of the following pairs will have a higher E2:E1 product ratio and explain why.

(a) 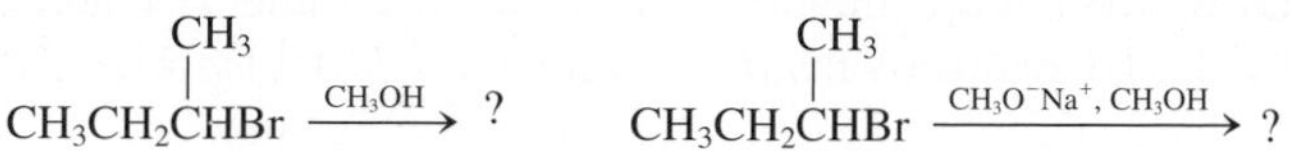 $CH_3CH_2\overset{\displaystyle CH_3}{\underset{}{C}}HBr \xrightarrow{CH_3OH} ?$ $CH_3CH_2\overset{\displaystyle CH_3}{\underset{}{C}}HBr \xrightarrow{CH_3O^-Na^+,\ CH_3OH} ?$

(b) 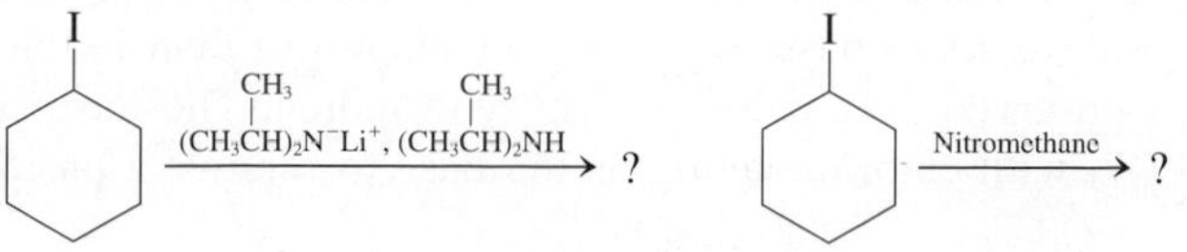 $\xrightarrow{(CH_3CH)_2N^-Li^+,\ (CH_3CH)_2NH} ?$ $\xrightarrow{Nitromethane} ?$

TABLE 7-4	**Likely Mechanisms by Which Haloalkanes React with Nucleophiles (Bases)**			
	Type of nucleophile (base)			
Type of haloalkane	**Poor nucleophile (e.g., H_2O)**	**Weakly basic, good nucleophile (e.g., I^-)**	**Strongly basic, unhindered nucleophile (e.g., CH_3O^-)**	**Strongly basic, hindered nucleophile (e.g., $(CH_3)_3CO^-$)**
Methyl	No reaction	S_N2	S_N2	S_N2
Primary				
Unhindered	No reaction	S_N2	S_N2	E2
Branched	No reaction	S_N2	E2	E2
Secondary	Slow S_N1, E1	S_N2	E2	E2
Tertiary	S_N1, E1	S_N1, E1	E2	E2

CHAPTER INTEGRATION PROBLEM

a. 2-Bromo-2-methylpropane (*tert*-butyl bromide) reacts readily in nitromethane with chloride and iodide ions.
1. Write the structures of the substitution products, and write the complete mechanism by which one of them is formed.
2. Assume equal concentrations of all reactants, and predict the relative rates of these two reactions.
3. Which reaction will give more elimination? Write its mechanism.

SOLUTION
1. We begin by analyzing the participating species and then recognizing just what kind of reaction is likely to take place. The substrate is a haloalkane with a good leaving group attached to a *tertiary* carbon. According to Table 7-4, displacement by the S_N2 mechanism is not an option, but S_N1, E1, and E2 processes are possibilities. Chloride and iodide are good nucleophiles and weak bases, suggesting that substitution should predominate to give $(CH_3)_3CCl$ and $(CH_3)_3CI$ as the products, respectively (Section 7-8). The mechanism (S_N1) is as shown in Section 7-2 except that, subsequent to initial ionization of the C–Br bond to give the carbocation, halide ion attacks at carbon to give the final product directly. The very polar nitromethane is a good solvent for S_N1 reactions (Section 7-4).
2. We learned in Section 7-4 that different nucleophilic power has no effect on the rates of unimolecular processes. The rates should be (and, experimentally, are) identical.
3. This part requires a bit more thought. According to Table 7-4 and Section 7-8, elimination by the E1 pathway always accompanies S_N1 displacement. However, increasing the base strength of the nucleophile may "turn on" the E2 mechanism, increasing the proportion of elimination product. Referring to Tables 6-4 and 6-7, we see that chloride is more basic (and less nucleophilic) than iodide. More elimination is indeed observed with chloride than with iodide. The mechanism is as shown in Figure 7-7, with chloride acting as the base to remove a proton from the carbocation.

b. The table in the margin presents data for the reactions that take place when the chloro compound shown here is dissolved in propanone (acetone) containing varying quantities of water and sodium azide, NaN_3:

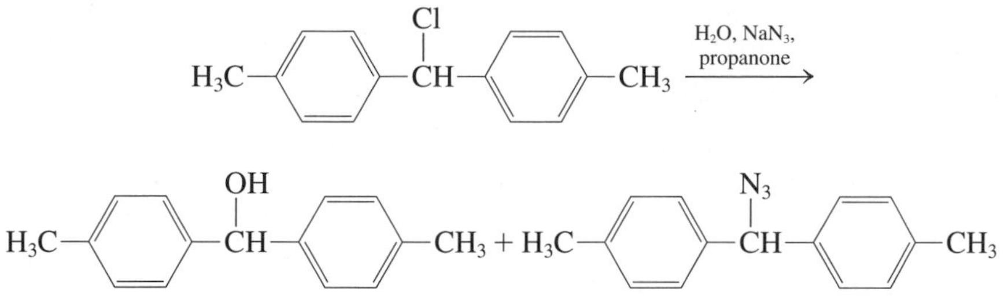

% H_2O	[N_3^-]	% RN_3	k_{rel}
10	0 M	0	1
10	0.05 M	60	1.5
15	0.05 M	60	7
20	0.05 M	60	22
50	0.05 M	60	*
50	0.10 M	75	*
50	0.20 M	85	*
50	0.50 M	95	*

*Too fast to measure.

In the table, % H_2O is the percentage of water by volume in the solvent, [N_3^-] is the initial concentration of sodium azide, % RN_3 is the percentage of organic azide in the product mixture (the remainder is the alcohol), and k_{rel} is the relative rate constant for the reaction, derived from the rate at which the starting material is consumed. The initial concentration of substrate is 0.04 M in all experiments. Answer the following questions.

   1. Describe and explain the effects of changing the percentage of H_2O on the rate of the reaction and on the product distribution.
   2. Do the same for the effects of changing [N_3^-]. Additional information: The reaction rates shown are the same when other ions—for example, Br^- or I^-—are used instead of azide.

SOLUTION

1. We begin by examining the data in the table, specifically lines 2-5, which compare reactions in the presence of different amounts of water at constant azide concentration. The rate of substitution increases rapidly as the proportion of water goes up, but the ratio of the two products stays the same: 60% azide and 40% alcohol. These two results suggest that the only effect of increasing the amount of water is to make the solvent environment more polar, thereby speeding up the initial ionization of the substrate. Even when the proportion of water is only 10%, it is present in great excess and is trapping carbocations as fast as it can, relative to the rate that azide ion reacts with the same intermediates (Section 7-2).

2. We note from lines 1 and 2 in the table that the reaction rate rises by about 50% when NaN_3 is added. Without further information, we might assume that this effect is a consequence of the occurrence of the S_N2 mechanism. Were that to be the case, however, other anions should affect the rate differently. But we were told that bromide and iodide, far more powerful nucleophiles, affect the measured rate in exactly the same way as does azide. We can explain this observation only by assuming that displacement is entirely by the S_N1 mechanism, and added ions affect the rate only by increasing the polarity of the solution and speeding up ionization (Section 7-4).

   In lines 5–8 of the table we note that increasing the amount of azide ion increases the amount of azide-containing product that is formed. At the higher concentrations, azide, a better nucleophile than water, is better able to complete for reaction with the carbocation intermediate.

NEW REACTIONS

1. Bimolecular Substitution—S$_N$2 (Sections 6-3 through 6-10, 7-5)

Primary and secondary substrates only

$$H_3C, H, CH_2CH_3\text{—}C\text{—}I \xrightarrow{:Nu^-} Nu\text{—}C(CH_3)(H)(CH_2CH_3) + I^-$$

Direct backside displacement with 100% inversion of configuration

2. Unimolecular Substitution—S$_N$1 (Sections 7-1 through 7-5)

Secondary and tertiary substrates only

$$CH_3\text{—}CBr(CH_3)(CH_3) \xrightarrow{-Br^-} CH_3\text{—}C^+(CH_3)(CH_3) \xrightarrow{:Nu^-} CH_3\text{—}CNu(CH_3)(CH_3)$$

Through carbocation: Chiral systems are racemized

3. Unimolecular Elimination—E1 (Section 7-6)

Secondary and tertiary substrates only

$$CH_3\text{—}CCl(CH_3)(CH_3) \xrightarrow{-Cl^-} CH_3\text{—}C^+(CH_3)(CH_3) \xrightarrow{:B^-} CH_2\!\!=\!\!C(H_3C)(CH_3) + BH$$

Through carbocation

4. Bimolecular Elimination—E2 (Section 7-7)

$$CH_3CH_2CH_2I \xrightarrow{:B^-} CH_3CH\!\!=\!\!CH_2 + BH + I^-$$

Simultaneous elimination of leaving group and neighboring proton

IMPORTANT CONCEPTS

1. Secondary haloalkanes undergo slow and tertiary haloalkanes fast **unimolecular substitution** in polar media. When the solvent serves as the nucleophile, the process is called **solvolysis.**
2. The slowest, or rate-determining, step in unimolecular substitution is dissociation of the C–X bond to form a **carbocation** intermediate. Added strong nucleophile changes the product but not the reaction rate.
3. Carbocations are stabilized by **hyperconjugation:** Tertiary are the most stable, followed by secondary. Primary and methyl cations are too unstable to form in solution.
4. **Racemization** often results when unimolecular substitution takes place at a chiral carbon.
5. **Unimolecular elimination** to form an alkene accompanies substitution in secondary and tertiary systems.
6. High concentrations of strong base may bring about **bimolecular elimination.** Expulsion of the leaving group accompanies removal of a hydrogen from the neighboring carbon by the base. The stereochemistry indicates an *anti* conformational arrangement of the hydrogen and the leaving group.
7. Substitution is favored by unhindered substrates and small, less basic nucleophiles.
8. Elimination is favored by hindered substrates and bulky, more basic nucleophiles.

PROBLEMS

16. What is the major substitution product of each of the following solvolysis reactions?

(a) CH$_3$$\overset{\displaystyle CH_3}{\underset{\displaystyle CH_3}{\overset{|}{\underset{|}{C}}}}$Br $\xrightarrow{\text{CH}_3\text{CH}_2\text{OH}}$
(b) (CH$_3$)$_2$$\overset{\displaystyle Br}{\overset{|}{C}}CH_2CH_3$ $\xrightarrow{\text{CF}_3\text{CH}_2\text{OH}}$

(c) [cyclopentane with CH$_3$CH$_2$Cl substituent] $\xrightarrow{\text{CH}_3\text{OH}}$
(d) [cyclohexyl]$\overset{\displaystyle Br}{\underset{\displaystyle CH_3}{\overset{|}{\underset{|}{C}}}}$—CH$_3$ $\xrightarrow{\overset{\displaystyle O}{\overset{\|}{\text{HCOH}}}}$

(e) CH$_3$$\overset{\displaystyle CH_3}{\underset{\displaystyle CH_3}{\overset{|}{\underset{|}{C}}}}$Cl $\xrightarrow{\text{D}_2\text{O}}$
(f) CH$_3$$\overset{\displaystyle CH_3}{\underset{\displaystyle CH_3}{\overset{|}{\underset{|}{C}}}}$Cl $\xrightarrow{\text{[cyclohexyl } \overset{H}{\underset{OD}{}}\text{]}}$

17. Write the two major substitution products of the reaction shown in the margin. **(a)** Write a mechanism to explain the formation of each of them. **(b)** Monitoring the reaction mixture reveals that an *isomer* of the starting material is generated as an intermediate. Draw its structure and explain how it is formed.

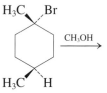

18. Give the two major substitution products of the following reaction.

[Newman projection: OSO$_2$CH$_3$, H$_3$C, C$_6$H$_5$, H$_3$C, C$_6$H$_5$, H] $\xrightarrow{\text{CH}_3\text{CH}_2\text{OH}}$

19. How would each reaction in Problem 16 be affected by the addition of each of the following substances to the solvolysis mixture?
(a) H$_2$O   **(b)** KI
(c) NaN$_3$ **(d)** CH$_3$CH$_2$OCH$_2$CH$_3$ (**Hint:** Low polarity.)

20. Rank the following carbocations in decreasing order of stability.

[three cyclopentane carbocation structures: H CH$_3$ with +, H CH$_2$$^+$, CH$_3$ with +]

21. Rank the compounds in each of the following groups in order of decreasing rate of solvolysis in aqueous propanone (acetone).

(a) CH$_3$$\overset{\displaystyle CH_3}{\overset{|}{C}}HCH_2CH_2$Cl   CH$_3$$\overset{\displaystyle CH_3}{\overset{|}{C}}HCH\underset{\displaystyle Cl}{\underset{|}{C}}H_3$ CH$_3$$\overset{\displaystyle CH_3}{\underset{\displaystyle Cl}{\overset{|}{\underset{|}{C}}}}CH_2CH_3$

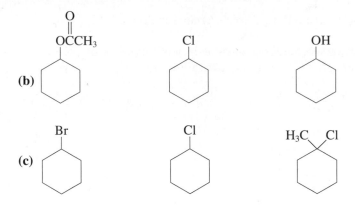

22. Give the products of the following substitution reactions. Indicate whether they arise through the S_N1 or the S_N2 process. Formulate the detailed mechanisms of their generation.

(a) $(CH_3)_2CHOSO_2CF_3 \xrightarrow{CH_3CH_2OH}$ **(b)** [cyclopentane with CH₃ and Br substituents] $\xrightarrow{\text{Excess } CH_3SH, \, CH_3OH}$

(c) $CH_3CH_2CH_2CH_2Br \xrightarrow{(C_6H_5)_3P, \, DMSO}$

(d) $CH_3CH_2CHClCH_2CH_3 \xrightarrow{NaI, \, propanone \, (acetone)}$

23. Give the product of each of the following substitution reactions. Which of these transformations should proceed faster in a polar, aprotic solvent (such as propanone or DMSO) than in a polar, protic solvent (such as water or CH_3OH)? Explain your answer on the basis of the mechanism that you expect to be operating in each case.
(a) $CH_3CH_2CH_2Br + Na^+ \, {}^-CN \longrightarrow$
(b) $(CH_3)_2CHCH_2I + Na^+N_3^- \longrightarrow$
(c) $(CH_3)_3CBr + HSCH_2CH_3 \longrightarrow$
(d) $(CH_3)_2CHOSO_2CH_3 + HOCH(CH_3)_2 \longrightarrow$

24. Propose a synthesis of (R)-$CH_3CHN_3CH_2CH_3$, starting from (R)-2-chloro-butane.

25. Two substitution reactions of (S)-2-bromobutane are shown here. Show their stereochemical outcomes.

$$(S)\text{-}CH_3CH_2CHBrCH_3 \xrightarrow{H\overset{O}{\overset{\|}{C}}OH}$$

$$(S)\text{-}CH_3CH_2CHBrCH_3 \xrightarrow{H\overset{O}{\overset{\|}{C}}O^-Na^+, \, DMSO}$$

26. Write all possible E1 products of each reaction in Problem 16.

27. Write the products of the following elimination reactions. Specify the predominant mechanism (E1 or E2) and formulate it in detail.

(a) $(CH_3CH_2)_3CBr \xrightarrow{NaNH_2, \ NH_3}$

(b) $CH_3CH_2CH_2CH_2Cl \xrightarrow{KOC(CH_3)_3, \ (CH_3)_3COH}$

(c) 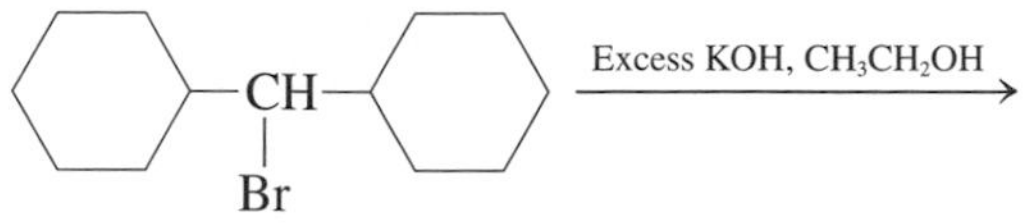 $\xrightarrow{Excess \ KOH, \ CH_3CH_2OH}$

(d) 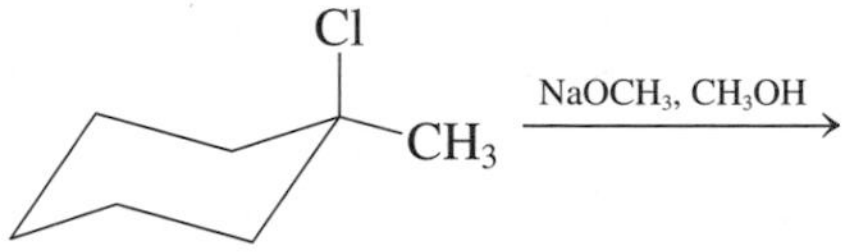 $\xrightarrow{NaOCH_3, \ CH_3OH}$

28. Predict the major product(s) that should form from reaction between 1-bromobutane and each of the following substances. By which reaction mechanism is each formed—S_N1, S_N2, E1, or E2? If it appears that a reaction will either not take place or be exceedingly slow, write "no reaction." Assume that each reagent is present in large excess. The solvent for each reaction is given.
(a) KCl in DMF   **(b)** KI in DMF   **(c)** KCl in CH_3NO_2
(d) NH_3 in CH_3CH_2OH **(e)** $NaOCH_2CH_3$ in CH_3CH_2OH **(f)** CH_3CH_2OH
(g) $KOC(CH_3)_3$ in $(CH_3)_3COH$ **(h)** $(CH_3)_3P$ in CH_3OH **(i)** CH_3CO_2H

29. Predict the major product(s) and mechanism(s) for reaction between 2-bromobutane (*sec*-butyl bromide) and each of the reagents in Problem 28.

30. Predict the major product(s) and mechanism(s) for reaction between 2-bromo-2-methylpropane (*tert*-butyl bromide) and each of the reagents in Problem 28.

31. Three reactions of 2-chloro-2-methylpropane are shown here. **(a)** Write the major product of each transformation. **(b)** Compare the rates of the three reactions. Assume identical solution polarities and reactant concentrations. Explain mechanistically.

$$(CH_3)_3CCl \xrightarrow{H_2S, \ CH_3OH}$$

$$(CH_3)_3CCl \xrightarrow{CH_3\overset{\overset{\displaystyle O}{\|}}{C}O^-K^+, \ CH_3OH}$$

$$(CH_3)_3CCl \xrightarrow{CH_3O^-K^+, \ CH_3OH}$$

32. Give the major product(s) of the following reactions. Indicate which of the following mechanism(s) is in operation: S_N1, S_N2, E1, or E2. If no reaction takes place, write "no reaction."

(a) 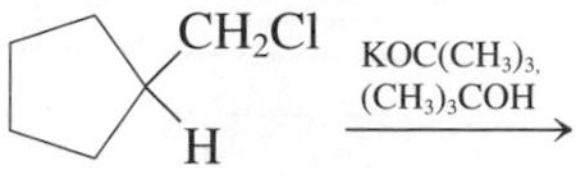 $\xrightarrow[(CH_3)_3COH]{KOC(CH_3)_3,}$

(b) 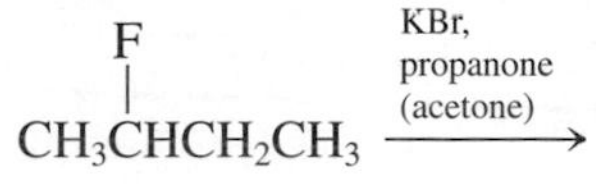 $CH_3\overset{\overset{\displaystyle F}{|}}{C}HCH_2CH_3 \xrightarrow[\substack{propanone \\ (acetone)}]{KBr,}$

(c) $H_3C\overset{\overset{\displaystyle CH_2CH_3}{|}}{\underset{\underset{\displaystyle H}{|}}{C}}Br \xrightarrow{H_2O}$

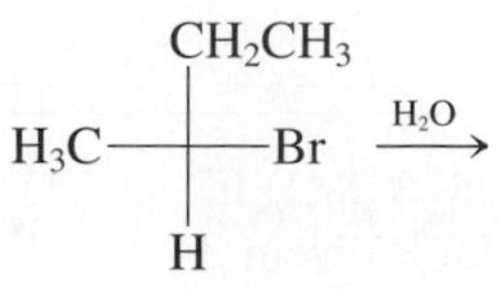

(d) 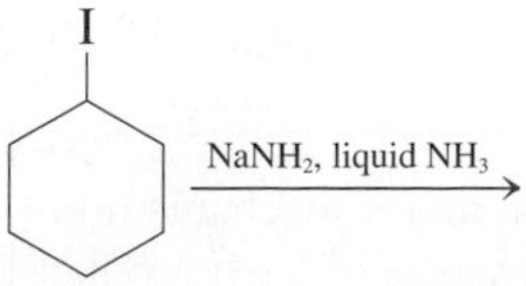 $\xrightarrow{NaNH_2, \ liquid \ NH_3}$

(e) $(CH_3)_2CHCH_2CH_2CH_2Br$ $\xrightarrow{NaOCH_2CH_3,\ CH_3CH_2OH}$

(f) 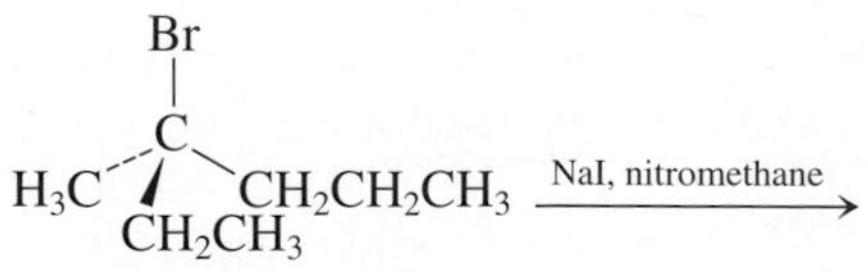 $H_3C\overset{\displaystyle \underset{|}{C}}{\diagdown}CH_2CH_2CH_3$ with Br above C and CH_2CH_3 below $\xrightarrow{NaI,\ nitromethane}$

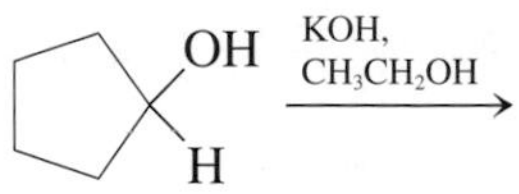

(g) cyclopentane ring with OH and H $\xrightarrow[CH_3CH_2OH]{KOH,}$

(h) $Cl-$ cyclohexane ring $-CH_2CH_2CH_2Br$ $\xrightarrow[\substack{KCN,\\CH_3OH}]{Excess}$

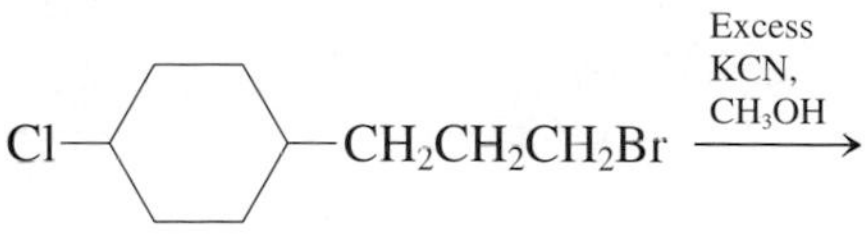

(i) 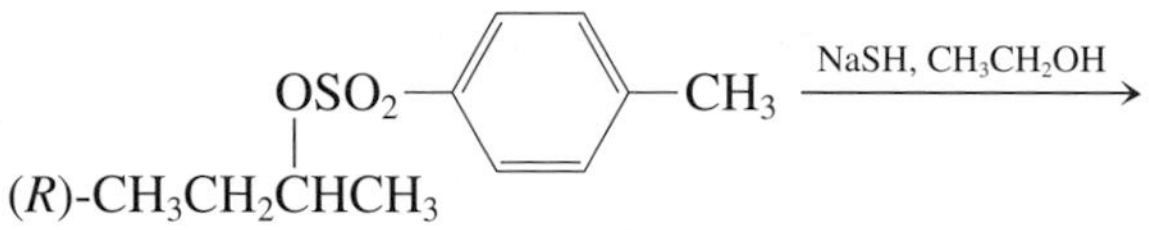 (R)-$CH_3CH_2\overset{\underset{|}{OSO_2}-\text{benzene}-CH_3}{CH}CH_3$ $\xrightarrow{NaSH,\ CH_3CH_2OH}$

(j) 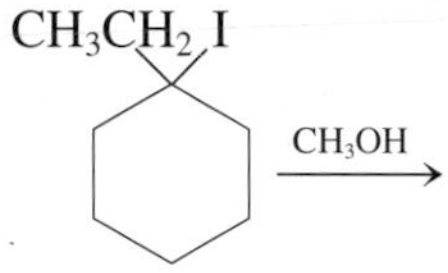 cyclohexane ring with CH_3CH_2 and I $\xrightarrow{CH_3OH}$

(k) $(CH_3)_3C\overset{\underset{|}{Br}}{C}HCH_3$ $\xrightarrow{KOH,\ CH_3CH_2OH}$

(l) CH_3CH_2Cl $\xrightarrow{\overset{\displaystyle\overset{O}{\|}}{CH_3COH}}$

33. Consider the reaction shown here. Will it proceed by substitution or by elimination? What factors determine the most likely mechanism? Write the expected product. (**Hint:** Draw the chair conformation of the substrate or, better yet, make a model.)

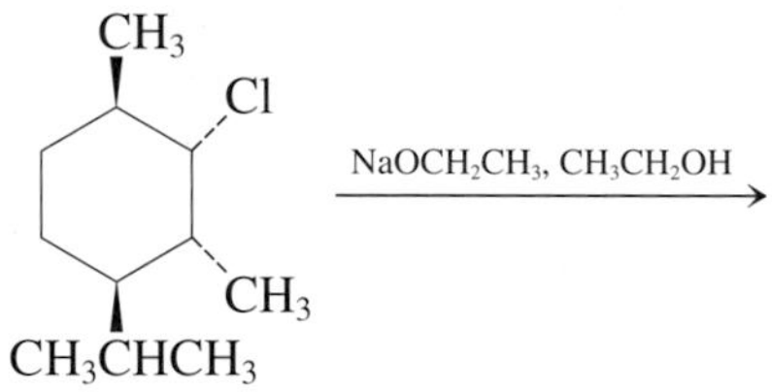

cyclohexane ring with CH_3, Cl, CH_3, and CH_3CHCH_3 substituents $\xrightarrow{NaOCH_2CH_3,\ CH_3CH_2OH}$

34. Fill the blanks in the following table with the major product(s) of the reaction of each haloalkane with the reagents shown.

Haloalkane	Reagent			
	H_2O	$NaSeCH_3$	$NaOCH_3$	$KOC(CH_3)_3$
CH_3Cl	___	___	___	___
$CH_3CH_2CH_2Cl$	___	___	___	___
$(CH_3)_2CHCl$	___	___	___	___
$(CH_3)_3CCl$	___	___	___	___

35. Indicate the major mechanism(s) (simply specify S_N2, S_N1, E2, or E1) required for the formation of each product that you wrote in Problem 34.

36. For each of the following reactions, indicate whether the reaction would work well, poorly, or not at all. Formulate alternative products, if appropriate.

(a) $CH_3CH_2CHCH_3$ $\xrightarrow{\text{NaOH, propanone (acetone)}}$ $CH_3CH_2CHCH_3$
  $\quad\quad\quad\quad\;\;|$ $|$
  $\quad\quad\quad\quad\;\;Br$ OH

(b) $\overset{\displaystyle H_3C}{\underset{\displaystyle |}{CH_3CHCH_2Cl}}$ $\xrightarrow{\text{CH}_3\text{OH}}$ $\overset{\displaystyle H_3C}{\underset{\displaystyle |}{CH_3CHCH_2OCH_3}}$

(c)

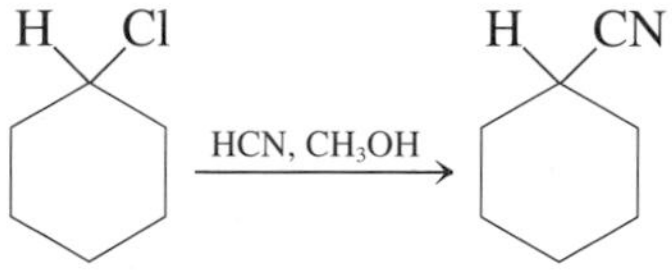

$\xrightarrow{\text{HCN, CH}_3\text{OH}}$

(d) 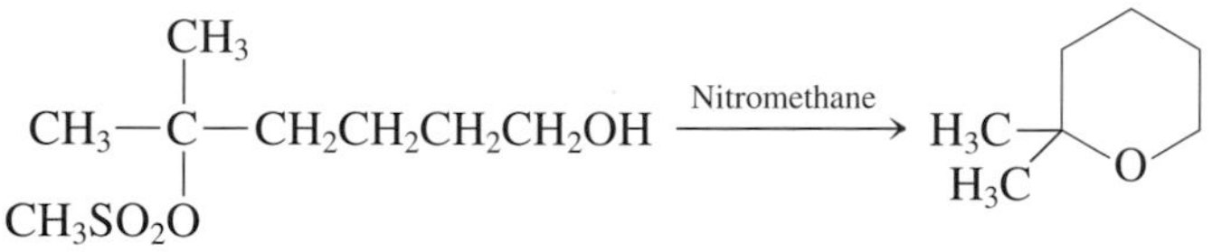 $CH_3-\overset{\displaystyle CH_3}{\underset{\displaystyle CH_3SO_2O}{\overset{|}{\underset{|}{C}}}}-CH_2CH_2CH_2CH_2OH$ $\xrightarrow{\text{Nitromethane}}$

(e)

$\xrightarrow{\text{NaSCH}_3,\ \text{CH}_3\text{OH}}$

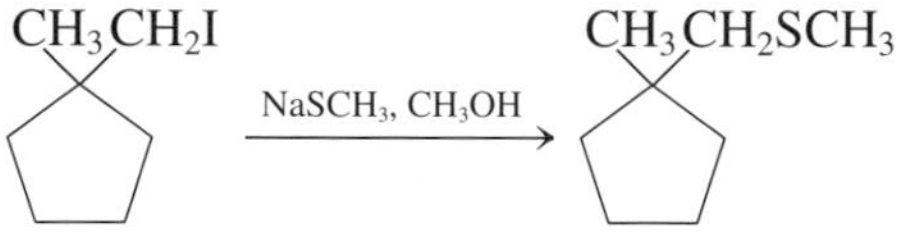

(f) $CH_3CH_2CH_2Br$ $\xrightarrow{\text{NaN}_3,\ \text{CH}_3\text{OH}}$ $CH_3CH_2CH_2N_3$

(g) $(CH_3)_3CCl$ $\xrightarrow{\text{NaI, nitromethane}}$ $(CH_3)_3CI$

(h) $(CH_3CH_2)_2O$ $\xrightarrow{\text{CH}_3\text{I}}$ $(CH_3CH_2)_2\overset{+}{O}CH_3 + I^-$

(i) CH_3I $\xrightarrow{\text{CH}_3\text{OH}}$ CH_3OCH_3

(j) $(CH_3CH_2)_3COCH_3$ $\xrightarrow{\text{NaBr, CH}_3\text{OH}}$ $(CH_3CH_2)_3CBr$

(k) $\overset{\displaystyle CH_3}{\underset{\displaystyle |}{CH_3CHCH_2CH_2Cl}}$ $\xrightarrow{\text{NaOCH}_2\text{CH}_3,\ \text{CH}_3\text{CH}_2\text{OH}}$ $\overset{\displaystyle CH_3}{\underset{\displaystyle |}{CH_3CHCH{=}CH_2}}$

(l) $CH_3CH_2CH_2CH_2Cl$ $\xrightarrow{\text{NaOCH}_2\text{CH}_3,\ \text{CH}_3\text{CH}_2\text{OH}}$ $CH_3CH_2CH{=}CH_2$

37. Propose syntheses of the following molecules from the indicated materials. Make use of any other reagents or solvents that you need. In some cases, there may be no alternative but to employ a reaction that results in a mixture of products. If so, use reagents and conditions that will maximize the yield of the desired material (compare Problem 39 in Chapter 6).

(a) $CH_3CH_2CHICH_3$, from butane   **(b)** $CH_3CH_2CH_2CH_2I$, from butane
(c) $(CH_3)_3COCH_3$, from methane and 2-methylpropane   **(d)** Cyclohexene, from cyclohexane

(e) Cyclohexanol, from cyclohexane   **(f)** 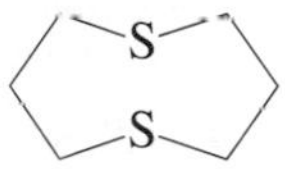 , from 1,3-dibromopropane

38. [(1-Bromo-1-methyl)ethyl]benzene, shown in the margin, undergoes solvolysis in a unimolecular, strictly first order process. The reaction rate for [RBr] = 0.1 M RBr in 9 : 1 propanone (acetone) : water is measured to be 2×10^{-4} mol L^{-1} s^{-1}. **(a)** Calculate the rate constant k from these data. What is the product of this reaction? **(b)** In the presence of 0.1 M LiCl, the rate is found to increase to 4×10^{-4} mol L^{-1} s^{-1}, although the reaction still remains strictly first order. Calculate the new rate constant k_{LiCl} and suggest an explanation. **(c)** When 0.1 M LiBr is present instead of LiCl, the measured rate

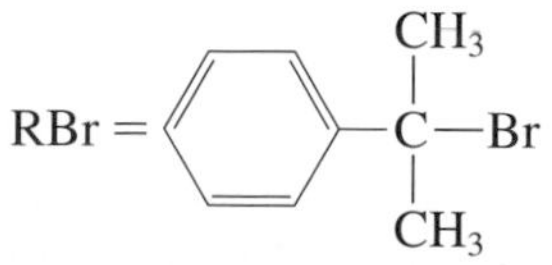

$$RBr = \text{}$$

drops to 1.6×10^{-4} mol L^{-1} s^{-1}. Explain this observation, and write the appropriate chemical equations to describe the reactions.

39. The stabilities of three cyclic cations illustrated here differ greatly. Predict their order of stability and provide a rationalization for your assignments.

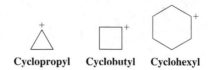

Cyclopropyl   Cyclobutyl   Cyclohexyl

40. Match each of the following transformations to the correct reaction profile shown here, and draw the structures of the species present at all points on the energy curves marked by capital letters.

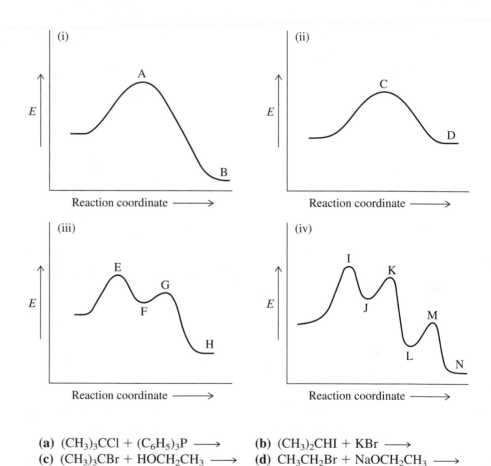

(a) $(CH_3)_3CCl + (C_6H_5)_3P \longrightarrow$ **(b)** $(CH_3)_2CHI + KBr \longrightarrow$
(c) $(CH_3)_3CBr + HOCH_2CH_3 \longrightarrow$ **(d)** $CH_3CH_2Br + NaOCH_2CH_3 \longrightarrow$

41. Formulate the structure of the most likely product of the following reaction of 4-chloro-4-methyl-1-pentanol in neutral polar solution.

$$\underset{\overset{|}{\text{Cl}}}{(CH_3)_2C}CH_2CH_2CH_2OH \longrightarrow HCl + C_6H_{12}O$$

In strongly *basic* solution, the starting material again converts into a molecule with the molecular formula $C_6H_{12}O$, but with a completely different structure. What is it? Explain the difference between the two results.

42. The following reaction can proceed through both E1 and E2 mechanisms.

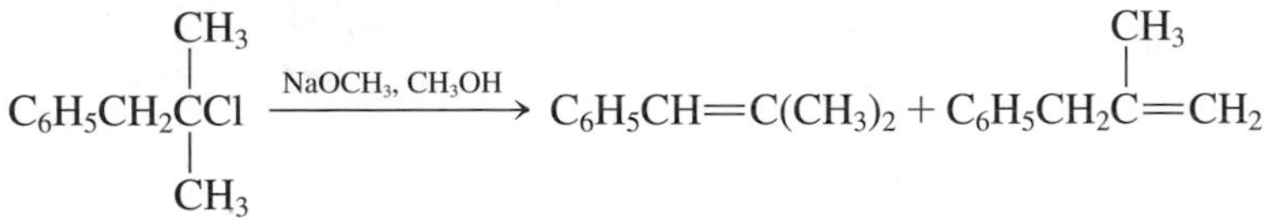

$$C_6H_5CH_2\overset{\overset{\displaystyle CH_3}{|}}{\underset{\underset{\displaystyle CH_3}{|}}{C}}Cl \xrightarrow{\text{NaOCH}_3,\ \text{CH}_3\text{OH}} C_6H_5CH{=}C(CH_3)_2 + C_6H_5CH_2\overset{\overset{\displaystyle CH_3}{|}}{C}{=}CH_2$$

The E1 rate constant $k_{E1} = 1.4 \times 10^{-4}\ s^{-1}$ and the E2 rate constant $k_{E2} = 1.9 \times 10^{-4}\ L\ mol^{-1}\ s^{-1}$; 0.02 M haloalkane. **(a)** What is the predominant elimination mechanism with 0.5 M $NaOCH_3$? **(b)** What is the predominant elimination mechanism with 2.0 M $NaOCH_3$? **(c)** At what concentration of base does exactly 50% of the starting material react by an E1 route and 50% by an E2 pathway?

43. When 2-methyl-2-propanol is shaken with concentrated aqueous HBr, 2-bromo-2-methylpropane (margin) rapidly forms, which is the reverse of S_N1 hydrolysis (Section 7-1). Propose a detailed mechanism for this process.

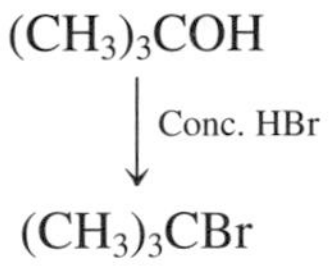

$(CH_3)_3COH$

$\downarrow$ Conc. HBr

$(CH_3)_3CBr$

44. Give the mechanism and major product for the reaction of a secondary haloalkane in a polar aprotic solvent with the following nucleophiles. The pK_a value of the conjugate acid of the nucleophile is given in parentheses.
(a) N_3^- (4.6)   **(b)** H_2N^- (35)   **(c)** NH_3 (9.5)
(d) HSe^- (3.7)   **(e)** F^- (3.2)   **(f)** $C_6H_5O^-$ (9.9)
(g) PH_3 (−12)   **(h)** NH_2OH (6.0)   **(i)** NCS^- (−0.7)

45. Cortisone is an important steroidal anti-inflammatory agent. Cortisone can be synthesized efficiently from the alkene shown here.

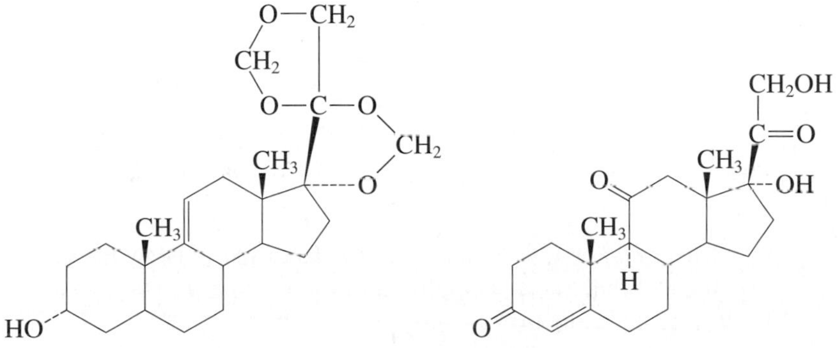

Alkene   **Cortisone**

Of the following three chlorinated compounds, two give reasonable yields of the alkene shown above by E2 elimination with base, but one does not. Which one does not work well, and why? What does it give during attempted E2 elimination? (**Hint:** Consider the geometry of each system.)

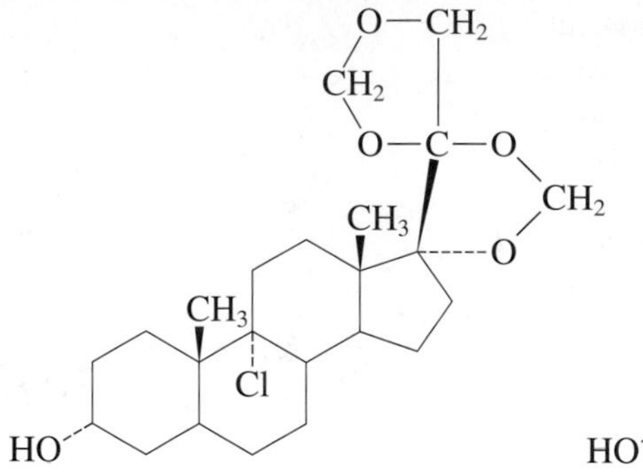

A 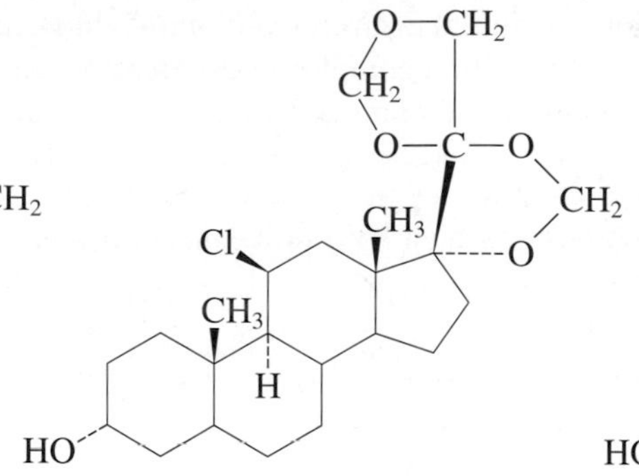 **B** 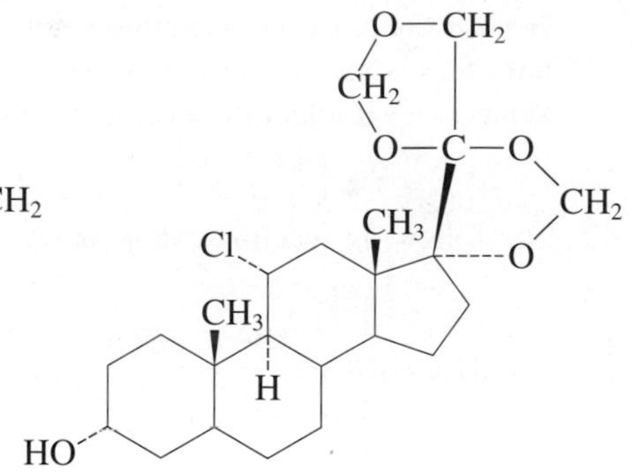 **C**

46. The chemistry of derivatives of *trans*-decalin is of interest because this ring system is part of the structure of steroids. Make models of the brominated systems (i and ii) to help you answer the following questions.

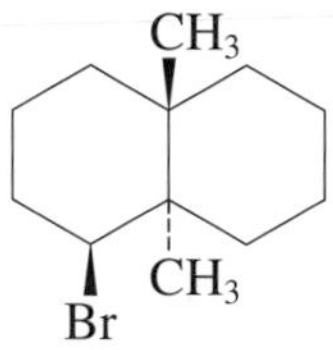

i 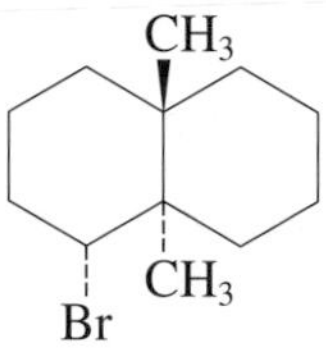 **ii**

(a) One of the molecules undergoes E2 reaction with $NaOCH_2CH_3$ in CH_3CH_2OH considerably faster than does the other. Which molecule is which? Explain. **(b)** The following deuterated analogs of systems i and ii react with base to give the products shown.

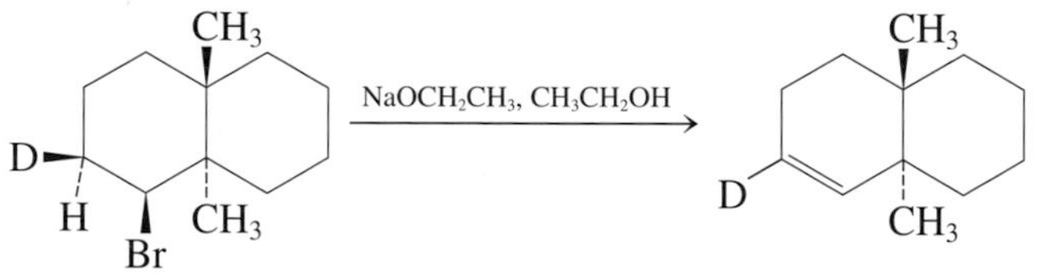

i-deuterated (All D retained)

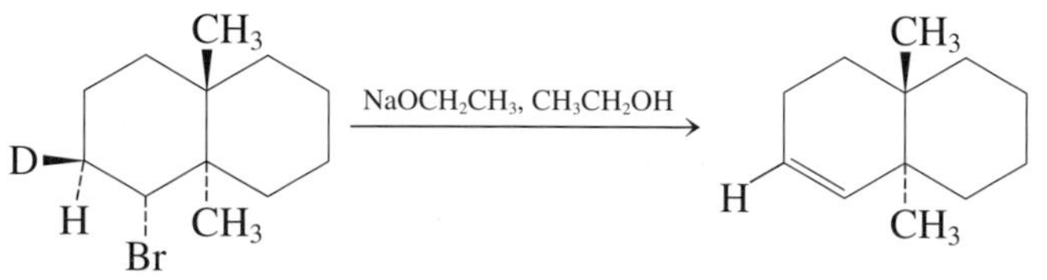

ii-deuterated (All D lost)

Specify whether *anti* or *syn* eliminations have taken place. Draw the conformations that the molecules must adopt for elimination to occur. Does your answer to (b) help you in solving (a)?

Team Problem

47. Consider the general substitution-elimination reactions of the bromoalkanes.

$$R{-}Br \xrightarrow{\text{Nu/Base}} R{-}Nu + \text{alkene}$$

How do the reaction mechanisms and product formation differ when the structure of the substrate and reaction conditions change? To begin to unravel the nuances of bimolecular and unimolecular substitution and elimination reactions, focus on the treatment of bromoalkanes A through D under conditions (a) through (e). Divide the problem evenly among yourselves so that each of you tackles the questions of reaction mechanism(s) and qualitative distribution of product(s), if any. Reconvene to discuss your conclusions and come to a consensus. When you are explaining a reaction mechanism to the rest of the team, use curved arrows to show the flow of electrons. Label the stereochemistry of starting materials and products as R or S, as appropriate.

(a) NaN_3, DMF   **(b)** LDA, DMF   **(c)** NaOH, DMF   **(d)** $CH_3CO^- Na^+$, CH_3COH **(e)** CH_3OH

Preprofessional Problems

48. Which of the following haloalkanes will undergo hydrolysis most rapidly?
  (a) $(CH_3)_3CF$ **(b)** $(CH_3)_3CCl$ **(c)** $(CH_3)_3CBr$ **(d)** $(CH_3)_3CI$

49. The reaction

$$(CH_3)_3CCl \xrightarrow{CH_3O^-} \underset{CH_3}{\overset{CH_3}{}}C=CH_2$$

is an example of which of the following processes?
  (a) E1   **(b)** E2   **(c)** S_N1 **(d)** S_N2

50. In this transformation

$$A \xrightarrow{H_2O, \text{ propanone (acetone)}} CH_3CH_2C(CH_3)_2$$
$$\underset{OH}{|}$$

what is the best structure for A?

  (a) $BrCH_2CH_2CH(CH_3)_2$ **(b)** $CH_3CH_2\overset{CH_3}{\underset{CH_3}{|}}CBr$

  (c) $CH_3CH_2\overset{CH_3}{\underset{CH_2Br}{|}}CH$ **(d)** $CH_3CHCH(CH_3)_2$ with Br

51. Which of the following isomeric carbocations is the most stable?

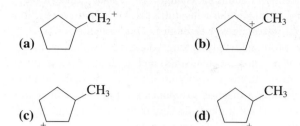

(a)

(b)

(c)

(d)

52. Which reaction intermediate is involved in the following reaction?

$$2\text{-methylbutane} \xrightarrow{\text{Br}_2,\ h\nu} 2\text{-bromo-3-methylbutane}$$
(not the major product)

(a) A secondary radical   (b) A tertiary radical
(c) A secondary carbocation   (d) A tertiary carbocation

Hydroxy Functional Group

Properties of the Alcohols and Strategy in Synthesis

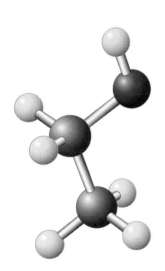

What is your first thought when you hear the word "alcohol"? Undoubtedly, whether pleasant or not, it is connected in some way to ethanol, as contained in alcoholic beverages. The euphoric effects of (limited) ethanol consumption have been known and purposely used for thousands of years. This is perhaps not surprising, because ethanol is naturally generated by the fermentation of carbohydrates. For example, the addition of yeast to an aqueous sugar solution leads to the evolution of CO_2 and the formation of ethanol.

$$C_6H_{12}O_6 \xrightarrow{\text{Yeast enzymes}} 2\ CH_3CH_2OH\ +\ 2\ CO_2$$

Sugar   **Ethanol**

Ethanol is a member of a large family of compounds called **alcohols.** This chapter introduces you to some of their chemistry. From Chapter 2, we know that alcohols have carbon backbones bearing the substituent OH, the **hydroxy** group. They may be viewed as derivatives of water in which one hydrogen has been replaced by an alkyl group. Replacement of the second hydrogen gives an **ether** (Chapter 9).

Alcohols are abundant in nature and varied in structure (see, e.g., Section 4-7). Simple alcohols are used as solvents; others aid in the synthesis of more complex molecules. They are a good example of how functional groups shape the properties and applications of organic compounds. Our discussion will begin with the naming

H—O—H
Water

CH_3—O—H
Methanol
(An alcohol)

CH_3—O—CH_3
Methoxymethane
(Dimethyl ether)
(An ether)

The ancient art of brewing beer by the yeast fermentation of cereal grains is now practiced on a large scale and controlled by computers.

of alcohols, followed by a brief description of their structures and other physical properties, particularly in comparison with those of the alkanes and haloalkanes. Finally, their preparation will introduce us to strategies for efficiently synthesizing new organic compounds.

8-1 Naming the Alcohols

Like other compounds, alcohols may have both systematic and common names. Systematic nomenclature treats alcohols as derivatives of alkanes. The ending *-e* of the alkane is replaced by **-ol.** Thus, an alkane is converted into an **alkanol.** For example, the simplest alcohol is derived from methane. It is methanol. Ethanol stems from ethane, propanol from propane, and so on. In more complicated, branched systems, the name of the alcohol is based on the longest chain *containing the OH substituent*— not necessarily the longest chain in the molecule.

A methyl heptanol   A methyl propyl octanol

To locate positions along the chain, number each carbon atom beginning from the end closest to the OH group. The names of other substituents along the chain can then be added to the alkanol stem as prefixes. Complex alkyl appendages are named according to the IUPAC rules for hydrocarbons (Section 2-3).

$$\overset{3}{C}H_3\overset{2}{C}H_2\overset{1}{C}H_2OH$$

1-Propanol

$$\overset{1}{C}H_3\overset{2}{\underset{\underset{OH}{|}}{C}}\overset{3}{H}\overset{3}{C}H_2\overset{4}{C}H_2\overset{5}{C}H_3$$

2-Pentanol

2,2,5-Trimethyl-3-hexanol

Cyclic alcohols are called **cycloalkanols.** Here the carbon carrying the functional group automatically receives the number 1.

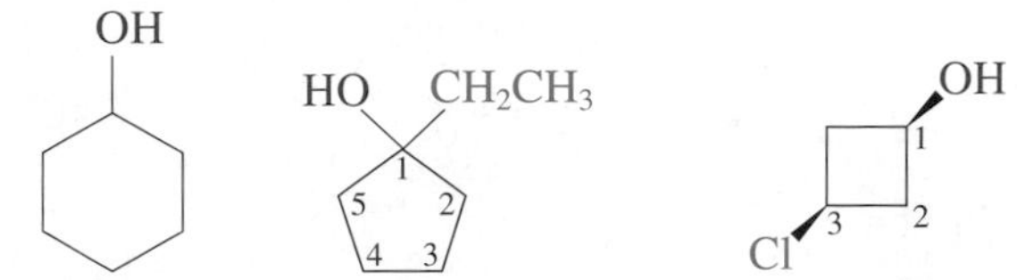

OH

Cyclohexanol   1-Ethylcyclopentanol   *cis*-3-Chlorocyclobutanol

When named as a substituent, the OH group is called *hydroxy.* Like haloalkanes, alcohols can be classified as primary, secondary, or tertiary.

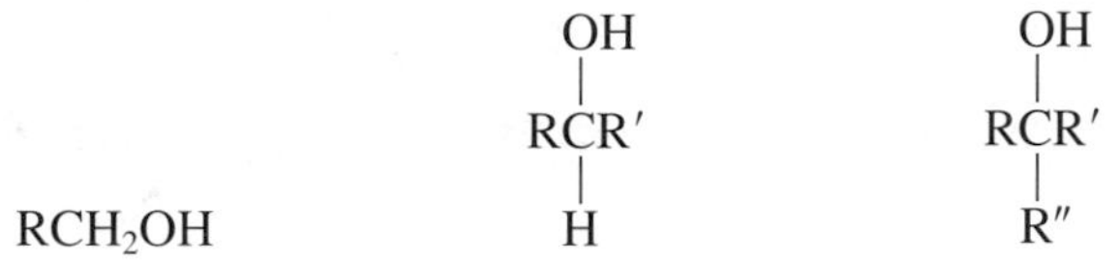

RCH₂OH
A primary alcohol

OH
RCR′
H
A secondary alcohol

OH
RCR′
R″
A tertiary alcohol

EXERCISE 8-1

Draw the structures of the following alcohols. **(a)** (*S*)-3-Methyl-3-hexanol; **(b)** *trans*-2-bromocyclopentanol; **(c)** 2,2-dimethyl-1-propanol (neopentyl alcohol).

EXERCISE 8-2

Name the following compounds.

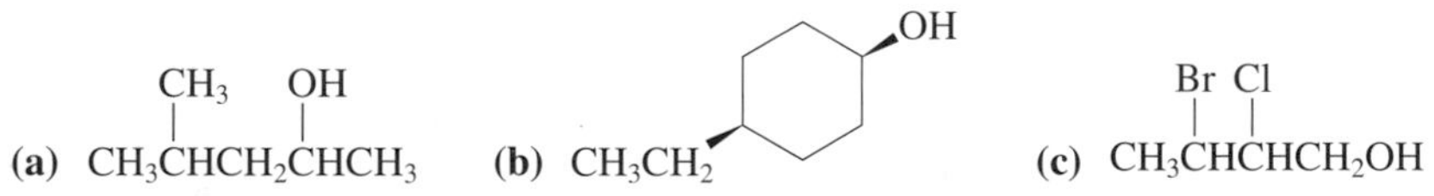

(a) CH₃CHCH₂CHCH₃ (CH₃, OH)   **(b)** CH₃CH₂ (cyclohexane, OH)   **(c)** CH₃CHCHCH₂OH (Br Cl)

In common nomenclature, the name of the alkyl group is followed by the word *alcohol,* written separately. Common names are found in the older literature; although it is best not to use them, we should be able to recognize them.

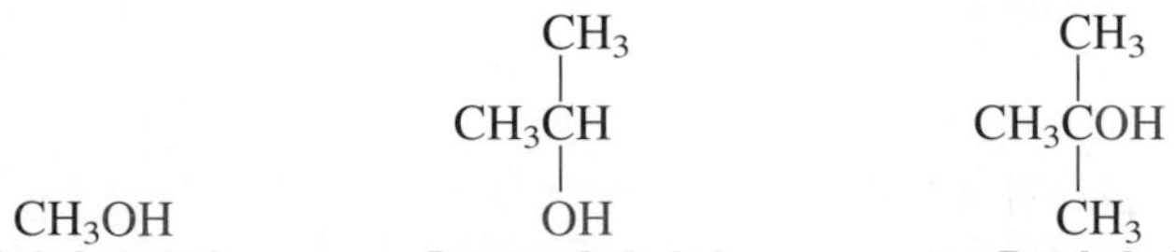

CH₃OH
Methyl alcohol

CH₃CH (CH₃, OH)
Isopropyl alcohol

CH₃COH (CH₃, CH₃)
***tert*-Butyl alcohol**

In summary, alcohols can be named as alkanols (IUPAC) or alkyl alcohols. In IUPAC nomenclature, the name is derived from the chain bearing the hydroxy group, whose position is given the lowest possible number.

8-2 Structural and Physical Properties of Alcohols

The hydroxy functional group strongly shapes the physical characteristics of the alcohols. It affects their molecular structure and allows them to enter into hydrogen bonding. As a result, it raises their boiling points and increases their solubilities in water.

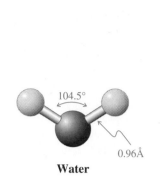

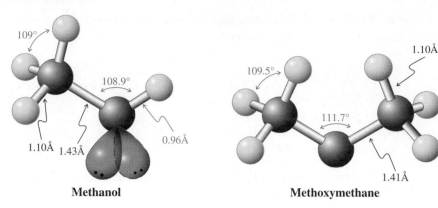

Water   Methanol   Methoxymethane

FIGURE 8-1 ──────

The similarity in structure of water, methanol, and methoxymethane. The oxygens are approximately sp^3 hybridized, so all three structures exhibit nearly tetrahedral bond angles around the heteroatom. Remember that the oxygen bears two lone electron pairs in two nonbonding sp^3 hybrid orbitals (shown in the center for methanol).

The structure of alcohols resembles that of water

Figure 8-1 shows how closely the structure of methanol resembles those of water and of methoxymethane (dimethyl ether). In all three, the oxygen atoms are roughly sp^3 hybridized and their bond angles nearly tetrahedral. The minor differences are due to the steric effect produced by the replacement of hydrogen atoms by alkyl groups. The O–H bond is considerably shorter than the C–H bond, in part because of the high electronegativity of oxygen relative to that of carbon. Consistent with this bond shortening is the order of bond strengths: $DH^\circ_{O-H} = 104$ kcal mol^{-1}; $DH^\circ_{C-H} = 98$ kcal mol^{-1}.

The electronegativity of oxygen causes an unsymmetrical distribution of charge in alcohols. This effect polarizes the O–H bond so that the hydrogen has a partial positive charge and gives rise to a molecular dipole (Section 1-3) similar to that observed for water.

TABLE 8-1	Physical Properties of Alcohols and Selected Analogous Haloalkanes and Alkanes				
Compound	IUPAC name	Common name	Melting point (°C)	Boiling point (°C)	Solubility in H_2O at 23°C
CH_3OH	Methanol	Methyl alcohol	−97.8	65.0	Infinite
CH_3Cl	Chloromethane	Methyl chloride	−97.7	−24.2	0.74 g/100 mL
CH_4	Methane		−182.5	−161.7	3.5 mL (gas)/100 mL
CH_3CH_2OH	Ethanol	Ethyl alcohol	−114.7	78.5	Infinite
CH_3CH_2Cl	Chloroethane	Ethyl chloride	−136.4	12.3	0.447 g/100 mL
CH_3CH_3	Ethane		−183.3	−88.6	4.7 mL (gas)/100 mL
$CH_3CH_2CH_2OH$	1-Propanol	Propyl alcohol	−126.5	97.4	Infinite
$CH_3CH_2CH_3$	Propane		−187.7	−42.1	6.5 mL (gas)/100 mL
$CH_3CH_2CH_2CH_2OH$	1-Butanol	Butyl alcohol	−89.5	117.3	8.0 g/100 mL
$CH_3(CH_2)_4OH$	1-Pentanol	Pentyl alcohol	−79	138	2.2 g/100 mL

Bond and Molecular Dipoles of Water and Methanol

Hydrogen bonding raises the boiling points and water solubilities of alcohols

In Section 6-2 we invoked the polarity of the haloalkanes to explain why their boiling points are higher than those of the corresponding nonpolar alkanes. The polarity of alcohols is similar to that of the haloalkanes. Does this mean that the boiling points of haloalkanes and alcohols should correspond? Inspection of Table 8-1 shows that they do not: Alcohols have unusually high boiling points, much higher than those of comparable alkanes and haloalkanes.

The explanation lies in hydrogen bonding. Hydrogen bonds may form between the oxygen atoms of one alcohol molecule and the hydroxy hydrogen atoms of another. Alcohols build up an extensive network of these interactions (Figure 8-2). Although hydrogen bonds are longer and much weaker ($DH° \sim$ 5–6 kcal mol^{-1}) than the covalent O–H linkage ($DH° \sim$ 104 kcal mol^{-1}), so many of them form that their combined strength makes it difficult for molecules to escape the liquid. The result is a higher boiling point.

FIGURE 8-2

Hydrogen bonding in an aqueous solution of methanol. The molecules form a complex three-dimensional array, and only one layer is depicted here. Pure water (in ice) tends to arrange itself in cyclic hexamer units (top left); neat small alcohols prefer a cyclic tetramer structure (bottom right).

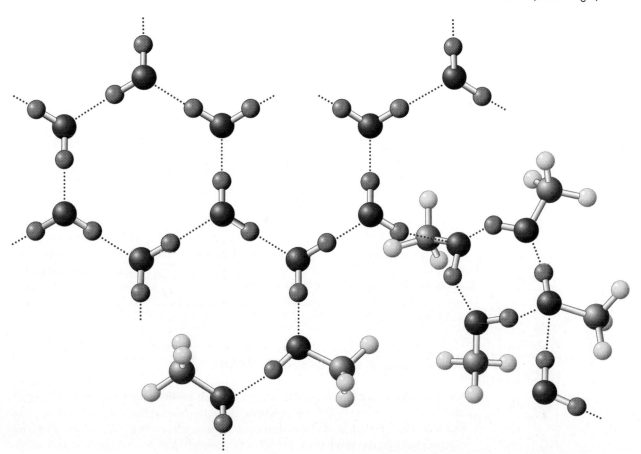

Methanol

1-Pentanol

FIGURE 8-3 ————————————————————————————————
The hydrophobic (green) and hydrophilic (red) parts of methanol and 1-pentanol (space-filling models). The polar functional group dominates the physical properties of methanol: The molecule is completely soluble in water but only partially so in hexane. Conversely, the increased size of the hydrophobic part in the higher alcohol leads to infinite solubility in hexane but reduced solubility in water (Table 8-1).

The effect is even more pronounced in water, which has two hydrogens available for hydrogen bonding (see Figure 8-2). This phenomenon explains why water, with a molecular weight of only 18, has a boiling point of 100°C. Without this property, water would be a gas at ordinary temperatures. Considering the importance of water in all living organisms, imagine how the absence of liquid water would have affected the development of life on our planet.

Hydrogen bonding in water and alcohols is responsible for another property: Many alcohols are appreciably water soluble (Table 8-1). This behavior contrasts with that of the nonpolar alkanes, which are poorly solvated by this medium. Because of their characteristic insolubility in water, alkanes are said to be **hydrophobic** (*hydro,* Greek, water; *phobos,* Greek, fear). So are most alkyl chains. Conversely, the OH group and other polar substituents, such as COOH and NH_2, are **hydrophilic:** They enhance water solubility.

As the values in Table 8-1 show, the larger the alkyl (hydrophobic) part of an alcohol, the lower its solubility in water. At the same time, the alkyl part increases the solubility of the alcohol in nonpolar solvents (Figure 8-3). (For example, alcohols are effective at removing greasy deposits from the tape heads of cassette decks and VCRs.) The "waterlike" structure of the lower alcohols, particularly methanol and ethanol, also makes them excellent solvents for polar compounds and even salts. It is not surprising, then, that alcohols are popular solvents in the S_N2 reaction (Chapter 6).

In summary, the oxygen in alcohols (and ethers) is tetrahedral and sp^3 hybridized. The covalent O–H bond is shorter and stronger than the C–H bond. Because of the electronegativity of the oxygen, alcohols, like water and ethers, exhibit appreciable molecular polarity. The hydroxy hydrogen enters into hydrogen bonding with other alcohol molecules. These properties lead to a substantial increase in the boiling points and in the solubilities of alcohols in polar solvents relative to those of the alkanes and haloalkanes.

8-3 Alcohols as Acids and Bases

Many applications of the alcohols depend on their ability to act both as acids and as bases. (See the review of these concepts in Section 6-8.) Thus, deprotonation gives alkoxide ions. We shall see how structural features affect their pK_a values. The lone electron pairs on oxygen render alcohols basic as well, and protonation results in alkyloxonium ions.

The acidity of alcohols resembles that of water

The acidity of alcohols in water is expressed by the equilibrium constant K.

$$\text{RO\overset{\frown}{H} + H_2O} \overset{K}{\rightleftharpoons} \text{H}_3\text{O}^+ + \text{RO}^-$$

Alkoxide
ion

Making use of the constant concentration of water (55 mol L^{-1}; Section 6-8), we derive a new equilibrium constant K_a.

$$K_a = K[\text{H}_2\text{O}] = \frac{[\text{H}_3\text{O}^+][\text{RO}^-]}{[\text{ROH}]} \text{ mol } L^{-1}, \text{ and } pK_a = -\log K_a$$

Table 8-2 lists the pK_a values of several alcohols and related compounds. A comparison of these values with those given in Table 2-6 for mineral and other strong acids shows that alcohols, like water, are fairly weak acids. Their acidity is far greater, however, than that of alkanes and haloalkanes.

Why are alcohols acidic, whereas alkanes and haloalkanes are not? The answer lies in the relatively strong electronegativity of the oxygen to which the proton is attached, which stabilizes the negative charge of the alkoxide ion.

To drive the equilibrium between alcohol and alkoxide to the side of the conjugate base, it is necessary to use a base *stronger* than the alkoxide formed (i.e., a base derived from a conjugate acid *weaker* than the alcohol; see also Section 9-1). An example is the reaction of sodium amide, $NaNH_2$, with methanol to furnish sodium methoxide and ammonia.

$$\text{CH}_3\text{O\overset{\frown}{H} + Na}^+ \text{ }^-\text{NH}_2 \overset{K}{\rightleftharpoons} \text{Na}^+ \text{ }^-\text{OCH}_3 + \text{NH}_3$$

$pK_a = 15.5$ $pK_a = 35$
Sodium   Sodium
amide   methoxide

This equilibrium lies well to the right ($K \sim 10^{35-15.5} = 10^{19.5}$), because methanol is a much stronger acid than is ammonia, or, conversely, because amide is a much stronger base than is methoxide.

TABLE 8-2	pK_a Values of Alcohols and Related Compounds in Water		
Compound	**pK_a**	**Compound**	**pK_a**
H_2O	15.7	HOCl	7.53
CH_3OH	15.5	$ClCH_2CH_2OH$	14.3
CH_3CH_2OH	15.9	CF_3CH_2OH	12.4
$(CH_3)_2CHOH$	17.1	$CF_3CH_2CH_2OH$	14.6
$(CH_3)_3COH$	18	$CF_3CH_2CH_2CH_2OH$	15.4
H_2O_2	11.64		

It is sometimes sufficient to generate alkoxides in less than stoichiometric equilibrium concentrations. For this purpose, we may add an alkali metal hydroxide to the alcohol.

$$\mathrm{CH_3CH_2OH} + \mathrm{Na^+\ \bar{}OH} \underset{}{\overset{K}{\rightleftharpoons}} \mathrm{CH_3CH_2O^-Na^+} + \mathrm{H_2O}$$

$\mathrm{p}K_\mathrm{a} = 15.9 \qquad\qquad\qquad\qquad\qquad\qquad\qquad\qquad \mathrm{p}K_\mathrm{a} = 15.7$

With this base present, approximately one-half of the alcohol will exist as the alkoxide, if we assume equimolar concentrations of starting materials. If the alcohol is the solvent (i.e., present in large excess), however, all of the base will exist in the form of the alkoxide.

EXERCISE 8-3

Which of the following bases are strong enough to cause essentially complete deprotonation of methanol? The $\mathrm{p}K_\mathrm{a}$ of the conjugate acid is given in parentheses. **(a)** KCN (9.2); **(b)** $\mathrm{CH_3CH_2CH_2CH_2Li}$ (50); **(c)** $\mathrm{CH_3CO_2Na}$ (4.7); **(d)** $\mathrm{LiN[CH(CH_3)_2]_2}$ (LDA, 40); **(e)** KH (38); **(f)** $\mathrm{CH_3SNa}$ (10).

Steric disruption and inductive effects control the acidity of alcohols

Table 8-2 shows a large variation in the $\mathrm{p}K_\mathrm{a}$ values of the alcohols. A closer look at the first column reveals that the acidity decreases ($\mathrm{p}K_\mathrm{a}$ increases) from methanol to primary, secondary, and finally tertiary systems.

Relative $\mathrm{p}K_\mathrm{a}$ Values of Alcohols (in Solution)

$\mathrm{CH_3OH}$ < primary < secondary < tertiary

Strongest acid   **Weakest acid**

This ordering has been ascribed to steric disruption of solvation and to hydrogen bonding in the alkoxide (Figure 8-4). Because solvation and hydrogen bonding stabilize the negative charge on oxygen, interference with these processes leads to an increase in $\mathrm{p}K_\mathrm{a}$.

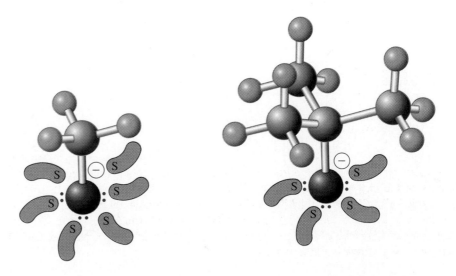

FIGURE 8-4

The smaller methoxide is better solvated than is the larger tertiary butoxide ion. (S, solvent molecules).

The second column in Table 8-2 reveals another contribution to the pK_a of alcohols: The presence of halogens increases acidity. Recall that the carbon of the C–X bond is positively polarized as a result of the high electronegativity of X (Sections 1-3 and 6-2). Electron withdrawal by the halogen also causes atoms farther away to be slightly positively charged. This phenomenon of transmission of charge, both negative and positive, through the σ bonds in a chain of atoms is called an **inductive effect.** Here it stabilizes the negative charge on the alkoxide oxygen by electrostatic attraction. The inductive effect in alcohols increases with the number of electronegative groups but decreases with distance from the oxygen.

Inductive Effect of the Chlorine in 2-Chloroethoxide

$$Cl—CH_2—CH_2—\overset{..}{\underset{..}{O}}:^-$$

EXERCISE 8-4

Rank the following alcohols in order of increasing acidity.

EXERCISE 8-5

Which side of the following equilibrium reaction is favored (assume equimolar concentrations of starting materials)?

$$(CH_3)_3CO^- + CH_3OH \rightleftharpoons (CH_3)_3COH + CH_3O^-$$

The lone electron pairs on oxygen make alcohols basic

Alcohols may also be basic, although weakly so. Very strong acids are required to protonate the OH group, as indicated by the low pK_a values (strong acidity) of their conjugate acids, the alkyloxonium ions (Table 8-3). Molecules that may be both acids and bases are called **amphoteric** (*ampho,* Greek, both).

The amphoteric nature of the hydroxy functional group characterizes the chemical reactivity of alcohols. In strong acids, they exist as alkyloxonium ions, in neutral media as alcohols, and in strong bases as alkoxides.

Alcohols Are Amphoteric

In summary, alcohols are amphoteric. They are acidic by virtue of the electronegativity of the oxygen and are converted into alkoxides by strong bases. In solution, the steric bulk of branching inhibits solvation of the alkoxide, thereby raising the pK_a of the corresponding alcohol. Electron-withdrawing substituents close to the functional group lower the pK_a. Alcohols are also weakly basic and can be protonated by strong acids to furnish alkyloxonium ions.

TABLE 8-3	pK_a Values of Four Protonated Alcohols	
Compound	**pK_a**
$CH_3\overset{+}{O}H_2$	−2.2
$CH_3CH_2\overset{+}{O}H_2$	−2.4
$(CH_3)_2CH\overset{+}{O}H_2$	−3.2
$(CH_3)_3C\overset{+}{O}H_2$	−3.8

8-4 Industrial Sources of Alcohols: Carbon Monoxide and Ethene

Let us now turn to the preparation of alcohols. We start in this section with methods of special importance in industry. Subsequent sections will take up procedures that are used more generally in the synthetic laboratory to introduce the hydroxy functional group into a wide range of organic molecules.

Methanol is made on a large scale (more than 11 billion pounds [5 billion kg] in the United States in 1995) from a pressurized mixture of CO and H_2 called **synthesis gas.** The reaction makes use of a catalyst consisting of copper, zinc oxide, and chromium(III) oxide.

$$CO + 2\,H_2 \xrightarrow{\text{Cu-ZnO-Cr}_2\text{O}_3,\ 250°\text{C},\ 50–100\ \text{atm}} CH_3OH$$

Changing the catalyst to rhodium or ruthenium leads to 1,2-ethanediol (ethylene glycol), an important industrial chemical that is the principal component of automobile antifreeze.

$$2\,CO + 3\,H_2 \xrightarrow{\text{Rh or Ru, pressure, heat}} \underset{\substack{| \quad\ \ | \\ OH \quad OH}}{CH_2-CH_2}$$

1,2-Ethanediol
(Ethylene glycol)

Other reactions that would permit the selective formation of a given alcohol from synthesis gas are the focus of much current research, because synthesis gas is readily available by the gasification of coal in the presence of water.

$$\text{Coal} \xrightarrow{\text{Air, H}_2\text{O, }\Delta} x\,CO + y\,H_2$$

Ethanol is prepared in large quantities by fermentation of sugars or by the phosphoric acid–catalyzed hydration of ethene (ethylene). The United States produced about 620 million pounds (280 million kg) of ethanol in 1995. The hydration (and other addition reactions) of alkenes are considered in detail in Chapter 12.

$$CH_2{=}CH_2 + HOH \xrightarrow{\text{H}_3\text{PO}_4,\ 300°\text{C}} \underset{\substack{| \quad\ \ | \\ H \quad\ \ OH}}{CH_2-CH_2}$$

In summary, the industrial preparation of methanol and 1,2-ethanediol proceeds by reduction of carbon monoxide with hydrogen, and that of ethanol by the acid-catalyzed hydration of ethene (ethylene).

8-5 Synthesis of Alcohols by Nucleophilic Substitution

On a smaller than industrial scale, we can prepare alcohols from a wide variety of starting materials. For example, conversions of haloalkanes into alcohols by S_N2 and S_N1 processes featuring hydroxide and water, respectively, as nucleophiles were described in Chapters 6 and 7. These methods are not as widely used as one might think, however, because the required halides are often accessible only from the corresponding alcohols (Chapter 9). They also suffer from the usual drawbacks of nucleophilic substitution: Bimolecular elimination can be a major side reaction of hindered systems, and tertiary halides form carbocations that may undergo E1 reactions. Nevertheless, a number of alcohol syntheses have made use of such simple nucleophilic substitution reactions, two of which are shown here.

Alcohols by Nucleophilic Substitution

NaOH, H$_2$O, pyridine
S_N2

Pyridine =

95%

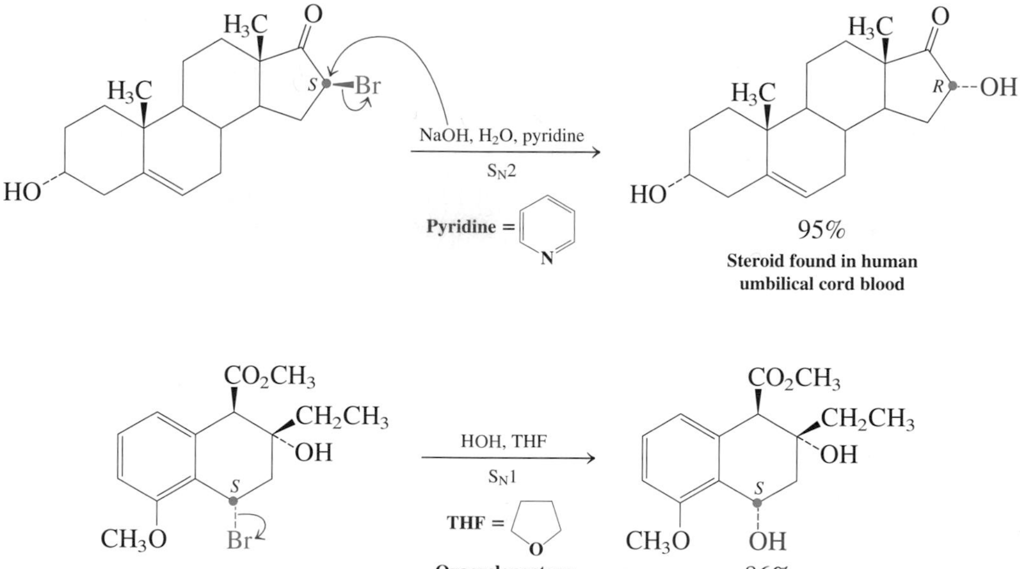

Steroid found in human
umbilical cord blood

HOH, THF
S_N1

THF =

Oxacyclopentane
(Tetrahydrofuran)

86%

Precursor for syntheses
of antitumor antibiotics

In the first example, stereochemical inversion was achieved by an S_N2 reaction with hydroxide. The second reaction is a solvolysis that follows the S_N1 mechanism. Stereoselectivity in this case is observed because the bottom face of the molecule is considerably less sterically hindered than the top. Both nucleophilic substitutions use organic **cosolvents** (pyridine and THF, respectively) to improve the solubility of the substrates.

EXERCISE 8-6

Show how you might convert the following haloalkanes into alcohols.
(a) Bromoethane; **(b)** chlorocyclohexane; **(c)** 3-chloro-3-methylpentane.

A way around the problem of elimination in S_N2 reactions of oxygen nucleophiles with secondary or sterically encumbered, branched primary substrates is the use of less basic functional equivalents of water, such as acetate (Section 6-9). The resulting alkyl acetate can then be converted into the desired alcohol by aqueous hydroxide. In a second step, the carbonyl oxygen bond is broken, thereby leaving the alkoxy residue unchanged. We shall consider this reaction, known as *ester hydrolysis*, in Chapter 20.

Alcohols from Haloalkanes by Acetate Substitution—Hydrolysis

STEP 1. Acetate formation (S_N2 reaction)

$$\underset{\textbf{1-Bromo-3-methylpentane}}{CH_3CH_2\overset{\overset{\displaystyle CH_3}{|}}{C}HCH_2CH_2\!-\!Br} + CH_3\overset{\overset{\displaystyle O}{\|}}{C}O^-Na^+ \xrightarrow{\text{DMF, 80°C}} \underset{95\%}{\underset{\textbf{3-Methylpentyl acetate}}{CH_3CH_2\overset{\overset{\displaystyle CH_3}{|}}{C}HCH_2CH_2O\overset{\overset{\displaystyle O}{\|}}{C}CH_3}} + Na^+Br^-$$

STEP 2. Conversion into the alcohol (ester hydrolysis)

$$CH_3CH_2\overset{\overset{\displaystyle CH_3}{|}}{C}HCH_2CH_2O\!+\!\overset{\overset{\displaystyle O}{\|}}{C}CH_3 + Na^+\ ^-OH \xrightarrow[\underset{-CH_3\overset{O}{C}O^-Na^+}{}]{H_2O} \underset{\underset{\textbf{3-Methyl-1-pentanol}}{85\%}}{CH_3CH_2\overset{\overset{\displaystyle CH_3}{|}}{C}HCH_2CH_2OH}$$

In summary, alcohols may be prepared from haloalkanes by nucleophilic substitution, provided the haloalkane is readily available and side reactions such as elimination do not interfere.

8-6 Synthesis of Alcohols: Oxidation–Reduction Relation Between Alcohols and Carbonyl Compounds

This section describes an important synthesis of alcohols: reduction of aldehydes and ketones. Later, we shall see that these compounds may be converted into alcohols by addition of organometallic reagents, with concomitant formation of a new carbon–carbon bond. Because of the versatility of aldehydes and ketones in synthesis, we shall also illustrate their preparation by alcohol oxidation.

Oxidation and reduction have special meanings in organic chemistry

We can readily recognize inorganic oxidation and reduction processes as the loss and gain of electrons, respectively. With organic compounds, it is often less clear whether electrons are being gained or lost in a reaction. Hence, organic chemists find it more useful to define oxidation and reduction in other terms. A process that adds electronegative atoms such as halogen or oxygen to, or removes hydrogen from, a molecule constitutes an **oxidation;** conversely, the removal of halogen or oxygen or the addition of hydrogen is defined as **reduction.** You can readily visualize this definition in the step-by-step oxidation of methane, CH_4, to carbon dioxide, CO_2.

Step-by-Step Oxidation of CH_4 to CO_2

$$CH_4 \xrightarrow{+O} CH_3OH \xrightarrow{-2H} H_2C=O \xrightarrow{+O} H\overset{\overset{\displaystyle O}{\|}}{C}OH \xrightarrow{-2H} CO_2$$

This definition of an oxidation–reduction relation allows us to connect alcohols to aldehydes and ketones. Addition of two hydrogen atoms to the double bond of a carbonyl group constitutes reduction to the corresponding alcohol. Aldehydes give primary alcohols; ketones give secondary alcohols. The reverse process, removal of hydrogen to furnish carbonyl compounds, is an example of oxidation. Together, these processes are referred to as **redox reactions.**

The Redox Relation Between Alcohols and Carbonyl Compounds

Aldehyde   Primary alcohol   Ketone   Secondary alcohol

Reduction of carbonyl compounds can be carried out either by addition of molecular hydrogen or by exposure to hydride reagents.

Alcohols can be made by catalytic hydrogenation of aldehydes and ketones

The addition of gaseous hydrogen to a double bond, called **hydrogenation,** requires a catalyst (see Chemical Highlight 3-1) and, in many cases, must be carried out under high pressure in order to proceed at a useful rate. Most such hydrogenations utilize **heterogeneous catalysts** (*héteros,* Greek, other; *génos,* Greek, kind) that are insoluble in the reaction solvent. (In contrast, soluble catalysts are called **homogeneous.**) The reaction takes place on the surface of the suspended particles, which typically consist of finely divided metals such as platinum, palladium, or nickel, often deposited on a supporting material, such as carbon, to maximize the surface area. Catalytic hydrogenation is also an important reaction of alkenes; its mechanism will be discussed in Chapter 12.

Hydrogenation of an Aldehyde   **Hydrogenation of a Ketone**

3-Methylbutanal   3-Methyl-1-butanol   Cyclohexanone   Cyclohexanol

Alcohols can form by hydride reduction of the carbonyl group

The electrons in the **carbonyl group,** C=O, are not distributed evenly between the two component atoms. Because oxygen is more electronegative than carbon, the carbon of a carbonyl group is electrophilic, the oxygen nucleophilic. This polarization can be represented by a charge-separated resonance form.

Polar Character of the Carbonyl Function

Because the carbonyl carbon is electrophilic, nucleophilic hydride, H^-, may be delivered to it by **hydride reagents.** Two such commercial reagents are sodium boro-

CHEMICAL HIGHLIGHT **8-1** **Biological Oxidation and Reduction**

Alcohols are metabolized by oxidation to carbonyl compounds. In biological systems, ethanol is converted into acetaldehyde by the cationic oxidizing agent *nicotinamide adenine dinucleotide* (abbreviated as NAD^+; for its structure, see Chapter 25). The process is catalyzed by the enzyme *alcohol dehydrogenase.* (The latter also catalyzes the reverse process, reduction of aldehydes and ketones to alcohols; see Problems 47 and 48 at the end of this chapter.) When the two enantiomers of 1-deuterioethanol were subjected to the enzyme, the

biochemical oxidation was found to be stereospecific, NAD^+ removing only the hydrogen marked by the solid arrowhead in the first reaction below from C1 of the alcohol (see Box 25-4).

Other alcohols are similarly oxidized biochemically. The relatively high toxicity of methanol ("wood alcohol") is due largely to its oxidation to formaldehyde. The latter specifically interferes with a system responsible for the transfer of one-carbon fragments between nucleophilic sites in biomolecules.

$$ \underset{\textbf{(}S\textbf{)-1-Deuterioethanol}}{\overset{\displaystyle CH_3}{\underset{\displaystyle D}{\underset{\displaystyle H}{\overset{|}{\underset{|}{C}}}}}\!\!-\!OH \; + \; NAD^+ \quad \xrightarrow[-NAD\text{--}H]{\text{Alcohol}\atop \text{dehydrogenase}} \quad \underset{CH_3}{\overset{O}{\overset{\|}{C}}}\!\!\diagdown\! D $$

$$ \underset{\textbf{(}R\textbf{)-1-Deuterioethanol}}{\overset{\displaystyle CH_3}{\underset{\displaystyle H}{\underset{\displaystyle D}{\overset{|}{\underset{|}{C}}}}}\!\!-\!OH \; + \; NAD^+ \quad \xrightarrow[-NAD\text{--}D]{\text{Alcohol}\atop \text{dehydrogenase}} \quad \underset{CH_3}{\overset{O}{\overset{\|}{C}}}\!\!\diagdown\! H $$

The capability of alcohols to undergo enzymatic oxidation makes them important relay stations in metabolism. One of the most important functions of the metabolic degradation of the food that we eat is its controlled "burning" (i.e., combustion; see Section 3-10) to release the heat and chemical energy required to run our bodies. Another is the selective introduction of functional groups, especially hydroxy groups, into unfunctionalized molecules or parts of molecules—in other words, alkanes or alkyl substituents. The *cytochrome* proteins are crucial biomolecules that enable nature to accomplish this task. These molecules are present in almost all living cells and emerged almost 1.5 billion years ago, before the development of plants and animals as separate species. Cytochrome P-450 (see Section 22-9) uses O_2 to accomplish the direct hydroxylation of organic molecules. In the liver,

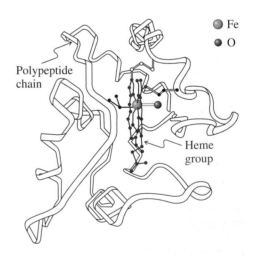

● Fe

● O

Polypeptide chain

Heme group

Cytochrome model.

this process serves to detoxify substances foreign to the body (xenobiotic), many of which are the medicines that we take. Often, the primary effect of hydroxylation is simply to impart greater water solubility, thereby accelerating the excretion of a drug and thus preventing its accumulation to toxic levels.

Selective hydroxylation is important in steroid synthesis (Section 4-7). For example, progesterone is converted by triple hydroxylation at C17, C21, and C11 into cortisol. Not only does the protein pick specific positions as targets for functionalization with complete stereoselectivity, but it also controls the sequence in which these reactions take place. You can get an inkling of the origin of this selectivity when you inspect the cytochrome model shown on the opposite page.

The active site is an Fe atom tightly held by a strongly bound heme group (see Section 26-8) embedded in the cloak of a polypeptide (protein) chain. The Fe center binds O_2 to generate an $Fe-O_2$ species, which is then reduced to H_2O and the Fe=O unit shown. This oxide reacts as a radical (Section 3-4) with R–H, producing an Fe–OH intermediate in the presence of R · . The carbon-based radical then abstracts OH to furnish the alcohol.

The steric and electronic environment provided by the polypeptide mantle allows substrates, such as progesterone, to approach the active iron site only in very specific orientations, leading to preferential oxidation at only certain positions, such as C17, C21, and C11.

hydride, $NaBH_4$, and lithium aluminum hydride, $LiAlH_4$. These two compounds possess the advantage of higher solubility in common organic solvents than do simpler analogs such as LiH and NaH.

General Hydride Reductions of Aldehydes and Ketones

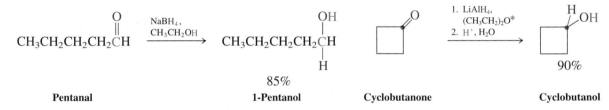

The chemistry of these reagents is dominated by the hydridic (H^-) character of the hydrogen atoms. Reduction of a carbonyl group is achieved by addition of a hydride to carbon and a proton to oxygen.

Examples of Hydride Reductions of Aldehydes and Ketones

$$CH_3CH_2CH_2CH_2\overset{\displaystyle O}{\overset{\|}{C}}H \xrightarrow[CH_3CH_2OH]{NaBH_4} CH_3CH_2CH_2CH_2\overset{\displaystyle OH}{\underset{\displaystyle H}{\overset{|}{C}}}H$$

Pentanal **1-Pentanol** 85%

Cyclobutanone $\xrightarrow[\text{2. } H^+, H_2O]{\begin{array}{l}\text{1. } LiAlH_4,\\(CH_3CH_2)_2O^*\end{array}}$ Cyclobutanol 90%

Although free hydride ion is a powerful base that is immediately protonated by protic solvents [see Exercise 8-3(e)], attachment to boron in BH_4^- moderates its re-

---

*The numbers refer to reagents that are used *sequentially*. Thus, in the first reaction, the substrate at the left of the arrow is treated with the reagents listed after the number 1. The product of this transformation then undergoes a reaction with the reagents listed after the number 2, and so on. The last reaction gives the final product shown on the right.

activity considerably, thus allowing $NaBH_4$ to be used in solvents such as ethanol. When an aldehyde or a ketone is exposed to borohydride, the reagent donates an H^- to the carbonyl carbon, with simultaneous protonation of the carbonyl oxygen by the solvent. The ethoxide by-product combines with the remaining boron fragment, giving ethoxyborohydride.

Mechanism of $NaBH_4$ Reduction

$$Na^+ \ H_3\bar{B}-H \quad \rangle C=O \quad H-OCH_2CH_3 \longrightarrow H-\underset{|}{\overset{|}{C}}-OH \ + \ Na^+ \ H_3\bar{B}OCH_2CH_3$$

Ethanol solvent   **Product alcohol**   **Sodium ethoxyborohydride**

The resulting ethoxyborohydride may attack three more carbonyl substrates before all the hydride atoms of the original reagent have been used up. As a result, one equivalent of borohydride is capable of reducing *four* equivalents of aldehyde or ketone to alcohol. The boron reagent is finally converted into tetraethoxyborate, $^-B(OCH_2CH_3)_4$.

Lithium aluminum hydride is more reactive than sodium borohydride, because its hydrogens are less strongly bound to the metal and more negatively polarized (Table 1-2). They are thus much more basic (as well as nucleophilic) and are attacked vigorously by water and alcohols to give hydrogen gas. Reductions utilizing lithium aluminum hydride are therefore carried out in aprotic solvents, such as ethoxyethane (diethyl ether).

Reaction of Lithium Aluminum Hydride with Protic Solvents

$$LiAlH_4 \ + \ 4\,CH_3OH \ \xrightarrow{\text{Fast}} \ LiAl(OCH_3)_4 \ + \ 4\,H{-}H\uparrow$$

Addition of lithium aluminum hydride to an aldehyde or ketone initially furnishes alkoxyaluminum hydride, which continues to deliver H^- to three more carbonyl groups, thus reducing a total of four equivalents of aldehyde or ketone. Addition of aqueous acid (aqueous, or HOH, work-up) consumes excess reagent, hydrolyzes the tetraalkoxyaluminate, and releases the product alcohol.

Mechanism of $LiAlH_4$ Reduction

$$Li^+ \ H_3\bar{Al}-H \quad \rangle C=\overset{..}{O} \ -\!\!\!\rightarrow \ H-\underset{|}{\overset{|}{C}}-O\bar{Al}H_3 \ Li^+ \xrightarrow{\begin{array}{c}\text{Repeat three times:}\\ \text{React with three more } \rangle C=O\end{array}}$$

Lithium alkoxyaluminum hydride

$$(H-\underset{|}{\overset{|}{C}}-O)_4 Al^- \ Li^+ \xrightarrow{\text{HOH work-up}} 4 \ H-\underset{|}{\overset{|}{C}}-OH \ + \ Al(OH)_3 \ + \ LiOH$$

Lithium tetraalkoxy-aluminate   **Product alcohol**

EXERCISE 8-9

Formulate reductions that would give rise to the following alcohols. **(a)** 1-Decanol; **(b)** 4-methyl-2-pentanol; **(c)** cyclopentylmethanol; **(d)** 1,4-cyclohexanediol.

$$CrO_3 + H_2O$$

$$\Big\updownarrow \text{pH} > 6$$

$$CrO_4^{2-}$$

$$\Big\updownarrow \text{pH} = 2\text{–}6$$

$$HCrO_4^- + Cr_2O_7^{2-}$$

$$\Big\updownarrow \text{pH} < 1$$

$$H_2CrO_4$$

Alcohol synthesis by reduction can be reversed: chromium reagents

We have seen several methods for synthesizing alcohols from aldehydes and ketones by reduction with hydrogen or hydride reagents. The reverse is also possible: Alcohols may be oxidized to produce aldehydes and ketones. A useful reagent for this purpose is a transition metal in a high oxidation state: chromium(VI). In this form, chromium has a yellow-orange color. On exposure to an alcohol, the Cr(VI) species is reduced to the deep green Cr(III). The reagent is usually supplied as a dichromate salt ($K_2Cr_2O_7$ or $Na_2Cr_2O_7$) or as CrO_3. Oxidation of secondary alcohols to ketones is often carried out in aqueous acid solution, in which all of the chromium reagents are generating varying amounts of chromic acid, H_2CrO_4, depending on pH.

Oxidation of a Secondary Alcohol to a Ketone with Aqueous Cr(VI)

$$\xrightarrow{Na_2Cr_2O_7, \ H_2SO_4, \ H_2O}$$

96%

EXERCISE 8-10

Write a balanced equation for the preceding redox process. The inorganic products are $Cr_2(SO_4)_3$ and Na_2SO_4. (**Hint:** Review the appropriate section of your general chemistry textbook on balancing half reactions.)

Under these conditions, primary alcohols tend to *overoxidize* to carboxylic acids, as shown for 1-propanol.

$$CH_3CH_2CH_2OH \xrightarrow[H_2SO_4, \ H_2O]{K_2Cr_2O_7,} \underset{\textbf{Propanal}}{CH_3CH_2\overset{\displaystyle O}{\overset{\|}{C}}H} \xrightarrow{\text{Overoxidation}} \underset{\textbf{Propanoic acid}}{CH_3CH_2\overset{\displaystyle O}{\overset{\|}{C}}OH}$$

In the absence of water, however, aldehydes are not susceptible to overoxidation. Therefore, a water-free form of Cr(VI) has been developed by reaction of CrO_3 with HCl, followed by the addition of the organic base pyridine. The result is the oxidizing agent **pyridinium chlorochromate,** abbreviated as $pyH^+ \ CrO_3Cl^-$ or just **PCC** (margin), which gives excellent yields of aldehydes on exposure to primary alcohols in dichloromethane solvent.

$$\overset{+}{N}H \ CrO_3Cl^-$$

Pyridinium chlorochromate (PCC or $pyH^+CrO_3Cl^-$)

PCC Oxidation of a Primary Alcohol to an Aldehyde

$$CH_3(CH_2)_8CH_2OH \xrightarrow{pyH^+ \ CrO_3Cl^-, \ CH_2Cl_2} CH_3(CH_2)_8\overset{\displaystyle O}{\overset{\|}{C}}H$$

92%

PCC oxidation conditions are often also used with secondary alcohols, because the relatively nonacidic reaction conditions minimize side reactions (e.g., carbocation formation; Sections 7-2, 7-3, and 9-3) and often give better yields than does the aqueous chromate method. Tertiary alcohols are unreactive toward oxidation by Cr(VI), because they do not carry hydrogens next to the OH function and therefore cannot readily form a carbon–oxygen double bond.

Chromic esters are intermediates in alcohol oxidation

What is the mechanism of the chromium(VI) oxidation of alcohols? The first step is formation of an intermediate called a **chromic ester;** the oxidation state of chromium stays unchanged in this process.

Chromic Ester Formation from an Alcohol

$$RCH_2\overset{..}{\underset{..}{O}}H + HO\overset{..}{\underset{..}{O}}{-}\overset{\overset{\displaystyle :O:}{\|}}{\underset{\underset{\displaystyle :O:}{\|}}{Cr}}{-}\overset{..}{\underset{..}{O}}H \rightleftharpoons RCH_2\overset{..}{\underset{..}{O}}{-}\overset{\overset{\displaystyle :O:}{\|}}{\underset{\underset{\displaystyle :O:}{\|}}{Cr}}{-}\overset{..}{\underset{..}{O}}H + H_2\overset{..}{\underset{..}{O}}$$

Chromic acid   Chromic ester

The next step in alcohol oxidation is equivalent to an E2 reaction. Here water (or pyridine, in the case of PCC) acts as a mild base, removing the proton next to the alcohol oxygen; $HCrO_3^-$ functions as a leaving group. The donation of an electron pair to chromium changes its oxidation state by two units, yielding Cr(IV).

Aldehyde Formation from a Chromic Ester

$$R{-}\overset{\overset{\displaystyle H}{|}}{\underset{\underset{\displaystyle H}{|}}{C}}{-}\overset{..}{O}{:}\overset{\overset{\displaystyle :\overset{..}{O}\quad\overset{..}{O}H}{\diagdown\;\diagup}}{\underset{}{Cr^{VI}}}\overset{..}{O}{:} \longrightarrow \underset{R}{\overset{H}{\diagdown}}C{=}\overset{..}{O}{:} + H_3\overset{+}{O}{:} + {}^-O_3\overset{IV}{Cr}H$$

$$H_2\overset{..}{\underset{..}{O}}$$

In contrast with the kinds of E2 reactions considered so far, this elimination furnishes a carbon–oxygen instead of a carbon–carbon double bond. The Cr(IV) species formed undergoes a redox reaction with itself to Cr(III) and Cr(V), and the latter may function as an oxidizing agent independently. Eventually all Cr(VI) is reduced to Cr(III).

EXERCISE 8-11

Formulate a synthesis of each of the following carbonyl compounds from the corresponding alcohol.

(a) $CH_3CH_2\overset{\overset{\displaystyle O}{\|}}{C}CH(CH_3)_2$ (b) [cyclobutane with H and CHO]   (c) CH_3CH_2—[cyclohexanone ring]—CH_3

To summarize, reductions of aldehydes and ketones constitute general syntheses of primary and secondary alcohols, respectively. Either catalytic hydrogenation or hydride reagents may be employed in these processes. The reverse reactions, oxidations of primary alcohols to aldehydes and secondary alcohols to ketones, are achieved with chromium(VI) reagents. Use of pyridinium chlorochromate (PCC) prevents overoxidation of primary alcohols to carboxylic acids.

8-7 Organometallic Reagents: Sources of Nucleophilic Carbon for Alcohol Synthesis

The reduction of aldehydes and ketones with hydride reagents is a useful way of synthesizing alcohols. This approach would be even more powerful if, instead of hydride,

---

The color change from Cr(VI) (orange) in the presence of alcohols to Cr(III) (green) is used in a preliminary determination of the ethanol level in the breath (and therefore blood) of suspected alcohol-intoxicated persons, especially drivers. (For the physiological effects of ethanol, see Section 9-11.) If this test is positive, it is taken as a justification by law enforcement officers to administer a more accurate blood or urine screening. It works because of the diffusion of blood alcohol through the lung into the breath, with a measured distribution ratio of roughly 2100:1 (i.e., 2100 ml of breath contains as much ethanol as 1 ml of blood). In the simplest version of this test, the participant is asked to blow into a tube containing $K_2Cr_2O_7$ and H_2SO_4 supported on powdered silica gel (SiO_2) for a duration of 10–20 seconds (as indicated by the inflation of a plastic bag attached at the end of the tube). Any alcohol present in the breath is oxidized to acetic (ethanoic) acid, a reaction signaled by the progressive color change from orange to green along the tube, according to the following balanced equation.

$$2 \underset{\text{Orange}}{K_2Cr_2O_7} + 8 H_2SO_4 + 3 CH_3CH_2OH \longrightarrow$$

$$2 \underset{\text{Green}}{Cr_2(SO_4)_3} + 2 K_2SO_4 + 3 CH_3\overset{\displaystyle O}{\overset{\displaystyle \|}{C}}OH + 11 H_2O$$

If green develops beyond the halfway mark, a blood alcohol concentration greater than 0.08% is indicated, which is considered a criminal offense in many countries. A more sophisticated version of this simple procedure uses spectrophotometric techniques to quantify the extent of oxidation, the so-called *breath analyzer test,* and more recent developments include the use of mini-gas chromatographs, electrochemical

Kit for testing alcohol in breath.

analyzers, and infrared spectrometers (Section 11-5). Some people claim that a breath analyzer can be tricked into producing a "false negative" by smoking, chewing coffee beans, eating garlic, or ingesting chlorophyll preparations beforehand: *These claims are false.*

---

we could use a source of *nucleophilic carbon.* Attack by a carbon nucleophile on a carbonyl group would give an alcohol and simultaneously form a carbon–carbon bond. *We would thus have constructed a product with more carbon atoms and, therefore, more complexity than that present in the starting materials.*

To achieve such transformations, we need to find a way of making carbon-based nucleophiles, R^-. This section will describe how this goal can be reached. Metals, particularly lithium and magnesium, act on haloalkanes to generate new compounds, called **organometallic reagents,** in which a carbon atom of an organic group is bound to a metal. These species are strong bases and good nucleophiles, and they are extremely useful in organic syntheses.

Alkyllithium and alkylmagnesium reagents are prepared from haloalkanes

Organometallic compounds of lithium and magnesium are most conveniently prepared by direct reaction of a haloalkane with the metal suspended in ethoxyethane (diethyl ether) or oxacyclopentane (tetrahydrofuran, THF). The reactivity of the haloalkanes increases in the order $Cl < Br < I$; fluorides are not normally used as starting materials in these reactions.

Alkyllithium Synthesis

$$CH_3Br + 2\ Li \xrightarrow{(CH_3CH_2)_2O,\ 0°–10°C} CH_3Li + LiBr$$

**Methyl-
lithium**

Alkylmagnesium (Grignard) Synthesis

**1-Methylethyl-
magnesium iodide**

Organomagnesium compounds, RMgX, are also called **Grignard reagents,** named after their discoverer, F. A. Victor Grignard.* They form in reactions starting with primary, secondary, and tertiary haloalkanes (and, as we shall see in later chapters, with haloalkenes and halobenzenes).

Alkyllithium compounds and Grignard reagents are rarely isolated; they are formed in solution and used immediately in the desired reaction. Sensitive to air and moisture, they must be prepared and handled under rigorously air-free and dry conditions.

The formulas RLi and RMgX oversimplify the true structures of these reagents. Thus, as written, the metal ions are highly electron deficient. To make up the desired electron octet, the metals function as Lewis acids (Section 2-9) and attach themselves to the Lewis basic solvent molecules. For example, alkylmagnesium halides are stabilized by bonding to two ether molecules. The solvent is said to be **coordinated** to the metal. When the structures of Grignard reagents are written, this coordination is rarely shown. It is crucial, however, because their formation is very difficult in its absence.

Grignard Reagents Are Coordinated to Solvent

$$CH_3CH_2\overset{..}{\underset{..}{O}}CH_2CH_3$$
$$\downarrow$$
$$R\!-\!X + Mg \xrightarrow{(CH_3CH_2)_2O} R\!-\!Mg\!-\!X$$
$$\uparrow$$
$$CH_3CH_2\overset{..}{\underset{..}{O}}CH_2CH_3$$

*Professor François Auguste Victor Grignard (1871–1935), University of Lyon, France, Nobel Prize 1912 (chemistry).

The alkylmetal bond is strongly polar

Alkyllithium and alkylmagnesium reagents have strongly polarized carbon–metal bonds; the strongly electropositive metal (Table 1-2) is the positive end of the dipole. The degree of polarization is sometimes referred to as "percentage of ionic bond character." The carbon–lithium bond, for example, has about 40% ionic character and the carbon–magnesium bond 35%. Such systems react chemically as if they contained a negatively charged carbon. To symbolize this behavior, we can show the carbon–metal bond with a resonance form that places the full negative charge on the carbon atom: a **carbanion.**

**Carbon–Metal Bond
in Alkyllithium and Alkylmagnesium Compounds**

$$\left[\quad -\overset{|}{\underset{|}{C}}\overset{\delta^-}{} \overset{\delta^+}{M} \quad \longleftrightarrow \quad -\overset{|}{\underset{|}{C}}{:}^- M^+ \quad \right]$$

Polarized   **Charge separated**

M = metal

The preparation of alkylmetals from haloalkanes illustrates an important principle in synthetic organic chemistry: **reverse polarization.** In a haloalkane, the presence of the electronegative halogen turns the carbon into an electrophilic center. On treatment with a metal, the $C^{\delta+}$–$X^{\delta-}$ unit is converted into $C^{\delta-}$–$M^{\delta+}$. In other words, the direction of polarization is reversed. Metallation has turned an electrophilic carbon into a nucleophilic center.

The alkyl group in alkylmetals is strongly basic

Carbanions are very strong bases. In fact, alkylmetals are much more basic than are amides or alkoxides, because carbon is considerably less electronegative than either nitrogen or oxygen (Table 1-2) and much less capable of supporting a negative charge. Recall (Table 2-6, Section 2-9) that alkanes are *extremely* weak acids: The pK_a of methane is estimated to be 50. It is not surprising, therefore, that carbanions are such strong bases: They are, after all, the *conjugate bases of alkanes.* Their basicity makes organometallic reagents moisture sensitive and incompatible with OH or similarly acidic functional groups. In the presence of water, hydrolysis—often violent—furnishes the metal hydroxide and an alkane. The outcome of this transformation is predictable on purely electrostatic grounds.

$$\overset{\delta^-}{R}\text{—}\overset{\delta^+}{M} + \overset{\delta^+}{H}\text{—}\overset{\delta^-}{OH} \longrightarrow R\text{—}H + M\text{—}OH$$

Organo-   **Alkane**   **Metal**
metal   **hydroxide**

Hydrolysis of an Organometallic Reagent

$$\overset{\displaystyle CH_3}{\underset{\displaystyle |}{}} \qquad\qquad\qquad \overset{\displaystyle CH_3}{\underset{\displaystyle |}{}}$$

$$CH_3CH_2CHCH_2CH_2MgBr + HOH \longrightarrow CH_3CH_2CHCH_2CH_2H + BrMgOH$$

                                                         100%

3-Methylpentylmagnesium   **3-Methylpentane**
bromide

The metallation–hydrolysis sequence affords a means by which a haloalkane can be converted into an alkane. A more direct way of achieving the same goal is the reaction of a haloalkane with the powerful hydride donor lithium aluminum hydride, an S_N2 displacement of halide by H^-. The less reactive $NaBH_4$ is incapable of performing this substitution.

$$CH_3(CH_2)_7CH_2-Br \xrightarrow[-LiBr]{LiAlH_4, (CH_3CH_2)_2O} CH_3(CH_2)_7CH_2-H$$

1-Bromononane   **Nonane**

A useful application of metallation–hydrolysis is the introduction of hydrogen isotopes, such as deuterium, into a molecule by exposure of the organometallic compound to labeled water.

**Introduction of Deuterium by Reaction
of an Organometallic Reagent
with D₂O**

$$(CH_3)_3CCl \xrightarrow[\text{2. D}_2\text{O}]{\text{1. Mg}} (CH_3)_3CD$$

EXERCISE 8-12

Show how you would prepare monodeuteriocyclohexane from cyclohexane.

In summary, haloalkanes can be converted into organometallic compounds of lithium or magnesium (Grignard reagents) by reaction with the respective metals in ether solvents. In these compounds, the alkyl group is negatively polarized, a charge distribution opposite that found in the haloalkane. Although the alkyl–metal bond is to a large extent covalent, the carbon attached to the metal behaves as a strongly basic carbanion, exemplified by its ready protonation.

8-8 Organometallic Reagents in the Synthesis of Alcohols

Among the most useful applications of organometallic reagents of magnesium and lithium are those in which the alkyl group reacts as a nucleophile. Like the hydrides, these reagents can attack the carbonyl group of an aldehyde or ketone to produce an alcohol. The difference is that a new carbon–carbon bond is formed in the process.

**Alcohol Syntheses from
Aldehydes, Ketones, and Organometallics**

M=Li or MgX

Following the flow of electrons can help us understand the reaction. In the first step, the nucleophilic alkyl group in the organometallic compound attacks the carbonyl carbon. As an electron pair from the alkyl group shifts to generate the new carbon–carbon linkage, it "pushes" two electrons from the double bond onto the oxygen, thus producing a metal alkoxide. The addition of a dilute aqueous acid furnishes the alcohol by hydrolyzing the metal–oxygen bond, another example of aqueous work-up.

The reaction of organometallic compounds with *formaldehyde* results in *primary alcohols.*

NOTE: Sequential arrows denote the execution of two (or more) steps in sequence. In the present examples, they indicate that first the Grignard addition reaction is carried out in the ether solvent, followed by acidic aqueous work-up.

Formation of a Primary Alcohol from a Grignard Reagent and Formaldehyde

$$CH_3CH_2CH_2CH_2MgBr \ + \ H_2C{=}O \xrightarrow{(CH_3CH_2)_2O} \xrightarrow{H^+, H_2O} CH_3CH_2CH_2CH_2\overset{\overset{\displaystyle H}{|}}{\underset{\underset{\displaystyle H}{|}}{C}}OH$$

Butylmagnesium bromide   Formaldehyde   93%
                                                       1-Pentanol

However, *aldehydes* other than formaldehyde convert into *secondary alcohols.*

Formation of a Secondary Alcohol from a Grignard Reagent and an Aldehyde

$$CH_3CH_2CH_2CH_2MgBr \ + \ CH_3\overset{\overset{\displaystyle O}{\|}}{C}H \xrightarrow{(CH_3CH_2)_2O} \xrightarrow{H^+, H_2O} CH_3CH_2CH_2CH_2\overset{\overset{\displaystyle CH_3}{|}}{\underset{\underset{\displaystyle H}{|}}{C}}OH$$

Butylmagnesium bromide   Acetaldehyde   78%
                                                  2-Hexanol

Ketones furnish *tertiary alcohols.*

Formation of a Tertiary Alcohol from a Grignard Reagent and a Ketone

$$CH_3CH_2CH_2CH_2MgBr \ + \ CH_3\overset{\overset{\displaystyle O}{\|}}{C}CH_3 \xrightarrow{THF} \xrightarrow{H^+, H_2O} CH_3CH_2CH_2CH_2\overset{\overset{\displaystyle CH_3}{|}}{\underset{\underset{\displaystyle CH_3}{|}}{C}}OH$$

Butylmagnesium bromide   Propanone (Acetone)   95%
                                                   2-Methyl-2-hexanol

---

EXERCISE 8-13

Write a synthetic scheme for the conversion of 2-bromopropane, $(CH_3)_2CHBr$, into 2-methyl-1-propanol, $(CH_3)_2CHCH_2OH$.

EXERCISE 8-14

Propose efficient syntheses of the following products from starting materials containing no more than four carbons.

(a) $CH_3(CH_2)_4OH$

(b) $CH_3CH_2CH_2\overset{\overset{\displaystyle OH}{|}}{C}HCH_2CH_2CH_3$

(c) $\overset{\overset{\displaystyle C(CH_3)_3}{|}}{\underset{}{\boxed{}}}\!\!-OH$

(d) $CH_3CH_2CH_2\overset{\overset{\displaystyle OH}{|}}{\underset{\underset{\displaystyle CH_3}{|}}{C}}CH_2CH_3$

Although the nucleophilic addition of alkyllithium and Grignard reagents to the carbonyl group provides us with a powerful C–C bond–forming transformation, such nucleophilic attack is too slow on haloalkanes and related electrophiles, such as those encountered in Chapter 6. This kinetic problem is what enables us to make the organometallic reagents described in Section 8-7: The product alkylmetal does not attack the haloalkane from which it is made. To achieve such coupling reactions, we must resort to copper-based reagents (Section 13-10).

In summary, alkyllithium and alkylmagnesium reagents add to aldehydes and ketones to give alcohols in which the alkyl group of the organometallic reagent has formed a bond to the original carbonyl carbon.

8-9 Complex Alcohols: An Introduction to Synthetic Strategy

The reactions introduced so far are part of the "vocabulary" of organic chemistry; unless we know the vocabulary, we cannot speak the language of organic chemistry. These reactions allow us to manipulate molecules and interconvert functional groups, so it is important to become familiar with these transformations—their types, the reagents used, the conditions under which they occur (especially when the conditions are crucial to the success of the process), and the limitations of each type.

This task may seem monumental, one that will require much memorization. But *it is made easier by an understanding of the reaction mechanisms.* We already know that reactivity can be predicted from a small number of factors, such as electronegativity, Coulombic forces, and bond strengths. Let us see how organic chemists apply this understanding to devise useful synthetic strategies.

Let us begin with a few examples in which we predict reactivity on mechanistic grounds. Then we shall turn to synthesis—the making of molecules. How do chemists develop new synthetic methods, and how can we make a "target" molecule as efficiently as possible? The two topics are closely related. The second, known as **total synthesis,** usually requires a series of reactions. In studying these tasks, therefore, we will also be reviewing much of the reaction chemistry that we have considered so far.

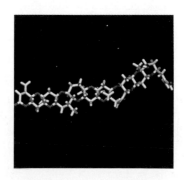

Mechanisms help in predicting the outcome of a reaction

First, recall how we predict the outcome of a reaction. What are the factors that let a particular mechanism go forward? Here are three examples.

How to Predict the Outcome of a Reaction on Mechanistic Grounds

$$ICH_2CH_2CH_2Br \overset{I^-}{\longleftarrow\!\!\!\times\!\!\!} FCH_2CH_2CH_2Br \overset{I^-}{\longrightarrow} FCH_2CH_2CH_2I$$

Not formed

Brevetoxin, the molecule on the front cover, was synthesized in 1994 after 12 years of effort in 83 steps and an overall yield of 0.043% (average yield for each step: 91%) from the simple sugar 2-deoxyribose (see Chapter 24).

Explanation. Bromide is a better leaving group than fluoride.

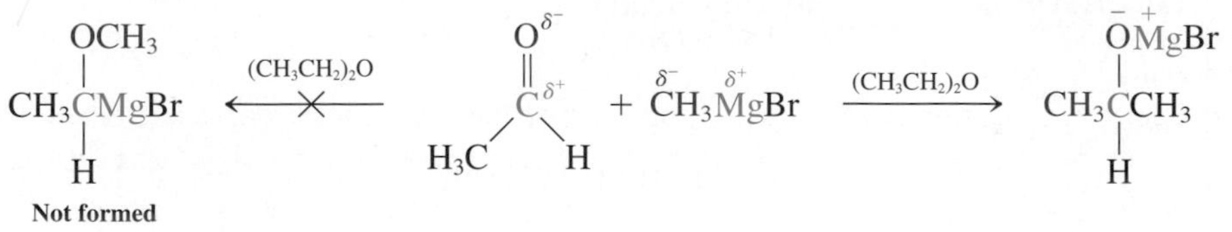
OCH3
CH3CMgBr
H
Not formed
(CH3CH2)2O
O
C
H3C
H
+ CH3MgBr
(CH3CH2)2O
OMgBr
CH3CCH3
H

Explanation. The positively polarized carbonyl carbon forms a bond to the negatively polarized alkyl group of the organometallic reagent.

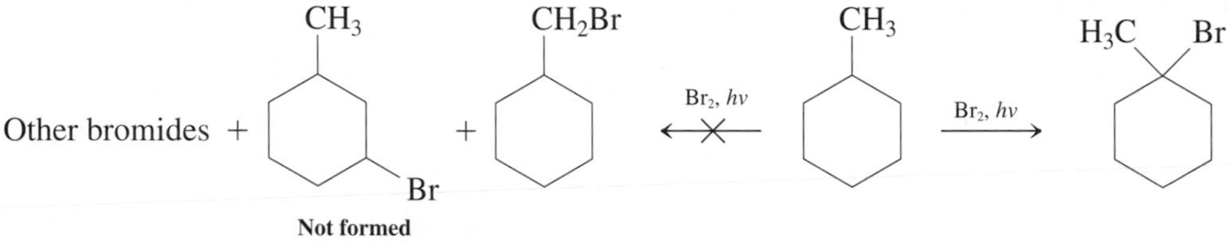
Other bromides +
CH3
Br
Not formed
+
CH2Br
Br2, hv
CH3
Br2, hv
H3C
Br

Explanation. The tertiary C–H bond is weaker than a primary or secondary C–H bond, and Br_2 is quite selective in radical halogenations.

EXERCISE 8-15

Predict and explain the outcome of each of the following reactions on mechanistic grounds.

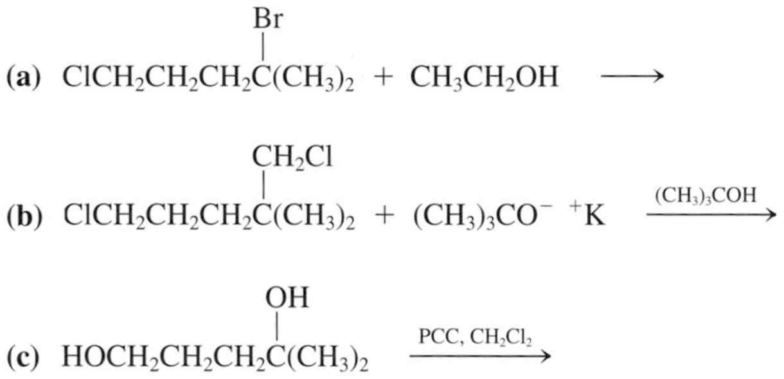
(a) ClCH2CH2CH2C(CH3)2 + CH3CH2OH
Br
(b) ClCH2CH2CH2C(CH3)2 + (CH3)3CO− +K
CH2Cl
(CH3)3COH
(c) HOCH2CH2CH2C(CH3)2
OH
PCC, CH2Cl2

New reactions lead to new synthetic methods

New reactions are found by design or by accident. For example, consider how two different students might discover the reactivity of a Grignard reagent with a ketone to give an alcohol. The first student, knowing about electronegativity and the electronic makeup of ketones, would predict that the nucleophilic alkyl group of the Grignard species should attach itself to the electrophilic carbonyl carbon. This student would be pleased by the successful outcome of the experiment, verifying chemical principles in practice. The second student, with less knowledge, might attempt to dilute a particularly concentrated solution of a Grignard reagent with what one might conceive to be a perfectly good polar solvent: propanone (acetone). A violent reaction would immediately reveal that this notion is incorrect, and further investigation would uncover the powerful potential of the reagent in alcohol synthesis.

When a reaction has been discovered, it is important to show its scope and its limitations. For this purpose, many different substrates are tested, side products (if any) noted, new functional groups subjected to the reaction conditions, and mechanistic studies carried out. Should these investigations prove the new reaction to be generally applicable, it will be added as a new synthetic method to the organic chemist's arsenal.

Because a reaction leads to a very specific change in a molecule, it is frequently useful to emphasize this "molecular alteration." A simple example is the addition of a Grignard reagent to formaldehyde. What is the structural change in this transformation? A one-carbon unit is added to an alkyl group. The method is valuable because it allows a straightforward one-carbon extension, also called a *homologation*.

Even though our synthetic vocabulary at this stage is relatively limited, we already have quite a number of molecular alterations at our disposal. For example, bromoalkanes are excellent starting points for numerous transformations.

R—M
Alkyl group
+
H_2C=O
One-carbon unit
↓
R—CH_2—OH

Each one of the products in the scheme can enter into further transformations of its own, thereby leading to more complicated products.

When we ask, "What good is a reaction? What sort of structures can we make by applying it?" we address a problem of *synthetic methodology*. Let us ask a different question. Suppose that we want to prepare a specific target molecule. How would we go about devising an efficient route to it? How do we find suitable starting materials? The problem with which we are dealing now is *total synthesis*.

Organic chemists want to make complex molecules for specific purposes. For example, certain compounds might have valuable medicinal properties but are not readily available from natural sources. Biochemists need a particular isotopically labeled molecule to trace metabolic pathways. Physical organic chemists frequently design novel structures to study. There are many reasons for the total synthesis of organic molecules.

Whatever the final target, a successful synthesis is characterized by brevity and high overall yield. The starting materials should be readily available, preferably commercially, and inexpensive. Moreover, safety and environmental concerns demand that, ideally, the reagents used be relatively nontoxic and easy to handle.

Retrosynthetic analysis simplifies synthesis problems

Many compounds that are commercially available and inexpensive are also small, containing six or fewer carbon atoms. Therefore, the most frequent task facing the synthetic planner is that of building up a larger, complicated molecule from smaller, simple fragments. The best approach to the preparation of the target is to work its synthesis *backward* on paper, an approach called **retrosynthetic analysis*** (*retro,* Latin, backward). In this analysis, strategic carbon–carbon bonds are "broken" at points at which their formation seems possible. This way of thinking backward may seem strange to you at first, because you are accustomed to learning reactions in a forward way—for example, "A plus B *gives* C." Retrosynthesis requires that you think of this process in the reverse manner—for example, "C is *derived* from A plus B."

Why retrosynthesis? The answer is that, in any "building" of a complex framework from simple building blocks, the number of possibilities of adding pieces increases drastically when going forward and includes myriad "dead end" options. In contrast, in working backward, complexity decreases and unworkable solutions are minimized. A simple analogy is a jigsaw puzzle: It is clearly easier to dismantle step by step than it is to assemble. For example, a retrosynthetic analysis of the synthesis of 3-hexanol from two three-carbon units would suggest its formation from a propyl organometallic compound and propanal.

Retrosynthetic Analysis of 3-Hexanol Synthesis from Two Three-Carbon Fragments

$$\overset{\text{OH}}{\underset{|}{\text{CH}_3\text{CH}_2\text{CH}_2\text{CHCH}_2\text{CH}_3}} \Longrightarrow \text{CH}_3\text{CH}_2\text{CH}_2\text{MgBr} + \overset{\text{O}}{\overset{||}{\text{HCCH}_2\text{CH}}}$$

Propylmagnesium bromide   **Propanal**

The double-shafted arrow indicates the so-called **strategic disconnection.** We recognize that the bond "broken" in this analysis, that between C3 and C4 in the product, is one that we can construct by using a transformation that we know, $\text{CH}_3\text{CH}_2\text{CH}_2\text{MgBr} + \text{CH}_3\text{CH}_2\text{CHO}$. In this case, only one reaction is necessary to achieve the connection; in others, it might require several steps. Two alternate, but inferior, retrosyntheses of 3-hexanol are

$$\overset{\text{OH}}{\underset{|}{\text{CH}_3\text{CH}_2\text{CH}_2\text{CHCH}_2\text{CH}_3}} \Longrightarrow \text{NaBH}_4 + \overset{\text{O}}{\overset{||}{\text{CH}_3\text{CH}_2\text{CH}_2\text{CCH}_2\text{CH}_3}}$$

$$\overset{\text{OH}}{\underset{|}{\text{CH}_3\text{CH}_2\text{CH}_2\text{CHCH}_2\text{CH}_3}} \Longrightarrow \overset{\text{O}}{\overset{||}{\text{NaOCCH}_3}} + \overset{\text{Br}}{\underset{|}{\text{CH}_3\text{CH}_2\text{CH}_2\text{CHCH}_2\text{CH}_3}}$$

They are not as good as the first because they do not significantly *simplify* the target structure: No carbon–carbon bonds are "broken."

Retrosynthetic analysis aids in alcohol construction

Let us apply retrosynthetic analysis to the preparation of a tertiary alcohol, 4-ethyl-4-nonanol. Because of their steric encumbrance and hydrophobic nature, this alcohol and its homologs have important industrial applications as cosolvents and additives

---

*Pioneered by Professor Elias J. Corey (b. 1928), Harvard University, Nobel Prize 1990 (chemistry).

in certain polymerization processes (Section 12-13). There are two steps to follow at each stage of the process. First, we identify all possible strategic disconnections, "breaking" all bonds that can be formed by reactions that we know. Second, we evaluate the relative merits of these disconnections, seeking the one that best simplifies the target structure. The strategic bonds in 4-ethyl-4-nonanol are those around the functional group. There are three disconnections leading to simpler precursors. Path *a* cleaves the ethyl group from C4, suggesting as the starting materials for its construction ethylmagnesium bromide and 4-nonanone. Cleavage *b* is an alternative possibility leading to a propyl Grignard reagent and 3-octanone as precursors. Finally, disconnection *c* reveals a third synthesis route derived from the addition of pentylmagnesium bromide to 3-hexanone.

Partial Retrosynthetic Analysis of the Synthesis of 4-Ethyl-4-nonanol

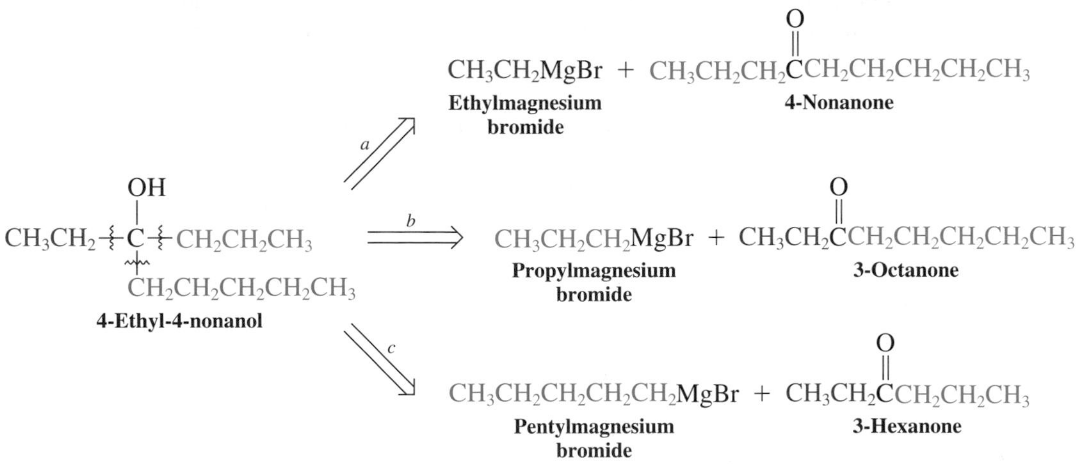

Evaluation reveals that pathway *c* is best: The necessary building blocks are almost equal in size, containing five and six carbons; thus, this disconnection provides the greatest simplification in structure.

EXERCISE 8-16

Apply retrosynthetic analysis to 4-ethyl-4-nonanol, disconnecting the carbon–*oxygen* bond. Does this lead to an efficient synthesis? Explain.

Can we pursue either of the fragments arising from disconnection by pathway *c* to even simpler starting materials? Yes; recall (Section 8-6) that ketones are obtained from the oxidation of secondary alcohols by Cr(VI) reagents. We may therefore envision preparation of 3-hexanone from the corresponding alcohol, 3-hexanol.

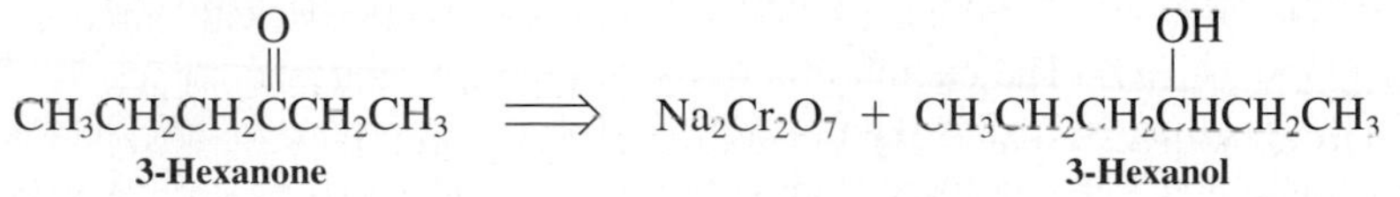

Because we earlier identified an efficient disconnection of the latter into three-carbon fragments, we are now in a position to present our complete synthetic scheme (see page 312):

Synthesis of 4-Ethyl-4-nonanol

$$\underset{\textbf{Propanal}}{CH_3CH_2CH} \overset{O}{\Vert} \quad \xrightarrow[\text{2. H}^+,\text{ H}_2\text{O}]{\text{1. CH}_3\text{CH}_2\text{CH}_2\text{MgBr, (CH}_3\text{CH}_2)_2\text{O}} \quad \underset{\textbf{3-Hexanol}}{CH_3CH_2\overset{OH}{\underset{|}{CH}}CH_2CH_2CH_3} \xrightarrow{\text{Na}_2\text{Cr}_2\text{O}_7,\text{ H}_2\text{SO}_4,\text{ H}_2\text{O}}$$

$$\underset{\textbf{3-Hexanone}}{CH_3CH_2\overset{O}{\overset{\Vert}{C}}CH_2CH_2CH_3} \quad \xrightarrow[\text{2. H}^+,\text{ H}_2\text{O}]{\text{1. CH}_3\text{CH}_2\text{CH}_2\text{CH}_2\text{CH}_2\text{MgBr, (CH}_3\text{CH}_2)_2\text{O}} \quad \begin{array}{c} OH \\ | \\ CH_3CH_2\overset{}{C}CH_2CH_2CH_3 \\ | \\ CH_2CH_2CH_2CH_2CH_3 \\ \textbf{4-Ethyl-4-nonanol} \end{array}$$

This example illustrates a very powerful general sequence for the construction of complex alcohols: first, Grignard or organolithium addition to an aldehyde to give a secondary alcohol; then oxidation to a ketone; and finally, addition of another organometallic reagent to give a tertiary alcohol.

Utility of Alcohol Oxidations in Synthesis

$$\underset{}{R\overset{O}{\overset{\Vert}{CH}}} \xrightarrow[\text{2. H}^+,\text{ H}_2\text{O}]{\text{1. R'MgBr, (CH}_3\text{CH}_2)_2\text{O}} \begin{array}{c} OH \\ | \\ R\,CH \\ | \\ R' \end{array} \xrightarrow{\text{CrO}_3,\text{ H}^+,\text{ H}_2\text{O}} R\overset{O}{\overset{\Vert}{C}}R' \xrightarrow[\text{2. H}^+,\text{ H}_2\text{O}]{\text{1. R''MgBr, (CH}_3\text{CH}_2)_2\text{O}} \begin{array}{c} OH \\ | \\ R\,CR' \\ | \\ R'' \end{array}$$

EXERCISE 8-17

Write an economical retrosynthetic analysis of 3-cyclobutyl-3-octanol. (**Hint:** Consult the Chapter Integration Problem.)

EXERCISE 8-18

Show how you would prepare 2-methyl-2-propanol from methane as the only organic starting material.

Watch out for pitfalls in planning syntheses

There are several considerations to keep in mind when practicing synthetic chemistry that will help to avoid designing unsuccessful or low-yielding approaches to a target molecule. First, *try to minimize the total number of transformations required to convert the initial starting material into the desired product.* This point is so important that, in some cases, it is worthwhile to accept a low-yield step if it allows a significant shortening of the synthetic sequence. For example (with the assumption that all starting materials are of comparable cost), a seven-step synthesis in which each step has an 85% yield is less productive than a four-step synthesis with three yields at 95% and one at 45%. The overall efficiency in the first sequence comes to ($0.85 \times 0.85 \times 0.85 \times 0.85 \times 0.85 \times 0.85 \times 0.85$) $\times$ 100 = 32%, whereas the second synthesis, in addition to being three steps shorter, gives ($0.95 \times 0.95 \times 0.95 \times 0.45$) $\times$ 100 = 39%. In these examples, all steps take place consecutively, a procedure called **linear synthesis.** In general, it is better to approach complex targets through two or more concurrent routes, as long as the overall number of steps is the same, a strategy described as **convergent synthesis.** Although a simple overall-yield calculation is not possible for a convergent strategy, you can readily convince yourself of its increased efficiency by comparing the actual *amounts* of starting materials required by the two approaches to make the same amount of product. In the following example, 10 g of a product H is prepared in three steps (50% each) by a linear sequence $A \rightarrow B \rightarrow C \rightarrow H$ and a convergent one starting from D and F, respectively, through E and G. If we assume (for the sake of simplicity) that the molecular weights of these compounds are all the same, the first preparation requires 80 g and, the second only (a combined) 40 g of starting materials.

$$
\begin{array}{ccccccc}
\text{A} & \xrightarrow{50\%} & \text{B} & \xrightarrow{50\%} & \text{C} & \xrightarrow{50\%} & \text{H} \\
80\text{ g} & & 40\text{ g} & & 20\text{ g} & & 10
\end{array}
$$

Linear synthesis of H

$$
\begin{array}{ccc}
\text{D} & \xrightarrow{50\%} & \text{E} \\
20\text{ g} & & 10\text{ g}
\end{array} \searrow
$$
$$
\xrightarrow{50\%} \quad \text{H} \quad 10\text{ g}
$$
$$
\begin{array}{ccc}
\text{F} & \xrightarrow{50\%} & \text{G} \\
20\text{ g} & & 10\text{ g}
\end{array} \nearrow
$$

Convergent synthesis of H

Second, *do not use reagents whose molecules have functional groups that would interfere with the desired reaction.* For example, treating a hydroxyaldehyde with a Grignard reagent leads to an acid–base reaction, destroying the Grignard, and not to carbon–carbon bond formation.

$$
\underset{\overset{|}{CH_3}}{HOCH_2CH_2\overset{OH}{\underset{|}{CH}}} \quad \xleftarrow{\quad\times\quad} \quad HOCH_2CH_2\overset{O}{\overset{||}{CH}} + CH_3MgBr \longrightarrow BrMgOCH_2CH_2\overset{O}{\overset{||}{CH}} + \overset{H}{\underset{|}{CH_3}}
$$

A possible solution to this problem would be to add two equivalents of Grignard reagent: *one* to react with the acidic hydrogen as shown, the *other* to achieve the desired addition to the carbonyl group.

Do not try to make a Grignard reagent from a bromoketone. Such a reagent is not stable and will, as soon as it is formed, decompose by reacting with its own carbonyl group (in the same or another molecule).

Third, *take into account any mechanistic and structural constraints affecting the reactions under consideration.* For example, radical brominations are more selective than chlorinations. Keep in mind the structural limitations on nucleophilic reactions, and do not forget the lack of reactivity of the 2,2-dimethyl-1-halopropanes (neopentyl halides). Although sometimes difficult to recognize, many haloalkanes have "neopentyl-like" structures and are similarly unreactive. Nevertheless, such systems do form organometallic reagents and may be further functionalized in this manner.

Neopentyl-like Hindered Haloalkanes

For example, treatment of the Grignard reagent made from 1-bromo-2,2-dimethyl-propane with formaldehyde leads to the corresponding alcohol.

$$(CH_3)_3CCH_2Br \xrightarrow[\substack{1.\ Mg \\ 2.\ CH_2=O \\ 3.\ H^+,\ H_2O}]{} (CH_3)_3CCH_2CH_2OH$$

1-Bromo-2,2-dimethylpropane   **3,3-Dimethyl-1-butanol**

Tertiary halides, if incorporated into a more complex framework, also are sometimes difficult to recognize. Remember that tertiary halides do not undergo S_N2 reactions but eliminate in the presence of bases.

Expertise in synthesis, as in many other aspects of organic chemistry, develops largely from practice. Planning the synthesis of complex molecules requires a review of the reactions and mechanisms covered in earlier sections. The knowledge thus acquired can then be applied to the solution of synthetic problems.

CHAPTER INTEGRATION PROBLEM

Tertiary alcohols are important additives in some industrial processes utilizing Lewis acidic metal compounds (Sections 2-9 and 6-4) as catalysts. The alcohol provides the metal with a sterically protecting and hydrophobic environment (see Figure 8-3; see also Chemical Highlight 8-1), which ensures solubility in organic solvents, longer lifetimes, and selectivity in substrate activation. Preparation of these tertiary alcohols typically follows the synthetic principles outlined in Section 8-9.

Starting from cyclohexane and using any other building blocks containing four or fewer carbons, in addition to any necessary reagents, formulate a synthesis of tertiary alcohol A.

A

SOLUTION

Before we start a random trial and error approach to solving this problem, it is better to take an inventory of what is given. First, we are given cyclohexane, and we note that this unit shows up as a substituent in tertiary alcohol A. Second, a total of seven additional carbons appears in the product, so our synthesis will require some additional stitching together of smaller fragments because we cannot use compounds containing more than four carbons. Third, target A is a tertiary alcohol, which should be amenable to the retrosynthetic analysis introduced in Section 8-9 (M = metal):

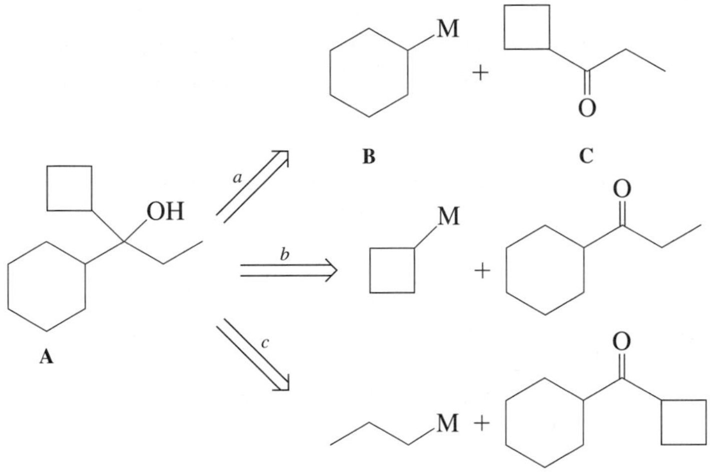

Approach *a* is clearly the one of choice because it breaks up tertiary alcohol A into evenly sized fragments B and C.

Having chosen route *a* as the most suitable way to find direct precursors to A, our analysis proceeds to working backward further: What are the appropriate precursors to B and C, respectively? The former is readily traced by retrosynthesis to our starting hydrocarbon, cyclohexane: The precursor to the organometallic compound B must be a halocyclohexane, which in turn can be made from cyclohexane by free-radical halogenation:

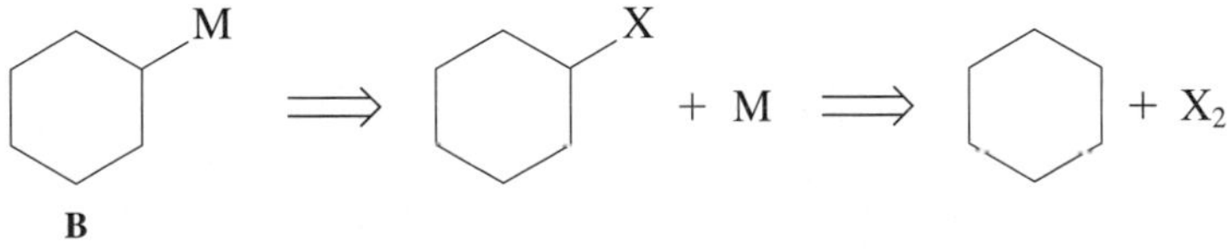

Ketone C must be broken into two smaller components; the best breakdown would be a "four + three" carbon combination: It is the most evenly sized solution, and it suggests the use of cyclobutyl intermediates. Because the only C–C disconnection that we know at this stage is that of alcohols, the first retrosynthetic step from C is its precursor alcohol (plus a chromium oxidant). Further retrosynthesis then provides the required starting pieces D and E.

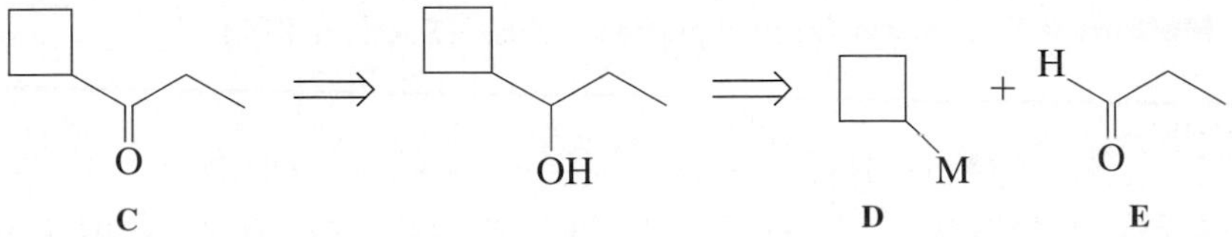

Now we can write the detailed synthetic scheme in a forward mode, with cyclohexane and pieces D and E as our starting materials:

A final note: In this and subsequent synthetic exercises in this book, retrosynthetic analysis requires that you have command of all reactions not only in a forward fashion (i.e., starting material + reagent → product) but also in reverse (i.e., product ← starting material + reagent). The two sets address two different questions. The first asks: What are all the possible products that I can make from my starting material in the presence of all the reagents that I know? The second asks: What are all the conceivable starting materials that, with the appropriate reagents, will lead to my product? The two types of schematic summaries of reactions that you will see at the end of this and subsequent chapters emphasize this point.

NEW REACTIONS

1. Acid-Base Properties of Alcohols (Section 8-3)

Alkyloxonium ion   **Alcohol**   **Alkoxide**

Acidity: $RO-H \approx HO-H > H_2N-H > H_3C-H$
Basicity: $RO^- \approx HO^- < H_2N^- < H_3C^-$

Industrial Preparation of Alcohols

2. Synthesis Gas (Section 8-4)

$$Coal \xrightarrow{\text{Air, } H_2O, \Delta} x\,CO + y\,H_2$$

Synthesis gas

3. Methanol Synthesis from Synthesis Gas (Section 8-4)

$$CO + 2\,H_2 \xrightarrow{\text{Cu-ZnO-Cr}_2O_3,\ 250°C,\ 50\text{–}100\ \text{atm}} CH_3OH$$

4. Ethanol by Hydration of Ethene (Section 8-4)

$$CH_2{=}CH_2 + HOH \xrightarrow{H_3PO_4,\ 300°C} CH_3CH_2OH$$

Laboratory Preparation of Alcohols

5. Nucleophilic Displacement of Halides and Other Leaving Groups by Hydroxide (Section 8-5)

$$RCH_2X + HO^- \xrightarrow[S_N2]{H_2O} RCH_2OH + X^-$$

X = halide, sulfonate
Primary, secondary (tertiary undergoes elimination)

$$\underset{\overset{|}{R'}}{RCHX} + CH_3\overset{O}{\overset{||}{C}}O^- \xrightarrow{S_N2} \underset{\overset{|}{R'}}{RCHO}\overset{O}{\overset{||}{C}}CH_3 \xrightarrow[\text{Ester hydrolysis}]{HO^-} \underset{\overset{|}{R'}}{RCHOH}$$

$$\underset{\overset{|}{R''}}{\overset{\overset{R}{|}}{R'CX}} \xrightarrow[S_N1]{H_2O, \text{ propanone (acetone)}} \underset{\overset{|}{R''}}{\overset{\overset{R}{|}}{R'COH}}$$

Best method for tertiary

6. Catalytic Hydrogenation of Aldehydes and Ketones (Section 8-6)

$$\overset{O}{\overset{||}{RCH}} \xrightarrow{H_2, Pt} RCH_2OH \qquad \overset{O}{\overset{||}{RCR'}} \xrightarrow{H_2, \text{ catalyst}} \underset{\overset{|}{H}}{\overset{\overset{OH}{|}}{RCR'}}$$

Aldehyde   Primary alcohol   Ketone   Secondary alcohol

7. Reduction of Aldehydes and Ketones by Hydrides (Section 8-6)

$$\overset{O}{\overset{||}{RCH}} \xrightarrow{NaBH_4, CH_3CH_2OH} RCH_2OH \qquad \overset{O}{\overset{||}{RCR'}} \xrightarrow{NaBH_4, CH_3CH_2OH} \underset{\overset{|}{H}}{\overset{\overset{OH}{|}}{RCR'}}$$

$$\overset{O}{\overset{||}{RCH}} \xrightarrow[\text{2. } H^+, H_2O]{\text{1. } LiAlH_4, (CH_3CH_2)_2O} RCH_2OH \qquad \overset{O}{\overset{||}{RCR'}} \xrightarrow[\text{2. } H^+, H_2O]{\text{1. } LiAlH_4, (CH_3CH_2)_2O} \underset{\overset{|}{H}}{\overset{\overset{OH}{|}}{RCR'}}$$

Aldehyde   Primary alcohol   Ketone   Secondary alcohol

Oxidation of Alcohols

8. Chromium Reagents (Section 8-6)

$$RCH_2OH \xrightarrow{PCC, CH_2Cl_2} \overset{O}{\overset{||}{RCH}} \qquad \underset{}{\overset{\overset{OH}{|}}{RCHR'}} \xrightarrow{Na_2Cr_2O_7, H_2SO_4} \overset{O}{\overset{||}{RCR'}}$$

Primary alcohol   Aldehyde   Secondary alcohol   Ketone

Organometallic Reagents

9. Reaction of Metals with Haloalkanes (Section 8-7)

$$RX + Li \xrightarrow{(CH_3CH_2)_2O} RLi$$
Alkyllithium reagent

$$RX + Mg \xrightarrow{(CH_3CH_2)_2O} RMgX$$
Grignard reagent

R cannot contain acidic groups such as O–H or electrophilic groups such as C=O

10. Hydrolysis (Section 8-7)

$$RLi \quad or \quad RMgX + H_2O \longrightarrow RH$$
$$RLi \quad or \quad RMgX + D_2O \longrightarrow RD$$

11. Addition of Organometallic Compounds to Aldehydes and Ketones (Section 8-8)

$$RLi \quad or \quad RMgX + CH_2{=}O \longrightarrow RCH_2OH$$
Formaldehyde **Primary alcohol**

$$RLi \quad or \quad RMgX + R'\overset{O}{\overset{\|}{C}}H \longrightarrow R\overset{OH}{\underset{H}{\overset{|}{C}}}R'$$
Aldehyde **Secondary alcohol**

$$RLi \quad or \quad RMgX + R'\overset{O}{\overset{\|}{C}}R'' \longrightarrow R\overset{OH}{\underset{R''}{\overset{|}{C}}}R'$$
Ketone **Tertiary alcohol**

Aldehyde or ketone cannot contain other groups that react with
organometallic reagents such as O–H or other C=O groups

12. Alkanes from Haloalkanes and Lithium Aluminum Hydride (Section 8-7)

$$RX + LiAlH_4 \xrightarrow{(CH_3CH_2)_2O} RH$$

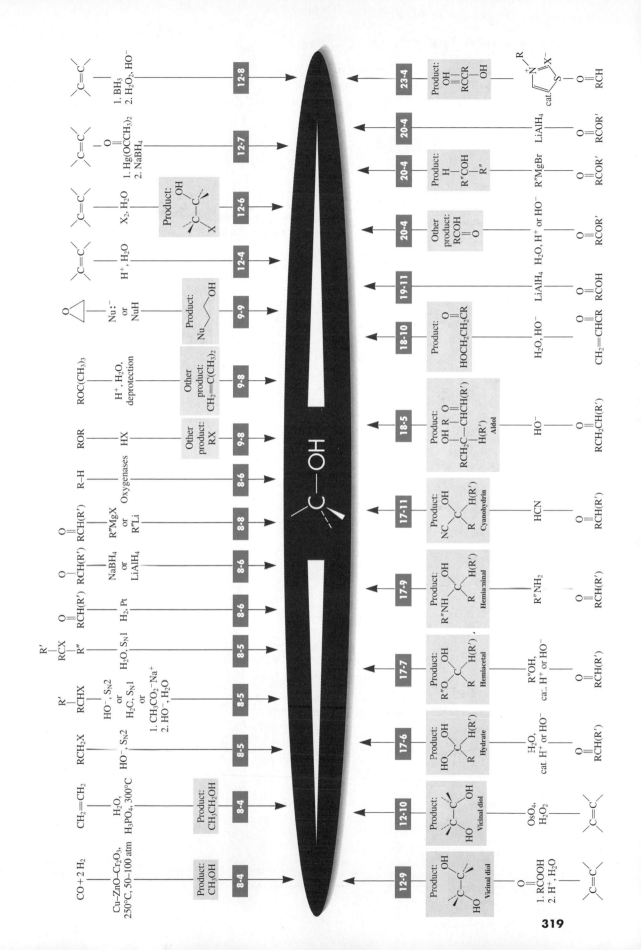

Preparation of Alcohols | section number

319

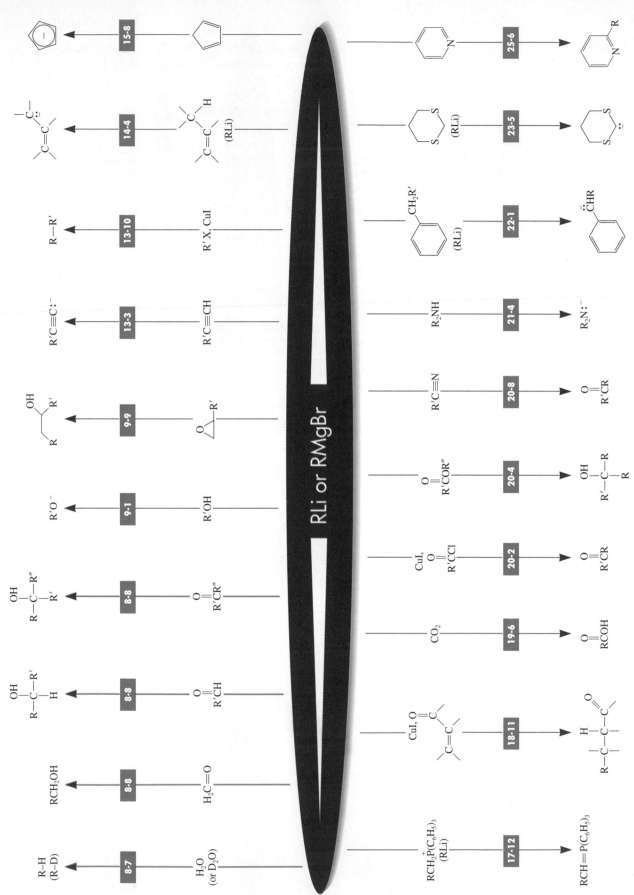

Reactions of Alkyllithium and Grignard Reagents   section number

RLi or RMgBr

320

IMPORTANT CONCEPTS

1. Alcohols are **alkanols** in IUPAC nomenclature. The stem containing the functional group gives the alcohol its name. Alkyl and halo substituents are added as prefixes.
2. Like water, alcohols have a **polarized** and short O–H bond. The proton is **hydrophilic** and enters into **hydrogen bonding.** Consequently, alcohols have unusually high boiling points and, in many cases, appreciable water solubility. The alkyl part of the molecule is **hydrophobic.**
3. Again like water, alcohols are **amphoteric:** They are both acidic and basic. Complete deprotonation to an **alkoxide** takes place with bases whose conjugate acids are considerably weaker than the alcohol. Protonation gives an **alkyloxonium ion.** In solution, the order of acidity is primary > secondary > tertiary alcohol. Electron-withdrawing substituents increase the acidity (and reduce the basicity).

4. The **hydrogenation** of aldehydes and ketones furnishes alcohols and requires a catalyst.
5. The conversion of the electrophilic alkyl group in a haloalkane, $C^{\delta+}-X^{\delta-}$, into its nucleophilic analog in an **organometallic compound,** $C^{\delta-}-M^{\delta+}$, is an example of **reverse polarization.**
6. The carbon atom in the **carbonyl group,** C=O, of an aldehyde or a ketone is electrophilic and therefore subject to attack by nucleophiles, such as hydride in **hydride reagents** or alkyl in organometallic compounds. Subsequent to aqueous work-up, the products of such transformations are alcohols.
7. The **oxidation** of alcohols to aldehydes and ketones by chromium(VI) reagents opens up important synthetic possibilities based on further reactions with organometallic reagents.
8. **Retrosynthetic analysis** aids in planning the synthesis of complex organic molecules by identifying strategic bonds that may be constructed in an efficient sequence of reactions.

PROBLEMS

19. Name the following alcohols according to the IUPAC nomenclature system. Indicate stereochemistry (if any) and label the hydroxy groups as primary, secondary, or tertiary.

(a) $CH_3CH_2CHCH_3$ with OH

(b) $CH_3CHCH_2CHCH_2CH_3$ with Br and OH

(c) $HOCH_2CH(CH_2CH_2CH_3)_2$

(d) structure with CH_2Cl, H, C, H_3C, OH

(e) cyclobutane with CH_2CH_3 and OH

(f) decalin-type ring with OH and Br

(g) $C(CH_2OH)_4$

(h) Fischer projection: CH_2OH / H—OH / H—OH / CH_2OH

(i) cyclopentane with OH and CH_2CH_2OH

(j) CH_3CH_2—C(Cl)(CH_3)—CH_2OH

20. Draw the structures of the following alcohols. **(a)** 2-(Trimethylsilyl)ethanol; **(b)** 1-methylcyclopropanol; **(c)** 3-(1-methylethyl)-2-hexanol; **(d)** (*R*)-2-pentanol; **(e)** 3,3-dibromocyclohexanol.

21. Rank each group of compounds in order of increasing boiling point. **(a)** Cyclohexane, cyclohexanol, chlorocyclohexane; **(b)** 2,3-dimethyl-2-pentanol, 2-methyl-2-hexanol, 2-heptanol.

22. Explain the order of water solubilities for the compounds in each of the following groups. **(a)** Ethanol > chloroethane > ethane; **(b)** methanol > ethanol > 1-propanol.

23. 1,2-Ethanediol exists to a much greater extent in the *gauche* conformation than does 1,2-dichloroethane. Explain. Would you expect the *gauche : anti* conformational ratio of 2-chloroethanol to be similar to that of 1,2-dichloroethane or more like that of 1,2-ethanediol?

24. Rank the compounds in each group in order of decreasing acidity.
(a) $CH_3CHClCH_2OH$, $CH_3CHBrCH_2OH$, $BrCH_2CH_2CH_2OH$
(b) $CH_3CCl_2CH_2OH$, CCl_3CH_2OH, $(CH_3)_2CClCH_2OH$
(c) $(CH_3)_2CHOH$, $(CF_3)_2CHOH$, $(CCl_3)_2CHOH$

25. Write an appropriate equation to show how each of the following alcohols acts as, first, a base, and, second, an acid in solution. How do the base and acid strengths of each compare with those of methanol? **(a)** $(CH_3)_2CHOH$; **(b)** CH_3CHFCH_2OH; **(c)** CCl_3CH_2OH.

26. Given the pK_a values of -2.2 for $CH_3\overset{+}{O}H_2$ and 15.5 for CH_3OH, calculate the pH at which **(a)** methanol will contain exactly equal amounts of $CH_3\overset{+}{O}H_2$ and CH_3O^-; **(b)** 50% CH_3OH and 50% $CH_3\overset{+}{O}H_2$ will be present; **(c)** 50% CH_3OH and 50% CH_3O^- will be present.

27. Do you expect hyperconjugation to be important in the stabilization of alkyloxonium ions (e.g., $R\overset{+}{O}H_2$, $R_2\overset{+}{O}H$)? Explain your answer.

28. Evaluate each of the following possible alcohol syntheses as being good (the desired alcohol is the major or only product), not so good (the desired alcohol is a minor product), or worthless. (**Hint:** Refer to Section 7-9 if necessary.)

(a) $CH_3CH_2CH_2CH_2Cl$ $\xrightarrow{\text{H}_2\text{O, CH}_3\overset{\text{O}}{\underset{||}{\text{C}}}\text{CH}_3}$ $CH_3CH_2CH_2CH_2OH$

(b) CH_3OSO_2—⟨benzene ring⟩—CH_3 $\xrightarrow{\text{HO}^-,\ \text{H}_2\text{O},\ \Delta}$ CH_3OH

(c) ⟨cyclohexane ring with I⟩ $\xrightarrow{\text{HO}^-,\ \text{H}_2\text{O},\ \Delta}$ ⟨cyclohexane ring with OH⟩

(d)
$$\underset{\text{CH}_3\overset{\displaystyle |}{\overset{\displaystyle \text{I}}{\text{CH}}}\text{CH}_2\text{CH}_2\text{CH}_3}{} \xrightarrow{\text{H}_2\text{O, }\Delta} \underset{\text{CH}_3\overset{\displaystyle |}{\overset{\displaystyle \text{OH}}{\text{CH}}}\text{CH}_2\text{CH}_2\text{CH}_3}{}$$

(e)
$$\underset{\text{CH}_3\overset{\displaystyle |}{\overset{\displaystyle \text{CN}}{\text{CH}}}\text{CH}_3}{} \xrightarrow{\text{HO}^-\text{, H}_2\text{O, }\Delta} \underset{\text{CH}_3\overset{\displaystyle |}{\overset{\displaystyle \text{OH}}{\text{CH}}}\text{CH}_3}{}$$

(f) $CH_3OCH_3 \xrightarrow{\text{HO}^-\text{, H}_2\text{O, }\Delta} CH_3OH$

(g) 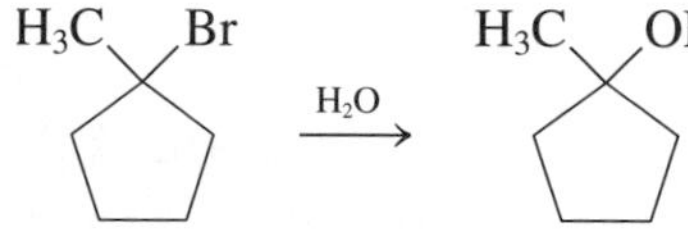
$$\xrightarrow{\text{H}_2\text{O}}$$

(h)
$$\underset{\text{CH}_3\overset{\displaystyle |}{\overset{\displaystyle \text{CH}_3}{\text{CH}}}\text{CH}_2\text{Cl}}{} \xrightarrow{\text{HO}^-\text{, H}_2\text{O, }\Delta} \underset{\text{CH}_3\overset{\displaystyle |}{\overset{\displaystyle \text{CH}_3}{\text{CH}}}\text{CH}_2\text{OH}}{}$$

29. Give the major product(s) of each of the following reactions. Aqueous work-up steps (when necessary) have been omitted.

(a) $CH_3CH{=}CHCH_3 \xrightarrow[\text{(Hint: See Section 8-4)}]{\text{H}_3\text{PO}_4\text{, H}_2\text{O, }\Delta}$

(b)
$$\underset{\text{CH}_3\overset{\displaystyle \text{O}}{\overset{\displaystyle \|}{\text{C}}}\text{CH}_2\text{CH}_2\overset{\displaystyle \text{O}}{\overset{\displaystyle \|}{\text{C}}}\text{CH}_3}{} \xrightarrow{\text{H}_2\text{, Pt}}$$

(c) 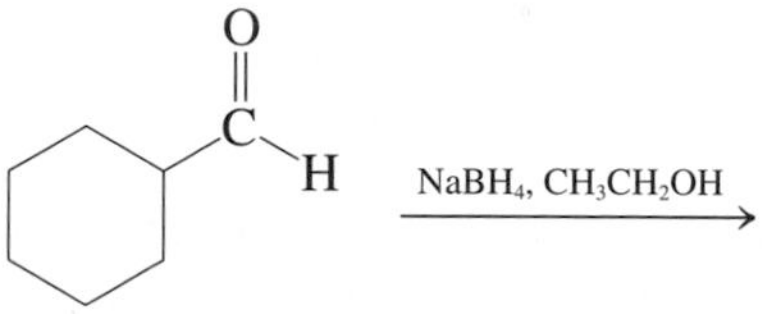
$\xrightarrow{\text{NaBH}_4\text{, CH}_3\text{CH}_2\text{OH}}$

(d) 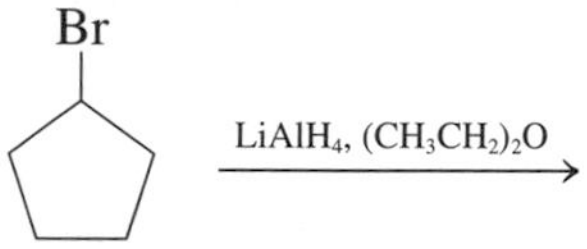
$\xrightarrow{\text{LiAlH}_4\text{, (CH}_3\text{CH}_2)_2\text{O}}$

(e) 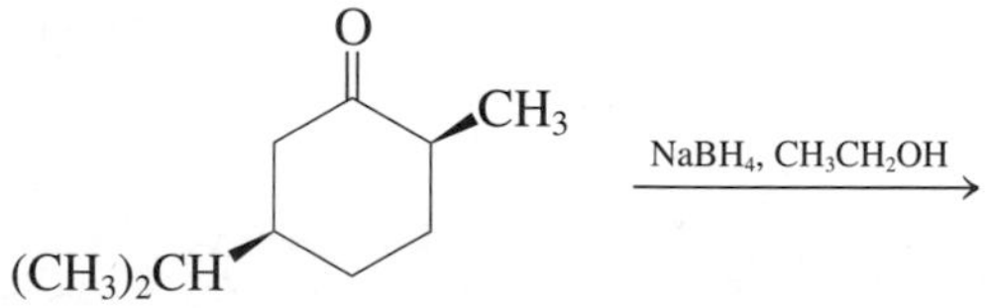
$\xrightarrow{\text{NaBH}_4\text{, CH}_3\text{CH}_2\text{OH}}$

(f) 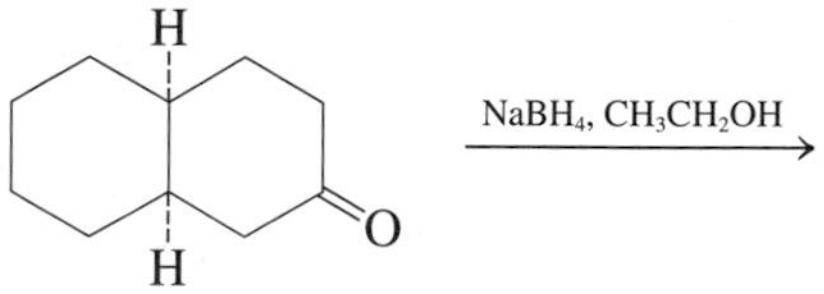
$\xrightarrow{\text{NaBH}_4\text{, CH}_3\text{CH}_2\text{OH}}$

30. What is the direction of the following equilibrium? (**Hint:** The pK_a for H_2 is about 38.)

$$H^- + H_2O \rightleftharpoons H_2 + HO^-$$

31. Formulate the product of each of the following reactions. The solvent in each case is $(CH_3CH_2)_2O$.

(a)
$$\underset{\text{CH}_3\text{CH}}{\overset{\displaystyle \text{O}}{\overset{\displaystyle \|}{}}} \xrightarrow[\text{2. H}^+\text{, H}_2\text{O}]{\text{1. LiAlD}_4}$$

(b)
$$\underset{\text{CH}_3\text{CH}}{\overset{\displaystyle \text{O}}{\overset{\displaystyle \|}{}}} \xrightarrow[\text{2. D}^+\text{, D}_2\text{O}]{\text{1. LiAlH}_4}$$

(c) $CH_3CH_2I \xrightarrow{\text{LiAlD}_4}$

32. Give the major product(s) of each of the following reactions [after work-up with aqueous acid in (d), (f), and (h)].

(a) $CH_3(CH_2)_5\overset{\overset{\displaystyle Cl}{\displaystyle |}}{C}HCH_3$ $\xrightarrow{Mg, (CH_3CH_2)_2O}$

(b) Product of (a) $\xrightarrow{D_2O}$

(c) 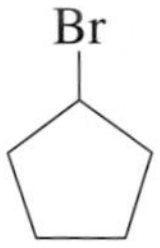 $\xrightarrow{Li, (CH_3CH_2)_2O}$

(d) Product of (c) 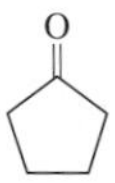 $\longrightarrow$

(e) $CH_3CH_2CH_2Cl + Mg$ $\xrightarrow{(CH_3CH_2)_2O}$

(f) Product of (e) + $\longrightarrow$

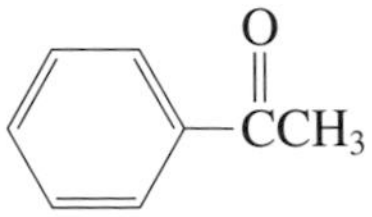

(g) 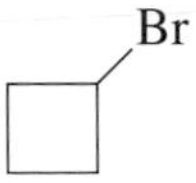 $+ 2\ Li$ $\xrightarrow{(CH_3CH_2)_2O}$

(h) 2 mol product of (g) + 1 mol
$$CH_3\overset{\overset{\displaystyle O}{\displaystyle \|}}{C}CH_2CH_2\overset{\overset{\displaystyle O}{\displaystyle \|}}{C}CH_3 \longrightarrow$$

33. The common practice of washing laboratory glassware with propanone (acetone) can lead to unintended consequences. For example, a student plans to carry out the preparation of methylmagnesium iodide, CH_3MgI, which he will add to benzaldehyde, C_6H_5CHO. What compound is he intending to synthesize after aqueous work-up? Using his freshly washed glassware, he carries out the procedure and finds that he has produced an unexpected tertiary alcohol as a product. What substance did he make? How did it form?

34. Give the major product(s) of each of the following reactions (after aqueous work-up). The solvent in each case is ethoxyethane (diethyl ether).

(a) 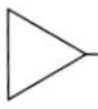 $-MgBr + H\overset{\overset{\displaystyle O}{\displaystyle \|}}{C}H$ $-$

(b) $CH_3\overset{\overset{\displaystyle CH_3}{\displaystyle |}}{C}HCH_2MgCl + CH_3\overset{\overset{\displaystyle O}{\displaystyle \|}}{C}H$ $\longrightarrow$

(c) $C_6H_5CH_2Li + C_6H_5\overset{\overset{\displaystyle O}{\displaystyle \|}}{C}H$ $\longrightarrow$

(d) $CH_3\overset{\overset{\displaystyle MgBr}{\displaystyle |}}{C}HCH_3 +$ $\longrightarrow$

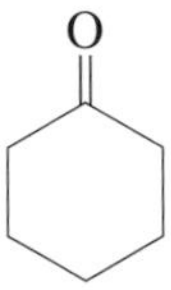

(e) 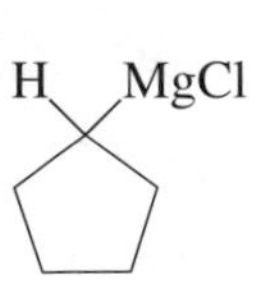 $+ CH_3CH_2\overset{\overset{\displaystyle O\diagdown\ \diagup H}{\displaystyle C}}{C}HCH_2CH_3$ $\longrightarrow$

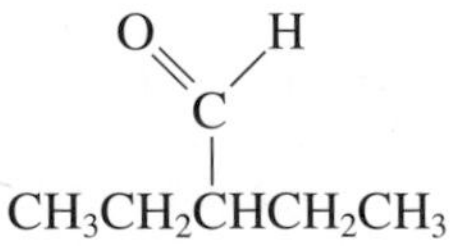

35. Write the structures of the products of reaction of ethylmagnesium bromide, CH_3CH_2MgBr, with each of the following carbonyl compounds. Identify any reaction that gives more than one stereoisomeric product, and indicate whether you would expect the products to form in identical or in differing amounts.

(a)

(b) [structure: 3-methylbutanal]

(c) [structure: 2-methylbutanal]

(d) [structure: 2-butanone]

(e) [structure: 2-pentanone]

(f) [structure: 3-hexanone]

(g) [structure: 3-methyl-2-butanone]

(h) [structure: 2-methylcyclohexanone, H down]

(i) [structure: 2,2-dimethylcyclohexanone]

(j) [structure: 2,6-dimethylcyclohexanone]

36. Give the expected major product of each of the following reactions. PCC is the abbreviation for pyridinium chlorochromate (Section 8-6).

(a) $CH_3CH_2CH_2OH$ $\xrightarrow{Na_2Cr_2O_7,\ H_2SO_4,\ H_2O}$

(b) $(CH_3)_2CHCH_2OH$ $\xrightarrow{PCC,\ CH_2Cl_2}$

(c) [cyclohexane with H and CH₂OH] $\xrightarrow{Na_2Cr_2O_7,\ H_2SO_4,\ H_2O}$

(d) [cyclohexane with H and CH₂OH] $\xrightarrow{PCC,\ CH_2Cl_2}$

(e) [cyclohexane with H and OH] $\xrightarrow{PCC,\ CH_2Cl_2}$

37. Give the expected major product of each of the following reaction *sequences.* PCC refers to pyridinium chlorochromate.

(a) $(CH_3)_2CHOH$ $\xrightarrow{\substack{1.\ CrO_3,\ H_2SO_4,\ H_2O \\ 2.\ CH_3CH_2MgBr,\ (CH_3CH_2)_2O \\ 3.\ H^+,\ H_2O}}$

(b) $CH_3CH_2CH_2CH_2Cl$ $\xrightarrow{\substack{1.\ ^-OH,\ H_2O \\ 2.\ PCC,\ CH_2Cl_2 \\ 3.\ \text{[cyclopentyl]}-Li,\ (CH_3CH_2)_2O \\ 4.\ H^+,\ H_2O}}$

(c) Product of (b) $\xrightarrow{\substack{1.\ CrO_3,\ H_2SO_4,\ H_2O \\ 2.\ LiAlD_4,\ (CH_3CH_2)_2O \\ 3.\ H^+,\ H_2O}}$

38. Unlike Grignard and organolithium reagents, organometallic compounds of the most electropositive metals (Na, K, etc.) react rapidly with haloalkanes. As a result, attempts to convert RX into RNa or RK by reaction with the corresponding metal lead to alkanes by a reaction called *Wurtz* coupling.

$$2\ RX + 2\ Na \longrightarrow R\text{—}R + 2\ NaX$$

which is the result of

$$R\text{—}X + 2\,Na \longrightarrow R\text{—}Na + NaX$$

followed rapidly by

$$R\text{—}Na + R\text{—}X \longrightarrow R\text{—}R + NaX$$

When it was still in use, the Wurtz coupling reaction was employed mainly for the preparation of alkanes by the coupling of two identical alkyl groups (e.g., equation 1 below). Suggest a reason why Wurtz coupling might not be a useful method for coupling two *different* alkyl groups (equation 2).

$$2\,CH_3CH_2CH_2Cl + 2\,Na \longrightarrow CH_3CH_2CH_2CH_2CH_2CH_3 + 2\,NaCl \qquad (1)$$

$$CH_3CH_2Cl + CH_3CH_2CH_2Cl + 2\,Na \longrightarrow CH_3CH_2CH_2CH_2CH_3 + 2\,NaCl \qquad (2)$$

39. The reaction of two equivalents of Mg with 1,4-dibromobutane produces compound A. The reaction of A with two equivalents of CH_3CHO (acetaldehyde), followed by work-up with dilute aqueous acid, produces compound B, having the formula $C_8H_{18}O_2$. What are the structures of A and B?

40. Suggest the best synthetic route to each of the following simple alcohols, using in each case a simple alkane as your initial starting molecule. What are some disadvantages of beginning syntheses with alkanes?
    (a) Methanol   (b) Ethanol   (c) 1-Propanol
    (d) 2-Propanol   (e) 1-Butanol   (f) 2-Butanol
    (g) 2-Methyl-2-propanol

41. For each alcohol in Problem 40, suggest (if possible) a synthetic route that starts with, first, an aldehyde and, second, a ketone.

42. Outline the best method for preparing each of the following compounds from an appropriate alcohol.

(a) [cyclopentanone structure]

(b) $CH_3CH_2CH_2CH_2COOH$

(c) [cyclohexanecarbaldehyde structure]

(d) $CH_3\underset{\underset{O}{\|}}{\overset{CH_3}{\overset{|}{C}H}}CCH_3$

(e) $CH_3\overset{\overset{O}{\|}}{C}H$

43. Suggest three different syntheses of 2-methyl-2-hexanol. Each route should utilize one of the following starting materials. Then use any number of steps and any other reagents needed.

(a) [acetone structure]   (b) [2-pentanone structure]   (c) [butanal structure]

44. Devise three different syntheses of 3-octanol starting with **(a)** a ketone; **(b)** an aldehyde; **(c)** an aldehyde different from that employed in (b).

45. Propose sensible synthetic schemes for the preparation of each of the following compounds, using only the organic starting material(s) indicated. Use any organic solvents or inorganic reagents necessary.

(a) [structure] from ethane and propane
OH

(b) [structure] from butane, ethane, and formaldehyde $\left(\begin{array}{c}O\\||\\HCH\end{array}\right)$

46. Waxes are naturally occurring esters (alkyl alkanoates) containing long, straight alkyl chains. Whale oil contains the wax 1-hexadecyl hexadecanoate, as shown in the margin. How would you synthesize this wax, using an S_N2 reaction?

$$CH_3(CH_2)_{14}\overset{O}{\overset{||}{C}}O(CH_2)_{15}CH_3$$
1-Hexadecyl hexadecanoate

47. The B vitamin commonly known as niacin is used by the body to synthesize the coenzyme nicotinamide adenine dinucleotide (NAD^+; Chapter 25). In the presence of a variety of enzyme catalysts, the reduced form of this substance (NADH) acts as a biological hydride donor, capable of reducing aldehydes and ketones to alcohols, according to the general formula

$$\overset{O}{\overset{||}{RCR}} + NADH + H^+ \xrightarrow{Enzyme} \overset{OH}{\overset{|}{RCHR}} + NAD^+$$

The COOH functional group of carboxylic acids is not reduced. Write the products of the NADH reduction of each of the molecules below.

(a) $CH_3\overset{O}{\overset{||}{CH}} + NADH \xrightarrow{Alcohol\ dehydrogenase}$

(b) $CH_3\overset{O\ O}{\overset{||\ ||}{CCOH}}$ + NADH $\xrightarrow{Lactate\ dehydrogenase}$
**2-Oxopropanoic acid
(Pyruvic acid)** — **Lactic acid**

(c) $HOCCH_2\overset{O\ \ \ O\ O}{\overset{||\ \ \ ||\ ||}{CCOH}}$ + NADH $\xrightarrow{Malate\ dehydrogenase}$
**2-Oxobutanedioic acid
(Oxaloacetic acid)** — **Malic acid**

48. Reductions by NADH (Problem 47) are stereospecific, with the stereochemistry of the product controlled by an enzyme (see Chemical Highlight 8-1). The common forms of lactate and malate dehydrogenases produce exclusively the *S* stereoisomers of lactic and malic acids, respectively. Draw these stereoisomers.

49. Chemically modified steroids have become increasingly important in medicine. Give the possible product(s) of the following reactions. In each case, identify the major stereoisomer formed on the basis of delivery of the attacking reagent from the less hindered side of the substrate molecule. (**Hint:** Make models and refer to Section 4-7.)

(a)

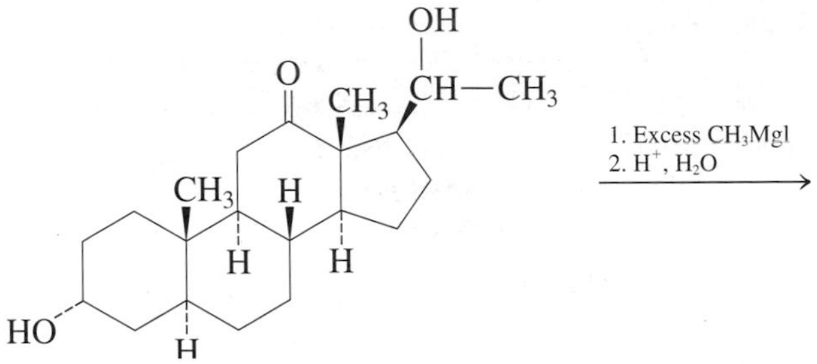
OH
O
CH_3
CH—CH_3
CH_3
H
H
H
HO
H
1. Excess CH_3MgI
2. H^+, H_2O

(b)

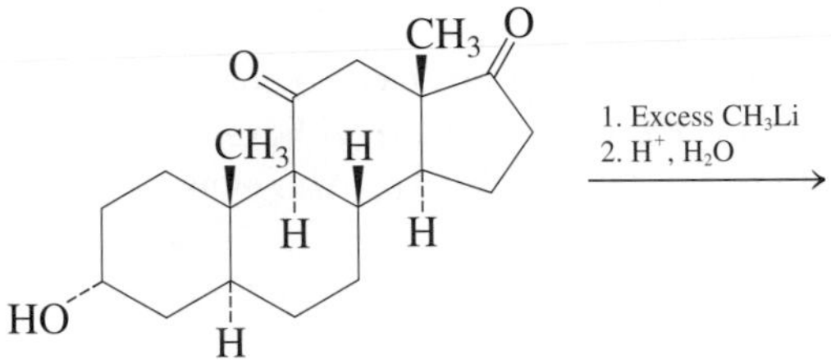
CH_3
O
O
CH_3
H
H
H
HO
H
1. Excess CH_3Li
2. H^+, H_2O

Team Problem

50. Your team has been asked to devise a synthesis of the tertiary alcohol 2-cyclohexyl-2-butanol, A. Your laboratory is well stocked with the usual organic and inorganic reagents and solvents. An inventory check reveals that there are many appropriate bromoalkanes and alcohols on hand. As a group, analyze alcohol A retrosynthetically and propose all possible strategic disconnections. Check the inventory to see if a particular route is feasible in regard to available starting materials. Then divide the proposed routes evenly among yourselves to evaluate the merits or pitfalls of these strategies. Write a detailed synthetic plan based on your chosen retrosynthesis for the synthesis of 2-cyclohexyl-2-butanol. Reconvene to defend or reject these plans. Finally, take into consideration the prices of your starting materials. Which one of your routes to A is the cheapest?

Target Molecule

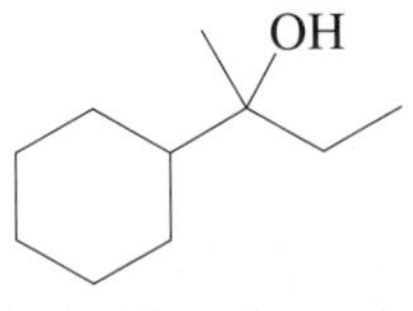
OH

2-Cyclohexyl-2-butanol
A

Inventory (Price)

2-Bromobutane (\$31/kg)	Cyclohexanol (\$14/kg)
Bromocyclohexane (\$50/kg)	1-Cyclohexylethanol (\$120/25 g)
Bromoethane (\$20/kg)	Cyclohexylmethanol (\$42/100 g)
Bromomethane (\$400/kg)	(Bromomethyl)cyclohexane (\$86/100 g)
2-Butanol (\$23/kg)	

Preprofessional Problems

51. A compound known to contain only C, H, and O gives the following upon micro-
analysis (atomic weights: C = 12.0, H = 1.00, O = 16.0): 52.1% C, 13.1% H.
It is found to have a boiling point of 78°C. Its structure is
(a) CH_3OCH_3 (b) CH_3CH_2OH (c) $HOCH_2CH_2CH_2CH_2OH$
(d) $HOCH_2CH_2CH_2OH$ (e) none of these

52. The compound whose structure is $(CH_3)_2CHCH_2CHCH_2CH_3$ is best named (IUPAC):
$$\underset{\displaystyle OH}{|}$$

(a) 2-methyl-4-hexanol (b) 5-methyl-3-hexanol (c) 1,4,4-trimethyl-2-hexanol (d) 1-isopropyl-2-hexanol

53. In this transformation, what is the best structure for "A"?

54. Ester hydrolysis is best illustrated by

Further Reactions of Alcohols and the Chemistry of Ethers

9

The alkoxy alcohol polyoxyethylene is a nonionic detergent used in washing machines. The polar end combines the ether and alcohol functional groups and interacts with water through hydrogen bonding, imparting water solubility. The nonpolar alkyl chain is responsible for dissolving grease and soil from contaminated fiber.

Sodium reacts with water under vigorous hydrogen evolution.

Do you remember the fizzing that occurred when you (or your teacher) dropped a pellet of sodium into water? The violent reaction that you observed was due to the conversion of the metal into NaOH and H_2 gas. Alcohols, which can be regarded as "alkylated water" (Section 8-2), undergo the same reaction, albeit less vigorously, to give NaOR and H_2. This chapter will describe this and further transformations of the hydroxy substituent and other derived functional groups.

Figure 9-1 depicts a variety of reaction modes available to alcohols. Usually at least one of the four bonds marked a, b, c, or d is cleaved. In Chapter 8 we learned that oxidation to aldehydes and ketones breaks bonds a and d. We found that the use of this reaction in combination with additions of organometallic reagents provides us with the means of preparing alcohols of considerable structural diversity. To further explore the reactions of alcohols, we shall start by reviewing their acidic and basic properties. Deprotonation at bond a furnishes alkoxides (Section 8-3), which are valuable both as strong bases and as nucleophiles (Section 7-8). Strong acids transform alcohols into alkyloxonium ions (Section 8-3), converting the OH into a good leaving group. Subsequently, bond b may break, thereby leading to substitution; or elimination can take place by cleavage of bonds b and c. We shall see that the carbocation intermediates arising from acid treatment of secondary and tertiary alcohols have a varied chemistry. An introduction to the preparation of esters and their applications in synthesis is followed by the chemistry of ethers and sulfur compounds. Alcohols, ethers, and their sulfur analogs occur widely in nature and have numerous applications in industry and medicine.

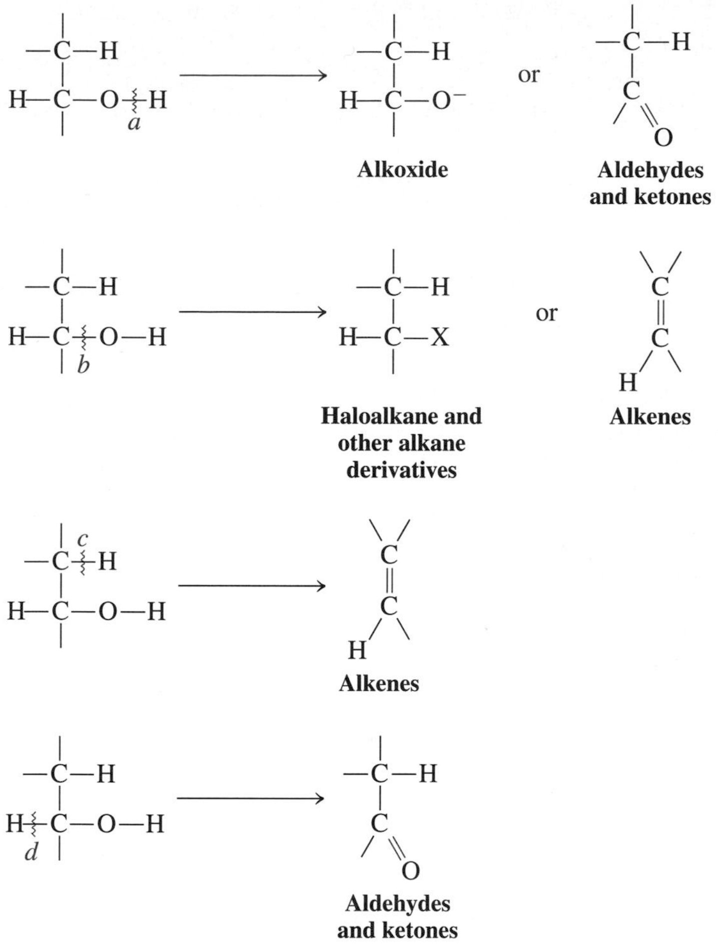

FIGURE 9-1

Four typical reaction modes of alcohols. In each mode, one or more of the four bonds marked *a–d* are cleaved (wavy line denotes bond cleavage): (*a*) deprotonation by base; (*b*) protonation by acid followed by uni- or bimolecular substitution; (*b, c*) elimination; and (*a, d*) oxidation.

9-1 Reactions of Alcohols with Base: Preparation of Alkoxides

As described in Section 8-3, alcohols can be both acids and bases. In this section, we shall review the methods by which the hydroxy group of alcohols is deprotonated to furnish their conjugate bases, the alkoxides.

Strong bases are needed to deprotonate alcohols completely

To remove a proton from the OH group of an alcohol (Figure 9-1, cleavage of bond *a*), we must use a base stronger than the alkoxide. Examples include lithium diisopropylamide, butyllithium, and alkali metal hydrides, such as potassium hydride, KH. Such hydrides are particularly useful because the only by-product of the reaction is hydrogen gas.

Three Ways of Making Methoxide from Methanol

$$\underset{\text{p}K_a = 15.5}{CH_3OH} \; + \; \underset{}{Li^+ \; ^-NCH(CH_3)_2} \; \overset{K = 10^{24.5}}{\rightleftharpoons} \; CH_3O^-Li^+ \; + \; \underset{\text{p}K_a = 40}{HNCH(CH_3)_2}$$

(with $CH(CH_3)_2$ substituents shown above the nitrogen atoms)

$$\underset{\text{p}K_a = 15.5}{CH_3OH} \; + \; CH_3CH_2CH_2CH_2Li \; \overset{K = 10^{34.5}}{\rightleftharpoons} \; CH_3O^-Li^+ \; + \; \underset{\text{p}K_a = 50}{CH_3CH_2CH_2CH_2H}$$

$$\underset{\text{p}K_a = 15.5}{CH_3OH} \; + \; K^+H^- \; \overset{K = 10^{22.5}}{\rightleftharpoons} \; CH_3O^-K^+ \; + \; \underset{\text{p}K_a = 38}{H-H}$$

EXERCISE 9-1

Considering the $\text{p}K_a$ data in Table 2-6, would you use sodium cyanide as a reagent to convert methanol into sodium methoxide? Explain your answer. (**Hint:** See Section 2-9.)

Alkali metals also deprotonate alcohols—but by reduction of H⁺

Another common way of obtaining alkoxides is by the reaction of alcohols with alkali metals, such as lithium. Such metals reduce water—in some cases, violently—to yield alkali metal hydroxides and hydrogen gas. When the more reactive metals (sodium, potassium, and cesium) are exposed to water in air, the hydrogen generated can ignite spontaneously or even detonate.

$$2\,H-OH \; + \; 2\,M\,(Li, Na, K, Cs) \; \longrightarrow \; 2\,M^{+\,-}OH \; + \; H_2$$

The alkali metals act similarly on the alcohols to give alkoxides, but the transformation is less vigorous. Here are two examples.

Alkoxides from Alcohols and Alkali Metals

$$2\,CH_3CH_2OH \; + \; 2\,Na \; \longrightarrow \; 2\,CH_3CH_2O^-Na^+ \; + \; H_2$$
$$2\,(CH_3)_3COH \; + \; 2\,K \; \longrightarrow \; 2\,(CH_3)_3CO^-K^+ \; + \; H_2$$

The reactivity of the alcohols used in this process decreases with increasing substitution, methanol being most reactive and tertiary alcohols least reactive.

Relative Reactivity of ROH with Alkali Metals

$$R = CH_3 > \text{primary} > \text{secondary} > \text{tertiary}$$

2-Methyl-2-propanol reacts so slowly that it can be used to safely destroy potassium residues in the laboratory.

What are alkoxides good for? We have already seen that they can be useful reagents in organic synthesis. For example, the reaction of hindered alkoxides with haloalkanes gives elimination.

$$CH_3CH_2CH_2CH_2Br \; \xrightarrow{(CH_3)_3CO^-K^+, \; (CH_3)_3COH} \; CH_3CH_2CH=CH_2 \; + \; (CH_3)_3COH \; + \; K^+Br^-$$

Less branched alkoxides attack primary haloalkanes by the S_N2 reaction to give ethers. This method is described in Section 9-6.

In summary, the treatment of an alcohol with a strong base or an alkali metal gives an alkoxide. Let us now inspect the protonation of alcohols with strong acids.

9-2 Reactions of Alcohols with Strong Acids: Alkyloxonium Ions in Substitution and Elimination Reactions of Alcohols

We have seen that heterolytic cleavage of the O–H bond in alcohols is readily achieved with strong bases. Can we break the C–O linkage (bond *b*, Figure 9-1) as easily? Yes, but now we need acid. Recall (Section 2-9) that water has a high pK_a: It is a weak acid. Consequently, hydroxide, its conjugate base, is an exceedingly poor leaving group. *For alcohols to undergo substitution or elimination reactions, the OH must first be converted into a better leaving group.*

Haloalkanes from primary alcohols and HX: water can be a leaving group

The simplest way of turning the hydroxy substituent in alcohols into a good leaving group is to protonate the oxygen to form an alkyloxonium ion. Recall (Section 8-3) that this process ties up one of the oxygen lone pairs in bonding to a proton. The positive charge therefore resides on the oxygen atom. *Protonation turns OH from a bad leaving group into a good one, neutral water.*

Alkyloxonium ions derived from primary alcohols are subject to nucleophilic attack. For example, the alkyloxonium ion resulting from the treatment of 1-butanol with HBr undergoes displacement by bromide to form 1-bromobutane. The originally nucleophilic (red) oxygen is protonated by the electrophilic proton (blue) to give the alkyloxonium ion, containing an electrophilic (blue) carbon and H_2O as a leaving group (green). In the subsequent S_N2 reaction, bromide acts as a nucleophile.

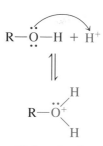

Alkyloxonium ion

Primary Bromoalkane Synthesis from an Alcohol

$$CH_3CH_2CH_2CH_2OH \; + \; HBr \; \longrightarrow \; CH_3CH_2CH_2CH_2\overset{+}{O}H_2 \; + \; Br^- \; \longrightarrow$$
$$CH_3CH_2CH_2CH_2Br \; + \; H_2O$$

Iodoalkanes also can be made in this fashion; however, primary chloroalkanes cannot, because chloride ion is too weak a nucleophile. The conversion of a primary alcohol into the corresponding chloroalkane requires other methods (Section 9-4). In general, the acid-catalyzed S_N2 reaction of *primary* alcohols with HBr or HI is a good way of preparing simple *primary* haloalkanes. What about secondary and tertiary alcohols?

Secondary and tertiary alcohols undergo carbocation reactions with acids: S_N1 and E1

Alkyloxonium ions derived from secondary and tertiary alcohols, in contrast with their primary counterparts, lose water with increasing ease to give the corresponding carbocations. The reason for this difference in behavior is the difference in carbocation stability in the procession from primary to secondary to tertiary (Section 7-5). Primary carbocations are too high in energy to be accessible under ordinary laboratory conditions, whereas secondary and tertiary carbocations are generated with increasing ease. Thus, primary alkyloxonium ions undergo only S_N2 reactions, whereas their secondary

Iodoalkane Synthesis

$$HO(CH_2)_6OH \; + \; 2 \; HI$$
1,6-Hexanediol

$$\downarrow$$

$$I(CH_2)_6I \; + \; 2 \; H_2O$$
$$85\%$$
1,6-Diiodo-hexane

and tertiary relatives enter into S$_N$1 and E1 processes. When good nucleophiles are present, we observe S$_N$1 products. Synthetically, this fact is exploited in the preparation of *tertiary* haloalkanes from *tertiary* alcohols in the presence of excess concentrated aqueous hydrogen halide. The product forms in minutes at room temperature. The mechanism is precisely the reverse of that of solvolysis (Section 7-2).

Conversion of 2-Methyl-2-propanol into 2-Bromo-2-methylpropane

$$(CH_3)_3COH \ + \ \underset{\text{Excess}}{HBr} \ \rightleftharpoons \ (CH_3)_3CBr \ + \ H_2O$$

Mechanism of the S$_N$1 Reaction of Tertiary Alcohols with Hydrogen Halides

$$(CH_3)_3C-\ddot{O}H \ + \ H-Br \ \rightleftharpoons \ (CH_3)_3C-\overset{+}{\ddot{O}}H_2 \ + \ Br^-$$

$$\rightleftharpoons \ H_2O \ + \ (CH_3)_3\overset{+}{C} \ + \ Br^- \ \rightleftharpoons \ H_2O \ + \ (CH_3)_3C-Br$$

The reason for the success of this process is that relatively low temperatures are required to generate the tertiary carbocation, thus largely preventing competing E1 reactions (Section 7-6). Indeed, at higher temperatures (or in the absence of good nucleophiles) elimination becomes dominant.* This explains why protonated *secondary* alcohols show the most complex behavior in the presence of HX, following S$_N$2, S$_N$1, and E1 pathways: They are relatively hindered, compared with their primary counterparts (i.e., retarded S$_N$2 reactivity), and relatively slow in forming carbocations, compared with their tertiary counterparts (e.g., retarded S$_N$1 reactivity); review Section 7-9.

The E1 reaction, here called **dehydration** because it results in the loss of a molecule of water (Figure 9-1, breaking bonds *b* and *c*; see also Sections 9-3 and 9-7), is one of the methods for the synthesis of alkenes (Section 11-9). Rather than the "nucleophilic" acids HBr and HI, so called because the conjugate base is a good nucleophile, "nonnucleophilic" acids, such as H$_3$PO$_4$ or H$_2$SO$_4$, are employed.

Alcohol Dehydration by the E1 Mechanism

OH

$$\xrightarrow[-H_2O]{H_2SO_4, \ 130°-140°C}$$

87%

Cyclohexanol   Cyclohexene

Mechanism of the Dehydration of Cyclohexanol

---

*The preference for elimination at high temperatures has its origin in its relatively large positive entropy (Section 2-8) as two molecules (alkene + H$_2$O) result from one (alcohol), and the $\Delta S°$ term in the expression $\Delta G° = \Delta H° - T\Delta S°$ increases with temperature (see also Exercise 7-14).

In the example, the counterion of the acid is the poor nucleophile HSO_4^- and proton loss from the intermediate carbocation is observed. Dehydration of tertiary alcohols is even easier, often occurring at slightly above room temperature.

EXERCISE 9-2

Write the structure of the product that you expect from the reaction of 4-methyl-1-pentanol with concentrated aqueous HI. Give the mechanism of its formation.

EXERCISE 9-3

Write the structure of the products that you expect from the reaction of 1-methylcyclohexanol with **(a)** concentrated HCl and **(b)** concentrated H_2SO_4. Compare and contrast the mechanisms of the two processes.

In summary, treatment of alcohols with strong acid leads to protonation to give alkyloxonium ions, which, when primary, undergo S_N2 reactions in the presence of good nucleophiles. Alkyloxonium ions from secondary or tertiary alcohols convert into carbocations, which furnish products of substitution and elimination (dehydration).

$$ROH \xrightarrow{H^+} R \overset{+}{-}O \overset{H}{\underset{H}{\big\langle}} $$

$$\xrightarrow[\text{R= prim}]{X^-,\ S_N2} RX + H_2O$$

$$\xrightarrow[\text{R= sec, tert}]{-H_2O} R^+ \overset{\xrightarrow{X^-,\ S_N1} RX}{\underset{\xrightarrow{-H^+,\ E1} \text{Alkene}}{}}$$

9-3 Carbocation Rearrangements

Carbocations can rearrange by both hydride and alkyl shifts to become new carbocations, which can then undergo S_N1 and E1 reactions. Unless there is a thermodynamic driving force toward one product, the result is likely to be a complex mixture.

Hydride shifts give new S_N1 products

Treatment of 2-propanol with hydrogen bromide gives 2-bromopropane, as expected. However, exposure of the more highly substituted secondary alcohol, 3-methyl-2-butanol, to the same reaction conditions produces a surprising result. The expected S_N1 product, 2-bromo-3-methylbutane, is only a minor component of the reaction mixture, and the major product is 2-bromo-2-methylbutane.

Hydride Shift in the S_N1 Reaction of an Alcohol with HBr

$$\underset{\substack{\text{3-Methyl-2-}\\\text{butanol}}}{\overset{\displaystyle H \quad OH}{\underset{\displaystyle H_3C \quad H}{CH_3C-CCH_3}}} \xrightarrow{\text{HBr, 0°C}} \underset{\substack{\text{2-Bromo-3-}\\\text{methylbutane}\\\text{(Normal product)}}}{\overset{\displaystyle H \quad Br}{\underset{\displaystyle H_3C \quad H}{CH_3C-CCH_3}}} + \underset{\substack{\text{2-Bromo-2-}\\\text{methylbutane}\\\text{(Rearranged product)}}}{\overset{\displaystyle Br \quad H}{\underset{\displaystyle H_3C \quad H}{CH_3C-CCH_3}}} + H-OH$$

Minor product **Major product**

Normal S_N1 Reaction of an Alcohol (No Rearrangement)

$$\underset{OH}{CH_3CHCH_3} + HBr$$

$$\Big\downarrow {0°C}$$

$$\underset{Br}{CH_3CHCH_3} + H-OH$$

What is the mechanism of this transformation? The answer is that *carbocations can undergo rearrangement* by **hydride shifts,** in which the hydrogen (yellow) moves *with both electrons* from its original position to the neighboring carbon atom. Initially, protonation of the alcohol followed by loss of water gives the expected secondary carbocation. A shift of the tertiary hydrogen to the electron-deficient neighbor then generates a tertiary cation, *which is more stable.* This species is finally trapped by bromide ion to give the rearranged S_N1 product.

Mechanism of Carbocation Rearrangement

STEP 1. Protonation   STEP 2. Loss of water

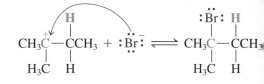

Color is used to indicate the electrophilic (blue), nucleophilic (red), and leaving-group (green) character of the reacting centers. Therefore, a color may "switch" from one group or atom to another as the reaction proceeds.

STEP 3. Hydride shift   STEP 4. Trapping by bromide

The details of the transition state of the observed hydride shift are shown schematically in Figure 9-2. A simple rule to remember when executing hydride shifts in carbocations is that *the hydrogen and the positive charge formally exchange places* between the two neighboring carbon atoms participating in the reaction.

Hydride shifts of carbocations are generally very fast—faster than S_N1 or E1. In part, this rapidity is due to hyperconjugation, which labilizes the C–H bond (Section 7-5 and Figure 9-2B). They are particularly favored when the new carbocation is more stable than the original one, as in the example depicted in Figure 9-2.

EXERCISE 9-4

2-Methylcyclohexanol on treatment with HBr gives 1-bromo-1-methylcyclohexane. Explain by a mechanism.

EXERCISE 9-5

Predict the major product from the following reactions.

(a) 2-Methyl-3-pentanol + H_2SO_4, CH_3OH solvent   (b) [structure] + HCl

---

Primary carbocations are too unstable to be formed by rearrangement. However, carbocations of comparable stability—for example, secondary–secondary or

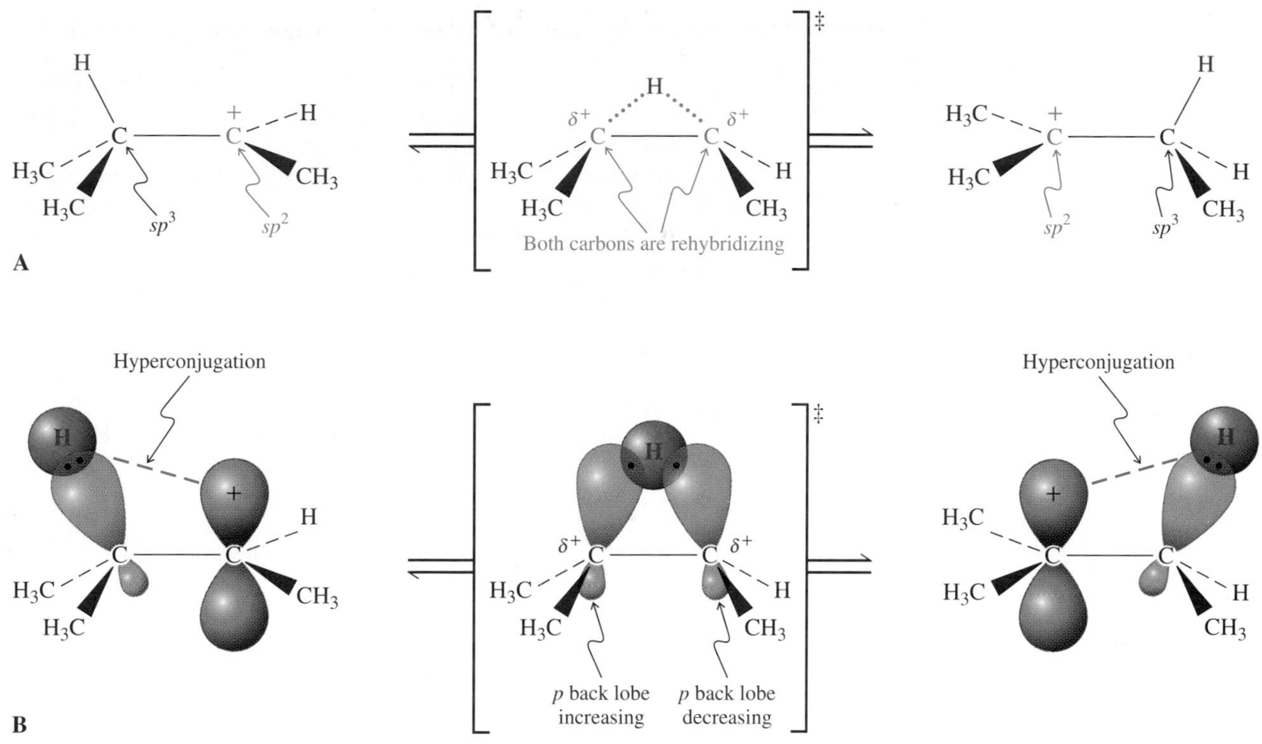

A

Hyperconjugation

Both carbons are rehybridizing

Hyperconjugation

p back lobe increasing *p* back lobe decreasing

B

FIGURE 9-2

The rearrangement of a carbocation by a hydrogen shift: (A) dotted-line notation; (B) orbital picture. Note that the migrating hydrogen and the positive charge exchange places. In addition, you can see how hyperconjugation labilizes the C–H bond by effecting some electron transfer into the empty neighboring π orbital.

tertiary–tertiary—equilibrate readily. In this case, any added nucleophile will trap all carbocations present, furnishing mixtures of products.

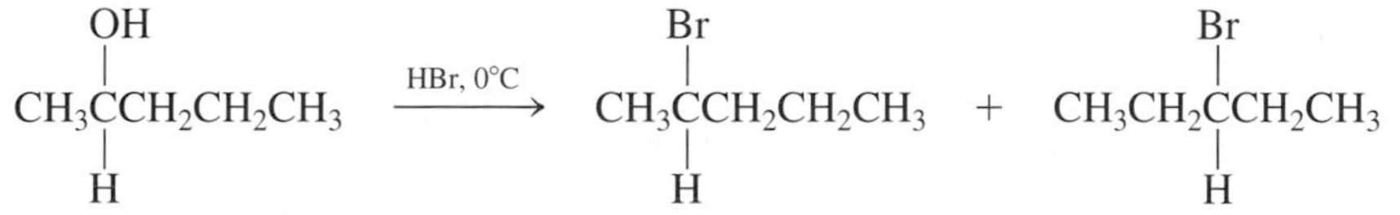

Carbocation rearrangements take place regardless of the nature of the precursor to the carbocation: alcohols (this section), haloalkanes (Chapter 7), and alkyl sulfonates (Section 6-7). For example, ethanolysis of 2-bromo-3-ethyl-2-methylpentane gives the two possible tertiary ethers.

Rearrangement in Solvolysis of a Haloalkane

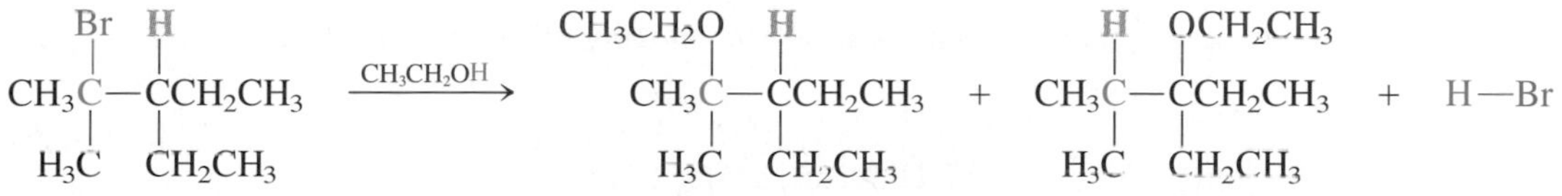

EXERCISE 9-6

Give a mechanism for the preceding reaction. Then predict the outcome of the reaction of 2-chloro-4-methylpentane with methanol. (**Hint:** Try two successive hydride shifts to the most stable carbocation.)

Carbocation rearrangements also give new E1 products

How does the rearrangement of intermediates affect the outcome of reactions under conditions that favor elimination? At elevated temperatures and in relatively nonnucleophilic media, rearranged carbocations yield alkenes by the E1 mechanism (Section 9-2). For example, treatment of 2-methyl-2-pentanol with sulfuric acid at 80°C gives the same major alkene product as that formed when the starting material is 4-methyl-2-pentanol. The conversion of the latter alcohol includes a hydride shift of the initial carbocation followed by deprotonation.

Rearrangement in E1 Elimination

$$
\begin{array}{ccc}
\underset{\displaystyle \text{2-Methyl-2-pentanol}}{\underset{\displaystyle \overset{\textstyle CH_3}{|}}{\overset{\textstyle OH}{\underset{|}{\overset{|}{CH_3C}}}-CH_2CH_2CH_3}}
& \xrightarrow[-H_2O]{H_2SO_4,\ 80°C}
& \underset{\displaystyle \text{Major product}}{\underset{\displaystyle \text{2-Methyl-2-pentene}}{\underset{\displaystyle H_3C}{\overset{\displaystyle H_3C}{C=C}}\overset{\displaystyle CH_2CH_3}{\underset{\displaystyle H}{}}}}
& \xleftarrow[\substack{-H_2O \\ \textbf{With} \\ \textbf{rearrangement}}]{H_2SO_4,\ 80°C}
& \underset{\displaystyle \text{4-Methyl-2-pentanol}}{\underset{CH_3\quad H}{\overset{H\qquad OH}{CH_3C-CH_2CCH_3}}}
\end{array}
$$

EXERCISE 9-7

(a) Give mechanisms for the preceding E1 reactions. **(b)** Treatment of 4-methylcyclohexanol with hot acid gives 1-methylcyclohexene. Explain by a mechanism. (**Hint:** Consider multiple H shifts.)

Other carbocation rearrangements are due to alkyl shifts

Carbocations, particularly when lacking suitable (secondary and tertiary) hydrogens next to the positively charged carbon, can undergo another mode of rearrangement, known as **alkyl group migration** or **alkyl shift.**

Rearrangement by Alkyl Shift in S$_N$1 Reaction

$$
\underset{\displaystyle \text{3,3-Dimethyl-2-butanol}}{\underset{H_3C\quad H}{\overset{H_3C\quad CH_3}{CH_3C-COH}}}
\xrightarrow[-HOH]{HBr}
\underset{\displaystyle \substack{94\% \\ \text{2-Bromo-2,3-dimethylbutane}}}{\underset{H_3C\quad H}{\overset{Br\quad CH_3}{CH_3C-CCH_3}}}
$$

As in the hydride shift, the migrating group takes its electron pair with it to form a bond to the neighboring carbocation. *The moving alkyl group and the positive charge formally exchange places.*

Mechanism of Alkyl Shift

$$
\underset{H_3C\quad H}{\overset{H_3C\quad CH_3}{CH_3C-COH}} + H^+
\underset{+H_2O}{\overset{-H_2O}{\rightleftharpoons}}
\underset{H_3C\quad H}{\overset{H_3C}{CH_3C-CCH_3}}
\xrightarrow{CH_3\ shift}
\underset{H_3C\quad H}{\overset{CH_3}{CH_3C-CCH_3}}
\underset{-Br^-}{\overset{+Br^-}{\rightleftharpoons}}
\underset{H_3C\quad H}{\overset{Br\quad CH_3}{CH_3C-CCH_3}}
$$

The rates of alkyl and hydride shifts are comparable when leading to carbocations of similar stability. However, either type of migration is faster when furnishing tertiary carbocations relative to those ending in their secondary counterparts. This explains why, in the preceding discussion of hydride shifts, alkyl group rearrangement was not observed: The less substituted cation would have been the result. Exceptions to this observation are found only if there are other compelling reasons for the preference of alkyl migration, such as electronic stabilization or steric relief (see Chapter Integration Problem and Problem 9-51).

EXERCISE 9-8

At higher temperatures, 3,3-dimethyl-2-butanol gives three products of E1 reaction, one derived from the carbocation present prior to rearrangement, the other two from that formed after alkyl shift has taken place. Give the structures of these elimination products.

Primary alcohols may undergo rearrangement

Treatment of a primary alcohol with HBr or HI normally produces the corresponding haloalkane through S_N2 reaction of the alkyloxonium ion (Section 9-2). However, it is possible in some cases to observe alkyl and hydride shifts to primary carbons bearing leaving groups, even though primary carbocations are not formed in solution. For example, treating 2,2-dimethyl-1-propanol (neopentyl alcohol) with strong acid causes rearrangement, despite the fact that a primary carbocation cannot be an intermediate.

Rearrangement in a Primary Substrate

$$CH_3CCH_2OH \xrightarrow[-H-OH]{HBr, \Delta} CH_3CCH_2CH_3$$

2,2-Dimethyl-1-propanol
(Neopentyl alcohol)

2-Bromo-2-methylbutane

In this case, after protonation to form the alkyloxonium ion, steric hindrance interferes with direct displacement by bromide (Section 6-10). Instead, water leaves *at the same time* as a methyl group migrates from the neighboring carbon, thus bypassing formation of a primary carbocation.

Mechanism of Concerted Alkyl Shift

$$CH_3CCH_2\overset{..}{\underset{..}{O}}H \rightleftharpoons CH_3\overset{+}{C}-CH_2-\overset{..}{\underset{..}{O}}H_2 \xrightarrow{-H_2\overset{..}{O}} CH_3\overset{+}{C}CH_2CH_3 \underset{-:\overset{..}{Br}:}{\overset{+:\overset{..}{Br}:}{\rightleftharpoons}} CH_3CCH_2CH_3$$

Rearrangements of primary substrates are relatively difficult processes, usually requiring elevated temperatures and long reaction times.

In summary, another mode of reactivity of carbocations, in addition to regular S_N1 and E1 processes, is rearrangement by hydride or alkyl shifts. In such rearrangements, the migrating group delivers its bonding electron pair to a positively charged carbon neighbor, exchanging places with the charge. Rearrangement may lead to a more stable cation—as in the conversion of a secondary cation into a tertiary one. Primary alcohols also can undergo rearrangement, but they do so by concerted pathways and not through the intermediacy of primary cations.

Nucleophile: red
Electrophile: blue
Leaving group: green

9-4 Organic and Inorganic Esters from Alcohols

Among the most useful reactions of alcohols is their conversion into esters. This term commonly refers to **organic esters,** also called **carboxylates** or **alkanoates** (Table 2-1). They are formally derived from organic (carboxylic) acids by replacement of the acidic hydrogen with an alkyl group. **Inorganic esters** are the analogous derivatives of inorganic acids.

| A carboxylic acid | Chromic acid | Phosphoric acid | A sulfonic acid |
| (An organic acid) | (Inorganic acids) | | |

| A carboxylate ester | A chromate ester | A phosphate ester | A sulfonate ester |
| (An organic ester) | (Inorganic esters) | | |

A preparation of organic esters by reaction of haloalkanes with alkanoate ions was presented in Sections 6-9 and 8-5. We have also considered the role of chromate esters in the oxidation of alcohols to aldehydes and ketones (Section 8-6). Living systems make use of alkyl phosphates in many ways, including the conversion of alcohols into leaving groups. Others will be outlined in Chapters 20 and 26. Sulfonates were briefly introduced as substrates in nucleophilic displacement reactions in Section 6-8. Here we shall see how these compounds can be made directly by reactions of alcohols with carboxylic acids or a variety of inorganic reagents.

Alcohols react with carboxylic acids to give organic esters

Alcohols react with carboxylic acids in the presence of catalytic amounts of a strong inorganic acid, such as H_2SO_4 or HCl, to give esters and water, a process called **esterification.** Starting materials and products in this transformation form an equilibrium that can be shifted in either direction. The formation and reactions of organic esters will be presented in detail in Chapters 19 and 20.

Esterification

| Acetic acid | Ethanol solvent | Ethyl acetate |

Haloalkanes can be made from alcohols through inorganic esters

Because of the difficulties and complications that can be encountered in the acid-catalyzed conversions of alcohols into haloalkanes (Section 9-2), several alternatives have been developed relying on a number of inorganic reagents capable of changing the hydroxy function into a good leaving group under milder conditions. Thus, primary and secondary alcohols react with phosphorus tribromide, a readily available commercial compound, to give bromoalkanes and phosphorous acid. This method constitutes a general way of making bromoalkanes from alcohols. All three bromine atoms are transferred from phosphorus to alkyl groups.

Bromoalkane Synthesis by Using PBr$_3$

3-Pentanol   Phosphorus tribromide   +   PBr$_3$ $\xrightarrow{\text{(CH}_3\text{CH}_2)_2\text{O}}$ 3   3-Bromopentane   +   H$_3$PO$_3$

47%

3-Pentanol   Phosphorus tribromide   3-Bromopentane   Phosphorous acid

What is the mechanism of action of PBr$_3$? In the first step, the alcohol and the phosphorus reagent form a protonated inorganic ester, a derivative of phosphorous acid.

STEP 1

$$RCH_2\overset{..}{\underset{..}{O}}H \ + \ \overset{Br}{\underset{Br}{P}}{-}Br \ \longrightarrow \ RCH_2\overset{+}{\underset{H}{O}}{-}PBr_2 \ + \ Br^-$$

Next, HOPBr$_2$, a good leaving group, is displaced (S$_N$2) by the bromide generated in step 1, finally producing the haloalkane.

STEP 2

$$:\!\overset{..}{\underset{..}{Br}}\!:^- \ + \ RCH_2{-}\overset{+}{\underset{H}{\overset{..}{O}}}{-}PBr_2 \ \longrightarrow \ RCH_2\overset{..}{\underset{..}{Br}}: \ + \ H\overset{..}{O}PBr_2$$

This method of haloalkane synthesis is especially efficient because HOPBr$_2$ continues to react successively with two more molecules of alcohol, converting them into haloalkane as well.

$$2\,RCH_2\overset{..}{\underset{..}{O}}H \ + \ H\overset{..}{O}PBr_2 \ \longrightarrow \longrightarrow \ 2\,RCH_2\overset{..}{\underset{..}{Br}}: \ + \ H_3PO_3$$

What if, instead of a bromoalkane, we want the corresponding iodoalkane? The required phosphorus triiodide, PI$_3$, is best generated in the reaction mixture in which it will be used, because it is a reactive species. We do this by adding red elemental phosphorus and elemental iodine to the alcohol (margin). The reagent is consumed as soon as it is formed.

A chlorinating agent commonly used to convert alcohols into chloroalkanes is thionyl chloride, SOCl$_2$. Simply warming an alcohol in its presence results in the evolution of SO$_2$ and HCl and the formation of the chloroalkane.

Chloroalkane Synthesis with SOCl$_2$

$$CH_3CH_2CH_2OH \ + \ SOCl_2 \ \longrightarrow \ CH_3CH_2CH_2Cl \ + \ O{=}S{=}O \ + \ HCl$$

91%

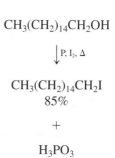

$$CH_3(CH_2)_{14}CH_2OH$$
$$\downarrow {\scriptstyle P,\,I_2,\,\Delta}$$
$$CH_3(CH_2)_{14}CH_2I$$
$$85\%$$
$$+$$
$$H_3PO_3$$

Mechanistically, the alcohol RCH_2OH again first forms an inorganic ester, RCH_2O_2SCl. The chloride ion created in this process then acts as a nucleophile and attacks the ester, an S_N2 reaction yielding one molecule each of SO_2 and HCl.

Mechanism of Substitution by Using Thionyl Chloride

The reaction works better in the presence of an amine, which neutralizes the hydrogen chloride generated. One such reagent is *N,N*-diethylethanamine (triethylamine), which forms the corresponding ammonium hydrochloride under these conditions.

$(CH_3CH_2)_3N:$

N,N-**Diethylethanamine**
(Triethylamine)
+
HCl
↓
$(CH_3CH_2)_3\overset{+}{N}H\ Cl^-$

Alkyl sulfonates are versatile substrates for substitution reactions

The inorganic esters in the reactions of $SOCl_2$ are special examples of leaving groups derived from sulfur-based acids. They are related to the sulfonates (Section 6-8). Alkyl sulfonates contain excellent leaving groups and can be readily prepared from the corresponding sulfonyl chlorides and an alcohol. A mild base such as pyridine or another amine is often added to remove the HCl formed.

Synthesis of an Alkyl Sulfonate

| 2-Methyl-1-propanol | Methanesulfonyl chloride (Mesyl chloride) | Pyridine | 2-Methylpropyl methanesulfonate (2-Methylpropyl mesylate) | Pyridinium hydrochloride |

Unlike the inorganic esters derived from phosphorus tribromide and thionyl chloride, alkyl sulfonates are often crystalline solids that can be isolated and purified before further reaction. They then can be used in reactions with a variety of nucleophiles to give the corresponding products of nucleophilic substitution.

Sulfonate Intermediates in Nucleophilic Displacement of the Hydroxy Group in Alcohols

The displacement of sulfonate groups by halide ions (Cl^-, Br^-, I^-) readily yields the corresponding haloalkanes, particularly with primary and secondary systems, in which

S_N2 reactivity is good. In addition, however, alkyl sulfonates allow replacement of the hydroxy group by *any* good nucleophile: They are not limited to halides alone, as was the case with hydrogen, phosphorus, and thionyl halides.

Substitution Reactions of Alkyl Sulfonates

$$CH_3CH_2CH_2O\overset{\displaystyle O}{\underset{\displaystyle O}{\overset{\|}{\underset{\|}{S}}}}CH_3 \;+\; I^- \;\longrightarrow\; CH_3CH_2CH_2I \;+\; {}^-O\overset{\displaystyle O}{\underset{\displaystyle O}{\overset{\|}{\underset{\|}{S}}}}CH_3$$

90%

$$CH_3-\underset{\underset{\displaystyle CH_3}{|}}{\overset{\overset{\displaystyle H}{|}}{C}}-O-\underset{\underset{\displaystyle O}{\|}}{\overset{\overset{\displaystyle O}{\|}}{S}}-\!\!\!\!\bigcirc\!\!\!\!-CH_3 \;+\; CH_3CH_2S^- \;\longrightarrow\; CH_3\underset{\underset{\displaystyle CH_3}{|}}{CH}SCH_2CH_3 \;+$$

85%

EXERCISE 9-9

What is the product of the reaction sequence shown in the margin?

EXERCISE 9-10

Supply reagents with which you would prepare the following haloalkanes from the corresponding alcohols.

(a) $I(CH_2)_6I$ **(b)** $(CH_3CH_2)_3CCl$ **(c)**

In summary, alcohols react with carboxylic acids by loss of water to furnish organic esters and with inorganic halides, such as PBr_3, $SOCl_2$, and RSO_2Cl, by loss of HX to produce inorganic esters. These inorganic esters contain good leaving groups in nucleophilic substitutions that are, for example, displaced by halide ions to give the corresponding haloalkanes.

9-5 Names and Physical Properties of Ethers

Ethers have been mentioned on several occasions in preceding chapters (see Table 2-1). We now introduce this class of compounds more systematically. This section gives the rules for naming ethers and describes some of their physical properties.

In the IUPAC system, ethers are alkoxyalkanes

The IUPAC system for naming **ethers** treats them as alkanes that bear an alkoxy substituent—that is, as alkoxyalkanes. The smaller substituent is considered part of the alkoxy group, and the larger defines the stem.

The alkoxyalkanes may be thought of as derivatives of alcohols in which the hydroxy proton has been replaced by an alkyl. Their common names are based on this picture: The names of the two alkyl groups are followed by the word ether. Hence, CH_3OCH_3 is dimethyl ether, $CH_3OCH_2CH_3$ is ethyl methyl ether, and so forth.

$CH_3OCH_2CH_3$
Methoxyethane

$$CH_3CH_2\overset{\displaystyle CH_3}{\overset{|}{\underset{\underset{\displaystyle CH_3}{|}}{\ddot{O}}}}CH_3$$
2-Ethoxy-2-methylpropane

cis-1-Ethoxy-2-methoxycyclopentane

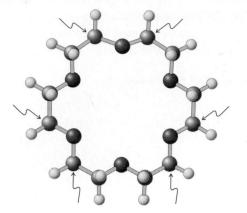

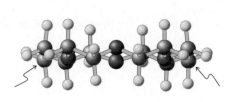

In this perspective, six of the H atoms are masked by the attached carbons (marked by arrows).

In this perspective, two of the O atoms are masked by the attached carbons (marked by arrows).

FIGURE 9-3

The crownlike structural arrangement of 18-crown-6.

Ethers are generally fairly unreactive (except for strained cyclic derivatives; see Section 9-9) and are therefore frequently used as solvents in organic reactions. Some of these ether solvents are cyclic; they may even contain several ether functions. All have common names.

Ether Solvents and Their Names

$CH_3CH_2OCH_2CH_3$
Ethoxyethane
(Diethyl ether)

**1,4-Dioxacyclo-
hexane
(1,4-Dioxane)**

$CH_3OCH_2CH_2OCH_3$
**1,2-Dimethoxyethane
(Glycol dimethyl ether,
glyme)**

**Oxacyclopentane
(Tetrahydrofuran,
THF)**

Cyclic ethers are members of a class of cycloalkanes in which one or more carbons have been replaced by a *heteroatom*—in this case, oxygen. (A **heteroatom** is defined as any atom except carbon and hydrogen.) Cyclic compounds of this type, called **heterocycles,** are discussed more fully in Chapter 25.

The simplest system for naming cyclic ethers is based on the **oxacycloalkane** stem, in which the prefix *oxa* indicates the replacement of carbon by oxygen in the ring. Thus, three-membered cyclic ethers are oxacyclopropanes (other names used are oxiranes, epoxides, and ethylene oxides), four-membered systems are oxacyclobutanes, and the next two higher homologs are oxacyclopentanes (tetrahydrofurans) and oxacyclohexanes (tetrahydropyrans). The compounds are numbered by starting at the oxygen and proceeding around the ring.

Cyclic polyethers which contain multiple ether functional groups based on the 1,2-ethanediol unit are called **crown ethers,** so named because the molecules adopt a crownlike conformation in the crystalline state and, presumably, in solution. The polyether 18-crown-6 is shown in Figure 9-3. The number 18 refers to the total number of atoms in the ring, and 6 to the number of oxygens.

The absence of hydrogen bonding affects the physical properties of ethers

The molecular formula of simple alkoxyalkanes is $C_nH_{2n+2}O$, identical with that of the alkanols. However, because of the absence of hydrogen bonding, the boiling points

TABLE 9-1	Boiling Points of Ethers and the Isomeric 1-Alkanols				
Ether	Name	Boiling point (°C)	1-Alkanol	Boiling point (°C)
CH_3OCH_3	Methoxymethane (Dimethyl ether)	−23.0	CH_3CH_2OH	78.5
$CH_3OCH_2CH_3$	Methoxyethane (Ethyl methyl ether)	10.8	$CH_3CH_2CH_2OH$	82.4
$CH_3CH_2OCH_2CH_3$	Ethoxyethane (Diethyl ether)	34.5	$CH_3(CH_2)_3OH$	117.3
$(CH_3CH_2CH_2CH_2)_2O$	1-Butoxybutane (Dibutyl ether)	142	$CH_3(CH_2)_7OH$	194.5

of ethers are much lower than those of the corresponding isomeric alcohols (Table 9-1). The two smallest members of the series are water miscible, but ethers become less water soluble as the hydrocarbon residues increase in size. For example, methoxymethane is completely water soluble, whereas ethoxyethane forms only an approximately 10% aqueous solution.

Polyethers solvate metal ions: crown ethers and ionophores

As in the alcohols, the oxygen in ethers is (Lewis) basic (Section 2-9); that is, its lone pairs can coordinate to electron-deficient metals, such as the magnesium in Grignard reagents (Section 8-7). This solvating power, by virtue of the presence of the lone electronic pairs on oxygen, is particularly striking in polyethers, where several such oxygens may surround metal ions. Crown ethers have the remarkable capacity of strongly binding to cations, such as those found in ordinary salts, and rendering them soluble in organic solvents. For example, potassium permanganate, a deep-violet, completely insoluble solid in benzene, is readily dissolved in that solvent if 18-crown-6 is added to it. This solution is useful because it allows oxidations with potassium permanganate to be carried out in organic solvents. Dissolution is possible by effective solvation of the metal ion by the six crown oxygens.

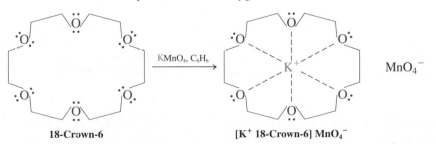

18-Crown-6   [K⁺ 18-Crown-6] MnO₄⁻

Space-filling model of the cation [K⁺ 18-crown-6].

The size of the "cavity" in the crown ether can be tailored to allow for the selective binding of only certain cations—namely, those whose ionic radius is best ac-

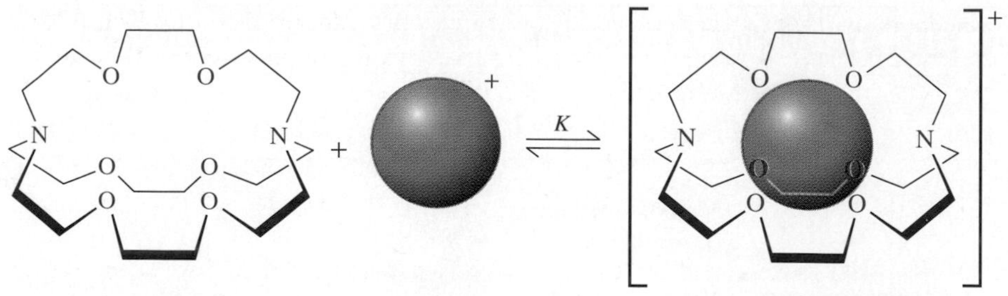

A cryptand   **A cation**   **A cryptate complex**

FIGURE 9-4

The binding of a cation by a polycyclic ether (cryptand) to form a complex (cryptate). The system shown selectively binds the potassium ion, with a binding constant of $K = 10^{10}$. The order of selectivity is $K^+ > Rb^+ > Na^+ > Cs^+ > Li^+$. The binding constant for lithium is about 100. Thus, the total range within the series of alkali metals spans eight orders of magnitude.

commodated by the polyether. This concept has been extended successfully into three dimensions by the synthesis of polycyclic ethers, also called **cryptands** (*kryptos,* Greek, hidden), which are highly selective in alkali and other metal binding (Figure 9-4). The significance of these compounds was recognized by the award of the Nobel Prize (in chemistry) in 1987, shared by Cram, Lehn, and Pedersen.*

Crown ethers and cryptands are often called **ion transport agents** and are part of the general class of **ionophores** (-*phoros,* Greek, bearing, hence "ion bearing"), compounds that organize themselves around cations by coordination. In general, the result of this interaction is that the polar hydrophilic nature of the ion is masked by a hydrophobic shell, hence making the ion much more soluble in nonpolar solvents. In nature, ionophores can transport ions through hydrophobic cell membranes. The ion balance between the inside and outside of the cell is carefully regulated to ensure cell survival, and therefore any undue disruption causes cell destruction. This property is put to medicinal use in fighting invading organisms with polyether antibiotics. However, because ion transport affects nerve transmission, some naturally occurring ionophores are also deadly neurotoxins.

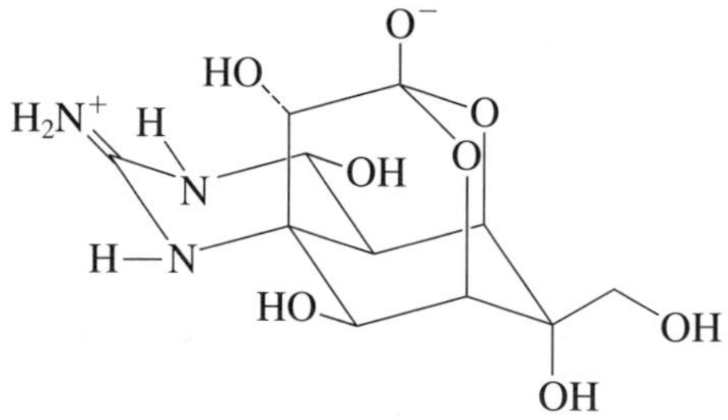

Tetrodotoxin
(The nerve toxin in puffer fish or fugu)

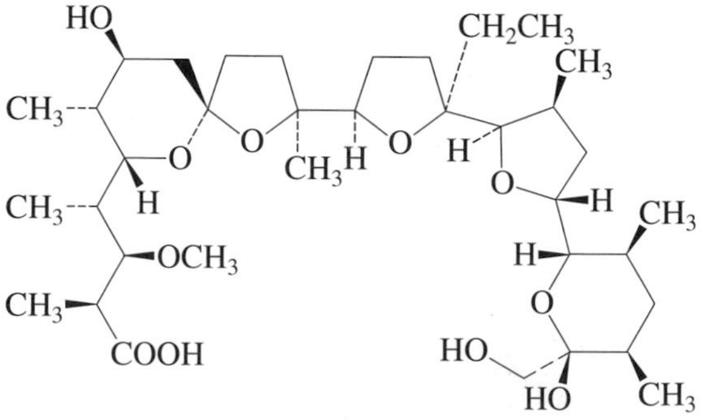

Monensin
(Antibiotic from a *Streptomyces* strain)

In summary, acyclic ethers can be named as alkoxyalkanes or as alkyl ethers. Their cyclic counterparts are called oxacycloalkanes. Ethers have lower boiling points than

*Professor Donald J. Cram (b. 1919), University of California at Los Angeles; Professor Jean-Marie Lehn (b. 1939), University of Strasbourg and Collège de France, Paris; Dr. Charles J. Pedersen (1904–1989), E. I. du Pont de Nemours & Company, Wilmington, Delaware.

do alcohols of comparable size, because they cannot hydrogen bond to each other. The Lewis basicity of the oxygen lone pairs, particularly in polyethers, allows efficient metal ion complexation and hence solubilization.

9-6 Williamson Ether Synthesis

Alkoxides are excellent nucleophiles. This section describes their use in the most common method for the preparation of ethers.

Ethers are prepared by S_N2 reactions

The simplest way to synthesize an ether is to have an alkoxide react with a primary haloalkane or a sulfonate ester under typical S_N2 conditions (Chapter 6). This approach is known as the **Williamson* ether synthesis.** The alcohol from which the alkoxide is derived can be used as the solvent (if inexpensive), but other polar molecules, such as dimethyl sulfoxide (DMSO) or hexamethylphosphoric triamide (HMPA), are often better (Table 6-5).

Williamson Ether Syntheses

$$CH_3CH_2CH_2CH_2O^-Na^+ \quad + \quad ClCH_2CH_2CH_2CH_3 \xrightarrow[-Na^+Cl^-]{\substack{CH_3CH_2CH_2CH_2OH, \ 14 \ h \\ or \ DMSO, \ 9.5 \ h}} CH_3CH_2CH_2CH_2OCH_2CH_2CH_2CH_3$$

60% (butanol solvent)
95% (DMSO solvent)
1-Butoxybutane

$$\text{(cyclopentane)}\ \underset{\overset{|}{H}}{O^-Na^+} \quad + \quad CH_3(CH_2)_{15}CH_2OSO_2CH_3 \xrightarrow{DMSO} \text{(cyclopentane)}\ \underset{\overset{|}{H}}{OCH_2(CH_2)_{15}CH_3} \quad + \quad Na^+{}^-O_3SCH_3$$

91%
Cyclopentoxyheptadecane

Nucleophile: red
Electrophile: blue
Leaving group: green

Because alkoxides are strong bases, their use in ether synthesis is restricted to primary unhindered alkylating agents; otherwise, a significant amount of E2 product is formed (Section 7-8).

EXERCISE 9-11

Write Williamson syntheses for the following ethers. **(a)** 1-ethoxybutane (two ways); **(b)** 2-methoxypentane (Are there two ways that work well?); **(c)** propoxycyclohexane; **(d)** 1,4-diethoxybutane.

Cyclic ethers can be prepared by intramolecular Williamson synthesis

The Williamson ether synthesis is also applicable to the preparation of cyclic ethers, starting from halo alcohols. Figure 9-5 (see page 348) depicts the reaction of hydroxide ion with a bromoalcohol. The black curved lines denote the chain of carbon

*Professor Alexander W. Williamson (1824–1904), University College, London.

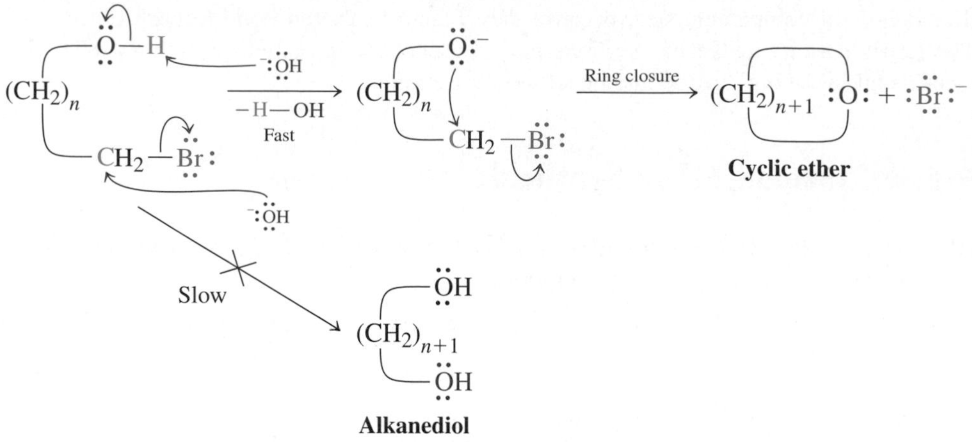

Alkanediol

FIGURE 9-5

The mechanism of cyclic ether synthesis from a bromoalcohol and hydroxide ion (upper reactions). A competing but slower side reaction, direct displacement of bromide by hydroxide, is also shown (lower reaction). The curved lines denote a chain of carbon atoms. The Williamson synthesis of cyclic ethers includes an intramolecular nucleophilic substitution.

atoms linking the functional groups. The mechanism consists of initial formation of a bromoalkoxide by fast proton transfer to the base, followed by ring closure to furnish the cyclic ether. The latter process is an example of an intramolecular displacement. Cyclic ether formation is much faster than the side reaction shown in Figure 9-5; that is, direct displacement of bromide by hydroxide to give a diol. In those cases in which the intermolecular S_N2 reaction competes with its intramolecular counterpart, the former can be effectively suppressed by using high dilution conditions, which drastically reduce the rate of bimolecular processes (Section 2-8).

Intramolecular Williamson synthesis allows the preparation of cyclic ethers of various sizes, including small rings.

Cyclic Ether Synthesis

The blue dot indicates the carbon atom in the product that corresponds to the site of ring closure in the starting material.

$$HOCH_2CH_2Br \ + \ HO^- \longrightarrow$$

Oxacyclopropane
(Oxirane, ethylene oxide)

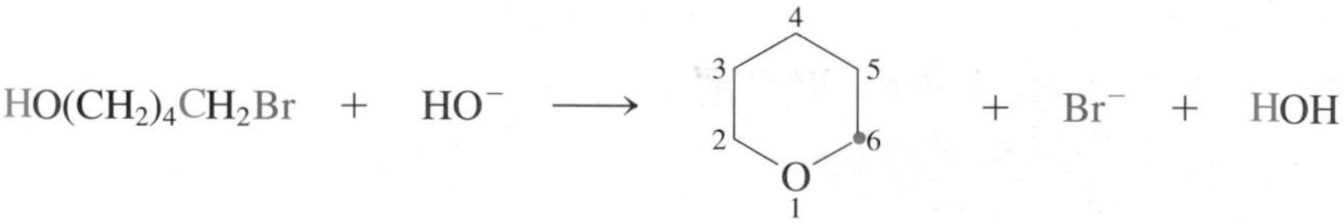

$$HO(CH_2)_4CH_2Br \ + \ HO^- \longrightarrow$$

Oxacyclohexane
(Tetrahydropyran)

EXERCISE 9-12

The product of the reaction of 5-bromo-3,3-dimethyl-1-pentanol with hydroxide is a cyclic ether. Suggest a mechanism for its formation.

CHEMICAL HIGHLIGHT 9-1 | Chemiluminescence of 1,2-Dioxacyclobutanes

A 2-bromohydroperoxide

3,3,4,4-Tetramethyl-1,2-dioxacyclobutane
(A 1,2-dioxetane)

Propanone
(Acetone)

An intramolecular Williamson-type reaction of a special kind is that in which a 2-bromohydroperoxide is the reactant. The peroxide product is a 1,2-dioxacyclobutane (1,2-dioxetane). This species is unusual because it decomposes to the corresponding carbonyl compounds with emission of light *(chemiluminescence)*. Dioxacyclobutanes seem to be responsible for the *bioluminescence* of certain species in nature. Terrestrial organisms, such as the firefly, the glowworm, and certain click beetles, are well known to emit light. However, most bioluminescent species live in the ocean; they range from

microscopic bacteria and plankton to fish. The emitted light serves many purposes and seems to be important in courtship and communication, sex differentiation, finding prey, and hiding from or scaring off predators.

An example of a chemiluminescent molecule in nature is firefly luciferin. The base oxidation of this molecule furnishes a dioxacyclobutanone intermediate that decomposes in a manner analogous to that of 3,3,4,4-tetramethyl-1,2-dioxacyclobutane to give a complex heterocycle, carbon dioxide, and emitted light.

Firefly luciferin

1,2-Dioxacyclobutanone intermediate

$+ \ CO_2 + h\nu$

Firefly with luminous tail glowing.

Ring size controls the speed of cyclic ether formation

A comparison of the relative rates of cyclic ether formation reveals a surprising fact: Three-membered rings form quickly, about as fast as five-membered rings. Six-membered ring systems, four-membered rings, and the larger oxacycloalkanes are generated more slowly. What effects are at work here? The answer includes both entropy factors and ring strain.

Relative Rates of Cyclic Ether Formation

$$k_3 \geq k_5 > k_6 > k_4 \geq k_7 > k_8$$
$k_n = \text{reaction rate}, \ n = \text{ring size}$

The preparation of an oxacyclopropane from a 2-bromoalcohol is favored by entropy, because nucleophile and leaving group are as close to each other as possible. Therefore, even though ring strain is at its worst in this case, the transition-state energy is relatively small, because a favorable entropy contribution allows relatively rapid ring construction. What about four-membered rings? Oxacyclobutanes are generated much more slowly. The entropy factor here is considerably worse, because the two reacting centers are separated by an extra methylene group. The ring strain, however, is still considerable. The net result is a very low relative rate of formation. In contrast, the synthesis of five-membered ring ethers is easy. Although the reacting centers are even farther apart, the strain is much less. Proceeding to oxacyclohexanes, strain no longer plays a role, but the entropy factor gets worse. This trend continues for the larger rings, their rates of formation suffering, in addition, from eclipsing, *gauche,* and transannular strain.

The intramolecular Williamson synthesis is stereospecific

The Williamson ether synthesis proceeds with inversion of configuration at the carbon bearing the leaving group, in accord with expectations based on an S_N2 mechanism. The attacking nucleophile approaches the electrophilic carbon from the opposite side of the leaving group. Only one conformation of the haloalkoxide can undergo efficient substitution. For example, oxacyclopropane formation requires an *anti* arrangement of the nucleophile and the leaving group. The alternative two *gauche* conformations cannot give the product (Figure 9-6).

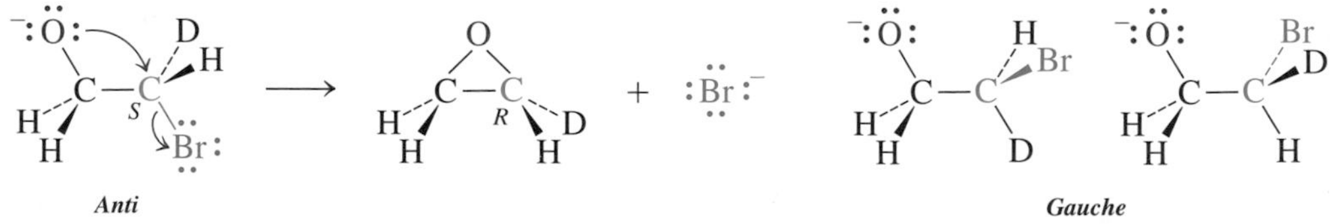

FIGURE 9-6

Only the *anti* conformation of a 2-bromoalkoxide allows for oxacyclopropane formation. The two *gauche* conformers cannot undergo intramolecular backside attack at the bromine-bearing carbon.

EXERCISE 9-13

(1*R*,2*R*)-2-Bromocyclopentanol reacts rapidly with sodium hydroxide to yield an optically inactive product. In contrast, the (1*S*,2*R*) isomer is much less reactive. Explain. (**Hint:** Refer to Figure 9-6.)

In summary, ethers are prepared by the Williamson synthesis, an S_N2 reaction of an alkoxide with a haloalkane. This reaction works best with primary halides or sulfonates that do not undergo ready elimination. Cyclic ethers are formed by the intramolecular version of this method. The relative rates of ring closure in this case are highest for three- and five-membered rings.

9-7 Synthesis of Ethers: Alcohols and Mineral Acid

An even simpler, albeit less selective, route to ethers is the reaction of a strong inorganic acid (e.g., H_2SO_4) with an alcohol. Protonation of the OH group in one alcohol generates water as a leaving group. Nucleophilic displacement of this leaving group by a second alcohol then results in the corresponding alkoxyalkane.

Alcohols give ethers by both S$_N$2 and S$_N$1 mechanisms

We have learned that treating primary alcohols with HBr or HI furnishes the corresponding haloalkanes through intermediate alkyloxonium ions (Section 9-2). However, when strong nonnucleophilic acids—such as sulfuric acid—are used at elevated temperatures, the main products are ethers.

Ether Synthesis from a Primary Alcohol with Strong Acid

$$2 \text{ CH}_3\text{CH}_2\text{OH} \xrightarrow{\text{H}_2\text{SO}_4,\ 130°\text{C}} \text{CH}_3\text{CH}_2\text{OCH}_2\text{CH}_3 + \text{HOH}$$

In this reaction, the strongest nucleophile present in solution is the unprotonated starting alcohol. As soon as one alcohol molecule has been protonated, nucleophilic attack begins, the ultimate products being an ether and water.

Mechanism of Ether Synthesis

Only symmetric ethers can be prepared by this method.

At even higher temperatures (see footnote on page 334), elimination of water to generate an alkene is observed. This reaction proceeds by an E2 mechanism (Sections 7-7 and 11-9), in which the neutral alcohol serves as the base that attacks the alkyloxonium ion.

Alkene Synthesis from an Alcohol and Strong Acid at High Temperature

$$\underset{\displaystyle \text{CH}_3\overset{\displaystyle \overset{\text{H}}{|}}{\text{C}}\text{HCH}_2\text{OH}}{} \xrightarrow{\text{H}_2\text{SO}_4,\ 180°\text{C}} \text{CH}_3\text{CH}{=}\text{CH}_2 + \text{HOH}$$

Secondary and tertiary ethers can also be made by acid treatment of secondary and tertiary alcohols. However, in these cases, a carbocation is formed initially and is then trapped by an alcohol (S$_N$1), as described in Section 9-2.

$$2\ \text{CH}_3\overset{\displaystyle \overset{\text{OH}}{|}}{\underset{\displaystyle \underset{\text{H}}{|}}{\text{C}}}\text{CH}_3 \xrightarrow{\text{H}^+,\ 40°\text{C}} (\text{CH}_3)_2\text{CHOCH}(\text{CH}_3)_2 + \text{HOH} + \text{H}^+$$

2-Propanol

75%

2-(1-Methylethoxy)propane
(Diisopropyl ether)

The major side reaction follows the E1 pathway (Sections 9-2, 9-3, and 11-9), which, again, becomes dominant at higher temperatures.

It is harder to synthesize ethers containing two different alkyl groups, because mixing two alcohols in the presence of an acid usually results in mixtures of all three possible products. However, mixed ethers containing one tertiary and one primary or secondary alkyl substituent can be prepared in good yield in the presence of dilute acid. Under these conditions, the much more rapidly formed tertiary carbocation is trapped by the other alcohol.

Synthesis of a Mixed Ether

$$
\underset{\underset{CH_3}{|}}{\overset{\overset{CH_3}{|}}{CH_3\overset{|}{C}OH}} + CH_3CH_2OH \xrightarrow[-\ HOH]{15\%\ aqueous\ NaHSO_4,\ 40°C} \underset{\underset{CH_3}{|}}{\overset{\overset{CH_3}{|}}{CH_3\overset{|}{C}OCH_2CH_3}}
$$

80%

2-Ethoxy-2-methylpropane

Write mechanisms for the following two reactions. **(a)** 1,4-Butanediol + H$^+$ → oxacyclopentane (tetrahydrofuran); **(b)** 5-methyl-1,5-hexanediol + H$^+$ → 2,2-dimethyloxacyclohexane (2,2-dimethyltetrahydropyran).

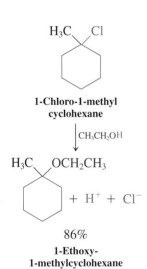

1-Chloro-1-methyl cyclohexane

CH$_3$CH$_2$OH

+ H$^+$ + Cl$^-$

86%

1-Ethoxy-1-methylcyclohexane

Ethers also form by alcoholysis

As we know, tertiary and secondary ethers may also form by the alcoholysis of the corresponding haloalkanes or alkyl sulfonates (Section 7-1). The starting material is simply dissolved in an alcohol until the S$_N$1 process is complete.

There are several ways of constructing an ether from an alcohol and a haloalkane. Which approach would you choose for the preparation of **(a)** 2-methyl-2-(1-methylethoxy)butane; **(b)** 1-methoxy-2,2-dimethylpropane? [**Hint:** The product for (a) is a tertiary ether, that for (b) is a neopentyl ether.]

In summary, ethers can be prepared by treatment of alcohols with acid through S$_N$2 and S$_N$1 pathways, with alkyloxonium ions or carbocations as intermediates, and by alcoholysis of secondary or tertiary haloalkanes or alkyl sulfonates.

9-8 Reactions of Ethers

As mentioned earlier, ethers are normally rather inert. They do, however, react slowly with oxygen by radical mechanisms to form hydroperoxides and peroxides. Because peroxides can decompose explosively, extreme care should be taken with samples of ethers that have been exposed to air for several days. This section describes a more preparatively useful reaction, cleavage by strong acid.

Peroxides from Ethers

$$2\ \underset{|}{\overset{|}{ROCH}}\ +\ O_2\ \longrightarrow\ 2\ \underset{|}{\overset{|}{ROC}}\!-\!O\!-\!OH\ \longrightarrow\ \underset{|}{\overset{|}{ROC}}\!-\!O\!-\!O\!-\!\underset{|}{\overset{|}{COR}}$$

An ether
hydroperoxide   An ether peroxide

The oxygen in ethers, like that in alcohols, may be protonated to generate alkyl-oxonium ions. The subsequent reactivity of these ions depends on the alkyl substituents. With *primary* groups and strong nucleophilic acids such as HBr, S_N2 displacement takes place.

Ether Cleavage with HBr

$$CH_3CH_2OCH_2CH_3 \xrightarrow{\text{HBr}} CH_3CH_2Br\ +\ CH_3CH_2OH$$

Ethoxyethane   Bromoethane   Ethanol

Mechanism of Ether Cleavage

$$CH_3CH_2\ddot{O}CH_2CH_3 \underset{}{\overset{H^+}{\rightleftharpoons}} CH_3CH_2\!-\!\overset{H}{\underset{CH_2CH_3}{\overset{|}{O}^+}} \underset{}{\overset{:\ddot{B}r:}{\rightleftharpoons}} CH_3CH_2Br\ +\ H\ddot{O}CH_2CH_3$$

Alkyloxonium ion

The alcohol formed as the second product may in turn be attacked by additional HBr to give more of the bromoalkane.

EXERCISE 9-16

Treatment of methoxymethane with hot concentrated HI gives two equivalents of iodomethane. Suggest a mechanism.

EXERCISE 9-17

Reaction of oxacyclohexane (tetrahydropyran; shown in the margin) with hot concentrated HI gives 1,5-diiodopentane. Give a mechanism for this reaction.

Ethers containing *tertiary* alkyl groups transform even in dilute acid to give intermediate tertiary carbocations, which are either trapped by nucleophiles (S_N1) or deprotonated (E1); see Chemical Highlight 9-2.

Oxacyclohexane
(Tetrahydropyran)

$$\text{(propyl-O-tert-butyl ether)} \xrightarrow{\text{H}_2\text{SO}_4,\ \text{H}_2\text{O, 50°C}} \text{(propanol)}\!-\!OH\ +\ \text{(isobutylene)}$$

Oxonium ions derived from *secondary* ethers can transform by either S_N2 or S_N1 (E1) reactions, depending on the system and conditions (Section 7-9 and Table 7-2). For example, 2-ethoxypropane is protonated by aqueous HI and then converted into 2-propanol and iodoethane by selective attack by iodide at the less hindered primary center.

$$\text{2-Ethoxypropane} \xrightarrow{\text{HI, H}_2\text{O}} \overset{OH}{\text{2-Propanol}}\ +\ \text{Iodoethane}$$

2-Ethoxypropane   2-Propanol   Iodoethane

CHEMICAL HIGHLIGHT 9-2   **Protecting Groups in Synthesis**

$$ROH \xrightarrow[-H_2O]{(CH_3)_3COH, H^+} ROC(CH_3)_3 \xrightarrow[\substack{\text{Carry out reactions on R by}\\\text{using Grignard reagents,}\\\text{oxidizing agents, etc.}}]{} R'OC(CH_3)_3 \xrightarrow{H^+, H_2O} R'OH$$

Protection step   **Protected alcohol**   **R changed into R′**   **Deprotection**

To remove the potentially complicating presence of a reactive functionality, chemists use *protecting groups.* That is, a functional group is temporarily converted into an unreactive form *(protection)* to prevent its interference with transformations to be carried out elsewhere in the molecule. Subsequently, the original unit is regenerated *(deprotection).* For example, one protected form of an alcohol is a 1,1-dimethylethyl ether, readily obtained from an acid-catalyzed reaction with 2-methyl-2-propanol (*tert*-butyl alcohol) (Section 9-7). The protected functional group is now inert to base, organometallic reagents, and oxidizing and reducing agents. Reactions of such species may therefore be completed without interference. The protecting group is removed by dilute aqueous acid as shown above.

Another method of alcohol protection is esterification (Section 9-4).

One application of these protecting groups is in the synthesis of the sex hormone testosterone (Section 4-7) from the cholesterol-derived starting material shown on the facing page. Natural sources of steroid hormones are far too limited to meet the needs of medicine and research; these molecules must be synthesized. In our case, selective reduction of the carbonyl group at C17 and oxidation of the hydroxy function at C3 are required.

Formation of the 1,1-dimethylethyl (*tert*-butyl) ether at C3 is followed by reduction at C17. A second protection step at C17 is esterification (Section 9-4). Esters are stable in dilute acid, which hydrolyzes tertiary ethers. This strategy allows the hydroxy group at C3 to be freed and oxidized to a carbonyl, while that at C17 remains protected. Exposure to strong acid finally converts the product of the sequence shown here into testosterone.

EXERCISE 9-18

Show how you would achieve the following interconversion (the dashed arrow indicates that several steps are required). (**Hint:** You need to protect the OH function.)

$$BrCH_2CH_2CH_2OH \dashrightarrow DCH_2CH_2CH_2OH$$

In summary, ethers are cleaved by (strong) acids. Protonation of an ether containing methyl or primary alkyl groups gives an alkyloxonium ion that is subject to S_N2 attack by nucleophiles. Carbocation formation follows protonation when secondary and tertiary groups are present, leading to S_N1 and E1 products.

9-9 Reactions of Oxacyclopropanes

Although ordinary ethers are relatively inert, the strained ring in oxacyclopropanes undergoes a variety of ring-opening reactions with nucleophiles. This section presents details of these processes.

Nucleophilic ring opening of oxacyclopropanes by S_N2 is regioselective and stereospecific

Oxacyclopropane is subject to bimolecular ring opening by anionic nucleophiles. Because of the symmetry of the substrate, substitution occurs to the same extent at either carbon. The reaction proceeds by nucleophilic attack, with the ether oxygen functioning as an intramolecular leaving group.

This S_N2 transformation is unusual for two reasons. First, alkoxides are usually very bad leaving groups. Second, the leaving group does not actually "leave"; it stays bound to the molecule. The driving force is the release of strain as the ring opens.

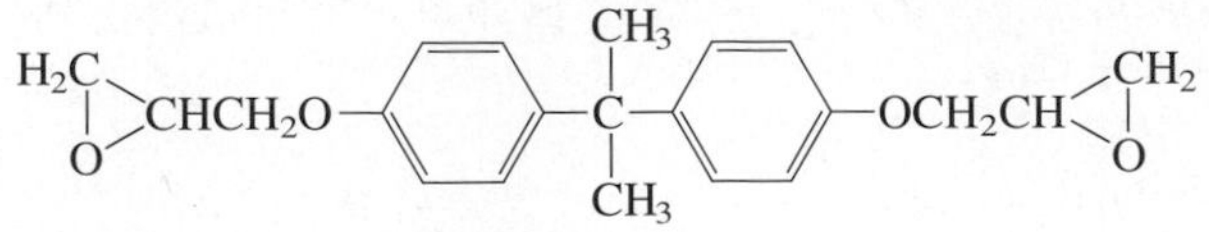

DGEBA epoxy resin
(Diglycidyl ether of bisphenol A)

The reactivity of oxacyclopropanes has been exploited in the development of the *epoxy resins* (a term derived from epoxide, a common name for oxacyclopropane). A typical example is DGEBA.

Epoxy resins undergo a transformation called *curing* upon heating or exposure to a variety of reagents, usually acids or bases. The result is an extremely hard solid. Most resins contain two oxacyclopropane functional groups. The base-catalyzed curing process begins with ring opening by hydroxide. The resulting alkoxide units then attack other molecules of the resin, a process called *cross-linking*. Each ring opening liberates another alkoxide, which cross-links with another molecule of resin, ultimately providing a material of very high molecular weight. The process is shown schematically here.

When curing takes place in contact with a solid containing surface hydroxy groups (glass is a good example), *surface bonding* is accomplished through covalent linkages: The resin acts as a strong adhesive.

Curing of an Epoxy Resin

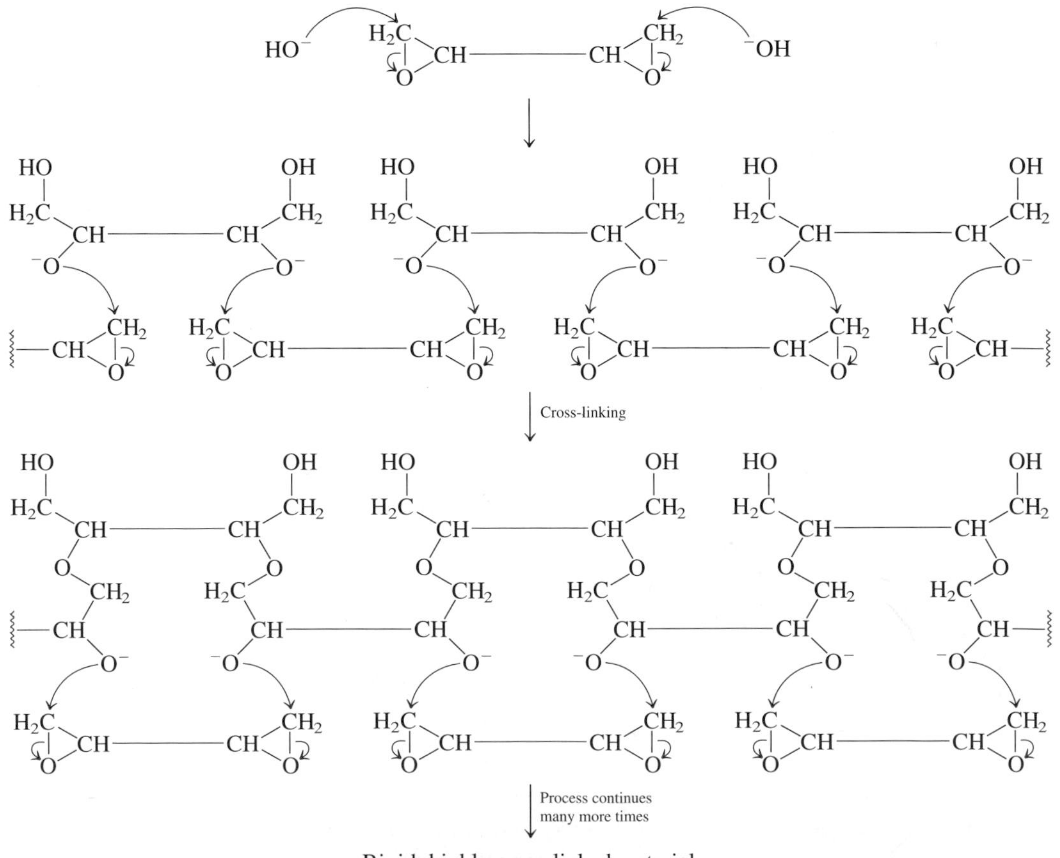

Cross-linking

Process continues
many more times

Rigid, highly cross-linked material

What is the situation with unsymmetric systems? Consider, for example, the reaction of 2,2-dimethyloxacyclopropane with methoxide. There are *two* possible reaction sites: at the primary carbon (*a*), to give 1-methoxy-2-methyl-2-propanol; and at the tertiary carbon (*b*), to yield 2-methoxy-2-methyl-1-propanol. Evidently, this system transforms solely through path *a*.

**Nucleophilic Ring Opening of an Unsymmetrically
Substituted Oxacyclopropane**

1-Methoxy-
2-methyl-2-propanol

$\xleftarrow[a]{CH_3O^-, CH_3OH}$

2,2-Dimethyloxacyclopropane

$\xrightarrow[b]{CH_3O^-, CH_3OH}$ ✗

2-Methoxy-
2-methyl-1-propanol
(Not formed)

Is this result surprising? No, because, as we know, if there is more than one possibility, S_N2 attack will be at the *less* substituted carbon center (Section 6-10). This selectivity in the nucleophilic opening of substituted oxacyclopropanes is referred to as **regioselectivity,** because, of two possible and similar "regions," the nucleophile attacks only one.

In addition, when the ring opens at a stereocenter, inversion is observed. Thus, we find that the rules of nucleophilic substitution developed for simple alkyl derivatives also apply to strained cyclic ethers.

Hydride and organometallic reagents convert strained ethers into alcohols

The highly reactive lithium aluminum hydride is able to open the rings of oxacyclopropanes, a reaction leading to alcohols. Ordinary ethers, lacking the ring strain of oxacyclopropanes, do not react with $LiAlH_4$. The reaction also proceeds by the S_N2 mechanism. Thus, in unsymmetric systems, the hydride attacks the less substituted side; when the latter constitutes a stereocenter, inversion is observed.

**Ring Opening of an Oxacylopropane
by Lithium Aluminum Hydride**

$\xrightarrow[2. \ H^+, H_2O]{1. \ LiAlH_4, (CH_3CH_2)_2O}$

Less hindered

Inversion on Oxaycyclopropane Opening

$\xrightarrow[2. \ H^+, H_2O]{1. \ LiAlD_4}$

99.4%

D and OH are trans, not cis.

In this example, $LiAlD_4$ reacts as a source of D^-.

EXERCISE 9-19

What oxacyclopropane would give 3-hexanol after treatment with LiAlH$_4$ followed by acidic aqueous work-up? (**Hints:** Apply retrosynthetic analysis, as described in Section 8-9. There are two possible answers, but one will give a mixture of 3-hexanol with an isomer.)

In contrast with haloalkanes (Section 8-8), oxacyclopropanes are sufficiently reactive electrophiles to be attacked by organometallic compounds. Thus, Grignard reagents and alkyllithium compounds undergo 2-hydroxyethylation by ether ring opening, following the S$_N$2 mechanism.

Oxacyclopropane Ring Opening by a Grignard Reagent: 2-Hydroxyethylation

$$H_2C\overset{O}{\diagdown}CH_2 \ + \ CH_3CH_2CH_2CH_2MgBr \ \xrightarrow[]{THF} \ \xrightarrow[]{H^+, H_2O} \ CH_3CH_2CH_2CH_2CH_2CH_2OH$$
$$62\%$$
$$\textbf{1-Hexanol}$$

EXERCISE 9-20

Propose an efficient synthesis of 3,3-dimethyl-1-butanol from starting materials containing no more than four carbons. (**Hint:** Consider the product retrosynthetically as a 2-hydroxyethylated tertiary butyl.)

Acids catalyze oxacyclopropane ring opening

Ring opening of oxacyclopropanes is also catalyzed by acids. The reaction in this case proceeds through initial cyclic alkyloxonium ion formation followed by ring opening as a result of nucleophilic attack.

Acid-Catalyzed Ring Opening of Oxacyclopropane

$$H_2C\overset{O}{\diagdown}CH_2 \ + \ CH_3OH \ \xrightarrow[]{H_2SO_4} \ HOCH_2CH_2OCH_3$$
$$\textbf{2-Methoxyethanol}$$

Mechanism of Acid-Catalyzed Ring Opening

Is this reaction also regioselective and stereospecific, like the anionic nucleophilic opening of oxacyclopropanes discussed first? Yes, but the details are different. Thus, acid-catalyzed methanolysis of 2,2-dimethyloxacyclopropane proceeds by exclusive ring opening at the *more* hindered carbon.

Acid-Catalyzed Ring Opening of 2,2-Dimethyloxacyclopropane

2,2-Dimethyloxacyclopropane   **2-Methoxy-2-methyl-1-propanol**

Why is the more hindered position attacked? Protonation at the oxygen of the ether generates a reactive intermediate alkyloxonium ion with substantially polarized oxygen–carbon bonds. This polarization places partial positive charges on the ring carbons. Because alkyl groups act as electron donors (Section 7-5), more positive charge is located on the tertiary than on the primary carbon.

Mechanism of Acid-Catalyzed Ring Opening
of 2,2-Dimethyloxacyclopropane by Methanol

1-Methoxy-2-methyl-2-propanol   **2-Methoxy-2-methyl-1-propanol**
        (Not formed)

This uneven charge distribution counteracts steric hindrance: Methanol is attracted by Coulombic forces more to the tertiary than to the primary center. Although the result is clear-cut in this example, it is less so in cases in which the two carbons are not quite as different. For example, mixtures of isomeric products are formed by acid-catalyzed ring opening of 2-methyloxacyclopropane.

Why do we not simply write free carbocations as intermediates in the acid-catalyzed ring openings? The reason is that inversion is observed when the reaction takes place at a stereocenter. Like the reaction of oxacyclopropanes with anionic nucleophiles, the acid-catalyzed process includes backside displacement—in this case, on a highly polarized cyclic alkyloxonium ion.

EXERCISE 9-21

Predict the major product of ring opening of 2,2-dimethyloxacyclopropane on treatment with each of the following reagents. (a) $LiAlH_4$, then H^+, H_2O; (b) $CH_3CH_2CH_2MgBr$, then H^+, H_2O; (c) CH_3SNa in CH_3OH; (d) dilute HCl in CH_3CH_2OH; (e) concentrated aqueous HBr.

In summary, although ordinary ethers are relatively inert, the ring in oxacyclopropanes can be opened both regioselectively and stereospecifically. For anionic nucleophiles, the usual rules of bimolecular nucleophilic substitution hold: Attack is at the less hindered carbon center, which undergoes inversion. Acid catalysis, however, changes the regioselectivity: Attack is at the more hindered center. Hydride and organometallic reagents behave like other anionic nucleophiles, furnishing alcohols by an S_N2 pathway.

9-10 Sulfur Analogs of Alcohols and Ethers

We conclude our study of the chemistry of alcohols and ethers by looking at some of their sulfur analogs, in which a sulfur heteroatom replaces the oxygen in the functional group. In this section, we shall see what difference this substitution makes.

The sulfur analogs of alcohols and ethers are thiols and sulfides

The sulfur analogs of alcohols, R–SH, are called **thiols** in the IUPAC system (*theion,* Greek, brimstone—an older name for sulfur). The ending *thiol* is added to the alkane stem to yield the alkanethiol name. The SH group is referred to as **mercapto,** and its location is indicated by numbering the longest chain, as in alkanol nomenclature. The mercapto functional group has lower precedence than hydroxy.

CH$_3$SH
Methanethiol

CH$_3$CH$_2$CHCH$_2$SH
4 3 2 1
CH$_3$ (above)
2-Methyl-1-butanethiol

CH$_3$CH$_2$CHCH$_2$CH$_3$
:SH (above)
3-Pentanethiol

:SH
Cyclohexanethiol

HSCH$_2$CH$_2$OH
2-Mercaptoethanol

The sulfur analogs of ethers (common name, thioethers) are called **sulfides,** as in alkyl ether nomenclature. The RS group is named **alkylthio,** the RS$^-$ group **alkanethiolate.**

CH$_3$SCH$_2$CH$_3$
Ethyl methyl sulfide

CH$_3$
CH$_3$CS(CH$_2$)$_6$CH$_3$
CH$_3$
1,1-Dimethylethyl heptyl sulfide

CH$_3$S:$^-$
Methanethiolate ion

Thiols are less hydrogen bonded and more acidic than alcohols

Sulfur, because of its large size, its diffuse orbitals, and the relatively nonpolarized S–H bond (Table 1-2), does not enter into hydrogen bonding very efficiently. Thus, the boiling points of thiols are not as abnormally high as those of alcohols; rather, their volatilities lie close to those of the analogous haloalkanes (Table 9-2).

Partly because of the relatively weak S–H bond, thiols are also more acidic than water, with pK_a values ranging from 9 to 12. They can therefore be more readily deprotonated by hydroxide and alkoxide ions.

Acidity of Thiols

$$RSH \ + \ HO:^- \ \rightleftharpoons \ RS:^- \ + \ HOH$$
p$Ka = 9{-}12$

Thiols and sulfides react much like alcohols and ethers

Many reactions of thiols and sulfides resemble those of their oxygen analogs. The sulfur in these compounds is even more nucleophilic than the oxygen in alcohols and ethers. Therefore, thiols and sulfides are readily made through nucleophilic attack by RS$^-$ or HS$^-$ on haloalkanes. A large excess of the HS$^-$ is used in the preparation of thiols to ensure that the product does not react with the starting halide to give the dialkyl sulfide.

TABLE 9-2

Comparison of the Boiling Points of Thiols, Haloalkanes, and Alcohols

Compound	Boiling point (°C)
CH$_3$SH	6.2
CH$_3$Br	3.6
CH$_3$Cl	−24.2
CH$_3$OH	65.0
CH$_3$CH$_2$SH	37
CH$_3$CH$_2$Br	38.4
CH$_3$CH$_2$Cl	12.3
CH$_3$CH$_2$OH	78.5

$$\underset{\text{Excess}}{\overset{\overset{\displaystyle CH_3}{|}}{CH_3CHBr}} + Na^+{}^-SH \xrightarrow{CH_3CH_2OH} \underset{\text{2-Propanethiol}}{\overset{\overset{\displaystyle CH_3}{|}}{CH_3CHSH}} + Na^+Br^-$$

Sulfides are prepared in an analogous way by alkylation of thiols in the presence of base, such as hydroxide. The base generates the alkanethiolate, which reacts with the haloalkane by an S_N2 process. Because of the strong nucleophilicity of thiolates, there is no competition from hydroxide in this displacement.

Sulfides by Alkylation of Thiols

$$RSH + R'Br \xrightarrow{NaOH} RSR' + NaBr + H_2O$$

The nucleophilicity of sulfur also explains the ability of sulfides to attack haloalkanes to furnish **sulfonium ions.**

$$\underset{H_3C}{\overset{H_3C}{>}}\ddot{S}\colon + CH_3\ddot{-}\ddot{I}\colon \longrightarrow \underset{H_3C}{\overset{H_3C}{>}}\overset{+}{S}-CH_3 + \colon\ddot{I}\colon^-$$

95%
Trimethylsulfonium iodide

Sulfonium salts are subject to nucleophilic attack at carbon, the sulfide functioning as the leaving group.

$$H\ddot{O}\colon^- + CH_3\overset{+}{-}\ddot{S}(CH_3)_2 \longrightarrow H\ddot{O}CH_3 + \ddot{S}(CH_3)_2$$

Soldiers wearing chemical protective gear.

EXERCISE 9-22

(a) Sulfide A is a powerful poison known as "mustard gas," a devastating chemical warfare agent used in the First World War and again in the eight-year war between Iraq and Iran in the 1980s. The specter of chemical and biological weapons loomed again during the Persian Gulf war of 1990–1991 and a medical condition known as "Gulf war syndrome" has been, at times, ascribed to the suspected exposure of ground troops to chemical and perhaps biological agents during the campaign. The Geneva protocol of 1925 explicitly bans the use of chemical and biological weapons. The 1982 Biological and 1993 (ratified by the United States in 1997) Chemical Weapons Conventions ban possession of such materials, but there is great concern about compliance and enforcement. One of the problems is the relative ease with which such toxic chemicals can be produced, as highlighted in this problem. Propose a synthesis of A starting with oxacyclopropane. (**Hint:** Your retrosynthetic analysis should proceed through the diol precursor to A.) **(b)** Its mechanism of action is believed to include sulfonium salt B, which is thought to react with nucleophiles in the body. How is compound B formed and how would it react with nucleophiles?

$$\underset{\text{A}}{ClCH_2CH_2SCH_2CH_2Cl}$$

$$\underset{\text{B}}{ClCH_2CH_2S\overset{\overset{\displaystyle CH_2}{|}}{\underset{\underset{\displaystyle CH_2}{|}}{\diagup\diagdown}}}\,Cl^-$$

Valence-shell expansion of sulfur accounts for the special reactivity of thiols and sulfides

As a third-row element with *d* orbitals, sulfur's valence shell can expand to accommodate more electrons than are allowed by the octet rule (Section 1-4). We have already seen that, in some of its compounds, sulfur is surrounded by 10 or even 12

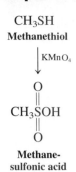

CH₃SH
Methanethiol

↓ KMnO₄

$$CH_3\overset{O}{\underset{O}{\overset{\|}{\underset{\|}{S}}}}OH$$

**Methane-
sulfonic acid**

valence electrons, and this capacity enables sulfur compounds to undergo reactions inaccessible to the corresponding oxygen analogs. For example, oxidation of thiols with strong oxidizing agents, such as hydrogen peroxide or potassium permanganate, gives the corresponding sulfonic acids. In this way, methanethiol is converted into methanesulfonic acid. Sulfonic acids react with PCl₅ to give sulfonyl chlorides, which are used in sulfonate synthesis, as discussed in Section 9-4.

Milder oxidation of thiols, by the use of iodine, results in the formation of **disulfides,** which are readily reduced back to thiols by alkali metals.

The Thiol-Disulfide Redox Reaction

Oxidation

$$2\ CH_3CH_2CH_2SH\ +\ I_2\ \longrightarrow\ CH_3CH_2CH_2SSCH_2CH_2CH_3\ +\ 2\ HI$$
    1-Propanethiol   **Dipropyl disulfide**

Reduction

$$CH_3CH_2CH_2SSCH_2CH_2CH_3\ \xrightarrow[\text{2. H}^+,\ \text{H}_2\text{O}]{\text{1. Li, liquid NH}_3}\ CH_3CH_2CH_2SH$$

Reversible disulfide formation from thiols is an important biological process. Many proteins and peptides contain free SH groups that form bridging disulfide linkages. Nature exploits this mechanism to link amino acid chains. By thus helping to control the shape of enzymes in three dimensions, the mechanism makes biocatalysis far more efficient and selective.

Amino acid chain   Amino acid chain   **Disulfide bridge**

$$CH_3\ddot{S}CH_3$$
**Dimethyl
sulfide**

↓ H₂O₂

$$CH_3\overset{:O:}{\overset{\|}{S}}CH_3$$
**Dimethyl
sulfoxide
(DMSO)**

↓ H₂O₂

$$CH_3\overset{:O:}{\underset{:O:}{\overset{\|}{\underset{\|}{S}}}}CH_3$$
**Dimethyl
sulfone**

Sulfides are readily oxidized to **sulfones,** a transformation proceeding through the intermediacy of a **sulfoxide.** For example, oxidation of dimethyl sulfide first gives dimethyl sulfoxide, which subsequently furnishes dimethyl sulfone. Dimethyl sulfoxide has already been mentioned as a highly polar nonprotic solvent of great use in organic chemistry, particularly in nucleophilic substitutions (see Section 6-9 and Table 6-5).

In summary, the naming of thiols and sulfides is related to the system used for alcohols and ethers. Thiols are more volatile, more acidic, and more nucleophilic than alcohols. Thiols and sulfides can be oxidized, the former to disulfides or sulfonic acids and the latter to sulfoxides and sulfones.

9-11 Physiological Properties and Uses of Alcohols and Ethers

Modern industrial *methanol* synthesis uses the catalytic reduction of carbon monoxide with hydrogen at high pressures and temperatures (Section 8-4). Methanol is sold as a solvent for paint and other materials, as a fuel for camp stoves and soldering torches, and as a synthetic intermediate. It is highly poisonous, and ingestion or chronic exposure may lead to blindness. Death from ingestion of as little as 30 mL has been reported. It is sometimes added to commercial ethanol to render it unfit for consumption (denatured alcohol). The toxicity of methanol is thought to be due to metabolic oxidation to formaldehyde, $CH_2=O$, which interferes with the physicochemical processes of vision. Further oxidation to formic acid, HCOOH, causes acidosis, an unusual lowering of the blood pH. This condition disrupts oxygen transport in the blood and leads eventually to coma.

Methanol has been studied as a possible precursor of gasoline. For example, certain zeolite catalysts (Section 3-3) allow the conversion of methanol into a mixture of hydrocarbons, ranging in length from four-carbon chains to ten-carbon ones, with a composition that, on distillation, yields largely gasoline (see Table 3-3).

$$ n\,CH_3OH \xrightarrow{\text{Zeolite, 340}°\text{–375}°\text{C}} \underset{67\%}{C_nH_{2n+2}} + \underset{6\%}{C_nH_{2n}} + \underset{27\%}{\text{aromatics}} $$

Ethanol—diluted by various amounts of flavored water—is an alcoholic beverage. It is classified pharmacologically as a general depressant, because it induces a non-selective, reversible depression of the central nervous system. Approximately 95% of the alcohol ingested is metabolized in the body (usually in the liver) to products that are eventually transformed into carbon dioxide and water. Although high in calories, ethanol has little nutritional value.

The rate of metabolism of most drugs in the liver increases with their concentration, but this is not true for alcohol, which is degraded linearly with time. An adult metabolizes about 10 mL of pure ethanol per hour, roughly the ethanol content of a cocktail, a shot of spirits, or a can of beer. As few as two or three drinks—depending on a person's weight, the ethanol content of the drink, and the speed with which it is consumed—can produce a level of alcohol in the blood that is more than 0.08%, a concentration at or above the legal limit beyond which the operation of a motorized vehicle is prohibited in much of the United States.

Ethanol is poisonous. Its lethal concentration in the bloodstream has been estimated at 0.4%. Its effects include progressive euphoria, disinhibition, disorientation, and decreased judgment (drunkenness), followed by general anesthesia, coma, and death. It dilates the blood vessels, producing a "warm flush," but it actually decreases body temperature. Although long-term ingestion of moderate amounts (the equivalent of about two beers a day) does not appear to be harmful, larger amounts can be the cause of a variety of physical and psychological disorders, usually described by the general term *alcoholism*. These disorders include hallucinations, psychomotor agitation, liver diseases, dementia, gastritis, and psychological dependence.

Interestingly, a near-toxic dose of ethanol is applied in cases of acute methanol or 1,2-ethanediol (ethylene glycol) poisoning. This treatment prevents the metabolism of the more toxic alcohols and allows their excretion before damaging concentrations of secondary products can accumulate.

Ethanol destined for human consumption is prepared by fermentation of sugars and starch (rice, potatoes, corn, wheat, flowers, fruit, etc.; Chapter 24). Fermentation is catalyzed by enzymes in a multistep sequence that converts carbohydrates into ethanol and carbon dioxide.

A cirrhotic liver due to excessive alcohol consumption becomes characteristically knobby and light brown in appearance.

$$ \underset{\textbf{Starch}}{(C_6H_{10}O_5)_n} \xrightarrow{\text{Enzymes}} \underset{\textbf{Glucose}}{C_6H_{12}O_6} \xrightarrow{\text{Enzymes}} \underset{\textbf{Ethanol}}{2\,CH_3CH_2OH} + 2\,CO_2 $$

Commercial alcohol not intended as a beverage is made industrially by hydration of ethene (Section 8-4). It is used, for example, as a solvent in perfumes, varnishes, and shellacs and as a synthetic intermediate, as demonstrated in earlier equations. Interest in ethanol production has surged recently because of its potential as a gasoline additive ("gasohol").

2-Propanol is toxic but (unlike methanol) is not absorbed through the skin. It is popular as a rubbing alcohol and used as a solvent and a cleaning agent.

1,2-Ethanediol (ethylene glycol) is prepared by oxidation of ethene to oxacyclopropane, followed by hydrolysis, in quantities exceeding 5.2 billion pounds (2.6 million tons) in the United States per year. Its low melting point ($-11.5°C$), its high boiling point ($198°C$), and its complete miscibility with water make it a useful antifreeze. Its toxicity is similar to that of other simple alcohols.

$$CH_2{=}CH_2 \xrightarrow{\text{Oxidation}} \triangle \xrightarrow{\text{H}_2\text{O}} HOCH_2CH_2OH$$

Ethene   **Oxacyclopropane**   **1,2-Ethanediol**
(Ethylene)   **(Ethylene oxide)**   **(Ethylene glycol)**

1,2,3-Propanetriol (glycerol, glycerine), $HOCH_2CHOHCH_2OH$, is a viscous greasy substance, soluble in water, and nontoxic. It is obtained by alkaline hydrolysis of triglycerides, the major component of fatty tissue. The sodium and potassium salts of the long-alkyl-chain acids produced from fats ("fatty acids;" Chapter 19) are sold as soaps.

$$
\begin{array}{c}
\underset{\text{O}}{\overset{\text{O}}{\parallel}} \\
CH_2OCR \\
\mid \\
CHOCR \\
\mid \\
CH_2OCR
\end{array}
\xrightarrow{\text{H}_2\text{O, NaOH}}
\begin{array}{c}
CH_2OH \\
\mid \\
HCOH \\
\mid \\
CH_2OH
\end{array}
+ \; RCO^-Na^+
$$

Triglyceride   **1,2,3-Propanetriol**   **Soap**
("Fat")   **(Glycerol, glycerine)**

R = long alkyl chain

Phosphoric esters of 1,2,3-propanetriols (phosphoglycerides; Section 20-4) are primary components of cell membranes.

1,2,3-Propanetriol is used in lotions and other cosmetics, as well as in medicinal preparations. Treatment with nitric acid gives a trinitrate ester known as nitroglycerine, an extremely powerful explosive. The explosive potential of this substance results from its shock-induced, highly exothermic decomposition to gaseous products (N_2, CO_2, H_2O gas, O_2), raising temperatures to more than $3000°C$ and creating pressures higher than 2000 atmospheres in a fraction of a second.

Cholesterol is an important steroid alcohol (Section 4-7).

Ethoxyethane (diethyl ether) was at one time used as a general anesthetic. It produces unconsciousness by depressing central nervous system activity. Because of adverse effects such as irritation of the respiratory tract and extreme nausea, its use has been discontinued, and 1-methoxypropane (methyl propyl ether, "neothyl") and other compounds have replaced it in such applications. Ethoxyethane and other ethers are explosive when mixed with air.

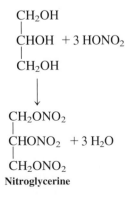

$$
\begin{array}{c}
CH_2OH \\
\mid \\
CHOH \\
\mid \\
CH_2OH
\end{array}
+ \; 3\,HONO_2
$$

$$\downarrow$$

$$
\begin{array}{c}
CH_2ONO_2 \\
\mid \\
CHONO_2 \\
\mid \\
CH_2ONO_2
\end{array}
+ \; 3\,H_2O
$$

Nitroglycerine

Oxacyclopropane (oxirane, ethylene oxide) is an industrial chemical intermediate (1996 U.S. production, 7.2 billion pounds [3.6 million tons]) and a fumigating agent for seeds and grains. In nature, oxacyclopropane derivatives control insect metamorphosis (see Chemical Highlight 12-2) and are formed in the course of enzyme-catalyzed oxidation of aromatic hydrocarbons, often leading to highly **carcinogenic** (cancer-causing) products (see Section 16-7).

Many *natural products,* some of which are quite active physiologically, contain alcohol and ether groups. For example, morphine is a powerful analgesic. Its synthetic acetate derivative, heroin, is a widely abused street drug. Tetrahydrocannabinol is the main active ingredient in marijuana *(Cannabis),* whose mood-altering effects have been known for thousands of years. In 1996, two states in the United States, California and Arizona, voted to legalize marijuana for medical purposes, on the basis of the finding that it relieves patients afflicted with cancer, AIDS, multiple sclerosis, and other diseases from the effects of nausea, pain, and loss of appetite.

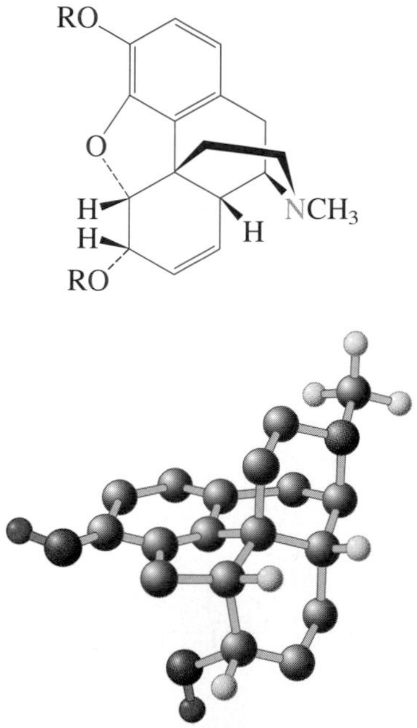

Morphine
(R = H)
Heroin

$$\left(R = \underset{\underset{O}{\parallel}}{C}CH_3 \right)$$

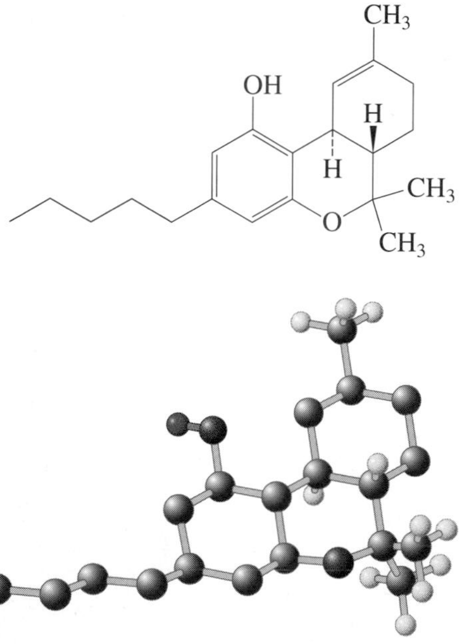

Tetrahydrocannabinol

The lower thiols and sulfides are most notorious for their foul smell. *Ethanethiol* is detectable by its odor even when diluted in 50 million parts of air. The major volatile components of the skunk's defensive spray are *3-methyl-1-butanethiol, trans-2-butene-1-thiol,* and *trans-2-butenyl methyl disulfide.*

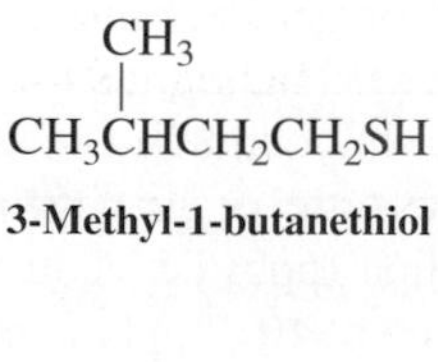

3-Methyl-1-butanethiol

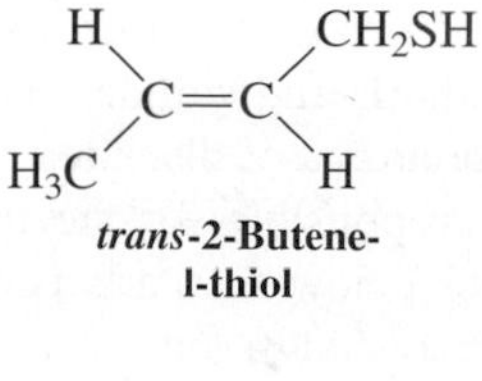

trans-**2-Butene-**
1-thiol

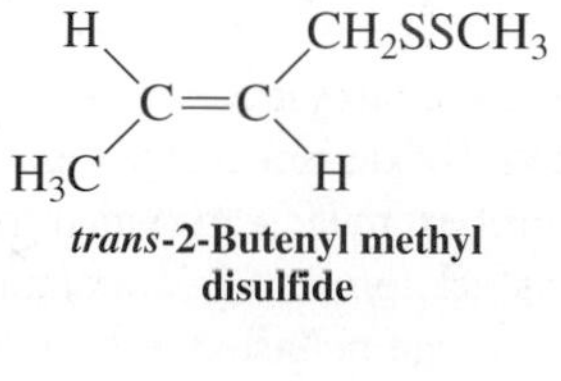

trans-**2-Butenyl methyl**
disulfide

CHEMICAL HIGHLIGHT 9-4   **Garlic and Sulfur**

What a culinary delight it is to augment your meal with the flavorful components of the genus *Allium:* garlic, onion, leeks, chives, scallions, and shallots! The odorants in all of these foods are based on the same element: sulfur. What is surprising is that, in many cases, the compounds giving rise to the desirable odor are not actually present in the intact plants but are biosynthesized on crushing, frying, or boiling the "starting material." For example, a clove of garlic does not itself smell, and uncut onions neither are flavorful nor bring tears to your eyes. In regard to garlic, crushing the clove releases so-called *allinase enzymes* that convert sulfoxide precursors into intermediate sulfenic acids, which subsequently dimerize with the loss of water to the flavorants, such as allicin. Garlic generates a host of other compounds in this way, all containing the functional groups of

these compounds as chemical warfare agents against invading organisms. In China, a significant reduction in gastric cancer risk has been noted to parallel the consumption of garlic. Garlic lowers cholesterol levels, it has cardiovascular properties, and it inhibits blood platelet aggregation. Among the most notable "negative" effects of garlic is bad breath, originating from the lungs by way of the blood and not, as you might have thought, from garlic traces in your mouth. Indeed, ingested garlic can persist in your urine for 3 to 4 days. Allicin is readily absorbed through the skin (a property that it shares with dimethyl sulfoxide). Thus, it has been claimed that rubbing garlic on the foot soon leads to the taste of garlic in the mouth, a claim confirmed by one of the authors of your text.

Component of intact garlic → (Allinase enzymes)

A sulfenic acid $\xrightarrow{-H_2O}$ **Allicin** (A flavorant)

The flavor of garlic, leeks, and onions is due to the extrusion of volatile sulfur compounds upon cutting.

sulfides, RSR′, sulfoxides, RSR′, and disulfides, RSSR′. Interestingly, some of these compounds are medicinally active. For example, allicin is a powerful antibacterial. Before modern antibiotics became available, garlic preparations were used in the treatment of typhus, cholera, dysentery, and tuberculosis. It is likely that the garlic plant uses

---

(R,S)-2-(4-Methyl-3-cyclohexenyl)-2-propanethiol

Strangely enough, when highly diluted, sulfur compounds can have a rather pleasant odor. For example, the smell of freshly chopped onions or garlic is due to the presence of low molecular weight thiols and sulfides (Chemical Highlight 9-4). Dimethyl sulfide is a component of the aroma of black tea. The compound 2-(4-methyl-3-cyclohexenyl)-2-propanethiol is responsible for the unique taste of grapefruit in which it is present in concentrations below the parts per billion (ppb; i.e., 1 in 10^9) range. It can be tasted at even lower concentrations, at a dilution of 10^{-4} ppb. In other words, you can notice the presence of 1 mg of this compound when it is dissolved in 10 million liters of water!

Many beneficial drugs contain sulfur in their molecular framework. Particularly well known are the *sulfonamides,* or *sulfa drugs,* powerful antibacterial agents (Section 15-11):

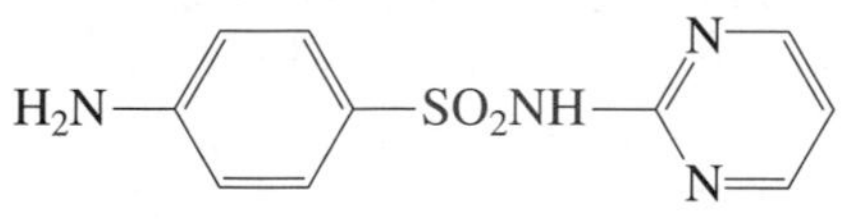

Sulfadiazine

(An antibacterial drug)

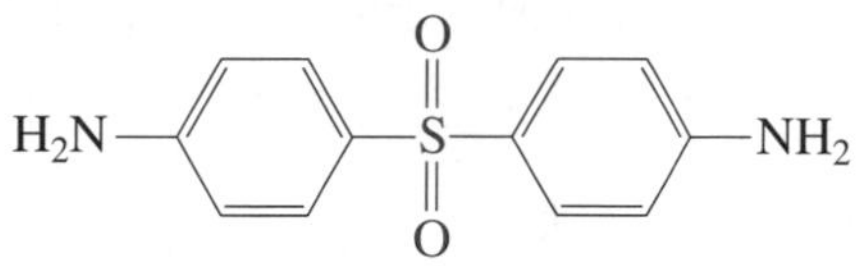

Diaminodiphenylsulfone

(An antileprotic drug)

To summarize, alcohols and ethers have various uses, both as chemical raw materials and as medicinal agents. Many of their derivatives can be found in nature; others are readily synthesized.

CHAPTER INTEGRATION PROBLEM

Treatment of alcohol A with acidic methanethiol gives sulfide B. Explain by a mechanism.

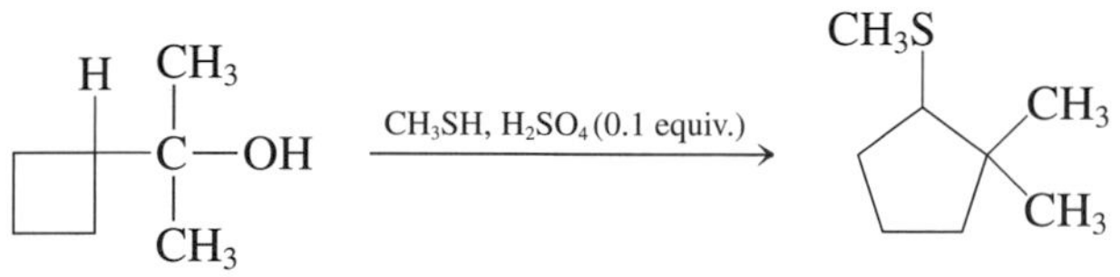

SOLUTION

This is an example of a mechanistic, rather than a synthetic, problem. In other words, to solve it, we cannot add any reagents as in a multistep synthetic sequence; what we see is what we have to work with. Let us take an inventory of what we have:

1. The (tertiary) alcohol function disappears and the (secondary) thioether unit (from CH_3SH) is introduced.
2. The four-membered ring turns into cyclopentyl.
3. The molecular formula of compound A, $C_7H_{14}O$, turns into that of compound B, $C_8H_{16}S$.

Focusing on the alkyl part attached to the respective functional group, we can rewrite these changes as $C_7H_{13}–OH \rightarrow C_7H_{13}–SCH_3$.

4. The reaction medium contains catalytic acid in the presence of a tertiary alcohol.

What can we conclude from this information? We have a carbocation rearrangement (Section 9-3) in which the strained cyclobutane ring undergoes expansion to a substituted cyclopentane. The initial carbocation must derive from a protonation–water loss sequence (Section 9-2), as applied to A, and product B must be formed through S_N1 capture of the rearranged cation by CH_3SH.

We can now begin to formulate a step-by-step description of these thoughts.

Step 1. The hydroxy group is protonated and leaves as H_2O.

Step 2. The tertiary carbocation undergoes ring expansion by alkyl shift (the migrating carbon is represented by a dot).

Step 3. The new carbocation is trapped by the relatively (compared with water) nucleophilic sulfur of CH_3SH, followed by proton loss to give product B (Section 9-10).

Visually, the most difficult step to follow in this sequence is step 2, because it consists of a fairly extreme topological change: The migrating carbon "drags" with it the appended chain, which is part of the ring. A good way of eliminating confusion is to label the "action pieces" in your molecule, as was done in the scheme for step 2, and to keep in mind the "bare bones" of an alkyl (or H) shift. Thus, only three key atoms take part: the cationic center, which will receive the migrating group; the neighboring carbon, which will become charged; and the migrating atom. A simple aid in remembering the basic feature of a carbocation rearrangement is the slogan: "The charge and the migrating center trade places."

Finally, you will have noted that rearrangement step 2 converts a tertiary into a secondary carbocation. The driving force is the release of ring strain in going from a four- (26.3 kcal mol^{-1} strain) to a five-membered ring (6.5 kcal mol^{-1} strain; Section 4-2). In our particular case, the secondary carbocation is trapped by the highly nucleophilic sulfur of the thiol before it rearranges further by methyl migration to the tertiary counterpart, a potential source of other products, which are not observed.

NEW REACTIONS

1. Alkoxides from Alcohols (Sections 8-3, 9-1)

Using strong bases

$$ROH \quad \underset{}{\overset{\text{Strong base}}{\rightleftharpoons}} \quad RO^-$$

Examples of strong bases: $Li^+ {}^- N[CH(CH_3)_2]_2$; $CH_3CH_2CH_2CH_2Li$; K^+H^-

Using alkali metals

$$ROH + M \longrightarrow RO^- M^+ + \tfrac{1}{2}H_2$$

M = Li, Na, K

Substitution Reactions of Alcohols

2. Using Hydrogen Halides (Sections 8-3, 9-2, 9-3)

Primary ROH $\xrightarrow{\text{Conc. HX}}$ RX

X = Br or I (S_N2 mechanism)

Secondary or tertiary ROH $\xrightarrow{\text{Conc. HX}}$ RX

X = Cl, Br, or I (S_N1 mechanism)

3. Using Phosphorus Reagents (Section 9-4)

$$3\ ROH + PBr_3 \longrightarrow 3\ RBr + H_3PO_3$$
$$6\ ROH + 2\ P + 3\ I_2 \longrightarrow 6\ RI + 2\ H_3PO_3$$

S_N2 mechanism with primary and secondary ROH.
Less likelihood of carbocation rearrangements than with HX

4. Using Sulfur Reagents (Section 9-4)

$$ROH + SOCl_2 \xrightarrow{\text{N(CH}_2\text{CH}_3)_3} RCl + SO_2 + (CH_3CH_2)_3\overset{+}{N}H\ Cl^-$$

$$ROH + R'SO_2Cl \longrightarrow ROSO_2R' \xrightarrow{\text{Nu}^-,\ \text{DMSO}} RNu + R'SO_3^-$$

Alkyl
sulfonate

Carbocation Rearrangements in Alcohols

5. Carbocation Rearrangements by Alkyl and Hydride Shifts (Section 9-3)

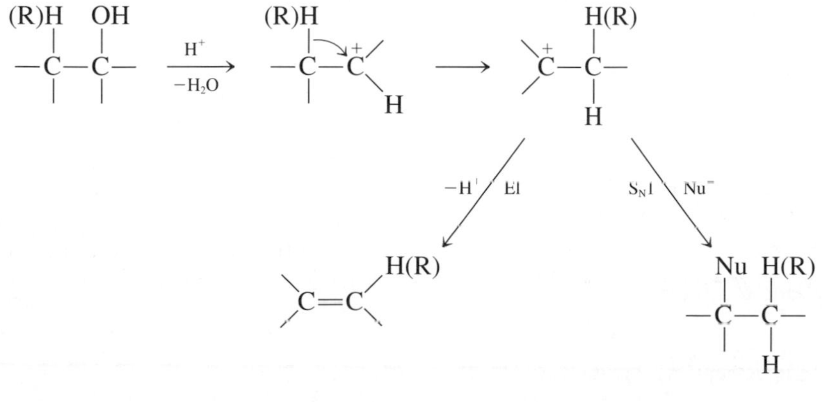

6. Concerted Alkyl Shifts from Primary Alcohols (Section 9-3)

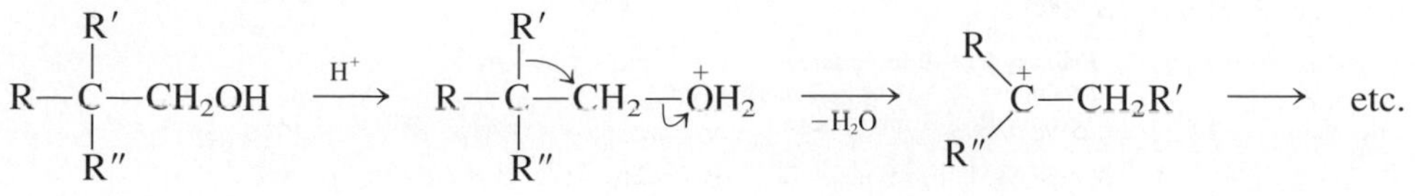

Elimination Reactions of Alcohols

7. Dehydration with Strong Nonnucleophilic Acid (Sections 9-2, 9-3, 9-7, 11-9)

$$\underset{\substack{\text{H OH} \\ | \ \ | \\ -\text{C}-\text{C}- \\ | \ \ |}}{} \xrightarrow{\text{H}_2\text{SO}_4,\ \text{heat}} \quad \text{C}=\text{C} \quad + \quad \text{H}_2\text{O}$$

Carbocation rearrangements may occur

Temperature required
Primary ROH: $170°–180°C$ (E_2 mechanism)
Secondary ROH: $100°–140°C$ (usually E1)
Teritary ROH: $25°–80°C$ (E1 mechanism)

Preparation of Ethers

8. Williamson Synthesis (Section 9-6)

$$\text{ROH} \xrightarrow{\text{NaH, DMSO}} \text{RO}^-\text{Na}^+ \xrightarrow[\text{S}_\text{N}2]{\text{R}'\text{X, DMSO}} \text{ROR}'$$

R′ must be methyl or primary
ROH can be primary or secondary (tertiary alkoxides usually give E2 products, unless R′ = methyl)
Ease of intramolecular version forming cyclic ethers: $k_3 \geq k_5 > k_6 > k_4 \geq k_7 > k_8$
(k_n = reaction rate, n = ring size)

9. Mineral Acid Method (Section 9-7)

Primary alcohols

$$\text{RCH}_2\text{OH} \xrightarrow{\text{H}^+,\ \text{low temperature}} \text{RCH}_2\overset{+}{\text{OH}}_2 \xrightarrow[-\text{H}_2\text{O}]{\text{RCH}_2\text{OH, }130°–140°C} \text{RCH}_2\text{OCH}_2\text{R}$$

Secondary alcohols

$$\underset{\substack{\text{OH} \\ | \\ \text{RCHR}}}{} \xrightarrow[-\text{H}_2\text{O}]{\text{H}^+} \underset{\substack{\text{R} \\ \\ \text{R}}}{\text{CH}}-\text{O}-\underset{\substack{\text{R} \\ \\ \text{R}}}{\text{CH}} \quad + \quad \text{E1 products}$$

Tertiary alcohols

$$\text{R}_3\text{COH} \quad + \quad \text{R}'\text{OH} \xrightarrow[\text{S}_\text{N}1,\ -\text{H}_2\text{O}]{\text{NaHSO}_4,\ \text{H}_2\text{O}} \text{R}_3\text{C}-\text{OR}' \quad + \quad \text{E1 products}$$

R′ = (mainly) primary

Reactions of Ethers

10. Cleavage by Hydrogen Halides (Section 9-8)

$$\text{ROR} \xrightarrow{\text{Conc. HX}} \text{RX} + \text{ROH} \xrightarrow{\text{Conc. HX}} 2\,\text{RX}$$

X = Br or I
Primary R: $S_\text{N}2$ mechanism
Secondary R: $S_\text{N}1$ or $S_\text{N}2$ mechanism
Tertiary R: $S_\text{N}1$ mechanism

11. Nucleophilic Opening of Oxacyclopropanes (Sections 9-9, 25-2)

Anionic nucleophiles

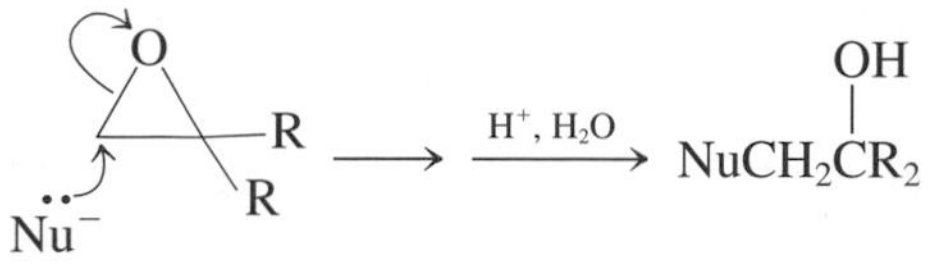

Examples of Nu⁻: HO⁻, RO⁻, RS⁻

Acid-catalyzed opening

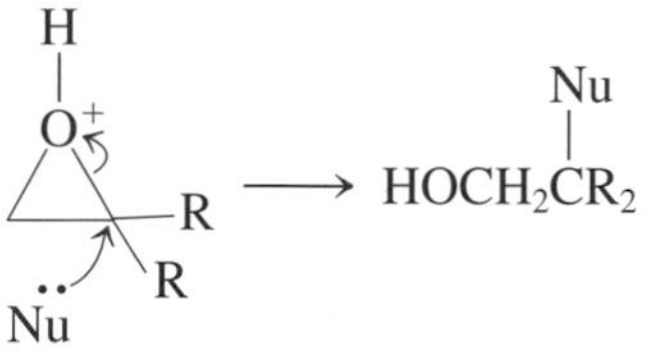

Examples of Nu: H_2O, ROH, halide

12. Nucleophilic Opening of Oxacyclopropane by Lithium Aluminum Hydride (Section 9-9)

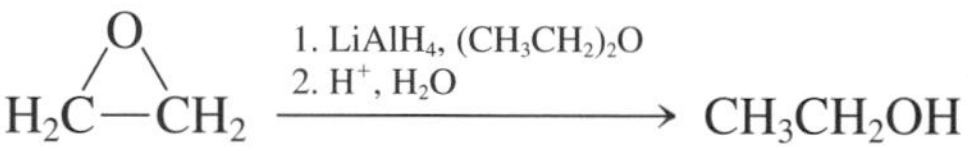

13. Nucleophilic Opening of Oxacyclopropane by Organometallic Compounds (Section 9-9)

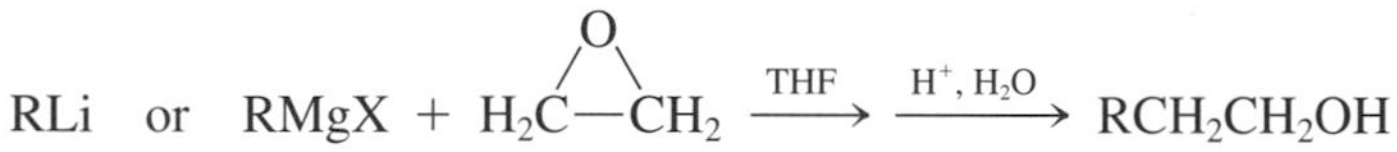

Sulfur Compounds

14. Preparation of Thiols and Sulfides (Section 9-10)

$$RX \ + \ HS^- \ \longrightarrow \ RSH$$
$$\text{Excess} \qquad \qquad \text{Thiol}$$

$$RSH \ + \ R'X \ \xrightarrow{\text{Base}} \ RSR'$$
$$\text{Alkyl sulfide}$$

15. Acidity of Thiols (Section 9-10)

$$RSH \ + \ HO^- \ \rightleftharpoons \ RS^- \ + \ H_2O \qquad pK_a(RSH) = 9-12$$
Acidity of RSH > H_2O ~ ROH

16. Nucleophilicity of Sulfides (Section 9-10)

$$R_2\ddot{S} \ + \ R'X \ \longrightarrow \ R_2\overset{+}{S}R' \ X^-$$
$$\text{Sulfonium salt}$$

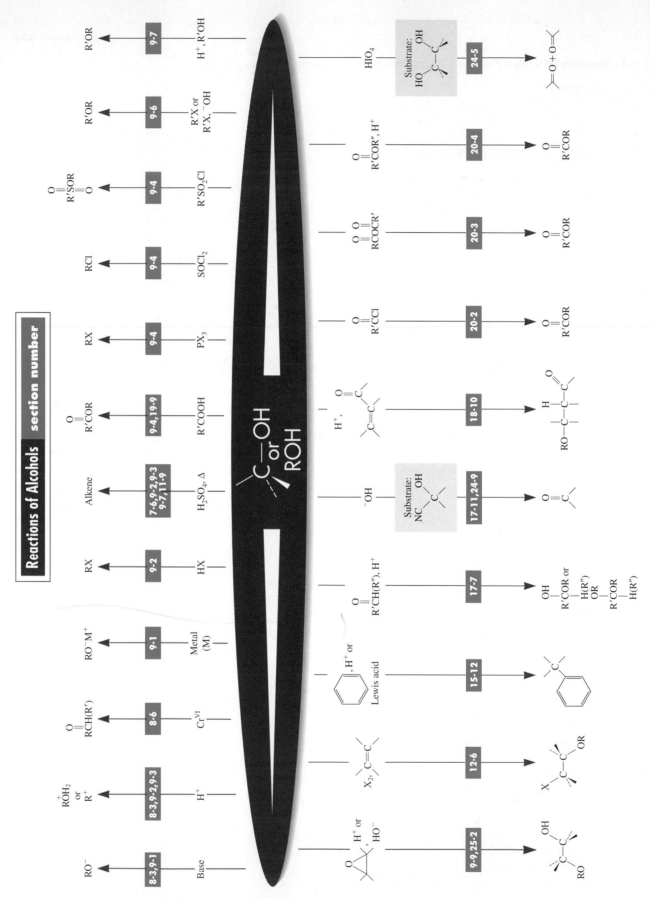

Reactions of Alcohols   section number

17. Oxidation of Thiols (Section 9-10)

$$RSH \xrightarrow{\text{KMnO}_4 \text{ or } \text{H}_2\text{O}_2} RSO_3H$$

Alkanesulfonic acid

$$RSH \underset{\text{Li, liquid NH}_3}{\overset{\text{I}_2}{\rightleftharpoons}} RS\!-\!SR$$

Dialkyl disulfide

18. Oxidation of Sulfides (Section 9-10)

$$RSR' \xrightarrow{\text{H}_2\text{O}_2} \overset{\displaystyle O}{\overset{\displaystyle \|}{R\ddot{S}R'}} \xrightarrow{\text{H}_2\text{O}_2} \overset{\displaystyle O}{\underset{\displaystyle O}{\overset{\displaystyle \|}{\underset{\displaystyle \|}{RSR'}}}}$$

Dialkyl sulfoxide   **Dialkyl sulfone**

IMPORTANT CONCEPTS

1. The reactivity of ROH with alkali metals to give **alkoxides** and hydrogen follows the order R = CH_3 > primary > secondary > tertiary.

2. In the presence of acid and a nucleophilic counterion, primary alcohols undergo S_N2 reactions. Secondary and tertiary alcohols tend to form **carbocations** in the presence of acid, capable of **E1** and **S_N1** product formation, before and after **rearrangement.**

3. **Carbocation rearrangements** take place by **hydride** and **alkyl** group **shifts.** They usually result in interconversion of secondary carbocations or conversion of a secondary into a tertiary carbocation. Primary **alkyloxonium** ions can rearrange by a concerted process consisting of loss of water and simultaneous hydride or alkyl shift to give secondary or tertiary carbocations.

4. Synthesis of primary and secondary haloalkanes can be achieved with less risk of rearrangement by methods using **inorganic esters.**

5. Ethers are prepared by either the **Williamson ether synthesis** or by reaction of alcohols with strong nonnucleophilic acids. The first method is best when S_N2 reactivity is high. In the latter case, elimination (dehydration) is a competing process at higher temperatures.

6. **Crown ethers** and **cryptands** are examples of **ionophores,** polyethers that coordinate around metal ions, thus rendering them soluble in hydrophobic media.

7. Whereas nucleophilic **ring opening of oxacyclopropanes** by anions is at the less substituted ring carbon according to the rules of the S_N2 reaction, acid-catalyzed opening favors the more substituted carbon, because of charge control of nucleophilic attack.

8. Sulfur has more diffuse orbitals than does oxygen. In **thiols,** the S–H bond is less polarized than the O–H bond in alcohols, thus leading to **diminished hydrogen bonding.** Because the S–H bond is also weaker than the O–H bond, the **acidity** of thiols is **greater** than that of alcohols.

9. **Note on color use:** Throughout the main parts of the text, beginning in Chapter 6, reacting species in mechanisms and most examples of new transformations are color coded **red** for **nucleophiles, blue** for **electrophiles,** and **green** for **leaving groups.** Color coding is *not* used in exercises, summaries of new reactions, or chapter-end problems.

PROBLEMS

23. On which side of the equation do you expect each of the following equilibria to lie (left or right)?

(a) $(CH_3)_3COH + K^{+\,-}OH \rightleftharpoons (CH_3)_3CO^-K^+ + H_2O$

(b) $CH_3OH + NH_3 \rightleftharpoons CH_3O^- + NH_4^+$ ($pK_a = 9.2$)

(c) 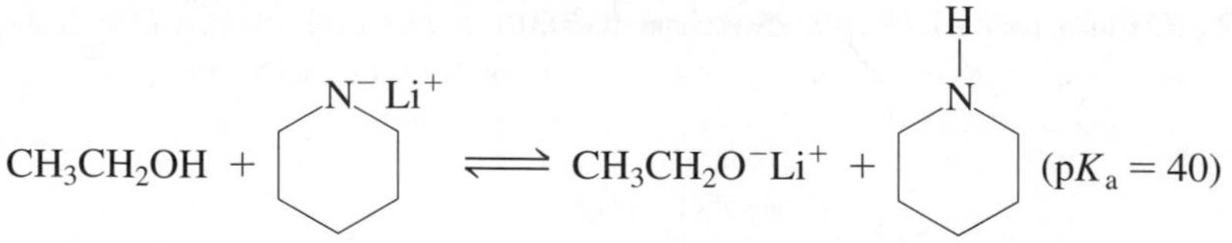

(d) $NH_3(pK_a = 35) + Na^+H^- \rightleftharpoons Na^{+\,-}NH_2 + H_2\,(pK_a \sim 38)$

24. Give the expected major product of each of the following reactions.

(a) $CH_3CH_2CH_2OH \xrightarrow{\text{Conc. HI}}$ **(b)** $(CH_3)_2CHCH_2CH_2OH \xrightarrow{\text{Conc. HBr}}$

(c) 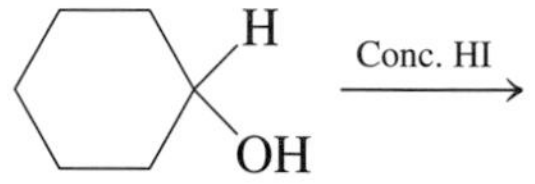 **(d)** $(CH_3CH_2)_3COH \xrightarrow{\text{Conc. HCl}}$

25. For each of the following alcohols, write the structure of the alkyloxonium ion produced after protonation by strong acid; if the alkyloxonium ion is capable of losing water readily, write the structure of the resulting carbocation; and, if the carbocation obtained is likely to be susceptible to rearrangement, write the structures of all new carbocations that might be reasonably expected to form.

(a) $CH_3CH_2CH_2OH$ **(b)** $CH_3CH(OH)CH_3$ **(c)** $CH_3CH_2CH_2CH_2OH$

(d) $(CH_3)_2CHCH_2OH$ **(e)** $(CH_3)_3CCH_2CH_2OH$ **(f)** 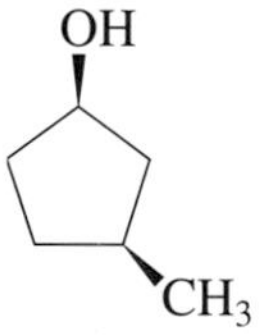

(g) 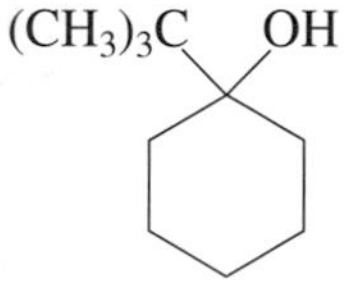 **(h)** 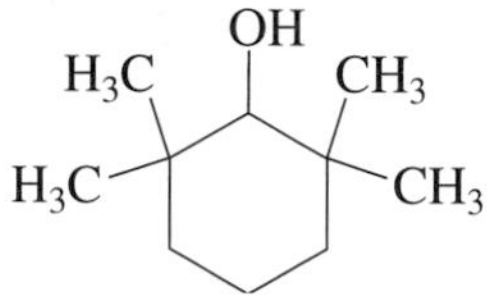

26. Write all products of the reaction of each of the alcohols in Problem 25 with concentrated H_2SO_4 under elimination conditions.

27. Write all sensible products of the reaction of each of the alcohols in Problem 25 with concentrated aqueous HBr.

28. Give detailed mechanisms and final products for the reaction of 3-methyl-2-pentanol with each of the reagents that follow.

(a) NaH **(b)** Concentrated HBr
(c) PBr_3 **(d)** $SOCl_2$
(e) Concentrated H_2SO_4 at 130°C **(f)** Dilute H_2SO_4 in $(CH_3)_3COH$

29. Primary alcohols are often converted into bromides by reaction with NaBr in H_2SO_4. Explain how this transformation works and why it might be considered a superior method to that using concentrated aqueous HBr.

$$CH_3CH_2CH_2CH_2OH \xrightarrow{NaBr, H_2SO_4} CH_3CH_2CH_2CH_2Br$$

30. What are the most likely product(s) of each of the following reactions?

(a) 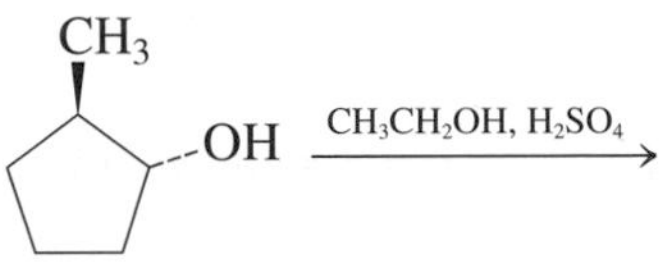 —OH $\xrightarrow{CH_3CH_2OH, H_2SO_4}$

(b) 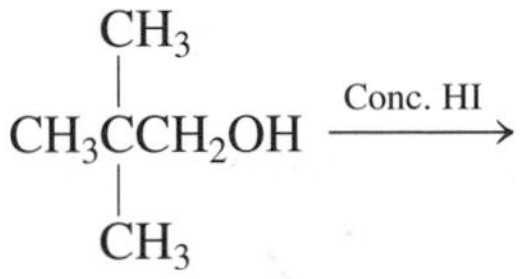 CH_3CCH_2OH (with CH_3 groups) $\xrightarrow{Conc.\ HI}$

(c) cyclohexane-CH_2OH $\xrightarrow{Conc.\ H_2SO_4,\ 180°C}$

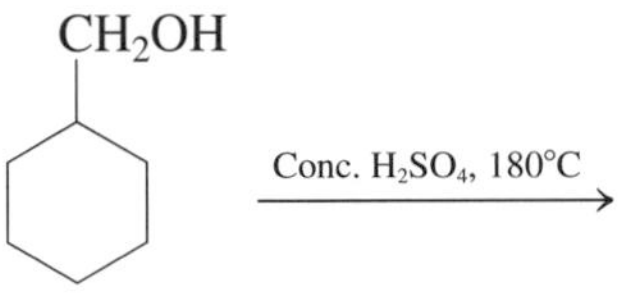

(d) 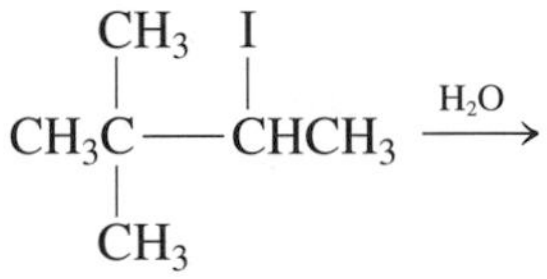 $CH_3C(CH_3)(CH_3)—CHICH_3$ $\xrightarrow{H_2O}$

31. Give the expected main product of the reaction of each of the alcohols in Problem 25 with PBr_3. Compare the results with those of Problem 27.

32. Give the expected product(s) of the reaction of 1-pentanol with each of the following reagents.
  (a) $K^{+-}OC(CH_3)_3$
  (b) Sodium metal
  (c) CH_3Li
  (d) Concentrated HI
  (e) Concentrated HCl
  (f) FSO_3H
  (g) Concentrated H_2SO_4 at 130°C
  (h) Concentrated H_2SO_4 at 180°C
  (i) $CH_3SO_2Cl, (CH_3CH_2)_3N$
  (j) PBr_3
  (k) $SOCl_2$
  (l) $K_2Cr_2O_7 + H_2SO_4 + H_2O$
  (m) PCC, CH_2Cl
  (n) $(CH_3)_3COH + H_2SO_4$ (as catalyst)

33. Give the expected product(s) of the reaction of *trans*-3-methylcyclopentanol with each of the reagents in Problem 32.

34. Suggest a good synthetic method for preparing each of the following haloalkanes from the corresponding alcohols.

(a) $CH_3CH_2CH_2Cl$ **(b)** $CH_3CH_2CHCH_2Br$ (with CH_3) **(c)** 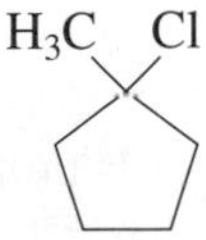 **(d)** $CH_3CHCH(CH_3)_2$ (with I)

35. Name each of the following molecules according to IUPAC.

(a) $(CH_3)_2CHOCH_2CH_3$ **(b)** $CH_3OCH_2CH_2OH$ **(c)**

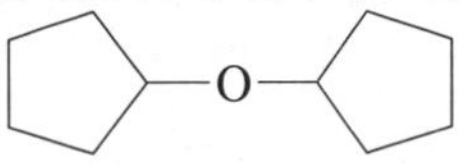

(d) $(ClCH_2CH_2)_2O$ **(e)** 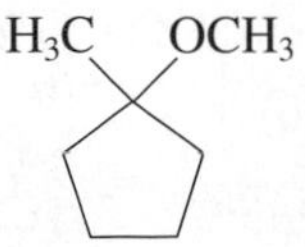 **(f)** 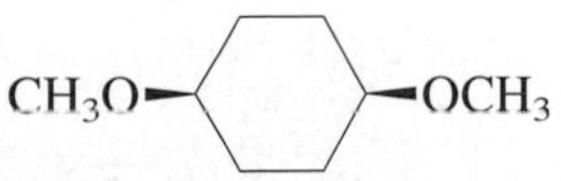 CH_3O——OCH_3 **(g)** CH_3OCH_2Cl

36. Explain why the boiling points of ethers are lower than those of the isomeric alcohols. Would you expect the relative water solubilities to differ in a similar way?

37. Suggest the best syntheses for each of the following ethers. Use alcohols or haloalkanes or both as your starting materials.

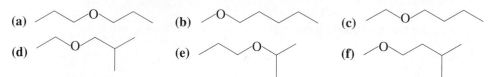

(a) **(b)** **(c)**

(d) **(e)** **(f)**

38. Write the expected major product(s) of each of the following attempted ether syntheses.

(a) $CH_3CH_2CH_2Cl$ + $CH_3CH_2\overset{\displaystyle O^-}{\underset{|}{C}}HCH_2CH_3$ $\xrightarrow{\text{DMSO}}$

(b) $CH_3CH_2CH_2O^-$ + $CH_3CH_2\overset{\displaystyle Cl}{\underset{|}{C}}HCH_2CH_3$ $\xrightarrow{\text{HMPA}}$

(c) $\overset{H_3C \quad O^-}{\bigcirc}$ + CH_3I $\xrightarrow{\text{DMSO}}$

(d) $(CH_3)_2CHO^-$ + $(CH_3)_2CHCH_2CH_2Br$ $\xrightarrow{(CH_3)_2CHOH}$

(e) $\bigcirc\overset{H}{\underset{O^-}{}}$ + $\overset{H}{\underset{Cl}{}}\bigcirc$ $\xrightarrow{\text{Cyclohexanol}}$

(f) $\bigcirc\overset{CH_3}{\underset{}{C}}-O^-$ + CH_3CH_2I $\xrightarrow{\text{DMSO}}$

39. For each synthesis proposed in Problem 38 that is not likely to give a good yield of ether product, suggest an alternative synthesis beginning with suitable alcohols or haloalkanes that will give a superior result. (**Hint:** See Problem 16 in Chapter 7.)

40. Write the product(s) of reaction of each of the following molecules with NaOH in dilute solution in DMSO.

(a) $CH_3\overset{OH}{\underset{|}{C}}HCH_2CH_2\overset{Cl}{\underset{|}{C}}H_2$ **(b)** **(c)**

41. Propose efficient syntheses for each of the following ethers, using haloalkanes or alcohols as starting materials.

(a) $CH_3CH_2CHOCH_2CH_3$ (with CH_3 on the central carbon)

(b) 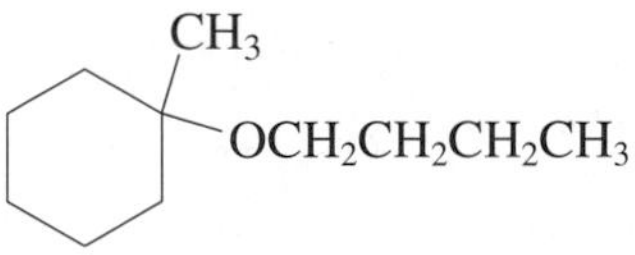 [cyclohexane with CH_3 and $OCH_2CH_2CH_2CH_3$ substituents]

(c) 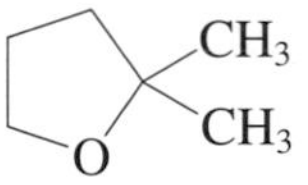 [tetrahydrofuran ring with two CH_3 groups at the 2-position]

(d) 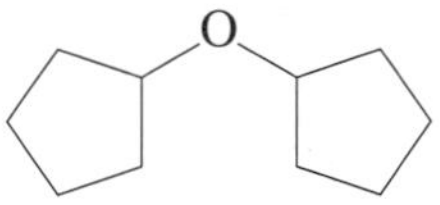 [two cyclopentyl groups joined by O]

42. Give the major product(s) of each of the following reactions.

(a) $CH_3CH_2OCH_2CH_2CH_3$ $\xrightarrow{\text{Excess conc. HI}}$

(b) $CH_3OCH(CH_3)_2$ $\xrightarrow{\text{Excess conc. HBr}}$

(c) $CH_3OCH_2CH_2OCH_3$ $\xrightarrow{\text{Excess conc. HI}}$

(d) 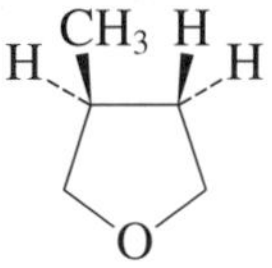 [oxolane ring with H, CH_3, H, H substituents] $\xrightarrow{\text{Excess conc. HBr}}$

(e) 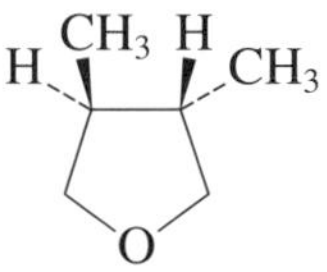 [oxolane ring with H, CH_3, H, CH_3 substituents] $\xrightarrow{\text{Excess conc. HBr}}$

(f) 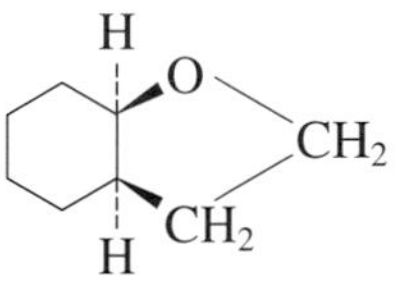 [bicyclic oxacyclopropane fused to cyclohexane with CH_2, CH_2] $\xrightarrow{\text{Excess conc. HBr}}$

43. Give the expected major product of reaction of 2,2-dimethyloxacyclopropane with each of the following reagents.
(a) Dilute H_2SO_4 in CH_3OH (b) Na^+ $^-OCH_3$ in CH_3OH (c) Dilute, aqueous HBr
(d) Concentrated HBr   (e) CH_3MgI, then H^+, H_2O (f) C_6H_5Li, then H^+, H_2O

44. Give the major product(s) of each of the following reactions. (**Hint:** The strained oxacyclobutanes react like oxacyclopropanes.)

(a)  [oxacyclopropane] $\xrightarrow{\text{Na}^+\text{NH}_2, \text{NH}_3}$

(b) 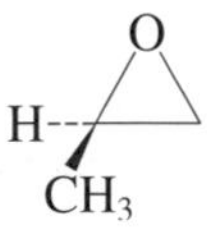 H--[oxacyclopropane with CH_3] $\xrightarrow{\text{Na}^+\text{SCH}_2CH_3, \text{CH}_3CH_2OH}$

(c) 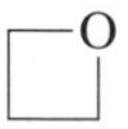 [oxacyclobutane] $\xrightarrow{\text{Excess conc. HBr}}$

(d) 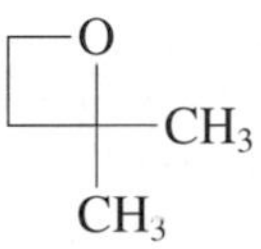 [oxacyclobutane with CH_3, CH_3] $\xrightarrow{\text{Dilute HCl in CH}_3\text{OH}}$

(e) 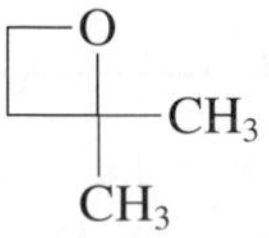 [oxacyclobutane with CH_3, CH_3] $\xrightarrow{\text{Na}^+ {}^-\text{OCH}_3 \text{ in CH}_3\text{OH}}$

(f) 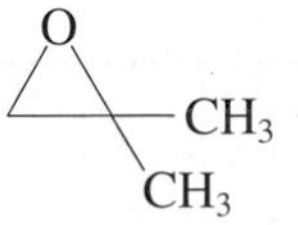 [oxacyclopropane with CH_3, CH_3] $\xrightarrow{\text{1. LiAlD}_4, (CH_3CH_2)_2O \quad 2. H^+, H_2O}$

(g) [oxacyclopropane with CH_3, CH_3] $\xrightarrow{\text{1. (CH}_3)_2\text{CHMgCl, (CH}_3\text{CH}_2)_2\text{O} \quad 2. H^+, H_2O}$

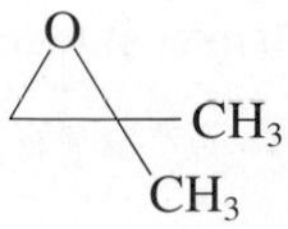

(h) [oxacyclopropane] $\xrightarrow{\text{1. [cyclopentyl]--Li, (CH}_3\text{CH}_2)_2\text{O} \quad 2. H^+, H_2O}$

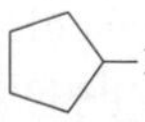

45. For each alcohol in Problem 40 of Chapter 8, suggest a synthetic route that starts with an oxacyclopropane (if possible).

46. Give the major product(s) expected from each of the reactions shown below. Watch stereochemistry.

(a) 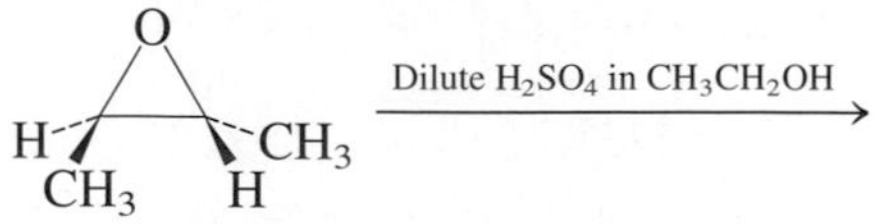 $\xrightarrow{\text{Dilute } H_2SO_4 \text{ in } CH_3CH_2OH}$ (b) 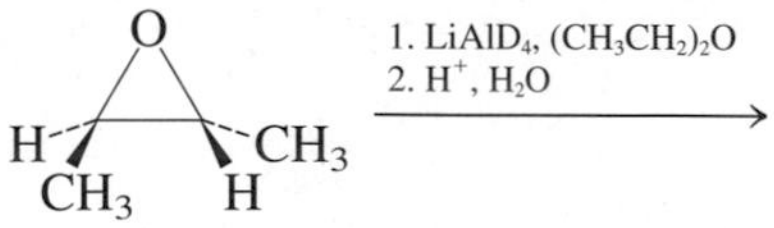 $\xrightarrow[\text{2. } H^+, H_2O]{\text{1. LiAlD}_4, (CH_3CH_2)_2O}$

47. Name each of the following compounds according to IUPAC.

(a) 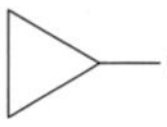 $-CH_2SH$ (b) $CH_3CH_2\overset{\overset{\displaystyle CH_3}{|}}{C}HSCH_3$ (c) $CH_3CH_2CH_2SO_3H$ (d) CF_3SO_2Cl

48. In each of the following pairs of compounds, indicate which is the stronger acid and which is the stronger base. (a) CH_3SH, CH_3OH; (b) HS^-, HO^-; (c) H_3S^+, H_2S.

49. Give reasonable products for each of the following reactions.

(a) $ClCH_2CH_2CH_2CH_2Cl$ $\xrightarrow{\text{One equivalent Na}_2S}$

(b) 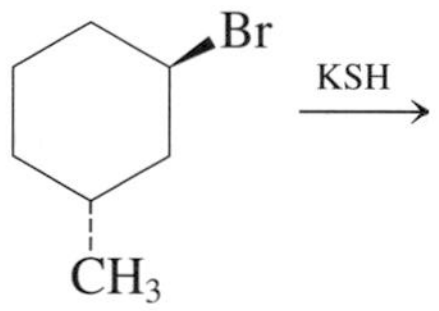 $\xrightarrow{\text{KSH}}$ (c) 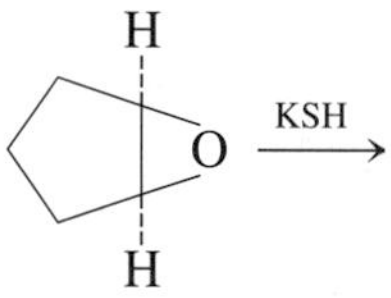 $\xrightarrow{\text{KSH}}$ (d) 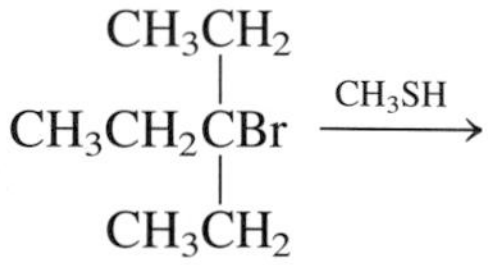 $CH_3CH_2\overset{\overset{\displaystyle CH_3CH_2}{|}}{\underset{\underset{\displaystyle CH_3CH_2}{|}}{C}}Br$ $\xrightarrow{CH_3SH}$

(e) 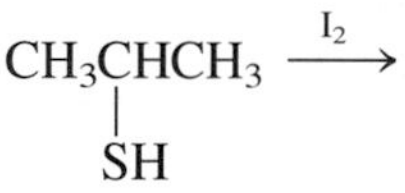 $CH_3\overset{\overset{\displaystyle }{|}}{\underset{\underset{\displaystyle SH}{|}}{C}}HCH_3$ $\xrightarrow{I_2}$ (f) 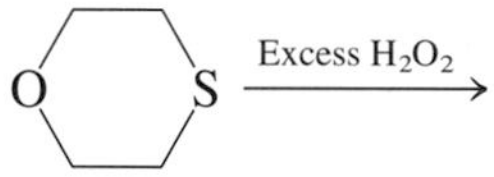 $\xrightarrow{\text{Excess } H_2O_2}$

50. Give the structures of compounds A, B, and C (with stereochemistry) from the information in the following scheme. (**Hint:** A is acyclic.) To what compound class does the product belong?

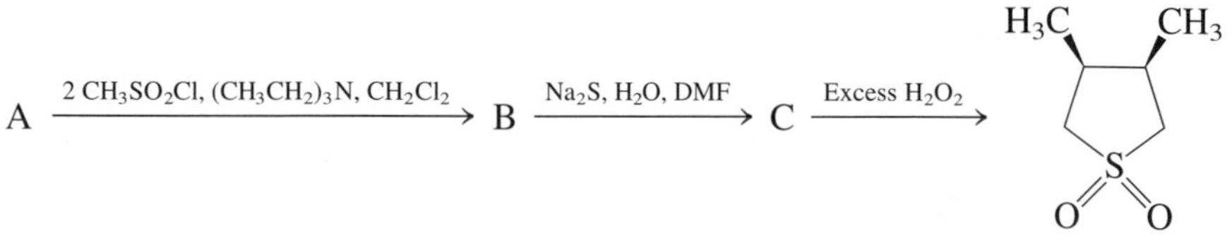

A $\xrightarrow{2\ CH_3SO_2Cl,\ (CH_3CH_2)_3N,\ CH_2Cl_2}$ B $\xrightarrow{Na_2S,\ H_2O,\ DMF}$ C $\xrightarrow{\text{Excess } H_2O_2}$

$C_6H_{14}O_2$ $C_8H_{18}S_2O_6$ $C_6H_{12}S$

51. In an attempt to make 1-chloro-1-cyclobutylpentane, the following reaction sequence was employed. The actual product isolated, however, was not the desired molecule but an isomer of it. Suggest a structure for the product and give a mechanistic explanation for its formation. (**Hint:** See Chapter Integration Problem.)

52. What is the product of the reaction shown in the margin? (Pay attention to stereochemistry at the reacting centers.) What is the kinetic order of this reaction?

53. Propose syntheses of the following molecules, choosing reasonable starting materials on the basis of the principles of synthetic strategy introduced in preceding chapters, particularly in Section 8-9. Suggested positions for carbon–carbon bond formation are indicated by wavy lines.

$CH_3CH_2CH \dashv CH_2CH_2SO_3H$

(a)

(b) $CH_3CH_2CH_2 \dashv C \dashv CHO$ with CH_3 and CH_2CH_3 substituents

54. Give efficient syntheses of each of the following compounds, beginning with the indicated starting material.
(a) *trans*-1-Bromo-2-methylcyclopentane, from *cis*-2-methylcyclopentanol
(b) 3-Cyanopentane, from 3-pentanol
(c) 3-Chloro-3-methylhexane, from 3-methyl-2-hexanol

(d) , from 2-bromoethanol (two equivalents)

55. Compare the following methods of alkene synthesis from a general primary alcohol. State the advantages and disadvantages of each one.

56. Sugars, being polyhydroxylic compounds (Chapter 24), undergo reactions characteristic of alcohols. In one of the later steps in glycolysis (the metabolism of glucose), one of the glucose metabolites with a remaining hydroxy group, 2-phosphoglyceric acid, is converted into 2-phosphoenolpyruvic acid. This reaction is catalyzed by the enzyme enolase in the presence of a Lewis acid such as Mg^{2+}. **(a)** How would you classify this reaction? **(b)** What is the possible role of the Lewis acidic metal ion?

$$HOCH_2-\underset{\underset{OPO_3{}^{2-}}{|}}{CH}-COOH \xrightarrow{\text{Enolase, Mg}^{2+}} CH_2=C\underset{CO_2H}{\overset{OPO_3{}^{2-}}{<}}$$

2-Phospho-
glyceric acid

2-Phosphoenol-
pyruvic acid

57. The formidable-looking molecule 5-methyltetrahydrofolic acid (abbreviated 5-methyl-FH$_4$) is the product of sequences of biological reactions that convert carbon atoms from a variety of simple molecules, such as formic acid and the amino acid histidine, into methyl groups.

5-Methyltetrahydrofolic acid
(5-Methyl-FH₄)

The simplest synthesis of 5-methyltetrahydrofolic acid is from tetrahydrofolic acid (FH$_4$) and trimethylsulfonium ion, a reaction carried out by microorganisms in the soil.

FH₄   Trimethylsulfonium ion   5-Methyl-FH₄

(a) Can this reaction be reasonably assumed to proceed through a nucleophilic substitution mechanism? Write the mechanism, using the "electron pushing" arrow notation. **(b)** Identify the nucleophile, the nucleophilic and electrophilic atoms participating in the reaction, and the leaving group. **(c)** On the basis of the concepts presented in Sections 6-8, 6-9, 9-2, and 9-9, are all the groups that you identified in (b) behaving in a reasonable way in this reaction? Does it help to know that species such as H$_3$S$^+$ are very strong acids (e.g., pK_a of CH$_3$SH$_2{}^+$ is -7)?

58. The role of 5-methyl-FH$_4$ (Problem 57) in biology is to serve as a donor of methyl groups to small molecules. The synthesis of the amino acid methionine from homocysteine is perhaps the best-known example.

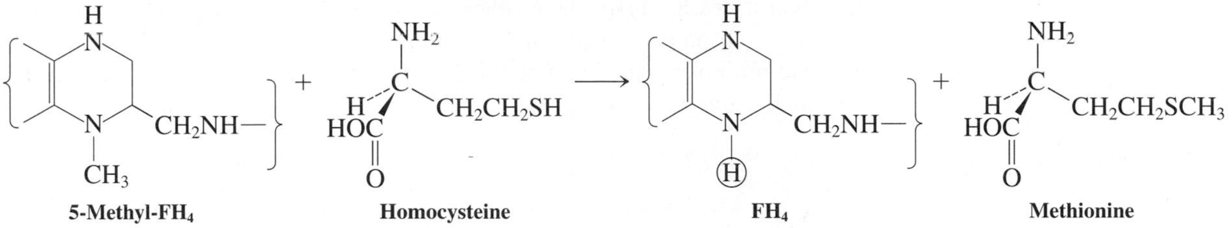

5-Methyl-FH₄   **Homocysteine**   **FH₄**   **Methionine**

For this problem, answer the same questions that were posed in Problem 57. The pK_b of the circled hydrogen in FH₄ is 5. Does this cause a problem with any feature of your mechanism? In fact, methyl transfer reactions of 5-methyl-FH₄ require a proton source. Review the material in Section 9-2, especially the subsection titled "Haloalkanes from primary alcohols and HX." Then suggest a useful role for a proton in the reaction illustrated here.

59. Epinephrine (adrenalin) is produced in your body in a two-step process that accomplishes the transfer of a methyl group from methionine (Problem 58) to norepinephrine (see reactions 1 and 2 below). **(a)** Explain in detail what is going on mechanistically in these two reactions, and analyze the role played by the molecule of ATP. **(b)** Would you expect methionine to react directly with norepinephrine? Explain. **(c)** Propose a laboratory synthesis of epinephrine from norepinephrine.

REACTION 1

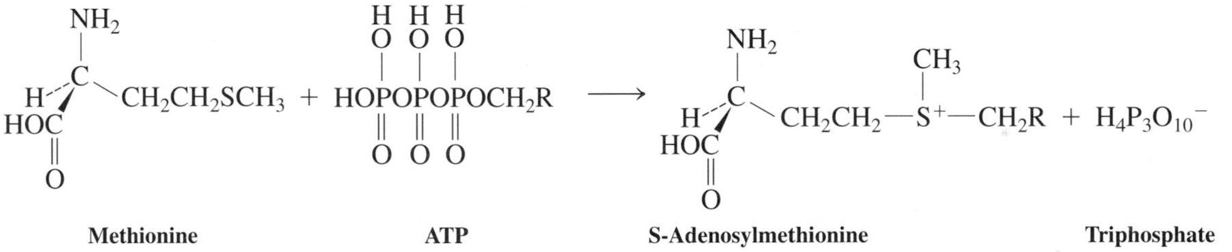

Methionine   **ATP**   **S-Adenosylmethionine**   **Triphosphate**

REACTION 2

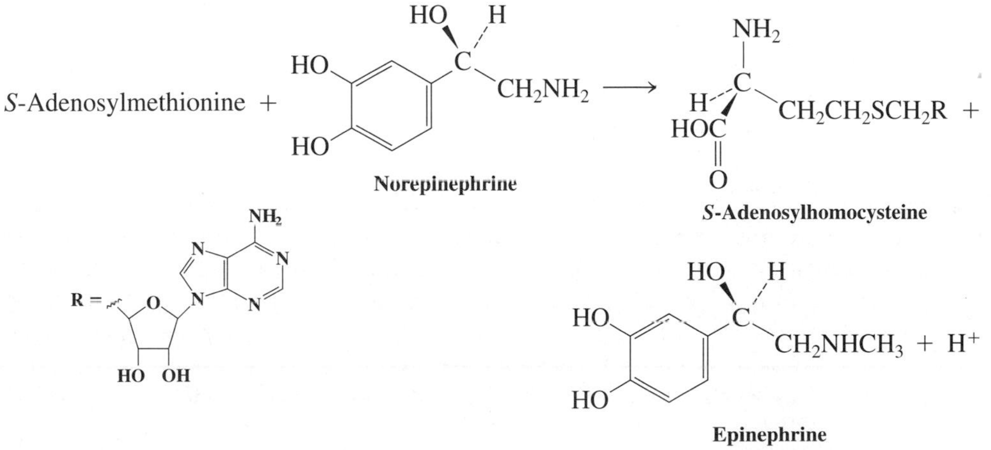

S-Adenosylmethionine +

Norepinephrine   **S-Adenosylhomocysteine**

Epinephrine

60. (a) Only the trans isomer of 2-bromocyclohexanol can react with sodium hydroxide to form an oxacyclopropane-containing product. Explain the lack of reactivity of the cis isomer. (**Hint:** Draw the available conformations of both the cis and trans isomers around the C1–C2 bonds [compare Figure 4-12]. Use models if necessary.) **(b)** The synthesis of some oxacyclopropane-containing steroids has been achieved by use of a two-step procedure starting with steroidal bromoketones. Suggest suitable reagents for accomplishing a conversion such as the following one.

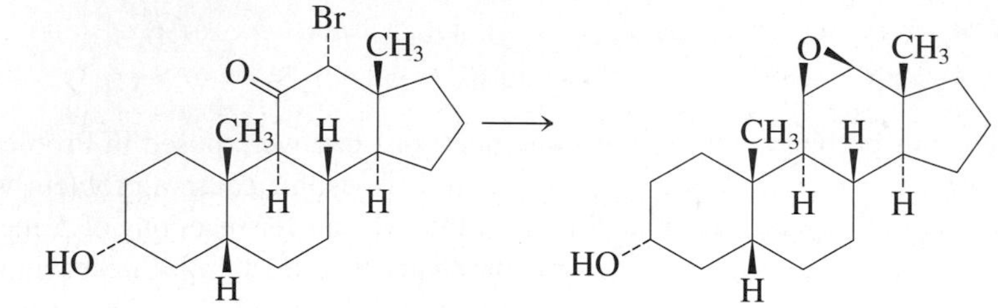

(c) Do any of the steps in your proposed sequence have specific stereochemical requirements for the success of the oxacyclopropane-forming step?

61. Freshly cut garlic contains allicin, a compound responsible for the true garlic odor (see Chemical Highlight 9-4). Propose a short synthesis of allicin, starting with 3-chloropropene.

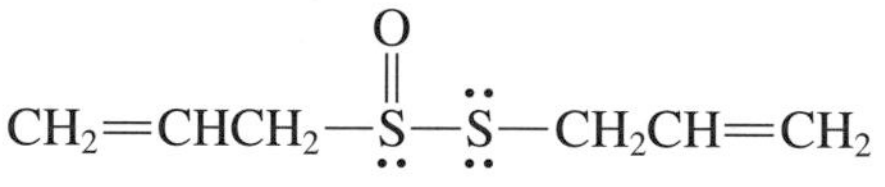

Allicin

Team Problem

62. There are four diastereomers (A–D) of (4*S*)-2-bromo-4-phenylcyclohexanol. As a team, formulate their structures and draw each diastereomer in the most stable chair conformation (see Table 4-3; the $\Delta G°$ value for axial versus equatorial C_6H_5 is 2.9 kcal mol^{-1}). Divide your team into equal groups to consider the outcome of the reaction of each isomer with base ($^-$OH).

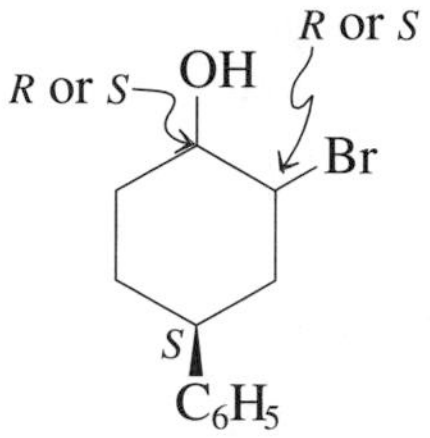

Diastereomers A–D of (4*S*)-2-bromo-4-phenylcyclohexanol

Note: C_6H_5 equals

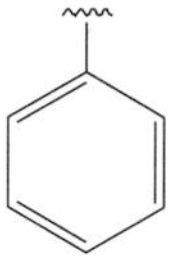

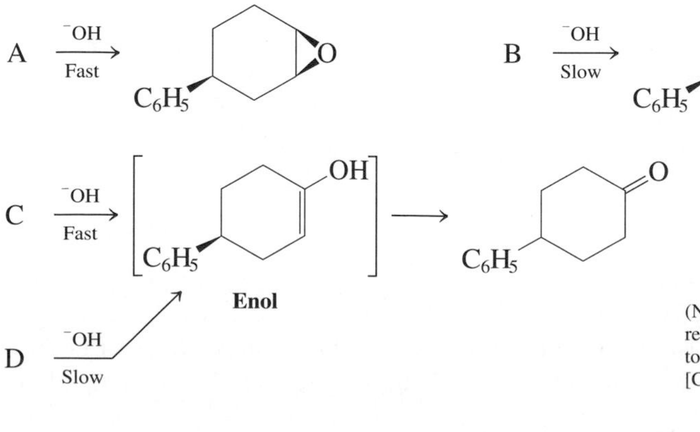

(Note: Enols are unstable with respect to isomerization to the corresponding ketone [Chapters 13 and 18].)

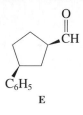

E

(a) Using the curved-arrow formalism (Section 6-4), show the flow of electrons in the attack of the base on the various cyclohexane conformers. Reconvene and present your mechanisms to your teammates, justifying the structural assignments of A–D. Find an explanation for the qualitative rate differences and the divergent course of the reactions of A and B versus C and D.

(b) When compounds A–D are exposed to conditions favoring bromide dissociation in the presence of Ag^+ salts (to accelerate heterolysis with formation of insoluble AgBr), A, C, and D give the same products as those obtained on treatment with base. Discuss the mechanism as a group.

(c) Curiously, compound B traverses another pathway under the conditions described in (b); that is, rearrangement to the aldehyde E. Discuss a possible mechanism for this ring contraction. (**Hint:** Keep in mind the principles outlined in Section 9-3. The mechanism proceeds through a hydroxycation. What is the driving force for its formation?)

Preprofessional Problems

63. The compound whose structure is

$$ (C_7H_{14}O) $$

is best named (IUPAC)

(a) 3,5-dimethylcyclopentyl ether   **(b)** 3,5-dimethylcyclopentane-oxo

(c) *cis*-3,5-dimethyloxacyclohexane   **(d)** *trans*-3,5-dimethyloxacyclohexane

64. The first step in the detailed mechanism for the dehydration of 1-propanol with concentrated H_2SO_4 would be

(a) loss of OH^- **(b)** formation of a sulfate ester   **(c)** protonation of the alcohol

(d) loss of H^+ by the alcohol   **(e)** elimination of H_2O by the alcohol

65. Identify the nucleophile in the following reaction:

$$ RX + H_2O \longrightarrow ROH + H^+X^- $$

(a) X^- **(b)** H^+ **(c)** H_2O **(d)** ROH   **(e)** RX

66. Which is the method of choice for preparing the ether $(CH_3CH_2)_3COCH_3$?

(a) $CH_3Br + (CH_3CH_2)_3CO^-K^+$ **(b)** $(CH_3CH_2)_3COH + CH_3MgBr$

(c) $(CH_3CH_2)_3CMgBr + CH_3OH$ **(d)** $(CH_3CH_2)_3CBr + CH_3O^-K^+$

Using Nuclear Magnetic Resonance Spectroscopy to Deduce Structure

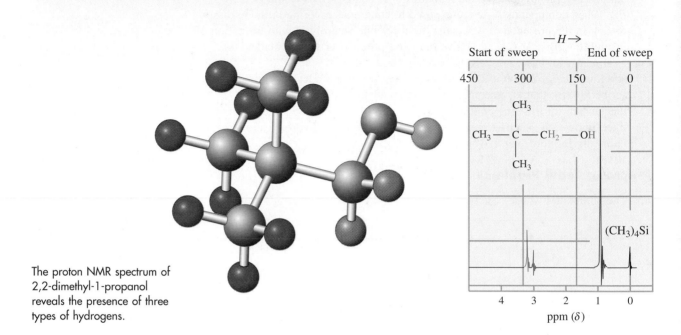

The proton NMR spectrum of 2,2-dimethyl-1-propanol reveals the presence of three types of hydrogens.

Now that we have studied a variety of organic reactions and functional groups, we should be able to plan the synthesis of a reasonably complicated organic molecule in the laboratory. But how can we identify a molecule that we have synthesized? And how can we ascertain its structure? How do we know, for example, that the Grignard reagent that we have used has really converted a ketone into the desired alcohol? Would it not be nice if we could, by some technique, observe the presence of certain nuclei in a molecule, their relative numbers, the nature of their environment (their neighbors), and how they are connected to other atoms?

This chapter will supply answers to these questions. We shall see that a tool known as nuclear magnetic resonance (NMR) spectroscopy provides a kind of "blueprint" of a molecule's structure. Not only does this method allow us to identify organic molecules, but it can be applied even to the imaging (magnetic resonance imaging, MRI) of whole body organs, with powerful applications in medical diagnosis.

We shall begin with a brief consideration of how classical physical measurements and chemical tests can help in the elucidation of structure, before dealing with NMR.

10-1 Physical and Chemical Tests

Let us imagine that you have run a reaction yielding an unidentified compound. To study our sample, we must first purify it—by chromatography, distillation, or recrystallization. We can then compare its melting point, refractive index, and other physical properties with data for known compounds. If our measurements match values in the literature (or appropriate handbooks), we can be reasonably certain of the identity and structure of our molecule. Yet many substances made in the laboratory

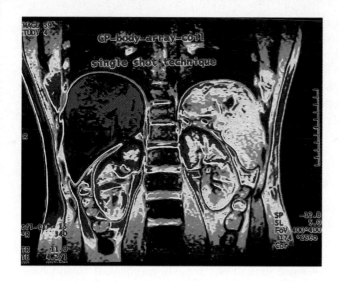

MRI of section through a human abdomen. The lower spinal vertebrae are seen running from top to bottom (center). The kidneys are visible at the lower center around the spine. Above them are the liver (left, dark colored) and stomach and spleen (right).

are new: No published data are available. We need ways to determine their structures *ourselves.*

Elemental analysis will reveal the sample's gross chemical composition. Tests of the chemistry of the compound can then help us to identify its functional groups. For example, we saw in Section 1-9 that we can distinguish between methoxymethane and ethanol on the basis of their physical properties.

The problem becomes considerably more difficult for larger molecules, which vary far more in structure. What if a reaction gave us an alcohol of molecular formula $C_7H_{16}O$? A test with sodium metal would reveal a hydroxy functional group—but not an unambiguous structure. In fact, there are many possibilities, only three of which are shown here.

Three Structural Possibilities for an Alcohol $C_7H_{16}O$

$$CH_3(CH_2)_5CH_2OH \qquad \underset{\underset{CH_3}{|}}{\overset{\overset{CH_3}{|}}{CH_3CCH_2CH_2CH_2OH}} \qquad \underset{\underset{CH_3}{|}}{\overset{\overset{CH_2CH_3}{|}}{CH_3CCH_2CH_2OH}}$$

EXERCISE 10-1

Write the structures of several secondary and tertiary alcohols having the molecular formula $C_7H_{16}O$.

To differentiate between these alternatives, a modern organic chemist makes use of another tool: spectroscopy.

10-2 Defining Spectroscopy

Spectroscopy is a technique for analyzing molecules based on differences in how they absorb radiation. Although there are many types of spectroscopy, four are used most often in organic chemistry: (1) nuclear magnetic resonance (NMR) spectroscopy; (2) infrared (IR) spectroscopy; (3) ultraviolet (UV) spectroscopy; and (4) mass

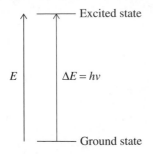

FIGURE 10-1 ——————

The energy difference, ΔE, between the ground state and the excited state of a molecule is overcome by incident radiation of frequency ν matched exactly to equal ΔE. [ν, frequency of absorbed radiation; h (Planck's constant) $= 6.626 \times 10^{-27}$ erg s (1 erg $= 10^{-7}$ J).]

spectrometry (MS). The first, **NMR spectroscopy,** probes the structure in the vicinity of individual nuclei, particularly hydrogens and carbons, and provides the most detailed information regarding the construction of a molecule.

We begin with a simple overview of spectroscopy as it relates to NMR, IR, and UV. Then we describe how a spectrometer works. Finally, we consider the principles and applications of NMR spectroscopy in more detail. We shall return to the other major forms of spectroscopy in Chapters 11, 14, and 20.

Molecules undergo distinctive excitations

Organic molecules absorb electromagnetic radiation in discrete "packets" of energy, or **quanta.** Absorption occurs only when radiation supplying exactly the right packet reaches the compound under investigation. If the frequency of the incident radiation is ν, the packet has energy $\Delta E = h\nu$ (Figure 10-1).

The absorbed energy causes some kind of electronic or mechanical "motion" in the molecule, a process called **excitation.** This motion also is quantized; and, because a molecule can undergo many different kinds of excitation, each kind of motion requires its own distinctive energy. X-rays, for example, which are a form of high-energy radiation, can promote electrons in atoms from inner shells to outer ones; this change, called an **electronic transition,** requires energy higher than 300 kcal mol^{-1}. Ultraviolet radiation and visible light, in contrast, excite valence-shell electrons, typically from a filled bonding molecular orbital to an unfilled antibonding one (see Figure 1-12); here the energy needed ranges from 40 to 300 kcal mol^{-1}. Infrared radiation causes vibrational excitation of a compound's bonds ($\Delta E = 2$–10 kcal mol^{-1}), whereas quanta of microwave radiation cause bond rotations to occur ($\Delta E = \sim 10^{-4}$ kcal mol^{-1}). Finally, radio waves can produce changes in the alignment of nuclear magnetism in a magnetic field ($\Delta E = \sim 10^{-6}$ kcal mol^{-1}); in the next section we see how this phenomenon is the basis of nuclear magnetic resonance spectroscopy.

Figure 10-2 depicts the various forms of radiation, the energy (ΔE) related to each form, and the corresponding wavelengths. Remember that radiation increases

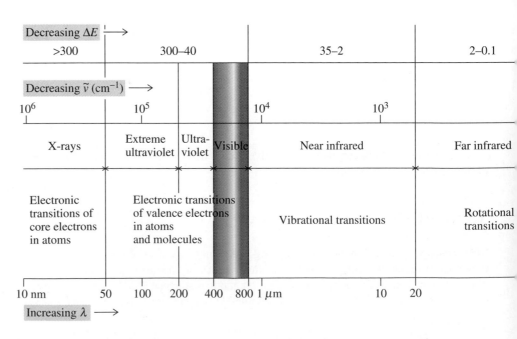

in energy with increasing frequency (ν) or wavenumber ($\tilde{\nu}$) but *decreasing* wavelength (λ).*

EXERCISE 10-2

What type of radiation (in wavelength, λ) would be minimally required to initiate the radical chlorination of methane? [**Hint:** The initiation step requires the breaking of the Cl–Cl bond (see Section 3-4).]

A spectrometer records the absorption of radiation

As illustrated in Figure 10-1, the absorption of a quantum of radiation by a molecule brings about a transition from its (normal) ground state to an excited state. Spectroscopy is a procedure by which these absorptions can be mapped by instruments called **spectrometers.**

Figure 10-3 (see page 388) shows how the spectrometer operates. It contains a source of electromagnetic radiation with a frequency in the region of interest, such as the visible, infrared, or radio. The apparatus is designed so that radiation of a specific wavelength range (e.g., UV, IR, NMR, etc.) passes through the sample. The frequency of this incident light is changed continuously, and its intensity (relative to a reference beam) is measured at a detector and recorded on calibrated paper. In the absence of absorption, the sweep of radiation appears as a straight line, the **baseline.** Whenever the sample absorbs incident light, however, the resulting change in intensity at the detector registers as a **peak,** or deviation from the baseline. The resulting pattern, or **plot,** is the **spectrum** (Latin, appearance, apparition) of the sample.

In summary, electromagnetic radiation is absorbed in discrete quanta of incident energy measurable by spectroscopy. Spectrometers allow the scanning of light of

---

*Wavenumber, defined as $\tilde{\nu} = 1/\lambda$, is the number of waves per centimeter. This quantity is related to (but should not be confused with) frequency, $\nu = c/\lambda$ [in cycles per second, or hertz (Hz), named for a German physicist, Heinrich Rudolf Hertz (1857–1894)], in which c = velocity of light $= 3 \times 10^{10}$ cm s^{-1}. A simple conversion between ΔE (kcal mol^{-1}) and λ (nm) is given by the equation $\Delta E = 28,600/\lambda$.

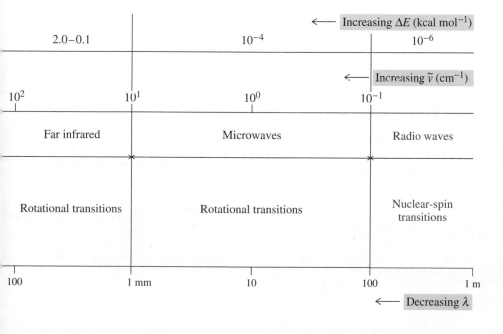

FIGURE 10-2
The spectrum of electromagnetic radiation. The top line is an energy scale, in units of kilocalories per mole, increasing from right to left. The next line contains the corresponding wavenumbers, $\tilde{\nu}$, in units of reciprocal centimeters. The types of radiation associated with the principal types of spectroscopy and the transitions induced by each type are given in the middle. A wavelength scale is at the bottom (λ, in units of nanometers, 1 nm = 10^{-9} m; micrometers, 1 μm = 10^{-6} m; millimeters, mm; and meters, m).

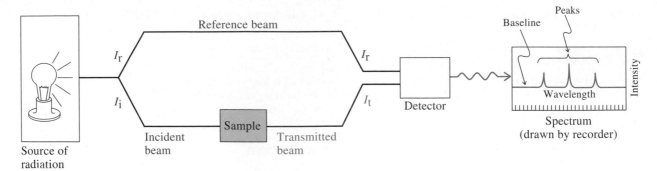

FIGURE 10-3 _____
General diagram of a spectrometer. The source beam is split into reference and incident beams of equal intensity ($I_r = I_i$). The incident beam traverses the sample and emerges as the transmitted beam (I_t). When no absorption occurs, its intensity, I_t, will equal I_i and hence I_r. When $I_t = I_r$, the detector notes no difference and a straight line (zero or baseline) is drawn by the recorder. If absorption occurs, then $I_t \neq I_r$ and a peak is recorded. The resulting diagram is called a spectrum.

varying wavelengths as it passes through a sample of the molecule under investigation, resulting in a plot of the absorptions occurring at certain energies: the spectrum.

10-3 Proton Nuclear Magnetic Resonance

Nuclear magnetic resonance spectroscopy requires low-energy radiation in the radio-frequency range. This section presents the principles behind this technique.

Nuclear spins can be excited by the absorption of radio waves

Many atomic nuclei have the property of spinning around an axis and are therefore said to have a **nuclear spin.** One of those nuclei is hydrogen, written as ^1H (the hydrogen isotope of mass 1) to differentiate it from other isotopes [deuterium (^2H), tritium (^3H)]. Let us consider the simplest form of hydrogen, the proton. Because the proton is positively charged, its spinning motion creates (as does any moving charged particle) a magnetic field. The net result is that a proton may be viewed as a tiny bar magnet floating freely in solution or in space. When the proton is exposed to an *external* magnetic field of strength H_0, it may have one of two orientations: It may be aligned either with H_0, an energetically favorable choice, or (unlike a normal bar magnet) against H_0, a move that costs energy. The two possibilities are designated the α and β **spin states,** respectively (Figure 10-4).

FIGURE 10-4 _____
(A) Single protons (H) act as tiny bar magnets. The direction of the magnetic field created by nuclear spin is indicated by the arrow. (B) In a magnetic field, H_0, the nuclear spins align with (α) or against (β) the field.

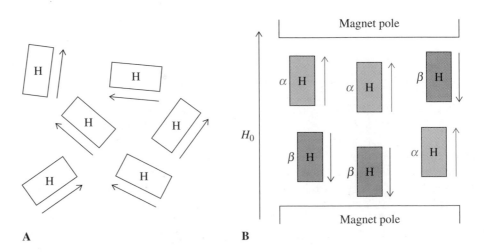

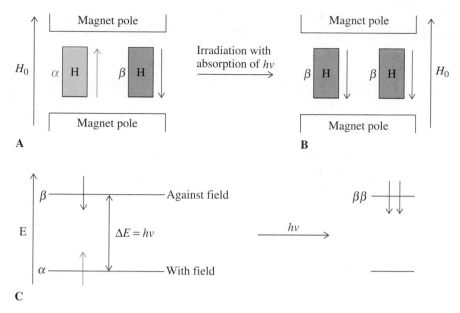

FIGURE 10-5
(A) When protons are exposed to an external magnetic field, two spin states differing in energy by $\Delta E = h\nu$ are generated. (B) Irradiation with energy of the right frequency, ν, causes absorption, "flipping" the nuclear spin of a proton from the α to the β state (resonance). (C) An energy diagram, showing a proton (denoted by only an arrow) gaining the energy $\Delta E = h\nu$ and undergoing "spin flip" from the α to the β state.

These two energetically different states afford the necessary condition for spectroscopy. Irradiation of the sample with a source of just the right frequency to bridge the difference in energy between the α and β states produces **resonance,** an absorption of energy as an α proton "flips" to the β spin state. This phenomenon is illustrated for a pair of protons in Figure 10-5. After excitation, the nuclei relax and return to their original states by a variety of pathways (which will not be discussed here), all of which release the absorbed energy as heat. At resonance, therefore, there is continuous excitation and relaxation.

The difference in energy, ΔE, between spin states α and β depends directly on the external field strength, H_0. The stronger the external field, the larger the difference in energy: The absorption frequency, ν, is proportional to H_0. Thus, the resonance frequency for hydrogen nuclei is 90 MHz (megahertz, millions of hertz) at $H_0 = 21{,}150$ gauss* (abbreviated G), 180 MHz at $H_0 = 42{,}300$ G, 300 MHz at $H_0 = 70{,}500$ G, and so on.

How much energy must be expended for the spin of a proton to flip from α to β? Because $\Delta E_{\beta-\alpha} = h\nu$, we can calculate how much. The amount is very small, that is, $\Delta E_{\beta-\alpha}$ at 90 MHz is on the order of 9×10^{-6} kcal mol^{-1}. Equilibration between the two states is fast, and typically only slightly more than one-half of all proton nuclei in a magnetic field will adopt the α state, the remainder having a β spin.

Many nuclei undergo magnetic resonance

Hydrogen is not the only nucleus capable of magnetic resonance. Table 10-1 (see page 390) lists a number of nuclei responsive to NMR and of importance in organic chemistry, as well as several that lack NMR activity. In general, nuclei composed of an odd number of protons, such as ^{1}H (and its isotopes), ^{14}N, ^{19}F, and ^{31}P, or an odd number of neutrons, such as ^{13}C, show magnetic behavior. On the other hand, when *both* the proton and the neutron counts are even, as in ^{12}C or ^{16}O, the nuclei are nonmagnetic.

*Karl Friedrich Gauss (1777–1855), German mathematician.

TABLE 10-1	NMR Activity and Natural Abundance of Selected Nuclei					
Nucleus	NMR activity	Natural abundance (%)		Nucleus	NMR activity	Natural abundance (%)
^1H	Active	99.985		^{16}O	Inactive	99.759
^2H (D)	Active	0.015		^{17}O	Active	0.037
^3H (T)	Active	0		^{18}O	Inactive	0.204
^{12}C	Inactive	98.89		^{19}F	Active	100
^{13}C	Active	1.11		^{31}P	Active	100
^{14}N	Active	99.63		^{35}Cl	Active	75.53
^{15}N	Active	0.37		^{37}Cl	Active	24.47

Abbreviations: D, deuterium; T, tritium.

On exposure to equal magnetic fields, *different NMR-active nuclei will resonate at different values of ν*. For example, if we were to scan a hypothetical spectrum of a sample of chlorofluoromethane, CH_2ClF in a 21,250-G magnet,* we would observe six absorptions corresponding to the six NMR-active nuclei in the sample: the highly abundant ^1H, ^{19}F, ^{35}Cl, and ^{37}Cl and the much less plentiful ^{13}C (1.11%) and ^2H (0.015%), as shown in Figure 10-6.

FIGURE 10-6
A hypothetical NMR spectrum of CH_2ClF at 21,150 G. Because each NMR-active nucleus resonates at a characteristic frequency, six lines are observed. We show them here with similar heights for simplicity. Actually, special techniques are required to observe signals of less abundant nuclei such as ^2H and ^{13}C. In addition, it would not normally be possible to observe all these lines on a single instrument: Most NMR spectrometers are not capable of scanning such a wide range of frequencies.

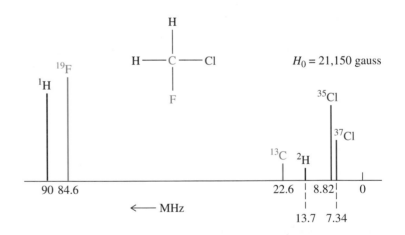

High-resolution NMR spectroscopy can differentiate nuclei of the same element

Consider now the NMR spectrum of chloro(methoxy)methane (chloromethyl methyl ether), $ClCH_2OCH_3$. A sweep at 21,250 G from 0 to 90 MHz would give one peak for each element present (Figure 10-7A). However, modern instrumentation enables

*For comparison, the maximum intensity of Earth's magnetic field anywhere on its surface is about 0.7 G.

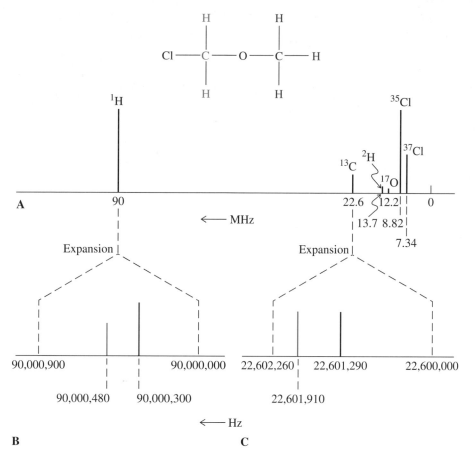

FIGURE 10-7 ──────

High resolution can reveal additional peaks in an NMR spectrum. (A) At low resolution, the spectrum of $ClCH_2OCH_3$ at 21,250 G shows six peaks for the six NMR active isotopes present in the molecule. Signals for carbon, oxygen, and deuterium are relatively small, to reflect their relatively low natural abundance. (^{12}C and ^{16}O are NMR inactive; see Table 10-1.) (B) At high resolution, the hydrogen spectrum shows two peaks for the two sets of hydrogens (one shown in blue in the structure, the other in red). Note that the high-resolution sweep covers only 0.001% of that at low resolution. (C) The high-resolution ^{13}C spectrum (see Section 10-9) shows peaks for the two different carbon atoms in the molecule.

us to view a small part of the spectrum in the immediate vicinity of a peak in great detail. With this technique, called **high-resolution NMR spectroscopy,** we may study the hydrogen resonance from 90,000,000 to 90,000,900 Hz. We find that what appeared to be only one peak in that region actually consists of two peaks that were not resolved at first (Figure 10-7B). Similarly, the high-resolution ^{13}C spectrum measured in the vicinity of 22.6 MHz shows two peaks (Figure 10-7C). These absorptions reveal the presence of *two* types of hydrogens and carbons, respectively. *Because high-resolution NMR spectroscopy distinguishes both hydrogen and carbon atoms in different structural environments, it is a powerful tool for elucidating structures.* The organic chemist uses NMR spectroscopy more often than any other spectroscopic technique.

Proton NMR spectrometers operate in two different modes: continuous wave and Fourier transform

NMR spectroscopy has been routinely available since about 1960. The first mass-produced proton NMR spectrometers used magnets at 14,100 G, which set the resonance frequency of hydrogen nuclei at about 60 MHz. Recent advances in technology have led to the widespread use of NMR spectrometers at higher field strength, which yield hydrogen resonance frequencies as high as 750 MHz and greatly improved resolution as a result. The magnets employed to achieve the corresponding high field strengths are superconducting at liquid helium temperature ($-269°C$, or

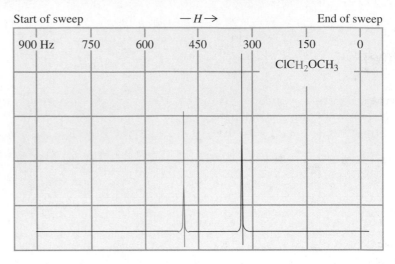

FIGURE 10-8

90-MHz ^{1}H NMR spectrum of chloro(methoxy)methane. The zero-hertz line is set at exactly 90 MHz at the right-hand side of the spectral paper. The strength of the applied magnetic field (H) increases from left to right.

4 K). This book depicts spectra at either 90 MHz or 300 MHz, depending on complexity.

NMR spectra may be recorded in either of two ways, depending on the design of the instrument. In the *continuous wave* (CW) spectrometers, the method is based on the general principles outlined in Section 10-2, in which a spectrum is obtained by scanning over a frequency range. For NMR, there are two possible variants: Either the magnetic field is held constant while the radio frequency is scanned or the magnetic field is varied while the radio frequency is held constant. The spectra obtained are indistinguishable. For convenience, all are calibrated in hertz (s^{-1}), as if frequency had been varied at constant H_0. An actual ^{1}H NMR spectrum of $ClCH_2OCH_3$ is shown in Figure 10-8.

The current state-of-the-art instruments use a different technique for obtaining the same spectra: *Fourier transform* (FT).* Rather than "sweeping" through the spectral region of interest, this method exposes the sample to one or multiple radio frequency pulses such that all the nuclei are excited at the same time and the data stored by computer. The advantages are much faster spectral accumulation (seconds instead of minutes) and the possibility of improving the signal-to-noise ratio by computer averaging of multiple spectra. An additional technical point is that FT spectra are initially measured as a function of time (i.e., the decay of the initial absorptions with time is recorded). Hence the data have to be processed by computer, through a mathematical treatment called Fourier transformation, to convert them into the standard appearance that is of use to organic chemists. Although these details are of no concern to us, it should be emphasized that FT NMR has greatly enhanced the utility of NMR as an analytical tool (see also Section 10-9).

*Invented by Professor Richard R. Ernst (b. 1933), Swiss Federal Institute of Technology (ETH), Zürich, Switzerland, Nobel Prize 1991 (chemistry); named for Baron (Jean Baptiste) Joseph Fourier (1768–1830), French mathematician.

CHEMICAL HIGHLIGHT 10-1 | Recording an NMR Spectrum

The sample to be studied (a few milligrams) is usually dissolved in a solvent (0.3–0.5 mL) preferably not containing any atoms that themselves absorb in the NMR range under investigation. Typical solvents are tetrachloromethane (CCl_4) or deuterated ones such as trichlorodeuteriomethane (deuteriochloroform), $CDCl_3$; hexadeuteriopropanone (hexadeuterioacetone), CD_3COCD_3; hexadeuteriobenzene, C_6D_6; and octadeuteriooxacyclopentane (octadeuteriotetrahydrofuran), C_4D_8O. The solution is transferred into an NMR sample container (a cylindrical precision-bore glass tube), which is inserted into the magnet. (The left-hand photograph below shows the magnet housing and the inlet for the sample.) To make sure that all molecules in the sample are rapidly averaged with respect to their position in the magnetic field, the NMR tube is rapidly spun by an air jet. In a continuous-wave instrument, energy from a radio-frequency (RF) generator is emitted by one of two coils of wire surrounding the sample cavity. The second coil is connected to an RF receiver and detects any energy absorption by the sample, converting it into a plot on a recorder as the spectrum is scanned (see diagram). In a higher-field Fourier-transform instrument, a superconducting magnet (cooled by liquid helium) is employed, and the pulsed signals are sent to a receiver that, in conjunction with a computer, elaborates them to the standard spectrum.

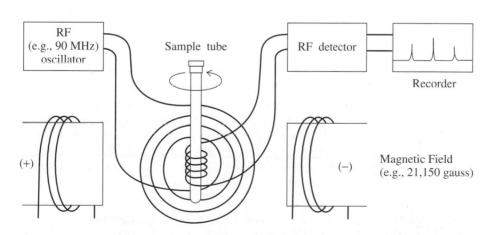

In summary, certain nuclei, such as ^1H and ^{13}C, can be viewed as tiny atomic magnets that, when exposed to a magnetic field, can align with it (α) or against it (β). These two states are of unequal energy, a condition giving rise to nuclear magnetic resonance spectroscopy. At resonance, radio-frequency radiation is absorbed by the nucleus to effect α-to-β transitions (excitation). The β state relaxes to the α state by giving off a small amount of heat. The resonance frequency, which is characteristic of the nucleus and its environment, is proportional to the strength of the external magnetic field.

10-4 Using NMR Spectra to Analyze Molecular Structure: The Proton Chemical Shift

Why do the two different groups of hydrogens in chloro(methoxy)methane give rise to distinct NMR peaks? How does the molecular structure affect the position of an NMR signal? This section will answer these questions.

We shall see that the position of an NMR absorption, also called the chemical shift, depends on the electron density around the hydrogen, which in turn is controlled by the molecular structure in the vicinity of the observed nucleus (its "structural environment"). Therefore, *the NMR chemical shifts of the hydrogens in a molecule will be important clues for determining its exact makeup.*

The position of an NMR signal depends on the nucleus's electronic environment

The high-resolution ^1H NMR spectrum of chloro(methoxy)methane depicted in Figure 10-8 reveals that the two kinds of hydrogens give rise to two separate resonance absorptions. What is the origin of this effect? It is the differing electronic environments of the respective hydrogen nuclei. A free proton is essentially unperturbed by electrons. Organic molecules, however, contain covalently bonded hydrogen nuclei, *not* free protons, and the electrons in these bonds affect nuclear magnetic resonance absorptions.*

Bound hydrogens are surrounded by orbitals whose electron density varies, depending on the polarity of the bond, the hybridization of the attached atom, and the presence of electron-donating or -withdrawing groups. When a nucleus surrounded by electrons is exposed to a magnetic field of strength H_0, these electrons move in such a way as to generate a small **local magnetic field, h_{local},** *opposing* H_0. As a consequence, the total field strength near the hydrogen nucleus is *reduced,* and the nucleus is thus said to be **shielded** from H_0 by its electron cloud (Figure 10-9). The degree of shielding depends on the amount of electron density surrounding the nucleus. Adding electrons increases shielding; their removal causes **deshielding.**

What is the effect of shielding on the relative position of an NMR absorption? Using the currently customary procedure of keeping the radio frequency constant and varying the magnetic field, we find that a higher external field strength is required to overcome the shielding effect and cause resonance. Spectra are recorded with field

---

*In discussions of NMR, the terms *proton* and *hydrogen* are frequently (albeit incorrectly) interchanged. "Proton NMR" and "protons in molecules" are used even in reference to covalently bound hydrogen.

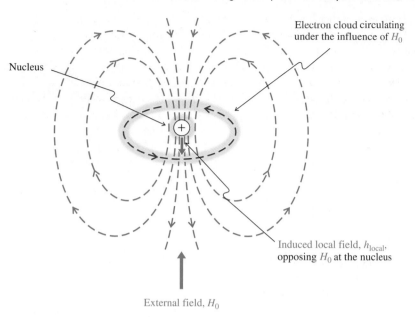

FIGURE 10-9

The external field, H_0, causes an electronic current of the bonding electrons around a hydrogen nucleus, which in turn generates a local magnetic field opposing H_0 (you may recognize this as Lenz's law, named for a Russian physicist, Heinrich Friedrich Emil Lenz [1804–1865]).

strength increasing from left to right, so the phrase "a shift upfield" means that the signal (peak) appears farther *to the right* (Figure 10-10).

Because each chemically distinct hydrogen has a unique electronic environment, it gives rise to a characteristic resonance. Likewise, *chemically equivalent hydrogens*

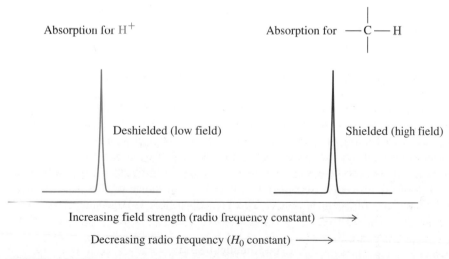

FIGURE 10-10

Effect of shielding on the absorption of a covalently bound proton. At constant radio frequency of energy $h\nu$, the free proton resonates at H_0. Shielding decreases the value of the field around a given nucleus to $H_0 - h_{local}$. Hence, to "match" $h\nu$, the external field must be increased by an amount equal to h_{local}. The net result is a displacement of the corresponding signal to the right-hand side of the spectrum. Conversely, at constant H_0, to "match" the magnetic field around the nucleus ($H_0 - h_{local}$), the radio frequency $h\nu$ is decreased.

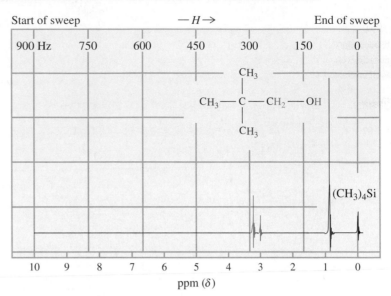

FIGURE 10-11 _____
90-MHz ^{1}H NMR spectrum of 2,2-dimethyl-1-propanol (containing a little tetramethylsilane as an internal standard) in CCl_4. Three peaks are observed for the three sets of different hydrogens. (The scale at the bottom indicates the chemical shift in δ, the distance from tetramethylsilane, to be defined in the next subsection.)

show peaks at the same position. This is shown for the NMR spectrum of 2,2-di-methyl-1-propanol in Figure 10-11: There are three absorptions—one (most shielded) for the nine equivalent methyl hydrogens of the tertiary butyl group, another for the OH, and a third (most deshielded) for the CH_2 hydrogens.

The chemical shift describes the position of an NMR peak

In which manner are spectral data reported? As noted earlier, most hydrogen absorptions in 90-MHz ^{1}H NMR fall within a range of 900 Hz. Rather than record the exact frequency of each resonance, we measure it relative to an internal standard, the compound tetramethylsilane, $(CH_3)_4Si$ (b.p. 26.5°C). Its 12 equivalent hydrogens are shielded relative to those in most organic molecules, a situation resulting in a resonance line conveniently removed from the usual spectral range. The position of the NMR absorptions of a compound under investigation can then be measured (in hertz) relative to the internal standard. In this way, the signals of, for example, 2,2-dimethyl-1-propanol (Figure 10-11) are reported as being located 78, 258, and 287 Hz downfield from $(CH_3)_4Si$.

A problem with these numbers, however, is that *they vary with the strength of the applied magnetic field.* Because field strength and resonance frequency are directly proportional, doubling or tripling the field strength will double or triple the distance (in hertz) of the observed peaks relative to $(CH_3)_4Si$. To prevent this complication and to be able to compare reported literature spectra at different field strengths, we standardize the measured frequency by dividing the distance to $(CH_3)_4Si$ (in hertz) by the frequency of the spectrometer. This procedure yields a *field-independent* number, the **chemical shift δ.**

> **The Chemical Shift**
>
> $$\delta = \frac{\text{distance of peak from } (CH_3)_4Si \text{ in hertz}}{\text{spectrometer frequency in megahertz}} \text{ ppm}$$

The chemical shift is reported in units of parts per million (ppm)—for hydrogen, usually to three significant figures. For $(CH_3)_4Si$, δ is defined as 0.00. The NMR spectrum of 2,2-dimethyl-1-propanol in Figure 10-11 would then be reported in the following format: ^{1}H NMR (90 MHz, CCl_4) $\delta = 0.87, 2.87, 3.19$ ppm.

EXERCISE 10-3

With a 300-MHz NMR instrument, the three signals for 2,2-dimethyl-1-propanol are recorded at 261, 861, and 957 Hz, respectively, downfield from $(CH_3)_4Si$. Recalculate the δ values and compare them with those obtained at 90 MHz.

Functional groups have characteristic chemical shifts

The reason that NMR is such a valuable analytical tool is that it can identify certain types of hydrogens in a molecule. Each type has a characteristic chemical shift

---

TABLE 10-2	**Typical Hydrogen Chemical Shifts in Organic Molecules**	
Type of hydrogen[a]	**Chemical shift δ in ppm**	
Primary alkyl, RCH_3	0.8–1.0	Alkane and alkanelike hydrogens
Secondary alkyl, RCH_2R'	1.2–1.4	
Tertiary alkyl, R_3CH	1.4–1.7	
Allylic (next to a double bond), $R_2C{=}C\overset{CH_3}{\underset{R'}{\diagdown}}$	1.6–1.9	Hydrogens adjacent to unsaturated functional groups
Benzylic (next to a benzene ring), $ArCH_2R$	2.2–2.5	
Ketone, $RCCH_3$ with $\overset{\|}{O}$	2.1–2.6	
Alkyne, $RC{\equiv}CH$	1.7–3.1	
Chloroalkane, RCH_2Cl	3.6–3.8	Hydrogens adjacent to electronegative atoms
Bromoalkane, RCH_2Br	3.4–3.6	
Iodoalkane, RCH_2I	3.1–3.3	
Ether, RCH_2OR'	3.3–3.9	
Alcohol, RCH_2OH	3.3–4.0	
Terminal alkene, $R_2C{=}CH_2$	4.6–5.0	Alkene hydrogens
Internal alkene, $R_2C{=}CH$ with R'	5.2–5.7	
Aromatic, ArH	6.0–9.5	
Aldehyde, RCH with $\overset{\|}{O}$	9.5–9.9	
Alcoholic hydroxy, ROH	0.5–5.0	(variable)
Thiol, RSH	0.5–5.0	(variable)
Amine, RNH_2	0.5–5.0	(variable)

[a]R, R′, alkyl groups; Ar, aromatic group (not argon).

depending on its structural environment. The hydrogen chemical shifts typical of standard organic structural units are listed in Table 10-2 (see page 397). It is important to be familiar with the chemical shift ranges for the structural types discussed so far: alkanes, haloalkanes, ethers, alcohols, aldehydes, and ketones. Others will be discussed in more detail in subsequent chapters.

Note that the absorptions of the alkane hydrogens occur at relatively high field ($\delta = 0.8$–1.7 ppm). A hydrogen close to an electron-withdrawing group or atom (such as a halogen or oxygen) is shifted to relatively lower field: Such substituents *deshield* their neighbors. Table 10-3 shows how adjacent heteroatoms affect the chemical shifts of a methyl group. The more electronegative the atom, the more deshielded are the methyl hydrogens relative to methane. Several such substituents exert a cumulative effect, as seen in the series of the three chlorinated methanes shown in the margin. The deshielding influence of electron-withdrawing groups diminishes rapidly with distance.

Cumulative Deshielding in Chloromethanes

CH_3Cl
$\delta = 3.05$ ppm

CH_2Cl_2
$\delta = 5.30$ ppm

$CHCl_3$
$\delta = 7.27$ ppm

$$CH_3-CH_2-CH_2-Br$$
$\delta = 1.06 \quad 1.81 \quad 3.47$ ppm

1.73 ppm

$$CH_3-\overset{\displaystyle CH_3}{\underset{\displaystyle H}{\overset{|}{\underset{|}{C}}}}-Br$$

4.21 ppm

EXERCISE 10-4

Explain the assignment of the ^1H NMR signals of chloro(methoxy)methane (Figure 10-8). (**Hint:** Consider the number of electronegative neighbors to each type of hydrogen.)

As noted in Table 10-2, hydroxy, mercapto, and amino hydrogens absorb over a range of frequencies. In spectra of such samples, the absorption peak of the proton attached to the heteroatom may be relatively broad. This variability of chemical shift is due to hydrogen bonding and depends on temperature, concentration, and the presence of other hydrogen-bonding species such as water (i.e., moisture). In simple terms,

TABLE 10-3	The Deshielding Effect of Electronegative Atoms	
CH_3X	Electronegativity of X (from Table 1-2)	Chemical shift δ (ppm) of CH_3 group
CH_3F	4.0	4.26
CH_3OH	3.4	3.40
CH_3Cl	3.2	3.05
CH_3Br	3.0	2.68
CH_3I	2.7	2.16
CH_3H	2.2	0.23

different degrees of hydrogen bonding alter the electronic environment of the hydrogen nuclei. When such line broadening is observed, it is usually diagnostic of the presence of OH, SH, or NH groups.

In summary, the various hydrogen atoms present in an organic molecule can be recognized by their characteristic NMR peaks at certain chemical shifts δ. An electron-poor environment is deshielded and leads to low-field (high-δ) absorptions, whereas an electron-rich environment results in the opposite (shielded or high-field peaks). The chemical shift δ is measured in parts per million by dividing the difference between the measured resonance and that of the internal standard, tetramethylsilane, $(CH_3)_4Si$, in hertz by the spectrometer frequency in megahertz. The NMR spectra for the OH groups of alcohols, the SH groups of thiols, and the NH_2 groups of amines exhibit characteristically broad peaks with concentration- and moisture-dependent δ values.

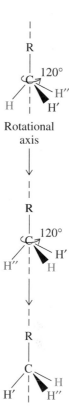

10-5 Tests for Chemical Equivalence

In the NMR spectra presented so far, two or more hydrogens occupying positions that are chemically equivalent give rise to only *one* NMR absorption. It can be said, in general, that *chemically equivalent protons have the same chemical shift.* However, we shall see that it is not always easy to identify chemically equivalent nuclei. We shall resort to the symmetry operations presented in Chapter 5 to help us decide on the expected NMR spectrum of a specific compound.

Molecular symmetry helps establish chemical equivalence

To establish chemical equivalence, we have to recognize the symmetry of molecules and their substituent groups. As we know, one form of symmetry is the presence of a mirror plane (Section 5-1, Figure 5-4). Another is rotational equivalence. For example, Figure 10-12 demonstrates how two successive 120° rotations of a methyl group allow each hydrogen to occupy the position of either of the other two without effecting any structural change. Thus, in a rapidly rotating methyl group, all hydrogens are equivalent and should have the same chemical shift. We shall see shortly that this is indeed the case.

FIGURE 10-12
Rotation of a methyl group as a test of symmetry.

Application of the principles of rotational or mirror symmetry or both allows the assignment of equivalent nuclei in other compounds (Figure 10-13; see page 400).

EXERCISE 10-5

How many NMR absorptions would you expect for **(a)** 2,2,3,3-tetramethylbutane; **(b)** $CH_3OCH_2CH_2OCH_2CH_2OCH_3$; **(c)** oxacyclopropane?

Conformational interconversion may result in equivalence on the NMR time scale

Let us look more closely at two more examples, chloroethane and cyclohexane. The first should have two NMR peaks because it has the two sets of equivalent hydrogens; the second has twelve chemically equivalent hydrogen nuclei and is expected to show only one absorption. However, are these expectations really justified? Consider the possible conformations of these two molecules (Figure 10-14).

Begin with chloroethane. The most stable conformation is the staggered arrangement, in which one of the methyl hydrogens (H_{b_3}) is located *anti* with respect to the chlorine atom. We expect this particular nucleus to have a chemical shift different

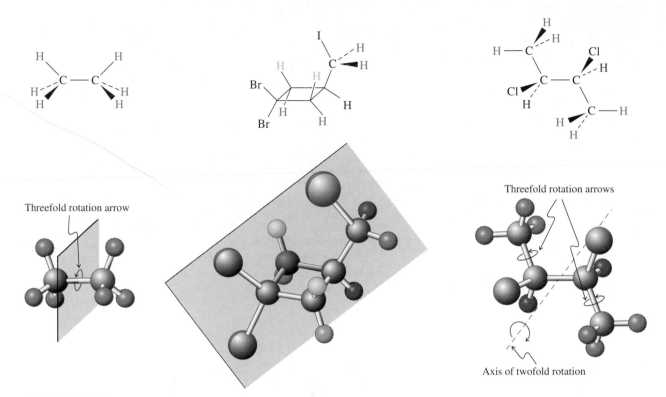

FIGURE 10-13

The recognition of rotational and mirror symmetry in organic molecules allows the identification of chemical-shift-equivalent hydrogens. The different colors distinguish between nuclei giving rise to separate absorptions with distinct chemical shifts.

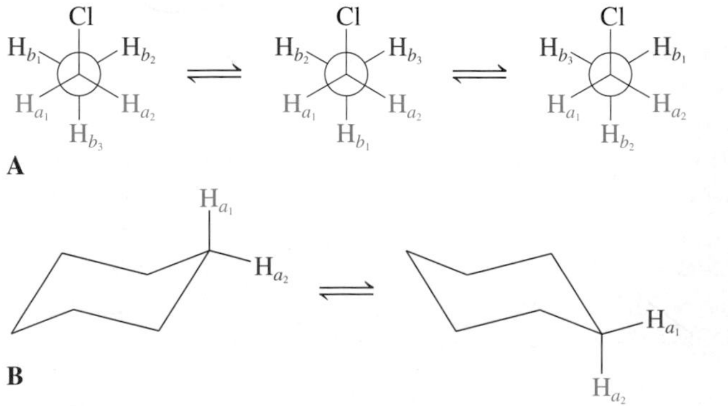

FIGURE 10-14

(A) Newman projections of chloroethane. H_{b_3} is located *anti* to the chlorine substituent and is therefore not in the same environment as H_{b_1} and H_{b_2}. However, fast rotation averages all the methyl hydrogens on the NMR time scale. (B) In any given conformation of cyclohexane, the axial hydrogens are different from the equatorial ones. However, conformational flip is rapid on the NMR time scale, so only one average signal is observed. Colors are used here to distinguish between environments and thus to indicate specific chemical shifts.

CHEMICAL HIGHLIGHT 10-2 | Magnetic Resonance Imaging in Medicine

After NMR spectroscopy was introduced to organic chemistry in the late 1960s, it did not take long for physicists and chemists to ask whether the technique could be used in medical diagnosis. In particular, if a spectrum is viewed as a sort of image of a molecule, why not image sections of the human (or animal) body? The answer emerged between the early 1970s and mid-1980s and relied not on the usual information embedded in chemical shifts, integration, and spin–spin splitting, but on a different phenomenon: *proton relaxation times.* Thus, the rate at which a hydrogen that has been induced to undergo the $\alpha \rightarrow \beta$ spin flip "relaxes" back to the α state is not constant but depends on the environment. Relaxation times can range from milliseconds to seconds and affect the line shapes of the corresponding signal. In the body, the hydrogens of water attached to the surface of biological molecules have been found to relax faster than those in the free fluid. In addition, there are slight differences depending on the nature of the tissue or structure to which the water is bound. For example, water in some cancerous tumors has a shorter relaxation time

Diagnostic testing whole body MRI.

than that in healthy cells. These differences can be used to image the inside of the human body by *magnetic resonance imaging,* or *MRI.* In this application, a patient's entire body is placed within the poles of a large electromagnet, and proton NMR spectra are collected and computer processed to give a series of cross-sectional plots of signal intensity. These cross-sectional plots are combined to produce a three-dimensional image of tissue proton density, as shown here.

Because most of the signal is due to water, variations from normal water-density patterns may be detected and used in diagnosis. Improvements in the late 1980s shortened the time needed for analysis from minutes to seconds, permitting direct viewing of blood flow, kidney secretion, and other medically relevant phenomena. MRI is particularly helpful in detecting abnormalities not readily found by CAT (computerized axial tomography*) scans and conventional X-rays. Unlike other imaging methods, the technique is *noninvasive,* requiring neither ionizing radiation nor the injection of radioactive substances for visualization.

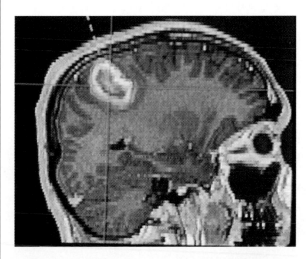

MRI brain scan revealing a tumor. This scan was used for computer-aided brain laser surgery. The dotted green line is the surgical point of attack; the patient's face is at right. Many such scans are programmed into a powerful microscope which provides a 3-D virtual reality map of the tumor that in turn allows the surgeon to cut and destroy malignant tissue with extreme precision.

*Tomography is a method for taking pictures of a specific plane of an object.

CH₃CH₂Cl
b   *a*

from the two *gauche* hydrogens (H_{b_1} and H_{b_2}). In fact, however, the NMR spectrometer cannot resolve that difference, because the fast rotation of the methyl group averages the signals for H_b. This rotation is said to be "fast on the NMR time scale." The resulting absorption appears at an average δ of the two signals expected for H_b.

In theory, it should be possible to slow the rotation in chloroethane by cooling the sample. In practice, "freezing" the rotation is very difficult to do, because the activation barrier to rotation is only a few kilocalories per mole. We would have to cool the sample to about $-180°C$, at which point most solvents would crystallize—and ordinary NMR spectroscopy would not be possible.

A similar situation is encountered for cyclohexane. Here fast conformational isomerism causes the axial hydrogens to be in equilibrium with the equatorial ones on the NMR time scale (Figure 10-14B); so, at room temperature, the NMR spectrum shows only one sharp line at $\delta = 1.36$ ppm. However, in contrast with that for chloroethane, the process is slow enough at $-90°C$ that, instead of a single absorption, two are observed: one for the six axial hydrogens at $\delta = 1.12$ ppm; the other for the six equatorial hydrogens at $\delta = 1.60$ ppm. The conformational isomerization in cyclohexane is frozen on the NMR time scale at this temperature because the activation barrier to ring flip is much higher ($E_a = 10.8$ kcal mol^{-1}; Section 4-3) than the barrier to rotation in chloroethane.

In general, the lifetime of a species in such an equilibrium must be on the order of about a second to allow its resolution by NMR. If this period decreases substantially, an average spectrum is obtained. Such time- and temperature-dependent NMR phenomena are in fact used by chemists to estimate the activation parameters of chemical processes.

In summary, the properties of symmetry, particularly mirror images and rotations, help to establish the chemical-shift equivalence or nonequivalence of the hydrogens in organic molecules. Those structures that undergo rapid conformational changes on the NMR time scale show only averaged spectra at room temperature. Examples are rotamers and ring conformers. In some cases, these processes may be "frozen" at low temperatures to allow distinct absorptions to be observed.

10-6 Integration

So far we have looked only at the *position* of NMR peaks. We shall see in this section that another useful feature of NMR spectroscopy is its ability to measure the relative integrated intensity of a signal, which is proportional to the relative number of nuclei giving rise to that absorption.

Integration reveals the number of hydrogens responsible for an NMR peak

The more hydrogens of one kind there are in a molecule, the more intense the corresponding NMR absorption relative to the other signals. By measuring the area under a peak (the "integrated area") and comparing it with the corresponding peak areas of other signals, we can quantitatively estimate the nuclear ratios. For example, in the spectrum of 2,2-dimethyl-1-propanol (Figure 10-11), three signals are observed, with relative areas of $9:2:1$.

To simplify this measurement, the spectrometer has an electronic feature called **integration.** At the push of a button, the instrument switches into the *integrator mode,* which allows a rescanning of the spectrum to record the intensity ratio of the observed absorptions. This spectral integration is plotted as follows: Whenever a point in the scan is reached at which there is an absorption peak, the pen in the integrator mode moves vertically upward a distance proportional to the intensity of the peak. It then again moves horizontally until the next peak is reached and so forth. A ruler can be used to measure the distance by which the horizontal line is displaced at every peak. *The relative sizes of these displacements furnish the ratio of hydrogens giving rise to the various signals.* Figure 10-15 depicts the ^{1}H NMR spectra of 2,2-dimethyl-1-propanol and 1,2-dimethoxyethane, including plots of the integration. Many higher-field instruments also provide digital readouts of integrated peak intensity automatically. Therefore, in the subsequent spectra, these values will be given above the corresponding signals.

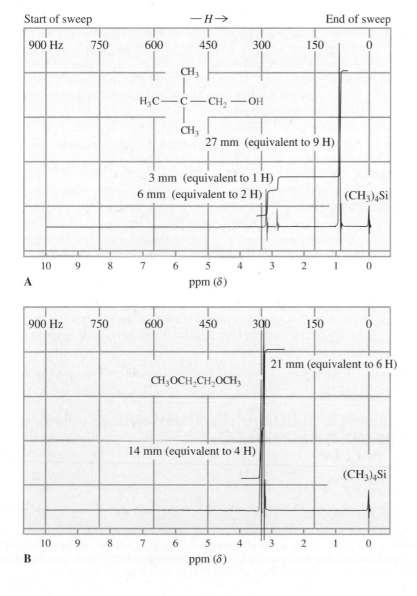

FIGURE 10-15 ⸺

Integrated 90-MHz spectra of (A) 2,2-dimethyl-1-propanol and (B) 1,2-dimethoxyethane, in CCl_4 with added $(CH_3)_4Si$. In (A), the integrated areas measured by a ruler are 6:3:27 (in mm). (Note that the integrator converts units of area into units of distance.) Normalization through division by the smallest number gives a peak ratio of 2:1:9. Note that the integration gives only *ratios,* not absolute values for the number of hydrogens present in the sample. Thus, in (B), the integrated peak ratio is ~3:2, yet the compound contains hydrogens in a ratio of 6:4.

Chemical shifts and peak integration can be used to determine structure

Consider the three products obtained in the monochlorination of 1-chloropropane, $CH_3CH_2CH_2Cl$. All have the same molecular formula $C_3H_6Cl_2$ and very similar physical properties (such as boiling points).

$$CH_3CH_2CH_2Cl \xrightarrow[-HCl]{Cl_2, \ hv, \ 100°C} CH_3CH_2CHCl_2 \ + \ CH_3CHClCH_2Cl \ + \ ClCH_2CH_2CH_2Cl$$

10%	27%	14%
1,1-Dichloropropane	**1,2-Dichloropropane**	**1,3-Dichloropropane**
(b.p. 87°–90°C)	(b.p. 96°C)	(b.p. 120°C)

NMR spectroscopy clearly distinguishes all three isomers. 1,1-Dichloropropane contains three types of nonequivalent hydrogens, a situation giving rise to three NMR signals in the ratio of $3:2:1$. The single hydrogen absorbs at relatively low field ($\delta = 5.93$ ppm) because of the cumulative deshielding effect of the two halogen atoms; the others absorb at relatively high field ($\delta = 1.01$ and 2.34 ppm).

1,2-Dichloropropane also shows three sets of signals associated with CH_3, CH_2, and CH groups. In contrast, however, their chemical shifts are quite different: Each of the last two groups now bears a halogen atom and gives rise to a low-field signal as a result ($\delta = 3.68$ ppm for the CH_2,* and $\delta = 4.17$ ppm for the CH). Only one signal, shown by integration to represent three hydrogens and therefore the CH_3 group, is at relatively high field ($\delta = 1.70$ ppm).

Finally, 1,3-dichloropropane shows only two peaks ($\delta = 3.71$ and 2.25 ppm) in a relative ratio of $2:1$, a pattern clearly distinct from those of the other two isomers. By this means, the structures of the three products are readily assigned by a simple measurement.

EXERCISE 10-6

Chlorination of chlorocyclopropane gives three compounds of molecular formula $C_3H_4Cl_2$. Draw their structures and describe how you would differentiate them by ^1H NMR. (**Hint:** Look for symmetry. Use the deshielding effect of chlorine and integration.)

To summarize, the NMR spectrometer in the integration mode records the relative areas under the various peaks, values that represent the relative numbers of hydrogens giving rise to these absorptions. This information, in conjunction with chemical shift data, can be used for structure elucidation—for example, in the identification of isomeric compounds.

10-7 Spin–Spin Splitting: The Effect of Nonequivalent Neighboring Hydrogens

The high-resolution NMR spectra presented so far have rather simple line patterns—single sharp peaks, also called **singlets.** The compounds giving rise to these spectra

---

*This description is somewhat simplified. The two hydrogens of this CH_2 group are not equivalent, because the presence of the adjacent stereocenter at C2 renders the molecule chiral and devoid of mirror symmetry. Indeed, on a high-field instrument with greater resolving power, these hydrogens have measurably different chemical shifts ($\delta = 3.62$ and 3.74).

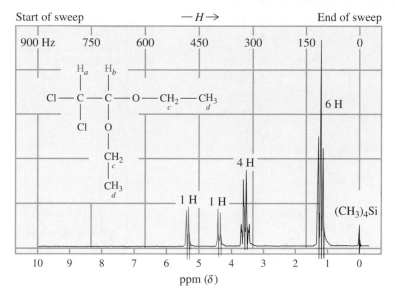

FIGURE 10-16 ————
Spin–spin splitting in the 90-MHz spectrum of 1,1-dichloro-2,2-diethoxyethane in CCl_4. The splitting patterns include two doublets, one triplet, and one quartet for the four types of protons. These multiplets reveal the effect of adjacent hydrogens. Note: The relative assignments of H_a and H_b are not obvious (see Table 10-3) and can be made only on consideration of additional data.

have one feature in common: In each compound, nonequivalent hydrogens are separated by at least one carbon or oxygen atom. These examples were chosen for good reason, because neighboring hydrogen nuclei can complicate the spectrum as the result of a phenomenon called **spin–spin splitting** or **spin–spin coupling.**

Figure 10-16 shows that the NMR spectrum of 1,1-dichloro-2,2-diethoxyethane has four absorptions, characteristic of four sets of hydrogens (H_a–H_d). Instead of single peaks, they adopt more complex patterns called **multiplets:** two two-peak absorptions, or **doublets** (blue and green), one of four peaks, **quartet** (black), and one of three, **triplet** (red). The detailed appearance of these multiplets depends on the number and kind of hydrogen atoms directly adjacent to the nuclei giving rise to the absorption.

In conjunction with chemical shifts and integration, spin–spin splitting frequently helps us arrive at a complete structure for an unknown compound. How can it be understood?

One neighbor splits the signal of a resonating nucleus into a doublet

Let us first consider the two doublets of relative integration 1, assigned to the two single hydrogens H_a and H_b. The splitting of these peaks is explained by the behavior of nuclei in an external magnetic field: They are like tiny magnets aligned with (α) or against (β) the field. The energy difference between the two states is minuscule (see Section 10-3), and at room temperature their populations are nearly equal. In the case under consideration, this means that there are two magnetic types of H_a—approximately half next to an H_b in the α state, the other half with a neighboring H_b in its β state. Conversely, H_b has two types of neighboring H_a—half of them in the α, half in the β state. What are the consequences of this phenomenon in the NMR spectrum?

A proton of type H_a that has as its neighbor an H_b aligned *with* the field is exposed to a total magnetic field that is strengthened by the addition of that due to the α spin of H_b. To achieve resonance for this type of H_a, a smaller external field strength is required than that necessary for H_a in the absence of a perturbing neighbor. A peak

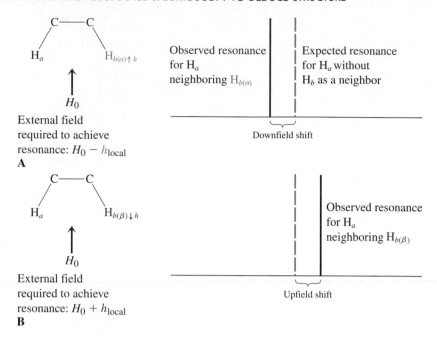

FIGURE 10-17

The effect of a hydrogen nucleus on the chemical shift of its neighbor is an example of spin–spin splitting. Two peaks are generated, because the hydrogen under observation has two types of neighbors. (A) When the neighboring nucleus H_b is in its α state, it contributes a local field h_{local}, so the external field needed for H_a to achieve resonance is reduced from H_0 to $H_0 - h_{local}$, thereby resulting in a downfield shift of the H_a peak. (B) When the neighboring nucleus is in its β state, its local field opposes the external one, and as a result an increase in the external field from H_0 to $H_0 + h_{local}$ is necessary for resonance to occur. The H_a peak is shifted to higher field.

at lower field than that expected is indeed observed (Figure 10-17A). However, this absorption is due to only half the H_a protons. The other half has H_b in its β state as a neighbor. Because H_b in its β state is aligned *against* the external field, the strength of the local field around H_a in this case is *diminished*. To achieve resonance, the external field H_0 has to be increased; an upfield shift is observed (Figure 10-17B).

Because the local contribution of H_b to H_0, whether positive or negative, is of the same magnitude, the downfield shift of the hypothetical signal equals the upfield shift. The single absorption expected for a neighbor-free H_a is said to be *split* by H_b into a doublet. Integration of each peak of this doublet shows a 50% contribution of each hydrogen. The chemical shift of H_a is reported as the center of the doublet (Figure 10-18).

The signal for H_b is subject to similar considerations. This hydrogen also has two types of hydrogens as neighbors—$H_{a(\alpha)}$ and $H_{a(\beta)}$. Consequently its absorption lines appear in the form of a doublet. So, in NMR jargon, H_b is split by H_a. The amount of this mutual splitting is equal; that is, the distance (in hertz) between the individual peaks making up each doublet is identical. This distance is termed the **coupling constant, J.** In our example, $J_{ab} = 7$ Hz (Figure 10-18). Because the coupling con-

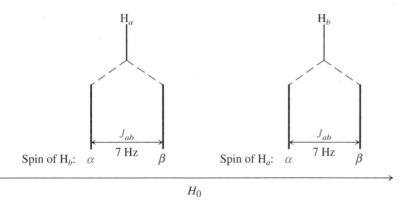

FIGURE 10-18

Spin–spin splitting between H_a and H_b in 1,1-dichloro-2,2-diethoxyethane. The coupling constant J_{ab} is the same for both doublets. The chemical shift is reported as the center of the doublet in the following format: $\delta_{H_a} = 5.36$ ppm (d, $J = 7$ Hz, 1 H), $\delta_{H_b} = 4.39$ ppm [d, $J = 7$ Hz, 1 H, in which "d" stands for the splitting pattern (doublet) and the last entry refers to the integrated value of the absorption].

stant is related only to magnetic field contributions by neighboring nuclei, it is *independent of the external field strength*. Coupling constants *remain unchanged* regardless of the field strength of the NMR instrument being used.

Spin–spin splitting is generally observed only between hydrogens that are immediate neighbors, bound either to the same carbon atom [**geminal coupling** (*geminus*, Latin, twin)] or to two adjacent carbons [**vicinal coupling** (*vicinus*, Latin, neighbor)]. Hydrogen nuclei separated by more than two carbon atoms are usually too far apart to exhibit appreciable coupling.

Note also that *the NMR signals of nuclei with equivalent chemical shifts do not exhibit spin–spin splitting.* Thus, the NMR spectrum of ethane, CH_3–CH_3, consists of a *single line* at $\delta = 0.85$ ppm. Splitting is observed only between nuclei with *different* chemical shifts.

Local-field contributions from more than one hydrogen are additive

How do we handle nuclei with two or more neighboring hydrogens? It turns out that we must consider the effect of each neighbor separately. Let us return to the spectrum of 1,1-dichloro-2,2-diethoxyethane shown in Figure 10-16. In addition to the two doublets assignable to H_a and H_b, this spectrum records a triplet due to the methyl protons H_d and a quartet assignable to the methylene hydrogens H_c. Because these two nonequivalent sets of nuclei are next to each other, vicinal coupling is observed as expected. However, compared with the peak patterns for H_a and H_b, those for H_c and H_d are considerably more complicated. They can be understood by expanding on the explanation used for the mutual coupling of H_a and H_b.

Consider first the triplet whose chemical shift and integrated value allow it to be assigned to the hydrogens H_d of the two methyl groups. Instead of one peak, we observe three, in the approximate ratio 1:2:1. The splittings must be due to coupling to the adjacent methylene groups—but how?

The three equivalent methyl hydrogens in each ethoxy group have two equivalent methylene hydrogens as their neighbors, and each of these methylene hydrogens may

Coupling Between Close-Lying Hydrogens

H_a $\quad$ H_b

J_{ab}, geminal coupling,
variable 0–18 Hz

H_a $\quad$ H_b

J_{ab}, vicinal coupling,
typically 6–8 Hz

H_a $\quad$ H_b

J_{ab}, 1,3-coupling,
usually negligible

FIGURE 10-19 _____

Nucleus H_d is represented by a three-peak NMR pattern because of the presence of three magnetically nonequivalent neighbor combinations: $H_{c(\alpha\alpha)}$, $H_{c(\alpha\beta \text{ and } \beta\alpha)}$, and $H_{c(\beta\beta)}$. The chemical shift of the absorption is reported as that of the center line of the triplet: $\delta_{H_d} = 1.23$ ppm (t, $J = 8$ Hz, 6 H, in which "t" stands for triplet).

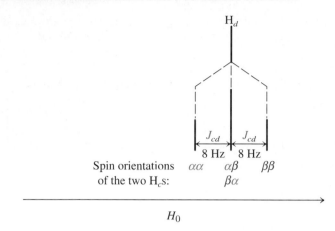

adopt the α or β spin orientation. Thus, each H_d may "see" its two H_c neighbors as an $\alpha\alpha$, $\alpha\beta$, $\beta\alpha$, or $\beta\beta$ combination (Figure 10-19). Those methyl hydrogens that are adjacent to the first possibility, $H_{c(\alpha\alpha)}$, are exposed to a twice-strengthened local field and give rise to a lower-field absorption. In the $\alpha\beta$ or $\beta\alpha$ combination, one of the H_c nuclei is aligned with the external field and the other is opposed to it. The net result is no net local-field contribution at H_d. In these cases, a spectral peak should appear at a chemical shift identical with the one expected if there were no coupling between H_c and H_d. Moreover, because *two* equivalent combinations of neighboring H_cs [$H_{c(\alpha\beta)}$ and $H_{c(\beta\alpha)}$] contribute to this signal [instead of only one, as did $H_{c(\alpha\alpha)}$ to the first peak], its approximate height should be double that of the first peak, as observed. Finally, H_d may have the $H_{c(\beta\beta)}$ combination as its neighbor. In this case, the local field subtracts from the external field, and an upfield peak of relative intensity 1 is produced. The resulting pattern for H_d is a *1:2:1 triplet* with a total integration corresponding to six hydrogens (because there are two methyl groups). The coupling constant J_{cd}, measured as the distance between each pair of adjacent peaks, is 8 Hz.

The quartet observed in Figure 10-16 for H_c can be analyzed in the same manner (Figure 10-20). This nucleus is exposed to four different types of H_d proton combinations as neighbors: one in which all protons are aligned with the field [$H_{d(\alpha\alpha\alpha)}$]; three equivalent arrangements in which one H_d is opposed to the external field and

FIGURE 10-20 _____

Splitting of H_c into a quartet by the various spin combinations of H_d. The chemical shift of the quartet is reported as its midpoint: $\delta_{H_c} = 3.63$ ppm (q, $J = 8$ Hz, 4 H, in which "q" stands for quartet).

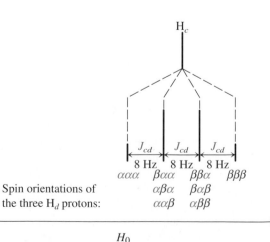

the other two are aligned with it [H$_{d(\beta\alpha\alpha,\ \alpha\beta\alpha,\ \alpha\alpha\beta)}$]; another set of three equivalent arrangements in which only one proton remains aligned with the field [H$_{d(\beta\beta\alpha,\ \beta\alpha\beta,\ \alpha\beta\beta)}$]; and a final possibility in which all H$_d$s oppose the external magnetic field [H$_{d(\beta\beta\beta)}$]. The resulting spectrum is predicted—and observed—to consist of a *1:3:3:1 quartet* (integrated intensity 4). The coupling constant J_{cd} is identical with that measured in the triplet for H$_d$ (8 Hz).

In many cases, spin–spin splitting is given by the *N* + 1 rule

We can summarize our analysis so far by a set of simple rules:

1. Equivalent nuclei located adjacent to one neighboring hydrogen resonate as a *doublet*.
2. Equivalent nuclei located adjacent to two hydrogens of a second set of equivalent nuclei resonate as a *triplet*.
3. Equivalent nuclei located adjacent to a set of three equivalent hydrogens resonate as a *quartet*.

Table 10-4 shows the expected splitting patterns for nuclei adjacent to *N* equivalent neighbors. The NMR signals of these nuclei *split into N + 1 peaks,* a result known as the *N* + **1 rule.** Their relative ratio is given by a mathematical mnemonic device called Pascal's* triangle. Each number in this triangle is the sum of the two numbers closest to it in the line above. The splitting patterns of two common alkyl groups,

*Blaise Pascal (1623–1662), French mathematician, physicist, and religious philosopher.

TABLE 10-4	NMR Splittings of a Set of Hydrogens with *N* Equivalent Neighbors and Their Integrated Ratios (Pascal's Triangle)		
Equivalent neighboring (*N*) hydrogens	Number of peaks (*N* + 1)	Name for peak pattern (abbreviation)	Integrated ratios of individual peaks
0	1	Singlet (s)	1
1	2	Doublet (d)	1:1
2	3	Triplet (t)	1:2:1
3	4	Quartet (q)	1:3:3:1
4	5	Quintet (quin)	1:4:6:4:1
5	6	Sextet (sex)	1:5:10:10:5:1
6	7	Septet (sep)	1:6:15:20:15:6:1

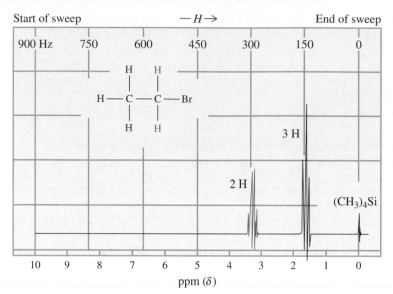

FIGURE 10-21 _____
The 90-MHz NMR spectrum of bromoethane in CCl_4 illustrates the $N + 1$ rule. The methylene group, which has three equivalent neighbors, appears as a quartet at $\delta = 3.24$ ppm, $J = 7$ Hz. The methyl hydrogens, which have two equivalent neighbors, absorb as a triplet at $\delta = 1.58$ ppm, $J = 7$ Hz.

ethyl and 1-methylethyl (isopropyl), are shown in Figures 10-21 and 10-22, respectively. In both spectra, integrations of the multiplets reveal the relative number of hydrogens responsible for each multiplet.

It is important to remember that nonequivalent nuclei mutually split one another. In other words, the observation of one split absorption necessitates the presence of another split signal in the spectrum. Moreover, the coupling constants for these patterns must be the same. Some frequently encountered multiplets and the corresponding structural units are shown in Table 10-5.

EXERCISE 10-7

Predict the NMR spectra of (**a**) ethoxyethane (diethyl ether); (**b**) 1,3-dibromopropane; (**c**) 2-methyl-2-butanol; (**d**) 1,1,2-trichloroethane. Specify approximate chemical shifts, relative abundance (integration), and multiplicities.

FIGURE 10-22 _____
90-MHz NMR spectrum of 2-iodopropane in CCl_4: $\delta = 4.12$ (sep, $J = 7.5$ Hz, 1 H), 1.82 (d, $J = 7.5$ Hz, 6 H). The six equivalent nuclei on the two methyl groups give rise to a septet for the tertiary hydrogen ($N + 1$ rule). Note that the outer peaks of the septet are of such small intensity that they are difficult to see in the spectrum recorded on scale. It is therefore frequently advisable to "blow up" split peaks in intensity to clarify some of their features. Such an enlargement is shown in the inset: The septet for the tertiary hydrogen has been rerecorded at higher sensitivity.

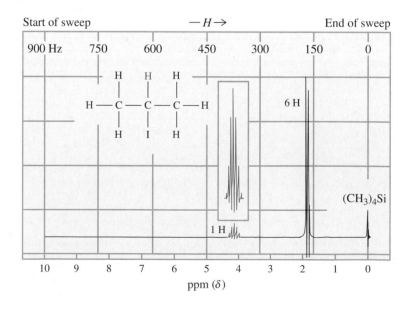

TABLE **10-5**	Frequently Observed Spin–Spin Splittings in Common Alkyl Groups	

Splitting Pattern for H_a	Structure	Splitting Pattern for H_b

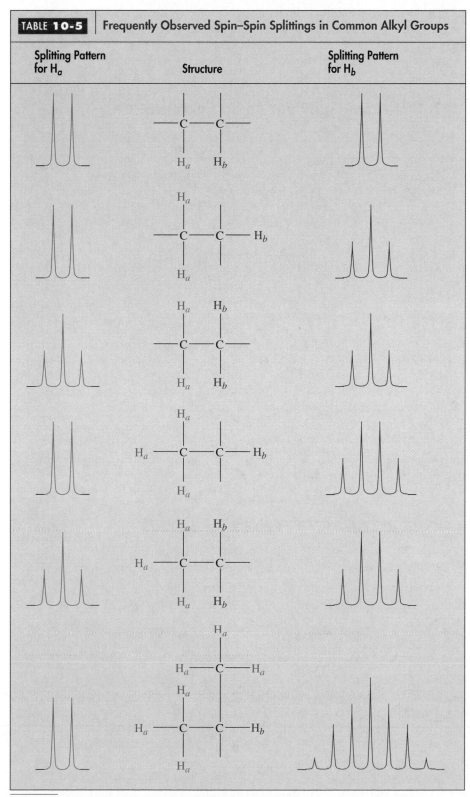

To summarize, spin–spin splitting occurs between vicinal and geminal nonequivalent hydrogens. Usually N equivalent neighbors will split the absorption of the observed hydrogen into $N + 1$ peaks, their relative intensities being in accord with Pascal's triangle. The common alkyl groups give rise to characteristic NMR patterns.

10-8 Spin–Spin Splitting: Some Complications

The rules governing the appearance of split peaks outlined in Section 10-7 are somewhat idealized. There are cases in which, because of a relatively small difference in δ between two absorptions, more complicated patterns (complex multiplets) are observed that are not interpretable without the use of computers. Moreover, the $N + 1$ rule may not be applicable in a direct way if two or more types of neighboring hydrogens are coupled to the resonating nucleus with fairly different coupling constants. Finally, the hydroxy proton may appear as a singlet (see Figure 10-11) even if vicinal hydrogens are present. Let us look in turn at each of these complications.

Close-lying peak patterns may give rise to non-first-order spectra

A careful look at the spectra shown in Figures 10-16, 10-21, and 10-22 reveals that the relative intensities of the splitting patterns do not conform to the idealized peak ratios expected from consideration of Pascal's triangle: The patterns are not completely symmetrical, but skewed. Specifically, the two multiplets of two mutually coupled hydrogens are skewed toward each other such that the intensity of the lines facing each other is slightly larger than expected. The exact intensity ratios dictated by Pascal's triangle and the $N + 1$ rule are observed only when the difference between the resonance frequencies of coupled protons is much larger than their coupling constant: $\Delta \nu \gg J$. Under these circumstances, the spectrum is said to be **first order.*** However, when these differences become smaller, the expected peak pattern is subject to increasing distortion.

In extreme cases, the simple rules devised in Section 10-7 do not apply any more, the resonance absorptions assume more complex shapes, and the spectra are said to be **non-first order.** Although such spectra can be simulated with the help of computers, this treatment is beyond the scope of the present discussion.

Particularly striking examples of non-first-order spectra are those of compounds containing alkyl chains. Figure 10-23 shows an NMR spectrum of octane, which is not first order, because all nonequivalent hydrogens (there are four types) have very similar chemical shifts. All methylenes absorb as one broad multiplet. In addition, there is a highly distorted triplet for the terminal methyl groups.

Non-first-order spectra arise when $\Delta \nu \sim J$, so it should be possible to "improve" the appearance of a multiplet by measuring a spectrum at higher field because the resonance frequency is proportional to the external field strength, whereas the coupling constant J is independent of field (Section 10-7).

Improved field strength has a dramatic effect on the spectrum of 2-chloro-1-(2-chloroethoxy)ethane (Figure 10-24). In this compound, the deshielding effect of the oxygen is about equal to that of the chlorine substituent. As a consequence, the two sets of methylene hydrogens give rise to very close lying peak patterns. The re-

---

*This expression derives from the term *first-order theory;* that is, one that takes into account only the most important variables and terms of a system.

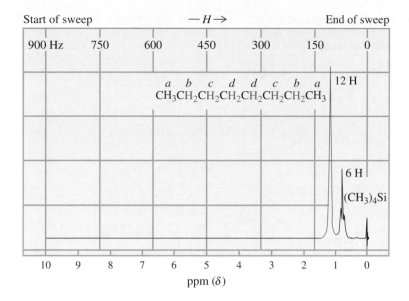

FIGURE 10-23
90-MHz NMR spectrum of octane in CCl₄. Compounds containing alkyl chains often display such non-first-order patterns.

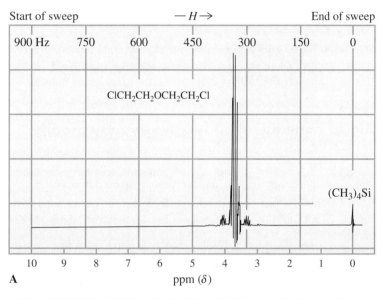

A

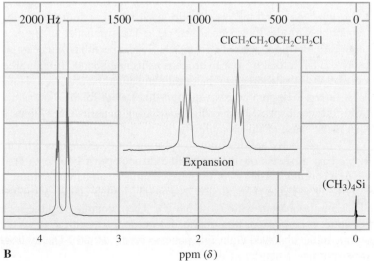

B

FIGURE 10-24
The effect of increased field strength on a non-first-order NMR spectrum: 2-chloro-1-(2-chloroethoxy)ethane at (A) 90 MHz; (B) 500 MHz in CCl₄. At high field strength, the complex multiplet observed at 90 MHz is simplified into two slightly distorted triplets, as might be expected for two mutually coupled CH₂ groups.

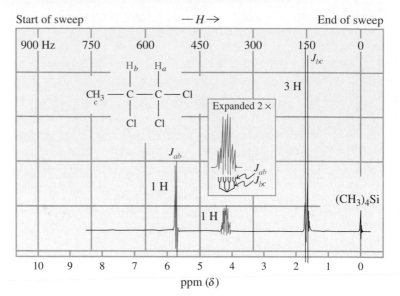

FIGURE 10-25
90-MHz NMR spectrum of 1,1,2-trichloropropane in CCl_4. Nucleus H_b gives rise to a quartet of doublets at $\delta = 4.18$ ppm: eight peaks.

sulting absorption has a symmetric shape but is very complicated, exhibiting more than 32 peaks of various intensities. However, recording the NMR spectrum with a 500-MHz spectrometer (Figure 10-24B) produces a first-order pattern.

Coupling to nonequivalent neighbors may modify the $N + 1$ rule

When hydrogens are coupled to two sets of nonequivalent neighbors, complicated splitting patterns may result. The spectrum of 1,1,2-trichloropropane illustrates this point (Figure 10-25). In this compound, the hydrogen at C2 is located between a methyl and a $CHCl_2$ group, and it is coupled to the hydrogens of each group independently.

Let us analyze the spectrum in detail. We first notice two doublets, one at low field ($\delta = 5.69$ ppm, $J = 3.6$ Hz, 1 H) and one at high field ($\delta = 1.64$ ppm, $J = 6.8$ Hz, 3 H). The low-field absorption is assignable to the hydrogen at C1 (H_a), adjacent to two deshielding halogens; and the methyl hydrogens (H_c) resonate as expected at highest field. In accord with the $N + 1$ rule, each signal is split into a doublet because of coupling with the hydrogen at C2 (H_b). The resonance of the latter, however, is quite different in appearance from what we expect. The nucleus giving rise to this absorption has a total of four hydrogens as its neighbors: H_a and three H_c. Application of the $N + 1$ rule suggests that a quintet should be observed. However, the signal for H_b at $\delta = 4.18$ ppm consists of *eight* lines, with relative intensities that do not conform to those expected for ordinary splitting patterns (see Tables 10-4 and 10-5). What is the cause of this complexity?

The $N + 1$ rule strictly applies only to splitting by *equivalent* neighbors. In this molecule, we have two sets of different adjacent nuclei that couple to H_b *with different coupling constants*. The effect of these couplings can be understood, however, if we apply the $N + 1$ rule sequentially. The methyl group causes a splitting of the H_b resonance into a quartet, with $J_{bc} = 6.8$ Hz. Then, coupling by the hydrogen at C1 further splits *each peak* in this quartet into a doublet, with $J_{ab} = 3.6$ Hz, both splits resulting in the observed eight-line pattern (Figure 10-26). The hydrogen at C2 is said to be split into a quartet of doublets.

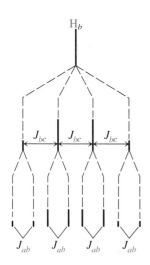

FIGURE 10-26
The splitting pattern for H_b in 1,1,2-trichloropropane follows the sequential $N + 1$ rule. Each of the four lines arising from coupling to the methyl group is further split into a doublet by the hydrogen on C1.

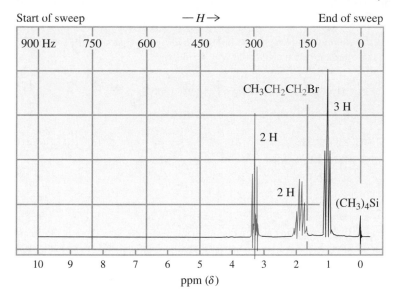

FIGURE 10-27
90-MHz NMR spectrum of 1-bromopropane in CCl₄.

The hydrogens on C2 of 1-bromopropane also couple to two nonequivalent sets of neighbors. In this case, however, the resulting splitting pattern appears to conform with the $N + 1$ rule, and a (slightly distorted) sextet is observed (Figure 10-27). The reason is that the coupling constants to the two different groups are very similar, about 6–7 Hz. Although an analysis similar to that given earlier for 1,1,2-trichloropropane would lead us to predict as many as 12 lines in this signal (a quartet of triplets), the nearly equal coupling constants cause many of the lines to overlap, thus simplifying the pattern (Figure 10-28). The hydrogens in many simple alkyl derivatives display similar coupling constants and, therefore, spectra that are in accord with the $N + 1$ rule.

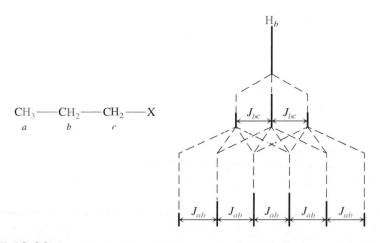

FIGURE 10-28
Splitting pattern expected for H_b in a propyl derivative when $J_{ab} \sim J_{bc}$. Several of the peaks coincide, giving rise to a deceptively simple spectrum: a sextet.

EXERCISE 10-8

Predict the coupling patterns for the boldface hydrogens, first according to the $N + 1$ rule and then according to the sequential $N + 1$ rule, in:

(a) $BrCH_2CH_2CH_2Cl$

(b) $CH_3CHCHCl_2$
  |
  OCH_3

(c) $Cl_2CHCHCH_2CH_3$
  |
  CH_2CH_3

(d) [ring structure: O, S with CH_3 **H**]

(e) $(CH_3)_2CHCH_2OH$

EXERCISE 10-9

In Section 10-6 you learned how to distinguish between the three monochlorination products of 1-chloropropane—1,1-, 1,2-, and 1,3-dichloropropane—by using just chemical shift data and chemical integration. Would you also expect to tell them apart on the basis of their coupling patterns?

Fast proton exchange decouples hydroxy hydrogens

With our knowledge of vicinal coupling, let us now return to the NMR of alcohols. We note in the NMR spectrum of 2,2-dimethyl-1-propanol (Figure 10-11) that the OH absorption appears as a single peak, devoid of any splitting. This is curious, because the hydrogen is adjacent to two others, which should cause its appearance as a triplet. The CH_2 hydrogens that show up as a singlet should in turn appear as a doublet with the same coupling constant. Why, then, do we not observe spin–spin splitting? It is because the weakly acidic OH hydrogens are rapidly transferring both between alcohol molecules and to traces of water on the NMR time scale at room temperature. As a consequence of this process, the NMR spectrometer sees only an average signal for the OH hydrogen. No coupling is visible, because the binding time of the proton to the oxygen is too short (about 10^{-5} s). It follows that the CH_2 nuclei are similarly uncoupled, a condition resulting in the observed singlet.

Rapid Proton Exchange Averages the Local Magnetic Field Experienced by the OH Hydrogen

And That Experienced by Its Neighbors

Absorptions of this type are said to be **decoupled** by **fast proton exchange.** The exchange may be slowed by removal of traces of water or acid or by cooling. In these

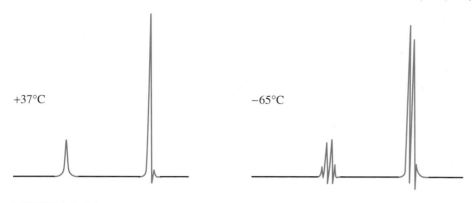

FIGURE 10-29

Temperature dependence of spin–spin splitting in methanol. The singlets at 37°C illustrate the effect of fast proton exchange in alcohols. *(After H. Günther, NMR-Spektroskopie, Georg Thieme Verlag, Stuttgart, 1973.)*

cases, the OH bond retains its integrity long enough (more than 1 s) for coupling to be observed on the NMR time scale. An example is shown in Figure 10-29 for methanol. At 37°C, two singlets are observed, corresponding to the two types of hydrogens, both devoid of spin–spin splitting. However, at −65°C, the expected coupling pattern is detectable: a quartet and a doublet.

Rapid magnetic exchange "self-decouples" chlorine, bromine, and iodine nuclei

All the halogen nuclei are magnetic. Therefore, the ^1H NMR spectra of haloalkanes would be expected to exhibit spin–spin splitting owing to their presence (in addition to the normal H–H coupling). In practice, only fluorine exhibits this effect, in much the same way as the proton does but with much larger J values. Thus, for example, the ^1H NMR spectrum of CH_3F exhibits a doublet with $J = 81$ Hz. Because fluoroorganic compounds are relatively rare, we will not deal with their NMR spectroscopy any further.

Turning to the other halides, inspection of the spectra of the haloalkanes depicted in Figures 10-16, 10-21, 10-22, 10-24, 10-25, and 10-27 (fortunately) reveals the *absence* of any visible spin–spin splitting by these nuclei. The reason for this observation lies in their relatively fast internal magnetic equilibration on the NMR time scale, precluding their recognition by the adjacent hydrogens as having differing alignments with respect to the external magnetic field H_0. They "self-decouple," in contrast with the "exchange-decoupling" exhibited by the hydroxy protons.

In summary, the peak patterns in many NMR spectra are not first order, because the differences between the chemical shifts of nonequivalent hydrogens are close to the values of the corresponding coupling constants. Use of higher-field NMR instruments may improve the appearance of such spectra. Coupling to nonequivalent hydrogen neighbors occurs separately, with different coupling constants. In some cases, they are sufficiently dissimilar to allow for an analysis of the multiplets. In many simple alkyl derivatives, they are sufficiently similar ($J = 6$–7 Hz) that the spectra observed are simplified to those predicted in accordance with the $N + 1$ rule. Vicinal coupling through the oxygen in alcohols is not frequently observed, because of decoupling by fast proton exchange.

10-9 Carbon-13 Nuclear Magnetic Resonance

Proton nuclear magnetic resonance is a powerful method for determining organic structures because most organic compounds contain hydrogens. Of even greater potential utility is NMR spectroscopy of carbon. After all, by definition, *all* organic compounds contain this element. In combination with ^{1}H NMR, it has become the most important analytical tool in the hands of the organic chemist. This section will summarize some of the essential aspects of this technique.

Carbon NMR utilizes an isotope in low natural abundance: ^{13}C

Carbon NMR is possible. However, there is a complication: The most abundant isotope of carbon, carbon-12, is not detectable by NMR. Fortunately, another isotope, carbon-13, is present in nature at a level of about 1.11%. Its behavior in the presence of a magnetic field is the same as that of hydrogen. One might therefore expect it to give spectra very similar to those observed in ^{1}H NMR spectroscopy. This expectation turns out to be only partly correct, because of several important (and very useful) differences between the two types of NMR techniques.

Carbon-13 NMR (^{13}C NMR) spectra used to be much more difficult to record than hydrogen spectra, not only because of the low natural abundance of the nucleus under observation, but also because of the much weaker magnetic resonance of ^{13}C. Thus, under comparable conditions, ^{13}C signals are about 1/6000 as strong as those for hydrogen. These problems have been overcome with the advent of routine Fourier transform NMR (FT NMR; Section 10-3), which allows the rapid recording of ^{13}C NMR spectra.

One advantage of the low abundance of ^{13}C is the absence of carbon–carbon coupling. Just like hydrogens, two adjacent carbons, if magnetically nonequivalent (as they are in, e.g., bromoethane), split each other. In practice, however, such splitting is not observed. Why? Because coupling can occur only if two ^{13}C isotopes come to lie next to each other. With the abundance of ^{13}C in the molecule at 1.11%, this event has a very low probability (roughly 1% of 1%; i.e., 1 in 10,000). Most ^{13}C nuclei are surrounded by only ^{12}C nuclei, which, having no spin, do not give rise to spin–spin splitting. This feature simplifies ^{13}C NMR spectra appreciably, reducing the problem of their analysis to a determination of the coupling patterns to any attached hydrogens.

Figure 10-30 depicts the ^{13}C NMR spectrum of bromoethane (for the ^{1}H NMR spectrum, see Figure 10-21). The chemical shift δ is defined as in ^{1}H NMR and is determined relative to an internal standard, normally the carbon absorption in $(CH_3)_4Si$. The chemical-shift range of carbon is much larger than that of hydrogen. For most organic compounds, it covers a distance of about 200 ppm, in contrast with the relatively narrow spectral "window" (10 ppm) of hydrogen. Figure 10-30 reveals the relative complexity of the absorptions caused by extensive ^{13}C–H spin–spin splittings. Not surprisingly, directly bound hydrogens are most strongly coupled ($\sim$125–200 Hz). Coupling tapers off however, with increasing distance from the ^{13}C nucleus under observation, such that the geminal coupling constant $J_{^{13}C-C-H}$ is in the range of only 0.7–6.0 Hz.

EXERCISE 10-10

Predict the ^{13}C NMR spectral pattern of 1-bromopropane (for the ^{1}H NMR spectrum, see Figure 10-27). (**Hint:** Use the sequential $N + 1$ rule.)

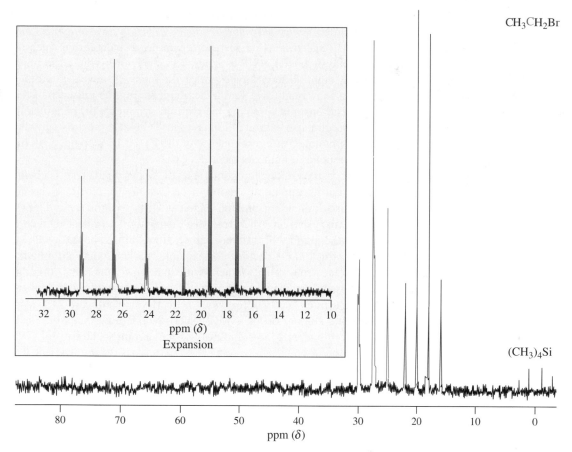

CH₃CH₂Br

(CH₃)₄Si

FIGURE 10-30

62.8-MHz ^{13}C NMR spectrum of bromoethane, showing the complexity of ^{13}C–H coupling. There is an upfield quartet (δ = 18.3 ppm, J = 126 Hz) and a downfield triplet (δ = 26.6 ppm, J = 151 Hz) resonance for the two carbon atoms. Note the large chemical-shift range. Tetramethylsilane, defined to be located at δ = 0 ppm as in ^{1}H NMR, absorbs as a quartet (J = 118 Hz; the two outside peaks are barely visible) because of coupling of each carbon to three equivalent hydrogens. The inset shows a part of the spectrum expanded horizontally to reveal the fine splitting of each of the main peaks that is due to coupling of each ^{13}C with protons on the neighboring carbon.

Hydrogen decoupling gives single lines

A technique that completely removes ^{13}C–H coupling is called **broad-band hydrogen** (or **proton**) **decoupling.** This method employs a strong, broad radio-frequency signal that covers the resonance frequencies of all the hydrogens and is applied at the same time as the ^{13}C spectrum is recorded. For example, in a magnetic field of 58,750 G, carbon-13 resonates at 62.8 MHz, hydrogen at 250 MHz. To obtain a proton-decoupled carbon spectrum at this field strength, we irradiate the sample at both frequencies. The first radio-frequency signal is used to produce carbon magnetic resonance. Simultaneous exposure to the second signal causes all the hydrogens to undergo rapid α–β spin flips, fast enough to average their local magnetic field contributions. The net result is the absence of coupling. Use of this technique simplifies the ^{13}C NMR spectrum of bromoethane to two single lines, as shown in Figure 10-31 (see page 420).

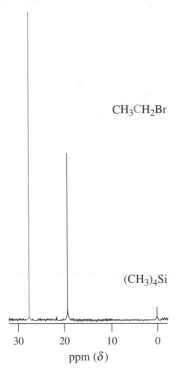

FIGURE 10-31

This 62.8-MHz ^{13}C NMR spectrum of bromoethane was recorded with broad-band decoupling at 250 MHz. All lines simplify to singlets, including the absorption for $(CH_3)_4Si$.

The power of proton decoupling becomes evident when spectra of relatively complex molecules are recorded. *Every magnetically distinct carbon gives only one single peak in the ^{13}C NMR spectrum.* Consider, for example, a hydrocarbon such as methylcyclohexane. Analysis by ^{1}H NMR is made very difficult by the close-lying chemical shifts of the eight different types of hydrogens. However, a proton-decoupled ^{13}C spectrum shows only five peaks, clearly depicting the presence of the five different types of carbons and revealing the twofold symmetry in the structure (Figure 10-32). These spectra also exhibit a limitation in ^{13}C NMR spectroscopy: Integration is not usually possible. As a consequence of the FT NMR method, peak intensities no longer correspond to numbers of nuclei.

Table 10-6 shows that carbon, like hydrogen (Table 10-2), has characteristic chemical shifts depending on its structural environment. As in ^{1}H NMR, electron-withdrawing groups cause deshielding, and the chemical shifts go up in the order primary < secondary < tertiary carbon. Apart from the diagnostic usefulness of such δ values, a knowledge of the number of different carbon atoms in the molecule can be an aid to structural identification. Consider, for example, the analytical differentiation of methylcyclohexane from other isomers with the same molecular formula, C_7H_{14}. Many of the possibilities have a different number of nonequivalent carbons incorporated in their structures and therefore give distinctly different carbon spectra (find some with the same number of ^{13}C NMR peaks). Notice how much the (lack of) symmetry in a molecule affects the complexity of the carbon spectrum.

Number of ^{13}C Peaks in Some C_7H_{14} Isomers

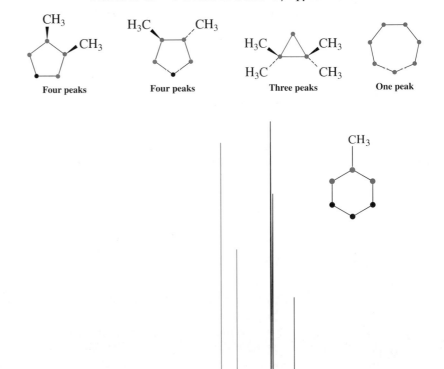

FIGURE 10-32

62.8-MHz ^{13}C NMR spectrum of methylcyclohexane with hydrogen decoupling (in C_6D_6). Each of the five magnetically different types of carbon in this compound gives rise to a distinct peak: δ = 23.1, 26.7, 26.8, 33.1, and 35.8 ppm.

TABLE 10-6 Typical ^{13}C NMR Chemical Shifts	
Type of carbon	**Chemical shift δ (ppm)**
Primary alkyl, RCH_3	5–20
Secondary alkyl, RCH_2R'	20–30
Tertiary alkyl, R_3CH	30–50
Quaternary alkyl, R_4C	30–45
Allylic, $R_2C{=}CCH_2R'$ with R''	20–40
Chloroalkane, RCH_2Cl	25–50
Bromoalkane, RCH_2Br	20–40
Ether or alcohol, RCH_2OR' or RCH_2OH	50–90
Carboxylic acids, RCOOH	170–180
Aldehyde or ketone, $R\overset{O}{\overset{\|}{C}}H$ or $R\overset{O}{\overset{\|}{C}}R'$	190–210
Alkene, aromatic, $R_2C{=}CR_2$	100–150
Alkyne, $RC{\equiv}CR$	65–95

EXERCISE 10-11

How many peaks would you expect in the proton-decoupled ^{13}C NMR spectra of the following compounds? (**Hint:** Look for symmetry.)

(a) 2,2-Dimethyl-1-propanol

(b)

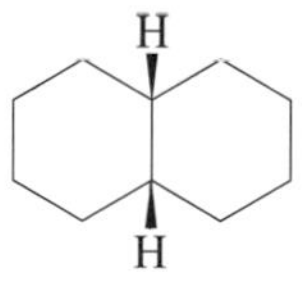

(c)

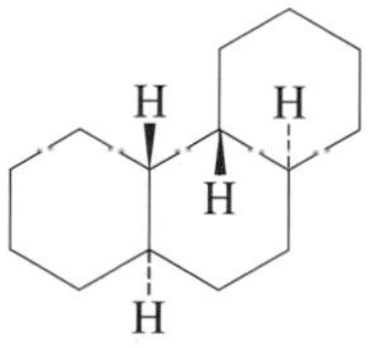

(d)

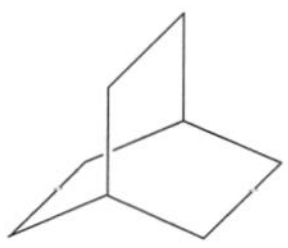

Advances in FT NMR are greatly aiding structure elucidation: DEPT ^{13}C NMR

The FT technique for the measurement of NMR spectra is extremely versatile, allowing data to be collected and presented in a variety of ways, each providing information about the structure of molecules. Most recent advances are due to the development of sophisticated time-dependent pulse sequences, including the application of two simultaneous but independent frequencies (*two-dimensional NMR,* or *2D-NMR).* With these methods, it is now possible to establish coupling (and therefore bonding or, at least, close proximity) between close-lying hydrogens *(homonuclear correlation)* or connected carbon and hydrogen atoms *(heteronuclear correlation)*. Thus, ^1H and ^{13}C NMR allow the determination of molecular connectivity by measuring the magnetic effect of neighboring atoms on one another along a carbon chain.

An example of such a pulse sequence, now routine in the research laboratory, is the **DEPT ^{13}C NMR** spectrum **(distortionless enhanced polarization transfer)**, which tells you what type of carbon gives rise to a specific signal in the normal ^{13}C spectrum: CH_3, CH_2, CH, or $C_{quaternary}$. It avoids the complications arising from a proton-coupled ^{13}C NMR spectrum (see Figure 10-30), particularly overlapping multiplets of close-lying carbon signals. The DEPT experiment consists of three spectra: the normal broad-band decoupled spectrum, a second spectrum with a pulse sequence (DEPT-90) that *reveals signals only of carbons bound to one hydrogen (CH)*, and, finally, a third spectrum (DEPT-135) that produces normal CH_3 and CH signals, but *negative absorptions for CH_2*, and no peaks for quaternary carbons. Figure 10-33 depicts a series of such spectra for limonene (see also Chapter 5 Opening and Chemical Highlight 11-3).

The first spectrum (Figure 10-33A) depicts the expected number of lines (10) and groups them into the six alkyl and four alkenyl carbon signals at high and at low field, respectively. The second spectrum (Figure 10-33B) directly identifies the CH carbons. These show up again in Figure 10-33C, with additional positive lines that must be

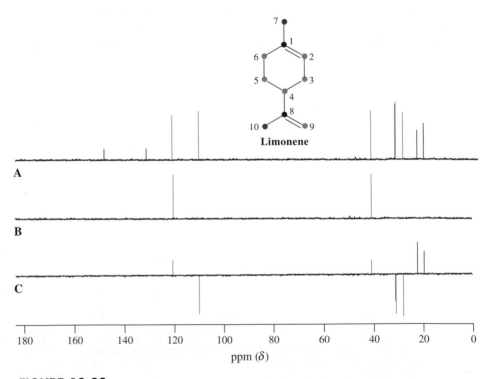

FIGURE 10-33

The DEPT ^{13}C NMR experiment with limonene: (A) broad-band decoupled spectrum revealing the six alkyl carbon signals at high field (20–40 ppm) and the four alkenyl carbon signals at low field (108–150 ppm; see Table 10-6); (B) DEPT-90 spectrum in which only the two CH signals for C2 and C4 appear; (C) DEPT-135 spectrum with positive absorptions for CH_3 (C7 and C10) and CH (C2 and C4) and negative peaks for CH_2 (C3, C5, C6, and C9) but no signals for C1 and C8.

due to CH$_3$ carbons, whereas the negative absorptions identify the CH$_2$ groups. Subtracting all the lines in Figure 10-33C from those in Figure 10-33A leaves the quaternary carbon signals.

CHEMICAL HIGHLIGHT 10-3

Structural Characterization of Medicinal Agents from Marine Sources

The sea is a rich source of pharmaceutically useful substances. The rare 25-carbon compound manoalide (a *sesterterpene;* see Section 4-7) was isolated from a sponge in 1977, but only 58 mg of the compound was obtained. As a result, elemental analysis, chemical tests, and any other procedures that would destroy the tiny amount of available material could not be used in structural elucidation. Instead, a combination of spectroscopic techniques (IR, UV, MS, and NMR) was employed, leading in 1980 to the structure shown here for manoalide, which was later confirmed by total synthesis.

The characterization of manoalide relied in part on ^{1}H and ^{13}C NMR data. For example, five signals for protons on double bonds or on carbons between two oxygens appear between $\delta = 4.8$ and 6.2 ppm (see Table 10-2). ^{13}C NMR shows the C=O carbon at $\delta = 172.3$ ppm and clearly distinguishes between alkene carbons ($\delta > 115$ ppm), the carbon adjacent to one ether oxygen ($\delta = 63.3$ ppm), the carbons between two oxygens ($\delta = 91.7$ and 99.1 ppm), and the remaining tetrahedral carbons (δ between 15 and 45 ppm; see Table 10-6), all of which were resolved and identified by the number of attached hydrogens. Although such a structural elucidation may seem extraordinary, improvements in instrumentation since 1980 now allow comparable procedures to be carried out on *micrograms* of a substance.

Clinical trials in 1990 revealed manoalide to be a potent drug with outstanding ability to block the release of the enzymes responsible for pain and inflammation; it thus constitutes a potential treatment for the symptoms of arthritis and muscular dystrophy. Additional aspects of the structural determination of manoalide will be presented in Chapter 20 (Chemical Highlights 20-1 and 20-4).

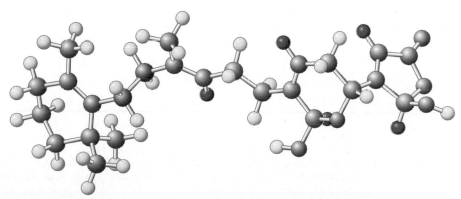

Manoalide

For the remainder of the text, whenever the depiction or description of a ^{13}C NMR spectrum includes spectral assignments to CH_3 CH_2, CH, or $C_{quaternary}$ carbons, they are based on a DEPT experiment.

We can apply ^{13}C NMR spectroscopy to the problem of the monochlorination of 1-chloropropane

In Section 10-6 we learned how we could distinguish between the three isomers of dichloropropane arising from the chlorination of 1-chloropropane by ^{1}H NMR chemical shifts and integration. Exercise 10-9 addressed the use of spin–spin splitting patterns as a complementary means of solving this problem. How does ^{13}C NMR fare in this task? Our prediction is straightforward: both 1,1- and 1,2-dichloropropane should exhibit three carbon signals each, but spaced significantly differently because the former has the two electron-withdrawing chlorine atoms located on the same carbon (hence no chlorines on the remaining two), whereas the latter bears one each on C1 and C2. In contrast, 1,3-dichloropropane would be clearly distinct from the other two isomers because of its symmetry: Only two lines should be observed. The experimental data are shown in the margin and confirm our expectations. The specific assignments of the two deshielded chlorine-bearing carbons in 1,2-dichloropropane (as indicated in the margin) can be made on the basis of the DEPT technique: The signal at 49.5 ppm appears inverted in DEPT-135 (CH_2), and that at 55.8 ppm is the only observable absorption in the DEPT-90 experiment.

You can see from this example how ^{1}H NMR and ^{13}C NMR spectroscopy complement each other. ^{1}H NMR spectra provide an estimate of the electronic environment (i.e., electron rich versus electron poor) of a hydrogen nucleus under observation (δ), a measure of its relative abundance (integration), and an indication of how many neighbors (and their number of types) it has (spin–spin splitting). Proton-decoupled ^{13}C NMR provides the total number of chemically distinct carbons, their electronic environment (δ), and, in the DEPT mode, even the quantity of their attached hydrogens. Application of both techniques to the solution of a structural problem is not unlike the methods used to solve a crossword puzzle. The horizontal entries (such as the data provided by ^{1}H NMR spectroscopy) have to fit the vertical ones (i.e., the corresponding ^{13}C NMR information) to provide the correct answer.

CH$_3$CH$_2$CHCl$_2$

10.1   34.9   73.2 ppm

CH$_3$CHClCH$_2$Cl

22.4   55.8   49.5 ppm

ClCH$_2$CH$_2$CH$_2$Cl

42.2   35.6 ppm

EXERCISE 10-12

Are bicyclic compounds A and B shown below readily distinguished by their proton-decoupled ^{13}C NMR spectra? Would DEPT spectra be of use in solving this problem?

H$_3$C   H   CH$_3$

H$_3$C   H   CH$_3$

A   B

In summary, ^{13}C NMR requires FT techniques because of the low natural abundance of the carbon-13 isotope and its intrinsically lower sensitivity in this experiment. ^{13}C–^{13}C coupling is not observed, because the scarcity of the isotope in the sample renders the likelihood of neighboring ^{13}C nuclei negligible. ^{13}C–H coupling can be measured but is usually removed by broad-band proton decoupling, providing single lines for each distinct carbon atom in the molecule under investigation. The ^{13}C NMR chemical-shift range is large, about 200 ppm for organic structures. ^{13}C NMR spectra cannot usually be integrated, but the DEPT experiment allows the identification of each signal as arising from CH_3, CH_2, CH, or $C_{quaternary}$ units, respectively.

CHAPTER INTEGRATION PROBLEM

A researcher executed the following reaction sequence in a preparation of (*S*)-2-chlorobutane:

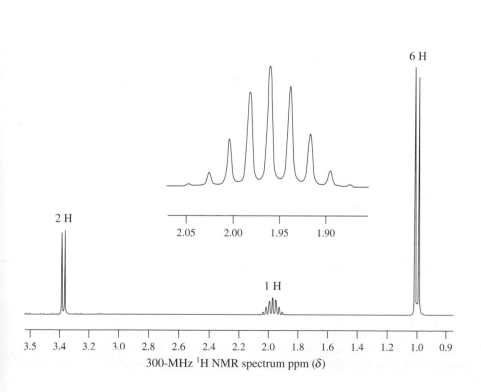

(*R*)-2-Methyloxacyclopropane   (*S*)-2-Chlorobutane

Careful preparative gas chromatography of the reaction product (b.p. 68.2°C) allowed the separation of a very small amount of another compound, C_4H_9Cl (b.p. 68.5°C), which was optically inactive and exhibited the NMR spectra depicted below. What is the structure of this compound and how could it have been formed?

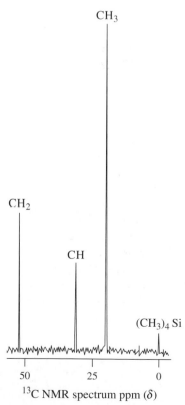

300-MHz ^1H NMR spectrum ppm (δ) ^{13}C NMR spectrum ppm (δ)

SOLUTION

As we have done for other synthetic and mechanistic problems (see Chapter Integration Problems in Chapters 8 and 9) it is important to take an inventory of what information is available and what it means (in general terms) before delving into the details of the case in hand. First, we have a reaction scheme in which an oxacyclopropane is treated with CH_3Li (Section 9-9) and the resulting alcohol with $SOCl_2$ (Section 9-4) to provide the chlorobutane with inversion of configuration. Somewhere along this sequence there is an opportunity to generate another product, albeit in only small amounts. Second, the side product has the molecular formula C_4H_9Cl, the same as 2-chlorobutane. In other words, it is an isomer. Third, although the compound contains four carbon atoms, the ^{13}C NMR spectrum (assignments by DEPT) shows only three lines at about $\delta = 20$ (CH_3), 31 (CH), and 52 (CH_2) ppm. The first two appear in the alkyl region, and the third must be due to the unique carbon bearing the chlorine (Table 10-6). Thus, one of the alkyl carbon peaks is due to two equivalent carbon nuclei. Fourth, the ^1H NMR spectrum (300 MHz) reveals three types of hydrogens at about $\delta = 1.0$, 1.9, and 3.4 ppm. As with the carbon spectrum, the most deshielded hydrogens must be those attached to the chlorine-bearing carbon (Section 10-4). The integrated values for the three proton signals are 6, 1, and 2, respectively, corresponding to the total number of nine hydrogens present (Section 10-6). Finally, there is spin–spin splitting. Both the highest and lowest field signals are doublets, with almost identical J values. This means that each of these sets of protons (6 + 2) has a single neighbor (the remaining H). That hydrogen shows up in the middle as a nine-line pattern, as expected by the $N + 1$ rule ($N = 6 + 2$; Section 10-7).

Now let us combine this information in a structural assignment. As with most puzzles, one can arrive at the answer in several ways. In NMR spectral problems, it is often best to start with the formulation of partial structures, as dictated by the ^1H NMR spectrum, and use the other information as corroborating evidence. Thus, the high-field doublet integrating for 6 H indicates a $(CH_3)_2CH–$ substructure. Similarly, the low-field counterpart points to $–CH_2CH–$. Combining the two provides $–CH_2CH(CH_3)_2$ and, adding the Cl atom, the solution: the achiral (hence optically inactive) $ClCH_2CH(CH_3)_2$. This assignment is confirmed by the ^{13}C NMR spectrum, the highest-field line being due to the presence of the two equivalent methyl carbons. The center line is due to the tertiary, the most deshielded absorption due to the chlorine-bearing carbon (Table 10-6). You can confirm your solution in another manner, taking advantage of the relatively small molecular formula; thus, there are only four possible chlorobutane isomers: $CH_3CH_2CH_2CH_2Cl$, $CH_3CHCH_2CH_3$ (our major

$$\overset{|}{Cl}$$

product), $(CH_3)_2CHCH_2Cl$ (the minor product), and $(CH_3)_3CCl$. They differ drastically in their ^1H and ^{13}C NMR spectra with respect to number of signals, chemical shifts, integration, and multiplicities. (Verify.)

The second aspect of this problem is mechanistic. How do we get 1-chloro-2-methylpropane from the preceding reaction sequence? The answer presents itself on retrosynthetic analysis, using the reagents given in our initial scheme:

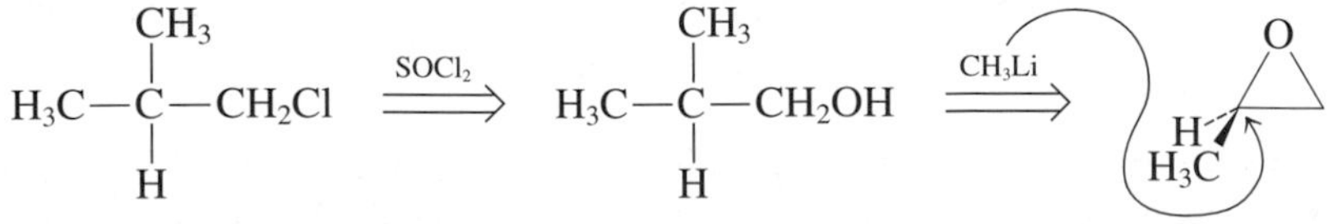

Therefore, the minor product is the result of nucleophilic ring opening of the oxacyclopropane by attack at the more hindered position, usually neglected because less favored.

IMPORTANT CONCEPTS

1. **NMR** is the most important spectroscopic tool in the elucidation of the **structures** of organic molecules.

2. **Spectroscopy** is possible because molecules exist in various energetic forms, those at lower energy being convertible into states of higher energy by absorption of discrete quanta of **electromagnetic radiation.**

3. NMR is possible because certain nuclei, especially ^{1}H and ^{13}C, when exposed to a strong magnetic field, align **with** it (α) or **against** it (β). The α-to-β transition can be effected by radio-frequency radiation, leading to **resonance** and a spectrum with characteristic absorptions. The higher the external field strength, the higher the resonance frequency. For example, a magnetic field of 21,150 G causes hydrogen to absorb at 90 MHz.

4. **High-resolution NMR** allows for the differentiation of hydrogen and carbon nuclei in different chemical environments. Their characteristic positions in the spectrum are measured as the **chemical shift, δ,** in ppm from an internal standard, tetramethylsilane.

5. The chemical shift is highly dependent on the presence (causing **shielding**) or absence (causing **deshielding**) of electron density. The former results in relatively high-field [to the right, toward $(CH_3)_4Si$] peaks, the latter in low-field ones. Therefore, electron-donor substituents shield, and electron-withdrawing components deshield. The pro-

tons on the heteroatoms of alcohols, thiols, and amines show variable chemical shifts and often appear as broad peaks because of hydrogen bonding and exchange.

6. Chemically equivalent hydrogens and carbons have the same chemical shift. Equivalence is best established by the application of **symmetry** operations, such as those using **mirror planes** and **rotations.**

7. The number of hydrogens giving rise to a peak is measured by **integration.**

8. The number of hydrogen neighbors of a nucleus is given by the **spin–spin splitting** pattern of its NMR resonance, following the **$N + 1$ rule.** Equivalent hydrogens show no mutual spin–spin splitting.

9. When the chemical-shift difference between coupled hydrogens is comparable to their coupling constant, **non-first-order spectra** with complicated patterns are observed.

10. When the constants for coupling to nonequivalent types of neighboring hydrogens are different, the **$N + 1$ rule** is applied **sequentially.**

11. **Carbon NMR** utilizes the low-abundance ^{13}C isotope. Carbon–carbon coupling is not observed in ordinary ^{13}C spectra. Carbon–hydrogen coupling can be removed by proton decoupling, thereby simplifying most ^{13}C spectra to a collection of single peaks. DEPT ^{13}C NMR allows the assignment of absorptions to CH_3, CH_2, CH, and quaternary carbons, respectively.

PROBLEMS

13. Where on the chart presented in Figure 10-2 would the following be located: AM radio waves ($\nu \sim 1$ MHz = 1000 kHz = 10^6 Hz = 10^6 s^{-1}, or cycles s^{-1}); FM and TV broadcast frequencies ($\nu \sim 100$ MHz = 10^8 s^{-1})?

14. Convert each of the following quantities into the specified units. **(a)** 1050 cm^{-1} into λ, in μm; **(b)** 510 nm (green light) into ν, in s^{-1} (cycles s^{-1}, or hertz); **(c)** 6.15 μm into $\tilde{\nu}$, in cm^{-1}; **(d)** 2250 cm^{-1} into ν, in s^{-1} (Hz).

15. Convert each of the following quantities into energies, in kcal mol^{-1}. **(a)** A bond rotation of 750 wavenumbers (cm^{-1}); **(b)** a bond vibration of 2900 wavenumbers (cm^{-1}); **(c)** an electronic transition of 350 nm (ultraviolet light, capable of sunburn); **(d)** the broadcast frequency of the audio signal of TV channel 6 (87.25 MHz); **(e)** a "hard" X-ray with a 0.07-nm wavelength.

16. Calculate to three significant figures the amount of energy absorbed by a hydrogen when it undergoes an α-to-β spin flip in the field of **(a)** a 21,150-G magnet (ν = 90 MHz); **(b)** a 117,500-G magnet (ν = 500 MHz).

17. Sketch a hypothetical low-resolution NMR spectrum, showing the positions of the resonance peaks for all magnetic nuclei for each of the following molecules. Assume an external magnetic field of 21,150 G. How would the spectra change if the magnetic field were 84,600 G?

(a) $CHCl_3$ (chloroform)   **(b)** $CFCl_3$ (Freon 11)   **(c)** $CF_3\overset{\displaystyle Cl}{\underset{\displaystyle Br}{|\atop|}}CH$ (Halothane)

18. If the NMR spectra of the molecules in Problem 17 were recorded by using high resolution for each nucleus, what differences would be observed?

19. The ^1H NMR spectrum of $CH_3COCH_2C(CH_3)_3$, 4,4-dimethyl-2-pentanone, taken at 90 MHz shows signals at the following positions: 92, 185, and 205 Hz, downfield from tetramethylsilane. **(a)** What are the chemical shifts (δ) of these signals? **(b)** What would their positions be in hertz, relative to tetramethylsilane, if the spectrum were recorded at 60 MHz? At 360 MHz? **(c)** Assign each signal to a set of hydrogens in the molecule.

20. Which hydrogens in the following molecules exhibit the more downfield signal relative to $(CH_3)_4Si$ in the NMR experiment? Explain.

(a) $(CH_3)_2O$ or $(CH_3)_3N$ **(b)** $CH_3\overset{\displaystyle O}{\overset{\displaystyle ||}{C}}CH_3$ **(c)** $CH_3CH_2CH_2OH$ **(d)** $(CH_3)_2S$ or $(CH_3)_2S{=}O$

                                                   ↑ or ↑   ↑ or ↑

21. How many signals would be present in the ^1H NMR spectrum of each of the following molecules? What would the *approximate* chemical shift be for each of these signals? Ignore spin–spin splitting in this problem and in Problem 22.

(a) $CH_3CH_2CH_2CH_3$ **(b)** $CH_3\overset{\displaystyle CHCH_3}{\underset{\displaystyle Br}{|}}$ **(c)** $HOCH_2\overset{\displaystyle CH_3}{\underset{\displaystyle CH_3}{\overset{|}{C}}}Cl$ **(d)** $CH_3\overset{\displaystyle CH_3}{\overset{|}{C}}HCH_2CH_3$

(e) $CH_3\overset{\displaystyle CH_3}{\underset{\displaystyle CH_3}{\overset{|}{C}}}NH_2$ **(f)** $CH_3CH_2CH(CH_2CH_3)_2$ **(g)** $CH_3OCH_2CH_2CH_3$ **(h)** $\begin{matrix} H_2C{-}CH_2 \\ | \quad\quad | \\ H_2C{-}C \\ \quad\quad\quad \diagdown O \end{matrix}$

(i) $CH_3CH_2{-}\overset{\displaystyle O}{\overset{\diagup\diagup}{C}}\diagdown_H$ **(j)** $CH_3\overset{\displaystyle CH_3O}{\overset{|}{C}}H{-}\overset{\displaystyle CH_3}{\underset{\displaystyle CH_3}{\overset{|}{C}}}{-}CH_3$

22. For each compound in each of the following groups of isomers, indicate the number of signals in the ^1H NMR spectrum, the *approximate* chemical shift of

each signal, and the integration ratios for the signals. Finally, indicate whether all the isomers in each group can be distinguished from one another by these three pieces of information alone.

(a) 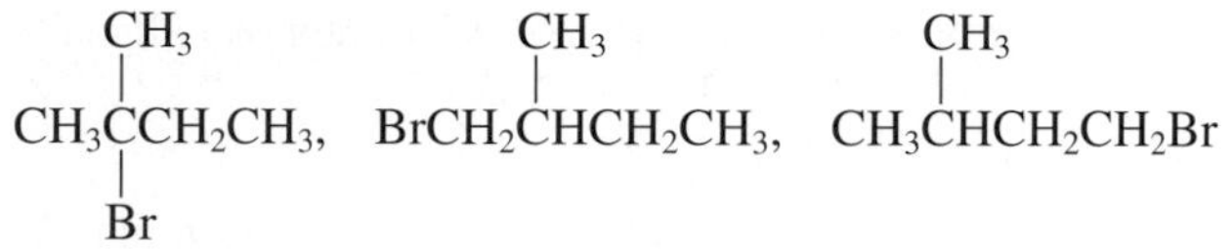

(b) 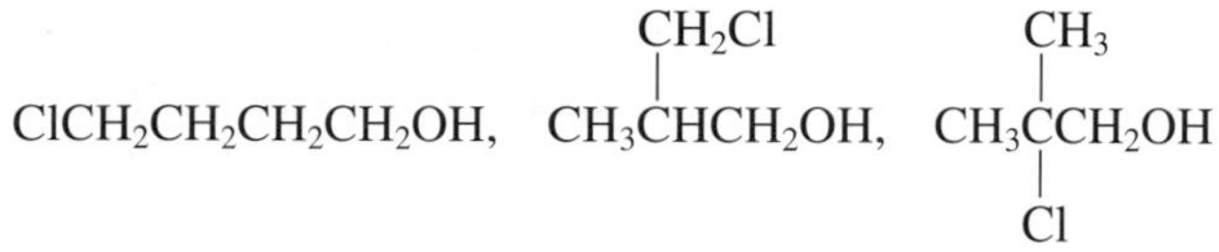

(c) 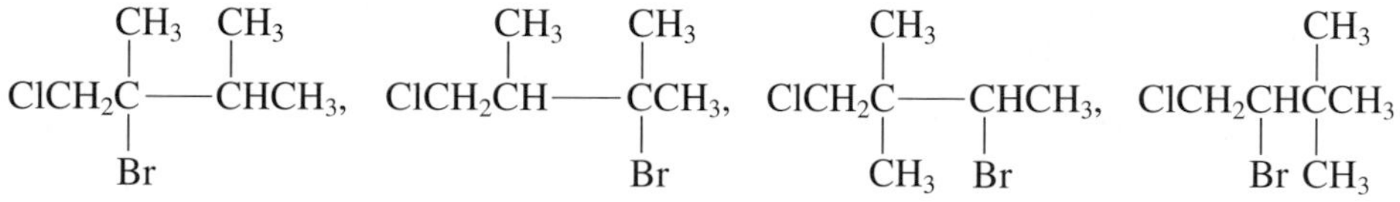

23. 90-MHz ^{1}H NMR spectra for two haloalkanes are shown below. Propose structures for these compounds that are consistent with the spectra. **(a)** $C_5H_{11}Cl$, spectrum A; **(b)** $C_4H_8Br_2$, spectrum B.

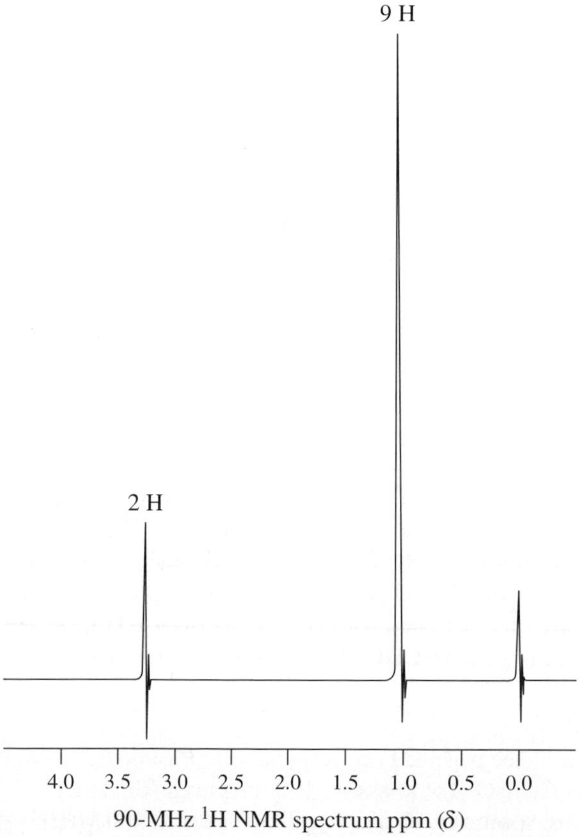

A

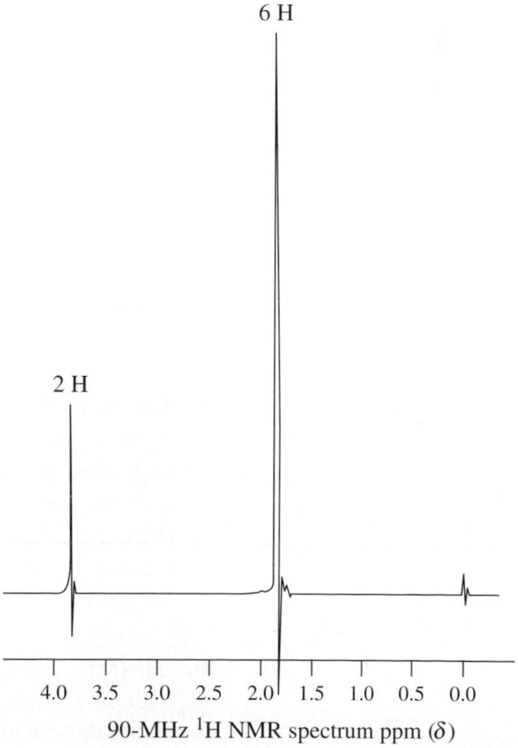

B

24. The following ^{1}H NMR signals are for three molecules with ether functional groups. All the signals are singlets (single, sharp peaks). Propose structures for these compounds. (a) $C_3H_8O_2$, $\delta = 3.3$ and 4.4 ppm (ratio $3:1$); (b) $C_4H_{10}O_3$, $\delta = 3.3$ and 4.9 ppm (ratio $9:1$); (c) $C_5H_{12}O_2$, $\delta = 1.2$ and 3.1 ppm (ratio $1:1$). Compare and contrast these spectra with that of 1,2-dimethoxyethane (Figure 10-15B).

25. (a) The ^{1}H NMR spectrum of a ketone with the molecular formula $C_6H_{12}O$ has $\delta = 1.2$ and 2.1 ppm (ratio $3:1$). Propose a structure for this molecule.
(b) Each of two isomeric molecules related to the ketone in (a) has the molecular formula $C_6H_{12}O_2$. Their ^{1}H NMR spectra are described as follows: isomer 1, $\delta = 1.5$ and 2.0 ppm (ratio $3:1$); isomer 2, $\delta = 1.2$ and 3.6 ppm (ratio $3:1$). All signals in these spectra are singlets. Propose structures for these compounds. To what compound class do they belong?

26. Below are shown three $C_4H_8Cl_2$ isomers on the left and three sets of ^{1}H NMR data on the right. Match the structures to the proper spectral data. (**Hint:** You may find it helpful to sketch the spectra on a piece of scratch paper.)

(a) CH$_3$CH$_2$CHCH$_2$ (Cl Cl)   (i) $\delta = 1.5$ (d, 6 H) and 4.1 (quin, 2 H) ppm

(b) CH$_3$CHCHCH$_3$ (Cl Cl)   (ii) $\delta = 1.6$ (d, 3 H), 2.1 (q, 2 H), 3.6 (t, 2 H), and 4.2 (sex, 1 H) ppm

(c) CH$_3$CHCH$_2$CH$_2$ (Cl Cl)   (iii) $\delta = 1.0$ (t, 3 H), 1.9 (quin, 2 H), 3.6 (d, 2 H), and 3.9 (quin, 1 H) ppm

27. Predict the spin–spin splitting that you would expect to observe in the NMR spectra of each compound in Problem 21. (**Reminder:** Hydrogens attached to oxygen and nitrogen do not normally exhibit spin–spin splitting.)

28. Predict the spin–spin splitting that you would expect to observe in the NMR spectra of each compound in Problem 22.

29. The ^{1}H NMR chemical shifts are given for each of the following compounds. As best you can, assign each signal to the proper group of hydrogens in the molecule, and sketch a spectrum for each compound, incorporating spin–spin splitting whenever appropriate. (a) Cl_2CHCH_2Cl, δ 4.0 and 5.8 ppm; (b) $CH_3CHBrCH_2CH_3$, δ 1.0, 1.7, 1.8, and 4.1 ppm; (c) $CH_3CH_2CH_2COOCH_3$, δ 1.0, 1.7, 2.3, and 3.6 ppm; (d) $ClCH_2CHOHCH_3$, δ 1.2, 3.0, 3.4, and 3.9 ppm.

30. ^{1}H NMR spectra C through F (see page 431) correspond to four isomeric alcohols with the molecular formula $C_5H_{12}O$. Try to assign their structures. To clarify some of the splitting patterns, spectra D, E, and F were recorded at 300 MHz; the insets in spectra D and F show expansions of several useful signals.

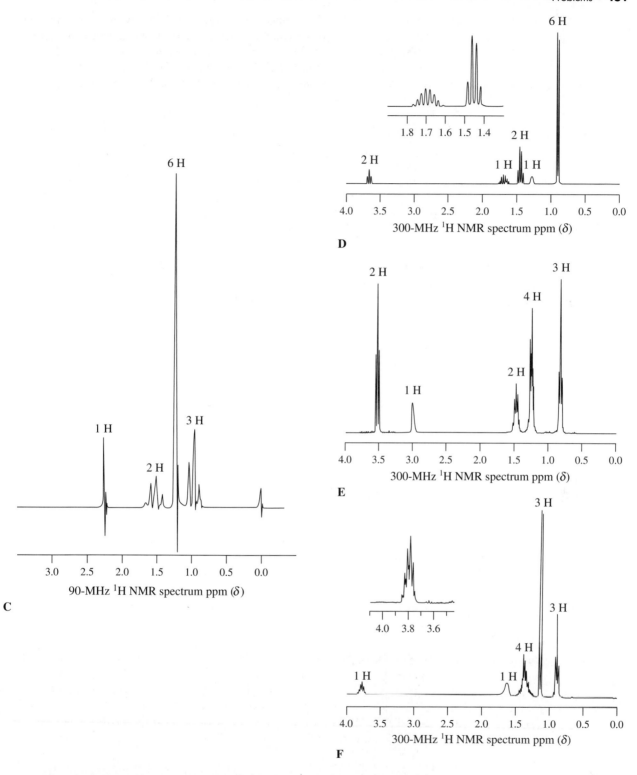

C 90-MHz ¹H NMR spectrum ppm (δ)

D 300-MHz ¹H NMR spectrum ppm (δ)

E 300-MHz ¹H NMR spectrum ppm (δ)

F 300-MHz ¹H NMR spectrum ppm (δ)

31. Sketch ¹H NMR spectra for the following compounds. Estimate chemical shifts (see Section 10-4) and show the proper multiplets for peaks that exhibit spin–spin coupling. **(a)** $CH_3CH_2OCH_2Br$; **(b)** $CH_3OCH_2CH_2Br$; **(c)** $CH_3CH_2CH_2OCH_2CH_2CH_3$; **(d)** $CH_3CH(OCH_3)_2$.

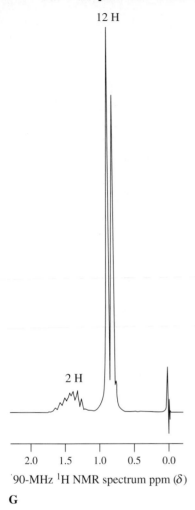

12 H

2 H

2.0   1.5   1.0   0.5   0.0

90-MHz ^{1}H NMR spectrum ppm (δ)

G

32. A hydrocarbon with the formula C_6H_{14} gives rise to 90-MHz ^{1}H NMR spectrum G (at left). What is its structure? This molecule has a structural feature similar to that of another compound whose spectrum is illustrated in this chapter. What molecule is that? Explain the similarities and differences in the spectra of the two.

33. Treatment of the alcohol corresponding to NMR spectrum D in Problem 30 with hot concentrated HBr yields a substance with the formula $C_5H_{11}Br$. Its ^{1}H NMR spectrum exhibits signals at $\delta = 1.0$ (t, 3 H), 1.2 (s, 6 H), and 1.6 (q, 2 H) ppm. Explain. (**Hint:** See NMR spectrum C in Problem 30.)

34. The ^{1}H NMR spectrum of 1-chloropentane is shown at 60 MHz (spectrum H, see below) and 500 MHz (spectrum I, see page 433). Explain the differences in appearance of the two spectra, and assign the signals to specific hydrogens in the molecule.

35. Can the three isomeric pentanes be distinguished unambiguously from their broad-band proton-decoupled ^{13}C NMR spectra *alone?* Can the five isomeric hexanes be distinguished in this way?

36. Predict the ^{13}C NMR spectra of the compounds in Problem 21, with and without proton decoupling.

37. Rework Problem 22 as it pertains to ^{13}C NMR spectroscopy.

38. How would the DEPT ^{13}C spectra of the compounds discussed in Problems 21 and 22 differ in appearance from the ordinary ^{13}C spectra?

39. From each group of three molecules, choose the one whose structure is most consistent with the proton-decoupled ^{13}C NMR data. Explain your choices. **(a)** $CH_3(CH_2)_4CH_3$, $(CH_3)_3CCH_2CH_3$, $(CH_3)_2CHCH(CH_3)_2$; $\delta = 19.5$ and 33.9 ppm. **(b)** 1-Chlorobutane, 1-chloropentane, 3-chloropentane; $\delta = 13.2$,

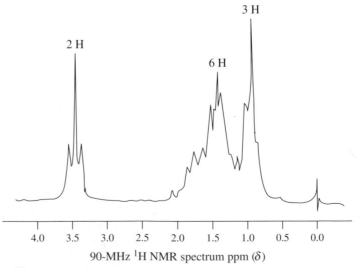

3 H

2 H

6 H

4.0   3.5   3.0   2.5   2.0   1.5   1.0   0.5   0.0

90-MHz ^{1}H NMR spectrum ppm (δ)

H

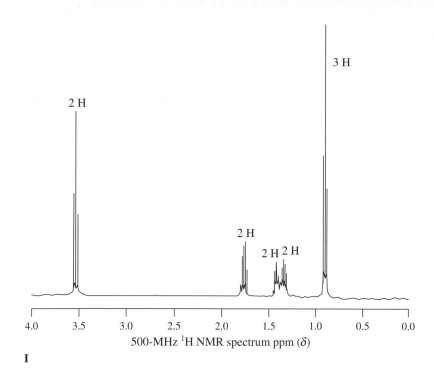

I

500-MHz ^{1}H NMR spectrum ppm (δ)

20.0, 34.6, and 44.6 ppm. **(c)** Cyclopentanone, cycloheptanone, cyclononanone; δ = 24.0, 30.0, 43.5, and 214.9 ppm. **(d)** ClCH$_2$CHClCH$_2$Cl, CH$_3$CCl$_2$CH$_2$Cl, CH$_2$=CHCH$_2$Cl; δ = 45.1, 118.3, and 133.8 ppm. (**Hint:** Consult Table 10-6.)

40. Propose a reasonable structure for each of the following molecules on the basis of the given molecular formula and of the ^{1}H and proton-decoupled ^{13}C NMR data. **(a)** C$_7$H$_{16}$O, spectra J and K (see below); **(b)** C$_8$H$_{18}$O$_2$, spectra L and M (see page 434). The ^{13}C signals marked with asterisks (*) give inverted peaks in the ^{13}C DEPT-135 spectra.

300-MHz ^{1}H NMR spectrum ppm (δ)

J

^{13}C NMR spectrum ppm (δ)

K

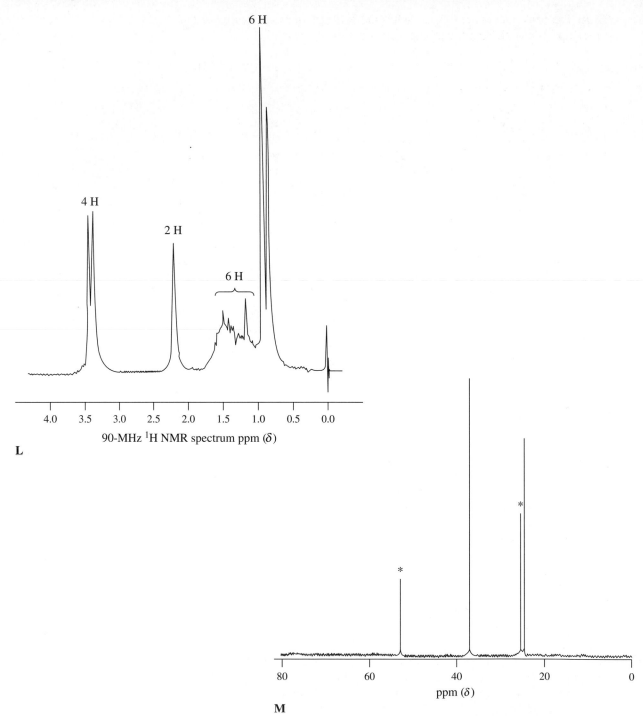

6 H

4 H

2 H

6 H

90-MHz ^{1}H NMR spectrum ppm (δ)

L

*

*

ppm (δ)

M

41. The 300-MHz ^{1}H NMR spectrum of cholesteryl benzoate (see Section 4-7) is shown as spectrum N (see page 435). Although complex, it contains a number of distinguishing features. Analyze the absorptions marked by integrated values. The inset is an expansion of the signal at $\delta = 4.85$ ppm and exhibits an approximately first-order splitting pattern. How would you describe this pattern?

(**Hint:** The peak patterns at δ = 2.5, 4.85, and 5.4 ppm are simplified by the occurrence of chemical shift and/or coupling constant equivalencies.)

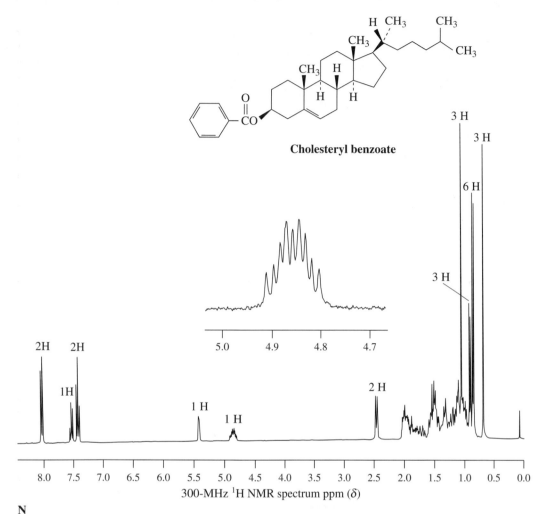

Cholesteryl benzoate

300-MHz ^{1}H NMR spectrum ppm (δ)

N

42. The terpene α-terpineol has the molecular formula $C_{10}H_{18}O$ and is a constituent of pine oil. As the -ol ending in the name indicates, it is an alcohol. Use its ^{1}H NMR spectrum (spectrum O, see page 436) to deduce as much as you can about the structure of α-terpineol. [**Hints:** (1) α-Terpineol has the 1-methyl-4-(1-methylethyl)cyclohexane framework also found in a number of other terpenes (e.g., carvone, Problem 33 of Chapter 5). (2) In your analysis of spectrum O, concentrate on the most obvious features (peaks at δ = 1.1, 1.6, and 5.3 ppm) and use chemical shifts, integrations, and splitting patterns (if any) to help you.]

43. Study of the solvolysis of derivatives of menthol [5-methyl-2-(1-methyl-ethyl)cyclohexanol] has greatly enhanced our understanding of these types of reactions. Heating the 4-methylbenzenesulfonate ester of the isomer shown here in 2,2,2-trifluoroethanol (a highly ionizing solvent of low nucleophilicity) leads to two products with the molecular formula $C_{10}H_{18}$. (**a**) The major product displays ten different signals in its ^{13}C NMR spectrum. Two of them are at

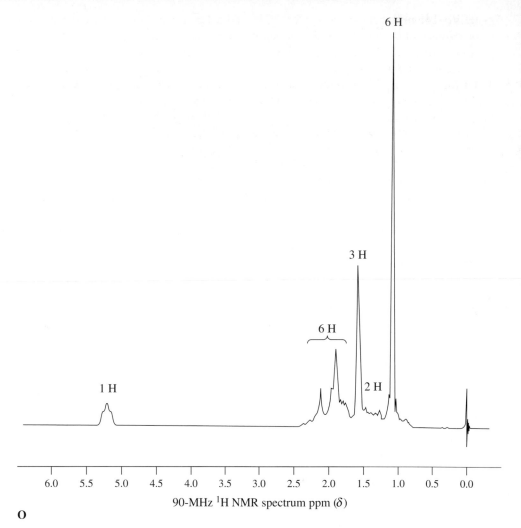

90-MHz ^{1}H NMR spectrum ppm (δ)

O

relatively low field, about $\delta = 120$ and 145 ppm, respectively. The ^{1}H NMR spectrum exhibits a multiplet near $\delta = 5$ ppm (1 H); all other signals are up-field of $\delta = 3$ ppm. Identify this compound. **(b)** The minor product gives only seven ^{13}C signals. Again, two are at low field ($\delta \sim 125$ and 140 ppm), but, in contrast with the ^{1}H NMR data on the major isomer, there are no signals at lower field than $\delta = 3$ ppm. Identify this product and explain its formation mechanistically. **(c)** When the solvolysis is carried out starting with the ester labeled with deuterium at C2, the ^{1}H spectrum of the resulting major product isomer in (a) reveals a significant reduction of the intensity of the signal at $\delta = 5$ ppm, a result indicating the *partial* incorporation of deuterium at the po-sition associated with this peak. How might this result be explained? [**Hint:** The answer lies in the mechanism of formation of the minor product in (b).]

Team Problem

44. Your team is faced with a puzzle. Four isomeric compounds, A–D, with the molecular formula C_4H_9BrO react with KOH to produce E–G with the molecular formula C_4H_8O. Molecules A and B yield compounds E and F, respectively. The NMR spectra of compounds C and D are identical, and both furnish the same product, G. Although some of the starting materials are optically active, none of the products are. Moreover, each of E, F, and G displays only two ^1H NMR signals of varying chemical shifts, none of them located between $\delta\,4.6$ and 5.7 ppm. Both the respective resonances of E and G are complex, whereas F exhibits two singlets. Proton-decoupled ^{13}C NMR spectra for E and G show only two peaks, whereas F exhibits three. Using this spectral information, work together to determine which isomers of C_4H_9BrO will yield the respective isomers of C_4H_8O. When you have matched reactant and product, divide the task of predicting the proton and carbon NMR spectra of E, F, and G among yourselves. Estimate the ^1H and ^{13}C chemical shifts for all, and predict the respective DEPT spectra.

Preprofessional Problems

45. The molecule $(CH_3)_4Si$ (tetramethylsilane) is used as an internal standard in ^1H NMR spectroscopy. One of the following properties makes it especially useful. Which one?
(a) Highly paramagnetic   **(b)** Highly colored   **(c)** Highly volatile   **(d)** Highly nucleophilic

46. One of the following compounds will show a doublet as part of its ^1H NMR spectrum. Which one?
(a) CH_4 **(b)** $ClCH(CH_3)_2$ **(c)** $CH_3CH_2CH_3$ **(d)** $H_2C—CH_2$ with a C bearing Br Br

47. In the ^1H NMR spectrum of 1-fluorobutane, the most deshielded hydrogens are those bound to
(a) C_4 **(b)** C_3 **(c)** C_2 **(d)** C_1

48. One of the following compounds will have one peak in its ^1H NMR spectrum and two peaks in its ^{13}C NMR spectrum. Which one?

(a)   **(b)** (triangle)   **(c)** $CH_3—CH_3$ **(d)** $CH_3CHCHCH_3$ with Cl on each central carbon   **(e)** (ring with two O and two F)

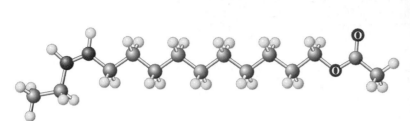

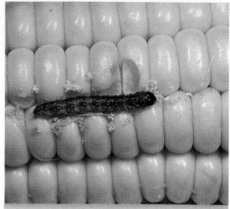

cis-11-Tetradecenyl acetate is a powerful sex attractant (pheromone) for the European corn-borer beetle. Isomeric compounds in which the double bond is shifted just one carbon either way or possesses trans instead of cis geometry are much weaker attractants or are biologically inactive altogether.

What differentiates solid shortening from liquid cooking oil? Remarkably, the *only* significant structural difference is that the liquid contains carbon–carbon double bonds and the solid does not. Cooking oils are derivatives of alkenes, the simplest organic compounds containing multiple bonds. In this chapter and in Chapter 12, we will investigate the properties, generation, and reactivity of alkenes. In the preceding several chapters, we focused on systems containing single-bonded functional groups. We found that, under appropriate conditions, these molecules may undergo elimination to form alkenes. In this chapter we will return to these processes and explore some additional features that relate to their behavior.

Just as alkenes may be prepared from single-bonded compounds by elimination, they can be converted back into single-bonded substances by the complementary process of addition. We shall see in Chapter 12 how alkenes can thus serve as intermediaries in many synthetic interconversions. They are also useful starting materials for plastics, synthetic fibers, construction materials, and a great many other industrially useful substances, as well as being valuable sources of them. For example, addition reactions of many gaseous alkenes give oils as products, which is why this class of compounds used to be called olefins (from *oleum facere,* Latin, to make oil). Indeed, "margarine" is a shortened version of the original name, oleomargarine, for this product.* Because alkenes can undergo addition reactions, they are described as **unsaturated** compounds. In contrast, alkanes, which possess the maximum number of single bonds and thus are inert with respect to addition, are referred to as **saturated.**

---

*The name margarine itself originates indirectly from the Greek, *margaron,* pearl, and directly from margaric acid, the common name given to one of the constituent fatty acids of margarine, heptadecanoic acid, because of the shiny, "pearly" crystals it forms.

We begin with the names and physical properties of the alkenes, and show how we evaluate the relative stability of their isomers. A review of elimination reactions allows us to further our discussion of alkene preparation. We also introduce a second type of spectroscopy—infrared (IR) spectroscopy—and show how it complements NMR by providing organic chemists with a method to help determine the presence or absence of functional groups in a molecule.

11-1 Naming the Alkenes

A carbon–carbon double bond is the characteristic functional group of the alkenes. Their general formula is C_nH_{2n}, the same as that for the cycloalkanes.

Like other organic compounds, some alkenes are still known by common names, in which the *-ane* ending of the corresponding alkanes is replaced by **-ylene.** Substituent names are added as prefixes.

Common Names of Typical Alkenes

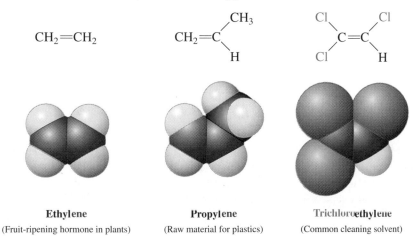

Ethylene	Propylene	Trichloroethylene
(Fruit-ripening hormone in plants)	(Raw material for plastics)	(Common cleaning solvent)

In IUPAC nomenclature, the simpler ending **-ene** is used instead of -ylene, as in ethene and propene. More complicated systems require adaptations and extensions of the rules for naming alkanes (Section 2-3).

RULE 1. To name the stem, find the longest chain that *includes* the functional group—in this case, *both* carbons making up the double bond. The molecule may have still longer carbon chains, but ignore them.

$$CH_2\!=\!CHCHCH_2CH_3$$

A methylpentene

$$CH_2\!=\!CHCH(CH_2)_4CH_3$$

A propyloctene

(Not a hexene or a nonane derivative)

$$CH_3CH_2CH_2CH_2C\!=\!CCH_2CH_2CH_2CH_3$$

An ethylmethyldecene

(Not a pentene or a heptene or an octene derivative)

RULE 2. Indicate the location of the double bond in the main chain by number, starting at the end *closer* to the double bond. (Cycloalkenes do not require the numerical prefix, but the carbons making up the double bond are assigned the numbers 1 and 2.) Alkenes that have the same molecular formula but differ in the location of the double bond (such as 1-butene and 2-butene) are called **double-bond isomers.** A 1-alkene is also referred to as a **terminal alkene;** the others are called **internal.** Note that alkenes are easily depicted in line notation.

$$\overset{1}{C}H_2 \overset{2}{=}\overset{3}{C}H\overset{4}{C}H_2CH_3$$

1-Butene

(A terminal alkene; not 3-butene)

$$\overset{1}{C}H_2\overset{2}{C}H\overset{3}{=}\overset{4}{C}HCH_3$$

2-Butene

(An internal alkene and a double-bond isomer of 1-butene)

2-Pentene

(Not 3-pentene)

Cyclohexene

RULE 3. Add substituents and their positions to the alkene name as prefixes. If the alkene stem is symmetric, begin from the end that gives the first substituent along the chain the lowest possible number.

$$\overset{1}{C}H_2\overset{2}{=}\overset{3}{C}H\overset{|}{C}H\overset{4}{C}H_2\overset{5}{C}H_3$$
$$\overset{}{\underset{CH_3}{}}$$

3-Methyl-1-pentene

3-Methylcyclohexene

(Not 6-methylcyclohexene)

$$\overset{1}{C}H_3\overset{2}{C}H\overset{3}{C}H\overset{3}{=}\overset{4}{C}H\overset{5}{C}H_2\overset{6}{C}H_3$$
$$\overset{}{\underset{CH_3}{}}$$

2-Methyl-3-hexene

(Not 5-methyl-3-hexene)

EXERCISE 11-1

Name the following two alkenes.

(a) **(b)**

RULE 4. Identify any stereoisomers. In a 1,2-disubstituted ethene, the two substituents may be on the same side of the molecule or on opposite sides. The first stereochemical arrangement is called cis, and the second trans, in analogy to the cis-trans names of the disubstituted cycloalkanes (Section 4-1). Two alkenes of the same molecular formula differing only in their stereochemistry are called cis-trans isomers and are examples of diastereomers: stereoisomers that are not mirror images of each other.

cis

trans

$$H_3C \quad CH_3$$
$$\underset{H}{\overset{}{}}C=C\underset{H}{\overset{}{}}$$

cis-2-Butene

$$H_3C \quad H$$
$$\underset{H}{\overset{}{}}C=C\underset{CH_3}{\overset{}{}}$$

trans-2-Butene

$$CH_3\overset{\underset{Cl}{|}}{C}H \quad CH_3$$
$$\underset{H}{\overset{}{}}C=C\underset{H}{\overset{}{}}$$

4-Chloro-cis-2-pentene

Name the following three alkenes.

(a)

$$CH_3 \quad Cl \diagdown \quad Cl \diagup C=C \diagup \diagdown H \quad H$$

(b)

(c) Br

In the smaller substituted cycloalkenes, the double bond can exist only in the cis configuration. The trans arrangement is prohibitively strained (as building a model reveals). However, in larger cycloalkenes, trans isomers are stable.

3-Fluoro-1-methylcyclopentene   **1-Ethyl-2,4-dimethylcyclohexene**   *trans*-**Cyclodecene**

(In both cases, only the cis isomer is stable)

RULE 5. Use a more general method, the *E,Z* system, to label more complex diastereomers. The labels *cis* and *trans* cannot be applied when there are three or more different substituents attached to the double-bond carbons. An alternative system for naming such alkenes has been adopted by IUPAC: the ***E,Z* system.** In this convention, the sequence rules devised for establishing priority in *R,S* names (Section 5-3) are applied separately to the two groups on each double-bonded carbon. When the two groups of higher priority are on opposite sides, the molecule is of the *E* configuration (E from *entgegen*, German, opposite). When the two substituents of higher priority appear on the same side, the molecule is a *Z* isomer (Z from *zusammen*, German, together).

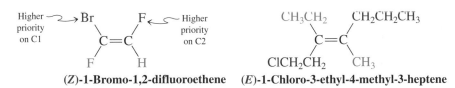

(Z)-1-Bromo-1,2-difluoroethene   **(E)-1-Chloro-3-ethyl-4-methyl-3-heptene**

Name the following three alkenes.

(a)

$$D \diagdown \quad D \diagup C=C \diagup \diagdown H_3C \quad H$$

(b)

$$F \diagdown \quad OCH_3 \diagup C=C \diagup \diagdown H_3C \quad CH_2CH_3$$

(c) Cl

RULE 6. Give the hydroxy functional group precedence over the double bond in numbering a chain. Alcohols containing double bonds are named as **alkenols,** and the stem incorporating both functions is numbered to give the carbon bearing the OH group the lowest possible assignment. Note that the last *e* in alkene is dropped in the naming of alkenols.

$$\overset{3}{CH_2}=\overset{2}{CH}\overset{1}{CH_2}OH$$

2-Propen-1-ol

(Not 1-propen-3-ol)

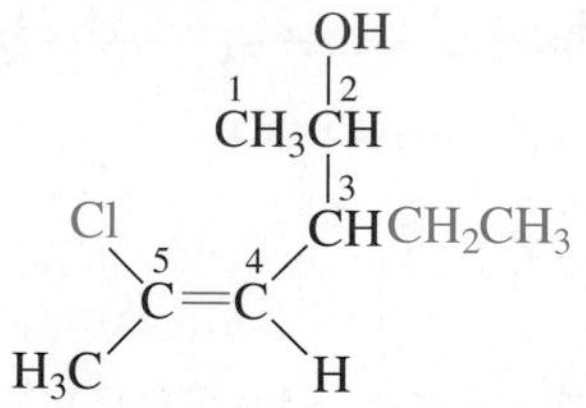

(Z)-5-Chloro-3-ethyl-4-hexen-2-ol

(The two stereocenters are unspecified)

EXERCISE 11-4

Draw the structures of the following molecules. **(a)** *trans*-3-Penten-1-ol; **(b)** 3-cyclo-hexenol.

RULE 7. Substituents containing a double bond are named **alkenyl;** for example, ethenyl (common name, vinyl), 2-propenyl (allyl), and *cis*-1-propenyl.

$$CH_2=CH— \qquad CH_2=CH—CH_2—$$

Ethenyl

(Vinyl)

2-Propenyl

(Allyl)

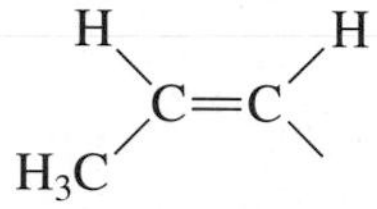

cis-**1-Propenyl**

As usual, the numbering of a substituent chain begins at the point of attachment to the basic stem.

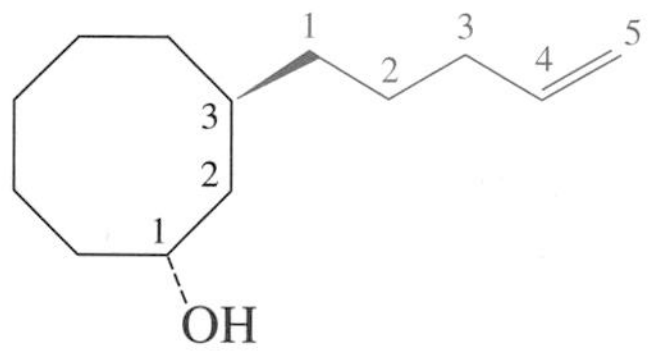

trans-**3-(4-Pentenyl)cyclooctanol**

EXERCISE 11-5

(a) Draw the structure of *trans*-2-ethenylcyclopropanol. **(b)** Name the structure shown below.

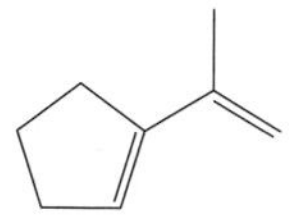

11-2 Structure and Bonding in Ethene: The Pi Bond

The carbon–carbon double bond in alkenes has special electronic and structural features. This section describes the hybridization of the carbon atoms in this functional group, the nature of its two bonds, defined as σ and π, and their relative strengths. We consider ethene, the simplest of the alkenes.

The double bond consists of sigma and pi components

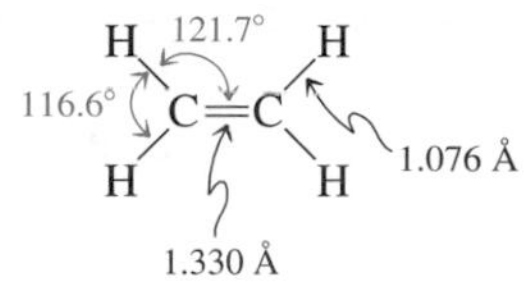

Ethene is planar, with two trigonal carbon atoms and bond angles close to 120° (Figure 11-1). Therefore, both carbon atoms are best described as being sp^2 hybridized (Section 1-8; Figure 1-21). Two sp^2 hybrids on each carbon atom overlap with hydrogen 1s orbitals to form four σ bonds. This leaves one unused sp^2 hybrid orbital on each carbon. The carbon–carbon σ bond is formed by combining these sp^2 hybrid orbitals. In addition, the carbon atoms possess two 2p orbitals, which are in parallel alignment with each other and are close enough to overlap (Figure 11-2A). This second type of bonding interaction, called a π **bond,** is typical of the double bonds in alkenes. The electrons in a π bond are delocalized over both carbons *above and below* the molecular plane, as indicated in Figure 11-2B.

FIGURE 11-1
Molecular structure of ethene.

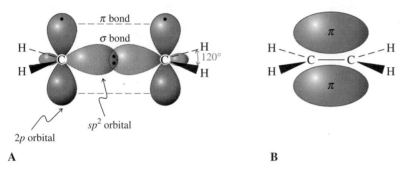

A

B

FIGURE 11-2
A molecular-orbital picture of the double bond in ethene. The σ carbon–carbon bond is made by sp^2–sp^2 overlap. The pair of p orbitals perpendicular to the ethene molecular plane overlap to form the additional π bond. For clarity, this overlap is indicated in (A) by the dashed green lines; the orbital lobes are shown artificially separated. Another way of presenting the π bond is depicted in (B), in which the "π electron cloud" is above and below the molecular plane.

The pi bond in ethene is relatively weak

As our orbital picture shows, the double bond is made up of two different types of bonds: a σ bond and a π bond. How much does each contribute to the total double-bond strength? We know from Section 1-7 that bonds are made by overlap of orbitals and that their relative strengths depend on the effectiveness of this overlap. Therefore, we can expect overlap in a σ bond to be considerably better than that in a π bond, because the sp^2 orbitals lie along the internuclear axis (Figure 11-2). This situation is illustrated in energy-level-interaction diagrams (Figures 11-3 and 11-4) analogous to those used to describe the bonding in the hydrogen molecule (Figures 1-11 and 1-12). Figure 11-5 summarizes our predictions of the relative energies of the molecular orbitals that make up the double bond in ethene.

Thermal isomerization allows us to measure the strength of the pi bond

How do these predictions of the π-bond strength compare with experimental values? We can measure the energy required to interconvert the cis form of a substituted

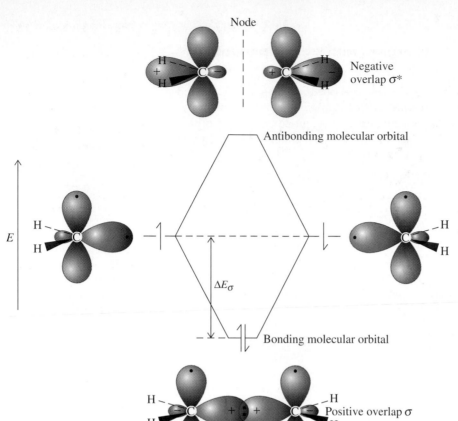

Node

Negative overlap σ*

Antibonding molecular orbital

E

ΔE_σ

Bonding molecular orbital

Positive overlap σ

FIGURE 11-3

Overlap between two sp^2 hybrid orbitals (containing one electron each) determines the relative strength of the σ bond of ethene. In-phase interaction between regions of the wave function having the *same* sign reinforces bonding (compare in-phase overlap of waves, Figure 1-4B) and creates a *bonding molecular orbital.* [Recall: These signs do *not* refer to charges; the + designations are arbitrarily chosen (see Figure 1-11).] Both electrons end up occupying this orbital and have a high probability of being located near the internuclear axis. The orbital stabilization energy, ΔE_σ, corresponds to the σ-bond strength. The out-of-phase interaction, between regions of *opposite* sign (compare Figure 1-4C), results in an unfilled *antibonding molecular orbital* (designated σ^*) with a node.

FIGURE 11-4

Compare this picture of the formation of the π bond in ethene with Figure 11-3. In-phase interaction between two parallel p orbitals (containing one electron each) results in positive overlap and a filled bonding π orbital. The representation of this orbital indicates the probability of finding the electrons between the carbons above and below the molecular plane. Because π overlap is less effective than σ, the stabilization energy, ΔE_π, is smaller than ΔE_σ. The π bond is therefore weaker than the σ bond. The out-of-phase interaction results in the antibonding molecular orbital π^*.

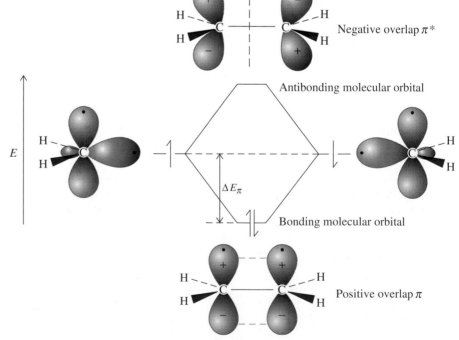

Node

Negative overlap π*

Antibonding molecular orbital

E

ΔE_π

Bonding molecular orbital

Positive overlap π

alkene—say, 1,2-dideuterioethene—with its trans isomer. In this process, called **thermal isomerization,** the two *p* orbitals making up the π bond are rotated 180°. At the midpoint of this rotation—90°—the π (but not the σ) bond has been broken (Figure 11-6). Thus, the activation energy for the reaction can be roughly equated with the π energy of the double bond.

Experimentally, thermal isomerization requires fairly high temperatures (400°–500°C) to take place at measurable rates. The activation energy is 65 kcal mol^{-1}, a value usually assigned to the strength of the π bond. At temperatures below 300°C, most double bonds are configurationally stable; that is, cis stays cis and trans remains trans. The strength of the double bond as a whole in ethene—in other words, the energy required for dissociation into two methylene fragments—is estimated to be 173 kcal mol^{-1}. Consequently, the σ bond in this molecule amounts to about 108 kcal mol^{-1} (Figure 11-7). The bond between an alkyl substituent or a hydrogen atom and the alkenyl carbon is also strong in comparison with the analogous bonds in alkanes (Table 3-2). To a large extent, this effect is due to the improved overlap between the relatively compact sp^2 hybrids and the substituent orbitals. As a con-

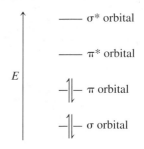

Antibonding orbitals: π^*, σ^*
Bonding orbitals: π, σ

FIGURE 11-5
Energy ordering of the molecular orbitals making up the double bond. The four electrons occupy only bonding orbitals.

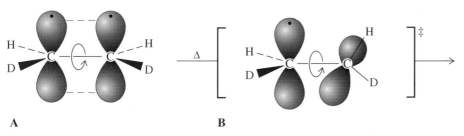

A   B   C

FIGURE 11-6
Thermal isomerization of *cis*-dideuterioethene to the trans isomer requires breaking the π bond. The reaction proceeds from starting material (A) through rotation around the C–C bond until it reaches the point of highest energy, the transition state (B). At this stage, the two *p* orbitals used to construct the π bond are perpendicular to each other. Further rotation in the same direction results in a product in which the two deuterium atoms are trans (C).

Remember:
The symbol ‡ in Figure 11-6 denotes a transition state.

sequence, radical reactions of alkenes do not take place by abstraction of the strongly bound alkenyl (vinyl) hydrogen. In fact, most of the chemistry of the double bond is characterized by the reactivity of the weaker bond: the π bond (Chapter 12).

In summary, the characteristic hybridization scheme for the double bond of an alkene accounts for its physical and electronic features. This hybridization also explains the formation of a strong σ, as well as a weaker π, bond; stable cis and trans isomers; and the strength of the alkenyl–substituent bond. It gives rise to a planar double bond, incorporating trigonal carbon atoms.

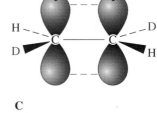

FIGURE 11-7
Approximate bond strengths in an alkene (in kcal mol^{-1}). Note the relative weakness of the π bond.

TABLE 11-1 Comparison of Melting Points of Alkenes and Alkanes	
Compound	Melting point (°C)
Butane	−138
trans-2-Butene	−106
cis-2-Butene	−139
Pentane	−130
trans-2-Pentene	−135
cis-2-Pentene	−180
Hexane	−95
trans-2-Hexene	−133
cis-2-Hexene	−141
trans-3-Hexene	−115
cis-3-Hexene	−138

11-3 Physical Properties of Alkenes

Does the carbon–carbon double bond alter the physical properties of alkenes relative to those of alkanes? The boiling points of alkenes are very similar to those of the corresponding alkanes. Like their alkane counterparts, ethene, propene, and the butenes are gases at room temperature. Melting points, however, depend in part on the packing of molecules in the crystal lattice, a function of molecular shape. The double bond in cis-disubstituted alkenes imposes a U-shaped bend in the molecule that disrupts packing and reduces the melting point, usually below that of either the corresponding alkane or isomeric trans alkene (Table 11-1). A cis double bond is responsible for the sub-room-temperature melting point of vegetable oil. The functional group also affects other physical properties of alkenes, including polarity and acidity.

Depending on their structure, alkenes may exhibit weak dipolar character. Why? Bonds between alkyl groups and an alkenyl carbon are polarized in the direction of the sp^2 hybridized atom, because the degree of s character in an sp^2 hybrid orbital is greater than that in an sp^3. Electrons in orbitals with increased s character are held closer to the nucleus than those in orbitals containing more p character. This effect makes the sp^2 carbon relatively electron withdrawing (although much less so than electronegative atoms such as O and Cl) and creates a dipole along the substituent–alkenyl carbon bond.

Often, particularly in cis-disubstituted alkenes, a net molecular dipole is the result of the two individual dipoles. In trans-disubstituted alkenes, such dipoles are small, because the polarizations of individual local bonds are opposed and tend to cancel each other.

Polarization in Alkenes

Another consequence of the electron-attracting character of the sp^2 carbon is the increased acidity of the alkenyl hydrogen. Whereas ethane has an approximate pK_a of 50, ethene is somewhat more acidic with a pK_a of 44. Even so, ethene is a very poor source of protons compared with other compounds, such as the carboxylic acids or alcohols.

Acidity of the Ethenyl Hydrogen

Ethenyllithium (vinyllithium) is not generally prepared by direct deprotonation of ethene but rather from chloroethene (vinyl chloride) by metallation (Section 8-7).

$$CH_2{=}CHCl \quad + \quad 2\ Li \quad \xrightarrow{(CH_3CH_2)_2O} \quad \underset{60\%}{CH_2{=}CHLi} \quad + \quad LiCl$$

On treatment of ethenyllithium with propanone (acetone) followed by aqueous work-up, a colorless liquid is obtained in 74% yield. Propose a structure.

To summarize, the presence of the double bond does not significantly affect the boiling points of alkenes, but it does give rise to weakly polar bonds and increased acidity of alkenyl hydrogens relative to those in alkanes.

11-4 Nuclear Magnetic Resonance of Alkenes

The double bond exerts characteristic effects on the ^1H and ^{13}C chemical shifts of alkenes (see Tables 10-2 and 10-6). We shall see how to make use of this information in structural assignments.

The pi electrons exert a deshielding effect on alkenyl hydrogens

Figure 11-8 shows the ^1H NMR spectrum of *trans*-2,2,5,5-tetramethyl-3-hexene. Only two signals are observed, one for the 18 equivalent methyl hydrogens and one for the 2 alkenyl protons. The absorptions appear as singlets because the methyl hydrogens are too far away from the alkenyl hydrogens to produce detectable coupling. The low-field resonance of the latter ($\delta = 5.30$ ppm) is typical of hydrogen atoms bound to alkenyl carbons. Terminal alkenyl hydrogens ($RR'C{=}CH_2$) appear at $\delta = 4.6–5.0$ ppm, their internal counterparts ($RCH{=}CHR'$) at $\delta = 5.2–5.7$ ppm.

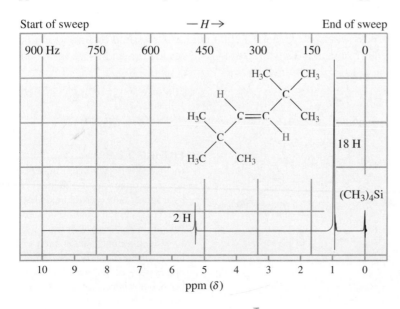

FIGURE 11-8
90-MHz ^1H NMR spectrum of *trans*-2,2,5,5-tetramethyl-3-hexene in CCl_4, illustrating the deshielding effect of the π bond in alkenes. It reveals two sharp singlets for two sets of hydrogens: the 18 methyl hydrogens at $\delta = 0.97$ ppm and 2 highly deshielded alkenyl protons at $\delta = 5.30$ ppm.

CHEMICAL HIGHLIGHT 11-1 | Prostaglandins

The prostaglandins (PGs) are a family of extremely potent hormonelike substances with many biological functions, including muscle stimulation, inhibition of platelet aggregation, lowering of blood pressure, enhancement of inflammatory reactions, and induction of labor in childbirth. Indeed, the anti-inflammatory effects of aspirin (see Chemical Highlight 22-3) are due to its ability to suppress prostaglandin biosynthesis. Illustrated here are three members of the PG family, of which the most biologically active is that labeled PGE_2.

The ^{1}H NMR spectra of these PGEs are quite complicated, with many overlapping absorptions. In contrast, ^{13}C NMR permits rapid identification of PG derivatives, merely by counting the peaks in three chemical shift ranges. For example, PGE_2 is readily distinguished by the presence of two signals near $\delta = 70$ ppm for two alcohol carbons, four alkene ^{13}C resonances between $\delta = 125$ and 140 ppm, and two carbonyls above $\delta = 170$ ppm.

Why is deshielding so pronounced for alkenyl hydrogens? Although the electron-withdrawing character of the sp^2-hybridized carbon is partly responsible, another phenomenon is more important: *the movement of the electrons in the π bond.* When subjected to an external magnetic field perpendicular to the double-bond axis, these electrons enter into a circular motion. This motion induces a local magnetic field that *reinforces* the external field at the edge of the double bond (Figure 11-9). As a consequence, less external magnetic field strength is required to bring the alkenyl hydrogens into resonance: They are strongly deshielded (Section 10-4).

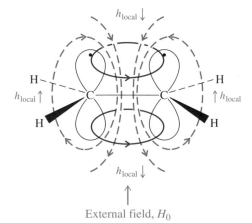

FIGURE 11-9

Movement of electrons in the π bond causes pronounced deshielding of alkenyl hydrogens. An external field, H_0, induces a circular motion of the π electrons (shown in red) above and below the plane of a double bond. This motion in turn induces a local field (shown in green) that opposes H_0 at the center of the double bond but reinforces it in the regions occupied by the alkenyl hydrogens.

The hydrogens on methyl groups attached to alkenyl carbons resonate at about $\delta = 1.6$ ppm (see Table 10-2). Explain the deshielding of these hydrogens relative to hydrogens on methyl groups in alkanes. (**Hint:** Try to apply the principles in Figure 11-9.)

Cis coupling through the double bond is different from trans

When a double bond is not symmetrically substituted, the alkenyl hydrogens will be nonequivalent, a situation leading to observable spin–spin coupling such as that shown in the spectra of *cis*- and *trans*-3-chloropropenoic acid (Figure 11-10). Note that the coupling constant for the hydrogens situated cis ($J = 9$ Hz) is different from that for the hydrogens arranged trans ($J = 14$ Hz). Table 11-2 (see page 450) gives the magnitude of the various possible couplings around a double bond. Although the range of J_{cis} overlaps that of J_{trans}, within a set of isomers the first is always smaller than the second. In this way cis and trans isomers can be readily distinguished.

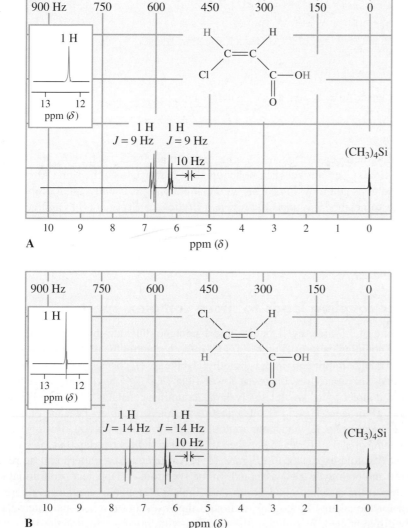

FIGURE 11-10

90-MHz ^{1}H NMR spectra of (A) *cis*-3-chloropropenoic acid and (B) the corresponding trans isomer, each in CCl_4. The two alkenyl hydrogens are nonequivalent and coupled. The carboxylic acid proton ($-CO_2H$) resonates at $\delta = 12.35$ ppm and is shown in the inset.

TABLE 11-2	**Coupling Constants Around a Double Bond**		
		J (Hz)	
Type of coupling	**Name**	**Range**	**Typical**
(structure: cis C=C with two H)	Vicinal, cis	6–14	10
(structure: trans C=C with two H)	Vicinal, trans	11–18	16
(structure: C=C with geminal H, H)	Geminal	0–3	2
(structure: C=C with H)	None	4–10	6
(structure: H–C=C–C–H allylic)	Allylic, (1,3)-cis or -trans	0.5–3.0	2
(structure: –C–C=C–C– long range)	(1,4)- or long-range	0.0–1.6	1

Coupling between hydrogens on adjacent carbon atoms, such as J_{cis} and J_{trans}, is called **vicinal.** Coupling between nonequivalent hydrogens on the same carbon atom is referred to as **geminal.** In alkenes, geminal coupling is usually small (Table 11-2). Coupling to neighboring alkyl hydrogens (**allylic,** see Section 11-1) and across the double bond (**1,4** or **long-range**) also is possible, sometimes giving rise to complicated spectral patterns. Thus, the simple rule devised for saturated systems, discounting coupling between hydrogens farther than two intervening atoms apart, does not hold for alkenes.

Further coupling leads to more complex spectra

The spectra of 3,3-dimethyl-1-butene and 1-pentene illustrate the potential complexity of the coupling patterns (Figure 11-11). In both spectra, the alkenyl hydrogens appear as complex multiplets. In 3,3-dimethyl-1-butene, H_a located on the more highly substituted carbon atom resonates at lower field ($\delta = 5.81$ ppm) and in the form of a double doublet with two relatively large coupling constants ($J_{ab} = 18$ Hz, $J_{ac} = 10.5$ Hz). Hydrogens H_b and H_c also absorb as double doublets because of their respective coupling to H_a and their mutual coupling ($J_{bc} = 1.5$ Hz). Because of the small chemical-shift difference between them, their signals overlap, but, as shown in the inset (twofold expansion) in Figure 11-11A, the coupling pattern can be readily analyzed and both hydrogens assigned. In the spectrum of 1-pentene, additional coupling due to the attached alkyl group (see Table 11-2) creates a relatively complex pattern for the alkenyl hydrogens, although the two sets (terminal and internal) are clearly differentiated. In addition, the electron-withdrawing effect of the sp^2 carbon

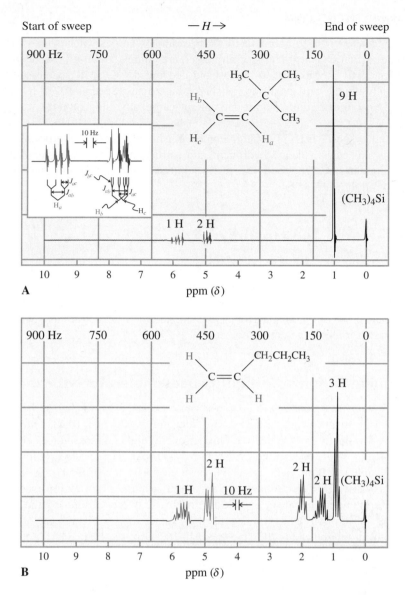

FIGURE 11-11
90-MHz ^{1}H NMR spectra of (A) 3,3-dimethyl-1-butene and (B) 1-pentene, each in CCl_4.

and the movement of the π electrons (Figure 11-9) cause a slight deshielding (shift to lower field) of the directly attached (allylic) CH_2 group. The magnitude of the coupling between these hydrogens and the neighboring alkenyl hydrogen is about the same (6–7 Hz) as the coupling with the two CH_2 hydrogens on the other side. As a result, the multiplet for this allylic CH_2 group appears as a quartet, in accordance with the simple $N + 1$ rule.

EXERCISE 11-8

Ethyl 2-butenoate (ethyl crotonate), $CH_3CH{=}CHCO_2CH_2CH_3$, in CCl_4 has the following ^{1}H NMR spectrum: $\delta = 6.95$ (dq, $J = 16, 6.8$ Hz, 1 H), 5.81 (dq, $J = 16, 1.7$ Hz, 1 H), 4.13 (q, $J = 7$ Hz, 2 H), 1.88 (dd, $J = 6.8, 1.7$ Hz, 3 H), and 1.24 (t, $J = 7$ Hz, 3 H) ppm; dd denotes a doublet of doublets, dq a doublet of quartets. Assign the various hydrogens and indicate whether the double bond is substituted cis or trans (consult Table 11-2).

Alkenyl carbons are deshielded in ^{13}C NMR

The carbon NMR absorptions of the alkenes also are highly revealing. Relative to alkanes, the corresponding alkenyl carbons (with similar substituents) absorb at about

TABLE 11-3 Comparison of ^{13}C NMR Absorptions of Alkenes with the Corresponding Alkane Carbon Chemical Shifts (in ppm)

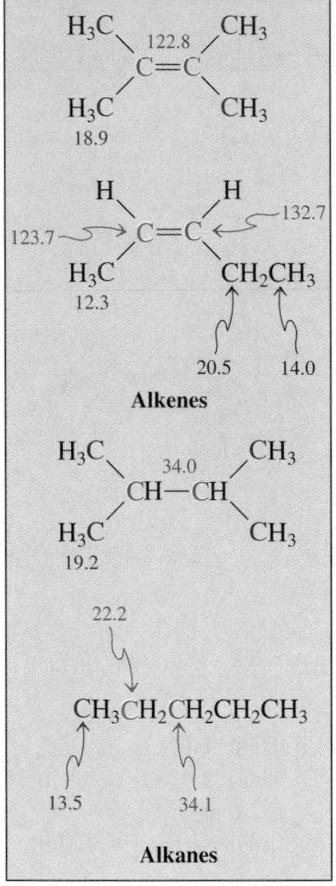

$$H_3C\diagdown \underset{C}{\overset{122.8}{}}\diagup CH_3$$
$$H_3C\diagup \overset{C}{} \diagdown CH_3$$
18.9

H   H
123.7 ↘ C=C ↙ 132.7
H_3C CH_2CH_3
12.3

20.5   14.0

Alkenes

$$H_3C\diagdown \underset{CH-CH}{\overset{34.0}{}}\diagup CH_3$$
$$H_3C\diagup \diagdown CH_3$$
19.2

22.2

$CH_3CH_2CH_2CH_2CH_3$

13.5   34.1

Alkanes

100-ppm lower field (see Table 10-6). Two examples are shown in Table 11-3, in which the carbon chemical shifts of an alkene are compared with those of its saturated counterpart. Recall that, in broad-band decoupled ^{13}C NMR spectroscopy, all magnetically unique carbons absorb as sharp single lines (Section 10-9). It is therefore very easy to determine the presence of sp^2 carbons by this method.

In summary, NMR is highly effective in establishing the presence of double bonds in organic molecules. Alkenyl hydrogens and carbons are strongly deshielded. The order of coupling is $J_{\text{gem}} < J_{\text{cis}} < J_{\text{trans}}$. In addition, various coupling constants are typical for allylic substituents.

11-5 Infrared Spectroscopy

Another method of identifying carbon–carbon double bonds and other functional groups is **infrared (IR) spectroscopy,** which measures the vibrational excitation of atoms around the bonds that connect them. The position of the absorption lines depends on the types of functional groups present, and the IR spectrum as a whole is a unique "fingerprint" of the entire molecule.

Absorption of infrared light causes molecular vibrations

In nuclear magnetic resonance, radio waves cause nuclear spins to change their alignment with the magnetic field ($\Delta E \sim 10^{-6}$ kcal mol^{-1}; Section 10-2). Ultraviolet–visible spectroscopy is performed with higher-energy light, which induces electronic transitions $\Delta E \sim 40$–300 kcal mol^{-1}; Section 14-11). At energies slightly lower than those of visible radiation, light causes **vibrational excitation** of the bonds in a molecule. This part of the electromagnetic spectrum is the infrared region (see Figure 10-2). The intermediate range, or **middle infrared,** is most useful to the organic chemist. IR absorption bands are described by either the wavelength, λ, of the absorbed light in micrometers (10^{-6} m; $\lambda \sim 2.5$–16.7 μm; see Figure 10-2) or its reciprocal value, called wavenumber, $\tilde{v}$ (in units of cm^{-1}; $\tilde{v} = 1/\lambda$). Thus, a typical infrared spectrum ranges from 600 to 4000 cm^{-1}, and the energy changes associated with absorption of this radiation range from 1 to 10 kcal mol^{-1}.

Figure 10-3 describes a simple IR spectrometer. Modern systems use sophisticated rapid-scan techniques and are linked with computers. This equipment allows for data storage, spectra manipulation, and computer library searches, so that unknown compounds can be matched with stored spectra.

Vibrational excitation can be envisioned simply by thinking of two atoms, A and B, linked by a bond as two weights on a spring that stretches and compresses at a certain frequency, v (Figure 11-12). In this picture, the frequency of the vibrations

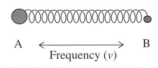

A   ⟵   ⟶   B
Frequency (v)

FIGURE 11-12

Two unequal weights on an oscillating ("vibrating") spring: a model for vibrational excitation of a bond.

between two atoms depends both on the strength of the bond between them and on their atomic weights. In fact, it is governed by Hooke's* law, just like the motion of a spring.

Hooke's Law and Vibrational Excitation

$$\tilde{v} = k \sqrt{f \frac{(m_1 + m_2)}{m_1 m_2}}$$

$\tilde{v}$ = vibrational frequency in wavenumbers (cm^{-1})
k = constant
f = force constant, indicating the strength of the spring (bond)
m_1, m_2 = masses of attached weights (atoms)

This equation might lead us to expect every individual bond in a molecule to show one specific absorption band in the infrared spectrum. However, in practice, an interpretation of the entire infrared spectrum is considerably more complex, because molecules that absorb infrared light undergo not only stretching, but also various bending motions (Figure 11-13), as well as combinations of the two. The bending vibrations are mostly of weaker intensity, they overlap with other absorptions, and they may show complicated patterns. The practicing organic chemist can, however, find some good use for IR spectroscopy for two reasons: The vibrational bands of many

*Professor Robert Hooke (1635–1703), physicist at Gresham College, London.

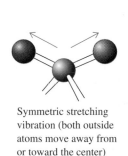

Symmetric stretching vibration (both outside atoms move away from or toward the center)

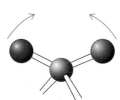

Symmetric bending vibration in a plane (scissoring)

Symmetric bending vibration out of a plane (twisting)

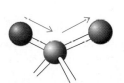

Asymmetric stretching vibration (as one atom moves toward the center, the other moves away)

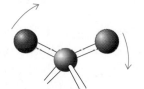

Asymmetric bending vibration in a plane (rocking)

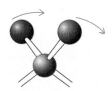

Asymmetric bending vibration out of a plane (wagging)

FIGURE 11-13

Various vibrational modes around tetrahedral carbon. The motions are labeled symmetric and asymmetric stretching or bending, scissoring, rocking, twisting, and wagging.

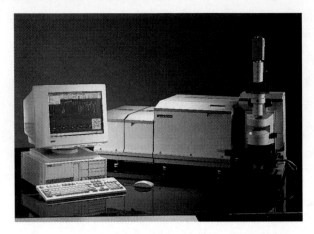

Modern infrared spectrophotometer.

functional groups appear at characteristic wavenumbers, and the entire infrared spectrum may be used as a unique fingerprint of a compound.

Functional groups have typical infrared absorptions

Table 11-4 lists the characteristic stretching wavenumber values for the bonds (shown in red) in some common structural units. Notice how most absorb in the region above 1500 cm^{-1}. We shall show the IR spectra typical of new functional groups when we introduce each of the corresponding compound classes in subsequent chapters.

Figures 11-14 and 11-15 show the IR spectra of pentane and hexane. Above 1500 cm^{-1}, we see only the C–H stretching absorptions typical of alkanes, in the

TABLE 11-4	Characteristic Infrared Stretching Wavenumber Ranges of Organic Molecules		
Bond or functional group	$\tilde{v}\,(\text{cm}^{-1})$	**Bond or functional group**	$\tilde{v}\,(\text{cm}^{-1})$
RO—H (alcohols)	3200–3650	RC≡N (nitriles)	2220–2260
RCO—H (carboxylic acids), with C=O	2500–3300	RCH, RCR′ (aldehydes, ketones), with C=O	1690–1750
R_2N—H (amines)	3250–3500	RCOR′ (esters), with C=O	1735–1750
RC≡C—H (alkynes)	3260–3330	RCOH (carboxylic acids), with C=O	1710–1760
C=C (alkenes), with H	3050–3150	C=C (alkenes)	1620–1680
—C—H (alkanes)	2840–3000	RC—OR′ (alcohols, ethers)	1000–1260
RC≡CH (alkynes)	2100–2260		

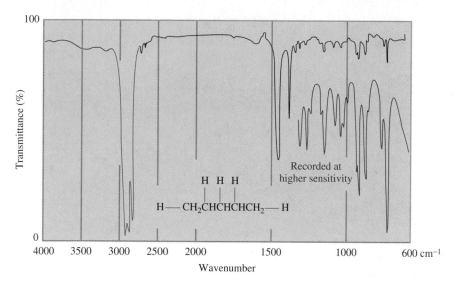

FIGURE 11-14

IR spectrum of pentane. Note the format: Wavenumber is plotted (decreasing from left to right) against percentage of transmittance. A 100% transmittance means *no* absorption; therefore "peaks" in an IR spectrum point *downward*. The spectrum shows absorbances at $\tilde{\nu}_{C-H \text{ stretch}} = 2960, 2930,$ and 2870 cm^{-1}; $\tilde{\nu}_{C-H \text{ bend}} = 1460, 1380,$ and 730 cm^{-1}. The region between 600 and 1300 cm^{-1} is also shown recorded at higher sensitivity (in red), revealing details of the pattern in the fingerprint region.

range from 2840 to 3000 cm^{-1}; no other functional groups are present, and the spectra of the two alkanes are very similar in this region. However, in the range between 600 and 1500 cm^{-1}, called the **fingerprint region,** they show differences in fine details, which become clearer at higher recorder sensitivity. Bands at approximately

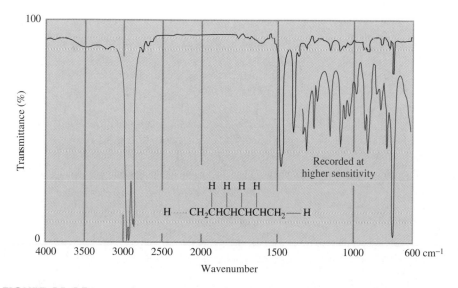

FIGURE 11-15

IR spectrum of hexane. Comparison with that of pentane (Figure 11-14) shows that the location and appearance of the major bands are very similar, but the two fingerprint regions show significant differences at higher recorder sensitivity (in red).

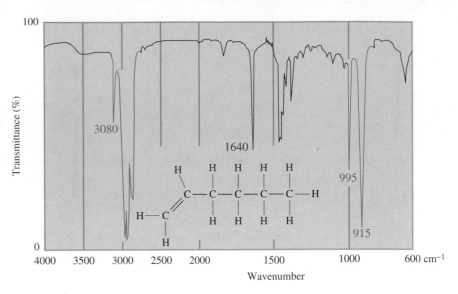

FIGURE 11-16

IR spectrum of 1-hexene: $\tilde{\nu}_{C_{sp^2}-H \text{ stretch}} = 3080 \text{ cm}^{-1}$; $\tilde{\nu}_{C=C \text{ stretch}} = 1640 \text{ cm}^{-1}$; $\tilde{\nu}_{C_{sp^2}-H \text{ bend}} = 995$ and 915 cm^{-1}.

1460, 1380, and 730 cm^{-1} result from bending motions and are common to all saturated hydrocarbons.

Figure 11-16 shows the IR spectrum of 1-hexene. A characteristic feature of alkenes when compared with alkanes is the stronger C_{sp^2}–H bond, which should therefore have a peak at higher energy in the IR spectrum. Indeed, as Figure 11-16 shows, there

The Garlic Story: Infrared Spectroscopy in Food Chemistry

You may have noticed that Table 11-4 omits infrared data for some functional groups, such as haloalkanes. This material is omitted because the IR bands for the stretching motions of C–X bonds lie in the fingerprint region of the spectrum, where assignment of individual absorptions is difficult. However, many sulfur-containing functional groups are readily identified by IR. The S=O bond in allicin, the principal volatile compound obtained from crushed garlic (see Chemical Highlight 9-4), gives rise to an intense band at 1080 cm^{-1}. Allicin is unstable: At room temperature, the 1080-cm^{-1} IR absorption

disappears in less than 24 h. The major decomposition product, which also smells like garlic, was identified spectroscopically. Two likely candidates, 2-propene-1-thiol and its corresponding disulfide, were synthesized and found to have similar spectra, with alkene C=C and alkenyl C–H bands at 1630 and 3070 cm^{-1}, respectively. However, the thiol also shows an absorption for the S–H group at 2535 cm^{-1}. The absence of this band in the IR spectrum of the allicin decomposition product led to its identification as the disulfide, not the thiol.

Allicin
IR: 1080 cm^{-1}

2-Propene-1-thiol
IR: 2535 cm^{-1}

2-Propenyl disulfide
IR: No S=O or S—H bands

is a sharp spike at 3080 cm^{-1}, due to this stretching mode, at slightly higher wavenumber than the remainder of the C–H stretching absorptions. According to Table 11-4, the C=C stretching band should appear between about 1620 and 1680 cm^{-1}. Figure 11-16 shows a relatively strong and sharp band at 1640 cm^{-1} assigned to this vibration. The other strong peaks are the result of bending motions. For example, the two signals at 915 and 995 cm^{-1} are typical of a terminal alkene.

Two other strong bending modes may be used as a diagnostic tool for the substitution pattern in alkenes. One mode results in a single band at 890 cm^{-1} and is characteristic of 1,1-dialkylethenes; the other gives a sharp peak at 970 cm^{-1} and is produced by the C$_{sp^2}$–H bending mode of a trans double bond. In conjunction with NMR (Section 11-4), the presence or absence of such bands allows for fairly certain structural assignments of specifically substituted double bonds.

The O–H stretching absorption is the most characteristic band in the IR spectra of alcohols (Chapters 8 and 9), appearing as a readily recognizable strong, broad peak over a wide range (3200–3650 cm^{-1}, Figure 11-17). The broadness of this peak is due to hydrogen bonding to other alcohol molecules or to water. Dry alcohols in dilute solution exhibit sharp, narrow bands in the range from 3620 to 3650 cm^{-1}. In contrast, the C–X bonds of haloalkanes (Chapters 6 and 7) possess IR stretching frequencies at energies too low (<800 cm^{-1}) to be generally useful for characterization.

Approximate Infrared Frequencies of Strong Bending Modes for Alkenes

R, H / C=C / H, H
915, 995 cm⁻¹

R, H / C=C / R, H
890 cm⁻¹

R, H / C=C / H, R
970 cm⁻¹

EXERCISE 11-9

Three alkenes with the formula C$_4$H$_8$ exhibit the following IR absorptions: alkene A, 964 cm^{-1}; alkene B, 908 and 986 cm^{-1}; alkene C, 890 cm^{-1}. Assign a structure to each alkene.

In summary, the presence of specific functional groups can be ascertained by infrared spectroscopy. Infrared light causes the vibrational excitation of bonds in molecules. Strong bonds and light atoms vibrate at relatively high stretching frequencies measured in wavenumbers (reciprocal wavelengths). Conversely, weak bonds and heavy atoms absorb at lower wavenumbers, as would be expected from Hooke's law.

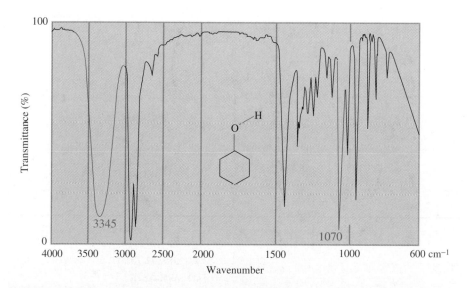

FIGURE 11-17

IR spectrum of cyclohexanol: $\tilde{\nu}_{\text{O–H stretch}} = 3345$ cm^{-1}; $\tilde{\nu}_{\text{C–O}} = 1070$ cm^{-1}. Note the broad O–H peak.

Because of the variety of stretching and bending modes, infrared spectra usually show complicated patterns. However, these patterns are diagnostic fingerprints for particular compounds. The presence of variously substituted alkenes may be detected by stretching signals at about 3080 (C–H) and 1640 (C=C) cm^{-1} and by bending modes between 890 and 990 cm^{-1}. Alcohols show a characteristic band for the OH group in the range from 3200 to 3650 cm^{-1}.

11-6 Degree of Unsaturation: Another Aid to Identifying Molecular Structure

NMR and IR spectroscopy are important tools for determining the structure of an unknown. However, concealed in the molecular formula of every compound is an additional piece of information that can make the job easier. This is the **degree of unsaturation,** defined as the *sum of the numbers of rings and π bonds* present in the molecule. Table 11-5 illustrates the relation between molecular formula, structure, and degree of unsaturation for several hydrocarbons.

As Table 11-5 shows, each increase in the degree of unsaturation corresponds to a decrease of *two* hydrogens in the molecular formula. Therefore, starting with the general formula for acyclic alkanes (saturated; degree of unsaturation = 0), C_nH_{2n+2} (Section 2-2), the degree of unsaturation may be determined for any hydrocarbon merely by comparing the actual number of hydrogens present with the number required for the molecule to be saturated, namely, $2n + 2$, where n = the number of carbon atoms present. For example, what is the degree of unsaturation in a hydrocarbon of the formula C_5H_8? A *saturated* compound with five carbons has the formula C_5H_{12} (C_nH_{2n+2}, with $n = 5$). Because C_5H_8 is four hydrogens short of being

TABLE 11-5	Degree of Unsaturation as a Key to Structure	
Formula	**Representative structures**	**Degree of unsaturation**
C_6H_{14}		0
C_6H_{12}	(one π bond) (one ring)	1
C_6H_{10}	(two π bonds) (one π bond + one ring) (two rings)	2
C_6H_8	(three π bonds) (two π bonds + one ring) (one π bond + two rings)	3

saturated, the degree of unsaturation is 4/2 = 2. All molecules with this formula contain a combination of rings and π bonds adding up to two.

The presence of heteroatoms may affect the calculation. Let us compare the molecular formulas of several saturated compounds: ethane, C_2H_6, and ethanol, C_2H_6O, have the same number of hydrogen atoms; chloroethane, C_2H_5Cl has one less, ethanamine, C_2H_7N, one more. The number of hydrogens required for saturation is reduced by the presence of halogen, increased when nitrogen is present, and unaffected by oxygen. We can generalize the procedure for determination of the degree of unsaturation from a molecular formula as follows.

Some C_5H_8 Hydrocarbons

(Two π bonds)

(One π bond + one ring)

(Two rings)

STEP 1. Determine from the number of carbons (n_C), halogens (n_X), and nitrogens (n_N) in the molecular formula the number of hydrogens required for the molecule to be saturated, H_{sat}.

$$H_{sat} = 2n_C + 2 - n_X + n_N \qquad \text{(Oxygen and sulfur are disregarded.)}$$

STEP 2. Compare H_{sat} with the actual number of hydrogens in the molecular formula, H_{actual}, to determine the degree of unsaturation.

$$\text{Degree of unsaturation} = (H_{sat} - H_{actual})/2$$

EXERCISE 11-10

Calculate the degree of unsaturation indicated by each of the following molecular formulas. **(a)** C_5H_{10}; **(b)** $C_9H_{12}O$; **(c)** C_8H_7ClO; **(d)** $C_8H_{15}N$; **(e)** $C_4H_8Br_2$.

EXERCISE 11-11

Spectroscopic data for three compounds with the molecular formula C_5H_8 are given below; m denotes a complex multiplet. Assign a structure to each compound. (**Hint:** One is acyclic; the others each contain one ring.) **(a)** IR 910, 1000, 1650, 3100 cm^{-1}; ^{1}H NMR δ = 2.79 (t, J = 8 Hz), 4.8–6.2 (m) ppm, integrated intensity ratio of the signals = 1:3. **(b)** IR 900, 995, 1650, 3050 cm^{-1}; ^{1}H NMR δ = 0.5–1.5 (m), 4.8–6.0 (m) ppm, integrated intensity ratio of the signals = 5:3. **(c)** IR 1611, 3065 cm^{-1}; ^{1}H NMR δ = 1.5–2.5 (m), 5.7 (m) ppm, integrated intensity ratio of the signals = 3:1. Is there more than one possibility?

In summary, the degree of unsaturation is equal to the sum of the numbers of rings and π bonds in a molecule. Calculation of this parameter makes solving structure problems from spectroscopic data easier.

11-7 Relative Stability of Double Bonds: Heats of Hydrogenation

We have seen several possible substitution patterns around a double bond, depending on the number of substituents and their positions. Is there a difference in the relative thermodynamic stability of these molecular arrangements? For example, are the three isomeric butenes—1-butene, *cis*-2-butene, and *trans*-2-butene—equally stable? The heat of hydrogenation, an important measure of relative energy content, provides an answer.

The fat molecules in butter and hard (stick) margarines are highly saturated, whereas those in vegetable oils have a high proportion of *cis*-alkene functions. Partial hydrogenation of the latter affords soft (tub) margarine.

The heat of hydrogenation is a measure of stability

How can we establish the relative energy of any compound? One method was presented in Section 3-10: measuring heat of combustion. The greater the energy content of the molecule, the more energy is released in this process. Another possibility, particularly applicable to the alkenes, is to measure the heat of another reaction: *hydrogenation of the double bond.*

When an alkene and hydrogen gas are mixed in the presence of a catalyst (usually palladium or platinum), the two hydrogen atoms in H_2 add to the double bond to give the saturated alkane (see Section 12-2), much like the hydrogenation of a carbonyl compound to give an alcohol (Section 8-6). The heat of this reaction, or **heat of hydrogenation,** can be measured accurately and is about -30 kcal mol^{-1} per double bond: The reaction is very exothermic.

Hydrogenation of an Alkene

$$\diagdown C = C \diagup \quad + \quad H-H \quad \xrightarrow{\text{Pd or Pt}} \quad -\overset{|}{\underset{|}{C}} - \overset{|}{\underset{|}{C}} - \\ H \quad H$$

$\Delta H° \sim -30$ kcal mol^{-1}

Hydrogenation of each butene isomer leads to the same product: butane. If their respective energy contents are equal, their heats of hydrogenation should also be equal; however, as the following reactions illustrate, they are not.

1-Butene $+$ H_2 $\xrightarrow{\text{Pt}}$ Butane $\quad \Delta H° = -30.3$ kcal mol^{-1}

cis-2-Butene $+$ H_2 $\xrightarrow{\text{Pt}}$ Butane $\quad \Delta H° = -28.6$ kcal mol^{-1}

trans-2-Butene $+$ H_2 $\xrightarrow{\text{Pt}}$ Butane $\quad \Delta H° = -27.6$ kcal mol^{-1}

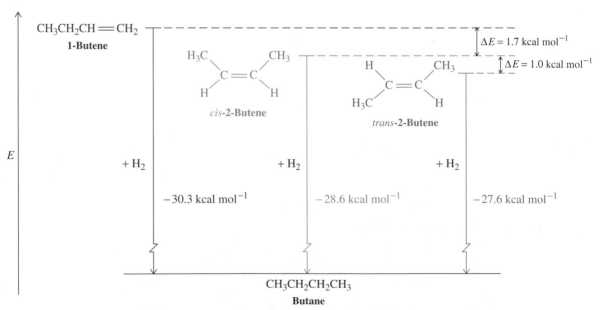

FIGURE 11-18

The relative energy contents of the butene isomers, as measured by their heats of hydrogenation, tell us their relative stabilities. Diagram is not drawn to scale.

The most heat is evolved by hydrogenation of the terminal double bond; the next most exothermic reaction is that with *cis*-2-butene; and finally the trans isomer gives off the least heat. Therefore, the thermodynamic stability of the butenes must increase in the order 1-butene < *cis*-2-butene < *trans*-2-butene (Figure 11-18).

Highly substituted alkenes are most stable; trans isomers are more stable than cis

The results of the preceding hydrogenation reactions may be generalized: The relative stability of the alkenes increases with increasing substitution, and trans isomers are usually more stable than their cis counterparts. The first trend is due in part to hyperconjugation. Just as the stability of a radical increases with increasing alkyl substitution (Section 3-2), the p orbitals of a π bond can be stabilized by alkyl substituents. The second finding is easily understood by looking at molecular models. In cis-disubstituted alkenes, the substituent groups frequently crowd each other.

Relative Stabilities of the Alkenes

$$CH_2{=}CH_2 \quad < \quad RCH{=}CH_2 \quad < \quad \underset{H}{\overset{R}{>}}C{=}C\underset{H}{\overset{R}{<}} \quad < \quad \underset{R}{\overset{H}{>}}C{=}C\underset{H}{\overset{R}{<}} \quad < \quad \underset{R}{\overset{R}{>}}C{=}C\underset{H}{\overset{R}{<}} \quad < \quad \underset{R}{\overset{R}{>}}C{=}C\underset{R}{\overset{R}{<}}$$

$$\text{(cis)} \qquad\qquad \text{(trans)}$$

Least stable   **Most stable**

This steric interference is energetically disadvantageous and absent in the corresponding trans isomers (Figure 11-19).

EXERCISE 11-12

Rank the following alkenes in order of stability of the double bond to hydrogenation (order of $\Delta H°$ of hydrogenation): 2,3-dimethyl-2-butene, *cis*-3-hexene, *trans*-4-octene, and 1-hexene.

Cycloalkenes are exceptions to the generalization that trans alkenes are more stable than their cis isomers. In the small- and medium-ring members of this class of compounds (Section 4-2), the trans isomers are much more strained (Section 11-1). The smallest isolated simple trans cycloalkene is *trans*-cyclooctene. It is 9.2 kcal mol^{-1} less stable than the cis isomer and has a highly twisted structure.

Structure of
***trans*-Cyclooctene**

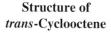

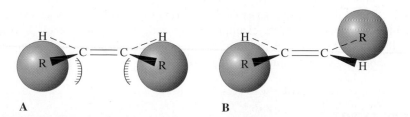

A   B

FIGURE 11-19

(A) Steric congestion in cis-disubstituted alkenes and (B) its absence in trans alkenes explain the greater stability of the latter isomers.

Alkene A hydrogenates to compound B with an estimated release of 65 kcal mol^{-1}, more than double the values of the hydrogenations shown in Figure 11-18. Explain.

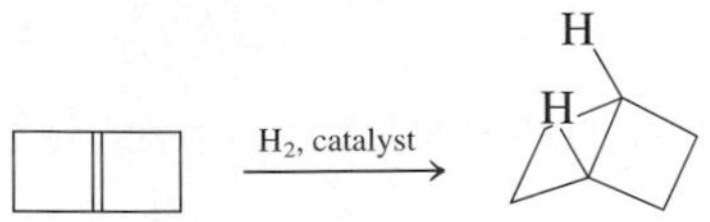

$$\text{A} \quad \xrightarrow{H_2, \text{ catalyst}} \quad \text{B} \qquad \Delta H^\circ \;=\; -65 \text{ kcal mol}^{-1}$$

In summary, the relative energies of isomeric alkenes can be estimated by measuring their heats of hydrogenation. The more energetic alkene has a higher ΔH° of hydrogenation. Stability increases with increasing substitution because of hyperconjugation. Trans alkenes are more stable than their cis isomers because of steric hindrance. Exceptions are the small- and medium-ring cycloalkenes, in which cis substitution is more stable than trans because of ring strain.

11-8 Preparation of Alkenes from Haloalkanes and Alkyl Sulfonates: Bimolecular Elimination Revisited

With the physical aspects of alkene structure and stability as a background, let us now concern ourselves with the various ways in which alkenes can be made. The most general approach is by *elimination,* in which two adjacent groups on a carbon framework are removed. The E2 reaction (Section 7-7) is the most common laboratory source of alkenes. Another method of alkene synthesis, the dehydration of alcohols, is described in Section 11-9.

General Elimination

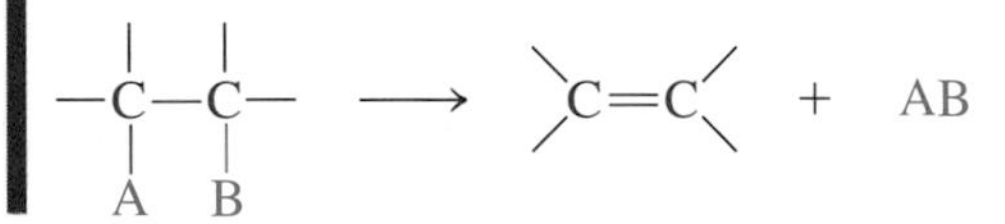

$$-\overset{|}{\underset{\underset{A}{|}}{C}}-\overset{|}{\underset{\underset{B}{|}}{C}}- \quad \longrightarrow \quad \overset{}{\underset{}{>}}C{=}C\overset{}{\underset{}{<}} \;+\; AB$$

Regioselectivity in E2 reactions depends on the base

In Chapter 7 we discussed how haloalkanes (or alkyl sulfonates) in the presence of strong base can undergo elimination of the elements of HX with simultaneous formation of a carbon–carbon double bond.

With many substrates, elimination can be in more than one direction, thus giving rise to double-bond isomers. In such cases, can we control which hydrogen is attacked—that is, the *regioselectivity* of the reaction (Section 9-9)? The answer is yes, to a limited extent. A simple example is the dehydrobromination of 2-bromo-2-methylbutane. Reaction with sodium ethoxide in hot ethanol furnishes mainly 2-methyl-2-butene, but also some 2-methyl-1-butene.

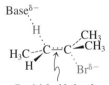

Partial double-bond character leading to trisubstituted double bond

A

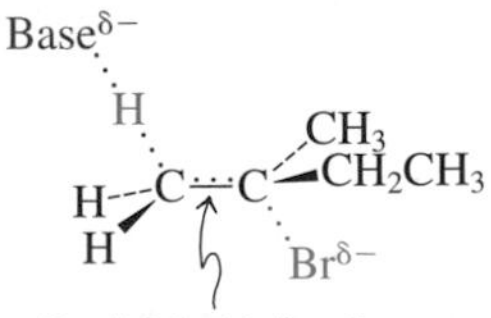

Partial double-bond character leading to terminal double bond

B

FIGURE 11-20

The two transition states leading to products in the dehydrobromination of 2-bromo-2-methylbutane. Transition state A is preferred over transition state B because there are more substituents around the partial double bond (Saytzev rule).

E2 Reaction of 2-Bromo-2-methylbutane with Ethoxide

$$CH_3CH_2\overset{\overset{\displaystyle CH_3}{|}}{\underset{\underset{\displaystyle Br}{|}}{C}}CH_3 \xrightarrow[-HBr]{CH_3CH_2O^-Na^+, CH_3CH_2OH, 70°C}$$

2-Bromo-
2-methylbutane

$$\underset{H}{\overset{H_3C}{>}}C=C\underset{CH_3}{\overset{CH_3}{<}} \quad + \quad \underset{H_3C}{\overset{CH_3CH_2}{>}}C=CH_2$$

70%   30%

2-Methyl-   2-Methyl-
2-butene   1-butene

In our example, the first alkene contains a trisubstituted double bond, so it is thermodynamically more stable than the second. Indeed, many eliminations are regioselective in this way, with the thermodynamically preferred product predominating. This result can be explained by analysis of the transition state of the reaction (Figure 11-20). Elimination of the elements of hydrogen bromide proceeds through attack by the base on one of the vicinal hydrogens situated *anti* to the leaving group. In the transition state, there is partial C–H bond rupture, partial C–C double-bond formation, and partial cleavage at C–Br (compare Figure 7-8). The transition state leading to 2-methyl-2-butene is slightly more stabilized than that generating 2-methyl-1-butene (Figure 11-21A). The more stable product is formed faster because *the structure of the transition state of the reaction resembles that of the products to some extent.* Elimination reactions of this type that lead to the more highly substituted alkene are said to follow the **Saytzev* rule:** The double bond preferentially forms between the carbon that contained the leaving group and *the most highly substituted adjacent carbon* that bears a hydrogen.

A different product distribution is obtained when a more hindered base is used in the same reaction; more of the thermodynamically *less* favored terminal alkene is generated.

*Alexander M. Saytzev (also spelled Zaitsev or Saytzeff; 1841–1910), Russian chemist.

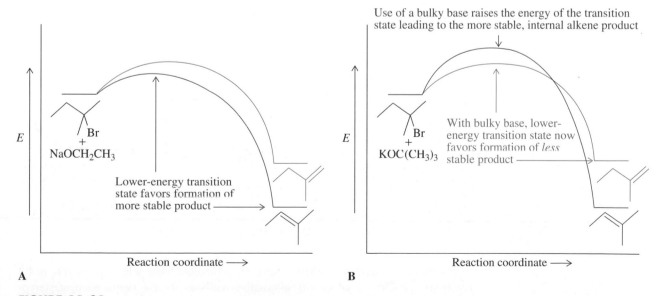

FIGURE 11-21

Potential energy diagrams for E2 reactions of 2-bromo-2-methylbutane with (A) sodium ethoxide (Saytzev rule) and (B) potassium *tert*-butoxide (Hofmann rule).

E2 Reaction of 2-Bromo-2-methylbutane with *tert*-Butoxide, a Hindered Base

$$\text{CH}_3\text{CH}_2\overset{\overset{\displaystyle \text{CH}_3}{|}}{\underset{\underset{\displaystyle \text{Br}}{|}}{\text{C}}}\text{CH}_3 \xrightarrow[\;-\text{HBr}\;]{(\text{CH}_3)_3\text{CO}^-\text{K}^+,\ (\text{CH}_3)_3\text{COH}}$$

$$\underset{27\%}{\overset{\displaystyle \text{H}_3\text{C}\quad\quad\ \text{CH}_3}{\underset{\displaystyle \text{H}\quad\quad\quad \text{CH}_3}{\text{C}=\text{C}}}} \quad + \quad \underset{73\%}{\overset{\displaystyle \text{CH}_3\text{CH}_2}{\underset{\displaystyle \text{H}_3\text{C}}{\text{C}=\text{CH}_2}}}$$

To see why, we again examine the transition state. Removal of a secondary hydrogen (from C3 in the starting bromide) is sterically more difficult than abstracting one of the more exposed methyl hydrogens. When the base used is bulky, as in our example, the energy of the transition state leading to the more stable product is increased by steric interference relative to that leading to the less substituted isomer; thus, the latter becomes the major product (Figure 11-21B). An E2 reaction that generates the thermodynamically less favored isomer is said to follow the **Hofmann rule,** named after the chemist* who investigated a series of eliminations that proceeded with this particular mode of regioselectivity (Chapter 21).

EXERCISE 11-14

When the following reaction is carried out with *tert*-butoxide in 2-methyl-2-propanol (*tert*-butyl alcohol), two products, A and B, are formed in the ratio 23:77. When ethoxide in ethanol is used, this ratio changes to 82:18. What are products A and B, and how do you explain the difference in their ratios in the two experiments?

$$\underset{\underset{\displaystyle \text{H}\ \ \ \overset{|}{\text{O}_3\text{S}}\!-\!\!\!\bigcirc\!\!\!-\text{CH}_3}{}}{\overset{\overset{\displaystyle \text{H}_3\text{C}\ \ \ \ \text{H}}{|\quad\ \ \ |}}{\text{CH}_3\text{C}-\text{CCH}_3}} \quad \xrightarrow{\text{Base, solvent}} \quad \text{A} \ + \ \text{B}$$

E2 reactions often favor trans over cis

Depending on the structure of the alkyl substrate, the E2 reaction can also lead to cis, trans alkene mixtures, in some cases with selectivity. For example, treatment of 2-bromopentane with sodium ethoxide furnishes 51% *trans*- and only 18% *cis*-2-pentene, the remainder of the product being the terminal regioisomer. The outcome of this and related reactions appears to be controlled again to some extent by the relative thermodynamic stabilities of the products, the most stable trans double bond being formed preferentially.

Stereoselective Dehydrobromination of 2-Bromopentane

$$\text{CH}_3\text{CH}_2\text{CH}_2\overset{\overset{\displaystyle \text{CH}_3}{|}}{\underset{\underset{\displaystyle \text{H}}{|}}{\text{C}}}\text{Br} \xrightarrow[\;-\text{HBr}\;]{\substack{\text{CH}_3\text{CH}_2\text{O}^-\text{Na}^+,\\ \text{CH}_3\text{CH}_2\text{OH}}}$$

$$\underset{51\%}{\overset{\displaystyle \text{CH}_3\text{CH}_2\quad\quad\ \text{H}}{\underset{\displaystyle \text{H}\quad\quad\quad\ \text{CH}_3}{\text{C}=\text{C}}}} \ + \ \underset{18\%}{\overset{\displaystyle \text{CH}_3\text{CH}_2\quad\quad\ \text{CH}_3}{\underset{\displaystyle \text{H}\quad\quad\quad\ \text{H}}{\text{C}=\text{C}}}} \ + \ \underset{31\%}{\text{CH}_3\text{CH}_2\text{CH}_2\text{CH}=\text{CH}_2}$$

Unfortunately from a synthetic viewpoint, complete trans selectivity is rare in E2 reactions. Chapter 13 deals with alternative methods for the preparation of stereochemically pure cis and trans alkenes.

*Professor August Wilhelm von Hofmann (1818–1892), University of Berlin.

Some E2 processes are stereospecific

Recall (Section 7-7) that the preferred transition state of elimination places the proton to be removed and the leaving group *anti* with respect to each other. Thus, before an E2 reaction takes place, bond rotation to such a conformation will occur. This fact has additional consequences when reaction may lead to *Z* or *E* stereoisomers. For example, the E2 reaction of the two diastereomers of 2-bromo-3-methylpentane to give 3-methyl-2-pentene is stereospecific. Both the (*R,R*) and the (*S,S*) isomer yield *exclusively* the (*E*) isomer of the alkene. Conversely, the (*R,S*) and (*S,R*) diastereomers give only the (*Z*) alkene. (Build models!)

Stereospecificity in the E2 Reaction of 2-Bromo-3-methylpentane

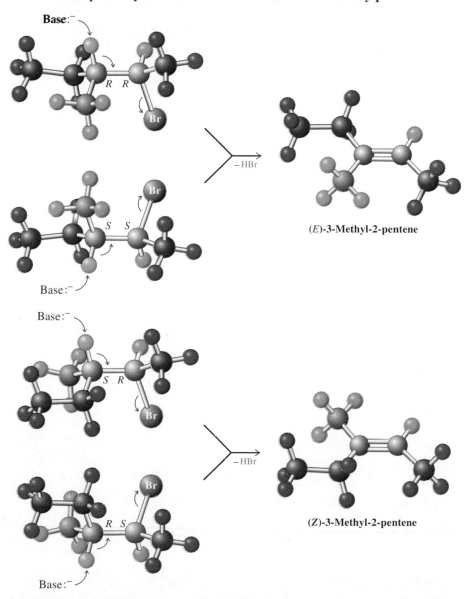

(*E*)-3-Methyl-2-pentene

(*Z*)-3-Methyl-2-pentene

As shown in these three-dimensional structures, *anti* elimination of HBr dictates the eventual configuration around the double bond. The reaction is stereospecific, one diastereomer (and its mirror image) producing only one stereoisomeric alkene, the other diastereomer furnishing the opposite configuration.

EXERCISE 11-15

Which diastereomer of 2-bromo-3-deuteriobutane gives (*E*)-2-deuterio-2-butene, and which diastereomer gives the *Z* isomer?

In summary, alkenes are most generally made by E2 reactions. Usually, the thermodynamically more stable internal alkenes are formed faster than the terminal isomers (Saytzev rule). The reaction may be stereoselective, producing greater quantities of trans isomers than their cis counterparts from racemic starting materials. It is also stereospecific, certain haloalkane diastereomers furnishing only one of the two possible stereoisomeric alkenes. Bulky bases may lead to more of the products with thermodynamically less stable (e.g., terminal) double bonds (Hofmann rule).

11-9 Preparation of Alkenes by Dehydration of Alcohols

Treatment of alcohols with mineral acid at elevated temperatures results in alkene formation by loss of water, a process called **dehydration,** which proceeds by E1 or E2 pathways (Chapters 7 and 9).

The usual way in which to dehydrate an alcohol is to heat it in the presence of sulfuric or phosphoric acid at relatively high temperatures (120°–170°C).

Acid-Mediated Dehydration of Alcohols

$$-\overset{|}{\underset{H}{C}}-\overset{|}{\underset{OH}{C}}- \xrightarrow{Acid, \Delta} \;\; \diagup C=C \diagdown \;\; + \;\; HOH$$

Another, sometimes more efficient, dehydration procedure requires the use of aluminum oxide (alumina), Al_2O_3, as a Lewis acid catalyst. In this case, vapors of the alcohol are passed over alumina powder at a temperature ranging from 350° to 400°C, and the alkene is collected in a trap at the exit of the reaction vessel.

The ease of elimination of water from alcohols increases with increasing substitution of the hydroxy-bearing carbon.

Relative Reactivity of Alcohols (ROH) in Dehydration Reactions

R = primary < secondary < tertiary

$$CH_3CH_2OH \xrightarrow[-HOH]{Conc. H_2SO_4, 170°C} CH_2=CH_2$$

$$CH_3\overset{HO}{\underset{H}{C}}-\overset{H}{\underset{H}{C}}CH_3 \xrightarrow[-HOH]{50\% H_2SO_4, 100°C} CH_3CH=CHCH_3 \;\; + \;\; CH_2=CHCH_2CH_3$$
80% Trace

$$(CH_3)_3COH \xrightarrow[-HOH]{Dilute H_2SO_4, 50°C} H_2C=C\diagup^{CH_3}_{CH_3}$$
100%

CHEMICAL HIGHLIGHT 11-3 — Acid-Catalyzed Dehydration of α-Terpineol

Acid-catalyzed dehydration is not generally useful for preparative purposes, because product mixtures often form. An example is the dehydration of α-terpineol, a naturally occurring, unsaturated terpene alcohol (see below and Section 4-7) with a pleasant lilac fragrance, isolated from pine oil. Fortunately, in this case the product mixture also has a pleasing aroma, due in part to the presence of *limonene,* a major constituent of lemon and orange oils and *α-terpinene,* which also possesses a lemon fragrance. Indeed, most of us have encountered this hydrocarbon mixture—it is used in the manufacture of scented soaps.

α-Terpineol → 33% H₂SO₄, 1 h, 100°C, −H₂O →

α-Terpineol	Terpinolene	Limonene
	15%	9%

α-Terpinene	Isoterpinolene	γ-Terpinene
28.5%	18.5%	15%

Gift basket of terpene-scented soaps.

Secondary and tertiary alcohols dehydrate by the unimolecular elimination pathway (E1), discussed in Sections 7-6 and 9-2. Protonation of the weakly basic hydroxy oxygen forms an alkyloxonium ion, now containing water as a good potential leaving group. Loss of H_2O supplies the respective secondary or tertiary carbocations, and deprotonation furnishes the alkene. The reaction is subject to all the side reactions of which carbocations are capable, particularly hydrogen and alkyl shifts (Section 9-3).

Dehydration with Rearrangement

$$CH_3\overset{CH_3}{\underset{H}{C}}-CH_2-\overset{OH}{\underset{H}{C}}CH_3 \xrightarrow[-H_2O]{H_2SO_4, \Delta} \quad \overset{H_3C}{\underset{H_3C}{}}C=C\overset{H}{\underset{CH_2CH_3}{}} \quad + \quad CH_3\overset{CH_3}{\underset{H}{C}}CH=CHCH_3 \quad + \quad \text{other minor isomers}$$

54%　　　　　　　8%

EXERCISE 11-16

Referring to Sections 7-6 and 9-3, write a mechanism for the preceding reaction.

Typically, the thermodynamically most stable alkene or alkene mixture results from unimolecular dehydration in the presence of acid. Thus, whenever possible, the most highly substituted system is generated; if there is a choice, trans-substituted alkenes predominate over the cis isomers. For example, acid-catalyzed dehydration of 2-butanol furnishes the equilibrium mixture of butenes, consisting of 74% *trans*-2-butene, 23% of the cis isomer, and only 3% 1-butene.

Treatment of primary alcohols with mineral acids at elevated temperatures also leads to alkenes; for example, ethanol gives ethene and 1-propanol yields propene (Section 9-7).

$$CH_3CH_2CH_2OH \xrightarrow{\text{Conc. } H_2SO_4,\ 180°C} CH_3CH{=}CH_2$$

The mechanism of this reaction begins with the initial protonation of oxygen. Then, attack by hydrogen sulfate ion or another alcohol molecule effects bimolecular elimination of the elements of H_3O^+ (see Section 7-7).

EXERCISE 11-17

(a) Propose a mechanism for the formation of propene from 1-propanol on treatment with hot concentrated H_2SO_4. (b) Propene is also formed when propoxypropane (dipropyl ether) is subjected to the same conditions. Explain.

$$CH_3CH_2CH_2OCH_2CH_2CH_3 \xrightarrow{\text{Conc. } H_2SO_4,\ 180°C} 2\ CH_3CH{=}CH_2\ +\ H_2O$$

In summary, alkenes can be made by dehydration of alcohols. Secondary and tertiary systems proceed through the intermediacy of carbocations, whereas primary alcohols can undergo E2 reactions from the intermediate alkyloxonium ions. All systems are subject to rearrangement and thus frequently give mixtures.

Chapter Integration Problem

Acid-catalyzed dehydration of 2-methyl-2-pentanol (dilute H_2SO_4, 50°C) gives one major and one minor product. Both have the formula C_6H_{12}. Spectroscopic data for these compounds are as follows:

1. **Major product:** IR 1660 and 3080 cm^{-1}; ^1H NMR $\delta = 0.91$ (t, $J = 7$ Hz, 3 H), 1.60 (s, 3 H), 1.70 (s, 3 H), 1.98 (quin, $J = 7$ Hz, 2 H), and 5.08 (t, $J = 7$ Hz, 1 H) ppm.
2. **Minor product:** IR 1640 and 3090 cm^{-1}; ^1H NMR $\delta = 0.92$ (t, $J = 7$ Hz, 3 H), 1.4 (sex, $J = 7$ Hz, 2 H), 1.74 (s, 3 H), 2.02 (t, $J = 7$ Hz, 2 H), 4.78 (s, 2 H) ppm.

Deduce the structures of the products, suggest mechanisms for their formation, and discuss why the major product forms in the greater amount.

Solution

First, we write the structure of the starting material (alcohol nomenclature, Section 8-1) and what we know about the reaction:

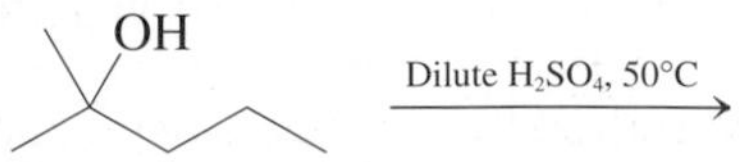

2-Methyl-2-pentanol

This is an acid-catalyzed reaction of a tertiary alcohol (Section 11-9). Even though we may know enough to be able to make a sensible prediction, let us proceed by interpretation of the spectra first, and see if the answer that we get is consistent with our expectations.

For the major product, peaks at 1660 and 3080 cm^{-1} in the IR spectrum are in the ranges for the alkene C=C and C—H bond-stretching frequencies (1620–1680 and 3050–3150 cm^{-1}, Table 11-3). Armed with this information, we turn to the NMR spectrum and immediately look for signals in the region characteristic of alkene hydrogens, $\delta = 4.6$–5.7 ppm (Table 10-2 and Section 11-4). Indeed, we find one at $\delta = 5.08$ ppm. The information following the chemical shift position (t, $J = 7$ Hz, 1 H) tells us that the signal is split into a triplet with a coupling constant of 7 Hz and has a relative integrated intensity corresponding to one hydrogen. The rest of the spectrum shows three signals with integrations of 3 H each; it is usually reasonable to assume that simple signals with intensities of 3 between $\delta = 0$–4 ppm indicate the presence of methyl groups, unless information to the contrary appears. Two of the methyl groups are singlets, and the other appears as a triplet. Finally, a signal integrating for 2 H at $\delta = 1.98$ ppm is split into five lines. Assuming that a CH$_2$ group gives rise to the latter signal, we have the following fragments to consider in piecing together a structure: CH=C, 3 CH$_3$, and CH$_2$. The sum of atoms in these fragments is C$_6$H$_{12}$, in agreement with the formula given for the product and giving us confidence that we are on the right track. There is only a limited number of ways in which to connect these pieces into a proper structure:

$$\underset{\text{CH}_3-\text{CH}=\overset{\displaystyle \overset{\text{CH}_3}{|}}{\text{C}}-\text{CH}_2-\text{CH}_3}{} \text{ (ignoring stereochemistry) and } \underset{\text{CH}_3-\text{CH}_2-\text{CH}=\overset{\displaystyle \overset{\text{CH}_3}{|}}{\text{C}}-\text{CH}_3}{}$$

We can use either our chemical knowledge or, again, turn to the spectroscopy to determine which is correct. With the use of the splitting patterns in the NMR spectrum, a quick decision may be made. Using the $N + 1$ rule (Section 10-7), we see that, under ideal conditions, an NMR signal will be split by N neighboring hydrogens into $N + 1$ lines. In the first structure, the alkenyl hydrogen is neighbor to a CH$_3$ group and should therefore appear with $3 + 1 = 4$ lines, a quartet. The actual spectrum shows this signal as a triplet. Furthermore, this structure contains three methyl groups, but they have 0, 1, and 2 neighboring hydrogens, respectively, and should appear as one singlet, one doublet, and one triplet, again, in disagreement with the actual spectrum. In contrast, the second structure fits: the alkenyl hydrogen has two neighbors, consistent with the observed triplet splitting, and two of the three CH$_3$ groups are on an alkene carbon lacking a neighboring hydrogen and should be singlets. Bringing our chemical knowledge into play, we note that this correct structure has the same carbon connectivity as that of the starting material, whereas the incorrect one would have required a rearrangement for its formation.

Turning to the minor product, we use the same logic: Again, the IR spectrum shows an alkene C=C stretch (at 1640 cm^{-1}) and an alkenyl C–H stretch at 3090 cm^{-1}. The NMR spectrum contains a singlet at $\delta = 4.78$, integrating for 2 H: two alkenyl hydrogens. It also shows two methyl signals, as well as two others integrating 2 H for each. Therefore, we have 2 CH$_3$, 2 CH$_2$, and 2 alkenyl H, adding again (with the two alkenyl carbons) to C$_6$H$_{12}$. Even though many possible combinations can be devised at this point, we can make use of NMR splitting information to rapidly close in on the answer. One of the CH$_3$ groups is a singlet, meaning that it must be attached to a carbon lacking hydrogens. If we look at the preceding array of fragments, this latter carbon can only be an alkenyl carbon. Thus, we have CH$_3$–C=C, in which the bold carbon is *not* attached to hydrogen. Therefore, by process of elimination, both alkenyl hydrogens must be attached to the *other* alkenyl carbon: $CH_3\text{--}\overset{|}{C}\text{==}CH_2$. The remaining fragments can be attached to this piece in only one way, giving the final structure: $CH_3\text{---}CH_2\text{---}CH_2$

$\quad CH_3\text{---}\overset{|}{C}\text{==}CH_2$. Thus we can complete the equation stated at the beginning of the problem as follows:

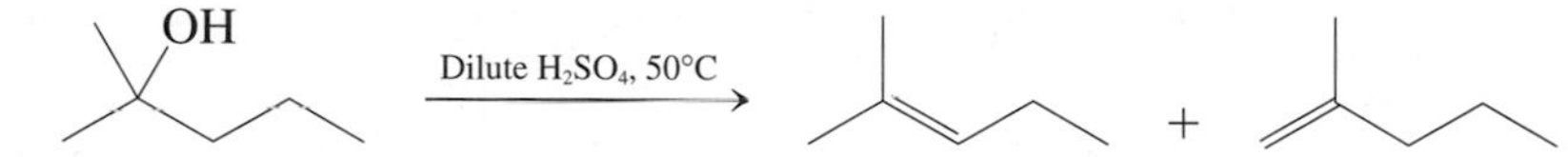

2-Methyl-2-pentanol   Dilute H_2SO_4, 50°C   Major product   +

Is this result in accord with our chemical expectations? Let us consider the mechanism (Section 11-9). Dehydration of a secondary or a tertiary alcohol under acidic conditions begins with protonation of the oxygen atom, giving a good potential leaving group (water). Departure of the leaving group results in a carbocation, and removal of a proton from a neighboring carbon atom (most likely by a second molecule of alcohol acting as a Lewis base) gives the alkene, overall an E1 process:

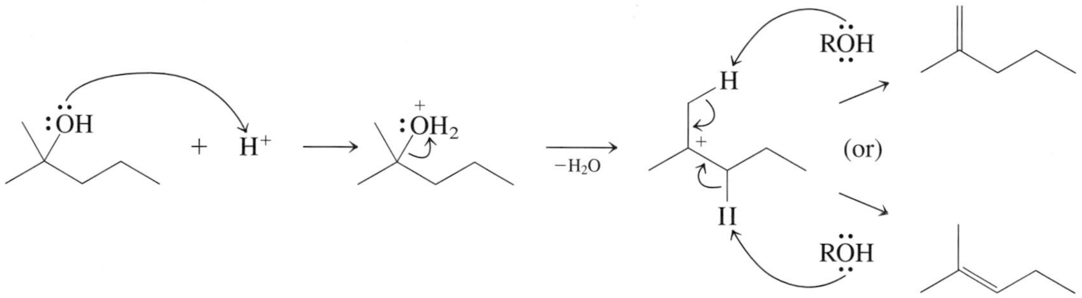

The major product is the one with the more substituted double bond, the thermodynamically more stable alkene (Sections 11-7 and 11-9), as is typical in E1 dehydrations.

NEW REACTIONS

1. Hydrogenation of Alkenes (Section 11-7)

$$C=C \quad + \quad H_2 \quad \xrightarrow{\text{Pd or Pt}} \quad -\overset{|}{\underset{H}{C}}-\overset{|}{\underset{H}{C}}- \qquad \Delta H° \sim -30 \text{ kcal mol}^{-1}$$

Order of stability of the double bond

$$\underset{R}{\overset{R}{>}}C=CH_2 \quad < \quad \underset{H}{\overset{R}{>}}C=C\underset{H}{\overset{R}{<}} \quad < \quad \underset{H}{\overset{R}{>}}C=C\underset{R}{\overset{H}{<}} \quad < \quad \text{more substituted alkene}$$

Preparation of Alkenes

2. From Haloalkanes, E2 with Unhindered Base (Section 11-8)

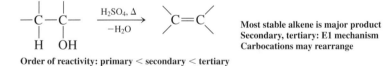

$$\underset{\underset{X}{|}}{\overset{\overset{H}{|}}{-C}}\overset{|}{\underset{|}{C}}-CH_3 \xrightarrow[-HX]{CH_3CH_2O^-Na^+,\ CH_3CH_2OH} \overset{}{\underset{}{C}}=C\overset{CH_3}{\underset{}{}}$$

More substituted
(more stable) alkene

Saytzev rule

3. From Haloalkanes, E2 with Sterically Hindered Base (Section 11-8)

$$\underset{\underset{X}{|}}{\overset{\overset{H}{|}}{-C}}\overset{|}{\underset{|}{C}}-CH_3 \xrightarrow[-HX]{(CH_3)_3CO^-K^+,\ (CH_3)_3COH} \overset{H}{\underset{C=CH_2}{C}}$$

Less substituted
(less stable) alkene

Hofmann rule

4. Stereochemistry of E2 Reaction (Section 11-8)

$$\overset{H}{\underset{R'\ R}{C}}-\overset{R''}{\underset{X}{C}}R''' \xrightarrow[-HX]{Base} \overset{R'}{\underset{R}{C}}=\overset{R''}{\underset{R'''}{C}}$$

Anti elimination

5. Dehydration of Alcohols (Section 11-9)

$$-\overset{|}{\underset{H}{C}}-\overset{|}{\underset{OH}{C}}- \xrightarrow[-H_2O]{H_2SO_4,\ \Delta} \overset{}{C}=C\overset{}{}$$

Most stable alkene is major product
Secondary, tertiary: E1 mechanism
Carbocations may rearrange

Order of reactivity: primary < secondary < tertiary

IMPORTANT CONCEPTS

1. **Alkenes** are **unsaturated** molecules. Their IUPAC names are derived from alkanes, the longest chain incorporating the double bond serving as the stem. **Double-bond isomers** include **terminal, internal, cis,** and **trans** arrangements. Tri- and tetrasubstituted alkenes are named according to the **E,Z system,** in which the *R,S* priority rules apply.

2. The double bond is composed of a σ and a π part. The former is obtained by overlap of the two sp^2 hybrid lobes on carbon, the latter by interaction of the two remaining p orbitals. The π **bond** is weaker (~65 kcal mol^{-1}) than its σ counterpart (~108 kcal mol^{-1}) but strong enough to allow for the existence of stable cis and trans isomers.

3. The functional group in the alkenes is flat, sp^2 **hybridization** being responsible for the possibility of creating dipoles and for the relatively high acidity of the alkenyl hydrogen.

4. Alkenyl hydrogens and carbons appear at **low field** in ^{1}H NMR ($\delta = 4.6$–5.7 ppm) and ^{13}C NMR ($\delta -$ 100–140 ppm) experiments, respectively. J_{trans} is larger than J_{cis}, $J_{geminal}$ is very small, and $J_{allylic}$ variable but small.

5. **Infrared spectroscopy** measures **vibrational excitation.** The energy of the incident radiation ranges from about 1 to 10 kcal mol^{-1} ($\lambda \sim 2.5$– 16.7 μm; $\tilde{\nu} \sim 600$–4000 cm^{-1}). Characteristic peaks are observed for certain functional groups, a conse-

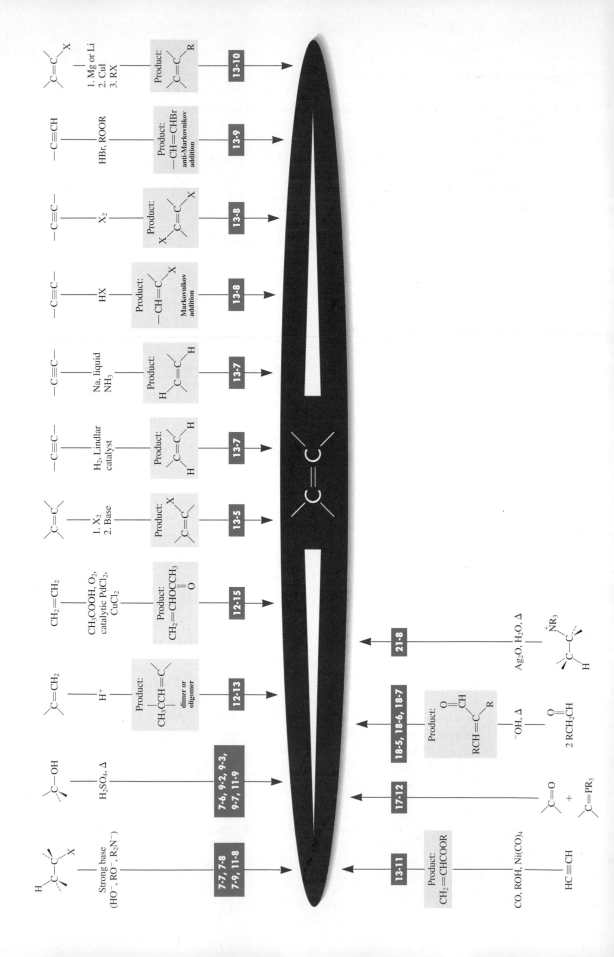

Preparation of Alkenes section number

472

quence of stretching, bending, and other modes of vibration, and their combination. Moreover, each molecule exhibits a characteristic infrared spectral pattern, called a **fingerprint.**

6. Alkanes show IR bands characteristic of C–H bonds in the range from 2840 to 3000 cm^{-1}. The C=C stretching absorption for alkenes is in the range from 1620 to 1680 cm^{-1}, that for the alkenyl C–H bond about 3100 cm^{-1}. Bending modes sometimes give useful peaks below 1500 cm^{-1}. Alcohols are usually characterized by a broad peak for the O–H stretch between 3200 and 3650 cm^{-1}.

7. **Degree of unsaturation** (number of rings + number of π bonds) is calculated from the molecular formula by using the equation

$$\text{Degree of unsaturation} = (H_{sat} - H_{actual})/2$$

where $H_{sat} = 2n_C + 2 - n_X + n_N$ (disregard oxygen and sulfur).

8. The relative stability of isomeric alkenes can be established by comparing **heats of hydrogenation.** It decreases with decreasing substitution; trans isomers are more stable than cis.

9. Elimination of haloalkanes (and other alkyl derivatives) may follow the **Saytzev rule** (nonbulky base, internal alkene formation) or the **Hofmann rule** (bulky base, terminal alkene formation). Trans alkenes as products predominate over cis alkenes. Elimination is **stereospecific,** as dictated by the *anti* transition state.

10. **Dehydration** of alcohols in the presence of strong acid usually leads to a mixture of products, with the most stable alkene being the major constituent.

PROBLEMS

18. Name each of the following molecules in accord with the IUPAC system of nomenclature.

19. Assign structures to the following molecules on the basis of the indicated ^{1}H NMR spectra on the following two pages. Consider stereochemistry, where applicable.
(a) C_4H_7Cl, NMR spectrum A; (b) $C_5H_8O_2$, NMR spectrum B;
(c) $C_6H_{11}I$, NMR spectrum C; (d) another $C_6H_{11}I$, NMR spectrum D;
(e) $C_3H_4Cl_2$, NMR spectrum E.

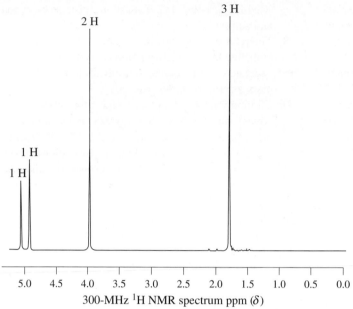

1 H

1 H

2 H

3 H

5.0 4.5 4.0 3.5 3.0 2.5 2.0 1.5 1.0 0.5 0.0

300-MHz ^{1}H NMR spectrum ppm (δ)

A

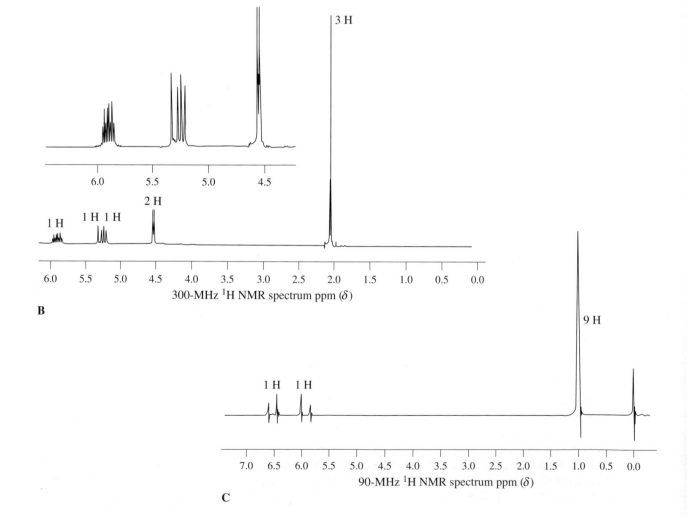

6.0 5.5 5.0 4.5

3 H

1 H

1 H 1 H

2 H

6.0 5.5 5.0 4.5 4.0 3.5 3.0 2.5 2.0 1.5 1.0 0.5 0.0

300-MHz ^{1}H NMR spectrum ppm (δ)

B

9 H

1 H 1 H

7.0 6.5 6.0 5.5 5.0 4.5 4.0 3.5 3.0 2.5 2.0 1.5 1.0 0.5 0.0

90-MHz ^{1}H NMR spectrum ppm (δ)

C

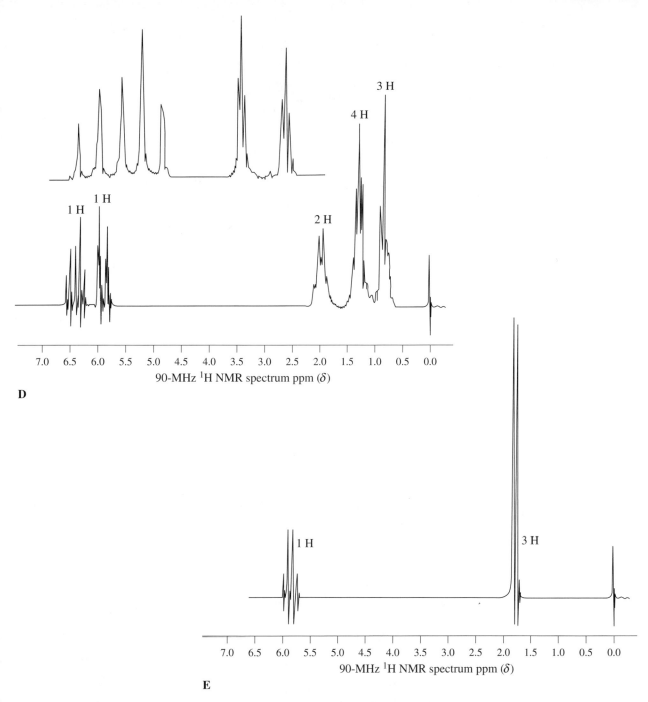

20. Explain the splitting patterns in ¹H NMR spectrum D in detail. Inset is
δ = 5.7–6.7 region expanded fivefold.

21. For each of the pairs of alkenes on page 476, indicate whether measurements of
polarity alone would be sufficient to distinguish the compounds from one another.
Where possible, predict which compound would be more polar.

(a) 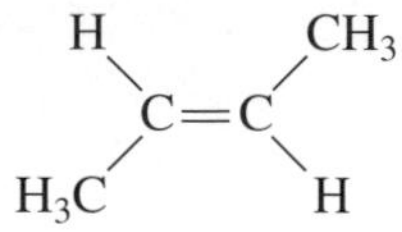
$$\begin{array}{c} H \\ \diagdown \\ C=C \\ \diagup \quad \diagdown \\ H_3C \qquad H \end{array} \begin{array}{c} CH_3 \end{array}$$ and $CH_3CH_2CH=CH_2$

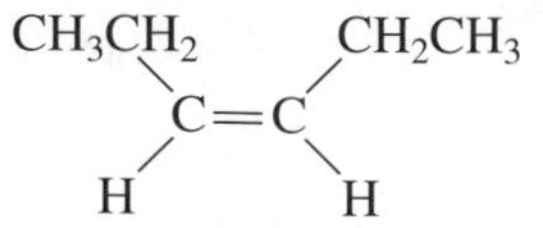

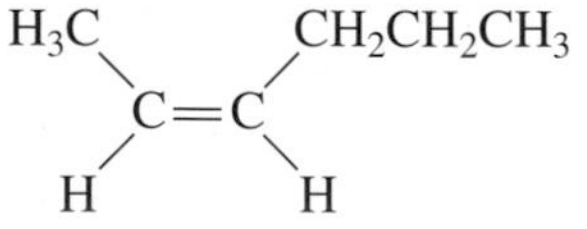

(b) 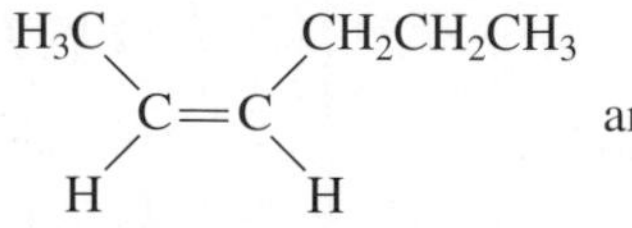
$$\begin{array}{c} H_3C \qquad CH_2CH_2CH_3 \\ \diagdown \quad \diagup \\ C=C \\ \diagup \quad \diagdown \\ H \qquad H \end{array}$$ and $$\begin{array}{c} CH_3CH_2 \qquad CH_2CH_3 \\ \diagdown \quad \diagup \\ C=C \\ \diagup \quad \diagdown \\ H \qquad H \end{array}$$

(c)
$$\begin{array}{c} H_3C \qquad CH_2CH_2CH_3 \\ \diagdown \quad \diagup \\ C=C \\ \diagup \quad \diagdown \\ H \qquad H \end{array}$$ and $$\begin{array}{c} H \qquad CH_2CH_3 \\ \diagdown \quad \diagup \\ C=C \\ \diagup \quad \diagdown \\ CH_3CH_2 \qquad H \end{array}$$

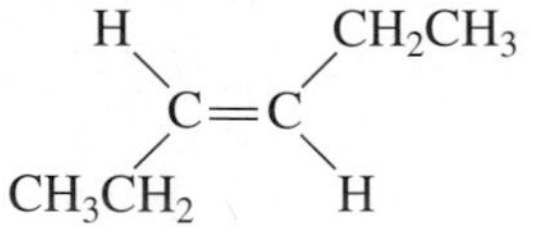

22. The molecular formulas and ^{13}C NMR data (in ppm) for several compounds are given here. The type of carbon, as revealed from DEPT spectra, is specified in each case. Deduce a structure for each compound. **(a)** C_4H_6: 30.2 (CH$_2$), 136.0 (CH); **(b)** C_4H_6O: 18.2 (CH$_3$), 134.9 (CH), 153.7 (CH), 193.4 (CH); **(c)** C_4H_8: 13.6 (CH$_3$), 25.8 (CH$_2$), 112.1 (CH$_2$), 139.0 (CH); **(d)** $C_5H_{10}O$: 17.6 (CH$_3$), 25.4 (CH$_3$), 58.8 (CH$_2$), 125.7 (CH), 133.7 (C$_{quaternary}$); **(e)** C_5H_8: 15.8 (CH$_2$), 31.1 (CH$_2$), 103.9 (CH$_2$), 149.2 (C$_{quaternary}$); **(f)** C_7H_{10}: 25.2 (CH$_2$), 41.9 (CH), 48.5 (CH$_2$), 135.2 (CH). (**Hint:** This one is difficult. The molecule has one double bond. How many rings must it have?)

23. Data from both ordinary and DEPT ^{13}C NMR spectra for several compounds with the formula C_5H_{10} are given here. Deduce a structure for each compound. **(a)** 25.3 (CH$_2$); **(b)** 13.3 (CH$_3$), 17.1 (CH$_3$), 25.5 (CH$_3$), 118.7 (CH), 131.7 (C$_{quaternary}$); **(c)** 12.0 (CH$_3$), 13.8 (CH$_3$), 20.3 (CH$_2$), 122.8 (CH), 132.4 (CH).

24. From the Hooke's law equation, would you expect the C–X bonds of common haloalkanes (X = Cl, Br, I) to have IR bands at higher or lower wavenumbers than are typical for bonds between carbon and lighter elements (e.g., oxygen)?

25. Convert each of the following IR frequencies into micrometers.
    (a) 1720 cm^{-1} (C=O) **(b)** 1650 cm^{-1} (C=C) **(c)** 3300 cm^{-1} (O—H)
    (d) 890 cm^{-1} (alkene bend) **(e)** 1100 cm^{-1} (C—O) **(f)** 2260 cm^{-1} (C≡N)

26. Match each of the following structures with the IR data that correspond best. Abbreviations: w, weak; m, medium; s, strong; br, broad. **(a)** 905 (s), 995 (m), 1040 (m), 1640 (m), 2850–2980 (s), 3090 (m), 3400 (s, br) cm^{-1}; **(b)** 2840 (s), 2930 (s) cm^{-1}; **(c)** 1665 (m), 2890–2990 (s), 3030 (m) cm^{-1}; **(d)** 1040 (m), 2810–2930 (s), 3300 (s, br) cm^{-1}.

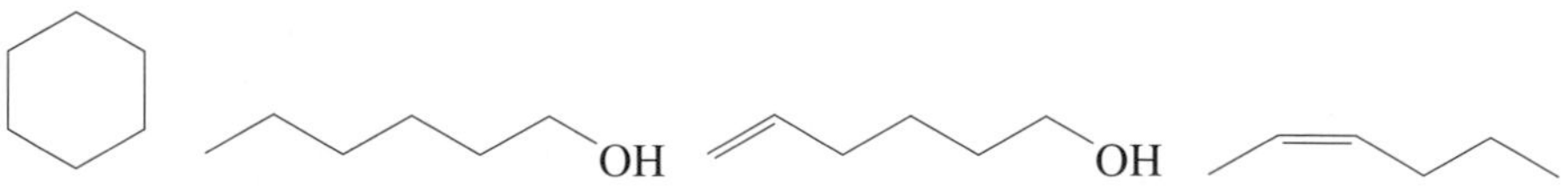

A   B   C   D

27. You have just entered the chemistry stockroom to look for several isomeric bromopentanes. There are three bottles on the shelf marked $C_5H_{11}Br$, but their labels have fallen off. The NMR machine is broken, so you devise the following experiment in an attempt to determine which isomer is in which bottle: You first treat a sample of the contents in each bottle with NaOH in aqueous ethanol, and then you determine the IR spectrum of each product or product mixture. Here are the results:

(i) $C_5H_{11}Br$ isomer in bottle A $\xrightarrow{\text{NaOH}}$ IR bands at 1660, 2850–3020, and 3350 cm^{-1}

(ii) $C_5H_{11}Br$ isomer in bottle B $\xrightarrow{\text{NaOH}}$ IR bands at 1670 and 2850–3020 cm^{-1}

(iii) $C_5H_{11}Br$ isomer in bottle C $\xrightarrow{\text{NaOH}}$ IR bands at 2850–2960 and 3350 cm^{-1}

(a) What do the data tell you about each product or product mixture?
(b) Suggest possible structures for the contents of each bottle.

28. Determine the molecular formulas corresponding to each of the following structures. For each structure, calculate the number of degrees of unsaturation from the molecular formula, and evaluate whether your calculations agree with the structures shown.

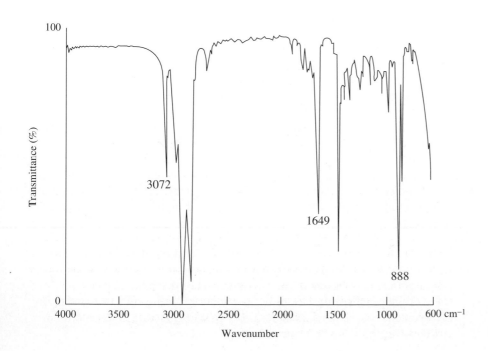

29. Calculate the degree of unsaturation that corresponds to each of the following molecular formulas. (a) C_7H_{12} (use the result in Problem 30); (b) $C_8H_7NO_2$; (c) C_6Cl_6; (d) $C_{10}H_{22}O_{11}$; (e) $C_6H_{10}S$; (f) $C_{18}H_{28}O_2$.

30. A hydrocarbon with the molecular formula C_7H_{12} exhibits the following spectroscopic data: ^1H NMR $\delta = 1.3$ (m, 2 H), 1.7 (m, 4 H), 2.2 (m, 4 H), and 4.8 (quin, $J = 3$ Hz, 2 H) ppm; ^{13}C NMR $\delta = 26.8, 28.7, 35.7, 106.9,$ and 149.7 ppm. The IR spectrum is shown below (spectrum F). Hydrogenation

100

Transmittance (%)

3072

1649

888

0
4000   3500   3000   2500   2000   1500   1000   600 cm^{-1}

Wavenumber

furnishes a product with the molecular formula C_7H_{14}. Suggest a structure for the compound consistent with these data.

31. The isolation of a new form of molecular carbon, C_{60}, was reported in 1990. The substance has the shape of a soccer ball of carbon atoms and possesses the nickname "buckyball" (you don't want to know the IUPAC name). Hydrogenation produces a hydrocarbon with the molecular formula $C_{60}H_{36}$. How many degrees of unsaturation are present in C_{60}? In $C_{60}H_{36}$? Does the hydrogenation result place limits on the numbers of π bonds and rings in "buckyball"? (More on C_{60} is found in Chemical Highlight 15-1.)

32. Place the alkenes in each group in order of increasing stability of the double bond and increasing heat of hydrogenation.

33. Write the structures of as many simple alkenes as you can that, upon catalytic hydrogenation with H_2 over Pt, will give as the product **(a)** 2-methylbutane; **(b)** 2,3-dimethylbutane; **(c)** 3,3-dimethylpentane; **(d)** 1,1,4-trimethylcyclohexane. In each case in which you have identified more than one alkene as an answer, rank the alkenes in order of stability.

34. The reaction between 2-bromobutane and sodium ethoxide in ethanol gives rise to three E2 products. What are they? Predict their relative amounts.

35. What key structural feature distinguishes haloalkanes that give more than one stereoisomer on E2 elimination (e.g., 2-bromobutane, Problem 34) from those that give only a single isomer exclusively (e.g., 2-bromo-3-methylpentane, Section 11-8)?

36. Write the most likely major product(s) of each of the following haloalkanes with sodium ethoxide in ethanol or potassium *tert*-butoxide in 2-methyl-2-propanol (*tert*-butyl alcohol). **(a)** Chloromethane; **(b)** 1-bromopentane; **(c)** 2-bromopentane; **(d)** 1-chloro-1-methylcyclohexane; **(e)** (1-bromoethyl)-cyclopentane; **(f)** (2R,3R)-2-chloro-3-ethylhexane; **(g)** (2R,3S)-2-chloro-3-ethylhexane; **(h)** (2S,3R)-2-chloro-3-ethylhexane.

37. Draw Newman projections of the four stereoisomers of 2-bromo-3-methyl-pentane in the conformation required for E2 elimination. (See the structures labeled "Stereospecificity in the E2 Reaction of 2-Bromo-3-methylpentane" on p. 465.) Are the reactive conformations also the most stable conformations? Explain.

38. Referring to the answer to Problem 26 of Chapter 7, predict (qualitatively) the relative amounts of isomeric alkenes that are formed in the elimination reactions shown.

39. Referring to the answers to Problem 26 of Chapter 9, predict (qualitatively) the relative yields of all the alkenes formed in each reaction.

40. Compare and contrast the major products of dehydrohalogenation of 2-chloro-4-methylpentane with **(a)** sodium ethoxide in ethanol and **(b)** potassium *tert*-butoxide in 2-methyl-2-propanol (*tert*-butyl alcohol). Write the mechanism of each process. Next consider the reaction of 4-methyl-2-pentanol with concentrated H_2SO_4 at 130°C, and compare its product(s) and the mechanism of its (their) formation with those from the dehydrohalogenations in (a) and (b). (**Hint:** The dehydration gives as its major product a molecule that is not observed in the dehydrohalogenations.)

41. Referring to Problem 45 of Chapter 7, write the structure of the alkene that you would expect to be formed as the major product from E2 elimination of each of the chlorinated steroids shown.

42. In the bacterium *Escherichia coli,* an enzyme has been discovered that catalyzes the dehydration of a thioester derivative of $(-)$-3-hydroxydecanoic acid to give a mixture of the corresponding thioester derivatives of *trans*-2-decenoic acid and *cis*-3-decenoic acid:

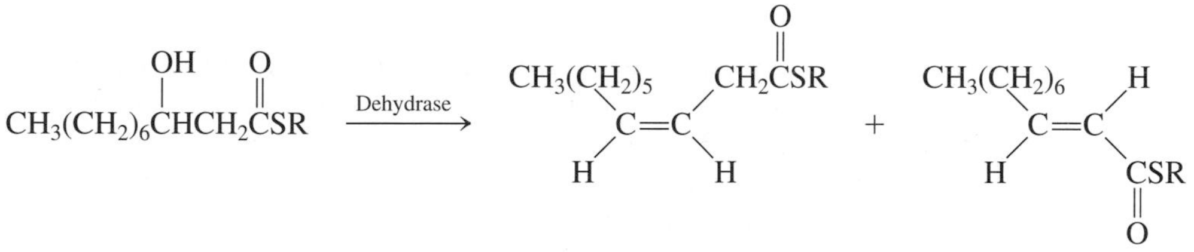

How does this result compare with those that can be expected from simple acid-catalyzed dehydration (e.g., H_2SO_4 and heat)?

43. The heat of combustion of ethane is -372.8 kcal mol^{-1}, that of ethene is -337.2 kcal mol^{-1}, and for H_2, -68.3 kcal mol^{-1}. Use these data to calculate the heat of hydrogenation of ethene. Is your calculation consistent with your expectations based on the discussion of alkene stabilities in Section 11-7? Using an energy input–energy output calculation for $\Delta H°$ (Section 3-4), do you get a reasonable value for the strength of the π bond, as discussed in Section 11-2? What errors arise in this type of estimation?

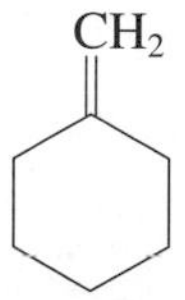

44. 1-Methylcyclohexene is more stable than methylenecyclohexane (A, in the margin), but methylenecyclopropane (B) is more stable than 1-methyl-cyclopropene. Explain.

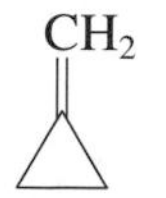

A   B

45. Give the products of bimolecular elimination from each of the following isomeric halogenated compounds.

(a)

Br   H
| |
C₆H₅ — — C₆H₅
| |
H   CH₃

(b)

Br   H
| |
C₆H₅ — — C₆H₅
| |
H₃C H

One of these compounds undergoes elimination 50 times as fast as the other. Which compound is it? Why? (**Hint:** See Problem 36.)

46. Explain in detail the differences between the mechanisms giving rise to the following two experimental results.

CH₃ ... Cl, CH(CH₃)₂ → Na⁺ ⁻OCH₂CH₃, CH₃CH₂OH → CH₃ ... CH(CH₃)₂ 100%

CH₃ ... Cl, CH(CH₃)₂ → Na⁺ ⁻OCH₂CH₃, CH₃CH₂OH → CH₃ ... CH(CH₃)₂ 25% + CH₃ ... CH(CH₃)₂ 75%

47. You have just been named president of the famous perfume company, Scents "R" Us. Searching for a hot new item to market, you run across a bottle labeled only $C_{10}H_{20}O$, which contains a liquid with a wonderfully sweet rose aroma. You want more, so you set out to elucidate its structure. Do so from the following data. (i) ¹H NMR: clear signals at $\delta = 0.94$ (d, $J = 7$ Hz, 3 H), 1.63 (s, 3 H), 1.71 (s, 3 H), 3.68 (t, $J = 7$ Hz, 2 H), 5.10 (t, $J = 6$ Hz, 1 H) ppm; the other 8 H have overlapping absorptions in the range $\delta = 1.3$–2.2 ppm. (ii) ¹³C NMR (¹H decoupled): $\delta = 60.7, 125.0, 130.9$ ppm; seven other signals are upfield of $\delta = 40$ ppm. (iii) IR: $\tilde{v} = 1640$ and 3350 cm⁻¹. (iv) Oxidation with buffered PCC (Section 8-6) gives a compound with the molecular formula $C_{10}H_{18}O$. Its spectra show the following changes compared with the starting material: ¹H NMR: signal at $\delta = 3.68$ ppm is gone, but a new signal is seen at $\delta = 9.64$ ppm; ¹³C NMR: signal at $\delta = 60.7$ ppm is gone, replaced by one at $\delta = 202.1$ ppm; IR: loss of signal at $\tilde{v} = 3350$ cm⁻¹; new peak at $\tilde{v} = 1728$ cm⁻¹. (v) Hydrogenation gives $C_{10}H_{22}O$, identical with that formed on hydrogenation of the natural product geraniol.

Geraniol

48. Using the information in Table 11-4, match up each set of the following IR signals with one of these naturally occurring compounds: camphor; menthol; chrysanthemic ester; epiandrosterone. You can find the structures of the natural products in Section 4-7. (**a**) 3355 cm⁻¹; (**b**) 1630, 1725, 3030 cm⁻¹; (**c**) 1730, 3410 cm⁻¹; (**d**) 1738 cm⁻¹.

49. Identify compounds A, B, and C from the following information and explain the chemistry that is taking place. Reaction of the alcohol shown in the margin with 4-methylbenzenesulfonyl chloride in pyridine produced A ($C_{15}H_{20}O_3S$). Reaction of A with lithium diisopropylamide (LDA, Section 7-8) produces a single product, B (C_8H_{12}), which displays in its ^1H NMR a two-proton multiplet at about $\delta = 5.6$ ppm. If, however, compound A is treated with NaI before the reaction with LDA, two products are formed: B and an isomer, C, whose NMR shows a multiplet at $\delta = 5.2$ ppm that integrates as only one proton.

50. The *citric acid cycle* is a series of biological reactions that play a central role in cell metabolism. The cycle includes dehydration reactions of both malic and citric acids, yielding fumaric and aconitic acids, respectively (all common names). Both proceed strictly by enzyme-catalyzed *anti*-elimination mechanisms.

(a) In each dehydration, only the hydrogen identified by an asterisk is removed, together with the OH group on the carbon below. Write the structures for fumaric and aconitic acids as they are formed in these reactions. Make sure that the stereochemistry of each product is clearly indicated. **(b)** Specify the stereochemistry of each of these products, using either cis-trans or *E,Z* notation, as appropriate. **(c)** Isocitric acid (shown in the margin) also is dehydrated by aconitase. How many stereoisomers can exist for isocitric acid? Remembering that this reaction proceeds through *anti* elimination, write the structure of a stereoisomer of isocitric acid that will give on dehydration the same isomer of aconitic acid that is formed from citric acid. Label the chiral carbons in this isomer of isocitric acid, using *R,S* notation.

OH
|
$HO_2CCHCHCH_2CO_2H$
|
CO_2H
Isocitric acid

Team Problem

51. The following data indicate that the dehydration of certain amino acid derivatives is stereospecific.

Divide the task of analyzing these data among yourselves to determine the nature of the stereocontrolled eliminations. Assign the absolute configuration (*R,S*) to compounds 1a–1d and the *E,Z* configuration to compounds 2a–2d. Draw a Newman projection of the active conformation of each starting compound (1a–1d). As a team, apply your understanding of this information to determine the absolute configuration of the unassigned stereocenter (marked by an asterisk) in compound 3, which was dehydrated to afford compound 4, an intermediate in an approach to the synthesis of compound 5, an antitumor agent.

	R^1	R^2
a	CH_3	H
b	H	CH_3
c	$CH(CH_3)_2$	H
d	H	$CH(CH_3)_2$

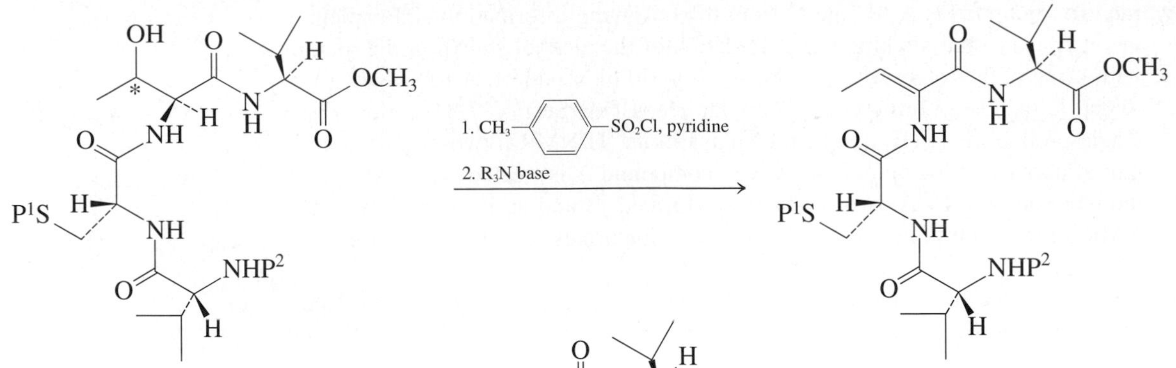

3

(P¹ and P² are protecting groups)

1. CH₃⟨⟩—SO₂Cl, pyridine

2. R₃N base

4

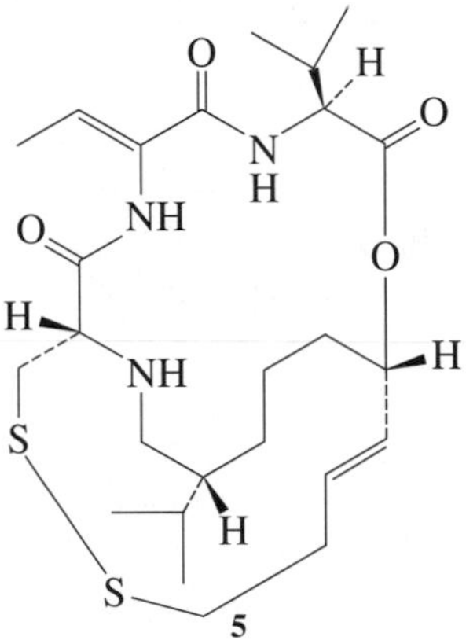

5

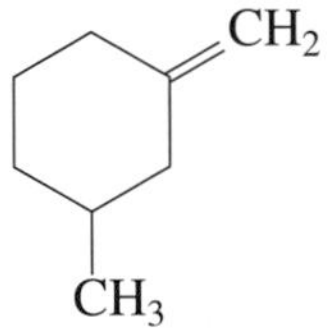

A

Preprofessional Problems

52. What is the empirical formula of compound A?
(a) C_8H_{14}; (b) C_8H_{16}; (c) C_8H_{12}; (d) C_4H_7

53. What is the degree of unsaturation in cyclobutane?
(a) Zero; (b) one; (c) two; (d) three

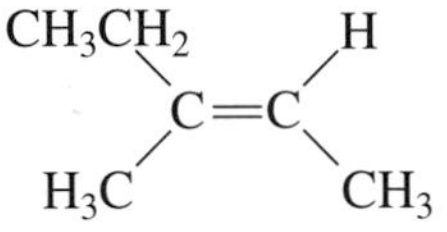

B

54. What is the IUPAC name for compound B (see margin)?
(a) (E)-2-Methyl-3-pentene; (b) (E)-3-methyl-2-pentene; (c) (Z)-2-methyl-3-pentene; (d) (Z)-3-methyl-2-pentene

55. Which of the following molecules would have the lowest heat of hydrogenation?

(a) 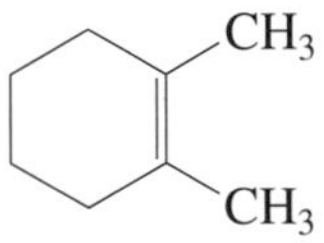 ⟨⟩ (b) 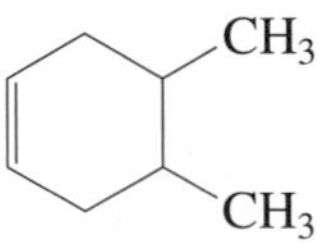 ⟨⟩ (c) 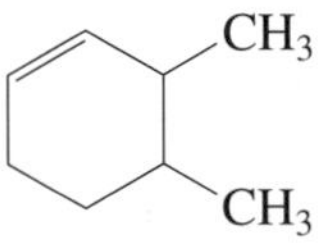 ⟨⟩ (d) 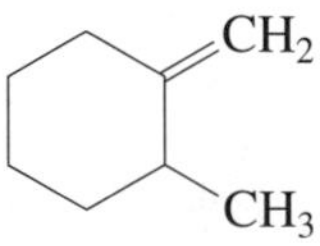 ⟨⟩

56. A certain hydrocarbon containing eight carbons was found to have two degrees of unsaturation but no absorption bands in the IR spectrum at 1640 cm⁻¹. The best structure for this compound is:

(a) 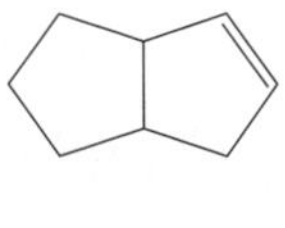 ⟨⟩ (b) 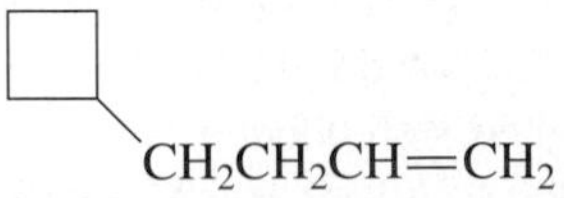 ⟨⟩ $CH_2CH_2CH=CH_2$ (c)  ⟨⟩ (d) 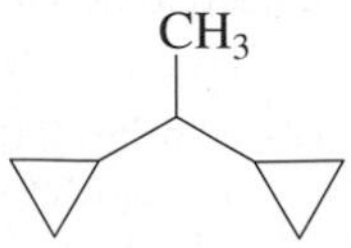 ⟨⟩

Reactions of Alkenes

Polymerization of ethene (ethylene) with the use of special metal-based catalysts generates extraordinarily tear-resistant polyethylene sheets.

Take a look around your room. Can you imagine how different it would appear if every polymer-derived material (including everything made of plastic) were to be removed? Polymers have had an enormous effect indeed on modern society. The chemistry of alkenes underlies our ability to produce polymeric materials of diverse structure, strength, elasticity, and function. In the later sections of this chapter, we shall investigate the processes that give rise to such substances. They are, however, only a subset of the varied reaction types that alkenes undergo. Addition reactions constitute the largest group and lead to saturated products. Through addition, we can take advantage of the fact that the alkene functional group bridges *two* carbons, and we can elaborate on the molecular structure at both ends. Fortunately for us, most additions to the π bond are exothermic: They are almost certain to take place if a mechanistic pathway can be found.

Beyond our simple ability to add to the double bond, additional features further enhance the usefulness and versatility of the addition process. Many alkenes possess defined stereochemistry (*E* and *Z*) and, as we shall see in our discussions, *many of their addition reactions proceed in a stereochemically defined manner.* By combining these facts with the realization that additions to unsymmetrical alkenes may also take place regioselectively, we shall find that we have the ability to exert a large measure of control over the course that an alkene addition follows and, consequently, on the structure of the product that is formed. This control has been exquisitely refined in applications toward the synthesis of enantiomerically pure pharmaceuticals, a phenomenon of the 1990s (see Chemical Highlights 5-4 and 12-3).

We shall begin with a discussion of hydrogenation, focusing on the details of the catalytic activation. Then we turn to the largest class of addition processes, those in which electrophiles such as protons, halogens, and metal ions are added. Other additions that will contribute further to our synthetic repertoire include hydroboration, several oxidation processes (which can lead to complete rupture of the double bond if desired), and radical reactions. Each of these transformations takes us in a different direction; the Reaction Summary Road Map at the end of the chapter provides an overview of the interconversions leading to and from this versatile compound class.

12-1 Why Addition Reactions Proceed: Thermodynamic Feasibility

The carbon–carbon π bond is relatively weak, and the chemistry of the alkenes is largely governed by its reactions. The most common transformation is **addition** of a reagent A–B to give a saturated compound. In the process, the A–B bond is broken, and A and B form single bonds to carbon. Thus, the *thermodynamic feasibility* of this process depends on the strength of the π bond, the dissociation energy DH°_{A-B}, and the strengths of the newly formed bonds of A and B.

Addition to the Alkene Double Bond

$$\text{C=C} \quad + \quad \text{A—B} \quad \xrightarrow{\Delta H^{\circ}\,=\,?} \quad \overset{\overset{\text{A}}{|}}{\underset{|}{\text{C}}}-\overset{\overset{\text{B}}{|}}{\underset{|}{\text{C}}}$$

Recall that we can *estimate* the ΔH° of such reactions by subtracting the combined strength of the bonds made from that of the bonds broken (Section 3-4):

$$\Delta H^{\circ} = (DH^{\circ}_{\pi\ bond} + DH^{\circ}_{A-B}) - (DH^{\circ}_{C-A} + DH^{\circ}_{C-B})$$

in which C stands for carbon.

Table 12-1 gives the DH° values (obtained by using the data from Table 3-1 and Section 3-5 and by equating the strength of the π bond to 65 kcal mol^{-1}) and the estimated ΔH° values for various additions to ethene. In all the examples, the combined strength of the bonds formed exceeds, sometimes significantly, that of the bonds broken. Therefore, thermodynamically, *additions to alkenes should proceed to products with release of energy.*

EXERCISE 12-1

Calculate the ΔH° for the addition of H_2O_2 to ethene to give 1,2-ethanediol (ethylene glycol) ($DH^{\circ}_{HO-OH} = 49$ kcal mol^{-1}).

12-2 Catalytic Hydrogenation

The simplest reaction of the double bond is its saturation with hydrogen. As discussed in Section 11-7, this reaction allows us to estimate the relative stability of substituted alkenes from their heats of hydrogenation. The process requires a catalyst, which may be either heterogeneous or homogeneous—that is, either insoluble or soluble in the reaction medium.

Hydrogenation takes place on the surface of a heterogeneous catalyst

The hydrogenation of an alkene to an alkane, although exothermic, will not take place even at elevated temperatures. Ethene and hydrogen can be heated in the gas phase

TABLE 12-1	Estimated $\Delta H°$ (all values in kcal mol^{-1}) for Additions to Ethene[a]				
$CH_2{=}CH_2$	$+$	$A{-}B$	$\longrightarrow$	$\begin{array}{cc} A & B \\ \mid & \mid \\ H-C-C-H \\ \mid & \mid \\ H & H \end{array}$	
$DH°_{\pi\,bond}$		$DH°_{A-B}$		$DH°_{A-C} \quad DH°_{B-C}$	$\sim \Delta H°$

Hydrogenation

$CH_2{=}CH_2$	$+$	$H{-}H$	$\longrightarrow$	$\begin{array}{cc} H & H \\ \mid & \mid \\ CH_2-CH_2 \end{array}$	-27
65		104		98 $\quad$ 98	

Bromination

$CH_2{=}CH_2$	$+$	$Br{-}Br$	$\longrightarrow$	$\begin{array}{cc} Br & Br \\ \mid & \mid \\ H-C-C-H \\ \mid & \mid \\ H & H \end{array}$	-25
65		46		~68 $\quad$ ~68	

Hydrochlorination

$CH_2{=}CH_2$	$+$	$H{-}Cl$	$\longrightarrow$	$\begin{array}{cc} H & Cl \\ \mid & \mid \\ H-C-C-H \\ \mid & \mid \\ H & H \end{array}$	-10
65		103		~98 $\quad$ 80	

Hydration

$CH_2{=}CH_2$	$+$	$H{-}OH$	$\longrightarrow$	$\begin{array}{cc} H & OH \\ \mid & \mid \\ H-C-C-H \\ \mid & \mid \\ H & H \end{array}$	-6
65		119		~98 $\quad$ 92	

[a] These values are only estimates: They do not take into account the changes in C–C σ-bond strength that accompany changes in hybridization. (Compare the value for hydrogenation in the table with that calculated in Problem 42 of Chapter 11.)

Commercial apparatus for catalytic hydrogenation. The metal container at the left encloses a glass reaction bottle that contains both substrate and catalyst. Hydrogen is admitted into the bottle from the tank in the back, and the gauges measure H_2 uptake. A motorized mechanism shakes the reaction vessel, dispersing the insoluble catalyst. (Courtesy Parr Instrument Company, Moline, IL.)

to 200°C for prolonged periods without any measurable change. However, as soon as a catalyst is added, hydrogenation proceeds even at room temperature at a steady rate. The catalysts frequently are the same as those used for the catalytic hydrogenation of carbonyl compounds to alcohols (Section 8-6): insoluble materials such as palladium (e.g., dispersed on carbon, Pd-C), platinum (Adams's* catalyst, PtO_2, which is converted into colloidal platinum metal in the presence of hydrogen), and nickel (finely dispersed, as in a preparation called Raney[†] nickel, Ra-Ni).

The major function of the catalyst is the activation of hydrogen to generate metal-bound hydrogen on the catalyst surface (Figure 12-1). Without the metal, thermal

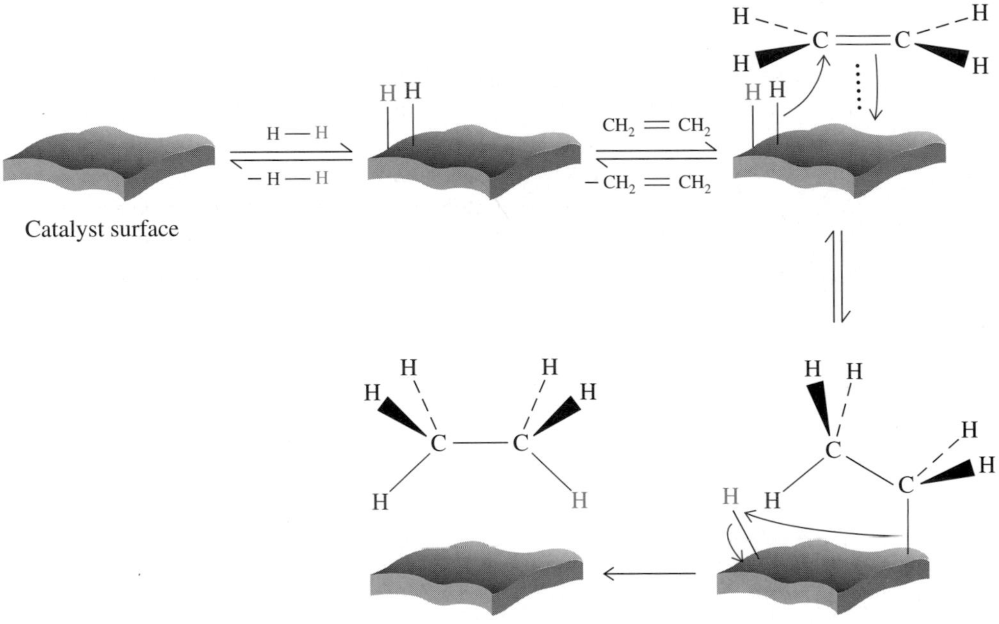

FIGURE 12-1

In the catalytic hydrogenation of ethene to produce ethane, the hydrogens bind to the catalyst surface and are then delivered to the carbons of the surface-adsorbed alkene.

cleavage of the strong H–H bond is energetically prohibitive. Solvents commonly used in such hydrogenations include methanol, ethanol, acetic acid, and ethyl acetate, as shown in the following example.

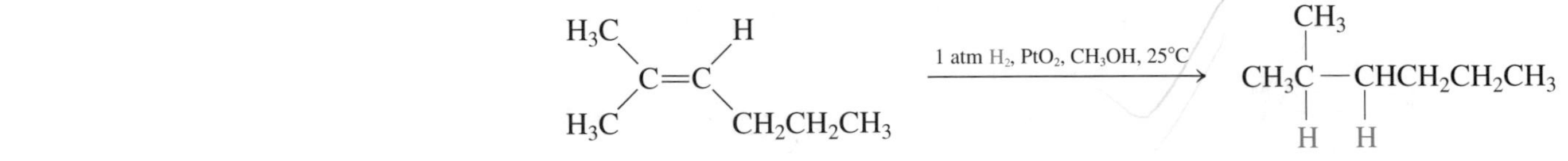

2-Methyl-2-hexene

1 atm H_2, PtO_2, CH_3OH, 25°C

100%
2-Methylhexane

Hydrogenation is stereospecific

An important feature of catalytic hydrogenation is *stereospecificity*. The two hydrogen atoms are added to the same face of the double bond (*syn* addition). For example, 1-ethyl-2-methylcyclohexene is hydrogenated over platinum to give specifically

*Professor Roger Adams (1889–1971), University of Illinois at Urbana-Champaign.
[†]Dr. Murray Raney (1885–1966), Raney Catalyst Company, South Pittsburg, Tennessee.

cis-1-ethyl-2-methylcyclohexane. Addition of hydrogen can be from above or from below the molecular plane with equal probability. Therefore, each stereocenter is generated as both image and mirror image, and the product is racemic.

1-Ethyl-2-methylcyclohexene

$\xrightarrow{\text{H}_2,\ \text{PtO}_2,\ \text{CH}_3\text{CH}_2\text{OH},\ 25°C}$

82%
cis-1-Ethyl-2-methylcyclohexane
(Racemic)

However, if steric hindrance inhibits hydrogenation on one side of a ring, addition will take place exclusively at the *less hindered* side. Thus, Pt-catalyzed hydrogenation of the bicyclic alkene car-3-ene, a constituent of turpentine, gives only one saturated product, with the common name *cis*-carane. The prefix cis indicates that the methyl group and the cyclopropane ring are on the same side of the cyclohexane ring. This result shows that hydrogen has been added only from the less hindered face of the double bond, opposite the three-membered ring, thus pushing the methyl group cis to that ring. (Make a model and use a tabletop to represent the catalyst surface, as in Figure 12-1.)

$\xrightarrow{\text{100 atm H}_2,\ \text{PtO}_2,\ \text{CH}_3\text{CH}_2\text{OH},\ 25°C}$

98%

not

Car-3-ene   *cis*-**Carane**

EXERCISE 12-2

Catalytic hydrogenation of (*S*)-2,3-dimethyl-1-pentene gives only one optically active product. Show the product and explain the result. [**Hint:** Does addition of H₂ either (1) create a new stereocenter or (2) affect any of the bonds around the stereocenter already present?]

CHEMICAL HIGHLIGHT 12-1

Optically Active Amino Acids by Asymmetric Hydrogenation

Amino acids, the building blocks of proteins (see Chapter 26), are synthesized in the laboratory and in industry in optically active form by catalytic hydrogenation of achiral, nitrogen-substituted alkenes called *enamides*. The process uses a homogeneous (soluble) catalyst, consisting of a metal such as rhodium or ruthenium and an optically active *phosphine ligand,* which binds to the metal. Bulky groups on the phosphorus atom direct the hydrogenation to one face of the alkene with high selectivity, thereby furnishing a single enantiomer.

Most of the ligands used in such *asymmetric hydrogenations* are available in both enantiomeric forms; therefore, either enantiomer of a product may be prepared. For example, hydrogenation of (Z)-2-acetamido-3-phenylpropenoic acid by using a rhodium catalyst containing the ligand "(R,R)-Degphos" produces the (S)-phenylalanine derivative shown here with greater than 99% stereoselectivity.

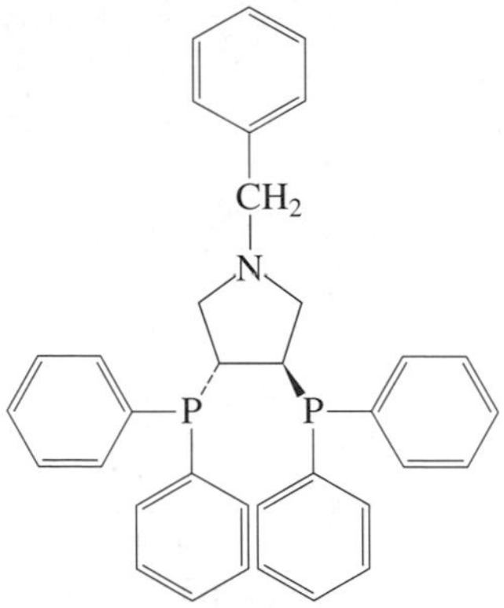

(R,R)-Degphos
(A phosphine ligand)

(S)-Phenylalanine is used in the manufacture of the artificial sweetener aspartame. Its R enantiomer, which is present in the antibiotic gramicidin S, is equally readily obtained by hydrogenation of the same enamide with Rh in the presence of the enantiomeric ligand (S,S)-Degphos.

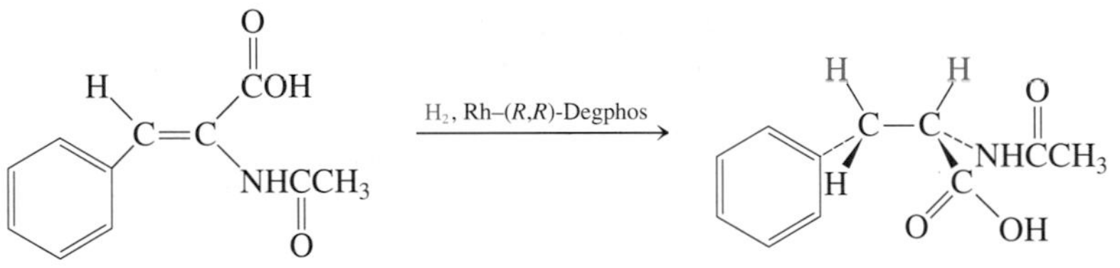

**(Z)-2-Acetamido-
3-phenylpropenoic acid**
(An enamide)

H_2, Rh–(R,R)-Degphos

(S)-N-Acetylphenylalanine

In summary, hydrogenation of the double bond in alkenes requires a catalyst. This transformation occurs stereospecifically by *syn* addition, and, when there is a choice, from the least hindered side of the molecule.

12-3 Nucleophilic Character of the Pi Bond: Electrophilic Addition of Hydrogen Halides

As noted earlier, the π electrons of a double bond are not as strongly bound as those of a σ bond. The electron cloud above and below the molecular plane of the alkene is polarizable and subject to attack by electron-deficient species, just as are the lone electron pairs in typical Lewis bases. Halogens and the mercuric ion are examples of electrophilic species that attack the π electrons of the double bond. As in hydrogenation, additions occur, but by different mechanisms. These transformations, called *electrophilic additions,* can be regioselective and stereospecific. We begin this section with the simplest of all electrophiles, the proton.

Electrophilic attack by protons gives carbocations

The proton of a strong acid may add to a double bond to yield a carbocation. The transition state for the process is the same as that formulated for the deprotonation step in the E1 reaction (Figure 7-7). In the absence of a good nucleophile capable of trapping the carbocation, rearrangement may occur. However, in the presence of such a nucleophile, particularly at low temperatures, the carbocation is intercepted to give the product of an **electrophilic addition** to the double bond. For example, treatment of alkenes with hydrogen halides leads to the corresponding haloalkanes.

Electrophilic Addition of HX to Alkenes

In a typical experiment, the gaseous hydrogen halide, which may be HCl, HBr, or HI, is bubbled through pure or dissolved alkene. Alternatively, HX can be added in a solvent, such as acetic acid. Aqueous work-up furnishes the haloalkane in high yield.

The Markovnikov rule predicts regioselectivity in electrophilic additions

Are additions of HX to unsymmetric alkenes regioselective? To answer this question, let us consider the reaction of propene with hydrogen chloride. Two products are possible: 2-chloropropane and 1-chloropropane. However, the only product observed is 2-chloropropane.

Regioselective Electrophilic Addition to Propene

$$CH_3CH=CH_2 \xrightarrow{HCl} CH_3CHCH_2 \quad \text{but no} \quad CH_3CHCH_2$$

Less substituted   2-Chloropropane   1-Chloropropane

Similarly, reaction of 2-methylpropene with hydrogen bromide gives only 2-bromo-2-methylpropane, and 1-methylcyclohexene combines with HI to furnish only 1-iodo-1-methylcyclohexane.

Less substituted   Less substituted

We can see from these examples that, if the carbon atoms participating in the double bond are not equally substituted, *the proton from the hydrogen halide attaches itself to the less substituted carbon.* As a consequence, the halogen ends up at the more substituted carbon. This phenomenon, referred to as the **Markovnikov* rule,** can be explained by what we know about the mechanism of electrophile additions of protons to alkenes. The key is the relative stability of the resulting carbocations.

Consider the hydrochlorination of propene. The regiochemistry of the reaction is determined in the first step, in which the proton attacks the π system to give an intermediate carbocation. Carbocation generation is rate determining; once it occurs, reaction with chloride proceeds quickly. Let us look at the crucial first step in more detail. The proton may attack either of the two carbon atoms of the double bond. Addition to the internal carbon leads to the primary propyl cation.

Protonation of Propene at C2

$$H_3C\overset{H}{\diagdown}C{=}C\overset{H}{\diagup}{}_H \quad H^+ \quad \longrightarrow \quad \left[H_3C\overset{H}{\diagdown}C\overset{\delta^+}{\cdots}C\overset{H}{\diagup}{}_H \atop H^{\delta^+} \right]^{\ddagger} \quad \longrightarrow \quad CH_3CH_2CH_2{}^+$$

TS-1

Primary carbocation
(Not observed)

In contrast, protonation at the terminal carbon results in the formation of the secondary 1-methylethyl (isopropyl) cation.

Protonation of Propene at C1

$$H_3C\overset{H}{\diagdown}C{=}C\overset{H}{\diagup}{}_H \quad H^+ \quad \longrightarrow \quad \left[H_3C\overset{H}{\diagdown}\overset{\delta^+}{C}{\cdots}C\overset{H}{\diagup}{}_H \atop H^{\delta^+} \right]^{\ddagger} \quad \longrightarrow \quad CH_3\overset{+}{C}HCH_3$$

TS-2

Secondary carbocation
(Favored)

The second species is more stable and, because the structure of the transition state for protonation resembles that of the resulting cation, is formed considerably faster. Figure 12-2 is a potential-energy diagram of this situation.

On the basis of this analysis, we can rephrase the empirical Markovnikov rule: HX adds to unsymmetric alkenes in such a way that *the initial protonation gives the more stable carbocation.* For alkenes that are similarly substituted at both sp^2 carbons, product mixtures are to be expected, because carbocations of comparable stability are formed. In analogy with other carbocation reactions (e.g., S_N1, Section 7-3), when addition to an achiral alkene generates a chiral product, the latter is obtained as a racemic mixture.

EXERCISE 12-3

Predict the outcome of the addition of HBr to **(a)** 1-hexene; **(b)** *trans*-2-pentene; **(c)** 2-methyl-2-butene; **(d)** 4-methylcyclohexene. How many isomers can be formed in each case?

---

*Professor Vladimir V. Markovnikov (1838–1904), University of Moscow, formulated his rule in 1869.

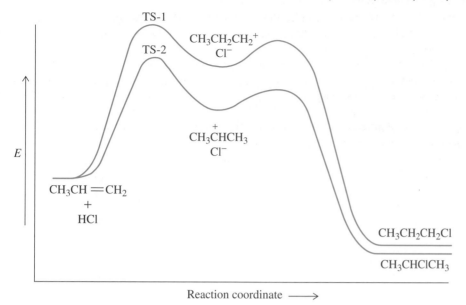

FIGURE 12-2 Potential-energy diagram for the two possible modes of HCl addition to propene. Transition state 1 (TS-1), which leads to the higher-energy primary propyl cation, is less favored than transition state 2 (TS-2), which gives the 1-methylethyl (isopropyl) cation.

EXERCISE 12-4

Draw a potential-energy diagram for the reaction in (c) of Exercise 12-3.

In summary, additions of hydrogen halides to alkenes are electrophilic reactions that begin with protonation of the double bond to give a carbocation. Trapping of the carbocation by halide ion gives the final product. The Markovnikov rule predicts the regioselectivity of hydrohalogenation to haloalkanes.

12-4 Alcohol Synthesis by Electrophilic Hydration: Thermodynamic Control

So far, we have seen attack on the double bond by a proton, followed by nucleophilic attachment of its halide counterion to the intermediate carbocation. Can other nucleophiles participate? Upon exposure of an alkene to an *aqueous* solution of an acid such as sulfuric acid, which has a poorly nucleophilic counterion, *water* acts as the nucleophile to trap the carbocation formed by initial protonation. Overall, the elements of water add to the double bond, an **electrophilic hydration.** The addition follows the Markovnikov rule in that H^+ adds to the less substituted carbon, and the OH group ends up at the more substituted one.

This addition process is the reverse of the acid-induced elimination of water from alcohols (dehydration, Section 11-9). Its mechanism is the same in reverse, as illustrated in the hydration of 2-methylpropene, a reaction of industrial importance leading to 2-methyl-2-propanol (*tert*-butyl alcohol).

Electrophilic Hydration

$$\underset{\textbf{2-Methylpropene}}{\begin{array}{c} H_3C \\ \diagdown \\ \\ H_3C \diagup \end{array} C=CH_2} \quad \xrightarrow{\text{50\% HOH, H}_2\text{SO}_4} \quad \underset{\begin{array}{c} 92\% \\ \textbf{2-Methyl-2-propanol} \end{array}}{\begin{array}{c} H \\ | \\ H_3C \qquad CH_2 \\ \diagdown \quad \diagup \\ C \\ \diagup \quad \diagdown \\ H_3C \qquad OH \end{array}}$$

Mechanism of the Hydration of 2-Methylpropene

$$H_3C \overset{H_3C}{\underset{H_3C}{C}} = CH_2 \quad \underset{-H^+}{\overset{H^+}{\rightleftharpoons}} \quad \overset{H_3C}{\underset{CH_3}{C}} \overset{CH_3}{\underset{+}{C}} \quad \underset{-H\ddot{O}H}{\overset{+H\ddot{O}H}{\rightleftharpoons}} \quad \overset{CH_3}{\underset{CH_3}{CH_3C}} - \overset{H}{\underset{H}{\ddot{O}}} \quad \underset{+H^+}{\overset{-H^+}{\rightleftharpoons}} \quad \overset{CH_3}{\underset{CH_3}{CH_3C}} - \ddot{O}H$$

Alkene hydration and alcohol dehydration are equilibrium processes

Hydration–Dehydration Equilibrium

$$RCH{=}CH_2 \ + \ H_2O$$

$$\Updownarrow \ \text{Catalytic } H^+$$

$$\underset{OH}{\overset{|}{RCHCH_3}}$$

In the mechanism of alkene hydration, *all the steps are reversible.* The proton acts only as a catalyst and is not consumed in the overall reaction. Indeed, without the acid, hydration would not occur; alkenes are stable in neutral water. The presence of acid, however, establishes an equilibrium between alkene and alcohol. This equilibrium can be driven toward the alcohol by using low reaction temperatures and a large excess of water. Conversely, we have seen (Section 11-9) that treating the alcohol with *concentrated* acid favors dehydration, especially at elevated temperatures.

$$\text{Alcohol} \quad \underset{\text{H}_2\text{SO}_4,\text{ excess H}_2\text{O, low temperature}}{\overset{\text{Conc. H}_2\text{SO}_4,\text{ high temperature}}{\rightleftharpoons}} \quad \text{alkene} \ + \ H_2O$$

EXERCISE 12-5

Treatment of 2-methylpropene with catalytic deuterated sulfuric acid (D_2SO_4) in D_2O gives $(CD_3)_3COD$. Explain by a mechanism.

The reversibility of alkene protonation leads to alkene equilibration

Section 11-9 explained that the acid-catalyzed dehydration of alcohols gives mixtures of alkenes in which the more stable isomers predominate. Equilibrating carbocation rearrangements are partly responsible for these results. However, another pathway is **reversible protonation** of the initial alkene products arising by E1 pathways.

Recall the dehydration of 2-butanol. Loss of water from the protonated alcohol gives the 2-butyl carbocation. Proton loss from this cation can then give any of three observed products: 1-butene, *cis*-2-butene, or *trans*-2-butene. However, under the strongly acidic conditions, a proton can re-add to the double bond. As noted earlier, this addition will be highly regioselective (Markovnikov rule) and will regenerate the same secondary carbocation that was formed in the initial dehydration. Because this cation may again lose a proton to give any of the same three alkene isomers, the net effect is *interconversion* of the isomers to an *equilibrium mixture,* in which the thermodynamically most stable isomer is the major component. This system is therefore an example of a reaction that is under *thermodynamic control.*

Mechanism of Thermodynamic Control in Acid-Catalyzed Dehydrations

STEP 1. Proton loss from the terminal carbon of the carbocation gives 1-butene.

$$\underset{H}{\overset{OH}{\underset{|}{\overset{|}{CH_3CH_2CCH_3}}}} \quad \underset{-H_2O}{\overset{H^+}{\rightleftharpoons}} \quad \underset{H}{\overset{+}{\underset{|}{\overset{|}{CH_3CH_2CCH_3}}}} \quad \underset{}{\overset{-H^+}{\rightleftharpoons}} \quad CH_3CH_2CH{=}CH_2$$

STEP 2. Reprotonation regenerates the carbocation, which eventually leads to a more stable internal alkene.

$$CH_3CH_2CH\!=\!CH_2 \quad \overset{H^+}{\rightleftharpoons} \quad CH_3CH_2\overset{+}{C}CH_3 \quad \overset{-H^+}{\rightleftharpoons} \quad CH_3CH\!=\!CHCH_3$$
$$\underset{H}{\overset{|}{}}$$

By this procedure, less stable alkenes may be catalytically converted into their more stable isomers.

Acid-Catalyzed Equilibration of Alkenes

Terminal

$$\underset{H}{\overset{(CH_3)_3C}{}}C\!=\!C\underset{H}{\overset{C(CH_3)_3}{}} \quad \xrightarrow{\text{Catalytic } H^+} \quad \underset{H}{\overset{(CH_3)_3C}{}}C\!=\!C\underset{C(CH_3)_3}{\overset{H}{}}$$

Cis   **Trans**

Catalytic H^+

EXERCISE 12-6

Write a mechanism for this rearrangement. What is the driving force for the reaction?

$$\underset{H_3C}{} \xrightarrow{H^+} \underset{H_3C}{\overset{CH_2CH_3}{}}$$

Internal

In summary, the carbocation formed by addition of a proton to an alkene may be trapped by water to give an alcohol, the reverse of alkene synthesis by alcohol dehydration. Reversible protonation equilibrates alkenes in the presence of acid, thereby forming a thermodynamically controlled mixture of isomers.

12-5 Electrophilic Addition of Halogens to Alkenes

Reagents that do not appear to contain electrophilic atoms can attack double bonds electrophilically. An example is the halogenation of alkenes, which proceeds with addition of two halogen atoms to the double bond to give a vicinal dihalide. The reaction works best for chlorine and bromine. Fluorine reacts too violently (often explosively) with alkenes, whereas diiodide formation is virtually thermoneutral and is not generally observed.

Halogenation of Alkenes

$$\underset{\overset{|}{}}{\overset{\diagup}{}}C\!=\!C\overset{\diagdown}{\underset{\overset{|}{}}{}} \quad \xrightarrow{X-X} \quad \underset{X}{\overset{X}{}}C-C\overset{\diagup}{\underset{X}{}}$$

Vicinal dihalide

X = Cl, Br

EXERCISE 12-7

Calculate (as in Table 12-1) the $\Delta H°$ values for the addition of F_2 and I_2 to ethene. (For $DH°_{X_2}$, see Section 3-5.)

Bromine addition is particularly easy to recognize because bromine solutions immediately change from red to colorless when exposed to an alkene. This phenomenon is sometimes used to test for unsaturation. The reaction of bromine with saturated systems is a much slower radical process and requires initiation by heat or light (Section 3-5).

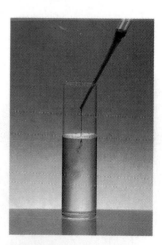

Addition of bromine to an alkene results in almost immediate loss of the red-brown color of Br_2 as the reaction takes place.

Halogenations are best carried out at room temperature or, with cooling, in inert halogenated solvents such as the halomethanes.

Electrophilic Halogen Addition of Br$_2$ to 1-Hexene

$$CH_3(CH_2)_3CH=CH_2 \xrightarrow{Br-Br, CCl_4} CH_3(CH_2)_3CHCH_2Br$$

$$\underset{Br}{|}$$

90%

1-Hexene   **1,2-Dibromohexane**

Halogen additions to double bonds may seem to be similar to hydrogenations. However, their mechanism is quite different, as revealed by the stereochemistry of bromination; similar arguments hold for the other halogens.

Bromination takes place through *anti* addition

What is the stereochemistry of bromination? Are the two bromine atoms added from the same side of the double bond (as in catalytic hydrogenation) or from opposite sides? Let us examine the bromination of cyclohexene. Double addition on the same side should give *cis*-1,2-dibromocyclohexane; the alternative would result in *trans*-1,2-dibromocyclohexane. The second alternative is borne out by experiment—only ***anti* addition** is observed. Because *anti* addition to the two reacting carbon atoms can take place with equal probability in two possible ways—in either case, from both above and below the π bond—the product is racemic.

Anti Bromination of Cyclohexene

83%

Racemic *trans*-1,2-dibromocyclohexane

With acyclic alkenes, the reaction is also cleanly stereospecific. For example, *cis*-2-butene is brominated to furnish a racemic mixture of (2*R*,3*R*)- and (2*S*,3*S*)-2,3-dibromobutane; *trans*-2-butene results in the meso diastereomer.

Stereospecific 2-Butene Bromination

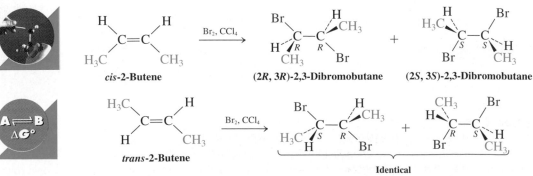

cis-2-Butene   (2*R*, 3*R*)-2,3-Dibromobutane   (2*S*, 3*S*)-2,3-Dibromobutane

trans-2-Butene

Identical

meso-2 3-Dibromobutane

Cyclic bromonium ions explain the stereochemistry

How does bromine attack the electron-rich double bond even though it does not appear to contain an electrophilic center? The answer lies in the polarizability of the Br–Br bond, which is prone to heterolytic cleavage on reaction with a nucleophile. The π electron cloud of the alkene is nucleophilic and attacks one end of the bromine molecule with simultaneous displacement of the second bromine atom as bromide ion in an S_N2-like process. What is the product of this process? We might expect a carbocation, in analogy with the proton additions discussed in Sections 12-3 and 12-4. However, the intermediacy of a carbocation is inconsistent with the stereospecificity observed. For example, if the first step of addition of bromine to cyclohexene were to give a carbocation, the bromide ion released in this process could attack the positively charged carbon atom from either the same or the opposite side of the ring and should give a mixture of *cis*- and *trans*-1,2-dibromocyclohexanes. However, as shown previously, *only* the trans product is obtained. How can we explain this result?

The stereochemistry of bromination is explained if we propose that initial attack of bromine on the double bond gives a cyclic **bromonium ion,** in which the bromine bridges both carbon atoms of the original double bond to form a three-membered ring (Figure 12-3). The structure of this ion is rigid, and it may be attacked by bromide ion only on the side opposite the bridging bromine atom. The three-membered ring is thus opened stereospecifically (compare nucleophilic ring opening of oxacyclopropanes in Section 9-9). The leaving group is the bridging bromine. In symmetric bromonium ions, attack is equally probable at either carbon atom, thereby giving the racemic (or meso) products observed.

Nucleophilic Opening of a Cyclic Bromonium Ion

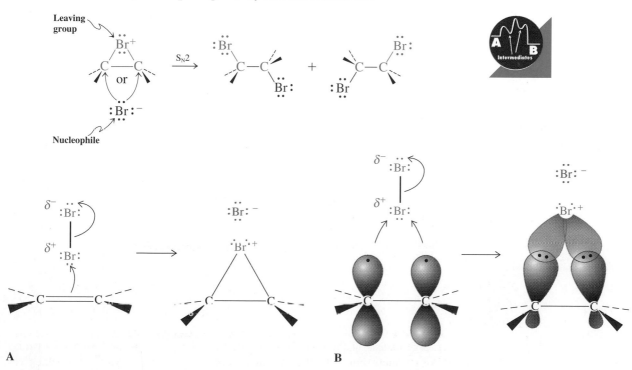

FIGURE 12-3

(A) Electron-pushing picture of cyclic bromonium ion formation. The alkene (red) acts as a nucleophile to displace bromide ion (green) from bromine. The molecular bromine behaves as if it were strongly polarized, one atom as a bromine cation, the other as an anion.
(B) Molecular-orbital picture of bromonium ion formation.

EXERCISE 12-8

Draw the intermediate in the bromination of cyclohexene, using the following conformational picture. Show why the product is racemic. What can you say about the initial conformation of the product?

Conformational flip in cyclohexene

In summary, halogens add, as electrophiles, to alkenes to produce intermediate bridged halonium ions. These intermediates are opened stereospecifically by the halide ions displaced in the initial step to give overall *anti* addition to the double bond.

12-6 The Generality of Electrophilic Addition

A number of electrophile–nucleophile combinations are capable of addition to double bonds, the reactions resulting in a wide variety of useful products.

The bromonium ion can be trapped by other nucleophiles

The creation of a bromonium ion in alkene brominations suggests that, in the presence of other nucleophiles, competition might be observed in the trapping of the intermediate. Indeed, this is possible. For example, bromination of cyclopentene in the presence of excess chloride ion (added as a salt) gives a mixture of *trans*-1,2-dibromocyclopentane and *trans*-1-bromo-2-chlorocyclopentane.

Competitive Trapping of a Cyclic Bromonium Ion
(Although all products are racemic, only one enantiomer is shown in each case)

A large excess of the competing nucleophile may prevent the formation of mixtures of products. For example, bromination of cyclopentene in water as solvent gives the vicinal bromoalcohol (common name, bromohydrin) exclusively. In this case, the bromonium ion is attacked by water. The net transformation is the *anti* addition of Br and OH to the double bond. The other product formed is HBr. The corresponding chloroalcohols (chlorohydrins) can be made from chlorine in water through the intermediacy of a chloronium ion.

Bromoalcohol (Bromohydrin) Synthesis

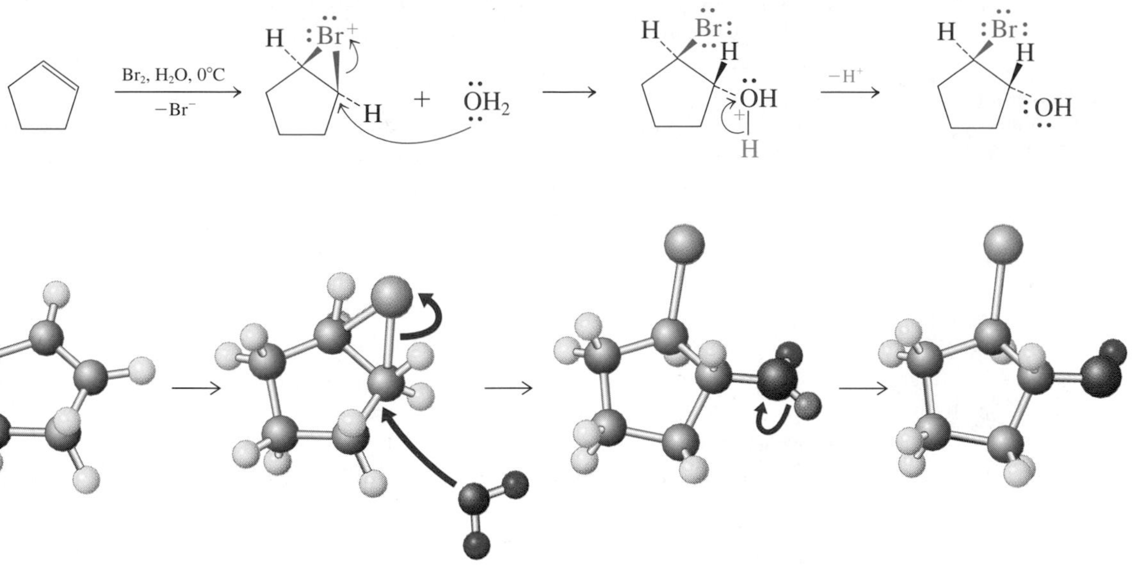

trans-2-
Bromocyclopentanol

EXERCISE 12-9

Write the expected product from the reaction of **(a)** *trans*-2-butene and **(b)** *cis*-2-pentene with aqueous chlorine. Show the stereochemistry clearly.

Vicinal haloalcohols undergo intramolecular ring closure in the presence of base to give oxacyclopropanes (Section 9-9) and are therefore useful intermediates in organic synthesis.

Oxacyclopropane Formation from an Alkene Through the Haloalcohol

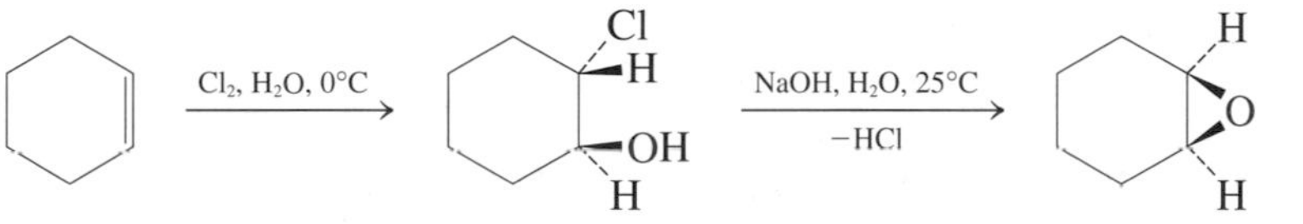

70–73% 70–73%

If alcohol is used as a solvent instead of water in these halogenations, the corresponding vicinal haloethers are produced, as shown in the margin.

Vicinal Haloether Synthesis

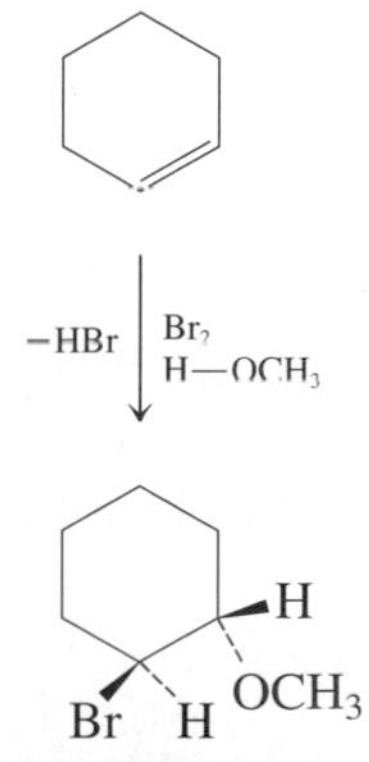

76%

***trans*-1-Bromo-2-methoxycyclohexane**

Halonium ion opening can be regioselective

In contrast with dihalogenations, mixed additions to double bonds can pose regiochemical problems. Is the addition of Br and OR to an unsymmetric double bond selective? The answer is yes. For example, 2-methylpropene is converted by aqueous bromine into only 1-bromo-2-methyl-2-propanol; none of the alternative regioisomer, 2-bromo-2-methyl-1-propanol, is formed.

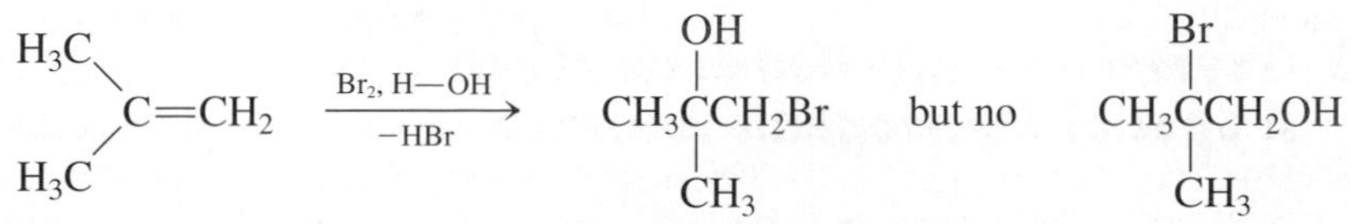

82% yield

1-Bromo-2-methyl-2-propanol

2-Bromo-2-methyl-1-propanol

The electrophilic halogen in the product always becomes linked to the less substituted carbon of the original double bond, and the subsequently added nucleophile attaches to the more highly substituted center.

How can this be explained? The situation is very similar to the acid-catalyzed nucleophilic ring opening of oxacyclopropanes (Section 9-9), in which the intermediate contains a protonated oxygen in the three-membered ring. In both reactions, *the nucleophile attacks the more highly substituted carbon of the ring, because this carbon is more positively polarized than the other.*

Regioselective Opening of the Bromonium Ion
Formed from 2-Methylpropene

Greater
δ^+ here

$(CH_3)_2C \overset{\frown}{\underset{}{}} CH_2 \quad + \quad H-\overset{..}{\underset{..}{O}}H \quad \longrightarrow \quad (CH_3)_2C-CH_2\overset{..}{\underset{..}{Br}}: \quad \xrightarrow{-H^+} \quad CH_3\overset{CH_3}{\underset{|}{C}}-CH_2\overset{..}{Br}$

Attack at more substituted
carbon of bromonium ion

$H-\overset{+}{\underset{|}{O}}:^+$ $:OH$

H

A simple rule of thumb is that electrophilic additions of unsymmetric reagents of this type add in a regiochemical sense that is Markovnikov-like, the electrophilic unit emerging at the less substituted carbon of the double bond. Mixtures are formed only when the two carbons are not sufficiently differentiated [see (b) of Exercise 12-10].

EXERCISE 12-10

What are the product(s) of the following reactions?

(a) $CH_3CH{=}CH_2 \xrightarrow{Cl_2, \, CH_3OH}$ **(b)** [cyclohexene ring with H_3C substituent] $\xrightarrow{Br_2, \, H_2O}$

EXERCISE 12-11

What would be a good alkene precursor for a racemic mixture of (2R,3R)- and (2S,3S)-2-bromo-3-methoxypentane? What other isomeric products might you expect to find from the reactions that you propose?

Alkenes in general can undergo stereo- and regiospecific addition reactions with reagents of the type A–B, in which the A–B bond is polarized such that A acts as the electrophile A^+, B as the nucleophile B^-. Table 12-2 shows how such reagents add to 2-methylpropene.

In summary, halonium ions are subject to stereospecific and regioselective ring opening in a manner mechanistically very similar to the nucleophilic opening of protonated oxacyclopropanes. Halonium ions can be trapped by halide ions, water, and alcohols to give vicinal dihaloalkanes, haloalcohols, and haloethers, respectively. The principle of electrophilic additions can be applied to any reagent A–B containing a polarized or polarizable bond.

12-7 Oxymercuration–Demercuration: A Special Electrophilic Addition

The last example in Table 12-2 is an electrophilic addition of a mercuric salt to an alkene. The reaction is called mercuration, and the resulting compound is an alkylmercury derivative, from which the mercury can be removed in a subsequent step. One

TABLE 12-2 Reagents A—B That Add to Alkenes by Electrophilic Attack

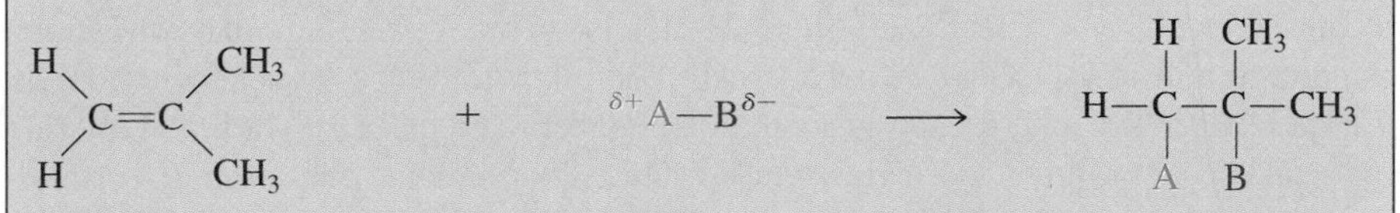

Name	Structure	Addition product to 2-methylpropene
Bromine chloride	Br—Cl	$BrCH_2CCl(CH_3)_2$
Cyanogen bromide	Br—CN	$BrCH_2C(CN)(CH_3)_2$
Iodine chloride	I—Cl	$ICH_2CCl(CH_3)_2$
Sulfenyl chlorides	RS—Cl	$RSCH_2CCl(CH_3)_2$
Mercuric salts	XHg—X[a]	$XHgCH_2CX(CH_3)_2$

[a]X here denotes acetate.

particularly useful reaction sequence is **oxymercuration–demercuration,** in which mercuric acetate acts as the reagent. In the first step (oxymercuration), treatment of an alkene with this species in the presence of water leads to the corresponding addition product.

Oxymercuration

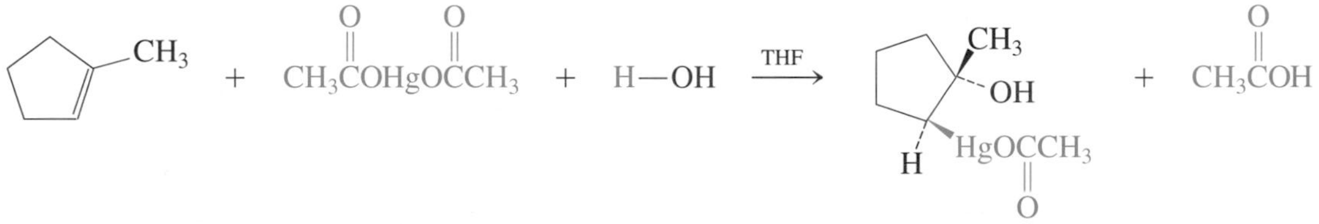

Mercuric acetate **Alkylmercuric acetate**

In the subsequent demercuration, the mercury-containing substituent can be replaced by hydrogen through treatment with sodium borohydride in base. The net result is hydration of the double bond to give an alcohol.

Demercuration

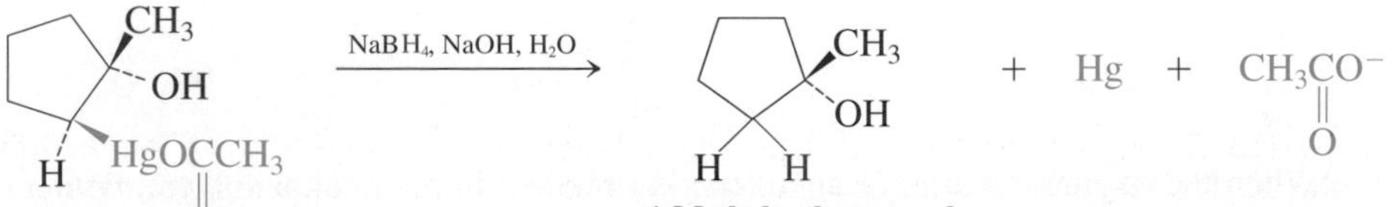

1-Methylcyclopentanol

Oxymercuration is usually *anti* stereospecific; it is also regioselective. This outcome implies a mechanism similar to that for the electrophilic addition reactions discussed so far. The mercury reagent may initially dissociate to give an acetate ion and a cationic mercury species. The latter then attacks an alkene double bond, furnishing a mercurinium ion, probably with a structure similar to that of a cyclic bromonium ion. The water that is present attacks the more substituted carbon (Markovnikov rule regioselectivity) to give an alkylmercuric acetate intermediate. Replacement of mercury by hydrogen (demercuration) is achieved by sodium borohydride reduction through a complex and only incompletely understood mechanism. It is not a stereospecific process.

The alcohol obtained after demercuration is the same as the product of Markovnikov hydration (Section 12-4) of the starting material. However, oxymercuration–demercuration is a valuable alternative to acid-catalyzed hydration, because no carbocation is involved; therefore *oxymercuration–demercuration is not susceptible to the rearrangements that commonly occur under acidic conditions.*

Mechanism of Oxymercuration–Demercuration

STEP 1. Dissociation

$$CH_3\overset{O}{\overset{\|}{C}}OHg\overset{O}{\overset{\|}{O}}CCH_3 \;\rightleftharpoons\; CH_3\overset{O}{\overset{\|}{C}}O^- \;+\; {}^+Hg\overset{O}{\overset{\|}{O}}CCH_3$$

STEP 2. Electrophilic attack

$$\text{C}{=}\text{C} \;+\; {}^+Hg\overset{O}{\overset{\|}{O}}CCH_3 \;\longrightarrow\; \overset{\overset{+}{Hg}O\overset{O}{\overset{\|}{C}}CH_3}{\text{C}{-}\text{C}}$$

Mercurinium ion

STEP 3. Nucleophilic opening

Ether Synthesis by Oxymercuration–Demercuration

![1-Hexene structure]

1-Hexene

1. Hg(OCCH₃)₂,
   CH₃OH
2. NaBH₄,
   NaOH, H₂O

![2-Methoxyhexane structure with OCH₃ and H]

65%

2-Methoxyhexane

$$\overset{\overset{+}{Hg}O\overset{O}{\overset{\|}{C}}CH_3}{\text{C}{-}\text{C}} \;+\; \text{H}{-}\overset{..}{\underset{..}{O}}\text{H} \;\xrightarrow{-H^+}\; \overset{CH_3\overset{O}{\overset{\|}{C}}OHg}{\underset{\underset{OH}{|}}{-\text{C}{-}\text{C}-}}$$

Alkylmercuric acetate

STEP 4. Reduction

$$\overset{CH_3\overset{O}{\overset{\|}{C}}OHg}{\underset{\underset{OH}{|}}{-\text{C}{-}\text{C}-}} \;\xrightarrow{NaBH_4,\ NaOH,\ H_2O}\; \overset{H}{\underset{\underset{OH}{|}}{-\text{C}{-}\text{C}-}} \;+\; Hg \;+\; CH_3\overset{O}{\underset{O}{\overset{\|}{C}}}{}^-$$

When the oxymercuration of an alkene is executed in an alcohol solvent, demercuration gives an ether, as shown in the margin.

Explain the result shown below.

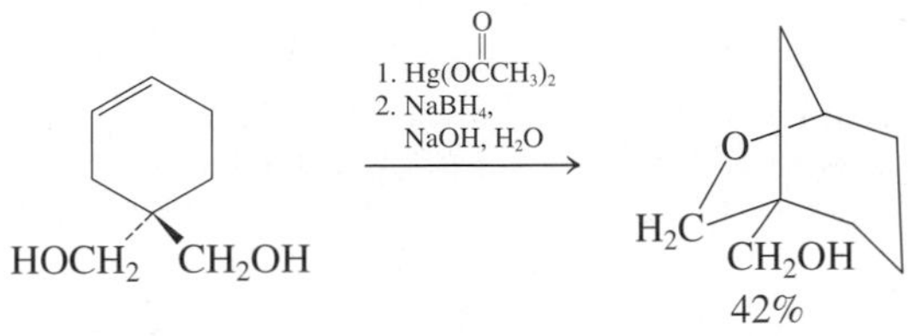

In summary, oxymercuration–demercuration is a synthetically useful method for converting alkenes regioselectively (following the Markovnikov rule) into alcohols or ethers. Carbocations are not involved; therefore, rearrangements do not occur.

CHEMICAL HIGHLIGHT 12-2

Synthesis of a Juvenile Hormone Analog

An application of the oxymercuration–demercuration reaction to the synthesis of an analog of juvenile hormone is shown here. Juvenile hormone is a substance that controls the larval metamorphosis in insects. It is produced by the male wild silk moth *Hyalophora cecropia L.,* and its presence prevents the maturation of insect larvae. The compound itself and modified analogs have been proposed as potential agents in insect control. Unfortunately, however, the activity of this analog is only 1/500 that of the natural compound. This example is noteworthy because the reaction can be controlled so that it will take place only at the least hindered electron-rich double bond.

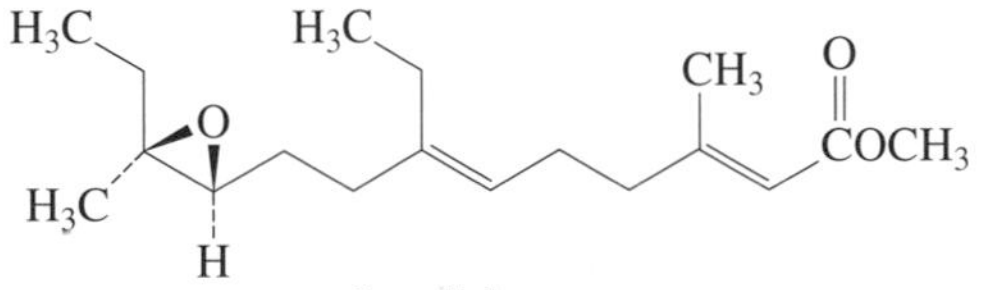

Juvenile hormone

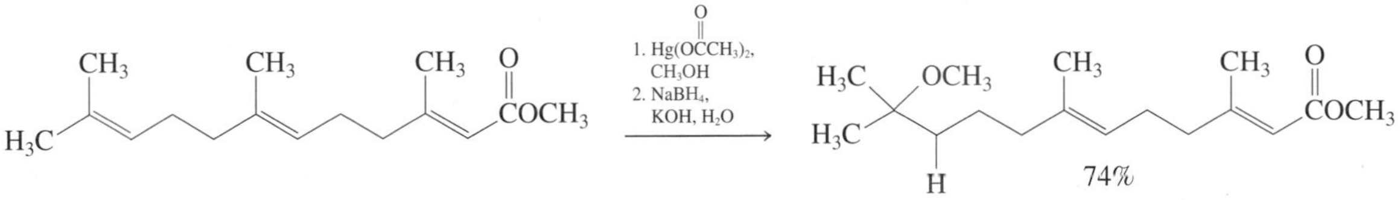

Analog of juvenile hormone

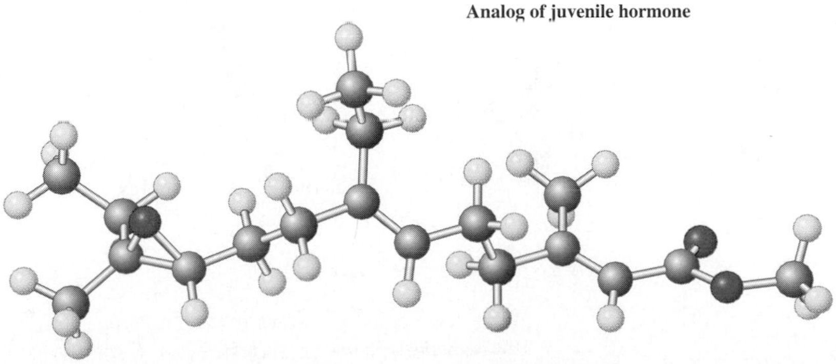

12-8 Hydroboration–Oxidation: A Stereospecific Anti-Markovnikov Hydration

This section deals with a reaction that seems to lie mechanistically somewhere between hydrogenation and electrophilic addition: the hydroboration of double bonds. The resulting alkylboranes can be oxidized to alcohols.

The boron–hydrogen bond adds across double bonds

Borane, BH_3, adds to double bonds without catalytic activation, a reaction called **hydroboration** by its discoverer, H. C. Brown.*

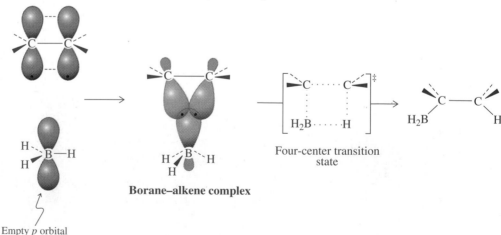

Hydroboration of Alkenes

Borane   **An alkylborane**   **A trialkylborane**

Borane (which by itself exists as a dimer, B_2H_6) is commercially available in ether and oxacyclopentane (tetrahydrofuran, THF). In these solutions, borane exists as a Lewis acid-base complex with the ether oxygen (see Sections 2-9 and 9-5), an aggregate that allows the boron to have an electron octet (for the molecular-orbital picture of BH_3, see Figure 1-17).

How does the B–H unit add to the π bond? Because the π bond is electron rich and borane electron poor, it is reasonable to formulate an initial Lewis acid-base complex similar to that of a bromonium ion (Figure 12-3), requiring the participation of the empty p orbital on BH_3. Subsequently, one of the hydrogens is transferred by means of a four-center transition state to one of the alkene carbons, while the boron shifts to the other. The stereochemistry of the addition is *syn*. All three B–H bonds are reactive in this way.

Borane–THF complex

Mechanism of Hydroboration

Borane–alkene complex   **Four-center transition state**

Empty p orbital

*Professor Herbert C. Brown (b. 1912), Purdue University, West Lafayette, Indiana, Nobel Prize 1979 (chemistry).

Hydroboration is not only stereospecific (*syn* addition), but also regioselective. Unlike the electrophilic additions described previously, steric more than electronic factors primarily control the regioselectivity: The boron binds to the less hindered (less substituted) carbon. The reactivity of the trialkylboranes resulting from these hydroborations are of special interest to us.

The oxidation of alkylboranes gives alcohols

Trialkylboranes can be oxidized with basic aqueous hydrogen peroxide to furnish alcohols in which the hydroxy function has replaced the boron atom. The net result of the two-step sequence, **hydroboration–oxidation,** is the addition of the elements of water to a double bond. In contrast with the hydrations described in Sections 12-4 and 12-7, those using borane proceed with the opposite regioselectivity: in this sequence, the OH group ends up at the *less* substituted carbon, an example of **anti-Markovnikov addition.**

Regioselectivity of Hydroboration

$$3 \ RCH = CH_2 \ + \ BH_3$$
$$\downarrow$$
$$(RCH_2CH_2)_3B$$

Hydroboration–Oxidation Sequence

$$3 \ RCH = CHR \xrightarrow{BH_3, \ THF} (RCH_2CHR)_3B \xrightarrow{H_2O_2, \ NaOH, \ H_2O} 3 \ RCH_2\overset{\overset{\displaystyle R}{|}}{C}HOH$$

$$(CH_3)_2CHCH_2CH = CH_2 \xrightarrow[\text{2. } H_2O_2, \ NaOH, \ H_2O]{\text{1. } BH_3, \ THF} (CH_3)_2CHCH_2CH_2CH_2OH$$
$$80\%$$

4-Methyl-1-pentene   **4-Methyl-1-pentanol**

In the mechanism of alkylborane oxidation, the nucleophilic hydroperoxide ion attacks the electron-poor boron atom. The resulting species undergoes a rearrangement in which an alkyl group migrates with its electron pair—and with *retention* of configuration—to the neighboring oxygen atom, thus expelling a hydroxide ion in the process.

Mechanism of Alkylborane Oxidation

This process is repeated until all three alkyl groups have migrated to oxygen atoms, finally forming a triakyl borate $(RO)_3B$. This inorganic ester is then hydrolyzed by base to the alcohol and sodium borate.

$$(RO)_3B \ + \ 3 \ NaOH \xrightarrow{H_2O} Na_3BO_3 \ + \ 3 \ ROH$$

Because borane addition to double bonds and subsequent oxidation are so selective, this sequence allows the stereospecific and regioselective synthesis of alcohols from alkenes.

**A Stereospecific and Regioselective Alcohol Synthesis
by Hydroboration–Oxidation**

1-Methylcyclopentene

$\xrightarrow{\text{BH}_3,\ \text{THF}}$

$\xrightarrow{\text{H}_2\text{O}_2,\ \text{NaOH},\ \text{H}_2\text{O}}$

86%

trans-2-Methylcyclopenta...

EXERCISE 12-13

Give the products of hydroboration–oxidation of (**a**) propene and (**b**) (*E*)-3-methyl-2-pentene. Show the stereochemistry clearly.

In summary, hydroboration–oxidation constitutes another method for hydrating alkenes. The initial addition is *syn* and regioselective, the boron shifting to the less hindered carbon. Oxidation of alkyl boranes with basic hydrogen peroxide gives anti-Markovnikov alcohols with retention of configuration of the alkyl group.

Peroxycarboxylic Acids

$$R-\overset{\overset{\displaystyle O}{\|}}{C}-\overset{\delta^-}{O}-\overset{\delta^+}{O}H$$

A peroxycarboxylic acid

$$\text{CH}_3\overset{\overset{\displaystyle O}{\|}}{\text{C}}\text{OOH}$$

**Peroxyethanoic
(peracetic) acid**

meta-Chloro-
peroxybenzoic acid
(MCPBA)

Magnesium
monoperoxyphthalate
(MMPP)

12-9 Oxacyclopropane Synthesis: Oxidation by Peroxycarboxylic Acids

This section and Sections 12-10 and 12-11 describe how electrophilic oxidizing agents are capable of delivering oxygen atoms to the π bond, thereby producing oxacyclopropanes, vicinal *syn* and *anti* diols, and carbonyl compounds by complete cleavage of the double bond. Let us start with oxacyclopropane formation, a process that can be used to synthesize vicinal *anti* diols.

Peroxycarboxylic acids deliver oxygen atoms to double bonds

The OH group in peroxycarboxylic acids, RCOOH (with $\overset{\displaystyle O}{\|}$ on the carbonyl carbon), contains an electrophilic oxygen. These compounds react with alkenes by adding this oxygen to the double bond to form oxacyclopropanes. The other product of the reaction is a carboxylic acid. The transformation is of value because, as we know, oxacyclopropanes are versatile synthetic intermediates (Section 9-9). It proceeds at room temperature in an inert solvent, such as chloroform, dichloromethane, or benzene. This reaction is commonly referred to as **epoxidation,** a term derived from *epoxide,* one of the older common names for oxacyclopropanes. A popular peroxycarboxylic acid for use in the research laboratory is *meta*-chloroperoxybenzoic acid (MCPBA). For large-scale and industrial purposes, however, the somewhat shock-sensitive MCPBA has been replaced by magnesium monoperoxyphthalate (MMPP).

Oxacyclopropane Formation: Epoxidation of a Double Bond

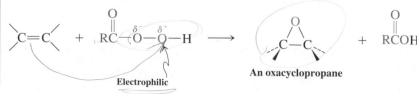

An oxacyclopropane

Electrophilic

$$CH_3CH_2CH{=}CH_2 \quad + \qquad \qquad \xrightarrow{\ CHCl_3\ } \quad CH_3CH_2CH{-}CH_2$$

90%

1-Butene   *meta*-**Chloroperoxybenzoic acid**   **Ethyloxacyclopropane**
                                    (MCPBA)

The transfer of oxygen is stereospecifically *syn,* the stereochemistry of the starting alkene being retained in the product. For example, *trans*-2-butene gives *trans*-2,3-dimethyloxacyclopropane; conversely, the *cis*-2-butene yields *cis*-2,3-dimethyloxacyclopropane.

$$\xrightarrow{\text{MCPBA, CH}_2\text{Cl}_2}$$

85%

trans-**2-Butene**   *trans*-**2,3-Dimethyloxacyclopropane**

What is the mechanism of this oxidation? It is related to but not quite identical with electrophilic halogenation (Section 12-5). In epoxidation, we can write a cyclic transition state in which the electrophilic oxygen is added to the π bond at the same time as the peroxycarboxylic acid proton is transferred to its own carbonyl group, releasing a molecule of carboxylic acid, a good leaving group. The two new C–O bonds in the oxacyclopropane product are formally derived from the electron pairs of the alkene π bond and of the cleaved O–H linkage.

Mechanism of Oxacyclopropane Formation

EXERCISE 12-14

Outline a short synthesis of *trans*-2-methylcyclohexanol from cyclohexene. (**Hint:** Review the reactions of oxacyclopropanes in Section 9-9.)

In accord with the electrophilic mechanism, the reactivity of alkenes to peroxycarboxylic acids increases with alkyl substitution, allowing for selective oxidations.

Relative Rates of Oxacyclopropane Formation (Epoxidation)

$$CH_2{=}CH_2 < RCH{=}CH_2 < RCH{=}CHR \sim R_2C{=}CH_2 < R_2C{=}CHR < R_2C{=}CR_2$$

| 1 | 24 | 500 | 500 | 6500 | Very fast |

For example,

$$\xrightarrow{\ CH_3COOH\ (1\ equivalent),\ CHCl_3,\ 10°C\ }$$

86%

Hydrolysis of oxacyclopropanes furnishes the products of *anti* dihydroxylation of an alkene

Treatment of oxacyclopropanes with water in the presence of catalytic acid or base leads to ring opening to the corresponding vicinal diols. These reactions follow the mechanisms described in Section 9-9: The nucleophile (water or hydroxide) attacks the side opposite the oxygen in the three-membered ring, so the net result of the oxidation–hydrolysis sequence constitutes an *anti* **dihydroxylation** of an alkene. In this way, *trans*-2-butene gives *meso*-2,3-butanediol, whereas *cis*-2-butene furnishes the racemic mixture of the 2*R*,3*R* and 2*S*,3*S* enantiomers.

Vicinal *Anti* Dihydroxylation of Alkenes

$$C=C \xrightarrow[\text{2. H}^+, \text{H}_2\text{O}]{\text{1. MCPBA, CH}_2\text{Cl}_2} \quad \underset{\text{OH}}{\overset{\text{HO}}{C-C}}$$

Synthesis of Isomers of 2,3-Butanediol

$$\underset{\substack{\text{H} \qquad \text{CH}_3 \\ \textit{trans-}\text{2-Butene}}}{\overset{\text{H}_3\text{C} \qquad \text{H}}{C=C}} \xrightarrow[\text{2. H}^+, \text{H}_2\text{O}]{\text{1. MCPBA, CH}_2\text{Cl}_2} \quad \underset{\substack{\text{H}_3\text{C} \quad \text{H} \qquad \text{OH} \\ \textit{meso-}\text{2,3-Butanediol}}}{\overset{\text{HO} \qquad \text{H}}{C-C{\text{CH}_3}}}$$

$$\underset{\substack{\text{H}_3\text{C} \qquad \text{CH}_3 \\ \textit{cis-}\text{2-Butene}}}{\overset{\text{H} \qquad \text{H}}{C=C}} \xrightarrow[\text{2. H}^+, \text{H}_2\text{O}]{\text{1. MCPBA, CH}_2\text{Cl}_2} \quad \underset{\substack{\text{H} \quad R \quad R \quad \text{OH} \\ \text{CH}_3 \qquad \text{OH} \\ (2R,3R)\text{-2,3-Butanediol}}}{\overset{\text{HO} \qquad \text{H}}{C-C{\text{CH}_3}}} + \underset{\substack{\text{HO} \qquad \text{CH}_3 \\ (2S,3S)\text{-2,3-Butanediol}}}{\overset{\text{H}_3\text{C} \qquad \text{OH}}{C-C}}$$

EXERCISE 12-15

Give the products obtained by treating the following alkenes with MCPBA and then aqueous acid. **(a)** 1-Hexene; **(b)** cyclohexene; **(c)** *cis*-2-pentene.

In summary, peroxycarboxylic acids supply oxygen atoms to convert alkenes into oxacyclopropanes (epoxidation). Oxidation–hydrolysis reactions with peroxycarboxylic acids furnish vicinal diols in a stereospecifically *anti* manner.

12-10 Vicinal *Syn* Dihydroxylation with Osmium Tetroxide

Osmium tetroxide reacts with alkenes in a two-step process to give the corresponding vicinal diols in a stereospecifically *syn* manner.

Vicinal *Syn* Dihydroxylation with Osmium Tetroxide

$$C=C \xrightarrow[\text{2. H}_2\text{S}]{\text{1. OsO}_4, \text{THF, 25°C}} \quad \underset{}{\overset{\text{HO} \qquad \text{OH}}{C-C}}$$

The process leads initially to an isolable cyclic ester, which is reductively hydrolyzed with H$_2$S or bisulfite, NaHSO$_3$. For example,

$$\text{OsO}_4, \text{THF}, 25°\text{C}, 48 \text{ h} \longrightarrow \quad \xrightarrow{\text{H}_2\text{S}} \quad 90\%$$

What is the mechanism of this transformation? The initial reaction of the π bond with osmium tetroxide constitutes a concerted addition in which three electron pairs move simultaneously to give a cyclic ester containing Os(VI). This process can be viewed as an electrophilic attack on the alkene: Two electrons flow from the alkene onto the metal, which is reduced [Os(VIII) →Os(VI)]. For steric reasons, the product can form only in a way that introduces the two oxygen atoms on the *same* face of the double bond—*syn*. This intermediate is usually not isolated but converted upon reductive work-up into the free diol.

Mechanism of the Osmium Tetroxide Oxidation of Alkenes

Because OsO$_4$ is expensive and highly toxic, a newer modification calls for the use of only catalytic quantities of the osmium reagent and stoichiometric amounts of another oxidizing agent such as H$_2$O$_2$, which serves to reoxidize reduced osmium.

An older reagent for vicinal *syn* dihydroxylation of alkenes is potassium permanganate, KMnO$_4$. Although this reagent functions in a manner similar mechanistically to OsO$_4$, it is less useful preparatively, because of a tendency to give poorer yields of diols owing to overoxidation. Potassium permanganate solutions, which are deep purple, are useful as a test for alkenes, however: Upon reaction, the purple reagent immediately turns colorless, and a brown precipitate of its reduction product, MnO$_2$, is observed to form.

Potassium Permanganate Test for Alkene Double Bonds

$$\underset{\textbf{Dark purple}}{\text{C=C}} + \text{KMnO}_4 \longrightarrow \underset{}{\overset{\text{HO} \quad \text{OH}}{\text{C—C}}} + \underset{\substack{\textbf{Brown} \\ \textbf{precipitate}}}{\text{MnO}_2}$$

EXERCISE 12-16

The stereochemical consequences of the vicinal *syn* dihydroxylation of alkenes are complementary to those of vicinal *anti* dihydroxylation. Show the products (indicate stereochemistry) of the vicinal *syn* dihydroxylation of *cis*- and *trans*-2-butene.

To summarize, osmium tetroxide, either stoichiometrically or catalytically together with a second oxidizing agent, converts alkenes into *syn*-1,2-diols. A similar reaction

Enantioselective Dihydroxylation in the Synthesis of Antitumor Drugs

In the decade of the 1990s, there has been a significant shift in the way in which new pharmaceuticals are synthesized. Before this time, most of the available methods for the preparation of chiral molecules in enantiomerically pure form were unsuitable for synthesis on an industrial scale. Racemic mixtures, which were much less expensive to prepare, were generally synthesized, although in many cases only one of the enantiomers in such mixtures possessed the desired activity (see Chemical Highlight 5-4). However, fundamental conceptual advances in catalysis have changed this situation. Some of the most useful examples are a series of highly enantioselective oxidation reactions of double bonds developed by K. B. Sharpless.* These processes include a dihydroxylation, which is catalyzed by OsO_4 in the presence of a ligand constructed from two molecules of an enantiomerically pure natural alkaloid called cinchona and uses an inexpensive Fe^{3+}-based stoichiometric oxidizing agent.

This method has been applied by E. J. Corey (Section 8-9) to the enantioselective synthesis of

ovalicin, a member of a class of drugs called antiangiogenesis agents: They inhibit the growth of new blood vessels and therefore have the capability to cut off the blood supply to solid tumors. Ovalicin possesses nearly the activity of the most potent existing antiangiogenesis agents but is much more chemically stable as well as being nontoxic, noninflammatory, and potentially amenable to oral administration. The key to the synthesis is the enantioselective *syn* dihydroxylation of the achiral cyclohexene derivative shown below. The role of the chiral ligand is to coordinate to the OsO_4 and provide a pocket into which the substrate can enter in only one spatial orientation. In this respect, it bears the characteristics of many enzymes, biological catalysts that function in essentially the same way (see Chemical Highlight 5-5 and Chapter 26). The presence of the bulky ester group attached to the oxygen at the "top" of the structure helps to restrict the substrate to only one possible orientation in the chiral pocket of the ligand. In the absence of the chiral ligand, a racemic mixture forms.

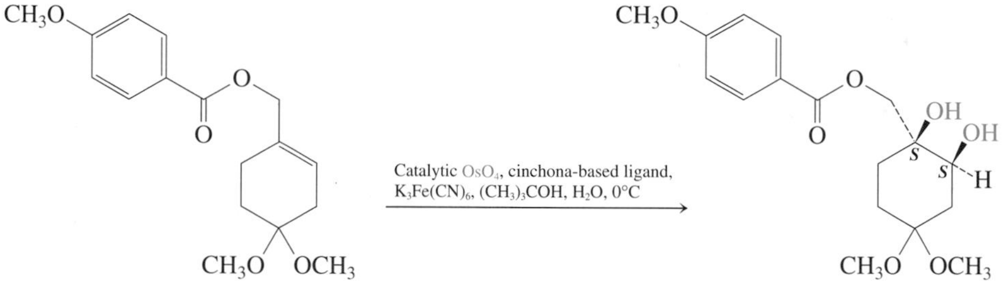

Catalytic OsO_4, cinchona-based ligand, $K_3Fe(CN)_6$, $(CH_3)_3COH$, H_2O, 0°C

93%
(99% optically pure)

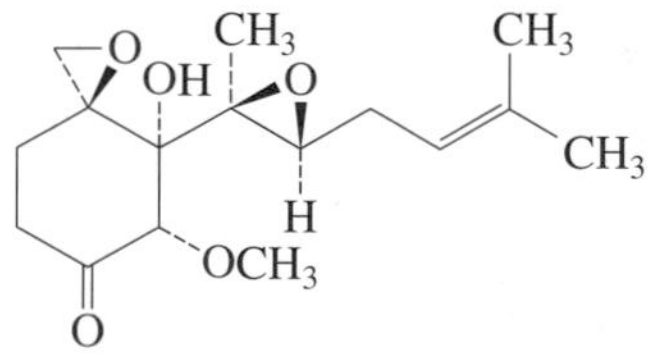

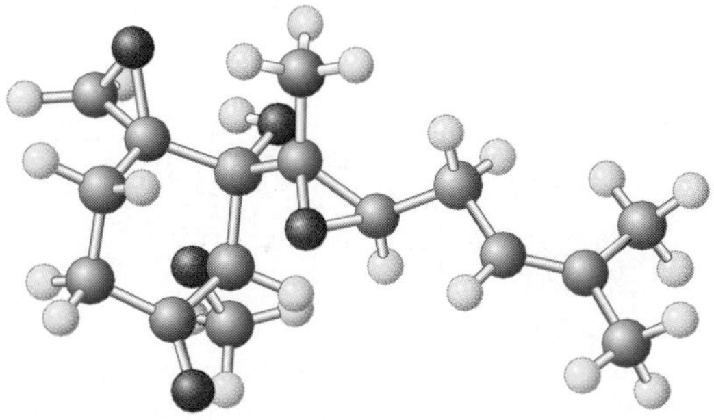

(–)-Ovalicin

*Professor K. Barry Sharpless (b. 1941), Scripps Research Institute, La Jolla, California.

of purple potassium permanganate is accompanied by decolorization, a result making it a useful test for the presence of double bonds.

12-11 Oxidative Cleavage: Ozonolysis

Although oxidation of alkenes with osmium tetroxide breaks only the π bond, other reagents may rupture the σ bond as well. The most general and mildest method of oxidatively cleaving alkenes into carbonyl compounds is through the reaction with ozone, **ozonolysis.**

Ozone, O_3, is produced in the laboratory in an instrument called an ozonator, in which an arc discharge generates 3–4% ozone in a dry oxygen stream. The gas mixture is passed through a solution of the alkene in methanol or dichloromethane. The first isolable intermediate is a species called an **ozonide,** which is reduced directly in a subsequent step by exposure to zinc in acetic acid or by reaction with dimethyl sulfide. The net result of the ozonolysis–reduction sequence is the cleavage of the molecule at the carbon–carbon double bond; oxygen becomes attached to each of the carbons that had originally been doubly bonded.

Ozone is a blue gas that condenses to a dark blue, highly unstable liquid.

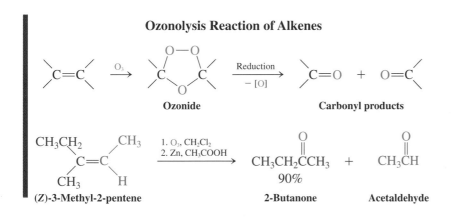

Ozonolysis Reaction of Alkenes

Ozonide

Carbonyl products

(Z)-3-Methyl-2-pentene   2-Butanone   Acetaldehyde

90%

The mechanism of ozonolysis proceeds through initial electrophilic addition of ozone to the double bond, a transformation that yields the so-called **molozonide.** In this reaction, as in several others already presented, six electrons move in concerted fashion in a cyclic transition state. The molozonide is unstable and breaks apart into a carbonyl fragment and a carbonyl oxide fragment through another cyclic six-electron rearrangement. Recombination of the two fragments as shown yields the ozonide.

Mechanism of Ozonolysis

STEP 1. Molozonide formation and cleavage

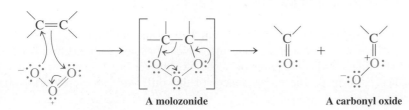

A molozonide   A carbonyl oxide

STEP 2. Ozonide formation and reduction

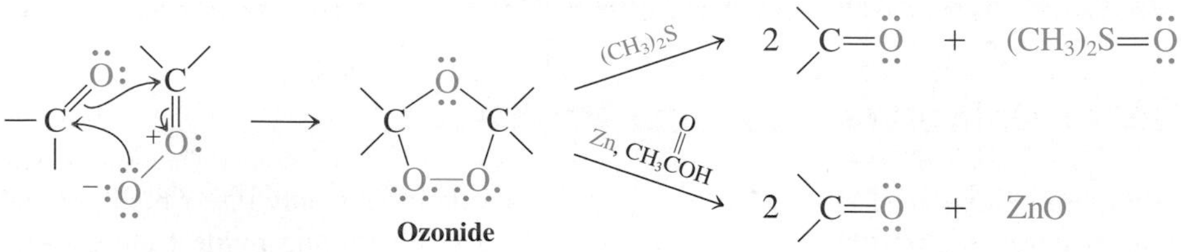

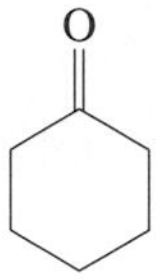

Ozonide

An unknown hydrocarbon of the molecular formula $C_{12}H_{20}$ exhibited an ^1H NMR spectrum with a complex multiplet of signals between 1.0 and 2.2 ppm. Ozonolysis of this compound gave two equivalents of cyclohexanone, whose structure is shown in the margin. What is the structure of the unknown?

Give the products of the following reactions.

What is the structure of the following starting material?

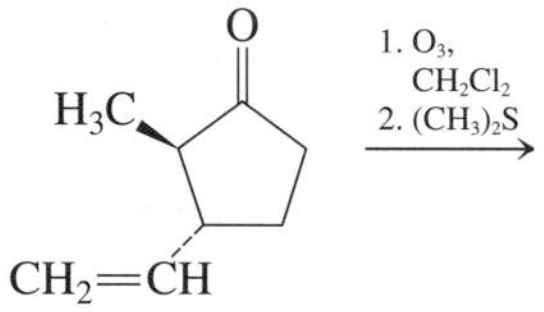

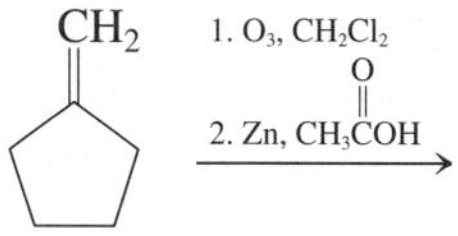

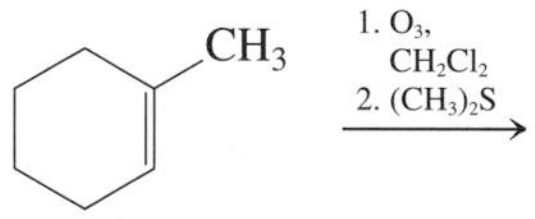

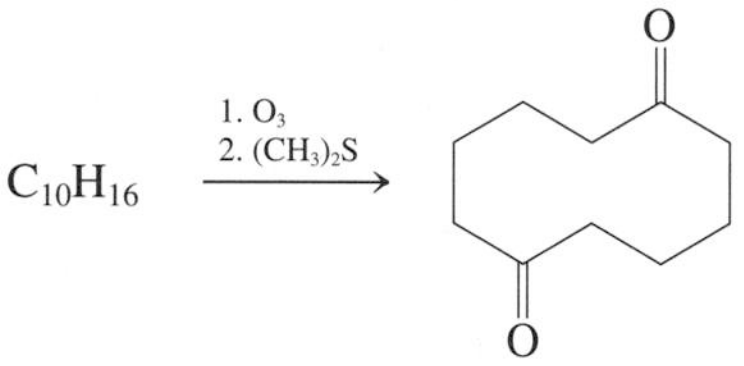

$$C_{10}H_{16} \quad \xrightarrow[\text{2. }(CH_3)_2S]{\text{1. }O_3}$$

In summary, ozonolysis followed by reduction yields aldehydes and ketones. Mechanistically, the reactions presented in Sections 12-9 through 12-11 can be related in that initial attack by an electrophilic oxidizing agent on the π bond leads to its rupture. Unlike the reaction sequences studied in Sections 12-9 and 12-10, however, ozonolysis causes cleavage of both the π and the σ bonds.

12-12 Radical Additions: Anti-Markovnikov Product Formation

In this section, all radicals and single atoms are shown in green, as in Chapter 3.

This section deals with another mode of reactivity of the double bond: radical addition. In contrast with electrophilic reagents, which consume both electrons of the π bond on addition, a radical requires only one electron for bond formation so that an alkyl radical is formed. The consequences of this difference are anti-Markovnikov products.

Hydrogen bromide can add to alkenes in anti-Markovnikov fashion: a change in mechanism

When freshly distilled 1-butene is exposed to hydrogen bromide, clean Markovnikov addition to give 2-bromobutane is observed. This result is in accord with the ionic mechanism for electrophilic addition of HBr discussed in Section 12-3. Curiously, the same reaction, when carried out with a sample of 1-butene that has been exposed to air, proceeds much faster and gives an entirely different result. In this case, we isolate 1-bromobutane, formed by anti-Markovnikov addition.

This change caused considerable confusion in the early days of alkene chemistry, because one researcher would obtain only one hydrobromination product, whereas another would obtain a different product or mixtures from a seemingly identical reaction. The mystery was solved by Kharasch* in the 1930s, when it was discovered that the culprits responsible for anti-Markovnikov additions were radicals formed from peroxides, ROOR, in alkene samples that had been stored in the presence of air.

The mechanism of the addition reaction under these conditions is not an ionic sequence; rather, it is a much faster **radical chain sequence.** The initiation steps are, first, the homolytic cleavage of the weak RO–OR bond ($DH° \sim 39$ kcal mol^{-1}) and, then, reaction of the resulting alkoxy radical with hydrogen bromide. The driving force for the second (exothermic) step is the formation of the strong O–H bond. The bromine atom so generated initiates chain propagation by attacking the double bond. One of the π electrons combines with the unpaired electron on the bromine atom to form the carbon–bromine bond. The other π electron remains on carbon, giving rise to a radical.

The halogen atom's attack is *regioselective,* creating the relatively more stable secondary radical rather than the primary one. This result is reminiscent of the ionic additions of hydrogen bromide (Section 12-3), except that the roles of the proton and bromine are reversed. In the ionic mechanism, a proton attacks first to generate the more stable carbocation, which is then trapped by bromide ion. *In the radical mechanism, a bromine atom is the attacking species,* creating the more stable radical center. The latter subsequently reacts with HBr by abstracting a hydrogen and regenerating the chain-carrying bromine atom. Both propagation steps are exothermic, and the reaction proceeds rapidly. As usual, termination is by radical combination or by some other removal of the chain carriers (Section 3-4).

Markovnikov Addition of HBr

$$CH_3CH_2CH{=}CH_2$$
(Freshly distilled)

$\downarrow$ HBr 24 h

$$\underset{\text{Br}}{CH_3CH_2\overset{|}{C}HCH_2H}$$
90%
Markovnikov product
(by *ionic* mechanism)

Anti-Markovnikov Addition of HBr

$$CH_3CH_2CH{=}CH_2$$
(Exposed to oxygen)

$\downarrow$ HBr 4 h

$$\underset{\text{H}}{CH_3CH_2\overset{|}{C}HCH_2Br}$$
65%
Anti-Markovnikov product
(by *radical* mechanism)

Mechanism of Radical Hydrobromination

INITIATION STEPS

$$\ddot{R\ddot{O}}{-}\ddot{O}R \overset{\Delta}{\longrightarrow} 2\,R\ddot{O}\cdot \qquad \Delta H° \sim +39 \text{ kcal mol}^{-1}$$

$$R\ddot{O}\cdot + H\ddot{Br}\!: \overset{\Delta}{\longrightarrow} R\ddot{O}H + :\ddot{Br}\cdot \qquad \Delta H° \sim -17 \text{ kcal mol}^{-1}$$

PROPAGATION STEPS

$$\underset{CH_3CH_2}{\overset{H}{\diagdown}}C{=}CH_2 + :\ddot{Br}\cdot \longrightarrow CH_3CH_2\dot{C}H{-}CH_2\ddot{Br}: \qquad \Delta H° \sim -3 \text{ kcal mol}^{-1}$$
$$\text{Secondary radical}$$

*Professor Morris S. Kharasch (1895–1957), University of Chicago.

$$CH_3CH_2\overset{\displaystyle\cdot}{C}HCH_2Br \;+\; H\!:\!\overset{\cdot\cdot}{\underset{\cdot\cdot}{Br}}\!: \;\longrightarrow\; CH_3CH_2\overset{\displaystyle\overset{H}{\mid}}{C}HCH_2\overset{\cdot\cdot}{\underset{\cdot\cdot}{Br}}\!: \;+\; :\!\overset{\cdot\cdot}{\underset{\cdot\cdot}{Br}}\!\cdot \quad \Delta H^\circ \sim -7.5 \text{ kcal mol}^{-1}$$

Are radical additions general?

Hydrogen chloride and hydrogen iodide do not give anti-Markovnikov addition products to alkenes; in both cases, one of the propagating steps is endothermic and consequently so slow that the chain reaction terminates. As a result, HBr is the *only* hydrogen halide that adds to an alkene under radical conditions to give anti-Markovnikov products. Additions of HCl and HI proceed by ionic mechanisms only to give normal Markovnikov products regardless of the presence or absence of radicals. Other reagents, however, do undergo successful radical additions to alkenes. Examples are thiols and some of the halomethanes.

Other Radical Additions to Alkenes

$$CH_3CH{=}CH_2 \;+\; CH_3CH_2SH \;\xrightarrow{\text{ROOR}}\; CH_3\overset{\displaystyle\overset{\;}{\underset{\underset{H}{\mid}}{C}}}{}HCH_2SCH_2CH_3$$

| **Ethanethiol** | **Ethyl propyl sulfide** |

$$CH_3(CH_2)_5CH{=}CH_2 \;+\; HCCl_3 \;\xrightarrow{\text{ROOR}}\; CH_3(CH_2)_5\overset{\displaystyle\overset{H}{\mid}}{C}HCH_2CCl_3$$
$$22\%$$

1,1,1-Trichlorononane

In these examples, the initiating alkoxy radical abstracts a hydrogen from the substrate to yield a chain carrier, because of the strength of the resulting OH bond. A typical example is trichloromethane (chloroform).

$$RO\cdot \;+\; CHCl_3 \;\longrightarrow\; ROH \;+\; \underset{\textbf{Chain carrier}}{\cdot CCl_3} \quad \text{not} \quad RO\cdot \;+\; Cl{-}CHCl_2 \;\longrightarrow\; ROCl \;+\; \cdot CHCl_2$$

Bis(1,1-dimethylethyl) peroxide

(Di-*tert*-butyl peroxide)

Dibenzoyl peroxide

Bis(1,1-dimethylethyl) peroxide (di-*tert*-butyl peroxide) and dibenzoyl peroxide are commercially available initiators for such radical addition reactions.

EXERCISE 12-20

Ultraviolet irradiation of a mixture of 1-octene and diphenylphosphine, $(C_6H_5)_2PH$, furnishes 1-(diphenylphosphino)octane by radical addition. Write a plausible mechanism for this reaction.

$$(C_6H_5)_2PH \;+\; H_2C{=}CH(CH_2)_5CH_3 \;\xrightarrow{h\nu}\; (C_6H_5)_2P{-}CH_2{-}CH_2(CH_2)_5CH_3$$

Anti-Markovnikov additions are synthetically useful because their products complement those obtained from ionic additions. This type of regiochemical control is an important feature in the development of new synthetic methods.

In summary, radical initiators alter the mechanism of the addition of HBr to alkenes from ionic to radical chain. The consequence of this change is anti-Markovnikov regioselectivity. Other species, most notably thiols and some halomethanes, are capable of undergoing similar reactions.

12-13 Dimerization, Oligomerization, and Polymerization of Alkenes

Is it possible for alkenes to react with one another? Indeed it is, but only in the presence of an appropriate catalyst—for example, an acid, a radical, a base, or a transition metal. In this reaction the unsaturated centers of the alkene monomer (*monos,* Greek, single; *meros,* Greek, part) are linked to form dimers, trimers, **oligomers** (*oligos,* Greek, few, small), and ultimately **polymers** (*polymeres,* Greek, of many parts), substances of great industrial importance.

Polymerization

Monomers $\longrightarrow$ Polymer

Carbocations attack pi bonds

Treatment of 2-methylpropene with hot aqueous sulfuric acid gives two dimers: 2,4,4-trimethyl-1-pentene and 2,4,4,-trimethyl-2-pentene. This transformation is possible because 2-methylpropene can be protonated under the reaction conditions to furnish the 1,1-dimethylethyl (*tert*-butyl) cation. This species can attack the electron-rich double bond of 2-methylpropene with formation of a new carbon–carbon bond. Electrophilic addition proceeds according to the Markovnikov rule to generate the more stable carbocation. Subsequent deprotonation in each of two directions furnishes a mixture of the two observed products.

Dimerization of 2-Methylpropene

$$CH_2{=}C(CH_3)_2 \;+\; CH_2{=}C(CH_3)_2 \;\xrightarrow{\;H^+\;}\; CH_3\overset{\overset{\displaystyle CH_3}{|}}{\underset{\underset{\displaystyle CH_3}{|}}{C}}CH_2\overset{\overset{\displaystyle CH_3}{|}}{C}{=}CH_2 \;+\; CH_3\overset{\overset{\displaystyle CH_3}{|}}{\underset{\underset{\displaystyle CH_3}{|}}{C}}CH{=}C(CH_3)_2$$

2,4,4-Trimethyl-1-pentene   **2,4,4-Trimethyl-2-pentene**

Mechanism of Dimerization of 2-Methylpropene

$$CH_2{=}C\overset{CH_3}{\underset{CH_3}{\big\langle}} \;\xrightarrow{\;H^+\;}\; CH_3\overset{+}{C}\overset{CH_3}{\underset{CH_3}{\big\langle}} \quad CH_2{=}C\overset{CH_3}{\underset{CH_3}{\big\langle}} \;\longrightarrow$$

$$CH_3\overset{\overset{\displaystyle CH_3}{|}}{\underset{\underset{\displaystyle CH_3}{|}}{C}}{-}\overset{\overset{\displaystyle H}{|}}{\underset{\underset{\displaystyle H}{|}}{C}}{-}\overset{+}{C}\overset{CH_3}{\underset{\underset{\underset{\displaystyle H}{|}}{CH_2}}{}} \;\xrightarrow{\;-H^+\;}\; CH_3\overset{\overset{\displaystyle CH_3}{|}}{\underset{\underset{\displaystyle CH_3}{|}}{C}}CH_2\overset{\overset{\displaystyle CH_3}{|}}{C}{=}CH_2 \;+\; CH_3\overset{\overset{\displaystyle CH_3}{|}}{\underset{\underset{\displaystyle CH_3}{|}}{C}}CH{=}C(CH_3)_2$$

Steroid Synthesis in Nature

A remarkable series of intramolecular alkene couplings occur in nature as part of the biosynthetic pathway to steroids, including cholesterol and the powerfully biologically active mammalian sex hormones (Section 4-7). In this process, a molecule called *squalene* is first enzymatically oxidized to the oxacyclopropane squalene oxide. Enzymatic acid-catalyzed ring opening is followed by the sequential formation of four carbon–carbon bonds. Each bond-forming step is mechanistically related to alkene oligomerization. Further conversion leads to lanosterol, a biological precursor of cholesterol. Very similar *(biomimetic)* reactions have been carried out in the laboratory. These processes, which are highly regioselective and stereospecific, are excellent methods for synthesizing many steroids.

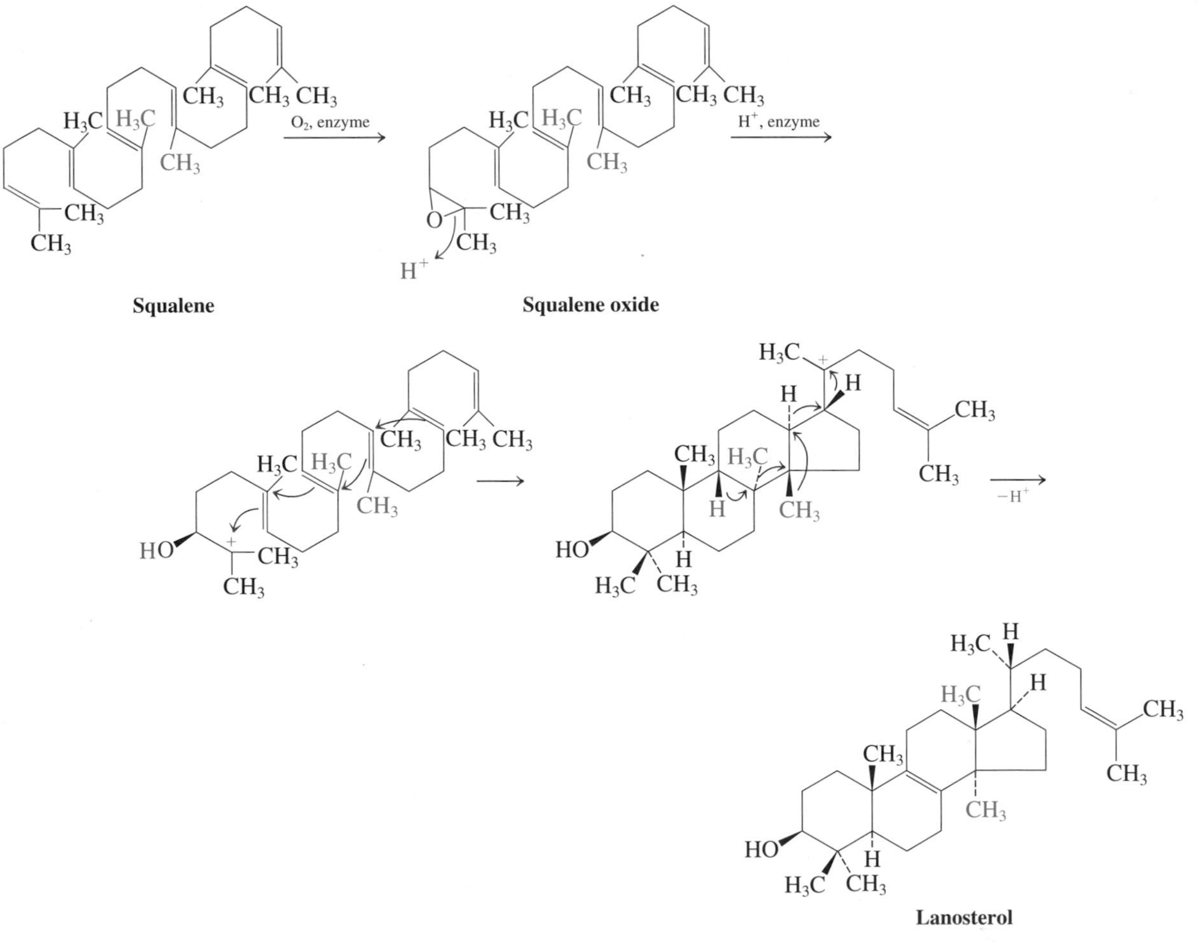

Squalene

Squalene oxide

Lanosterol

Repeated attack can lead to oligomerization and polymerization

The two dimers of 2-methylpropene tend to react further with the starting alkene. For example, when 2-methylpropene is treated with mineral acid under more stringent conditions, trimers, tetramers, pentamers, and so forth, are formed by repeated electrophilic attack of intermediate carbocations on the double bond. This process, which leads to alkane chains of intermediate length, is called **oligomerization.**

Oligomerization of the 2-Methylpropene Dimers

$$CH_3\underset{\underset{CH_3}{|}}{\overset{\overset{CH_3}{|}}{C}}CH=C(CH_3)_2 \quad + \quad CH_3\underset{\underset{CH_3}{|}}{\overset{\overset{CH_3}{|}}{C}}CH_2\overset{CH_2}{\underset{CH_3}{C}} \quad \xrightarrow{H^+} \quad CH_3\underset{\underset{CH_3}{|}}{\overset{\overset{CH_3}{|}}{C}}CH_2\overset{CH_3}{\underset{CH_3}{\overset{+}{C}}} \quad + \quad CH_2=C\overset{CH_3}{\underset{CH_3}{}} \quad \longrightarrow$$

$$CH_3\underset{\underset{CH_3}{|}}{\overset{\overset{CH_3}{|}}{C}}CH_2\underset{\underset{CH_3}{|}}{\overset{\overset{CH_3}{|}}{C}}CH_2\overset{CH_3}{\underset{CH_3}{\overset{+}{C}}} \quad + \quad CH_2=C\overset{CH_3}{\underset{CH_3}{}} \quad \longrightarrow \quad CH_3\underset{\underset{CH_3}{|}}{\overset{\overset{CH_3}{|}}{C}}CH_2\underset{\underset{CH_3}{|}}{\overset{\overset{CH_3}{|}}{C}}CH_2\underset{\underset{CH_3}{|}}{\overset{\overset{CH_3}{|}}{C}}CH_2\overset{CH_3}{\underset{CH_3}{\overset{+}{C}}} \quad etc.$$

At higher temperatures, the oligomerization of alkenes continues to give polymers containing many subunits.

Polymerization of 2-Methylpropene

$$n\ CH_2=C(CH_3)_2 \quad \xrightarrow{H^+,\ 200°C} \quad H-(CH_2-\underset{\underset{CH_3}{|}}{\overset{\overset{CH_3}{|}}{C}})_{n-1}-CH_2\overset{CH_3}{\underset{}{C}}=CH_2$$

Poly(2-methylpropene)
(Polyisobutylene)

In summary, catalytic acid causes alkene–alkene additions to occur, a process forming dimers, trimers, oligomers containing several components, and finally polymers, which are composed of a great many alkene subunits.

12-14 Synthesis of Polymers

Many alkenes are suitable monomers for polymerization. Polymerization is exceedingly important in the chemical industry, because many polymers have desirable properties, such as durability, inertness to many chemicals, elasticity, transparency, and electrical and thermal resistance.

Although the production of polymers has contributed to pollution—many of them are not biodegradable—they have varied uses as synthetic fibers, films, pipes, coatings, and molded articles. Polymers are also being increasingly used as coatings for medical implants. Names such as polyethylene, poly(vinyl chloride) (PVC), Teflon, polystyrene, Orlon, and Plexiglas (Table 12-3) have become household words.

Acid-catalyzed polymerizations, such as that described for poly(2-methylpropene), are carried out with H_2SO_4, HF, and BF_3 as the initiators. Because they proceed through the intermediacy of carbocations, they are also called *cationic polymerizations.* Other mechanisms of polymerizations are *radical, anionic,* and *metal catalyzed.*

Radical polymerizations lead to commercially useful materials

An example of **radical polymerization** is that of ethene in the presence of an organic peroxide at high pressures and temperatures. The reaction proceeds by a mechanism

TABLE 12-3	Common Polymers and Their Monomers				
Monomer	**Structure**	**Polymer (common name)**	**Structure**	**Uses**	
Ethene	$H_2C{=}CH_2$	Polyethylene	$-(CH_2CH_2)_n-$	Food storage bags, containers	
Chloroethene (vinyl chloride)	$H_2C{=}CHCl$	Poly(vinyl chloride) (PVC)	$-(CH_2CH)_n-$ Cl	Pipes, vinyl fabrics	
Tetrafluoroethene	$F_2C{=}CF_2$	Teflon	$-(CF_2CF_2)_n-$	Nonstick cookware	
Ethenylbenzene (styrene)	$\overset{}{\bigcirc}$—CH$=CH_2$	Polystyrene	$-(CH_2CH)_n-$	Foam packing material	
Propenenitrile (acrylonitrile)	$H_2C{=}C\overset{H}{\underset{C{\equiv}N}{}}$	Orlon	$-(CH_2CH)_n-$ CN	Clothing, synthetic fabrics	
Methyl 2-methyl-propenoate (methyl methacrylate)	$H_2C{=}C\overset{CH_3}{\underset{\underset{O}{COCH_3}}{}}$	Plexiglass	$-(CH_2C)_n-$ CH$_3$ CO$_2$CH$_3$	Impact-resistant paneling	
2-Methylpropene (isobutylene)	$H_2C{=}C\overset{CH_3}{\underset{CH_3}{}}$	Elastol	$-(CH_2C)_n-$ CH$_3$ CH$_3$	Oil-spill cleanup	

Radical Polymerization of Ethene

$$n\ CH_2{=}CH_2$$

$$\Big\downarrow \begin{array}{l} \text{ROOR,} \\ \text{1000 atm,} \\ >100°C \end{array}$$

$$-(CH_2{-}CH_2)_n{-}$$

Polyethene (Polyethylene)

that, in its initial stages, resembles that of the radical addition to alkenes (Section 12-12). The peroxide initiators cleave into alkoxy radicals, which begin polymerization by addition to the double bond of ethene. The alkyl radical thus created attacks the double bond of another ethene molecule, furnishing another radical center, and so on. Termination of the polymerization can be by dimerization, disproportionation of the radical, or other radical-trapping reactions.

Mechanism of Radical Polymerization of Ethene

INITIATION STEPS

$$RO{-}OR \longrightarrow RO\cdot$$

$$RO\cdot \ + \ CH_2{=}CH_2 \longrightarrow ROCH_2{-}\overset{\cdot}{C}H_2$$

PROPAGATION STEPS

$$ROCH_2CH_2\cdot \ + \ CH_2{=}CH_2 \longrightarrow ROCH_2CH_2CH_2CH_2\cdot$$

$$ROCH_2CH_2CH_2CH_2\cdot \ \xrightarrow{(n-1)\ CH_2{=}CH_2} \ RO{-}(CH_2CH_2)_n{-}CH_2CH_2\cdot$$

CHEMICAL HIGHLIGHT 12-5   **Polymers in the Cleanup of Oil Spills**

Poly(2-methylpropene) (polyisobutylene) is the principal ingredient in a product called Elastol, which has been demonstrated to be an effective agent in the cleanup of oil spills. When Elastol is sprayed on an oil slick, its polymer chains, which normally wrap tightly about each other, unravel and mix with the oil. The oil is bound by the polymer into a viscous mat that can be skimmed off the water surface. An especially valuable feature of this cleanup method is

Poly(2-methylpropene)
(Polyisobutylene)

that the oil may be recovered from the polymer by running the mixture through a special type of pump. The process was successfully demonstrated on an oil spill in the New Haven, Connecticut, harbor in 1990.

Polyethene (polyethylene) produced in this way does not have the expected linear structure. *Branching* occurs by abstraction of a hydrogen along the growing chain by another radical center followed by chain growth originating from the new radical. The average molecular weight of polyethene is almost 1 million.

Polychloroethene [poly(vinyl chloride), PVC] is made by similar radical polymerization. Interestingly, the reaction is regioselective. The peroxide initiator and the intermediate chain radicals add only to the unsubstituted end of the monomer, because the radical center formed next to chlorine is relatively stable. Thus, PVC has a very regular *head-to-tail structure* of molecular weight in excess of 1.5 million. Although PVC itself is fairly hard and brittle, it can be softened by addition of carboxylic acid esters (Section 20-4), called **plasticizers** (*plastikos,* Greek, to form). The resulting elastic material is used in "vinyl leather," plastic covers, and garden hoses.

Branching in
Polyethene
(Polyethylene)

$$\sim\sim\sim CH_2\overset{\displaystyle H}{\underset{\displaystyle \underset{\displaystyle CH_2}{|}}{\overset{\displaystyle |}{C}}}CH_2CH_2\sim\sim\sim$$

$$CH_2$$

$$CH_2$$

$$CH_2=CHCl \xrightarrow{\text{ROOR}} -(CH_2CH)_{\overline{n}}- \\ \qquad\qquad\qquad\qquad | \\ \qquad\qquad\qquad\qquad Cl$$

Polychloroethene
[Poly(vinyl chloride)]

Exposure to chloroethene (vinyl chloride) has been linked to the incidence of a rare form of liver cancer (angiocarcinoma). The Occupational Safety and Health Administration (OSHA) has set limits to human exposure of less than an average of 1 ppm per 8-hr working day per worker.

An iron compound, $FeSO_4$, in the presence of hydrogen peroxide promotes the radical polymerization of propenenitrile (acrylonitrile). **Polypropenenitrile** (polyacrylonitrile), $-(CH_2CHCN)n-$, also known as Orlon, is used to make fibers. Similar polymerizations of other monomers furnish Teflon and Plexiglas.

EXERCISE 12-21

Saran Wrap is made by radical copolymerization of 1,1-dichloroethene and chloroethene. Propose a structure.

Anionic polymerizations require initiation by bases

Anionic polymerizations are initiated by strong bases such as alkyllithiums, amides, alkoxides, and hydroxide. For example, methyl 2-cyanopropenoate (methyl α-cyanoacrylate) polymerizes rapidly in the presence of even small traces of hydroxide. When spread between two surfaces, it forms a tough solid film that cements the surfaces together. For this reason, commercial preparations of this monomer are marketed as Super Glue.

What accounts for this ease of polymerization? When the base attacks the methylene group of α-cyanoacrylate, it generates a carbanion whose negative charge is located next to the nitrile and ester groups, both of which are strongly electron withdrawing. The anion is stabilized because the nitrogen and oxygen atoms polarize their multiple bonds in the sense $^{\delta+}C\equiv N^{\delta-}$ and $^{\delta+}C=O^{\delta-}$ and because the charge can be delocalized by resonance.

Anionic Polymerization of Super Glue (Methyl α-Cyanoacrylate)

Metal-catalyzed polymerizations produce highly regular chains

An important **metal-catalyzed polymerization** is that initiated by Ziegler-Natta* catalysts. They are typically made from titanium tetrachloride and a trialkylaluminum, such as triethylaluminum, $Al(CH_2CH_3)_3$. The system polymerizes alkenes, particularly ethene, at relatively low pressures with remarkable ease and efficiency.

Although we shall not consider the mechanism here, two features of Ziegler-Natta polymerization are the regularity with which substituted alkane chains are constructed from substituted alkenes, such as propene, and the high linearity of the chains. The polymers that result possess higher density and much greater strength than those obtained from radical polymerization. An example of this contrast is found in the prop-

*Professor Karl Ziegler (1898–1973), Max Planck Institute for Coal Research, Mülheim, Germany, Nobel Prize 1963 (chemistry); Professor Giulio Natta (1903–1979), Polytechnic Institute of Milan, Nobel Prize 1963 (chemistry).

erties of polyethene (polyethylene) prepared by the two methods. The chain branching that occurs during radical polymerization of ethene results in a flexible, transparent material *(low-density polyethylene)* used for food storage bags, whereas the Ziegler-Natta method produces a tough, chemically resistant plastic *(high-density polyethylene)* that may be molded into containers.

In summary, alkenes are subject to attack by carbocations, radicals, anions, and transition metals to give polymers. In principle, any alkene can function as a monomer. The intermediates are usually formed according to the rules that govern the stability of charges and radical centers.

12-15 Ethene: An Important Industrial Feedstock

Ethene serves as a case study of the significance of alkenes in industrial chemistry. This monomer is important for polyethene production. More than 26 billion pounds (12 million tons) of this polymer were produced in the United States in 1996. Currently, the major source of ethene is the pyrolysis of petroleum or hydrocarbons, such as ethane, propane, other alkanes, and cycloalkanes, derived from natural gas. Temperatures range from 750° to 900°C and the yields of ethene from 20 to 30%. Cracking of larger alkanes typically proceeds through C–C bond breaking to alkyl radicals, the further fragmentation of which eliminates ethene (Section 3-3). In 1996, 49 billion pounds (22.3 million tons) of ethene was made in this way. This amount is equivalent to about 17% of the total production of organic chemicals.

Apart from its direct use as a monomer, ethene is the starting material for many other industrial chemicals. For example, acetaldehyde is obtained in the reaction of ethene with water in the presence of a palladium(II) catalyst, air, and $CuCl_2$. The initially formed product, ethenol (vinyl alcohol), is unstable and spontaneously rearranges to the aldehyde (see Chapters 13 and 18). The catalytic conversion of ethene into acetaldehyde is also known as the *Wacker process*. A similar reaction, but using acetic acid instead of water, furnishes ethenyl acetate (vinyl acetate).

$$CH_2{=}CHOCCH_3$$
$$\overset{\displaystyle O}{\overset{\displaystyle \|}{}}$$

Ethenyl acetate
(Vinyl alcohol)

The Wacker Process

$$CH_2{=}CH_2 \xrightarrow{\text{H}_2\text{O, O}_2, \text{ catalytic PdCl}_2, \text{ CuCl}_2} CH_2{=}CHOH \longrightarrow CH_3\overset{\displaystyle O}{\overset{\displaystyle \|}{C}}H$$

Ethenol   **Acetaldehyde**
(Vinyl alcohol)
(Unstable)

Chloroethene (vinyl chloride) is made from ethene by a chlorination–dehydrochlorination sequence. Because chlorine is relatively expensive, an indirect process that uses HCl in the presence of oxygen and $CuCl_2$ has been developed. These conditions lead to the same intermediate, 1,2-dichloroethane, which is converted into the desired product by elimination of HCl.

Chloroethene (Vinyl Chloride) Synthesis

$$CH_2{=}CH_2 \xrightarrow{\text{Cl}_2} \underset{\underset{\displaystyle \text{Cl}}{\displaystyle |}}{CH_2}{-}\underset{\underset{\displaystyle \text{Cl}}{\displaystyle |}}{CH_2} \xrightarrow[-\text{HCl}]{\Delta} CH_2{=}CHCl$$

1,2-Dichloroethane   **Chloroethene**
(Vinyl chloride)

CHEMICAL HIGHLIGHT 12-6

Polymer-Supported Synthesis of Chemical Libraries

In the pharmaceutical industry, the major bottleneck to the discovery of new drugs has been the time required to prepare new organic molecules as candidates for biological testing. Perhaps the best example is the decades-long, worldwide effort to synthesize relatives of penicillin for examination as potential antibiotics in the ongoing battle against drug resistance that continually arises in infectious microorganisms. Thousands of compounds have been prepared in this effort, and every single synthesis has required laborious separation and purification procedures to generate just one pure candidate for examination.

It has become clear in the 1990s that this kind of effort simply takes too long: The number of different compounds that must be screened to uncover a single good drug candidate can be very large, and synthesis one-at-a-time is out of the question. As a result, strategies have been developed for *simultaneous parallel* synthesis and testing of large numbers of related compounds, so-called *chemical libraries.* These syntheses are typically *not* done in solution, where the product of each synthetic step must be separated from by-products, excess reagents, and solvents. Instead, the initial reaction substrates are covalently attached to insoluble polymers (symbolized by P in the scheme), which are then exposed to the necessary reagents while suspended in the reaction solvents. Substrates are chemically transformed into products while anchored to the solid polymer, and at the end of each step a simple filtration removes all non-polymer-bound substances. After the desired sequence of chemical processes has been completed, cleavage of the linkage between the insoluble polymer and the final product gives the latter in high purity. The adaptation of this strategy to the simultaneous parallel synthesis of a nine-compound chemical library can be shown in schematic form.

The sequence consists of just two reaction steps. However, in each step, three variants (which, e.g., may be three reagents that contain different alkyl substituents) give the nine permutations of final products shown at the bottom of the scheme. Four variants in each of two steps would give 16 products; in each of three steps, 64. Libraries of up to millions(!) of compounds have been produced by this method, which has been given the name *combinatorial synthesis,* because the products derive from multiple combinations of a few structurally different units. Thus, the need for large numbers of compounds for pharmaceutical testing is finally being met, and an entirely new style for doing organic synthesis has been developed.

Oxidation of ethene with oxygen in the presence of silver furnishes oxacyclo-propane (ethylene oxide), the hydrolysis of which gives 1,2-ethanediol (ethylene glycol) (Section 9-11). Hydration of ethene gives ethanol (Section 9-11).

$$CH_2{=}CH_2 \xrightarrow{\text{O}_2,\ \text{catalytic Ag}} \underset{\substack{\textbf{Oxacyclopropane}\\ \textbf{(Ethylene oxide)}}}{H_2C{-}CH_2} \xrightarrow{\text{H}^+,\ \text{H}_2\text{O}} \underset{\substack{\textbf{1,2-Ethanediol}\\ \textbf{(Ethylene glycol)}}}{\overset{\text{OH}\quad\text{OH}}{CH_2{-}CH_2}}$$

Table 12-4 (see page 523) gives an idea of the sizable amount of ethene-derived raw materials produced in the United States in 1995.

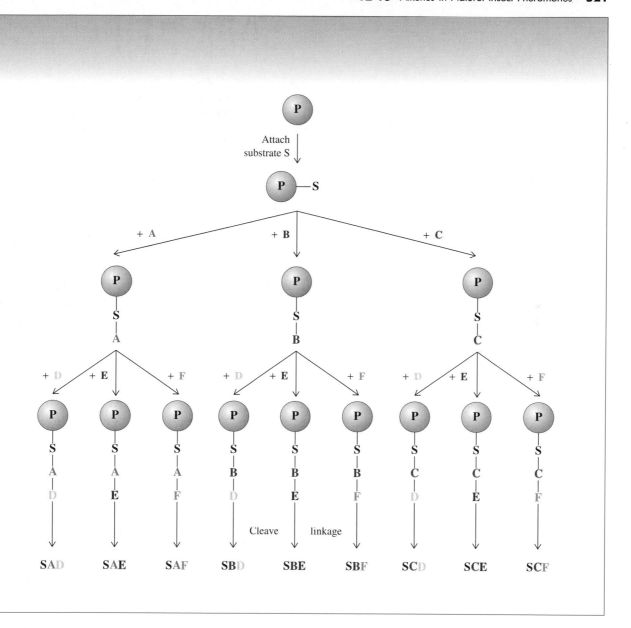

In summary, ethene is a valuable source of various industrial raw materials, particularly monomers, ethanol, and 1,2-ethanediol (ethylene glycol).

12-16 Alkenes in Nature: Insect Pheromones

Many natural products contain π bonds; several were mentioned in Sections 4-7 and 9-11. This section describes a specific group of naturally occurring alkenes, the **insect pheromones** (*pherein*, Greek, to bear; *hormon*, Greek, to stimulate).

Insect Pheromones

European vine moth

Japanese beetle

Male boll weevil

Defense pheromone of larvae of chrysomelid beetle

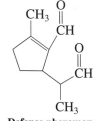

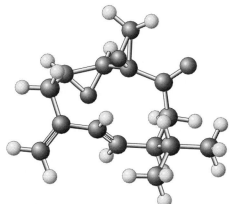

American cockroach

California red scale

Pheromones are chemical substances used for communication within a living species. There are sex, trail, alarm, and defense pheromones, to mention a few. Many insect pheromones are simple alkenes; they are isolated by extraction of certain parts of the insect and separation of the resulting product mixture by chromatographic techniques. Often only minute quantities of the bioactive compound can be obtained, in which case the synthetic organic chemist can play a very important role in the design and execution of total syntheses. Interestingly, the specific activity of a pheromone frequently depends on the configuration around the double bond (e.g., *E* or *Z*), as well as on the absolute configuration of any chiral centers present (*R,S*) *and* the composition of isomer mixtures. For example, the sex attractant for the male silkworm moth, 10-*trans*-12-*cis*-hexadecadien-1-ol (known as bombykol), is 10 billion times more ac-

TABLE 12-4	Chemicals Made from Ethene	
Chemical		**Tons**
1,2-Dichloroethane (ethylene dichloride)		8632
Ethenylbenzene (styrene)		5693
Ethylbenzene		6828
Chloroethene (vinyl chloride)		7588
Oxacyclopropane (ethylene oxide)		3811
1,2-Ethanediol (ethylene glycol)		2615
Acetic acid[a]		2342
Ethanol		313
Ethenyl acetate (vinyl acetate)		1446

[a]Also made by other processes.

tive in eliciting a response than the 10-*cis*-12-*trans* isomer, and 10 *trillion* times more so than the trans, trans compound.

Bombykol

Pheromone research affords an important opportunity for achieving pest control. Minute quantities of sex pheromones can be used per acre of land to confuse male insects about the location of their female partners. These pheromones can thus serve as lures in traps to effectively remove insects without spraying crops with large amounts of other chemicals. It is clear that organic chemists in collaboration with insect biologists will make important contributions in this area in the years to come.

CHAPTER INTEGRATION PROBLEM

Compare and contrast the addition reactions of each of the following reagents with (*E*)-3-methyl-3-hexene: H_2 (catalyzed by PtO_2), HBr, dilute aqueous H_2SO_4, Br_2 in CCl_4, mercuric acetate in H_2O, and B_2H_6 in THF. Consider regiochemistry and stereochemistry. Which of these reactions can be used to synthesize alcohols? In what respects do the resulting alcohols differ?

SOLUTION

First, we need to identify the starting material's structure. Recall (Section 11-1) that the designation "*E*" describes the stereoisomer in which the two highest-priority groups (according to Cahn-Ingold-Prelog guidelines, Section 5-3) are on opposite sides of the double bond (i.e., *trans* to each other). In 3-methyl-3-hexene, the two ethyl groups are of highest priority. We therefore are starting with the compound shown in the margin.

Now let us consider the behavior of this alkene toward our list of reagents. In each case, it may be necessary to choose between two different regiochemical and two different stereochemical modes of addition. We recognize this situation because the starting alkene is substituted differently at each carbon, a regiochemical characteristic, and it has a defined (*E*) stereochemistry. To answer such questions *completely* correctly, it is essential to consider the *mechanism* of each reaction—that is, to *think* mechanistically.

Thus, the addition of H_2 with PtO_2 as a catalyst is an example of catalytic hydrogenation. Because the same kind of atom (hydrogen) adds to each of the alkene carbons, regiochemistry is not a consideration. Stereochemistry may be one, however. Catalytic hydrogenation is a *syn* addition, in which both hydrogen atoms attach to the same face of the alkene π bond (Section 12-2). If we view the alkene in a plane perpendicular to the plane of the page (Figure 12-1), addition will be on the top face 50% of the time and on the bottom face 50% of the time:

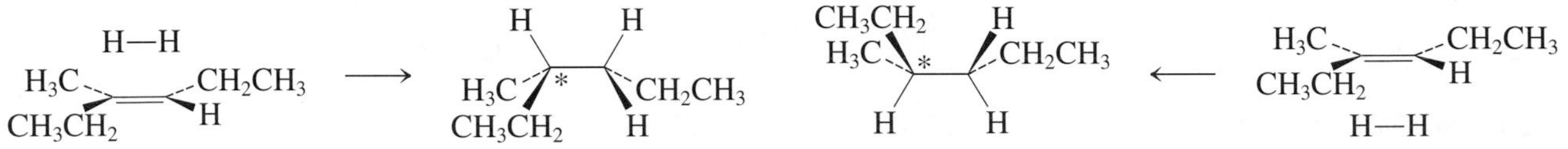

Racemic mixture

The addition generates one stereocenter (marked in each of the two products in the center of the scheme by an asterisk); therefore each product molecule is chiral (Section 5-1). Because the products form in equal amounts, the consequence is a racemic mixture of (*R*)- and (*S*)-3-methylhexane.

The next two reactions, with HBr and with aqueous H_2SO_4, begin with addition of the electrophile H^+ (Sections 12-3 and 12-4). These processes generate carbocations and will follow the regiochemical guideline known as Markovnikov's rule: Addition will be by attachment of H^+ to the less substituted alkenyl carbon to give the more stable carbocation. The latter is trapped by any available nucleophile—Br^- in the case of HBr, and H_2O in the case of aqueous H_2SO_4. Both steps proceed without stereoselectivity and, because the carbocation formed is already tertiary, rearrangement to a more stable carbocation is not possible. So we have the following result:

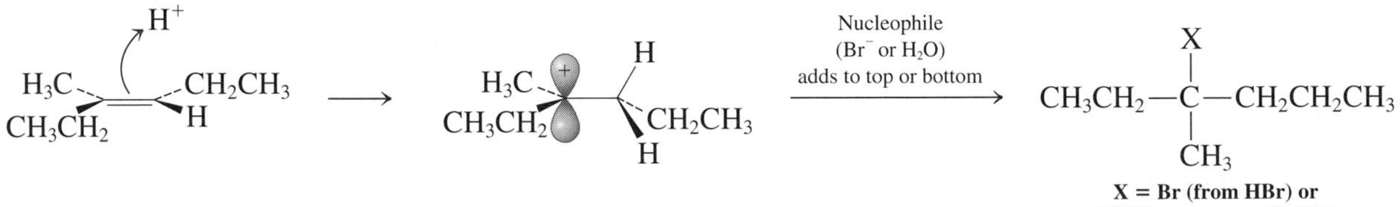

X = Br (from HBr) or
OH (from aqueous H_2SO_4)
Racemic mixture

The next two are examples of additions of electrophiles that form bridged cationic intermediates: A cyclic bromonium ion in the first case (Section 12-5), and a cyclic mercurinium ion in the second (Section 12-7). Additions will therefore proceed stereospecifically *anti,* because the cyclic ion can be attacked only from the direction opposite the location of the bridging electrophile (Figure 12-3). With Br_2, identical atoms add to both alkene carbons, so regiochemistry is not a consideration. In oxymercuration, the nucleophile is water, and it will add to the most substituted alkene carbon, because the latter is tertiary carbocation-like. Stereochemistry must be considered because addition of the electrophile occurs with equal probability from the top and from the bottom and stereocenters are created. So we have:

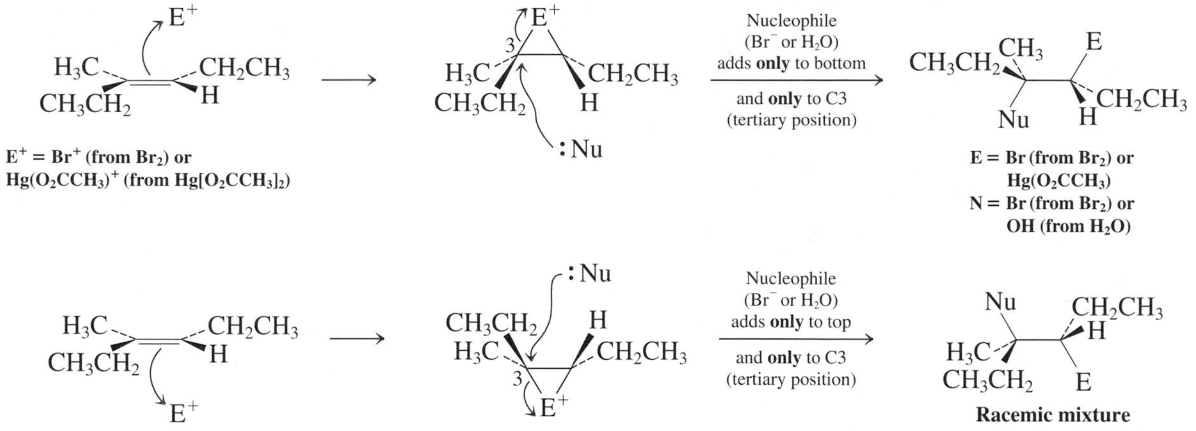

$E^+ = Br^+$ (from Br_2) or $Hg(O_2CCH_3)^+$ (from $Hg[O_2CCH_3]_2$)

E = Br (from Br_2) or $Hg(O_2CCH_3)$
N = Br (from Br_2) or OH (from H_2O)

Racemic mixture

Equal amounts—a racemic mixture—of the two chiral, enantiomerically related products are formed.

Finally, we come to hydroboration. Again, both regio- and stereoselectivity must be considered. As in hydrogenation, the stereochemistry is *syn;* unlike the preceding electrophilic additions, the regiochemistry is anti-Markovnikov: Boron attaches to the *less* substituted alkene carbon:

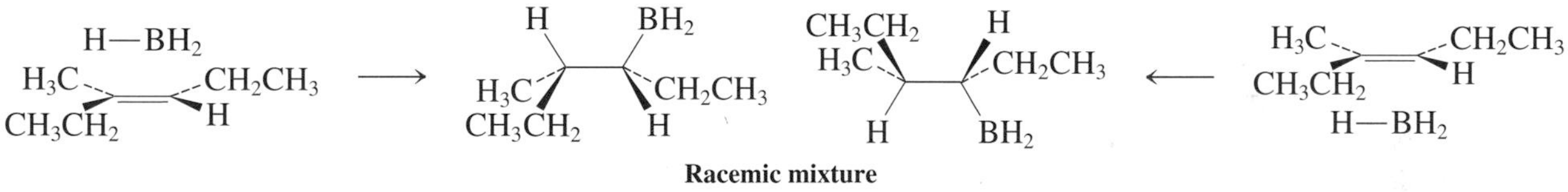

Racemic mixture

For simplicity, addition of only one of the B–H bonds is shown. The picture strongly resembles that of hydrogenation, but, because of the unsymmetric nature of the reagent, both alkene carbons are now transformed into stereocenters.

Three of the six reactions are well suited for the synthesis of alcohols: acid-catalyzed hydration, oxymercuration (after reduction of the C–Hg bond by $NaBH_4$), and hydroboration (by oxidation of the C–B bond by H_2O_2). Compare the alcohol from hydration (shown earlier) with that from demercuration of the oxymercuration product:

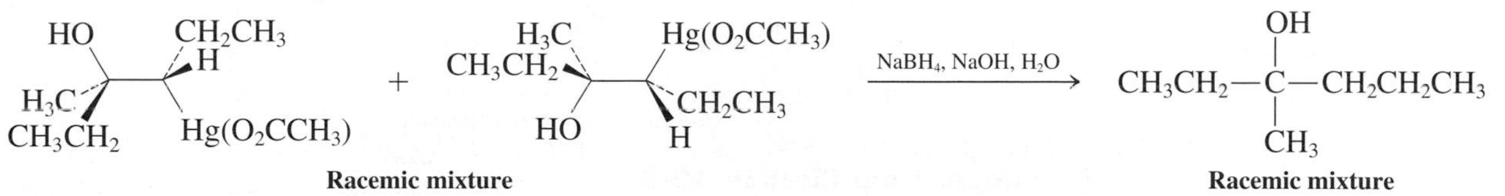

Racemic mixture

Racemic mixture

They are the same. Had a rearrangement taken place during hydration, this outcome might not have been the case (Section 9-3).

Oxidation of the hydroboration product gives a different, regioisomeric alcohol having two stereocenters, also as a racemic mixture (the boron atoms are shown without the additional alkyl groups):

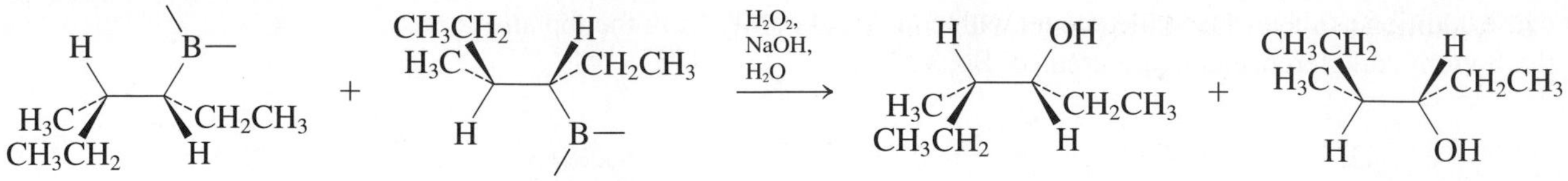

Racemic mixture   **Racemic mixture**

NEW REACTIONS

1. General Addition to Alkenes (Section 12-1)

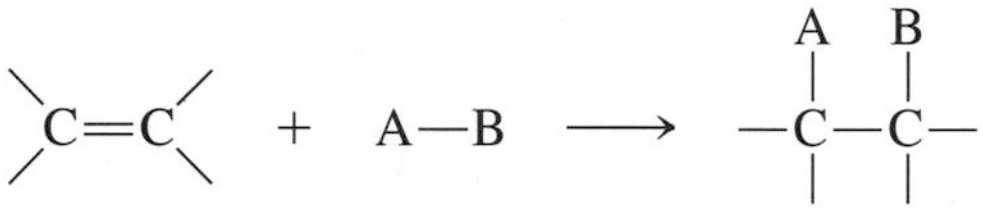

2. Hydrogenation (Section 12-2)

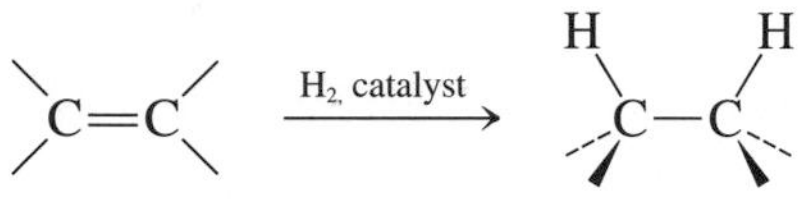

Syn addition

Typical catalysts: PtO$_2$, Pd-C, Ra-Ni

Electrophilic Additions

3. Hydrohalogenation (Section 12-3)

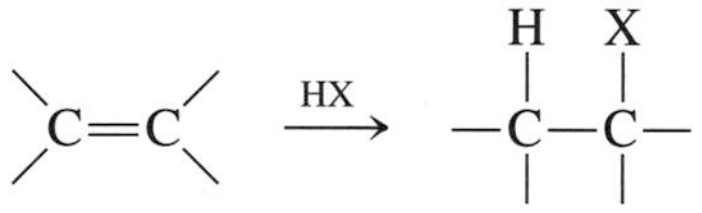

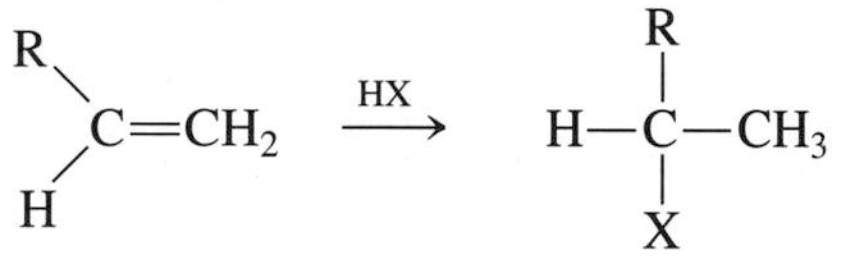

Regiospecific
(Markovnikov rule)

Through more stable carbocation

4. Hydration (Section 12-4)

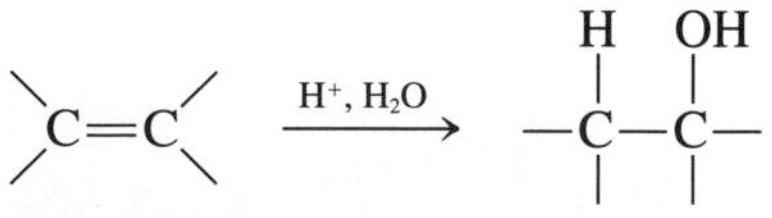

Through more stable carbocation

5. Halogenation (Section 12-5)

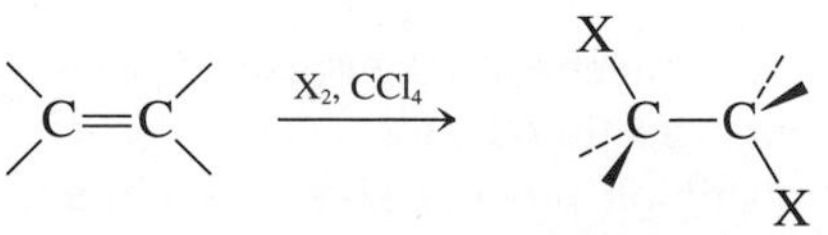

Stereospecific (*anti*)

X$_2$ = Cl$_2$ or Br$_2$, but not I$_2$

6. Vicinal Haloalcohol Synthesis (Section 12-6)

$$\text{C=C} \xrightarrow{\text{X}_2,\ \text{H}_2\text{O}} \overset{\text{X}}{\underset{\text{OH}}{\text{C—C}}}$$

OH attaches to more substituted carbon

7. Vicinal Haloether Synthesis (Section 12-6)

$$\text{C=C} \xrightarrow{\text{X}_2,\ \text{ROH}} \overset{\text{X}}{\underset{\text{OR}}{\text{C—C}}}$$

OR attaches to more substituted carbon

8. General Electrophilic Additions (Section 12-6, Table 12-2)

$$\text{C=C} \xrightarrow{\text{AB}} \overset{\overset{+}{\text{A}}}{\text{C—C}} \xrightarrow{\text{B}^-} \overset{\text{A}}{\underset{\text{B}}{\text{C—C}}}$$

A = electropositive, B = electronegative
B attaches to more substituted carbon

9. Oxymercuration–Demercuration (Section 12-7)

$$\text{C=C} \xrightarrow[\text{2. NaBH}_4,\ \text{NaOH, H}_2\text{O}]{\text{1. Hg(OCCH}_3)_2,\ \text{H}_2\text{O}} \overset{\text{H}\quad\text{OH}}{\text{—C—C—}}$$

Initial addition is *anti*, through mercurinium ion

$$\text{C=C} \xrightarrow[\text{2. NaBH}_4,\ \text{NaOH, H}_2\text{O}]{\text{1. Hg(OCCH}_3)_2,\ \text{ROH}} \overset{\text{H}\quad\text{OR}}{\text{—C—C—}}$$

OH or OR attaches to more substituted carbon

10. Hydroboration (Section 12-8)

$$\text{C=C} + \text{BH}_3 \xrightarrow{\text{THF}} (\overset{|}{\underset{\text{H}}{\text{—C}}}\overset{|}{\underset{}{\text{—C—}}})_3\text{B}$$

$$\overset{\text{R}}{\underset{\text{H}}{\text{C=CH}_2}} + \text{BH}_3 \xrightarrow{\text{THF}} (\text{RCH}_2\text{CH}_2)_3\text{B}$$
Regiospecific

B attaches to less substituted carbon

$$+ \text{BH}_3 \xrightarrow{\text{THF}}$$

Stereospecific (*syn*)
and anti-Markovnikov

11. Hydroboration–Oxidation (Section 12-8)

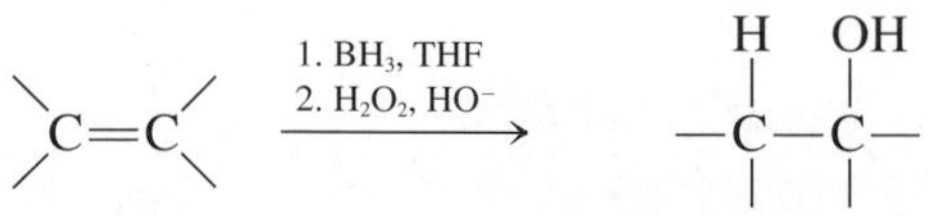

$$\overset{1.\ BH_3,\ THF}{\underset{2.\ H_2O_2,\ HO^-}{\longrightarrow}}$$

Stereospecific (*syn*)
and anti-Markovnikov

OH attaches to less substituted carbon

Oxidation

12. Oxacyclopropane Formation (Section 12-9)

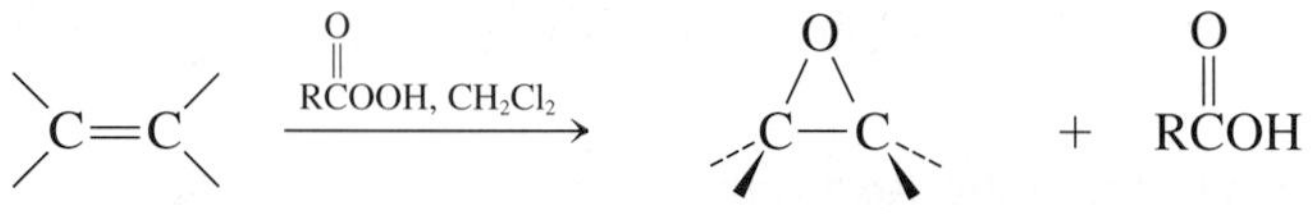

$$\overset{RCOOH,\ CH_2Cl_2}{\longrightarrow}$$

Stereospecific (*syn*)

13. Vicinal *Anti* Dihydroxylation (Section 12-9)

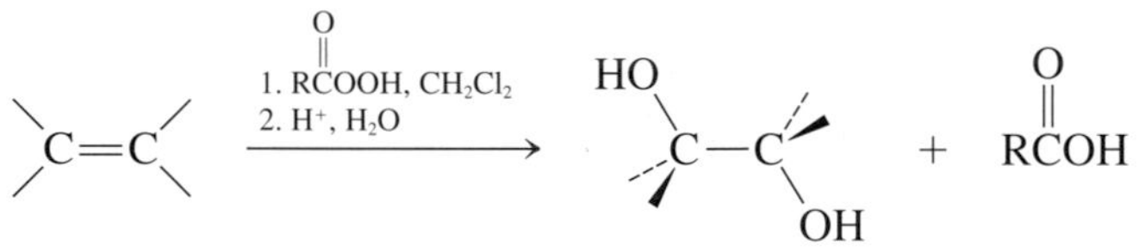

$$\overset{1.\ RCOOH,\ CH_2Cl_2}{\underset{2.\ H^+,\ H_2O}{\longrightarrow}}$$

14. Vicinal *Syn* Dihydroxylation (Section 12-10)

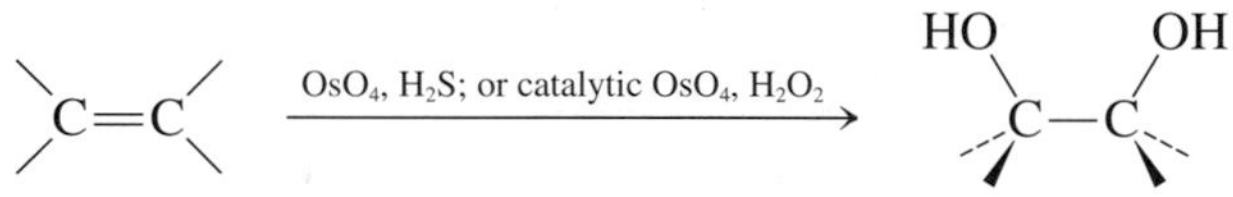

$$\overset{OsO_4,\ H_2S;\ or\ catalytic\ OsO_4,\ H_2O_2}{\longrightarrow}$$

Through cyclic intermediates

15. Ozonolysis (Section 12-11)

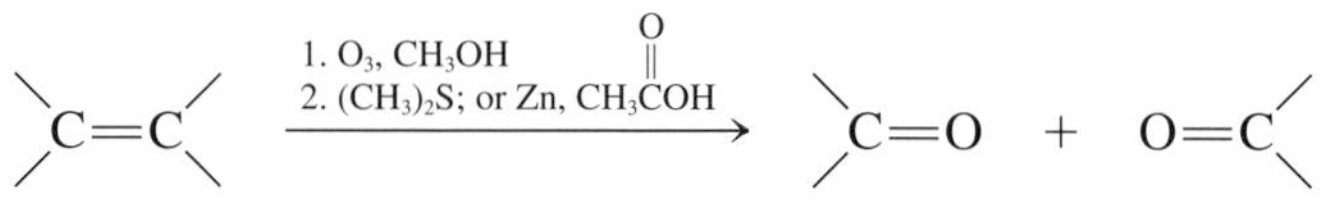

$$\overset{1.\ O_3,\ CH_3OH}{\underset{2.\ (CH_3)_2S;\ or\ Zn,\ CH_3COH}{\longrightarrow}}$$

Through molozonide and ozonide intermediates

Radical Additions

16. Radical Hydrobromination (Section 12-12)

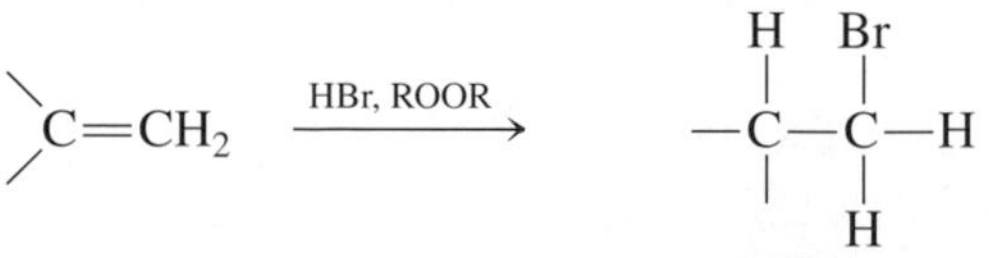

$$\overset{HBr,\ ROOR}{\longrightarrow}$$

Anti-Markovnikov
Does not occur with HCl or HI

17. Other Radical Additions (Section 12-12)

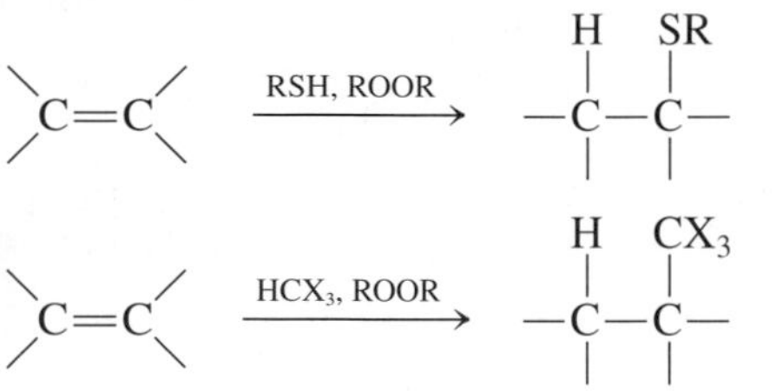

$$\text{C=C} \xrightarrow{\text{RSH, ROOR}} -\overset{H}{\underset{|}{C}}-\overset{SR}{\underset{|}{C}}-$$

$$\text{C=C} \xrightarrow{\text{HCX}_3,\ \text{ROOR}} -\overset{H}{\underset{|}{C}}-\overset{CX_3}{\underset{|}{C}}-$$

Anti-Markovnikov

Monomers and Polymers

18. Dimerization, Oligomerization, and Polymerization (Sections 12-13 and 12-14)

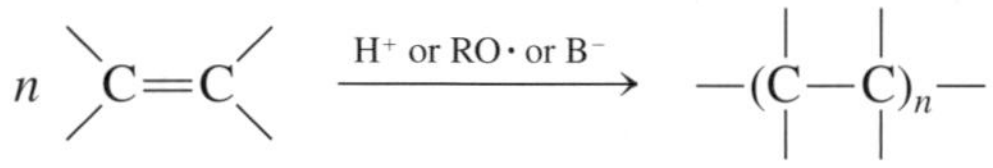

$$n \ \text{C=C} \xrightarrow{\text{H}^+ \text{ or RO} \cdot \text{ or B}^-} -(\overset{|}{\underset{|}{C}}-\overset{|}{\underset{|}{C}})_n-$$

Ziegler-Natta polymerization

$$n\ CH_2{=}CH_2 \xrightarrow{TiCl_4,\ AlR_3} -(CH_2CH_2)_n-$$

19. Ethene in Industrial Processes (Section 12-15)

Synthesis by cracking

$$R-CH_2-CH_2-R \xrightarrow{\Delta} CH_2{=}CH_2 + R-R + \text{other hydrocarbons}$$

Wacker process

$$CH_2{=}CH_2 \xrightarrow{H_2O,\ O_2,\ \text{catalytic PdCl}_2,\ CuCl_2} CH_3\overset{O}{\overset{\|}{C}}H$$

Chloroethene (vinyl chloride) synthesis

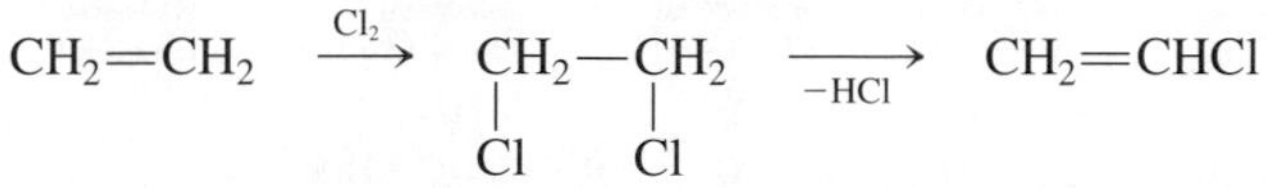

$$CH_2{=}CH_2 \xrightarrow{Cl_2} \underset{Cl\quad Cl}{CH_2-CH_2} \xrightarrow{-HCl} CH_2{=}CHCl$$

Oxacyclopropane (ethylene oxide) and 1,2-ethanediol (ethylene glycol) synthesis

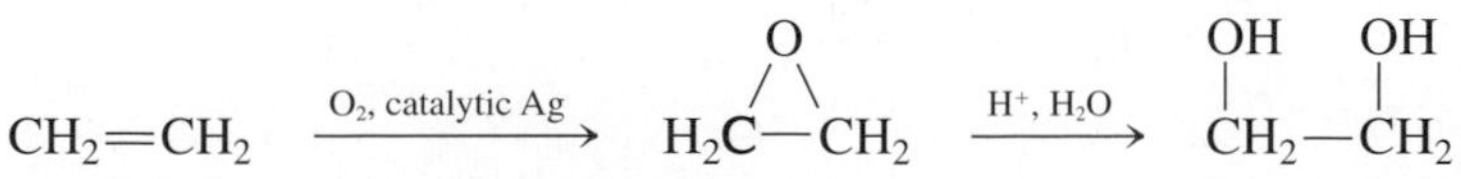

$$CH_2{=}CH_2 \xrightarrow{O_2,\ \text{catalytic Ag}} H_2C\overset{O}{\overset{\triangle}{-}}CH_2 \xrightarrow{H^+,\ H_2O} \underset{OH\quad OH}{CH_2-CH_2}$$

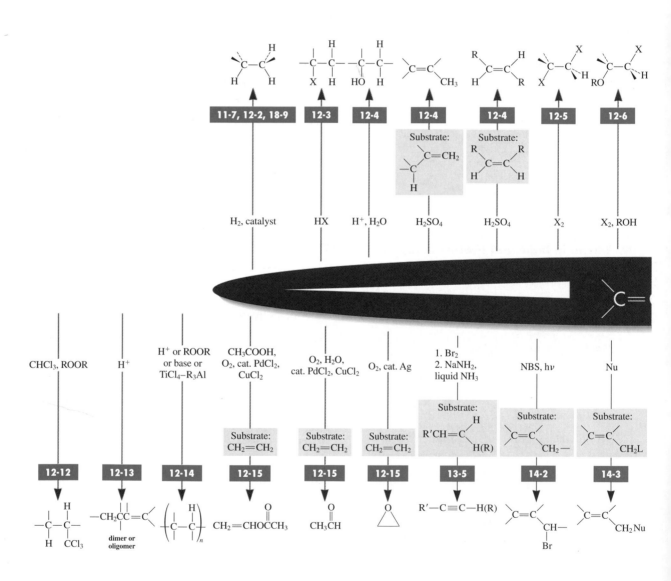

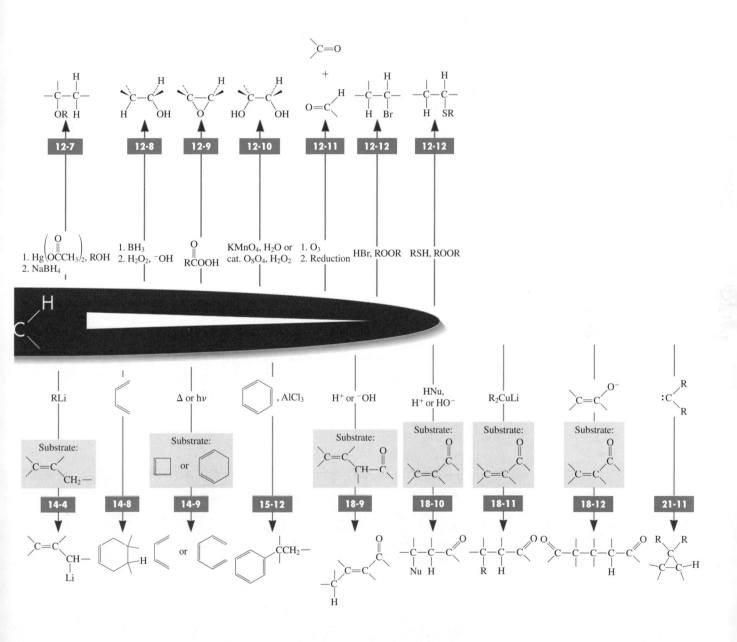

14-4 **14-8** **14-9** **15-12** **18-9** **18-10** **18-11** **18-12** **21-11**

12-7 **12-8** **12-9** **12-10** **12-11** **12-12** **12-12**

1. Hg(OCCH₃)₂, ROH
2. NaBH₄

1. BH₃
2. H₂O₂, ⁻OH

RCOOH

KMnO₄, H₂O or
cat. OₛO₄, H₂O₂

1. O₃
2. Reduction

HBr, ROOR

RSH, ROOR

RLi   Δ or hν   , AlCl₃   H⁺ or ⁻OH   HNu, H⁺ or HO⁻   R₂CuLi   :C⟨R R⟩

531

IMPORTANT CONCEPTS

1. The reactivity of the double bond manifests itself in exothermic **addition** reactions leading to **saturated** products.
2. The **hydrogenation** of alkenes is immeasurably slow unless a **catalyst** capable of splitting the strong H–H bond is used. Possible catalysts are palladium on carbon, platinum (as PtO_2), and Raney nickel. Addition of hydrogen is subject to steric control, the least hindered face of the least substituted double bond frequently being attacked preferentially.
3. As a Lewis base, the π bond is subject to attack by acid and **electrophiles,** such as H^+, X_2, and Hg^{2+}. If the initial intermediate is a free **carbocation,** the more highly substituted carbocation is formed. Alternatively, a **cyclic onium ion** is generated subject to nucleophilic ring opening at the more substituted carbon. The former case leads to control of regiochemistry **(Markovnikov rule),** the latter to control of both regio- and stereochemistry.
4. Mechanistically, **hydroboration** lies between hydrogenation and electrophilic addition. The first step is π complexation to the electron-deficient boron, whereas the second is a concerted transfer of the hydrogen to carbon. **Hydroboration–oxidation** results in the **anti-Markovnikov hydration** of alkenes.
5. **Peroxycarboxylic acids** may be thought of as containing an electrophilic oxygen atom, transferable to alkenes to give **oxacyclopropanes.** The process is often called epoxidation.
6. Osmium tetroxide acts as an electrophilic oxidant of alkenes; in the course of the reaction, the oxidation state of the metal is reduced by two units. Addition takes place in a concerted *syn* manner through cyclic six-electron transition states to give vicinal diols.
7. **Ozonolysis** followed by reduction yields carbonyl compounds derived by cleavage of the double bond.
8. In **radical chain additions** to alkenes, the chain carrier adds to the π bond to create the more highly substituted radical. This method allows for the anti-Markovnikov hydrobromination of alkenes, as well as the addition of thiols and some halomethanes.
9. Alkenes react with themselves through initiation by charged species, radicals, or some transition metals to give **polymers.** The initial attack at the double bond yields a reactive intermediate that perpetuates carbon–carbon bond formation.

PROBLEMS

22. With the help of the $DH°$ values given in Tables 3-1 and 3-4, calculate the $\Delta H°$ values for addition of each of the following molecules to ethene, using 65 kcal mol^{-1} for the carbon–carbon π bond strength.
    (a) Cl_2
    (b) IF ($DH° = 67$ kcal mol^{-1})
    (c) IBr ($DH° = 43$ kcal mol^{-1})
    (d) HF
    (e) HI
    (f) HO—Cl ($DH° = 60$ kcal mol^{-1})
    (g) Br—CN ($DH° = 83$ kcal mol^{-1}; $DH°$ for C_{sp^3}—CN $= 124$ kcal mol^{-1})
    (h) CH_3S—H ($DH° = 88$ kcal mol^{-1}; $DH°$ for C_{sp^3}—S $= 60$ kcal mol^{-1})

23. Give the expected major product of catalytic hydrogenation of each of the following alkenes. Clearly show and explain the stereochemistry of the resulting molecules.

(a) CH_3 $(CH_3)_2CH$ H   (b) CH_3 (c)   H H_2C H

24. Would you expect the catalytic hydrogenation of a small-ring cyclic alkene such as cyclobutene to be more or less exothermic than that of cyclohexene? (**Hint:** Which has more bond-angle strain, cyclobutene or cyclobutane?)

25. Give the expected major product from the reaction of each alkene with (i) peroxide-free HBr and (ii) HBr in the presence of peroxides.
(**a**) 1-Hexene; (**b**) 2-methyl-1-pentene; (**c**) 2-methyl-2-pentene;
(**d**) (Z)-3-hexene; (**e**) cyclohexene.

26. Give the product of addition to Br_2 to each alkene in Problem 25. Pay attention to stereochemistry.

27. What alcohol would be obtained from treatment of each alkene in Problem 25 with aqueous sulfuric acid? Would any of these alkenes give a different product upon oxymercuration–demercuration? On hydroboration–oxidation?

28. Give the reagents and conditions necessary for each of the following transformations and comment on the thermodynamics of each.
(**a**) Cyclohexanol → cyclohexene; (**b**) cyclohexene → cyclohexanol;
(**c**) chlorocyclopentane → cyclopentene; (**d**) cyclopentene → chlorocyclopentane.

29. (**a**) Give the product of addition of methylselenenyl chloride, CH_3SeCl (polarized $^{\delta+}Se–Cl^{\delta-}$), to propene. Explain the regiochemistry. (**b**) Reaction of CH_3SeCl with cyclohexene is a stereospecific *anti* addition. Explain mechanistically.

30. Formulate the product(s) that you would expect from each of the following reactions. Show stereochemistry clearly.

(**a**) 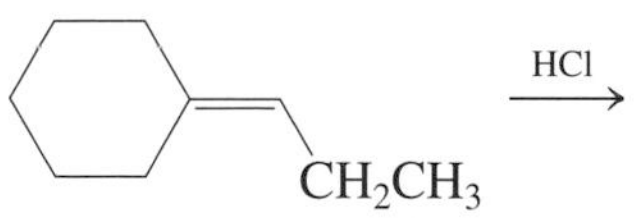 (**b**) *trans*-3-Heptene $\xrightarrow{Cl_2}$ (**c**) 1-Ethylcyclohexene $\xrightarrow{Br_2, H_2O}$

(**d**) Product of (c) $\xrightarrow{NaOH, H_2O}$ (**e**) 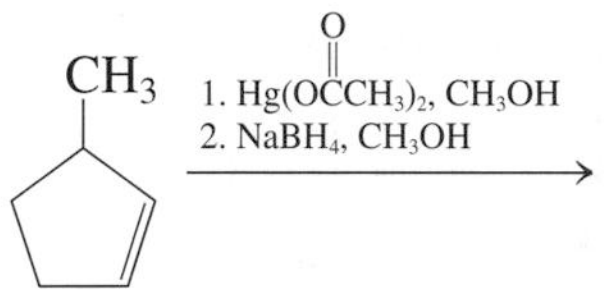 (**f**) *cis*-2-Butene $\xrightarrow{Br_2, \text{excess } Na^+N_3^-}$

(**g**)

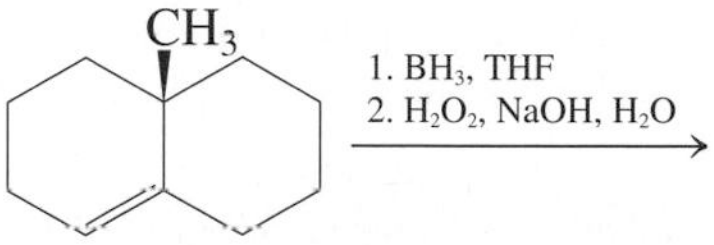

31. Show how you would synthesize each of the following molecules from an alkene of appropriate structure (your choice).

(**a**) 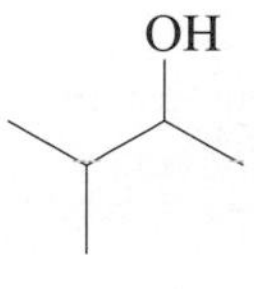 (**b**) 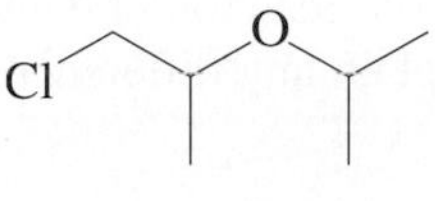 (**c**) 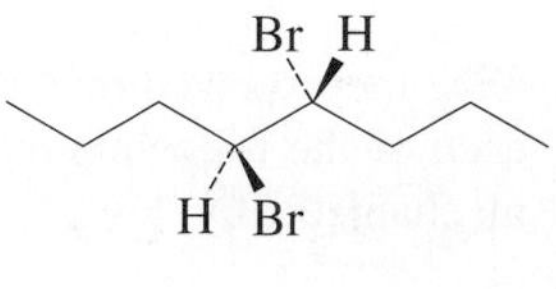 (*meso*—4R,5S—isomer)

(d) + (Racemate of 4*R*,5*R* and 4*S*,5*S* isomers)

(e)   **(f)** (More challenging. **Hint:** See Section 12-6.)

32. Propose efficient methods for accomplishing each of the following transformations. Most will require more than one step.

(a)   **(b)** (*meso*—2*R*,3*S*—isomer)

(c) + (Racemate of 2*R*,3*R* and 2*S*,3*S* isomers)

(d)

33. Give the expected product of reaction of 2-methyl-1-pentene with each of the following reagents.
    (a) H_2, PtO_2, CH_3CH_2OH (b) D_2, Pd-C, CH_3CH_2OH
    (c) BH_3, THF then NaOH + H_2O_2 (d) HCl
    (e) HBr   (f) HBr + peroxides
    (g) HI + peroxides   (h) H_2SO_4 + H_2O
    (i) Cl_2 (j) ICl
    (k) Br_2 + CH_3CH_2OH (l) CH_3SH + peroxides
    (m) MCPBA, CH_2Cl_2 (n) OsO_4, then H_2S
    (o) O_3, then Zn + $CH_3\overset{O}{\overset{\|}{C}}OH$ (p) $Hg(O\overset{O}{\overset{\|}{C}}CH_3)_2$ + H_2O, then $NaBH_4$
    (q) $CHBr_3$ + peroxides   (r) Catalytic H_2SO_4 + heat

34. What are the products of reaction of (*E*)-3-methyl-3-hexene with each of the reagents in Problem 33?

35. Write the expected products of reaction of 1-ethylcyclopentene with each of the reagents in Problem 33.

36. Give the expected major product from reaction of 3-methyl-1-butene with each of the following reagents. Explain any differences in the products mechanistically.

$$\overset{O}{\underset{||}{}}$$

(a) 50% aqueous H_2SO_4; (b) $Hg(OCCH_3)_2$ in H_2O, followed by $NaBH_4$;
(c) BH_3 in THF, followed by NaOH and H_2O_2.

37. Answer the question posed in Problem 36 for cyclohexylethene.

38. Give the expected major product of reaction of magnesium monoperoxy-phthalate (MMPP) with each alkene. In each case, also give the structure of the material formed upon hydrolysis in aqueous acid of the initial product.
(a) 1-Hexene; (b) (Z)-3-ethyl-2-hexene; (c) (E)-3-ethyl-2-hexene;
(d) (E)-3-hexene; (e) 1,2-dimethylcyclohexene.

39. Give the expected major product of reaction of OsO_4, followed by H_2S, with each alkene in Problem 38.

40. Give the expected major product of reaction of CH_3SH in the presence of per-oxides with each alkene in Problem 38.

41. ^1H NMR spectrum A corresponds to a molecule with the formula C_3H_5Cl. The compound shows significant IR bands at 730 (see Problem 24 of Chapter 11), 930, 980, 1630, and 3090 cm^{-1}. (a) Deduce the structure of the molecule. (b) Assign each NMR signal to a hydrogen or group of hydrogens. (c) The "doublet" at $\delta = 4.05$ ppm has $J = 6$ Hz. Is this in accord with your assignment in (b)? (d) This "doublet," on fivefold expansion, becomes a doublet of triplets (inset, spectrum A), with $J \sim 1$ Hz for the triplet splittings. What is the origin of this triplet splitting? Is it reasonable in light of your assignment in (b)?

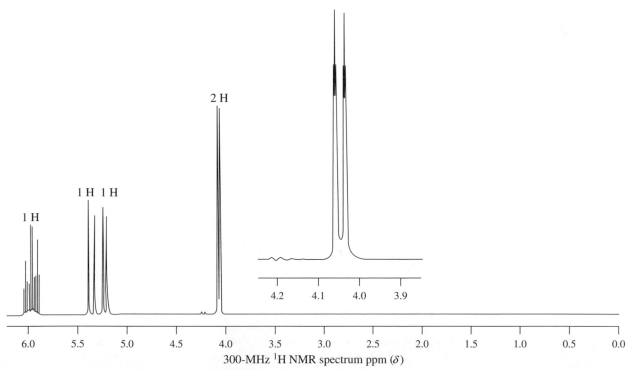

300-MHz ^1H NMR spectrum ppm (δ)

A

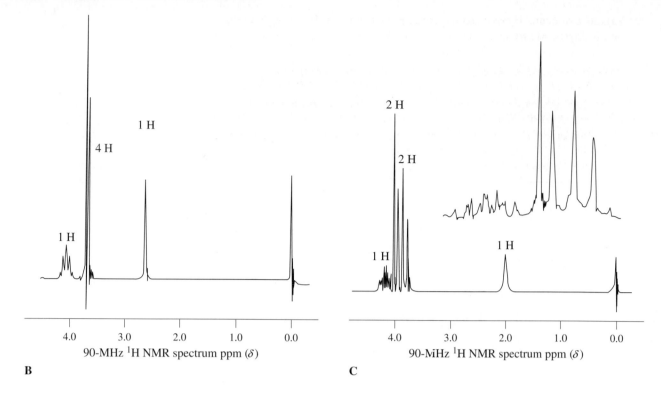

B

C

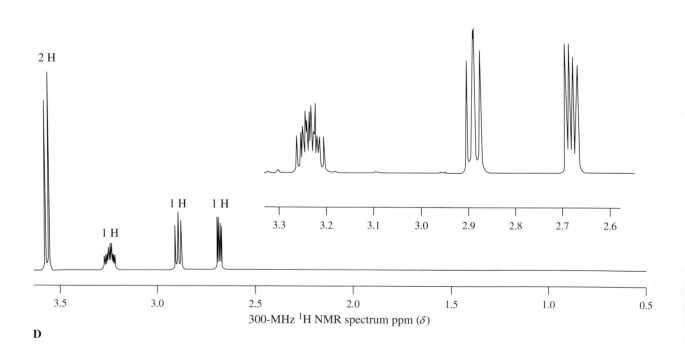

D

42. Reaction of C_3H_5Cl (Problem 41, spectrum A) with Cl_2 in H_2O gives rise to two products, both $C_3H_6Cl_2O$, whose spectra are shown in B and C. Reaction of either of these products with KOH yields the same molecule C_3H_5ClO (spectrum D). The insets show expansions of some of the multiplets. IR spectrum D reveals bands at 720 and 1260 cm^{-1} and the absence of signals between 1600 and 1800 cm^{-1} and between 3200 and 3700 cm^{-1}. **(a)** Deduce the structures of the compounds giving rise to spectra B, C, and D. **(b)** Why does reaction of the starting chloride compound with Cl_2 in H_2O give two isomeric products? **(c)** Write mechanisms for the formation of the product C_3H_5ClO from both isomers of $C_3H_6Cl_2O$.

43. ^1H NMR spectrum E corresponds to a molecule with the formula C_4H_8O. Its IR spectrum has important bands at 945, 1015, 1665, 3095, and 3360 cm^{-1}. **(a)** Determine the structure of the unknown. **(b)** Assign each NMR and IR signal. **(c)** Explain the splitting patterns for the signals at $\delta = 1.3$, 4.3, and 5.9 ppm (see inset for tenfold expansion).

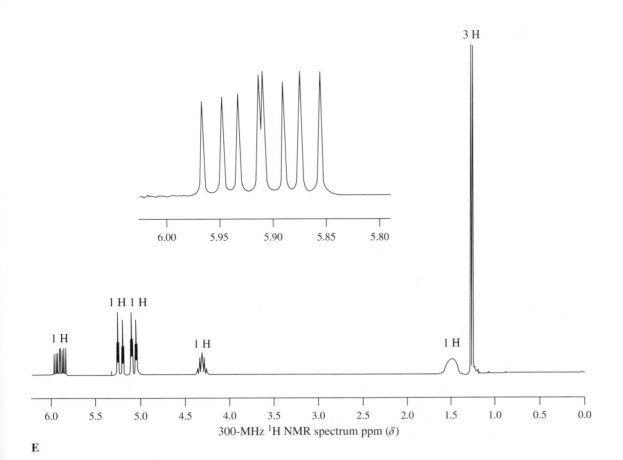

300-MHz ^1H NMR spectrum ppm (δ)

E

44. Reaction of the compound corresponding to spectrum E with $SOCl_2$ produces a chloroalkane, C_4H_7Cl, whose NMR spectrum is almost identical with spectrum E, except that the broad signal at $\delta = 1.5$ ppm is absent. Its IR spectrum E shows bands at 700 (Problem 24, Chapter 11), 925, 985, 1640, and 3090 cm^{-1}. Treatment with H_2 over PtO_2 results in C_4H_9Cl (spectrum F). Its IR spectrum reveals the absence of all the bands quoted for its precursor, except for the signal at 700 cm^{-1}. Identify these two molecules.

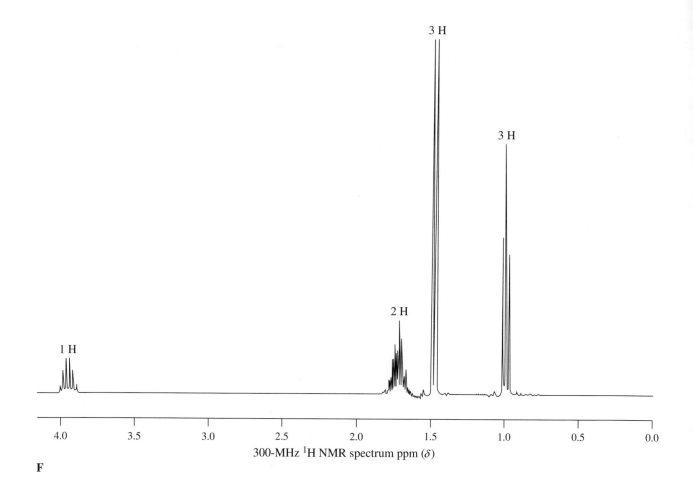

3 H

3 H

2 H

1 H

4.0   3.5   3.0   2.5   2.0   1.5   1.0   0.5   0.0

300-MHz ^{1}H NMR spectrum ppm (δ)

F

45. Give the structure of an alkene that will give the following carbonyl compounds upon ozonolysis followed by reduction with $(CH_3)_2S$.

(a) CH_3CHO only

(b) CH_3CHO and CH_3CH_2CHO

(c) $(CH_3)_2C{=}O$ and $H_2C{=}O$

(d) $CH_3CH_2\overset{\overset{\displaystyle O}{\|}}{C}CH_3$ and CH_3CHO

(e) cyclopentanone and CH_3CH_2CHO

46. Plan syntheses of each of the following compounds, utilizing retrosynthetic-analysis techniques. Starting compounds are given in parentheses. However, other simple alkanes or alkenes also may be used, as long as you include at least one carbon–carbon bond-forming step in each synthesis.

(a) 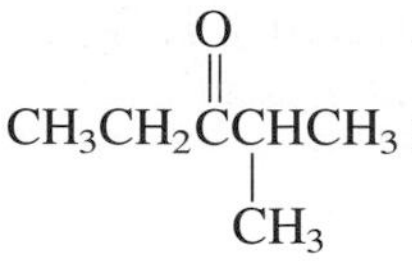 $CH_3CH_2\overset{\displaystyle O}{\overset{\|}{C}}CHCH_3$ (propene)
$\quad\quad\quad\underset{CH_3}{|}$

(b) 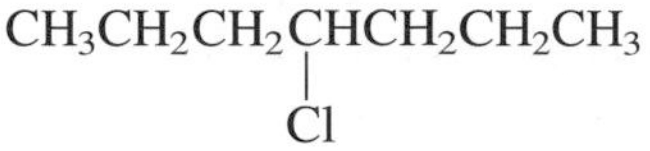 $CH_3CH_2CH_2CHCH_2CH_2CH_3$ (propene, again)
$\quad\quad\quad\quad\quad\quad\underset{Cl}{|}$

(c) 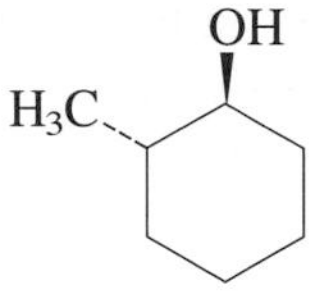 (cyclohexene)

47. Show how you would convert cyclopentane into each of the following molecules.

(a) *cis*-1,2-Dideuteriocyclopentane **(b)** *trans*-1,2-Dideuteriocyclopentane

(c) 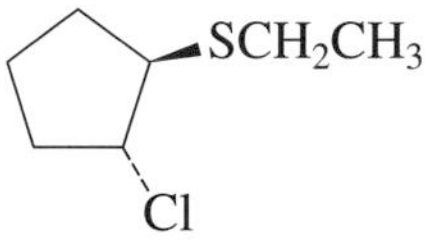 **(d)** 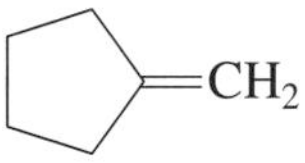 **(e)** 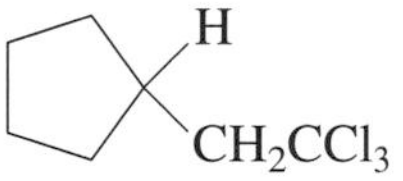 **(f)**

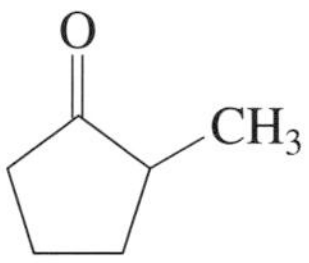

(g) 1,2-Dimethylcyclopentene **(b)** *trans*-1,2-Dimethyl-1,2-cyclopentanediol

48. Give the expected major product(s) of each of the following reactions.

(a) 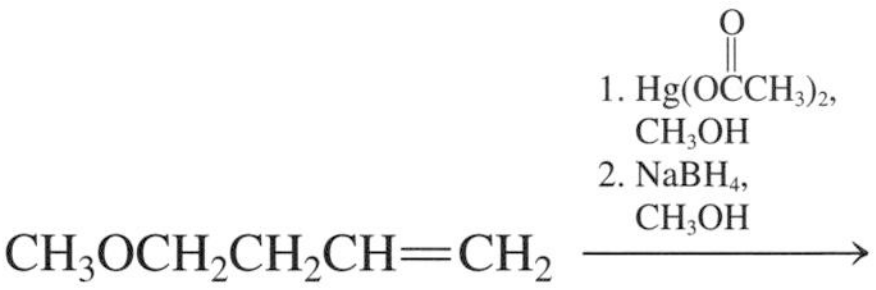 $CH_3OCH_2CH_2CH{=}CH_2$ $\xrightarrow{\begin{array}{l}1.\ Hg(OCCH_3)_2,\\ \ \ \ CH_3OH\\ 2.\ NaBH_4,\\ \ \ \ CH_3OH\end{array}}$

(b) 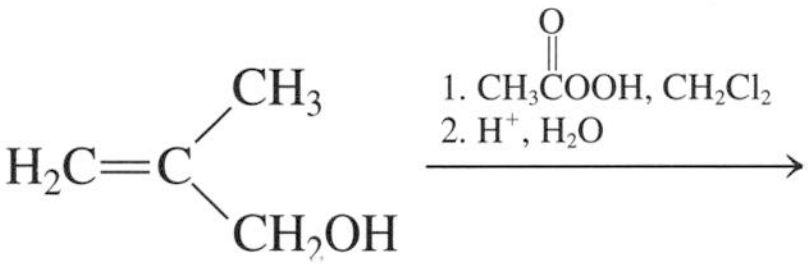 $H_2C{=}\underset{\displaystyle CH_2OH}{\overset{\displaystyle CH_3}{C}}$ $\xrightarrow{\begin{array}{l}1.\ CH_3COOH,\ CH_2Cl_2\\ 2.\ H^+,\ H_2O\end{array}}$

(c) $\xrightarrow{\text{Conc. HI}}$

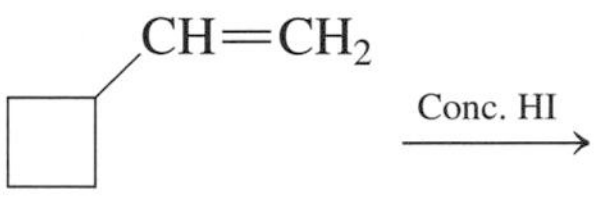

(d) 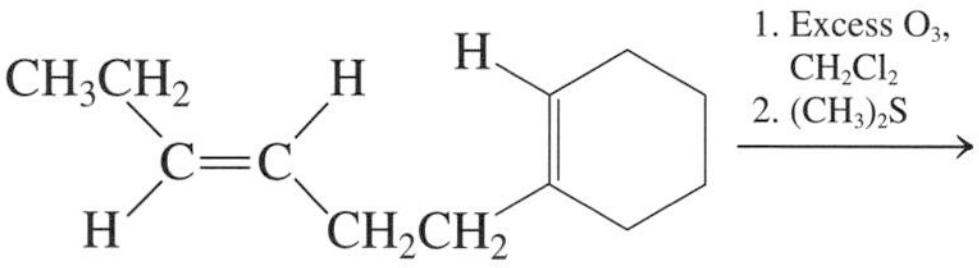 $\underset{\displaystyle H}{\overset{\displaystyle CH_3CH_2}{}}C{=}C\underset{\displaystyle CH_2CH_2}{\overset{\displaystyle H}{}}$ $\xrightarrow{\begin{array}{l}1.\ Excess\ O_3,\\ \ \ \ CH_2Cl_2\\ 2.\ (CH_3)_2S\end{array}}$

(e) 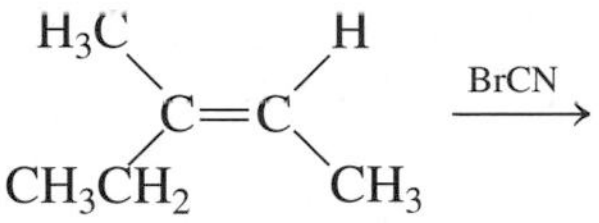 $\underset{\displaystyle CH_3CH_2}{\overset{\displaystyle H_3C}{}}C{=}C\underset{\displaystyle CH_3}{\overset{\displaystyle H}{}}$ $\xrightarrow{\text{BrCN}}$

(f) 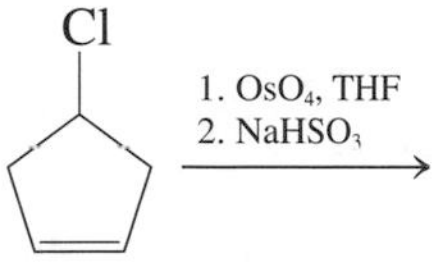 $\xrightarrow{\begin{array}{l}1.\ OsO_4,\ THF\\ 2.\ NaHSO_3\end{array}}$

(g) $CH_3CH{=}CH_2$ $\xrightarrow{\text{Catalytic HF}}$

(h) $CH_2{=}CHNO_2$ $\xrightarrow{\text{Catalytic KOH}}$
(**Hint:** Draw Lewis structures for the NO_2 group.)

49. (*E*)-5-Hepten-1-ol reacts with the following reagents to give products with the indicated formulas. Determine their structures and explain their formation by detailed mechanisms. **(a)** HCl, $C_7H_{14}O$ (no Cl!); **(b)** Cl_2, $C_7H_{13}ClO$ (IR: 740 cm^{-1}; nothing between 1600 and 1800 cm^{-1} and between 3200 and 3700 cm^{-1}).

50. When a cis alkene is mixed with a small amount of I_2 in the presence of heat or light, it isomerizes to some trans alkene. Propose a detailed mechanism to account for this observation.

51. Treatment of α-terpineol (Chapter 10, Problem 42) with aqueous mercuric acetate followed by sodium borohydride reduction leads predominantly to an isomer of the starting compound ($C_{10}H_{18}O$) instead of a hydration product. This isomer is the chief component in oil of eucalyptus and, appropriately enough, is called eucalyptol. It is popularly used as a flavoring for otherwise foul-tasting medicines because of its pleasant spicy taste and aroma. Deduce a structure for eucalyptol on the basis of sensible mechanistic chemistry and the following proton-decoupled ^{13}C NMR data. (**Hint:** The IR spectrum shows nothing between 1600 and 1800 cm^{-1} or between 3200 and 3700 cm^{-1}!)

$$
\text{CH}_3
$$

1. $Hg(OCCH_3)_2, H_2O$
2. $NaBH_4, H_2O$

$\longrightarrow$ eucalyptol, ^{13}C NMR: $\delta = 22.8, 27.5,$
($C_{10}H_{18}O$) $\qquad$ $28.8, 31.5,$
$\qquad$ $32.9, 69.6,$
$\qquad$ and 73.5 ppm

$(CH_3)_2COH$

α-**Terpineol**

52. Both borane and MCPBA react highly selectively with molecules, such as limonene, that contain double bonds in very different environments. Predict the products of reaction of limonene with (**a**) one equivalent of BH_3 in THF, followed by basic aqueous H_2O_2, and (**b**) one equivalent of MCPBA in CH_2Cl_2. Explain your answers.

CH_3

$H_3C \qquad CH_2$

Limonene

53. Oil of marjoram contains a pleasant, lemon-scented substance, $C_{10}H_{16}$ (compound G). Upon ozonolysis, G forms two products. One of them, H, has the formula $C_8H_{14}O_2$ and can be independently synthesized in the following way.

H_3C

$H_3C \qquad CH_2Br$

1. $Mg, (CH_3CH_2)_2O$
$\qquad$ O
2. $H_2C—CHCH_3$

$\longrightarrow C_8H_{16}O \xrightarrow{PCC, CH_2Cl_2} C_8H_{14}O \xrightarrow{\begin{array}{l}1.\ BH_3,\ THF\\2.\ H_2O_2,\ NaOH\\3.\ PCC,\ CH_2Cl_2\end{array}} H$
$\qquad\qquad\qquad$ I $\qquad\qquad\qquad\qquad$ J

From this information, propose reasonable structures for compounds G through J.

54. Humulene and α-caryophyllene alcohol are terpene constituents of carnation extracts. The former is converted into the latter by acid-catalyzed hydration in one step. Write a mechanism. (**Hint:** Follow the labeled carbon atoms retrosynthetically. The mechanism includes carbocation-induced cyclizations and hydrogen and alkyl-group migrations. This is difficult.)

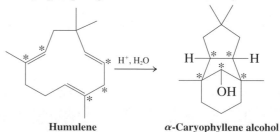

Humulene $\qquad\qquad\qquad$ α-**Caryophyllene alcohol**

55. Caryophyllene ($C_{15}H_{24}$) is an unusual sesquiterpene familiar to you as a major cause of the odor of cloves. Determine its structure from the following information. (**Caution:** The structure is totally different from that of α-caryophyllene alcohol in Problem 54.)

REACTION 1

$$\text{Caryophyllene} \xrightarrow{\text{H}_2,\ \text{Pd–C}} C_{15}H_{28}$$

REACTION 2

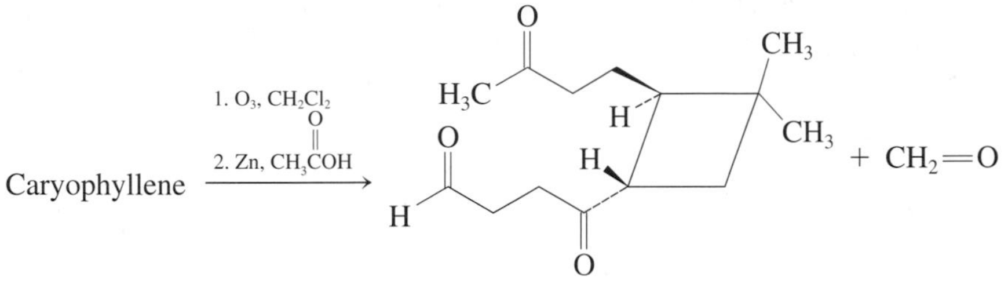

REACTION 3

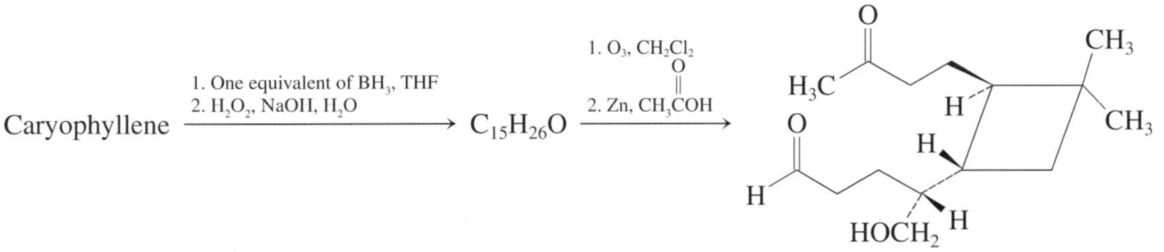

An isomer, isocaryophyllene, gives the same products as caryophyllene upon hydrogenation and ozonolysis. Hydroboration–oxidation of isocaryophyllene gives a $C_{15}H_{26}O$ product isomeric to the one shown in reaction 3; however, ozonolysis converts this compound into the same final product shown. In what way do caryophyllene and its isomer differ?

56. Juvenile hormone (JH, Chemical Highlight 12-2) has been synthesized in several ways. Two intermediates are shown in (a) and (b). Propose completions for syntheses of JH that start with each of them. Your synthesis for (a) should be stereospecific. Also for (a), note that the double bond between C10 and C11 is the most reactive toward electrophilic reagents (compare the synthesis of the JH analog in Chemical Highlight 12-2).

(a)

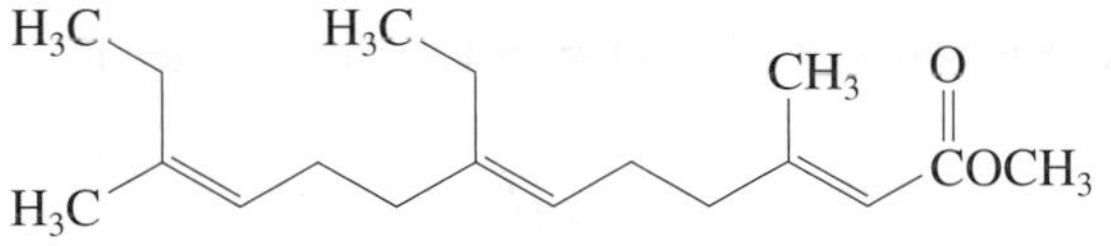

(b)

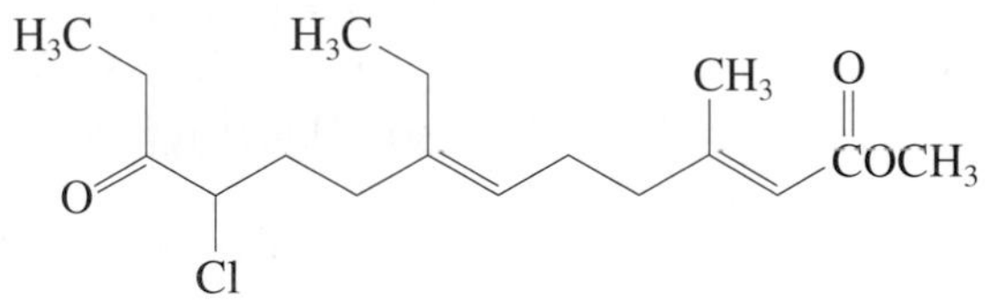

Team Problem

57. The selectivity of hydroboration increases with increasing bulkiness of the borane reagent.

**Bis(1,2-dimethylpropyl)borane
(Disiamylborane)**

9-BBN

(a) For example, 1-pentene is selectively hydroborated in the presence of *cis*- and *trans*-2-pentene when treated with bis(1,2-dimethylpropyl)borane (disiamylborane) or with 9-borabicyclo[3.3.1]nonane, 9-BBN. Divide the task of formulating the structure of the starting alkene used in preparing both of these bulky borane reagents among yourselves. Make models to visualize the features of these reagents that direct the structural selectivity.

(b) In an enantioselective approach to making secondary alcohols, two equivalents of one enantiomer of α-pinene are treated with BH_3. The resulting borane reagent is treated with *cis*-2-butene followed by basic hydrogen peroxide to yield optically active 2-butanol.

Share your model kits to make a model of α-pinene and the resulting borane reagent. Discuss what is directing the enantioselectivity of this hydroboration–oxidation reaction. What products besides 2-butanol result from the oxidation step?

Preprofessional Problems

58. A chiral compound, C_5H_8, upon simple catalytic hydrogenation yields an achiral compound, C_5H_{10}. What is the best name for the former? (a) 1-Methylcyclobutene; (b) 3-methylcyclobutene; (c) 1,2-dimethylcyclopropene; (d) cyclopentene.

59. A chemist reacted 300 g of 1-butene with excess Br_2 (in CCl_4) at 25°C. He isolated 418 g of 1,2-dibromobutane. What is the percent yield? (Atomic weights: C = 12.0, H = 1.00, Br = 80.0.) (a) 26; (b) 36; (c) 46; (d) 56; (e) 66.

60. *Trans*-3-hexene and *cis*-3-hexene differ in one of the following ways. Which one? (a) Products of hydrogenation; (b) products of ozonolysis; (c) products of Br_2 addition in CCl_4; (d) products of hydroboration–oxidation; (e) products of combustion.

61. Which reaction intermediate is believed to be part of the reaction shown?

$$RCH{=}CH_2 \xrightarrow{\text{HBr, ROOR}} RCH_2CH_2Br$$

(a) Radical; (b) carbocation; (c) oxacyclopropane; (d) bromonium ion.

62. When 1-pentene is treated with mercuric acetate, followed by sodium borohydride, which of the following compounds is the resulting product? (a) 1-Pentyne; (b) pentane; (c) 1-pentanol; (d) 2-pentanol.

Alkynes

The Carbon–Carbon Triple Bond

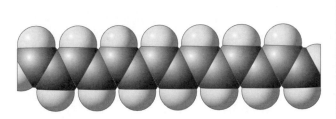

This shiny, flexible foil is polyacetylene (polyethyne), made by polymerization of gaseous ethyne on the walls of the reaction vessel shown. Polyacetylene was the first electrically conducting organic polymer ever made. It is used in the foil packaging for computer components to dissipate static electrical charge.

Will the characteristics of alkynes, compounds containing carbon–carbon triple bonds, resemble the properties and behavior of alkenes, their double-bonded cousins? In this chapter we shall see that, like alkenes, alkynes have numerous uses in a variety of modern settings. For example, the polymer derived from the parent compound, ethyne (HC≡CH), can be fashioned into electrically conductive sheets usable in lightweight all-polymer batteries. Ethyne is also a substance with a relatively high energy content, a property exploited in oxyacetylene torches. A variety of alkynes, both naturally occurring and synthetic, are useful in medicine for their antibacterial, antiparasitic, and antifungal activity.

Because the –C≡C– functional group contains two π linkages (which are mutually perpendicular; recall Figure 1-21), its reactivity is much like that of the double bond. For example, like alkenes, alkynes are electron rich and subject to attack by electrophiles. Many of the alkenes that serve as monomers for the production of polymeric fabrics, elastics, and plastics are prepared by electrophilic addition reactions to ethyne and other alkynes. Alkynes can be prepared by elimination reactions similar to those that are used to generate alkenes, and they are likewise more stable when the multiple bond is internal rather than terminal. A further and useful feature is that the alkynyl hydrogen is much more acidic than its alkenyl or alkyl counterparts, a property that permits easy deprotonation by strong bases. The resulting alkynyl anions are valuable nucleophilic reagents in synthesis.

We begin with a consideration of the naming and structural characteristics of the alkynes. Subsequent sections cover the spectroscopic and chemical properties of the compound class.

13-1 Naming the Alkynes

Common Names for Alkynes

$HC\equiv CH$

Acetylene

$CH_3C\equiv CCH_3$

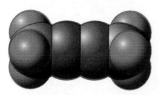

Dimethylacetylene

$CH_3CH_2CH_2C\equiv CH$

Propylacetylene

A carbon–carbon triple bond is the functional group characteristic of the **alkynes.** The general formula for the alkynes is C_nH_{2n-2}, the same as that for the cycloalkenes. The common names for many alkynes are still in use, including *acetylene,* the common name of the smallest alkyne, C_2H_2. Other alkynes are treated as its derivatives— for example, the alkylacetylenes.

The IUPAC rules for naming alkenes (Section 11-1) also apply to alkynes, the ending **-yne** replacing *-ene.* A number indicates the position of the triple bond in the main chain.

$$HC\equiv CH \qquad CH_3C\equiv CCH_3 \qquad \overset{1}{C}H_3\overset{2}{C}\equiv\overset{3}{C}\overset{4}{C}H\overset{5}{C}H_2\overset{6}{C}H_3 \qquad \overset{4}{C}H_3\overset{3}{C}\overset{2}{C}\equiv\overset{1}{C}H$$

with Br on C4 of the third structure and CH_3 groups on C3 of the fourth structure

Ethyne   **2-Butyne**   **4-Bromo-2-hexyne**   **3,3-Dimethyl-1-butyne**

Alkynes having the general structure $RC\equiv CH$ are **terminal,** whereas those with the structure of $RC\equiv CR'$ are **internal.**

Substituents bearing a triple bond are **alkynyl** groups. Thus, the substituent $-C\equiv CH$ is named **ethynyl;** its homolog $-CH_2C\equiv CH$ is **2-propynyl** (propargyl). Like alkanes and alkenes, alkynes can be depicted in straight-line notation.

trans-**1,2-Diethynylcyclohexane**   **2-Propynylcyclopropane**   **2-Propyn-1-ol**
                                  (Propargylcyclopropane)   **(Propargyl alcohol)**

In IUPAC nomenclature, a hydrocarbon containing both double and triple bonds is called an **alkenyne.** The chain is numbered starting from the end closest to either of the functional groups. When a double bond and a triple bond are at equidistant positions from either terminus, the *double* bond is given the lower number. Alkynes incorporating the hydroxy function are named **alkynols.** Note the omission of the final *e* of -ene in -enyne and of -yne in -ynol. The OH group takes precedence over both double and triple bonds in the numbering of a chain.

$$\overset{6}{C}H_3\overset{5}{C}H_2\overset{4}{C}H=\overset{3}{C}H\overset{2}{C}\equiv\overset{1}{C}H \qquad \overset{1}{C}H_2=\overset{2}{C}H\overset{3}{C}H_2\overset{4}{C}\equiv\overset{5}{C}H \qquad \text{5-Hexyn-2-ol structure with OH on C2}$$

3-Hexen-1-yne   **1-Penten-4-yne**   **5-Hexyn-2-ol**

(Not 3-hexen-5-yne)   (Not 4-penten-1-yne)   (Not 1-hexyn-5-ol)

EXERCISE 13-1

Give the IUPAC names for **(a)** all the alkynes of composition C_6H_{10};

(b)

$$\underset{H}{\overset{H_3C}{>}}C-C\equiv CH$$
with $CH=CH_2$ substituent

(c) all butynols. Remember to include and designate stereoisomers.

13-2 Properties and Bonding in the Alkynes

The nature of the triple bond helps explain the physical and chemical properties of the alkynes. In molecular-orbital terms, we shall see that the carbons are *sp* hybridized, and the four singly filled *p* orbitals form two perpendicular π bonds.

Alkynes are relatively nonpolar

Alkynes have boiling points very similar to those of the corresponding alkenes and alkanes. Ethyne is unusual in that it has no boiling point at atmospheric pressure; rather it sublimes at −84°C. Propyne (b.p. −23.2°C) and 1-butyne (b.p. 8.1°C) are gases, whereas 2-butyne is barely a liquid (b.p. 27°C) at room temperature. The medium-sized alkynes are distillable liquids. Care must be taken in the handling of alkynes: They polymerize very easily—frequently with violence. Ethyne explodes under pressure but can be shipped in pressurized gas cylinders that contain propanone (acetone) and a porous filler such as pumice.

Ethyne is linear and has strong, short bonds

In ethyne, the two carbons are *sp* hybridized (Figure 13-1A). One of the hybrid orbitals on each carbon overlaps with hydrogen, and a σ bond between the two carbon atoms results from mutual overlap of the remaining *sp* hybrids. The two perpendicular *p* orbitals on each carbon contain one electron each. These two sets overlap to form two perpendicular π bonds (Figure 13-1B). Because π bonds are diffuse, the distribution of electrons in the triple bond resembles a cylindrical cloud (Figure 13-1C). As a consequence of hybridization and the two π interactions, the strength of the triple bond is about 229 kcal mol^{-1}, considerably stronger than either the carbon–carbon double or single bonds (margin). As with alkenes, however, the alkyne π bonds are much weaker than the σ component of the triple bond, a feature that gives rise to much of its chemical reactivity. The C–H bond-dissociation energy of terminal alkynes is also substantial: 131 kcal mol^{-1}.

Dissociation Energies of C–C Bonds

$$HC\equiv CH$$

$$DH° = 229 \text{ kcal mol}^{-1}$$

$$H_2C\!\!\not\equiv\!\!CH_2$$

$$DH° = 173 \text{ kcal mol}^{-1}$$

$$H_3C\!\!\not\equiv\!\!CH_3$$

$$DH° = 90 \text{ kcal mol}^{-1}$$

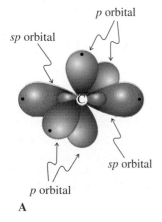

p orbital
sp orbital
sp orbital
p orbital

A

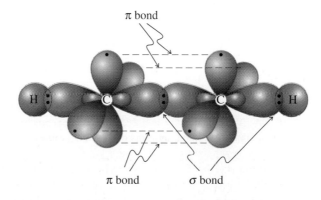

π bond
H
C
C
H
π bond   σ bond

B

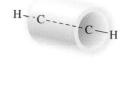

H⸱⸱⸱C⸱⸱⸱⸱C—H

C

FIGURE 13-1

(A) Molecular-orbital picture of *sp*-hybridized carbon, showing the two perpendicular *p* orbitals. (B) The triple bond in ethyne: The orbitals of two *sp*-hybridized CH fragments overlap to create a σ bond and two π bonds. (C) The two π bonds produce a cylindrical electron cloud around the molecular axis of ethyne.

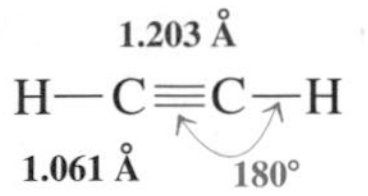

1.203 Å

$$H-C\equiv C-H$$

1.061 Å   180°

FIGURE 13-2 ——————
Molecular structure of ethyne.

**Deprotonation
of 1-Alkynes**

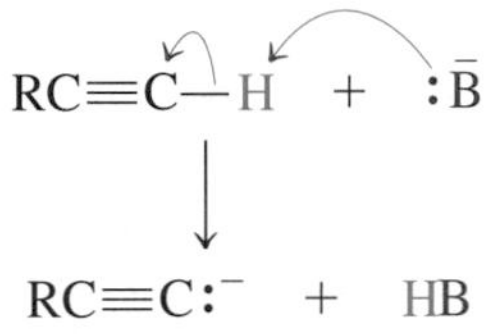

$$RC\equiv C\overset{\frown}{-}H + :\overset{..}{B}$$

$$\downarrow$$

$$RC\equiv C:^- + HB$$

Because both carbon atoms in ethyne are *sp* hybridized, its structure is linear (Figure 13-2). The carbon–carbon bond length is 1.20 Å, shorter than that of a double bond (1.33 Å, Figure 11-1). The carbon–hydrogen bond also is short, again because of the relatively large degree of *s* character in the *sp* hybrids used for bonding to hydrogen. The electrons in these orbitals (and in the bonds that they form by overlapping with other orbitals) reside relatively close to the nucleus and produce shorter (and stronger) bonds.

Terminal alkynes are remarkably acidic

In Section 2-9 you learned that the strength of an acid, H–A, increases with increasing electronegativity, or electron-attracting capability, of atom A. Is the electronegativity of an atom the same in all structural environments? The answer is *no*: Electronegativity varies with hybridization. Electrons in *s* orbitals are more strongly attracted to an atomic nucleus than are electrons in *p* orbitals. As a consequence, an atom with hybrid orbitals high in *s* character (e.g., *sp*, with 50% *s* and 50% *p* character) will be slightly more electronegative than the same atom with hybrid orbitals with less *s* character (*sp³*, 25% *s* and 75% *p* character). The relatively high *s* character in the carbon hybrid orbitals of terminal alkynes makes them more acidic than alkanes and alkenes. The pK_a of ethyne, for example, is 25, remarkably low compared with that of ethene and ethane.

Relative Acidities of Alkanes, Alkenes, and Alkynes

$$H_3C-CH_3 < H_2C=CH_2 < HC\equiv CH$$

Hybridization:	sp^3	sp^2	sp
pK_a:	50	44	25

This property is useful, because strong bases such as sodium amide in liquid ammonia, alkyllithiums, and Grignard reagents can deprotonate terminal alkynes to the corresponding **alkynyl anions.** These species react as bases and nucleophiles, much like other carbanions (Section 13-6).

Deprotonation of a Terminal Alkyne

$$CH_3CH_2C\equiv CH + CH_3CH_2CH_2CH_2Li \xrightarrow{(CH_3CH_2)_2O} CH_3CH_2C\equiv CLi + CH_3CH_2CH_2\overset{\displaystyle H}{\overset{|}{C}H_2}$$

EXERCISE 13-2

Strong bases other than those mentioned here for the deprotonation of alkynes were introduced earlier. Two examples are potassium *tert*-butoxide and lithium diisopropylamide (LDA). Would either (or both) of these compounds be suitable for making ethynyl anion from ethyne? Explain, in terms of their pK_a values.

In summary, the characteristic hybridization scheme for the triple bond of an alkyne controls its physical and electronic features. It is responsible for strong bonds, the linear structure, and the relatively acidic alkynyl hydrogen.

13-3 Spectroscopy of the Alkynes

Alkenyl hydrogens (and carbons) are deshielded and give rise to relatively low-field NMR signals compared with those in saturated alkanes (Section 11-4). In contrast,

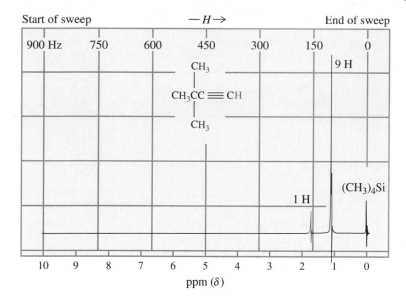

Start of sweep $-H\rightarrow$ End of sweep

FIGURE 13-3 _____

90-MHz ^{1}H NMR spectrum
of 3,3-dimethyl-1-butyne in
CCl_4, showing the high-field
position ($\delta = 1.74$ ppm) of
the signal due to the alkynyl
hydrogen.

FIGURE 13-4 _____

Electron circulation in the
presence of an external
magnetic field generates
local magnetic fields that
cause the characteristic
chemical shifts of alkenyl and
alkynyl hydrogens.
(A) Alkenyl hydrogens are
located in a region of space
where h_{local} reinforces H_0.
Therefore, a smaller applied
field is necessary for
resonance and these protons
are relatively deshielded.
(B) Electron circulation in an
alkyne generates a local field
that *opposes* H_0 in the
vicinity of the alkynyl
hydrogen. A higher field is
required for resonance
because of this shielding
effect.

alkynyl hydrogens have chemical shifts at relatively high field, much closer to those in alkanes. Similarly, the *sp*-hybridized carbons absorb in a range between that recorded for alkenes and alkanes. Alkynes, especially terminal ones, are also readily identified by IR spectroscopy.

The NMR absorptions of alkyne hydrogens show a characteristic shielding

Unlike alkenyl hydrogens, which are deshielded and give ^{1}H NMR absorptions at $\delta = 4.6$–5.7 ppm, protons bound to *sp*-hybridized carbon atoms are found at $\delta = 1.7$–3.1 ppm (Table 10-2). For example, in the NMR spectrum of 3,3-dimethyl-1-butyne, the alkynyl hydrogen absorbs at $\delta = 1.74$ ppm (Figure 13-3).

Why is the terminal alkyne hydrogen so shielded? Like the π electrons of an alkene, those in the triple bond enter into a circular motion when an alkyne is subjected to an external magnetic field (Figure 13-4). However, the cylindrical distribution of these

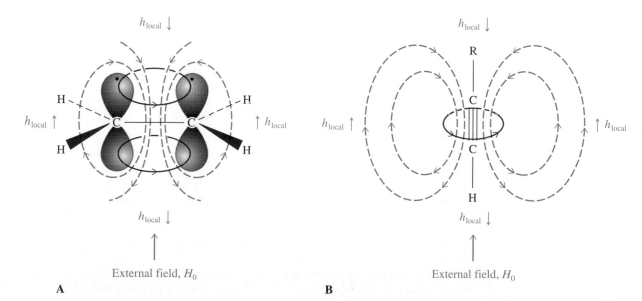

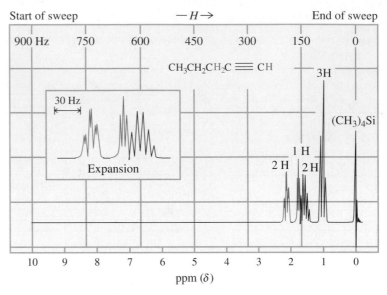

FIGURE 13-5

90-MHz ^{1}H NMR spectrum of 1-pentyne in CCl_4, showing coupling between the alkynyl (green) and propargylic (blue) hydrogens. See inset for more detail.

electrons (Figure 13-1C) now allows the major direction of this motion to be perpendicular to that in alkenes and to generate a local magnetic field that *opposes* H_0 in the vicinity of the alkyne hydrogen. The result is a strong *shielding* effect that cancels the deshielding tendency of the electron-withdrawing *sp*-hybridized carbon and gives rise to a relatively high field chemical shift.

The triple bond transmits spin–spin coupling

Long-Range Coupling in Alkynes

$$-\overset{\displaystyle H}{\underset{\displaystyle |}{C}}-C\equiv C-H$$

$$J = 2\text{–}4\text{ Hz}$$

The alkyne functional group transmits coupling so well that the terminal hydrogen is split by the hydrogens across the triple bond, even though they are separated from them by three carbons. This result is an example of long-range coupling. The coupling constants are small and range from about 2 to 4 Hz. Figure 13-5 shows the NMR spectrum of 1-pentyne. The alkynyl hydrogen signal at $\delta = 1.71$ ppm is a triplet ($J = 2.5$ Hz) because of coupling to the two equivalent hydrogens at C3, which appear at $\delta = 2.07$ ppm. The latter, in turn, give rise to a doublet of triplets, representing coupling to the two hydrogens at C4 ($J = 6$ Hz) as well as that at C1 ($J = 2.5$ Hz).

EXERCISE 13-3

Predict the first-order splitting pattern in the ^{1}H NMR spectrum of 3-methyl-1-butyne.

The ^{13}C NMR chemical shifts of alkyne carbons are distinct from those of the alkanes and alkenes

Carbon-13 NMR spectroscopy also is useful in deducing the structure of alkynes. For example, the triple-bonded carbons in alkyl-substituted alkynes absorb in the narrow range of $\delta = 65$–95 ppm, quite separate from the chemical shifts of analogous alkane ($\delta = 5$–45 ppm) and alkene ($\delta = 100$–150 ppm) carbon atoms (Table 10-6).

Typical Alkyne ^{13}C NMR Chemical Shifts

HC≡CH   HC≡CCH$_2$CH$_2$CH$_2$CH$_3$ CH$_3$CH$_2$C≡CCH$_2$CH$_3$

δ = 71.9   68.6   84.0   18.6 31.1 22.4 14.1   81.1 15.6   13.2 ppm

Terminal alkynes give rise to two characteristic infrared absorptions

Infrared spectroscopy is helpful in identifying terminal alkynes. Characteristic stretching bands appear for the alkynyl hydrogen at 3260–3330 cm^{-1} and for the C≡C triple bond at 2100–2260 cm^{-1} (Figure 13-6). Such data are especially useful when ^{1}H NMR spectra are complex and difficult to interpret. However, the band for the C≡C stretching vibration in internal alkynes is often weak, thus reducing the value of IR spectroscopy for characterizing these systems.

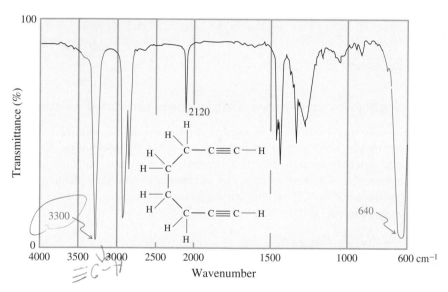

FIGURE 13-6

IR spectrum of 1,7-octadiyne: $\tilde{\nu}_{Csp-H\ stretch}$ = 3300 cm^{-1}; $\tilde{\nu}_{C≡C\ stretch}$ = 2120 cm^{-1}; $\tilde{\nu}_{Csp-H\ bend}$ = 640 cm^{-1}.

In summary, the cylindrical π cloud around the carbon–carbon triple bond induces local magnetic fields that lead to NMR chemical shifts for alkynyl hydrogens at higher fields than those of alkenyl protons. Long-range coupling is observed through the C≡C linkage. Infrared spectroscopy provides a useful complement to NMR data, displaying characteristic bands for the C≡C and ≡C–H bonds of terminal alkynes.

13-4 Stability of the Triple Bond

We are almost ready to address the preparation and reactions of alkynes. As with alkenes, a brief examination of the relative stabilities of these compounds gives a useful context in which to consider their chemistry (recall Section 11-7). We can gain such information by studying comparative heats of combustion and hydrogenation.

The high temperatures required for welding are attained by combustion of ethyne (acetylene).

Alkynes are high-energy compounds

Alkynes have a high energy content. Because of this property, alkynes react in many cases with considerable release of energy. For example, when shocked by high pressure or exposed to catalytic amounts of copper, ethyne explosively decomposes to carbon and hydrogen. On combustion, it releases 311 kcal of energy per mole. This energy is put to practical use in acetylene torches employed in welding, which requires very hot flames (more than 2500°C).

Combustion of Ethyne

$$HC \equiv CH \;+\; 2.5\,O_2 \;\longrightarrow\; 2\,CO_2 \;+\; H_2O \qquad \Delta H° = -311 \text{ kcal mol}^{-1}$$

As discussed in Problem 33 of Chapter 3, the flame temperature is related to both the heat of combustion and the quantity of exhaust gas produced. In the combustion of ethyne, the heat released is distributed among only three molecules of product gas per molecule of ethyne consumed. Therefore, each product molecule is heated to a higher temperature than in the case of propane, for instance, another commonly used heating source. The heat of combustion of the latter is 530.6 kcal mol^{-1}, but it is distributed among *seven* molecules of product gas (3 CO_2 and 4 H_2O) per molecule of propane, a situation resulting in a cooler flame.

Internal alkynes are more stable than terminal alkynes

Although heats of combustion can be used to determine relative stabilities of alkyne isomers, it is more convenient to compare their heats of hydrogenation, just as we did for alkenes. In the presence of the usual catalysts (platinum or palladium on charcoal), hydrogenation of alkynes proceeds by addition of two molar equivalents of hydrogen to produce alkanes. Thus, both butyne isomers hydrogenate to furnish butane but, on doing so, release different amounts of heat.

$$CH_3CH_2C \equiv CH \;+\; 2\,H_2 \;\xrightarrow{\text{Catalyst}}\; CH_3CH_2CH_2CH_3 \qquad \Delta H° = -69.9 \text{ kcal mol}^{-1}$$

$$CH_3C \equiv CCH_3 \;+\; 2\,H_2 \;\xrightarrow{\text{Catalyst}}\; CH_3CH_2CH_2CH_3 \qquad \Delta H° = -65.1 \text{ kcal mol}^{-1}$$

Relative Stabilities of the Alkynes

$$RC \equiv CH < RC \equiv CR'$$

The results show that the internal π system is more stable than the terminal one. As with the alkenes, the stabilizing influence is hyperconjugation (Sections 3-2 and 11-7).

EXERCISE 13-4

Are the heats of hydrogenation of the butynes consistent with the notion that alkynes are high-energy compounds? Explain. (**Hint:** Compare these values with the heats of hydrogenation of alkene double bonds.)

In summary, alkynes are highly energetic compounds. Internal isomers are more stable than terminal ones, as shown by the relative heats of hydrogenation.

13-5 Preparation of Alkynes by Double Elimination

The two basic methods used to prepare alkynes are double elimination from 1,2-dihaloalkanes and alkylation of alkynyl anions. This section deals with the first method, which provides a synthetic route to alkynes from alkenes; Section 13-6 addresses the second, which converts terminal alkynes into more complex, internal ones.

Alkynes are prepared from dihaloalkanes by elimination

As discussed in Section 11-8, alkenes can be prepared by E2 reactions of haloalkanes. Application of this principle to alkyne synthesis suggests that treatment of vicinal dihaloalkanes with two equivalents of strong base should result in double elimination to furnish a triple bond.

Double Elimination from Dihaloalkanes to Give Alkynes

    Vicinal
    dihaloalkane

Indeed, addition of 1,2-dibromohexane (prepared by bromination of 1-hexene, Section 12-5) to sodium amide in liquid ammonia followed by evaporation of solvent and aqueous work-up gives 1-hexyne.

$$CH_3CH_2CH_2CH_2\underset{\underset{Br}{|}}{CH}-CH_2Br \xrightarrow[-2\ HBr]{\substack{1.\ 3\ NaNH_2,\ liquid\ NH_3 \\ 2.\ H_2O}} CH_3CH_2CH_2CH_2C\equiv CH$$

Three equivalents of $NaNH_2$ are necessary in the preparation of a terminal alkyne because, as the latter forms, its acidic terminal hydrogen (Section 13-2) immediately protonates an equivalent amount of base. The inorganic products of this reaction are the protonated base (i.e., NH_3), which is evaporated, and NaBr, which is removed in the work-up. Eliminations in liquid ammonia are usually carried out at its boiling point, $-33°C$.

Because vicinal dihaloalkanes are readily available from alkenes by halogenation, this sequence, called **halogenation–double dehydrohalogenation,** is a ready means of converting alkenes into the corresponding alkynes.

**A Halogenation–Double Dehydrohalogenation
Used in Alkyne Synthesis**

1. Br_2, CCl_4
2. $NaNH_2$, liquid NH_3
3. H_2O

53%

1,5-Hexadiene   **1,5-Hexadiyne**

EXERCISE 13-5

Illustrate the use of halogenation–double dehydrohalogenation in the synthesis of the alkynes **(a)** 2-pentyne; **(b)** 1-octyne; **(c)** 2-methyl-3-hexyne.

Haloalkenes are intermediates in alkyne synthesis by elimination

Dehydrohalogenation of dihaloalkanes proceeds through the intermediacy of haloalkenes, also called **alkenyl halides.** Although mixtures of *E*- and *Z*-haloalkenes are in principle possible, with diastereomerically pure vicinal dihaloalkanes only one product is formed because elimination proceeds stereospecifically *anti* (Section 11-8).

EXERCISE 13-6

Give the structure of the bromoalkene involved in the bromination–double dehydrobromination of *cis*-2-butene to 2-butyne. Do the same for the trans isomer. (**Hint:** Refer to Section 12-5 for useful information, and use models.)

The stereochemistry of the intermediate haloalkene is of no consequence when the sequence is used for alkyne synthesis. Both *E*- and *Z*-haloalkenes eliminate with base to give the same alkyne.

In summary, alkynes are made from vicinal dihaloalkanes by double elimination. Alkenyl halides are intermediates, being formed stereospecifically in the first elimination.

13-6 Preparation of Alkynes by Alkylation of Alkynyl Anions

Alkynes can also be prepared from other alkynes. The reaction of terminal alkynyl anions with alkylating agents, such as primary haloalkanes, oxacyclopropanes, aldehydes, or ketones, results in carbon–carbon bond formation. As we know (Section 13-2), such anions are readily prepared from terminal alkynes by deprotonation with strong bases (mostly alkyllithium reagents, sodium amide in liquid ammonia, or Grignard reagents). Alkylation with primary haloalkanes is typically done in liquid ammonia or in ether solvents. These solvents sometimes contain 1,2-ethanediamine, $H_2NCH_2CH_2NH_2$; N,N,N',N'-tetramethyl-1,2-ethanediamine, $(CH_3)_2NCH_2CH_2N(CH_3)_2$; or HMPA as cosolvents to increase the nucleophilic power of the anion (Section 6-9). The process is unusual, because ordinary alkyl organometallic compounds are unreactive in the presence of haloalkanes. Alkynyl anions are an exception, however (see also Section 13-10).

Alkylation of an Alkynyl Anion

85%
1-Pentynylcyclohexane

Attempted alkylation of alkynyl anions with secondary and tertiary halides leads to E2 products because of the strongly basic character of the nucleophile (recall Section 7-8). Ethyne itself may be alkylated in a series of steps through the selective formation of the monoanion to give mono- and dialkyl-derivatives. The ethynyllithium-1,2-ethanediamine complex is also commercially available.

Alkynyl anions react with other carbon electrophiles such as oxacyclopropanes and carbonyl compounds in the same manner as do other organometallic reagents (Sections 8-8 and 9-9).

Reactions of Alkynyl Anions

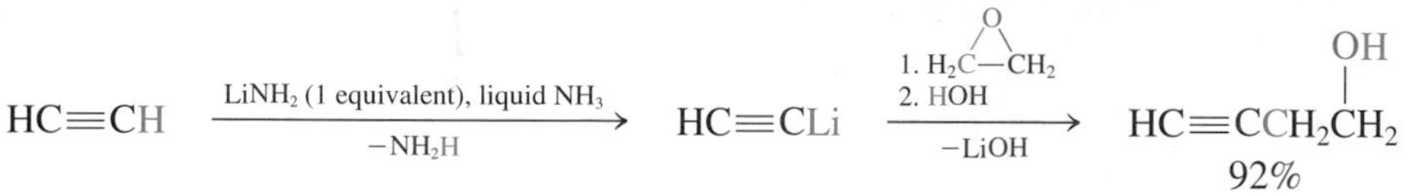

$$HC{\equiv}CH \xrightarrow[-NH_2H]{\text{LiNH}_2 \text{ (1 equivalent), liquid NH}_3} HC{\equiv}CLi \xrightarrow[-\text{LiOH}]{\begin{array}{l}1.\ H_2C{-}CH_2 \text{ (O)}\\ 2.\ HOH\end{array}} HC{\equiv}CCH_2CH_2\overset{OH}{|}$$

92%

3-Butyn-1-ol

$$CH_3C{\equiv}CH \xrightarrow[-CH_3CH_2H]{CH_3CH_2MgBr, (CH_3CH_2)_2O, 20°C} CH_3C{\equiv}CMgBr \xrightarrow[2.\ H_2O]{1.\ \text{(cyclopentanone)}}$$

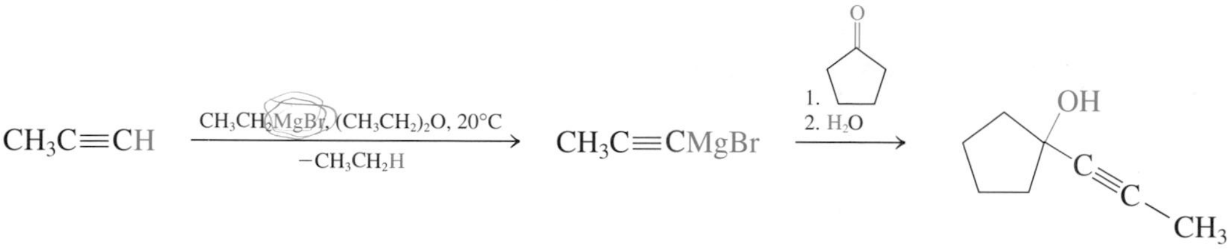

66%

1-(1-Propynyl)cyclopentanol

EXERCISE 13-7

Suggest efficient and short syntheses of these two compounds. (**Hint:** Review Section 8-9.)

(a) 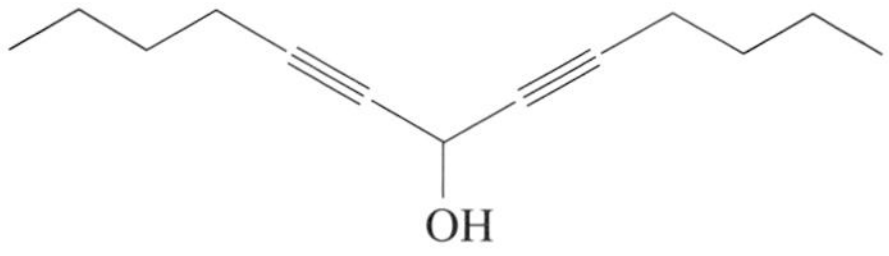 [structure of a diyne with OH group] from [structure of a terminal alkyne]

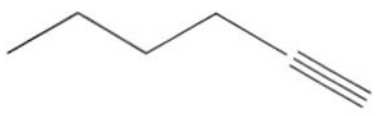

(b) 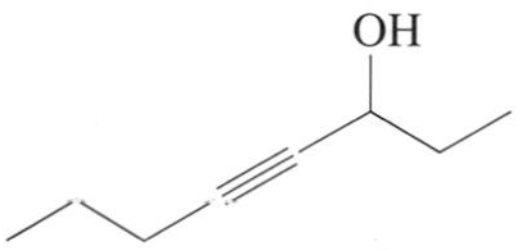 [structure with OH group and internal alkyne] from ethyne

---

EXERCISE 13-8

3-Butyn-2-ol is an important raw material in the pharmaceutical industry. It is the starting point for the synthesis of a variety of medicinally valuable alkaloids (Section 25-8), steroids (Section 4-7), and prostaglandins (Chemical Highlights 11-1 and 18-5, Section 18-14), as well as vitamins E (Section 22-9) and K. Propose a short synthesis of 3-butyn-2-ol by using the techniques outlined in this section.

In summary, alkynes can be prepared from other alkynes by alkylation with primary haloalkanes, oxacyclopropanes, or carbonyl compounds. Ethyne itself can be alkylated in a series of steps.

13-7 Reduction of Alkynes: The Relative Reactivity of the Two Pi Bonds

Now we turn from the preparation of alkynes to the characteristic reactions of the triple bond. In many respects, alkynes are like alkenes, except for the availability of *two* π bonds. Thus, alkynes can undergo additions, such as hydrogenation and electrophilic attacks.

Addition of Reagents A–B to Alkynes

$$R-C\equiv C-R \xrightarrow{\text{A–B}} \underset{A}{\overset{R}{C}}=\underset{B}{\overset{R}{C}} \text{ or } \underset{A}{\overset{R}{C}}=\underset{R}{\overset{B}{C}} \xrightarrow{\text{A–B}} A-\underset{A}{\overset{R}{\underset{|}{C}}}-\underset{B}{\overset{R}{\underset{|}{C}}}-B \text{ or } A-\underset{B}{\overset{R}{\underset{|}{C}}}-\underset{A}{\overset{R}{\underset{|}{C}}}-B$$

In this section we introduce two new hydrogen addition reactions: step-by-step hydrogenation and dissolving-metal reduction by sodium to give cis and trans alkenes, respectively.

Cis alkenes can be synthesized by catalytic hydrogenation

Alkynes can be hydrogenated under the same conditions used to hydrogenate alkenes. Typically, platinum or palladium on charcoal is suspended in a solution containing the alkyne, and the mixture is exposed to a hydrogen atmosphere. Under these conditions, the triple bond is saturated completely.

Complete Hydrogenation of Alkynes

$$CH_3CH_2CH_2C\equiv CCH_2CH_3 \xrightarrow{H_2, \text{ Pt}} CH_3CH_2CH_2CH_2CH_2CH_2CH_3$$

3-Heptyne   100%   **Heptane**

Hydrogenation is a step-by-step process that may be stopped at the intermediate alkene stage by the use of modified catalysts, such as the **Lindlar* catalyst.** This catalyst is palladium that has been precipitated on calcium carbonate and treated with lead acetate and quinoline. The surface of the metal rearranges to a less active configuration than that of palladium on carbon so that only the first π bond of the alkyne is hydrogenated. As with catalytic hydrogenation of alkenes (Section 12-2), the addition of H_2 is a *syn* process. As a result, this method affords a stereoselective synthesis of cis alkenes from alkynes.

Hydrogenations with Lindlar Catalyst

$$\text{Lindlar catalyst: 5\% Pd–CaCO}_3, \text{Pb(OCCH}_3)_2,$$

Quinoline

$$\text{3-Heptene} \xrightarrow{H_2, \text{ Lindlar catalyst, 25°C}} \text{cis-3-Heptene}$$

100%

3-Heptene   *cis*-**3-Heptene**

*Dr. Herbert H. M. Lindlar (b. 1909), Hoffman-La Roche Ltd., Basel.

EXERCISE 13-9

Write the structure of the product expected from the following reaction.

$$\xrightarrow{\text{H}_2,\ \text{Lindlar catalyst, 25°C}}$$

EXERCISE 13-10

The perfume industry makes considerable use of naturally occurring substances such as those obtained from rose and jasmine extracts. In many cases, the quantities of fragrant oils available by natural product isolation are so small that it is necessary to synthesize them. Examples are found in the olfactory components of violets, which include *trans*-2-*cis*-6-nonadien-1-ol and the corresponding aldehyde. An important intermediate in their large-scale synthesis is *cis*-3-hexen-1-ol, whose industrial preparation is described as "a closely guarded secret." Using the methods in this and the preceding sections, propose a synthesis from 1-butyne.

With a method for the construction of cis alkenes at our disposal, we might ask: Can we modify the reduction of alkynes to give only trans alkenes? The answer is yes, with a different reducing agent and through a different mechanism.

Sequential one-electron reductions of alkynes produce trans alkenes

When, instead of catalytically activated hydrogen, we use *sodium metal* dissolved in liquid ammonia (dissolving-metal reduction) as the reagent for the reduction of alkynes, we obtain trans alkenes as the products. For example, 3-heptyne is reduced to *trans*-3-heptene in this way. Unlike sodium amide in liquid ammonia, which functions as a strong base, elemental sodium in liquid ammonia acts as a powerful electron donor (i.e., a reducing agent).

$$\xrightarrow[\text{2. H}_2\text{O}]{\text{1. Na, liquid NH}_3}$$

86%

3-Heptyne   ***trans*-3-Heptene**

In the first step of the mechanism of this reduction, the π framework of the triple bond accepts one electron to give a radical anion. This anion is protonated by the ammonia solvent to give an alkenyl radical that is further reduced by accepting another electron to give an alkenyl anion. This species is again protonated to give the product alkene, which is stable to further reduction. The trans stereochemistry of the final alkene is set in the first two steps of the mechanism, which give rise preferentially to the less sterically hindered trans alkenyl radical. Under the reaction conditions (liquid NH$_3$, −33°C), the second one-electron transfer takes place faster than cis-trans equilibration of the radical. This type of reduction typically provides >98% stereochemically pure trans alkene.

Mechanism of the Reduction of Alkynes by Sodium in Liquid Ammonia

STEP 1. One-electron transfer

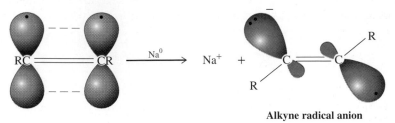

A

STEP 2. First protonation

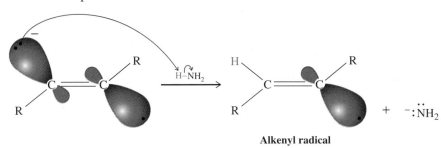

B

STEP 3. Second one-electron transfer

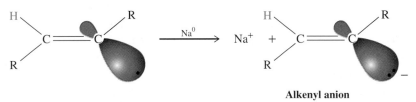

C

STEP 4. Second protonation

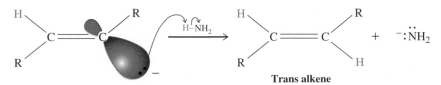

D

EXERCISE 13-11

When 1,7-undecadiyne (11 carbons) was treated with a mixture of sodium *and* sodium amide in liquid ammonia, only the internal bond was reduced to give *trans*-7-undecen-1-yne. Explain. (**Hint:** What reaction takes place between sodium amide and a terminal alkyne? Note that the pK_a of NH_3 is 35.)

In summary, alkynes are very similar in reactivity to alkenes, except that they have two π bonds, both of which may be saturated by addition reactions. Hydrogenation of the first π bond, which gives cis alkenes, is best achieved by using the Lindlar catalyst. Alkynes are converted into trans alkenes by treatment with sodium in liquid ammonia, a process that includes two successive one-electron reductions.

13-8 Electrophilic Addition Reactions of Alkynes

As a center of high electron density, the triple bond is readily attacked by electrophiles. This section describes the results of three such processes: addition of hydrogen halides, reaction with halogens, and hydration. The hydration is catalyzed by mercury(II) ions. As is the case in electrophilic additions to unsymmetrical alkenes (Section 12-3), the Markovnikov rule is followed in transformations of terminal alkynes: The electrophile adds to the terminal (less substituted) carbon atom.

Addition of hydrogen halides forms haloalkenes and geminal dihaloalkanes

The addition of hydrogen bromide to 2-butyne yields (Z)-2-bromobutene.

Addition of a Hydrogen Halide to an Internal Alkyne

$$CH_3C \equiv CCH_3 \xrightarrow{\text{HBr, Br}^-} \underset{\text{(Z)-2-Bromobutene}}{\underset{60\%}{\overset{\displaystyle H_3C}{\underset{\displaystyle}{}} C = C \overset{\displaystyle CH_3}{\underset{\displaystyle Br}{}}}}$$

The stereochemistry of this type of addition is frequently (but not always) *anti,* particularly in the presence of excess bromide ion. Another molecule of hydrogen bromide added to the bromoalkene gives the geminal dihaloalkane with regioselectivity in accord with the Markovnikov rule.

$$\overset{H}{\underset{H_3C}{}} C = C \overset{CH_3}{\underset{Br}{}} \xrightarrow{\text{HBr}} \underset{\underset{\text{2,2-Dibromobutane}}{90\%}}{CH_3\overset{H}{\underset{}{C}}H\overset{Br}{\underset{Br}{C}}CH_3}$$

The addition of hydrogen halides to terminal alkynes also proceeds in accord with the Markovnikov rule.

Addition to a Terminal Alkyne

$$CH_3C \equiv CH \xrightarrow{\text{HI, }-70°C} \underset{35\%}{\overset{I}{\underset{H_3C}{}} C = C \overset{H}{\underset{H}{}}} + \underset{65\%}{CH_3\overset{I}{\underset{I}{C}} - \overset{H}{\underset{H}{C}} - H}$$

It is usually difficult to limit such reactions to addition of a single molecule of HX.

CHEMICAL HIGHLIGHT 13-1   **Synthesis of a Sex Pheromone**

We have seen that pheromones may be used as chemical lures to aid in eradication of pests (Section 12-16). The sequential one-electron reduction of alkynes is applied to the synthesis of the sex pheromone of the spruce budworm, as shown in the scheme below. The hydroxy group in the starting material, 10-bromo-1-decanol, is initially protected as an ether by acid-catalyzed addition to 2-methylpropene. This step is necessary to prevent the protonation of the organolithium reagent employed in the next step. 1-Butynyllithium then displaces the bromine, the oxygen is deprotected, and the resulting alkynol is reduced to the trans alkenol. Oxidation of the alcohol function gives the pheromone.

$$HO(CH_2)_{10}Br \xrightarrow[\substack{\text{Protection} \\ \text{(Sections 9-7} \\ \text{and 12-4)}}]{(CH_3)_2C=CH_2, H^+} (CH_3)_3CO(CH_2)_{10}Br \xrightarrow[-LiBr]{LiC \equiv CCH_2CH_3, THF, HMPA}$$

10-Bromo-1-decanol

$$(CH_3)_3CO(CH_2)_{10}C \equiv CCH_2CH_3 \xrightarrow[\substack{-(CH_3)_2C=CH_2 \\ \text{Deprotection} \\ \text{(Section 9-8)}}]{H^+, H_2O} HO(CH_2)_{10}C \equiv CCH_2CH_3 \xrightarrow[\text{Reduction}]{Na, \text{ liquid } NH_3}$$

11-Tetradecyn-1-ol

trans-**11-Tetradecen-1-ol**   **Sex pheromone of the spruce budworm**

(reaction: PCC, CH_2Cl_2, Oxidation (Section 8-6))

EXERCISE 13-12

Write a step-by-step mechanism for the addition of HBr twice to 2-butyne to give 2,2-dibromobutane. Show clearly the structure of the intermediate formed in each step.

Halogenation also takes place once or twice

Electrophilic addition of halogen to alkynes proceeds through the intermediacy of isolable vicinal dihaloalkenes, the products of a single *anti* addition. Reaction with additional halogen gives tetrahaloalkanes. For example, halogenation of 3-hexyne gives the expected (*E*)-dihaloalkene and the tetrahaloalkane.

$$CH_3CH_2C \equiv CCH_2CH_3 \xrightarrow{Br_2, CH_3COOH, LiBr}$$

3-Hexyne

(E)-3,4-Dibromo-3-hexene, 99%

$$\xrightarrow{Br_2, CCl_4}$$

3,3,4,4-Tetrabromohexane, 95%

EXERCISE 13-13

Give the products of addition of one and two molecules of Cl_2 to 1-butyne.

Mercuric ion–catalyzed hydration of alkynes furnishes ketones

In a process analogous to the hydration of alkenes, water can be added to alkynes in a Markovnikov sense to give alcohols—in this case **enols,** in which the hydroxy group is attached to a double-bond carbon. As mentioned in Section 12-15, enols spontaneously rearrange to the isomeric carbonyl compounds. This process, called **tautomerism,** interconverts two isomers by simultaneous proton and double bond shifts. The enol is said to **tautomerize** to the carbonyl compound, and the two species are called **tautomers** (*tauto,* Greek, the same; *meros,* Greek, part). In this way, alkynes are ultimately converted into ketones. The reaction is catalyzed by Hg(II) ions.

Hydration of Alkynes

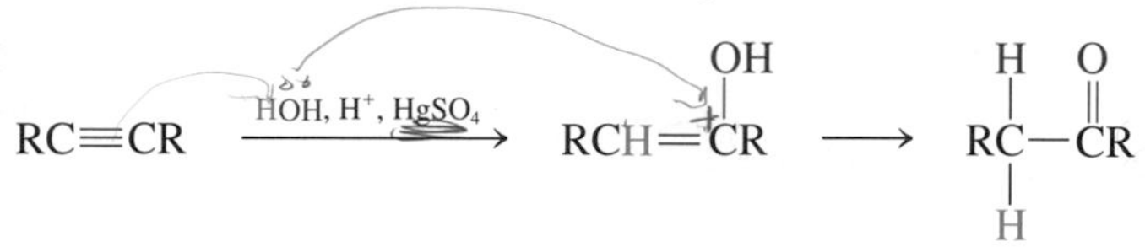
RC≡CR
HOH, H⁺, HgSO₄
OH
RCH=CR
H O
RC—CR
H

Enol **Ketone**

Hydration follows Markovnikov's rule: Terminal alkynes give methyl ketones.

Hydration of a Terminal Alkyne

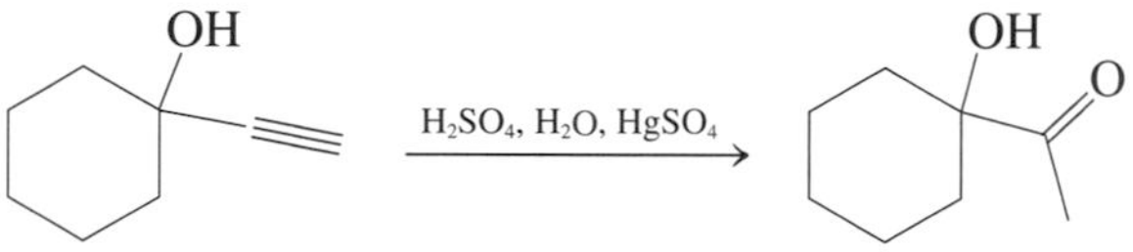
OH
H₂SO₄, H₂O, HgSO₄
OH
O

91%

Symmetric internal alkynes give a single carbonyl compound; unsymmetric systems lead to a mixture of ketones.

Hydration of Internal Alkynes

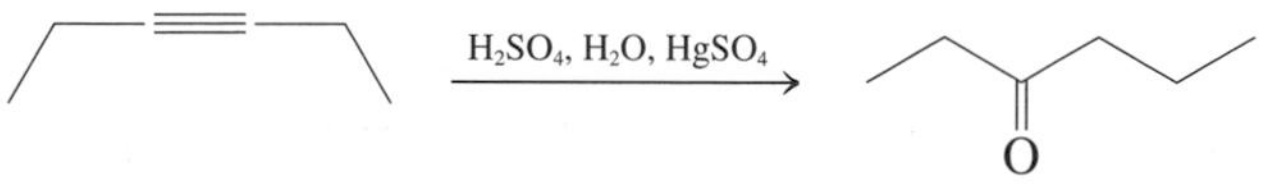
H₂SO₄, H₂O, HgSO₄
O

80%

Only possible product

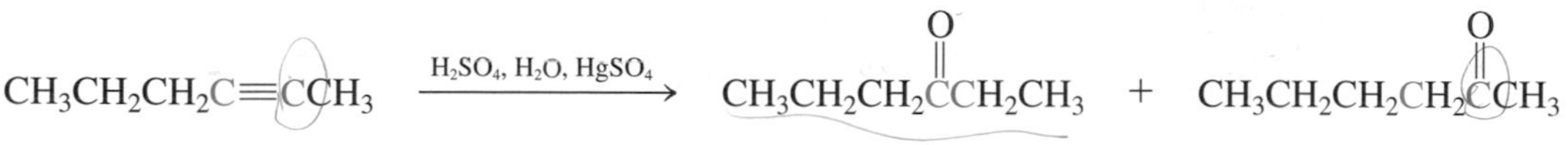
CH₃CH₂CH₂C≡CCH₃
H₂SO₄, H₂O, HgSO₄
O
CH₃CH₂CH₂CCH₂CH₃
+
O
CH₃CH₂CH₂CH₂CCH₃

EXERCISE 13-14

Give the products of mercuric ion–catalyzed hydration of **(a)** ethyne; **(b)** propyne; **(c)** 1-butyne; **(d)** 2-butyne; **(e)** 2-methyl-3-hexyne.

EXERCISE 13-15

Propose a synthetic scheme that will convert compound A into B (see margin). [**Hint:** Consider a route that proceeds through the alkynyl alcohol $(CH_3)_2C(OH)C{\equiv}CH$.]

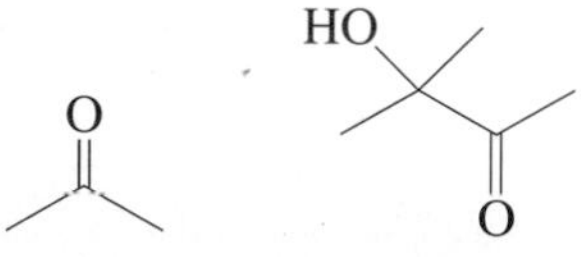
O
HO
O

A **B**

To summarize, alkynes can react with electrophiles such as hydrogen halides and halogens either once or twice. Terminal alkynes transform in accord with the Markovnikov rule. Mercuric ion–catalyzed hydration furnishes enols, which convert into ketones by a process called tautomerism.

13-9 Anti-Markovnikov Additions to Triple Bonds

This section describes two methods by which addition to terminal alkynes can be carried out in an anti-Markovnikov manner.

Radical addition of HBr gives 1-bromoalkenes

As with alkenes, hydrogen bromide can add to triple bonds by a radical mechanism in an anti-Markovnikov fashion if light or other radical initiators are present. Both *syn* and *anti* additions are observed.

$$CH_3(CH_2)_3C\equiv CH \xrightarrow{\text{HBr, ROOR}} CH_3(CH_2)_3CH=CHBr$$
$$74\%$$

1-Hexyne   *cis-* and *trans-***1-Bromo-1-hexene**

Aldehydes result from hydroboration–oxidation of terminal alkynes

Terminal alkynes are hydroborated in a regioselective, anti-Markovnikov fashion, the boron attacking the less hindered carbon. However, with borane itself, this reaction leads to hydroboration of both π bonds. To stop at the alkenylborane stage, less reactive bulky borane reagents, such as dicyclohexylborane, are used.

Hydroboration of a Terminal Alkyne

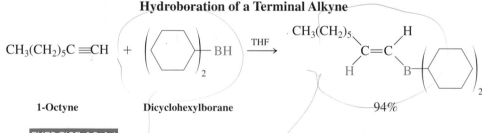

1-Octyne   **Dicyclohexylborane**   94%

Dicyclohexylborane is made by a hydroboration reaction. What are the starting materials for its preparation?

Like alkylboranes (Section 12-8), alkenylboranes can be oxidized to the corresponding alcohols—in this case, to terminal enols that spontaneously rearrange to aldehydes.

$$CH_3(CH_2)_5C\equiv CH \xrightarrow[\text{2. H}_2\text{O}_2, \text{HO}^-]{\text{1. Dicyclohexylborane}} \left[\begin{array}{c} CH_3(CH_2)_5 \quad\quad H \\ C=C \\ H \quad\quad OH \end{array} \right] \xrightarrow{\text{Tautomerism}} CH_3(CH_2)_5\overset{\displaystyle H}{\underset{\displaystyle H}{C}}-\overset{\displaystyle O}{C}H$$

$$70\%$$

Enol   **Octanal**

Give the products of hydroboration–oxidation of (**a**) ethyne; (**b**) 1-propyne; (**c**) 1-butyne.

Outline a synthesis of the following molecule from 3,3-dimethyl-1-butyne.

$$(CH_3)_3CCH_2\overset{\overset{\displaystyle O}{\|}}{C}H$$

In summary, HBr in the presence of peroxides undergoes anti-Markovnikov addition to terminal alkynes to give 1-bromoalkenes. Hydroboration–oxidation with sterically hindered boranes furnishes intermediate enols that tautomerize to the final product aldehydes.

13-10 Chemistry of Alkenyl Halides and Cuprate Reagents

We have encountered haloalkenes as intermediates in both the preparation of alkynes by double dehydrohalogenation and the addition of hydrogen halides to triple bonds. This section presents some chemistry of these systems and introduces a new type of organometallic reagent for carbon–carbon bond formation, the organocuprates.

Alkenyl halides do not undergo S$_N$2 or S$_N$1 reactions

Unlike haloalkanes, alkenyl halides are relatively unreactive toward nucleophiles. Although we have seen that, with strong bases, alkenyl halides undergo elimination reactions to give alkynes, they do not react with weak bases and relatively nonbasic nucleophiles, such as iodide. Similarly, S$_N$1 reactions do not normally take place, because the intermediate alkenyl cations are species of high energy.

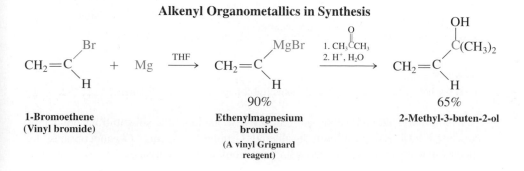

Does not take place

Ethenyl (vinyl) cation

Does not take place

Alkenyl halides, however, can react through the intermediate formation of alkenyl organometallics (see Exercise 11-6). These species allow access to a variety of specifically substituted alkenes.

Alkenyl Organometallics in Synthesis

1-Bromoethene (Vinyl bromide) + Mg $\xrightarrow{\text{THF}}$ **Ethenylmagnesium bromide (A vinyl Grignard reagent)** 90% $\xrightarrow[\text{2. H}^+, \text{H}_2\text{O}]{\text{1. CH}_3\text{CCH}_3}$ **2-Methyl-3-buten-2-ol** 65%

Organocuprates form carbon–carbon bonds

Many important molecules contain double bonds that bear simple alkyl groups (Section 12-9). In principle, these compounds may be derived by the direct alkylation of organometallic reagents. Unfortunately, however, RLi and RMgX molecules do not generally react with haloalkanes (Section 8-8; except when R is an alkynyl group; Section 13-6). Fortunately, another type of organometallic reagent, an organocuprate, is very effective. **Organocuprates** have the empirical formula R_2CuLi, and they may be prepared directly by adding two equivalents of an organolithium reagent to one of cuprous iodide, CuI.

Cuprate Synthesis

$$2\,RLi \;+\; CuI \;\longrightarrow\; \underset{\textbf{An organocuprate}}{R_2CuLi} \;+\; LiI$$

For example,

$$\underset{\textbf{Ethenyllithium}}{2\,CH_2{=}CHLi} \;+\; CuI \;\longrightarrow\; \underset{\substack{\textbf{Lithium diethenylcuprate}\\ \textbf{(An organocuprate reagent)}}}{(CH_2{=}CH)_2CuLi} \;+\; LiI$$

Lithium alkenylcuprates undergo so-called **coupling reactions** with haloalkanes to give substituted alkene products. Primary iodides are best, although other primary and secondary haloalkanes also give satisfactory results.

Reaction of Organocuprates with Haloalkanes

$$R_2CuLi \;+\; R'X \;\longrightarrow\; R{-}R' \;+\; LiX$$

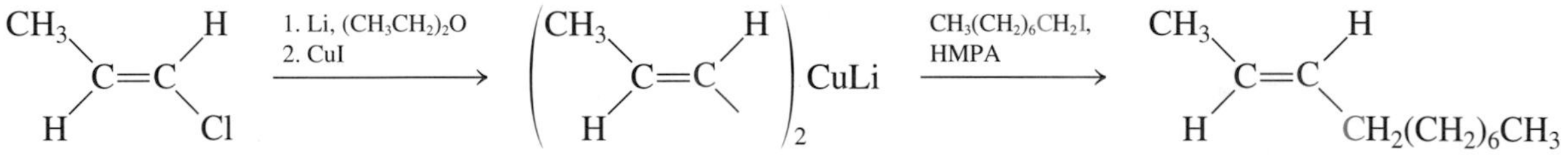

90%

Lithium bis(*trans*-1-propenyl)cuprate

***trans*-2-Undecene**

Such couplings are also accomplished by using alkylcuprate species.

$$CH_3I \xrightarrow[\text{2. CuI}]{\text{1. Li, }(CH_3CH_2)_2O} \underset{\substack{\textbf{Lithium}\\ \textbf{dimethylcuprate}}}{(CH_3)_2CuLi} \xrightarrow[(CH_3CH_2)_2O,\ 0^\circ C]{CH_3(CH_2)_8CH_2I,} \underset{\substack{90\%\\ \textbf{Undecane}}}{CH_3(CH_2)_8CH_2CH_3}$$

EXERCISE 13-19

Show how you would prepare (*Z*)-3-methyl-2-heptene from *cis*-2-butene.

In summary, alkenyl halides are unreactive in nucleophilic substitutions but can be converted into alkenyllithium or alkenyl Grignard reagents. Organocuprates are species that undergo efficient coupling reactions with haloalkanes.

13-11 Ethyne as an Industrial Starting Material

Ethyne was once one of the four or five major starting materials in the chemical industry for two reasons: Addition reactions to one of the π bonds produce useful alkene monomers (Section 12-14), and it has a high heat content. Its industrial use has declined because of the availability of cheap ethene, propene, butadiene, and other hydrocarbons through oil-based technology. However, in the twenty-first century, oil reserves are expected to dwindle to the point that other sources of energy will have to be developed. One such source is coal. There are currently no known processes for converting coal directly into the aforementioned alkenes; ethyne, however, can be produced from coal and hydrogen or from coke (a coal residue obtained after removal of volatiles) and limestone through the formation of calcium carbide. Consequently, it may once again become an important industrial raw material.

Production of ethyne from coal requires high temperatures

The high energy content of ethyne requires the use of production methods that are costly in energy. One process for making ethyne from coal uses hydrogen in an arc reactor at temperatures as high as several thousand degrees Celsius.

$$ \text{Coal} \ + \ \text{H}_2 \ \xrightarrow{\Delta} \ \text{HC}\equiv\text{CH} \ + \ \text{nonvolatile salts} $$
33% conversion

The oldest large-scale preparation of ethyne proceeds through calcium carbide. Limestone (calcium oxide) and coke are heated to about 2000°C, which results in the desired product and carbon monoxide.

$$ \underset{\text{Coke}}{3\,\text{C}} \ + \ \underset{\text{Lime}}{\text{CaO}} \ \xrightarrow{2000°\text{C}} \ \underset{\text{Calcium carbide}}{\text{CaC}_2} \ + \ \text{CO} $$

The calcium carbide is then treated with water at ambient temperatures to give ethyne and calcium hydroxide.

$$ \text{CaC}_2 \ + \ 2\,\text{H}_2\text{O} \ \longrightarrow \ \text{HC}\equiv\text{CH} \ + \ \text{Ca(OH)}_2 $$

Vivid demonstration of the combustion of ethyne, generated by the addition of water to calcium carbide.

Ethyne is a source of valuable monomers for industry

Ethyne chemistry underwent important commercial development in the 1930s and 1940s in the laboratories of Badische Anilin and Sodafabriken (BASF) in Ludwigshafen, Germany. Ethyne under pressure was brought into reaction with carbon monoxide, carbonyl compounds, alcohols, and acids in the presence of catalysts to give a multitude of valuable raw materials to be used in further transformations. For example, nickel carbonyl catalyzes the addition of carbon monoxide and water to ethyne to give propenoic (acrylic) acid. Similar exposure to alcohols or amines instead of water results in the corresponding acid derivatives. All of the products are valuable monomers (see Section 12-14).

Industrial Chemistry of Ethyne

$$HC \equiv CH \ + \ CO \ + \ H_2O \ \xrightarrow{\text{Ni(CO)}_4, \ 100 \ \text{atm}, \ >250°C} \ \begin{array}{c} H \\ \diagdown \\ \diagup \\ H \end{array} C = CHCOOH$$

Propenoic acid
(Acrylic acid)

$$HC \equiv CH \ + \ CO \ + \ CH_3OH \ \xrightarrow{\text{Ni(CO)}_4, \ \Delta} \ \begin{array}{c} H \\ \diagdown \\ \diagup \\ H \end{array} C = CH\overset{\overset{\displaystyle O}{\|}}{C}OCH_3$$

80%
Methyl propenoate
(Methyl acrylate)

Polymerization of propenoic (acrylic) acid and its derivatives produces materials of considerable utility. The polymeric esters (**polyacrylates**) are tough, resilient, and flexible polymers that have replaced natural rubber (see Chapter 14) in many applications. Poly(ethyl acrylate) is used for O-rings, valve seals, and related purposes in automobiles. Other polyacrylates are found in biomedical and dental appliances, such as dentures.

The addition of formaldehyde to ethyne is achieved with high efficiency by using copper acetylide as a catalyst.

$$HC \equiv CH \ + \ CH_2 = O \ \xrightarrow{\text{Cu}_2\text{C}_2–\text{SiO}_2, \ 125°C, \ 5 \ \text{atm}} \ HC \equiv CCH_2OH \ \text{ or } \ HOCH_2C \equiv CCH_2OH$$

2-Propyn-1-ol **2-Butyne-1,4-diol**
(Propargyl alcohol)

The resulting alcohols are useful synthetic intermediates. For example, 2-butyne-1,4-diol is a precursor for the production of oxacyclopentane (tetrahydrofuran, one of the solvents most frequently employed for Grignard and organolithium reagents) by hydrogenation, followed by acid-catalyzed dehydration.

Oxacyclopentane (Tetrahydrofuran) Synthesis

$$HOCH_2C \equiv CCH_2OH \ \xrightarrow{\text{Catalyst, } H_2} \ HO(CH_2)_4OH \ \xrightarrow[-H_2O]{\substack{\text{H}_3\text{PO}_4, \ \text{pH 2,} \\ 260°–280°C, \ 90–100 \ \text{atm}}} \ \square$$

99%
Oxacyclopentane
(Tetrahydrofuran, THF)

A number of technical processes have been developed in which reagents $^{\delta+}A–B^{\delta-}$ in the presence of a catalyst add to the triple bond. For example, the catalyzed addition of hydrogen chloride gives chloroethene (vinyl chloride), and addition of hydrogen cyanide produces propenenitrile (acrylonitrile). Nearly 1.7 million tons of acrylonitrile were made in the United States in 1996.

Addition Reactions of Ethyne

$$HC\equiv CH \quad + \quad HCl \quad \xrightarrow{\text{Hg}^{2+},\ 100°-200°C} \quad \underset{H}{\overset{H}{>}}C=CHCl$$

Chloroethene
(Vinyl chloride)

$$HC\equiv CH \quad + \quad HCN \quad \xrightarrow{\text{Cu}^+,\ \text{NH}_4\text{Cl},\ 70°-90°C,\ 1.3\ \text{atm}} \quad \underset{H}{\overset{H}{>}}C=CHCN$$

80–90%

Propenenitrile
(Acrylonitrile)

Poly(vinyl chloride) is extensively used in the construction industry for water and sewer pipes.

Each year, the United States produces more than 200,000 tons of **acrylic fibers,** polymers containing at least 85% propenenitrile (acrylonitrile). Their applications include clothing (Orlon), carpets, and insulation. Copolymers of acrylonitrile and 10–15% vinyl chloride have fire-retardant properties and are used in children's sleepwear.

In summary, ethyne was once, and may again be in the future, a valuable industrial feedstock because of its ability to react with a large number of substrates to yield useful monomers and other compounds having functional groups. It can be made from coal and H_2 at high temperatures or it can be liberated from calcium carbide by hydrolysis. Some of the industrial reactions that it undergoes are carbonylation, addition of formaldehyde, and addition reactions with HX.

13-12 Naturally Occurring and Physiologically Active Alkynes

Although alkynes are not very abundant in nature, they do exist in some plants and other organisms. The first such substance to be isolated, in 1826, was dehydromatricaria ester, from the chamomile flower. More than a thousand such compounds are now known, and some of them are physiologically active. For example, some naturally occurring ethynylketones, such as capillin, an oil found in the chrysanthemum, have fungicidal activity.

$$CH_3C\equiv C-C\equiv C-C\equiv C\underset{H}{\overset{}{>}}C=C\underset{COCH_3}{\overset{H}{<}}$$

Dehydromatricaria ester

$$CH_3C\equiv C-C\equiv C-\overset{O}{\overset{\|}{C}}-\bigcirc$$

Capillin
(Active against skin fungi)

The alkyne ichthyothereol is the active ingredient of a poisonous substance used by the Indians of the Lower Amazon River Basin in arrowheads. It causes convulsions in mammals. Two enyne functional groups are incorporated in the compound hystrionicotoxin. It is one of the substances isolated from the skin of "poison arrow frog," a highly colorful species of the genus *Dendrobates.* The frog secretes this compound and similar ones as defensive venoms and mucosal-tissue irritants against both mammals and reptiles. How the alkyne units are constructed biosynthetically is not clear.

"Poison arrow" frog.

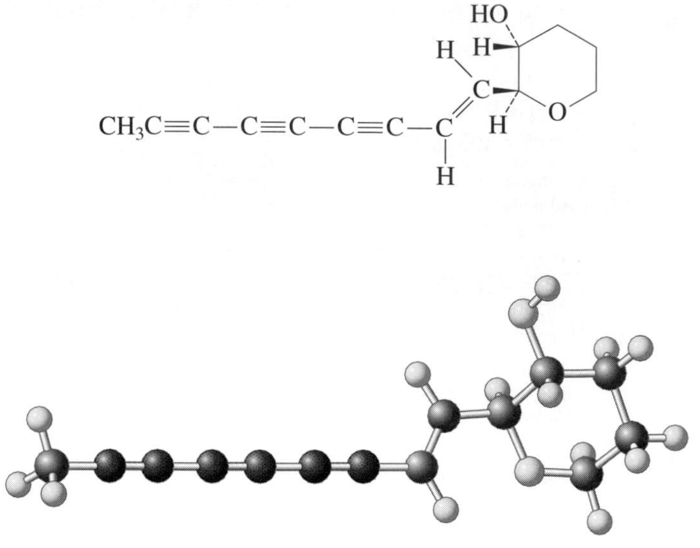

CH₃C≡C−C≡C−C≡C−C

Ichthyothereol
(A convulsant)

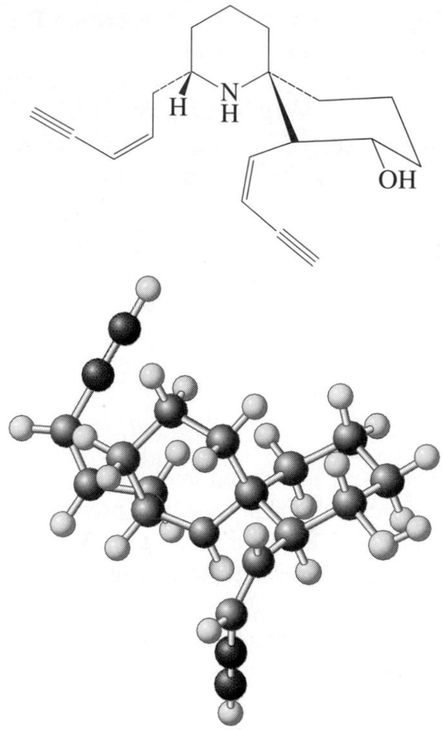

Hystrionicotoxin

Many drugs have been modified by synthesis to contain alkyne substituents, because such compounds are frequently more readily absorbed by the body, less toxic, and more active than the corresponding alkenes or alkanes. For example, 3-methyl-1-pentyn-3-ol is available as a nonprescription hypnotic, and several other alkynols are similarly effective.

Highly reactive enediyne (–C≡C–CH=CH–C≡C–) and trisulfide (RSSSR) functional groups characterize a class of naturally occurring antibiotic-antitumor agents discovered in the late 1980s, such as calicheamicin and esperamicin.

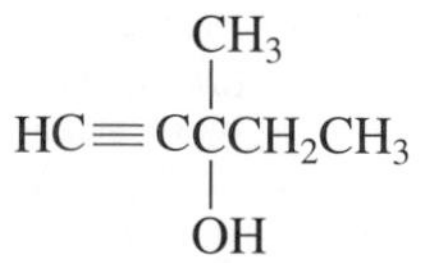

$$HC \equiv C\underset{\underset{OH}{|}}{\overset{\overset{CH_3}{|}}{C}}CH_2CH_3$$

3-Methyl-1-pentyn-3-ol
(Hypnotic)

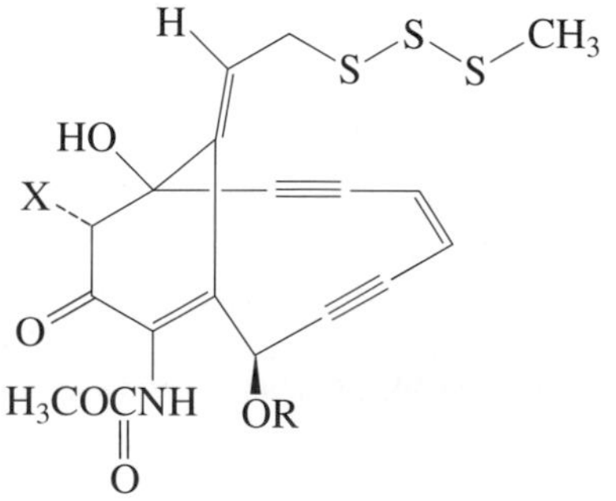

Calicheamicin (X = H)
Esperamicin (X = OR′)

R and R′ = sugars (Chapter 24)

Ethynyl estrogens, such as 17-ethynylestradiol, are considerably more potent birth control agents than are the naturally occurring hormones (see Section 4-7). The diaminoalkyne tremorine induces symptoms characteristic of Parkinson's disease:

spasms of uncontrolled movement. Interestingly, a simple cyclic homolog of tremorine acts as a muscle relaxant and counteracts the effect of tremorine. Compounds that cancel the physiological effects of other compounds are called **antagonists** (*antagonizesthai,* Greek, to struggle against). Finally, ethynyl analogs of amphetamine have been prepared in a search for alternative, more active, more specific, and less addictive central nervous system stimulants.

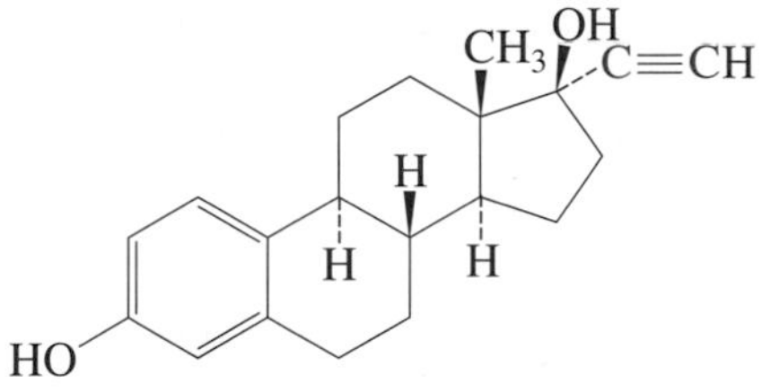

17-Ethynylestradiol

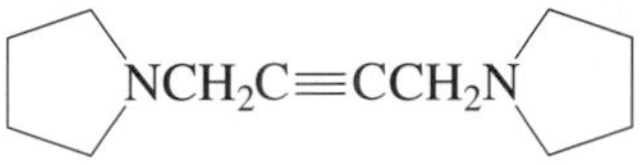

Tremorine

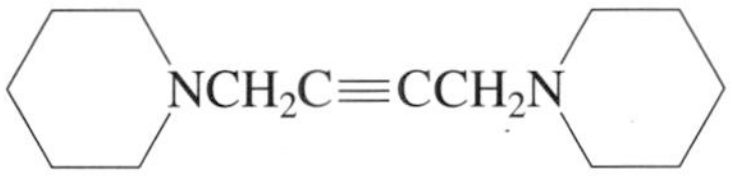

Tremorine antagonist

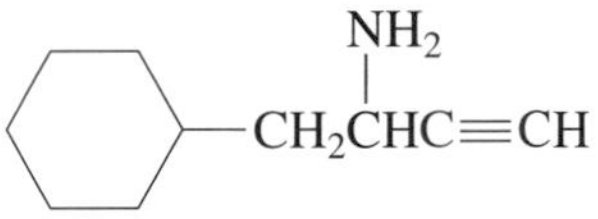

An amphetamine analog
(**Active in the central nervous system**)

In summary, the alkyne unit is present in a number of physiologically active natural and synthetic compounds.

CHAPTER INTEGRATION PROBLEM

Propose an efficient synthesis of 2,7-dimethyl-4-octanone, using organic building blocks containing no more than four carbon atoms.

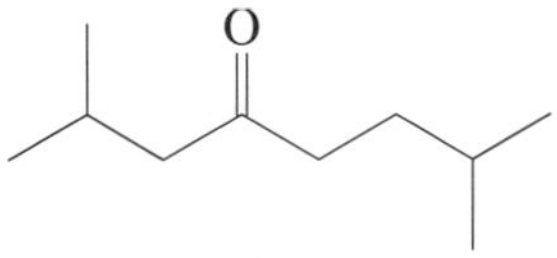

2,7-Dimethyl-4-octanone

SOLUTION

We begin by analyzing the problem retrosynthetically (Section 8-9). What methods do we know for the preparation of ketones? We can oxidize alcohols (Section 8-6). Is this a productive line of analysis? If we look at the corresponding alcohol precursor, we can imagine synthesizing it by addition of an appropriate organometallic reagent to an aldehyde to form either bond *a* or bond *b* (Section 8-9):

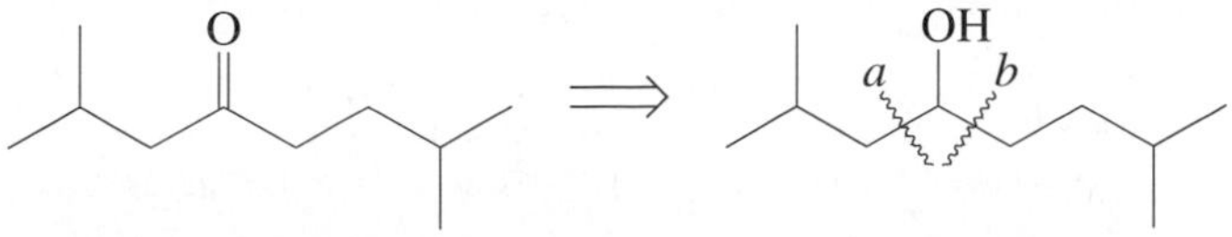

Target molecule   **Alcohol precursor**

Let us count carbon atoms in the fragments necessary for each of these synthetic pathways. To make bond *a*, we need to add a four-carbon organometallic to a six-carbon aldehyde. The bond *b* alternative would employ 2 five-carbon building blocks. Remember the restriction that only four-carbon starting materials are allowed. From this point of view, neither of the preceding options is overly attractive. We will shortly look again at route *a* but not route *b*. Do you see why? The latter would require initial construction of 2 five-carbon pieces, whereas the former needs formation of only 1 six-carbon unit from fragments containing four carbons or fewer.

Now let us consider a second, fundamentally different ketone synthesis—hydration of an alkyne (Section 13-8). Either of two precursors, 2,7-dimethyl-3-octyne and 2,7-dimethyl-4-octyne, will lead to the target molecule. As shown here, however, only the latter, *symmetrical* alkyne undergoes hydration to give just one ketone, regardless of the initial direction of addition.

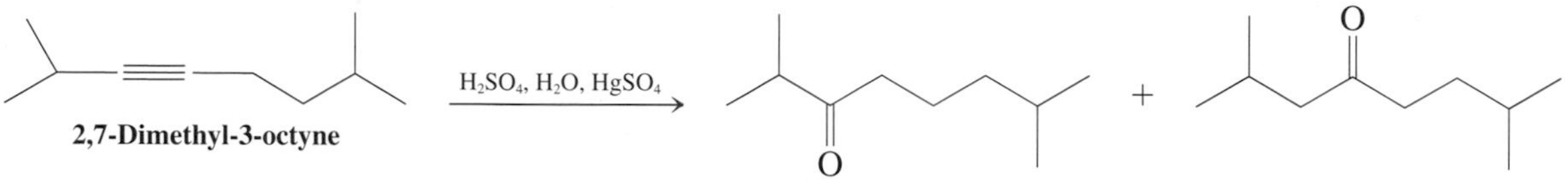

2,7-Dimethyl-3-octyne $\xrightarrow{H_2SO_4,\ H_2O,\ HgSO_4}$

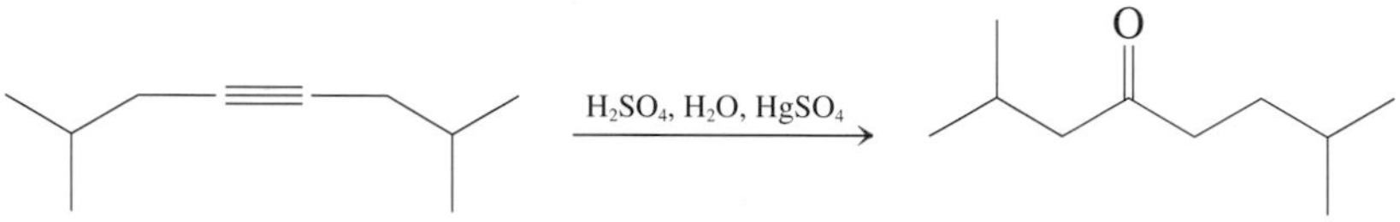

2,7-Dimethyl-4-octyne $\xrightarrow{H_2SO_4,\ H_2O,\ HgSO_4}$

With 2,7-dimethyl-4-octyne appearing to be the precursor of choice, we continue by investigating its synthesis from building blocks of four carbons or fewer. The alkylation of terminal alkynes (Section 13-6) affords us a method of bond formation that divides the molecule into three suitable fragments, shown in the following analysis:

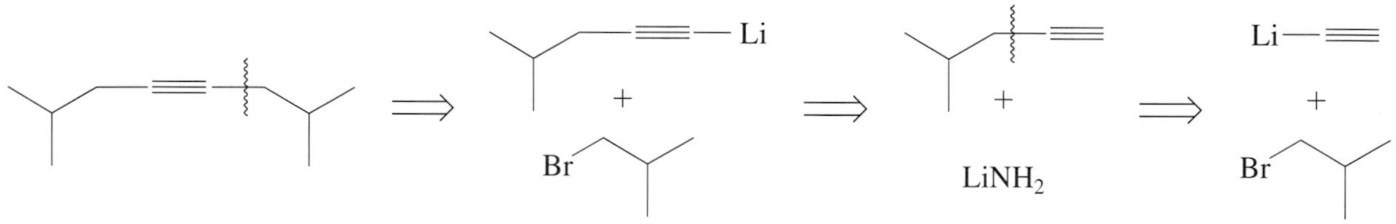

The synthesis follows directly:

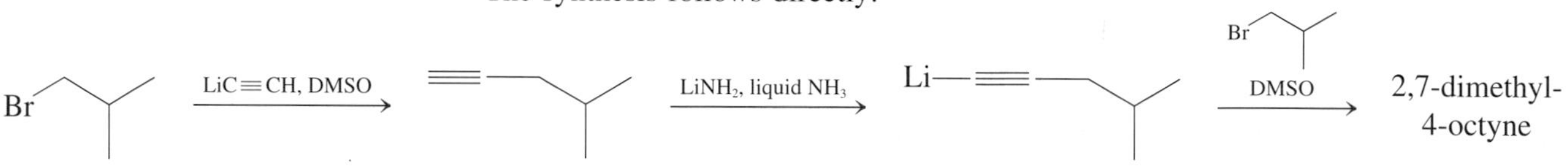

$\xrightarrow{LiC\equiv CH,\ DMSO}$ $\xrightarrow{LiNH_2,\ liquid\ NH_3}$ $\xrightarrow{DMSO}$ 2,7-dimethyl-4-octyne

Although this three-step synthesis is the most efficient answer, a related approach derives from our earlier consideration of ketone synthesis with the use of an alcohol. It, too, proceeds through an alkyne. Construction of bond *a* of the target molecule, shown earlier, requires addition of an organometallic reagent to a six-carbon aldehyde, which, in turn, may be produced through hydroboration–oxidation (Section 13-9) of the terminal alkyne shown in the preceding scheme.

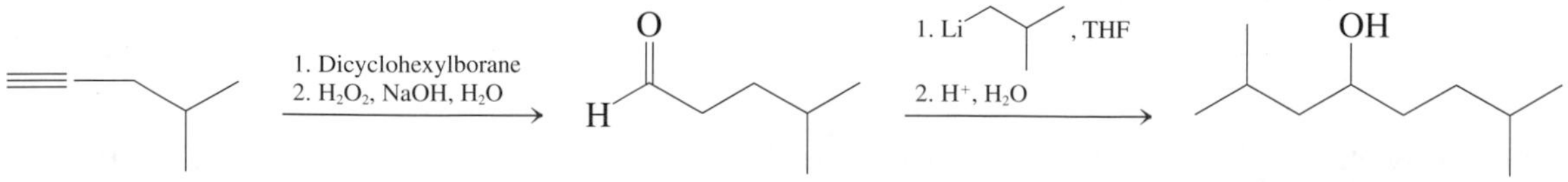

$\xrightarrow[\text{2. } H_2O_2,\ NaOH,\ H_2O]{\text{1. Dicyclohexylborane}}$ $\xrightarrow[\text{2. } H^+,\ H_2O]{\text{1. Li} \quad , THF}$

Oxidation of this alcohol by using a Cr(VI) reagent (Section 8-6) completes a synthesis that is just slightly longer than the optimal one described first.

NEW REACTIONS

1. Acidity of l-Alkynes (Section 13-2)

$$RC{\equiv}CH \quad + \quad :B^- \quad \rightleftharpoons \quad RC{\equiv}C:^- \quad + \quad BH$$

pK_a ~ 25

Base (B): $NaNH_2$–liquid NH_3; RLi–$(CH_3CH_2)_2O$; RMgX–THF

Preparation of Alkynes

2. Double Elimination from Dihaloalkanes (Section 13-5)

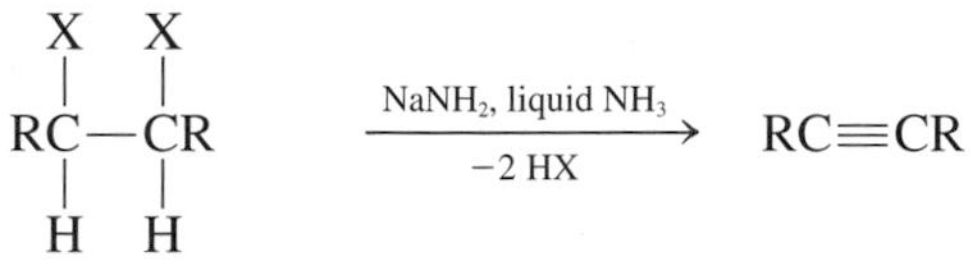

Vicinal dihaloalkane

3. From Alkenes by Halogenation-Dehydrohalogenation (Section 13-5)

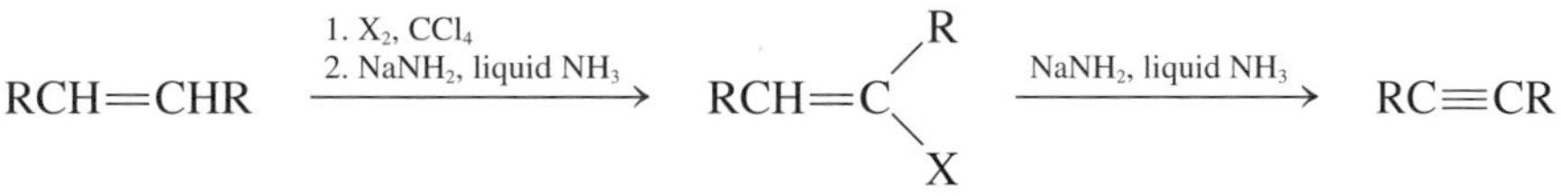

Alkenyl halide intermediate

Conversion of Alkynes into Other Alkynes

4. Alkylation of Alkynyl Anions (Section 13-6)

$$RC{\equiv}CH \xrightarrow[]{\text{1. } NaNH_2\text{, liquid } NH_3 \quad \text{2. } R'X} RC{\equiv}CR'$$

S_N2 reaction: R′ must be primary

5. Alkylation with Oxacyclopropane (Section 13-6)

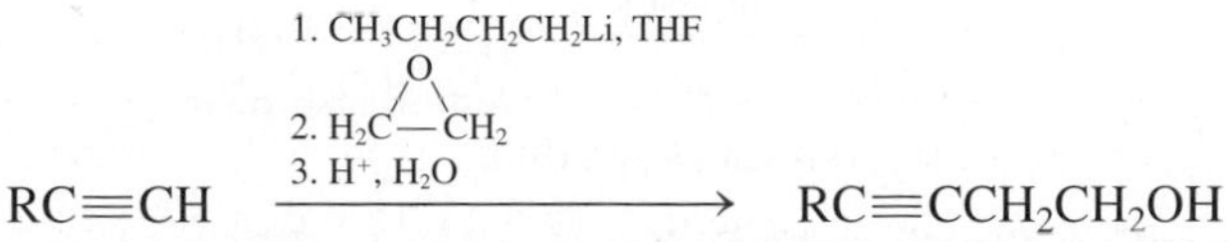

Attack takes place at less substituted carbon in unsymmetric oxacyclopropanes

6. Alkylation with Carbonyl Compounds (Section 13-6)

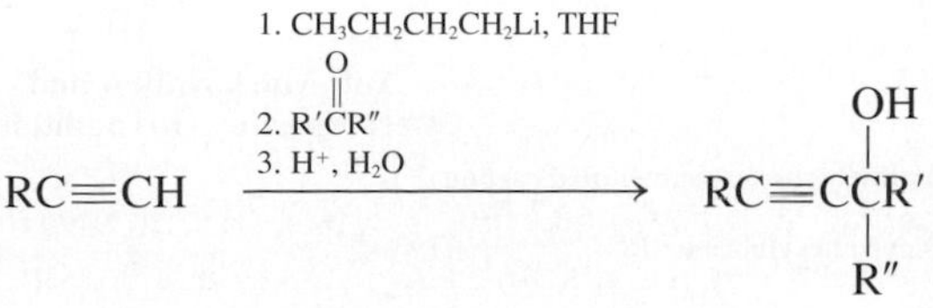

Reactions of Alkynes

7. Hydrogenation (Section 13-7)

$$RC\equiv CR \xrightarrow{\text{Catalyst, H}_2} RCH_2CH_2R \qquad \Delta H° \sim -70 \text{ kcal mol}^{-1}$$

Catalysts: Pt, Pd–C

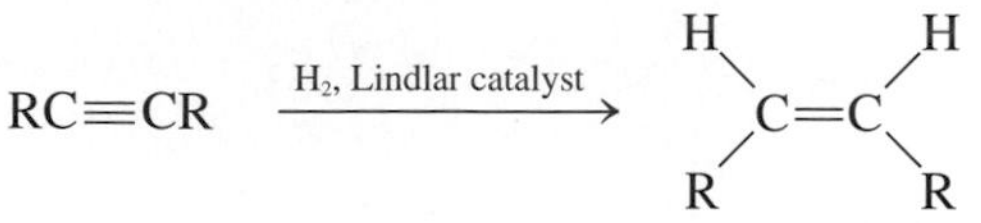

$$RC\equiv CR \xrightarrow{\text{H}_2,\ \text{Lindlar catalyst}} \underset{\text{Cis alkene}}{\overset{\displaystyle HH}{\underset{\displaystyle RR}{C=C}}} \qquad \Delta H° \sim -40 \text{ kcal mol}^{-1}$$

Cis alkene

8. Reduction with Sodium in Liquid Ammonia (Section 13-7)

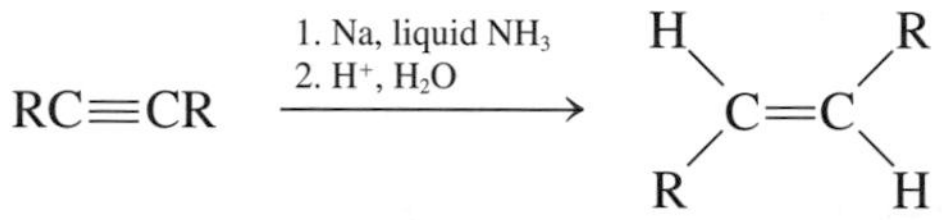

$$RC\equiv CR \xrightarrow[\text{2. H}^+,\ \text{H}_2\text{O}]{\text{1. Na, liquid NH}_3} \overset{\displaystyle HR}{\underset{\displaystyle RH}{C=C}}$$

Trans alkene

9. Electrophilic (and Markovnikov) Additions: Hydrohalogenation, Halogenation, and Hydration (Section 13-8)

$$RC\equiv CR \xrightarrow{\text{HX}} RCH=CXR \xrightarrow{\text{HX}} RCH_2CX_2R$$

Geminal dihaloalkane

$$RC\equiv CH \xrightarrow{\text{2 HX}} RCX_2CH_3$$

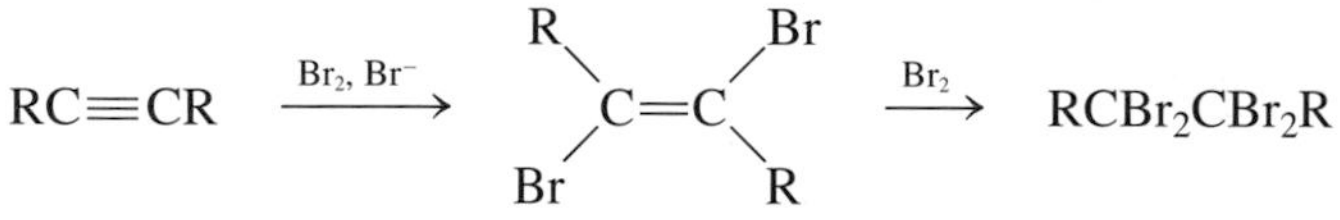

$$RC\equiv CR \xrightarrow{\text{Br}_2,\ \text{Br}^-} \overset{\displaystyle RBr}{\underset{\displaystyle BrR}{C=C}} \xrightarrow{\text{Br}_2} RCBr_2CBr_2R$$

Mainly trans

$$RC\equiv CR \xrightarrow{\text{Hg}^{2+},\ \text{H}_2\text{O}} RCH_2\overset{\displaystyle O}{\overset{\displaystyle \|}{C}}R$$

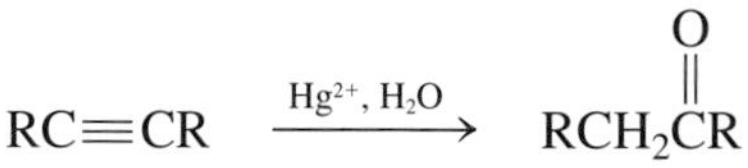

10. Radical Addition of Hydrogen Bromide (Section 13-9)

$$RC\equiv CH \xrightarrow{\text{HBr, ROOR}} RCH=CHBr$$

Anti-Markovnikov

Br attaches to less substituted carbon

11. Hydroboration (Section 13-9)

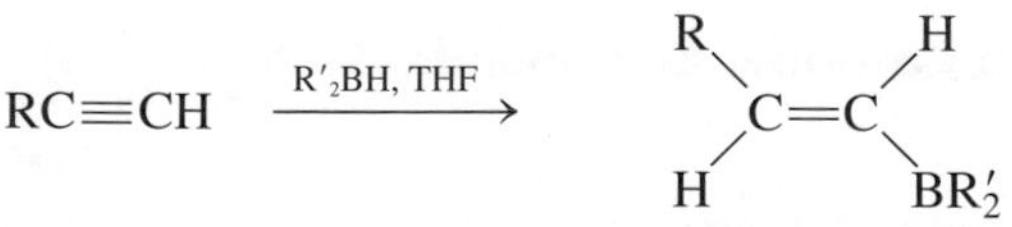

$$RC\equiv CH \xrightarrow{\text{R}'_2\text{BH, THF}} \overset{\displaystyle RH}{\underset{\displaystyle HBR'_2}{C=C}}$$

Anti-Markovnikov and stereospecific (*syn*) addition

B attaches to less substituted carbon

Dicyclohexylborane (R′ = ⬡)

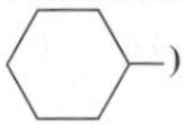

12. Oxidation of Alkenylboranes (Section 13-9)

$$\underset{\substack{R \\ }}{\overset{\substack{H \\ }}{C}}\!\!=\!\!\underset{\substack{H \\ }}{\overset{\substack{B— \\ }}{C}} \xrightarrow{H_2O_2,\ HO^-} \left[\underset{\substack{R \\ }}{\overset{\substack{H \\ }}{C}}\!\!=\!\!\underset{\substack{H \\ }}{\overset{\substack{OH \\ }}{C}}\right] \xrightarrow[\text{Tautomerism}]{} RCH_2\overset{\displaystyle O}{\overset{\|}{C}}H$$

Enol

Organometallic Reagents

13. Alkenyl Organometallics (Section 13-10)

$$\underset{\substack{R' \\ }}{\overset{\substack{R \\ }}{C}}\!\!=\!\!\underset{\substack{R'' \\ }}{\overset{\substack{X \\ }}{C}} \xrightarrow{Mg,\ THF} \underset{\substack{R' \\ }}{\overset{\substack{R \\ }}{C}}\!\!=\!\!\underset{\substack{R'' \\ }}{\overset{\substack{MgX \\ }}{C}}$$

14. Coupling Reactions of Organocuprates (Section 13-10)

$$R\!-\!X \xrightarrow[\substack{2.\ CuI}]{\substack{1.\ Li,\ (CH_3CH_2)_2O}} R_2CuLi \xrightarrow{R'\!-\!X,\ (CH_3CH_2)_2O} R\!-\!R'$$

R = alkyl or alkenyl   **Lithium
diorganocuprate**

15. Industrial Preparation and Uses of Ethyne (Sections 13-4 and 13-11)

Preparation

Directly from coal $+ H_2$, Δ; or from coke $+ CaO \longrightarrow CaC_2 \xrightarrow{H_2O} HC\equiv CH$

Combustion (acetylene torch)

$$HC\equiv CH\ +\ 2.5\ O_2 \longrightarrow 2\ CO_2\ +\ H_2O \qquad \Delta H° = -311\ kcal\ mol^{-1}$$

Industrial chemistry

$$C_2H_2\ +\ CO\ +\ H_2O \xrightarrow{Ni(CO)_4} CH_2\!\!=\!\!CHCO_2H$$

$$C_2H_2\ +\ CH_2O \xrightarrow{Cu_2C_2} HOCH_2C\equiv CH\ +\ HOCH_2C\equiv CCH_2OH$$

Additions

$$C_2H_2\ +\ HX \longrightarrow CH_2\!\!=\!\!CHX$$
$$C_2H_2\ +\ HCN \longrightarrow CH_2\!\!=\!\!CHCN$$
Catalyzed by transition metals

IMPORTANT CONCEPTS

1. The rules for **naming alkynes** are essentially the same as those formulated for alkenes. Molecules with both double and triple bonds are called **alkenynes,** the double bond receiving the lower number if both are at equivalent positions. Hydroxy groups are given precedence in numbering alkynyl alcohols (**alkynols).**

2. The **electronic structure** of the triple bond reveals two π bonds, perpendicular to each other, and a σ bond, formed by two overlapping sp hybrid orbitals.

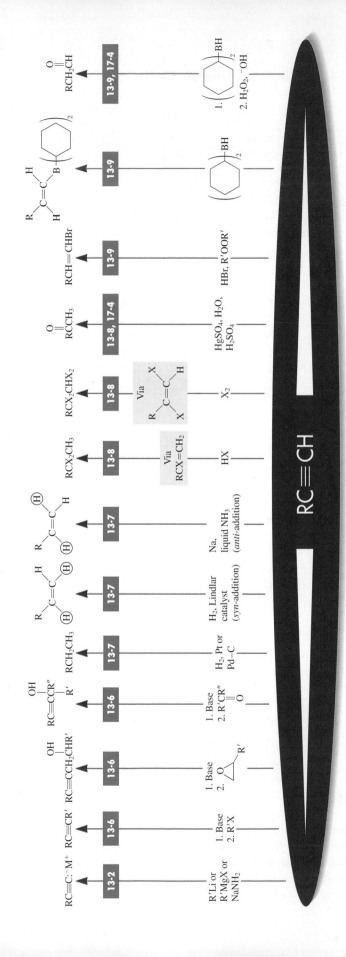

Reactions of Alkynes | section number

$RC \equiv CH$

13-2 → $RC \equiv C:^- M^+$

$R'Li$ or $R'MgX$ or $NaNH_2$

13-6 → $RC \equiv CR'$

1. Base
2. R'X

13-6 → $RC \equiv CCH_2CHR'$ | OH

1. Base
2. O (epoxide with R')

13-6 → $RC \equiv CCR''$ | OH, R'

1. Base
2. R'CR'' (O)

13-7 → RCH_2CH_3

H_2, Pt or Pd–C

13-7 → R—C=C—(H), H, R (H) (syn)

H_2, Lindlar catalyst (*syn*-addition)

13-7 → R—C=C, R, (H) H (H) (anti)

Na, liquid NH_3 (*anti*-addition)

13-8 → RCX_2CH_3

Via $RCX=CH_2$

HX

13-8 → RCX_2CHX_2

Via R—C=C—X, X, H, X

X_2

13-8, 17-4 → $RCCH_3$ | O

$HgSO_4$, H_2O, H_2SO_4

13-9 → $RCH=CHBr$

HBr, R'OOR'

13-9 → R—C=C—B(cyclohexyl)$_2$, H, H

BH (dicyclohexylborane)$_2$

13-9, 17-4 → RCH_2CH | O

1. BH (dicyclohexylborane)$_2$
2. H_2O_2, ^-OH

The strength of the triple bond is about 229 kcal mol^{-1}; that of the alkynyl C–H bond is 131 kcal mol^{-1}. Triple bonds form **linear structures** with respect to other attached atoms, with short C–C (1.20 Å) and C–H (1.06 Å) bonds.

3. The high *s* character of the terminal alkyne carbon makes the bound hydrogen relatively **acidic** (p$K_a \sim$ 25).

4. The **chemical shift** of the alkynyl hydrogen is low (δ = 1.7–3.1 ppm) compared with that of alkenyl hydrogens because of the shielding effect of an induced electron current around the molecular axis caused by the external magnetic field. The triple bond allows for long-range coupling. **IR spectroscopy** indicates the presence of the C≡C and ≡C–H bonds of terminal alkynes through bands at 2100–2260 cm^{-1} and 3260–3330 cm^{-1}, respectively.

5. Internal alkynes are more stable than the isomeric terminal alkynes by about 4–5 kcal mol^{-1}.

6. The **elimination** reaction with vicinal dihaloalkanes proceeds regioselectively and stereospecifically to give alkenyl halides.

7. Selective *syn* **dihydrogenation** of alkynes is possible with Lindlar catalyst, the surface of which is less active than palladium on carbon and therefore not capable of hydrogenating alkenes. Selective *anti* **hydrogenation** is possible with sodium metal dissolved in liquid ammonia because simple alkenes cannot be reduced by one-electron transfer. The stereochemistry is set by the greater stability of a trans disubstituted alkenyl radical intermediate.

8. To stop the **hydroboration** of terminal alkynes at the alkenylboron intermediate stage, modified dialkylboranes—particularly dicyclohexylborane—are used. Oxidation of the resulting alkenylboranes produces enols that are unstable with respect to rearrangement to carbonyl compounds **(tautomerism).**

9. **Organocuprate reagents** undergo coupling reactions with haloalkanes.

PROBLEMS

20. Name each of the compounds below, using the IUPAC system of nomenclature.

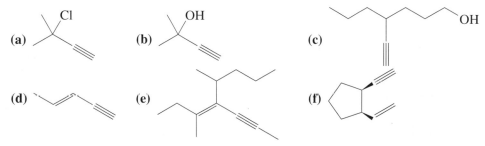

(a) (b) (c)

(d) (e) (f)

21. Compare C–H bond strengths in ethane, ethene, and ethyne. Reconcile these data with hybridization, bond polarity, and acidity of the hydrogen.

22. Compare the C2–C3 bonds in propane, propene, and propyne. Should they be any different with respect to either bond length or bond strength? If so, how should they vary?

23. Predict the order of acid strengths in the following series of cationic species: $CH_3CH_2NH_3{}^+$, $CH_3CH{=}NH_2{}^+$, $CH_3C{\equiv}NH^+$. (**Hint:** Look for an analogy among hydrocarbons [Section 13-2].)

24. The heats of combustion for three compounds with the molecular formula C_5H_8 are as follows: cyclopentene, ΔH_{comb} = -1027 kcal mol^{-1}; 1,4-pentadiene, ΔH_{comb} = -1042 kcal mol^{-1}; and 1-pentyne, ΔH_{comb} = -1052 kcal mol^{-1}. Explain in terms of relative stability and bond strengths.

25. Rank in order of decreasing stability.
(**a**) 1-Heptyne and 3-heptyne

(**b**)

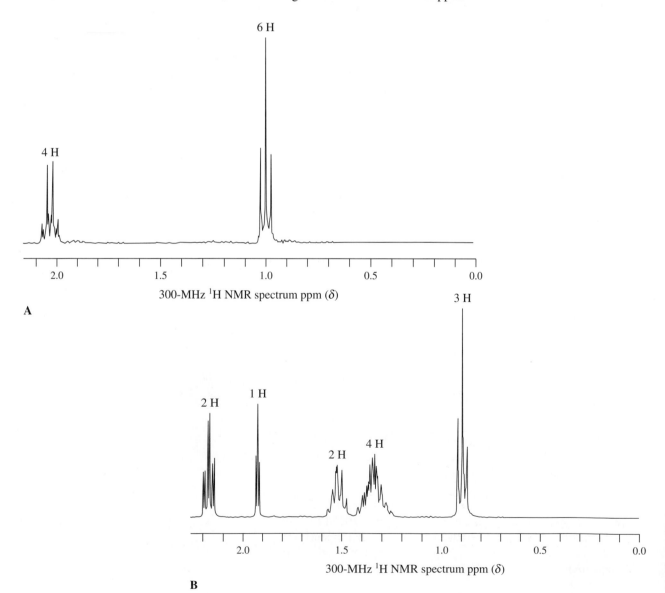

(**Hint:** Make a model of the third structure. Is there anything unusual about its triple bond?)

26. Deduce structures for each of the following. (**a**) Molecular formula C_6H_{10}; NMR spectrum A; no strong IR bands between 2100 and 2300 or 3250 and 3350 cm^{-1}. (**b**) Molecular formula C_7H_{12}; NMR spectrum B; IR bands at about 2120 and 3330 cm^{-1}. (**c**) Molecular formula C_5H_8O; NMR and IR spectra C (see the facing page). The inset in NMR spectrum C provides better resolution of the signals between 1.6 and 2.4 ppm.

6 H

4 H

300-MHz ^1H NMR spectrum ppm (δ)

A

3 H

2 H

1 H

2 H

4 H

300-MHz ^1H NMR spectrum ppm (δ)

B

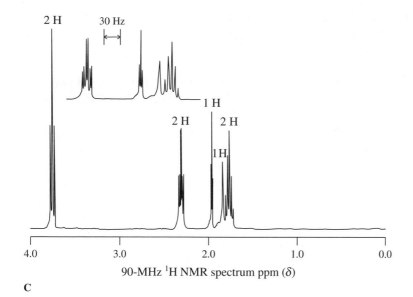

C

90-MHz ^{1}H NMR spectrum ppm (δ)

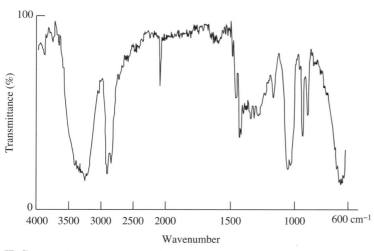

Wavenumber

IR-C

27. The IR spectrum of 1,8-nonadiyne displays a strong, sharp band at 3300 cm^{-1}. What is the origin of this absorption? Treatment of 1,8-nonadiyne with NaNH$_2$, then with D$_2$O, leads to the incorporation of two deuterium atoms, leaving the molecule unchanged otherwise. The IR spectrum reveals that the peak at 3300 cm^{-1} has disappeared, but a new one is present at 2580 cm^{-1}. **(a)** What is the product of this reaction? **(b)** What new bond is responsible for the IR absorption at 2580 cm^{-1}? **(c)** Using Hooke's law, calculate the approximate expected position of this new band from the structure of the original molecule and its IR spectrum. Assume that k and f have not changed.

28. Write the expected product(s) of each of the following reactions.

(a) CH$_3$CH$_2$CHCHCH$_2$Cl $\xrightarrow[\text{liquid NH}_3]{\text{3 NaNH}_2,}$

(with CH$_3$ on the first CH and Cl on the second CH)

(b) CH$_3$OCH$_2$CH$_2$CH$_2$CHCHCH$_3$ $\xrightarrow[\text{liquid NH}_3]{\text{2 NaNH}_2,}$

(with Br Br on the last two CH groups)

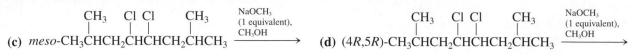

(c) *meso*-CH₃CHCH₂CHCHCH₂CHCH₃ $\xrightarrow{\substack{\text{NaOCH}_3 \\ \text{(1 equivalent),} \\ \text{CH}_3\text{OH}}}$ **(d)** (4R,5R)-CH₃CHCH₂CHCHCH₂CHCH₃ $\xrightarrow{\substack{\text{NaOCH}_3 \\ \text{(1 equivalent),} \\ \text{CH}_3\text{OH}}}$

29. Write the expected major product of reaction of 1-propynyllithium, $CH_3C{\equiv}C^-Li^+$, with each of the following molecules in THF.

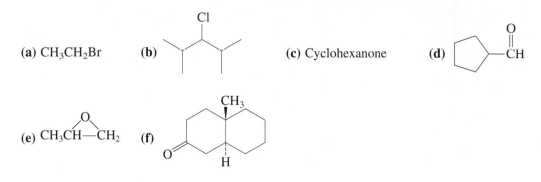

(a) CH_3CH_2Br **(b)**   **(c)** Cyclohexanone   **(d)**

(e) $CH_3CH{-}CH_2$ **(f)**

30. Which of the following methods is best suited as a high-yield synthesis of

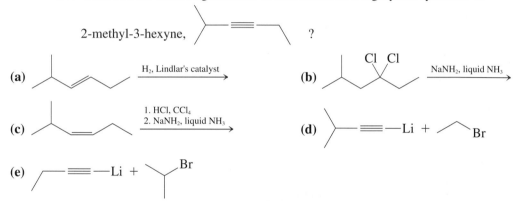

2-methyl-3-hexyne,   ?

(a) $\xrightarrow{\text{H}_2, \text{ Lindlar's catalyst}}$ **(b)** $\xrightarrow{\text{NaNH}_2, \text{ liquid NH}_3}$

(c) $\xrightarrow{\substack{1.\ \text{HCl, CCl}_4 \\ 2.\ \text{NaNH}_2, \text{ liquid NH}_3}}$ **(d)**

(e)

31. Propose reasonable syntheses of each of the following alkynes, using the principles of retrosynthetic analysis. Each alkyne functional group in your synthetic target should come from a *separate* molecule, which may be any two-carbon compound (e.g., ethyne, ethene, ethanal).

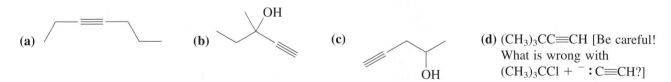

(a)   **(b)**   **(c)**   **(d)** $(CH_3)_3CC{\equiv}CH$ [Be careful! What is wrong with $(CH_3)_3CCl + {}^-{:}C{\equiv}CH?$]

32. Draw the structure of (R)-4-deuterio-2-hexyne. Propose a suitable retro-S_N2 precursor of this compound.

33. Give the expected major product of the reaction of propyne with each of the following reagents. **(a)** D_2, Pd–CaCO₃, Pb(O₂CCH₃)₂, quinoline; **(b)** Na, ND₃; **(c)** 1 equivalent HI; **(d)** 2 equivalents HI; **(e)** 1 equivalent Br₂; **(f)** 1 equivalent ICl; **(g)** 2 equivalents ICl; **(h)** H_2O, HgSO₄, H_2SO_4; **(i)** dicyclohexylborane, then NaOH, H_2O_2.

34. What are the products of the reactions of dicyclohexylethyne with the reagents in Problem 33?

35. Give the products of the reactions of your first two answers to Problem 34 with each of the following reagents. **(a)** H_2, Pd–C, CH_3CH_2OH; **(b)** Br_2, CCl_4; **(c)** BH_3, THF, then NaOH, H_2O_2; **(d)** MCPBA, CH_2Cl_2; **(e)** OsO_4, then H_2S.

36. Propose several syntheses of *cis*-3-heptene, beginning with each of the following molecules. Note in each case whether your proposed route gives the desired compound as a major or minor final product. **(a)** 3-Chloroheptane; **(b)** 4-chloroheptane; **(c)** 3,4-dichloroheptane; **(d)** 3-heptanol; **(e)** 4-heptanol; **(f)** *trans*-3-heptene; **(g)** 3-heptyne.

37. Give the products of each reaction or reaction sequence.

(a) $(CH_3)_2CHBr + 2\,Li \xrightarrow{(CH_3CH_2)_2O}$

(b) Product of (a) (2 equivalents) + CuI $\longrightarrow$

(c) Product of (b) + $CH_3\overset{\overset{\displaystyle CH_3}{|}}{C}HCH_2CH_2Br \longrightarrow$

(d) $CH_3CH_2CH_2C{\equiv}CCH_2CH_2CH_3 + HBr$ (1 equivalent) + Br^- (excess) $\longrightarrow$

(e) Product of (d) $\xrightarrow[\text{2. CuI (0.5 equivalents)}]{\text{1. Li, }(CH_3CH_2)_2O}$

(f) Product of (e) + $H_2C{=}CHCH_2Cl \longrightarrow$

38. Propose reasonable syntheses of each of the following molecules, using an alkyne at least once in each synthesis.

(a) Br Cl

(b) I I

(c) *meso*-2,3-Dibromobutane

(d) Racemic mixture of (2*R*,3*R*)- and (2*S*,3*S*)-2,3-dibromobutane

(e)
$$CH_3$$
Br——Cl
H——Cl
$$CH_3$$

(f)

(g) OH   HO

(h) H   O

(i)

(j)

39. Propose a reasonable structure for calcium carbide, CaC_2, on the basis of its chemical reactivity (Section 13-11). What might be a more systematic name for it?

40. Propose *two different* syntheses of linalool, a terpene found in cinnamon, sassafras, and orange flower oils. Start with the eight-carbon ketone shown here and use ethyne as your source of the necessary additional two carbons in both syntheses.

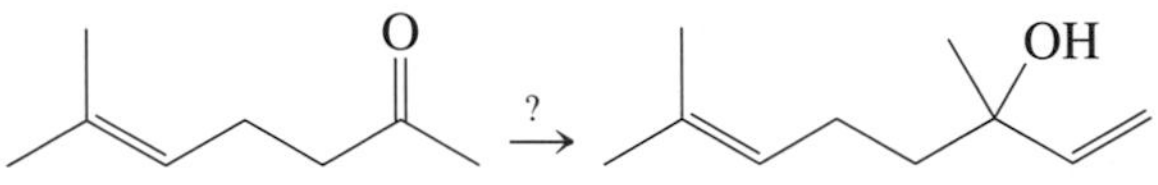

Linalool

41. The synthesis of chamaecynone, the essential oil of the Benihi tree, requires the conversion of a chloroalcohol into an alkynyl ketone. Propose a synthetic strategy to accomplish this task.

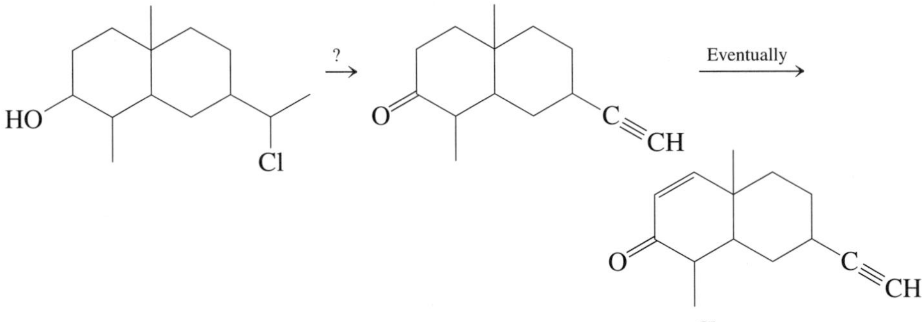

Chamaecynone

42. Synthesis of the sesquiterpene bergamotene proceeds from the alcohol shown here. Suggest a sequence to complete the synthesis.

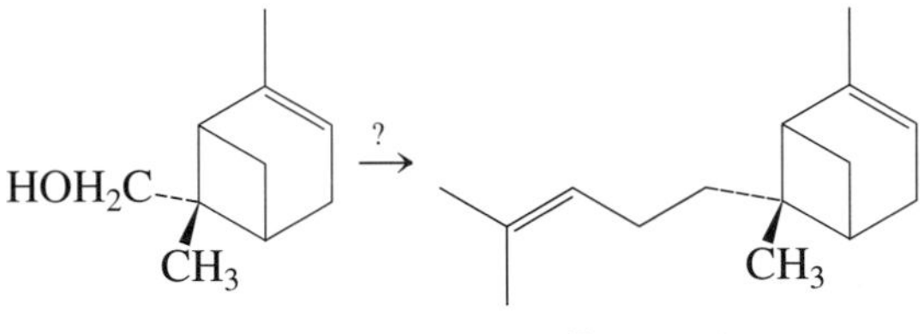

Bergamotene

43. An unknown molecule displays ^1H NMR and IR spectra D (see the facing page). Reaction with H_2 in the presence of the Lindlar catalyst gives a compound that, after ozonolysis and treatment with Zn in aqueous acid, gives rise to one equivalent of $CH_3\overset{O}{\overset{||}{C}}CH$ and two of $H\overset{O}{\overset{||}{C}}H$. What was the structure of the original molecule?

44. Formulate a plausible mechanism for the hydration of ethyne in the presence of mercuric chloride. (**Hint:** Review the hydration of alkenes catalyzed by mercuric ion, Section 12-7.)

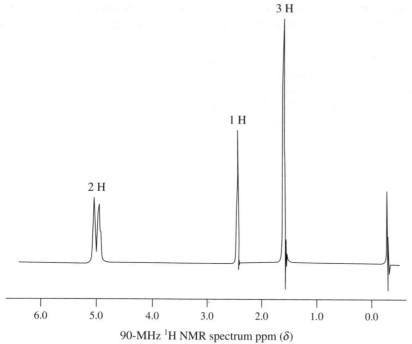

3 H

1 H

2 H

90-MHz ¹H NMR spectrum ppm (δ)

D

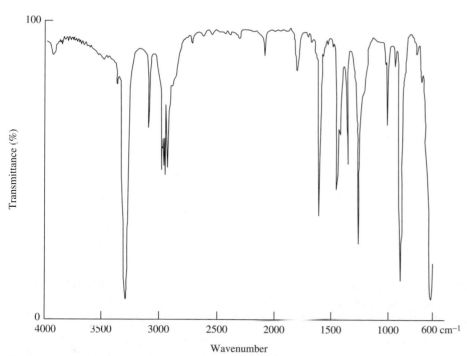

Transmittance (%)

100

0

4000   3500   3000   2500   2000   1500   1000   600 cm⁻¹

Wavenumber

IR-D

45. A synthesis of the sesquiterpene farnesol requires the conversion of a dichloro compound into an alkynol, as shown below. Suggest a way of achieving this transformation. (**Hint:** Devise a conversion of the starting compound into a terminal alkyne.)

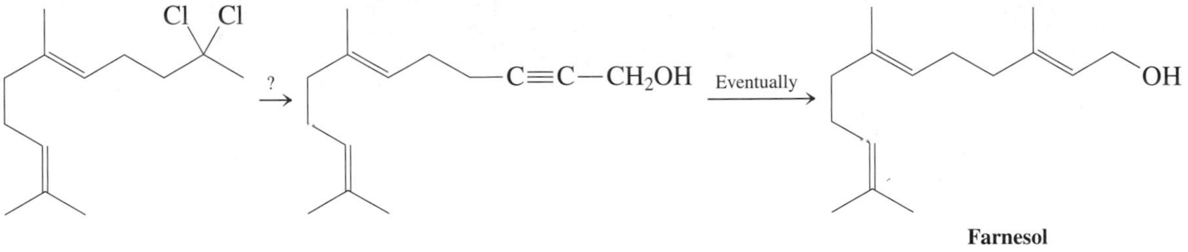

Farnesol

Team Problem

46. Your team is studying the problem of an intramolecular ring closure of enediyne systems important in the total synthesis of dynemicin A, which exhibits potent antitumor activity.

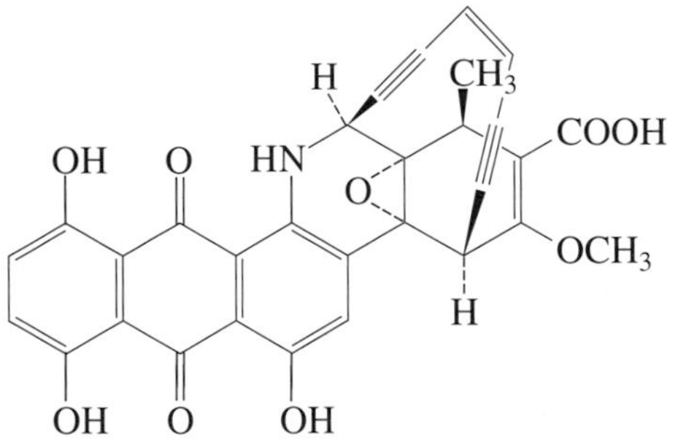

Dynemicin A

One research group tried the following approaches to effect this process. Unfortunately, all were unsuccessful. Divide the schemes among yourselves and assign structures to compounds A through D. (Note: R′ and R″ are protecting groups.)

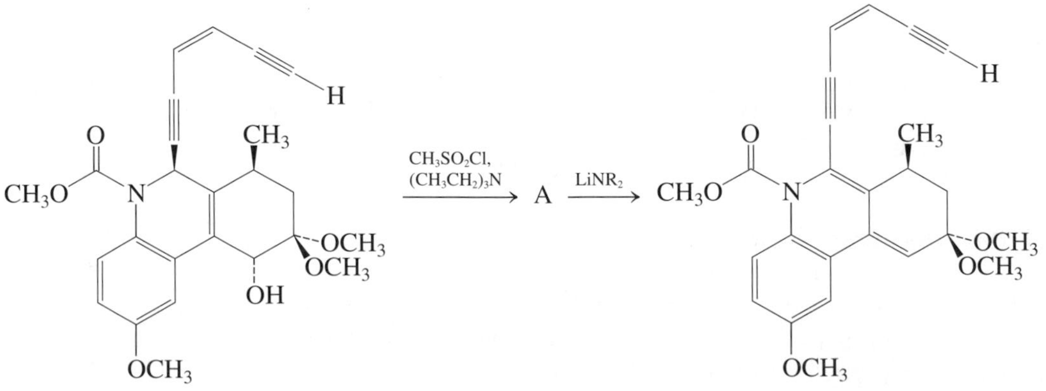

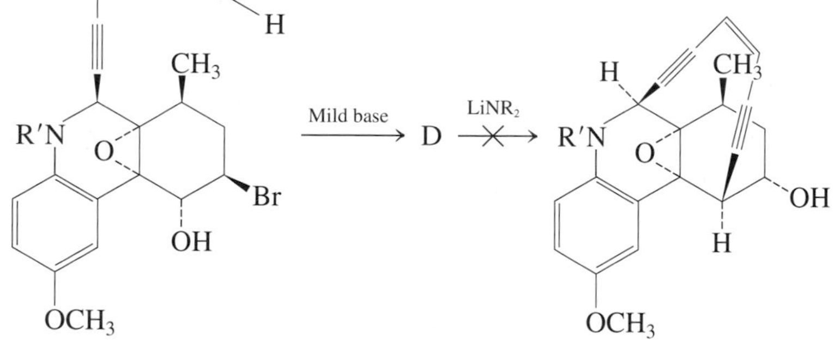

A successful model study (shown here) provided an alternative strategy toward the completion of the total synthesis.

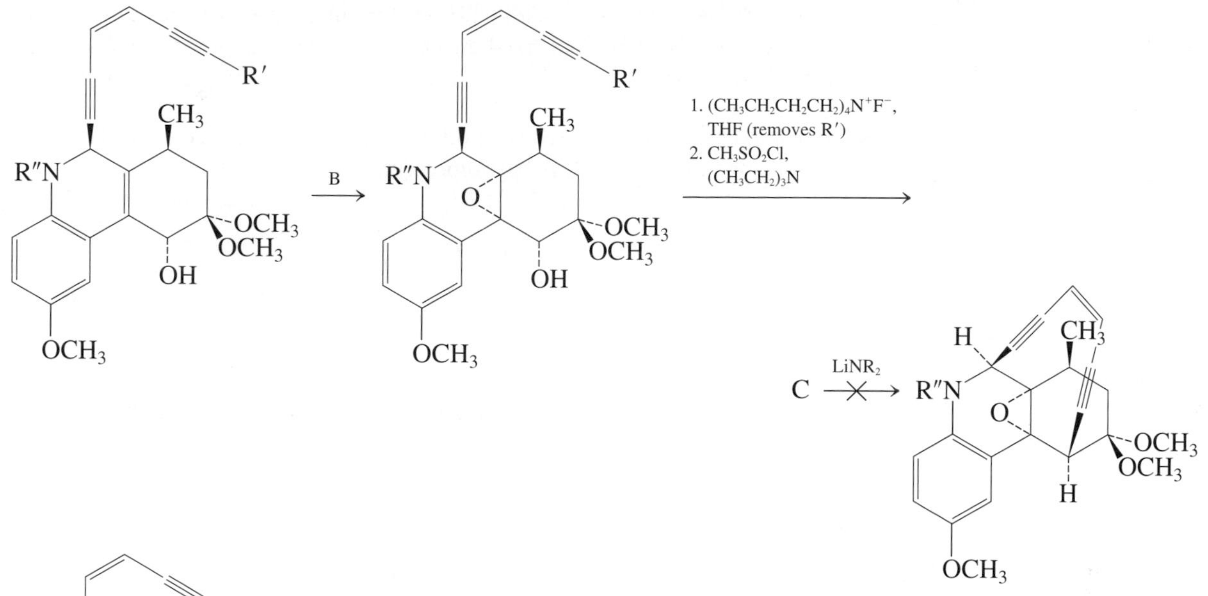

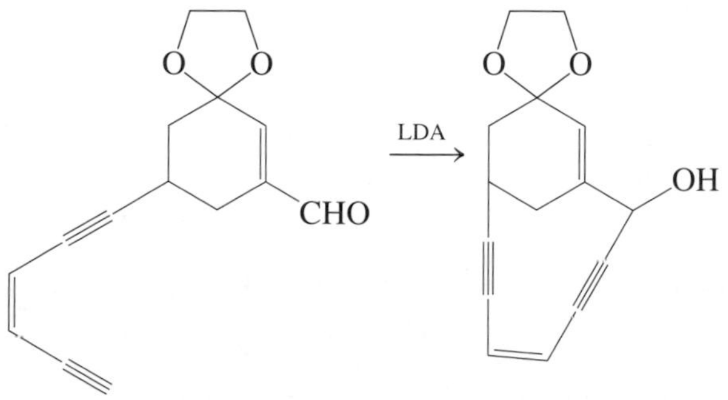

Discuss the advantages of this approach and apply it to the appropriate compound in approaches 1 through 3.

Preprofessional Problems

47. The compound whose structure is H–C≡C(CH$_2$)$_3$Cl is best named (IUPAC)
(a) 4-chloro-1-pentyne; (b) 5-chloropent-1-yne; (c) 4-pentyne-1-chloroyne;
(d) 1-chloropent-4-yne.

48. A nucleophile made by deprotonation of propyne is
(a) $^-:CH_2CH_3$; (b) $^-:HC=CH_2$; (c) $^-:C\equiv CH$; (d) $^-:C\equiv CCH_3$;
(e) $^-:HC=CHCH_3$.

49. When cyclooctyne is treated with dilute, aqueous sulfuric acid and $HgSO_4$, a
new compound results. It is best represented as

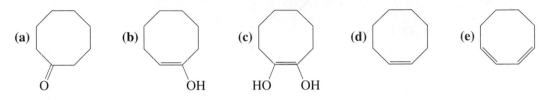

50. From the choices shown below, pick the one which best describes the structure
of compound A.

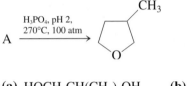

(a) $HOCH_2CH(CH_2)_2OH$ (b) $HOCH_2CHCH_2OH$
             $|$ $|$
             CH_3 CH_3

(c) $HC\equiv CCHCH_2OH$ (d) $HC\equiv CCH_2CHCH_2OH$
             $|$ $|$
             CH_3 CH_3

51. From the choices shown below, pick the one which best describes the structure
of compound A.

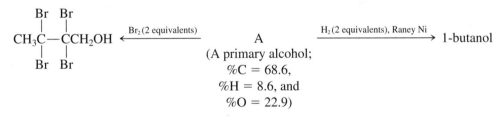

(a) $CH_2=CHCH_2CH_2OH$ (b) ▷—CH_2OH (c) $CH_3C\equiv CCH_2OH$

(d) $CH_3CH=CH—CH=CHOH$ (e)   ⬜
                                                  OH

Delocalized Pi Systems

Investigation by Ultraviolet and Visible Spectroscopy

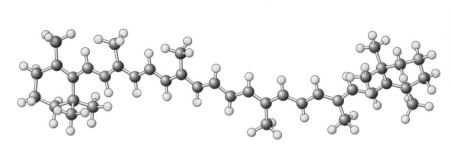

Persimmons, carrots, and autumn leaves contain carotene, whose conjugated system of 11 double bonds absorbs light in the blue region of the visible spectrum. Removal of the blue frequencies from the light reflected by these items results in their orange coloration.

We live in a universe of color. Our ability to perceive and distinguish thousands of hues and shades of color is tied to the ability of molecules to absorb different frequencies of visible light. In turn, this molecular property is frequently a consequence of the presence of multiple π bonds. In the preceding three chapters we introduced the topic of compounds containing carbon–carbon π bonds, the products of overlap between two adjacent parallel p orbitals. We found that addition reactions to these chemically versatile systems provided entries both to relatively simple products, of use in synthesis, and to more complex products, including polymers—substances that have affected modern society enormously. In this chapter we shall expand further on all these themes by studying compounds in which *three or more* parallel p orbitals participate in π-type overlap. The electrons in such orbitals are therefore shared by three or more atomic centers and are said to be **delocalized.**

Our discussion begins with the 2-propenyl system—also called allyl—containing three interacting p orbitals. We then proceed to compounds that contain *several* double bonds: dienes and higher analogs. The latter in particular give rise to some of the most widely used polymers in the modern world, found in everything from automobile tires to the plastic cases around the computers used to prepare the manuscript for this textbook.

The special situation of alternating double and single bonds gives rise to **conjugated** dienes, trienes, and so forth, possessing more extended delocalization of their π electrons. These substances illustrate new modes of reactivity, including thermal and photochemical cycloadditions and ring closures, which are among the most powerful methods for the synthesis of cyclic compounds such as steroidal pharmaceuticals. These processes exemplify a fundamentally new mechanism class: *pericyclic re-actions*, the last such class to be considered in this book, giving us an opportunity to review all the major types of organic processes covered in this course. We continue with a first look at the unusual cyclic triene *benzene*, which is at the heart of much organic and bioorganic chemistry and will be the focus of several subsequent chapters. We conclude the chapter with a discussion of light absorption by molecules with delocalized π systems, the basis of ultraviolet and visible spectroscopy.

14-1 Overlap of Three Adjacent *p* Orbitals: Electron Delocalization in the 2-Propenyl (Allyl) System

What is the effect of a neighboring double bond on the reactivity of a carbon center? Three key observations answer this question.

Dissociation Energies of Various C–H Bonds

$CH_2{=}CHCH_2{-}\!\!\!\!\!\!\!\!\!\frac{}{}\,H$
DH° = 87 kcal mol⁻¹

$(CH_3)_3C{-}\!\!\!\!\!\!\!\!\!\frac{}{}\,H$
DH° = 93 kcal mol⁻¹

$(CH_3)_2CH{-}\!\!\!\!\!\!\!\!\!\frac{}{}\,H$
DH° = 94.5 kcal mol⁻¹

$CH_3CH_2{-}\!\!\!\!\!\!\!\!\!\frac{}{}\,H$
DH° = 98 kcal mol⁻¹

OBSERVATION 1. The primary carbon–hydrogen bond in propene is relatively weak, only 87 kcal mol⁻¹.

A comparison with the values found for other hydrocarbons (see margin) shows that it is even weaker than a tertiary C–H bond. *Evidently, the 2-propenyl radical enjoys some type of special stability.*

OBSERVATION 2. In contrast with saturated primary haloalkanes, 3-chloropropene dissociates relatively fast under S_N1 (solvolysis) conditions and undergoes rapid unimolecular substitution through a carbocation intermediate.

This finding clearly contradicts our expectations (recall Section 7-5). *It appears that the cation derived from 3-chloropropene is somehow more stable than other primary carbocations.* By how much? The ease of formation of the 2-propenyl cation in solvolysis reactions has been found to be roughly equal to that of a secondary carbocation.

OBSERVATION 3. The pK_a of propene is about 40.

Thus, propene is considerably more acidic than propane ($pK_a \sim 50$), and *the formation of the propenyl anion by deprotonation appears to be unusually favored.*

How can we explain these three observations?

Delocalization stabilizes 2-propenyl (allyl) intermediates

Each of the preceding three processes generates a reactive carbon center—a radical, a carbocation, or a carbanion, respectively—that is adjacent to the π framework of a double bond. This arrangement seems to impart special stability. Why? The reason is electron delocalization: Each species may be described by a pair of equivalent contributing resonance forms. These three-carbon intermediates have been given the name **allyl** (followed by the appropriate term: radical, cation, or anion). The activated carbon is called **allylic.**

Resonance Representation of Delocalization in the 2-Propenyl (Allyl) System

Radical

Cation

Anion

> Remember that resonance forms are *not* isomers but partial molecular representations. The true structure (the resonance hybrid) is derived by their superposition, better represented by the dotted-line drawings at the right of the classical picture.

The 2-propenyl (allyl) pi system is represented by three molecular orbitals

The stabilization of the 2-propenyl (allyl) system by resonance can also be described in terms of molecular orbitals. Each of the three carbons is sp^2 hybridized and bears a *p* orbital perpendicular to the molecular plane (Figure 14-1). *Make a model: The structure is symmetric, with equal C–C bond lengths.*

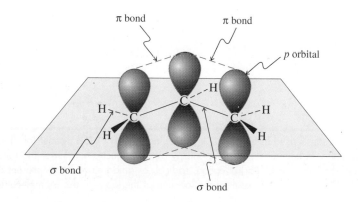

FIGURE 14-1

The three *p* orbitals in the 2-propenyl (allyl) group overlap, giving a symmetric structure with delocalized electrons. The σ framework is shown as black lines.

Two nodes

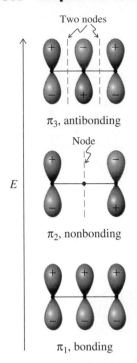

π_3, antibonding

Node

π_2, nonbonding

E

π_1, bonding

FIGURE 14-2

The three π molecular orbitals of 2-propenyl (allyl), obtained by combining the three adjacent atomic p orbitals.

Ignoring the σ framework, we can combine the three p orbitals mathematically to give three π molecular orbitals. This process is analogous to mixing two atomic orbitals to give two molecular orbitals describing a π bond (Figures 11-2 and 11-4), except that there is now a third atomic orbital. Of the three resulting molecular orbitals, one (π_1) is bonding and has no nodes, one (π_2) is *nonbonding* (in other words, it has the same energy as a noninteracting p orbital) and has one node, and one (π_3) is antibonding, with two nodes, as shown in Figure 14-2. We can use the Aufbau principle to fill up the π molecular orbitals with the appropriate number of electrons for the 2-propenyl cation, radical, and anion (Figure 14-3). The cation, with a total of two electrons, contains only one filled orbital, π_1. For the radical and the anion, we place one or two additional electrons, respectively, into the nonbonding orbital, π_2. The total π-electron energy of each system is lower (more favorable) than that expected from three noninteracting p orbitals—essentially because π_1 is greatly stabilized and filled in all cases, whereas the antibonding level, π_3, stays empty throughout.

The resonance formulations for the three 2-propenyl species indicate that it is mainly the two *terminal* carbons that accommodate the charges in the ions or the odd electron in the radical. The molecular-orbital picture is consistent with this view: The three structures differ only in the number of electrons present in molecular orbital π_2, which possesses a node passing through the central carbon; therefore, very little of the electron excess or deficiency will show up at this position.

Partial Electron Density Distribution in the 2-Propenyl (Allyl) System

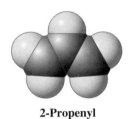

2-Propenyl (Allyl)

In summary, allylic radicals, cations, and anions are unusually stable. In Lewis terms, this stabilization is readily explained by electron delocalization. In a molecular-orbital description, the three interacting p orbitals form three new molecular orbitals: One is considerably lower in energy than the p level, another one stays the same, and a third moves up. Because only the first two are populated with electrons, the total π energy of the system is lowered.

E

π_3 ———   ———   ———

π_2 ———   —↿—   —↿⇂—

π_1 —↿⇂—   —↿⇂—   —↿⇂—

+   ·   —

FIGURE 14-3

The Aufbau principle is used to fill up the π molecular orbitals of 2-propenyl (allyl) cation, radical, and anion. In each case, the total energy of the π electrons is lower than that of three noninteracting p orbitals. Partial cation, radical, or anion character is present at the end carbons in these systems, a result of the location of the lobes in the π_2 molecular orbital.

14-2 Radical Allylic Halogenation

A consequence of delocalization is that resonance-stabilized allylic intermediates can readily participate in reactions of unsaturated molecules. For example, although halogens can add to alkenes to give the corresponding vicinal dihalides (Section 12-5), the course of this reaction is changed when the halogen is present only in low concentrations. These conditions favor radical chain mechanisms and lead to **radical allylic substitution.***

Radical Allylic Halogenation

$$CH_2{=}CHCH_3 \xrightarrow{\ X_2 \text{ (low conc.)}\ } CH_2{=}CHCH_2X \ + \ HX$$

A reagent frequently used in allylic brominations in the laboratory is *N*-bromobutanimide (*N*-bromosuccinimide, NBS, Section 3-8) suspended in tetrachloromethane. This species is nearly insoluble in CCl_4 and is a steady source of very small amounts of bromine formed by reaction with traces of HBr.

NBS as a Source of Bromine

N-Bromobutanimide
(N-Bromosuccinimide, NBS)   **Butanimide**

Example:

85%
3-Bromocyclohexene

The bromine reacts with the alkene by a radical chain mechanism (Section 3-4). The process is initiated by light or by traces of radical initiators that cause dissociation of Br_2 into bromine atoms. Propagation of the chain involves abstraction of a weakly bound allylic hydrogen by Br·.

DH° = 87 kcal mol⁻¹   *DH°* ~ 87 kcal mol⁻¹

*The explanation for this change requires a detailed kinetic analysis that is beyond the scope of this book. Suffice it to say that at low bromine concentrations the competing addition processes are reversible, and substitution wins out.

The resonance-stabilized radical may then react with Br_2 at either end of the allylic system to furnish an allylic bromide and regenerate $Br\cdot$, which continues the chain. Thus, alkenes that form unsymmetric allylic radicals can give mixtures of products upon treatment with NBS. For example,

$$CH_2\!\!=\!\!CH(CH_2)_5CH_3 \xrightarrow[-HBr]{NBS,\ h\nu} \left[CH_2\!\!=\!\!CH\overset{\cdot}{C}H(CH_2)_4CH_3 \longleftrightarrow \cdot CH_2CH\!\!=\!\!CH(CH_2)_4CH_3 \right]$$

1-Octene

$$\overset{Br}{\underset{|}{CH_2\!\!=\!\!CHCH(CH_2)_4CH_3}} \quad + \quad BrCH_2CH\!\!=\!\!CH(CH_2)_4CH_3$$

28%   72%

3-Bromo-1-octene   **1-Bromo-2-octene**

(Mixture of *cis* and *trans*)

EXERCISE 14-1

Ignoring stereochemistry, give all the isomeric bromoheptenes that are possible from NBS treatment of *trans*-2-heptene.

Allylic chlorinations are important in industry because chlorine is relatively cheap. For example, 3-chloropropene (allyl chloride) is made commercially by the gas-phase chlorination of propene at 400°C. It is a building block for the synthesis of DGEBA epoxy resin (Chemical Highlight 9-3) and many other useful substances.

$$CH_3CH\!\!=\!\!CH_2 \quad + \quad Cl_2 \xrightarrow{400°C} ClCH_2CH\!\!=\!\!CH_2 \quad + \quad HCl$$

3-Chloropropene
(Allyl chloride)

EXERCISE 14-2

Predict the outcome of the allylic bromination of the following substrates with NBS (1 equivalent).

(a) Cyclohexene   **(b)**   **(c)** 1-Methylcyclohexene

The biochemical degradation of unsaturated molecules frequently involves radical abstraction of allylic hydrogens by oxygen-containing species. Such processes will be discussed in Chapter 22.

To summarize, under radical conditions, alkenes containing allylic hydrogens enter into allylic halogenation. A particularly good reagent for allylic bromination is *N*-bromobutanimide (*N*-bromosuccinimide, NBS).

14-3 Nucleophilic Substitution of Allylic Halides: Kinetic and Thermodynamic Control

As our example of 3-chloropropene in Section 14-1 shows, allylic halides dissociate readily to produce allylic cations. These can be trapped by nucleophiles at either end in S_N1 reactions. Allylic halides also readily undergo S_N2 transformations.

Allylic halides undergo S$_N$1 reactions

The ready dissociation of allylic halides has important chemical consequences. Different allylic halides may give identical products upon solvolysis if they dissociate to the same allylic cation. For example, the hydrolysis of either 1-chloro-2-butene or 3-chloro-1-butene results in the same alcohol mixture. The reason is the intermediacy of the same allylic cation.

Hydrolysis of Isomeric Allylic Chlorides

$$CH_3CH{=}CHCH_2Cl \xrightarrow[-Cl^-]{} \left[\begin{array}{c} \overset{4}{C}H_3\overset{3}{C}H{=}\overset{2}{C}H\overset{1}{C}H_2{}^+ \\ \updownarrow \\ CH_3\overset{+}{C}HCH{=}CH_2 \\ {}_{4}\quad{}_{3}\quad{}_{2}\quad{}_{1} \end{array} \right] \xleftarrow[-Cl^-]{} \begin{array}{c} Cl \\ | \\ CH_3CHCH{=}CH_2 \end{array}$$

1-Chloro-2-butene	Allylic cation	3-Chloro-1-butene

$$\Big\downarrow HOH$$

$$CH_3CH{=}CHCH_2OH \quad + \quad \begin{array}{c} OH \\ | \\ CH_3CHCH{=}CH_2 \end{array} \quad + \quad H^+$$

Minor   Major

2-Buten-1-ol   **3-Buten-2-ol**

EXERCISE 14-3

Hydrolysis of (*R*)-3-chloro-1-butene gives exclusively racemic alcohols. Explain. (**Hint:** Review Section 7-3.)

Interestingly, at room temperature, the major product from this hydrolysis reaction is 3-buten-2-ol. The thermodynamically *less* stable terminal double bond is created. Why? The less stable isomer must be forming faster, through a *lower-energy transition state.* We say that the reaction is under **kinetic control.**

At higher temperatures or after longer reaction times, 2-buten-1-ol, the *more* stable product, starts to predominate. By studying the equilibration of the two isomers, we find out why: The less stable but kinetically favored product forms first, *but its formation is reversible.* Under the acidic conditions of the reaction (HCl is a product), the allylic alcohols can revert back to the intermediate carbocation (Section 9-2), a process that is accelerated by higher temperatures. Through this cation, the isomeric alcohols eventually equilibrate to a mixture in which *the ratio of the products is determined by their relative energies,* and the more stable isomer is the major product obtained. Under these conditions, we say that the reaction is under **thermodynamic control.** Thus, although the less stable isomer, which forms faster, predominates early in the course of the reaction or at lower temperatures (kinetic control), the more stable isomer is the major product observed after longer reaction times or at higher temperatures (thermodynamic control).

Kinetic Compared with Thermodynamic Control

$$\underset{\substack{\text{Less stable product,}\\\text{predominates when}\\\text{reaction time is}\\\text{short or at low}\\\text{temperatures}}}{\overset{\overset{\text{OH}}{|}}{CH_3CHCH=CH_2}} + H^+ \underset{\substack{\text{Kinetic}\\\text{control (fast,}\\\text{irreversible}\\\text{at low}\\\text{temperature)}}}{\rightleftharpoons} \begin{bmatrix} CH_3CH=CHCH_2^+ \\ \updownarrow \\ CH_3\overset{+}{C}HCH=CH_2 \end{bmatrix} \underset{\substack{\text{Thermo-}\\\text{dynamic}\\\text{control}\\\text{(slow)}}}{\rightleftharpoons} \underset{\substack{\text{More stable product,}\\\text{predominates when reaction}\\\text{time is long or at}\\\text{higher temperatures}}}{CH_3CH=CHCH_2OH} + H^+$$

$$+ \quad HOH$$

The potential-energy diagram in Figure 14-4 shows the lower activation barrier associated with the formation of the less stable product. At higher temperatures and longer reaction times, the (relatively slow) generation of the thermodynamically favored isomer becomes competitive.

Why does the formation of the less stable product have a lower activation barrier? The reason is that the allylic cation in this system is unsymmetric; its positive charge is *unequally* distributed between carbons 1 and 3. More positive charge resides at C3, the more substituted carbon, a condition leading to faster attack by the nucleophile (in this case, water) at this position.

EXERCISE 14-4

Treatment of 3-buten-2-ol with cold hydrogen bromide gives 1-bromo-2-butene and 3-bromo-1-butene in a 15:85 ratio. After heating, this ratio changes to give mainly 1-bromo-2-butene. Explain.

Allylic halides can also undergo S$_N$2 reactions

S$_N$2 reactions of allylic halides with good nucleophiles (Section 6-9) are faster than those of the corresponding saturated haloalkanes. Overlap between the double bond

FIGURE 14-4 ———

Kinetic control compared with thermodynamic control in the reaction of 1-methyl-2-propenyl cation with water. At low temperatures, the lower activation-energy barrier leading to the formation of 3-buten-2-ol results in predominant formation of this isomer (i.e., $k_1 > k_2$). At higher temperatures, the reverse process (governed by k_{-1}) becomes competitive, regenerating the cation, which slowly forms more and more of the thermodynamic product, 2-buten-1-ol. (Alkyloxonium ion intermediates have been omitted from the diagram for clarity.)

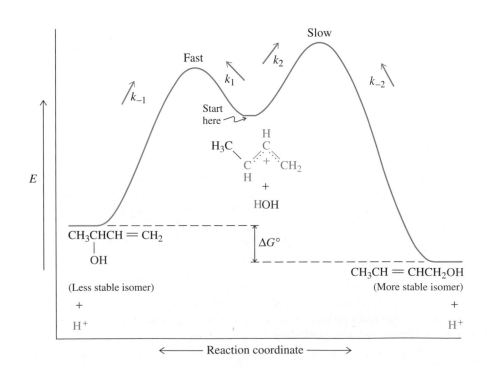

and the p orbital in the transition state of the displacement (see Figure 6-4) stabilizes the latter, resulting in a relatively low activation barrier.

The S_N2 Reactions of 3-Chloro-1-propene and 1-Chloropropane

			Relative rate
$CH_2{=}CHCH_2Cl \ + \ I^-$	$\xrightarrow{\text{Propanone (acetone), 50°C}}$	$CH_2{=}CHCH_2I \ + \ Cl^-$	73
$CH_3CH_2CH_2Cl \ + \ I^-$	$\xrightarrow{\text{Propanone (acetone), 50°C}}$	$CH_3CH_2CH_2I \ + \ Cl^-$	1

EXERCISE 14-5

The solvolysis of 3-chloro-3-methyl-1-butene in acetic acid at 25°C gives initially a mixture containing mostly the structurally isomeric chloride with some of the acetate. After a longer period of time, no allylic chloride remains, and the acetate is the only product present. Explain this result.

| 3-Chloro-3-methyl-1-butene | 1-Chloro-3-methyl-2-butene | 3-Methyl-2-butenyl acetate |

The unusually high reactivity of allylic compounds in both S_N1 and S_N2 displacements make them highly versatile intermediates in synthesis, and they are compounds with which most organic chemists are quite familiar. As such, one physical property of many volatile allylic derivatives deserves comment: They typically possess some of the most pungent and penetrating odors this side of thiols (Sections 9-10 and 9-11). For (just one) example, allyl isothiocyanate, $CH_2{=}CHCH_2N{=}C{=}S$, is such a powerful *lachrymator*—it makes you cry—that its actual smell is difficult to describe in words. It has also been used as a war gas. You can experience its effects in a more benign setting if you wish: It is the substance responsible for the "hot" in horseradish.

In summary, allylic halides undergo both S_N1 and S_N2 reactions. At low temperatures (conditions favoring kinetic control), unimolecular substitution in unsymmetric systems results in nucleophilic attack at the more substituted allylic position, an outcome due to the greater presence of partial positive charge in the intermediate carbocation. Prolonged heating and longer reaction times lead to a predominance of the more stable substitution product (thermodynamic control). With good nucleophiles, allylic halides undergo S_N2 reaction more rapidly than the corresponding saturated substrates.

14-4 Allylic Organometallic Reagents: Useful Three-Carbon Nucleophiles

Propene is appreciably more acidic than propane because of the relative stability of the conjugated carbanion that results from deprotonation (Section 14-1). Therefore, allylic lithium reagents can be made from propene derivatives by proton abstraction by an alkyllithium. The process is facilitated by *N,N,N′,N′*-tetramethylethane-1,2-diamine (tetramethylethylenediamine, TMEDA), a good solvating agent (see Section 13-6).

Allylic Deprotonation

$$CH_3CH_2CH_2CH_2Li \ + \ H_2C\!\!=\!\!C\overset{CH_3}{\underset{CH_2}{\diagdown}} \quad \xrightarrow{\text{(CH}_3\text{)}_2\text{NCH}_2\text{CH}_2\text{N(CH}_3\text{)}_2 \ \text{(TMEDA)}} \quad H_2C\!\!=\!\!C\overset{CH_3}{\underset{CH_2Li}{\diagdown}} \ + \ CH_3CH_2CH_2CH_2\!\!-\!\!H$$

Another way of producing an allylic organometallic is Grignard formation. For example,

$$CH_2\!\!=\!\!CHCH_2Br \quad \xrightarrow{\text{Mg, THF, 0°C}} \quad CH_2\!\!=\!\!CHCH_2MgBr$$
3-Bromo-1-propene **2-Propenylmagnesium bromide**

Like their alkyl counterparts (Section 8-8), allylic lithium and Grignard reagents can function as nucleophiles.

$$CH_2\!\!=\!\!CHCH_2MgBr \ + \ CH_3\overset{O}{\overset{\|}{C}}CH_3 \quad \xrightarrow[\text{(CH}_3\text{CH}_2\text{)}_2\text{O}]{} \quad \xrightarrow[\text{H}^+, \ \text{H}_2\text{O}]{} \quad CH_2\!\!=\!\!CHCH_2\overset{OH}{\underset{CH_3}{\overset{|}{\underset{|}{C}}}}CH_3$$
85%
2-Methyl-4-penten-2-ol

EXERCISE 14-6

Show how to accomplish the following conversion in as few steps as possible.

In summary, alkenes tend to be deprotonated at the allylic position, a reaction resulting in the corresponding delocalized anions. Allylic Grignard reagents are made from the corresponding halides. Like their alkyl analogs, allyl organometallics function as nucleophiles.

14-5 Two Neighboring Double Bonds: Conjugated Dienes

Now that we have caught a glimpse of the consequences of delocalization over three atoms, it should be interesting to learn what happens if we go one step further. Let us consider the addition of a fourth *p* orbital, which results in two double bonds separated by one single bond: a **conjugated diene** (*conjugatio,* Latin, union). In these compounds, delocalization again results in stabilization, as measured by heats of hydrogenation. Extended π overlap is also revealed in their molecular and electronic structures and in their chemistry.

Hydrocarbons with two double bonds are named dienes

Conjugated dienes have to be contrasted with their **nonconjugated** isomers, in which the two double bonds are separated by saturated carbons, and the **allenes** (or **cumulated** dienes), in which the π bonds share a single *sp* hybridized carbon and are perpendicular to each other (Figure 14-5).

The names of conjugated and nonconjugated dienes are derived from those of the alkenes in a straightforward manner. The longest chain incorporating both double bonds is found and then numbered to indicate the positions of the functional and substituent groups. If necessary, cis-trans or *E,Z* prefixes indicate the geometry around the double bonds. Cyclic dienes are named accordingly.

The Simplest Conjugated and Nonconjugated Dienes

$CH_2\!=\!CH\!-\!CH\!=\!CH_2$

1,3-Butadiene

$CH_2\!=\!CHCH_2CH\!=\!CH_2$

1,4-Pentadiene

$CH_2\!=\!C\!=\!CH_2$

1,2-Propadiene
(Allene)

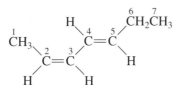

trans-**1,3-Pentadiene**　　*cis*-**2**-*trans*-**4-Heptadiene**　　(*Z*)-**4**-Bromo-**1,3-pentadiene**

cis-**1,4-Heptadiene**　　**1,3-Cyclohexadiene**　　**1,4-Cycloheptadiene**
(A nonconjugated diene)　　　　　　　　　　　(A nonconjugated cyclic diene)

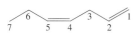

EXERCISE 14-7

Suggest names or draw structures, as appropriate, for the following compounds.

(a) ... Br　　　　**(b)** H_3C CH_3 ... CH_3

(c) *cis*-3,6-Dimethyl-1,4-cyclohexadiene　　**(d)** *cis,cis*-1,4-Dibromo-1,3-butadiene

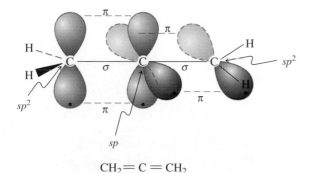

$CH_2\!=\!C\!=\!CH_2$

FIGURE 14-5
The two π bonds of an allene share a single carbon and are perpendicular to each other.

Conjugated dienes are more stable than nonconjugated dienes

The preceding sections noted that delocalization of electrons makes the allylic system especially stable. Does a conjugated diene have the same property? If so, that stability should be manifest in its heat of hydrogenation. We know that the heat of hydrogenation of a terminal alkene is about -30 kcal mol^{-1} (see Section 11-7). A compound containing two *noninteracting* (i.e., separated by one or more saturated carbon atoms) terminal double bonds should exhibit a heat of hydrogenation roughly twice this value, about -60 kcal mol^{-1}. Indeed, catalytic hydrogenation of either 1,5-hexadiene or 1,4-pentadiene releases just about that amount of energy.

Heat of Hydrogenation of Nonconjugated Alkenes

$$CH_3CH_2CH=CH_2 \ + \ H_2 \ \xrightarrow{Pt} \ CH_3CH_2CH_2CH_3 \qquad \Delta H° = -30.3 \text{ kcal mol}^{-1}$$

$$CH_2=CHCH_2CH_2CH=CH_2 \ + \ 2\,H_2 \ \xrightarrow{Pt} \ CH_3(CH_2)_4CH_3 \qquad \Delta H° = -60.5 \text{ kcal mol}^{-1}$$

$$CH_2=CHCH_2CH=CH_2 \ + \ 2\,H_2 \ \xrightarrow{Pt} \ CH_3(CH_2)_3CH_3 \qquad \Delta H° = -60.8 \text{ kcal mol}^{-1}$$

When the same experiment is carried out with the conjugated diene, 1,3-butadiene, *less* energy is produced.

Heat of Hydrogenation of 1,3-Butadiene

$$CH_2=CH-CH=CH_2 \ + \ 2\,H_2 \ \xrightarrow{Pt} \ CH_3CH_2CH_2CH_3 \qquad \Delta H° = -57.1 \text{ kcal mol}^{-1}.$$

The difference, about 3.5 kcal mol^{-1}, is due to a stabilizing interaction between the two double bonds, as illustrated in Figure 14-6. It is referred to as the **resonance energy** of 1,3-butadiene.

EXERCISE 14-8

The heat of hydrogenation of *trans*-1,3-pentadiene is -54.2 kcal mol^{-1}, 6.6 kcal mol^{-1} less than that of 1,4-pentadiene, even lower than that expected from the resonance energy of 1,3-butadiene. Explain. (**Hint:** See Section 11-7.)

FIGURE 14-6

The difference between the heats of hydrogenation of two molecules of 1-butene (a terminal monoalkene) and one molecule of 1,3-butadiene (a doubly terminal conjugated diene) reveals their relative stabilities. The value, about 3.5 kcal mol^{-1}, is a measure of the stabilization of the latter due to conjugation.

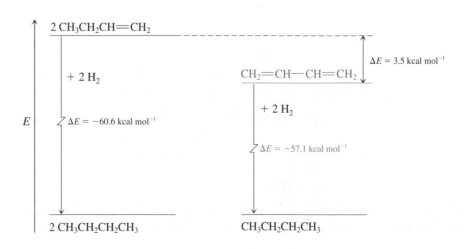

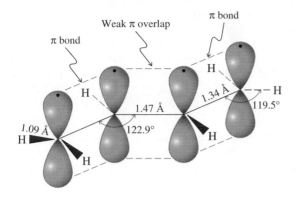

$E_a = 3.9 \text{ kcal mol}^{-1}$

s-**cis**

$\Delta H° = -2.8 \text{ kcal mol}^{-1}$

s-**trans**

FIGURE 14-7

(A) The structure of 1,3-butadiene. The central bond is shorter than that in an alkane (1.54 Å for the central C–C bond in butane). The p orbitals aligned perpendicularly to the molecular plane form a contiguous interacting array. (B) 1,3-Butadiene can exist in two planar conformations.

Conjugation in 1,3-butadiene results from overlap of the pi bonds

How do the two double bonds in 1,3-butadiene interact? The answer lies in the alignment of their π systems, an arrangement that permits the p orbitals on C2 and C3 to overlap (Figure 14-7A). The π interaction that results is weak but nevertheless amounts to a few kilocalories per mole because the π electrons are delocalized over the system of four p orbitals.

Besides adding stability to the diene, this π interaction also raises the barrier to rotation about the single bond. An inspection of models shows that the molecule can adopt two possible extreme coplanar conformations. In one, designated *s*-**cis,** the two π bonds lie on the same side of the C2–C3 axis; in the other, called *s*-**trans,** the π bonds are on opposite sides (Figure 14-7B). The prefix *s* refers to the fact that the bridge between C2 and C3 constitutes a *single* bond. The *s*-cis form is almost 3 kcal mol^{-1} less stable than the *s*-trans conformation because of the steric interference between the two hydrogens on the inside of the diene unit.*

EXERCISE 14-9

The dissociation energy of the central C–H bond in 1,4-pentadiene is only 77 kcal mol^{-1}. Explain. (**Hint:** See Sections 14-1 and 14-2 and draw the product of H atom abstraction.)

The π-electronic structure of 1,3-butadiene may be described by constructing four molecular orbitals out of the four p atomic orbitals (Figure 14-8) (see page 596). Because the four π electrons occupy solely the two bonding levels, the energy of the system is lower than that of four noninteracting p orbitals.

*The *s*-cis conformation is very close in energy to a nonplanar conformation in which the two double bonds are *gauche* (Section 2-7). Whether the *s*-cis or the *gauche* conformation is more stable remains a subject of controversy.

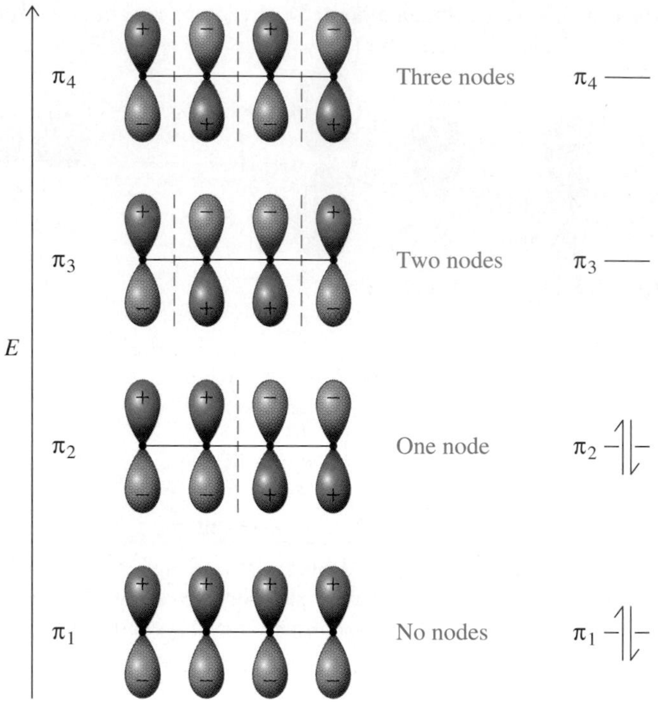

π_4 Three nodes π_4 ——

π_3 Two nodes π_3 ——

π_2 One node π_2

π_1 No nodes π_1

E

FIGURE 14-8 ————
A π-molecular-orbital description of 1,3-butadiene. Its four electrons are placed into the two lowest π (bonding) orbitals, π_1 and π_2.

In summary, dienes are named according to the rules formulated for ordinary alkenes. Conjugated dienes are more stable than dienes containing two isolated double bonds, as measured by their heats of hydrogenation; the difference is called their resonance energy. Conjugation is manifested in the molecular structure of 1,3-butadiene, which has a relatively short central carbon–carbon bond with a small barrier to rotation of about 4 kcal mol^{-1}. The two rotamers *s*-trans and *s*-cis differ in energy by about 3 kcal mol^{-1}. The molecular-orbital picture of the π system in 1,3-butadiene shows two bonding and two antibonding orbitals. The four electrons are placed in the first two bonding levels.

14-6 Electrophilic Attack on Conjugated Dienes

Does the structure of conjugated dienes affect their reactivity? Although more stable thermodynamically than dienes with isolated double bonds, conjugated dienes are actually *more reactive* kinetically in the presence of electrophiles and other reagents. 1,3-Butadiene, for example, readily absorbs 1 mol of gaseous hydrogen chloride. Two isomeric addition products are formed: 3-chloro-1-butene and 1-chloro-2-butene.

$$CH_2=CH-CH=CH_2 \ + \ HCl \ \xrightarrow{25°C} \ \underset{\substack{80\% \\ \textbf{3-Chloro-1-butene}}}{HCH_2-\overset{\displaystyle \overset{Cl}{|}}{CH}-CH=CH_2} \ + \ \underset{\substack{20\% \\ \textbf{1-Chloro-2-butene}}}{HCH_2-CH=CH-CH_2Cl}$$

The generation of the first product is readily understood in terms of ordinary alkene chemistry. It is the result of a Markovnikov addition to one of the double bonds. But what about the second product?

The presence of 1-chloro-2-butene is explained by the reaction mechanism. Initial protonation at C1 gives the thermodynamically most favored allylic cation.

Protonation of 1,3-Butadiene

$$\underset{\substack{\text{Primary nondelocalized cation}\\\text{not formed}}}{\overset{+}{C}H_2-\overset{\overset{\displaystyle H}{|}}{C}H-CH=CH_2} \quad \overset{H^+}{\underset{\substack{\text{Attack}\\\text{at C2}}}{\longleftarrow\!\!\!\times\!\!\!-}} \quad \overset{1}{C}H_2=\overset{2}{C}H-\overset{3}{C}H=\overset{4}{C}H_2 \quad \overset{H^+}{\underset{\substack{\text{Attack}\\\text{at C1}}}{\longrightarrow}}$$

$$\left[\ \overset{\overset{\displaystyle H}{|}}{C}H_2-\overset{+}{C}H\!\!\curvearrowright\!\!CH=CH_2 \quad \longleftrightarrow \quad \overset{\overset{\displaystyle H}{|}}{C}H_2-CH=CH-\overset{+}{C}H_2\ \right]$$

Delocalized allylic cation
formed exclusively

is the same as $CH_3CH\overset{\overset{\displaystyle H}{\underset{\displaystyle C}{}}}{\underset{\ }{\overset{\cdots\overset{+}{\ }\cdots}{}}}CH_2$

This cation can be trapped by chloride in two possible ways to form the two observed products: At the terminal carbon, it yields 1-chloro-2-butene; and, at the internal carbon, it furnishes 3-chloro-1-butene. The 1-chloro-2-butene is said to result from *1,4-addition* of hydrogen chloride to butadiene, because reaction has taken place at C1 and C4 of the original diene. The other product arises by the normal 1,2-addition. Many electrophilic additions to dienes give rise to product mixtures by both modes of addition.

Nucleophilic Trapping of the Allylic Cation Formed
on Protonation of 1,3-Butadiene

$$\underset{\substack{\text{Cl}\\ \textbf{1,2-Addition}}}{CH_3\overset{|}{C}HCH=CH_2} \quad \overset{Cl^-}{\underset{\substack{\text{Attack at}\\\text{internal carbon}}}{\longleftarrow}} \quad CH_3CH\overset{\overset{\displaystyle H}{\underset{\displaystyle C}{}}}{\underset{\ }{\overset{\cdots\overset{+}{\ }\cdots}{}}}CH_2 \quad \overset{Cl^-}{\underset{\substack{\text{Attack at}\\\text{terminal carbon}}}{\longrightarrow}} \quad \underset{\textbf{1,4-Addition}}{CH_3CH=CHCH_2Cl}$$

Example:

$$CH_2=CH-CH=CH_2 \ + \ Br-Br \ \xrightarrow{CCl_4} \ \underset{\underset{\substack{\textbf{54\%}\\\textbf{3,4-Dibromo-1-butene}}}{}}{\overset{\ }{C}H_2-\overset{\ }{C}H-CH=CH_2} \ + \ \underset{\underset{\substack{\textbf{46\%}\\\textbf{1,4-Dibromo-2-butene}}}{}}{\overset{\ }{C}H_2-CH=CH-\overset{\ }{C}H_2}$$

(with Br on the 1 and 2 positions in the first product, Br Br; and Br on the 1 and 4 positions in the second product, Br ... Br)

Conjugated dienes also function as monomers in polymerizations induced by electrophiles, radicals, and other initiators (see Sections 12-13 and 12-14) and, as such, will be discussed in Section 14-10.

EXERCISE 14-10

Conjugated dienes can be made by the methods applied to the preparation of ordinary alkenes. Propose syntheses of **(a)** 2,3-dimethylbutadiene from 2,3-dimethyl-1,4-butanediol; **(b)** 1,3-cyclohexadiene from cyclohexane.

EXERCISE 14-11

Write the products of 1,2-addition and 1,4-addition of **(a)** HBr and **(b)** DBr to 1,3-cyclohexadiene. What is unusual about the products of 1,2- and 1,4-addition of HX to unsubstituted cyclic 1,3-dienes?

In summary, conjugated dienes are electron rich and are attacked by electrophiles to give intermediate allylic cations on the way to 1,2- and 1,4-addition products.

14-7 Delocalization Among More Than Two Pi Bonds: Extended Conjugation and Benzene

What happens if a molecule contains more than two conjugated double bonds? Are cyclic conjugated systems different from their acyclic analogs? This section will begin to answer these questions.

Extended pi systems are thermodynamically stable but kinetically reactive

When more than two double bonds are in conjugation, the molecule is called an **extended π system.** An example is 1,3,5-hexatriene, the next higher double-bond

---

CHEMICAL HIGHLIGHT 14-1   **Use of Sorbic Acid in Making Wine**

Wine making is truly both an art and a science. Wines made in the United States contain fewer additives than do most processed foods and beverages, but those that are permitted serve very specific purposes. Wine does not support the growth of any microorganisms known to be harmful to people; however, bacteria and yeast can still render it unfit to drink and subject to rapid spoilage. A small amount (about 0.001%) of sulfur dioxide, SO_2, usually introduced as sulfite, inhibits the growth of bacteria but is less effective in controlling fungi that give rise to off-odors and unpleasant flavors. Sorbic acid (IUPAC name: *trans,trans*-2,4-hexadienoic acid), available from the berries of the mountain ash tree and synthesized for commercial use since 1954, is the fungicide of choice for wines as well as many foods, but it has no antibacterial capability. Sulfur dioxide and sorbic acid are thus complementary in their wine-preserving ability: Without SO_2, bacteria cause reduction of the acid to sorbyl alcohol (*trans,trans*-2,4-hexadien-1-ol), which under the acidic conditions in wine converts into the foul-smelling ether *trans*-5-ethoxy-1,3-hexadiene (recall odors of allylic derivatives, Section 14-3). Sorbic acid was approved by the Food and Drug Administration in 1971. In 1982, the Environmental Protection Agency allowed its use on whole grain and other agricultural products as a fungistat.

Vintage wine, well preserved by the combination of sulfur dioxide and sorbic acid.

Sorbic acid $\xrightarrow{\text{Malolactic bacteria}}$ Sorbyl alcohol $\xrightarrow{\text{H}^+, \text{CH}_3\text{CH}_2\text{OH}}$ *trans*-5-Ethoxy-1,3-hexadiene

homolog of 1,3-butadiene. This compound is quite reactive and readily polymerizes, particularly in the presence of electrophiles. Despite its reactivity as a delocalized π system, it is also relatively stable thermodynamically.

The increased reactivity of this extended π system is due to the low activation barriers for electrophilic additions, which proceed through highly delocalized carbocations. For example, the bromination of 1,3,5-hexatriene produces a substituted pentadienyl cation intermediate that can be described by three resonance structures.

Bromination of 1,3,5-Hexatriene

$$CH_2\!\!=\!\!CH\!\!-\!\!CH\!\!=\!\!CH\!\!-\!\!CH\!\!=\!\!CH_2 \xrightarrow{\ Br_2\ } \left[\begin{array}{c} BrCH_2\!\!-\!\!\overset{+}{CH}\!\!-\!\!CH\!\!=\!\!CH\!\!-\!\!CH\!\!=\!\!CH_2 \\ \updownarrow \\ BrCH_2\!\!-\!\!CH\!\!=\!\!CH\!\!-\!\!\overset{+}{CH}\!\!-\!\!CH\!\!=\!\!CH_2 \\ \updownarrow \\ BrCH_2\!\!-\!\!CH\!\!=\!\!CH\!\!-\!\!CH\!\!=\!\!CH\!\!-\!\!\overset{+}{CH}_2 \end{array} \right] + \ Br^- $$

1,3,5-Hexatriene

Br
|
BrCH₂CHCH=CHCH=CH₂   +   BrCH₂CH=CHCHCH=CH₂   +   BrCH₂CH=CHCH=CHCH₂Br

$$BrCH_2\overset{\displaystyle Br}{\overset{|}{C}}HCH\!\!=\!\!CHCH\!\!=\!\!CH_2 \quad + \quad BrCH_2CH\!\!=\!\!CH\overset{\displaystyle Br}{\overset{|}{C}}HCH\!\!=\!\!CH_2 \quad + \quad BrCH_2CH\!\!=\!\!CHCH\!\!=\!\!CHCH_2Br$$

5,6-Dibromo-1,3-hexadiene   **3,6-Dibromo-1,4-hexadiene**   **1,6-Dibromo-2,4-hexadiene**

(A 1,2-addition product)   **(A 1,4-addition product)**   **(A 1,6-addition product)**

The final mixture is the result of 1,2-, 1,4-, and 1,6-additions, the last product being the most favored thermodynamically, because it retains an internal conjugated diene system.

EXERCISE 14-12

On treatment with two equivalents of bromine, 1,3,5-hexatriene has been reported to give moderate amounts of 1,2,5,6-tetrabromo-3-hexene. Write a mechanism for the formation of this product.

Some highly extended π systems are found in nature. An example is β-carotene (see Chapter Opening), the orange coloring agent in carrots, and its biological degradation product, vitamin A. Compounds of this type can be very reactive because there

β-Carotene

Vitamin A

are many potential sites for attack by reagents that add to double bonds. In contrast, some cyclic conjugated systems may be considerably less reactive, depending on the number of π electrons (see Chapter 15). The most striking example of this effect is benzene, the cyclic analog of 1,3,5-hexatriene.

Benzene and Its Resonance Structures

Benzene

Benzene, a conjugated cyclic triene, is unusually stable

Cyclic conjugated systems are special cases. The most common examples are the cyclic triene C_6H_6, better known as benzene, and its derivatives (Chapters 15, 16, and 22). In contrast with hexatriene, benzene is unusually stable both thermodynamically and kinetically, because of its special electronic makeup (see Chapter 15). That benzene is unusual can be seen by drawing its resonance forms: There are two *equally* contributing Lewis structures. Benzene does not readily undergo addition reactions typical of unsaturated systems, such as catalytic hydrogenation, hydration, halogenation, and oxidation. In fact, because of its low reactivity, benzene can be used as a solvent in organic reactions.

Benzene Is Unusually Unreactive

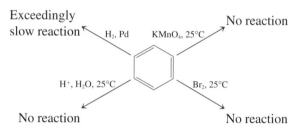

In the chapters that follow, we shall see that the unusual lack of reactivity of benzene is related to the number of π electrons present in its cyclically conjugated array—namely, six. The next section introduces a reaction that is made possible only because its transition state benefits from six-electron cyclic overlap.

To summarize, acyclic extended conjugated systems show increasing reactivity because of the many sites open to attack by reagents and the ease of formation of delocalized intermediates. In contrast, the cyclohexatriene benzene is unusually unreactive.

14-8 A Special Transformation of Conjugated Dienes: Diels-Alder Cycloaddition

Conjugated double bonds participate in more than just the reactions typical of the alkenes, such as electrophilic addition. This section describes a process in which conjugated dienes and alkenes combine to give substituted cyclohexenes. In this transformation, known as Diels-Alder cycloaddition, the nuclei at the ends of the diene add to the alkene double bond, thereby closing a ring. The new bonds form simultaneously and stereospecifically.

The cycloaddition of dienes to alkenes gives cyclohexenes

When a mixture of 1,3-butadiene and ethene is heated in the gas phase, a remarkable reaction takes place in which cyclohexene is formed by the simultaneous generation of two new carbon–carbon bonds. This is the simplest example of the **Diels-Alder***

*Professor Otto P. H. Diels (1876–1954), University of Kiel, Germany, Nobel Prize 1950 (chemistry); Professor Kurt Alder (1902–1958), University of Köln, Germany, Nobel Prize 1950 (chemistry).

reaction, in which a conjugated diene adds to an alkene to yield cyclohexene derivatives. The Diels-Alder reaction is in turn a special case of the more general class of **cycloaddition reactions** between π systems, the products of which are called **cycloadducts.** In the Diels-Alder reaction, an assembly of four conjugated atoms containing four π electrons reacts with a double bond containing two π electrons. For that reason, the reaction is also called a [4 + 2]cycloaddition.

Diels-Alder Cycloaddition of Ethene and 1,3-Butadiene

1,3-Butadiene	**Ethene**	**Cyclohexene**
(Four π electrons)	(Two π electrons)	(A cycloadduct)

The prototype reaction of butadiene and ethene actually does not work very well and gives only low yields of cyclohexene. It is much better to use an *electron-poor alkene* with an *electron-rich diene.* Substitution of the alkene with electron-attracting groups and of the diene with electron-donating groups therefore creates excellent reaction partners.

The trifluoromethyl group, for example, is inductively (Section 8-3) electron attracting owing to its highly electronegative fluorine atoms. The presence of such a substituent enhances the Diels-Alder reactivity of an alkene. Conversely, alkyl groups are electron donating because of hyperconjugation (Sections 7-5, 11-3, 11-7, and 12-9); their presence increases electron density and is beneficial to dienes in the Diels-Alder reaction.

Other alkenes have substituents that interact with double bonds by resonance. For example, carbonyl-containing groups and nitriles are good electron acceptors. Double bonds containing such substituents are electron poor because of the contribution of resonance forms that place a positive charge on an alkene carbon atom.

An electron-poor alkene

An electron-rich diene

Groups That Are Electron Withdrawing by Resonance

EXERCISE 14-13

Classify each of the following alkenes as electron poor or electron rich, relative to ethene. Explain your assignments.

(a) H_2C=$CHCH_2CH_3$ (b)   (c)   (d)

EXERCISE 14-14

The double bond in nitroethene, H_2C=$CHNO_2$, is electron poor, and that in methoxyethene, H_2C=$CHOCH_3$, is electron rich. Explain, using resonance structures.

Examples of reaction partners that undergo efficient Diels-Alder cycloaddition are 2,3-dimethyl-1,3-butadiene and propenal (acrolein).

2,3-Dimethyl-   Propenal   90%
1,3-butadiene   (Acrolein)   Diels-Alder cycloadduct

The carbon–carbon double bond in the cycloadduct is electron rich and sterically hindered. It therefore does not participate in further Diels-Alder reaction with additional diene.

The parent 1,3-butadiene, without additional substituents, is electron rich enough to undergo cycloadditions with electron-poor alkenes.

Ethyl propenoate   94%
(Ethyl acrylate)

In cycloadditions, the substituted ethene is frequently called the **dienophile** ("diene loving") to contrast it with the diene. Many typical dienes and dienophiles have common names, owing to their widespread use in Diels-Alder syntheses of substances of industrial and pharmaceutical importance (Table 14-1).

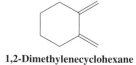

1,2-Dimethylenecyclohexane

EXERCISE 14-15

Formulate the products of [4 + 2]cycloaddition of tetracyanoethene with
(a) 1,3-butadiene; (b) cyclopentadiene; (c) 1,2-dimethylenecyclohexane (see margin).

The Diels-Alder reaction is concerted

The Diels-Alder reaction takes place in one step. Both new carbon–carbon single bonds and the new π bond form simultaneously, just as the three π bonds in the start-

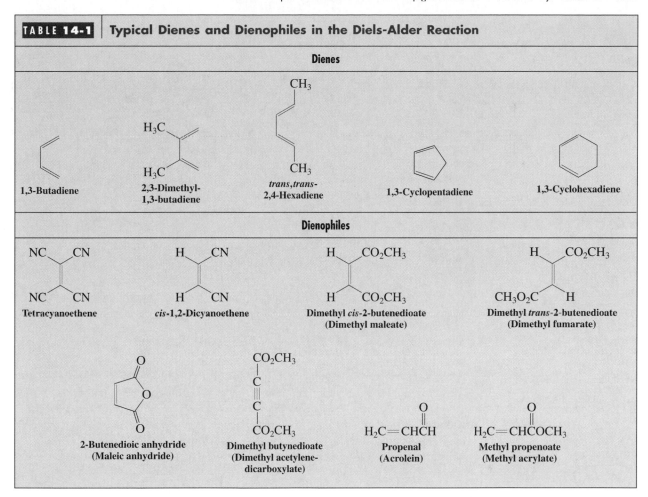

TABLE 14-1	**Typical Dienes and Dienophiles in the Diels-Alder Reaction**

Dienes

1,3-Butadiene   2,3-Dimethyl-1,3-butadiene   *trans,trans*-2,4-Hexadiene   1,3-Cyclopentadiene   1,3-Cyclohexadiene

Dienophiles

Tetracyanoethene   *cis*-1,2-Dicyanoethene   Dimethyl *cis*-2-butenedioate (Dimethyl maleate)   Dimethyl *trans*-2-butenedioate (Dimethyl fumarate)

2-Butenedioic anhydride (Maleic anhydride)   Dimethyl butynedioate (Dimethyl acetylene-dicarboxylate)   Propenal (Acrolein)   Methyl propenoate (Methyl acrylate)

ing materials break. As mentioned earlier (Section 6-4), one-step reactions, in which bond breaking happens at the same time as bond making, are *concerted*. The concerted nature of this transformation can be depicted in either of two ways: by a dotted circle, representing the six delocalized π electrons, or by electron-pushing arrows. Just as six-electron cyclic overlap stabilizes benzene (Section 14-7), the Diels-Alder process benefits from the presence of such an array in its transition state.

Two Pictures of the Transition State of the Diels-Alder Reaction

Dotted-line picture   or   Electron-pushing picture

A molecular-orbital representation (Figure 14-9) clearly shows bond formation by overlap of the p orbitals of the dienophile with the terminal p orbitals of the diene. While these four carbons rehybridize to sp^3, the remaining two internal diene p orbitals give rise to the new π bond.

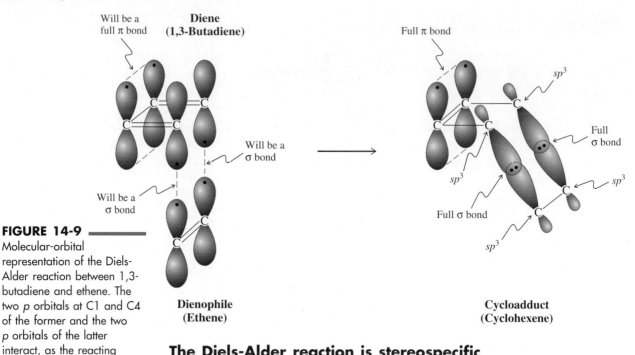

FIGURE 14-9

Molecular-orbital representation of the Diels-Alder reaction between 1,3-butadiene and ethene. The two *p* orbitals at C1 and C4 of the former and the two *p* orbitals of the latter interact, as the reacting carbons rehybridize to *sp*³ to maximize overlap in the two resulting new single bonds. At the same time, π overlap between the two *p* orbitals on C2 and C3 of the diene increases to create a full double bond.

The Diels-Alder reaction is stereospecific

As a consequence of the concerted mechanism, the Diels-Alder reaction is *stereospecific*. For example, reaction of 1,3-butadiene with dimethyl *cis*-2-butenedioate (dimethyl maleate, a cis alkene) gives dimethyl *cis*-4-cyclohexene-1,2-dicarboxylate. *The stereochemistry at the original double bond of the dienophile is retained in the product.* In the complementary reaction, dimethyl *trans*-2-butenedioate (dimethyl fumarate, a trans alkene) gives the trans adduct.

In the Diels-Alder Reaction,
the Stereochemistry of the Dienophile Is Retained

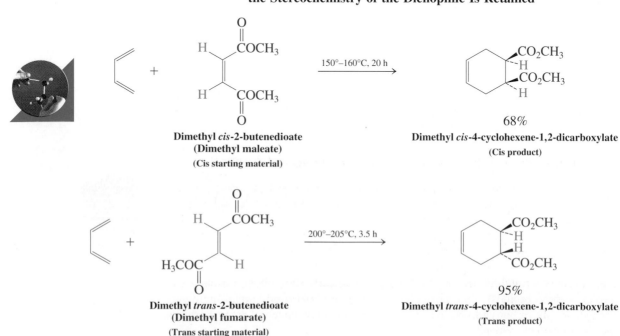

Similarly, *the stereochemistry of the diene also is retained.*

**In the Diels-Alder Reaction,
the Stereochemistry of the Diene Is Retained**

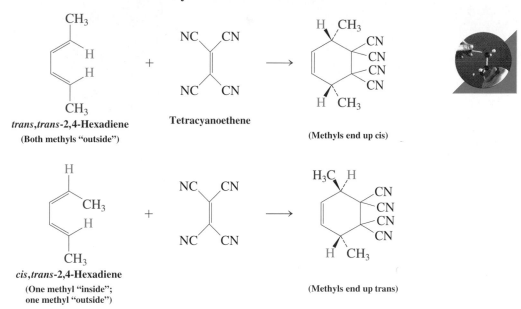

trans,trans-**2,4-Hexadiene**
(Both methyls "outside")

Tetracyanoethene

(Methyls end up cis)

cis,trans-**2,4-Hexadiene**
(One methyl "inside";
one methyl "outside")

(Methyls end up trans)

EXERCISE 14-16

Add structures of the missing products or starting materials to the following schemes.

(a) H₃C + NC—CN ⟶ ?

(b) ? + ? ⟶

EXERCISE 14-17

cis,trans-2,4-Hexadiene reacts very sluggishly in [4 + 2]cycloadditions; the trans,trans isomer does so much more rapidly. Explain. (**Hint:** The Diels-Alder reaction requires the *s*-cis arrangement of the diene [Figure 14-9, and see Figure 14-7].)

Diels-Alder cycloadditions follow the endo rule

The Diels-Alder reaction is stereospecific, not only with respect to the substitution pattern of the original double bonds, but also with respect to the orientation of the starting materials relative to each other. Consider the reaction of 1,3-cyclopentadiene with dimethyl *cis*-2-butenedioate. Two products are conceivable, one in which the two ester substituents on the bicyclic frame are on the same side (cis) as the methylene bridge, the other in which they are on the side opposite (trans) from the bridge. The first is called the **exo adduct,** the second the **endo adduct** (*exo,* Greek, outside; *endo,* Greek, within). The terms refer to the position of groups in bridged systems. Exo substituents are placed cis with respect to the shorter bridge; endo substituents are positioned trans to this bridge.

Exo and Endo Cycloadditions to Cyclopentadiene

The Diels-Alder reaction usually proceeds with *endo selectivity; the endo product is preferred kinetically,* a result referred to as the **endo rule.** This preference appears to originate from an attractive interaction between the π system of the diene and that of the unsaturated substituents on the dienophile.

The Endo Rule

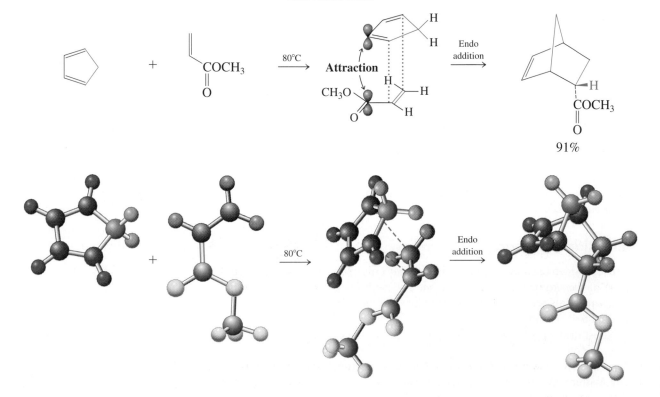

Methyl propenoate

Endo product

The relative stereochemistry of the product of a general Diels-Alder reaction that follows the endo rule is illustrated below.

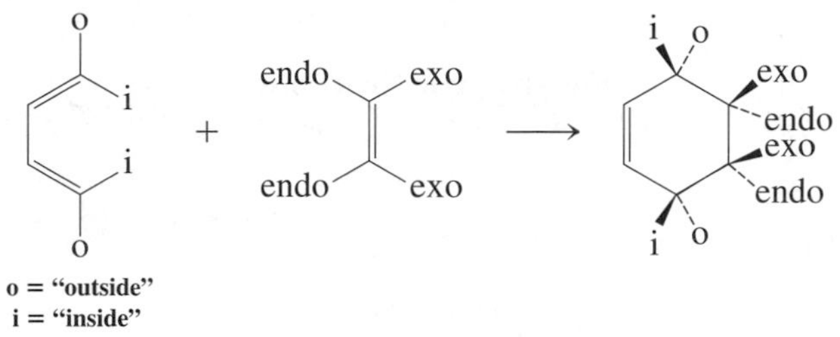

o = "outside"
i = "inside"

EXERCISE 14-18

Predict the products of the following reactions (show the stereochemistry clearly). **(a)** *trans,trans*-2,4-Hexadiene with methyl propenoate; **(b)** *trans*-1,3-pentadiene with 2-butenedioic anhydride (maleic anhydride); **(c)** 1,3-cyclopentadiene with dimethyl *trans*-2-butenedioate (dimethyl fumarate).

EXERCISE 14-19

The Diels-Alder reaction can also occur in an intramolecular fashion. Draw the two transition states leading to products in the following reaction.

65   :   35
75% (combined yield)

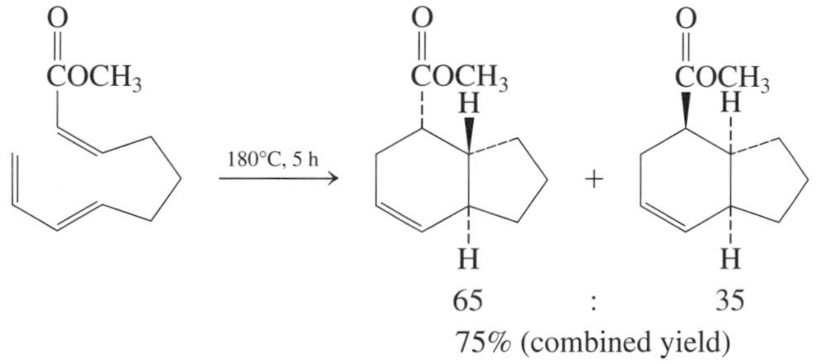

In summary, the Diels-Alder reaction is a cycloaddition that proceeds best between electron-rich 1,3-dienes and electron-poor dienophiles to furnish cyclohexenes. It is stereospecific with respect to the stereochemistry of the double bonds and with respect to the arrangements of the substituents on diene and dienophile: It follows the endo rule.

14-9 Electrocyclic Reactions

The Diels-Alder reaction couples the ends of two separate π systems. Can rings be formed by the linkage of the termini of a *single* conjugated di-, tri-, or polyene? Yes, and this section will describe the conditions under which such ring closures, called **electrocyclic reactions,** take place. Cycloadditions and electrocyclic reactions belong to a class of transformations called **pericyclic** (*peri,* Greek, around), because they exhibit transition states with a cyclic array of nuclei and electrons.

Electrocyclic transformations are driven by heat or light

Let us consider first the conversion of 1,3-butadiene into cyclobutene. This process is endothermic because of ring strain. Indeed, the reverse reaction, ring *opening* of cyclobutene, occurs readily upon heating. However, ring *closure* of *cis*-1,3,5-hexatriene to 1,3-cyclohexadiene is exothermic and takes place thermally. Is it possible to drive these transformations in the thermally disfavored directions? Again, the answer is yes. Irradiation with ultraviolet light (**photochemical conditions**) provides sufficient energy to reverse both processes.

Electrocyclic Reactions

cis-1,3,5-Hexatriene $\rightleftharpoons$ **1,3-Cyclohexadiene** $\Delta H° = -14.5 \text{ kcal mol}^{-1}$

Cyclobutene $\rightleftharpoons$ **1,3-Butadiene** $\Delta H° = -9.7 \text{ kcal mol}^{-1}$

EXERCISE 14-20

When heated, benzocyclobutene (A) in the presence of dimethyl *trans*-2-butenedioate (B) gives compound C. Explain. (**Hint:** Combine an electrocyclic with a Diels-Alder reaction.)

A + B $\xrightarrow{180°C}$ C

Electrocyclic reactions are concerted and stereospecific

Like the Diels-Alder cycloaddition, electrocyclic reactions are concerted and stereospecific. Thus, the thermal isomerization of *cis*-3,4-dimethylcyclobutene gives only *cis,trans*-2,4-hexadiene.

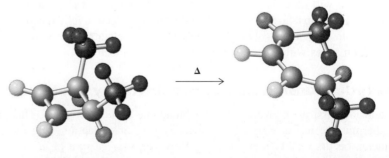

cis-**3,4-Dimethylcyclobutene**   *cis,trans*-**2,4-Hexadiene**

Heated *trans*-3,4-dimethylcyclobutene, however, opens to *trans,trans*-2,4-hexadiene.

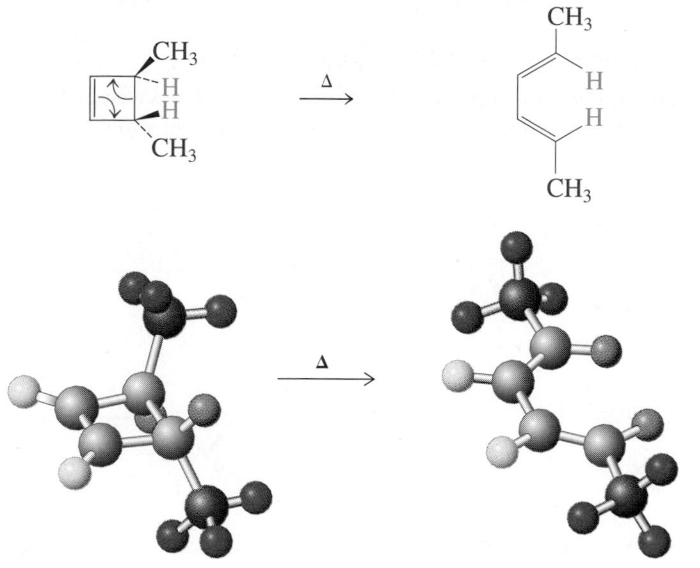

CH$_3$

H

H

CH$_3$

$\xrightarrow{\Delta}$

CH$_3$

H

H

CH$_3$

trans-3,4-Dimethylcyclobutene $\xrightarrow{\Delta}$ **trans,trans-2,4-Hexadiene**

Figure 14-10 takes a closer look at these processes. As the bond between carbons C3 and C4 in the cyclobutene is broken, these carbon atoms must rehybridize from sp^3 to sp^2 and rotate to permit overlap between the emerging p orbitals and those originally present. In thermal cyclobutene ring opening, the carbon atoms are found

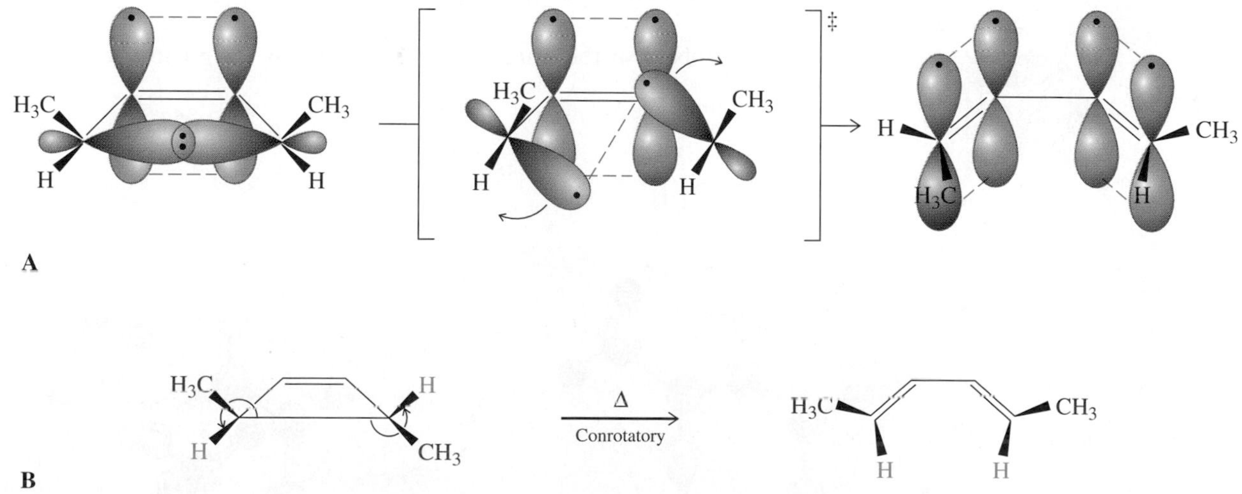

A

H$_3$C   CH$_3$

H   H

$\longrightarrow$

H$_3$C   CH$_3$

H   H

$\xrightarrow{\ddagger}$

H   CH$_3$

H$_3$C   H

B

H$_3$C   H

H   CH$_3$

$\xrightarrow[\text{Conrotatory}]{\Delta}$

H$_3$C   CH$_3$

H   H

FIGURE 14-10

(A) Conrotatory ring opening of *cis*-3,4-dimethylcyclobutene. Both reacting carbons rotate clockwise. The sp^3 hybrid lobes in the ring change into p orbitals, the carbons becoming sp^2 hybridized. Overlap of these p orbitals with those already present in the cyclobutene starting material creates the two double bonds of the cis,trans diene. (B) Similar conrotatory opening of *trans*-3,4-dimethylcyclobutene proceeds to the trans,trans diene.

to rotate *in the same direction,* either both clockwise or both counterclockwise. This mode of reaction is called a **conrotatory** process.

Fascinatingly, the photochemical closure **(photocyclization)** of butadiene to cyclobutene proceeds with stereochemistry exactly *opposite* that observed in the thermal opening. In this case, the products arise by rotation of the two reacting carbons in opposite directions. In other words, if one rotates clockwise, the other does so counterclockwise. This mode of movement is called **disrotatory** (Figure 14-11).

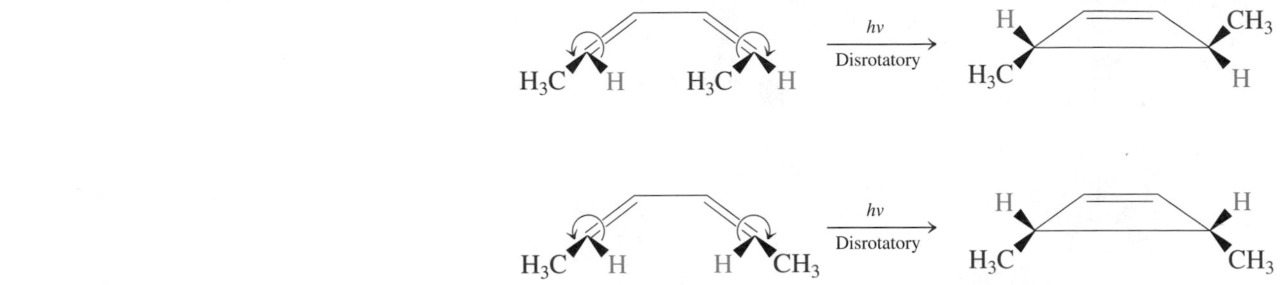

FIGURE 14-11

Disrotatory photochemical ring closure of *cis,trans-* and *trans,trans-*2,4-hexadiene. In the disrotatory mode, one carbon rotates clockwise, the other counterclockwise.

Can these observations be generalized? Let us look at the stereochemistry of the *cis*-1,3,5-hexatriene–cyclohexadiene interconversion. Surprisingly, the six-membered ring is formed thermally by the disrotatory mode, as can be shown by using derivatives. For example, heated *trans,cis,trans-*2,4,6-octatriene gives *cis*-5,6-dimethyl-1,3-cyclohexadiene, and *cis,cis,trans-*2,4,6-octatriene converts into *trans*-5,6-dimethyl-1,3-cyclohexadiene, both disrotatory closures.

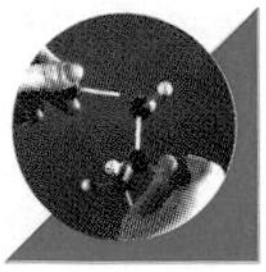

Stereochemistry of the Thermal 1,3,5-Hexatriene Ring Closure

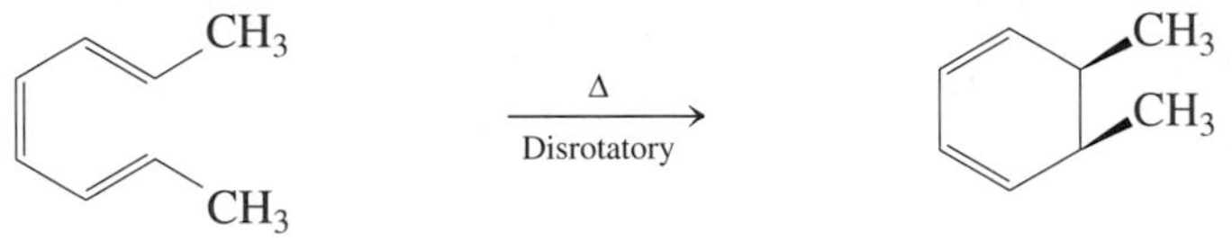

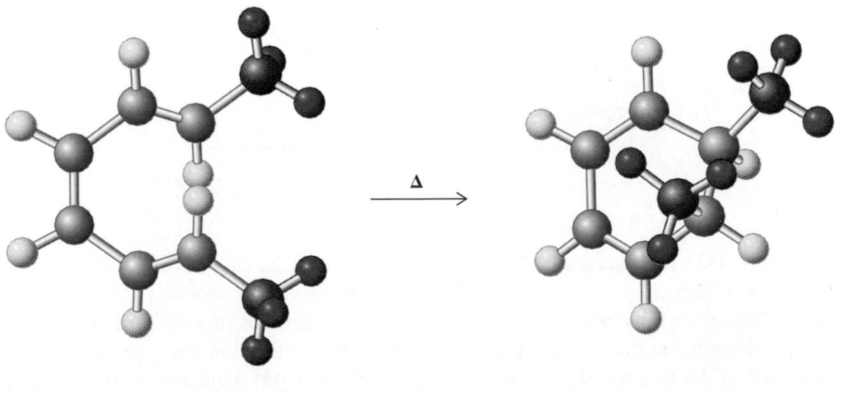

*trans,cis,trans-***2,4,6-Octatriene**   *cis-***5,6-Dimethyl-1,3-cyclohexadiene** *cis,cis,trans*-2,4,6-Octatriene   *trans*-5,6-Dimethyl-1,3-cyclohexadiene

In contrast, the corresponding photochemical reactions occur in conrotatory fashion.

Stereochemistry of the Photochemical 1,3,5-Hexatriene Ring Closure

This stereocontrol is observed in many other electrocyclic transformations and is governed by the symmetry properties of the relevant π molecular orbitals. The **Woodward-Hoffmann* rules** describe these interactions and predict the stereochemical outcome of all electrocyclic reactions as a function of the number of electrons taking part in the process and whether the reaction is carried out photochemically or thermally. A complete treatment of this subject is best left to a more advanced course in organic chemistry.

EXERCISE 14-21

Photolysis of ergosterol gives provitamin D_2, a precursor of vitamin D_2 (an antirachitic agent, shown in the margin). Is the ring opening conrotatory or disrotatory?

Ergosterol   **Provitamin D_2**   **Vitamin D_2**

*Professor Robert B. Woodward (1917–1979), Harvard University, Cambridge, Massachusetts, Nobel Prize 1965 (chemistry); Professor Roald Hoffmann (b. 1937), Cornell University, Ithaca, New York, Nobel Prize 1981 (chemistry).

CHEMICAL HIGHLIGHT **14-2**

An Extraordinary Electrocyclic Reaction of Anticancer Agents

Heating of 3-hexene-1,5-diyne to 200°C induces an electrocyclization (Bergman* reaction) related to the ring closure of 1,3,5-hexatrienes. However, the extra electrons in the two triple bonds mean that, instead of a cyclohexadiene, a reactive intermediate known as a 1,4-benzene diradical is generated.

Electrocyclization of *cis*-3-Hexene-1,5-diyne

1,4-Benzene diradical

The enediyne-containing antitumor antibiotics calicheamicin and esperamicin (Section 13-12)

***

*Professor Robert G. Bergman (b. 1942), University of California at Berkeley.

exhibit impressive biological activity (4000 times as great as that of the clinical drug adriamycin). A sequence of enzyme-catalyzed steps (shown for calicheamicin below) cleaves an S–S bond and adds sulfur to the double bond at C9. The change in hybridization from sp^2 to sp^3 at C9 "pinches" the molecule, bringing C2 and C7 within bonding distance. (Make a model.) Bergman-type electrocyclization ensues (below $-10°C$), a reaction leading to a benzene diradical.

It is these radical centers that are thought to decompose the DNA (Chapter 26) of tumor cells by hydrogen abstraction. Ironically, the enzyme system that activates the antitumor molecule usually *protects* cells from damage by radicals (see Chapter 25). Thus, these anticancer agents destroy tumor cells *with the tumor cells' own defense systems*. Although the enediyne-to-benzene diradical transformation was discovered in the laboratory in 1973, the finding that such a remarkable process occurs in nature was, to say the least, unexpected.

Calicheamycin (R = sugar)

In summary, conjugated dienes and hexatrienes are capable of (reversible) electrocyclic ring closures to cyclobutenes and 1,3-cyclohexadienes, respectively. The diene–cyclobutene system prefers thermal conrotatory and photochemical disrotatory modes. The triene–cyclohexadiene system reacts in the opposite way, proceeding through thermal disrotatory and photochemical conrotatory rearrangements. The stereochemistry of such electrocyclic reactions is governed by the Woodward-Hoffmann rules.

14-10 Polymerization of Conjugated Dienes: Rubber

Like simple alkenes (Section 12-14), conjugated dienes can be polymerized. The elasticity of the resulting materials has led to their use as synthetic rubbers. The biochemical pathway to natural rubber features an activated form of the five-carbon unit 2-methyl-1,3-butadiene (isoprene, see Section 4-7), which is an important building block in nature.

1,3-Butadiene can form cross-linked polymers

When 1,3-butadiene is polymerized at C1 and C2, it yields a polyethenylethene (polyvinylethylene).

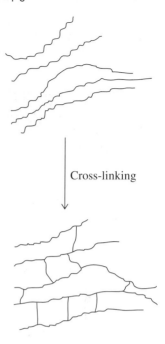

1,2-Polymerization of 1,3-Butadiene

$$2n\ CH_2=CH-CH=CH_2 \xrightarrow{\text{Initiator}} \begin{array}{cc} CH_2 & CH_2 \\ \| & \| \\ CH & CH \\ | & | \\ -(CH-CH_2-CH-CH_2)_n- \end{array}$$

Alternatively, polymerization at C1 and C4 gives either *trans*-polybutadiene, *cis*-polybutadiene, or a mixed polymer.

Cross-linking

1,4-Polymerization of 1,3-Butadiene

$$n\ CH_2=CH-CH=CH_2 \xrightarrow{\text{Initiator}} -(CH_2-CH=CH-CH_2)_n-$$
cis- or *trans-***Polybutadiene**

FIGURE 14-12
Cross-linking underlies the elasticity of polybutadiene chains in rubber.

Butadiene polymerization is unique in that the product itself may be unsaturated. The double bonds in this initial polymer may be linked by further treatment with added chemicals, such as radical initiators, or by radiation. In this way, **cross-linked polymers** arise, in which individual chains have been connected into a more rigid framework (Figure 14-12). Cross-linking generally increases the density and hardness of such materials. It also greatly affects a property characteristic of butadiene polymers: **elasticity.** The individual chains in *most* polymers can be moved past each other, a process allowing for their molding and shaping. In cross-linked systems, however, such deformations are rapidly reversible: The chain snaps back to its original shape (more or less). Such elasticity is characteristic of rubbers.

Synthetic rubbers are derived from poly-1,3-dienes

Polymerization of 2-methyl-1,3-butadiene (isoprene, Section 4-7) by a Ziegler-Natta catalyst (Section 12-14) results in a synthetic rubber *(polyisoprene)* of almost 100% Z configuration. Similarly, 2-chloro-1,3-butadiene furnishes an elastic, heat- and oxygen-resistant polymer called neoprene, with trans chain double bonds. More than 2.5 million tons of synthetic rubber are produced in the United States every year.

$$n\ H_2C=\overset{\overset{\displaystyle CH_3}{|}}{C}-CH=CH_2 \xrightarrow{\text{TiCl}_4,\ \text{AlR}_3} \begin{array}{c} H_3C \diagdown \qquad \diagup H \\ C=C \\ -(H_2C \diagup \qquad \diagdown CH_2)_n- \end{array}$$

2-Methyl-1,3-butadiene   **(Z)-Polyisoprene**

$$n\ H_2C=\overset{\overset{\displaystyle Cl}{|}}{C}-CH=CH_2 \xrightarrow{\text{TiCl}_4,\ \text{AlR}_3} \begin{array}{c} Cl \diagdown \qquad \diagup CH_2)_n- \\ C=C \\ -(H_2C \diagup \qquad \diagdown H \end{array}$$

2-Chloro-1,3-butadiene   **Neoprene**

Latex, the precursor of natural rubber, is drained from the bark of *Hevea brasiliensis*.

This popular flexible flashlight is held together by ABS copolymer, the same material used in the construction of computer housings, camera bodies, automobile grilles and interiors, and a host of other products where toughness and durability are important.

Natural *Hevea* rubber is a 1,4-polymerized (*Z*)-poly(2-methyl-1,3-butadiene), similar in structure to polyisoprene. To increase its elasticity, it is treated with hot elemental sulfur in a process called **vulcanization** (*Vulcanus,* Latin, the Roman god of fire), which creates sulfur cross-links. This reaction was discovered by Goodyear* in 1839. One of the earliest and most successful uses of the product, Vulcanite, was the manufacture of dentures that could be molded for good fit. Before the 1860s, false teeth were embedded in animal bone, ivory, or metal. The swollen appearance of George Washington's lips (visible on U.S. currency) is attributed to his ill-fitting ivory dentures. Nowadays, dentures are made from acrylics (Section 13-11). Rubber remains an indispensable component of many commercial products, including tires (the major use), shoes, rainwear, and other clothing incorporating elastic fibers.

Copolymers in which the double bonds of 1,3-butadiene undergo polymerization with the double bonds of other alkenes have assumed increasing importance in recent years. By varying the proportions of the different monomers in the polymerization mix, the properties of the final product may be "tuned" over a considerable range. One such substance is a three-component copolymer of propenenitrile, 1,3-butadiene, and ethenylbenzene, known as ABS (for *a*crylonitrile/*b*utadiene/*s*tyrene copolymer). The diene imparts the rubberlike property of flexibility, whereas the nitrile hardens the polymer. The result is a highly versatile material that may be fashioned into sheets or molded into virtually any shape. Its strength and ability to tolerate deformation and stress have allowed its utilization in everything from clock mechanisms to camera and computer cases to automobile bodies and bumpers.

Polyisoprene is the basis of natural rubber

How is rubber made in nature? Plants construct the polyisoprene framework of natural rubber by using as a building block 3-methyl-3-butenyl pyrophosphate (isopentenyl pyrophosphate). This molecule is an ester of pyrophosphoric acid and 3-methyl-3-buten-1-ol. An enzyme equilibrates a small amount of this material with the 2-butenyl isomer, an allylic pyrophosphate.

Biosynthesis of the Two Isomers of 3-Methylbutenyl Pyrophosphate

$$H_2C=CCH_2CH_2OH$$
$$CH_3$$
3-Methyl-3-buten-1-ol

$$HO-\overset{\overset{O}{\|}}{\underset{\underset{OH}{|}}{P}}-OH \quad + \quad HO-\overset{\overset{O}{\|}}{\underset{\underset{OH}{|}}{P}}-OH$$
Phosphoric acid

$-HOH$

$$HO-\overset{\overset{O}{\|}}{\underset{\underset{OH}{|}}{P}}-O-\overset{\overset{O}{\|}}{\underset{\underset{OH}{|}}{P}}-OH$$
Pyrophosphoric acid

$-HOH$

$$H_2C=CCH_2CH_2O-\overset{\overset{O}{\|}}{\underset{\underset{OH}{|}}{P}}-O-\overset{\overset{O}{\|}}{\underset{\underset{OH}{|}}{P}}-OH$$
$$CH_3$$
3-Methyl-3-butenyl pyrophosphate

$\overset{Enzyme}{\rightleftharpoons}$

$$\overset{H_3C}{\underset{H_3C}{}}C=CHCH_2O-\overset{\overset{O}{\|}}{\underset{\underset{OH}{|}}{P}}-O-\overset{\overset{O}{\|}}{\underset{\underset{OH}{|}}{P}}-OH$$
3-Methyl-2-butenyl pyrophosphate

*Charles Goodyear (1800–1860), American inventor, Washington, D.C.

Although the subsequent processes are enzymatically controlled, they can be formulated simply in terms of familiar mechanisms (OPP = pyrophosphate).

Mechanism of Natural Rubber Synthesis

STEP 1. Ionization to stabilized (allylic) cation

STEP 2. Electrophilic attack

STEP 3. Proton loss

Geranyl pyrophosphate

STEP 4. Second oligomerization

Farnesyl pyrophosphate

In the first step, ionization of the allylic pyrophosphate gives an allylic cation. Attack by a molecule of 3-methyl-3-butenyl pyrophosphate, followed by proton loss, yields a dimer called geranyl pyrophosphate. Repetition of this process leads to natural rubber.

Many natural products are composed of 2-methyl-1,3-butadiene (isoprene) units

Many natural products are derived from 3-methyl-3-butenyl pyrophosphate, including the terpenes first discussed in Section 4-7. Indeed, the structures of terpenes can be dissected into five-carbon units connected as in 2-methyl-1,3-butadiene. Their structural diversity can be attributed to the multiple ways in which 3-methyl-3-butenyl pyrophosphate can couple. The monoterpene geraniol and the sesquiterpene farnesol, two of the most widely distributed substances in the plant kingdom, form by hydrolysis of their corresponding pyrophosphates.

Geraniol   **Farnesol**

Coupling of two molecules of farnesyl pyrophosphate leads to squalene, a biosynthetic precursor of the steroid nucleus (Chemical Highlight 12-4).

⟶ steroids

Squalene

Bicyclic substances, such as camphor, a chemical used in mothballs, nasal sprays, and muscle rubs, are built up from geranyl pyrophosphate by enzymatically controlled electrophilic carbon–carbon bond formations.

Camphor Biosynthesis from Geranyl Pyrophosphate

Cis, trans isomerization ⟶

−OPP⁻ ⟶

⟶

Geranyl pyrophosphate

is the same as ⟶ ⟶ ⟶

Camphor

Other higher terpenes are constructed by similar cyclization reactions.

To summarize, 1,3-butadiene polymerizes in a 1,2 or 1,4 manner to give polybutadienes with various amounts of cross-linking and therefore variable elasticity. Synthetic rubber can be made from 2-methyl-1,3-butadiene and contains varied numbers of E and Z double bonds. Natural rubber is constructed by isomerization of 3-methyl-3-butenyl pyrophosphate to the 2-butenyl system, ionization, and electrophilic (step-by-step) polymerization. Similar mechanisms account for the incorporation of 2-methyl-1,3-butadiene (isoprene) units into the polycyclic structure of terpenes.

Sunlight is split into its component colors (the visible spectrum) by raindrops to give a rainbow.

14-11 Electronic Spectra: Ultraviolet and Visible Spectroscopy

Section 10-2 explained that organic molecules may absorb radiation at various wavelengths. Spectroscopy is possible because the phenomenon is restricted to quanta of defined energies, $h\nu$, to effect specific excitations with energy change ΔE.

$$\Delta E = h\nu = \frac{hc}{\lambda} \qquad (c = \text{velocity of light})$$

This section will consider a form of spectroscopy that requires electromagnetic radiation of relatively high energy, within the wavelength ranges from 200 to 400 nm, called **ultraviolet spectroscopy,** and from 400 to 800 nm, **visible spectroscopy** (see Figure 10-2). Both are useful for investigating the electronic structures of unsaturated molecules and for measuring the extent of their conjugation.

A UV-visible spectrometer is constructed according to the general scheme in Figure 10-3. As in NMR, samples are usually dissolved in solvents that do not absorb in the spectral region under scrutiny. Examples are ethanol, methanol, and cyclohexane, none of which have absorption peaks above 200 nm. The events triggered by electromagnetic radiation at the UV and visible wavelengths, the excitation of electrons from filled bonding (and nonbonding) to unfilled antibonding molecular orbitals, are recorded as **electronic spectra.**

Ultraviolet and visible light give rise to electronic excitations

Consider the bonds in an average molecule: We can safely assume that, except for lone pairs, all electrons occupy bonding molecular orbitals. The compound is said to be in its **ground electronic state.** Electronic spectroscopy is possible because ultraviolet radiation and visible light have sufficient energy to transfer many such electrons to antibonding orbitals, thereby creating an **excited electronic state** (Figure 14-13). Dissipation of the absorbed energy may occur in the form of a chemical reaction (cf. Section 14-9), as emission of light (fluorescence, phosphorescence), or simply as emission of heat.

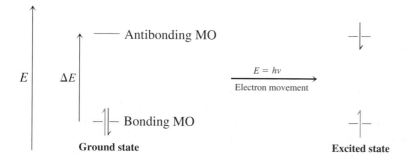

FIGURE 14-13

Electronic excitation by transfer of an electron from a bonding to an antibonding orbital converts a molecule from its ground electronic state into an excited state.

Organic σ bonds have a large gap between bonding and antibonding orbitals. To excite electrons in such bonds requires wavelengths much below the practical range (<200 nm). As a result, the technique has found its major uses in the study of π systems, in which filled and unfilled orbitals are much closer in energy. Excitation of such electrons gives rise to $\pi \rightarrow \pi^*$ **transitions.** Nonbonding (n) electrons are even more readily promoted through $n \rightarrow \pi^*$ **transitions** (Figure 14-14). Because the number of π molecular orbitals is equal to that of the component p orbitals, the simple

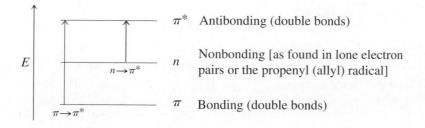

FIGURE 14-14

Electronic transitions in a simple π system. The wavelength of radiation required to cause them is revealed as a peak in the ultraviolet or visible spectrum.

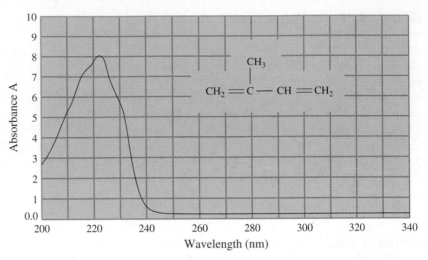

FIGURE 14-15

Ultraviolet spectrum of 2-methyl-1,3-butadiene in methanol, $\lambda_{max} = 222.5$ nm ($\epsilon = 10,800$). The two indentations at the sides of the main peak are called shoulders.

picture presented in Figure 14-14 is rapidly complicated by extending conjugation: The number of possible transitions skyrockets and with it the complexity of the spectra.

A typical UV spectrum is that of 2-methyl-1,3-butadiene (isoprene), shown in Figure 14-15. The position of a peak is defined by the wavelength at its maximum, a λ_{max} value (in nanometers). Its height is reported as the **molar extinction coefficient** or **molar absorptivity, ϵ,** that is characteristic of the molecule. The value of ϵ is calculated by dividing the measured peak height (absorbance, A) by the molar concentration, C, of the sample (assuming a standard cell length of 1 cm).

$$\epsilon = \frac{A}{C}$$

The value of ϵ can range from less than a hundred to several hundred thousand. It provides a good estimate of the efficiency of light absorption. Electronic spectral peaks are frequently broad, as in Figure 14-16, not the sharp lines typical of many NMR spectra.

FIGURE 14-16

The energy gap between the highest occupied and lowest unoccupied molecular orbitals (HOMO and LUMO) decreases along the series ethene, the 2-propenyl (allyl) radical, and butadiene. Excitation therefore requires less energy and is observed at longer wavelengths.

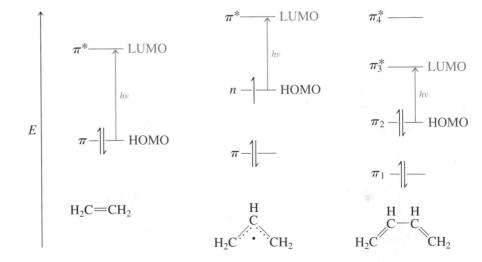

Electronic spectra tell us the extent of delocalization

Electronic spectra often indicate the size and degree of delocalization in an extended π system. The more double bonds there are in conjugation, the longer is the wavelength for the lowest energy excitation (and the more peaks will appear in the spectrum). For example, ethene absorbs at $\lambda_{max} = 171$ nm, and an unconjugated diene, such as 1,4-pentadiene, absorbs at $\lambda_{max} = 178$ nm. A conjugated diene, such as 1,3-butadiene, absorbs at much lower energy ($\lambda_{max} = 217$ nm). Further extension of the conjugated system leads to corresponding incremental increases in the λ_{max} values, as shown in Table 14-2. The hyperconjugation of alkyl groups and the improved π overlap in rigid, planar cyclic systems both seem to contribute. Beyond 400 nm (in the visible range), molecules become colored, first yellow, then orange, red, violet, and finally blue green. For example, the contiguous array of 11 double bonds in β-carotene is responsible for its characteristic intense orange appearance.

Why should larger conjugated π systems have more readily accessible and lower-energy excited states? The answer is pictured in Figure 14-16. As the overlapping *p*-orbital array gets longer, the energy gap between filled and unfilled orbitals gets smaller, and more bonding and antibonding orbitals are available to give rise to additional electronic excitations.

Finally, conjugation in cyclopolyenes is governed by a separate set of rules, to be introduced in the next two chapters. Just compare the electronic spectra of benzene (Figure 15-6) with that of azulene (Figure 14-17) and then compare both with the data in Table 14-2.

EXERCISE 14-22

Each substitution of a hydrogen by an alkyl group at an sp^2 carbon in a conjugated system causes the wavelength associated with the lowest energy $\pi \rightarrow \pi^*$ transition to increase by 5 nm. Use this fact and the measured value of λ_{max} for 1,3-butadiene to calculate λ_{max} for 2-methylbutadiene and 2,5-dimethyl-2,4-hexadiene. Compare your results with the measured values in Table 14-2.

In summary, UV and visible spectroscopy can be used to detect electronic excitations in conjugated molecules. With an increasing number of molecular orbitals, there is an increasing variety of possible transitions and hence number of absorption bands. The band of longest wavelength is typically associated with the movement of an electron from the highest occupied to the lowest unoccupied molecular orbital. Its energy decreases with increasing conjugation.

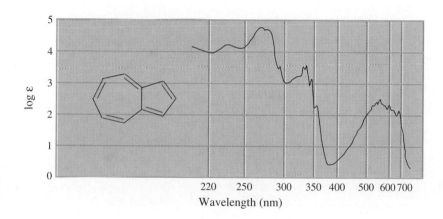

FIGURE 14-17

UV-visible spectrum of azulene in cyclohexane. The absorbance is plotted as log ϵ to compress the scale. The horizontal axis, representing wavelength, also is nonlinear.

Alkene structure	Name	λ_{max} (nm)	ϵ
	Ethene	171	15,500
	1,4-Pentadiene	178	Not measured
	1,3-Butadiene	217	21,000
	2-Methyl-1,3-butadiene	222.5	10,800
	trans-1,3,5-Hexatriene	268	36,300
	trans,trans-1,3,5,7-Octatetraene	330	Not measured
	2,5-Dimethyl-2,4-hexadiene	241.5	13,100
	1,3-Cyclopentadiene	239	4,200
	1,3-Cyclohexadiene	259	10,000
	A steroid diene	282	Not measured
	A steroid triene	324	Not measured
	A steroid tetraene	355	Not measured
(For structure, see Section 14-7.)	β-Carotene (Vitamin A precursor)	497 (orange appearance)	133,000
	Azulene, a cyclic conjugated hydrocarbon	696 (blue-violet appearance)	150

14-12 A Summary of Organic Reaction Mechanisms

Although we are only just past the halfway point in our survey of organic chemistry, with the completion of Chapter 14 we have in fact now seen examples of each of the three major classes of organic transformations: radical, polar, and pericyclic processes. This section summarizes all of the individual mechanism types that we have so far encountered in each of these reaction classes.

Radical reactions follow chain mechanisms

Radical reactions begin with the generation of a reactive odd-electron intermediate by means of an initiation step and convert starting materials into products through a chain of propagation steps. We have seen both **radical substitution** (Chapter 3) processes and **radical additions** (Chapter 12). Substitution is capable of introducing a functional group into a previously unfunctionalized molecule; radical addition is an example of functional-group interconversion. These individual subcategories are summarized in Table 14-3.

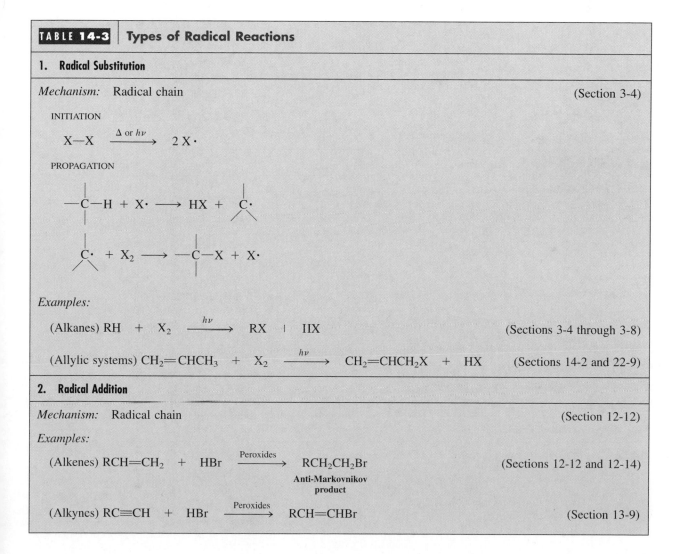

TABLE 14-3	**Types of Radical Reactions**

1. Radical Substitution

Mechanism: Radical chain   (Section 3-4)

INITIATION

$$X\!-\!X \xrightarrow{\ \Delta\ or\ h\nu\ } 2\,X\cdot$$

PROPAGATION

$$-\overset{|}{\underset{|}{C}}\!-\!H + X\cdot \longrightarrow HX + \overset{|}{C}\cdot$$

$$\overset{|}{C}\cdot + X_2 \longrightarrow -\overset{|}{\underset{|}{C}}\!-\!X + X\cdot$$

Examples:

(Alkanes) $RH + X_2 \xrightarrow{\ h\nu\ } RX + HX$ (Sections 3-4 through 3-8)

(Allylic systems) $CH_2\!=\!CHCH_3 + X_2 \xrightarrow{\ h\nu\ } CH_2\!=\!CHCH_2X + HX$ (Sections 14-2 and 22-9)

2. Radical Addition

Mechanism: Radical chain   (Section 12-12)

Examples:

(Alkenes) $RCH\!=\!CH_2 + HBr \xrightarrow{\ Peroxides\ } RCH_2CH_2Br$ (Sections 12-12 and 12-14)

Anti-Markovnikov product

(Alkynes) $RC\!\equiv\!CH + HBr \xrightarrow{\ Peroxides\ } RCH\!=\!CHBr$ (Section 13-9)

Polar reactions constitute the largest class of organic transformations

Interactions of polarized, or charged, species lead to the greatest variety in organic chemistry and to the largest number of reaction mechanism types: the typical chemistry of the organic functional groups. Two mechanisms each for **substitution** and **elimination** were first presented in Chapters 6 and 7. Both unimolecular and bimolecular pathways were found to be possible for each mechanism, depending on the structure of the substrate and, in some cases, the reaction conditions. With the introduction of functional groups containing π bonds, we encountered polar **addition reactions** in two distinct forms: nucleophilic in Chapter 8 and electrophilic in Chapter 12. These processes are summarized in Table 14-4.

TABLE 14-4	**Types of Polar Reactions**

1. Bimolecular Nucleophilic Substitution

Mechanism: Concerted backside displacement (S_N2) (Sections 6-4 and 6-5)

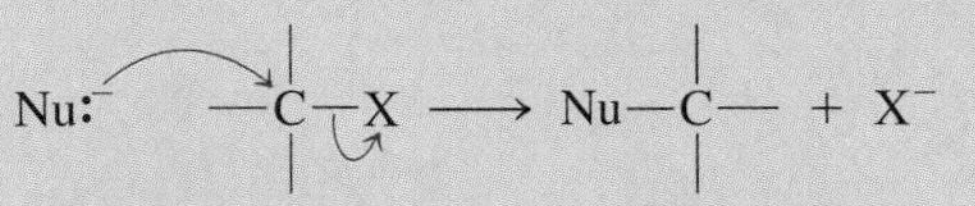

$$Nu: \quad -\overset{|}{\underset{|}{C}}-X \longrightarrow Nu-\overset{|}{\underset{|}{C}}- + X^-$$

Example:

$$HO^- + CH_3Cl \longrightarrow CH_3OH + Cl^-$$
100% Inversion at a stereocenter (Sections 6-3 through 6-9)

2. Unimolecular Nucleophilic Substitution

Mechanism: Carbocation formation–nucleophilic attack (S_N1; usually accompanied by E1) (Section 7-2)

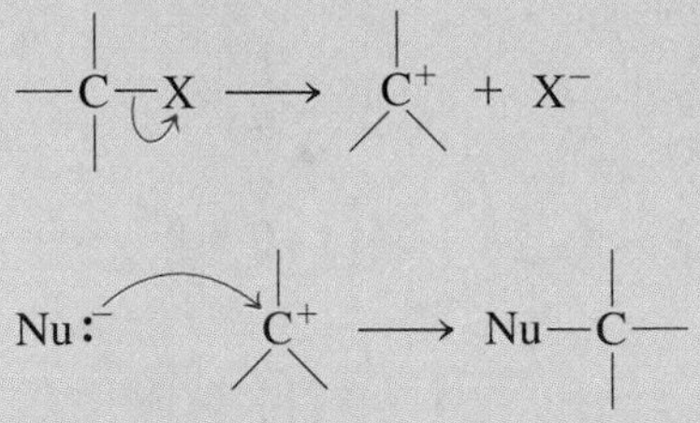

$$-\overset{|}{\underset{|}{C}}-X \longrightarrow \overset{+}{C} + X^-$$

$$Nu: \quad \overset{+}{C} \longrightarrow Nu-\overset{|}{\underset{|}{C}}-$$

Example:

$$H_2O + (CH_3)_3CCl \longrightarrow (CH_3)_3COH + HCl$$
Racemization at a stereocenter (Sections 7-2 through 7-5)

3. Bimolecular Elimination

Mechanism: Concerted deprotonation–π bond formation–expulsion of leaving group (E2) (Section 7-7)

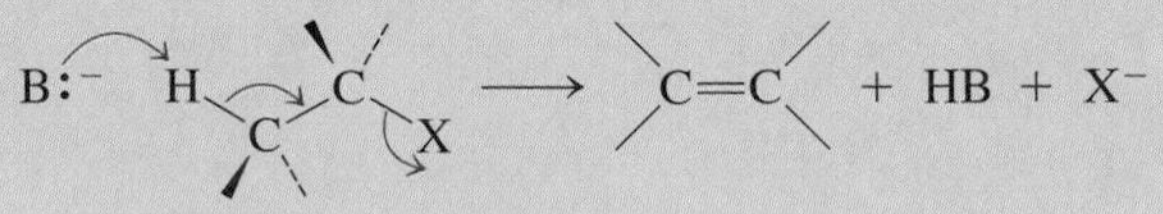

$$B:^- \quad H-\overset{}{\underset{C}{C}}\overset{}{\underset{X}{C}} \longrightarrow C=C + HB + X^-$$

TABLE 14-4	**Types of Polar Reactions** *(continued)*

Example:

$$CH_3CH_2O^- + CH_3CHClCH_3 \longrightarrow CH_3CH_2OH + CH_3CH=CH_2 + X^-$$
Anti transition state preferred (Sections 7-7 and 11-8)

4. Unimolecular Elimination

Mechanism: Carbocation formation–deprotonation and π bond formation (E1) accompanies S_N1 (Section 7-6)

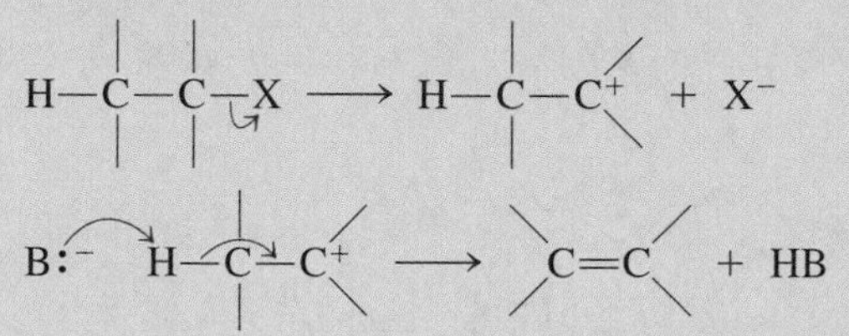

Example:

$$(CH_3)_3CCl \xrightarrow{H_2O} (CH_3)_2C=CH_2 + HCl$$
(Sections 7-6 and 11-9)

5. Nucleophilic Addition

Mechanism: Nucleophilic addition–protonation (Sections 8-6 and 8-8)

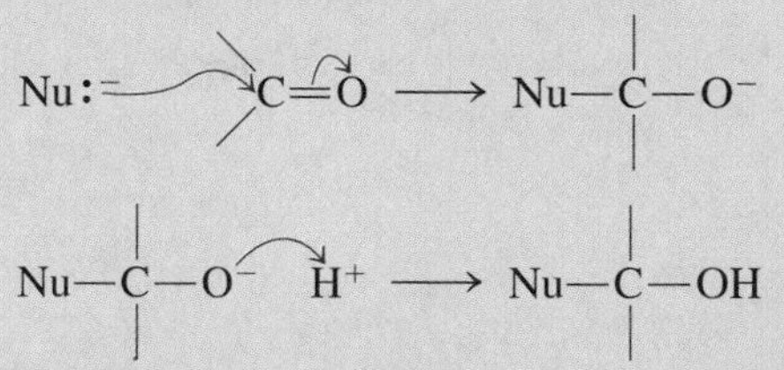

Examples:

(Hydride reagents) $NaBH_4 + (CH_3)_2C=O \longrightarrow (CH_3)_2CHOH$ (Section 8-6)

(Organometallic reagents) $RMgX + (CH_3)_2C=O \longrightarrow R-\overset{\underset{|}{CH_3}}{\underset{\underset{|}{CH_3}}{C}}-OH$ (Sections 8-8 and 14-4)

6. Electrophilic Addition

Mechanism: Electrophilic addition–nucleophilic attack (Sections 12-3 and 12-5)

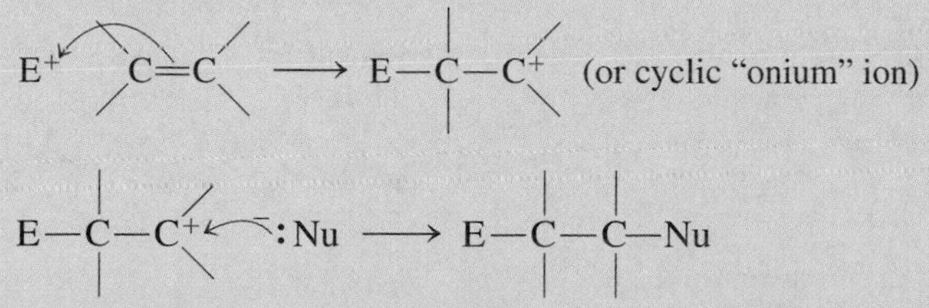

Example:

(Alkenes) $RCH=CH_2 + HBr \longrightarrow RCH\overset{\underset{|}{Br}}{}CH_3$
Markovnikov product (Sections 12-3 through 12-7)

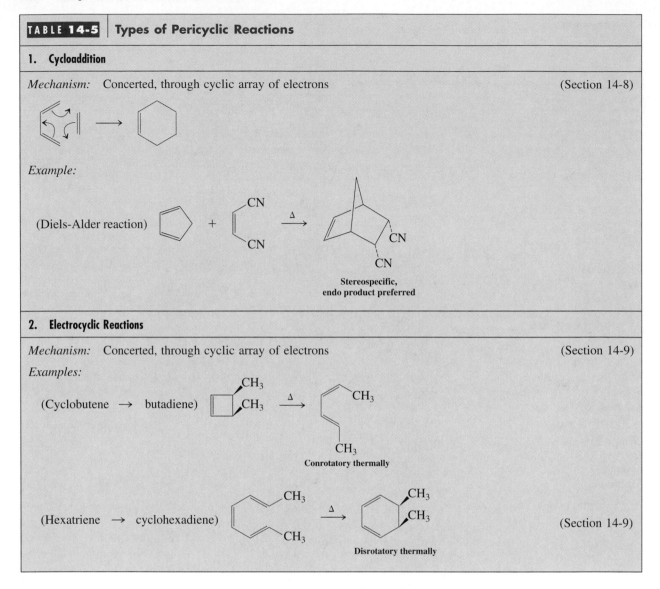

TABLE 14-5 | **Types of Pericyclic Reactions**

1. Cycloaddition

Mechanism: Concerted, through cyclic array of electrons (Section 14-8)

Example:

(Diels-Alder reaction)

**Stereospecific,
endo product preferred**

2. Electrocyclic Reactions

Mechanism: Concerted, through cyclic array of electrons (Section 14-9)

Examples:

(Cyclobutene → butadiene)

Conrotatory thermally

(Hexatriene → cyclohexadiene) (Section 14-9)

Disrotatory thermally

Pericyclic reactions lack intermediates

The final class of reactions are those characterized by cyclic transition states in which there is continuous overlap of a cyclic array of orbitals. These processes take place in a single step, without the intervention of any intermediate species. They may combine multiple components to introduce new rings, as in the Diels-Alder and other **cycloaddition reactions,** or they may take the form of ring-opening and ring-closing processes, the **electrocyclic reactions.** Examples are presented in Table 14-5.

CHAPTER INTEGRATION PROBLEM

a. Propose a reasonable mechanism for the transformation of *trans,trans*-2,4-hexadien-1-ol (sorbyl alcohol) into *trans*-5-ethoxy-1,3-hexadiene, as described in Chemical Highlight 14-1.

SOLUTION

This process takes place in acidic ethanol solution (otherwise known as wine). We begin by inspection of the structures of the starting and final molecules to find that (1) an alcohol has become an ether and (2) the double bonds have moved. Let us consider relevant information that we have seen concerning these functional groups. The conversion of two alcohols into an ether in the presence of acid was introduced in Section 9-7. Protonation of one alcohol molecule gives an alkyloxonium ion, whose leaving group (water) may be replaced in either an S_N2 or an S_N1 fashion by a molecule of a second alcohol:

$$CH_3CH_2\overset{+}{O}H_2 \ + \ CH_3CH_2OH \ \xrightarrow{S_N2} \ CH_3CH_2OCH_2CH_3$$

$$\underset{\underset{CH_3}{|}}{\overset{\overset{CH_3}{|}}{CH_3\overset{+}{C}OH_2}} \ + \ CH_3CH_2OH \ \xrightarrow{S_N1} \ \underset{\underset{CH_3}{|}}{\overset{\overset{CH_3}{|}}{CH_3COCH_2CH_3}}$$

Can either of these processes be adapted to the present question? Sorbyl alcohol is an allylic alcohol, and we have just seen (Section 14-3) that allylic *halides* readily undergo both S_N2 and S_N1 displacement reactions. A good deal of Chapter 9 dealt with comparing and contrasting the behavior of haloalkanes with that of alcohols: protonation of the OH group of an alcohol opens the way to both substitution and elimination chemistry. We should therefore expect that allylic alcohols will be similarly reactive. We need next to consider whether the S_N2 or the S_N1 mechanism is more appropriate.

The fact that the double bonds have moved in the course of the reaction is a helpful piece of information. Referring again to Section 14-3, we note that the allylic cation that results from dissociation of a leaving group in the first step of an S_N1 reaction is delocalized, and nucleophiles can therefore attach at more than one site. Let us explore this line of reasoning by examining the carbocation derived from protonation and loss of water from sorbyl alcohol:

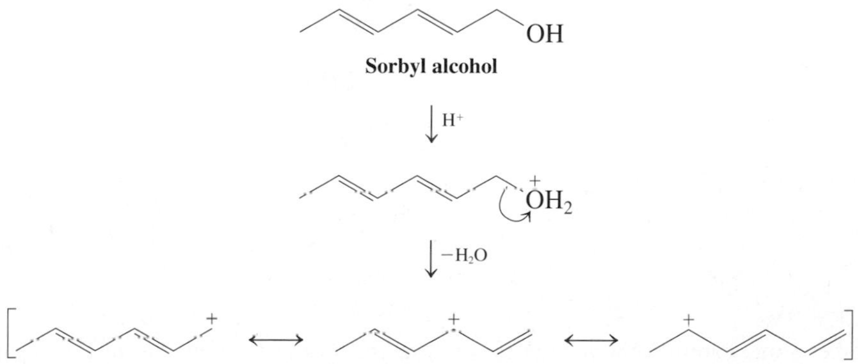

Sorbyl alcohol

Like an allylic cation, this carbocation is delocalized. However, the initial system is a more extended conjugated one, and the resulting cation is a hybrid of three, rather than two, contributing resonance forms. As in electrophilic addition to a conjugated triene (Section 14-7), an incoming nucleophile has the option of attaching to any of three positions. In this particular case, attachment of ethanol to a secondary carbon, where a relatively high fraction of the positive charge resides (Section 14-3), is the predominant (kinetic) result.

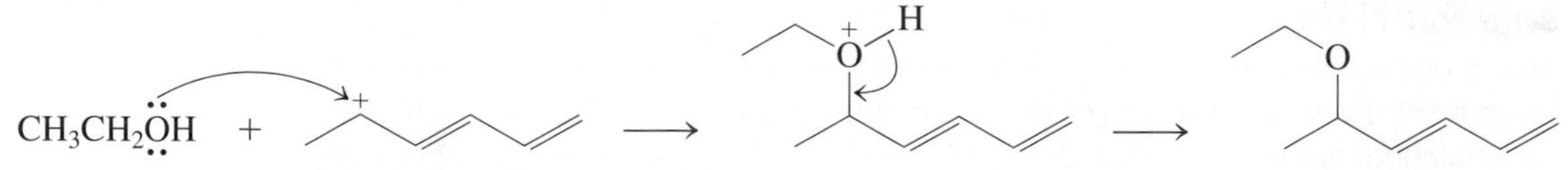

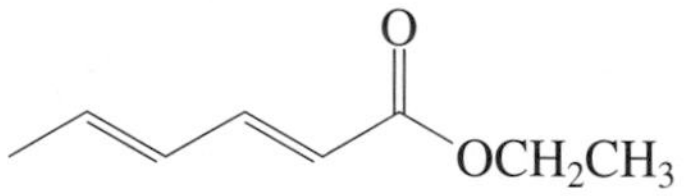

$$CH_3CH_2\ddot{O}H \ + \quad \longrightarrow \quad \longrightarrow$$

Ethyl *trans, trans*-2,4-hexadienoate
(Ethyl sorbate)

b. Esters of sorbic acid undergo Diels-Alder reactions. Predict the major product to arise from the heating of ethyl sorbate with 2-butenedioic anhydride (maleic anhydride, see structure, Table 14-1). Consider carefully all stereochemical aspects of the process.

SOLUTION

The Diels-Alder reaction is a cycloaddition between a diene, in this case the ethyl sorbate, and a dienophile, usually an electron-poor alkene (Section 14-8). We should first have a look at the two reaction partners to visualize the new bonding connections that this process will effect. To do so, we need to rotate about the single bond between carbon 3 and carbon 4 of ethyl sorbate to place its double bonds in the necessary conformation. We do this carefully, so as not to mistakenly change the stereochemistry of the double bonds: They both start out trans and must be that way after we have finished. Next, we connect the ends of the diene system to the carbon atoms of the dienophile's alkene function (dotted lines below), giving us the connectivity of the product, but without stereochemistry specified:

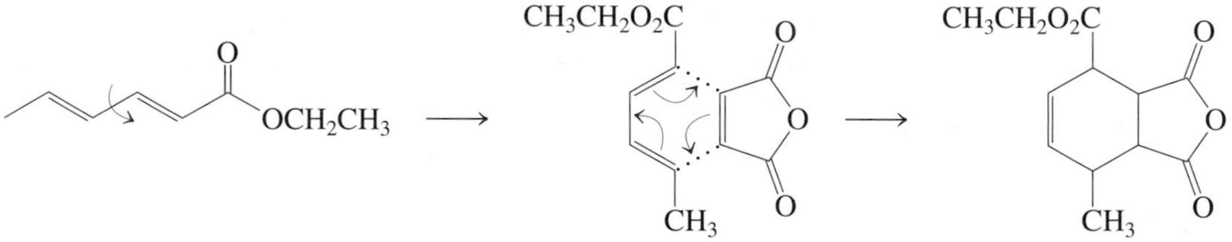

To complete the problem, we must finally take into account two details of the Diels-Alder reaction: (1) stereochemical relations in the components are retained in the course of the reaction; and (2) unsaturated substituents on the double bond of the dienophile prefer to locate themselves underneath the diene system (in endo positions) in the course of the cycloaddition. Making use of illustrations similar to those in the text, we can visualize the process as follows:

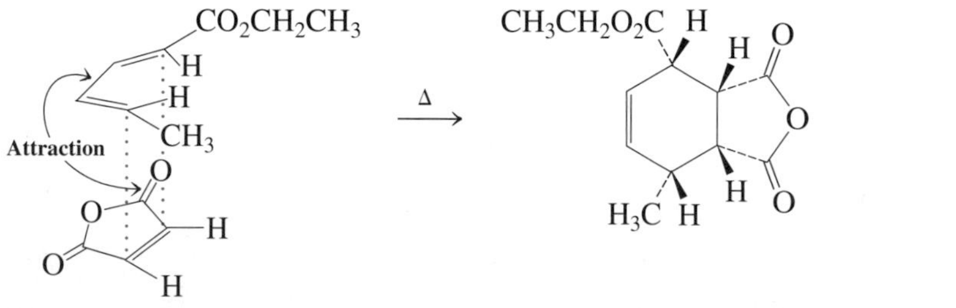

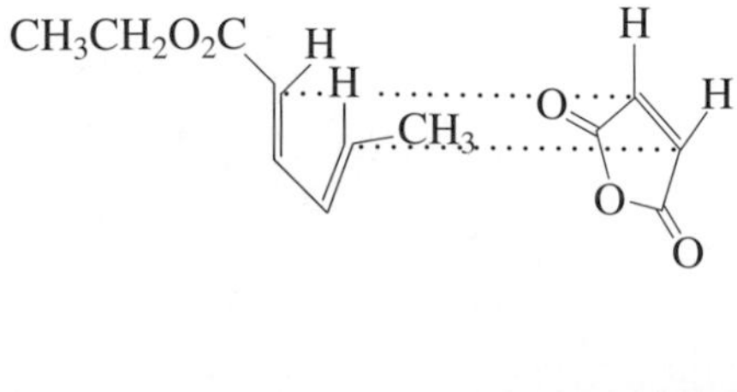

It takes some careful examination to see how the spatial arrangement of the groups in the sketch of the transition state translates into the final positions in the product. As you work on this aspect of the problem, look in particular at each carbon taking part in the formation of the two new single bonds [dotted lines in transition state (left side of reaction)]. Examine the positions of the substituents on these carbons relative to the two new bonds. A view that may be useful for this purpose is one in which the whole picture is rotated by 90° (see margin). It is reasonably clear from this perspective that the four hydrogen atoms will end up all cis to each other in the newly formed cyclohexene ring.

NEW REACTIONS

1. Radical Allylic Halogenations (Section 14-2)

$$RCH_2CH{=}CH_2 \xrightarrow[\substack{DH° \text{ of allylic C–H bond } \sim 87 \text{ kcal mol}^{-1}}]{NBS,\ CCl_4,\ hv} \underset{\overset{|}{Br}}{RCHCH{=}CH_2} + RCH{=}CHCH_2Br$$

2. Thermodynamic Compared with Kinetic Control in S_N1 Reactions of Allylic Derivatives (Section 14-3)

$$CH_3CH{=}CHCH_2X \xleftarrow{\text{Slow}} CH_3CH{=}CH\overset{+}{C}H_2 + X^- \xrightleftharpoons{\text{Fast}} \underset{\overset{|}{X}}{CH_3CHCH{=}CH_2}$$

More stable product **Stability of primary allylic cation ~ ordinary secondary cation** **Less stable product**

(Formation reversible at higher temperature)

3. S_N2 Reactivity of Allylic Halides (Section 14-3)

$$CH_2{=}CHCH_2X + Nu{:}^- \xrightarrow{\text{Propanone or DMSO}} CH_2{=}CHCH_2Nu + X^-$$

Faster than ordinary primary halides

4. Allylic Grignard Reagents (Section 14-4)

$$CH_2{=}CHCH_2Br \xrightarrow{Mg,\ (CH_3CH_2)_2O} CH_2{=}CHCH_2MgBr$$

Can be used in additions to carbonyl compounds

5. Allyllithium Reagents (Section 14-4)

$$RCH_2CH{=}CH_2 \xrightarrow{CH_3CH_2CH_2CH_2Li,\ TMEDA} R\ddot{C}HCH{=}CH_2\ Li^+$$

pK_a of allylic C–H bond ~ 40

6. Hydrogenation of Conjugated Dienes (Section 14-5)

$$CH_2{=}CH{-}CH{=}CH_2 \xrightarrow{H_2,\ Pd\text{-}C,\ CH_3CH_2OH} CH_3CH_2CH_2CH_3 \quad \Delta H° = -57.1 \text{ kcal mol}^{-1}$$

but compare

$$CH_2{=}CH{-}CH_2{-}CH{=}CH_2 \xrightarrow{H_2,\ Pd\text{-}C,\ CH_3CH_2OH} CH_3(CH_2)_3CH_3 \quad \Delta H° = -60.8 \text{ kcal mol}^{-1}$$

7. Electrophilic Reactions of 1,3-Dienes: 1,2- and 1,4-Addition (Section 14-6)

$$CH_2{=}CH{-}CH{=}CH_2 \xrightarrow{HX} \underset{\overset{|}{X}}{CH_2{=}CHCHCH_3} + XCH_2CH{=}CHCH_3$$

$$CH_2{=}CH{-}CH{=}CH_2 \xrightarrow{X_2} \underset{\overset{|}{X}}{CH_2{=}CHCHCH_2X} + XCH_2CH{=}CHCH_2X$$

8. Diels-Alder Reaction (Concerted and Stereospecific, Endo Rule) (Section 14-8)

A = electron acceptor
Requires *s*-cis diene; better with electron-poor dienophile

9. Electrocyclic Reactions (Section 14-9)

$$\xrightarrow[\text{Conrotatory}]{\Delta}$$

$$\xrightarrow[\text{Disrotatory}]{\Delta}$$

$$\xrightarrow[\text{Disrotatory}]{h\nu}$$

$$\underset{\text{Conrotatory}}{\overset{h\nu}{\rightleftharpoons}}$$

10. Polymerization of 1,3-Dienes (Section 14-10)

1,2-Polymerization

$$2n \ CH_2{=}CH{-}CH{=}CH_2 \xrightarrow{\text{Initiator}}$$

$$\begin{array}{cc} CH_2 & CH_2 \\ \| & \| \\ CH & CH \\ | & | \\ -(CH{-}CH_2{-}CH{-}CH_2)_n{-} \end{array}$$

1,4-Polymerization

$$n \ CH_2{=}CH{-}CH{=}CH_2 \xrightarrow{\text{Initiator}} -(CH_2{-}CH{=}CH{-}CH_2)_n{-}$$
$$\text{Cis or trans}$$

11. 3-Methyl-3-butenyl Pyrophosphate as a Biochemical Building Block (Section 14-10)

$$\underset{\substack{\textbf{3-Methyl-3-butenyl} \\ \textbf{pyrophosphate}}}{CH_2{=}\overset{\overset{\textstyle CH_3}{|}}{C}{-}CH_2CH_2OPP} \underset{\text{Enzyme}}{\rightleftharpoons} (CH_3)_2C{=}CHCH_2OPP \longrightarrow \underset{\textbf{Allylic cation}}{(CH_3)_2C{=}CHCH_2^+} \ + \ \underset{\substack{\textbf{Pyrophosphate} \\ \textbf{ion}}}{{}^-OPP}$$

C–C bond formation

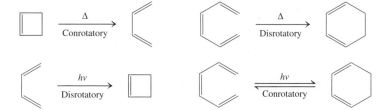

IMPORTANT CONCEPTS

1. The 2-propenyl (**allyl**) system is stabilized by **resonance.** Its molecular-orbital description shows the presence of three π molecular levels: one bonding, one nonbonding, and one antibonding. Its structure is symmetric, any charges or odd electrons being equally distributed between the two end carbons.

2. The chemistry of the 2-propenyl (allyl) cation is subject to both **thermodynamic and kinetic control.** Nucleophilic trapping may occur more rapidly at an internal carbon that bears relatively more positive charge, giving the thermodynamically less stable product. The kinetic product may rearrange to its thermodynamic isomer by dissociation followed by eventual thermodynamic trapping.

3. The stability of allylic radicals allows **radical halogenations** of alkenes at the allylic position.

4. The S_N2 reaction of allylic halides is accelerated by orbital overlap in the transition state.

5. The special stability of allylic anions allows **allylic deprotonation** by a strong base, such as butyllithium–TMEDA.

6. 1,3-Dienes reveal the effects of **conjugation** by their resonance energy and a relatively short internal bond (1.47 Å).

7. Electrophilic attack on 1,3-dienes leads to the preferential formation of allylic cations.

8. **Extended conjugated systems** are reactive because they have many sites for attack and the resulting intermediates are stabilized by resonance.

9. Benzene has special stability because of **cyclic delocalization.**

10. The **Diels-Alder reaction** is a concerted stereospecific **cycloaddition reaction** of an *s*-cis diene to a dienophile; it leads to cyclohexene derivatives. It follows the **endo rule.**

11. Conjugated dienes and trienes equilibrate with their respective cyclic isomers by concerted and stereospecific **electrocyclic reactions.**

12. **Polymerization** of 1,3-dienes results in 1,2- or 1,4-additions to give polymers that are capable of further **cross-linking.** Synthetic rubbers can be synthesized in this way. Natural rubber is made by electrophilic carbon–carbon bond formation involving biosynthetic five-carbon cations derived from 3-methyl-3-butenyl pyrophosphate.

13. **Ultraviolet and visible spectroscopy** gives a way of estimating the extent of conjugation in a molecule. Peaks in **electronic spectra** are usually broad and are reported as λ_{max} (nm). Their relative intensities are given by the **molar absorptivity** (extinction coefficient) ϵ.

PROBLEMS

23. Draw all resonance forms and a representation of the appropriate resonance hybrid for each of the following species.

(a)   (b)   (c)   (d)   (e)

24. For each species in Problem 23, indicate the resonance form that is the major contributor to the resonance hybrid. Explain your choices.

25. Illustrate by means of appropriate structures (including all relevant resonance forms) the initial species formed by (a) breaking the weakest C–H bond in 1-butene; (b) treating 4-methylcyclohexene with a powerful base (e.g., butyllithium–TMEDA); (c) heating a solution of 3-chloro-1-methylcyclopentene in aqueous ethanol.

26. Rank primary, secondary, tertiary, and allylic radicals in order of decreasing stability. Do the same for the corresponding carbocations. Do the results indicate something about the relative ability of hyperconjugation and resonance to stabilize radical and cationic centers?

27. Give the major product(s) of each of the following reactions. If more than one product forms, indicate which is kinetic (i.e., formed fastest at low temperature and short reaction time) and which is thermodynamic (i.e., formed in highest yield at higher temperature after longer reaction times).

(a) 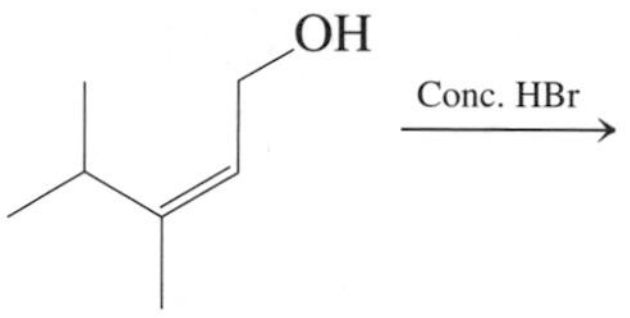 Conc. HBr →

(b) 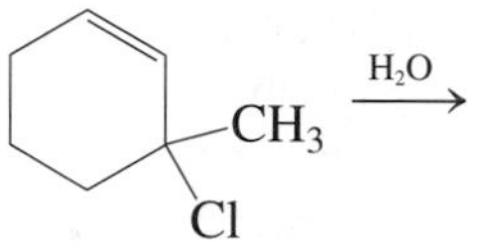 H_2O →

(c) 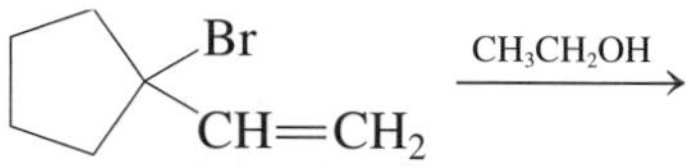 CH_3CH_2OH →

(d) 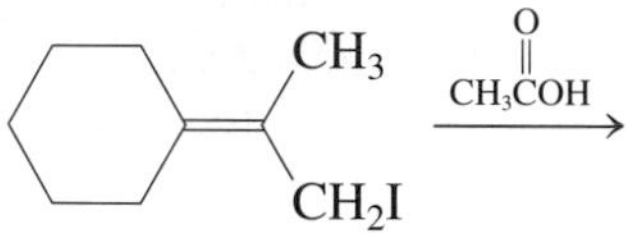 CH_3COH →

(e) 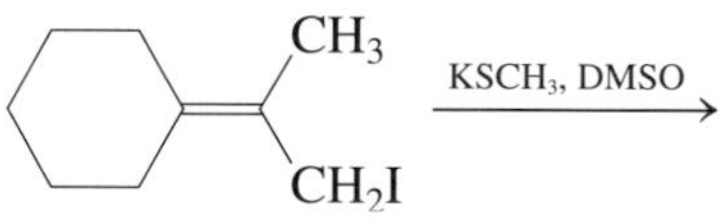 $KSCH_3$, DMSO →

(f) 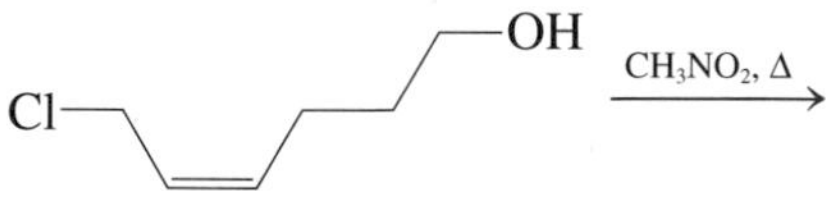 CH_3NO_2, Δ →

28. Formulate detailed mechanisms for the reactions in Problem 27 (a, c, e, f).

29. Rank primary, secondary, tertiary, and (primary) allylic chlorides in approximate order of (a) decreasing S_N1 reactivity; (b) decreasing S_N2 reactivity.

30. Rank the following six molecules in approximate order of decreasing S_N1 reactivity and decreasing S_N2 reactivity.

(a)

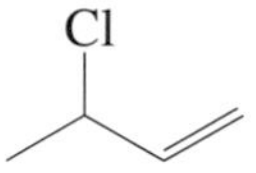

(b)

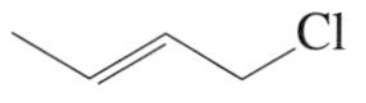

(c)

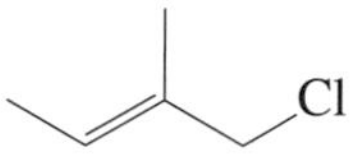

(d)

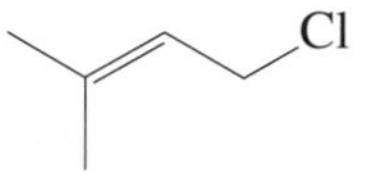

(e)

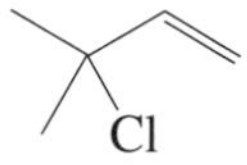

(f)

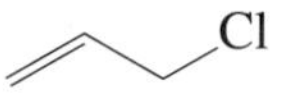

31. How would you expect the S_N2 reactivities of simple saturated primary, secondary, and tertiary chloroalkanes to compare with the S_N2 reactivities of the compounds in Problem 30? Make the same comparison for S_N1 reactivities.

32. Give the major product(s) of each of the following reactions.

(a) 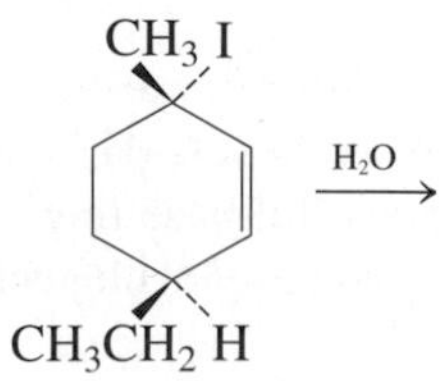 H_2O →

(b) 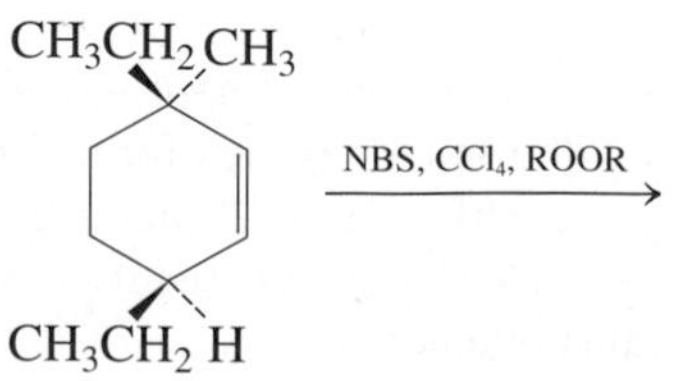 NBS, CCl_4, ROOR →

(c) 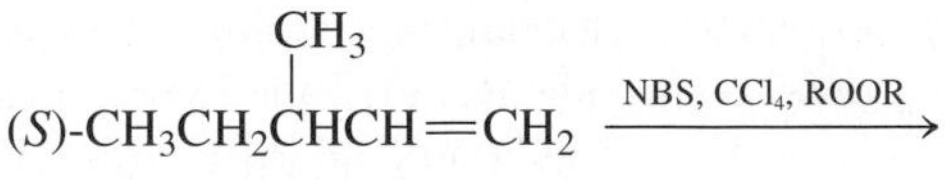 (S)-CH$_3$CH$_2$CHCH=CH$_2$ $\xrightarrow{\text{NBS, CCl}_4,\text{ ROOR}}$

with CH$_3$ substituent

(d)

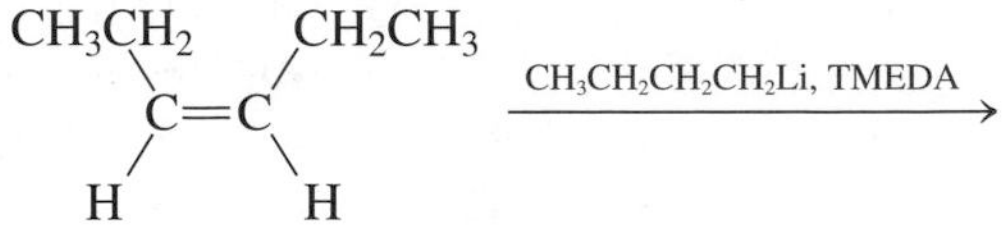

(CH$_3$CH$_2$)(CH$_2$CH$_3$)C=C(H)(H) $\xrightarrow{\text{CH}_3\text{CH}_2\text{CH}_2\text{CH}_2\text{Li, TMEDA}}$

(e) Product of (d) 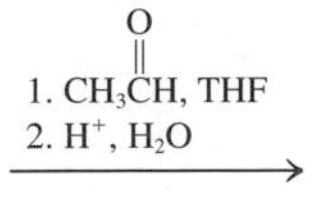 $\xrightarrow[\text{2. H}^+,\text{ H}_2\text{O}]{\text{1. CH}_3\overset{\text{O}}{\overset{\|}{\text{CH}}},\text{ THF}}$

(f)

(CH$_3$)$_2$C=CH—C(H)(CH$_3$)(Br) $\xrightarrow{\text{KSCH}_3,\text{ DMSO}}$

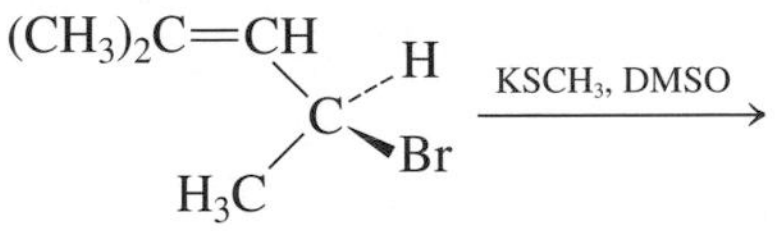

33. The following reaction sequence gives rise to two isomeric products. What are they? Explain the mechanism of their formation.

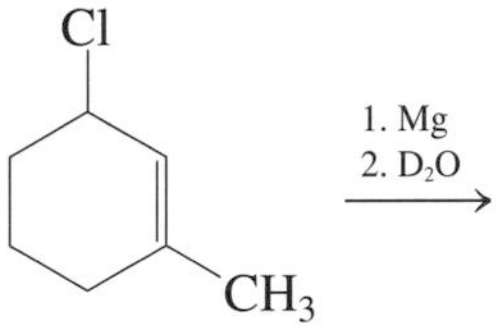

Cl-substituted cyclohexene with CH$_3$ $\xrightarrow[\text{2. D}_2\text{O}]{\text{1. Mg}}$

34. Starting with cyclohexene, propose a reasonable synthesis of the cyclohexene derivative shown in the margin.

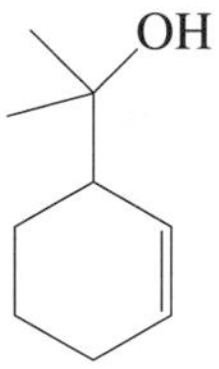

2-(cyclohex-2-enyl)propan-2-ol (OH)

35. Give a systematic name to each of the following molecules.

(a)

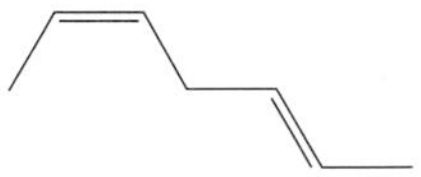

(b) 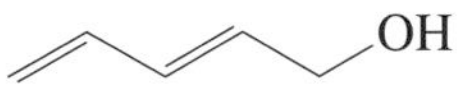 structure with OH

(c) 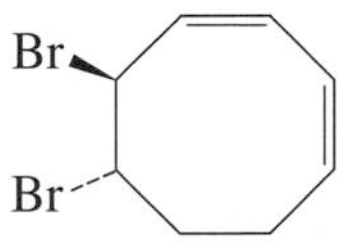 cyclooctadiene with two Br

(d) 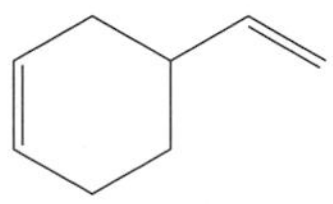 cyclohexene with vinyl group

36. Compare the allylic bromination reactions of 1,3-pentadiene and 1,4-pentadiene. Which should be faster? Which is more energetically favorable? How do the product mixtures compare?

CH$_2$=CH—CH=CH—CH$_3$ $\xrightarrow{\text{NBS, ROOR, CCl}_4}$

CH$_2$=CH—CH$_2$—CH=CH$_2$ $\xrightarrow{\text{NBS, ROOR, CCl}_4}$

37. Compare the addition of H$^+$ to 1,3-pentadiene and 1,4-pentadiene (see Problem 36). Draw the structures of the products. Draw a qualitative reaction profile showing both dienes and both proton addition products on the same graph. Which diene adds the proton faster? Which one gives the more stable product?

38. What products would you expect from the electrophilic addition of each of the following reagents to 1,3-cycloheptadiene? **(a)** HI; **(b)** Br$_2$ in H$_2$O; **(c)** IN$_3$; **(d)** H$_2$SO$_4$ in CH$_3$CH$_2$OH.

39. Give the products of the reaction of *trans*-1,3-pentadiene with each of the reagents in Problem 38.

40. What are the products of reaction of 2-methyl-1,3-pentadiene with each of the reagents in Problem 38?

41. Give the products expected from reaction of deuterium iodide (DI) with **(a)** 1,3-cycloheptadiene; **(b)** *trans*-1,3-pentadiene; **(c)** 2-methyl-1,3-pentadiene. In what way does the observable result of reaction of DI differ from that of reaction of HI with these same substrates [compare with Problems 38(a), 39(a), and 40(a)]?

42. Arrange the following carbocations in order of decreasing stability. Draw all possible resonance forms for each of them.

(a) $CH_2{=}CH{-}\overset{+}{C}H_2$ (b) $CH_2{=}\overset{+}{C}H$ (c) $CH_3\overset{+}{C}H_2$

(d) $CH_3{-}CH{=}CH{-}\overset{+}{C}H{-}CH_3$ (e) $CH_2{=}CH{-}CH{=}CH{-}\overset{+}{C}H_2$

43. Sketch the molecular orbitals for the pentadienyl system in order of ascending energy (see Figures 14-2 and 14-8). Indicate how many electrons are present, and in which orbitals, for **(a)** the radical; **(b)** the cation; **(c)** the anion (see Figures 14-3 and 14-8). Draw all reasonable resonance forms for any one of these three species.

44. Dienes may be prepared by elimination reactions of substituted allylic compounds. For example,

Propose detailed mechanisms for each of these 2-methyl-1,3-butadiene (isoprene) syntheses.

45. Give the structures of all possible products of the acid-catalyzed dehydration of vitamin A (Section 14-7).

46. Propose a synthesis of each of the following molecules by Diels-Alder reactions.

47. Propose an efficient synthesis of the cyclohexenol in the margin, beginning exclusively with acyclic starting materials and employing sound retrosynthetic analysis strategy. (**Hint:** A Diels-Alder reaction may be useful, but be sure to take note of structural features in dienes and dienophiles that permit Diels-Alder reactions to work well [Section 14-8].)

Dimethyl azodicarboxylate

48. Dimethyl azodicarboxylate (see margin) takes part in the Diels-Alder reaction as a dienophile. Write the structure of the product of cycloaddition of this molecule with each of the following dienes. **(a)** 1,3-Butadiene; **(b)** *trans,trans*-2,4-hexadiene; **(c)** 5,5-dimethoxycyclopentadiene; **(d)** 1,2-dimethylenecyclohexane.

49. Bicyclic diene A reacts readily with appropriate alkenes by the Diels-Alder reaction, whereas diene B is totally unreactive. Explain.

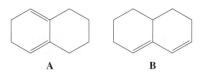

A   B

50. Formulate the expected product of each of the following reactions.

(a) H₃CO $\xrightarrow{hv}$ CH₃O

(b) $\xrightarrow{hv}$ (with D, H, D, H substituents)

(c) H₃C, CH₃ ... H, CH₃, H₃C, H $\xrightarrow{\Delta}$

(d) $\xrightarrow{\Delta}$

51. Give abbreviated structures of each of the following compounds. **(a)** (*E*)-1,4-Poly-2-methyl-1,3-butadiene [(*E*)-1,4-polyisoprene]; **(b)** 1,2-poly-2-methyl-1, 3-butadiene (1,2-polyisoprene); **(c)** 3,4-poly-2-methyl-1,3-butadiene (3,4-polyisoprene); **(d)** copolymer of 1,3-butadiene and ethenylbenzene (styrene, $C_6H_5CH=CH_2$, SBR, used in automobile tires); **(e)** copolymer of 1,3-butadiene and propenenitrile (acrylonitrile, $CH_2=CHCN$, latex); **(f)** copolymer of 2-methyl-1,3-butadiene (isoprene) and 2-methylpropene (butyl rubber, for inner tubes).

52. The structure of the terpene limonene is shown in the margin. Identify the two 2-methyl-1,3-butadiene (isoprene) units in limonene. **(a)** Treatment of isoprene with catalytic amounts of acid leads to a variety of oligomeric products, one of which is limonene. Devise a detailed mechanism for the acid-catalyzed conversion of two molecules of isoprene into limonene. Take care to use sensible intermediates in each step. **(b)** Two molecules of isoprene may also be converted into limonene by a completely different mechanism, which takes place in the strict absence of catalysts of any kind. Describe this mechanism. What is the name of the reaction?

CH₃

Limonene (H₃C, CH₂)

53. The carbocation derived from geranyl pyrophosphate (Section 14-10) is the biosynthetic precursor of not only camphor, but also limonene (Problem 52) and α-pinene (Chapter 4, Problem 32). Formulate mechanisms for the formation of the latter two compounds.

54. What is the longest wavelength electronic transition in each of the following species? Use molecular-orbital designations such as $n\rightarrow\pi^*$, $\pi_1\rightarrow\pi_2$, in your answer. (**Hint:** Prepare a molecular-orbital energy diagram such as that in Figure 14-16 for each.) **(a)** 2-Propenyl (allyl) cation; **(b)** 2-propenyl (allyl) radical; **(c)** formaldehyde, $H_2C=O$; **(d)** N_2; **(e)** pentadienyl anion (Problem 43); **(f)** 1,3,5-hexatriene.

55. Ethanol, methanol, and cyclohexane are commonly used solvents for UV spectroscopy because they do not absorb radiation of wavelength longer than 200 nm. Why not?

56. The ultraviolet spectrum of a 2×10^{-4} M solution of 3-penten-2-one exhibits a $\pi \rightarrow \pi^*$ absorption at 224 nm with $A = 1.95$ and an $n \rightarrow \pi^*$ band at 314 nm with $A = 0.008$. Calculate the molar absorptivities (extinction coefficients) for these bands.

57. In a published synthetic procedure, propanone (acetone) is treated with ethenyl (vinyl) magnesium bromide, and the reaction mixture is then neutralized with strong aqueous acid. The product exhibits ^{1}H NMR spectrum A. What is its structure?

When the reaction mixture is (improperly) allowed to remain in contact with aqueous acid for too long, a mixture of products is obtained. This mixture gives rise to ^{1}H NMR spectrum B. New signals are present at $\delta = 1.77$ (several lines), 4.10 (a doublet with $J = 8$ Hz), and 5.45 (a broad triplet with $J = 8$ Hz) ppm with relative intensities $6:2:1$. What is the structure of the new compound? How did it get there?

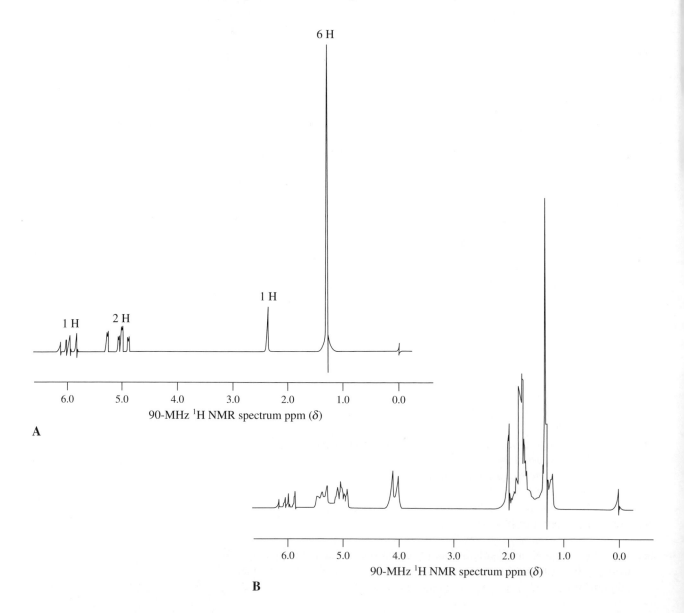

58. Farnesol is a molecule that makes flowers smell good (lilacs, for instance). Treatment with hot concentrated H_2SO_4 converts farnesol first into bisabolene and finally into cadinene, a compound of the essential oils of junipers and cedars. Propose detailed mechanisms for these conversions.

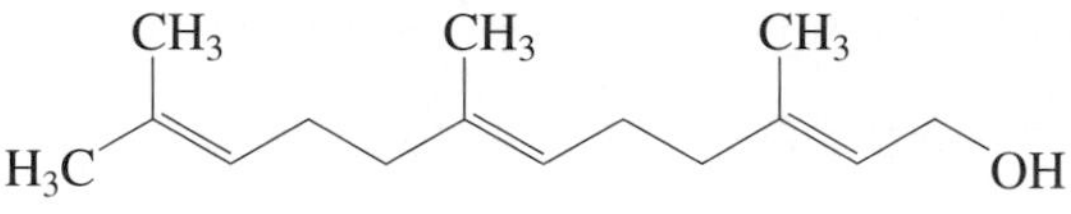

Farnesol

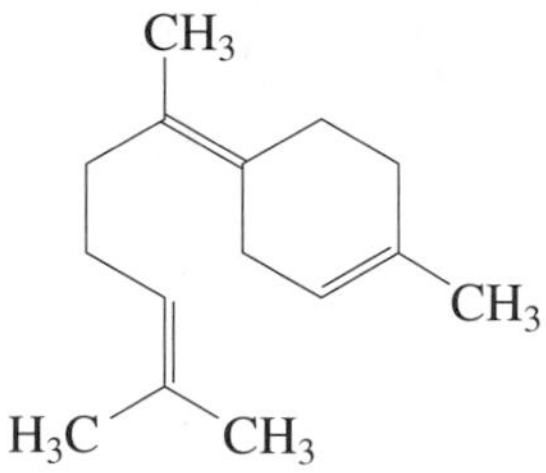

Bisabolene

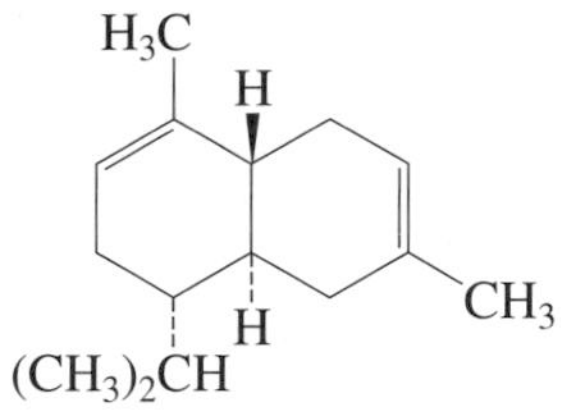

Cadinene

59. The ratio of 1,2- to 1,4-addition of Br_2 to 1,3-butadiene (Section 14-6) is temperature dependent. Identify the kinetic and thermodynamic products, and explain your choices.

60. Diels-Alder cycloaddition of 1,3-butadiene with the cyclic dienophile shown in the margin takes place at only one of the two carbon–carbon double bonds in the latter to give a single product. Give its structure and explain your answer. Watch stereochemistry.

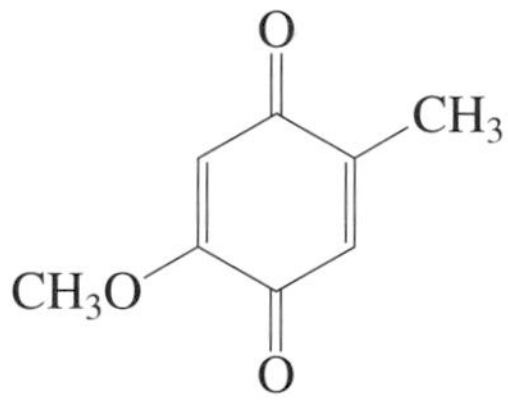

   This transformation was the initial step in the total synthesis of cholesterol (Section 4-7), completed by R. B. Woodward (see Section 14-9) in 1951. This achievement, monumental for its time, revolutionized synthetic organic chemistry.

Team Problem

61. As a team consider the following historic preparation of a tris(1,1-dimethylethyl) derivative of Dewar benzene, B, by the photochemical isomerisation of 1,2,4-tris(1,1-dimethylethyl)benzene by van Tamelen and Pappas (1962). B does not revert to A via either a thermal or photochemical electrocyclic mechanism. Formulate a mechanism for the conversion of A to B and explain the kinetic robustness of B with respect to the regeneration of A.

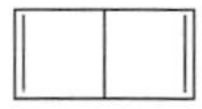

Dewar benzene

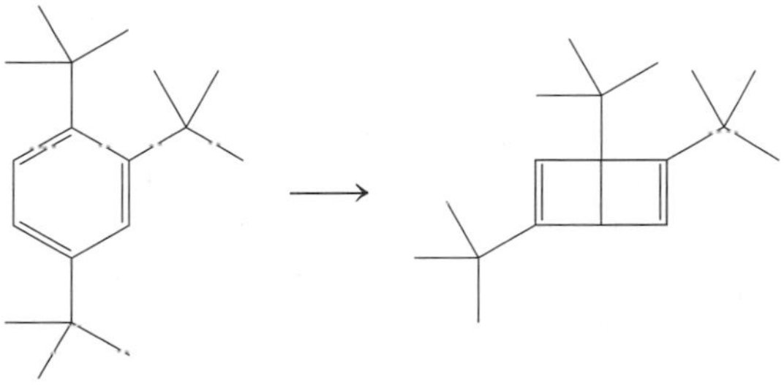

A

B

Preprofessional Problems

62. How many nodes are present in the LUMO (lowest unoccupied molecular orbital) of 1,3-butadiene?
   (a) Zero; **(b)** one; **(c)** two; **(d)** three; **(e)** four

63. Arrange the following three chlorides in decreasing order of S_N1 reactivity.

$$CH_3CH_2CH_2Cl \qquad H_2C{=}CHCHCH_3 \qquad CH_3CH_2CHCH_3$$

          Cl   Cl

A   B   C

(a) $A > B > C$; **(b)** $B > C > A$; **(c)** $B > A > C$; **(d)** $C > B > A$.

64. When cyclopentadiene is treated with tetracyanoethene, a new product results. Its most likely structure is

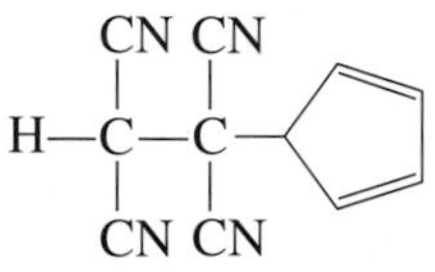

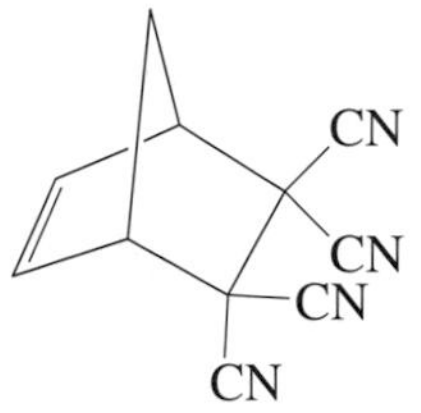

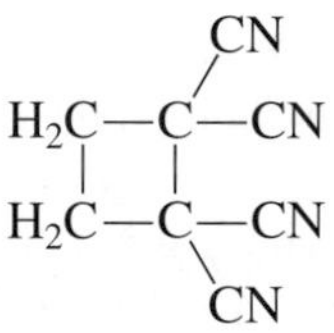

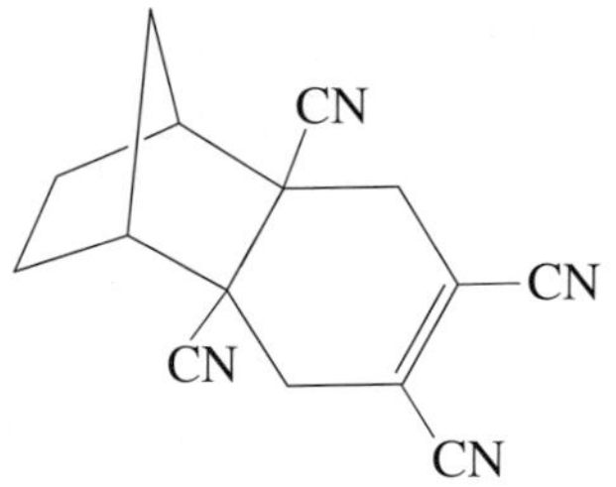

65. Which common analytical method will most clearly and rapidly distinguish A from B?

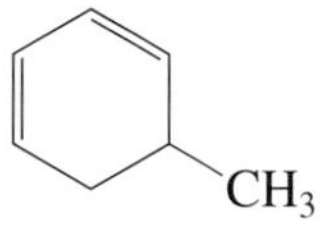

(a) IR spectroscopy; **(b)** UV spectroscopy; **(c)** combustion analysis; **(d)** visible spectroscopy.

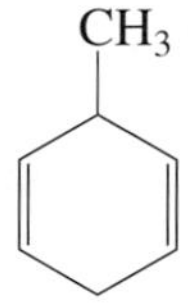

Benzene, Other Cyclic Polyenes, and Electrophilic Aromatic Substitution

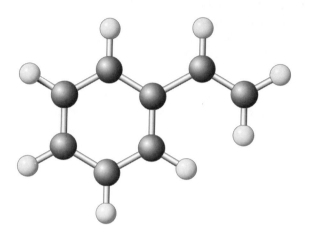

Polystyrene, in the form of Styrofoam, is a resilient polymer of low thermal conductivity used in packing materials and various molded products. Foaming is induced by the addition of volatiles (such as hexane) during the polymerization process of ethenylbenzene (styrene) and subsequent heating. The photo shows a volunteer sculpting a Teddy bear for a float in a Thanksgiving Day parade.

During the early 19th century, whales supplied the oil (from whale fat, or "blubber") used to illuminate the streets of London and other cities. Eager to determine its components, in 1825 the English scientist Michael Faraday* heated whale oil to obtain a colorless liquid (b.p. 80.1°C, m.p. 5.5°C) that had the empirical formula CH. This compound posed a problem for the theory that carbon had to have four valences to other atoms, and it was of particular interest because of its unusual stability and chemical inertness. The molecule was named **benzene** (Section 14-7), and it was eventually shown to have the molecular formula C_6H_6. This hydrocarbon has four degrees of unsaturation (see Table 11-5), satisfied by a 1,3,5-cyclohexatriene structure (one ring, three double bonds) first proposed by Kekulé[†] and Loschmidt,[‡] yet it does not exhibit the kind of reactivity expected for a conjugated triene (Section 14-7). As a consequence, various investigators proposed several incorrect structures, such as Dewar benzene, Claus benzene, Ladenburg[§] prismane, and benzvalene. Since then, Dewar benzene, prismane, and benzvalene (but not Claus benzene) have in fact been synthesized. These compounds are unstable and isomerize to benzene in very exothermic transformations.

*Professor Michael Faraday (1791–1867), Royal Institute of Chemistry, London.
[†]Professor F. August Kekulé; see Section 1-4.
[‡]Professor Josef Loschmidt (1821–1895), University of Vienna, Austria.
[§]Named for Professor James Dewar (1842–1923), Cambridge University, England; Professor Adolf Carl Ludwig Claus (1838–1900), University of Freiburg, Germany; and Professor Albert Ladenburg (1842–1911), University of Breslau, Germany, respectively.

Proposed Benzene Structures

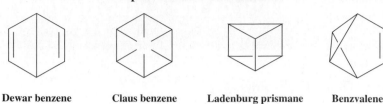

Dewar benzene   **Claus benzene**   **Ladenburg prismane**   **Benzvalene**

Although relatively inert, benzene is not completely unreactive, however. For example, it will be attacked by bromine, albeit only in the presence of catalytic amounts of a Lewis acid, such as iron tribromide, $FeBr_3$ (Section 15-10). Surprisingly, the result is not addition but substitution to furnish bromobenzene.

Benzene   **Bromobenzene**   **Addition product
(Substitution product)   not formed**

The formation of only one product of monobromination is nicely consistent with the sixfold symmetry of the structure of benzene. Further reaction with bromine introduces a second halogen atom to give the three isomeric 1,2-, 1,3-, and 1,4-dibromobenzenes.

1,2-Dibromobenzene   **1,3-Dibromo-**   **1,4-Dibromo-**
(the same as 1,6-dibromobenzene)   **benzene**   **benzene**

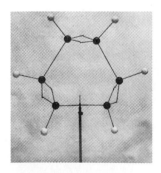

Kekulé's original model of
benzene.

Historically, the observation of only one 1,2-dibromobenzene product posed another puzzle. If the molecule has a cyclohexatriene topology with alternating single and double bonds, two isomers should have been formed: 1,2- and 1,6-dibromobenzene, in which the two substituents are connected by double or single carbon–carbon bonds, respectively. Kekulé solved this puzzle brilliantly by boldly proposing that benzene should be viewed as a set of two rapidly equilibrating (he used the word "oscillating") cyclohexatriene *isomers,* which would render 1,2- and 1,6-dibromobenzene indistinguishable. We now know that this notion was not quite right. In terms of modern electronic theory, benzene is a single compound, best described by two equivalent cyclohexatrienic *resonance* forms (Section 14-7).

Why does the cyclic array of π electrons in benzene impart unusual stability? How can.we quantify this observation? Considering what we know about the diagnostic power of NMR (Section 11-4) and electronic spectroscopy (Section 14-11) in establishing the presence of delocalized π systems, will there be special effects for benzene? This chapter will answer these questions. We look first at the system of naming substituted benzenes, second at the electronic and molecular structure of the parent molecule, and third at the evidence for its unusual stabilizing energy, the **aromaticity** of benzene. This aromaticity and the special structure of benzene affect its spectral properties and reactivity. Subsequently, we shall see what happens when two or more benzene rings are fused to give more extended π systems. Do these compounds also enjoy the special stability and properties of benzene? Is aromaticity unique to six-membered rings, or do other cyclic polyenes possess this property? Finally, you will learn the mechanism by which benzene is brominated, as shown earlier, and that this transformation is only one example of a class of reactions that introduce substituents into the benzene ring: electrophilic aromatic substitution.

15-1 Naming the Benzenes

Many derivatives of benzene were originally called **aromatic compounds** because of their strong aroma. Benzene, even though its odor is not particularly pleasant, is viewed as the "parent" aromatic molecule. Whenever the symbol for the benzene ring with its three double bonds is written, it should be understood to represent only one of a pair of contributing resonance forms. Alternatively, the ring is sometimes drawn as a regular hexagon with an inscribed circle.

Many monosubstituted benzenes are named by adding a substituent prefix to the word benzene.

is the same as

Fluorobenzene   Nitrobenzene   (1,1-Dimethylethyl)benzene
(*tert*-Butylbenzene)

There are three possible arrangements of disubstituted benzenes. These arrangements are designated by the prefixes **1,2- (ortho,** or **o-,** Greek, straight) for adjacent substituents, **1,3 (meta-,** or **m-,** Greek, transposed) for 1,3-disubstitution, and **1,4 (para-,** or **p-,** Greek, beyond) for 1,4-disubstitution. The substituents are listed in alphabetical order.

1,2-Dichlorobenzene   1-Bromo-3-nitrobenzene   1-Ethyl-4-(1-methylethyl)benzene
(*o*-Dichlorobenzene)   (*m*-Bromonitrobenzene)   (*p*-Ethylisopropylbenzene)

To name tri- and more highly substituted derivatives, the six carbons of the ring are numbered with the lowest set of locants, and the substituents are labeled accordingly, as in cyclohexane nomenclature.

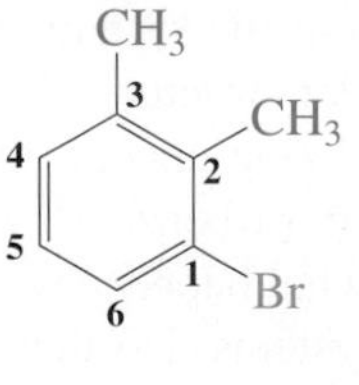

1-Bromo-2,3-dimethylbenzene 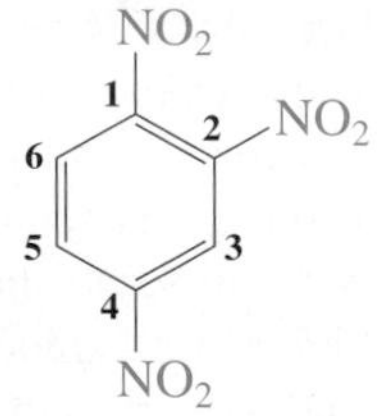   **1,2,4-Trinitrobenzene** 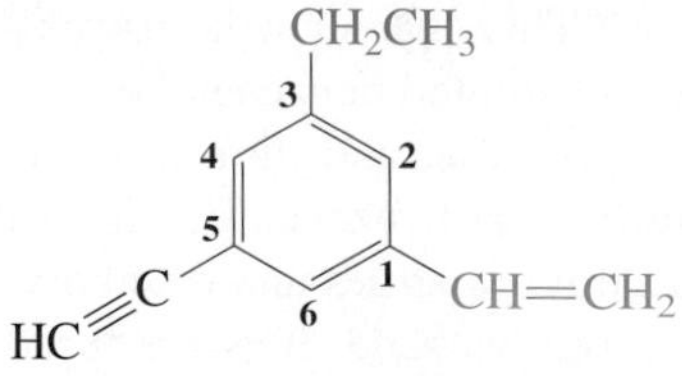   **1-Ethenyl-3-ethyl-5-ethynylbenzene**

The following benzene derivatives will be encountered in this book.

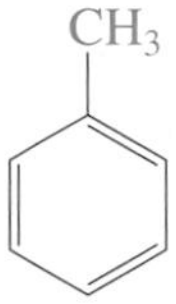

Methylbenzene
(Toluene)

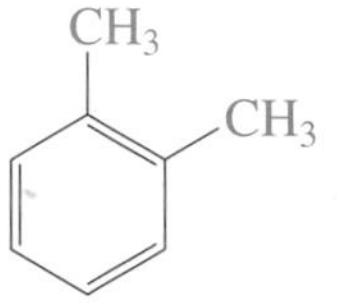

1,2-Dimethylbenzene
(o-Xylene)

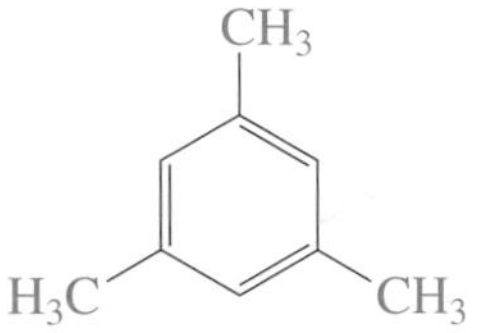

1,3,5-Trimethylbenzene
(Mesitylene)

(Common industrial and laboratory solvents)

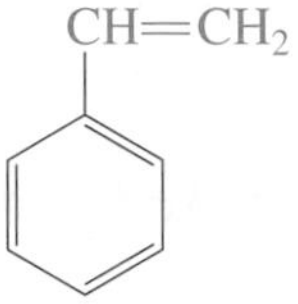

Ethenylbenzene
(Styrene)

(Used in polymer manufacture)

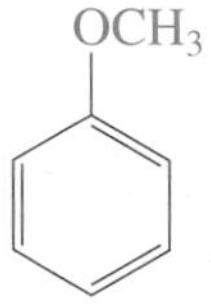

Methoxybenzene
(Anisole)

(Used in perfume)

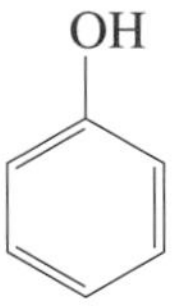

Benzenol
(Phenol)

(An antiseptic and anesthetic used in sore throat sprays)

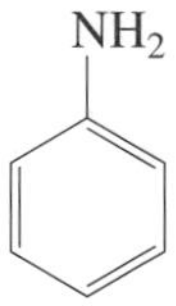

Benzenamine
(Aniline)

(Used in dye manufacture)

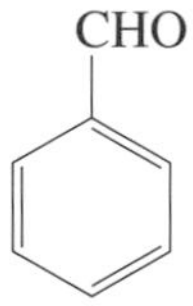

Benzenecarbaldehyde
(Benzaldehyde)

(An artificial flavoring)

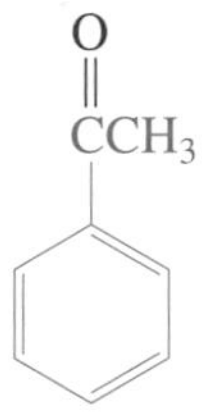

1-Phenylethanone
(Acetophenone)

(A hypnotic drug)

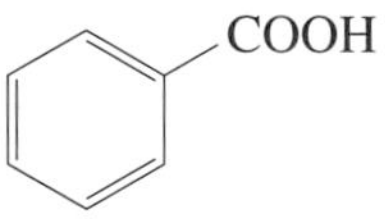

Benzenecarboxylic acid
(Benzoic acid)

(A food preservative)

We shall employ IUPAC nomenclature for all but three systems. In accord with the indexing preferences of *Chemical Abstracts,* the common names phenol, benzaldehyde, and benzoic acid will be used in place of their systematic counterparts.

Ring-substituted derivatives of such compounds are named by numbering the ring positions or by using the prefixes *o-*, *m-*, and *p-*. The substituent that gives the compound its base name is placed at carbon 1.

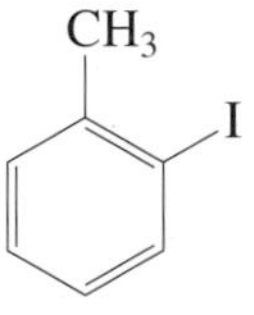

1-Iodo-2-methylbenzene
(o-Iodotoluene)

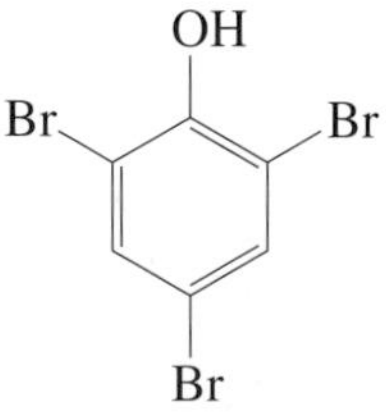

2,4,6-Tribromophenol

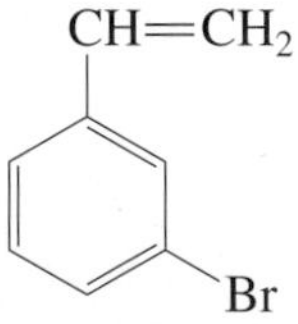

1-Bromo-3-ethenylbenzene
(m-Bromostyrene)

A number of the common names for aromatic compounds refer to their fragrance and natural sources. Several of them have been accepted by IUPAC. As before, a consistent logical naming of these compounds will be adhered to as much as possible, with common names mentioned in parentheses.

Aromatic Flavoring Agents

Methyl 2-**hydroxy**-
benzoate
(Methyl salicylate,
oil of wintergreen flavor)

4-**Hydroxy**-3-methoxy-
benzaldehyde
(Vanillin, vanilla flavor)

5-**Methyl**-2-(1-methyl-
ethyl)phenol
(Thymol, thyme flavor)

The generic term for substituted benzenes is **arene.** An arene as a substituent is referred to as an **aryl group,** abbreviated **Ar.** The parent aryl substituent is **phenyl,** C_6H_5. The $C_6H_5CH_2-$ group, which is related to the 2-propenyl (allyl) substituent (Sections 14-1 and 22-1), is called **phenylmethyl (benzyl).**

Phenylmethanol
(Benzyl alcohol)

EXERCISE 15-1

Write systematic and common names of the following substituted benzenes.

(a) **(b)** **(c)**

trans-1-(4-Bromophenyl)-
2-methylcyclohexane

EXERCISE 15-2

Draw the structures of **(a)** (1-methylbutyl)benzene; **(b)** 1-ethenyl-4-nitrobenzene (*p*-nitrostyrene); **(c)** 2-methyl-1,3,5-trinitrobenzene (2,4,6-trinitrotoluene—the explosive TNT).

EXERCISE 15-3

The following names are incorrect. Write the correct form. **(a)** 3,5-Dichlorobenzene; **(b)** *o*-aminophenyl fluoride; **(c)** *p*-fluorobromobenzene.

In summary, simple singly substituted benzenes are named by placing the substituent name before the word benzene. For more highly substituted systems, 1,2-, 1,3-, and 1,4- or ortho, meta, and para prefixes indicate the positions of disubstitution or the ring is numbered and the so-labeled substituents are named in alphabetical order. Many simple substituted benzenes have common names.

15-2 Structure and Resonance Energy of Benzene: A First Look at Aromaticity

Benzene is unusually unreactive: At room temperature, benzene is inert to acids, H_2, Br_2, and $KMnO_4$, reagents that readily add to conjugated alkenes (Section 14-6). This section explains why that is so: The cyclic six-electron arrangement imparts a special stability in the form of a large resonance energy (Section 14-7). We shall first review the evidence for the structure of benzene and then estimate its resonance energy by comparing its heat of hydrogenation with those of model systems that lack cyclic conjugation, such as 1,3-cyclohexadiene.

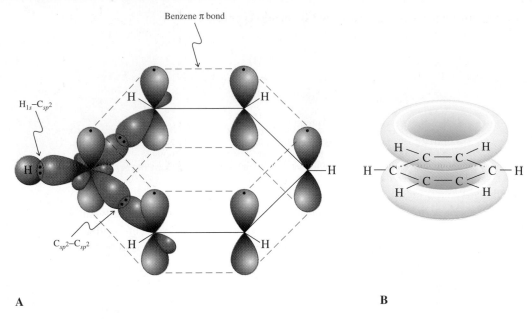

Benzene π bond

H_{1s}–C_{sp^2}

C_{sp^2}–C_{sp^2}

A

B

FIGURE 15-1

Orbital picture of the bonding in benzene. (A) The σ framework is depicted as straight lines except for the bonding to one carbon, in which the p orbital and the sp^2 hybrids are shown explicitly. (B) The six overlapping p orbitals in benzene form a π electron cloud located above and below the molecular plane.

The benzene ring contains six equally overlapping *p* orbitals

The electronic structure of the benzene ring is shown in Figure 15-1. All carbons are sp^2 hybridized, and each p orbital overlaps to an equal extent with its two neighbors. The consequently delocalized electrons form a π *cloud* above and below the ring.

According to this picture, the benzene molecule should be a completely symmetric hexagon with equal C–C bond lengths. Such is in fact the experimentally determined structure (Figure 15-2), which reveals the absence of alternation between single and double bonds. This type of alternation would be expected if benzene were a conjugated triene, a "cyclohexatriene." The C–C bond length in benzene is 1.39 Å, *between* the values found for the single (1.47 Å) and the double bond (1.34 Å) in 1,3-butadiene (Figure 14-7).

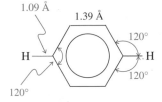

1.09 Å

1.39 Å

120°

120°

120°

H

H

FIGURE 15-2

The molecular structure of benzene. All six C–C bonds are equal; all bond angles are 120°.

Benzene is especially stable: heats of hydrogenation

A way to establish the relative stability of a series of alkenes is to measure their heats of hydrogenation (Sections 11-7 and 14-5). We may carry out a similar experiment with benzene, relating its heat of hydrogenation to those of 1,3-cyclohexadiene and cyclohexene. These molecules are conveniently compared because hydrogenation changes all three into cyclohexane.

The hydrogenation of cyclohexene is exothermic by -28.6 kcal mol^{-1}, a value expected for the hydrogenation of a cis double bond (Section 11-7). The heat of hydrogenation of 1,3-cyclohexadiene ($\Delta H^\circ = -54.9$ kcal mol^{-1}) is slightly less than double that of cyclohexene, because of the resonance stabilization in a conjugated diene (Section 14-5); the energy of that stabilization is $(2 \times 28.6) - 54.9 = 2.3$ kcal mol^{-1}.

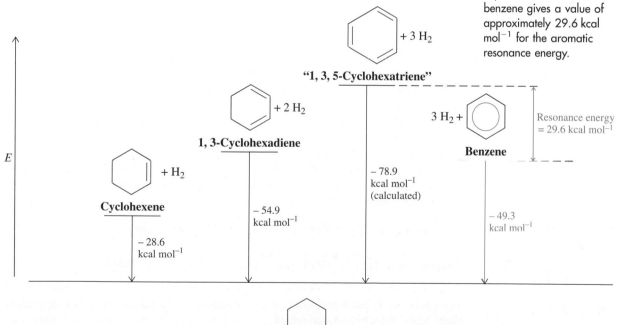

Armed with these numbers, we can calculate the expected heat of hydrogenation of benzene, as though it were simply composed of three double bonds like that of cyclohexene.

FIGURE 15-3

Heats of hydrogenation provide a measure of benzene's unusual stability. Experimental values for cyclohexene and 1,3-cyclohexadiene allow us to estimate the heat of hydrogenation for the hypothetical "1,3,5-cyclohexatriene." Comparison with the experimental $\Delta H°$ for benzene gives a value of approximately 29.6 kcal mol^{-1} for the aromatic resonance energy.

Now let us look at the experimental data. Although benzene is hydrogenated only with difficulty (Section 14-7), special catalysts carry out this reaction, so the heat of hydrogenation of benzene can be measured: $\Delta H° = -49.3$ kcal mol^{-1}, much less than the -78.9 kcal mol^{-1} predicted.

Figure 15-3 summarizes these results. It is immediately apparent that benzene is *much* more stable than a cyclic triene containing alternating single and double bonds.

The difference is the **resonance energy** of benzene, about 30 kcal mol^{-1}. Other terms used to describe this quantity are *delocalization energy, aromatic stabilization,* or simply the **aromaticity** of benzene. The original meaning of the word *aromatic* has changed with time, now referring to a thermodynamic property rather than to odor.

To summarize, the structure of benzene is a regular hexagon made up of six sp^2-hybridized carbons. The C–C bond length is between those of a single and a double bond. The electrons occupying the p orbitals form a π cloud above and below the plane of the ring. The structure of benzene can be represented by two equally contributing cyclohexatriene resonance forms. Hydrogenation to cyclohexane releases about 30 kcal mol^{-1} less energy than is expected on the basis of nonaromatic models. This difference is the resonance energy of benzene.

15-3 Pi Molecular Orbitals of Benzene

This section compares the six π molecular orbitals of benzene with those of 1,3,5-hexatriene, the open-chain analog. Both sets are the result of the contiguous overlap of six p orbitals, yet the cyclic system differs considerably from the acyclic one. A comparison of the energies of the bonding orbitals in these two compounds shows that cyclic conjugation of three double bonds is better than acyclic conjugation.

Cyclic overlap modifies the energy of benzene's molecular orbitals

Figure 15-4 compares the π molecular orbitals of benzene with those of 1,3,5-hexatriene. The acyclic triene follows a pattern similar to that of 1,3-butadiene (Figure 14-8), but with two more molecular orbitals: The orbitals all have different energies, the number of nodes increasing in the procession from π_1 to π_6. The picture for benzene is different in all respects: different orbital energies, two sets of degenerate (equal energy) orbitals, and completely different nodal patterns.

Is the cyclic π system more stable than the acyclic one? To answer this question, we have to compare the combined energies of the three filled bonding orbitals in both. Figure 15-5 (see page 646) shows the answer: *The cyclic π system is stabilized relative to the acyclic one.* On going from 1,3,5-hexatriene to benzene, two of the bonding orbitals (π_1 and π_3) are lowered in energy and one is raised; the effect of the latter is more than offset by the former.

Inspection of the signs of the wave functions at the terminal carbons of 1,3,5-hexatriene (Figure 15-4) explains why the orbital energies change in this way: Linkage of C1 and C6 causes p orbital overlap that is in phase for π_1 and π_3 but out of phase for π_2.

Some reactions have aromatic transition states

The structure of benzene accounts in a simple way for several reactions that proceed readily by what has seemed to be a complicated, concerted movement of three electron pairs: the Diels-Alder reaction (Section 14-8), osmium tetroxide addition to alkenes (Section 12-10), and the first step of ozonolysis (Section 12-11). In all three processes, there is a transition state with cyclic overlap of six electrons in π orbitals (or orbitals with π character). This electronic arrangement is similar to that in ben-

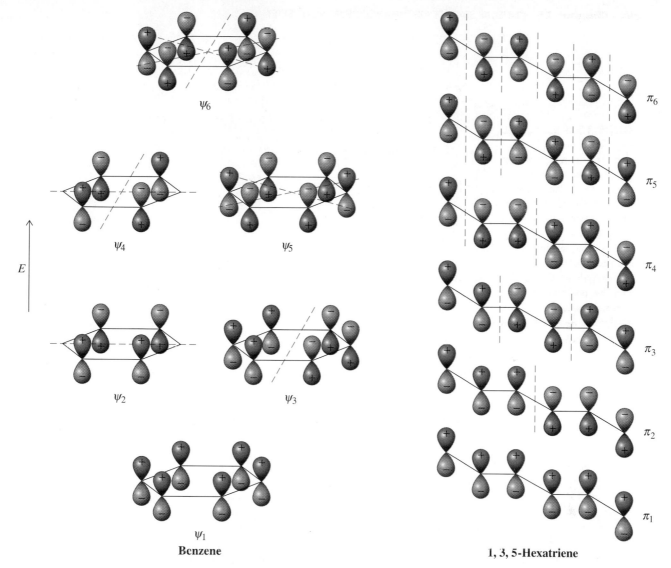

Benzene

1, 3, 5-Hexatriene

E

ψ_6

ψ_4 ψ_5

ψ_2 ψ_3

ψ_1

π_6

π_5

π_4

π_3

π_2

π_1

FIGURE 15-4

Pi molecular orbitals of benzene compared with those of 1,3,5-hexatriene. The orbitals are shown at equal size for simplicity. Favorable overlap (bonding) takes place between orbital lobes of equal sign. A sign change is indicated by a node (dashed line). As the number of nodes increases, so does the energy of the orbitals. Note that benzene has two sets of degenerate (equal energy) orbitals, the lower energy set occupied (ψ_2, ψ_3), the other not (ψ_4, ψ_5), as shown in Figure 15-5.

zene and is energetically more favored than the alternative, sequential bond breaking and bond making. Such transition states are called aromatic.

Aromatic Transition States

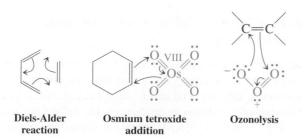

Diels-Alder reaction

Osmium tetroxide addition

Ozonolysis

645

FIGURE 15-5

Energy levels of the π molecular orbitals in benzene and 1,3,5-hexatriene. In both, the six π electrons fill the three bonding molecular orbitals. In benzene, two of them are lower in energy and one is higher than the corresponding orbitals in 1,3,5-hexatriene. Overall, the cumulative energy loss outweighs energy gain. Thus, the net result is a more stable system of six π electrons.

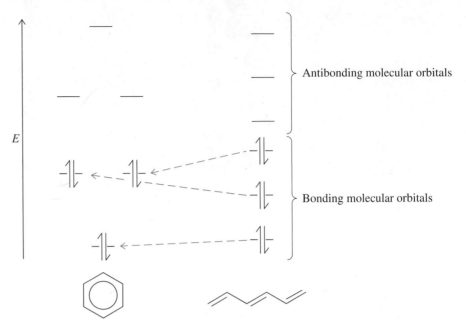

EXERCISE 15-4

If benzene were a cyclohexatriene, 1,2-dichloro- and 1,2,4-trichlorobenzene should each exist as two isomers. Draw them.

EXERCISE 15-5

The thermal ring opening of cyclobutene to 1,3-butadiene is exothermic by about 10 kcal mol^{-1} (Section 14-9). Conversely, the same reaction for benzocyclobutene, A, to compound B (shown in the margin) is *endothermic* by the same amount. Explain.

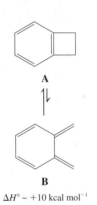

A

B

$\Delta H^{\circ} \sim +10 \text{ kcal mol}^{-1}$

To summarize, two of the filled π molecular orbitals of benzene are lower in energy than are those in 1,3,5-hexatriene. Benzene is therefore stabilized by considerably more resonance energy than is its acyclic analog. A similar orbital structure also stabilizes aromatic transition states.

15-4 Spectral Characteristics of the Benzene Ring

Is the unique nature of benzene and its derivatives revealed in their spectra? This section shows that the electronic arrangement in these molecules gives rise to characteristic ultraviolet spectra. The hexagonal structure is also manifest in specific infrared bands. What is most striking, cyclic delocalization causes induced ring currents in NMR spectroscopy, resulting in unusual deshielding of hydrogens attached to the aromatic ring. Moreover, the different coupling constants of 1,2- (ortho), 1,3- (meta), and 1,4- (para) hydrogens in substituted benzenes indicate the substitution pattern.

The UV-visible spectrum of benzene reveals its electronic structure

Cyclic delocalization in benzene gives rise to a characteristic arrangement of energy levels for its molecular orbitals (Figure 15-4). In particular, the energetic gap between bonding and antibonding orbitals is relatively large (Figure 15-5). Is this manifested in its electronic spectrum? As shown in Section 14-11, the answer is yes: Compared with the spectra of the acyclic trienes, smaller λ_{max} values are expected for benzene

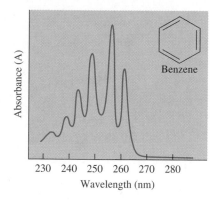

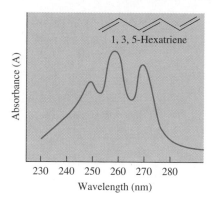

FIGURE 15-6

The distinctive ultraviolet spectra of benzene: $\lambda_{max}(\epsilon) = 234(30), 238(50), 243(100),$ $249(190), 255(220), 261(150)$ nm; and 1,3,5-hexatriene: $\lambda_{max}(\epsilon) = 247(33,900),$ $258(43,700), 268(36,300)$ nm. The extinction coefficients of the absorptions of 1,3,5-hexatriene are very much larger than those of benzene; therefore the spectrum at the right was taken at lower concentration.

and its derivatives. You can verify this effect in Figure 15-6: The highest wavelength absorption for benzene occurs at 261 nm, closer to that of 1,3-cyclohexadiene (259 nm, Table 14-2) than to that of 1,3,5-hexatriene (268 nm).

The ultraviolet and visible spectra of aromatic compounds vary with the introduction of substituents; this phenomenon has been exploited in the tailored synthesis of dyes (Section 22-11). Simple substituted benzenes absorb between 250 and 290 nm. For example, the water-soluble 4-aminobenzoic acid (*p*-aminobenzoic acid, or PABA), $4\text{-}H_2N\text{-}C_6H_4\text{-}COOH$, has a λ_{max} at 289 nm, with a rather high extinction coefficient of 18,600. Because of this property, it is used in suntan lotions, in which it filters out the dangerous ultraviolet radiation emanating from the sun in this wavelength region. Some individuals cannot tolerate PABA because of an allergic skin reaction, and many recent sun creams use other compounds for protection from the sun ("PABA-free").

Suntan lotions are applied to protect the skin from potential cancer-causing high-energy UV radiation. They contain substances, such as PABA, that absorb light in the damaging region of the electromagnetic spectrum.

The infrared spectrum reveals substitution patterns in benzene derivatives

The infrared spectra of benzene and its derivatives show characteristic bands in three regions. The first is at 3030 cm^{-1} for the phenyl–hydrogen stretching mode. The second ranges from 1500 to 2000 cm^{-1} and includes aromatic ring C–C stretching vibrations. Finally, a useful set of bands due to C–H out-of-plane bending motions is found between 650 and 1000 cm^{-1}.

Typical Infrared C–H Out-of-Plane Bending Vibrations for Substituted Benzenes (cm^{-1})

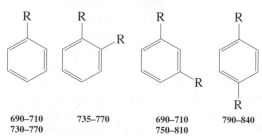

| 690–710 | 735–770 | 690–710 | 790–840 |
| 730–770 | | 750–810 | |

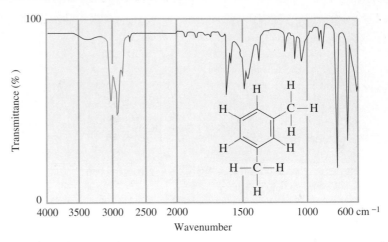

FIGURE 15-7

The infrared spectrum of 1,3-dimethylbenzene (*m*-xylene). There are two C–H stretching absorptions, one due to the aromatic bonds (3030 cm^{-1}), the other to saturated C–H bonds (2920 cm^{-1}). The two bands at 690 and 765 cm^{-1} are typical of 1,3-disubstituted benzenes.

Their precise location indicates the specific substitution pattern. For example, 1,2-dimethylbenzene (*o*-xylene) has this band at 738 cm^{-1}, the 1,4 isomer at 793 cm^{-1}; the 1,3 isomer (Figure 15-7) shows two absorptions in this range, at 690 and 765 cm^{-1}.

The NMR spectra of benzene derivatives show the effects of an electronic ring current

^{1}H NMR is a powerful spectroscopic technique for the identification of benzene and its derivatives. The cyclic delocalization of the aromatic ring gives rise to unusual deshielding, which causes the ring hydrogens to resonate at very low field ($\delta \sim$ 6.5–8.5 ppm), even lower than the already rather deshielded alkenyl hydrogens ($\delta \sim$ 4.6–5.7 ppm, see Section 11-4).

The ^{1}H NMR spectrum of benzene, for example, exhibits a sharp singlet for the six equivalent hydrogens at $\delta = 7.27$ ppm. How can this strong deshielding be explained? In a simplified picture, the cyclic π system with its delocalized electrons may be compared to a loop of conducting metal. When such a loop is exposed to a perpendicular magnetic field (H_0), an electric current (called a **ring current**) flows in the loop, in turn generating a new local magnetic field (h_{local}). This induced field opposes H_0 on the inside of the loop (Figure 15-8), but it reinforces H_0 on the outside—just where the hydrogens are located. This reinforcement results in a local field in their vicinity equal to $H_0 + h_{local}$. So, to cause resonance at constant radio frequency ν, the applied field strength has to be reduced to ($H_0 - h_{local}$): The nuclei are deshielded. This effect is strongest close to the ring and diminishes rapidly with increasing distance from it. Thus, benzylic nuclei are deshielded only about 0.4 to 0.8 ppm more than their allylic counterparts, and hydrogens farther away from the π system have chemical shifts that do not differ much from each other and are similar to those in the alkanes.

Whereas benzene exhibits a sharp singlet in its NMR spectrum, substituted derivatives may have more complicated patterns. For example, introduction of one substituent renders the hydrogens positioned ortho, meta, and para nonequivalent and subject to mutual coupling. An example is the NMR spectrum of bromobenzene, in which the sig-

Chemical Shifts of Allylic and Benzylic Hydrogens

CH_2=CH—CH_3

Allylic:
δ 1.68 ppm

—CH_3

Benzylic:
δ 2.35 ppm

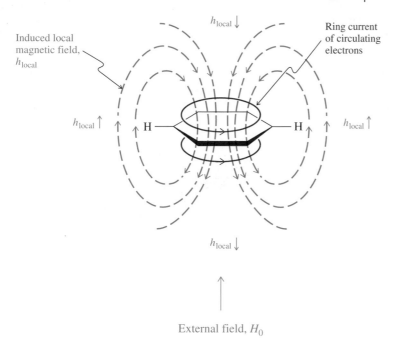

FIGURE 15-8
The π electrons of benzene may be compared to those in a loop of conducting metal. Exposure of this loop of electrons to an external magnetic field H_0 causes them to circulate like an electric current. This "ring current" generates a local field, reinforcing H_0 on the outside of the ring. Thus, the hydrogens resonate at a lower applied field strength.

nal for the ortho hydrogens is shifted downfield and those for the meta and para hydrogens occur slightly upfield, relative to that of benzene. Moreover, all the protons are coupled to one another, thus giving rise to a complicated spectral pattern (Figure 15-9).

Figure 15-10 shows the NMR spectrum of 4-(N,N-dimethylamino)benzaldehyde. The large chemical shift difference between the two sets of hydrogens on the ring results in a first-order pattern of two doublets. The observed coupling constant is 9 Hz, a typical splitting between ortho protons.

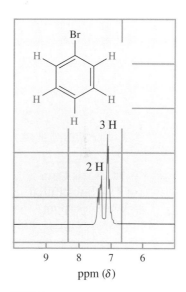

FIGURE 15-9
A part of the 90-MHz ^1H NMR spectrum of bromobenzene in CCl_4, a non-first-order spectrum.

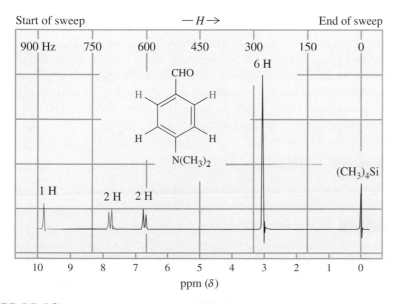

FIGURE 15-10
90-MHz ^1H NMR spectrum of 4-(N,N-dimethylamino)benzaldehyde (p-dimethylaminobenzaldehyde) in $CDCl_3$. In addition to the two aromatic doublets ($J = 9$ Hz), there are singlets for the methyl ($\delta = 3.07$ ppm) and the formyl hydrogens ($\delta = 9.80$ ppm).

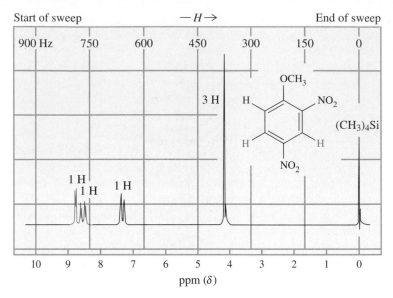

FIGURE 15-11

^{1}H NMR spectrum of 1-methoxy-2,4-dinitrobenzene (2,4-dinitroanisole) in CDCl$_3$.

A first-order spectrum revealing all three types of coupling is shown in Figure 15-11. 1-Methoxy-2,4-dinitrobenzene (2,4-dinitroanisole) bears three ring hydrogens with different chemical shifts and distinct splittings. The hydrogen ortho to the methoxy group appears at $\delta = 7.30$ ppm as a doublet with a 9-Hz ortho coupling. The hydrogen flanked by the two nitro groups (at $\delta = 8.73$ ppm) also appears as a doublet, with a small (3-Hz) meta coupling. Finally, we find the remaining ring hydrogen absorbing at $\delta = 8.50$ ppm as a double doublet because of simultaneous coupling to the two other ring protons. Para coupling between the hydrogens at C3 and C6 is too small (<1 Hz) to be resolved; it is evident as a slight broadening of the resonances of these protons.

In contrast with ^{1}H NMR, ^{13}C NMR chemical shifts in benzene derivatives are dominated by hybridization and substituent effects. As a result of the large chemical shift range for carbon nuclei (200 ppm), the contribution from the ring current (only a few parts per million) becomes relatively less significant. Therefore, benzene carbons exhibit chemical shifts similar to those in alkenes, between 120 and 135 ppm when unsubstituted (see margin). Benzene itself exhibits a single line at $\delta = 128.7$ ppm.

^{13}C NMR Data of Two Substituted Benzenes (ppm)

H$_3$C ← 21.3

137.8

129.3

128.5

125.6

NO$_2$

148.3

123.4

129.5

134.7

EXERCISE 15-6

Can the three isomeric trimethylbenzenes be distinguished solely on the basis of the number of peaks in their proton-decoupled ^{13}C NMR spectra? Explain.

EXERCISE 15-7

A hydrocarbon has the molecular formula C$_{10}$H$_{14}$. The spectral data for this compound are as follows: ^{1}H NMR (90 MHz) $\delta = 7.02$ (broad s, 4 H), 2.82 (septet, $J = 7.0$ Hz, 1 H), 2.28 (s, 3 H), and 1.22 ppm (d, $J = 7.0$ Hz, 6 H); ^{13}C NMR $\delta = 21.3$, 24.2, 38.9, 126.6, 128.6, 134.8, and 145.7 ppm; IR $\tilde{\nu} = 3030$, 2970, 2880, 1515, 1465, and 813 cm^{-1}; UV $\lambda_{max}(\epsilon) = 265(450)$ nm. What is its structure?

In summary, benzene and its derivatives can be recognized and structurally characterized by their spectral data. Electronic absorptions take place between 250 and 290 nm. The infrared vibrational bands are found at 3030 cm^{-1} (C$_{aromatic}$–H), from 1500 to 2000 cm^{-1} (C–C), and from 650 to 1000 cm^{-1} (C–H out-of-plane bending).

Most informative is NMR, with low-field resonances for the aromatic hydrogens and carbons. Coupling is largest between the ortho hydrogens, smaller in their meta and para counterparts.

15-5 Polycyclic Benzenoid Hydrocarbons

What happens if several benzene rings are fused to give a more extended π system? Molecules of this class are called **polycyclic benzenoid** or **polycyclic aromatic hydrocarbons (PAHs).** In these structures two or more benzene rings share two or more carbon atoms. Do these compounds also enjoy the special stability of benzene? The next two sections will show that they largely do.

There is no simple system for naming these structures; so we shall use their common names. The fusion of one benzene ring to another results in a compound called naphthalene. Further fusion can occur in a linear manner to give anthracene, tetracene, pentacene, and so on, a series called the **acenes. Angular fusion** results in phenanthrene, which can be further annulated to a variety of other benzenoid polycycles.

Naphthalene   **Anthracene**   **Tetracene (Naphthacene)**   **Phenanthrene**

Each has its own numbering system around the periphery. A quaternary carbon is given the number of the preceding carbon in the sequence followed by the letters a, b, and so on, depending on how close it is to that carbon.

EXERCISE 15-8

Name the following compounds or draw their structures.
(a) 2,6-Dimethylnaphthalene   **(b)** 1-Bromo-6-nitrophenanthrene   **(c)** 9,10-Diphenylanthracene

(d)   **(e)**

15-6 Fused Benzenoid Hydrocarbons: Naphthalene and the Tricyclic Systems

In contrast with benzene, which is a liquid, naphthalene is a colorless crystalline material with a melting point of 80°C. It is probably best known as a moth repellent and insecticide, although in these capacities it has been partly replaced by chlorinated compounds such as 1,4-dichlorobenzene (*p*-dichlorobenzene).

Is naphthalene still aromatic? Does it share benzene's delocalized electronic structure and thermodynamic stability? Its spectral properties suggest strongly that it does. Particularly revealing are the ultraviolet and NMR spectra.

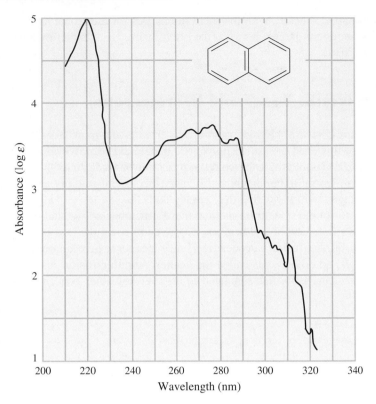

FIGURE 15-12

Extended π conjugation in naphthalene is manifest in its UV spectrum (measured in 95% ethanol). The complexity and location of the absorptions are typical of extended π systems.

Naphthalene is aromatic: a look at spectra

The ultraviolet spectrum of naphthalene (Figure 15-12) shows a pattern typical of an extended conjugated system, with peaks at wavelengths as long as 320 nm. On the basis of this observation, we conclude that the electrons are delocalized more extensively than in benzene (Section 15-2 and Figure 15-6). Thus, it appears that the added four π electrons enter into efficient overlap with those of the attached benzene ring. In fact, it is possible to draw several resonance forms.

Resonance Forms in Naphthalene

Alternatively, the continuous overlap of the 10 p orbitals can be shown as in Figure 15-13.

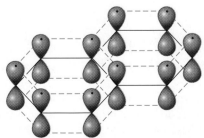

FIGURE 15-13

Orbital picture of naphthalene, showing its extended overlap of p orbitals.

According to these representations, the structure of naphthalene should be symmetric, with planar and almost hexagonal benzene rings and two perpendicular mirror planes bisecting the molecule. X-ray crystallographic measurements confirm this prediction (Figure 15-14). The C–C bonds deviate only slightly in length from those in benzene (1.39 Å), and they are clearly different from pure single (1.54 Å) and double bonds (1.33 Å).

Further evidence of aromaticity is found in the ^{1}H NMR spectrum of naphthalene (Figure 15-15). Two symmetric multiplets can be observed at δ = 7.40 and 7.77 ppm, characteristic of aromatic hydrogens deshielded by the ring-current effect of the π electron loop (see Section 15-4, Figure 15-8). Coupling in the naphthalene nucleus is very similar to that in substituted benzenes: J_{ortho} = 7.5 Hz, J_{meta} = 1.4 Hz, and J_{para} = 0.7 Hz. The ^{13}C NMR spectrum shows three lines with chemical shifts that are in the range of other benzene derivatives (see margin). Thus, on the basis of structural and spectral criteria, naphthalene is aromatic.

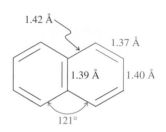

FIGURE 15-14
The molecular structure of naphthalene. The bond angles within the rings are 120°.

EXERCISE 15-9

A substituted naphthalene $C_{10}H_8O_2$ gave the following spectral data: ^{1}H NMR δ = 5.20 (broad s, 2 H), 6.92 (dd, J = 7.5 Hz and 1.4 Hz, 2 H), 7.00 (d, J = 1.4 Hz, 2 H), and 7.60 (d, J = 7.5 Hz, 2 H) ppm; ^{13}C NMR δ = 107.5, 115.3, 123.0, 129.3, 136.8, and 155.8 ppm; IR $\tilde{\nu}$ = 3100 cm^{-1}. What is its structure? (**Hints:** Review the values for $J_{ortho, meta, para}$ in benzene [Section 15-4]. The number of NMR peaks is less than maximum.)

^{13}C NMR Data of Naphthalene (ppm)

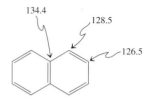

Most fused benzenoid hydrocarbons are aromatic

These properties of naphthalene hold for most of the other polycyclic benzenoid hydrocarbons. It appears that the cyclic delocalization in the individual benzene rings is not significantly perturbed by the fact that they have to share at least one π bond. Linear and angular fusion of a third benzene ring onto naphthalene results in the systems anthracene and phenanthrene. Although isomeric and seemingly very similar, they

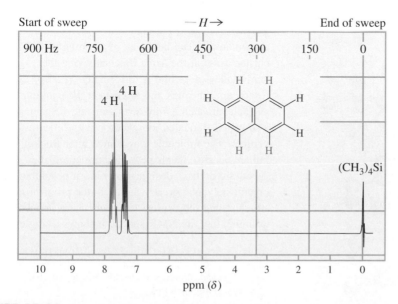

FIGURE 15-15
90-MHz ^{1}H NMR spectrum of naphthalene in CCl_4 reveals the characteristic deshielding due to a π-electronic ring current.

The Allotropes of Carbon: Graphite, Diamond, and Fullerenes

Elements can exist in several forms, called *allotropes,* depending on conditions and modes of synthesis. Thus, elemental carbon can arrange in more than 40 configurations, most of them amorphous (i.e., noncrystalline). For example, *coke* is the solid residue left after distillation of crude petroleum (Section 3-3) and coal (Section 13-11). You all know the *soot* that is generated on incomplete combustion of organic materials. *Carbon black* is an important commercial product, synthesized by heating gaseous hydrocarbons to nearly 1000°C in the absence of air, and has many uses, including the ink used in printing. *Activated carbon* is highly porous and capable of absorbing organic trace materials in air (air purifiers, gas masks) and water (water filters). You probably know best two crystalline modifications of carbon, *graphite* and *diamond.* Graphite, the most stable carbon allotrope, is a completely fused polycyclic benzenoid π system, consisting of layers arranged in an open honeycomb pattern and 3.35Å apart. The fully delocalized nature of these sheets (all carbons are sp^2-hybridized) gives rise to their black color and conductive capability. Graphite's lubricating property is the result of the ready mutual sliding of its component planes. The "lead" of pencils is graphitic carbon and the black pencil marks left on a sheet of paper consist of rubbed-off layers of the element.

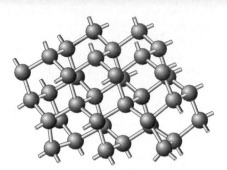

Diamond

In the colorless diamond, the carbon atoms (all sp^3-hybridized) form an insulating network of cross-linked cyclohexane chair conformers. Diamond is the densest and hardest (least deformable) material known. It is also less stable than graphite, by 0.45 kcal/g C atom, and transforms into graphite at high temperatures or when subjected to high-energy radiation, a little-appreciated fact in the jewelry business.

A spectacular discovery was made in 1985 by Curl, Kroto, and Smalley* (for which they received the Nobel Prize in 1996): *buckminsterfullerene,* C_{60}, a new, spherical allotrope of carbon in the shape of a soccer ball. They discovered that laser evaporation of graphite generated a variety of carbon clusters in the gas phase, the most abundant of which contained 60 carbon atoms. The best way of assembling such a cluster while at the same time satisfying the tetravalency of carbon is to formally "roll up" 20 fused benzene rings and to connect the dangling valencies in such a way as to generate 12 pentagons: a so-called truncated icosahedron with 60 equivalent vertices. The molecule was named after Buckminster Fuller[†] because its shape is reminiscent of the "geodesic domes" designed by him. It is soluble in organic solvents, greatly aiding in the proof of its structure and the exploration of its chemistry. For example, the ^{13}C NMR spectrum shows a single line

*Professor Robert F. Curl (b. 1933), Rice University, Houston, Texas; Professor Harold W. Kroto (b. 1939), University of Sussex, England; Professor Richard E. Smalley (b. 1943), Rice University, Houston, Texas.

[†]Richard Buckminster Fuller (1895–1983), American architect, inventor, and philosopher.

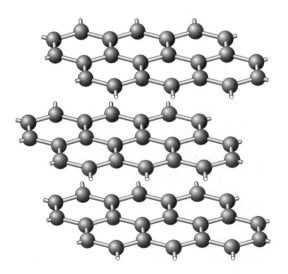

Graphite

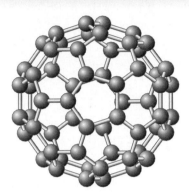

Buckminsterfullerene

An example of a geodesic dome (of the type whose design was pioneered by Buckminster Fuller) forms part of the entrance to EPCOT Center, Disney World, Florida. It is 180 feet high with a diameter of 165 feet.

at δ 142.7 ppm, in the expected range (Sections 15-4 and 15-6). Because of its curvature, the constituent benzene rings in C_{60} are strained and the energy content relative to graphite is 10.16 kcal/g C atom. This strain is manifested in a rich chemistry, including electrophilic, nucleophilic, radical, and concerted addition reactions (Chapter 14). The enormous interest spurred by the discovery of C_{60} rapidly led to a number of exciting developments, such as the design of multigram synthetic methods (commercial material sells at about $30.00 per gram); the isolation of many other larger carbon clusters, dubbed "fullerenes," such as the rugby-ball-shaped C_{70}; chiral systems (e.g., as in C_{76}); isomeric forms; fullerenes encaging host atoms, such as He and metal nuclei ("endohedral fullerenes"); and the synthesis of conducting salts (e.g., Cs_3C_{60}, which becomes superconducting at 40 K). Moreover, reexamination of the older literature and newer studies have revealed that C_{60} and other fullerenes are produced simply on incomplete combustion of organic matter under certain conditions or by varied heat treatments of soot and therefore have probably been "natural products" on our planet since early in its formation. From a

materials point of view, perhaps most useful has been the synthesis of graphitic tubules, so-called *nanotubes,* based on the fullerene motif. Nanotubes promise to be even harder than diamond, yet elastic, and to show unusual magnetic and electrical (metallic) properties. They also function as a molecular "packaging material" for other structures, such as metal catalysts and even biomolecules. Thus, carbon in the fullerene modification has taken center stage in the new field of nanotechnology, aimed at the construction of devices at the molecular level.

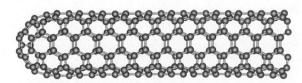

Carbon nanotube

have different thermodynamic stabilities: Anthracene is about 6 kcal mol^{-1} less stable than phenanthrene even though both are aromatic. Enumeration of the various resonance forms of the molecules explains why. Anthracene has only four, and only two contain two fully aromatic benzene rings (red in the structures shown here). Phenanthrene has five, three of which incorporate two aromatic benzenes, one even three benzenes.

Resonance in Anthracene

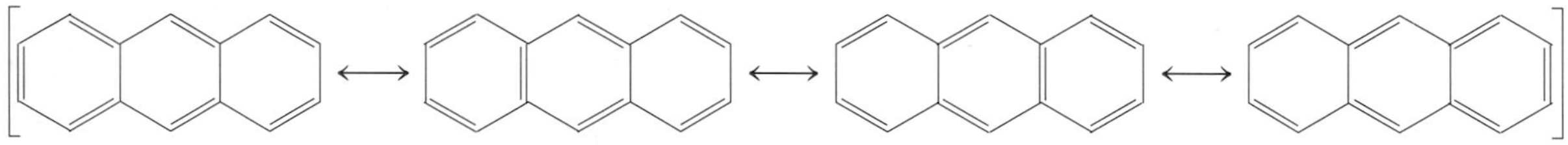

Resonance in Phenanthrene

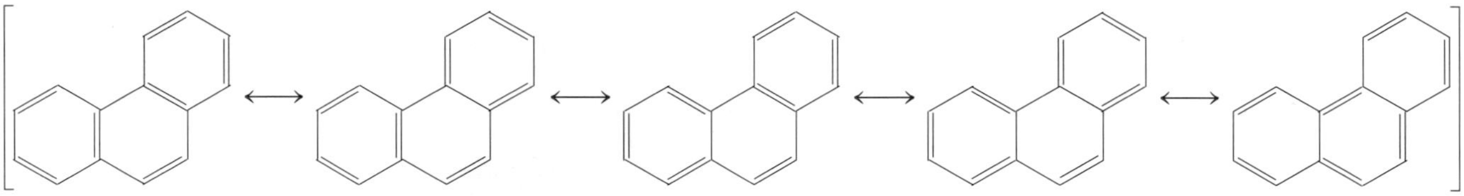

EXERCISE 15-10

Draw all the possible resonance forms of tetracene (naphthacene, Section 15-5). What is the maximum number of completely aromatic benzene rings in these structures?

In summary, the physical properties of naphthalene are typical of an aromatic system. Its UV spectrum reveals that there is extensive delocalization of all π electrons, its molecular structure shows bond lengths and bond angles very similar to those in benzene, and its ^{1}H NMR spectrum reveals deshielded ring hydrogens indicative of an aromatic ring current. Other polycyclic benzenoid hydrocarbons have similar properties and are considered aromatic.

15-7 Other Cyclic Polyenes: Hückel's Rule

Is the special stability and reactivity associated with cyclic delocalization unique to benzene and polycyclic benzenoids, or do other cyclic π systems have similar properties? The answers are no and yes, respectively, but not in a straightforward manner. Thus, we will see that other cyclic conjugated polyenes can be aromatic, but only if they contain $4n + 2$ π electrons ($n = 0, 1, 2, 3...$). In contrast, $4n$ π circuits may be *destabilized* by conjugation, or are **antiaromatic.** This pattern is known as **Hückel's* rule.** Nonplanar systems in which cyclic overlap is disrupted sufficiently to impart alkenelike properties are classified as **nonaromatic.** Let us look at some members of this series, starting with 1,3-cyclobutadiene.

*Professor Erich Hückel (1896–1984), University of Marburg, Germany.

1,3-Cyclobutadiene, the smallest cyclic polyene, is antiaromatic

1,3-Cyclobutadiene, a $4n$ π system ($n = 1$), is an air sensitive and extremely reactive molecule (much more so than its congeners 1,3-butadiene or cyclobutene) not only devoid of aromatic properties, but also destabilized by π overlap. Consequently, unlike the symmetrical benzene in which two cyclohexatriene forms are in *resonance,* cyclobutadiene consists of a pair of rapidly equilibrating rectangular structural *isomers,* interconverting through a symmetrical transition state.

1,3-Cyclobutadiene Is Unsymmetrical

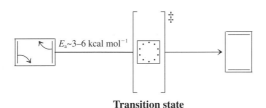

Transition state

Free 1,3-Cyclobutadiene can be prepared and observed only at very low temperatures. The reactivity of cyclobutadiene can be seen in its rapid Diels-Alder reactions, in which it can act as both diene (shown in red) and dienophile (blue).

EXERCISE 15-11

1,3-Cyclobutadiene dimerizes at temperatures as low as $-200°C$ to give the two products shown. Explain mechanistically.

Substituted cyclobutadienes are less reactive because of steric protection, particularly if the substituents are bulky; they have been used to probe the spectroscopic features of the cyclic system of four π electrons. For example, in the ^1H NMR spectrum of 1,2,3-tris(1,1-dimethylethyl)cyclobutadiene (1,2,3-tri-*tert*-butylcyclobutadiene), the ring hydrogen resonates at $\delta = 5.38$ ppm, at much higher field than expected for an aromatic system. This and other properties of cyclobutadiene show that it is quite unlike benzene.

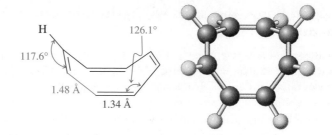

FIGURE 15-16 _____
The molecular structure of 1,3,5,7-cyclooctatetraene. Note the alternating single and double bonds of this nonplanar, nonaromatic molecule.

$4 \text{ HC} \equiv \text{CH}$

Ni(CN)$_2$,
70°C,
15–25 atm

70%
1,3,5,7-Cyclooctatetraene

1,3,5,7-Cyclooctatetraene is nonplanar and nonaromatic

Let us now examine the properties of the next *higher* cyclic polyene analog of benzene, 1,3,5,7-cyclooctatetraene, another $4n$ π cycle ($n = 2$). Is it antiaromatic, like 1,3-cyclobutadiene? First prepared in 1911 by Willstätter,* this substance is now readily available from a special reaction, the nickel-catalyzed cyclotetramerization of ethyne. It is a yellow liquid (b.p. 152°C) that is stable if kept cold but polymerizes when heated. It is oxidized by air, catalytically hydrogenated to cyclooctane, and subject to electrophilic additions and to cycloadditions. This chemical reactivity is diagnostic of a normal polyene (Section 14-7).

Spectral and structural data confirm this assessment. Thus, the ^{1}H NMR spectrum shows a sharp singlet at $\delta = 5.68$ ppm, typical of an alkene. The molecular-structure determination reveals that cyclooctatetraene is actually *nonplanar* and tubshaped (Figure 15-16). The double bonds are nearly orthogonal (perpendicular), and they alternate with single bonds. Conclusion: The molecule is nonaromatic.

EXERCISE 15-12

On the basis of the molecular structure of 1,3,5,7-cyclooctatetraene, would you describe its double bonds as conjugated (i.e., does it exhibit extended π overlap)? Would it be correct to draw two resonance forms for this molecule, as we do for benzene? (**Hint:** Build molecular models of the two forms of 1,2-dimethylcyclooctatetraene.)

EXERCISE 15-13

Cyclooctatetraene A exists in equilibrium with less than 0.05% of a bicyclic isomer B, which is trapped by Diels-Alder cycloaddition to 2-butenedioic anhydride (maleic anhydride, Table 14-1) to give compound C.

What is isomer B? Show a mechanism for the A → B → C interconversion. (**Hint:** Work backward from C to B and review Section 14-9.)

*Professor Richard Willstätter (1872–1942), Technical University, Munich, Germany, Nobel Prize 1915 (chemistry).

Only cyclic conjugated polyenes containing 4n + 2 π electrons are aromatic

Unlike cyclobutadiene and cyclooctatetraene, certain higher cyclic conjugated polyenes are aromatic. All of them have two properties in common: They contain $4n + 2$ π electrons and they are sufficiently planar to allow for delocalization.

The first such system was prepared in 1956 by Sondheimer;* it was 1,3,5,7,9,11,13,15,17-cyclooctadecanonaene, containing 18 π electrons ($4n + 2, n = 4$). To avoid the use of such cumbersome names, Sondheimer introduced a simpler system of naming cyclic conjugated polyenes. He named completely conjugated monocyclic hydrocarbons $(CH)_N$ as **[N]annulenes,** in which N denotes the ring size. Thus, cyclobutadiene would be called [4]annulene; benzene, [6]annulene; cyclooctatetraene, [8]annulene. The first almost unstrained aromatic system in the series after benzene is [18]annulene.

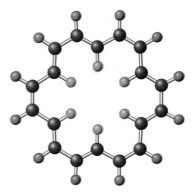

[18]Annulene
(1,3,5,7,9,11,13,15,17-Cyclooctadecanonaene)

EXERCISE 15-14

The three isomers of [10]annulene shown here have been prepared and exhibit nonaromatic behavior. Why? (**Hint:** Build models!)

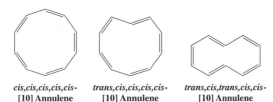

cis,cis,cis,cis,cis-
[10] Annulene

trans,cis,cis,cis,cis-
[10] Annulene

trans,cis,trans,cis,cis-
[10] Annulene

---

[18]Annulene contains delocalized electrons, is fairly planar, and shows little alternation of the single and double bonds. Like benzene, therefore, it can be described by a set of two equivalent resonance forms. In accord with its aromatic character, the molecule is relatively stable and undergoes electrophilic aromatic substitution. It also exhibits a benzenelike ring-current effect in ^1H NMR (see Problems 50 and 51).

Since the preparation of [18]annulene, many other annulenes have been made: Those with $4n$ π electrons, such as cyclobutadiene and cyclooctatetraene, are either anti- or nonaromatic, but those with $4n + 2$ electrons, such as benzene and [18]annulene, are aromatic. This behavior had been predicted earlier by the theoretical chemist Hückel, who formulated this $4n + 2$ rule in 1931.

---

*Professor Franz Sondheimer (1926–1981), University College, London.

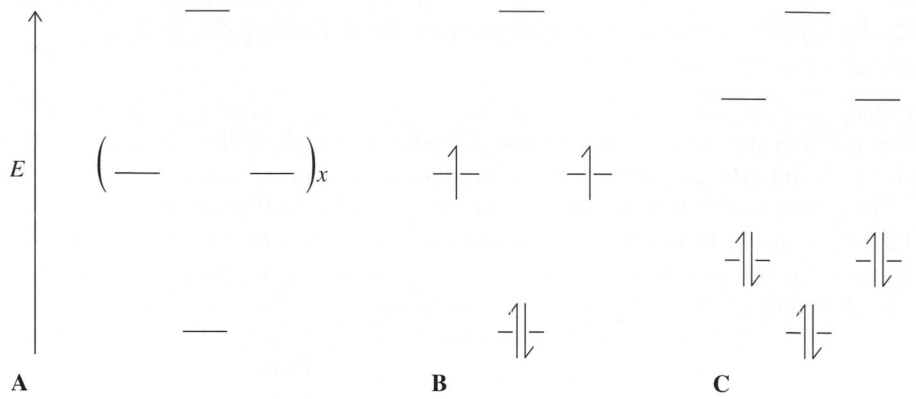

FIGURE 15-17

(A) Hückel's $4n + 2$ rule is based on the regular pattern of the π molecular orbitals in cyclic conjugated polyenes. The energy levels are equally spaced, and only the highest and lowest ones are nondegenerate. (B) Molecular-orbital levels in 1,3-cyclobutadiene. Four π electrons are not enough to result in a closed shell (in other words, doubly filled molecular orbitals), so the molecule is not aromatic. (C) The six π electrons in benzene produce a closed-shell configuration.

Hückel's rule expresses the regular molecular-orbital patterns in cyclic conjugated polyenes. The p orbitals mix to give an equal number of π molecular orbitals, as shown in Figure 15-17. All levels are composed of degenerate pairs, except for the lowest bonding and highest antibonding orbitals. A closed-shell system is possible only if all bonding molecular orbitals are occupied (see Section 1-7); that is, only if there are $4n + 2$ π electrons. On the other hand, $4n$ π cycles always contain a pair of singly occupied orbitals, an unfavorable electronic arrangement.

EXERCISE 15-15

On the basis of Hückel's rule, label the following molecules as aromatic or antiaromatic. **(a)** [30]Annulene; **(b)** [16]annulene; **(c)** *trans*-15,16-dihydropyrene; **(d)** the deep blue (see Figure 14-16) azulene; **(e)** S-indacene.

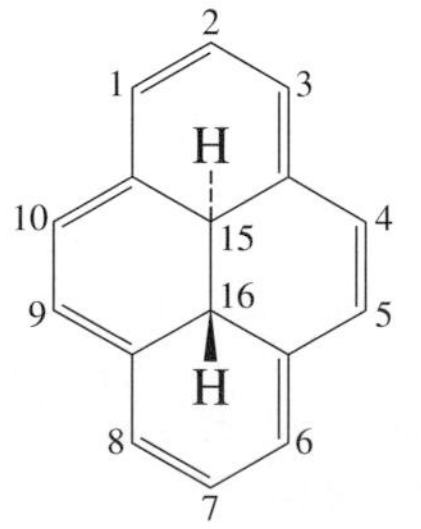

***Trans*-15,16-dihydropyrene** 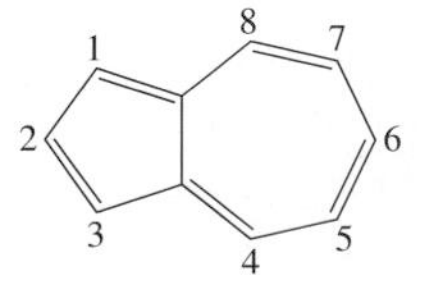   **Azulene** 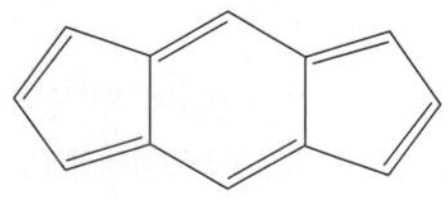   **S-indacene**

In summary, cyclic conjugated polyenes are aromatic if their π electron count is $4n + 2$. This number corresponds to a completely filled set of bonding molecular orbitals. Conversely, $4n$ π systems have open-shell, antiaromatic structures that are unstable, are reactive, and lack aromatic ring-current effects in ^{1}H NMR. Finally, when steric constraints impose nonplanarity, cyclic polyenes behave as nonaromatic alkenes.

15-8 Hückel's Rule and Charged Molecules

Hückel's rule also applies to charged molecules, as long as cyclic delocalization can occur. This section shows how charged aromatic systems can be prepared.

The cyclopentadienyl anion and the cycloheptatrienyl cation are aromatic

1,3-Cyclopentadiene is unusually acidic [$pK_a \sim 16$; comparable to alcohols (Section 8-3)] because the cyclopentadienyl anion resulting from deprotonation contains a delocalized, aromatic system of six π electrons. The negative charge is equally distributed over all five carbon atoms. For comparison, the pK_a of propene is 40.

Aromatic Cyclopentadienyl Anion

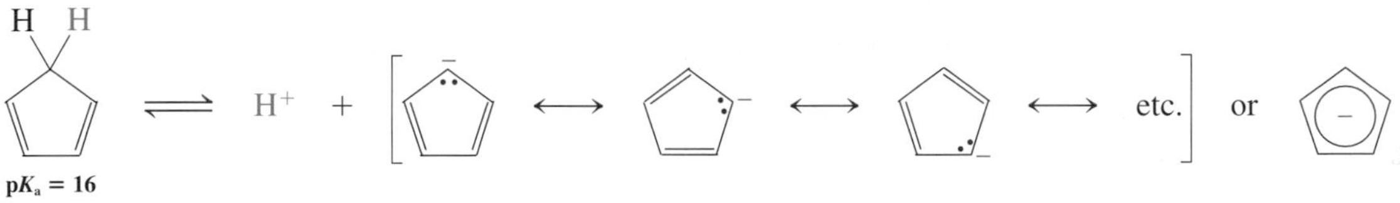

In contrast, the cyclopentadienyl cation, a system of four π electrons, can be produced only at low temperature and is extremely reactive.

When 1,3,5-cycloheptatriene is treated with bromine, a stable salt is formed, cycloheptatrienyl bromide. In this molecule, the organic cation contains six delocalized π electrons, and the positive charge is equally distributed over seven carbons. Even though a carbocation, the system is remarkably unreactive, as expected for an aromatic system. In contrast, the cycloheptatrienyl anion is antiaromatic, as indicated by the much lower acidity of cycloheptatriene ($pK_a = 39$) compared with that of cyclopentadiene.

Aromatic Cycloheptatrienyl Cation

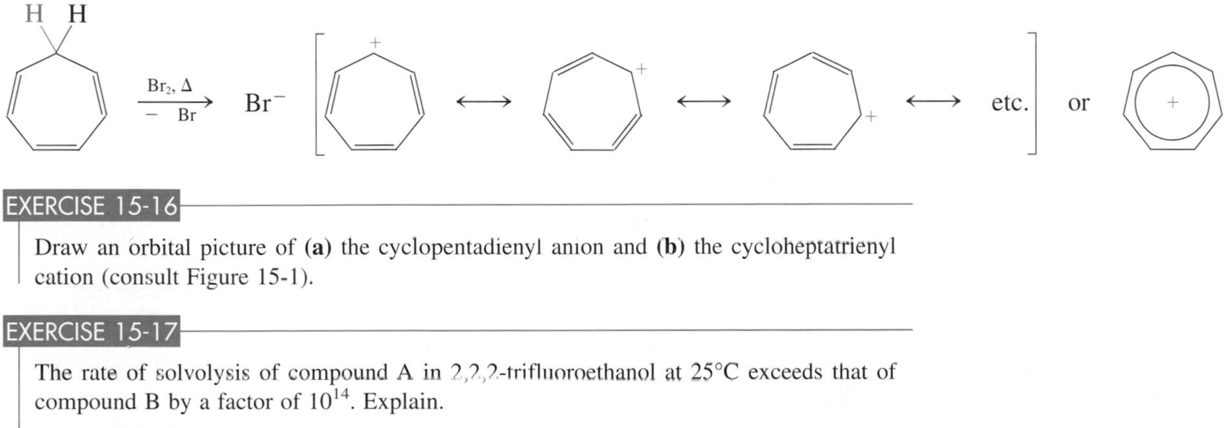

EXERCISE 15-16

Draw an orbital picture of **(a)** the cyclopentadienyl anion and **(b)** the cycloheptatrienyl cation (consult Figure 15-1).

EXERCISE 15-17

The rate of solvolysis of compound A in 2,2,2-trifluoroethanol at 25°C exceeds that of compound B by a factor of 10^{14}. Explain.

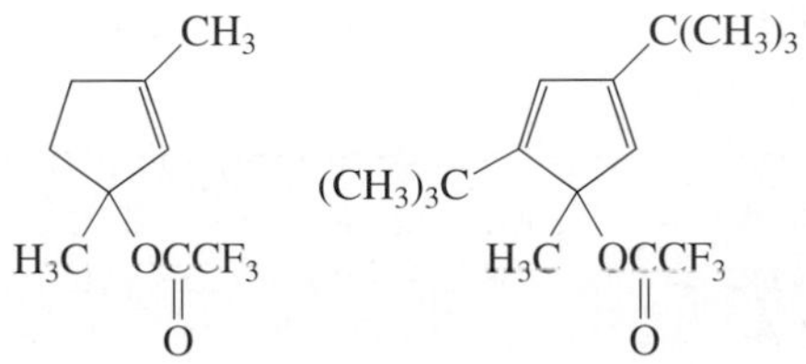

On the basis of Hückel's rule, label the following molecules aromatic or antiaromatic. **(a)** Cyclopropenyl cation; **(b)** cyclononatetraenyl anion; **(c)** cycloundecapentaenyl anion.

Nonaromatic cyclic polyenes can form aromatic dianions and dications

Cyclic systems of $4n$ π electrons can be converted into their aromatic counterparts by two-electron oxidations and reductions. For example, cyclooctatetraene is reduced by alkali metals to the corresponding aromatic dianion. This species is planar, contains fully delocalized electrons, and is relatively stable. It also exhibits an aromatic ring current in ^{1}H NMR.

Nonaromatic Cyclooctatetraene Forms an Aromatic Dianion

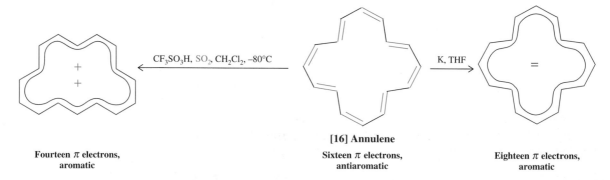

Eight π electrons,   Ten π electrons,
nonaromatic   aromatic

Similarly, [16]annulene can be either reduced to its dianion or oxidized to its dication, both products being aromatic. On formation of the dication, the configuration of the molecule changes.

Aromatic [16]Annulene Dication and Dianion from Antiaromatic [16]Annulene

CF$_3$SO$_3$H, SO$_2$, CH$_2$Cl$_2$, –80°C   K, THF

[16] Annulene

Fourteen π electrons,   Sixteen π electrons,   Eighteen π electrons,
aromatic   antiaromatic   aromatic

The triene A can be readily deprotonated twice to give the stable dianion B. However, the neutral analog of B, the tetraene C (pentalene), is extremely unstable. Explain.

A   Base   B   C

EXERCISE 15-20

Azulene [see Exercise 15-15(d)] is readily attacked by electrophiles at C1, by nucleophiles at C4. Explain. (**Hint:** Formulate the resonance forms of the resulting cation and anion, respectively, and assess their cyclic conjugation in light of Hückel's rule.)

In summary, charged species may be aromatic, provided they exhibit cyclic delocalization and obey the $4n + 2$ rule.

15-9 Synthesis of Benzene Derivatives: Electrophilic Aromatic Substitution

In this section we explore the reactivity of benzene, the prototype aromatic compound. The stability of benzene makes it relatively unreactive. As a result, its chemical transformations require special conditions and proceed through new pathways.

Benzene undergoes substitution reactions with electrophiles

Benzene can be attacked by electrophiles. In contrast with the corresponding reactions of alkenes, this reaction results in *substitution* of hydrogens—**electrophilic aromatic substitution**—*not addition* to the ring.

Electrophilic Aromatic Substitution

$$\text{(benzene)} + E^+X^- \text{ (Electrophile)} \longrightarrow \text{(substituted benzene)} + H^+X^-$$

Under the conditions employed for these processes, nonaromatic conjugated polyenes would rapidly polymerize. However, the stability of the benzene ring allows it to survive. Let us begin with the general mechanism of electrophilic aromatic substitution.

Electrophilic aromatic substitution in benzene proceeds by addition of the electrophile and then by proton loss

The mechanism of electrophilic aromatic substitution has two steps. First, the electrophile E^+ attacks the benzene nucleus, much as it would attack an ordinary double bond.* The cationic intermediate thus formed then loses a proton to regenerate the aromatic ring.

*We use the phrase "an electrophile attacks the benzene nucleus" somewhat arbitrarily, as the "attack" is mutual—that is, one could also say that the benzene double bond attacks the electrophile, a description that would conform with the actual flow of electrons. However, the former use of words depicts the electrophile as the "aggressor," in tune with its electron-deficient nature, and the benzene ring as the "victim," reflecting its relative stability.

Mechanism of Electrophilic Aromatic Substitution

STEP 1. Electrophilic attack

STEP 2. Proton loss

The first step is not favored thermodynamically. Although charge is delocalized in the cationic intermediate, the formation of the C–E bond results in an sp^3-hybridized carbon in the ring, which interrupts cyclic conjugation: *The intermediate is not aromatic* (Figure 15-18). However, the next step, loss of the proton at the sp^3-hybridized carbon, regenerates the aromatic ring. This process is more favored than nucleophilic trapping by the anion that accompanies E^+. The latter would give a nonaromatic addition product. The overall substitution is exothermic, the new bonds being stronger than the old ones.

Figure 15-19 depicts a potential energy diagram in which the first step is rate determining, a kinetic finding that applies to most electrophiles that we will encounter. The subsequent loss of proton is much faster than initial electrophilic attack because it leads to the aromatic product in an exothermic step, which furnishes the driving force for the overall sequence.

The following sections look more closely at the most common reagents employed in this transformation and the details of the mechanism.

EXERCISE 15-21

Review Section 12-1 and explain why addition reactions to benzene do not take place. (Calculate the $\Delta H°$ for some additions.)

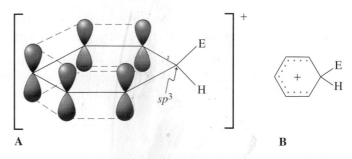

FIGURE 15-18

(A) Orbital picture of the cationic intermediate resulting from attack by an electrophile on the benzene ring. Aromaticity is lost because cyclic conjugation is interrupted by the sp^3-hybridized carbon. The four electrons in the π system are not shown. (B) Dotted-line notation to indicate delocalized nature of the charge in the cation.

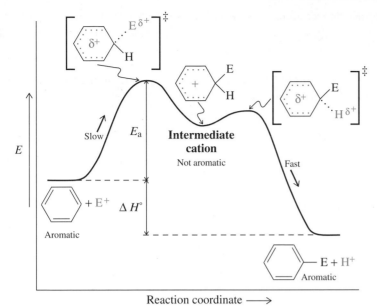

FIGURE 15-19

Potential-energy diagram describing the course of the reaction of benzene with an electrophile. The first transition state is rate determining. Proton loss is relatively fast. The overall rate of the reaction is determined by E_a, the exothermicity by $\Delta H°$.

In summary, the general mechanism of electrophilic aromatic substitution begins with electrophilic attack by E^+ to give an intermediate, charge-delocalized but nonaromatic cation in a rate-determining step. Subsequent fast proton loss regenerates the (now substituted) aromatic ring.

15-10 Halogenation of Benzene: The Need for a Catalyst

An example of electrophilic aromatic substitution is halogenation. Benzene is normally unreactive in the presence of halogens, because halogens are not electrophilic enough to disrupt its aromaticity. However, the halogen may be activated by Lewis acidic catalysts, such as ferric halides (FeX_3) or aluminum halides (AlX_3), to become a much more powerful electrophile.

How does this activation work? The characteristic of Lewis acids is their ability to accept electron pairs. When a halogen such as bromine is exposed to $FeBr_3$, the two molecules combine as Lewis acid and base.

Bromination of Benzene

Br_2, $FeBr_3$

Br

Bromobenzene

Activation of Bromine by the Lewis Acid $FeBr_3$

$$:\overset{..}{\underset{..}{Br}}-\overset{..}{\underset{..}{Br}}:\curvearrowright FeBr_3 \longrightarrow \left[:\overset{..}{\underset{..}{Br}}-\overset{..}{\underset{..}{\overset{+}{Br}}}-\overset{-}{F}eBr_3 \longleftrightarrow \overset{..}{\underset{..}{\overset{+}{Br}}}=\overset{..}{\underset{..}{Br}}-\overset{-}{F}eBr_3\right]$$

In this complex, the Br–Br bond is polarized, thereby imparting electrophilic character to the bromine atoms. Electrophilic attack on benzene is at the terminal bromine, allowing the other bromine atom to depart with the good leaving group $FeBr_4^-$. In terms of electron flow, you can also view this process as a nucleophilic substitution of $[Br_2FeBr_3]$ by the benzene double bond, not unlike an S_N2 reaction.

Electrophilic Attack on Benzene by Activated Bromine

The $FeBr_4^-$ formed in this step now functions as a base, abstracting a proton from the cyclohexadienyl cation intermediate. This transformation not only furnishes the two products of the reaction, bromobenzene and hydrogen bromide, but also regenerates the $FeBr_3$ catalyst.

Bromobenzene Formation

A quick calculation confirms the exothermicity of the electrophilic bromination of benzene. A phenyl–hydrogen bond (approximately 112 kcal mol^{-1}, Table 15-1) and a bromine molecule (46 kcal mol^{-1}) are lost in the process. Counterbalancing this loss is the formation of a phenyl–bromine bond ($DH° = 81$ kcal mol^{-1}) and an H–Br bond ($DH° = 87.5$ kcal mol^{-1}). Thus, the overall reaction is exothermic by $168.5 - 158 = 10.5$ kcal mol^{-1}.

As in the radical halogenation of alkanes (Section 3-7), the exothermicity of aromatic halogenation decreases down the periodic table. Fluorination is so exothermic that direct reaction of fluorine with benzene is explosive. Chlorination, on the other hand, is controllable but requires the presence of an activating catalyst, such as aluminum chloride or ferric chloride. The mechanism of this reaction is identical with that of bromination. Finally, electrophilic iodination with iodine is endothermic and thus not normally possible.

TABLE 15-1

Strengths ($DH°$) of Bonds A–B (kcal mol^{-1})

A	B	$DH°$
F	F	37
Cl	Cl	58
Br	Br	46
I	I	36
F	C_6H_5	126
Cl	C_6H_5	96
Br	C_6H_5	81
I	C_6H_5	65
C_6H_5	H	112
F	H	135.8
Cl	H	103.2
Br	H	87.5
I	H	71.3

EXERCISE 15-22

When benzene is dissolved in D_2SO_4, its ^1H NMR absorption at $\delta = 7.27$ ppm disappears and a new compound is formed having a molecular weight of 84. What is it? Propose a mechanism for its formation.

EXERCISE 15-23

Professor G. Olah* and his colleagues exposed benzene to the especially strong acid system HF-SbF_5 in an NMR tube and observed a new ^1H NMR spectrum with absorptions at $\delta = 5.69$ (2 H), 8.22 (2 H), 9.42 (1 H), and 9.58 (2 H) ppm. Propose a structure for this species.

In summary, the halogenation of benzene increases in exothermicity from I_2 (endothermic) to F_2 (explosive). Chlorinations and brominations are achieved with the

*Professor George A. Olah (b. 1927), University of Southern California, Los Angeles, Nobel Prize 1994 (chemistry).

help of Lewis acid catalysts, which polarize the X–X bond and activate the halogen by increasing its electrophilic power.

15-11 Nitration and Sulfonation of Benzene

In two other typical electrophilic substitutions of benzene, the electrophiles are the nitronium ion (NO_2^+), leading to nitrobenzene, and sulfur trioxide (SO_3), giving benzenesulfonic acid.

Benzene is subject to electrophilic attack by the nitronium ion

To bring about nitration of the ring at moderate temperatures, it is not sufficient just to treat benzene with concentrated nitric acid. Because the nitrogen in the nitrate group of HNO_3 has no electrophilic power, it must somehow be activated. Addition of concentrated sulfuric acid serves this purpose, protonating the nitric acid. Loss of water then yields the **nitronium ion,** NO_2^+, a strong electrophile.

Nitration of Benzene

HNO₃, H₂SO₄

NO₂

Nitrobenzene

Activation of Nitric Acid by Sulfuric Acid

Nitric acid

Nitronium ion

Electrophilic attack on benzene proceeds by the positively charged nitrogen.

Mechanism of Aromatic Nitration

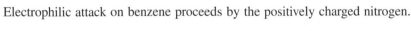

Nitrobenzene

Sulfonation is reversible

Concentrated sulfuric acid does not sulfonate benzene at room temperature. However, a more reactive form, called *fuming sulfuric acid,* permits electrophilic attack by SO_3. Commercial fuming sulfuric acid is made by adding about 8% of sulfur trioxide, SO_3, to the concentrated acid. Because of the strong electron-withdrawing effect of the three oxygens, the sulfur in SO_3 is electrophilic enough to attack benzene directly. Subsequent proton transfer results in the sulfonated product, benzenesulfonic acid.

Mechanism of Aromatic Sulfonation

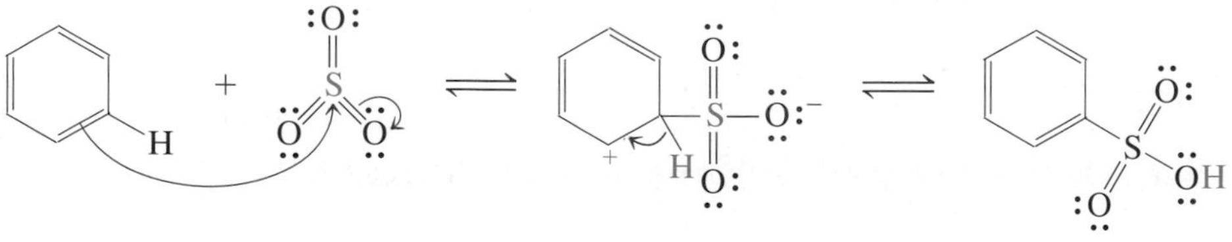

Benzenesulfonic acid

Aromatic sulfonation is readily reversible. The reaction of sulfur trioxide with water to give sulfuric acid is so exothermic that heating benzenesulfonic acid in dilute aqueous acid completely reverses sulfonation.

Sulfonation of Benzene

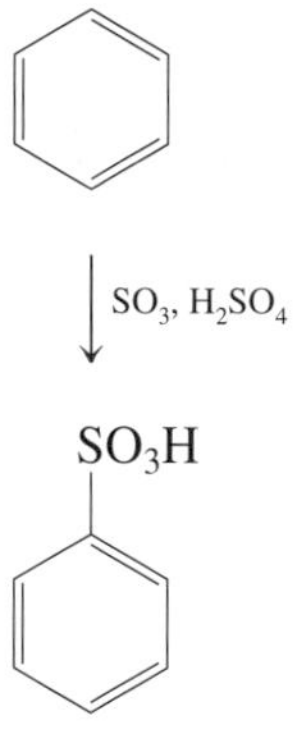

SO₃, H₂SO₄

Benzenesulfonic acid

Hydration of SO₃

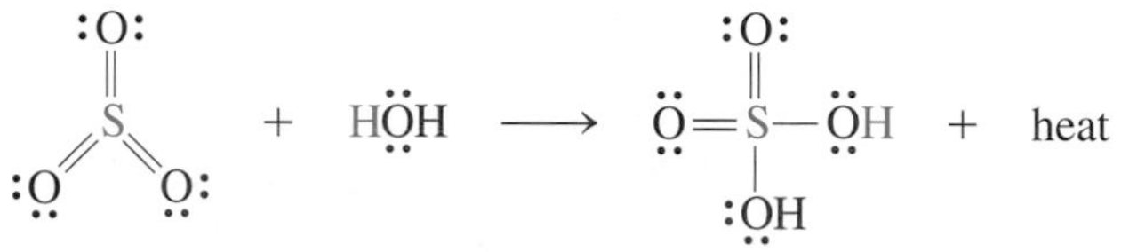

Reverse Sulfonation

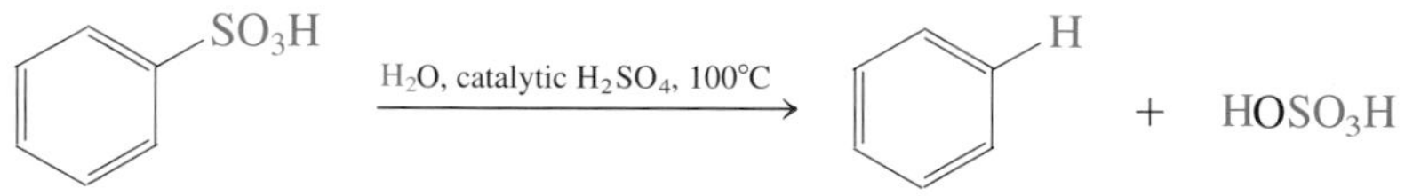

The reversibility of sulfonation may be used to control further aromatic substitution processes. The ring carbon containing the substituent is blocked from attack, and electrophiles are directed to other positions. Thus, the sulfonic acid group can be introduced to serve as a *directing blocking group* and then removed by reverse sulfonation. Synthetic applications of this strategy will be discussed in Section 16-5.

Benzenesulfonic acids have important uses

Sulfonation of substituted benzenes is used in the synthesis of *detergents*. Thus, long-chain branched alkylbenzenes are sulfonated to the corresponding sulfonic acids, then converted into their sodium salts. Because such detergents are not readily biodegradable, they have been replaced by more environmentally acceptable alternatives. We shall examine this class of compounds in Chapter 19.

Aromatic Detergent Synthesis

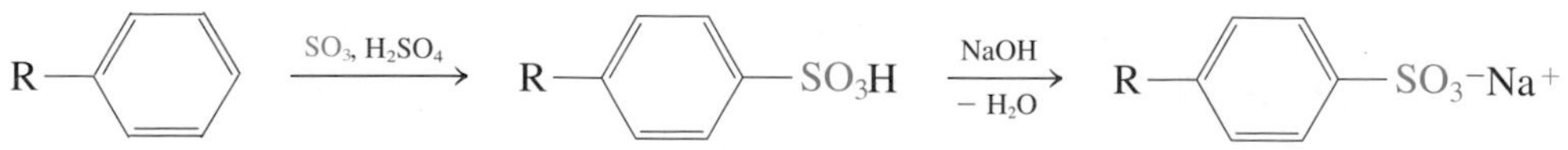

R = branched alkyl group

Another application of sulfonation is to the manufacture of dyes, because the sulfonic acid group imparts water solubility (Chapter 22).

Sulfonyl chlorides, the acid chlorides of sulfonic acids (see Section 9-4), are usually prepared by reaction of the sodium salt of the acid with PCl₅ or SOCl₂.

Preparation of Benzenesulfonyl Chloride

Sulfonyl chlorides are frequently employed in synthesis. For example, recall that the hydroxy group of an alcohol may be turned into a good leaving group by conversion of the alcohol into the 4-methylbenzenesulfonate (*p*-toluenesulfonate, tosylate; Sections 6-8 and 9-4).

Sulfonyl chlorides are important precursors of **sulfonamides,** many of which are chemotherapeutic agents, such as the *sulfa drugs* discovered in 1932 (Section 9-11). Sulfonamides are derived from the reaction of a sulfonyl chloride with an amine. Sulfa drugs specifically contain the 4-aminobenzenesulfonamide (sulfanilamide) function. Their mode of action is to interfere with the bacterial enzymes that help to synthesize folic acid (Chemical Highlight 25-4), thereby depriving them of an essential nutrient and thus causing bacterial cell death.

Sulfa Drugs

General structure

Sulfamethoxazole
(Gantanol)
(Antibacterial, used to treat urinary infections)

Sulfalene
(Kelfizina)
(Antilepral)

Sulfadiazine
(Antimalarial)

About 15,000 sulfa derivatives have been synthesized and screened for antibacterial activity; some have become new drugs. With the advent of newer generations of antibiotics, the medicinal use of sulfa drugs has greatly diminished, but their discovery was a milestone in the systematic development of medicinal chemistry.

EXERCISE 15-24

Formulate mechanisms for **(a)** the reverse of sulfonation; **(b)** the hydration of SO_3.

In summary, nitration of benzene requires the generation of the nitronium ion, NO_2^+, which functions as the active electrophile. The nitronium ion is formed by the loss of water from protonated nitric acid. Sulfonation is achieved with fuming sulfuric acid, in which sulfur trioxide, SO_3, is the electrophile. Sulfonation is reversed by hot aqueous acid. Benzenesulfonic acids are used in the preparation of detergents, dyes, compounds containing leaving groups, and sulfa drugs.

15-12 Friedel-Crafts Alkylation

None of the electrophilic substitutions mentioned so far have led to carbon–carbon bond formation, one of the primary challenges in organic chemistry. In principle, such reactions could be carried out with benzene in the presence of a sufficiently electrophilic carbon-based electrophile. This section introduces the first of two such transformations, the **Friedel-Crafts* reactions.** The secret to the success of both processes is the use of a Lewis acid, usually aluminum chloride. In the presence of this reagent, haloalkanes attack benzene to form alkylbenzenes.

In 1877, Friedel and Crafts discovered that a haloalkane reacts with benzene in the presence of an aluminum halide. The resulting products are the alkylbenzene and hydrogen halide. This reaction, which can be carried out in the presence of other Lewis acid catalysts, is called the **Friedel-Crafts alkylation** of benzene.

Friedel-Crafts Alkylation

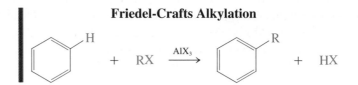

*Professor Charles Friedel (1832–1899), Sorbonne, Paris; Professor James M. Crafts (1839–1917), Massachusetts Institute of Technology, Cambridge, MA.

First experimental notes from Professor Crafts' notebook concerning the reaction of benzene with CH_3I (written ICH^3) in the presence of $AlCl_3$ (written Al^2Cl^6), dated May 21, 1877. [*Ecole Nationale Supérieure de Chimie de Paris, founded by Charles Friedel and Henri Moissan.*]

The reactivity of the haloalkane increases with the polarity of the C—X bond in the order RI < RBr < RCl < RF. Typical Lewis acids are (in the order of increasing activity) BF_3, $SbCl_5$, $FeCl_3$, $AlCl_3$, and $AlBr_3$.

Friedel-Crafts Alkylation of Benzene with Chloroethane

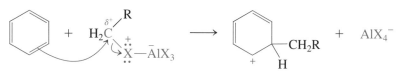

$$CH_3CH_2Cl \ + \ \underset{}{\text{(benzene)}} \ \xrightarrow{\text{AlCl}_3,\ 25°C} \ \underset{\substack{28\% \\ \textbf{Ethylbenzene}}}{\text{(ethylbenzene)}} \ + \ HCl$$

With primary halides, the reaction begins with coordination of the Lewis acid to the halogen of the haloalkane, much as in the activation of halogens in electrophilic halogenation. This coordination places a partial positive charge on the halogen-bearing carbon, rendering it more electrophilic. Attack on the benzene ring is followed by proton loss in the usual manner, giving the observed product.

Mechanism of Friedel-Crafts Alkylation with Primary Haloalkanes

STEP 1. Haloalkane activation

$$RCH_2\text{—}\overset{..}{\underset{..}{X}}\text{:} \ + \ AlX_3 \ \rightleftharpoons \ \overset{\delta+}{R}\overset{+}{C}H_2 \text{:} \overset{..}{\underset{..}{X}} \text{:} \overset{-}{A}lX_3$$

STEP 2. Electrophilic attack

STEP 3. Proton loss

With secondary and tertiary halides, free carbocations are usually formed as intermediates; these species attack the benzene ring in the same way as the cation NO_2^+.

EXERCISE 15-25

Write a mechanism for the formation of (1,1-dimethylethyl)benzene (*tert*-butylbenzene) from 2-chloro-2-methylpropane (*tert*-butyl chloride), benzene, and catalytic $AlCl_3$.

Intramolecular Friedel-Crafts alkylations can be used to fuse a new ring onto the benzene nucleus.

An Intramolecular Friedel-Crafts Alkylation

$$\xrightarrow[\text{— HCl}]{\text{AlCl}_3,\ CS_2\ \text{and}\ CH_3NO_2\ \text{(solvents)},\ 25°C,\ 72\ h}$$

31%
Tetralin
(Common name)

Friedel-Crafts alkylations can be carried out with any starting material, such as an alcohol or alkene, that functions as a precursor to a carbocation (Sections 9-2 and 12-3).

Friedel-Crafts Alkylations Using Other Carbocation Precursors

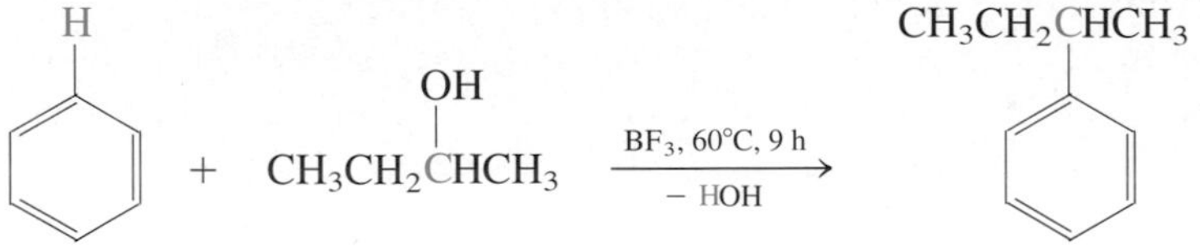

36%
(1-Methylpropyl)benzene

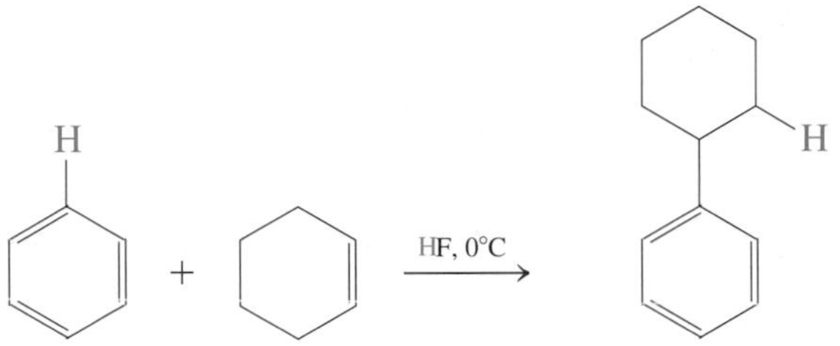

62%
Cyclohexylbenzene

EXERCISE 15-26

In 1995, more than 5.6 billion pounds (2.5 billion kilograms) of (1-methylethyl)benzene (isopropylbenzene or cumene), an important industrial intermediate in the manufacture of phenol (Chapter 22), was synthesized in the United States from propene and benzene in the presence of phosphoric acid. Write a mechanism for its formation in this reaction.

In summary, the Friedel-Crafts alkylation produces carbocations (or their equivalents) capable of electrophilic aromatic substitution by formation of aryl–carbon bonds. Haloalkanes, alkenes, and alcohols can be used to achieve aromatic alkylation in the presence of a Lewis or mineral acid.

15-13 Limitations of Friedel-Crafts Alkylations

The alkylation of benzenes under Friedel-Crafts conditions is accompanied by two important side reactions: One is *polyalkylation;* the other, *carbocation rearrangement.* Both cause the yield of the desired products to diminish and lead to mixtures that are difficult to separate.

Consider first polyalkylation. Benzene reacts with 2-bromopropane in the presence of $FeBr_3$ as a catalyst to give products of both single and double substitution. The yields are low because of the formation of many by-products.

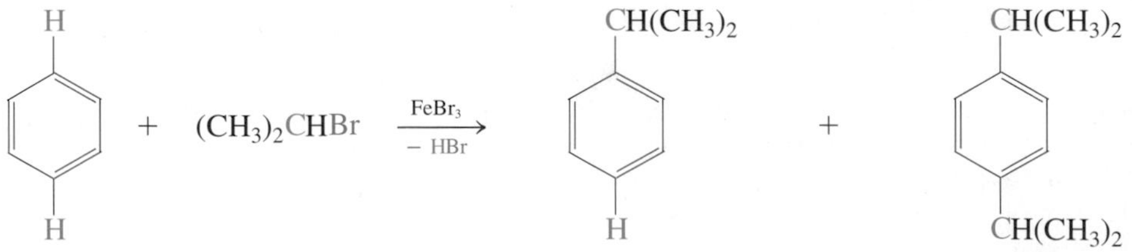

25%
(1-Methylethyl)benzene
(Isopropylbenzene)

15%
1,4-Bis(1-methylethyl)benzene
(*p*-Diisopropylbenzene)

The electrophilic aromatic substitutions that we studied in Sections 15-10 and 15-11 can be stopped at the monosubstitution stage. Why do Friedel-Crafts alkylations have the problem of multiple electrophilic substitution? It is because the substituents differ in electronic structure (a subject discussed in more detail in Chapter 16). Bromination, nitration, and sulfonation introduce an electron-withdrawing group into the benzene ring, which renders the product *less* susceptible than the starting material to electrophilic attack. In contrast, an alkylated benzene is more electron rich than unsubstituted benzene and thus *more* susceptible to electrophilic attack.

EXERCISE 15-27

Treatment of benzene with chloromethane in the presence of aluminum chloride results in a complex mixture of tri-, tetra-, and pentamethylbenzenes. One of the components in this mixture crystallizes out selectively: m.p. = 80°C; molecular formula = $C_{10}H_{14}$; ^{1}H NMR $\delta = 2.27$ (s, 12 H) and 7.15 (s, 2 H) ppm; ^{13}C NMR $\delta = 19.2$, 131.2, and 133.8 ppm. Draw a structure for this product.

The second complication in aromatic alkylation is skeletal rearrangement (Section 9-3). For example, the attempted propylation of benzene with 1-bromopropane and $AlCl_3$ produces (1-methylethyl)benzene.

The starting haloalkane rearranges to the thermodynamically favored 1-methylethyl (isopropyl) cation in the presence of the Lewis acid.

Rearrangement of 1-Bromopropane to 1-Methylethyl (Isopropyl) Cation

EXERCISE 15-28

Attempted alkylation of benzene with 1-chlorobutane in the presence of $AlCl_3$ gave not only the expected butylbenzene, but also, as a major product, (1-methylpropyl)benzene. Write a mechanism for this reaction.

Because of these limitations, Friedel-Crafts alkylations are used rarely in synthetic chemistry. Is there a way to improve this process? It would require an electrophilic carbon species that could not rearrange and that would, moreover, deactivate the ring to prevent further substitution. There is such a species—an acylium cation—and it is used in the second Friedel-Crafts reaction, the topic of the next section.

In summary, Friedel-Crafts alkylation suffers from overalkylation and skeletal rearrangements by both hydrogen and alkyl shifts.

15-14 Friedel-Crafts Alkanoylation (Acylation)

The second electrophilic aromatic substitution that forms carbon–carbon bonds is **Friedel-Crafts alkanoylation, or acylation.** This reaction proceeds through the intermediacy of **acylium cations,** with the general structure $RC{\equiv}O{:}^+$. This section will describe how these ions readily attack benzene to form ketones.

Even though it is unstable and reactive, HCO^+ is a fundamental small organic molecule that is (relatively) abundant in such diverse environments as hot flames and cold interstellar space. It has been detected in the gas surrounding the comet Hale-Bopp, a spectacular visitor to Earth's skies in 1997.

Friedel-Crafts Alkanoylation

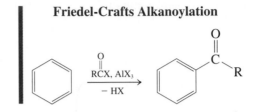

Friedel-Crafts alkanoylation employs alkanoyl chlorides

Benzene reacts with alkanoyl (acyl) halides in the presence of an aluminum halide to give 1-phenylalkanones (phenyl ketones). An example is the preparation of 1-phenylethanone (acetophenone) from benzene and acetyl chloride, by using aluminum chloride as the Lewis acid.

**Friedel-Crafts Alkanoylation of Benzene
with Acetyl Chloride**

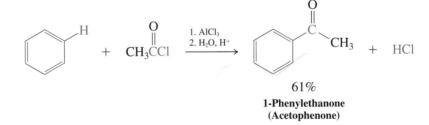

61%
**1-Phenylethanone
(Acetophenone)**

Alkanoyl (acyl) chlorides are reactive derivatives of carboxylic acids. They are readily formed from the acids by reaction with $SOCl_2$. (We shall explore this process in detail in Chapter 19.)

Preparation of an Alkanoyl (Acyl) Chloride

$$\underset{\text{RCOH}}{\overset{O}{\parallel}} \quad + \quad SOCl_2 \quad \longrightarrow \quad \underset{\text{RCCl}}{\overset{O}{\parallel}} \quad + \quad SO_2 \quad + \quad HCl$$

Alkanoyl halides react with Lewis acids to produce acylium ions

Acylium cations are formed by the reaction of alkanoyl halides with aluminum chloride. The Lewis acid initially coordinates to the carbonyl oxygen. This complex is in equilibrium with an isomer in which the aluminum chloride is bound to the halogen. Dissociation then produces the acylium ion, which is stabilized by resonance.

Acylium Ion Generation

$$RC\overset{\displaystyle :O:}{\underset{\displaystyle}{\|}}\overset{}{C}-\ddot{X}: \ + \ \overline{AlCl_3} \ \rightleftharpoons \ RC\overset{+:O\ \ \overline{AlCl_3}}{\|}-\ddot{X}: \ \rightleftharpoons \ RC\overset{:O:}{\|}-\overset{+}{\ddot{X}}-\overline{AlCl_3} \ \rightleftharpoons$$

$$[RC\overset{+}{\equiv}\overset{}{O}: \ \longleftrightarrow \ R\overset{+}{C}=\ddot{\overset{..}{O}}] \ + \ :\overset{..}{\ddot{X}}\overline{AlCl_3}$$
Acylium ion

Sometimes carboxylic anhydrides are used in alkanoylation in place of the halides. These molecules react with Lewis acids in a similar way.

Acylium Ions from Carboxylic Anhydrides

$$RC\overset{:O:\ \ :O:}{\underset{}{\|\ \ \ \ \|}}-\overset{}{\ddot{O}}-CR \ + \ \overline{AlCl_3} \ \rightleftharpoons \ RC\overset{+:O\ \ \ \ :O:\ \ \overline{AlCl_3}}{\|\ \ \ \ \ \ \ \|}-\ddot{O}-CR \ \rightleftharpoons \ RC\overset{:O:\ \ \ \ :O:}{\|\ \ \ \ \ \ \|}-\overset{+}{\underset{\overline{AlCl_3}}{\ddot{O}}}-CR \ \rightleftharpoons$$

$$[RC\overset{+}{\equiv}\overset{}{O}: \ \longleftrightarrow \ R\overset{+}{C}=\ddot{\overset{..}{O}}] \ + \ Cl_3Al\overset{-\ \ ..\ \ :O:}{\underset{}{\ddot{O}}}\overset{\|}{C}R$$

Acylium ions undergo electrophilic aromatic substitution

The acylium ion is sufficiently electrophilic to attack benzene by the usual aromatic substitution mechanism.

Electrophilic Alkanoylation

(reaction scheme: benzene with H + $R\overset{+}{-}C\overset{}{=}\overset{..}{O}:$ → arenium intermediate with H, CR, and $\overset{+}{:O:}$ → product 1-phenylalkanone with $\overset{O}{\|}\overset{}{C}R$ + H^+)

Because the newly introduced alkanoyl substituent is electron withdrawing (see Sections 14-8 and 16-1), it deactivates the ring and protects it from further substitution. The effect is accentuated by the formation of a strong complex between the aluminum chloride catalyst and the carbonyl function of the product ketone.

Lewis Acid Complexation with 1-Phenylalkanones

(reaction scheme: phenyl ketone with $\overset{:O:}{\|}\overset{}{C}R$ + $\overline{AlCl_3}$ $\rightleftharpoons$ phenyl ketone with $\overset{+:O\ \ \overline{AlCl_3}}{\|}\overset{}{C}R$)

This complexation removes the $AlCl_3$ from the reaction mixture and necessitates the use of *at least one full equivalent* of the Lewis acid to allow the reaction to go to completion. Aqueous work-up is necessary to liberate the ketone from its aluminum chloride complex, as illustrated by the following examples.

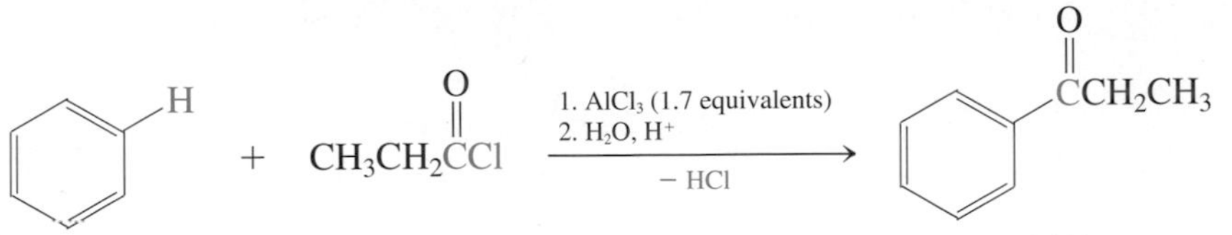

84%
**1-Phenyl-1-propanone
(Propiophenone)**

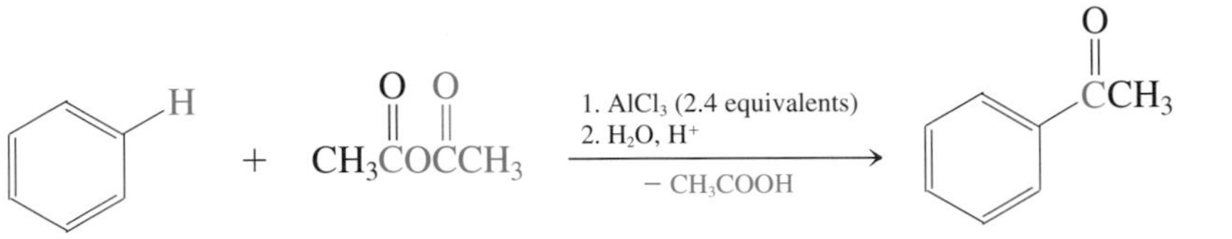

85%
**1-Phenylethanone
(Acetophenone)**

EXERCISE 15-29

The simplest alkanoyl chloride, formyl chloride, H–C–Cl, is unstable, decomposing to HCl and CO upon attempted preparation. Therefore, direct Friedel-Crafts formylation of benzene is impossible. An alternative process, the Gattermann-Koch reaction, enables the introduction of the formyl group, –CHO, into the benzene ring by treatment with CO under pressure, in the presence of HCl and Lewis acid catalysts. For example, methylbenzene (toluene) can be formylated at the para position in this way in 51% yield. The electrophile in this process was observed directly for the first time in 1997 by treating CO with HF-SbF$_5$ under high pressure: ^{13}C NMR: δ 139.5 ppm; IR: $\tilde{\nu} = 2110$ cm^{-1}. What is the structure of this species and the mechanism of its reaction with methylbenzene? Explain the spectral data. (**Hints:** Draw the Lewis structure of CO and proceed by considering the species that may arise in the presence of acid. The comparative spectral data for free CO are ^{13}C NMR: δ 181.3 ppm; IR: $\tilde{\nu} = 2143$ cm^{-1}.)

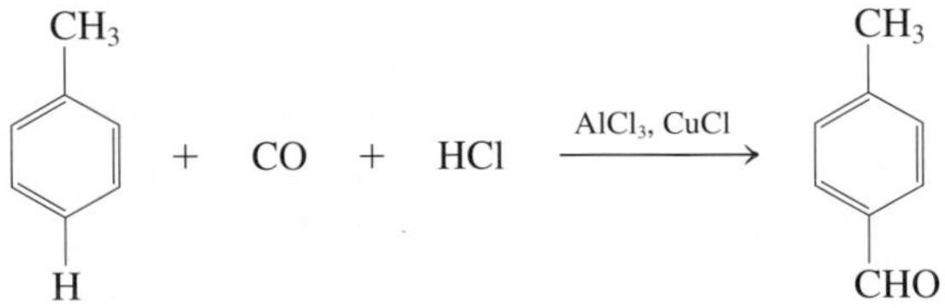

In summary, the problems of Friedel-Crafts alkylation (multiple substitution and carbocation rearrangements) are avoided in Friedel-Crafts alkanoylations, in which an alkanoyl halide or carboxylic acid anhydride is the reaction partner, in the presence of a Lewis acid. The intermediate acylium cations undergo electrophilic aromatic substitution to yield the corresponding aromatic ketones.

CHAPTER INTEGRATION PROBLEM

The insecticide DDT (see Chemical Highlight 3-3) has been made in ton quantities by treatment of chlorobenzene with 2,2,2-trichloroacetaldehyde in the (required) presence of concentrated H_2SO_4. Formulate a mechanism for this reaction.

| Chlorobenzene | 2,2,2-Trichloro-acetaldehyde | | DDT |

SOLUTION

Let us first take an inventory of what is given:

1. The product is composed of two subunits derived from chlorobenzene and one subunit derived from the aldehyde.
2. In the process, the starting materials, two chlorobenzenes and one trichloroacetaldehyde, which amount to a combined atomic total of $C_{14}H_{11}Cl_5O$, turn into DDT, with the molecular formula $C_{14}H_9Cl_5$. Conclusion: The elements of H_2O are extruded, perhaps as water.
3. Topologically, the transformation constitutes a replacement of an aromatic hydrogen by an alkyl carbon substituent, strongly implicating the occurrence of a Friedel-Crafts alkylation (Section 15-12).

We can now address the details of a possible mechanism. Friedel-Crafts alkylations require positively polarized or cationic carbon electrophiles. In our case, inspection of the product clearly suggests that the carbonyl carbon in the aldehyde is the electrophilic aggressor. This carbon is positively polarized to begin with, because of the presence of the electron withdrawing oxygen and chlorine substituents. A nonoctet dipolar resonance form (Section 1-5) illustrates this point.

Activation of 2,2,2-Trichloroacetaldehyde as an Electrophile

In the presence of strong acid, protonation of the negatively polarized oxygen of the neutral species generates a positively charged intermediate in which the electrophilic character of the carbonyl carbon is further accentuated in the resonance form depicting a hydroxy carbocation. The stage is now set for the first of two electrophilic aromatic substitution steps (Section 15-12).

First Electrophilic Aromatic Substitution Step

The product of this step is an alcohol, which can be readily converted into the corresponding carbocation by acid (Section 9-2), partly because the ensuing charge is resonance stabilized by the adjacent benzene ring (*benzylic resonance*, Section 22-1, related to *allylic resonance*, Sections 14-1 and 14-3). The cation subsequently enables the second electrophilic aromatic substitution step to provide DDT.

Alcohol Activation and Second Electrophilic Aromatic Substitution Step

Finally, you may have wondered why this synthesis does not lead to a mixture of ortho, meta, and para products, but rather only para substitution is observed. The topic of directing effects in substituted benzenes undergoing electrophilic substitution is considered in the next chapter. Chlorine as a substituent is an ortho, para director (Section 16-3), because the positive charge in the cationic intermediates of *o,p*-electrophilic attack can be stabilized by resonance with the chlorine lone electron pairs. This is not the case when such attack is at the meta position. (Draw the resonance forms of the corresponding intermediates fully.) In the DDT synthesis, steric hindrance suffered by the relatively large electrophile makes ortho substitution less favorable, leading to the observed result.

NEW REACTIONS

1. Hydrogenation of Benzene (Section 15-2)

$$\Delta H^\circ = -49.3 \text{ kcal mol}^{-1}$$
Resonance energy: $\sim -30 \text{ kcal mol}^{-1}$

Electrophilic Aromatic Substitution

2. Chlorination, Bromination, Nitration, and Sulfonation (Sections 15-10 and 15-11)

$$C_6H_6 \xrightarrow{X_2,\ FeX_3} C_6H_5X \ + \ HX \qquad \text{X = Cl, Br}$$

$$C_6H_6 \xrightarrow{HNO_3,\ H_2SO_4} C_6H_5NO_2 \ + \ H_2O$$

$$C_6H_6 \underset{H_2SO_4,\ H_2O,\ \Delta}{\overset{SO_3,\ H_2SO_4}{\rightleftharpoons}} C_6H_5SO_3H \qquad \textbf{Reversible}$$

3. Benzenesulfonyl Chlorides (Section 15-11)

$$C_6H_5SO_3Na \ + \ PCl_5 \longrightarrow C_6H_5SO_2Cl \ + \ POCl_3 \ + \ NaCl$$

4. Friedel-Crafts Alkylation (Section 15-12)

$$C_6H_6 \ + \ RX \xrightarrow{AlCl_3} C_6H_5R \ + \ HX \ + \ \text{overalkylated product}$$

R⁺ is subject to carbocation rearrangements

Intramolecular

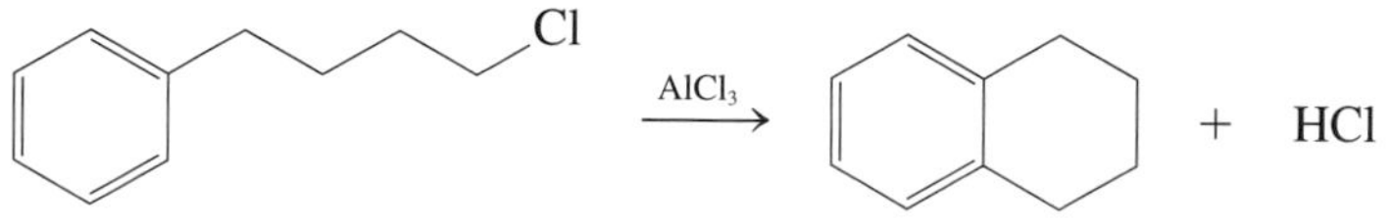

Alcohols and alkenes as substrates

$$C_6H_6 \ + \ \underset{\underset{OH}{|}}{RCHR'} \xrightarrow[-H_2O]{BF_3,\ 60°C} \underset{\underset{R'}{|}}{C_6H_5CHR}$$

$$C_6H_6 \ + \ RCH{=}CH_2 \xrightarrow{HF,\ 0°C} \underset{\underset{R}{|}}{C_6H_5CHCH_3}$$

5. Friedel-Crafts Alkanoylation (Section 15-14)

Alkanoyl halides

$$C_6H_6 \ + \ R\overset{O}{\overset{\|}{C}}Cl \xrightarrow[2.\ H_2O]{1.\ AlCl_3} C_6H_5\overset{O}{\overset{\|}{C}}R \ + \ HCl$$

Requires at least one full equivalent of Lewis acid

Anhydrides

$$C_6H_6 \ + \ CH_3\overset{O}{\overset{\|}{C}}O\overset{O}{\overset{\|}{C}}CH_3 \xrightarrow[2.\ H_2O]{1.\ AlCl_3} C_6H_5\overset{O}{\overset{\|}{C}}CH_3 \ + \ CH_3COOH$$

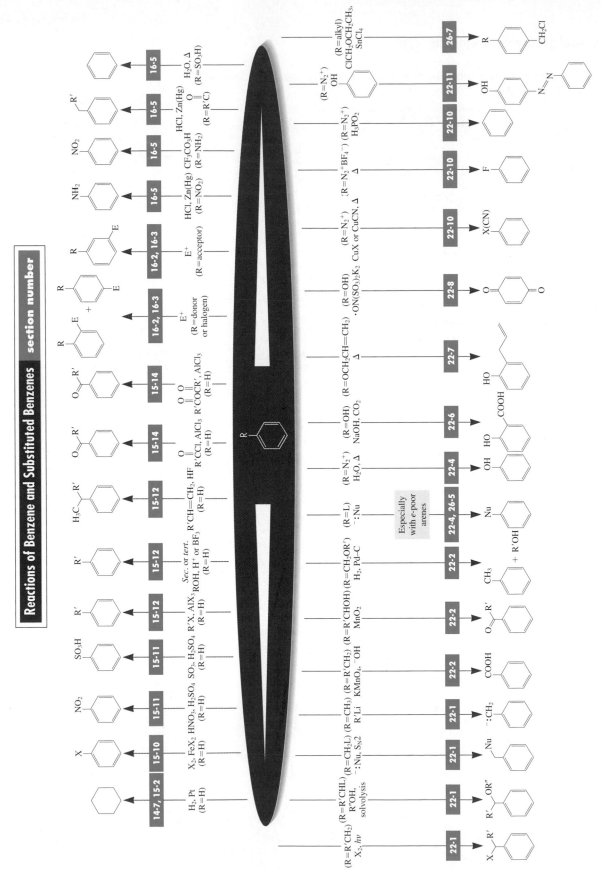

Reactions of Benzene and Substituted Benzenes | **section number**

680

IMPORTANT CONCEPTS

1. Substituted benzenes are named by adding prefixes or suffixes to the word *benzene.* Disubstituted systems are labeled as 1,2-, 1,3-, and 1,4- or **ortho, meta,** and **para,** depending on the location of the substituents. Many benzene derivatives have common names, sometimes used as bases for naming their substituted analogs. As a substituent, an aromatic system is called **aryl;** the parent aryl substituent, C_6H_5, is called **phenyl;** its homolog $C_6H_5CH_2$ is named **phenylmethyl (benzyl).**

2. Benzene is not a cyclohexatriene but a delocalized cyclic system of six π electrons. It is a **regular hexagon** of six sp^2-hybridized carbons. All six p orbitals overlap equally with their neighbors. Its unusually low heat of hydrogenation indicates a **resonance energy** or **aromaticity** of about 30 kcal mol^{-1}. The stability imparted by aromatic delocalization is also evident in the transition state of some reactions, such as the Diels-Alder cycloaddition and ozonolysis.

3. The special structure of benzene gives rise to unusual UV, IR, and NMR spectral data. ^1H NMR spectroscopy is particularly diagnostic because of the unusual **deshielding** of aromatic hydrogens by an **induced ring current.** Moreover, the substitution pattern is revealed by examination of the *o, m,* and *p* coupling constants.

4. The **polycyclic benzenoid hydrocarbons** are composed of linearly or angularly fused benzene rings. The simplest members of this class of compounds are naphthalene, anthracene, and phenanthrene.

5. In these molecules, benzene rings **share** two (or more) carbon atoms, whose π electrons are delocalized over the entire ring system. Thus, naphthalene shows some of the properties characteristic of the aromatic ring in benzene: The electronic spectra reveal extended conjugation, ^1H NMR exhibits

deshielding ring-current effects, and there is little bond alternation.

6. Benzene is the smallest member of the class of aromatic cyclic polyenes following **Hückel's 4n + 2 rule.** Most of the $4n$ π systems are relatively reactive **anti-** or **nonaromatic** species. Hückel's rule also extends to aromatic charged systems, such as the cyclopentadienyl anion, cycloheptatrienyl cation, and cyclooctatetraene dianion.

7. The most important reaction of benzene is **electrophilic aromatic substitution.** The rate-determining step is addition by the electrophile to give a delocalized hexadienyl cation in which the aromatic character of the original benzene ring has been lost. Fast deprotonation restores the aromaticity of the (now substituted) benzene ring. Exothermic substitution is preferred over endothermic addition. The reaction can lead to halo- and nitrobenzenes, benzenesulfonic acids, and alkylated and alkanoylated derivatives. When necessary, Lewis acid (chlorination, bromination, Friedel-Crafts reaction) or mineral acid (nitration, sulfonation) catalysts are applied. These enhance the electrophilic power of the reagents or generate strong, positively charged electrophiles.

8. **Sulfonation** of benzene is a **reversible** process. The sulfonic acid group is removed by heating with dilute aqueous acid.

9. **Benzenesulfonic acids** are precursors of benzenesulfonyl chlorides. The chlorides react with alcohols to form sulfonic esters containing useful **leaving groups** and with amines to give sulfonamides, some of which are medicinally important.

10. In contrast with other electrophilic substitutions, **Friedel-Crafts alkylations** activate the aromatic ring to further electrophilic substitution, leading to product mixtures.

PROBLEMS

30. Name each of the following compounds by using the IUPAC system and, if possible, a reasonable common alternative. (**Hint:** The order of functional group precedence is –COOH > –CHO > –OH > –NH_2.)

(a) COOH ... Cl   (b) OCH_3 ... NO_2 (c) OH CHO

(d)

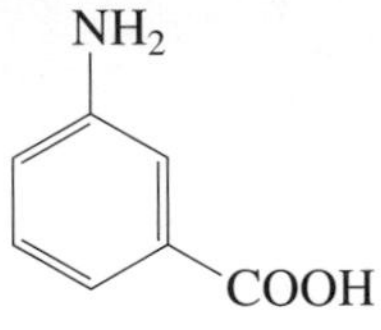

(e)

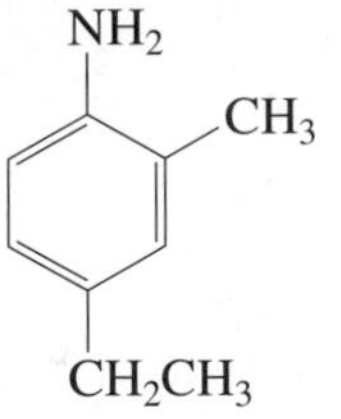

(f)

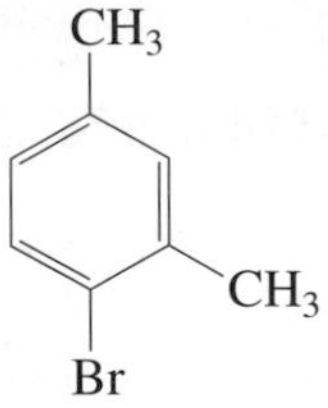

(g)

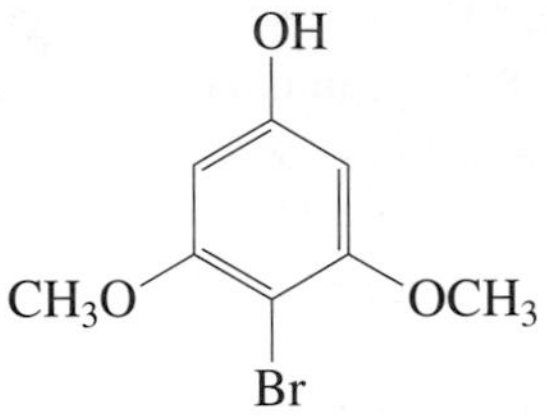

(h)

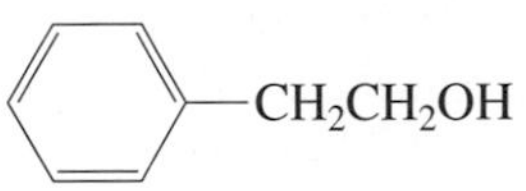

(i)

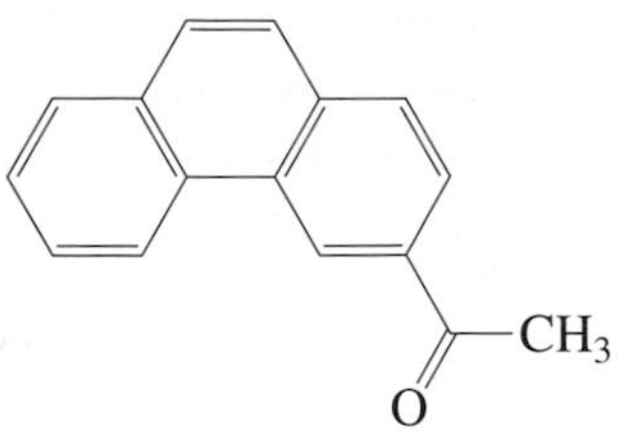

31. Give a proper IUPAC name for each of the following commonly named substances.

(a) 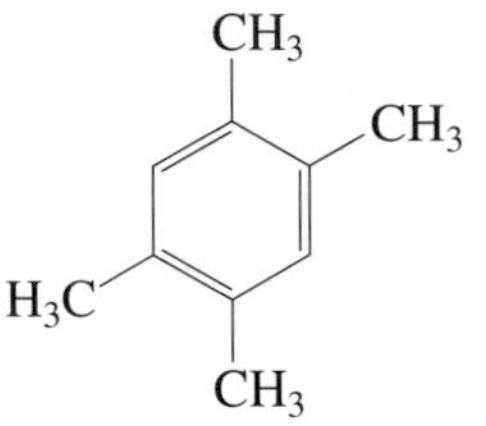
Durene

(b) 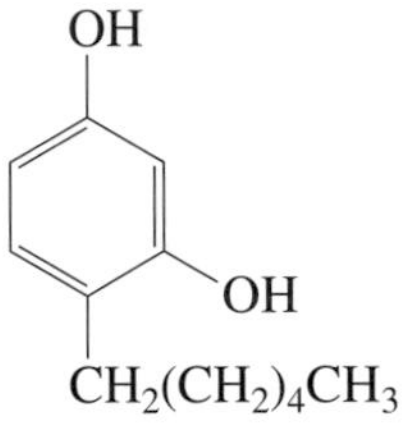
Hexylresorcinol

(c) 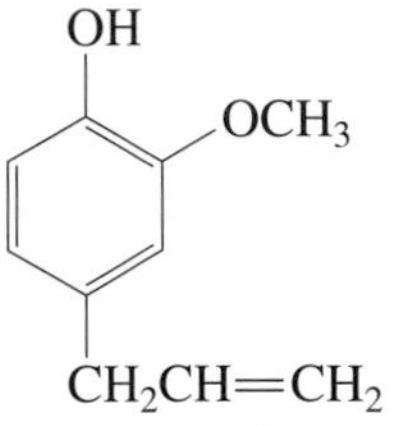
Eugenol

32. Draw the structure of each of the following compounds. If the name itself is incorrect, give a correct systematic alternative. **(a)** *o*-Chlorobenzaldehyde; **(b)** 2,4,6-trihydroxybenzene; **(c)** 4-nitro-*o*-xylene; **(d)** *m*-isopropylbenzoic acid; **(e)** 4,5-dibromoaniline; **(f)** *p*-methoxy-*m*-nitroacetophenone.

33. The complete combustion of benzene is exothermic by approximately -789 kcal mol^{-1}. What would this value be if benzene lacked aromatic stabilization?

34. The ^{1}H NMR spectrum of naphthalene shows two multiplets (Figure 15-15). The upfield absorption ($\delta = 7.40$ ppm) is due to the hydrogens at C2, C3, C6, and C7, and the downfield multiplet ($\delta = 7.77$ ppm) is due to the hydrogens at C1, C4, C5, and C8. Why do you suppose the latter hydrogens are more deshielded than the former?

35. Complete hydrogenation of 1,3,5,7-cyclooctatetraene is exothermic by -101 kcal mol^{-1}. Hydrogenation of cyclooctene proceeds with $\Delta H^\circ = -23$ kcal mol^{-1}. Are these data consistent with the description of cyclooctatetraene presented in the chapter?

36. Which of the following structures qualify as being aromatic, according to Hückel's rule?

(a) 
(b) 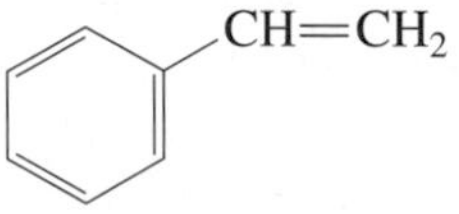
(c) 
(d)

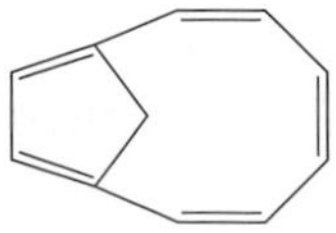

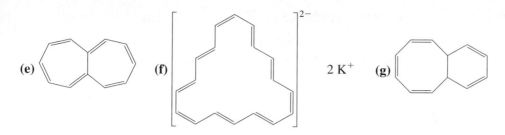

(e)　(f) 2 K⁺ (g)

37. Following are spectroscopic and other data for several compounds. Propose a structure for each of them. (a) Molecular formula = $C_6H_4Br_2$. ¹H NMR spectrum A. ¹³C NMR: 3 peaks. IR: $\tilde{\nu} = 745$ (s, broad) cm⁻¹. UV: $\lambda_{max}(\epsilon) =$ 263(150), 270(250), and 278(180) nm. (b) Molecular formula = C_7H_7BrO. ¹H NMR spectrum B. ¹³C NMR: 7 peaks. IR: $\tilde{\nu} = 765$(s) and 680(s) cm⁻¹. (c) Molecular formula = $C_9H_{11}Br$. ¹H NMR spectrum C. ¹³C NMR: $\delta = 20.6$ (CH_3), 23.6 (CH_3), 124.2 ($C_{quaternary}$), 129.0 (CH), 136.0 ($C_{quaternary}$), and 137.7 ($C_{quaternary}$) ppm.

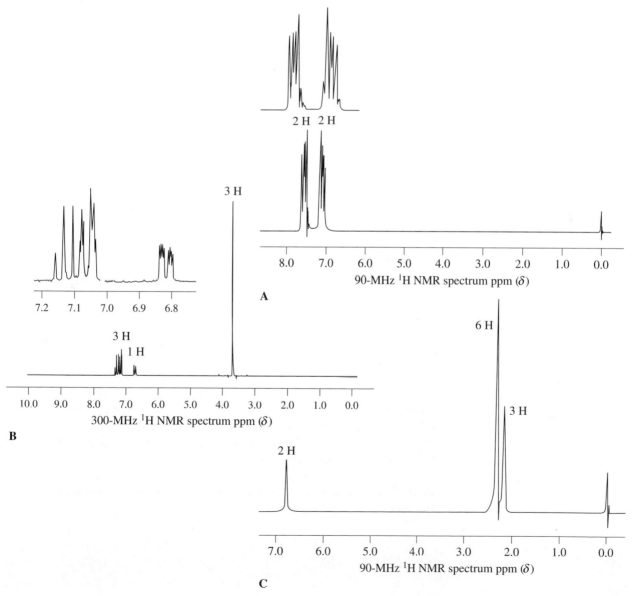

38. (a) Is it possible to distinguish the three isomers of dimethoxybenzene solely on the basis of the number of peaks in their proton-decoupled ^{13}C NMR spectra? Explain. **(b)** How many different isomers of dimethoxynaphthalene exist? How many peaks should each one exhibit in the proton-decoupled ^{13}C NMR spectrum?

39. The species resulting from the addition of benzene to $HF\text{-}SbF_5$ (Exercise 15-23) shows the following ^{13}C NMR absorptions: $\delta = 52.2(CH_2)$, $136.9(CH)$, $178.1(CH)$, and $186.6(CH)$ ppm. The signals at $\delta = 136.9$ and $\delta = 186.6$ are twice the intensity of the other peaks. Assign this spectrum.

40. Write the expected major product that should form on addition of each of the following reagent mixtures to benzene.
(a) $Cl_2 + AlCl_3$
(b) $T_2O + T_2SO_4$ (T = tritium, ^3H)
(c) $ICl + FeCl_3$ (Careful! $DH°_{ICl} = 50$ kcal mol^{-1}. Is this reaction exothermic?)
(d) N_2O_5 (which tends to dissociate into NO_2^+ and NO_3^-)
(e) $(CH_3)_2C{=}CH_2 + H_3PO_4$
(f) $(CH_3)_3CCH_2CH_2Cl + AlCl_3$
(g) $(CH_3)_2\overset{\overset{\displaystyle Br}{|}}{C}CH_2CH_2\overset{\overset{\displaystyle Br}{|}}{C}(CH_3)_2 + AlBr_3$
(h) $H_3C\!-\!\!\bigcirc\!\!-\!COCl + AlCl_3$

41. Write mechanisms for reactions (c) and (f) in Problem 40.

42. Propose a mechanism for the direct chlorosulfonylation of benzene (in the margin), an alternative synthesis of benzenesulfonyl chloride.

43. Benzene reacts with sulfur dichloride, SCl_2, in the presence of $AlCl_3$ to give diphenyl sulfide, $C_6H_5\text{–}S\text{–}C_6H_5$. Propose a mechanism for this process.

44. (a) 3-Phenylpropanoyl chloride, $C_6H_5CH_2CH_2COCl$, reacts with $AlCl_3$ to give a single product with the formula C_9H_8O and an ^1H NMR spectrum with signals at $\delta = 2.53$ (t, $J = 8$ Hz, 2 H), 3.02 (t, $J = 8$ Hz, 2 H), and 7.2–7.7 (m, 4 H) ppm. Propose a structure and a mechanism for the formation of this product.
(b) The product of the process described in (b) is subjected to the following reaction sequence: (1) $NaBH_4$, CH_3CH_2OH; (2) Conc. H_2SO_4, 100°C; (3) H_2, Pd-C, CH_3CH_2OH. The final product exhibits five resonance lines in its proton-decoupled ^{13}C NMR spectrum. What is the structure of the substance formed after each of the steps in this sequence?

45. The text states that alkylated benzenes are more susceptible to electrophilic attack than is benzene itself. Draw a graph like that in Figure 15-19 to show

+
2 ClSO₃H

↓

SO₂Cl

+
HCl
+
H₂SO₄

how the energy profile of electrophilic substitution of methylbenzene (toluene) would differ quantitatively from that of benzene.

46. Like haloalkanes, haloarenes are readily converted into organometallic reagents, which are sources of nucleophilic carbon.

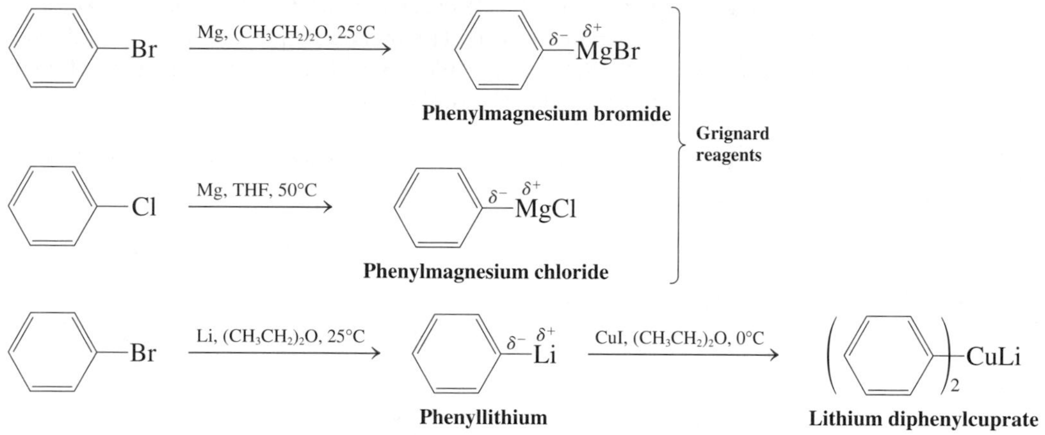

C_6H_5—Br $\xrightarrow{\text{Mg, (CH}_3\text{CH}_2)_2\text{O, 25°C}}$ C_6H_5—$\overset{\delta-}{}\overset{\delta+}{\text{MgBr}}$

Phenylmagnesium bromide

C_6H_5—Cl $\xrightarrow{\text{Mg, THF, 50°C}}$ C_6H_5—$\overset{\delta-}{}\overset{\delta+}{\text{MgCl}}$

Phenylmagnesium chloride

} Grignard reagents

C_6H_5—Br $\xrightarrow{\text{Li, (CH}_3\text{CH}_2)_2\text{O, 25°C}}$ C_6H_5—$\overset{\delta-}{}\overset{\delta+}{\text{Li}}$ $\xrightarrow{\text{CuI, (CH}_3\text{CH}_2)_2\text{O, 0°C}}$ $(C_6H_5)_2$CuLi

Phenyllithium　　　　　　　　　　**Lithium diphenylcuprate**

The chemical behavior of these reagents is very similar to that of their alkyl counterparts. Write the main product of each of the following sequences.

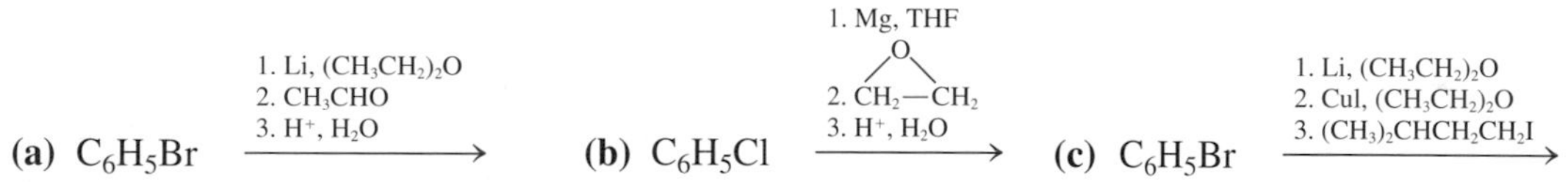

(a) C_6H_5Br $\xrightarrow[\substack{\text{1. Li, (CH}_3\text{CH}_2)_2\text{O} \\ \text{2. CH}_3\text{CHO} \\ \text{3. H}^+, \text{H}_2\text{O}}]{}$

(b) C_6H_5Cl $\xrightarrow[\substack{\text{1. Mg, THF} \\ \text{2. CH}_2\!-\!\text{CH}_2 \text{ (epoxide)} \\ \text{3. H}^+, \text{H}_2\text{O}}]{}$

(c) C_6H_5Br $\xrightarrow[\substack{\text{1. Li, (CH}_3\text{CH}_2)_2\text{O} \\ \text{2. CuI, (CH}_3\text{CH}_2)_2\text{O} \\ \text{3. (CH}_3)_2\text{CHCH}_2\text{CH}_2\text{I}}]{}$

47. Give efficient syntheses of the following compounds, beginning with benzene. **(a)** 1-Phenyl-1-heptanol; **(b)** 2-phenyl-2-butanol; **(c)** 1-phenyloctane. (**Hint:** Use a method from Problem 46. Why will Friedel-Crafts alkylation not work?)

48. Because of cyclic delocalization, structures A and B shown here for *o*-dimethylbenzene (*o*-xylene) are simply two resonance forms of the same molecule. Can the same be said for the two dimethylcyclooctatetraene structures C and D? Explain.

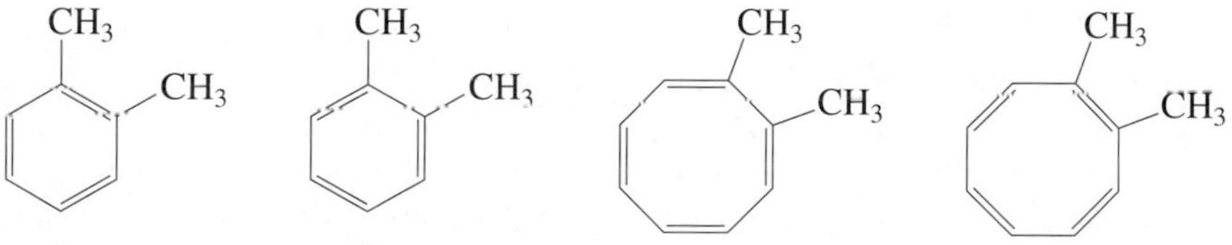

A　　　　**B**　　　　**C**　　　　**D**

49. The energy levels of the 2-propenyl (allyl) and cyclopropenyl π systems (see margin) are compared qualitatively in the diagram below. **(a)** Draw the three molecular orbitals of each system, using plus and minus signs and dotted lines to indicate bonding overlap and nodes, as in Figure 15-4. Does either of these systems possess degenerate molecular orbitals? **(b)** How many π electrons would give rise to the maximum stabilization of the cyclopropenyl system, relative to 2-propenyl (allyl)? (Compare Figure 15-5, for benzene.) Draw Lewis structures for both systems with this number of π electrons and any appropriate atomic charges. **(c)** Could the cyclopropenyl system drawn in (b) qualify as being "aromatic"? Explain.

**2-Propenyl
(Allyl)**

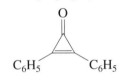

Cyclopropenyl

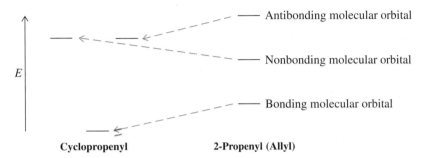

50. 2,3-Diphenylcyclopropenone (see structure in the margin) forms an addition product with HBr that exhibits the properties of an ionic salt. Suggest a structure for this product and a reason for its existence as a stable entity.

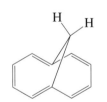

2,3-Diphenylcyclopropenone

51. (a) The ^{1}H NMR spectrum of [18]annulene shows two signals, at $\delta = 9.28$ (12 H) and -2.99 (6 H) ppm. The negative chemical shift value refers to a resonance *upfield* (to the *right*) of $(CH_3)_4Si$. Explain this spectrum. (**Hint:** Consult Figure 15-8.) **(b)** The unusual molecule 1,6-methano[10]annulene (shown in the margin) exhibits two sets of signals in the ^{1}H NMR spectrum at $\delta = 7.10$ (8 H) and -0.50 (2 H) ppm. Is this result a sign of aromatic character?

1,6-Methano[10]annulene

52. The ^{1}H NMR spectrum of the most stable isomer of [14]annulene shows two signals, at $\delta = -0.61$ (4 H) and 7.88 (10 H) ppm. Two possible structures for [14]annulene are shown here. How do they differ? Which one corresponds to the NMR spectrum described?

A   B

53. Explain the following reaction and the indicated stereochemical result mechanistically.

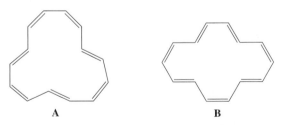

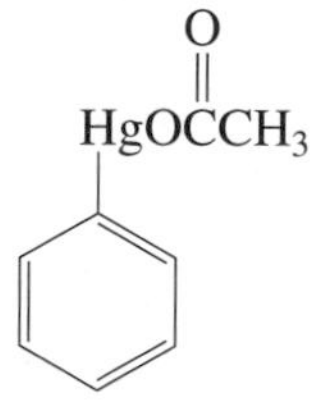

Phenylmercury acetate

54. Metal-substituted benzenes have a long history of use in medicine. Before antibiotics were discovered, phenylarsenic derivatives were the only treatment for a number of diseases. Phenylmercury compounds continue to be used as fungicides and antimicrobial agents to the present day. On the basis of the general principles explained in this chapter and your knowledge of the characteristics of compounds of Hg^{2+} (see Section 12-7), propose a sensible synthesis of phenylmercury acetate (shown in the margin).

Team Problem

55. As a team, discuss the following complementary experimental results as they pertain to the mechanism of electrophilic aromatic substitution.
   (a) A solution of HCl and benzene is colorless and does not conduct electricity, whereas a solution of HCl and $AlCl_3$ and benzene is colored and does conduct electricity.
   (b) The following are ^{13}C NMR chemical shifts for the species below:

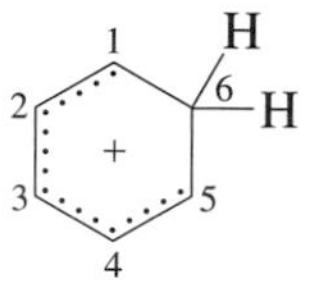

C1 and C5: 186.6 ppm
C3: 178.1 ppm
C2 and C4: 136.9 ppm
C6: 52.2 ppm

   (c) The relative rates of chlorination of the following compounds are as shown.

Compound	Relative Rate
Benzene	0.0005
Methylbenzene	0.157
1,4-Dimethylbenzene	1.00
1,2-Dimethylbenzene	2.1
1,2,4-Trimethylbenzene	200
1,2,3-Trimethylbenzene	340
1,2,3,4-Tetramethylbenzene	2000
1,2,3,5-Tetramethylbenzene	240,000
Pentamethylbenzene	360,000

   (d) When 1,3,5-trimethylbenzene is treated with fluoroethane and one equivalent of BF_3 at $-80°C$, an isolable solid salt with a melting point of $-15°C$ is produced. Heating the salt results in 1-ethyl-2,4,6-trimethylbenzene.

Preprofessional Problems

56. *o*-Iodoaniline is the common name of which of the following compounds?

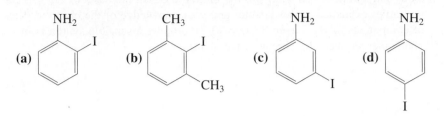

57. The species that is *not* aromatic according to Hückel's rule is:

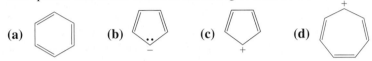

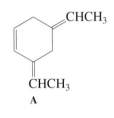

58. When compound A (shown in the margin) is treated with dilute mineral acid, an isomerization takes place. Which of the following compounds is the new isomer formed?

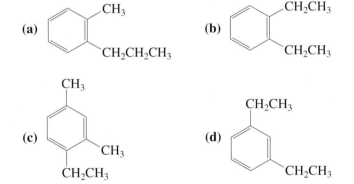

59. Which set of reagents will best carry out the conversion shown?

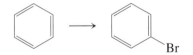

(a) HBr, peroxides; (b) Br_2, $FeBr_3$; (c) Br_2 in CCl_4; (d) KBr.

60. One of the compounds shown here contains carbon–carbon bonds that are 1.39Å long. Which one?

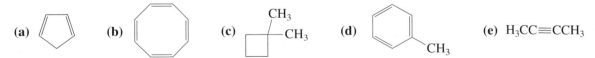

Substituents Control Regioselectivity

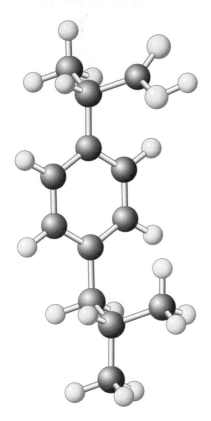

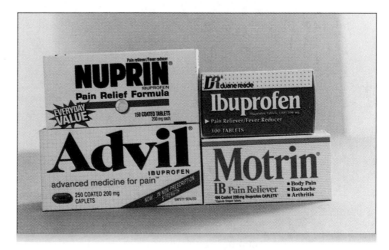

Ibuprofen, prepared industrially by selective electrophilic aromatic substitutions of benzene, is a widely used nonsteroidal anti-inflammatory drug (brand names Advil, Motrin, Nuprin, and many others) for the treatment of inflammation and to relieve pain or discomfort caused by arthritis, soft tissue injuries, and fever.

A t some time in your life you have probably ingested at least one of the painkillers aspirin, acetaminophen, naproxen, or ibuprofen, perhaps better known under one of their respective brand names Bayer Aspirin, Tylenol, Naprosyn, and Advil. Aspirin, acetaminophen, and ibuprofen are ortho or para disubstituted benzenes; naproxen, a disubstituted naphthalene. How are such compounds made? The answer is by electrophilic aromatic substitution.

2-Acetyloxybenzoic acid
(Aspirin)

N-(4-Hydroxyphenyl)acetamide
(Acetaminophen)

2-[2-(6-Methoxynaphthyl)]-
propanoic acid
(Naproxen)

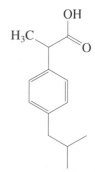

2-[4-(2-Methylpropyl)-
phenyl]propanoic acid
(Ibuprofen)

Chapter 15 described the use of this transformation in the preparation of mono-substituted benzenes. This chapter will analyze the effect of a thus-introduced first substituent on the reactivity and regioselectivity (orientation) of a subsequent electrophilic substitution reaction. Specifically, we shall see that substituents on benzene can be grouped into (1) **activators** (electron donors), which generally direct a second electrophilic attack to the **ortho** and **para positions,** and (2) **deactivators** (electron acceptors), which generally direct electrophiles to the **meta positions.** Armed with this knowledge, we will learn to devise strategies toward the synthesis of substituted arenes, such as the aforementioned analgesics.

16-1 Activation or Deactivation by Substituents on a Benzene Ring

Section 14-8 introduced you to the effect that substituents have on the efficiency of the Diels-Alder reaction. We learned there that electron donors on the diene and acceptors on the dienophile were beneficial to the outcome of the cycloaddition. Chapter 15 revealed another manifestation of these effects: We saw that the introduction of electron-withdrawing substituents into the benzene ring (e.g., as in nitration) caused further electrophilic aromatic substitution to slow down, whereas the incorporation of donors, as in the Friedel Crafts alkylation, caused it to accelerate. What are the factors that contribute to the activating or deactivating nature of substituents in these processes? How do they make a monosubstituted benzene more or less susceptible to further electrophilic attack?

The electronic influence of any substituent is determined by an interplay of two effects that, depending on the structure of the substituent, may operate simultaneously: **induction** and **resonance.** *Induction* occurs through the *σ framework,* tapers off rapidly with distance, and is mostly governed by the relative electronegativity of atoms and the so-induced polarization of bonds (Table 1-2 and 8-2). *Resonance* takes place through *π bonds,* is therefore longer range, and is particularly strong in charged systems (Section 1-5, Chapter 14).

Let us look at both of these effects of typical groups introduced by electrophilic aromatic substitution, starting with inductive donors and acceptors. Thus, simple alkyl groups, such as methyl, are donating by virtue of their hyperconjugating σ frame, a phenomenon that we encountered earlier (Sections 7-5, 11-3, 11-7, 12-9). On the other hand, trifluoromethyl (by virtue of its electronegative fluorines) is electron withdrawing. Similarly, directly bound heteroatoms, such as N, O, and the halogens (by virtue of their relative electronegativity), as well as positively polarized atoms, such as those in carbonyl, cyano, nitro, and sulfonyl functions, are inductively electron withdrawing.

Inductive Effects of Some Substituents on the Benzene Ring

Donors D

Acceptors A

D = –CH$_3$, other alkyl

A = –CF$_3$, –NR$_2$, –OR, –X (–F, –Cl, –Br, –I),

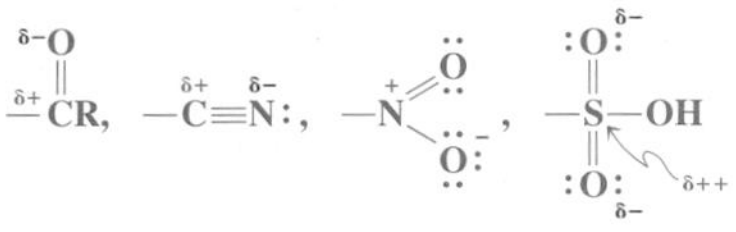

Now let us look at resonating substituents. Resonance donors bear at least one electron pair capable of delocalization into the benzene ring. Therefore, such groups as –NR$_2$, –OR, and the halogens belong in that category. You will note that, inductively, these groups are electron withdrawing; in other words, here the two phenomena, induction and resonance, are opposing each other. Which one wins out? The answer depends on the relative electronegativity of the heteroatoms (Table 1-2) and on the ability of their respective p orbitals to overlap with the aromatic π system. For amino and alkoxy groups, resonance overrides induction. For the halogens, the interplay of the two effects makes them weak overall electron acceptors.

Resonance Donation to Benzene

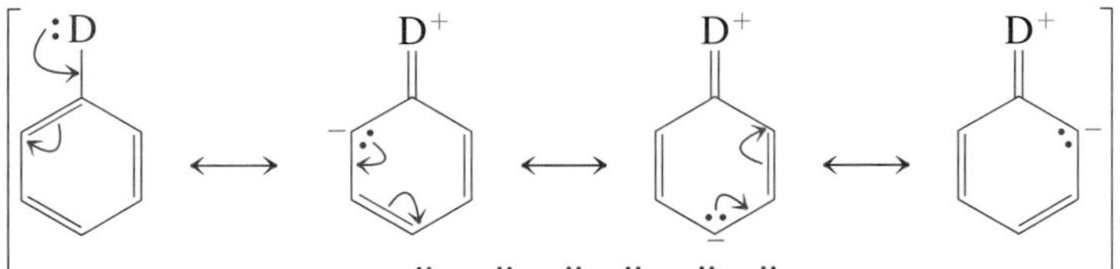

D = –ÑR$_2$, –ÖR, –F̈:, –C̈l:, –B̈r:, –Ï:

Finally, groups bearing a polarized double or triple bond, whose positive (δ^+) end is attached to the benzene nucleus, such as carbonyl, cyano, nitro, and sulfonyl, are resonance withdrawing.

Resonance Acceptance from Benzene

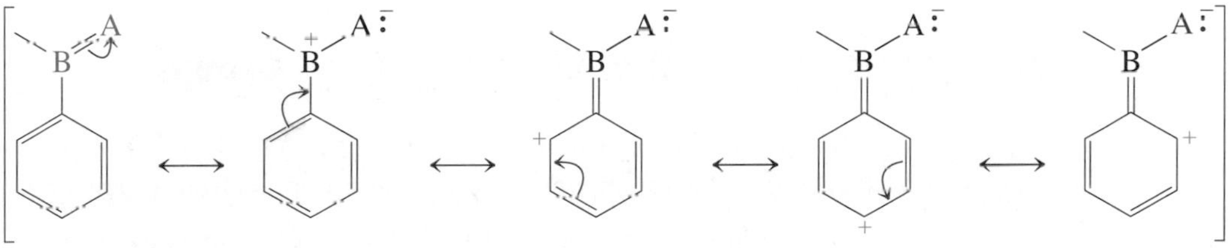

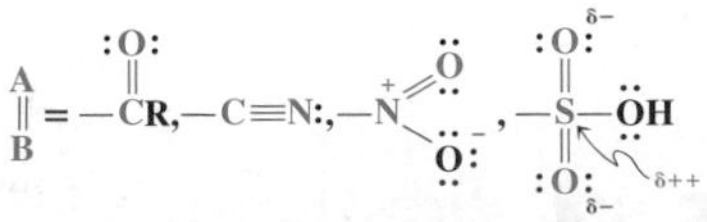

Note that, here, resonance reinforces induction.

EXERCISE 16-1

Explain the ^{1}H NMR spectral assignments in Figures 15-10 and 15-11. (**Hint:** Draw resonance structures involving the substituents to the benzene ring.)

EXERCISE 16-2

The ^{13}C NMR spectrum of benzenol (phenol), C_6H_5OH, shows four lines at $\delta = 116.1$ (C2), 120.8 (C4), 130.5 (C3), and 155.6 (C1) ppm. Explain these assignments. (**Hint:** The ^{13}C chemical shift for benzene is $\delta = 128.7$ ppm.)

How do we know whether a substituent functions as a donor or acceptor? In electrophilic aromatic substitution, the answer is simple. Because the attacking species is an electrophile, the more electron rich the arene, the faster the reaction. Conversely, the more electron poor the arene, the slower the reaction. Hence, electron donors activate the ring, acceptors deactivate.

Relative Rates of Nitration of C_6H_5R

R =	OH	>	CH$_3$	>	H	>	Cl	>	CO$_2$CH$_2$CH$_3$	>	CF$_3$	>	NO$_2$
	1000		25		1		0.033		0.0037		2.6×10^{-5}		6×10^{-8}

EXERCISE 16-3

Specify whether the benzene rings in the compounds below are activated or deactivated.

(a) CH$_2$CH$_3$ ... CH$_2$CH$_3$ (b) NO$_2$ ... CH$_3$ (c) COOH ... CF$_3$ (d) OCH$_3$ ... N(CH$_3$)$_2$

In summary, when considering the effect of substituents on the reactivity of the benzene nucleus, we have to analyze the contributions that occur by induction and resonance. We can group these substituents into two classes: (1) electron donors, which accelerate electrophilic aromatic substitutions relative to benzene, and (2) electron acceptors, which retard them.

We are now ready to tackle the question of the regioselectivity (orientation) of electrophilic aromatic substitution of substituted benzenes. What controls the position that an electrophile will attack?

16-2 Directing Inductive Effects of Alkyl Groups

We begin with the electrophilic substitution reactions of alkyl-substituted benzenes, such as methylbenzene (toluene), in which the methyl group is electron donating by induction.

Groups that donate electrons by induction are activating and direct ortho and para

Electrophilic bromination of methylbenzene (toluene) is considerably faster than the bromination of benzene itself. The reaction is also regioselective: It results mainly in para (60%) and ortho (40%) substitutions, with virtually no meta product.

Electrophilic Bromination of Methylbenzene (Toluene)
Gives Ortho and Para Substitution

40%   <1%   60%

1-Bromo-2-methylbenzene   **1-Bromo-3-methylbenzene**   **1-Bromo-4-methylbenzene**
(*o*-Bromotoluene)   (*m*-Bromotoluene)   (*p*-Bromotoluene)

Is bromination a special case? The answer is no; nitration, sulfonation, and Friedel-Crafts reactions of the alkylbenzene give similar results—mainly ortho and para substitutions (see also Table 16-2). Evidently, the nature of the attacking electrophile has little influence on the observed orientation; it is the methyl group that matters. Because there is virtually no meta product, we say that the activating methyl substituent is **ortho and para directing.**

Can we explain this selectivity by a mechanism? Let us inspect the possible resonance forms of the cations formed after the electrophile, E^+, has attacked the ring in the first, and rate-determining, step.

Ortho, Meta, and Para Attack on Methylbenzene (Toluene)

Ortho attack (E^+ = electrophile)

**Most significant
resonance contributor**

More stable cation

Meta attack

Less stable cation

Para attack

**Methylbenzene
(Toluene)**

**(1,1-Dimethylethyl)-
benzene
(*tert*-Butylbenzene)**

Most significant
resonance contributor

More stable cation

The alkyl group is electron donating by induction (Section 16-1). Attacks at the ortho and para positions produce an intermediate carbocation in which one of the resonance forms places the positive charge *next* to the alkyl substituent, rendering it tertiary carbocation-like (Section 7-5). Because the alkyl group can donate electron density to stabilize the positive charge, that resonance form is a more important contributor to the resonance hybrid than the others, in which the positive charge is at an unsubstituted carbon. Meta attack, however, produces an intermediate in which *none* of the resonance forms benefit from such direct stabilization. Thus, electrophilic attack on a carbon located ortho or para to the methyl (or another alkyl group) leads to a cationic intermediate that is more stable than the one derived from attack at the meta carbon. The transition state leading to the more stable intermediate is of relatively low energy (Hammond postulate, Section 3-5), and is therefore reached relatively rapidly.

Why are the two favored products, ortho and para, not formed in equal amounts? Frequently, the answer is steric effects. Thus, attack ortho to an existing substituent, especially when it is bulky, by an electrophile (again, especially when bulky) is sterically more encumbered than attack para. Therefore, para products often predominate over their ortho isomers. In the bromination of methylbenzene (toluene), this predominance is small. However, similar halogenation of 1,1-dimethylethyl (*tert*-butyl) benzene results in a much larger para:ortho ratio (~ 10:1).

Groups that withdraw electrons inductively are deactivating and meta directing

The strongly electronegative fluorine atoms in (trifluoromethyl)benzene make the trifluoromethyl group inductively electron withdrawing (Section 16-1). In this case, reaction with electrophiles is very sluggish. Under stringent conditions, substitution does take place—but *only* at the meta positions: The trifluoromethyl group is deactivating and **meta directing.**

**Electrophilic Nitration of (Trifluoromethyl)benzene
Gives Meta Substitution**

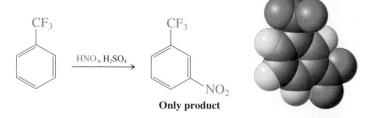

Only product

Once again, the explanation lies in the various resonance forms for the cation produced by ortho, meta, and para attack.

Ortho, Meta, and Para Attack on (Trifluoromethyl)benzene

Ortho attack

Strongly destabilized cation

Poor
resonance contributor

Meta attack

Less destabilized cation

Para attack

Poor
resonance contributor

Strongly destabilized cation

The presence of an inductively electron-withdrawing substituent *de*stabilizes the carbocations resulting from electrophilic attack at *all* positions in the ring. However, ortho and para attack are even less favored than meta attack for the same reasons that they are relatively favored with methylbenzene (toluene): In each case, one of the resonance forms in the intermediate cation places the positive charge next to the substituent. This structure is stabilized by an electron-donating group, but it is *destabilized* by an *electron-withdrawing* substituent—removing electron density from a positively charged center is energetically unfavored. Meta attack avoids this situation. The destabilizing inductive effect is still felt in the meta intermediate, but to a lesser extent. Therefore, the trifluoromethyl group retards substitution but, when it takes place, it directs the electrophile meta, or, more accurately, *away* from the ortho and para carbons.

EXERCISE 16-4

Rank these compounds in order of decreasing activity in electrophilic substitution.

(a)

(b)

(c)

(d)

EXERCISE 16-5

Electrophilic bromination of an equimolar mixture of methylbenzene (toluene) and (trifluoromethyl)benzene with one equivalent of bromine gives only 1-bromo-2-methylbenzene and 1-bromo-4-methylbenzene. Explain.

In summary, electron-donating substituents inductively activate the benzene ring and direct electrophiles ortho and para; their electron-accepting counterparts deactivate the benzene ring and direct electrophiles to the meta positions.

16-3 Directing Effects of Substituents in Conjugation with the Benzene Ring

What is the influence of substituents whose electrons are in conjugation with those of the benzene ring? We can answer this question by again comparing the resonance forms of the intermediates formed by the various modes of electrophilic attack.

Groups that donate electrons by resonance activate and direct ortho and para

Benzene rings bearing the groups –NH₂ and –OH are strongly activated. For example, halogenations of benzenamine (aniline) and phenol not only take place in the absence of catalysts but also are difficult to stop at single substitution. The reactions proceed very rapidly and, as in inductive activation (Section 16-2), furnish exclusively *ortho- and para-substituted products.*

Electrophilic Brominations of Benzenamine (Aniline) and Phenol Give Ortho and Para Substitution

Benzenamine (Aniline)	**2,4,6-Tribromobenzenamine (2,4,6-Tribromoaniline)** 100%	**Phenol**	**2,4,6-Tribromophenol** 100%

Better control of monosubstitution is attained with modified amino and hydroxy substituents, such as in *N*-phenylacetamide (acetanilide) and methoxybenzene (anisole). These groups are, again, ortho and para directing.

Electrophilic Nitration of *N*-Phenylacetamide (Acetanilide)

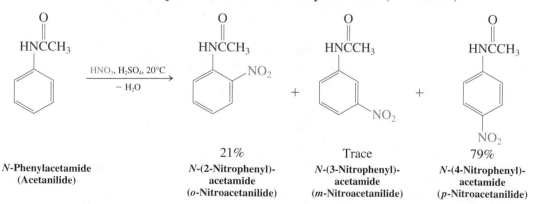

N-Phenylacetamide (Acetanilide)	*N*-(2-Nitrophenyl)-acetamide (*o*-Nitroacetanilide) 21%	*N*-(3-Nitrophenyl)-acetamide (*m*-Nitroacetanilide) Trace	*N*-(4-Nitrophenyl)-acetamide (*p*-Nitroacetanilide) 79%

Both the activated nature of these compounds and the observed regioselectivity upon electrophilic substitution can be explained by writing resonance forms for the various intermediate cations.

Ortho, Meta, and Para Attack on Benzenamine (Aniline)

Ortho attack

Meta attack

Para attack

Because nitrogen is more electronegative than carbon, the amino group in benzenamine (aniline) is inductively electron withdrawing (Section 16-1). However, the lone electron pair on the nitrogen atom may participate in resonance, thereby stabilizing the intermediate cations resulting from ortho and para (but not meta) substitutions. This resonance contribution outweighs the inductive effect. The result is a much reduced activation barrier for ortho or para attack. Consequently, benzenamine (aniline) is strongly activated toward electrophilic substitution relative to benzene itself, and the reaction is highly regioselective as well.

EXERCISE 16-6

Formulate resonance forms for the various modes of electrophilic attack on methoxybenzene (anisole).

EXERCISE 16-7

In strongly acidic solution, benzenamine (aniline) becomes quite unreactive to electrophilic attack, and increased meta substitution is observed. Explain. (**Hint:** The nitrogen atom in benzenamine may behave as a base. How would you classify the resulting substituent within the framework of the discussion in Section 16-1?)

CHEMICAL HIGHLIGHT 16-1

Explosive Nitroarenes: TNT and Picric Acid

Complete ortho, para nitration of methylbenzene (toluene) or phenol (benzenol) furnishes the corresponding trinitro derivatives, both of which are powerful explosives (see Chemical Highlight 4-1): *TNT* and *picric acid.* Both compounds have a long history as military and industrial explosives, not the least because of their extraordinary ease of preparation.

2-Methyl-1,3,5-trinitro-benzene
(2,4,6-Trinitrotoluene, TNT)

2,4,6-Trinitrophenol
(Picric acid)

The Sands Hotel collapses as it is imploded early Tuesday morning, November 26, 1996, in Las Vegas.

Indeed, picric acid can be made by simply treating crushed aspirin tablets (Chemical Highlight 22-3) with KNO_3 and H_2SO_4, a recipe available on the World Wide Web. These and similar preparations are **extremely dangerous** and have caused considerable injury and suffering (not to speak of material damage) among those foolish enough to attempt them.

TNT has become such a standard, particularly in military uses, that the destructive power of other explosives, especially in bombs, is often compared to that of an equivalent of TNT. For example, the first atomic bomb detonated on July 16, 1945, in New Mexico had the equivalent power of 19,000 tons of TNT. The device exploded over Hiroshima, Japan, which killed more than 140,000 people, had the power of 13,000 tons of TNT. Although these numbers appear huge, comparison with the hydrogen bomb with the destructive equivalent of 10 million tons of TNT dwarfs them. For further calibration, all of the explosions of World War II combined amounted to the equivalent of "only" 2 million tons of TNT.

Picric acid has some commercial applications other than that as an explosive—in matches, in the leather industry, in electric batteries, and in colored glass. Its name is due to the unusually high acidity of its hydroxy group (pK_a 0.38; Section 22-3), which is increased beyond that of acetic acid (pK_a 4.7) and even hydrogen fluoride (pK_a 3.2; Table 2-6) by the electron-withdrawing effect of the three nitro groups.

In modern commercial applications, particularly mining and building demolition, TNT and picric acid have been replaced by nitroglycerine (Section 9-11).

Groups that withdraw electrons by resonance deactivate and direct meta

Several groups *deactivate* the benzene ring by resonance (Section 16-1). An example is the carboxy group in benzoic acid, C_6H_5COOH. Nitration of benzoic acid takes place at only about 1/1000th the rate of benzene nitration and gives predominantly meta substitution. The COOH group is deactivating and, as in inductive deactivation (Section 16-2), *meta directing.*

Electrophilic Meta-Nitration of Benzoic Acid

18.5%	80%	1.5%
2-Nitrobenzoic acid	**3-Nitrobenzoic acid**	**4-Nitrobenzoic acid**
(*o*-Nitrobenzoic acid)	(*m*-Nitrobenzoic acid)	(*p*-Nitrobenzoic acid)

Let us see how conjugation with the COOH function affects the resonance forms of the cations resulting from electrophilic attack on benzoic acid.

Ortho, Meta, and Para Attack on Benzoic Acid

Ortho attack

Strongly destabilized cation

Meta attack

None are poor
Less destabilized cation

Para attack

Strongly destabilized cation

Attack at the meta position avoids placing the positive charge next to the electron-withdrawing carboxy group, whereas ortho and para attacks necessitate the formulation of poor resonance contributors. The carbocation resulting from meta substitution is still destabilized by the substituent, however, and substitution is slower than in benzene itself.

In summary, it appears that deactivating groups, whether operating by induction or resonance, direct incoming electrophiles to the meta position, whereas activating groups direct to the ortho and para carbons. This statement is true for all classes of substituents except one—the halogens.

EXERCISE 16-8

Electrophilic nitration of nitrobenzene gives almost exclusively 1,3-dinitrobenzene. Formulate the (poor) resonance forms of the intermediate cations resulting from attack by NO_2^+ at the ortho and para positions that explain this result.

There is always an exception: halogen substituents, although deactivating, direct ortho and para

Halogen substituents inductively withdraw electron density (Section 16-1); however, they are donors by resonance. On balance, the inductive effect wins out, rendering haloarenes *deactivated*. Nevertheless, the electrophilic substitution that does take place is mainly at the *ortho and para positions*.

**Electrophilic Bromination of Bromobenzene Results
in Ortho- and Para-Dibromobenzene**

Br—Br, FeBr₃
— HBr

13%
**1,2-Dibromobenzene
(*o*-Dibromobenzene)**

2%
**1,3-Dibromobenzene
(*m*-Dibromobenzene)**

85%
**1,4-Dibromobenzene
(*p*-Dibromobenzene)**

The competition between resonance and inductive effects explains this seemingly contradictory reactivity. Again, we must examine the resonance forms for the various possible intermediates.

Ortho, Meta, and Para Attack on a Halobenzene

Ortho attack

**Strong
contributor**

More stable cation

Meta attack

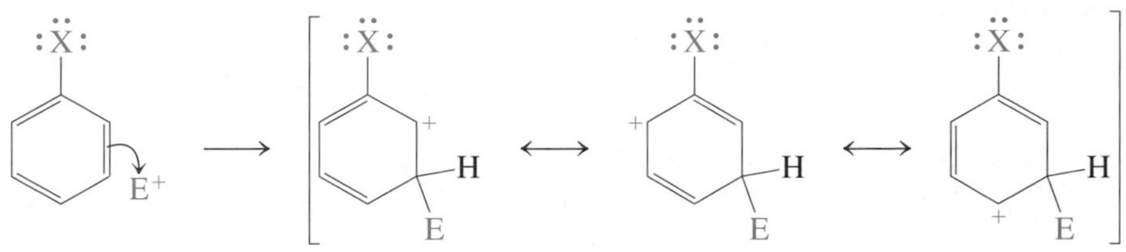

Less stable cation

Para attack

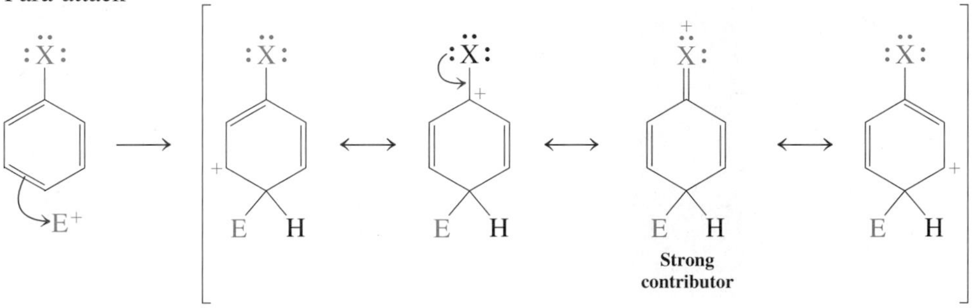

Strong contributor

More stable cation

Note that ortho and para attack lead to resonance forms in which the positive charge is placed next to the halogen substituent. Although this might be expected to be unfavorable, because the halogen is inductively electron withdrawing, resonance with the lone electron pairs allows the charge to be delocalized. Therefore, ortho and para substitutions become the preferred modes of reaction. The inductive effect of the halogen is still strong enough to make all three possible cations less stable than the one derived from benzene itself. Therefore, we have the unusual result that halogens are *ortho and para directing,* but *deactivating.*

This section completes the survey of the regioselectivity of electrophilic attack on monosubstituted benzenes, summarized in Table 16-1. Table 16-2 ranks various substituents by their activating power and lists the product distributions obtained on electrophilic nitration of the benzene ring.

EXERCISE 16-9

Explain why (a) $-NO_2$, (b) $-\overset{+}{N}R_3$, and (c) $-SO_3H$ are meta directing. (d) Why should phenyl be activating and ortho and para directing (Table 16-1)? (**Hint:** Draw resonance forms for the appropriate cationic intermediates of electrophilic attack on phenylbenzene [diphenyl].)

EXERCISE 16 10

Compound A is an intermediate in one of the syntheses of ibuprofen (see page 689). Propose a synthetic route to compound A, starting with (2-methylpropyl)benzene. (**Hint:** You need to introduce the cyano group by nucleophilic substitution.)

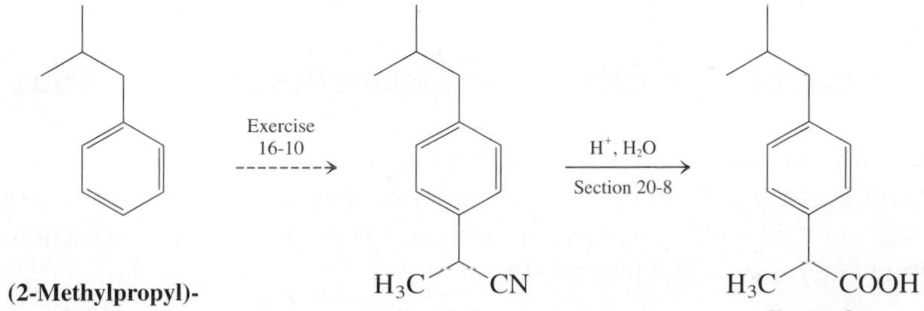

Exercise 16-10

H^+, H_2O
Section 20-8

(**2-Methylpropyl)-benzene**

H_3C　CN
A

H_3C　COOH
Ibuprofen

| TABLE 16-1 | Effects of Substituents in Electrophilic Aromatic Substitution |

Ortho and para directors	Meta directors
Moderate and strong activators	Strong deactivators

$$-\ddot{N}H_2 \quad -\ddot{N}HR \quad -\ddot{N}R_2 \quad -\ddot{N}HCR \quad\quad -NO_2 \quad -CF_3 \quad -\overset{+}{N}R_3 \quad -C\overset{O}{\overset{\|}{O}}H$$

$$-\ddot{O}H \quad -\ddot{O}R$$

Weak activators

Alkyl, phenyl

Weak deactivators

$$-\ddot{F}: \quad -\ddot{C}l: \quad -\ddot{B}r: \quad -\ddot{I}:$$

(Meta directors, lower row)

$$-\overset{O}{\overset{\|}{C}}OR \quad -\overset{O}{\overset{\|}{C}}R \quad -SO_3H \quad -C\equiv N$$

| TABLE 16-2 | Relative Rates and Orientational Preferences in the Nitration of Some Monosubstituted Benzenes, RC_6H_5 |

R	Relative rate	Percentage of isomer		
		Ortho	Meta	Para
OH	1000	40	<2	58
CH_3	25	58	4	38
H	1			
CH_2Cl	0.71	32	15.5	52.5
I	0.18	41	<0.2	59
Cl	0.033	31	<0.2	69
$CO_2CH_2CH_3$	0.0037	24	72	4
CF_3	2.6×10^{-5}	6	91	3
NO_2	6×10^{-8}	5	93	2
$\overset{+}{N}(CH_3)_3$	1.2×10^{-8}	0	89	11

16-4 Electrophilic Attack on Disubstituted Benzenes

Do the rules developed so far in this chapter predict the reactivity and regioselectivity of still higher substitution? We shall see that they do, provided we take into account the individual effect of each substituent. Let us investigate the reactions of disubstituted benzenes with electrophiles.

The strongest activator wins out

In trying to predict the regioselectivity of electrophilic substitution of disubstituted benzenes, we have to apply the same considerations that were necessary to understand the directing effects of one substituent (Sections 16-1 through 16-3). This may seem rather difficult at first, because the two substituents either may be ortho, para or may be meta directing, and they can be placed around the ring in three possible ways, 1,2, 1,3, or 1,4. However, the problem becomes much simpler when you remember that ortho, para directors are activators and therefore accelerate attack of the electrophile at the ortho, para positions relative to benzene. In contrast, meta directors are deactivators and exert regiocontrol by retarding ortho, para substitution more than meta substitution. Taking these electronic effects into account, in conjunction with steric considerations, allows us to formulate a set of simple guidelines that enable us to foretell the outcome of most such reactions.

GUIDELINE 1. The most powerful activator controls the position of attack.

(There are two equivalent
ortho positions to OH,
attack at only one of
which is marked.)

GUIDELINE 2. Experimentally, we can rank the directing power of substituents into three groups:

$$NR_2, OR \quad > \quad X, R \quad > \quad \text{meta directors}$$
$$I \qquad\qquad II \qquad\qquad III$$

Members of the higher-ranking groups override the effect of members of lower rank (guideline 1). However, substituents within each group compete to give isomer mixtures (except when such substituents direct to the same position or when symmetry ensures the generation of a single product).

(There are four
equivalent ortho
positions to the two Br
substituents, only one
of which is marked.)

(There are four pairwise
equivalent, respective
ortho, para positions to
the two CH₃O substituents,
only one set of which
is marked.)

(There are four
pairwise equivalent,
respective meta positions
to the two COOH
substituents, only two
of which are marked.)

GUIDELINE 3. In such cases where product mixtures are predicted on the basis of guidelines 1 and 2, you may typically discount those that constitute ortho attack to bulky groups or to two substituents.

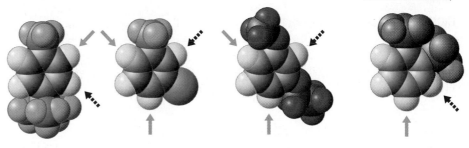

(There are two equivalent ortho positions to CH₃, only one of which is marked.)

(There are four pairwise equivalent, respective meta positions to the two SO₃H substituents, only two of which are marked.)

GUIDELINE 4. Guidelines 1 through 3 are applicable to more highly substituted benzenes for which the decreasing number of reactive positions simplifies the number of possible products.

EXERCISE 16-11

Predict the result of mononitration of

(a)   (b)   (c)   (d)

EXERCISE 16-12

The German chemist Wilhelm Körner (1893–1925) observed in 1874 that each of the three dibromobenzenes A, B, and C furnished a different number of tribromobenzenes on further bromination, allowing him to assign their respective structures. Try to do the same on the basis of the following results.

(i)   A gives two tribromobenzenes in comparable amounts.
(ii) B gives three tribromobenzenes, one of them in minor quantities.
(iii) C gives only one tribromobenzene.

The food preservative BHT (*tert*-butylated hydroxytoluene) has the structure shown here. Suggest a synthesis starting from 4-methylphenol (*p*-cresol).

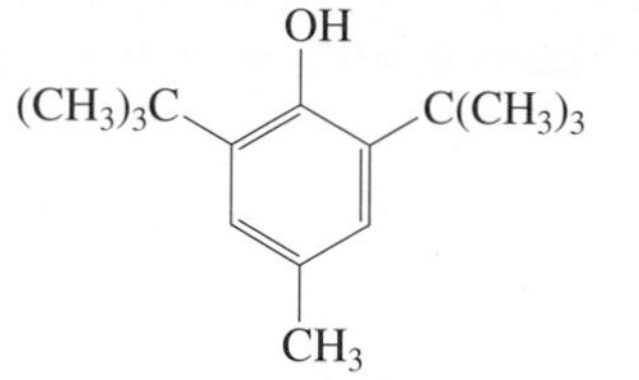

OH

(CH₃)₃C⟶⟵C(CH₃)₃

CH₃

4-Methyl-2,6-bis(1,1-dimethylethyl)phenol
(2,6-Di-*tert*-butyl-4-methylphenol)

In summary, electrophilic aromatic substitution of multiply substituted benzenes is controlled by the strongest activator and, to a certain extent, by steric effects. The greatest product selectivity appears when there is only one dominant activator or when the substitution pattern minimizes the number of isomers that can be formed.

16-5 Synthetic Strategies Toward Substituted Benzenes

The synthesis of substituted benzenes requires planning to ensure that a specific substitution pattern is obtained. How do we approach the problem of targeting a product in which this pattern seems incompatible with the directing sense of the substituents? For example, how can we make a meta-substituted benzenamine (aniline) or an ortho, para-substituted nitrobenzene? To solve problems such as these, we need to know a couple of synthetic "tricks." Among them are the chemical interconversions of ortho, para with meta directors, such as nitro ⇌ amino or carbonyl ⇌ methylene, some additional knowledge about the practicality of certain electrophilic substitutions, and the employment of reversible blocking strategies of certain positions with sulfonic acid groups (–SO₃H).

We can change the sense of the directing power of substituents

The easiest way to introduce a nitrogen substituent into an arene is by nitration. Yet many desired substituted benzenes have amino functions. Moreover, the nitro group is a meta director, limiting its use in the preparation of ortho, para–substituted systems. A solution to these problems is provided by the availability of simple reagents that reversibly convert –NO₂, a meta director, into –NH₂, an ortho, para director. Thus, the nitro group can be reduced to the amino function by either catalytic hydrogenation or exposure to acid in the presence of active metals such as iron or zinc amalgam. The reverse oxidation reaction employs trifluoroperacetic acid.

Interconversion of Nitro (Meta Director) with Amino (Ortho, Para Director)

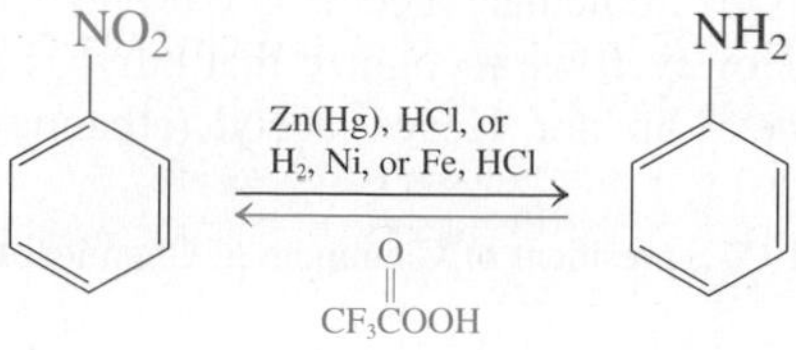

NO₂ ⟶ NH₂

Zn(Hg), HCl, or
H₂, Ni, or Fe, HCl
⇌
O
‖
CF₃COOH

For an example of the application of this capability in synthetic strategy, consider the preparation of 3-bromobenzenamine. Direct bromination of benzenamine (aniline) leads to complete ortho and para substitution (Section 16-3) and is therefore useless. However, bromination of nitrobenzene allows preparation of 3-bromonitrobenzene, which can be converted into the required target molecule by reduction. The outcome is a benzene in which two ortho, para directors emerge positioned meta to each other.

3-Bromobenzenamine

EXERCISE 16-14

Would nitration of bromobenzene be a useful alternative way to begin a synthesis of 3-bromobenzenamine?

EXERCISE 16-15

Propose a synthesis from benzene of 3-aminobenzenesulfonic acid [metanilic acid, used in the synthesis of azo dyes (Section 22-11), such as "Metanil yellow," and certain sulfa drugs (Section 15-11)].

EXERCISE 16-16

Use the methods just presented to devise a synthesis of 4-nitrobenzenesulfonic acid from benzene. (**Hint:** Sulfonation proceeds selectively para to activating groups, because the reagent is sterically hindered and the process reversible [Section 15-11].)

Another example of the interconversion of the directing ability of a substituent on a benzene ring is the redox reaction alkanoyl $\rightleftharpoons$ alkyl. Thus, the carbonyl group in alkanoylarenes can be completely reduced by using palladium-catalyzed hydrogenation or treatment with zinc amalgam in concentrated HCl (**Clemmensen* reduction**). Conversely, the methylene group next to the aromatic ring in an alkylarene is susceptible to oxidation to the carbonyl function with the use of CrO_3, H_2SO_4 (Chapter 22-2).

Interconversion of Alkanoyl (Meta Director) with Alkyl (Ortho, Para Director)

How is this useful? Consider the preparation of 1-chloro-3-ethylbenzene from benzene. Retrosynthetic analysis (Section 8-9) indicates that neither chlorobenzene nor ethylbenzene is suitable as the immediate precursor to product, because each substituent is ortho, para directing. However, we recognize that ethyl is a retrosynthetic product of acetyl (ethanoyl), a meta director. Hence, acetyl (ethanoyl) benzene, readily pre-

*E. C. Clemmensen (1876–1941), president of Clemmensen Chemical Corporation, Newark, New Jersey.

pared by Friedel-Crafts alkanoylation, is a perfect relay point, because it can be chlorinated meta, and the carbonyl group, after it has completed its directing job, reduced.

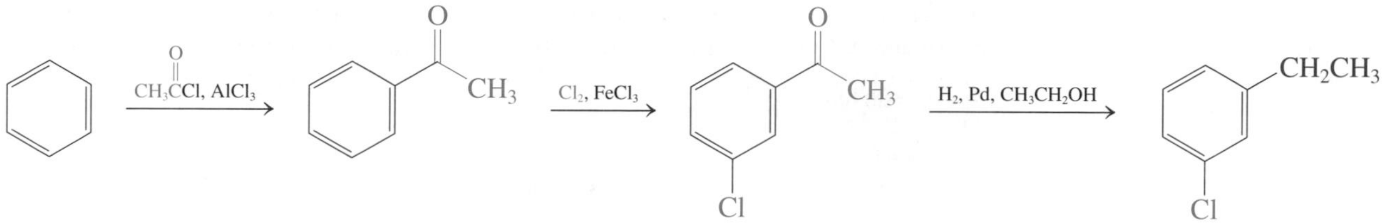

1-Chloro-3-ethylbenzene

EXERCISE 16-17

Propose a synthesis of 1-chloro-3-propylbenzene, starting from propylbenzene.

The ready reduction of alkanoyl- to alkylarenes also provides a way to synthesize alkylbenzenes without the complication of alkyl group rearrangement and overalkylation. For example, butylbenzene is best synthesized by the sequence of Friedel-Crafts alkanoylation with butanoyl chloride, followed by Clemmensen reduction.

Synthesis of Butylbenzene Without Rearrangement

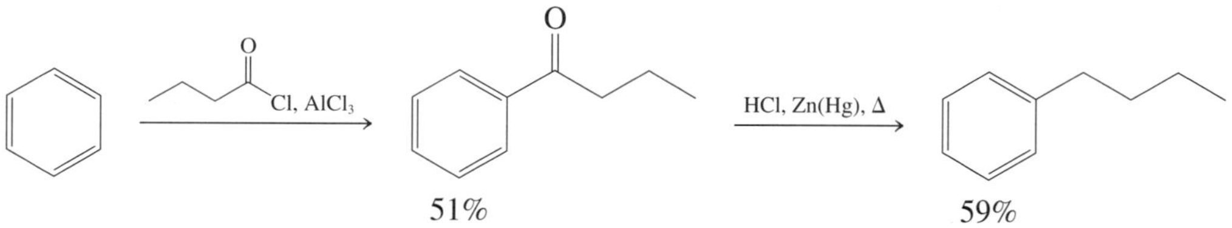

51%   59%

The more direct alternative, Friedel-Crafts butylation of benzene, fails because of the formation of the rearranged product (1-methylpropyl)benzene (*sec*-butylbenzene; Section 15-13) and di- and trialkylation.

EXERCISE 16-18

Give an efficient synthesis of (2-methylpropyl)benzene (isobutylbenzene, the starting material for the preparation of ibuprofen; see Exercise 16-10), starting from benzene. (**Hint:** What would you expect as the major monosubstitution product of Friedel-Crafts alkylation of benzene with 1-chloro-2-methylpropane [isobutyl chloride]?)

Friedel-Crafts electrophiles do not attack strongly deactivated benzene rings

Let us examine possible syntheses of 1-(3-nitrophenyl)ethanone (*m*-nitroacetophenone). Because both groups are meta directors, two possibilities appear available: nitration of 1-phenylethanone or Friedel-Crafts acetylation of nitrobenzene. However, in practice, only the first route succeeds.

Successful and Unsuccessful Syntheses of 1-(3-Nitrophenyl)ethanone
(*m*-Nitroacetophenone)

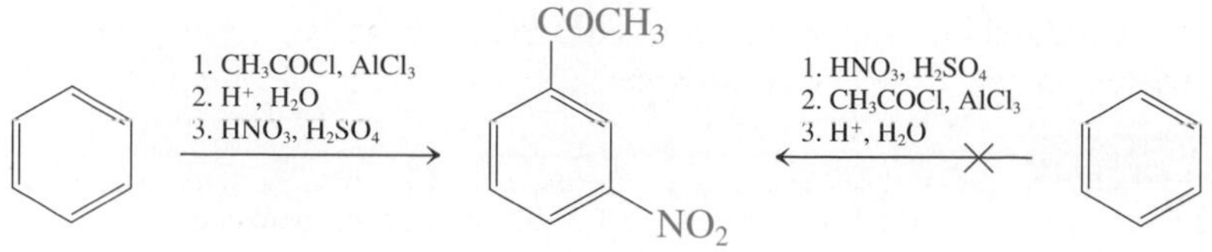

The failure of the second route results from a combination of factors. One is the extreme deactivation of the nitrobenzene ring. Another is the relatively low electrophilicity of the acylium ion, at least compared with other electrophiles in aromatic substitution. As a general rule, neither Friedel-Crafts alkylations nor alkanoylations take place with benzene derivatives strongly deactivated by meta-directing groups.

EXERCISE 16-19

Propose a synthesis of 5-propyl-1,3-benzenediamine, starting from benzene. (**Hint:** Consider carefully the order in which you introduce the groups.)

Reversible sulfonation allows the efficient synthesis of ortho-disubstituted benzenes

A problem of another sort arises in the attempt to prepare an *o*-disubstituted benzene, even when one of the groups is an ortho, para director. Although appreciable amounts of ortho isomers may form in electrophilic substitutions of benzenes containing such groups, the para isomer is the major product in most such cases (Sections 16-2 and 16-3). Suppose you required an efficient synthesis of 1-(1,1-dimethylethyl)-2-nitrobenzene [*o*-(*t*-butyl)nitrobenzene]? Direct nitration of (1,1-dimethylethyl)benzene (*t*-butylbenzene) is unsatisfactory.

A Poor Synthesis of 1-(1,1-Dimethylethyl)-2-nitrobenzene
[*o*-(*t*-Butyl)nitrobenzene]

	16%	11%	73%
	1-(1,1-Dimethylethyl)-2-nitrobenzene [*o*-(*t*-Butyl)nitrobenzene]	1-(1,1-Dimethylethyl)-3-nitrobenzene [*m*-(*t*-Butyl)nitrobenzene]	1-(1,1-Dimethylethyl)-4-nitrobenzene [*p*-(*t*-Butyl)nitrobenzene]

A clever solution makes use of reversible sulfonation (Section 15-11) as a blocking procedure. Both the substituent *and* the electrophile are sterically bulky; hence, (1,1-dimethylethyl)benzene is sulfonated almost entirely para, blocking this carbon from further electrophilic attack. Nitration now can occur only ortho to the alkyl group. Heating in aqueous acid removes the blocking group and completes the synthesis.

Reversible Sulfonation as a Blocking Procedure

EXERCISE 16-20

Suggest a synthetic route from benzene to 1,3-dibromo-2-nitrobenzene.

Protection strategies moderate the activating power of amine and hydroxy groups

We noted in Section 16-3 that electrophilic attack on benzenamine (aniline) and phenol are sometimes difficult to stop at the stage of monosubstitution, because of the highly activating nature of the NH_2 and OH groups. In addition, with these Lewis basic functions (Section 2-9), complications can arise by direct attack on the heteroatom. To prevent these problems, protecting groups are used: acetyl (ethanoyl) for benzenamine, as in *N*-phenylacetamide (acetanilide; Section 20-2), and methyl for phenol, as in methoxybenzene. Deprotection is achieved by basic or acidic hydrolyses, respectively.

In this way, selective halogenation, nitration, and Friedel-Crafts reactions are accomplished. For example, the synthesis of 2-nitrobenzenamine (*o*-nitroaniline) employs this protection strategy in conjunction with sulfonation to block the para position.

Protecting the Oxygen Atom in Phenol

OH

Phenol

Conc. HI ⇵ NaOH, CH_3I

OCH_3

Methoxybenzene

A Synthesis of 2-Nitrobenzenamine (*o*-Nitroaniline) Through Protected Benzenamine (Aniline)

Benzenamine (Aniline) → CH₃CCl, pyridine (Section 20-2) → *N*-Phenylacetamide (Acetanilide) → SO₃, conc. H₂SO₄ → → HNO₃ → → 1. H⁺, H₂O, Δ 2. ⁻OH, H₂O → 2-Nitrobenzenamine (*o*-Nitroaniline)

EXERCISE 16-21

Apply the strategy just discussed to a synthesis of 4-acetyl-2-chlorophenol, starting with phenol.

In summary, by careful choice of the sequence in which new groups are introduced, it is possible to devise specific syntheses of multiply substituted benzenes. Such strategies may require changing the directing power of substituents, modifying their activating ability, and reversibly blocking positions on the ring.

16-6 Reactivity of Polycyclic Benzenoid Hydrocarbons

In this section we will use resonance forms to predict the regioselectivity and reactivity of polycyclic aromatic molecules (Section 15-5), using naphthalene as an example. Some biological implications of the reactivity of these substances will be explored in Section 16-7.

Naphthalene is activated toward electrophilic substitution

The aromatic character of naphthalene is manifest in its reactivity: It undergoes electrophilic substitution rather than addition. For example, treatment with bromine, even in the absence of a catalyst, results in smooth conversion into 1-bromonaphthalene.

The mild conditions required for this process reveal that naphthalene is activated with respect to electrophilic aromatic substitution.

$$\xrightarrow[{- \text{HBr}}]{\text{Br}-\text{Br, CCl}_4, \, \Delta}$$

75%
1-Bromonaphthalene

Other electrophilic substitutions also are readily achieved and, again, are highly selective for reaction at C1. For example,

$$\xrightarrow[{- \text{H}_2\text{O}}]{\text{HNO}_3, \text{CH}_3\text{COOH}, \, \Delta}$$

Major   **Minor**
1-Nitronaphthalene   **2-Nitronaphthalene**

The highly delocalized nature of the intermediate explains the ease of attack. The cation can be nicely pictured as a hybrid of five resonance forms.

Electrophilic Reactivity of Naphthalene: Attack at C1

However, attack of an electrophile at C2 also produces a cation that may be described by five contributing resonance forms.

Electrophilic Attack on Naphthalene at C2

Why, then, do electrophiles prefer to attack naphthalene at C1 rather than at C2? Closer inspection of the resonance contributors for the two cations reveals an important difference: Attack at C1 allows *two* of the resonance forms of the intermediate to keep an intact benzene ring, with the full benefit of aromatic cyclic delocalization. Attack at C2 allows only *one* such structure, so the resulting carbocation is less stable, and the transition state leading to it less energetically favorable. Because the first step in electrophilic aromatic substitution is rate determining, attack is therefore faster at C1 than at C2.

Electrophiles attack substituted naphthalenes regioselectively

The rules of orientation in electrophilic attack on monosubstituted benzenes extend easily to naphthalenes. *The ring carrying the substituent is the one most affected:* An activating group usually directs the incoming electrophile to the same ring, a deactivating group directs it away. For example, 1-naphthalenol (1-naphthol) undergoes electrophilic nitration at C2 and C4.

Nitration of 1-Naphthalenol (1-Naphthol)

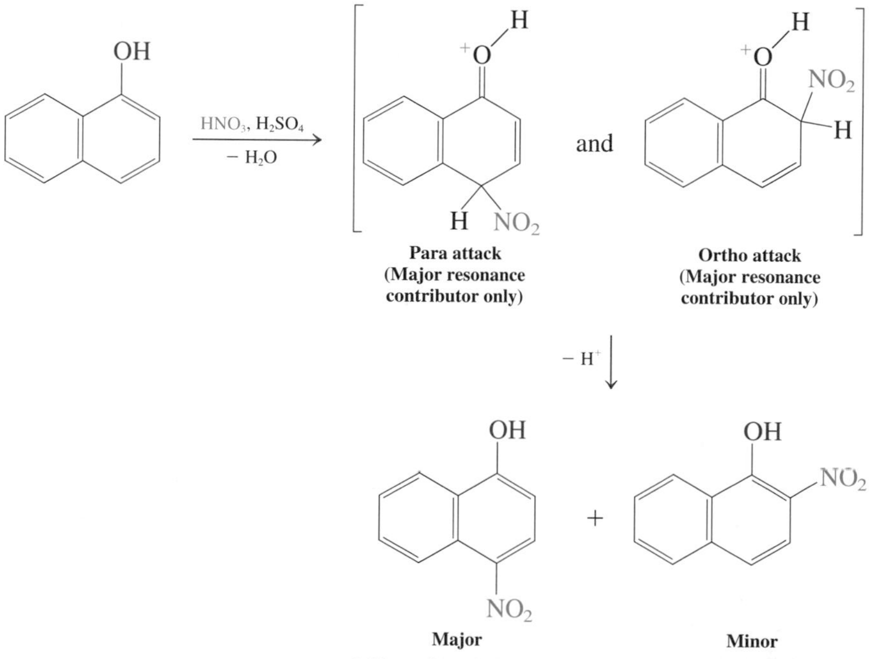

Para attack
(Major resonance
contributor only)

Ortho attack
(Major resonance
contributor only)

Major
**4-Nitro-1-naphthalenol
(4-Nitro-1-naphthol)**

Minor
**2-Nitro-1-naphthalenol
(2-Nitro-1-naphthol)**

Deactivating groups in one ring usually direct electrophilic substitutions to the other ring and preferentially in the positions C5 and C8.

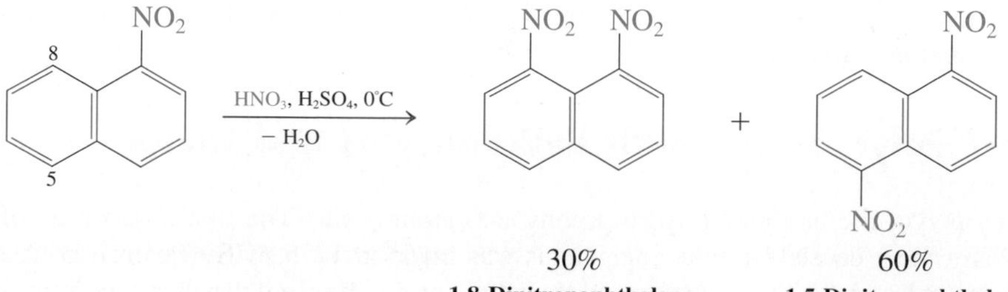

30%
1,8-Dinitronaphthalene

60%
1,5-Dinitronaphthalene

EXERCISE 16-22

On the basis of the relative viability of the sets of resonance forms arising from initial electrophilic attack, predict the position of electrophilic aromatic nitration in **(a)** 1-ethyl-naphthalene; **(b)** 2-nitronaphthalene; **(c)** 5-methoxy-1-nitronaphthalene.

Resonance structures aid in predicting the regioselectivity of larger polycyclic aromatic hydrocarbons

The same principles of resonance, steric considerations, and directing power of substituents apply to larger polycyclic systems, derived from naphthalene by additional benzofusion, such as anthracene and phenanthrene (Section 15-5). For example, the site of preferred electrophilic attack on phenanthrene is C9 (10), because the dominant resonance contributor to the resulting cation retains two intact, delocalized benzene rings, whereas all the other forms require disruption of the aromaticity of either one or two of those rings.

Electrophilic Attack on Phenanthrene

Major contributor

Similar considerations pertain to anthracene and the higher benzologs of phenanthrene and anthracene in an evaluation of the relative ease of electrophilic attack at various positions.

EXERCISE 16-23

Formulate a resonance form of the cation derived by electrophilic attack at C9 of phenanthrene in which the aromaticity of *all* benzene rings is disrupted.

EXERCISE 16-24

Electrophilic protonation of anthracene exhibits the following relative rates: $k(C9) : k(C1) : k(C2) \approx 11{,}000 : 7 : 1$. Explain. (For the numbering of the anthracene skeleton, see Section 15-5.)

In summary, naphthalene is activated with respect to electrophilic aromatic substitution; favored attack takes place at C1. Electrophilic attack on a substituted naphthalene takes place on an activated ring and away from a deactivated ring, with regioselectivity in accordance with the general rules developed for electrophilic aromatic substitution of benzene derivatives. Similar considerations apply to the higher polycyclic aromatic hydrocarbons.

16-7 Polycyclic Aromatic Hydrocarbons and Cancer

Forest fires generate major environmental pollutants, including benzo[*a*]pyrene.

Many polycyclic benzenoid hydrocarbons are carcinogenic. The first observation of human cancer caused by such compounds was made in 1775 by Sir Percival Pott, a surgeon at London's St. Bartholomew's hospital, who recognized that chimney sweeps were prone to scrotal cancer. Since then, a great deal of research has gone into identifying which polycyclic benzenoid hydrocarbons have this physiological property and how their structures correlate with activity. A particularly well studied molecule is benzo[a]pyrene, a widely distributed environmental pollutant. It is produced in the combustion of organic matter, such as automobile fuel and oil (for domestic heating and industrial power generation), in incineration of refuse, in forest fires, in burning cigarettes and cigars, and even in roasting meats. The annual release into the atmosphere in the United States alone has been estimated at 3000 tons.

Carcinogenic Benzenoid Hydrocarbons

Benzo[a]pyrene   Benz[a]anthracene   Dibenz[a,h]anthracene

What is the mechanism of carcinogenic action of benzo[a]pyrene? An oxidizing enzyme (an *oxidase*) of the liver converts the hydrocarbon into the oxacyclopropane at C7 and C8. Another enzyme *(epoxide hydratase)* catalyzes the hydration of the product to the trans diol. Further oxidation then results in the ultimate carcinogen, a new oxacyclopropane at C9 and C10.

Enzymatic Conversion of Benzo[a]pyrene into the Ultimate Carcinogen

Benzo[a]pyrene
oxacyclopropane

7,8-Dihydrobenzo[a]pyrene-
trans-7,8-diol   Carcinogen

What makes the compound carcinogenic? It is believed that the amine nitrogen in guanine, one of the bases in DNA (see Chapter 26), attacks the three-membered ring as a nucleophile. The so-altered guanine disrupts the DNA double helix, leading to a mismatch in DNA replication.

The Carcinogenic Event

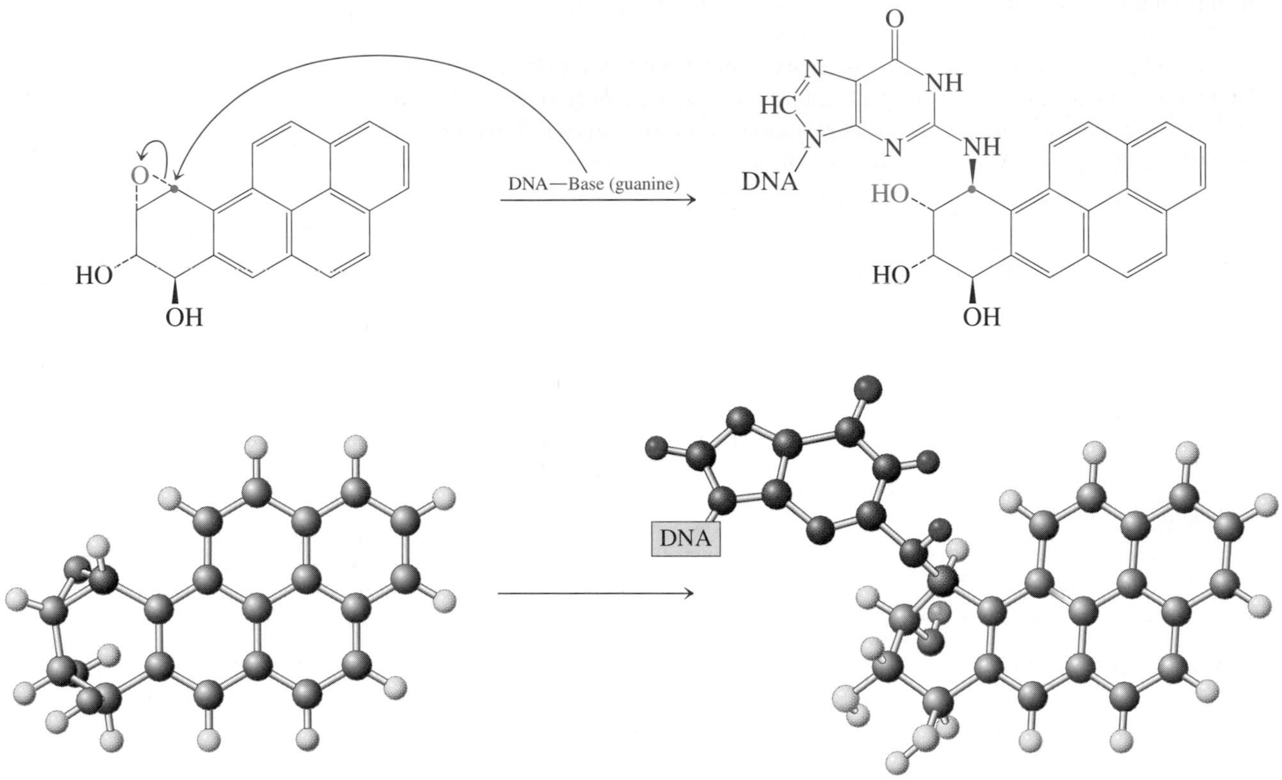

**Carcinogenic
Alkylating Agents and
Sites of Reactivity**

$BrCH_2CH_2Br$

1,2-Dibromoethane

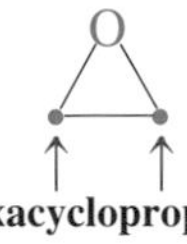

Oxacyclopropane

$ClCH_2OCH_3$

**Chloro(methoxy)methane
(Chloromethyl methyl ether)**

This change can lead to an alteration (mutation) of the genetic code, which may then generate a line of rapidly and indiscriminately proliferating cells typical of cancer. Not all mutations are carcinogenic; in fact, most of them lead to the destruction of only the one affected cell. Exposure to the carcinogen simply increases the likelihood of a carcinogenic event.

Notice that the carcinogen acts as an alkylating agent on DNA. This observation implies that other alkylating agents could also be carcinogenic, and indeed that is found to be the case. The Occupational Safety and Health Administration (OSHA) has published lists of carcinogens and probable carcinogens that include simple alkylating agents such as 1,2-dibromoethane and oxacyclopropane (see Problem 38 of Chapter 1).

The discovery of carcinogenicity in a number of organic compounds necessitated their replacement in synthetic applications. Both 1- and 2-naphthalenamines (naphthyl amines) were once widely used in the synthesis of dyes because of the brilliant colors of many of their derivatives (azo dyes; see Section 22-11). These substances were discovered to be carcinogens many years ago, a finding leading to the development both of synthetic routes that avoided their use as intermediates and of new dyes with completely unrelated structures. A more recent example is chloro(methoxy)-methane ($ClCH_2OCH_3$, chloromethyl methyl ether), once a commonly used reagent for the protection of alcohols by ether formation. The discovery of carcinogenicity in this alkylating agent in the 1970s resulted in the development of several less hazardous reagents.

CHAPTER INTEGRATION PROBLEM

Specifically substituted, functionalized benzenamines (anilines) are important synthetic intermediates in medicinal chemistry and the dye industry. Propose a selective synthesis of 5-chloro-2-methoxy-1, 3-benzenediamine, B, from methoxybenzene, A.

SOLUTION

Our retrosynthetic analysis (Section 8-9) answers the question: What are the feasible bond disconnections that lead to a simpler precursor to B directly, *a, b,* or *c?* The answer is none. Retrosynthetic step *a* proposes a transformation that is impossible to achieve with our present repertory of reactions and, in fact, very difficult even with special reagents (requiring a source of "CH_3O^+"). Quite apart from this problem, step *a* appears unwise because it cleaves a bond that is given in the starting material. Step *c* is a reverse electrophilic chlorination (Section 15-10), feasible in principle but not in practice because no selectivity for the desired position at C5 can be expected. Thus, while CH_3O directs para (and hence to C5) as wanted, the amino groups activate the other positions ortho, para (and hence to C4 and C6) even more (Section 16-3), ruling out step *c* as a good option, at least as such.

What about disconnection *b?* While direct electrophilic amination of arenes is (like alkoxylation or hydroxylation) not viable, we know that we can achieve it indirectly through a nitration-reduction sequence (Section 16-5). Thus, the problem reduces to a nitration of 1-chloro-4-methoxybenzene. Will it proceed with the desired regioselectivity? Guideline 2 in Section 16-4 answers in the affirmative. This analysis provides 1-chloro-4-methoxybenzene as a new relay point, available, in addition to its ortho isomer, by chlorination of methoxybenzene.

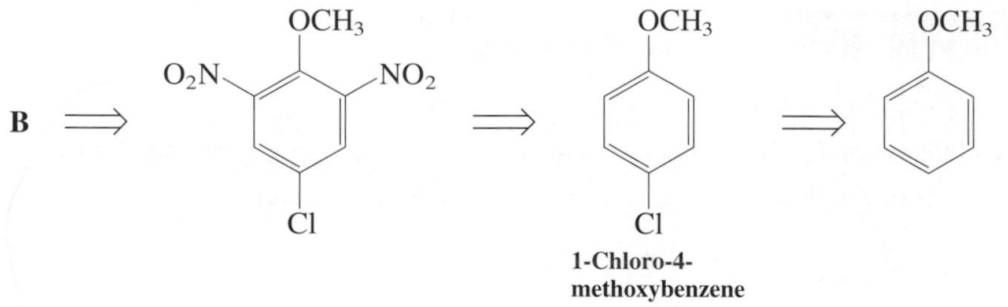

1-Chloro-4-
methoxybenzene

Therefore, a reasonable solution to our synthetic problem would be as follows below, shown as a proper synthetic scheme, with reagents, synthetic intermediates, and forward arrows in place.

SOLUTION SYNTHETIC SCHEME 1

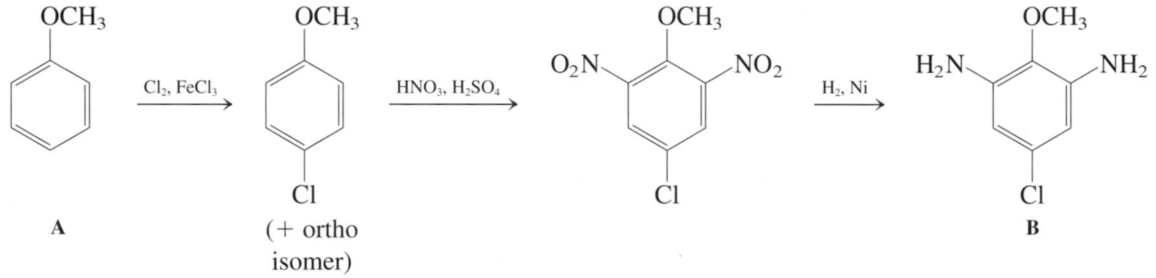

As a purist, you might not be satisfied with the lack of regiochemical control of the first step, which not only cuts down on yields but also requires a cumbersome separation. Consideration of a blocking strategy employing a sulfonation may help (Section 16-5). SO_3 is more bulky than chlorine and will furnish exclusively the para substitution product of electrophilic attack on methoxybenzene. Blocking the para position allows selective dinitration ortho to methoxy. After deblocking, the chlorine can be introduced, as shown below.

SOLUTION SYNTHETIC SCHEME 2

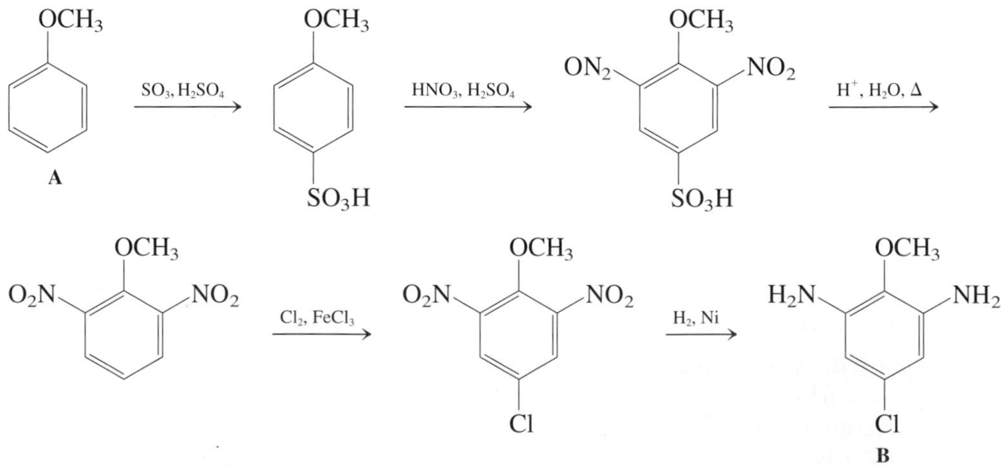

This sequence requires two additional steps and, in practice, yields, ease of experimentation, including workup, disposal cost of waste, and the value and availability of starting materials will determine which route to take, Scheme 1 or Scheme 2.

NEW REACTIONS

Electrophilic Substitution of Substituted Benzenes

1. Ortho- and Para-Directing Groups (Sections 16-1 through 16-3)

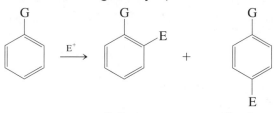

Ortho isomer   **Para isomer**

(Usually predominates)

G = NH₂, OH; strongly activating
= NHCOR, OR; moderately activating
= alkyl, aryl; weakly activating
= halogen; weakly deactivating

2. Meta-Directing Groups (Sections 16-1 through 16-3)

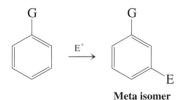

Meta isomer

G = $\overset{+}{N}(CH_3)_3$, NO₂, CF₃, C≡N, SO₃H; very strongly deactivating
= CHO, COR, COOH, COOR, CONH₂; strongly deactivating

Synthetic Planning: Switching and Blocking of Directing Power

3. Interconversion of Nitro and Amino Groups (Section 16-5)

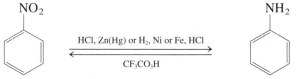

HCl, Zn(Hg) or H₂, Ni or Fe, HCl

CF₃CO₃H

Meta directing   **Ortho, para directing**

4. Interconversion of Alkanoyl and Alkyl (Section 16-5)

H₂, Pd, CH₃CH₂OH or
Zn(Hg), HCl, Δ

CrO₃, H₂SO₄, H₂O

Meta directing   **Ortho, para directing**

5. Blocking by Sulfonation (Section 16-5)

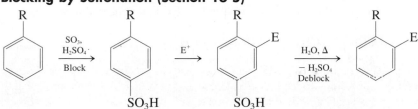

SO₃,
H₂SO₄
Block

E⁺

H₂O, Δ
− H₂SO₄
Deblock

6. Moderation of Strong Activators by Protection (Section 16-5)

7. Electrophilic Aromatic Substitution of Naphthalene (Section 16-6)

IMPORTANT CONCEPTS

1. Substituents on the benzene ring can be divided into two classes: those that **activate** the ring by **electron donation** and those that **deactivate** it by **electron withdrawal.** The mechanisms of donation and withdrawal are based on **induction** or **resonance.** These effects may operate simultaneously to either reinforce or oppose each other. Amino and alkoxy substituents are strongly activating, alkyl and phenyl groups weakly so; nitro, trifluoromethyl, sulfonyl, oxo, nitrile, and cationic groups are strongly deactivating, whereas halogens are weakly so.

2. **Activators** direct electrophiles **ortho** and **para;** de-**activators** direct **meta,** although at a much lower rate. The exceptions are the **halogens,** which deactivate but direct **ortho** and **para.**

3. When there are several substituents, the strongest activator (or weakest deactivator) controls the regioselectivity of attack, the extent of control decreasing in the following order:

$$NR_2, OR > X_2, R > \text{meta directors}$$

4. Strategies for the **synthesis** of highly substituted benzenes rely on the **directing power** of the substituents, the synthetic ability to **change** the **sense of direction** of these substituents by chemical manipulation, and the use of **blocking** and **protecting** groups.

5. **Naphthalene** undergoes preferred electrophilic substitution at **C1** because of the relative stability of the intermediate carbocation.

6. Electron-donating substituents on one of the naphthalene rings direct electrophiles to the same ring, ortho and para. Electron-withdrawing substituents direct electrophiles away from that ring; substitution is mainly at C5 and C8.

7. The actual carcinogen derived from benzo[*a*]pyrene appears to be an oxacyclopropanediol in which C7 and C8 bear hydroxy groups and C9 and C10 are bridged by oxygen. This molecule alkylates one of the nitrogens of one of the DNA bases, thus causing mutations.

PROBLEMS

25. Rank the compounds in each of the following groups in order of decreasing reactivity toward electrophilic substitution. Explain your rankings.

(a)

(b)

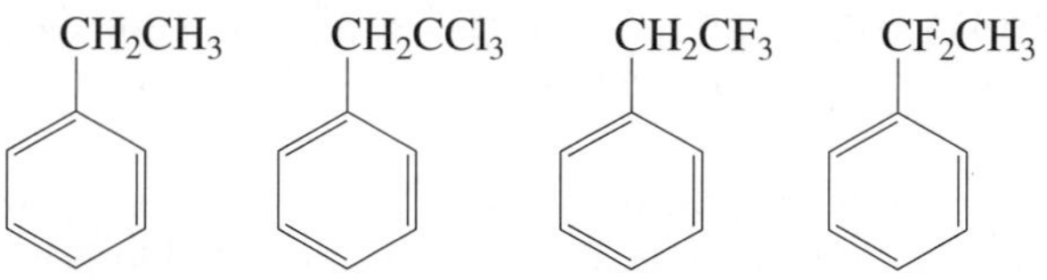

CH_2CH_3 CH_2CCl_3 CH_2CF_3 CF_2CH_3

(c)

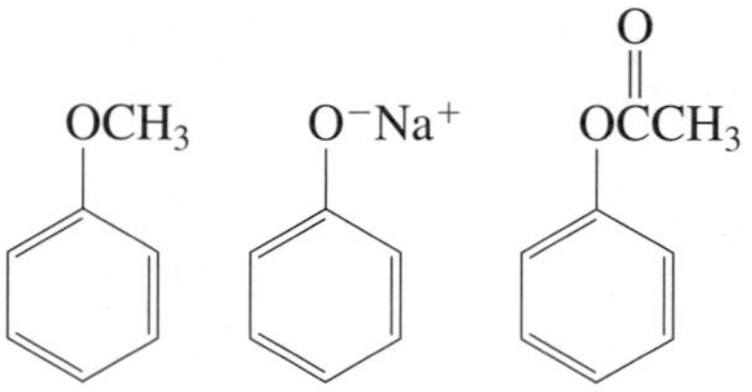

OCH_3 O^-Na^+ $\overset{\overset{\displaystyle O}{\parallel}}{OCCH_3}$

26. Specify whether you expect the benzene rings in the following compounds to be activated or deactivated.

(a) 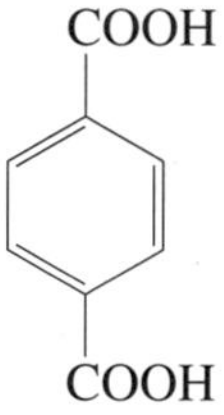 COOH / COOH (1,4-dicarboxybenzene)

(b) 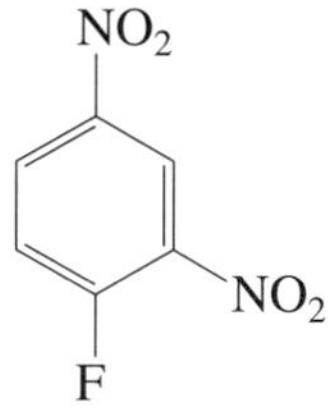 NO$_2$ / NO$_2$ / F

(c) 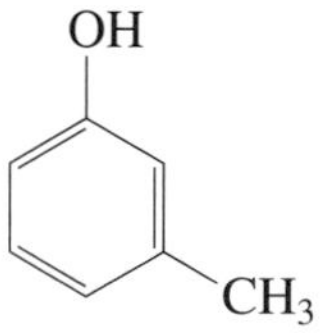 OH / CH$_3$

(d) 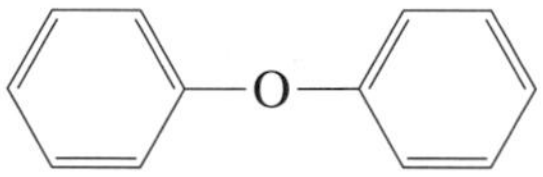 phenyl–O–phenyl

(e) 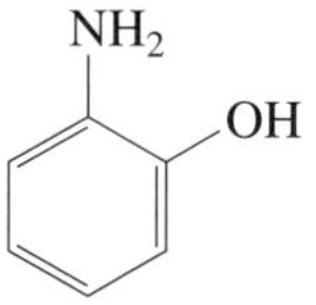 NH$_2$ / OH

(f) SO$_3$H / NO$_2$

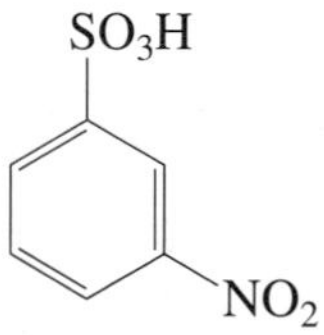

(g) 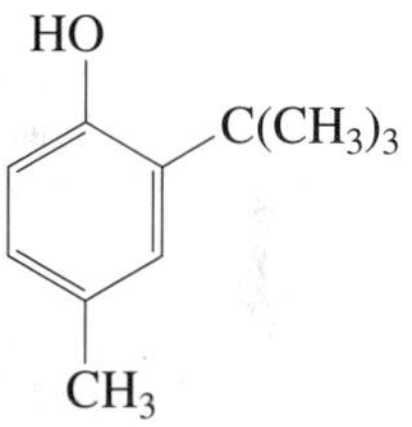 HO / C(CH$_3$)$_3$ / CH$_3$

27. Rank the compounds in each of the following groups in order of decreasing reactivity toward electrophilic aromatic substitution. Explain your answers.

(a) 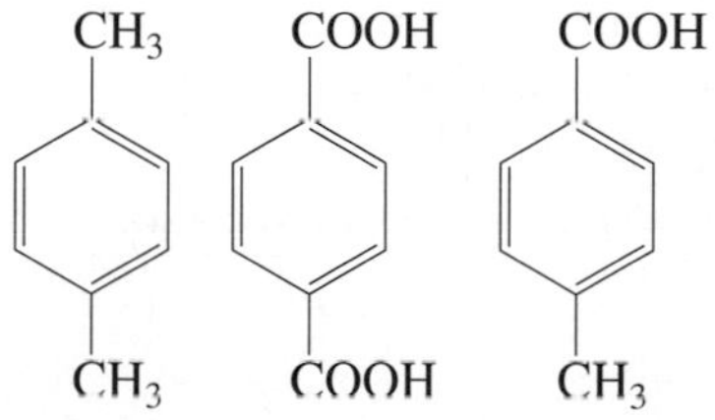 CH$_3$/CH$_3$ COOH/COOH   COOH/CH$_3$

(b) 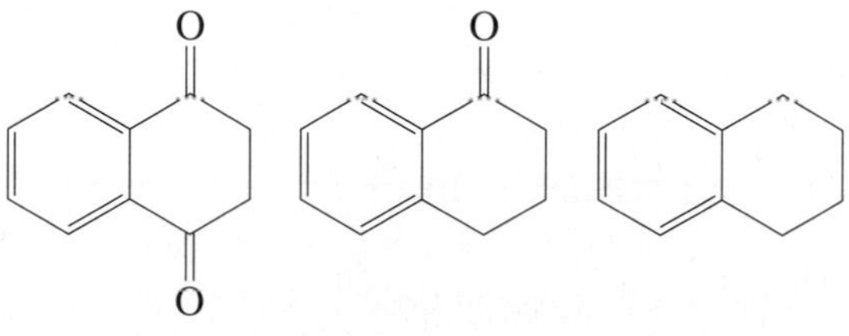 three bicyclic ketone structures

28. Write the structure(s) of the major product(s) that you expect from each of the following electrophilic aromatic substitutions. **(a)** Nitration of methylbenzene (toluene); **(b)** sulfonation of methylbenzene (toluene); **(c)** nitration of 1,1-dimethylethylbenzene (*tert*-butylbenzene); **(d)** sulfonation of 1,1-dimethylethylbenzene (*tert*-butylbenzene).

In what way does changing the substrate structures from methylbenzene (toluene) in (a) and (b) to 1,1-dimethylethylbenzene (*tert*-butylbenzene) in (c) and (d) affect the expected product distributions?

29. Write the structure(s) of the major product(s) that you expect from each of the following electrophilic aromatic substitutions.

    (a) Bromination of trifluoromethylbenzene; (b) nitration of methoxybenzene (anisole); (c) chlorination of benzoic acid; (d) sulfonation of chlorobenzene.

30. Draw appropriate resonance forms to explain the deactivating meta-directing character of the SO_3H group in benzenesulfonic acid.

31. Do you agree with the following statement? "Strongly electron-withdrawing substituents on benzene rings are meta directing because they deactivate the meta positions less than they deactivate the ortho and para positions." Explain your answer.

32. Give the expected major product(s) of each of the following electrophilic substitution reactions.

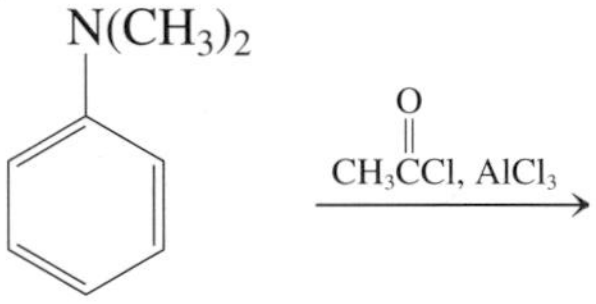

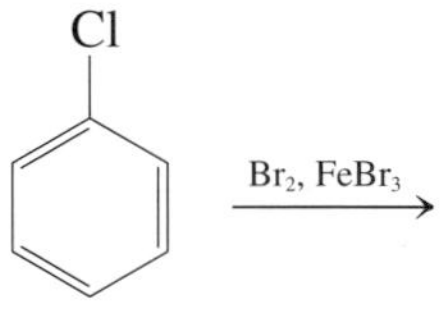

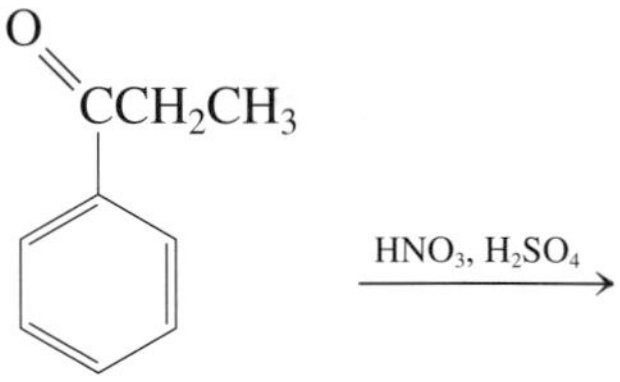

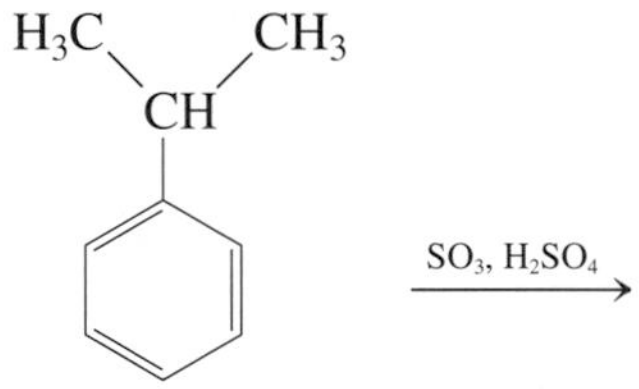

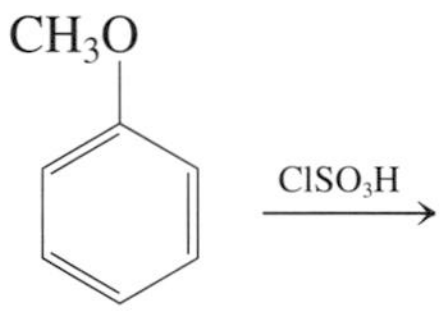

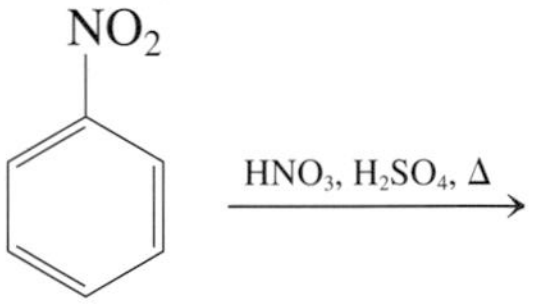

33. Give the expected major product(s) of each of the following reactions.

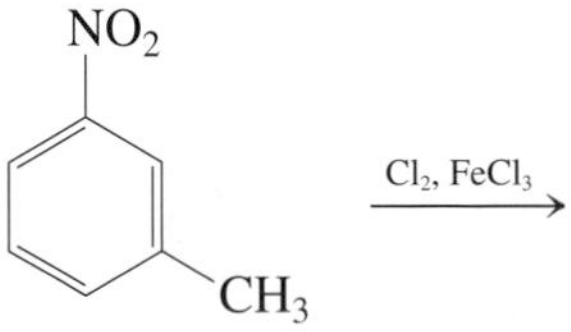

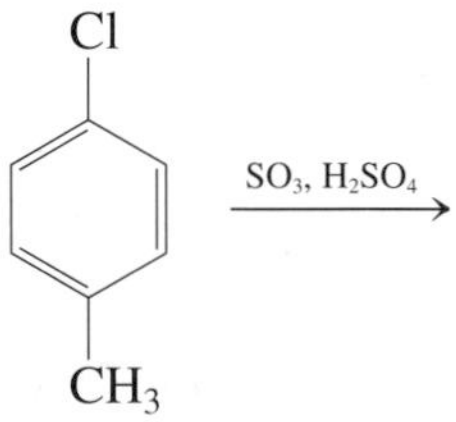

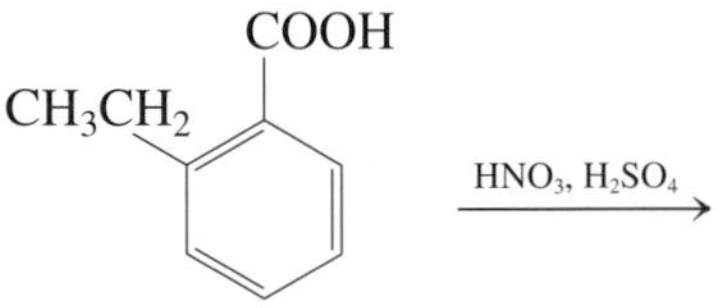

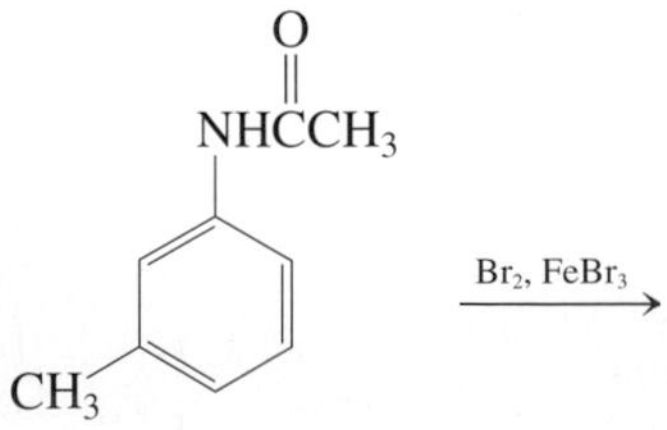

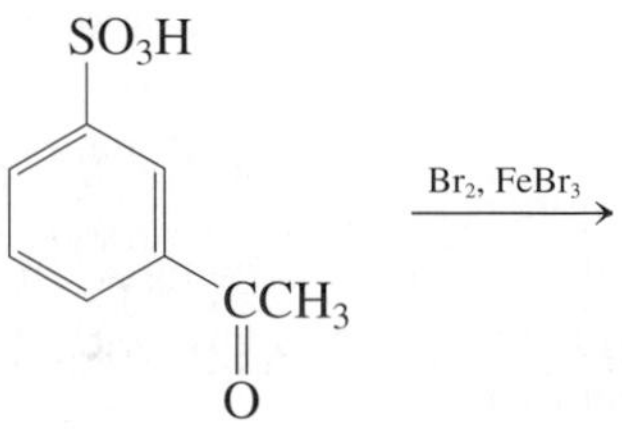

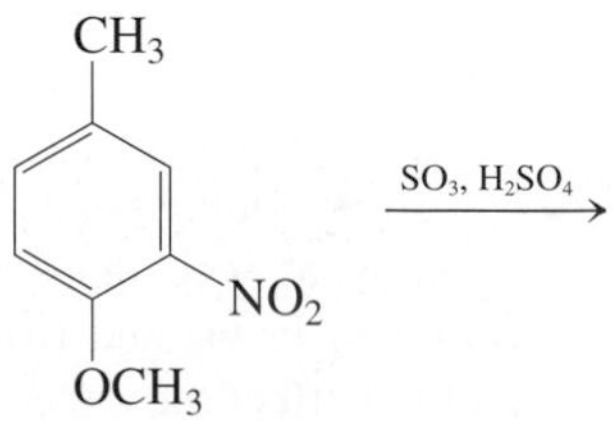

(g) $\xrightarrow{\text{HNO}_3, \text{H}_2\text{SO}_4}$

(h) $\xrightarrow{\text{Cl}_2, \text{FeCl}_3}$

(i) $\xrightarrow{\text{CH}_3\text{Cl}, \text{AlCl}_3}$

34. (a) When a mixture containing one mole each of the three dimethylbenzenes (*o*-, *m*-, and *p*-xylene) is treated with one mole of chlorine in the presence of a Lewis acid catalyst, one of the three hydrocarbons is monochlorinated in 100% yield, whereas the other two remain completely unreacted. Which isomer reacts? Explain the differences in reactivity. **(b)** The same experiment carried out on a mixture of the following three trimethylbenzenes gives a similar outcome. Answer the questions posed in (a) for this mixture of compounds.

1,2,3-Trimethylbenzene   1,2,4-Trimethylbenzene   1,3,5-Trimethylbenzene

35. Propose a reasonable synthesis of each of the following multiply substituted arenes from benzene.

(a) **(b)** **(c)** **(d)**

(e) **(f)** **(g)** **(h)**

36. (4-Methoxyphenyl)methanol (anisyl alcohol) (see margin) contributes both to the flavor of licorice and to the fragrance of lavender. Propose a synthesis of this compound from methoxybenzene (anisole). (**Hint:** Consider your range of options for alcohol synthesis. If necessary, refer to Problem 45 of Chapter 15.)

(4-Methoxyphenyl)methanol
(Anisyl alcohol)

37. The NMR and IR spectra for four unknown compounds A through D are presented below and on the next two pages. Possible empirical formulas for them (not in any particular order) are C_6H_5Br, C_6H_6BrN, and $C_6H_5Br_2N$ (one of these formulas is used twice—two of the unknowns are isomers). Propose a structure and suggest a synthesis of each unknown from benzene.

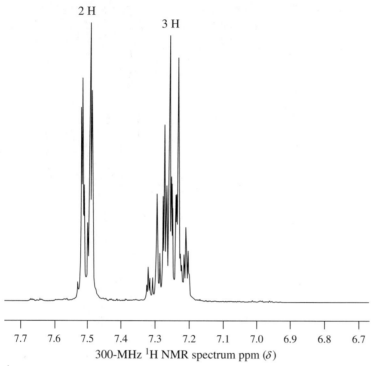

300-MHz ^{1}H NMR spectrum ppm (δ)

A

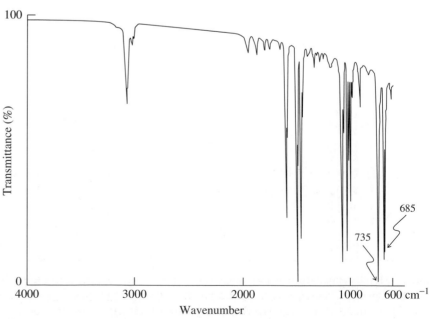

IR-A

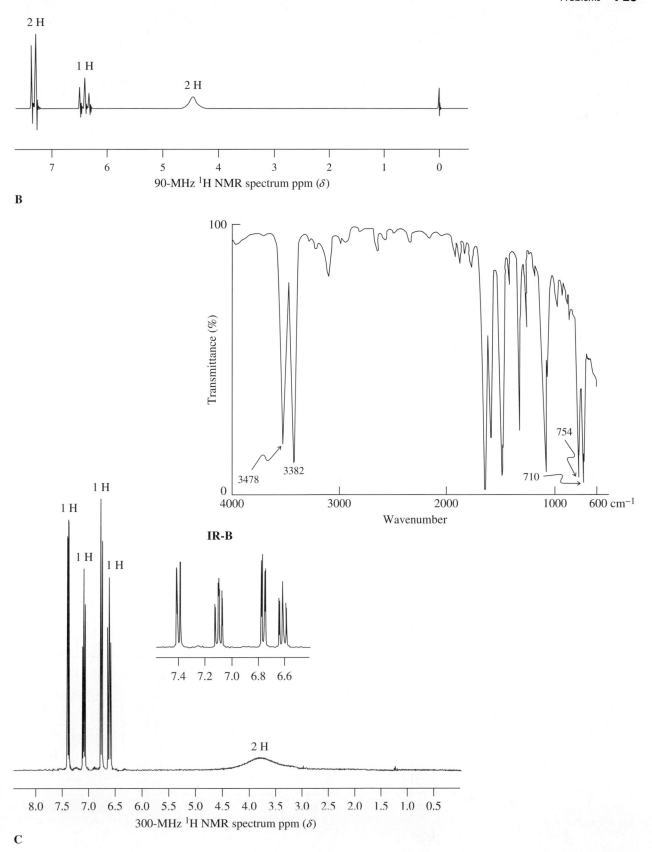

2 H

1 H

2 H

90-MHz ^{1}H NMR spectrum ppm (δ)

B

100

Transmittance (%)

3478

3382

754

710

0

4000   3000   2000   1000   600 cm^{-1}

Wavenumber

IR-B

1 H

1 H

1 H

1 H

7.4   7.2   7.0   6.8   6.6

2 H

8.0 7.5 7.0 6.5 6.0 5.5 5.0 4.5 4.0 3.5 3.0 2.5 2.0 1.5 1.0 0.5

300-MHz ^{1}H NMR spectrum ppm (δ)

C

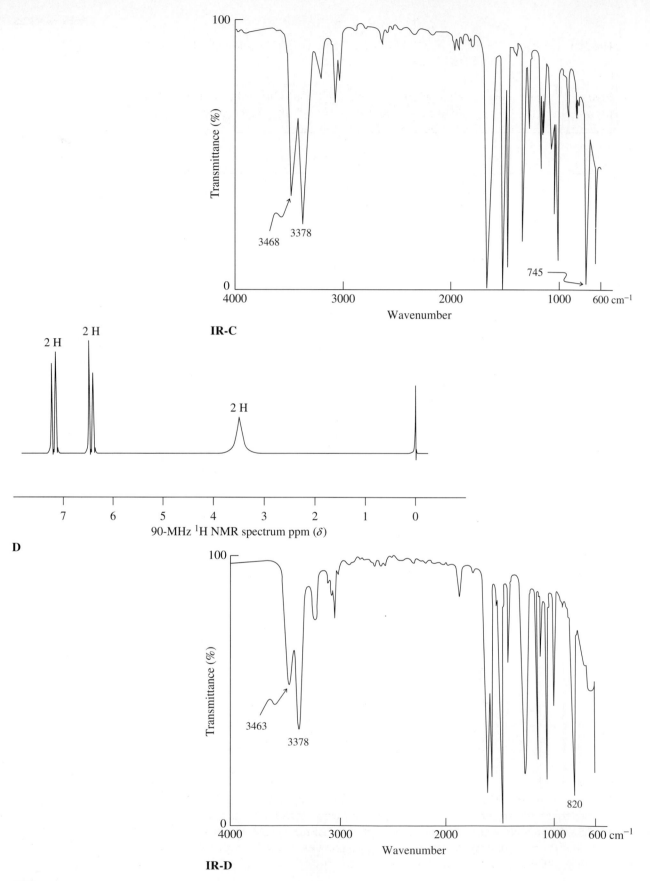

100

Transmittance (%)

3468 3378

745

0
4000   3000   2000   1000   600 cm⁻¹
Wavenumber

IR-C

2 H   2 H

2 H

7   6   5   4   3   2   1   0
90-MHz ¹H NMR spectrum ppm (δ)

D

100

Transmittance (%)

3463

3378

820

0
4000   3000   2000   1000   600 cm⁻¹
Wavenumber

IR-D

38. Catalytic hydrogenation of naphthalene over Pd-C results in rapid addition of two moles of H_2. Propose a structure for this product.

39. Predict the major mononitration product of each of the following disubstituted naphthalenes. **(a)** 1,3-Dimethylnaphthalene; **(b)** 1-chloro-5-methoxynaphthalene; **(c)** 1,7-dinitronaphthalene; **(d)** 1,6-dichloronaphthalene.

40. Write the expected product(s) of each of the following reactions.

(a) 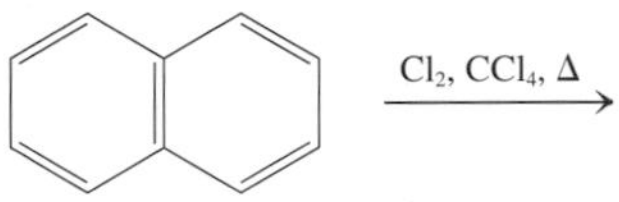 $\xrightarrow{Cl_2,\ CCl_4,\ \Delta}$

(b) 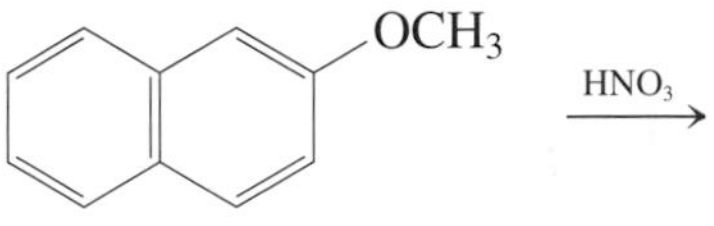 $\xrightarrow{HNO_3}$

(c) 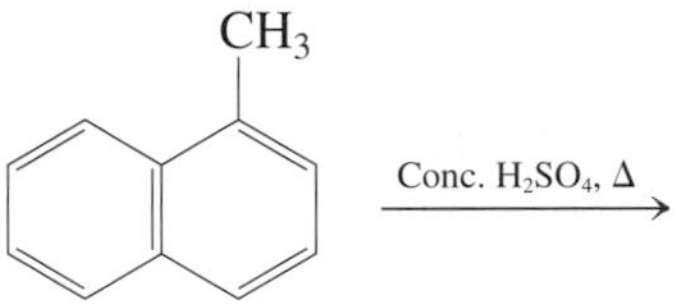 $\xrightarrow{Conc.\ H_2SO_4,\ \Delta}$

(d) 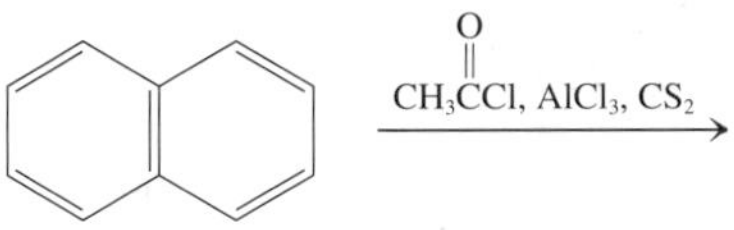 $\xrightarrow{CH_3CCl,\ AlCl_3,\ CS_2}$

(e) 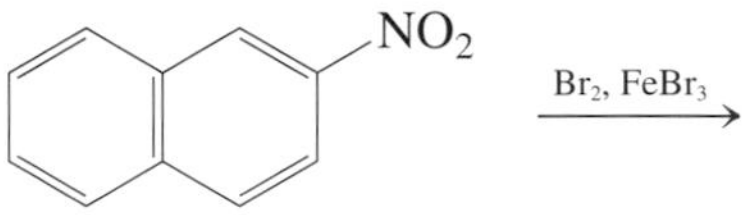 $\xrightarrow{Br_2,\ FeBr_3}$

41. Electrophilic substitution on the benzene ring of benzenethiol (thiophenol, C_6H_5SH) is not possible. Why? What do you think happens when benzenethiol is allowed to react with an electrophile? (**Hint:** Review Section 9-10.)

42. Although methoxy is a strongly activating (and ortho, para-directing) group, the meta positions in methoxybenzene (anisole) are actually slightly *deactivated* toward electrophilic substitution relative to benzene. Explain.

43. Predict the result of mononitration of

(a)

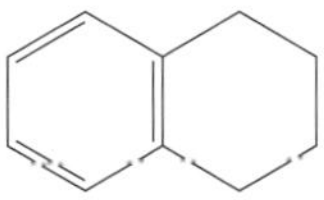

(b)

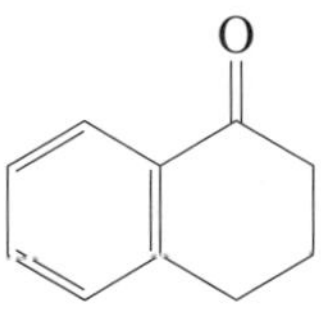

(c)

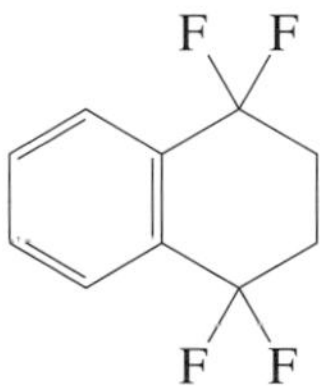

(d) 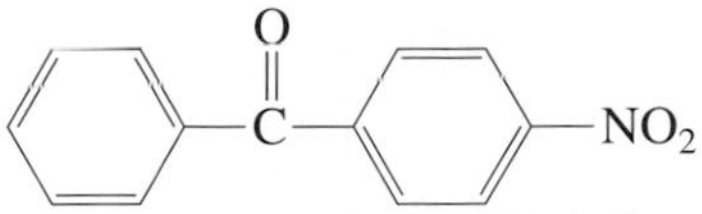 ...—NO_2

(e) 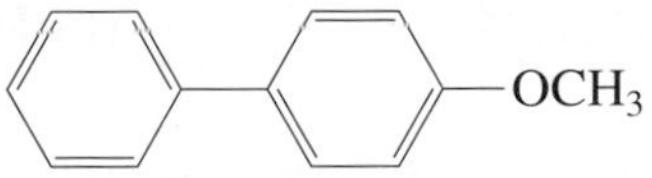 ...—OCH_3

44. The *nitroso* group, –NO, as a substituent on a benzene ring acts as an ortho, para-directing group but is deactivating. Use the Lewis structure of the nitroso group and its inductive and resonance interactions with the benzene ring to explain this finding. (**Hint:** Consider possible similarities to another type of substituent that is ortho, para-directing but deactivating.)

45. Typical conditions for nitrosation are illustrated in the following equation. Propose a detailed mechanism for this reaction.

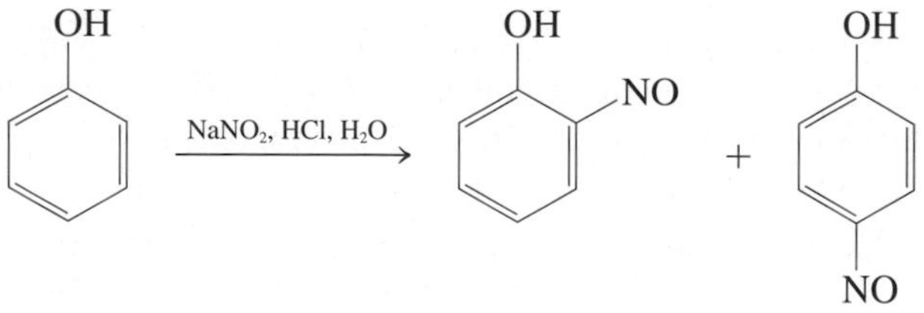

Team Problem

46. Polystyrene (polyethenylbenzene) is a familiar polymer used in the manufacture of foam cups and packing beads (see Chapter 15 Opening). One could, in principle, synthesize polystyrene by cationic polymerization with acid. However, this approach is unsuccessful because of the formation of dimer A.

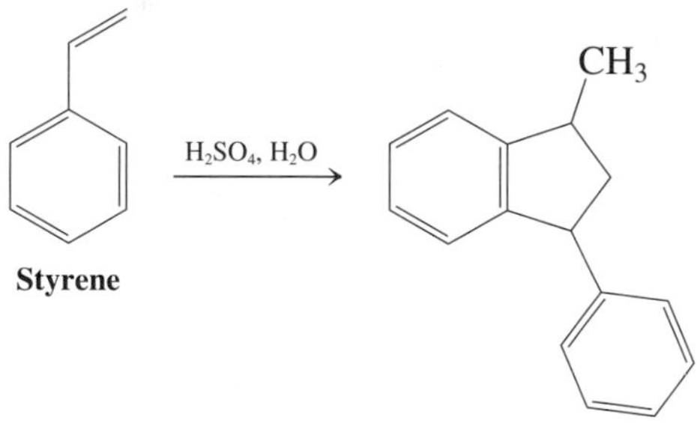

Styrene

A

Split your group into two teams. The first should formulate a mechanism for the cationic polymerization of styrene using acid; the second should do so for the formation of A. Reconvene and compare your notes. At which stage is the normal polymerization sequence diverted to generate A?

Preprofessional Problems

47. Which of the following reactions is an electrophilic aromatic substitution?

(a) $C_6H_{12} \xrightarrow{\text{Se, 300°}} C_6H_6$

(b) $C_6H_5CH_3 \xrightarrow{\text{Cl}_2, h\nu} C_6H_5CH_2Cl$

(c) $C_6H_6 + (CH_3)_2CHOH \xrightarrow{\text{BF}_3, 60°C} C_6H_5CH(CH_3)_2$

(d) $C_6H_5Br \xrightarrow{\text{Mg, ether}} C_6H_5MgBr$

48. The intermediate cation A in the sequence $C_6H_6 + E^+ \rightarrow A \rightarrow C_6H_5E + H^+$ is best shown as

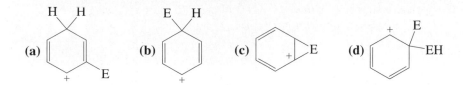

(a)   (b)   (c)   (d)

49. The ^{1}H NMR spectrum of an unknown compound shows absorptions at (*no multiplicities given*) $\delta = 7.3$ (5 H), 2.3 (1 H), and 0.9 (6 H) ppm. One of the following five structures satisfies these data. Which one? (**Hint:** The ^{1}H NMR signal for ethane is at $\delta = 0.9$ ppm, that for benzene at $\delta = 7.3$ ppm.)

(a) $C_6H_5CH_2CH_2CH_3$

(b)

(c) $C_6H_5CH(CH_3)_2$

(d)

(e)

50. A volume of 135 mL of benzene was treated with excess Cl_2 and $AlCl_3$ to yield 50 mL of chlorobenzene. Given the atomic weights of C = 12.0, H = 1.00, and Cl = 35.5, and the densities of benzene, 0.78 g mL^{-1}, and chlorobenzene, 1.10 g mL^{-1}, the percent yield is closest to: (**a**) 15; (**b**) 26; (**c**) 35; (**d**) 46; (**e**) 55.

51. Among the following choices, the group that *activates* the benzene ring toward electrophilic aromatic substitution is (**a**) –NO$_2$; (**b**) –CF$_3$; (**c**) –CO$_2$H; (**d**) –OCH$_3$; (**e**) –Br.

Aldehydes and Ketones

The Carbonyl Group

Propanone, commonly known as acetone, is one of the most widely used solvents in industry, the laboratory, and the home.

:O:
‖
C

Carbonyl group

:O:
‖
C
R   H

An aldehyde

:O:
‖
C
R   R′

A ketone

Have you ever sensed an odor that suddenly brought back a long-forgotten memory? If so, you have experienced a phenomenon unique to our sense of smell— it is a primitive sense and the only one for which the related sensory nerves are part of the brain itself. These nerves respond to both the shape and the presence of polar functional groups in volatile molecules, and prominent among the organic compounds with the most potent and varied odors are those possessing the carbon–oxygen double bond, the **carbonyl group.**

In this chapter and the next, we focus on two classes of carbonyl compounds: **aldehydes,** in which the carbon atom of the carbonyl group is bound to at least one hydrogen atom, and **ketones,** in which it is bound to two carbons. These compounds exist throughout nature, contributing to the flavors and aromas of many foods and assisting in the biological functions of a number of enzymes. In addition, industry makes considerable use of aldehydes and ketones, both as reagents and as solvents in synthesis. Indeed, the carbonyl group is frequently considered to be the most important function in organic chemistry.

After explaining how to name aldehydes and ketones, this chapter will look at their structures and physical properties. Like alcohols, carbonyl groups are weak Lewis bases, because the oxygen bears two lone electron pairs. Also, the carbon–oxygen double bond is polarized, thereby making the carbonyl carbon electrophilic. The remainder of the chapter will show how these properties shape the chemistry of this versatile functional group.

17-1 Naming the Aldehydes and Ketones

The carbonyl function is the highest ranking group we have encountered so far for the purposes of naming. The aldehyde function takes precedence over that of ketones. For historical reasons, the simpler aldehydes often retain their common names. These names are derived from the common name of the corresponding carboxylic acid, with the ending *-ic acid* replaced by **-aldehyde.**

Formic acid   **Formaldehyde**   **Acetic acid**   **Acetaldehyde**   *o*-**Bromo-benzoic acid**   *o*-**Bromo-benzaldehyde**

Many ketones also have common names, which consist of the names of the substituent groups followed by the word *ketone.* Dimethyl ketone, the simplest example, is a common solvent best known as **acetone.** Phenyl ketones have common names ending in **-phenone.**

CH_3CCH_3 $CH_3CCH_2CH_3$ $CH_3CH_2CCH_2CH_3$

Dimethyl ketone   **Ethyl methyl ketone**   **Diethyl ketone**   **Benzophenone**
(Acetone)

IUPAC names treat aldehydes as derivatives of the alkanes, with the ending *-e* replaced by **-al.** An alkane thus becomes an **alkanal.** Methanal, the systematic name of the simplest aldehyde, is thus derived from methane, ethanal from ethane, propanal from propane, and so forth. However, *Chemical Abstracts* retains the common names for the first two, and so will we. We number the substituent chain beginning with the carbonyl carbon.

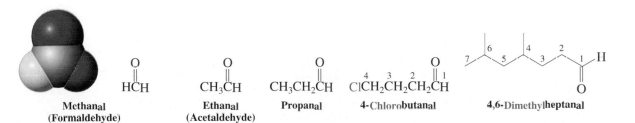

HCH   **CH₃CH**   **CH₃CH₂CH**   **ClCH₂CH₂CH₂CH**

Methanal   **Ethanal**   **Propanal**   **4-Chlorobutanal**   **4,6-Dimethylheptanal**
(Formaldehyde)   **(Acetaldehyde)**

Notice that the names parallel those of the 1-alkanols (Section 8-1), except that the position of the aldehyde carbonyl group does not have to be specified; *its carbon is defined as* C1.

Aldehydes not readily named after alkanes are instead called **carbaldehydes.** The parent aromatic aldehyde, for instance, is benzenecarbaldehyde, although its common name, benzaldehyde, is still widely used and is accepted by *Chemical Abstracts.*

Cyclohexanecarbaldehyde

4-Hydroxy-3-methoxy-benzenecarbaldehyde
(4-Hydroxy-3-methoxy-benzaldehyde)

Ketones are called **alkanones,** the ending *-e* of the alkane replaced with **-one.** We can see, for example, why acetone should be called propanone. The carbonyl carbon is assigned the lowest possible number in the chain, regardless of the presence of other substituents or the OH, C=C, or C≡C functional groups. Aromatic ketones are named as aryl-substituted alkanones. Ketones, unlike aldehydes, may also be part of a ring, an arrangement giving compounds called **cycloalkanones.**

2-Pentanone

4-Chloro-**6-**methyl**-3-**heptanone **2,2-**Dimethylcyclopentanone **1-**Phenylethanone
(Acetophenone)

Notice that we assign the number 1 to the carbonyl carbon when it is in a ring.

Aldehydes and Ketones with Other Functional Groups

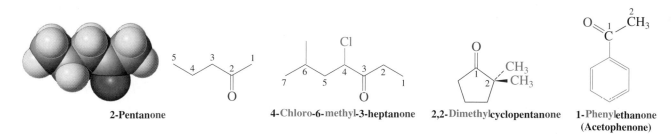

$$CH_3CCH_2CH=CHCH_2CCH_3$$
$$CH_3$$

7-Hydroxy**-7-**methyl**-4-**octen**-2-**one

Propynal

5-Bromo**-3-**ethynylcycloheptanone

(Note that the *e* in *-ene* and *-yne* is dropped in *enone* and *ynal*)

4-Formylcyclohexane-carboxylic acid

3-Oxobutanal

The systematic name for the fragment RC– is **alkanoyl,** although the older term **acyl** is widely used. Both the IUPAC and *Chemical Abstracts* retain the common names **formyl** for HC– and **acetyl** for CH₃C–. The term **oxo** denotes the location of a ketone carbonyl group when it is present together with an aldehyde function.

Name or draw the structures of the following compounds.

(a) 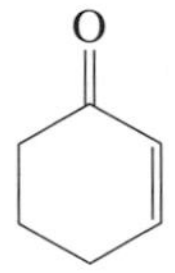 [structure: cyclohexenone]

(b) 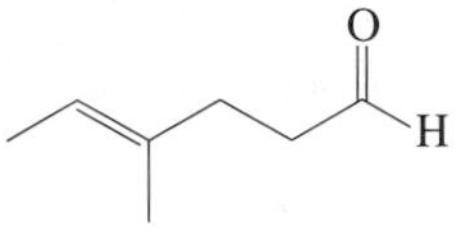 [structure: aldehyde chain with H]

(c) 4-Octyn-3-one

(d) 3-Hydroxybutanal   (e) 4-Bromocyclohexanone

There are various ways of drawing aldehydes and ketones. As usual, condensed formulas or the zigzag notation may be used. Note that the condensed formulas for aldehydes are written as RCHO, and *never* as RCOH, to prevent confusion with the hydroxy group of alcohols.

Various Ways of Writing Aldehyde and Ketone Structures

Butanal: $CH_3CH_2CH_2CH$ (with =O) $CH_3CH_2CH_2CHO$ [zigzag with O and H]

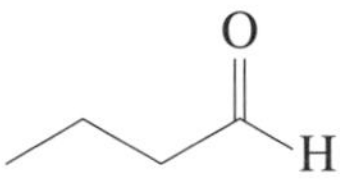

Not a hydroxy group

2-Butanone: $CH_3CH_2CCH_3$ (with =O) $CH_3CH_2COCH_3$ [zigzag with O]

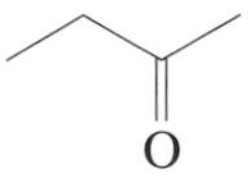

In summary, aldehydes and ketones are named systematically as alkanals and alkanones. The carbonyl group takes precedence over the hydroxy function and C—C double and triple bonds in numbering. With these rules, the usual guidelines for numbering the stem and labeling the substituents are followed.

17-2 Structure of the Carbonyl Group

If we think of the carbonyl group as an oxygen analog of the alkene functional group, we can correctly predict its molecular-orbital description, the structures of aldehydes and ketones, and some of their physical properties. However, the alkene and carbonyl double bonds do differ considerably in reactivity because of the electronegativity of oxygen and its two lone pairs of electrons.

The carbonyl group contains a short, strong, and very polar bond

Both the carbon and the oxygen of the carbonyl group are sp^2 hybridized. They therefore lie in the same plane as the two additional groups on carbon, with bond angles

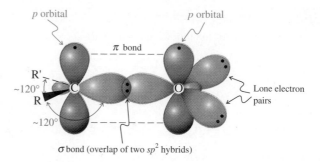

FIGURE 17-1

Molecular-orbital picture of the carbonyl group. The sp^2 hybridization and the orbital arrangement are similar to those of ethene (Figure 11-2). However, both the two lone electron pairs and the electronegativity of the oxygen atom modify the properties of the functional group.

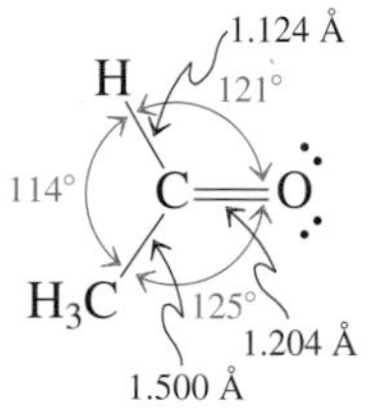

FIGURE 17-2

Molecular structure of acetaldehyde.

approximating 120°. Perpendicular to the molecular frame are two p orbitals, one on carbon and one on oxygen, making up the π bond (Figure 17-1).

Figure 17-2 shows some of the structural features of acetaldehyde. As expected, the molecule is planar, with a trigonal carbonyl carbon and a short carbon–oxygen bond, indicative of its double-bond character. Not surprisingly, this bond is also rather strong, ranging from 175 to 180 kcal mol^{-1}.

Comparison with the electronic structure of an alkene double bond reveals two important differences. First, the oxygen atom bears two lone electron pairs located in two sp^2 hybrid orbitals. Second, oxygen is more electronegative than carbon. This property causes an appreciable polarization of the carbon–oxygen double bond, with a partial positive charge on carbon and an equal amount of negative charge on oxygen. Thus, the carbon is electrophilic, the oxygen nucleophilic and slightly basic. This polarization can be described either by a polar resonance structure for the carbonyl moiety or by partial charges. As we have seen (Section 16-1), the partial positive charge on the carbonyl carbon renders alkanoyl groups electron withdrawing.

Descriptions of a Carbonyl Group

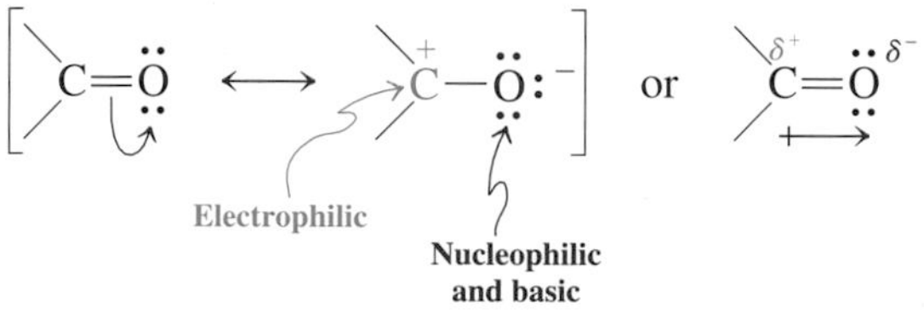

Electrophilic

Nucleophilic and basic

Polarization alters the physical constants of aldehydes and ketones

The polarization of the carbonyl functional group makes the boiling points of aldehydes and ketones higher than those of hydrocarbons of similar molecular weight (Table 17-1). Because of their polarity, the smaller carbonyl compounds such as acetaldehyde and propanone (acetone) are completely miscible with water. As the hydrophobic hydrocarbon part of the molecule increases in size, however, water solubility decreases. Carbonyl compounds with more than six carbons are rather insoluble in water.

To summarize, the carbonyl group in aldehydes and ketones is an oxygen analog of the carbon–carbon double bond. However, the electronegativity of oxygen polarizes the π bond, thereby rendering the alkanoyl substituent electron withdrawing. The

TABLE 17-1	**Boiling Points of Aldehydes and Ketones**	
Formula	**Name**	**Boiling point (°C)**
HCHO	Formaldehyde	−21
CH_3CHO	Acetaldehyde	21
CH_3CH_2CHO	Propanal (propionaldehyde)	49
CH_3COCH_3	Propanone (acetone)	56
$CH_3CH_2CH_2CHO$	Butanal (butyraldehyde)	76
$CH_3CH_2COCH_3$	Butanone (ethyl methyl ketone)	80
$CH_3CH_2CH_2CH_2CHO$	Pentanal	102
$CH_3COCH_2CH_2CH_3$	2-Pentanone	102
$CH_3CH_2COCH_2CH_3$	3-Pentanone	102

arrangement of bonds around the carbon and oxygen is planar, a consequence of sp^2 hybridization.

17-3 Spectroscopic Properties of Aldehydes and Ketones

What are the characteristics of the spectra of carbonyl compounds? In ^{1}H NMR spectroscopy, the formyl hydrogen of the aldehydes is very strongly deshielded, appearing between 9 and 10 ppm, a chemical shift that is unique for this class of compounds. The reason for this effect is twofold. First, the movement of the π electrons, like that in alkenes (Section 11-4), causes a local magnetic field, which strengthens the external field. Second, the charge on the positively polarized carbon exerts an additional deshielding effect. Figure 17-3 shows the ^{1}H NMR spectrum of propanal with the formyl hydrogen resonating at $\delta = 9.89$ ppm, split into a triplet ($J = 2$ Hz) because of a small coupling to the neighboring hydrogens. The latter are also slightly

^{1}H NMR Deshielding in Aldehydes and Ketones

$$RCH_2\overset{\overset{\displaystyle O}{\|}}{C}H$$

$\delta \sim 2.5 \quad \sim 9.8$ **ppm**

$$R\overset{\displaystyle C}{\underset{\displaystyle H}{|}}\overset{\overset{\displaystyle O}{\|}}{C}CH_3$$

$\delta \sim 2.6 \quad \sim 2.0$ **ppm**

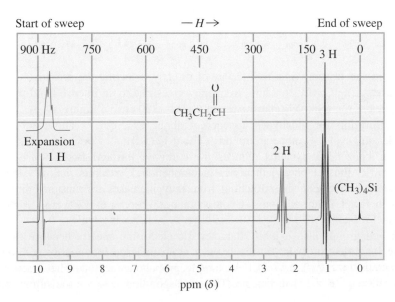

FIGURE 17-3 _____
90-MHz ^{1}H NMR spectrum of propanal in CCl_4. The formyl hydrogen (at $\delta = 9.89$ ppm) is strongly deshielded.

FIGURE 17-4 ⎯⎯⎯

^{13}C NMR spectrum of cyclohexanone at 75.5 MHz in CDCl$_3$. The carbonyl carbon at 211.8 ppm is strongly deshielded relative to the other carbons. Because of symmetry, the molecule exhibits only four peaks; the three methylene carbon resonances absorb at increasingly lower field the closer they are to the carbonyl group.

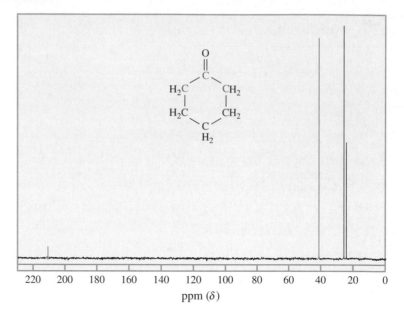

deshielded relative to alkane hydrogens because of the electron-withdrawing character of the carbonyl group. This effect is also seen in the ^{1}H NMR spectra of ketones: The α-hydrogens normally appear in the region δ = 2.0–2.8 ppm.

Carbon-13 NMR spectra are diagnostic of both aldehydes *and* ketones because of the characteristic chemical shift of the carbonyl carbon. Partly because of the electronegativity of the directly bound oxygen, the carbonyl carbons in aldehydes and ketones appear at even lower field (~200 ppm) than do the sp^2-hybridized carbon atoms of alkenes (Section 11-4). The carbons next to the carbonyl group also are deshielded relative to those located farther away. The ^{13}C NMR spectrum of cyclohexanone is shown in Figure 17-4.

^{13}C NMR Chemical Shifts of Typical Aldehydes and Ketones

$$CH_3-\overset{\overset{\displaystyle O}{\|}}{CH} \qquad CH_3-CH_2-\overset{\overset{\displaystyle O}{\|}}{CH} \qquad CH_3\overset{\overset{\displaystyle O}{\|}}{C}CH_3 \qquad CH_3\overset{\overset{\displaystyle O}{\|}}{C}-CH_2-CH_2-CH_3$$

δ = 31.2   199.6 **ppm**   δ = 5.2   36.7   201.8 **ppm**   δ = 30.2   205.1 **ppm**   δ = 29.3   206.6   45.2   17.5   13.5 **ppm**

Infrared spectroscopy is a useful way of directly detecting the presence of a carbonyl group. The C=O stretching frequency gives rise to an intense band that typically appears in a relatively narrow range (1690–1750 cm^{-1}; Figure 17-5). The carbonyl absorption for aldehydes appears at about 1735 cm^{-1}. Those for acyclic alkanones and cyclohexanone are found at about 1715 cm^{-1}. Conjugation with either alkene or benzene π systems reduces the carbonyl infrared frequency by about 30–40 cm^{-1}; thus 1-phenylethanone (acetophenone) exhibits an IR band at 1680 cm^{-1}. Conversely, the stretching frequency increases for carbonyl groups in rings with fewer than six atoms: Cyclopentanone absorbs at 1745 cm^{-1}, cyclobutanone at 1780 cm^{-1}.

Carbonyl groups also exhibit characteristic electronic spectra because the nonbonding lone electron pairs on the oxygen atom undergo low-energy $n \rightarrow \pi^*$ and $\pi \rightarrow \pi^*$ transitions (Figure 17-6). For example, propanone (acetone) shows an $n \rightarrow \pi^*$ band at 280 nm (ε = 15) in hexane. The corresponding $\pi \rightarrow \pi^*$ transition appears

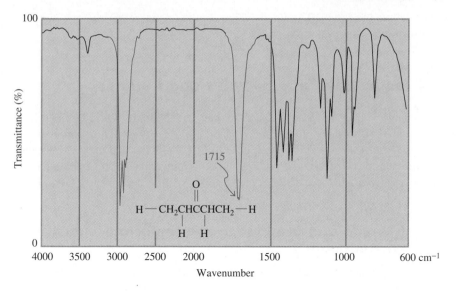

FIGURE 17-5

IR spectrum of 3-pentanone; $\tilde{\nu}_{C=O \text{ stretch}} = 1715 \text{ cm}^{-1}$.

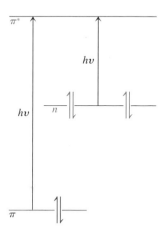

FIGURE 17-6

The $\pi \rightarrow \pi^*$ and $n \rightarrow \pi^*$ transitions in propanone (acetone).

at about 190 nm ($\epsilon = 1100$). Conjugation with a carbon–carbon double bond shifts absorptions to longer wavelengths. For example, the electronic spectrum of 3-buten-2-one, CH_2=$CHCOCH_3$, has peaks at 324 nm ($\epsilon = 24$, $n \rightarrow \pi^*$) and 219 nm ($\epsilon = 3600$, $\pi \rightarrow \pi^*$).

Electronic Transitions of Propanone (Acetone) and 3-Buten-2-one

$$CH_3\overset{\overset{\displaystyle O}{\|}}{C}CH_3$$

Propanone
(Acetone)

$\lambda_{max}(\epsilon) = 280(15) \qquad n \rightarrow \pi^*$
$\qquad\qquad 190(1100) \qquad \pi \rightarrow \pi^*$

$$CH_2=CH\overset{\overset{\displaystyle O}{\|}}{C}CH_3$$

3-Buten-2-one

$\lambda_{max}(\epsilon) = 324(24) \qquad n \rightarrow \pi^*$
$\qquad\qquad 219(3600) \qquad \pi \rightarrow \pi^*$

EXERCISE 17-2

How would you use spectroscopy to differentiate the compounds from each other in each of the following pairs? Indicate the spectroscopic method and the features in the spectra that would be most useful in each case.

(a) $CH_3CH_2CH_2CH_2OH$ and $CH_3CH_2CH_2CHO$ **(b)** $CH_3COCH_2CH_3$ and $CH_3CH_2CH_2CHO$
(c) CH_3CH=$CHCH_2CHO$ and CH_3CH_2CH=$CHCHO$ **(d)** 2-pentanone and 3-pentanone

EXERCISE 17-3

An unknown C_4H_6O exhibited the following spectral data: ^1H NMR (CCl_4) $\delta = 2.03$ (dd, $J = 6.7, 1.6$ Hz, 3 H), 6.06 (ddq, $J = 16.1, 7.7, 1.6$ Hz, 1 H), 6.88 (dq, $J = 16.1, 6.7$ Hz, 1 H), 9.47 (d, $J = 7.7$ Hz, 1 H) ppm; ^{13}C NMR (CCl_4) $\delta = 18.4, 132.8, 152.1, 191.4$ ppm; UV $\lambda_{max}(\epsilon) = 220(15,000)$ and 314(32) nm. Suggest a structure.

To summarize the spectroscopic features of aldehydes and ketones, the NMR spectra of the formyl hydrogens and the carbonyl carbons show strong deshielding. The carbon–oxygen double bond gives rise to a strong infrared band at about 1715 cm^{-1}, which is shifted to lower frequency by conjugation and to higher frequency in small rings. Finally, the ability of nonbonding electrons to be excited into the π^* molecular orbitals causes the carbonyl group to exhibit characteristic, relatively long wavelength UV absorptions.

17-4 Preparation of Aldehydes and Ketones

Several ways to prepare aldehydes and ketones have already been described in connection with the chemistry of other functional groups (see the Reaction Summary Road Map on pages 768–769). This section introduces some industrial preparations and reviews the methods that we have studied, pointing out special features and additional examples. Other routes to aldehydes and ketones will be described in later chapters.

Formaldehyde and propanone are important industrial carbonyl compounds

Formaldehyde is a product of the incomplete combustion of organic compounds and is present in wood smoke. Smoked meats are preserved both by the bactericidal effect of the surface coating of formaldehyde and by the antioxidant properties of phenolic substances also present in smoke (Section 22-9).

The most important aldehyde in industry is formaldehyde; the most important ketone is propanone (acetone). About 8 billion pounds of formaldehyde are made yearly in the United States by oxidation of methanol.

$$CH_3OH \xrightarrow{\text{O}_2,\ 600°–650°C,\ \text{catalytic Ag}} CH_2\text{=}O$$

In aqueous solution (formalin), it has applications as a disinfectant and a fungicide. Its greatest use is in the preparation of phenolic resins (Section 22-6).

Propanone (acetone) is a valuable by-product of the cumene hydroperoxide process (Section 22-4) and is sold as a solvent and starting material for the production of other industrial materials. Annual production in the United States alone is more than 2 billion pounds.

Butanal is made by a process called hydroformylation, in which propene is exposed to synthesis gas (CO + H$_2$; Section 8-4) in the presence of a cobalt or rhodium catalyst. The reaction can also be used to prepare other aldehydes.

$$CH_3CH\text{=}CH_2\ +\ CO\ +\ H_2\ \xrightarrow{\text{Co or Rh, }\Delta,\ \text{pressure}}\ CH_3CH_2CH_2CHO$$

Four methods allow laboratory syntheses of aldehydes and ketones

Table 17-2 summarizes four approaches to synthesizing aldehydes and ketones. First, we have seen (Section 8-6) that *oxidation* of alcohols by chromium (VI) reagents gives carbonyl compounds. Secondary alcohols give ketones and primary alcohols give aldehydes, but, in the latter case, only in the absence of water, to prevent overoxidation to carboxylic acids. The chromium oxidant is selective even in the presence of alkene and alkyne units.

Selective Alcohol Oxidation

$$\underset{\text{3-Octyn-2-ol}}{\overset{\overset{\displaystyle OH}{|}}{CH_3CHC\text{≡}C(CH_2)_3CH_3}} \xrightarrow{\text{CrO}_3,\ \text{H}_2\text{SO}_4,\ \text{propanone (acetone), 0°C}} \underset{\substack{\text{80\%} \\ \text{3-Octyn-2-one}}}{\overset{\overset{\displaystyle O}{\|}}{CH_3CC\text{≡}C(CH_2)_3CH_3}}$$

TABLE **17-2**	Syntheses of Aldehydes and Ketones
Reaction	**Illustration**
1. Oxidation of alcohols	$-CH_2OH$ $\xrightarrow{PCC, CH_2Cl_2}$ $-\overset{\displaystyle O}{\overset{\|}{C}}H$
2. Ozonolysis of alkenes	$\overset{\diagdown}{\diagup}C=C\overset{\diagup}{\diagdown}$ $\xrightarrow[\text{2. } (CH_3)_2S]{\text{1. } O_3, CH_2Cl_2}$ $\overset{\diagdown}{\diagup}C=O + O=C\overset{\diagup}{\diagdown}$
3. Hydration of alkynes	$-C\equiv C-$ $\xrightarrow{H_2O, H^+, Hg^{2+}}$ $-\overset{\displaystyle O}{\overset{\|}{C}}-CH_2-$
4. Friedel-Crafts alkanoylation	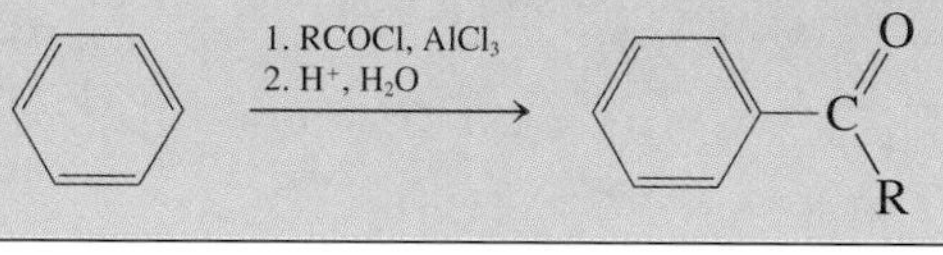 ⬡ $\xrightarrow[\text{2. } H^+, H_2O]{\text{1. RCOCl, AlCl}_3}$ ⬡$-\overset{\displaystyle O}{\overset{\|}{C}}\diagdown_R$

Use of PCC (CrO₃ + Pyridine + HCl) to Oxidize a Primary Alcohol to an Aldehyde

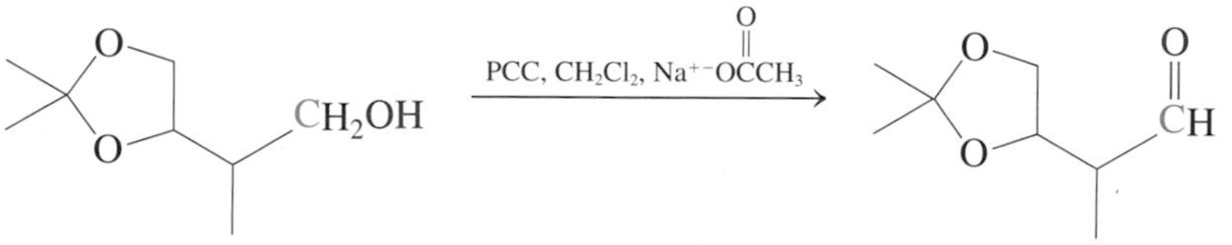

$\xrightarrow{PCC, CH_2Cl_2, Na^{+-}OCCH_3}$

85%

Overoxidation of aldehydes in the presence of water is due to hydration to a 1,1-diol (Section 17-6). Oxidation of this diol leads to the carboxylic acid.

Water Causes the Overoxidation of Primary Alcohols

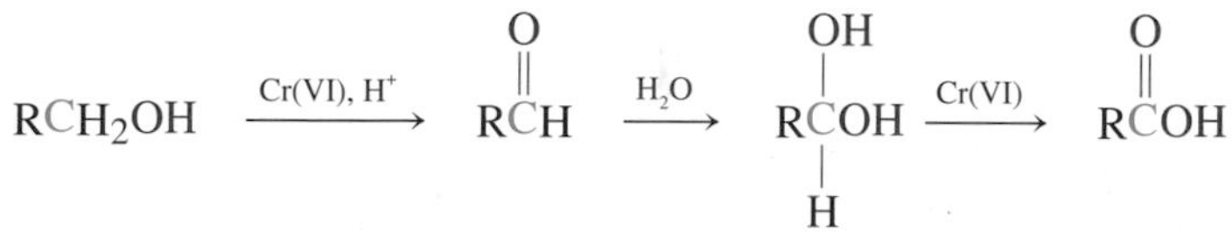

$$RCH_2OH \xrightarrow{Cr(VI), H^+} R\overset{\displaystyle O}{\overset{\|}{C}}H \xrightarrow{H_2O} R\overset{\displaystyle OH}{\underset{\displaystyle H}{\overset{\|}{C}}}OH \xrightarrow{Cr(VI)} R\overset{\displaystyle O}{\overset{\|}{C}}OH$$

Another mild reagent that specifically oxidizes allylic alcohols (Section 14-3) is manganese dioxide. Ordinary alcohols are not attacked at room temperature, as shown below in the selective oxidation to form a steroid found in the adrenal gland.

Selective Allylic Oxidations with Manganese Dioxide

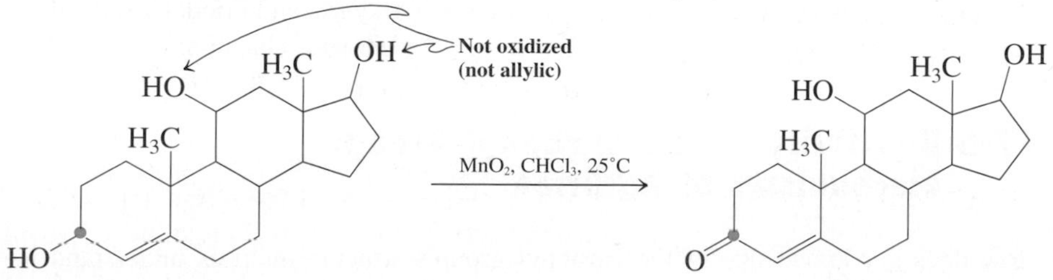

Not oxidized (not allylic)

$\xrightarrow{MnO_2, CHCl_3, 25°C}$

62%

EXERCISE 17-4

Design a synthesis of cyclohexyl 1-propynyl ketone starting from cyclohexane. You may use any other reagents.

Ozonolysis

1. O₃
2. Reducing agent

O   O
‖   ‖
CH₃C(CH₂)₄CH
85%

The second method of preparation that we studied is the oxidative cleavage of carbon–carbon double bonds—*ozonolysis* (Section 12-11). Exposure to ozone followed by treatment with a mild reducing agent, such as catalytically activated hydrogen or dimethyl sulfide, cleaves alkenes to give aldehydes and ketones.

Third, *hydration* of the carbon–carbon triple bond yields enols that tautomerize to carbonyl compounds (Sections 13-8 and 13-9). In the presence of mercuric ion, addition of water follows Markovnikov's rule to furnish ketones.

Markovnikov Hydration of Alkynes

$$RC\equiv CH \xrightarrow{HOH, H^+, Hg^{2+}} \left[\begin{array}{c} HO \quad\quad H \\ \diagdown\diagup \\ C=C \\ \diagup\quad\quad\diagdown \\ R \quad\quad H \end{array} \right] \longrightarrow \begin{array}{c} O \\ \| \\ RCCH_3 \end{array}$$

Anti-Markovnikov addition is observed in hydroboration–oxidation.

Anti-Markovnikov Hydration of Alkynes

$$RC\equiv CH \xrightarrow{\left(\bigcirc\right)_2 BH} \begin{array}{c} R \quad\quad H \\ \diagdown\diagup \\ C=C \\ \diagup\quad\quad\diagdown \\ H \quad\quad B{-}\left(\bigcirc\right)_2 \end{array} \xrightarrow{H_2O_2, HO^-} \left[\begin{array}{c} R \quad\quad H \\ \diagdown\diagup \\ C=C \\ \diagup\quad\quad\diagdown \\ H \quad\quad OH \end{array} \right] \longrightarrow \begin{array}{c} O \\ \| \\ RCH_2CH \end{array}$$

Finally, Section 15-14 discussed the synthesis of aryl ketones by *Friedel-Crafts alkanoylation,* a form of electrophilic aromatic substitution. The following example furnishes an industrially useful perfume additive.

Friedel-Crafts Alkanoylation (Acylation)

In summary, four methods of synthesizing aldehydes and ketones are oxidation of alcohols, oxidative cleavage of alkenes, hydration of alkynes, and Friedel-Crafts alkanoylation. Many other approaches will be considered in later chapters.

17-5 Reactivity of the Carbonyl Group: Mechanisms of Addition

How does our knowledge of the carbonyl group's structure help us understand its characteristic reactions? We shall see that the carbon–oxygen double bond is prone to additions, just like the π bond in alkenes. Being highly polar, however, the C=O function is predisposed toward attack by nucleophiles at carbon and electrophiles at oxygen. This section begins the discussion of the chemistry of the carbonyl group in aldehydes and ketones.

There are three regions of reactivity in aldehydes and ketones

Aldehydes and ketones contain three regions at which most reactions take place: the Lewis basic oxygen, the electrophilic carbonyl carbon, and the adjacent carbon.

Regions of Reactivity in Aldehydes and Ketones

The remainder of this chapter concerns the first two areas of reactivity, both of which lead to addition to the carbonyl π bond. This section illustrates two distinct ways in which this addition is achieved: catalytic hydrogenation and ionic addition. Other reactions, centered instead on the acidic hydrogen of the adjacent carbon, will be the subject of Chapter 18.

The carbonyl pi bond undergoes hydrogenation

Like carbon–carbon π bonds, the carbonyl group is susceptible to catalytic hydrogenation (Section 8-6), which results in the formation of alcohols. Aldehydes and ketones are usually more sluggish than alkenes in this reaction, requiring pressure or elevated temperature to proceed at a useful rate.

Selective Enone Hydrogenation

This difference in reactivity can be exploited in selective hydrogenations of the carbon–carbon π bond in unsaturated aldehydes (enals) or ketones (enones).

The catalytic hydrogenations of aldehydes and ketones are addition reactions that proceed through surface-bound intermediates (see Section 12-2). Additions to the carbon–oxygen double bond may also be achieved through ionic mechanisms that exploit the dipolar nature of the functional group.

The carbonyl group undergoes ionic additions

Polar reagents add to the dipolar carbonyl group according to Coulomb's law (Section 1-2). Nucleophiles bond to the carbon, electrophiles to the oxygen. Sections 8-6 and 8-8 described a number of such additions by organometallic and hydride reagents to give alcohols. Table 17-3 reviews these processes.

Ionic Additions to the Carbonyl Group

$$\overset{\delta^-}{O} \quad\quad \overset{OX}{} $$

$$\overset{\delta^-}{\underset{}{O}} \quad\quad\quad\quad\quad \overset{OX}{}$$

$$\overset{\delta^+}{C} + \overset{\delta^+}{X}-Y^{\delta^-} \longrightarrow -\overset{|}{\underset{|}{C}}- $$
$$\qquad\qquad\qquad\qquad\qquad\qquad Y$$

The hydride reagents $NaBH_4$ and $LiAlH_4$ reduce carbonyl groups but not carbon–carbon double bonds. These reagents therefore convert unsaturated aldehydes and ketones into unsaturated alcohols, with a selectivity opposite that exhibited in catalytic hydrogenation.

$$\xrightarrow{\begin{array}{l}1.\ LiAlH_4,\ (CH_3CH_2)_2O,\ -10°C\\ 2.\ H^+,\ H_2O\end{array}}$$

90%

Because the nucleophilic reagents illustrated in Table 17-3 are strong bases, their additions are irreversible. This section and Sections 17-6 through 17-9 consider ionic additions of less basic nucleophiles Nu–H, such as water, alcohols, thiols, and amines. These processes are not strongly exothermic but instead establish equilibria that can be pushed in either direction by the appropriate choice of reaction conditions.

TABLE 17-3	**Additions of Hydride and Organometallic Reagents to Aldehydes and Ketones**	
Reaction	**Illustration**	
1. Aldehyde + hydride reagent	$RCHO \xrightarrow{NaBH_4,\ CH_3CH_2OH} RCH_2OH$	Primary alcohol
2. Ketone + hydride reagent	$R_2CO \xrightarrow{NaBH_4,\ CH_3CH_2OH} R_2CHOH$	Secondary alcohol
3. Formaldehyde + Grignard reagent	$H_2CO \xrightarrow{R'MgX,\ (CH_3CH_2)_2O} R'CH_2OH^a$	Primary alcohol
4. Aldehyde + Grignard reagent	$RCHO \xrightarrow{R'MgX,\ (CH_3CH_2)_2O} R'RCHOH^a$	Secondary alcohol
5. Ketone + Grignard reagent	$R_2CO \xrightarrow{R'MgX,\ (CH_3CH_2)_2O} R'R_2COH^a$	Tertiary alcohol

[a]After aqueous work-up.

What is the mechanism of the ionic addition of these milder reagents to the carbon–oxygen double bond? Two pathways can be formulated: nucleophilic addition–protonation and electrophilic protonation–addition. The first, which begins with nucleophilic attack, takes place under neutral or, more commonly, basic conditions. As the nucleophile approaches the electrophilic carbon, the latter rehybridizes, and the electron pair of the π bond moves over to the oxygen, thereby producing an alkoxide ion. Subsequent protonation, usually from a protic solvent such as water or alcohol, yields the final addition product.

Nucleophilic Addition–Protonation (Basic Conditions)

Alkoxide ion

Note that the new Nu–C bond is made up entirely of the electron pair of the nucleophile. The entire transformation is reminiscent of an S_N2 reaction. In that process, a leaving group is displaced. Here, it is an electron pair that is displaced from a shared position between carbon and oxygen to one solely on the oxygen atom. Additions of strongly basic nucleophiles to carbonyl groups typically follow the nucleophilic addition–protonation pathway.

The second mechanism predominates under acidic conditions and begins with electrophilic attack. In the first step, protonation of the carbonyl oxygen occurs, facilitated by the polarization of the C=O bond and the presence of lone electron pairs on the oxygen atom. The latter is only weakly basic, as shown by the pK_a of the conjugate acid, which ranges from about -7 to -8. Thus, in a dilute acidic medium, in which many carbonyl addition reactions are carried out, most of the carbonyl compound will stay unprotonated. However, the small amount of protonated material behaves like a very reactive carbon electrophile. Nucleophilic attack by the nucleophile completes the addition process and shifts the first, unfavorable equilibrium.

Electrophilic Protonation–Addition (Acidic Conditions)

Protonated carbonyl group
$pK_a \sim -8$

The electrophilic protonation–addition mechanism is best suited for reactions of relatively weakly basic nucleophiles. The acidic conditions are incompatible with strongly basic nucleophiles, because protonation of the nucleophile occurs.

In summary, there are three regions of reactivity in aldehydes and ketones. The first two are the two atoms of the carbonyl group and are the subject of the remainder of this chapter. The third is the adjacent carbon. The reactivity of the carbonyl group is governed by addition processes. Catalytic hydrogenation gives alcohols. Nucleophilic organometallic and hydride reagents result in irreversible alcohol formation (after protonation). Ionic additions of NuH (Nu = OH, OR, SR, NR₂) are reversible and may begin with nucleophilic attack at the carbonyl carbon, followed by electrophilic trapping of the alkoxide anion so generated. Alternatively, in acidic media, protonation precedes the addition of the nucleophile.

17-6 Addition of Water to Form Hydrates

This section and Sections 17-7 and 17-8 introduce the reactions of aldehydes and ketones with water and alcohols. These compounds attack the carbonyl group through the mechanisms just outlined, utilizing both acid and base catalysis.

Water hydrates the carbonyl group

Water is capable of attacking the carbonyl function of aldehydes and ketones. The transformation can be catalyzed by either acid or base and leads to equilibration with the corresponding **geminal diols**, $RC(OH)_2R'$, also called **carbonyl hydrates.**

Hydration of the Carbonyl Group

$$\text{C}=\text{O} \ + \ \text{HOH} \ \underset{K}{\overset{\text{H}^+ \text{ or HO}^-}{\rightleftharpoons}} \ \text{C}-\text{OH}$$

Geminal diol

In the base-catalyzed mechanism, the hydroxide functions as the nucleophile. Water then protonates the intermediate adduct, a hydroxy alkoxide, to give the product diol and to regenerate the catalyst.

Mechanism of Base-Catalyzed Hydration

$$\text{C}=\ddot{\text{O}} \ + \ ^-\!\!:\!\ddot{\text{O}}\text{H} \ \rightleftharpoons \ \underset{\text{HO}}{\text{C}}-\ddot{\text{O}}:^- \ \overset{\text{H}\ddot{\text{O}}\text{H}}{\rightleftharpoons} \ \underset{\text{HO}}{\text{C}}-\ddot{\text{O}}\text{H} \ + \ \text{HO}:^-$$

Hydroxy alkoxide   **Geminal diol**

In the acid-catalyzed mechanism, the sequence of events is reversed. Here, initial protonation facilitates nucleophilic attack by the weak nucleophile water. Subsequently, the catalytic proton is lost to reenter the catalytic cycle.

Mechanism of Acid-Catalyzed Hydration

$$\text{C}=\ddot{\text{O}} \ + \ \text{H}^+ \ \rightleftharpoons \ \text{C}=\overset{+}{\ddot{\text{O}}}-\text{H} \ \overset{\text{H}_2\ddot{\text{O}}}{\rightleftharpoons} \ \underset{\overset{+}{\text{H}_2\text{O}}:}{\text{C}}-\ddot{\text{O}}\text{H} \ \rightleftharpoons \ \underset{\text{HO}}{\text{C}}-\ddot{\text{O}}\text{H} \ + \ \text{H}^+$$

Protonated carbonyl   **Geminal diol**

Hydration is reversible

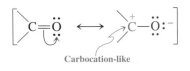

Carbocation-like

As these equations indicate, hydrations of aldehydes and ketones are reversible. The equilibrium lies to the left for ketones and to the right for formaldehyde and aldehydes bearing inductively electron-withdrawing substituents. For ordinary aldehydes, the equilibrium constant approaches unity.

How can these trends be explained? Look again at the resonance structures of the carbonyl group, as described in Section 17-2 and repeated here in the margin. In the dipolar resonance form, the carbon possesses carbocation character. Therefore, alkyl groups, which stabilize carbocations (Section 7-5), stabilize carbonyl compounds as well. Conversely, inductively electron-withdrawing substituents, such as CCl_3 and CF_3, increase positive charge at the carbonyl carbon and destabilize carbonyl compounds. The stabilities of the product diols are affected to a lesser extent by substituents. As a net result, therefore, relative to hydration of formaldehyde, the hydration reactions of aldehydes and ketones are progressively more endothermic, whereas hydrations of carbonyl compounds containing electron-withdrawing groups are more exothermic. These thermodynamic effects are paralleled by differences in kinetic reactivity: Carbonyl compounds containing electron-withdrawing groups are the most electrophilic and the most reactive, followed in turn by formaldehyde, other aldehydes, and finally ketones. We shall see that the same trends also govern other addition reactions.

Equilibrium Constants K for the Hydration of Typical Carbonyl Compounds

$$Cl_3CCH \quad K > 10^4$$

$$H_2C{=}O \quad K > 10^3$$

$$CH_3CH \quad K \sim 1$$

$$CH_3CCH_3 \quad K < 10^{-2}$$

Relative Reactivities of Carbonyl Groups

Electron-withdrawing CCl_3 group generates additional positive charge at carbonyl carbon

Electron-donating CH_3 groups reduce positive charge at carbonyl carbon

---

EXERCISE 17-5

(a) Rank in order of increasing favorability of hydration: Cl_3CCH, Cl_3CCCH_3, Cl_3CCCCl_3. (b) Treatment of propanone (acetone) with $H_2{}^{18}O$ results in the formation of ${}^{18}O$ labeled propanone, CH_3CCH_3. Explain.

In summary, the carbonyl group of the aldehydes and ketones is hydrated by water. Aldehydes are more reactive than ketones. Electron-withdrawing substituents render the carbonyl carbon more electrophilic. Hydration is an equilibrium process that may be catalyzed by both acids and bases.

17-7 Addition of Alcohols to Form Hemiacetals and Acetals

In this section we shall see that alcohols add to carbonyl groups in much the same manner as water does. Both acids and bases catalyze the process. In addition, acids catalyze the further transformation of the initial addition product to furnish acetals by replacement of the hydroxy group with an alkoxy substituent.

Aldehydes and ketones form hemiacetals reversibly

Not surprisingly, alcohols also undergo addition to aldehydes and ketones, by a mechanism virtually identical with that outlined for water. The adducts formed are called **hemiacetals** (*hemi,* Greek, half) because they are intermediates on the way to acetals.

Hemiacetal Formation

$$
\underset{\substack{R \diagdown \\ H}}{\overset{O}{\parallel}}{C} \; + \; R'OH \; \rightleftharpoons \; R{-}\overset{\overset{\textstyle H}{\overset{\textstyle |}{O}}}{\underset{\underset{\textstyle H}{|}}{C}}{-}OR' \qquad\qquad \underset{\substack{R \diagdown \\ R}}{\overset{O}{\parallel}}{C} \; + \; R'OH \; \rightleftharpoons \; R{-}\overset{\overset{\textstyle H}{\overset{\textstyle |}{O}}}{\underset{\underset{\textstyle R}{|}}{C}}{-}OR'
$$

A hemiacetal

A hemiacetal

Like hydration, these addition reactions are governed by equilibria that usually favor the starting carbonyl compound. Hemiacetals, like hydrates, are therefore usually not isolable. Exceptions are those formed from reactive carbonyl compounds such as formaldehyde or 2,2,2-trichloroacetaldehyde. Hemiacetals are also isolable from hydroxy aldehydes and ketones when cyclization leads to the formation of relatively strain free five- and six-membered rings.

Intramolecular Hemiacetal Formation

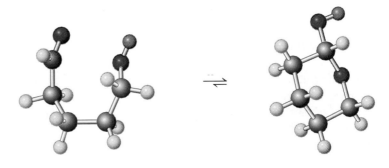

Glucose

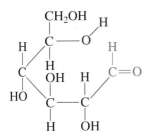

0.003%

Aldehyde form

$$\Big\updownarrow \; \text{H}^+ \text{ or HO}^-$$

> 99%

Cyclic hemiacetal
(Two stereoisomers)

5-Hydroxypentanal   **A cyclic hemiacetal, stable**

Intramolecular hemiacetal formation is common in sugar chemistry (Chapter 24). For example, glucose, the most common simple sugar in nature, exists as an equilibrium mixture of an acyclic pentahydroxyaldehyde and two stereoisomeric cyclic hemiacetals. The latter forms constitute more than 99% of the mixture in aqueous solution.

Acids catalyze acetal formation

In the presence of excess alcohol, the *acid*-catalyzed reaction of aldehydes and ketones proceeds beyond the hemiacetal stage. Under these conditions, the hydroxy function of the initial adduct is replaced by another alkoxy unit derived from the alcohol. The resulting compounds are called **acetals. (Ketal** is an older term for an acetal derived from a ketone.)

Acetal Synthesis

$$\underset{\underset{\displaystyle \text{RCR}}{\|}}{O} \quad + \quad 2\,R'OH \quad \underset{}{\overset{H^+}{\rightleftharpoons}} \quad \underset{\displaystyle R}{\overset{\displaystyle OR'}{\underset{|}{\overset{|}{R-C-OR'}}}} \quad + \quad H_2O$$

An acetal

The net change is the replacement of the carbonyl oxygen by two alkoxy groups and the formation of one equivalent of water.

Let us examine the mechanism of this transformation for an aldehyde. The initial reaction is the ordinary acid-catalyzed addition of the first molecule of alcohol. The resulting hemiacetal can be protonated at the hydroxy group, changing this substituent into water, a good leaving group. On loss of water, the resulting carbocation is stabilized by resonance with a lone electron pair on oxygen. A second molecule of alcohol now adds to the electrophilic carbon, initially giving the protonated acetal, which is then deprotonated to the final product.

Mechanism of Acetal Formation

STEP 1. Hemiacetal generation

STEP 2. Acetal generation

Acetal

Each step is reversible; the entire sequence, starting from the carbonyl compound and ending with the acetal, is an equilibrium process. The overall equilibrium may be shifted in either direction by manipulating the reaction conditions: toward acetal by using excess alcohol or by continually removing water from the reaction medium; toward aldehyde or ketone by using excess water. The latter process is called **acetal hydrolysis.**

In summary, the reaction of alcohols with aldehydes and ketones results in hemi-acetal formation. This process, like hydration, is reversible and is catalyzed by both acids and bases. Hemiacetals are converted into relatively stable acetals by acid and excess alcohol.

17-8 Acetals as Protecting Groups

In the conversion of an aldehyde or ketone to an acetal, the reactive carbonyl function is transformed into a relatively unreactive ether moiety. Because acetalization is reversible, this process amounts to one of *masking,* or *protecting,* the carbonyl group. Such protection is necessary when selective reactions (for example, with nucleophiles) are required at another functional site of the same molecule to which the presence of the unprotected carbonyl would be detrimental. This section will describe the execution of such protecting strategies.

Cyclic acetal formation protects carbonyl groups from attack by nucleophiles

Diols, such as 1,2-ethanediol, are particularly effective acetalization reagents compared to normal alcohols. Their use converts aldehydes and ketones into *cyclic* acetals which are generally more stable than their acyclic counterparts. The reason lies in part in their relatively more favorable (or, better, less unfavorable) entropy of formation (Section 2-8). Thus, as shown here, two molecules of starting material (carbonyl compound and diol) are converted into two molecules of product (acetal and water) in this process. In contrast, acetal formation with a normal alcohol (Section 17-7) uses up *three* molecules (carbonyl compound and two equivalents of alcohol) to result in the same number of product molecules.

Cyclic Acetalization

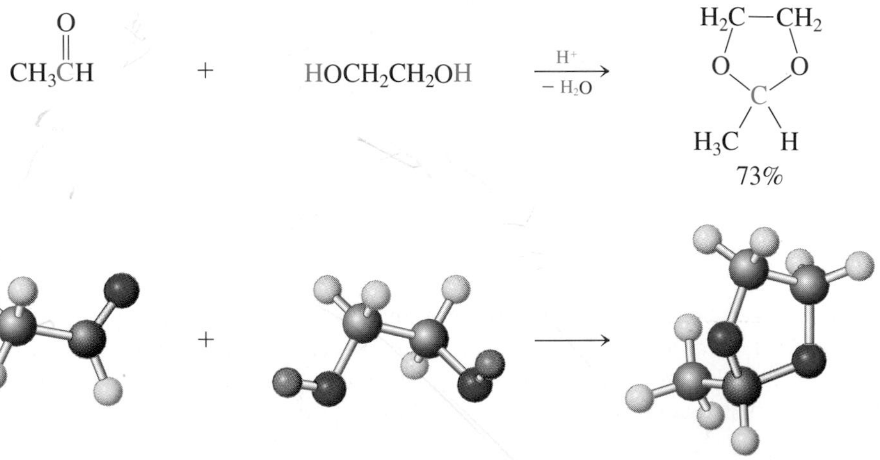

A cyclic acetal

Cyclic acetals are nevertheless readily hydrolyzed in the presence of excess acidic water, but are not attacked by many basic, organometallic, and hydride reagents. These properties make them most useful as **protecting groups** for the carbonyl function in aldehydes and ketones. An example is the alkylation of an alkynyl anion with 3-iodopropanal 1,2-ethanediol acetal.

Use of a Protected Aldehyde in Synthesis

$$CH_3(CH_2)_3C\equiv C^-Li^+ \quad + \quad ICH_2CH_2-\overset{\overset{\displaystyle H}{|}}{\underset{}{C}}\overset{O}{\underset{O}{\diagdown}} \quad \xrightarrow[-\text{ LiI}]{}$$

1-Hexynyllithium   **3-Iodopropanal**
                                  1,2-ethanediol acetal

$$CH_3(CH_2)_3C\equiv C-CH_2CH_2-\overset{\overset{\displaystyle H}{|}}{\underset{O}{C}}\overset{O}{\diagdown} \quad \xrightarrow[-\text{ HOCH}_2\text{CH}_2\text{OH}]{H^+, H_2O} \quad CH_3(CH_2)_3C\equiv C-CH_2CH_2-CHO$$

                                                      Deprotection

        70%   **90%**
4-Nonynal 1,2-ethanediol acetal   **4-Nonynal**

When the same alkylation is attempted with unprotected 3-iodopropanal, the alkynyl anion attacks the carbonyl group.

EXERCISE 17-6

Suggest a convenient way of converting compound A into B.

   A   **B**

If a diol can be used to protect a carbonyl function, then a carbonyl compound should be able to protect a diol. This is indeed the case. For example, propanone (acetone) blocks the acidic sites in vicinal diols in the form of an acetal. The protection of diols as propanone (acetone) acetals is an important reaction in sugar chemistry (Chapter 24).

Protection of a Vicinal Diol as the Propanone (Acetone) Acetal

6-Bromo-1,2-hexanediol   **75%**

Thiols react with the carbonyl group to form thioacetals

Thiols, the sulfur analogs of alcohols (see Section 9-10), react with aldehydes and ketones by a mechanism identical with the one described for alcohols. Instead of a proton catalyst, a Lewis acid, such as BF_3 or $ZnCl_2$, is often used. The reaction produces the sulfur analogs of acetals, **thioacetals.**

Cyclic Thioacetal Formation from a Ketone

95%
A cyclic thioacetal

These sulfur derivatives are stable in aqueous acid, a medium that hydrolyzes ordinary acetals. The difference in reactivity may be useful in synthesis when it is necessary to differentiate two different carbonyl groups in the same molecule. Hydrolysis of thioacetals is carried out by using mercuric chloride in aqueous acetonitrile. The driving force is the formation of insoluble mercuric sulfides.

CHEMICAL HIGHLIGHT 17-1 | **Protecting Groups in Vitamin C Synthesis**

$$\xrightarrow[\text{$-2\,H_2O$}]{\text{$H^+, 2\,CH_3CCH_3$}}$$

Sorbose

Virtually all vitamin C sold is synthetic material. At an early stage of the synthesis, the primary alcohol group at C1 of the sugar sorbose must be selectively oxidized to a carboxylic acid function, without affecting the OH groups at carbons 3 through 6. Sorbose, like glucose (Section 17-7), exists mostly as a cyclic hemiacetal from the addition of the hydroxy group at C5 to the C2 carbonyl. On acid-catalyzed reaction with excess propanone, the hydroxy groups at C2 and C3 are blocked as a five-membered cyclic acetal and those at C4 and C6 are protected as a six-membered acetal ring. Potassium permanganate then converts the unprotected CH_2OH (C1) into CO_2H.

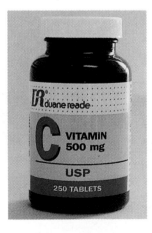

Unlike most mammals, humans lack the ability to synthesize vitamin C and must obtain this necessary nutrient from dietary sources.

Vitamin C

Thioacetals are **desulfurized** to the corresponding hydrocarbon by treatment with Raney nickel (Section 12-2). Thioacetal generation followed by desulfurization is used to convert a carbonyl into a methylene group.

Thioacetal Hydrolysis

EXERCISE 17-7

Suggest a possible synthesis of cyclodecane from [structure]. (**Hint:** See Section 12-11.)

In summary, cyclic acetals are good protecting groups for the carbonyl function and for diols. Thiols show similar reactivity. The formation of thioacetals is usually catalyzed by Lewis acids. Their hydrolysis requires the presence of mercuric salts. Thioacetals are reduced by Raney nickel to hydrocarbons.

17-9 Nucleophilic Addition of Ammonia and Its Derivatives

Ammonia and the amines may be regarded as nitrogen analogs of water and alcohols. Do they add to aldehydes and ketones? In fact, they do, giving products corresponding to those just studied. We shall see one important difference, however. The products of addition of amines and their derivatives lose water, furnishing either of two new derivatives of the original carbonyl compounds: imines and enamines.

Ammonia   **Primary amine**

Ammonia and primary amines form imines

On exposure to an amine, aldehydes and ketones form **hemiaminals,** the nitrogen analogs of hemiacetals. Hemiaminals of primary amines readily lose water to form a carbon–nitrogen double bond. This function is called an **imine** (an older name is **Schiff* base**) and is the nitrogen analog of a carbonyl group.

Imine Formation from Amines and Aldehydes or Ketones

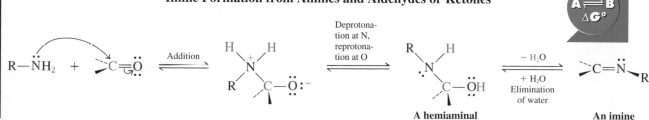

A hemiaminal   An imine

The mechanism of the elimination of water from the hemiaminal is the same as that for the decomposition of a hemiacetal to the carbonyl compound and alcohol. It begins with protonation of the hydroxy group. (Protonation of the more basic nitrogen just leads back to the carbonyl compound.) Dehydration follows, and then deprotonation of the intermediate **iminium ion.**

*Professor Hugo Schiff (1834–1915), University of Florence, Italy.

Mechanism of Hemiaminal Dehydration

Hemiaminal   Iminium ion   Imine

Processes such as imine formation from a primary amine and an aldehyde or ketone, in which two molecules are joined with the elimination of water, are called **condensations.** Imine formation is reversible, and the usual measures have to be employed to shift the equilibrium in the desired direction.

Condensation of a Ketone with a Primary Amine

$$RNH_2 \ + \ O=C \overset{R'}{\underset{R''}{\diagup}} \ \rightleftharpoons \ RN=C \overset{R'}{\underset{R''}{\diagup}} \ + \ H_2O$$

CHEMICAL HIGHLIGHT **17-2** | **Imines in Biological Transformations**

The related molecules pyridoxal and pyridoxamine assist in the interconversion of carbonyl groups and primary amine functions in biology. Pyridoxamine undergoes enzyme-catalyzed condensation with the carbonyl group of 2-oxocarboxylic acids to produce an imine. Rearrangement furnishes a new

imine, which hydrolyzes to pyridoxal and a 2-aminocarboxylic acid (an *amino acid*). In the forward direction, this process synthesizes several of the naturally occurring amino acids; in the reverse mode, it aids in their metabolism.

Pyridoxamine   **2-Oxocarboxylic acid**   Rearrangement

Pyridoxal   **2-Aminocarboxylic acid**

Examples:

$$CH_3\overset{\displaystyle O}{\overset{\displaystyle \|}{C}}H \ + \ H_2NCH_2CH_2CH_2CH_3 \ \xrightarrow{KOH} \ CH_3CH{=}NCH_2CH_2CH_2CH_3 \ + \ H_2O$$
83%

$$CH_3\overset{\displaystyle O}{\overset{\displaystyle \|}{C}}CH_3 \ + \quad\underset{}{\text{(cyclohexylamine with } NH_2\text{)}} \quad\xrightarrow{H^+}\quad \underset{H_3C}{\overset{H_3C}{>}}C{=}N{-}\text{(cyclohexyl)} \ + \ H_2O$$
95%

EXERCISE 17-8

Reagent A has been used with aldehydes to prepare crystalline imidazolidine derivatives such as B, for the purpose of their isolation and structural identification. Write a mechanism for the formation of B. (**Hint:** Notice that the product is a nitrogen analog of an acetal. Develop a mechanism analogous to that for acetal formation [Section 17-7], using the two amine groups in the starting diamine in place of two alcohol molecules.)

A
N,N′-Diphenyl-1,2-ethanediamine

$H_3C\overset{\displaystyle O}{\overset{\displaystyle \|}{}}\,\,H$
+   H₃C—CHO $\xrightarrow[-\,H_2O]{CH_3OH,\ H^+}$

B
2-Methyl-1,3-diphenyl-1,3-diazacyclopentane
(2-Methyl-1,3-diphenylimidazolidine)
(m.p. 102°C)

Special imines aid in identification of aldehydes and ketones

Several amine derivatives condense with aldehydes and ketones to form imine products that are highly crystalline and often have sharp melting points. For example, *hydroxylamine*, H_2NOH, as its hydrochloride, condenses with aldehydes to **oximes.**

A Typical Oxime Synthesis

$$CH_3(CH_2)_5\overset{\overset{\displaystyle O}{\|}}{C}H \ + \ H_3\overset{+}{N}OH \ Cl^- \ \xrightarrow[-\ H_2O]{H^+} \ \underset{H}{\overset{CH_3(CH_2)_5}{\diagdown}}C{=}NOH$$

Hydroxylamine	93%
hydrochloride	**Heptanal oxime**
	(m.p. 57°C)

Derivatives of *hydrazine, H₂NNH₂*—in particular, *phenylhydrazine, C₆H₅NHNH₂*, and *2,4-dinitrophenylhydrazine, 2,4-(NO₂)₂C₆H₃NHNH₂*—condense to the corresponding **hydrazones.**

Formation of a 2,4-Dinitrophenylhydrazone from a Ketone

2,4-Dinitrophenylhydrazine

85%
Cyclopentanone 2,4-dinitrophenylhydrazone
(m.p. 142°C)

Finally, the hydrazine derivative *semicarbazide* reacts with aldehydes and ketones to give **semicarbazones.**

A Semicarbazone Synthesis

Semicarbazide

90%
Cyclohexanecarbaldehyde semicarbazone
(m.p. 173°C)

Melting points have been tabulated for imine derivatives of thousands of aldehydes and ketones. Before spectroscopic methods became routinely available, chemists used comparisons of these melting points to determine whether an aldehyde or ketone of unknown structure was identical with a previously characterized compound. For example, the distillation of castor oil yields a compound with the formula $C_7H_{14}O$. Its boiling point of about 150°C is similar to those recorded for heptanal, 2-heptanone, and 4-heptanone. The unknown is readily identified as heptanal by preparing and determining the melting points of its semicarbazone (109°C) and 2,4-dinitrophenylhydrazone (108°C). These values match those tabulated for the derivatives of the alde-hyde and are quite different from those recorded for the two ketones: The respective derivatives of 2-heptanone exhibit melting points of 127° and 89°C, and those of 4-heptanone, 133° and 75°C.

Condensations with secondary amines give enamines

The condensations of amines described so far are possible only for primary deriva-tives, because the nitrogen of the amine has to supply both of the protons necessary to form water. Reaction with a secondary amine therefore takes a different course. After the initial addition, water is eliminated by deprotonation at *carbon* to produce an **enamine.** This functional group incorporates both the *ene* function of an alkene and the *amino* group of an amine.

$$\underset{\text{Secondary amine}}{\underset{|}{\overset{R}{\diagdown}}\overset{..}{\underset{R'}{N}}\overset{H}{\diagup}}$$

Enamine Formation

$$\underset{O}{\overset{\parallel}{CH_3CH_2CCH_2CH_3}} + \overset{H}{\underset{}{\diagdown}}N\diagup \rightleftharpoons \underset{CH_2CH_3}{\overset{OH}{\underset{|}{CH_3CH_2-C-N}}}\diagup \underset{+ H_2O}{\overset{- HOH}{\rightleftharpoons}} \underset{CH_2CH_3}{\overset{N}{CH_3CH=C}}$$

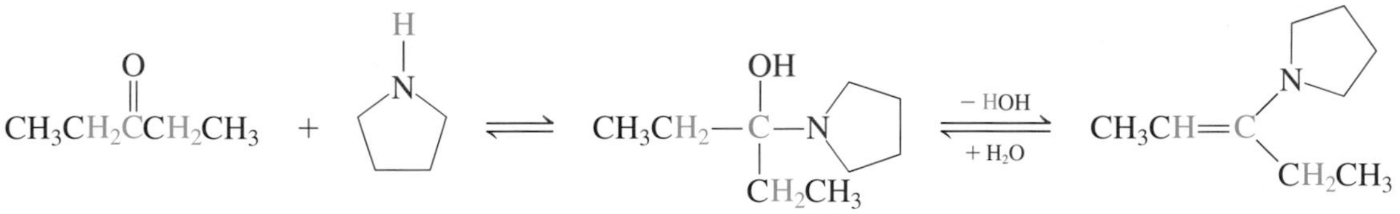

90%
An enamine

Enamine formation is reversible, and hydrolysis occurs readily in the presence of acidic water. Enamines are useful substrates in alkylations (Section 18-4).

EXERCISE 17-9

Write the products of the following reactions.

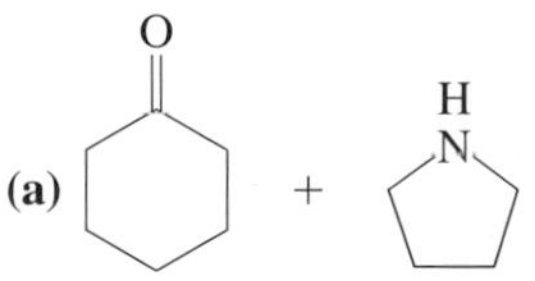

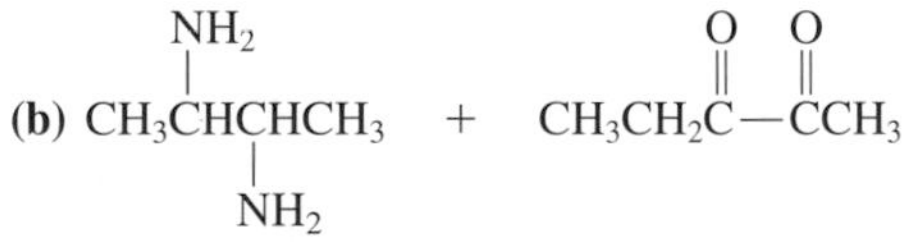

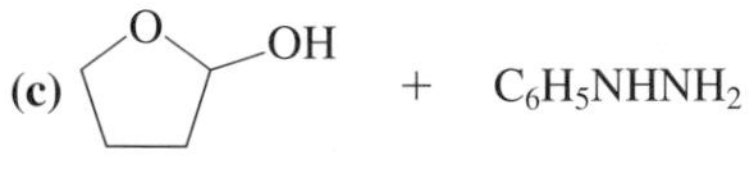

(a) (cyclohexanone) + (pyrrolidine N—H)

(b) $CH_3\underset{\underset{NH_2}{|}}{CHCHCH_3}$ + $\underset{O \quad O}{\overset{\parallel \quad \parallel}{CH_3CH_2C-CCH_3}}$
      with NH_2

(c) (tetrahydrofuran with O and OH) + $C_6H_5NHNH_2$

(**Hint:** See Section 17-7.)

In summary, amines attack aldehydes and ketones to form imines by condensa-tion. Hydroxylamine gives oximes, hydrazines lead to hydrazones, and semicarbazide results in semicarbazones. Secondary amines react with aldehydes and ketones to give enamines.

17-10 Deoxygenation of the Carbonyl Group

In Section 17-5 we reviewed methods by which carbonyl compounds are reduced to alcohols. Reduction of the C=O group to CH_2 (**deoxygenation**) also is possible. Two ways in which this may be achieved are Clemmensen reduction (Section 16-5) and

Synthesis of a Hydrazone

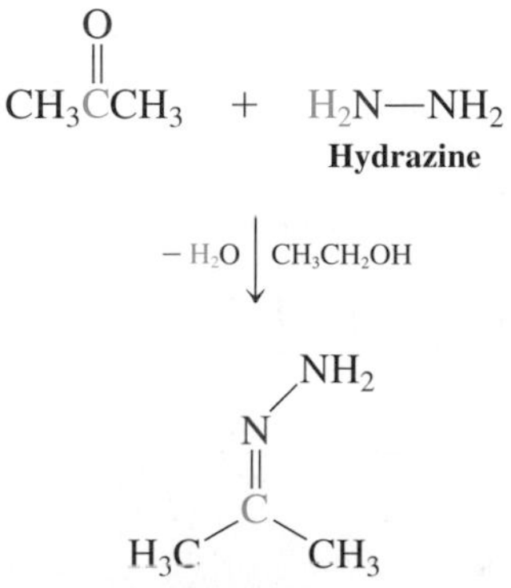

$$\underset{O}{\overset{\parallel}{CH_3CCH_3}} + H_2N-NH_2$$
Hydrazine

$$\downarrow \overset{CH_3CH_2OH}{\underset{- H_2O}{}}$$

$$\underset{\underset{\text{hydrazone}}{\text{Propanone (acetone)}}}{\overset{\overset{NH_2}{\diagup}}{\underset{H_3C}{\overset{\parallel}{\underset{}{C}}}\overset{N}{\diagdown}CH_3}}$$

thioacetal formation followed by desulfurization (Section 17-8). This section presents a third method for deoxygenating aldehydes and ketones—the Wolff-Kishner reduction.

Strong base converts simple hydrazones into hydrocarbons

Condensation of hydrazine itself with aldehydes and ketones produces simple hydrazones (see margin, page 753). These derivatives undergo decomposition with evolution of nitrogen when treated with base at elevated temperatures. The product of this reaction, called the **Wolff-Kishner* reduction,** is the corresponding hydrocarbon.

Wolff-Kishner Reduction

$$
\underset{\substack{\|\\ RCR'}}{\overset{\substack{NH_2 \\ N}}{}} \quad + \quad NaOH \quad \xrightarrow{\text{(HOCH}_2\text{CH}_2)_2\text{O, }180°-200°C} \quad RCH_2R' \quad + \quad N_2
$$

The mechanism of nitrogen elimination includes a sequence of base-mediated hydrogen shifts. The base first removes a proton from the hydrazone to give the corresponding delocalized anion. Reprotonation may occur on nitrogen, regenerating the starting material, or on carbon, thereby furnishing an intermediate azo compound (Section 22-11) leading on to the product. The base removes another proton from the azo nitrogen to generate a new anion, which rapidly decomposes irreversibly, with the extrusion of nitrogen gas. The very basic alkyl anion so formed is immediately protonated in the aqueous medium to the hydrocarbon.

Mechanism of Nitrogen Elimination in the Wolff-Kishner Reduction

In practice, the Wolff-Kishner reduction is carried out without isolating the intermediate hydrazone. An 85% aqueous solution of hydrazine (hydrazine hydrate) is added to the carbonyl compound in a high-boiling alcohol solvent such as diethylene glycol (HOCH$_2$CH$_2$OCH$_2$CH$_2$OH, b.p. 245°C) or triethylene glycol (HOCH$_2$CH$_2$OCH$_2$CH$_2$OCH$_2$CH$_2$OH, b.p. 285°C) containing NaOH or KOH, and the mixture is heated. Aqueous work-up yields the pure hydrocarbon.

*Professor Ludwig Wolff (1857–1919), University of Jena, Germany; Professor N. M. Kishner (1867–1935), University of Moscow.

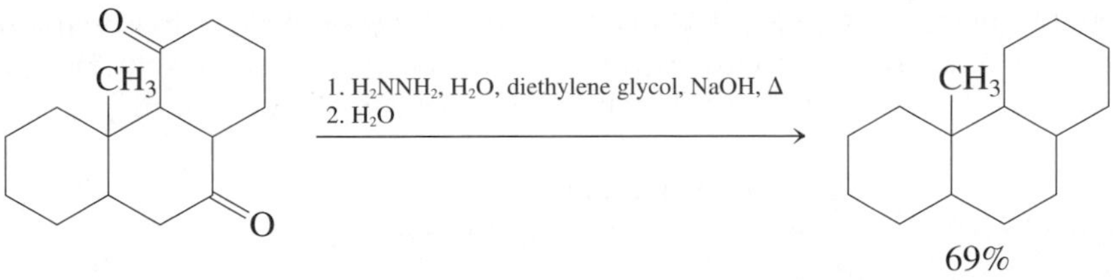

69%

Wolff-Kishner reduction complements the Clemmensen and thioacetal desulfurization methods of deoxygenating aldehydes and ketones. Thus, the Clemmensen reduction is unsuitable for compounds containing acid-sensitive groups, and hydrogenation of multiple bonds can accompany desulfurization with hydrogen and Raney nickel. Such functional groups are generally not affected by Wolff-Kishner conditions.

Wolff-Kishner reduction aids in alkylbenzene synthesis

We have already seen that the products of Friedel-Crafts alkanoylations may be converted into alkyl benzenes by using Clemmensen reduction. Wolff-Kishner deoxygenation also is frequently employed for this purpose and is particularly useful for acid-sensitive, base-stable substrates.

Wolff-Kishner Reduction of a Friedel-Crafts Alkanoylation Product

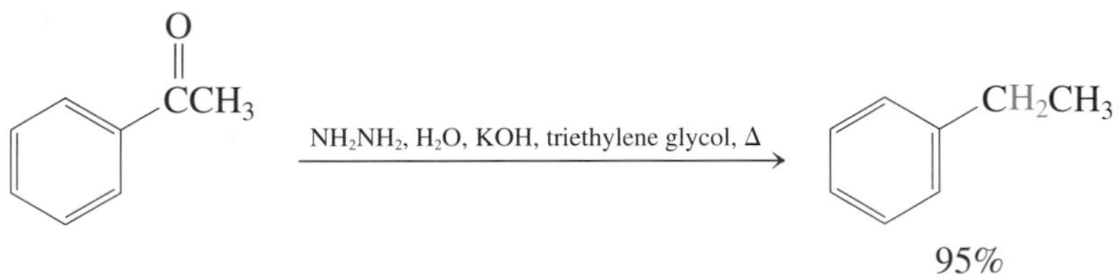

95%

EXERCISE 17-10

Propose a synthesis of hexylbenzene from hexanoic acid.

In summary, the Wolff-Kishner reduction is the decomposition of a hydrazone by base, the second part of a method of deoxygenating aldehydes and ketones. It complements Clemmensen and thioacetal desulfurization procedures.

17-11 Addition of Hydrogen Cyanide to Give Cyanohydrins

Besides alcohols and amines, several other nucleophilic reagents may attack the carbonyl group. Particularly important are carbon nucleophiles, because new carbon–carbon bonds can be made in this way. Section 8-8 explained that organometallic compounds, such as Grignard and alkyllithium reagents, add to aldehydes and ketones to produce alcohols. This section and Section 17-12 deal with the behavior of carbon nucleophiles that are not organometallic reagents—the additions of cyanide ion and of a new class of compounds called ylides.

Hydrogen cyanide adds reversibly to the carbonyl group to form hydroxy alkanenitrile adducts commonly called **cyanohydrins.** The equilibrium may be shifted toward the adduct by the use of liquid HCN as solvent. However, it is dangerous to use such large amounts of HCN, which is volatile and highly toxic. Alternatively, HCN may be generated in situ from a cyanide salt by the slow addition of an acid.

Cyanohydrin Formations

$$CH_3CH + HCN \rightleftharpoons CH_3CCN$$

70%
2-Hydroxypropanenitrile
(Acetaldehyde cyanohydrin)

$$+ \; Na^{+-}CN \xrightarrow[- \; NaCl]{Conc. \; HCl}$$

60%
1-Hydroxycyclohexanecarbonitrile
(Cyclohexanone cyanohydrin)

The mechanism of cyanohydrin formation begins with nucleophilic attack by cyanide ion and ends with protonation at oxygen.

Mechanism of Cyanohydrin Formation

$$:N \equiv C:^- \; + \; \overset{}{>}C = \overset{..}{O}: \rightleftharpoons \underset{NC}{\overset{|}{>}C - \overset{..}{O}:^-} \underset{- \; H^+}{\overset{+ \; H^+}{\rightleftharpoons}} \underset{NC}{\overset{|}{>}C - \overset{..}{O}H}$$

The reaction is readily reversed by the addition of base, which shifts the equilibrium to the free cyanide side by removing protons from the equation.

$$CH_3CCN \rightleftharpoons CH_3CH + HCN \xrightarrow[Shifts \; equilibrium]{HO^-} CH_3CH + HOH + {}^-CN$$

EXERCISE 17-11

Rank the following carbonyl compounds in order of thermodynamic favorability of HCN addition: propanone (acetone), formaldehyde, 3,3-dimethyl-2-butanone, acetaldehyde. (**Hint:** See Section 17-6.)

We shall see in subsequent chapters (Sections 19-6 and 26-2) that cyanohydrins are useful intermediates because the nitrile group can be modified by further reaction.

In summary, the carbonyl group in aldehydes and ketones can be attacked by carbon-based nucleophiles. Organometallic reagents give alcohols and cyanide gives cyanohydrins.

17-12 Addition of Phosphorus Ylides: The Wittig Reaction

Another useful reagent in nucleophilic additions contains a carbanion that is stabilized by an adjacent, positively charged phosphorus group. Such a species is called a **phosphorus ylide,** and its attack on aldehydes and ketones is called the **Wittig***

$$\overset{R}{\underset{H}{>}} \overset{..}{\overset{+}{C}} - \overset{+}{P}(C_6H_5)_3$$

Ylide

*Professor Georg Wittig (1897–1987), University of Heidelberg, Germany, Nobel Prize 1979 (chemistry).

reaction. The Wittig reaction is a powerful method for the selective synthesis of alkenes from aldehydes and ketones.

Deprotonation of phosphonium salts gives phosphorus ylides

Phosphorus ylides are most conveniently prepared from haloalkanes by a two-step sequence; the first step is the nucleophilic displacement of halide by triphenylphosphine to furnish an alkyltriphenylphosphonium salt.

Phosphonium Salt Synthesis

$$(C_6H_5)_3P: \ + \ \underset{R}{CH_2}\text{---}\ddot{\ddot{X}}: \ \xrightarrow{C_6H_6} \ RCH_2\overset{+}{P}(C_6H_5)_3 \ :\ddot{\ddot{X}}:^-$$

Triphenyl-   An alkyltriphenyl-
phosphine   phosphonium
              halide

The positively charged phosphorus atom renders any neighboring proton acidic. In the second step, deprotonation by bases, such as alkoxides, sodium hydride, or butyl-lithium, gives the ylide. Ylides can be isolated, although they are usually only generated in solutions that are subsequently treated with other reagents.

Ylide Formation

$$RCH_2\overset{+}{P}(C_6H_5)_3 \ X^- \ + \ CH_3CH_2CH_2CH_2Li \ \xrightarrow{THF} \ \left[\begin{array}{c} R\overset{\cdot\cdot}{\overline{CH}}\text{---}\overset{+}{P}(C_6H_5)_3 \\ \updownarrow \\ RCH{=}P(C_6H_5)_3 \end{array} \right] \ + \ CH_3CH_2CH_2CH_2H \ + \ LiX$$

Ylide

Notice that we may formulate a second resonance structure for the ylide by delocalizing the negative charge onto the phosphorus. In this form, the valence shell on phosphorus has been expanded and a carbon–phosphorus double bond is present.

The Wittig reaction forms carbon–carbon double bonds

When an ylide is exposed to an aldehyde or ketone, their reaction ultimately furnishes an alkene by coupling the ylide carbon with that of the carbonyl. The other product of this transformation is triphenylphosphine oxide.

The Wittig Reaction

$$\underset{\substack{\text{Aldehyde} \\ \text{or ketone}}}{\Big\rangle C{=}O} \ + \ \underset{\text{Ylide}}{(C_6H_5)_3P{=}C\Big\langle} \ \longrightarrow \ \underset{\text{Alkene}}{\Big\rangle C{=}C\Big\langle} \ + \ \underset{\text{Triphenylphosphine oxide}}{(C_6H_5)_3P{=}O}$$

$$\underset{\substack{O \\ \|}}{CH_3CH_2CH_2\overset{O}{\overset{\|}{C}}H} \ + \ CH_3CH_2\overset{CH_3}{\overset{|}{C}}{=}P(C_6H_5)_3 \ \xrightarrow[- (C_6H_5)_3PO]{(CH_3CH_2)_2O, \ 10°C} \ CH_3CH_2CH_2CH{=}\overset{CH_3}{\overset{|}{C}}CH_2CH_3$$

66%
3-Methyl-3-heptene

The Wittig reaction is a valuable addition to our synthetic arsenal because it forms carbon–carbon double bonds. In contrast with eliminations (Sections 11-8 and 11-9), it gives rise to alkenes in which the position of the newly formed double bond is unambiguous. Compare, for example, two syntheses of 2-ethyl-1-butene, one by the Wittig reaction, the other by elimination.

Comparison of Two Syntheses of 2-Ethyl-1-butene

By Wittig reaction

$$CH_3CH_2\overset{\overset{\displaystyle O}{\|}}{C}CH_2CH_3 \ + \ CH_2{=}P(C_6H_5)_3 \ \longrightarrow \ CH_3CH_2\overset{\overset{\displaystyle CH_2}{\|}}{C}CH_2CH_3 \ + \ (C_6H_5)_3P{=}O$$
(Only one isomer)

By elimination

$$CH_3CH_2\overset{\overset{\displaystyle CH_3}{|}}{\underset{\underset{\displaystyle Br}{|}}{C}}CH_2CH_3 \ \xrightarrow{\text{Base}} \ CH_3CH_2\overset{\overset{\displaystyle CH_2}{\|}}{C}CH_2CH_3 \ + \ CH_3CH_2\overset{\overset{\displaystyle CH_3}{|}}{C}{=}CHCH_3$$
(Mixture of isomers)

What is the mechanism of the Wittig reaction? The negatively polarized carbon in the ylide is nucleophilic and can attack the carbonyl group. The result is a **phosphorus betaine,*** a dipolar species of the kind called a *zwitterion* (*Zwitter,* German, hybrid). The betaine is short lived and rapidly forms a neutral **oxaphosphacyclobutane (oxaphosphetane),** characterized by a four-membered ring containing phosphorus and oxygen. This substance then decomposes to the product alkene and triphenylphosphine oxide. The driving force for the last step is the formation of the very strong phosphorus–oxygen double bond.

Mechanism of the Wittig Reaction

A phosphorus betaine

**An oxaphosphacyclobutane
(Oxaphosphetane)**

Wittig reactions can be carried out in the presence of ether, ester, halogen, alkene, and alkyne functions. However, they are only sometimes stereoselective, and mixtures of *Z* and *E* alkenes may form.

*Betaine is the name of an amino acid, $(CH_3)_3\overset{+}{N}CH_2COO^-$, which is found in beet sugar (*beta,* Latin, beet) and exists as a zwitterion.

CHEMICAL HIGHLIGHT 17-3 | **The Wittig Reaction in Synthesis**

$$CH_3CH_2CH_2C \equiv CCH_2Br \xrightarrow{P(C_6H_5)_3} CH_3CH_2CH_2C \equiv CCH_2\overset{+}{P}(C_6H_5)_3\ Br^- \xrightarrow{CH_3CH_2O^-Na^+}$$

$$CH_3CH_2CH_2C \equiv CCH = P(C_6H_5)_3 \xrightarrow{CH_3CH_2O_2C(CH_2)_8CHO} CH_3CH_2CH_2C \equiv CCH = CH(CH_2)_8CO_2CH_2CH_3$$

Cis and trans isomers

The Wittig reaction is employed extensively in total synthesis. For example, the total synthesis of the pheromone bombykol (see Section 12-16), the sex attractant of the silkworm moth, employs a Wittig reaction to construct the molecular backbone, as shown above.

The Wittig reaction has also found extensive application in industry. The chemical company

Badische Anilin und Soda Fabriken (BASF) in Germany synthesizes vitamin A_1 (Section 14-7) by a Wittig reaction in the crucial step. In this case, the reaction is stereoselective, giving only the trans alkene.

BASF Vitamin A_1 Synthesis

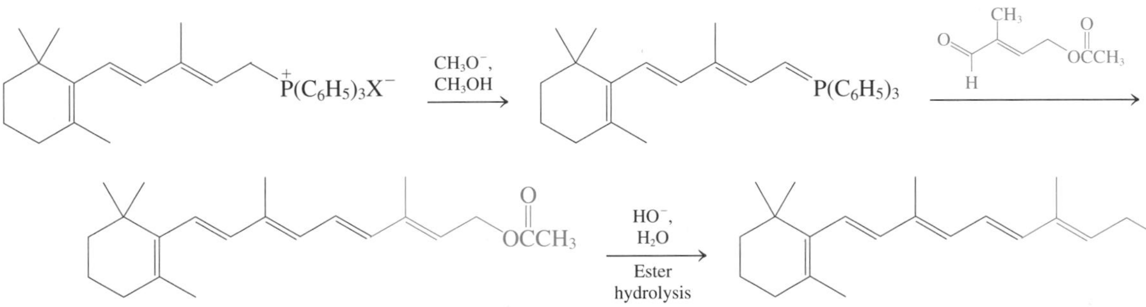

Vitamin A_1

$$CH_3CH_2CH_2CH = P(C_6H_5)_3\ +\ CH_3(CH_2)_4CHO \xrightarrow[-(C_6H_5)_3PO]{THF}$$

$$CH_3(CH_2)_2CH = CH(CH_2)_4CH_3$$
$$70\%$$
$$\textbf{(cis : trans ratio = 6 : 1)}$$

EXERCISE 17-12

Propose syntheses of 3-methylenecyclohexene from (**a**) 2-cyclohexenone and (**b**) 3-bromo-cyclohexene, using Wittig reactions.

EXERCISE 17-13

Propose a synthesis of the following dienone from the indicated starting materials. (**Hint:** Make use of a protecting group [Section 17-8].)

$$\overset{O}{\overset{\|}{CH_3CCH_2CH_2CH}} = CHCH - CH_2 \quad \text{from} \quad \overset{O}{\overset{\|}{CH_3CCH_2CH_2CH_2Br}} \quad \text{and} \quad \overset{O}{\overset{\|}{HCCH}} = CH_2$$

EXERCISE 17-14

Develop concise synthetic routes from starting material to product. You may use any material in addition to the given compound (more than one step will be required).

(a) ⬡ – – –> $CH_2\text{=}CH(CH_2)_4CH\text{=}CH_2$ (**Hint:** See Section 12-11.)

(b) – – –> (**Hint:** See Section 14-8.)

To summarize, phosphorus ylides add to aldehydes and ketones to give betaines that decompose by forming carbon–carbon double bonds. The Wittig reaction affords a means of synthesizing alkenes from carbonyl compounds and haloalkanes by way of the corresponding phosphonium salts.

17-13 Oxidation by Peroxycarboxylic Acids: The Baeyer-Villiger Oxidation

When ketones are treated with peroxycarboxylic acids (Section 12-9), the result is an oxidation of the carbonyl function to an ester, a transformation called **Baeyer-Villiger* oxidation.** The mechanism of this reaction starts by nucleophilic addition of the hydroperoxy end of the peracid to the carbonyl group to generate a reactive peroxide analog of a hemiacetal. This unstable adduct decomposes through a cyclic transition state in which an alkyl group shifts from the original carbonyl carbon to oxygen to give an ester.

Baeyer-Villiger Oxidation

$$CH_3\overset{O}{\overset{\|}{C}}CH_2CH_3 \xrightarrow{CF_3COOH, CH_2Cl_2} CH_3\overset{O}{\overset{\|}{C}}OCH_2CH_3$$
$$\text{72\%}$$

2-Butanone   **Ethyl acetate**

$$R''\overset{O}{\overset{\|}{C}}R' \;+\; R\overset{O}{\overset{\|}{C}}OOH \longrightarrow$$

Ketone   **Peroxycarboxylic acid**   **Ester**

Cyclic ketones are converted into cyclic esters. Attack is at the carbonyl rather than at the carbon–carbon double bond (Section 12-9). Unsymmetric ketones, such as that in the following reaction, can in principle lead to two different esters. Why is only one observed? The answer is that some substituents migrate more easily than others. Experiments have established their relative ease of migration, or **migratory aptitude.**

*Professor Johann Friedrich Wilhelm Adolf von Baeyer (1835–1917), University of Munich, Nobel Prize 1905 (chemistry); Victor Villiger (1868–1934), BASF, Ludwigshafen, Germany.

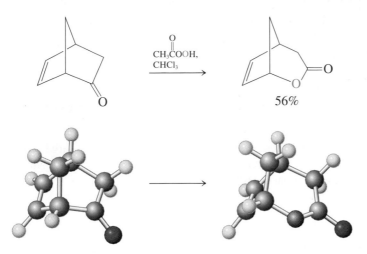

The ordering suggests that the migrating carbon possesses carbocationic character in the transition state for rearrangement.

Migratory Aptitudes in the Baeyer-Villiger Reaction

Methyl < primary < phenyl ~ secondary < cyclohexyl < tertiary

EXERCISE 17-15

Predict the outcome of the following oxidations with a peroxycarboxylic acid.

(a)   **(b)**   **(c)**

In summary, ketones can be oxidized with peroxycarboxylic acids to give esters; with unsymmetric ketones, the esters can be formed selectively by migration of only one of the substituents.

17-14 Oxidative Chemical Tests for Aldehydes

Although the advent of NMR and other spectroscopy has made chemical tests for functional groups a rarity, they are still used in special cases in which other analytical tests may fail. Two characteristic simple tests for aldehydes will again turn up in the discussion of sugar chemistry in Chapter 24; they make use of the ready oxidation of aldehydes to carboxylic acids. The first is **Fehling's* test,** in which cupric ion is the oxidant. In a basic medium, the precipitation of red cuprous oxide indicates the presence of an aldehyde function.

Fehling's Test

$$\overset{\overset{\text{O}}{\|}}{\text{RCH}} + \text{Cu}^{2+} \xrightarrow[\text{H}_2\text{O}]{\text{NaOH, tartrate (see Chemical Highlight 5-3),}} \underset{\textbf{Brick-red}}{\text{Cu}_2\text{O}} + \overset{\overset{\text{O}}{\|}}{\text{RCOH}}$$

*Professor Hermann C. von Fehling (1812–1885), Polytechnic School of Stuttgart, Germany.

The presence of an aldehyde group in a molecule is signified by the formation of a brick-red precipitate of copper(I) oxide in the Fehling test.

The Tollens test detects the presence of readily oxidized functional groups in organic molecules, such as aldehydes, by the rapid deposition of a silver mirror on the glass walls of the tube.

The second is **Tollens's* test,** in which a solution of silver ion precipitates a silver mirror on exposure to an aldehyde.

Tollens's Test

$$\underset{\text{RCH}}{\overset{O}{\parallel}} \;+\; Ag^+ \;\xrightarrow{NH_3, H_2O}\; Ag \;+\; \underset{\text{RCOH}}{\overset{O}{\parallel}}$$

Mirror

The Fehling and Tollens tests are not commonly used in large-scale syntheses. However, the Tollens reaction is employed industrially to produce shiny silver mirrors on glass surfaces, such as the insides of Thermos bottles.

CHAPTER INTEGRATION PROBLEM

a. Propose a synthesis of norethynodrel, the major component in one common oral contraceptive preparation (Enovid, compare Chemical Highlight 4-2). Use as your starting compound the following nortestosterone derivative (left-hand structure).

Starting compound

Norethynodrel

SOLUTION

Analyze the problem retrosynthetically (Section 8-9). The target molecule is identical with the starting compound in all but one respect: the additional substituent on the cyclopentane ring (at C17). We know that we cannot add groups directly to secondary alcohols to make tertiary alcohols. However, addition reactions of organometallic reagents to *ketones* afford tertiary alcohols (Section 8-8). The necessary organometallic reagent for our purpose is $Li^+ \; {}^-C\equiv CH$, an alkynyl anion (Section 13-6). Let us see if we can devise a plan based on this chemistry.

The final step in the synthesis would appear to be addition of $LiC\equiv CH$ to a C17 ketone, but, as the following reaction shows, there is a major problem: The precursor molecule contains a *second carbonyl group* in the six-membered ring at the lower left (at C3). In such a process, it would be impossible to prevent the alkynyl anion from adding indiscriminately to C3 as well as C17.

1. $LiC\equiv CH$
2. H^+, H_2O

*Professor Bernhard C. G. Tollens (1841–1918), University of Göttingen, Germany.

What to do? A similar problem arose in the synthesis of testosterone described in Chemical Highlight 9-2. It was solved by the use of protecting groups. We can do something similar here. The underlying conceptual problem is that *we need to avoid having both C3 and C17 exist as carbonyls at the same time in **any** molecule in our synthesis.* Failure to do so will doom a synthetic plan. With this concept in mind, we can imagine a modified ending to our synthesis, in which the carbon–oxygen double bond at C3 is protected in a form unreactive toward organometallic reagents during the previously shown step. After the addition reaction has been completed, the protecting group may be removed. An acetal derived from 1,2-ethanediol should do nicely (Section 17-8):

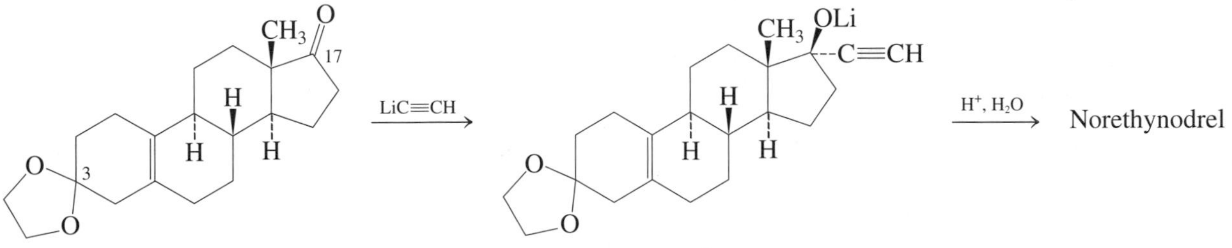

In this scheme, the aqueous acid does double duty: It protonates the alkoxide that results from alkynyllithium addition, and it hydrolyzes the acetal to a ketone.

How does the synthesis begin? The C17 hydroxy group in the original starting material must be oxidized to a carbonyl. However, we cannot perform that oxidation in the presence of the carbonyl function at C3, because we would then violate the aforedescribed principle: We would create a molecule with both carbonyls present at the same time and the need to later react one and not the other with a reagent that cannot discriminate between them. With these considerations in mind, it becomes clear that protection of C3 must come first, before oxidation at C17:

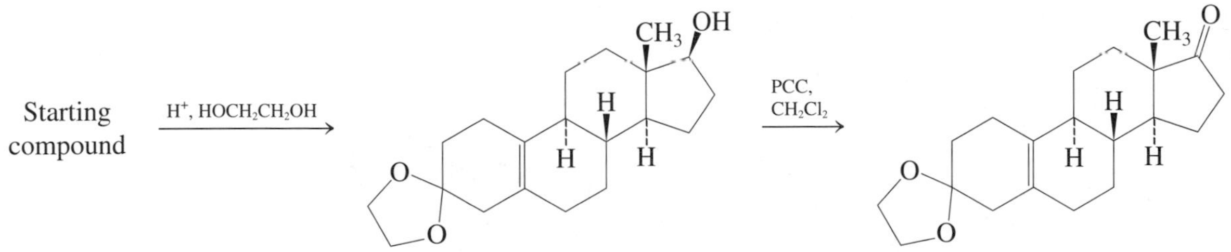

b. Why is pyridinium chlorochromate (PCC) in CH_2Cl_2 a better choice than $K_2Cr_2O_7$ for an oxidant in the second step?

SOLUTION

What could go wrong with $K_2Cr_2O_7$? Consider the conditions under which it is used: typically in aqueous sulfuric acid (Section 8-6). Therefore, this reagent carries the risk of hydrolysis of the acetal group under the acidic conditions. PCC is a neutral, nonaqueous reagent, ideal for substrates containing acid-sensitive functions.

c. Addition of alkynyllithium to the carbonyl at C17 is stereoselective to give the tertiary alcohol shown. Why? Would addition to a carbonyl at C3 also be stereoselective?

SOLUTION

What is the usual origin of stereoselectivity in reactions such as this? Steric hindrance in the immediate vicinity of the reacting center is the most common cause. The methyl substituent at the position adjacent to C17 is axial with respect to the six-membered ring and sterically hinders approach of the alkynyl reagent to the top face of the C17 carbonyl carbon (see Section 4-7). Addition from below is therefore strongly favored. There is no comparable steric hindrance in the vicinity of C3 to give rise to any preference for addition from either side.

NEW REACTIONS

Synthesis of Aldehydes and Ketones

1. Oxidation of Alcohols (Section 17-4)

$$\underset{\text{RCHOH}}{\overset{\text{R}'}{|}} \xrightarrow{\text{CrO}_3,\ \text{H}_2\text{SO}_4} \underset{\text{RCR}'}{\overset{\text{O}}{\|}} \qquad \text{RCH}_2\text{OH} \xrightarrow{\text{PCC, CH}_2\text{Cl}_2} \underset{\text{RCH}}{\overset{\text{O}}{\|}}$$

Stable to oxidizing agent

Allylic oxidation

$$\underset{\text{C}=\text{C}}{\overset{\overset{\text{H}\quad\text{OH}}{\diagdown\ /}}{\underset{\diagup}{\text{C}}}} \xrightarrow{\text{MnO}_2,\ \text{CHCl}_3} \underset{\text{C}=\text{C}}{\overset{\overset{\text{O}}{\|}}{\underset{\diagup}{\text{C}}}}$$

2. Ozonolysis of Alkenes (Section 17-4)

$$\overset{\diagdown}{\underset{\diagup}{\text{C}}}=\overset{\diagup}{\underset{\diagdown}{\text{C}}} \xrightarrow[\text{2. Zn, CH}_3\text{CO}_2\text{H}]{\text{1. O}_3,\ \text{CH}_2\text{Cl}_2} \overset{\diagdown}{\underset{\diagup}{\text{C}}}{=}\text{O} \ + \ \text{O}{=}\overset{\diagup}{\underset{\diagdown}{\text{C}}}$$

3. Hydration of Alkynes (Section 17-4)

$$\text{RC}\equiv\text{CH} \xrightarrow{\text{H}_2\text{O, Hg}^{2+},\ \text{H}_2\text{SO}_4} \underset{\text{RCCH}_3}{\overset{\text{O}}{\|}}$$

4. Friedel-Crafts Alkanoylation (Section 17-4)

$$\text{C}_6\text{H}_6 \ + \ \underset{\text{RCCl}}{\overset{\text{O}}{\|}} \xrightarrow[\text{2. H}^+,\ \text{H}_2\text{O}]{\text{1. AlCl}_3} \underset{\text{C}_6\text{H}_5\text{CR}}{\overset{\text{O}}{\|}} \ + \ \text{HCl}$$

Reactions of Aldehydes and Ketones

5. Hydrogenation (Section 17-5)

$$\underset{\text{RCR}'}{\overset{\text{O}}{\|}} \xrightarrow{\text{H}_2,\ \text{Raney Ni, pressure}} \underset{\underset{\text{H}}{\overset{|}{\text{RCR}'}}}{\overset{\text{OH}}{\overset{|}{}}}$$

Selectivity

$$\text{RCH}{=}\text{CHCH}_2\text{CH}_2\underset{\overset{\|}{\text{CR}'}}{\overset{\text{O}}{}} \xrightarrow{\text{H}_2,\ \text{Pt}} \text{R(CH}_2)_4\underset{\overset{\|}{\text{CR}'}}{\overset{\text{O}}{}}$$

6. Reduction by Hydrides (Section 17-5)

$$\underset{\text{RCH}}{\overset{\text{O}}{\|}} \xrightarrow{\text{NaBH}_4,\ \text{CH}_3\text{CH}_2\text{OH}} \text{RCH}_2\text{OH} \qquad \underset{\text{RCR}'}{\overset{\text{O}}{\|}} \xrightarrow[\text{2. H}^+,\ \text{H}_2\text{O}]{\text{1. LiAlH}_4,\ (\text{CH}_3\text{CH}_2)_2\text{O}} \underset{\text{H}}{\overset{\text{OH}}{\underset{|}{\text{RCR}'}}}$$

Selectivity

$$\underset{\text{RCH}=\text{CHCR}'}{\overset{\text{O}}{\|}} \xrightarrow[\text{2. H}^+,\ \text{H}_2\text{O}]{\text{1. LiAlH}_4,\ (\text{CH}_3\text{CH}_2)_2\text{O}} \underset{\text{H}}{\overset{\text{OH}}{\underset{|}{\text{RCH}=\text{CHCR}'}}}$$

7. Addition of Organometallic Compounds (Section 17-5)

$$\text{RLi or RMgX} \quad + \quad \underset{\textbf{Formaldehyde}}{\text{CH}_2=\text{O}} \xrightarrow{\text{THF}} \underset{\textbf{Primary alcohol}}{\text{RCH}_2\text{OH}}$$

$$\text{RLi or RMgX} \quad + \quad \underset{\textbf{Aldehyde}}{\overset{\text{O}}{\underset{}{\overset{\|}{\text{R}'\text{CH}}}}} \xrightarrow{\text{THF}} \underset{\textbf{Secondary alcohol}}{\overset{\text{OH}}{\underset{\text{H}}{\underset{|}{\text{RCR}'}}}}$$

$$\text{RLi or RMgX} \quad + \quad \underset{\textbf{Ketone}}{\overset{\text{O}}{\overset{\|}{\text{R}'\text{CR}''}}} \xrightarrow{\text{THF}} \underset{\textbf{Tertiary alcohol}}{\overset{\text{OH}}{\underset{\text{R}''}{\underset{|}{\text{RCR}'}}}}$$

8. Addition of Water and Alcohols—Hemiacetals (Sections 17-6 and 17-7)

$$\underset{\text{RCR}'}{\overset{\text{O}}{\|}} \underset{\text{H}_2\text{O, H}^+ \text{ or HO}^-}{\rightleftharpoons} \underset{\underset{\textbf{(A geminal diol)}}{\textbf{Carbonyl hydrate}}}{\overset{\text{OH}}{\underset{\text{OH}}{\underset{|}{\text{RCR}'}}}} \qquad \underset{\text{RCR}'}{\overset{\text{O}}{\|}} \underset{\text{R}''\text{OH, H}^+ \text{ or HO}^-}{\rightleftharpoons} \underset{\textbf{Hemiacetal}}{\overset{\text{OH}}{\underset{\text{OR}''}{\underset{|}{\text{RCR}'}}}} \qquad K_{\text{eq}}:\ \underset{}{\overset{\text{O}}{\overset{\|}{\text{R}-\text{C}-\text{R}}}} < \underset{}{\overset{\text{O}}{\overset{\|}{\text{R}-\text{C}-\text{H}}}} < \text{H}_2\text{C}=\text{O}$$

Intramolecular addition

$$\text{HO}\diagdown\diagup\diagdown\underset{\text{R}}{\overset{\text{O}}{\overset{\|}{\diagdown}}} \xrightarrow{\text{H}^+ \text{ or HO}^-} \underset{\textbf{Cyclic hemiacetal}}{\overset{\text{HO}\quad\text{R}}{\bigcirc\text{O}}}$$

9. Acid-Catalyzed Addition of Alcohols—Acetals (Sections 17-7 and 17-8)

$$\underset{}{\overset{\text{O}}{\overset{\|}{\text{RCR}'}}} \quad + \quad 2\,\text{R}''\text{OH} \underset{}{\overset{\text{H}^+}{\rightleftharpoons}} \underset{\underset{\textbf{Acetal}}{\text{OR}''}}{\overset{\text{OR}''}{\underset{|}{\text{RCR}'}}} \quad + \quad \text{H}_2\text{O}$$

Cyclic acetals

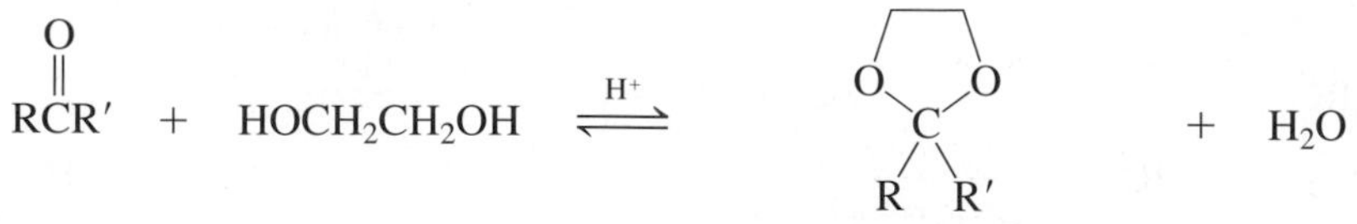

$$RCR' \ + \ HOCH_2CH_2OH \ \underset{}{\overset{H^+}{\rightleftharpoons}} \ \ + \ H_2O$$

Ketone

**Ketone, protected as
cyclic acetal**
(Stable to base, LiAlH$_4$, RMgX)

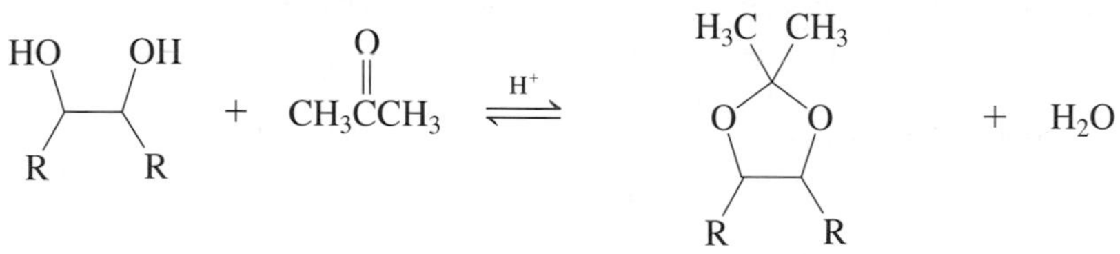

$$\text{Diol} \ + \ CH_3CCH_3 \ \underset{}{\overset{H^+}{\rightleftharpoons}} \ \ + \ H_2O$$

Diol

**Diol, protected as
propanone (acetone) acetal**

10. Thioacetals (Section 17-8)

Formation

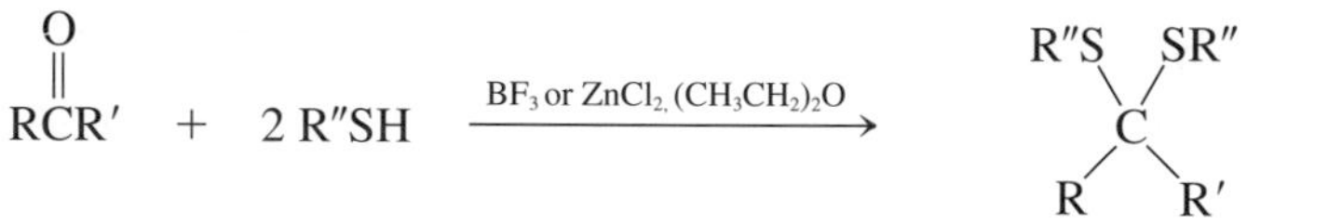

$$RCR' \ + \ 2\ R''SH \ \xrightarrow{\ BF_3 \text{ or } ZnCl_2, \ (CH_3CH_2)_2O\ }$$

**(Stable to aqueous acid,
base, LiAlH$_4$, RMgX)**

Hydrolysis

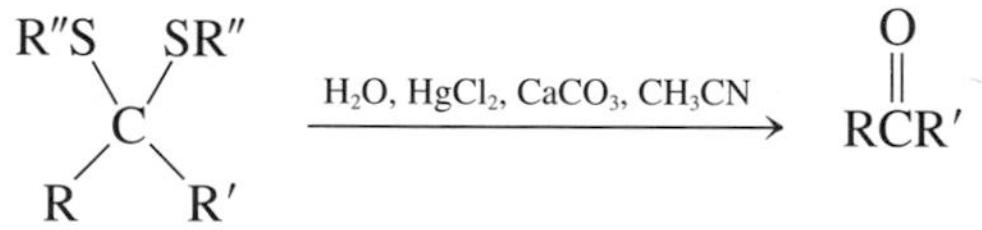

$$\xrightarrow{\ H_2O,\ HgCl_2,\ CaCO_3,\ CH_3CN\ } \ RCR'$$

11. Raney Nickel Desulfurization (Section 17-8)

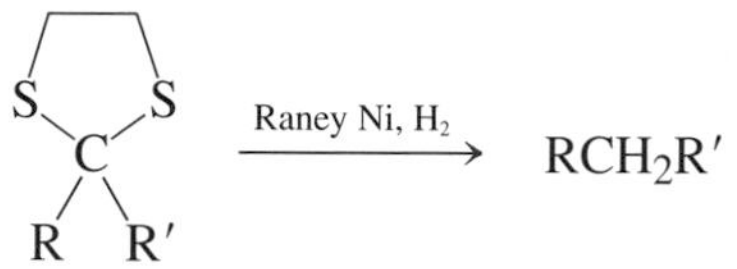

$$\xrightarrow{\text{Raney Ni, } H_2} \ RCH_2R'$$

12. Addition of Amine Derivatives (Section 17-9)

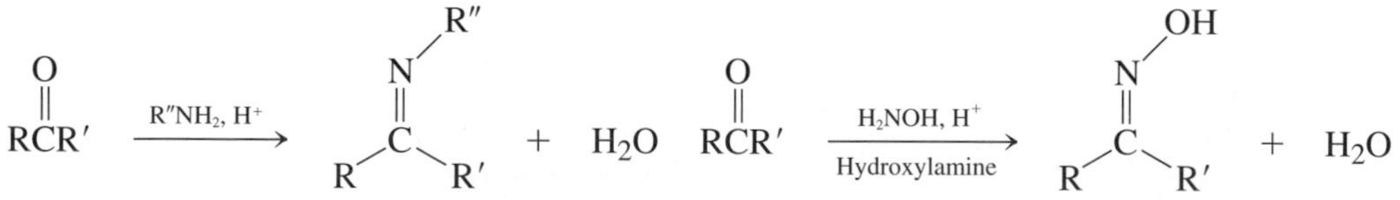

$$RCR' \ \xrightarrow{R''NH_2,\ H^+} \ \ + \ H_2O$$

Imine

$$RCR' \ \xrightarrow[\text{Hydroxylamine}]{H_2NOH,\ H^+} \ \ + \ H_2O$$

Oxime

$$RCR' \ \xrightarrow[\substack{\text{2,4-Dinitrophenyl-}\\\text{hydrazine}}]{} \ \ + \ H_2O$$

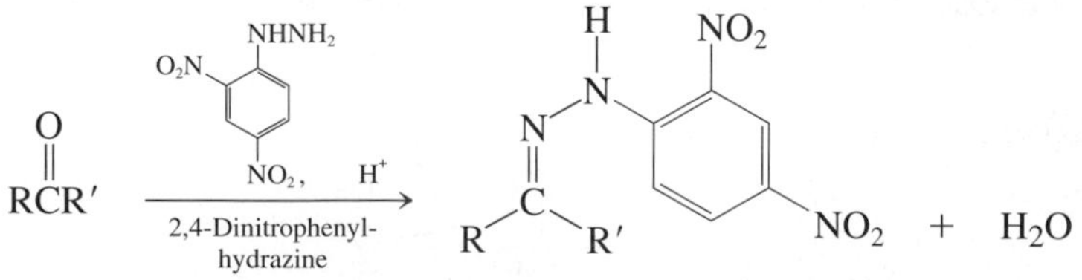

2,4-Dinitrophenylhydrazone

$$\underset{\text{RCR'}}{\overset{\text{O}}{\|}} \xrightarrow[\text{Semicarbazide}]{\text{H}_2\text{NNHCNH}_2, \text{ H}^+} \quad \underset{\text{R}\quad\text{R'}}{\overset{\text{NHCNH}_2}{\underset{\text{C}}{\overset{\text{N}}{\|}}}} \quad + \quad \text{H}_2\text{O}$$

Semicarbazone

13. Enamines (Section 17-9)

$$\underset{\text{RCH}_2\text{CR'}}{\overset{\text{O}}{\|}} + \underset{\substack{\text{R''}\\\text{R'''}}}{\overset{\text{R''}}{\diagup}}\text{NH} \rightleftharpoons \underset{\text{R'}}{\overset{\text{R'''}}{\text{RCH}=\text{C}}}\overset{\text{N}-\text{R''}}{\diagup} + \text{H}_2\text{O}$$

Secondary amine   **Enamine**

14. Wolff-Kishner Reduction (Section 17-10)

$$\underset{\text{RCR'}}{\overset{\text{O}}{\|}} \xrightarrow{\text{H}_2\text{NNH}_2, \text{H}_2\text{O}, \text{HO}^-, \Delta} \text{RCH}_2\text{R'}$$

15. Cyanohydrins (Section 17-11)

$$\underset{\text{RCR'}}{\overset{\text{O}}{\|}} + \text{HCN} \rightleftharpoons \underset{\text{R}\quad\text{R'}}{\overset{\text{HO}\quad\text{CN}}{\underset{\text{C}}{\diagdown\diagup}}}$$

Cyanohydrin

16. Wittig Reaction (Section 17-12)

$$\text{R''CH}_2\text{X} + \text{P}(\text{C}_6\text{H}_5)_3 \xrightarrow{\text{C}_6\text{H}_6} \text{R''CH}_2\overset{+}{\text{P}}(\text{C}_6\text{H}_5)_3 \ \text{X}^-$$

Triphenylphosphine   **Phosphonium halide**

Works with primary or secondary haloalkanes

$$\text{R''CH}_2\overset{+}{\text{P}}(\text{C}_6\text{H}_5)_3 \ \text{X}^- \xrightarrow{\text{Base}} \text{R''CH}=\text{P}(\text{C}_6\text{H}_5)_3$$

Ylide

$$\underset{\text{RCR'}}{\overset{\text{O}}{\|}} + \text{R''CH}=\text{P}(\text{C}_6\text{H}_5)_3 \xrightarrow{\text{THF}} \underset{\text{R'}}{\overset{\text{R}}{\diagup}}\text{C}=\text{CHR''} + (\text{C}_6\text{H}_5)_3\text{P}=\text{O}$$

(Not always stereoselective)

17. Baeyer-Villiger Oxidation (Section 17-13)

$$\underset{\text{RCR'}}{\overset{\text{O}}{\|}} + \underset{\text{R''COOH}}{\overset{\text{O}}{\|}} \xrightarrow{\text{CH}_2\text{Cl}_2} \underset{\text{RCOR'}}{\overset{\text{O}}{\|}} + \underset{\text{R''COH}}{\overset{\text{O}}{\|}}$$

Ketone   **Ester**

Migratory aptitudes in Baeyer-Villiger oxidation

Methyl < primary < phenyl ~ secondary < cyclohexyl < tertiary

18. Test for Aldehydes (Section 17-14)

$$\underset{\text{RCH}}{\overset{\text{O}}{\|}} \xrightarrow{\text{Cu}^{2+} \text{ or Ag}^+} \underset{\text{RCOH}}{\overset{\text{O}}{\|}} + \text{Cu}_2\text{O} \quad \text{or} \quad \text{Ag}$$

Red precipitate   **Mirror**

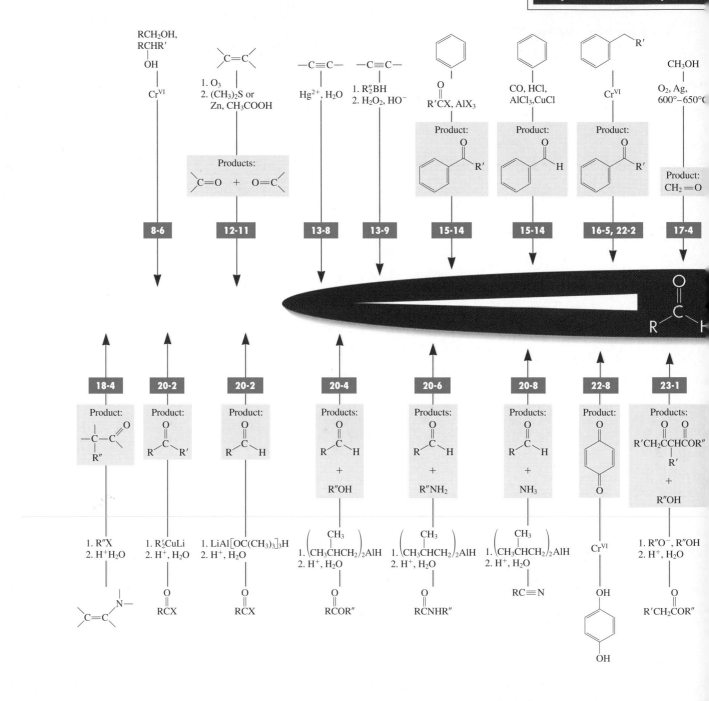

RCH₂OH,
RCHR′
|
OH

Cr^{VI}

8-6

1. O₃
2. (CH₃)₂S or
   Zn, CH₃COOH

Products:

12-11

Hg²⁺, H₂O

13-8

1. R″₂BH
2. H₂O₂, HO⁻

13-9

O
||
R′CX, AlX₃

Product:

15-14

CO, HCl,
AlCl₃,CuCl

Product:

15-14

Cr^{VI}

Product:

16-5, 22-2

CH₃OH

O₂, Ag,
600°–650°C

Product:
CH₂=O

17-4

18-4

Product:

1. R″X
2. H⁺H₂O

20-2

Product:
O
||
R—C—R′

1. R′₂CuLi
2. H⁺, H₂O

O
||
RCX

20-2

Product:
O
||
R—C—H

1. LiAl[OC(CH₃)₃]₃H
2. H⁺, H₂O

O
||
RCX

20-4

Products:
O
||
R—C—H
+
R″OH

1. (CH₃CHCH₂)₂AlH
     |
     CH₃
2. H⁺, H₂O

O
||
RCOR″

20-6

Products:
O
||
R—C—H
+
R″NH₂

1. (CH₃CHCH₂)₂AlH
     |
     CH₃
2. H⁺, H₂O

O
||
RCNHR″

20-8

Products:
O
||
R—C—H
+
NH₃

1. (CH₃CHCH₂)₂AlH
     |
     CH₃
2. H⁺, H₂O

RC≡N

22-8

Product:
O

Cr^{VI}

OH

OH

23-1

Products:
O   O
|| ||
R′CH₂CCHCOR″
        |
        R′
+
R″OH

1. R″O⁻, R″OH
2. H⁺, H₂O

O
||
R′CH₂COR″

768

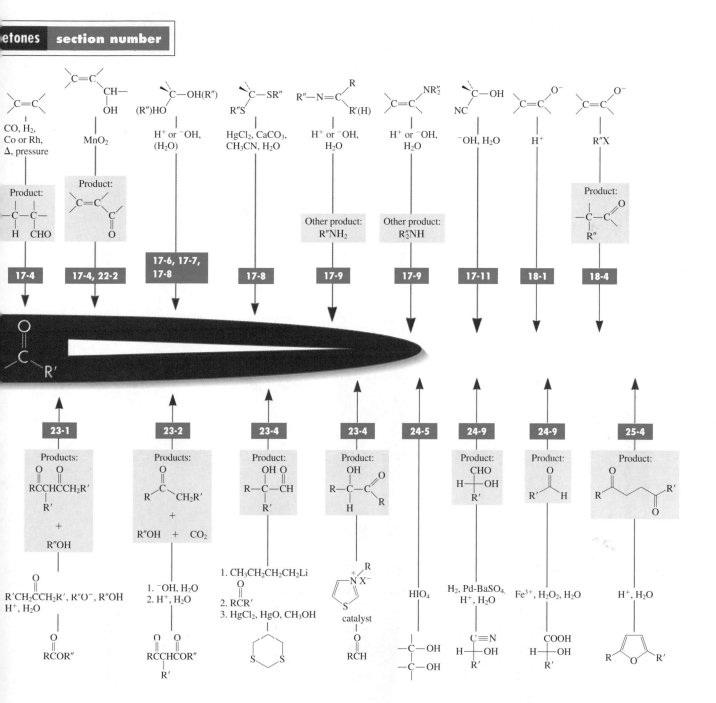

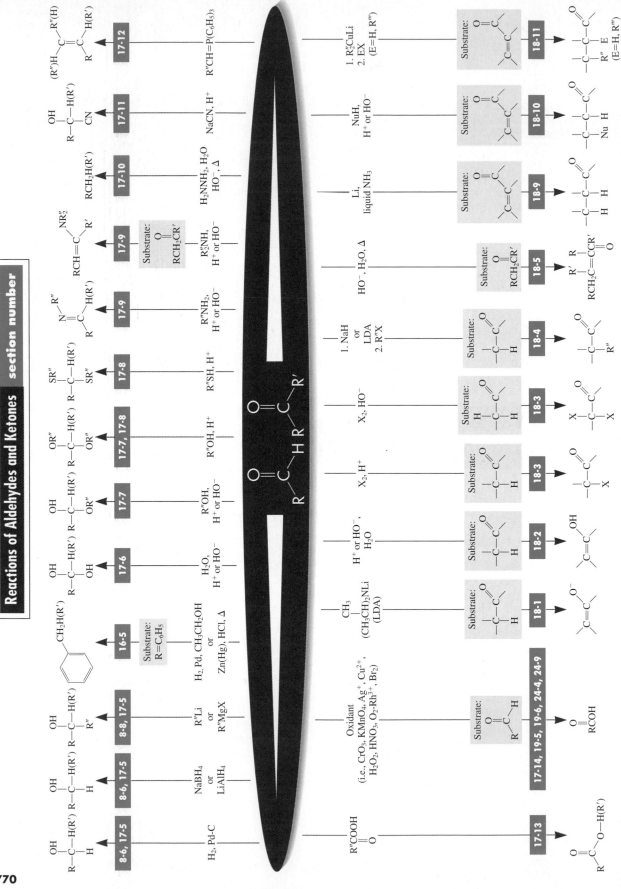

Reactions of Aldehydes and Ketones section number

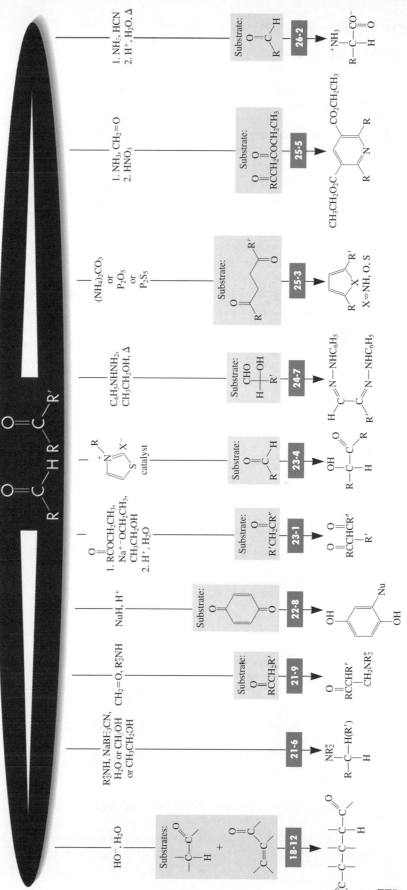

771

IMPORTANT CONCEPTS

1. The **carbonyl group** is the functional group of the aldehydes **(alkanals)** and **ketones (alkanones).** It has precedence over the hydroxy, alkenyl, and alkynyl groups in the naming of molecules.

2. The carbon–oxygen double bond and its two attached nuclei in aldehydes and ketones form a plane. The C=O unit is **polarized,** with a partial negative charge on oxygen and a partial positive charge on carbon.

3. The ^{1}H NMR **spectra** of aldehydes exhibit a peak at $\delta \sim 9.8$ ppm. The carbonyl carbon absorbs at ~ 200 ppm. Aldehydes and ketones exhibit strong infrared bands in the region 1690–1750 cm^{-1}; this absorption is due to the stretching of the C=O bond. Because of the availability of low-energy $n \rightarrow \pi^*$ transitions, the electronic spectra of aldehydes and ketones have relatively long wavelength bands.

4. The carbon–oxygen double bond undergoes **catalytic hydrogenation** and **ionic additions.** The catalysts for the former are heterogeneous transition metal surfaces; for the latter, acid or base.

5. The reactivity of the carbonyl group increases with increasing **electrophilic character** of the carbonyl carbon. Therefore, aldehydes are more reactive than ketones.

6. Primary amines undergo **condensation** reactions with aldehydes and ketones to imines; secondary amines condense to enamines.

7. The combination of Friedel-Crafts alkanoylation and Wolff-Kishner or Clemmensen reduction allows synthesis of alkylbenzenes free of the limitations of Friedel-Crafts alkylation.

8. The **Wittig reaction** is an important carbon–carbon bond-forming reaction that produces alkenes directly from aldehydes and ketones.

9. The reaction of **peroxycarboxylic acids** with the carbonyl group of ketones produces **esters.**

PROBLEMS

16. Draw structures and provide IUPAC names for each of the following compounds. **(a)** Methyl ethyl ketone; **(b)** ethyl isobutyl ketone; **(c)** methyl *tert*-butyl ketone; **(d)** diisopropyl ketone; **(e)** acetophenone; **(f)** *m*-nitroacetophenone.

17. Name or draw the structure of each of the following compounds.

(a) $(CH_3)_2CHCCH(CH_3)_2$ **(b)** **(c)**

(d) **(e)** **(f)**

(g) (*Z*)-2-Acetyl-2-butenal   **(h)** *trans*-3-Chlorocyclobutanecarbaldehyde

18. The following spectroscopic data are for two carbonyl compounds with the formula $C_8H_{12}O$. Suggest a structure for each compound. The letter "m" stands for the appearance of this particular part of the spectrum as an uninterpretable

multiplet. **(a)** ¹H NMR: δ = 1.6 (m, 4 H), 2.15 (s, 3 H), 2.19 (m, 4 H), and 6.78 (t, 1 H) ppm. ¹³C NMR: δ = 21.8, 22.2, 23.2, 25.0, 26.2, 139.8, 140.7, and 198.6 ppm. **(b)** ¹H NMR: δ = 0.94 (t, 3 H), 1.48 (sex, 2 H), 2.21 (q, 2 H), 5.8–7.1 (m, 4 H), and 9.56 (d, 1 H) ppm. ¹³C NMR: δ = 13.6, 21.9, 35.2, 129.0, 135.2, 146.7, 152.5, and 193.2 ppm.

19. The compounds described in Problem 18 have very different ultraviolet spectra. One has $\lambda_{max}(\epsilon)$ = 232(13,000) and 308(1450) nm, whereas the other has $\lambda_{max}(\epsilon)$ = 272(35,000) nm and a weaker absorption near 320 nm (this value is hard to determine accurately because of the intensity of the stronger absorption). Match the structures that you determined in Problem 18 to these UV spectral data. Explain the spectra in terms of the structures.

20. Indicate which reagent or combination of reagents is best suited for each of the following reactions.

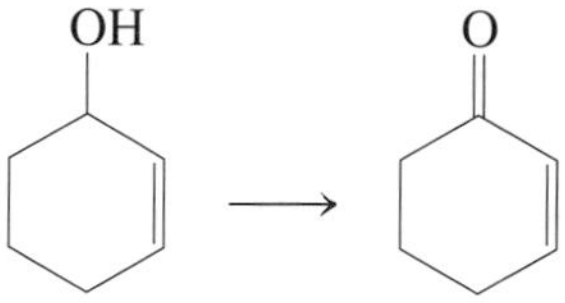

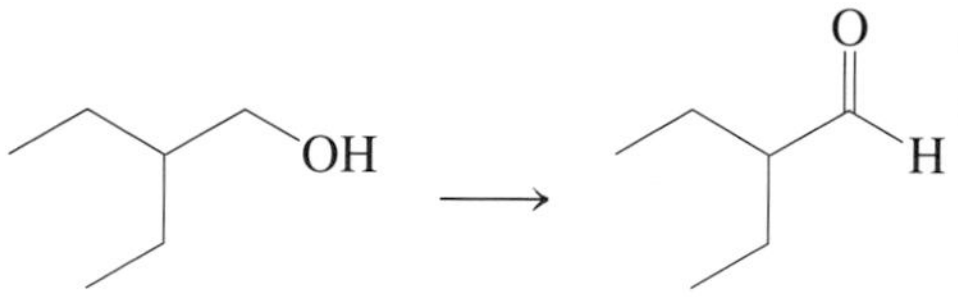

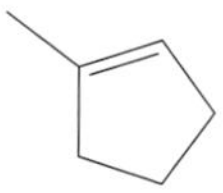

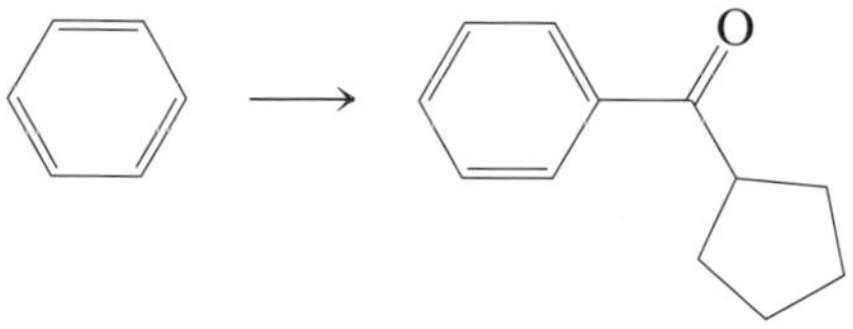

21. Write the expected products of ozonolysis (followed by mild reduction—e.g., by Zn) of each of the following molecules.

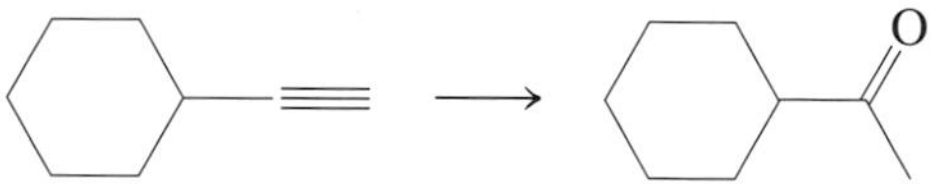

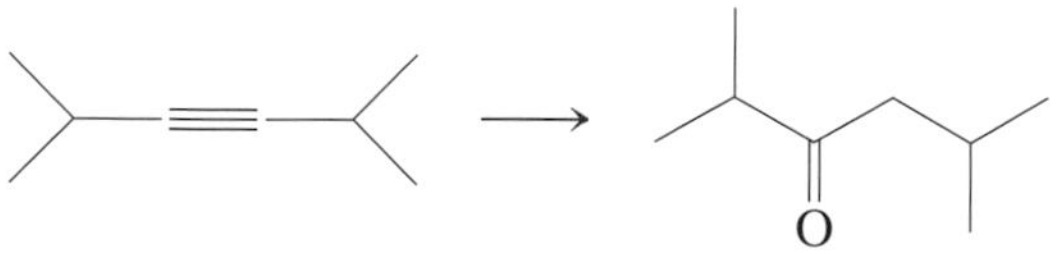

(a) $CH_3CH_2CH_2CH=CH_2$ **(b)** 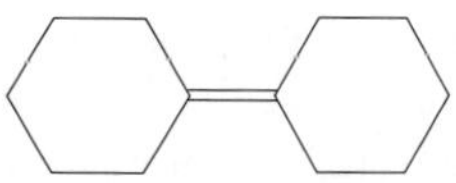 **(c)** 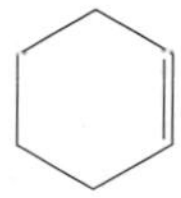 **(d)**

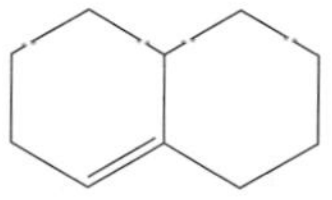

22. For each of the following groups, rank the molecules in decreasing order of reactivity toward addition of a nucleophile to the most electrophilic sp^2-hybridized carbon.

(a) $(CH_3)_2C=O$, $(CH_3)_2C=NH$, $(CH_3)_2C=\overset{+}{O}H$ **(b)** $CH_3\overset{\overset{O}{\|}}{C}CH_3$, $CH_3\overset{\overset{O\ O}{\|\ \|}}{C}CCH_3$, $CH_3\overset{\overset{O\ O\ O}{\|\ \|\ \|}}{C}CCCH_3$

(c) $BrCH_2COCH_3$, CH_3COCH_3, CH_3CHO, $BrCH_2CHO$

23. Give the expected products of reaction of butanal with each of the following reagents.
   (a) H_2 (1 equivalent), Pd, CH_3CH_2OH
   (b) $LiAlH_4$, $(CH_3CH_2)_2O$, then H^+, H_2O
   (c) CH_3CH_2MgBr, $(CH_3CH_2)_2O$, then H^+, H_2O

24. Give the expected products of reaction of 2-pentanone with each of the reagents in Problem 23.

25. Give the expected products of reaction of 4-acetylcyclohexene with each of the reagents in Problem 23.

26. Give the expected product(s) of each of the following reactions.

(a) 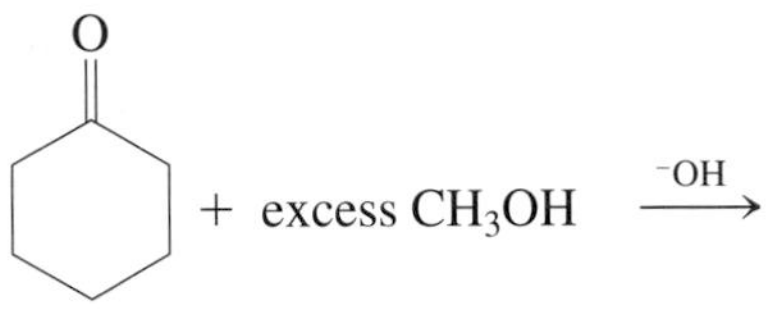 + excess CH_3OH $\xrightarrow{^-OH}$

(b) 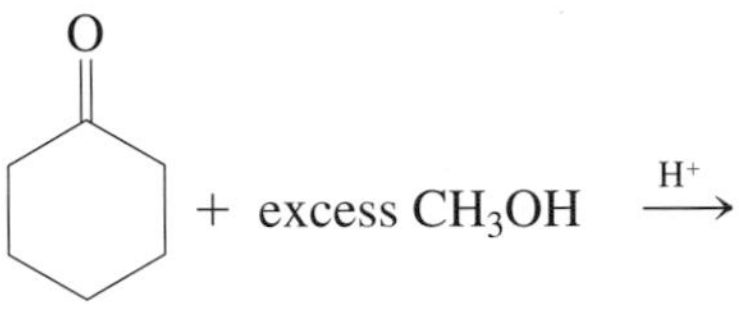 + excess CH_3OH $\xrightarrow{H^+}$

(c) 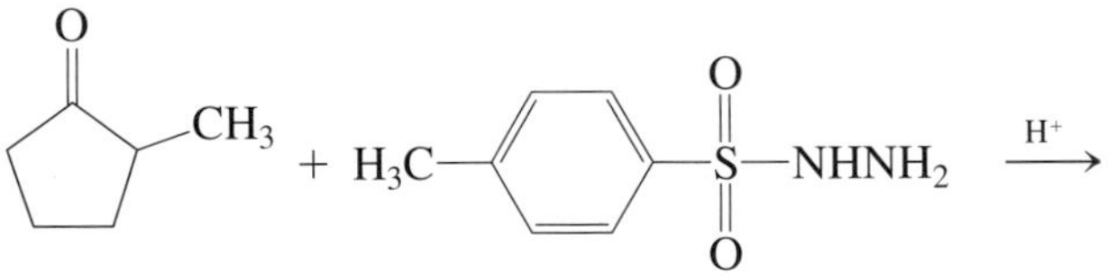 + H_3C- $-S-NHNH_2$ $\xrightarrow{H^+}$

(d) $CH_3\overset{O}{\overset{\|}{C}}CH_3$ + $HOCH_2\overset{OH}{\overset{|}{C}H}CH_2CH_2CH_3$ $\xrightarrow{H^+}$

(e) 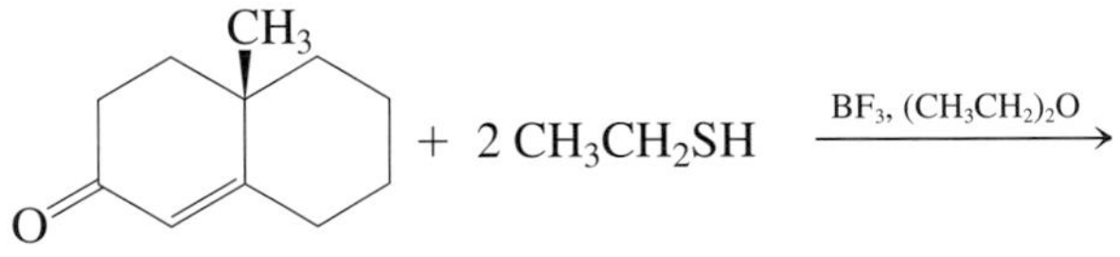 + $2\ CH_3CH_2SH$ $\xrightarrow{BF_3,\ (CH_3CH_2)_2O}$

(f) 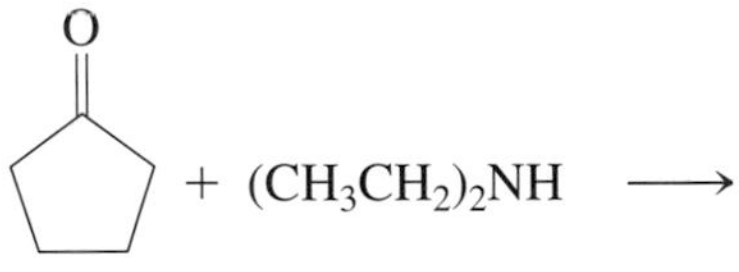 + $(CH_3CH_2)_2NH$ $\longrightarrow$

27. Formulate detailed mechanisms for **(a)** the formation of the hemiacetal of acetaldehyde and methanol under both acid- and base-catalyzed conditions and **(b)** the formation of the intramolecular hemiacetal of 5-hydroxypentanal (Section 17-7), again under both acid- and base-catalyzed conditions.

28. Formulate the mechanism of the BF_3-catalyzed reaction of CH_3SH with butanal (Section 17-8).

29. Overoxidation of primary alcohols to carboxylic acids is caused by the water present in the usual aqueous acidic Cr(VI) reagents. The water adds to the initial aldehyde product to form a hydrate, which is further oxidized (Section 17-6). In view of these facts, explain the following two observations. **(a)** Water adds to ketones to form hydrates, but no overoxidation follows the conversion of a secondary alcohol into a ketone. **(b)** Successful oxidation of primary alcohols to aldehydes by the water-free PCC reagent requires that the alcohol be added slowly to the Cr(VI) reagent. If, instead, the PCC is added *to the alcohol,* a new side reaction forms an ester. This is illustrated for 1-butanol.

$$CH_3CH_2CH_2CH_2OH \xrightarrow{PCC,\ CH_2Cl_2} CH_3CH_2CH_2\overset{O}{\overset{\|}{C}}OCH_2CH_2CH_2CH_3$$

(c) Give the products expected from reaction of 3-phenyl-1-propanol and water-free CrO_3 (1) when the alcohol is added to the oxidizing agent and (2) when the oxidant is added to the alcohol.

30. Explain the results of the following reactions by means of mechanisms.

(a)

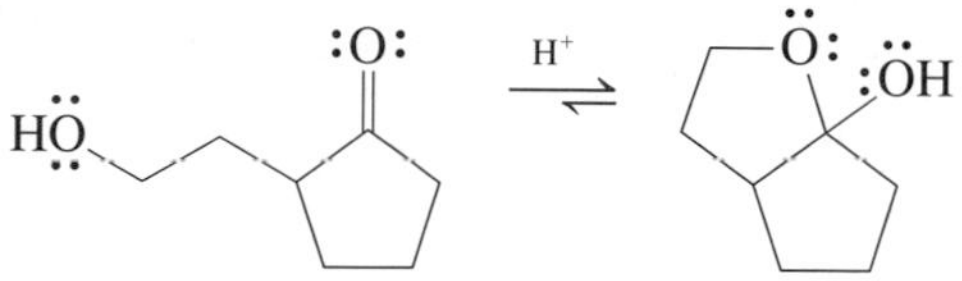

(b)

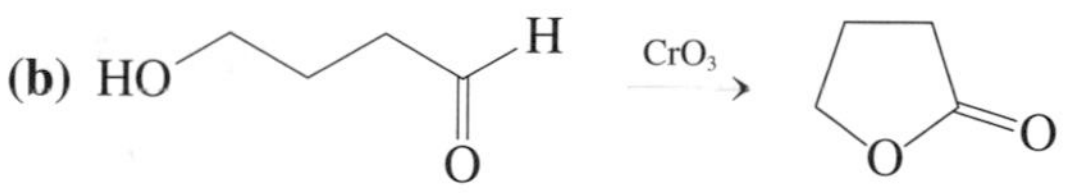

(c)

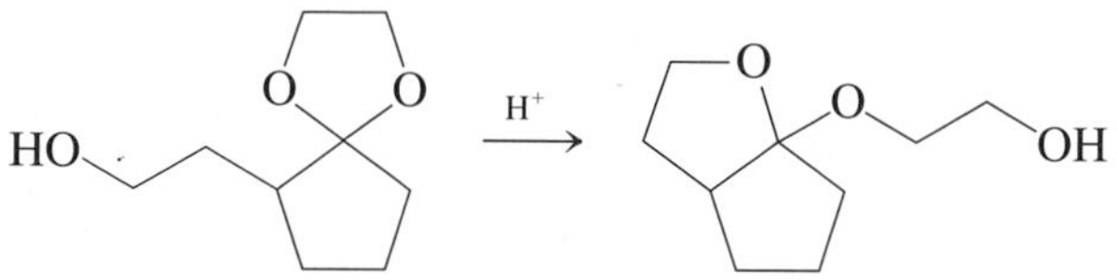

(d) Explain why hemiacetal formation may be catalyzed by either acid or base, but acetal formation is catalyzed only by acid, not by base.

31. Formulate a plausible mechanism for the following reaction. The product is a precursor of *mediquox* (shown in the margin), an agent used to treat respiratory infections in chickens (no, we are not making this up).

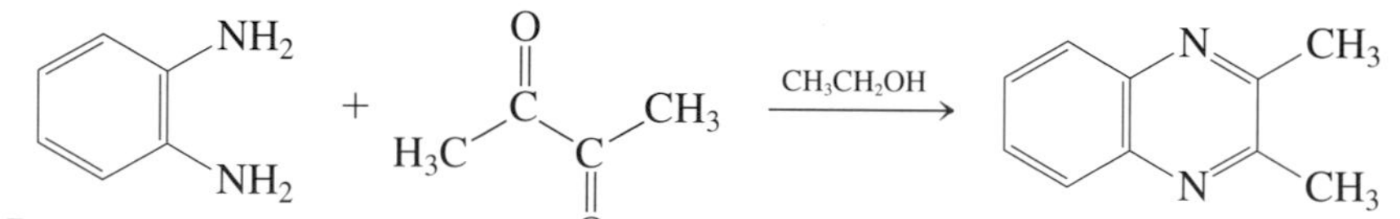

Benzene-1,2-diamine

2,3-Dimethylquinoxaline

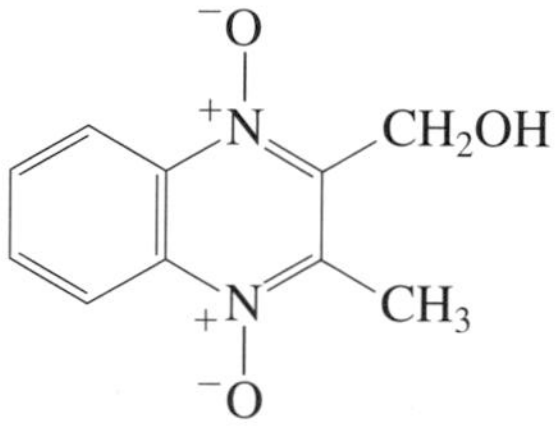

Mediquox

32. The formation of imines, oximes, hydrazones, and related derivatives from carbonyl compounds is reversible. Write a detailed mechanism for the acid-catalyzed hydrolysis of cyclohexanone semicarbazone to cyclohexanone and semicarbazide.

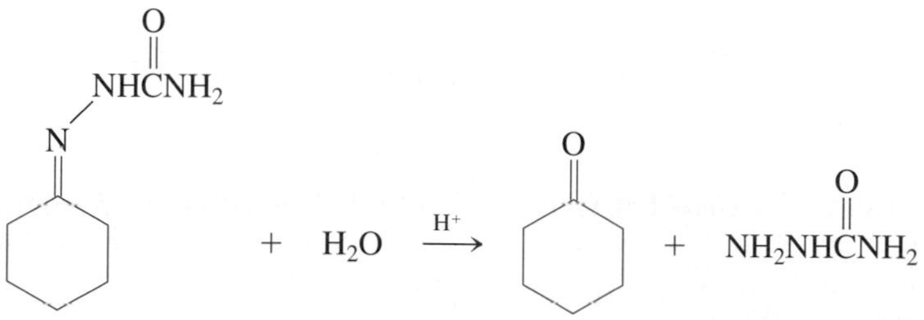

33. Propose reasonable syntheses of each of the following molecules, beginning with the indicated starting material.

(a)

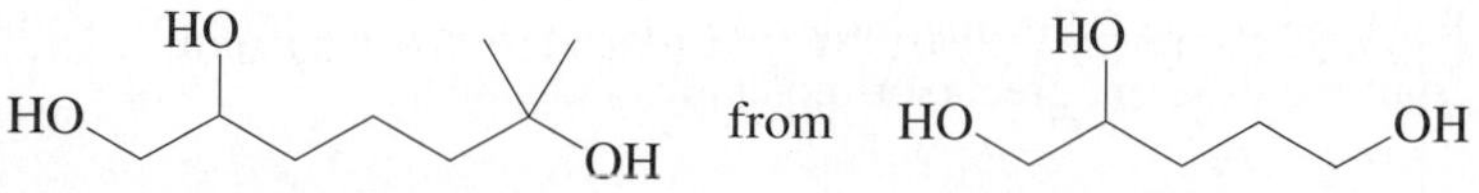

(b) $C_6H_5N{=}C(CH_2CH_3)_2$ from 3-pentanol

(c) 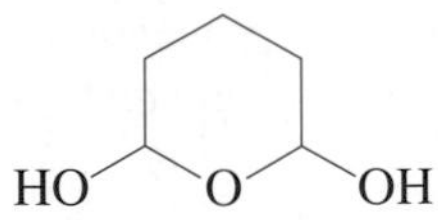 from 1,5-pentanediol

(d) 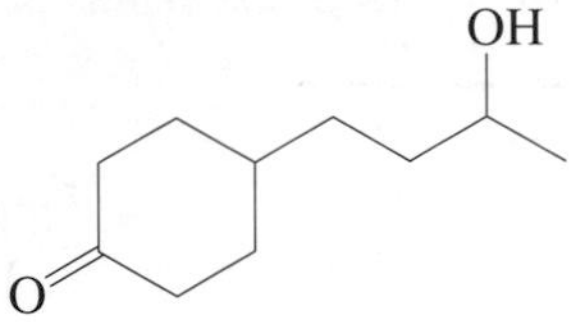 from

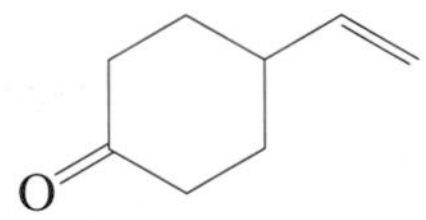

34. The UV absorptions and colors of 2,4-dinitrophenylhydrazone derivatives of aldehydes and ketones depend sensitively on the structure of the carbonyl compound. Suppose that you are asked to identify the contents of three bottles whose labels have fallen off. The labels indicate that one bottle contained butanal, one contained *trans*-2-butenal, and one contained *trans*-3-phenyl-2-propenal. The 2,4-dinitrophenylhydrazones prepared from the contents of the bottles have the following characteristics.

Bottle 1: m.p. 187°–188°C; λ_{max} = 377 nm; orange color
Bottle 2: m.p. 121°–122°C; λ_{max} = 358 nm; yellow color
Bottle 3: m.p. 252°–253°C, λ_{max} = 394 nm; red color

Match up the hydrazones with the aldehydes (*without* first looking up the melting points of these derivatives), and explain your choices. (**Hint:** See Section 14-11.)

35. Indicate the reagent(s) best suited to effect these transformations.

(a)

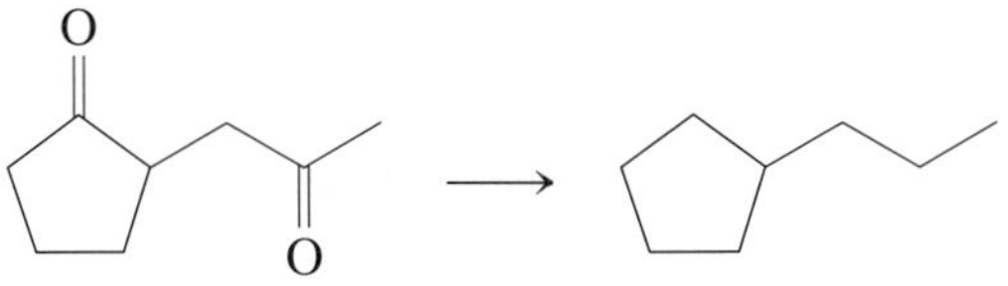

(b) $CH_3CH{=}CHCH_2CH_2\overset{O}{\overset{\|}{C}}H \longrightarrow CH_3CH_2CH_2CH_2CH_2\overset{O}{\overset{\|}{C}}H$

(c) $CH_3CH{=}CHCH_2CH_2\overset{O}{\overset{\|}{C}}H \longrightarrow CH_3CH{=}CHCH_2CH_2CH_2OH$

(d)

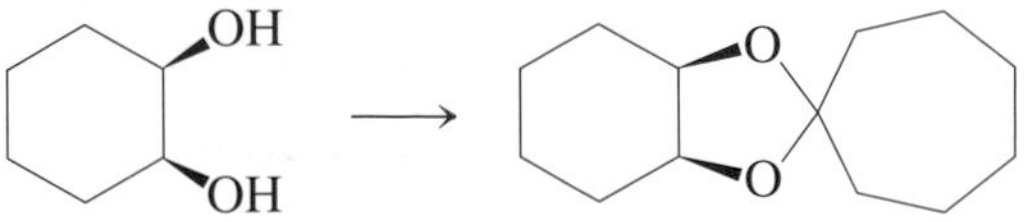

36. For each of the following molecules, propose *two* methods of synthesis from the different precursor molecules indicated.

(a) $CH_3CH{=}CHCH_2CH(CH_3)_2$ from (1) an aldehyde and (2) a different aldehyde

(b)  from (1) a dialdehyde and (2) a diketone

37. Three isomeric ketones with the molecular formula $C_7H_{14}O$ are converted into heptane by Clemmensen reduction. Compound A gives a single product on Baeyer-Villiger oxidation; compound B gives two different products in very different yields; compound C gives two different products in virtually a $1:1$ ratio. Identify A, B, and C.

38. Give the product(s) of reaction of hexanal with each of the following reagents.

(a) $HOCH_2CH_2OH$, H^+ **(b)** $LiAlH_4$, then H^+, H_2O **(c)** NH_2OH, H^+

(d) NH_2NH_2, KOH, heat   **(e)** $(CH_3)_2CHCH_2CH{=}P(C_6H_5)_3$ **(f)** , H^+

(g) Ag^+, NH_3, H_2O **(h)** CrO_3, H_2SO_4, H_2O **(i)** HCN

39. Give the product(s) of reaction of cycloheptanone with each of the reagents in Problem 38.

40. Formulate a detailed mechanism for the Baeyer-Villiger oxidation of the ketone shown in the margin (refer to Exercise 17-15).

41. Give the two theoretically possible Baeyer-Villiger products from each of the following compounds. Indicate which one is formed preferentially.

(a)   **(b)**   **(c)**

(d)   **(e)** $C_6H_5CCH_3$

42. Propose efficient syntheses of each of the following molecules, beginning with the indicated starting materials.

(a)   from   **(b)**   from

(c) from ClCH$_2$CH$_2$CH$_2$OH

43. Explain the fact that, although hemiacetal formation between methanol and cyclohexanone is thermodynamically disfavored, addition of methanol to cyclopropanone goes essentially to completion:

44. The rate of the reaction of NH$_2$OH with aldehydes and ketones is very sensitive to pH. It is very low in solutions more acidic than pH 2 or more basic than pH 7. It is highest in moderately acidic solution (pH ~ 4). Suggest explanations for these observations.

45. Compound D, formula C$_8$H$_{14}$O, is converted by CH$_2$=P(C$_6$H$_5$)$_3$ into compound E, C$_9$H$_{16}$. Treatment of compound D with LiAlH$_4$ yields *two* isomeric products F and G, both C$_8$H$_{16}$O, in unequal yield. Heating either F or G with concentrated H$_2$SO$_4$ produces H, with the formula C$_8$H$_{14}$. Ozonolysis of H produces a keto aldehyde after Zn–H$^+$, H$_2$O treatment. Oxidation of this keto aldehyde with aqueous Cr(VI) produces

Identify compounds D through H. Pay particular attention to the stereochemistry of D.

46. In 1862, it was discovered that cholesterol (for structure, see Section 4-7) is converted into a new substance named coprostanol by the action of bacteria in the human digestive tract. Make use of the following information to deduce the structure of coprostanol. Identify the structures of unknowns J through M as well. (i) Coprostanol, on treatment with Cr(VI) reagents, gives compound J, UV $\lambda_{max}(\epsilon)$ = 281(22) nm and IR = 1710 cm^{-1}. (ii) Exposure of cholesterol to H$_2$ over Pt results in compound K, a stereoisomer of coprostanol. Treatment of K with the Cr(VI) reagent furnishes compound L, which has a UV peak very similar to that of compound J, $\lambda_{max}(\epsilon)$ = 258(23) nm, and turns out to be a stereoisomer of J. (iii) Careful addition of Cr(VI) reagent to cholesterol produces M: UV $\lambda_{max}(\epsilon)$ = 286(109) nm. Catalytic hydrogenation of M over Pt also gives L.

47. Three reactions that include compound M (see Problem 46) are described here. Answer the questions that follow. **(a)** Treatment of M with catalytic amounts of acid in ethanol solvent causes isomerization to compound N: UV $\lambda_{max}(\epsilon)$ = 241(17,500) and 310(72) nm. Propose a structure for N. **(b)** Hydrogenation of compound N (H$_2$-Pd, ether solvent) produces compound J (Problem 46). Is this the result that you would have predicted, or is there something unusual about it? **(c)** Wolff-Kishner reduction of compound N (H$_2$NNH$_2$, H$_2$O, HO$^-$, Δ) leads to 3-cholestene. Propose a mechanism for this transformation.

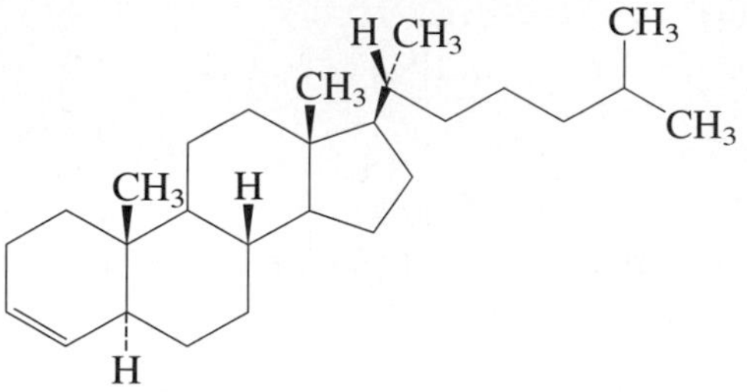

3-Cholestene

Team Problem

48. In acidic methanol, 3-oxobutanal is transformed into a compound with the molecular formula $C_6H_{12}O_3$.

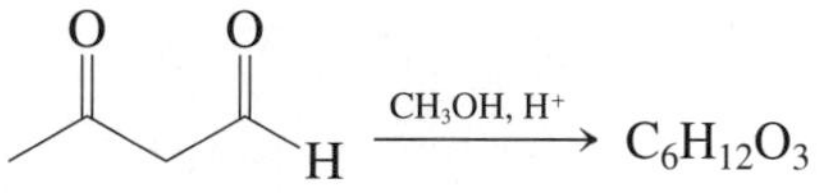

3-Oxobutanal

As a group, analyze the following ^{1}H NMR and IR spectral data: ^{1}H NMR (CCl_4) $\delta = 2.19$ (s, 3 H), 2.75 (d, 2 H), 3.38 (s, 6 H), 4.89 (t, 1 H) ppm; IR 1715 cm^{-1}. Consider the chemical shifts, the splitting patterns, and the integrations of the signals in the NMR spectrum and discuss possible fragments that could give rise to the observed multiplicities. Use the IR information to assign the functional group that exists in the new molecule. Present an explanation for your structural determination, including reference to the spectral data, and suggest a detailed mechanism for the formation of the new compound.

Preprofessional Problems

49. In the transformation shown here, which of the following compounds is most likely to be compound A (use IUPAC name)? **(a)** 5-Octyn-7-one; **(b)** 5-octyn-2-one; **(c)** 3-octyn-2-one; **(d)** 2-octyn-3-one.

$$3\text{-octyn-2-ol} \xrightarrow{\text{CrO}_3,\text{H}_2\text{SO}_4,\text{propanone}} A$$

50. The reaction demonstrates **(a)** resonance; **(b)** tautomerism; **(c)** conjugation; **(d)** deshielding.

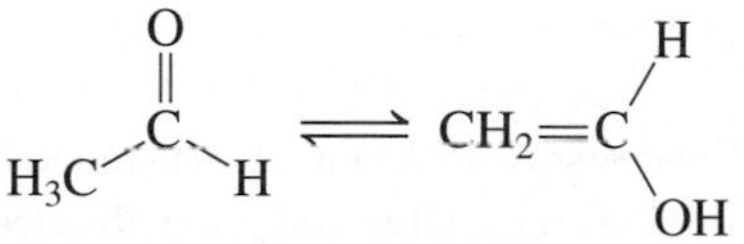

51. Which of the following reagents converts benzenecarbaldehyde (benzaldehyde) into an oxime? **(a)** $H_2NNHC_6H_5$; **(b)** H_2NNH_2; **(c)** O_3; **(d)** H_2NOH; **(e)** $CH_3CH(OH)_2$.

52. In the IR spectrum of 3-methyl-2-butanone, the most intense absorption is at **(a)** 3400 cm^{-1}, owing to an OH stretching mode; **(b)** 1700 cm^{-1}, owing to a $C{=}O$ stretching mode; **(c)** 2000 cm^{-1}, owing to a CH stretching mode; **(d)** 1500 cm^{-1}, owing to the rocking of an isopropyl group.

Enols and Enones

α,β-Unsaturated Alcohols, Aldehydes, and Ketones

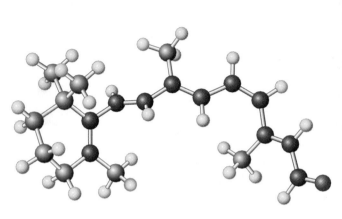

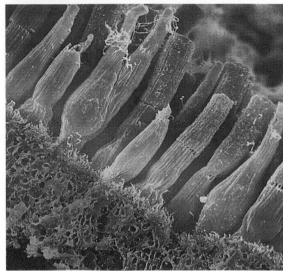

Photomicrograph of rod and cone cells in the retina. All known image-resolving eyes and, indeed, all known visual systems in nature use a single molecule, *cis*-retinal, for light detection. Absorption of a photon isomerizes the cis double bond to trans, and the accompanying massive change in overall molecular geometry is responsible for triggering the nerve impulse that is perceived as vision.

Have another look at the Chapter Opening above. The very fact that you can see it is made possible by the chemistry alluded to in the caption: Photons impinge on the chromophore (Section 14-1) of *cis*-retinal bound to a protein to cause cis-trans isomerization. The associated conformational changes result, within picoseconds, in a nerve impulse that is translated by your brain into "vision" (Chemical Highlight 18-4). The crucial chemical feature of retinal that makes this process possible is the (in this case extended) conjugative communication between the carbonyl group and the adjacent π system. This chapter will show you that the carbonyl group (much like an ordinary C—C double bond; Section 14-1) activates adjacent C—H and C=C bonds even in much simpler systems, because of resonance. After you have absorbed the material that follows, you will be able to "look" at this page with quite a different perspective!

The last chapter examined two sites of reactivity in aldehydes and ketones. We saw in Sections 17-5 through 17-12 that the carbonyl oxygen is easily attacked by electrophiles (usually protons), its carbon by nucleophiles. We turn now to a third site of reactivity, the carbon *next to* the carbonyl group, also called the **α-carbon.** Hydrogens on α-carbons are rendered acidic by the neighboring carbonyl group; loss of such **α-hydrogens** leads to either of two electron-rich species: unsaturated alcohols, called enols, or their corresponding anions, known as enolate ions. Both species are important nucleophilic intermediates in reactions of carbonyl compounds. We shall see that they can attack electrophiles such as protons, alkylating agents, halogens, and even other carbonyl carbons.

We begin by introducing the chemistry of enolates and enols. Especially important will be a reaction between enolate ions and carbonyl compounds, the aldol condensation. This process is widely used to form carbon–carbon bonds both in the laboratory and in nature. Among the possible products of aldol condensation are α,β-unsaturated aldehydes and ketones, which contain conjugated carbon–carbon and carbon–oxygen π bonds. As expected, electrophilic additions may take place at either one. However, more significantly, α,β-unsaturated carbonyl compounds are also subject to ionic attack, a reaction affecting the *entire* conjugated system.

18-1 Acidity of Aldehydes and Ketones: Enolate Ions

The pK_a values of aldehyde and ketone α-hydrogens range from 19 to 21, much lower than the pK_a values of ethene (44) or ethyne (25) but higher than those of alcohols (15–18). Strong bases can therefore remove an α-hydrogen. The anions that result are known as **enolate ions** or simple **enolates.**

Deprotonation of a Carbonyl Compound

Why are aldehydes and ketones relatively acidic? We know that acid strength is enhanced by stabilization of the conjugate base (Section 2-9). In the enolate ion, the inductive effect of the positively polarized carbonyl carbon strongly stabilizes the negative charge at the α-position. Further stabilization is provided by delocalization of charge onto the electronegative oxygen, as described by the resonance forms just pictured. An example of enolate formation is the deprotonation of cyclohexanone by lithium diisopropylamide (LDA; Section 7-8).

Enolate Preparation

EXERCISE 18-1

Identify the most acidic hydrogens in each of the following molecules. Give the structure of the enolate ion arising from deprotonation. **(a)** Acetaldehyde; **(b)** propanal; **(c)** propanone; **(d)** 4-heptanone; **(e)** cyclopentanone.

Each resonance form contributes to the characteristics of the enolate ion and thus to the chemistry of carbonyl compounds. The resonance hybrid possesses partial negative charges on both carbon and oxygen; as a result, it is nucleophilic and may attack electrophiles at either position. A species that can react at two different sites to

Resonance hybrid

give two different products is called **ambident** ("two fanged": from *ambi,* Latin, both; *dens,* Latin, tooth). The enolate ion is thus an ambident anion. Its carbon atom is normally the site of reaction, undergoing nucleophilic substitution with S_N2 substrates such as suitable haloalkanes. Because this reaction attaches an alkyl group to the reactive carbon, it is called **alkylation** (more specifically, C-alkylation) and, as we shall see in Section 18-4, is a powerful method for carbon–carbon bond formation of ketones. For example, alkylation of cyclohexanone enolate with 3-chloropropene takes place at carbon. Alkylation at oxygen (O-alkylation) is uncommon, although oxygen is typically the site of *protonation.* The product of protonation is an unsaturated alcohol, called an **alkenol** (or **enol** for short). Enols are unstable and rapidly isomerize back to the original ketones (recall Section 13-8).

Ambident Behavior of Cyclohexanone Enolate Ion

62%
2-(2-Propenyl)cyclo-
hexanone

Cyclohexanone
enolate ion

Cyclo-
hexanone
enol

Cyclo-
hexanone

EXERCISE 18-2

Give the products of reaction of cyclohexanone enolate with **(a)** iodoethane (reacts by C-alkylation) and **(b)** chlorotrimethylsilane, $(CH_3)_3Si–Cl$ (reacts by O-silylation).

In summary, the hydrogens on the carbon next to the carbonyl group in aldehydes and ketones are acidic, with pK_a values ranging from 19 to 21. Deprotonation leads to the corresponding enolate ions, which may attack electrophilic reagents at either oxygen or carbon. Protonation at oxygen gives enols.

18-2 Keto–Enol Equilibria

We have seen that protonation of an enolate at oxygen leads to an enol. The enol, an unstable isomer of an aldehyde or ketone, rapidly converts into the carbonyl system: It **tautomerizes** (Section 13-8). These isomers are called **enol** and **keto tautomers.** This section begins by discussing factors affecting their equilibria, in which the keto form usually predominates. It then describes the mechanism of tautomerism and its chemical consequences.

An enol equilibrates with its keto form in acidic or basic solution

Enol–keto tautomerism proceeds by either acid or base catalysis. Base simply removes the proton from the enol oxygen, reversing the initial protonation. Subsequent (and slower) C-protonation furnishes the thermodynamically more stable keto form.

Base-Catalyzed Enol–Keto Equilibration

$$\underset{\textbf{Enol form}}{\text{C=C}\overset{\ddot{O}\text{H}}{}} + :B^- \rightleftharpoons \left[\underset{\textbf{Enolate ion}}{\text{C=C}\overset{\ddot{O}:^-}{} \longleftrightarrow \text{C}-\text{C}\overset{\ddot{O}:}{}} \right] + BH \rightleftharpoons \underset{\textbf{Keto form}}{-\overset{H}{\underset{|}{\text{C}}}-\text{C}\overset{\ddot{O}:}{}} + :B^-$$

In the acid-catalyzed process, the enol form is protonated at the double bond to give the resonance-stabilized carbocation next to oxygen. This species is simply a protonated carbonyl function. Its deprotonation then gives the keto form.

Acid-Catalyzed Enol–Keto Equilibration

$$\underset{\textbf{Enol form}}{\text{C=C}\overset{\ddot{O}\text{H}}{}} + H^+ \rightleftharpoons \left[\underset{\textbf{Protonated carbonyl system}}{-\overset{H}{\underset{|}{\text{C}}}-\overset{+}{\text{C}}\overset{\ddot{O}\text{H}}{} \longleftrightarrow -\overset{H}{\underset{|}{\text{C}}}-\text{C}\overset{+\ddot{O}\text{H}}{}} \right] \rightleftharpoons \underset{\textbf{Keto form}}{-\overset{H}{\underset{|}{\text{C}}}-\text{C}\overset{\ddot{O}:}{}} + H^+$$

Both the acid- and base-catalyzed enol–keto interconversions occur relatively fast in solution, whenever there are traces of the required catalysts. Remember that although the keto form (usually) predominates, the enol to keto form conversion is reversible, and the mechanisms by which the keto form equilibrates with its enol counterpart are the exact reverse of the above two schemes.

Substituents can shift the keto–enol equilibrium

The equilibrium constants for the conversion of the keto into the enol forms are very small for ordinary aldehydes and ketones, only traces of enol being present. However, relative to its keto form, the enol of acetaldehyde is about a hundred times as stable as the enol of propanone (acetone), because the less substituted aldehyde carbonyl is less stable than the more substituted ketone carbonyl.

Keto–Enol Equilibra

$$\underset{}{\text{H}-\text{CH}_2\overset{O}{\overset{\|}{\text{C}}}\text{H}} \underset{K = 6 \times 10^{-7}}{\rightleftharpoons} \text{H}_2\text{C=C}\overset{\text{OH}}{\underset{\text{H}}{}} \qquad \Delta G^\circ \sim +8.5 \text{ kcal mol}^{-1}$$

Acetaldehyde ⇌ **Ethenol (Vinyl alcohol)**

$$\underset{\substack{\textbf{Propanone}\\\textbf{(Acetone)}}}{\text{H}-\text{CH}_2\overset{O}{\overset{\|}{\text{C}}}\text{CH}_3} \underset{K = 5 \times 10^{-9}}{\rightleftharpoons} \underset{\textbf{2-Propenol}}{\text{H}_2\text{C=C}\overset{\text{OH}}{\underset{\text{CH}_3}{}}} \qquad \Delta G^\circ \sim +11.3 \text{ kcal mol}^{-1}$$

Enol formation leads to deuterium exchange and stereoisomerization

What are some consequences of enol formation by tautomerism? One is that treatment of a ketone with traces of acid or base in D_2O solvent leads to the complete exchange of *all* the α-hydrogens.

Hydrogen–Deuterium Exchange of Enolizable Hydrogens

$$\underset{\textbf{2-Butanone}}{CH_3\overset{\overset{\textstyle O}{\|}}{C}CH_2CH_3} \xrightarrow{D_2O,\ DO^-} \underset{\textbf{1,1,1,3,3-Pentadeuterio-2-butanone}}{CD_3\overset{\overset{\textstyle O}{\|}}{C}CD_2CH_3}$$

This reaction can be conveniently followed by ^1H NMR, because the signal for these hydrogens slowly disappears as each one is sequentially replaced by deuterium. In this way, the number of α-hydrogens present in a molecule can be readily determined.

EXERCISE 18-3

Formulate mechanisms for the base- and acid-catalyzed replacement of a single α-hydrogen in propanone by deuterium from D_2O.

EXERCISE 18-4

Write the products (if any) of deuterium incorporation by the treatment of the following compounds with D_2O–NaOD.

(a) Cycloheptanone

(b) 2,2-Dimethylpropanal

(c) 3,3-Dimethyl-2-butanone

(d) O

CHO

EXERCISE 18-5

(a) The ^1H NMR spectrum of cyclobutanone consists of a quintet at $\delta = 2.00$ ppm and a triplet at $\delta = 3.13$ ppm. Assign the signals in this spectrum to the appropriate hydrogens in the molecule. (b) What would you expect to observe in the ^1H NMR spectrum upon treatment of cyclobutanone with D_2O–NaOD?

Another consequence of enol formation, or **enolization,** is the ease with which stereoisomers at α-carbons interconvert. For example, treatment of cis-2,3-disubstituted cyclopentanones with mild base furnishes the corresponding trans isomers. The latter are more stable for steric reasons.

Base-Catalyzed Isomerism of an α-Substituted Ketone

$$\overset{\overset{\textstyle O}{\|}}{\underset{CH_3}{\bigcirc}}CH_2CH=CH_2 \underset{\longleftarrow}{\overset{10\%\ KOH,\ CH_3CH_2OH}{\longrightarrow}} \overset{\overset{\textstyle O^-K^+}{\|}}{\underset{CH_3}{\bigcirc}}CH_2CH=CH_2 \rightleftharpoons \overset{\overset{\textstyle O}{\|}}{\underset{\underset{>95\%}{CH_3}}{\bigcirc}}CH_2CH=CH_2$$

The reaction proceeds through the enolate ion, in which the α-carbon is planar and, hence, no longer a stereocenter. Reprotonation from the side cis to the 3-methyl group results in the trans diastereomer (Section 5-5).

Another consequence of enolization is the difficulty of maintaining optical activity in a compound whose only stereocenter is an α-carbon. Why? As the (achiral) enol converts back into the keto form, a racemic mixture of R and S enantiomers is produced. For example, at room temperature, optically active 3-phenyl-2-butanone racemizes with a half-life of minutes in basic ethanol.

Racemization of Optically Active 3-Phenyl-2-butanone

(*S*)-3-Phenyl-2-butanone	Achiral	(*R*)-3-Phenyl-2-butanone

EXERCISE 18-6

Bicyclic ketone A rapidly equilibrates with a stereoisomer upon treatment with base, but ketone B does not. Explain.

In summary, aldehydes and ketones are in equilibrium with their enol forms, which are roughly 10 kcal mol^{-1} less stable. Keto–enol equilibration is catalyzed by acid or base. Enolization allows for easy H–D exchange in D$_2$O and causes stereoisomerization at stereocenters next to the functional group.

18-3 Halogenation of Aldehydes and Ketones

This section examines a reaction of the carbonyl group that can proceed through the intermediacy of either enols or enolate ions—halogenation. Aldehydes and ketones react with halogens at the α-carbon. In contrast with deuteration, the extent of halogenation depends on whether acid or base catalysis has been used.

In the presence of acid, halogenation usually stops after the first halogen has been introduced, as shown in the following example.

Acid-Catalyzed α-Halogenation of Ketones

$$H-CH_2CCH_3 \xrightarrow{\text{Br–Br, CH}_3\text{CO}_2\text{H, H}_2\text{O, 70°C}} BrCH_2CCH_3 + HBr$$

44%

Bromopropanone
(Bromoacetone)

The rate of the acid-catalyzed halogenation is independent of the halogen concentration, an observation suggesting a rate-determining first step involving the carbonyl substrate. This step is enolization. The halogen then attacks the double bond to give an intermediate oxygen-stabilized halocarbocation. Subsequent deprotonation of this species furnishes the product.

Mechanism of the Acid-Catalyzed Bromination of Propanone (Acetone)

STEP 1. Enolization

$$CH_3\overset{:O:}{\overset{\|}{C}}CH_3 \overset{H^+}{\rightleftharpoons} H_2C=\underset{CH_3}{\overset{\overset{..}{O}H}{C}}$$

STEP 2. Halogen attack

$$H_2C=\underset{CH_3}{\overset{\overset{..}{O}H}{C}} \quad\longrightarrow\quad \left[\ H_2C-\underset{\underset{Br}{|}}{\overset{+}{C}}\underset{CH_3}{\overset{\overset{..}{O}H}{}} \quad\longleftrightarrow\quad H_2C-\underset{\underset{Br}{|}}{C}\underset{CH_3}{\overset{\overset{+}{O}H}{}}\ \right] \quad+\quad Br^-$$

$$Br-Br$$

STEP 3. Deprotonation

$$BrCH_2\overset{\overset{+}{:O}}{\overset{\|}{C}}CH_3 \quad\longrightarrow\quad BrCH_2\overset{:O:}{\overset{\|}{C}}CH_3 \quad+\quad H^+$$

Why is further halogenation retarded? The answer lies in the requirement for enolization. To repeat halogenation, the halo carbonyl compound must enolize again by the usual acid-catalyzed mechanism. However, the electron-withdrawing power of the halogen makes protonation, the initial step in enolization, *more difficult* than in the original carbonyl compound.

Halogenation Slows Down Enolization

Less basic than unsubstituted ketone

$$BrCH_2\overset{:O:}{\overset{\|}{C}}CH_3 \underset{\text{Electron withdrawing}}{\overset{H^+}{\rightleftharpoons}} BrCH_2\overset{\overset{+}{:O}\,H}{\overset{\|}{C}}CH_3$$

**Mechanism
of Halogenation
of an Enolate Ion**

$$\underset{R}{\overset{:\overset{..}{O}:^-}{C}}=CH_2 \ +\ Br-Br$$

$$\downarrow$$

$$R\overset{:O:}{\overset{\|}{C}}CH_2Br \ +\ Br^-$$

More acidic
than unsubstituted
ketone

Therefore, the singly halogenated product is not attacked by additional halogen until the starting aldehyde or ketone has been used up.

Base-mediated halogenation is entirely different. It proceeds instead by the formation of an enolate ion, which then attacks the halogen. Here the reaction continues until it *completely* halogenates the same α-carbon, leaving unreacted starting material (when insufficient halogen is employed). Why is base-catalyzed halogenation so difficult to stop at the stage of monohalogenation? The electron-withdrawing power of the halogen increases the acidity of the remaining α-hydrogens, accelerating further enolate formation and hence further halogenation.

Haloform Reaction: A Test for Methyl Ketones

The base-mediated halogenation of methyl ketones can proceed beyond complete halogenation of the methyl group. The trihalomethyl substituent acts as a leaving group, giving rise eventually to a carboxylic acid and a trihalomethane. The process is called the *haloform reaction,* after the common name of the product.

Mechanism of the Haloform Reaction

$$
\overset{:\ddot{O}:}{\underset{||}{RCCBr_3}} + ^{-}:\ddot{O}H \longrightarrow \overset{:\ddot{O}:^{-}}{\underset{\underset{:\ddot{O}H}{|}}{RC-CBr_3}} \longrightarrow
$$

$$
\overset{:O:}{\underset{||}{RC\ddot{O}H}} + ^{-}:CBr_3 \longrightarrow \overset{:O:}{\underset{||}{RC\ddot{O}:^{-}}} + HCBr_3 \xrightarrow[H_2O]{H^{+},}
$$

$$
\overset{O}{\underset{||}{RCOH}} + HCBr_3
$$

When the halogen is iodine, triiodomethane (iodoform) precipitates as a yellow solid. Its formation, called the *iodoform reaction,* is a

$$
\text{qualitative test for the } R\overset{O}{\overset{||}{C}}CH_3 \text{ structural unit.}
$$

An Iodoform Reaction

Iodoform, by the way, is a topical disinfectant. It was responsible for the "smell of the doctor's office" familiar to patients of past generations because of its widespread use and characteristic odor.

Write the products of the acid- and base-catalyzed bromination of cyclohexanone.

In summary, halogenation of aldehydes and ketones in acid can proceed selectively to the monohalocarbonyl compounds. In base, *all* α-hydrogens are replaced before another molecule of starting material is attacked.

18-4 Alkylation of Aldehydes and Ketones

We have seen that enolates may be generated from ketones and aldehydes by treatment with base. This section presents examples of enolate formation in which the base is sodium hydride, NaH. When formed, the enolate of a ketone may be alkylated, providing a general way to introduce an alkyl substituent next to the carbonyl group. We shall compare this process with the similar alkylation of another intermediate, called an enamine.

Alkylation of enolates can be difficult to control

Ketone enolates typically alkylate at the α-carbon. When a ketone possesses only a single α-hydrogen, high yields of the alkylation product can be obtained. (Aldehydes cannot be alkylated in this way; their enolates undergo an alternative process, which will be described in the next section.)

Alkylation of a Ketone

$$C_6H_5CCH(CH_3)_2 \xrightarrow[\substack{-\text{ H–H,} \\ -\text{ NaBr}}]{\substack{1.\ \text{NaH, benzene, }\Delta \\ 2.\ (CH_3)_2C=CHCH_2Br}} C_6H_5CC(CH_3)_2$$

2-Methyl-1-phenyl-	2,2,5-Trimethyl-1-phenyl-
1-propanone	4-hexen-1-one

88%

In other cases, prevention of dialkylation is a problem. Under the reaction conditions, a monoalkylated ketone might become deprotonated by the starting enolate and could therefore undergo further alkylation.

Single and Double Alkylations of a Ketone

1. NaH, $CH_3OCH_2CH_2OCH_3$
2. CH_3I

   $-\text{ H–H, } -\text{ NaI}$

27% + 38%

Another complication arises in the alkylation of unsymmetric ketones: Both α positions are subject to electrophilic attack, leading to two different products.

Alkyation of an Unsymmetric Ketone

NaH, CH_3I,
$CH_3OCH_2CH_2OCH_3$

36% + 52%

(1 : 1 cis : trans)

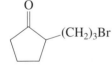

EXERCISE 18-8

The reaction of the compound shown in the margin with base gives three isomeric products $C_8H_{12}O$. What are they? (**Hint:** Try intramolecular alkylations.)

EXERCISE 18-9

C-alkylation of cyclohexanone enolate with 3-chloropropene (Section 18-1) is much faster than the corresponding reaction with 1-chloropropane. Explain. (**Hint:** See Section 14-3.) What product(s) would you expect from reaction of cyclohexanone enolate with (**a**) 2-bromopropane and (**b**) 2-bromo-2-methylpropane? (**Hint:** See Chapter 7.)

Enamines afford an alternative route for the alkylation of aldehydes and ketones

**Azacyclopentane
(Pyrrolidine)**

Section 17-9 showed that the reaction of secondary amines such as azacyclopentane (pyrrolidine) with aldehydes or ketones produces enamines. As the following resonance forms indicate, the nitrogen substituent renders the enamine carbon–carbon double bond electron rich. Furthermore, the dipolar resonance contribution gives rise to significant nucleophilicity at carbon, even though the enamine is neutral. As a result, electrophiles may attack at this position. Let us see how this attack may be used to synthesize alkylated aldehydes and ketones.

Exposure of enamines to haloalkanes results in alkylation at carbon to produce an iminium salt. On aqueous work-up, iminium salts hydrolyze by a mechanism that is the reverse of the one formulated for imine formation in Section 17-9. The results are a new alkylated aldehyde or ketone and the original secondary amine.

Resonance in Enamines

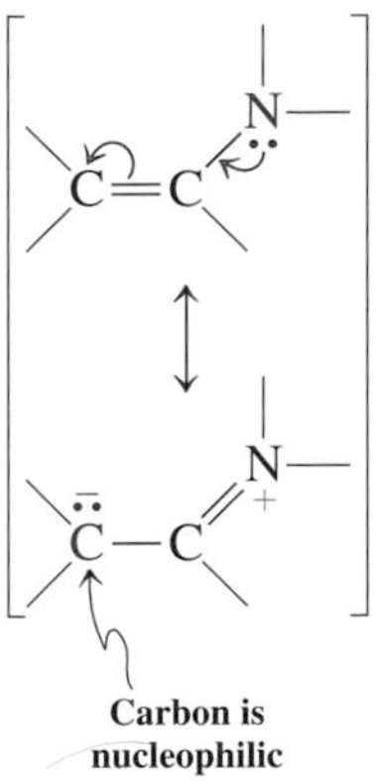

Carbon is nucleophilic

Alkylation of an Enamine

Alkylated at α-carbon

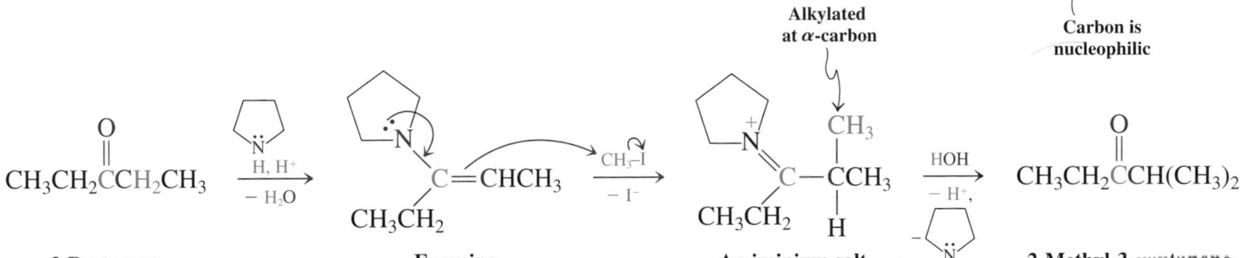

3-Pentanone   Enamine   An iminium salt   2-Methyl-3-pentanone

Formulate the mechanism for the final step of the sequence just shown: hydrolysis of the iminium salt.

How does the alkylation of an enamine compare with the alkylation of an enolate? Enamine alkylation is far superior, because it minimizes double or multiple alkylation. It has the additional advantage that it can be used to prepare alkylated aldehydes, as shown here.

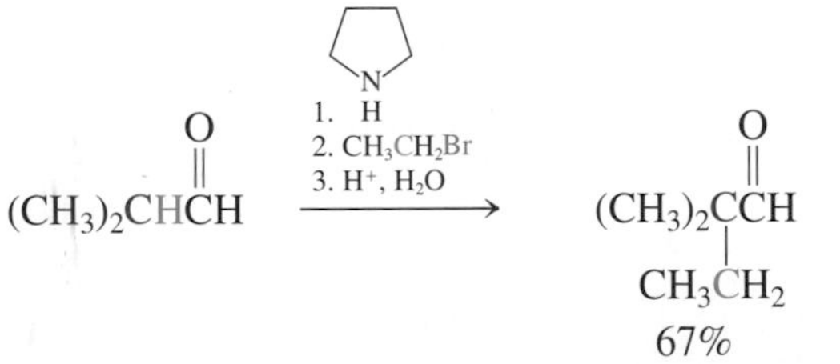

2-Methylpropanal   2,2-Dimethylbutanal

Alkylations of the enolate of ketone A are very difficult to stop before dialkylation occurs, as illustrated here. Show how you would use an enamine to prepare monoalkylated ketone B.

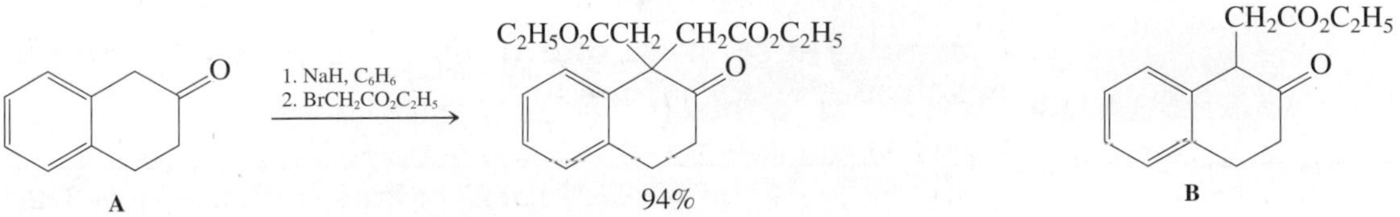

A   94%   B

In summary, enolates give rise to alkylated derivatives upon exposure to haloalkanes. In these reactions, control of the extent and the position of alkylation, when there is a choice, may be a problem. Enamines derived from aldehydes and ketones undergo alkylation to the corresponding iminium salts, which can hydrolyze to the alkylated carbonyl compounds.

18-5 Attack by Enolates on the Carbonyl Function: Aldol Condensation

Among the most frequently employed carbon–carbon bond forming strategies is the attack of an enolate ion at a carbonyl carbon. The product of this process is a hydroxy carbonyl compound. Subsequent elimination of water leads to α,β-unsaturated aldehydes and ketones. The next three sections describe these reactions in detail. Chemical Highlights 18-2 and 18-3 illustrate them in a biological context.

CHEMICAL HIGHLIGHT 18-2 | **Aldol Condensations in Nature**

Aldol condensations take place in natural systems. For example, collagen fibers are strengthened by chemical cross-linking of aldehyde units through aldol condensations. Collagen is the most abundant fibrous protein in mammals, being the major fibrous component of skin, bone, tendon, cartilage, and teeth. One of its functions is to hold cells together in discrete units. Its structure is basically a staggered array of *tropocollagen* molecules, which consist of triply stranded helical polypeptide chains (Chapter 26).

Cross-linking (see Sections 12-14 and 14-10 for cross-linking of polymers) of these chains takes place by aldol condensations catalyzed by enzymes. First, lysine residues (Section 26-1) in the chain are enzymatically oxidized to aldehyde derivatives. Then, an aldol condensation connects two chains.

The extent of cross-linking depends on the function of the tissue. For example, the collagen in the Achilles' tendon of rats is highly cross-linked, but that of the more flexible tail tendon is less so.

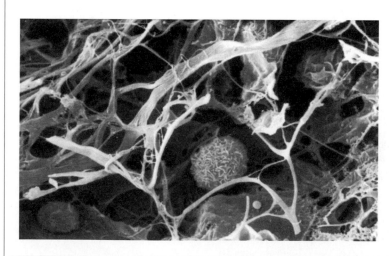

A photomicrograph of collagen fibers intertwined about fibroblasts, the cells that secrete tropocollagen.

Aldehydes undergo base-catalyzed condensations

Addition of a small amount of dilute aqueous sodium hydroxide to acetaldehyde at low temperature initiates the conversion of the aldehyde into a dimer, 3-hydroxybutanal. Upon heating, this hydroxyaldehyde dehydrates to give the final condensation product, the α,β-unsaturated aldehyde *trans*-2-butenal (crotonaldehyde). This reaction is an example of the **aldol condensation**.

Aldol Condensation Between Two Molecules of Acetaldehyde

$$
\underset{H_3C}{\overset{H}{\diagdown}}C{=}O \;+\; H_2C\overset{O}{\overset{\|}{C}}H \xrightarrow{\text{NaOH, H}_2\text{O, 5°C}} CH_3\underset{\underset{H}{|}}{\overset{\overset{OH}{|}}{C}}-\underset{\underset{H}{|}}{CH}\overset{O}{\overset{\|}{C}}H \xrightarrow{\Delta} \underset{H_3C}{\overset{H}{\diagdown}}\overset{\beta}{C}{=}\overset{\alpha}{\underset{H}{C}}\overset{\overset{O}{\|}}{CH} \;+\; H_2O
$$

$$\text{3-Hydroxybutanal} \qquad\qquad \substack{\textit{trans}\text{-2-Butenal}\\ \text{(Crotonaldehyde)}}$$

(An α,β-unsaturated aldehyde)

H—N
H—C—(CH₂)₂—CH₂—CH₂—$\overset{+}{N}H_3$ $H_3\overset{+}{N}$—CH₂—CH₂—(CH₂)₂—C—H   N—H
O=C   C=O

Lysine residues

↓ Oxidation

H—N
H—C—(CH₂)₂—CH₂—C=O   O=C—CH₂—(CH₂)₂—C—H   N—H
O=C   C=O

Aldehyde derivatives

−H₂O ↓

H—N   H
H—C—(CH₂)₂—CH₂—C=C—(CH₂)₂—C—H   N—H
O=C   C=O
                   O=C—H

Aldol cross-link

The aldol condensation is general for aldehydes and, as we shall see, sometimes succeeds with ketones as well. We shall first describe its mechanism before turning to its uses in synthesis.

The aldol condensation highlights the two most important facets of carbonyl group reactivity: enolate formation and nucleophilic attack at a carbonyl carbon. The base sets up an equilibrium between a small amount of the aldehyde and its corresponding enolate ion. Because the ion is surrounded by excess aldehyde, its nucleophilic α-carbon can then attack the carbonyl group of another molecule of aldehyde. Protonation of the resulting alkoxide furnishes 3-hydroxybutanal, which has the common name *aldol* (from *ald*ehyde alcoh*ol*).

Mechanism of Aldol Formation

STEP 1. Enolate generation

$$HC\overset{:O:}{\overset{\|}{}}CH_2-H \;+\; {}^-:\ddot{O}H \;\rightleftharpoons\; H_2C=C\overset{:\ddot{O}:^-}{\underset{H}{}} \;+\; H\ddot{O}H$$

Small equilibrium concentration of enolate

STEP 2. Nucleophilic attack

$$CH_3\overset{:O:}{\overset{\|}{C}}H \quad CH_2=C\overset{\ddot{O}:^-}{\underset{H}{}} \;\rightleftharpoons\; CH_3\overset{{}^-:\ddot{O}:}{\overset{}{C}}-CH_2\overset{:O:}{\overset{\|}{C}}H$$

STEP 3. Protonation

$$CH_3\overset{:\ddot{O}:^-}{\underset{H}{\overset{|}{C}}}-CH_2\overset{:O:}{\overset{\|}{C}}H \;+\; H\ddot{O}H \;\rightleftharpoons\; CH_3\overset{:\ddot{O}H}{\underset{H}{\overset{|}{C}}}-CH_2\overset{:O:}{\overset{\|}{C}}H \;+\; H\ddot{O}:^-$$

50–60%

**3-Hydroxybutanal
(Aldol)**

Note that hydroxide ion functions as a catalyst in this reaction. The last two steps of the sequence drive the initially unfavorable equilibrium toward product, but the overall reaction is not very exothermic. The aldol is formed in 50–60% yield and does not react further when its preparation is carried out at low temperature (5°C).

EXERCISE 18-12

Give the structure of the hydroxyaldehyde product of aldol condensation at 5°C of each of the following aldehydes. **(a)** Propanal; **(b)** butanal; **(c)** 2-phenylacetaldehyde; **(d)** 3-phenylpropanal.

EXERCISE 18-13

Can benzaldehyde undergo aldol condensation? Why or why not?

At elevated temperature, the aldol is converted into its enolate ion, which eliminates hydroxide ion to yield the final product. The net result of this second sequence is a hydroxide-catalyzed dehydration of the aldol.

Mechanism of Dehydration

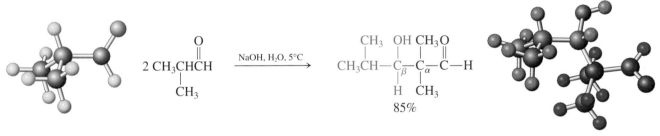

What makes the aldol condensation synthetically useful? It couples two functionalities, with carbon–carbon bond formation, to furnish two new functional groups. Thus, depending on temperature, its outcome is a β-hydroxycarbonyl or an α,β-unsaturated carbonyl compound. For example, at low temperature,

$$2 \ CH_3CHCH \ \xrightarrow{NaOH, \ H_2O, \ 5°C} \ CH_3CH-\underset{\beta}{C}-\underset{\alpha}{C}-C-H$$

85%

2-Methylpropanal   **3-Hydroxy-2,2,4-trimethylpentanal**

At higher temperature, however,

$$\xrightarrow{K_2CO_3, \ H_2O, \ \Delta}$$

80%

Heptanal   **(Z)-2-Pentyl-2-nonenal**

EXERCISE 18-14

Give the structure of the α,β-unsaturated aldehyde product of aldol condensation at higher temperature of each of the aldehydes in Exercise 18-12.

Ketones can undergo aldol condensation

So far, only aldehydes have been discussed as substrates in the aldol condensation. What about ketones? Treatment of propanone (acetone) with base does indeed lead to some 4-hydroxy-4-methyl-2-pentanone, but the conversion is poor because of an unfavorable equilibrium with starting material.

Aldol Formation from Propanone (Acetone)

$$\underset{\text{94\%}}{CH_3\overset{\overset{\textstyle O}{\|}}{C}CH_3} \underset{HO^-}{\rightleftharpoons} \underset{\underset{\text{6\%}}{}}{CH_3\underset{\underset{CH_3}{|}}{\overset{\overset{\textstyle OH}{|}}{C}}-CH_2\overset{\overset{\textstyle O}{\|}}{C}CH_3}$$

4-Hydroxy-4-methyl-2-pentanone

$$CH_3\underset{\underset{CH_3}{|}}{\overset{\overset{\textstyle OH}{|}}{C}}-CH_2\overset{\overset{\textstyle O}{\|}}{C}CH_3$$

↓ NaOH, H$_2$O, Δ

$$\underset{H_3C}{\overset{H_3C}{>}}C=CH\overset{\overset{\textstyle O}{\|}}{C}CH_3$$

80%
4-Methyl-3-penten-2-one

+

H$_2$O (removed)

What explains the lesser driving force of the aldol reaction of ketones? Because the carbonyl bond in a ketone is somewhat stronger (by about 3 kcal mol^{-1}) than that in an aldehyde, aldol addition of ketones is endothermic. To drive the reaction forward, we can extract the product alcohol continuously from the reaction mixture as it is formed. Alternatively, under more vigorous conditions, dehydration and removal of water move the equilibrium toward the α,β-unsaturated ketone (see margin).

EXERCISE 18-15

Formulate the mechanism for the aldol addition of propanone. This process is reversible. Propose a mechanism for the conversion of 4-hydroxy-4-methyl-2-pentanone into propanone in the presence of $^-$OH, an example of a *retro-aldol reaction.*

In summary, treatment of enolizable aldehydes with catalytic base leads to β-hydroxy aldehydes at low temperature and to α,β-unsaturated aldehydes upon heating. The reaction proceeds by enolate attack on the carbonyl function. Aldol addition to a ketone carbonyl group is energetically unfavorable. To drive the aldol condensation of ketones to product, special conditions have to be used, such as removal of the water or the aldol formed in the reaction.

18-6 Crossed Aldol Condensation

What happens if we try to carry out an aldol condensation between the enolate of one aldehyde and the carbonyl carbon of another? In such a situation, called **crossed aldol condensation,** mixtures ensue, because enolates of both aldehydes are present and may react with the carbonyl groups of either starting compound. For example, a 1:1 mixture of acetaldehyde and propanal gives the four possible aldol addition products in comparable amounts.

Nonselective Crossed Aldol Reaction of Acetaldehyde and Propanal

(All four reactions occur simultaneously)

$$CH_3\overset{\overset{\textstyle O}{\|}}{C}H \; + \; CH_3CH_2\overset{\overset{\textstyle O}{\|}}{C}H \longrightarrow CH_3\underset{\underset{H}{|}}{\overset{\overset{\textstyle OH}{|}}{C}}-\underset{\underset{CH_3}{|}}{C}H\overset{\overset{\textstyle O}{\|}}{C}H$$

3-Hydroxy-2-methylbutanal

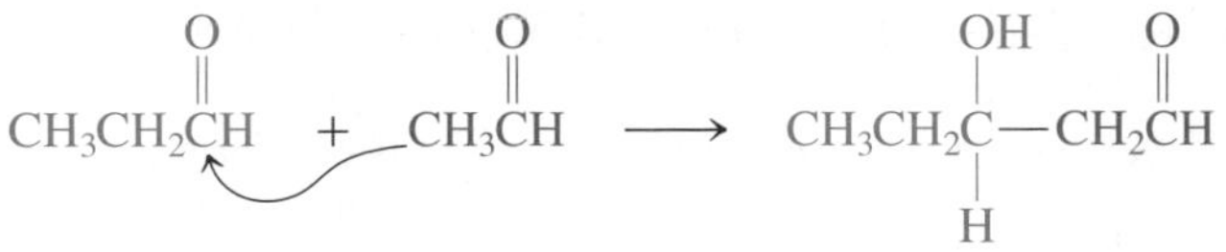

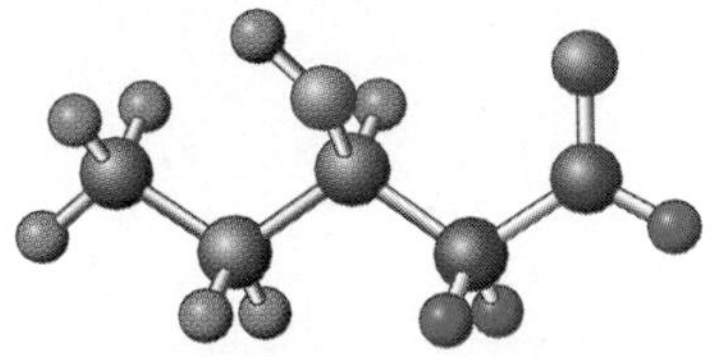

$$CH_3CH_2CH + CH_3CH \longrightarrow CH_3CH_2\underset{\underset{H}{|}}{\overset{\overset{OH}{|}}{C}}-CH_2CH$$

3-Hydroxypentanal

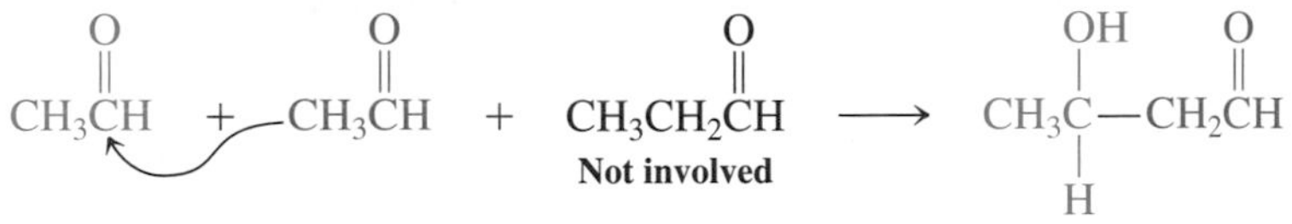

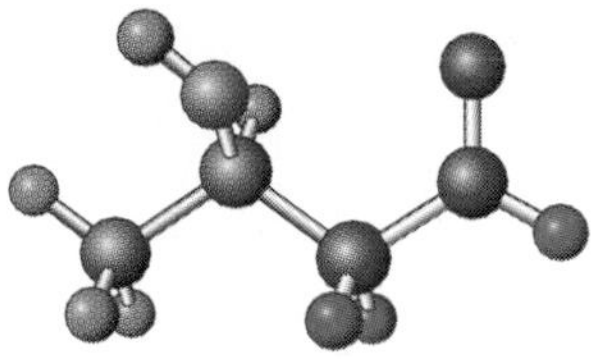

$$CH_3CH + CH_3CH + CH_3CH_2CH \longrightarrow CH_3\underset{\underset{H}{|}}{\overset{\overset{OH}{|}}{C}}-CH_2CH$$
Not involved

3-Hydroxybutanal

$$CH_3CH_2CH + CH_3CH_2CH + CH_3CH \longrightarrow CH_3CH_2\underset{\underset{H}{|}}{\overset{\overset{OH}{|}}{C}}-\underset{\underset{CH_3}{|}}{CH}CH$$
Not involved

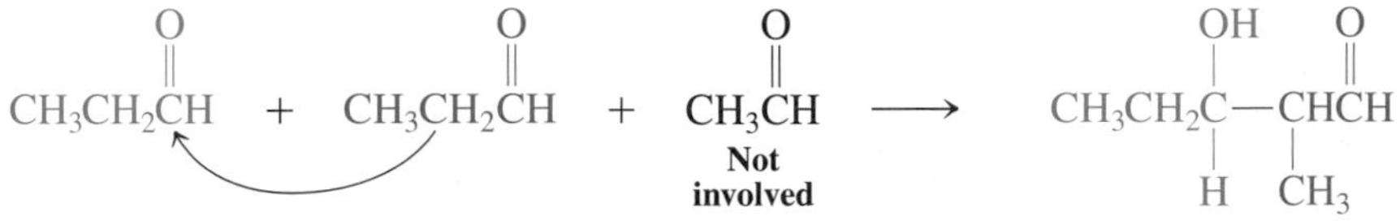

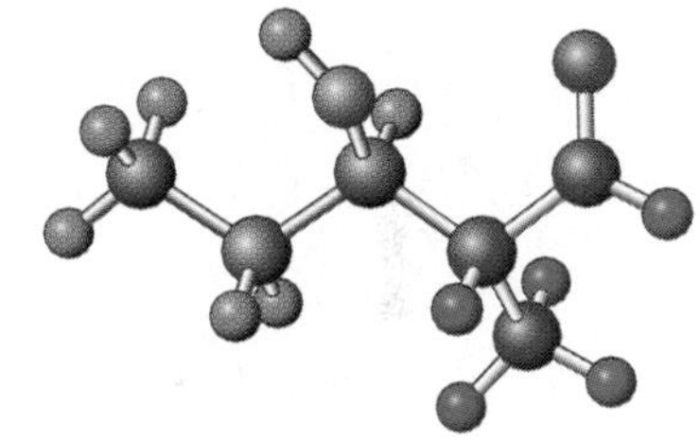

3-Hydroxy-2-methylpentanal

CHEMICAL HIGHLIGHT 18-3   Enzymes in Crossed Aldol Condensations

The biosynthesis of sugars uses crossed aldol condensations (catalyzed by an enzyme appropriately named *aldolase*) to construct carbon–carbon bonds. A protein-bound primary amine (a substituent on the amino acid lysine) first condenses with the carbonyl group of 1,3-dihydroxypropanone (1,3-dihydroxyacetone) monophosphate to form an iminium salt.

Enzyme-catalyzed deprotonation gives an enamine, whose nucleophilic carbon attacks the carbonyl group of 2,3-dihydroxypropanal (glyceraldehyde) 3-phosphate, in a reaction that constitutes a nitrogen analog of a crossed aldol condensation. Hydrolysis of the resulting iminium salt furnishes the six-carbon sugar *fructose*.

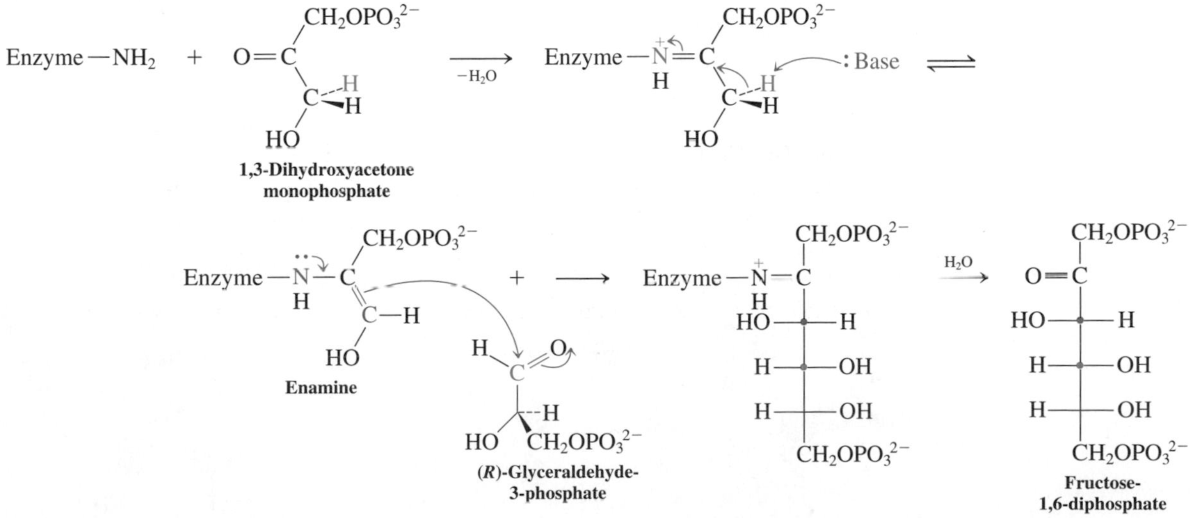

1,3-Dihydroxyacetone monophosphate

Enamine

(R)-Glyceraldehyde-3-phosphate

Fructose-1,6-diphosphate

Can we ever efficiently synthesize a single aldol product from the reaction of two different aldehydes? We can, when one of the aldehydes has *no enolizable hydrogens,* because two of the four possible condensation products cannot form. We add the enolizable aldehyde slowly to an excess of the nonenolizable reactant in the presence of base. As soon as the enolate of the addend is generated, it is trapped by the other aldehyde.

A Successful Crossed Aldol Condensation

$$\underset{\substack{\text{2,2-Dimethyl-}\\\text{propanal}\\\text{(No }\alpha\text{-hydrogens)}}}{\underset{\underset{\text{CH}_3}{|}}{\overset{\overset{\text{CH}_3}{|}}{\text{CH}_3\text{CCHO}}}} + \underset{\substack{\text{Propanal}\\\textbf{Added slowly}}}{\text{CH}_3\text{CH}_2\text{CHO}} \xrightarrow{\text{NaOH, H}_2\text{O, }\Delta} \underset{\text{2,4,4-Trimethyl-2-pentenal}}{\underset{\underset{\text{CH}_3}{|}}{\overset{\overset{\text{CH}_3\ \ \text{CH}_3}{|\ \ \ \ \ |}}{\text{CH}_3\text{CCH}=\text{CCHO}}}} + \text{H}_2\text{O}$$

The product of reaction (a) of Exercise 18-16 derives its name—cinnamaldehyde—from the flavor that it imparts to these popular items.

EXERCISE 18-16

Show the likely products of the following aldol condensations.

(a) [benzaldehyde structure with CH=O] + CH₃CHO   (b) 2 [cyclohexanecarbaldehyde structure, CHO] (reacts with itself)

(c) CH₂=CHCHO + CH₃CH₂CHO

In summary, crossed aldol condensations furnish product mixtures unless one of the reaction partners cannot enolize.

18-7 Intramolecular Aldol Condensation

It is possible to carry out aldol condensation between an enolate ion and a carbonyl group *in the same molecule;* such a reaction is called an **intramolecular aldol condensation.** This section describes the utility of this type of reaction in the synthesis of cyclic compounds.

Treatment of a dilute solution of hexanedial with aqueous base results in the formation of a cyclic product. As soon as an enolate ion is generated at one of the termini of the molecule, it attacks the carbonyl carbon at the other.

$$\underset{}{\overset{\overset{\text{O}}{\|}\ \ \ \ \ \ \ \ \ \ \ \ \ \ \ \ \overset{\text{O}}{\|}}{\text{HCCH}_2\text{CH}_2\text{CH}_2\text{CH}_2\text{CH}}} \xrightarrow[-\text{H}_2\text{O}]{\text{KOH, H}_2\text{O}} \text{[1-cyclopentenecarbaldehyde, 62\%]}$$

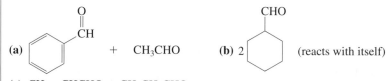

Hexanedial   1-Cyclopentenecarbaldehyde

Attack at the carbonyl carbon of a different molecule (*intermolecular* aldol condensation) is minimized by the low concentration of the dialdehyde and the favorable kinetics of five-membered-ring formation.

Intramolecular ketone condensations are a ready source of cyclic and bicyclic α,β-unsaturated ketones. Usually the least strained ring is generated, typically one that is five or six membered. Thus, reaction of 2,5-hexanedione results in condensation between an enolate at C1 and the C5 carbonyl group. An alternative pathway, attack of a C3 enolate on C5 to give a strained three-membered ring, is not observed.

Intramolecular Aldol Condensation of a Dione

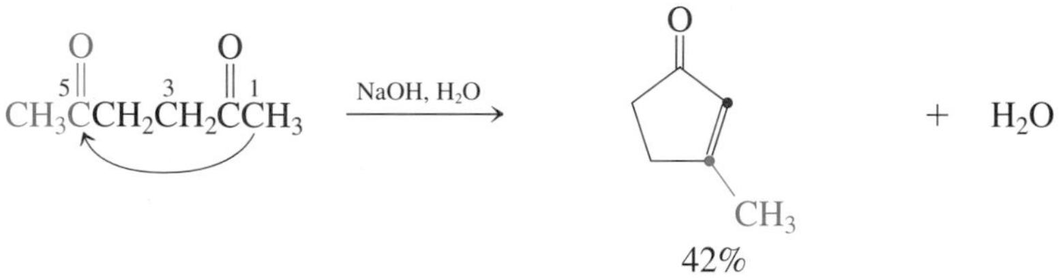

$$ \underset{\text{2,5-Hexanedione}}{\overset{O}{\underset{5}{CH_3}}\overset{}{\underset{}{C}}\overset{}{CH_2}\overset{}{\underset{3}{C}}H_2\overset{O}{\underset{1}{C}}CH_3} \xrightarrow{\text{NaOH, H}_2\text{O}} \underset{\underset{\text{3-Methyl-2-cyclopentenone}}{42\%}}{\overset{O}{\overset{}{}}\quad CH_3} \quad + \quad H_2O $$

Why is intramolecular aldol condensation of ketones so readily achieved when the intermolecular analog is so difficult? Part of the answer lies in the thermodynamics of the process. We know that reaction of two ketone molecules suffers from an unfavorable enthalpy contribution because of the loss of a strong carbonyl bond (Section 18-5). In addition, the process must overcome the loss of entropy arising from the conversion of two molecules into one, as the initial hydroxyketone product forms. In contrast, the intramolecular transformation merely converts an acyclic substrate into a cyclic product, a process "costing" much less in entropic terms and therefore more favorable thermodynamically.

EXERCISE 18-17

Predict the outcome of intramolecular aldol condensations of the following compounds.

(a) Cyclodecane-1,5-dione

(b) $C_6H_5\overset{O}{\overset{\|}{C}}(CH_2)_2\overset{O}{\overset{\|}{C}}CH_3$

(c) 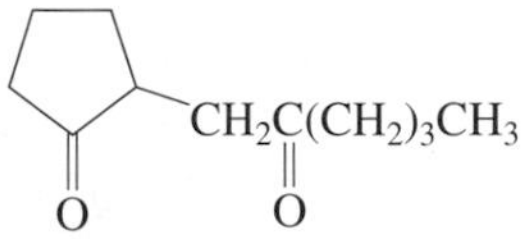 [structure: cyclopentanone ring with $CH_2C(CH_2)_3CH_3$ substituent and adjacent ketone]

(d) 2,7-Octanedione

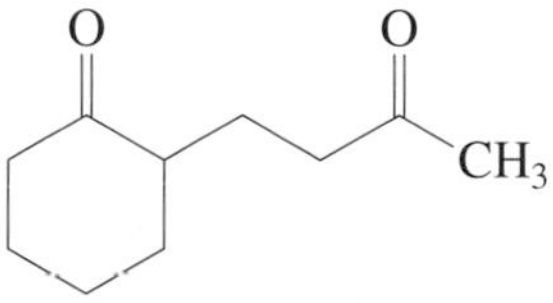

EXERCISE 18-18

The intramolecular aldol condensation of 2-(3-oxobutyl)cyclohexanone (margin) can in principle lead to four different compounds (ignoring stereochemistry). Draw them and suggest which one would be the most likely to form. (**Hint:** Build models!)

[structure in margin]

2-(3-Oxobutyl)cyclohexanone

EXERCISE 18-19

Prepare the following compounds from any starting material, using aldol reactions in the crucial step. (**Hint:** The last preparation requires a double aldol addition.)

(a) 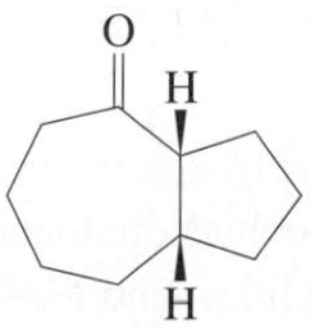 [bicyclic ketone structure with H substituents]

(b) 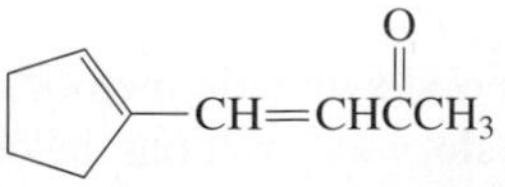 [cyclopentene structure with $-CH=CHCCH_3$ side chain]

(c) 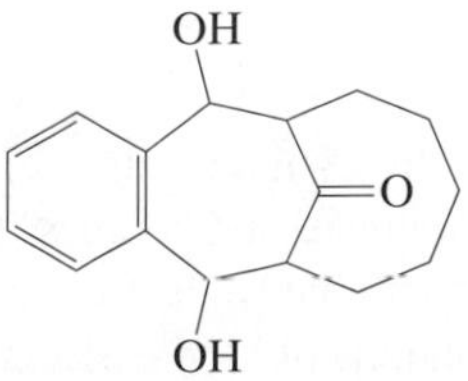 [tricyclic structure with OH groups and =O]

In summary, intramolecular aldol condensation succeeds with both aldehydes and ketones. It can be highly selective and gives the least strained cycloalkenones.

18-8 Other Preparations of α,β-Unsaturated Aldehydes and Ketones

Cyclopentanone

1. Cl₂, CCl₄
2. Na₂CO₃

73%

2-Cyclopentenone

Sections 18-5 through 18-7 showed how α,β-unsaturated aldehydes and ketones are prepared by the aldol condensation. Are there other synthetic routes to these compounds? This section will present several, beginning with the one in the margin. We have seen how a halogen atom may be introduced next to the carbonyl function (Section 18-3). Subsequent base-mediated dehydrohalogenation (Sections 7-7 and 11-8) affords the carbon–carbon double bond.

Another way of introducing the double bond is by a Wittig reaction (Section 17-12) of carbonyl-substituted ylides, which are stabilized by resonance.

$$\left[(C_6H_5)_3P=CH-\overset{\overset{\displaystyle :O:}{\|}}{C}H \;\longleftrightarrow\; (C_6H_5)_3\overset{+}{P}-\overset{\overset{\displaystyle :O:}{\|}}{\underset{..}{C}}H-\overset{\overset{\displaystyle :O:}{\|}}{C}H \;\longleftrightarrow\; (C_6H_5)_3\overset{+}{P}-CH=\overset{\overset{\displaystyle :\ddot{O}:^-}{|}}{C}H \right]$$

A stabilized ylide

Such **stabilized ylides** are comparatively unreactive and can be readily isolated and stored. However, they do react with aldehydes to form the corresponding α,β-unsaturated aldehydes, as the following example shows.

Wittig Synthesis of an α,β-Unsaturated Aldehyde

$(C_6H_5)_3P=CHCH{=}{O}$ +

Heptanal

$\xrightarrow[- (C_6H_5)_3PO]{(CH_3CH_2)_2O,\ \Delta}$

81%

trans-2-Nonenal

A fourth way to prepare α,β-unsaturated aldehydes and ketones is by oxidation of allylic alcohols with manganese dioxide, MnO₂ (Section 17-4). Vitamin A, for example, can be oxidized in this way to all-*trans*-retinal, a molecule of importance in the chemistry of vision (see Chemical Highlight 18-4).

Vitamin A

$\xrightarrow[\text{propanone (acetone)}]{MnO_2,}$

80%

All-*trans*-retinal

In summary, this section reviewed synthetic methods of preparing α,β-unsaturated aldehydes and ketones. They are aldol condensations, halogenation–dehydrohalogenation of saturated aldehydes and ketones, Wittig reactions with stabilized ylides, and MnO₂ oxidations of allylic alcohols.

CHEMICAL HIGHLIGHT 18-4

Reactions of Unsaturated Aldehydes in Nature: The Chemistry of Vision

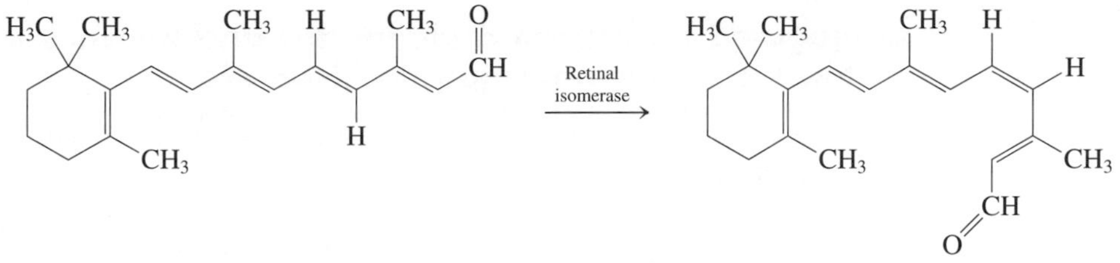

trans-Retinal

cis-Retinal

Vitamin A (retinol) is an important nutritional factor in vision. Vitamin A deficiency causes night blindness. *Trans*-retinal is present in the light-receptor cells of the human eye but, before it can fulfill its biological function, it has to be isomerized by an enzyme, *retinal isomerase,* to give *cis*-retinal. This molecule fits well into the active site of a protein called *opsin* (approximate molecular weight 38,000). As shown below, *cis*-retinal reacts with one of the amine substituents of opsin to form the imine *rhodopsin,* the light-sensitive chemical unit in the eye. The electronic spectrum of rhodopsin, with a λ_{max} at 506 nm ($\epsilon = 40,000$), indicates the presence of a protonated imine group.

When a photon strikes rhodopsin, the *cis*-retinal part isomerizes extremely rapidly, in only picoseconds (10^{-12} s), to the trans isomer. This isomerization induces a tremendous geometric change, which appears to severely disrupt the snug fit of the original molecule in the protein cavity. Within nanoseconds (10^{-9} s), a series of new intermediates form from this photoproduct, accompanied by conformational changes in the protein structure, followed by eventual hydrolysis of the ill-fitting retinal unit. This sequence initiates a nerve impulse perceived by us as light. The *trans*-retinal is then reisomerized to the cis form by retinal isomerase and reforms rhodopsin, ready for another photon. What is extraordinary about this mechanism is its sensitivity, which allows the eye to register as little as one photon impinging on the retina. Curiously, all known visual systems in nature, even though they might have a completely different evolutionary history, use the retinal system for visual excitation. Evidently, this molecule offers an optimal solution to the problem of vision.

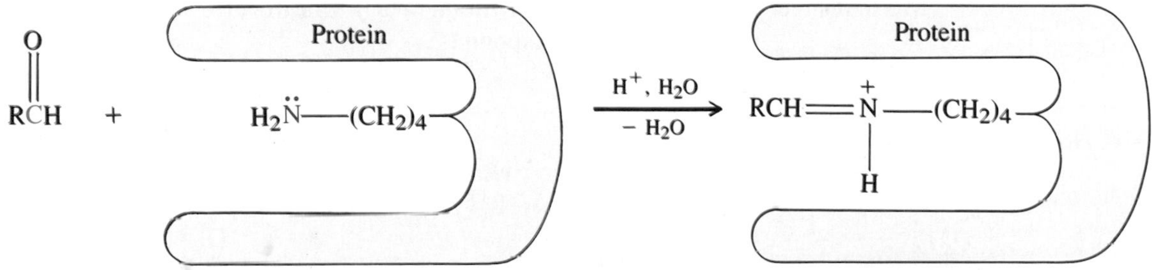

cis-Retinal   **Opsin**   **Rhodopsin**

18-9 Properties of α,β-Unsaturated Aldehydes and Ketones

How do the properties of α,β-unsaturated aldehydes and ketones compare with the individual characteristics of their two types of double bonds? We shall find that their chemistry in some situations is a simple composite of the two and under other circumstances involves the α,β-unsaturated carbonyl, or **enone,** functional group as a whole. As later chapters will indicate, this complex reactivity is quite typical of molecules with two functional groups, or **difunctional compounds.**

Conjugated unsaturated aldehydes and ketones are more stable than their unconjugated isomers

Like conjugated dienes (Section 14-5), α,β-unsaturated aldehydes and ketones are stabilized by electron delocalization.

Resonance Forms of 2-Butenal

$$\left[CH_3CH{=}CH{-}\overset{\overset{\displaystyle :\ddot{O}:}{\|}}{C}H \longleftrightarrow CH_3CH{=}CH{-}\overset{\overset{\displaystyle :\ddot{O}:^-}{|}}{C}\overset{+}{H} \longleftrightarrow CH_3\overset{+}{C}H{-}CH{=}\overset{\overset{\displaystyle :\ddot{O}:^-}{|}}{C}H \right]$$

Thus, unconjugated β,γ-unsaturated carbonyl compounds rearrange readily to their conjugated isomers. The carbon–carbon double bond is said to "move into conjugation" with the carbonyl group, as the following example shows.

Isomerization of a β,γ-Unsaturated Carbonyl Compound to a Conjugated System

$$\overset{\gamma}{C}H_2{=}\overset{\beta}{C}HCH_2\overset{\overset{\displaystyle O}{\|}}{C}H \xrightarrow{\text{H}^+ \text{ or HO}^-,\ \text{H}_2\text{O}} CH_3\overset{\beta}{C}H{=}\overset{\alpha}{C}H\overset{\overset{\displaystyle O}{\|}}{C}H$$

3-Butenal **2-Butenal**

The isomerization can be acid or base catalyzed. In the base-catalyzed reaction, the intermediate is the conjugated dienolate ion, which is reprotonated at the carbon terminus.

Mechanism of Base-Mediated Isomerization of β,γ-Unsaturated Carbonyl Compounds

$$CH_2{=}CHCH_2\overset{\overset{\displaystyle :O:}{\|}}{C}H \ + \ H\ddot{O}:^- \ \rightleftharpoons$$

Protonation here gives conjugated aldehyde

$$\left[CH_2{=}CH{-}\overset{..}{\underset{..}{C}}H{-}\overset{\overset{\displaystyle :\ddot{O}:}{\|}}{C}H \longleftrightarrow \overset{+}{C}H_2{=}CH{-}CH{=}\overset{\overset{\displaystyle :\ddot{O}:^-}{|}}{C}H \longleftrightarrow :\overset{-}{C}H_2{-}CH{=}CH{-}\overset{\overset{\displaystyle :O:}{\|}}{C}H \right] \ + \ H\overset{..}{\underset{..}{O}}H \ \rightleftharpoons$$

Dienolate ion

$$CH_3CH{=}CH\overset{\overset{\displaystyle :O:}{\|}}{C}H \ + \ H\ddot{O}:^-$$

EXERCISE 18-20

Propose a mechanism for the *acid*-catalyzed isomerization of 3-butenal to 2-butenal. (**Hint:** An intermediate is 1,3-butadien-1-ol.)

α,β-Unsaturated aldehydes and ketones undergo the reactions typical of their component functional groups

α,β-Unsaturated aldehydes and ketones undergo many reactions that are perfectly predictable from the known chemistry of the carbon–carbon and carbon–oxygen double bonds. For example, hydrogenation by palladium on carbon gives the saturated carbonyl compound (Section 17-5).

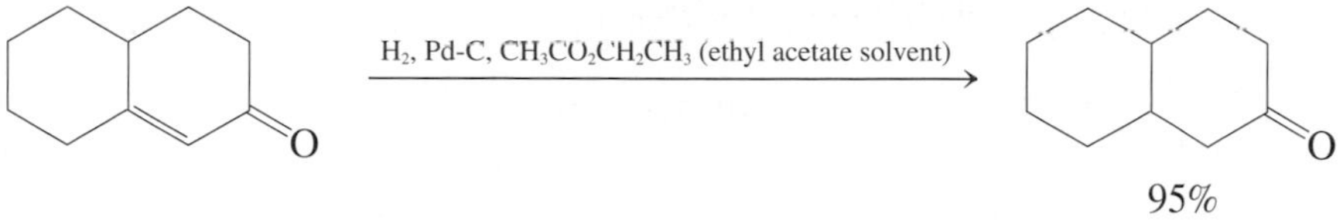

$$\xrightarrow{\text{H}_2,\ \text{Pd-C, CH}_3\text{CO}_2\text{CH}_2\text{CH}_3\ (\text{ethyl acetate solvent})}$$

95%

The double bond of some conjugated enones and enals (unsaturated aldehydes) can undergo "conjugate reduction" by a reducing system described earlier in the conversion of alkynes into trans alkenes: an alkali metal in liquid ammonia (Section 13-7). The mechanisms of these reductions are similar and include two one-electron transfers and two protonations. The method allows for the selective reduction of the conjugated double bond in the presence of unconjugated ones.

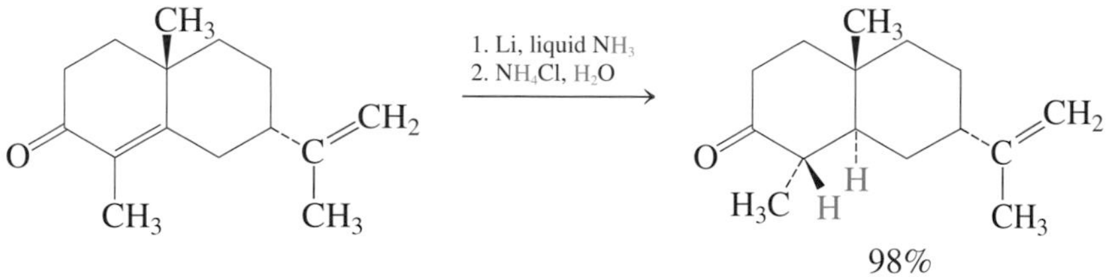

$$\xrightarrow{\begin{array}{l}\text{1. Li, liquid NH}_3\\ \text{2. NH}_4\text{Cl, H}_2\text{O}\end{array}}$$

98%

The carbon–carbon π system also undergoes electrophilic additions. For example, bromination furnishes a dibromocarbonyl compound (compare Section 12-5).

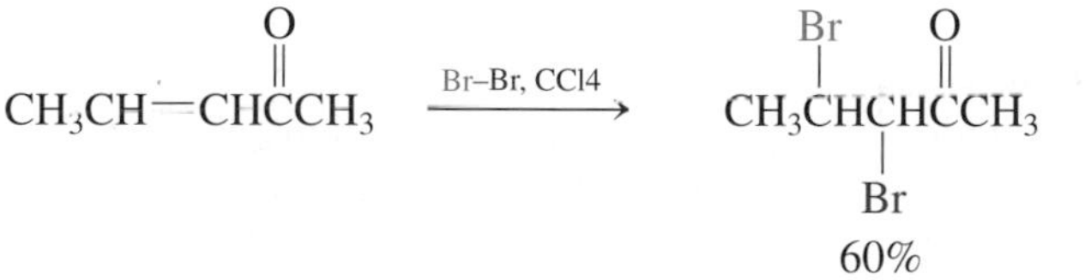

$$\text{CH}_3\text{CH}=\text{CHCCH}_3 \xrightarrow{\text{Br–Br, CCl4}} \text{CH}_3\text{CHCHCCH}_3$$

3-Penten-2-one

3,4-Dibromo-2-pentanone

60%

The carbonyl function undergoes the usual addition reactions (Section 17-9). Thus, nucleophilic addition of amines results in the expected condensation products (see, however, the next section).

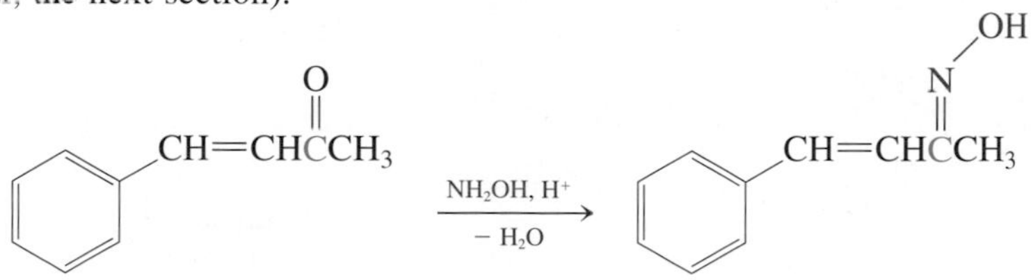

$$\xrightarrow[-\ \text{H}_2\text{O}]{\text{NH}_2\text{OH, H}^+}$$

4-Phenyl-3-buten-2-one

Oxime
(m.p. 115°C)

EXERCISE 18-21

Propose a synthesis of 1-pentanol starting from propanal.

In summary, α,β-unsaturated aldehydes and ketones are more stable than their non-conjugated counterparts. Either base or acid catalyzes interconversion of the isomeric systems. Reactions typical of alkenes and carbonyl compounds are also characteristic of α,β-unsaturated aldehydes and ketones.

18-10 Conjugate Additions to α,β-Unsaturated Aldehydes and Ketones

We now show how the conjugated carbonyl group of α,β-unsaturated aldehydes and ketones can enter into reactions that involve the entire functional system. These reactions are 1,4-additions of the type encountered with 1,3-butadiene (Section 14-6). The reactions proceed by acid-catalyzed, radical, or nucleophilic addition mechanisms, depending on the reagents.

The entire conjugated system takes part in 1,4-additions

Addition reactions in which only one of the π bonds of a conjugated system takes part are classified as 1,2-additions (compare Section 14-6). Examples are the additions of Br_2 to the carbon–carbon double bond and NH_2OH to the carbon–oxygen double bond of α,β-unsaturated aldehydes and ketones considered in the preceding section.

1,2-Addition of a Polar Reagent A–B to a Conjugated Enone

However, several reagents add to the conjugated π system in a 1,4-manner, a result also called **conjugate addition.** In these transformations, the nucleophilic part of a reagent attaches itself to the β-carbon, and the electrophilic part (commonly, a proton) binds to the carbonyl oxygen. The initial product is an enol, which subsequently rearranges to its keto form. Thus, the end result *appears* to be that of 1,2-addition of HNu to the carbon-carbon double bond.

1,4-Addition of a Polar Reagent A–B to a Conjugated Enone

Oxygen and nitrogen nucleophiles undergo conjugate additions

Water, alcohols, amines, and similar nucleophiles undergo 1,4-additions, as the following examples (hydration and amination) show. Although these reactions can be catalyzed by acid or base, the products are usually formed faster and in higher yields with base. These processes are readily reversed at elevated temperatures.

$$CH_2=CHCCH_3 \quad \xrightarrow{\text{HOH, Ca(OH)}_2,\ -5^\circ\text{C}} \quad HOCH_2-CHCCH_3$$

3-Buten-2-one   **4-Hydroxy-2-butanone**

$$\underset{\text{H}_3\text{C}}{\overset{\text{H}_3\text{C}}{\diagdown}}C=CHCCH_3 \quad \xrightarrow{\text{CH}_3\text{N}\diagup\overset{\text{H}}{\diagdown}\text{H},\ \text{H}_2\text{O}} \quad (CH_3)_2\overset{CH_3NH}{\underset{H}{C}}-\overset{O}{\overset{\|}{C}}HCCH_3$$

4-Methyl-3-penten-2-one   75%
                                       4-Methyl-4-(methylamino)-2-pentanone

What determines whether 1,2- or 1,4-additions will take place? With the nucleophiles shown, both processes are reversible. Usually the 1,4-products form because they are carbonyl compounds and are more stable than the species arising from 1,2-addition (hydrates, hemiacetals, and hemiaminals, Sections 17-6, 17-7, and 17-9). Exceptions include amine derivatives such as hydroxylamine, semicarbazide, or the hydrazines, for which 1,2-addition eventually leads to an imine product whose precipitation out of solution drives the equilibrium.

The mechanism of the base-catalyzed addition to conjugated aldehydes and ketones is direct nucleophilic attack at the β-carbon to give the enolate ion, which is subsequently protonated.

Mechanism of Base-Catalyzed Hydration of α,β-Unsaturated Aldehydes and Ketones

$$HO:^- + \quad \overset{\text{:O:}}{\underset{}{C=C-C}} \quad \rightleftharpoons \quad HOC-C=C \quad \overset{\text{HOH}}{\rightleftharpoons} \quad HOC-CHC- + \ ^-\text{:OH}$$

<u>**EXERCISE 18-22**</u>

Treatment of 3-chloro-2-cyclohexenone with sodium methoxide in methanol gives 3-methoxy-2-cyclohexenone. Write the mechanism of this reaction. (**Hint:** Start with a conjugate addition.)

Hydrogen cyanide also undergoes conjugate addition

Treatment of a conjugated aldehyde or ketone with cyanide in the presence of acid may result in attack by cyanide at the β-carbon, in contrast with cyanohydrin formation (Section 17-11). This transformation proceeds through a 1,4-addition pathway. The reaction includes protonation of the oxygen, then nucleophilic β-attack, and finally enol–keto tautomerization.

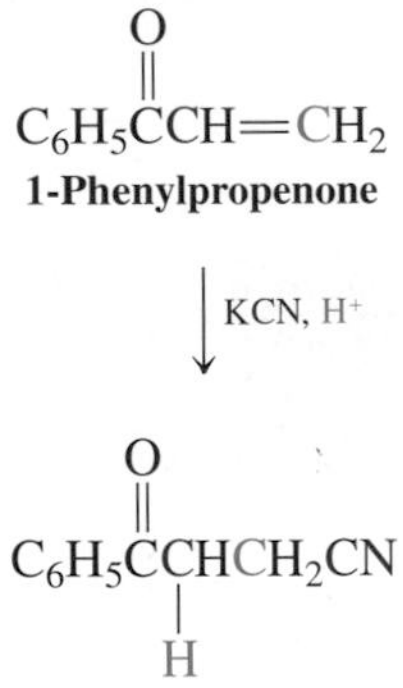

$$\underset{\substack{\text{1-Phenylpropenone}}}{C_6H_5\overset{\overset{\displaystyle O}{\|}}{C}CH=CH_2}$$

$\downarrow$ KCN, H⁺

$$\underset{\substack{\text{67\%}\\\text{4-Oxo-4-phenyl-}\\\text{butanenitrile}}}{C_6H_5\overset{\overset{\displaystyle O}{\|}}{C}\underset{\underset{\displaystyle H}{|}}{C}HCH_2CN}$$

EXERCISE 18-23

Formulate the mechanism of the acid-catalyzed 1,4-addition of cyanide to 1-phenyl-propenone (see margin).

In summary, α,β-unsaturated aldehydes and ketones are synthetically useful building blocks in organic synthesis because of their ability to undergo 1,4-additions. Hydrogen cyanide addition leads to β-cyano carbonyl compounds; oxygen and nitrogen nucleophiles also can add to the β-carbon.

18-11 1,2- and 1,4-Additions of Organometallic Reagents

Organometallic reagents may add to the α,β-unsaturated carbonyl function in either 1,2 or 1,4 fashion. Organolithium reagents, for example, react preferentially by direct nucleophilic attack at the carbonyl carbon.

Exclusive 1,2-Addition of an Organolithium

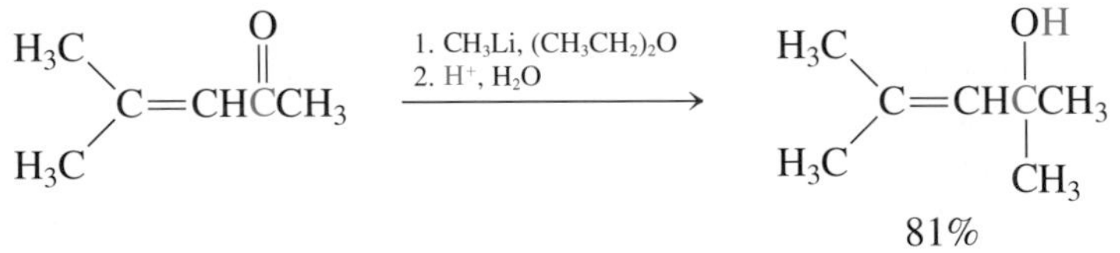

4-Methyl-3-penten-2-one → 2,4-Dimethyl-3-penten-2-ol, 81%
(1. CH₃Li, (CH₃CH₂)₂O; 2. H⁺, H₂O)

Conversely, cuprates give only products of conjugate addition.

Exclusive 1,4-Addition of a Lithium Organocuprate

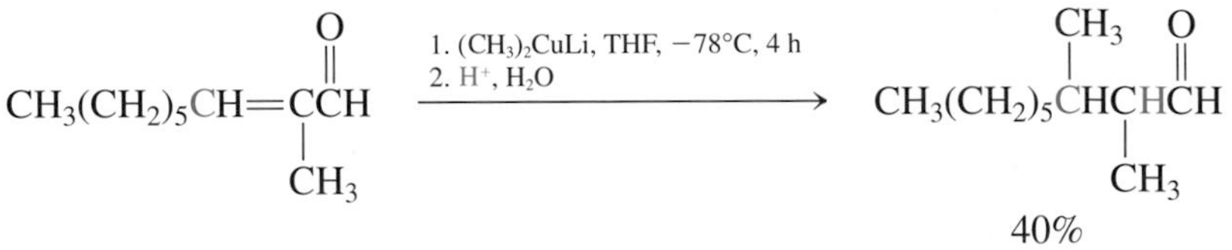

2-Methyl-2-nonenal → 2,3-Dimethylnonanal, 40%
(1. (CH₃)₂CuLi, THF, −78°C, 4 h; 2. H⁺, H₂O)

The copper-mediated 1,4-addition reactions are thought to proceed through complex electron-transfer mechanisms. The first isolable intermediate is an enolate ion, which can be trapped by alkylating species, as shown in Section 18-4. Conjugate addition followed by alkylation constitutes a useful sequence for α,β-dialkylation of unsaturated aldehydes and ketones.

α,β-Dialkylation of Unsaturated Carbonyl Compounds

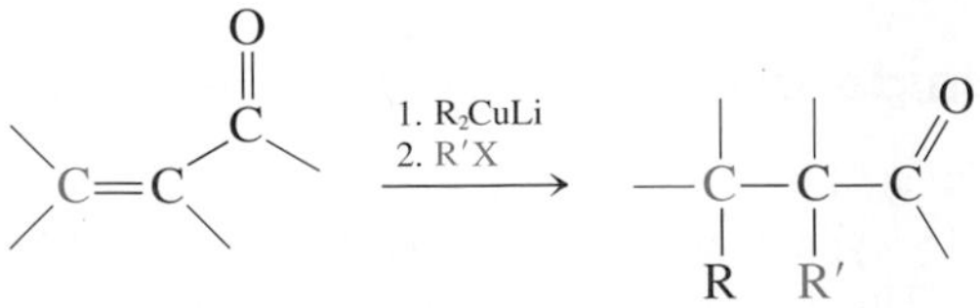

(1. R₂CuLi; 2. R′X)

The following example illustrates this reaction.

1. (CH$_3$CH$_2$CH$_2$CH$_2$)$_2$CuLi, THF
2. CH$_3$I

→

84%, 4 : 1
***trans*-** and ***cis*-3-Butyl-2-methylcyclohexanone**

Reactions of ordinary Grignard reagents with α,β-unsaturated aldehydes and ketones give 1,2-addition, 1,4-addition, or both, depending on the structures of the reacting species and conditions.

EXERCISE 18-24

Show how you might synthesize the following compounds from 3-methyl-2-cyclohexenone. (**Hints:** Work backward; the last step in (b) is an intramolecular aldol condensation.)

(a)   **(b)**

In summary, organolithium reagents add 1,2 and organocuprate reagents add 1,4 to α,β-unsaturated carbonyl systems. The latter process initially gives rise to a β-substituted enolate, which upon exposure to haloalkanes furnishes α,β-dialkylated aldehydes and ketones.

18-12 Conjugate Additions of Enolate Ions: Michael Addition and Robinson Annulation

Like other nucleophiles, enolate ions undergo conjugate additions to α,β-unsaturated aldehydes and ketones, in a reaction known as the **Michael* addition.** This transformation works best with enolates derived from β-dicarbonyl compounds (Chapter 21), but it also occurs with simpler systems, as the following examples show.

Michael Addition

$$CH_3CCH_2CCH_3 \ + \ CH_2{=}CHCH \xrightarrow{\text{Pyridine}} $$

27%

$$+ \ CH_2{=}CHCC_6H_5 \xrightarrow{CH_3CH_2O^-K^+,\ CH_3CH_2OH,\ (CH_3CH_2)_2O}$$

64%

---

*Professor Arthur Michael (1853–1942), Harvard University, Cambridge, MA.

CHEMICAL HIGHLIGHT 18-5

Prostaglandins: α,β-Dialkylation in Synthesis

The α,β-dialkylation procedure has been exploited in the total synthesis of *prostaglandins,* powerful physiologically active compounds (see Chemical Highlight 11-1 and Section 19-14). They appear to regulate a remarkable variety of bodily functions, including those of the endocrine, reproductive, nervous, digestive, hemostatic, respiratory, cardiovascular, and renal systems. Because of these properties, they are potential drugs in the treatment of hypertension, asthma, fever, inflammations, and ulcers. One of the commercially available prostaglandins induces labor in pregnant women. Others have applications in animal breeding by controlling the day in which the animal goes into heat.

A synthesis of prostaglandin $PGF_2\alpha$, developed by Stork,* includes two conjugate additions as well as an aldol reaction.

*Professor Gilbert Stork (b. 1921), Columbia University, New York.

A Prostaglandin Synthesis
(R, R′, R″ are protecting groups; C_5H_{11} = pentyl)

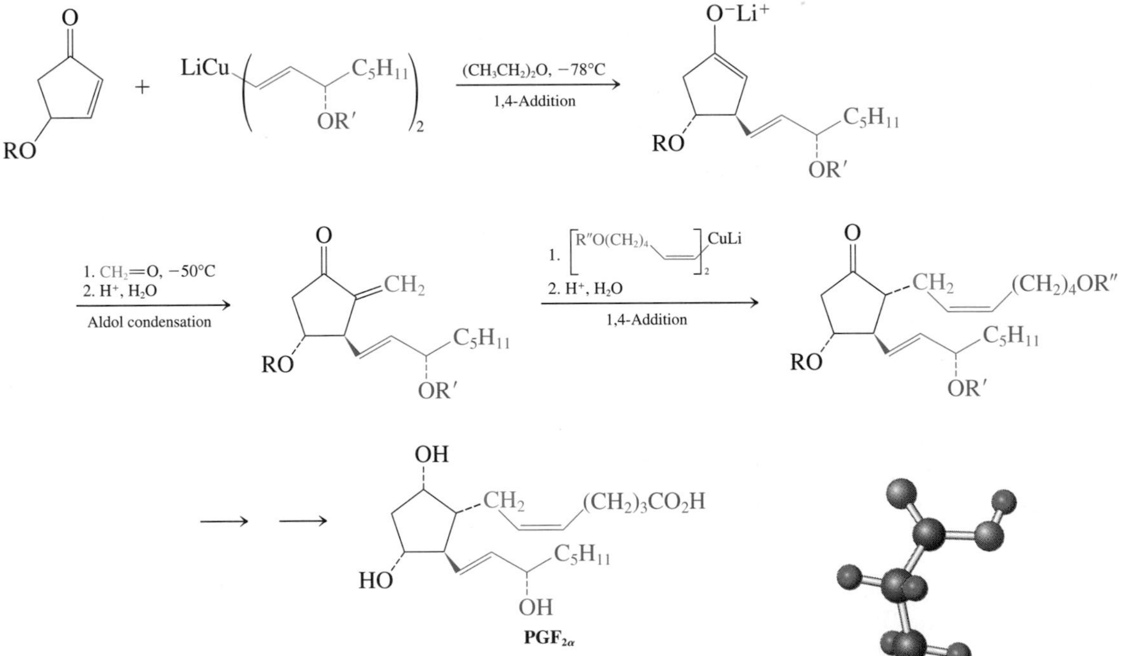

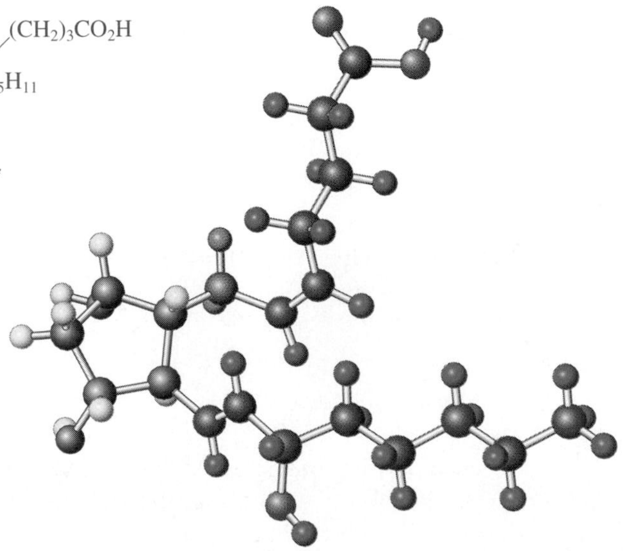

$PGF_{2\alpha}$

The mechanism of the Michael addition includes nucleophilic attack by the enolate ion on the β-carbon of the unsaturated carbonyl compound (the Michael "acceptor"), followed by protonation of the resulting enolate.

Mechanism of the Michael Addition

As the mechanism indicates, the reaction works because of the nucleophilic potential of the α-carbon of an enolate and the electrophilic reactivity of the β-carbon of an α,β-unsaturated carbonyl compound.

With some Michael acceptors, such as 3-buten-2-one, the products of the initial addition are capable of a subsequent intramolecular aldol condensation, which creates a new ring.

Michael Addition Followed by Intramolecular Condensation

The synthetic sequence of a Michael addition followed by an intramolecular aldol condensation is also called a **Robinson* annulation.**

Robinson Annulation

The Robinson annulation has found extensive use in the synthesis of polycyclic ring systems, including steroids and other natural products containing six-membered rings.

*Sir Robert Robinson (1886–1975), Oxford University, England, Nobel Prize 1947 (chemistry).

EXERCISE 18-25

Propose syntheses of the following compounds by Michael or Robinson reactions.

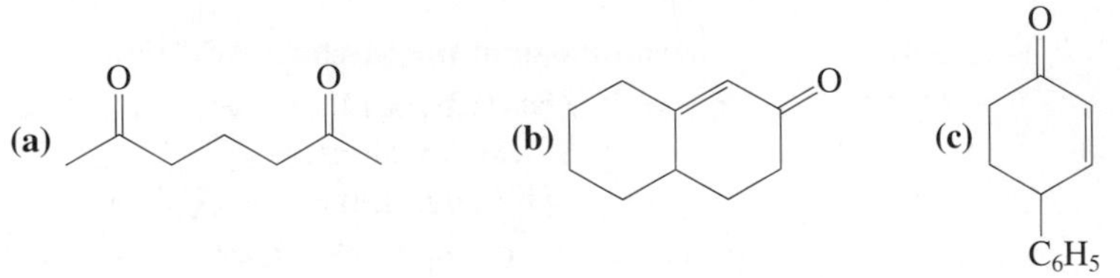

(a) (b) (c)

In summary, the Michael addition results in the conjugate addition of an enolate ion to give dicarbonyl compounds. The Robinson annulation reaction combines a Michael addition with a subsequent intramolecular aldol condensation to produce new cyclic enones.

CHAPTER INTEGRATION PROBLEM

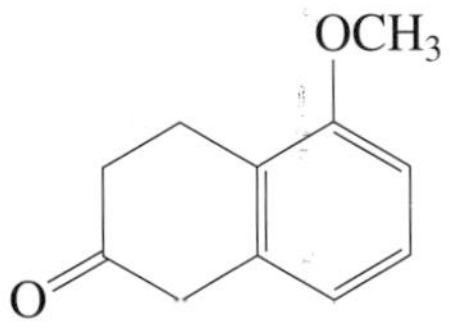

The Robinson annulation sequence is a powerful method for the construction of six-membered rings. It is therefore no surprise that this sequence has been widely applied in steroid synthesis. Beginning with the bicyclic ketone shown in the margin, propose a synthesis that makes use of one or more Robinson annulations for the steroid shown here.

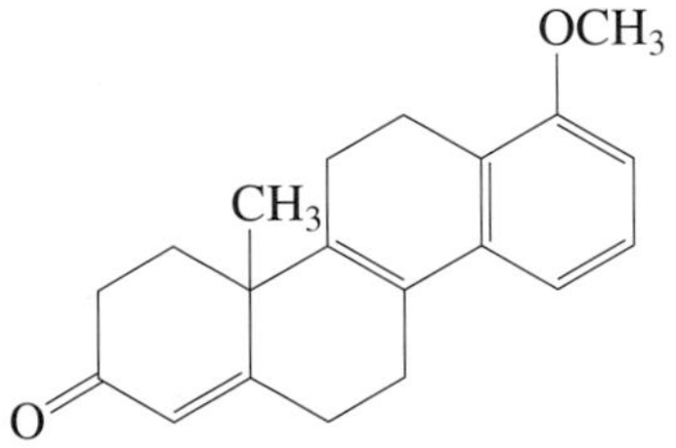

SOLUTION

The Robinson annulation combines Michael addition to an α,β-unsaturated ketone with intramolecular aldol condensation (Section 18-12) to afford a cyclohexenone. Retrosynthetic analysis (Section 8-9) of the target molecule leads to the disconnection of two bonds in ring A: the carbon–carbon double bond, by a retro-aldol condensation, and a single bond by a retro-Michael addition. The Robinson annulation described for the construction of ring A is closely related to the example in Section 18-12, which condenses 2-methylcyclohexanone with 3-buten-2-one.

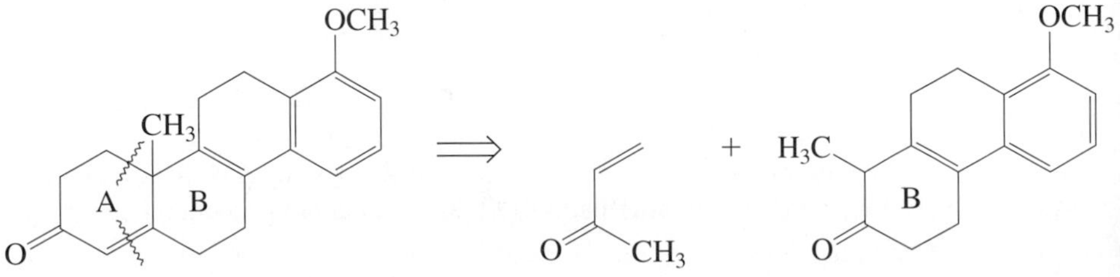

At this stage, we have simplified the target from a tetracyclic to a tricyclic molecule. However, the latter is a β,γ-unsaturated ketone, *not* an obvious product of Robinson annulation for the formation of ring B. Postponing consideration of this problem for the moment, can we formulate a Robinson annulation that transforms the initial ketone into a tricyclic product with the *same molecular skeleton* as that of the product of the last reaction? Yes we can: In fact, we can construct a double-bond isomer:

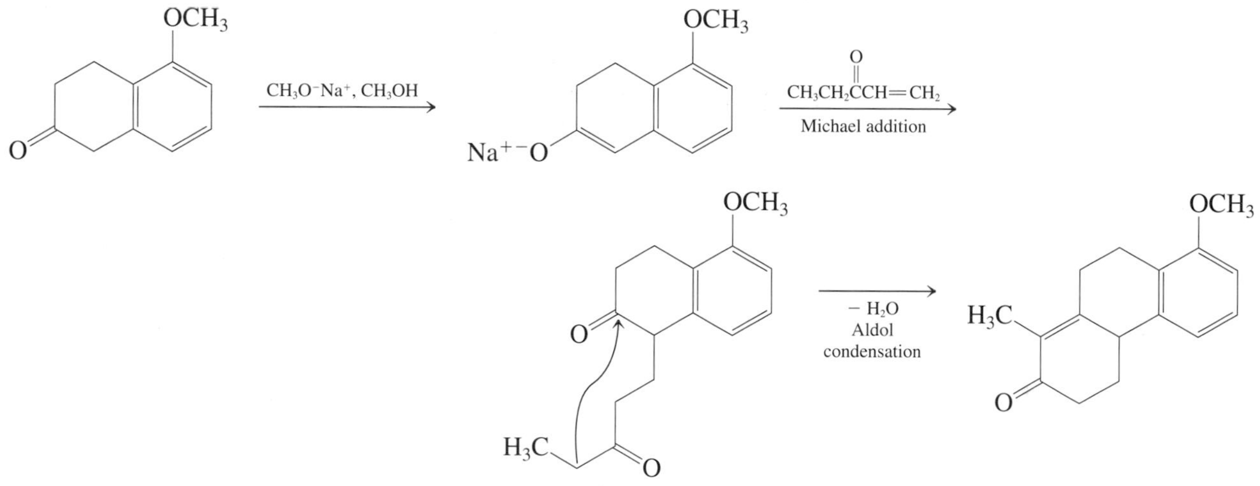

We are quite close to a complete solution. All that remains is to connect the tricyclic ketone that we have just prepared with the necessary second Robinson sequence. To do so, we *think mechanistically,* asking ourselves, "What is the structure of the enolate needed to initiate the annulation?" It is shown at the right in the following reaction, and we note that it is an *allylic* enolate anion (Section 14-4), for which we can write a second resonance contributor. The latter reveals that it may be derived by deprotonation of the tricyclic ketone in the preceding reaction at the allylic (γ) position.

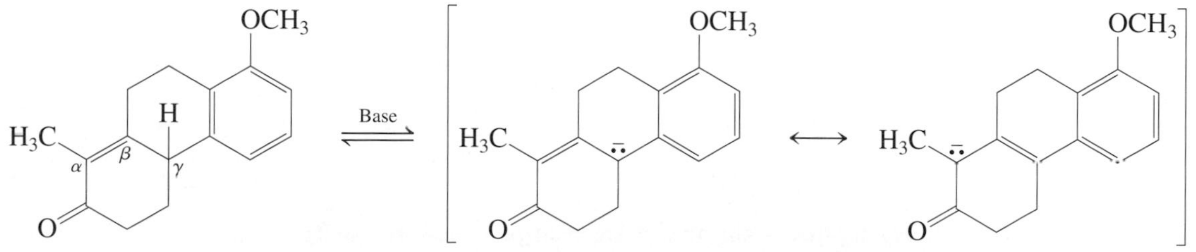

Resonance-stablilized allylic enolate anion

What does this realization mean in practical terms? We can carry out the needed second Robinson annulation starting directly from the preceding α,β-unsaturated ketone; it is not necessary to first prepare the β,γ-unsaturated ketone revealed in our initial retrosynthetic disconnection, because they give the same enolate upon deprotonation. The complete synthesis is shown on the next page.

A final note: You may have recognized that the allylic anion discussed above is also benzylic and hence enjoys additional resonance stabilization by conjugation with the benzene ring. Benzylic resonance will be discussed in detail in Section 22-1.

NEW REACTIONS

Synthesis and Reactions of Enolates and Enols

1. Enolate Ions (Section 18-1)

Enolate ion

2. Keto–Enol Equilibria (Section 18-2)

3. Hydrogen–Deuterium Exchange (Section 18-2)

4. Stereoisomerization (Section 18-2)

5. Halogenation (Section 18-3)

$$RCH_2CR' \xrightarrow[-HX]{X_2,\ H^+} RCHCR'$$

with structures showing C=O groups and X substituent.

6. Enolate Alkylation (Section 18-4)

$$RCH=C \begin{smallmatrix} O^- \\ R' \end{smallmatrix} \xrightarrow[-X^-]{R''X} RCHCR'$$

S_N2 reaction: R″X must be methyl or primary halide

7. Enamine Alkylation (Section 18-4)

$$\ce{C=C} \xrightarrow{R'X} \ce{-C-C} X^- \xrightarrow{H^+,\ H_2O} \ce{-C-C} + R_2NH$$

S_N2 reaction: R′X must be methyl, primary, or secondary halide

8. Aldol Condensations (Sections 18-5 through 18-7)

$$2\ RCH_2CH \underset{}{\overset{HO^-}{\rightleftharpoons}} RCH_2C-CHCH \underset{}{\overset{HO^-,\ \Delta}{\rightleftharpoons}} RCH_2CH=C \begin{smallmatrix} CHO \\ R \end{smallmatrix} + H_2O$$

Aldol adduct   **Condensation product**

Crossed aldol condensation (one aldehyde not enolizable)

$$RCH + R'CH_2CH \xrightarrow[-H_2O]{HO^-,\ \Delta} RCH=C \begin{smallmatrix} R' \\ CHO \end{smallmatrix}$$

Ketones

$$RCCH_2R' \overset{HO^-}{\rightleftharpoons} RC-CH-CR \xrightarrow[\text{Drive equilibrium}]{\Delta,\ -H_2O} RC=C-CR$$

Intramolecular aldol condensation

$$(CH_2)_n \xrightarrow[-H_2O]{HO^-} (CH_2)_n$$

Unstrained rings preferred

9. Synthesis of α,β-Unsaturated Aldehydes and Ketones (Sections 18-8 and 18-9)

Aldol condensation (see preceding reactions)

Bromination–dehydrobromination of aldehydes and ketones

$$RCH_2CH_2CR' \xrightarrow[\text{2. Base}]{\text{1. Br}_2,\ \text{H}^+} RCH=CHCR'$$

Wittig reaction with stabilized ylides

$$RCH + (C_6H_5)_3P=CHCR' \xrightarrow{\text{THF}} \underset{H}{\overset{R}{C}}=CHCR' + (C_6H_5)_3P=O$$

Oxidation of allylic alcohols

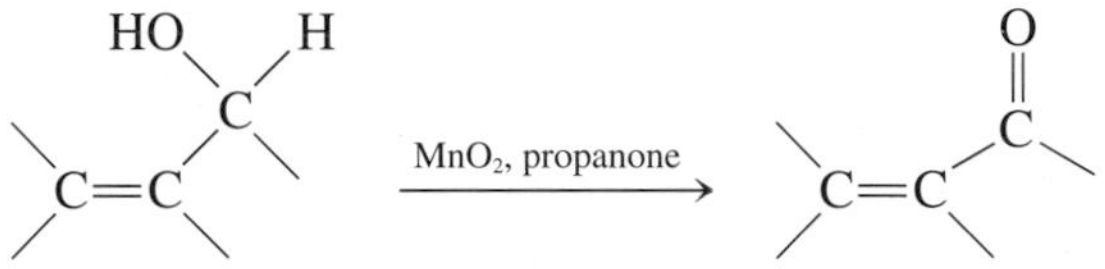

$$\overset{HO\quad H}{\underset{C=C}{C}} \xrightarrow{\text{MnO}_2,\ \text{propanone}} \underset{C=C}{\overset{O}{C}}$$

Isomerization of β,γ-unsaturated aldehydes and ketones to conjugated carbonyl compounds

$$RCH=CHCH_2CH \xrightarrow{\text{H}^+ \text{ or HO}^-,\ \text{H}_2\text{O}} RCH_2CH=CHCH$$

Reactions of α,β-Unsaturated Aldehydes and Ketones

10. Reductions (Section 18-9)

Hydrogenation

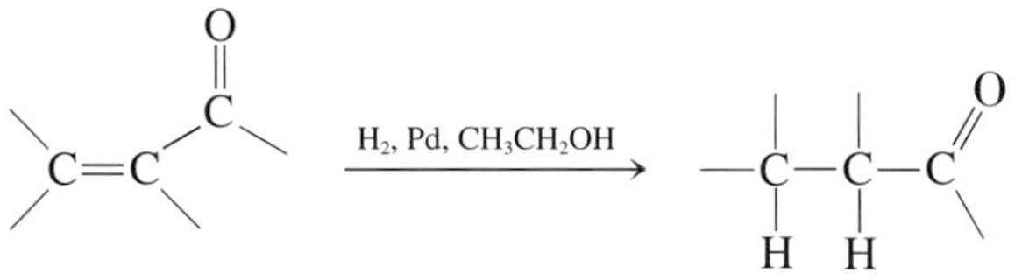

$$\underset{C=C}{\overset{O}{C}} \xrightarrow{\text{H}_2,\ \text{Pd},\ \text{CH}_3\text{CH}_2\text{OH}} \underset{H\ \ H}{-C-C-C}$$

One-electron transfer reduction

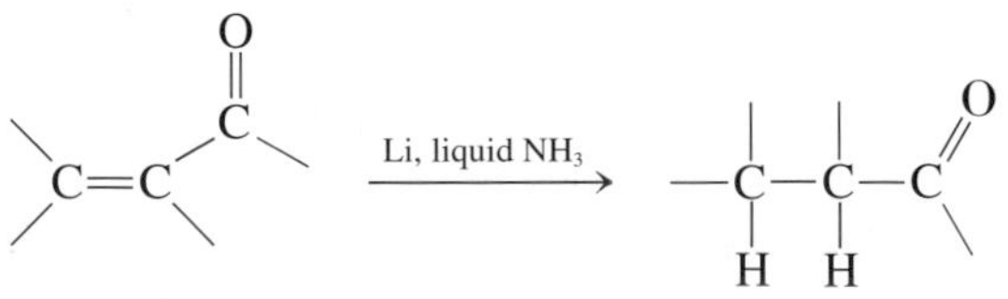

$$\underset{C=C}{\overset{O}{C}} \xrightarrow{\text{Li, liquid NH}_3} \underset{H\ \ H}{-C-C-C}$$

11. Addition of Halogen (Section 18-9)

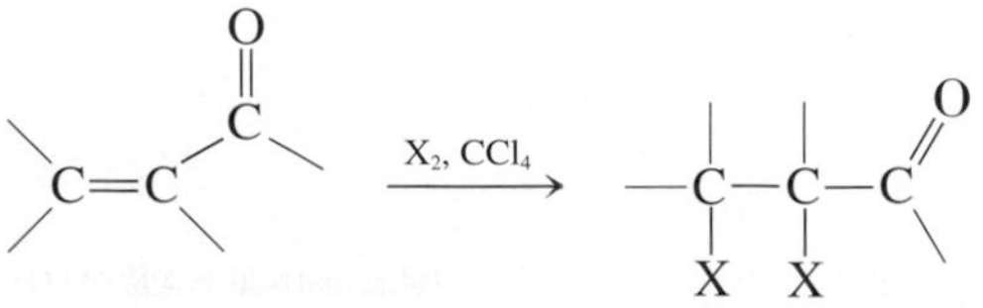

$$\underset{C=C}{\overset{O}{C}} \xrightarrow{\text{X}_2,\ \text{CCl}_4} \underset{X\ \ X}{-C-C-C}$$

12. Condensations with Amine Derivatives (Section 18-9)

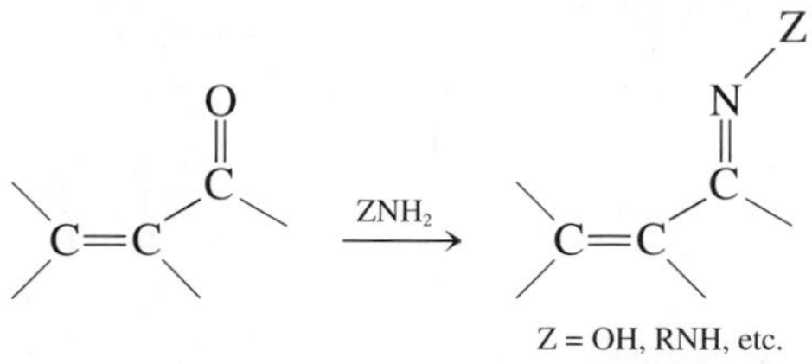

Z = OH, RNH, etc.

Conjugate Additions to α,β-Unsaturated Aldehydes and Ketones

13. Hydrogen Cyanide Addition (Section 18-10)

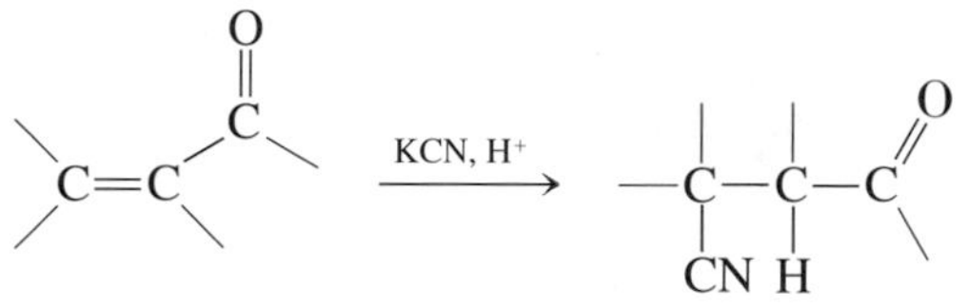

14. Water, Alcohols, Amines (Section 18-10)

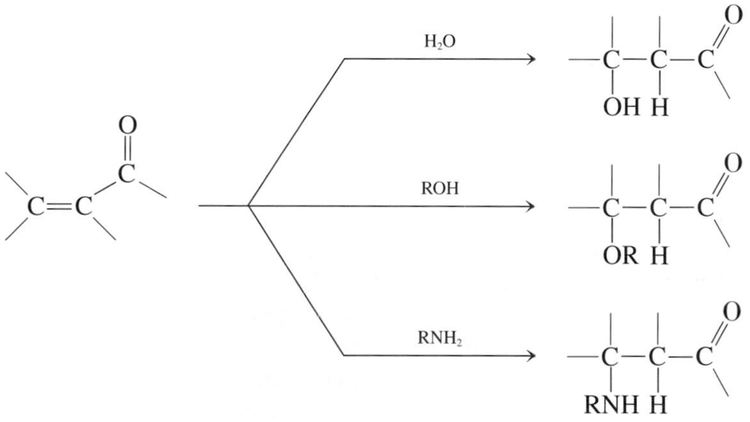

15. Organometallic Reagents (Section 18-11)

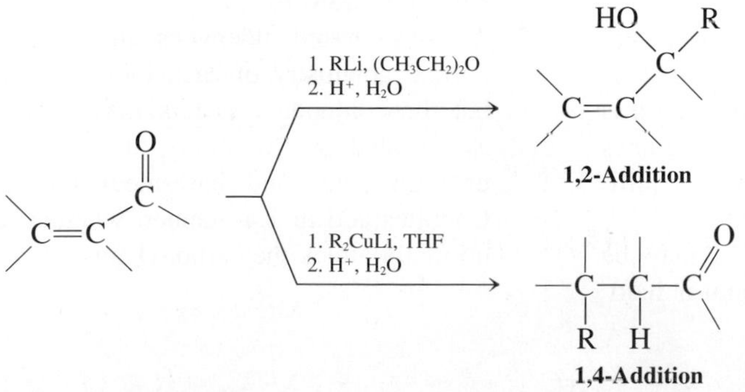

1,2-Addition

1,4-Addition

Additions of RMgX may be 1,2 or 1,4, depending on reagent and substrate structure

Cuprate additions followed by enolate alkylations

16. Michael Addition (Section 18-12)

17. Robinson Annulation (Section 18-12)

IMPORTANT CONCEPTS

1. Hydrogens next to the carbonyl group (**α-hydrogens**) are acidic because of the electron-withdrawing nature of the functional group and because the resulting **enolate ion** is resonance stabilized.

2. Electrophilic attack on enolates can occur at both the α-carbon and the oxygen. Haloalkanes usually prefer the former. Protonation of the latter leads to **enols.**

3. **Enamines** are neutral analogs of enolates. They can be β-alkylated to give iminium cations that hydrolyze to aldehydes and ketones on aqueous work-up.

4. Aldehydes and ketones are in equilibrium with their tautomeric enol forms; the **enol–keto conversion** is catalyzed by acid or base. This equilibrium allows for facile α-deuteration and stereochemical equilibration.

5. **α-Halogenation** of carbonyl compounds may be acid or base catalyzed. With acid, the enol is halogenated by attack at the double bond; subsequent renewed enolization is slowed down by the halogen substituent. With base, the enolate is attacked at carbon, and subsequent enolate formation is accelerated by the halogens introduced.

6. Enolates are nucleophilic and reversibly attack the carbonyl carbon of an aldehyde or a ketone in the **aldol condensation;** they also attack the β-carbon of an α,β-unsaturated carbonyl compound in the **Michael addition.**

7. α,β-Unsaturated aldehydes and ketones show the normal chemistry of each individual double bond, but the conjugated system may react as a whole, as revealed by the ability of these compounds to undergo acid- and base-mediated **1,4-additions.** Cuprates add in 1,4-manner, whereas alkyllithiums normally attack the carbonyl function.

PROBLEMS

26. Underline the α-carbons and circle the α-hydrogens in each of the following structures.

(a) $CH_3CH_2\overset{\displaystyle O}{\overset{\displaystyle \|}{C}}CH_2CH_3$

(b) $CH_3\overset{\displaystyle O}{\overset{\displaystyle \|}{C}}CH(CH_3)_2$

(c)

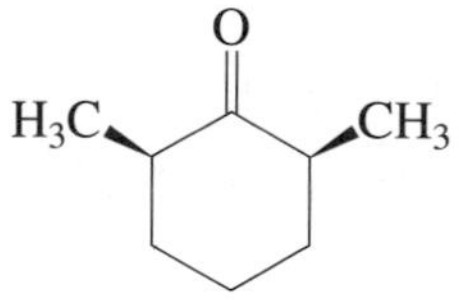

(d)

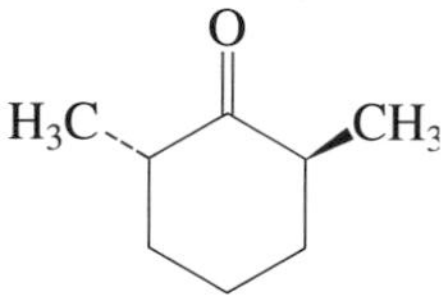

(e)

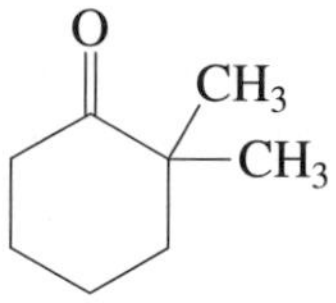

(f)

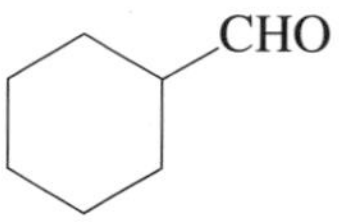

(g) $(CH_3)_3C\overset{\displaystyle O}{\overset{\displaystyle \|}{C}}H$

(h) $(CH_3)_3CCH_2\overset{\displaystyle O}{\overset{\displaystyle \|}{C}}H$

27. Write the structures of every enol and enolate ion that can arise from each of the carbonyl compounds illustrated in Problem 26.

28. What product(s) would form if each carbonyl compound in Problem 26 were treated with **(a)** alkaline D_2O; **(b)** 1 equivalent of Br_2 in acetic acid; **(c)** excess Cl_2 in aqueous base?

29. Describe the experimental conditions that would be best suited for the efficient synthesis of each of the following compounds from the corresponding nonhalogenated ketone.

(a) $C_6H_5\overset{\displaystyle Br}{\overset{\displaystyle |}{C}}H\overset{\displaystyle O}{\overset{\displaystyle \|}{C}}CH_3$

(b)

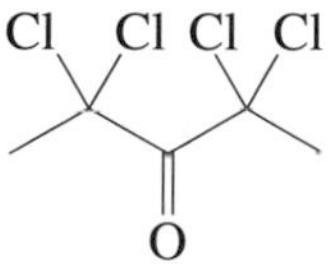

(c)

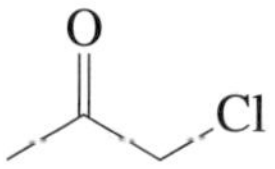

30. Propose a mechanism for the following reaction. (**Hint:** Take note of all of the products that are formed and base your answer on the mechanism for acid-catalyzed bromination of propanone [acetone] shown in Section 18-3.)

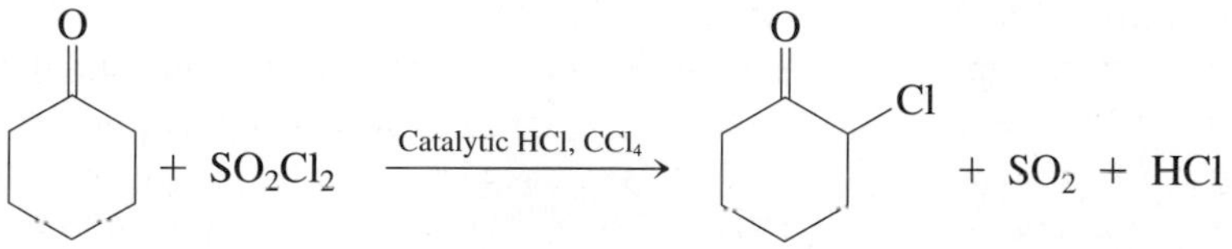

31. Give the product(s) that would be expected on reaction of 3-pentanone with 1 equivalent of LDA, followed by addition of 1 equivalent of

(a) CH_3CH_2Br

(b) $(CH_3)_2CHCl$

(c) 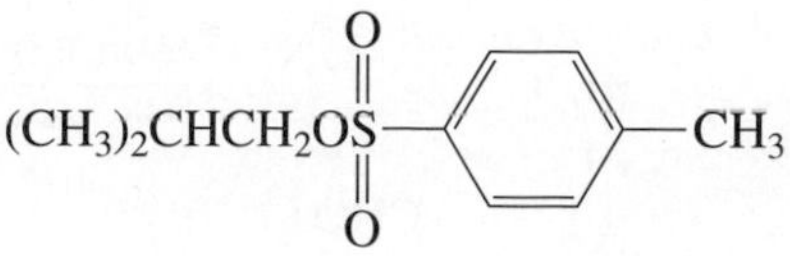 $(CH_3)_2CHCH_2O\overset{\displaystyle O}{\underset{\displaystyle O}{\overset{\displaystyle \|}{\underset{\displaystyle \|}{S}}}}\!-\!\!\!\bigcirc\!\!\!-\!CH_3$

(d) $(CH_3)_3CCl$

32. Give the product(s) of the following reaction sequences.

(a) CH₃CHO
1. H, H⁺
2. (CH₃)₂C=CHCH₂Cl
3. H⁺, H₂O
⟶

(b) [phenyl]—CH₂CHO
1. H, H⁺
2. [phenyl]—CH₂Br
3. H⁺, H₂O
⟶

33. The problem of double compared with single alkylation of ketones by iodomethane and base is mentioned in Section 18-4. Write a detailed mechanism showing how some double alkylation occurs even when only 1 equivalent each of the iodide and base is used. Suggest a reason why the use of the enamine alkylation procedure solves this problem.

34. Would the use of an enamine instead of an enolate improve the likelihood of successful alkylation of a ketone by a secondary haloalkane?

35. Formulate a mechanism for the acid-catalyzed hydrolysis of the pyrrolidine enamine of cyclohexanone (shown in the margin).

36. Write the structures of the aldol condensation products of **(a)** pentanal; **(b)** 3-methylbutanal; **(c)** cyclopentanone.

37. Write the structures of the expected major products of crossed aldol condensation at elevated temperature between excess benzaldehyde and **(a)** 1-phenylethanone (acetophenone—see Section 17-1 for structure); **(b)** propanone (acetone); **(c)** 1,1-dimethylcyclopentanone.

38. Formulate a detailed mechanism for the reaction that you wrote in (c) of Problem 37.

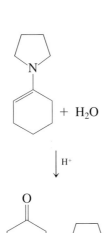

39. Give the likely products of each of the following aldol addition reactions.

(a) 2 [phenyl]—CH₂CHO
$\xrightarrow{\text{NaOH, H}_2\text{O}}$

(b) [phenyl]—CHO + (CH₃)₂CHCHO
$\xrightarrow{\text{NaOH, H}_2\text{O}}$

(c) H—C(=O)—C(CH₃)(CH₃)—CH₂CH₂—C(=O)—CH₃

(d) [bicyclic with CHO and CHO]
$\xrightarrow{\text{NaOH, H}_2\text{O}}$

40. Describe how you would prepare each of the following compounds, using an aldol condensation.

(a)

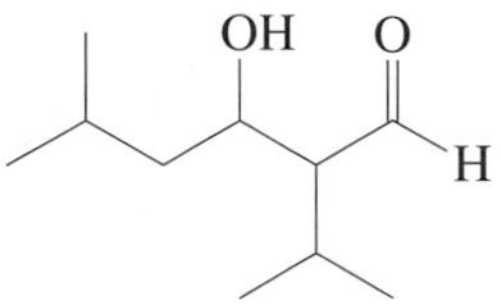

(b)

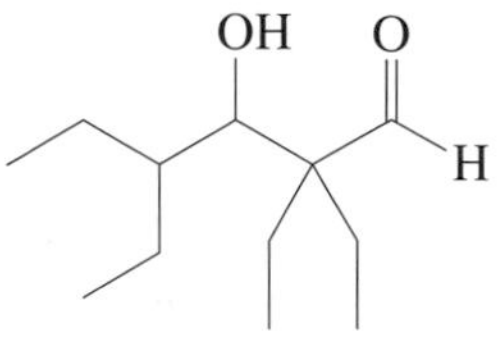

(c)

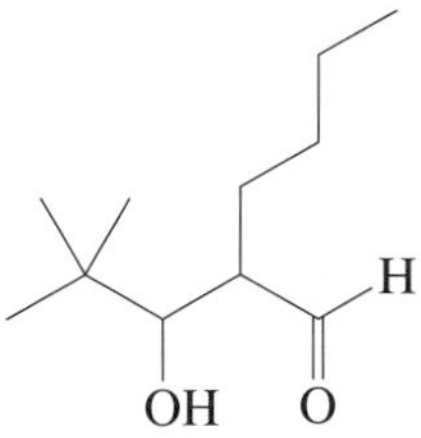

(d)

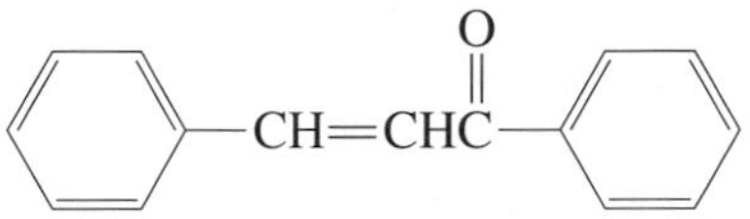

(e)

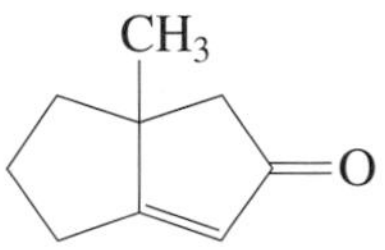

(f)

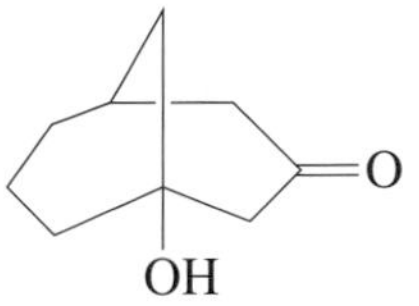

41. Aldol condensations may be catalyzed by acids. Suggest a role for H^+ in the acid-catalyzed version. (**Hint:** Consider what kind of nucleophile might exist in acidic solution, where enolate ions are *unlikely* to be present.)

42. A fresh salad may contain the following compounds: 2-hexenal (aroma of tomatoes), 3-octen-2-one (flavor of mushrooms), and 2-nonenal and 2,4-nonadienal (flavor and aroma of cucumbers). In Section 18-8, the synthesis of 2-nonenal was illustrated. Write similar sequences for the preparation of these other naturally occurring compounds from saturated aldehydes.

43. Four general synthetic routes to α,β-unsaturated aldehydes and ketones are described in Section 18-8. For each of the following three compounds, select the route that, you feel, might be especially useful and practical.

(a)

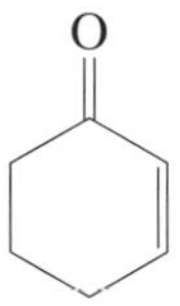

(b) 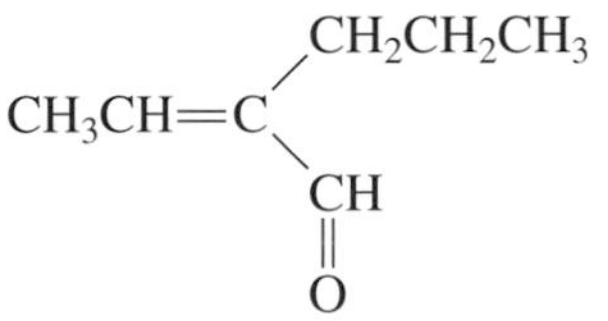 $CH_3CH=C$

(c)

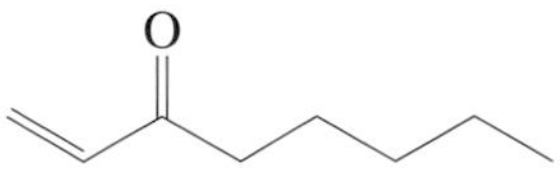

44. For each carbonyl compound listed in Problem 43, write the expected major product of reaction with each of the following reagents.

(a) H_2, Pd, CH_3CH_2OH
(b) $LiAlH_4$, $(CH_3CH_2)_2O$
(c) Cl_2, CCl_4
(d) KCN, H^+, H_2O
(e) CH_3Li, $(CH_3CH_2)_2O$
(f) $(CH_3CH_2CH_2CH_2)_2CuLi$, THF
(g) NH_2NHCNH_2, CH_3CH_2OH
 $\overset{\|}{O}$
(h) $(CH_3CH_2CH_2CH_2)_2CuLi$, followed by treatment with $CH_2=CHCH_2Cl$ in THF

45. Give the expected product(s) of each of the following reactions.

(a) $C_6H_5\overset{\overset{\displaystyle O}{\|}}{C}CH_2CH_2CH_3$ $\xrightarrow[\text{2. } CH_3CH_2Br, \text{ HMPA}]{\text{1. LDA, THF}}$

(b) 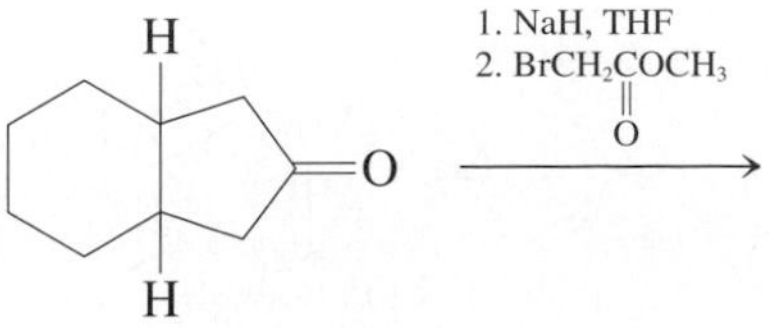 $\xrightarrow[\text{2. } BrCH_2COCH_3]{\text{1. NaH, THF}}$

(c)

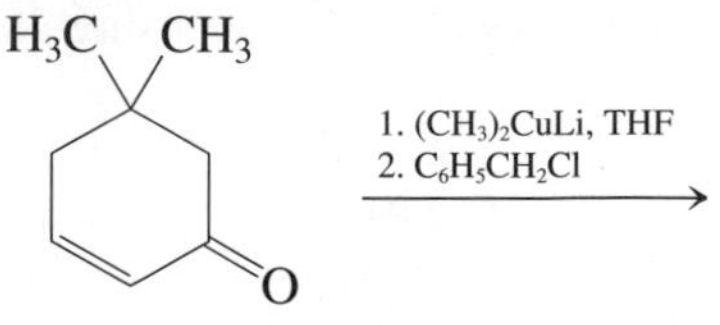

$$\text{H}_3\text{C} \quad \text{CH}_3$$

$$\xrightarrow[\text{2. C}_6\text{H}_5\text{CH}_2\text{Cl}]{\text{1. (CH}_3)_2\text{CuLi, THF}}$$

(d)

$$\begin{array}{c} \text{CH}_3 \\ (\text{CH}_2)_4\text{Br} \end{array}$$

$$\xrightarrow{\text{LDA, THF}}$$

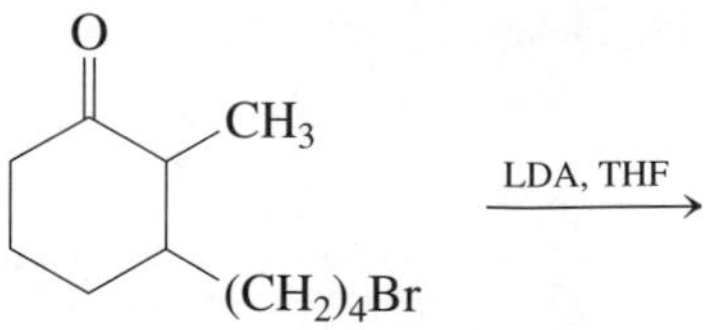

46. Write the products of each of the following reactions after aqueous work-up.

(a) $\text{C}_6\text{H}_5\overset{\text{O}}{\underset{\|}{\text{C}}}\text{CH}_3 + \text{CH}_2{=}\text{CHCC}_6\text{H}_5 \xrightarrow{\text{LDA, THF}}$

(b) 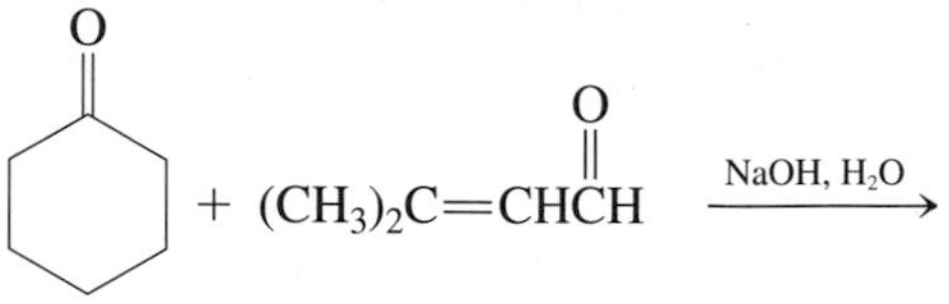 [cyclohexanone] $+ (\text{CH}_3)_2\text{C}{=}\text{CHCH} \xrightarrow{\text{NaOH, H}_2\text{O}}$

(c) [cyclopentenone]
$$\xrightarrow[\text{2. CH}_2{=}\text{CHCCH}_3]{\text{1. (CH}_2{=}\text{CH})_2\text{CuLi, THF}}$$

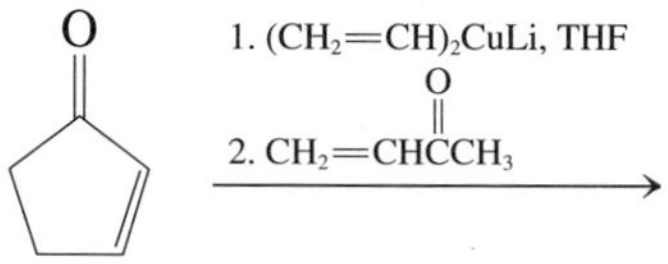

(d) 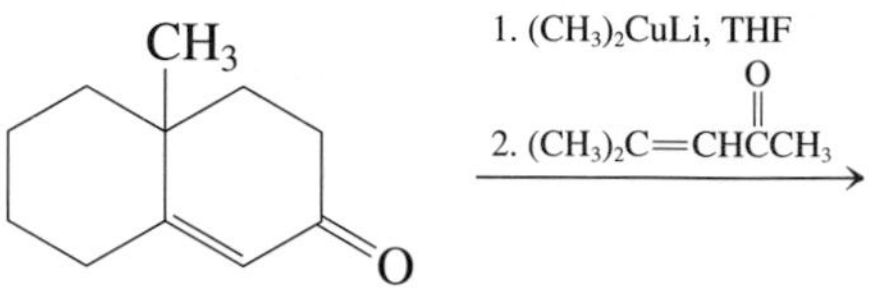 [octalone with CH$_3$]
$$\xrightarrow[\text{2. (CH}_3)_2\text{C}{=}\text{CHCCH}_3]{\text{1. (CH}_3)_2\text{CuLi, THF}}$$

(e) Write the results that you expect from base treatment of the products of reactions (c) and (d).

47. Write the final products of the following reaction sequences.

(a) 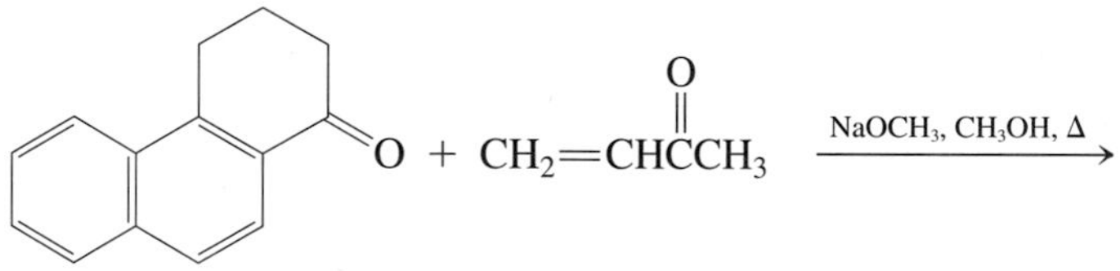 [phenanthrenone] $+ \text{CH}_2{=}\text{CHCCH}_3 \xrightarrow{\text{NaOCH}_3, \text{CH}_3\text{OH}, \Delta}$

(b) 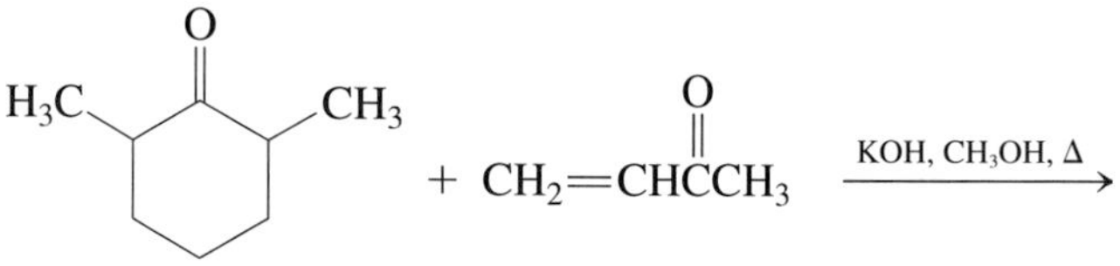 [dimethyl cyclohexanone] $+ \text{CH}_2{=}\text{CHCCH}_3 \xrightarrow{\text{KOH, CH}_3\text{OH}, \Delta}$

(c) 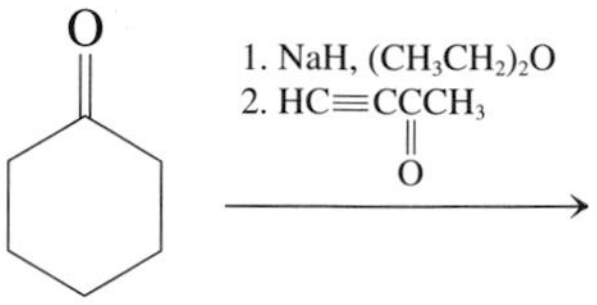 [cyclohexanone]
$$\xrightarrow[\text{2. HC}{\equiv}\text{CCCH}_3]{\text{1. NaH, (CH}_3\text{CH}_2)_2\text{O}}$$

(d) Write a detailed mechanism for reaction sequence (c). (**Hint:** Treat the 3-butyn-2-one reagent as a Michael acceptor in the first step.)

48. Propose syntheses of the following compounds by using Michael additions followed by aldol condensations (i.e., Robinson annulation). Each of the compounds shown has been instrumental in one or more total syntheses of steroidal hormones.

(a) 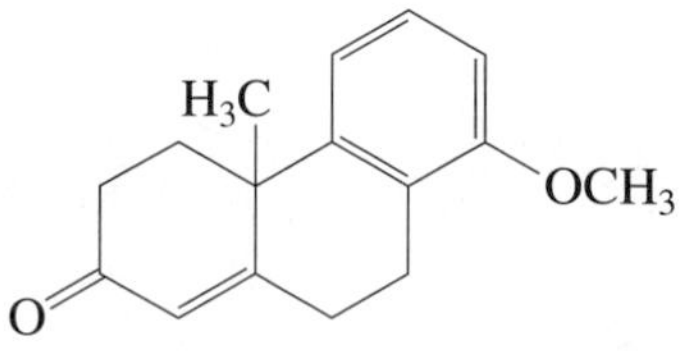 [tricyclic structure with H$_3$C and OCH$_3$]

(b) 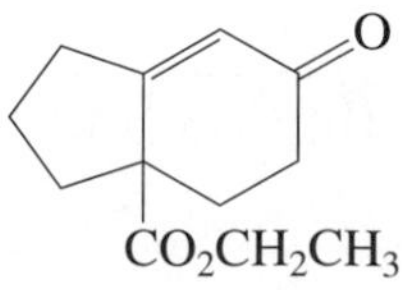 [bicyclic structure with CO$_2$CH$_2$CH$_3$]

(c)

(d)

49. Would you expect addition of HCl to the double bond of 3-buten-2-one (shown in the margin) to follow Markovnikov's rule? Explain your answer by a mechanistic argument.

$$CH_3\overset{\displaystyle O}{\overset{\displaystyle \|}{C}}CH=CH_2$$

3-Buten-2-one

50. Using the following information, propose structures for each of these compounds. **(a)** $C_5H_{10}O$, NMR spectrum A, UV $\lambda_{max}(\epsilon) = 280(18)$ nm; **(b)** C_5H_8O, NMR spectrum B, UV $\lambda_{max}(\epsilon) = 220(13,200), 310(40)$ nm; **(c)** C_6H_{12}, NMR spectrum C, UV $\lambda_{max}(\epsilon) = 189(8,000)$ nm; **(d)** $C_6H_{12}O$, NMR spectrum D, UV $\lambda_{max}(\epsilon) = 282(25)$ nm. (See the next page for spectra C and D.)

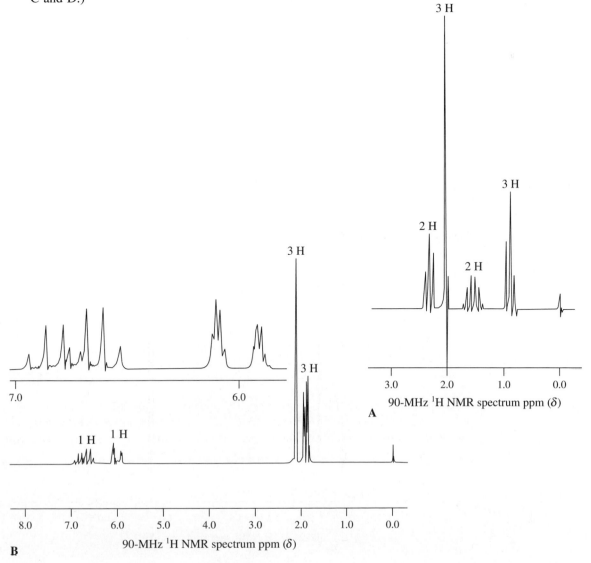

90-MHz ^{1}H NMR spectrum ppm (δ)

A

90-MHz ^{1}H NMR spectrum ppm (δ)

B

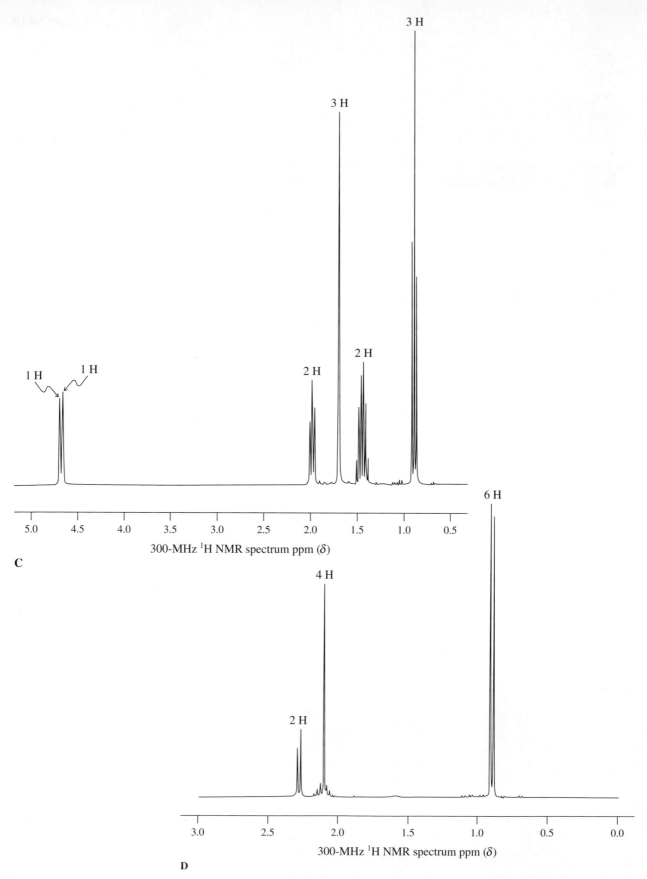

C

300-MHz ^{1}H NMR spectrum ppm (δ)

D

300-MHz ^{1}H NMR spectrum ppm (δ)

820

Next, for each of the following reactions, name an appropriate reagent for the indicated interconversion. (The letters refer to the compounds giving rise to NMR spectra A through D.) **(e)** A → C; **(f)** B → D; **(g)** B → A.

51. Treatment of cyclopentane-1,3-dione with iodomethane in the presence of base leads mainly to a mixture of three products.

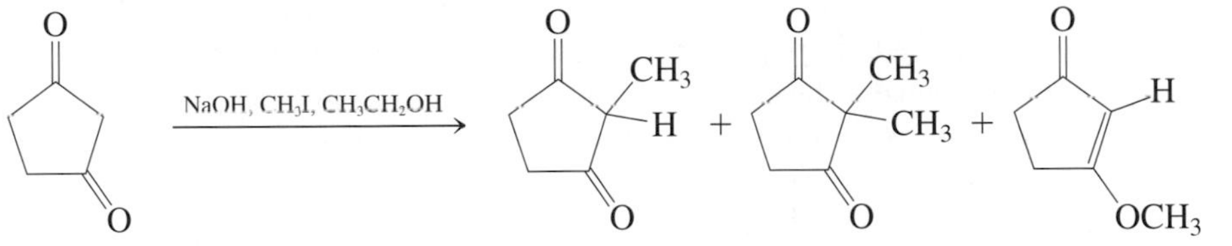

(a) Give a mechanistic description of how these three products are formed.
(b) Reaction of product C with a cuprate reagent results in loss of the methoxy group. For example,

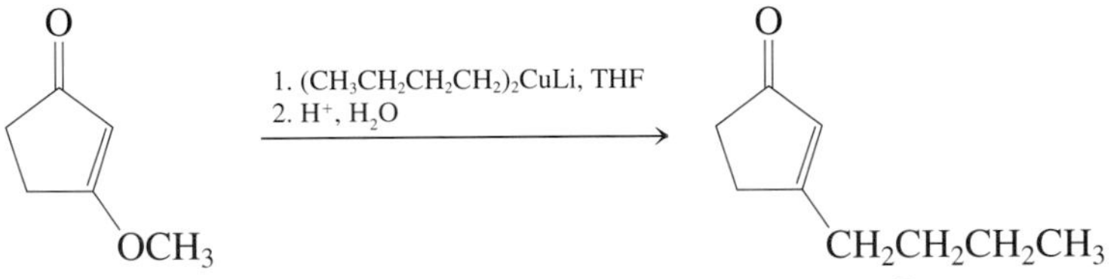

Suggest a mechanism for this reaction, which is another synthetic route to enones substituted at the β-carbon. (**Hint:** See Exercise 18-22.)

52. A somewhat unusual synthesis of cortisone-related steroids includes the following two reactions.

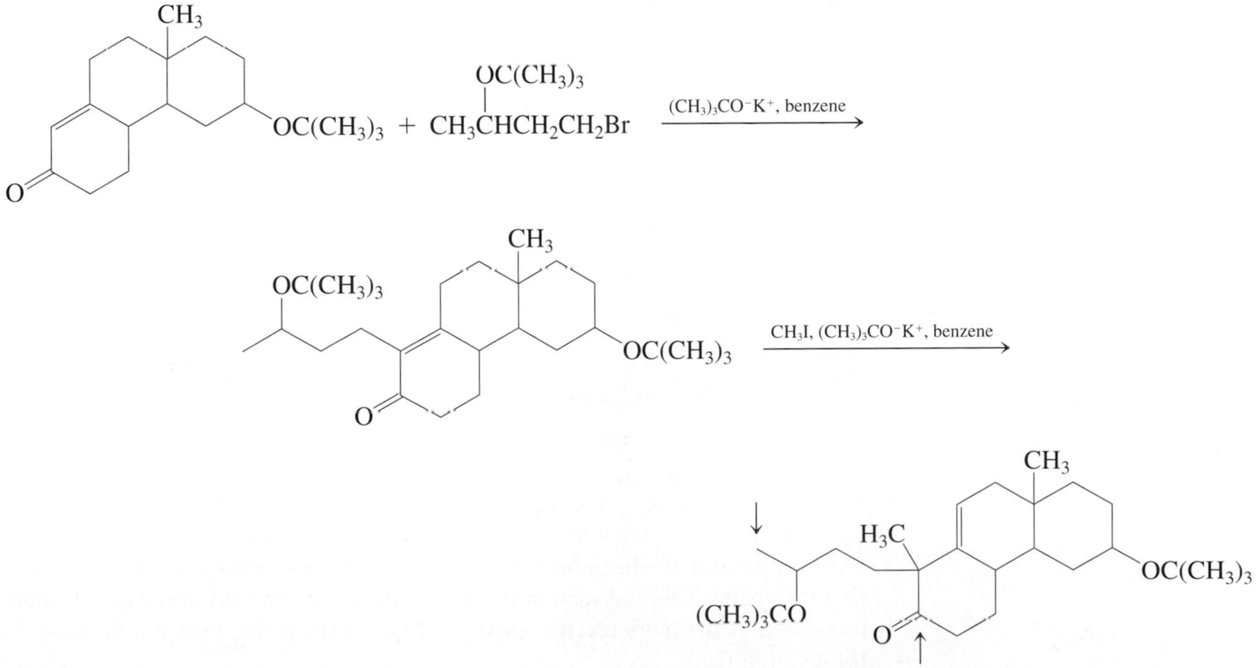

(a) Propose mechanisms for these two transformations. Be careful in choosing the initial site of deprotonation in the starting enone. The alkenyl hydrogen, in particular, is *not acidic enough* to be the one initially removed by base in this reaction.

(b) Propose a sequence of reactions that will connect the carbons marked by arrows in the third structure shown to form another six-membered ring.

53. The following steroid synthesis contains modified versions of two key types of reactions presented in this chapter. Identify these reaction types and give detailed mechanisms for each of the transformations shown.

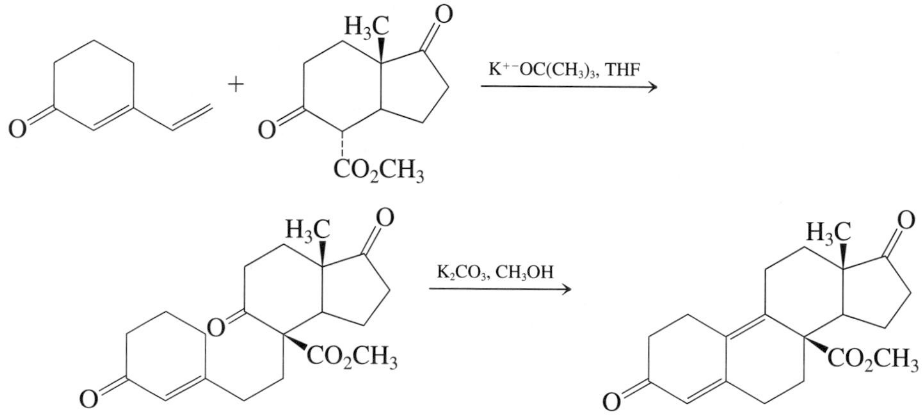

54. Devise reasonable plans for carrying out the following syntheses. Ignore stereochemistry in your strategies.

(a) 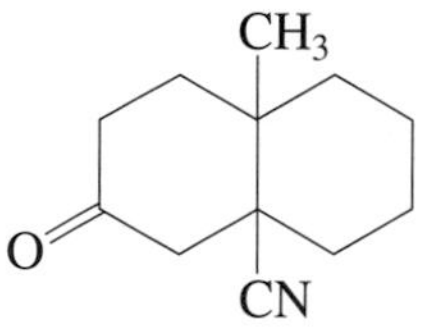 , starting from cyclohexanone

(b) 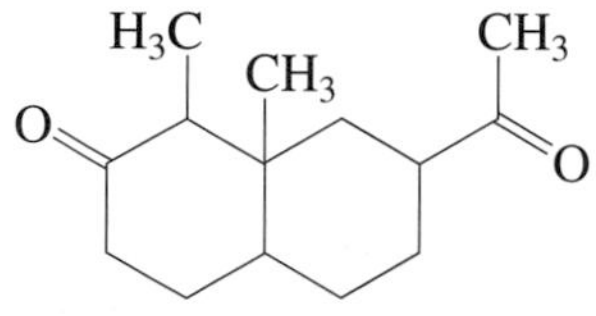 , starting from 2-cyclohexenone

(**Hint:** Prepare 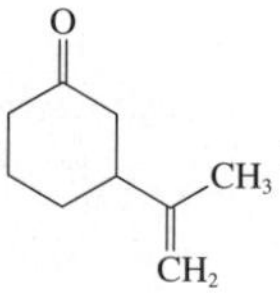 in your first step.)

55. Write reagents (a, b, c, d, e) where they have been omitted from the following synthetic sequence. Each letter may correspond to one or more reaction steps. This sequence is the beginning of a synthesis of germanicol, a naturally occurring triterpene. The diol used in the step between (a) and (b) provides selective protection of the more reactive carbonyl group. [**Hint:** See Problem 52 when formulating (b).]

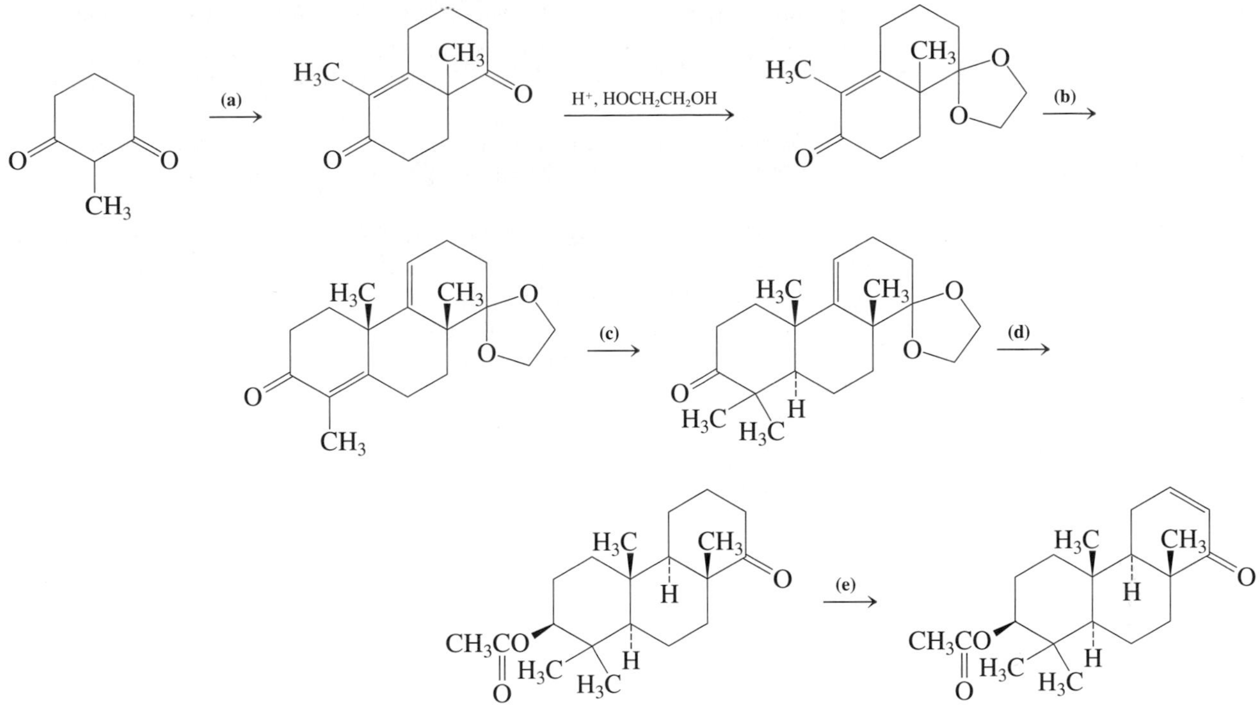

(a)

$H^+, HOCH_2CH_2OH$

(b)

(c)

(d)

(e)

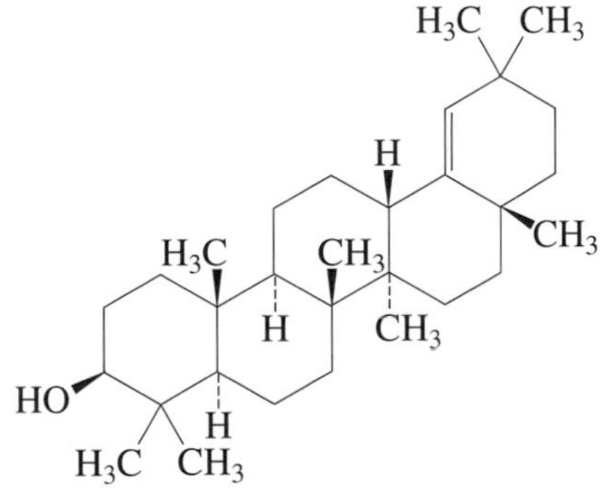

Germanicol

56. (a) The enzymatic oxidation of a lysine group to give an aldehyde is described in Section 18-5 (Chemical Highlight 18-2). What sort of intermediate(s) might appear in this oxidation? (**Hints:** Refer to Problem 36 of Chapter 4 and to Sections 3-10 and 17-9.) **(b)** A similar enzyme-catalyzed oxidation is the first step in the metabolism of amphetamine (see margin), which takes place in the endoplasmic reticulum of the liver. Write the structures of both the final product of this oxidation and the intermediate that immediately precedes its formation.

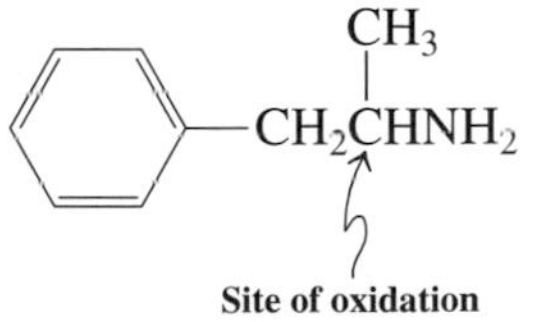

Site of oxidation

Amphetamine

Team Problem

57. When 2-methylcyclopentanone is treated with the bulky base triphenylmethyllithium under the two sets of conditions shown, the two possible enolates are generated in differing ratios. Why is this so?

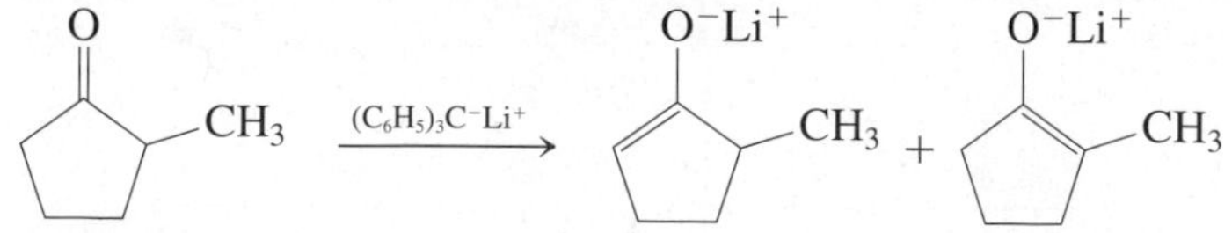

Conditions A: Ketone added to excess base 72% 28%
Conditions B: Excess ketone added to base 6% 94%

To tackle this problem, you have to invoke the principles of kinetic versus thermodynamic control (review Sections 11-8, 14-3, and 18-2); that is, which enolate is formed faster and which one is more stable? Divide your team so that one group considers conditions A and the other conditions B. Use curved arrows to show the flow of electrons leading to each enolate. Then assess whether your set of conditions is subject to enolate equilibration (thermodynamic control) or not (kinetic control). Reconvene to discuss these issues and draw a qualitative potential energy diagram depicting the progress of deprotonation at the two α sites.

Preprofessional Problems

58. When 3-methyl-1,3-diphenyl-2-butanone is treated with excess D_2O in the presence of catalytic acid, some of its hydrogens are replaced by deuterium. How many? **(a)** One; **(b)** two; **(c)** three; **(d)** six; **(e)** eight.

59. How would you best classify the following reaction? **(a)** Wittig reaction; **(b)** cyanohydrin formation; **(c)** conjugate addition; **(d)** aldol addition.

$$H_2C{=}CHCH{=}\overset{+}{N}H_2 \xrightarrow{NC^-} NCCH_2CH{=}CH\ddot{N}H_2$$

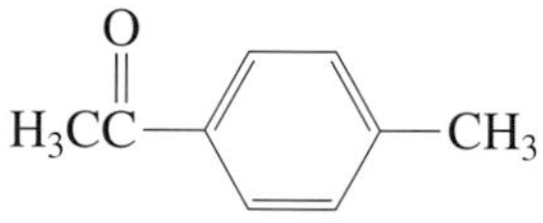

60. The aqueous hydroxide-promoted reaction of the compound shown in the margin with $(CH_3)_3CCHO$ yields exclusively one compound. Which one is it?

(a) 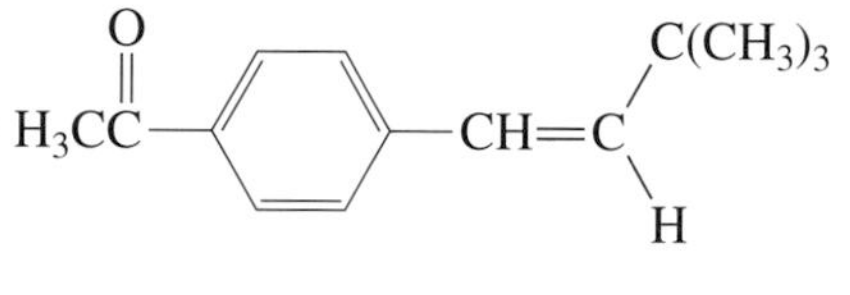

(b) 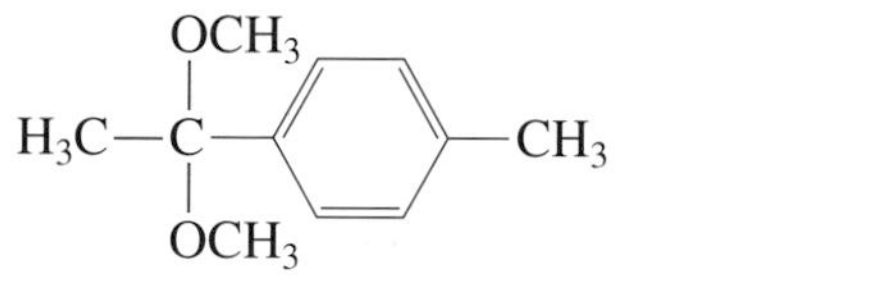

(c) 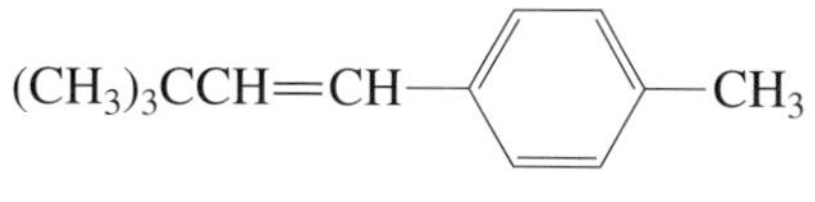

(d) 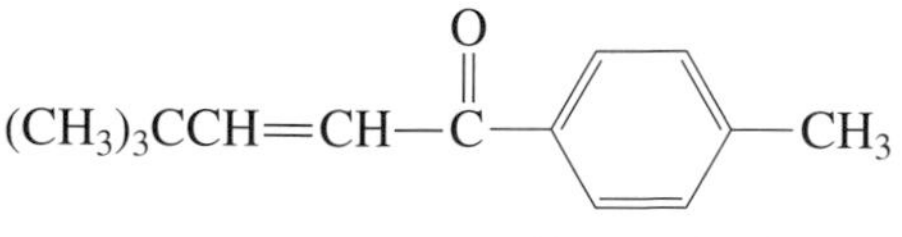

61. The ^{1}H NMR spectrum of 2,4-pentanedione indicates the presence of an enol tautomer of the dione. What is its most likely structure?

(a) 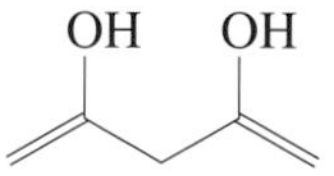

(b) 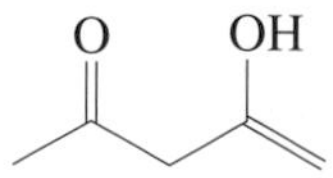

(c) 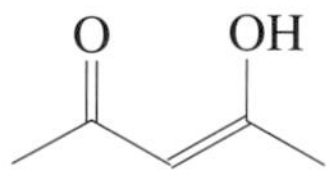

(d) 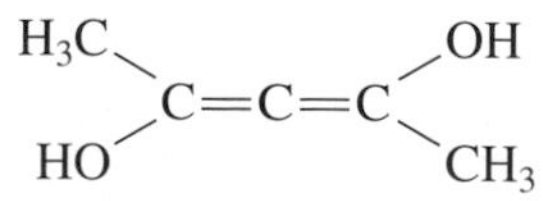

Carboxylic Acids

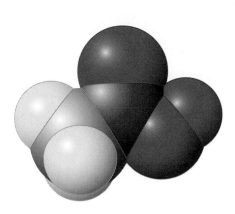

Vinegar's sharp odor and sour taste arise from the presence of acetic acid, the product of ethanol oxidation (the Latin word *acetum* and the French words *vin aigre* both mean "sour wine"). Acetic acid is generated by a particular strain of yeast used in the preparation of sourdough bread; it also is nature's building block for long-chain fatty acids, steroids, and many other complex molecules.

What types of compounds are the end products of biological metabolism? We inhale oxygen and exhale carbon dioxide, one product of the oxidation of organic compounds (Chapter 3). **Carboxylic acids** also are major products of biochemistry generated under oxidizing conditions and are thus widespread throughout living systems on Earth. You are all familiar with the bottle of wine that has turned to vinegar, the result of enzymatic oxidation of ethanol to acetic acid. Carboxylic acids are characterized by the presence of the **carboxy group,** a functional group containing a hydroxy unit attached to a carbonyl carbon. This substituent is usually written COOH or CO_2H; both of these representations will be used in this chapter.

Carboxylic acids not only are widely distributed in nature, but are also important industrial chemicals. For example, besides being the most important building block in the assembly of complex biological molecules, acetic acid is an industrial commodity produced in very large quantities.

Much of the reactivity of carboxylic acids can be anticipated if we view them as hydroxy carbonyl derivatives. Thus, the hydroxy hydrogen is acidic, the oxygens are basic and nucleophilic, and the carbonyl carbon is subject to nucleophilic attack.

$$\begin{array}{c} O \\ \| \\ -COH \end{array}$$

Carboxy group

Electrophilic — :O: — **Basic**

R — δ⁺ — C — :O: — H — Acidic

This chapter first introduces the system of naming carboxylic acids and some of their physical and spectroscopic characteristics. We then examine their acidity and basicity, two properties that are strongly influenced by the interaction between the electron-withdrawing carbonyl group and the hydroxy function. Methods for the preparation of the carboxy group are considered next, followed by a survey of its reactivity. The latter will feature a new substitution mechanism, *addition–elimination*, for the replacement of the hydroxy group by other nucleophiles, such as halide, alkoxide, and amide. The chemistry of the carboxylic acid derivatives, which result from these transformations, is the subject of Chapter 20.

19-1 Naming the Carboxylic Acids

Like other organic compounds, many carboxylic acids have common names that are used frequently in the literature. These names often indicate the natural sources from which the acids were first isolated (Table 19-1). *Chemical Abstracts* retains the common names for the two simplest, **formic** and **acetic** acids. The carboxy function (and those of its derivatives) takes precedence in naming over any other discussed so far.

Order of Precedence of Functional Groups

$$\underset{\substack{\text{Carboxylic} \\ \text{acid}}}{\text{RCOH}} > \underset{\text{Anhydride}}{\text{RCOCR}} > \underset{\text{Ester}}{\text{RCOR}'} > \underset{\substack{\text{Alkanoyl} \\ \text{halide}}}{\text{RCX}} > \underset{\text{Amide}}{\text{RCNR}_2'} > \underset{\text{Nitrile}}{\text{RCN}} > \underset{\text{Aldehyde}}{\text{RCH}} > \underset{\text{Ketone}}{\text{RCR}'} > \underset{\text{Alcohol}}{\text{ROH}} > \underset{\text{Thiol}}{\text{RSH}} > \underset{\text{Amine}}{\text{RNH}_2}$$

TABLE 19-1 | **Names and Natural Sources of Carboxylic Acids**

	Structure	IUPAC name	Common name	Natural source
	HCOOH	Methanoic acid	Formic acid[a]	From the "destructive distillation" of ants (*formica,* Latin, ant)
	CH₃COOH	Ethanoic acid	Acetic acid[a]	Vinegar (*acetum,* Latin, vinegar)
	CH₃CH₂COOH	Propanoic acid	Propionic acid	Dairy products (*pion,* Greek, fat)
	CH₃CH₂CH₂COOH	Butanoic acid	Butyric acid	Butter (particularly if rancid) (*butyrum,* Latin, butter)
	CH₃(CH₂)₃COOH	Pentanoic acid	Valeric acid	Valerian root
	CH₃(CH₂)₄COOH	Hexanoic acid	Caproic acid	Odor of goats (*caper,* Latin, goat)

[a]Used by *Chemical Abstracts.*

The IUPAC system derives the names of the carboxylic acids by replacing the ending *-e* in the name of the alkane by **-oic acid.** The alkanoic acid stem is numbered by assigning the number 1 to the carboxy carbon and labeling any substituents along the longest chain incorporating the CO_2H group accordingly.

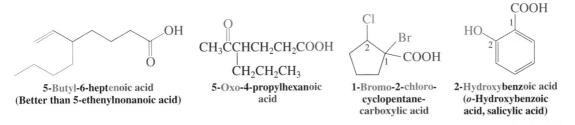

(R)-2-Bromopropanoic acid
(**α-Bromopropionic acid**)

$CH_2=CHCOOH$

Propenoic acid
(**Acrylic acid**)

(2R,3S)-Dimethylpentanoic acid
(**αR,βS-Dimethylvaleric acid**)

In multiply functionalized carboxylic acids, the main chain is chosen to include other functional groups as much as possible. Saturated cyclic acids are named as **cycloalkanecarboxylic acids.** Their aromatic counterparts are the **benzoic acids.** In these compounds, the carbon attached to the carboxy functional group is Cl.

5-Butyl-6-heptenoic acid
(**Better than 5-ethenylnonanoic acid**)

$CH_3\overset{O}{\overset{\|}{C}}CHCH_2CH_2COOH$
$\quad\quad\quad |$
$\quad\quad CH_2CH_2CH_3$

5-Oxo-4-propylhexanoic acid

1-Bromo-2-chloro-cyclopentane-carboxylic acid

2-Hydroxybenzoic acid
(**o-Hydroxybenzoic acid, salicylic acid**)

2-Hydroxybenzoic (salicylic) acid has been known for over 2000 years as an analgesic (painkilling) folk remedy present in willow bark (*salix,* Latin, willow). The acetic acid ester of salicylic acid is better known as aspirin (Chemical Highlight 22-3).

Dicarboxylic acids should be referred to as **dioic acids.**

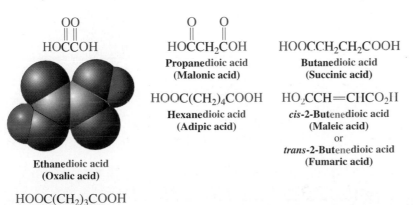

$HO\overset{O\quad O}{\overset{\|\quad\|}{C}}\!\!COH$

Ethanedioic acid
(**Oxalic acid**)

$HOOC(CH_2)_3COOH$

Pentanedioic acid
(**Glutaric acid**)

$HO\overset{O}{\overset{\|}{C}}CH_2\overset{O}{\overset{\|}{C}}OH$

Propanedioic acid
(**Malonic acid**)

$HOOCCH_2CH_2COOH$

Butanedioic acid
(**Succinic acid**)

$HOOC(CH_2)_4COOH$

Hexanedioic acid
(**Adipic acid**)

$HO_2CCH=CHCO_2H$

cis-**2-Butenedioic acid**
(**Maleic acid**)
or
trans-**2-Butenedioic acid**
(**Fumaric acid**)

Human beings have known since prehistoric times that chewing on the bark of the white willow relieves pain.

Their common names again derive from their natural origins. For example, butanedioic (succinic) acid was discovered in the distillate from amber (*succinum,* Latin, amber), and *trans*-2-butenedioic (fumaric) acid is found in the plant *Fumaria,* which was burned in ancient times to ward off evil spirits (*fumus,* Latin, smoke).

EXERCISE 19-1

Give systematic names or write the structure, as appropriate, of the following compounds.

(a) 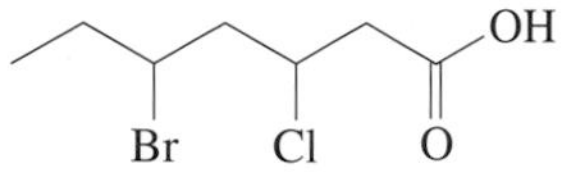 [structure]

(b) 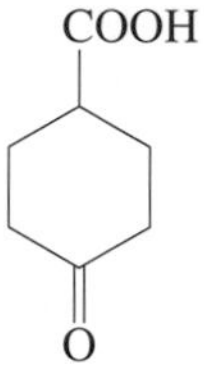 [structure]

(c) 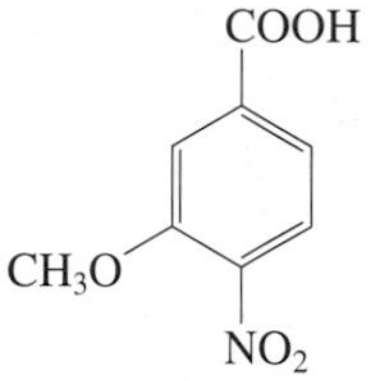 [structure]

(d) 2,2-Dibromohexanedioic acid   (e) 4-Hydroxypentanoic acid
(f) 4-(1,1-Dimethylethyl)benzoic acid

In summary, the systematic naming of the carboxylic acids is based on the alkanoic acid stem. Cyclic derivatives are called cycloalkanecarboxylic acids, their aromatic counterparts are referred to as benzoic acids, and dicarboxylic systems are labeled alkanedioic acids.

19-2 Structural and Physical Properties of Carboxylic Acids

What is the structure of a typical carboxylic acid? Do carboxylic acids have characteristic physical properties? This section answers these questions, beginning with the structure of formic acid. We shall see that carboxylic acids exist mainly as hydrogen-bonded dimers.

Formic acid is planar

The molecular structure of formic acid is shown in Figure 19-1. It is roughly planar, as expected for a "hydroxyformaldehyde," with an approximately trigonal carbonyl

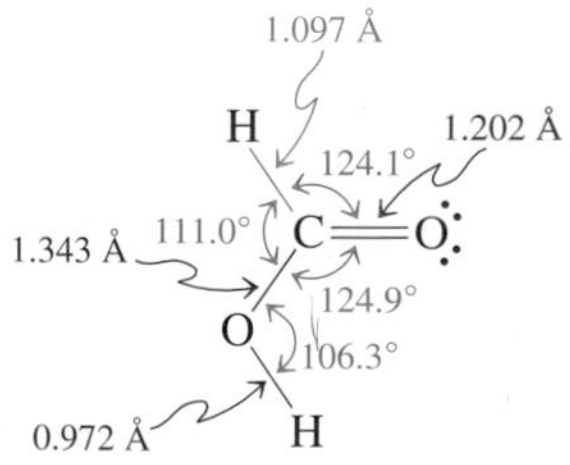

FIGURE 19-1
Molecular structure of formic acid. It is planar, with a roughly equilateral trigonal arrangement about the carbonyl carbon.

carbon. (See the structure of methanol, Figure 8-1, and that of acetaldehyde, Figure 17-2.) These structural characteristics are found in carboxylic acids in general.

The carboxy group is polar and forms hydrogen-bonded dimers

The carboxy function derives strong polarity from the carbonyl double bond and the hydroxy group, which forms hydrogen bonds to other polarized molecules, such as water, alcohols, and other carboxylic acids. Not surprisingly, therefore, the lower carboxylic acids (up to butanoic acid) are completely soluble in water. As neat liquids and even in fairly dilute solutions (in aprotic solvents), carboxylic acids exist to a large extent as hydrogen-bonded dimers, each O–H···O interaction ranging in strength from about 6 to 8 kcal mol^{-1}.

Carboxylic Acids Form Dimers Readily

$$2 \ RCOOH \longrightarrow$$

Two hydrogen bonds

Table 19-2 shows that carboxylic acids have relatively high melting and boiling points, because they form hydrogen bonds in the solid as well as the liquid state.

TABLE 19-2	Melting and Boiling Points of Functional Alkane Derivatives with Various Chain Lengths	
Derivative	Melting point (°C)	Boiling point (°C)
CH_4	−182.5	−161.7
CH_3Cl	−97.7	−24.2
CH_3OH	−97.8	65.0
HCHO	−92.0	−21.0
HCOOH	8.4	100.6
CH_3CH_3	−183.3	−88.6
CH_3CH_2Cl	−136.4	12.3
CH_3CH_2OH	−114.7	78.5
CH_3CHO	−121.0	20.8
CH_3COOH	16.7	118.2
$CH_3CH_2CH_3$	−187.7	−42.1
$CH_3CH_2CH_2Cl$	−122.8	46.6
$CH_3CH_2CH_2OH$	−126.5	97.4
CH_3COCH_3	−95.0	56.5
CH_3CH_2CHO	−81.0	48.8
CH_3CH_2COOH	−20.8	141.8

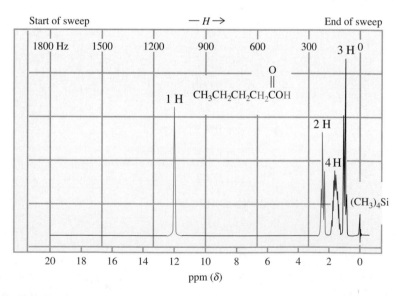

(E)-3-Methyl-2-hexenoic acid

Carboxylic acids, especially those possessing relatively low molecular weights and correspondingly high volatility, exhibit characteristically strong odors. For example, the presence of butanoic acid contributes to the characteristic strong flavor of many cheeses, and (E)-3-methyl-2-hexenoic acid was identified in 1991 as the principal compound responsible for the smell of human sweat.

In summary, the carboxy function is planar and contains a polarizable carbonyl group. Carboxylic acids exist as hydrogen-bonded dimers and exhibit unusually high melting and boiling points.

19-3 NMR and IR Spectroscopy of Carboxylic Acids

The polarizable double bond and the hydroxy group also shape the spectra of carboxylic acids. In this section, we show how both NMR and IR methods are used to characterize the carboxy group.

The carboxy hydrogen and carbon are deshielded

As in aldehydes and ketones, the hydrogens positioned on the carbon next to the carbonyl group are slightly deshielded in ^{1}H NMR spectra. The effect diminishes rapidly with increasing distance from the functional group. The hydroxy proton resonates at very low field ($\delta = 10$–13 ppm). As in the NMR spectra of alcohols, its chemical shift varies strongly with concentration, solvent, and temperature, because of the strong ability of the OH group to enter into hydrogen bonding. The ^{1}H NMR spectrum of pentanoic acid is shown in Figure 19-2.

FIGURE 19-2

90-MHz ^{1}H NMR spectrum of pentanoic acid in CCl_4. The scale has been expanded to 20 ppm, to allow the signal for the acid proton at $\delta = 11.83$ ppm to be shown. The methylene hydrogens at C2 absorb at the next lowest field as a triplet ($\delta = 2.25$ ppm, $J = 7$ Hz), followed by a four-hydrogen multiplet for the next two sets of methylenes. The methyl group appears as a distorted triplet at highest field ($\delta = 0.90$ ppm, $J = 6$ Hz).

^{1}H NMR Chemical Shifts of Alkanoic Acids

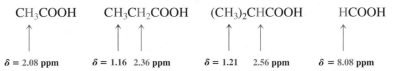

CH$_3$COOH	CH$_3$CH$_2$COOH	(CH$_3$)$_2$CHCOOH	HCOOH
δ = 2.08 ppm	δ = 1.16 2.36 ppm	δ = 1.21 2.56 ppm	δ = 8.08 ppm

The ^{13}C NMR chemical shifts of carboxylic acids also are similar to those of the aldehydes and ketones, with moderately deshielded carbons next to the carbonyl group and the typically low field carbonyl absorptions. However, the amount of deshielding is smaller, because the positive polarization of the carboxy carbon is somewhat attenuated by the presence of the extra OH group.

Typical ^{13}C NMR Chemical Shifts of Alkanoic Acids

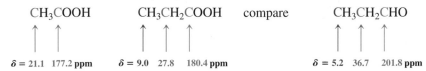

CH$_3$COOH	CH$_3$CH$_2$COOH	compare	CH$_3$CH$_2$CHO
δ = 21.1 177.2 ppm	δ = 9.0 27.8 180.4 ppm		δ = 5.2 36.7 201.8 ppm

This attenuation is best understood by writing resonance forms. Recall from Section 17-2 that aldehydes and ketones are described by two such representations. The dipolar Lewis structure indicates the polarization of the C=O bond. Even though it is a minor contributor (because of the lack of an octet on carbon), it explains the deshielding of the carbonyl carbon. In carboxylic acids, however, the corresponding dipolar form contributes less to the resonance hybrid: The hydroxy oxygen can donate an electron pair to give a third arrangement in which the carbon and both oxygen atoms have octets. The degree of positive charge on the carbonyl carbon and therefore its deshielding are greatly reduced.

Resonance in Aldehydes and Ketones

The contribution of the second resonance form, though minor, explains the strong deshielding of the carbonyl and adjacent carbons

Resonance in Carboxylic Acids

The third resonance form explains the attenuated deshielding of the carbonyl carbon relative to that of aldehydes and ketones

---

EXERCISE 19-2

A foul-smelling carboxylic acid with b.p. 164°C gave the following NMR data: ^{1}H NMR (CCl$_4$) δ = 1.00 (t, J = 7.4 Hz, 3 H), 1.65 (sextet, J = 7.5 Hz, 2 H), 2.31 (t, J = 7.4 Hz, 2 H), and 11.68 (s, 1 H) ppm; ^{13}C NMR (CS$_2$) δ = 13.4, 18.5, 36.3, and 179.6 ppm. Assign a structure to it.

The carboxy group shows two important IR bands

The carboxy group consists of a carbonyl group and an attached hydroxy substituent. Consequently, both characteristic stretching frequencies are seen in the infrared

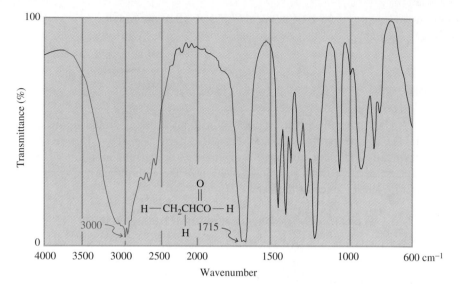

FIGURE 19-3
IR spectrum of propanoic acid:
$\tilde{\nu}_{\text{O-H stretch}} = 3000 \text{ cm}^{-1}$;
$\tilde{\nu}_{\text{C=O stretch}} = 1715 \text{ cm}^{-1}$.
The peaks associated with
these stretching vibrations are
broad because of hydrogen
bonding.

spectrum. The O–H bond gives rise to a very broad band at a lower wavenumber (2500–3300 cm^{-1}) than is observed for alcohols, largely because of strong hydrogen bonding. The IR spectrum of propanoic acid is shown in Figure 19-3.

In summary, the NMR signals for carboxylic acids reveal highly deshielded acid protons and carbonyl carbons and moderately deshielded nuclei next to the functional group. The infrared spectrum shows characteristic bands at about 2500–3300 (O–H) and 1710 (C=O) cm^{-1}.

19-4 Acidic and Basic Character of Carboxylic Acids

Like alcohols (Section 8-3), carboxylic acids exhibit both acidic and basic character: Deprotonation to carboxylate ions is relatively easy, protonation more difficult.

Carboxylic acids are relatively strong acids

Carboxylic acids have much lower pK_a values than alcohols do, even though the relevant hydrogen in each case is that of a hydroxy group.

Carboxylic Acids Dissociate Readily

$$\underset{\substack{K_a \sim 10^{-4}-10^{-5}, \\ pK_a \sim 4-5}}{\text{RCOH}} + \text{H}_2\text{O} \rightleftharpoons \underset{\text{Carboxylate ion}}{\text{RCO:}^-} + \text{HOH}_2^+$$

Why do carboxylic acids dissociate to a greater extent than do alcohols? The difference is that the hydroxy substituent of a carboxylic acid is attached to a carbonyl group, whose positively polarized carbonyl carbon exerts a powerful electron-withdrawing inductive effect. In addition, the carboxylate ion is significantly stabilized by resonance, much as is the enolate ion formed by deprotonation of the α-carbon in aldehydes and ketones (Section 18-1).

Resonance in Carboxylate and Enolate Ions

(B = base)

Carboxylate ion

$$B:^- \ + \ RC\overset{:O:}{\underset{..}{\overset{\|}{C}}}OH \ \rightleftharpoons \ BH \ + \ \left[\ RC\overset{:O:}{-}\ddot{O}: \ \longleftrightarrow \ RC\overset{:\ddot{O}:^-}{\underset{+}{-}}\ddot{O}: \ \longleftrightarrow \ RC\overset{:\ddot{O}:}{=}\ddot{O} \ \right] \quad pK_a \sim 4\text{–}5$$

Enolate ion

$$B:^- \ + \ R'C\overset{:O:}{\overset{\|}{C}}CH_2R \ \rightleftharpoons \ BH \ + \ \left[\ R'C\overset{:O:}{-}\ddot{C}HR \ \longleftrightarrow \ R'C\overset{:\ddot{O}:^-}{\underset{+}{-}}\ddot{C}HR \ \longleftrightarrow \ R'C\overset{:\ddot{O}:}{=}CHR \ \right] \quad pK_a \sim 19\text{–}21$$

In contrast with enolates, two of the three resonance forms in carboxylate ions are equivalent (Section 1-5). As a result, carboxylates are symmetric, with equal carbon–oxygen bond lengths (1.26 Å), in between the lengths typical of the carbon–oxygen double (1.20 Å) and single (1.34 Å) bonds in the corresponding acids (see Figure 19-1).

Electron-withdrawing substituents increase the acidity of carboxylic acids

As with alcohols, the inductive effects of electron-withdrawing groups close to the carboxy function increase the acidity. Table 19-3 shows the pK_a values of selected carboxylic acids. Note that two or three such groups on the α-carbon can result in carboxylic acids that are nearly as strong as typical mineral acids.

EXERCISE 19-3

In the following sets of acids, rank the components in order of *decreasing* acidity.

(a) CH_3CH_2COOH $CH_3\overset{Br}{\underset{|}{C}}HCOOH$ CH_3CBr_2COOH

(b) $CH_3\overset{F}{\underset{|}{C}}HCH_2COOH$ $CH_3\overset{Br}{\underset{|}{C}}HCH_2COOH$

(c)

The dioic acids have two pK_a values, one for each CO_2H group. In ethanedioic (oxalic) and propanedioic (malonic) acids, the first pK_a is lowered by the electron-withdrawing inductive effect of the second carboxy group on the first. In the higher dioic acids, the two pK_a values are close to those of monocarboxylic acids.

TABLE 19-3 pK$_a$ Values of Various Carboxylic and Other Acids			
Compound	**pK$_a$**	**Compound**	**pK$_a$**
Alkanoic acids		**Benzoic acids**	
CH_3COOH	4.76	$4\text{-}CH_3C_6H_4COOH$	4.36
$ClCH_2COOH$	2.87	C_6H_5COOH	4.20
$Cl_2CHCOOH$	1.25	$4\text{-}ClC_6H_4COOH$	3.98
Cl_3CCOOH	0.65		
F_3CCOOH	0.23	**Other acids**	
$CH_3CH_2CH_2COOH$	4.82	H_3PO_4	2.15 (first pK$_a$)
$CH_3CH_2CHClCOOH$	2.84	HNO_3	−1.3
$CH_3CHClCH_2COOH$	4.06	HCl	−2.2
$ClCH_2CH_2CH_2COOH$	4.52	H_2SO_4	−5.2 (first pK$_a$)
		H_2O	15.7
		CH_3OH	15.5
Dioic acids			
$HOOCCOOH$	1.23, 4.19		
$HOOCCH_2COOH$	2.83, 5.69		
$HOOCCH_2CH_2COOH$	4.16, 5.61		
$HOOC(CH_2)_4COOH$	4.43, 5.41		

The relatively strong acidity of carboxylic acids means that their corresponding **carboxylate salts** are readily made by treatment of the acid with base, such as NaOH, Na_2CO_3, or $NaHCO_3$. These salts are named by specifying the metal and replacing the ending *-ic acid* in the acid name by **-ate.** Thus, $HCOO^-Na^+$ is called sodium formate; $CH_3COO^-Li^+$ is named lithium acetate; and so forth. Carboxylate salts are much more water soluble than the corresponding acids, because the polar anionic group is readily solvated.

Carboxylate Salt Formation

$$CH_3CCH_2CH_2COOH \xrightarrow{\text{NaOH, H}_2\text{O}} CH_3CCH_2CH_2COO^-Na^+ \;+\; HOH$$

4,4-Dimethylpentanoic acid
(Slightly water soluble)

Sodium 4,4-dimethylpentanoate
(Water soluble)

Carboxylic acids may be protonated on the carbonyl oxygen

The lone electron pairs on both of the oxygen atoms in the carboxy group can, in principle, be protonated, just as alcohols are protonated by strong acids to give alkyloxonium ions (Section 8-3). It has been found that it is the carbonyl oxygen that is more basic. Why? Protonation at the carbonyl oxygen gives a species whose positive charge is delocalized by resonance. The alternative possibility, resulting from protonation at the hydroxy group, does not benefit from such stabilization.

Protonation of a Carboxylic Acid

Note that protonation is nevertheless very difficult, as shown by the very high acidity ($pK_a \sim -6$) of the conjugate acid. We shall see, however, that such protonations are important in many reactions of the carboxylic acids and their derivatives.

EXERCISE 19-4

The pK_a of protonated propanone (acetone) is -7.2 and that of protonated acetic acid is -6.1. Explain.

In summary, carboxylic acids are acidic because the polarized carbonyl carbon is strongly electron withdrawing, and deprotonation gives resonance-stabilized anions. Electron-withdrawing groups increase acidity, although this effect wears off rapidly with increasing distance from the carboxy group. Protonation is difficult, but possible, and occurs on the carbonyl oxygen to give a resonance-stabilized cation.

19-5 Carboxylic Acid Synthesis in Industry

Carboxylic acids are useful reagents and synthetic precursors. The two simplest ones are manufactured on a large scale industrially. Formic acid is employed in the tanning process in the manufacture of leather and in the preparation of latex rubber. It is synthesized efficiently by the reaction of powdered sodium hydroxide with carbon monoxide under pressure. This transformation proceeds by nucleophilic addition followed by protonation.

Formic Acid Synthesis

$$NaOH \; + \; CO \; \xrightarrow{150^\circ C, \; 100 \; psi} \; HCOO^- Na^+ \; \xrightarrow{H^+, \; H_2O} \; HCOOH$$

There are three important industrial preparations of acetic acid: ethene oxidation through acetaldehyde (Section 12-15); air oxidation of butane; and carbonylation of methanol. The mechanisms of these reactions are complex.

Acetic Acid by Oxidation of Ethene

$$CH_2{=}CH_2 \; \xrightarrow[\text{Wacker process}]{O_2, \; H_2O, \; \text{catalytic PdCl}_2 \text{ and CuCl}_2} \; CH_3CHO \; \xrightarrow{O_2, \; \text{catalytic Co}^{3+}} \; CH_3COOH$$

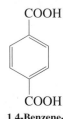

COOH

COOH

**1,4-Benzene-
dicarboxylic acid
(Terephthalic acid)**

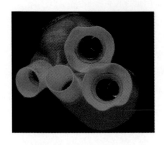

Polyethylene terephthalate is
Dacron, used in the Jarvik
artificial heart. As a thin film,
it is Mylar.

Acetic Acid by Oxidation of Butane

$$CH_3CH_2CH_2CH_3 \xrightarrow{\text{O}_2,\ \text{catalytic Co}^{3+},\ 15\text{–}20\ \text{atm},\ 180°C} CH_3COOH$$

Acetic Acid by Carbonylation of Methanol

$$CH_3OH \xrightarrow[\text{Monsanto process}]{\text{CO, catalytic Rh}^{3+},\ \text{I}_2,\ 30\text{–}40\ \text{atm},\ 180°C} CH_3COOH$$

Annual production of acetic acid in the United States alone exceeds 2 million tons (2×10^9 kg). It is used to manufacture monomers for polymerization, such as ethenyl acetate (vinyl acetate; Section 12-15), as well as pharmaceuticals, dyes, and pesticides. Two dicarboxylic acids in large-scale chemical production are hexanedioic (adipic) acid, used in the manufacture of nylon (see Section 21-12), and 1,4-benzenedicarboxylic (terephthalic) acid, whose polymeric esters with diols are fashioned into plastic sheets, films, and bottles for soft drinks.

19-6 Methods for Introducing the Carboxy Functional Group

The oxidation of primary alcohols and aldehydes to carboxylic acids by aqueous Cr(VI) was described in Sections 8-6 and 17-4. This section presents two additional reagents suitable for this purpose. It is also possible to introduce the carboxy function by adding a carbon atom to a haloalkane. This transformation can be achieved in either of two ways: the carbonation of organometallic reagents or the preparation and hydrolysis of nitriles.

Oxidation of primary alcohols and of aldehydes furnishes carboxylic acids

Primary alcohols oxidize to aldehydes, which in turn may readily oxidize further to the corresponding carboxylic acids.

Carboxylic Acids by Oxidation

$$RCH_2OH \xrightarrow{\text{Oxidation}} \underset{\displaystyle RCH}{\overset{\displaystyle O}{\|}} \xrightarrow{\text{Oxidation}} \underset{\displaystyle RCOH}{\overset{\displaystyle O}{\|}}$$

In addition to aqueous CrO_3, $KMnO_4$ and nitric acid (HNO_3) are frequently used in this process. Because it is one of the cheapest strong oxidants, nitric acid is often chosen for large-scale and industrial applications.

$$2\ HNO_3 + \underset{\textbf{3-Chloropropanal}}{ClCH_2CH_2\overset{\overset{\displaystyle O}{\|}}{C}H} \xrightarrow{25°C} \underset{\underset{\textbf{3-Chloropropanoic acid}}{79\%}}{ClCH_2CH_2\overset{\overset{\displaystyle O}{\|}}{C}OH} + 2\ NO_2 + H_2O$$

Give the products of nitric acid oxidation of (**a**) pentanal; (**b**) 1,6-hexanediol;
(**c**) 4-(hydroxymethyl)cyclohexanecarbaldehyde.

Organometallic reagents react with carbon dioxide to give carboxylic acids

Organometallic reagents attack carbon dioxide (usually in the solid form known as "dry ice") much as they would attack aldehydes or ketones. The product of this **carbonation** process is a carboxylate salt, which upon protonation by aqueous acid yields the carboxylic acid.

Carbonation of Organometallics

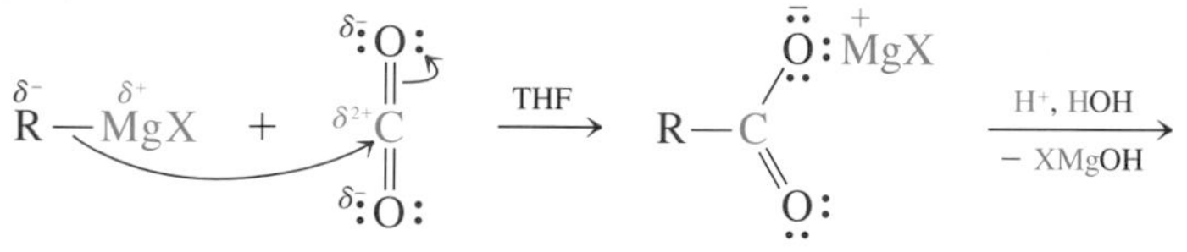

$$RLi + CO_2 \xrightarrow{THF} RCOO^-Li^+ \xrightarrow[-\ LiOH]{H^+,\ HOH} RCOOH$$

Recall that an organometallic reagent is usually made from the corresponding haloalkane: $RX + Mg \rightarrow RMgX$. Hence, carbonation of the latter allows the two-step transformation of RX into RCOOH, the carboxylic acid with one more carbon. For example,

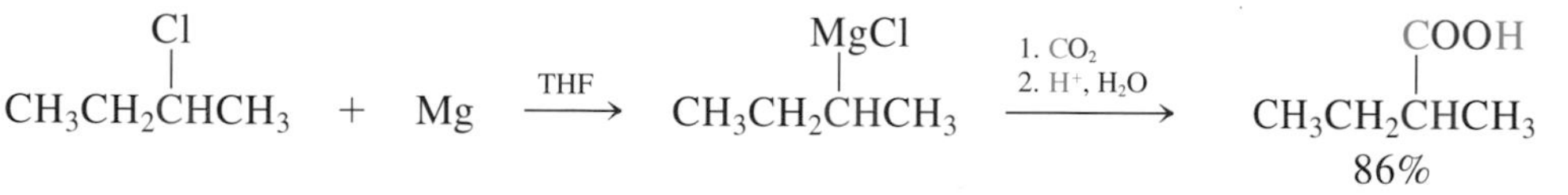

2-Chlorobutane

2-Methylbutanoic acid

86%

Nitriles hydrolyze to carboxylic acids

Another method for converting a haloalkane into a carboxylic acid with an additional carbon atom is through the preparation and hydrolysis of a nitrile, $RC\equiv N$. Recall (Section 6-3) that cyanide ion, $^-:C\equiv N:$, is a good nucleophile and may be used to synthesize nitriles through S_N2 reactions. Hydrolysis of the nitrile in hot acid or base furnishes the corresponding carboxylic acid (and ammonia).

Carboxylic Acids from Haloalkanes Through Nitriles

$$RX \xrightarrow[-\ X^-]{^-CN} RC\equiv N \xrightarrow[2.\ H^+,\ H_2O]{1.\ HO^-} RCOOH + NH_3$$

The mechanism of this reaction will be described in detail in Section 20-8.

Carboxylic acid synthesis with the use of nitrile hydrolysis is preferable to Grignard carbonation when the substrate contains groups that react with organometallic reagents, such as the hydroxy, carbonyl, and nitro functionalities.

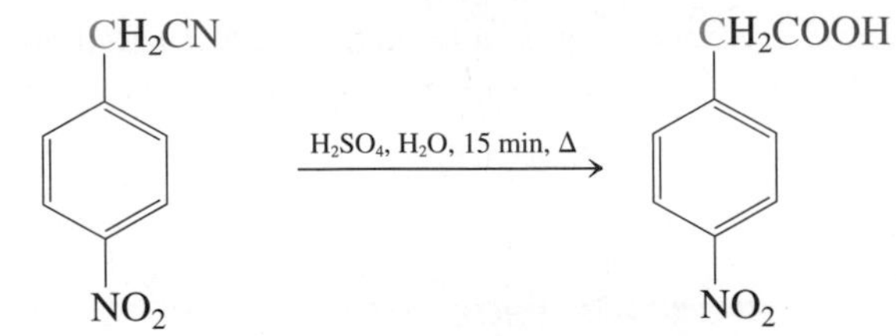

CH₂CN → CH₂COOH with NO₂ substituents

$$\text{CH}_2\text{CN} \xrightarrow{\text{H}_2\text{SO}_4,\ \text{H}_2\text{O},\ 15\ \text{min},\ \Delta} \text{CH}_2\text{COOH}$$

(4-Nitrophenyl)-
ethanenitrile

(4-Nitrophenyl)-
acetic acid
95%

EXERCISE 19-6

Using chemical equations, show how you would convert each of the following halogenated compounds into a carboxylic acid with one additional carbon atom. If more than one method can be used successfully, show all. If not, give the reason behind your choice of approach. **(a)** 1-Chloropentane; **(b)** iodocyclopentane; **(c)** 4-bromobutanoic acid (**Hint:** See Sections 8-7 and 8-9); **(d)** chloroethene (**Hint:** See Section 13-10); **(e)** bromocyclopropane (**Hint:** See Problem 45 of Chapter 6).

Hydrolysis of the nitrile group in a cyanohydrin, prepared by addition of HCN to an aldehyde or a ketone (Section 17-11), is a general route to 2-hydroxycarboxylic acids, which possess antiseptic properties.

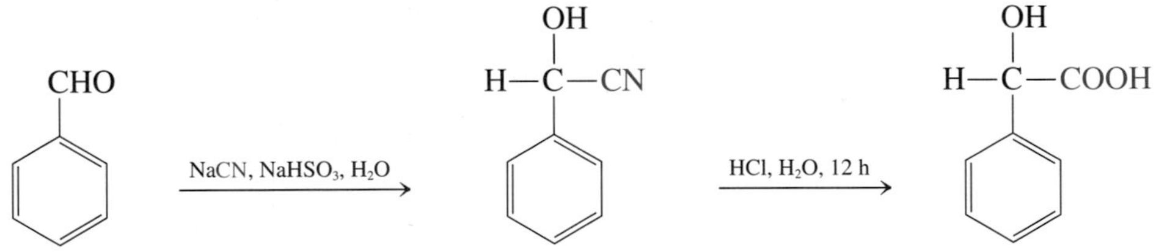

Benzaldehyde
$\xrightarrow{\text{NaCN, NaHSO}_3,\ \text{H}_2\text{O}}$
2-Hydroxy-2-phenyl-
ethanenitrile
(Mandelonitrile)
$\xrightarrow{\text{HCl, H}_2\text{O, 12 h}}$
2-Hydroxy-2-phenyl-
acetic acid
(Mandelic acid)
46%

EXERCISE 19-7

Suggest ways to effect the following conversions (more than one step will be required).

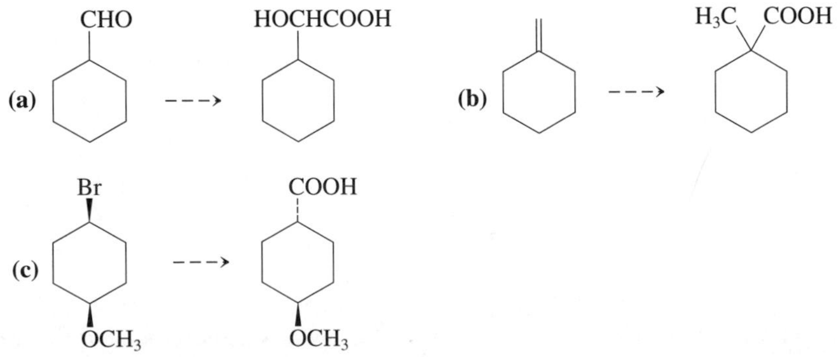

(a) CHO ---→ HOCHCOOH

(b) ---→ H₃C COOH

(c) Br, OCH₃ ---→ COOH, OCH₃

In summary, several reagents oxidize primary alcohols and aldehydes to carboxylic acids. A haloalkane is transformed into the carboxylic acid containing one additional carbon atom either by conversion into an organometallic reagent and carbonation or by displacement of halide by cyanide followed by nitrile hydrolysis.

19-7 Substitution at the Carboxy Carbon: The Addition–Elimination Mechanism

Carboxylic acids show reactivity similar to that of aldehydes and ketones (Section 17-5): The carbonyl carbon is subject to nucleophilic attack, and the corresponding oxygen is the target of electrophiles. After addition occurs, however, the carboxy OH, like that in alcohols, may be converted into a leaving group (Section 9-2). Expulsion of the latter results in a net substitution process and gives rise to new carbonyl compounds. This section introduces this process and general mechanisms by which it takes place for both carboxylic acids and their derivatives.

The carbonyl carbon is attacked by nucleophiles

Carbonyl carbons are electrophilic and potentially can be attacked by nucleophiles. This type of reactivity is observed in the carboxylic acids and the **carboxylic acid derivatives,** substances with the general formula RCOL (L stands for leaving group).

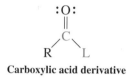

Carboxylic acid derivative

Carboxylic Acid Derivatives

$$\underset{\substack{\text{Alkanoyl}\\\text{halide}}}{\overset{\overset{\textstyle O}{\|}}{RCX}} \qquad \underset{\text{Anhydride}}{\overset{\overset{\textstyle O}{\|}\,\overset{\textstyle O}{\|}}{RCOCR}} \qquad \underset{\text{Ester}}{\overset{\overset{\textstyle O}{\|}}{RCOR'}} \qquad \underset{\text{Amide}}{\overset{\overset{\textstyle O}{\|}}{RCNR_2'}}$$

In contrast with the addition products of aldehydes and ketones (Sections 17-5 through 17-7), the intermediate formed upon attack of a nucleophile on a carboxy carbon can decompose *by eliminating a leaving group.* The overall result is substitution of the nucleophile for the leaving group through a process called **addition–elimination.** The species formed first in this transformation contains (in contrast with both starting material and product) a tetrahedral carbon center. It is therefore called a **tetrahedral intermediate.**

Color code reminder:
Nucleophile—red
Electrophile—blue
Leaving group—green

Nucleophilic Substitution by Addition–Elimination

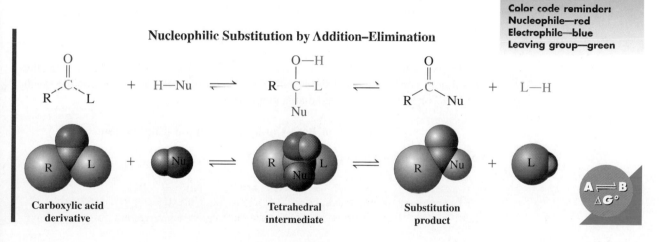

Carboxylic acid derivative   Tetrahedral intermediate   Substitution product

Substitution by addition–elimination is the most important pathway for the formation of carboxylic acid derivatives as well as for their interconversion—that is,

$$\overset{O}{\underset{\parallel}{RCL}} \rightarrow \overset{O}{\underset{\parallel}{RCL'}}.$$ The remainder of this section and those that follow will describe how such derivatives are prepared from carboxylic acids, and Chapter 20 will explore their properties and chemistry.

Addition–elimination is catalyzed by acid or base

Addition–elimination reactions benefit from acid catalysis. The acid functions in two ways: First, it protonates the carbonyl oxygen, activating the carbonyl group toward nucleophilic attack (Section 17-5). Second, protonation of L makes it a better leaving group (recall Sections 6-8 and 9-2).

Mechanism of Acid-Catalyzed Addition–Elimination

STEP 1. Protonation

STEP 2. Addition–elimination

STEP 3. Deprotonation

In base-catalyzed addition–elimination, the base (denoted as $:B^-$) ensures the maximum concentration of the negatively charged (deprotonated) nucleophile (such as HO^- or RO^-) when it is the attacking species.

Mechanism of Base-Catalyzed Addition–Elimination

STEP 1. Deprotonation of NuH

$$Nu-H \ + \ ^-:B \ \rightleftharpoons \ ^-:Nu \ + \ BH$$

STEP 2. Addition–elimination

STEP 3. Regeneration of catalyst

(Alternatively, ⁻:L may act as a base in step 1)

Substitution in carboxylic acids is complicated by a poor leaving group and the acidic proton

Can we apply the general addition–elimination process to the conversion of carboxylic acids into their derivatives? We must first overcome two problems. First, we know that the hydroxide ion is a poor leaving group (Section 6-8). Second, the carboxy proton is acidic, and most nucleophiles are bases (Section 6-9). Therefore, the desired nucleophilic attack (path *a* in the following equation) can encounter interference from an acid-base reaction (path *b*). Indeed, if a nucleophile is very basic, such as alkoxide, formation of the carboxylate ion will be essentially irreversible, and nucleophilic addition to the carbonyl carbon becomes very difficult.

Competing Reactions of a Carboxylic Acid with a Nucleophile

| Tetrahedral intermediate | (path *a*) | | (path *b*) | Carboxylate ion |

Esterification

$$RCOOH + R'OH$$

$$\xrightarrow{H^+}$$

$$RCOOR' + H_2O$$

On the other hand, with less basic nucleophiles, especially under acidic conditions, the ready reversibility of carboxylate formation may permit nucleophilic addition to compete and ultimately lead to substitution through the addition–elimination mechanism. A typical example is the **esterification** of a carboxylic acid (Section 9-4), in which an alcohol and a carboxylic acid react to yield an ester and water. The nucleophile, an alcohol, is a weak base, and acid is present to protonate both the carbonyl oxygen, activating it toward nucleophilic addition, and the carboxy OH, converting it into a better leaving group, water. The sections that follow will examine this and other carboxy substitutions in detail.

In summary, nucleophilic attack on the carbonyl group of carboxylic acid derivatives is a key step in substitution by addition–elimination. Either acid or base catalysis may be observed. For carboxylic acids, the process is complicated by the poor leaving group (hydroxide) and competitive deprotonation of the acid by the nucleophile, acting as a base. With less basic nucleophiles, addition can occur.

19-8 Carboxylic Acid Derivatives: Alkanoyl (Acyl) Halides and Anhydrides

With this section we begin a survey of the preparation of carboxylic acid derivatives. Replacement of the hydroxy group in RCOOH by halide gives rise to **alkanoyl (acyl) halides;** substitution by alkanoate (RCOO⁻) furnishes **carboxylic anhydrides.** *Both processes first require transformation of the hydroxy functionality into a better leaving group.*

Alkanoyl (acyl) halides are formed by using inorganic derivatives of carboxylic acids

$$\overset{O}{\underset{\|}{\text{RCX}}}$$

Alkanoyl (acyl) halide

The conversion of carboxylic acids into alkanoyl (acyl) halides employs the same reagents used in the synthesis of haloalkanes from alcohols (Section 9-4), $SOCl_2$ and PBr_3. The problem to be solved in both cases is identical—changing a poor leaving group (OH) into a good one.

Alkanoyl (Acyl) Halide Synthesis

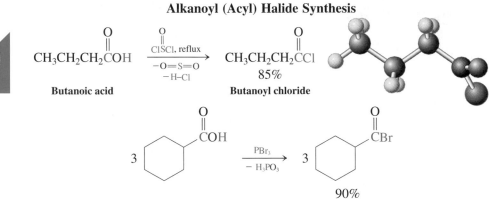

$$CH_3CH_2CH_2\overset{O}{\underset{\|}{C}}OH \xrightarrow[\substack{-O=S=O \\ -H-Cl}]{ClSCl, \text{ reflux}} CH_3CH_2CH_2\overset{O}{\underset{\|}{C}}Cl$$

Butanoic acid — Butanoyl chloride 85%

(These reactions fail with formic acid, HCOOH, because formyl chloride, HCOCl, and formyl bromide, HCOBr, are unstable. See Exercise 15-29.)

The mechanisms governing these transformations are reminiscent of those in the reactions of alcohols. Initially, an inorganic derivative of the acid is generated. The newly formed substituent is a good leaving group, facilitating acid-catalyzed addition–elimination, with halide as the nucleophile.

Mechanism of Alkanoyl (Acyl) Chloride Formation with Thionyl Chloride ($SOCl_2$)

STEP 1. Activation

STEP 2. Addition

Tetrahedral intermediate

STEP 3. Elimination

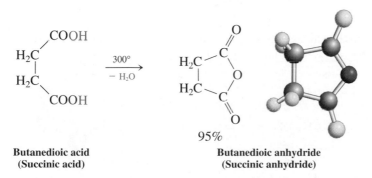

The mechanism of alkanoyl bromide formation using phosphorus tribromide (PBr$_3$) is similar (see Section 9-4).

Acids combine with alkanoyl halides to produce anhydrides

The electronegative power of the halogens in alkanoyl halides activates the carbonyl function to attack by other, even weak, nucleophiles (Chapter 20). For example, treatment of alkanoyl halides with carboxylic acids results in **carboxylic anhydrides.**

$$\underset{\text{Carboxylic anhydride}}{\text{RCOCR}}$$

$$\underset{\text{Butanoic acid}}{\text{CH}_3\text{CH}_2\text{CH}_2\overset{\text{O}}{\overset{\|}{\text{C}}}\text{OH}} \;+\; \underset{\text{Butanoyl chloride}}{\text{Cl}\overset{\text{O}}{\overset{\|}{\text{C}}}\text{CH}_2\text{CH}_2\text{CH}_3} \quad \xrightarrow[-\text{HCl}]{\Delta,\,8\,\text{h}}$$

$$\underset{85\%}{\text{CH}_3\text{CH}_2\text{CH}_2\overset{\text{O}}{\overset{\|}{\text{C}}}\text{O}\overset{\text{O}}{\overset{\|}{\text{C}}}\text{CH}_2\text{CH}_2\text{CH}_3}$$

Butanoic anhydride

As the name indicates, carboxylic anhydrides are formally derived from the corresponding acids by loss of water. Although carboxylic acid dehydration is not a general method for anhydride synthesis, cyclic examples may be prepared by heating dicarboxylic acids. A condition for the success of this transformation is that the ring closure lead to a five- or six-membered ring product.

Cyclic Anhydride Formation

$$\begin{array}{c}\text{H}_2\text{C}\diagup^{\text{COOH}}\\ |\\ \text{H}_2\text{C}\diagdown_{\text{COOH}}\end{array} \quad \xrightarrow[-\,\text{H}_2\text{O}]{300°} \quad \begin{array}{c}\text{H}_2\text{C}-\overset{\text{O}}{\overset{\|}{\text{C}}}\\ |\qquad\;\;\diagup\text{O}\\ \text{H}_2\text{C}-\underset{\|}{\text{C}}\\ \qquad\quad\text{O}\end{array}$$

$$95\%$$

| **Butanedioic acid** | **Butanedioic anhydride** |
| **(Succinic acid)** | **(Succinic anhydride)** |

Because the halogen in the alkanoyl halide and the RCO$_2$ substituent in the anhydride are good leaving groups and because they activate the adjacent carbonyl function, these carboxylic acid derivatives are useful synthetic intermediates for the preparation of other compounds, a topic to be discussed in Sections 20-2 and 20-3.

EXERCISE 19-8

Suggest two preparations, starting from carboxylic acids or their derivatives, for each of the following compounds.

$$\text{(a)} \ \underset{\substack{\| \quad \|\\ O \quad O}}{CH_3COCCH_2CH_3} \qquad \text{(b)} \ \underset{\substack{\ | \quad \|\\ H_3C \quad O}}{CH_3CHCCl}$$

EXERCISE 19-9

Propose a mechanism for the thermal formation of butanedioic anhydride from the dioic acid.

In summary, the hydroxy group in COOH can be replaced by halogen by using the same reagents used to convert alcohols into haloalkanes—$SOCl_2$ and PBr_3. The resulting alkanoyl (acyl) halides are sufficiently reactive to be attacked by carboxylic acids to generate carboxylic anhydrides. Cyclic examples of the latter may be made from dicarboxylic acids by thermal dehydration.

19-9 Carboxylic Acid Derivatives: Esters

Esters have the general formula $\overset{\scriptsize O}{\underset{\scriptsize \|}{RCOR'}}$. Their widespread occurrence in nature and many practical uses make them perhaps the most important of the carboxylic acid derivatives. This section describes two methods by which esters are made—the mineral acid-catalyzed reaction of carboxylic acids with alcohols and the synthesis of methyl esters using diazomethane.

Carboxylic acids react with alcohols to form esters

When a carboxylic acid and an alcohol are mixed together, no reaction takes place. However, upon addition of catalytic amounts of a mineral acid, such as sulfuric acid or HCl, the two components combine in an equilibrium process to give an ester and water (Section 9-4).

Acid-Catalyzed Esterification

$$\underset{\substack{\| \\ O}}{RCOH} \ + \ R'OH \ \underset{}{\overset{H^+}{\rightleftharpoons}} \ \underset{\substack{\| \\ O}}{RCOR'} \ + \ H_2O$$

  Carboxylic acid   **Alcohol**   **Ester**

Esterification is not very exothermic. How can the equilibrium be shifted toward the ester product? One way is to use an excess of either of the two starting materials; another is to remove the ester or the water product from the reaction mixture. In practice, esterifications are most often achieved by using the alcohol as a solvent.

$$\underset{\substack{\| \\ O}}{CH_3COH} \ + \ CH_3OH \ \underset{-H_2O}{\overset{H_2SO_4, \ \Delta}{\rightleftharpoons}} \ \underset{\substack{\| \\ O}}{CH_3COCH_3}$$
85%

Acetic acid   **Solvent**   **Methyl acetate**

EXERCISE 19-10

Show the products of the acid-catalyzed reaction of each of the following pairs of compounds. **(a)** Methanol + pentanoic acid; **(b)** formic acid + 1-pentanol; **(c)** cyclohexanol + benzoic acid; **(d)** 2-bromoacetic acid + 3-methyl-2-butanol.

The opposite of esterification is **ester hydrolysis.** This reaction is carried out under the same conditions as esterification, but, to shift the equilibrium, an excess of water is used in a water-miscible solvent.

$$CH_3CH_2CH_2CH_2\underset{\underset{\displaystyle CH_3}{|}}{\overset{\overset{\displaystyle CH_3}{|}}{C}}COOCH_2CH_3 \xrightarrow{H_2SO_4,\ HOH,\ propanone\ (acetone),\ \Delta} CH_3CH_2CH_2CH_2\underset{\underset{\displaystyle CH_3}{|}}{\overset{\overset{\displaystyle CH_3}{|}}{C}}COOH \ + \ CH_3CH_2OH$$

Ethyl 2,2-dimethylhexanoate   85%
                                                     2,2-Dimethylhexanoic acid

Give the products of acid-catalyzed hydrolysis of each of the following esters.

(a) $CH_3(CH_2)_3C\equiv CCH_2CH_2COOCH(CH_3)_2$ (b)

(c) $CH_3CH_2\underset{\underset{\displaystyle CH_3}{|}}{C}HCH_2OOC$—⬠

Esterification proceeds through acid-catalyzed addition–elimination

The presence of the acid catalyst features prominently in the mechanism of ester formation: It causes the carbonyl function to undergo nucleophilic attack by the alcohol (step 2) and the hydroxy group to leave as water (step 3).

Mechanism of Acid-Catalyzed Esterification and Ester Hydrolysis

STEP 1. Protonation of the carboxy group

STEP 2. Attack by methanol

Tetrahedral intermediate

Relay point:
← **Can go back to starting material**
Can go forward to product →

STEP 3. Elimination of water

Initially, protonation of the carbonyl oxygen gives a delocalized carbocation (step 1). Now the carbonyl carbon is susceptible to nucleophilic attack by methanol. Proton loss from the initial adduct furnishes the tetrahedral intermediate (step 2). This species is a crucial relay point, because it can react in either of *two* ways in the presence of the mineral acid catalyst. First, it can lose methanol by the reverse of steps 1 and 2. This process, beginning with protonation of the methoxy oxygen, leads back to the carboxylic acid. The second possibility, however, is protonation at either hydroxy oxygen, leading to elimination of water and to the ester (step 3). All the steps are reversible; therefore, either addition of excess alcohol or removal of water favors esterification by shifting the equilibria in steps 2 and 3, respectively. Ester hydrolysis proceeds by the reverse of the sequence and is favored by aqueous conditions.

EXERCISE 19-12

Give the mechanism of esterification of a general carboxylic acid RCOOH with methanol in which the alcohol oxygen is labeled with the ^{18}O isotope ($CH_3{}^{18}OH$). Does the labeled oxygen appear in the ester or in the water product?

Hydroxy acids may undergo intramolecular esterification to lactones

When hydroxy carboxylic acids are treated with catalytic amounts of mineral acid, cyclic esters—or **lactones**—may form. This process is called **intramolecular esterification** and is favorable for formation of five- and six-membered rings.

Formation of a Lactone

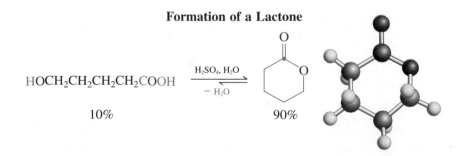

$$HOCH_2CH_2CH_2CH_2COOH \xrightleftharpoons[- H_2O]{H_2SO_4,\ H_2O}$$

10%   90%

EXERCISE 19-13

Explain the following result by a mechanism. (**Hint:** See Section 17-7.)

Carboxylic acids form esters by other processes

In addition to acid-catalyzed esterification, other reactions can transform carboxylic acids into esters. Section 8-5 described one, the nucleophilic substitution of halo-alkanes with carboxylate ions. Another method is used only on a small scale for the specific conversion of an acid into its methyl ester. This transformation employs **diazomethane,** CH_2N_2, a highly reactive, toxic, and explosive gas, and is driven by the production of gaseous nitrogen (see Section 21-11).

$$CH_3C\equiv CCOOH + CH_2N_2 \xrightarrow{(CH_3CH_2)_2O} CH_3C\equiv CCOOCH_2H + N_2$$
$$80\%$$

2-Butynoic acid   **Methyl 2-butynoate**

In summary, carboxylic acids react with alcohols to form esters, as long as a mineral acid catalyst is present. This reaction is only slightly exothermic, and its equilibrium may be shifted in either direction by the choice of reaction conditions. The reverse of ester formation is ester hydrolysis. The mechanism of esterification is acid-catalyzed addition of alcohol to the carbonyl group followed by acid-catalyzed dehydration. Intramolecular ester formation results in lactones, favored only when five- or six-membered rings are produced. Esters can be formed from carboxylic acids by other mechanisms—for example, the reaction of carboxylate ions with (primary) haloalkanes, and, for methyl esters, that of carboxylic acids with diazomethane.

19-10 Carboxylic Acid Derivatives: Amides

As we saw earlier (Section 17-9), amines also are capable of attacking the carbonyl function. When the latter is that of a carboxylic acid, the product is a **carboxylic amide,** the last major class of carboxylic acid derivatives. The mechanism is again addition–elimination but is complicated by acid-base chemistry.

$$\overset{O}{\underset{\|}{RCNR_2'}}$$
Carboxylic amide

Amines react with carboxylic acids as bases and as nucleophiles

Nitrogen lies to the left of oxygen in the periodic table. Therefore, amines (Chapter 21) are better bases as well as better nucleophiles than alcohols (Section 6-9). To synthesize carboxylic amides, therefore, we must address the problem discussed in Section 19-7: interference from competing acid-base reaction. Indeed, exposure of a carboxylic acid to an amine initially forms an ammonium carboxylate salt, in which the negatively charged carboxylate is very resistant to nucleophilic attack.

Ammonium Salts from Carboxylic Acids

Notice that salt formation, though very favorable, is nonetheless reversible. Upon heating, a slower but thermodynamically favored reaction between the acid and the amine can take place. The acid and the amine are removed from the equilibrium, and eventually salt formation is completely reversed. In this second mode of reaction, the nitrogen acts as a nucleophile and attacks the carbonyl carbon. Completion of an addition–elimination sequence leads to the amide.*

Formation of an Amide from an Amine and a Carboxylic Acid

$$CH_3CH_2CH_2\overset{O}{\overset{\|}{C}}OH \quad + \quad (CH_3)_2NH \xrightarrow[-H_2O]{155°C} \quad CH_3CH_2CH_2\overset{O}{\overset{\|}{C}}N(CH_3)_2$$

84%

N,N-Dimethylbutanamide

Mechanism of Amide Formation

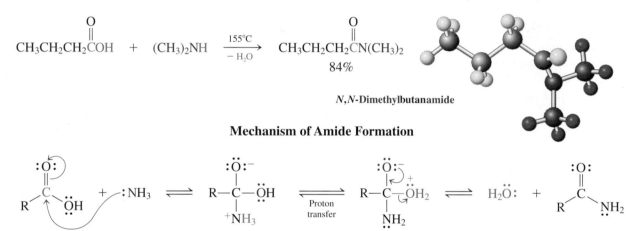

Tetrahedral intermediates   **An amide**

Amide formation is reversible. Thus, treatment of amides with hot acidic or basic water regenerates the component carboxylic acids and amines.

Dicarboxylic acids react with amines to give imides

Dicarboxylic acids may react twice with the amine nitrogen of ammonia or of a primary amine. This sequence gives rise to **imides,** the nitrogen analogs of cyclic anhydrides.

$$\begin{array}{c} CH_2COOH \\ | \\ CH_2COOH \end{array} \xrightarrow{NH_3} \begin{array}{c} CH_2COO^-NH_4{}^+ \\ | \\ CH_2COO^-NH_4{}^+ \end{array} \xrightarrow[-NH_3]{290°C}$$

Butanedioic acid

83%

**Butanimide
(Succinimide)**

Recall the use of *N*-halobutanimides in halogenations (Sections 3-8 and 14-2).

*Remember not to confuse the names of carboxylic amides with those of the alkali salts of amines, also called amides (e.g., lithium amide, LiNH$_2$).

Amino acids cyclize to lactams

In analogy to hydroxycarboxylic acids, some amino acids undergo cyclization to the corresponding cyclic amides, called **lactams** (Section 20-7).

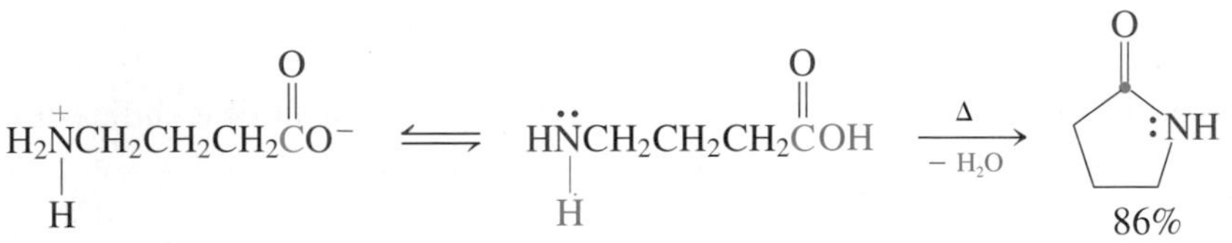

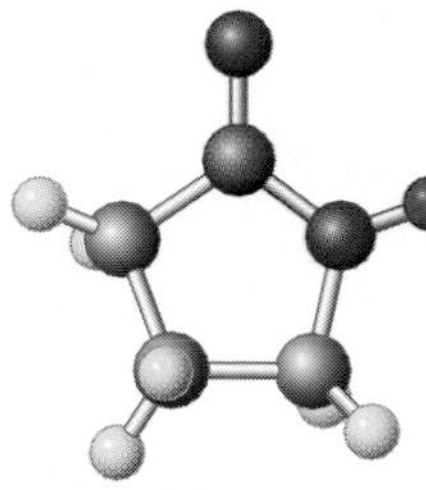

A lactam

The penicillin and cephalosporin classes of antibiotics derive their biological activity from the presence of a lactam function (see Chemical Highlight 20-2).

EXERCISE 19-14

Formulate a detailed mechanism for the formation of butanimide from butanedioic acid and ammonia.

In summary, amines react with carboxylic acids to form amides by an addition–elimination process that begins with nucleophilic attack by the amine on the carboxy carbon. Amide formation is complicated by reversible deprotonation of the carboxylic acid by the basic amine to give an ammonium salt.

19-11 Reduction of Carboxylic Acids by Lithium Aluminum Hydride

Lithium aluminum hydride is capable of reducing carboxylic acids all the way to the corresponding primary alcohols, which are obtained upon aqueous acidic work-up.

Reduction of a Carboxylic Acid

$$RCOOH \xrightarrow[\text{2. H}^+, \text{H}_2\text{O}]{\text{1. LiAlH}_4, \text{THF}} RCH_2OH$$

Example:

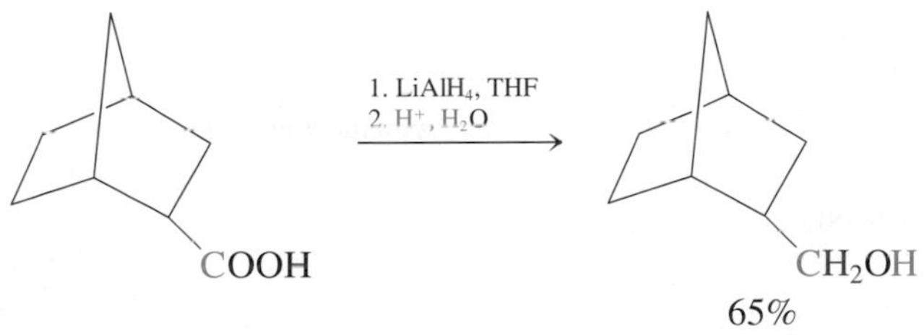

Although the exact mechanism of this transformation is not completely understood, it is clear that the hydride reagent first acts as a base, forming the lithium salt of the acid and hydrogen gas. Carboxylate salts are generally resistant to attack by nucleophiles. Yet, despite the negative charge, lithium aluminum hydride is so reactive that it is capable of donating two hydrides to the carbonyl function of the carboxylate. The product of this sequence, a simple alkoxide, gives the alcohol after protonation.

EXERCISE 19-15

Propose synthetic schemes that produce compound B from compound A.

(a) CH₃CH₂CH₂CN   CH₃CH₂CH₂CH₂OH   **(b)** ▷—CH₂COOH   ▷—CH₂CD₂OH

       A   **B**   **A**   **B**

In summary, the nucleophilic reactivity of lithium aluminum hydride is sufficiently great to effect the reduction of carboxylates to primary alcohols.

19-12 Bromination Next to the Carboxy Group: The Hell-Volhard-Zelinsky Reaction

Like aldehydes and ketones, alkanoic acids can be monobrominated at the α-carbon by exposure to Br_2. The addition of a trace amount of PBr_3 is necessary to get the reaction started. Because the highly corrosive PBr_3 is difficult to handle, it is often generated in the reaction flask (in situ). This is achieved by the addition of a little elemental (red) phosphorus to the mixture of starting materials; it is converted into PBr_3 instantaneously by the bromine present.

A Hell-Volhard-Zelinsky* Reaction

$$CH_3CH_2CH_2CH_2COOH \xrightarrow{\text{Br–Br, trace P}} \underset{\substack{80\% \\ \textbf{2-Bromopentanoic acid}}}{CH_3CH_2CH_2\overset{\overset{\displaystyle Br}{|}}{C}HCOOH} + HBr$$

The alkanoyl bromide, formed by the reaction of PBr_3 with the carboxylic acid (Section 19-8), is subject to rapid acid-catalyzed enolization. The enol is brominated to give the 2-bromoalkanoyl bromide. The latter then undergoes an exchange reaction with unreacted acid to furnish the product bromoacid and another molecule of alkanoyl bromide, which reenters the reaction cycle.

Mechanism of the Hell-Volhard-Zelinsky Reaction

STEP 1. Alkanoyl bromide formation

$$3 \; RCH_2\overset{\overset{\displaystyle O}{\|}}{C}OH \;+\; PBr_3 \longrightarrow \underset{\textbf{Alkanoyl bromide}}{3 \; RCH_2\overset{\overset{\displaystyle O}{\|}}{C}Br} \;+\; H_3PO_3$$

STEP 2. Enolization

$$RCH_2\overset{\overset{\displaystyle O}{\|}}{C}Br \underset{}{\overset{H^+}{\rightleftharpoons}} \underset{\textbf{Enol}}{RCH=C\overset{\displaystyle OH}{\underset{\displaystyle Br}{\big<}}}$$

*Professor Carl M. Hell (1849–1926), University of Stuttgart, Germany; Professor Jacob Volhard (1834–1910), University of Halle, Germany; Professor Nicolai D. Zelinsky (1861–1953), University of Moscow.

STEP 3. Bromination

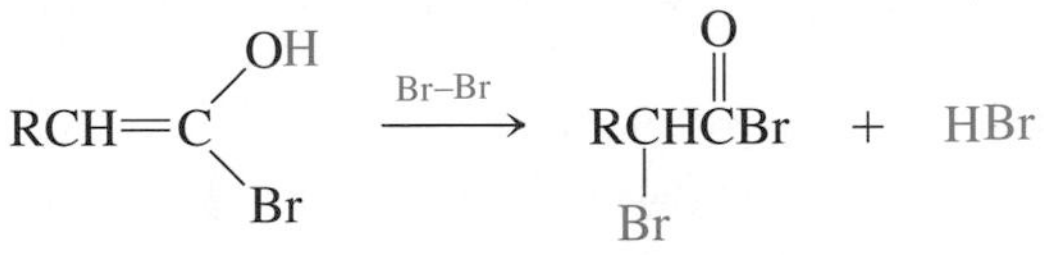

$$RCH=C\begin{smallmatrix}OH\\\\Br\end{smallmatrix} \xrightarrow{Br-Br} RCHCBr \;+\; HBr$$

STEP 4. Exchange

$$\underset{Br}{RCHCBr} \;+\; RCH_2COH \;\rightleftharpoons\; \underset{Br}{RCHCOH} \;+\; \underset{\substack{\textbf{Reenters}\\\textbf{step 2}}}{RCH_2CBr}$$

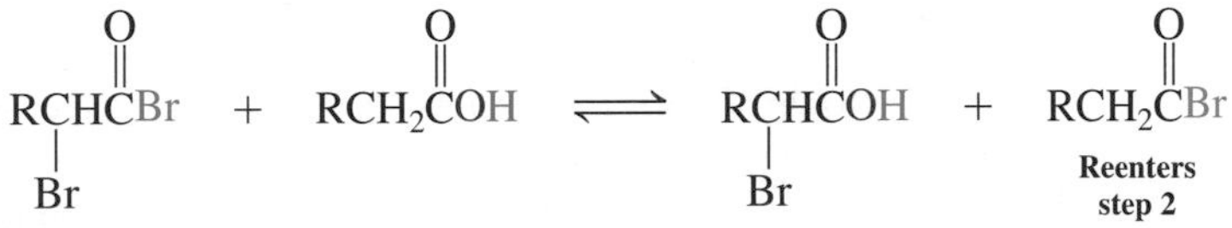

EXERCISE 19-16

Formulate detailed mechanisms for steps 2 and 3 of the Hell-Volhard-Zelinsky reaction. (**Hints:** Review Sections 18-2 for step 2 and 18-3 for step 3.)

The bromocarboxylic acids formed in the Hell-Volhard-Zelinsky reaction can be converted into other 2-substituted derivatives. For example, treatment with aqueous base gives 2-hydroxy acids, whereas amines yield α-amino acids (Chapter 26). An example of the latter is the synthesis of racemic 2-aminohexanoic acid (norleucine, a naturally occurring but rare amino acid) shown here.

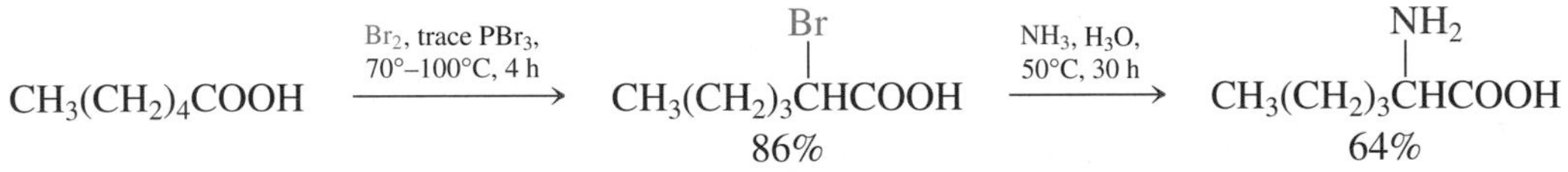

$$CH_3(CH_2)_4COOH \xrightarrow[70°-100°C,\,4\,h]{Br_2,\,trace\,PBr_3,} \underset{86\%}{CH_3(CH_2)_3\overset{Br}{CHCOOH}} \xrightarrow[50°C,\,30\,h]{NH_3,\,H_3O,} \underset{64\%}{CH_3(CH_2)_3\overset{NH_2}{CHCOOH}}$$

Hexanoic acid   **2-Bromohexanoic acid**   **2-Aminohexanoic acid**
(Norleucine)

In summary, with trace amounts of phosphorus (or phosphorus tribromide), carboxylic acids are brominated at C2 (the Hell-Volhard-Zelinsky reaction). The transformation proceeds through 2-bromoalkanoyl bromide intermediates.

19-13 Biological Activity of Carboxylic Acids

Considering the variety of reactions that carboxylic acids can undergo, it is no wonder that they are very important, not only as synthetic intermediates in the laboratory, but also in biological systems. This section will provide a glimpse of the enormous structural and functional diversity of natural carboxylic acids. A discussion of amino acids will be deferred to Chapter 26.

As Table 19-1 indicates, even the simplest carboxylic acids are abundant in nature. Formic acid is present not only in ants, where it functions as an alarm pheromone, but also in plants. For example, one reason why human skin hurts after it touches the stinging nettle is that formic acid is deposited in the wounds.

Acetic acid is formed through the enzymatic oxidation of ethanol produced by fermentation. Vinegar is the term given to the dilute (ca. 4–12%) aqueous solution thus generated in ciders, wines, and malt extracts. Louis Pasteur in 1864 established the involvement of bacteria in the oxidation stage of this ancient process.

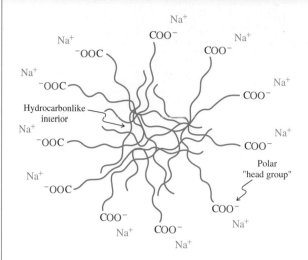

The sodium and potassium salts of long-chain carboxylic acids have the property of aggregating as spherical clusters called *micelles in* aqueous solution. In such aggregates, the hydrophobic alkyl chains (Section 8-2) of the acids attempt to occupy the same region in space because of their attraction to one another by London forces (Section 2-4) and their tendency to avoid exposure to polar water. As shown above, the polar, water-solvated carboxylate "head groups" form a spherical shell around the hydrocarbonlike center.

Because these carboxylate salts also create films on aqueous surfaces, they act as soaps. The polar groups stick into the water while the alkyl chains assemble into a hydrophobic layer. This construction reduces the surface tension of the water, permitting it to permeate cloth and other fabrics and giving rise to the foaming typical of soaps. Cleansing is accomplished by dissolving ordinarily water-insoluble materials (oils, fats) in the hydrocarbon interior of the micelles.

One problem with carboxylate soaps is that they form curdlike precipitates with the ions present in hard water (e.g., Mg^{2+} and Ca^{2+}). Detergents based on alkanesulfonates, $RSO_3^- Na^+$, and alkyl sulfates, $ROSO_3^- Na^+$, avoid this drawback but have caused pollution problems in lakes and streams because branches in their alkyl chains made early versions *nonbiodegradable.* The microorganisms associated with normal sewage treatment procedures are capable of breaking down only straight-chain systems.

The recent widespread introduction of energy- and water-saving front-loading washing machines has given rise to the need for detergents of very high solubility and cleansing power but reduced foaming (sudsing). Novel sulfonates containing ether linkages have been developed to meet this demand.

Certain steroid *bile acids* such as *cholic acid* (Section 4-7) also have surfactant or detergentlike properties and are found in the bile duct. These substances are released into the upper intestinal tract where they emulsify water-insoluble fats through the formation of micelles. Hydrolytic enzymes can then digest the dispersed fat molecules.

Cholic acid

False-color scanning electron micrographs of the collar of a pure cotton shirt before washing (left) and after washing (right). Before washing, the threads and subthreads are covered with grime and skin flakes. (Magnified 70 times.)

Fatty acids are derived from coupling of acetic acid

Acetic acid exhibits diverse biological activities, ranging from a defense pheromone in some ants and scorpions to the primary building block for the biosynthesis of more naturally occurring organic compounds than any other single precursor substance. For example, 3-methyl-3-butenyl pyrophosphate, the crucial precursor in the buildup of the terpenes (Section 14-10), is made by the enzymatic conversion of three molecules of CH_3COOH into an intermediate called mevalonic acid. Further reactions degrade the system to the five-carbon (isoprene) unit of the product.

$$3\ CH_3-COOH \xrightarrow{\text{Enzymes}} CH_3-\overset{\overset{\displaystyle CH_2-COOH}{|}}{\underset{\underset{\displaystyle CH_2-CH_2OH}{|}}{C}}OH \xrightarrow{\text{Enzymes}} CH_3-C$$

Mevalonic acid   **3-Methyl-3-butenyl pyrophosphate**

A conceptually more straightforward mode of multiple coupling is found in the biosynthesis of **fatty acids.** This class of compounds derives its name from its source, the natural **fats,** which are esters of long-chain carboxylic acids (see Section 20-4). Hydrolysis or **saponification** (so called because the corresponding salts form soaps—see Chemical Highlight 19-1; *sapo,* Latin, soap) yields the corresponding fatty acids.

CHEMICAL HIGHLIGHT 19-2 | Trans Fatty Acids and Your Health

More than 90% of the double bonds in naturally occurring unsaturated fatty acids possess the cis configuration, contributing to the reduced melting temperatures of vegetable oils compared with saturated fats (Section 11-3). Exposure of vegetable oil to catalytic hydrogenation conditions produces solid margarine. However, this process does not hydrogenate all of the double bonds: A significant fraction of cis double bonds are merely isomerized to trans by the catalyst and remain in the final solid product. For example, synthetic hard (stick) margarine contains about 35% saturated fatty acids (SFAs) and 12% trans fatty acids (TFAs). For comparison, natural butter consists of more than 50% SFAs but only 3–4% TFAs. Soft (tub) margarines, whose exposure to catalytic hydrogenation conditions is less than that of hard margarine, possess about 15% saturated acids and 5% TFAs.

What, if any, are the health consequences of TFAs in the human diet? It has long been suspected that TFAs are not metabolized by the body in the same way as their cis counterparts; in the 1960s and 1970s,

this suspicion was confirmed by studies that indicated that TFAs in foods greatly affect lipid metabolism. Perhaps the most alarming finding was that TFAs accumulate in cell membranes and increase the levels of **low-density lipoproteins** (**LDLs,** popularly but imprecisely known as "bad cholesterol") in the bloodstream while reducing **high-density lipoprotein** levels (**HDLs,** the so-called good cholesterol). Follow-up studies have focused on possible links between dietary TFAs and increased risk of heart disease and, more recently, breast cancer, but they have not led to easily interpreted results, in part because foods containing TFAs contain saturated fatty acids as well. The current prevailing opinion is that TFAs behave similarly to SFAs in their overall health effects. Because TFAs are in general a very minor component of a typical diet, the American Heart Association's recommendation to limit fat consumption to no more than 30% of total caloric intake is still considered reasonable advice for most healthy people.

The most important of them are between 12 and 22 carbons long and may contain cis carbon–carbon double bonds.

Fatty Acids

$$CH_3(CH_2)_{14}COOH$$

$$CH_3(CH_2)_7 \quad (CH_2)_7COOH$$
$$C=C$$
$$H \qquad H$$

Hexadecanoic acid   ***cis*-9-Octadecenoic acid**
(Palmitic acid)   **(Oleic acid)**

In accord with their biosynthetic origin, fatty acids consist mostly of even-numbered carbon chains. A very elegant experiment demonstrated that linear coupling occurred in a highly regular fashion. In it, singly labeled radioactive (^{14}C) acetic acid was fed to several organisms. The resulting fatty acids were labeled only at every other carbon atom.

$$CH_3{}^{14}COOH \xrightarrow{\text{Organism}} CH_3{}^{14}CH_2CH_2{}^{14}CH_2CH_2{}^{14}CH_2CH_2{}^{14}CH_2CH_2{}^{14}CH_2CH_2{}^{14}CH_2CH_2{}^{14}CH_2CH_2{}^{14}COOH$$

Labeled hexadecanoic (palmitic) acid

The mechanism of chain formation is very complex, but the following scheme provides a general idea of the process. A key player is the mercapto group of an important biological relay compound called coenzyme A (abbreviated HSCoA; Figure 19-4). This function binds acetic acid in the form of a **thiolester** called acetyl CoA. Carboxylation transforms some of this thiolester into malonyl CoA. The two acyl groups are then transferred to two molecules of **acyl carrier protein.** Coupling occurs with loss of CO_2 to furnish a 3-oxobutanoic thiolester.

Mechanism of Coupling of Acetic Acid Units

STEP 1. Thiolester formation

$$CH_3COOH \;+\; HSCoA \longrightarrow \overset{\displaystyle O}{\overset{\|}{CH_3C}}SCoA \;+\; HOH$$

Acetic acid   **Coenzyme A**   **Acetyl coenzyme A**

STEP 2. Carboxylation

$$\overset{\displaystyle O}{\overset{\|}{CH_3C}}SCoA \;+\; CO_2 \xrightarrow{\text{Acetyl CoA carboxylase}} \overset{\displaystyle O \qquad O}{\overset{\| \qquad \|}{HOCCH_2C}}SCoA$$

Malonyl CoA

STEP 3. Acetyl and malonyl group transfers

$$\overset{\displaystyle O}{\overset{\|}{CH_3C}}SCoA \;+\; HS\!-\!\boxed{\text{protein}} \longrightarrow \overset{\displaystyle O}{\overset{\|}{CH_3C}}S\!-\!\boxed{\text{protein}} \;+\; HSCoA$$

Acyl carrier protein

$$\overset{\displaystyle O \qquad O}{\overset{\| \qquad \|}{HOCCH_2C}}SCoA \;+\; HS\!-\!\boxed{\text{protein}} \longrightarrow \overset{\displaystyle O \qquad O}{\overset{\| \qquad \|}{HOCCH_2C}}S\!-\!\boxed{\text{protein}} \;+\; HSCoA$$

Acyl carrier protein

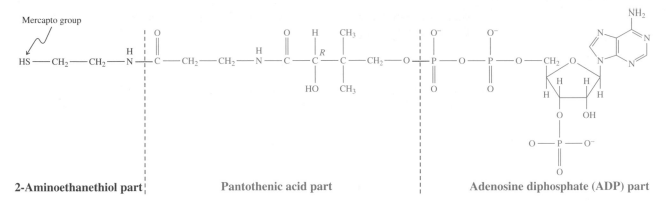

Mercapto group

2-Aminoethanethiol part   **Pantothenic acid part**   **Adenosine diphosphate (ADP) part**

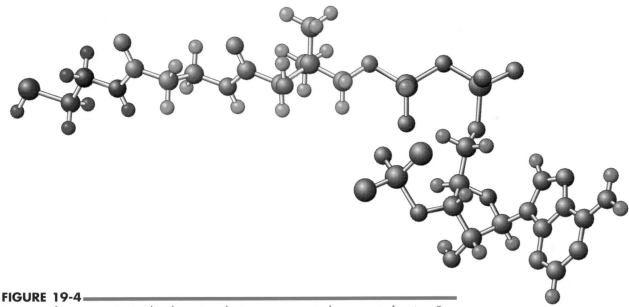

FIGURE 19-4
Structure of coenzyme A. For this discussion, the important part is the mercapto function. For convenience, the molecule may be abbreviated HSCoA.

STEP 4. Coupling

$$\underset{\text{HOCCH}_2\text{CS}}{\overset{\text{O}\quad\text{O}}{\|\quad\|}}\boxed{\text{protein}} \xrightarrow[-\text{CO}_2]{\overset{\text{O}}{\overset{\|}{\text{CH}_3\text{CS}}}\boxed{\text{protein}}} \underset{\text{CH}_3\text{CCH}_2\text{CS}}{\overset{\text{O}\quad\text{O}}{\|\quad\|}}\boxed{\text{protein}}$$

A 3-oxobutanoic thiolester

Reduction of the ketone function to a methylene group follows. The resulting butanoic thiolester repeatedly undergoes a similar sequence of reactions elongating the chain, always by two carbons. The eventual product is a long-chain alkanoyl group, which is removed from the protein by hydrolysis.

$$\text{CH}_3\text{CH}_2\text{-(CH}_2\text{CH}_2)_n\text{-CH}_2\overset{\text{O}}{\overset{\|}{\text{C}}}\text{-S-}\boxed{\text{protein}} \xrightarrow{\text{H}_2\text{O}} \text{CH}_3\text{CH}_2\text{-(CH}_2\text{CH}_2)_n\text{-CH}_2\overset{\text{O}}{\overset{\|}{\text{C}}}\text{-OH} + \text{HS-}\boxed{\text{protein}}$$

Fatty acid

Arachidonic acid is a biologically important unsaturated fatty acid

Naturally occurring *unsaturated* fatty acids are capable of undergoing further transformations leading to a variety of unusual structures. An example is arachidonic acid, which is the biological precursor to a multitude of important chemicals in the human body, such as prostaglandins (Chemical Highlight 11-1), thromboxanes, prostacyclins, and leukotrienes.

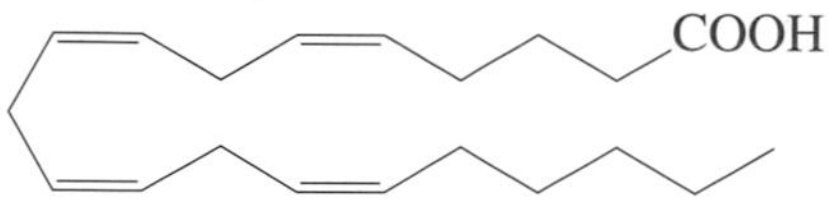

Arachidonic acid

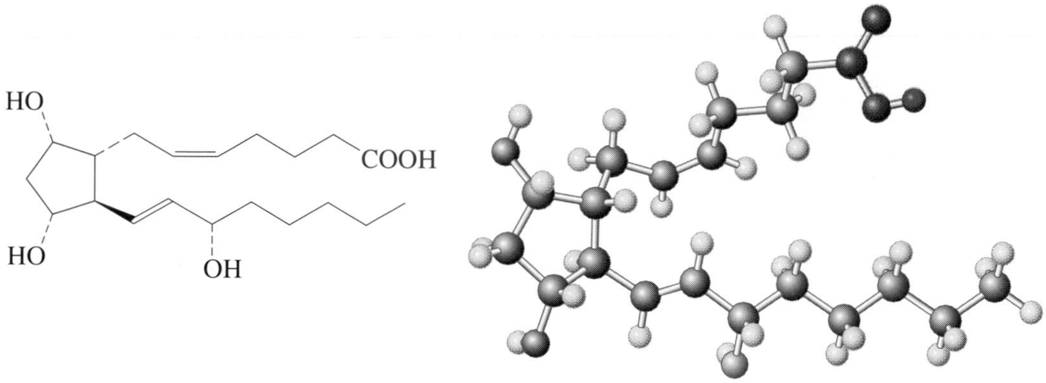

Prostaglandin F$_{2\alpha}$
(Induces labor, abortion, menstruation)

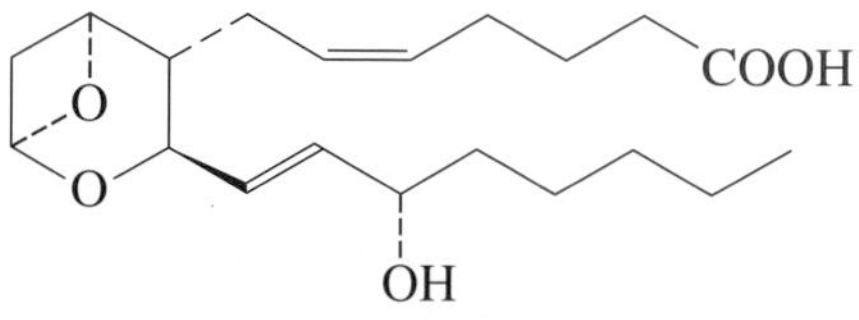

Thromboxane A$_2$

(Contracts smooth muscles;
aggregates blood platelets)

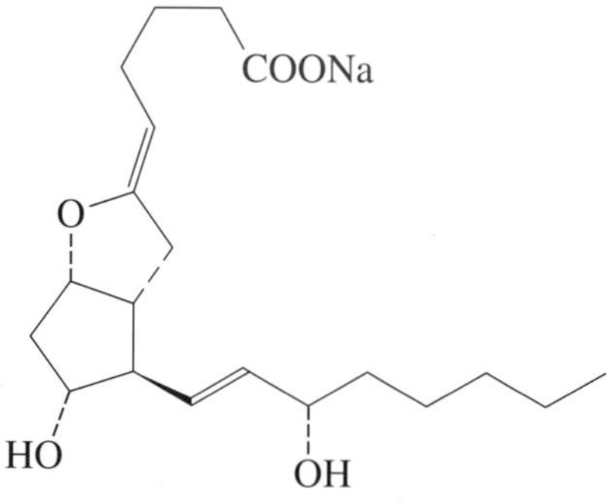

Prostacyclin I$_2$, sodium salt

(The most potent natural inhibitor of platelet
aggregation; vasodilator, used in heart-bypass
operations, in kidney patients, and in others)

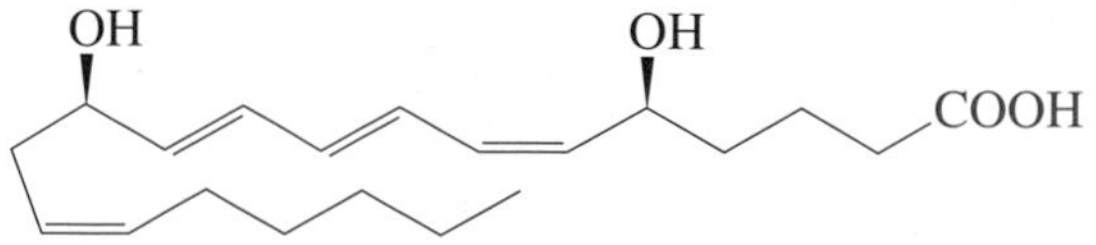

Leukotriene B$_4$

(Potent chemotactic factor;
e.g., causes cell migrations)

CHEMICAL HIGHLIGHT 19-3   Biodegradable Polyester Plastics

With applications of polymers and plastics becoming increasingly important in modern society, the problem of plastic waste disposal grows ever more severe. Many landfills are approaching capacity with these highly degradation resistant substances (Section 12-14). As a consequence, efforts to develop commercially acceptable polymers that degrade under appropriate environmental conditions are underway.

Biodegradable polymers and plastics would be suitable for nonreusable items such as plastic bags, films, and bottles. A recently developed and commercialized example is poly(β-hydroxybutyrate-co-β-hydroxyvalerate) (PHBV), a copolymer of 3-hydroxybutanoic acid and 3-hydroxypentanoic acid

mixture of 3-oxobutanoic and 3-oxopentanoic thiolesters, whose ketone carbonyl functions are then reduced by NADH to hydroxy groups (see Chapter 8, Problem 47). Finally, an enzyme-mediated esterification process couples hydroxy and carboxy functions of the two hydroxyesters in random order to generate the copolymer.

The properties of PHBV copolymers vary according to the ratio of 3-hydroxybutanoic acid, which leads to a stiffer polymer, to 3-hydroxypentanoic acid, which imparts flexibility. PHBV is used in specialty packaging, orthopedic devices, and even controlled drug release. In the drug-release application, the polymer encapsulates a

in which the individual monomer units are connected by ester linkages. PHBV copolymers are produced not in the conventional manner but by fermentation of mixtures containing both acetic and propanoic acids by the bacterium *Alcaligenes eutropus*. Both acids are converted into thiolesters with coenzyme A. Condensation with malonyl CoA then yields a

pharmaceutical, which is released only after the polymer coating has degraded sufficiently. When PHBV is used for such medical applications, it degrades by ester hydrolysis to 3-hydroxybutanoic and 3-hydroxypentanoic acids, which are natural products of human metabolism. PHBV is also subject to bacterial degradation upon disposal in the environment.

Each of these compounds possesses powerful biological properties. For example, some of the prostaglandins are responsible for the tissue inflammation associated with rheumatoid arthritis. The salicylate derivative aspirin (Chemical Highlight 22-3) is capable of combating such symptoms by inhibiting the conversion of arachidonic acid into prostaglandins. In the early 1990s, asthma, a condition that afflicts some 15 million Americans, was recognized to be a chronic inflammatory disease. Recent research has shown that the leukotrienes mediate airway inflammation, and the thromboxanes contribute to bronchoconstriction. These findings have led to the use of anti-inflammatory agents such as the **corticosteroids** (Section 4-7) as antiasthmatics, because they are capable of blocking the biosynthesis of arachidonic acid itself.

Aspirin

EXERCISE 19-17

Identify the arachidonic acid backbone in each of its four derivatives just shown.

Nature also produces complex polycyclic carboxylic acids

Many biologically active natural products that have carboxy groups as substituents of complex polycyclic frames have physiological potential that derives from other sites in the molecule. In these compounds the function of the carboxy group may be to impart water solubility, to allow for salt formation or ion transport, and to enable micellar-type aggregation. Two examples are gibberellic acid, one of a group of plant growth-promoting substances manufactured by fermentation, and lysergic acid, a major product of hydrolyzed extracts of ergot, a fungal parasite that lives on grasses, including rye. Many lysergic acid derivatives possess powerful psychotomimetic activity. In the Middle Ages, thousands who ate rye bread contaminated by ergot experienced the poisonous effects characteristic of these compounds: hallucinations, convulsions, delirium, epilepsy, and death. The synthetic lysergic acid diethylamide (LSD) is one of the most powerful hallucinogens known. The effective oral dose for humans is only about 0.05 mg.

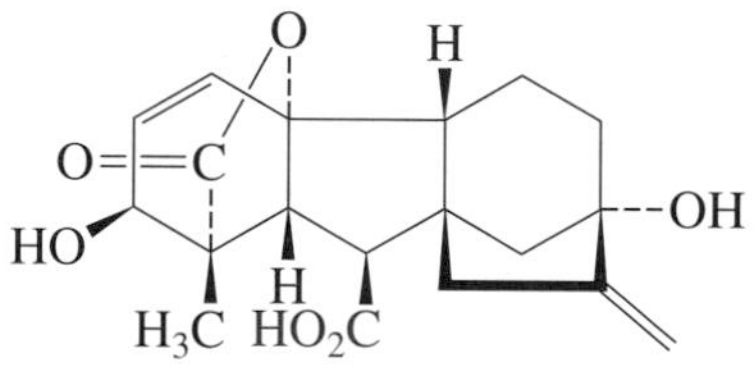

Gibberellic acid

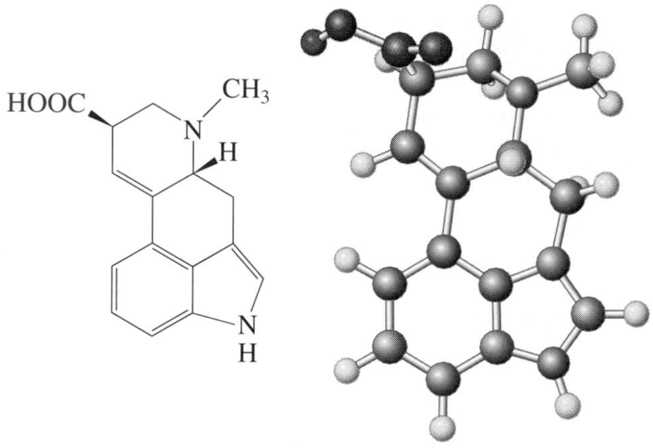

Lysergic acid

In summary, the numerous naturally occurring carboxylic acids are structurally and functionally diverse.

CHAPTER INTEGRATION PROBLEM

Lactones with rings of more than six members may be synthesized if ring strain and transannular interactions are minimized (see Sections 4-2 through 4-5, and 9-6). One of the most important commercial musk fragrances is a dilactone, ethylene brassylate. Its synthesis, illustrated here, begins with acid-catalyzed reaction of tridecanedioic acid with 1,2-ethanediol, which yields a substance A with the molecular formula $C_{15}H_{28}O_5$. Under the conditions in which it forms, substance A converts into a polymer (lower-left structure). Strong heating reverses polymer formation, regenerating substance A. In a slower process, A transforms into the final macrocyclic product (lower-right structure), which is distilled from the reaction mixture, thereby shifting the equilibria in its favor.

Commercial Synthesis of a Macrocyclic Musk

Tridecanedioic acid
(Brassylic acid)

1,2-Ethanediol
(Ethylene glycol)

$$\left[\text{C(CH}_2)_{11}\text{COCH}_2\text{CH}_2\text{O}\right]_n \quad \underset{}{\overset{\text{H}^+}{\rightleftharpoons}} \quad \text{C}_{15}\text{H}_{28}\text{O}_5 \quad \underset{}{\overset{\text{H}^+, \Delta}{\rightleftharpoons}}$$

Polymer   **A**   **Ethylene brassylate**

a. Suggest a structure for compound A.

SOLUTION

What information do we possess regarding this compound? We have its molecular formula and the structures both of its immediate precursors and of two of its products. It is possible to arrive at the solution by using any of these clues. For example, the polymeric product is constructed of repeating monomer units, each of which contains an ethylene glycol ester of brassylic acid, just missing the OH from the carboxy group at the left and the H from the hydroxy group at the right. Indeed, the formula of this unit is $C_{15}H_{26}O_4$, or structure A less one molecule of water. If we simply write its structure and then add the missing elements of a molecule of water, we get A, the monoester shown.

Ethylene glycol monoester of
brassylic acid (A)

To use another approach, the molecular formulas of the two starting materials add up to $C_{15}H_{30}O_6$, or A + H_2O. Therefore, A is the product of a reaction that combines brassylic acid and ethylene glycol, and *releases* a molecule of water in the process, exactly what is observed in esterification.

b. Propose mechanisms for the reactions that interconvert structure A with both the polymer and the macrocycle.

SOLUTION

Both reactions are examples of esterifications. Polymer formation begins with ester formation between the hydroxy group of one molecule of structure A and the carboxy group of another. We follow the pattern of addition–elimination (Section 19-9), making use of acid catalysis *as specified in the original equation.* Remember: The answers to mechanism problems should make use of only those chemical species actually specified in the reaction. The mechanistic sequence of steps is adapted directly from the text examples: (1) protonation of the carbonyl carbon to be attacked, (2) nucleophilic addition of a hydroxy oxygen to form the tetrahedral intermediate, (3) protonation of one hydroxy group of the intermediate to form a good leaving group, (4) departure of water, and (5) proton loss from oxygen to yield the final product.

Tetrahedral intermediate

The product of this process is a dimer; repetition of such ester formation many times at both ends of this molecule eventually gives rise to the polyester observed in the reaction. The same steps take place in reverse to convert the polymer back into monoester A upon heating.

What about the mechanism for macrocyclic lactone formation? We realize that the process must be intramolecular to form a ring (compare Sections 9-6 and 17-7). We can write exactly the same sequence of steps as we did for esterification, but we use the free hydroxy group at one end of molecule A to attack the carboxy carbon at the other end.

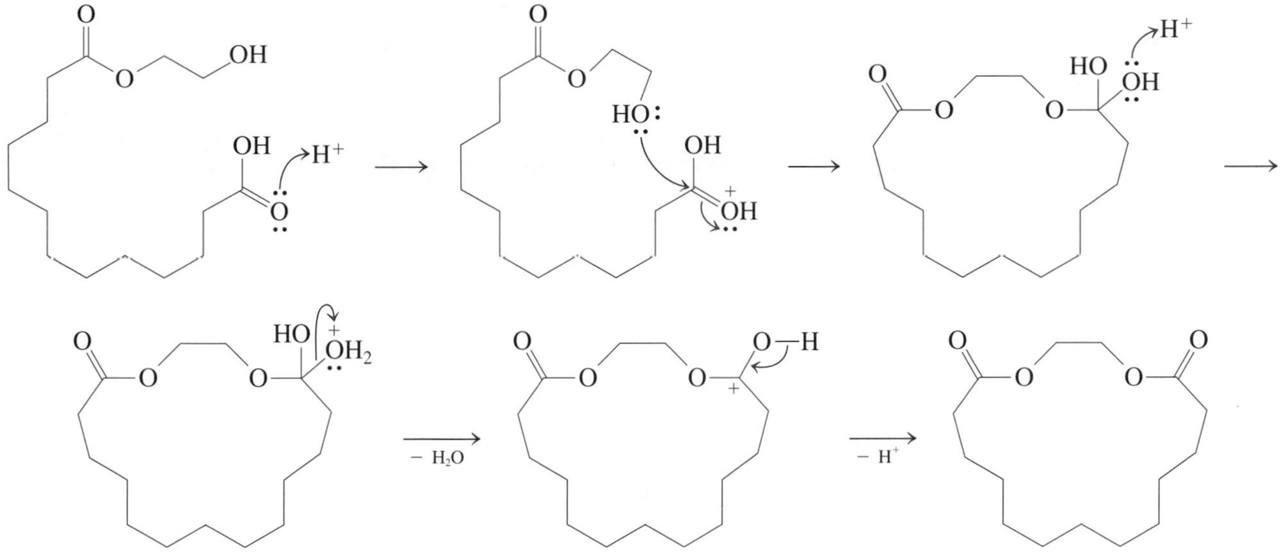

Macrocyclic lactone synthesis is a topic of considerable interest to the pharmaceutical industry because this function constitutes the basic framework of many medicinally valuable compounds. Examples include erythromycin A, a widely used *macrolide antibiotic,* and tacrolimus (FK-506), a powerful *immunosuppressant* that shows great promise in controlling the rejection of transplanted organs in human patients.

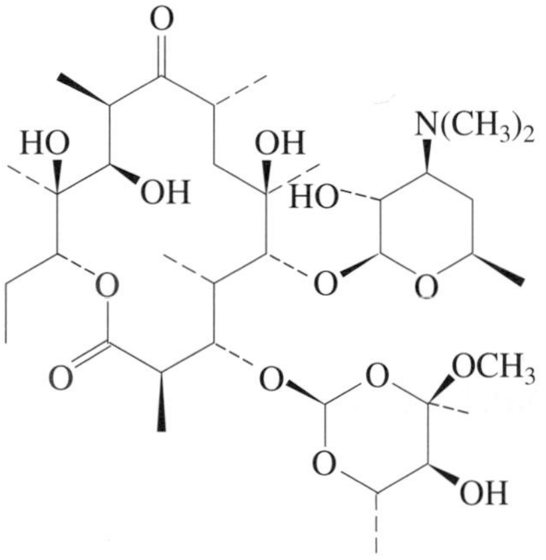

Erythromycin A

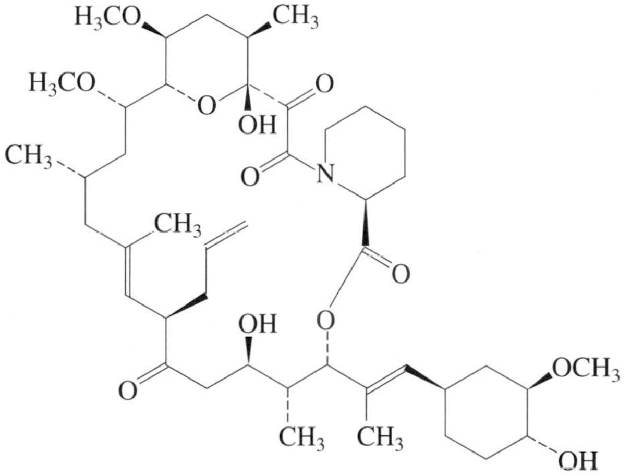

Tacrolimus (FK-506)

NEW REACTIONS

1. Acidity of Carboxylic Acids (Section 19-4)

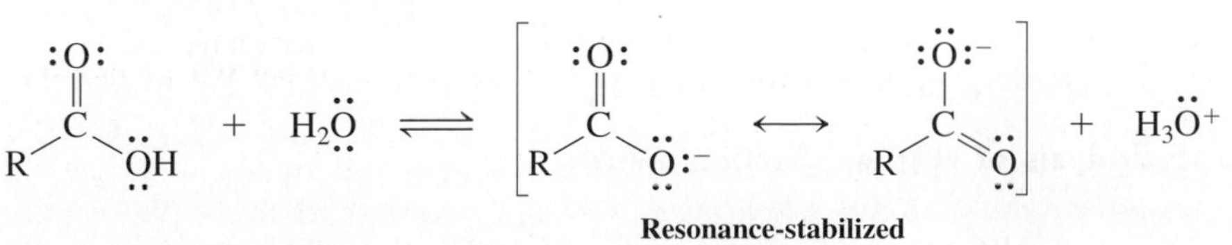

$K_a = 10^{-4} – 10^{-5}, \text{p}K_a \sim 4–5$

Resonance-stabilized carboxylate ion

Salt formation

$$RCOOH \ + \ NaOH \ \longrightarrow \ RCOO^-Na^+ \ + \ H_2O$$

Also with Na$_2$CO$_3$, NaHCO$_3$

2. Basicity of Carboxylic Acids (Section 19-4)

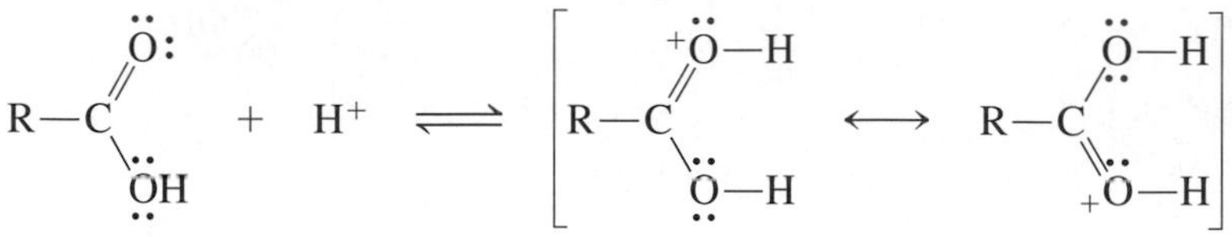

**Resonance-stabilized
protonated carboxylic acid**

Preparation of Carboxylic Acids

3. Formic Acid (Section 19-5)

$$CO \ + \ NaOH \ \xrightarrow{\Delta, \ 100 \ psi} \ HCOO^-Na^+ \ \xrightarrow{H^+, \ H_2O} \ HCOOH$$

4. Acetic Acid (Section 19-5)

Ethene oxidation

$$CH_2{=}CH_2 \ \xrightarrow{O_2, \ H_2O, \ catalytic \ Pd^{2+} \ and \ Cu^{2+}} \ CH_3CHO \ \xrightarrow{O_2, \ catalytic \ Co^{3+}} \ CH_3COOH$$

Butane oxidation

$$CH_3CH_2CH_2CH_3 \ \xrightarrow{O_2, \ catalytic \ Co^{3+}, \ \Delta, \ pressure} \ CH_3COOH$$

Carbonylation of methanol

$$CH_3OH \ \xrightarrow{CO, \ catalytic \ Rh^{3+}, \ I_2, \ \Delta, \ pressure} \ CH_3COOH$$

5. Oxidation of Primary Alcohols and Aldehydes (Section 19-6)

$$RCH_2OH \ \xrightarrow{Oxidizing \ agent} \ RCOOH$$

Oxidizing agents: aqueous CrO$_3$, KMnO$_4$, HNO$_3$

$$RCHO \ \xrightarrow{Oxidizing \ agent} \ RCOOH$$

Oxidizing agents: aqueous CrO$_3$, KMnO$_4$, Ag$^+$, H$_2$O$_2$, HNO$_3$

6. Carbonation of Organometallic Reagents (Section 19-6)

$$RMgX \ + \ CO_2 \ \xrightarrow{THF} \ RCOO^-{}^+MgX \ \xrightarrow{H^+, \ H_2O} \ RCOOH$$

$$RLi \ + \ CO_2 \ \xrightarrow{THF} \ RCOO^-Li^+ \ \xrightarrow{H^+, \ H_2O} \ RCOOH$$

7. Hydrolysis of Nitriles (Section 19-6)

$$RC{\equiv}N \ \xrightarrow{H_2O, \ \Delta, \ H^+ \ or \ HO^-} \ RCOOH \ + \ NH_3$$

Reactions of Carboxylic Acids

8. Nucleophilic Attack at the Carbonyl Group (Section 19-7)

Base-catalyzed addition–elimination

$$\underset{\text{L = leaving group}}{\overset{\overset{\displaystyle O}{\|}}{RCL}} \ + \ :Nu^- \ \xrightarrow[\text{Addition}]{} \ \underset{\substack{\text{Tetrahedral} \\ \text{intermediate}}}{R-\overset{\overset{\displaystyle O^-}{|}}{\underset{\underset{\displaystyle Nu}{|}}{C}}-L} \ \xrightarrow[\text{Elimination}]{} \ \overset{\overset{\displaystyle O}{\|}}{RCNu} \ + \ L^-$$

Acid-catalyzed addition–elimination

$$\overset{\overset{\displaystyle O}{\|}}{RCL} \ \xrightarrow{H^+} \ \overset{\overset{\displaystyle {}^+O-H}{\|}}{RCL} \ \xrightarrow{:NuH} \ \underset{\substack{\text{Tetrahedral} \\ \text{intermediate}}}{R-\overset{\overset{\displaystyle OH}{|}}{\underset{\underset{\displaystyle {}^+NuH}{|}}{C}}-L} \ \xrightleftharpoons{-H^+}$$

$$R-\overset{\overset{\displaystyle OH}{|}}{\underset{\underset{\displaystyle Nu}{|}}{C}}-L \ \xrightleftharpoons{H^+} \ R-\overset{\overset{\displaystyle OH}{|}}{\underset{\underset{\displaystyle Nu}{|}}{\overset{+}{C}}-\underset{\displaystyle H}{L}} \ \xrightleftharpoons{-HL} \ \overset{\overset{\displaystyle {}^+O-H}{\|}}{RCNu} \ \xrightleftharpoons{-H^+} \ \overset{\overset{\displaystyle O}{\|}}{RCNu}$$

Derivatives of Carboxylic Acids

9. Alkanoyl Halides (Section 19-8)

$$RCOOH \ + \ SOCl_2 \ \longrightarrow \ \underset{\substack{\text{Alkanoyl} \\ \text{chloride}}}{\overset{\overset{\displaystyle O}{\|}}{RCCl}} \ + \ SO_2 \ + \ HCl$$

$$3\,RCOOH \ + \ PBr_3 \ \longrightarrow \ 3\,\underset{\substack{\text{Alkanoyl} \\ \text{bromide}}}{\overset{\overset{\displaystyle O}{\|}}{RCBr}} \ + \ H_3PO_3$$

10. Carboxylic Anhydrides (Section 19-8)

$$RCOOH \ + \ \overset{\overset{\displaystyle O}{\|}}{RCCl} \ \xrightarrow{\Delta} \ \underset{\text{Anhydride}}{\overset{\overset{\displaystyle O \quad O}{\| \quad \|}}{RCOCR}} \ + \ HCl$$

Cyclic anhydrides

Best for five- or six-membered rings

11. Esters (Section 19-9)

Acid-catalyzed esterification

$$RCO_2H \ + \ R'OH \ \underset{K \sim 1}{\overset{H^+}{\rightleftharpoons}} \ \overset{\overset{\displaystyle O}{\displaystyle \|}}{R}COR' \ + \ H_2O$$

Cyclic esters (lactones)

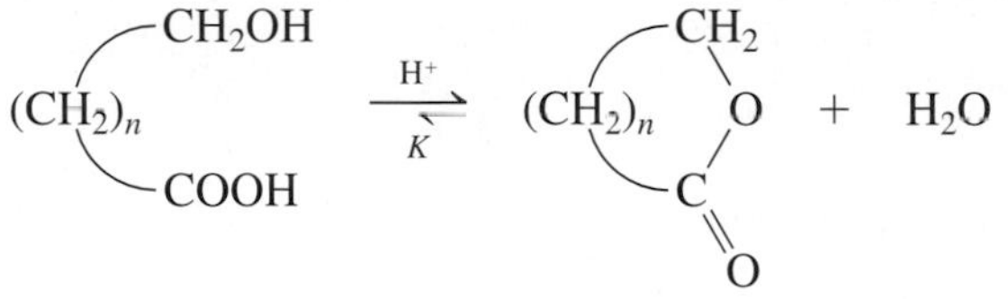

Lactone

$K > 1$ for five- and six-membered rings

Nucleophilic displacements with carboxylate ions

$$RCO_2^-Na^+ \ + \ R'X \ \xrightarrow{\text{HMPA}} \ RCO_2R' \ + \ NaX$$

Diazomethane reaction

$$RCOOH \ + \ CH_2N_2 \ \xrightarrow{(CH_3CH_2)_2O} \ RCOOCH_3 \ + \ N_2$$

12. Carboxylic Amides (Section 19-10)

$$RCOOH \ + \ R'NH_2 \ \longrightarrow \ RCOO^- \ + \ R'NH_3^+ \ \overset{\Delta}{\rightarrow} \ \overset{\overset{\displaystyle O}{\displaystyle \|}}{R}CNHR' \ + \ H_2O$$

Imides

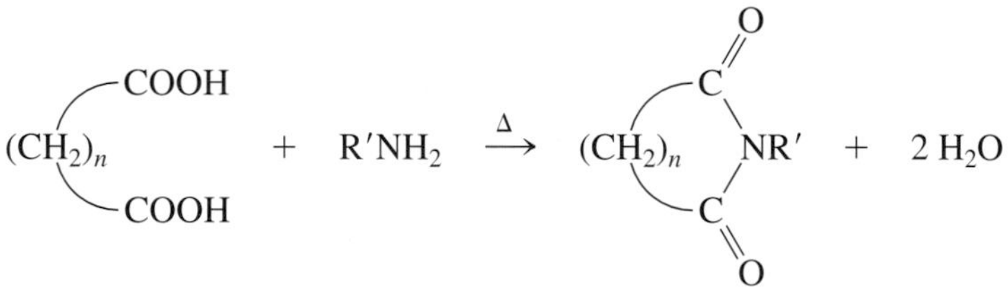

Cyclic amides (lactams)

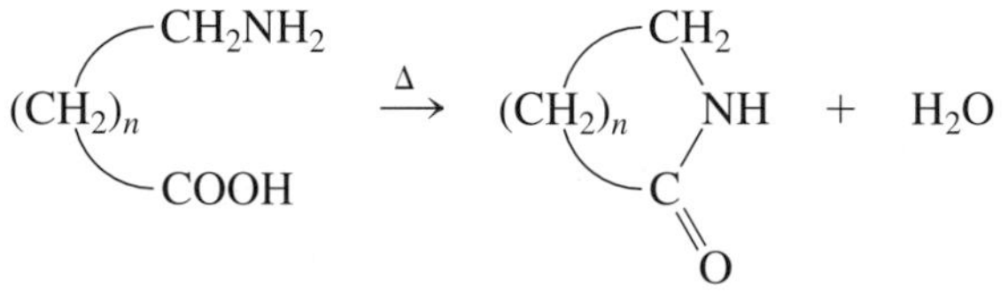

Lactam

13. Reduction with Lithium Aluminum Hydride (Section 19-11)

$$RCOOH \ \xrightarrow[\text{2. H}^+, \text{ H}_2\text{O}]{\text{1. LiAlH}_4, \text{ (CH}_3\text{CH}_2)_2\text{O}} \ RCH_2OH$$

14. Bromination: Hell-Volhard-Zelinsky Reaction (Section 19-12)

$$RCH_2COOH \ \xrightarrow{Br_2, \text{ trace P}} \ \overset{\overset{\displaystyle Br}{\displaystyle |}}{R}CHCOOH$$

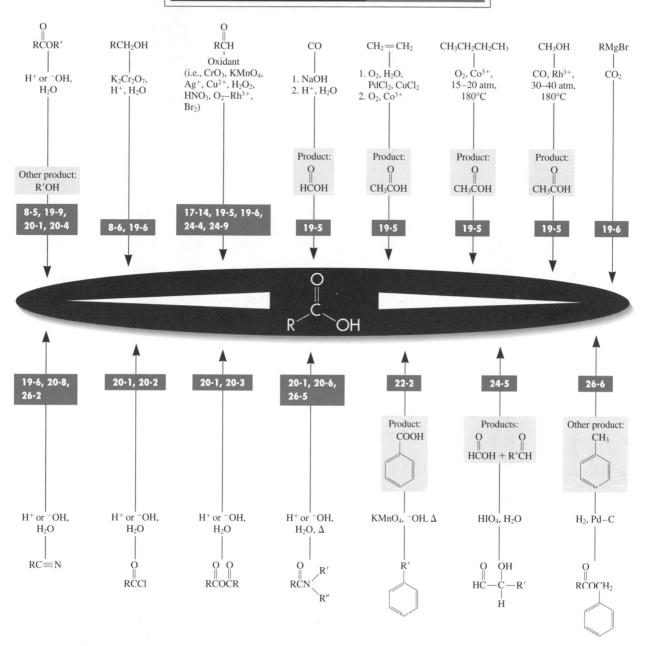

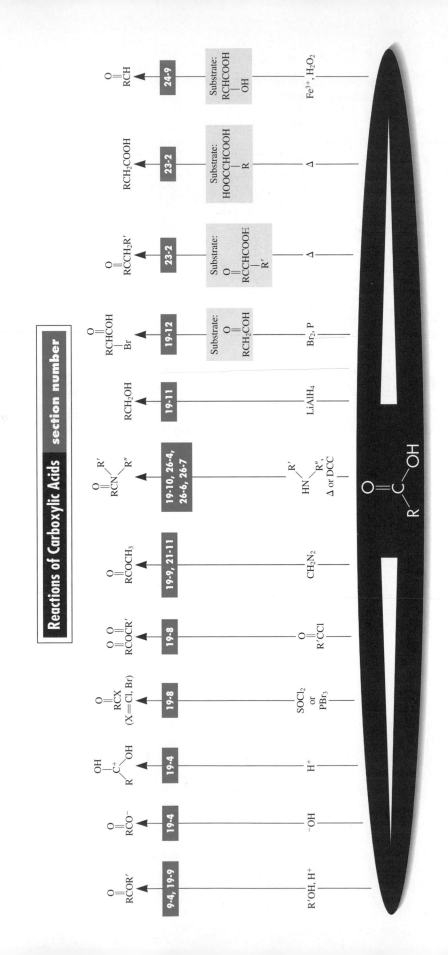

Reactions of Carboxylic Acids | section number

IMPORTANT CONCEPTS

1. **Carboxylic acids** are named as **alkanoic acids.** The carbonyl carbon is numbered 1 in the longest chain incorporating the carboxy group. Dicarboxylic acids are called **alkanedioic acids.** Cyclic and aromatic systems are called **cycloalkanecarboxylic** and **benzoic acids,** respectively. In these systems the ring carbon bearing the carboxy group is assigned the number 1.

2. The **carboxy group** is approximately **trigonally planar.** Except in very dilute solution, carboxylic acids form dimers by hydrogen bonding.

3. The carboxylic acid **proton chemical shift** is variable but relatively **high** ($\delta = 10$–13), because of hydrogen bonding. The **carbonyl carbon** is also relatively **deshielded** but not as much as in aldehydes

and ketones, because of the resonance contribution of the hydroxy group. The carboxy function shows two important infrared bands, one at about $1710 \ \text{cm}^{-1}$ for the C=O bond and a very broad band between 2500 and $3300 \ \text{cm}^{-1}$ for the O–H group.

4. The carbonyl group in carboxylic acids is subject to addition by **nucleophiles** to give an unstable **tetrahedral intermediate.** This intermediate may decompose by elimination of the hydroxy group to give a **carboxylic acid derivative.**

5. **Lithium aluminum hydride** is a strong enough nucleophile to add to the carbonyl group of carboxylate ions. This process allows the reduction of carboxylic acids to **primary alcohols.**

PROBLEMS

18. Name (IUPAC or *Chemical Abstracts* system) or draw the structure of each of the following compounds.

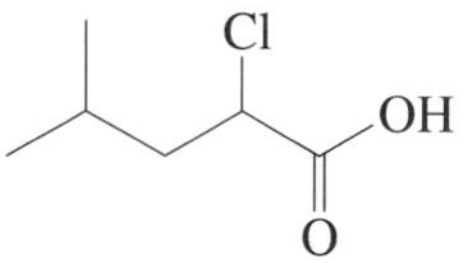

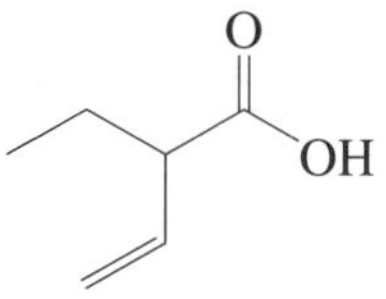

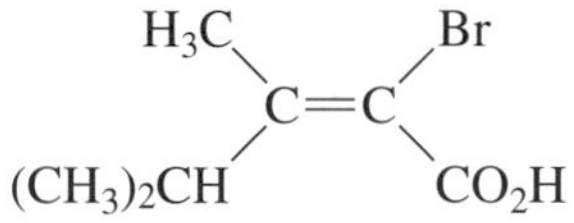

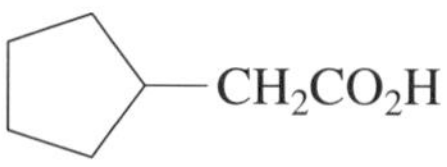

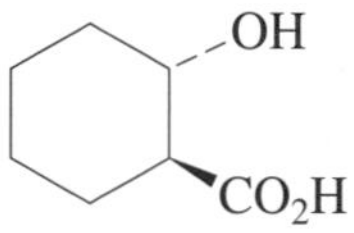

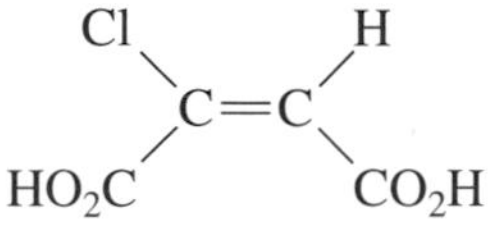

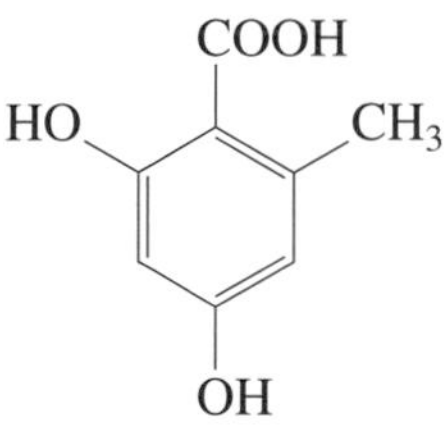

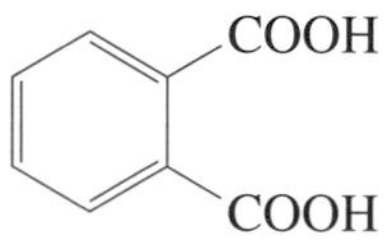

(i) 4-aminobutanoic acid (also known as GABA, a critical participant in brain biochemistry); **(j)** *meso*-2,3-dimethylbutanedioic acid; **(k)** 2-oxopropanoic acid (pyruvic acid); **(l)** *trans*-2-formylcyclohexanecarboxylic acid; **(m)** (*Z*)-3-phenyl-2-butenoic acid; **(n)** 1,8-naphthalenedicarboxylic acid.

19. Rank the group of molecules shown here in decreasing order of boiling point and solubility in water. Explain your answers.

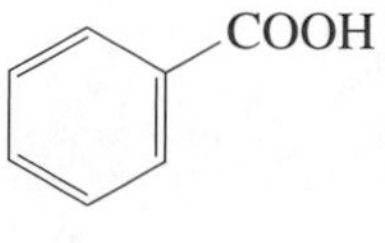

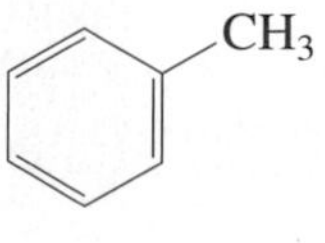

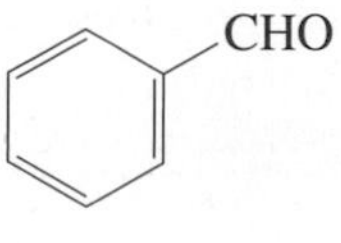

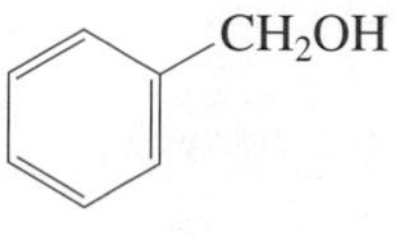

20. Rank each of the following groups of organic compounds in order of decreasing acidity.

(a) $CH_3CH_2CO_2H$, $CH_3\overset{\overset{\displaystyle O}{\|}}{C}CH_2OH$, $CH_3CH_2CH_2OH$

(b) $BrCH_2CO_2H$, $ClCH_2CO_2H$, FCH_2CO_2H

(c) $CH_3\overset{\overset{\displaystyle Cl}{|}}{C}HCH_2CO_2H$, $ClCH_2CH_2CH_2CO_2H$, $CH_3CH_2\overset{\overset{\displaystyle Cl}{|}}{C}HCO_2H$

(d) CF_3CO_2H, CBr_3CO_2H, $(CH_3)_3CCO_2H$

(e) ⬡—COOH, O_2N—⬡—COOH, O_2N—⬡(NO_2)—COOH, CH_3O—⬡—COOH

21. (a) An unknown compound A has the formula $C_7H_{12}O_2$ and infrared spectrum A (page 869). To which class does this compound belong? (b) Use the other spectra (NMR-B, page 869, and F, page 871; IR-D, E, and F, pages 870–871) and spectroscopic and chemical information in the reaction sequence to determine the structures of compound A and the other unknown substances B through F. References are made to relevant sections of earlier chapters, but do not look them up before you have tried to solve the problem without the extra help. (c) Another unknown compound, G, has the formula $C_8H_{14}O_4$ and the NMR and IR spectra labeled G (page 872). Propose a structure for this molecule. (d) Compound G may be readily synthesized from B. Propose a sequence that accomplishes this efficiently. (e) Propose a completely different sequence from that shown in (b) for the conversion of C into A. (f) Finally, construct a synthetic scheme that is the reverse of that shown in (b); namely, the conversion of A into B.

C_6H_{10}
B

$\xrightarrow[\text{Section 12-7}]{\begin{array}{l}\text{1. Hg(O}\overset{\overset{\displaystyle O}{\|}}{C}CH_3)_2, H_2O \\ \text{2. NaBH}_4\end{array}}$

$C_6H_{12}O$
C

$\xrightarrow[\text{Section 8-6}]{\begin{array}{l}CrO_3, H_2SO_4, \\ \text{propanone (acetone), 0°C}\end{array}}$

^{13}C NMR: δ = 22.1
24.5
126.2 ppm

^1H NMR-B

^{13}C NMR: δ = 24.4
25.9
35.5
69.5 ppm

$C_6H_{10}O$
D

$\xrightarrow[\text{Section 17-12}]{CH_2=P(C_6H_5)_3}$ C_7H_{12}
E

$\xrightarrow[\text{Section 12-8}]{\begin{array}{l}\text{1. BH}_3, \text{THF} \\ \text{2. HO}^-, H_2O_2\end{array}}$ $C_7H_{14}O$
F

$\xrightarrow[\text{Section 8-6}]{Na_2Cr_2O_7, H_2O, H_2SO_4}$ **A**

^{13}C NMR: δ = 23.8
26.5
40.4
208.5 ppm

IR-D

IR-E

IR-F
^1H NMR-F

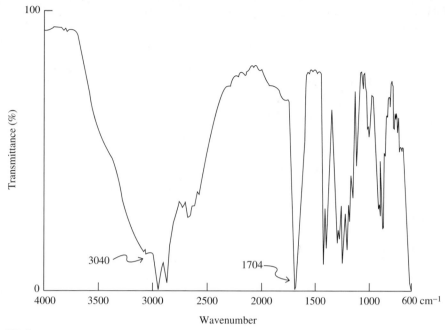

IR-A

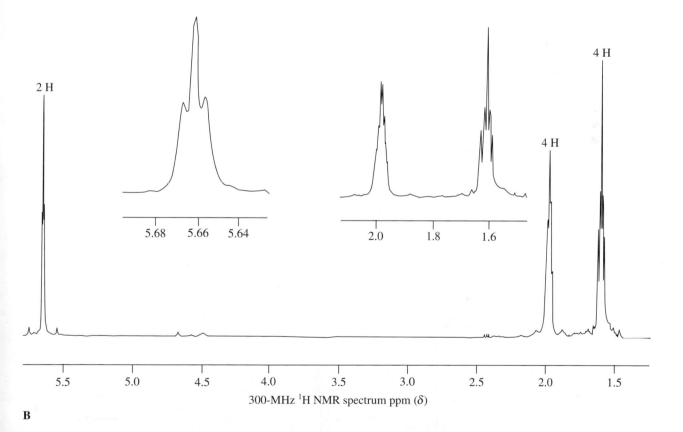

300-MHz ^{1}H NMR spectrum ppm (δ)

B

22. Give the products of each of the following reactions.

(a) $(CH_3)_2CH_2CH_2CO_2H + SOCl_2 \longrightarrow$ 　　　　(b) $(CH_3)_2CH_2CH_2CO_2H + CH_3COBr \longrightarrow$

(c) 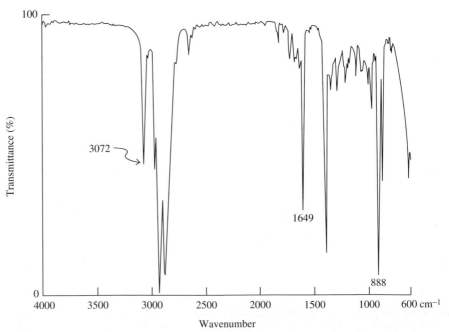 + CH₃CH₂OH $\xrightarrow{H^+}$ **(d)** CH₃O—⟨⟩—COOH + NH₃ ⟶

(e) Product of (d), heated strongly   **(f)** Phthalic acid [Problem 18(h)], heated strongly

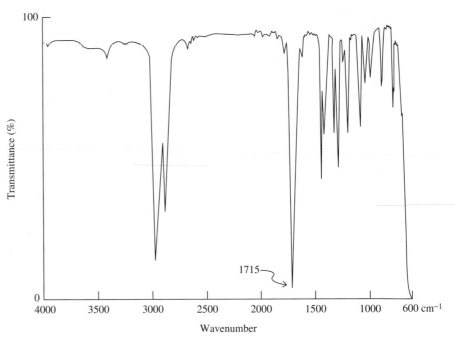

1715

IR-D

3072

1649

888

IR-E

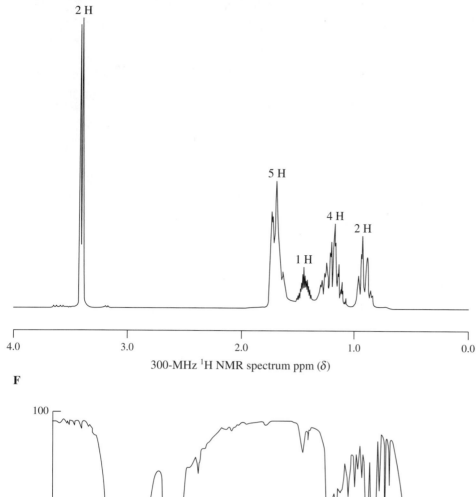

2 H

5 H

1 H

4 H

2 H

4.0   3.0   2.0   1.0   0.0

300-MHz ^{1}H NMR spectrum ppm (δ)

F

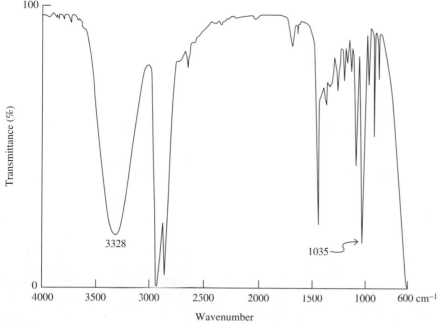

100

Transmittance (%)

3328

1035

0

4000   3500   3000   2500   2000   1500   1000   600 cm^{-1}

Wavenumber

IR-F

23. When 1,4- and 1,5-dicarboxylic acids, such as butanedioic (succinic) acid
(Section 19-8), are treated with SOCl$_2$ or PBr$_3$ in attempted preparations of the
dialkanoyl halides, the corresponding cyclic anhydrides are obtained. Explain
mechanistically.

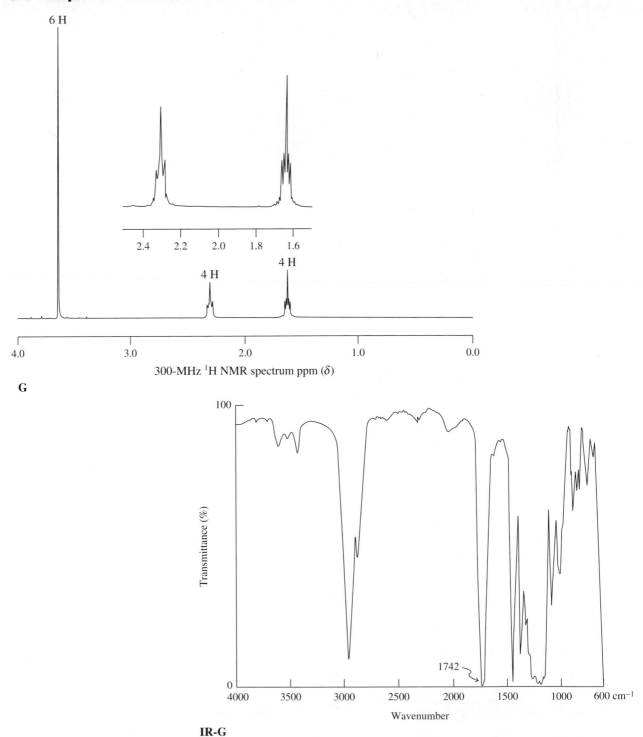

6 H

2.4 2.2 2.0 1.8 1.6

4 H

4 H

4.0   3.0   2.0   1.0   0.0

300-MHz ^{1}H NMR spectrum ppm (δ)

G

100

Transmittance (%)

1742

0
4000   3500   3000   2500   2000   1500   1000   600 cm^{-1}

Wavenumber

IR-G

24. Fill in suitable reagents to carry out the following transformations.

(a) $(CH_3)_2CHCH_2CHO \longrightarrow (CH_3)_2CHCH_2CO_2H$ **(b)** ⬠—CHO $\longrightarrow$ ⬠—$\overset{\overset{\displaystyle OH}{|}}{C}HCO_2H$

(c) 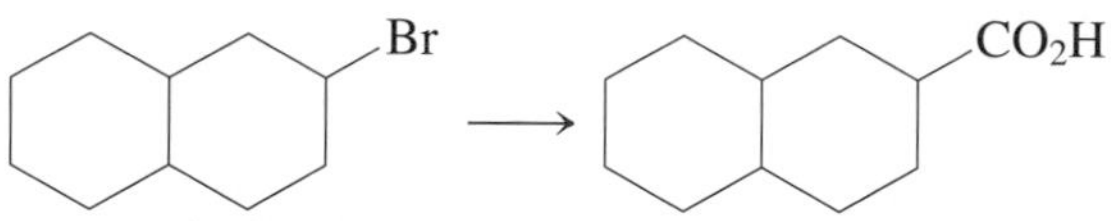 Br → CO₂H

(d)

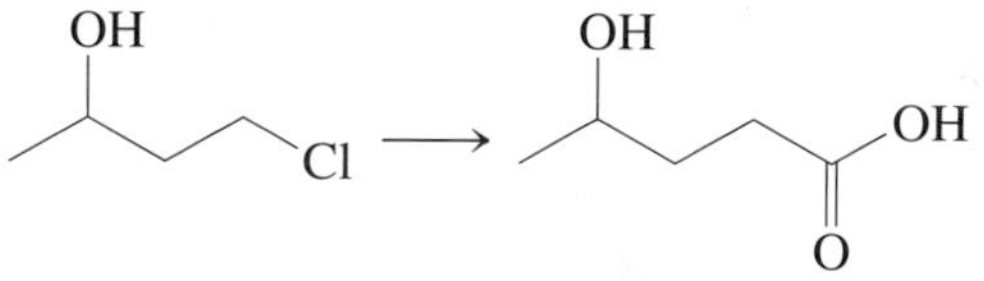

(e) 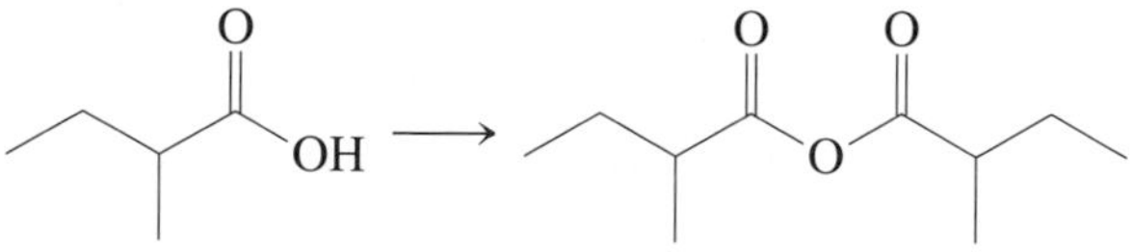 →

(f) $(CH_3)_3CCO_2H \longrightarrow (CH_3)_3CCO_2CH(CH_3)_2$

(g) 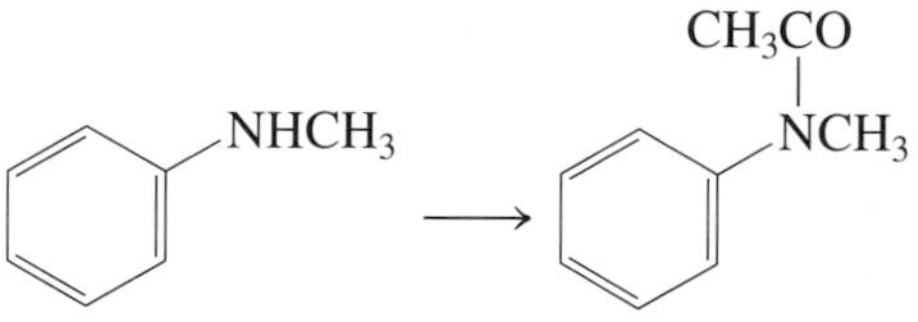 →

25. Propose syntheses of each of the following carboxylic acids that employ at least one reaction that forms a carbon–carbon bond.

(a) $CH_3CH_2CH_2CH_2CH_2CH_2CO_2H$ **(b)** $CH_3\overset{\underset{\displaystyle |}{OH}}{C}HCH_2CO_2H$ **(c)** $H_3C-\overset{\underset{\displaystyle CH_3}{\overset{\displaystyle |}{|}}}{\underset{\displaystyle CH_3}{C}}-CO_2H$

26. **(a)** Write a mechanism for the esterification of propanoic acid with ^{18}O-labeled ethanol. Show clearly the fate of the ^{18}O label. **(b)** Acid-catalyzed hydrolysis of an unlabeled ester with ^{18}O-labeled water ($H_2{}^{18}O$) leads to incorporation of some ^{18}O into *both* oxygens of the carboxylic acid product. Explain by a mechanism. (**Hint:** You must use the fact that all steps in the mechanism are reversible.)

27. Give the products of reaction of propanoic acid with each of the following reagents.

(a) $SOCl_2$ **(b)** PBr_3 **(c)** CH_3CH_2COBr + pyridine
(d) $(CH_3)_2CHOH$ + HCl   **(e)** CH_2N_2 **(f)** KOH, CH_3CH_2I

(g) 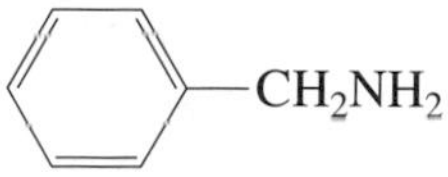 —CH_2NH_2 **(h)** Product of (g), heated strongly   **(i)** $LiAlH_4$, then H^+, H_2O

(j) Br_2, P

28. Give the product of reaction of cyclopentanecarboxylic acid with each of the reagents in Problem 27.

29. Suggest a preparation of hexanoic acid from pentanoic acid.

30. Give reagents and reaction conditions that would allow efficient conversion of 2-methylbutanoic acid into **(a)** the corresponding alkanoyl chloride; **(b)** the corresponding methyl ester; **(c)** the corresponding ester with 2-butanol; **(d)** the anhydride; **(e)** the *N*-methylamide;

(f) $CH_3CH_2\overset{\displaystyle CH_3}{\underset{\displaystyle |}{C}}HCH_2OH$

(g) $CH_3CH_2\overset{\displaystyle Br}{\underset{\displaystyle \underset{\displaystyle CH_3}{|}}{\overset{\displaystyle |}{C}}}CO_2H$

4-Pentenoic acid

31. Treatment of 4-pentenoic acid (margin) with Br_2 in the presence of dilute aqueous base yields a nonacidic compound with the formula $C_5H_7BrO_2$.
(a) Suggest a structure for this compound and a mechanism for its formation.
(b) Can you find a second, isomeric product whose formation is also mechanistically reasonable? **(c)** Discuss the issues that contribute to determining which of the two is the major product in this reaction. (**Hint:** Review Section 9-6.)

32. Show how the Hell-Volhard-Zelinsky reaction might be used in the synthesis of each of the following compounds, beginning in each case with a simple monocarboxylic acid. Write detailed mechanisms for all the reactions in *one* of your syntheses.

(a) $CH_3CH_2\underset{\displaystyle \underset{\displaystyle NH_2}{|}}{C}HCO_2H$

(b) benzene ring—$\underset{\displaystyle \underset{\displaystyle CO_2H}{|}}{C}HCO_2H$

(c) structure with O, OH, OH

(d) $HO_2CCH_2SSCH_2CO_2H$

(e) $(CH_3CH_2)_2NCH_2CO_2H$

(f) $(C_6H_5)_3\overset{+}{P}\underset{\displaystyle \underset{\displaystyle CH_3}{|}}{C}HCO_2H\ Br^-$

33. Although the original Hell-Volhard-Zelinsky reaction is limited to the synthesis of bromocarboxylic acids, the chloro and iodo analogs may be prepared by modifications of the procedure. Thus, alkanoyl chlorides may be converted into their α-chloro and α-bromo derivatives by reaction with *N*-chloro and *N*-bromobutanimide (*N*-chloro- and *N*-bromosuccinimide, NCS and NBS: Sections 3-8 and 14-2), respectively. Reaction with I_2 gives α-iodo compounds. Suggest a mechanism for any one of these processes.

$$C_6H_5CH_2CH_2COCl$$

$\xrightarrow{\text{NCS, HCl, SOCl}_2,\ 70°C} C_6H_5CH_2CHClCOCl$
84%

$\xrightarrow{\text{NBS, HBr, SOCl}_2,\ 70°C} C_6H_5CH_2CHBrCOCl$
71%

$\xrightarrow{\text{I}_2,\ \text{HI, SOCl}_2,\ 85°C} C_6H_5CH_2CHICOCl$
75%

34. Propose a mechanism for the exchange between an alkanoyl bromide and a carboxylic acid, as it occurs in step 4 of the Hell-Volhard-Zelinsky reaction. (**Hint:** Refer to the mechanisms presented in Section 19-8.)

35. How would you expect the acidity of acetamide to compare with that of acetic acid? With that of propanone (acetone)? Which protons in acetamide are the most acidic? Where would you expect acetamide to be protonated by very strong acid?

$$\underset{\text{Acetamide}}{CH_3\overset{\displaystyle O}{\overset{\displaystyle \|}{C}}NH_2}$$

36. Attempted CrO_3 oxidation of 1,4-butanediol to butanedioic acid results in significant yields of "γ-butyrolactone." Explain mechanistically.

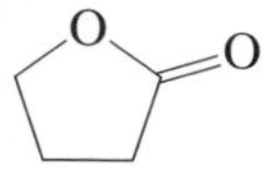

γ-Butyrolactone

37. Following the general mechanistic scheme given in Section 19-7, write detailed mechanisms for each of the following substitution reactions. (Note: These transformations are part of Chapter 20, but you should be able to solve the problem without looking ahead.)

(a)

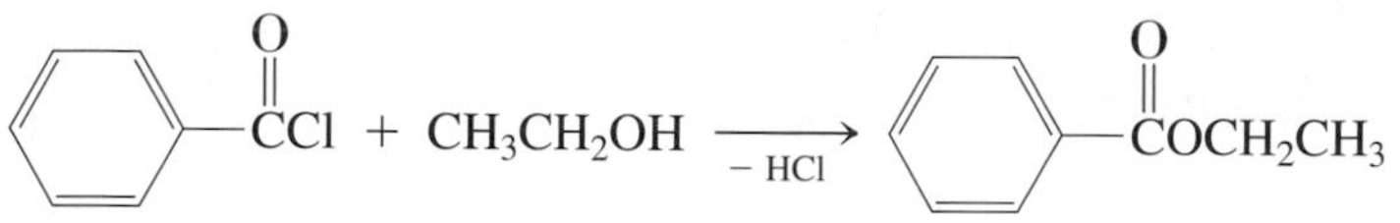

(b) $CH_3\overset{O}{\overset{\|}{C}}NH_2 + H_2O \xrightarrow{H^+} CH_3\overset{O}{\overset{\|}{C}}OH + \overset{+}{N}H_4$

38. Suggest structures for the products of each reaction in the following synthetic sequence.

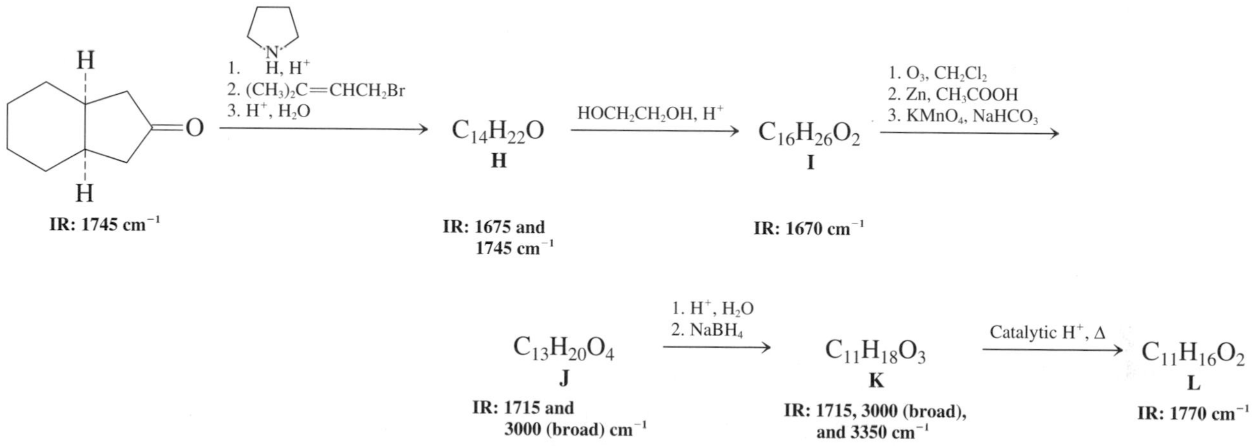

39. S_N2 reactions of simple carboxylate ions with haloalkanes in aqueous solution generally do not give good yields of esters. **(a)** Explain why this is so. **(b)** Reaction of 1-iodobutane with sodium acetate gives an excellent yield of ester if carried out in acetic acid (as shown here). Why is acetic acid a better solvent for this process than water?

$$\underset{\textbf{Iodobutane}}{CH_3CH_2CH_2CH_2I} + \underset{\textbf{Sodium acetate}}{CH_3\overset{O}{\overset{\|}{C}}O^-Na^+} \xrightarrow{CH_3CO_2H,\ 100°C} \underset{\substack{95\% \\ \textbf{Butyl acetate}}}{CH_3CH_2CH_2CH_2O\overset{O}{\overset{\|}{C}}CH_3} + Na^+I^-$$

(c) The reaction of 1-iodobutane with sodium dodecanoate proceeds surprisingly well in aqueous solution, much better than the reaction with sodium acetate (see the following equation). Explain this observation. (**Hint:** Sodium dodecanoate is a soap and forms micelles in water. See Chemical Highlight 19-1.)

$$CH_3CH_2CH_2CH_2I + CH_3(CH_2)_{10}CO_2^-Na^+ \xrightarrow{H_2O} CH_3(CH_2)_{10}\overset{O}{\overset{\|}{C}}OCH_2CH_2CH_2CH_3$$

40. (a) Diazomethane, CH_2N_2, is usually represented as a resonance hybrid of two contributing Lewis structures. Write them. **(b)** Propose a mechanism for the formation of a methyl ester from diazomethane and a carboxylic acid.

41. The *iridoids* are a class of monoterpenes with powerful and varied biological activities. They include insecticides, agents of defense against predatory insects, and animal attractants. The following reaction sequence is a synthesis of neonepetalactone, one of the nepetalactones, which are primary constituents of catnip. Use the information given to deduce the structures that have been left out, including that of neonepetalactone itself.

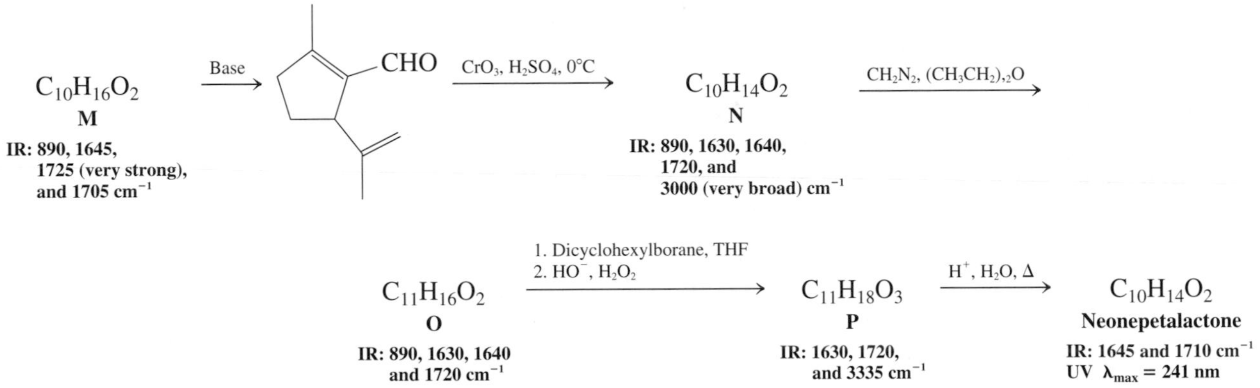

42. Propose *two* possible mechanisms for the following reaction. (**Hint:** Consider the possible sites of protonation in the molecule and the mechanistic consequences of each.) Devise an isotope-labeling experiment that might distinguish your two mechanisms.

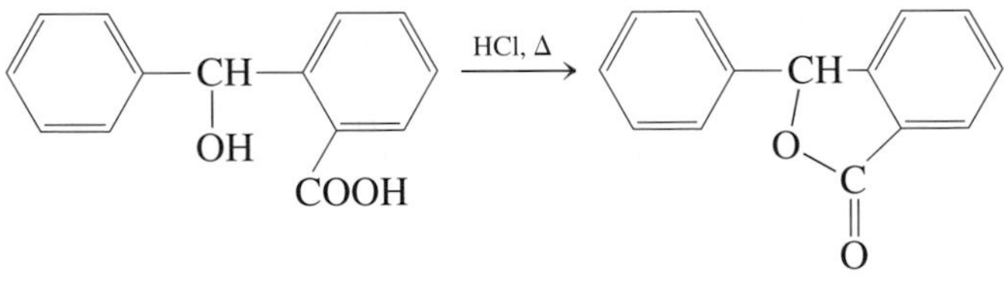

43. Propose a short synthesis of 2-butynoic acid, $CH_3C{\equiv}CCO_2H$, starting from propyne. (**Hint:** Review Section 13-2.)

44. The benzene rings of many compounds in nature are prepared by a biosynthetic pathway similar to that operating in fatty acid synthesis. Acetyl units are coupled, but the ketone functions are not reduced. The result is a *polyketide thiolester*, which forms rings by intramolecular aldol condensation.

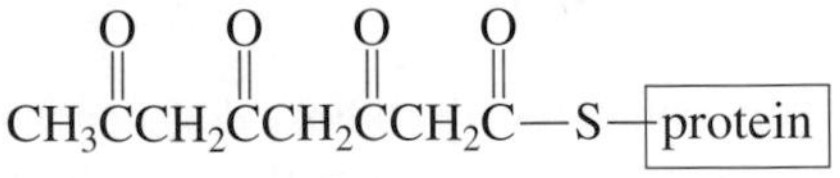

Polyketide thiolester

o-Orsellinic acid [for structure, see Problem 18(g)] is a derivative of salicylic acid and is prepared biosynthetically from the polyketide thiolester shown. Explain how this transformation might take place. Hydrolysis of the thiolester to give the free carboxylic acid is the last step.

Team Problem

45. Section 19-9 showed you that 4- and 5-hydroxy acids can undergo acid catalyzed intramolecular esterification to produce the corresponding lactone in good yield. Consider the following two examples of lactonization. Divide the analysis of the reaction sequences among your group. Propose reasonable mechanisms to explain the respective product formation.

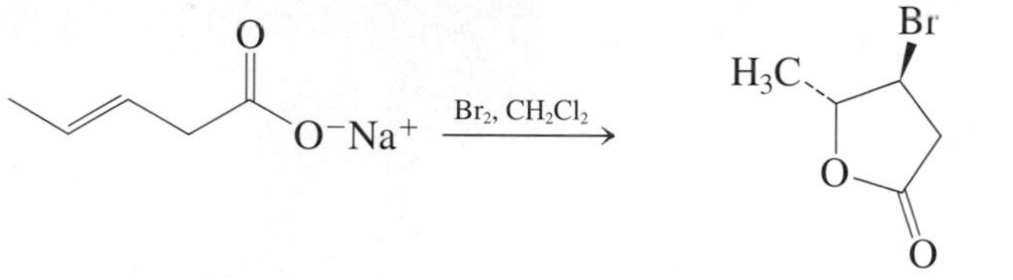

(Note stereochemistry!)

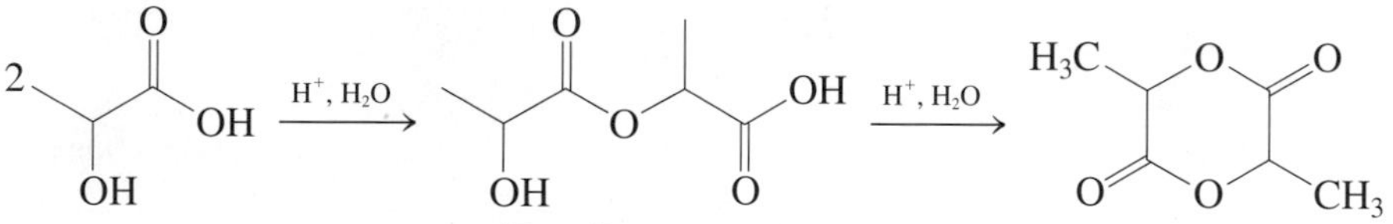

(Two diastereomers) **(Cis and trans isomers)**

Reconvene to present your mechanisms to each other.

Preprofessional Problems

46. What is the IUPAC name of the compound shown?

(a) (*E*)-3-Methyl-2-hexenoic acid
(b) (*Z*)-3-Methyl-2-hexenoic acid
(c) (*E*)-3-Methyl-3-hexenoic acid
(d) (*Z*)-3-Methyl-3-hexenoic acid

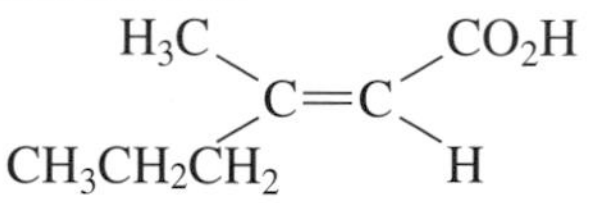

47. Select the acid with the highest K_a (i.e., lowest pK_a).

(a) H_3CCO_2H **(b)**

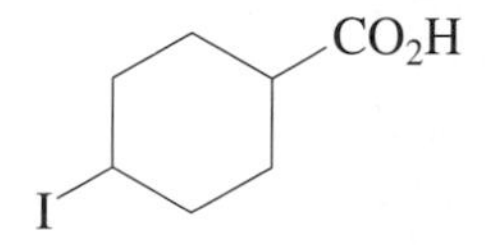

(c)

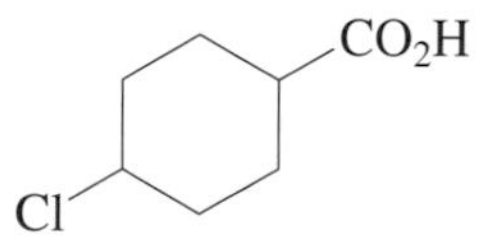

(d) Cl_2CHCO_2H

48. The acid whose structure is shown in the margin can best be prepared via one of these sequences. Which one?

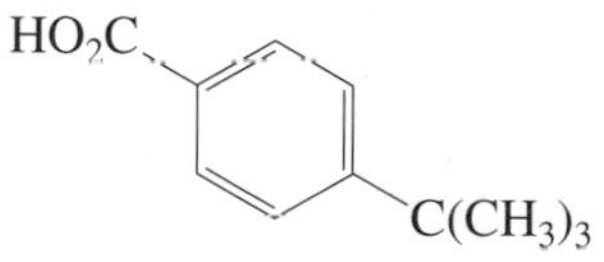

(a) $H_3CBr + Br_3CCO_2H \xrightarrow{K} \xrightarrow{\text{benzene}}$

(b) $(CH_3)_3CI \xrightarrow{\text{Mg, ether}} \xrightarrow{CO_2} \xrightarrow{H^+, H_2O\text{(work-up)}}$

(c)

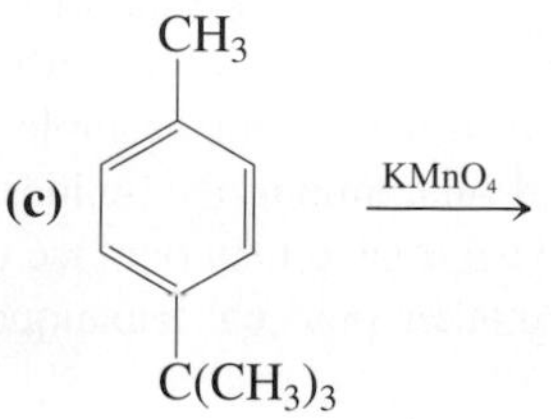

(d)

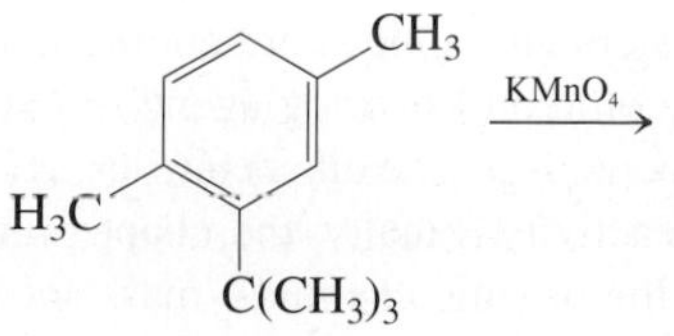

Carboxylic Acid Derivatives and Mass Spectrometry

The sturdiness of the amide linkage is amply demonstrated by the properties of poly(*p*-phenylene terephthalamide), known commercially as Kevlar. Used in bulletproof vests and body armor, just 20 layers of Kevlar fabric stitched together are capable of stopping the bullet from a 9-mm handgun traveling at 1200 feet per second. The enforced planarity of the benzene rings combines with the restricted rotation of the amide linkage (Section 20-6) to make Kevlar some 16 times as stiff as nylon (Section 21-12), the synthetic polyamide formerly used in such applications.

$$\text{R}\overset{\displaystyle :\!O\!:}{\underset{}{\overset{\|}{C}}}\text{L}$$

$$\text{L} = \text{X, O}\overset{O}{\overset{\|}{C}}\text{R, OR, NR}_2$$

Carboxylic acid derivatives

When complex organic systems are constructed in nature, they are frequently made by the combination of carboxy carbonyl groups and the heteroatoms of alcohols, amines, or thiols. Why? The addition–elimination processes introduced in Chapter 19 provide mechanistic pathways of relatively low activation energy for the interconversion of variously substituted carboxylic acid derivatives, many of which play central roles in biology (Chapter 26). The chemist finds these compounds similarly useful, as we shall see in this chapter, which deals with the chemistry of four major carboxylic acid derivatives: halides, anhydrides, esters, and amides. Each has a substituent, L, that can function as a leaving group in substitution reactions. We already know, for instance, that displacement of the halide in RCX (where the carbon bears a doubly bonded O) by a carboxy group leads to anhydrides.

We begin with a comparison of the structures, properties, and relative reactivities of carboxylic acid derivatives. We then explore the chemistry of each type of compound. Halides and anhydrides are valuable reagents in the synthesis of other carbonyl compounds. Esters and amides are enormously important in nature; for example, the esters include common flavoring agents, waxes, fats, and oils; among the amides we find urea and penicillin. The alkanenitriles, $\text{RC}{\equiv}\text{N}$, are also treated here because they have similar reactivity. Finally, the chapter introduces another physical technique of great value to the organic chemist—mass spectrometry.

20-1 Relative Reactivities, Structures, and Spectra of Carboxylic Acid Derivatives

Carboxylic acid derivatives undergo substitution reactions with nucleophiles, such as water, organometallic compounds, and hydride reducing agents. These transformations proceed through the familiar (often acid- or base-catalyzed) addition–elimination sequence (Section 19-7).

Addition–Elimination in Carboxylic Acid Derivatives

The relative reactivities of the substrates follow a consistent order: Alkanoyl (acyl) halides are most reactive, followed by anhydrides, then esters, and finally the amides, which are least reactive.

Relative Reactivities of Carboxylic Acid Derivatives in Nucleophilic Addition–Elimination with Water

The observed order of reactivity depends directly on the effect of the leaving group, L, on the carbonyl function. In the carboxylic derivatives, just as in the acids themselves (Section 19-3), lone pairs on the carboxy substituent L can be delocalized onto the carbonyl oxygen.

Resonance in Carboxylic Acid Derivatives

The dipolar resonance contributor on the right-hand side is most important when L is NR_2, because nitrogen is the least electronegative atom in the series. In esters, this resonance form is somewhat less important, because oxygen is more electronegative than nitrogen. Nonetheless, resonance wins out (see also Section 16-1) and both the NR_2 and the OR groups strongly stabilize amides and esters, respectively. On the other hand, anhydrides are more reactive than esters, since the lone pairs on the central oxygen in anhydrides are shared by two carbonyl groups, diminishing resonance stabilization. Finally, alkanoyl (acyl) halides are the least stable for either or both of two reasons: relative electronegativity (F, Cl; Table 1-2) and poor overlap of the large halogen p orbitals (Cl, Br, I; Section 1-6) with the relatively small $2p$ lobes of the carbonyl carbon.

EXERCISE 20-1

Phosgene, phenylmethyl chloroformate (see Section 26-6), bis(1,1-dimethylethyl) dicarbonate (see Section 26-6), dimethyl carbonate, and urea are derivatives of carbonic acid H_2CO_3 (made by dissolving CO_2 in water). Rank them in order of decreasing reactivity in nucleophilic addition-eliminations.

$\overset{\text{O}}{\overset{\|}{\text{ClCCl}}}$	$\overset{\text{O}}{\overset{\|}{\text{C}_6\text{H}_5\text{CH}_2\text{OCCl}}}$	$\overset{\text{O O}}{\overset{\|\;\|}{(\text{CH}_3)_3\text{COCOCOC}(\text{CH}_3)_3}}$	$\overset{\text{O}}{\overset{\|}{\text{CH}_3\text{OCOCH}_3}}$
Phosgene	**Phenylmethyl chloroformate**	**Bis(1,1-dimethylethyl) dicarbonate**	**Dimethyl carbonate**

$\overset{\text{O}}{\overset{\|}{\text{H}_2\text{NCNH}_2}}$
Urea

The greater the resonance, the shorter the C–L bond

The extent of resonance can be observed directly in the structures of carboxylic acid derivatives. In the procession from alkanoyl halides to esters and amides, the C–L bond becomes progressively shorter, owing to increased double-bond character (Table 20-1). The NMR spectra of amides reveal that rotation about this bond has become restricted. For example, *N,N*-dimethylformamide at room temperature exhibits *two* singlets for the two methyl groups, because rotation about the C–N bond is very slow on the NMR time scale. The evidence points to considerable π overlap between the lone pair on nitrogen and the carbonyl carbon, as a result of the increased importance of the dipolar resonance form in amides. The measured barrier to this rotation is about 21 kcal mol^{-1}.

Slow Rotation in *N, N*-Dimethylformamide

$E_a = 21$ kcal mol^{-1}

L	Bond length (Å) in R–L	Bond length (Å) in RC–L
Cl	1.78	1.79 (not shorter)
OCH$_3$	1.43	1.36 (shorter by 0.07 Å)
NH$_2$	1.47	1.36 (shorter by 0.11 Å)

TABLE 20-1 C–L Bond Lengths in RCL Compared with R–L Single Bond Distances

---

EXERCISE 20-2

The methyl group in the ^{1}H NMR spectrum of 1-acetyl 2-phenylhydrazide, shown in the margin, exhibits two singlets at $\delta = 2.02$ and 2.10 ppm at room temperature. On heating to 100°C in the NMR probe, the same compound gives rise to only one signal in that region. Explain.

$$CH_3CNHNHC_6H_5$$
1-Acetyl 2-phenylhydrazide

Infrared spectroscopy can also be used to probe resonance in carboxylic acid derivatives. The dipolar resonance structure weakens the C=O bond and causes a corresponding decrease in the carbonyl stretching frequency (Table 20-2).

The ^{13}C NMR signals of the carbonyl carbons in carboxylic acid derivatives are less sensitive to polarity differences and fall in a narrow range near 170 ppm.

^{13}C NMR Chemical Shifts of the Carbonyl Carbon in Carboxylic Acid Derivatives

CH$_3$CCl	CH$_3$COCCH$_3$	CH$_3$COH	CH$_3$COCH$_3$	CH$_3$CNH$_2$
$\delta = 170.3$	166.9	177.2	170.7	172.6 ppm

TABLE 20-2 Carbonyl Stretching Frequencies of Carboxylic Acid Derivatives RCL

L	$\tilde{\nu}_{C=O}$ (cm^{-1})	
Cl	1790–1815	
OCR	1740–1790 1800–1850	Two bands are observed, corresponding to asymmetric and symmetric stretching motions
OR	1735–1750	
NR$_2'$	1650–1690	

Spectroscopic Characterization of the Lactone Ring in Manoalide

The identification of the lactone ring in the anti-inflammatory natural product manoalide, isolated from a marine sponge (Chemical Highlight 10-3), depended largely on IR and UV spectroscopy.

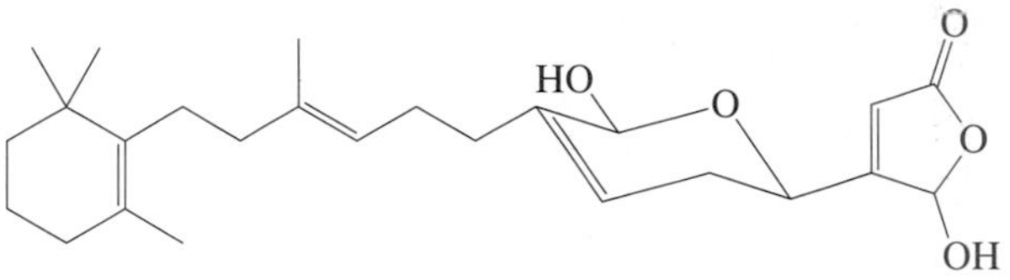

Manoalide

The signal at $\delta = 172.3$ ppm in the ^{13}C NMR spectrum revealed that a carboxylic acid derivative was present but did not reveal which type. The IR spectrum showed a carbonyl stretching frequency at 1765 cm^{-1}. Although this value is higher than usual for an ester, recall that the incorporation of a carbonyl group in a ring containing fewer than six atoms raises the value of $\tilde{\nu}_{C=O}$ (Section 17-3). The observed frequency is in the range common for γ-butyrolactone and its derivatives.

The presence of unsaturation in the lactone ring of manoalide was suggested by the UV spectrum. Saturated esters normally exhibit λ_{max} values in the range 200–210 nm, but conjugation with a double bond shifts this band into the range 215–250 nm (see Section 14-11). The λ_{max} value for manoalide is 227 nm.

Support for the presence of an α,β-unsaturated five-membered lactone ring with an alkyl substituent at C3 and a hydroxy group at C4 relied on comparison with the spectra of known substances, such as the simpler lactone shown below, which displays an IR band at 1760 cm^{-1}.

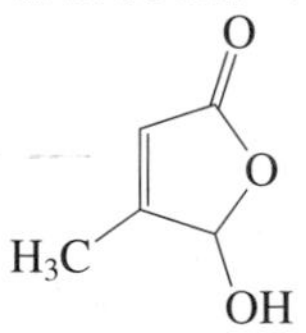

Carboxylic acid derivatives are basic and acidic

The extent of resonance in carboxylic acid derivatives is also seen in their basicity (protonation at the carbonyl oxygen) and acidity (enolate formation). In all cases, protonation requires strong acid, but it gets easier as the electron-donating ability of the L group increases. Protonation is important in acid-catalyzed nucleophilic addition–elimination reactions.

Protonation of a Carboxylic Acid Derivative

A relatively strong contribution of this resonance form stabilizes the protonated species

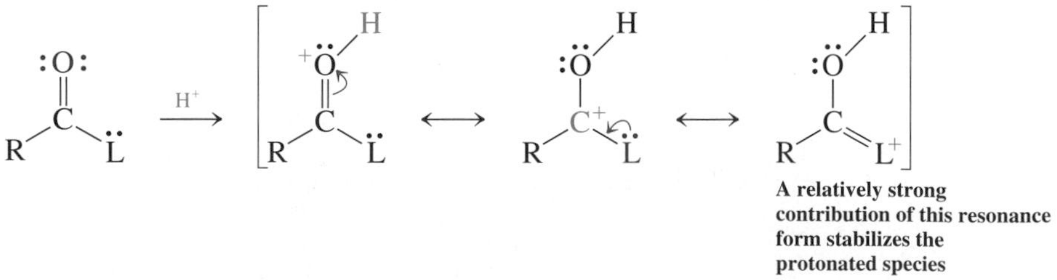

EXERCISE 20-3

Acetyl chloride is a much weaker base than acetamide. Explain, using resonance structures.

For related reasons, the acidity of the hydrogens next to the carbonyl group increases along the series. The acidity of a ketone lies between those of an alkanoyl chloride and an ester.

**Acidities of α-Hydrogens in Carboxylic Acid Derivatives
in Comparison with Propanone (Acetone)**

$$\underset{\sim 30}{CH_3\overset{O}{\overset{\|}{C}}N(CH_3)_2} \quad < \quad \underset{\sim 25}{CH_3\overset{O}{\overset{\|}{C}}OCH_3} \quad < \quad \underset{\sim 20}{CH_3\overset{O}{\overset{\|}{C}}CH_3} \quad < \quad \underset{\sim 16}{CH_3\overset{O}{\overset{\|}{C}}Cl}$$

pK_a

EXERCISE 20-4

Can you think of a reaction from a previous chapter that takes advantage of the relatively high acidity of the α-hydrogens in alkanoyl (acyl) halides?

In summary, the relative electronegativity and the size of L in R$\overset{O}{\overset{\|}{C}}$L controls the extent of resonance of the lone electron pair(s) and the relative reactivity of a carboxylic acid derivative in nucleophilic addition–elimination reactions. This effect manifests itself structurally and spectroscopically, as well as in the relative acidity and basicity of the α-hydrogen and the carbonyl oxygen, respectively.

20-2 Chemistry of Alkanoyl Halides

The alkano**yl halides**, R$\overset{O}{\overset{\|}{C}}$X, are named after the alkano*ic acid* from which they are derived. The halides of cycloalkane*carboxylic acids* are called cycloalkane**carbonyl halides.**

Alkanoyl halides undergo addition–elimination reactions in which nucleophiles displace the halide leaving group.

Addition–Elimination Reactions of Alkanoyl Halides

$$R\overset{:\overset{..}{O}:}{\overset{\|}{\underset{..}{C}}}X: \quad + \quad :Nu^- \quad \longrightarrow \quad R-\overset{:\overset{..}{O}:^-}{\underset{\underset{Nu}{|}}{\overset{|}{C}}}-\overset{..}{X}: \quad \longrightarrow \quad R\overset{:O:}{\overset{\|}{C}}Nu \quad + \quad :\overset{..}{\underset{..}{X}}:^-$$

Figure 20-1 shows a variety of nucleophilic reagents and the corresponding products. It is because of this wide range of reactivity that alkanoyl halides are useful synthetic intermediates.

$$CH_3\overset{O}{\overset{\|}{C}}Cl$$

Acetyl chloride

$$\underset{\text{3-Methylbutanoyl bromide}}{\overset{CH_3}{\overset{|}{CH_3CHCH_2}}\overset{O}{\overset{\|}{C}}Br}$$

Pentanoyl fluoride

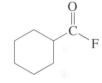

**Cyclohexanecarbonyl
fluoride**

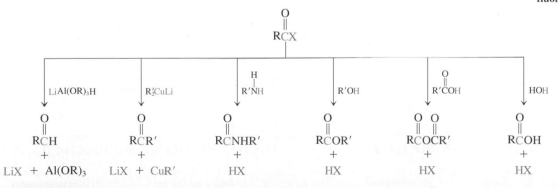

FIGURE 20-1

Nucleophilic addition–elimination reactions of alkanoyl halides.

Let us consider these transformations one by one (except for anhydride formation, which was covered in Section 19-8). Examples will be restricted to alkanoyl chlorides, which are the most readily accessible, but their transformations can be generalized to a considerable extent to the other alkanoyl halides.

Water hydrolyzes alkanoyl chlorides to carboxylic acids

Alkanoyl chlorides react with water, often violently, to give the corresponding carboxylic acids and hydrogen chloride. This transformation is a simple example of addition–elimination.

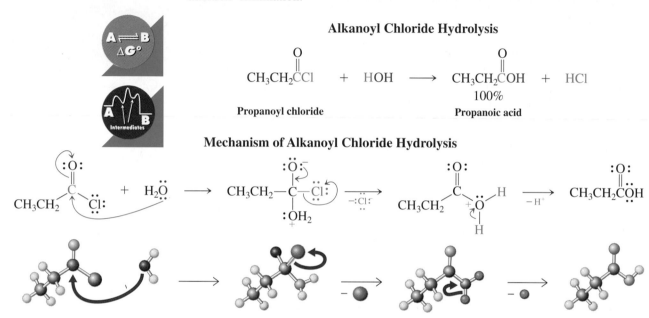

Alkanoyl Chloride Hydrolysis

$$CH_3CH_2\overset{\overset{\textstyle O}{\|}}{C}Cl \;+\; HOH \longrightarrow CH_3CH_2\overset{\overset{\textstyle O}{\|}}{C}OH \;+\; HCl$$

Propanoyl chloride   Propanoic acid (100%)

Mechanism of Alkanoyl Chloride Hydrolysis

Alcohols convert alkanoyl chlorides into esters

The analogous reaction of alkanoyl chlorides with alcohols is a highly effective way of producing esters. A base such as an alkali metal hydroxide, pyridine, or a tertiary amine is usually added to neutralize the HCl by-product. Because alkanoyl chlorides are readily made from the corresponding carboxylic acids (Section 19-8), the sequence RCOOH → RCOCl → RCOOR′ is a good method for esterification. By maintaining neutral conditions, this preparation avoids the equilibrium problem of acid-catalyzed ester formation (Section 19-9).

Ester Synthesis from Carboxylic Acids Through Alkanoyl Chlorides

$$RCOH \xrightarrow[-\,HCl]{SOCl_2} RCCl \xrightarrow[-\,HCl]{R'OH,\ base} RCOR'$$

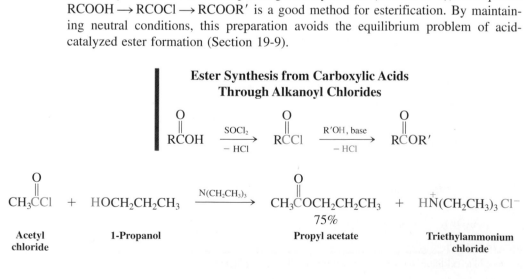

$$CH_3\overset{\overset{\textstyle O}{\|}}{C}Cl \;+\; HOCH_2CH_2CH_3 \xrightarrow{N(CH_2CH_3)_3} CH_3\overset{\overset{\textstyle O}{\|}}{C}OCH_2CH_2CH_3 \;+\; H\overset{+}{N}(CH_2CH_3)_3\ Cl^-$$

Acetyl chloride   1-Propanol   Propyl acetate (75%)   Triethylammonium chloride

$$H_3C-\overset{\overset{\displaystyle O}{\|}}{C}-OC(CH_3)_3$$

1,1-Dimethylethyl acetate
(*tert*-Butyl acetate)

EXERCISE 20-5

You have learned that 2-methyl-2-propanol (*tert*-butyl alcohol) dehydrates in the presence of acid (Section 9-2). Suggest a synthesis of 1,1-dimethylethyl acetate (*tert*-butyl acetate, shown in the margin) from acetic acid. Avoid conditions that might dehydrate the alcohol.

Amines convert alkanoyl chlorides into amides

Secondary and primary amines, as well as ammonia, convert alkanoyl chlorides into amides. Again, the hydrogen chloride formed is neutralized by added base (which can be excess amine).

Benzoyl chloride + NH₃ (Excess) →(H₂O) Benzamide (86%) + NH₄⁺Cl⁻

$$CH_2=CHCCl + 2\ CH_3NH_2 \xrightarrow[\text{5°C}]{\text{Benzene,}} CH_2=CHCNHCH_3 + CH_3NH_3^+Cl^-$$

Propenoyl chloride   68%   *N*-**Methylpropenamide**

The mechanism of this transformation is, again, addition–elimination, beginning with attack of the nucleophilic amine nitrogen at the carbonyl carbon.

Mechanism of Amide Formation from Alkanoyl Chlorides

Note that, in the last step, a proton must be lost from nitrogen to give the amide. Consequently, tertiary amines (which have no hydrogens on nitrogen) cannot form amides.

EXERCISE 20-6

Some amide preparations from alkanoyl halides require a primary or secondary amine that is too expensive to use also as the base to neutralize the hydrogen halide. Suggest a solution to this problem.

Organometallic reagents convert alkanoyl chlorides into ketones

Organometallic reagents attack the carbonyl group of alkanoyl chlorides to give the corresponding ketones. However, the latter are themselves prone to further attack by the relatively unselective organolithium and Grignard reagents to give alcohols (see Section 8-8). Ketone formation is best achieved by using diorganocuprates (Section 13-10), which are more selective than RLi or RMgX and do not add to the product ketone.

$$2 \quad \overset{\overset{\displaystyle O}{\|}}{\underset{\bigcirc}{C}Cl} \quad + \quad [(CH_3)_2C{=}CH]_2CuLi \quad \longrightarrow \quad 2 \quad \overset{\overset{\displaystyle O}{\|}}{\underset{\bigcirc}{C}CH{=}C(CH_3)_2} \quad + \quad LiCl \quad + \quad (CH_3)_2C{=}CHCu$$

70%

Reduction of alkanoyl chlorides results in aldehydes

We can convert an alkanoyl chloride into an aldehyde by hydride reduction. In this transformation, we again face a selectivity problem: Sodium borohydride and lithium aluminum hydride convert aldehydes into alcohols. To prevent such overreduction, we must modify LiAlH$_4$ by letting it react first with three molecules of 2-methyl-2-propanol (*tert*-butyl alcohol; see Section 8-6). This treatment neutralizes three of the reactive hydride atoms, leaving one behind that is nucleophilic enough to attack an alkanoyl chloride but not the resulting aldehyde.

Reductions by Modified Lithium Aluminum Hydride

Preparation of reagent

$$LiAlH_4 \quad + \quad 3\,(CH_3)_3COH \quad \longrightarrow \quad LiAl[OC(CH_3)_3]_3H \quad + \quad 3\,H{-}H$$

Lithium tri(*tert*-butoxy)aluminum hydride

Reduction

$$\overset{\overset{\displaystyle O}{\|}}{R{C}Cl} \quad + \quad LiAl[OC(CH_3)_3]_3H \quad \xrightarrow[\text{2. H}^+, \text{H}_2\text{O}]{\text{1. Ether solvent}} \quad \overset{\overset{\displaystyle O}{\|}}{R{C}H} \quad + \quad LiCl \quad + \quad Al[OC(CH_3)_3]_3$$

$$\underset{\textbf{2-Butenoyl chloride}}{CH_3CH{=}CH\overset{\overset{\displaystyle O}{\|}}{C}Cl} \quad \xrightarrow[\text{2. H}^+, \text{H}_2\text{O}]{\text{1. LiAl[OC(CH}_3)_3]_3\text{H, (CH}_3\text{OCH}_2\text{CH}_2)_2\text{O, } -78°\text{C}} \quad \underset{\substack{48\% \\ \textbf{2-Butenal}}}{CH_3CH{=}CH\overset{\overset{\displaystyle O}{\|}}{C}H}$$

EXERCISE 20-7

Prepare the following compounds from butanoyl chloride.

(a) [structure: chain with C(=O)OH]

(b) [structure: chain with C(=O)O–cyclohexyl]

(c) [structure: chain with C(=O)N(CH$_3$)$_2$]

(d) [structure: chain with C(=O) ketone]

(e) [structure: chain with C(=O)H aldehyde]

In summary, alkanoyl chlorides are attacked by a variety of nucleophiles, the reactions leading to new carboxylic acid derivatives, ketones, and aldehydes by addition–

elimination mechanisms. The reactivity of alkanoyl halides makes them useful synthetic relay points on the way to other carbonyl derivatives.

20-3 Chemistry of Carboxylic Anhydrides

Carboxylic anhydrides, $\overset{\displaystyle O \quad O}{\overset{\displaystyle \| \quad \|}{RCOCR}}$, are named by simply adding the term *anhydride* to the acid name (or names, in regard to mixed anhydrides). This method also applies to cyclic derivatives.

$$\overset{\displaystyle O \quad O}{\overset{\displaystyle \| \quad \|}{CH_3COCCH_3}}$$

**1,2-Benzenedicarboxylic anhydride
(Phthalic anhydride)**

Acetic anhydride

**2-Butenedioic anhydride
(Maleic anhydride)**

**Pentanedioic anhydride
(Glutaric anhydride)**

$$\overset{\displaystyle O \quad O}{\overset{\displaystyle \| \quad \|}{CH_3COCCH_2CH_3}}$$
**Acetic propanoic anhydride
(A mixed anhydride)**

The reactions of carboxylic anhydrides with nucleophiles, although less vigorous, are completely analogous to those of the alkanoyl halides. The leaving group is a carboxylate instead of a halide ion.

Nucleophilic Addition–Elimination of Anhydrides

$$RC\overset{\ddot{O}}{\underset{}{\|}}\overset{..}{O}\overset{\ddot{O}}{\underset{}{\|}}CR + :NuH \longrightarrow RC\overset{:\ddot{O}:^-}{\underset{+NuH}{\|}}\overset{..}{O}\overset{:O:}{\underset{}{\|}}CR \longrightarrow RC\overset{:O:}{\underset{}{\|}}\overset{+}{NuH} + {}^-:\overset{:O:}{\underset{}{\|}}OCR \longrightarrow RCNu\overset{:O:}{\underset{}{\|}} + HOCR\overset{:O:}{\underset{}{\|}}$$

$$\overset{\displaystyle O \quad O}{\overset{\displaystyle \| \quad \|}{CH_3COCCH_3}} \xrightarrow{\text{HOH}} CH_3\overset{\displaystyle O}{\overset{\displaystyle \|}{C}}OH + HO\overset{\displaystyle O}{\overset{\displaystyle \|}{C}}CH_3$$
Acetic anhydride   100%   **Acetic acid**

Propanoic anhydride $\xrightarrow{\text{CH}_3\text{OH}}$ **Methyl propanoate** 83% + **Propanoic acid**

In every addition–elimination reaction except hydrolysis, the carboxylic acid side product is usually undesired and is removed by work-up with basic water. Cyclic anhydrides undergo similar nucleophilic addition–elimination reactions that lead to ring opening.

Nucleophilic Ring Opening of Cyclic Anhydrides

$$\xrightarrow{\text{CH}_3\text{OH, 100°C}}$$

HOCCH$_2$CH$_2$COCH$_3$

96%

Butanedioic (succinic) anhydride

EXERCISE 20-8

Treatment of butanedioic (succinic) anhydride with ammonia at elevated temperatures leads to a compound $C_4H_5NO_2$. What is its structure?

EXERCISE 20-9

Formulate the mechanism for the reaction of acetic anhydride with methanol in the presence of sulfuric acid.

In summary, anhydrides react with nucleophiles in the same way as alkanoyl halides do, except that the leaving group is a carboxylate ion. Cyclic anhydrides furnish dicarboxylic acid derivatives.

20-4 Chemistry of Esters

As mentioned in Section 19-9, esters, RCOR′, constitute perhaps the most important class of carboxylic derivatives. They are particularly widespread in nature, contributing especially to the pleasant flavors and aromas of many flowers and fruits. Esters undergo typical carbonyl chemistry but with reduced reactivity relative to that of either alkanoyl halides or carboxylic anhydrides. This section begins with a discussion of ester nomenclature, which includes a sampling of naturally occurring esters. Descriptions of the transformations that esters undergo with a variety of nucleophilic reagents follow.

Esters are alkyl alkanoates

Esters are named as alkyl alkanoates. The ester grouping, –COR, as a substituent is called **alkoxycarbonyl.** A cyclic ester is called a **lactone** (the common name, Section 19-9); the systematic name is **oxa-2-cycloalkanone** (Section 25-1). Its name is preceded by α, β, γ, δ, and so forth, depending on ring size.

CH$_3$COCH$_3$

Methyl acetate

CH$_3$CH$_2$COCH$_2$CH$_3$

Ethyl propanoate
(Ethyl propionate)

CH$_3$COCH$_2$CH$_2$CHCH$_3$

CH$_3$

3-Methylbutyl acetate
(Isopentyl acetate)

(A component of banana flavor)

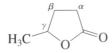

2-Methylpropyl propanoate
(Isobutyl propionate)

(A component of rum flavor)

3-Methylbutyl pentanoate
(Isopentyl valerate)

(A component of apple flavor)

Methyl 2-aminobenzoate
(Methyl anthranilate)

(A component of grape flavor)

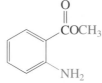

γ-Valerolactone
(Better: 5-methyloxa-2-cyclopentanone)

β-Propiolactone

(This compound is a carcinogen
and is systematically
called oxa-2-cyclobutanone;
see Section 25-1)

γ-Butyrolactone
(Better: oxa-2-cyclopentanone)

In addition to their prevalence in plants, esters play biological roles in the animal kingdom. Section 12-16 included several examples of esters that function as insect pheromones. Perhaps the most bizarre of these esters is (Z)-7-dodecenyl acetate, a component of the pheromone mixture in several species of moths. This same compound was recently found to also be the mating pheromone of the elephant (who said nature has no sense of humor?). Section 20-5 will describe a number of more conventional biological functions of esters.

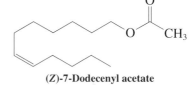

(Z)-7-Dodecenyl acetate

EXERCISE 20-10

Name the following esters.

(a) ![structure] **(b)** $CH_3OCCH_2CH_2COCH_3$ **(c)** $CH_2{=}CHCO_2CH_3$

In industry, lower esters, such as ethyl acetate (b.p. 77°C) and butyl acetate (b.p. 127°C), are used as solvents. For example, butyl butanoate has replaced the ozone-depleting trichloroethane as a cleaning solvent in the manufacture of electronic components such as computer chips. Higher nonvolatile esters are used as softeners (called plasticizers; see Section 12-14) for brittle polymers—in flexible tubing (e.g., Tygon tubing), rubber pipes, and upholstery.

Esters hydrolyze to carboxylic acids

Esters undergo nucleophilic substitution reactions by means of addition–elimination pathways, albeit with reduced reactivity relative to halides and anhydrides. Thus, in contrast to the latter, catalysis by acid or base becomes a frequent necessity. For example, esters are cleaved to carboxylic acids and alcohols in the presence of excess

An elephant, decidedly unimpressed by the fact that it uses the same molecule as a mating pheromone as that of several species of moth.

water and strong acid, and the reaction requires heating to proceed at a reasonable rate. The mechanism of this transformation is the reverse of acid-catalyzed esterification (Section 19-9).

Formulate a mechanism for the acid-catalyzed hydrolysis of γ-butyrolactone.

Strong *bases* also promote ester hydrolysis through an addition–elimination mechanism (Sections 19-7 and 20-1). The base (B) converts the poor nucleophile water into the negatively charged and more highly nucleophilic hydroxide ion.

$$B:^- \quad + \quad H{-}OH \quad \longrightarrow \quad ^-{:}OH \quad + \quad B{-}H$$

Ester hydrolysis is frequently achieved by using hydroxide itself, in at least stoichiometric amounts, as the base.

Methyl 3-methylbutanoate → [1. KOH, H₂O, CH₃OH, Δ 2. H⁺, H₂O] → 3-Methylbutanoic acid + CH₃OH

Methyl 3-methylbutanoate   100%   **3-Methylbutanoic acid**

Mechanism of Base-Mediated Ester Hydrolysis

$$RC\ddot{O}CH_3 + {:}\ddot{O}H \;\rightleftharpoons\; R{-}\overset{:O:^-}{\underset{:\ddot{O}H}{C}}{-}\ddot{O}CH_3 \;\rightleftharpoons\; RC\ddot{O}{-}H + {:}\ddot{O}CH_3 \;\longrightarrow\; RC\ddot{O}{:}^- + H\ddot{O}CH_3$$

Carboxylate ion

Unlike acid-catalyzed hydrolysis, which is reversible, the base-mediated process (saponification; see Section 19-13) is driven essentially to completion by the last step, in which the stoichiometric base converts the acid into a carboxylate ion. The carboxylic acid product is obtained by acidic aqueous work-up.

Formulate a mechanism for the base-mediated hydrolysis of γ-butyrolactone.

Transesterification takes place with alcohols

Transesterification

$$\underset{RC\overset{O}{\parallel}OR'}{} + R''OH$$

$$\updownarrow \; H^+ \text{ or } ^-OR''$$

$$\underset{RC\overset{O}{\parallel}OR''}{} + R'OH$$

Esters react with alcohols in an acid- or base-catalyzed transformation called **transesterification.** It allows for the direct conversion of one ester into another without proceeding through the free acid. Like esterification, transesterification is a reversible reaction. To shift the equilibrium, a large excess of the alcohol is usually employed, sometimes in the form of solvent.

$$C_{17}H_{35}C\overset{O}{\parallel}OCH_2CH_3 + CH_3OH \xrightarrow{H^+ \text{ or } ^-OCH_3} C_{17}H_{35}C\overset{O}{\parallel}OCH_3 + CH_3CH_2OH$$

                                                                                    90%

Ethyl octadecanoate   **Solvent**   **Methyl octadecanoate**

Lactones are opened to hydroxy esters by transesterification.

γ-Butyrolactone   3-Bromopropanol   3-Bromopropyl 4-hydroxybutanoate

80%

The mechanisms of transesterification by acid and base are straightforward extensions of the mechanisms of the corresponding hydrolyses to the carboxylic acids. Thus, acid-catalyzed transesterification begins with protonation of the carbonyl oxygen, followed by nucleophilic attack of the alcohol on the carbonyl carbon. In contrast, in base the alcohol is first deprotonated, and the resulting alkoxide ion then adds to the ester carbonyl group.

EXERCISE 20-13

Formulate mechanisms for the acid- and base-catalyzed transesterifications of γ-butyrolactone by 3-bromopropanol.

Amines convert esters into amides

Amines, which are more nucleophilic than alcohols, readily transform esters into amides. No catalyst is needed, but heating is required.

Amide Formation from Methyl Esters

$$RCOCH_3 + R'NH_2 \xrightarrow{\Delta} RCNHR' + CH_3OH$$

$$CH_3(CH_2)_7CH=CH(CH_2)_7COCH_3 + CH_3(CH_2)_{11}NH_2 \xrightarrow{230°C}$$
Methyl 9-octadecenoate   **1-Dodecanamine**

$$CH_3(CH_2)_7CH=CH(CH_2)_7CNH(CH_2)_{11}CH_3 + CH_3OH$$
69%
***N*-Dodecyl-9-octadecenamide**

Grignard reagents transform esters into alcohols

Esters can be converted into alcohols by using *two* equivalents of a Grignard reagent. In this way, ordinary esters are transformed into tertiary alcohols, whereas formic esters furnish secondary alcohols.

Alcohols from Esters and Grignard Reagents

$$CH_3CH_2COCH_2CH_3 + 2\ CH_3CH_2CH_2MgBr \xrightarrow[-CH_3CH_2OH]{\substack{1.\ (CH_3CH_2)_2O \\ 2.\ H^+,\ H_2O}} CH_3CH_2CCH_2CH_2CH_3$$
$$CH_2CH_2CH_3$$
OH
69%

Ethyl propanoate   **Propylmagnesium bromide**   **4-Ethyl-4-heptanol**

$$
\underset{\substack{\text{Methyl formate}}}{\overset{\displaystyle O}{\overset{\|}{HCOCH_3}}} + \underset{\substack{\text{Butylmagnesium}\\\text{bromide}}}{2\,CH_3CH_2CH_2CH_2MgBr} \xrightarrow[\substack{-\,CH_3OH}]{\substack{1.\,(CH_3CH_2)_2O\\2.\,H^+,\,H_2O}} \underset{\substack{\text{5-Nonanol}}}{\overset{\displaystyle OH}{\overset{\displaystyle |}{\underset{\displaystyle CH_2CH_2CH_2CH_3}{\underset{\displaystyle |}{HCCH_2CH_2CH_2CH_3}}}}}
$$

85%

The reaction begins with addition of the organometallic to the carbonyl function in the usual manner to give the magnesium salt of a hemiacetal (Section 17-7). At room temperature, rapid elimination results in the formation of an intermediate ketone (or aldehyde, from formates). The resulting carbonyl group then immediately adds a second equivalent of Grignard reagent. Subsequent acidic aqueous work-up leads to the observed alcohol.

Mechanism of the Alcohol Synthesis from Esters and Grignard Reagents

$$
\underset{R}{\overset{:O:}{\overset{\|}{C}}}{\overset{}{\underset{OCH_3}{}}} + R'-MgBr \longrightarrow R-\underset{R'}{\overset{O-MgBr}{\overset{|}{\underset{|}{C}}}}-OCH_3 \xrightarrow{-\,CH_3OMgBr}
$$

$$
\underset{\substack{R\qquad R'\\\text{(Cannot be stopped}\\\text{at this stage)}}}{\overset{O}{\overset{\|}{C}}} \xrightarrow{R'MgBr} R-\underset{R'}{\overset{OMgBr}{\overset{|}{\underset{|}{C}}}}-R' \xrightarrow{H^+,\,H_2O} R-\underset{R'}{\overset{OH}{\overset{|}{\underset{|}{C}}}}-R'
$$

$$
\underset{\substack{\text{Methyl benzoate}}}{\overset{\displaystyle O}{\overset{\|}{C_6H_5COCH_3}}}
$$

EXERCISE 20-14

Propose a synthesis of triphenylmethanol, $(C_6H_5)_3COH$, beginning with methyl benzoate (shown in the margin) and bromobenzene.

Esters are reduced by hydride reagents to give alcohols or aldehydes

The reduction of esters to alcohols by $LiAlH_4$ requires 0.5 equivalent of the hydride, because only two hydrogens are needed per ester function.

Reduction of an Ester to an Alcohol

$$
\underset{\substack{CH_3}}{\overset{\displaystyle O}{N\overset{\|}{CHCOCH_2CH_3}}} \xrightarrow[\substack{-\,CH_3CH_2OH}]{\substack{1.\,LiAlH_4\ (0.5\ \text{equivalent}),\,(CH_3CH_2)_2O\\2.\,H^+,\,H_2O}} \underset{\substack{CH_3\\90\%}}{NCHCH_2OH}
$$

A milder reducing agent allows the reaction to be stopped at the aldehyde oxidation stage. Such a reagent is bis(2-methylpropyl)aluminum hydride (diisobutylaluminum hydride) when used at low temperatures in toluene.

Reduction of an Ester to an Aldehyde

$$\underset{\textbf{Ethyl 2-methylpropanoate}}{\overset{\overset{\displaystyle H_3C\quad O}{\underset{\displaystyle |\quad\ |}{}}}{CH_3CHCOCH_2CH_3}} + \underset{\substack{\textbf{Bis(2-methylpropyl)aluminum hydride} \\ \textbf{(Diisobutylaluminum hydride, DIBAL)}}}{\overset{\overset{\displaystyle CH_3}{\underset{\displaystyle |}{}}}{(CH_3CHCH_2)_2AlH}} \xrightarrow[-\,CH_3CH_2OH]{\substack{1.\ Toluene,\ -60°C \\ 2.\ H^+,\ H_2O}} \underset{\textbf{2-Methylpropanal}}{\overset{\overset{\displaystyle CH_3}{\underset{\displaystyle |}{}}}{CH_3CHCHO}}$$

Esters form enolates that can be alkylated

The acidity of the α-hydrogens in esters is sufficiently high that **ester enolates** are formed by treatment of esters with strong base at low temperatures. Ester enolates react like ketone enolates, undergoing alkylations.

Alkylation of an Ester Enolate

$$\underset{\substack{\\ pK_a \sim 25}}{\overset{\overset{\displaystyle O}{\underset{\displaystyle \|}{}}}{CH_3COCH_2CH_3}} \xrightarrow{LDA,\ THF,\ -78°C} \underset{\textbf{Ethyl acetate}}{CH_2=C\overset{\displaystyle O^-Li^+}{\underset{\displaystyle OCH_2CH_3}{}}} \xrightarrow[-\,LiBr]{\substack{CH_2=CHCH_2-Br, \\ HMPA}} \underset{\underset{\textbf{Ethyl 4-pentenoate}}{97\%}}{\overset{\overset{\displaystyle O}{\underset{\displaystyle \|}{}}}{CH_2=CHCH_2CH_2COCH_2CH_3}}$$

The pK_a of esters is about 25. Consequently, ester enolates exhibit the typical side reactions of strong bases: E2 processes (especially with secondary, tertiary, and β-branched halides) and deprotonations. The most characteristic reaction of ester enolates is the Claisen condensation, in which the enolate attacks the carbonyl carbon of another ester. This process will be considered in Chapter 23.

EXERCISE 20-15

Give the products of the reaction of ethyl cyclohexanecarboxylate with the following compounds or under the following conditions (and followed by acidic aqueous work-up, if necessary). **(a)** H^+, H_2O; **(b)** HO^-, H_2O; **(c)** CH_3O^-, CH_3OH; **(d)** NH_3, Δ; **(e)** $2\ CH_3MgBr$; **(f)** $LiAlH_4$; **(g)** 1. LDA, 2. CH_3I.

In summary, esters are named as alkyl alkanoates. Many of them have pleasant odors and are present in nature. They are less reactive than alkanoyl halides or carboxylic anhydrides and therefore often require the presence of acid or base to transform. With water, esters hydrolyze to the corresponding carboxylic acids or carboxylates; with alcohols, they undergo transesterification and, with amines at elevated temperatures, they furnish amides. Grignard reagents add twice to give tertiary alcohols (or secondary alcohols, from formates). Lithium aluminum hydride reduces esters all the way to alcohols, whereas bis(2-methylpropyl)aluminum (diisobutylaluminum) hydride allows the process to be stopped at the aldehyde stage. With LDA, it is possible to form ester enolates, which can be alkylated by electrophiles.

20-5 Esters in Nature: Waxes, Fats, Oils, and Lipids

Esters are essential components in the cells of all living organisms. This section introduces several of the most common types of natural esters with brief descriptions of their biological functions.

Waxes are simple esters, whereas fats and oils are more complex

Esters made up of long-chain carboxylic acids and long-chain alcohols constitute **waxes,** which form hydrophobic (Section 8-3) and insulating coatings on the skin and fur of animals, the feathers of birds, and the fruits and leaves of many plants. Spermaceti and beeswax are liquids or very soft solids at room temperature and are used as lubricants. Sheep's wool provides wool wax, which when purified yields lanolin, a widely used cosmetic base. The leaves of a Brazilian palm are the source of carnauba wax, a tough and water-resistant mixture of several solid esters. Carnauba wax is highly valued for its ability to take and maintain a high gloss and is used as a floor and automobile wax.

$$CH_3(CH_2)_{14}\overset{\displaystyle O}{\overset{\|}{C}}O(CH_2)_{15}CH_3$$

Hexadecyl hexadecanoate
(Cetyl palmitate)
(Wax from the sperm whale)

$$CH_3(CH_2)_n\overset{\displaystyle O}{\overset{\|}{C}}O(CH_2)_mCH_3$$

n = 24, 26; *m* = 29, 31

Beeswax

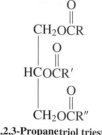

CH₂OH
|
CHOH
|
CH₂OH

**1,2,3-Propanetriol
(Glycerol)**

$$\begin{array}{l}
CH_2O\overset{\displaystyle O}{\overset{\|}{C}}R \\[4pt]
HCO\overset{\displaystyle O}{\overset{\|}{C}}R' \\[4pt]
CH_2O\overset{\displaystyle O}{\overset{\|}{C}}R''
\end{array}$$

**1,2,3-Propanetriol triester
(Triglyceride)**

Triesters of 1,2,3-propanetriol (glycerol) with long-chain carboxylic acids constitute **fats** and **oils** (Sections 11-3 and 19-13; Chemical Highlight 19-2). They are also called **triglycerides.** The acids in fats and oils (**fatty acids**) are typically unbranched and contain an even number of carbon atoms; if unsaturated, the double bonds are usually cis. Fats are biological energy reserves, which are stored in the body's tissues until their metabolism leads ultimately to CO_2 and water. Oils have a similar function in plant seeds. As food components, fats and oils serve as solvents for food flavors and colors and contribute to a sense of "fullness" after eating, because they leave the stomach relatively slowly. Saturated fats containing hexadecanoic (palmitic), tetradecanoic (myristic), and dodecanoic (lauric) acids have been implicated as dietary factors in atherosclerosis (hardening of the arteries). Fortunately for lovers of chocolate, cocoa butter is mainly a low-melting triglyceride that contains two molecules of octadecanoic (stearic) acid—which, despite being saturated, does not contribute to elevated low-density lipoprotein (LDL) levels or atherosclerosis—and one molecule of (Z)-9-octadecenoic (oleic) acid. The latter is also the chief fatty acid component of olive oil, the major source of dietary fat among the Greeks, who have very low rates of heart disease. Research reported in 1996 revealed that chocolate (especially the dark variety) contains an assortment of phenolic antioxidants that strongly inhibit radical oxidation of lipids and subsequent cell damage (Section 22-9).

A "chocoholic's" dream.

Lipids are biomolecules soluble in nonpolar solvents

Extraction of biological material with nonpolar solvents gives a wide assortment of compounds that includes terpenes and steroids (Section 4-7), fats and oils, and a variety of other low-polarity substances collectively called **lipids** (*lipos,* Greek, fat). Lipid fractions include **phospholipids,** important components of cell membranes, which are derived from carboxylic acids and phosphoric acid. In the **phosphoglycerides,** glycerol is esterified with two adjacent fatty acids and a phosphate unit which bears another substituent derived from a low molecular weight alcohol such as choline, $[HOCH_2CH_2N(CH_3)_3]^+ \, ^-OH$. The substance shown here is an example of a **lecithin,** a lipid found in the brain and nervous system.

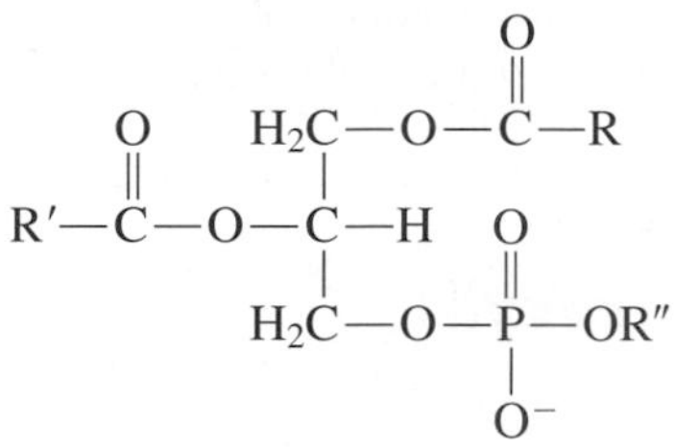

A phosphoglyceride

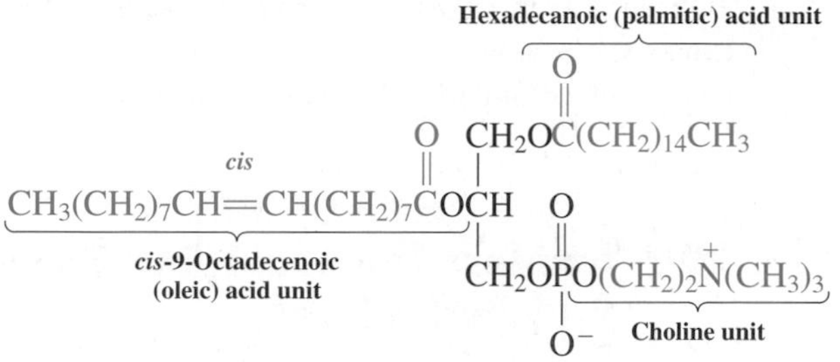

Hexadecanoic (palmitic) acid unit

Palmitoyloleoylphosphatidylcholine
(A lecithin)

cis-9-Octadecenoic
(oleic) acid unit

Choline unit

(Z)-9-Octadecenamide

Compounds such as these play roles in the transmission of nerve impulses and exhibit other biological effects. For example, a lipid containing the amide analog of oleic acid [(*Z*)-9-octadecenamide] was recently identified as an essential sleep-inducing agent in the brain; indeed, persons with lipid-restricted diets have difficulty experiencing deep sleep.

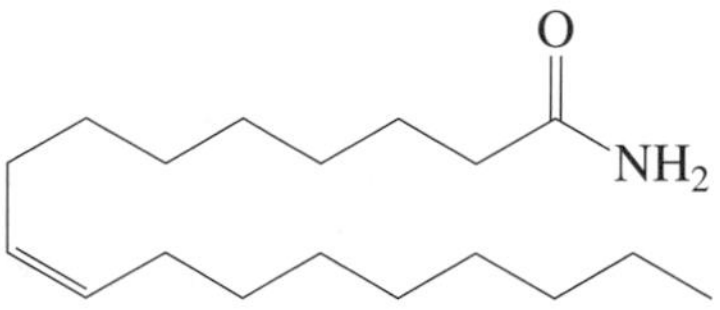

Because these molecules carry two long hydrophobic fatty acid chains and a polar head group (the phosphate and choline substituent), they are capable of forming micelles in aqueous solution (see Chemical Highlight 19-1). In the micelles, the phosphate unit is solubilized by water, and the ester chains are clustered inside the hydrophobic micellar sphere (Figure 20-2A).

Phosphoglycerides can also aggregate in a different way: They may form sheets called a **lipid bilayer** (Figure 20-2B). This capability is significant because, whereas micelles are usually limited in size (<200 Å in diameter), bilayers may be as much as 1 mm (10^7 Å) in length. This property makes them ideal constituents of cell membranes, which act as permeability barriers regulating molecular transport into and out of the cell. Lipid bilayers are relatively stable molecular assemblies. The forces that drive their formation are similar to those at work in micelles: London interactions (Section 2-4) between the hydrophobic alkane chains, and Coulombic and solvation forces between the polar head groups and between these polar groups and water.

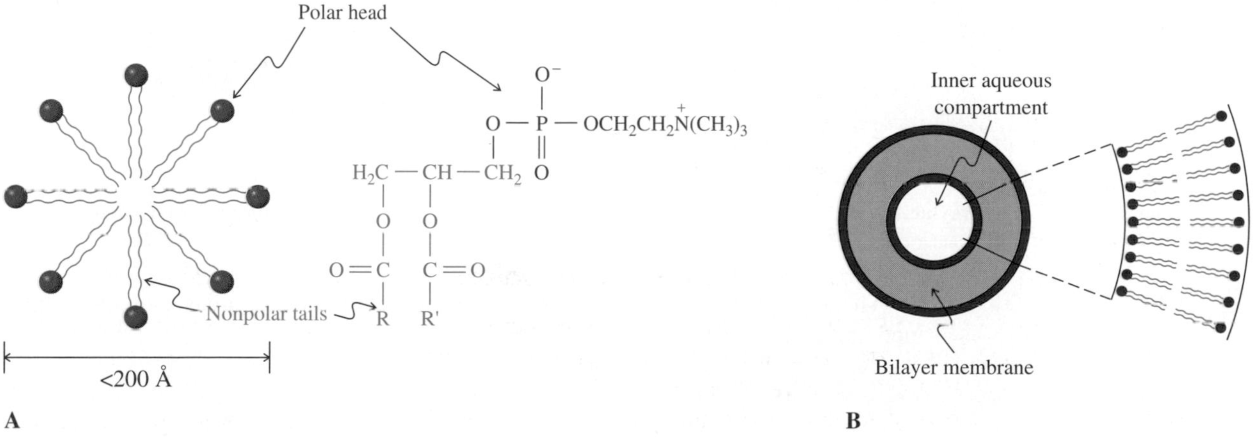

Polar head

Nonpolar tails

<200 Å

A

Inner aqueous compartment

Bilayer membrane

B

FIGURE 20-2

Phospholipids are substituted esters essential to the structures of cell membranes. These molecules aggregate to form (A) a micelle or (B) a lipid bilayer. The polar head groups and nonpolar tails in phospholipids drive these aggregations. (After *Biochemistry*, 4th ed., by Lubert Stryer. W. H. Freeman and Company. Copyright © 1975, 1981, 1988, 1995).

In summary, waxes, fats, and oils are naturally occurring, biologically active esters. Lipids are a broad class of biological molecules soluble in nonpolar solvents. They include glycerol-derived triesters, which are components of cell membranes.

20-6 Amides: The Least Reactive Carboxylic Acid Derivatives

Among all carboxylic acid derivatives, the amides, $\overset{O}{\overset{\|}{RCNR'_2}}$, are the least susceptible to nucleophilic attack. After a brief introduction to amide naming, this section describes their reactions.

Amides are named alkanamides, cyclic amides are lactams

Amides are called **alkanamides,** the ending *-e* of the alkane stem having been replaced by **-amide.** In common names, the ending *-ic* of the acid name is replaced by the **-amide** suffix. In cyclic systems, the ending *-carboxylic acid* is replaced by **-carboxamide.** Substituents on the nitrogen are indicated by the prefix *N-*, or *N,N-*, depending on the number of groups. Accordingly, there are primary, secondary, and tertiary amides.

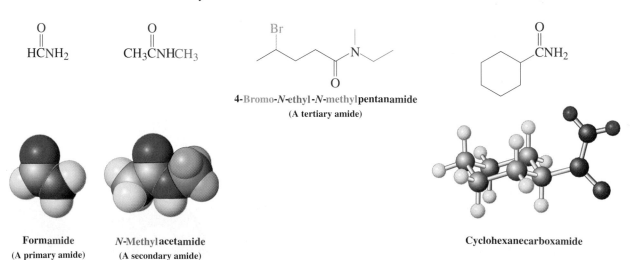

$$\overset{O}{\overset{\|}{HCNH_2}} \qquad \overset{O}{\overset{\|}{CH_3CNHCH_3}}$$

4-Bromo-*N*-ethyl-*N*-methylpentanamide
(A tertiary amide)

Formamide
(A primary amide)

N-Methylacetamide
(A secondary amide)

Cyclohexanecarboxamide

There are several amide derivatives of carbonic acid, H_2CO_3: ureas, carbamic acids, and carbamic esters (urethanes).

A urea   A carbamic acid   A carbamic ester
(Urethane)

Cyclic amides are called **lactams** (Section 19-10)—the systematic name is aza-2-cycloalkanones (Section 25-1)—and the rules for naming them follow those used for lactones. The penicillins are annulated β-lactams.

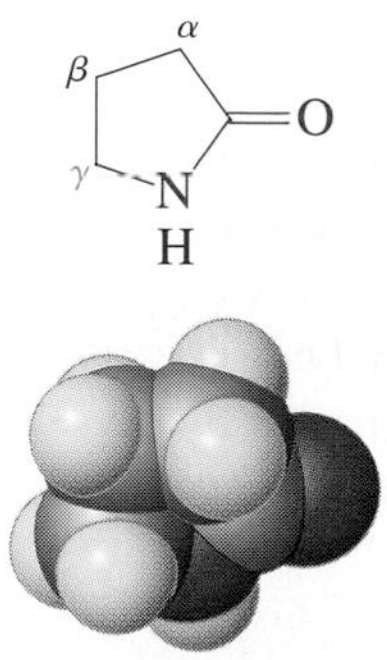

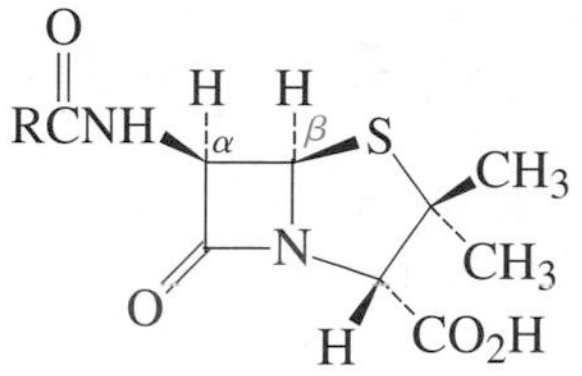

Penicillin

(A β-lactam derivative)

γ-Butyrolactam
(Systematic name:
aza-2-cyclopentanone)

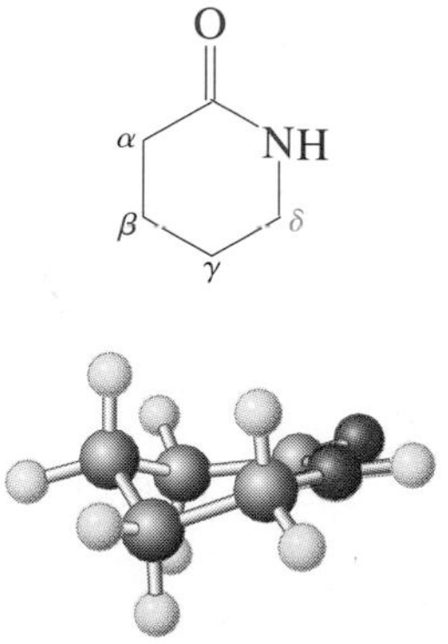

δ-Valerolactam
(Systematic name:
aza-2-cyclohexanone)

Urethane-based polymers form lightweight, tough, abrasion-resistant, high-performance materials ideal for use in a variety of sporting goods.

Amides are essential to biochemistry; amide groups constitute the linkages between the amino acid subunits that make up the biopolymers called proteins (Chapter 26). Many simpler amides possess varied forms of biological activity. For example, anandamide, the amide of arachidonic acid (Section 19-13) with 2-aminoethanol, has been found to bind to the same receptor in the brain as does tetrahydrocannabinol (Section 9-11), the active ingredient in marijuana. Anandamide has been isolated from chocolate, suggesting that people who claim to be "addicted to chocolate" may really know what they are talking about.

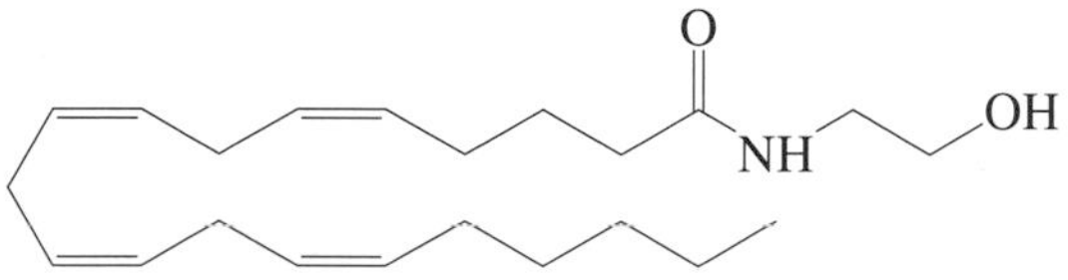

Anandamide

Amide hydrolysis requires strong heating in concentrated acid or base

The amides are the least reactive of the carboxylic acid derivatives, in part because they are strongly stabilized by delocalization of the nitrogen lone pair (Section 20-1). As a consequence, their nucleophilic addition–eliminations require relatively harsh conditions. For example, hydrolysis to the corresponding carboxylic acid occurs only on prolonged heating in strongly acidic or basic water by addition–elimination mechanisms. Acid hydrolysis liberates the amine in the form of an ammonium salt.

Acid Hydrolysis of an Amide

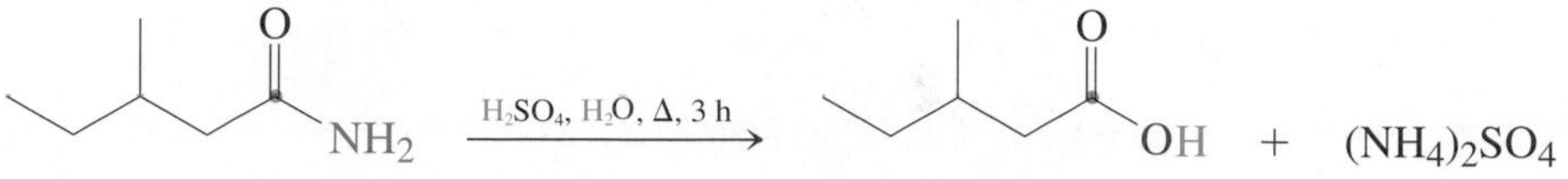

$$\xrightarrow{H_2SO_4,\ H_2O,\ \Delta,\ 3\ h}$$

3-Methylpentanamide → **3-Methylpentanoic acid** + $(NH_4)_2SO_4$

CHEMICAL HIGHLIGHT 20-2

Penicillin: An Antibiotic Containing a β-Lactam Ring

The discovery of penicillin as a powerful broad-spectrum antibiotic was one of the milestones in medicinal chemistry. As with many such advances, serendipity played a major role. In 1928, the Scottish bacteriologist Alexander Fleming* noted that several *Staphylococcus* cultures set aside on a laboratory bench had been contaminated by microorganisms from the laboratory air. A green mold, *Penicillium notatum,* had grown in some places, and the *Staphylococcus* in its vicinity was disintegrating. The substance causing this antibiotic activity was called penicillin.

Penicillin became available in pure form only about 10 years later. Many different penicillins have subsequently been synthesized with different R groups. *Penicillin G* has a phenylmethyl (benzyl, $C_6H_5CH_2$) group attached to the amide function; *ampicillin* has a phenylaminomethyl ($C_6H_5CHNH_2$) substituent. Structurally and biologically related are the *cephalosporins,* important antibiotics that are frequently active when the penicillins are not.

Cephalosporin C

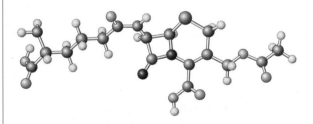

*Sir Alexander Fleming (1881–1955), St. Mary's Hospital, University of London, Nobel Prize 1945 (physiology or medicine).

The strained β-lactam ring is responsible for the antibiotic activity of these drugs. Because ring strain is relieved on opening, β-lactams are unusually reactive compared with ordinary amides. The enzyme *transpeptidase,* which catalyzes a crucial cross-linking reaction in the biosynthesis of bacterial cell walls, accepts penicillin as a substrate. The penicillin then alkanoylates a nucleophilic oxygen of the enzyme, rendering it inactive. Cell wall construction stops, and the organism soon dies. The reaction is the reverse of amide formation from esters (Section 20-4).

Penicillin in Action

Transpeptidase + penicillin

↓

Inactivated enzyme

Some bacteria are resistant to penicillin because they produce an enzyme, *penicillinase,* that hydrolyzes the β-lactam ring before it can attach itself to transpeptidase. The rate of this hydrolysis depends on the structure of the β-lactam. Cephalosporins are not affected by penicillinase. Nevertheless, the continual emergence of new, antibiotic-resistant bacterial strains, resulting from the frequently indiscriminate prescription of penicillin and other antibiotics, has spurred intensive ongoing efforts to discover novel, more active, and more selective systems.

Base hydrolysis initially furnishes the carboxylate salt and the amine. Acidic aqueous work-up then produces the acid.

Base Hydrolysis of an Amide

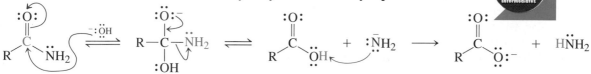

Mechanism of Hydrolysis of Amides by Aqueous Base

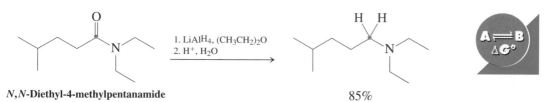

Amides can be reduced to amines or aldehydes

In contrast with carboxylic acids and esters, the reaction of amides with lithium aluminum hydride produces amines instead of alcohols.

Reduction of an Amide to an Amine

N,N-Diethyl-4-methylpentanamide

1. LiAlH₄, (CH₃CH₂)₂O
2. H⁺, H₂O

85%
N,N-Diethyl-4-methylpentanamine

EXERCISE 20-16

What product would you expect from LiAlH₄ reduction of the compound depicted in the margin?

The mechanism of reduction begins with hydride addition, which gives a tetrahedral intermediate. Elimination of an aluminum alkoxide leads to an **iminium ion.** Addition of a second hydride gives the final amine product.

Mechanism of Amide Reduction by LiAlH₄

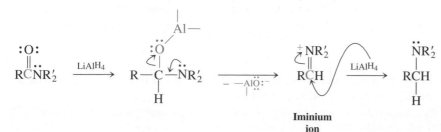

Iminium ion

Reduction of amides by bis(2-methylpropyl)aluminum hydride (diisobutylaluminum hydride) furnishes aldehydes. Recall that esters also are converted into aldehydes by this reagent (Section 20-4).

Reduction of an Amide to an Aldehyde

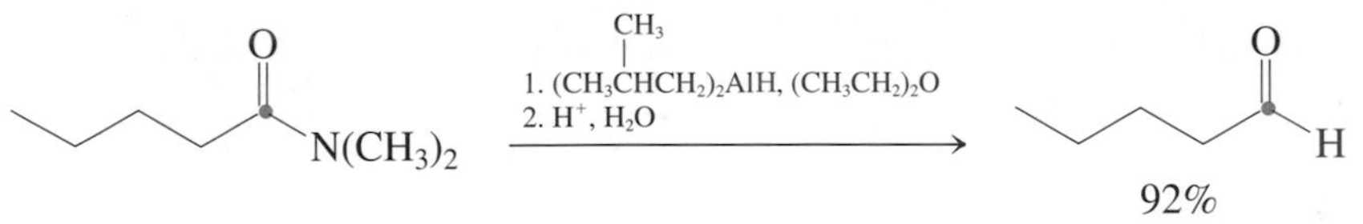

N,N-Dimethylpentanamide

1. $(CH_3CHCH_2)_2AlH$, $(CH_3CH_2)_2O$
2. H^+, H_2O

92%
Pentanal

EXERCISE 20-17

Treatment of amide A with $LiAlH_4$, followed by acidic aqueous work-up, gave compound B. Explain. (**Hint:** Review Sections 17-8 and 17-9.)

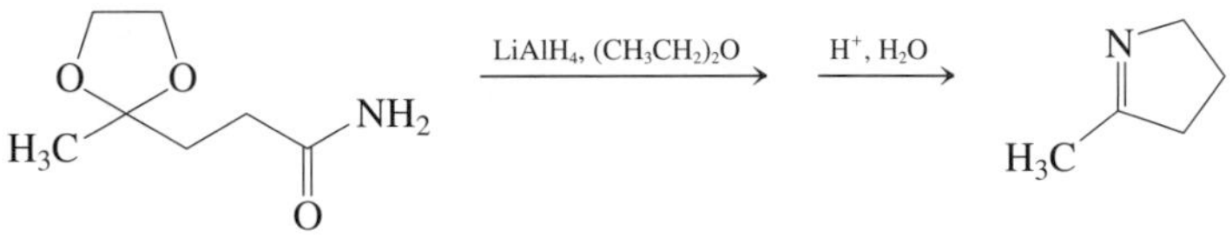

$LiAlH_4$, $(CH_3CH_2)_2O$ H^+, H_2O

A

B

In summary, carboxylic amides are named as alkanamides or as lactams if cyclic. They can be hydrolyzed to carboxylic acids by acid or base and reduced to amines by lithium aluminum hydride. Reduction by bis(2-methylpropyl)aluminum hydride (diisobutylaluminum hydride) stops at the aldehyde stage.

20-7 Amidates and Their Halogenation: The Hofmann Rearrangement

Hydrogens on both the carbon and the nitrogen atoms next to the amide carbonyl group are acidic. Removal of the NH hydrogen, which has a pK_a of about 22, with base leads to an **amidate ion.** The CH proton is less acidic, with a pK_a of about 30 (Section 20-1); therefore, deprotonation of the α-carbon, leading to an **amide enolate,** is more difficult.

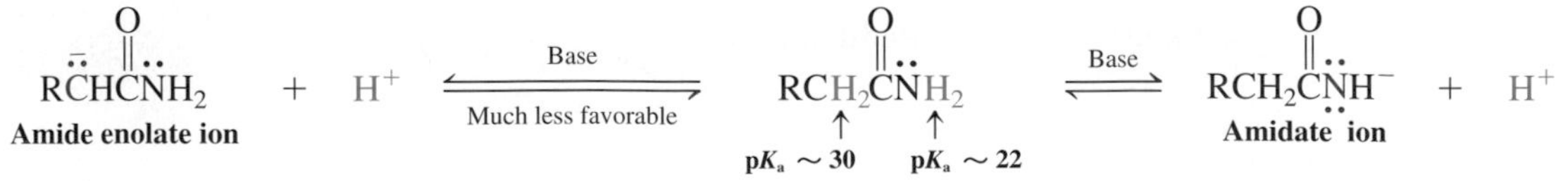

$\overset{\cdot\cdot}{R}\overset{\cdot\cdot}{CH}\overset{O}{\overset{\|}{C}}NH_2$ + H^+ ⇌ (Much less favorable) $RCH_2\overset{O}{\overset{\|}{C}}NH_2$ ⇌ (Base) $RCH_2\overset{O}{\overset{\|}{C}}\overset{\cdot\cdot}{N}H^-$ + H^+

Amide enolate ion $pK_a \sim 30$ $pK_a \sim 22$ Amidate ion

Practically speaking, therefore, a proton may be removed from carbon only with tertiary amides, in which the nitrogen is blocked.

The amidate ion formed by deprotonation of a primary amide is a synthetically useful nucleophile. This section will focus on one of its reactions, the Hofmann rearrangement.

The pK_a of 1,2-benzenedicarboximide (phthalimide, A) is 8.3, considerably lower than the pK_a of benzamide (B). Why?

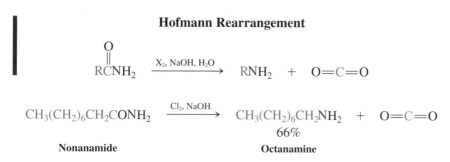

A   B

In the presence of base, primary amides undergo a special halogenation reaction, the **Hofmann*** **rearrangement.** In it, the carbonyl group is expelled from the molecule to give a primary amine with one carbon fewer in the chain.

Hofmann Rearrangement

$$\underset{RCNH_2}{\overset{O}{\|}} \xrightarrow{X_2, \text{ NaOH, } H_2O} RNH_2 \ + \ O{=}C{=}O$$

$$CH_3(CH_2)_6CH_2CONH_2 \xrightarrow{Cl_2, \text{ NaOH}} CH_3(CH_2)_6CH_2NH_2 \ + \ O{=}C{=}O$$
$$\phantom{CH_3(CH_2)_6CH_2CONH_2 \xrightarrow{Cl_2, \text{ NaOH}} } 66\%$$

Nonanamide   **Octanamine**

The Hofmann rearrangement begins with deprotonation of nitrogen to form an amidate ion (step 1). Halogenation of the nitrogen follows, much like the α-halogenation of aldehyde and ketone enolates (step 2; see Section 18-3). Subsequently, the second proton on the nitrogen is abstracted by additional base to give an N-haloamidate, which spontaneously eliminates halide (steps 3 and 4). The species formed contains an uncharged nitrogen atom surrounded by only an electron sextet. Such intermediates, called **nitrenes,** are highly reactive and short lived. In the Hofmann rearrangement, the acyl nitrene undergoes a 1,2-shift of an alkyl group (step 5) to give an **isocyanate,** R—N=C=O, a nitrogen analog of carbon dioxide, O=C=O. The *sp*-hybridized carbonyl carbon in the isocyanate is highly electrophilic and is attacked by water to produce an unstable **carbamic acid.** Finally, the carbamic acid decomposes to carbon dioxide and the amine (step 6).

Mechanism of the Hofmann Rearrangement

STEP 1. Amidate formation

$$\underset{R}{\overset{:O:}{\overset{\|}{C}}}\!\!\!-\!\!NH_2 \ + \ {}^-\!:\!\ddot{O}H \ \rightleftharpoons \ \underset{R}{\overset{:O:}{\overset{\|}{C}}}\!\!\!-\!\!\ddot{N}H^- \ + \ H\ddot{O}H$$

STEP 2. Halogenation

$$\underset{R}{\overset{O}{\overset{\|}{C}}}\!\!\!-\!\!\ddot{N}H^- \ + \ :\!\ddot{X}\!-\!\ddot{X}: \ \longrightarrow \ \underset{R}{\overset{O}{\overset{\|}{C}}}\!\!\!-\!\!\ddot{N}\!-\!H \ + \ :\!\ddot{X}\!:^- $$
$$ \underset{}{:\!\ddot{X}\!:}$$

*This is the Hofmann of the Hofmann rule of E2 reactions (Section 11-8).

STEP 3. *N*-Haloamidate formation

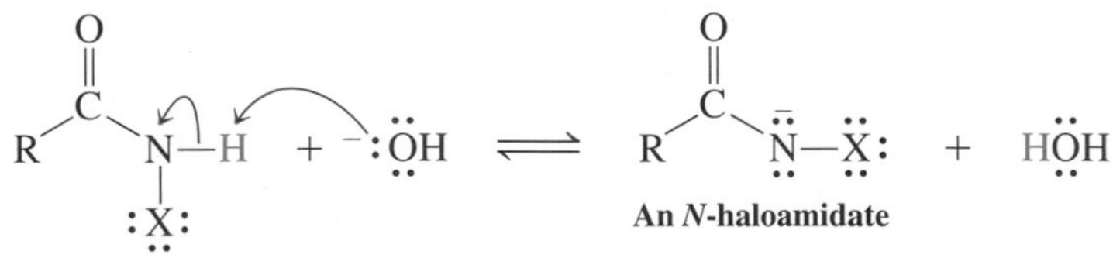

An *N*-haloamidate

STEP 4. Halide elimination

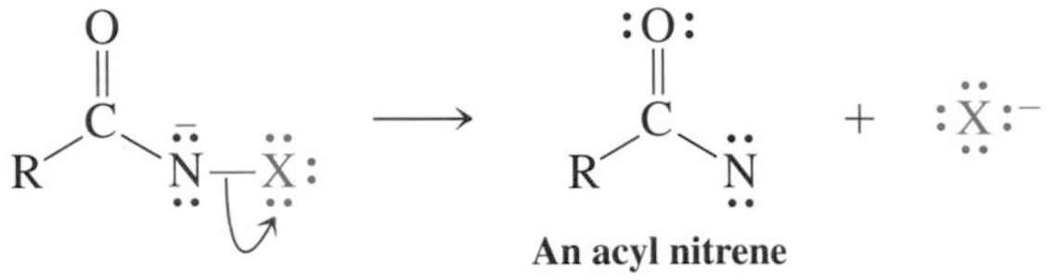

An acyl nitrene

STEP 5. Rearrangement

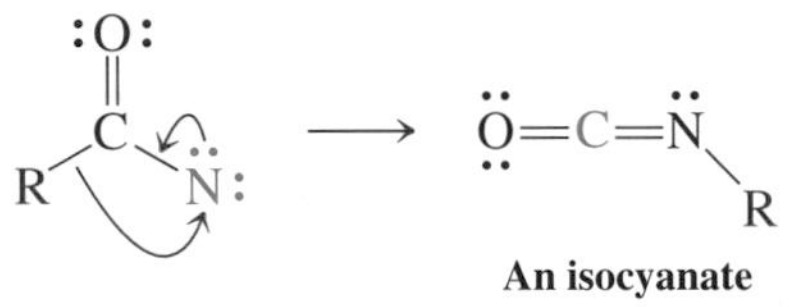

An isocyanate

CHEMICAL HIGHLIGHT 20-3

Methyl Isocyanate and the Bhopal Tragedy

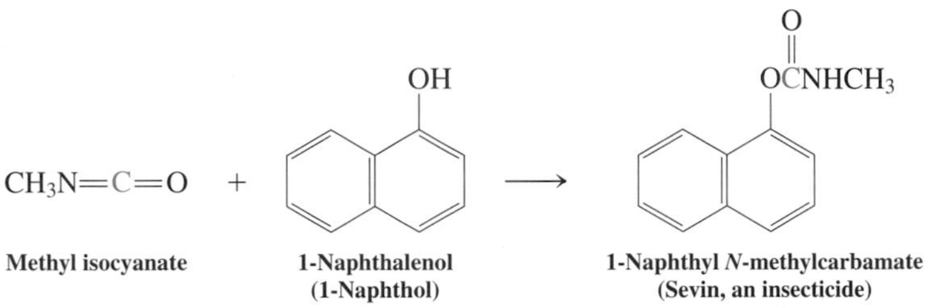

CH₃N=C=O + [1-Naphthalenol structure] ⟶ [Sevin structure]

Methyl isocyanate **1-Naphthalenol (1-Naphthol)** **1-Naphthyl *N*-methylcarbamate (Sevin, an insecticide)**

The reaction of methyl isocyanate with various alcohols and amines is used in the industrial preparation of several powerful herbicides and insecticides.

Consumption of methyl isocyanate in the United States has been estimated at 30 to 35 million pounds per year. In late 1984 in the city of Bhopal, India, a massive leak of this substance, used in the preparation of the insecticide Sevin, resulted in the deaths of more than 2000 people; at least 300,000 more were exposed to it. This catastrophe, the worst chemical industrial accident in history, led to a complete reappraisal of the safety measures for the handling of large quantities of toxic chemicals.

The toxicity of the isocyanate function derives from its rapid reaction with nucleophilic sites in biological molecules. Indiscriminate attack on the hydroxy, amino, and thiol groups in, for example, peptides and proteins inactivates them with respect to their biological functions. Other substances, which would be similarly affected by such attack, include small molecules taking part in the transmission of nerve impulses and various aspects of cell regulation.

STEP 6. Hydration to the carbamic acid and decomposition

$$\underset{R}{\overset{..}{N}}=C=\overset{..}{\underset{..}{O}} \quad \xrightarrow{H_2O} \quad \underset{R}{\overset{H}{\underset{|}{N}}}-\overset{\overset{:O:}{\|}}{C}-\overset{..}{\underset{..}{O}}H \quad \longrightarrow \quad R\overset{..}{N}H_2 \;+\; CO_2$$

A carbamic acid

EXERCISE 20-19

Write a detailed mechanism for the addition of water to an isocyanate under basic conditions and for the decarboxylation of the resulting carbamic acid.

EXERCISE 20-20

Suggest a sequence by which you could convert ester A into amine B.

COOCH₃   NH₂

A   --->   B

In summary, treatment of primary and secondary amides with base leads to deprotonation at nitrogen, giving amidate ions. Bases abstract protons from the α-carbon of tertiary amides. In the Hofmann rearrangement, amides react with halogens in base to furnish amines with one carbon fewer. In the course of this process an alkyl shift takes place, which converts an acyl nitrene into an isocyanate.

20-8 Alkanenitriles: A Special Class of Carboxylic Acid Derivatives

Nitriles, $RC\equiv N$, are considered derivatives of carboxylic acids because the nitrile carbon is in the same oxidation state as the carboxy carbon and because nitriles are readily converted into other carboxylic acid derivatives. This section describes the rules for naming nitriles, the structure and bonding in the nitrile group, and some of its spectral characteristics. Then it compares the chemistry of the nitrile group with that of other carboxylic acid derivatives.

In IUPAC nomenclature, nitriles are named from alkanes

A systematic way of naming this class of compounds is as **alkanenitriles.** In common names, the ending *-ic acid* of the carboxylic acid is usually replaced with **-nitrile.** The chain is numbered like those of carboxylic acids. Similar rules apply to dinitriles derived from dicarboxylic acids. The substituent CN is called **cyano.** Cyanocycloalkanes are called cycloalkane**carbonitriles.** We will retain the common name benzonitrile (from benzoic acid), rather than use the systematic benzenecarbonitrile.

$CH_3C\equiv N$

**Ethanenitrile
(Acetonitrile)**

$CH_3CH_2C\equiv N$
**Propanenitrile
(Propionitrile)**

$$\overset{\overset{\displaystyle CH_3}{|}}{CH_3CHCH_2C}\equiv N$$
3-Methylbutanenitrile

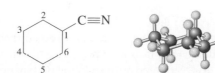

$CH_2C\equiv N$
$|$
$CH_2C\equiv N$

Butanedinitrile (Succinonitrile)

2
3 ⌇ C≡N
   ⌇ 1
4   6
   5

Cyclohexanecarbonitrile

$$N\equiv CCH_2\overset{\overset{\displaystyle O}{\|}}{C}OCH_2CH_3$$
Ethyl cyanoacetate

FIGURE 20-3

(A) Molecular-orbital picture of the nitrile group, showing the *sp* hybridization of both atoms in the C≡N function. (B) Molecular structure of ethanenitrile (acetonitrile), which is similar to that of the corresponding alkyne.

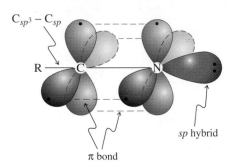

A   B

The C≡N bond in nitriles resembles the C≡C bond in alkynes

In the nitriles, both atoms in the functional group are *sp* hybridized, and there is a lone electron pair on nitrogen occupying an *sp* hybrid orbital pointing away from the molecule along the C–N axis. The hybridization and structure of the nitrile functional group very much resemble those of the alkynes (Figure 20-3; see also Figures 13-1 and 13-2).

In the infrared spectrum, the C≡N stretching vibration appears at about 2250 cm^{-1}, in the same range as the C≡C absorption, but much more intense. The ^{1}H NMR spectra of nitriles indicate that protons near the nitrile group are deshielded about as much as those in other carboxylic acid and alkyne derivatives (Table 20-3).

The ^{13}C NMR absorption for the nitrile carbon appears at lower field ($\delta \sim$ 112–126 ppm) than that of the alkynes ($\delta \sim$ 65–85 ppm), because nitrogen is more electronegative than carbon.

TABLE 20-3 ^{1}H NMR Chemical Shifts of Substituted Methanes CH$_3$X	
X	δ_{CH_3} **(ppm)**
—H	0.23
—Cl	3.06
—OH	3.39
—CHO	2.18
—COOH	2.08
—CONH$_2$	2.02
—C≡N	1.98
—C≡CH	1.80

EXERCISE 20-21

1,3-Dibromopropane was treated with sodium cyanide in dimethyl sulfoxide-d_6 and the mixture monitored by ^{13}C NMR. After a few minutes, four new intermediate peaks appeared, one of which was located well downfield from the others at δ = 117.6 ppm. Subsequently, another three peaks began growing at δ = 119.1, 22.6, and 17.6 ppm, at the expense of the signals of starting material and the intermediate. Explain.

Nitriles undergo hydrolysis to carboxylic acids

As mentioned in Section 19-6, nitriles can be hydrolyzed to give the corresponding carboxylic acids. The conditions are usually stringent, requiring concentrated acid or base at high temperatures.

$$N≡C(CH_2)_4C≡N \xrightarrow{\text{H}^+, \text{H}_2\text{O, 300°C}} HOOC(CH_2)_4COOH$$
$$97\%$$

Hexanedinitrile   **Hexanedioic acid**
(Adiponitrile)   **(Adipic acid)**

The mechanisms of these reactions proceed through the intermediate amide and include addition–elimination steps.

In the acid-catalyzed process, initial protonation on nitrogen facilitates nucleophilic attack by water. Loss of a proton from oxygen furnishes a neutral intermediate, which is a tautomer of an amide. A second protonation on the nitrogen occurs, to be followed again by deprotonation of oxygen and formation of the amide. Hydrolysis of the latter proceeds by the usual addition–elimination pathway.

Mechanism of the Acid-Catalyzed Hydrolysis of Nitriles

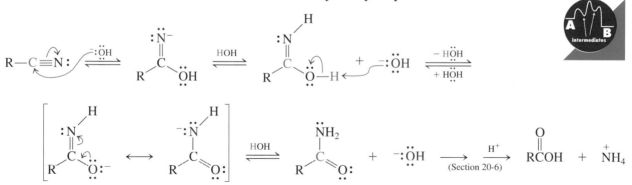

In base-catalyzed nitrile hydrolysis, direct attack of hydroxide gives the anion of the amide tautomer, which protonates on nitrogen. The remaining proton on the oxygen is then removed by base, and a second *N*-protonation gives the amide. Hydrolysis is completed as described in Section 20-6.

Mechanism of the Base-Catalyzed Hydrolysis of Nitriles

Organometallic reagents attack nitriles to give ketones

Strong nucleophiles, such as organometallic reagents, add to nitriles to give anionic imine salts. Work-up with acidic water gives the neutral imine, which is rapidly hydrolyzed to the ketone (Section 17-9).

Ketone Synthesis from Nitriles

CH_3CN
Ethanenitrile
(Acetonitrile)

1. $CH_3(CH_2)_3CH_2MgBr$, THF
2. H^+, H_2O

$CH_3\overset{O}{\overset{\|}{C}}(CH_2)_4CH_3$
44%
2-Heptanone

Reduction of nitriles by hydride reagents leads to aldehydes and amines

As in its reactions with esters and amides, bis(2-methylpropyl)aluminum (diisobutyl-aluminum) hydride (DIBAL) adds to a nitrile only once to give an imine derivative. Aqueous hydrolysis then produces an aldehyde.

Aldehyde Synthesis from Nitriles

$$R-C\equiv N \ + \ R_2'AlH \ \longrightarrow \ R-C\overset{N-AlR_2'}{\underset{H}{\Big\Vert}} \ \xrightarrow{H^+, H_2O} \ \underset{R}{\overset{O}{\Vert}}\underset{}{C}\underset{}{H}$$

85%

Treatment of nitriles with strong hydride reducing agents results in double hydride addition, giving the amine on aqueous work-up. The best reagent for this purpose is lithium aluminum hydride.

$$CH_3CH_2CH_2C\equiv N \ \xrightarrow[\text{2. }H^+, H_2O]{\text{1. LiAlH}_4} \ CH_3CH_2CH_2CH_2NH_2$$

$$\qquad\qquad\qquad\qquad\qquad\qquad\qquad\qquad\qquad\qquad 85\%$$

Butanenitrile   **Butanamine**

EXERCISE 20-22

The reduction of a nitrile by $LiAlH_4$ to give an amine adds four hydrogen atoms to the C–N triple bond: two from the reducing agent and two from the water in the aqueous work-up. Formulate a mechanism for this transformation.

Like the triple bond of alkynes (Section 13-7), the nitrile group is hydrogenated by catalytically activated hydrogen. The result is the same as that of reduction by lithium aluminum hydride—amine formation. All four hydrogens are from the hydrogen gas.

$$CH_3CH_2CH_2C\equiv N \ \xrightarrow{H_2, \ PtO_2, \ CH_3CH_2OH, \ CHCl_3} \ CH_3CH_2CH_2CH_2NH_2$$

$$\qquad\qquad\qquad\qquad\qquad\qquad\qquad\qquad\qquad\qquad 96\%$$

Butanenitrile   **Butanamine**

EXERCISE 20-23

Show how you would prepare the following compounds from pentanenitrile.

(a) $CH_3(CH_2)_3COOH$ (b) $CH_3(CH_2)_3\overset{O}{\overset{\Vert}{C}}(CH_2)_3CH_3$

(c) $CH_3(CH_2)_3\overset{O}{\overset{\Vert}{C}}H$ (d) $CH_3(CH_2)_3CD_2ND_2$

In summary, nitriles are named as alkanenitriles. Both atoms making up the C–N unit are *sp* hybridized, the nitrogen bearing a lone pair in an *sp*-hybrid orbital. The nitrile stretching vibration appears at 2250 cm^{-1}, the ^{13}C NMR absorption at about 120 ppm. Acid- or base-catalyzed hydrolysis of nitriles gives carboxylic acids, and organometallic reagents (RLi, RMgBr) add to give ketones after hydrolysis. With bis(2-methylpropyl)aluminum hydride (diisobutylaluminum hydride, DIBAL), addition and hydrolysis furnishes aldehydes, whereas LiAlH$_4$ or catalytically activated hydrogen converts the nitrile function into the amine.

20-9 Measuring the Molecular Weight of Organic Compounds: Mass Spectrometry

In the various examples and problems dealing with structure determinations of organic compounds, we have so far always been given the molecular formula of the "unknown." How is this information obtained? Elemental analysis (Section 1-9) gives us an *empirical* formula, which tells us the *ratios* of the different elements in a molecule. However, empirical and molecular formulas are not necessarily identical. For example, elemental analysis of benzene merely reveals the presence of carbon and hydrogen atoms in a 1:1 ratio; it does not tell us that the molecule contains six of each. To reach this conclusion, the chemist turns to one last important physical technique used to characterize organic molecules: **mass spectrometry,** which measures molecular weights. This section begins with a description of the apparatus used and the physical principle on which it is based. Subsequently, we will consider a process by which molecules fragment under the conditions required for molecular weight measurement, giving rise to characteristic recorded patterns called **mass spectra.**

The mass spectrometer distinguishes ions by weight

Mass spectrometry is not a form of spectroscopy in the conventional sense, because no radiation is absorbed (Section 10-2). A sample of an organic compound is introduced into an inlet chamber (Figure 20-4, upper right). It is vaporized and a small quantity is allowed to leak into the source chamber of the spectrometer. Here the neutral molecules (M) pass through a beam of high-energy (usually 70 eV, or about 1600 kcal mol^{-1}) electrons. On electron impact, some of the molecules eject an electron to form the corresponding radical cation, M$^{+\cdot}$, called the **parent** or **molecular ion.** Most organic molecules undergo only a single ionization.

Ionization of a Molecule on Electron Impact

$$\text{M} \quad + \quad e\,(70\,\text{eV}) \quad \longrightarrow \quad \text{M}^{+\cdot} \quad + \quad 2\,e$$

Neutral molecule	Ionizing beam	Radical cation (Molecular ion)	

As charged particles, the molecular ions are next accelerated to high velocity by an electric field. (Molecules that were not ionized remain in the source chamber to be pumped away.) The accelerated M$^{+\cdot}$ ions are then subjected to a magnetic field, which deflects them from a linear to a circular path. The curvature of this path is a function of the strength of the magnetic field. Much as in an NMR spectrometer (Section 10-3), the strength of the magnetic field can be varied and therefore adjusted to give the exact degree of curvature to the path of the ions necessary to direct them through the collector slit to the collector, where they are detected and counted.

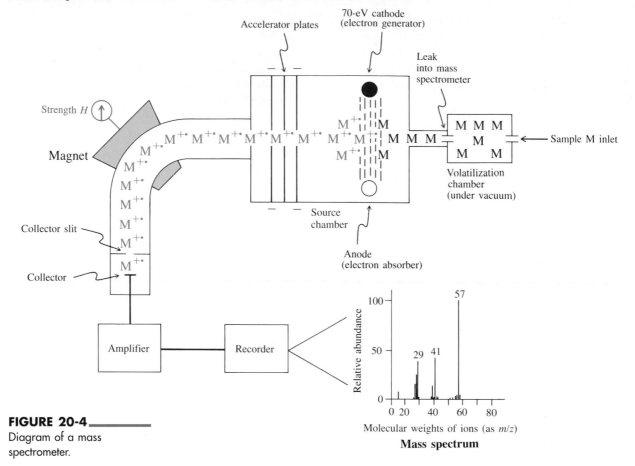

FIGURE 20-4
Diagram of a mass spectrometer.

Molecular Weights of Organic Molecules

CH_4
$m/z = 16$

CH_3OH
$m/z = 32$

$$\underset{m/z = 74}{CH_3\overset{\overset{\displaystyle O}{\|}}{C}OCH_3}$$

Because lighter species are deflected more than heavier ones, the strength of the field necessary to direct the ions through the slit to the collector is a function of the mass of $M^{+\cdot}$ and therefore of the original molecule M. Thus, at a given magnetic field strength, *only ions of a specific mass can pass through the collector slit.* All others will collide with the internal walls of the instrument. Finally, the arrival of ions at the collector is translated electronically into a signal and recorded on a chart. The chart plots mass-to-charge ratio, m/z (on the abscissa), versus peak height (on the ordinate), the latter being a measure of the relative number of ions with a given m/z ratio. Because only singly charged species are normally formed, $z = 1$, and m/z equals the mass of the ion being detected.

EXERCISE 20-24

Three unknown compounds containing only C, H, and O gave rise to the following molecular weights. Draw as many reasonable structures as you can. **(a)** $m/z = 46$; **(b)** $m/z = 30$; **(c)** $m/z = 56$.

Molecular ions undergo fragmentation

Mass spectrometry gives information not only about the molecular ion, but also about its component structural parts. Because the energy of the ionizing beam far exceeds that required to break typical organic bonds, some of the molecular ions break apart

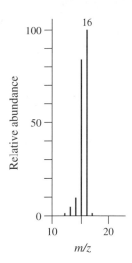

Tabulated Spectrum

m/z	Relative abundance (%)	Molecular or fragment ion
17	1.1	$(M + 1)^{+\cdot}$
16	100.0 (base peak)	$M^{+\cdot}$ (parent ion)
15	85.0	$(M - 1)^{+}$
14	9.2	$(M - 2)^{+\cdot}$
13	3.9	$(M - 3)^{+}$
12	1.0	$(M - 4)^{+\cdot}$

FIGURE 20-5

Mass spectrum of methane. At the left is the spectrum actually recorded; at the right is the tabulated form, the largest peak (base peak) being defined as 100%. For methane, the base peak at $m/z = 16$ is due to the parent ion. Fragmentation gives rise to peaks of lower mass.

into virtually all possible combinations of neutral and ionized fragments. This **fragmentation** gives rise to a number of additional mass-spectral peaks, *all of lower mass* than the molecular ion from which they are derived. The spectrum that results is called the **mass-spectral fragmentation pattern.** The most intense peak in the spectrum is called the **base peak.** Its relative intensity is defined to be 100, and the intensities of all other peaks are described as a percentage of the intensity of the base peak. The base peak in a mass spectrum may be the molecular ion peak or it may be the peak of one of the fragment ions.

For example, the mass spectrum of methane contains, in addition to the molecular ion peak, lines for CH_3^{+}, $CH_2^{+\cdot}$, CH^{+}, and $C^{+\cdot}$ (Figure 20-5). These lines are formed by the processes shown in the margin. The relative abundances of these species, as indicated by the heights of the peaks, give a useful indication of the relative ease of their formation. It can be seen that the first C–H bond is cleaved readily, the $m/z = 15$ peak reaching 85% of the abundance of the molecular ion, which in this case is the base peak. The breaking of additional C–H bonds is increasingly difficult, and the corresponding ions have lower relative abundance. Section 20-10 considers the process of fragmentation in more detail and reveals how fragmentation patterns may be used as an aid to molecular structure determination.

Mass spectra reveal the presence of isotopes

An unusual feature in the mass spectrum of methane is a small (1.1%) peak at $m/z = 17$; it is designated $(M + 1)^{+\cdot}$. How is it possible to have an ion present that has an extra mass unit? The answer lies in the fact that carbon is not isotopically pure. About 1.1% of natural carbon is the ^{13}C isotope (see Table 10-1), giving rise to the additional peak. In the mass spectrum of ethane, the height of the $(M + 1)^{+\cdot}$ peak, at $m/z = 31$, is about 2.2% that of the parent ion. The reason for this finding is statistical. The chance of finding a ^{13}C atom in a compound containing two carbons is double that expected of a one-carbon molecule. For a three-carbon moiety, it would be threefold, and so on. A mass spectrum of the eighteen-carbon steroid estrone (see Section 4-7) is shown in Figure 20-6. The height of the $(M + 1)^{+\cdot}$ peak is 18 times 1.1% of the height of $M^{+\cdot}$. There is a simple rule of thumb for organic compounds that lack isotopically impure heteroatoms: The $(M + 1)^{+\cdot}$ peak has a height (relative to $M^{+\cdot}$) of n times 1.1%, where n is the number of carbon atoms in the molecule.

Fragmentation of Methane in the Mass Spectrometer

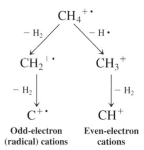

Odd-electron (radical) cations   Even-electron cations

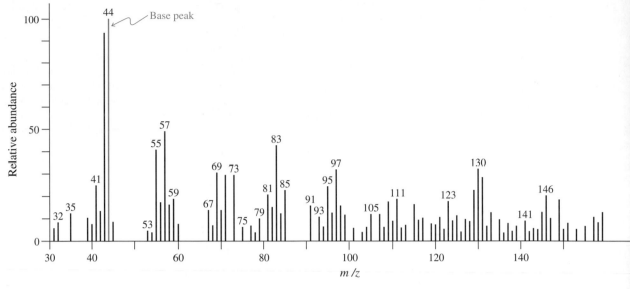

FIGURE 20-6

Mass spectrum (above and on the facing page) of the female sex hormone estrone. The molecule contains 18 carbon atoms and thus the $(M + 1)^{+ \bullet}$ peak height is predicted to be approximately 18 times 1.1% of the intensity of the $M^{+ \bullet}$ peak, a value close to that observed (18 × 1.1% × 62% = 12.3%). Note the extensive and complex fragmentation pattern. The base peak is found at $m/z = 44$.

Other elements, too, have naturally occurring higher isotopes: Hydrogen (deuterium, ^2H, about 0.015% abundance), nitrogen (0.366% ^{15}N), and oxygen (negligible ^{17}O, 0.204% ^{18}O) are examples. These isotopes also contribute to the intensity of peaks at masses higher than $M^{+ \bullet}$, but less so than ^{13}C.

Fluorine and iodine are isotopically pure. However, chlorine (75.53% ^{35}Cl; 24.47% ^{37}Cl) and bromine (50.54% ^{79}Br; 49.46% ^{81}Br) each exist as a mixture of two isotopes and give rise to readily identifiable isotopic patterns. For example, the mass spectrum of 1-bromopropane (Figure 20-7) shows two peaks of nearly equal inten-

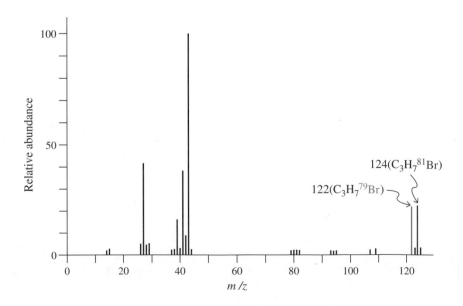

FIGURE 20-7

Mass spectrum of 1-bromopropane. Note the nearly equal heights of the peaks at $m/z = 122$ and 124, owing to the almost equal abundances of the two bromine isotopes.

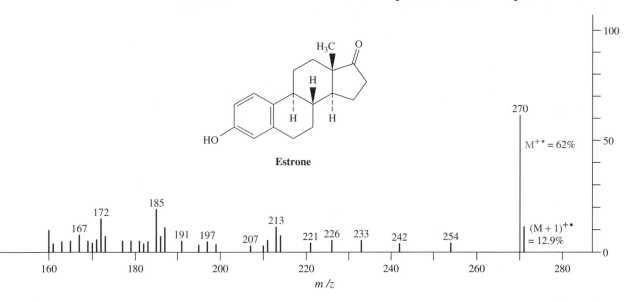

Estrone

$M^{+\bullet} = 62\%$

$(M + 1)^{+\bullet} = 12.9\%$

m/*z*

sity at *m/z* = 122 and 124, and only a very small peak at 123, which is the sum of the average atomic weights of the elements in the periodic table. Why? The true isotopic composition of the molecule is a nearly $1:1$ mixture of $CH_3CH_2CH_2{}^{79}Br$ and $CH_3CH_2CH_2{}^{81}Br$. Similarly, the spectra of monochloroalkanes exhibit ions two mass units apart in a $3:1$ intensity ratio, because of the presence of about 75% $R^{35}Cl$ and 25% $R^{37}Cl$. Peak patterns such as these are useful in revealing the presence of chlorine or bromine.

EXERCISE 20-25

What peak pattern do you expect for the molecular ion of dibromomethane?

EXERCISE 20-26

Nonradical compounds containing C, H, and O have even molecular weights, those containing C, H, O, and an odd number of N atoms have odd molecular weights, but those with an even number of N atoms are even again. Explain.

In summary, molecules can be ionized by an electron beam at 70 eV to give radical cations that are accelerated by an electric field and then separated by the different deflections that they undergo in a magnetic field. In a mass spectrometer, this effect is used to measure the molecular weights of molecules. The molecular ion is usually accompanied by less massive fragments and isotopic "satellites" owing to the presence of less abundant isotopes. In some cases, such as with Cl and Br, more than one isotope may be present in substantial quantities.

20-10 Fragmentation Patterns of Organic Molecules

On electron impact, molecules dissociate at weaker bonds first and then at the stronger ones. The resulting fragments may themselves fall apart into smaller pieces. The detection of this fragmentation pattern by the mass spectrometer provides clues to the structure of the parent molecule.

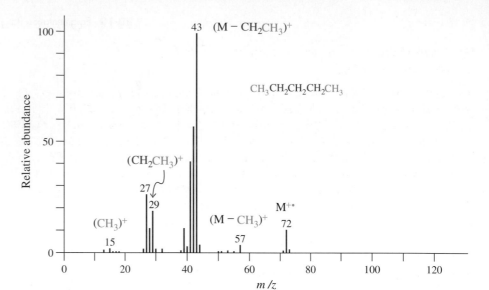

FIGURE 20-8

Mass spectrum of pentane, revealing that all C–C bonds in the chain have been ruptured.

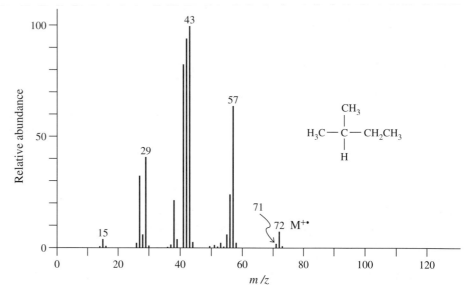

FIGURE 20-9

Mass spectrum of 2-methylbutane. The peaks at $m/z = 43$ and 57 result from preferred fragmentation around C2 to give secondary carbocations.

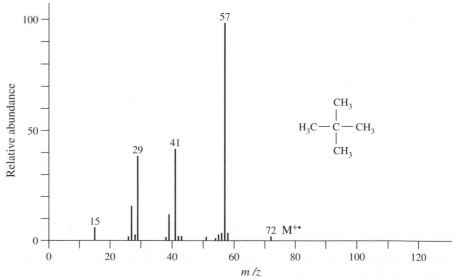

FIGURE 20-10

Mass spectrum of 2,2-dimethylpropane. Only a very weak molecular ion peak is seen, because the fragmentation to give a tertiary cation is favored.

Fragmentation is more likely at a highly substituted center

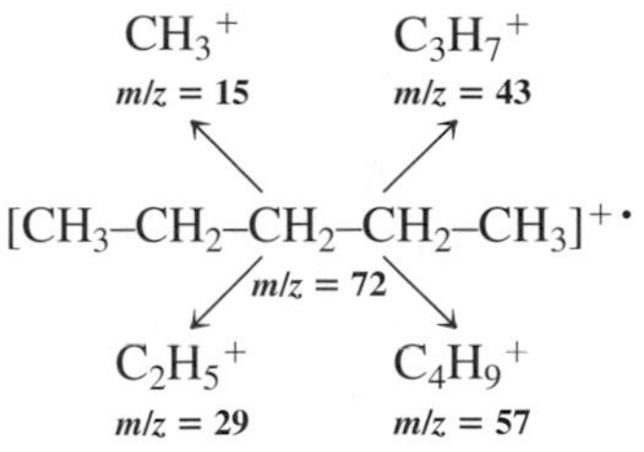

The relative ease of bond dissociation can be seen in the mass spectra of the isomeric hydrocarbons pentane, 2-methylbutane, and 2,2-dimethylpropane (Figures 20-8, 20-9, and 20-10). In each case, the molecular ion ($m/z = 72$) produces a relatively small peak, but the spectra are otherwise very different for the three compounds. Pentane fragments by more or less indiscriminate C—C bond breaking (Figure 20-8; see margin). Thus, in addition to the molecular ion signal, there is a line at $m/z = 57$ (M − CH$_3$)$^+$ followed by peaks indicating progressive loss of CH$_2$ units: $m/z = 43$ (M − CH$_2$CH$_3$)$^+$, and $m/z = 29$ (CH$_3$CH$_2$)$^+$. These fragment peaks are surrounded by clusters of smaller lines because of the presence of ^{13}C (M + 1)$^{+\cdot}$ and the loss of hydrogens [(M − 1)$^+$, (M − 2)$^+$, etc.].

The mass spectrum of 2-methylbutane (Figure 20-9) shows a pattern similar to that of pentane; however, the relative intensities of the various peaks differ. Thus, there is a larger (M − 1)$^+$ peak at $m/z = 71$, and strong (M − alkyl)$^+$ signals at $m/z = 57$ and 43, because of the relative stability of the cations produced by preferred fragmentation at the more highly substituted center at C2.

Preferred Fragmentation of 2-Methylbutane

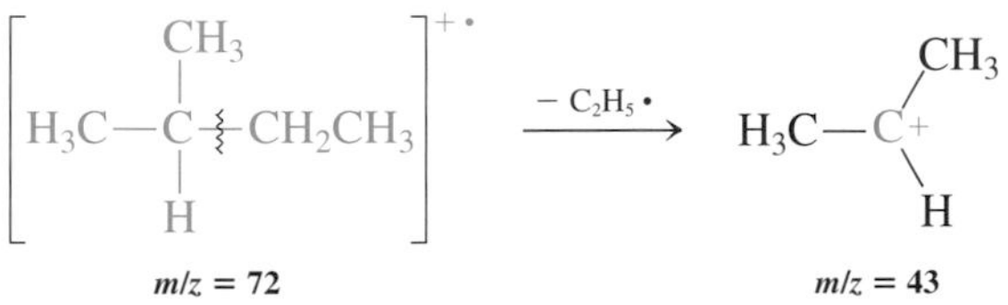

The preference for fragmentation at a highly substituted center is even more pronounced in the mass spectrum of 2,2-dimethylpropane (Figure 20-10). Here, loss of a methyl radical from the molecular ion produces the 1,1-dimethylethyl (*tert*-butyl) cation as the base peak at $m/z = 57$. This fragmentation is so easy that the molecular ion is barely visible. The spectrum also reveals peaks at $m/z = 41$ and 29, the result of complex structural reorganizations, such as the carbocation rearrangements considered in Section 9-3.

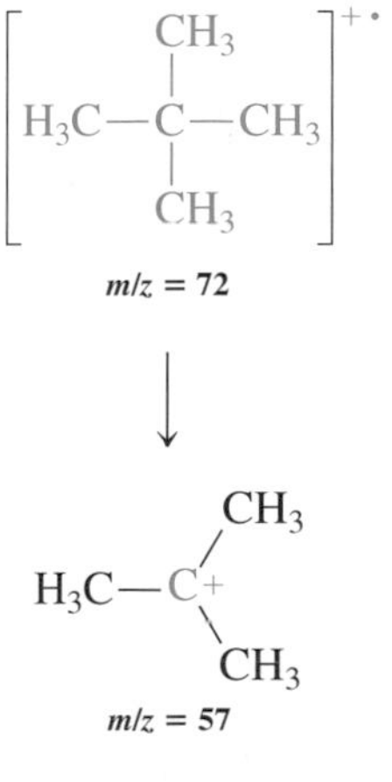

Fragmentations also help to identify functional groups

Particularly easy fragmentation of relatively weak bonds is also seen in the mass spectra of the haloalkanes. The fragment ion (M − X)$^+$ is frequently the base peak in these spectra. A similar phenomenon is observed in the mass spectra of alcohols, which eliminate water to give a large (M − H$_2$O)$^{+\cdot}$ peak 18 mass units below the parent ion (Figure 20-11). The bonds to the C–OH group also readily dissociate in a process called **α cleavage,** leading to resonance-stabilized hydroxycarbocations:

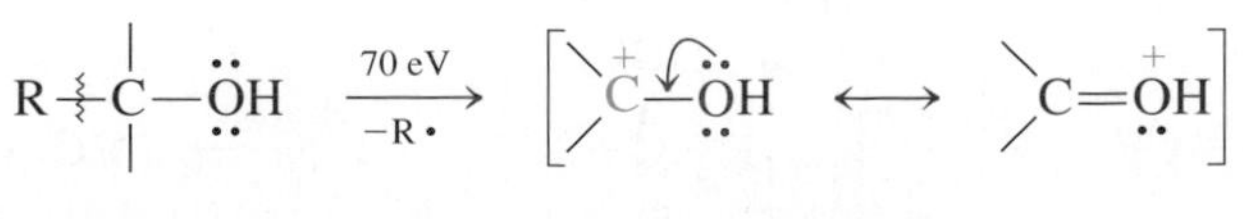

The strong peak at $m/z = 31$ in the mass spectrum of 1-butanol is due to the hydroxymethyl cation, $^+$CH$_2$OH, which arises from α cleavage.

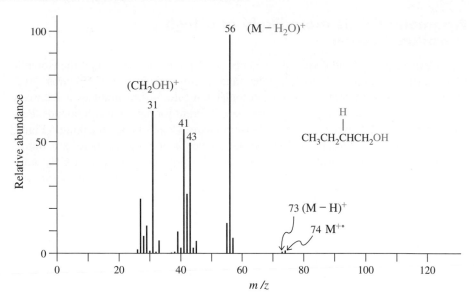

FIGURE 20-11
Mass spectrum of 1-butanol. The parent ion, at $m/z = 74$, gives rise to a small peak because of ready loss of water to give the ion at $m/z = 56$. Other fragment ions due to α cleavage are propyl ($m/z = 43$), 2-propenyl (allyl) ($m/z = 41$), and hydroxymethyl ($m/z = 31$).

Alcohol Fragmentation by Dehydration and α Cleavage

$$\left[\begin{array}{c} \text{HO} \quad \text{H} \\ | \quad \quad | \\ \text{R}-\text{C}-\text{CHR}' \\ | \\ \text{H} \end{array}\right]^{+\cdot} \longrightarrow [\text{RCH}=\text{CHR}']^{+\cdot} + \text{H}_2\text{O}$$

$$\textbf{M}^{+\cdot} \qquad\qquad\qquad (\textbf{M} - \textbf{18})^{+\cdot}$$

$$-\,\text{R}\cdot \swarrow \qquad -\,\text{H}\cdot \downarrow \qquad \searrow\, -\,\text{R}'\text{CH}_2\cdot$$

$$\left[\begin{array}{c} \text{HO} \\ | \\ \text{C} \\ \diagup \quad \diagdown \\ \text{H} \quad\quad \text{CH}_2\text{R}' \end{array}\right]^{+} \quad \left[\begin{array}{c} \text{HO} \\ | \\ \text{C} \\ \diagup \quad \diagdown \\ \text{R} \quad\quad \text{CH}_2\text{R}' \end{array}\right]^{+} \quad \left[\begin{array}{c} \text{HO} \\ | \\ \text{C} \\ \diagup \quad \diagdown \\ \text{R} \quad\quad \text{H} \end{array}\right]^{+}$$

EXERCISE 20-27

Try to predict the appearance of the mass spectrum of 3-methyl-3-heptanol.

The fragmentation patterns of carbonyl compounds are often useful in structural identifications. For example, the mass spectra of the isomeric ketones 2-pentanone, 3-pentanone, and 3-methyl-2-butanone (Figure 20-12) reveal very clean and distinct fragment ions. The predominant decomposition pathway is α cleavage, which severs an alkyl bond to the carbonyl function to give the corresponding **acylium cation** and an alkyl radical.

α Cleavage of Carbonyl Compounds

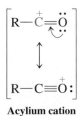

$$\begin{bmatrix} \text{R}-\overset{+}{\text{C}}=\overset{\cdot\cdot}{\text{O}} \\ \text{\Large\textasciicircum}\;\cdot\cdot \\ \updownarrow \\ \text{R}-\text{C}\equiv\overset{+}{\text{O}}: \end{bmatrix}$$

Acylium cation

$$:\!\overset{+}{\text{O}}\!\equiv\text{CR}' \quad\xleftarrow[-\,\text{R}\cdot]{\alpha\text{ cleavage}}\quad \left[\text{R}\!-\!\!\!\left.\begin{array}{c} :\text{O}: \\ \| \\ \text{C} \end{array}\right.\!\!\!-\!\text{R}'\right]^{+\cdot} \quad\xrightarrow[-\,\text{R}'\cdot]{\alpha\text{ cleavage}}\quad \text{RC}\equiv\overset{+}{\text{O}}:$$

The acylium cation forms easily because of resonance stabilization. These fragment ions allow the gross composition of the two alkyl groups in a ketone to be read

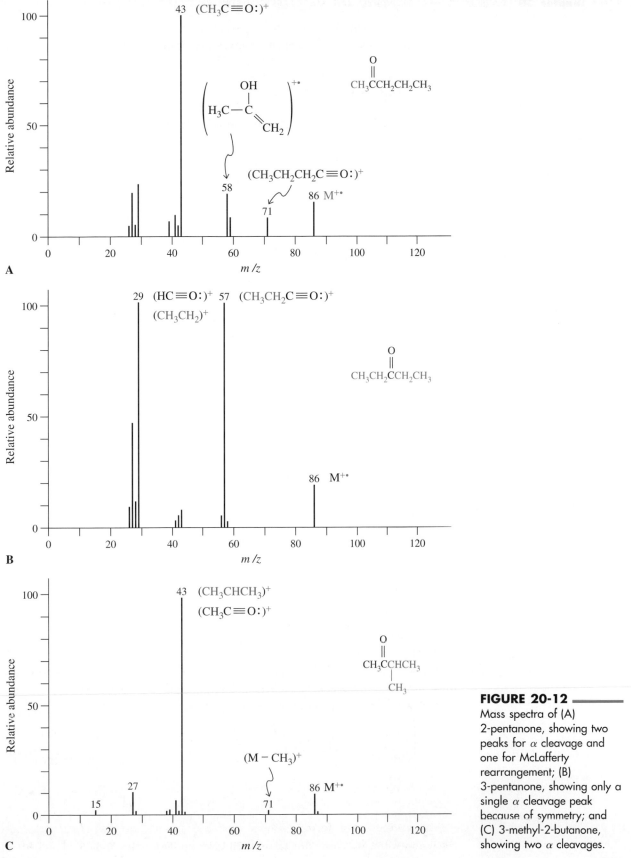

FIGURE 20-12
Mass spectra of (A) 2-pentanone, showing two peaks for α cleavage and one for McLafferty rearrangement; (B) 3-pentanone, showing only a single α cleavage peak because of symmetry; and (C) 3-methyl-2-butanone, showing two α cleavages.

from the spectrum. In this way, 2-pentanone is readily differentiated from 3-pentanone: α cleavage of 2-pentanone gives two acylium ions, at $m/z = 43$ and 71, but 3-pentanone gives only one, at $m/z = 57$. (The $m/z = 29$ peak in the mass spectrum of the latter is due partly to $CH_3CH_2^+$ and partly to $HC\equiv O^+$, which arises by loss of C_2H_4 from $CH_3CH_2C\equiv O^+$.)

α Cleavage in 2-Pentanone

$$:\overset{+}{O}\equiv CCH_2CH_2CH_3 \quad \longleftarrow \quad H_3C\overset{\displaystyle :O:}{\underset{}{\overset{\|}{C}}}\!\!\!-\!\!\!\overset{}{C}\!\!\!-\!\!\!CH_2CH_2CH_3 \quad \longrightarrow \quad CH_3C\equiv\overset{+}{O}:$$

$m/z = 71$ $m/z = 86$ $m/z = 43$

2-Pentanone

α Cleavage in 3-Pentanone

$$CH_3CH_2\!\!-\!\!\overset{\displaystyle :O:}{\overset{\|}{C}}\!\!-\!\!CH_2CH_3 \quad \longrightarrow \quad CH_3CH_2C\equiv\overset{+}{O}:$$

$m/z = 86$ $m/z = 57$

3-Pentanone

Can 2-pentanone be distinguished from 3-methyl-2-butanone? Not by the observation of α cleavage—in both molecules, the substituent groups are CH_3 and C_3H_7. However, comparison of the mass spectra of the two compounds (Figure 20-12A and C) reveals an additional prominent peak for 2-pentanone at $m/z = 58$, signifying the loss of a molecular fragment of weight $m/z = 28$. This fragment is absent from the spectra of both other isomers and is characteristic of the presence of hydrogens located gamma to the carbonyl group. Compounds with this structural feature and with sufficient flexibility to allow the γ-hydrogen to be close to the carbonyl oxygen decompose by the **McLafferty* rearrangement.** In this reaction, the molecular ion of the starting ketone splits into two pieces (a neutral fragment and a radical cation) in a unimolecular process.

McLafferty Rearrangement

$$\left[\begin{array}{c} \overset{\gamma}{RHC}\diagdown^{\displaystyle H} \\ \overset{|}{\underset{\beta}{}}\!H_2C \diagdown \quad O \\ \quad\quad \diagup\!\!\! \diagdown \\ \quad\quad\quad \underset{\alpha}{CH_2}\quad C\diagdown R' \end{array} \right]^{+\,\bullet} \longrightarrow \quad \overset{RCH}{\underset{CH_2}{\overset{\|}{}}} \quad + \quad \left[\begin{array}{c} \quad OH \\ \quad | \\ \quad C \\ H_2C\diagup \diagdown R' \end{array} \right]^{+\,\bullet}$$

The McLafferty rearrangement yields an alkene and the enol form of a new ketone. In the case of 2-pentanone, ethene and the enol of 2-propanone (acetone) are produced; the radical cation of the latter is observed with $m/z = 58$.

$$\left[\begin{array}{c} O \quad\quad\quad H \\ \| \quad\quad\quad | \\ CH_3\overset{}{C}CH_2CH_2CH_2 \\ \quad\alpha \quad\; \beta \quad\; \gamma \end{array} \right]^{+\,\bullet} \longrightarrow \left[\begin{array}{c} \quad\quad OH \\ \quad\quad | \\ H_3C\!\!-\!\!C \\ \quad\quad \diagdown \\ \quad\quad\; CH_2 \end{array} \right]^{+\,\bullet} + \quad CH_2\!\!=\!\!CH_2$$

$m/z = 86$ $m/z = 58$

---

*Professor Fred W. McLafferty (b. 1923), Cornell University, Ithaca, New York.

The Contribution of Mass Spectrometry to the Characterization of Manoalide

Completion of the structural elucidation of manoalide (Chemical Highlights 10-3 and 20-1) required that information from several sources, including mass spectrometry, be reconciled. The parent ion in the mass spectrum was too weak for an exact mass determination with the instruments of the late 1970s (such limitations no longer exist). However, a strong peak at M − 18, corresponding to loss of water, was found to have the exact $m/z = 398.2459$. The composition fitting this mass is $C_{25}H_{34}O_4$ ($m/z = 398.2457$); extrapolation gives a formula of $C_{25}H_{36}O_5$ for manoalide itself. The base peak at $m/z = 137$ was identified by analogy with similar compounds as the $C_{10}H_{17}$ fragment arising from cleavage of an allylic C–C bond.

$C_{10}H_{17}{}^+$

NMR spectroscopic and mass spectrometric information revealed the presence of an acyclic C_5H_8 (isoprene) unit (see Section 14-10). The remaining fragment was identified as follows. Selective protection of one OH group followed by PCC oxidation of the other (Section 8-6) gave a product with new bands in the IR spectrum at $1725 \ cm^{-1}$ and in the UV spectrum with $\lambda_{max} = 311$ nm. Only an α,β-unsaturated, six-membered lactone is consistent with these data, establishing the final piece of the puzzle as an unsaturated cyclic hemiacetal (see below).

The potential therapeutic value of manoalide arises from a rare mode of biological activity: It is one of only a tiny number of nonsteroidal substances that inhibits the release of arachidonic acid (Section 19-13). Manoalide inactivates the enzyme that cleaves arachidonic acid from the phospholipid ester (Section 20-5) by which it is transported through cell membranes. Recall that arachidonic acid is the biochemical precursor to both the prostaglandins, which contribute to inflammation, and the leukotrienes, the overproduction of which is associated with uncontrolled cell division. When these substances are present in excess in the outer layers of the skin, they give rise to abnormalities such as the lesions due to psoriasis. Other nonsteroidal anti-inflammatory drugs (NSAIDs), such as aspirin, are effective as prostaglandin but not leukotriene inhibitors. By blocking the release of arachidonic acid, manoalide impedes both prostaglandin *and* leukotriene biosynthesis, without the undesirable side effects of the usual steroid treatments.

1. $(CH_3CO)_2O$, pyridine
2. PCC, CH_2Cl_2

IR 1725 cm^{-1}
UV λ_{max} 211 nm

Neither 3-pentanone nor 3-methyl-2-butanone possesses a γ-hydrogen; therefore, neither is able to undergo McLafferty rearrangement.

EXERCISE 20-28

How would you tell the difference between **(a)** 3-methyl-2-pentanone and 4-methyl-2-pentanone, and **(b)** 2-ethylcyclohexanone and 3-ethylcyclohexanone, using only mass spectrometry?

Similar rearrangements and α cleavages can be seen in the mass spectra of aldehydes and carboxylic acid derivatives.

EXERCISE 20-29

Interpret the labeled peaks in the mass spectra of pentanal, pentanoic acid, and methyl pentanoate shown in Figure 20-13.

In summary, fragmentation patterns can be interpreted for structural elucidation. For example, the radical cations of alkanes cleave to form the most stable positively charged fragments, haloalkanes fragment by rupture of the carbon–halogen bond, alcohols readily dehydrate and undergo α cleavage, and carbonyl compounds decompose by both α cleavage and McLafferty rearrangement.

20-11 High-Resolution Mass Spectrometry

Consider substances with the following molecular formulas: C_7H_{14}, $C_6H_{10}O$, $C_5H_6O_2$, and $C_5H_{10}N_2$. All possess the same **integral mass;** that is, to the nearest integer, all four would be expected to exhibit a parent ion at $m/z = 98$. However, the atomic weights of the elements are composites of the masses of their naturally occurring isotopes, *which are not integers.* Thus, if we use the atomic masses for the most abundant isotopes of C, H, O, and N (Table 20-4) to calculate the **exact mass** corresponding to each of the aforementioned molecular formulas, we see significant differences.

TABLE 20-4 Exact Masses of Several Common Isotopes	
Isotope	**Mass**
^1H	1.00783
^{12}C	12.00000
^{14}N	14.0031
^{16}O	15.9949
^{32}S	31.9721
^{35}Cl	34.9689
^{37}Cl	36.9659
^{79}Br	78.9183
^{81}Br	80.9163

Exact Masses of Four Compounds with $m/z = 98$

C_7H_{14}	$C_6H_{10}O$	$C_5H_6O_2$	$C_5H_{10}N_2$
98.1096	98.0732	98.0368	98.0845

Can we use mass spectrometry to differentiate between these species? Yes. Modern **high-resolution mass spectrometers** are capable of distinguishing between ions that differ in mass by as little as a few thousandths of a mass unit. We can therefore measure the exact mass of any parent or fragment ion. By comparing this experimentally determined value with that calculated for each species possessing the same integral mass, computers can assign a molecular formula to the unknown ion. High-resolution mass spectrometry is now the most widely used method for determining the molecular formulas of unknowns.

EXERCISE 20-30

Choose the molecular formula that matches the exact mass. **(a)** $m/z = 112.0888$, C_8H_{16}, $C_7H_{12}O$, or $C_6H_8O_2$; **(b)** $m/z = 86.1096$, C_6H_{14}, $C_4H_6O_2$, or $C_4H_{10}N_2$.

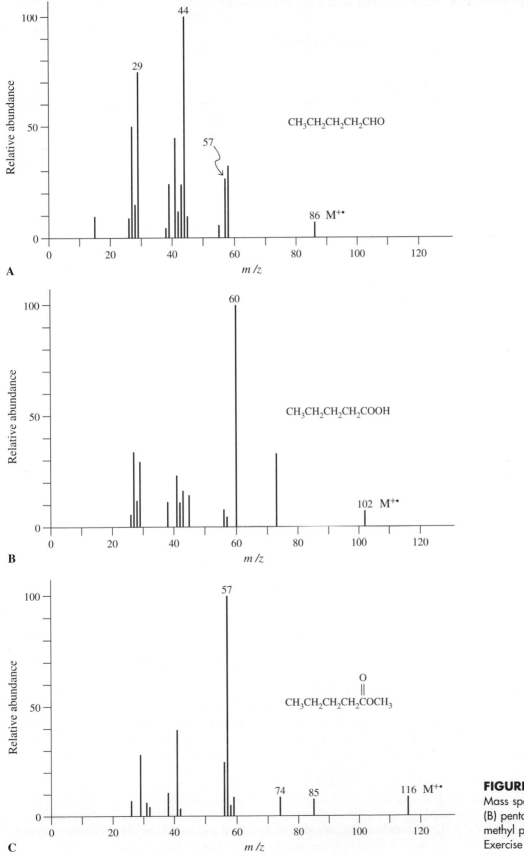

FIGURE 20-13

Mass spectra of (A) pentanal; (B) pentanoic acid; and (C) methyl pentanoate. (See Exercise 20-29.)

Clovene

CHAPTER INTEGRATION PROBLEM

Clovene is a sesquiterpene that arises from acid-catalyzed rearrangement of caryophyllene (Problem 55 of Chapter 12—odor of cloves). The following sequence of transformations constitutes a part of a total synthesis of clovene, but the reagents have been omitted. Supply reasonable reagents and reaction conditions for each transformation, some of which require more than one step. In some cases, you may need to refer to earlier chapters, especially Chapters 17 through 19, for information.

SOLUTION

How do we analyze this sort of synthesis problem? We can begin by characterizing each transformation: Identify exactly what has changed. When we have this information, we can refer to the reactions that we know and determine whether more than one step is needed or whether one step is sufficient to carry out the process. Let us proceed in the order shown above.

a. A molecule containing methoxycarbonyl and hydroxy groups converts into a cyclic ester—a lactone. A change from one ester into another would be transesterification (Section 20-4), a reversible process catalyzed equally well by acid or base. How do we shift the equilibrium in the desired direction? Note that methanol is the by-product of the process; it has a low boiling point (65°C, Table 8-1) and may be driven off by heating the reaction mixture. Therefore, (a) is catalytic H^+.

b. The lactone converts into a diol. One could effect this change in two steps: Hydrolysis would cleave the lactone into a hydroxycarboxylic acid (Section 20-4); subsequent reduction by $LiAlH_4$ would furnish the diol (Section 19-11). Much easier is $LiAlH_4$ reduction of the lactone, which would lead to the diol directly. Notice (Section 20-4) that, in the course of these reduction processes alkoxides are the initial products, requiring acidic work-up. Therefore, (b) is 1. $LiAlH_4$, ether (a typical solvent for such reductions) and 2. H^+, H_2O.

c. This step requires some thought. There are two hydroxy groups to start with, and we want to oxidize only one. The usual methods based on, for example, Cr(VI) will not distinguish between them and cannot be used. However, one of the two –OH groups (the one on the ring) is *allylic,* and (Sections 17-4 and

18-9) can be oxidized *selectively* to an α,β-unsaturated ketone by MnO_2, leaving the other unchanged. So, (c) is MnO_2, propanone (acetone).

d. Now we are back to more straightforward chemistry—oxidation of a primary alcohol to a carboxylic acid. Any Cr(VI) reagent in water will do, such as $K_2Cr_2O_7$ in aqueous acid (Sections 8-6 and 19-6). Thus (d) is $K_2Cr_2O_7$, H_2SO_4, H_2O.

e. Two functional group changes are evident: The enone C–C double bond has been reduced and the carboxy group esterified. We have seen reductions of enones to saturated ketones (Section 18-9); catalytic hydrogenation is the simplest applicable method (a metal–NH_3 reduction is unnecessary, because selectivity is not relevant). Methyl ester formation may be achieved in one of several ways (Sections 19-8, 19-9, and 20-2): via the alkanoyl halide (made with $SOCl_2$) and CH_3OH, by direct acid-catalyzed reaction with CH_3OH, or by reaction with diazomethane, CH_2N_2. The last method is easiest on a small scale. Does the order of steps matter? Neither acids nor esters are affected by catalytic hydrogenation conditions, and esterifications may be carried out in the presence of either enones or ordinary ketones, so the answer is no. For (e), then, 1. H_2, Pd–C, CH_3CH_2OH and 2. CH_2N_2, ether.

f. Here is a curious pair of changes: The carbonyl moiety is protected as a cyclic acetal, and the ester function is hydrolyzed back to the –COOH group, which raises the question of why the latter was esterified in the first place. Most likely, the chemists working on this project tried to form the cyclic acetal on the molecule containing the free carboxy substituent and ran into difficulty. Protection of the carbonyl requires acid catalysis and 1,2-ethanediol (ethylene glycol, Section 17-8). Perhaps complications due to the formation of esters with the –COOH group were observed. In any event, esterification of the latter before acetalization apparently solved the problem. Thus, for (f), 1. $HOCH_2CH_2OH$, H^+ and 2. ^-OH, H_2O (ester hydrolysis *with base,* Section 20-4, to prevent hydrolysis of the acetal unit).

g. The ring carbonyl group has been deprotected, and the carboxy group transformed into a ketone. For the first time in the synthesis, a new carbon–carbon bond has been introduced, between the carboxy carbon and an ethyl group. This process will most likely require the use of an organometallic reagent, from which the ring carbonyl has been protected in sequence (f). From which carboxylic acid derivatives can ketones be prepared? Alkanoyl halides (Section 20-2) and nitriles (Section 20-8). Since you don't know how to convert a carboxylic acid into its nitrile, we choose the following sequence for (g): 1. $SOCl_2$ (gives halide), 2. $(CH_3CH_2)_2CuLi$, ether (converts halo into ethylcarbonyl), and 3. H^+, H_2O (deprotects ring carbonyl, Section 17-8).

h. Another carbon–carbon bond is made, forming a ring. We trace the pattern of carbon atoms before and after this step to identify which ones were connected (see margin). They are the ring carbonyl carbon and the CH_2 of the ethyl group, α to the side-chain carbonyl. We have here an intramolecular aldol condensation (Section 18-7), readily achieved by aqueous base. So (h), the final step, is ^-OH, H_2O.

Analysis of Step (h) Aldol Condensation

NEW REACTIONS

1. Order of Reactivity of Carboxylic Acid Derivatives (Section 20-1)

Esters and amides require acid or base catalysts to react with weak nucleophiles.

2. Basicity of the Carbonyl Oxygen (Section 20-1)

$$L = \text{leaving group}$$

Basicity increases with increasing contribution of resonance structure C

3. Enolate Formation (Section 20-1)

Acidity of the neutral derivative generally increases with decreasing contribution of resonance structure C in the anion

Reactions of Alkanoyl Halides

4. Water (Section 20-2)

$$\underset{\text{RCX}}{\overset{O}{\|}} + H_2O \longrightarrow \underset{\substack{\text{RCOH} \\ \text{Carboxylic} \\ \text{acid}}}{\overset{O}{\|}} + HX$$

5. Carboxylic Acids (Sections 19-8 and 20-2)

$$\underset{\text{RCX}}{\overset{O}{\|}} + R'CO_2H \longrightarrow \underset{\substack{\text{RCOCR'} \\ \text{Carboxylic} \\ \text{anhydride}}}{\overset{O\ \ O}{\|\ \ \|}} + HX$$

6. Alcohols (Section 20-2)

$$\underset{\text{RCX}}{\overset{O}{\|}} + R'OH \longrightarrow \underset{\substack{\text{RCOR'} \\ \text{Ester}}}{\overset{O}{\|}} + \underset{\substack{\text{(Removed with pyridine,} \\ \text{triethylamine, or other base)}}}{HX}$$

7. Amines (Section 20-2)

$$\underset{\text{RCX}}{\overset{O}{\|}} + R'NH_2 \longrightarrow \underset{\substack{\text{RCNHR'} \\ \text{Amide}}}{\overset{O}{\|}} + \underset{\substack{\text{(Removed with pyridine,} \\ \text{triethylamine, excess R'NH}_2, \\ \text{or other base)}}}{HX}$$

8. Cuprate Reagents (Section 20-2)

$$\underset{\text{RCX}}{\overset{O}{\|}} \xrightarrow[\text{2. H}^+, \text{H}_2\text{O}]{\text{1. R}_2'\text{CuLi, THF}} \underset{\substack{\text{RCR'} \\ \text{Ketone}}}{\overset{O}{\|}} + R'Cu + LiX$$

9. Hydrides (Section 20-2)

$$\underset{RCX}{\overset{O}{\parallel}} \xrightarrow[\text{2. H}^+\text{, H}_2\text{O}]{\text{1. LiAl[OC(CH}_3)_3]_3\text{H, (CH}_3\text{CH}_2)_2\text{O}} \underset{\text{Aldehyde}}{\overset{O}{\underset{RCH}{\parallel}}} + \text{LiX} + \text{Al[OC(CH}_3)_3]_3$$

Reactions of Carboxylic Acid Anhydrides

10. Water (Section 20-3)

$$\underset{RCOCR}{\overset{O\ \ O}{\parallel\ \ \parallel}} + \text{H}_2\text{O} \longrightarrow \underset{\text{Carboxylic acid}}{2\ \overset{O}{\underset{RCOH}{\parallel}}}$$

11. Alcohols (Section 20-3)

$$\underset{RCOCR}{\overset{O\ \ O}{\parallel\ \ \parallel}} + \text{R}'\text{OH} \longrightarrow \underset{\text{Ester}}{\overset{O}{\underset{RCOR'}{\parallel}}} + \overset{O}{\underset{RCOH}{\parallel}}$$

12. Amines (Section 20-3)

$$\underset{RCOCR}{\overset{O\ \ O}{\parallel\ \ \parallel}} + \text{R}'\text{NH}_2 \longrightarrow \underset{\text{Amide}}{\overset{O}{\underset{RCNHR'}{\parallel}}} + \overset{O}{\underset{RCOH}{\parallel}}$$

Reactions of Esters

13. Water (Ester Hydrolysis) (Sections 19-9 and 20-4)

Acid catalysis

$$\underset{RCOR'}{\overset{O}{\parallel}} + \text{H}_2\text{O} \xrightarrow{\text{Catalytic H}^+} \underset{\text{Carboxylic acid}}{\overset{O}{\underset{RCOH}{\parallel}}} + \text{R}'\text{OH}$$

Base catalysis

$$\underset{RCOR'}{\overset{O}{\parallel}} + \underset{\text{1 equivalent}}{\ ^-\text{OH}} \xrightarrow{\text{H}_2\text{O}} \underset{\text{Carboxylate ion}}{\overset{O}{\underset{RCO^-}{\parallel}}} + \text{R}'\text{OH}$$

14. Alcohols (Transesterification) and Amines (Section 20-4)

$$\underset{RCOR'}{\overset{O}{\parallel}} + \text{R}''\text{OH} \xrightarrow{\text{H}^+\text{ or }^-\text{OR}''} \underset{\text{Ester}}{\overset{O}{\underset{RCOR''}{\parallel}}} + \text{R}'\text{OH}$$

$$\underset{RCOR'}{\overset{O}{\parallel}} + \text{R}''\text{NH}_2 \xrightarrow{\text{Heat}} \underset{\text{Amide}}{\overset{O}{\underset{RCNHR''}{\parallel}}} + \text{R}'\text{OH}$$

15. Organometallic Reagents (Section 20-4)

$$\underset{\text{RCOR''}}{\overset{\text{O}}{\|}} \xrightarrow[\text{2. H}^+,\text{H}_2\text{O}]{\text{1. 2 R'MgX, (CH}_3\text{CH}_2)_2\text{O}} \underset{\underset{\text{R'}}{|}}{\overset{\text{OH}}{|}} \underset{\text{Tertiary alcohol}}{R-\overset{|}{\underset{|}{C}}-R'} \quad + \quad \text{R''OH}$$

Methyl formate

$$\underset{\text{HCOCH}_3}{\overset{\text{O}}{\|}} \xrightarrow[\text{2. H}^+,\text{H}_2\text{O}]{\text{1. 2 R'MgX, (CH}_3\text{CH}_2)_2\text{O}} \underset{\underset{\text{R'}}{|}}{\overset{\text{OH}}{|}} \underset{\text{Secondary alcohol}}{H-\overset{|}{\underset{|}{C}}-R'} \quad + \quad \text{CH}_3\text{OH}$$

16. Hydrides (Section 20-4)

$$\underset{\text{RCOR'}}{\overset{\text{O}}{\|}} \xrightarrow[\text{2. H}^+,\text{H}_2\text{O}]{\text{1. LiAlH}_4, (\text{CH}_3\text{CH}_2)_2\text{O}} \text{RCH}_2\text{OH}$$

$$\underset{\text{RCOR'}}{\overset{\text{O}}{\|}} \xrightarrow[\text{2. H}^+,\text{H}_2\text{O}]{\text{1. (CH}_3\overset{\overset{\text{CH}_3}{|}}{\text{CHCH}_2)_2}\text{AlH, toluene, } -60°\text{C}} \underset{\text{RCH}}{\overset{\text{O}}{\|}}$$

17. Enolates (Section 20-4)

$$\underset{\text{RCH}_2\text{COR'}}{\overset{\text{O}}{\|}} \xrightarrow{\text{LDA, THF}} \left[\underset{\overset{-}{\text{RCH}}-\text{COR'}}{\overset{:\ddot{\text{O}}:}{\|}} \longleftrightarrow \underset{\text{RCH}=\text{COR'}}{\overset{^-:\ddot{\text{O}}:}{|}} \right] \xrightarrow{\text{R''X}} \underset{\text{RCHCOR'}}{\overset{\text{R'' O}}{\underset{|}{\|}}}$$

Ester enolate ion

Reactions of Amides

18. Water (Section 20-6)

$$\underset{\text{RCNHR'}}{\overset{\text{O}}{\|}} + \text{H}_2\text{O} \xrightarrow{\text{H}^+,\Delta} \underset{\underset{\substack{\text{Carboxylic}\\\text{acid}}}{\text{RCOH}}}{\overset{\text{O}}{\|}} + \text{R'}\overset{+}{\text{N}}\text{H}_3$$

$$\underset{\text{RCNHR'}}{\overset{\text{O}}{\|}} + \text{H}_2\text{O} \xrightarrow{\text{HO}^-,\Delta} \underset{\text{RCO}^-}{\overset{\text{O}}{\|}} + \text{R'NH}_2$$

19. Hydrides (Section 20-6)

$$\underset{\text{RCNHR'}}{\overset{\text{O}}{\|}} \xrightarrow[\text{2. H}^+,\text{H}_2\text{O}]{\text{1. LiAlH}_4, (\text{CH}_3\text{CH}_2)_2\text{O}} \underset{\text{Amine}}{\text{RCH}_2\text{NHR'}}$$

$$\underset{\text{RCNHR'}}{\overset{\text{O}}{\|}} \xrightarrow[\text{2. H}^+,\text{H}_2\text{O}]{\text{1. (CH}_3\overset{\overset{\text{CH}_3}{|}}{\text{CHCH}_2)_2}\text{AlH, (CH}_3\text{CH}_2)_2\text{O}} \underset{\text{Aldehyde}}{\overset{\text{O}}{\underset{\text{RCH}}{\|}}}$$

20. Enolates and Amidates (Section 20-7)

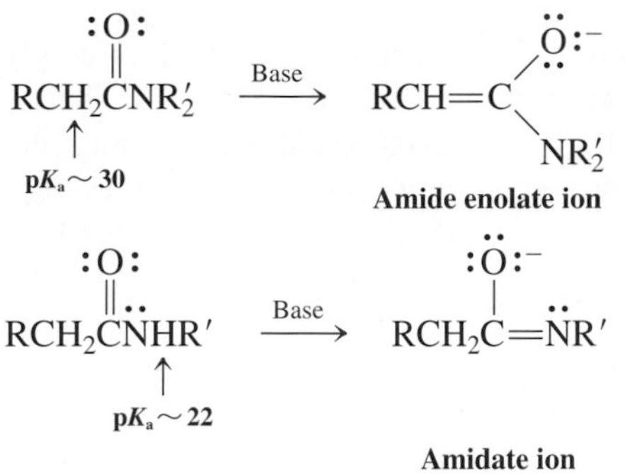

$$\underset{\substack{\uparrow \\ pK_a \sim 30}}{RCH_2\overset{\overset{\displaystyle :O:}{\|}}{C}NR_2'} \quad \xrightarrow{\text{Base}} \quad RCH\!=\!\overset{\overset{\displaystyle \ddot{O}:^-}{\diagdown}}{C}\!\underset{NR_2'}{}$$

Amide enolate ion

$$\underset{\substack{\uparrow \\ pK_a \sim 22}}{RCH_2\overset{\overset{\displaystyle :O:}{\|}}{C}\ddot{N}HR'} \quad \xrightarrow{\text{Base}} \quad RCH_2\overset{\overset{\displaystyle :\ddot{O}:^-}{|}}{C}\!=\!\ddot{N}R'$$

Amidate ion

21. Hofmann Rearrangement (Section 20-7)

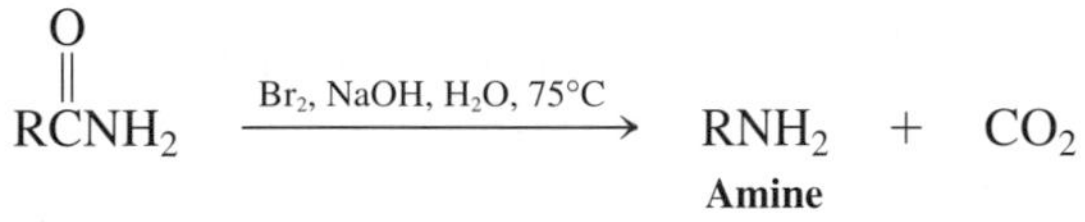

$$R\overset{\overset{\displaystyle O}{\|}}{C}NH_2 \quad \xrightarrow{Br_2,\ NaOH,\ H_2O,\ 75°C} \quad \underset{\textbf{Amine}}{RNH_2} \quad + \quad CO_2$$

Reactions of Nitriles

22. Water (Section 20-8)

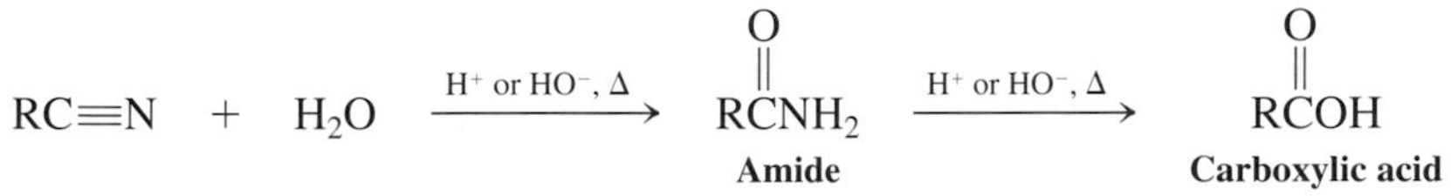

$$RC\!\equiv\!N \;+\; H_2O \quad \xrightarrow{H^+\ or\ HO^-,\ \Delta} \quad \underset{\textbf{Amide}}{R\overset{\overset{\displaystyle O}{\|}}{C}NH_2} \quad \xrightarrow{H^+\ or\ HO^-,\ \Delta} \quad \underset{\textbf{Carboxylic acid}}{R\overset{\overset{\displaystyle O}{\|}}{C}OH}$$

23. Organometallic Reagents (Section 20-8)

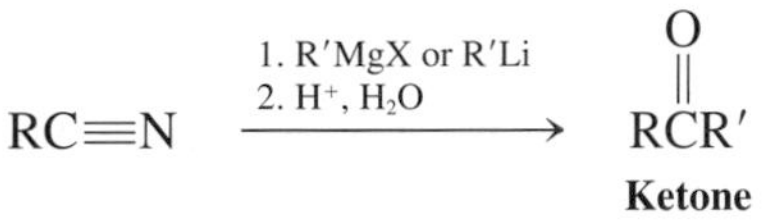

$$RC\!\equiv\!N \quad \xrightarrow[\text{2. } H^+,\ H_2O]{\text{1. } R'MgX\ or\ R'Li} \quad \underset{\textbf{Ketone}}{R\overset{\overset{\displaystyle O}{\|}}{C}R'}$$

24. Hydrides (Section 20-8)

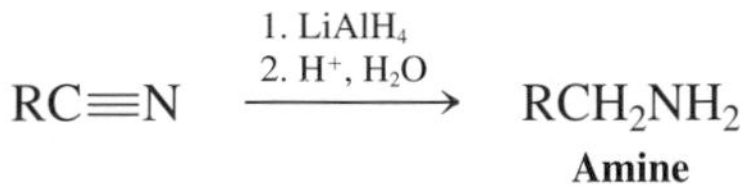

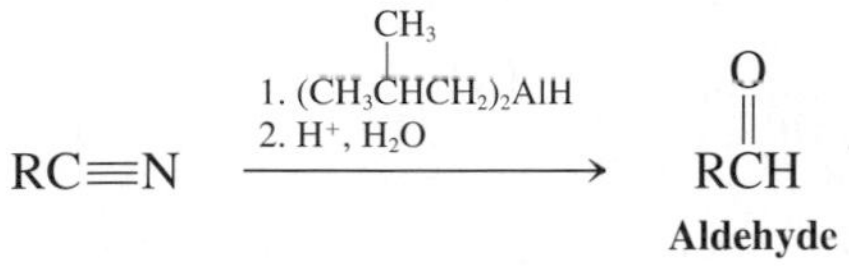

$$RC\!\equiv\!N \quad \xrightarrow[\text{2. } H^+,\ H_2O]{\text{1. } LiAlH_4} \quad \underset{\textbf{Amine}}{RCH_2NH_2}$$

$$RC\!\equiv\!N \quad \xrightarrow[\text{2. } H^+,\ H_2O]{\text{1. } (CH_3\overset{\overset{\displaystyle CH_3}{|}}{C}HCH_2)_2AlH} \quad \underset{\textbf{Aldehyde}}{R\overset{\overset{\displaystyle O}{\|}}{C}H}$$

25. Catalytic Hydrogenation (Section 20-8)

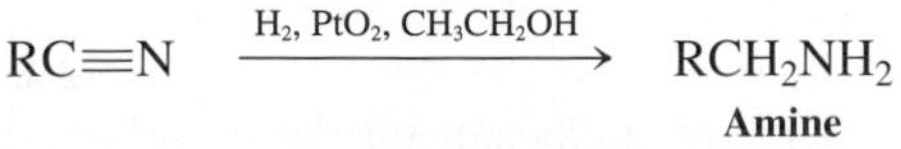

$$RC\!\equiv\!N \quad \xrightarrow{H_2,\ PtO_2,\ CH_3CH_2OH} \quad \underset{\textbf{Amine}}{RCH_2NH_2}$$

IMPORTANT CONCEPTS

1. The **electrophilic reactivity** of the carbonyl carbon in **carboxylic acid derivatives** is weakened by good electron-donating substituents. This effect, measurable by IR spectroscopy, is responsible not only for the decrease in the reactivity with nucleophiles and acid, but also for the increased basicity along the series: alkanoyl halides–anhydrides–esters–amides. Electron donation by resonance from the nitrogen in amides is so pronounced that there is **hindered rotation** about the amide bond on the NMR time scale.

2. **Carboxylic acid derivatives** are named as **alkanoyl halides, carboxylic anhydrides, alkyl alkanoates, alkanamides,** and **alkanenitriles,** depending on the functional group.

3. **Carbonyl stretching frequencies** in the IR spectra are diagnostic of the carboxylic acid derivatives: Alkanoyl chlorides absorb at $1790–1815\ cm^{-1}$, anhydrides at $1740–1790$ and $1800–1850\ cm^{-1}$, esters at $1735–1750\ cm^{-1}$, and amides at $1650–1690\ cm^{-1}$.

4. Carboxylic acid derivatives generally react with **water** (under acid or base catalysis) to hydrolyze to the corresponding carboxylic acid; they combine with **alcohols** to give esters and with **amines** to furnish amides. With **Grignard** and other **organometallic reagents,** they form ketones; esters may react further to form the corresponding alcohols. Reduction by **hydrides** gives products in various oxidation states: aldehydes, alcohols, or amines.

5. Long-chain esters are the constituents of animal and plant **waxes.** Triesters of glycerol are contained in natural **oils** and **fats.** Their hydrolysis gives **soaps. Triglycerides** containing phosphoric acid ester subunits belong to the class of **phospholipids.** Because they carry a highly polar head group and hydrophobic tails, phospholipids form **micelles** and **lipid bilayers.**

6. **Transesterification** can be used to convert one ester into another.

7. The functional group of **nitriles** is somewhat similar to that of the alkynes. The two component atoms are *sp* hybridized. The IR stretching vibration appears at about $2250\ cm^{-1}$. The hydrogens next to the cyano group are deshielded in ^1H NMR. The ^{13}C NMR absorptions for nitrile carbons are at relatively low field ($\delta \sim 112–126$ ppm), a consequence of the electronegativity of nitrogen.

8. **Mass spectrometry** is a technique for ionizing molecules and separating the resulting ions magnetically by molecular weight. Because the ionizing beam has high energy, the ionized molecules also **fragment** into smaller particles, all of which are separated and recorded as the **mass spectrum** of a compound. The presence of certain elements (such as Cl, Br) can be detected by their isotopic patterns. The presence of fragment-ion signals in mass spectra can be used to deduce the structure of a molecule.

9. **High-resolution mass spectral data** allow determination of molecular formulas from **exact mass** values.

PROBLEMS

31. Name (IUPAC system) or draw the structure of each of the following compounds.

(a) [structure: (CH₃)₂CHCH₂C(=O)I]

(b) [structure: cyclopentane with COCl and CH₃ substituents]

(c) CF_3COCCF_3 (with two C=O)

(d) [structure: phenyl–C(=O)C(=O)CH₂CH₃]

(e) $(CH_3)_3CCOCH_2CH_3$ (with C=O)

(f) CH_3CNH–phenyl (with C=O)

(g) Propyl butanoate
(h) Butyl propanoate
(i) 2-Chloroethyl benzoate
(j) *N,N*-Dimethylbenzamide
(k) 2-Methylhexanenitrile
(l) Cyclopentanecarbonitrile

32. (a) Use resonance forms to explain in detail the relative order of acidity of carboxylic acid derivatives, as presented in Section 20-1. (b) Do the same, but use an argument based on inductive effects.

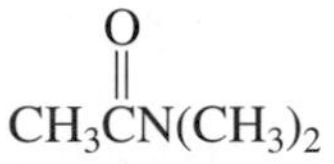

33. In each of the following pairs of compounds, decide which possesses the indicated property to the greater degree. **(a)** Length of C–X bond: acetyl fluoride or acetyl chloride. **(b)** Acidity of the boldface H: **CH$_2$**(COCH$_3$)$_2$ or **CH$_2$**(COOCH$_3$)$_2$. **(c)** Reactivity toward addition of a nucleophile: (i) an amide or (ii) an imide (as shown in the margin). **(d)** High-energy infrared carbonyl stretching frequency: ethyl acetate or ethenyl acetate.

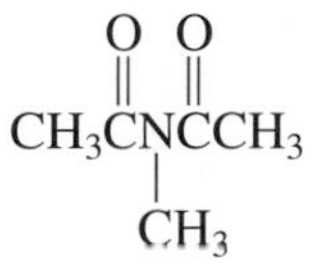

34. Give the product(s) of each of the following reactions.

i

ii

(a)

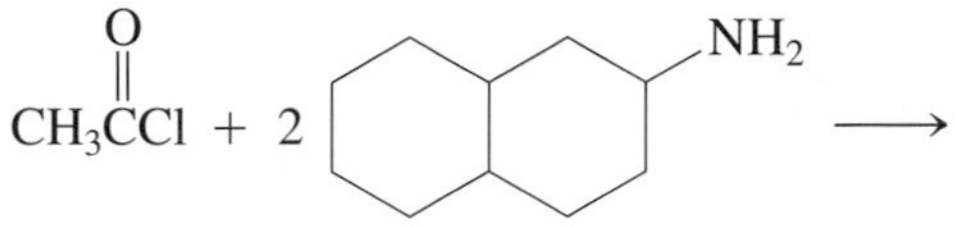

(b)

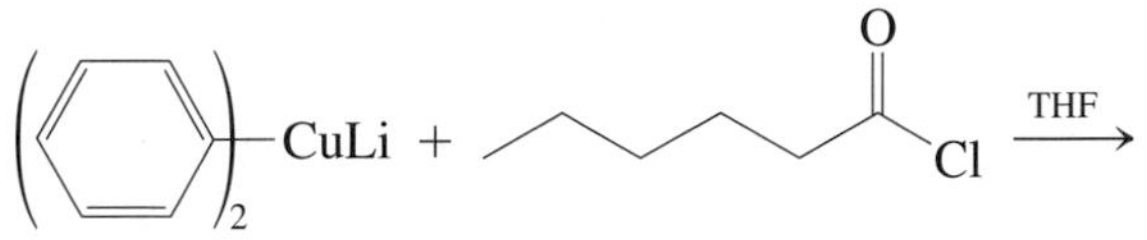

(c)

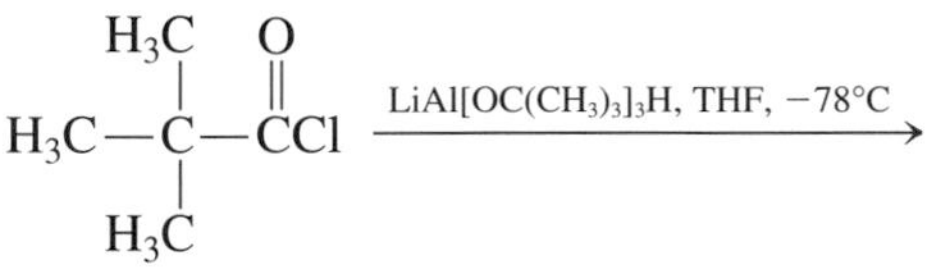

(d)

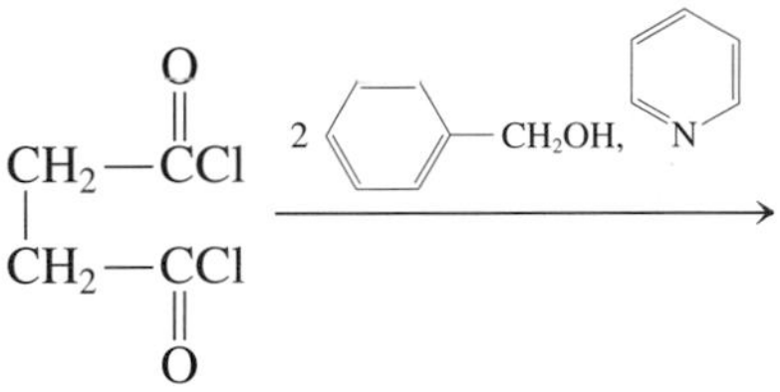

(e)

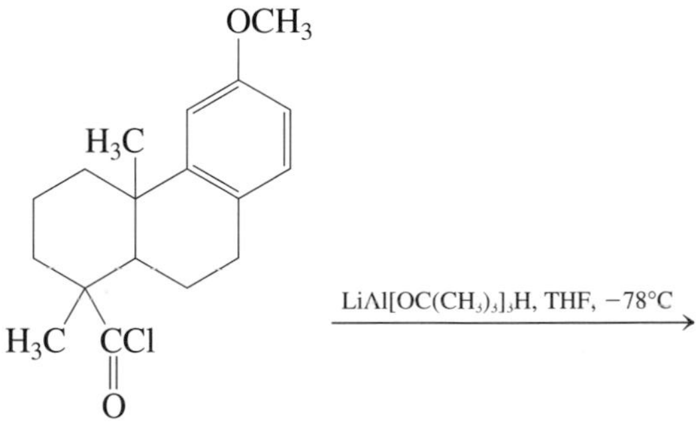

35. Formulate a mechanism for the reaction of acetyl chloride with 1-propanol shown on page 884.

36. Give the product(s) of the reactions of acetic anhydride with each of the following reagents. Assume in all cases that the reagent is present in excess.

(a) (CH$_3$)$_2$CHOH

(b) NH$_3$

(c) 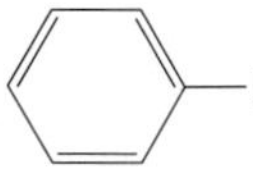 —MgBr, THF; then H$^+$, H$_2$O

(d) LiAlH$_4$, (CH$_3$CH$_2$)$_2$O; then H$^+$, H$_2$O

37. Give the product(s) of the reaction of butanedioic (succinic) anhydride with each of the reagents in Problem 36.

38. Formulate a mechanism for the reaction of butandoic (succinic) anhydride with methanol shown on page 888.

39. Give the products of reaction of methyl pentanoate with each of the following reagents under the conditions shown.

(a) NaOH, H_2O, heat; then H^+, H_2O
(b) $(CH_3)_2CHCH_2CH_2OH$ (excess), H^+
(c) $(CH_3CH_2)_2NH$, heat
(d) CH_3MgI (excess), $(CH_3CH_2)_2O$; then H^+, H_2O
(e) $LiAlH_4$, $(CH_3CH_2)_2O$; then H^+, H_2O
(f) $[(CH_3)_2CHCH_2]_2AlH$, toluene, low temperature; then H^+, H_2O

40. Give the products of reaction of γ-valerolactone (5-methyloxa-2-cyclopentanone, Section 20-4) with each of the reagents in Problem 39.

41. Draw the structure of each of the following compounds. **(a)** β-Butyrolactone; **(b)** β-valerolactone; **(c)** δ-valerolactone; **(d)** β-propiolactam; **(e)** α-methyl-δ-valerolactam; **(f)** *N*-methyl-γ-butyrolactam.

42. Formulate a mechanism for the acid-catalyzed transesterification of ethyl 2-methylpropanoate (ethyl isobutyrate) into the corresponding methyl ester. Your mechanism should clearly illustrate the catalytic role of the proton.

43. Formulate a mechanism for the reaction of methyl 9-octadecenoate with 1-dodecanamine shown on page 891.

44. Give the product of each of the following reactions.

(a)
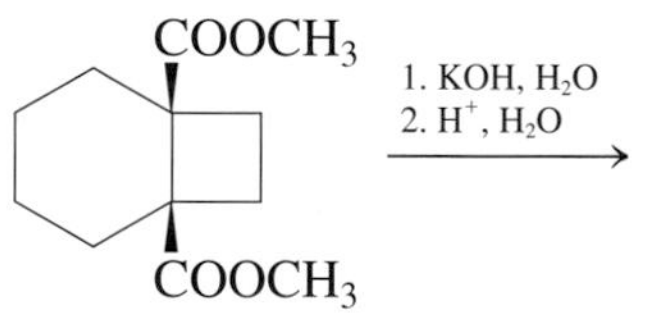
COOCH3
1. KOH, H2O
2. H+, H2O
COOCH3

(b)
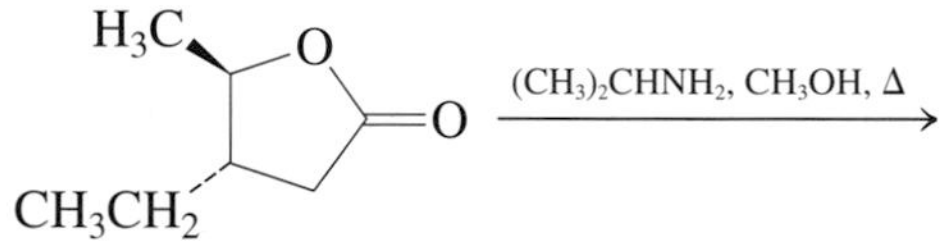
H3C
O
O
(CH3)2CHNH2, CH3OH, Δ
CH3CH2

(c)
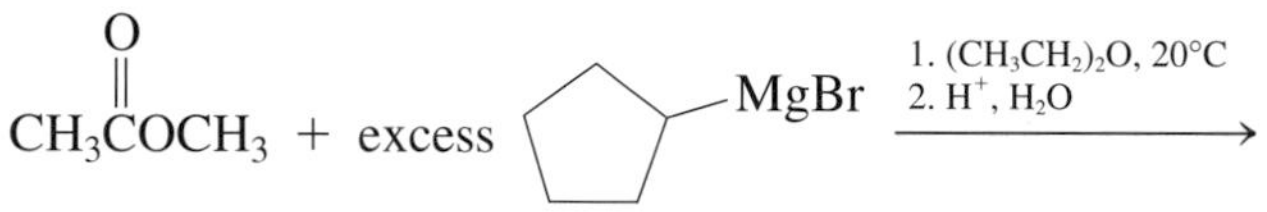
O
CH3COCH3 + excess
MgBr
1. (CH3CH2)2O, 20°C
2. H+, H2O

(d)
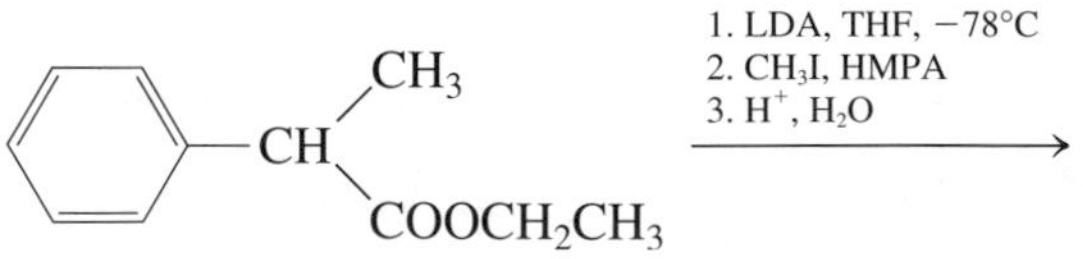
CH3
CH
COOCH2CH3
1. LDA, THF, −78°C
2. CH3I, HMPA
3. H+, H2O

(e)
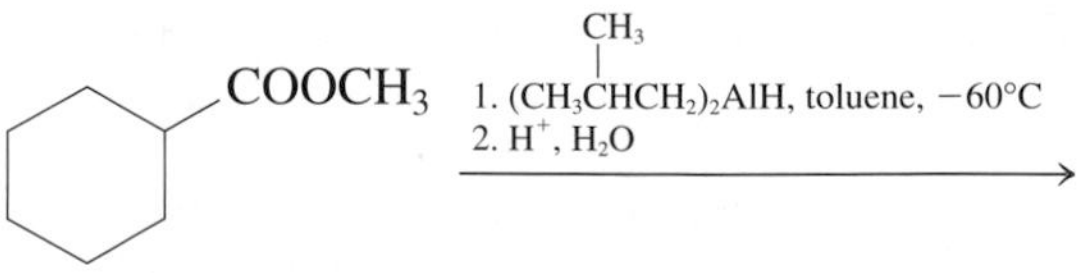
COOCH3
CH3
1. (CH3CHCH2)2AlH, toluene, −60°C
2. H+, H2O

45. A useful synthesis of certain types of diols includes the reaction of a "bis-Grignard" reagent with a lactone:

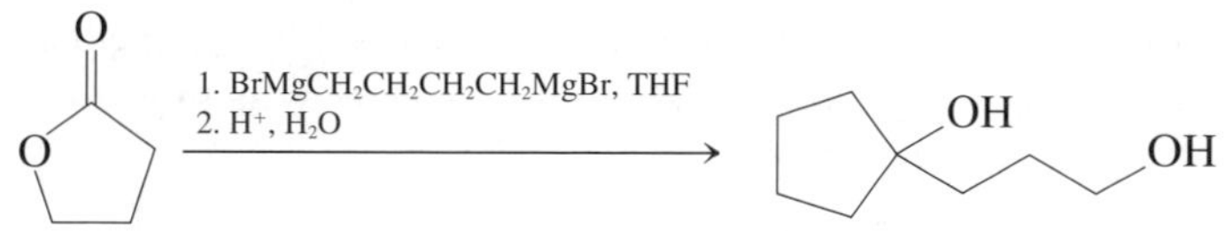

(a) Formulate a mechanism for this transformation. **(b)** Show how you would apply this general method to the synthesis of diols A and B.

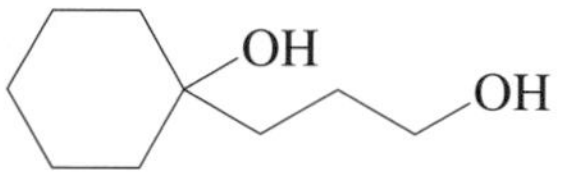

A 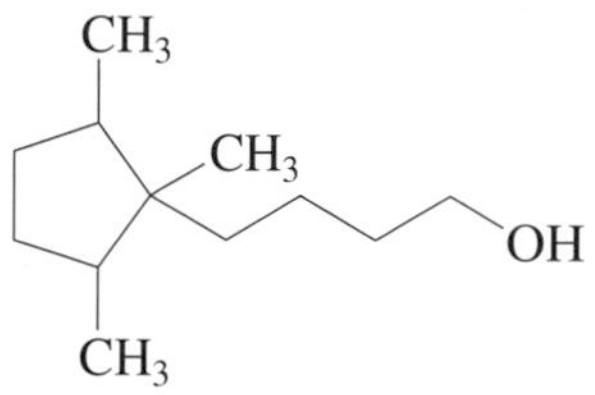   B

46. Formulate a mechanism for the formation of acetamide, $CH_3\overset{\displaystyle O}{\overset{\|}{C}}NH_2$, from methyl acetate and ammonia.

47. Give the products of the reactions of pentanamide with the reagents given in Problem 39(a, e, f). Repeat for *N,N*-dimethylpentanamide.

48. Formulate a mechanism for the acid-catalyzed hydrolysis of 3-methylpentanamide shown on page 897. (**Hint:** Use as a model the mechanism for general acid-catalyzed addition–elimination presented in Section 19-7.)

49. What reagents would be necessary to carry out the following transformations?
(a) Cyclohexanecarbonyl chloride → pentanoylcyclohexane; **(b)** 2-butenedioic (maleic) anhydride → (*Z*)-butene-1,4-diol; **(c)** 3-methylbutanoyl bromide → 3-methylbutanal; **(d)** benzamide → 1-phenylmethanamine; **(e)** propanenitrile → 3-hexanone; **(f)** methyl propanoate → 4-ethyl-4-heptanol.

50. Nylon-6,6 was the first completely synthetic fiber, initially prepared in the mid-1930s at DuPont. It is a copolymer whose subunits are alternating molecules of 1,6-hexanediamine and hexanedioic (adipic) acid, linked by amide functions. **(a)** Write a structural representation of Nylon-6,6. **(b)** Hexanedinitrile (adiponitrile) may be used as a common precursor to both 1,6-hexanediamine and hexanedioic acid. Show the reactions necessary to convert the dinitrile into both the diamine and the diacid. **(c)** Formulate a mechanism for the conversion into the diacid under the reaction conditions that you chose in (b).

51. On treatment with strong base followed by protonation, compounds A and B undergo cis–trans isomerization, but compound C does not. Explain.

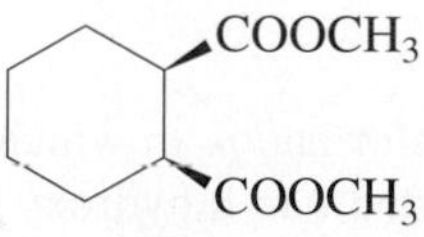

A 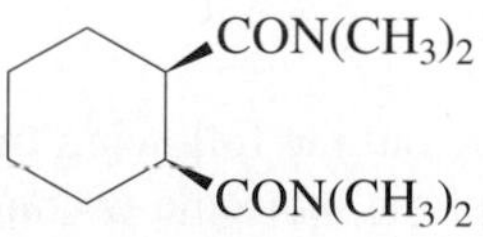   B 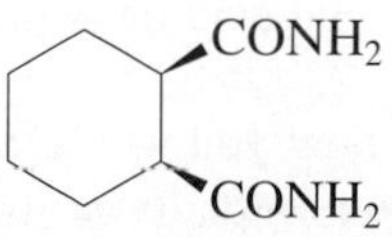   C

52. 2-Aminobenzoic (anthranilic) acid is prepared from 1,2-benzenedicarboxylic anhydride (phthalic anhydride) by using the two reactions shown here. Explain these processes mechanistically.

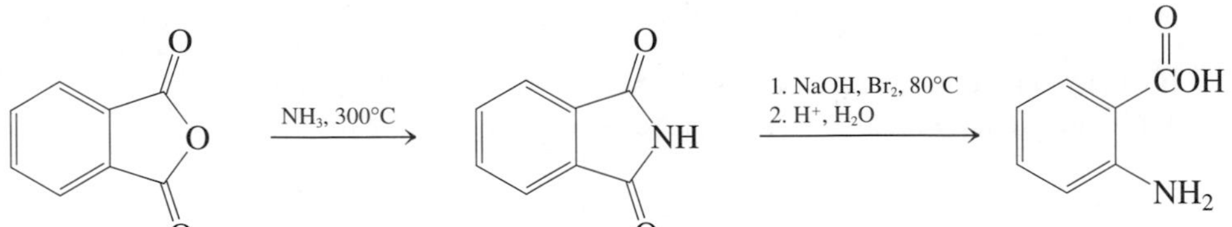

1,2-Benzenedicarboxylic anhydride (Phthalic anhydride) → (NH₃, 300°C) → 1,2-Benzenedicarboximide (Phthalimide) → (1. NaOH, Br₂, 80°C; 2. H⁺, H₂O) → 2-Aminobenzoic acid (Anthranilic acid)

53. On the basis of the reactions presented in this chapter, write reaction summary charts for esters and amides similar to the chart for alkanoyl halides (Figure 20-1). Compare the number of reactions for each of the compound classes. Is this information consistent with your understanding of the relative reactivity of each of the functional groups?

54. Show how you might synthesize chlorpheniramine, a powerful antihistamine used in several decongestants, from each of carboxylic acids A and B. Use a different carboxylic amide in each synthesis.

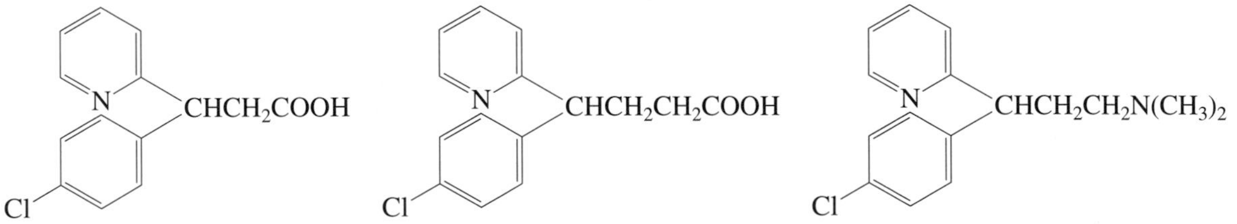

A — CHCH₂COOH; B — CHCH₂CH₂COOH; Chlorpheniramine — CHCH₂CH₂N(CH₃)₂

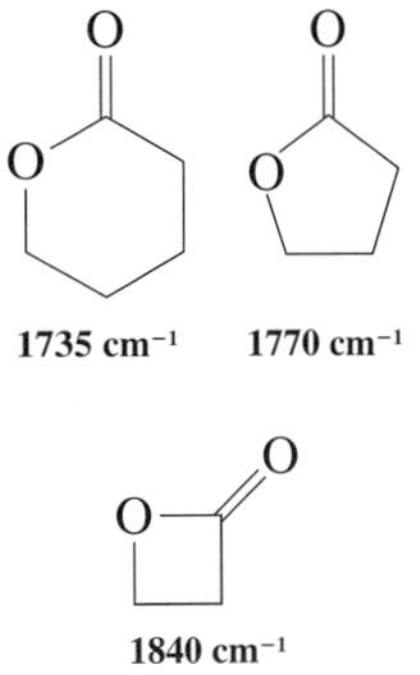

55. Although esters typically have carbonyl stretching frequencies at about 1740 cm⁻¹ in the infrared spectrum, the corresponding band for lactones can vary greatly with ring size. Three examples are shown in the margin. Propose an explanation for the IR bands of these smaller-ring lactones.

1735 cm⁻¹ 1770 cm⁻¹ 1840 cm⁻¹

56. On completing a synthetic procedure, every chemist is faced with the job of cleaning glassware. Because the compounds present may be dangerous in some way or have unpleasant properties, a little serious chemical thinking is often beneficial before "doing the dishes." Suppose that you have just completed a synthesis of hexanoyl chloride, perhaps to carry out the reaction in Problem 34(b); first, however, you must clean the glassware contaminated with this alkanoyl halide. *Both hexanoyl chloride and hexanoic acid have terrible odors.* **(a)** Would cleansing the glassware with soap and water be a good idea? Explain. **(b)** Suggest a more pleasant alternative, based on the chemistry of alkanoyl halides and the physical properties (particularly the odors) of the various carboxylic acid derivatives.

57. Show how you would carry out the following transformation in which the ester function at the lower left of the molecule is converted into a hydroxy group but that at the upper right is preserved. (**Hint:** Do not try ester hydrolysis. Look carefully at how the ester groups are linked to the steroid and consider an approach based on transesterification.)

58. The removal of the C17 side chain of certain steroids is a critical element in the synthesis of a number of hormones, such as testosterone, from steroids in the relatively readily available pregnane family.

Pregnan-3α-ol-20-one   **Testosterone**

How would you carry out the comparable transformation, shown in the margin, of acetylcyclopentane into cyclopentanol? (Note: In this and subsequent synthetic problems you may need to use reactions from several areas of carbonyl chemistry discussed in Chapters 17–20.)

59. Propose a synthetic sequence to convert carboxylic acid A into the naturally occurring sesquiterpene α-curcumene.

A   **α-Curcumene**

60. Propose a synthetic scheme for the conversion of lactone A into amine B, a precursor to the naturally occurring monoterpene C.

A   **B**   **C**

61. Propose a synthesis of β-selinene, a member of a very common family of sesquiterpenes, beginning with the alcohol shown here. Use a nitrile in your synthesis. Inspection of a model may help you choose a way to obtain the desired stereochemistry. (Is the 1-methylethenyl group axial or equatorial?)

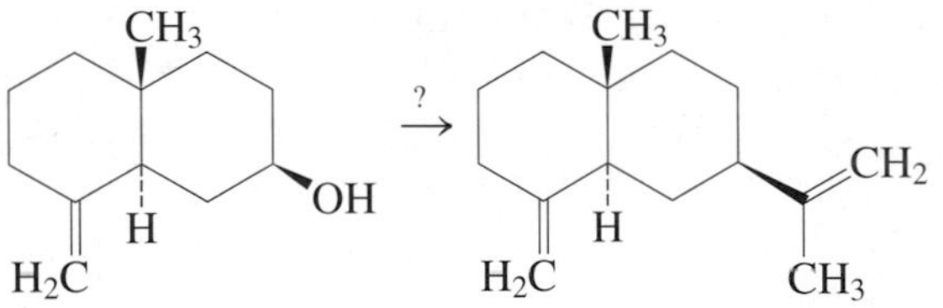
CH3
H
OH
H2C
?
CH3
CH2
H
H2C
CH3

***β*-Selinene**

62. Give the structure of the product of the first of the following reactions, and then propose a scheme that will ultimately convert it into the methyl-substituted ketone at the end of the scheme. This example illustrates a common method for the introduction of "angular methyl groups" into synthetically prepared steroids. (**Hint:** It will be necessary to protect the carbonyl function.)

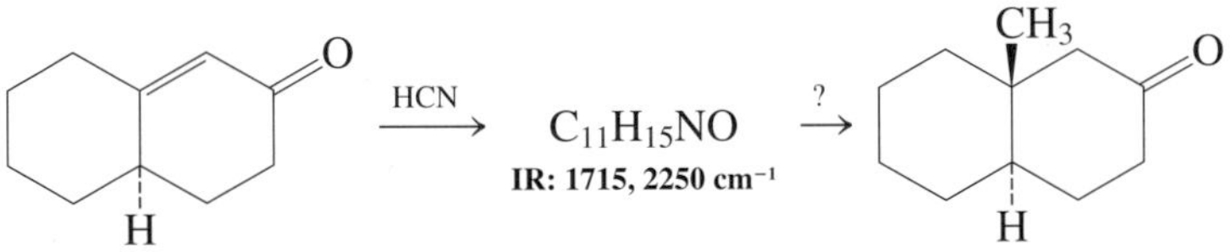
O
H
HCN
C11H15NO
IR: 1715, 2250 cm−1
?
CH3
O
H

63. Assign as many peaks as you can in the mass spectrum of 1-bromopropane (Figure 20-7).

64. The following table lists selected mass-spectral data for three isomeric alcohols with the formula $C_5H_{12}O$. On the basis of the peak positions and intensities, suggest structures for each of the three isomers. A dash means that the peak is very weak or absent entirely.

Relative Peak Intensities

m/z	Isomer A	Isomer B	Isomer C
88 M^+	—	—	—
87 $(M - 1)^+$	2	2	—
73 $(M - 15)^+$	—	7	55
70 $(M - 18)^+$	38	3	3
59 $(M - 29)^+$	—	—	100
55 $(M - 15 - 18)^+$	60	17	33
45 $(M - 43)^+$	5	100	10
42 $(M - 18 - 28)^+$	100	4	6

65. Following are spectroscopic and analytical characteristics of two unknown compounds. Propose a structure for each. **(a)** Empirical formula: $C_8H_{16}O$. ^1H NMR: $\delta = 0.90$ (t, 3 H), 1.0–1.6 (m, 8 H), 2.05 (s, 3 H), and 2.25 (t, 2 H) ppm. IR: 1715 cm^{-1}. UV: $\lambda_{max}(\epsilon) = 280(15)$ nm. MS: $m/z = 128$ ($M^{+\cdot}$); intensity of $(M + 1)^+$ peak is 9% of $M^{+\cdot}$ peak; important fragments are at $m/z = 113$ $(M - 15)^+$, $m/z = 85$ $(M - 43)^+$, $m/z = 71$ $(M - 57)^+$, $m/z = 58$ $(M - 70)^+$ (the second largest peak), and $m/z = 43$ $(M - 85)^+$ (the base peak). **(b)** Empirical formula: C_5H_8. ^1H NMR: spectrum B. ^{13}C NMR: $\delta = 20.5$ (CH_3), 23.8 (CH_3), 28.0 (CH_2), 30.6 (CH_2), 30.9 (CH_2), 41.2 (CH), 108.4 (CH_2), 120.8 (CH), 133.2 ($C_{quaternary}$), and 149.7 ($C_{quaternary}$) ppm. IR: significant bands at 3060 (medium), 3010 (medium), 1680 (weak), 1646 (medium), and 880 (very strong) cm^{-1}. UV: $\lambda_{max} < 200$ nm. MS: $m/z = 136$ ($M^{+\cdot}$); intensity of $(M + 1)^+$ peak is 11% of $M^{+\cdot}$ peak; important fragments are at $m/z = 121$ $(M - 15)^+$, $m/z = 95$ $(M - 41)^+$, $m/z = 68$ $(M - 68)^+$ (the base peak), and $m/z = 41$ $(M - 95)^+$.

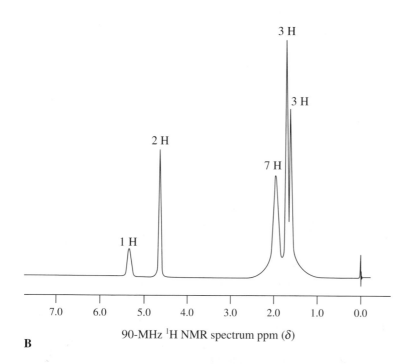

90-MHz ^1H NMR spectrum ppm (δ)

B

66. Spectroscopic data for two carboxylic acid derivatives are given in NMR-C and NMR-D. Identify these compounds, which may contain C, H, O, N, Cl, and Br but no other elements. **(a)** ^1H NMR: spectrum C (one signal has been amplified to reveal all peaks in the multiplet). IR: 1728 cm^{-1}. High-resolution mass spectrum: m/z for the molecular ion is 116.0837. See table for important MS fragmentation peaks. **(b)** ^1H NMR: spectrum D. IR: 1739 cm^{-1}. High-resolution mass spectrum: The intact molecule gives two peaks with almost equal intensity: $m/z = 179.9786$ and 181.9766. See table for important MS fragmentation peaks. (For spectra and tables, see page 934.)

Mass Spectrum of Unknown C	
m/z	Intensity relative to base peak (%)
116	0.5
101	12
75	26
57	100
43	66
29	34

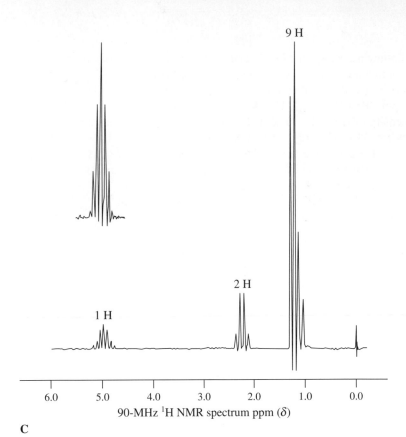

C

Mass Spectrum of Unknown D	
m/z	Intensity relative to base peak (%)
182	13
180	13
109	78
107	77
101	3
29	100

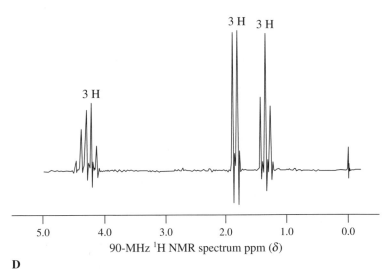

D

Team Problem

67. Friedel-Crafts alkanoylations are best carried out with alkanoyl halides, but other carboxylic acid derivatives undergo this process, too, such as carboxylic anhydrides or esters. These reagents may have some drawbacks, however, the subject of this problem.

Before you start, discuss as a group the mechanisms for forming acylium ions from alkanoyl halides and carboxylic anhydrides in Section 15-14. Then divide your group in two and analyze the outcome of the following two reactions. Use the NMR spectral data given to confirm your product assignments. (**Hint:** D is formed via C.)

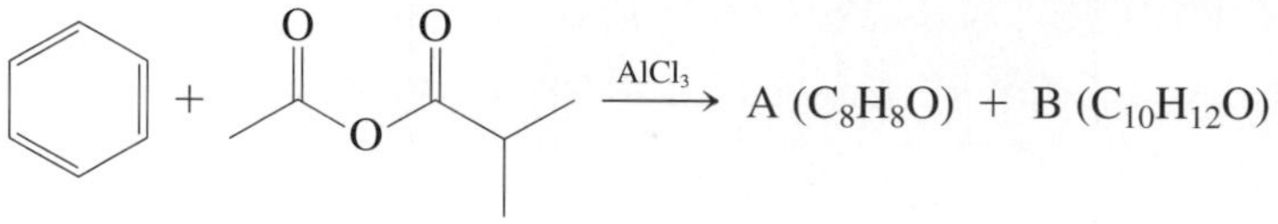

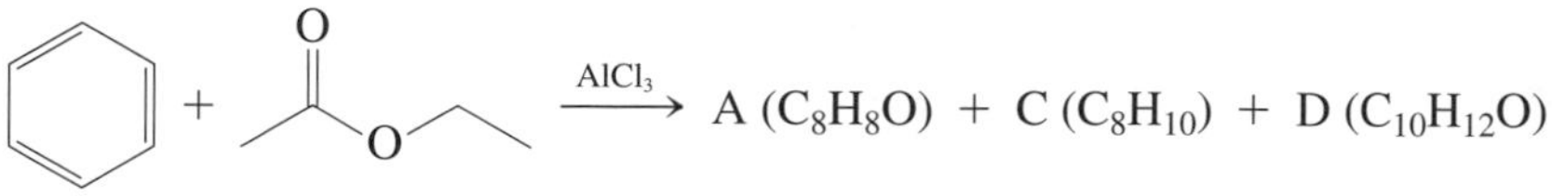

$$\text{benzene} + \text{anhydride} \xrightarrow{\text{AlCl}_3} \text{A } (C_8H_8O) + \text{B } (C_{10}H_{12}O)$$

$$\text{benzene} + \text{ester} \xrightarrow{\text{AlCl}_3} \text{A } (C_8H_8O) + \text{C } (C_8H_{10}) + \text{D } (C_{10}H_{12}O)$$

Compound A: ^{1}H NMR δ 2.60 (s, 3H), 7.40-7.50 (m, 2H), 7.50-7.60 (m, 1H), 7.90-8.00 (m, 2H).

Compound B: ^{1}H NMR δ 2.22 (d, 6H), 3.55 (sep, 1H), 7.40-7.50 (m, 2H), 7.50-7.60 (m, 1H), 7.90-8.00 (m, 2H).

Compound C: ^{1}H NMR δ 2.20 (t, 3H), 2.64 (q, 2H), 7.10-7.30 (m, 5H).

Compound D: ^{1}H NMR δ 1.25 (t, 3H), 3.57 (s, 3H), 2.70 (q, 2H), 7.20 (d, 2H), 7.70 (d, 2H).

Reconvene to share your solutions. Then specifically address the nature of the complications ensuing when using the reagents shown. Finally, altogether, consider the following reaction sequence. Again, use a mechanistic approach to arrive at the structures of the products.

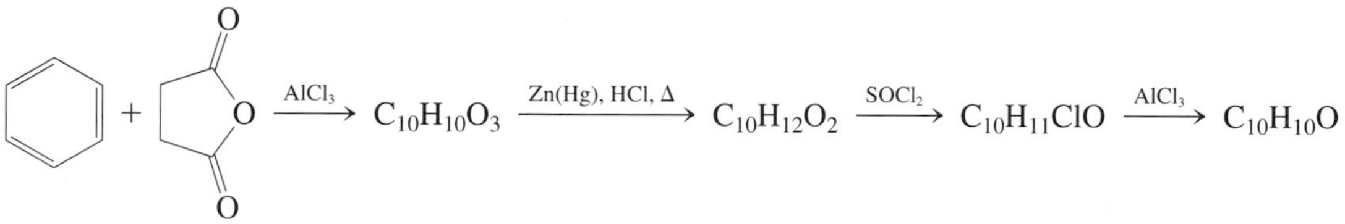

$$\text{benzene} + \text{anhydride} \xrightarrow{\text{AlCl}_3} C_{10}H_{10}O_3 \xrightarrow{\text{Zn(Hg), HCl, }\Delta} C_{10}H_{12}O_2 \xrightarrow{\text{SOCl}_2} C_{10}H_{11}ClO \xrightarrow{\text{AlCl}_3} C_{10}H_{10}O$$

Preprofessional Problems

68. What is the IUPAC name of the compound shown in the margin?
(a) Isopropyl 2-fluoro-3-methylbutanoate; (b) 2-fluoroisobutanoyl 2-propanoate; (c) 1-methylethyl 2-fluorobutyrate; (d) 2-fluoroisopropyl isopropanoate; (e) 1-methylethyl 2-fluoro-2-methylpropanoate.

$(CH_3)_2CCO_2CH(CH_3)_2$ (with F substituent)

69. Saponification of $(CH_3)_2CHC^{18}OCH_2CH_2CH_3$ with aqueous NaOH will give
(a) $(CH_3)_2CHCO_2{}^-Na^+ + CH_3CH_2CH_2{}^{18}OH$;

(b) $(CH_3)_2CHC^{18}O{}^-\ Na^+ + CH_3CH_2CH_2OH$; (c) $(CH_3)_2CHOCH_2CH_2CH_3 + C\equiv^{18}O$; (d) $(CH_3)_2CHCHO + CH_3CH_2CH_2{}^{18}OH$.

70. The best description for compound A (see margin) is (a) an amide; (b) a lactam; (c) an ether; (d) a lactone.

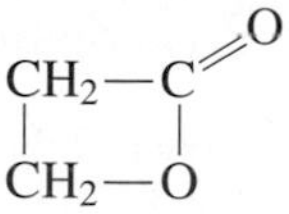

$$\begin{array}{c} CH_2-C{\diagup}^{\displaystyle O} \\ | \qquad\quad | \\ CH_2-O \end{array}$$

A

71. Which species is formed in the mass spectrometric fragmentation shown in the margin?

(a) $:\overset{..}{\underset{..}{Cl}}\cdot{}^+$; (b) $:\overset{..}{\underset{..}{Cl}}{}^+$; (c) $CH_3\cdot{}^+$; (d) $CH_3:{}^-$

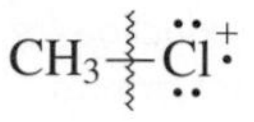

$CH_3 {-}\!\!\!{|}\ \overset{..}{\underset{..}{Cl}}\cdot{}^+$

Functional Groups Containing Nitrogen

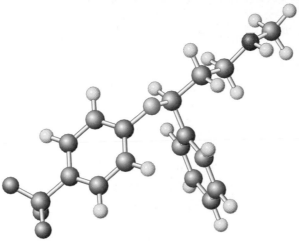

Prozac [fluoxetin; *R,S-N*-methyl-3-(4-trifluoromethylphenoxy)-3-phenyl-1-propanamine] is one of the most widely prescribed drugs in the treatment of depression. It has been estimated that, since its introduction in 1987, more than 20 million people have used this compound worldwide.

The Scream (Edvard Munch)

Although higher organisms cannot activate nitrogen by reduction to ammonia, some microorganisms do. Thus, the nodules in the root system of the soybean are the sites of nitrogen reduction by *Rhizobium* bacteria.

Our atmosphere is composed of about one-fifth oxygen, O_2, and nearly four-fifths nitrogen, N_2. We are fully aware of the importance of the oxygen part: We need it to breathe, and nature uses it abundantly in water, alcohols, ethers, and many other organic and inorganic molecules. What about the nitrogen component? Unlike O_2, which is ultimately the reactive ingredient in biological oxidations, N_2 itself is relatively inert. However, in its reduced form of ammonia, NH_3, and its organic derivatives, the amines, it plays as active a role in nature as oxygen. Thus, amines and other nitrogen-bearing compounds are among the most abundant organic molecules. As components of the amino acids, peptides, proteins, and alkaloids, they are essential to biochemistry. Many, such as the neurotransmitters, possess powerful physiological activity; related substances have found medicinal uses as decongestants, anesthetics, sedatives, and stimulants (Chemical Highlight 21-1). Similar activity is found in cyclic amines in which nitrogen is part of a ring, the nitrogen heterocycles (Chapter 25).

In many respects, the chemistry of the amines is analogous to that of the alcohols and ethers (Chapters 8 and 9). For example, all amines are basic (although primary and secondary ones can also behave as acids), they form hydrogen bonds, and they act as nucleophiles in substitution reactions. However, there are some differences in reactivity, because nitrogen is less electronegative than oxygen. Thus, primary and secondary amines are less acidic and form weaker hydrogen bonds than alcohols and ethers, and they are more basic and more nucleophilic. This chapter will show that these properties underlie their physical and chemical characteristics and give us a variety of ways to synthesize amines.

21-1 Naming the Amines

Amines are derivatives of ammonia, in which one (primary), two (secondary), or three (tertiary) of the hydrogens have been replaced by alkyl or aryl groups. Therefore, amines are related to ammonia in the same sense as ethers and alcohols are related to water. You will note, however, that the designations primary, secondary, and tertiary (see margin) are used in a different way. In alcohols, ROH, the nature of the R group defines this designation; in amines, the number of R substituents on nitrogen does so.

The system for naming amines is confused by the variety of common names in the literature. Probably the best way to name aliphatic amines is that used by *Chemical Abstracts*—that is, as **alkanamines,** in which the name of the alkane stem is modified by replacing the ending *-e* by **-amine.** The position of the functional group is indicated by a prefix designating the carbon atom to which it is attached, as in the alcohols (Section 8-1).

$$CH_3NH_2 \qquad CH_3CHCH_2NH_2 \qquad H_2N \quad H$$

$$\text{CH}_3$$

Methanamine   2-Methyl-1-propanamine   (*R*)-*trans*-3-Penten-2-amine

Substances with two amine functions are **diamines,** two examples of which are 1,4-butanediamine and 1,5-pentanediamine. Their contribution to the smell of dead fish and rotting flesh leads to their descriptive common names, putrescine and cadaverine, respectively.

$$H_2N \qquad NH_2 \qquad H_2N \qquad NH_2$$

1,4-Butanediamine   **1,5-Pentanediamine**
(Putrescine)   **(Cadaverine)**

The aromatic amines, or anilines, are called **benzenamines** (Section 15-1). For secondary and tertiary amines, the largest alkyl substituent on nitrogen is chosen as the alkanamine stem, and the other groups are named by using the letter *N*-, followed by the name of the additional substituent(s).

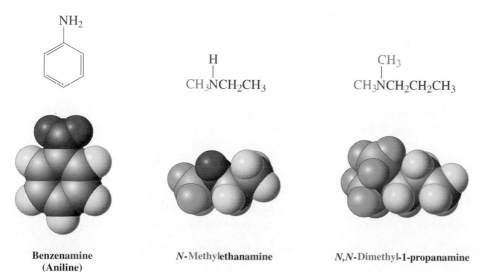

$$NH_2$$

$$\text{H} \qquad\qquad\qquad CH_3$$

$$CH_3NCH_2CH_3 \qquad\qquad CH_3NCH_2CH_2CH_3$$

Benzenamine   *N*-Methylethanamine   *N,N*-Dimethyl-1-propanamine
(Aniline)

Ammonia

Primary amine

Secondary amine

Tertiary amine

CHEMICAL HIGHLIGHT 21-1

Physiologically Active Amines and Weight Control

A large number of physiologically active compounds owe their activity to the presence of amino groups. Several simple examples are well-known prescription or illegal drugs, such as adrenaline, Benzedrex, Urotropine, amphetamine, mescaline, and Prozac (see illustration at the beginning of the chapter). A recurring pattern in many (but not all) of these compounds is the 2-phenylethanamine (β-phenethylamine) unit—that is, a nitrogen connected by a two-carbon chain to a benzene nucleus (highlighted in green where applicable). This structural feature appears to be crucial for the binding to brain receptor sites responsible for neurotransmitter action at certain nerve terminals that (among other functions) control appetite and muscular activity, on the one hand, and (potentially addictive) euphoric stimulation, on the other.

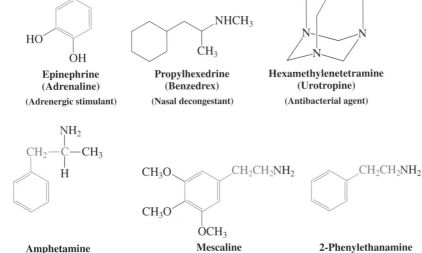

Epinephrine (Adrenaline)
(Adrenergic stimulant)

Propylhexedrine (Benzedrex)
(Nasal decongestant)

Hexamethylenetetramine (Urotropine)
(Antibacterial agent)

Amphetamine
(Antidepressant, central nervous system stimulant. The *N*-methyl derivative, methamphetamine, is a dangerous, addictive drug known as speed, crank, crystal, or ice.)

Mescaline
(Hallucinogen)

2-Phenylethanamine
(β-Phenethylamine)

Controlling weight through exercise: the New York City marathon.

The problem of selective targeting of these sites by molecular design has played a major role in the development of drugs that control weight and combat obesity. This effort has taken several directions. Thus, whereas many of the "diet pills" rely on their anorectic (appetite-suppressing) effect, newer products, such as sibutramine (approved by the FDA in 1998), which has no stimulating side effects, act by increasing satiety. In other words, you feel hungry, but you are more quickly satisfied after you start eating, or feel full longer. Another approach to controlling weight is to increase metabolic rate (and hence body temperature), a goal most simply and commonly attained by exercising. An arguably less taxing alternative is a thermogenic drug. Caffeine is such a compound, but its enhancing effect on the metabolic rate is short lived and is followed by a period of actually depressing it. A potential solution, a drug in a long line of trials and (like many other compounds in drug development) simply known by its number, CL 316243, is currently being tested.

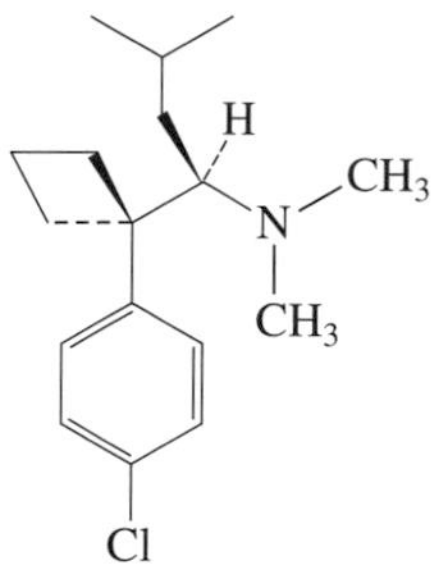

**Sibutramine
(Meridia)**

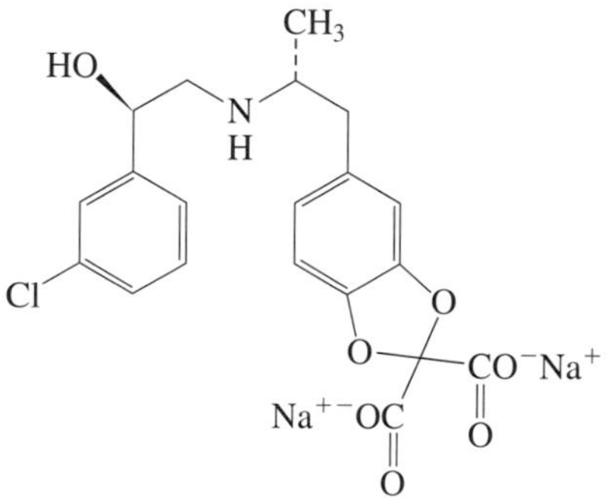

CL 316243

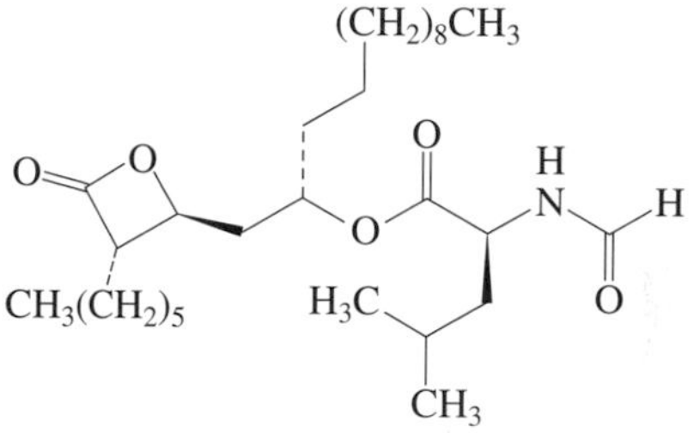

**Orlistat
(Xenical)**

Whereas thermogenic drugs help control body weight by "burning off fat," the exact opposite— namely, inhibition of calorific intake by preventing metabolism—is being attempted with yet another class of drugs. An example is orlistat, a molecule that retards the enzymatic breakdown of fat in the gut, thus allowing food to be excreted undigested. Alternatively, a wholly nonabsorbable fat, Olestra, based on a sucrose (Chapter 24) polyester, is currently used in some foods, especially snack products such as potato chips.

You may question the validity of this type of drug research, because it appears to cater to the glutton or the "fashionable" in the search of the perfect body. Not so. Do not forget that obesity is a major health hazard, causing chronic diseases, such as cardiovascular and respiratory problems, hypertension, diabetes, and certain cancers, and that it shortens the life span of those so afflicted. It appears to be mostly the result of a metabolic (perhaps in some cases genetic) disposition that cannot otherwise be controlled.

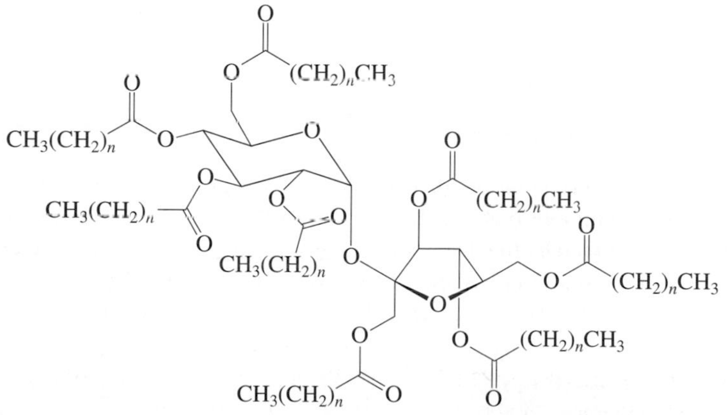

Olestra

CH_3NH_2
Methylamine

$(CH_3)_3N$
Trimethylamine

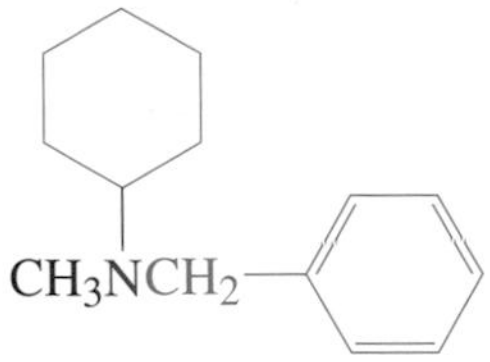

CH_3NCH_2—

Benzylcyclohexylmethylamine

An alternative way to name amines treats the functional group, called **amino-,** as a substituent of the alkane stem. This procedure is analogous to naming alcohols as hydroxyalkanes.

$CH_3CH_2NH_2$ $(CH_3)_2NCH_2CH_2CH_3$
Aminoethane   *N,N*-**Dimethylamino**propane

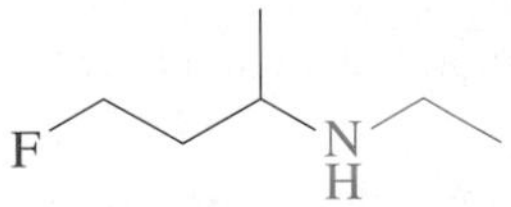

3-(*N*-**Ethylamino**)-1-fluorobutane

Many common names are based on the term **alkylamine** (see margin), as in the naming of alkyl alcohols.

Name each of the following molecules twice, first as an alkanamine, then as an alkyl amine.

(a) 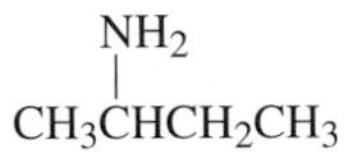 $CH_3CHCH_2CH_3$ (with NH_2 on second carbon)   (b) 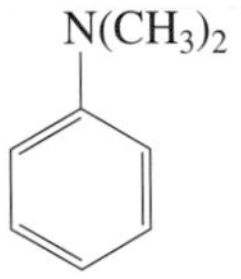 (phenyl with $N(CH_3)_2$)   (c) 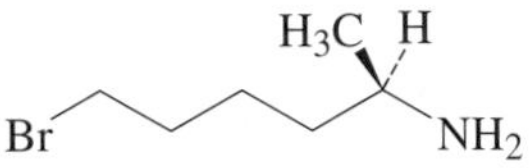 Br—...—NH_2 (with H_3C, H)

EXERCISE 21-2

Draw structures for the following compounds (common name in parentheses). **(a)** 2-Propynamine (propargylamine); **(b)** (*N*-2-propenyl)phenylmethanamine (*N*-allylbenzylamine); **(c)** *N*,2-dimethyl-2-propanamine (*tert*-butylmethylamine).

In summary, there are several systems for naming amines. *Chemical Abstracts* uses names of the type alkanamine and benzenamine. Alternatives are the terms aminoalkane, aniline, and alkylamine.

21-2 Structural and Physical Properties of Amines

Let us now look at some of the structural and physical characteristics of simple amines. Amines adopt a tetrahedral geometry around the heteroatom, but this arrangement is not rigid because of a rapid isomerization process called inversion. This section also compares the polarity and hydrogen-bonding ability of amines with that of alcohols.

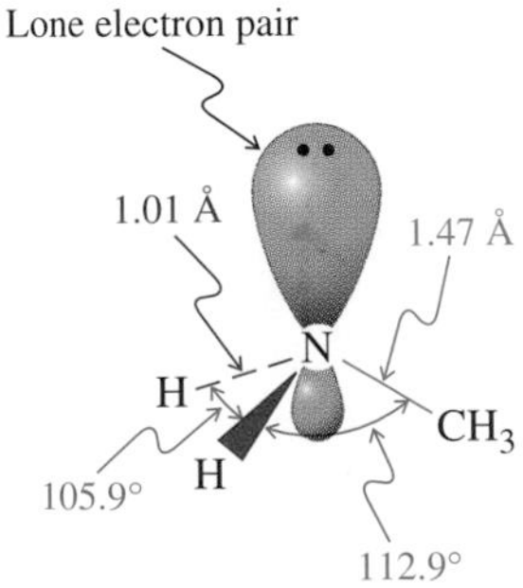

The alkanamine molecule is tetrahedral

The nitrogen orbitals in amines are very nearly sp^3 hybridized (see Section 1-8), forming an approximately tetrahedral arrangement. Three vertices of the tetrahedron are occupied by the three substituents, the fourth by the lone electron pair. The term **pyramidal** is often used to describe the geometry adopted by the nitrogen and its three substituents. Figure 21-1 depicts the structure of methanamine (methylamine).

Lone electron pair
1.01 Å   1.47 Å
H   N   CH₃
105.9°   H
112.9°

FIGURE 21-1
Nearly tetrahedral structure of methanamine (methylamine).

EXERCISE 21-3

The bonds to nitrogen in methanamine (methylamine; see Figure 21-1) are slightly longer than the bonds to oxygen in the structure of methanol (Figure 8-1). Explain. (**Hint:** See Table 1-2.)

The tetrahedral geometry around an amine nitrogen suggests that it should be chiral if it bears three different substituents, the lone electron pair serving as the fourth. The image and mirror image of such a compound are not superimposable, by analogy with carbon-based stereocenters (Section 5-1). This fact is illustrated with the simple chiral alkanamine *N*-methylethanamine (ethylmethylamine).

Image and Mirror Image of *N*-Methylethanamine (Ethylmethylamine)

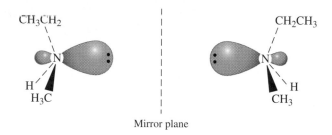

Mirror plane

However, samples of the amine prove *not* to be optically active. Why? Amines are not configurationally stable at nitrogen, because of rapid isomerization by a process called **inversion.** The molecule passes through a transition state incorporating an sp^2-hybridized nitrogen atom, as illustrated in Figure 21-2. The barrier to this motion in ordinary small amines has been measured by spectroscopic techniques and found to be between 5 and 7 kcal mol^{-1}. It is therefore impossible to keep an enantiomerically pure, simple di- or trialkylamine from racemizing at room temperature when the nitrogen atom is the only stereocenter.

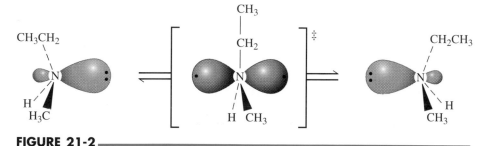

FIGURE 21-2

Inversion at nitrogen rapidly interconverts the two enantiomers of *N*-methylethanamine (ethylmethylamine). Thus, the compound exhibits no optical activity.

Amines form weaker hydrogen bonds than alcohols do

Because alcohols readily form hydrogen bonds (Section 8-2), they have unusually high boiling points. In principle, so should amines, and indeed Table 21-1 (see page 942) bears out this expectation. However, because amines form weaker hydrogen bonds* than alcohols do, their boiling points are lower and their solubility in water less. In general, the boiling points of the amines lie between those of the corresponding alkanes and alcohols. The smaller amines are soluble in water and in alcohols because they can form hydrogen bonds to the solvent. If the hydrophobic part of an amine exceeds six carbons, the solubility in water decreases rapidly; the larger amines are essentially insoluble in water.

*Whereas *all* amines can act as proton acceptors in hydrogen bonding, only primary and secondary amines can function as proton donors, because tertiary amines lack such protons.

TABLE 21-1	Physical Properties of Amines, Alcohols, and Alkanes					
Compound	Melting point (°C)	Boiling point (°C)	Compound	Melting point (°C)	Boiling point (°C)
CH_4	−182.5	−161.7	$(CH_3)_2NH$	−93	7.4
CH_3NH_2	−93.5	−6.3	$(CH_3)_3N$	−117.2	2.9
CH_3OH	−97.5	65.0			
			$(CH_3CH_2)_2NH$	−48	56.3
CH_3CH_3	−183.3	−88.6	$(CH_3CH_2)_3N$	−114.7	89.3
$CH_3CH_2NH_2$	−81	16.6			
CH_3CH_2OH	−114.1	78.5	$(CH_3CH_2CH_2)_2NH$	−40	110
			$(CH_3CH_2CH_2)_3N$	−94	155
$CH_3CH_2CH_3$	−187.7	−42.1			
$CH_3CH_2CH_2NH_2$	−83	47.8	NH_3	−77.7	−33.4
$CH_3CH_2CH_2OH$	−126.2	97.4	H_2O	0	100

To summarize, amines adopt an approximately tetrahedral structure in which the lone electron pair occupies one vertex of the tetrahedron. They can, in principle, be chiral at nitrogen but are difficult to maintain in enantiomerically pure form because of fast nitrogen inversion. Amines have boiling points higher than those of alkanes of similar size. Their boiling points are lower than those of the analogous alcohols because of weaker hydrogen bonding, and their water solubility is between that of comparable alkanes and alcohols.

21-3 Spectroscopy of the Amine Group

Primary and secondary amines can be recognized by infrared spectroscopy because they exhibit a characteristic broad N–H stretching absorption in the range between 3250 and 3500 cm^{-1}. Primary amines show two strong peaks in this range, whereas secondary amines give rise to only a very weak single line. Primary amines also show a band near 1600 cm^{-1} that is due to a scissoring motion of the NH_2 group (Section 11-5, Figure 11-13). Tertiary amines do not give rise to such signals, because they do not have a hydrogen that is bound to nitrogen. Figure 21-3 shows the infrared spectrum of cyclohexanamine.

Nuclear magnetic resonance spectroscopy also may be useful for detecting the presence of amino groups. Amine hydrogens resonate to give often broadened peaks, like the OH signal in the NMR spectra of alcohols. Their chemical shift depends mainly on the rate of exchange of protons with water in the solvent and the degree of hydrogen bonding. Figure 21-4 shows the ^{1}H NMR spectrum of azacyclohexane (piperidine), a cyclic secondary amine. The amine hydrogen appears at $\delta = 1.29$ ppm, and there are two other sets of signals, at $\delta = 1.52$ and 2.73 ppm. The absorption at lowest field can be assigned to the hydrogens neighboring the nitrogen, which are deshielded by the nearby electronegative nitrogen atom.

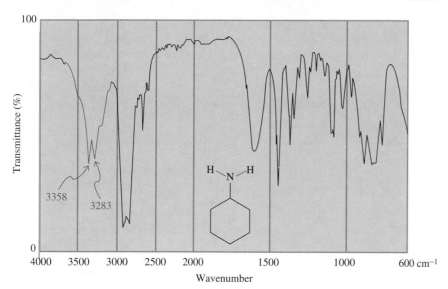

FIGURE 21-3

Infrared spectrum of cyclohexanamine. The amine exhibits two strong peaks between 3250 and 3500 cm^{-1}, characteristic of the N–H stretching absorptions of the primary amine functional group. Note also the broad band near 1600 cm^{-1}, which results from scissoring motions of the N–H bonds.

EXERCISE 21-4

Would you expect the hydrogens next to the heteroatom in an amine, RCH_2NH_2, to be more or less deshielded than those in an alcohol, RCH_2OH? Explain. (**Hint:** See Exercise 21-3.)

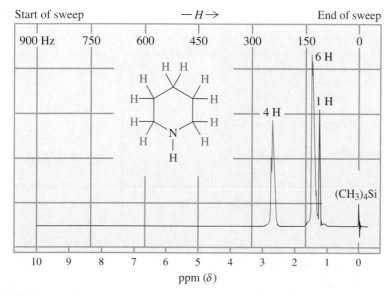

FIGURE 21-4

90 MHz ^{1}H NMR spectrum of azacyclohexane (piperidine) in dry CCl_4. Like the OH hydrogen signal in alcohols, the amine NH peak may appear almost anywhere in the normal hydrogen chemical-shift range. Here $\delta = 1.29$ ppm, and the NH absorption is sharp, because of the use of dry solvent.

The ^{13}C NMR spectra of amines show a similar trend: Carbons directly bound to nitrogen resonate at considerably lower field than do the carbon atoms in alkanes. However, as in the hydrogen spectra (Exercise 21-4), nitrogen is less deshielding than oxygen.

^{13}C Chemical Shifts in Various Amines (ppm)

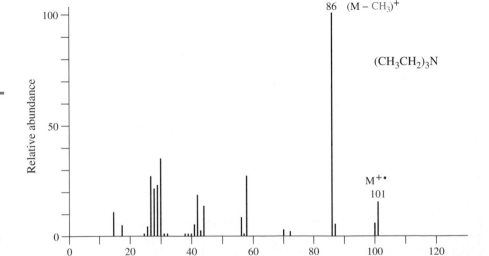

Mass spectrometry readily establishes the presence of nitrogen in an organic compound. Unlike carbon, which is tetravalent, nitrogen is trivalent. Because of these valence requirements and because nitrogen has an even atomic weight (14), molecules incorporating one nitrogen (or any odd number of nitrogens) have an *odd* molecular weight (recall Exercise 20-26). For example, the mass spectrum of *N,N*-diethylethanamine (triethylamine) shows the peak for the molecular ion at $m/z = 101$ (Figure 21-5). The base peak, at $m/z = 86$, is caused by the loss of a methyl group by α-cleavage (Section 20-10). Such a fragmentation is favored because it results in a resonance-stabilized **iminium ion.**

Mass-Spectral Fragmentation of *N,N*-Diethylethanamine

$$(CH_3CH_2)_2\overset{\bullet+}{N}CH_2 \!\!\!-\!\!\!\!\!| CH_3 \longrightarrow CH_3\cdot + \left[(CH_3CH_2)_2\overset{+}{N}\!\!=\!\!CH_2 \longleftrightarrow (CH_3CH_2)_2\overset{\bullet\bullet}{N}\!\!-\!\!\overset{+}{C}H_2 \right]$$

$m/z = 101$
N,N-**Diethylethanamine**
(Triethylamine)

$m/z = 86$
Iminium ion

The rupture of the C–C bond next to nitrogen is frequently so easy that the molecular ion cannot be observed. For example, in the mass spectrum of 1-hexanamine,

FIGURE 21-5

Mass spectrum of *N,N*-diethylethanamine (triethylamine), showing a molecular ion peak at $m/z = 101$. In general, molecules incorporating one nitrogen atom have an odd molecular weight. The base peak is due to loss of a methyl group, resulting in an iminium ion with $m/z = 86$.

the molecular ion ($m/z = 101$) is barely visible; the dominating peak corresponds to the methyleneiminium fragment $[CH_2{=}NH_2]^+$ ($m/z = 30$).

EXERCISE 21-5

What approximate spectral data (IR, NMR, m/z) would you expect for *N*-ethyl-2,2-dimethylpropanamine, shown in the margin?

N-**Ethyl-2,2-dimethyl-propanamine**

In summary, the IR stretching absorption of the N–H bond ranges between 3250 and 3500 cm^{-1}; the corresponding ^{1}H NMR peak is often broad and can be found at variable δ. The electron-withdrawing nitrogen deshields neighboring carbons and hydrogens, although to a lesser extent than the oxygen in alcohols and ethers. The mass spectra of simple alkanamines that contain only one nitrogen atom have odd-numbered molecular ion peaks, because of the trivalent character of nitrogen. Fragmentation occurs in such a way as to produce resonance-stabilized iminium ions.

21-4 Acidity and Basicity of Amines

Like the alcohols (Section 8-3), amines are both basic and acidic. Because nitrogen is less electronegative than oxygen, the acidity of amines is about 20 orders of magnitude less than that of comparable alcohols. Conversely, the lone pair is much more available for protonation, thereby causing amines to be better bases.

Acidity and Basicity of Amines

Amine acting as an acid: RNH–H + $^-$:B $\underset{}{\overset{K_a}{\rightleftharpoons}}$ RNH$^-$ + HB

Amine acting as a base: RNH$_2$ + HA $\underset{}{\overset{K_b}{\rightleftharpoons}}$ RNH$_2^+$–H + $^-$:A

Amines are very weak acids

We have seen evidence that amines are much less acidic than alcohols: Amide ions, R$_2$N$^-$, are used to deprotonate alcohols (Section 9-1). The equilibrium of this proton transfer is strongly shifted to the side of the alkoxide ion. The high value of the equilibrium constant, about 10^{20}, is due to the strong basicity of amide ions, which is manifested, in turn, in the low acidity of amines. The pK_a of ammonia and alkanamines is on the order of 35.

Acidity of Amines

RNH–H + H$_2$O $\underset{}{\overset{K_a}{\rightleftharpoons}}$ RNH$^-$ + H$_2$OH$^+$ $\qquad K_a = \dfrac{[\text{RNH}^-][\text{H}_2\text{OH}^+]}{[\text{RNH}_2]} \sim 10^{-35}$

Amine (Weak acid)   Amide ion (Strong base)   p$K_a \sim 35$

CHEMICAL HIGHLIGHT `21-2`

Separation of Amines from Other Organic Compounds by Aqueous Extraction Techniques

We can exploit the basicity of amines and, more generally, the acid-base properties of a number of functional groups in their purification by aqueous extraction of the corresponding salts. These procedures make use of the fact that salts are, in general, water soluble, whereas their neutral conjugate acids or bases are not. Thus, for example, the basic amines can be extracted from organic solvents into an *acidic* aqueous phase in the form of their ammonium salts. Conversely, acidic compounds, such as carboxylic acids, are removed from an organic medium by extraction with *basic* water, in which they are soluble as carboxylates. These processes can then be reversed by respective basification or acidification of the aqueous layers, ultimately allowing the separation of basic, neutral, and acidic components of a mixture, a problem encountered frequently in the isolation of natural products or organic synthesis.

A flow chart that describes such a separation procedure is shown on page 947.

First, a solution of the mixture in an organic solvent is treated with mild aqueous base, such as NaHCO$_3$, thereby converting any carboxylic acid that may be present into its sodium salt. This salt, being water soluble, moves from the organic into the aqueous layer. After the layers are separated, the carboxylic acid may be brought out of the water

Extraction of solutes from an organic into an aqueous layer (and the reverse) is accomplished by shaking the mixture in a separating funnel and then letting the two phases separate, as shown. Physical separation of the layers occurs by careful draining through the stopcock opening at the bottom.

The deprotonation of amines requires extremely strong bases, such as alkyllithium reagents. For example, lithium diisopropylamide, the special sterically hindered base used in some bimolecular elimination reactions (Section 7-8), is made in the laboratory by treatment of *N*-(1-methylethyl)-2-propanamine (diisopropylamine) with butyllithium.

Preparation of LDA

$$
\underset{\substack{\textit{N}\text{-(1-Methylethyl)-}\\ \text{2-propanamine}\\ \text{(Diisopropylamine)}}}{\overset{\displaystyle \text{H}_3\text{C} \quad\; \text{CH}_3}{\underset{\overset{|}{\text{H}}}{\underset{|}{\text{CH}_3\text{CHNCHCH}_3}}}} \quad
\xrightarrow[{-\ \text{CH}_3\text{CH}_2\text{CH}_2\text{CH}_2\text{H}}]{\text{CH}_3\text{CH}_2\text{CH}_2\text{CH}_2\text{Li}} \quad
\underset{\substack{\textbf{Lithium}\\ \textbf{diisopropylamide, LDA}}}{\overset{\displaystyle \overset{\text{Li}^+}{\text{H}_3\text{C} \quad\; \text{CH}_3}}{\underset{|}{\underset{|}{\text{CH}_3\text{CHNCHCH}_3}}}}
$$

solution by acidification with strong mineral acid, such as HCl. Next, the organic solution that remains is treated with acid, thus converting any amine present into the corresponding water-soluble ammonium salt, which moves into the water phase.

Separation of the layers gives an organic solution, which contains the neutral substances. The amine is removed from water solution by treatment with strong base, such as NaOH.

Separation of Acidic (Carboxylic Acid), Basic (Amine), and Neutral (Haloalkane) Organic Compounds by Aqueous Extraction

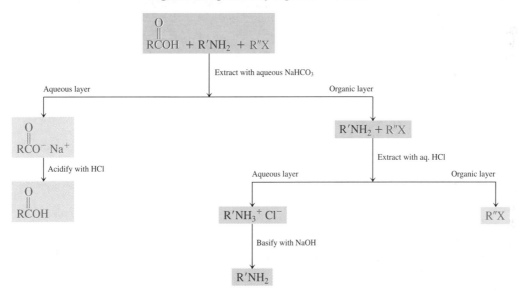

An alternative synthesis of amide ions is the treatment of amines with alkali metals. Alkali metals dissolve in amines (albeit relatively slowly) with the evolution of hydrogen and the formation of amine salts (much like they dissolve in water and alcohol, furnishing H_2 and metal hydroxides or alkoxides, Section 9-1). For example, sodium amide can be made in liquid ammonia from sodium metal in the presence of catalytic amounts of Fe^{3+}, which facilitates electron transfer to the amine. In the absence of such a catalyst, sodium simply dissolves in ammonia (labeled "Na, liquid NH_3") to form a strongly reducing solution (Section 13-7).

Preparation of Sodium Amide

$$2\,Na\ +\ 2\,NH_3$$

$$\downarrow \text{Catalytic } Fe^{3+}$$

$$2\,NaNH_2\ +\ H_2$$

Amines are moderately basic

Amines deprotonate water to a *small* extent to form ammonium and hydroxide ions. Thus, amines are more strongly basic than alcohols but not nearly as basic as alkoxides; their pK_b values (Section 2-9) are about 4.

Basicity of Amines

$$RNH_2 + HOH \underset{K_b}{\rightleftharpoons} \overset{\overset{H}{|}}{RNH_2^+} + HO:^- \qquad K_b = \frac{[RNH_2][HO:^-]}{[RNH_2]} \sim 10^{-4}$$

Amine   Ammonium $pK_b \sim 4$
                    ion

Alkanamines are slightly more basic than ammonia, because of the electron-donating property of alkyl groups (Sections 7-5 and 16-1), which stabilizes the positive charge on the ammonium nitrogen, but they are less basic than hydroxide ion ($pK_b = -1.7$).

pK_b Values of a Series of Simple Amines

NH_3	CH_3NH_2	$(CH_3)_2NH$	$(CH_3)_3N$
$pK_b = 4.76$	3.38	3.27	4.21

The protonation of amines gives ammonium salts. These salts can be primary, secondary, or tertiary, depending on the number of substituents on nitrogen.

$$RNH_2 + H^+Cl^- \longrightarrow RNH_3^+Cl^-$$
Primary ammonium chloride

$$R_2NH + H^+Br^- \longrightarrow R_2NH_2^+Br^-$$
Secondary ammonium bromide

$$R_3N: + H^+I^- \longrightarrow R_3NH^+I^-$$
Tertiary ammonium iodide

$CH_3NH_3^+Cl^-$
Methylammonium chloride

Ammonium salts are named by attaching the substituent names to the ending *-ammonium* followed by the name of the anion.

You will note that, contrary to expectation, the pK_b values of the alkanamines do not decrease in a regular way (i.e., the amine basicities do not increase) with increasing alkyl substitution. This finding is due to increasing steric disruption of solvation (Section 8-3) of the ammonium ions resulting on protonation. This effect counteracts the inductive donor properties of the alkyl groups. Indeed, in the gas phase, the basicity trend is as expected: (pK_b) $NH_3 > CH_3NH_2 > (CH_3)_2NH > (CH_3)_3N$.

The nitrogen lone pairs in arenamines and imines are less available for protonation

The basicity of amines is strongly affected by the electronic environment around the nitrogen. For example, benzenamine (aniline) is considerably less basic than its saturated analog cyclohexanamine (and other primary amines). There are two reasons for this effect. One is the sp^2 hybridization of the aromatic carbon attached to nitrogen in benzenamine, which renders it relatively electron withdrawing (Sections 11-3

$:NH_2$

$pK_b = 9.37$

Benzenamine
(Aniline)

and 13-2), in this way making the nitrogen lone pair less available for protonation. The second is resonance: The lone pair is said to be "tied up" by resonance with the aromatic π system (Section 16-1).

As might be expected, the hybridization at nitrogen itself also drastically affects basicity in the order: $:NH_3 > R_2C=\ddot{N}R' > RC\equiv N:$, a phenomenon that we already encountered in the discussion of the relative acidity of alkanes, alkenes, and alkynes (Section 13-2). Thus, imines have pK_b values estimated to be on the order of 7 to 9; nitriles are even less basic ($pK_b > 20$).

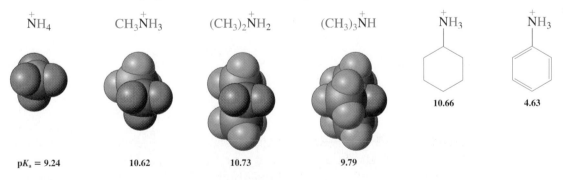

$pK_b = 3.34$

Cyclohexanamine

Ammonium ions are weakly acidic

It is useful to view the basicity of the amines as a measure of the acidity of their conjugate acids (Section 2-9), the ammonium ions. These species are stronger acids than water or alcohols but much weaker than carboxylic acids.

Acidity of Ammonium Ions

$$RNH_2 \xrightarrow{H} + H_2\ddot{O} \overset{K_a}{\rightleftharpoons} R\ddot{N}H_2 + H_2\overset{+}{\ddot{O}}H \qquad K_a = \frac{[R\ddot{N}H_2][H_2\overset{+}{\ddot{O}}H]}{[R\overset{|+}{\underset{H}{N}H_2}]} \sim 10^{-10}$$

$$pK_a \sim 10$$

The pK_b value is easily converted into the pK_a of the conjugate acid by using the relation $pK_a + pK_b = 14$ (Section 2-9).

pK_a Values of a Series of Simple Ammonium Ions*

$\overset{+}{N}H_4$	$CH_3\overset{+}{N}H_3$	$(CH_3)_2\overset{+}{N}H_2$	$(CH_3)_3\overset{+}{N}H$	$\overset{+}{N}H_3$	$\overset{+}{N}H_3$
				10.66	4.63
$pK_a = 9.24$	10.62	10.73	9.79		

In summary, amines are poor acids and require alkyllithium reagents or alkali metal treatment to form amide ions. In contrast, they are good bases, although they are weaker than hydroxide ion.

---

*A confusing practice in the literature is to refer to the pK_a value of an ammonium ion as being that of the neutral amine. In the statement "the pK_a of methanamine is 10.62," what is meant is the pK_a of the methylammonium ion. The pK_a of methanamine is actually 35.

21-5 Synthesis of Amines by Alkylation

Amines can be synthesized by alkylating nitrogen atoms. Several such procedures take advantage of an important property of the nitrogen in many compounds: It is *nucleophilic*.

Amines can be derived from other amines

As nucleophiles, amines react with haloalkanes to give ammonium salts (Section 6-3). Unfortunately, this reaction is not clean, because the resulting amine product usually undergoes further alkylation. How does this complication arise?

Consider the alkylation of ammonia with bromomethane. When this transformation is carried out with equimolar quantities of starting materials, the weakly acidic product (methylammonium bromide), as soon as it is formed, (reversibly) donates a proton to the starting, weakly basic ammonia. The small quantities of methanamine generated in this way then compete effectively with the ammonia for the alkylating agent, and this further methylation generates a dimethylammonium salt.

The process does not stop there, either. This salt can donate a proton to either of the other two nitrogen bases present, furnishing *N*-methylmethanamine (dimethylamine). This compound constitutes yet another nucleophile competing for bromomethane; its further reaction leads to *N,N*-dimethylmethanamine (trimethylamine) and, eventually, to tetramethylammonium bromide, a *quaternary* ammonium salt. The final outcome is a mixture of alkylammonium salts and alkanamines.

Methylation of Ammonia

FIRST ALKYLATION. Gives primary amine

$$H_3N: \ + \ CH_3Br \ \longrightarrow \ CH_3\overset{+}{N}H_3 \ Br^-$$

Methylammonium bromide

$$CH_3\overset{+}{N}H_2 \ Br^- \ + \ :NH_3 \ \rightleftharpoons \ CH_3\overset{..}{N}H_2 \ + \ H\overset{+}{N}H_3 \ Br^-$$
$$|$$
$$H$$

Methanamine
(Methylamine)

SUBSEQUENT ALKYLATION. Gives secondary, tertiary, and quaternary amines or ammonium salts, respectively

$$CH_3\overset{..}{N}H_2 \ \xrightarrow[\substack{- \ Amine \\ hydrobromide}]{CH_3Br} \ (CH_3)_2\overset{..}{N}H \ \xrightarrow[\substack{- \ Amine \\ hydrobromide}]{CH_3Br} \ (CH_3)_3N: \ \xrightarrow{CH_3Br} \ (CH_3)_4N^+ \ Br^-$$

N-Methylmethanamine   **N,N-Dimethylmethanamine**   **Tetramethylammonium**
(Dimethylamine)   **(Trimethylamine)**   **bromide**

The mixture of products obtained on treatment of haloalkanes with ammonia or amines is a serious drawback that limits the usefulness of direct alkylation in synthesis. As a result, indirect alkylation methods are frequently applied.

EXERCISE 21-6

Like other amines, benzenamine (aniline) can be benzylated with chloromethylbenzene (benzyl chloride), $C_6H_5CH_2Cl$. In contrast with the reaction with alkanamines, which proceeds at room temperature, this transformation requires heating to between 90° and 95°C. Explain. (**Hint:** See Section 21-4.)

Indirect alkylation is more effective

Controlled alkylation of amines requires a nitrogen-containing nucleophile that will undergo reaction *only once*. For example, cyanide ion, ^-CN, turns primary and secondary haloalkanes into nitriles, which can be subsequently reduced to the corresponding amines (Section 20-8). This sequence allows the conversion $RX \rightarrow RCH_2NH_2$. Note, however, that this method introduces an additional carbon into the haloalkane framework.

Conversion of a Haloalkane into the Homologous Amine by Cyanide Displacement–Reduction

$$RX + {}^-CN \longrightarrow RC{\equiv}N + X^-$$

$$RC{\equiv}N \xrightarrow{\text{LiAlH}_4 \text{ or H}_2\text{, metal catalyst}} RCH_2NH_2$$

$$\text{Br(CH}_2)_8\text{Br} + \text{NaCN} \xrightarrow[-\,2\text{NaBr}]{\text{DMSO}} \underset{93\%}{\text{NC(CH}_2)_8\text{CN}} \xrightarrow{\text{H}_2\text{, Raney Ni, 100 atm}} \underset{80\%}{\text{H}_2\text{NCH}_2(\text{CH}_2)_8\text{CH}_2\text{NH}_2}$$

1,8-Dibromooctane   **Decanedinitrile**   **1,10-Decanediamine**

To convert a haloalkane selectively into the corresponding amine without additional carbons requires a modified nitrogen nucleophile which should be unreactive after the first alkylation. Such a nucleophile is the **azide ion**, N_3^-, which reacts with haloalkanes to furnish **alkyl azides**. These azides in turn are reduced by catalytic hydrogenation (Pd-C) or by lithium aluminum hydride to the primary amines.

$$R{-}N_3$$
Alkyl azide

Azide Displacement–Reduction

3-Cyclopentylpropyl azide   91%

3-Cyclopentylpropanamine   89%

A nonreductive approach to synthesizing primary amines uses the (commercially available) anion of 1,2-benzenedicarboximide (phthalimide), the imide of 1,2-benzenedicarboxylic (phthalic) acid. This process is also known as the **Gabriel*** **synthesis.** Because the nitrogen in the imide is adjacent to two carbonyl functions, the acidity of the NH group ($pK_a = 8.3$) is much greater than that of an ordinary amide ($pK_a = 22$, Section 20-7). Deprotonation can therefore be achieved with as mild a base as carbonate ion, the resulting anion being monoalkylated in good yield. The amine subsequently can be liberated by acidic hydrolysis, initially as the ammonium salt. Base treatment of the salt then produces the free amine.

*Professor Siegmund Gabriel (1851–1924), University of Berlin.

Gabriel Synthesis of a Primary Amine

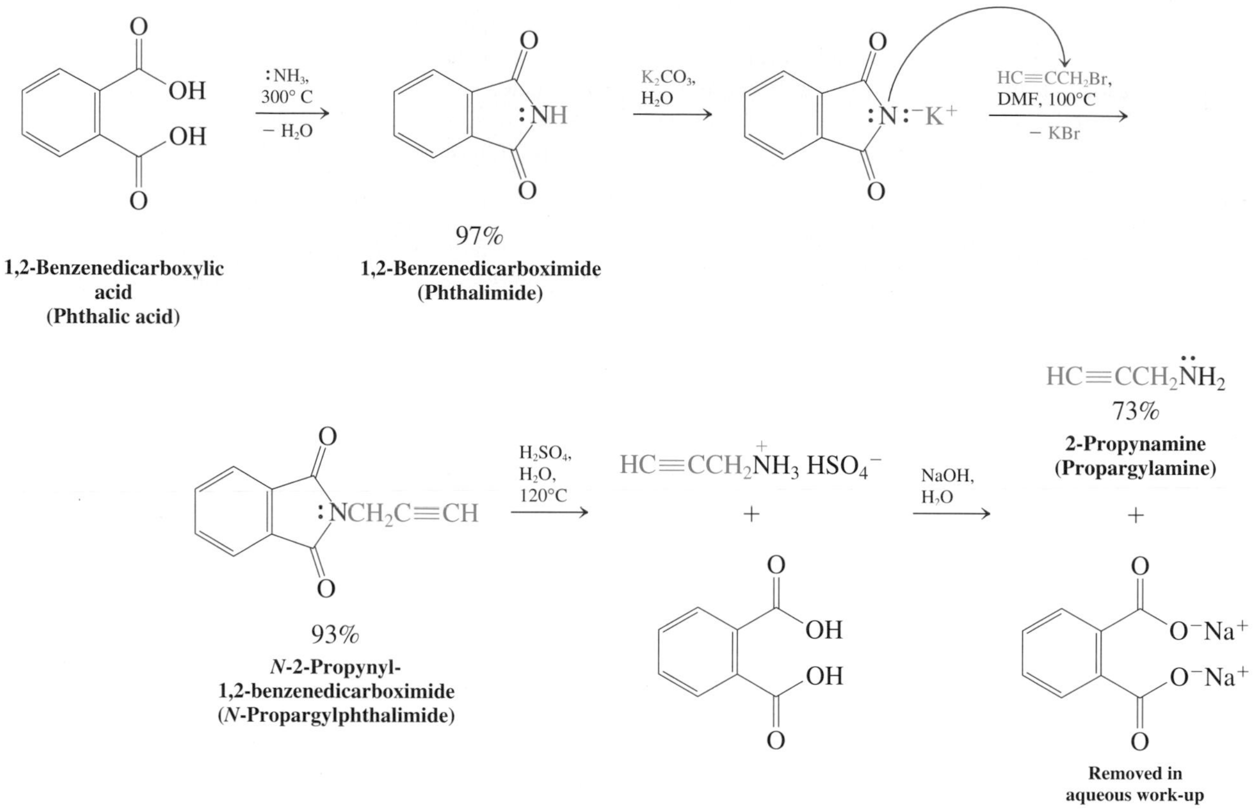

1,2-Benzenedicarboxylic
acid
(Phthalic acid)

97%
1,2-Benzenedicarboximide
(Phthalimide)

93%
N-2-Propynyl-
1,2-benzenedicarboximide
(*N*-Propargylphthalimide)

HC≡CCH₂N̈H₂
73%
2-Propynamine
(Propargylamine)

Removed in
aqueous work-up

The cleavage of an *N*-alkyl-1,2-benzenedicarboximide (*N*-alkyl phthalimide) is frequently
carried out with base or with hydrazine, H_2NNH_2. The respective products of these two
treatments are the 1,2-benzenedicarboxylate A or the hydrazide B. Write mechanisms for
these two transformations. (**Hint:** Review Section 20-6.)

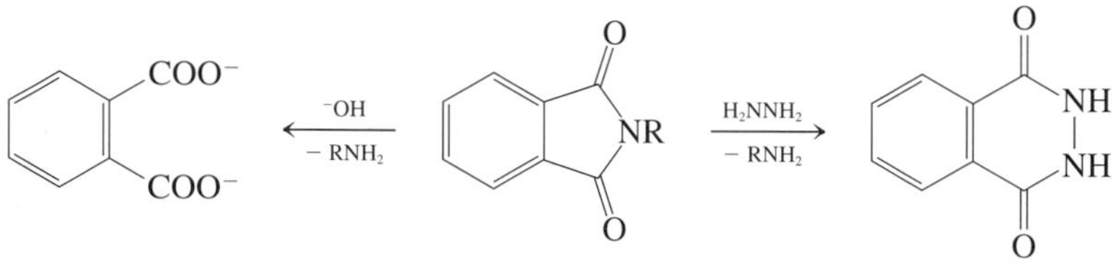

A

B

Show how you would apply the Gabriel method to the synthesis of each of the following amines. (**a**) 1-Hexanamine; (**b**) 3-methylpentanamine; (**c**) cyclohexanamine;
(**d**) $H_2NCH_2CO_2H$, the amino acid glycine. (**Hint:** During the last of these syntheses, the
carboxylic acid group should be protected as an ester. Can you see why?) For each of
these four syntheses, would the azide displacement–reduction method be equally good,
better, or worse?

To summarize, amines can be made from ammonia or other amines by simple alkylation, but this method gives mixtures and poor yields. It is better to use step-by-step

methods that employ nitrile and azide groups, or protected systems, such as 1,2-benzenedicarboxylic imide (phthalimide) in the Gabriel synthesis.

21-6 Synthesis of Amines by Reductive Amination

Another method of amine synthesis, called **reductive amination** of aldehydes and ketones, begins by the condensation of amines with carbonyl compounds to produce imines (Section 17-9). Like the carbon–oxygen double bond in aldehydes and ketones, the carbon–nitrogen double bond in imines can then be reduced by catalytic hydrogenation or by hydride reagents. Reductive amination may be performed on either aldehydes or ketones.

Reductive Amination of a Ketone

This reaction succeeds because of the selectivity of the reducing agents: catalytically activated hydrogen or sodium cyanoborohydride, $Na^{+-}BH_3CN$. Both react faster with the imine double bond than with the carbonyl group under the conditions employed. With $Na^{+-}BH_3CN$, the conditions are relatively acidic (pH = 2-3), which allows for effective activation of the imine double bond by protonation on nitrogen and thus facilitates hydride attack at carbon. The relative stability of the modified borohydride reagent at such low pH (at which $NaBH_4$ hydrolyzes; Section 8-6) is due to the presence of the electron-withdrawing cyano group, which renders the hydrogens less basic (hydridic). In a typical procedure, the carbonyl component and the amine are allowed to equilibrate with the imine and water in the presence of the reducing agent.

Amine Synthesis by Reductive Amination

Reductive aminations with secondary amines give the corresponding *N,N*-dialkylamino derivatives.

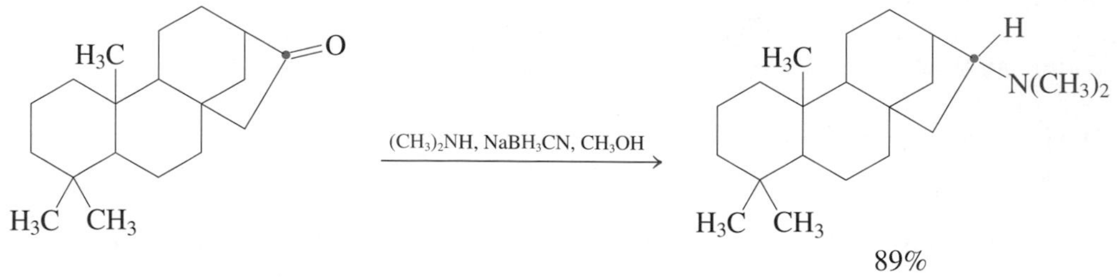
H3C
O
(CH3)2NH, NaBH3CN, CH3OH
H3C CH3
H
N(CH3)2
89%

The reaction proceeds through the intermediacy of an iminium ion, which is reduced by addition of H^- from cyanoborohydride.

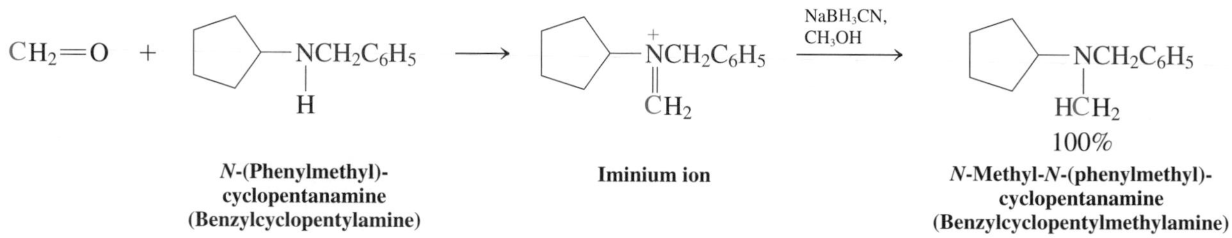
CH2=O +
NCH2C6H5
H
NCH2C6H5
CH2
NaBH3CN, CH3OH
NCH2C6H5
HCH2
100%
N-(Phenylmethyl)-cyclopentanamine (Benzylcyclopentylamine)
Iminium ion
N-Methyl-N-(phenylmethyl)-cyclopentanamine (Benzylcyclopentylmethylamine)

EXERCISE 21-9

Formulate a mechanism for the reductive amination with the secondary amine shown in the preceding example.

EXERCISE 21-10

Explain the following transformation by a mechanism. (**Hint:** The mechanism proceeds through two consecutive reductive aminations.)

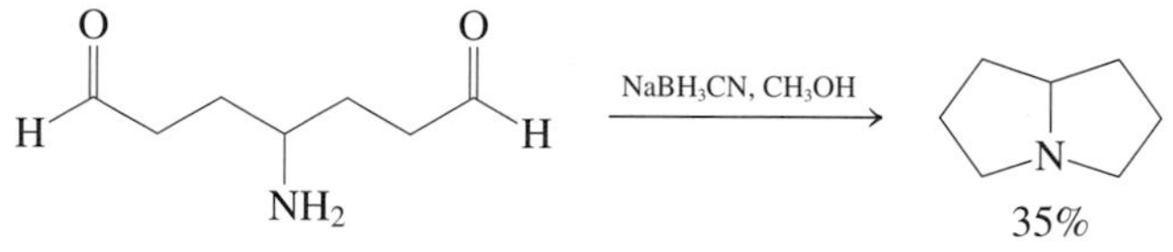
O
H
NH2
O
H
NaBH3CN, CH3OH
N
35%

In summary, reductive amination furnishes alkanamines by reductive condensation of amines with aldehydes and ketones.

21-7 Synthesis of Amines from Carboxylic Amides

Carboxylic amides can be versatile precursors of amines (Section 20-6). Recall that amides are readily available by reaction of alkanoyl halides with amines. Reduction with lithium aluminum hydride then converts them into amines.

Utility of Amides in Amine Synthesis

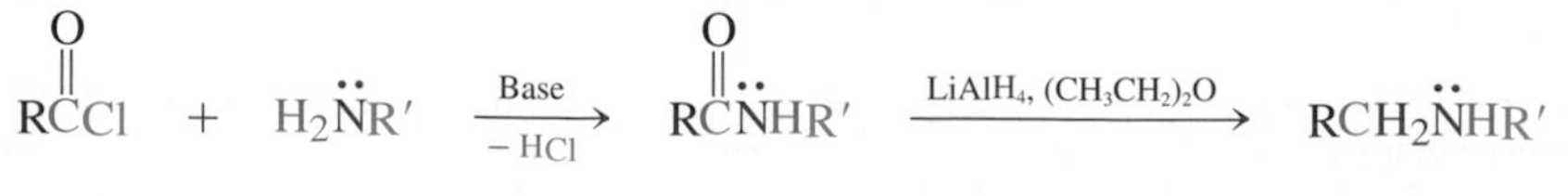
O
RCCl + H2NR'
Base
− HCl
O
RCNHR'
LiAlH4, (CH3CH2)2O
RCH2NHR'

Primary amides can also be turned into amines by oxidation with bromine or chlorine in the presence of sodium hydroxide—in other words, by the Hofmann rearrangement (Section 20-7). Recall that in this transformation the carbonyl group is extruded as carbon dioxide, so the resulting amine bears one carbon fewer than the starting material.

Amines by Hofmann Rearrangement

$$RCNH_2 \xrightarrow{\text{Br}_2, \text{ NaOH, H}_2\text{O}} RNH_2 \ + \ O{=}C{=}O$$

with O double-bonded above the carbonyl carbon of $RCNH_2$.

EXERCISE 21-11

Suggest synthetic methods for the preparation of *N*-methylhexanamine from hexanamine (two syntheses) and from *N*-hexylmethanamide (*N*-hexylformamide).

In summary, amides can be reduced to amines by treatment with lithium aluminum hydride. The Hofmann rearrangement converts amides into amines with loss of the carbonyl group.

21-8 Quaternary Ammonium Salts: Hofmann Elimination

Much like the protonation of alcohols, which turns the –OH into the better leaving group $^+\text{OH}_2$ (Section 9-2), you could envisage that protonation of amines would render the resulting ammonium salts subject to nucleophilic attack. In practice, however, amines are not sufficiently good leaving groups (they are more basic than water) to partake in substitution reactions. Nevertheless, they can function as such in the **Hofmann* elimination,** in which a tetraalkylammonium salt is converted to an alkene with base.

Section 21-5 described how peralkylation of amines leads to the corresponding quaternary alkylammonium salts. These species are unstable in the presence of strong base, because of a bimolecular elimination reaction that furnishes alkenes (Section 7-7). The base attacks the hydrogen in the β-position with respect to the nitrogen, and a trialkylamine departs as a neutral leaving group.

Bimolecular Elimination of Quaternary Ammonium Ions

$$\overset{\displaystyle \ \ \ \overset{+}{N}R_3}{\underset{\underset{\displaystyle ^-{:}\overset{..}{O}H}{\uparrow}}{\underset{\displaystyle H}{C{-}C}}} \longrightarrow \quad C{=}C \ + \ HOH \ + \ :NR_3$$

Alkene

In the procedure of Hofmann elimination, the amine is first completely methylated with excess iodomethane **(exhaustive methylation)** and then treated with wet silver oxide (a source of HO^-) to produce the ammonium hydroxide. Heating degrades this salt to the alkene. When more than one regioisomer is possible, Hofmann elimination,

---

*This is the Hofmann of the Hofmann rule for E2 reactions (Section 11-8) and the Hofmann rearrangement (Section 20-7).

in contrast with most E2 processes, tends to give less substituted alkenes as the major products.

Hofmann Elimination of Butanamine

$$CH_3CH_2CH_2CH_2NH_2 \xrightarrow{\text{Excess } CH_3I,\ K_2CO_3,\ H_2O} CH_3CH_2CH_2CH_2\overset{+}{N}(CH_3)_3\ I^- \xrightarrow[-AgI]{Ag_2O,\ H_2O}$$

Butanamine   **Butyltrimethylammonium
iodide**

**Butyltrimethylammonium
hydroxide**   **1-Butene**

$$\xrightarrow{\Delta} CH_3CH_2CH{=}CH_2 \ + \ HOH \ + \ N(CH_3)_3$$

EXERCISE 21-12

Give the structures of the possible alkene products of the Hofmann elimination of **(a)** N-ethylpropanamine (ethylpropylamine) and **(b)** 2-butanamine.

The Hofmann elimination of amines has been used to elucidate the structure of nitrogen-containing natural products, such as alkaloids (Section 25-8). Each sequence of exhaustive methylation and Hofmann elimination cleaves one C–N bond. Repeated cycles allow the heteroatom to be precisely located, particularly if it is part of a ring. In this case, the first carbon–nitrogen bond cleavage opens the ring.

N-Methylazacycloheptane   **N,N-Dimethyl-5-hexenamine**   **1,5-Hexadiene**

EXERCISE 21-13

Why is exhaustive *methyl*ation and not, say, ethylation used in Hofmann eliminations for structure elucidation? (**Hint:** Look for other possible elimination pathways.)

EXERCISE 21-14

An unknown amine of the molecular formula $C_7H_{13}N$ has a ^{13}C NMR spectrum containing only three lines of $\delta = 21.0$, 26.8, and 47.8 ppm. Three cycles of Hofmann elimination are required to form 3-ethenyl-1,4-pentadiene (trivinylmethane) and its double-bond isomers (as side products arising from base-catalyzed isomerization). Propose a structure for the unknown.

In summary, quaternary methyl ammonium salts, synthesized by amine methylation, undergo bimolecular elimination in the presence of base to give alkenes.

21-9 Mannich Reaction: Alkylation of Enols by Iminium Ions

In the aldol reaction, an enolate ion attacks the carbonyl group of an aldehyde or ketone (Section 18-5) to furnish a β-hydroxycarbonyl product. A process that is quite analogous is the **Mannich* reaction.** Here, however, it is an enol that functions as the nucleophile and an iminium ion, derived by condensation of a second carbonyl component with amines, as the substrate. The outcome is a β-aminocarbonyl product.

To differentiate the reactivity of the three components of the Mannich reaction, it is usually carried out with a ketone, a more reactive (see Section 17-6) aldehyde (often formaldehyde, CH_2=O), and the amine in an alcohol solvent containing HCl. These conditions give the hydrochloride salt of the product. The free amine, called a **Mannich base,** can be obtained on treatment with base.

Mannich Reaction

$$+ \quad CH_2{=}O \quad + \quad (CH_3)_2NH \quad \xrightarrow{\text{HCl, CH}_3\text{CH}_2\text{OH, } \Delta} \quad \begin{array}{c} CH_2\overset{+}{N}(CH_3)_2 \; Cl^- \\ | \\ H \end{array}$$

85%
Salt of Mannich base

$$\begin{array}{c} CH_3 \\ | \\ CH_3CHCH{=}O \end{array} \quad + \quad CH_2{=}O \quad + \quad CH_3NH_2 \quad \xrightarrow[\text{2. HO}^-, \text{H}_2\text{O}]{\text{1. HCl, CH}_3\text{CH}_2\text{OH, } \Delta} \quad \begin{array}{c} CH_3 \\ | \\ CH_3C{-}CH{=}O \\ | \\ CH_2NHCH_3 \end{array}$$

2-Methylpropanal

70%
2-Methyl-2-(N-methyl-aminomethyl)propanal
Mannich base

The mechanism of this process starts with iminium ion formation between the aldehyde (e.g., formaldehyde) and the amine, on the one hand, and enolization of the ketone, on the other. As soon as it is formed, the enol undergoes nucleophilic attack on the electrophilic iminium carbon, and the resulting species converts to the Mannich salt by proton transfer from the carbonyl oxygen to the amino group.

Mechanism of the Mannich Reaction

STEP 1. Iminium ion formation

$$CH_2{=}O \quad + \quad (CH_3)_2\overset{+}{N}H_2\,Cl^- \quad \longrightarrow \quad CH_2{=}\overset{+}{N}(CH_3)_2\,Cl^- \quad + \quad H_2O$$

STEP 2. Enolization

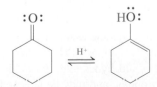

$$:O: \qquad \overset{\cdot\cdot}{H\overset{\cdot\cdot}{O}:}$$

$$\underset{H^+}{\rightleftharpoons}$$

---

*Professor Carl Mannich (1877–1947), University of Berlin.

STEP 3. Carbon–carbon bond formation

STEP 4. Proton transfer

Salt of Mannich base

The following example shows the Mannich reaction in natural product synthesis. In this instance, one ring is formed by condensation of the amino with one carbonyl group. Mannich reaction of the resulting iminium salt with the enol form of the other carbonyl function follows. The product has the framework of retronecine, an alkaloid that is present in many shrubs and is hepatotoxic (causes liver damage) to grazing livestock.

Mannich Reaction in Synthesis

$$\xrightarrow[- \text{H}_2\text{O}]{\text{H}^+} \qquad \xrightarrow{- \text{H}^+} \qquad 52\%$$

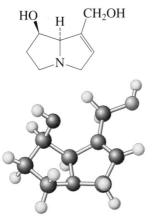

Retronecine

EXERCISE 21-15

Write the products of each of the following Mannich reactions. **(a)** Ammonia + formaldehyde + cyclopentanone; **(b)** 1-hexanamine + formaldehyde + 2-methylpropanal; **(c)** N-methylmethanamine + formaldehyde + propanone; **(d)** cyclohexanamine + formaldehyde + cyclohexanone.

EXERCISE 21-16

β-Dialkylamino alcohols and their esters are useful local anesthetics. Suggest a synthesis of the anesthetic Tutocaine hydrochloride, beginning with 2-butanone.

$$\text{H}_2\text{N}-\!\!\!\!\bigcirc\!\!\!\!-\overset{\overset{\text{O}}{\|}}{\text{C}}\text{OCHCHCH}_2\overset{+}{\text{N}}\text{H}(\text{CH}_3)_2 \ \text{Cl}^-$$
$$\qquad\qquad\qquad\quad \text{H}_3\text{C} \ \ \text{CH}_3$$

Tutocaine hydrochloride

In summary, condensation of aldehydes (e.g., formaldehyde) with amines furnishes iminium ions, which are electrophilic and may be attacked by the enols of ketones (or other aldehydes) in the Mannich reaction. The products are β-aminocarbonyl compounds.

21-10 Nitrosation of Amines: *N*-Nitrosamines and Diazonium Ions

Amines react with nitrous acid, through nucleophilic attack on the **nitrosyl cation,** NO^+. The product depends very much on whether the reactant is an alkanamine or a benzenamine (aniline) and on whether it is primary, secondary, or tertiary. This section deals with alkanamines; aromatic amines will be considered in the next chapter.

To generate NO^+, we must first prepare the unstable nitrous acid by the treatment of sodium nitrite with aqueous HCl. In such an acid solution, an equilibrium is established with the nitrosyl cation. (Compare this sequence with the preparation of the nitronium cation from nitric acid; Section 15-11.)

Nitrosyl Cation from Nitrous Acid

| Sodium nitrite | Nitrous acid | | Nitrosyl cation | |

The nitrosyl cation is electrophilic and is attacked by amines to form an *N*-nitrosammonium salt.

N-Nitrosammonium
salt

The course of the reaction now depends on whether the amine nitrogen bears zero, one, or two hydrogens. *Tertiary N*-nitrosammonium salts are stable only at low temperatures and decompose on heating to give a mixture of compounds. *Secondary N*-nitrosammonium salts are simply deprotonated to furnish the relatively stable **N-nitrosamines** as the major products.

$$R_2N—N=O$$
N-Nitrosamine

88–90%

N-Nitroso**dimethylamine**

Similar treatment of *primary* amines initially gives the analogous monoalkyl-*N*-nitrosamines. However, these products are unstable because of the remaining proton on the nitrogen. By a series of hydrogen shifts, they first rearrange to the corresponding diazohydroxides. Then protonation, followed by loss of water, gives highly reactive **diazonium ions,** $R—N_2^+$. When R is a secondary or a tertiary alkyl group, these ions lose molecular nitrogen, N_2, and form the corresponding carbocations, which may rearrange, deprotonate, or undergo nucleophilic trapping (Section 9-3) to yield the observed mixtures of compounds.

CHEMICAL HIGHLIGHT 21-3

Carcinogenicity of *N*-Nitrosodialkanamines and Cured Meats

N-Nitrosodialkanamines are notoriously potent carcinogens in a variety of animals. Although there is no direct evidence, they are suspected of causing cancer in humans as well. Most nitrosamines appear to cause liver cancer, but certain of them are very organ specific in their carcinogenic potential (bladder, lungs, esophagus, nasal cavity, etc.).

Their mode of carcinogenic action appears to begin with enzymatic oxidation of one of the α-positions, which allows eventual formation of an unstable monoalkyl-*N*-nitrosamine. This compound then decomposes to a carbocation that, as a powerful electrophile, is thought to attack one of the bases in DNA to inflict the kind of genetic damage that seems to lead to cancerous cell behavior.

Nitrosamines have been detected in a variety of cured meats, such as smoked fish, frankfurters (*N*-nitrosodimethylamine), and fried bacon [*N*-nitrosoazacyclopentane (*N*-nitrosopyrrolidine); see Exercise 25-7].

Curing is a process used for centuries to preserve meat. Initially, such treatment was based on salting with sodium chloride, the effect of which was direct or indirect (through drying) prevention of bacterial growth. Near the turn of the century, salting with sodium nitrate, $NaNO_3$, was found to have the desirable side effect of producing an appetizing pink coloration and special flavor in so-treated meat. Later, the origin of this effect was traced to sodium nitrite, $NaNO_2$, produced from $NaNO_3$ by bacterial action during processing. Hence today, $NaNO_2$ is used for curing. It inhibits the growth of bacteria responsible

for botulism, retards development of rancidity and off-odors during storage, and preserves the flavor of added spices and smoke (if the meat is smoked). It evolves nitric oxide, NO, which forms a red complex with the iron in myoglobin. (The oxygen complex of myoglobin gives blood its characteristic color, see Section 26-8). The level of nitrite in foods is closely regulated (<200 ppm), because it is itself toxic (fatal dose: 22–23 mg/kg body weight) and because it may convert natural amines present in the stomach into nitrosamines. However, to place these facts into perspective, less than 10% of our nitrate (nitrite) intake is (on average) from cured meats. The remainder comes from natural sources—namely, vegetables such as spinach, beets, radishes, celery, and cabbages.

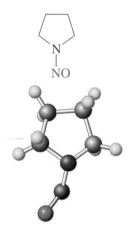

N-Nitroso**azacyclopentane**
(*N*-**Nitrosopyrrolidine**)

Mechanism of Decomposition of Primary *N*-Nitrosamines

STEP 1. Rearrangement to a diazohydroxide

$$
\underset{H}{\overset{R}{\diagdown}}\ddot{N}-\ddot{N}=\ddot{O} \;\underset{-H^+}{\overset{+\,H^+}{\rightleftharpoons}}\; \left[\underset{H}{\overset{R}{\diagdown}}\ddot{N}-\ddot{N}=\overset{+}{\ddot{O}}H \;\longleftrightarrow\; \underset{H}{\overset{R}{\diagdown}}\overset{+}{N}=\ddot{N}-\ddot{O}H \right] \;\underset{+H^+}{\overset{-H^+}{\rightleftharpoons}}\; R-\ddot{N}=\ddot{N}-\ddot{O}H
$$

Diazohydroxide

STEP 2. Loss of water to give a diazonium ion

$$R-\ddot{\underset{\cdot\cdot}{N}}=\ddot{N}-\ddot{\underset{\cdot\cdot}{O}}H \underset{-H^+}{\overset{+H^+}{\rightleftharpoons}} R-\overset{\frown}{N}=\ddot{N}-\overset{+}{\ddot{O}}H_2 \underset{+H_2O}{\overset{-H_2O}{\rightleftharpoons}} R-\overset{+}{N}\equiv N:$$

Diazonium cation

STEP 3. Nitrogen loss to give a carbocation

$$R\overset{\frown}{\underset{-N_2}{-N}}\equiv N: \xrightarrow{\quad} R^+ \longrightarrow \text{product mixtures}$$

EXERCISE 21-17

The result shown in the margin was reported in 1991 and addresses the applicability of the mechanism just shown to the decomposition of diazonium ions with *primary* R groups. Does the same mechanism operate here?

In summary, nitrous acid attacks amines, thereby causing *N*-nitrosation. Secondary amines give *N*-nitrosamines, which are notorious for their carcinogenicity. *N*-Nitrosamines derived from primary amines decompose through carbocations to a variety of products.

$$CH_3CH_2CH_2-C\overset{H}{\underset{NH_2}{\diagup D}}$$

Pure *R* enantiomer

↓ NaNO₂, HCl, H₂O

$$D\overset{H}{\underset{HO}{\diagdown}}C-CH_2CH_2CH_3$$

100%
Pure *S* enantiomer

21-11 Diazomethane, Carbenes, and Cyclopropane Synthesis

The nitrosyl cation also attacks the nitrogen of *N*-methylamides. The products, *N*-methyl-*N*-nitrosamides, are precursors to useful synthetic intermediates.

Nitrosation of an N-Methylamide

$$\underset{\text{\textit{N}-Methylamide}}{\overset{\displaystyle O}{\overset{\|}{\underset{}{R\text{C}}}\text{NHCH}_3}} \xrightarrow[-H^+]{NO^+} \underset{\text{\textit{N}-Methyl-\textit{N}-nitrosamide}}{\overset{\displaystyle O}{\overset{\|}{\underset{\displaystyle \underset{}{NO}}{R\text{C}\underset{|}{\text{N}}\text{CH}_3}}}}$$

Base treatment of *N*-methyl-*N*-nitrosamides gives diazomethane

N-Methyl-*N*-nitrosamides are converted into **diazomethane,** CH_2N_2, on treatment with aqueous base.

Making Diazomethane

$$\underset{\substack{\displaystyle | \\ N=O \\ \textit{N}\text{-Methyl-}\textit{N}\text{-nitrosourea}}}{\overset{\displaystyle O}{\overset{\|}{CH_3 N C NH_2}}} \xrightarrow{KOH, H_2O, (CH_3CH_2)_2O, 0°C} \underset{\textbf{Diazomethane}}{CH_2=\overset{+}{N}=\ddot{N}:^-} + NH_3 + K_2CO_3 + H_2O$$

Diazomethane is used in the synthesis of methyl esters from carboxylic acids (Section 19-9). However, it is exceedingly toxic and highly explosive in the gaseous state (b.p. $-24°C$) and in concentrated solutions. It is therefore usually generated in dilute ether solution and immediately allowed to react with the acid. This method is very mild and permits esterification of molecules possessing acid- and base-sensitive functional groups, as shown in the following example.

$$CH_2N_2, (CH_3CH_2)_2O, CH_3OH$$

75%

Diazomethane forms methylene, which converts alkenes into cyclopropanes

$R_2C:$

Carbene

On exposure to light, heat, or catalytic copper metal, diazomethane extrudes N_2 to give the highly reactive species **methylene,** $H_2C:$, the simplest **carbene.**

$$H_2\overset{..}{C}-\overset{+}{N}\equiv N: \xrightarrow{h\nu \text{ or } \Delta \text{ or } Cu} H_2C: + :N\equiv N:$$

Methylene

When methylene is generated in the presence of compounds containing double bonds, addition takes place to furnish cyclopropanes, usually stereospecifically.

Methylene Additions to Double Bonds

40%

Bicyclo[4.1.0]heptane

50–70%

cis-**Diethylcyclopropane**

EXERCISE 21-18

Diazomethane is the simplest member of the class of compounds called *diazoalkanes* or *diazo compounds,* $R_2C=N_2$. When diazo compound A is irradiated in heptane solution at $-78°C$, it gives a hydrocarbon, C_4H_6, exhibiting three signals in ^1H NMR and two signals in ^{13}C NMR spectroscopy, all in the aliphatic region. Suggest a structure for this molecule.

$$CH_2=CHCH_2CH=\overset{+}{N}=\overset{..}{N}:^-$$

A

Halogenated carbenes and carbenoids also give cyclopropanes

Cyclopropanes may also be synthesized from halogenated carbenes, which are prepared from halomethanes. For example, treatment of trichloromethane (chloroform) with a strong base causes an unusual elimination reaction in which both the proton and the leaving group are removed from the same carbon. The product is dichlorocarbene, which gives cyclopropanes when generated in the presence of alkenes.

Dichlorocarbene from Chloroform and Its Trapping by Cyclohexene

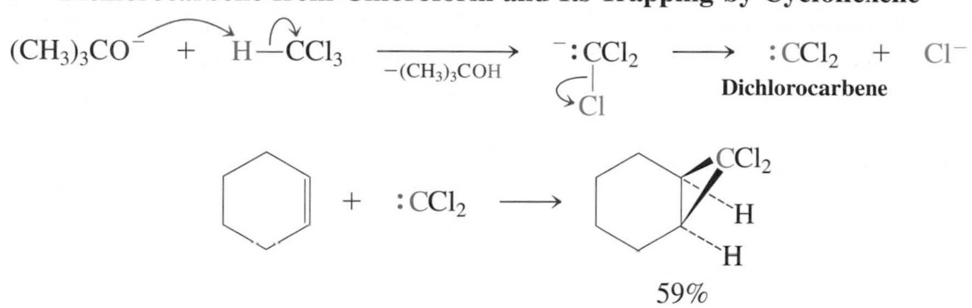

$$(CH_3)_3CO^- \ + \ H-CCl_3 \ \xrightarrow[-(CH_3)_3COH]{} \ ^-:CCl_2 \ \underset{\curvearrowleft Cl}{} \ \longrightarrow \ :CCl_2 \ + \ Cl^-$$

Dichlorocarbene

Cyclohexene $+ \ :CCl_2 \ \longrightarrow$ bicyclic dichlorocyclopropane product

59%

In another route to cyclopropanes, diiodomethane is treated with zinc powder (usually activated with copper) to generate ICH_2ZnI, called the **Simmons-Smith* reagent.** This species is an example of a **carbenoid,** or carbenelike substance, because, like carbenes, it also converts alkenes into cyclopropanes stereospecifically. Use of the Simmons-Smith reagent in cyclopropane synthesis avoids the hazards associated with diazomethane preparation.

Simmons-Smith Reagent in Cyclopropane Synthesis

$$\underset{H_3C}{\overset{H}{>}}C=C\underset{H}{\overset{CH_3}{<}} \ + \ CH_2I_2 \ \xrightarrow[-\text{Metal iodide}]{Zn-Cu, \ (CH_3CH_2)_2O} \ H\cdots\underset{H_3C}{\overset{CH_2}{\underset{|}{C}}}-\underset{H}{\overset{|}{C}}\cdots CH_3$$

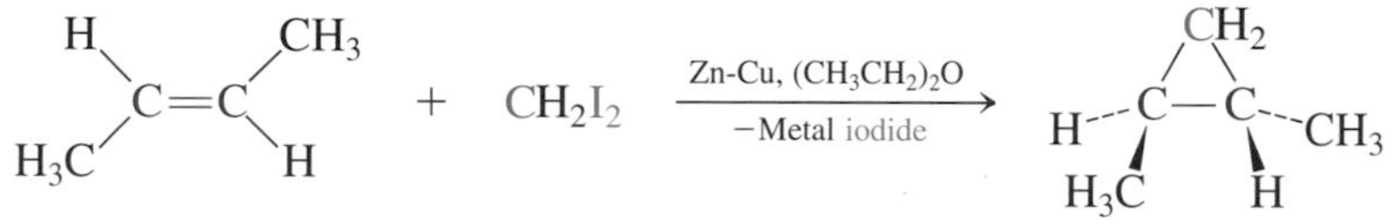

An impressive example of the use of the Simmons-Smith reagent in the construction of natural products is the highly unusual, potent antifungal agent FR-900848 obtained in 1990 from a fermentation broth of *Streptoverticillium fervens* and first synthesized in 1996. Its most noteworthy feature is the fatty acid residue, which contains five cyclopropanes, four of which are contiguous and all of which were made by Simmons-Smith cyclopropanations.

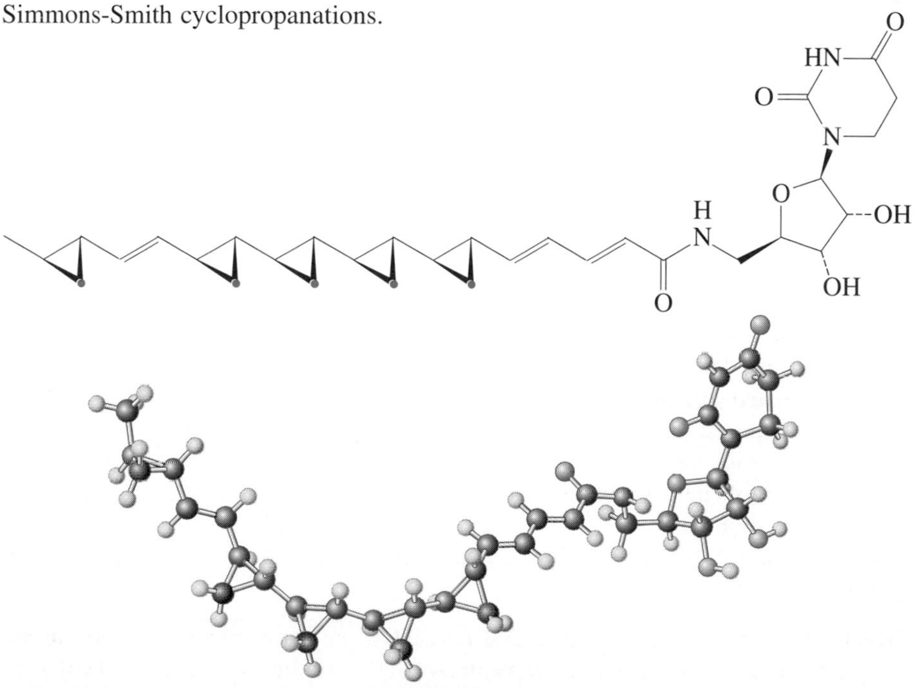

FR-900848

*Dr. Howard E. Simmons (1929–1997) and Dr. Ronald D. Smith (b. 1930), both with E. I. du Pont de Nemours and Company, Wilmington, Delaware.

To summarize, *N*-methylnitrosamides release diazomethane on treatment with hydroxide. Diazomethane is a useful synthetic intermediate in the methyl esterification of carboxylic acids and as a methylene source for forming cyclopropanes from alkenes. Halogenated carbenes, which are formed by dehydrohalogenation of halomethanes, and the Simmons-Smith reagent, a carbenoid arising from the reaction of diiodomethane with zinc, also convert alkenes into cyclopropanes.

21-12 Amines in Industry: Nylon

In addition to their significance in medicine (Chemical Highlight 21-1), the amines have numerous industrial applications. This section deals with one commercially important amine, 1,6-hexanediamine (hexamethylenediamine, HMDA), needed in the manufacture of nylon. This compound is polycondensed with hexanedioic (adipic) acid to produce Nylon 6,6, out of which hosiery, gears, and millions of tons of textile fiber are made.

Polycondensation of Adipic Acid with HMDA

**Hexanedioic acid
(Adipic acid)**

+

**1,6-Hexanediamine
(Hexamethylenediamine)**

Double salt

270°C, 250 psi
$-H_2O$
Polymerization

Nylon 6,6

Nylon is used in the manufacture of many products, including rock-climbing ropes and gear.

Nylon 6,6 is a polyamide formed by condensation of the acid with the diamine under pressure. The high demand for nylon (U.S. production in 1995: 1.2 million tons) stimulated the development of several ingenious cheap syntheses of the monomeric precursors. Originally, Wallace Carothers* at the du Pont Company made the diamine from hexanedioic (adipic) acid. The diacid was turned into hexanedinitrile (adiponitrile) by treatment with ammonia. Finally, catalytic hydrogenation furnished the diamine.

$$HOC(CH_2)_4COH \xrightarrow[-4\,H_2O]{NH_3,\,\Delta} N{\equiv}C(CH_2)_4C{\equiv}N \xrightarrow{H_2,\,Ni,\,130°C,\,2000\,psi} H_2N(CH_2)_6NH_2$$

**Hexanedinitrile
(Adiponitrile)**

**1,6-Hexanediamine
(Hexamethylene-
diamine)**

*Dr. Wallace H. Carothers (1896–1937), E. I. du Pont de Nemours and Company, Wilmington, Delaware.

Later, a still shorter hexanedinitrile synthesis was discovered that used 1,3-butadiene as a starting material. Chlorination of butadiene furnished a mixture of 1,2- and 1,4-dichlorobutene (Section 14-6). This mixture can be directly converted into only the required 3-hexenedinitrile with sodium cyanide in the presence of cuprous cyanide. Selective hydrogenation of the alkenyl double bond then furnishes the desired product.

Hexanedinitrile (Adiponitrile) from 1,3-Butadiene

$$CH_2 = CH - CH = CH_2 \xrightarrow{\text{Cl}-\text{Cl}} ClCH_2CH = CHCH_2Cl \ + \ ClCH_2\overset{\overset{\displaystyle Cl}{|}}{C}HCH = CH_2 \xrightarrow[-NaCl]{\text{CuCN, NaCN}}$$

$$NCCH_2CH = CHCH_2CN \xrightarrow{\text{H}_2,\ \text{catalyst}} NC(CH_2)_4CN$$
3-Hexenedinitrile

In the mid-1960s, Monsanto developed a process that uses a more expensive starting material, but takes just one step: the electrolytic hydrodimerization of propenenitrile (acrylonitrile).

Electrolytic Hydrodimerization of Propenenitrile (Acrylonitrile)

$$2\ CH_2 = CHC \equiv N \ + \ 2\ e \ + \ 2\ H^+ \longrightarrow N \equiv CCH_2CH_2 - CH_2CH_2C \equiv N$$

To counter Monsanto's challenge, du Pont devised yet another synthesis, again starting with 1,3-butadiene, but now eliminating the consumption of chlorine, removing the toxic-waste problems of the disposal of copper salts, and using cheaper hydrogen cyanide rather than sodium cyanide. The synthesis is based on the conceptually simplest approach: direct regioselective anti-Markovnikov addition of two molecules of hydrogen cyanide to butadiene. A transition metal catalyst, such as iron, cobalt, or nickel, is needed to effect this regiocontrol. Typically also required are Lewis acids and phosphines, usually triphenylphosphine, $P(C_6H_5)_3$.

In summary, 1,6-hexanediamine, one of the monomeric components of Nylon 6,6, has been made industrially in a variety of ways, each synthesis improving on previous ones in terms of economy, selectivity, waste disposal, and ease of execution.

Hydrogen Cyanide Addition to 1,3-Butadiene

$$CH_2 = CHCH = CH_2$$
$$+$$
$$2\ HCN$$
$$\downarrow \text{Catalyst}$$
$$NC(CH_2)_4CN$$

CHAPTER INTEGRATION PROBLEM

On the basis of the synthetic methods for amines provided in this chapter, carry out retrosynthetic analyses of Prozac, with 4-trifluoromethylphenol and benzene as your starting materials.

Prozac
**[*R*,*S*-*N*-Methyl-3-(4-trifluoro-
phenoxy)-3-phenyl-1-propanamine]**

SOLUTION

As for any synthetic problem, you can envisage many possible solutions. However, the constraints of given starting materials, convergence, and practicality rapidly narrow the number of available options. Thus, it is clear that the 4-trifluorophenoxy-group is best introduced by Williamson ether synthesis (Section 9-6) on an appropriate benzylic halide, compound A (Sections 15-1 and 22-1), with the use of our first starting compound, 4-trifluoromethylphenol.

Retrosynthetic Step to A

The task therefore reduces to devising routes to A, utilizing the second starting material, benzene, and suitable building blocks. We know that the best way to introduce an alkyl chain into benzene is by Friedel-Crafts alkanoylation (Section 15-14). This reaction would also supply a carbonyl function, as in compound B, which could be readily converted [by reduction to the alcohol (Section 8-6) and conversion of the hydroxy into a good leaving group, X (Chapter 9)] into compound A (or something similar). On paper, an attractive alkanoylating reagent would be $\text{ClCCH}_2\text{CH}_2\text{NHCH}_3$.

However, this reagent carries two functional groups that would react with each other either intermolecularly to produce a polyamide (see Section 21-12) or intramolecularly to produce a β-lactam (Section 20-6). This problem could be circumvented by using a leaving group in place of the amino function, the latter to be introduced by one of the methods in Section 21-5.

Retrosynthesis of A

Consideration of cyanide as a building block (Section 21-5) opens up another avenue of retrosynthetic disconnection of compound B (and therefore compound A) through compound C.

Retrosynthesis of B

The required compound D could be envisaged to arise from ethanoyl (acetyl) benzene (made by Friedel-Crafts acetylation of benzene), followed by acid-catalyzed halogenation of the ketone (Section 18-3). Reduction of the nitrile group in compound C could be carried out with concomitant carbonyl conversion to give a primary amine version of compound A; namely, E. *N*-Methylation might be most conveniently accomplished by reductive amination (although, as we shall see in Chapter 22, the benzylic position may be sensitive to the reductive conditions employed).

Retrosynthesis of E

C	E	Phenyloxacyclo-propane	

Intermediate E can be approached by using nucleophilic cyanide in a different manner—namely, by attack at the less hindered position of phenyloxacyclopropane. The latter would arise from phenylethene (styrene) by oxacyclopropanation (Section 12-9), and phenylethene could in turn be readily made from ethanoyl (acetyl) benzene by reduction–dehydration.

Inspection of the synthetic methods described in Sections 21-5 through 21-8 may give you further ideas about how to tackle the synthesis of Prozac, but they are all variations on the schemes formulated so far. However, consideration of the Mannich reaction (Section 21-9) provides a more fundamental alternative, attractive because of its more highly convergent nature to a derivative of compound B with the intact amino function in place. Indeed, this is the commercial route used by Eli Lilly to the final drug.

Retro-Mannich-Synthesis of B (X=CH₃NH)

$$+ \quad CH_2{=}O \quad + \quad CH_3NH_2$$

B

NEW REACTIONS

1. Acidity of Amines and Amide Formation (Section 21-4)

$$RNH_2 \;+\; H_2O \;\overset{K}{\rightleftharpoons}\; R\overset{-}{N}H \;+\; H_3O^+ \qquad K_a \sim 10^{-35}$$

$$R_2NH \;+\; CH_3CH_2CH_2CH_2Li \;\rightleftharpoons\; R_2N^-Li^+ \;+\; CH_3CH_2CH_2CH_3$$

Lithium
dialkylamide

$$2\,NH_3 \;+\; 2\,Na \;\xrightarrow{\text{Catalytic Fe}^{3+}}\; 2\,NaNH_2 \;+\; H_2$$

2. Basicity of Amines (Section 21-4)

$$RNH_2 + H_2O \underset{K}{\rightleftharpoons} \overset{+}{R}NH_3 + HO^- \qquad K_b \sim 10^{-4}$$

$$\overset{+}{R}NH_3 + H_2O \underset{K}{\rightleftharpoons} RNH_2 + H_3O^+ \qquad K_a \sim 10^{-10}$$

Salt formation

$$RNH_2 + HCl \longrightarrow R\overset{+}{N}H_3\ Cl^-$$

Alkylammonium chloride

General for primary, secondary, and tertiary amines

Preparation of Amines

3. Amines by Alkylation (Section 21-5)

$$R\overset{..}{N}H_2 + R'X \longrightarrow R\overset{R'}{\underset{}{\overset{|+}{N}}}H_2\ X^-$$

General for primary, secondary, and tertiary amines

Drawback: Multiple alkylation

$$R\overset{R'}{\underset{}{\overset{|+}{N}}}H_2\ X^- + R'X \longrightarrow \longrightarrow \longrightarrow R\overset{+}{N}R'_3\ X^-$$

4. Primary Amines from Nitriles (Section 21-5)

$$RX + {}^-CN \xrightarrow[-X^-]{\overset{DMSO}{S_N2}} RCN \xrightarrow{LiAlH_4 \text{ or } H_2,\ catalyst} RCH_2NH_2$$

R limited to methyl, primary, and secondary alkyl groups

5. Primary Amines from Azides (Section 21-5)

$$RX + N_3^- \xrightarrow[-X^-]{\overset{CH_3CH_2OH}{S_N2}} RN_3 \xrightarrow{\overset{1.\ LiAlH_4,\ (CH_3CH_2)_2O}{2.\ H_2O}} RNH_2$$

R limited to methyl, primary, and secondary alkyl groups

6. Primary Amines by Gabriel Synthesis (Section 21-5)

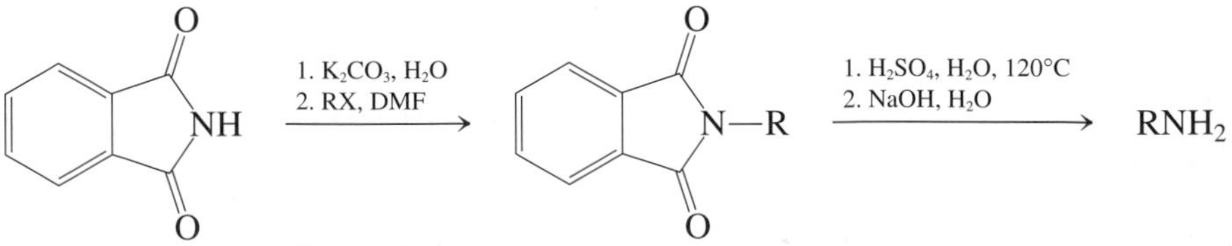

$$\xrightarrow[\text{2. RX, DMF}]{\text{1. K}_2\text{CO}_3,\ \text{H}_2\text{O}}$$

$$\xrightarrow[\text{2. NaOH, H}_2\text{O}]{\text{1. H}_2\text{SO}_4,\ \text{H}_2\text{O, 120°C}} RNH_2$$

R limited to methyl, primary, and secondary alkyl groups

7. Amines by Reductive Amination (Section 21-6)

$$\underset{RCR'}{\overset{O}{\overset{\|}{}}} \xrightarrow{NH_3,\ NaBH_3CN,\ H_2O\ or\ alcohol} R\overset{NH_2}{\underset{H}{\overset{|}{-C-}}}R'$$

Reductive methylation with formaldehyde

$$R_2NH + CH_2{=}O \xrightarrow{NaBH_3CN,\ CH_3OH} R_2NCH_3$$

8. Amines from Carboxylic Amides (Section 21-7)

$$RC(=O)N(R')R'' \xrightarrow[2.\ H_2O]{1.\ LiAlH_4,\ (CH_3CH_2)_2O} RCH_2N(R')R''$$

9. Hofmann Rearrangement (Section 21-7)

$$RC(=O)NH_2 \xrightarrow{Br_2,\ NaOH,\ H_2O} RNH_2 + CO_2$$

Reactions of Amines

10. Hofmann Elimination (Section 21-8)

$$RCH_2CH_2NH_2 \xrightarrow{\text{Excess } CH_3I,\ K_2CO_3} RCH_2CH_2\overset{+}{N}(CH_3)_3\ I^- \xrightarrow[-\ AgI]{Ag_2O,\ H_2O}$$

$$RCH_2CH_2\overset{+}{N}(CH_3)_3\ {}^-OH \xrightarrow{\Delta} RCH{=}CH_2 + N(CH_3)_3 + H_2O$$

11. Mannich Reaction (Section 21-9)

$$RC(=O)CH_2R' + CH_2{=}O + (CH_3)_2NH \xrightarrow[2.\ HO^-]{1.\ HCl} RC(=O)CH(CH_2N(CH_3)_2)R'$$

12. Nitrosation of Amines (Section 21-10)

Tertiary amines

$$R_3N \xrightarrow{NaNO_2,\ H^+X^-} R_3\overset{+}{N}{-}NO\ X^-$$

Tertiary *N*-nitrosammonium salt

Secondary amines

$$R(R')NH \xrightarrow{NaNO_2,\ H^+} R(R')N{-}N{=}O$$

***N*-Nitrosamine**

Primary amincs

$$RNH_2 \xrightarrow{NaNO_2,\ H^+} RN{=}NOH \xrightarrow[-\ H_2O]{H^+} RN_2^+ \xrightarrow[-\ N_2]{} R^+ \longrightarrow \text{mixture of products}$$

13. Diazomethane (Section 21-11)

$$CH_3N(N{=}O)C(=O)R \xrightarrow{KOH} CH_2{-}\overset{+}{N}{=}N{:}^-$$

Reactions of diazomethane

$$RCOH + CH_2N_2 \longrightarrow RCOCH_3 + N_2$$

$$R\overset{}{\diagup}\overset{}{\diagdown}R' + CH_2N_2 \xrightarrow{\textit{hv} \text{ or } \Delta \text{ or Cu}} \overset{\triangle}{\underset{R \quad R'}{}} \quad \textbf{Stereospecific}$$

Other sources of carbenes or carbenoids

$$CHCl_3 \xrightarrow{\text{Base}} :CCl_2$$

$$CH_2I_2 \xrightarrow{\text{Zn-Cu}} ICH_2ZnI$$

14. Nylon 6,6 (Section 21-12)

$$\underset{\substack{\textbf{Hexanedioic acid} \\ \textbf{(Adipic acid)}}}{HOC(CH_2)_4COH} + \underset{\substack{\textbf{1,6-Hexanediamine} \\ \textbf{(Hexamethylene-} \\ \textbf{diamine)}}}{H_2N(CH_2)_6NH_2} \xrightarrow{-H_2O} \underset{\textbf{Nylon 6,6}}{-\left[NH(CH_2)_6NHC(CH_2)_4C\right]_n-}$$

IMPORTANT CONCEPTS

1. **Amines** can be viewed as derivatives of **ammonia,** just as ethers and alcohols can be regarded as derivatives of water.

2. *Chemical Abstracts* names amines as **alkanamines** (and **benzenamines**), alkyl substituents on the nitrogen being designated as *N*-alkyl. Another system is based on the label aminoalkane. Common names are based on the label alkylamine.

3. The **nitrogen** in amines is sp^3 **hybridized,** the nonbonding electron pair functioning as the equivalent of a substituent. This tetrahedral arrangement **inverts rapidly** through a planar transition state.

4. The **lone electron pair** in amines is less tightly held than in alcohols and ethers, because nitrogen is **less electronegative** than oxygen. The consequences are a diminished capability for hydrogen bonding, higher basicity and nucleophilicity, and lower acidity.

5. **Infrared spectroscopy** helps to differentiate between primary and secondary amines. **Nuclear magnetic resonance** spectroscopy indicates the presence of nitrogen-bound hydrogens; both hydrogen and carbon atoms are **deshielded** in the vicinity of the nitrogen. **Mass spectra** are characterized by **iminium ion** fragments.

6. Indirect methods, such as displacements with azide or cyanide, or reductive amination, are superior to direct alkylation of ammonia for the **synthesis** of amines.

7. The **NR₃** group in a quaternary amine, $R'-\overset{+}{N}R_3$, is a **good leaving group** in E2 reactions; this enables the Hofmann elimination to take place.

8. The **nucleophilic reactivity** of amines manifests itself in reactions with electrophilic carbon, as in haloalkanes, aldehydes, ketones, and carboxylic acids and their derivatives.

9. **Carbenes** and carbenoids are useful for the synthesis of **cyclopropanes** from alkenes.

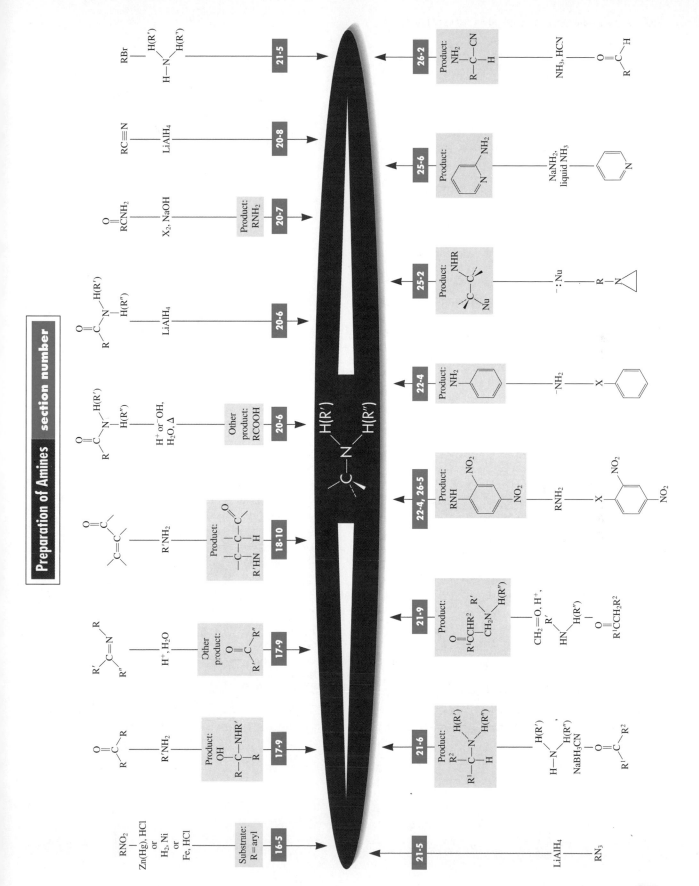

Preparation of Amines   section number

Reactions of Amines | section number

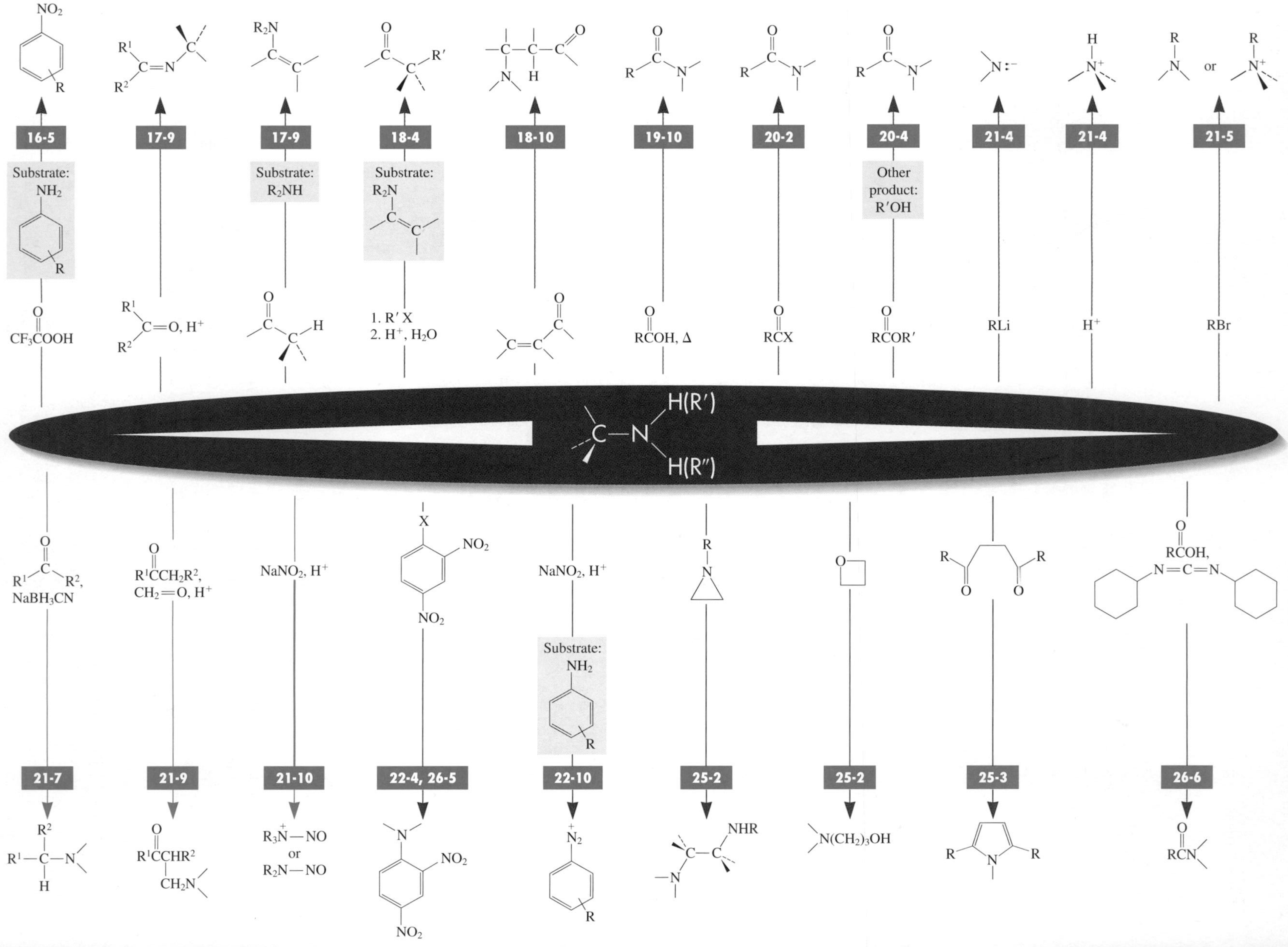

21-5 ← RBr

21-4 ← H⁺

21-4 ← RLi

20-4 ← RCOR'   Other product: R'OH

20-2 ← RCX

19-10 ← RCOH, Δ

18-10 ←

18-4 ← 1. R'X 2. H⁺, H₂O   Substrate: R₂N

17-9 ← Substrate: R₂NH

17-9 ← R¹C=O, H⁺ / R²

16-5 ← CF₃COOH   Substrate: NH₂

26-6 → RCN   RCOH, N=C=N

25-3 →

25-2 → N(CH₂)₃OH

25-2 → NHR

22-10 → N₂⁺   NaNO₂, H⁺   Substrate: NH₂

22-4, 26-5 →

21-10 → R₃N⁺—NO or R₂N—NO   NaNO₂, H⁺

21-9 → R¹CCHR² CH₂N   R¹CCH₂R², CH₂=O, H⁺

21-7 → R²—C—H / R¹   R¹CR², NaBH₃CN

PROBLEMS

19. Give at least two names for each of the following amines.

(a) [structure: pentane chain with NH₂ on C3]

(b) [structure: (CH₃)₂CH-NH-CH₃]

(c) [structure: benzene ring with -NH₂ and Cl]

(d) [structure: benzene ring with -N(CH₃)CH₂CH₂CH₃]

(e) $(CH_3)_3N$

(f) $CH_3CCH_2CH_2N(CH_3)_2$ with O double bonded to second carbon

(g) [structure: cyclopentyl-N(CH₃)-chain with Cl and methyl branch]

(h) $(CH_3CH_2)_2NCH_2CH{=}CH_2$

20. Give structures that correspond to each of the following names.
(a) *N,N*-Dimethyl-3-cyclohexenamine; **(b)** *N*-ethyl-2-phenylethylamine;
(c) 2-aminoethanol; **(d)** *m*-chloroaniline.

21. As mentioned in Section 21-2, the inversion of nitrogen requires a change of hybridization. **(a)** What is the approximate energy difference between pyramidal nitrogen (sp^3 hybridized) and trigonal planar nitrogen (sp^2 hybridized) in ammonia and simple amines? (**Hint:** Refer to the E_a of inversion.) **(b)** Compare the nitrogen atom in ammonia with the carbon atom in each of the following species: methyl cation, methyl radical, and methyl anion. Compare the most stable geometries and the hybridizations of each of these species. Using fundamental notions of orbital energies and bond strengths, explain the similarities and differences among them.

22. Use the following NMR- and mass-spectral data to identify the structures of two unknown compounds, A and B.
> **A:** ^1H NMR $\delta = 0.92$ (t, $J = 6$ Hz, 3 H), 1.32 (broad s, 12 H),
>   2.28 (broad s, 2 H), and 2.69 (t, $J = 7$ Hz, 2 H) ppm.
>   Mass spectrum m/z (relative intensity) = 129(0.6) and 30(100).
> **B:** ^1H NMR $\delta = 1.00$ (s, 9 H), 1.17 (s, 6 H), 1.28 (s, 2 H),
>   and 1.42 (s, 2 H) ppm.
>   Mass spectrum m/z (relative intensity) = 129(0.05),
>   114(3), 72(4), and 58(100).

23. The following spectroscopic data (^{13}C NMR and IR) are for several isomeric amines of the formula $C_6H_{15}N$. Propose a structure for each compound.
(a) ^{13}C NMR: $\delta = 23.7$ (CH_3) and 45.3 (CH) ppm. IR: 3300 cm^{-1}.
(b) ^{13}C NMR: $\delta = 12.6$ (CH_3) and 46.9 (CH_2) ppm. IR: no bands in 3250–3500 cm^{-1} range. **(c)** ^{13}C NMR: $\delta = 12.0$ (CH_3), 23.9 (CH_2), and 52.3 (CH_2) ppm. IR: 3280 cm^{-1}. **(d)** ^{13}C NMR: $\delta = 14.2$ (CH_3), 23.2 (CH_2), 27.1 (CH_2), 32.3 (CH_2), 34.6 (CH_2), and 42.7 (CH_2) ppm. IR: 1600 (broad), 3280, and 3365 cm^{-1}. **(e)** ^{13}C NMR: $\delta = 25.6$ (CH_3), 38.7 (CH_3), and 53.2 ($C_{quaternary}$) ppm. IR: no bands in 3250–3500 cm^{-1} region.

24. The following mass-spectral data are for two of the compounds in Problem 23. Match each mass spectrum with one of them. **(a)** m/z (relative intensity) = 101(8), 86(11), 72(79), 58(10), 44(40), and 30(100). **(b)** m/z (relative intensity) = 101(3), 86(30), 58(14), and 44(100).

25. Is a molecule with a high pK_b value a stronger or weaker base than a molecule with a low pK_b value? Explain by using a general equilibrium equation.

26. In which direction would you expect each of the following equilibria to lie?

(a) $NH_3 + {}^-OH \rightleftharpoons NH_2{}^- + H_2O$

(b) $CH_3NH_2 + H_2O \rightleftharpoons CH_3NH_3{}^+ + {}^-OH$

(c) $CH_3NH_2 + (CH_3)_3NH^+ \rightleftharpoons CH_3NH_3{}^+ + (CH_3)_3N$

27. How would you expect the following classes of compounds to compare with simple primary amines as bases and acids?

(a) Carboxylic amides; for example, CH_3CONH_2 **(b)** Imides; for example, $CH_3CONHCOCH_3$

(c) Enamines; for example, $CH_2{=}CHN(CH_3)_2$ **(d)** Benzenamines; for example,

28. Several functional groups containing nitrogen are considerably stronger bases than are ordinary amines. One is the amidine group found in DBN and DBU, both of which are widely used as bases in a variety of organic reactions.

Amidine group   **1,5-Diazabicyclo[4.3.0]non-5-ene (DBN)**   **1,8-Diazabicyclo[5.4.0]undec-7-ene (DBU)**

Another unusually strong organic base is guanidine, H_2NCNH_2. Indicate which nitrogen in each of these bases is the one most likely to be protonated and explain the enhanced strength of these bases relative to simple amines.

29. The following syntheses are proposed for amines. In each case, indicate whether the synthesis will work well, poorly, or not at all. If a synthesis will not work well, explain why.

(a) $CH_3CH_2CH_2CH_2Cl \xrightarrow[\text{2. LiAlH}_4,\ (CH_3CH_2)_2O]{\text{1. KCN, CH}_3CH_2OH} CH_3CH_2CH_2CH_2NH_2$

(b) $(CH_3)_3CCl \xrightarrow[\text{2. LiAlH}_4,\ (CH_3CH_2)_2O]{\text{1. NaN}_3,\ DMSO} (CH_3)_3CNH_2$

(c) [cyclohexane]−CONH₂ $\xrightarrow{\text{Br}_2,\ NaOH,\ H_2O}$ [cyclohexane]−NH₂

(d)

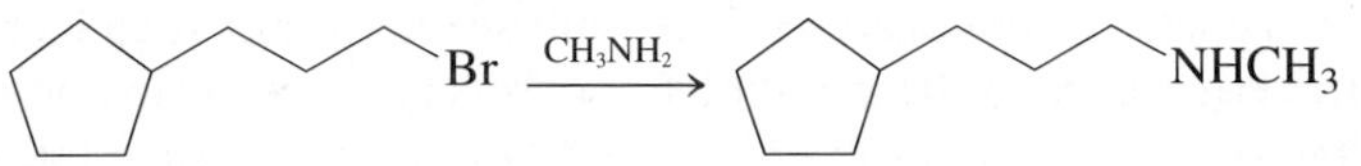

CH_3NH_2

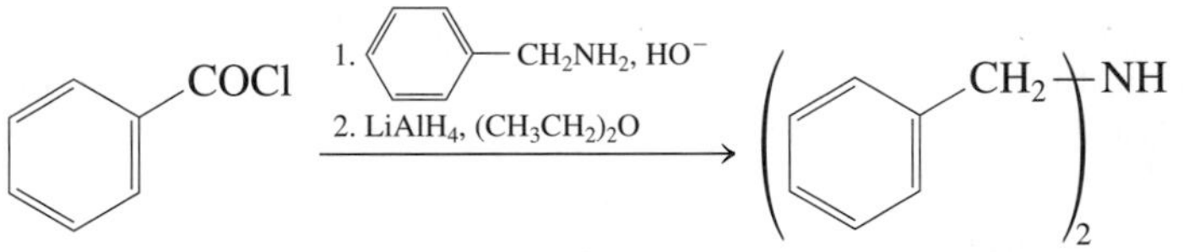

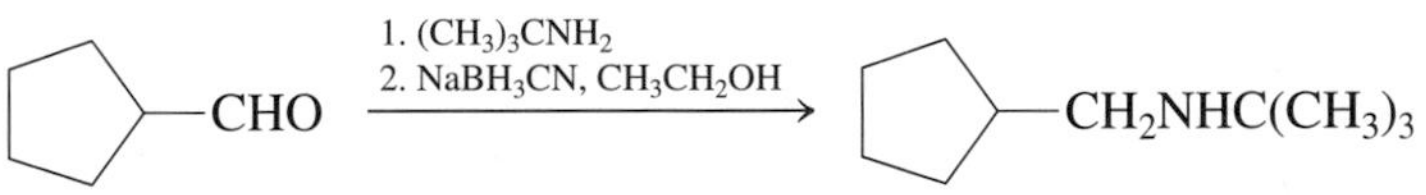

(e)

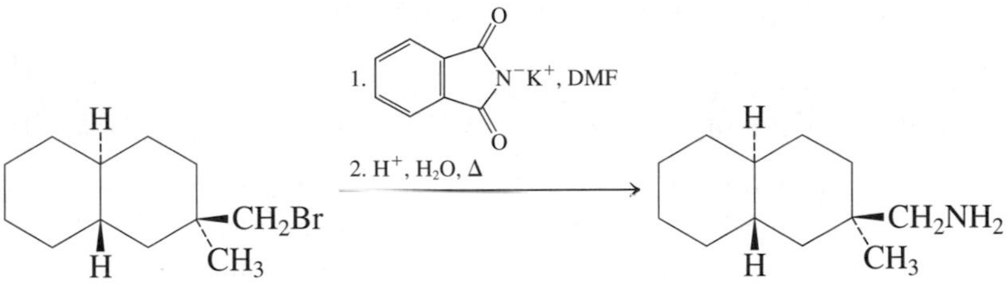

1. N^-K^+, DMF

2. H^+, H_2O, Δ

(f)

1. $-CH_2NH_2$, HO^-

2. $LiAlH_4$, $(CH_3CH_2)_2O$

(g)

1. $(CH_3)_3CNH_2$
2. $NaBH_3CN$, CH_3CH_2OH

(h) $H_2NCH_2CH_2CHO$

$\xrightarrow{NaBH_3CN,\ CH_3CH_2OH}$

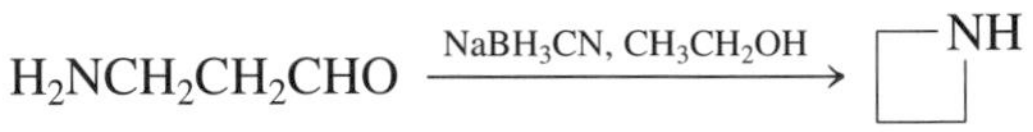

(i)

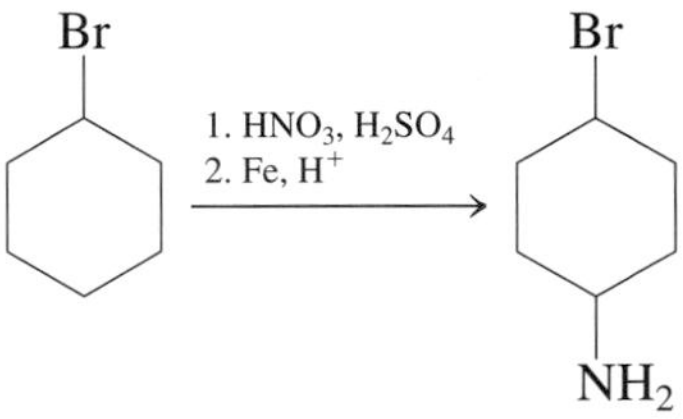

1. HNO_3, H_2SO_4
2. Fe, H^+

(j)

1. NaH, THF
2. CH_3I
3. $LiAlH_4$, $(CH_3CH_2)_2O$

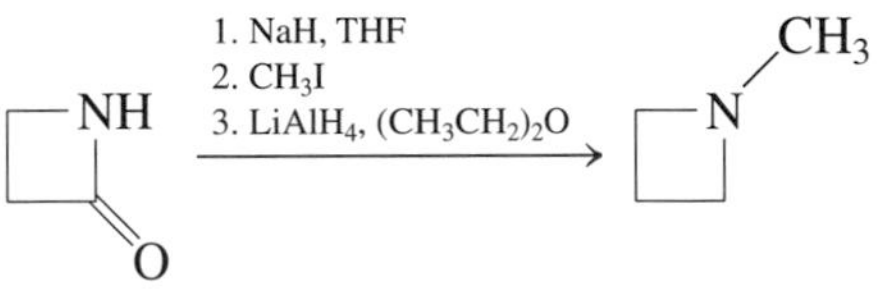

30. For each synthesis in Problem 29 that does not work well, propose an alternative synthesis of the final amine, starting either with the same material or with a material of similar structure and functionality.

31. Give the structures of all possible nitrogen-containing organic products that might be expected to form on reaction of chloroethane with ammonia. (**Hint:** Consider multiple alkylations.)

32. In the past several years, pseudoephedrine has gradually replaced phenylpropanolamine as the favored decongestant in over-the-counter cold remedies. (Pseudoephedrine is less likely to cause drowsiness.)

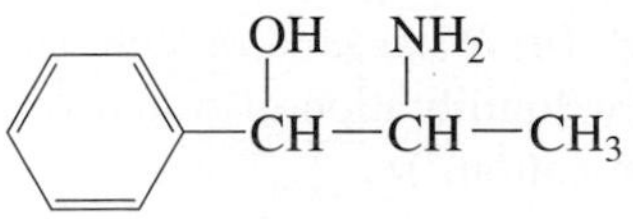

Phenylpropanolamine

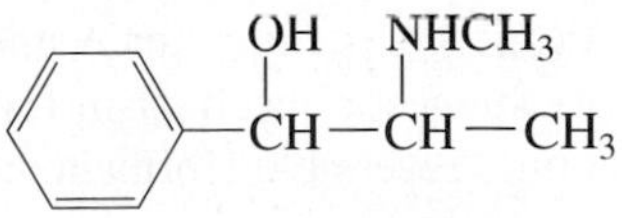

Pseudoephedrine

Suppose that you are the director of a major pharmaceutical laboratory with a huge stock of phenylpropanolamine on hand and the president of the company issues the order, "Pseudoephedrine from now on!" Analyze all your options, and propose the best solution that you can find for the problem.

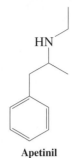

Apetinil

33. Apetinil, an appetite suppressant (i.e., diet pill; see Chemical Highlight 21-1), has the structure shown in the margin. Is it a primary, a secondary, or a tertiary amine? Propose an efficient synthesis of Apetinil from each of the following starting materials. Try to use a variety of methods.

(a) $C_6H_5CH_2COCH_3$ (b) $C_6H_5CH_2\overset{\overset{\displaystyle Br}{|}}{C}HCH_3$ (c) $C_6H_5CH_2\overset{\overset{\displaystyle CH_3}{|}}{C}HCOOH$

34. Suggest the best syntheses that you can for the following amines, beginning each with any organic compounds that do not contain nitrogen.
(a) Butanamine; (b) *N*-methylbutanamine; (c) *N,N*-dimethylbutanamine.

35. Give the structures of the possible alkene products of Hofmann elimination of each of the following amines. If a compound can be cycled through multiple eliminations, give the products of each cycle.

(a) [structure: phenyl–$CH(NH_2)CH_2CH_3$]

(b) [structure: cyclohexane with H_3C and NH_2 on same carbon]

(c) H_3C [structure: 2-methyl-1-methylpyrrolidine with N–CH_3]

(d) [structure: indoline with N–H]

(e) [structure: octahydroindolizine/quinolizidine with N]

36. Formulate a detailed mechanism for the Mannich reaction between 2-methyl-propanal, formaldehyde, and methanamine shown on page 957.

37. Reaction of the tertiary amine tropinone with (bromomethyl)benzene (benzyl bromide) gives not one but two quaternary ammonium salts, A and B.

[structure of tropinone with H_3C–N bridge and O ketone] + [benzene ring]—CH_2Br ⟶ A + B
$[C_{15}H_{20}NO]^+ Br^-$

Tropinone
$(C_8H_{13}NO)$

Compounds A and B are stereoisomers that are interconverted by base; that is, base treatment of either pure isomer leads to an equilibrium mixture of the two. (a) Propose structures for A and B. (b) What kind of stereoisomers are A and B? (c) Suggest a mechanism for the equilibration of A and B by base. (**Hint:** Think "reversible Hofmann elimination.")

38. Attempted Hofmann elimination of an amine containing a hydroxy group on the β-carbon gives an oxacyclopropane product instead of an alkene.

[structure: HO and NH_2 on H_2C—CH_2] $\xrightarrow[\text{3. } \Delta]{\begin{array}{l}\text{1. Excess } CH_3I\\\text{2. } Ag_2O, H_2O\end{array}}$ [structure: oxacyclopropane H_2C—CH_2 with O] + $(CH_3)_3N$

(a) Propose a sensible mechanism for this transformation. (b) Pseudoephedrine (see Problem 32) and ephedrine are closely related, naturally occurring compounds, as the similar names imply. In fact, they are stereoisomers. From the results of the following reactions, deduce the precise stereochemistries of ephedrine and pseudoephedrine.

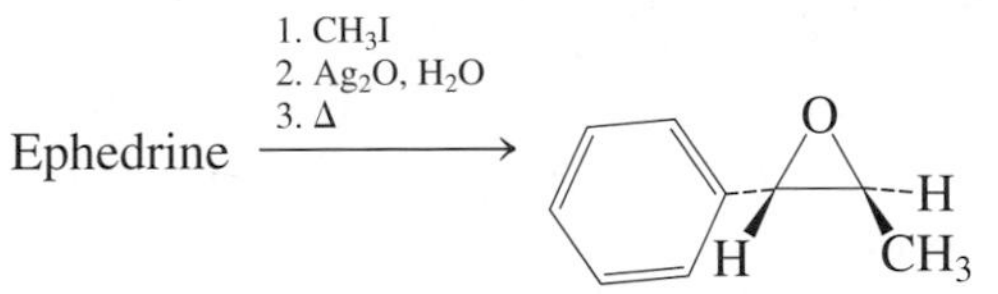

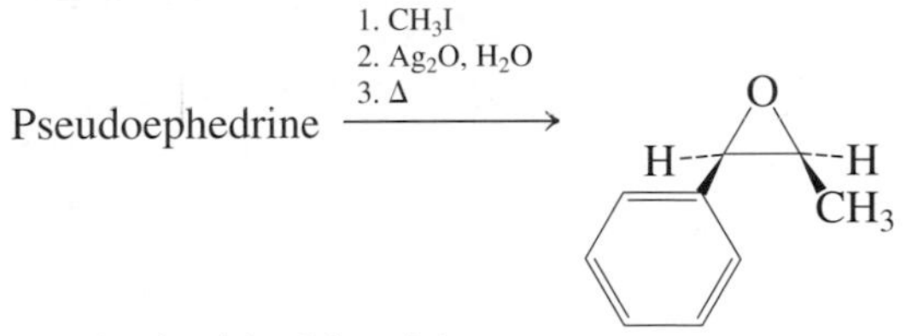

Ephedrine $\xrightarrow[\text{3. } \Delta]{\begin{array}{l}\text{1. CH}_3\text{I}\\\text{2. Ag}_2\text{O, H}_2\text{O}\end{array}}$

Pseudoephedrine $\xrightarrow[\text{3. } \Delta]{\begin{array}{l}\text{1. CH}_3\text{I}\\\text{2. Ag}_2\text{O, H}_2\text{O}\end{array}}$

39. Show how each of the following molecules might be synthesized by Mannich or Mannich-like reactions. (**Hint:** Work backward, identifying the bond made in the Mannich reaction.)

(a)

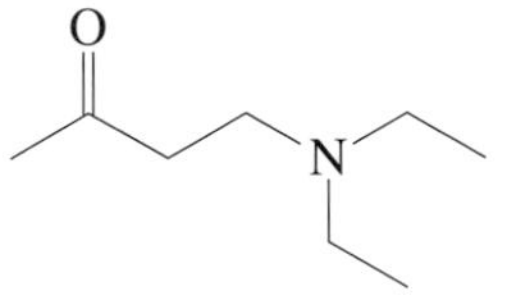

(b)

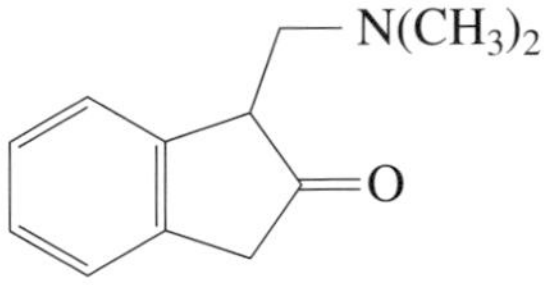

(c) H₃CCHCN
         |
        NH₂

(d)

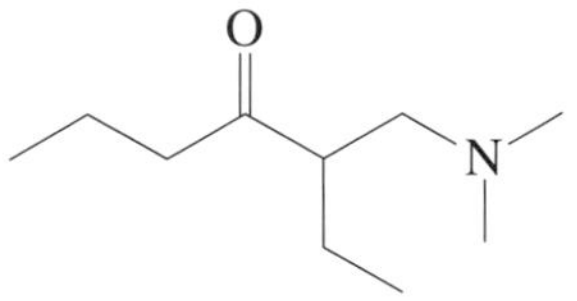

(e)

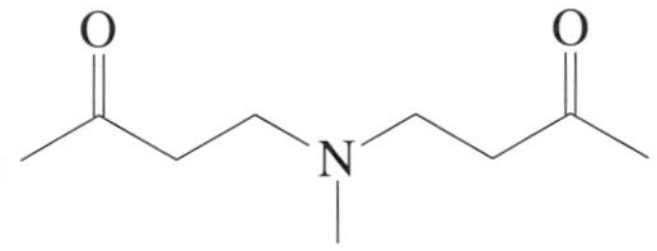

40. Tropinone (Problem 37) was first synthesized by Sir Robert Robinson (famous for the Robinson annulation reaction; Section 18-12), in 1917, by the following reaction. Show a mechanism for this transformation.

$+ \text{CH}_3\text{NH}_2 +$

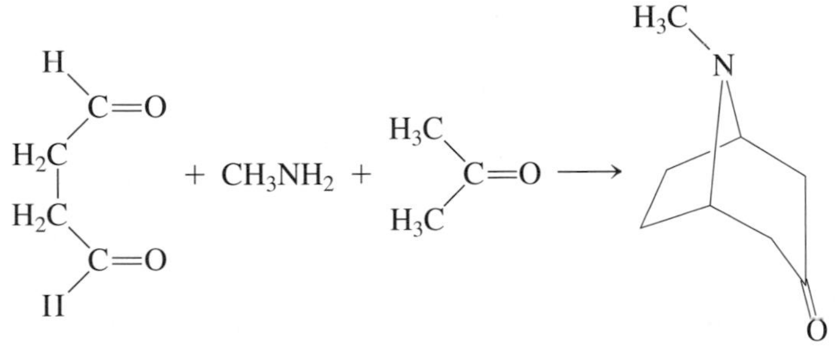

41. Illustrate a method for achieving the transformation shown in the margin, using combinations of reactions presented in Sections 21-8 and 21-9.

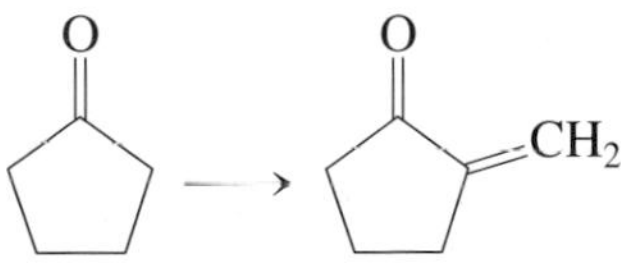

42. Give the expected product(s) of each of the following reactions.

(a) 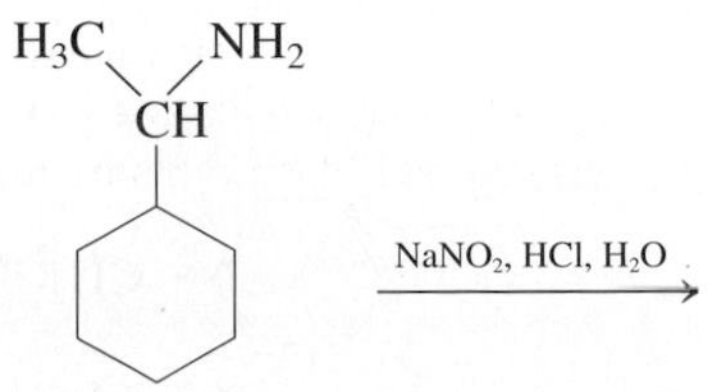 $\xrightarrow{\text{NaNO}_2, \text{HCl}, \text{H}_2\text{O}}$

(b) 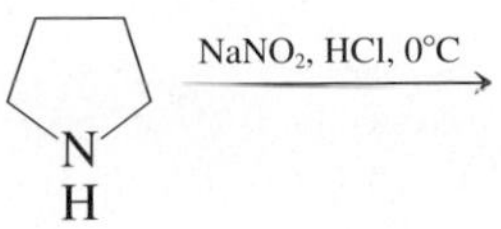 $\xrightarrow{\text{NaNO}_2, \text{HCl}, 0°\text{C}}$

43. Write the expected products of each of the following reactions.

(a) (*E*)-2-pentene + CHCl$_3$ $\xrightarrow{\text{KOC(CH}_3)_3,\ (\text{CH}_3)_3\text{COH}}$

(b) 1-Methylcyclohexene + CH$_2$I$_2$ $\xrightarrow{\text{Zn-Cu, (CH}_3\text{CH}_2)_2\text{O}}$

(c) Propene + CH$_2$N$_2$ $\xrightarrow{\text{Cu, }\Delta}$

(d) (*Z*)-1,2-diphenylethene + CHBr$_3$ $\xrightarrow{\text{KOC(CH}_3)_3,\ (\text{CH}_3)_3\text{COH}}$

(e) (*E*)-1,3-pentadiene + 2 CH$_2$I$_2$ $\xrightarrow{\text{Zn-Cu, (CH}_3\text{CH}_2)_2\text{O}}$

(f) CH$_2$=CHCH$_2$CH$_2$CH$_2$CHN$_2$ $\xrightarrow{hv}$

44. Reductive amination of *excess* formaldehyde with a primary amine leads to the formation of a *di*methylated tertiary amine as the product (see the following example). Propose an explanation.

(CH$_3$)$_3$CCH$_2$NH$_2$ + 2 CH$_2$=O $\xrightarrow{\text{NaBH}_3\text{CN, CH}_3\text{OH}}$ (CH$_3$)$_3$CCH$_2$N(CH$_3$)$_2$
84%

2,2-Dimethylpropanamine   **N,N,2,2-Tetramethylpropanamine**

45. Several of the natural amino acids are synthesized from 2-oxocarboxylic acids by an enzyme-catalyzed reaction with a special coenzyme called pyridoxamine. Use electron-pushing arrows to describe each step in the following synthesis of phenylalanine from phenylpyruvic acid.

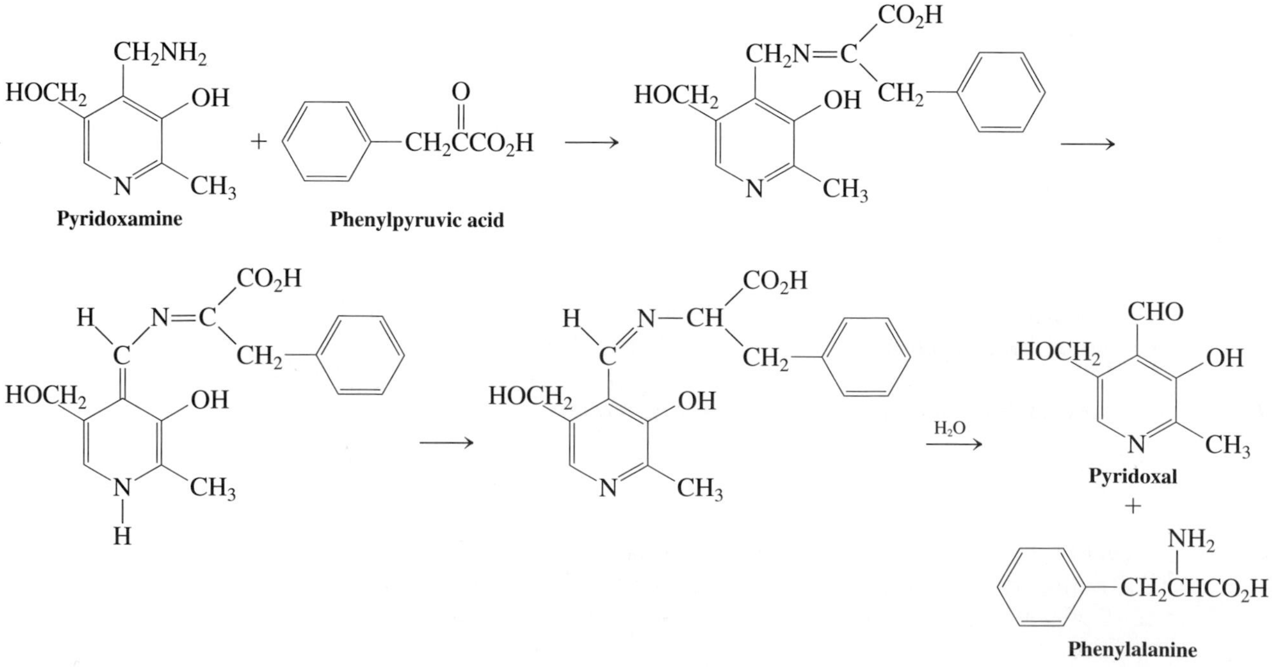

Pyridoxamine   Phenylpyruvic acid

Pyridoxal

Phenylalanine

46. From the following information, deduce the structure of coniine, an amine found in poison hemlock, which, deservedly, has a very bad reputation. IR: 3330 cm^{-1}. ^{1}H NMR: $\delta = 0.91$ (t, $J = 7$ Hz, 3 H), 1.33 (s, 1 H), 1.52 (m, 10 H), 2.70 (t, $J = 6$ Hz, 2 H), and 3.0 (m, 1 H) ppm. Mass spectrum: m/z (relative intensity) $= 127$ (M$^+$, 43), 84(100), and 56(20).

47. Pethidine, the active ingredient in the narcotic analgesic Demerol, was subjected to two successive exhaustive methylations with Hofmann eliminations, and then ozonolysis, with the following results:

(a) Propose a structure for pethidine based on this information.
(b) Propose a synthesis of pethidine that begins with ethyl phenylacetate and *cis*-1,4-dibromo-2-butene. (**Hint:** First prepare the dialdehyde ester shown in the margin, and then convert it into pethidine.)

48. Skytanthine is a monoterpene alkaloid with the following properties. Analysis: C$_{11}$H$_{21}$N. ^{1}H NMR: two CH$_3$ doublets ($J = 7$ Hz) at $\delta = 1.20$ and 1.33 ppm; one CH$_3$ singlet at $\delta = 2.32$ ppm; other hydrogens give rise to broad signals at $\delta = 1.3$–2.7 ppm. IR: no bands ≥ 3100 cm^{-1}. Deduce the structures of skytanthine and degradation products A, B, and C from this information.

49. Many alkaloids are synthesized in nature from a precursor molecule called norlaudanosoline which, in turn, appears to be derived from the condensation of amine A with aldehyde B. Formulate a mechanism for this transformation.

Note that a carbon–carbon bond is formed in the process. Name a reaction presented in this chapter that is closely related to this carbon–carbon bond formation.

A   B   Norlaudanosoline

Team Problem

50. Quaternary ammonium salts catalyze reactions between species dissolved in two immiscible phases, a phenomenon called phase-transfer catalysis. For example, heating a mixture of 1-chlorooctane dissolved in decane with aqueous sodium cyanide shows no sign of the S_N2 product, nonanenitrile. On the other hand, addition of a small amount of (phenylmethyl)triethylammonium chloride results in a rapid, quantitative reaction.

$$CH_3(CH_2)_7Cl + Na^{+-}CN \longrightarrow CH_3(CH_2)_7CN + Na^+Cl^-$$
                                                              100%

1-Chlorooctane   **(Phenylmethyl)-**   **Nonanenitrile**
                          triethylammonium
                          chloride

As a team, discuss possible answers to the following questions:

(a) What is the solubility of the catalyst in the two solvents?
(b) Why is the S_N2 reaction so slow without catalyst?
(c) How does the ammonium salt facilitate the reaction?

Preprofessional Problems

51. One of the following four amines is tertiary. Which one? (a) Propanamine; (b) *N*-methylethanamine; (c) *N,N*-dimethylmethanamine; (d) *N*-methylpropanamine.

52. Identify the best conditions for the following transformation:

$$CH_3CH_2\overset{\overset{\displaystyle O}{\|}}{C}NH_2 \longrightarrow CH_3CH_2NH_2 + CO_2$$

(a) H_2, metal catalyst; (b) excess CH_3I, K_2CO_3; (c) Br_2, NaOH, H_2O; (d) $LiAlH_4$, ether; (e) CH_2N_2, ether.

53. Rank the basicities of the following three nitrogen-containing compounds (most basic first):

$$NH_3 \qquad CH_3NH_2 \qquad (CH_3)_4N^+ \ NO_3^-$$

A　　　　　　　B　　　　　　　　　C

(a) A > B > C; **(b)** B > C > A; **(c)** C > A > B; **(d)** C > B > A;
(e) B > A > C.

54. Which of the following formulas best represents diazomethane?

(a) $CH_2 = \overset{+}{N} = \overset{..}{\underset{..}{N}}{}^{-}$ **(b)** $H - \overset{..}{N} = C = \overset{..}{N} - H$ **(c)** $\overset{-\ ..}{N} = C = \overset{+}{N} \diagdown{}_{H}^{H}$

(d) $:\overset{-}{C}H_2 - N \equiv \overset{+}{N}:$ **(e)** $CH_2 - \overset{+}{N} \equiv \overset{-}{N}:$

55. Use the following partial IR- and mass-spectral data to identify one of the structures among the selection given. IR spectrum: 3300 and 1690 cm^{-1}; mass spectrum: $m/z = 73$ (parent ion).

(a) $\overset{O}{\overset{\|}{H C}}N(CH_3)_2$ **(b)** [epoxide]$\diagup$NHCH$_3$ with H   **(c)** $H_2NCH_2C \equiv CCH_2NH_2$

(d) $CH_3CH_2\overset{O}{\overset{\|}{C}}NH_2$ **(e)** [oxetane ring]—NH$_2$ with H

Alkylbenzenes, Phenols, and Benzenamines

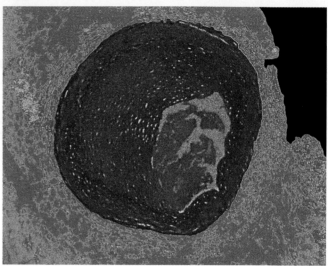

Aspirin, perhaps the most successful drug of all time, is an analgesic, antipyretic, anti-inflammatory, and antiplatelet agent. The pain-relieving property of its base structure has been known since the time of Hippocrates (460–377 B.C.), and it has been prescribed for the treatment of fever and rheumatism for a couple of centuries. More recent is its use in the prevention of heart attacks by reducing blood clotting in the coronary arteries (shown in the photograph).

Benzene (Chapter 15) used to be a common laboratory solvent until OSHA (U.S. Occupational Safety and Health Administration) placed it on its list of carcinogens. Chemists now use methylbenzene (toluene) instead, which has very similar solvating power but is *not* carcinogenic. Why not? The reason is the relatively high reactivity of the benzylic hydrogens that renders methylbenzene subject to fast metabolic degradation and extrusion from the body, unlike benzene, which can survive many days embedded in fatty and other tissues. Thus, the benzene ring, though itself quite unreactive because of its aromaticity, appears to activate neighboring bonds or, more generally, affects the chemistry of its substituents. You should not be too surprised by this finding, because it is complementary to the conclusions of Chapter 16. There we saw that substituents affect the behavior of benzene. Here we shall see the reverse.

How does the benzene ring modify the behavior of neighboring reactive centers? This chapter takes a closer look at the effect exerted by the ring on the reactivity of carbon, oxygen, and nitrogen substituents. We shall see that the interaction of the ring and the substituent allows the formation of compounds that are stabilized by resonance. After considering the special reactivity of aryl-substituted (benzylic) carbon atoms, we turn our attention to the preparation and reactions of phenols and benzenamines (anilines). These compounds are found widely in nature and are used in synthetic procedures as precursors to substances such as aspirin, dyes, and vitamins.

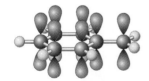

Phenylmethyl (benzyl) system

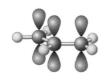

2-Propenyl (allyl) system

22-1 Reactivity at the Phenylmethyl (Benzyl) Carbon: Benzylic Resonance Stabilization

Methylbenzene is readily metabolized because one of the methyl C–H bonds is labile with respect to homolytic and heterolytic cleavage. The resulting **phenylmethyl (benzyl)** group, $C_6H_5CH_2$, may be viewed as a benzene ring whose π system overlaps with an extra p orbital. This interaction, called **benzylic resonance**, stabilizes adjacent radical, cationic, and anionic centers in much the same way that overlap of a π bond and a third p orbital stabilizes 2-propenyl (allyl) intermediates (Section 14-1).

Benzylic radicals are reactive intermediates in the halogenation of alkylbenzenes

We have seen that benzene will not react with chlorine or bromine unless a Lewis acid is added. The acid catalyzes halogenation of the ring (Section 15-10).

In contrast, heat or light allows attack by chlorine or bromine on methylbenzene (toluene) even in the absence of a catalyst. Analysis of the products shows that reaction takes place at the methyl group, *not* at the aromatic ring, and that excess halogen leads to multiple substitution.

(Bromomethyl)-benzene

(Chloromethyl)-benzene   (Dichloromethyl)-benzene   (Trichloromethyl)-benzene

Each substitution yields one molecule of hydrogen halide as a by-product.

As in the halogenation of alkanes (Sections 3-4 through 3-6) and the allylic halogenation of alkenes (Section 14-2), the mechanism of benzylic halogenation proceeds through radical intermediates. Heat or light induces dissociation of the halogen molecule into atoms. One of them abstracts a benzylic hydrogen, a reaction giving HX and a phenylmethyl (benzyl) radical. This intermediate reacts with another molecule of halogen to give the product, a (halomethyl)benzene, and another halogen atom, which propagates the chain process.

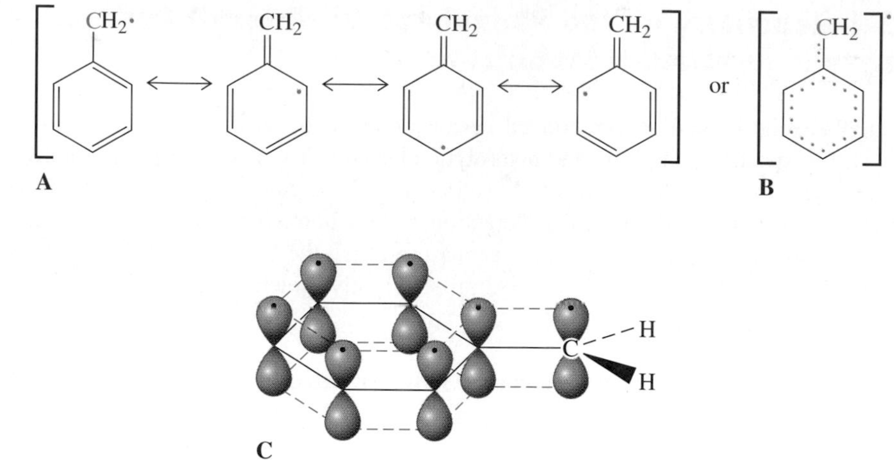

FIGURE 22-1

The benzene π system of the phenylmethyl (benzyl) radical enters into resonance with the adjacent radical center. The extent of delocalization may be depicted by (A) resonance structures, (B) dotted lines, or (C) orbitals.

Mechanism of Benzylic Halogenation

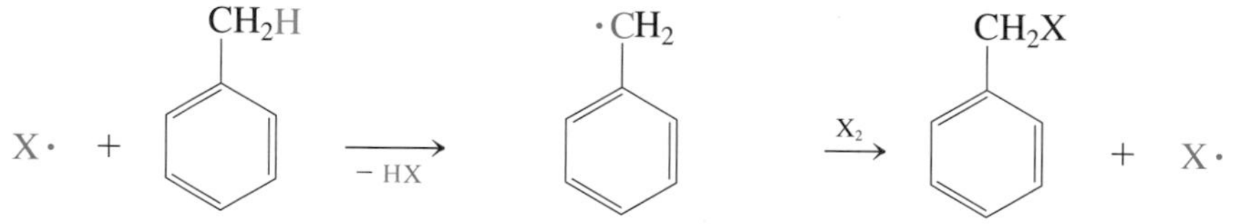

Phenylmethyl (benzyl) radical

What explains the ease of benzylic halogenation? The answer lies in the stabilization of the phenylmethyl (benzyl) radical by the phenomenon called benzylic resonance (Figure 22-1). As a consequence, the benzylic C–H bond is relatively weak ($DH° = 87$ kcal mol^{-1}); its cleavage is relatively favorable and proceeds with a low activation energy.

Inspection of the resonance structures in Figure 22-1 reveals why the halogen attacks only the *benzylic* position and not an aromatic carbon: Reaction at any but the benzylic carbon would destroy the aromatic character of the benzene ring.

EXERCISE 22-1

For each of the following compounds, draw the structure and indicate where radical halogenation is most likely to occur upon heating in the presence of Br$_2$. Then rank the compounds in approximate descending order of reactivity under bromination conditions. **(a)** Ethylbenzene; **(b)** 1,2-diphenylethane; **(c)** 1,3-diphenylpropane; **(d)** diphenylmethane; **(e)** (1-methylethyl)benzene.

Benzylic cations delocalize the positive charge

Reminiscent of the effects encountered in the corresponding allylic systems (Section 14-3), benzylic resonance can affect strongly the reactivity of benzylic halides and sulfonates in nucleophilic displacements. For example, the 4-methylbenzenesulfonate (tosylate) of 4-methoxyphenylmethanol (4-methoxybenzyl alcohol) undergoes ethanolysis rapidly via an S_N1 mechanism.

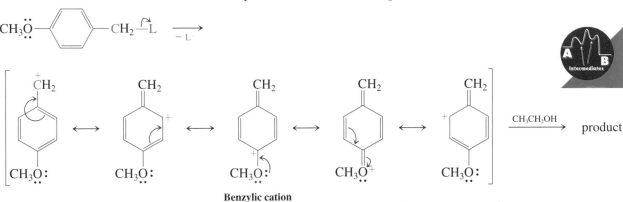

(4-Methoxyphenyl)methyl
4-methylbenzenesulfonate
(A primary benzylic tosylate)

1-(Ethoxymethyl)-4-methoxybenzene

The reason is the delocalization of the positive charge of the benzylic cation through the benzene ring, allowing for relatively facile dissociation of the starting sulfonate.

Mechanism of Benzylic Unimolecular Nucleophilic Substitution

Benzylic cation

A number of benzylic cations are stable enough to be isolable. For example, the X-ray structure of the 2-phenyl-2-propyl cation (as its SbF_6^- salt) was obtained in 1997 and shows the phenyl-C bond (1.41Å) to be intermediate in length between those of pure single (1.54 Å) and double bonds (1.33 Å), in addition to the expected planar framework and trigonal arrangement of all sp^2-carbons (Figure 22-2), as expected for a delocalized benzylic system.

EXERCISE 22-2

Which one of the two chlorides will solvolyze more rapidly: (1-chloroethyl)benzene, C_6H_5CHCl, or chloro(diphenyl)methane, $(C_6H_5)_2CHCl$? Explain your answer.
|
CH_3

The parent phenylmethyl (benzyl) cation is (in the gas phase) approximately as stable as the 1-methylethyl (isopropyl) cation. In solution nevertheless, phenylmethyl (benzyl) halides and sulfonates undergo preferential and unusually rapid S_N2 displacements, even under solvolytic conditions, but particularly in the presence of good

FIGURE 22-2
Structure of the 2-phenyl-2-propyl cation.

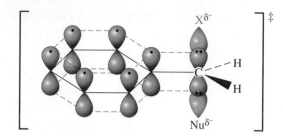

FIGURE 22-3
The benzene π system overlaps with the orbitals of the S_N2 transition state at a benzylic center. As a result, the transition state is stabilized, thereby lowering the activation barrier toward S_N2 reactions of (halomethyl)benzenes.

nucleophiles. The reasons for this observation are the lack of steric hindrance and the stabilization of the S_N2 transition state by overlap with the benzene π system (Figure 22-3).

$$\text{CH}_2\text{Br} + {}^-\text{CN} \xrightarrow{\text{S}_N2} \text{CH}_2\text{CN} + \text{Br}^-$$

81%

(Bromomethyl)benzene   **Phenylethanenitrile**
(Benzyl bromide)   **(Phenylacetonitrile)**
(~ 100 times faster than S_N2 reactions of primary bromoalkanes)

EXERCISE 22-3

Phenylmethanol (benzyl alcohol) is converted into (chloromethyl)benzene in the presence of hydrogen chloride much more rapidly than ethanol is converted into chloroethane. Explain.

Resonance in benzylic anions makes benzylic hydrogens relatively acidic

A negative charge adjacent to a benzene ring, as in phenylmethyl (benzyl) anion, is stabilized by conjugation in much the same way that the corresponding radical and cation are stabilized.

Resonance in Benzylic Anions

$$\text{CH}_3 \rightleftharpoons \left[{}^-\text{:CH}_2 \longleftrightarrow \text{CH}_2 \longleftrightarrow \text{CH}_2 \longleftrightarrow \text{CH}_2 \right] + \text{H}^+$$

$pK_a \sim 41$

The acidity of methylbenzene (toluene; $pK_a \sim 41$) is therefore considerably greater than that of ethane ($pK_a \sim 50$) and comparable to that of propene ($pK_a \sim 40$), which is deprotonated to produce the resonance-stabilized 2-propenyl (allyl) anion (Section 14-4). Consequently, methylbenzene (toluene) can be deprotonated by butyllithium to generate phenylmethyllithium.

Deprotonation of Methylbenzene

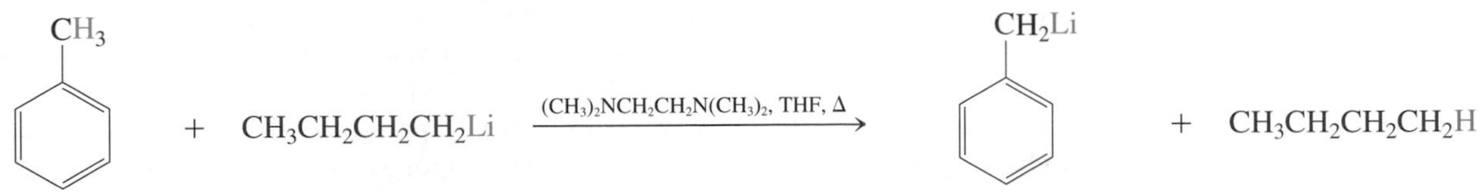

Methylbenzene
(Toluene)

$+$ $CH_3CH_2CH_2CH_2Li$ $\xrightarrow{(CH_3)_2NCH_2CH_2N(CH_3)_2,\ THF,\ \Delta}$

Phenylmethyllithium
(Benzyllithium)

$+$ $CH_3CH_2CH_2CH_2H$

EXERCISE 22-4

Which molecule in each of the following pairs is more reactive with the indicated reagents, and why?

(a) $(C_6H_5)_2CH_2$ or $C_6H_5CH_3$, with $CH_3CH_2CH_2CH_2Li$

(b)

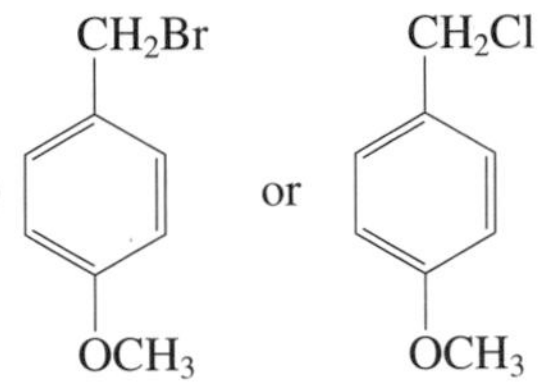

or , with $NaOCH_3$ in CH_3OH

(c)

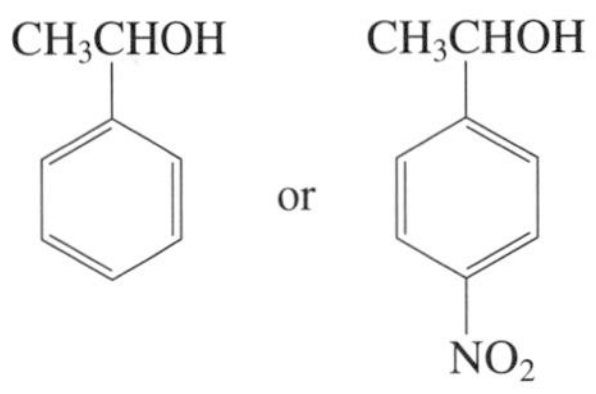

or , with HCl

In summary, benzylic radicals, cations, and anions are stabilized by resonance with the benzene ring. This effect allows for relatively easy radical halogenations, S_N1 and S_N2 reactions, and benzylic anion formation.

22-2 Benzylic Oxidations and Reductions

Because it is aromatic, the benzene ring is quite unreactive. It does undergo certain transformations—in particular, electrophilic aromatic substitution (Chapters 15 and 16)—but oxidations and reductions of the ring are difficult to achieve. However, *benzylic* positions are more susceptible to such transformations. This section describes how certain reagents oxidize and reduce alkyl substituents on the benzene ring.

Oxidation of alkyl-substituted benzenes leads to aromatic ketones and acids

Reagents such as hot $KMnO_4$ and $Na_2Cr_2O_7$ may oxidize alkylbenzenes all the way to benzoic acids. Benzylic carbon–carbon bonds are cleaved in this process, which usually requires at least one benzylic C–H bond to be present in the starting material (i.e., tertiary alkylbenzenes are inert).

Complete Benzylic Oxidations of Alkyl Chains

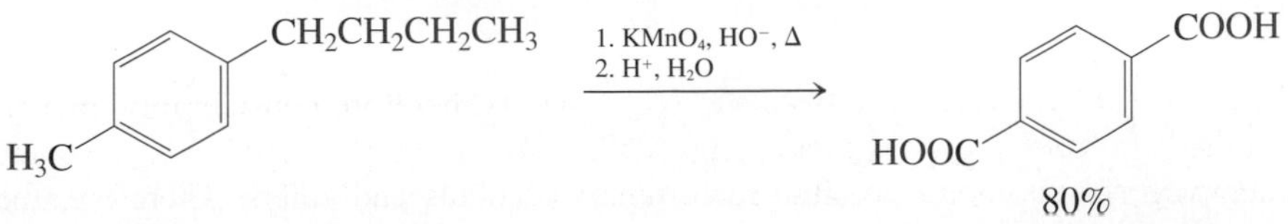

1. $KMnO_4$, HO^-, Δ
2. H^+, H_2O

80%

1-Butyl-4-methylbenzene

1,4-Benzenedicarboxylic acid
(Terephthalic acid)

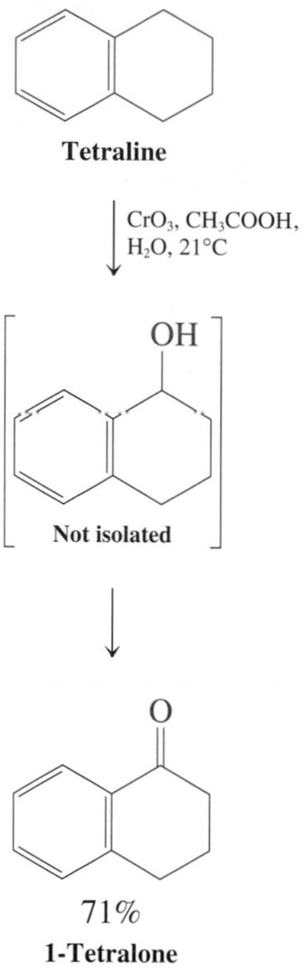

Tetraline

CrO₃, CH₃COOH,
H₂O, 21°C

OH

Not isolated

O

71%
1-Tetralone

The reaction proceeds through first the benzylic alcohol and then the ketone, at which stage it can be stopped under milder conditions (see margin and Section 16-5).

The special reactivity of the benzylic position is also seen in the mild conditions required for the oxidation of benzylic alcohols to the corresponding carbonyl compounds. For example, manganese dioxide, MnO_2, performs this oxidation selectively in the presence of other (nonbenzylic) hydroxy groups. (Recall that MnO_2 was used in the conversion of allylic alcohols into α,β-unsaturated aldehydes and ketones; see Section 18-8.)

Selective Oxidation of a Benzylic Alcohol with Manganese Dioxide

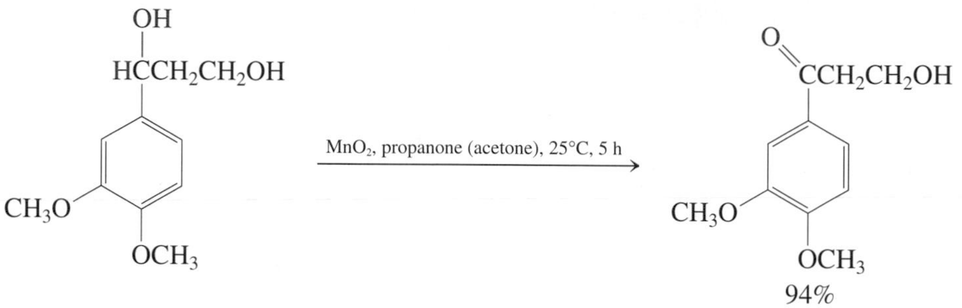

MnO_2, propanone (acetone), 25°C, 5 h

94%

Benzylic ethers are cleaved by hydrogenolysis

Exposure of benzylic alcohols or ethers to hydrogen in the presence of metal catalysts results in rupture of the reactive benzylic carbon–oxygen bond. This transformation is an example of **hydrogenolysis,** cleavage of a σ bond by catalytically activated hydrogen.

Cleavage of Benzylic Ethers by Hydrogenolysis

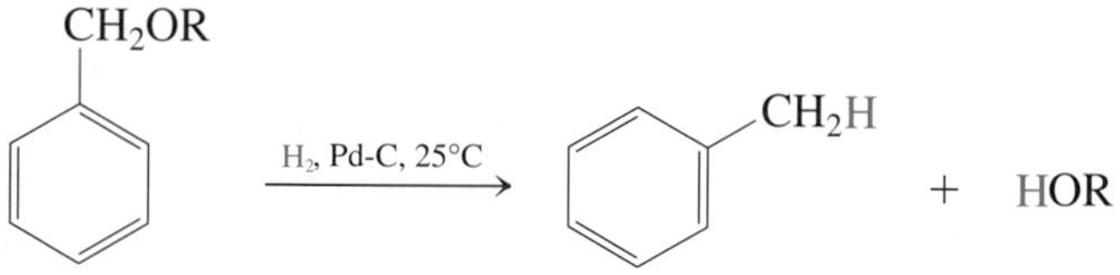

CH_2OR

H_2, Pd-C, 25°C

CH_2H + HOR

EXERCISE 22-5

Write synthetic schemes that would connect the following starting materials with their products.

(a) CH_2CH_3 ⇢ CH_3

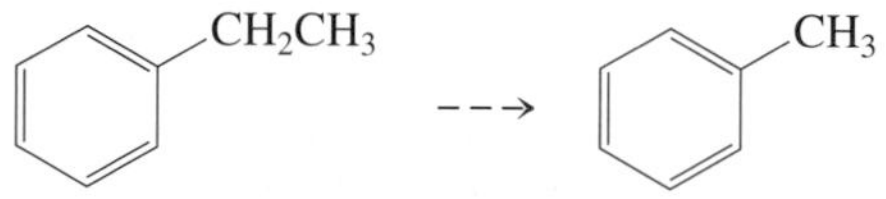

(b) ⇢

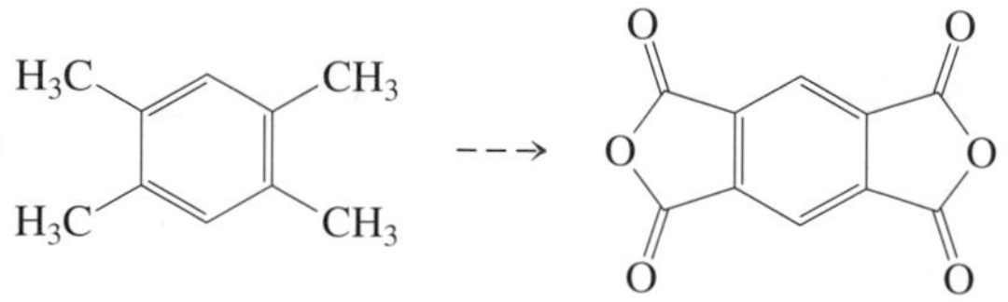

Hydrogenolysis is not possible for ordinary alcohols and ethers. Therefore, the phenylmethyl (benzyl) substituent is a valuable protecting group for hydroxy functions. The following scheme shows its use in part of a synthesis of a compound in

Phenylmethyl Protection in a Complex Synthesis

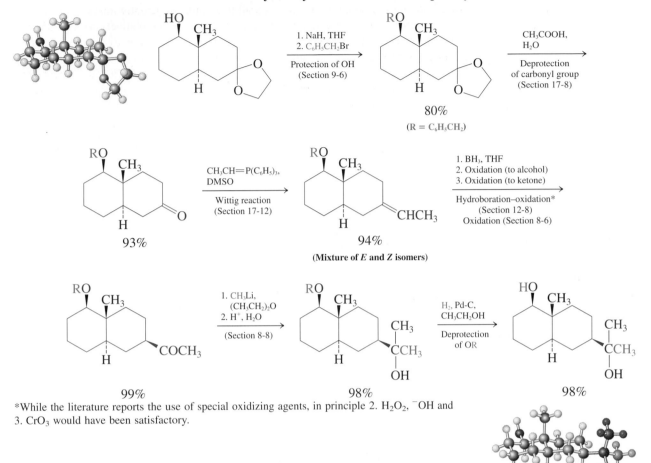

80%

(R = C₆H₅CH₂)

93%

94%

(Mixture of *E* and *Z* isomers)

99%

98%

98%

*While the literature reports the use of special oxidizing agents, in principle 2. H₂O₂, ⁻OH and 3. CrO₃ would have been satisfactory.

the eudesmane class of essential oils, which includes substances of importance in both medicine and perfumery.

Because the hydrogenolysis of the phenylmethyl (benzyl) ether in the final step occurs under neutral conditions, the tertiary alcohol function survives untouched. A tertiary butyl ether would have been a worse choice as a protecting group, because cleavage of its carbon–oxygen bond would have required strong acid (Section 9-8), which may cause dehydration (Section 9-2).

To summarize, benzylic oxidations of alkyl groups take place in the presence of permanganate or chromate; benzylic alcohols are converted into the corresponding ketones by manganese dioxide. The benzylic ether function can be cleaved by hydrogenolysis in a transformation that allows the phenylmethyl (benzyl) substituent to be used as a protecting group for the hydroxy function in alcohols.

22-3 Names and Properties of Phenols

Arenes substituted by hydroxy groups are called **phenols** (IUPAC name: benzenols; Section 15-1). The π system of the benzene ring overlaps with an occupied *p* orbital on the oxygen atom, a situation resulting in delocalization similar to that found in benzylic anions (Section 22-1). As one result of this extended conjugation, phenols possess an unusual, enolic structure. Recall that enols are usually unstable: They tautomerize easily to the corresponding ketones because of the relatively strong carbonyl bond (Section 18-2). Phenols, however, prefer the enol to the keto form because the aromatic character of the benzene ring is preserved.

Keto and Enol Forms of Phenol

2,4-Cyclohexadienone   Phenol

Phenols and their ethers are ubiquitous in nature; some derivatives have medicinal and herbicidal applications, whereas others are important industrial materials. This section first explains the names of these compounds. It then describes an important difference between phenols and alkanols—phenols are stronger acids because of the neighboring aromatic ring.

Phenols are hydroxyarenes

Bisphenol A

Phenol itself was formerly known as carbolic acid. It forms colorless needles (m.p. 41°C), has a characteristic odor, and is somewhat soluble in water. Aqueous solutions of it (or its methyl-substituted derivatives) are applied as disinfectants, but its main use is for the preparation of polymers (phenolic resins; Section 22-6). Total U.S. production of phenol in 1996 was 1.9 million tons. Pure phenol causes severe skin burns and is poisonous; deaths have been reported from the ingestion of as little as 1 g. Fatal poisoning may also result from absorption through the skin.

Substituted phenols are named as phenols, benzenediols, or benzenetriols, according to the system described in Section 15-1. These substances find uses in the photography, dyeing, and tanning industries. The compound bisphenol A (shown in the margin) is an important monomer (U.S. production in 1996, 0.74 million tons) in the synthesis of epoxyresins (see Chemical Highlight 9-3) and polycarbonates, materials widely employed in the manufacture of durable plastic materials, food packaging, dental sealants, and coatings inside beverage cans. Some recent reports of leaching of bisphenol A from these products and its activity as an estrogen mimic in mice have triggered an intensive effort in investigating its potential health hazards to humans.

Phenols containing the higher-ranking carboxylic acid functionality are called **hydroxybenzoic acids.** Many have common names. Phenyl ethers are named as **alkoxybenzenes.** As a substituent, C_6H_5O is called **phenoxy.**

| 4-Methylphenol (*p*-Cresol) | 4-Chloro-3-nitrophenol | 3-Hydroxybenzoic acid (*m*-Hydroxybenzoic acid) | 1,4-Benzenediol (Hydroquinone) | 1,2,3-Benzenetriol (Pyrogallol) |

Many examples, particularly those exhibiting physiological activity, are depicted in this book (e.g., see Chemical Highlights 5-4, 9-1, 21-1, 22-1, and 22-3, and Sections 4-7, 9-11, 15-1, 22-9, 24-12, 25-8, and 26-1). You are very likely to have ingested without knowing it the three phenol derivatives shown here.

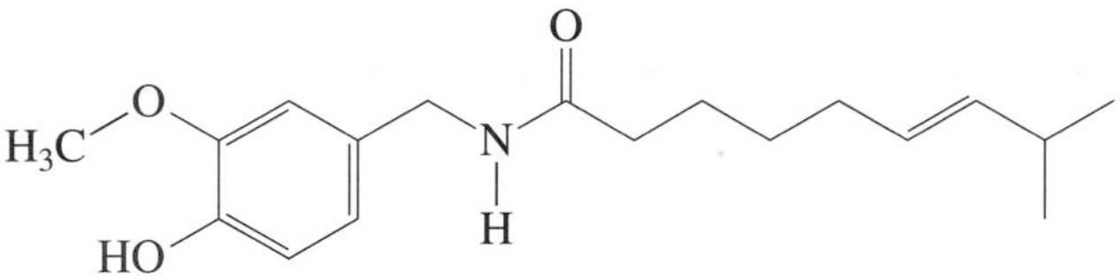

Capsaicin
(Active ingredient in hot pepper,
as in jalapeño or cayenne pepper)

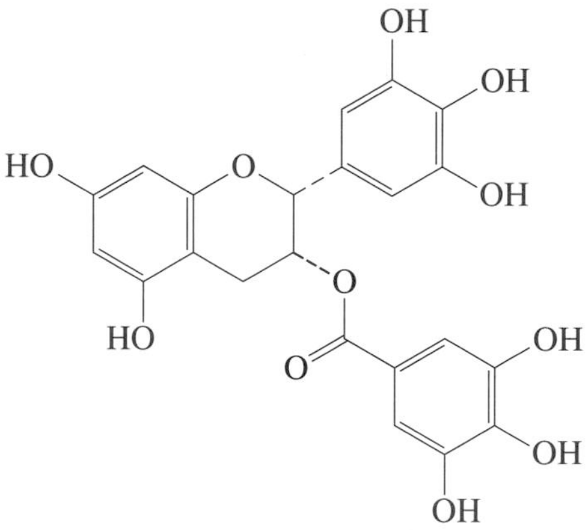

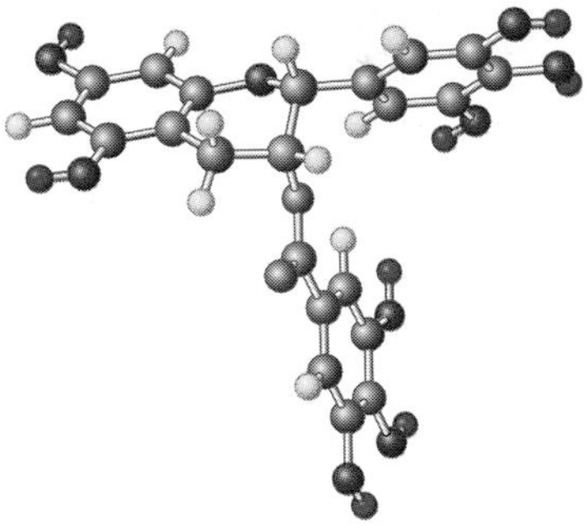

The hot pepper (Chili)
Capsicum frutescens (South America).

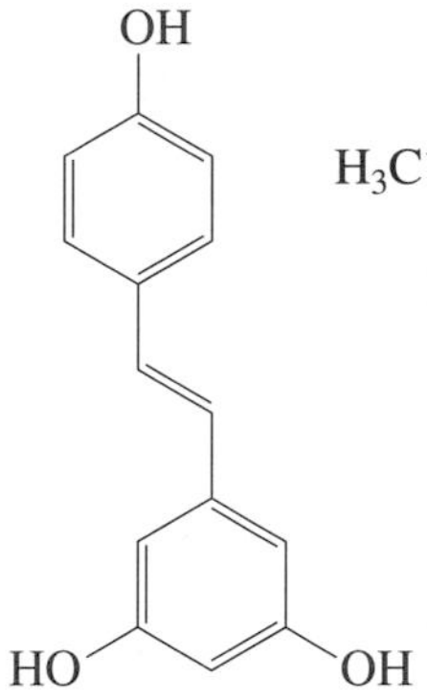

Resveratrol
(Cancer chemopreventive
from grapes)

Epigallocatechin-3 gallate
(Cancer chemopreventitive
from green tea)

Phenols are unusually acidic

Phenols have pK_a values that range from 8 to 10. Even though they are less acidic than carboxylic acids ($pK_a = 3$–5), they are stronger than alkanols ($pK_a = 16$–18). The reason is resonance: The negative charge in the conjugate base, called the **phenoxide ion,** is stabilized by delocalization into the ring.

Acidity of Phenol

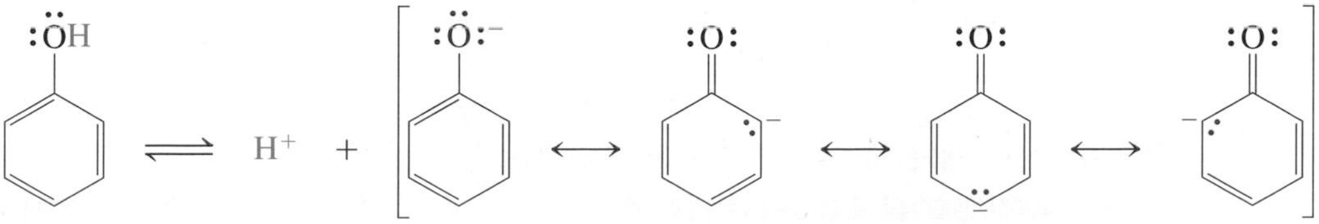

$pK_a \sim 10$ **Phenoxide ion**

The acidity of phenols is greatly affected by substituents that are capable of resonance. 4-Nitrophenol (*p*-nitrophenol), for example, has a pK_a of 7.15.

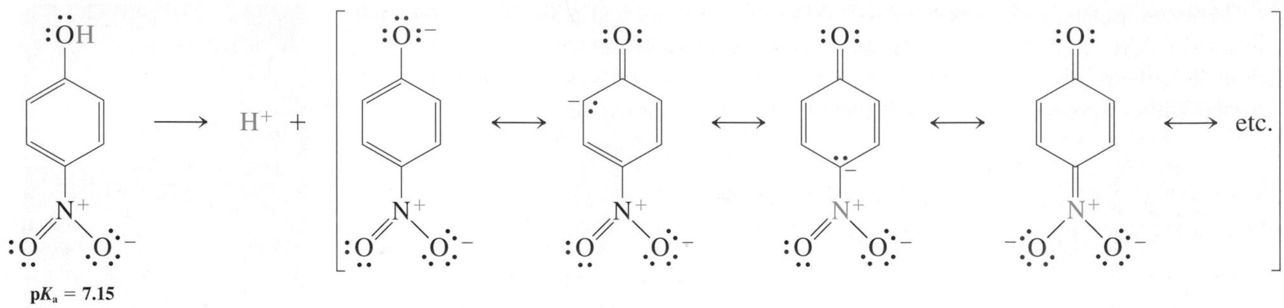

pK$_a$ = 7.15

The 2-isomer has similar acidity (pK$_a$ = 7.22), whereas nitrosubstitution at C3 results in a pK$_a$ of 8.39. Multiple nitration increases the acidity to that of carboxylic or even mineral acids. Electron-donating substituents have the opposite effect, raising the pK$_a$.

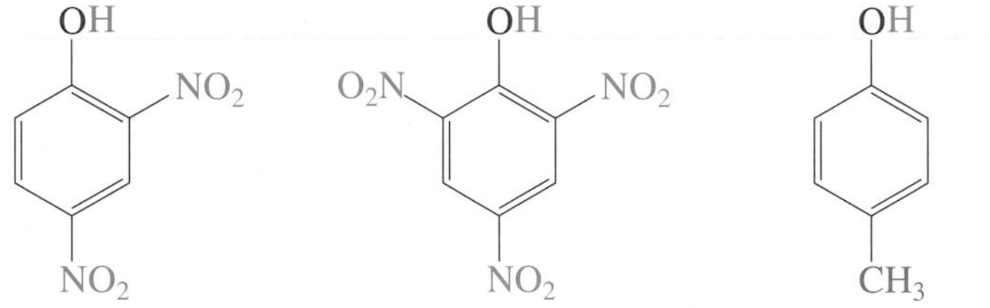

2,4-Dinitrophenol

pK$_a$ = **4.09**

**2,4,6-Trinitrophenol
(Picric acid)**

pK$_a$ = **0.25**

**4-Methylphenol
(*p*-Cresol)**

pK$_a$ = **10.26**

As Section 22-5 will show, the oxygen in phenol and its ethers is also weakly basic, in the case of ethers giving rise to acid-catalyzed cleavage.

EXERCISE 22-6

Why is 3-nitrophenol (*m*-nitrophenol) less acidic than its 2- and 4-isomers but more acidic than phenol itself?

EXERCISE 22-7

Rank in order of increasing acidity: phenol, A; 3,4-dimethylphenol, B; 3-hydroxybenzoic (*m*-hydroxybenzoic) acid, C; 4-(fluoromethyl)phenol [*p*-(fluoromethyl)phenol], D.

In summary, phenols exist in the enol form because of aromatic stabilization. They are named according to the rules for naming aromatic compounds explained in Section 15-1. Those derivatives bearing carboxy groups on the ring are called hydroxybenzoic acids. Phenols are acidic because the corresponding anions are resonance stabilized.

22-4 Preparation of Phenols: Nucleophilic Aromatic Substitution

Phenols are synthesized quite differently from the way in which ordinary substituted benzenes are made. Direct *electrophilic* addition of OH to arenes is difficult, because of the scarcity of reagents that generate an electrophilic hydroxy group, such as HO$^+$.

Instead, phenols are prepared by *nucleophilic* displacement of a leaving group from the arene ring by hydroxide, HO⁻, reminiscent of, but mechanistically quite different from, the synthesis of alkanols from haloalkanes. This section considers the ways in which this transformation may be achieved.

Nucleophilic aromatic substitution may follow an addition–elimination pathway

Treatment of 1-chloro-2,4-dinitrobenzene with hydroxide replaces the halogen with the nucleophile, furnishing the corresponding substituted phenol. Other nucleophiles, such as alkoxides or ammonia, may be similarly employed, forming alkoxyarenes and arenamines, respectively. Processes such as these, in which a group other than hydrogen is displaced from an aromatic ring, are called **ipso substitutions** (*ipso,* Latin, on itself). The products of these reactions are intermediates in the manufacture of useful dyes.

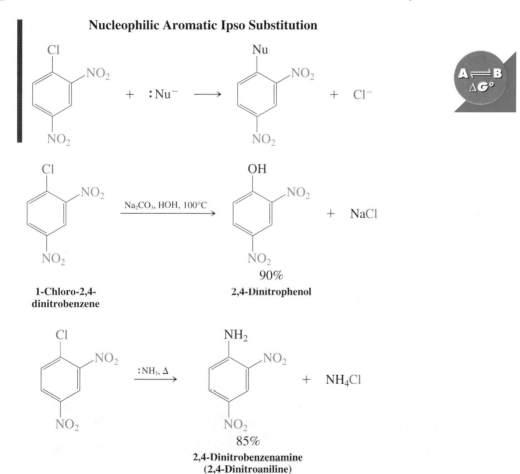

Nucleophilic Aromatic Ipso Substitution

1-Chloro-2,4-dinitrobenzene

2,4-Dinitrophenol
90%

2,4-Dinitrobenzenamine
(2,4-Dinitroaniline)
85%

The transformation is called **nucleophilic aromatic substitution.** The key to its success is the presence of one or more strongly electron-withdrawing groups on the benzene ring located ortho or para to the leaving group. Such substituents stabilize an intermediate anion by resonance. In contrast with the S_N2 reaction of haloalkanes, substitution in these reactions takes place by a *two-step mechanism,* an *addition–elimination sequence* similar to the mechanism of substitution of carboxylic acid derivatives (Chapter 20).

Mechanism of Nucleophilic Aromatic Substitution

STEP 1. Addition (facilitated by resonance stabilization)

The negative charge is strongly stabilized by resonance involving the ortho- and para-NO₂ groups

STEP 2. Elimination (only one resonance structure is shown)

In the first and rate-determining step, ipso attack by the nucleophile produces an anion with a highly delocalized charge, for which several resonance structures may be written, as shown. Note the ability of the negative charge to be delocalized into the electron-withdrawing groups. In contrast, such delocalization is *not* possible in 1-chloro-3,5-dinitrobenzene, in which these groups are located meta; so this compound does *not* undergo ipso substitution under the conditions employed.

Meta-NO₂ groups do *not* provide resonance stabilization of the negative charge

In the second step, the leaving group is expelled to regenerate the aromatic ring. The reactivity of haloarenes in nucleophilic substitutions increases with the nucleophilicity of the reagent and the number of electron-withdrawing groups on the ring, particularly if they are in the ortho and para positions.

CHEMICAL HIGHLIGHT 22-1 | Toxicity of Chlorophenols

2,4,5-Trichlorophenol
(2,4,5-TCP)

2,4,5-Trichlorophenoxyacetic acid
(2,4,5-T)

85%

The direct nucleophilic substitution of chloride in chloroarenes is a synthetic pathway to a number of herbicides, pesticides, and antibacterials. For example, as shown above, hydroxylation of 1,2,4,5-tetrachlorobenzene gives 2,4,5-trichlorophenol (2,4,5-TCP), an intermediate in the synthesis of 2,4,5-trichlorophenoxyacetic acid (2,4,5-T). This acid is a powerful herbicide of particular value in brush control. A 1:1 mixture of the butyl esters of 2,4,5-T and its 2,4-dichloro analog (2,4-D) was used in large amounts (estimated at more than 10 million gallons from 1965 to 1970) as a defoliant (code name, Agent Orange) during the Vietnam war.

These chemicals are toxic irritants. 2,4,5-T is notorious because of a much more toxic impurity that forms in small quantities in the course of its preparation: 2,3,7,8-tetrachlorodibenzo-*p*-dioxin (TCDD or, popularly but incorrectly, *dioxin*). Heating 2,4,5-T to between 500° and 600°C has been shown to produce TCDD, and thus extreme care has to be taken to control the reaction temperatures in the preparation of 2,4,5-T. TCDD can also be made directly from 2,4,5-trichlorophenol by coupling through double dehydrochlorination (below). The toxicity of TCDD (lethal dose, for test animals, in moles per kilogram of body weight) is about 500 times that of strychnine and more than 100,000 times that of sodium cyanide.

It is embryotoxic, teratogenic (causing deformations of the fetus), and a suspected carcinogen in humans. In smaller than lethal concentrations, it causes severe skin rashes and lesions (*chloracne*). In 1976, a runaway reaction in a chemical plant in Seveso, Italy, led to the accidental release of a cloud of overheated 2,4,5-trichlorophenol contaminated with TCDD. It is estimated that more than 130 pounds of the poison were vaporized, causing numerous deaths among animals and severe skin irritations in many humans.

Other examples of chlorinated phenols with physiological activity are 2,3,4,5,6-pentachlorophenol, a fungicide, and hexachlorophene (common name), a skin germicide formerly used in soaps and other toiletry products. Hexachlorophene was banned when it was discovered to cause brain damage.

2,3,4,5,6-Penta-
chlorophenol
(A fungicide)

Hexachlorophene
(A skin germicide)

$$2 \quad \xrightarrow[- \text{ 2 KCl, } - \text{ H}_2\text{O}]{\substack{0.5 \text{ M K}_2\text{CO}_3, \\ \text{Cu powder,} \\ 240°-250°C}}$$

2,3,7,8-Tetrachlorodibenzo-p-dioxin
(TCDD, or "Dioxin")

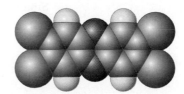

EXERCISE 22-8

Write the expected product of reaction of 1-chloro-2,4-dinitrobenzene with NaOCH₃ in boiling CH₃OH.

Another example of nucleophilic aromatic substitution is the conversion of arene-sulfonic acids into phenols by heating in molten NaOH. Phenol itself was at one time manufactured from sodium benzenesulfonate in this manner.

$SO_3^-Na^+$ + 2 NaOH $\xrightarrow{350°C}$ O^-Na^+ (Sodium salt of phenol) + Na_2SO_3 + H_2O

The reaction initially produces the phenoxide salt, which is subsequently protonated with HCl.

EXERCISE 22-9

Propose a mechanism for the following conversion. Considering that the first step is rate determining, draw a potential-energy diagram depicting the progress of the reaction. (**Hint:** This is a nucleophilic aromatic substitution.)

Haloarenes may react through benzyne intermediates

Haloarenes devoid of electron-withdrawing substituents do not undergo simple ipso substitution. However, at highly elevated temperatures and pressures, it is still possible to effect nucleophilic substitution. For example, if exposed to hot sodium hydroxide followed by neutralizing work-up, chlorobenzene furnishes phenol.

Chlorobenzene $\xrightarrow[\text{2. H}^+, \text{H}_2\text{O}]{\text{1. NaOH, H}_2\text{O, 340°C, 150 atm}}$ Phenol + NaCl

Similar treatment with potassium amide results in benzenamine (aniline).

Cl $\xrightarrow[\text{2. H}^+, \text{H}_2\text{O}]{\text{1. KNH}_2, \text{liquid NH}_3}$ NH₂ (Benzenamine (Aniline)) + KCl

It is tempting to assume that these substitutions follow a mechanism similar to that formulated for nucleophilic aromatic ipso substitution earlier in this section. However, when the last reaction is performed with radioactively labeled chlorobenzene (^{14}C at C1), a very curious result is obtained: Only half of the product is substituted at the labeled carbon; in the other half, the nitrogen is at the *neighboring* position.

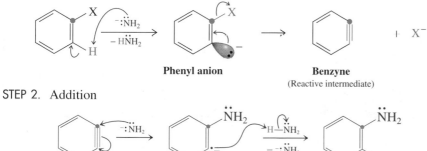

Chlorobenzene-1-^{14}C $\xrightarrow[\text{– KCl}]{\text{KNH}_2,\ \text{liquid NH}_3}$ Benzenamine-1-^{14}C   50%   +   Benzenamine-2-^{14}C   50%

Direct substitution does not seem to be the mechanism of these reactions. What, then, is the answer to this puzzle? A clue is the attachment of the incoming nucleophile *only* at the ipso or at the ortho position relative to the leaving group. This observation can be accounted for by an initial base-induced elimination of HX from the benzene ring, a process reminiscent of the dehydrohalogenation of haloalkenes to give alkynes (Section 13-5). In the present case, elimination is not a concerted process, but rather takes place in a sequential manner, deprotonation preceding the departure of the leaving group (step 1 of the mechanism shown). Both stages in step 1 are difficult, with the second being worse than the first. Why is that? With respect to the initial anion formation, recall (Section 11-3) that the acidity of $C_{sp}{}^2$–H is very low ($pK_a \sim 44$) and the same is true in general for phenyl hydrogens. The presence of the adjacent π system of benzene does not help, because the negative charge in the phenyl anion resides in an sp^2 orbital that is *perpendicular* to the π frame and is therefore incapable of resonance with the double bonds in the six-membered ring. Thus, deprotonation of the haloarene requires a strong base. It takes place ortho to the halogen, because the latter's inductive electron-withdrawing effect acidifies this position relative to the others.

Although deprotonation is not easy, the second stage of step 1, subsequent elimination of X$^-$, is even more difficult because of the highly strained structure of the resulting reactive species, called **1,2-dehydrobenzene** or **benzyne.**

Mechanism of Nucleophilic Substitution of Simple Haloarenes

STEP 1. Elimination

Phenyl anion   Benzyne
(Reactive intermediate)

STEP 2. Addition

Why is benzyne so strained? Recall that alkynes normally adopt a linear structure, a consequence of the *sp* hybridization of the carbons making up the triple bond (Section 13-2). Because of benzyne's cyclic structure, its triple bond is forced to be bent, rendering it unusually reactive. Thus, benzyne exists only as a reactive intermediate under these conditions, being rapidly attacked by any nucleophile present. For example (step 2), amide ion or even ammonia solvent can add to furnish the product benzenamine (aniline). Because the two ends of the triple bond are equivalent, addition can take place at either carbon, explaining the label distribution in the benzenamine obtained from ^{14}C-labeled chlorobenzene.

EXERCISE 22-10

1-Chloro-4-methylbenzene (*p*-chlorotoluene) is not a good starting material for the preparation of 4-methylphenol (*p*-cresol) by direct reaction with hot NaOH, because it forms a mixture of two products. Why does it do so, and what are the two products? Propose a synthesis from methylbenzene (toluene).

EXERCISE 22-11

Explain the regioselectivity observed in the following reaction. (**Hint:** Consider the effect of the methoxy group on the selectivity of attack by amide ion on the intermediate benzyne.)

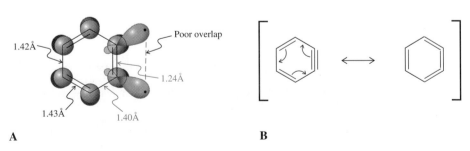

Benzyne is a strained cycloalkyne

Benzyne is too reactive to be isolated and stored in a bottle, but it can be observed spectroscopically under special conditions. Photolysis of benzocyclobutenedione at 77 K ($-196°$C) in frozen argon (m.p. = $-189°$C) produces a species whose IR and UV spectra are assignable to benzyne, formed by expulsion of two molecules of CO.

Although benzyne is usually represented as a cycloalkyne (Figure 22-4A), its triple bond exhibits an IR stretching frequency of 1846 cm^{-1}, intermediate between the values for normal double (ethene, 1655 cm^{-1}) and triple (ethyne, 1974 cm^{-1}) bonds. The ^{13}C NMR values for these carbons (δ = 182.7 ppm) are also atypical of pure triple bonds (Section 13-3), indicating a considerable contribution of the cumulated

Generation of Benzyne, a Reactive Intermediate

Benzocyclobutene-1,2-dione

hv, 77 K

+ 2 CO

1.42Å

Poor overlap

1.24Å

1.43Å

1.40Å

A

B

FIGURE 22-4

(A) The orbital picture of benzyne reveals that the six aromatic π electrons are located in orbitals that are perpendicular to the two additional hybrid orbitals making up the distorted triple bond. The latter overlap only poorly; therefore, benzyne is highly reactive.
(B) Resonance in benzyne.

triene (Section 14-5) resonance form (Figure 22-4B). The bond is weakened substantially by poor *p* orbital overlap in the plane of the ring.

Phenols are produced from arenediazonium salts

The most general laboratory procedure for making phenols is through **arenediazonium salts,** $ArN_2^+X^-$. Recall that primary alkanamines can be *N*-nitrosated but that the resulting species rearrange to diazonium ions, which are unstable—they lose nitrogen to give carbocations (Section 21-10). In contrast, primary benzenamines (anilines) are attacked by cold nitrous acid, in a reaction called **diazotization,** to give relatively stable arenediazonium salts.

Diazotization

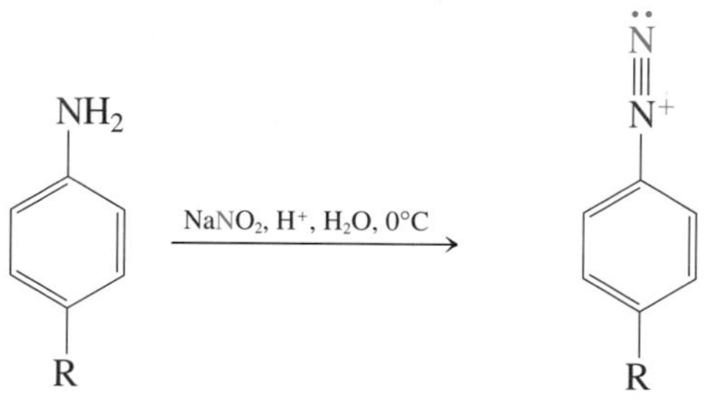

Arenediazonium ion

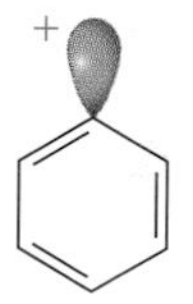

Phenyl cation

The reactions of these species will be discussed in detail in Section 22-10. When arenediazonium ions are heated, nitrogen is evolved and reactive **aryl cations** are produced (see Figure 22-5 in Section 22-10). These ions are trapped by water to give the desired phenols.

Decomposition of Arenediazonium Salts

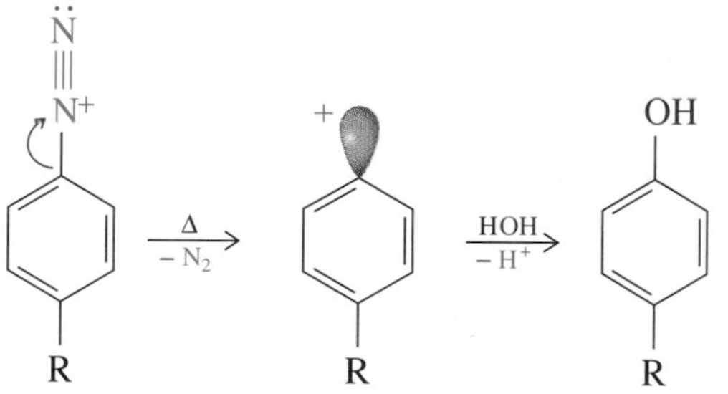

Aryl cation

ortho-Benzenediazoniumcarboxylate A (made by diazotization of 2-aminobenzoic acid, Problem 20-52) is explosive. When warmed in solution with *trans,trans*-2,4-hexadiene, it forms compound B. Explain by a mechanism. (**Hint:** Two other products are formed, both of which are gases.)

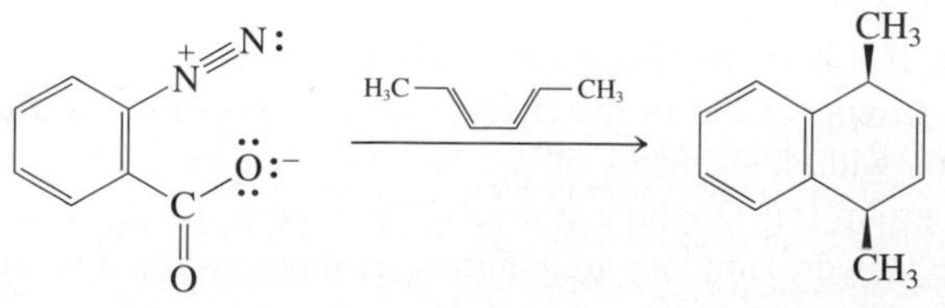

A   B

CHEMICAL HIGHLIGHT 22-2

Economics of Industrial Synthesis of Phenol: The Cumene Hydroperoxide Process

$$CH_3CH{=}CH_2 \; + \; \bigcirc \; + \; O_2 \; \dashrightarrow \; CH_3\overset{\displaystyle O}{\overset{\|}{C}}CH_3 \; + \; \text{(phenol, OH)}$$

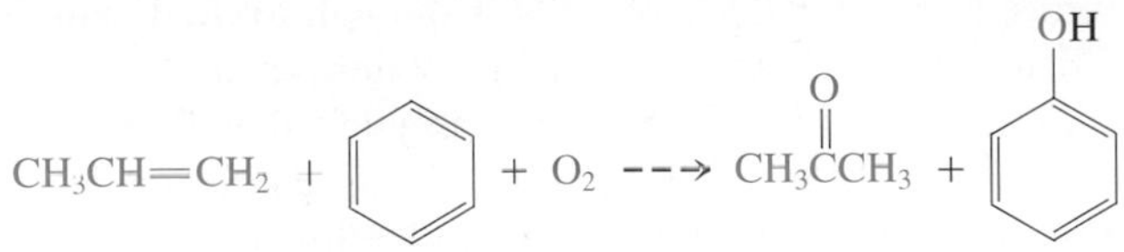

Another industrial preparation of phenol highlights the economic constraints on any process that has commercial significance. In this approach, called the cumene hydroperoxide process (shown above), benzene and propene are oxidized in a series of steps by air to phenol and propanone (acetone). Although the goal of the sequence is to make the former product, it is the sales potential of the ketone by-product that makes the process economically feasible.

The synthesis proceeds through several separate reactions. In the first, benzene is converted into 1-methylethylbenzene (isopropylbenzene, or cumene)

by Friedel-Crafts alkylation with propene under acidic conditions (Section 15-12). U.S. production in 1996 amounted to 5.6 billion pounds (2.5 million tons).

In the second reaction (shown below), the alkyl benzene is oxidized by air to the corresponding hydroperoxide. The ease with which this transformation takes place is due to the ready initiation of a radical chain process through the tertiary benzylic radical. In the final step, the hydroperoxide is treated with dilute acid to give the two products, phenol and propanone (acetone), by acid-catalyzed rearrangement.

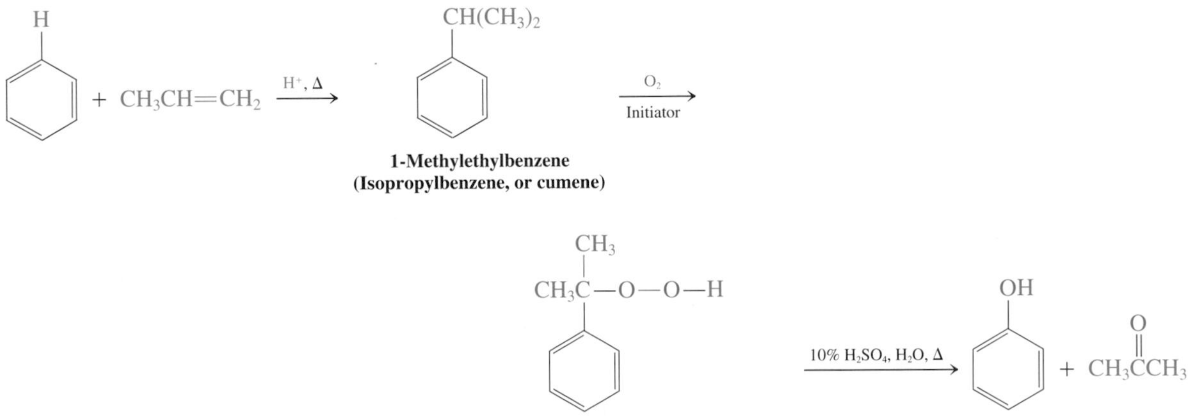

1-Methylethylbenzene
(Isopropylbenzene, or cumene)

1-Methyl-1-phenylethyl hydroperoxide
(Cumene hydroperoxide)

EXERCISE 22-13

Propose a synthesis of 4-(phenylmethyl)phenol (*p*-benzylphenol) from benzene.

In summary, if a benzene ring bears enough strongly electron-withdrawing substituents and a leaving group in the right place, nucleophilic addition to give an intermediate anion with delocalized charge becomes feasible, followed by elimination of the leaving group (nucleophilic aromatic ipso substitution). Phenols result when the nucleophile is hydroxide ion, arenamines (anilines) when it is ammonia, and alkoxyarenes when alkoxides are employed. Very strong bases are capable of eliminating HX from haloarenes to form the reactive intermediate benzynes, which are subject to nucleophilic attack to give substitution products. Finally, phenols may be prepared by decomposition of arenediazonium salts in water.

22-5 Alcohol Chemistry of Phenols

The phenol hydroxy group undergoes several of the reactions of alcohols (Chapter 9), such as protonation, Williamson ether synthesis, and esterification.

The oxygen in phenols is only weakly basic

Phenols are not only acidic but also weakly basic. They (and their ethers) can be protonated by strong acids to give the corresponding **phenyloxonium ions.** Thus, as with the alkanols, the hydroxy group imparts amphoteric character (Section 8-3). However, the basicity of phenol is even less than that of the alkanols, because the lone electron pairs on the oxygen are delocalized into the benzene ring (Sections 16-1 and 16-3). The pK_a values for phenyloxonium ions are, therefore, lower than those of alkyloxonium ions.

pK_a **Values of Methyl- and Phenyloxonium Ion**

$pK_a = -2.2$ $pK_a = -6.7$

Unlike secondary and tertiary alkyloxonium ions derived from alcohols, phenyloxonium derivatives do not dissociate to form phenyl cations, because such ions have too high an energy content (see Section 22-10). The phenyl–oxygen bond in phenols is very difficult to break. However, after protonation of alkoxybenzenes, the bond between the *alkyl* group and oxygen is readily cleaved in the presence of nucleophiles such as Br^- or I^- (e.g., from HBr or HI) to give phenol and the corresponding haloalkane.

3-Methoxybenzoic acid   **3-Hydroxybenzoic acid**
(*m*-**Methoxybenzoic acid**)   (*m*-**Hydroxybenzoic acid**)

EXERCISE 22-14

Why does cleavage of an alkoxybenzene by acid not produce a halobenzene and the alkanol?

Alkoxybenzenes are prepared by Williamson ether synthesis

The Williamson ether synthesis (Section 9-6) permits easy preparation of many alkoxybenzenes. The phenoxide ions obtained by deprotonation of phenols (Section 22-3) are good nucleophiles. They can displace the leaving groups from haloalkanes and alkyl sulfonates.

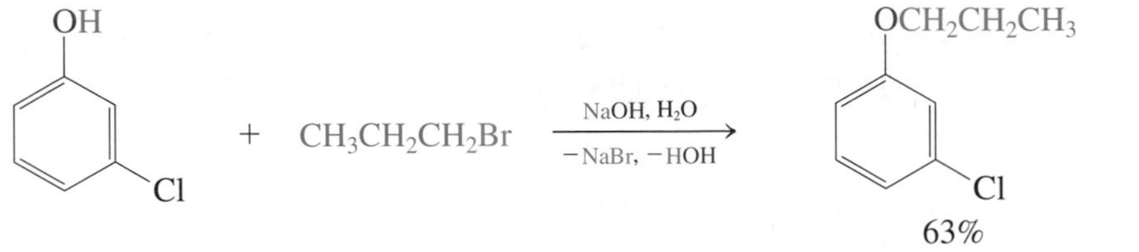
OH
Cl
+ $CH_3CH_2CH_2Br$
NaOH, H_2O
−NaBr, −HOH
$OCH_2CH_2CH_3$
Cl
63%

3-Chlorophenol
(*m*-Chlorophenol)

1-Chloro-3-propoxybenzene
(*m*-Chlorophenyl propyl ether)

Esterification leads to phenyl alkanoates

The reaction of a carboxylic acid with a phenol (Section 19-9) to form a phenyl ester is endothermic. Therefore, esterification requires an activated carboxylic acid derivative, such as an alkanoyl halide or a carboxylic anhydride.

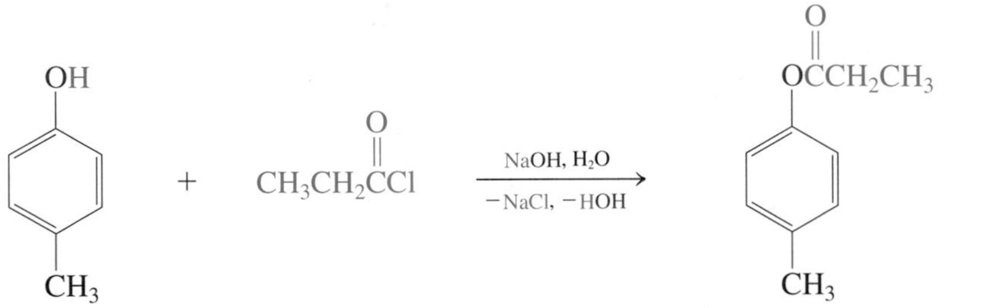
OH
CH_3
+
O
CH_3CH_2CCl
NaOH, H_2O
−NaCl, −HOH
O
$OCCH_2CH_3$
CH_3

4-Methylphenol
(*p*-Cresol)

Propanoyl chloride

4-Methylphenyl propanoate
(*p*-Methylphenyl propanoate)

EXERCISE 22-15

Explain why, in the preparation of Tylenol (Chemical Highlight 22-3), the amide is formed rather than the ester. (**Hint:** Review Section 6-9.)

In summary, the oxygen in phenols and alkoxybenzenes can be protonated even though it is less basic than the oxygen in the alkanols and alkoxyalkanes. Protonated phenols and their derivatives do not ionize to phenyl cations, but the ethers can be cleaved to phenols and haloalkanes by HX. Alkoxybenzenes are made by Williamson ether synthesis, aryl alkanoates by alkanoylation.

22-6 Electrophilic Substitution of Phenols

The aromatic ring in phenols is also a center of reactivity. The interaction between the OH group and the ring strongly activates the ortho and para positions toward elec-

CHEMICAL HIGHLIGHT 22-3   **Aspirin: A Phenyl Alkanoate Drug**

2-Hydroxybenzoic acid
(*o*-Hydroxybenzoic acid,
salicylic acid)

2-Acetyloxybenzoic acid
(*o*-Acetoxybenzoic acid,
acetylsalicylic acid, aspirin)

The year 1997 marked the 100th birthday of the synthesis of the acetic ester of 2-hydroxybenzoic acid (salicylic acid); namely, 2-acetyloxybenzoic acid (acetylsalicylic acid), better known as *aspirin* (see Section 19-13). Aspirin is the first drug that was clinically tested before it was marketed in 1899. More than 100 billion tablets of aspirin are taken by people throughout the world to relieve headaches and rheumatoid and other pain, to control fever, and to treat gout and arthritis. Production capacity in the United States alone is 10,000 tons per year.

Salicylic acid (also called spiric acid, hence the name aspirin ["a" for acetyl]) in extracts from the bark of the willow tree or from the meadowsweet plant had been used since ancient times (see beginning of this chapter) to treat pain, fever, and swelling. This acid was first isolated in pure form in 1829, then synthesized in the laboratory, and finally produced on a large scale in the 19th century and prescribed as an analgesic, antipyretic, and anti-inflammatory drug. Its bitter taste and side effects, such as mouth irritation and gastric bleeding, prompted the search for better derivatives that resulted in the discovery of aspirin.

In the body, aspirin functions as a precursor of salicylic acid. The latter irreversibly inhibits *cyclooxygenase,* an enzyme that causes the production

of prostaglandins (see Chemical Highlight 11-1 and Section 19-13), molecules that in turn are inflammatory and pain producing. In addition, one of them, thromboxane A_2, aggregates blood platelets, necessary for the clotting of blood when injury occurs. This same process is, however, undesirable inside arteries, causing heart attacks or brain strokes, depending on the location of the clot. Indeed, a large study conducted in the 1980s showed that aspirin lowered the risk of heart attacks in men by almost 50% and reduced the mortality rate during an actual attack by 23%.

Many other potential applications of aspirin are under investigation, such as in the treatment of pregnancy-related complications, viral inflammation in AIDS patients, dementia, Alzheimer's disease, and cancer. Despite its popularity, aspirin can have some serious side effects: it is toxic to the liver, prolongs bleeding, and causes gastric irritation. It is suspected as the cause of Reye's syndrome, a condition that leads to usually fatal brain damage. Because of some of these drawbacks, many other drugs compete with aspirin, particularly in the analgesics market, such as naproxen, ibuprofen, and acetaminophen (see the beginning of Chapter 16). Acetaminophen, better known as Tylenol, is prepared from 4-aminophenol by acetylation.

4-Aminophenol
(*p*-Aminophenol)

N-(4-Hydroxyphenyl)acetamide
[*N*-(*p*-Hydroxyphenyl)acetamide, acetaminophen, Tylenol]

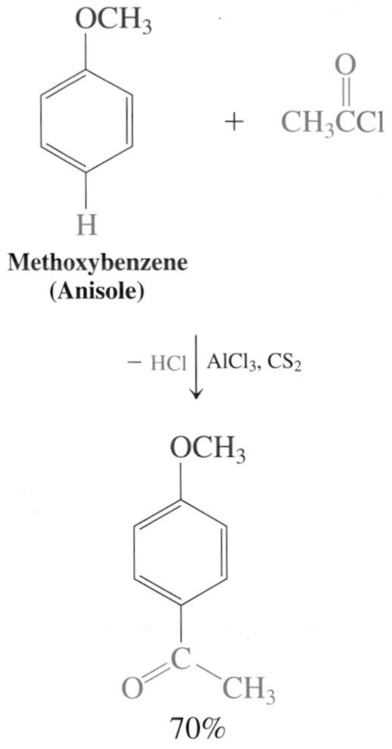

trophilic substitution (Sections 16-1 and 16-3). For example, even dilute nitric acid causes nitration.

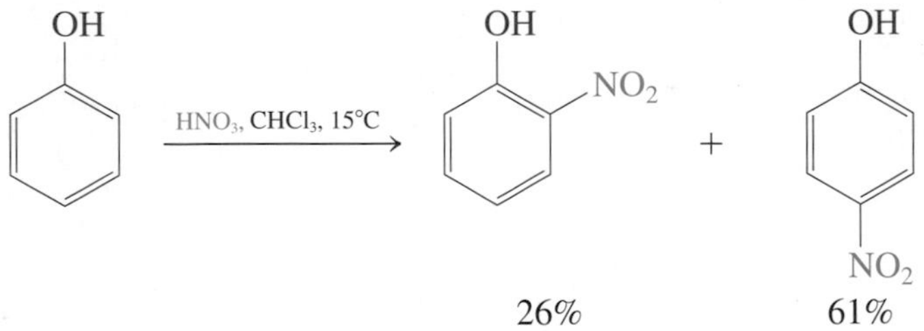

Methoxybenzene (Anisole)

$- HCl$ | $AlCl_3$, CS_2

70%
1-(4-Methoxyphenyl)ethanone (p-Methoxyacetophenone)

HNO$_3$, CHCl$_3$, 15°C

26%
2-Nitrophenol (o-Nitrophenol)

61%
4-Nitrophenol (p-Nitrophenol)

Friedel-Crafts alkanoylation (acylation) of phenols is complicated by ester formation and is better carried out on ether derivatives of phenol (Section 16-5), as shown in the margin.

Phenols are halogenated so readily that a catalyst is not required, and multiple halogenations are frequently observed (Section 16-3). As shown in the following reactions, tribromination occurs in water at 20°C, but the reaction can be controlled to produce the monohalogenation product through the use of a lower temperature and a less polar solvent.

Halogenation of Phenols

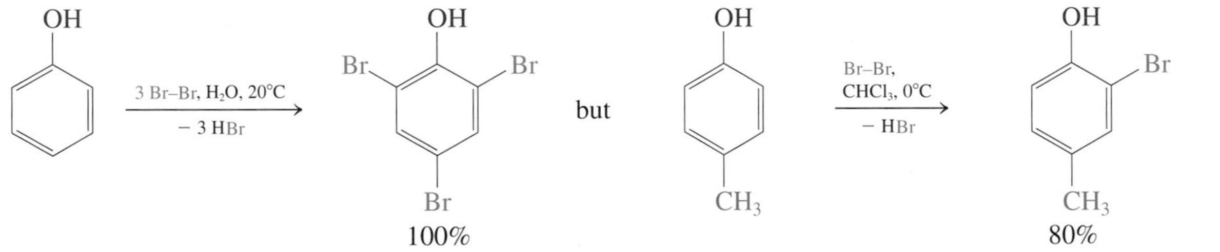

Phenol

3 Br–Br, H$_2$O, 20°C
$- 3$ HBr

100%
2,4,6-Tribromophenol

but

4-Methylphenol (p-Cresol)

Br–Br, CHCl$_3$, 0°C
$-$ HBr

80%
2-Bromo-4-methylphenol

Electrophilic attack at the para position is frequently dominant because of steric effects. However, it is normal to obtain mixtures resulting from both ortho and para substitutions, and their compositions are highly dependent on reagents and reaction conditions.

EXERCISE 22-16

Friedel-Crafts methylation of methoxybenzene (anisole) with chloromethane in the presence of AlCl$_3$ gives a 2:1 ratio of ortho:para products. Treatment of methoxybenzene with 2-chloro-2-methylpropane (*tert*-butyl chloride) under the same conditions furnishes only 1-methoxy-4-(1,1-dimethylethyl)benzene (*p-tert*-butylanisole). Explain. (**Hint:** Review Section 16-5.)

Under basic conditions, phenols can undergo electrophilic substitution, even with very mild electrophiles, through intermediate phenoxide ions. An industrially important application is the reaction with formaldehyde, which leads to *o*- and *p*-hydroxymethylation. Mechanistically, these processes may be considered enolate condensations, much like the aldol reaction (Section 18-5).

Hydroxymethylation of Phenol

The initial aldol products are unstable: They dehydrate on heating, giving reactive intermediates called **quinomethanes.**

o-Quinomethane   *p*-Quinomethane

Because quinomethanes are α,β-unsaturated carbonyl compounds, they may undergo Michael additions (Section 18-12) with excess phenoxide ion. The resulting phenols can be hydroxymethylated again and the entire process repeated. Eventually, a complex phenol–formaldehyde copolymer, also called a **phenolic resin** (e.g., Bakelite), is formed. Total production of these resins in the United States in 1996 exceeded 1.4 million tons. Their major uses are in plywood (45%), insulation (14%), molding compounds (9%), fibrous and granulated wood (9%), and laminates (8%).

Phenolic Resin Synthesis

In the **Kolbe* reaction,** phenoxide attacks carbon dioxide to furnish the salt of 2-hydroxybenzoic acid (*o*-hydroxybenzoic acid, salicylic acid, precursor to aspirin; see Chemical Highlight 22-3).

$$\text{OH} \quad + \quad CO_2 \quad \xrightarrow{\text{KHCO}_3,\ H_2O,\ \text{pressure}} \quad \text{OH} \quad \text{COO}^-\text{K}^+$$

EXERCISE 22-17

Formulate a mechanism for the Kolbe reaction. (**Hint:** This process is initially analogous to an aldol addition reaction; Section 18-5.)

EXERCISE 22-18

Hexachlorophene (see Chemical Highlight 22-1) is prepared in one step from 2,4,5-trichlorophenol and formaldehyde in the presence of sulfuric acid. How does this reaction proceed? (**Hint:** Formulate an acid-catalyzed hydroxymethylation for the first step.)

In summary, the benzene ring in phenols is subject to electrophilic aromatic substitution, particularly under basic conditions. Phenoxide ions can be hydroxymethylated and carbonated.

22-7 Claisen and Cope Rearrangements

At 200°C, 2-propenyloxybenzene (allyl phenyl ether) undergoes an unusual reaction that leads to the rupture of the allylic ether bond: The starting material rearranges to 2-(2-propenyl)phenol (*o*-allylphenol).

$$\underset{\substack{\textbf{2-Propenyloxybenzene}\\\textbf{(Allyl phenyl ether)}}}{\underset{\text{H}}{\overset{\text{O}-\text{CH}_2\text{CH}=\text{CH}_2}{\bigcirc}}} \quad \xrightarrow{\Delta} \quad \underset{\substack{\textbf{2-(2-Propenyl)phenol}\\\textbf{(}o\textbf{-Allylphenol)}\\75\%}}{\underset{\text{CH}_2\text{CH}=\text{CH}_2}{\overset{\text{OH}}{\bigcirc}}}$$

This transformation, called the **Claisen**[†] **rearrangement,** is another concerted reaction with a transition state that accommodates the movement of six electrons (Sections 14-8 and 15-3). The initial intermediate is a high-energy isomer, 6-(2-propenyl)-2,4-cyclohexadienone, which enolizes to the final product.

---

*Professor Adolph Wilhelm Hermann Kolbe (1818–1884), University of Leipzig, Germany.
[†]Professor Ludwig Claisen (1851–1930), University of Berlin, Germany.

Mechanism of the Claisen Rearrangement

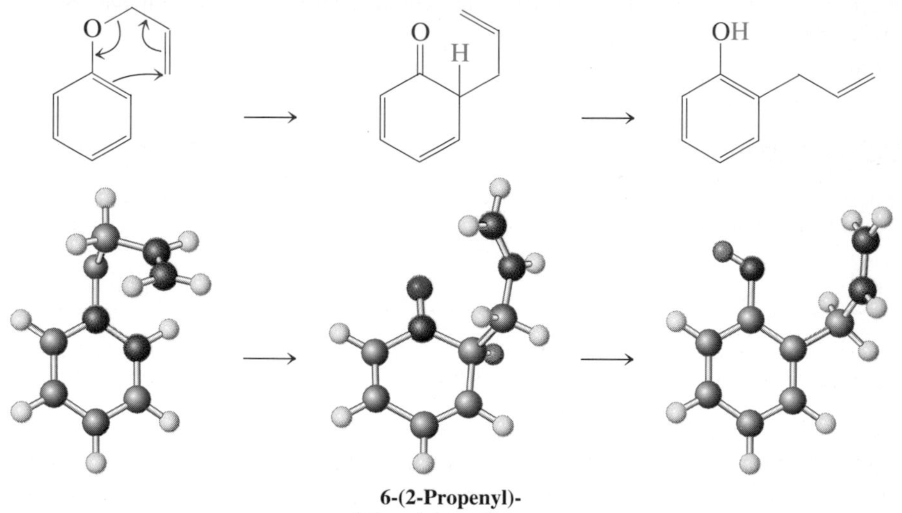

**6-(2-Propenyl)-
2,4-cyclohexadienone**

The Claisen rearrangement is general for other systems. With the nonaromatic 1-ethenyloxy-2-propene (allyl vinyl ether), it stops at the carbonyl stage because there is no driving force for enolization. This is called the **aliphatic Claisen rearrangement.**

Aliphatic Claisen Rearrangement

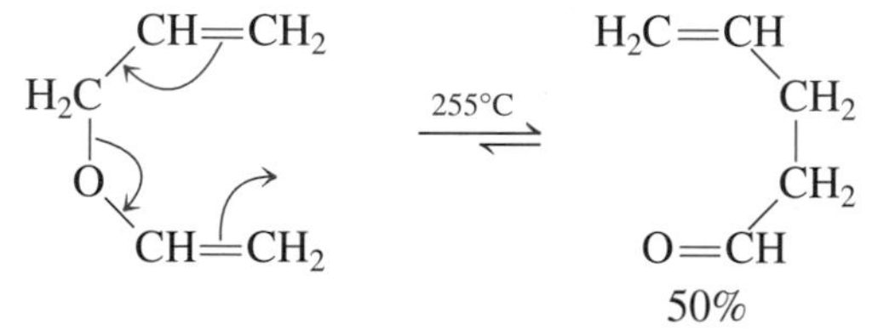

**1-Ethenyloxy-2-propene
(Allyl vinyl ether)**

255°C

4-Pentenal

50%

The carbon analog of the Claisen rearrangement is called the **Cope* rearrangement;** it takes place in compounds containing 1,5-diene units.

Cope Rearrangement

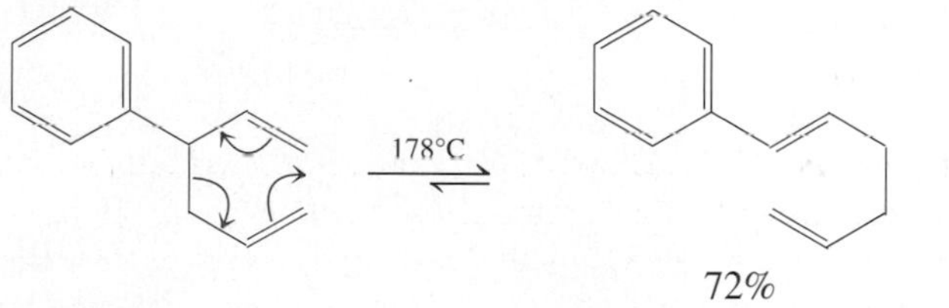

178°C

3-Phenyl-1,5-hexadiene

trans-1-Phenyl-1,5-hexadiene

72%

*Professor Arthur C. Cope (1909–1966), Massachusetts Institute of Technology, Cambridge.

Electrocyclic Reaction of
***cis*-1,3,5-Hexatriene**

Note that all of these rearrangements are related to the electrocyclic reactions that interconvert *cis*-1,3,5-hexatriene with 1,3-cyclohexadiene (see margin and Section 14-9). The only difference is the absence of a double bond connecting the terminal π bonds.

EXERCISE 22-19

Explain the following transformation by a mechanism. (**Hint:** The Cope rearrangement can be accelerated greatly if it leads to charge delocalization.)

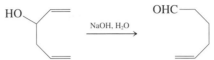

In summary, 2-propenyloxybenzene rearranges to 2-(2-propenyl)phenol (*o*-allylphenol) by an electrocyclic mechanism that moves six electrons (Claisen rearrangement). Similar concerted reactions are undergone by aliphatic unsaturated ethers (aliphatic Claisen rearrangement) and by hydrocarbons containing 1,5-diene units (Cope rearrangement).

22-8 Oxidation of Phenols: Cyclohexadienediones (Benzoquinones)

Phenols can be oxidized to carbonyl derivatives by one-electron transfer mechanisms, resulting in a new class of cyclic diketones, called **cyclohexadienediones (benzoquinones).**

Chemical Warfare in Nature: The Bombardier Beetle

The oxidizing power of 2,5-cyclohexadiene-1,4-diones (*p*-benzoquinones) is used by some arthropods, such as millipedes, beetles, and termites, as chemical defense agents. Most remarkable among these species is the bombardier beetle. It contains a cocktail of 1,4-benzenediol (hydroquinone), a number of its simple alkyl derivatives, and hydrogen peroxide in its defensive glands which, at times of an attack by a predator, is passed into a discharge compartment. Here, enzymes trigger an explosive reaction to furnish quinones and oxygen, which are audibly ejected onto the enemy at temperatures reaching 100°C with quite some accuracy and force.

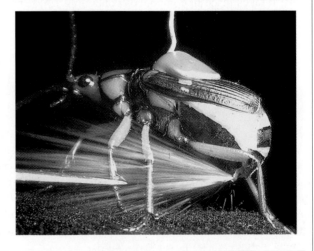

The bombardier beetle in action.

Cyclohexadienediones (benzoquinones) and benzenediols (hydroquinones and catechols) are redox couples

1,2- and **1,4-benzenediols** (**catechols** and **hydroquinones**) are oxidized to the corresponding diketones, 3,5-cyclohexadiene-1,2-diones and 2,5-cyclohexadiene-1,4-diones (*o*-benzoquinones and *p*-benzoquinones), by a variety of oxidizing agents, such as sodium dichromate or silver oxide. Yields can be variable when the resulting diones are reactive, as is the case with 3,5-cyclohexadiene-1,2-dione (*o*-benzoquinone), which partly decomposes under the conditions of its formation.

Cyclohexadienediones (Benzoquinones) from Oxidation of 1,2- and 1,4-Benzenediols

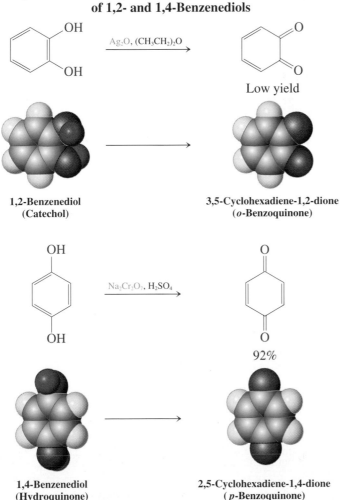

1,2-Benzenediol
(Catechol)

3,5-Cyclohexadiene-1,2-dione
(*o*-Benzoquinone)

1,4-Benzenediol
(Hydroquinone)

2,5-Cyclohexadiene-1,4-dione
(*p*-Benzoquinone)

The redox process that interconverts 1,4-benzenediol (hydroquinone) and 2,5-cyclohexadiene-1,4-dione (*p*-benzoquinone) can be visualized as a sequence of proton and electron transfers. Initial deprotonation gives a phenoxide ion, which is transformed into a **phenoxy radical** by one-electron oxidation. Proton dissociation from the remaining OH group furnishes a **semiquinone radical anion,** and a second one-electron oxidation step leads to the benzoquinone. All of the intermediate species in this sequence benefit from considerable resonance stabilization (two forms are shown for the semiquinone). We shall see in Section 22-9 that redox processes similar to those shown here occur widely in nature.

Redox Relation Between 2,5-Cyclohexadiene-1,4-dione
(*p*-Benzoquinone) and 1,4-Benzenediol (Hydroquinone)

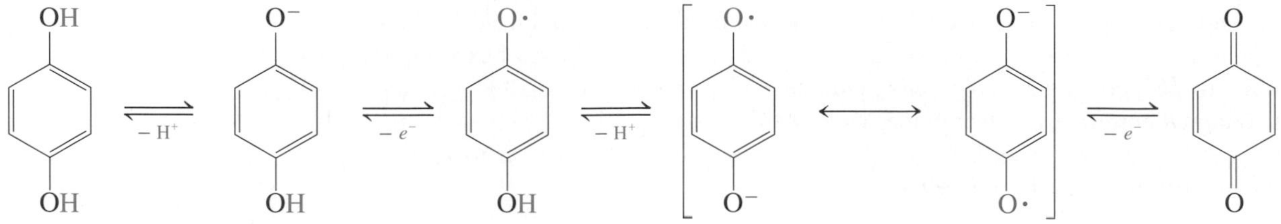

Phenoxide **Phenoxy** **Semiquinone**
ion **radical** **radical anion**

EXERCISE 22-20

Give a minimum of two additional resonance forms each for the phenoxide ion, phenoxy radical, and semiquinone radical anion shown in the preceding scheme.

The enone units in 2,5-cyclohexadiene-1,4-diones (*p*-benzoquinones) undergo conjugate and Diels–Alder additions

2,5-Cyclohexadiene-1,4-diones (*p*-benzoquinones) function as reactive α,β-unsaturated ketones in conjugate additions (see Section 18-10). For example, hydrogen chloride adds to give an intermediate hydroxy dienone that enolizes to the aromatic 2-chloro-1,4-benzenediol.

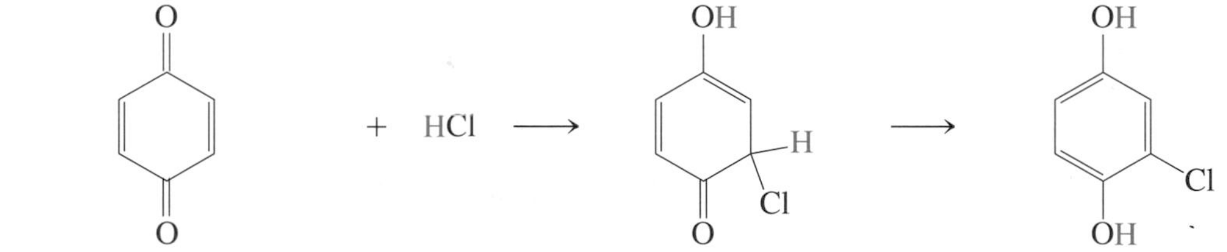

2,5-Cyclohexadiene-1,4-dione **6-Chloro-4-hydroxy-** **2-Chloro-1,4-benzenediol**
(*p*-Benzoquinone) **2,4-cyclohexadienone**

The double bonds also undergo cycloadditions to dienes (Section 14-8). The initial cycloadduct to 1,3-butadiene tautomerizes with acid to the aromatic system.

Diels–Alder Reactions
of 2,5-Cyclohexadiene-1,4-dione (*p*-Benzoquinone)

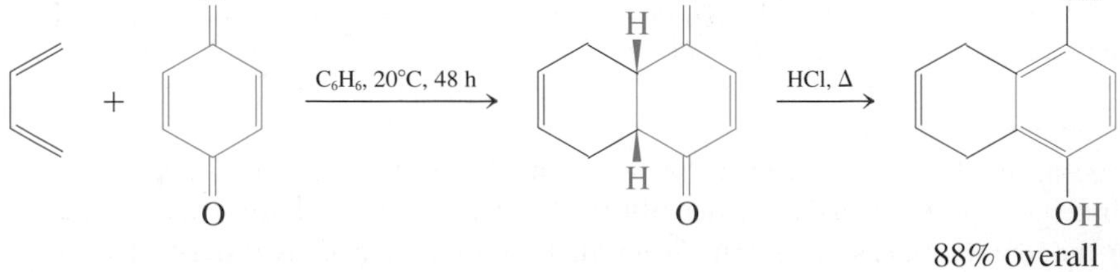

88% overall

Explain the following result by a mechanism. (**Hint:** Review Section 18-10.)

In summary, phenols are oxidized to the corresponding diones (benzoquinones). The diones enter into reversible redox reactions that yield the corresponding diols. They also undergo conjugate additions and Diels–Alder additions to the double bonds.

22-9 Oxidation–Reduction Processes in Nature

This section describes some chemical processes involving 1,4-benzenediols (hydroquinones) and 2,5-cyclohexadiene-1,4-diones (p-benzoquinones) that occur in nature. We begin with an introduction to the biochemical reduction of O_2. Oxygen can engage in reactions that cause damage to biomolecules. Naturally occurring **antioxidants** inhibit these transformations. The section concludes with a discussion of the properties of synthetic preservatives.

Ubiquinones mediate the biological reduction of oxygen to water

Nature makes use of the benzoquinone–hydroquinone redox couple in reversible oxidation reactions. These processes are part of the complicated cascade by which oxygen is used in biochemical degradations. An important series of compounds used for this purpose are the **ubiquinones** (a name coined to indicate their ubiquitous presence in nature), also collectively called **coenzyme Q** (**CoQ**, or simply **Q**). The ubiquinones are substituted 2,5-cyclohexadiene-1,4-dione (p-benzoquinone) derivatives bearing a side chain made up of 2-methylbutadiene units (isoprene; Sections 4-7 and 14-10). An enzyme system that utilizes NADH (Chemical Highlights 8-1 and 25-3) converts CoQ into its reduced form (QH_2).

Ubiquinones ($n = 6, 8, 10$)
(Coenzyme Q)

Reduced form of coenzyme Q
(Reduced Q, or QH_2)

QH_2 participates in a chain of redox reactions with electron-transporting iron-containing proteins called **cytochromes** (Chemical Highlight 8-1). The reduction of Fe^{3+} to Fe^{2+} in cytochrome b by QH_2 begins a sequence of electron transfers involving six different proteins. The chain ends with reduction of O_2 to water by addition of four electrons and four protons.

$$O_2 \ + \ 4\,H^+ \ + \ 4\,e^- \ \longrightarrow \ 2\,H_2O$$

Phenol derivatives protect cell membranes from oxidative damage

The biochemical conversion of oxygen into water includes several intermediates, including **superoxide,** $O_2^{-\cdot}$, the product of one-electron reduction, and **hydroxy radical,** $\cdot OH$, which arises from cleavage of H_2O_2. Both are highly reactive species capable of initiating reactions that damage organic molecules of biological importance. An example is the phosphoglyceride shown here, a cell-membrane component derived from the unsaturated fatty acid *cis,cis*-octadeca-9,12-dienoic acid (linoleic acid).

INITIATION STEP

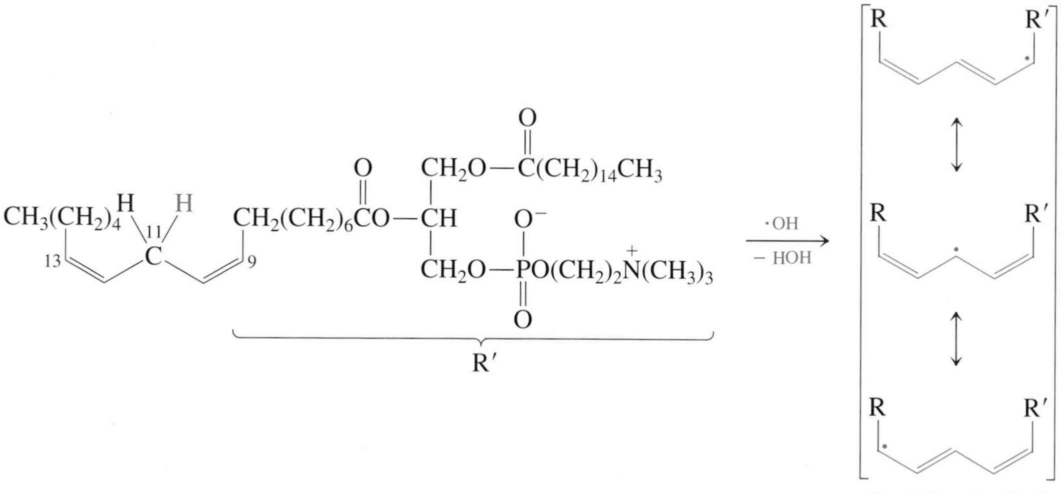

Pentadienyl radical

The doubly allylic hydrogens at C11 are readily abstracted by radicals such as $\cdot OH$ (Chapter 14).

PROPAGATION STEP 1

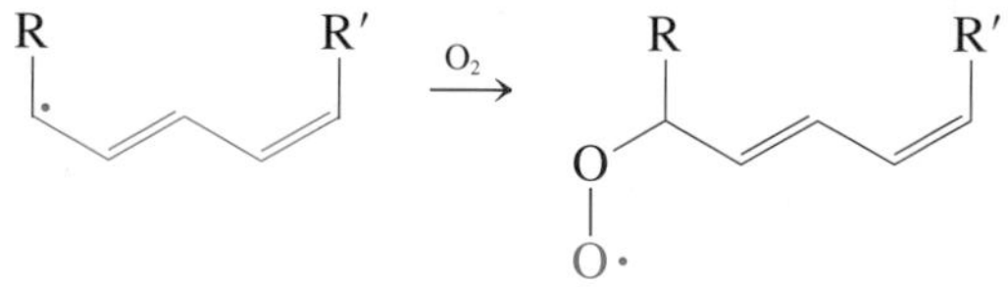

Peroxy radical

The resonance-stabilized pentadienyl radical combines rapidly with O_2 in the first of two propagation steps. Reaction occurs at either C9 or C13 (shown here), giving either of two peroxy radicals containing conjugated diene units.

PROPAGATION STEP 2

R   R'   R H H R'   R   R'   R   R'

O   11   O
O•   +   C   ⟶   O   +   •

                         OH   OH

Lipid hydroperoxide

In the second propagation step this species removes a hydrogen atom from C11 of another molecule of phosphoglyceride, or more generally, lipid (Section 20-5), thereby forming a new dienyl radical and a molecule of **lipid hydroperoxide.** The dienyl radical may then reenter propagation step 1. In this way, a large number of lipid molecules may be oxidized following just a single initiation event.

Numerous studies have confirmed that lipid hydroperoxides are toxic, their products of decomposition even more so. For example, loss of $\cdot$OH by cleavage of the relatively weak O–O bond gives rise to an alkoxy radical, which may decompose by breaking a neighboring C–C bond (β-scission), forming an unsaturated aldehyde.

β-Scission of a Lipid Alkoxy Radical

R   H R'   R   H R'   R   H

                O   O•   O
            OH

$\xrightarrow{-\cdot\text{OH}}$ $\xrightarrow{-\cdot\text{R}'}$

Alkoxy radical

Through related but more complex mechanisms, certain lipid hydroperoxides decompose to give unsaturated hydroxyaldehydes, such as *trans*-4-hydroxy-2-nonenal, as well as the dialdehyde propanedial (malondialdehyde). Molecules of these general types are partly responsible for the smell of rancid fats.

Both propanedial and the α,β-unsaturated aldehydes are extremely toxic, because they are highly reactive toward the proteins that are present in close proximity to the lipids in cell membranes. For example, both dials and enals are capable of reacting with nucleophilic amino and mercapto groups from two different parts of one protein or from two different protein molecules, and these reactions produce cross-linking (Section 14-10). Cross-linking severely inhibits protein molecules from carrying out their biological functions (Chapter 26).

HO

CH$_3$(CH$_2$)$_4$ H

                                    O

trans-**4-Hydroxy-2-nonenal**

O   O

H   H

**Propanedial
(Malondialdehyde)**

Cross-linking of Proteins by Reaction with Unsaturated Aldehydes

        O   protein$_1$—S   O   protein$_1$—S   N—protein$_2$

                          $\xrightarrow{\text{Protein}_1\text{—SH}}$ $\xrightarrow{\text{Protein}_2\text{—NH}_2}$

R   H   R   H   R   H

Processes such as these are thought by many to contribute to the development of emphysema, atherosclerosis (the underlying cause of several forms of heart disease and stroke), certain chronic inflammatory and autoimmune diseases, cancer, and, possibly, the process of aging itself.

Vitamin E
(α-Tocopherol)

R = branched C₁₆H₃₃ chain

Does nature provide the means for biological systems to protect themselves from such damage? A variety of naturally occurring antioxidant systems defend lipid molecules inside cell membranes from oxidative destruction. The most important is **vitamin E,** a reducing agent that possesses a long hydrocarbon chain (see Problem 30 of Chapter 2), a feature making it lipid soluble. Vitamin E contains a structure similar to that of 1,4-benzenediol (Section 22-8). The corresponding phenoxide ion is an excellent electron donor. The protective qualities of vitamin E rest on its ability to break the propagation chain of lipid oxidation by the reduction of radical species.

Reactions of Vitamin E with Lipid Hydroperoxy and Alkoxy Radicals

Vitamin E

$$- H^+$$

$$+ \quad \begin{array}{c} \text{lipid}-\text{O}\cdot \\ \text{or} \\ \text{lipid}-\text{O}-\text{O}\cdot \end{array} \quad \rightleftharpoons$$

Lipid radicals

Many dietary supplements advertise antioxidants.

α-Tocopheroxy radical

$$+ \quad \begin{array}{c} \text{lipid}-\text{O}^- \\ \text{or} \\ \text{lipid}-\text{O}-\text{O}^- \end{array} \quad \underset{+ H^+}{\rightleftharpoons} \quad \begin{array}{c} \text{lipid}-\text{OH} \\ \text{or} \\ \text{lipid}-\text{O}-\text{OH} \end{array}$$

In this process, lipid radicals are reduced and protonated. Vitamin E is oxidized to an α-tocopheroxy radical, which is relatively unreactive because of extensive delocalization and the steric hindrance of the methyl substituents. Vitamin E is regenerated at the membrane surface by reaction with water-soluble reducing agents such as **vitamin C.**

Regeneration of Vitamin E by Vitamin C

Vitamin C

$$+ \quad \longrightarrow \quad + \quad$$

Semidehydroascorbic acid

The product of vitamin C oxidation eventually decomposes to lower molecular weight water-soluble compounds, which are excreted by the body.

EXERCISE 22-22

Vitamin C is an effective antioxidant because its oxidation product semidehydroascorbic acid is stabilized by resonance. Give other resonance forms for this species.

Benzoquinones consume glutathione, an intracellular reducing agent

Virtually all living cells contain the substance **glutathione**, a peptide that incorporates a mercapto functional group (see Sections 9-10 and 26-4). The latter serves to reduce disulfide linkages in proteins to SH groups and to maintain the iron in hemoglobin in the 2+ oxidation state (Section 26-8). Glutathione also participates in the reduction of oxidants, such as hydrogen peroxide, H_2O_2, that may be present in the interior of the cell.

Glutathione

The molecule is converted into a disulfide (Section 9-10) in this process but is regenerated by an enzyme-mediated reduction.

Cyclohexadienediones (benzoquinones) and related compounds react irreversibly with glutathione in the liver by conjugate addition. Cell death results if glutathione depletion is extensive. Tylenol is an example of a substance that exhibits such liver toxicity at very high doses. Cytochrome P-450, a redox enzyme system in the liver, oxidizes Tylenol to an imine derivative of a cyclohexadienedione, which in turn consumes glutathione. Vitamin C is capable of reversing the oxidation.

Tylenol

2-(1,1-Dimethylethyl)-4-methoxyphenol
(**BHA**)

Synthetic analogs of vitamin E are preservatives

Synthetic phenol derivatives are widely used as antioxidants and preservatives in the food industry. Perhaps two of the most familiar are 2-(1,1-dimethylethyl)-4-methoxyphenol (butylated hydroxyanisole, or **BHA**) and 2,6-bis(1,1-dimethylethyl)-4-methylphenol (butylated hydroxytoluene, or **BHT**; see Exercise 16-13). For example, addition of BHA to butter increases its storage life from months to years. Both BHA and BHT function like vitamin E, reducing oxygen radicals and interrupting the propagation of oxidation processes.

In summary, oxygen-derived radicals are capable of initiating radical chain reactions in lipids, thereby leading to toxic decomposition products. Vitamin E is a naturally occurring phenol derivative that functions as an antioxidant to inhibit these

2,6-Bis(1,1-dimethylethyl)-4-methylphenol
(**BHT**)

processes within membrane lipids. Vitamin C and glutathione are biological reducing agents located in the intra- and extracellular aqueous environments. High concentrations of cyclohexadienediones (benzoquinones) can bring about cell death by consumption of glutathione; vitamin C can protect the cell by reduction of the benzoquinone. Synthetic food preservatives are structurally designed to mimic the antioxidant behavior of vitamin E.

22-10 Arenediazonium Salts

The final two sections of this chapter conclude our examination of functional group chemistry that has been modified by conjugation with a benzene ring. We focus now on nitrogen substituents and describe the chemistry of arenediazonium salts. As mentioned in Section 22-4, *N*-nitrosation of primary benzenamines (anilines) furnishes these salts, which can be used in the synthesis of phenols. Arenediazonium salts are stabilized by resonance and are converted into haloarenes, arenecarbonitriles, and other aromatic derivatives through replacement of nitrogen by the appropriate nucleophile.

Arenediazonium salts are stabilized by resonance

The reason for the stability of arenediazonium salts, relative to their alkane counterparts, is resonance and the high energy of the aryl cations formed by loss of nitrogen. One of the electron pairs making up the aromatic π system can be delocalized into the functional group, which results in charge-separated resonance structures containing a double bond between the benzene ring and the attached nitrogen.

Resonance in the Benzenediazonium Cation

At elevated temperatures (>50°C), nitrogen extrusion does take place, however, to form the very reactive phenyl cation. When this is done in aqueous solution, phenols are produced (Section 22-4).

Why is the phenyl cation so reactive? After all, it is a carbocation that is part of a benzene ring. Should it not be resonance stabilized, like the phenylmethyl (benzyl) cation? The answer is no, as may be seen in the molecular-orbital picture of the phenyl cation (Figure 22-5). The empty orbital associated with the positive charge is one of the sp^2 hybrids aligned in *perpendicular* fashion to the π framework that normally produces aromatic resonance stabilization. Hence, this orbital cannot overlap with the π bonds, and

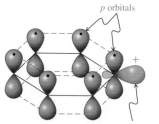

p orbitals

Empty sp^2 orbital

FIGURE 22-5

Orbital picture of the phenyl cation. The alignment of its empty sp^2 orbital is perpendicular to the six-π-electron framework of the aromatic ring. As a result, the positive charge is not stabilized by resonance.

the positive charge cannot be delocalized. We used the same picture to explain the difficulty in deprotonating benzene to the corresponding phenyl anion (Section 22-4).

Arenediazonium salts can be converted into other substituted benzenes

When arenediazonium salts are decomposed in the presence of nucleophiles other than water, the corresponding substituted benzenes are formed. For example, diazotization of arenamines (anilines) in the presence of hydrogen iodide results in the corresponding iodoarenes.

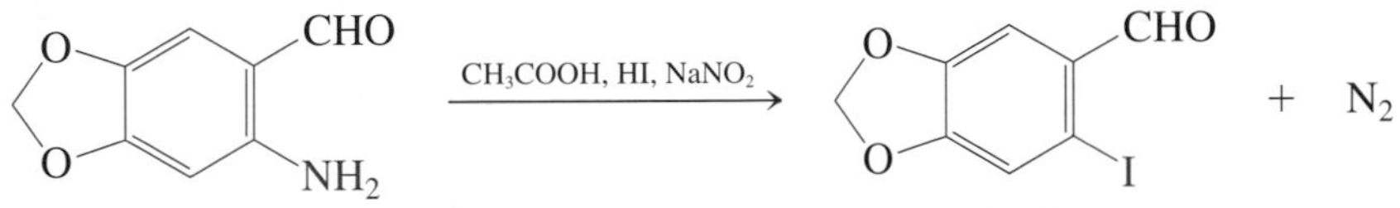

Attempts to obtain other haloarenes in this way are frequently complicated by side reactions. One solution to this problem is the **Sandmeyer* reaction,** which makes use of the fact that the exchange of the nitrogen substituent for halogen is considerably facilitated by the presence of cuprous [Cu(I)] salts. The detailed mechanism of this process is complex, and radicals are participants. Addition of cuprous cyanide, CuCN, to the diazonium salt in the presence of excess potassium cyanide gives aromatic nitriles.

Sandmeyer Reactions

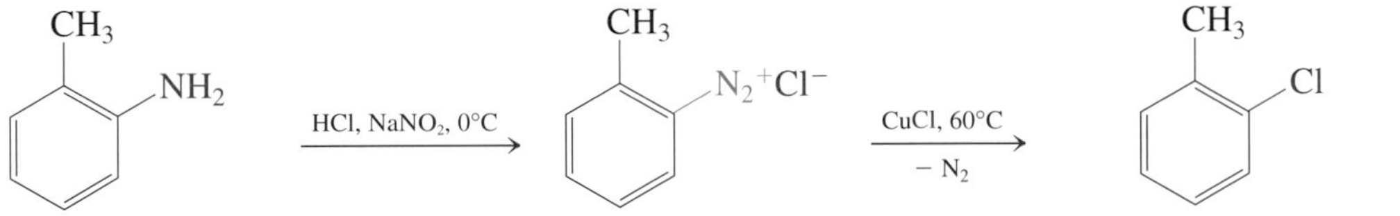

2-Methylbenzenamine
(*o*-Methylaniline)

1-Chloro-2-methylbenzene
(*o*-Chlorotoluene)

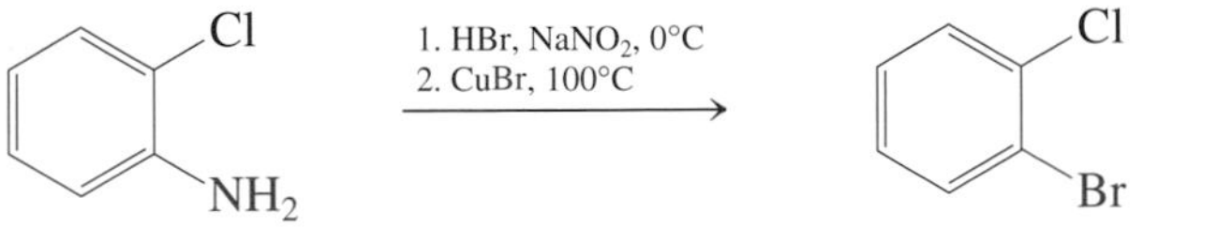

79% overall

2-Chlorobenzenamine
(*o*-Chloroaniline)

1-Bromo-2-chlorobenzene
(*o*-Bromochlorobenzene)

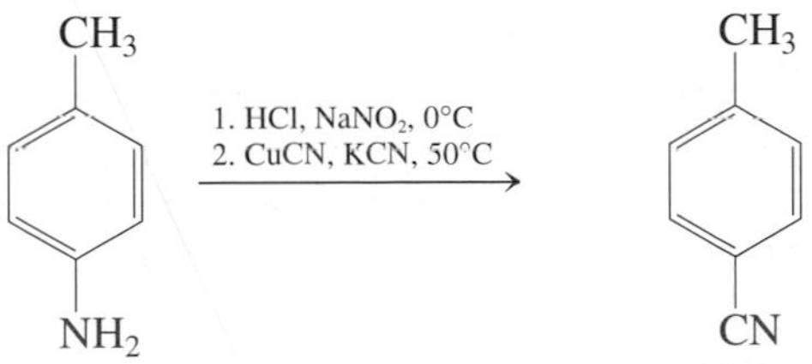

73%

4-Methylbenzonitrile
(*p*-Tolunitrile)

70%

*Dr. Traugott Sandmeyer (1854–1922), Geigy Company, Basel, Switzerland.

EXERCISE 22-23

Propose syntheses of the following compounds, starting from benzene.

(a) [structure: benzene ring with ethyl group and I]

(b) [structure: benzene ring with two CN groups, meta]

(c) [structure: benzene ring with OH and SO₃H, meta]

The diazonium group can be removed reductively by reducing agents. The sequence diazotization–reduction is a way to replace the amino group in arenamines (anilines) with hydrogen. The reducing agent employed is aqueous hypophosphorous acid, H_3PO_2. This method is especially useful in syntheses in which an amino group is used as a removable directing substituent in electrophilic aromatic substitution (Section 16-5).

Reductive Removal of a Diazonium Group

$$\text{(structure with } CH_3, NH_2, Br) \xrightarrow{NaNO_2, H^+, H_2O} \text{(structure with } CH_3, N_2^+, Br) \xrightarrow{H_3PO_2, H_2O, 25°C} \text{(structure with } CH_3, H, Br)$$

85%

1-Bromo-3-methylbenzene
(*m*-Bromotoluene)

Another application of diazotization in synthetic strategy is illustrated in the synthesis of 1,3-dibromobenzene (*m*-dibromobenzene). Direct electrophilic bromination of benzene is not feasible for this purpose; after the first bromine has been introduced, the second will attack ortho or para. What is required is a meta-directing substituent which can be transformed eventually into bromine. The nitro group is such a substituent. Double nitration of benzene furnishes 1,3-dinitrobenzene (*m*-dinitrobenzene). Reduction (Section 16-5) leads to the benzenediamine, which is then converted into the dihalo derivative.

Synthesis of 1,3-Dibromobenzene by Using a Diazotization Strategy

$$\text{benzene} \xrightarrow{HNO_3, H_2SO_4, \Delta} \text{(1,3-dinitrobenzene)} \xrightarrow{H_2, Pd} \text{(1,3-benzenediamine)} \xrightarrow[2.\ CuBr, 100°C]{1.\ NaNO_2, H^+, H_2O} \text{(1,3-dibromobenzene)}$$

EXERCISE 22-24

Propose a synthesis of 1,3,5-tribromobenzene from benzene.

To summarize, arenediazonium salts, which are more stable than alkanediazonium salts because of resonance, are starting materials not only for phenols, but also for haloarenes, arenecarbonitriles, and reduced aromatics by displacement of nitrogen gas. The intermediates in some of these reactions may be aryl cations, highly reactive because of the absence of any electronically stabilizing features, but other, more complicated mechanisms may be followed. The ability to transform arenediazonium salts in this way gives considerable scope to the regioselective construction of substituted benzenes.

22-11 Electrophilic Substitution with Arenediazonium Salts: Diazo Coupling

Being positively charged, arenediazonium ions are electrophilic. Although they are not very reactive in this capacity, they can accomplish electrophilic aromatic substitution when the substrate is an activated arene, such as phenol or benzenamine (aniline). This reaction, called **diazo coupling,** leads to highly colored compounds called **azo dyes.** For example, reaction of *N,N*-dimethylbenzenamine (*N,N*-dimethylaniline) with benzenediazonium chloride gives the brilliant orange dye Butter Yellow. This compound was once used as a food coloring agent but has been declared a suspect carcinogen by the Food and Drug Administration.

Dyes are important additives in the textile industry. Azo dyes, while still in widespread use, are becoming less attractive for this purpose because some have been found to degrade to carcinogenic benzenamines.

Diazo Coupling

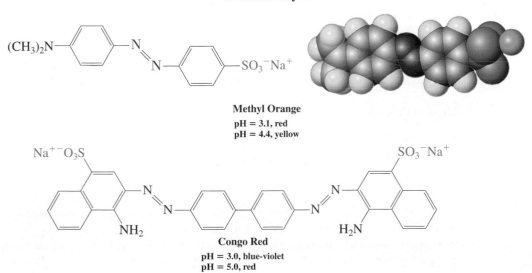

4-Dimethylaminoazobenzene
(*p*-Dimethylaminoazobenzene, Butter Yellow)

Dyes used in the clothing industry usually contain sulfonic acid groups that impart water solubility and allow the dye molecule to attach itself ionically to charged sites on the polymer framework of the textile.

Industrial Dyes

Methyl Orange
pH = 3.1, red
pH = 4.4, yellow

Congo Red
pH = 3.0, blue-violet
pH = 5.0, red

EXERCISE 22-25

Write the products of diazo coupling of benzenediazonium chloride with each of the following molecules. **(a)** Methoxybenzene; **(b)** 1-chloro-3-methoxybenzene; **(c)** 1-(dimethylamino)-4-(1,1-dimethylethyl)benzene. (**Hint:** Diazo couplings are quite sensitive to steric effects.)

In summary, arenediazonium cations attack activated benzene rings by diazo coupling, a process that furnishes azobenzenes, which are often highly colored.

CHAPTER INTEGRATION PROBLEM

5-Amino-2,4-dihydroxybenzoic acid A is a potential intermediate in the preparation of natural products of medicinal value (Section 22-3). Propose syntheses, starting from methylbenzene (toluene).

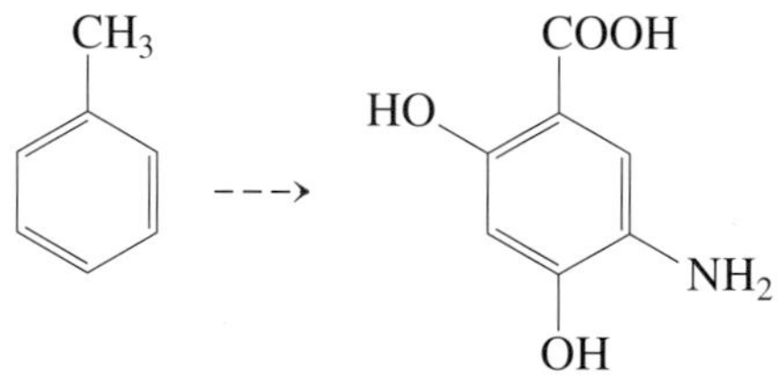

SOLUTION

This problem builds on the expertise that you gained (Chapter 16) in controlling the substitution patterns of target benzenes, but now with a greatly expanded range of reactions. The key is, again, recognition of the directing power of substituents, ortho, para or meta (Section 16-2), and their interconversion (Section 16-5).

Retrosynthetic analysis of compound A reveals one carbon-based substituent, the carboxy group, which one can be readily envisaged to be derivable from the methyl group in the starting material (by oxidation, Section 22-2). In the starting material, the carbon-based substituent is ortho, para directing, suggesting its use (retrosynthesis 1) in the introduction of the two hydroxy functions (as in compound B, through nitration–reduction–diazotization–hydrolysis; Sections 22-4 and 22-10). In compound A, it is meta directing and potentially utilizable (retrosynthesis 2) for the amination at C3 (as in compound C, through nitration–reduction).

Retrosynthesis 1

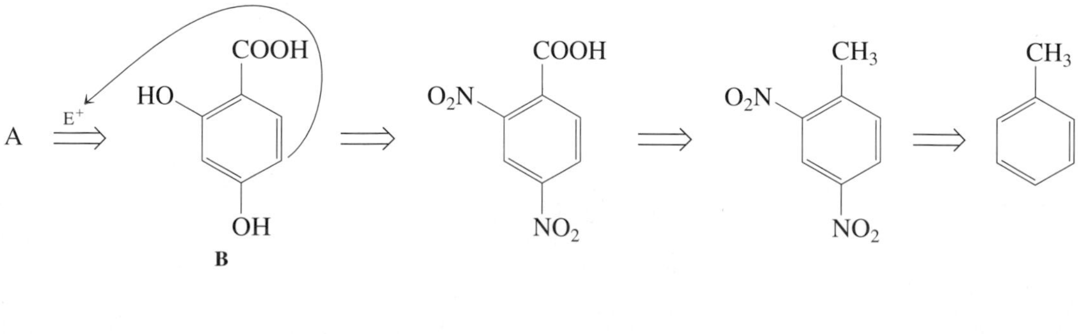

Retrosynthesis 2

The question is, Are compounds B and C effective precursors of compound A? The answer is yes. Nitration of compound B should take place at the desired position (C3 in the product), ortho and para, respectively, to the two hydroxy substituents, thus placing the nitrogen at its position in compound A. Electrophilic attack between the OH groups would be expected to be sterically hindered (Section 16-5). Conversely, the amino group in compound C, especially when protected as an amide, should direct electrophilic substitution to the less hindered ortho carbon and the para carbon, again yielding the desired pattern. The actual proposed synthetic schemes would be then as follows.

Synthesis 1

Synthesis 2

NEW REACTIONS

Benzylic Resonance

1. Radical Halogenation (Section 22-1)

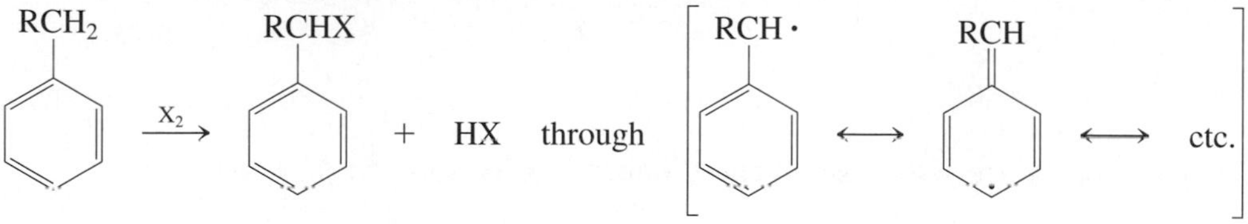

RCH_2—benzene $\xrightarrow{X_2}$ $RCHX$—benzene + HX through [$RCH\cdot$—benzene $\longleftrightarrow$ RCH=benzene radical] $\longleftrightarrow$ etc.

Benzylic radical

Requires heat, light, or a radical initiator

2. Solvolysis (Section 22-1)

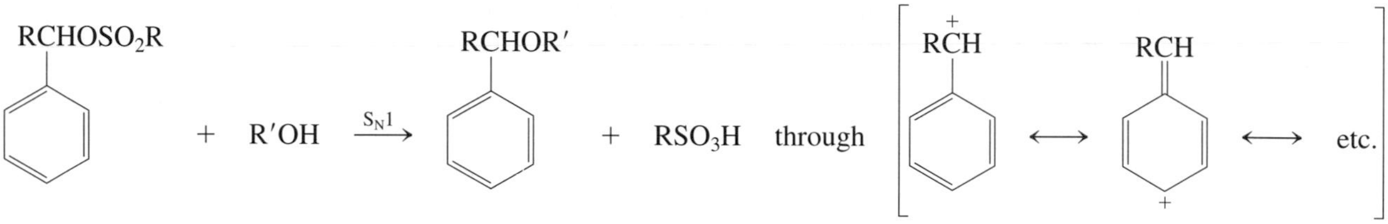

$RCHOSO_2R$—benzene + R'OH $\xrightarrow{S_N1}$ $RCHOR'$—benzene + RSO_3H through [$\overset{+}{RCH}$—benzene $\longleftrightarrow$ RCH=benzene cation] $\longleftrightarrow$ etc.

Benzylic cation

3. S$_N$2 Reactions of (Halomethyl)benzenes (Section 22-1)

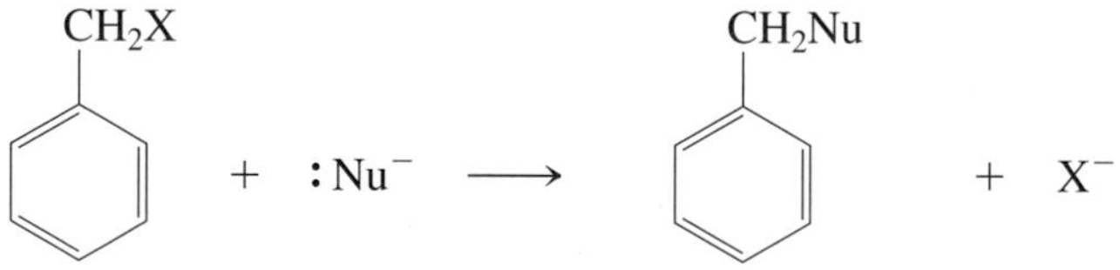

CH_2X—benzene + $:Nu^-$ $\longrightarrow$ CH_2Nu—benzene + X^-

Through delocalized
transition state

4. Benzylic Deprotonation (Section 22-1)

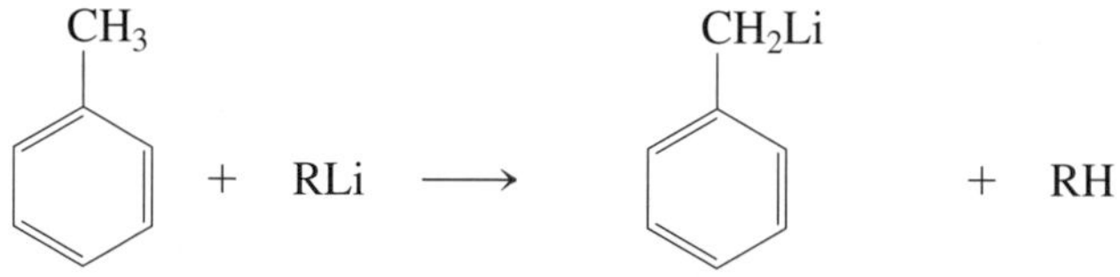

CH_3—benzene + RLi $\longrightarrow$ CH_2Li—benzene + RH

$pK_a \sim 41$

Phenylmethyllithium
(Benzyllithium)

Oxidation and Reduction Reactions on Aromatic Side Chains

5. Oxidation (Section 22-2)

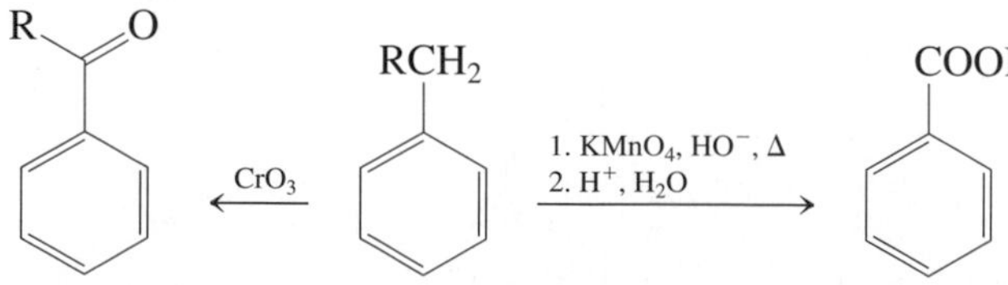

benzene—$\overset{O}{\underset{}{C}}$R $\xleftarrow{CrO_3}$ RCH_2—benzene $\xrightarrow[\text{2. } H^+, H_2O]{\text{1. } KMnO_4, HO^-, \Delta}$ COOH—benzene

Benzylic alcohols

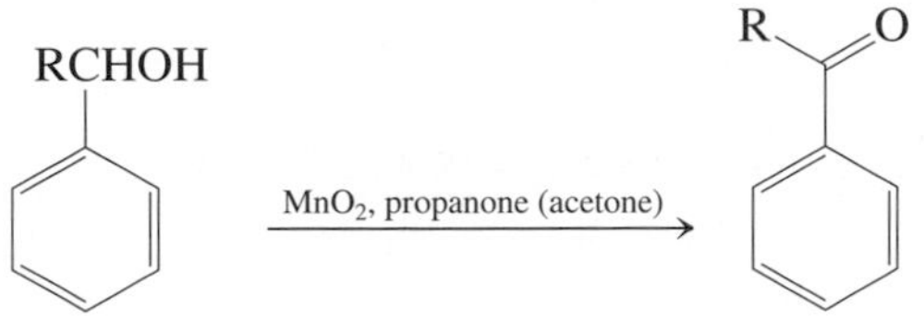

RCHOH (phenyl) $\xrightarrow{\text{MnO}_2,\text{ propanone (acetone)}}$ R–C(=O)–(phenyl)

6. Reduction by Hydrogenolysis (Section 22-2)

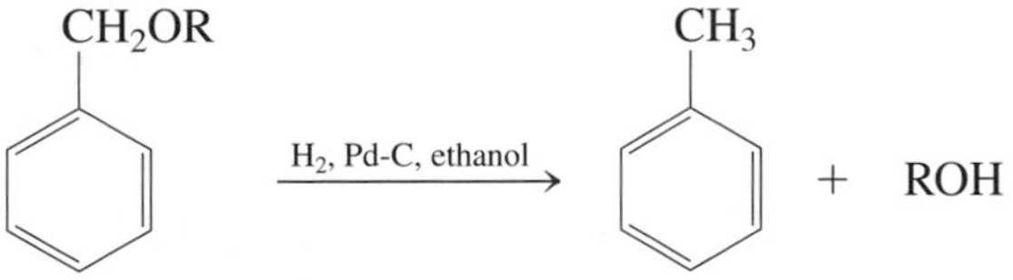

CH₂OR (phenyl) $\xrightarrow{\text{H}_2,\text{ Pd-C, ethanol}}$ CH₃ (phenyl) + ROH

$C_6H_5CH_2$ is a protecting group for ROH

Phenols and Ipso Substitution

7. Acidity (Section 22-3)

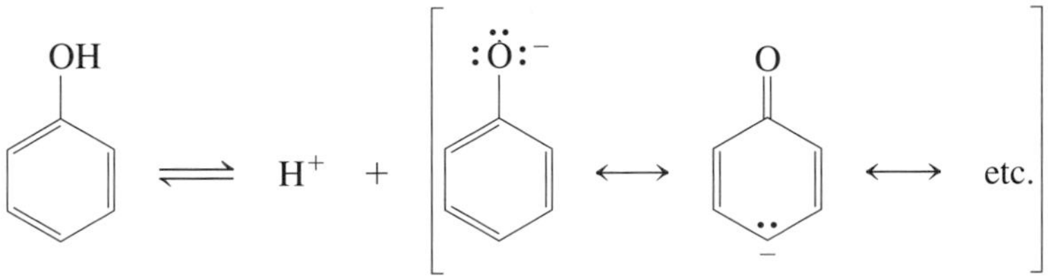

OH (phenyl) $\rightleftharpoons$ H⁺ + [:Ö:⁻ (phenoxide) ⟷ O (cyclohexadienone anion) ⟷ etc.]

$pK_a \sim 10$
Much stronger acid than simple alkanols

Phenoxide ion

8. Nucleophilic Aromatic Substitution (Section 22-4)

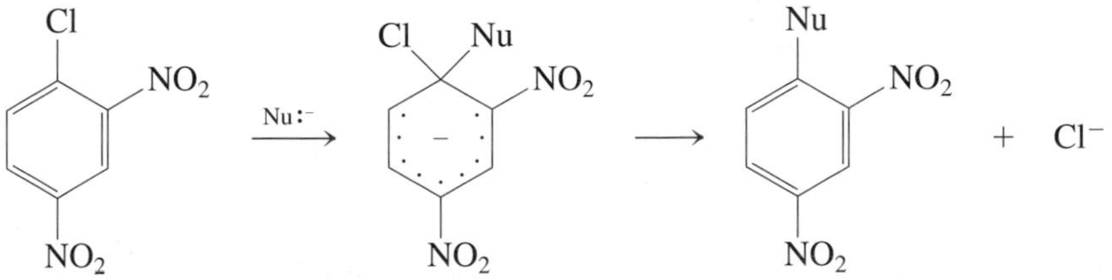

(Cl, NO₂, NO₂ substituted benzene) $\xrightarrow{\text{Nu:}^-}$ (Cl Nu, NO₂, NO₂ Meisenheimer complex) $\longrightarrow$ (Nu, NO₂, NO₂ substituted benzene) + Cl⁻

Nucleophile attacks at ipso position

9. Aromatic Substitution Through Benzyne Intermediates (Section 22-4)

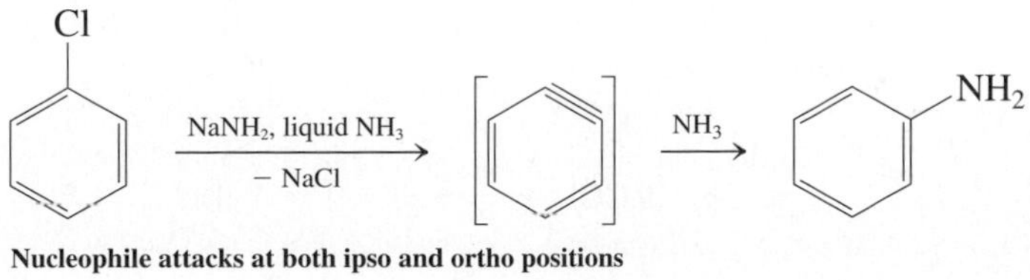

Cl (phenyl) $\xrightarrow[\text{– NaCl}]{\text{NaNH}_2,\text{ liquid NH}_3}$ [benzyne] $\xrightarrow{\text{NH}_3}$ NH₂ (phenyl)

Nucleophile attacks at both ipso and ortho positions

10. Preparation of Phenols by Nucleophilic Aromatic Substitution (Section 22-4)

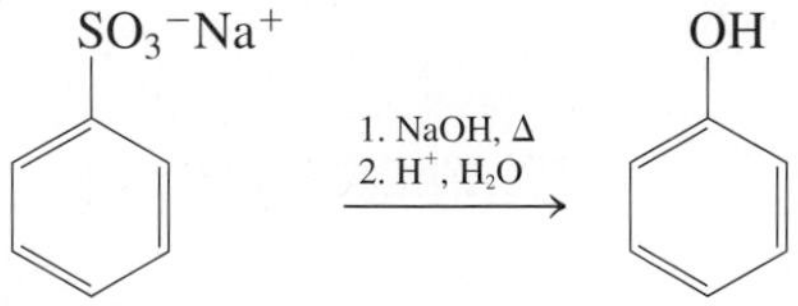

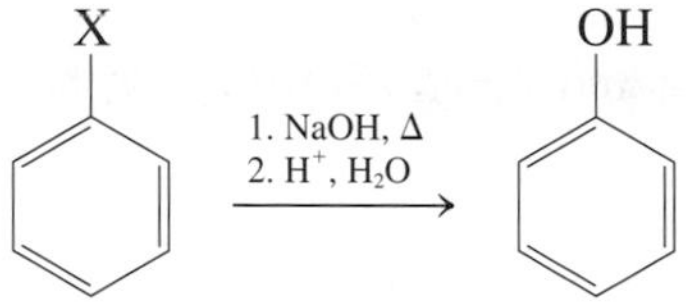

1. NaOH, Δ
2. H^+, H_2O

1. NaOH, Δ
2. H^+, H_2O

11. Arenediazonium Salt Hydrolysis (Section 22-4)

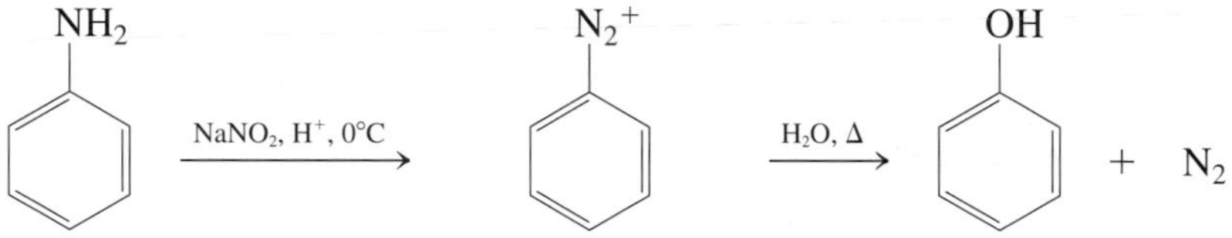

$NaNO_2$, H^+, 0°C

H_2O, Δ

$+$ N_2

Benzenediazonium
cation

Reactions of Phenols and Alkoxybenzenes

12. Ether Cleavage (Section 22-5)

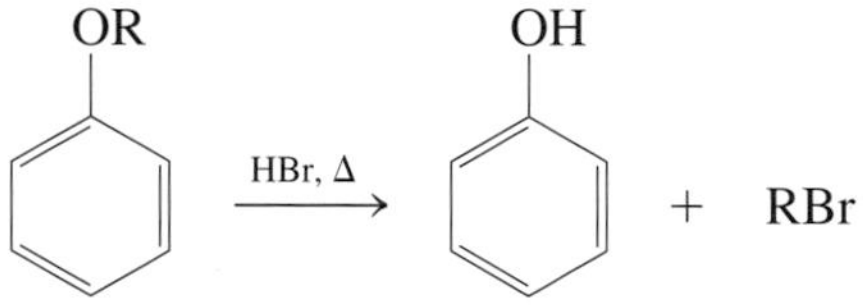

HBr, Δ

$+$ RBr

Aryl C–O bond is not cleaved

13. Ether Formation (Section 22-5)

$+$ RX

NaOH, H_2O

Alkoxybenzene

Williamson method (Section 9-6)

14. Esterification (Section 22-5)

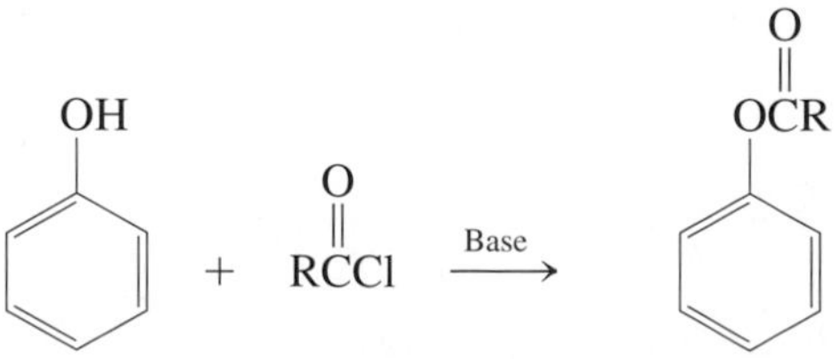

$+$ RCCl

Base

Phenyl alkanoate

15. Electrophilic Aromatic Substitution (Section 22-6)

16. Phenolic Resins (Section 22-6)

17. Kolbe Reaction (Section 22-6)

18. Claisen Rearrangement (Section 22-7)

Aromatic Claisen rearrangement

Aliphatic Claisen rearrangement

19. Cope Rearrangement (Section 22-7)

20. Oxidation (Section 22-8)

2,5-Cyclohexadiene-1,4-dione
(*p*-Benzoquinone)

21. Conjugate Additions to 2,5-Cyclohexadiene-1,4-diones (*p*-Benzoquinones) (Section 22-8)

22. Diels-Alder Cycloadditions to 2,5-Cyclohexadiene-1,4-diones (*p*-Benzoquinones) (Section 22-8)

23. Lipid Peroxidation (Section 22-9)

toxic substances such as
4-hydroxy-2-alkenals

24. Inhibition by Antioxidants (Section 22-9)

Vitamin E
(or BHA or BHT)

25. Vitamin C as an Antioxidant (Section 22-9)

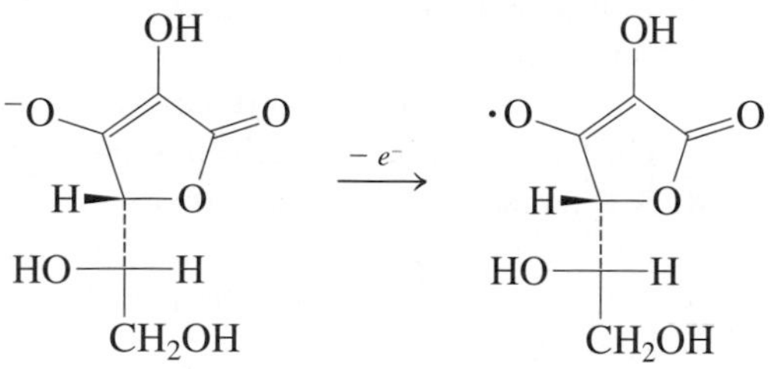

Semidehydroascorbic
acid

Arenediazonium Salts

26. Sandmeyer Reactions (Section 22-10)

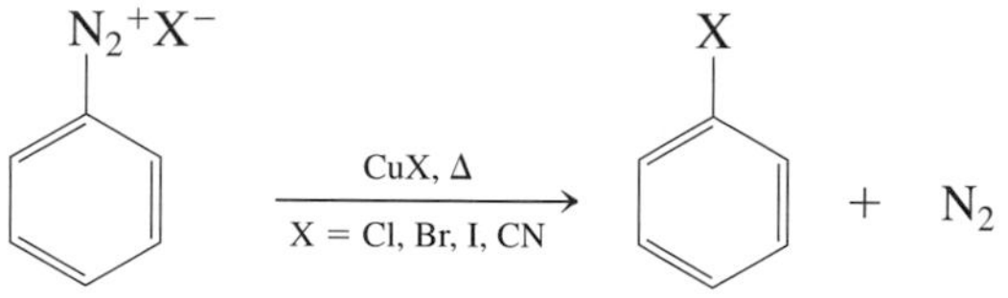

27. Reduction (Section 22-10)

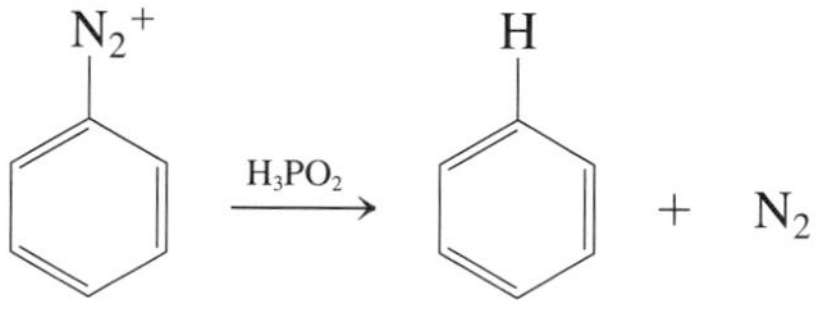

28. Diazo Coupling (Section 22-11)

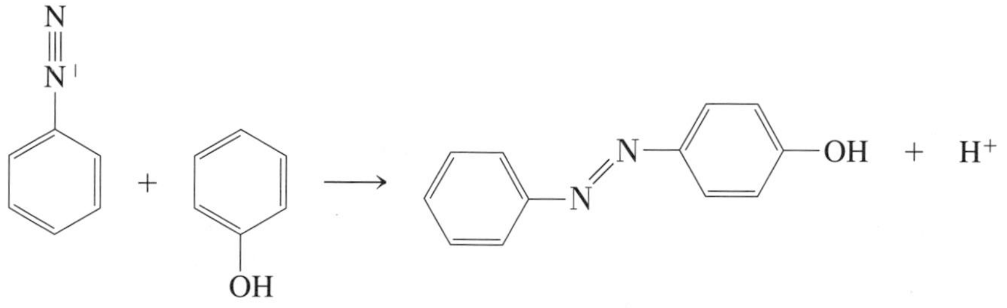

Azo compound

Occurs only with strongly activated rings

IMPORTANT CONCEPTS

1. Phenylmethyl and other **benzylic radicals, cations, and anions** are reactive intermediates stabilized by **resonance** of the resulting centers with a benzene π system.
2. **Nucleophilic aromatic ipso substitution** accelerates with the nucleophilicity of the attacking species and with the number of electron-withdrawing groups on the ring, particularly if they are located ortho or para to the point of attack.
3. **Benzyne** is destabilized by the strain of the two sp-hybridized carbons forming the triple bond.

4. **Phenols** are aromatic enols, undergoing reactions typical of the hydroxy group and the aromatic ring.

5. **Cyclohexadienediones** and benzenediols function as redox couples in the laboratory and in nature.

6. Vitamin E and the highly substituted phenol derivatives BHA and BHT function as inhibitors of the **radical-chain oxidation of lipids.** Vitamin C also is an antioxidant, capable of regenerating vitamin E at the surface of cell membranes.

7. **Arenediazonium ions** are stabilized by resonance but furnish reactive **aryl cations** whose positive charge cannot be delocalized into the aromatic ring.

8. The amino group can be used to direct electrophilic aromatic substitution, after which it is replaceable by diazotization and substitution, including reduction.

PROBLEMS

26. Give the expected major product(s) of each of the following reactions.

(a)

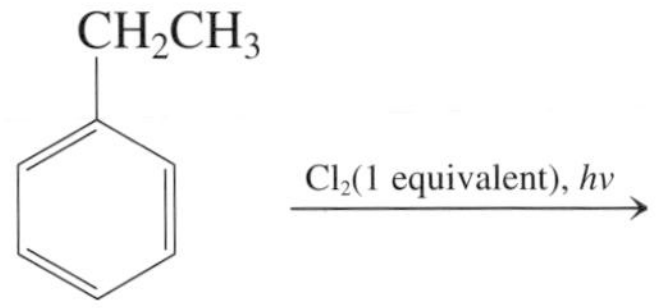

CH$_2$CH$_3$

$\xrightarrow{\text{Cl}_2(\text{1 equivalent}), hv}$

(b)

$\xrightarrow{\text{NBS (1 equivalent)}, hv}$

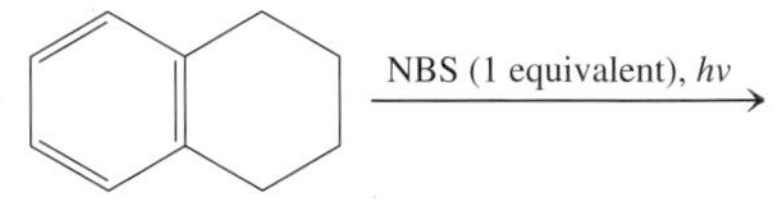

27. Formulate a mechanism for the reaction described in Problem 26(b).

28. Propose syntheses of each of the following compounds, beginning in each case with ethylbenzene. **(a)** (1-Chloroethyl)benzene; **(b)** 2-phenylpropanoic acid; **(c)** 2-phenylethanol; **(d)** 2-phenyloxacyclopropane.

29. Predict the order of relative stability of the three benzylic cations derived from chloromethylbenzene (benzyl chloride), 1-(chloromethyl)-4-methoxybenzene (4-methoxybenzyl chloride), and 1-(chloromethyl)-4-nitrobenzene (4-nitrobenzyl chloride). Rationalize your answer with the help of resonance structures.

30. By drawing appropriate resonance structures, illustrate why halogen atom attachment at the para position of phenylmethyl (benzyl) radical is unfavored compared with attachment at the benzylic position.

31. Triphenylmethyl radical, (C$_6$H$_5$)$_3$C ·, is stable at room temperature in dilute solution in an inert solvent, and salts of triphenylmethyl cation, (C$_6$H$_5$)$_3$C$^+$, can be isolated as stable crystalline solids. Propose explanations for the unusual stabilities of these species.

32. Give the expected products of the following reactions or reaction sequences.

(a)  BrCH$_2$CH$_2$CH$_2$—⟨ ⟩—CH$_2$Br $\xrightarrow{\text{H}_2\text{O}, \Delta}$

(b)

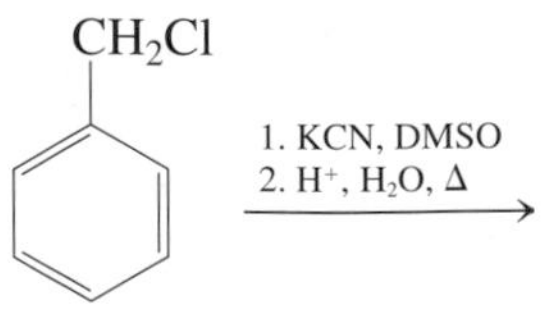

CH$_2$Cl

$\xrightarrow[\text{2. H}^+, \text{H}_2\text{O}, \Delta]{\text{1. KCN, DMSO}}$

(c)

$\xrightarrow{\begin{array}{l}\text{1. CH}_3\text{CH}_2\text{CH}_2\text{CH}_2\text{Li, (CH}_3)_2\text{NCH}_2\text{CH}_2\text{N(CH}_3)_2\text{, THF} \\ \text{2. C}_6\text{H}_5\text{CHO} \\ \text{3. H}^+, \text{H}_2\text{O}, \Delta\end{array}}$ C$_{16}$H$_{14}$

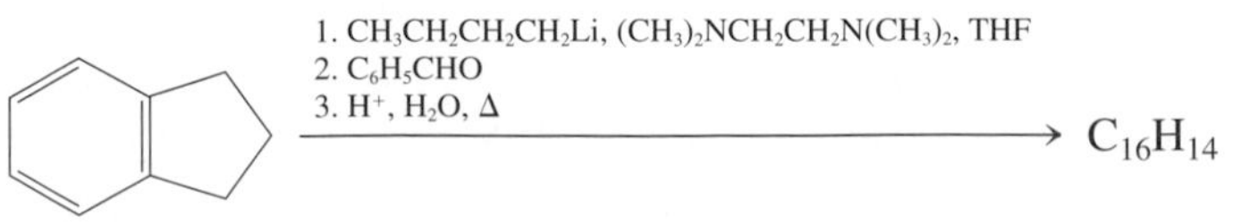

33. The hydrocarbon with the common name fluorene is acidic enough ($pK_a \sim 23$) to be a useful indicator in deprotonation reactions of compounds of greater acidity. Indicate the most acidic hydrogen(s) in fluorene. Draw resonance structures to explain the relative stability of its conjugate base.

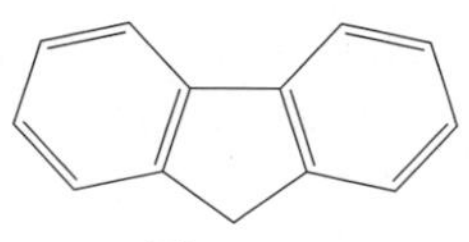

Fluorene

34. Outline a straightforward, practical, and efficient synthesis of each of the following compounds. Start with benzene or methylbenzene. Assume that the para isomer (but *not* the ortho isomer) may be separated efficiently from any mixtures of ortho and para substitution products.

(a) 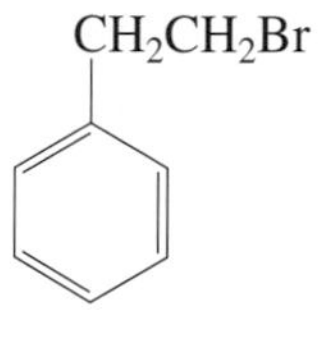 CH_2CH_2Br

(b) 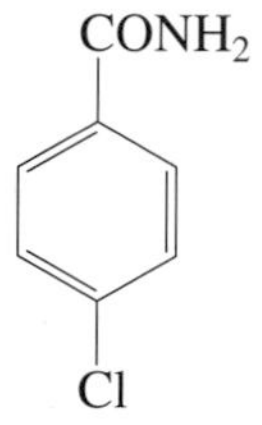 $CONH_2$ ... Cl

(c) 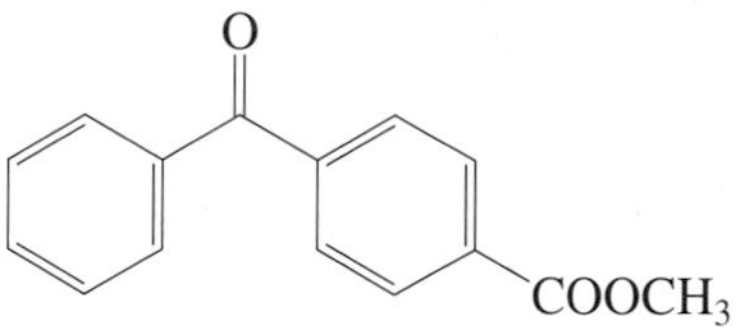 [diaryl ketone with $COOCH_3$]

(d) 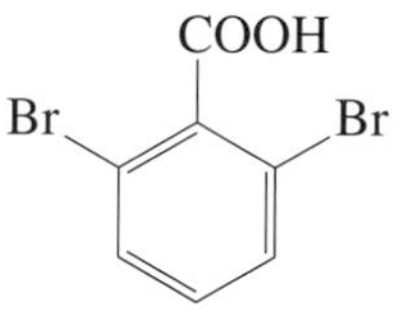 $COOH$, Br, Br

35. Rank the following compounds in descending order of reactivity toward hydroxide ion.

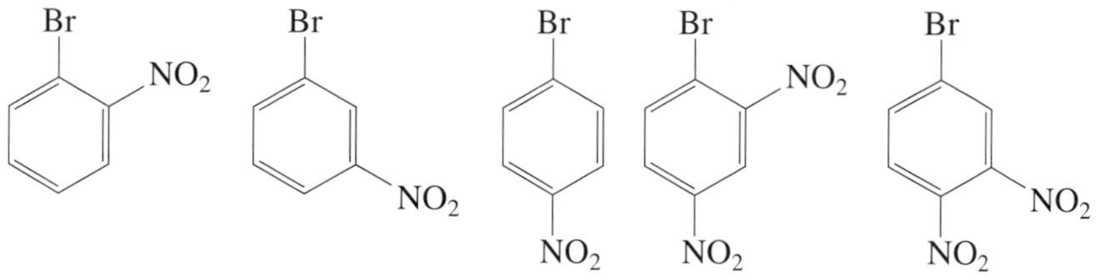

Br, NO_2

Br ... NO_2

Br ... NO_2

Br, NO_2 ... NO_2

Br ... NO_2, NO_2

36. Predict the main product(s) of the following reactions. In each case, describe the mechanism(s) in operation.

(a) 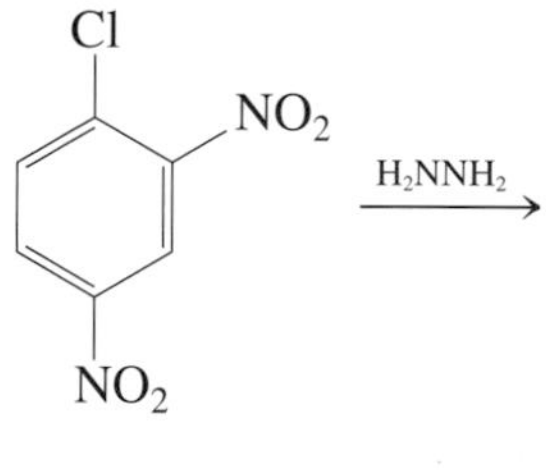 Cl, NO_2, NO_2 $\xrightarrow{H_2NNH_2}$

(b) 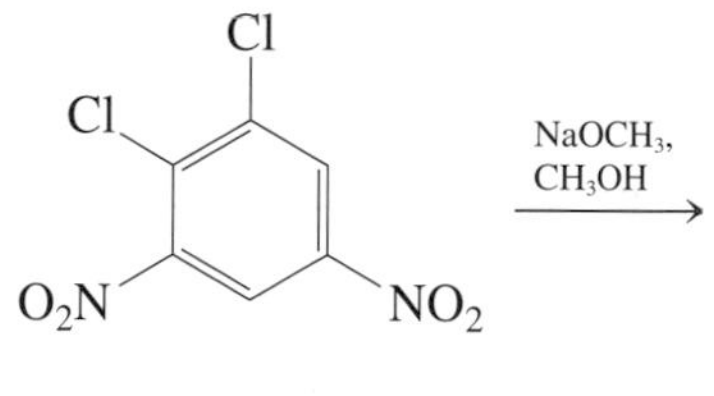 Cl, Cl, O_2N, NO_2 $\xrightarrow{NaOCH_3, \ CH_3OH}$

(c) 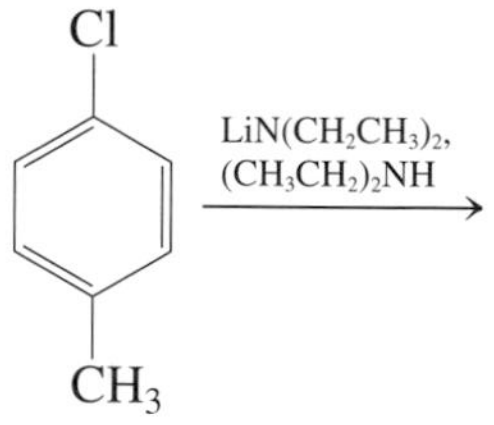 Cl, CH_3 $\xrightarrow{LiN(CH_2CH_3)_2, \ (CH_3CH_2)_2NH}$

37. Starting with benzenamine, propose a synthesis of aklomide, an agent used to treat certain exotic fungal and protozoal infections in veterinary medicine. Several intermediates are shown to give you the general route. Fill in the blanks that remain; each requires as many as three sequential reactions. (**Hint:** Review the oxidation of amino- to nitroarenes in Section 16-5.)

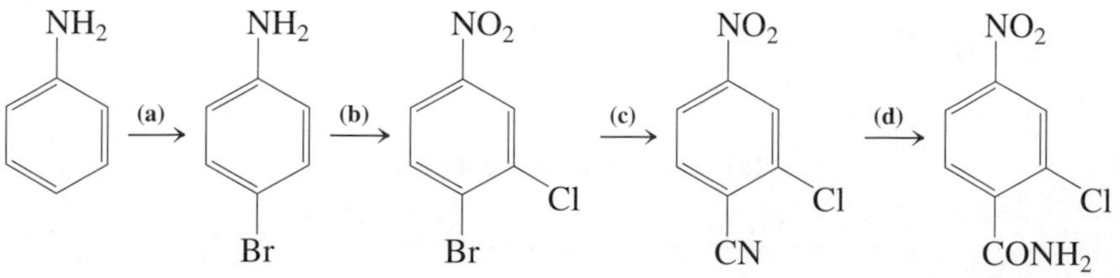

NH_2 $\xrightarrow{(a)}$ NH_2, Br $\xrightarrow{(b)}$ NO_2, Cl, Br $\xrightarrow{(c)}$ NO_2, Cl, CN $\xrightarrow{(d)}$ NO_2, Cl, $CONH_2$

Aklomide

38. Explain the mechanism of the following synthetic transformation. (**Hint:** Two equivalents of butyllithium are used.)

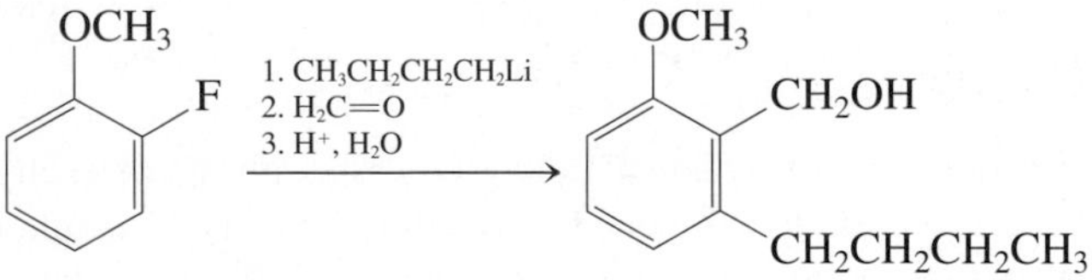

$$
\begin{array}{c}
\text{OCH}_3 \text{-substituted benzene with F} \\
\xrightarrow[\begin{array}{l}1.\ \text{CH}_3\text{CH}_2\text{CH}_2\text{CH}_2\text{Li}\\ 2.\ \text{H}_2\text{C}{=}\text{O}\\ 3.\ \text{H}^+,\ \text{H}_2\text{O}\end{array}]{}
\text{OCH}_3,\ \text{CH}_2\text{OH},\ \text{CH}_2\text{CH}_2\text{CH}_2\text{CH}_3
\end{array}
$$

39. In nucleophilic aromatic substitution reactions that proceed by the addition–elimination mechanism, fluorine is the most easily replaced halogen in spite of the fact that F^- is by far the worst leaving group among halide ions. For example, 1-fluoro-2,4-dinitrobenzene reacts much more rapidly with amines than does the corresponding chloro compound. Suggest an explanation. (**Hint:** Consider the effect of the identity of the halogen on the rate-determining step.)

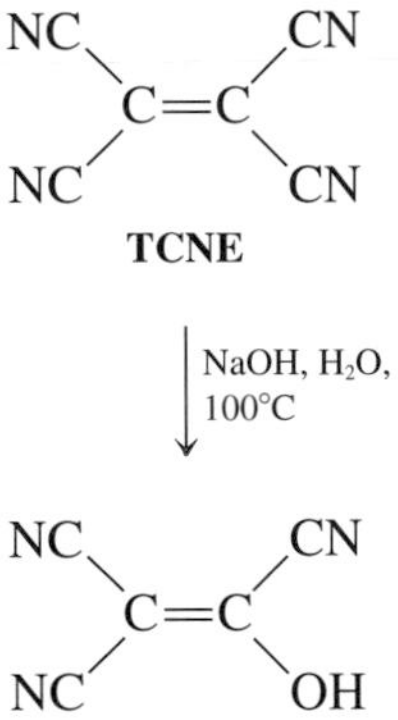

TCNE

NaOH, H_2O, 100°C

40. In a rather unusual reaction, tetracyanoethene (TCNE) is converted by boiling in aqueous base into tricyanoethenol (whose enol form is stabilized by the three nitrile groups). Suggest a mechanism for this transformation. (**Hint:** This reaction occurs for the same reason that one form of nucleophilic aromatic substitution takes place.)

41. Give the expected major product(s) of each of the following reactions and reaction sequences.

(a) 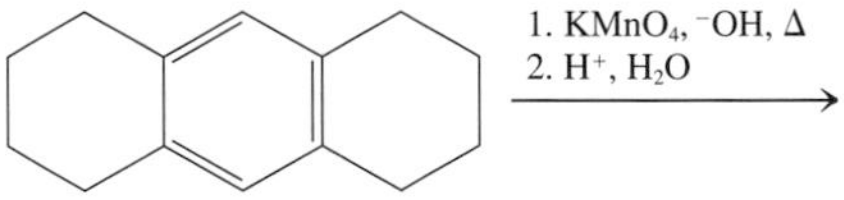
$$\xrightarrow[\begin{array}{l}1.\ \text{KMnO}_4,\ ^-\text{OH},\ \Delta\\ 2.\ \text{H}^+,\ \text{H}_2\text{O}\end{array}]{}$$

(b) 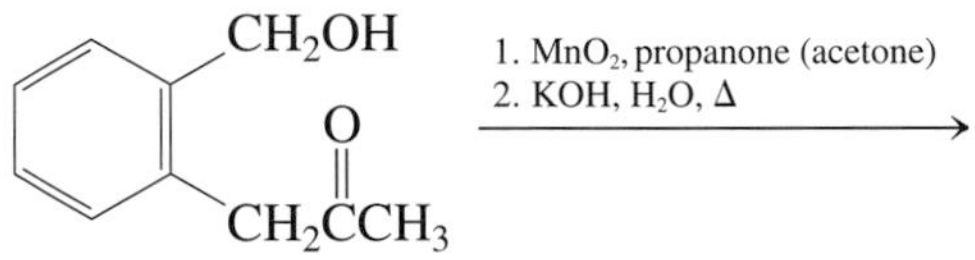
$$\xrightarrow[\begin{array}{l}1.\ \text{MnO}_2,\ \text{propanone (acetone)}\\ 2.\ \text{KOH},\ \text{H}_2\text{O},\ \Delta\end{array}]{}$$

(c) 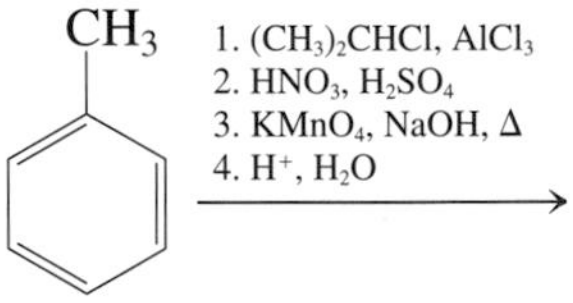
$$\xrightarrow[\begin{array}{l}1.\ (\text{CH}_3)_2\text{CHCl},\ \text{AlCl}_3\\ 2.\ \text{HNO}_3,\ \text{H}_2\text{SO}_4\\ 3.\ \text{KMnO}_4,\ \text{NaOH},\ \Delta\\ 4.\ \text{H}^+,\ \text{H}_2\text{O}\end{array}]{}$$

42. Rank the following compounds in order of descending acidity

(a) CH_3OH **(b)** CH_3COOH **(c)** 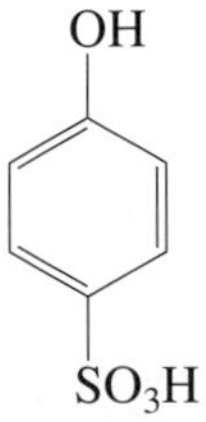 OH / SO_3H **(d)** 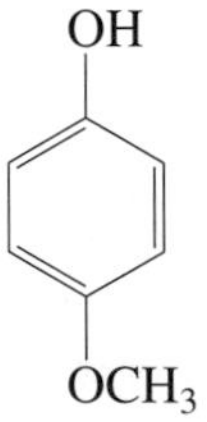 OH / OCH_3 **(e)** 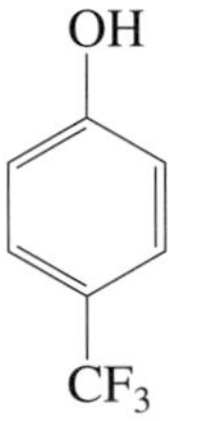 OH / CF_3 **(f)** 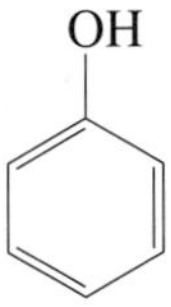 OH

43. Design a synthesis of each of the following phenols, starting with either benzene or any monosubstituted benzene derivative.

(a)  OH, CH_3 **(b)** 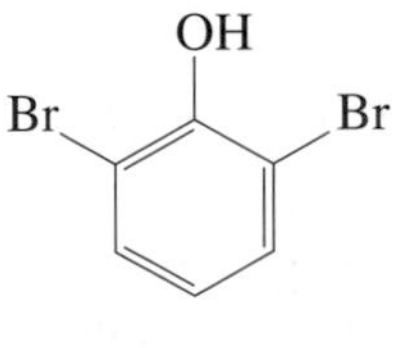 Br, OH, Br   **(c)** The three benzenediols   **(d)** 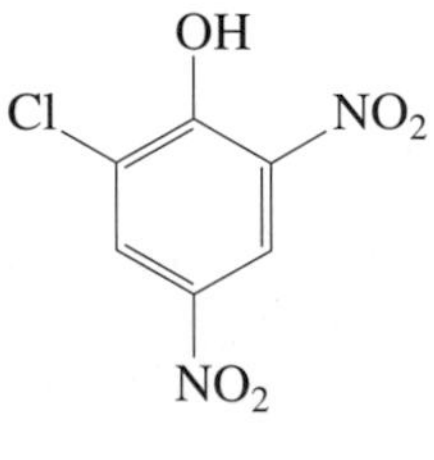 Cl, OH, NO_2, NO_2

44. Starting with benzene, propose syntheses of each of the following phenol derivatives.

(a)

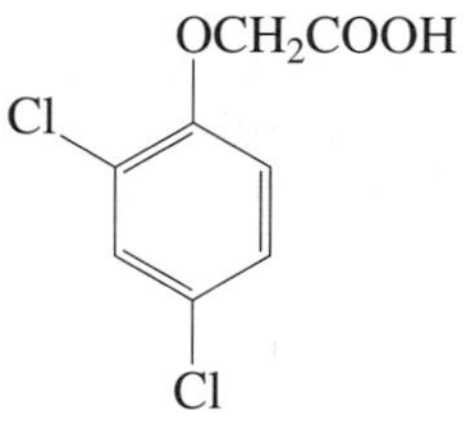

OCH₂COOH
Cl

Cl

The herbicide 2,4-D

(b)

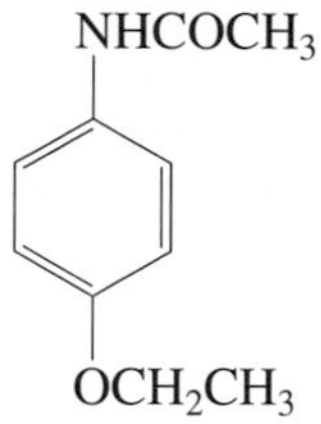

NHCOCH₃

OCH₂CH₃

Phenacetin
(The active ingredient
in Midol)

(c)

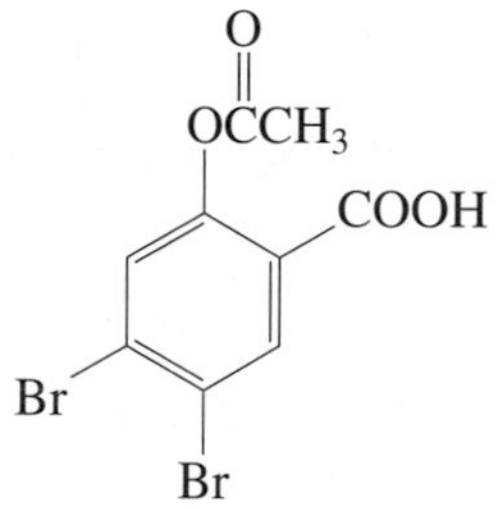

OCCH₃
O
COOH

Br

Br

Dibromoaspirin
(An experimental drug for the
treatment of sickle-cell anemia)

45. Name each of the following compounds.

(a)

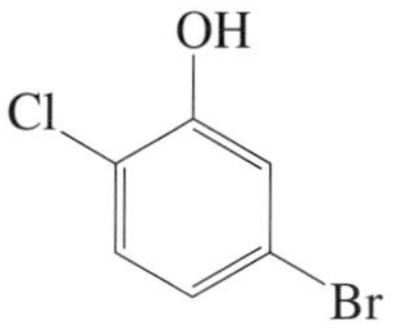

OH
Cl

Br

(b)

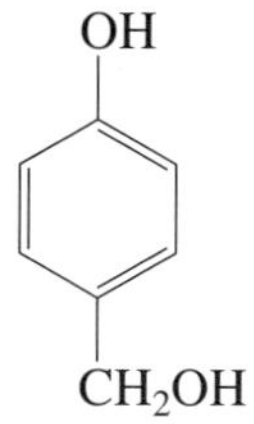

OH

CH₂OH

(c)

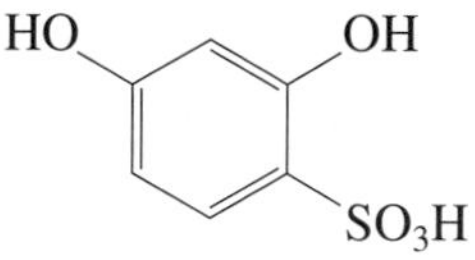

HO   OH

SO₃H

(d)

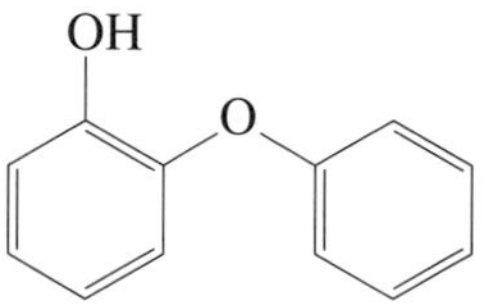

OH
O

(e)

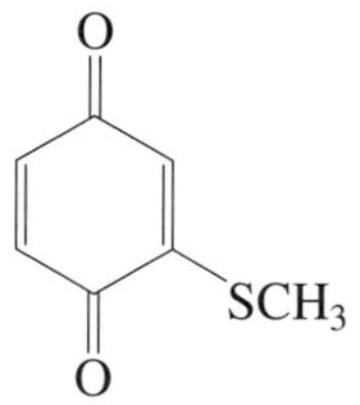

O

SCH₃

O

46. Give the expected product(s) of each of the following reaction sequences.

(a)

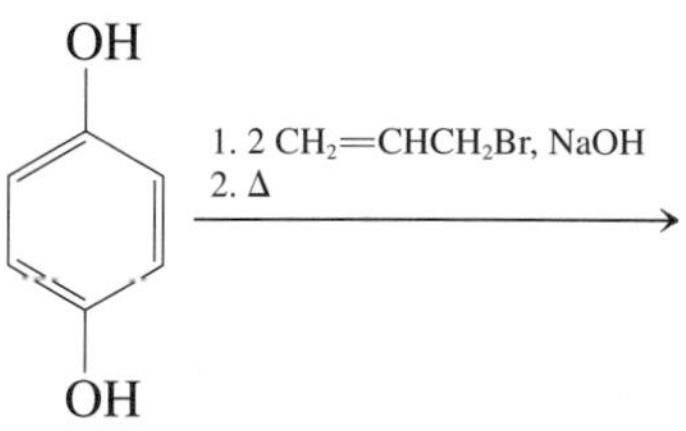

OH

OH

1. 2 CH₂=CHCH₂Br, NaOH
2. Δ

(b)

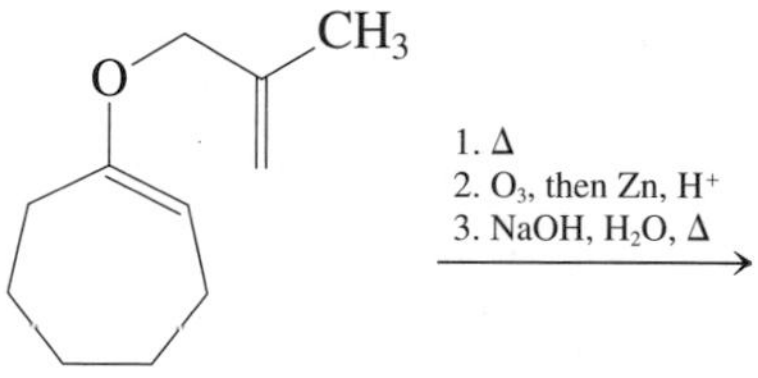

O   CH₃

1. Δ
2. O₃, then Zn, H⁺
3. NaOH, H₂O, Δ

(c)

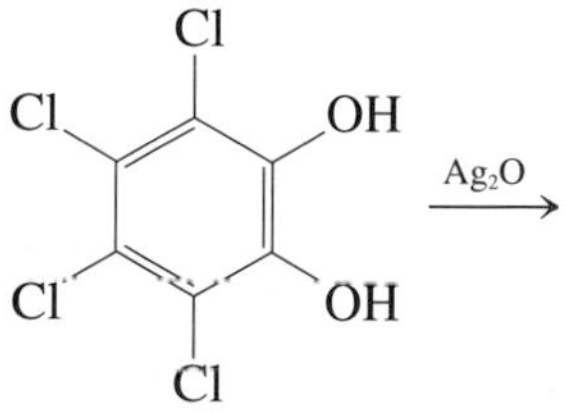

Cl   Cl
Cl   OH

Cl   OH
Cl

Ag₂O

(d)

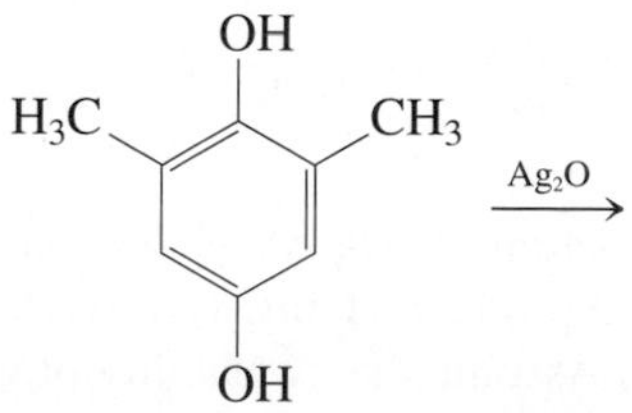

OH
H₃C   CH₃

OH

Ag₂O

(e)

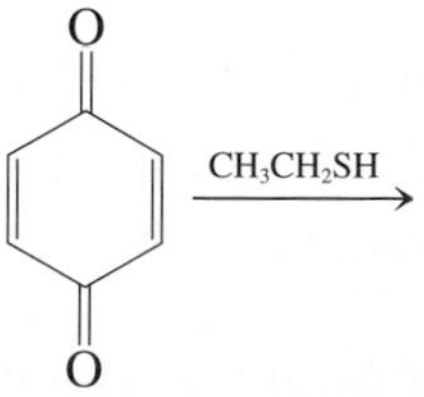

O

O

CH₃CH₂SH
(two possibilities)

(f)

O

O

+

⬡ →

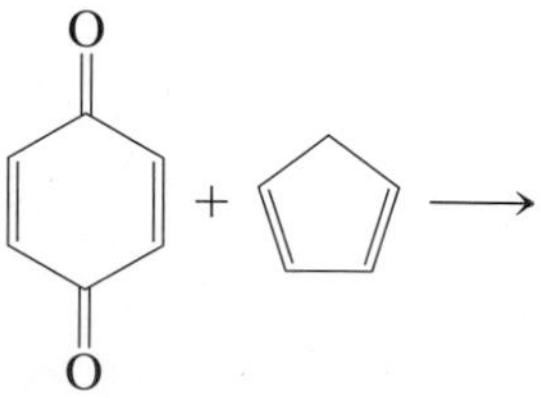

47. As a children's medicine, Tylenol has a major marketing advantage over aspirin: *Liquid Tylenol* preparations (essentially, Tylenol dissolved in flavored water) are stable, whereas comparable aspirin solutions are not. Explain.

48. Black-and-white photographic film contains tiny crystals of silver bromide. On exposure to light, the silver bromide becomes photoactivated. In this form, it is readily converted into particles of black silver metal by a reducing agent called a *developer*, an example of which is 4-(*N*-methylamino)phenol (Metol). After exposure and development, the film is washed with a *fixer*, such as ammonium thiosulfate, which dissolves and removes unreduced silver bromide. Give the structure of Metol and of the product of its oxidation by silver bromide. [**Hint:** The oxidation of Metol is closely related to that of 1,4-benzenediol (hydroquinone).]

49. Biochemical oxidation of aromatic rings is catalyzed by a group of liver enzymes called aryl hydroxylases. Part of this chemical process is the conversion of toxic aromatic hydrocarbons such as benzene into water-soluble phenols, which can be easily excreted. However, the primary purpose of the enzyme is to enable the synthesis of biologically useful compounds, such as the amino acid tyrosine.

Phenylalanine   Tyrosine

(a) Extrapolating from your knowledge of benzene chemistry, which of the following three possibilities seems most reasonable: The oxygen is introduced by electrophilic attack on the ring; the oxygen is introduced by free-radical attack on the ring; or the oxygen is introduced by nucleophilic attack on the ring? **(b)** It is widely suspected that oxacyclopropanes play a role in arene hydroxylation. Part of the evidence is the following observation: When the site to be hydroxylated is initially labeled with deuterium, a substantial proportion of the product still contains deuterium atoms, which have apparently migrated to the position ortho to the site of hydroxylation.

Suggest a plausible mechanism for the formation of the oxacyclopropane intermediate and its conversion into the observed product. (**Hint:** Hydroxylase converts O_2 into hydrogen peroxide, HO–OH.) Assume the availability of catalytic amounts of acids and bases, as necessary.

Note: In victims of the genetically transmitted disorder called phenylketonuria (PKU), the hydroxylase enzyme system described here does not function properly. Instead, phenylalanine in the brain is converted into 2-phenyl-2-oxopropanoic (phenylpyruvic) acid, the reverse of the process shown in Problem 45 of Chapter 21. The buildup of this compound in the brain can lead to severe retardation; thus people with PKU (which can be diagnosed at birth) must be restricted to diets low in phenylalanine.

50. A common application of the Cope rearrangement is in ring-enlargement sequences. Fill in the reagents and products missing from the following scheme, which illustrates the construction of a 10-membered ring.

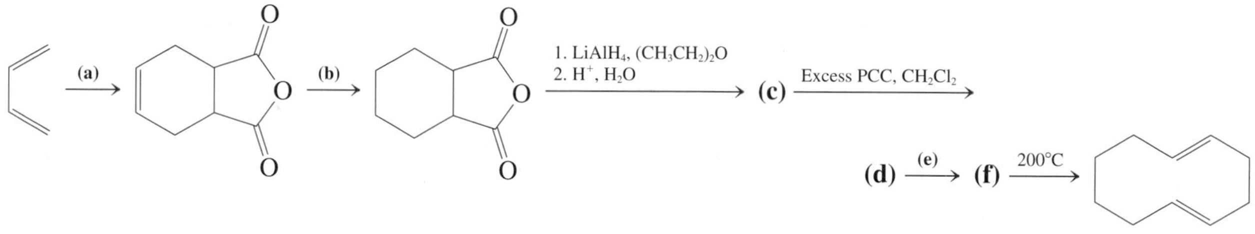

51. Formulate a complete mechanism for the diazotization of benzenamine (aniline) in the presence of HCl and NaNO$_2$ and a plausible mechanism (based on what you have learned in Section 22-10) for its subsequent conversion into iodobenzene by treatment with aqueous iodide ion (e.g., from K^+I^-).

52. Devise a synthesis of each of the following substituted benzene derivatives, starting from benzene.

(a) 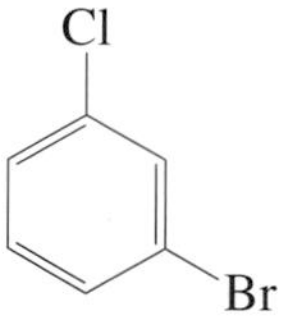

(b) 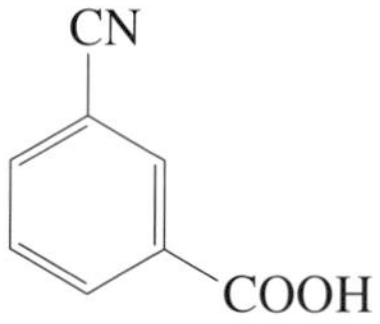

(c) 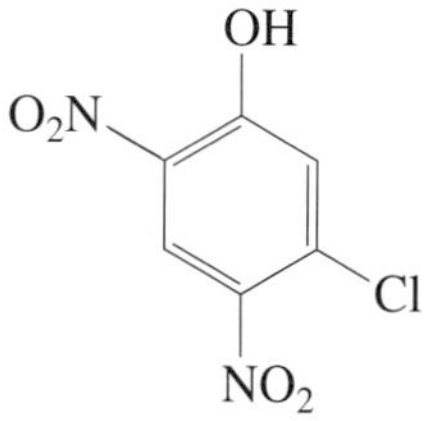

(d) 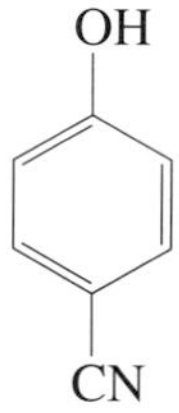

(e) 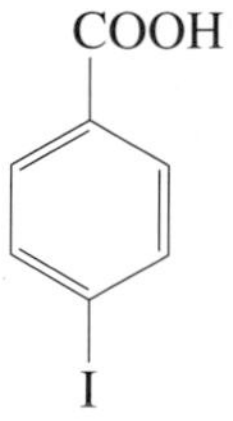

(f) 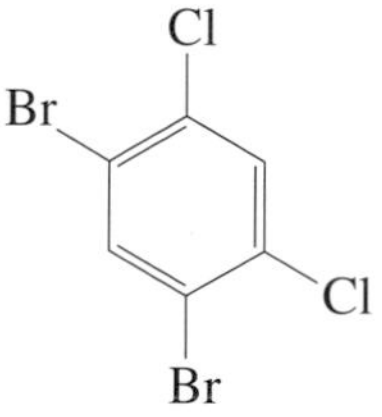

(g) 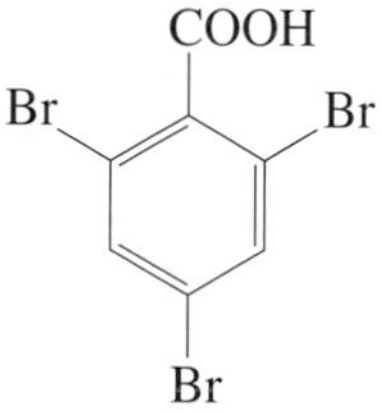

53. Write the most reasonable structure of the product of each of the following reaction sequences.

(a) 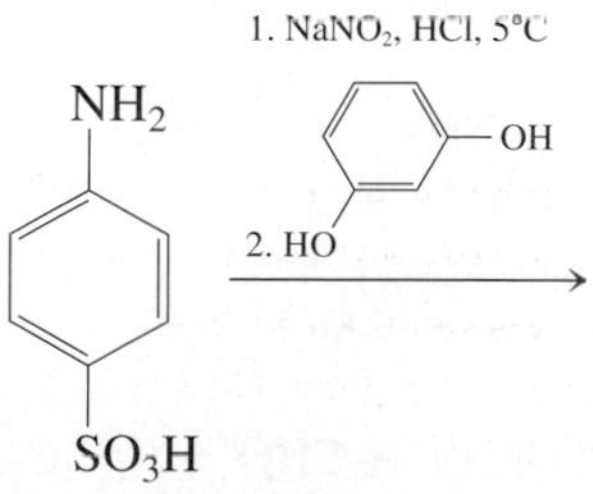

Golden Yellow

(b) 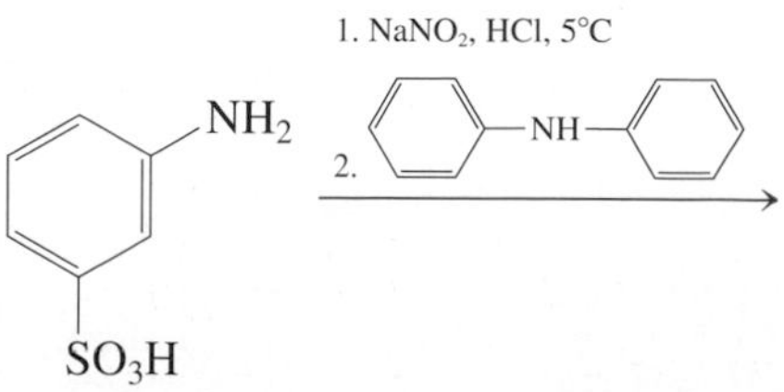

Metanil Yellow

For the following reaction, assume that electrophilic substitution occurs preferentially on the most activated ring.

(c)

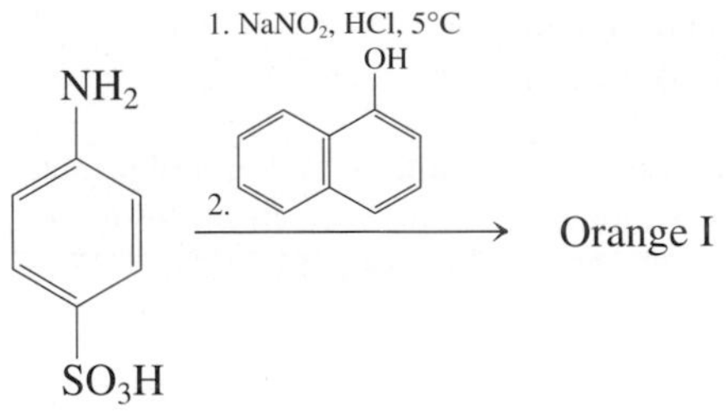

1. $NaNO_2$, HCl, 5°C

2.

$\longrightarrow$ Orange I

54. Show the reagents that would be necessary for the synthesis by diazo coupling of each of the following three compounds. For structures, see Section 22-11.

(a) Methyl Orange   (b) Congo Red

(c) Prontosil, 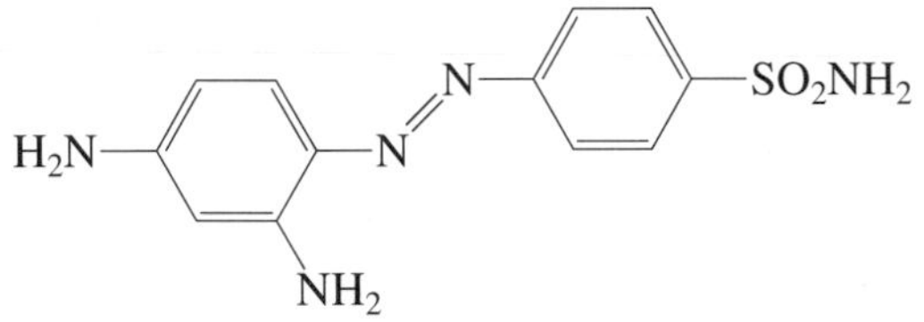 , which is

converted microbially into sulfanilamide, H_2N—⬡—SO_2NH_2

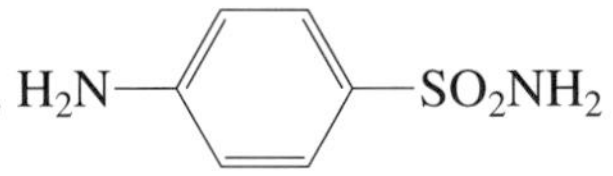

(The accidental discovery of the antibacterial properties of prontosil in the 1930s led indirectly to the development of sulfa drugs as antibiotics in the 1940s.)

55. (a) Give the key reaction that illustrates the inhibition of fat oxidation by the preservative BHT. (b) The extent to which fat is oxidized in the body can be determined by measuring the amount of *pentane* exhaled in the breath. Increasing the amount of vitamin E in the diet decreases the amount of pentane exhaled. Examine the processes described in Section 22-9 and identify one that could produce pentane. You will have to do some extrapolating from the specific reactions shown in the section.

56. The urushiols are the irritants in poison ivy and poison oak that give you rashes and make you itch upon exposure. Use the following information to determine the structures of urushiols I ($C_{21}H_{36}O_2$) and II ($C_{21}H_{34}O_2$), the two major members of this family of unpleasant compounds.

Urushiol II $\xrightarrow{H_2, \text{ Pd-C, } CH_3CH_2OH}$ urushiol I

Urushiol II $\xrightarrow{\text{Excess } CH_3I, \text{ NaOH}}$ $C_{23}H_{38}O_2$ $\xrightarrow[\text{2. Zn, } H_2O]{\text{1. } O_3, CH_2Cl_2}$

Dimethylurushiol II

$CH_3CH_2CH_2CH_2CH_2CH_2CHO$ + $C_{16}H_{24}O_3$

Aldehyde A

Synthesis of Aldehyde A

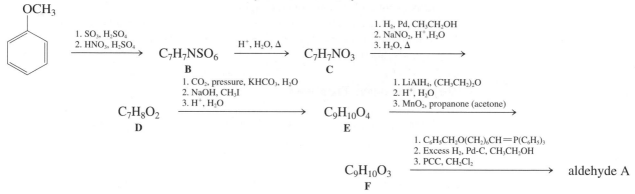

57. Is the site of reaction in the biosynthesis of norepinephrine from dopamine (see Chapter 5, Problem 52) consistent with the principles outlined in this chapter? Would it be easier or more difficult to duplicate this transformation nonenzymatically? Explain.

Team Problem

58. As a team, consider the following schemes that outline steps toward the total synthesis of taxodone D, a potential anticancer agent. For the first scheme, divide your team into two groups, one to discuss the best option for A to effect the initial reduction step, the second to assign a structure to B, using the partial spectral data provided.

^{1}H NMR spectrum:
$\delta = 5.99$ (dd, 1H), 6.50 (d, 1H) ppm;
IR spectrum: $\tilde{\nu} = 1720$ cm^{-1};
Mass spectrum: m/z 384 (M$^+$).

Reconvene to discuss both parts of the first scheme. Then, as a group, analyze the remainder of the synthesis shown in the second scheme below. Use the spectroscopic data to help you determine the structures of C and taxodone, D.

^{1}H NMR spectrum:
$\delta = 3.51$ (dd, 1H),
3.85 (d, 1H);
Mass spectrum:
m/z 400 (M$^+$).

Taxodone
^{1}H NMR spectrum:
$\delta = 6.55$ (d, 1H),
6.81 (s, 1H), no other alkenyl or aromatic signals;
IR spectrum: $\tilde{\nu} = 3610, 3500, 1628$ cm^{-1};
UV spectrum: λ_{max} (ε) = 316 (20,000) nm;
Mass spectrum: m/z 316 (M$^+$).

Propose a mechanism for the formation of D from C. (**Hints:** After ester hydrolysis, one of the phenolate oxygens can donate its electron pair through the benzene ring to effect a reaction at the para position. The product contains a carbonyl group in its enol form. For the interpretation of some of the spectral data, see Section 17-3.)

Preprofessional Problems

59. After chlorobenzene has been boiled in water for 2 h, which of the following organic compounds will be present in greatest concentration?

(a) C_6H_5OH (b) 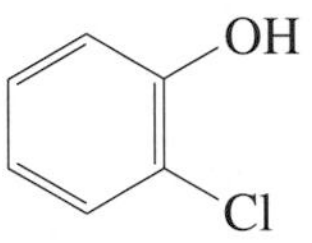   (c) 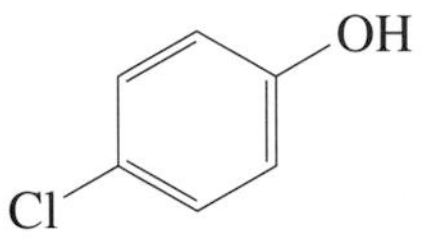   (d) C_6H_5Cl (e)

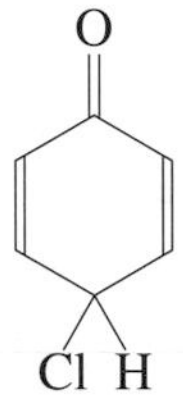

60. What are the products of the following reaction? $C_6H_5OCH_3 \xrightarrow{HI, \Delta} ?$

(a) $C_6H_5I + CH_3OH$ (b) $C_6H_5OH + CH_3I$ (c) $C_6H_5I + CH_3I$ (d) 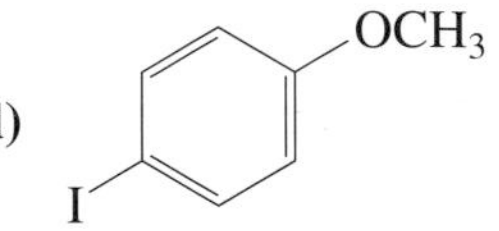 [structure] $+ \ H_2$

61. The transformation of 4-methylbenzenediazonium bromide to toluene is best carried out by using:

(a) H^+, H_2O (b) H_3PO_2, H_2O (c) $H_2O, \ ^-OH$ (d) $Zn, NaOH$

62. What is the principal product after the slurry obtained on treating benzenamine (aniline) with potassium nitrite and HCl at 0°C has been added to 4-ethylphenol?

(a) 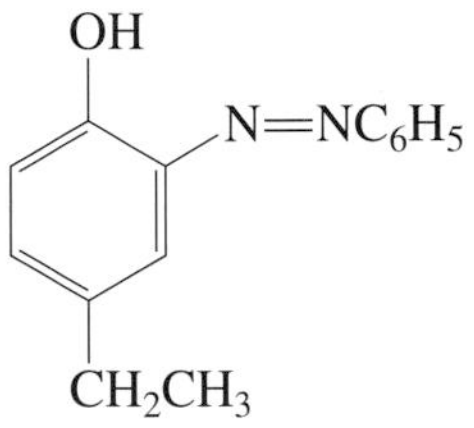   (b) 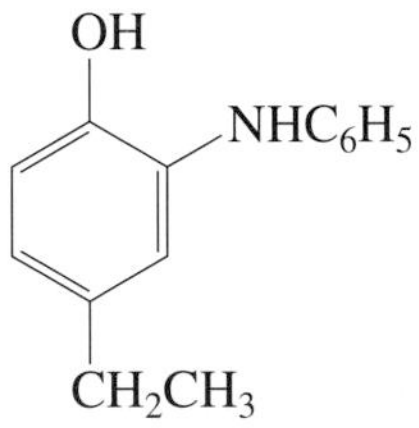   (c) 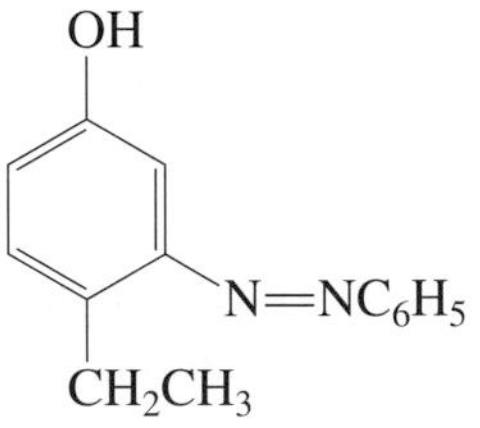   (d)

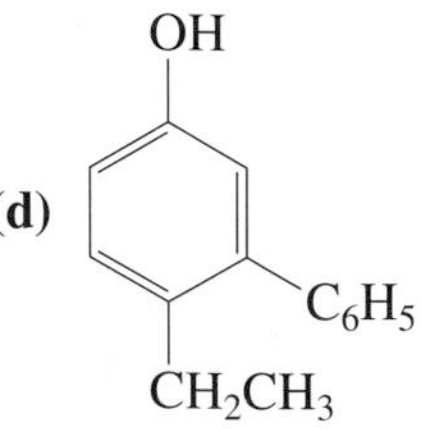

63. Examination of the ^1H NMR spectra of the following three isomeric nitrophenols reveals that one of them displays a hydroxy (phenolic) proton at substantially lower field than do the other two. Which one?

(a) 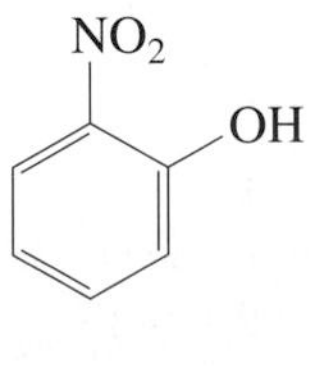   (b) 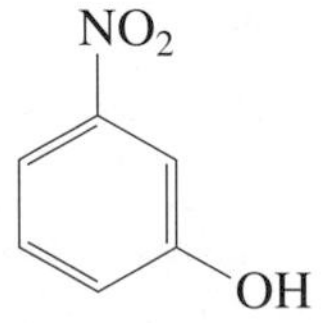   (c)

Syntheses of β-Dicarbonyl and α-Hydroxycarbonyl Compounds

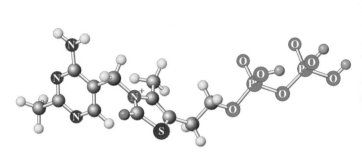

The conjugate base of thiamine pyrophosphate adds to carbonyl carbon atoms in a way that changes their electronic character from electrophilic to nucleophilic. In this way, a thiamine-derived two-carbon nucleophile mediates the biosynthesis of fructose, the principal sugar in bees' honey.

In Chapter 18 we began an expansion of our survey of carbonyl chemistry, learning that many techniques developed by the synthetic organic chemist were in fact based on natural processes for the construction of carbon–carbon bonds in biological systems. The aldol condensation (Sections 18-5 through 18-7) was one such process, a powerful method for converting aldehydes and ketones into β-hydroxycarbonyl compounds. In this chapter we shall examine first the related **Claisen condensation,** in which attack of an ester enolate on a carbonyl group generates a new carbon–carbon bond. We already saw its use in the biosynthesis of long-chain carboxylic acids (Section 19-13). The products of Claisen condensations are 1,3-dicarbonyl compounds, more commonly known as β-dicarbonyl compounds, which are important for their versatility in synthesis.

Examples of β-Dicarbonyl Compounds

CH$_3$CCH$_2$CCH$_3$ CH$_3$CCH$_2$COCH$_3$ HOCCH$_2$COH

2,4-Pentanedione   **Methyl 3-oxobutanoate (Methyl acetoacetate)**   **Propanedioic acid**
(Acetylacetone)   **(Malonic acid)**
(A β-diketone)   **(A β-ketoester)**   **(A β-dicarboxylic acid)**

Carbohydrates, polyhydroxycarbonyl compounds that will be the subjects of Chapter 24, are biosynthesized in several ways, of which aldol condensations are but one (see Chemical Highlight 18-3). In this chapter we learn another strategy, also employed in nature (Chemical Highlight 23-3) and adapted by synthetic chemists, which makes use of a new kind of nucleophile derived from aldehydes and ketones, an **alkanoyl** or **acyl anion equivalent.** These species illustrate how a normally electrophilic carbonyl carbon atom can be temporarily made nucleophilic and used to form a new carbon–carbon bond with the carbonyl carbon of another aldehyde or ketone molecule, giving α-hydroxyketones as the ultimate products.

23-1 β-Dicarbonyl Compounds: Claisen Condensations

Ester enolates undergo addition–elimination reactions with ester functions, furnishing β-keto esters. These transformations, known as **Claisen* condensations,** are the ester analogs of the aldol reaction (Section 18-5).

Claisen condensations form β-dicarbonyl compounds

Ethyl acetate reacts with a stoichiometric amount of sodium ethoxide to give ethyl 3-oxobutanoate (ethyl acetoacetate).

Claisen Condensation of Ethyl Acetate

$$\underset{\textbf{Ethyl acetate}}{CH_3\overset{O}{\overset{\|}{C}}OCH_2CH_3} \;+\; CH_3\overset{O}{\overset{\|}{C}}OCH_2CH_3 \;\xrightarrow[-\,CH_3CH_2OH]{Na^{+\,-}OCH_2CH_3,\ CH_3CH_2OH}\; \underset{\substack{75\% \\ \textbf{Ethyl 3-oxobutanoate} \\ \textbf{(Ethyl acetoacetate)}}}{CH_3\overset{O}{\overset{\|}{C}}CH_2\overset{O}{\overset{\|}{C}}OCH_2CH_3}$$

The Claisen condensation begins with formation of the ester enolate ion (step 1). Addition–elimination of this species to the carbonyl group of another ester molecule furnishes a 3-ketoester (steps 2 and 3). These steps are reversible; the overall process is *endothermic* through this stage. Indeed, the initial deprotonation of the ester ($pK_a \sim 25$) by ethoxide is particularly unfavorable (pK_a of ethanol = 15.9). Nonetheless, the equilibrium shifts to the product because, in step 4, the base deprotonates C2 of the 3-ketoester ($pK_a \sim 11$) essentially irreversibly. Work-up with aqueous acid (step 5) leads to protonation, completing the process.

Mechanism of the Claisen Condensation

STEP 1. Ester enolate formation

$$CH_3\overset{O}{\overset{\|}{C}}OCH_2CH_3 \xrightarrow{Na^{+\,-}OCH_2CH_3} Na^+ \left[\; ^-\!:\!CH_2\!-\!\overset{\ddot{O}:}{\overset{\|}{C}}\underset{\ddot{O}CH_2CH_3}{} \;\longleftrightarrow\; CH_2\!=\!\overset{\ddot{O}:^-}{\overset{\|}{C}}\underset{\ddot{O}CH_2CH_3}{} \; \right] \;+\; CH_3CH_2OH$$

STEP 2. Nucleophilic addition

$$CH_3\overset{:\ddot{O}:}{\overset{\|}{C}}OCH_2CH_3 \;+\; ^-\!:\!CH_2\overset{O}{\overset{\|}{C}}OCH_2CH_3 \;\rightleftharpoons\; CH_3\!-\!\underset{\underset{O}{\overset{\|}{C}}}{\overset{:\ddot{O}:^-}{\underset{|}{\overset{|}{C}}}}\!-\!OCH_2CH_3$$

$$CH_2\overset{O}{\overset{\|}{C}}OCH_2CH_3$$

*This is the Claisen of the Claisen rearrangement (Section 22-7).

STEP 3. Elimination

$$CH_3\overset{\overset{\displaystyle :\ddot{O}:^-}{|}}{\underset{\underset{\displaystyle O}{\overset{|}{CH_2COCH_2CH_3}}}{C-OCH_2CH_3}} \;\rightleftharpoons\; CH_3\overset{O}{\overset{||}{C}}CH_2\overset{O}{\overset{||}{C}}OCH_2CH_3 \;+\; {}^-:\ddot{O}CH_2CH_3$$

3-Ketoester

STEP 4. Deprotonation of ketoester drives equilibrium

$$\underset{\text{Acidic, }pK_a \sim 11}{CH_3\overset{O}{\overset{||}{C}}CH_2\overset{O}{\overset{||}{C}}OCH_2CH_3} \;+\; {}^-:\ddot{O}CH_2CH_3 \;\longrightarrow$$

$$\left[CH_3\overset{O}{\overset{||}{C}}\ddot{C}H-\overset{O}{\overset{||}{C}}OCH_2CH_3 \longleftrightarrow CH_3\overset{:\ddot{O}:^-}{\overset{|}{C}}=CH-\overset{O}{\overset{||}{C}}OCH_2CH_3 \longleftrightarrow CH_3\overset{O}{\overset{||}{C}}-CH=\overset{:\ddot{O}:^-}{\overset{|}{C}}OCH_2CH_3 \right] \;+\; CH_3CH_2OH$$

STEP 5. Protonation on aqueous work-up

$$CH_3\overset{O}{\overset{||}{C}}\overset{..}{\overset{-}{C}}H\overset{O}{\overset{||}{C}}OCH_2CH_3 \xrightarrow{\;H^+,\,H_2O\;} CH_3\overset{O}{\overset{||}{C}}CH_2\overset{O}{\overset{||}{C}}OCH_2CH_3$$

To prevent transesterification, both the alkoxide and the ester should be derived from the same alcohol.

EXERCISE 23-1

Give the products of Claisen condensation of **(a)** ethylpropanoate; **(b)** ethyl 3-methylbutanoate; **(c)** ethyl pentanoate. For each, the base is sodium ethoxide, the solvent ethanol.

Protons flanked by two carbonyl groups are acidic

Why is deprotonation of the 3-ketoester so favorable? The acidity of the hydrogens flanked by the two carbonyl groups is much enhanced by resonance stabilization of the corresponding anion. Table 23-1 lists the pK_a values of several *β*-dicarbonyl compounds and related systems, such as methyl cyanoacetate and propanedinitrile (malononitrile).

The importance of the final deprotonation step in the Claisen condensation is clearly apparent when the ester bears only one *α*-hydrogen. The product of a reaction with such an ester would be a 2,2-disubstituted 3-ketoester lacking the acidic protons necessary to drive the equilibrium. Hence, no Claisen condensation product is observed.

Failure of a Claisen Condensation

$$2\;(CH_3)_2CH\overset{O}{\overset{||}{C}}OCH_2CH_3 \underset{\xrightarrow{\;Na^+\,{}^-OCH_2CH_3,\;CH_3CH_2OH\;}}{\rightleftharpoons} (CH_3)_2CH\overset{O}{\overset{||}{C}}-\underset{\underset{\displaystyle CH_3}{\overset{|}{|}}}{\overset{\overset{\displaystyle CH_3}{|}}{C}}-\overset{O}{\overset{||}{C}}OCH_2CH_3 \;+\; CH_3CH_2OH$$

Note lack of acidic hydrogens

Ethyl 2-methylpropanoate   **Ethyl 2,2,4-trimethyl-3-oxopentanoate**

TABLE 23-1	pK_a Values for β-Dicarbonyl and Related Compounds	
Name	**Structure**	**pK_a**
2,4-Pentanedione (Acetylacetone)	$\overset{O}{\overset{\|}{CH_3C}}CH_2\overset{O}{\overset{\|}{C}}CH_3$	9
Methyl 2-cyanoacetate	$NCCH_2\overset{O}{\overset{\|}{C}}OCH_3$	9
Ethyl 3-oxobutanoate (Ethyl acetoacetate)	$CH_3\overset{O}{\overset{\|}{C}}CH_2\overset{O}{\overset{\|}{C}}OCH_2CH_3$	11
Propanedinitrile (Malonodinitrile)	$NCCH_2CN$	13
Diethyl propanedioate (Diethyl malonate)	$CH_3CH_2O\overset{O}{\overset{\|}{C}}CH_2\overset{O}{\overset{\|}{C}}OCH_2CH_3$	13

That this result is due to an unfavorable equilibrium can be demonstrated by treating a 2,2-disubstituted 3-ketoester with base: *complete reversal* of the Claisen condensation process (called **retro-Claisen condensation**) ensues, giving two molecules of simple ester through a mechanism that is the exact reverse of the forward reaction.

Reversal of a Claisen Condensation (Retro-Claisen Condensation)

$$CH_3CH_2\overset{..}{\underset{..}{O}}:^- \;+\; (CH_3)_2CHC\overset{:O:}{\overset{\|}{}}-\overset{CH_3}{\underset{\|}{C}}-\overset{:O:}{\overset{\|}{C}}OCH_2CH_3 \xrightarrow{CH_3CH_2\overset{..}{O}H} (CH_3)_2CHC-\overset{CH_3}{\underset{\|}{C}}-\overset{:O:}{\overset{\|}{C}}OCH_2CH_3 \longrightarrow$$

$$(CH_3)_2CHC\overset{:O:}{\overset{\|}{}}\overset{..}{\underset{..}{O}}CH_2CH_3 \;+\; (CH_3)_2C=C\overset{:\overset{..}{O}:^-}{\overset{\|}{}}OCH_2CH_3 \underset{\xrightarrow{CH_3CH_2\overset{..}{O}H}}{\rightleftharpoons} (CH_3)_2CHC\overset{:O:}{\overset{\|}{}}\overset{..}{\underset{..}{O}}CH_2CH_3 \;+\; CH_3CH_2\overset{..}{\underset{..}{O}}:^-$$

EXERCISE 23-2

Explain the following observation.

$$CH_3\overset{O}{\overset{\|}{C}}-\overset{CH_3}{\underset{\underset{CH_3}{|}}{C}}-COOCH_3 \xrightarrow[\text{2. H}^+,\text{ H}_2\text{O}]{\text{1. CH}_3\text{O}^-\text{Na}^+,\text{ CH}_3\text{OH}} CH_3\overset{O}{\overset{\|}{C}}CH_2COOCH_3 \;+\; 2\,(CH_3)_2CHCOOCH_3$$

Claisen condensations can have two different esters as reactants

Mixed Claisen condensations start with two different esters. Like crossed aldol condensations (Section 18-6), they are typically unselective and furnish product mixtures. However, a selective mixed condensation is possible when one of the reacting partners has no α-hydrogens, as in ethyl benzoate.

A Selective Mixed Claisen Condensation

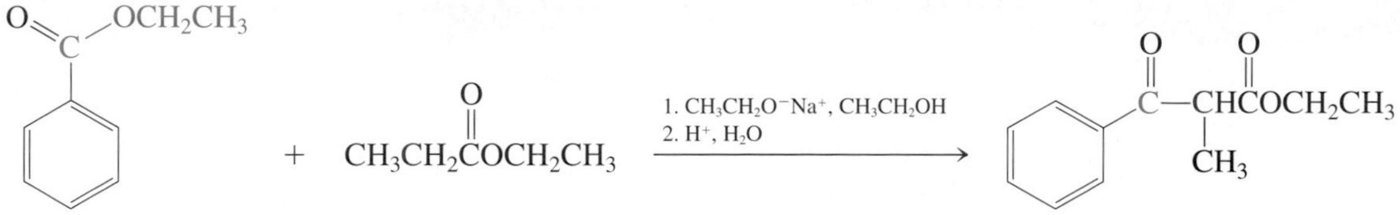

Ethyl benzoate

$+$ $CH_3CH_2COCH_2CH_3$

1. $CH_3CH_2O^-Na^+$, CH_3CH_2OH
2. H^+, H_2O

$\longrightarrow$

71%

Ethyl 2-methyl-3-oxo-3-phenylpropanoate

EXERCISE 23-3

Give all the Claisen condensation products that would result from treatment of a mixture of ethyl acetate and ethyl propanoate with sodium ethoxide in ethanol.

EXERCISE 23-4

Is the mixed Claisen condensation between ethyl formate and ethyl acetate likely to afford one major product? Explain and give the structure(s) of the product(s) you expect.

Intramolecular and double Claisen condensations result in cyclic compounds

The intramolecular version of the Claisen reaction, called the **Dieckmann* condensation,** produces cyclic 3-ketoesters. As expected (Section 9-6), it works best for the formation of five- and six-membered rings.

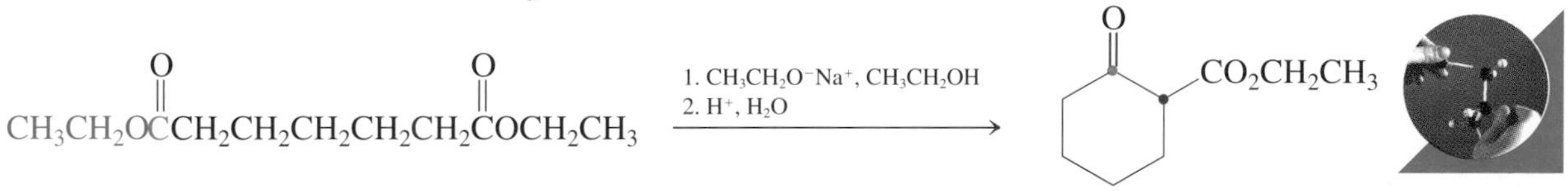

$CH_3CH_2OCCH_2CH_2CH_2CH_2CH_2COCH_2CH_3$

1. $CH_3CH_2O^-Na^+$, CH_3CH_2OH
2. H^+, H_2O

$\longrightarrow$

$CO_2CH_2CH_3$

60%

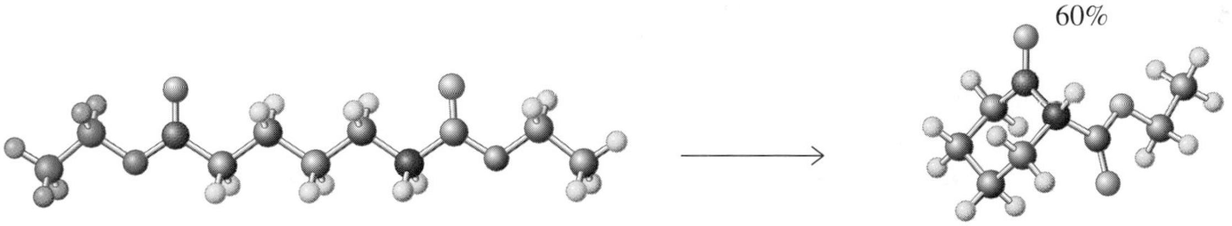

Diethyl heptanedioate

Ethyl 2-oxocyclohexanecarboxylate

Cyclic compounds can also be obtained by (intermolecular followed by intramolecular) **double Claisen condensations** with diesters such as diethyl ethanedioate (diethyl oxalate).

$CH_3CH_2OCCOCH_2CH_3$ $+$ $CH_3CH_2OCCH_2CH_2CH_2COCH_2CH_3$

1. $CH_3CH_2O^-Na^+$, CH_3CH_2OH
2. H^+, H_2O

$\longrightarrow$

$CH_3CH_2O_2C$

$CO_2CH_2CH_3$

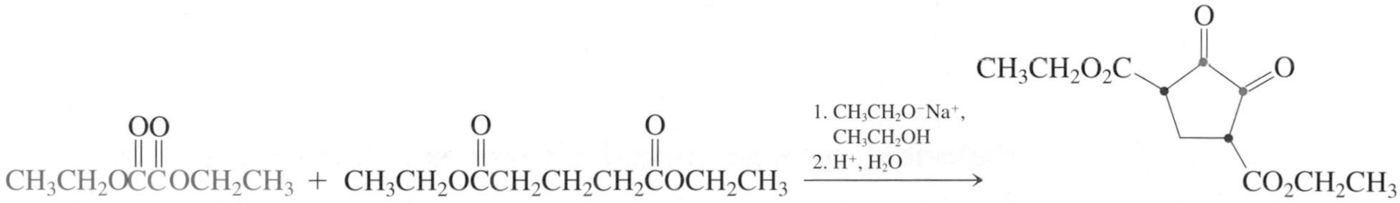

80%

Diethyl ethanedioate
(Diethyl oxalate)

Diethyl pentanedioate

Diethyl 4,5-dioxo-1,3-cyclopentanedicarboxylate

*Professor Walter Dieckmann (1869–1925), University of Munich, Germany.

| **Claisen Condensations in Biochemistry**

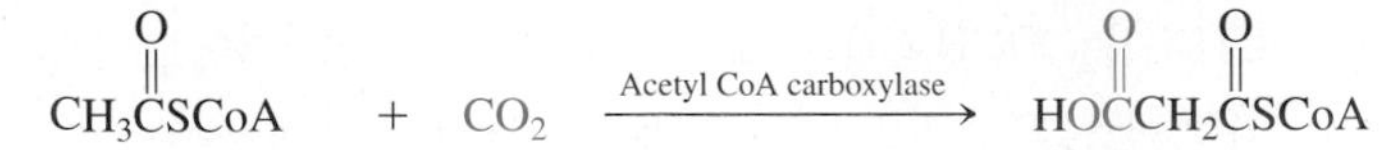

$$CH_3CSCoA \quad + \quad CO_2 \quad \xrightarrow{\text{Acetyl CoA carboxylase}} \quad HOCCH_2CSCoA$$

Acetyl coenzyme A   **Malonyl CoA**

The coupling processes that build fatty acid chains from thioesters of coenzyme A (Section 19-13) are forms of Claisen condensations. The carboxylation of acetyl CoA into malonyl CoA (shown above) is a variant in which the carbon of CO_2 rather than that of an ester carbonyl group is the site of nucleophilic attack.

The methylene group in the carboxylated species is much more reactive than the methyl group in acetyl thioesters and participates in a wide variety of Claisen-like condensations. Although these processes require enzyme catalysis, they may be formulated in simplified form, shown below.

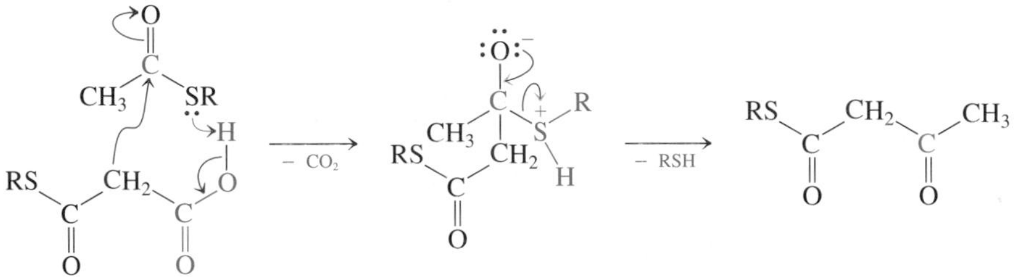

(**RSH** = acyl carrier protein; see Section 19-14)

Formulate a mechanism for the reaction of diethyl ethanedioate with diethyl pentanedioate.

Formulate a mechanism for the following reaction.

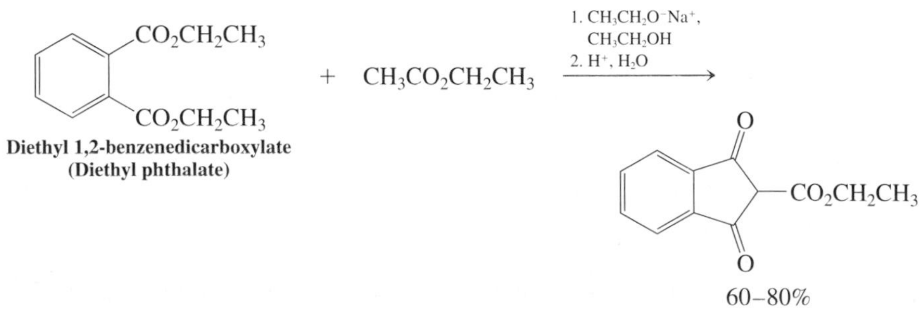

Diethyl 1,2-benzenedicarboxylate
(Diethyl phthalate)

$+ \quad CH_3CO_2CH_2CH_3 \quad \xrightarrow[\text{2. H}^+, \text{H}_2\text{O}]{\text{1. CH}_3\text{CH}_2\text{O}^-\text{Na}^+, \text{CH}_3\text{CH}_2\text{OH}}$

60–80%

Ketones undergo mixed Claisen reactions

Ketones can participate in the Claisen condensation. Because they are more acidic than esters, they are deprotonated before the ester has a chance to undergo self-condensation. The products (after acidic work-up) may be β-diketones, β-ketoaldehydes, or other β-dicarbonyl compounds. The reaction can be carried out with a variety of ketones and esters both inter- and intramolecularly.

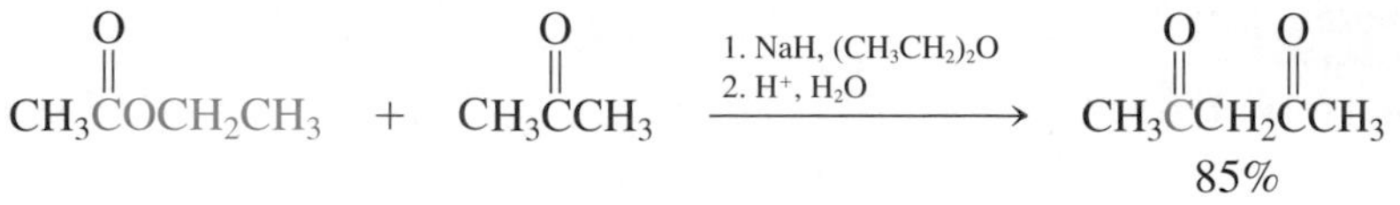

CH3COCH2CH3 + CH3CCH3
1. NaH, (CH3CH2)2O
2. H+, H2O
CH3CCH2CCH3
85%

EXERCISE 23-7

1,3-Cyclohexanedione can be prepared by an intramolecular mixed Claisen condensation between the ketone carbonyl and ester functions of a single molecule. What is the structure of this substrate molecule?

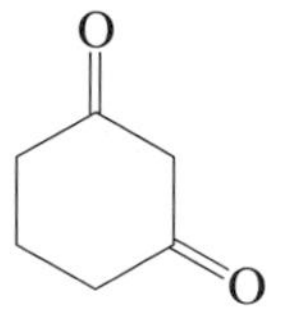

1,3-Cyclohexanedione

Retrosynthetic analysis clarifies the synthetic utility of the Claisen condensation

Having seen a variety of types of Claisen condensations, we may now ask how this process may be logically analyzed for synthetic use. Three facts are available to help us: (1) Claisen condensations always form 1,3-dicarbonyl compounds; (2) one of the reaction partners in a Claisen condensation must be an ester, whose alkoxide group is lost in the course of the condensation; and (3) the other reaction partner (the source of the nucleophilic enolate) must contain at least two acidic hydrogens on an α-carbon. In addition, if a mixed condensation is being considered, one reaction partner (the ester, if there is only one), should be incapable of self-condensation (e.g., it should lack α-hydrogens). If we are given the structure of a target molecule and wish to determine whether (and, if so, how) it can be made by a Claisen condensation, we must analyze it retrosynthetically with the preceding points in mind. For example, let us consider whether 2-benzoylcyclohexanone can be made by a Claisen condensation.

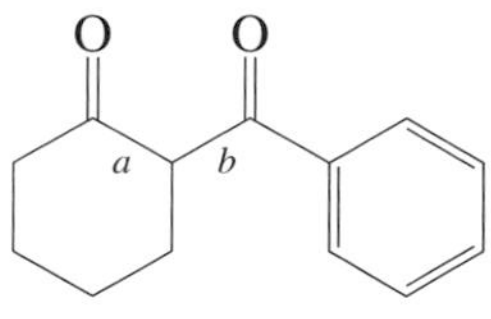

2-Benzoylcyclohexanone

It is a 1,3-dicarbonyl compound, meeting the first requirement. What bond forms in a Claisen condensation? By examining all the examples in this section, we find that the new bond in the product always connects one of the carbonyl groups of the 1,3-dicarbonyl moiety to the carbon atom *between* them. Our target molecule contains two such bonds, which we label *a* and *b*. As we continue our analysis by disconnecting each of these strategic bonds in turn, we must employ the second point: The carbonyl group at which the new carbon–carbon bond forms starts out as part of an ester function. Thus, working backward, *we must imagine reattaching an alkoxy group to this carbonyl carbon.* Thus:

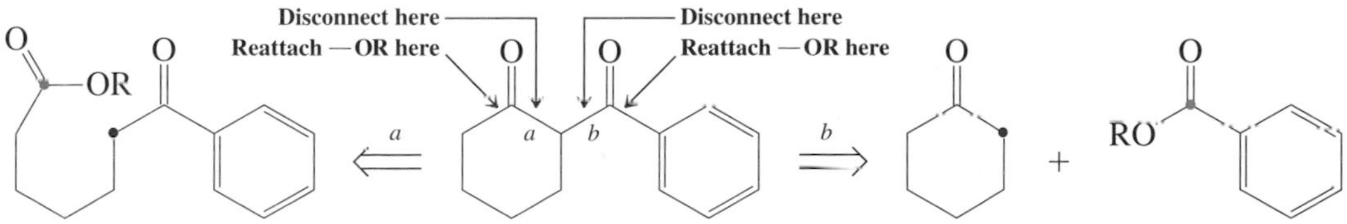

Disconnect here
Reattach —OR here
Disconnect here
Reattach —OR here
OR
a
b
RO
+

Disconnection of bond *a* reveals a ketoester, which undergoes intramolecular Claisen condensation in the forward direction, whereas disconnection of bond *b* gives cyclohexanone and a benzoic ester. Both condensations are quite feasible; however, the second is preferable because it constructs the target from two smaller pieces:

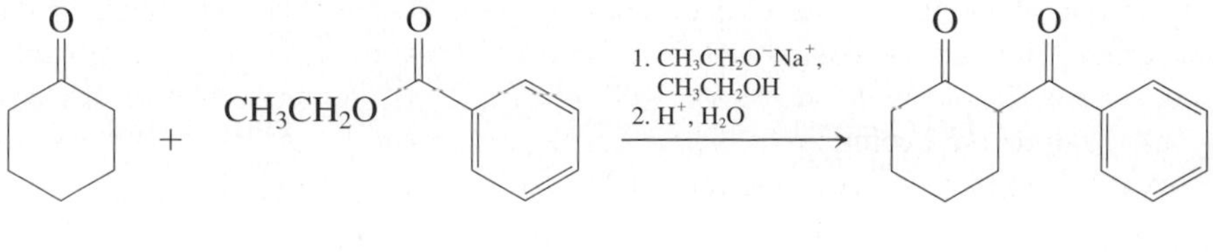

CH3CH2O
+
1. CH3CH2O−Na+, CH3CH2OH
2. H+, H2O

EXERCISE 23-8

Suggest syntheses of the following molecules by Claisen or Dieckmann condensations.

(a) [structure: cyclohexanone with $CCO_2CH_2CH_3$ substituent]

(b) CH_3CCH_2CH (with two C=O groups)

(c) [structure: cyclooctanone with $CO_2CH_2CH_3$ substituent]

(d) H_3C—[structure: cyclopentanone with acetyl group]

In summary, Claisen condensations are endothermic and therefore would not take place without a stoichiometric amount of base strong enough to deprotonate the resulting 3-ketoester. Mixed Claisen condensations between two esters are nonselective, unless they are intramolecular (Dieckmann condensation) or one of the components is devoid of α-hydrogens. Ketones also participate in selective mixed Claisen reactions because they are more acidic than esters.

23-2 β-Dicarbonyl Compounds as Synthetic Intermediates

Having seen how to prepare β-dicarbonyl compounds, let us explore their synthetic utility. This section will show that the corresponding anions are readily alkylated and that 3-ketoesters are hydrolyzed to the corresponding acids, which can be decarboxylated to give ketones or new carboxylic acids. These transformations open up versatile synthetic routes to other functionalized molecules.

β-Dicarbonyl anions are nucleophilic

The unusual acidity of β-ketocarbonyl compounds can be used to synthetic advantage, because the enolate ions obtained by deprotonation can be alkylated to give substituted derivatives. For example, in this way ethyl 3-oxobutanoate is readily converted into alkylated analogs.

β-Ketoester Alkylations

$CH_3CCHCOCH_2CH_3$ (with H) $\xrightarrow[\substack{- CH_3CH_2OH \\ - NaI}]{\substack{1.\ Na^+\ ^-OCH_2CH_3, \\ CH_3CH_2OH \\ 2.\ CH_3I}}$ $CH_3CCHCOCH_2CH_3$ (with CH_3) $\xrightarrow[\substack{- (CH_3)_3COH \\ - KBr}]{\substack{1.\ K^+\ ^-OC(CH_3)_3, \\ (CH_3)_3COH \\ 2.\ [PhCH_2Br]}}$ $CH_3C—C—COCH_2CH_3$ (with CH_3 and CH_2Ph)

	65%	77%
Ethyl 3-oxobutanoate	**Ethyl 2-methyl-3-oxobutanoate**	**Ethyl 2-methyl-2-(phenylmethyl)-3-oxobutanoate**

Other β-dicarbonyl compounds undergo similar reactions.

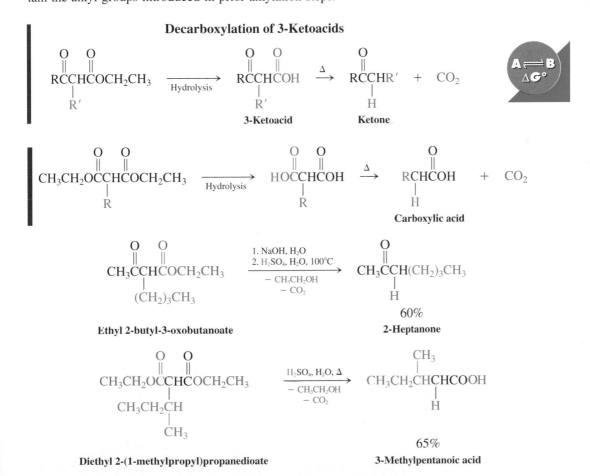

Diethyl propanedioate → **Diethyl 2-(1-methylpropyl)propanedioate**

Give a synthesis of 2,2-dimethyl-1,3-cyclohexanedione from methyl 5-oxohexanoate.

3-Ketoacids readily undergo decarboxylation

Hydrolysis of 3-ketoesters furnishes 3-ketoacids, which in turn readily undergo decarboxylation under mild conditions. The products, ketones and carboxylic acids, contain the alkyl groups introduced in prior alkylation steps.

Decarboxylation has a concerted mechanism with an aromatic transition state (Section 15-3), somewhat like that of the McLafferty rearrangement in mass spectrometry (Section 20-10).

Mechanism of Decarboxylation of 3-Ketoacids

Loss of CO_2 can occur readily only from the free carboxylic acid. If the ester is hydrolyzed under basic conditions, the resulting carboxylate salt is usually neutralized with acid to enable subsequent decarboxylation. Decarboxylation of substituted propanedioic (malonic) acids follows the same mechanism.

EXERCISE 23-10

Formulate a detailed mechanism for the decarboxylation of $CH_3CH(COOH)_2$ (methylmalonic acid).

The acetoacetic ester synthesis leads to methyl ketones

The combination of alkylation followed by ester hydrolysis and finally decarboxylation allows ethyl 3-oxobutanoate (ethyl acetoacetate) to be converted ultimately into 3-substituted or 3,3-disubstituted methyl ketones. This strategy is called the **acetoacetic ester synthesis.**

Acetoacetic Ester Synthesis

$$CH_3CCH_2COCH_2CH_3 \dashrightarrow CH_3C—C—COCH_2CH_3 \dashrightarrow CH_3CCH_{R}^{R'}$$

3,3-Disubstituted methyl ketone

Methyl ketones with either one or two substituent groups on C3 can be synthesized by using the acetoacetic sequence.

Syntheses of Substituted Methyl Ketones

$$CH_3CCH_2COCH_2CH_3 \xrightarrow[\substack{2.\ CH_3CH_2CH_2Br}]{\substack{1.\ NaOCH_2CH_3,\\ CH_3CH_2OH}} CH_3CCHCOCH_2CH_3$$

CH₂CH₂CH₂CH₃

72%

$$\xrightarrow[\substack{2.\ H_2SO_4,\\ H_2O,\ 100°C}]{\substack{1.\ NaOH,\ H_2O}} CH_3CCH_2—CH_2CH_2CH_2CH_3$$

60%
2-Heptanone

1. KOC(CH₃)₃, (CH₃)₃COH
2. CH₃CH₂CH₂CH₂I

$$CH_3CCCOCH_2CH_3$$

CH₃CH₂CH₂CH₂   CH₂CH₂CH₂CH₃

80%

$$\xrightarrow[\substack{2.\ HCl,\\ H_2O,\ 100°C}]{\substack{1.\ KOH,\\ H_2O,\ 100°C}} CH_3C—CH—CH_2CH_2CH_2CH_3$$

CH₂CH₂CH₂CH₃

64%
3-Butyl-2-heptanone

EXERCISE 23-11

Propose syntheses of the following ketones, beginning with ethyl 3-oxobutanoate (ethyl acetoacetate). **(a)** 2-Hexanone; **(b)** 2-octanone; **(c)** 3-ethyl-2-pentanone; **(d)** 4-phenyl-2-butanone.

The malonic ester synthesis furnishes carboxylic acids

Diethyl propanedioate (malonic ester) is the starting material for preparing 2-alkylated and 2,2-dialkylated acetic acids, a method called the **malonic ester synthesis.**

Malonic Ester Synthesis

$$CH_3CH_2OCCH_2COCH_2CH_3 \dashrightarrow CH_3CH_2OC-C-COCH_2CH_3 \dashrightarrow H-C-COOH$$

2,2-Dialkylated acetic acid

Like the acetoacetic ester route to ketones, the malonic ester synthesis can lead to carboxylic acids with either one or two substituents at C2.

Synthesis of a 2,2-Dialkylated Acetic Acid

1. NaOCH₂CH₃, CH₃CH₂OH
2. CH₃(CH₂)₉Br, 80°C
3. KOH, H₂O, CH₃CH₂OH, 80°C
4. H₂SO₄, H₂O, 180°C

$$CH_3CH_2OCCHCOCH_2CH_3 \longrightarrow CH_3(CH_2)_9CHCOOH$$

74%
2-Methyldodecanoic acid

**Diethyl 2-methylpropanedioate
(Diethyl methylmalonate)**

EXERCISE 23-12

(a) Give the structure of the product formed after each of the first three steps in the preceding synthesis of 2-methyldodecanoic acid. **(b)** How would you make the starting material, diethyl 2-methylpropanedioate?

The rules and limitations governing S_N2 reactions apply to the alkylation steps. Thus, tertiary haloalkanes exposed to β-dicarbonyl anions give mainly elimination products. However, the anions can successfully attack alkanoyl halides, α-bromoesters, α-bromoketones, and oxacyclopropanes.

EXERCISE 23-13

The first-mentioned compound in each of the following parts is treated with the subsequent series of reagents. Give the final products.

(a) CH₃CH₂O₂C(CH₂)₅CO₂CH₂CH₃: NaOCH₂CH₃; CH₃(CH₂)₃I; NaOH; and H⁺, H₂O, Δ

(b) CH₃CH₂O₂CCH₂CO₂CH₂CH₃: NaOCH₂CH₃; CH₃I; KOH; and H⁺, H₂O, Δ

(c) CH₃CCHCO₂CH₃: NaH, C₆H₆; C₆H₅CCl; and H⁺, H₂O, Δ

$$\overset{\displaystyle O}{\overset{\displaystyle \|}{}}$$

(d) $CH_3CCH_2CO_2CH_2CH_3$: $NaOCH_2CH_3$; $BrCH_2CO_2CH_2CH_3$; $NaOH$; and H^+, H_2O, Δ

(e) $CH_3CH_2CH(CO_2CH_2CH_3)_2$: $NaOCH_2CH_3$; $BrCH_2CO_2CH_2CH_3$; and H^+, H_2O, Δ

$$\overset{\displaystyle O}{\overset{\displaystyle \|}{}} \qquad\qquad\qquad\qquad \overset{\displaystyle O}{\overset{\displaystyle \|}{}}$$

(f) $CH_3CCH_2CO_2CH_2CH_3$: $NaOCH_2CH_3$; $BrCH_2CCH_3$; and H^+, H_2O, Δ

EXERCISE 23-14

Propose a synthesis of cyclohexanecarboxylic acid from diethyl propanedioate (malonate), $CH_2(CO_2CH_2CH_3)_2$, and 1-bromo-5-chloropentane, $Br(CH_2)_5Cl$.

In summary, β-dicarbonyl compounds such as ethyl 3-oxobutanoate (acetoacetate) and diethyl propanedioate (malonate) are versatile synthetic building blocks for elaborating more complex molecules. Their unusual acidity makes it easy to form the corresponding anions, which can be used in nucleophilic displacement reactions with a wide variety of substrates. Their hydrolysis produces 3-ketoacids that are unstable and undergo decarboxylation on heating.

23-3 β-Dicarbonyl Anion Chemistry: Michael Additions

Reaction of the stabilized anions derived from β-dicarbonyl compounds and related analogs (Table 23-1) with α,β-unsaturated carbonyl compounds leads to 1,4-additions. This transformation, an example of **Michael addition** (Section 18-12), is base-catalyzed and works with α,β-unsaturated ketones, aldehydes, nitriles, and carboxylic acid derivatives, all of which are termed **Michael acceptors.**

Michael Addition

$$CH_2(CO_2CH_2CH_3)_2 \quad + \quad \overset{\displaystyle O}{\overset{\displaystyle \|}{CH_2\!=\!CHCCH_3}} \quad \xrightarrow[\text{CH}_3\text{CH}_2\text{OH, } -10 \text{ to } 25°\text{C}]{\text{Catalytic CH}_3\text{CH}_2\text{O}^-\text{Na}^+,} \quad \overset{\displaystyle O}{\overset{\displaystyle \|}{(CH_3CH_2O_2C)_2CH\!-\!CH_2CH_2CCH_3}}$$

71%

Diethyl propanedioate (**Diethyl malonate**)   **3-Buten-2-one** (**Methyl vinyl ketone**) (**Michael acceptor**)   **Diethyl 2-(3-oxobutyl)propanedioate**

Why do stabilized anions undergo 1,4- rather than 1,2-addition to Michael acceptors? The latter process occurs but is reversible with relatively stable anionic nucleophiles because it leads to a relatively high energy alkoxide. Conjugate addition is favored thermodynamically because it produces a resonance-stabilized enolate ion.

EXERCISE 23-15

Formulate a detailed mechanism for the Michael addition process just depicted. Why is the base required in only catalytic amounts?

EXERCISE 23-16

Give the products of the following Michael additions [base in square brackets].

(a) $CH_3CH_2CH(CO_2CH_2CH_3)_2$ + $CH_2=CHCH$ (with C=O above) $[Na^{+-}OCH_2CH_3]$

(b) 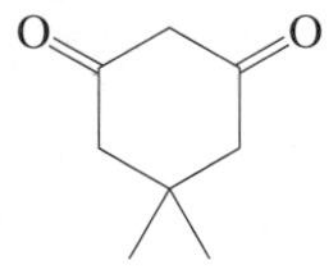 (structure) + $CH_2=CHC\equiv N$ $[Na^{+-}OCH_3]$

(c) 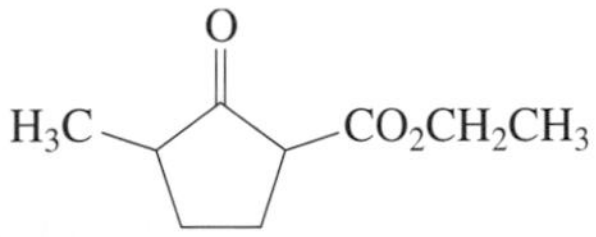 (structure) H_3C ... $CO_2CH_2CH_3$ + $CH_3CH=CHCO_2CH_2CH_3$ $[K^{+-}OCH_2CH_3]$

EXERCISE 23-17

Explain the following observation. (**Hint:** Consider proton transfer in the first Michael adduct.)

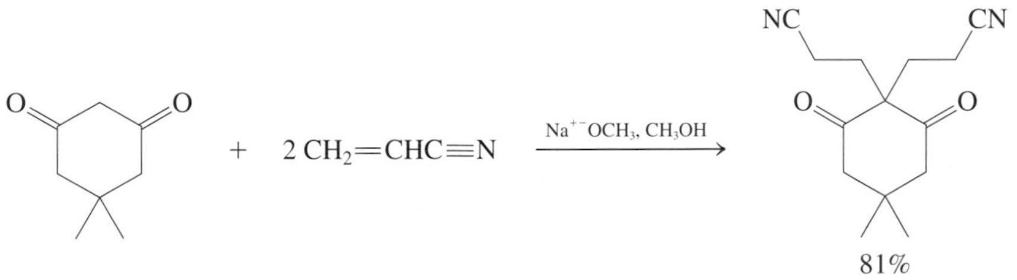

(structure) + $2\ CH_2=CHC\equiv N$ $\xrightarrow{Na^{+-}OCH_3,\ CH_3OH}$ (product structure) 81%

The following reaction is a useful synthetic application of Michael addition of anions of β-ketoesters to α,β-unsaturated ketones. The process leads to a diketone in which the enolate of one is positioned to form a six-membered ring on aldol condensation with the carbonyl group of the other. Recall (Section 18-12) that the synthesis of six-membered rings by Michael addition followed by aldol condensation is called **Robinson annulation.**

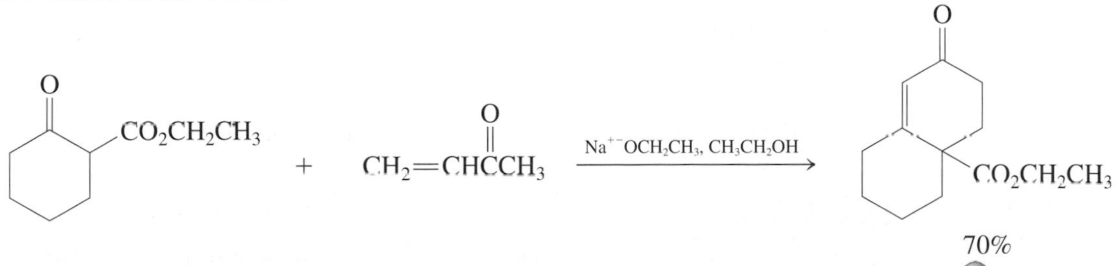

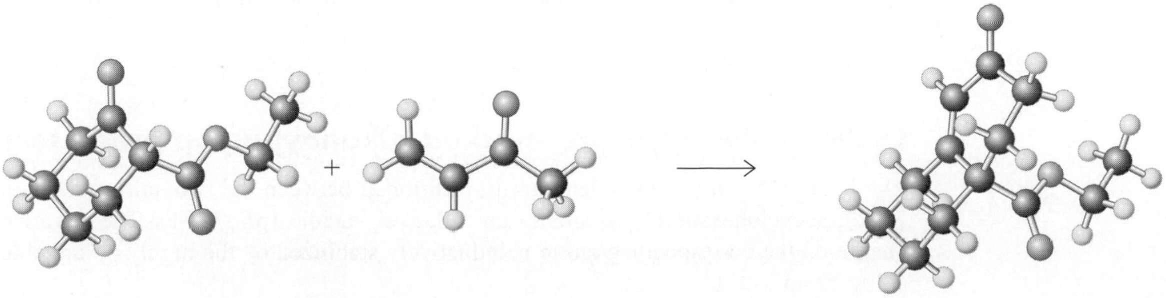

(structure) $CO_2CH_2CH_3$ + $CH_2=CHCCH_3$ $\xrightarrow{Na^{+-}OCH_2CH_3,\ CH_3CH_2OH}$ (product) $CO_2CH_2CH_3$

70%

EXERCISE 23-18

Formulate a detailed mechanism for the preceding transformation.

In summary, β-dicarbonyl anions, like ordinary enolate anions, undergo Michael additions to α,β-unsaturated carbonyl compounds. Addition of a β-ketoester to an enone gives a diketone, which can generate six-membered rings by intramolecular aldol condensation (Robinson annulation).

23-4 Alkanoyl (Acyl) Anion Equivalents: Preparation of α-Hydroxyketones

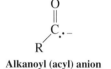
Alkanoyl (acyl) anion

Throughout our study of carbonyl compounds, we have been used to thinking of the carbonyl carbon atom as being electrophilic and its α-carbon, in the form of an enol or an enolate, as being nucleophilic. These tendencies govern the rich chemistry of the functional groups built around the C=O unit. However, although the array of transformations available to carbonyl compounds is large, it is not without limits. For example, we have no way of making a direct connection between two carbonyl carbon atoms, because both are electrophilic: Neither can serve as a nucleophilic electron source to attack the other. One may imagine a hypothetical carbonyl-derived nucleophilic species, such as an **alkanoyl (acyl) anion** (margin), which might add to an aldehyde or ketone to give an α-hydroxyketone, as follows.

A Plausible (but Unfeasible) Synthesis of α-Hydroxyketones

$$:\!O: \qquad :O\overset{..}{:} \qquad :O: :\overset{..}{O}:^{-} \qquad :O: :\overset{..}{O}H$$
$$\underset{RC:^{-}}{\overset{\|}{}} \xrightarrow{\underset{H\overset{..}{C}R}{}} \underset{RC-CHR}{\overset{\|\quad|}{}} \xrightarrow[-\,HO^{-}]{HOH} \underset{RC-CHR}{\overset{\|\quad|}{}}$$

The ability to generate such an anion would expand the versatility of carbonyl chemistry immensely, making a wide variety of 1,2-difunctionalized systems readily available, in analogy to the 1,3-difunctional products that we can so readily prepare by using aldol and Claisen condensations. Unfortunately, alkanoyl anions are high-energy species and cannot be generated readily for synthetic applications. Consequently, chemists have explored the construction of other chemical species containing negatively charged carbon atoms that can undergo addition reactions *and later be transformed into carbonyl groups.* These special nucleophiles, called **masked alkanoyl (acyl) anions** or **alkanoyl (acyl) anion equivalents,** are the subject of this section.

EXERCISE 23-19

Why are alkanoyl anions not formed by reaction of a base with an aldehyde? (**Hint:** See Sections 17-5 and 18-1.)

Cyclic dithioacetals are masked alkanoyl anion precursors

The hydrogens on the methylene group positioned between the two sulfur atoms in 1,3-dithiacyclohexane (1,3-dithiane) are relatively acidic ($pK_a = 31$). The negative charge on the corresponding anion is inductively stabilized by the highly polarizable sulfur atoms.

Deprotonation of 1,3-Dithiacyclohexane

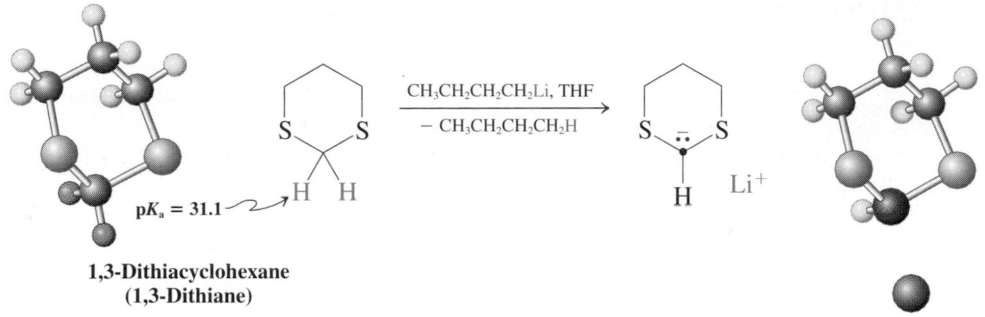
CH₃CH₂CH₂CH₂Li, THF
− CH₃CH₂CH₂CH₂H
S
S
H
H
S
S
H
Li⁺
pKₐ = 31.1

1,3-Dithiacyclohexane (1,3-Dithiane)

These anions add to aldehydes and ketones, thereby furnishing alcohols with an adjacent thioacetal function.

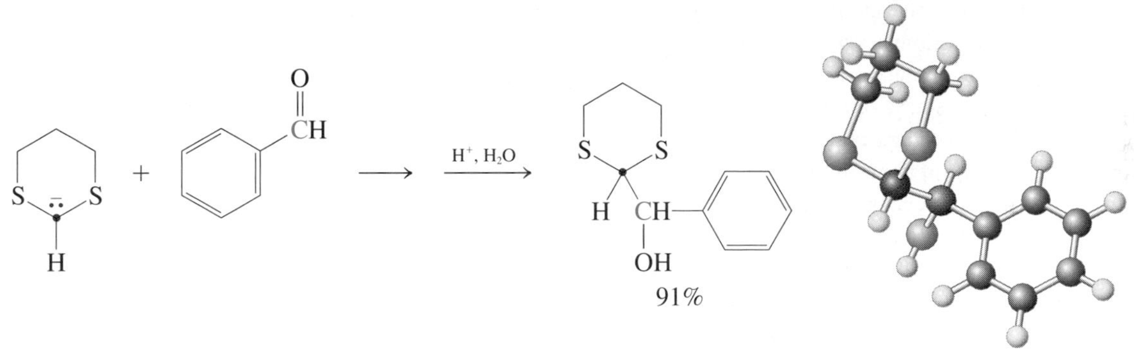
O
CH
S
S
H
H⁺, H₂O
S
S
H
CH
OH
91%

2-(1-Hydroxyphenylmethyl)-1,3-dithiacyclohexane

The thioacetals that result from such addition may be hydrolyzed by using mercuric salts (Section 17-8) to furnish the corresponding carbonyl compounds.

The synthesis of 1-acetyl-2-cyclohexen-1-ol illustrates the formation of a substituted 1,3-dithiacyclohexane from an aldehyde. Addition of the dithiacyclohexane anion to a ketone is followed by hydrolysis of the resulting thioacetal to give the product.

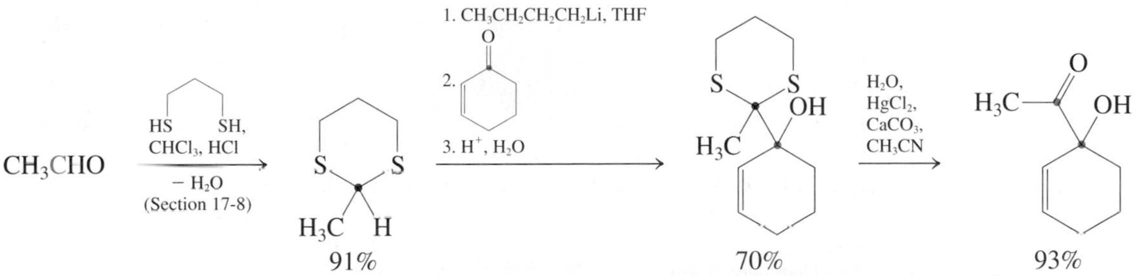
CH₃CHO
HS
SH,
CHCl₃, HCl
− H₂O
(Section 17-8)
S
S
H₃C
H
91%
1. CH₃CH₂CH₂CH₂Li, THF
O
2.
3. H⁺, H₂O
S
S
OH
H₃C
70%
H₂O,
HgCl₂,
CaCO₃,
CH₃CN
O
H₃C
OH
93%

1-Acetyl-2-cyclohexen-1-ol

In this synthesis, the electrophilic carbonyl carbon of the starting aldehyde is transformed into a *nucleophilic* atom, the negatively charged C2 of a 1,3-dithiacyclohexane anion. After the latter is added to the ketone, hydrolysis of the thioacetal function regenerates the original electrophilic carbonyl group. The sequence therefore employs the *reversal of the polarization* of this carbon atom to form the carbon–carbon bond. Reagents exhibiting reverse polarization greatly increase the strategies available to chemists in planning syntheses. We have in fact seen this strategy before: Conversion of a haloalkane into an organometallic (e.g., Grignard) reagent (Section 8-7) reverses the polarity of the functionalized carbon from electrophilic ($^{\delta+}C\text{–}X^{\delta-}$) to nucleophilic ($^{\delta-}C\text{–}M^{\delta+}$).

2-Oxopropanoic (Pyruvic) Acid, a Natural α-Ketoacid

$$CH_3\overset{\overset{\displaystyle O}{\|}}{C}Cl \;+\; Na^{+\,-}CN \xrightarrow[-\;NaCl]{}$$

$$CH_3\overset{\overset{\displaystyle O}{\|}}{C}CN \xrightarrow{\text{Conc. HCl, 0°C}} CH_3\overset{\overset{\displaystyle O}{\|}}{C}COOH$$
95%   100%

2-Oxopropanenitrile   **2-Oxopropanoic acid**
                                   (Pyruvic acid)

2-Oxopropanoic (pyruvic) acid is an α-ketoacid. It can be synthesized by the hydrolysis of 2-oxopropanenitrile, whose laboratory preparation from acetyl chloride and sodium cyanide is another example of the use of functional group polarity reversal in synthesis. In this case, the required carbon–carbon bond is formed by attack on the electrophilic carbonyl carbon atom in the alkanoyl chloride by the nucleophilic cyanide carbon. Hydrolysis gives the ketoacid, in which two electrophilic carbonyl carbon atoms are now directly bonded. The cyanide ion, $^-C{\equiv}N$, thus serves as a synthetic equivalent of the nonexistent "^-COOH" anion.

Pyruvic acid plays a central role in metabolism. It is an end product of the biochemical breakdown of glucose (Chapter 24) and, after loss of one molecule of CO_2, is converted into acetyl CoA (Section 19-13), the primary entry molecule for the **tricarboxylic acid (TCA) cycle,** also known as the **citric acid** or **Krebs* cycle.** The TCA cycle combines acetyl CoA with 2-oxobutanedioic (oxaloacetic) acid to form citric acid. In each turn of the cycle, two carbons of citric acid are oxidized to CO_2, regenerating oxaloacetic acid and giving rise in a coupled process to a molecule of **adenosine triphosphate (ATP),** the main energy source in cells. The TCA cycle generates two-thirds of the energy derived from food in higher organisms.

Under nonoxidative (anaerobic, or "oxygen debt") conditions, such as those found in muscle tissue undergoing extreme exertion, an alternate process drives the formation of energy-rich ATP: Pyruvic acid is *reduced* to (S)-$(+)$-2-hydroxypropanoic (lactic) acid by the enzyme *lactic acid dehydrogenase.* Excessive buildup of lactic acid in muscle tissue causes fatigue and cramps. Lactic acid is removed from the muscle both by slow diffusion into the bloodstream and by enzyme-catalyzed conversion back into pyruvic acid, after the condition of oxygen debt has been relieved. This oxygen debt is the reason why you breathe hard during and after physical exercise.

Upon crossing the finish line, the winner of the 1996 New York Marathon shows the combined effects of oxygen debt and lactic acid buildup in her muscles.

*Sir Hans Adolf Krebs (1900–1981), Oxford University, Nobel Prize 1953 (physiology or medicine).

$$\begin{array}{c}O{=}C{-}CO_2H\\ |\\ H_2C{-}CO_2H\end{array} \xrightarrow[\text{synthase}]{\substack{CH_3COSCoA,\\ \text{Citric acid}}} \begin{array}{c}H_2C{-}CO_2H\\ |\\ HO{-}C{-}CO_2H\\ |\\ H_2C{-}CO_2H\end{array}$$

2-Oxobutanedioic   **Citric acid**
(Oxaloacetic) acid

$$CH_3\overset{\overset{\displaystyle O}{\|}}{C}COOH \underset{\text{dehydrogenase}}{\overset{\text{Lactic acid}}{\rightleftharpoons}} \begin{array}{c}HO\\ H{\diagdown}\\ C{-}COOH\\ H_3C\end{array}$$

(S)-$(+)$-2-Hydroxy-propanoic (lactic) acid

While this section describes the application of dithiacyclohexane anions as masked alkanoyl anions to the preparation of α-hydroxyketones only, it is readily apparent that their alkylation with other electrophilic reagents allows for a general ketone synthesis (see Chapter Integration Problem).

EXERCISE 23-20

Formulate a synthesis of 2-hydroxy-2,4-dimethyl-3-pentanone, beginning with simple aldchydes and ketones and using a 1,3-dithiacyclohexane anion.

Thiazolium ions catalyze aldehyde coupling

Masked alkanoyl anions can be generated catalytically in the reaction of aldehydes with **thiazolium salts.** These salts are derived from thiazoles by alkylation at nitrogen. Thiazole is a heteroaromatic compound (Section 25-4) containing sulfur and nitrogen. Thiazolium salts have an unusual feature—a relatively acidic proton located between the two heteroatoms (at C2).

Thiazole

Thiazolium Cations Are Acidic

$pK_a \sim 17\text{–}19$
Thiazolium cation

$+ \ \text{HOH}$

Thiazolium salt

In the presence of thiazolium salts, aldehydes undergo conversion into α-hydroxyketones. An example of this process is the conversion of two molecules of butanal into 5-hydroxy-4-octanone. The catalyst is N-dodecylthiazolium bromide, which contains a long-chain alkyl substituent to improve its solubility in organic solvents.

Aldehyde Coupling

N-Dodecylthiazolium bromide

$$2 \ CH_3CH_2CH_2CH \xrightarrow{\hspace{2cm}} CH_3CH_2CH_2C-CHCH_2CH_2CH_3$$

76%

Butanal **5-Hydroxy-4-octanone**

The mechanism of this reaction begins with reversible addition of C2 in the deprotonated thiazolium salt to the carbonyl function of an aldehyde.

Mechanism of Thiazolium Ion Catalysis in Aldehyde Coupling

STEP 1. Deprotonation of thiazolium ion
STEP 2. Nucleophilic attack by catalyst

$+ \ ^-OH$

STEP 3. Masked alkanoyl anion formation

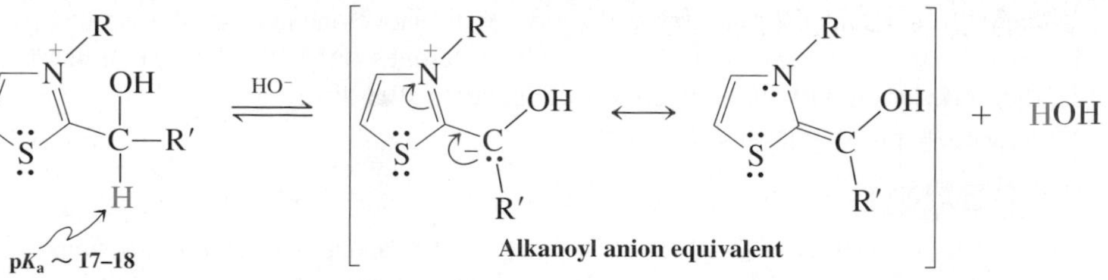

$pK_a \sim 17–18$

Alkanoyl anion equivalent

STEP 4. Nucleophilic attack on second aldehyde

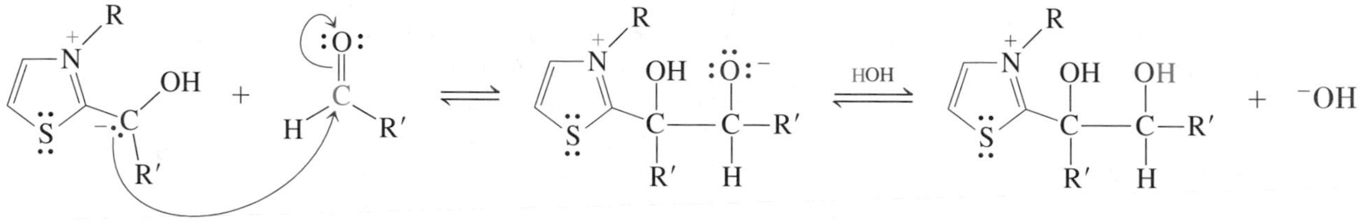

STEP 5. Liberation of α-hydroxyketone

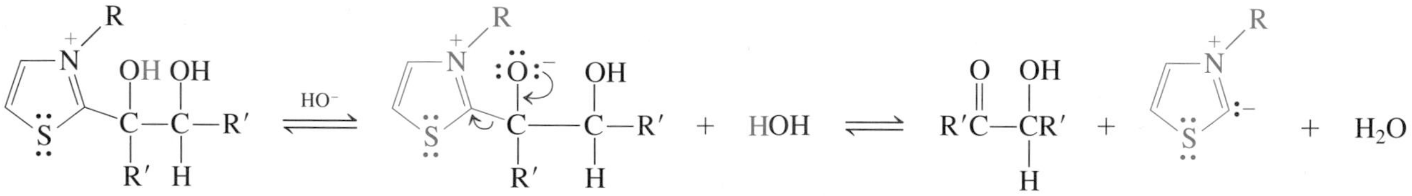

The product alcohol of step 2 is unique in that the thiazolium unit is a substituent. This group is electron withdrawing and increases the acidity of the adjacent proton. Deprotonation leads to an unusually stable masked alkanoyl anion. Nucleophilic attack by this anion on another molecule of aldehyde, followed by loss of the thiazolium substituent, liberates the α-hydroxyketone.

Comparison of the thiazolium method for synthesis of α-hydroxyketones with the use of dithiacyclohexane anions is instructive. Thiazolium salts have the advantage in that they are needed in only catalytic amounts. However, their use is limited to the synthesis of molecules R–C–CH–R in which the two R groups are identical. The dithiacyclohexane method is more versatile and can be used to prepare a much wider variety of substituted α-hydroxyketones.

EXERCISE 23-21

Which of the following compounds can be prepared by using thiazolium ion catalysts, and which are accessible only from 1,3-dithiacyclohexane anions? Formulate syntheses of at least two of these substances, one by each route.

(a) 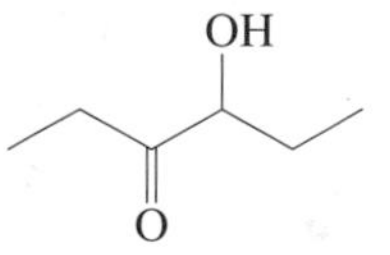   (b) 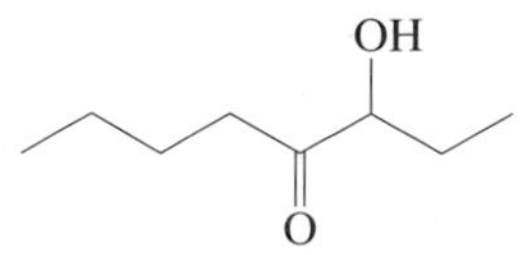   (c) 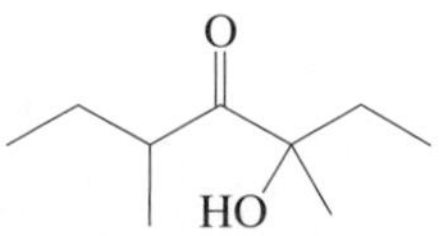   (d) 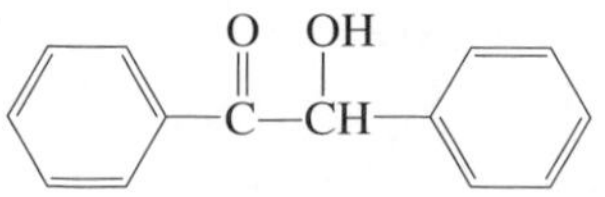   (e)

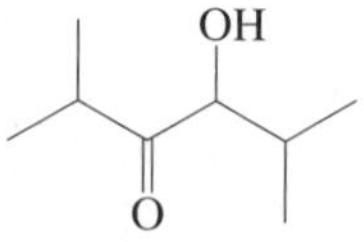

Thiamine: A Natural, Catalytically Active Thiazolium Ion

Thiamine
A=H

Thiamine pyrophosphate (TPP)

$$A = \begin{array}{c} O \\ \| \\ P \\ | \\ OH \end{array} - O - \begin{array}{c} O \\ \| \\ P \\ | \\ OH \end{array} - OH$$

The catalytic activity of thiazolium salts in aldehyde dimerization has an analogy in nature—the action of thiamine, or vitamin B_1 (see Chapter Opening). Thiamine, in the form of its pyrophosphate, is a coenzyme for several biochemical transformations that include intermediates of the type appearing in the synthesis of α-hydroxyketones. In the example below, the enzyme *transketolase* catalyzes the transfer of a (hydroxymethyl)carbonyl unit from the sugar xylulose (Chapter 24) to the sugar erythrose, thereby producing the new sugar fructose in the process. The active site of the enzyme (Chapter 26) contains a molecule of thiamine pyrophosphate (TPP).

Mechanistically, the deprotonated thiazolium ion first attacks the carbonyl group of the donor sugar (xylulose) to form an addition product, in a way completely analogous to addition to aldehydes. Because the donor sugar contains a hydroxy group next to the reaction site, this initial product can decompose by the reverse of the addition process to an aldehyde (glyceraldehyde) and a new thiamine intermediate. The latter species attacks another aldehyde (erythrose) to produce a new addition product. The catalyst then dissociates as thiamine pyrophosphate, releasing the new sugar molecule (fructose).

Transketolase-Catalyzed Biosynthesis of Fructose from Erythrose Using Thiamine Pyrophosphate

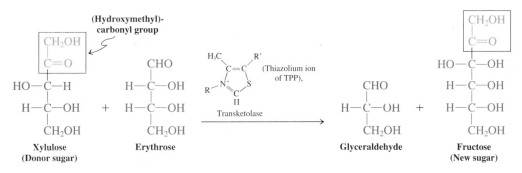

Sugar Activation

(*continued*)

CHEMICAL HIGHLIGHT **23-3** **Thiamine: A Natural, Catalytically Active Thiazolium Ion** *(continued)*

Removal of Old Aldehyde

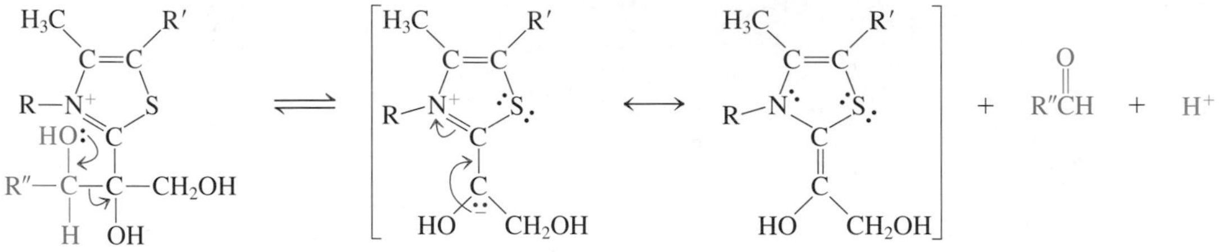

Introduction of New Aldehyde

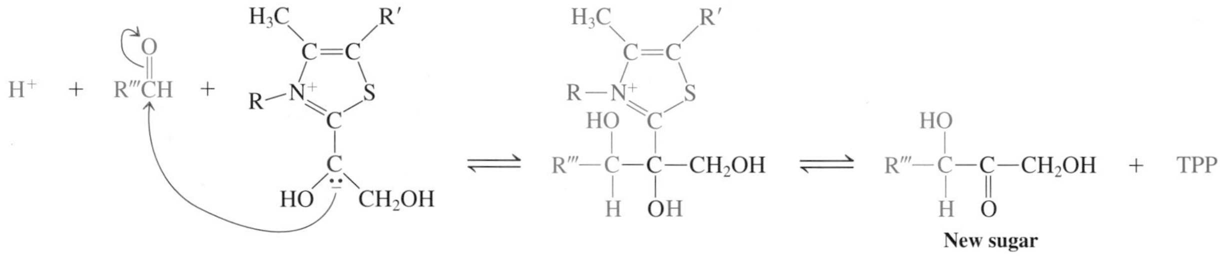

New sugar

In summary, α-hydroxyketones are available from addition of masked alkanoyl (acyl) anions to aldehydes and ketones. The conversion of aldehydes into the anions of the corresponding 1,3-dithiacyclohexanes (1,3-dithianes) illustrates the method of reverse polarization. The electrophilic carbon changes into a nucleophilic center, thereby allowing addition to an aldehyde or ketone carbonyl group. Thiazolium ions catalyze the dimerization of aldehydes, again through the transformation of the carbonyl carbon into a nucleophilic atom.

CHAPTER INTEGRATION PROBLEM

Propose syntheses of the two ketones in (a) and (b), using the methodology introduced in this chapter. Ground rules for these problems: You may use any organic compounds *provided that they contain no more than six carbons each.* Any inorganic reagents are allowable.

a. 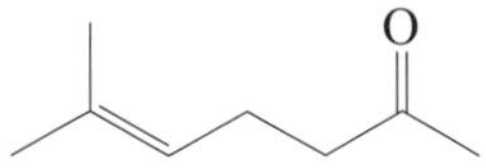 (an intermediate in the synthesis of the widely used perfume essence linalool—see Problem 40 of Chapter 13)

SOLUTION

Where do we start? We note that the target molecule is a 2-alkanone—a *methyl* ketone. As a result, we *may* be able to utilize the acetoacetic ester chemistry introduced in Section 23-2, which is specific in the preparation of methyl ketones. Recall that the acetoacetic ester synthesis affords ketones of the general formulas RCH_2COCH_3 and $RR'CHCOCH_3$, depending on whether we alkylate the starting ester once or twice. The desired product fits the first of these general formulas, so let us examine a retrosynthetic analysis along these lines. The strategic bond (Section 8-9) is the one between the R group and the C3 α-carbon of the ketone:

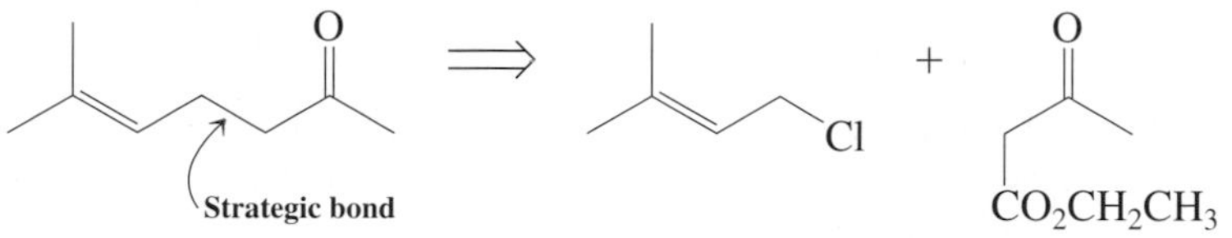

With a general strategy in hand, let us look at its details: A β-ketoester alkylation is necessary and we know (Section 18-4) that alkylations of enolates in general follow the S_N2 mechanism. Is the proposed substrate suitable for this process? We note that the leaving group, Cl, is located on a carbon atom that is both primary and allylic; hence this substance should be an excellent S_N2 substrate (Sections 6-10, 7-9, and 14-3). The final synthesis can therefore be laid out in a manner virtually identical with that of 2-heptanone in Section 23-2, replacing 1-bromobutane with the allylic chloride:

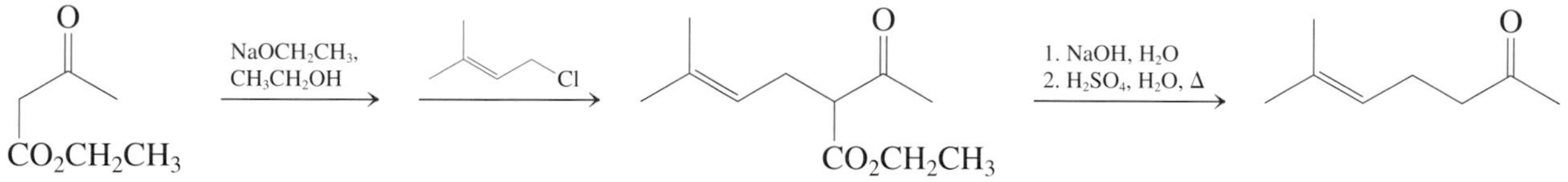

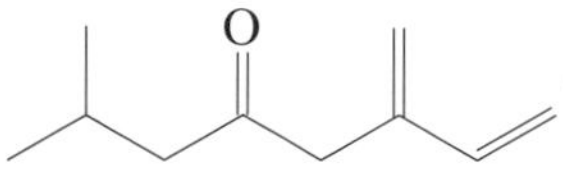

(an intermediate in the synthesis of the sex attractant of **b.** the bark beetle *Ips confusus*)

SOLUTION

We note immediately that this target molecule is not a methyl ketone; thus the acetoacetic ester synthesis is not conveniently applicable. We have made simple ketones before by adding organometallic reagents to aldehydes to give secondary alcohols, followed by oxidation (Section 8-9). By this route, the strategic bonds are those to the corresponding alcohol carbon, giving us two potential retrosynthetic disconnections, as shown here.

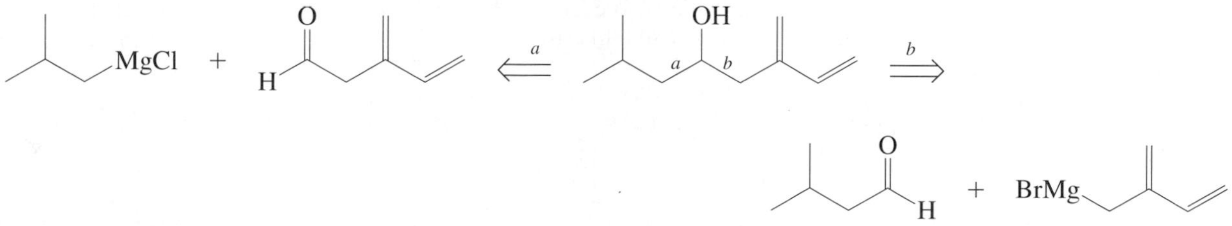

While the preceding provides perfectly good solutions, the problem requires us to employ methodology from the current chapter. Having ruled out acetoacetic ester chemistry, we have an alternative in the use of alkanoyl (acyl) anion equivalents, for which we saw examples applied to the synthesis of α-hydroxyketones (Section 23-4).

Can simple aldehydes and ketones be prepared with these equivalents as well? The answer is yes, by simple alkylation with RX, followed by hydrolysis. Thus,

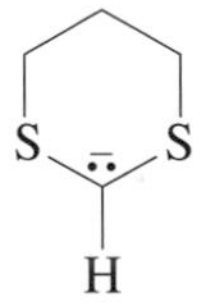 is the synthetic equivalent of 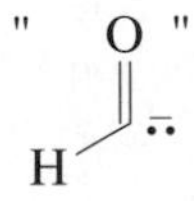 and 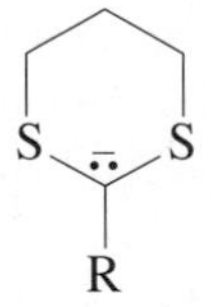 is the synthetic equivalent of 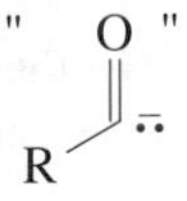

Turning to the problem at hand, we analyze retrosynthetically, identifying the two possible strategic bonds in the disubstituted 1,3-dithiacyclohexane corresponding to our target. The resulting monosubstituted dithianes are readily derived from the corresponding aldehydes by thioacetalization (Section 17-8):

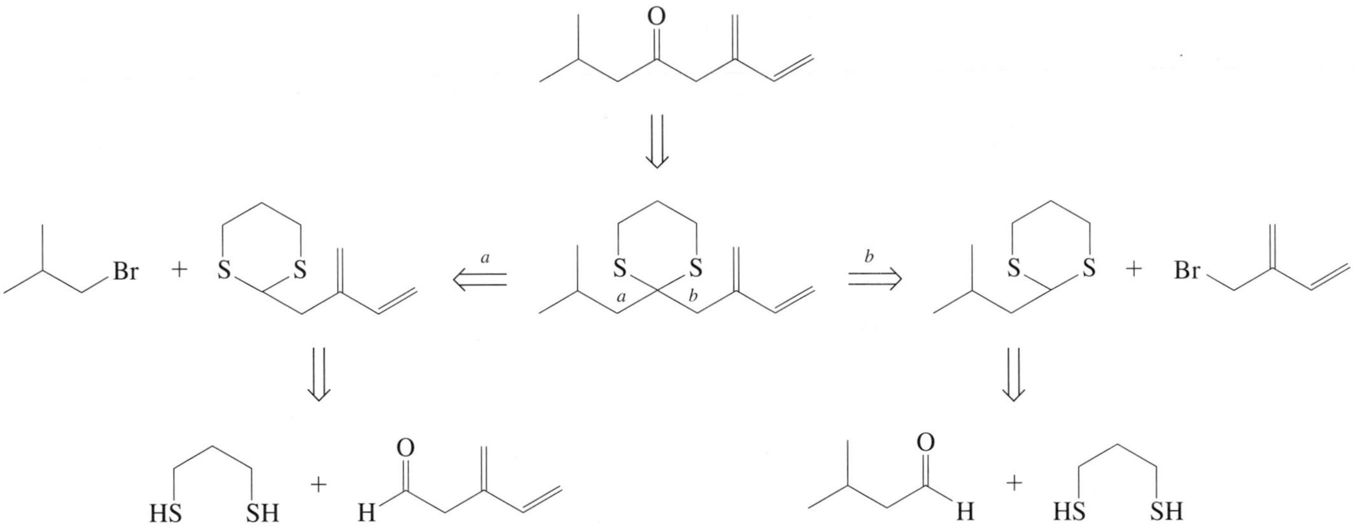

We have now to decide which strategic bond construction, *a* or *b*, is better before designing our final answer. Two considerations are important in this decision: (1) ease of bond formation and (2) size and structural and functional complexity of starting materials. Bond *a* forms in an S_N2 displacement of a branched, primary haloalkane—not the best situation, especially with strongly basic nucleophiles (Section 7-9), and dithiacyclohexanes have pK_a values in the 30s. Bond *b* employs a primary, allylic substrate—a far better alternative. Approach *b* is also preferable on the basis of our second consideration: two even-sized (five carbon units) pieces with separated functional groups. Our synthetic scheme for the solution is therefore as shown.

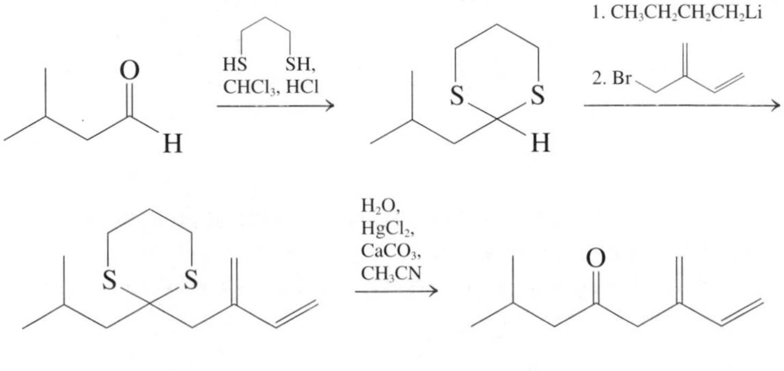

NEW REACTIONS

Synthesis of β-Dicarbonyl Compounds

1. Claisen Condensation (Section 23-1)

$$2\ CH_3COR \underset{\text{2. H}^+,\ H_2O}{\overset{\text{1. Na}^+\text{-OR, ROH}}{\rightleftharpoons}} CH_3CCH_2COR + ROH$$

Acidic

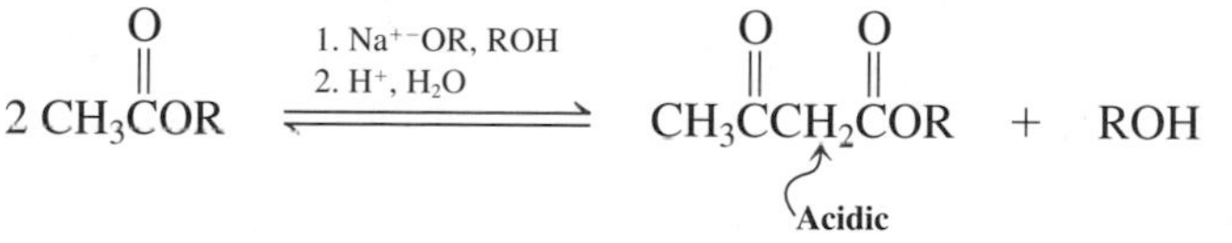

Endothermic reaction; equilibrium driven to anion of product by excess base

2. Dieckmann Condensation (Section 23-1)

$$(CH_2)_n \underset{\substack{-CO_2R \\ -CH_2CO_2R}}{} \xrightarrow[\text{2. H}^+,\ H_2O]{\text{1. Na}^+\text{-OR, ROH}} (CH_2)_n \underset{CHCO_2R}{\overset{C=O}{}} + ROH$$

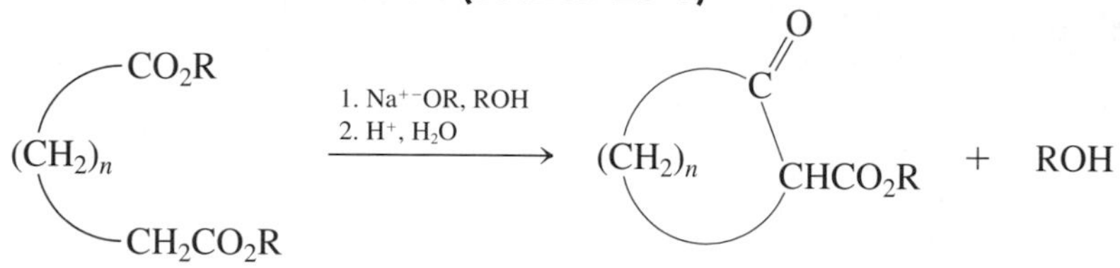

3. β-Diketone Synthesis (Section 23-1)

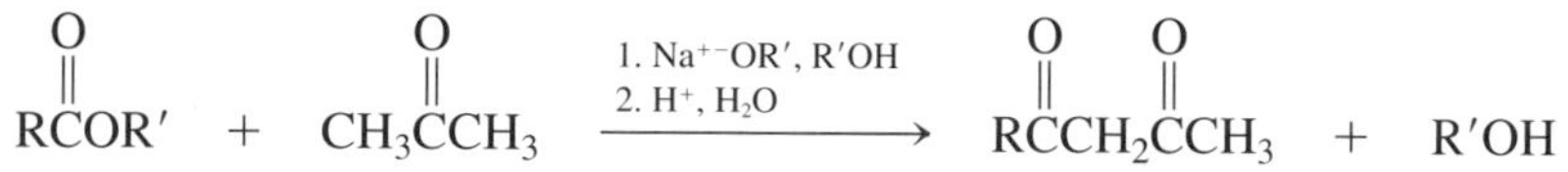

$$RCOR' + CH_3CCH_3 \xrightarrow[\text{2. H}^+,\ H_2O]{\text{1. Na}^+\text{-OR', R'OH}} RCCH_2CCH_3 + R'OH$$

Intramolecular

$$(CH_2)_n \underset{\substack{-CCH_3 \\ -CO_2R}}{} \xrightarrow[\text{2. H}^+,\ H_2O]{\text{1. Na}^+\text{-OR, ROH}} (CH_2)_n \underset{\substack{C=O \\ C=O}}{CH_2} + ROH$$

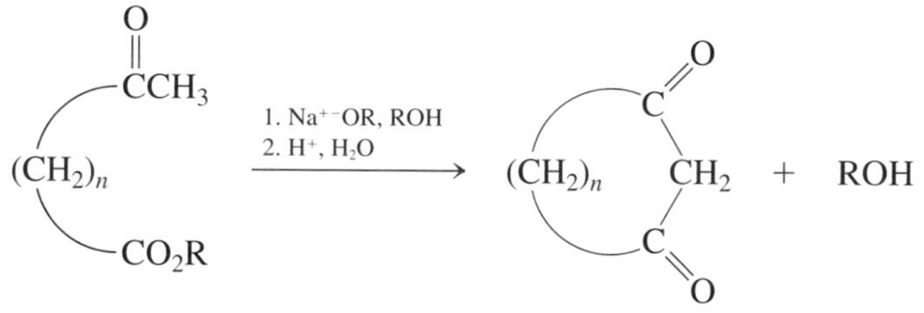

3-Ketoesters as Synthetic Building Blocks

4. Enolate Alkylation (Section 23-2)

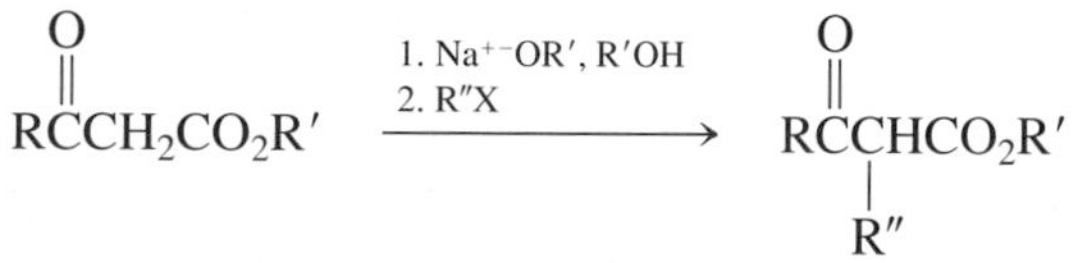

$$RCCH_2CO_2R' \xrightarrow[\text{2. R''X}]{\text{1. Na}^+\text{-OR', R'OH}} RCCHCO_2R' \atop R''$$

5. 3-Ketoacid Decarboxylation (Section 23-2)

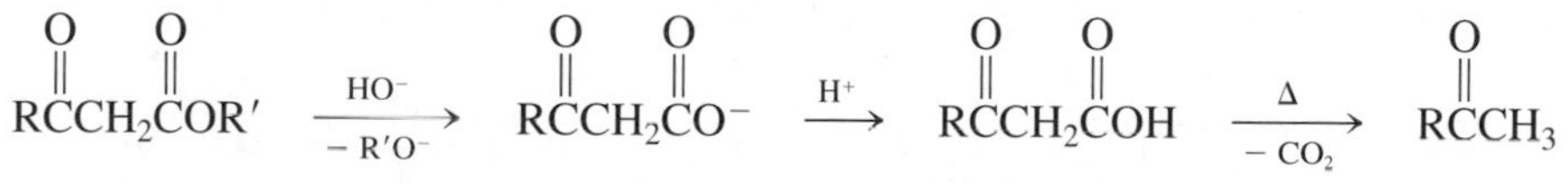

$$RCCH_2COR' \xrightarrow[-R'O^-]{HO^-} RCCH_2CO^- \xrightarrow{H^+} RCCH_2COH \xrightarrow[-CO_2]{\Delta} RCCH_3$$

6. Acetoacetic Ester Synthesis of Methyl Ketones (Section 23-2)

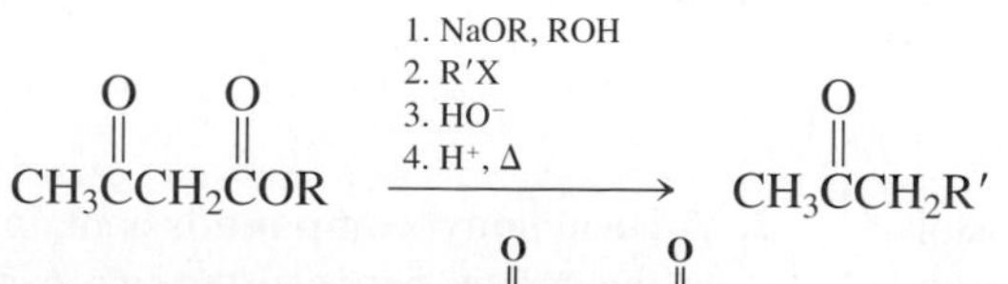

$$CH_3CCH_2COR \xrightarrow[\substack{\text{3. HO}^- \\ \text{4. H}^+,\ \Delta}]{\substack{\text{1. NaOR, ROH} \\ \text{2. R'X}}} CH_3CCH_2R'$$

R' = alkyl, alkanoyl (acyl), CH_2COR'', CH_2CR''
R'X = oxacyclopropane

7. Malonic Ester Synthesis of Carboxylic Acids (Section 23-2)

$$\underset{\text{ROCCH}_2\text{COR}}{\overset{\text{O} \quad\quad \text{O}}{\|\quad\quad\;\|}} \xrightarrow[\substack{\text{3. HO}^- \\ \text{4. H}^+,\, \Delta}]{\substack{\text{1. NaOR, ROH} \\ \text{2. R}'\text{X}}} \underset{\text{R}'\text{CH}_2\text{COH}}{\overset{\text{O}}{\|}}$$

R′ = alkyl, alkanoyl (acyl), CH_2COR'', CH_2CR''
R′X = oxacyclopropane

8. Michael Addition (Section 23-3)

Michael acceptor

Alkanoyl (Acyl) Anion Equivalents

9. 1,3-Dithiacyclohexane (1,3-Dithiane) Anions as Alkanoyl (Acyl) Anion Equivalents (Section 23-4)

$$R-\overset{\text{A}'}{\underset{\text{A}}{\text{C}}}:^- \quad \text{synthetic equivalent of} \quad \underset{R}{\overset{\text{O}}{\text{C}}}:^-$$

A = electron-withdrawing, conjugating, or polarizable group

10. Thiazolium Salts in Aldehyde Coupling (Section 23-4)

IMPORTANT CONCEPTS

1. The **Claisen condensation** is driven by the stoichiometric generation of a stable **β-dicarbonyl anion** in the presence of excess base.

2. **β-Dicarbonyl compounds** contain acidic hydrogens at the carbon between the two carbonyl groups because of the inductive electron-withdrawing effect of

the two neighboring carbonyl functions and because the anions resulting from deprotonation are resonance stabilized.

3. Although **mixed Claisen condensations** between esters are usually not selective, they can be so with certain substrates (nonenolizable esters, intramolecular versions, ketones).

4. 3-Ketoacids are unstable; they **decarboxylate** in a concerted process through an aromatic transition state. This property, in conjunction with the nucleophilic reactivity of 3-ketoester anions, allows the synthesis of substituted ketones and acids.

5. Because **alkanoyl (acyl) anions** are not directly available by deprotonation of aldehydes, they have to be made as masked reactive intermediates or stoichiometric reagents by transformations of functional groups.

PROBLEMS

22. Give the expected results of the reaction of each of the following molecules (or combinations of molecules) with excess $NaOCH_2CH_3$ in CH_3CH_2OH, followed by aqueous acidic work-up.

(a) $CH_3CH_2CH_2COOCH_2CH_3$

(b) $C_6H_5\overset{\overset{\displaystyle CH_3}{|}}{C}HCH_2COOCH_2CH_3$

(c) $C_6H_5CH_2\overset{\overset{\displaystyle CH_3}{|}}{C}HCOOCH_2CH_3$

(d) $CH_3CH_2O\overset{\overset{\displaystyle O}{||}}{C}(CH_2)_4\overset{\overset{\displaystyle O}{||}}{C}OCH_2CH_3$

(e) $CH_3CH_2O\overset{\overset{\displaystyle O}{||}}{C}\underset{\underset{\displaystyle CH_3}{|}}{C}H(CH_2)_4\overset{\overset{\displaystyle O}{||}}{C}OCH_2CH_3$

(f) $C_6H_5CH_2CO_2CH_2CH_3 + HCO_2CH_2CH_3$

(g) $C_6H_5CO_2CH_2CH_3 + CH_3CH_2CH_2CO_2CH_2CH_3$

(h)
$+ CH_3CH_2O\overset{\overset{\displaystyle O}{||}}{C}CH_2CH_2\overset{\overset{\displaystyle O}{||}}{C}OCH_2CH_3$

(i)
$+ CH_3CH_2O\overset{\overset{\displaystyle O}{||}}{C}-\overset{\overset{\displaystyle O}{||}}{C}OCH_2CH_3$

23. The following mixed Claisen condensation works best when one of the starting materials is present in large excess. Which of the two starting materials should be present in excess? Why? What side reaction will compete if the reagents are present in comparable amounts?

$$CH_3CH_2\overset{\overset{\displaystyle O}{||}}{C}OCH_3 + (CH_3)_2CH\overset{\overset{\displaystyle O}{||}}{C}OCH_3 \xrightarrow{\text{NaOCH}_3,\ \text{CH}_3\text{OH}} (CH_3)_2CH\overset{\overset{\displaystyle O}{||}}{C}\underset{\underset{\displaystyle CH_3}{|}}{C}H\overset{\overset{\displaystyle O}{||}}{C}OCH_3$$

24. Suggest a synthesis of each of the following β-dicarbonyl compounds by Claisen or Dieckmann condensations.

(a) 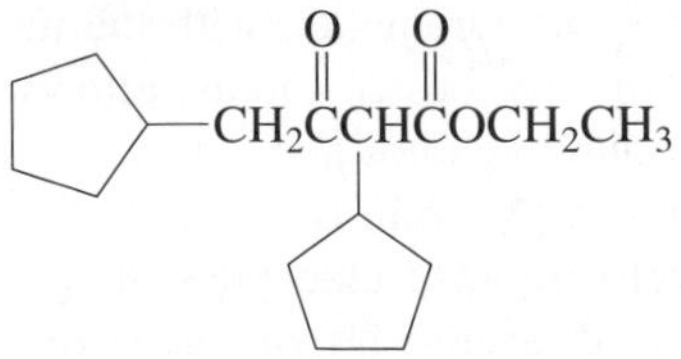 [structure: cyclopentyl-$CH_2CCHCOCH_2CH_3$ with two C=O and cyclopentyl substituent]

(b) $C_6H_5CCHCOCH_2CH_3$ with two C=O and C_6H_5

(c) H_3C [cyclohexanone structure] $CO_2CH_2CH_3$

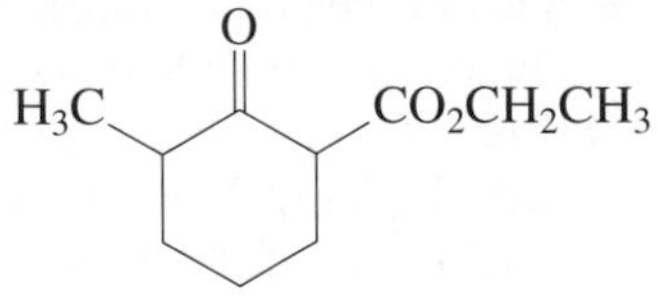

(d) 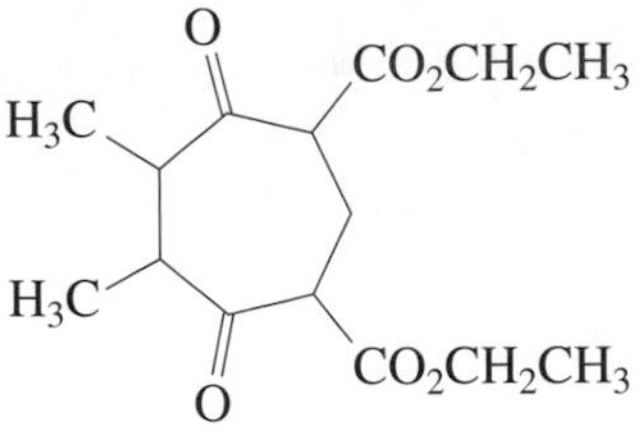 [cycloheptane structure with H_3C, $CO_2CH_2CH_3$, two C=O, H_3C, $CO_2CH_2CH_3$]

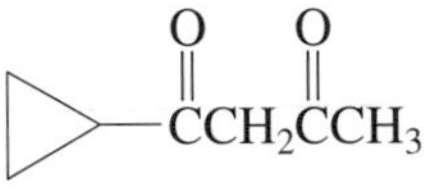

(e) $HCCCH_2COCH_2CH_3$ [three O's]

(f) $C_6H_5CCH_2CC_6H_5$ [two C=O]

(g) $CH_3CH_2OCCH_2COCH_2CH_3$ [two C=O]

(h) [cyclopropyl]$-CCH_2CCH_3$ [two C=O]

[margin structure: $HCCH_2CH$ with two C=O, labeled **Propanedial**]

25. Do you think that propanedial, shown in the margin, can be easily prepared by a simple Claisen condensation? Why or why not?

26. Devise a preparation of each of the following ketones by using the acetoacetic ester synthesis.

(a) 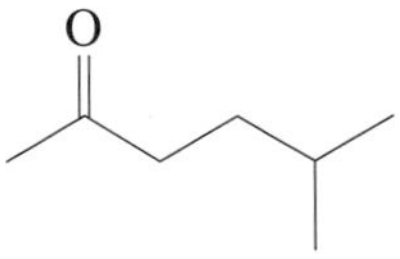 [structure: ketone with isobutyl group]

(b) 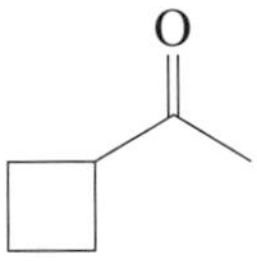 [structure: cyclobutyl methyl ketone]

(c) 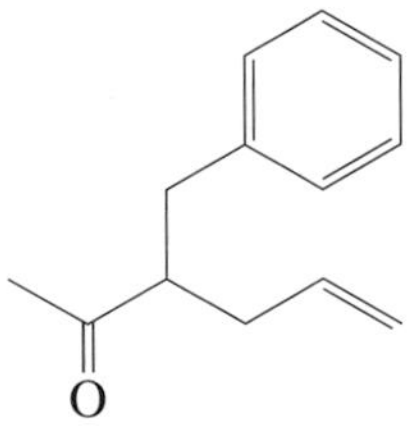 [structure: benzyl-substituted ketone with allyl group]

(d) 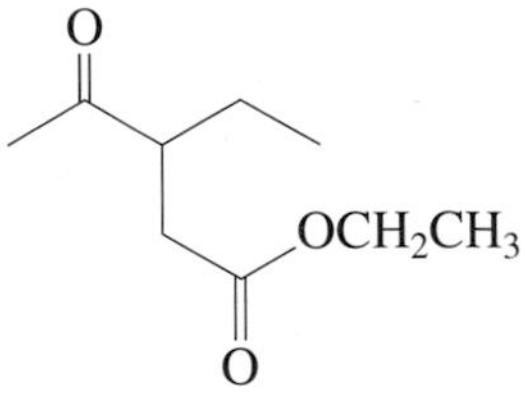 [structure: ketone with ethyl group and OCH_2CH_3 ester]

27. Devise a synthesis for each of the following four compounds by using the malonic ester synthesis.

(a) 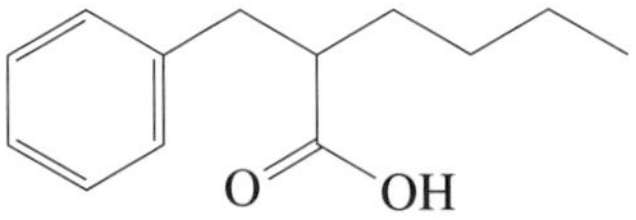 [structure: benzyl-substituted carboxylic acid with butyl chain, O OH]

(b) 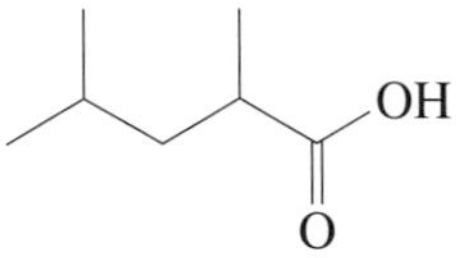 [structure: branched carboxylic acid, OH, O]

(c) H_2CCOOH / H_2CCOOH

(d) 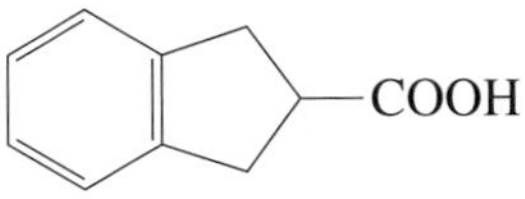 [indane structure]$-COOH$

28. Use the methods described in Section 23-3, with other reactions if necessary, to synthesize each of the following compounds. In each case, your starting materials should include one aldehyde or ketone and one β-dicarbonyl compound.

(a) 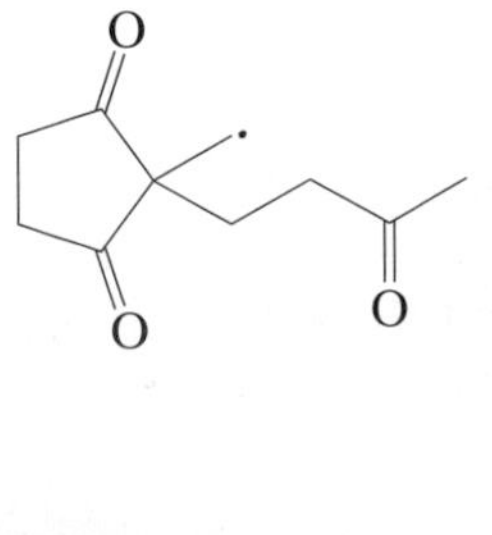 [structure: cyclopentane-1,3-dione with side chain ending in ketone]

(b) 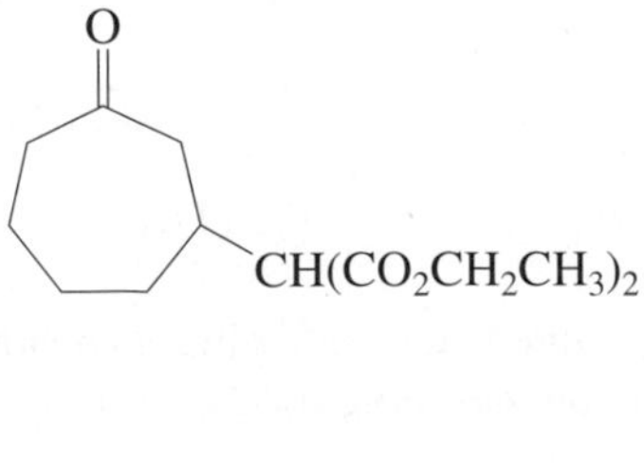 [structure: cycloheptanone with $CH(CO_2CH_2CH_3)_2$ substituent]

(c) 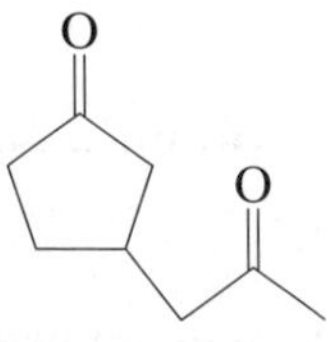 [structure: cyclopentanone with side chain ketone]

(**Hint:** A decarboxylation is necessary.)

29. Write out, in full detail, the mechanism of the Michael addition of malonic ester to 3-buten-2-one in the presence of ethoxide ion. Be sure to indicate all steps that are reversible. Does the overall reaction appear to be exo- or endothermic? Explain why only a catalytic amount of base is necessary.

30. Using the methods described in this chapter, design a multistep synthesis of each of the following molecules, making use of the indicated building blocks as the sources of all the carbon atoms in your final product.

(a) 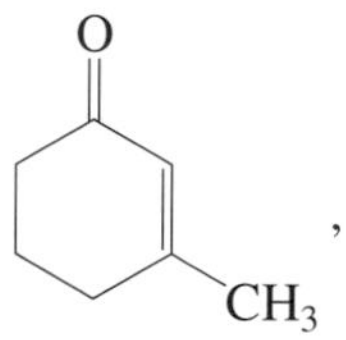 , from $CH_3CO_2CH_2CH_3$ and $CH_3\overset{O}{\overset{\|}{C}}CH=CH_2$

(b) 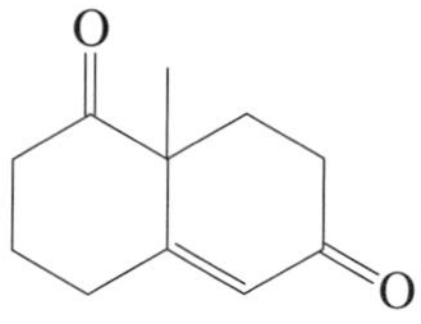 , from CH_3I, $CH_2(CO_2CH_2CH_3)_2$ and $CH_3\overset{O}{\overset{\|}{C}}CH=CH_2$ (**Hint:** First make 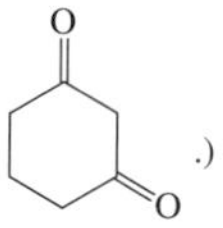 .)

(c) 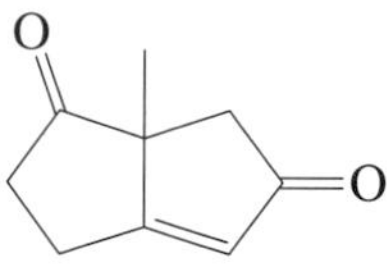 =O , from CH_3I, $CH_2(CO_2CH_2CH_3)_2$ and $BrCH_2\overset{O}{\overset{\|}{C}}CH_3$ (**Hint:** First make 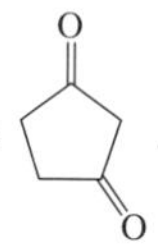 .)

31. Give the products of reaction of the following aldehydes with catalytic *N*-dodecylthiazolium bromide. **(a)** $(CH_3)_2CHCHO$; **(b)** C_6H_5CHO; **(c)** cyclohexanecarbaldehyde; **(d)** $C_6H_5CH_2CHO$.

32. Give the products of the following reactions.

(a) $C_6H_5CHO + HS(CH_2)_3SH \xrightarrow{BF_3}$ **(b)** Product of (a) + $CH_3CH_2CH_2CH_2Li \xrightarrow{THF}$

What are the results of reaction of the substance formed in (b) with each aldehyde in Problem 31, followed by hydrolysis in the presence of $HgCl_2$?

33. **(a)** On the basis of the following data, identify unknowns A, found in fresh cream prior to churning, and B, possessor of the characteristic yellow color and buttery odor of butter.
A: MS: m/z (relative abundance) = $88(M^{+\bullet}$, weak), 45(100), and 43(80).
  ^{1}H NMR: δ = 1.36 (d, J = 7 Hz, 3 H), 2.18 (s, 3 H), 3.73 (broad s, 1 H), 4.22 (q, J = 7 Hz, 1 H) ppm.
  IR: $\tilde{\nu}$ = 1718 and 3430 cm^{-1}.
B: MS: m/z (relative abundance) = 86(17) and 43(100).
  ^{1}H NMR: δ = 2.29 (s) ppm.
  IR: $\tilde{\nu}$ = 1708 cm^{-1}.
(b) What kind of reaction is the conversion of compound A into compound B? Does it make sense that this should take place in the churning of cream to make butter? Explain.
(c) Outline laboratory syntheses of compounds A and B, starting only with compounds containing two carbons. **(d)** The UV spectrum of compound A

has a λ_{max} at 271 nm, whereas that of compound B has a λ_{max} at 290 nm. (Extension of the latter absorption into the violet region of the visible spectrum is responsible for the yellow color of compound B.) Explain the difference in λ_{max}.

34. Write chemical equations to illustrate all primary reaction steps that can occur between a base such as ethoxide ion and a carbonyl compound such as acetaldehyde. Explain why the carbonyl carbon is not deprotonated to any appreciable extent in this system.

35. Nootkatone is found in grapefruit. Fill in the necessary steps in the following scheme to make nootkatone from 4-(1-methylethenyl)cyclohexanone.

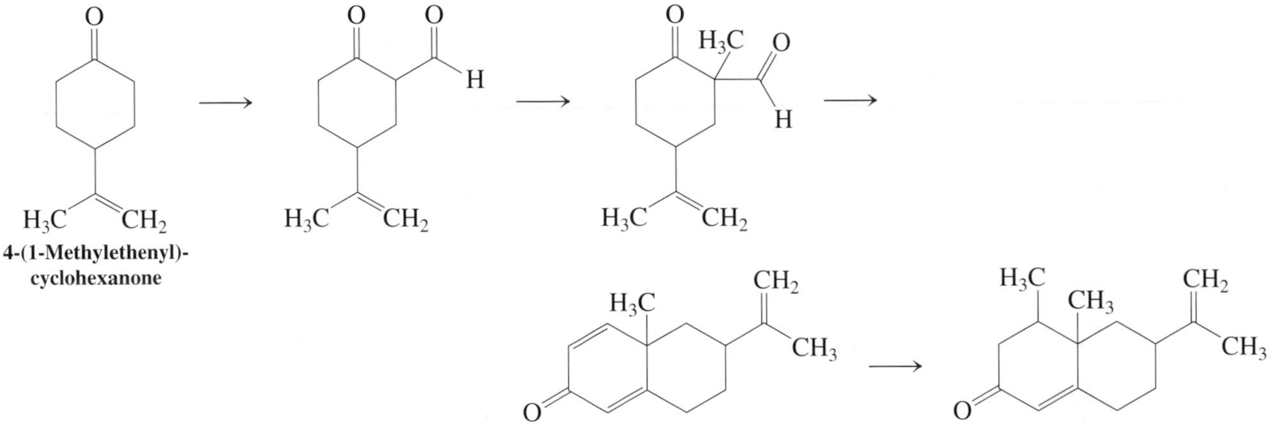

4-(1-Methylethenyl)-
cyclohexanone

Nootkatone

36. The following ketones cannot be synthesized by the acetoacetic ester method (why?), but they can be prepared by a modified version of it. The modification includes the preparation (by Claisen condensation) and use of an appropriate

$$\overset{\text{O}}{\overset{\|}{\text{R}}}\text{CCH}_2\overset{\text{O}}{\overset{\|}{\text{C}}}\text{OCH}_2\text{CH}_3,$$

3-ketoester, containing an R group that appears in the final product. Synthesize each of the following ketones. For each, show the structure and synthesis of the necessary 3-ketoester as well.

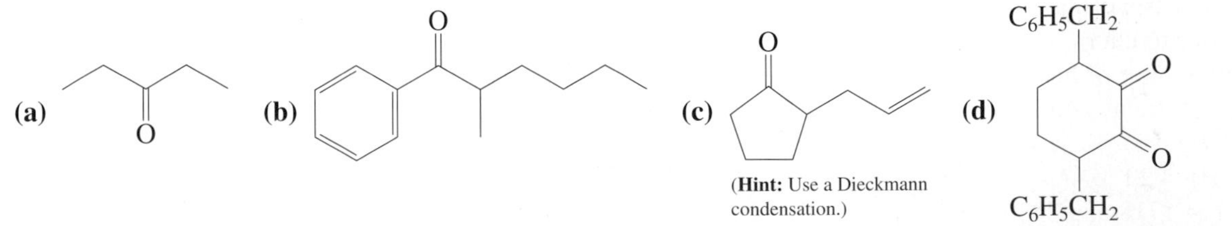

(a) (b) (c)

(**Hint:** Use a Dieckmann condensation.)

(d)

(**Hint:** Use a double Claisen condensation.)

37. Some of the most important building blocks for synthesis are very simple molecules. Although cyclopentanone and cyclohexanone are readily available commercially, an understanding of how they can be made from simpler molecules is instructive. The following are possible retrosynthetic analyses (Section 8-9) for both of these ketones. Using them as a guide, write out a synthesis of each ketone from the indicated starting materials.

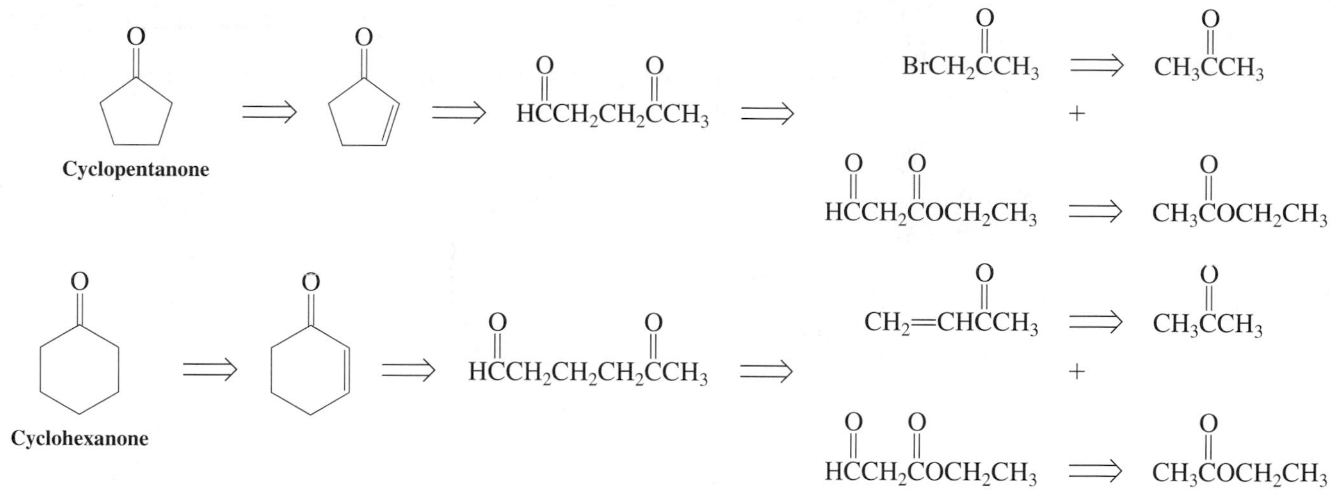

Cyclopentanone

Cyclohexanone

38. A short construction of the steroid skeleton (part of a total synthesis of the hormone estrone) is shown here. Formulate mechanisms for each of the steps. (**Hint:** A process similar to that taking place in the second step is presented in Problem 41 of Chapter 18.)

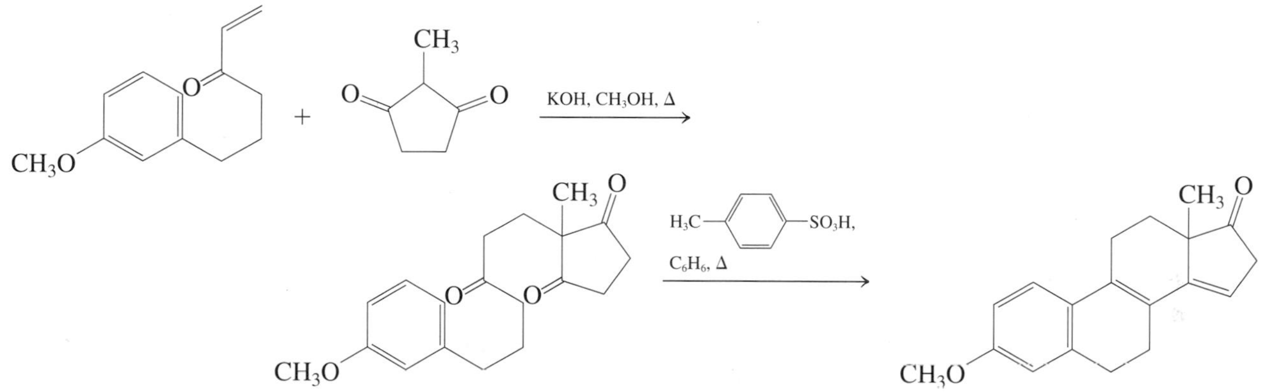

39. Using methods described in Section 23-4 (i.e., reverse polarization), propose a simple synthesis of each of the following molecules.

(a) CH_2=CHCHCCH$_2$C$_6$H$_5$ (b)   (c) CH$_3$CHCHCHO

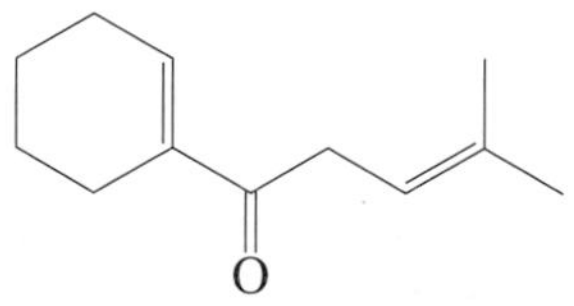

40. Propose a synthesis of ketone C, which was central in attempts to synthesize several antitumor agents. Start with aldehyde A, lactone B, and anything else you need.

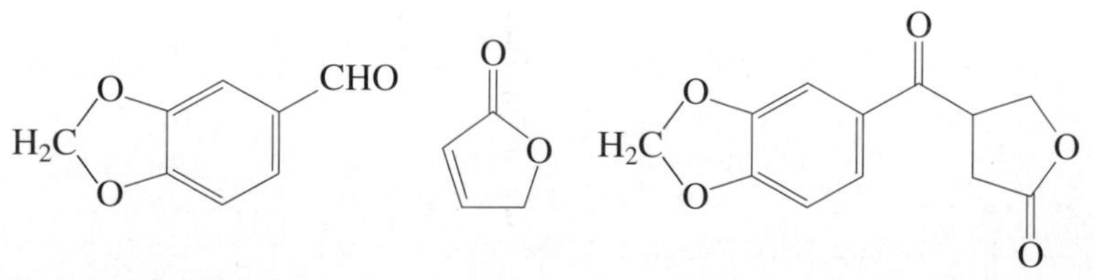

A   **B**   **C**

Team Problem

41. Split your team in two, each group to analyze one of the following reaction sequences by a mechanism (^{13}C = carbon-13 isotope).

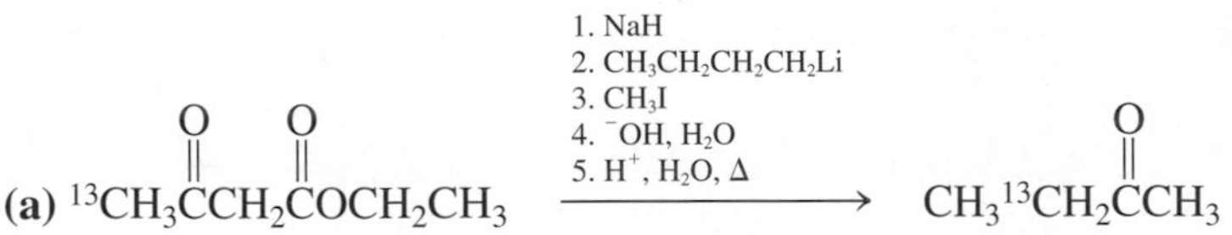

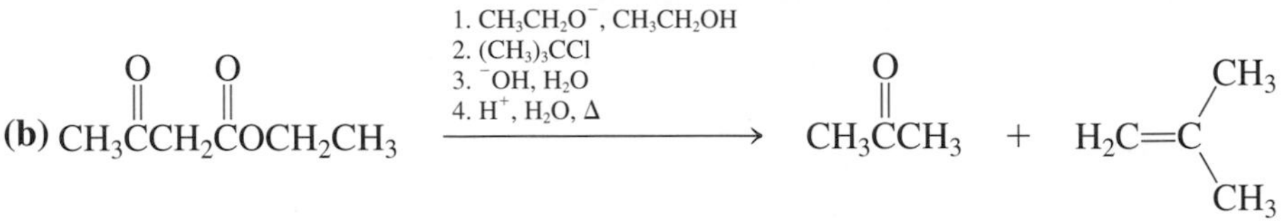

Reconvene and discuss your results. Specifically, address the position of the ^{13}C label in the product of (a) and the failure to obtain alkylation in (b).

As a complete team, also discuss the mechanism of the following transformation. (**Hint:** For the first step, a minimum of three equivalents of KNH_2 is required.)

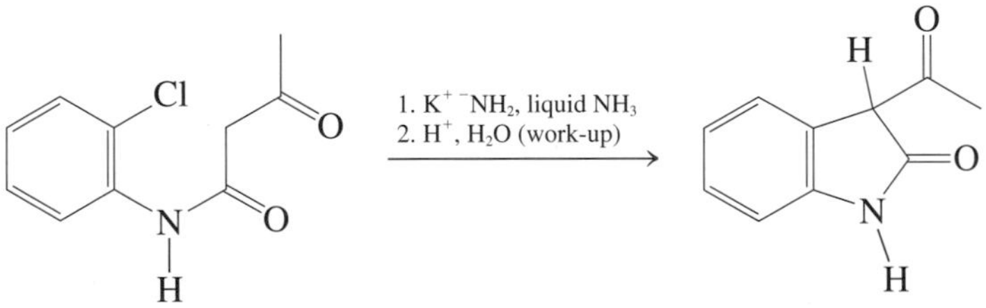

78%

Preprofessional Problems

42. Two of the following four compounds are more acidic than CH_3OH (i.e., two of these have K_a *greater* than methanol). Which ones?

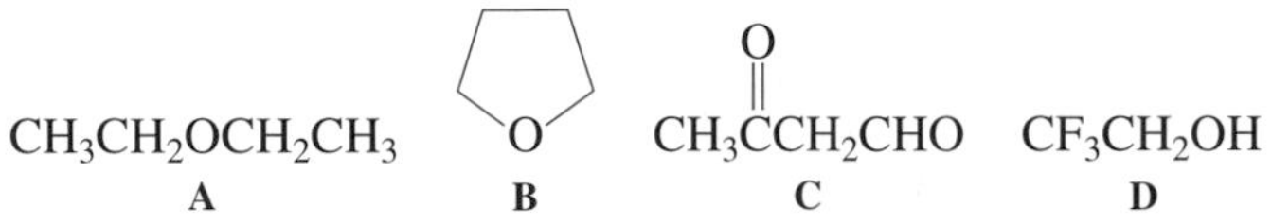

$$CH_3CH_2OCH_2CH_3 \qquad \text{B} \qquad CH_3CCH_2CHO \qquad CF_3CH_2OH$$

A   B   C   D

(a) A and B; **(b)** B and C; **(c)** C and D; **(d)** D and A; **(e)** D and B.

43. The reaction of ethyl butanoate with sodium ethoxide in CH_3CH_2OH gives

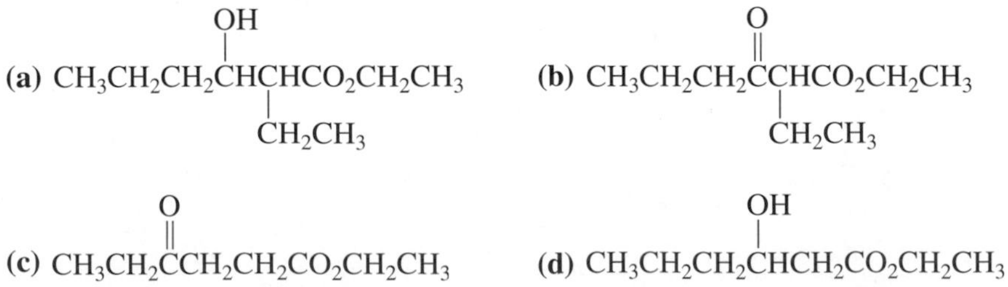

44. When acid A is heated to 230°C, CO_2 and H_2O are evolved and a new compound is formed. Which one?

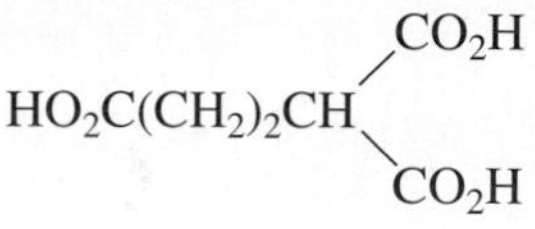

$$HO_2C(CH_2)_2CH \overset{CO_2H}{\underset{CO_2H}{}}$$

A

(a) 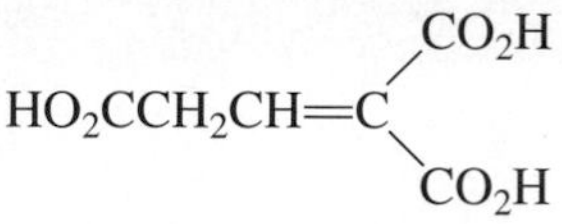 $HO_2CCH_2CH{=}C \overset{CO_2H}{\underset{CO_2H}{}}$

(b) $HO_2CCH_2CH_2CH_2CH_3$

(c) 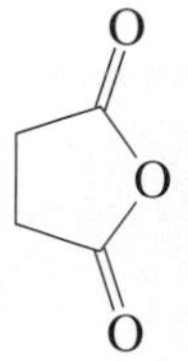 [structure: succinic anhydride]

(d) $CH_3CH_2CH(CO_2H)_2$

(e) 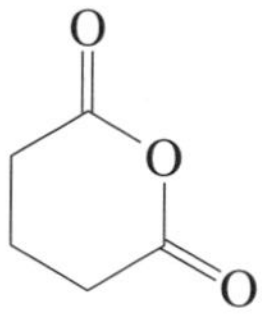 [structure: six-membered lactone ring]

45. A compound with m.p. = −22°C has a parent peak in its mass spectrum at $m/z = 113$. The ^1H NMR spectrum shows absorptions at $\delta = 1.2$ (t, 3H), 3.5 (s, 2H), and 4.2 (q, 2H). The IR spectrum exhibits significant bands at $\tilde{\nu} = 3000$, 2250, and 1750 cm^{-1}. What is its structure?

(a) 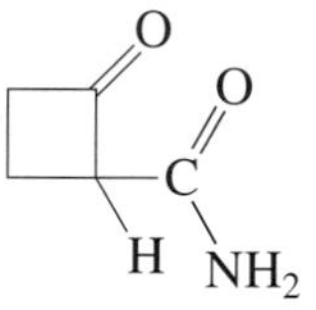 [structure: cyclobutanone with CHNH$_2$ and C=O substituent]

(b) [structure: cyclopentanone with CH=CH$_2$ and CN substituents]

(c) [structure: aziridine with OCH$_2$CH$_2$CH$_3$]

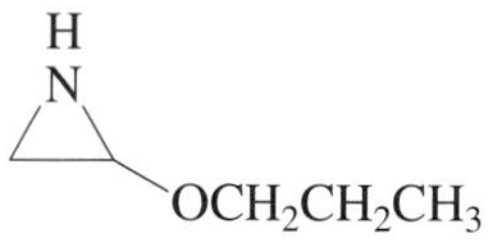

(d) 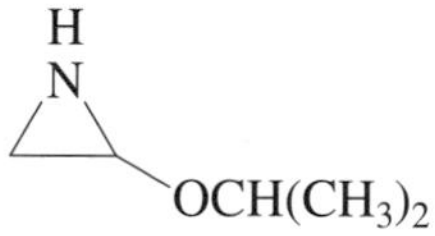 [structure: aziridine with OCH(CH$_3$)$_2$]

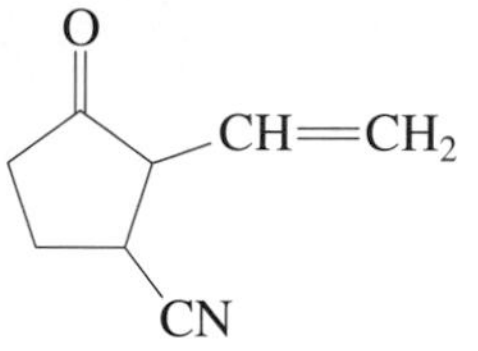

(e) $NCCH_2CO_2CH_2CH_3$

Carbohydrates

Polyfunctional Compounds in Nature

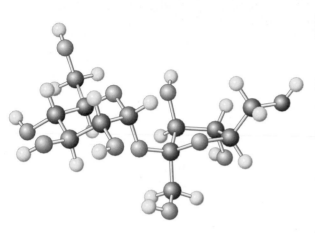

In everyday life, when we say the word "sugar" we usually refer to sucrose, the most widely occurring disaccharide in nature. Sucrose is prepared commercially in pure form in greater quantities than any other chemical substance.

Take a piece of bread and place it in your mouth. After a few minutes it will begin to taste distinctly sweet, as if you had added sugar to it. Indeed, in a way, this is what happened. The acid and enzymes in your saliva have cleaved the starch in the bread into its component units: glucose molecules. You all know glucose as dextrose or grape sugar. The polymer, starch, and its monomer, glucose, are two examples of carbohydrates.

Carbohydrates are most familiar to us as major contributors to our daily diets, in the form of sugars, fibers, and starches, such as bread, rice, and potatoes. In this capacity, they function as chemical energy-storage systems, being metabolized to water, carbon dioxide, and heat or other energy. Members of this class of compounds give structure to plants, flowers, vegetables, and trees. They also serve as building units of fats (Sections 19-13 and 20-5) and nucleic acids (Section 26-9). All are considered to be **polyfunctional,** because they possess multiple functional groups. Glucose, $C_6(H_2O)_6$, and many related simple members of this compound class form the building blocks of the complex carbohydrates and have the empirical formulas $C_n(H_2O)_n$, essentially hydrated carbon.

We shall first consider the structure and naming of the simplest carbohydrates—the sugars. We then turn our attention to their chemistry, which is governed by the presence of carbonyl and hydroxy functions along carbon chains of various lengths. We have already seen examples of the biosynthesis of carbohydrates (Chemical Highlights 18-3 and 23-3). Here we examine several methods for sugar synthesis and structural analysis. Finally, we describe a sampling of the many types of carbohydrates found in nature.

24-1 Names and Structures of Carbohydrates

The simplest carbohydrates are the sugars, or **saccharides.** As chain length increases, the increasing number of stereocenters gives rise to a multitude of diastereomers. Fortunately for chemists, nature deals mainly with only one of the possible series of enantiomers. Sugars are polyhydroxycarbonyl compounds, and so they form stable cyclic hemiacetals, which affords additional structural and chemical variety.

Sugars are classified as aldoses and ketoses

Carbohydrate is the general name for the monomeric (monosaccharides), dimeric (disaccharides), trimeric (trisaccharides), oligomeric (oligosaccharides), and polymeric (polysaccharides) forms of sugar (*saccharum,* Latin, sugar). A **monosaccharide,** or **simple sugar,** is an aldehyde or ketone containing at least two additional hydroxy groups. Thus, the two simplest members of this class of compounds are 2,3-dihydroxypropanal (glyceraldehyde) and 1,3-dihydroxypropanone (1,3-dihydroxyacetone). **Complex sugars** (Section 24-11) are those formed by the linkage of simple sugars through ether bridges.

Aldehydic sugars are classified as **aldoses;** those with a ketone function are called **ketoses.** On the basis of their chain length, we call sugars **trioses** (three carbons), **tetroses** (four carbons), **pentoses** (five carbons), **hexoses** (six carbons), and so on. Therefore, 2,3-dihydroxypropanal (glyceraldehyde) is an aldotriose, whereas 1,3-dihydroxypropanone is a ketotriose.

Glucose, also known as dextrose, blood sugar, or grape sugar (*glykys,* Greek, sweet), is a pentahydroxyhexanal and hence belongs to the class of aldohexoses. It occurs naturally in many fruits and plants and in concentrations ranging from 0.08 to 0.1% in human blood. A corresponding isomeric ketohexose is **fructose,** the sweetest natural sugar (some synthetic sugars are sweeter), which also is present in many fruits (*fructus,* Latin, fruit) and in honey. Another important natural sugar is the aldopentose **ribose,** a building block of the ribonucleic acids (Section 26-9).

$$
\begin{array}{c}
\text{CHO} \\
| \\
\text{H—C—OH} \\
| \\
\text{CH}_2\text{OH}
\end{array}
$$

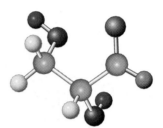

2,3-Dihydroxypropanal (Glyceraldehyde)

(An aldotriose)

$$
\begin{array}{c}
\text{CH}_2\text{OH} \\
| \\
\text{C=O} \\
| \\
\text{CH}_2\text{OH}
\end{array}
$$

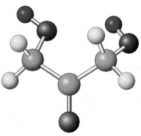

1,3-Dihydroxypropanone (1,3-Dihydroxyacetone)

(A ketotriose)

$$
\begin{array}{c}
\text{CHO} \\
| \\
\text{H—C—OH} \\
| \\
\text{HO—C—H} \\
| \\
\text{H—C—OH} \\
| \\
\text{H—C—OH} \\
| \\
\text{CH}_2\text{OH}
\end{array}
\qquad
\begin{array}{c}
\text{CH}_2\text{OH} \\
| \\
\text{C=O} \\
| \\
\text{HO—C—H} \\
| \\
\text{H—C—OH} \\
| \\
\text{H—C—OH} \\
| \\
\text{CH}_2\text{OH}
\end{array}
\qquad
\begin{array}{c}
\text{CHO} \\
| \\
\text{H—C—OH} \\
| \\
\text{H—C—OH} \\
| \\
\text{H—C—OH} \\
| \\
\text{CH}_2\text{OH}
\end{array}
$$

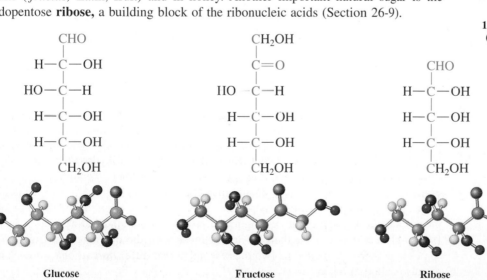

Glucose

(An aldohexose)

Fructose

(A ketohexose)

Ribose

(An aldopentose)

EXERCISE 24-1

To which class of sugars do the following monosaccharides belong?

	CHO	CH_2OH
CHO	HOCH	C=O
(a) HCOH	(b) HOCH	(c) HOCH
HCOH	HCOH	HCOH
CH_2OH	CH_2OH	CH_2OH
Erythrose	**Lyxose**	**Xylulose**

A **disaccharide** is derived from two monosaccharides by the formation of an ether (usually, acetal) bridge (Section 17-7). Hydrolysis regenerates the monosaccharides. Ether formation between a mono- and a disaccharide results in a trisaccharide, and repetition of this process eventually produces a natural polymer (polysaccharide). Polysaccharides constitute the framework of cellulose and starch (Section 24-12).

Most sugars are chiral and optically active

With the exception of 1,3-dihydroxypropanone, all the sugars mentioned so far contain at least one stereocenter. The simplest chiral sugar is 2,3-dihydroxypropanal (glyceraldehyde), with one asymmetric carbon. Its dextrorotatory form is found to be R and the levorotatory enantiomer S, as shown in the Fischer projections of the molecule. (See Section 5-4 for a discussion of this notation, which is used extensively to represent sugars.)

**Fischer Projections of the Two Enantiomers
of 2,3-Dihydroxypropanal (Glyceraldehyde)**

CHO

H———OH   is the same as   H—C◄OH   HO———H   is the same as   HO◄C◄H

CH_2OH CH_2OH CH_2OH CH_2O

(R)-(+)-**2,3-Dihydroxypropanal**   *(S)*-(−)-**2,3-Dihydroxypropanal**
[D-(+)-**Glyceraldehyde**]   [L-(−)-**Glyceraldehyde**]
$([\alpha]_D^{25°C} = +8.7)$ $([\alpha]_D^{25°C} = -8.7)$

Even though R and S nomenclature is perfectly satisfactory for naming sugars, an older system is still in general use. It was developed before the absolute configuration of sugars was established, and it relates all sugars to 2,3-dihydroxypropanal (glyceraldehyde). Instead of R and S, it uses the prefixes D for the (+) enantiomer of glyceraldehyde and L for the (−) enantiomer (Section 5-3). Those monosaccharides whose *highest-numbered stereocenter* (i.e., the one farthest from the aldehyde or keto group) has the same absolute configuration as that of D-(+)-2,3-dihydroxypropanal [D-(+)-glyceraldehyde] are then labeled D; those with the opposite configuration at that stereocenter are named L. Two diastereomers that differ *only at one stereocenter* are also called **epimers.**

Designation of a D and an L Sugar

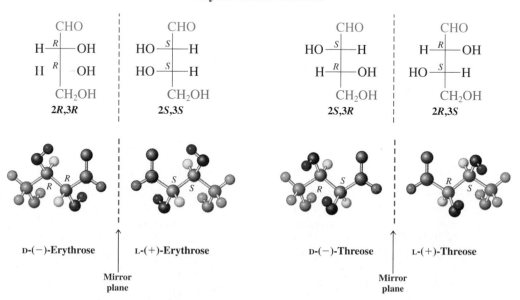

CHO
H——OH
HO——H
H——OH

Highest-numbered stereocenter
H——OH
CH₂OH
D-Aldose

CH₂OH
══O
H——OH
H——OH
HO——H
CH₂OH
L-Ketose

The D,L nomenclature divides the sugars into two groups. As the number of stereocenters increases, so does the number of stereoisomers. For example, the aldotetrose 2,3,4-trihydroxybutanal has two stereocenters and hence may exist as four stereoisomers: two diastereomers, each as a pair of enantiomers.

Like many natural products, these diastereomers have common names that are often used, mainly because the complexity of these molecules leads to long systematic names. This chapter will therefore deviate from our usual procedure of labeling molecules systematically. The isomer of 2,3,4-trihydroxybutanal with 2*R*,3*R* configuration is called erythrose; its diastereomer, threose. Note that each of these isomers has two enantiomers, one belonging to the family of the D sugars, its mirror image to the L sugars. The sign of the optical rotation is not correlated with the D and L label (just as in the *R,S* notation; see Section 5-3). For example, D-glyceraldehyde is dextrorotatory, but D-erythrose is levorotatory.

Diastereomeric 2,3,4-Trihydroxybutanals: Erythrose and Threose

CHO
H—*R*—OH
II *R*——OH
CH₂OH
2R,3R

D-(−)-Erythrose

CHO
HO—*S*—H
HO—*S*—H
CH₂OH
2S,3S

L-(+)-Erythrose

Mirror plane

CHO
HO—*S*—H
H—*R*—OH
CH₂OH
2S,3R

D-(−)-Threose

CHO
H—*R*—OH
HO—*S*—H
CH₂OH
2R,3S

L-(+)-Threose

Mirror plane

An aldopentose has three stereocenters and hence $2^3 = 8$ stereoisomers. There are $2^4 = 16$ such isomers in the group of aldohexoses. Why then use the D,L nomenclature even though it designates the absolute configuration of only one stereocenter?

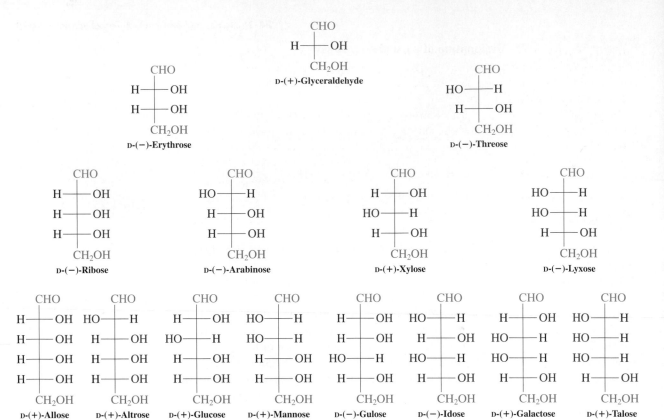

CHO
H——OH
CH₂OH
D-(+)-Glyceraldehyde

CHO
H——OH
H——OH
CH₂OH
D-(−)-Erythrose

CHO
HO——H
H——OH
CH₂OH
D-(−)-Threose

CHO
H——OH
H——OH
H——OH
CH₂OH
D-(−)-Ribose

CHO
HO——H
H——OH
H——OH
CH₂OH
D-(−)-Arabinose

CHO
H——OH
HO——H
H——OH
CH₂OH
D-(+)-Xylose

CHO
HO——H
HO——H
H——OH
CH₂OH
D-(−)-Lyxose

CHO
H——OH
H——OH
H——OH
H——OH
CH₂OH
D-(+)-Allose

CHO
HO——H
H——OH
H——OH
H——OH
CH₂OH
D-(+)-Altrose

CHO
H——OH
HO——H
H——OH
H——OH
CH₂OH
D-(+)-Glucose

CHO
HO——H
HO——H
H——OH
H——OH
CH₂OH
D-(+)-Mannose

CHO
H——OH
H——OH
HO——H
H——OH
CH₂OH
D-(−)-Gulose

CHO
HO——H
H——OH
HO——H
H——OH
CH₂OH
D-(−)-Idose

CHO
H——OH
HO——H
HO——H
H——OH
CH₂OH
D-(+)-Galactose

CHO
HO——H
HO——H
HO——H
H——OH
CH₂OH
D-(+)-Talose

FIGURE 24-1

D-Aldoses (up to the aldohexoses), their signs of rotation, and their common names.

CH₂OH
═O
CH₂OH
1,3-Dihydroxypropanone

CH₂OH
═O
H——OH
CH₂OH
D-(−)-Erythrulose

CH₂OH
═O
H——OH
H——OH
CH₂OH
D-(+)-Ribulose

CH₂OH
═O
HO——H
H——OH
CH₂OH
D-(+)-Xylulose

CH₂OH
═O
H——OH
H——OH
H——OH
CH₂OH
D-(+)-Psicose

CH₂OH
═O
HO——H
H——OH
H——OH
CH₂OH
D-(−)-Fructose

CH₂OH
═O
H——OH
HO——H
H——OH
CH₂OH
D-(+)-Sorbose

CH₂OH
═O
HO——H
HO——H
H——OH
CH₂OH
D-(−)-Tagatose

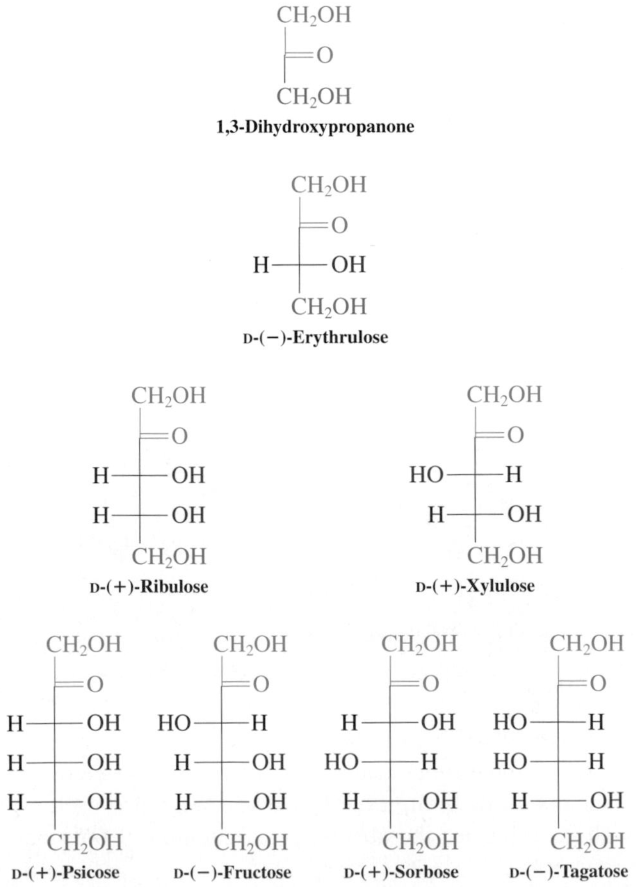

FIGURE 24-2

D-Ketoses (up to the ketohexoses), their signs of rotation, and their common names.

Probably because *almost all naturally occurring sugars have the D configuration.* Evidently somewhere in the structural evolution of the sugar molecules, nature "chose" only one configuration for one end of the chain. The amino acids are another example of such selectivity (Chapter 26).

Figure 24-1 shows Fischer projections of the series of D-aldoses up to the aldohexoses. To prevent confusion, chemists have adopted a standard way to draw these projections: The carbon chain extends vertically, and the aldehyde terminus is placed at the top. In this convention, the hydroxy group at the highest-numbered stereocenter (at the bottom) points to the right in all D sugars. Figure 24-2 shows the analogous series of ketoses.

EXERCISE 24-2

Give a systematic name for **(a)** D-(−)-ribose and **(b)** D-(+)-glucose. Remember to assign the *R* and *S* configuration at each stereocenter.

EXERCISE 24-3

Redraw the dashed-wedged line structure of sugar A (shown in the margin) as a Fischer projection and find its common name in Figure 24-1.

In summary, the simplest carbohydrates are sugars, which are polyhydroxy aldehydes (aldoses) and ketones (ketoses). They are classified as D when the highest-numbered stereocenter is *R*, L when it is *S*. Sugars related to each other by inversion at one stereocenter are called epimers. Most of the naturally occurring sugars belong to the D family.

24-2 Conformations and Cyclic Forms of Sugars

So far, the structures of monosaccharides have been represented by Fischer projections. Dashed-wedged line structures are more realistic, but we must be very careful when we convert a Fischer projection into the correct dashed-wedged line representation (and vice versa). This section describes this procedure and then continues with a discussion of the cyclic isomers that exist in solutions of simple sugars.

Fischer projections depict all-eclipsed conformations

Recall (Section 5-4) that the Fischer projection represents the molecule in an *all-eclipsed arrangement.* It can be translated into an all-eclipsed dashed-wedged line picture.

**Fischer Projection and Dashed-Wedged Line Structures
for D-(+)-Glucose**

| Fischer projection | All-eclipsed dashed-wedged line structure | All-staggered dashed-wedged line structure |

A molecular model will help you see that the all-eclipsed form actually possesses a roughly circular shape. Notice that the groups on the *right* of the carbon chain in the original Fischer projection now project *downward* in the dashed-wedged line structure. Rotation of two alternate internal carbons by 180° (C3 and C5 in this example) gives the all-staggered conformation.

Sugars form intramolecular hemiacetals

Sugars are hydroxycarbonyl compounds that should be capable of intramolecular hemiacetal formation (see Section 17-7). Indeed, glucose and the other hexoses, as well as the pentoses, exist as an equilibrium mixture with their cyclic hemiacetal isomers, in which the latter strongly predominate. In principle, any one of the five hydroxy groups could add to the carbonyl group of the aldehyde. However, three- and four-membered rings are too strained and, although five-membered ring formation is known, six-membered rings are usually the preferred product.

To correctly depict a sugar in its cyclic form, draw the dashed-wedged line representation of the all-eclipsed structure. Rotation of C5 places its hydroxyl group in position to form a six-membered cyclic hemiacetal by addition to the C1 aldehyde carbon. Similarly, a five-membered ring can be made by rotation of C4 to place its OH group in position to bond to C1. This procedure is general for all sugars in the D series.

Cyclic Hemiacetal Formation by Glucose

D-Glucose

D-Glucofuranose
(Less stable)

D-Glucopyranose
(More stable)

(Groups on the *right* in the original
Fischer projection [circled] *point downward*
in the cyclic hemiacetal except at C5, which
has been rotated)

EXERCISE 24-4

Draw the Fischer projection of L-(−)-glucose and illustrate its transformation into the corresponding six-membered cyclic hemiacetal.

The six-membered ring structure of a monosaccharide is called a **pyranose,** a name derived from *pyran,* a six-membered cyclic ether (see Sections 9-6 and 25-1). Sugars in the five-membered ring form are called **furanoses,** from *furan* (Section 25-3). In contrast with glucose, which exists primarily as the pyranose, fructose forms both fructopyranose and fructofuranose in a rapidly equilibrating 68:32 mixture.

Pyran   **Furan**

Cyclic Hemiacetal Formation by Fructose

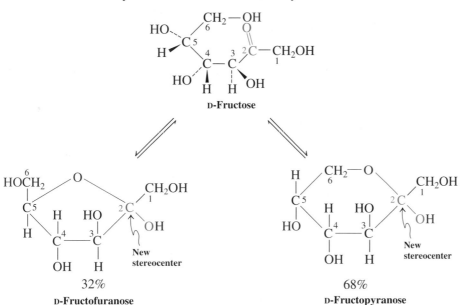

Note that, on cyclization, the carbonyl carbon turns into a new stereocenter. As a consequence, hemiacetal formation leads to *two* new compounds, two diastereomers differing in the configuration of the hemiacetal group. If that configuration is *S*, the sugar is labeled α; when it is *R*, it is called β. Hence, for example, glucose may form α- or β-glucopyranose or -furanose. Because this type of diastereomer formation is unique to sugars, such isomers have been given a separate name: **anomers.** The new stereocenter is called the **anomeric carbon.**

EXERCISE 24-5

The anomers α- and β-glucopyranose should form in equal amounts because they are enantiomers. True or false? Explain your answer.

Fischer, Haworth, and chair cyclohexane projections help depict cyclic sugars

How can we represent the stereochemistry of the cyclic forms of sugars? One approach uses Fischer projections. We simply draw elongated lines to indicate the bonds formed on cyclization, preserving the basic "grid" of the original formula.

Adapted Fischer Projections of Glucopyranoses

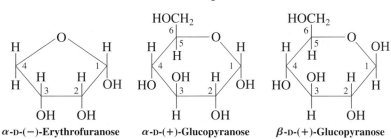

In the Fischer projection
of the α form, the
anomeric OH points
toward the *right*.
In the Fischer projection
of the β form, the
anomeric OH is on the
left.

α-D-(+)-Glucopyranose   β-D-(+)-Glucopyranose
(m.p. 146°C)   (m.p. 150°C)

Haworth* projections more accurately represent the real three-dimensional structure of the sugar molecule. The cyclic ether is written in line notation as a pentagon
or a hexagon, the anomeric carbon placed on the right, and the ether oxygen put on
top. The substituents located above or below the ring are attached to vertical lines. In
relating the Haworth projection to a three-dimensional structure, we understand the
ring bond at the bottom (between C2 and C3) to be *in front* of the plane of the paper, and the ring bonds containing the oxygen are understood to be in back.

Haworth Projections

Groups on the *right* in
the Fischer projection
point *downward* in the
Haworth formula.

α-D-(−)-Erythrofuranose   α-D-(+)-Glucopyranose   β-D-(+)-Glucopyranose

In a Haworth projection, the α anomer has the OH group at the anomeric carbon
pointing down, whereas the β anomer has it pointing up.

EXERCISE 24-6

Draw the structure of **(a)** α-D-fructofuranose; **(b)** β-D-glucofuranose; and
(c) β-D-arabinopyranose.

Haworth projections are used extensively in the literature, but here, to make use
of our knowledge of conformation (Sections 4-3 and 4-4), the cyclic forms of sugars
will be presented as envelope (for furanoses) or chair (for pyranoses) conformations.
As in Haworth notation, the ether oxygen will usually be placed top right, and the
anomeric carbon at the right vertex of the envelope or chair.

---

*Sir W. Norman Haworth (1883–1950), University of Birmingham, England, Nobel Prize 1937
(chemistry).

Conformational Pictures of Glucofuranose and -pyranose

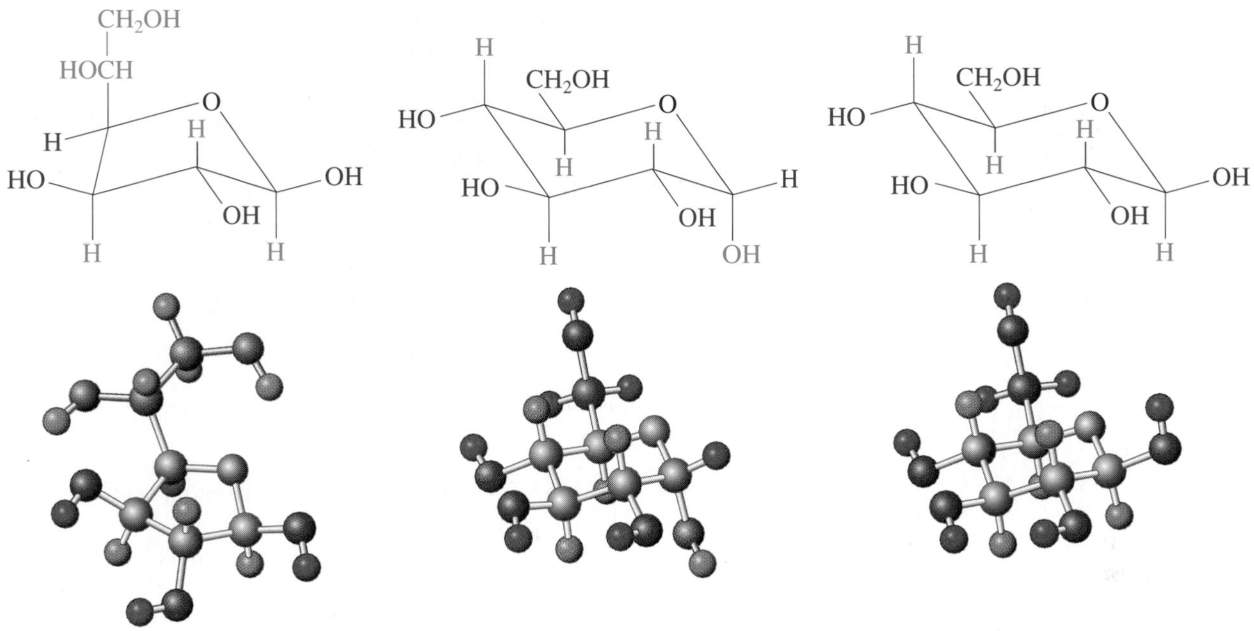

β-D-Glucofuranose

α-D-Glucopyranose

β-D-Glucopyranose

Although there are exceptions, *most aldohexoses adopt the chair conformation* that places the bulky hydroxymethyl group at the C5 terminus *in the equatorial position.* For glucose, this preference means that, in the α form, four of the five substituents can be equatorial, and one is forced to lie axial; in the β form, *all* substituents can be equatorial. This situation is unique for glucose; the other seven D aldohexoses (see Figure 24-1) contain one or more axial substituents.

EXERCISE 24-7

Using the values in Table 4-3, estimate the difference in free energy between the all-equatorial conformer of β-D-glucopyranose and that obtained by ring flip (assume that $\Delta G^\circ_{CH_2OH} = \Delta G^\circ_{CH_3} = 1.7$ kcal mol^{-1} and that the ring oxygen mimics a CH_2 group).

In summary, the hexoses and pentoses can take the form of five- or six-membered cyclic hemiacetals. These structures rapidly interconvert through the open-chain poly-hydroxyaldehyde or ketone, with the equilibrium usually favoring the six-membered (pyranose) ring.

24-3 Anomers of Simple Sugars: Mutarotation of Glucose

Glucose precipitates from concentrated solutions at room temperature to give crystals that melt at 146°C. Structural analysis by X-ray diffraction reveals that these crystals contain only the α-D-(+)-glucopyranose anomer (Figure 24-3). When crystalline α-D-(+)-glucopyranose is dissolved in water and its optical rotation measured immediately, a value $[\alpha]_D^{25°C} = +112$ is obtained. Curiously, this value decreases with

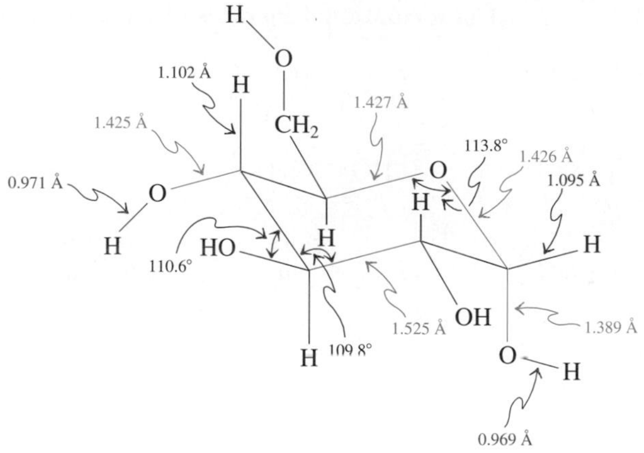

FIGURE 24-3
Structure of α-D-(+)-glucopyranose, with selected bond lengths and angles.

time until it reaches a constant +52.7. The process that gives rise to this effect is the *interconversion* of the α and β anomers.

In solution, the α-pyranose rapidly establishes an equilibrium (in a reaction that is catalyzed by acid and base; see Section 17-7) with a small amount of the open-chain aldehyde isomer, which in turn undergoes reversible ring closure to the β anomer.

Interconversion of Open-Chain and Pyranose Forms of D-Glucose

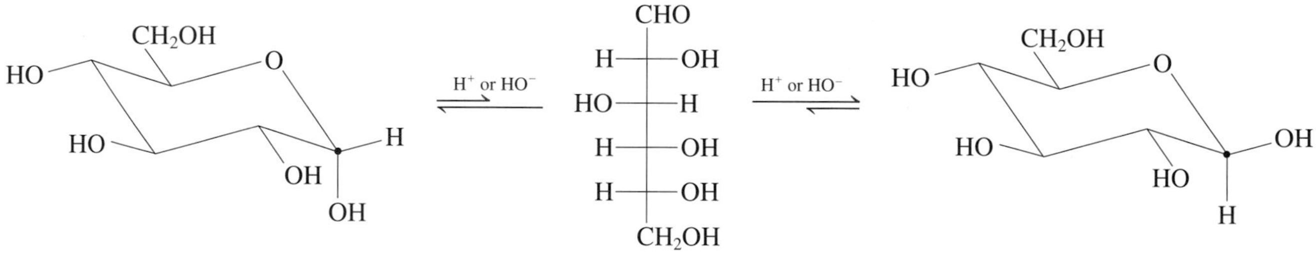

36.4% **α-D-(+)-Glucopyranose** ($[\alpha]_D^{25°C}$ = +112)

0.003% **Aldehyde form**

63.6% **β-D-(+)-Glucopyranose** ($[\alpha]_D^{25°C}$ = +18.7)

The β form has a considerably lower specific rotation (+18.7) than its anomer; therefore, the observed α value in solution decreases. Similarly, a solution of the pure β anomer [m.p. 150°C, obtainable by crystallizing glucose from acetic acid] gradually increases its specific rotation from +18.7 to +52.7. At this point, a final equilibrium has been reached, with 36.4% of the α anomer and 63.6% of the β anomer. The change in optical rotation observed when a sugar equilibrates with its anomer is called **mutarotation** (*mutare,* Latin, to change). Interconversion of α and β anomers is a general property of sugars. This includes all monosaccharides capable of existing as cyclic hemiacetals.

EXERCISE 24-8

An alternative mechanism for mutarotation bypasses the aldehyde intermediate and proceeds through oxonium ions. Formulate it.

EXERCISE 24-9

Calculate the equilibrium ratio of α- and β-glucopyranose (which has been given in the text) from the specific rotations of the pure anomers and the observed specific rotation at mutarotational equilibrium.

By using Table 4-3, estimate the difference in energy between α- and β-glucopyranose at room temperature (25°C). Then calculate it by using the equilibrium percentage.

In summary, the hemiacetal carbon (anomeric carbon) can have two configurations: α or β. In solution, the α and β forms of the sugars are in equilibrium with each other. The equilibration can be followed by starting with a pure anomer and observing the changes in specific rotation, a phenomenon also called mutarotation.

24-4 Polyfunctional Chemistry of Sugars: Oxidation to Carboxylic Acids

Simple sugars exist as various isomers: the open-chain structure and the α and β anomers of various cyclic forms. Because all of these isomers are equilibrated rapidly, the relative rates of their individual reactions with various reagents will determine the product distribution of a particular transformation. We can therefore divide the reactions of sugars into two groups, those of the linear form and those of the cyclic forms, although sometimes both react competitively. This section describes reactions of aldoses with oxidizing agents, which attack the open-chain form predominantly.

Fehling's and Tollens's tests detect reducing sugars

Because they are polyfunctional compounds, the open-chain monosaccharides undergo the reactions typical of each of their functional groups. For example, aldoses contain the oxidizable formyl group and therefore respond to the standard oxidation tests such as exposure to Fehling's or Tollens's solutions (Section 17-14). The α-hydroxy substituent in ketoses is similarly oxidized.

Results of Fehling's and Tollens's Tests on Aldoses and Ketoses

In these reactions, the aldoses are transformed into **aldonic acids,** ketoses into α-dicarbonyl compounds. Sugars that respond positively to these tests are called **reducing sugars.** All ordinary monosaccharides are reducing sugars.

Oxidation of aldoses can give mono- or dicarboxylic acids

Aldonic acids are made on a preparative scale by oxidation of aldoses with bromine in buffered aqueous solution (pH = 5–6). For example, D-mannose yields D-mannonic acid in this way. On subsequent evaporation of solvent from the aqueous solution of the aldonic acid, the γ-lactone (Section 20-4) forms spontaneously.

Aldonic Acid Preparation and Subsequent Dehydration to Give an Aldonolactone

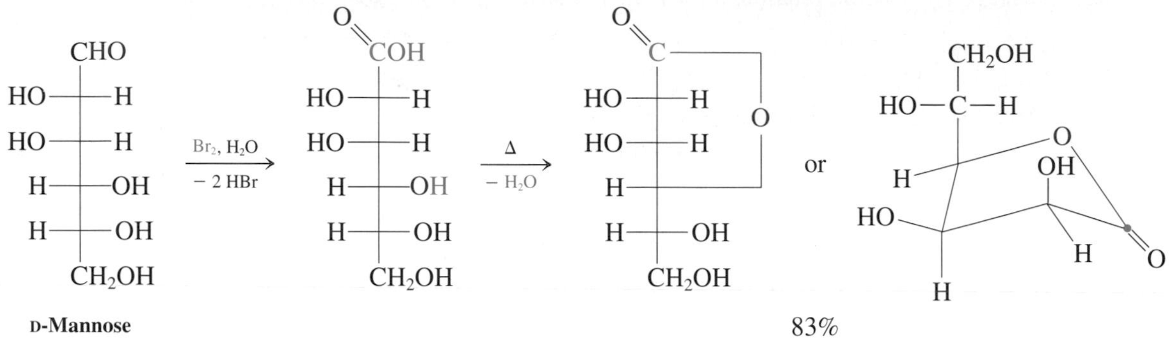

D-Mannose

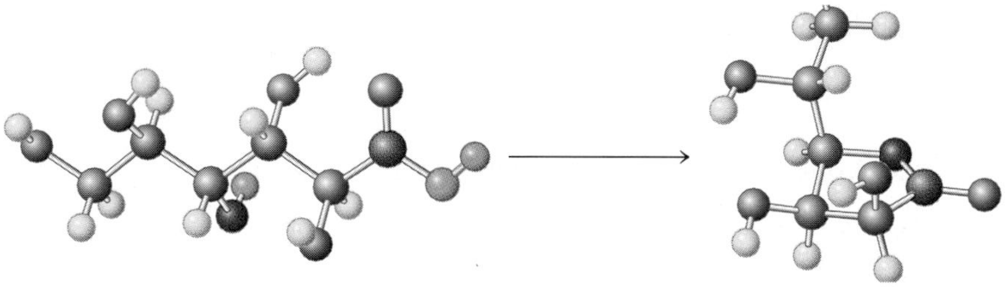

83%

D-Mannonic acid   D-Mannono-γ-lactone

More vigorous oxidation of an aldose leads to attack at the primary hydroxy function as well as at the formyl group. The resulting dicarboxylic acid is called an **aldaric**, or **saccharic, acid.** This oxidation can be achieved with warm dilute aqueous nitric acid (see Section 19-6). For example, D-mannose is converted into D-mannaric acid under these conditions.

Preparation of an Aldaric Acid

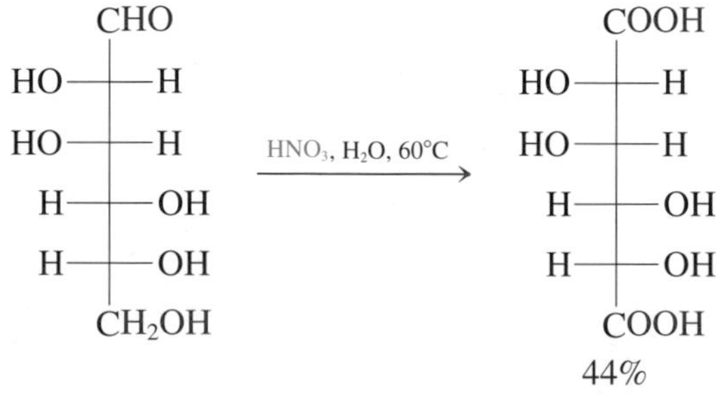

44%

D-Mannose   D-Mannaric acid

EXERCISE 24-11

The two sugars D-allose and D-glucose (Figure 24-1) differ in configuration only at C3. If you did not know which was which and you had samples of both, a polarimeter, and nitric acid at your disposal, how could you distinguish the two? (**Hint:** Write the products of oxidation.)

In summary, the chemistry of the sugars is largely that expected for carbonyl compounds containing several hydroxy substituents. Oxidation (by Br_2) of the formyl group of aldoses gives aldonic acids; more vigorous oxidation (by HNO_3) converts sugars into aldaric acids.

24-5 Oxidative Cleavage of Sugars

The methods for oxidation of sugars discussed so far leave the basic skeleton intact. A reagent that leads to C–C bond rupture is periodic acid, HIO_4. This compound oxidatively degrades vicinal diols to give carbonyl compounds.

Oxidative Cleavage of Vicinal Diols with Periodic Acid

cis-**1,2-Cyclohexanediol** → **Hexanedial** (77%)

The mechanism of this transformation proceeds through a cyclic **periodate ester,** which decomposes to give two carbonyl groups.

Mechanism of Periodic Acid Cleavage of Vicinal Diols

Cyclic periodate ester

Because most sugars contain several pairs of vicinal diols, oxidation with HIO_4 can give complex mixtures. Sufficient oxidizing agent causes complete degradation of the chain to one-carbon compounds, a technique that has been applied in the structural elucidation of sugars. For example, treatment of glucose with five equivalents of HIO_4 results in the formation of five equivalents of formic acid and one of formaldehyde. Similar degradation of the isomeric fructose consumes an equal amount of oxidizing agent, but the products are three equivalents of the acid, two of the aldehyde, and one of carbon dioxide.

Periodic Acid Degradation of Sugars

D-Glucose $\xrightarrow{5\ HIO_4}$ 5 HCOH (From C1–C5) + HCH (From C6)

D-Fructose $\xrightarrow{5\ HIO_4}$ 3 HCOH (From C3–C5) + 2 HCH (From C1 and C6) + CO_2 (From C2)

It is found that (1) the breaking of each C–C bond in the sugar consumes one molecule of HIO_4, (2) each aldehyde and secondary alcohol unit furnishes an equivalent of formic acid, and (3) the primary hydroxy function gives formaldehyde. The carbonyl group in ketoses gives CO_2. The number of equivalents of HIO_4 consumed reveals the size of the sugar molecule, and the ratios of products are important clues to the number and arrangement of hydroxy and carbonyl functions. Notice in particular that, after degradation, each carbon fragment retains the same number of attached hydrogen atoms as were present in the original sugar.

EXERCISE 24-12

Write the expected products (and their ratios), if any, of the treatment of the following compounds with HIO_4. **(a)** 1,2-Ethanediol (ethylene glycol); **(b)** 1,2-propanediol; **(c)** 1,2,3-propanetriol; **(d)** 1,3-propanediol; **(e)** 2,4-dihydroxy-3,3-dimethylcyclobutanone; **(f)** D-threose.

EXERCISE 24-13

Would degradation with HIO_4 permit the following sugars to be distinguished? Explain. (For structures, see Figures 24-1 and 24-2.) **(a)** D-Arabinose and D-glucose; **(b)** D-erythrose and D-erythrulose; **(c)** D-glucose and D-mannose.

In summary, oxidative cleavage with periodic acid degrades the sugar backbone to formic acid, formaldehyde, and CO_2. The ratio of these products depends on the structure of the sugar.

24-6 Reduction of Monosaccharides to Alditols

Aldoses and ketoses are reduced by the same types of reducing agents that convert aldehydes and ketones into alcohols. The resulting polyhydroxy compounds are called **alditols.** For example, D-glucose gives D-glucitol (older name, D-sorbitol) when treated with sodium borohydride. The hydride reducing agent traps the small amount of the open-chain form of the sugar, in this way shifting the equilibrium from the unreactive cyclic hemiacetal to the product.

Preparation of an Alditol

D-Glucose

D-Glucitol
(D-Sorbitol)

Many alditols are found in nature. D-Glucitol is present in red seaweed in concentrations as high as 14%, as well as in many berries (but not in grapes), in cherries, in plums, in pears, and in apples. It is prepared commercially from D-glucose by high-pressure hydrogenation or by its electrochemical reduction.

(a) Reduction of D-ribose with $NaBH_4$ gives a product without optical activity. Explain.
(b) Similar reduction of D-fructose gives two optically active products. Explain.

In summary, reduction of the carbonyl function in aldoses and ketoses (by $NaBH_4$) furnishes alditols.

24-7 Carbonyl Condensations with Amine Derivatives

As might be expected, the carbonyl function in aldoses and ketoses undergoes condensation reactions with amine derivatives (Section 17-9). For example, treatment of D-mannose with phenylhydrazine gives the corresponding **hydrazone,** D-mannose phenylhydrazone. Surprisingly, the reaction does not stop at this stage but can be induced to continue with additional phenylhydrazine (two extra equivalents). The final product is a double phenylhydrazone, also called an **osazone** (here, phenylosazone). In addition, one equivalent each of benzenamine (aniline), ammonia, and water is generated.

Phenylhydrazone and Phenylosazone Formation

	75%	95%
D-Mannose	D-Mannose phenylhydrazone	A phenylosazone

The mechanism of osazone synthesis is thought to incorporate tautomerism to a 2-keto amine, followed by a complex sequence of eliminations and condensations. Once formed, the osazones do not continue to react with excess phenylhydrazine but are stable under the conditions of the reaction.

Historically, the discovery of osazone formation marked a significant advance in the practical aspects of sugar chemistry. Sugars, like many other polyhydroxy compounds, are well known for their reluctance to crystallize from syrups. Their osazones, however, readily form yellow crystals with sharp melting points, thus simplifying the isolation and characterization of many sugars, particularly if they have been formed as mixtures or are impure.

Compare the structures of the phenylosazones of D-glucose, D-mannose, and D-fructose. Do you notice anything unusual?

In summary, one equivalent of phenylhydrazine converts a sugar into the corresponding phenylhydrazone. Additional hydrazine reagent causes oxidation of the center adjacent to the hydrazone function to furnish the osazone.

24-8 Ester and Ether Formation: Glycosides

Because of their multiple hydroxy groups, sugars can be converted into alcohol de-rivatives. This section explores the formation of simple esters and ethers of mono-saccharides as well as reactions that take place selectively at the anomeric carbon.

Sugars can be esterified and methylated

Esters can be prepared from monosaccharides by standard techniques (Sections 19-9, 20-2, and 20-3). Excess reagent will completely convert all hydroxy groups, including the hemiacetal function. For example, acetic anhydride transforms β-D-glucopyranose into the pentaacetate.

Complete Esterification of Glucose

β-D-Glucopyranose β-D-Glucopyranose pentaacetate

Williamson ether synthesis (Section 9-6) allows complete methylation.

Complete Methylation of a Pyranose

β-D-Ribopyranose β-D-Ribopyranose tetramethyl ether

Notice that the hemiacetal function at C1 is converted into an acetal group. The lat-ter can be selectively hydrolyzed back to the hemiacetal (see Section 17-7).

Selective Hydrolysis of a Sugar Acetal

D-Ribopyranose trimethyl ether
(Mixture of α and β forms)

^{18}F-Labeled Glucose as a Radiotracer: Imaging the Human Brain

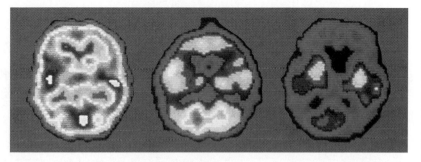

β-D-Mannopyranose
1,3,4,6-tetraacetate-2-triflate
(Tf = trifluoromethanesulfonyl group, CF$_3$SO$_2$)

2-Deoxy-2-[^{18}F]fluoro-D-glucose
(Only β anomer shown)

Positron emission tomography (PET) is a powerful medical imaging method for studying the movement of chemical substances through the organs of the body. Biochemically active substances are labeled with radioactive isotopes of common elements that disintegrate by positron emission. The PET scanner detects the photons that arise from positron-electron annihilation. The isotopes commonly used (^{11}C, ^{13}N, ^{15}O, and ^{18}F) have relatively short half-lives, and the photon bursts are easily detected even at low concentrations. Thus the method is very sensitive, and radiation doses to the body can be kept very low. The isotope ^{18}F is prepared readily from H$_2$^{18}O, has the longest and most convenient half-life of the group (110 min), and decays with the lowest energy into ^{18}O, making it the label of choice for many medical applications.

The most frequently used radiotracer molecule for PET scan studies is 2-deoxy-2-^{18}fluoro-D-glucose

([^{18}F]FDG). It is prepared by conversion of the 2-hydroxy group of suitably protected D-mannose into a good leaving group followed by S$_N$2 displacement with [^{18}F]fluoride ion. [^{18}F]FDG is transported into cells, as is glucose, but cannot undergo further metabolism and remains trapped in the cellular interior until ^{18}F decay occurs. Thus, by introduction of small quantities of [^{18}F]FDG to the body, the localization of glucose in the cells in which it is to be metabolized may be viewed by using the PET technique. False-color images depict regions of low glucose metabolism in blue and those of high glucose uptake in red and white. The three PET scans shown here reveal the relative glucose uptake in the brains of a healthy adult (left) and of patients with moderate (center) and severe (right) Alzheimer's disease. The latter two images indicate the degeneration of neural tissue as the disease progresses.

PET images of uptake of ^{18}F-labeled 2-deoxy-2-fluoroglucose in the brains of a healthy adult (left), and persons with moderate (center) and severe (right) Alzheimer's disease.

It is also possible to convert the hemiacetal unit of a sugar selectively into the acetal. For example, treatment of D-glucose with acidic methanol leads to the formation of the two methyl acetals. Sugar acetals are called **glycosides.** Thus, glucose forms **glucosides.**

Selective Preparation of a Glycoside (Sugar Acetal)

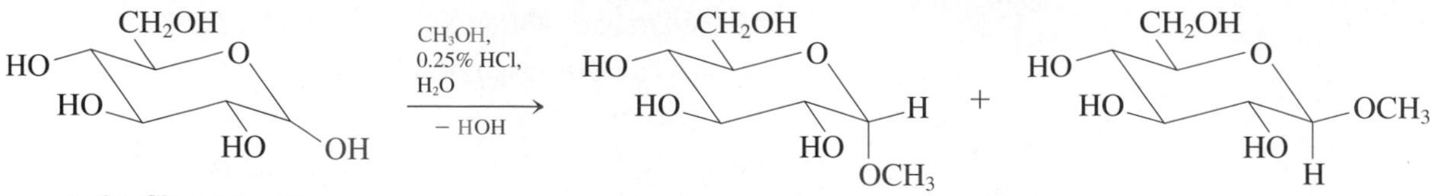

α- or β-D-Glucopyranose

Methyl α-D-glucopyranoside
(m.p. 166°C, $[\alpha]_D^{25°C} = +158$)

Methyl β-D-glucopyranoside
(m.p. 105°C, $[\alpha]_D^{25°C} = -33$)

Because glycosides contain a blocked anomeric carbon atom, they do not show mutarotation in the absence of acid, they test negatively to Fehling's and Tollens's reagents (they are *non*reducing sugars), and they are unreactive toward reagents that attack carbonyl groups. Such protection can be useful in synthesis and in structural analysis (see Exercise 24-16).

EXERCISE 24-16

The same mixture of glucosides is formed in the methylation of D-glucose with acidic methanol, regardless of whether you start with the α or β form. Why?

EXERCISE 24-17

Draw the structure of methyl α-D-arabinofuranoside.

EXERCISE 24-18

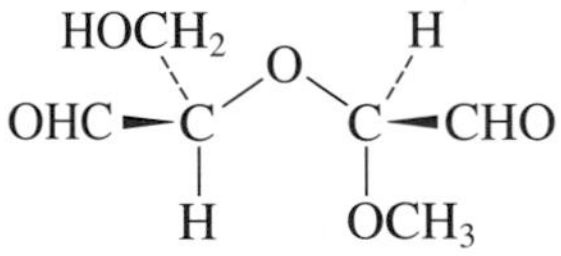

Methyl α-D-glucopyranoside consumes two equivalents of HIO_4 to give one equivalent each of formic acid and dialdehyde A (shown in the margin). An unknown aldopentose methyl furanose reacted with one equivalent of HIO_4 to give dialdehyde A, but no formic acid. Suggest a structure for the unknown. Is there more than one solution to this problem?

A

Neighboring hydroxy groups in sugars can be linked as cyclic ethers

The presence of neighboring pairs of hydroxy groups in the sugars allows for the formation of cyclic ether derivatives. For example, it is possible to synthesize five- or six-membered cyclic sugar acetals from the vicinal (and also from some β-diol) units by treating them with carbonyl compounds (Section 17-8).

Cyclic Acetal Formation from Vicinal Diols

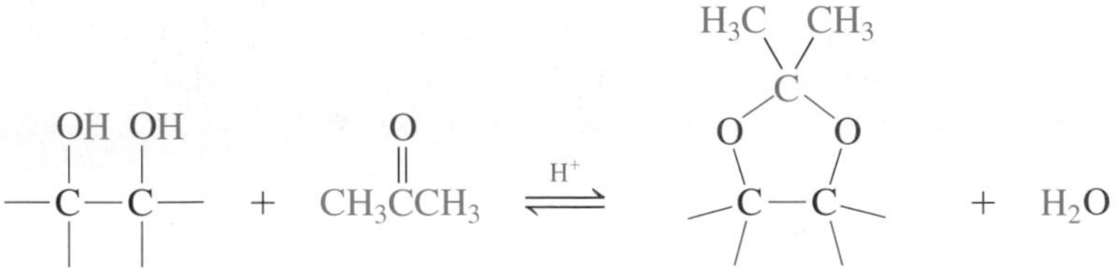

Propanone (acetone)
acetal

Such processes work best if the two OH groups are positioned cis to allow a relatively unrestrained ring to form. For example, excess acidic propanone (acetone) converts β-D-arabinopyranose into the double acetal.

Conversion of *cis*-Diols into Cyclic Acetals

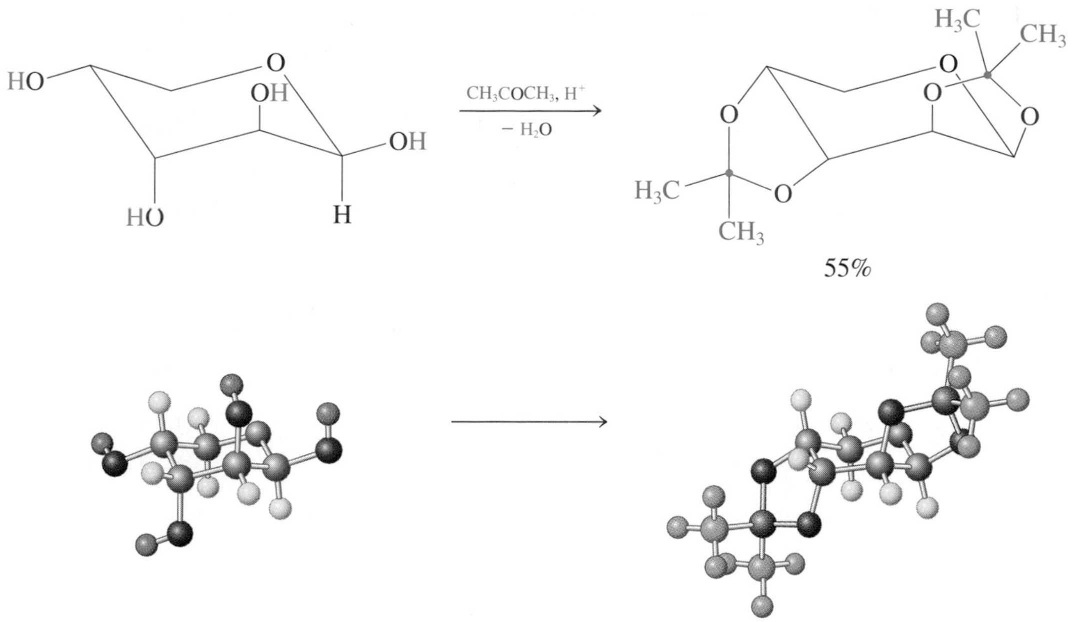

β-D-**Arabinopyranose**

β-D-**Arabinopyranose**
double acetal

Cyclic acetal and ester formation is often employed to protect selected alcohol functions. The remaining hydroxy groups can then be oxidized to carbonyl compounds, converted into leaving groups, or transformed by elimination.

EXERCISE 24-19

Suggest a synthesis of the compound shown in the margin from D-galactose. (**Hint:** Consider a strategy that uses protecting groups.)

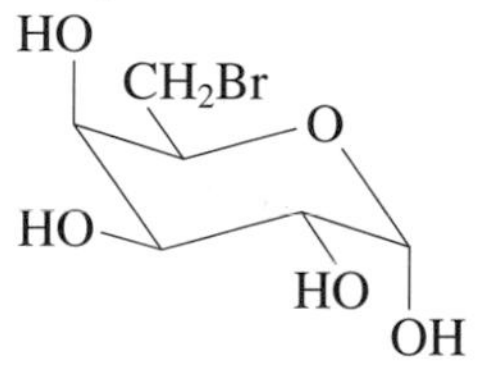

In summary, the various hydroxy groups of sugars can be esterified or converted into ethers. The hemiacetal unit can be selectively protected as the acetal, also called a glycoside. Finally, the various diol units in the sugar backbone can be linked as cyclic acetals, depending on steric requirements.

24-9 Step-by-Step Buildup and Degradation of Sugars

Larger sugars can be made from smaller ones and vice versa, by chain lengthening and chain shortening. These transformations can also be used to structurally correlate various sugars, a procedure applied by Fischer to prove the relative configuration of all the stereocenters in the aldoses shown in Figure 24-1.

Cyanohydrin formation and reduction lengthens the chain

To lengthen the chain of a sugar, an aldose is first treated with HCN to give the corresponding cyanohydrin (Section 17-11). Because this transformation forms a new stereocenter, two diastereomers appear. Separation of the diastereomers and partial reduction of the nitrile group by catalytic hydrogenation in aqueous acid then gives the aldehyde groups of the chain-extended sugars.

Sugar Chain Extension Through Cyanohydrins

STEP 1. Cyanohydrin formation

New stereocenter

$$\begin{array}{c} HC{=}O \\ H{-}{-}OH \\ HO{-}{-}H \end{array} \xrightarrow{\text{HCN}} \begin{array}{c} CN \\ H{-}{-}OH \\ H{-}{-}OH \\ HO{-}{-}H \end{array} + \begin{array}{c} CN \\ HO{-}{-}OH \\ H{-}{-}OH \\ HO{-}{-}H \end{array}$$

Two new diastereomeric nitriles

STEP 2. Reduction and hydrolysis (only one diastereomer is shown)

$$\begin{array}{c} C{\equiv}N \\ H{-}{-}OH \\ H{-}{-}OH \\ HO{-}{-}H \end{array} \xrightarrow[\text{H}^+,\ \text{H}_2\text{O},\ 4\ \text{atm}]{\text{H}_2,\ \text{Pd-BaSO}_4,} \begin{array}{c} HC{=}NH \\ H{-}{-}OH \\ H{-}{-}OH \\ HO{-}{-}H \end{array} \xrightarrow[-\ \text{NH}_2\text{H}]{\text{H}_2\text{O}} \begin{array}{c} HC{=}O \\ H{-}{-}OH \\ H{-}{-}OH \\ HO{-}{-}H \end{array}$$

Imine   Extended sugar

CHEMICAL HIGHLIGHT 24-2 | **Sugar Biochemistry**

In nature, carbohydrates are produced primarily by a reaction sequence called *photosynthesis*. In this process, sunlight impinging on the chlorophyll of green plants is absorbed, and the photochemical energy thus obtained is used to convert carbon dioxide and water into oxygen and the polyfunctional structure of carbohydrates.

Photosynthesis of Glucose in Green Plants

$$6\ CO_2 + 6\ H_2O \xrightarrow[\substack{\text{Released} \\ \text{metabolic} \\ \text{energy}}]{\substack{\text{Sunlight,} \\ \text{chlorophyll}}} C_6(H_2O)_6 + 6\ O_2$$

Glucose

The detailed mechanism of this transformation is complicated and takes many steps, the first of which is the absorption of one quantum of light by the extended π system (Chapter 14) of chlorophyll, shown below. Chemical Highlights 18-3 and 23-3 described two processes that construct and interconvert carbohydrates; Chemical Highlight 23-2

introduces a small part of the rich chemistry associated with the metabolism of glucose to produce biochemical energy. The cycle of photosynthesis and carbohydrate metabolism is a beautiful example of how nature reuses its resources. First, CO_2 and H_2O are consumed to convert solar energy into chemical energy and O_2. When an organism needs some of the stored energy, it is generated by conversion of carbohydrate into CO_2 and H_2O, using up roughly the same amount of O_2 originally liberated.

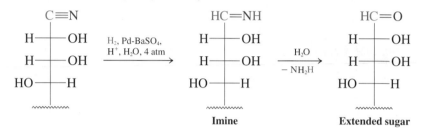

Chlorophyll a

In this hydrogenation, a modified palladium catalyst (similar to the Lindlar catalyst, Section 13-7) allows selective reduction of the nitrile to the imine, which hydrolyzes under the reaction conditions. The special catalyst is necessary to prevent the hydrogenation from proceeding all the way to the amine (Section 20-8). The preceding chain-lengthening sequence is an improved and shortened version of what is known as the **Kiliani-Fischer*** synthesis of chain-extended sugars. In the late 1800s, Kiliani demonstrated that addition of cyanide to an aldose gives the cyanohydrin, which, on hydrolysis (Sections 19-6 and 20-8), is converted into the chain-extended aldonic acid. Fischer subsequently succeeded in converting the latter into an aldose by lactonization (Sections 19-9 and 24-4) followed by reduction (Section 20-4; for Fischer's method, see Problem 47 at the end of this chapter).

EXERCISE 24-20

What are the products of chain extension of **(a)** D-erythrose and **(b)** D-arabinose?

Ruff degradation shortens the chain

Whereas the preceding approach synthesizes higher sugars, complementary strategies degrade higher sugars to lower sugars one carbon at a time. One of these strategies is the **Ruff†** **degradation.** This procedure removes the carbonyl group of an aldose and converts the neighboring carbon into the aldehyde function of the new sugar.

The Ruff degradation is an oxidative decarboxylation. The sugar is first oxidized to the aldonic acid by aqueous bromine. Exposure to hydrogen peroxide in the presence of ferric ion then leads to the loss of the carboxy group and oxidation of the new terminus to the aldehyde function of the lower aldose.

Ruff Degradation of Sugars

The oxidative decarboxylation takes place by means of two one-electron oxidations. The first gives a carboxy radical, which is unstable and rapidly loses CO_2. The resulting hydroxy-substituted radical is oxidized again to give the aldehyde.

Mechanism of Oxidative Decarboxylation

Ruff degradation gives low yields because of the sensitivity of the products to the reaction conditions. Nevertheless, the procedure is useful in structural elucidations

---

*Professor Heinrich Kiliani (1855–1945), University of Freiburg, Germany; Professor Emil Fischer (see Section 5-4).

†Professor Otto Ruff (1871–1939), University of Danzig, Germany.

(Exercise 24-21). Fischer originally carried out such studies to establish the relative configurations of the monosaccharides (the **Fischer proof**). The next section will describe some of the logic behind his approach to the problem.

EXERCISE 24-21

Ruff degradation of two D pentoses, A and B, gave two new sugars, C and D. Oxidation of C with HNO_3 gave *meso*-2,3-dihydroxybutanedioic (tartaric) acid, that of D resulted in an optically active acid. Oxidation of either A or B with HNO_3 furnished an optically active aldaric acid. What are compounds A, B, C, and D?

In summary, sugars can be made from other sugars by step-by-step one-carbon chain lengthening (cyanohydrin formation and reduction) or shortening (Ruff degradation).

24-10 Relative Configurations of the Aldoses: An Exercise in Structure Determination

Imagine that we have been presented with 14 jars, each filled with one of the tetroses, pentoses, and hexoses depicted in Figure 24-1 and each labeled with a name but *no structural formula*. How would we establish the structure of each substance?

This was essentially the challenge faced by Fischer in the late 19th century, when chemists had no sophisticated equipment such as spectrometers at their disposal. In an extraordinary display of scientific logic, Fischer showed that this problem could be solved by interpreting the results of a combination of carefully thought out synthetic manipulations, designed to convert the aldoses that he had available to him into one another and into related substances. Fischer made one assumption: The dextrorotatory enantiomer of 2,3-dihydroxypropanal (glyceraldehyde) has the D (not the L) configuration. This assumption was proved correct only after a special kind of X-ray analysis had been developed in 1950, long after Fischer's time. Fischer's was a lucky guess; otherwise, all the structures of the D sugars in Figure 24-1 would have had to be changed into their mirror images. At the time, however, it was more important to establish the *relative* configurations of all the stereoisomers—that is, to associate each unique sugar with a unique sequence of stereocenters.

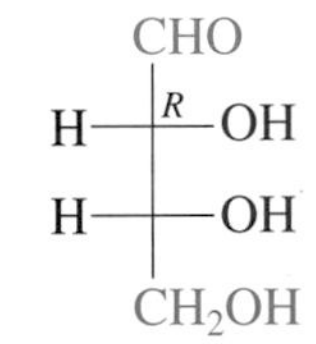

The structures of the four- and five-carbon aldoses can be determined from the optical activity of the corresponding aldaric acids

Given the structure of D-glyceraldehyde, we will now set out to unambiguously prove the structures of all the higher D aldoses. We will use Fischer's logic, although not his procedures, which were made more complicated by the unavailability of many of these aldoses in his time. (Consult Figure 24-1 as required while following these arguments.) First, we perform a chain-lengthening procedure on D-glyceraldehyde. We find that we obtain a mixture of D-erythrose and D-threose. Oxidation of the first with nitric acid gives *meso*-2,3-dihydroxybutanedioic (tartaric) acid; oxidation of the second leads to an *optically active* diastereomeric acid. Thus D-erythrose must have the structure shown, with both –OH groups on the same side in the Fischer representation, whereas D-threose must have the opposite stereochemistry at C2. Both must have the same (R) configuration at C3, having derived that carbon from the C2 stereocenter of our starting material, D-glyceraldehyde. Recall that this is the common structural feature of all D sugars. The difference is at C2: D-Erythrose is 2R; D-threose, 2S.

CHO

H—R—OH

H——OH

CH_2OH

**Structure of
D-erythrose**

↓ HNO_3

meso-tartaric acid

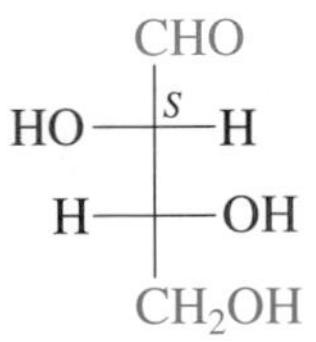

CHO

HO—S—H

H——OH

CH_2OH

**Structure of
D-threose**

↓ HNO_3

(*S,S*)-tartaric acid

Let us now use D-erythrose as our new starting material—because we know its structure—and lengthen the chain further.

$$
\begin{array}{c}
\text{CHO} \\
\text{H}\!\!-\!\!\!-\!\!\text{OH} \\
\text{H}\!\!-\!\!\!-\!\!\text{OH} \\
\text{CH}_2\text{OH}
\end{array}
$$

D-Erythrose

Chain extension

COOH	CHO	CHO	COOH
H——OH	H——OH	HO——H	HO——H
H——OH	H——OH	H——OH	H——OH
H——OH	H——OH	H——OH	H——OH
COOH	CH₂OH	CH₂OH	COOH

(HNO₃ ←) (→ HNO₃)

Meso — **Not optically active** / **Structure of D-ribose** / **Structure of D-arabinose** / **Optically active**

Again, two new sugars ensue (because we have again added a new stereocenter), two pentoses. We know their configuration at C3 and C4 (the same as that at C2 and C3 of their starting material) but not that at C2. Their oxidation again produces one optically inactive dicarboxylic acid and one that is active. The former must therefore have the structure of D-ribose, the latter that of D-arabinose.

A very similar train of thought leads to the unambiguous assignment of the structures of D-xylose (oxidized to a meso dioic acid) and D-lyxose (oxidized to an optically active dioic acid), derived synthetically from D-threose, whose structure we ascertained at the very beginning.

COOH	CHO	CHO	COOH
H——OH	H——OH	HO——H	HO——H
HO——H	HO——H	HO——H	HO——H
H——OH	H——OH	H——OH	H——OH
COOH	CH₂OH	CH₂OH	COOH

(HNO₃ ←) (→ HNO₃)

Meso **Not optically active** / **D-Xylose** / **D-Lyxose** / **Optically active**

Symmetry properties also define the structures of the six-carbon aldoses

We now know the structures of the four aldopentoses and can extend the chain of each of them. This process gives us four pairs of aldohexoses, each pair distinguished from the other by the unique sequence of stereocenters at C3, C4, and C5. The members of each pair differ only in their configuration at C2.

The structural assignment for the four sugars obtained from D-ribose and D-lyxose, respectively, is again accomplished by oxidation to the corresponding aldaric acids. Both D-allose and D-galactose give optically inactive oxidation products, in contrast with their counterparts D-altrose and D-talose, which give optically active dicarboxylic acids.

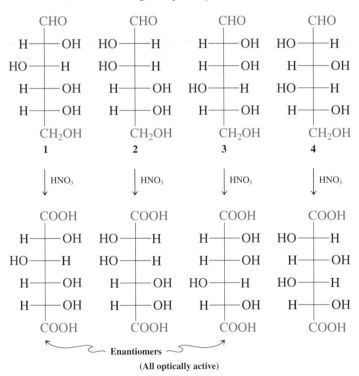

CHO	CHO	CHO	CHO
H——OH	H——OH	HO——H	HO——H
H——OH	HO——H	H——OH	HO——H
H——OH	HO——H	H——OH	HO——H
H——OH	H——OH	H——OH	H——OH
CH$_2$OH	CH$_2$OH	CH$_2$OH	CH$_2$OH
D-Allose	**D-Galactose**	**D-Altrose**	**D-Talose**
(From D-ribose)	(From D-lyxose)	(From D-ribose)	(From D-lyxose)
(Both give meso dicarboxylic acids)		(Both give optically active dicarboxylic acids)	

The structural assignment of the four remaining sugars cannot be based on the approach taken thus far, because all give optically active diacids on oxidation.

CHO	CHO	CHO	CHO
H——OH	HO——H	H——OH	HO——H
HO——H	HO——H	H——OH	H——OH
H——OH	H——OH	HO——H	HO——H
H——OH	H——OH	H——OH	H——OH
CH$_2$OH	CH$_2$OH	CH$_2$OH	CH$_2$OH
1	**2**	**3**	**4**

↓ HNO$_3$ ↓ HNO$_3$ ↓ HNO$_3$ ↓ HNO$_3$

COOH	COOH	COOH	COOH
H——OH	HO——H	H——OH	HO——H
HO——H	HO——H	H——OH	H——OH
H——OH	H——OH	HO——H	HO——H
H——OH	H——OH	H——OH	H——OH
COOH	COOH	COOH	COOH

Enantiomers

(All optically active)

However, it is found that the two carboxylic acids derived from sugars 1 and 3 are enantiomers—that is, mirror images of each other. This result is possible only if sugars 1 and 3 have the structures of D-glucose or D-gulose. This relation of the two aldaric acids can be verified by building molecular models.

We now proceed experimentally as follows. D-Arabinose is converted into a pair of new sugars, 1 and 2, by chain extension; D-xylose furnishes sugars 3 and 4. With these results in hand, the structural assignments fall into place. Sugar 1 must have the structure of D-glucose, and sugar 3 must have the structure of D-gulose. Therefore, sugar 2 is assigned the structure of D-mannose and sugar 4 that of D-idose.

EXERCISE 24-22

In the preceding discussion, we assigned the structures of D-ribose and D-arabinose by virtue of the fact that, on oxidation, the first gives a meso diacid, the second an optically active diastereomer. Could you arrive at the same result by ^{13}C NMR spectroscopy?

In summary, step-by-step one-carbon chain lengthening or shortening, in conjunction with the symmetry properties of the various aldaric acids, allows the stereochemical assignments of the aldoses.

24-11 Complex Sugars in Nature: Disaccharides

A substantial fraction of the natural sugars occur in dimeric, trimeric, higher oligomeric (between 2 and 10 sugar units), and polymeric forms. The sugar most familiar to us, so-called table sugar, is a dimer.

Sucrose is a disaccharide derived from glucose and fructose

Sucrose, ordinary table sugar, is one of the few natural chemicals consumed in unmodified form (water and NaCl are examples of others). Its average yearly consumption in the United States is about 150 pounds per person. Sugar is isolated from sugar cane and sugar beets, in which it is particularly abundant (about 14–20% by weight), although it is present in many plants in smaller concentrations. World production is about 100 billion tons a year, and there are countries (e.g., Cuba) whose entire economy depends on the world price of sucrose.

Sucrose has not been considered in this chapter so far, because it is not a simple monosaccharide; rather, it is a disaccharide composed of two units, glucose and fructose. The structure of sucrose can be deduced from its chemical behavior. Acidic hydrolysis splits it into glucose and fructose. It is a nonreducing sugar. It does not form an osazone. It does not undergo mutarotation. These findings suggest that the component monosaccharide units are linked by an acetal bridge connecting the two anomeric carbons; in this way, the two cyclic hemiacetal functions block each other. X-ray structural analysis confirms this hypothesis: Sucrose is a disaccharide in which the α-D-glucopyranose form of glucose is attached to β-D-fructofuranose in this way.

Sucrose, an α-D-glucopyranosyl-β-D-fructofuranose

Sucrose has a specific rotation of +66.5. Treatment with aqueous acid decreases the rotation until it reaches a value of −20. The same effect is observed with the enzyme invertase. The phenomenon, known as the **inversion of sucrose,** is related to mutarotation of monosaccharides. It includes three separate reactions: hydrolysis of the disaccharide to the component monosaccharides α-D-glucopyranose and β-D-fructofuranose; mutarotation of α-D-glucopyranose to the equilibrium mixture with the β form; and mutarotation of β-D-fructofuranose to the slightly more stable β-D-fructopyranose. Because the value for the specific rotation of fructose (−92) is more negative than the value for glucose (+52.7) is positive, the resulting mixture, sometimes called **invert sugar,** has a net negative rotation, *inverted* from that of the original sucrose solution.

Inversion of Sucrose

Sucrose

H^+, H_2O
or invertase

18%
α-D-Glucopyranose

+

32%
β-D-Glucopyranose

+

16%
β-D-Fructofuranose

+

34%
β-D-Fructopyranose

The noncaloric, edible fat substitute olestra (Chemical Highlight 21-1) consists of a mixture of sucroses esterified with seven or eight fatty acids obtained from vegetable oils, such as hexadecanoic (palmitic) acid. The fatty acids shield the sucrose core of olestra so effectively that it is completely immune to attack by digestive enzymes and passes through the digestive tract unchanged.

EXERCISE 24-23

Write the products (if any) of the reaction of sucrose with (a) excess $(CH_3)_2SO_4$, NaOH; (b) 1. H^+, H_2O, 2. $NaBH_4$; and (c) NH_2OH.

Acetals link the components of complex sugars

Sucrose contains an acetal linkage between the anomeric carbons of the component sugars. One could imagine other acetal linkages with other hydroxy groups. Indeed, **maltose** (malt sugar), which is obtained in 80% yield by enzymatic (amylase) degradation of starch (to be discussed later in this section), is a dimer of glucose in which the hemiacetal oxygen of one glucose molecule (in the α anomeric form) is bound to C4 of the second.

Note: The model is shown in its lowest energy rotameric form, which places the ether oxygen of glucose in front.

β-Maltose, an α-D-glucopyranosyl-β-D-glucopyranose

CHEMICAL HIGHLIGHT 24-3 | Carbohydrate-Derived Sugar Substitutes

Xylitol

Lactitol

In spite of the commercial success of the synthetic artificial sweeteners saccharine (Chemical Highlight 1-1) and aspartame (Section 26-4), these substances cannot completely replace sugar in products such as chocolate, because their physical properties are very different from those of sucrose. As a result, confections containing these sweeteners must be supplemented with other materials to give them perceived characteristics acceptable to the consumer. Alditols such as D-glucitol (D-sorbitol) are commonly used for such purposes. Sorbitol is about 60% as sweet as sucrose and used in hard candies and chewing gums. You may be surprised to discover that this substance is *not* a noncaloric sweetener; in fact, its caloric content is virtually the same as that of sucrose. Its value as a sugar replacement is twofold: The bacteria responsible for tooth decay are unable to consume sorbitol, and its metabolism is insulin-independent, furnishing mostly CO_2 instead of glucose. Thus, sorbitol is a suitable sweetener for use by diabetics.

The formulation of low-calorie chocolate is an unusual challenge. Much of the caloric content of chocolate comes from fat. However, reduction of the fat content compromises the smooth, "melt in your mouth" character of chocolate and results in an unappealing product. Sugar replacement with a mixture of low-calorie alditols, alditol derivatives, and other synthetic carbohydrates partly achieves the desired goal. Xylitol, one of the sweetest alditols, is easily incorporated into chocolate formulations but possesses two disadvantages: First, it has an endothermic heat of solution and so results in a cooling sensation in the mouth—suitable only for mint chocolates. Second, xylitol upsets the osmotic balance in the intestine, making it a laxative, not necessarily a desired property from the point of view of the true chocolate lover. A commercially workable solution is to use a combination of lactitol, which is less sweet than xylitol but possesses xylitol's disadvantages to a lesser degree, and polydextrose, a moderately sweet but very low calorie synthetic polymer consisting of glucopyranose units randomly crosslinked by α and β 1,2-, 1,3-, 1,4-, and 1,6-glycosidic bonds.

In this arrangement, one glucose retains its unprotected hemiacetal unit, with its distinctive chemistry. For example, maltose is a reducing sugar; it forms osazones, and it undergoes mutarotation. Maltose is hydrolyzed to two molecules of glucose by aqueous acid or by the enzyme maltase. It is about one-third as sweet as sucrose.

EXERCISE 24-24

Draw the structure of the initial product of β-maltose when it is subjected to **(a)** Br₂ oxidation; **(b)** phenylhydrazine (3 equivalents); **(c)** conditions that effect mutarotation.

Another common disaccharide is **cellobiose,** obtained by the hydrolysis of cellulose (to be discussed later in this section). Its chemical properties are almost identical with those of maltose, and so is its structure, which differs only in the stereochemistry at the acetal linkage—β instead of α.

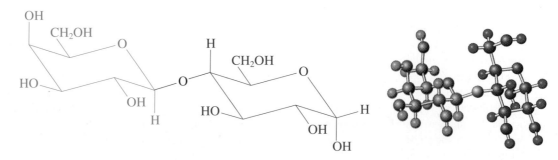

β-**Cellobiose, a** β-D-**glucopyranosyl-**β-D-**glucopyranose**

Aqueous acid cleaves cellobiose into two glucose molecules just as efficiently as it hydrolyzes maltose. However, enzymatic hydrolysis requires a different enzyme, β-glucosidase, which specifically attacks only the β-acetal bridge. In contrast, maltase is specific for α-acetal units of the type found in maltose.

After sucrose, the most abundant natural disaccharide is **lactose** (milk sugar). It is found in human and most animal milk (about 5% solution), constituting more than one-third of the solid residue remaining on evaporation of all volatiles. Its structure is made up of galactose and glucose units, connected in the form of a β-D-galactopyranosyl-α-D-glucopyranose. Crystallization from water furnishes only the α anomer.

Crystalline α-**lactose, a** β-D-**galactopyranosyl-**α-D-**glucopyranose**

In summary, sucrose is a dimer derived from linking α-D-glucopyranose with β-D-fructofuranose at the anomeric centers. It shows inversion of its optical rotation on hydrolysis to its mutarotating component sugars. The disaccharide maltose is a glucose dimer in which the components are linked by a carbon–oxygen bond between an α anomeric carbon of a glucose molecule and C4 of the second. Cellobiose is almost identical with maltose but has a β configuration at the acetal carbon. Lactose has a β-D-galactose linked to glucose in the same manner as in cellobiose.

24-12 Polysaccharides and Other Sugars in Nature

Polysaccharides are the polymers of monosaccharides. Their possible structural diversity exceeds that of alkene polymers (Sections 12-13 and 12-14), particularly in variations of chain length and branching. Nature, however, has been remarkably conservative in its construction of such macromolecules. The three most abundant natural polysaccharides, cellulose, starch, and glycogen, are derived from the same monomer—glucose.

Cellulose and starch are unbranched polymers

Cellulose is a poly-β-glucopyranoside linked at C4, containing about 3000 monomeric units and having a molecular weight of about 500,000. It is largely linear.

Cellulose

Individual strands of cellulose tend to align with one another and are connected by multiple hydrogen bonds. The development of so many hydrogen bonds is responsible for the highly rigid structure of cellulose and its effective use as the cell-wall material in organisms. Thus, cellulose is abundant in trees and other plants. Cotton fiber is almost pure cellulose, as is filter paper. Wood and straw contain about 50% of the polysaccharide.

Several derivatives of cellulose have commercial uses. Conversion of the free hydroxy groups into nitrate esters with nitric acid results in **nitrocellulose.** If the nitrate content is high, this material is explosive and is used in smokeless gunpowder. A lower nitrate content gives a polymer that was important as one of the first commercial plastics—celluloid. For a long time, nitrocellulose was used extensively in the photographic and film industries. Unfortunately, it is highly flammable and gradually decomposes; it is now used only rarely.

Although we normally associate cellulose with plants, this sea squirt produces cellulose for its outer wall, one of very few animals to do so.

Cellulose, which is insoluble in almost all solvents, can be made soluble by blocking the hydroxy groups as adducts to carbon disulfide, the sulfur analog of CO_2. The resulting functional group is called a **xanthate.** Subsequent treatment with acid regenerates the insoluble polymer; this process may be controlled to give fibers (rayon) or sheets (cellophane).

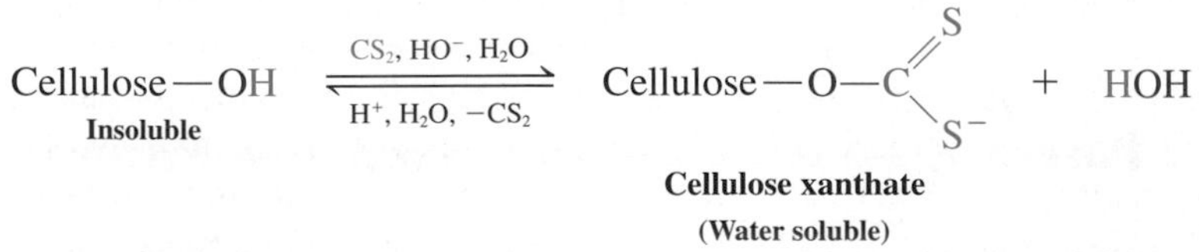

$$\text{Cellulose—OH} \underset{\text{H}^+, \text{H}_2\text{O}, -\text{CS}_2}{\overset{\text{CS}_2, \text{HO}^-, \text{H}_2\text{O}}{\rightleftharpoons}} \text{Cellulose—O—C} \begin{smallmatrix} \diagup\text{S} \\ \diagdown\text{S}^- \end{smallmatrix} + \text{HOH}$$

Insoluble
Cellulose xanthate
(Water soluble)

Like cellulose, **starch** is a polyglucose, but its subunits are connected by α-acetal linkages. It functions as a food reserve in plants and (like cellulose) is readily cleaved by aqueous acid into glucose. Major sources of starch are corn, potatoes, wheat, and rice. Hot water swells granular starch and allows the separation of the two major components: **amylose** (~20%) and **amylopectin** (~80%). Both are soluble in hot water, but the former is less soluble in cold water. Amylose contains a few hundred glucose units per molecule (molecular weight, 150,000–600,000). Its structure is different from that of cellulose, even though both polymers are unbranched. The difference in the stereochemistry at the anomeric carbons leads to the strong tendency of amylose to form a helical polymer arrangement (not the straight chain shown in the formula). Note that the disaccharide units in amylose are the same as those in maltose.

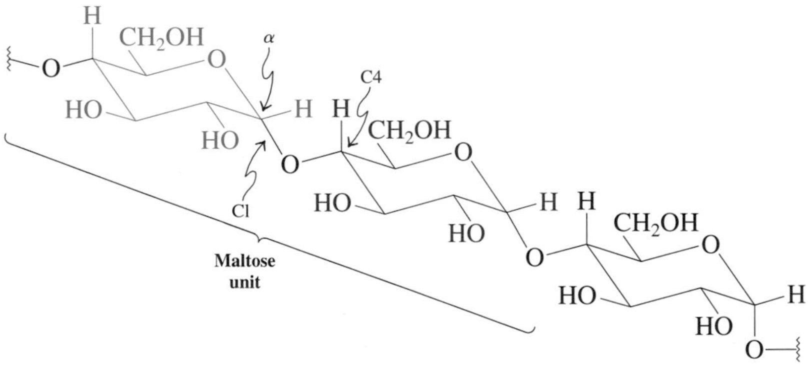

Amylose

In contrast with amylose, amylopectin is branched, mainly at C6, about once every 20 to 25 glucose units. Its molecular weight runs into the millions.

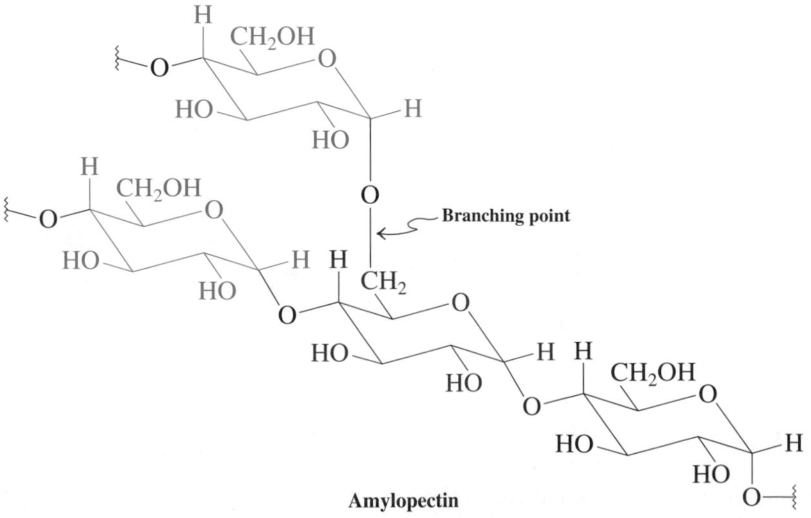

Amylopectin

Glycogen is a source of energy

Another polysaccharide similar to amylopectin but with greater branching (1 per 10 glucose units) and of much larger size (as much as 100 million molecular weight) is **glycogen.** This compound is of considerable biological importance because it is one of the major energy-storage polysaccharides in humans and animals and because it provides an immediate source of glucose between meals and during (strenuous) physical activity. It is accumulated in the liver and in rested skeletal muscle in relatively large amounts. The manner in which cells make use of this energy storage is a fascinating story in biochemistry.

A special enzyme, phosphorylase, first breaks glycogen down to give a derivative of glucose, α-D-glucopyranosyl 1-phosphate. This transformation takes place at one

Glycogen

H_3PO_4,
glycogen phosphorylase

α-D-Glucopyranosyl
1-phosphate

of the nonreducing terminal sugar groups of the glycogen molecule and proceeds step by step—one glucose molecule at a time. Because glycogen is so highly branched, there are many such end groups at which the enzyme can "nibble" away, making sure that, at a time of high-energy requirements, a sufficient amount of glucose becomes quickly available.

Phosphorylase cannot break α-1,6-glycosidic bonds. As soon as it gets close to such a branching point (in fact, as soon as it reaches a terminal residue four units away from that point), it stops (Figure 24-4). At this stage, a different enzyme comes into play, transferase, which can shift blocks of three terminal glucosyl residues from one branch to another. One glucose substituent remains at the branching point. Now a third enzyme is required to remove this last obstacle to obtaining a new straight chain. This enzyme is specific for the kind of bond at which cleavage is needed; it is α-1,6-glucosidase, also known as the debranching enzyme. When this enzyme has completed its task, phosphorylase can continue degrading the glucose chain until it reaches another branch, and so forth.

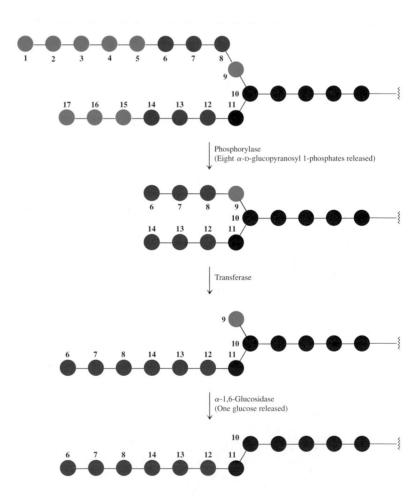

FIGURE 24-4

Steps in the degradation of a glycogen side chain. Initially, phosphorylase removes glucose units 1 through 5 and 15 through 17 step by step. The enzyme is now four sugar units away from a branching point (10). Transferase moves units 6 through 8 in one block and attaches them to unit 14. A third enzyme, α-1,6-glucosidase, debranches the system at glucose unit 10, by removing glucose 9. A straight chain has been formed and phosphorylase can continue its degradation job.

Modified sugars may contain nitrogen

Many of the naturally occurring sugars have a modified structure or are attached to some other organic molecule. There is a large class of sugars in which at least one of the hydroxy groups has been replaced by an amine function. They are called **gly-cosylamines** when the nitrogen is attached to the anomeric carbon and **amino deoxy sugars** when it is located elsewhere.

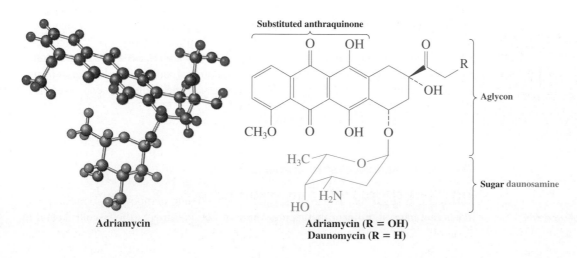

β-ᴅ-Glucopyranosylamine
(A glycosylamine)

2-Amino-2-deoxy-ᴅ-glucopyranose
(An amino deoxy sugar)

Glycosylamines are present in other important biological polymers: the nucleic acids (Section 26-9), which contain the genetic code and are responsible for protein biosynthesis. Ribonucleic acid is a polymer made of subunits called nucleotides, which are substituted glycosylamines. An example is uridylic acid.

Uridylic acid

When a sugar is attached by its anomeric carbon to the hydroxy group of another complex residue, it is called a **glycosyl group.** The remainder of the molecule (or the product after removal of the sugar by hydrolysis) is called the **aglycon.** An example is adriamycin, a member of the anthracycline family of antibiotics. Adriamycin and its deoxy analog daunomycin have been remarkably effective in the treatment of a wide variety of human cancers. They now constitute a cornerstone of combination cancer chemotherapy. The aglycon part of these systems is a linear tetracyclic frame-work incorporating an anthraquinone moiety (derived from anthracene; see Section 15-6). The amino sugar is called daunosamine.

Substituted anthraquinone

Aglycon

Sugar daunosamine

Adriamycin

Adriamycin (R = OH)
Daunomycin (R = H)

An unusual group of antibiotics, the aminoglycoside antibiotics, is based on oligosaccharide structures almost exclusively. Of particular therapeutic importance is streptomycin (an antituberculosis agent), isolated in 1944 from cultures of the mold *Streptomyces griseus*.

Streptomycin

The molecule consists of three subunits: the furanose streptose, the glucose derivative 2-deoxy-2-methylamino-L-glucose (an example of the rare L form), and streptidine, which is actually a hexasubstituted cyclohexane.

In summary, the polysaccharides cellulose, starch, and glycogen are polyglucosides. Cellulose consists of repeating dimeric cellobiose units. Starch may be regarded as a polymaltose derivative. Its occasional branching poses a challenge to enzymatic degradation, as does glycogen. Metabolism of these polymers first gives monomeric glucose, which is then further degraded. Finally, many sugars exist in nature in modified form or as simple appendages to other structures. Examples include amino sugars, glycosylamines, and adriamycin. The aminoglycoside antibiotics consist entirely of saccharide molecules, modified and unmodified.

The peels of these citrus fruits are rich in bioflavonoids.

CHAPTER INTEGRATION PROBLEM

Rutinose is a sugar that is part of several bioflavonoids, compounds found in many plants that have significant therapeutic value in maintaining cardiovascular health in general and the strength of the walls of blood vessels in particular. Rutin is one rutinose-containing bioflavonoid found in buckwheat and eucalyptus. Hesperidin is another, derived from the peels of lemons and oranges. Each contains rutinose bound to a tricyclic aglycon (Section 24-12).

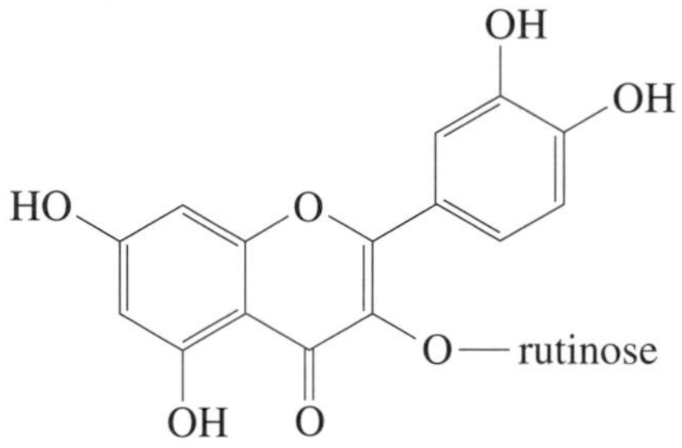

Rutin 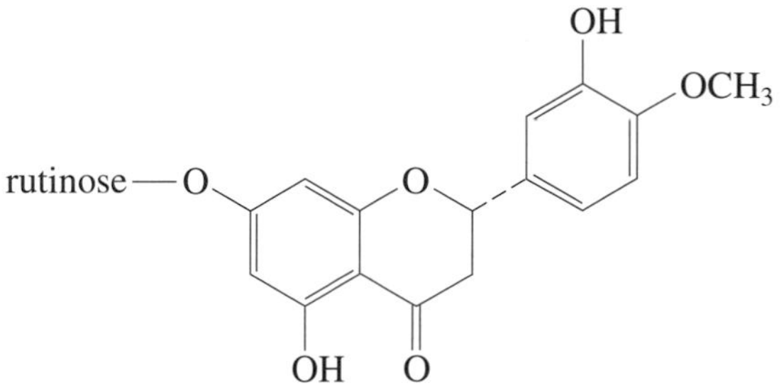   **Hesperidin**

Use the following information to deduce a structure for the sugar rutinose.
a. Rutinose is a reducing sugar that, upon acid hydrolysis, gives one equivalent each of D-glucose and a sugar A with the formula $C_6H_{12}O_5$. Sugar A reacts with four equivalents of HIO_4 to give four equivalents of formic acid and one equivalent of acetaldehyde. What can we conclude regarding sugar A at this stage?

SOLUTION

What does the result of HIO_4 degradation tell us? Each equivalent of HIO_4 cleaves a bond between two oxygen-bearing carbon atoms. As the examples in Section 24-5 illustrate, formic acid can arise from either a terminal formyl group or an internal secondary hydroxy group. Acetaldehyde is an unusual degradation product. Its formation implies a terminal methyl substituent, attached to a secondary hydroxy carbon. A logical reconstruction of sugar A from this information may look like this:

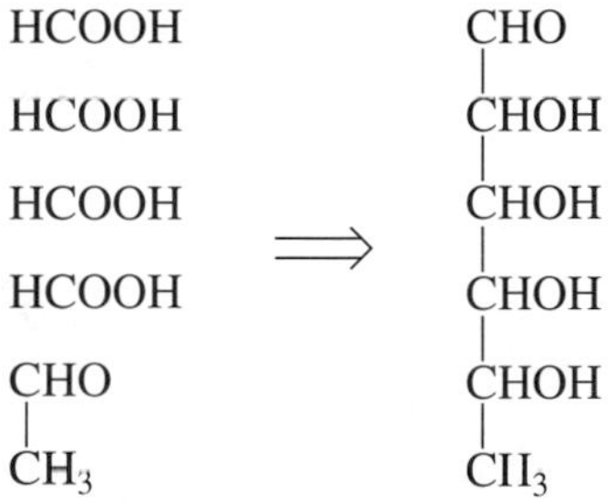

As a check, we note that our structure does indeed have the molecular formula $C_6H_{12}O_5$.

L-(−)-Mannose

b. Sugar A can be synthesized from L-(−)-mannose as shown in the margin. Step 3 (noted by an asterisk) is a special reaction that converts the terminal primary alcohol into a carboxylic acid group. What does this result reveal about the stereocenters in sugar A?

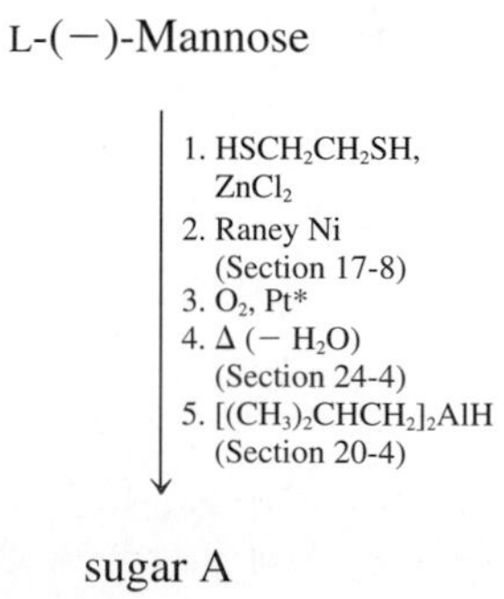

SOLUTION

We follow the reaction sequence step by step to obtain the full open-chain structure of sugar A:

1. $HSCH_2CH_2SH$, $ZnCl_2$
2. Raney Ni (Section 17-8)
3. O_2, Pt*
4. Δ (− H_2O) (Section 24-4)
5. $[(CH_3)_2CHCH_2]_2AlH$ (Section 20-4)

sugar A

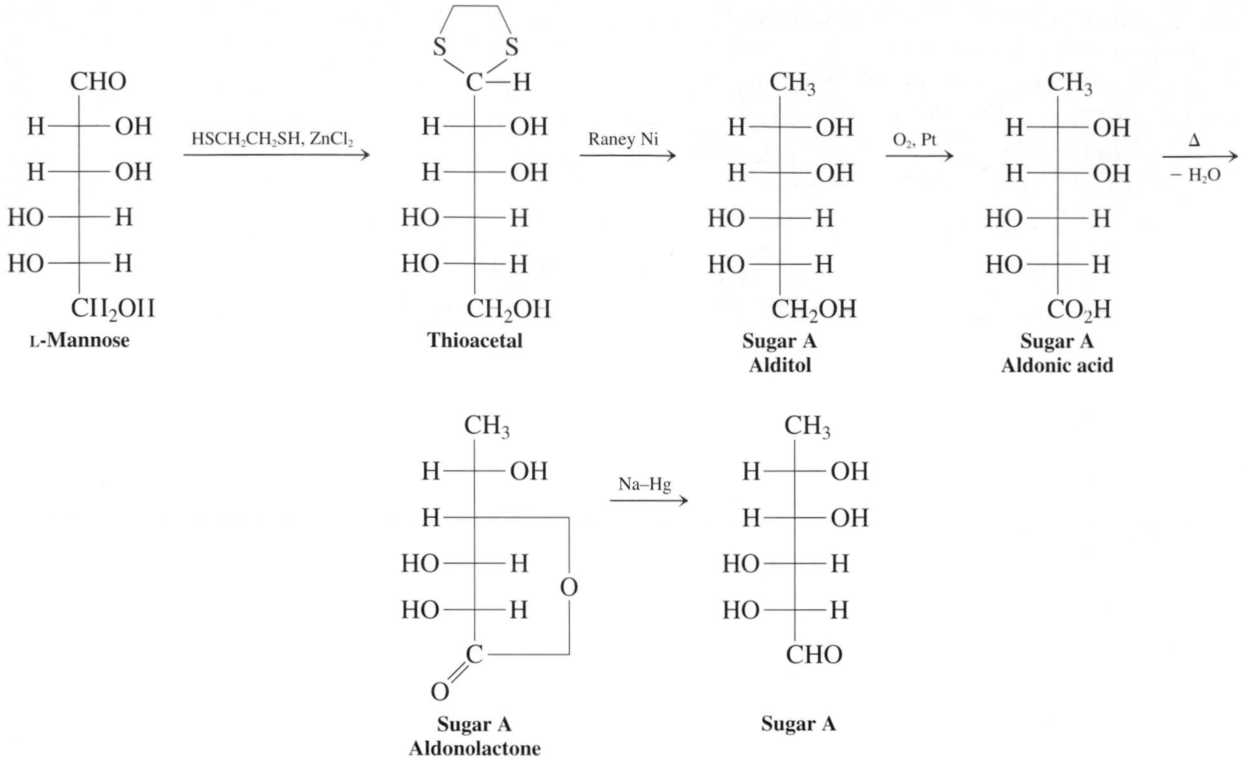

Reaction scheme (Fischer projections):

L-Mannose

$$\begin{array}{c} CHO \\ H\!-\!OH \\ H\!-\!OH \\ HO\!-\!H \\ HO\!-\!H \\ CH_2OH \end{array}$$

$\xrightarrow{HSCH_2CH_2SH, ZnCl_2}$

Thioacetal

$$\begin{array}{c} S\!-\!S \\ C\!-\!H \\ H\!-\!OH \\ H\!-\!OH \\ HO\!-\!H \\ HO\!-\!H \\ CH_2OH \end{array}$$

$\xrightarrow{Raney\ Ni}$

**Sugar A
Alditol**

$$\begin{array}{c} CH_3 \\ H\!-\!OH \\ H\!-\!OH \\ HO\!-\!H \\ HO\!-\!H \\ CH_2OH \end{array}$$

$\xrightarrow{O_2,\ Pt}$

**Sugar A
Aldonic acid**

$$\begin{array}{c} CH_3 \\ H\!-\!OH \\ H\!-\!OH \\ HO\!-\!H \\ HO\!-\!H \\ CO_2H \end{array}$$

$\xrightarrow[-\,H_2O]{\Delta}$

**Sugar A
Aldonolactone**

$$\begin{array}{c} CH_3 \\ H\!-\!OH \\ H \\ HO\!-\!H \\ HO\!-\!H \\ C\!=\!O \end{array}$$

$\xrightarrow{Na-Hg}$

Sugar A

$$\begin{array}{c} CH_3 \\ H\!-\!OH \\ H\!-\!OH \\ HO\!-\!H \\ HO\!-\!H \\ CHO \end{array}$$

Sugar A has the name 6-deoxy-L-mannose.

c. Complete methylation of rutinose with the use of excess dimethyl sulfate (Section 24-8) gives a heptamethylated derivative. On subsequent mild acid hydrolysis, one equivalent of 2,3,4-tri-*O*-methyl-D-glucose and one equivalent of the 2,3,4-tri-*O*-methyl derivative of sugar A are obtained. What possible structure(s) of rutinose is consistent with these data?

SOLUTION

Dimethyl sulfate treatment converts all free –OH groups into –OCH$_3$ moieties (Section 24-8). Therefore, we may conclude that rutinose possesses seven hydroxy substituents, and (at least) one must be part of a cyclic hemiacetal function. Recall that rutinose is a reducing sugar (Section 24-4). We may also conclude that *both* monosaccharide components of rutinose are in cyclic forms. Why? The open-chain forms of glucose and sugar A contain a total of *nine* –OH groups. For rutinose to possess only seven, two of the original nine must be built into a cyclic glycosidic acetal function linking the sugars, similar to the linkages in the disaccharides maltose, cellobiose, and lactose (Section 24-11).

Acid hydrolysis of the heptamethylated rutinose gives two *trimethylated* monosaccharides. Acid cleaves the glycoside linkage between the two sugars (Section 24-11), but where did the seventh methyl group go? It must have been attached to a hemiacetal oxygen, in the form of a methyl glycoside—an acetal—which we know is cleaved by acid much more easily than an ordinary methyl ether (Section 24-8). The six methyl groups (at carbons 2, 3, and 4 in the two resulting monosaccharides) must be bound to oxygen atoms that are *not* part of either the glycosidic linkage between the sugars or their rings. Only the oxygen atoms left unmethylated by dimethyl sulfate treatment can constitute these parts of rutinose. Thus, we may conclude that the oxygens at C5 in both methylated monosaccharide products are contained in pyranose (six-membered) rings, because five-membered rings would have contained the oxygens at C4 instead. We are almost finished. There remain three hydroxy groups that are candidates for the linkage between the sugars: the hemiacetal –OH at C1 of sugar A and either the C1 hemiacetal hydroxy or the C6 primary –OH of the glucose. To summarize:

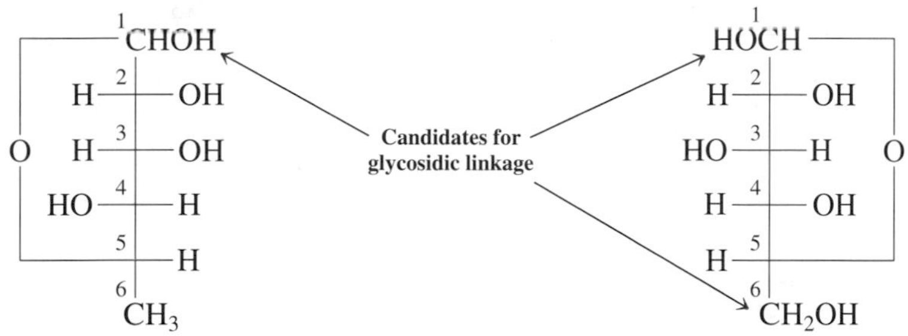

6-Deoxy-L-mannopyranose
Sugar A

Candidates for glycosidic linkage

D-Glucopyranose

In fact, we already have the answer: Rutinose is a *reducing sugar* and must therefore possess *at least one hemiacetal function.* The only option therefore is to assign the latter to the glucose moiety and the glycoside bond to C6-OH and C1 of sugar A. We conclude by redrawing the structures in more descriptive Haworth projections and chair conformations (the stereochemistry at C1 of sugar A is not defined) by following the procedures in Section 24-2. In the following three drawings, the glucopyranose moieties are below the sugar A plane. Because A is an L sugar, care is necessary to preserve the proper absolute stereochemistry. One method is to follow the same procedure as for D sugars: Place the ether linkage in back and rotate the Fischer projection 90° clockwise, such that groups originally on the right in the Fischer projection now point downward, and those on the left point upward. Indicators D and L refer to the configuration at C5. Whereas the substituent (C6) at this position points upward in a D sugar, with C5 inverted, this group must now point downward. The two structures below result. An equally correct alternative is to rotate the Fischer projection of an L sugar the opposite way—counterclockwise. In this manner the structure at the far right is obtained. Notice that the C6 methyl group now points *upward,* the anomeric carbon (C1) is to the *left,* and the structure of sugar A is presented in a view rotated by 180° about an axis perpendicular to the page, relative to the drawings at the right. Build models!

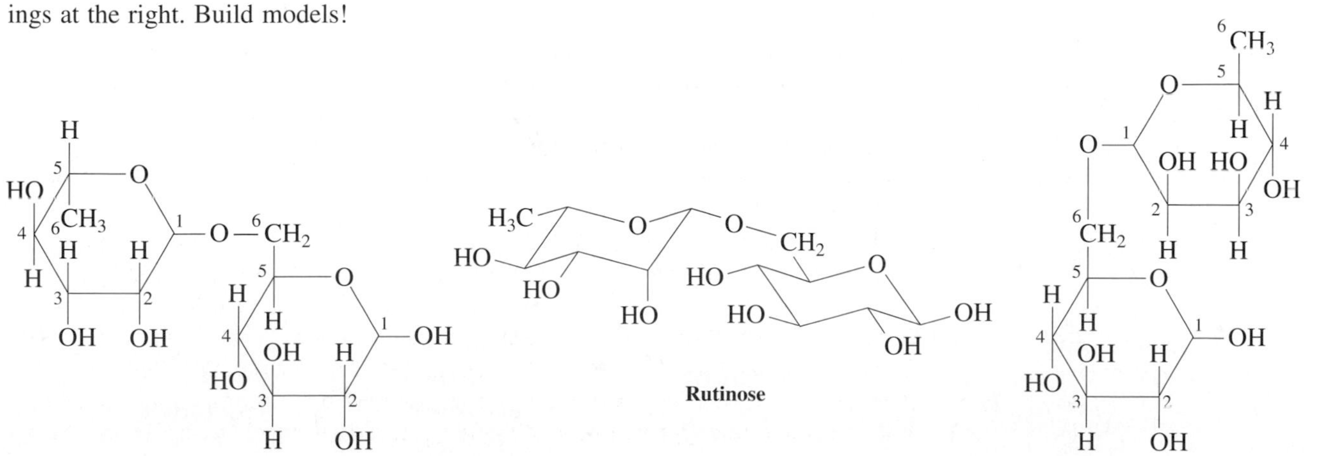

Rutinose

NEW REACTIONS

1. Cyclic Hemiacetal Formation in Sugars (Section 24-2)

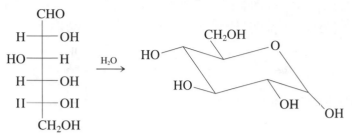

α- and β-Glucopyranoses

2. Mutarotation (Section 24-3)

α anomer
$([\alpha]_D^{25°C} = +112)$

(Equilibrium $[\alpha]_D^{25°C} = +52.7$)

β anomer
$([\alpha]_D^{25°C} = +18.7)$

3. Oxidation (Section 24-4, most nonessential H and OH substituents have been omitted)

Tests for reducing sugars

CHO

Cu^{2+}, OH^-, H_2O (Fehling's solution)
or Ag^+, NH_4OH, H_2O (Tollens's solution)

COOH

H——OH
CH_2OH

H——OH
CH_2OH

+ Cu_2O or Ag
Red   Silver
mirror

Aldonic acid synthesis

CHO

H——OH
CH_2OH

Br_2, H_2O

COOH

H——OH
CH_2OH

Aldonic acid

$- H_2O$

H——
CH_2OH

γ-**Lactone**

Aldaric acid synthesis

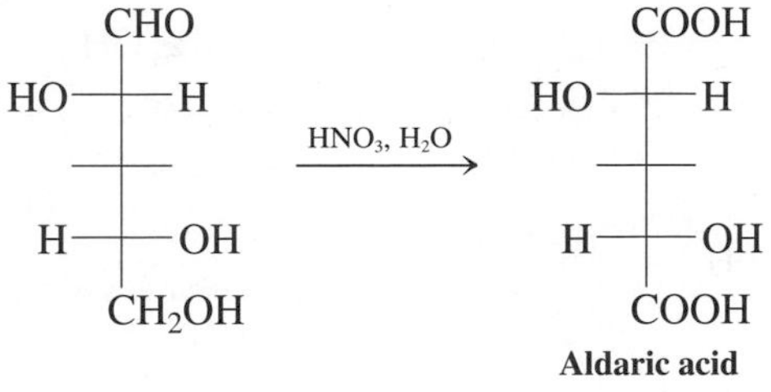

Aldaric acid

4. Sugar Degradation (Section 24-5)

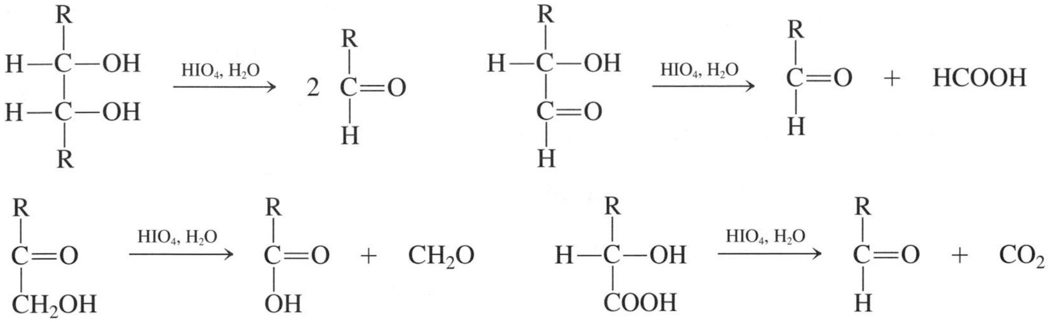

5. Reduction (Section 24-6)

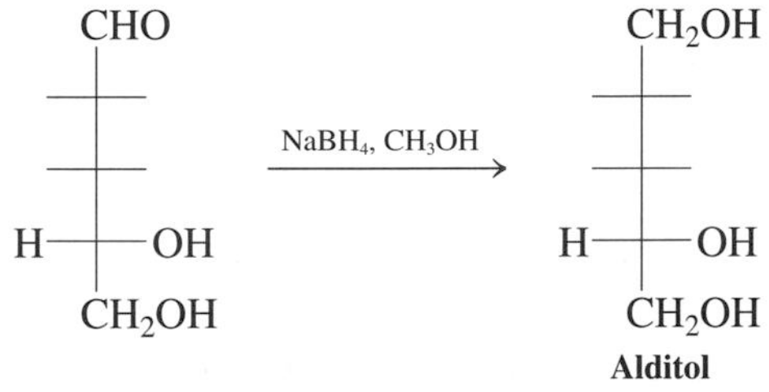

Alditol

6. Hydrazones and Osazones (Section 24-7)

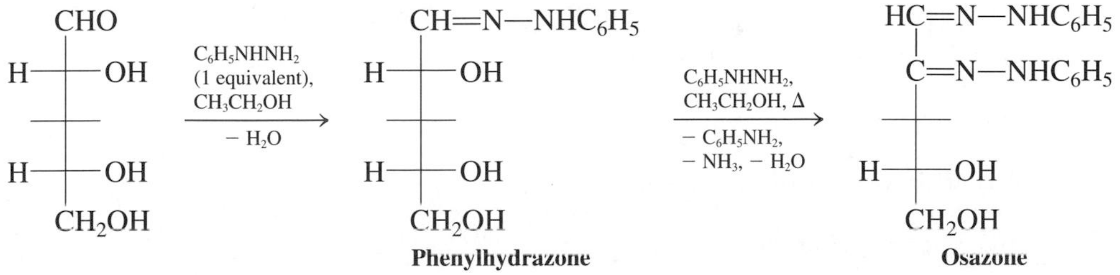

Phenylhydrazone **Osazone**

7. Esters (Section 24-8)

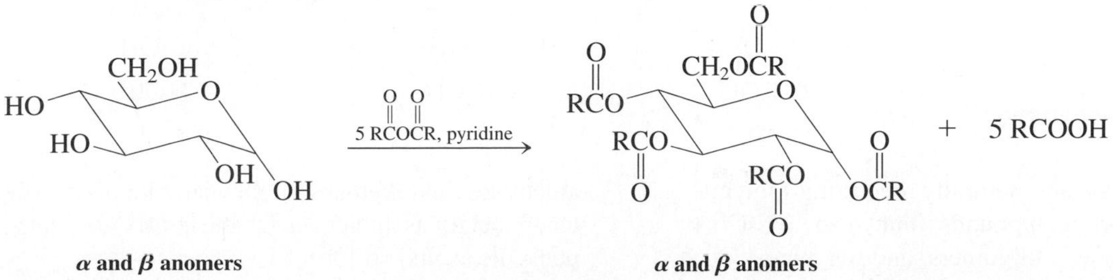

α and *β* anomers *α* and *β* anomers

8. Glycosides (Section 24-8)

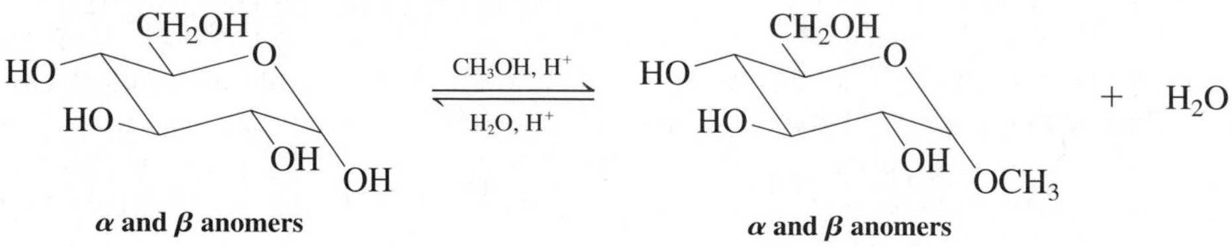

α and β anomers   α and β anomers

$$\text{CH}_3\text{OH, H}^+ \rightleftharpoons \text{H}_2\text{O, H}^+$$

+ H₂O

9. Ethers (Section 24-8)

α and β anomers   α and β anomers

5 (CH₃)₂SO₄, Na⁺⁻OH / − Na₂SO₄

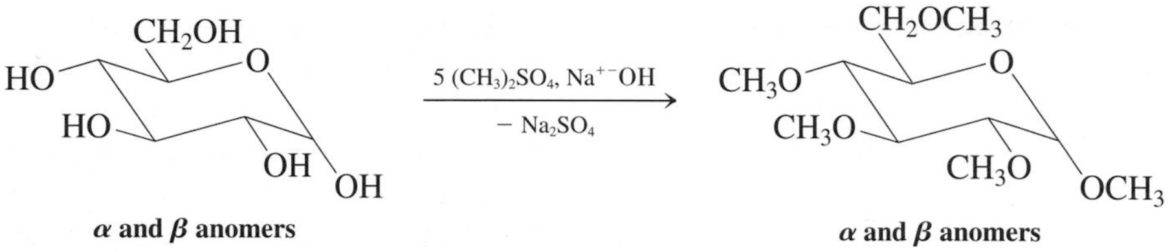

10. Cyclic Acetals (Section 24-8)

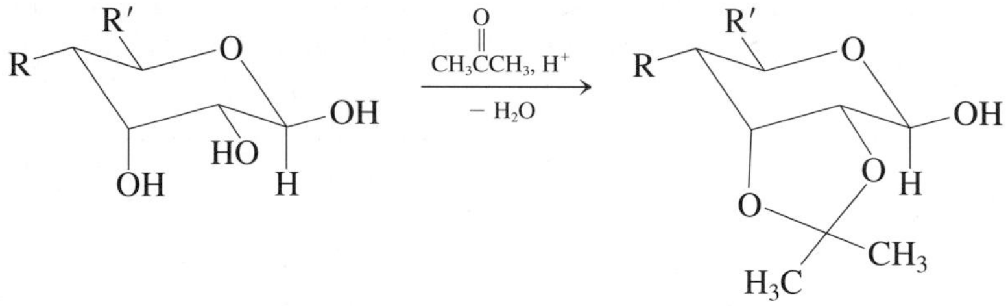

CH₃CCH₃, H⁺ / − H₂O

11. Chain Extension Through Cyanohydrins (Section 24-9)

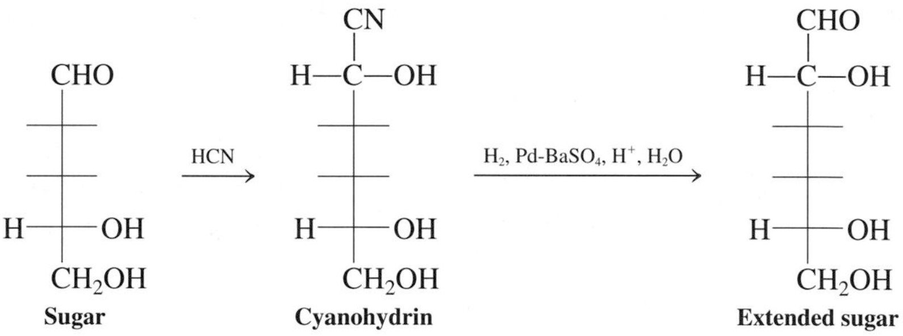

Sugar   Cyanohydrin   Extended sugar

HCN →   H₂, Pd-BaSO₄, H⁺, H₂O →

12. Ruff Degradation (Section 24-9)

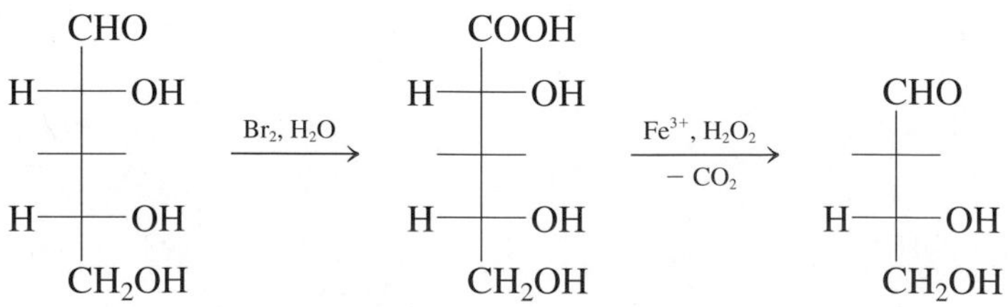

Br₂, H₂O →   Fe³⁺, H₂O₂ / − CO₂ →

IMPORTANT CONCEPTS

1. **Carbohydrates** are naturally occurring **polyhydroxycarbonyl** compounds that can exist as monomers, dimers, oligomers, and polymers.

2. **Monosaccharides** are called **aldoses** if they are aldehydes and **ketoses** if they are ketones. The chain length is indicated by the prefix tri-, tetr-, pent-, hex-, and so forth.

3. Most natural carbohydrates belong to the D **family;**

that is, the stereocenter farthest from the carbonyl group has the same configuration as that in (R)-$(+)$-2,3-dihydroxypropanal [D-$(+)$-glyceraldehyde].

4. The keto forms of carbohydrates exist in equilibrium with the corresponding five-membered (**furanoses**) or six-membered (**pyranoses**) cyclic hemiacetals. The new stereocenter formed by cyclization is called the **anomeric carbon,** and the two **anomers** are designated α and β.

5. **Haworth projections** of D sugars depict the cyclic ether in line notation as a pentagon or a hexagon, the anomeric carbon placed on the right and the ether oxygen at the top. The substituents located above or below the ring are attached to vertical lines. The ring bond at the bottom (between C2 and C3) is understood to be in front of the plane of the paper, and the ring bonds containing the oxygen are understood to be in back. The α anomer has the OH group at the anomeric carbon pointing downward, whereas the β anomer has it pointing upward.

6. Equilibration between anomers in solution gives rise to changes in the measured optical rotation called **mutarotation.**

7. The reactions of the saccharides are characteristic of carbonyl, alcohol, and hemiacetal groups. They include oxidation of the aldehyde to the carboxy function of **aldonic acids,** double oxidation to **aldaric acids,** oxidative cleavage of vicinal diol units,

reduction to **alditols,** condensations, esterifications, and acetal formations.

8. Sugars containing hemiacetal functions are called **reducing sugars,** because they readily reduce Tollens's and Fehling's solutions. Sugars in which the anomeric carbon is acetalized are nonreducing.

9. The synthesis of higher sugars is based on **chain lengthening,** the new carbon being introduced by cyanide ion. The synthesis of lower sugars relies on **Ruff chain shortening,** a terminal carbon being expelled as CO_2.

10. The **Fischer proof** uses the techniques of chain lengthening and shortening together with the symmetry properties of aldaric acids to determine the structures of the aldoses.

11. Di- and **higher saccharides** are formed by ether formation between monomers; the ether bridge usually includes at least one hemiacetal hydroxy group.

12. The change in optical rotation observed in aqueous solutions of sucrose, called the **inversion of sucrose,** is due to the equilibration of the starting sugar with the various cyclic and anomeric forms of its component monomers.

13. Many sugars contain modified backbones. Amino groups may have replaced hydroxy groups, there may be substituents of various complexity (**aglycons**), the backbone carbon atoms of a sugar may lack oxygens, and (rarely) the sugar may adopt the L configuration.

PROBLEMS

25. The designations D and L as applied to sugars refer to the configuration of the highest-numbered stereocenter. If the configuration of the highest-numbered stereocenter of D-ribose (Figure 24-1) is switched from D to L, is the product L-ribose? If not, what is the product? How is it related to D-ribose (i.e., what kind of isomers are they)?

26. To which classes of sugars do the following monosaccharides belong? Which are D and which are L?

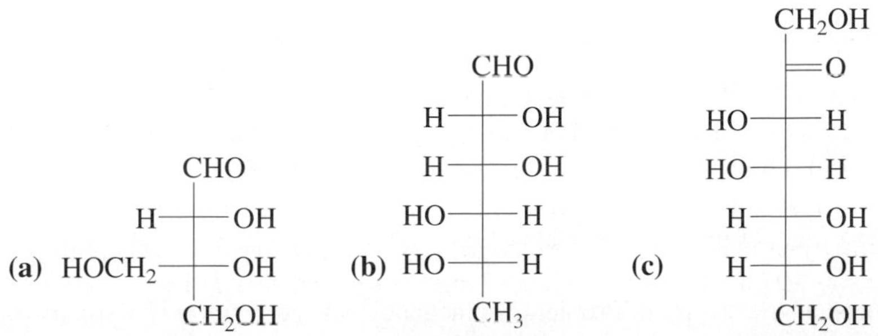

(a)

```
          CHO
      H ——┼—— OH
HOCH₂ ——┼—— OH
         CH₂OH
```
(+)-**Apiose**

(b)

```
          CHO
      H ——┼—— OH
      H ——┼—— OH
     HO ——┼—— H
     HO ——┼—— H
          CH₃
```
(−)-**Rhamnose**

(c)

```
         CH₂OH
            �═O
     HO ——┼—— H
     HO ——┼—— H
      H ——┼—— OH
      H ——┼—— OH
         CH₂OH
```
(+)-**Mannoheptulose**

27. Draw open-chain (Fischer-projection) structures for L-(+)-ribose and L-(−)-glucose (see Exercise 24-2). What are their systematic names?

28. Identify the following sugars, which are represented by unconventionally drawn Fischer projections. (**Hint:** It will be necessary to convert these projections into more conventional representations *without* inverting any of the stereocenters.)

(a)

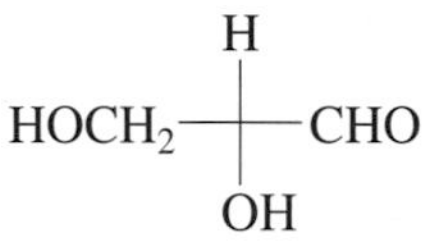

(b)

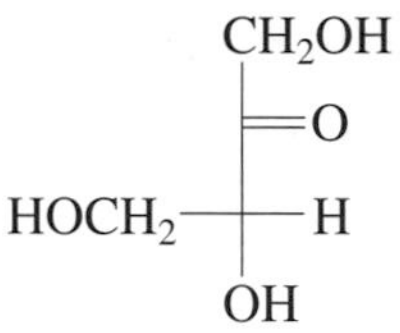

(c)

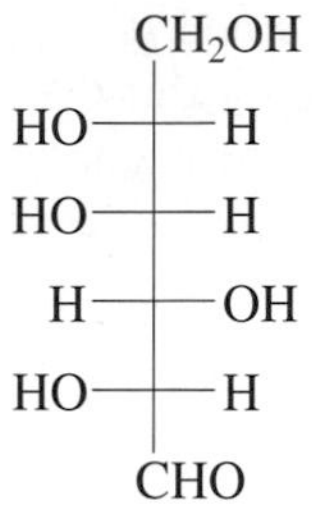

(d)

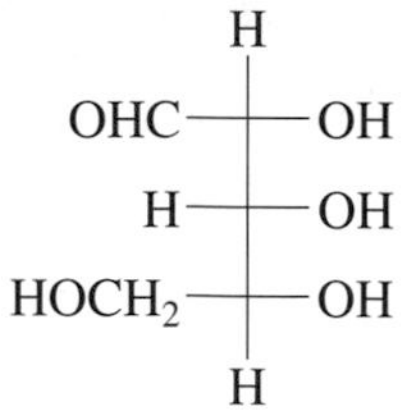

(e)

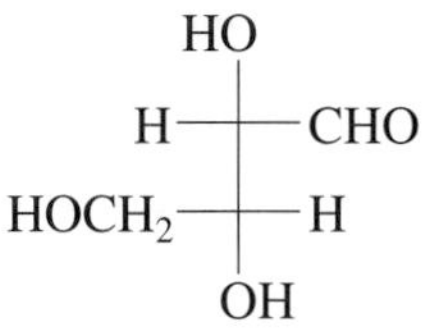

29. Redraw each of the following sugars in open-chain form as a Fischer projection, and find its common name.

(a)

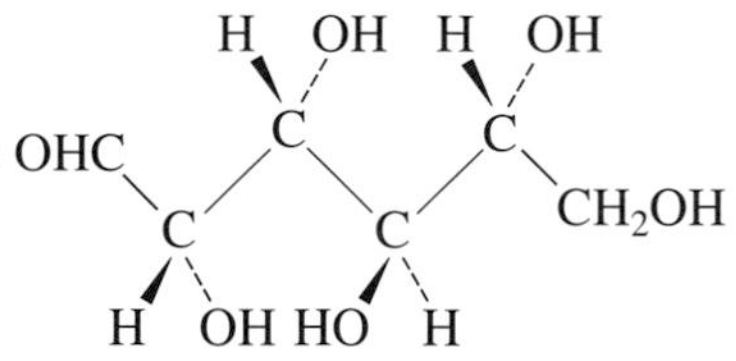

(b)

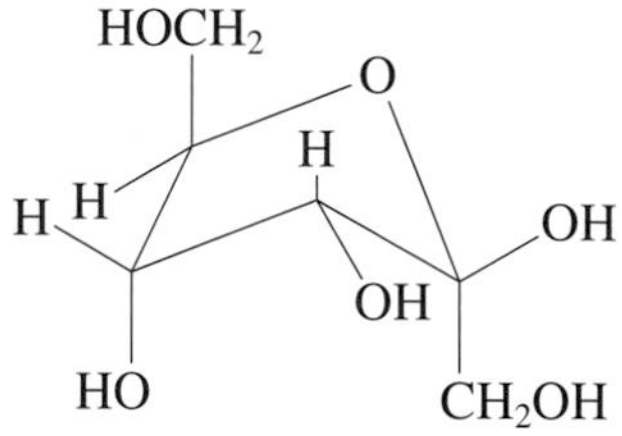

(c)

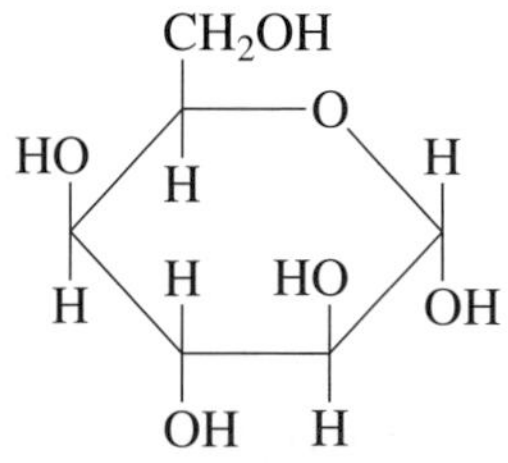

(d)

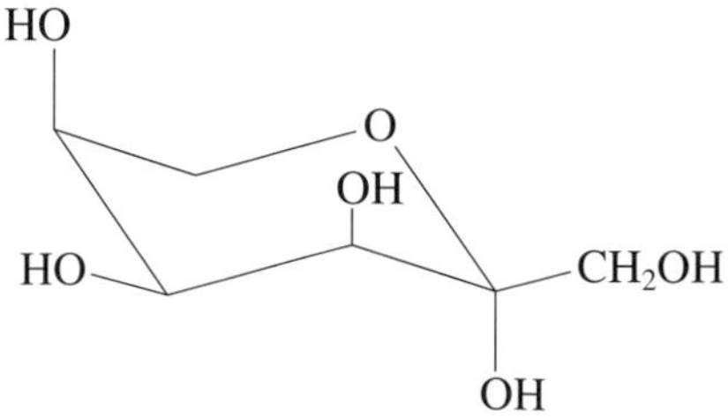

30. For each of the following sugars, draw all reasonable cyclic structures, using either Haworth or conformational formulas; indicate which structures are pyranoses and which are furanoses; and label α and β anomers.
    (a) (−)-Threose; (b) (−)-allose; (c) (−)-ribulose; (d) (+)-sorbose;
    (e) (+)-mannoheptulose (Problem 25).

31. Are any of the sugars in Problem 30 incapable of mutarotation? Explain your answer.

32. Draw the most stable pyranose conformation of each of the following sugars.
(a) α-D-Arabinose; (b) β-D-galactose; (c) β-D-mannose; (d) α-D-idose.

33. Write the expected products of the reaction of each of the following sugars
with (i) Br_2, H_2O; (ii) HNO_3, H_2O, 60°C; (iii) $NaBH_4$, CH_3OH; and
(iv) excess $C_6H_5NHNH_2$, CH_3CH_2OH, Δ. Find the common names of all
the products. (a) D-(−)-Threose; (b) D-(+)-xylose; (c) D-(+)-galactose.

34. Draw the Fischer projection of an aldohexose that will give the same osazone
as (a) D-(−)-idose and (b) L-(−)-altrose.

35. (a) Which of the aldopentoses (Figure 24-1) would give optically active
alditols upon reduction with $NaBH_4$? (b) Using D-fructose, illustrate the results
of $NaBH_4$ reduction of a ketose. Is the situation more complicated than reduction of an aldose? Explain.

36. Which of the following glucoses and glucose derivatives are capable of undergoing mutarotation? (a) α-D-Glucopyranose; (b) methyl α-D-glucopyranoside;
(c) methyl α-2,3,4,6-tetra-O-methyl-D-glucopyranoside (i.e., the tetramethyl
ether at carbon 2, 3, 4, and 6); (d) α-2,3,4,6-tetra-O-methyl-D-glucopyranose;
(e) α-D-glucopyranose 1,2-propanone acetal.

37. (a) Explain why the oxygen at C1 of an aldopyranose can be methylated so
much more easily than the other oxygens in the molecule. (b) Explain why the
methyl ether unit at C1 of a fully methylated aldopyranose can be hydrolyzed
so much more easily than the other methyl ether functions in the molecule.
(c) Write the expected product(s) of the following reaction.

$$\text{D-Fructose} \xrightarrow{CH_3OH,\ 0.25\%\ HCl,\ H_2O}$$

38. Of the four aldopentoses, two form diacetals readily when treated with excess
acidic propanone (acetone), but the other two form only monoacetals. Explain.

39. D-Sedoheptulose is a sugar that plays a role in a metabolic cycle (the *pentose
oxidation cycle*) that converts glucose into 2,3-dihydroxypropanal (glyceraldehyde) plus three equivalents of CO_2. Determine the structure of D-sedoheptulose from the following information.

$$\text{D-Sedoheptulose} \xrightarrow{6\ HIO_4} 4\ \overset{O}{\overset{\|}{HCOH}} + 2\ \overset{O}{\overset{\|}{HCH}} + CO_2$$

$$\text{D-Sedoheptulose} \xrightarrow{C_6H_5NHNH_2} \begin{array}{l}\text{an osazone identical with that formed}\\ \text{by another sugar, aldoheptose A}\end{array}$$

$$\text{Aldoheptose A} \xrightarrow{\text{Ruff degradation}} \text{aldohexose B}$$

$$\text{Aldohexose B} \xrightarrow{HNO_3,\ H_2O,\ \Delta} \text{an optically active product}$$

$$\text{Aldohexose B} \xrightarrow{\text{Ruff degradation}} \text{D-ribose}$$

40. Illustrate the results of chain elongation of D-talose through a cyanohydrin. How many products are formed? Draw them. After treatment with warm HNO_3, does the product(s) give optically active or inactive dicarboxylic acids?

41. **(a)** Write a detailed mechanism for the isomerization of β-D-fructofuranose from the hydrolysis of sucrose into an equilibrium mixture of the β-pyranose and β-furanose forms. **(b)** Although fructose usually appears as a furanose when it is part of a polysaccharide, in the pure crystalline form, fructose adopts a β-pyranose structure. Draw β-D-fructopyranose in its most stable conformation. In water at 20°C, the equilibrium mixture contains about 68% β-D-pyranose and 32% β-D-furanose. **(c)** What is the free-energy difference between the pyranose and furanose forms at this temperature? **(d)** Pure β-D-fructopyranose has $[\alpha]_D^{20°C} = -132$. The equilibrium pyranose–furanose mixture has $[\alpha]_D^{20°C} = -92$. Calculate $[\alpha]_D^{20°C}$ for pure β-D-fructofuranose.

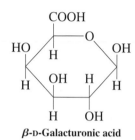

β-D-Galacturonic acid

42. Classify each of the following sugars and sugar derivatives as either reducing or nonreducing. **(a)** D-Glyceraldehyde; **(b)** D-arabinose; **(c)** β-D-arabinopyranose 3,4-propanone acetal; **(d)** β-D-arabinopyranose propanone diacetal; **(e)** D-ribulose; **(f)** D-galactose; **(g)** methyl β-D-galactopyranoside; **(h)** β-D-galacturonic acid (as shown in the margin); **(i)** β-cellobiose; **(j)** α-lactose.

43. Is α-lactose capable of mutarotation? Write an equation to illustrate your answer.

44. Trehalose, sophorose, and turanose are disaccharides. Trehalose is found in the cocoons of some insects, sophorose turns up in a few bean varieties, and turanose is an ingredient in low-grade honey made by bees with indigestion from a diet of pine tree sap. Identify among the following structures those that correspond to trehalose, sophorose, and turanose on the basis of the following information: (i) Turanose and sophorose are reducing sugars. Trehalose is nonreducing. (ii) On hydrolysis, sophorose and trehalose give two molecules each of aldoses. Turanose gives one molecule of an aldose and one molecule of a ketose. (iii) The two aldoses that constitute sophorose are anomers of each other.

(a)

(b)

(c)

(d)

45. **(a)** A mixture of (*R*)-2,3-dihydroxypropanal (D-glyceraldehyde) and 1,3-dihydroxypropanone (1,3-dihydroxyacetone) that is treated with aqueous NaOH rapidly yields a mixture of three sugars: D-fructose, D-sorbose, and racemic dendroketose (only one enantiomer is shown here). Explain this result by means of a detailed mechanism. **(b)** The same product mixture is also obtained if either the aldehyde or the ketone *alone* is treated with base. Explain. [**Hint:** Closely examine the intermediates in your answer to (a).]

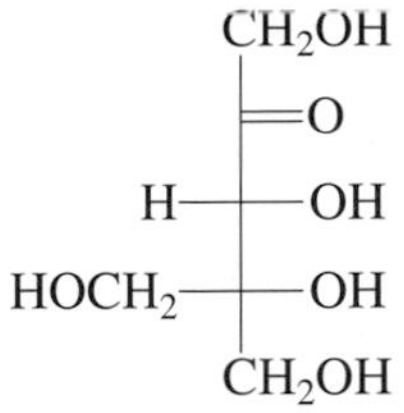

$$
\begin{array}{c}
CH_2OH \\
| \\
=O \\
H\!-\!\!-\!OH \\
HOCH_2\!-\!\!-\!OH \\
| \\
CH_2OH
\end{array}
$$

Dendroketose

46. Write or draw the missing reagents and structures (a) through (g). What is the common name of (g)?

D-(+)-Xylose $\xrightarrow{\text{(a)}}$ **(b)** $\xrightarrow{\text{(c)}}$ **(d)** $\xrightarrow{\text{NH}_3,\ \Delta}$ $C_5H_{11}NO_5$ $\xrightarrow{\text{Br}_2,\ \text{NaOH}}$

D-Xylonic acid   **Methyl D-xylonate**   (e)

$$CO_2 + C_4H_{11}NO_4 \xrightarrow{\Delta} NH_3 + C_4H_8O_4$$
             (f)   (g)

The preceding sequence (called the *Weerman degradation*) achieves the same end as what procedure described in this chapter?

47. Fischer's solution to the problem of sugar structures was actually much more difficult to achieve experimentally than Section 24-10 implies. For one thing, the only sugars that he could readily obtain from natural sources were glucose, mannose, and arabinose. (Erythrose and threose were, in fact, not then available at all, either naturally or synthetically.) His ingenious solution required a source of gulose so that he could make the critical comparison of dicarboxylic acids described at the end of the section. Unfortunately, gulose does not exist in nature; so Fischer had to make it. His synthesis, from glucose, was difficult because, at a key point, he got a troublesome mixture of products. Nowadays the following synthesis might be used.

Fill in the missing reagents and structures (a) through (g). Use Fischer projections for all structures. Follow the instructions and hints in parentheses.

D-(+)-Glucose $\xrightarrow{\text{(a)}}$ **(b)** $\xrightarrow[\substack{\text{(A special reaction}\\ \text{that oxidizes } only\\ \text{the primary hydroxy}\\ \text{at C6 into a}\\ \text{carboxylic group)}}]{\text{O}_2,\ \text{Pt}}$ **(c)** $\xrightarrow{\text{H}^+,\ \text{H}_2\text{O}}$

Methyl D-glucoside   **Methyl D-glucuronoside**

(Both isomers; write only one)

(d) $\xrightarrow{\text{NaBH}_4}$ **(e)** $\xrightarrow{\Delta}$ H_2O + **(f)** $\xrightarrow[\substack{\text{(Reduces lactones}\\ \text{to aldehydes)}}]{\text{Na-Hg}}$ **(g)**

D-Glucuronic acid   **Gulonic acid**   **Gulonolactone**   **Gulose**

(Write the open-chain form only)   (Write the open-chain form only)

Is the gulose that Fischer synthesized from D-glucose an L sugar or a D sugar? (Be careful. Fischer himself got the wrong answer at first, and that confused *everybody* for *years*.)

48. Vitamin C (ascorbic acid, Section 22-9) is present almost universally in the plant and animal kingdoms. (According to Linus Pauling, mountain goats biosynthesize from 12 to 14 g of it per day.) Animals produce it from D-glucose in the liver by the four-step sequence D-glucose → D-glucuronic acid (see Problem 47) → D-glucuronic acid γ-lactone → L-gulonic acid γ-lactone → vitamin C.

Vitamin C

The enzyme that catalyzes the last reaction, L-gulonolactone oxidase, is absent from humans, some monkeys, guinea pigs, and birds, presumably because of a defective gene resulting from a mutation that may have occurred some 60 million years ago. As a result, we have to get our vitamin C from food or make it in the laboratory. In fact, the ascorbic acid in almost all vitamin supplements is synthetic. An outline of one of the major commercial syntheses follows. Draw the missing reagents and products (a) through (f).

2-Keto-L-gulonic acid   **Keto form of vitamin C**

Team Problem

49. This problem is meant to encourage you to think as a team about how you might establish the structure of a simple disaccharide, with some additional information at your disposal. Consider D-lactose (Section 24-11) and assume you do not know its structure. You are given the knowledge that it is a disaccharide, linked in a β manner to the anomeric carbon of only one of the sugars, and you are given the structures of all of the aldohexoses (Table 24-1), as well as all of their possible methyl ethers. Deal with the following questions as a team or by dividing the work, before joint discussion, as appropriate.

1. Mild acid hydrolyzes your "unknown" to D-galactose and D-glucose. How much information can you derive from that result?

2. Propose an experiment that tells you that the two sugars are not connected through their respective anomeric centers.

3. Propose an experiment that tells you which one of the two sugars contains an acetal group used to bind the other. (**Hint:** The functional group chemistry of the monosaccharides described in the chapter can be applied to higher sugars as well. Specifically, consider Section 24-4 here.)

4. Making use of the knowledge of the structure of all possible methyl ethers of the component monosaccharides, design experiments that will tell you which (nonanomeric) hydroxy group is responsible for the disaccharide linkage.

5. Similarly, can you use this approach to distinguish between a furanose and pyranose structure for the component of the disaccharide that can mutarotate?

Preprofessional Problems

50. Most natural sugars have a stereocenter that is identical to that in (R)-2,3-dihydroxypropanal, shown as a Fischer projection in the margin. What is the (very popular) common name for this compound?
(**a**) D-(+)-Glyceraldehyde; (**b**) D-(−)-glyceraldehyde; (**c**) L-(+)-glyceraldehyde; (**d**) L-(−)-glyceraldehyde

51. What kind of sugar is the compound shown in the margin?
(**a**) An aldopentose; (**b**) a ketopentose; (**c**) an aldohexose; (**d**) a ketohexose

52. Which one of the following statements is *true* about the oxacyclohexane conformer of the sugar β-D-(+)-glucopyranose?
(**a**) One OH group is axial, but all remaining substituents are equatorial.
(**b**) The CH_2OH group is axial, but all remaining groups are equatorial.
(**c**) All groups are axial. (**d**) All groups are equatorial.

53. The methyl glycoside of mannose is made by treating the sugar with:
(**a**) $AlBr_3$, CH_3Br; (**b**) dilute aqueous CH_3OH; (**c**) CH_3OCH_3 and $LiAlH_4$;
(**d**) CH_3OH, HCl; (**e**) oxacyclopropane, $AlCl_3$

54. One of the statements below is correct about the sugar shown. Which one?

(**a**) It is a nonreducing sugar.   (**b**) It forms an osazone.
(**c**) It exists in two anomeric forms.   (**d**) It undergoes mutarotation.

Heterocycles

Heteroatoms in Cyclic Organic Compounds

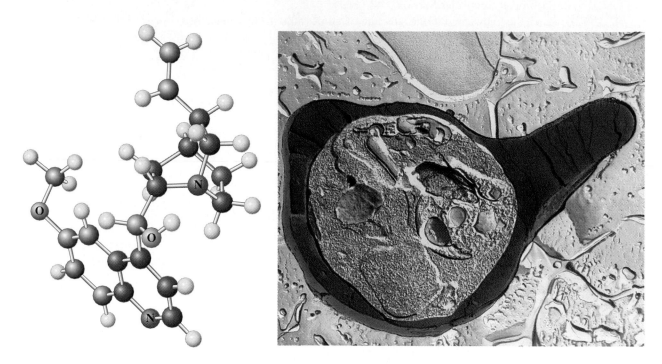

Quinine is an alkaloid isolated from the bark of the cinchona tree and constitutes the first effective agent in the treatment of malaria. It is best known as the origin of the bitter taste in tonic water. At right, malaria parasites (green) infect a blood cell.

Look at the list of the world's top ten prescription drugs (Table 25-1). What do their structures have in common? Apart from the presence of heteroatoms, they all contain at least one ring. Moreover, in all but two cases (fluoxetin and diclofenac), these rings are made up not solely of carbon, called **carbocycles** (Chapter 4), but also of heteroatoms, such as oxygen, nitrogen, and sulfur, and are thus called **heterocycles.**

This chapter describes the naming, syntheses, and reactions of some saturated and aromatic heterocyclic compounds in order of increasing ring size, starting with the heterocyclopropanes. Some of this chemistry is a simple extension of transformations presented earlier for carbocycles. However, the heteroatom often causes heterocyclic compounds to exhibit special chemical behavior.

Most physiologically active compounds owe their biological properties to the presence of heteroatoms, mainly in the form of heterocycles. A majority of the known natural products are heterocyclic. It is therefore not surprising that more than half the published chemical literature deals with such compounds, their synthesis, isolation,

TABLE 25-1 The World's Top Ten Prescription Drugs (Ranked by 1995 Sales) by Generic and, in Parentheses, Selected Brand Names[a]

1. Ranitidine (Zantac) Antiulcerative	2. Omeprazole (Prilosec) Antiulcerative
3. Amoxicillin (Amoxil, Trimox) Antibiotic	4. Nifedipine (Oxcord, Procardia XL) Antihypertensive
5. Enalapril (Vasotec) Antihypertensive	6. Fluoxetin (Prozac) Antidepressant
7. Diltiazem (Cardizem, Tiazac) Antihypertensive	8. Diclofenac (Cataflam, Voltaren) Antiinflammatory
9. Captopril (Capoten) Antihypertensive	10. Simvastatin (Sivastin, Zocor) Antihypercholesterolemic

[a]Total U.S. sales of pharmaceuticals topped \$85 billion in 1997, with omeprazole leading at \$2.3 billion.

Source: After H.-J. Quadbeck-Seeger, R. Faust, G. Knaus, and V. Siemeling, *Chemie Rekorde, Menschen, Märkte, Moleküle*, Wiley-VCH, Weinheim, 1997.

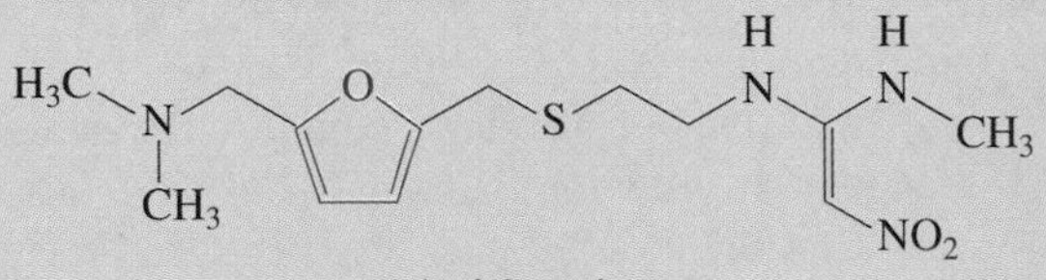

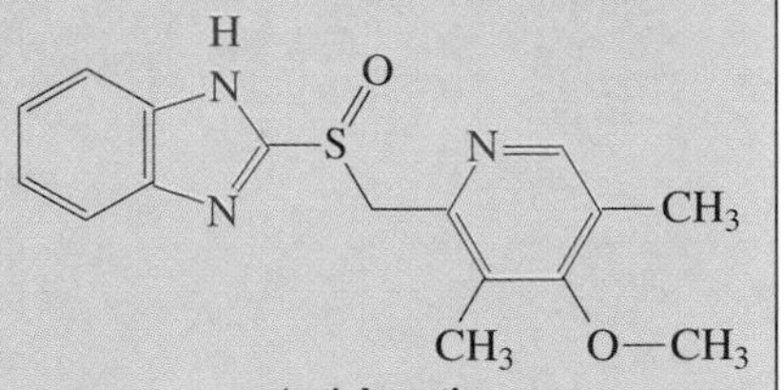

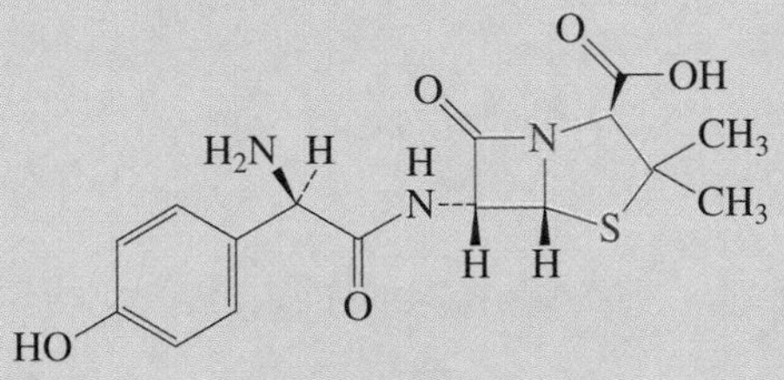

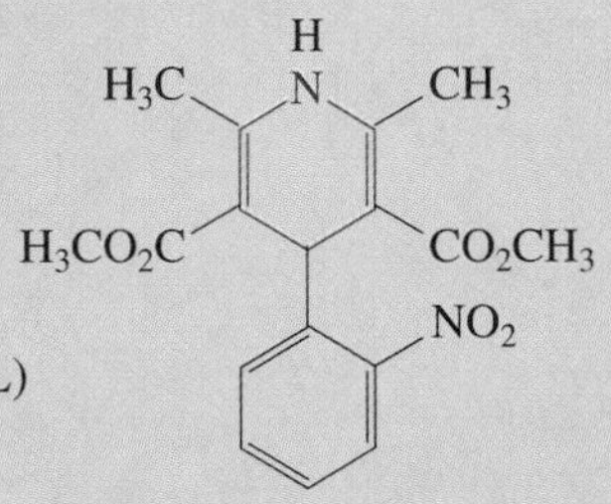

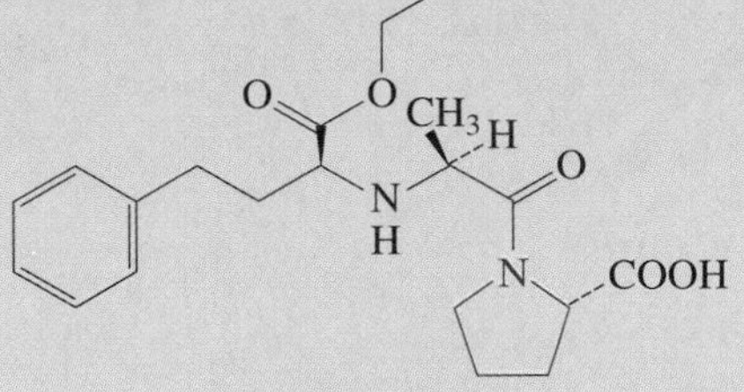

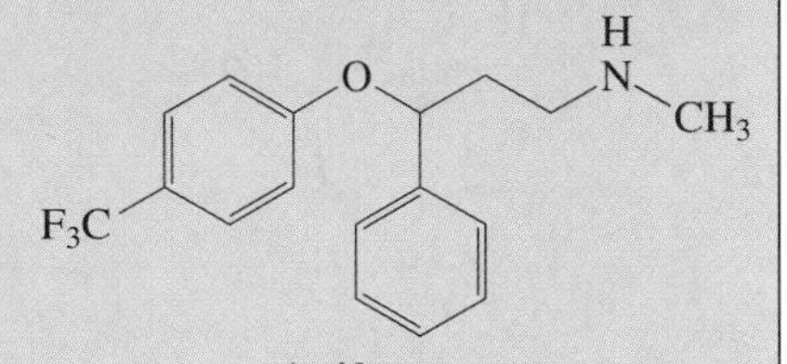

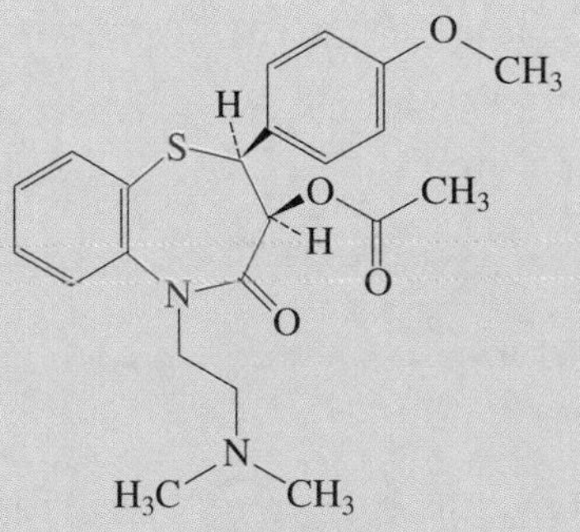

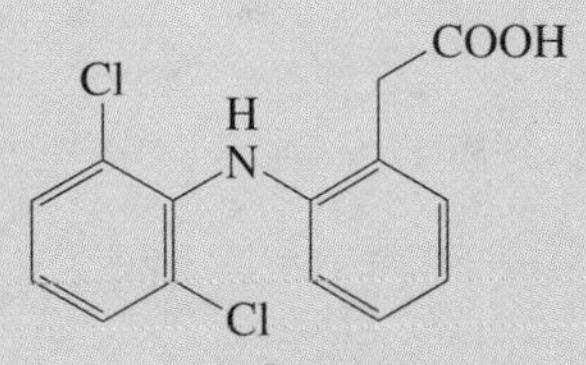

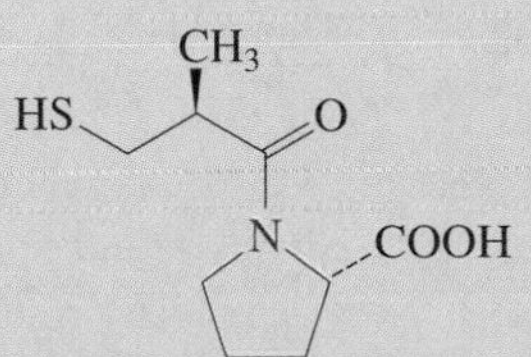

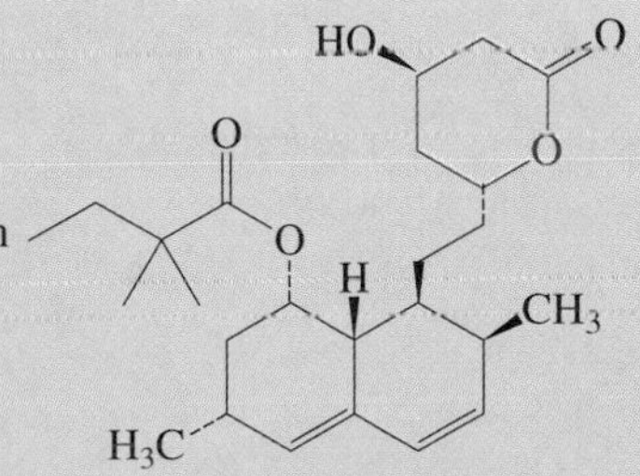

and interconversions. Indeed, we have already encountered many examples—the cyclic ethers (Section 9-6), acetals (Sections 17-8, 23-4, and 24-8), carboxylic acid derivatives (Chapters 19 and 20), and amines (Chapter 21). The bases in DNA, whose sequence stores hereditary information, are heterocycles (Section 26-9); so are many vitamins, such as B_1 (thiamine, Chemical Highlight 23-3), B_2 (riboflavin, Chemical Highlight 25-4), B_6 (pyridoxine), the spectacularly complex B_{12}, and vitamins C and E (Section 22-9). The structures of vitamins B_6 and B_{12}, as well as additional examples of heterocyclic systems and their varied uses, are depicted here.

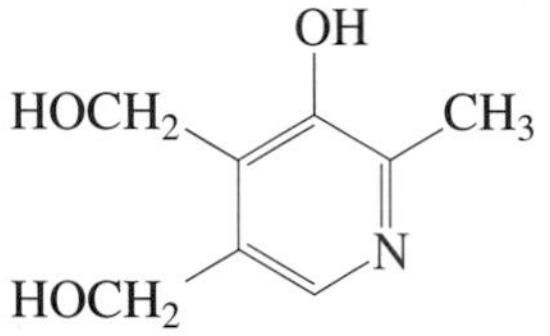

Pyridoxine, vitamin B_6

(Enzyme cofactor vitamin
with multiple functions)

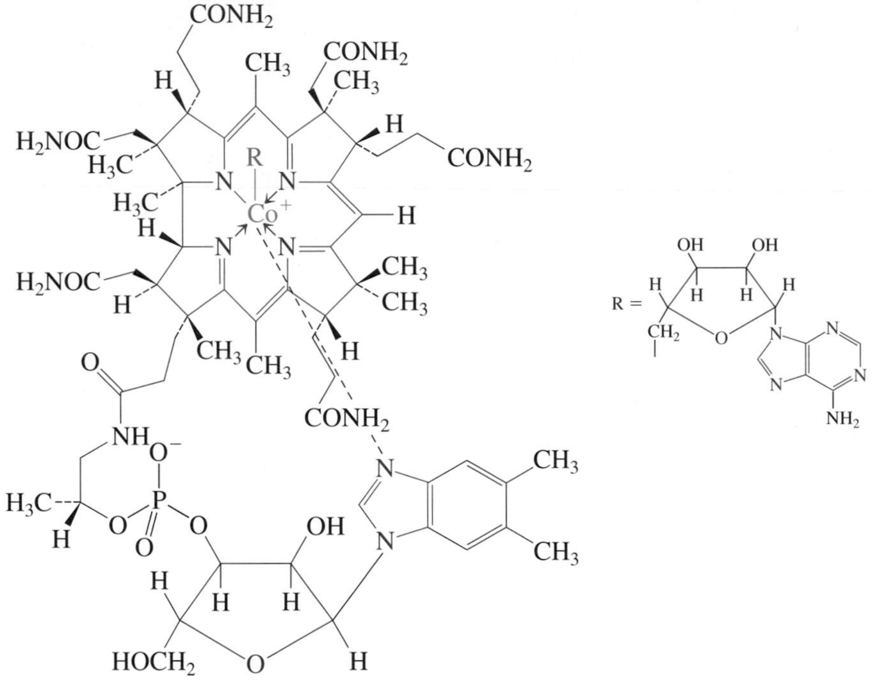

Vitamin B_{12}
(Cobalamin)

(Catalyzes biological rearrangements and methylations)

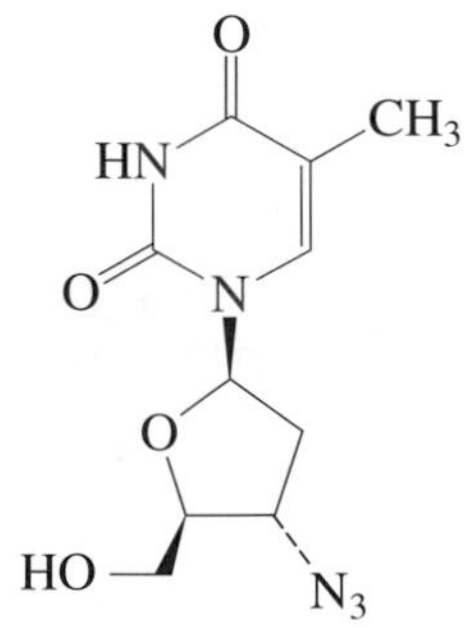

Zidovudine
(AZT)

(Antiviral AIDS drug,
see Chemical Highlight 26-4)

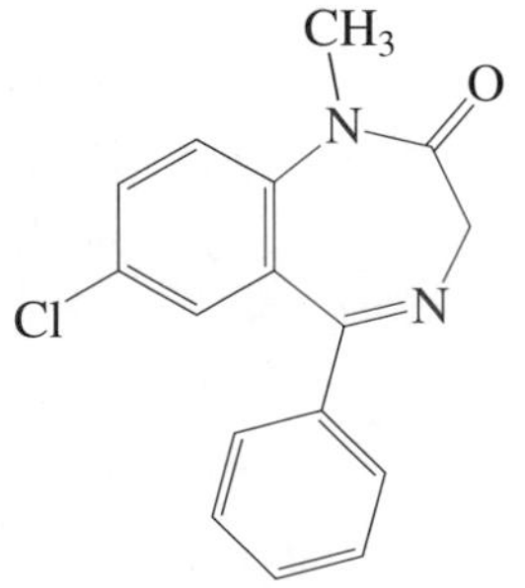

Diazepam
(Valium)

(Tranquilizer)

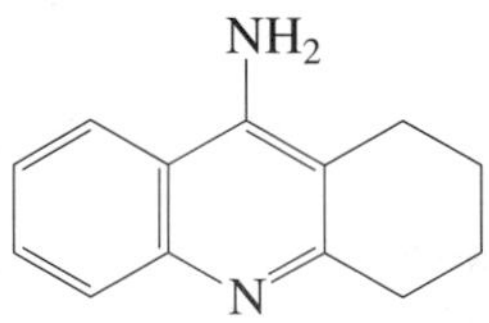

Tacrine
(Cognex)

(For the treatment of Alzheimer disease)

H₃C   CH₃

R

OH

OH

OH

Terfenadine (Seldane), R = CH₃
Fexofenadine (Allegra), R = COOH

(Antihistaminic allergy drugs;
terfenadine was withdrawn from the
market in 1998, in favor of fexofenadine, because of potential side
effects of the former)

CH_3CH_2S
O
O

N—N
S
N
CH₃
O
NHCH₃

Ethidimuron

(Antiphotosynthetic
herbicide, kills all
plants)

N
N
CH₃

8-Methylpyrrolo[1,2-*a*]-pyrazine
(Aroma of roasted meat)

OCH₃
N
N
Cl
N
N
H
SO₃H
CH=CH
HO₃S
N
H
N
N
OCH₃
N
Cl

Diaminostilbenedisulfonic acid "bleach"
(Optical whitener added to linen, paper, or plastics;
luminesces blue light to create the appearance of
intense whiteness)

25-1 Naming the Heterocycles

Like all the other classes of compounds that we have encountered, this one contains
many members bearing common names. Moreover, there are several competing systems for naming heterocycles that are sometimes confusing. We will adhere to the
simplest system. We regard saturated heterocycles as derivatives of the related carbocycles and use a prefix to denote the presence and identity of the heteroatom: **aza-**
for nitrogen, **oxa-** for oxygen, **thia-** for sulfur, **phospha-** for phosphorus, and so forth.
Other widely used names will be given in parentheses. The location of substituents
is indicated by numbering the ring atoms, starting with the heteroatom.

Oxacyclopropane
(Oxirane, ethylene oxide)

N
CH₃

N-Methylazacyclopropane
(N-Methylaziridine)

F   2   3

S
1

2-Fluorothiacyclopropane
(2-Fluorothiirane)

O

Oxacyclobutane
(Oxetane)

3-Ethylazacyclobutane
(3-Ethylazetidine)

2,2-Dimethylthiacyclobutane
(2,2-Dimethylthietane)

***trans*-3,4-Dibromooxacyclopentane**
(*trans*-3,4-Dibromotetrahydrofuran)

Azacyclopentane
(Pyrrolidine)

Thiacyclopentane
(Tetrahydrothiophene)

3-Methyloxacyclohexane
(3-Methyltetrahydropyran)

Azacyclohexane
(Piperidine)

3-Cyclopropylthiacyclohexane
(3-Cyclopropyltetrahydrothiopyran)

The common names of unsaturated heterocycles are so firmly entrenched in the literature that we shall use them here.

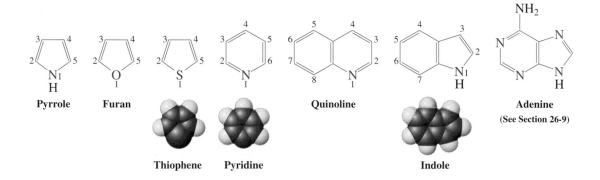

Pyrrole **Furan** **Thiophene** **Pyridine** **Quinoline** **Indole** **Adenine**
(See Section 26-9)

Name or draw the following compounds. **(a)** *trans*-2,4-Dimethyloxacyclopentane (*trans*-2,4-dimethyltetrahydrofuran); **(b)** *N*-ethylazacyclopropane;

(c)

O_2N — N — NO_2

(d)

Br

25-2 Nonaromatic Heterocycles

As illustrated by the chemistry of the oxacyclopropanes (Section 9-9), ring strain allows the three- and four-membered heterocycles to undergo nucleophilic ring opening readily. In contrast, the larger, unstrained systems are relatively inert to attack.

Ring strain makes heterocyclopropanes and heterocyclobutanes reactive

Heterocyclopropanes are relatively reactive because ring strain is released by nucleo-philic ring opening. Under basic conditions, this process gives rise to inversion at the less substituted center (Section 9-9).

2-Phenyloxacyclopropane $+$ CH_3O^- $\xrightarrow{CH_3OH}$ $\underset{\textbf{85\%}}{C_6H_5\overset{OH}{\underset{|}{C}}HCH_2OCH_3}$

2-Methoxy-1-phenylethanol

N-Ethyl-(2S,3S)-trans-2,3-dimethylazacyclopropane $\xrightarrow{70\% \ CH_3CH_2NH_2, \ H_2O, \ 120°C, \ 16 \ days}$ $\underset{\textbf{55\%}}{meso\text{-}N,N'\text{-Diethyl-2,3-butane-diamine}}$

Explain the following result by a mechanism. (**Hint:** Try a ring opening catalyzed by the Lewis acid and consider the options available to the resulting intermediate.)

$\xrightarrow{MgBr_2, \ (CH_3CH_2)_2O}$

100%

2-(Chloromethyl)oxacyclopropane reacts with hydrogen sulfide ion (HS^-) to give thiacy-clobutan-3-ol. Explain by a mechanism.

Isomeric cylindricines A and B, isolated in 1993, are the two main alkaloids (Section 25-8) present in extracts from the Australian marine plant *Clavelina cylindrica*. The two compounds equilibrate to a 3 : 2 mixture. Formulate a mechanism for this process. (**Hint:** Check Exercise 9-22.)

$\rightleftharpoons$

Cylindricine A   **Cylindricine B**

The reactivity of the four-membered heterocycloalkanes bears out expectations based on ring strain: They undergo ring opening, as do their three-membered cyclic

CHEMICAL HIGHLIGHT 25-1 | **Azacyclopropene Antibiotics**

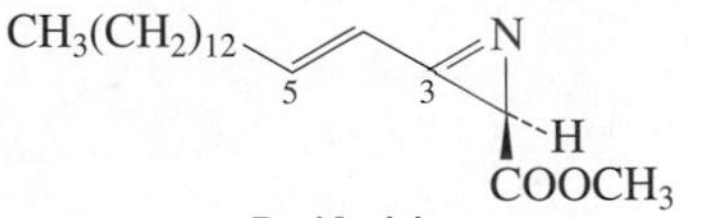

$$CH_3(CH_2)_{12} \overset{5}{\diagdown} \overset{3}{\diagup} N$$
$$\overset{H}{\underset{COOCH_3}{\diagdown}}$$
Dysidazirine

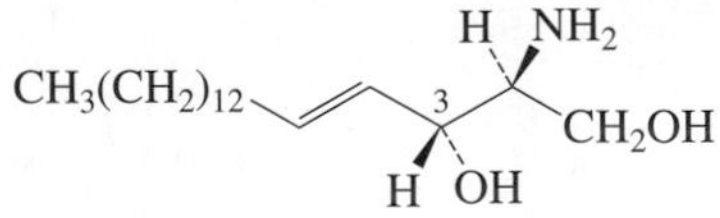

$$CH_3(CH_2)_{12} \diagdown \overset{H \ NH_2}{\underset{H \ OH}{\overset{3}{\diagdown}}} CH_2OH$$
D-Sphingosine

The sea is an abundant source of highly biologically active substances. One of the most unusual is dysidazirine, a natural product that was discovered in 1988 in a South Pacific species of sponge and contains an azacyclopropene ring. This substance is toxic toward certain strains of cancer cells and inhibits the growth of Gram-negative bacteria.

It is not known how dysidazirine is synthesized in nature, but one possible precursor is the amino alcohol D-sphingosine, which is a component of cell membranes. Enzyme-mediated oxidation of the secondary alcohol group at C3 of D-sphingosine to a ketone function could be a plausible first step in the transformation into dysidazirine. Intramolecular imine

formation between the carbonyl and amino groups would lead to the three-membered heterocyclic ring. Finally, oxidation and esterification at C1 would complete the synthesis.

Although the mechanism of antibiotic action of dysidazirine is not yet known, the chemical reactivity of its C–N double bond is greatly increased by the strain of the three-membered ring. Nucleophiles rapidly attack C3 and undergo conjugate addition to C5. Ring opening occurs under very mild conditions as well: Catalytic hydrogenation gives a mixture of the corresponding azacyclopropane and acyclic amino esters, as shown below.

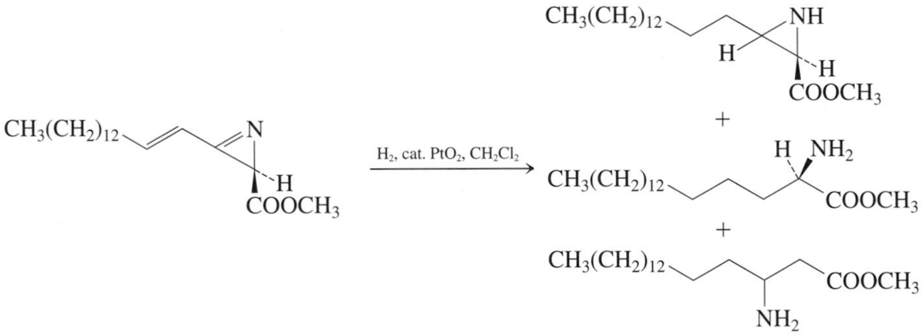

$$CH_3(CH_2)_{12} \diagdown \overset{N}{\underset{COOCH_3}{\overset{\diagdown}{\diagup}}} \xrightarrow{\text{H}_2, \text{ cat. PtO}_2, \text{ CH}_2\text{Cl}_2}$$

$$CH_3(CH_2)_{12} \diagdown \diagdown \overset{NH}{\underset{H \quad COOCH_3}{\overset{\diagdown}{\diagup}}}$$
$$+$$
$$CH_3(CH_2)_{12} \diagdown \diagdown \overset{H \ NH_2}{\underset{COOCH_3}{\diagdown}}$$
$$+$$
$$CH_3(CH_2)_{12} \diagdown \diagdown \diagdown \overset{}{\underset{NH_2}{\diagdown}} COOCH_3$$

counterparts, but more stringent reaction conditions are usually required. The reaction of oxacyclobutane with CH_3NH_2 is typical.

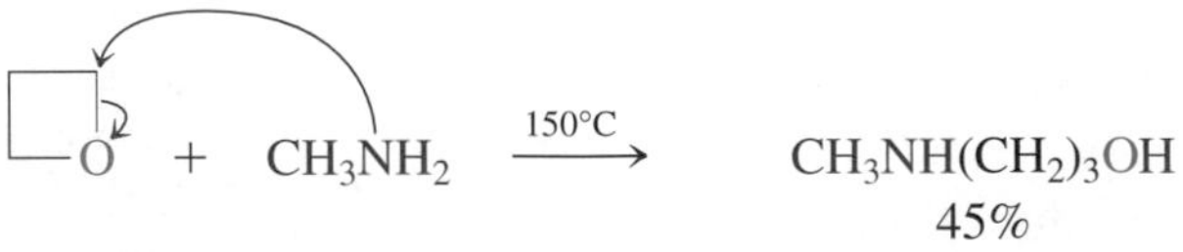

$$\square_O + CH_3NH_2 \xrightarrow{150°C} CH_3NH(CH_2)_3OH$$
$$45\%$$
N-Methyl-3-amino-1-propanol

The β-lactam antibiotics (penicillins and cephalosporins) function through related ring-opening processes (see Chemical Highlight 20-2).

EXERCISE 25-5

Treatment of thiacyclobutane with chlorine in $CHCl_3$ at $-70°C$ gives $ClCH_2CH_2CH_2SCl$ in 30% yield. Suggest a mechanism for this transformation. (**Hint:** The sulfur in sulfides is nucleophilic [Section 9-10].)

EXERCISE 25-6

2-Methyloxacyclobutane reacts with HCl to give two products. Write their structures.

Heterocyclopentanes and heterocyclohexanes are relatively unreactive

The unstrained heterocyclopentanes and heterocyclohexanes are relatively inert. Recall that oxacyclopentane (tetrahydrofuran, THF) is used as a solvent. However, the heteroatoms in five- and six-membered aza- and thiacycloalkanes allow these species to undergo their own characteristic transformations (see Sections 9-10, 17-8, 18-4, and Chapter 21). In general, ring opening by cleavage of a bond to the heteroatom does not occur unless the latter is first converted into a good leaving group.

CHEMICAL HIGHLIGHT 25-2 | Nicotine and Cancer

Nicotine

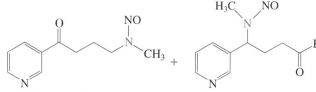

4-(*N*-Methyl-*N*-nitrosamino)-
1-(3-pyridyl)-1-butanone

+

4-(*N*-Methyl-*N*-nitrosamino)-
4-(3-pyridyl)butanal

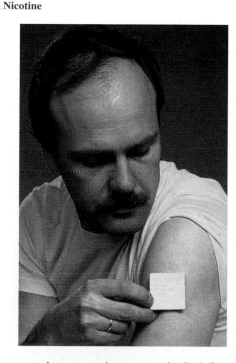

Nicotine patches are used to wean individuals from their addiction to cigarette smoking.

The mechanism by which nicotine in cigarette smoke is converted into highly carcinogenic species is gradually becoming more clearly understood. The initial step appears to be *N*-nitrosation of the azacyclopentane (pyrrolidine) nitrogen. Oxidation and ring opening (compare Chemical Highlight 21-3) take place giving a mixture of two *N*-nitrosodialkanamines (*N*-nitrosamines), each of which is a powerful carcinogen.

On protonation of the oxygen in a nitroso group, these substances become reactive alkylating agents, capable of transferring methyl groups to nucleophilic sites in biological molecules such as DNA, as shown below.

The diazohydroxide that remains decomposes through a diazonium ion to a carbocation, which may inflict additional molecular damage (Section 21-10).

Treatment of azacyclopentane (pyrrolidine) with sodium nitrite in acetic acid gives a liquid, b.p. 99°–100°C (15 mm Hg), that has the composition $C_4H_8N_2O$. Propose a structure for this compound. (**Hint:** Review Section 21-10.)

In summary, the reactivity of heterocyclopropanes and heterocyclobutanes results from the release of strain by ring opening. The five- and six-membered heterocycloalkanes are less reactive than their smaller-ring counterparts.

25-3 Structure and Properties of Aromatic Heterocyclopentadienes

Aromatic Heterocyclopentadienes

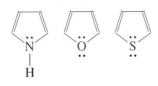

Pyrrole, furan, and **thiophene** are 1-hetero-2,4-cyclopentadienes. Each contains a butadiene unit bridged by a heteroatom bearing lone electron pairs. These systems contain delocalized π electrons in an aromatic six-electron framework. This section considers the structures and methods of preparation of these compounds.

Pyrrole, furan, and thiophene contain delocalized lone electron pairs

Cyclopentadienyl Anion

The electronic structure of the three heterocycles pyrrole, furan, and thiophene is similar to that of the cyclopentadienyl anion (Section 15-8). The cyclopentadienyl anion may be viewed as a butadiene bridged by a negatively charged carbon whose electron pair is delocalized over the other four carbons. The heterocyclic analogs contain a neutral atom in that place, again bearing lone electron pairs. One of these pairs is similarly delocalized, furnishing the two electrons needed to satisfy the $4n + 2$ rule (Section 15-7). To maximize overlap, the heteroatoms are hybridized sp^2 (Figure 25-1), the delocalized electron pair being assigned to the remaining p orbital. In pyrrole, the sp^2-hybridized nitrogen bears a hydrogen substituent in the plane of the mol-

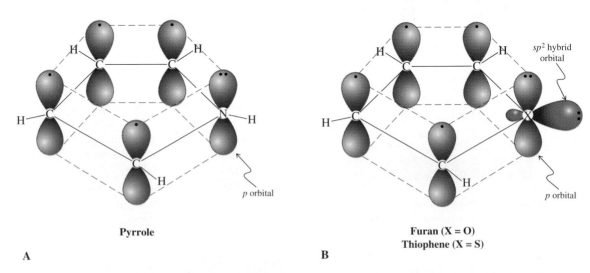

Pyrrole

A

Furan (X = O)
Thiophene (X = S)

B

FIGURE 25-1

Molecular-orbital pictures of (A) pyrrole and (B) furan (X = O), and thiophene (X = S). The heteroatom in each is sp^2 hybridized and bears one delocalized lone electron pair.

ecule. For furan and thiophene, the second lone electron pair is placed into one of the sp^2 hybrid orbitals, again in the plane and therefore with no opportunity to achieve overlap. This arrangement is much like that in the phenyl anion (Section 22-4).

The delocalization of the lone pair in the 1-hetero-2,4-cyclopentadienes can be described by charge-separated resonance forms, as shown for pyrrole.

Resonance Forms of Pyrrole

Notice that there are four dipolar forms in which a positive charge is placed on the heteroatom and a negative charge successively on each of the carbons. This picture suggests that the heteroatom should be relatively electron poor and the carbons relatively electron rich. Indeed, as we shall see, the reactivity of these compounds bears out that expectation.

EXERCISE 25-8

Azacyclopentane and pyrrole are both polar molecules. However, the dipole vectors in the two molecules point in opposite directions. What is the sense of direction of this vector in each structure? Explain your answer.

Pyrroles, furans, and thiophenes are prepared from γ-dicarbonyl compounds

Syntheses of the heterocyclopentadienes use a variety of cyclization strategies. A general approach is the **Paal-Knorr* synthesis** (for pyrroles) and its variations (for the other heterocycles). The target molecule is made from an enolizable γ-dicarbonyl compound that is treated with an amine derivative (for pyrroles) or P_2O_5 (for furans) or P_2S_5 (for thiophenes).

Cyclization of a γ-Dicarbonyl Compound to a 1-Hetero-2,4-Cyclopentadiene

$$R \overset{}{\underset{O\ \ O}{\diagdown}} R \quad \xrightarrow[{-H_2O}]{R'NH_2,\ or\ P_2O_5,\ or\ P_2S_5} \quad R\underset{X}{\diagdown} R$$

X = NR', O, S

$$CH_3\overset{O}{\overset{\|}{C}}CH_2CH_2\overset{O}{\overset{\|}{C}}CH_3 \ + \ (CH_3)_2CHNH_2 \quad \xrightarrow{CH_3COOH,\ \Delta,\ 17\ h} \quad H_3C\underset{\underset{(CH_3)_2CH}{N}}{\diagdown}CH_3$$

70%
N-(1-Methylethyl)-
2,5-dimethylpyrrole

---

*Professor Karl Paal (1860–1935), University of Erlangen, Germany; Professor Ludwig Knorr (1859–1921), University of Jena, Germany.

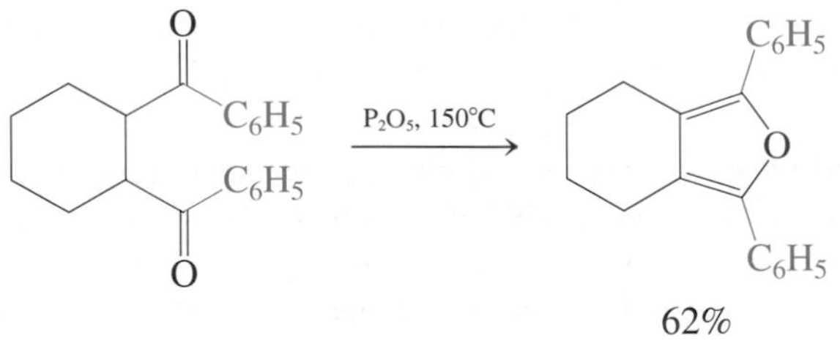

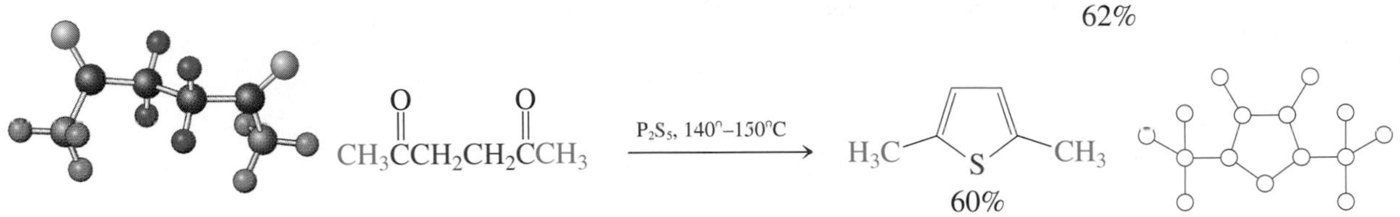

62%

$$CH_3\overset{O}{\overset{\|}{C}}CH_2CH_2\overset{O}{\overset{\|}{C}}CH_3 \xrightarrow{P_2S_5,\ 140^\circ-150^\circ C} H_3C \underset{S}{\diagdown} CH_3$$

60%

2,5-Dimethylthiophene

EXERCISE 25-9

Formulate a possible mechanism for the acid-catalyzed dehydration of 2,5-hexanedione to 2,5-dimethylfuran. (**Hint:** The crucial ring closure is accomplished by the oxygen of one carbonyl group attacking the carbon of the second.)

EXERCISE 25-10

4-Methylpyrrole-2-carboxylic acid (compound B) is the trail pheromone of the ant species *Atta texana*. A third of a milligram has been estimated to be sufficient to mark a path around Earth. Consequently, each ant carries only 3.3 ng (10^{-9} g). Propose a synthesis starting from 3-methylcyclobutene-1-carboxylic acid (compound A). (**Hint:** What dione is the retrosynthetic precursor to compound B, and how can you make it from compound A?)

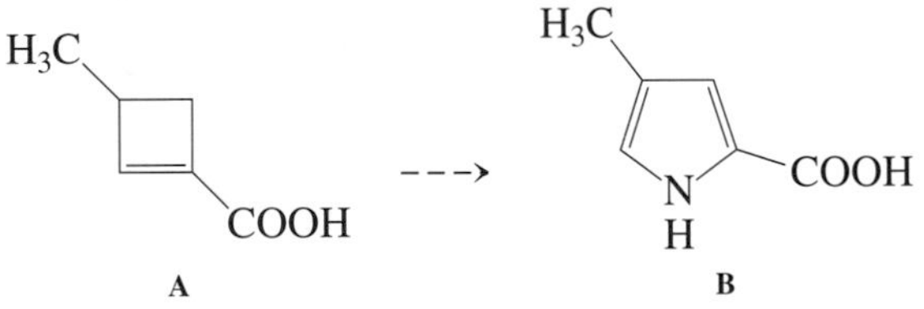

A   B

EXERCISE 25-11

The following equation is an example of another synthesis of pyrroles. Write a mechanism for this transformation. (**Hint:** Refer to Section 17-9.)

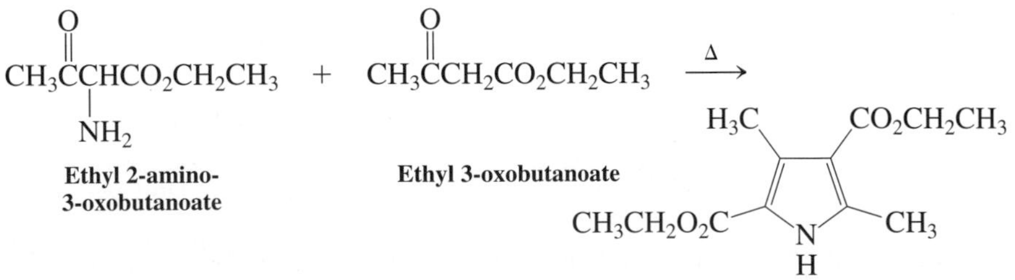

Ethyl 2-amino-3-oxobutanoate   **Ethyl 3-oxobutanoate**

Diethyl 3,5-dimethylpyrrole-2,4-dicarboxylate

In summary, pyrrole, furan, and thiophene contain delocalized aromatic π systems analogous to that of the cyclopentadienyl anion. A general method for the preparation of 1-hetero-2,4-cyclopentadienes is based on the cyclization of enolizable γ-dicarbonyl compounds.

25-4 Reactions of the Aromatic Heterocyclopentadienes

The reactivity of pyrrole, furan, and thiophene and their derivatives is largely governed by their aromaticity and based on the chemistry of benzene. This section describes some of their reactions, particularly electrophilic aromatic substitution, and introduces indole, a benzofused analog of pyrrole.

Pyrroles, furans, and thiophenes undergo electrophilic aromatic substitution

As expected for aromatic systems, the 1-hetero-2,4-cyclopentadienes undergo electrophilic substitution. There are two sites of possible attack—at C2 and at C3. Which one should be more reactive? An answer can be found by the same procedure used to predict the regioselectivity of electrophilic aromatic substitution of substituted benzenes (Chapter 16): enumeration of all the possible resonance forms for the two modes of reaction.

**Consequences of Electrophilic Attack at C2 and C3
in the Aromatic Heterocyclopentadienes**

Attack at C2

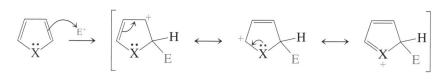

Attack at C3

Both modes benefit from the presence of the resonance-contributing heteroatom, but attack at C2 leads to an intermediate with an additional resonance form, thus indicating this position to be the preferred site of substitution. Indeed, such selectivity is generally observed. However, because C3 also is activated to electrophilic attack, mixtures of products can form, depending on conditions, substrates, and electrophiles.

Electrophilic Aromatic Substitution of Pyrrole, Furan, and Thiophene

2-Nitropyrrole 50%
3-Nitropyrrole 13%

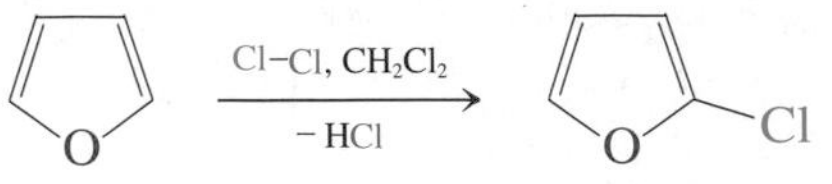

The relative nucleophilic reactivity of benzene and the three heterocycles increases in the order benzene $\ll$ thiophene $<$ furan $<$ pyrrole.

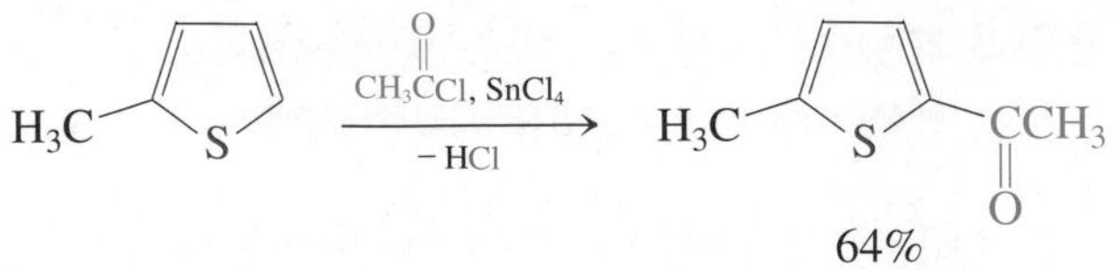

The monobromination of thiophene-3-carboxylic acid gives only one product. What is its structure and why is it the only product formed?

Because the lone electron pair on nitrogen is tied up by conjugation, pyrrole is extremely nonbasic. Very strong acid is required to effect protonation, which takes place not on nitrogen but on C2.

Protonation of Pyrrole

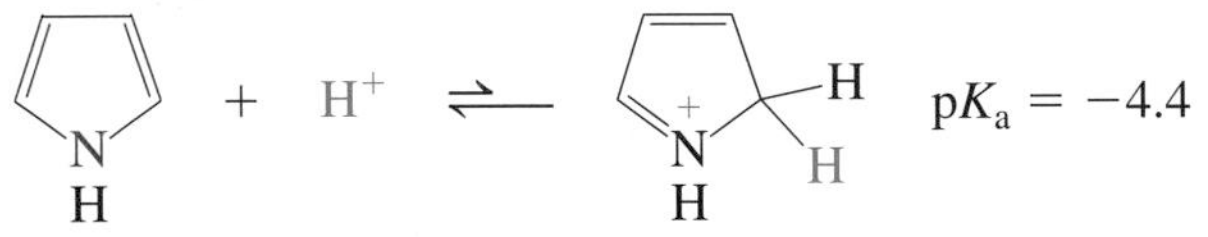

pyrrole + H^+ $\rightleftharpoons$ protonated pyrrole $pK_a = -4.4$

Explain why pyrrole is protonated on an α-carbon rather than on the nitrogen.

1-Hetero-2,4-cyclopentadienes can undergo ring opening and cycloaddition reactions

Furans can be hydrolyzed under mild conditions to γ-dicarbonyl compounds. The reaction may be viewed as the reverse of the Paal-Knorr-type synthesis of furans. Pyrrole polymerizes under these reaction conditions, whereas thiophene is stable.

Hydrolysis of a Furan to a γ-Dicarbonyl Compound

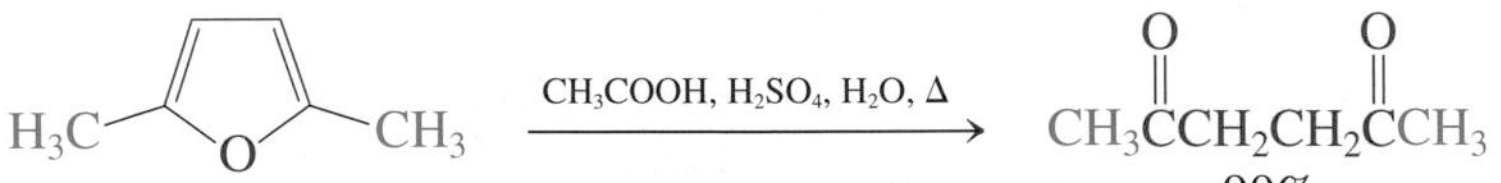

$$CH_3\overset{O}{\overset{\|}{C}}CH_2CH_2\overset{O}{\overset{\|}{C}}CH_3$$
90%
2,5-Hexanedione

Raney nickel desulfurization (Section 17-8) of thiophene derivatives results in sulfur-free acyclic saturated compounds.

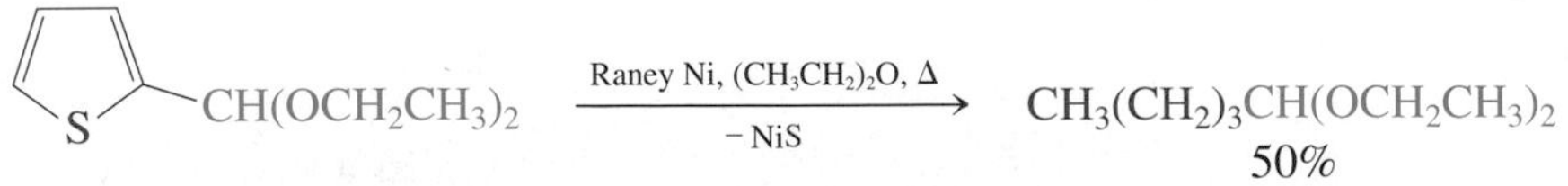

$$CH_3(CH_2)_3CH(OCH_2CH_3)_2$$
50%

The π system of furan (but not of pyrrole or thiophene) possesses sufficient diene character to undergo Diels-Alder cycloadditions (Section 14-8).

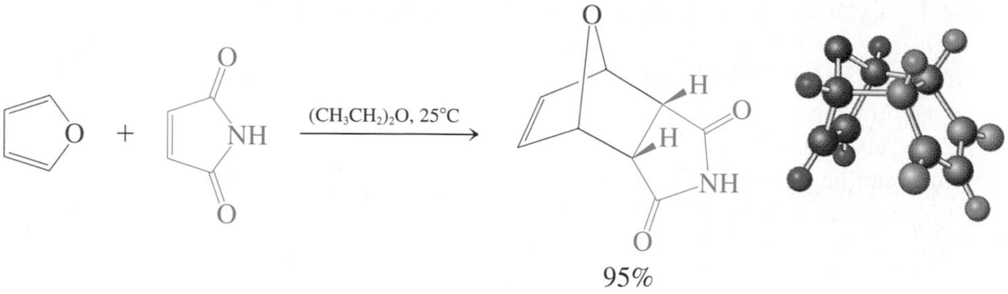

95%

Indole is a benzopyrrole

Indole is the most important *benzannulated* (fused-ring) derivative of the 1-hetero-2,4-cyclopentadienes. It forms part of many natural products, including the amino acid tryptophan (Section 26-1).

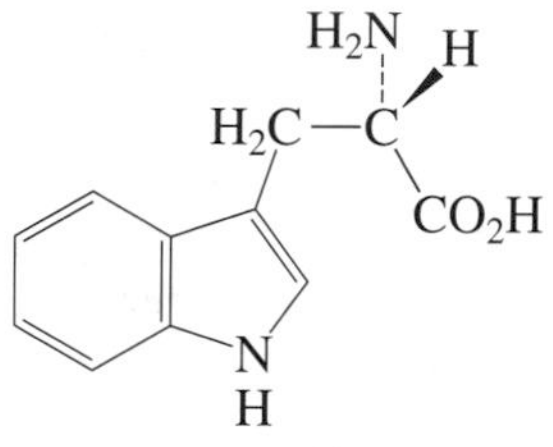

Indole is related to pyrrole in the same way that naphthalene is related to benzene. Its electronic makeup is indicated by the various possible resonance forms that can be formulated for the molecule. Although those resonance forms that disturb the cyclic six-π-electron system of the fused benzene ring are less important, they indicate the electron-donating effect of the heteroatom.

Tryptophan

Resonance in Indole

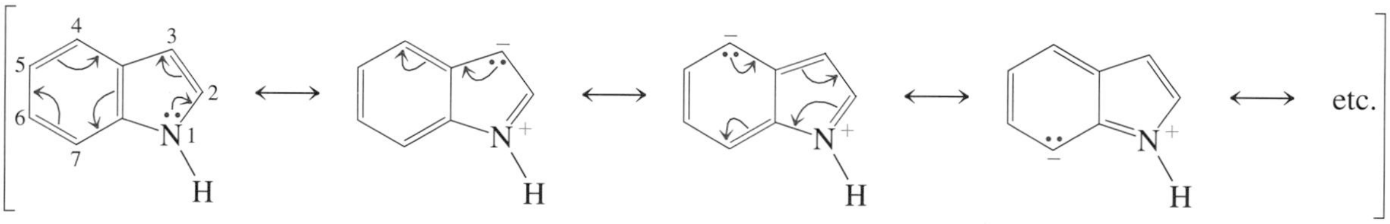

EXERCISE 25-14

Predict the preferred site of electrophilic aromatic substitution in indole. Explain your choice.

EXERCISE 25-15

Irradiation of compound A in ethoxyethane (diethyl ether) at $-100°C$ generates the enol form, B, of ethanoylbenzene (acetylbenzene) and a new compound, C, which isomerizes to indole on warming to room temperature.

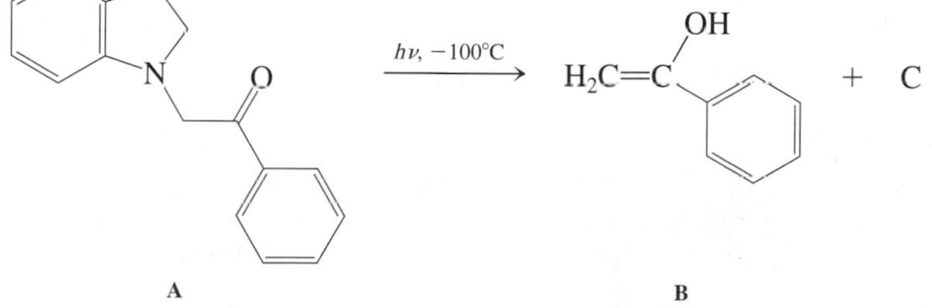

The ^1H NMR spectrum of compound C shows signals at $\delta = 3.79$ (d, 2 H) and 8.40 (t, 1 H) ppm in addition to four aromatic absorptions. Indole has peaks at $\delta = 6.34$ (d, 1 H), 6.54 (broad d, 1 H), and 7.00 (broad s, 1 H) ppm. What is compound C? (**Hint:** This photolysis proceeds by a mechanism similar to the mass spectral McLafferty rearrangement, Section 20-10.)

In summary, the donation of the lone electron pair on the heteroatom to the diene unit in pyrrole, furan, and thiophene makes the carbon atoms in these systems electron rich and therefore more susceptible to electrophilic aromatic substitution than those in benzene. Electrophilic attack is frequently favored at C2, but substitution at C3 is also observed, depending on conditions, substrates, and electrophiles. Some rings can be opened by hydrolysis or by desulfurization (for thiophenes). The diene unit in furan is reactive enough to undergo Diels-Alder cycloadditions. Indole is a benzopyrrole containing a delocalized π system.

25-5 Structure and Preparation of Pyridine: An Azabenzene

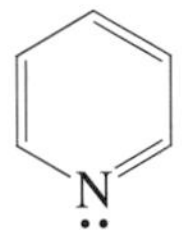

Pyridine

Pyridine can be regarded as a benzene derivative—an **azabenzene**—in which an sp^2-hybridized nitrogen atom replaces a CH unit. The pyridine ring is therefore aromatic, but its electronic structure is strongly perturbed by the presence of the electronegative nitrogen atom. This section describes the structure, spectroscopy, and preparation of this simple azabenzene.

Pyridine is a cyclic aromatic imine

Pyridine contains an sp^2-hybridized nitrogen atom like that in an imine (Section 17-9). In contrast with pyrrole, only one electron in the p orbital completes the aromatic π-electron arrangement of the aromatic ring; as in the phenyl anion, the lone electron pair is located in one of the sp^2 hybrid atomic orbitals in the molecular plane (Figure 25-2). Therefore, in pyridine, the nitrogen does not donate excess electron density to the remainder of the molecule. Quite the contrary: Because nitrogen is more electronegative than carbon (Table 1-2), it withdraws electron density from the ring, both inductively and by resonance.

Resonance in Pyridine

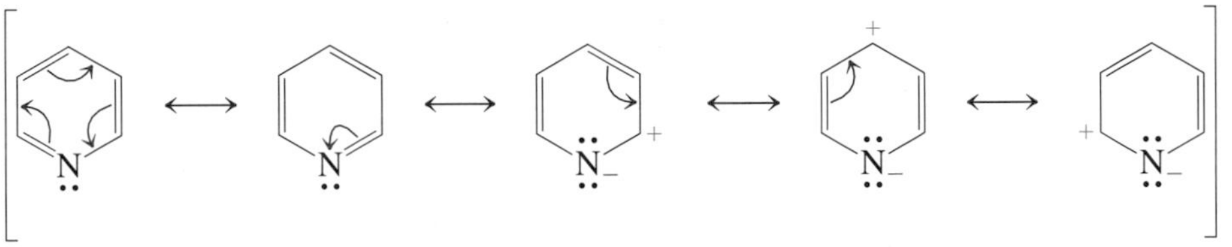

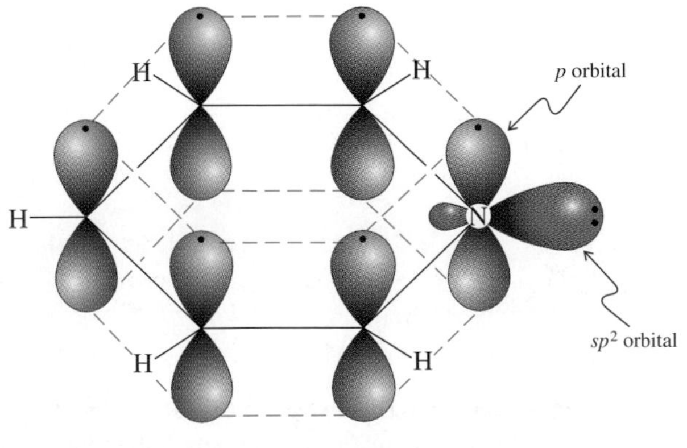

FIGURE 25-2

Molecular-orbital picture of pyridine. The lone electron pair on nitrogen is in an sp^2-hybridized orbital and is *not* part of the aromatic π system.

Azacyclohexane (piperidine) is a polar molecule. In which direction does its dipole vector point? Answer the same question for pyridine. Explain your answer.

Aromatic delocalization in pyridine is evident in the ^{1}H NMR spectrum, which reveals the presence of a ring current. The electron-withdrawing capability of the nitrogen is manifest in larger chemical shifts (more deshielding) at C2 and C4, as expected from the resonance picture.

^{1}H NMR Chemical Shifts (ppm) in Pyridine and Benzene

Because the lone pair on nitrogen is not tied up by conjugation (as it is in pyrrole, Exercise 25-13), pyridine is a weak base. (It is used as such in numerous organic transformations.) Compared with alkanamines (pK_a of ammonium salts ~ 10; Section 21-4), the pyridinium ion has a low pK_a, because the nitrogen is sp^2 and not sp^3 hybridized (see Section 11-3 for the effect of hybridization on acidity).

Pyridine is the simplest azabenzene. Some of its higher aza analogs are shown here. They behave like pyridine but show the increasing effect of aza substitution—in particular, increasing electron deficiency. Minute quantities of several 1,4-diazabenzene (pyrazine) derivatives are responsible for the characteristic odors of many vegetables. One drop of 2-methoxy-3-(1-methylethyl)-1,4-diazabenzene (2-isopropyl-3-methoxypyrazine) in a large swimming pool would be more than adequate to give the entire pool the odor of raw potatoes.

Pyridine Is a Weak Base

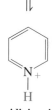

Pyridinium ion
pK_a = 5.29

| 1,2-Diazabenzene (Pyridazine) | 1,3-Diazabenzene (Pyrimidine) | 1,4-Diazabenzene (Pyrazine) | 1,2,3-Triazabenzene (1,2,3-Triazine) | 1,2,4-Triazabenzene (1,2,4-Triazine) |

| 1,3,5-Triazabenzene (1,3,5-Triazine) | 1,2,4,5-Tetraazabenzene (1,2,4,5-Tetrazine) | 2-Methoxy-3-(1-methylethyl)-1,4-diazabenzene (Potatoes) | 2-Methoxy-3-(2-methylpropyl)-1,4-diazabenzene (Green peppers) |

Pyridines are made by condensation reactions

Pyridine and simple alkylpyridines are obtained from coal tar. Many of the more highly substituted pyridines are in turn made by both electrophilic and nucleophilic substitution of the simpler derivatives.

Pyridines can be made by condensation reactions of acyclic starting materials such as carbonyl compounds with ammonia. The most general of these methods is the **Hantzsch* pyridine synthesis.** In this reaction, two molecules of a β-dicarbonyl compound, an aldehyde, and ammonia combine in several steps to give a substituted dihydropyridine, which is readily oxidized by nitric acid to the aromatic system.

Hantzsch Synthesis of 2,6-Dimethylpyridine

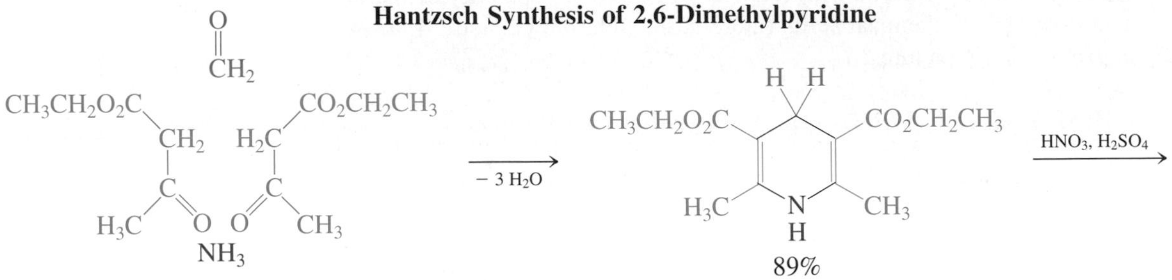

89%
Diethyl 1,4-dihydro-2,6-dimethyl-3,5-pyridinedicarboxylate

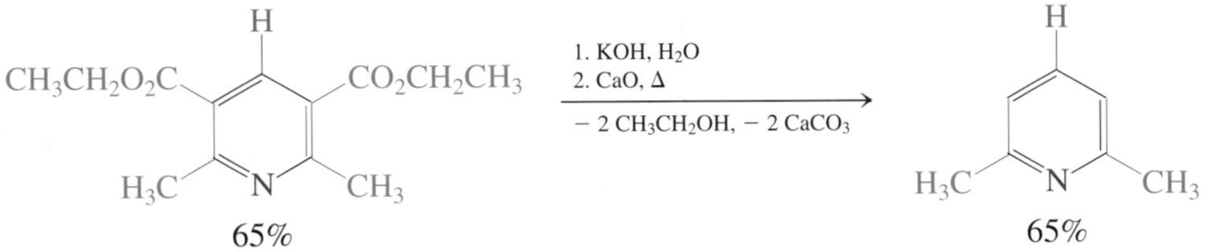

65%
Diethyl 2,6-dimethyl-3,5-pyridinedicarboxylate

65%
2,6-Dimethylpyridine

If the β-dicarbonyl compound is a 3-keto ester, the resulting product is a 3,5-pyridinedicarboxylic ester. Hydrolysis followed by pyrolysis of the calcium salt of the acid causes decarboxylation.

EXERCISE 25-17

What starting materials would you use in the Hantzsch synthesis of the following pyridines?

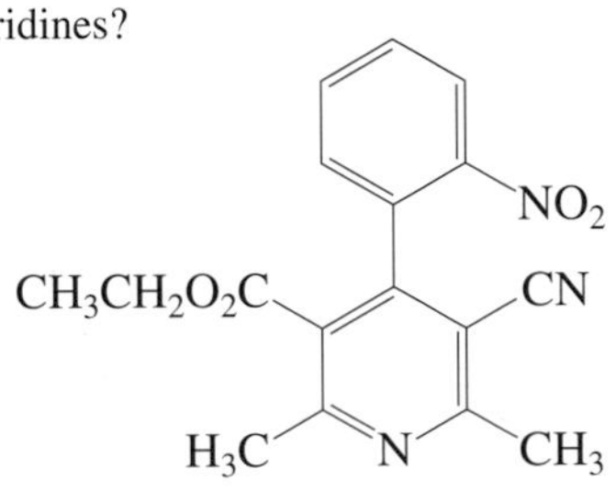

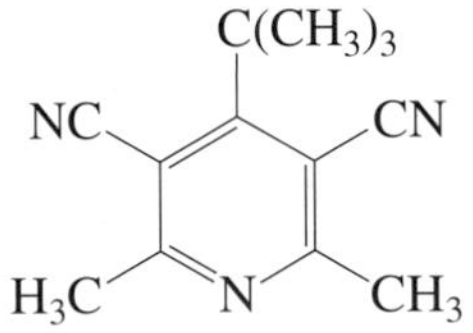

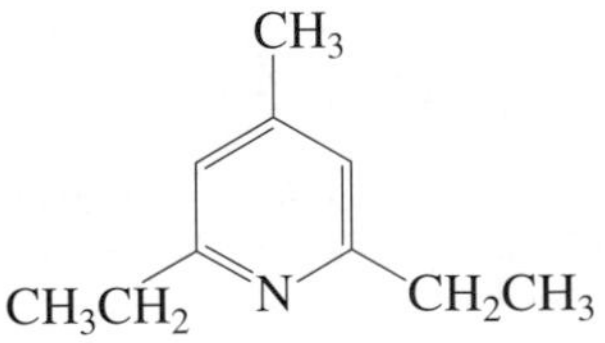

In summary, pyridines are aromatic but electron poor. The lone pair on nitrogen makes the heterocycle weakly basic. Pyridines are prepared by condensation of a β-dicarbonyl compound with ammonia and an aldehyde.

*Professor Arthur Hantzsch (1857–1935), University of Leipzig, Germany.

25-6 Reactions of Pyridine

The reactivity of pyridine derives from its dual nature as both an aromatic molecule and a cyclic imine. Both electrophilic and nucleophilic substitution processes may occur, leading to a variety of substituted derivatives.

Pyridine undergoes electrophilic aromatic substitution only under extreme conditions

Because the pyridine ring is electron poor, the system undergoes electrophilic aromatic substitution only with great difficulty, several orders of magnitude more slowly than benzene, and at C3 (see Section 15-9).

Electrophilic Aromatic Substitution of Pyridine

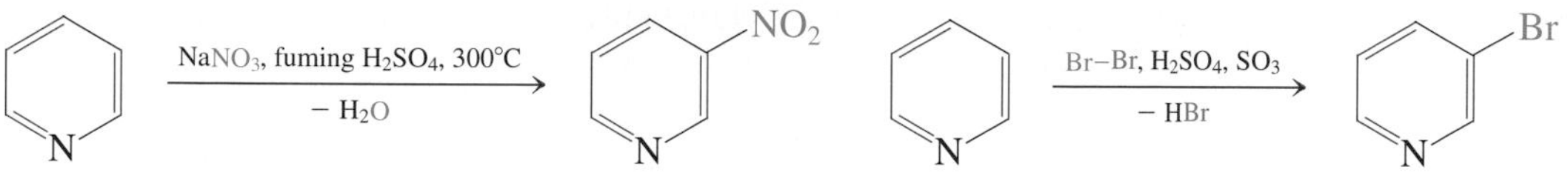

4.5%
3-Nitropyridine

86%
3-Bromopyridine

EXERCISE 25-18

Explain why electrophilic aromatic substitution of pyridine is at C3.

Activating substituents allow for milder conditions or improved yields.

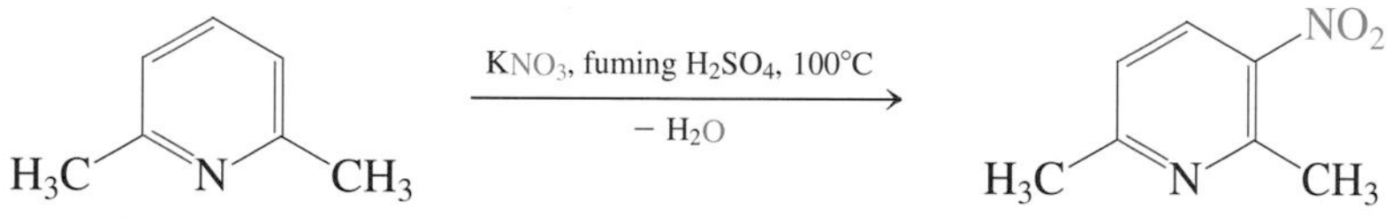

81%
2,6-Dimethyl-3-nitropyridine

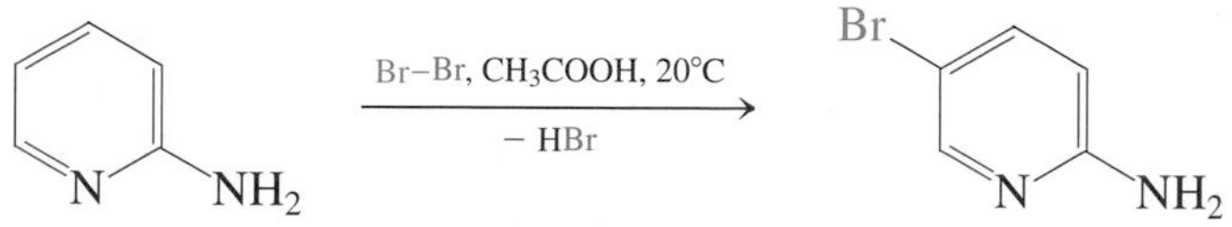

2,6-Dimethylpyridine

90%
2-Amino-5-bromopyridine

2-Aminopyridine

Pyridine undergoes nucleophilic substitution

Because the pyridine ring is relatively electron deficient, it undergoes nucleophilic substitution much more readily than does benzene (Section 22-4). Attack at C2 and C4 is preferred because it leads to intermediates in which the negative charge is on the nitrogen. An example of nucleophilic substitution of pyridine is the **Chichibabin* reaction,** in which the heterocycle is converted into 2-aminopyridine by treatment with sodium amide in liquid ammonia.

*Professor Alexei E. Chichibabin (1871–1945), University of Moscow, Russia.

Chichibabin Reaction

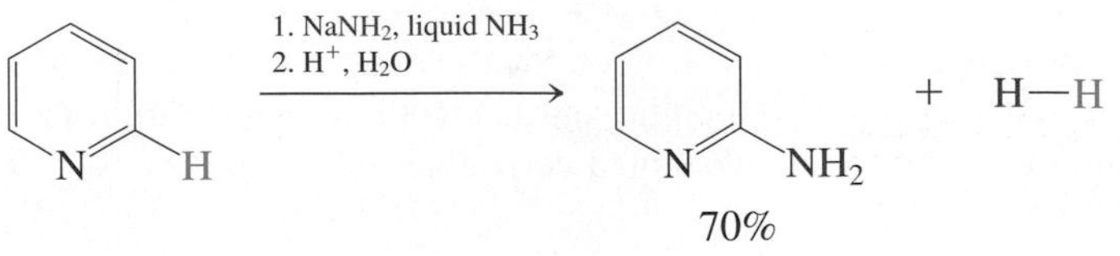

1. NaNH₂, liquid NH₃
2. H⁺, H₂O

70%
2-Aminopyridine

This reaction proceeds by the addition–elimination mechanism. The first step is attack by ⁻:N̈H₂ at C2, a process that resembles 1,2-addition to an imine function. Expulsion of a hydride ion, H:⁻, from C2 is followed by deprotonation of the amine nitrogen to give H₂ and a resonance-stabilized 2-pyridineamide ion. Protonation of the latter by aqueous work-up furnishes the final product. Note the contrast with *electrophilic* substitutions, which include *proton* loss, not expulsion of hydride as a leaving group.

Transformations related to the Chichibabin reaction take place when pyridines are treated with Grignard or organolithium reagents.

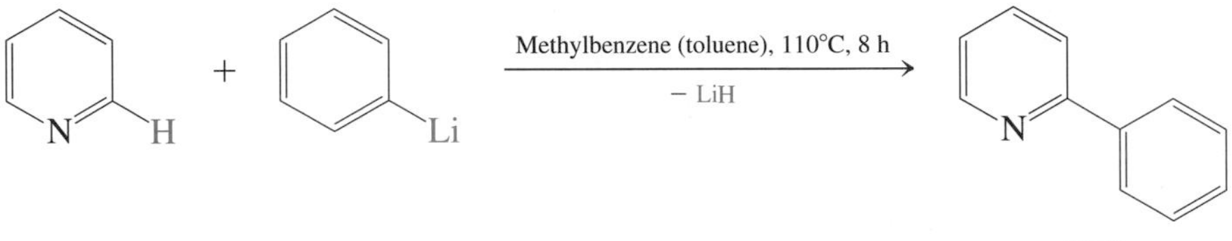

Methylbenzene (toluene), 110°C, 8 h
− LiH

49%
2-Phenylpyridine

In most nucleophilic substitutions of pyridines, halides are leaving groups, the 2- and 4-halopyridines being particularly reactive.

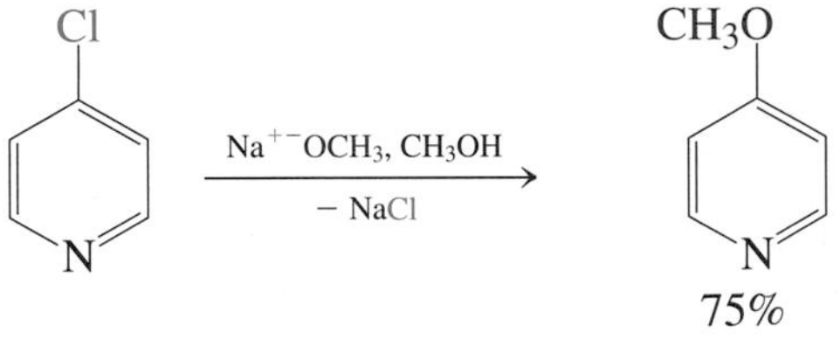

Na⁺⁻OCH₃, CH₃OH
− NaCl

75%
4-Methoxypyridine

EXERCISE 25-19

The relative rates of the reactions of 2-, 3-, and 4-chloropyridine with sodium methoxide in methanol are 3000:1:81,000. Explain.

EXERCISE 25-20

Propose a mechanism for the reaction of 4-chloropyridine with methoxide (see Exercise 25-19). (**Hint:** Think of the pyridine ring as containing an α,β-unsaturated imine function [see Sections 17-9 and 18-10].)

In summary, pyridine undergoes slow electrophilic aromatic substitution preferentially at C3. Nucleophilic substitution reactions occur more readily to expel hydride or another leaving group from either C2 or C4.

CHEMICAL HIGHLIGHT 25-3 | **Pyridinium Salts in Nature: Nicotinamide Adenine Dinucleotide**

Nicotinamide adenine dinucleotide

A complex pyridinium derivative, *nicotinamide adenine dinucleotide* (NAD^+) is an important biological oxidizing agent. The structure consists of a pyridine ring [derived from 3-pyridinecarboxylic (nicotinic) acid], two ribose molecules (Section 24-1) linked by a pyrophosphate bridge, and the heterocycle adenine (Section 26-9).

Most organisms derive their energy from the oxidation (removal of electrons) of fuel molecules, such as glucose or fatty acids; the ultimate oxidant (electron acceptor) is oxygen, which gives water. Such biological oxidations proceed through a cascade of electron-transfer reactions requiring the intermediacy of special redox reagents. NAD^+ is one such molecule. In the oxidation of a substrate, the pyridinium ring in NAD^+ undergoes a two-electron reduction with simultaneous protonation.

Reduction of NAD^+

NAD^+ is the electron acceptor in many biological oxidations of alcohols to aldehydes (including the conversion of vitamin A into retinal, Section 18-9). This reaction can be seen as a transfer of hydride from C1 of the alcohol to the pyridinium nucleus with simultaneous deprotonation.

Oxidation of Alcohols by NAD^+

25-7 Quinoline and Isoquinoline: The Benzopyridines

Quinoline

Isoquinoline

We can imagine the fusion of a benzene ring to pyridine in either of two ways, giving us **quinoline** and **isoquinoline** (1- and 2-azanaphthalene, according to our systematic nomenclature). Both are liquids with high boiling points. Many of their derivatives are found in nature or have been synthesized in the search for physiological activity. Like pyridine, quinoline and isoquinoline are readily available from coal tar.

As might be expected, because pyridine is electron poor compared with benzene, electrophilic substitutions on quinoline and isoquinoline take place at the *benzene* ring. As with naphthalene, substitution at the carbons next to the ring fusion predominates.

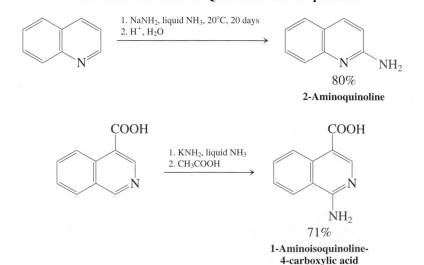

In contrast with electrophiles, nucleophiles prefer reaction at the electron-poor *pyridine* nucleus. These reactions are quite analogous to those with pyridine.

Quinoline and isoquinoline react with organometallic reagents exactly as pyridine does (Section 25-6). Give the products of their reaction with 2-propenylmagnesium bromide (allylmagnesium bromide).

The following structures are representative of higher aza-analogs of naphthalene.

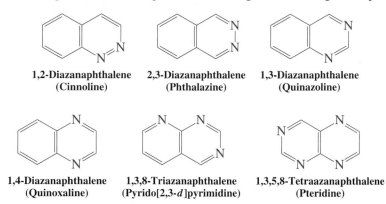

| 1,2-Diazanaphthalene
(Cinnoline) | 2,3-Diazanaphthalene
(Phthalazine) | 1,3-Diazanaphthalene
(Quinazoline) |

| 1,4-Diazanaphthalene
(Quinoxaline) | 1,3,8-Triazanaphthalene
(Pyrido[2,3-*d*]pyrimidine) | 1,3,5,8-Tetraazanaphthalene
(Pteridine) |

In summary, the azanaphthalenes quinoline and isoquinoline may be regarded as benzopyridines. Electrophiles attack the benzene ring, nucleophiles attack the pyridine ring.

25-8 Alkaloids: Physiologically Potent Nitrogen Heterocycles in Nature

The **alkaloids** are bitter-tasting, natural nitrogen-containing compounds found particularly in plants. The name is derived from their characteristic basic properties (alkali-like), which are induced by the lone electron pair of nitrogen.

As with acyclic amines (Chapter 21), the (Lewis) basic nature of the alkaloids, in conjunction with their particular three-dimensional architecture, gives rise to often potent physiological activity. We have already noted some examples of this behavior in the narcotics morphine and heroin (Section 9-11), the psychoactive lysergic acid and LSD (Section 19-13), and the antibiotic penicillins (Section 20-6).

Morphine (R = H)
Heroin (R = CH$_3$C=O)

Lysergic acid (X = OH)
Lysergic acid diethylamide, LSD
[X = (CH$_3$CH$_2$)$_2$N]

Penicillin

CHEMICAL HIGHLIGHT 25-4 | **Azanaphthalenes in Nature**

Xanthopterin
(Yellow butterfly and other insect pigment)

Leucopterin
(Colorless substance found in white butterfly wings)

1,3,5,8-Tetraaza-
naphthalene part

4-Aminobenzoic
acid part

(S)-2-Aminopentanedioic
(glutamic) acid part

Folic acid (X = OH, R = H)
Methotrexate (X = NH₂, R = CH₃)

The 1,3,5,8-tetraazanaphthalene (pteridine) ring system is present in a number of interesting natural products. Xanthopterin and leucopterin are insect pigments. Folic acid (Section 15-11) is a biologically important molecule incorporating a 1,3,5,8-tetraazanaphthalene (pteridine) ring, 4-aminobenzoic acid, and (S)-2-aminopentanedioic (glutamic) acid (Section 26-1). Folic acid is critical to the proper development of the nervous system in very early stages of pregnancy. A deficiency of this substance, which must be obtained from the diet, is associated with crippling and often fatal birth defects such as spina bifida ("open spine") and anencephaly (a failure of the brain to develop normally). The U.S. Public Health Service recommends that all women of childbearing age take 0.4 mg of folic acid daily.

Butterflies make spectacular use of pigments.

Nicotine (see also Chemical Highlight 25-2), present in dried tobacco leaves in 2–8% concentration, is the active ingredient in cigarettes and other tobacco products. Smoking and chewing tobacco have been connected with heart and lung disease and cancer, mainly a result of the presence of carcinogens, carbon monoxide, and other toxins. The mode of action of nicotine is complex: Ingestion of the molecule may stimulate or calm the user, and it may affect his or her mood, appetite, and cognition. There appears to be little doubt that nicotine is an addictive drug, and the debate about how to regulate its availability is ongoing.

Tetrahydrofolic acid functions as a biological carrier of one-carbon units. The reactive part of the molecule is at nitrogens N5 and N10, as shown below.

A derivative of folic acid, methotrexate, is sufficiently similar structurally that it can enter into some of the reactions of folic acid. It also acts as an inhibitor in some of the processes of cell division that are mediated by folic acid. As a result, it is a useful drug in cancer chemotherapy. Because cancer cells divide much more rapidly than normal cells, they are strongly affected by the presence of this compound.

Riboflavin (vitamin B_2) is a benzannulated analog of 1,3,5,8-tetraazanaphthalene (pteridine) bearing a ribose unit; it is found in animal and plant tissues.

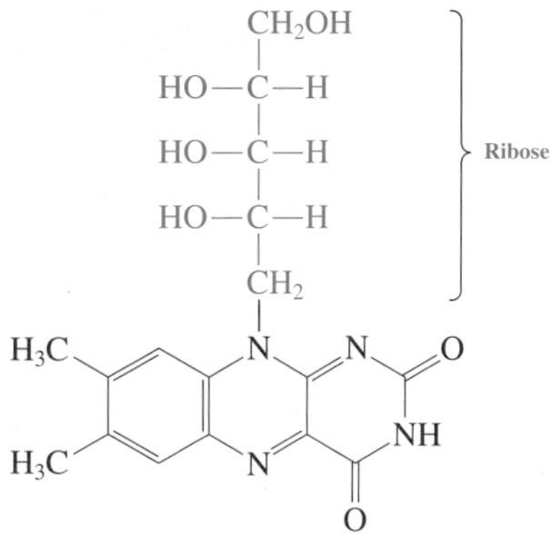

**Riboflavin
(Vitamin B_2)**

Tetrahydrofolic Acid as a Carrier of One-Carbon Units

Tetrahydrofolic acid

N-5,N-10-Methylenetetrahydrofolate

N-5-Methyltetrahydrofolate

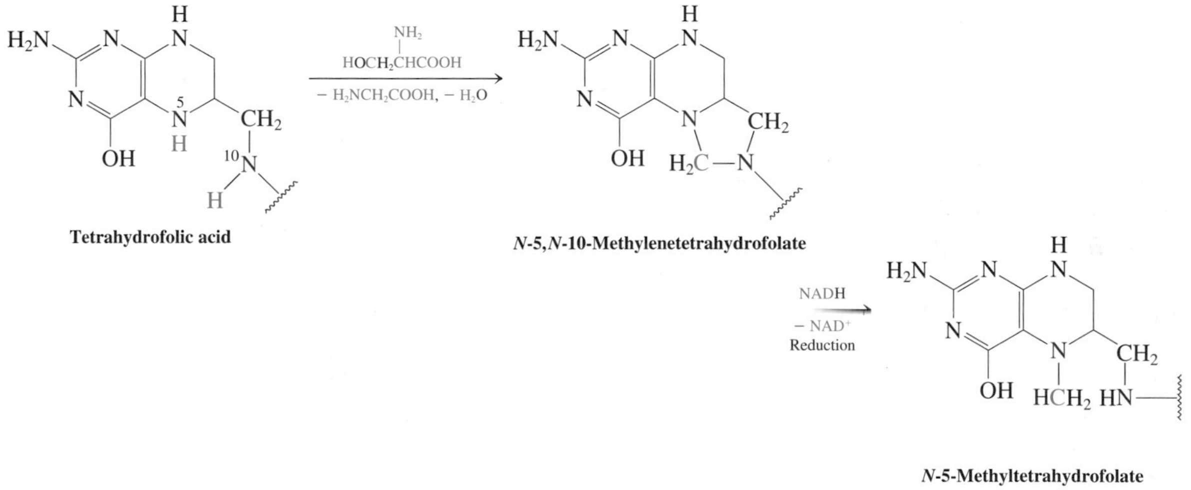

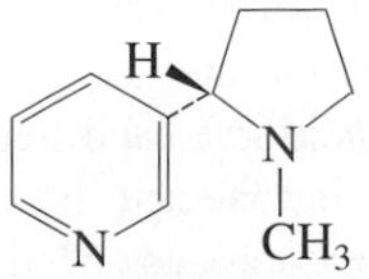

Nicotine

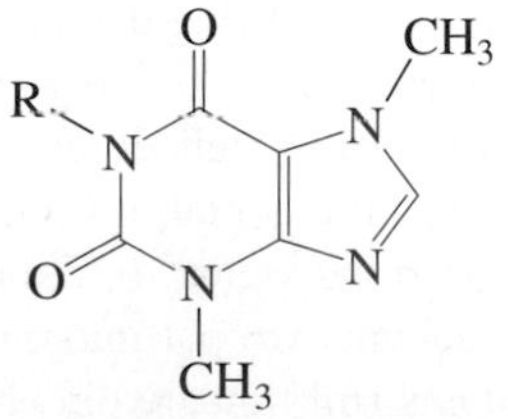

Caffeine (R = CH$_3$)

Theobromine (R = H)

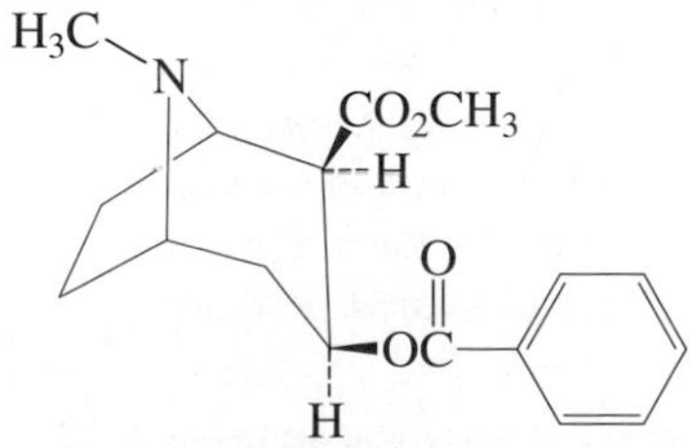

Cocaine

Even more stimulating than nicotine are caffeine and theobromine, present in coffee and tea or cocoa (chocolate), respectively. Perhaps the most dangerous stimulant is cocaine, extracted from the leaves of the coca shrub, which is cultivated mainly in South America for the purpose of illegal drug trafficking. Cocaine is shipped and sold in the form of the water-soluble hydrochloride salt ("street cocaine"), which may be ingested through the nasal passages by "snorting" or orally and intravenously. The actual alkaloid is known as "freebase" or "crack" and is inhaled by smoking. There are severe physical and psychological side effects of the drug, such as brain seizures, respiratory collapse, heart attack, paranoia, and depression. In 1996, a half million

---

CHEMICAL HIGHLIGHT **25-5** | **Nature Is Not Always Green: Natural Pesticides**

Many people believe that everything synthetic is somehow suspect and "bad" and that all of nature's chemicals are benign. As pointed out by Ames* and others, this is a misconception. Although we have seen that, indeed, many manufactured chemicals have problems with toxicity and adverse effects on the environment, people have managed to put most of these compounds to beneficial use. On the other side of the spectrum, nature's chemicals are not any different from synthetic ones. Nature has its own highly productive laboratory, which puts out compounds by the millions, many of which are highly toxic, as, for example, quite a few of the alkaloids found in plants. Consequently, there are numerous (sometimes lethal) cases of poisoning (especially of children) due to the accidental ingestion of plant material, the eating of green potatoes (exposed to sunlight, which increases their toxin level), the drinking of herbal teas, the consumption of "poison" mushrooms, and so forth. Abraham Lincoln's mother died from drinking milk from a cow that had grazed on the toxic snakeroot plant.

What is the purpose of these compounds in plant life? Plants cannot run away from predators and invading organisms, such as fungi, insects, animals, and humans, and they have no organs with which to defend themselves. Instead, they have developed an array of chemical weapons, "natural pesticides," with which to mount an effective defense strategy. Tens of thousands of these chemicals are now known. Americans consume about 1.5 g of natural pesticides

per person per day, in the form of vegetables, fruit, tea, coffee, and so forth—10,000 times more than their intake of synthetic pesticide residues. The concentration of these natural compounds ranges in the parts per million (ppm), orders of magnitude above the levels at which water pollutants (e.g., chlorinated hydrocarbons) and other synthetic pollutants (e.g., dioxin, Chemical Highlight 22-1) are usually measured (parts per billion, or ppb). Few of these plant toxins have been tested for carcinogenicity but, of those tested (in rodents), roughly half are carcinogenic, the same proportion as that of synthetic chemicals. Table 25-2 gives some examples of these compounds, their concentration, and occurrence in common foods.

Why, then, have we all not been exterminated by these poisons? One reason is that the level of our exposure to any one of these natural pesticides is very small. More importantly, we, like plants, have evolved to defend ourselves against this barrage of chemical projectiles. Thus, for starters, our first line of defense, the surface layers of the mouth, esophagus, stomach, intestine, skin, and lungs, is discarded once every few days as "cannon fodder." In addition, we have multiple detoxifying mechanisms, rendering ingested poisons nontoxic; we excrete a lot of material before it does any harm, our DNA has many ways of repairing damage, and, finally, our ability to smell and taste "repugnant" substances (such as the "bitter" alkaloids, rotten food, milk that is "off," eggs that smell of "sulfur") serves as an advance warning signal. In the final analysis, we each must judge what we put into our bodies, but the age-old wisdoms still hold: avoid anything in excess and maintain variety in your diet.

*Professor Bruce N. Ames (b. 1928), University of California at Berkeley.

TABLE 25-2	Carcinogenic Natural Plant Pesticides
Compound	**Plant food (concentration in ppm)**
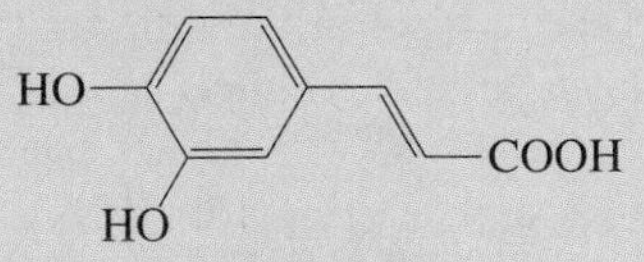 Caffeic acid	Apple, carrot, celery, grapes, lettuce, potato (50–200); basil, dill, sage, thyme, and other herbs (>1000); coffee (roasted beans, 1800)
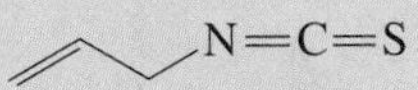 Allyl isothiocyanate (Section 14-3)	Cabbage (35–590); cauliflower (12–66), Brussels sprouts (110–1560); brown mustard (16,000–72,000); horseradish (4500)
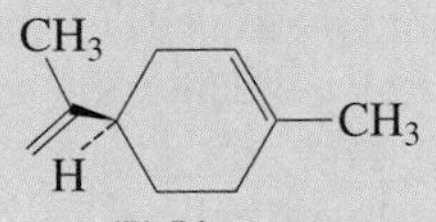 (R)-Limonene	Orange juice (31); black pepper (8000)
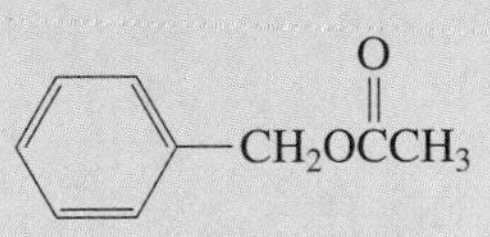 Benzyl acetate	Basil (82); jasmine tea (230); honey (15)

people received emergency treatment in the United States alone for cocaine over-doses. The compound has some good uses, nevertheless. For example, it functions as a very effective topical anesthetic in eye operations.

Quinine (see first page of this chapter), isolated from cinchona bark (as much as 8% concentration) is the oldest known effective antimalarial agent. A malaria attack consists of a chill accompanied or followed by a fever, which terminates in a sweat-ing stage. Such attacks may recur regularly. The name malaria is derived from the Italian *malo,* bad, and *aria,* air, referring to the old theory that the disease is caused by noxious effluent gases from marshland. However, malaria is actually caused by a protozoan parasite (*Plasmodium* species) transmitted by the bite of an infected fe-male mosquito of the genus *Anopheles*. It is estimated that from 300 to 500 million people are affected by this disease, killing 2 million each year, more than half of them children.

Strychnine is a powerful poison (the lethal dose in animals is about 5–8 mg kg^{-1}), the lethal ingredient of many a detective novel.

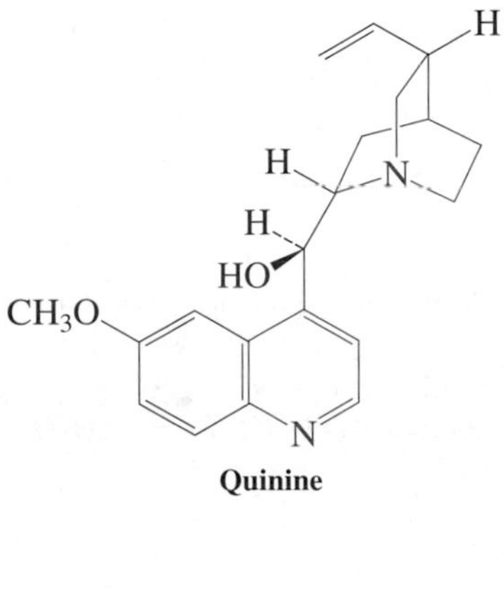

Quinine

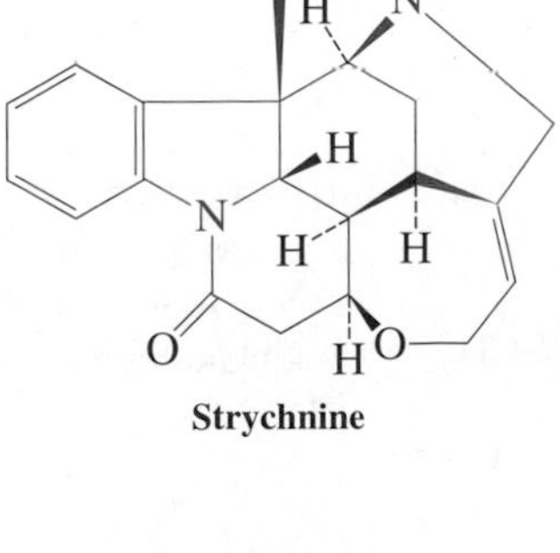

Strychnine

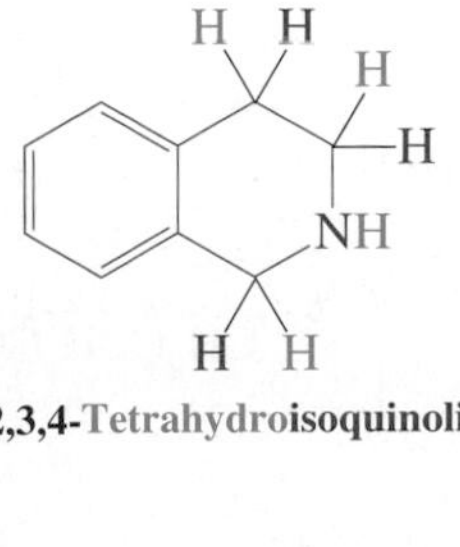

1,2,3,4-Tetrahydroisoquinoline

The isoquinoline and 1,2,3,4-tetrahydroisoquinoline nuclei are abundant among the alkaloids, and their derivatives are physiologically active, for example, as hallucinogens, central nervous system agents (depressants and stimulants), and hypotensives. Note that the pharmacophoric 2-phenylethanamine unit (see Chemical Highlight 21-1) is part of these nuclei and is also present in most of the other alkaloids considered in this section. (Find it in morphine, lysergic acid, quinine—there is a quirk here—and strychnine.)

In summary, the alkaloids are natural nitrogen-containing compounds, many of which are physiologically active.

CHAPTER INTEGRATION PROBLEM

As we have seen for the heterocyclopentadienes (Section 25-3) and pyridine (Section 25-5), heteroaromatic compounds can almost invariably be made by condensation reactions of carbonyl substrates with appropriate heterofunctions.

a. Write a plausible mechanism for the first step of the Hantzsch synthesis of 2,6-dimethylpyridine (Section 25-5) shown again here.

SOLUTION

By following the fate of the four components in the starting mixture, you can see that the ammonia has reacted at the two keto carbonyl carbons, presumably by imine and then enamine formation (Section 17-9), whereas the formaldehyde component has made bonds to the acidic methylene of the 3-oxobutanoates (Section 23-2), presumably initially by an aldol-like condensation process (Section 18-5). Let us formulate these steps one at a time.

Step 1. Aldol-like condensation of formaldehyde with ethyl 3-oxobutanoate

Step 2. Enamine formation of ammonia with ethyl 3-oxobutanoate

We note that step 1 furnished a Michael acceptor, whereas step 2 resulted in an enamine. The latter can react with the former, in analogy to the reaction of enolates, by Michael addition, in this case to a neutral oxoenamine.

Step 3. Michael addition of the enamine

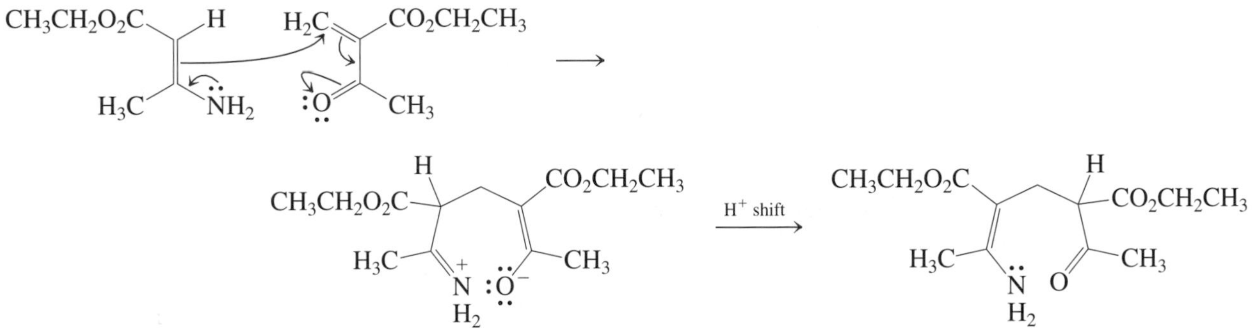

Oxoenamine

This species is now perfectly set up to undergo an intramolecular imine condensation of the 3,4-dihydropyridine, which tautomerizes (Sections 13-8 and 18-2) to the more stable 1,4-dihydro product.

Step 4. Intramolecular imine formation and tautomerization

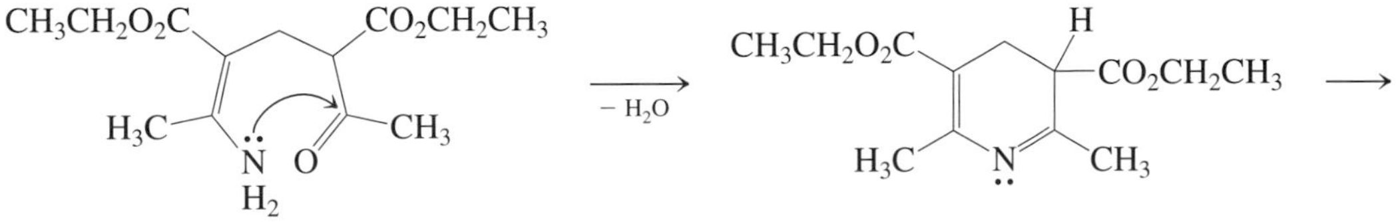

Oxoenamine

Diethyl 3,4-dihydro-2,6-dimethyl-3,5-pyridinedicarboxylate

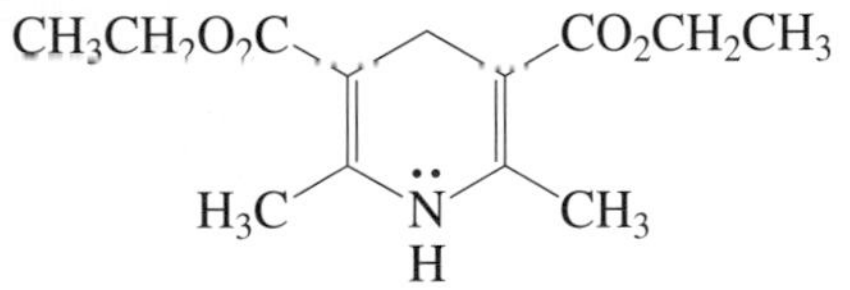

Diethyl 1,4-dihydro-2,6-dimethyl-3,5-pyridinedicarboxylate

b. On the basis of the discussion of (a) and Exercises 25-9 and 25-11, suggest some simple retrosynthetic (retrocondensation) reactions to indole, quinoline, and 1,4-diazanaphthalene (quinoxaline) from ortho-disubstituted benzenes.

Solution

You can view indole as a benzofused enamine. The enamine part is retrosynthetically connected to the corresponding enolizable carbonyl and amino function (Section 17-9).

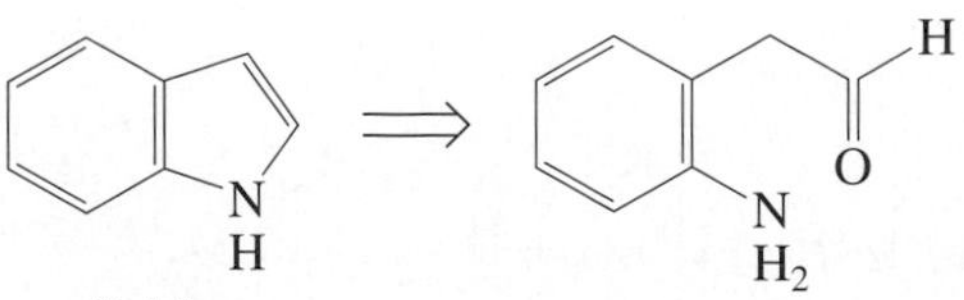

Indole

Quinoline can be viewed as a benzofused α,β-unsaturated imine. The imine nucleus is retrosynthetically opened to the corresponding amine and α,β-unsaturated carbonyl fragments (Section 17-9), which can be constructed by an aldol condensation (Section 18-5) employing acetaldehyde.

Quinoline

1,4-Diazanaphthalene can be dissected by retrosynthetic hydrolysis of the two imine functions (Section 17-9) to 1,2-benzenediamine and ethanedione (glyoxal).

**1,4-Diazanaphthalene
(Quinoxaline)**

NEW REACTIONS

1. Reactions of Heterocyclopropanes (Section 25-2)

2. Ring Opening of Heterocyclobutanes (Section 25-2)

Less reactive than heterocyclopropanes

3. Paal-Knorr Synthesis of 1-Hetero-2,4-cyclopentadienes (Section 25-3)

4. Reactions of 1-Hetero-2,4-cyclopentadienes (Section 25-4)

Electrophilic substitution

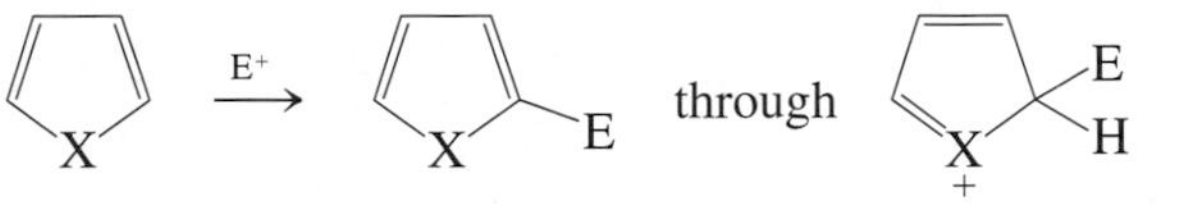

Main product 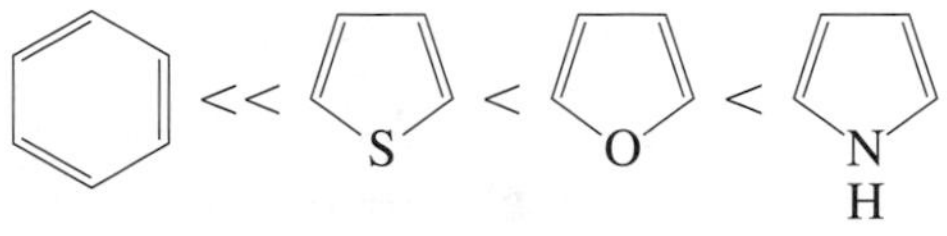   **Relative reactivity**

Ring opening

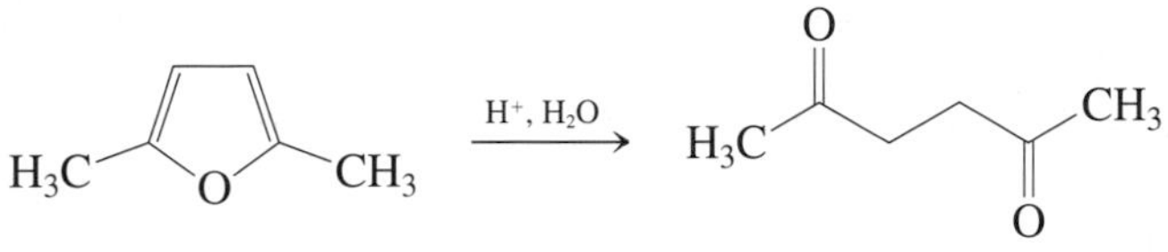

Cycloaddition

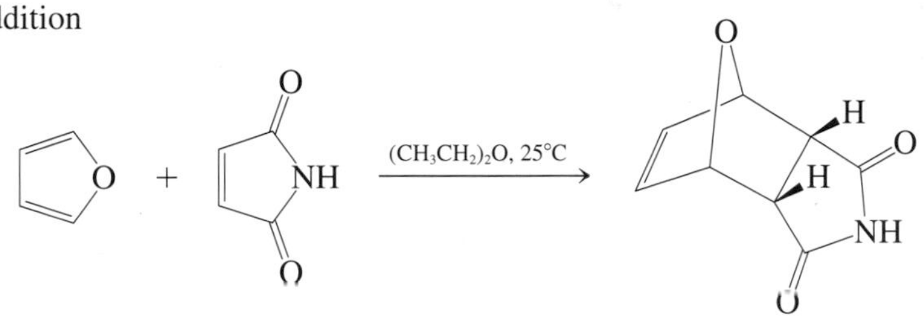

5. Hantzsch Synthesis of Pyridines (Section 25-5)

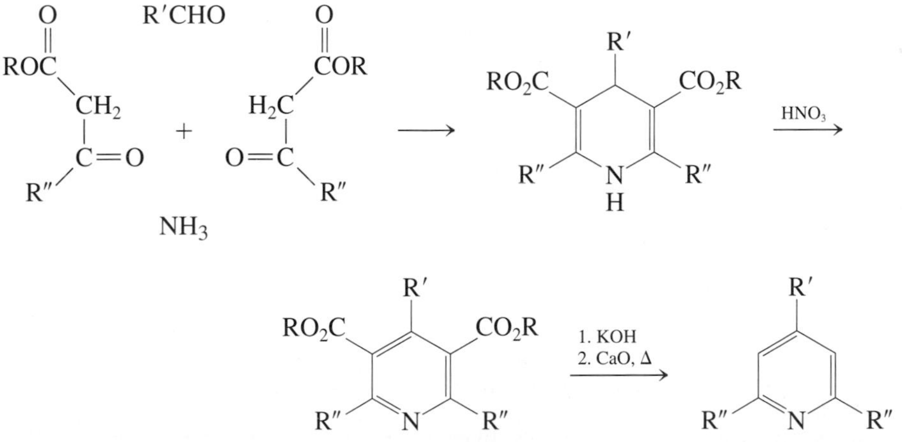

6. Reactions of Pyridine (Sections 25-5 and 25-6)

Protonation (Section 25-5)

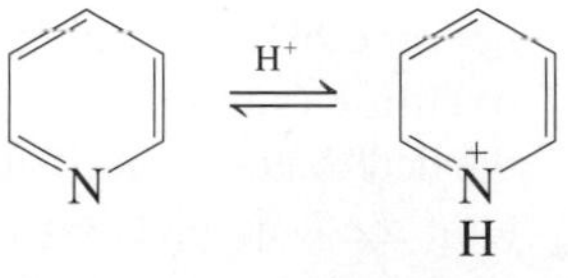

Pyridinium ion
$pK_a = 5.29$

Electrophilic substitution (Section 25-6)

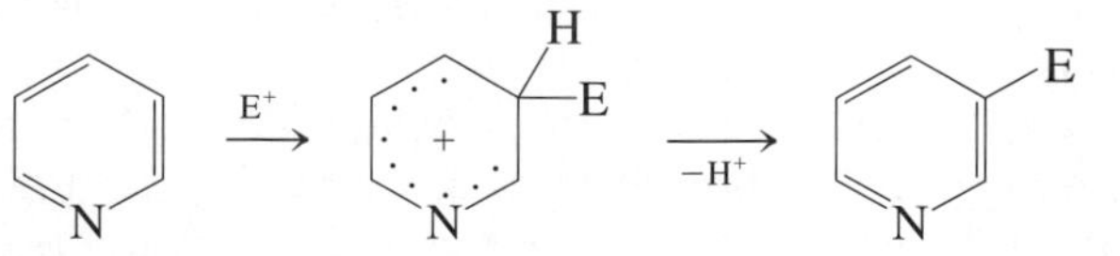

Ring is deactivated relative to benzene

Nucleophilic substitution (Section 25-6)

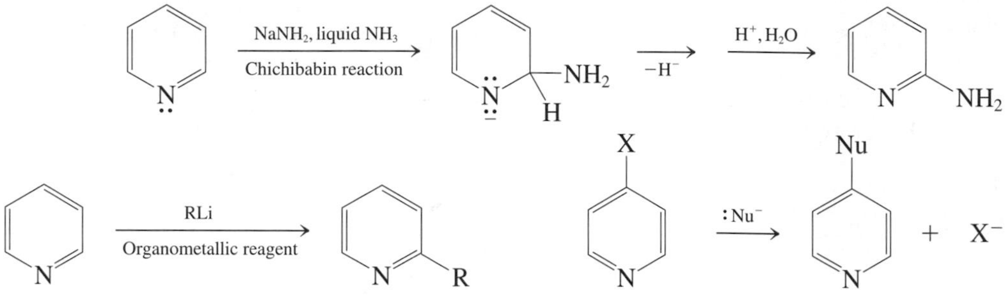

Halopyridine
(X = Br, Cl)

7. Reactions of Quinoline and Isoquinoline (Section 25-7)

Electrophilic substitution

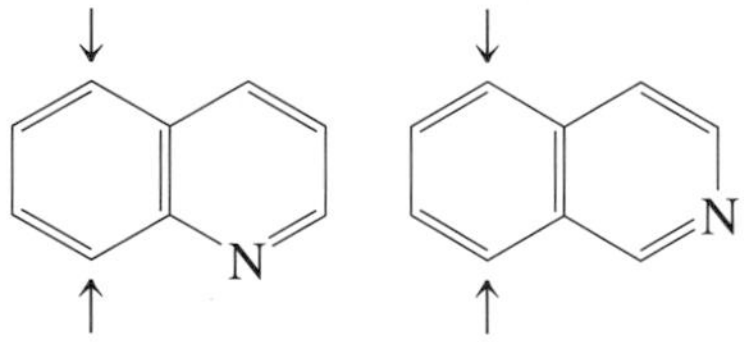

Nucleophilic substitution

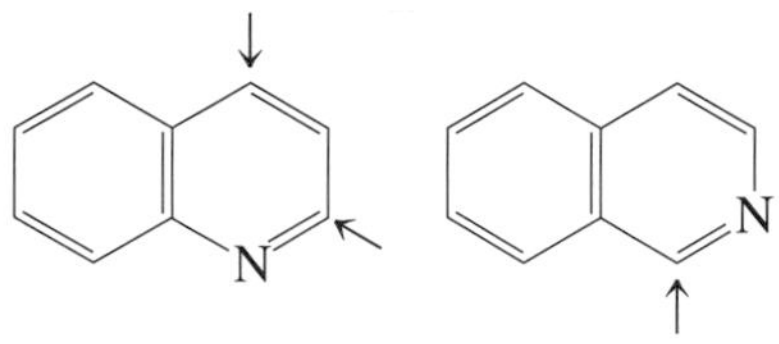

IMPORTANT CONCEPTS

1. The **heterocycloalkanes** can be named by using cycloalkane nomenclature. The prefix aza- for nitrogen, oxa- for oxygen, thia- for sulfur, and so forth, indicates the heteroatom. Other systematic and common names abound in the literature, particularly for the aromatic heterocycles.

2. The **strained three-** and **four-membered heterocycloalkanes** undergo **ring opening** with nucleophiles easily.

3. The **1-hetero-2,4-cyclopentadienes** are **aromatic** and have an arrangement of six π electrons, similar to that in the cyclopentadienyl anion. The heteroatom

is sp^2 hybridized, the p orbital contributing two electrons to the π system. As a consequence, the diene unit is electron rich and reactive in electrophilic aromatic substitutions.

4. Replacement of one (or more) of the CH units in benzene by an sp^2-hybridized nitrogen gives rise to **pyridine** (and other azabenzenes). The p orbital on the heteroatom contributes one electron to the π system; the lone electron pair is located in an sp^2 hybrid atomic orbital in the molecular plane. Azabenzenes are **electron poor,** because the electronegative nitrogen withdraws electron density from the ring by

induction and by resonance. Electrophilic aromatic substitution of azabenzenes is sluggish. Conversely, nucleophilic aromatic substitution occurs readily; this is shown by the Chichibabin reaction, substitutions by organometallic reagents next to the nitrogen, and the displacement of halide ion from halopyridines by nucleophiles.

5. The azanaphthalenes (benzopyridines) **quinoline** and **isoquinoline** contain an electron-poor pyridine ring, susceptible to nucleophilic attack, and an electron-rich benzene ring that enters into electrophilic aromatic substitution reactions, usually at the positions closest to the heterocyclic unit.

PROBLEMS

22. Name or draw the following compounds. (**a**) *cis*-2,3-Diphenyloxacyclopropane; (**b**) 3-azacyclobutanone; (**c**) 1,3-oxathiacyclopentane; (**d**) 2-butanoyl-1,3-dithiacyclohexane;

(**e**) 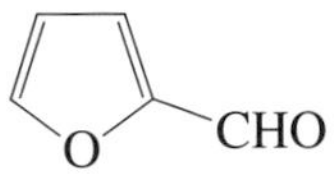 (**f**) 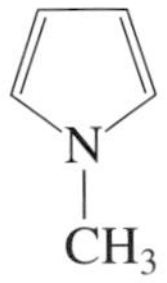 (**g**) 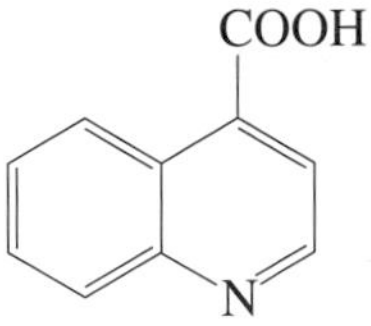 (**h**)

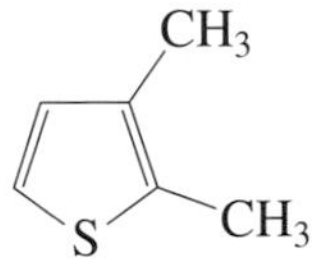

23. Give the expected product of each of the following reaction sequences.

(**a**)

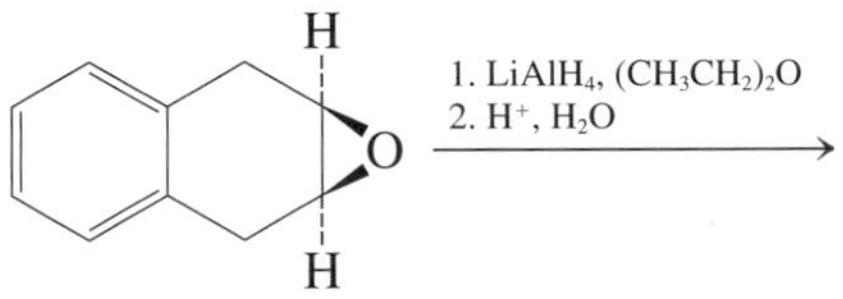

(**b**)

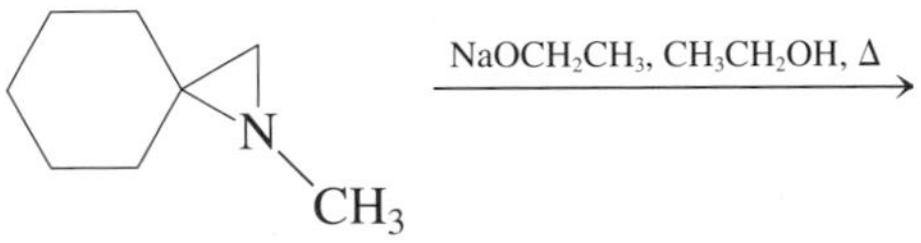

(**c**)

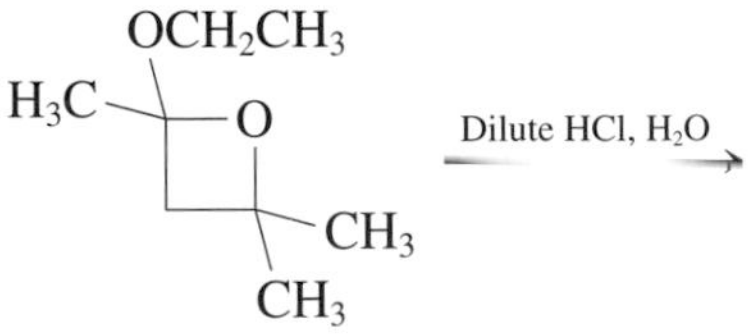

24. The penicillins are a class of antibiotics containing two heterocyclic rings that interfere with the construction of cell walls by bacteria. The interference results from reaction of the penicillin with an amino group of a protein that closes gaps that develop during construction of the cell wall. The insides of the cell leak out, and the organism dies. (**a**) Suggest a reasonable product for the reactions of penicillin G with the amino group of a protein (protein-NH$_2$). (**Hint:** First identify the most reactive electrophilic site in penicillin.)

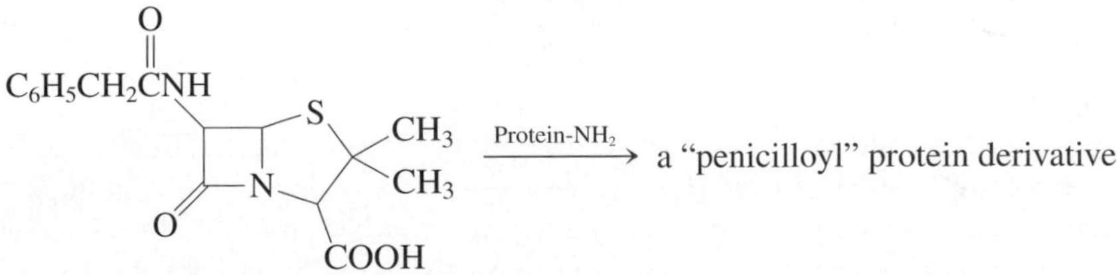

Penicillin G

(b) Penicillin-resistant bacteria secrete an enzyme (penicillinase) that catalyzes hydrolysis of the antibiotic faster than the antibiotic can attack the cell-wall proteins. Propose a structure for the product of this hydrolysis and suggest a reason why hydrolysis destroys the antibiotic properties of penicillin.

$$\text{Penicillin G} \xrightarrow{\text{H}_2\text{O penicillinase}} \text{penicilloic acid}$$

(Hydrolysis product; no antibiotic activity)

25. Propose reasonable mechanisms for the following transformations.

(a)

1. SnCl₄ (a Lewis acid), CH₂Cl₂
2. H⁺, H₂O

CH₂OH

(b)

1. CH₃CH₂CH₂CH₂Li, BF₃-O(CH₂CH₃)₂, THF
2. H⁺, H₂O

C₆H₅

OH

(c)

H₃C, H, O, CH₃

MgBr₂, CH₃COCH₃
(Hint: See Section 15-14.)

CH₃CO, H, CH₃, Br, H₃C, H

26. Rank the following compounds in increasing order of basicity: water, hydroxide, pyridine, pyrrole, ammonia.

27. Each of the heterocyclopentadienes in the margin contains more than one heteroatom. For each one, identify the orbitals occupied by all lone electron pairs on the heteroatoms and determine whether the molecule qualifies as aromatic. Are any of these heterocycles a stronger base than pyrrole?

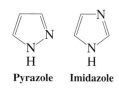

Pyrazole   **Imidazole**

Thiazole   **Isoxazole**

28. Give the product of each of the following reactions.

(a)

CH₃NH₂

(b)

P₂O₅, Δ

29. 1-Hetero-2,4-cyclopentadienes can be prepared by condensation of an α-dicarbonyl compound and certain heteroatom-containing diesters. Propose a mechanism for the following pyrrole synthesis.

$$\text{C}_6\text{H}_5\text{CCC}_6\text{H}_5 + \text{CH}_3\text{OCCH}_2\text{NCH}_2\text{COCH}_3 \xrightarrow{\text{NaOCH}_3,\ \text{CH}_3\text{OH},\ \Delta}$$

H₃C, C=O

H₅C₆, C₆H₅

CH₃OOC, COOCH₃

H₃C, C=O

How would you use a similar approach to synthesize 2,5-thiophenedicarboxylic acid?

30. Give the expected major product(s) of each of the following reactions. Explain how you chose the position of substitution in each case.

(a)  $\xrightarrow{\text{Cl}_2}$

(b) 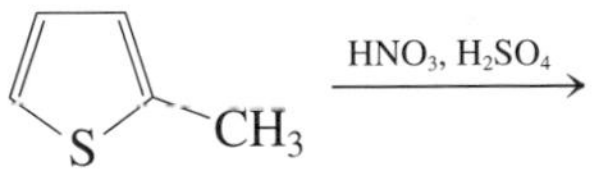 $\xrightarrow{\text{HNO}_3, \text{H}_2\text{SO}_4}$

(c) 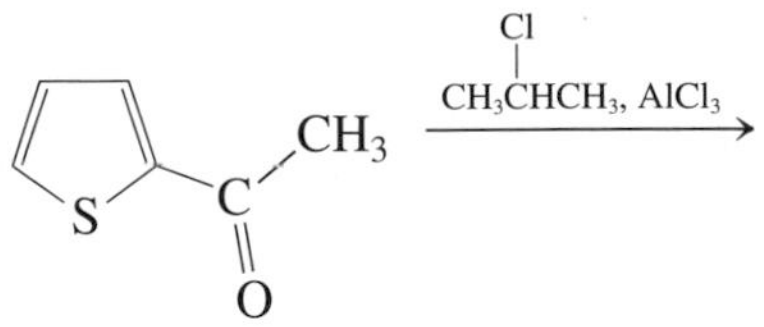 $\xrightarrow{\text{CH}_3\text{CHCH}_3, \text{AlCl}_3}$

(d) 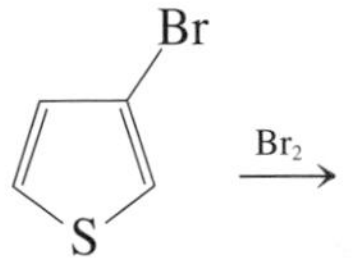 $\xrightarrow{\text{Br}_2}$

(e) 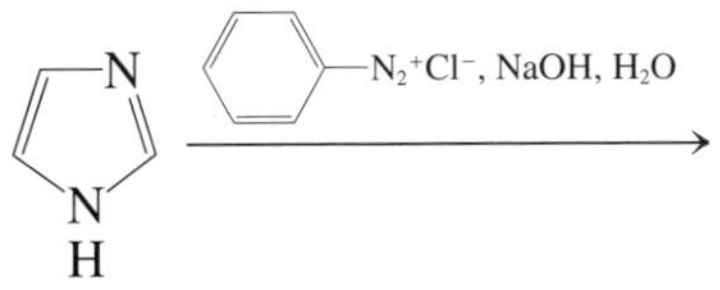 $\xrightarrow{\text{N}_2{}^+\text{Cl}^-, \text{NaOH}, \text{H}_2\text{O}}$

31. Give the products expected of each of the following reactions.

(a) 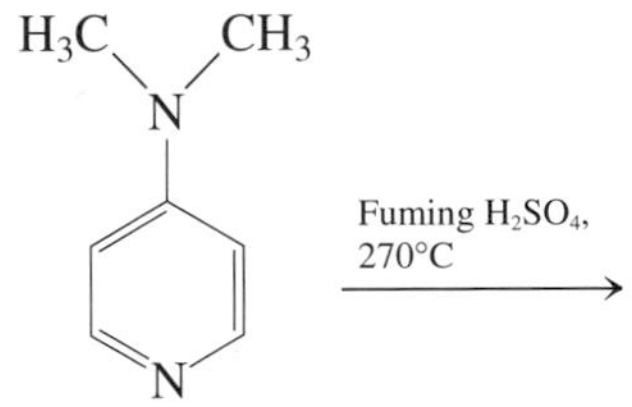 $\xrightarrow[270°C]{\text{Fuming H}_2\text{SO}_4,}$

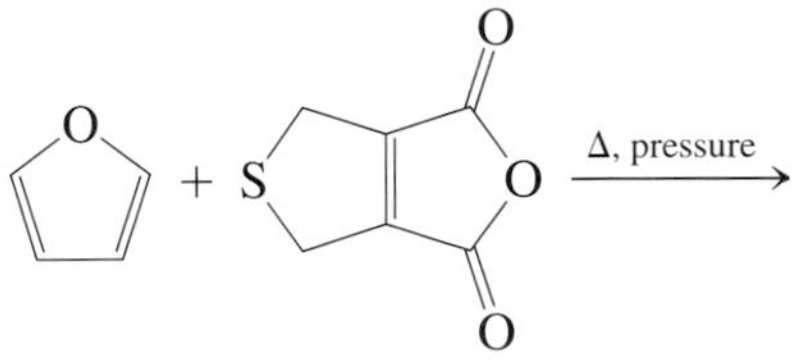

(b) $\xrightarrow{\Delta, \text{ pressure}}$

(c) $\xrightarrow{\text{KSH, CH}_3\text{OH}, \Delta}$

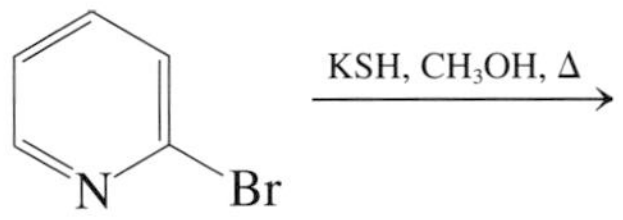

(d) 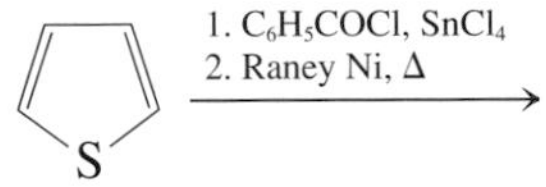 $\xrightarrow[\text{2. Raney Ni, } \Delta]{\text{1. C}_6\text{H}_5\text{COCl, SnCl}_4}$

(e) 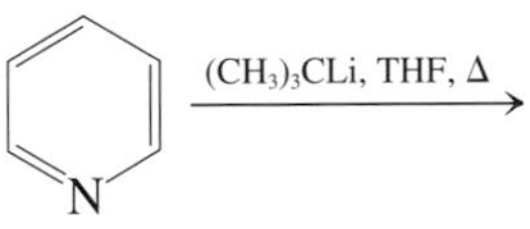 $\xrightarrow{(\text{CH}_3)_3\text{CLi, THF}, \Delta}$

32. Propose a synthesis of each of the following substituted heterocycles, using synthetic sequences presented in this chapter.

(a)

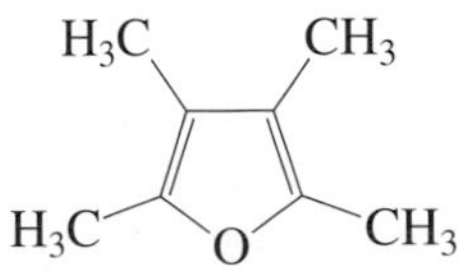

(b)

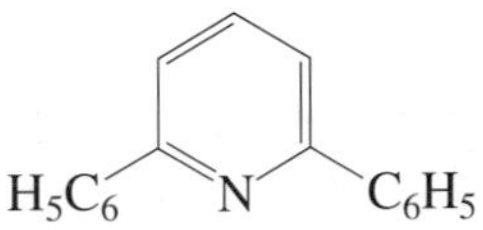

(c)

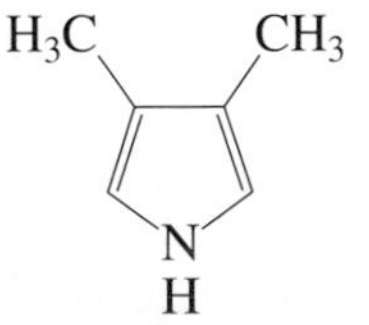

(d)

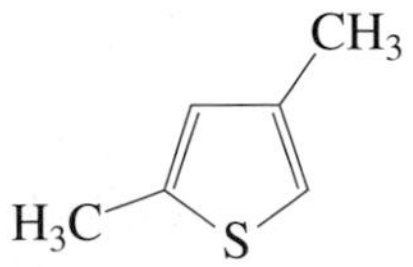

33. Chelidonic acid, a 4-oxacyclohexanone (common name, γ-pyrone), is found in a number of plants and is synthesized from propanone (acetone) and diethyl ethanedioate. Formulate a mechanism for this transformation.

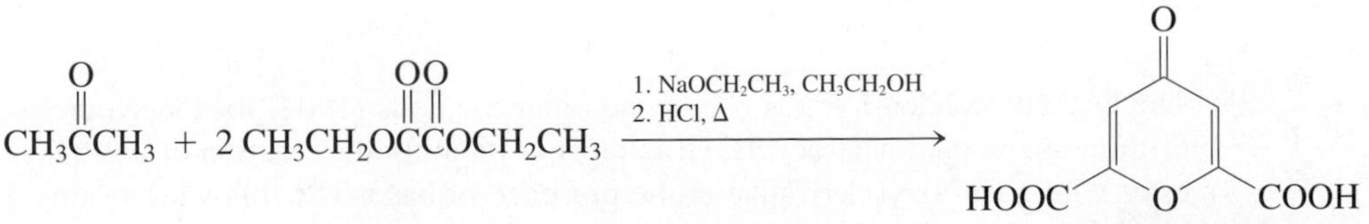

Chelidonic acid

34. Reserpine is a naturally occurring indole alkaloid with powerful tranquilizing and antihypertensive activity. Many such compounds possess a characteristic structural feature: one nitrogen atom at a ring fusion separated by two carbons from another nitrogen atom.

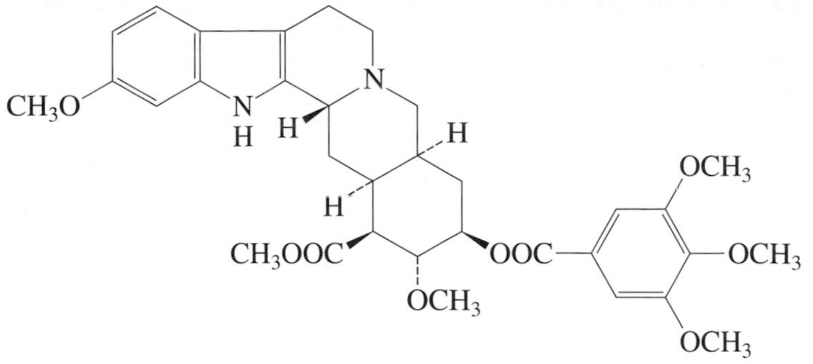

Reserpine

A series of compounds with modified versions of this structural feature have been synthesized and also shown to have antihypertensive activity, as well as antifibrillatory properties. One such synthesis is shown here. Name or draw the missing reagents and products (a) through (c).

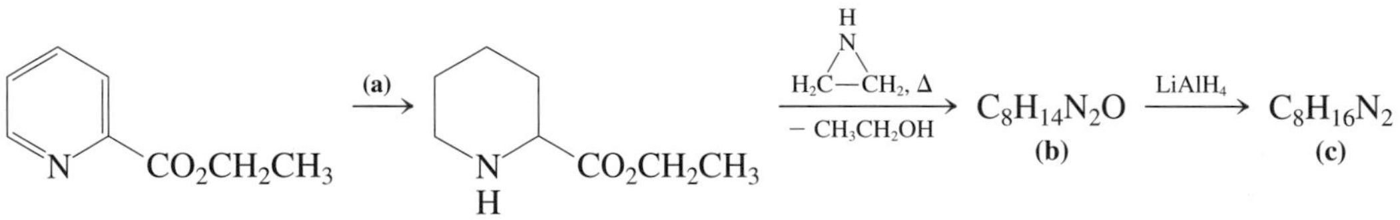

35. Starting with benzenamine (aniline) and pyridine, propose a synthesis for the antimicrobial sulfa drug sulfapyridine.

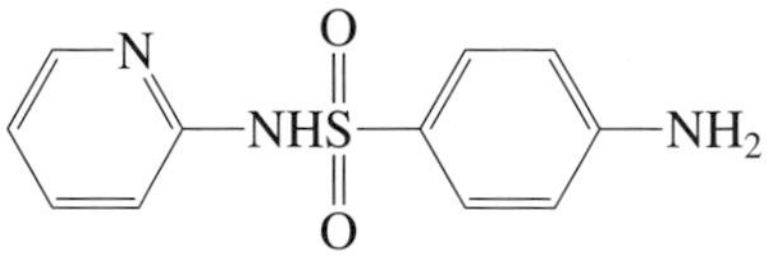

Sulfapyridine

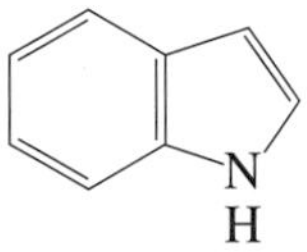

36. Derivatives of benzimidazole possess biological activity somewhat like that of indoles and purines (of which adenine, Section 25-1, is an example). Benzimidazoles are commonly prepared from benzene-1,2-diamine. Devise a short synthesis of 2-methylbenzimidazole from benzene-1,2-diamine.

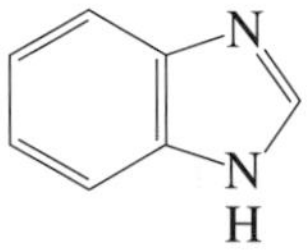

Indole

Benzimidazole

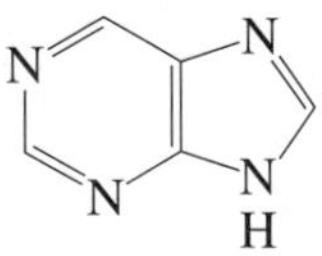

Purine

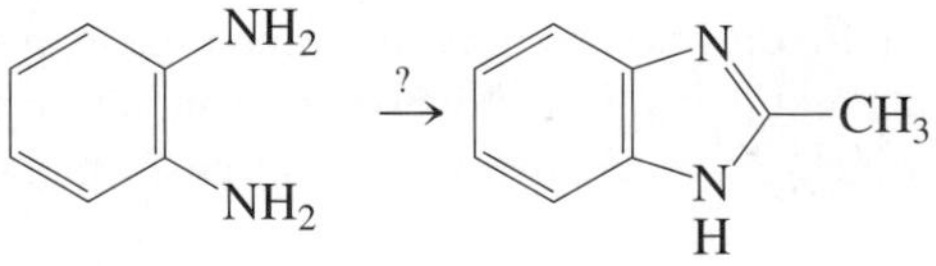

Benzene-1,2-diamine   **2-Methylbenzimidazole**

37. The Darzens condensation is one of the older methods (1904) for the synthesis of three-membered heterocycles. It is most commonly the reaction of a 2-halo ester with a carbonyl derivative in the presence of base. The following examples of the Darzens condensation show how it is applied to the synthesis of oxacyclopropane and azacyclopropane rings. Suggest a reasonable mechanism for each of these reactions.

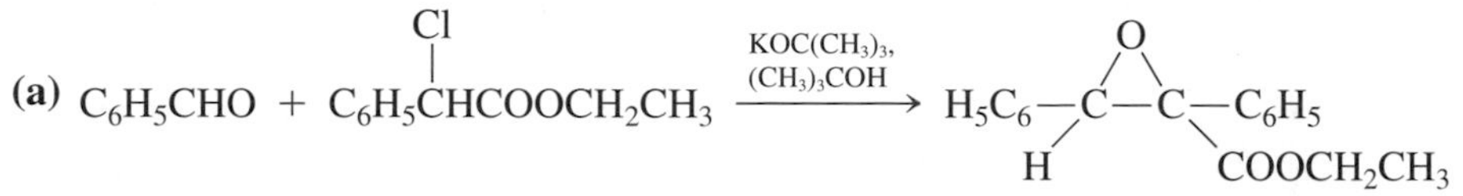

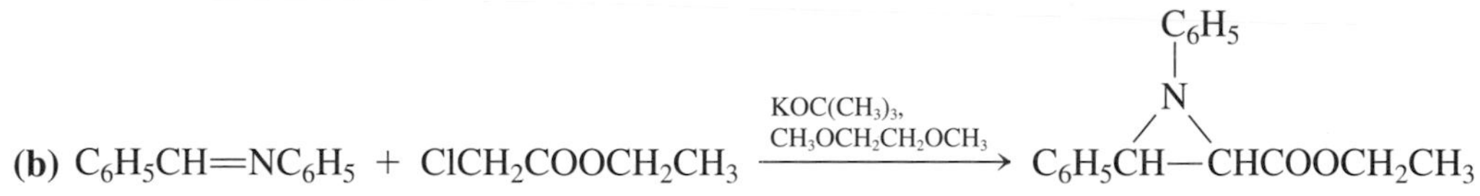

(a) C_6H_5CHO + $C_6H_5\overset{\underset{|}{Cl}}{C}HCOOCH_2CH_3$ $\xrightarrow[\text{(CH}_3)_3\text{COH}]{\text{KOC(CH}_3)_3,}$ $H_5C_6\overset{\underset{|}{}}{\underset{H}{C}}\overset{O}{\overset{\diagup\diagdown}{C}}\overset{\underset{|}{}}{\underset{COOCH_2CH_3}{C}}C_6H_5$

(b) $C_6H_5CH{=}NC_6H_5$ + $ClCH_2COOCH_2CH_3$ $\xrightarrow[\text{CH}_3\text{OCH}_2\text{CH}_2\text{OCH}_3]{\text{KOC(CH}_3)_3,}$ $C_6H_5\overset{\overset{\displaystyle C_6H_5}{\overset{\displaystyle |}{\overset{\displaystyle N}{\diagup\diagdown}}}}{CH{-}CH}COOCH_2CH_3$

38. **(a)** The compound shown in the margin, with the common name 1,3-dibromo-5,5-dimethylhydantoin, is useful as a source of electrophilic bromine (Br^+) for addition reactions. Give a more systematic name for this heterocyclic compound. **(b)** An even more remarkable heterocyclic compound (ii) is prepared by the following reaction sequence. Using the given information, deduce structures for compounds i and ii, and name the latter.

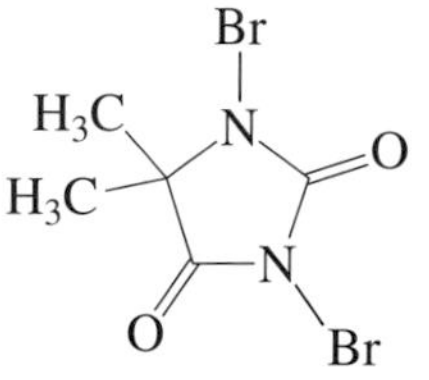

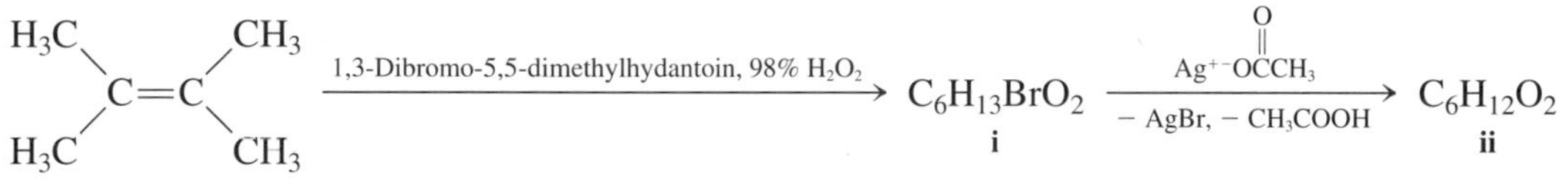

$$\underset{H_3C}{\overset{H_3C}{>}}C{=}C\underset{CH_3}{\overset{CH_3}{<}} \xrightarrow{\text{1,3-Dibromo-5,5-dimethylhydantoin, 98\% H}_2\text{O}_2} \underset{\textbf{i}}{C_6H_{13}BrO_2} \xrightarrow[-\text{ AgBr, }-\text{ CH}_3\text{COOH}]{Ag^+\ {}^-OCCH_3} \underset{\textbf{ii}}{C_6H_{12}O_2}$$

Heterocycle ii is a yellow, crystalline, sweet-smelling compound that decomposes on gentle heating into two molecules of propanone (acetone), one of which is formed directly in its $n \rightarrow \pi^*$ excited state (Sections 14-11 and 17-3). This electronically excited product is chemiluminescent.

$$\textbf{ii} \longrightarrow \underset{}{CH_3\overset{O}{\overset{||}{C}}CH_3} + \left[\underset{}{CH_3\overset{O}{\overset{||}{C}}CH_3}\right]^{n\rightarrow\pi^*} \longrightarrow h\nu + 2\ CH_3\overset{O}{\overset{||}{C}}CH_3$$

Heterocycles similar to compound ii are responsible for the chemiluminescence produced by a number of species (e.g., fireflies [see Chemical Highlight 9-1] and several deep-sea fish); they also serve as the energy sources in commercial chemiluminescent products.

39. Azacyclohexanes (piperidines) can be synthesized by reaction of ammonia with *cross-conjugated dienones:* ketones conjugated on both sides with double bonds. Propose a mechanism for the following synthesis of 2,2,6,6-tetramethyl-aza-4-cyclohexanone.

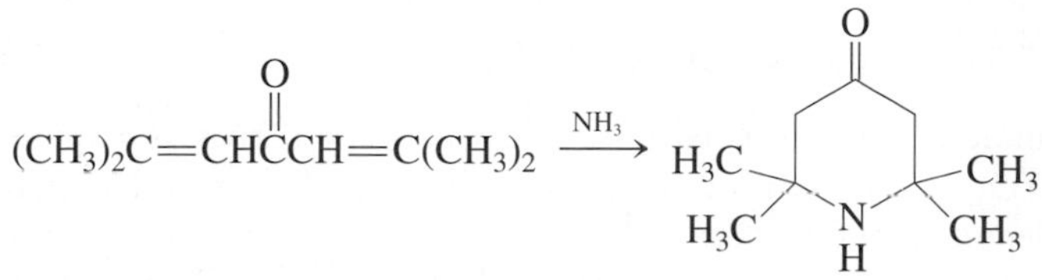

$$(CH_3)_2C{=}CH\overset{O}{\overset{||}{C}}CH{=}C(CH_3)_2 \xrightarrow{NH_3}$$

40. Compound A, C_8H_8O, exhibits ^1H NMR spectrum A (page 1152). On treatment with concentrated aqueous HCl, it is converted almost instantaneously into a compound that exhibits spectrum B (page 1152). What is compound A, and what is the product of its treatment with aqueous acid?

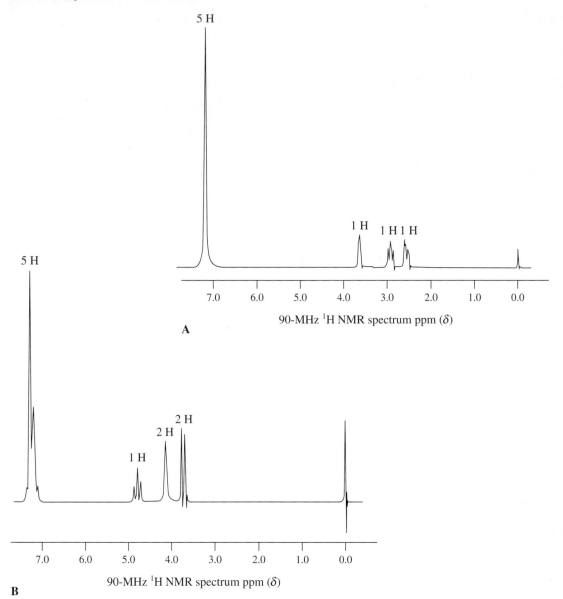

5 H

1 H   1 H 1 H

7.0   6.0   5.0   4.0   3.0   2.0   1.0   0.0

90-MHz ^{1}H NMR spectrum ppm (δ)

A

5 H

1 H

2 H

2 H

7.0   6.0   5.0   4.0   3.0   2.0   1.0   0.0

90-MHz ^{1}H NMR spectrum ppm (δ)

B

41. Heterocycle C, C_5H_6O, exhibits ^{1}H NMR spectrum C and is converted by H_2 and Raney nickel into compound D, $C_5H_{10}O$, with spectrum D. Identify compounds C and D. (Note: The coupling constants of the compounds in this problem and the next one are rather small; they are therefore not nearly as useful in structure elucidation as those around a benzene ring.)

42. The commercial synthesis of a useful heterocyclic derivative requires treatment of a mixture of aldopentoses (derived from corncobs, straw, etc.) with hot acid under dehydrating conditions. The product, E, has ^{1}H NMR spectrum E, shows a strong IR band at 1670 cm^{-1}, and is formed in nearly quantitative yield. Identify compound E and formulate a mechanism for its formation.

$$\text{Aldopentoses} \xrightarrow{\text{H}^+, \, \Delta} \underset{\mathbf{E}}{C_5H_4O_2}$$

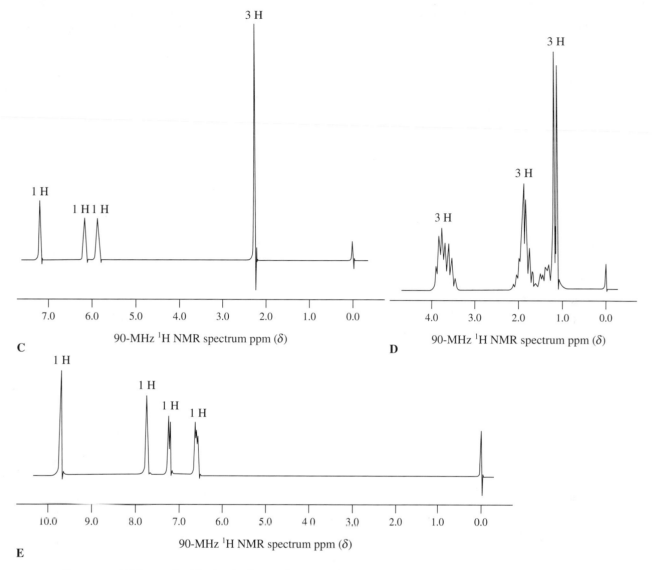

90-MHz ^{1}H NMR spectrum ppm (δ)

C

90-MHz ^{1}H NMR spectrum ppm (δ)

D

90-MHz ^{1}H NMR spectrum ppm (δ)

E

Compound E is a valuable synthetic starting material. The following sequence converts it into furethonium, which is useful in the treatment of glaucoma. What is the structure of furethonium?

$$E \xrightarrow[\text{2. Excess } CH_3I, (CH_3CH_2)_2O]{\text{1. } NH_3, NaBH_3CN} \text{furethonium}$$

43. Treatment of a 3-alkanoylindole with $LiAlH_4$ in $(CH_3CH_2)_2O$ reduces the carbonyl all the way to a CH_2 group. Explain by a plausible mechanism. (**Hint:** Direct S_N2 displacement of alkoxide by hydride is *not* plausible.)

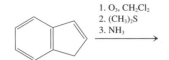

1. O₃, CH₂Cl₂
2. (CH₃)₂S
3. NH₃

44. The sequence in the margin is a rapid synthesis of one of the heterocycles in this chapter. Draw the structure of the product, which has ^{1}H NMR spectrum F.

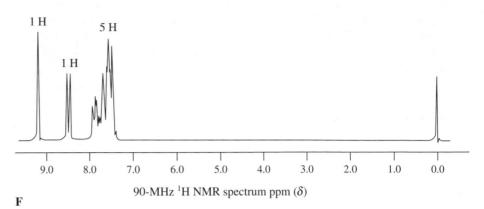

90-MHz ^{1}H NMR spectrum ppm (δ)

F

Team Problem

45. This problem introduces two literature syntheses of indole derivatives, and you are asked to come up with plausible mechanisms for them. Divide your team in two, each group concentrating on one of the methods.

Fischer Indole Synthesis of 2-Phenylindole

2-Phenylindole

In this procedure, a hydrazone of an enolizable aldehyde or ketone is heated in strong acid, causing ring closure with simultaneous expulsion of ammonia to furnish the indole nucleus. [**Hints:** The mechanism of the reaction proceeds in three stages: (1) an imine–enamine tautomerization (recall Section 17-9); (2) an electrocyclic reaction (a "diaza-Cope" rearrangement; recall Section 22-9); (3) another imine–enamine (in this case, benzenamine) tautomerization; (4) ring closure to the heterocycle; and (5) elimination of NH₃.]

Reissert Indole Synthesis of Ethyl Indole-2-carboxylate

2-Methylnitrobenzene
(*o*-Nitrotoluene)

$K^{+ \, -}OCH_2CH_3$

A 2-oxopropanoate ester
(A pyruvate ester)

H_2, Pt

$-CO_2CH_2CH_3$

Ethyl indole-2-carboxylate

In this sequence, a 2-methylnitrobenzene (*o*-nitrotoluene) is first converted into an ethyl 2-oxopropanoate (pyruvate, Chemical Highlight 23-2) ester, which, on reduction, is transformed into the target indole. [**Hints:** (1) The nitro group is essential for the success of the first step. Why? Does this step remind you of another reaction? Which one? (2) Which functional group is the target of the reduction step (recall Section 16-5)? (3) The ring closure to the heterocycle requires a condensation reaction.]

Preprofessional Problems

46. The proton decoupled ^{13}C NMR spectrum of pyridine will display how many peaks? (**a**) One; (**b**) two; (**c**) three; (**d**) four; (**e**) five.

47. Pyrrole is a much weaker base than azacyclopentane (pyrrolidine) for which of the following reasons? (**a**) The nitrogen in pyrrole is more electropositive than that in pyrrolidine; (**b**) pyrrole is a Lewis acid; (**c**) pyrrole has four electrons; (**d**) pyrrolidine can give up the proton on the nitrogen atom more readily than can pyrrole; (**e**) pyrrole is aromatic.

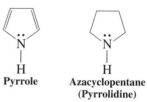

Pyrrole   **Azacyclopentane (Pyrrolidine)**

48. Which of the following compounds would you expect to be the major organic product of the two-step sequence shown here?

$$\text{(structure)} + (CH_3CH)_2NLi \xrightarrow{\text{THF}} \xrightarrow{CH_3CH_2I}$$

(**a**)   (**b**)   (**c**)   (**d**)

49. This reaction yields one main organic product. Which of the following compounds is it?

$$\text{2-Phenylthiophene} \xrightarrow{SnCl_4,\ CH_3\overset{O}{\overset{\|}{C}}Cl}$$

(**a**)   (**b**)   (**c**)   (**d**)

Amino Acids, Peptides, Proteins, and Nucleic Acids

Nitrogen-Containing Polymers in Nature

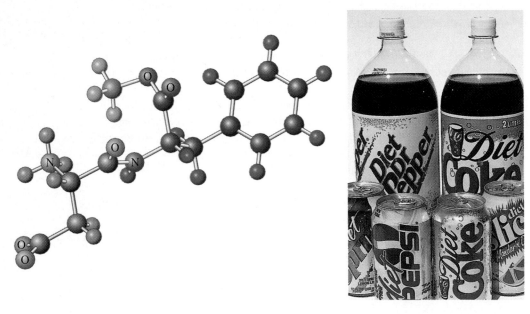

Aspartame (NutraSweet) is the artificial sweetener used in most diet drinks and other low-calorie products. It is 200 times as sweet as table sugar (sucrose) but contains only 4 calories per gram.

On page 1 of this text we defined organic chemistry as the chemistry of carbon-containing compounds. We then went on to point out that organic molecules constitute the chemical bricks of life. Indeed, a historical definition of organic chemistry restricts it to living organisms. What is life and how do we, as organic chemists, approach its study? A functional definition of life refers to it as a condition of matter that is manifested by growth, metabolism, reproduction, and evolution. The underlying basic processes are chemical, and researchers hope to decipher their complexity by investigating specific reactions or reaction sequences. The "whole," however, is much more complex than these individual pieces, as the latter interact by multiple feedback loops in a manner that is intractable in a deterministic way (i.e., such that each effect has a simple cause). This final chapter will provide you with a glimpse of that complexity by taking you from amino acids to their polymers, the polypeptides—in particular, the large natural polypeptides called **proteins**—and to their biological origin, **DNA.**

Proteins have an astounding diversity of functions in living systems. As **enzymes,** they catalyze transformations ranging in complexity from the simple hydration of carbon dioxide to the replication of entire chromosomes—great coiled strands of DNA, the genetic material in living cells. Enzymes can accelerate certain reactions many millionfold.

We have already encountered the protein rhodopsin, the photoreceptor that generates and transmits nerve impulses in retinal cells (Chemical Highlight 18-4). Other proteins serve for transport and storage. Thus, hemoglobin carries oxygen; iron is transported in the blood by transferrin and stored in the liver by ferritin. Proteins play a crucial role in coordinated motion, such as muscle contraction. They give mechanical support to skin and bone; they are the antibodies responsible for our immune protection; and they control growth and differentiation—that is, which part of the information stored in DNA is to be used at any given time.

We begin with the structure and preparation of the 20 most common amino acids, the building blocks of proteins. We then show how amino acids are linked by peptide bonds in the three-dimensional structure of hemoglobin and other polypeptides. Some proteins contain thousands of amino acids, but we shall see how to determine the sequence of amino acids in many polypeptides and synthesize these molecules in the laboratory. Finally, we consider how other polymers, the nucleic acids DNA and RNA, direct the synthesis of proteins in nature.

26-1 Structure and Properties of Amino Acids

Amino acids are carboxylic acids bearing an amine group. The most common of these in nature are the **2-amino acids,** or **α-amino acids,** which have the general formula $RCH(NH_2)COOH$; that is, the amino function is located at C2, the α-carbon. The R group can be alkyl or aryl, and it can contain hydroxy, amino, mercapto, sulfide, carboxy, guanidino, and imidazolyl groups. Because of the presence of both amino and carboxy functions, amino acids are both acidic and basic.

The stereocenter of common 2-amino acids has the *S* configuration

More than 500 amino acids exist in nature, but the proteins in all species, from bacteria to humans, consist mainly of only 20. Adult humans can synthesize all but eight, and two only in insufficient quantities. This group is often called the **essential amino acids** because they must be included in our diet. Although amino acids can be named in a systematic manner, they rarely are; so we shall use their common names. Table 26-1 lists the 20 most common amino acids, along with their structures, their pK_a values, and the three- and (the newer) one-letter codes that abbreviate their names. We shall see later how to use these codes to describe peptides conveniently.

Amino acids may be depicted by either dashed-wedged line structures or by Fischer projections.

How to Draw L-Amino Acids and Their Relation to the L-Sugars

C2, or α-carbon

H_2N — C — COOH
R — H
S (L)

H_2N — C — H
R

COOH
H_2N —— H
R

CHO
HO —— H
CH_2OH

(*S*)-2,3-Dihydroxypropanal
(L-Glyceraldehyde)

Dashed-wedged line structures   **Fischer projections**

| TABLE 26-1 | Natural (2S)-Amino Acids |

$$\begin{array}{c} COOH \\ | \\ H_2N - \!\!\!\!\!- H \\ | \\ R \end{array}$$

R	Name	Three-letter code	One-letter code	pK_a of α-COOH	pK_a of α-$^+NH_3$	pK_a of acidic function in R	Isoelectric point, pI
H	Glycine	Gly	G	2.3	9.6	—	6.0
Alkyl group							
CH_3	Alanine	Ala	A	2.3	9.7	—	6.0
$CH(CH_3)_2$	Valine[a]	Val	V	2.3	9.6	—	6.0
$CH_2CH(CH_3)_2$	Leucine[a]	Leu	L	2.4	9.6	—	6.0
$CHCH_2CH_3$ (S) / CH_3	Isoleucine[a]	Ile	I	2.4	9.6	—	6.0
H_2C—⬡	Phenylalanine[a]	Phe	F	1.8	9.1	—	5.5
Proline ring structure	Proline	Pro	P	2.0	10.6	—	6.3
Hydroxy-containing							
CH_2OH	Serine	Ser	S	2.2	9.2	—	5.7
$CHOH$ (R) / CH_3	Threonine[a]	Thr	T	2.1	9.1	—	5.6
H_2C—⬡—OH	Tyrosine	Tyr	Y	2.2	9.1	10.1	5.7
Amino-containing							
$CH_2\overset{O}{\overset{\|}{C}}NH_2$	Asparagine	Asn	N	2.0	8.8	—	5.4
$CH_2CH_2\overset{O}{\overset{\|}{C}}NH_2$	Glutamine	Gln	Q	2.2	9.1	—	5.7
$(CH_2)_4NH_2$	Lysine[a]	Lys	K	2.2	9.0	10.5[c]	9.7
$(CH_2)_3NH\overset{NH}{\overset{\|}{C}}NH_2$	Arginine[a]	Arg	R	2.2	9.0	12.5[c]	10.8

Proline R-group structure:
$$\begin{array}{c} COOH^b \\ | \\ HN - \!\!\!\!\!- H \\ \diagdown \quad | \\ \quad\;\; CH_2 \end{array}$$

| TABLE 26-1 | Natural (2S)-Amino Acids (continued) | | | | | | |

R	Name	Three-letter code	One-letter code	pK_a of α-COOH	pK_a of $\alpha^{\pm}NH_3$	pK_a of acidic function in R	Isoelectric point, pI
Amino-containing (continued)							
	Tryptophan[a]	Trp	W	2.8	9.4	—	5.9
	Histidine[a]	His	H	1.8	9.2	6.1[c]	7.6
Mercapto- or sulfide-containing							
CH_2SH	Cysteine[d]	Cys	C	2.0	10.3	8.2	5.1
$CH_2CH_2SCH_3$	Methionine[a]	Met	M	2.3	9.2	—	5.7
Carboxy-containing							
CH_2COOH	Aspartic acid	Asp	D	1.9	9.6	3.7	2.8
CH_2CH_2COOH	Glutamic acid	Glu	E	2.2	9.7	4.3	3.2

[a]Essential amino acids. [b]Entire structure. [c]pK_a of conjugate acid. [d] The stereocenter is R, because the CH_2SH substituent has higher priority than the COOH group.

In all but glycine, the simplest of the amino acids, C2 is a stereocenter and usually adopts the S configuration.

As in the names of the sugars (Section 24-1), an older amino acid nomenclature uses the prefixes D and L, which relate all the L-amino acids to (S)-2,3-dihydroxy-propanal (L-glyceraldehyde). As emphasized in the discussion of the natural D-sugars, a molecule belonging to the L family is not necessarily levorotatory. For example, both valine ($[\alpha]_D^{25°C} = +13.9$) and isoleucine ($[\alpha]_D^{25°C} = +11.9$) are dextrorotatory.

EXERCISE 26-1

Give the systematic names of alanine, valine, leucine, isoleucine, phenylalanine, serine, tyrosine, lysine, cysteine, methionine, aspartic acid, and glutamic acid.

EXERCISE 26-2

Draw dashed-wedged line structures for (S)-alanine, (S)-phenylalanine, (R)-phenylalanine, and (S)-proline.

EXERCISE 26-3

Among the amino acids in which R = alkyl (Table 26-1), proline can be readily distinguished from the others by IR spectroscopy. How? (**Hint:** Review Section 21-3.)

Amino acids are acidic and basic: zwitterions

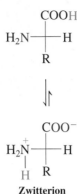

Zwitterion

Because of their two functional groups, the amino acids are both acidic and basic; that is, they are **amphoteric** (Section 8-3). In the solid state, the carboxylic acid group protonates the amine function, thus forming a **zwitterion.** This ammonium carboxylate form is favored because an ammonium ion is much less acidic ($pK_a \sim 10$–11) than a carboxylic acid ($pK_a \sim 2$–5). The highly polar zwitterionic structure allows amino acids to form particularly strong crystal lattices. Most of them therefore are fairly insoluble in organic solvents, and they decompose rather than melt when heated.

The structure of an amino acid in aqueous solution depends on the pH. Consider, for example, the simplest member of the series, glycine. The major form in neutral solution is the zwitterion. However, in strong acid (pH < 1), glycine exists predominantly as the cationic ammonium carboxylic acid, whereas strongly basic solutions (pH > 13) contain mainly the deprotonated 2-aminocarboxylate ion. These forms interconvert by acid-base equilibria (Section 2-9).

$$\overset{+}{H_2NCH_2COOH} \underset{H^+}{\overset{HO^-}{\rightleftharpoons}} \overset{+}{H_2NCH_2COO^-} \underset{H^+}{\overset{HO^-}{\rightleftharpoons}} H_2NCH_2COO^-$$
$$\underset{H}{} \qquad\qquad \underset{H}{}$$

| Predominates at pH < 1 | Predominates at pH ~ 6 | Predominates at pH > 13 |

Table 26-1 records pK_a values for each functional group of the amino acids. For glycine, the first value (2.4) refers to the equilibrium

$$\overset{+}{H_3NCH_2COOH} + H_2O \rightleftharpoons \overset{+}{H_3NCH_2COO^-} + \overset{+}{H_2OH}$$
$$pK_a = 2.4$$

$$K_1 = \frac{[\overset{+}{H_3NCH_2COO^-}][\overset{+}{H_2OH}]}{[\overset{+}{H_3NCH_2COOH}]} = 10^{-2.4}$$

Note that this pK_a is more than two units lower than that of an ordinary carboxylic acid (pK_a $CH_3COOH = 4.74$), an observation that is true for all the other α-aminocarboxy groups in Table 26-1. This difference is a consequence of the electron-withdrawing effect of the protonated amino group. The second pK_a value (9.8) describes the second deprotonation step.

$$\overset{+}{H_3NCH_2COO^-} + H_2O \rightleftharpoons H_2NCH_2COO^- + \overset{+}{H_2OH}$$
$$pK_a = 9.8$$

$$K_2 = \frac{[H_2NCH_2COO^-][\overset{+}{H_2OH}]}{[\overset{+}{H_3NCH_2COO^-}]} = 10^{-9.8}$$

At the isoelectric point, the charges are neutralized

The pH at which the extent of protonation equals that of deprotonation is called the **isoelectric pH** or the **isoelectric point (p*I*;** Table 26-1). At this pH, the concentration of the charge-neutralized zwitterionic form is at its greatest. For glycine and other

CHEMICAL HIGHLIGHT 26-1 | Arginine and Nitric Oxide in Biochemistry and Medicine

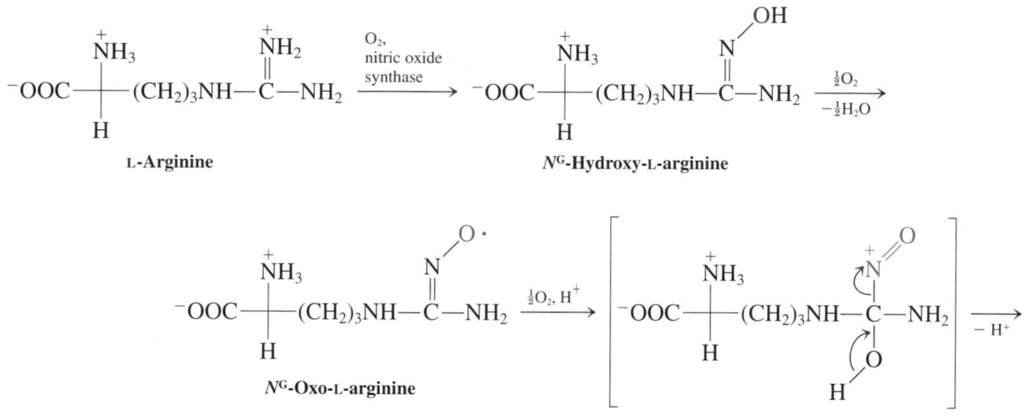

L-Arginine

N^G-Hydroxy-L-arginine

N^G-Oxo-L-arginine

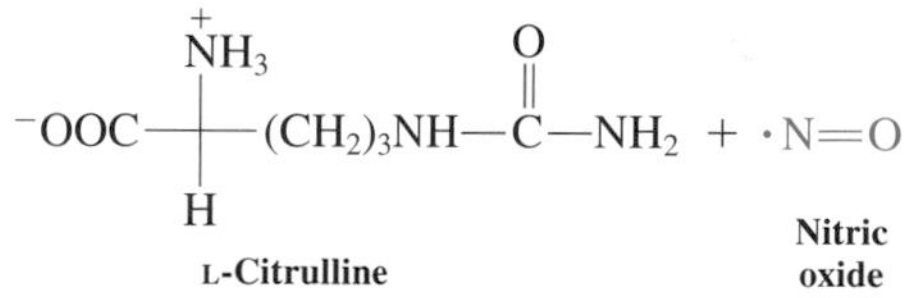

L-Citrulline

Nitric oxide

In the late 1980s and early 1990s, scientists made a series of startling discoveries. The simple but highly reactive and exceedingly toxic molecule nitric oxide, $:\overset{\cdot}{N}=\overset{\cdot\cdot}{O}:$, is synthesized in a wide variety of cells in mammals, including humans, and performs several critical biological functions. Macrophages (cells associated with the body's immune system) destroy bacteria and tumor cells by exposing them to nitric oxide, which is synthesized by the enzyme-catalyzed oxidation of arginine, as shown above.

Nitric oxide is released by cells on the inner walls of blood vessels and causes adjacent muscle fibers to relax. This 1987 discovery explains the effectiveness of nitroglycerin and other organic nitrates as treatments for angina and heart attacks, a nearly century old mystery: These substances are metabolically converted into NO, which dilates the blood vessels. More recent studies reveal a role for nitric oxide as a neurotransmitter in the brain. Paradoxically, NO is also a powerful neurotoxin

whose uncontrolled release may be responsible for the extensive cell destruction associated with vascular strokes and brain disorders such as Alzheimer's and Huntington's diseases. Indeed, after strokes in mice are initiated, administration of N^2-nitro-L-arginine, an inhibitor of NO synthesis, greatly reduces neuronal damage. Such studies are of great interest in the quest for effective therapies for these conditions.

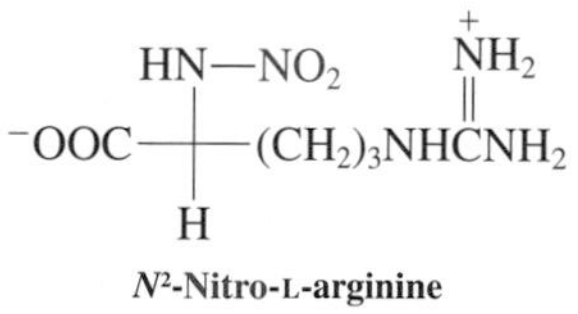

N^2-Nitro-L-arginine

amino acids devoid of additional acidic or basic groups, the value of pI is calculated from the expressions for K_1 and K_2, by setting $[H_3NCH_2COOH] = [H_2NCH_2COO^-]$. As can be verified readily, *the isoelectric point is the average of the two* pK_a *values* of the amino acid.

$$pI = \frac{pK_{COOH} + pK^+_{NH_2H}}{2} = \frac{2.4 + 9.8}{2} = 6.1$$

When the side chain of the acid bears an additional acidic or basic function, the pI is either decreased or increased, respectively, as one would expect. Table 26-1 shows seven entries in which this is the case, several of which are depicted in the following discussion, together with their assigned pK_a values.

Assignment of pK_a Values in Selected Amino Acids

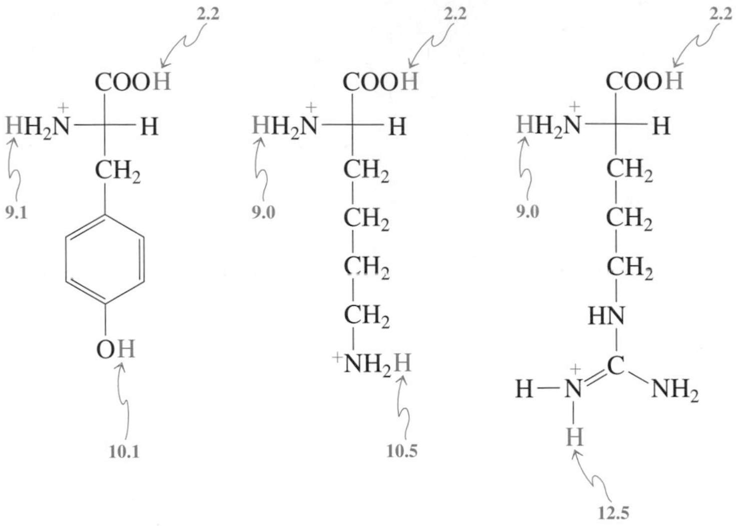

Tyrosine   **Lysine**   **Arginine**

Tyrosine bears the relatively nonacidic phenol function and its pI is 5.7.

Lysine has an additional amino group that can be protonated in a strongly acidic medium to furnish a dication. When the pH of the solution is raised, deprotonation of the carboxy group occurs first, to be followed by proton loss from the nitrogen at C2 and, finally, from the remote ammonium function. The isoelectric point is located halfway between the last two pK_a values, at pI = 9.7.

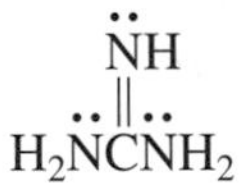

Arginine bears a substituent new to us: the relatively basic **guanidino** group,

Guanidine

, derived from the molecule guanidine (margin). The pK_a of its conjugate acid is ~13, three units larger than that of the ammonium ion (Section 21-4). Its pI is 10.8.

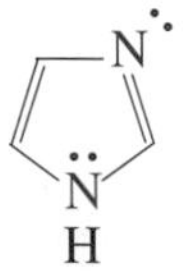

Imidazole

EXERCISE 26-4

Guanidine is found in turnip juice, mushrooms, corn germ, rice hulls, mussels, and earthworms. Its basicity is due to the formation of a highly resonance-stabilized conjugate acid. Draw its resonance forms. (**Hint:** Review Section 20-1.)

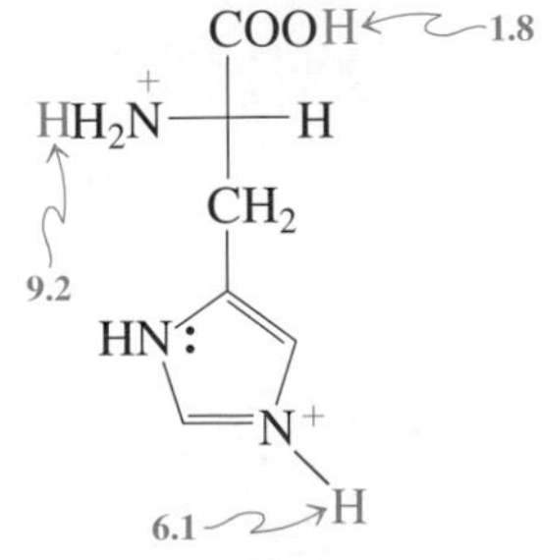

Histidine contains another new substituent, the basic **imidazole** ring (see Problem 27 of Chapter 25). In this aromatic heterocycle, one of the nitrogen atoms is hybridized as in pyridine, and the other is hybridized as in pyrrole.

EXERCISE 26-5

Draw an orbital picture of imidazole. (**Hint:** Use Figure 25-1 as a model.)

Histidine

The imidazole ring is relatively basic because the protonated species is stabilized by resonance.

Resonance in Protonated Imidazole

$pK_a = 7.0$

This resonance stabilization is related to that in amides (Sections 20-1 and 26-4). Imidazole is significantly protonated at physiological pH ($pI = 7.6$). It can therefore function as a proton acceptor and donor at the active site of a variety of enzymes (see, e.g., chymotrypsin, Section 26-4).

The amino acid cysteine bears a relatively acidic mercapto substituent ($pK_a = 8.2$, $pI = 5.1$). Recall that, apart from their acidic character, thiols can be oxidized to disulfides under mild conditions (Section 9-10). In nature, various enzymes are capable of oxidatively coupling and reductively decoupling the mercapto groups in the cysteines of proteins and peptides, thereby reversibly linking peptide strands (Section 9-10).

Aspartic acid and glutamic acid are amino dicarboxylic acids. At physiological pH, both of the carboxy functions are deprotonated, and the molecules exist as the zwitterionic anions aspartate and glutamate. Their pI values are ~3. Monosodium glutamate (MSG) is used as a flavor enhancer in various foods.

In summary, there are 20 elementary L-amino acids, all of which have common names. Unless there are additional acid-base functions in the side chain, their acid-base behavior is governed by two pK_a values, the lower one describing the deprotonation of the carboxy group. At the isoelectric point, the number of amino acid molecules with net zero charge is maximized. Some amino acids contain additional acidic or basic functions, such as hydroxy, amino, guanidino, imidazolyl, mercapto, and carboxy.

Aspartic acid

26-2 Synthesis of Amino Acids: A Combination of Amine and Carboxylic Acid Chemistry

Chapter 21 treated the chemistry of amines, Chapters 19 and 20 that of carboxylic acids and their derivatives. We use both in preparing 2-amino acids.

Hell-Volhard-Zelinsky bromination followed by amination converts carboxylic acids to 2-amino acids

What would be the quickest way of introducing a 2-amino substituent into a carboxylic acid? Section 19-12 pointed out that simple 2-functionalization of an acid is possible by the Hell-Volhard-Zelinsky bromination. Furthermore, the bromine in the product can be displaced by nucleophiles, such as ammonia. In these two steps, propanoic acid can be converted into racemic alanine.

Unfortunately, this approach suffers frequently from relatively low yields. A better synthesis utilizes Gabriel's procedure for the preparation of primary amines (Section 21-5).

The Gabriel synthesis can be adapted to produce amino acids

Recall that *N*-alkylation of 1,2-benzenedicarboxylic imide (phthalimide) anion followed by acid hydrolysis furnishes amines (Section 21-5). To prepare an amino acid instead, we can use diethyl 2-bromopropanedioate (diethyl 2-bromomalonate) in the first step of the reaction sequence. This alkylating agent is readily available from the bromination of diethyl propanedioate (malonate). Now the alkylation product can be hydrolyzed and decarboxylated. Hydrolysis of the imide group then furnishes an amino acid.

Gabriel Synthesis of Glycine

Potassium 1,2-benzene-dicarboxylic imide (Potassium phthalimide) + Diethyl 2-bromo-propanedioate (Diethyl 2-bromo-malonate) → 85% → Glycine

One of the advantages of this approach is the versatility of the initially formed 2-substituted propanedioate. This product can itself be alkylated, thus allowing for the preparation of a variety of substituted amino acids.

EXERCISE 26-6

Propose Gabriel syntheses of methionine, aspartic acid, and glutamic acid.

Amino acids are prepared from aldehydes by the Strecker synthesis

The crucial step in the **Strecker* synthesis** is a variation of the cyanohydrin formation from aldehydes and hydrogen cyanide (Section 17-11).

$$\underset{\text{RCH}}{\overset{\text{O}}{\|}} + \text{HCN} \rightleftharpoons \underset{\underset{\text{H}}{|}}{\overset{\text{OH}}{\underset{|}{\text{R}-\text{C}-\text{CN}}}}$$

Cyanohydrin

When the same reaction is carried out in the presence of ammonia, it is the intermediate imine that undergoes addition of hydrogen cyanide, to furnish the corresponding 2-amino nitriles. Subsequent acidic or basic hydrolysis results in the desired amino acids.

Strecker Synthesis of Alanine

$$\underset{\text{Acetaldehyde}}{\overset{\text{O}}{\underset{\text{CH}_3\text{CH}}{\|}}} \xrightarrow[-\text{H}_2\text{O}]{\text{NH}_3} \underset{\text{Imine}}{\overset{\text{NH}}{\underset{\text{CH}_3\text{CH}}{\|}}} \xrightarrow{\text{HCN}} \underset{\text{2-Aminopropanenitrile}}{\overset{\text{NH}_2}{\underset{\underset{\text{H}}{|}}{\text{H}_3\text{C}-\text{C}-\text{CN}}}} \xrightarrow{\text{H}^+,\text{H}_2\text{O},\Delta} \underset{\underset{\text{Alanine}}{55\%}}{\overset{^+\text{NH}_3}{\text{CH}_3\text{CHCOO}^-}}$$

EXERCISE 26-7

Propose Strecker syntheses of glycine (from formaldehyde) and methionine (from 2-propenal). (**Hint:** Review Section 18-10.)

In summary, racemic amino acids are made by the amination of 2-bromocarboxylic acids, applications of the Gabriel synthesis of amines, and the Strecker synthesis, which proceeds through an imine variation of the preparation of cyanohydrins, followed by hydrolysis.

26-3 Synthesis of Enantiomerically Pure Amino Acids

All the methods of the preceding section produce amino acids in racemic form. However, we noted that most of the amino acids in natural polypeptides have the *S* configuration. Thus, many synthetic procedures—in particular, peptide and protein syntheses—require enantiomerically pure compounds. To meet this requirement, either the racemic amino acids must be resolved (Section 5-8) or a single enantiomer must be prepared by enantioselective reactions.

A conceptually straightforward approach to the preparation of pure enantiomers of amino acids would be resolution of their diastereomeric salts. Typically, the amine group is first protected as an amide, and the resulting product is then treated with an optically active amine, such as the alkaloid brucine (Section 25-8). The two

*Professor Adolf Strecker (1822–1871), University of Würzburg, Germany.

diastereomers formed can be separated by fractional crystallization. Unfortunately, in practice, this method can be tedious and can suffer from poor yields.

Resolution of Racemic Valine

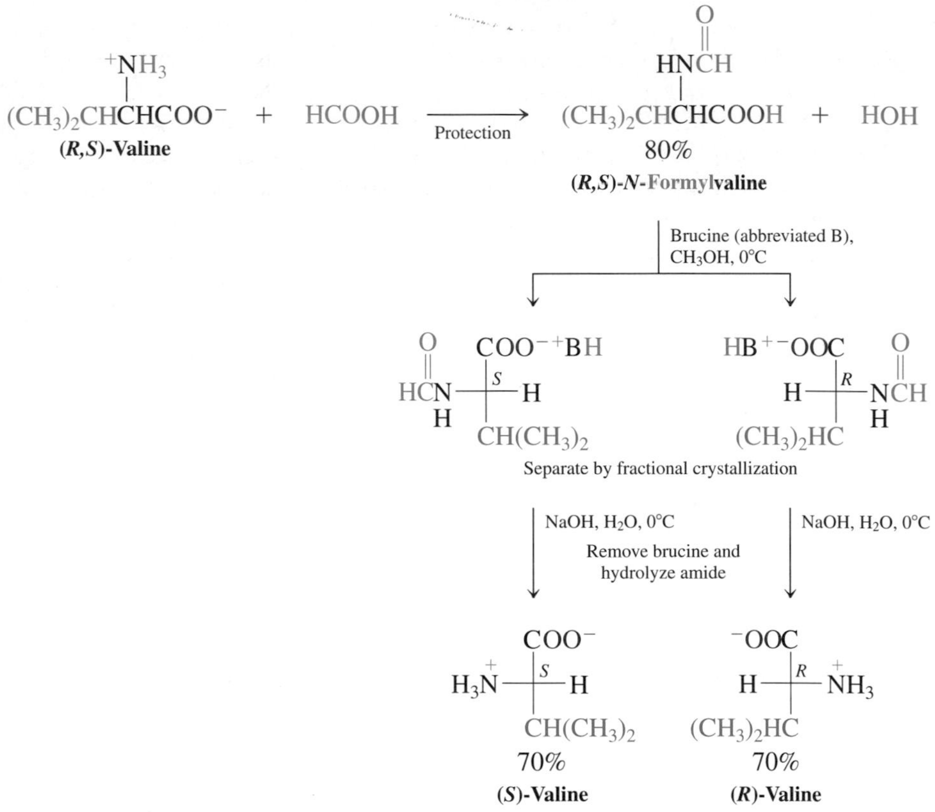

$$(CH_3)_2CHCHCOO^- \quad + \quad HCOOH \xrightarrow{\text{Protection}} (CH_3)_2CHCHCOOH \quad + \quad HOH$$

(R,S)-Valine   80%

(R,S)-N-Formylvaline

Brucine (abbreviated B), CH_3OH, 0°C

Separate by fractional crystallization

NaOH, H_2O, 0°C   Remove brucine and hydrolyze amide   NaOH, H_2O, 0°C

70%   70%

(S)-Valine   **(R)-Valine**

In an alternative approach, the stereocenter at C2 is formed enantioselectively, such as in enantioselective hydrogenations of α,β-unsaturated amino acids (Chemical Highlight 12-1). Nature makes use of this strategy in the biosynthesis of amino acids. Thus, the enzyme glutamate dehydrogenase converts the carbonyl group in 2-oxopentane-dioic acid into the amine substituent in (S)-glutamic acid by a biological reductive amination (for chemical reductive aminations, see Section 21-6). The reducing agent is NADH (Chemical Highlight 25-3).

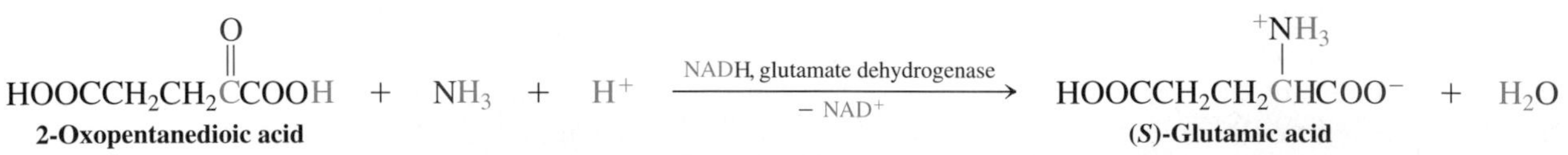

$$HOOCCH_2CH_2CCOOH \quad + \quad NH_3 \quad + \quad H^+ \xrightarrow[- \, NAD^+]{\text{NADH, glutamate dehydrogenase}} HOOCCH_2CH_2CHCOO^- \quad + \quad H_2O$$

2-Oxopentanedioic acid   **(S)-Glutamic acid**

(S)-Glutamic acid is the biosynthetic precursor of glutamine, proline, and arginine. Moreover, it functions to aminate other 2-oxo acids, with the help of another type of enzyme, a transaminase, making additional amino acids available.

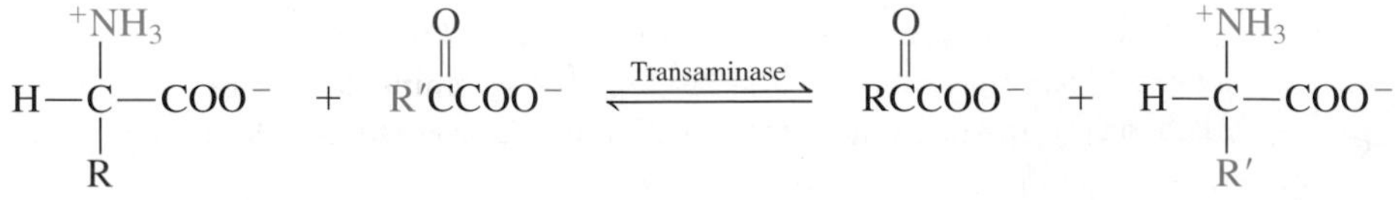

$$H-\overset{+NH_3}{\underset{R}{C}}-COO^- \quad + \quad R'CCOO^- \xrightleftharpoons{\text{Transaminase}} RCCOO^- \quad + \quad H-\overset{+NH_3}{\underset{R'}{C}}-COO^-$$

CHEMICAL HIGHLIGHT 26-2   **Synthesis of Optically Pure Amino Acids**

Catalytic asymmetric hydrogenation (Chemical Highlight 12-1, Section 12-2) is only one of many methods that have been developed to prepare optically pure amino acids. Another approach is the alkylation of glycine in the presence of optically active bases. Optically pure D-(4-chlorophenyl)alanine has been obtained by using the naturally occurring alkaloid cinchonine according to the scheme shown below.

Only a catalytic quantity of the alkaloid base is required for the first step of this process, which is carried out in a rapidly stirred mixture of dichloromethane and water. During alkylation, which

occurs in the CH_2Cl_2 layer, the base is converted into its conjugate acid, which is ionic and migrates into the water phase. The NaOH dissolved in the water phase regenerates the neutral alkaloid, which shifts back into the CH_2Cl_2 phase, thereby leading to alkylation of another molecule of substrate, and so forth. This cycle of reactions, which employs a catalyst that repeatedly shuttles back and forth between two immiscible solvents, is an example of *phase-transfer catalysis* (see also Problem 21-50). Alkylation gives an *optically enriched* intermediate, from which the pure major enantiomer is separated from residual racemic material by recrystallization.

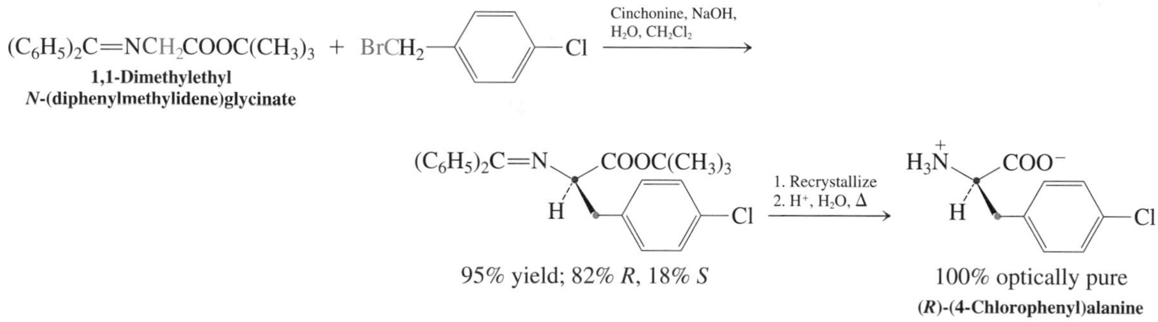

$(C_6H_5)_2C\text{=}NCH_2COOC(CH_3)_3$ + $BrCH_2$—〈benzene ring〉—Cl $\xrightarrow{\text{Cinchonine, NaOH,}\ H_2O,\ CH_2Cl_2}$

1,1-Dimethylethyl
***N*-(diphenylmethylidene)glycinate**

$(C_6H_5)_2C\text{=}N$, $COOC(CH_3)_3$
H 〈benzene ring〉—Cl
95% yield; 82% *R*, 18% *S*

$\xrightarrow[\text{2. }H^+,\ H_2O,\ \Delta]{\text{1. Recrystallize}}$

$H_3\overset{+}{N}$, COO^-
H 〈benzene ring〉—Cl
100% optically pure
(*R*)-(4-Chlorophenyl)alanine

In summary, optically pure amino acids can be obtained by resolution of a racemic mixture or by enantioselective formation of the C2 stereocenter.

26-4 Peptides and Proteins: Amino Acid Oligomers and Polymers

Amino acids are very versatile biologically because they can be polymerized. This section will describe the structure and properties of such **polypeptide** chains. These chains twist and fold in three dimensions to form the biologically active **proteins.**

Amino acids form peptide bonds

2-Amino acids are the monomer units in polypeptides. The polymer forms by repeated reaction of the carboxylic acid function of one amino acid with the amine group in another to make a chain of amides (Section 20-6). The amide linkage joining amino acids is also called a **peptide bond.**

Peptide bond

$$2n \text{ HN}-\underset{\underset{H}{|}}{\overset{\overset{R}{|}}{C}}-\text{COH} \longrightarrow -(\text{NH}-\underset{\underset{H}{|}}{\overset{\overset{R}{|}}{C}}-\overset{\overset{O}{\parallel}}{C}-\text{NH}-\underset{\underset{H}{|}}{\overset{\overset{R}{|}}{C}}-\overset{\overset{O}{\parallel}}{C})_n- + 2n \text{ H}_2\text{O}$$

2-Amino acid　　　　　　　　　**Polyamino acid**
(α-Amino acid)　　　　　　　　**(Polypeptide)**

The oligomers formed by linking amino acids in this way are called **peptides.** For example, two amino acids give rise to a **dipeptide,** three to a **tripeptide,** and so forth. The individual amino acid units forming the peptide are referred to as **residues.** In some proteins, two or more polypeptide chains are linked by disulfide bridges (Sections 9-10 and 26-1).

Of great importance for the structure of polypeptides is the fact that the peptide bond is fairly rigid at room temperature and planar, a result of the conjugation of the amide nitrogen lone electron pair with the carbonyl group (Section 20-1). The N–H hydrogen is almost always located trans to the carbonyl oxygen, and rotation about the C–N bond is slow because the C–N bond has partial double-bond character. The latter manifests itself in a relatively short bond (1.32 Å), between the length of a pure C–N single bond (1.47 Å, Figure 21-1) and that of a C–N double bond (1.27 Å). On the other hand, the bonds adjacent to the amide function enjoy free rotation. Thus, polypeptides are relatively rigid but nevertheless sufficiently mobile to adopt a variety of conformations. Hence, they may fold in many different ways. Most biological activity is due to such folded arrangements; straight chains are usually inactive.

Planarity Induced by Resonance in the Peptide Bond

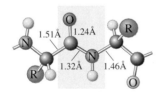

$$\left[\begin{array}{c} \overset{H}{\underset{|}{N}} \quad \overset{:\ddot{O}:}{\underset{\parallel}{C}} \quad \overset{H}{\underset{}{}} \overset{R}{\underset{}{}} \\ \quad \overset{}{\underset{\underset{R}{|}}{}} C \overset{}{\underset{\underset{H}{|}}{}} N \overset{}{\underset{\underset{H}{|}}{}} C \overset{}{\underset{\underset{O}{}}{}} C \end{array} \longleftrightarrow \begin{array}{c} \overset{H}{\underset{|}{N}} \quad \overset{:\ddot{O}:^-}{\underset{}{C}} \quad \overset{H}{\underset{}{}} \overset{R}{\underset{}{}} \\ \quad C \overset{+}{N} C \quad C \\ \overset{}{\underset{R \ H}{}} \quad \overset{}{\underset{H}{}} \quad \overset{}{\underset{O}{}} \end{array} \right]$$

Polypeptides are characterized by their sequence of amino acid residues

In drawing a polypeptide chain, the **amino end,** or **N-terminal amino acid,** is placed at the left, the **carboxy end,** or **C-terminal amino acid,** at the right. The configuration at the C2 stereocenters is usually presumed to be *S*.

How to Draw the Structure of a Tripeptide

$$\overset{+}{\text{H}_3}\text{N}-\underset{\underset{H}{|}}{\overset{\overset{R}{|}}{C}}-\overset{\overset{O}{\parallel}}{C}-\underset{\underset{H}{|}}{\overset{}{N}}-\underset{\underset{H}{|}}{\overset{\overset{R'}{|}}{C}}-\overset{\overset{O}{\parallel}}{C}-\underset{\underset{H}{|}}{\overset{}{N}}-\underset{\underset{H}{|}}{\overset{\overset{R''}{|}}{C}}-\text{COO}^-$$

Amino　　　　Amino　　　　Amino
acid 1　　　　acid 2　　　　acid 3

The chain incorporating the amide (peptide) bonds is called the **main chain,** the substituents R, R', and so forth, are the **side chains.**

CHEMICAL HIGHLIGHT 26-3

Glutathione and the Toxicity of Methyl Isocyanate

Examination of victims of the Bhopal tragedy (Chemical Highlight 20-3, Section 20-7) revealed that many of these people suffered extensive damage to *internal* tissues and organs, characteristic of exposure to methyl isocyanate. This finding was unexpected, because methyl isocyanate is so chemically reactive that none of it should have been capable of passing through the lungs and circulatory system to more remote locations in the body, such as the liver.

The solution to this mystery was found in 1992, when glutathione, one of whose roles is to protect

cells from damage by toxic agents (see Section 22-9), was found to be capable of carrying and releasing methyl isocyanate throughout the body. This insidious effect derives from the ability of methyl isocyanate to react *reversibly* with the mercapto group of the peptide (as shown below).

Intermediates such as *S*-(*N*-methylcarbamoyl)-glutathione are currently under investigation as possible mediators of the toxic effects of several other small nitrogen-containing organic compounds.

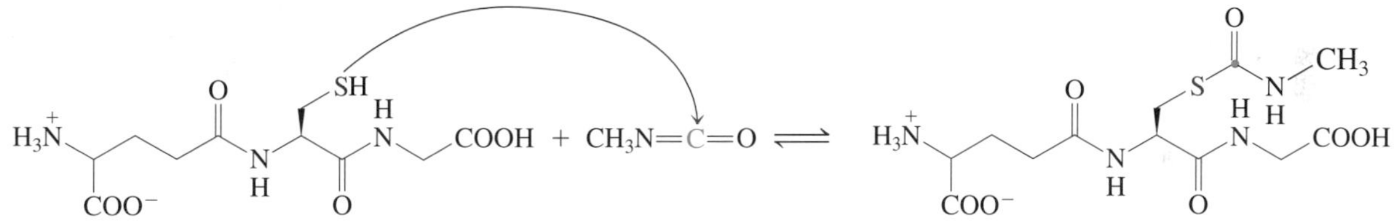

Glutathione   Methyl isocyanate   *S*-(*N*-Methylcarbamoyl)glutathione

The naming of peptides is straightforward. Starting from the amino end, the names of the individual residues are simply connected in sequence, each regarded as a substituent to the next amino acid, ending with the *C*-terminal residue. Because this procedure rapidly becomes cumbersome, the three-letter abbreviations listed in Table 26-1 are used for larger peptides.

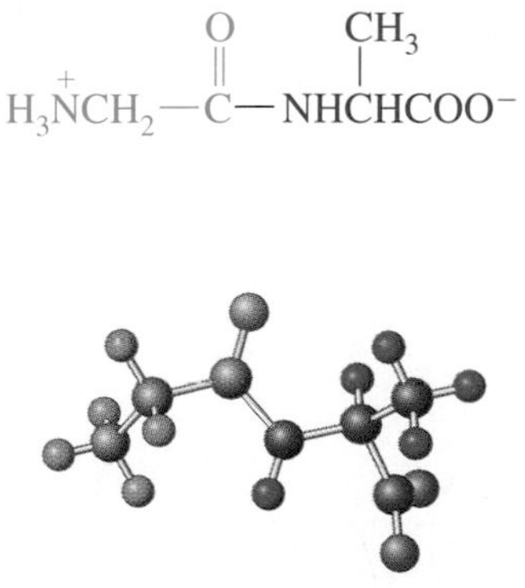

Glycylalanine
Gly-Ala

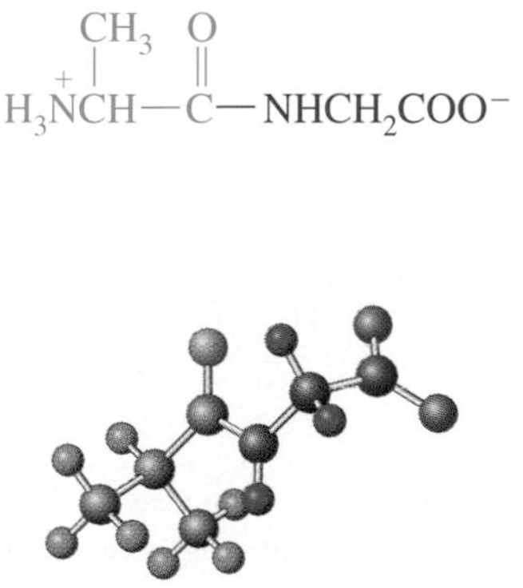

Alanylglycine
Ala-Gly

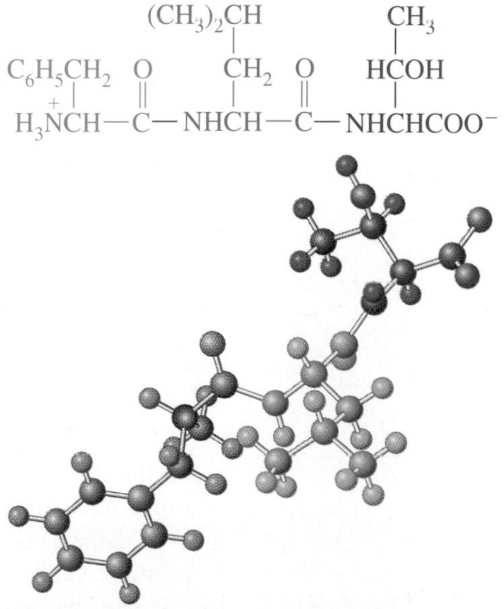

Phenylalanylleucylthreonine
Phe-Leu-Thr

Let us look at some examples of peptides and their structural variety. A dipeptide ester, aspartame, is a low-calorie artificial sweetener (NutraSweet; see first page of this chapter and the Chapter Integration Problem). In the three-letter notation, the ester end is denoted by OCH_3. Glutathione, a tripeptide, is found in all living cells, and in particularly high concentrations in the lens of the eye. It is unusual in that its glutamic acid residue is linked at the γ-carboxy group (denoted γ-Glu) to the rest of the peptide.

Aspartylphenylalanine methyl ester
Asp-Phe-OCH₃
(Aspartame)

γ-Glutamylcysteinylglycine
γ-Glu-Cys-Gly
(Glutathione)

It functions as a biological reducing agent by being readily oxidized enzymatically at the cysteine mercapto unit to the disulfide-bridged dimer.

Gramicidin S is a cyclic peptide antibiotic constructed out of two identical pentapeptides that have been joined head to tail. It contains phenylalanine in the *R* configuration and a rare amino acid, ornithine [Orn, a lower homolog (one less CH_2 group) of lysine]. In the short notation in which gramicidin S is shown in the margin, the sense in which the amino acids are linked (amino to carboxy direction) is indicated by arrows.

Although short notations are practical, they do not reveal some of the noncovalent bonding interactions in peptides. A dashed-wedged line picture of gramicidin S shows several important hydrogen bonds contributing (in addition to the peptide bonds) substantially to its stereochemical rigidity.

Orn → Leu
Val — (R)-Phe
Pro — Pro
(R)-Phe — Val
Leu ← Orn

Gramicidin S

Gramicidin S
(Dashed-wedged line notation)

Insulin illustrates the three-dimensional structure adopted by a complex sequence of amino acids (Figure 26-1). This protein hormone is a drug important in the treatment of diabetes because of its ability to regulate glucose metabolism.

Insulin contains 51 amino acid residues incorporated into two chains, denoted A and B. The chains are connected by two disulfide bridges, and there is an additional such linkage connecting the cysteine residues at positions 6 and 11 of the A chain,

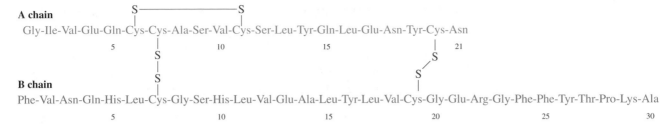

A chain

S————————S
| |
Gly-Ile-Val-Glu-Gln-Cys-Cys-Ala-Ser-Val-Cys-Ser-Leu-Tyr-Gln-Leu-Glu-Asn-Tyr-Cys-Asn

           5   S   10   15   S   21

B chain

Phe-Val-Asn-Gln-His-Leu-Cys-Gly-Ser-His-Leu-Val-Glu-Ala-Leu-Tyr-Leu-Val-Cys-Gly-Glu-Arg-Gly-Phe-Phe-Tyr-Thr-Pro-Lys-Ala

           5   10   15   20   25   30

FIGURE 26-1 —————————————————————————

Bovine (cattle) insulin is made up of two amino acid chains, linked by disulfide bridges. The amino (*N*-terminal) end is at the left in both chains.

causing it to loop. Both chains fold up in a way that minimizes steric interference and maximizes electrostatic, London, and hydrogen-bonding attractions. These forces give rise to a fairly condensed three-dimensional structure (Figure 26-2).

Because most synthetic methods give only low yields, a major source of insulin, until recently, has been the pancreas of slaughtered animals. An exciting new development is the efficient preparation of the protein by genetic-engineering methods. Huge vats of bacteria carrying the human insulin gene now generate enough material to treat thousands of diabetics throughout the world.

EXERCISE 26-8

Vasopressin, also known as antidiuretic hormone, controls the excretion of water from the body. Draw its complete structure. Note that there is an intramolecular disulfide bridge between the two cysteine residues.

S————————S
| |
Cys-Tyr-Phe-Gln-Asn-Cys-Pro-Arg-Gly-NH$_2$
Vasopressin

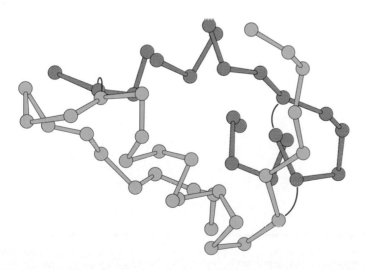

FIGURE 26-2 —————————————————————————

Three-dimensional structure of insulin. Residues in chain A are blue, those in B green. The disulfide bridges are indicated in red. (After *Biochemistry,* 4th ed., by Lubert Stryer, W. H. Freeman and Company. Copyright © 1975, 1981, 1988, 1995.)

Proteins fold into pleated sheets and helices: secondary and tertiary structure

Insulin and other polypeptide chains adopt well-defined three-dimensional structures. Whereas the sequence of amino acids in the chain defines the **primary structure,** the folding pattern of the chain induced by the spatial arrangement of close-lying amino acid residues gives rise to the **secondary structure** of the polypeptide. The secondary structure results mainly from the rigidity of the amide bond and from hydrogen and other noncovalent bonding along the chain(s). Two important arrangements are the pleated sheet, or β configuration, and the α helix.

In the **pleated sheet** (Figure 26-3), two chains line up with the amino groups of one peptide opposite the carbonyl groups of a second, thereby allowing hydrogen

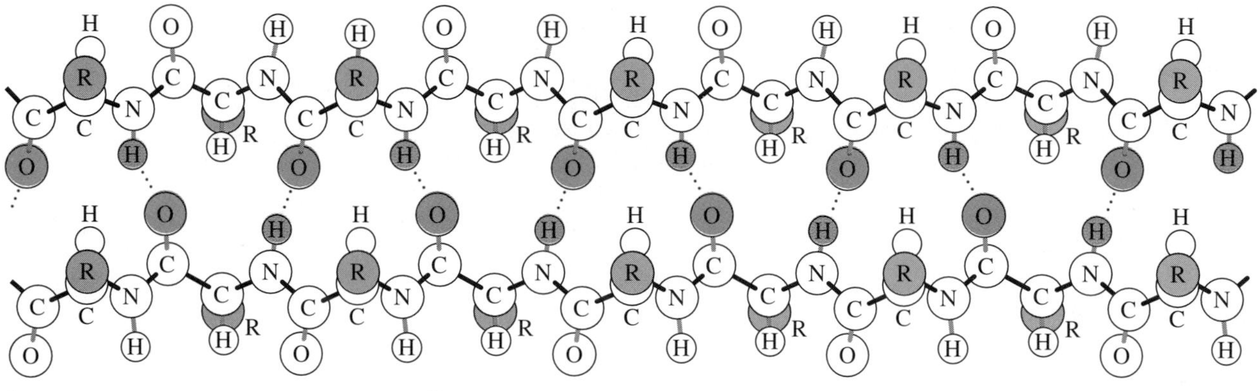

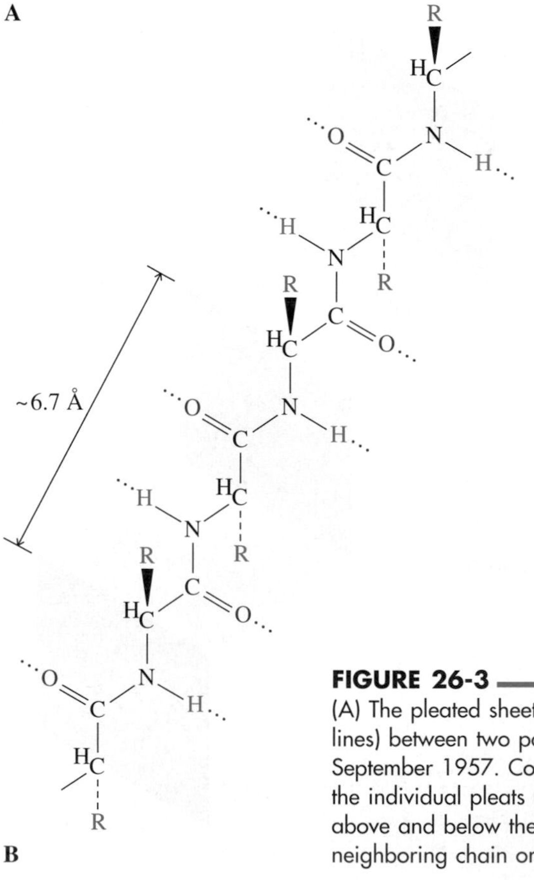

FIGURE 26-3

(A) The pleated sheet, or β configuration, which is held in place by hydrogen bonds (dotted lines) between two polypeptide strands. (After "Proteins," by Paul Doty, *Scientific American*, September 1957. Copyright © 1957, Scientific American, Inc.) (B) The peptide bonds define the individual pleats (shaded in yellow); the positions of the side chains, R, are alternately above and below the planes of the sheets. The dotted lines indicate hydrogen bonds to a neighboring chain or to water.

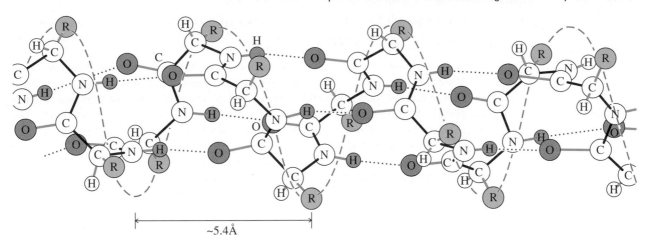

FIGURE 26-4

The α helix, in which the polymer chain is arranged as a right-handed spiral held rigidly in shape by intramolecular hydrogen bonds. (After "Proteins," by Paul Doty, *Scientific American*, September 1957. Copyright © 1957, Scientific American, Inc.)

bonds to form. Such bonds can also develop within a single chain if it loops back on itself. Multiple hydrogen bonding of this type can impart considerable rigidity to a system. The planes of adjacent amide linkages form a specific angle, a geometry that produces the observed pleated-sheet structure.

The **α helix** (Figure 26-4) allows for intramolecular hydrogen bonding between nearby amino acids in the chain: The carbonyl oxygen of each amino acid is interacting with the amide hydrogen four residues ahead. There are 3.6 amino acids per turn of the helix, two equivalent points in neighboring turns being about 5.4 Å apart.

Not all polypeptides adopt idealized structures such as these. If too much charge of the same kind builds up along the chain, charge repulsion will enforce a more random orientation. In addition, the rigid proline, because its amino nitrogen is also part of the substituent ring, can cause a kink or bend in an α helix.

Further folding, coiling, and aggregation of polypeptides is induced by distant residues in the chain end and give rise to their **tertiary structure.** A variety of forces, all arising from the R group, come into play to stabilize such molecules, including disulfide bridges, hydrogen bonds, London forces, and electrostatic attraction and repulsion. There are also **micellar effects** (Chemical Highlight 19-1 and Section 20-5): The polymer adopts a structure that maximizes exposure of polar groups to the aqueous environment, while minimizing exposure of hydrophobic groups (e.g., alkyl and phenyl). Pronounced folding is observed in the **globular proteins,** many of which perform chemical transport and catalysis (e.g., myoglobin and hemoglobin, Section 26-8). In the **fibrous proteins,** such as myosin (in muscle), fibrin (in blood clots), and α-keratin (in hair, nails, and wool), several α helices are coiled to produce a **superhelix** (Figure 26-5).

The tertiary structure of enzymes and transport proteins (proteins that carry molecules from place to place) usually gives rise to three-dimensional pockets, called **active sites.** The size and shape of the active site provide a highly specific "fit" for the **substrate,** the molecule on which the protein carries out its intended function. The inner surface of the pocket typically contains a specific arrangement of the side chains of polar amino acids that attracts functional groups in the substrate by hydrogen bonding or ionic interactions. In enzymes, the active site aligns functional groups and additional molecules in a way that promotes their reactions with the substrate.

20 Å

FIGURE 26-5

Idealized picture of a superhelix, a coiled coil.

An example is the active site of chymotrypsin, a mammalian digestive enzyme responsible for the degradation of proteins in food. Chymotrypsin accomplishes the hydrolysis of peptide bonds at body temperature and at physiological pH. Recall that ordinary amide hydrolysis requires much more drastic conditions (Section 20-6). Moreover, the enzyme also recognizes specific peptide linkages that are targeted for selective cleavage, such as the carboxy end of phenylalanine residues (see Section 26-5, Table 26-2). How does it do that?

A simplified picture of the action part of this large molecule (of dimensions $51 \times 40 \times 40$ Å) in this process is shown in the following scheme.

Peptide Hydrolysis in the Active Site of Chymotrypsin

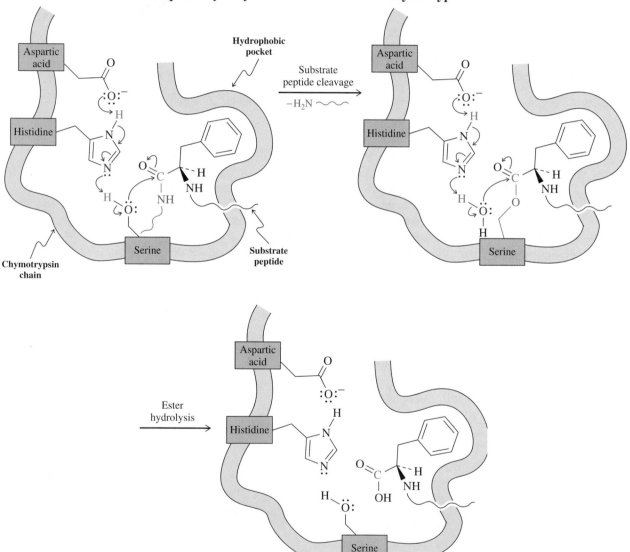

The enzyme has four important, close-lying parts, all of which work together to facilitate the hydrolysis reaction: a hydrophobic pocket, and the residues of aspartic acid, histidine, and serine. The hydrophobic pocket (see Section 8-2) helps to bind the polypeptide to be "digested" by attraction of the hydrophobic phenyl substituent of one of its component phenylalanine residues. With the latter held in this pocket, the three amino acid residues cooperate in a proton transfer relay sequence to effect nucleophilic addition–elimination (Sections 19-7 and 20-6) of the serine hydroxy group to the carbonyl function of Phe, releasing the amine part of the so-cleaved polypeptide. The remainder of the substrate is held by an ester linkage to the enzyme, positioned to undergo ester hydrolysis (Section 20-5) by a water molecule. This reaction is aided by a tandem proton transfer sequence similar to that used for hydrolyzing the peptide bond. With the link to the enzyme broken, the carboxy segment of the original substrate is now free to leave the intact active site of chymotrypsin, making room for another polypeptide.

EXERCISE 26-9

The scheme just shown omits the respective elimination steps of the two nucleophilic addition–elimination reactions. Show how the enzyme aids in accelerating them as well. (**Hint:** Draw the result of the "electron pushing" depicted in the first [or second] picture of the scheme and think about how a reversed electron and proton flow might help.)

Denaturation, or breakdown of the tertiary structure of a protein, usually causes precipitation of the protein and destroys its catalytic activity. Denaturation is caused by exposure to excessive heat or extreme pH values. Think, for example, of what happens to clear egg white when it is poured into a hot frying pan or to milk when it is added to lemon tea.

Some molecules, such as hemoglobin (Section 26-8), also adopt a **quaternary structure,** in which two or more polypeptide chains, each with its own tertiary structure, combine to form a larger assembly.

The three-dimensional constellation of polypeptides is a direct consequence of primary structure: In other words, the amino acid sequence specifies in which way the chain will coil, aggregate, and otherwise interact with internal and external molecular units. Therefore, knowledge of this sequence is of paramount importance in the understanding of protein structure and function. How to obtain this knowledge is the subject of the next section.

In summary, polypeptides are polymers of amino acids linked by amide bonds. Their amino acid sequences can be described in a shorthand notation using the three- or one-letter abbreviations compiled in Table 26-1. The amino end group is placed at the left, the carboxy end at the right. Polypeptides can be cyclic and can also be linked by disulfide and hydrogen bonds. The sequence of amino acids is the primary structure of a polypeptide, folding gives rise to its secondary structure, further folding and coiling produce its tertiary structure, and aggregation of several polypeptides results in the quaternary structure.

26-5 Determination of Primary Structure: Amino Acid Sequencing

Biological function in polypeptides and proteins requires a specific three-dimensional shape and arrangement of functional groups, which, in turn, necessitate a definite amino acid sequence. One "monkey wrench" residue in an otherwise normal protein can completely alter its behavior. For example, sickle-cell anemia, a potentially lethal condition, is the result of changing a *single* amino acid in hemoglobin (Section 26-8). The determination of the primary structure of a protein, called amino acid or polypeptide **sequencing,** can help us to understand the protein's mechanism of action.

In the late 1950s and early 1960s, amino acid sequences were discovered to be predetermined by DNA, the molecule containing our hereditary information (Section 26-9). Thus, through a knowledge of protein primary structure, we can learn how genetic material expresses itself. Functionally similar proteins in related species should, and do, have similar primary structures. The closer their sequences of amino acids, the more closely the species are related. Polypeptide sequencing therefore strikes at the heart of the question of the evolution of life itself. This section shows how chemical means, together with analytical techniques, allow us to obtain this information.

First, purify the polypeptide

The problem of polypeptide purification is an enormous one, and attempts at its solution consume many days in the laboratory. Several techniques can separate polypeptides on the basis of size, solubility in a particular solvent, charge, or ability to bind to a support. Although detailed discussions are beyond the scope of this book, we shall briefly describe some of the more widely used methods.

In **dialysis,** the polypeptide is separated from smaller fragments by filtration through a semipermeable membrane. A second method, **gel-filtration chromatography,** uses a carbohydrate polymer in the form of a column of beads as a support. Smaller molecules diffuse more easily into the beads, spending a longer time on the column than large ones do; thus they emerge from the column later than the large molecules. In **ion-exchange chromatography,** a charged support separates molecules according to the amount of charge that they carry. Another method based on electric charge is **electrophoresis.** A spot of the mixture to be separated is placed on a plate covered with a thin layer of chromatographic material (such as polyacrylamide) that is attached to two electrodes. When the voltage is turned on, positively charged species (e.g., polypeptides rich in protonated amine groups) migrate toward the cathode, negatively charged species (carboxy-rich peptides) toward the anode. The separating power of this technique is extraordinary. More than a thousand different proteins from one species of bacterium have been resolved in a single experiment.

Finally, **affinity chromatography** exploits the tendency of polypeptides to bind very specifically to certain supports by hydrogen bonds and other attractive forces. Peptides of differing sizes and shapes have differing retention times in a column containing such a support.

Second, determine which amino acids are present

When the polypeptide strand has been purified, the next step in structural analysis is to establish its composition. To determine which amino acids and how much of each is present in the polypeptide, the entire chain is degraded by amide hydrolysis (6 N HCl, 110°C, 24 h) to give a mixture of the free amino acids. The mixture is then separated and its composition recorded by an automated **amino acid analyzer.**

This instrument consists of a column bearing a negatively charged support, usually containing carboxylate or sulfonate ions. The amino acids pass through the column in slightly acidic solution. They are protonated to a greater or lesser degree, depending on their structure, and therefore are more or less retained on the column. This differential retention separates the amino acids, and they come off the column in a specific order, beginning with the most acidic and ending with the most basic. At the end of the column is a reservoir containing a special indicator. Each amino

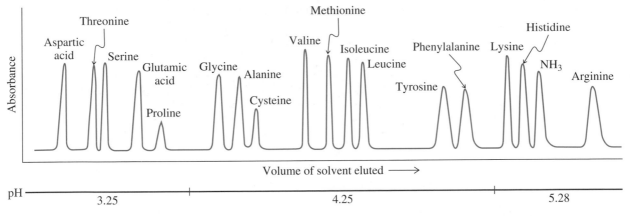

FIGURE 26-6

The result, recorded as a chromatogram, of separating various amino acids on an amino acid analyzer, using a polysulfonated ion-exchange resin. The more acidic products (e.g., aspartic acid) are generally eluted first. Ammonia is included for comparison.

acid produces a violet color whose intensity is proportional to the amount of that acid present and is recorded in a chromatogram (Figure 26-6). The area under each peak is a measure of the relative amount of a specific amino acid in the mixture.

The amino acid analyzer can readily establish the composition of a polypeptide. For example, the chromatogram of hydrolyzed glutathione (Section 26-4) gives three equal-sized peaks, corresponding to Glu, Gly, and Cys.

EXERCISE 26-10

Give the expected results of the amino acid analysis of the A chain in insulin (Figure 26-1).

Sequence the peptide from the amino (*N*-terminal) end

When we know the gross makeup of a polypeptide, we must determine the order in which the individual amino acids are bound to one another—the amino acid sequence.

Several different methods can reveal the identity of the residue at the amino end. Most exploit the uniqueness of the free amino substituent, which may enter into specific chemical reactions that serve to "tag" the *N*-terminal amino acid. One such procedure is the **Edman* degradation,** and the reagent used is phenyl isothiocyanate, $C_6H_5N{=}C{=}S$ (a sulfur analog of an isocyanate, Section 20-7).

Recall (Section 20-7) that isocyanates are very reactive with respect to nucleophilic attack, and the same is true of their sulfur analogs. In the Edman degradation, the terminal amino group adds to the isothiocyanate reagent to give a thiourea derivative (refer to Section 20-6 for the urea function). Mild acid causes extrusion of the tagged amino acid as a phenylthiohydantoin, leaving the remainder of the polypeptide unchanged. The phenylthiohydantoins of all amino acids are well known, so the *N*-terminal end of the original polypeptide can be readily identified. The new chain, carrying a new terminal amino acid, is now ready for another Edman degradation to tag the next residue, and so forth. The entire procedure has been automated to allow the routine identification of polypeptides containing 50 or more amino acids.

Beyond that number, the buildup of impurities becomes a serious impediment. The reason for this drawback is that each degradation round, though proceeding in high yield, is not completely quantitative, thus leaving small quantities of incompletely reacted peptide admixed with the new one. You can readily envisage that, with each step, this problem increases until the mixtures become intractable.

*Professor Pehr V. Edman (1916–1977), Max Planck Institute for Biochemistry, Martinsried, Germany.

Edman Degradation of the A Chain of Insulin

Phenylthiohydantoin derived from glycine

Phenylthiohydantoin derived from isoleucine

etc.

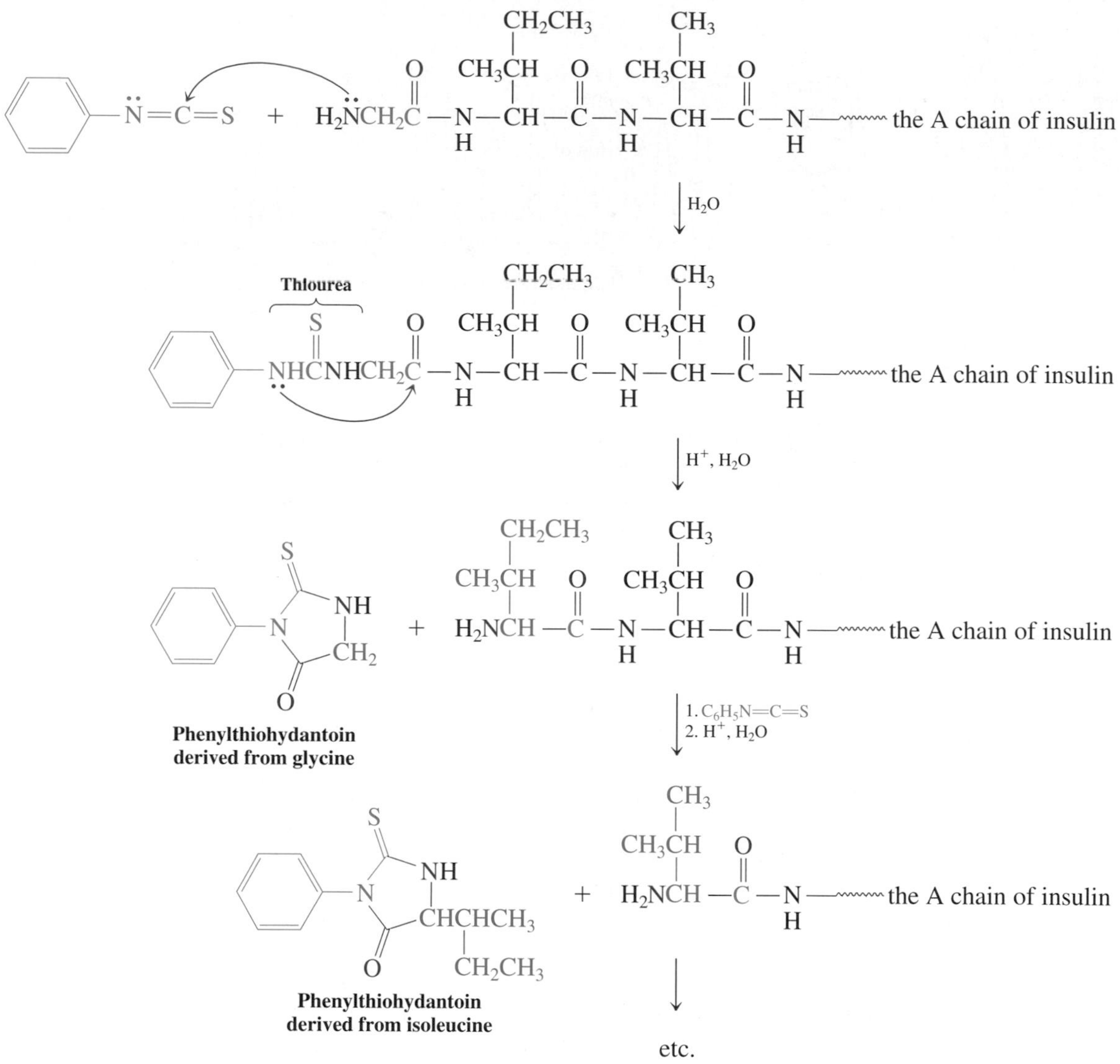

EXERCISE 26-11

Write a mechanism for the formation of glycine phenylthiohydantoin from the reaction of phenylisothiocyanate with glycine amide, $H_2NCH_2CONH_2$. (**Hint:** Note the red arrows in the scheme for the Edman degradation of the A chain of insulin.)

EXERCISE 26-12

Polypeptides can be cleaved into their component amino acid fragments by treatment with dry hydrazine. This method reveals the identity of the carboxy end. Explain.

The chopping up of longer chains is achieved with enzymes

The Edman procedure allows for the ready sequencing of only relatively short polypeptides. For longer ones (e.g., those with more than 50 residues), it is necessary to cleave the larger chains into shorter fragments in a selective and predictable manner.

TABLE 26-2	Specificity of Hydrolytic Enzymes in Polypeptide Cleavage	
Enzyme	**Site of cleavage**	
Trypsin	Lys, Arg, carboxy end	
Clostripain	Arg, carboxy end	
Chymotrypsin	Phe, Trp, Tyr, carboxy end	
Pepsin	Asp, Glu, Leu, Phe, Trp, Tyr, carboxy end	
Thermolysin	Leu, Ile, Val, amino end	

These cleavage methods rely mostly on hydrolytic enzymes. For example, trypsin, a digestive enzyme of intestinal liquids, cleaves polypeptides only at the carboxy end of arginine and lysine.

Selective Hydrolysis of the B Chain of Insulin by Trypsin

Phe-Val-Asn-Gln-His-Leu-Cys-Gly-Ser-His-Leu-Val-Glu-Ala-Leu-Tyr-Leu-Val-Cys-Gly-Glu-Arg-Gly-Phe-Phe-Tyr-Thr-Pro-Lys-Ala
                    5   10   15   20   25   30

$$\downarrow \text{Trypsin, } H_2O$$

Phe-Val-Asn-Gln-His-Leu-Cys-Gly-Ser-His-Leu-Val-Glu-Ala-Leu-Tyr-Leu-Val-Cys-Gly-Glu-Arg + Gly-Phe-Phe-Tyr-Thr-Pro-Lys + Ala

A more selective enzyme is clostripain, which cleaves only at the carboxy end of arginine. In contrast, chymotrypsin, which, like trypsin, is found in mammalian intestines, is less selective and cleaves at the carboxy end of phenylalanine (see Section 26-4), tryptophan, and tyrosine. Other enzymes have similar selectivity (Table 26-2). In this way, a longer polypeptide is broken down into several shorter ones, which may then be sequenced by the Edman procedure.

After a first enzymatic cleavage, sequences of segments of the polypeptide under investigation are determined, but the order in which they are linked is not. For this purpose, selective hydrolysis is carried out a second time, by using a different enzyme that provides pieces in which the connectivities broken under the first conditions are left intact, so-called *overlap peptides*. The solution is then found by literally "piecing" together the available information like a puzzle.

EXERCISE 26-13

A polypeptide containing 21 amino acids was hydrolyzed by thermolysin. The products of this treatment were Gly, Ile, Val-Cys-Ser, Leu-Tyr-Gln, Val-Glu-Gln-Cys-Cys-Ala-Ser, and Leu-Glu-Asn-Tyr-Cys-Asn. When the same polypeptide was hydrolyzed by chymotrypsin, the products were Cys-Asn, Gln-Leu-Glu-Asn-Tyr, and Gly-Ile-Val-Glu-Gln-Cys-Cys-Ala-Ser-Val-Cys-Ser-Leu-Tyr. Give the amino acid sequence of this molecule.

Protein sequencing is made possible by recombinant DNA technology

Despite the success of the techniques for polypeptide sequencing described so far, allowing for the structure elucidation of hundreds of proteins, their application to large systems (i.e., those containing more than 1000 residues) is an expensive, laborious, and time-consuming business. Progress in this field was significantly impeded until the advent of recombinant DNA technology (Section 26-11). As we shall see (Section 26-10, Table 26-3), the sequences of the four bases in DNA—adenine, thymine, guanine, and cytosine—are directly correlated to the amino acid sequences of the proteins encoded by genes or the corresponding messenger RNA. Modern developments have led to the rapid automated analysis of DNA, and the knowledge gained can be immediately translated into primary protein structure. In this way, tens of thousands of proteins have been sequenced in the past few years.

In summary, the structure of polypeptides is established by various degradation schemes. First, the polymer is purified; then the kind and relative abundance of the component amino acids are determined by complete hydrolysis and amino acid analysis. The *N*-terminal residues can be identified by Edman degradation. Repeated Edman degradation gives the sequence of shorter polypeptides, which are made from longer polypeptides by specific enzymatic hydrolysis. Finally, recombinant DNA technology has made primary structure analysis of larger proteins relatively easy.

26-6 Synthesis of Polypeptides: A Challenge in the Application of Protecting Groups

In a sense, the topic of peptide synthesis is a trivial one: Only one type of bond, the amide linkage, has to be made. The formation of this linkage was described in Section 19-10. Why discuss it further? This section shows that, in fact, achieving selectivity poses great problems, for which specific solutions have to be found.

Consider even as simple a target as the dipeptide glycylalanine. Just heating glycine and alanine to make the peptide bond by dehydration would result in a complex mixture of di-, tri-, and higher peptides with random sequences. Because the two starting materials can form bonds either to their own kind or to each other, there is no way to prevent random oligomerization.

An Attempt at the Synthesis of Glycylalanine by Thermal Dehydration

$$Gly \ + \ Ala \ \xrightarrow[- \ H_2O]{\Delta} \ Gly\text{-}Gly \ + \ Ala\text{-}Gly \ + \ \underset{\substack{\textbf{Desired} \\ \textbf{product}}}{Gly\text{-}Ala} \ + \ Ala\text{-}Ala \ + \ Gly\text{-}Gly\text{-}Ala \ + \ Ala\text{-}Gly\text{-}Ala \ etc.$$

Selective peptide synthesis requires protecting groups

To form peptide bonds selectively, the functional groups of the amino acids have to be protected. There are both amino- and carboxy-protecting groups.

The amino end is frequently blocked by a **phenylmethoxycarbonyl group** (abbreviated **carbobenzoxy** or **Cbz**), introduced by reaction of an amino acid with phenylmethyl chloroformate (benzyl chloroformate).

Protection of the Amino Group in Glycine

$H_3\overset{+}{N}CH_2COO^-$	$C_6H_5CH_2O\overset{\overset{\displaystyle O}{\|}}{C}Cl$	$C_6H_5CH_2O\overset{\overset{\displaystyle O}{\|}}{C}NHCH_2COOH$
Glycine	**Phenylmethyl chloroformate** (**Benzyl chloroformate**)	**Phenylmethoxycarbonylglycine** (**Carbobenzoxyglycine, Cbz-Gly**)

Reaction arrow: $\xrightarrow[\substack{- \ NaCl \\ - \ HOH}]{NaOH}$, 80%

The amino group is deprotected by hydrogenolysis (Section 22-2), which initially furnishes the carbamic acid as a reactive intermediate (Section 20-7). Decarboxylation occurs instantly to restore the amino function.

Deprotection of the Amino Group in Glycine

$$\text{C}_6\text{H}_5\text{—CH}_2\text{OCNHCH}_2\text{COOH} \xrightarrow{\text{H}_2, \text{Pd-C}} \text{C}_6\text{H}_5\text{—CH}_2\text{H} + \text{HOCNHCH}_2\text{COOH} \longrightarrow \text{CO}_2 + \overset{+}{\text{H}_3}\text{NCH}_2\text{COO}^-$$

Carbamic acid function

95%

Another amino-protecting group is **1,1-dimethylethoxycarbonyl (*tert*-butoxycarbonyl, Boc)**, introduced by reaction with bis(1,1-dimethylethyl) dicarbonate (di-*tert*-butyl dicarbonate).

Protection of the Amino Group in Amino Acids
as the Boc Derivative

$$\overset{+}{\text{H}_3}\text{NCHCOO}^- + (\text{CH}_3)_3\text{COCOCOC(CH}_3)_3 \xrightarrow[\substack{-\text{CO}_2, \\ -(\text{CH}_3)_3\text{COH}}]{(\text{CH}_3\text{CH}_2)_3\text{N}} (\text{CH}_3)_3\text{COCNHCHCOOH}$$

70–100%

Bis(1,1-dimethylethyl) dicarbonate (Di-*tert*-butyl dicarbonate)

1,1-Dimethylethoxy-carbonylamino acid (*tert*-Butoxycarbonylamino acid, Boc-amino acid)

Deprotection in this case is achieved by treatment with acid under conditions mild enough to leave peptide bonds untouched.

Deprotection of Boc-Amino Acids

$$(\text{CH}_3)_3\text{COCNHCHCOOH} \xrightarrow{\text{HCl or CF}_3\text{COOH}, 25°\text{C}} \overset{+}{\text{H}_3}\text{NCHCOO}^- + \text{CO}_2 + \text{CH}_2\text{=C(CH}_3)_2$$

EXERCISE 26-14

The mechanism of the deprotection of Boc-amino acids is different from that of the normal ester hydrolysis (Section 20-4): It proceeds through the intermediate 1,1-dimethylethyl (*tert*-butyl) cation. Formulate this mechanism.

The carboxy terminus of an amino acid is protected by the formation of a simple ester, such as methyl or ethyl. Deprotection results from treatment with base. Phenylmethyl (benzyl) esters can be cleaved by hydrogenolysis under neutral conditions (Section 22-2).

Peptide bonds are formed by using carboxy activation

With the ability to protect either end of the amino acid, we can synthesize peptides selectively by coupling an amino-protected unit with a carboxy-protected one. Because the protecting groups are sensitive to acid and base, the peptide bond must be formed under the mildest possible conditions. Special carboxy-activating reagents are used.

Perhaps the most general of these reagents is **dicyclohexylcarbodiimide (DCC).** The electrophilic reactivity of this molecule is similar to that of an isocyanate; it is ultimately hydrated to N,N'-dicyclohexylurea.

Peptide Bond Formation with Dicyclohexylcarbodiimide

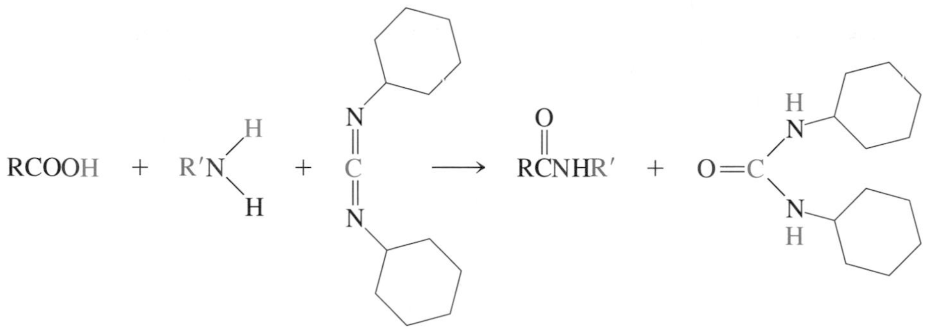

Dicyclohexyl-
carbodiimide

N,N'-Dicyclohexylurea

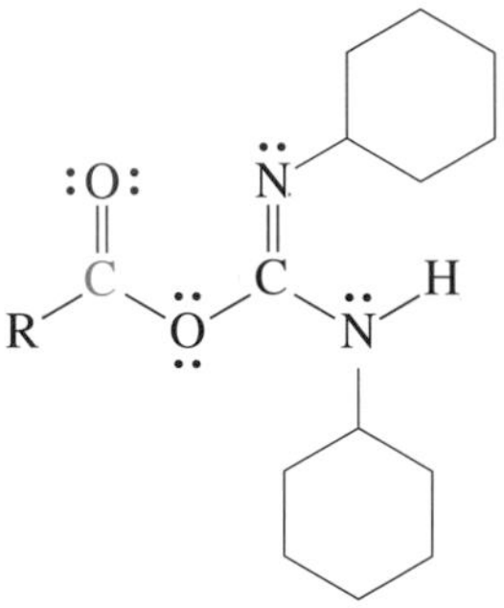

An *O*-acyl isourea

The role of DCC is to activate the carbonyl group of the acid to nucleophilic attack by the amine. This activation arises from the formation of an ***O*-acyl isourea,** in which the carbonyl group possesses reactivity similar to that in an anhydride (Section 20-3).

Armed with this knowledge, let us solve the problem of the synthesis of glycylalanine. We add amino-protected glycine to an alanyl ester in the presence of DCC. The resulting product is then deprotected to give the desired dipeptide.

Preparation of Gly-Ala

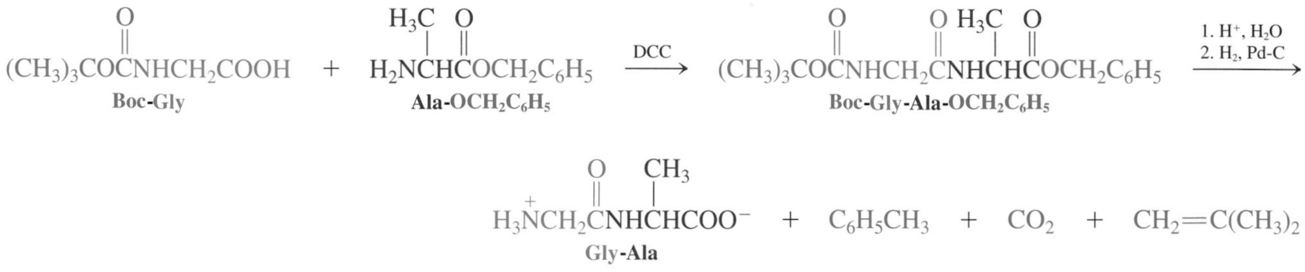

For the preparation of a higher peptide, deprotection of only one end is required, followed by renewed coupling, and so forth.

EXERCISE 26-15

Propose a synthesis of Leu-Ala-Val from the component amino acids.

In summary, polypeptides are made by coupling an amino-protected amino acid with another in which the carboxy end is protected. Typical protecting groups are readily cleaved esters and related functions. Coupling proceeds under mild conditions with dicyclohexylcarbodiimide as a dehydrating agent.

26-7 Merrifield Solid-Phase Peptide Synthesis

Polypeptide synthesis has been automated. This ingenious method, known as the **Merrifield* solid-phase peptide synthesis,** uses a solid support of polystyrene to anchor a peptide chain.

Polystyrene is a polymer (Section 12-14) whose subunits are derived from ethenylbenzene (styrene). Although beads of polystyrene are insoluble and rigid when dry, they swell considerably in certain organic solvents, such as dichloromethane. The swollen material allows reagents to move in and out of the polymer matrix easily. Thus, its phenyl groups may be functionalized by electrophilic aromatic substitution. For peptide synthesis, a form of Friedel-Crafts alkylation is used to chloromethylate a few percent of the phenyl rings in the polymer.

Electrophilic Chloromethylation of Polystyrene

Polystyrene   **Functionalized polystyrene**

EXERCISE 26-16

Formulate a plausible mechanism for the chloromethylation of the benzene rings in polystyrene. (**Hint:** Review Section 15-12.)

A dipeptide synthesis on chloromethylated polystyrene proceeds as follows.

Solid-Phase Synthesis of a Dipeptide

*Professor Robert B. Merrifield (b. 1921), Rockefeller University, New York, Nobel Prize 1984 (chemistry).

3. Coupling to protected
   second amino acid

$$(CH_3)_3COCNHCHCOOH, DCC$$

$$(CH_3)_3COCNHCHCNHCHCOCH_2\!\!-\!\!\bigcirc\!\!-\!\!\wwww\text{polystyrene chain}$$

4. Deprotection of
   amino terminus

$$CF_3CO_2H, CH_2Cl_2$$

$$H_2NCHCNHCHCOCH_2\!\!-\!\!\bigcirc\!\!-\!\!\wwww\text{polystyrene chain}$$

5. Disconnection of
   dipeptide from polymer

$$HF$$

$$\overset{+}{H_3}NCHCNHCHCOO^- \quad + \quad FCH_2\!\!-\!\!\bigcirc\!\!-\!\!\wwww\text{polystyrene chain}$$
Dipeptide

First an amino-protected amino acid is anchored on the polystyrene by nucleophilic substitution of the benzylic chloride by carboxylate. Deprotection is then followed by coupling with a second amino-protected amino acid. Renewed deprotection and final removal of the dipeptide by treatment with hydrogen fluoride complete the sequence.

The great advantage of solid-phase synthesis is the ease with which products can be isolated. Because all the intermediates are immobilized on the polymer, they can be purified by simple filtration and washing.

Obviously, it is not necessary to stop at the dipeptide stage. Repetition of the deprotection–coupling sequence leads to larger and larger peptides. Merrifield designed a machine that would carry out the required series of manipulations automatically, each cycle requiring only a few hours. In this way, the first total synthesis of the protein insulin was accomplished. More than 5000 separate operations were required to assemble the 51 amino acids in the two separate chains; thanks to the automated procedure, this took only several days.

Automated protein synthesis has opened up exciting possibilities. First, it is used to confirm the structure of polypeptides that have been analyzed by chain degradation and sequencing. Second, it can be used to construct unnatural proteins that might be more active and more specific than natural ones. Such proteins could be invaluable in the treatment of disease or in understanding biological function and activity.

In summary, solid-phase synthesis is an automated procedure in which a carboxy-anchored peptide chain is built up from amino-protected monomers by cycles of coupling and deprotection.

26-8 Polypeptides in Nature: Oxygen Transport by the Proteins Myoglobin and Hemoglobin

Two natural polypeptides function as oxygen carriers in vertebrates: the proteins myoglobin and hemoglobin. Myoglobin is active in the muscle, where it stores oxygen and releases it when needed. Hemoglobin is contained in red blood cells and facilitates

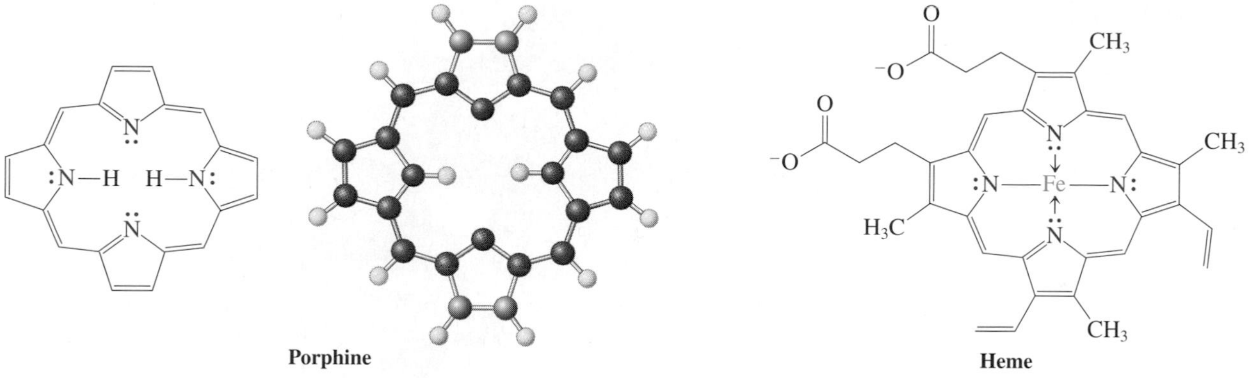

Porphine

Heme

FIGURE 26-7

Porphine is the simplest porphyrin. Note that the system forms an aromatic ring of 18 delocalized π electrons (indicated in red). A biologically important porphyrin is the heme group, responsible for binding oxygen. Two of the bonds to iron are dative (coordinate covalent), indicated by arrows.

oxygen transport. Without its presence, blood would be able to absorb only a fraction (about 2%) of the oxygen needed by the body.

How is the oxygen bound in these proteins? The secret of the oxygen-carrying ability of myoglobin and hemoglobin is a special nonpolypeptide unit, called a **heme group,** attached to the protein. Heme is a cyclic organic molecule (called a **porphyrin**) made of four linked, substituted pyrrole units surrounding an iron atom (Figure 26-7). The complex is red, giving blood its characteristic color.

The iron in the heme is attached to four nitrogens but can accommodate two additional groups above and below the plane of the porphyrin ring. In myoglobin, one of these groups is the imidazole ring of a histidine unit attached to one of the α-helical segments of the protein (Figure 26-8A). The other is most important for the

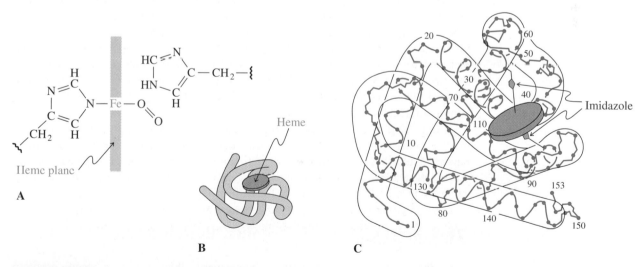

FIGURE 26-8

(A) Schematic representation of the active site in myoglobin, showing the iron atom in the heme plane bound to a molecule of oxygen and to the imidazole nitrogen atom of one histidine residue. (B) Schematic representation of the tertiary structures of myoglobin and its heme. (C) Secondary and tertiary structure of myoglobin. (After "The Hemoglobin Molecule," by M. F. Perutz, *Scientific American*, November 1964. Copyright © 1964, Scientific American, Inc.)

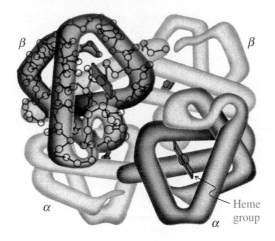

protein's function—bound oxygen. Close to the oxygen-binding site is a second imidazole of a histidine unit, which protects this side of the heme by steric hindrance. For example, carbon monoxide, which also binds to the iron in the heme group, and thus blocks oxygen transport, is prevented from binding as strongly as it normally would, because of the presence of the second imidazole group. Consequently, CO poisoning can be reversed by administering oxygen to a person who has been exposed to the gas. The two imidazole substituents in the neighborhood of the iron atom in the heme group are brought into close proximity by the unique folding pattern of the protein. The rest of the polypeptide chain serves as a mantle, shielding and protecting the active site from unwanted intruders and controlling the kinetics of its action (Figure 26-8B and C).

Myoglobin and hemoglobin offer excellent examples of the four structural levels in proteins. The primary structure of myoglobin consists of 153 amino acid residues of known sequence. Myoglobin has eight α-helical segments that constitute its secondary structure, the longest having 23 residues. The tertiary structure has the bends that give myoglobin its three-dimensional shape.

Hemoglobin contains four protein chains: two α *chains* of 141 residues each, and two β *chains* of 146 residues each. Each chain has its own heme group and a tertiary structure similar to that of myoglobin. There are many contacts between the chains; in particular, α_1 is closely attached to β_1, as is α_2 to β_2. These interactions give hemoglobin its quaternary structure (Figure 26-9).

The folding of the hemoglobin and myoglobin of several living species is strikingly similar even though the amino acid sequences differ. This finding implies that this particular tertiary structure is an optimal configuration around the heme group. The folding allows the heme to absorb oxygen as it is introduced through the lung, hang on to it as long as necessary for safe transport, and release it when required.

26-9 Biosynthesis of Proteins: Nucleic Acids

How does nature assemble proteins? The answer to this question is based on one of the most exciting discoveries in science, the nature and workings of the genetic code. All hereditary information is embedded in the **deoxyribonucleic acids (DNA).** The expression of this information in the synthesis of the many enzymes necessary for cell function is carried out by the **ribonucleic acids (RNA).** After the carbohydrates and polypeptides, the nucleic acids are the third major type of biological polymer. This section describes their structure and function.

Four heterocycles define the structure of nucleic acids

Considering the structural diversity of natural products, the structures of DNA and RNA are simple. All their components, called **nucleotides,** are polyfunctional, and it is one of the wonders of nature that evolution has eliminated all but a few specific combinations. Nucleic acids are polymers in which phosphate units link sugars, which bear various heterocyclic nitrogen **bases** (Figure 26-10).

In DNA, the sugar units are 2-deoxyriboses, and only four bases are present: **cytosine (C), thymine (T), adenine (A),** and **guanine (G).** The sugar characteristic of RNA is ribose, and again there are four bases, but the nucleic acid incorporates **uracil (U)** instead of thymine.

Nucleic Acid Sugars and Bases

2-Deoxyribose **Ribose**

Cytosine (C) **Thymine (T)** **Adenine (A)** **Guanine (G)** **Uracil (U)**

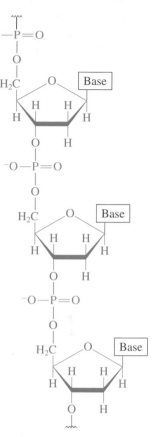

FIGURE 26-10
Part of a DNA chain. The base is a nitrogen heterocycle. The sugar is 2-deoxyribose.

EXERCISE 26-17

Even though the preceding structures do not (except for adenine) indicate so, cytosine, thymine, guanine, and uracil are aromatic, albeit somewhat less so than the corresponding azapyridines. Explain. (**Hint:** Recall the discussions about amide resonance in Sections 20-1 and 26-4.)

We construct a nucleotide from three components. First, we replace the hydroxy group at C1 in the sugar with one of the base nitrogens. This combination is called a **nucleoside.** Second, a phosphate substituent is introduced at C5. In this way we obtain the four nucleotides of both DNA and RNA. The positions on the sugars in nucleosides and nucleotides are designated 1′, 2′, and so forth, to distinguish them from the carbon atoms in the nitrogen heterocycles.

Adenosine
(A nucleoside)

Human male chromosomes, the structures in the cell that carry the genetic information encoded in their component DNA.

Synthetic Nucleic Acid Bases and Nucleosides in Medicine

5-Fluorouracil
(Fluracil)

9-[(2-Hydroxyethoxy)methyl]guanosine
(Acyclovir)

3′-Azido-3′-deoxythymidine
(Zidovudine, or AZT)

The central role played by nucleic acid replication in biology has been exploited in medicine. Many hundreds of synthetically modified bases and nucleosides have been prepared and their effects on nucleic acid synthesis investigated. Some of them in clinical use include 5-fluorouracil *(fluracil),* an anticancer agent, 9-[(2-hydroxyethoxy)methyl] guanosine *(acyclovir),* which is active against two strains of herpes simplex virus, and 3′-azido-3′-

deoxythymidine *(zidovudine, or AZT),* a drug that combats the AIDS virus.

Substances such as these may interfere with nucleic acid replication by masquerading as legitimate nucleic acid building blocks. The enzymes associated with this process are fooled into incorporating the drug molecule, and synthesis of the biological polymer cannot continue.

Nucleotides of DNA

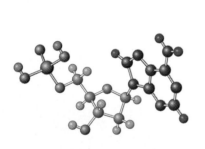

2′-Deoxyadenylic acid

2′-Deoxyguanidylic acid

2′-Deoxycytidylic acid

2′-Deoxythymidylic acid

Nucleotides of RNA

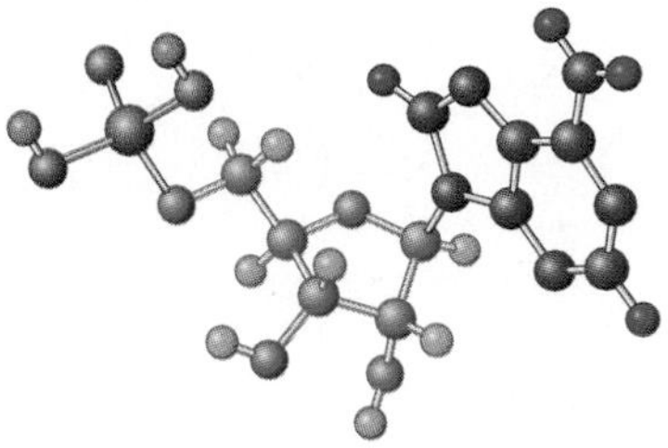

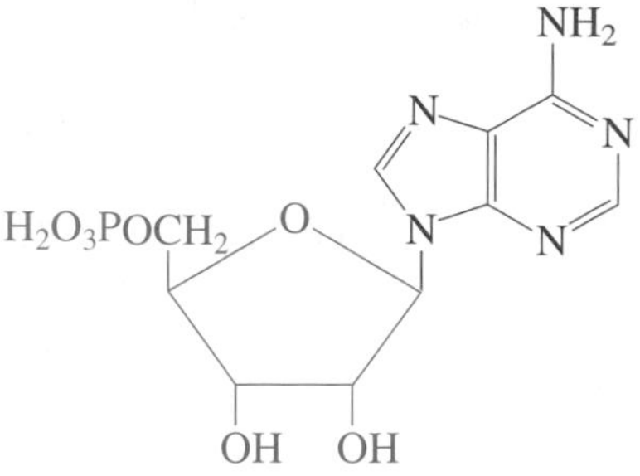

Adenylic acid

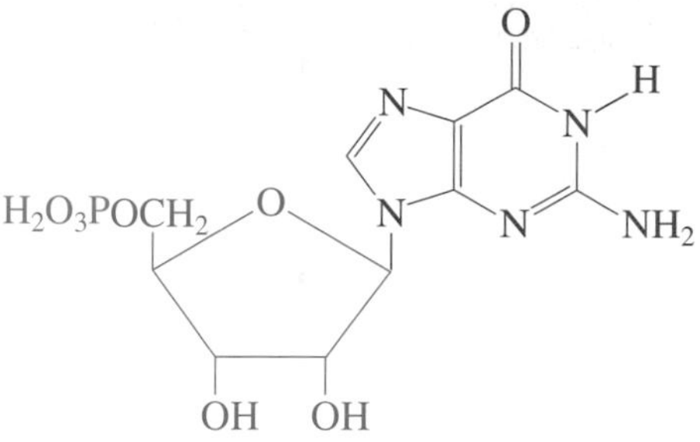

Guanidylic acid

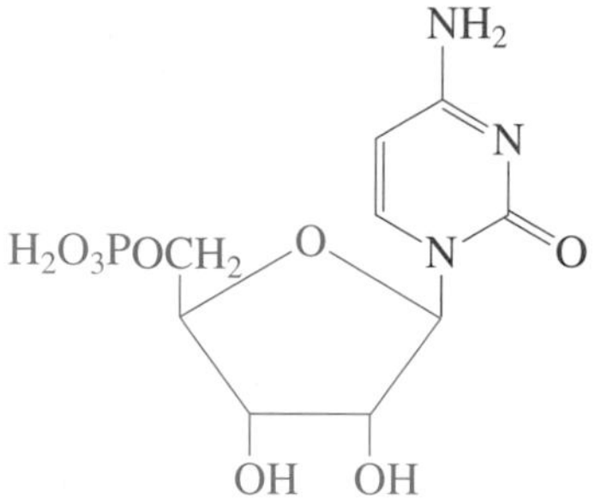

Cytidylic acid

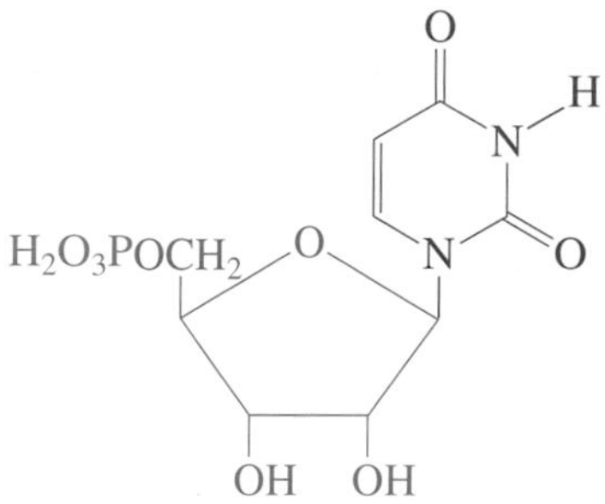

Uridylic acid

The polymeric chain shown in Figure 26-10 is then readily derived by repeatedly forming a phosphate ester bridge from C5′ (called the **5′ end**) of the sugar unit of one nucleotide to C3′ (the **3′ end**) of another.

In this polymer, the bases adopt the same role as that of the 2-substituent in the amino acids of a polypeptide: Their sequence varies from one nucleic acid to another and determines the fundamental biological properties of the system.

Nucleic acids form a double helix

Nucleic acids, especially DNA, can form extraordinarily long chains (as long as several centimeters) with molecular weights of as much as 150 billion. Like proteins, they adopt secondary and tertiary structures. In 1953, Watson and Crick* made their ingenious proposal that DNA is a double helix composed of two strands with complementary base sequences. A crucial piece of information was that, in the DNA of various species, the ratio of adenine to thymine, like that of guanine to cytosine, is always one to one. Watson and Crick concluded that two chains are held together by hydrogen bonding in such a way that adenine and guanine, respectively, in one chain always face thymine and cytosine in the other, and vice versa (Figure 26-11). Thus, if a piece of DNA in one strand has the sequence -A-G-C-T-A-C-G-A-T-C-, this entire segment is hydrogen bonded to a complementary strand running in the opposite direction, -T-C-G-A-T-G-C-T-A-G-, as shown.

$$\text{\tiny wwww}—A—G—C—T—A—C—G—A—T—C—\text{\tiny wwww}$$
$$\text{\tiny wwww}—T—C—G—A—T—G—C—T—A—G—\text{\tiny wwww}$$

*Professor James D. Watson (b. 1928), Harvard University, Cambridge, Massachusetts, Nobel Prize 1962 (medicine); Professor Francis H. C. Crick (b. 1916), Cambridge University, England, Nobel Prize 1962 (medicine).

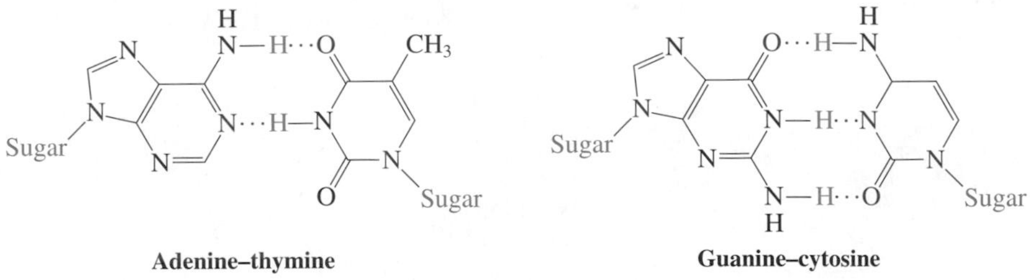

FIGURE 26-11

Hydrogen bonding between the base pairs adenine–thymine and guanine–cytosine. The components of each pair are always present in equal amounts.

Adenine–thymine

Guanine–cytosine

Because of other structural constraints, the arrangement that maximizes hydrogen bonding and minimizes steric repulsion is the double helix (Figure 26-12).

DNA replicates by unwinding and assembling new complementary strands

There is no restriction on the variety of sequences of the bases in the nucleic acids. Watson and Crick proposed that the specific base sequence of a particular DNA contained all genetic information necessary for the duplication of a cell and, indeed, the growth and development of the organism as a whole. Moreover, the double-helical structure suggested a way in which DNA might **replicate**—make exact copies of itself—and so pass on the genetic code. In this mechanism, each of the two strands of

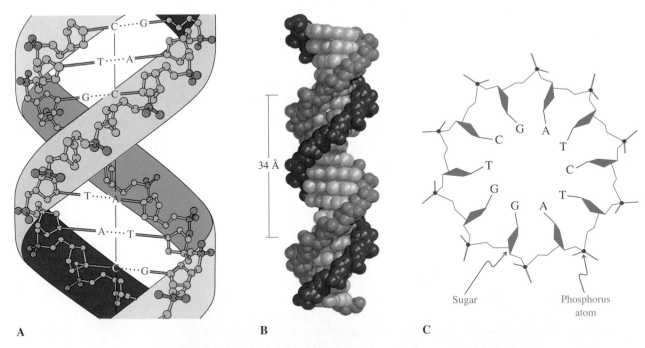

A
B
C

FIGURE 26-12

(A) The two nucleic acid strands of a DNA double helix are held together by hydrogen bonding between the complementary sets of bases. Note: The two chains run in opposite directions and all the bases are on the inside of the double helix. The diameter of the helix is 20 Å; base–base separation across the strands is ~ 3.4 Å; the helical turn repeats every 10 residues, or 34 Å. (B) Space-filling model of DNA double helix (green and red strands). (C) One of the strands of a DNA double helix, in a view down the axis of the molecule. The color scheme in (A) and (B) matches that in Figure 26-10. (After *Biochemistry,* 4th ed., by Lubert Stryer, W. H. Freeman and Company. Copyright © 1975, 1981, 1988, 1995.)

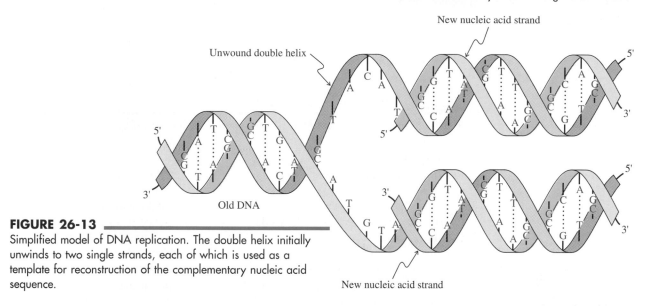

New nucleic acid strand

Unwound double helix

5'

3'

5'

3'

5'

Old DNA

5'

3'

3'

New nucleic acid strand

FIGURE 26-13

Simplified model of DNA replication. The double helix initially unwinds to two single strands, each of which is used as a template for reconstruction of the complementary nucleic acid sequence.

DNA functions as a template. The double helix partly unwinds, and enzymes called DNA polymerases then begin to assemble the new DNA by coupling nucleotides to one another in a sequence complementary to that in the template, always juxtaposing C to G and A to T (Figure 26-13). Eventually, two complete double helices are produced from the original. This process is at work throughout the entire human genetic material, or **genome**—some 3 billion base pairs—with an error frequency of less than 1 in 10 billion base pairs.

In summary, the nucleic acids DNA and RNA are polymers containing monomeric units called nucleotides. There are four nucleotides for each, varying only in the structure of the base: cytosine (C), thymine (T), adenine (A), and guanine (G) for DNA; cytosine, uracil (U), adenine, and guanine for RNA. The two nucleic acids differ also in the identity of the sugar unit: deoxyribose for DNA, ribose for RNA. DNA replication and RNA synthesis from DNA is facilitated by the complementary character of the base pairs A–T, G–C, and A–U. The double helix partly unwinds and functions as a template for replication.

Transmission electron microscope picture of replicating DNA. Unwinding of the two complementary strands generates a "bubble" that enlarges to form a Y-shaped molecule termed a replication fork.

26-10 Protein Synthesis Through RNA

The mechanism of duplicating the entire nucleotide sequence in DNA replication is used by nature and by chemists to obtain partial copies of the genetic code for various purposes. In nature, the most important application is the assembly of RNA, called **transcription,** which transcribes the parts of the DNA that contain the information (the **genes**) necessary to synthesize proteins in the cell. The process by which this transcribed information is decoded and used to construct proteins is called **translation.** The three key players in protein synthesis are the "DNA transcript" **messenger RNA (mRNA),** the "delivery unit" for the specific amino acids to be connected by peptide bonds, **transfer RNA (tRNA),** and the catalyst that enables amide bond formation—the **ribosome.**

Protein synthesis starts with mRNA, a transcript of a piece of a single strand of partly unwound DNA (Figure 26-14). Its chain is much shorter than that of DNA, and it does not stay bound to the DNA but breaks away as its synthesis is finished.

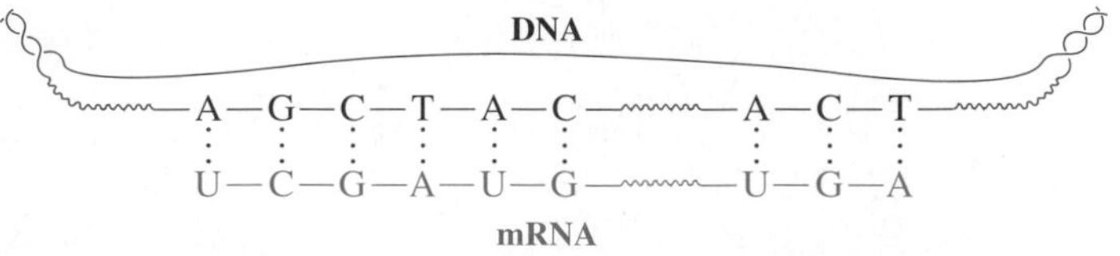

FIGURE 26-14

Simplified picture of messenger RNA synthesis from a single strand of (partly unwound) DNA.

The mRNA is the template responsible for the correct sequencing of the amino acid units in proteins. How does mRNA do that? Each sequence of three bases, called a **codon,** specifies a particular amino acid (Table 26-3). Simple permutation of this three-base code with a total of four bases gives $4^3 = 64$ possible distinct sequences. That number is more than enough, because only 20 different amino acids are needed for protein synthesis. This might seem like overkill, but consider that the next lower alternative—namely, a two-base code—would give only $4^2 = 16$ combinations, too few for the number of different amino acids found in natural proteins.

Codons do not overlap; in other words, the three bases specifying one amino acid are not part of another preceding or succeeding codon. Moreover, the "reading" of the base sequence is consecutive; each codon immediately follows the next, uninterrupted by genetic "commas" or "hyphens." Nature also makes full use of all 64 codons by allowing for several of them to describe the same amino acid (Table 26-3). Only tryptophan and methionine are characterized by single three-base codes. Some codons act as signals to initiate or terminate production of a polypeptide chain. Note that the initiator codon (AUG) is also the codon for methionine. Thus, if the codon AUG appears *after* a chain has been initiated, methionine will be produced. The complete base sequence of the DNA in a cell defines its **genetic code.**

Mutations in the base sequence of DNA can be caused by physical (radiation) or chemical (carcinogens; see, e.g., Section 16-7) interference. Mutations can either replace one base with another or add or delete one base or more. Here is some of the potential value of redundant codons. If, for example, the RNA sequence CCG (proline) were changed as a result of a DNA mutation to the RNA sequence CCC, proline would still be correctly synthesized.

EXERCISE 26-18

What is the sequence of an mRNA molecule produced from a DNA template strand with the following composition?   5'-ATTGCTCAGCTA-3'

EXERCISE 26-19

(a) What amino acid sequence is encoded in the following mRNA base sequence (starting at the left end)?   A-A-G-U-A-U-G-C-A-U-C-A-U-G-C-U-U-A-A-G-C
(b) Identify the mutation that would have to occur for Trp to be present in the resulting peptide.

With a copy of the requisite codons in hand, proteins are then synthesized along the mRNA template with the help of a set of other important nucleic acids, the tRNAs. These molecules are relatively small, containing from 70 to about 90 nucleotides. Each tRNA is specifically designed to carry one of the 20 amino acids for delivery to the mRNA in the course of protein buildup. The amino acid sequence

TABLE 26-3	Three-Base Code for the Common Amino Acids Used in Protein Synthesis				
Amino acid	**Base sequence**	**Amino acid**	**Base sequence**	**Amino acid**	**Base sequence**
Ala (A)	GCA	His (H)	CAC	Ser (S)	AGC
	GCC		CAU		AGU
	GCG				UCA
	GCU	Ile (I)	AUA		UCG
			AUC		UCC
Arg (R)	AGA		AUU		UCU
	AGG				
	CGA	Leu (L)	CUA	Thr (T)	ACA
	CGC		CUC		ACC
	CGG		CUG		ACG
	CGU		CUU		ACU
			UUA		
Asn (N)	AAC		UUG	Trp (W)	UGG
	AAU				
		Lys (K)	AAA	Tyr (Y)	UAC
Asp (D)	GAC		AAG		UAU
	GAU				
		Met (M)	AUG	Val (V)	GUA
Cys (C)	UGC				GUG
	UGU	Phe (F)	UUU		GUC
			UUC		GUU
Gln (Q)	CAA				
	CAG	Pro (P)	CCA	Chain initiation	AUG
			CCC		
Glu (E)	GAA		CCG	Chain termination	UGA
	GAG		CCU		UAA
					UAG
Gly (G)	GGA				
	GGC				
	GGG				
	GGU				

FIGURE 26-15 ———

Representation of the biosynthesis of the tripeptide Gly-Ala-Asn. The tRNAs carrying their specific amino acids line up their anticodons along the codons of mRNA, before ribosomal enzymes accomplish amide linkage.

~~~~~G—G—A—G—C—A—A—A—C~~~~~ mRNA **chain (with** codons**)**

C—C—U   C—G—U   U—U—G   **tRNAs (with anticodons)**

Gly   Ala   Asn

Amino acids **delivered by tRNAs**

encoded in mRNA is read codon by codon by a complementary three-base sequence on tRNA, called an **anticodon.** In other words, the individual tRNAs, each carrying its specific amino acid, line up along the mRNA strand in the correct order. At this stage, catalytic ribosomes (very large enzymes) containing their own RNA, facilitate peptide-bond formation (Figure 26-15). As the polypeptide chain grows longer, it begins to develop its characteristic secondary and tertiary structure (α helix, pleated sheets, etc.), helped by enzymes that form the necessary disulfide bridges. All of this happens with remarkable speed. It is estimated that a protein made up of approximately 150 amino acid residues can be biosynthesized in less than 1 minute. Clearly, nature still has the edge over the synthetic organic chemist, at least in this domain.

In summary, RNA is responsible for protein biosynthesis; each three-base sequence, or codon, specifies a particular amino acid. Codons do not overlap, and more than one codon can specify the same amino acid.

26-11 DNA Sequencing and Synthesis: Cornerstones of Gene Technology

The sheep Dolly, claimed to be the world's first clone of an adult mammal.

Molecular biology is undergoing a revolution because of our capability to decipher, reproduce, and alter the genetic code of organisms. Individual genes or other DNA sequences from a genome can be reproduced *(cloned),* often on a large scale. Genes of higher organisms can be expressed (i.e., made to start protein synthesis) in lower organisms or they can be modified to produce "unnatural" proteins. Some modified genes have been successfully reintroduced into the organism of their origin, causing changes in the physiology and biochemistry of the "starting" host. Much of this development is due to biochemical advances such as the discovery of enzymes that can selectively cut, join, or replicate DNA and RNA. For example, *restriction enzymes* cut long molecules into small defined fragments that can be joined by *DNA ligases* to other DNAs *(recombinant DNA technology). Polymerases* catalyze DNA replication, and selected DNA sequences are made on a large scale by the *polymerase chain reaction.* The list goes on and on and its contents are beyond the scope of organic chemistry.

The essential foundation for these discoveries has been the knowledge of the primary structure of nucleic acids and the development of methods for their synthesis.

DNA sequencing mimics polypeptide sequencing in concept

The sequencing of DNA can be accomplished by using both enzymatic and chemical methods. In analogy to protein analysis (Section 26-5), the unwieldy DNA chain is first cleaved at specific points by enzymes called **restriction endonucleases.** There are more than 200 such enzymes, providing access to a multitude of overlapping sequences. The pieces of DNA so obtained are then (again enzymatically) tagged at the $5'$-phosphate ends, not chemically as in the Edman degradation, but with a radioactive phosphorus (^{32}P) label, and not for structural identification, but for analytical detection in the next step. This step consists of subjecting the pure, labeled fragments in separate experiments to reagents that cleave only next to specific bases (i.e., only at the $5'$ side of, respectively, A, G, T, or C) and analyzing the products by chromatography with the use of radioactive detection (through a photographic plate). The patterns of products that are made and identified in this way reveal the base sequence. This method is partly automated and so efficient that thousands of base pairs can be identified in a day. In conjunction with other, biological sequencing strategies, it has made DNA sequencing the method of choice for the structural elucidation of the encoded proteins (see Section 26-5). In 1990, the U.S. Human Genome Project was begun, an international effort aimed at establishing the sequence of the 3 billion bases in human DNA and hence identifying the estimated associated 80,000 genes. Originally thought to take 15 years, it is now hoped that it will be completed shortly after the turn of the millennium. Apart from work on human DNA, the genetic makeup of several other organisms is being studied. Some spectacular milestones are the completion of the first nonviral whole genome to be sequenced (1.8 million base pairs, 1743 genes)—namely, of the bacterium *Haemophilus influenzae*—in 1995, followed by those of the human gut bacterium *Escherichia coli* and the cause of Lyme disease, *Borrelia burgdorferi,* both in 1997.

DNA synthesis is automated

You can order your custom-made oligonucleotide today and receive it tomorrow. The reason for this efficiency is the use of automated nucleotide coupling by so-called DNA synthesizers that operate on the same principle as that of the Merrifield polypeptide synthesis (Section 26-5): solid-phase attachment of the growing chain and the employment of protected nucleotide building blocks. Protection of the bases cytosine, adenine, and guanine is required at their amino functions (absent in thymine, which needs no protection), in the form of their amides.

Protected (Except for Thymine) DNA Bases

Cytosine (C) Thymine (T) Adenine (A) Guanine (G)

The sugar moiety is blocked at C5$'$ as a dimethoxytrityl [di(4-methoxyphenyl)phenylmethyl, DMT] ether, readily cleaved by mild acid through an S_N1 mechanism (Chapter 22, Section 22-1 and Problem 31), much like 1,1-dimethylethyl (*tert*-butyl) ethers

(Section 9-8, Chemical Highlight 9-2). To anchor the first so-protected nucleotide on the solid support, the C4′-OH is attached to an activated linker, a diester. Unlike the Merrifield medium of polystyrene, the solid used in oligonucleotide synthesis is surface-functionalized silica (SiO_2) bearing an amino substituent as a "hook." Coupling to the anchor nucleotide is by amide formation.

Anchoring the Protected Nucleotide on SiO₂

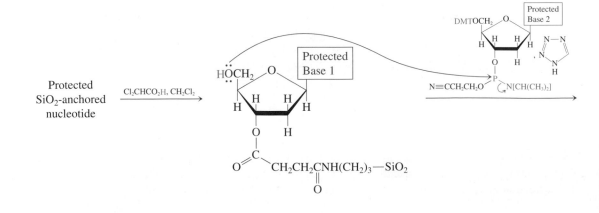

With the first nucleotide in place, we are ready to attach to it the second. For this purpose, the point of attachment, the 5′-OH, is deprotected with acid. Subsequent addition of a 3′-OH activated nucleotide effects coupling. The activating group is an unusual phosphoramidite [containing P(III)], which, as we shall see shortly, also serves as a masked phosphate [P(V)] for the final dinucleotide and is labile with respect to nucleophilic substitution, not unlike PBr_3 (recall Sections 9-4 and 19-8). The displacement reaction is base catalyzed and furnishes a phosphite derivative; the base is the, again unusual, aromatic heterocycle tetrazole, a tetrazacyclopentadiene related to pyrrole (Section 25-3) and imidazole (Section 26-1). Finally, the phosphorus is oxidized with iodine to the phosphate oxidation state.

Dinucleotide Synthesis: Deprotection, Coupling, and Oxidation

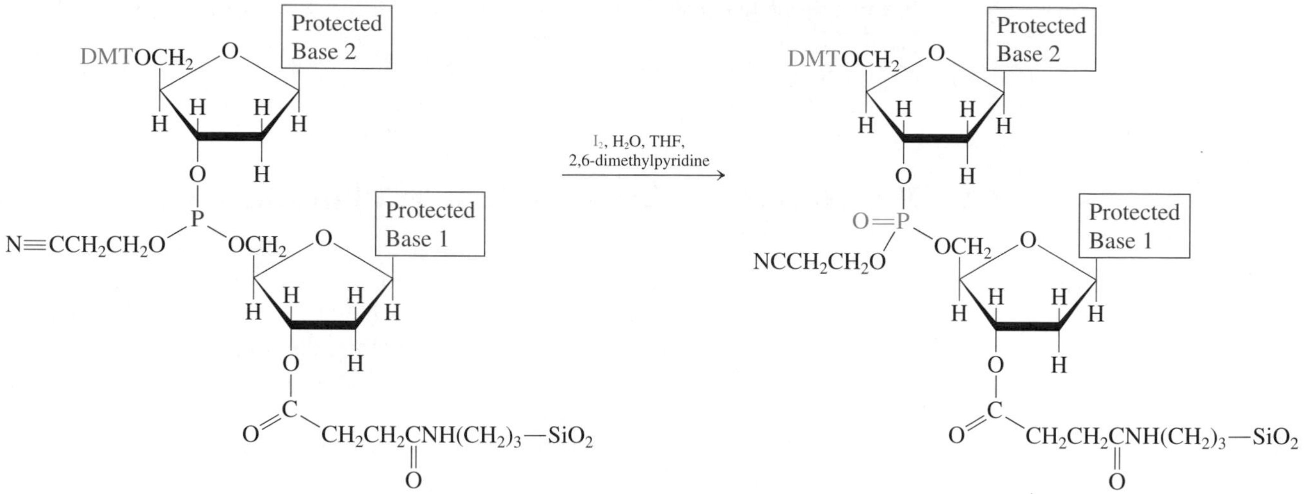

The sequence (1) DMT hydrolysis, (2) coupling, and (3) oxidation can be repeated multiple times in the synthesizer machine until the desired oligonucleotide is assembled in its protected and immobilized form. The final task is to remove the product from the silica and deprotect the DMT-bearing terminal sugar, all the bases, and the phosphate group, without cleaving any of the other bonds. Remarkably, this task can be done all at once with aqueous ammonia, as shown here for the dinucleotide made in the preceding scheme.

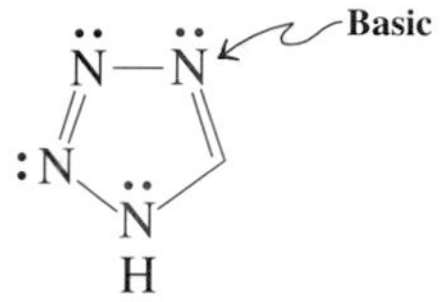

Tetrazole

Liberation of the Solid-Phase-Supported and Protected Dinucleotide

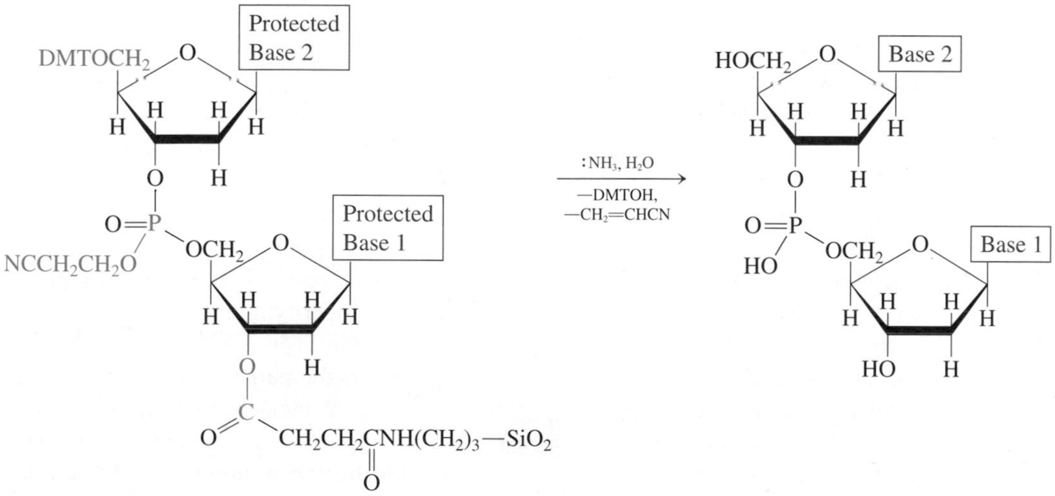

EXERCISE 26-20

Formulate mechanisms for all the hydrolysis reactions that effect deprotection in the scheme just shown.

EXERCISE 26-21

In the dinucleotide synthesis just described, the attachment of the first nucleotide to silica employs a 4-nitrophenyl ester as a leaving group. Why would the use of this substituent be advantageous?

The polymerase chain reaction (PCR) makes multiple copies of DNA

Cloning, the amplification of individual genes for the study of their sequence, expression, and regulation, revolutionized molecular biology in the mid-1970s. Cloning needs living cells to amplify inserted DNA. Therefore, the discovery of a procedure by Mullis* in 1984 that reproduces DNA segments in vitro by the millions, without the necessity of living cells, marked a stunning advance: the **polymerase chain reaction (PCR).** The key to this reaction is the ability of some DNA-copying enzymes to remain stable at temperatures as high as 95°C. The original enzyme used in PCR is Tag polymerase found in a bacterium, *Thermus aquaticus,* in a hot spring in Yellowstone National Park. Polymerases are the enzymes responsible for copying DNA after it has unwound (Figure 26-13). For this purpose, they need a supply of the four nucleotides and a short sequence of nucleotides to "prime" the reaction. In nature, a primase enzyme takes care of the priming and the entire DNA sequence behind the primer is reproduced (as the complementary strand, Section 26-9). In PCR, the primer consists of a short (20 bases) oligonucleotide strand complementary to a short piece of the DNA to be copied.

In practice, the PCR is carried out as shown schematically in Figure 26-16. The reaction flask is charged with the (double-stranded) DNA to be copied, the four nucleotides, the primer, and Tag polymerase. In the first step, the mixture is heated to 90°–95°C to induce the DNA to separate into two strands. Cooling to 54°C allows the primers to attach themselves to the individual DNA molecules. Raising the temperature to 72°C provides optimum (in fact, "native") conditions for Tag polymerase to add nucleotides to the primer along the attached DNA strand until the end, resulting in a complementary copy of the template. All of this takes only a few minutes and can be repeated in automated temperature-regulated vessels multiple times. Because the products of each cycle are again separated into their component strands to function as additional templates in the subsequent cycle, the amount of DNA produced with time increases exponentially: after *n* cycles, the quantity of DNA is 2^{n-1}. For example, after 20 cycles (less than 3 h), it can be about 1 million; after 32 cycles, 1 billion.

There are many practical applications of this technique. In medical diagnostics, bacteria and viruses (including HIV) are readily detected through the use of specific primers. Early cancer detection is made possible by identifying mutations in growth-control genes. In forensics and legal medicine, PCR has been used to identify the origin of blood, saliva, and other biological clues left by the perpetrator of a crime. Paternity and genealogy can be proved unambiguously. In molecular evolution, ancient DNA can be multiplied or even reconstructed by amplification of isolated fragments. In this way, some of the genes of a 2400-year-old Egyptian mummy have been deciphered. Even more impressively, the DNA of an approximately 30-million-year-old termite preserved in amber was amplified by PCR, revealing the presence of a gene for ribosomal RNA, thus providing insight into the evolution of this species. PCR constituted the scientific foundation for Michael Crichton's *Jurassic Park,* a novel in which dinosaurs are cloned from their ancient DNA. What will be next?

*Dr. Kary B. Mullis (b. 1945), La Jolla, California, Nobel Prize 1993 (chemistry).

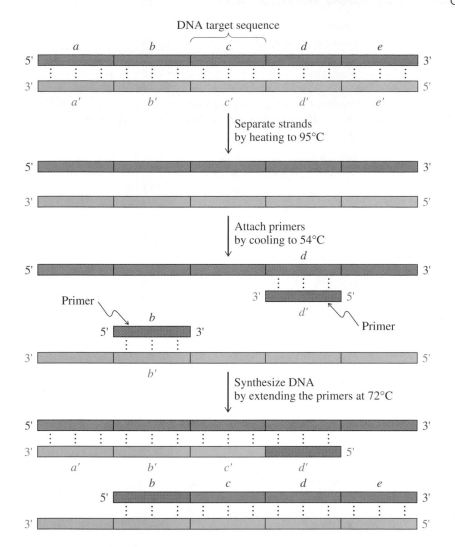

FIGURE 26-16 _____
Polymerase chain reaction
(PCR). A cycle consists of
three steps: strand
separation, attachment of
primers, and extension of
primers by DNA synthesis.
The reactions are carried out
in a closed vessel. The cycle
is driven by changes in
temperature. Sequences on
one strand of the original
DNA are denoted by *abcde*
and those on the
complementary strand by
a′b′c′d′e′. The primers are
shown in blue and the new
DNA extending from the
primers in the respective
complementary color to the
old strands (red or green).
(After *Biochemistry*,
4th ed., by Lubert Stryer,
W. H. Freeman and
Company. Copyright © 1975,
1981, 1988, 1995.)

Chapter Integration Problem

Aspartame (NutraSweet), Asp-Phe-OCH₃ (see first page of this chapter), appears to
be a simple target of synthesis. That this is not so becomes apparent when you try to
devise routes to it, starting simply from aspartic acid (Asp) and phenylalanine (Phe).
Analyze the problem and formulate various approaches to the synthesis of this
compound.

Aspartame

Solution

Simple retrosynthetic analysis dissects the molecule into its two amino acid components, Asp and Phe-OCH₃ (made by methyl esterification of Phe), which might be

envisaged to be coupled with DCC (Section 20-6). There is a problem, however: Asp has an additional β-carboxy group, bound to interfere. We are therefore facing the task of preparing a selectively carboxy protected Asp. We can think about accomplishing it by two strategies: (A) the direct protection of the Asp nucleus or, if strategy A is not successful, (B) a total synthesis of an appropriately derivatized Asp from simpler starting materials that avoids Asp altogether. The following strategies focus on practical solutions to our problem found in the literature, although you may come up with alternatives that might be even better.

A. Selective Protection of the Asp Nucleus

Are the two carboxy groups chemically sufficiently different to warrant the investigation of selective monoprotection of one of the carboxy groups in Asp? The answer is yes, perhaps. Thus, we have learned that an α-amino group causes an increase in the acidity of a carboxylic acid by about 2 pK units (Section 26-1). For Asp, pK_a (α-COOH) = 1.9, whereas pK_a (β-COOH) = 3.7 (Table 26-1). One might therefore be tempted to try a selective ammonium salt formation with the α-carboxylate function at carefully controlled pH, followed by thermolysis (Section 19-10). The problem is that, under the thermal conditions of the relatively slow amide formation, fast proton exchange takes place. It is better to hope for differing reactivity of the two carbonyl groups in addition–elimination reactions of an appropriate derivative. The electron-withdrawing effect of the α-amino group should manifest itself here in increased electrophilicity of the adjacent carbonyl carbon (Sections 17-6, 19-4, and 20-1), rendering the corresponding ester function more susceptible to basic hydrolysis. Indeed, this approach has been realized successfully with the N-Cbz protected (phenylmethyl) diester of Asp (prepared by standard procedures, see Section 26-6), as shown here.

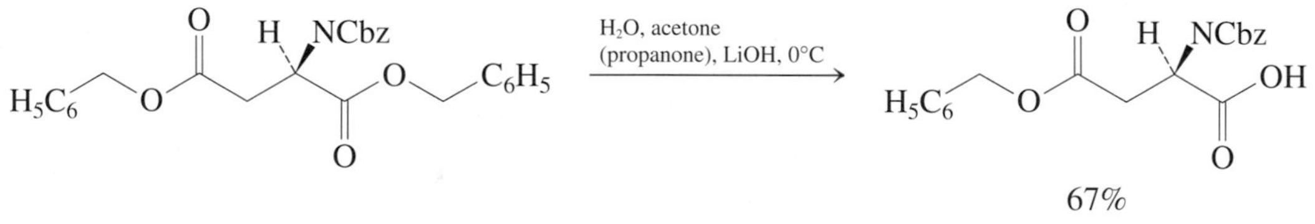

Cbz-Asp-(OCH₂C₆H₅)₂

The product was coupled with Phe-OCH₃ and DCC and then deprotected by catalytic hydrogenolysis (Section 26-6) to give aspartame.

A simplified synthesis relies on the potential to protect difunctional compounds as cyclic derivatives. For example, 1,2-diols are "masked" as cyclic acetals (Section 24-8), hydroxy acids as lactones (Section 19-9), amino acids as lactams (Section 19-10), and dicarboxylic acids as anhydrides (Section 19-8). The last two possibilities merit consideration as applied to Asp. However, direct lactam formation can be quickly ruled out, because of the complications of ring strain (although β-lactams have been used in the preparation of aspartame). This problem is absent with respect to dehydration to the five-membered ring anhydride. Because anhydrides are activated carboxylic acid derivatives (Section 20-3), the Asp anhydride can be coupled directly with Phe-OCH₃ without the help of added DCC. Nucleophilic attack of the amino end of Phe-OCH₃ occurs preferentially at the desired position, albeit not completely so; 19% of the product derives from peptide-bond formation at the β carboxy group of Asp.

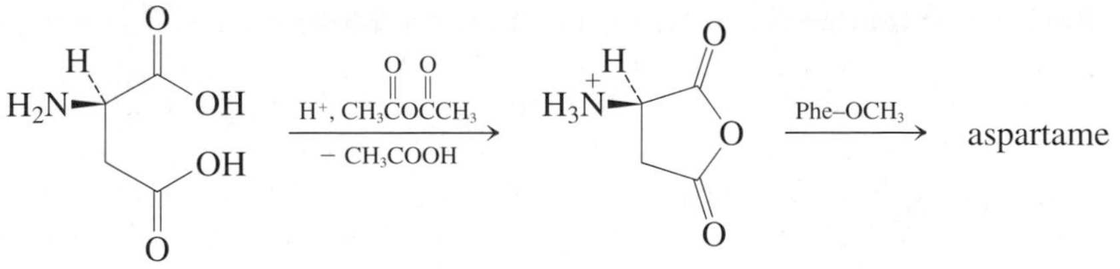

93%   61%

Asp anhydride

B. Total Syntheses of Carboxy-differentiated Asp Derivatives

The alternative to approach A is to start from scratch and build up the Asp framework by using the methods of Section 26-2 in such a way as to provide as a final product a selectively monoprotected carboxy derivative. There are many possible strategies for the solution of this problem, but, as you contemplate them, you will discover that finding an appropriate masked β-carboxy group that is not unmasked during the manipulations of the α-amino acid part is not easy. In the following Gabriel synthesis, recourse was made to the relatively innocuous 2-propenyl substituent, which is eventually elaborated by oxidative cleavage (Sections 12-10 and 24-5) to the free β-carboxy moiety.

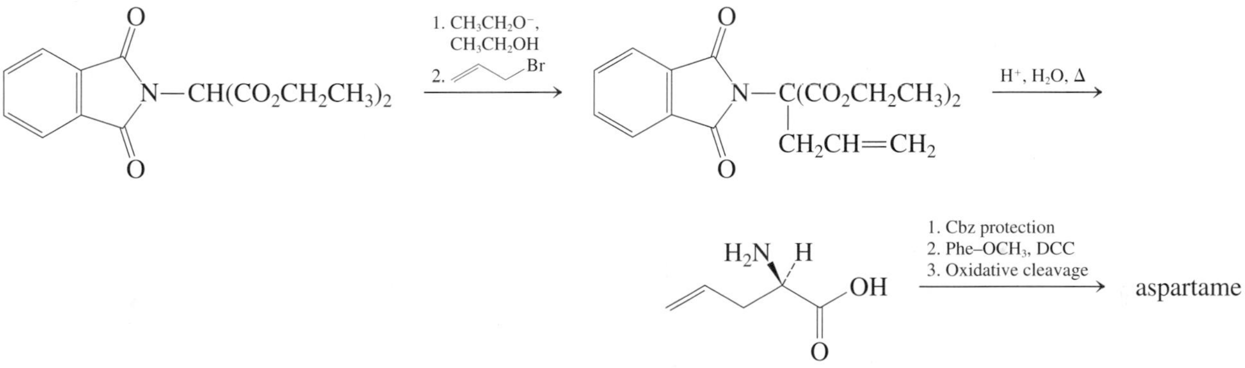

NEW REACTIONS

1. Acidity of Amino Acids (Section 26-1)

$H_3\overset{+}{N}CHCOOH$ $H_3\overset{+}{N}CHCOO^-$

$pK_a \sim 2\text{--}3$ $pK_a \sim 9\text{--}10$

Isoelectric point $pI = \dfrac{pK_{COOH} + pK_{NH_3}^+}{2}$

2. Strongly Basic Guanidino Group in Arginine (Section 26-1)

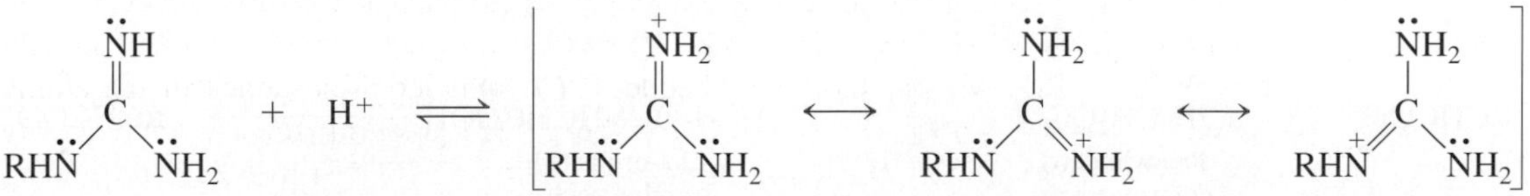

$pK_a \sim 13$

3. Basicity of Imidazole in Histidine (Section 26-1)

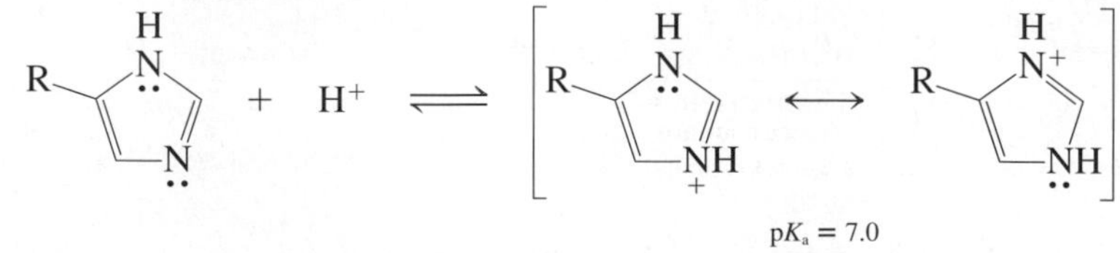

$$pK_a = 7.0$$

Preparation of Amino Acids

4. Hell-Volhard-Zelinsky Bromination Followed by Amination (Section 26-2)

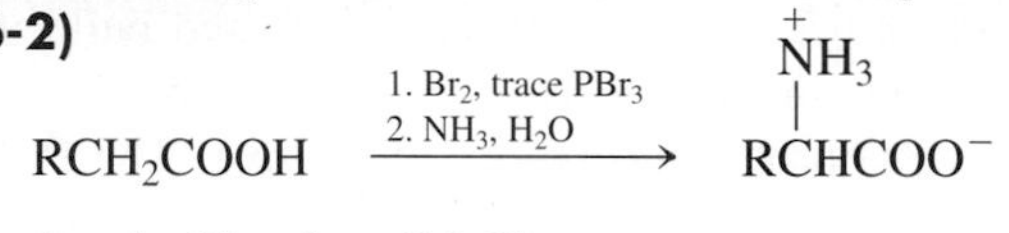

5. Gabriel Synthesis (Section 26-2)

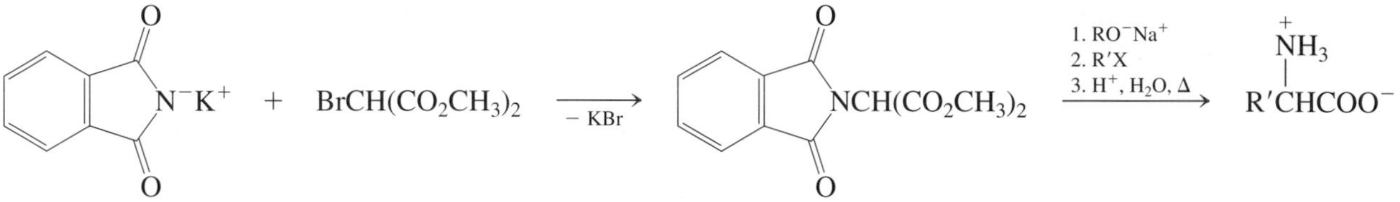

6. Strecker Synthesis (Section 26-2)

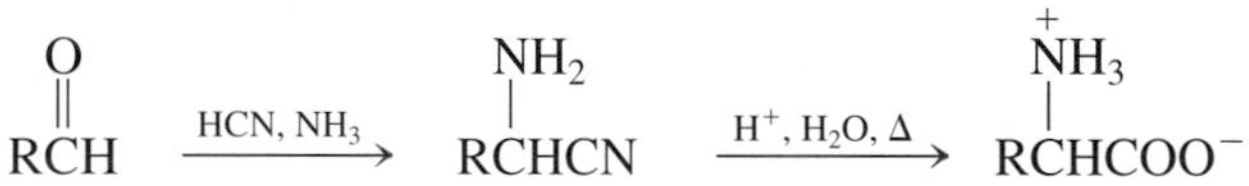

Polypeptide Sequencing

7. Hydrolysis (Section 26-5)

$$\text{Peptide} \xrightarrow{\text{6 N HCl, 110°C, 24 h}} \text{amino acids}$$

8. Edman Degradation (Section 26-5)

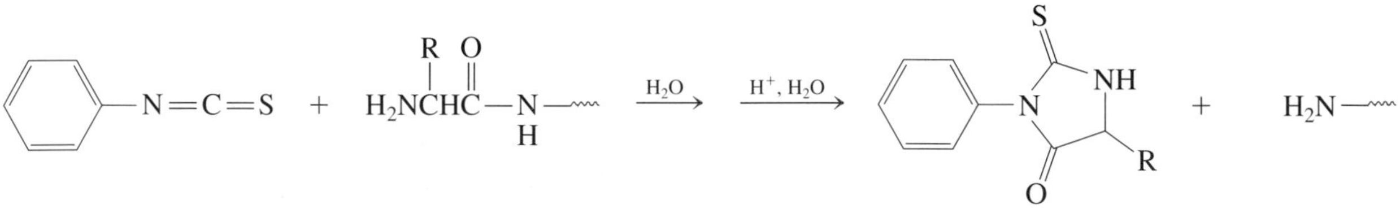

Phenylthiohydantoin   Lower polypeptide

Preparation of Polypeptides

9. Protecting Groups (Section 26-6)

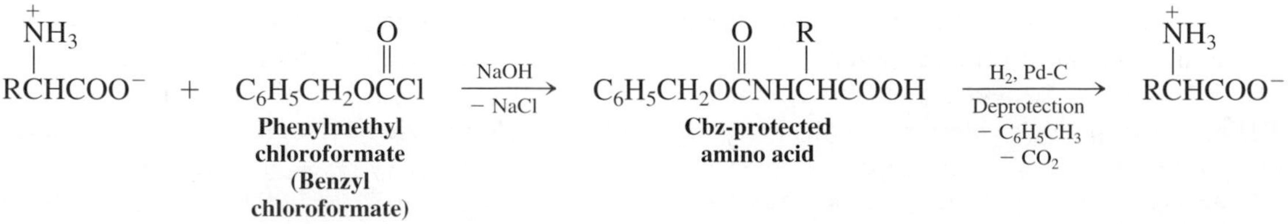

Phenylmethyl
chloroformate
(Benzyl
chloroformate)

Cbz-protected
amino acid

$$\underset{\substack{\text{RCHCOO}^- \\ \mid \\ \overset{+}{\text{NH}_3}}}{} \quad + \quad \underset{\substack{\text{Bis(1,1-Dimethylethyl)} \\ \text{dicarbonate} \\ \text{(Di-}\textit{tert}\text{-butyl dicarbonate)}}}{\text{(CH}_3\text{)}_3\text{COCOCOC(CH}_3\text{)}_3} \quad \xrightarrow{\text{(CH}_3\text{CH}_2\text{)}_3\text{N}} \quad \underset{\substack{\text{Boc-protected} \\ \text{amino acid}}}{\text{(CH}_3\text{)}_3\text{COCNHCHCOOH}} \quad \xrightarrow[\substack{\text{Deprotection} \\ -\text{CO}_2 \\ -\text{CH}_2=\text{C(CH}_3\text{)}_2}]{\text{H}^+, \text{H}_2\text{O}} \quad \underset{\substack{\mid \\ \overset{+}{\text{NH}_3}}}{\text{RCHCOO}^-}$$

10. Peptide-Bond Formation with Dicyclohexylcarbodiimide (Section 26-6)

$$\text{Cbz-Gly} \quad + \quad \text{Ala-OCH}_2\text{C}_6\text{H}_5 \quad + \quad \underset{\text{DCC}}{\text{C}_6\text{H}_{11}\text{N}=\text{C}=\text{NC}_6\text{H}_{11}} \quad \longrightarrow$$

$$\text{Cbz-Gly-Ala-OCH}_2\text{C}_6\text{H}_5 \quad + \quad \overset{\text{O}}{\overset{\|}{\text{C}_6\text{H}_{11}\text{NHCNHC}_6\text{H}_{11}}}$$

11. Merrifield Solid-Phase Synthesis (Section 26-7)

$$\text{(P)} \xrightarrow[\substack{-\text{CH}_3\text{CH}_2\text{OH}}]{\substack{\text{ClCH}_2\text{OCH}_2\text{CH}_3, \\ \text{SnCl}_4}} \text{(P)}-\text{CH}_2\text{Cl} \xrightarrow[\substack{2.\ \text{H}^+, \text{H}_2\text{O}}]{\substack{1.\ \text{(CH}_3\text{)}_3\text{COCNHCHCOO}^-}} \text{(P)}-\text{CH}_2\text{OC}-\text{CHNH}_2 \xrightarrow[\substack{2.\ \text{H}^+, \text{H}_2\text{O}}]{\substack{1.\ \text{(CH}_3\text{)}_3\text{COCNHCHCOOH}, \\ \text{DCC}}}$$

(P) = polystyrene

$$\text{(P)}-\text{CH}_2\text{OC}-\text{CHNHC}-\text{CHNH}_2 \xrightarrow{\text{HF}} \text{(P)}-\text{CH}_2\text{F} \quad + \quad \overset{+}{\text{H}_3}\text{NCHCNHCHCOO}^-$$

IMPORTANT CONCEPTS

1. **Polypeptides** are poly(amino acids) linked by **amide bonds.** Most natural polypeptides are made from only 19 different L-amino acids and glycine, all of which have common names and three- and one-letter abbreviations.

2. Amino acids are **amphoteric;** they can be protonated and deprotonated.

3. **Enantiomerically pure** amino acids can be made by classical fractional crystallization of diastereomeric derivatives or by enantioselective reactions of appropriate achiral precursors.

4. The **structures** of **polypeptides** are varied; they can be linear, cyclic, disulfide bridged, pleated sheet, α helical or superhelical, or disordered, depending on size, composition, hydrogen bonding, and electrostatic and London forces.

5. **Amino** and **nucleic acids** are **separated** mainly by virtue of their size- or charge-based differences in ability to bind to solid supports.

6. **Polypeptide sequencing** entails a combination of selective chain cleavage and amino acid analysis of the resulting shorter polypeptide fragments.

7. **Polypeptide synthesis** requires end-protected amino acids that are coupled by dicyclohexylcarbodiimide. The product can be selectively deprotected at either end to allow for further extension of the chain. The use of **solid supports,** as in the Merrifield synthesis, can be automated.

8. The proteins myoglobin and hemoglobin are polypeptides in which the amino acid chain envelops the active site, **heme.** The heme contains an iron atom that reversibly binds oxygen, allowing for oxygen uptake, transport, and delivery.

9. The **nucleic acids** are biological polymers made of phosphate-linked base-bearing sugars. Only four different bases and one sugar are used for DNA and RNA. Because the base pairs adenine–thymine, guanine–cytosine, and adenine–uracil pair up by particularly favorable **hydrogen bonding,** a nucleic acid can adopt a dimeric helical structure containing **complementary base sequences.** In DNA, this arrangement unwinds and functions as a template during **DNA replication** and **RNA synthesis.** In **protein synthesis,** each amino acid is specified by a set of three consecutive RNA bases, called a **codon.** Thus, the base sequence (**genetic code**) in

a strand of RNA translates into a specific amino acid sequence in a protein.

10. **DNA sequencing** relies on restriction enzymes, radioactive labeling, and specific chemical cleavage reactions to small fragments, analyzed by electrophoresis.

11. **DNA synthesis** employs silica as a support on which the growing oligonucleotide sequence is built up with the help of base, alcohol, and phosphite-phosphate **protecting groups.**

12. The **polymerase chain reaction (PCR)** makes multiple copies of DNA.

PROBLEMS

22. Draw stereochemically correct structural formulas for isoleucine and threonine (Table 26-1). What is a systematic name for threonine?

23. The abbreviation *allo* means *diastereomer* in amino acid terms. Draw allo-L-isoleucine and give it a systematic name.

24. Draw the structure that each of the following amino acids would have in aqueous solution at the indicated pH values. **(a)** Alanine at pH = 1, 7, and 12; **(b)** serine at pH = 1, 7, and 12; **(c)** lysine at pH = 1, 7, 9.5, and 12; **(d)** histidine at pH = 1, 5, 7, and 12; **(e)** cysteine at pH = 1, 7, 9, and 12; **(f)** aspartic acid at pH = 1, 3, 7, and 12; **(g)** arginine at pH = 1, 7, 12, and 14; **(h)** tyrosine at pH = 1, 7, 9.5, and 12.

25. Group the amino acids in Problem 24 according to whether they are **(a)** positively charged, **(b)** neutral, or **(c)** negatively charged at pH = 7.

26. Using either one of the methods in Section 26-2 or a route of your own devising, propose a reasonable synthesis of each of the following amino acids in racemic form. **(a)** Val; **(b)** Leu; **(c)** Pro; **(d)** Thr; **(e)** Lys.

27. **(a)** Illustrate the Strecker synthesis of phenylalanine. Is the product chiral? Does it exhibit optical activity? **(b)** It has been found that replacement of NH_3 by an optically active amine in the Strecker synthesis of phenylalanine leads to an excess of one enantiomer of the product. Assign the R or S configuration to each stereocenter in the following structures, and explain why the use of a chiral amine causes preferential formation of one stereoisomer of the final product.

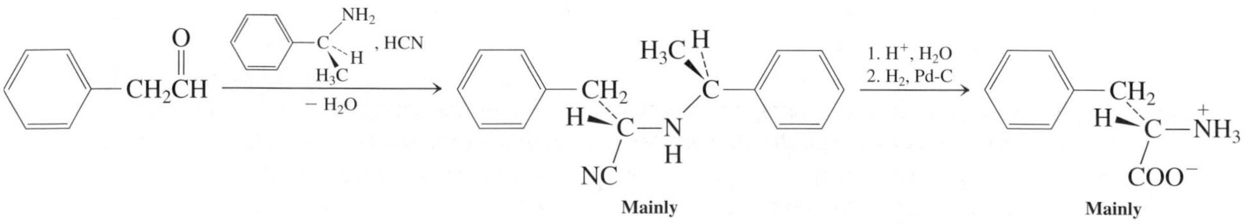

28. The antibacterial agent in garlic, allicin (Chemical Highlight 9-4, Problem 61 in Chapter 9), is synthesized from the unusual amino acid alliin by the action of the enzyme allinase. Because allinase is an extracellular enzyme, this process takes place only when garlic cells are crushed. Propose a reasonable

synthesis for the amino acid alliin. (**Hint:** Begin by designing a synthesis of an amino acid from Table 26-1 that is structurally related to alliin.)

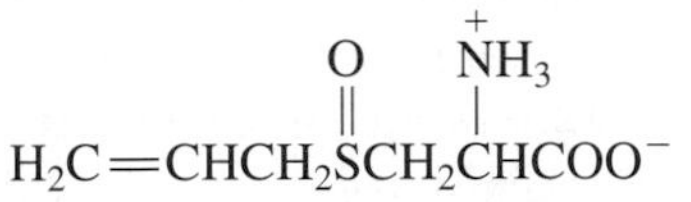

$$H_2C=CHCH_2\overset{\displaystyle O}{\underset{\displaystyle \|}{S}}CH_2\overset{\displaystyle \overset{+}{N}H_3}{\underset{\displaystyle |}{C}}HCOO^-$$

Alliin

29. Devise a procedure for separating a mixture of the four stereoisomers of isoleucine into its four components: $(+)$-isoleucine, $(-)$-isoleucine, $(+)$-alloisoleucine, and $(-)$-alloisoleucine (Problem 23). (Note: Alloisoleucine is much more soluble in 80% ethanol at all temperatures than is isoleucine.)

30. Identify each of the following structures as a dipeptide, tripeptide, and so forth, and point out all the peptide bonds.

(a) 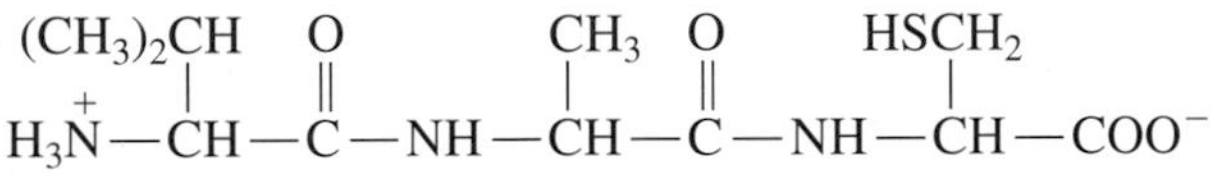 $\overset{(CH_3)_2CH}{H_3\overset{+}{N}-CH-\overset{O}{\overset{\|}{C}}-NH-\overset{CH_3}{CH}-\overset{O}{\overset{\|}{C}}-NH-\overset{HSCH_2}{CH}-COO^-}$

(b) $\overset{HOCH_2}{H_3\overset{+}{N}-CH-\overset{O}{\overset{\|}{C}}-NH-\overset{CH_2COO^-}{CH}-COO^-}$

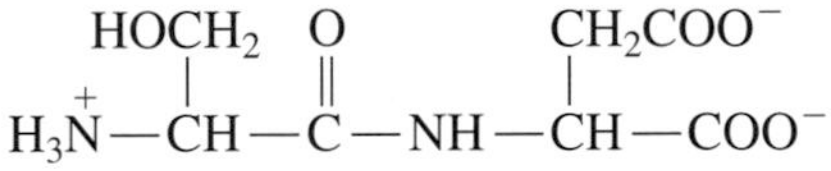

(c) $H_3\overset{+}{N}-CH-\overset{O}{\overset{\|}{C}}-NH-CH-\overset{O}{\overset{\|}{C}}-N-CH-\overset{O}{\overset{\|}{C}}-NH-CH-COO^-$

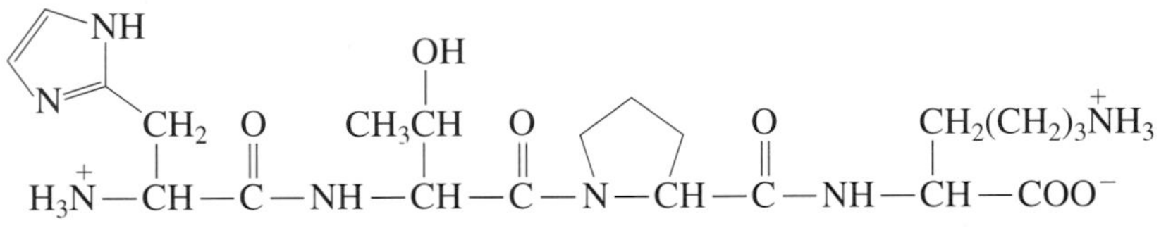

(d) 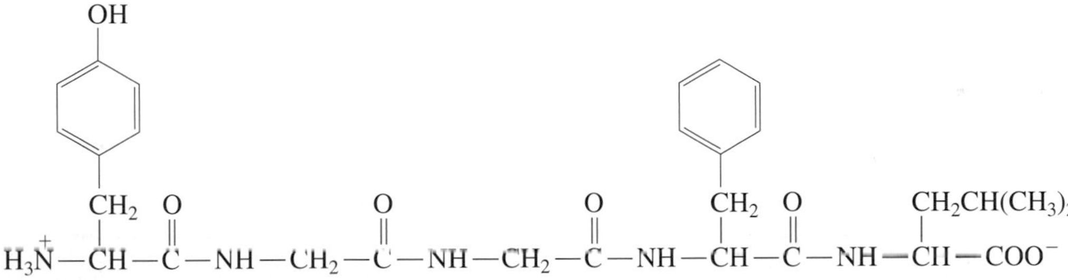 $H_3\overset{+}{N}-CH-\overset{O}{\overset{\|}{C}}-NH-CH_2-\overset{O}{\overset{\|}{C}}-NH-CH_2-\overset{O}{\overset{\|}{C}}-NH-CH-\overset{O}{\overset{\|}{C}}-NH-CH-COO^-$

31. Using the three-letter abbreviations for amino acids, write the peptide structures in Problem 30 in short notation.

32. Indicate which of the amino acids in Problem 24 and the peptides in Problem 30 would migrate in an electrophoresis apparatus at pH = 7
(**a**) toward the anode or (**b**) toward the cathode.

33. Silk is composed of β sheets whose polypeptide chains consist of the repeating sequence Gly-Ser-Gly-Ala-Gly-Ala. What characteristics of amino acid side chains appear to favor the β-sheet configuration? Do the illustrations of β-sheet structures (Figure 26-3) suggest an explanation for this preference?

34. Identify as many stretches of α helix as you can in the structure of myoglobin (Figure 26-8C). Prolines are located in myoglobin at positions 37, 88, 100, and 120. How does each of these prolines affect the tertiary structure of the molecule?

35. Of the 153 amino acids in myoglobin, 78 contain polar side chains (i.e., Arg, Asn, Asp, Gln, Glu, His, Lys, Ser, Thr, Trp, and Tyr). When myoglobin adopts its natural folded conformation, 76 of these 78 polar side chains (all but those of two histidines) project outward from its surface. Meanwhile, in addition to the two histidines, the interior of myoglobin contains only Gly, Val, Leu, Ala, Ile, Phe, Pro, and Met. Explain.

36. Explain the following three observations. **(a)** Silk, like most polypeptides with sheet structures, is water insoluble. **(b)** Globular proteins such as myoglobin generally dissolve readily in water. **(c)** Disruption of the tertiary structure of a globular protein (denaturation) leads to precipitation from aqueous solution.

37. In your own words, outline the procedure that might have been followed by the researchers who determined which amino acids were present in vasopressin (Exercise 26-8).

38. Write the products of a single Edman degradation of the peptides in Problem 30.

39. What would be the outcome of reaction of gramicidin S with phenyl isothiocyanate (Edman degradation)? (**Hint:** With which functional group does this substance react?)

40. The polypeptide bradykinin is a tissue hormone that can function as a potent pain-producing agent. By means of a single treatment with the Edman reagent, the *N*-terminal amino acid in bradykinin is identified as Arg. Incomplete acid hydrolysis of the intact polypeptide causes random cleavage of many bradykinin molecules into an assortment of peptide fragments that includes Arg-Pro-Pro-Gly, Phe-Arg, Ser-Pro-Phe, and Gly-Phe-Ser. Complete hydrolysis followed by amino acid analysis indicates a ratio of 3 Pro, 2 Phe, 2 Arg, and one each of Gly and Ser. Deduce the amino acid sequence of bradykinin.

41. The amino acid sequence of met-enkephalin, a brain peptide with powerful opiate-like biological activity, is Tyr-Gly-Gly-Phe-Met. What would be the products of step-by-step Edman degradation of met-enkephalin?

    The peptide shown in Problem 30(d) is leu-enkephalin, a relative of met-enkephalin with similar properties. How would the results of Edman degradation of leu-enkephalin differ from those of met-enkephalin?

42. Secreted by the pituitary gland, corticotropin is a hormone that stimulates the adrenal cortex. Determine its primary structure from the following information. (i) Hydrolysis by chymotrypsin produces six peptides: Arg-Trp, Ser-Tyr, Pro-Leu-Glu-Phe, Ser-Met-Glu-His-Phe, Pro-Asp-Ala-Gly-Glu-Asp-Gln-Ser-Ala-Glu-Ala-Phe, and Gly-Lys-Pro-Val-Gly-Lys-Lys-Arg-Arg-Pro-Val-Lys-Val-Tyr. (ii) Hydrolysis by trypsin produces free lysine, free arginine, and the following five peptides: Trp-Gly-Lys, Pro-Val-Lys, Pro-Val-Gly-Lys, Ser-Tyr-Ser-Met-Glu-His-Phe-Arg, and Val-Tyr-Pro-Asp-Ala-Gly-Glu-Asp-Gln-Ser-Ala-Glu-Ala-Phe-Pro-Leu-Glu-Phe.

43. Glucagon is a pancreatic hormone whose function opposes that of insulin: It causes an increase in glucose levels in the blood. It consists of a polypeptide chain with 29 amino acid units. Treatment of glucagon with thermolysin pro-

duces four fragments, including the tripeptide Val-Gln-Tyr, the tetrapeptide Leu-Met-Asn-Thr, a 9-amino-acid peptide A, and a 13-amino-acid peptide B. Carrying out a single Edman cycle on peptide A produces the phenylthio-hydantoin of leucine, and similar treatment of peptide B furnishes the phenyl-hydantoin of histidine.

Peptide A is not cleaved by chymotrypsin, but clostripain breaks it down into Leu-Asp-Ser-Arg, Ala-Gln-Asp-Phe, and a free Arg. Chymotrypsin cleaves peptide B into Ser-Lys-Tyr, Thr-Ser-Asp-Tyr, and His-Ser-Gln-Gly-Thr-Phe. **(a)** At this stage, how much do you know for certain about the structure of glucagon? What uncertainties still remain? **(b)** One of the products of trypsin hydrolysis of the intact glucagon molecule is the peptide Tyr-Leu-Asp-Ser-Arg. Does this help? **(c)** One product of chymotrypsin hydrolysis of the intact hormone is Leu-Met-Asn-Thr, the same tetrapeptide released by thermolysin. Now can you piece together the entire molecule?

44. Propose a synthesis of leu-enkephalin [see Problem 30(d)] from the component amino acids.

45. The following molecule is thyrotropin-releasing hormone (TRH). It is secreted by the hypothalamus, causing the release of thyrotropin from the pituitary gland, which, in turn, stimulates the thyroid gland. The thyroid produces hormones, such as thyroxine, that control metabolism in general.

The initial isolation of TRH required the processing of 4 tons of hypothalamic tissue, from which 1 mg of the hormone was obtained. Needless to say, it is a bit more convenient to synthesize TRH in the laboratory than to extract it from natural sources. Devise a synthesis of TRH from Glu, His, and Pro. Note that pyroglutamic acid is just the lactam of Glu and may be readily obtained by heating Glu to between 135° and 140°C.

46. (a) The structures illustrated for the four DNA bases (Section 26-9) represent only the most stable tautomers. Draw one or more alternative tautomers for each of these heterocycles (review tautomerism, Sections 13-8 and 18-2). **(b)** In certain cases, the presence of a small amount of one of these less stable tautomers can lead to an error in DNA replication or mRNA synthesis due to faulty base pairing. One example is the imine tautomer of adenine, which pairs with cytosine instead of thymine. Draw a possible structure for this hydrogen-bonded base pair (see Figure 26-11). **(c)** Using Table 26-3, derive a possible nucleic acid sequence for an mRNA that would code for the five amino acids in met-enkephalin (see Problem 42). If the mispairing described in (b) were at the first possible position in the synthesis of this mRNA sequence, what would be the consequence in the amino acid sequence of the peptide? (Ignore the initiation codon.)

47. Factor VIII is one of the proteins participating in the formation of blood clots. A defect in the gene whose DNA sequence codes for Factor VIII is responsible for classic hemophilia. Factor VIII contains 2332 amino acids. How many nucleotides are needed to code for its synthesis?

48. Hydroxyproline (Hyp), like many other amino acids that are not "officially" classified as essential, is nonetheless a very necessary biological substance. It constitutes about 14% of the amino acid content of the protein collagen. Collagen is the main constituent of skin and connective tissue. It is also present, together with inorganic substances, in nails, bones, and teeth. **(a)** The systematic name for hydroxyproline is (2S,4R)-4-hydroxyazacyclopentane-2-carboxylic acid. Draw a stereochemically correct structural formula for this amino acid. **(b)** Hyp is synthesized in the body in peptide-bound form from peptide-bound proline and O_2, in an enzyme-catalyzed process that requires vitamin C. In the absence of vitamin C, only a defective, Hyp-deficient collagen can be produced. Vitamin C deficiency causes scurvy, a condition characterized by bleeding of the skin and swollen, bleeding gums.

In the following reaction sequence, an efficient laboratory synthesis of hydroxyproline, fill in the necessary reagents (i) and (ii), and formulate detailed mechanisms for the steps marked with an asterisk.

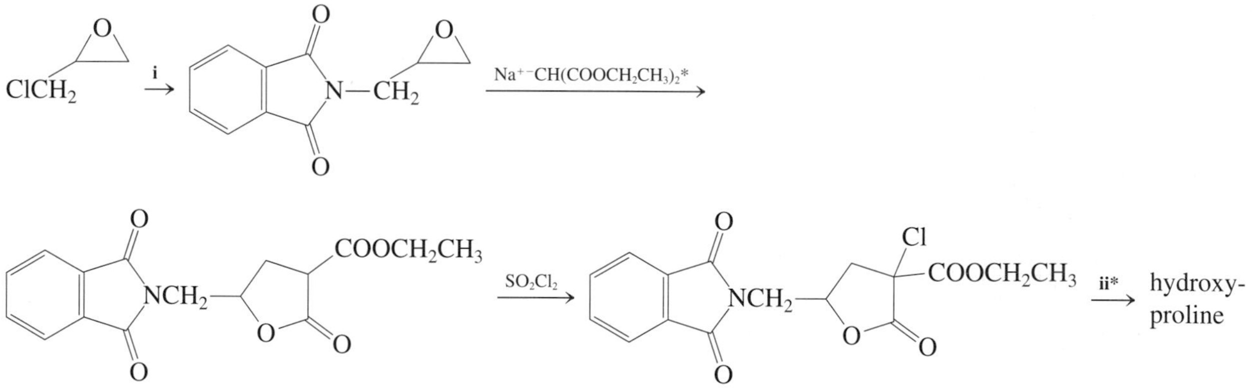

(c) Gelatin, which is partly hydrolyzed collagen, is rich in hydroxyproline and, as a result, is often touted as a remedy for split or brittle nails. Like most proteins, however, gelatin is almost completely broken down into individual amino acids in the stomach and small intestine before absorption. Is the free hydroxyproline thus introduced into the bloodstream of any use to the body in the synthesis of collagen? (**Hint:** Does Table 26-3 list a three-base code for hydroxyproline?)

49. Sickle-cell anemia is an often fatal genetic condition caused by a single error in the DNA gene that codes for the β chain of hemoglobin. The correct nucleic acid sequence (read from the mRNA template) begins with AUGGUGCACCU-GACUCCUGAGGAGAAG . . . , and so forth. **(a)** Translate this into the corresponding amino acid sequence of the protein. **(b)** The mutation that gives rise to the sickle-cell condition is replacement of the boldface A in the preceding sequence by U. What is the consequence of this error in the corresponding amino acid sequence? **(c)** This amino acid substitution alters the properties of the hemoglobin molecule—in particular, its polarity and its shape. Suggest reasons for both these effects. (Refer to Table 26-1 for amino acid structures and to Figure 26-8C for the structure of myoglobin, which is similar to that of hemoglobin. Note the location of the amino acid substitution in the tertiary structure of the protein.)

Team Problem

50. Amino acids can be used as enantiomerically pure starting materials in organic synthesis. Scheme I depicts the first steps in the synthesis of a reagent employed in the preparation of enantiomerically pure β-amino acids, such as that occurring in the side chain of taxol (Section 4-7). Scheme II features an ester of the same amino acid for the preparation of an unusual heterocyclic dipeptide used in the study of polypeptide conformations.

Scheme I: Synthesis of an enantiomerically pure reagent

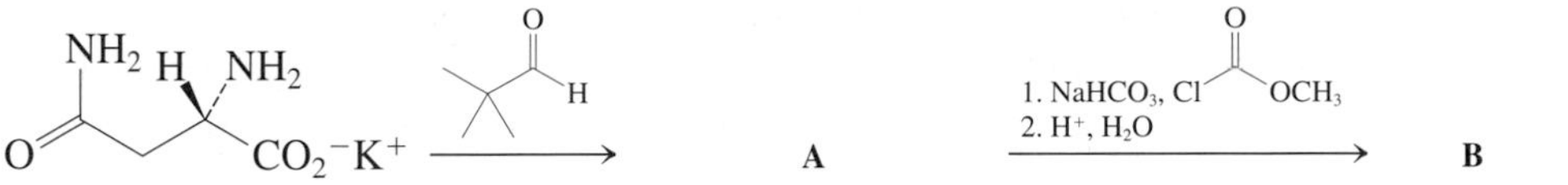

Potassium salt of asparagine

A

$C_9H_{15}N_2O_3^-$ K^+

Six-membered nitrogen heterocyle

1. NaHCO$_3$, Cl$\diagup$OCH$_3$
2. H$^+$, H$_2$O

B

$C_{11}H_{18}N_2O_5$

Consider the following questions:

1. There are two diastereomers of A formed in a 90:10 ratio. The major isomer is the one that can attain the most stable chair conformation. Are the two substituents on the ring cis or trans to each other? Label their positions as equatorial or axial.

2. Which nitrogen is nucleophilic and yields the carbamic ester (Section 20-6) B?

Scheme II: Synthesis of an unusual heterocyclic dipeptide

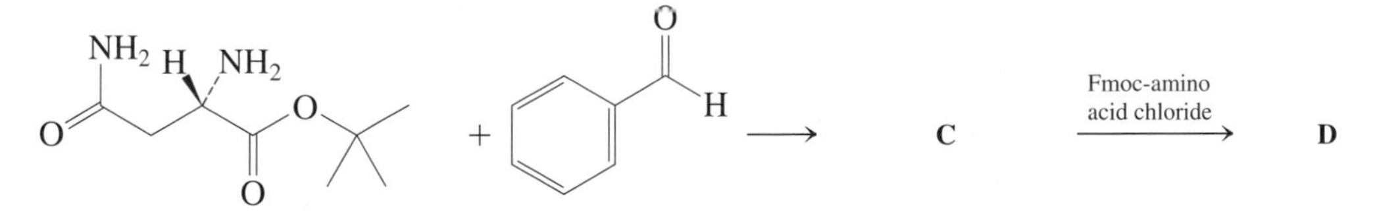

1,1-Dimethylethyl (tert-butyl) ester of asparagine

C

$C_{15}H_{20}N_2O_3$

Acyclic compound

Fmoc-amino acid chloride

D

Heterocyclic dipeptide

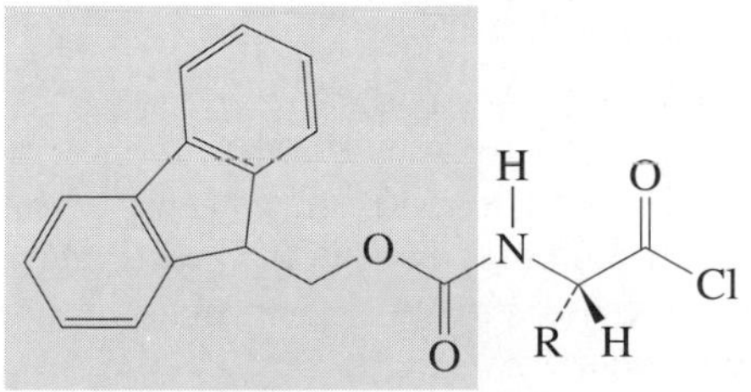

Fluorenylmethyloxycarbonyl(Fmoc)-amino acid chloride

The Fmoc protecting group (highlighted in the box) is used instead of Cbz or Boc with which you are familiar, because the amino *acid chloride* is necessary to make the new amide bond. Neither the Cbz nor the Boc group is stable under these conditions.

Consider the following questions:

1. What functional group is generated in C?
2. Where is the peptide bond in D? Circle it.

Reconvene to discuss the answers to the questions posed in each scheme and the structures you proposed for A–D.

Preprofessional Problems

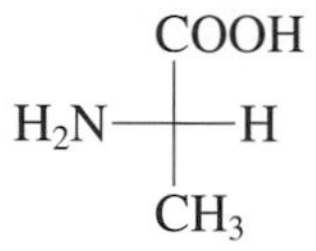

COOH

H_2N——H

CH_3

A

51. Structure A (shown in the margin) is that of a naturally occurring α-amino acid. Select its name from the following list. **(a)** Glycine; **(b)** alanine; **(c)** tyrosine; **(d)** cysteine.

52. The *primary structure* of a protein refers to: **(a)** cross-links with disulfide bonds; **(b)** presence of an α-helix; **(c)** the α-amino acid sequence in the polypeptide chain; **(d)** the orientation of the side chains in three-dimensional space.

53. Which one of the following five structures is a zwitterion?

(a) $^-O_2CCH_2\overset{O}{\overset{\|}{C}}NH_2$

(b) $^-O_2CCH_2CH_2CO_2^-$

(c) $H_3\overset{+}{N}CH_2CO_2^-$

(d) $CH_3(CH_2)_{16}CO_2^-\ K^+$

(e) $\left[H-\overset{O}{\overset{\diagup}{\underset{\ddot{\underset{..}{O}}:^-}{C}}} \longleftrightarrow H-\overset{\ddot{O}:^-}{\overset{\diagup}{\underset{O}{C}}} \right]$

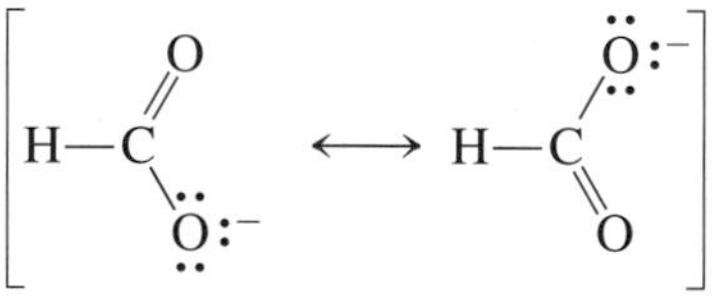

54. When an α-amino acid is dissolved in water and the pH of the solution adjusted to 12, which of the following species is predominant?

(a) $\overset{O}{\overset{\|}{RCHCOH}}$ | NH_2

(b) $\overset{O}{\overset{\|}{RCHCOH}}$ | $^+NH_3$

(c) $\overset{O}{\overset{\|}{RCHCO^-}}$ | $^+NH_3$

(d) $\overset{O}{\overset{\|}{RCHCO^-}}$ | NH_2

55. How many stereocenters are present in the small, naturally occurring protein glycylalanylalanine? **(a)** Zero; **(b)** one; **(c)** two; **(d)** three.

Answers to Exercises

CHAPTER 1

1-1

(a)

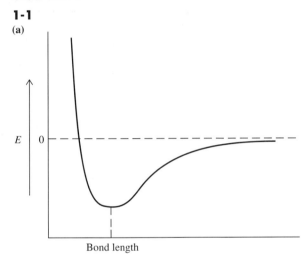

(b) Self-explanatory

1-2

Li^+ :Br:⁻ $[Na]_2^+$:Ö:²⁻ $Be^{2+}[:\ddot{F}:]_2^-$

$Al^{3+}[:\ddot{Cl}:]_3^-$ $Mg^{2+}:\ddot{S}:^{2-}$

1-3

:F:F: :F:C:F: with :F: above and :F: below   :Cl: above H:C:H and :Cl: below   H:P:H with H below

:Br:I: ⁻:Ö:H   H:N:H (with bar, H below)   ⁻:C:H with H above and H below

1-4

H→O←H   SC↔O   S↔O   I↔Br   H→C←H (with H above ↓ and H below ↑)

Cl←C→Cl (H above ↓, Cl below ↓)   H↔C→Cl (H above ↓, Cl below ↓)   H→C←H (H above ↓, Cl below ↓)

1-5

You can view NH_3 as being isoelectronic with H_3C^-, H_2O with H_2C^{2-}. Electron repulsion by the free electron pairs causes the bonding electrons to "bend away," giving rise to the respective pyramidal and bent structures.

1-6

H:Ï: H:C:C:C:H (with H H H above and H H H below)   H:C:Ö:H (with H above and H below)   H:S:S:H

Ö::Si::Ö   Ö::Ö   S::C::S

1-7

S::Ö   :F:Ö:F: H:Ö:Cl:Ö: (with + and − charges)   :F:B:N:H (with :F:H above, :F:H below, + and −)

H:C:Ö:H (with H below, + charge)   C::Ö with :Cl: above and :Cl: below   ⁻:C:::N: ⁻:C:::C:⁻

1-8

The geometry should be close to trigonal (counting the lone electron pair), with equal N–O bond lengths and one-half of a negative charge on each oxygen atom.

$$\left[:\ddot{O}\diagup\diagdown\ddot{O}:^- \quad \longleftrightarrow \quad {}^-:\ddot{O}\diagup\diagdown\ddot{O}: \right]$$
(with N below)

1-9

(a) $\left[{}^-:C\equiv\overset{+}{N}-\ddot{O}:^- \quad \longleftrightarrow \quad {}^{2-}\ddot{C}=\overset{+}{N}=\ddot{O} \right]$

The left-hand structure is preferred because the charges are more evenly distributed and the negative charge resides on the relatively electronegative oxygen.

(b) $\left[{}^-\ddot{N}=\ddot{O} \quad \longleftrightarrow \quad \ddot{N}-\ddot{O}:^- \right]$

The left-hand structure is preferred because the right-hand structure has no octet on nitrogen.

1-10

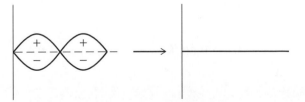

1-11

Draw according to the following electronic configurations:
S $(1s)^2(2s)^2(2p)^6(3s)^2(3p)^4$; P $(1s)^2(2s)^2(2p)^6(3s)^2(3p)^3$

1-12

The molecular-orbital picture is like that shown for the bonding in H_2 (Figure 1-11) and the energy splitting like that shown for He_2 in Figure 1-12B. However, the presence of only one antibonding electron but two bonding electrons results in net bonding.

He $\quad$ He$^+$

He$_2{}^+$

1-13

$CH_3{}^+$ or H:$\overset{+}{\underset{H}{C}}$:H $\qquad$ $CH_3{}^-$ or H:$\overset{\cdot\cdot}{\underset{H}{C}}$:H

No octet

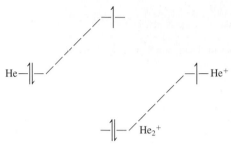

Empty 2*p* orbital
σ bond
σ bond
H
C
H
H
σ bond
sp^2 orbital $\quad$ 1*s* orbital

Trigonal, sp^2 hybridized, electron deficient like BH$_3$

Lone electron pair
1*s*
C
H
H
H
sp^3

Tetrahedral, sp^3 hybridized, closed shell

1-14

H H H H
H—C—C—C—C—H
H H H H

Butane

H H—C—H
H—C————C—H
H H—C—H
H

Isobutane

1-15

Self-explanatory. Note that the molecules are flexible and can adopt a variety of arrangements in space.

1-16

$CH_3CH_2CH_2CH_3$ $\qquad$ $CH_3\overset{CH_3}{\underset{}{CHCH_3}}$

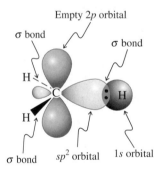 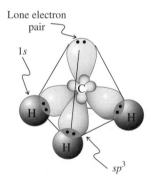

1-17

H H H H
H—C—C—C—H
H H H H

H CH$_3$
H—C—C—H
H H H H

CHAPTER 2

2-1

(a)

(b) Higher homologs:

$CH_3\overset{CH_3}{\underset{}{CH}}\!-\!\overset{CH_3}{\underset{}{CH}}CH_3$

$CH_3CH_2CH_2\overset{CH_3}{\underset{CH_3}{CH}}$ $\quad$ $CH_3CH_2\overset{CH_2}{\underset{CH_3}{CH}}$ $\quad$ $CH_3CH_2\overset{CH_3}{\underset{CH_3}{CCH_3}}$

Lower homologs:

$CH_3\overset{CH_3}{\underset{CH_3}{CH}}$ $\qquad$ $CH_3CH_2CH_2CH_3$

2-2

$CH_3\overset{CH_3}{\underset{}{CHCH_2CH_2CH_3}}$ $\qquad$ $CH_3\overset{CH_3}{\underset{CH_3}{CCH_3}}$

Isohexane $\qquad$ **Neopentane**

2-3

Sec
CH$_3$
$CH_3CCH_2CH_2CH_3$
Prim
H
Tert

2-4

Self-explanatory

2-5

H$_3$C H

H$_3$C H
H$_3$C H

2-Methylbutane $\qquad$ **2,3-Dimethylbutane**

2-6

In this example (graphed below), the energy difference between the two staggered conformers turns out to be quite small.

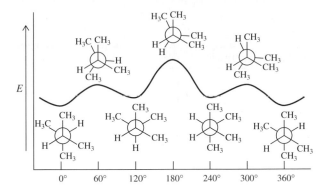

2-7

$0.9 = -1.36 \log K$ kcal mol^{-1} at 25°C

$K = 0.219$; $anti:gauche = 82:18$

$0.9 = -RT \ln K = -1.71 \log K$ kcal mol^{-1} at 100°C

$K = 0.297$; $anti:gauche = 77:23$

2-8

$\Delta G° = \Delta H° - T\Delta S°$

$\quad = 22.4$ kcal mol^{-1} − (298 deg ×

$\quad\quad\quad\quad$ 33.3 cal deg^{-1} mol^{-1})

$\quad = 12.5$ kcal mol^{-1}

The reaction is unfavorable at 25°C. At higher temperatures, $\Delta G°$ is less positive, eventually becoming negative. The crossover point, $\Delta G° = 0$, is reached at 400°C, when $\Delta H° = T\Delta S°$.

2-9

$\Delta G° = \Delta H° - T\Delta S°$

$\quad = -15.5$ kcal mol^{-1} − (298 deg ×

$\quad\quad\quad\quad$ −31.3 cal deg^{-1} mol^{-1})

$\quad = -6.17$ kcal mol^{-1}

The entropy is negative because two molecules are converted into one in this reaction.

2-10

After 50% conversion, only 1/2 of the molar concentration of starting materials is present. Hence, for first order: rate = k[A]. At 50% conversion, rate will be 1/2 initial rate. For second order: rate = k[A][B]. At 50% conversion, rate = k(1/2)[A](1/2)[B] = 1/4 initial rate.

2-11

(a) +6.17 kcal mol^{-1}

(b) $\Delta G° = 15.5 - (0.773 \times 31.3)$

$\quad\quad = -8.69$ kcal mol^{-1}

Hence, the dissociation equilibrium lies on the side of ethene and HCl at this high temperature, where the entropy factor overrides the $\Delta H°$ term.

2-12

$k = 10^{14}e^{-58.4/1.53} = 3.03 \times 10^{-3}$ s^{-1}

2-13

(a) HSO_3^- (b) ClO_3^- (c) HS^- (d) $(CH_3)_2O$

(e) SO_4^{2-}

2-14

(a) $(CH_3)_2NH$ (b) HS^- (c) $^+NH_4$ (d) $(CH_3)_2C\overset{+}{=}OH$

(e) CF_3CH_2OH

2-15

Phosphorous acid is stronger. It has the smaller pK_a value, which corresponds to a larger acid dissociation constant, K_a. $K_a(HNO_2) = 10^{-3.3}$; $K_a(H_3PO_3) = 10^{-1.3}$.

2-16

The Lewis structure for acetic acid is

$$H_3C-\overset{\displaystyle :O:}{\underset{\displaystyle}{C}}-\overset{..}{O}H$$

Protonation of the oxygen at the end of the double bond gives a structure with three contributing resonance forms:

[three resonance structures]

Protonation of the oxygen atom of the OH group gives a structure with only two contributing forms, the second of which is poor because of the two adjacent positively charged atoms:

[two resonance structures]

Protonation at the double-bonded oxygen is preferred.

CHAPTER 3

3-1

A simple answer would be that the strength of a bond depends not only on the size and energies of the orbitals, but also on Coulombic contributions. Thus, in the procession from N to O to F, nuclear charge increases in the core, allowing for more nuclear-core electronic attraction in binding to CH_3. The bond in question becomes more and more polar along the series.

3-2

First: $CH_3\text{-}C(CH_3)_3$ $\quad DH° = 84$ kcal mol^{-1}

Second: $CH_3\text{-}CH_3$ $\quad DH° = 90$ kcal mol^{-1}

3-3

$CH_3CH_3 + Cl_2 \xrightarrow{h\nu} CH_3CH_2Cl + HCl$

$\Delta H° = 98 + 58 - 80 - 103 = -27$ kcal mol^{-1}

Mechanism

Initiation

$Cl_2 \xrightarrow{h\nu} 2:\overset{..}{\underset{..}{Cl}}\cdot \quad \Delta H° = +58$ kcal mol^{-1}

Propagation

$$CH_3CH_3 \quad + \quad :\overset{..}{\underset{..}{Cl}}\cdot \quad \rightarrow \quad CH_3CH_2\cdot \quad + \quad H\overset{..}{\underset{..}{Cl}}:$$
$$\Delta H° = -5 \text{ kcal mol}^{-1}$$

$$CH_3CH_2\cdot \quad + \quad Cl_2 \quad \rightarrow \quad CH_3CH_2\overset{..}{\underset{..}{Cl}}: \quad + \quad :\overset{..}{\underset{..}{Cl}}\cdot$$
$$\Delta H° = -22 \text{ kcal mol}^{-1}$$

Termination

$$:\overset{..}{\underset{..}{Cl}}\cdot \quad + \quad :\overset{..}{\underset{..}{Cl}}\cdot \quad \rightarrow \quad Cl_2 \qquad \Delta H° = -58 \text{ kcal mol}^{-1}$$

$$CH_3CH_2\cdot \quad + \quad :\overset{..}{\underset{..}{Cl}}\cdot \quad \rightarrow \quad CH_3CH_2\overset{..}{\underset{..}{Cl}}:$$
$$\Delta H° = -80 \text{ kcal mol}^{-1}$$

$$CH_3CH_2\cdot + \cdot CH_2CH_3 \quad \longrightarrow CH_3CH_2CH_2CH_3$$
$$\Delta H° = -82 \text{ kcal mol}^{-1}$$

3-4

$$CH_4 + Cl_2 + Br_2 \quad \longrightarrow \quad CH_3Cl + CH_4 + Cl_2 + Br_2 + HCl$$
Cl_2 is more reactive than Br_2.

3-5

$$CH_3CH_2CH_2CH_3 \quad + \quad Cl_2 \quad \overset{h\nu}{\longrightarrow}$$
$$CH_3CH_2CH_2CH_2Cl \quad + \quad CH_3CH_2\overset{\underset{|}{Cl}}{C}HCH_3 \quad + \quad HCl$$

The ratio of primary to secondary product is calculated by multiplying the number of respective hydrogens in the starting material by their relative reactivity:

$$(6 \times 1):(4 \times 4) = 6:16 = 3:8$$

In other words, 2-chlorobutane : 1-chlorobutane = 8 : 3.

3-6

The starting compound has four distinct groups of hydrogens:

Four possible monochlorination products are therefore possible, corresponding to replacement of a hydrogen atom in each of these four groups by chlorine:

The table summarizes the determination of the relative amounts of each product that would be expected to form. In each case, multiply the number of hydrogens in the corresponding group by the re-

activity for the type of hydrogen in the group: primary for groups A and B, secondary for group C, and tertiary for group D. Division of the relative amount of each product by the sum gives the proportion of product, and normalization to a total of 100% gives its percentage yield.

Position (group)	Number of hydrogens	Relative reactivity	Relative yield	Percentage yield
A	3	1	3	10%
B	6	1	6	20%
C	4	4	16	53%
D	1	5	5	17%

3-7

$$CH_3CH_2CH_3$$
$$\uparrow \quad \uparrow$$

Will give mixture, because of competitive primary and secondary chlorination (indicated by arrows).

Only one type of C–H; so 2,2-dimethylpropane should give good selectivity.

Same situation as in 2,2-dimethylpropane.

Will give a bad mixture.

3-8

In this isomerization, a secondary hydrogen and a terminal methyl group in butane switch positions:

Hence,

$$\Delta H° = \text{(sum of the strengths of the bonds broken)}$$
$$\qquad - \text{(sum of the strengths of the bonds made)}$$
$$= (94.5 + 87) - (86 + 98)$$
$$= -2.5 \text{ kcal mol}^{-1}$$

CHAPTER 4

4-1

Aspects of ring strain and conformational analysis are discussed in Sections 4-2 through 4-5.

Note that the cycloalkanes are much less flexible than the straight-chain alkanes and thus have less conformational freedom. Cyclopropane must be flat and all hydrogens eclipsed. The higher cycloalkanes have increasing flexibility, more hydrogens attaining staggered positions and the carbon atoms of the ring eventually being able to adopt *anti* conformations.

4-2

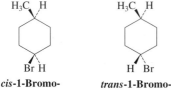

trans-1-Bromo-2-methylcyclohexane **cis-1-Bromo-3-methylcyclohexane** **trans-1-Bromo-3-methylcyclohexane**

cis-1-Bromo-4-methylcyclohexane **trans-1-Bromo-4-methylcyclohexane**

4-3

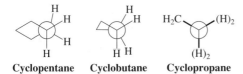

Trans   **Cis**

The cis isomer suffers from steric hindrance and has a larger heat of combustion (by about 1 kcal mol^{-1}).

4-4

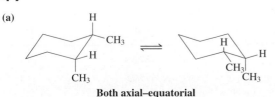

Cyclopentane   **Cyclobutane**   **Cyclopropane**

The respective C–H torsional angles are roughly 40°, 20°, and 0°.

4-5

log $K = -1.7/1.36 = -1.25$
  $K = 10^{-1.25} = 0.056$. Compare $K = 5/95 = 0.053$

4-6

(a) $\Delta G° =$ energy difference between an axial methyl and axial ethyl group: $1.75 - 1.70 =$ about 0.05 kcal mol^{-1}; that is, very small.
(b) Same as (a).
(c) $1.75 + 1.70 = 3.45$ kcal mol^{-1}

4-7

(a)

Both axial–equatorial

(b)

Diequatorial   **Diaxial**

(c)

Diequatorial   **Diaxial**

(d)

Both axial–equatorial

4-8

The substituent values in Table 4-3 are based on an axial substituent suffering from 1,3-diaxial interactions with two hydrogens only. In diaxial *cis*-1,3-dimethylcyclohexane, the *two methyl groups* affect each other in this way; that is, each methyl is proximal to one other axial methyl (and the remaining axial hydrogen), causing an additional 0.3 kcal mol^{-1} in strain.

0.3 kcal mol^{-1}

Similarly, *gauche* effects of neighboring substituents are disregarded when the values in Table 4-3 are applied in an additive manner. For diequatorial *cis*-1,2-dimethylcyclohexane, there is a *gauche* interaction equal to that in *gauche*-butane, 0.9 kcal mol^{-1}, which *destabilizes* this conformer relative to its diaxial counterpart in which the two methyl groups are located *anti* with respect to each other.

0.9 kcal mol^{-1}

4-9

trans-Decalin is fairly rigid. Full chair–chair conformational "flipping" is not possible. In contrast, the axial and equatorial positions in the cis isomer can be interchanged by conformational isomerization of both rings. The barrier to this exchange is small ($E_a = 14$ kcal mol^{-1}). Because one of the appended bonds is always axial, the cis isomer is less stable than the trans isomer by 2 kcal mol^{-1} (as measured by combustion experiments).

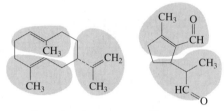

Ring flip in *cis*-decalin

4-10

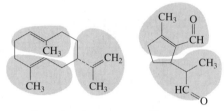

All equatorial

4-11

Sesquiterpene   **Monoterpene**

4-12

Chrysanthemic acid: $C=C$, —COOH, —COOR

Grandisol: $C=C$, —OH

Menthol: —OH

Camphor: $C=O$

β-Cadinene: $C=C$

Taxol: —OH, —O—, aromatic benzene rings, $C=O$, —COOR, —CONHR

CHAPTER 5

5-1

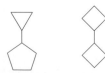

Cyclopropyl-   **Cyclobutyl-**
cyclopentane   **cyclobutane**

Both hydrocarbons have the same molecular formula: C_8H_{14}. Therefore, they are (constitutional) isomers.

5-2

There are several boat and twist-boat forms of methylcyclohexane, some of which are shown:

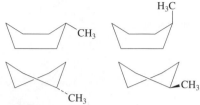

5-3

All are chiral. Note, however, that 2-methylbutadiene (isoprene) itself is achiral. Number of stereocenters: chrysanthemic acid, 2; grandisol, 2; menthol, 3; camphor, 2; β-cadinene, 3; taxol, 11; epiandrosterone, 7; cholesterol, 8; cholic acid, 11; cortisone, 6; testosterone, 6; estradiol, 5; progesterone, 6; norethindrone, 6; ethynylestradiol, 5; RU-486, 5.

5-4

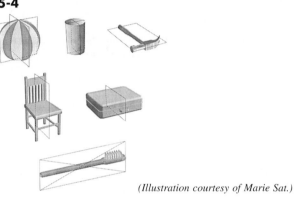

(Illustration courtesy of Marie Sat.)

5-5

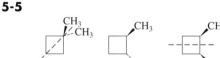

Mirror plane

Achiral   **Chiral**   **Achiral**

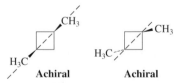

Achiral   **Achiral**

5-6

$$[\alpha] = \frac{6.65}{1 \times 0.1} = 66.5$$

The enantiomer of natural sucrose has $[\alpha] = -66.5$.

5-7

Optical purity (%)	Ratio (+/−)	$[\alpha]_{obs.}$
75	87.5/12.5	+17.3°
50	75/25	+11.6°
25	62.5/37.5	+5.8°

5-8

(a) $-CH_2Br > -CCl_3 > -CH_2CH_3 > -CH_3$

(b) cyclohexyl $> -\overset{\overset{CH_3}{|}}{C}HCH_3 > -CH_2\overset{\overset{CH_3}{|}}{C}HCH_3$

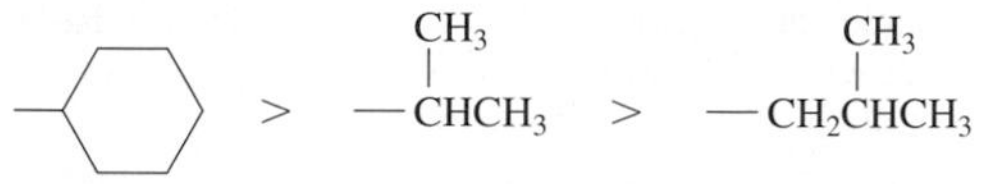

(c) $-C(CH_3)_3 > -\overset{\overset{CH_3}{|}}{C}HCH_2CH_3 > -CH_2\overset{\overset{CH_3}{|}}{C}HCH_3 > -CH_2CH_2CH_2CH_3$

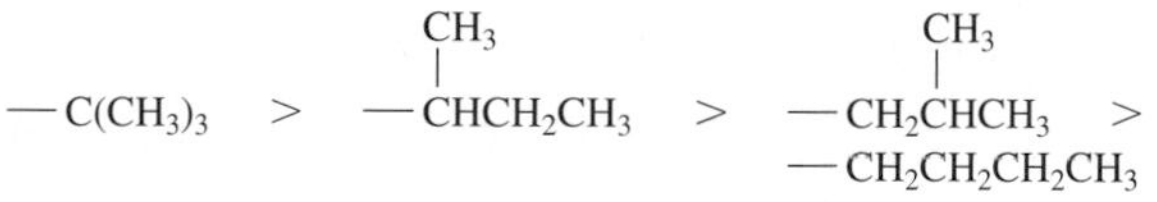

(d) $-\overset{\overset{Br}{|}}{C}HCH_3 > -\overset{\overset{Cl}{|}}{C}HCH_3 > -CH_2CH_2Br > -CH_2CH_3$

5-9

$(-)$-2-Bromobutane: R
$(+)$-2-Bromobutane: S
$(+)$-2-Aminopropanoic acid: S
$(-)$-2-Hydroxypropanoic acid: R

5-10

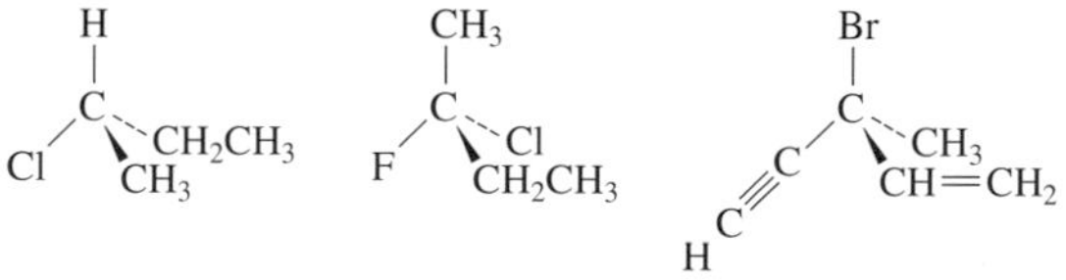

$S \qquad\qquad R \qquad\qquad S$

5-11

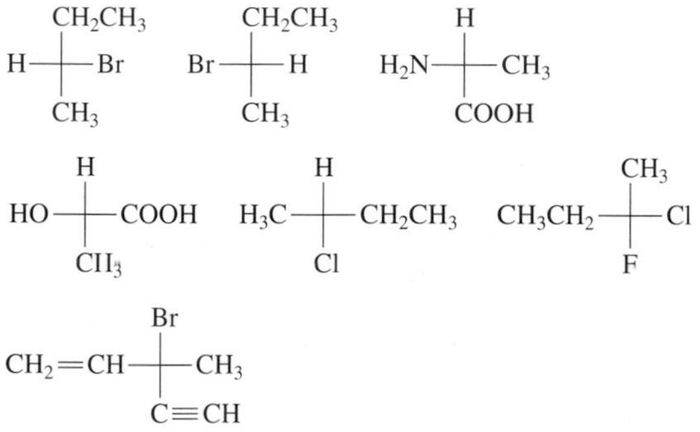

5-12

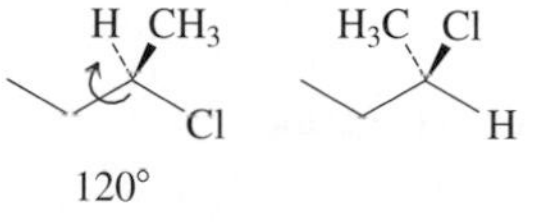

$120°$

5-13

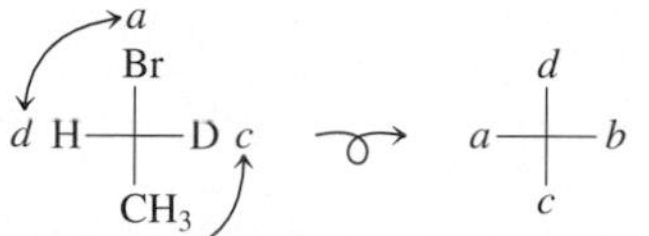

R

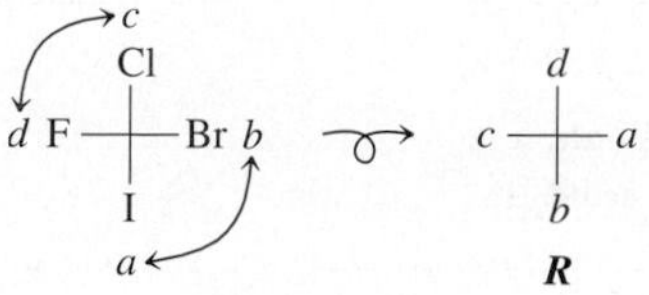

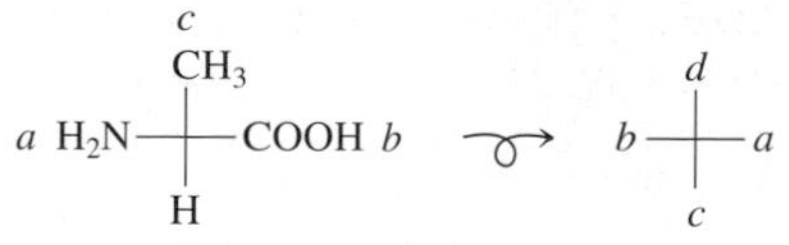

R

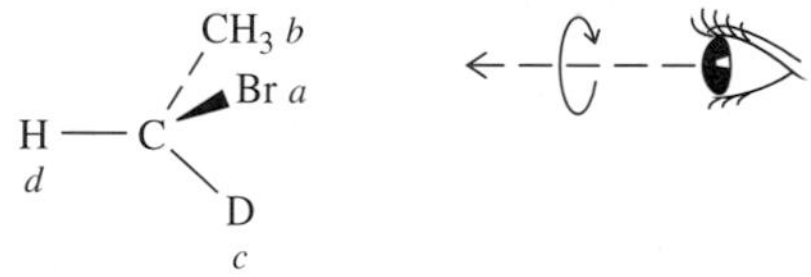

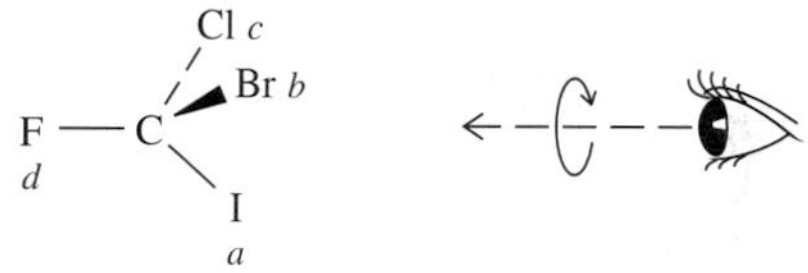

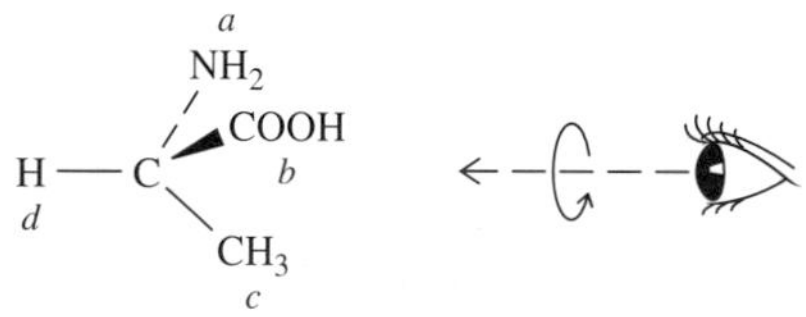

Placing the lowest priority substituent d at the top of a Fischer projection means that it is located behind the plane of the page, the place required for the correct assignment of absolute configuration by visual inspection.

5-15

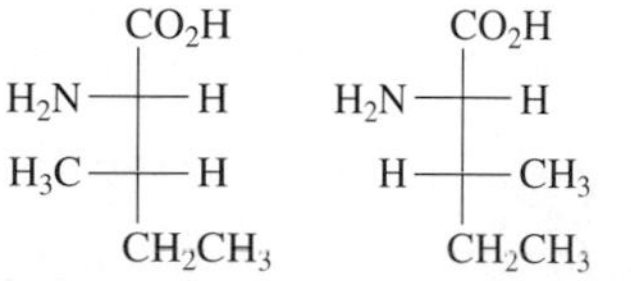

Isoleucine   **Alloisoleucine**

They are diastereomers.

5-16

1: (2S,3S)-2-Fluoro-3-methylpentane.
2: (2R,3S)-2-Fluoro-3-methylpentane.
3: (2R,3R)-2-Fluoro-3-methylpentane.
4: (2S,3S)-2-Fluoro-3-methylpentane.
1 and 2 are diastereomers; 1 and 3 are enantiomers; 1 and 4 are identical; 2 and 3 and 2 and 4 are diastereomers; 3 and 4 are enantiomers. With the inclusion of the mirror image of 2, there are four stereoisomers.

5-17

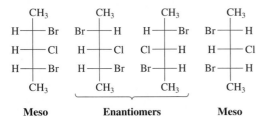

With the inclusion of the four mirror images, there are four enantiomeric pairs of diastereomers.

5-18

Meso   **Enantiomers**   **Meso**

5-19

Mirror plane

(a) Meso   **(b)** Chiral

(c) Meso   **(d)** Chiral

(e) Meso   **(f)** Chiral

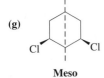

(g) Meso   **(h)** Chiral

5-20

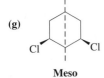

5-21

Almost any halogenation at C2 gives a racemate; the exception is bromination, which results in achiral 2,2-dibromobutane. In addition, bromination at C3 gives the two 2,3-dibromobutane diastereomers, one of which, 2R,3S, is meso.

5-22

Attack at Cl:

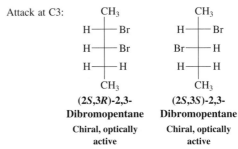

(R)-1,2-Dibromopentane
Chiral, optically active

Attack at C2:

2,2-Dibromopentane
Achiral

Attack at C3:

(2S,3R)-2,3-   (2S,3S)-2,3-
Dibromopentane   Dibromopentane
Chiral, optically Chiral, optically
active   active

Diastereomers, formed
in unequal amounts

Attack at C4:

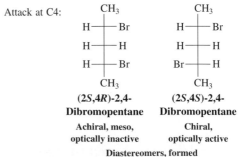

(2S,4R)-2,4-   (2S,4S)-2,4-
Dibromopentane   Dibromopentane
Achiral, meso,   Chiral,
optically inactive optically active

Diastereomers, formed
in unequal amounts

Attack at C5:

(S)-1,4-Dibromopentane
Chiral, optically active

CHAPTER 6

6-1

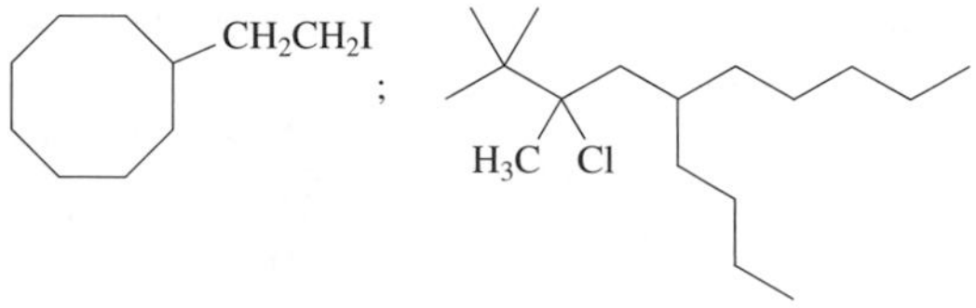

or

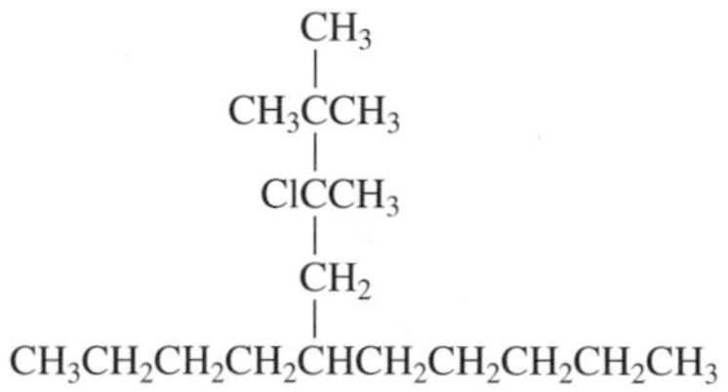

Note the similarity of the last structure to that of 6-(2-chloro-2,3,3-trimethylbutyl)undecane on p. 212. Why is it named so differently?

6-2

(a) $CH_3CH_2CH_2CH_2\ddot{\underset{..}{I}}:$

(b) $CH_3CH_2CH_2CH_2\ddot{\underset{..}{O}}CH_2CH_3$

(c) $CH_3CH_2CH_2CH_2\ddot{N}=\overset{+}{N}=\ddot{\underset{..}{N}}:^-$

(d) 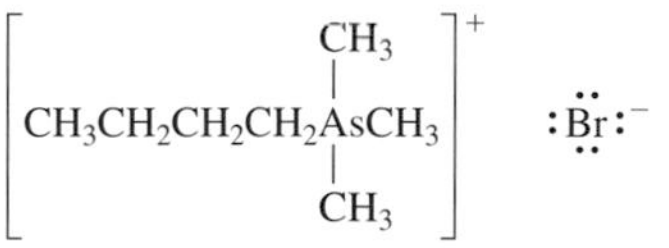
$$\left[CH_3CH_2CH_2CH_2\overset{\overset{\displaystyle CH_3}{|}}{\underset{\underset{\displaystyle CH_3}{|}}{As}}CH_3 \right]^+ \quad :\ddot{\underset{..}{Br}}:^-$$

(e) 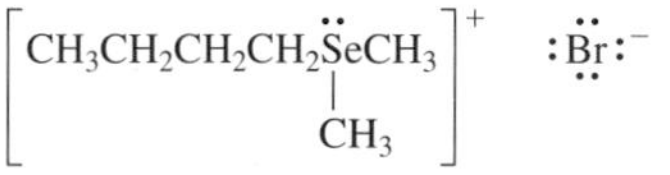
$$\left[CH_3CH_2CH_2CH_2\overset{..}{\underset{\underset{\displaystyle CH_3}{|}}{Se}}CH_3 \right]^+ \quad :\ddot{\underset{..}{Br}}:^-$$

6-3

(a) $CH_3I \quad + \quad :N(CH_3)_3$

(b) There are two approaches:

$CH_3\ddot{\underset{..}{S}}:^- \quad + \quad CH_3CH_2\ddot{\underset{..}{I}}:$ or $CH_3\ddot{\underset{..}{I}}: \quad + \quad CH_3CH_2\ddot{\underset{..}{S}}:^-$

6-4

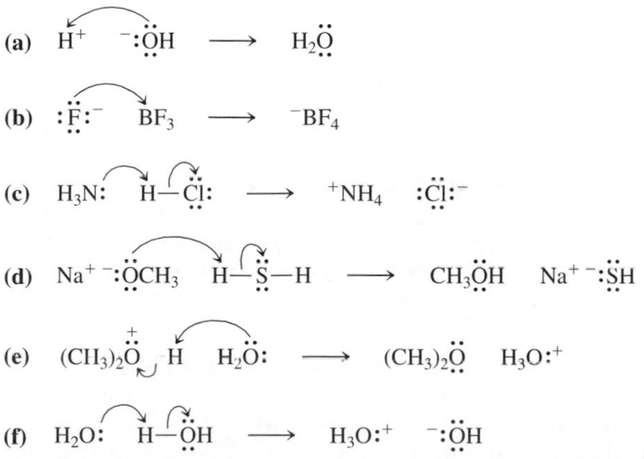

(a) $H^+ \quad ^-:\ddot{\underset{..}{O}}H \longrightarrow H_2\ddot{\underset{..}{O}}$

(b) $:\ddot{\underset{..}{F}}:^- \quad BF_3 \longrightarrow ^-BF_4$

(c) $H_3N: \quad H-\ddot{\underset{..}{Cl}}: \longrightarrow {}^+NH_4 \quad :\ddot{\underset{..}{Cl}}:^-$

(d) $Na^+ {}^-:\ddot{\underset{..}{O}}CH_3 \quad H-\ddot{\underset{..}{S}}-H \longrightarrow CH_3\ddot{\underset{..}{O}}H \quad Na^+ {}^-:\ddot{\underset{..}{S}}H$

(e) $(CH_3)_2\overset{+}{\ddot{O}} \, H \quad H_2\ddot{\underset{..}{O}}: \longrightarrow (CH_3)_2\ddot{\underset{..}{O}} \quad H_3O:^+$

(f) $H_2\ddot{\underset{..}{O}}: \quad H-\ddot{\underset{..}{O}}H \longrightarrow H_3O:^+ \quad {}^-:\ddot{\underset{..}{O}}H$

6-5

In mechanisms 1 and 3, oxygen is the nucleophile, and carbon is the electrophile. In mechanism 4, the carbon–carbon double bond is the nucleophile, and the proton the electrophile. In mechanism 2,

the dissociation, no external nucleophiles or electrophiles are pictured, but the carbon atom is electrophilic at the start and becomes more so after the chloride leaves.

6-6

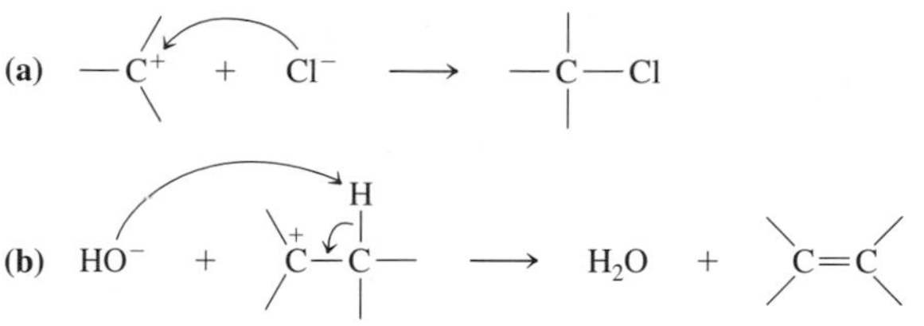

(a) $-\overset{|}{\underset{|}{C}}{}^+ + Cl^- \longrightarrow -\overset{|}{\underset{|}{C}}-Cl$

(b) $HO^- + \overset{+}{C}\overset{\overset{\displaystyle H}{|}}{C}- \longrightarrow H_2O + C=C$

6-7

$CH_3I + {}^-:\ddot{N}=\overset{+}{N}=\ddot{\underset{..}{N}}:^- \longrightarrow CH_3-\ddot{N}=\overset{+}{N}=\ddot{\underset{..}{N}}:^- + I^-$

$k = 3 \times 10^{-10}/10^{-4} = 3 \times 10^{-6} \text{ mol L}^{-1}\text{ s}^{-1}$

(a) Rate $= k[CH_3I][N_3^-] = (3 \times 10^{-6}) \times (2 \times 10^{-4})$
$= 6 \times 10^{-10} \text{ mol L}^{-1}\text{ s}^{-1}$

(b) $1.2 \times 10^{-9} \text{ mol L}^{-1}\text{ s}^{-1}$

(c) $2.7 \times 10^{-9} \text{ mol L}^{-1}\text{ s}^{-1}$

6-8

Frontside displacement

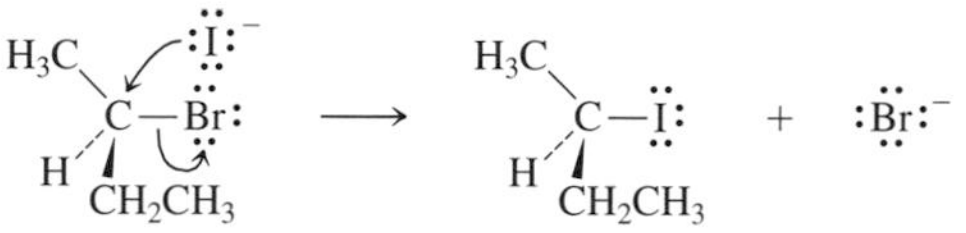

Backside displacement

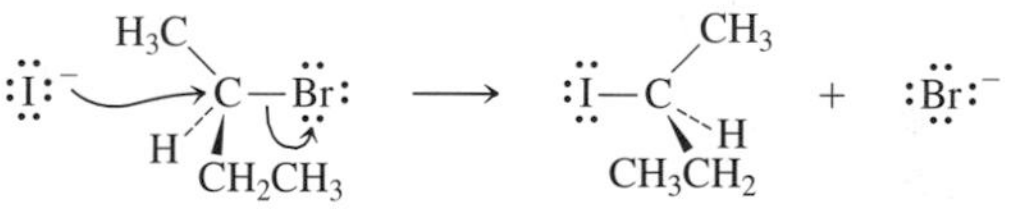

6-9

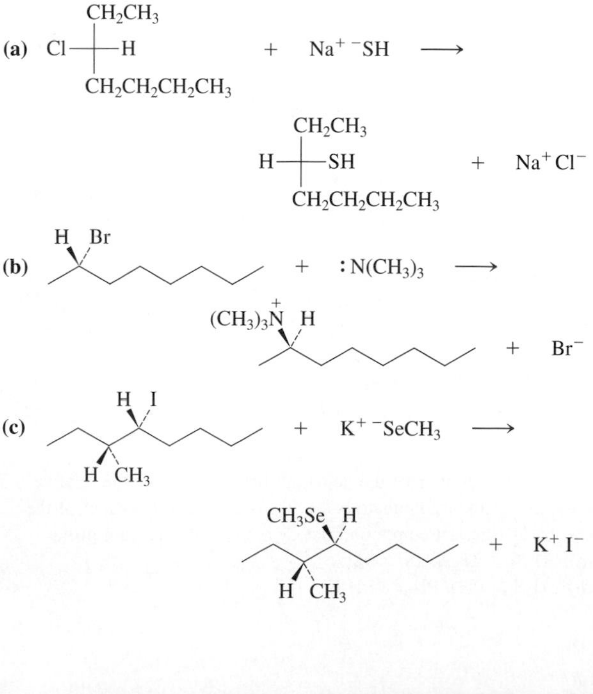

(a)
+ Na$^+$ $^-$SH $\longrightarrow$

+ Na$^+$Cl$^-$

(b)
+ :N(CH$_3$)$_3$ $\longrightarrow$

+ Br$^-$

(c)
+ K$^+$ $^-$SeCH$_3$ $\longrightarrow$

+ K$^+$I$^-$

6-10

Meso   Meso

Trans   Cis

6-11

6-12

(S)-2-Iodooctane $\xrightarrow{\text{I}^-}$ (R)-2-iodooctane; i.e., racemization.

6-13

2R,4R   2S,4S

2R,3R   2S,3R

All four components are diastereomers of their counterparts on p. 226.

6-14

I^- is a better leaving group than Cl^-. Hence the product is $Cl(CH_2)_6SeCH_3$.

6-15

The relative acidities of the acids are listed first, then the relative basicities of their conjugate bases. In each case, the weaker of the two bases (the last compound listed) is the better leaving group.
(a) $H_2Se > H_2S$, $HS^- > HSe^-$
(b) $H_2S > PH_3$, $PH_2^- > HS^-$

(c) $HClO_3 > HClO_2$, $ClO_2^- > ClO_3^-$
(d) $HBr > H_2Se$, $HSe^- > Br^-$
(e) $H_3O^+ > {}^+NH_4$, $NH_3 > H_2O$

6-16
(a) $^-OH > {}^-SH$ (b) $^-PH_2 > {}^-SH$

(c) $^-SeH > I^-$ (d)

The relative acidities of the respective conjugate acids follow the inverse order.

6-17
(a) $HS^- > H_2S$ (b) $CH_3S^- > CH_3SH$
(c) $CH_3NH^- > CH_3NH_2$ (d) $HSe^- > H_2Se$

6-18
(a) $CH_3S^- > Cl^-$ (b) $P(CH_3)_3 > S(CH_3)_2$
(c) $CH_3CH_2Se^- > Br^-$ (d) $H_2O > HF$

6-19
(a) $CH_3SeH > CH_3SH$ (b) $(CH_3)_2PH > (CH_3)_2NH$

6-20
(a) CH_3S^- (b) $(CH_3)_2NH$

6-21

The more reactive substrates are (a) [structure] and (b) $CH_3CH_2CH_2Br$.

6-22

CHAPTER 7

7-1
Compound A is a 2,2-dialkyl-1-halopropane (neopentyl halide) derivative. The carbon bearing the potential leaving group is primary but very hindered and therefore very unreactive with respect to any substitution reactions. Compound B is a 1,1-dialkyl-1-haloethane (*tert*-alkyl halide) derivative and undergoes solvolysis.

7-2
Bonds broken [R = $(CH_3)_3C$]: 67 (R–Br) + 119 (H–OH) =
$$186 \text{ kcal mol}^{-1}$$
Bonds made: 93 (R–OH) + 87 (H–Br) =
$$180 \text{ kcal mol}^{-1}$$
$$\Delta H^\circ = +6 \text{ kcal mol}^{-1}$$
By this calculation, the reaction should be endothermic. It still proceeds because of the excess water employed and the favorable solvation energies of the products.

7-3

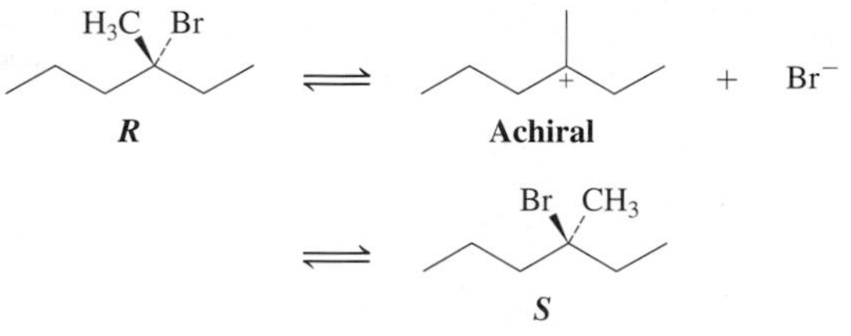

R Achiral   + Br^-

S

The molecule dissociates to the achiral tertiary carbocation. Recombination gives a 1:1 mixture of R and S product.

7-4

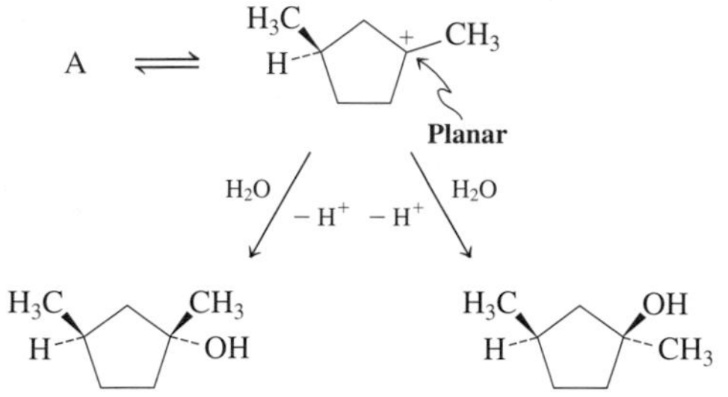

A ⇌

Planar

H_2O H_2O
$-H^+$ $-H^+$

7-5

$(CH_3)_3COSCH_3$ ⇌ $(CH_3)_3C^+$ + $CH_3SO_3^-$

F^- Faster Br^- Slower

$(CH_3)_3CF$ $(CH_3)_3CBr$

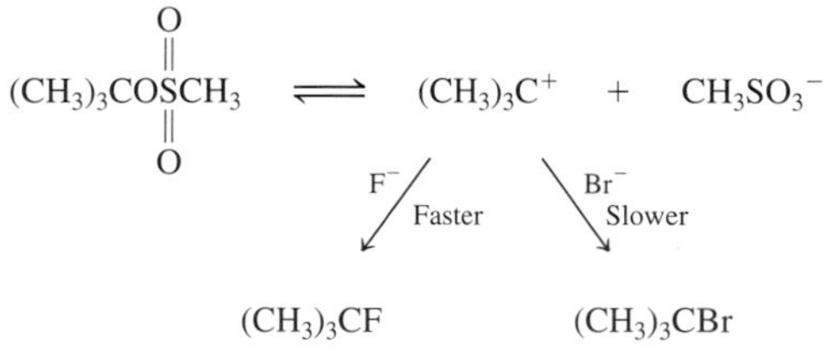

In polar aprotic solvents, fluoride, the stronger base, is also the better nucleophile.

7-6

(a) This is an S_N2 reaction that takes place with inversion.
(b) In a weakly nucleophilic protic solvent, mainly solvolysis takes place through the intermediacy of an achiral carbocation.

7-7

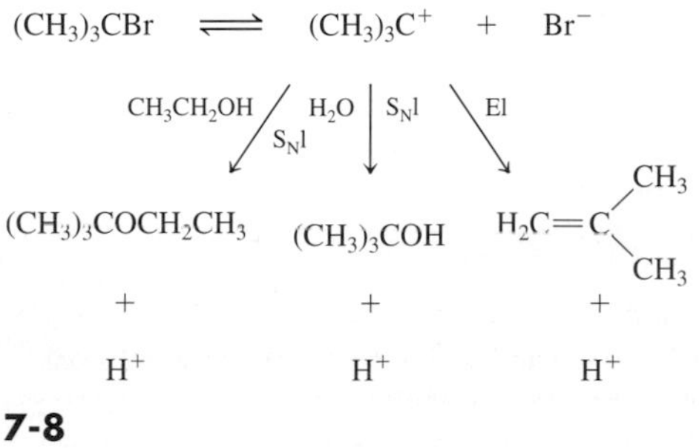

$(CH_3)_3CBr$ ⇌ $(CH_3)_3C^+$ + Br^-

CH_3CH_2OH / S_N1 H_2O | S_N1 $E1$

$(CH_3)_3COCH_2CH_3$ $(CH_3)_3COH$ $H_2C=C{\overset{CH_3}{\underset{CH_3}{}}}$

+   +   +

H^+ H^+ H^+

7-8

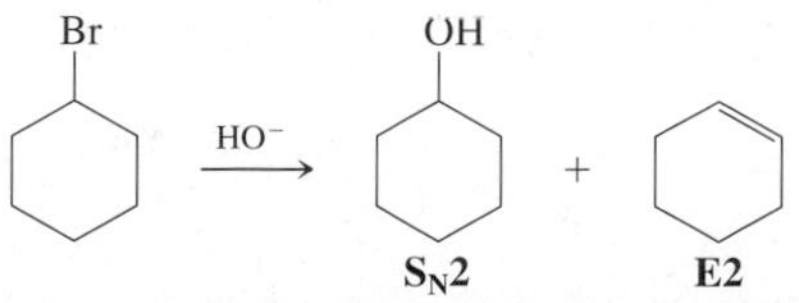

HO^-

S_N2 $E2$

7-9

$CH_2=CH_2$; no E2 possible; $CH_2=C(CH_3)_2$; no E2 possible.

7-10

I^- is a better leaving group, thus allowing for selective elimination of HI by E2.

7-11

The cis isomer undergoes E2; the trans compound, E1.

7-12

All chlorines are equatorial, lacking *anti* hydrogens.

7-13

(a) $N(CH_3)_3$, stronger base, worse nucleophile

(b) $(CH_3CH)_2N^-$, more hindered base (with CH₃ substituent shown above)

(c) Cl^-, stronger base, worse nucleophile (in protic solvent)

7-14

Thermodynamically, eliminations are usually favored by entropy, and the entropy term in $\Delta G° = \Delta H° - T\Delta S°$ is temperature dependent. Kinetically, eliminations have higher activation energies than do substitutions, so their rates rise more rapidly with increasing temperature (see Problem 33, Chapter 2).

7-15

(a) The second reaction will give more E2 product, because a stronger base is present.
(b) The first reaction will give E2 product, mainly because of the presence of a strong, hindered base, which is absent in the second reaction.

CHAPTER 8

8-1

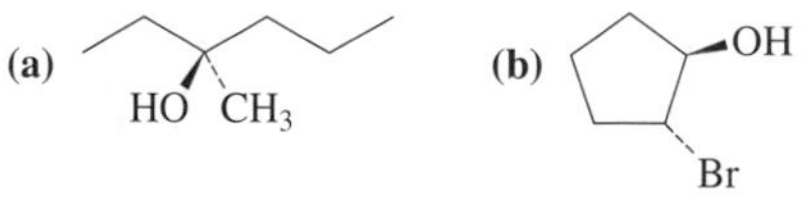

(a)   (b)

(c) $(CH_3)_3CCH_2OH$

8-2

(a) 4-Methyl-2-pentanol
(b) *cis*-4-Ethylcyclohexanol
(c) 3-Bromo-2-chloro-1-butanol

8-3

All the bases whose conjugate acids have pK_a values $\gg$ 15.5—i.e., $CH_3CH_2CH_2CH_2Li$, LDA, and KH.

8-4

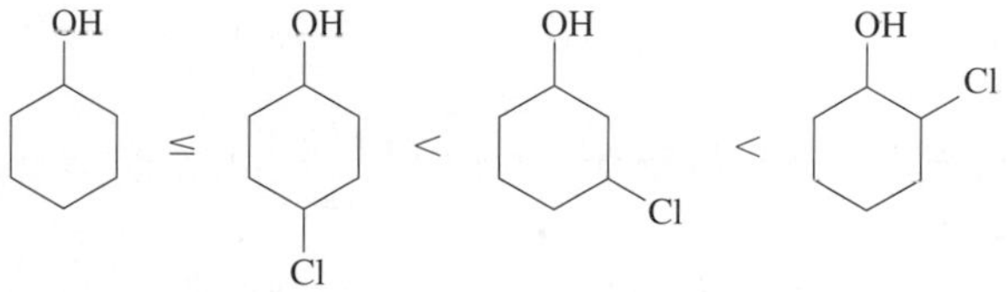

8-5

In condensed phase, $(CH_3)_3COH$ is a weaker acid than CH_3OH. The equilibrium lies to the right.

8-6

(a) NaOH, H_2O
(b) 1. CH_3CO_2Na, 2. NaOH, H_2O
(c) H_2O

8-7

(a)  +

(b) $CH_3CH_2CHCH_2CH_3$ (with OH)

(c) 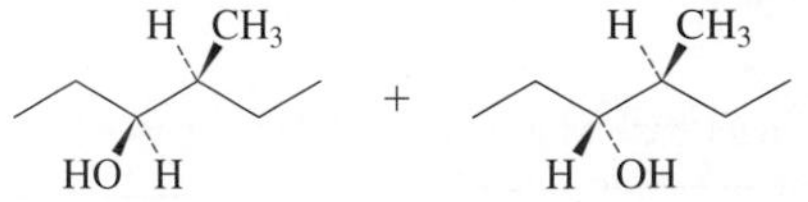 +

8-8

Less hindered

Hindered

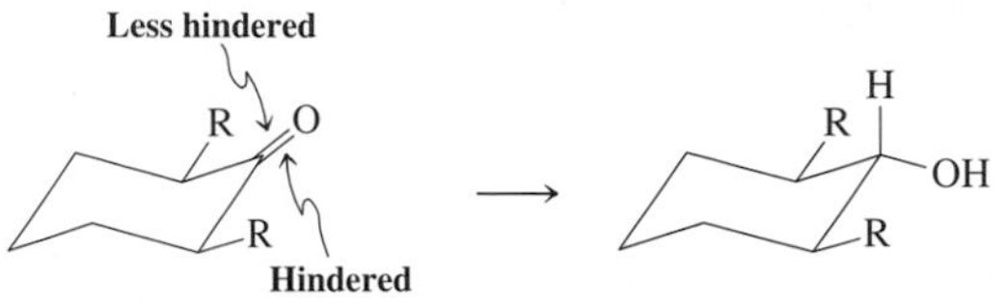

8-9

(a) $CH_3(CH_2)_8CHO + NaBH_4$ **(b)** 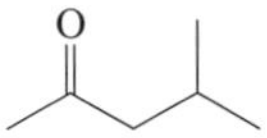 + $NaBH_4$

(c) 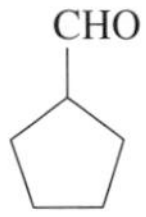 + $NaBH_4$ **(d)** 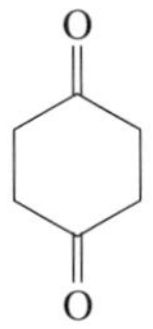 + $NaBH_4$

8-10

1. 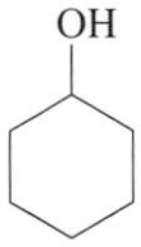 + $2 H^+$ + $2 e$

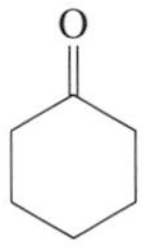

2. $Cr_2O_7^{2-} + 14 H^+ + 6 e \longrightarrow 2 Cr^{3+} + 7 H_2O$

Addition of (1) and (2) and balancing the electron count gives

3 + $Cr_2O_7^{2-}$ + $8 H^+ \longrightarrow$

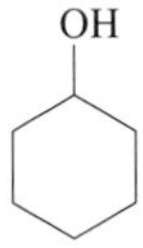

3 + $2 Cr^{3+}$ + $7 H_2O$

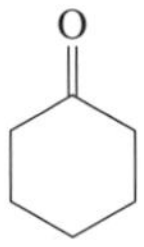

Adding the counterions results in

3 + $Na_2Cr_2O_7$ + $4 H_2SO_4 \longrightarrow$

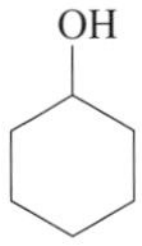

3 + $Cr_2(SO_4)_3$ + Na_2SO_4 + $7 H_2O$

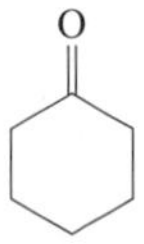

8-11

(a) $CH_3CH_2CHCH(CH_3)_2$ (with OH) + $Na_2Cr_2O_7$

(b) 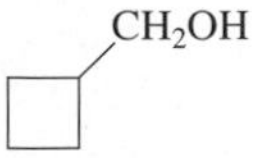 + PCC

(c) 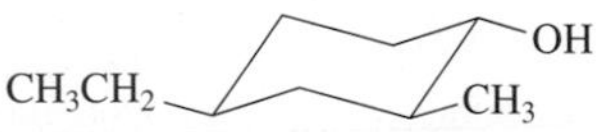 CH_3CH_2... CH_3 OH + $Na_2Cr_2O_7$

8-12

$\xrightarrow{Br_2, hv}$ (Br) $\xrightarrow{Mg}$ (MgBr) $\xrightarrow{D_2O}$ (D)

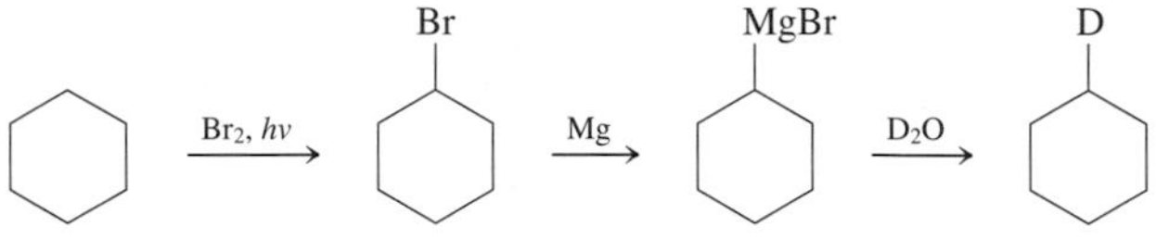

8-13

$(CH_3)_2CHBr \xrightarrow{Mg} (CH_3)_2CHMgBr \xrightarrow{CH_2=O} (CH_3)_2CHCH_2OH$

8-14

(a) $CH_3CH_2CH_2CH_2Li + CH_2=O$
(b) $CH_3CH_2CH_2MgBr + CH_3CH_2CH_2CHO$

(c) $(CH_3)_3CLi$ +

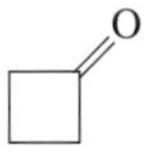

(d) $CH_3CH_2CH_2MgBr$ + $CH_3CCH_2CH_3$ (with O)

8-15
(a) Product:

$ClCH_2CH_2CH_2C(CH_3)_2$ By S_N1
(with OCH_2CH_3)

(b) Product:

$CH_2=CHCH_2C(CH_3)_2$ (with CH_2Cl) By E2 (hindered base)

The second chlorine is in a neopentyl position.

(c) Product:

$(CH_3)_2CCH_2CH_2CHO$ (with OH)

The second hydroxy function is tertiary.

8-16
The desired alcohol is tertiary and is therefore readily made from 4-ethylnonane by 1. Br_2, *hv*; 2. hydrolysis (S_N1). However, the starting hydrocarbon is itself complex and would require an elaborate synthesis. Thus, the retrosynthetic analysis by C–O disconnection is poor.

8-17

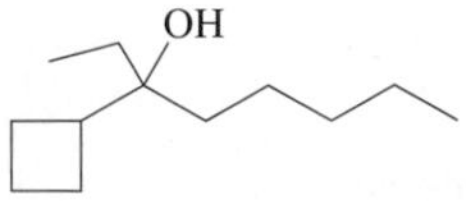

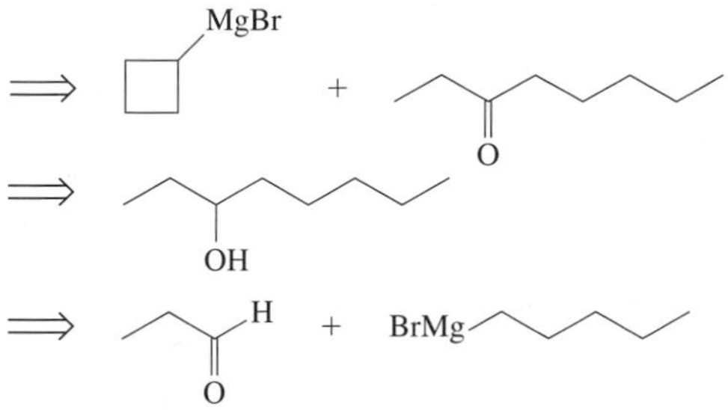

8-18

$$CH_4 \xrightarrow{Br_2, h\nu} CH_3Br \xrightarrow{Mg} CH_3MgBr$$

1. NaOH 2. PCC

1. $H_2C{=}O$ 2. PCC

$$H_2C{=}O \qquad CH_3CHO$$

$$CH_3CHO \xrightarrow[\substack{1.\ CH_3MgBr \\ 2.\ Na_2Cr_2O_7}]{} CH_3\overset{O}{\overset{\|}{C}}CH_3 \xrightarrow{CH_3MgBr} (CH_3)_3COH$$

CHAPTER 9

9-1

$$CH_3OH + \ ^-CN \xrightleftharpoons{K\ =\ 10^{-6.3}} CH_3O^- + HCN$$
$$pK_a = 15.5 \qquad\qquad pK_a = 9.2$$

Answer: No.

9-2

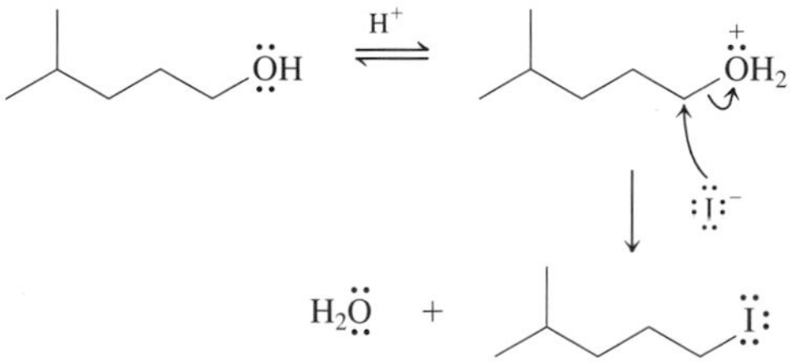

9-3

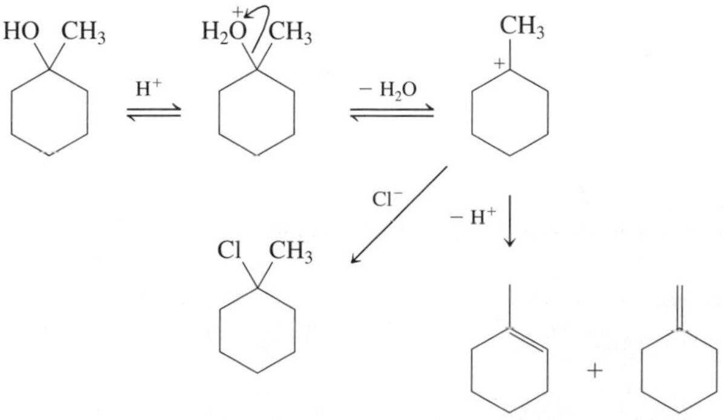

(a)

(Major) **(Minor)**

(b)

The tertiary carbocation is either trapped by the nucleophile (Cl^-) or undergoes E1. (SO_4^{2-} is a poor nucleophile.)

9-4

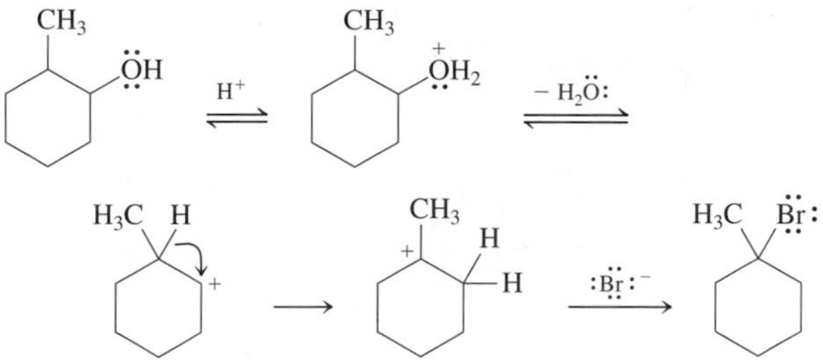

$$\xrightarrow{H^+} \qquad \xrightarrow{-\ H_2\ddot{O}:}$$

Secondary carbocation   **Tertiary carbocation**

9-5

(a)

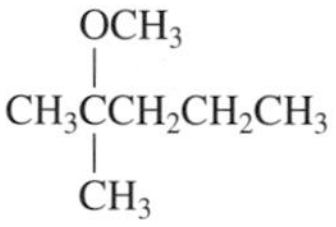

(b) $CH_3CH_2 \quad Cl$

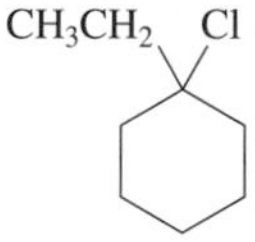

9-6

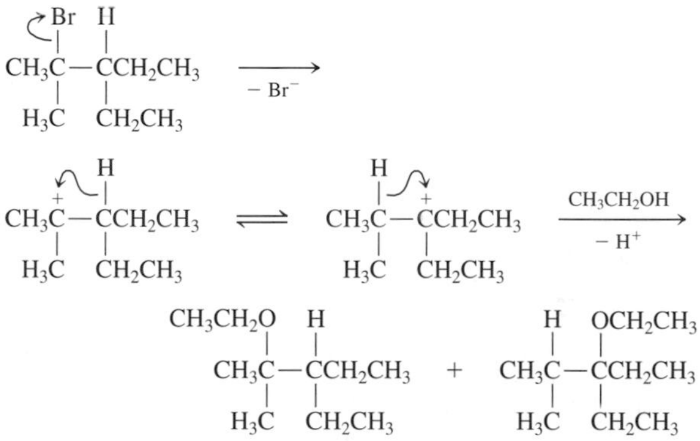

$$\xrightarrow{-\ Br^-}$$

$$\xrightleftharpoons{} \qquad \xrightarrow[-\ H^+]{CH_3CH_2OH}$$

Similarly,

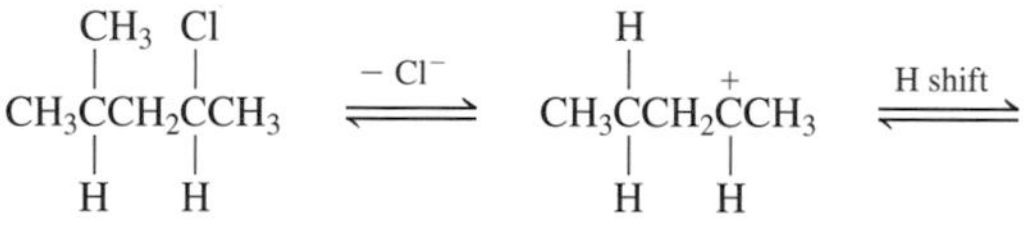

$$\xrightleftharpoons{-\ Cl^-} \qquad \xrightarrow{H\ shift}$$

Secondary carbocation

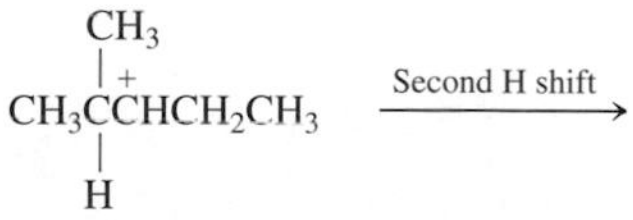

$$\xrightarrow{Second\ H\ shift}$$

Secondary carbocation

$$\xrightarrow[-\ H^+]{CH_3OH}$$

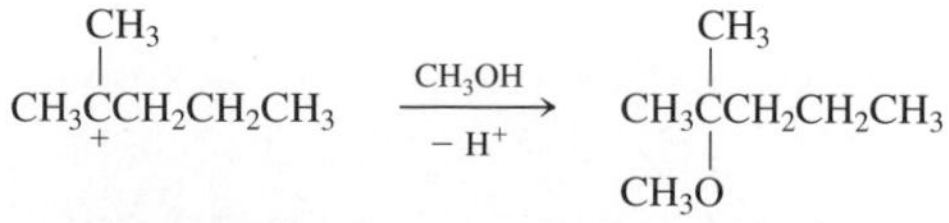

Tertiary carbocation

9-7

(a) $CH_3CCH_2CH_2CH_3$ (with OH and CH_3 substituents) $\xrightarrow[-H_2O]{\text{Straight E1}}$

$CH_3CCH_2CCH_3$ (with CH_3 and OH) $+ H^+$ $\underset{+H_2O}{\overset{-H_2O}{\rightleftharpoons}}$ $CH_3C-CHCCH_3$ $\xrightarrow{\text{H shift}}$

$CH_3C-CHCCH_3$ $\rightleftharpoons$ (alkene) $+ H^+$

(b) (4-methylcyclohexanol) $\xrightarrow[-H_2O]{H^+}$ (cation) $\xrightarrow{\text{H shift}}$ (cation) $\xrightarrow{\text{H shift}}$

(cation) $\xrightarrow{\text{H shift}}$ (cation) $\xrightarrow{-H^+}$ (1-methylcyclohexene)

9-8

$(CH_3)_3CCH=CH_2$, $CH_2=C(CH_3)CH(CH_3)_2$, and $(CH_3)_2C=C(CH_3)_2$

9-9

(cis-3-methylcyclohexanol) $\xrightarrow[-HCl]{CH_3SO_2Cl}$ (mesylate) $\xrightarrow[-CH_3SO_3Na]{NaI}$ (iodide)

9-10

(a) 1. CH_3SO_2Cl, 2. NaI
(b) HCl
(c) PBr_3

9-11

(a) 1. $CH_3CH_2I + CH_3CH_2CH_2CH_2O^-Na^+$,
2. $CH_3CH_2O^-Na^+ + CH_3CH_2CH_2CH_2I$

(b) Best is (2-pentanolate O^-Na^+) $+$ CH_3I

The alternative, $CH_3O^-Na^+$ $+$ (2-iodopentane), suffers from competing E2.

(c) (cyclohexanolate O^-Na^+) $+$ $CH_3CH_2CH_2Br$

(d) $Na^+ {}^-O$—(butanediolate)—$O^- Na^+ + CH_3CH_2OSO_2CH_3$

9-12

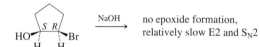

9-13

(1R,2R)-2-Bromocyclopentanol $\xrightarrow[\text{Fast}]{NaOH}$ (epoxide) **Meso**

(1R,2R)-2-Bromocyclopentanol
The nucleophilic oxygen and the leaving group are trans (anti)

(1S,2R)-2-Bromocyclopentanol $\xrightarrow{NaOH}$ no epoxide formation, relatively slow E2 and S_N2

(1S,2R)-2-Bromocyclopentanol
Here nucleophile and leaving group are cis (syn)

9-14

(a) HO—(chain)—$OH + H^+ \rightleftharpoons$ (protonated) $\xrightarrow{-H_2O}$ (oxocycle) $\rightleftharpoons$ (cycle) $+ H^+$

(b) (diol) $+ H^+ \underset{+H_2O}{\overset{-H_2O}{\rightleftharpoons}}$ (cyclic oxocarbenium) $+ H^+$ $\rightleftharpoons$ (product) $+ H^+$ $\longrightarrow$

9-15

(a) This ether is best synthesized by solvolysis:

$$CH_3CH_2CBr(CH_3) + CH_3COH(CH_3) \longrightarrow CH_3CH_2C(CH_3)_2-O-C(CH_3)_2H$$

Solvent

2-Methyl-2-(1-methylethoxy)butane

The alternative, an S_N2 reaction, would give elimination:

$$CH_3CH_2\underset{\underset{CH_3}{|}}{\overset{\overset{CH_3}{|}}{C}}O^- \quad + \quad CH_3\underset{\underset{H}{|}}{\overset{\overset{CH_3}{|}}{C}}Br \longrightarrow$$

$$CH_3CH=CH_2 \quad + \quad CH_3CH_2\underset{\underset{CH_3}{|}}{\overset{\overset{CH_3}{|}}{C}}OH$$

(b) This target is best prepared by an S_N2 reaction with a halomethane, because such an alkylating agent cannot undergo elimination. The alternative would be nucleophilic substitution of a 1-halo-2,2-dimethylpropane, a reaction that is normally too slow.

$$CH_3\underset{\underset{CH_3}{|}}{\overset{\overset{CH_3}{|}}{C}}CH_2O^- \quad + \quad CH_3Cl \longrightarrow$$

$$CH_3\underset{\underset{CH_3}{|}}{\overset{\overset{CH_3}{|}}{C}}CH_2OCH_3 \quad + \quad Cl^-$$

1-Methoxy-2,2-dimethylpropane

$$CH_3\underset{\underset{CH_3}{|}}{\overset{\overset{CH_3}{|}}{C}}CH_2Br \quad + \quad CH_3O^- \longrightarrow \text{slow reaction, impractical}$$

9-16

$$CH_3OCH_3 + 2 \; HI \overset{\Delta}{\longrightarrow} 2 \; CH_3I + H_2O$$

Mechanism

$$CH_3\ddot{O}CH_3 \;+\; H\ddot{I}: \;\rightleftharpoons\; CH_3\overset{\overset{H}{|+}}{\ddot{O}}CH_3 \;+\; :\ddot{I}:^-$$

$$:\ddot{I}:^- \;\;CH_3-\overset{\overset{H}{|}}{\underset{\cdot\cdot}{O}}{}^+\!-CH_3 \longrightarrow CH_3\ddot{I}: \;+\; H\ddot{O}CH_3$$

$$CH_3\ddot{O}H \;+\; H\ddot{I}: \;\rightleftharpoons\; CH_3\overset{\overset{H}{|+}}{\ddot{O}}H \;+\; :\ddot{I}:^-$$

$$:\ddot{I}:^- \;\;CH_3-\overset{\overset{H}{|}}{\underset{\cdot\cdot}{O}}{}^+\!-H \longrightarrow CH_3\ddot{I}: \;+\; H_2\ddot{O}$$

9-17

9-18

$$BrCH_2CH_2CH_2OH \xrightarrow[\substack{1.\,(CH_3)_3COH,\,H^+ \\ 2.\,Mg \\ 3.\,D_2O \\ 4.\,H^+,\,H_2O}]{} DCH_2CH_2CH_2OH$$

9-19

Gives product only   **Gives mixture**

9-20

$$(CH_3)_3CLi \quad + \quad \triangle\!O \longrightarrow \text{(neopentyl ethanol)} OH$$

9-21

(a) $(CH_3)_3COH$ **(b)** $CH_3CH_2CH_2CH_2C(CH_3)_2OH$
(c) $CH_3SCH_2C(CH_3)_2OH$
(d) $HOCH_2C(CH_3)_2OCH_2CH_3$
(e) $HOCH_2C(CH_3)_2Br$

9-22

(a)

$$\triangle\!O \;+\; HS^- \longrightarrow HO\frown S^- \xrightarrow{\triangle\!O}$$

$$\xrightarrow{H^+,\,H_2O} HO\frown S\frown OH \xrightarrow{SOCl_2}$$

$$Cl\frown S\frown Cl$$

(b) Intramolecular sulfonium salt formation

$$ClCH_2CH_2\underset{}{S}CH_2\overset{\frown}{C}H_2-\overset{\frown}{Cl} \longrightarrow ClCH_2CH_2-\overset{+}{\underset{}{S}}\!\!<\!\!\begin{array}{c}CH_2\\|\\CH_2\end{array} \quad Cl^-$$

Nucleophiles attack by ring opening

$$ClCH_2CH_2\overset{+}{\underset{}{S}}\!\!<\!\!\begin{array}{c}CH_2\\|\\CH_2\end{array}\!\!\frown:Nu \longrightarrow ClCH_2CH_2\ddot{S}CH_2CH_2Nu^+$$

CHAPTER 10

10-1

There are quite a number of isomers, e.g., several butanols, pentanols, hexanols, and heptanols. Examples include

$$CH_3\underset{\underset{H_3C}{|}}{\overset{\overset{H_3C\;\;\;CH_3}{\diagdown\;\;\;\diagup}}{C}}\!-\!\underset{\underset{CH_3}{|}}{\overset{\overset{CH_3}{|}}{C}}OH \qquad CH_3\underset{\underset{CH_3}{|}}{CH}CHCH_2CH_2OH \qquad CH_3(CH_2)_4\underset{\underset{CH_3}{\;}}{\overset{\overset{CH_3}{|}}{C}}HOH$$

2,3,3-Trimethyl-   **3,4-Dimethyl-**   **2-Heptanol**
2-butanol   **1-pentanol**

10-2

$$DH^\circ_{Cl_2} = 58 \text{ kcal mol}^{-1} = \Delta E$$
$$\Delta E = 28{,}600/\lambda$$
$$\lambda = 28{,}600/58 = 493 \text{ nm, in the ultraviolet-visible range}$$

10-3

$\delta = 261/300 = 0.87$ ppm; $\delta = 861/300 = 2.87$ ppm; $\delta = 957/300 = 3.19$ ppm; identical with the δ values measured at 90 MHz.

10-4

The methyl group resonates at higher field; the methylene hydrogens are relatively deshielded because of the cumulative electron-withdrawing effect of the two heteroatoms.

10-5

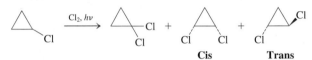

(a) $CH_3C - CCH_3$ One peak

(b) $CH_3OCH_2CH_2OCH_2CH_2OCH_3$ Three peaks

(c) 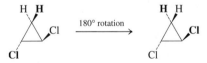   One peak

10-6

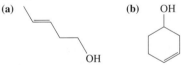

1,1-Dichlorocyclopropane shows only one signal for the four equivalent hydrogens. *Cis*-1,2-Dichlorocyclopropane exhibits three in the ratio 2:1:1. The lowest field absorption is due to the two equivalent hydrogens next to the chlorine atoms at C1 and C2. The two hydrogens at C3 are not equivalent: One lies cis to the chlorine atoms, the other trans. In contrast, the trans isomer reveals only two signals (integration ratio, 1:1). Now the hydrogens at C3 *are* equivalent, as shown by a 180° rotational symmetry operation:

10-7

The following δ values were recorded in CCl$_4$ solution. You could not have predicted them exactly, but how close did you get?
(a) $\delta = 3.38$ (q, $J = 7.1$ Hz, 4 H) and 1.12 (t, $J = 7.1$ Hz, 6 H) ppm
(b) $\delta = 3.53$ (t, $J = 6.2$ Hz, 4 H) and 2.34 (quin, $J = 6.2$ Hz, 2 H) ppm
(c) $\delta = 3.19$ (s, 1 H), 1.48 (q, $J = 6.7$ Hz, 2 H), 1.14 (s, 6 H), and 0.90 (t, $J = 6.7$ Hz, 3 H) ppm
(d) $\delta = 5.58$ (t, $J = 7$ Hz, 1 H) and 3.71 (d, $J = 7$ Hz, 2 H) ppm

10-8

(a) Quintet (quin); triplet of triplets (t,t); (b) quintet (quin); doublet of quartets (dq, or quartet of doublets, qd, which is the same); (c) sextet (sex); doublet of quintets (dquin); (d) octet (oct); triplet of triplets of quartets (ttq); (e) nonet (non); triplet of septets (tsep).

10-9

$CH_3-CH_2-CHCl_2$ $CH_3-CH-C-H$

t (7.0) dq   t (6.0)   d (6.44)   ddq   dd (10.82, 9.08)

dd (10.82, 4.71)

The hydrogens at Cl are not equivalent, because of the presence of the adjacent stereocenter (see footnote, page 404)

$ClCH_2CH_2CH_2Cl$

t   quin (6.0)

The experimental J values (Hz) are given in parentheses.

10-10

$H_3C-CH_2-CH_2-Br$

qt   ttq   tt

10-11

(a) 3;   (b) 3;   (c) 7;   (d) 2

10-12

For compound A, three lines, one of them at relatively high field (CH$_3$); DEPT would confirm CH$_3$ and two CH units. For compound B, three lines, no CH$_3$ absorption; DEPT would confirm the absence of CH$_3$ and the presence of two CH$_2$ units and one CH unit.

CHAPTER 11

11-1

(a) 2,3-Dimethyl-2-heptene   (b) 3-Bromocyclopentene

11-2

(a) *cis*-1,2-Dichloroethene
(b) *trans*-3-Heptene
(c) *cis*-1-Bromo-4-methyl-1-pentene

11-3

(a) (*E*)-1,2-Dideuterio-1-propene
(b) (*Z*)-2-Fluoro-3-methoxy-2-pentene
(c) (*E*)-2-Chloro-2-pentene

11-4

(a)   (b)

11-5

(a)

(b) (1-Methylethenyl)cyclopentene [(1-methylvinyl)cyclopentene]

11-6

$CH_2{=}CHLi + CH_3CCH_3 \longrightarrow CH_2{=}CHCCH_3$

The reaction of ethenyllithium (vinyllithium) with carbonyl compounds is like that of other alkyllithium organometallics.

11-7

The induced local magnetic field strengthens H_0 in the region occupied by the methyl hydrogens.

11-8

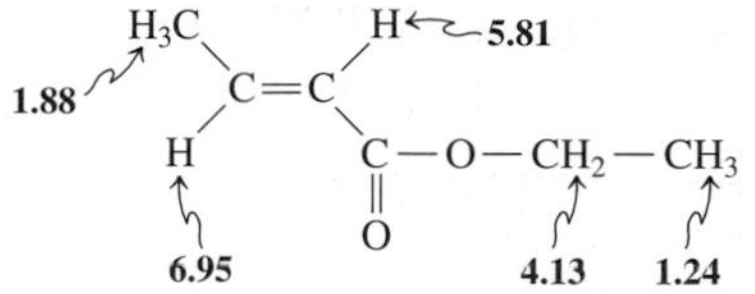

The trans coupling constant is 16.0 Hz. The couplings to the methyl group on the double bond conform with the values in Table 11-2.

11-9

Alkene A:

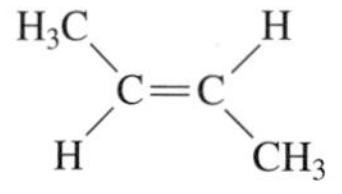

B: $CH_3CH_2CH=CH_2$

C:

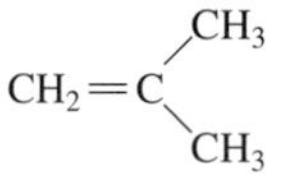

11-10

(a) $H_{sat} = 12$; degree of unsaturation = 1
(b) $H_{sat} = 20$; degree of unsaturation = 4
(c) $H_{sat} = 17$; degree of unsaturation = 5
(d) $H_{sat} = 19$; degree of unsaturation = 2
(e) $H_{sat} = 8$; degree of unsaturation = 0

11-11

(a)    (b)    (c)

Another possibility is H CH_2CH_3,

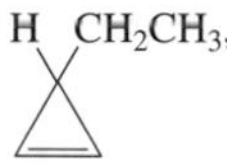

which should, however, exhibit a distinct methyl triplet signal as part of the high-field multiplet absorption.

11-12

1-Hexene < *cis*-3-hexene < *trans*-4-octene < 2,3-dimethyl-2-butene.

11-13

If you can make a model of alkene A (without breaking your plastic sticks), you will notice its extremely strained nature, much of which is released on hydrogenation. You can estimate the excess strain in A (relative to B) by subtracting the $\Delta H°$ of the hydrogenation of a "normal" tetrasubstituted double-bond (~ -27 kcal mol^{-1}) from the $\Delta H°$ of the A-to-B transformation: 38 kcal mol^{-1}.

11-14

$(CH_3)_2C=CHCH_3$ $(CH_3)_2CHCH=CH_2$
     A   **B**

Product B results from abstraction of the more accessible methyl group hydrogen at Cl, favored with the more hindered base.

11-15

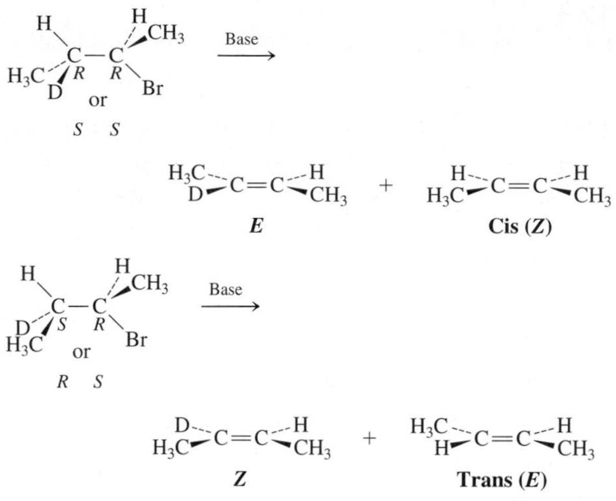

Note that, in the first case, a pair of isomers is formed with the configuration *opposite* that generated in the second. The E and Z isomers of 2-deuterio-2-butene are isotopically pure in each case; none of the protic 2-butene with the same configuration is generated. The protio-2-butenes are also pure, devoid of any deuterium.

11-16

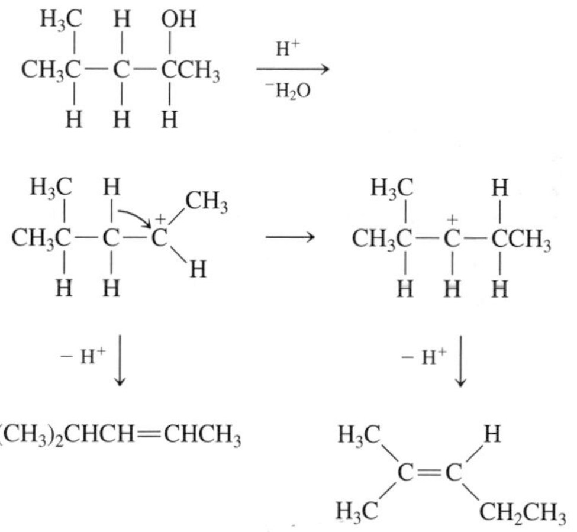

11-17

(a) 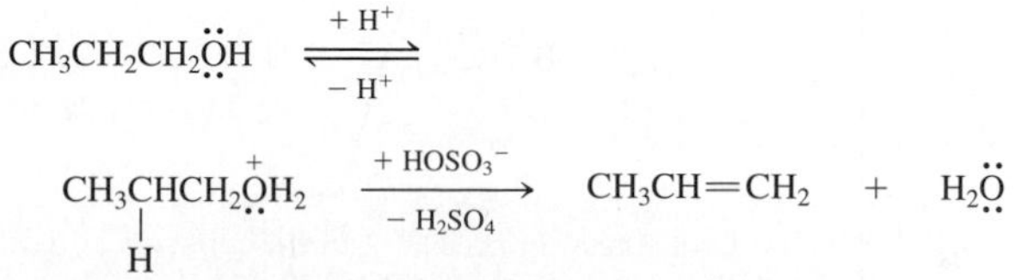 $CH_3CH_2CH_2\overset{..}{O}H$ $\underset{-H^+}{\overset{+H^+}{\rightleftharpoons}}$

$CH_3CH\overset{+}{C}H\overset{+}{O}H_2$ $\underset{-H_2SO_4}{\overset{+HOSO_3^-}{\longrightarrow}}$ $CH_3CH=CH_2$ + $H_2\overset{..}{\underset{..}{O}}$
  |
  H

(b) $CH_3CH_2CH_2OCH_2CH_2CH_3 \overset{H^+}{\rightleftharpoons} CH_3CH=CH_2$ + $CH_3CH_2CH_2OH$ in analogy to (a). The propanol may then be dehydrated as in (a).

CHAPTER 12

12-1

Estimating the strengths of the bonds broken and the bonds made

$$CH_2{=}CH_2 \quad + \quad HO{-}OH \quad \longrightarrow \quad \begin{array}{c} HO \quad OH \\ | \quad\quad | \\ H{-}C{-}C{-}H \\ | \quad\quad | \\ H \quad H \end{array}$$

$$\textbf{65}\textbf{49}\textbf{2 × (~92)}$$

gives $\Delta H° = -70$ kcal mol^{-1}. Even though very exothermic, this reaction requires a catalyst.

12-2

$$\begin{array}{c} CH_3CH_2 \quad\quad CH_3 \\ \underset{\underset{CH_3}{|}}{\overset{}{\underset{H \diagup S}{}C}}{-}\underset{\underset{CH_2}{}}{\overset{}{C}} \end{array} \quad\xrightarrow{\text{H}_2,\ \text{catalyst}}\quad$$

$$\begin{array}{c} CH_3CH_2 \\ \underset{\underset{CH_3}{|}}{\overset{}{\underset{H \diagup S}{}C}}{-}CH(CH_3)_2 \end{array}$$

Not a stereocenter

12-3

(a) Both enantiomers

(b) + Both enantiomers

(c) $(CH_3)_2\underset{\underset{}{\overset{\overset{Br}{|}}{C}}}CH_2CH_3$

(d) + Both cis and trans

12-4

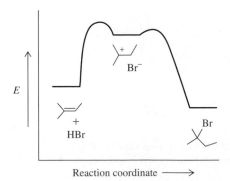

12-5

Protonation to the 1,1-dimethylethyl (*tert*-butyl) cation is reversible. With D$^+$, fast exchange of all hydrogens for deuterium will take place.

$$CH_2{=}C(CH_3)_2 \underset{-\,D^+}{\overset{+\,D^+}{\rightleftharpoons}} DCH_2\overset{+}{C}(CH_3)_2 \underset{+\,H^+}{\overset{-\,H^+}{\rightleftharpoons}}$$

$$DCH{=}C(CH_3)_2 \underset{-\,D^+}{\overset{+\,D^+}{\rightleftharpoons}} D_2CH\overset{+}{C}(CH_3)_2 \underset{+\,H^+}{\overset{-\,H^+}{\rightleftharpoons}}$$

$$D_2C{=}C(CH_3)_2 \underset{-\,D^+}{\overset{+\,D^+}{\rightleftharpoons}} D_3C\overset{+}{C}(CH_3)_2 \underset{+\,H^+}{\overset{-\,H^+}{\rightleftharpoons}}$$

$$\begin{array}{c} D_3C \\ \diagdown \\ C{=}CH_2 \\ \diagup \\ H_3C \end{array} \underset{-\,D^+}{\overset{+\,D^+}{\rightleftharpoons}} \quad \text{and so on} \quad {-}{-}\!\rightarrow$$

$$(CD_3)_3C^+ \underset{-\,D_2O}{\overset{D_2O}{\rightleftharpoons}} (CD_3)_3COD \quad + \quad D^+$$

12-6

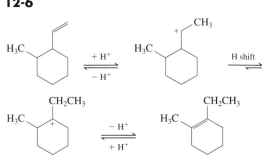

Tetrasubstituted alkene, most stable

12-7

$$CH_2{=}CH_2 \quad + \quad F{-}F \quad \longrightarrow \quad \begin{array}{c} F \quad\ F \\ | \quad\ | \\ CH_2{-}CH_2 \end{array}$$

$$\textbf{65}\textbf{37}\textbf{2 × (~107) kcal mol}^{-1}$$
$$\Delta H° = -112\ \text{kcal mol}^{-1}$$

$$CH_2{=}CH_2 \quad + \quad I{-}I \quad \longrightarrow \quad \begin{array}{c} I \quad\ I \\ | \quad\ | \\ CH_2{-}CH_2 \end{array}$$

$$\textbf{65}\textbf{36}\textbf{2 × (~53) kcal mol}^{-1}$$
$$\Delta H° = -5\ \text{kcal mol}^{-1}$$

12-8

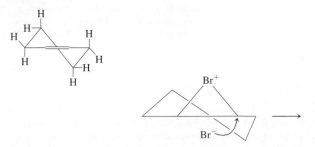

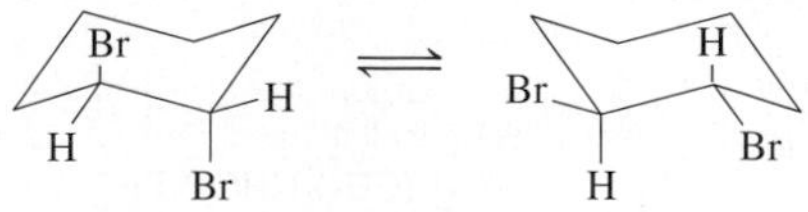

(1S,2S)-*trans*-1,2-Dibromocyclohexane

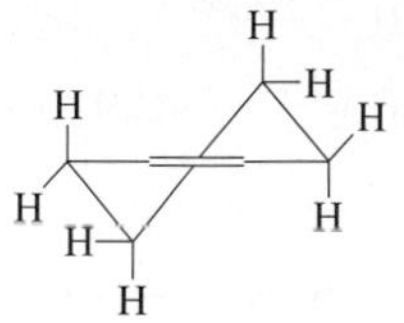

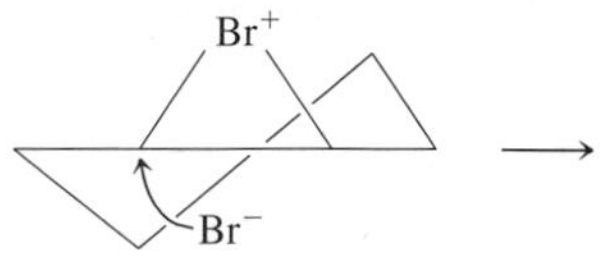

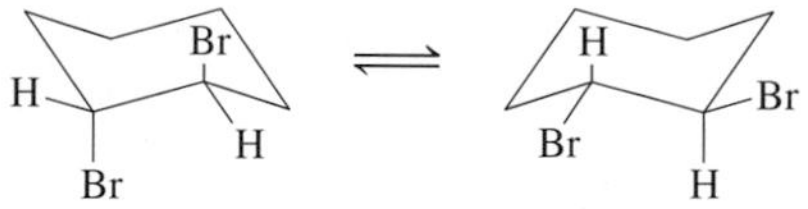

(1R,2R)-*trans*-1,2-Dibromocyclohexane

Anti addition to either conformation gives the trans-diaxial conformer initially.

12-9

(a) Only one diastereomer is formed (as a racemate):

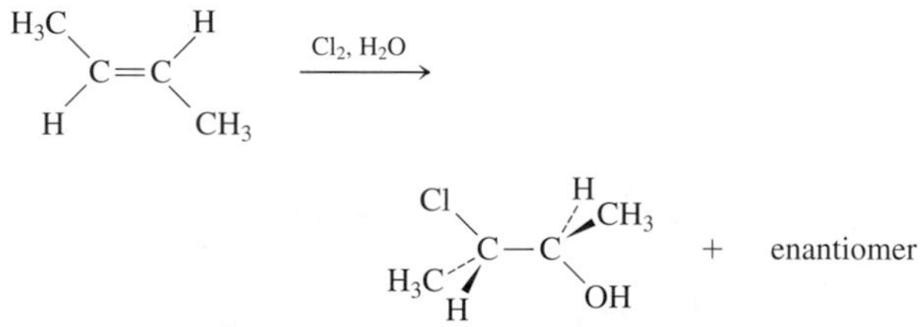

$$H_3C, H \quad C=C \quad H, CH_3 \quad \xrightarrow{Cl_2, H_2O}$$

Cl, CH₃ / H₃C—C—C—CH₃ / OH, H + enantiomer

(b) Two isomers are formed, but only one diastereomer of each (as racemates):

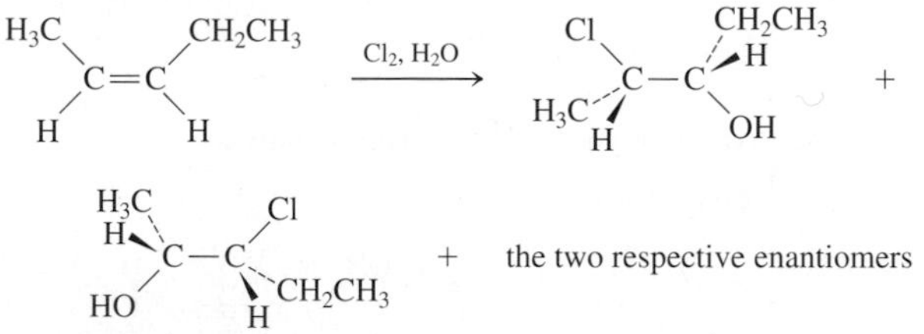

$$H_3C, CH_2CH_3 \quad C=C \quad H, H \quad \xrightarrow{Cl_2, H_2O}$$

Cl, CH₂CH₃ / H₃C—C—C—H / H, OH +

H₃C, H, Cl / C—C / HO, CH₂CH₃, H + the two respective enantiomers

12-10

(a) 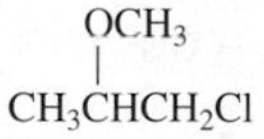 CH_3CHCH_2Cl with OCH₃ substituent (both enantiomers)

(b)

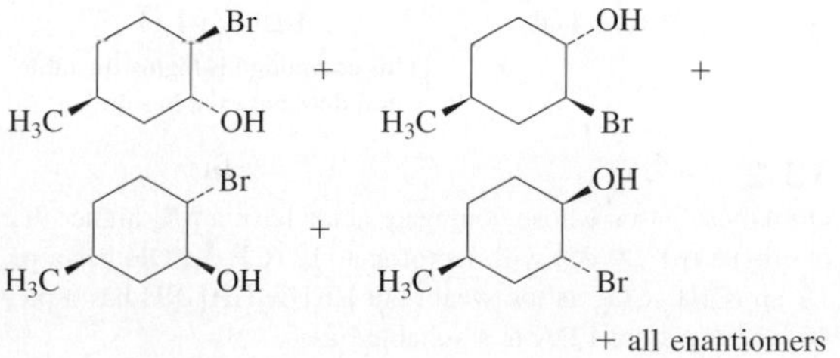

Br / H₃C—...—OH + OH / H₃C—...—Br +

Br / H₃C—...—OH + OH / H₃C—...—Br

+ all enantiomers

12-11

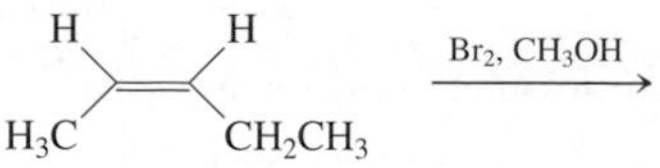

$$H, H \quad C=C \quad H_3C, CH_2CH_3 \quad \xrightarrow{Br_2, CH_3OH}$$

cis-2-Pentene

Br, H, CH₂CH₃ / H—C—C / H₃C (R) (R) OCH₃ + enantiomer

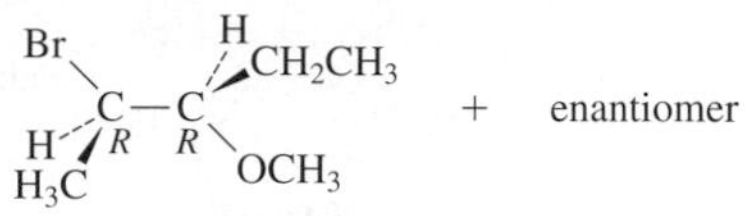

Opening of the bromonium ion can also give (3R,2R)- and (3S,2S)-3-bromo-2-methoxypentane.

12-12

Mercuration is followed by *intramolecular* trapping of the mercurinium ion by one of the hydroxy groups.

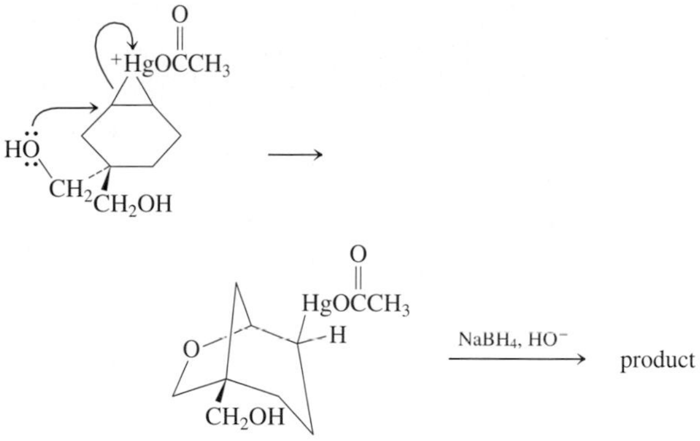

$$+HgOCCH_3 \text{ (with C=O)}$$

HO·· CH₂CH₂OH →

O—...—HgOCCH₃ (with C=O), H / CH₂OH $\xrightarrow{NaBH_4, HO^-}$ product

12-13

(a) $CH_3CH_2CH_2OH$

(b)

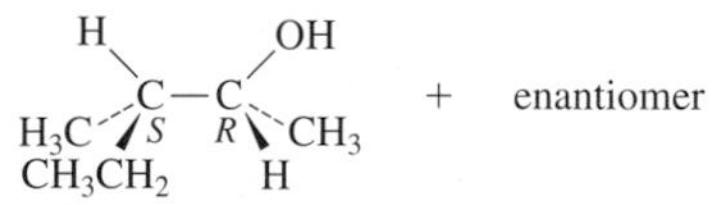

H, OH / H₃C—C—C—CH₃ / CH₃CH₂ (S) (R) H + enantiomer

12-14

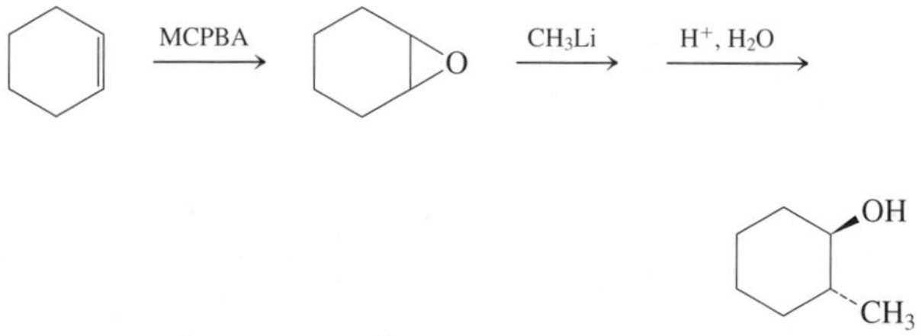

cyclohexene $\xrightarrow{MCPBA}$ epoxide $\xrightarrow{CH_3Li}$ $\xrightarrow{H^+, H_2O}$

cyclohexane with OH and CH₃

12-15

(a) pentyl chain with OH and HO, H + enantiomer

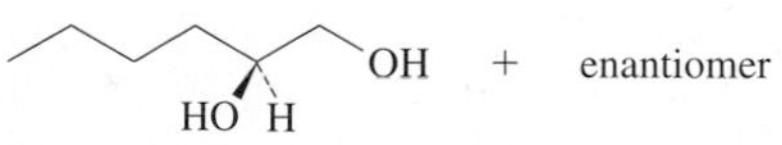

(b) cyclohexane with OH, OH + enantiomer

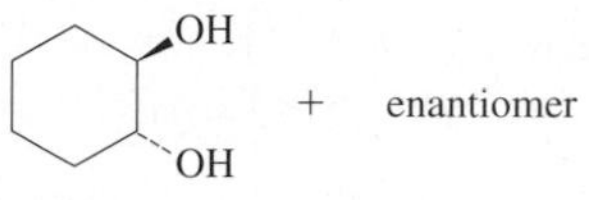

70%

(c)

HO, H, CH₂CH₃ / C—C / H, H₃C, OH + enantiomer

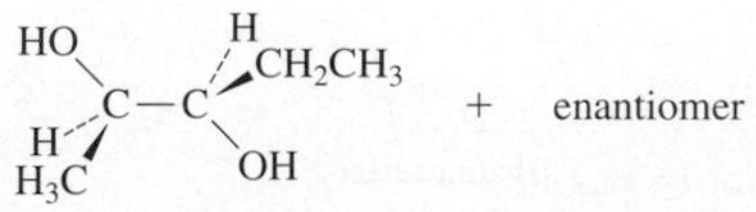

12-16

$$H_3C, CH_3 \atop H \quad C=C \quad H \xrightarrow{\text{H}_2\text{O}_2,\ \text{catalytic OsO}_4}$$

HO   OH
    C—C   same as   HO   H,CH₃
H₃C‖‖CH₃   C—C
   H   H   H₃C‖   OH
                                  H
Eclipsed   **Staggered**

Meso

$$H_3C, H \atop H \quad C=C \quad CH_3 \xrightarrow{\text{H}_2\text{O}_2,\ \text{catalytic OsO}_4}$$

HO   OH   HO   H₃C,H
    C—C   same as   C—C
H₃C‖   H   H₃C‖   OH
   H   CH₃   H
Eclipsed   **Staggered**

(R,R), (S,S)

12-17

C₁₂H₂₀

12-18

(a)
H₃C— (2-methyl cyclopentanone with CHO) + H₂C=O   **(b)** cyclopentanone + H₂C=O

(c)

12-19

Do not be fooled by the way structures are drawn.

is the same as

Therefore, the starting material is

12-20

Initiation

$$(\text{C}_6\text{H}_5)_2\text{PH} \xrightarrow{h\nu} (\text{C}_6\text{H}_5)_2\text{P}\cdot + \text{H}\cdot$$
Chain carrier

Propagation

$$\text{CH}_3(\text{CH}_2)_5\text{CH}=\text{CH}_2 + (\text{C}_6\text{H}_5)_2\text{P}\cdot \longrightarrow$$

$$\text{CH}_3(\text{CH}_2)_5\dot{\text{C}}\text{HCH}_2\text{P}(\text{C}_6\text{H}_5)_2$$
More stable radical

$$\text{CH}_3(\text{CH}_2)_5\dot{\text{C}}\text{HCH}_2\text{P}(\text{C}_6\text{H}_5)_2 + (\text{C}_6\text{H}_5)_2\text{PH} \longrightarrow$$

$$\text{CH}_3(\text{CH}_2)_5\text{CH}_2\text{CH}_2\text{P}(\text{C}_6\text{H}_5)_2 + (\text{C}_6\text{H}_5)_2\text{P}\cdot$$
Product

12-21

This is an irregular copolymer with both monomers incorporated in random numbers but regioselectively along the chain. Write a mechanism for its formation.

$$\left[(-\text{CH}_2\overset{\text{Cl}}{\underset{\text{Cl}}{\text{C}}})_m (-\text{CH}_2\overset{\text{H}}{\underset{\text{Cl}}{\text{C}}})_n - \right]$$

CHAPTER 13

13-1

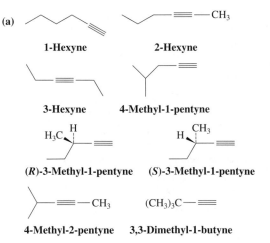

(a)

1-Hexyne   **2-Hexyne**

3-Hexyne   **4-Methyl-1-pentyne**

(R)-3-Methyl-1-pentyne   **(S)-3-Methyl-1-pentyne**

4-Methyl-2-pentyne   **3,3-Dimethyl-1-butyne**

(b) (R)-3-Methyl-1-penten-4-yne

(c)

3-Butyn-1-ol   **(S)-3-Butyn-2-ol**   **(R)-3-Butyn-2-ol**

2-Butyn-1-ol   **1-Butyn-1-ol**
(This compound is highly unstable
and does not exist in solution)

13-2

Only those bases whose conjugate acids have a pK_a higher than that of ethyne ($pK_a = 25$) will deprotonate it: $(\text{CH}_3)_3\text{COH}$ has a $pK_a \sim$ 18, so $(\text{CH}_3)_3\text{CO}^-$ is too weak; but $[(\text{CH}_3)_2\text{CH}]_2\text{NH}$ has a $pK_a \sim$ 40, and therefore LDA is a suitable base.

13-3

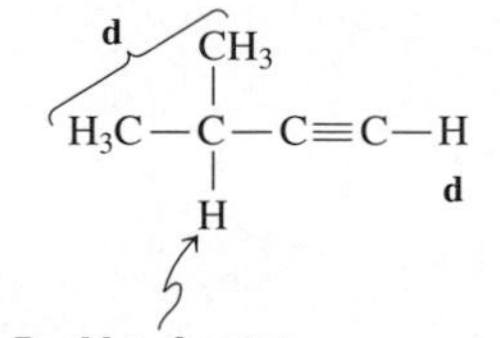

Doublet of septets
(or septet of doublets)

13-4

From the data in Section 11-7, we can calculate the heat of hydrogenation of the first π bond in the butynes.

$$CH_3CH_2C\equiv CH + H_2 \longrightarrow CH_3CH_2CH=CH_2$$
$$\Delta H° = -(69.9 - 30.3) = -39.6 \text{ kcal mol}^{-1}$$

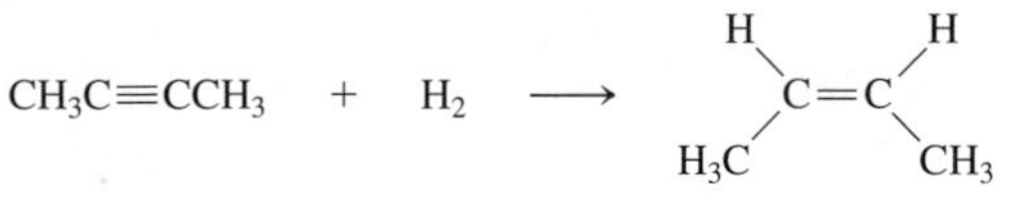

$$\Delta H° = -(65.1 - 28.6) = -36.5 \text{ kcal mol}^{-1}$$

In both cases, more heat is released than expected for a simple C–C double bond.

13-5

The starting materials in each case can be

(a)

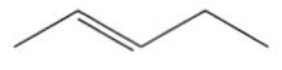

(b) 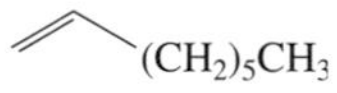 $(CH_2)_5CH_3$

(c)

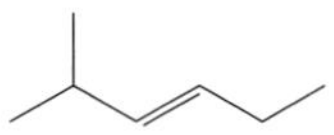

13-6

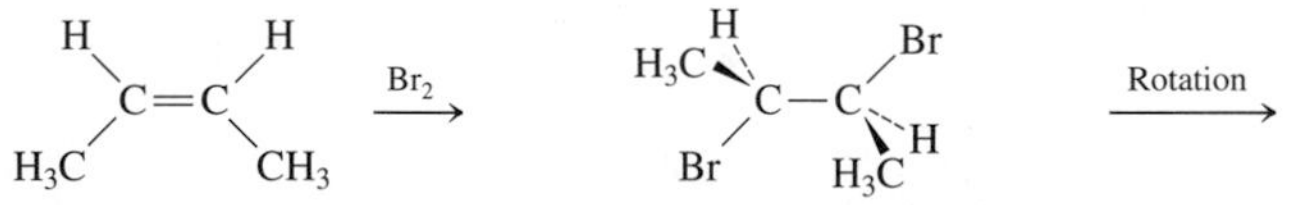

cis-**2-Butene**

(2S,3S)- [and
(2R,3R)] 2,3-Dibromobutane

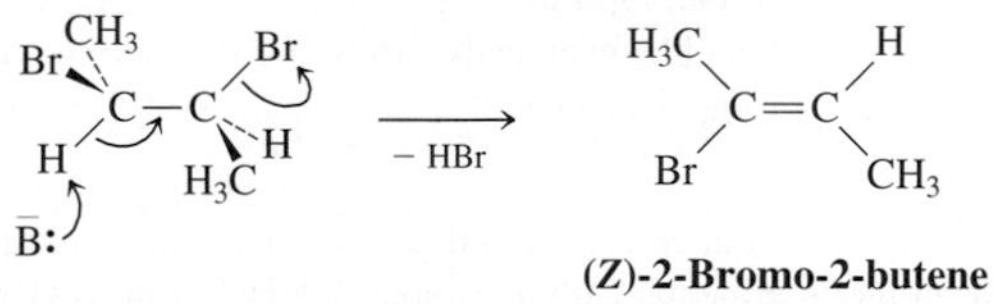

(Z)-2-Bromo-2-butene

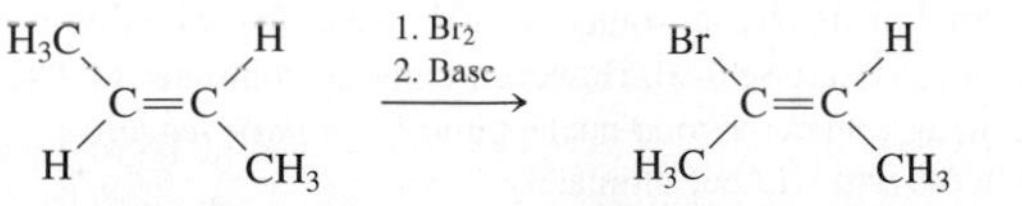

trans-**2-Butene**   **(E)-2-Bromo-2-butene**

13-7

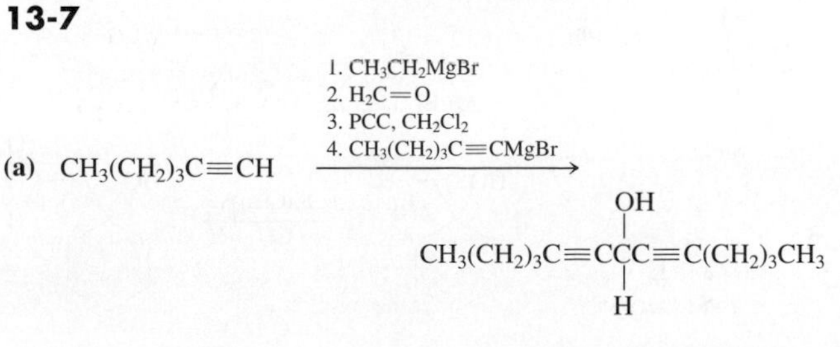

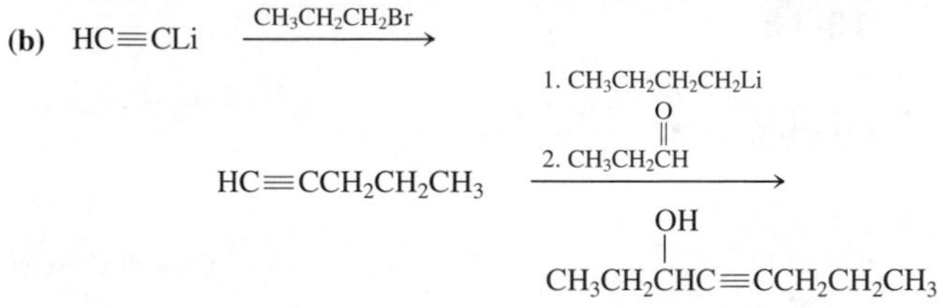

(a) $CH_3(CH_2)_3C\equiv CH$
1. CH_3CH_2MgBr
2. $H_2C=O$
3. PCC, CH_2Cl_2
4. $CH_3(CH_2)_3C\equiv CMgBr$

(b) $HC\equiv CLi$ $\xrightarrow{CH_3CH_2CH_2Br}$

$HC\equiv CCH_2CH_2CH_3$

1. $CH_3CH_2CH_2CH_2Li$
2. CH_3CH_2CH (with C=O)

$CH_3CH_2CHC\equiv CCH_2CH_2CH_3$ (with OH)

13-8

$\equiv$—Li   + CH_3CHO

13-9

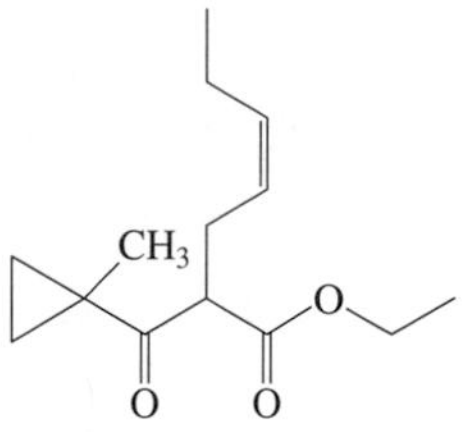

13-10

$CH_3CH_2C\equiv CH$
1. $CH_3CH_2CH_2CH_2Li$
2. (epoxide)
3. H_2, Lindlar cat.

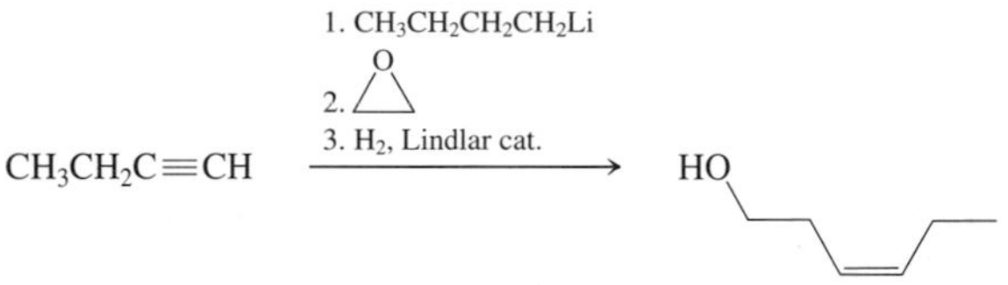

13-11

In the presence of sodium amide, the terminal alkyne unit is deprotonated. Electron transfer to a negatively charged alkynyl group is not favored.

$$CH_3(CH_2)_2C\equiv C(CH_2)_4C\equiv CH \xrightarrow{NaNH_2, \text{ liquid } NH_3}$$

$$CH_3(CH_2)_2C\equiv C(CH_2)_4C\equiv C:^- \xrightarrow{Na, \text{ liquid } NH_3} \xrightarrow{H^+, H_2O}$$

$$CH_3(CH_2)_2CH=CH(CH_2)_4C\equiv CH$$
trans
75%

13-12

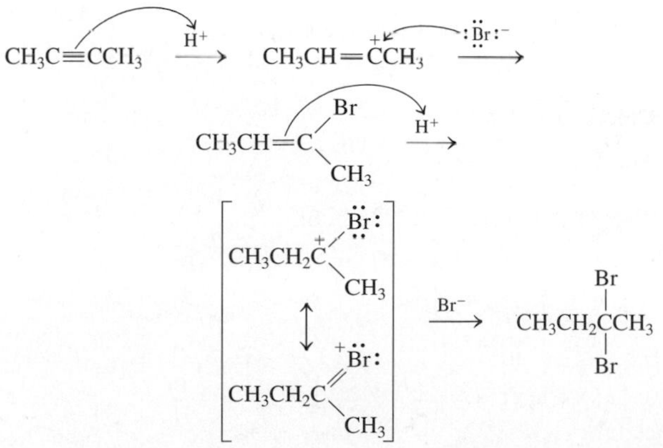

13-13

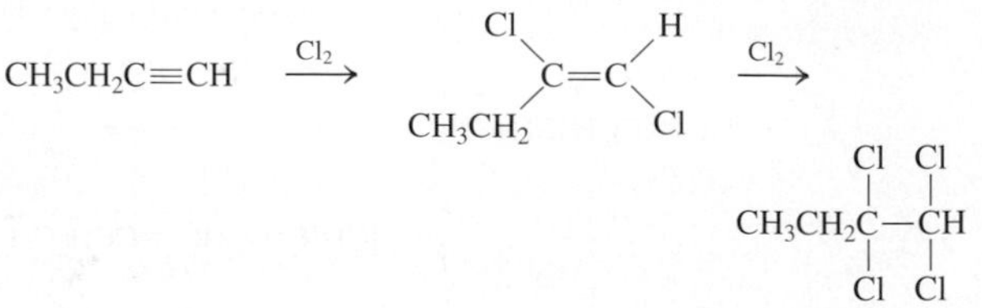

$$CH_3CH_2C\equiv CH \xrightarrow{Cl_2}$$ (product) $$\xrightarrow{Cl_2}$$ $$CH_3CH_2\overset{Cl\ Cl}{\underset{Cl\ Cl}{C-CH}}$$

13-14

(a) CH_3CHO (b) $CH_3\overset{O}{\overset{\|}{C}}CH_3$ (c) $CH_3CH_2\overset{O}{\overset{\|}{C}}CH_3$

(d) $CH_3CH_2\overset{O}{\overset{\|}{C}}CH_3$

(e)

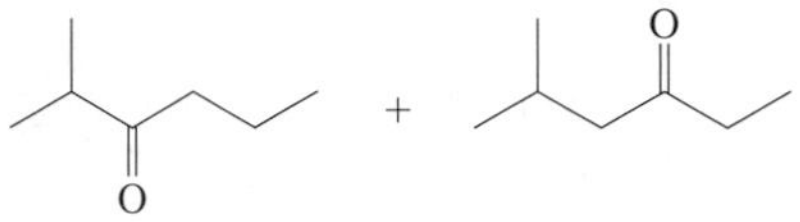

13-15

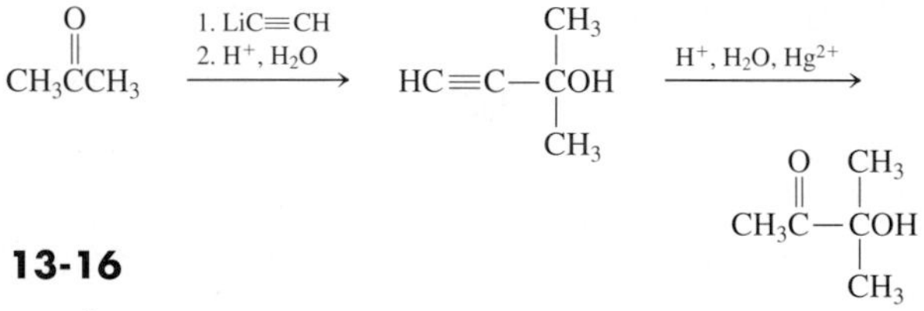

13-16

$$2\ \text{(cyclohexene)} + BH_3$$

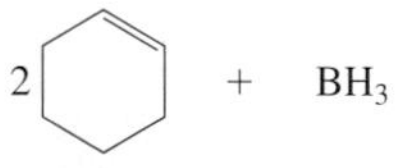

13-17

(a) CH_3CHO (b) CH_3CH_2CHO
(c) $CH_3CH_2CH_2CHO$

13-18

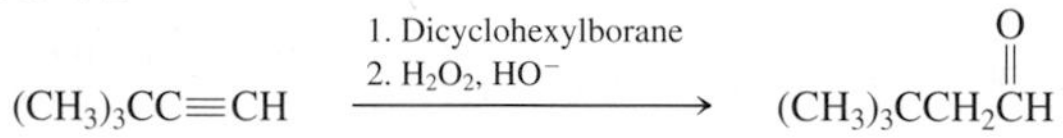

$$(CH_3)_3CC\equiv CH \xrightarrow[\text{2. H}_2\text{O}_2,\ \text{HO}^-]{\text{1. Dicyclohexylborane}} (CH_3)_3CCH_2\overset{O}{\overset{\|}{C}}H$$

13-19

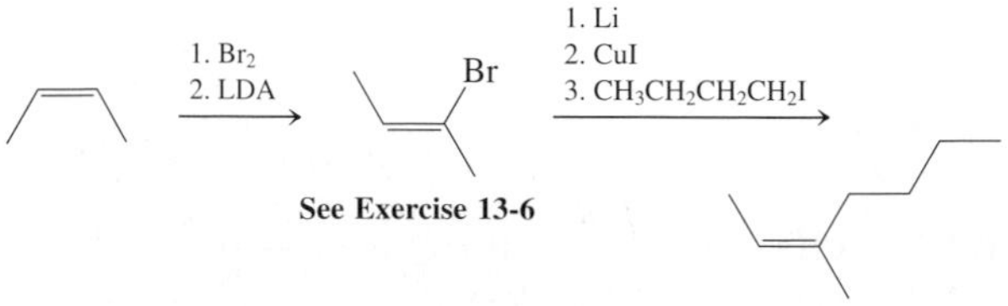

See Exercise 13-6

CHAPTER 14

14-1

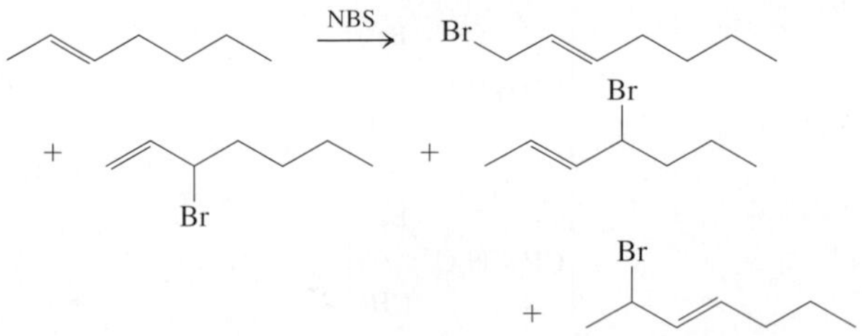

14-2

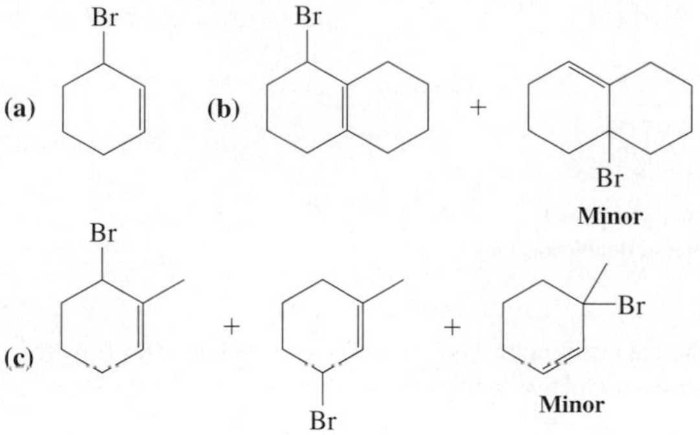

(a)

(b) +
Minor

(c) + +
Minor

Bromination at the primary allylic position is too slow.

14-3

The intermediate allylic cation is achiral.

14-4

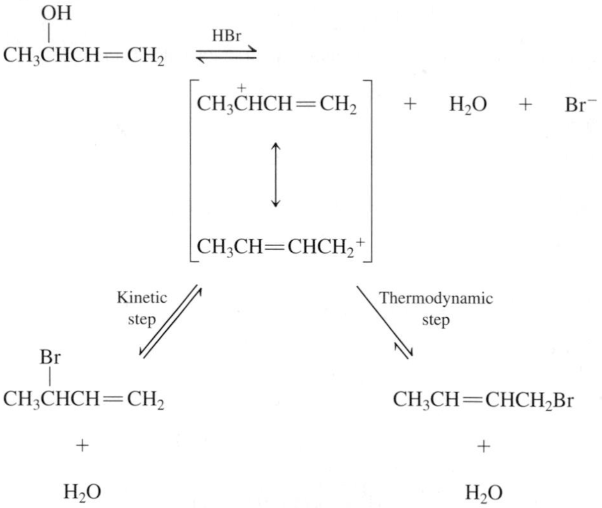

The intermediate allylic cation is trapped by bromide at C1 and C3 with a relative rate ratio of 15:85 (kinetic product ratio). This process is reversible, eventually furnishing the thermodynamic product: 1-bromo-2-butene.

14-5

Upon ionization, chloride ion does not immediately diffuse away from the intermediate allylic cation. Reattachment to give either starting material or its allylic isomer occurs—recall the reversibility of S_N1 reactions (Exercise 7-3). However, chloride continues to dissociate, allowing acetate, a good nucleophile *but a poor leaving group* (Table 6-4), to win out ultimately.

14-6

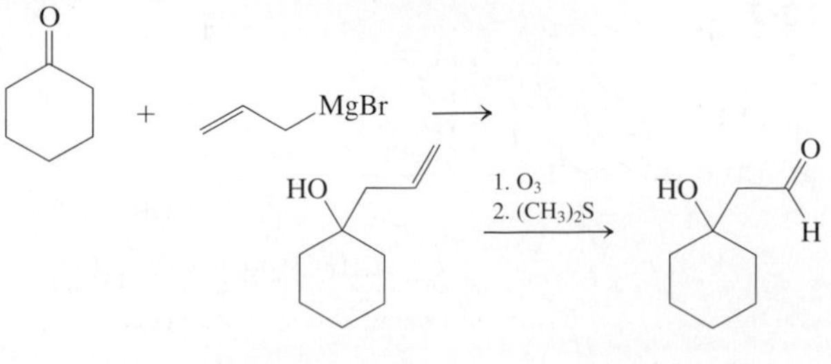

14-7

(a) 5-Bromo-1,3-cycloheptadiene

(b) (E)-2,3-Dimethyl-1,3-pentadiene

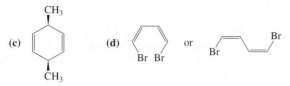

(c)   **(d)**   or

14-8

An internal trans double bond is more stable than a terminal double bond by about 2.7 kcal mol^{-1} (see Figure 11-18). This difference plus the expected resonance energy of 3.5 kcal mol^{-1} add up to 6.2 kcal mol^{-1}, pretty close to the observed value.

14-9

The product is the delocalized pentadienyl radical.

14-10

(a)   HOCH$_2$CHCHCH$_2$OH $\xrightarrow{\text{PBr}_3}$

BrCH$_2$CHCHCH$_2$Br $\xrightarrow{\text{(CH}_3)_3\text{CO}^-\text{K}^+,\ \text{(CH}_3)_3\text{COH}}$

(b)

$\xrightarrow[\text{– HBr}]{\text{Br}_2,\ h\nu}$ $\xrightarrow[\substack{\text{– CH}_3\text{OH,} \\ \text{– NaBr}}]{\text{CH}_3\text{O}^-\text{Na}^+}$ $\xrightarrow[\text{– HBr}]{\text{NBS}}$

$\xrightarrow[\text{– (CH}_3)_3\text{COH, – KBr}]{\text{(CH}_3)_3\text{CO}^-\text{K}^+,\ \text{(CH}_3)_3\text{COH}}$

14-11

(a)   Same product for both modes of addition

(b)   +   Both cis and trans

Addition of HX to unsubstituted cycloalka-1,3-dienes in either 1,2- or 1,4- manner gives the same product because of symmetry.

14-12

$\xrightarrow{\text{Br}_2}$

$+$ Br$^-$ $\longrightarrow$

$\xrightarrow{\text{Br}_2}$

$+$ Br$^-$ $\longrightarrow$

14-13

(a), (b) Electron rich, because alkyl groups are electron donors.
(c), (d) Electron poor, because the carbonyl group is electron withdrawing by induction and resonance and the fluoroalkyl group is so by induction only.

14-14

14-15

(a)

(b)

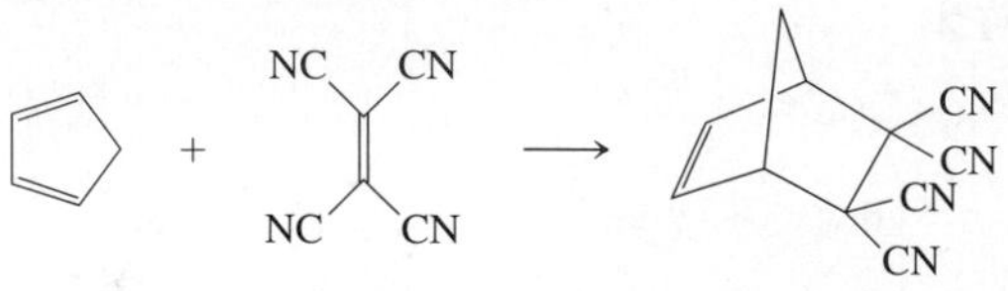

(c)

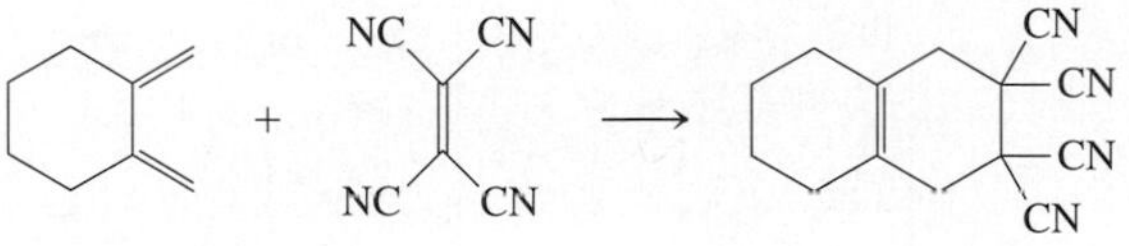

14-16

(a)

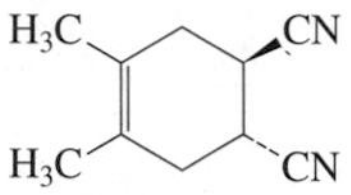

(b)

H$_3$C   CH$_3$ or   H$_3$C   CH$_3$ +   F F

CH$_3$

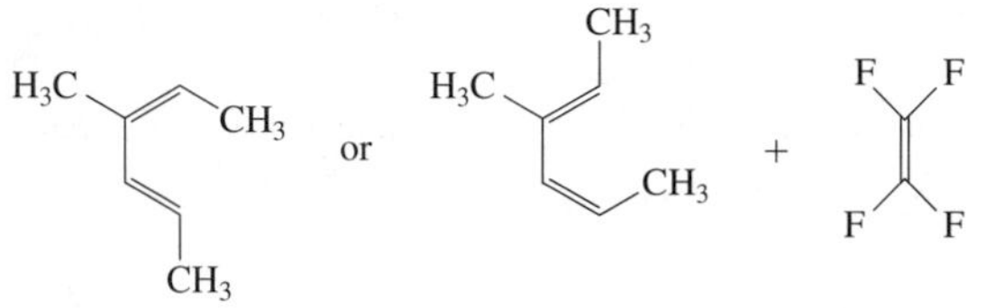

14-17

The cis,trans isomer cannot readily reach the *s*-cis conformation because of steric hindrance.

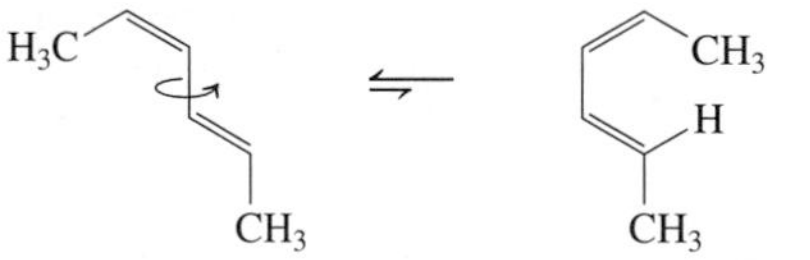

H$_3$C

CH$_3$ ⇌   CH$_3$ H

CH$_3$ CH$_3$

Sterically hindered

14-18

(a)

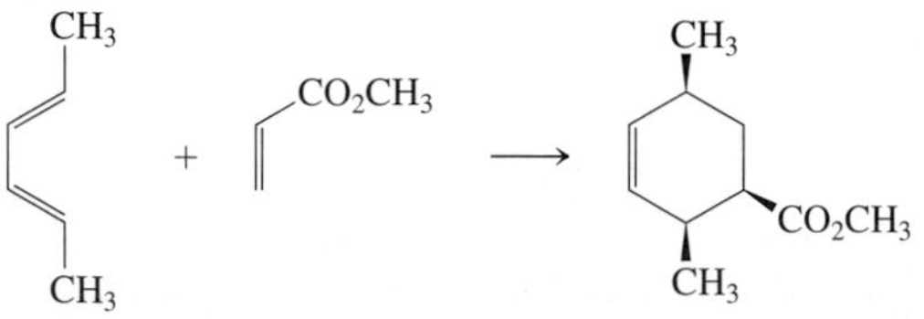

(b)

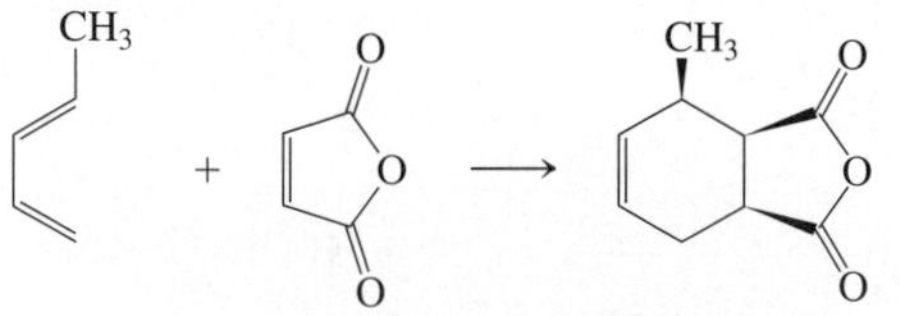

(c)

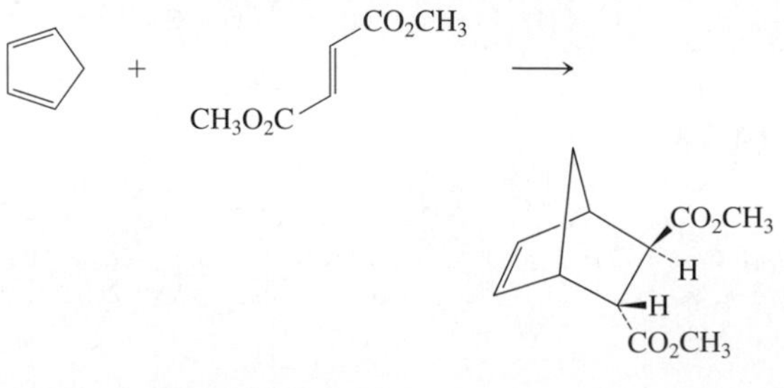

14-19

The first product is the result of exo addition; the second product, the outcome of endo addition.

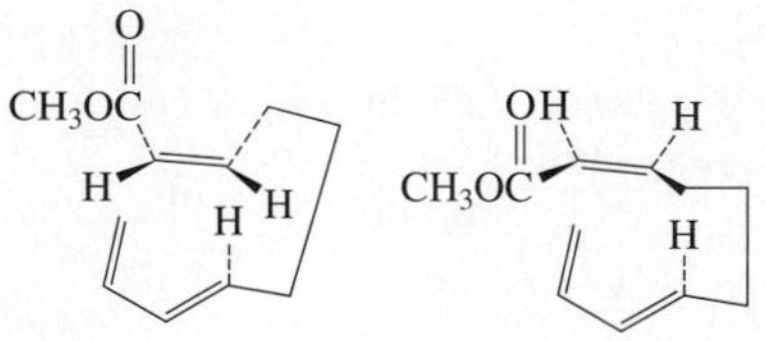

14-20

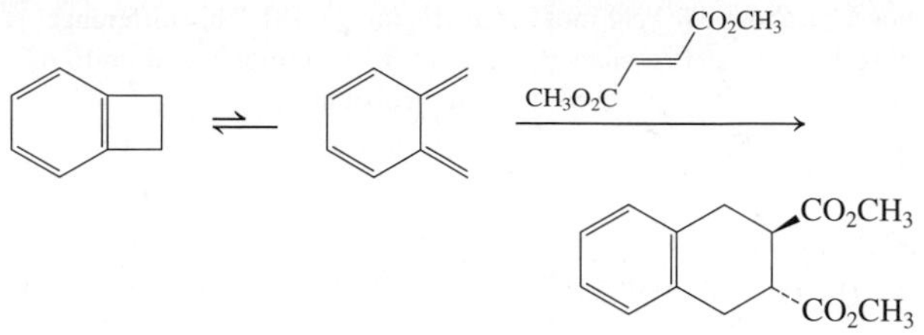

14-21

Conrotatory. Make a model.

14-22

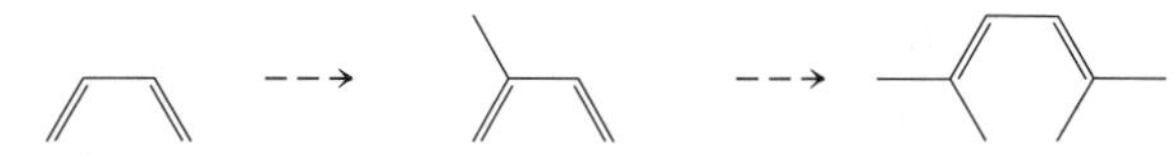

$\lambda_{max} = 217$ nm   Calculated 222 nm   Calculated 237 nm
                                   (measured 222.5 nm)   (measured 241.5 nm)

CHAPTER 15

15-1

(a) 1-Chloro-4-nitrobenzene (*p*-chloronitrobenzene)
(b) 1-Deuterio-2-methylbenzene (*o*-deuteriotoluene)
(c) 2,4-Dinitrophenol

15-2

(a)

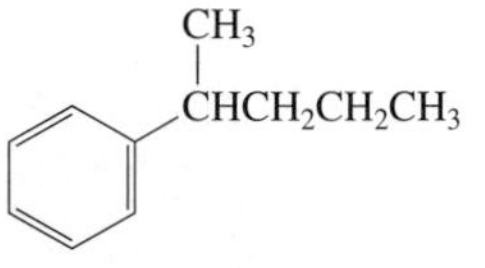

(b)

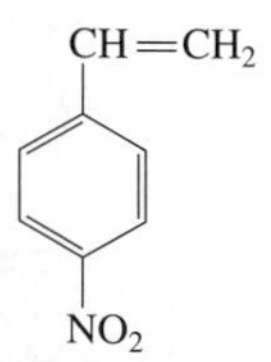

(c)

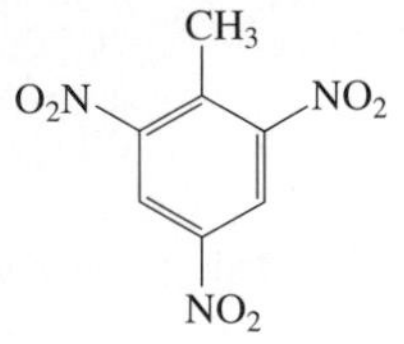

15-3

(a) 1,3-Dichlorobenzene (*m*-dichlorobenzene)
(b) 2-Fluorobenzenamine (*o*-fluoroaniline)
(c) 1-Bromo-4-fluorobenzene (*p*-bromofluorobenzene)

15-4

1,2-Dichlorobenzene

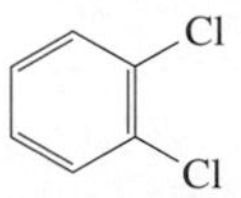

and

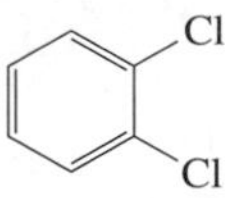

1,2,4-Trichlorobenzene

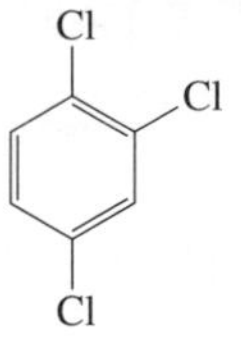

and

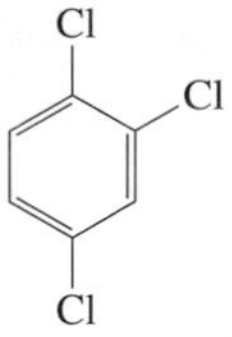

15-5

Compound B has lost its cyclic arrangement of six π electrons and therefore its aromaticity. Thus, ring opening is endothermic.

15-6

The unsymmetrically substituted 1,2,4-trimethylbenzene exhibits the maximum number, nine, of ^{13}C NMR lines. Symmetry reduces this count to six in 1,2,3- and three in 1,3,5-trimethylbenzene.

15-7

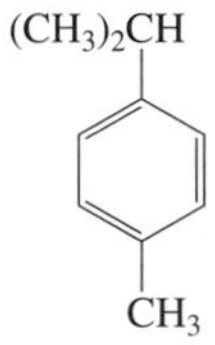

$(CH_3)_2CH$

CH_3

15-8

(a)

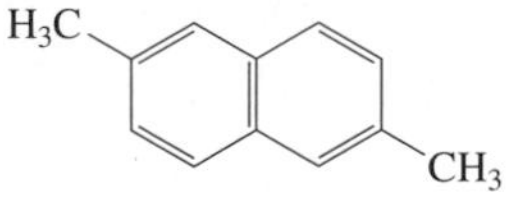

H_3C CH_3

(b)

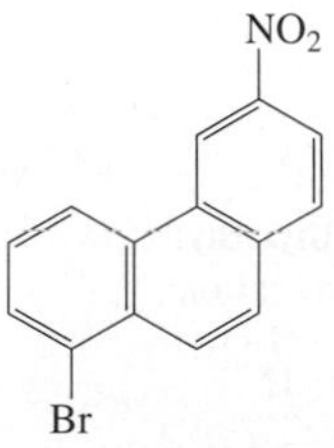

NO_2

Br

(c)

C_6H_5

C_6H_5

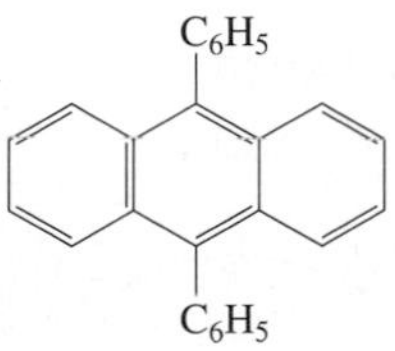

(d) 9-Bromophenanthrene

(e) 5-Nitro-2-naphthalenesulfonic acid

15-9

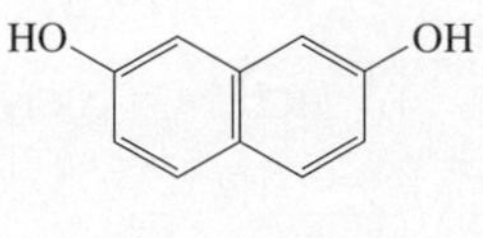

HO OH

15-10

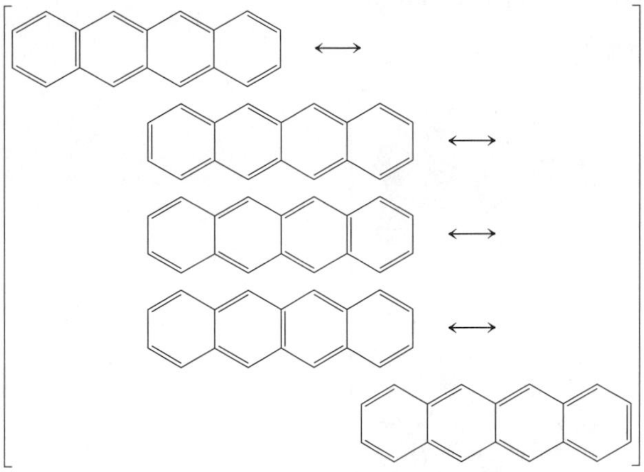

The maximum number of aromatic benzene Kekulé rings is two, in three of the resonance forms (the first, third, and fourth).

15-11

This is an unusual Diels-Alder reaction in which one molecule acts as a diene, the other as a dienophile. Make models.

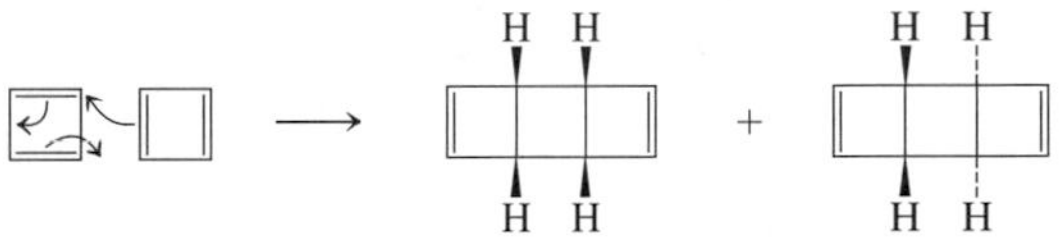

Endo product   **Exo product**

Note the surprising result of applying the general stereochemical scheme on p. 607 here (replace the i's in starting material and product by a bond).

15-12

No. Cyclooctatetraene has localized double bonds. Double-bond shift results in geometrical isomerization and not in a resonance form, as shown for 1,2-dimethylcyclooctatetraene.

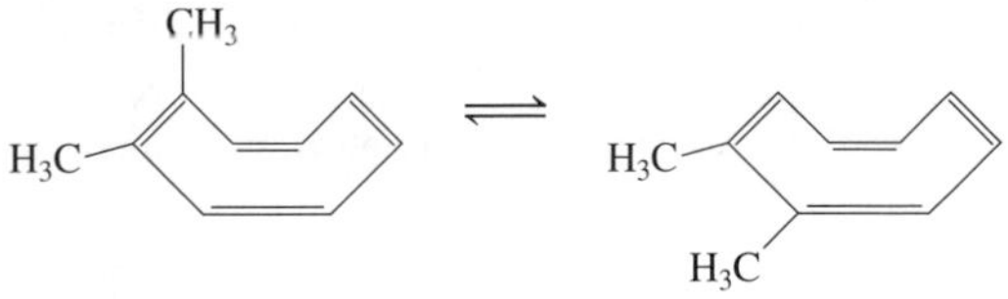

15-13

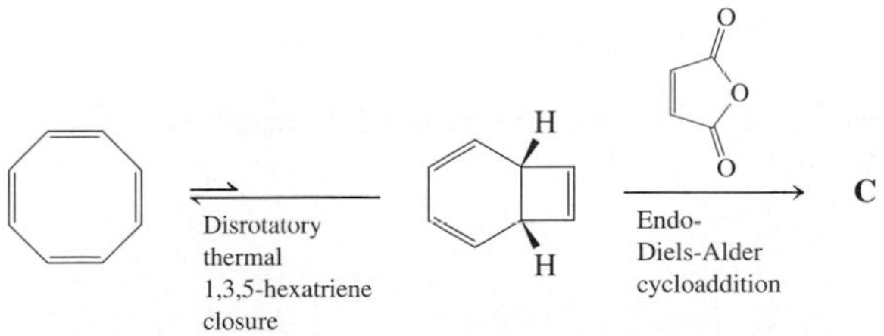

15-14

They are all nonplanar because of bond angle (e.g., a flat all-*cis*-[10]annulene requires C_{sp^2} bond angles of 144°, a considerable distortion from the normal value), eclipsing, and transannular strain (e.g., the two inside hydrogens of *trans,cis,trans,cis,cis*-[10]annulene occupy the same region in space).

15-15

(a), (c), (d) Aromatic; (b), (e) antiaromatic

15-16

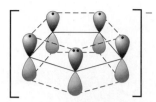

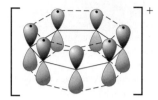

15-17

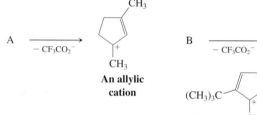

A $\xrightarrow{- CF_3CO_2^-}$ **An allylic cation**

B $\xrightarrow{- CF_3CO_2^-}$ **An antiaromatic cyclopentadienyl cation**

15-18

(a), (b) Aromatic; (c) antiaromatic

15-19

The dianion is an aromatic system of 10 π electrons, but pentalene has $4n$ π electrons.

15-20

Attack of E$^+$ at C1 gives the aromatic cycloheptatrienyl cation framework:

Attack of Nu$^-$ at C4 gives the aromatic cyclopentadienyl anion framework:

15-21

According to Table 12-1, common electrophilic additions to alkenes are exothermic by at most 27 kcal mol^{-1}. Because such additions to benzene would cause the loss of about 30 kcal mol^{-1} in resonance energy, they are not thermodynamically possible.

15-22

(reaction scheme)

Molecular weight = 84

15-23

$\xrightarrow{\text{HF, SbF}_5\text{, SO}_2\text{ClF,} \atop \text{SO}_2\text{F}_2\text{, } -129°C}$

H H δ = 5.69 ppm
H δ = 9.58 ppm
H δ = 8.22 ppm
H δ = 9.42 ppm

The NMR assignments correspond to the amount of charge expected at the various hexadienyl cation carbons on the basis of resonance.

15-24

(a) (reaction scheme with SO$_3$H ... $+ H^+ \rightleftharpoons$ ... $\xrightarrow{-H^+}$ ... $\rightleftharpoons$ ... $+ SO_3$)

(b) (reaction scheme) $+ HOH \longrightarrow$ ... $\xrightarrow{H^+ \text{ shift}}$

15-25

$(CH_3)_3CCl + AlCl_3 \longrightarrow (CH_3)_3C^+ + AlCl_4^-$

1,1-Dimethylethyl (*tert*-butyl) cation

$(CH_3)_3C^+ +$ (benzene) $\longrightarrow$ (cation with $-C(CH_3)_3$)

(cation with $-C(CH_3)_3$) $+ AlCl_4^- \longrightarrow$ (product with $-C(CH_3)_3$) $+ HCl + AlCl_3$

15-26

$CH_3CH{=}CH_2$ + H^+ →(Markovnikov addition)

$CH_3\overset{+}{C}HCH_3$ →(C_6H_6) [cyclohexadienyl cation with H and $CH(CH_3)_2$] →

[benzene ring]—$CH(CH_3)_2$ + H^+

15-27

[benzene ring with H_3C, CH_3, H_3C, CH_3]

1,2,4,5-Tetramethylbenzene (Durene)

15-28

$CH_3CH_2\overset{H}{C}HCH_2{-}Cl$ + $AlCl_3$ →

$CH_3CH_2\overset{+}{C}HCH_3$ + $AlCl_4^-$ →(C_6H_6) →

[benzene ring]—$\overset{CH_3}{C}HCH_2CH_3$ + H^+

15-29

$:\overset{-}{C}{\equiv}\overset{+}{O}:$ + H^+ ⇌ $[H{-}C{\equiv}\overset{+}{O}: \longleftrightarrow H{-}\overset{+}{C}{=}\overset{..}{O}:]$

Methanoyl (formyl) cation

[benzene ring with H_3C] + $H{-}\overset{+}{C}{=}\overset{..}{O}$ →

[cyclohexadienyl cation with H, CH, O, H_3C] → [benzene ring with H_3C and CHO] + H^+

The spectral data for the methanoyl cation indicate the dominant contribution of [H–C≡O : $^+$] in its resonance-structural description. This species can be viewed as a positively charged oxygen analog of ethyne ($\delta_{^{13}C}$ = 71.9 ppm.; $\tilde{\nu}_{C{\equiv}C}$ = 1974 cm^{-1}), the oxygen and the charge causing the observed relatively deshielded carbon resonance, in addition to the strengthening of the triple bond and the associated relatively high wavenumber band in the IR spectrum.

CHAPTER 16

16-1

4-(*N,N*-Dimethylamino)benzaldehyde: Formulation of dipolar resonance forms involving the electron-withdrawing carbonyl group reveals the presence of partial positive charges at its ortho (but not meta) carbons, reflected in the relative deshielding of the associated two (green) hydrogens. Conversely, formulation of dipolar reso-

nance forms involving the electron-donating amino substituent indicates the presence of partial negative charges at its ortho (but not meta) carbons, causing relative shielding of the corresponding (red) hydrogens.

1-Methoxy-2,4-dinitrobenzene: By a similar argument, the relatively deshielded hydrogens (blue and green) are those located ortho and para to the electron-withdrawing nitro functions, the third hydrogen (red) is placed ortho to the electron-donating methoxy substituent. The "extra" deshielding of the hydrogen (blue) at C3 may be ascribed to its relative proximity to the two nitro groups, maximizing their inductive effect on their common neighbor.

16-2

Relative to benzene, C1 is deshielded by the strong inductive withdrawing effect of the oxygen, C2 and C4 are shielded because resonance of the benzene π system with the oxygen lone electron pairs places partial negative charges at these positions, and C3 is essentially unaffected.

16-3

(a), (d) Activated; (b), (c) deactivated

16-4

(d) > (b) > (a) > (c)

16-5

Methylbenzene (toluene) is activated and will consume all of the electrophile before the latter has a chance to attack the deactivated ring of (trifluoromethyl)benzene.

16-6

Ortho attack

[resonance structures of ortho attack on anisole, $\overset{..}{O}CH_3$, with E, H]

Meta attack

[resonance structures of meta attack on anisole, $\overset{..}{O}CH_3$, with E, H]

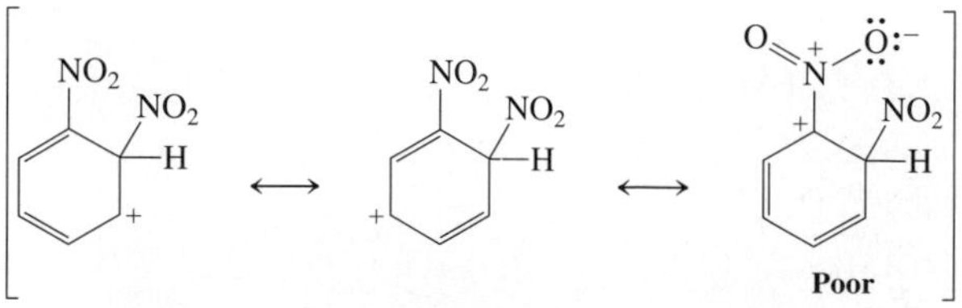

Para attack

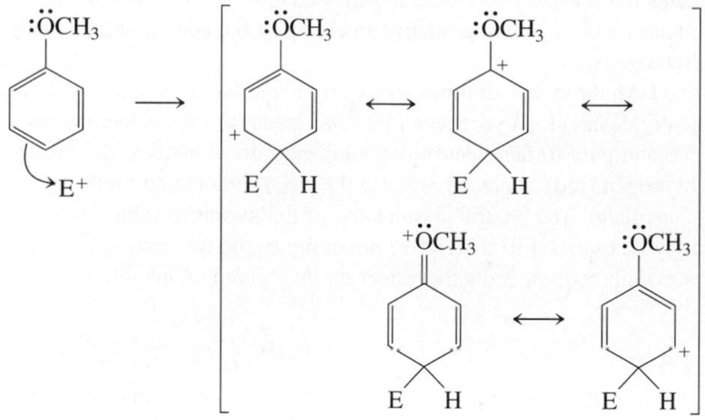

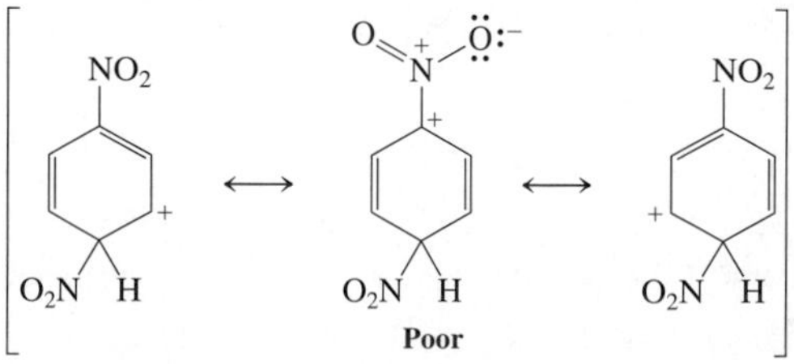

16-9

(a), (c) Electron withdrawing by resonance and induction.
(b) Electron withdrawing by induction.
(d) The phenyl substituent acts as a resonance donor.

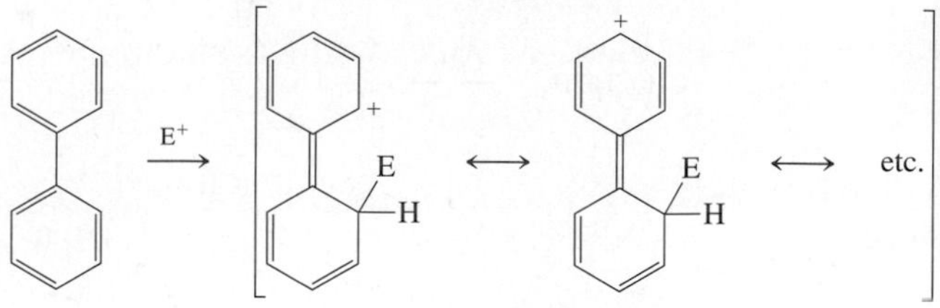

Total of six resonance structures

16-7

Benzenamine (aniline) is completely protonated in strong acid. The lone electron pair is no longer available for resonance with the ring. Hence, the ammonium substituent is an inductive deactivator and meta director.

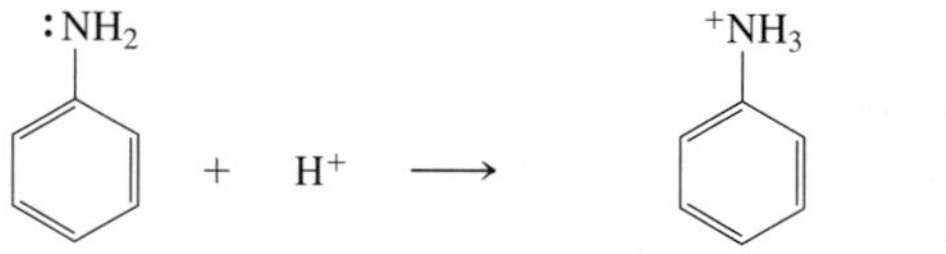

$pK_a = 4.60$

Benzenammonium ion
(Anilinium ion)

16-10

A

16-8

Ortho attack:

Poor

Para attack:

Poor

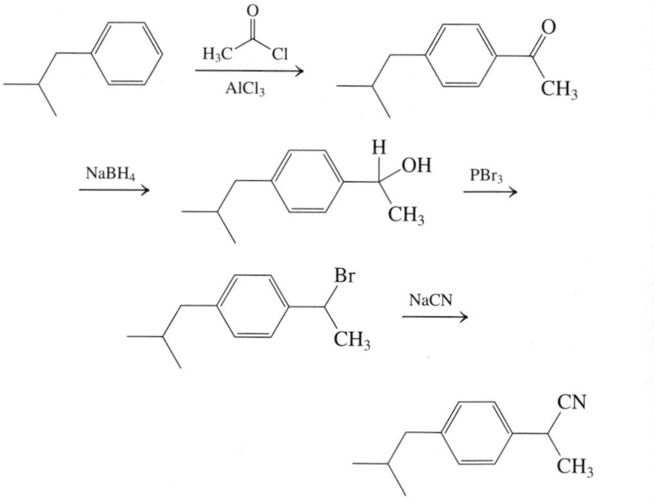

16-11

(a)

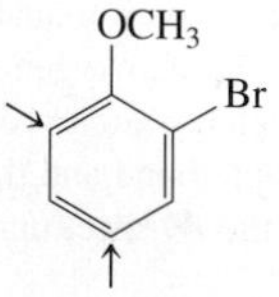

(b)

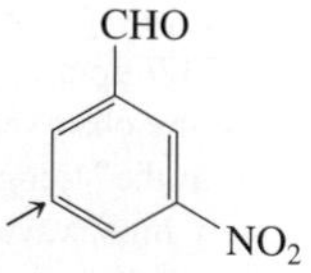

(c)

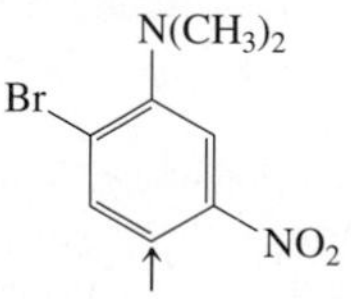

(d)

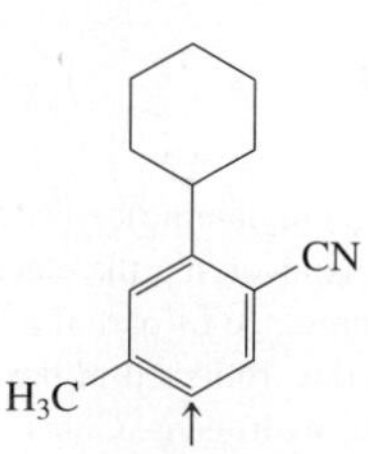

16-12

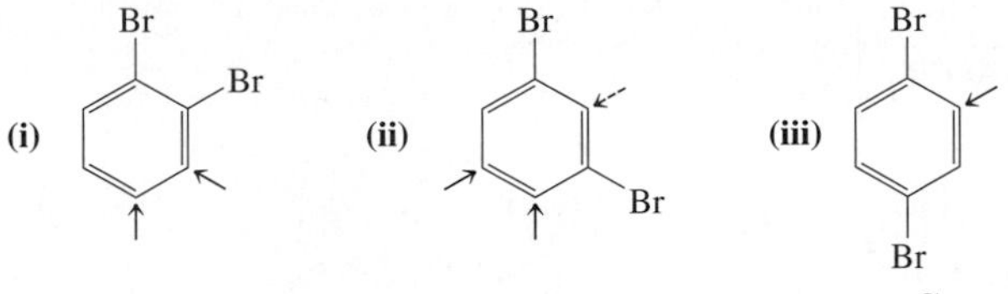

(i)
Br
Br
(ii)
Br
Br
(iii)
Br
Br

A **B** **C**

16-13

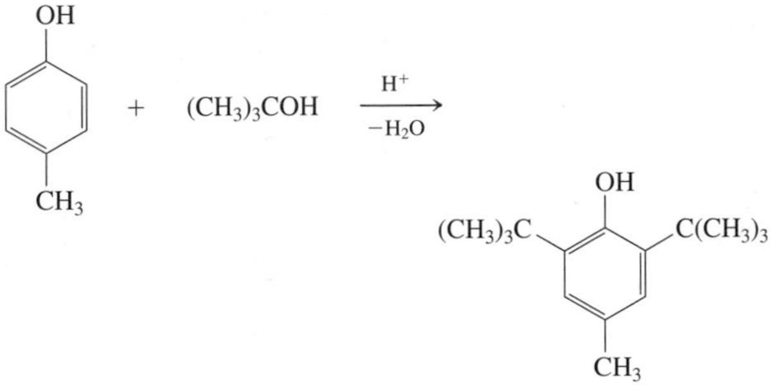

OH
CH3
+ (CH3)3COH
H+
−H2O
OH
(CH3)3C
C(CH3)3
CH3

16-14

No, because the nitrogen is introduced ortho and para to bromine only.

16-15

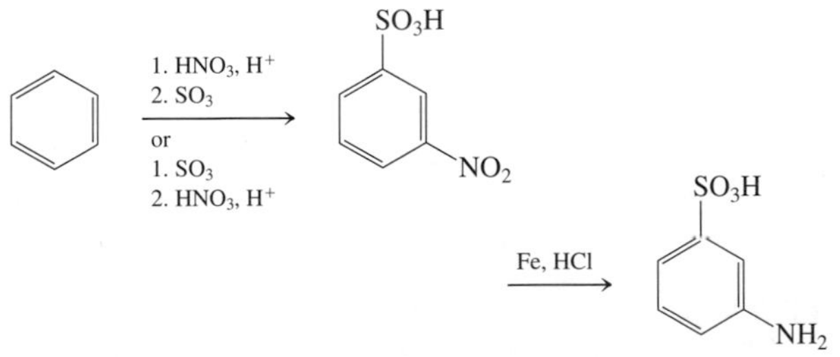

1. HNO3, H+
2. SO3
or
1. SO3
2. HNO3, H+
SO3H
NO2
Fe, HCl
SO3H
NH2

16-16

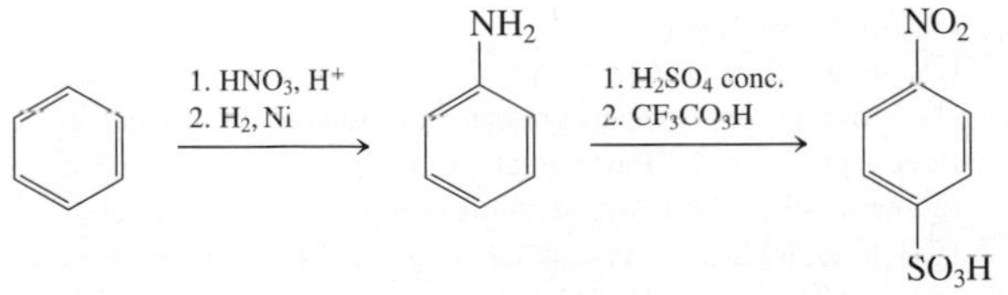

1. HNO3, H+
2. H2, Ni
NH2
1. H2SO4 conc.
2. CF3CO3H
NO2
SO3H

16-17

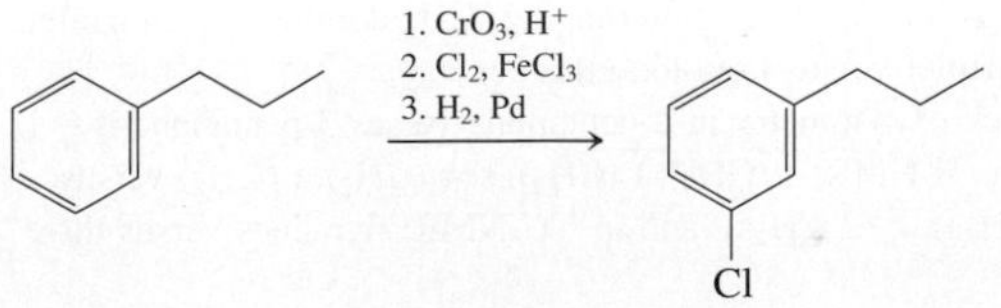

1. CrO3, H+
2. Cl2, FeCl3
3. H2, Pd
Cl

16-18

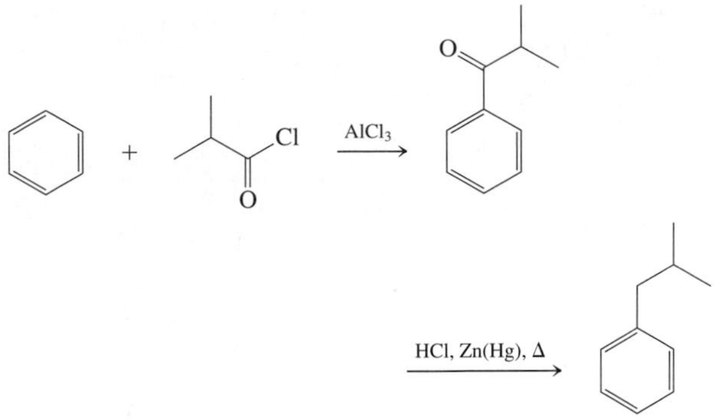

+
Cl
O
AlCl3
O
HCl, Zn(Hg), Δ

Direct Friedel-Crafts alkylation of benzene with 1-chloro-2-methylpropane gives (1,1-dimethylethyl)benzene (*tert*-butylbenzene) by rearrangement of the carbon electrophile (see Section 15-13).

16-19

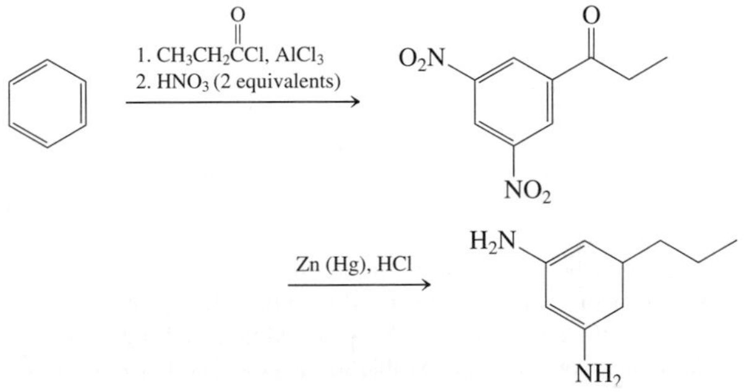

1. CH3CH2CCl, AlCl3
2. HNO3 (2 equivalents)
O2N
O
NO2
Zn (Hg), HCl
H2N
NH2

16-20

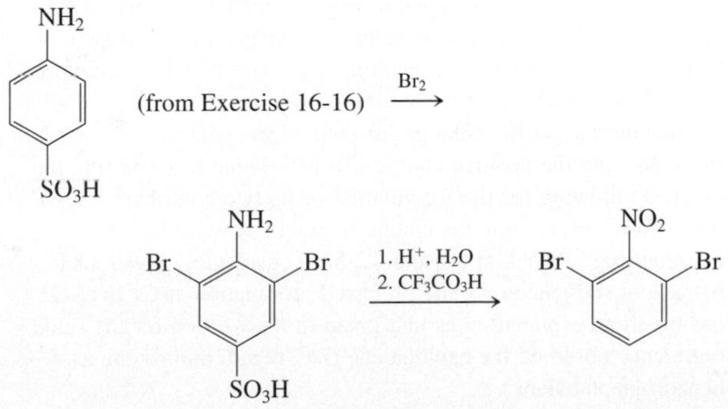

NH2
SO3H
(from Exercise 16-16)
Br2
NH2
Br
Br
SO3H
1. H+, H2O
2. CF3CO3H
NO2
Br
Br

16-21

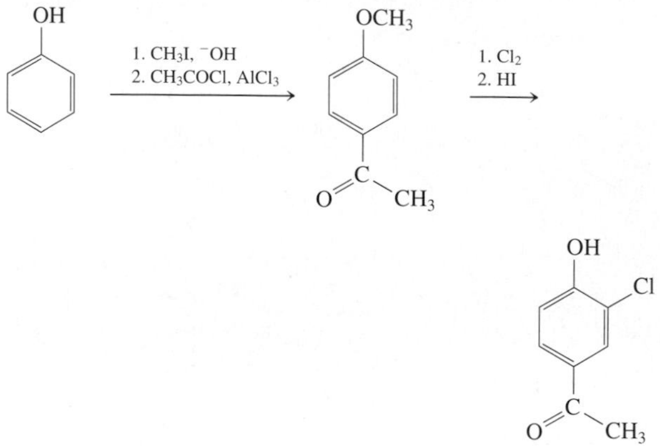

16-22

(a) At C4; (b) at C5 and C8; (c) at C6 and C8.

16-23

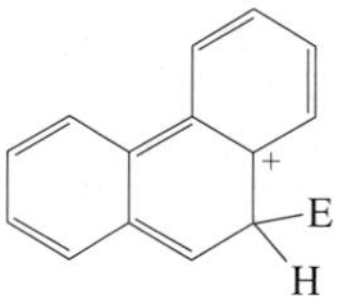

16-24

The answer is derived by inspection of the respective sets of resonance forms of the cations generated by protonation at the given positions; only the crucial ones are depicted here, although we recommend that you write all possible forms as an additional exercise.

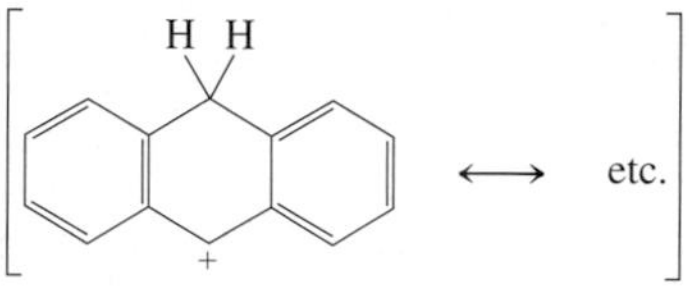

The essence of the analysis is, then, in each case, to count up the number of resonance contributors that contain intact benzene rings. You will recognize the important difference between protonation at C9 (which generates two separate benzene nuclei) and that at C1 and C2, which furnishes naphthalene fragments ("less aromatic" than two benzene rings, see Section 15-6). The [C9–H]$^+$ structure shown here already has four benzenoid resonance forms, even without moving the positive charge into one of the adjacent benzene rings. Moving the positive charge still leaves one benzene ring untouched (allowing for the formulation of its two contributing resonance forms every time the charge is moved around the other benzene ring). Attack at C1 and C2 has considerably fewer such benzenoid substructures. The preferred protonation at C1 over C2 has the same explanation as that given in this section for the same preference observed for naphthalene (i.e., regard anthracene as a benzo[*b*]naphthalene).

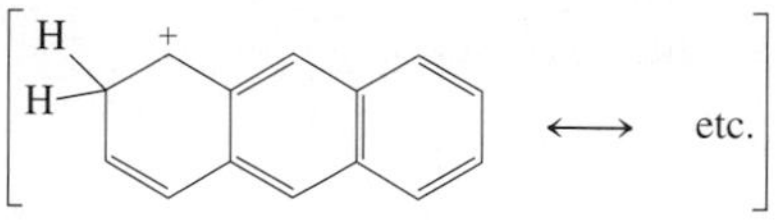

CHAPTER 17

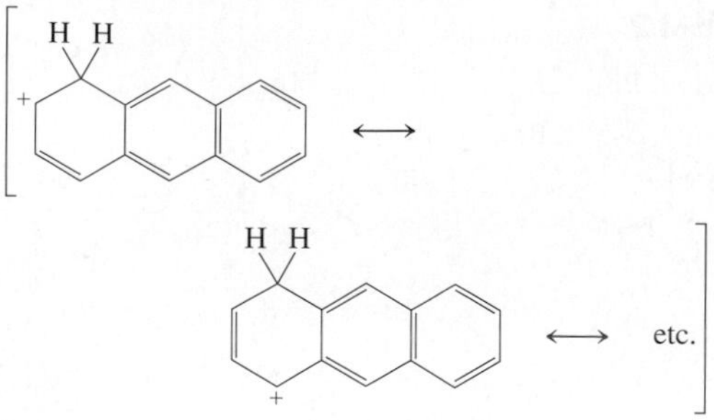

17-1

(a) 2-Cyclohexenone

(b) (*E*)-4-Methyl-4-hexenal

(c)

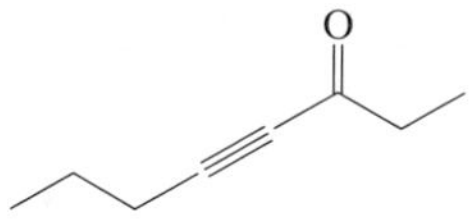

(d)

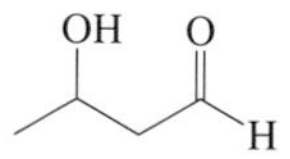

(e)

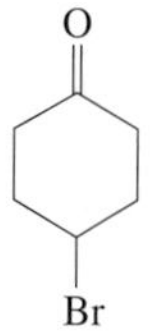

17-2

(a) ^{1}H NMR: chemical shift differences between $-$CH$_2$OH and $-$CHO.
^{13}C NMR: chemical shift differences between $-$CH$_2$OH and $-$CHO.
IR: $\tilde{v}_{OH}$ versus $\tilde{v}_{C=O}$.
UV: no absorption for alcohol, $\lambda_{max} \sim 280$ nm for aldehyde.

(b) ^{1}H NMR: absence versus presence of aldehyde resonance; three signals versus four signals; multiplicity differences are dramatic—i.e., the ketone exhibits a singlet (CH$_3$), triplet (CH$_3$), and quartet (CH$_2$), whereas the aldehyde shows a triplet (CH$_3$), sextet (CH$_2$ of C3), doublet of triplets (α-CH$_2$), and triplet (CHO).

(c) UV: $\lambda_{max} \sim 280$ (unconjugated carbonyl) versus 325 nm (conjugated carbonyl). ^{1}H NMR: most drastic differences would be the spin–spin splitting patterns, e.g. CH$_3$ doublet versus triplet, CHO triplet versus doublet, etc.

(d) The lack of symmetry in 2-pentanone versus 3-pentanone is evident in ^{1}H NMR: s (CH$_3$), t (CH$_3$), sex (CH$_2$), t (CH$_2$) versus t (2 CH$_3$s), q (2 CH$_2$s), and in ^{13}C NMR: five lines versus three lines.

17-3

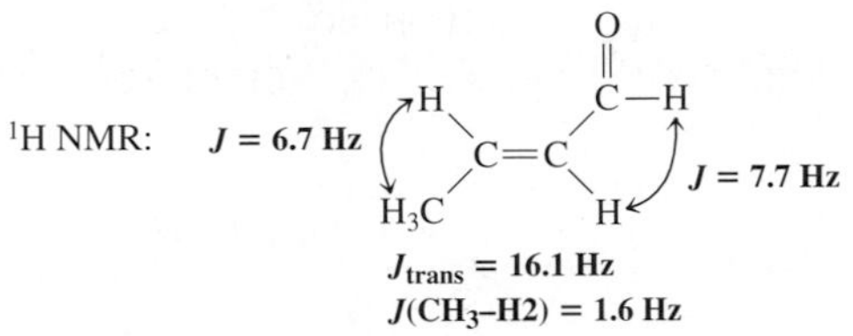

^{1}H NMR: $J = 6.7$ Hz

$J = 7.7$ Hz

$J_{trans} = 16.1$ Hz

$J(CH_3–H2) = 1.6$ Hz

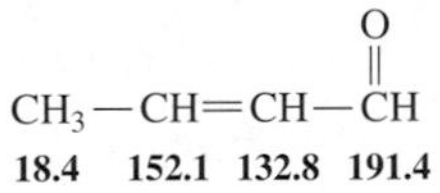

^{13}C NMR: $CH_3—CH=CH—\overset{O}{\overset{\|}{C}}H$
18.4   152.1   132.8   191.4

UV: Absorptions are typical for a conjugated enone.

17-4

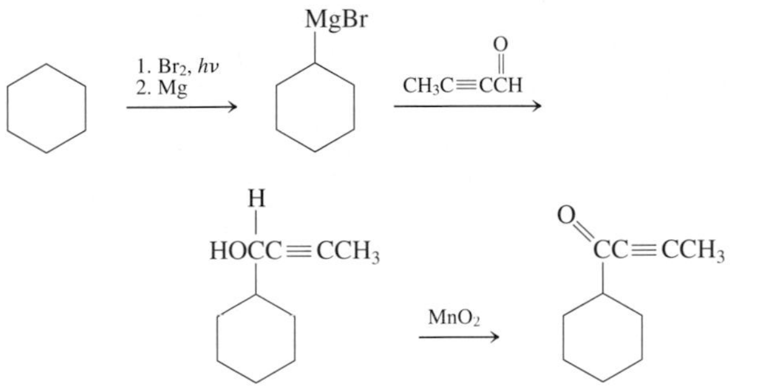

**Cyclohexyl
1-propynyl ketone**

17-5

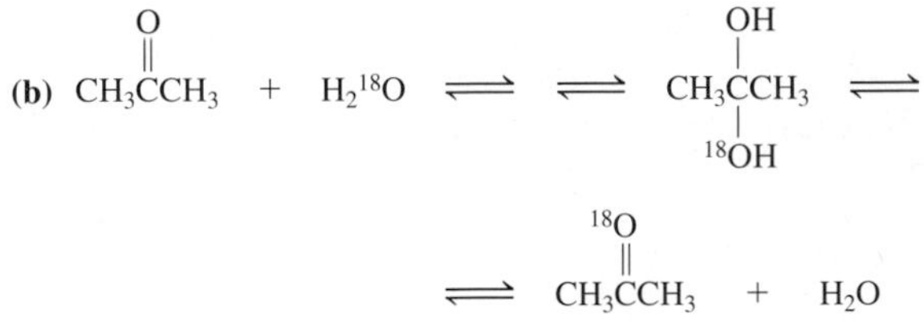

(a) Cl_3CCCH_3 < Cl_3CCH < Cl_3CCCCl_3

(b)

17-6

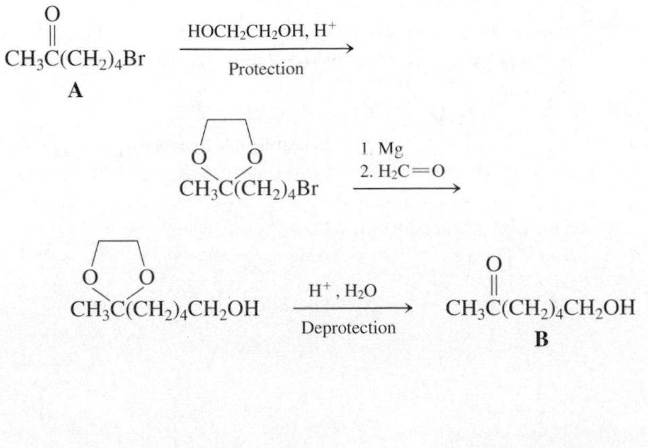

17-7

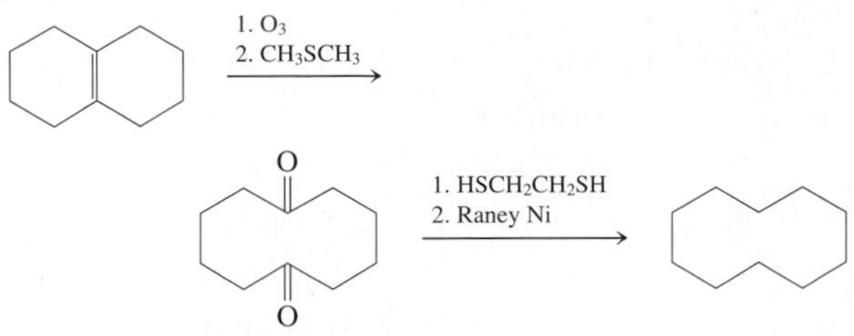

17-8

The mechanism of imidazolidine formation is similar to that formulated for cyclic acetal synthesis.

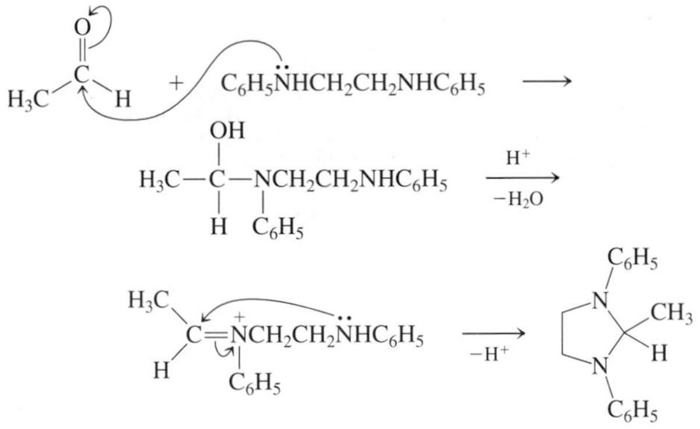

17-9

(a)

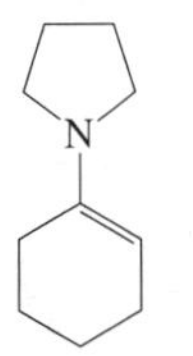

(b)

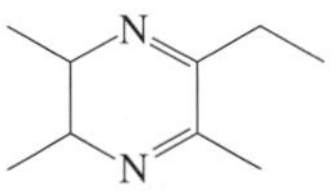

(c)

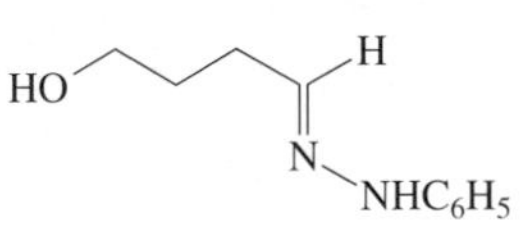

17-10

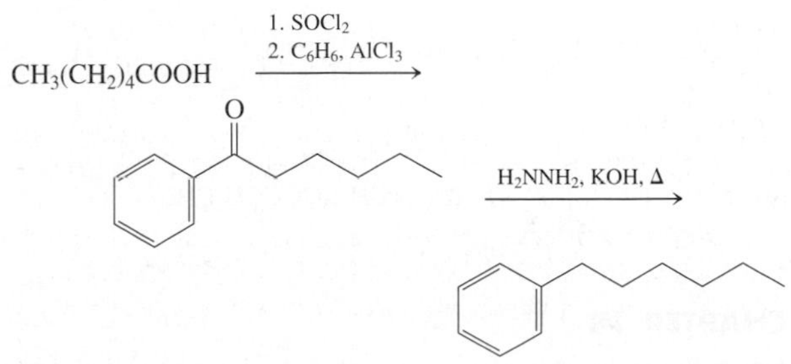

$CH_3(CH_2)_4COOH$

1. $SOCl_2$
2. C_6H_6, $AlCl_3$

H_2NNH_2, KOH, Δ

17-11

Formaldehyde > acetaldehyde > propanone > 3,3-dimethyl-2-butanone

17-12

(a) 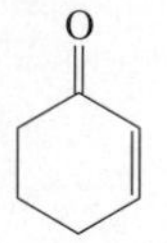 + $CH_2 = P(C_6H_5)_3$

(b) 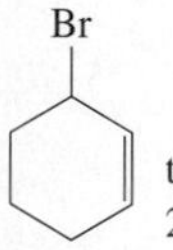 treated successively with 1. $P(C_6H_5)_3$,
2. $CH_3CH_2CH_2CH_2Li$, 3. $H_2C=O$ or
1. H_2O, 2. MnO_2, 3. $CH_2=P(C_6H_5)_3$.

17-13

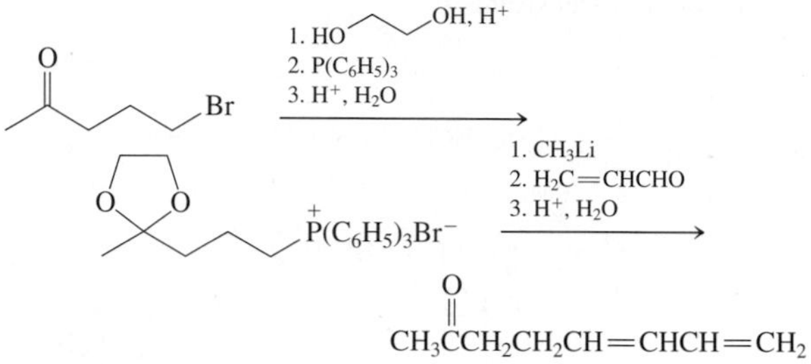

$$CH_3CCH_2CH_2CH=CHCH=CH_2$$

17-14

(a) 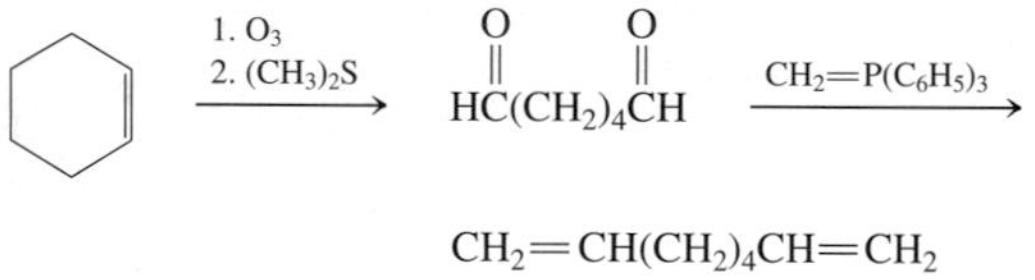 $\xrightarrow[\text{2. }(CH_3)_2S]{\text{1. }O_3}$ $HC(CH_2)_4CH$ (diketone) $\xrightarrow{CH_2=P(C_6H_5)_3}$

$$CH_2=CH(CH_2)_4CH=CH_2$$

(b)

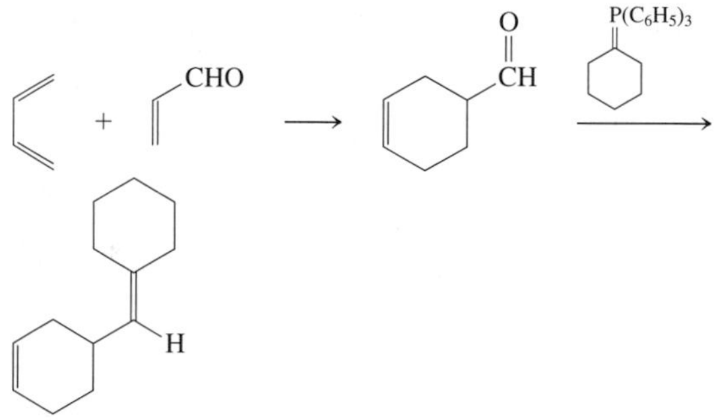

17-15

(a) $CH_2=CHCH_2CH_2OCCH_3$

(b)

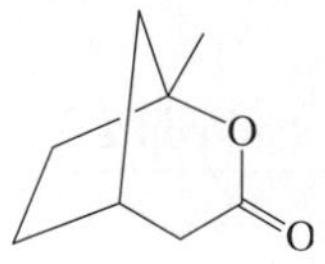

(c) $(CH_3)_3COCCH_2CH_3$

CHAPTER 18

18-1

(a) 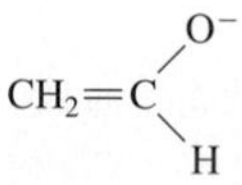 $CH_2=C\begin{smallmatrix}O^-\\|\\H\end{smallmatrix}$

(b) $CH_3CH=C\begin{smallmatrix}O^-\\|\\H\end{smallmatrix}$

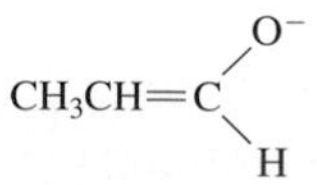

(c) $CH_2=C\begin{smallmatrix}O^-\\|\\CH_3\end{smallmatrix}$

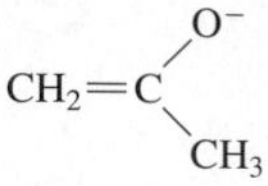

(d) 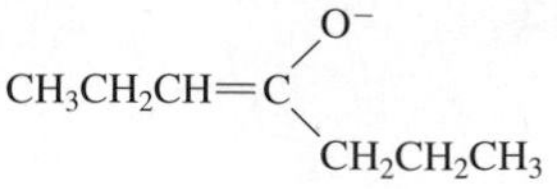 $CH_3CH_2CH=C\begin{smallmatrix}O^-\\|\\CH_2CH_2CH_3\end{smallmatrix}$

(e)

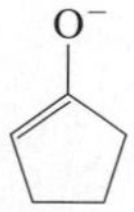

18-2

(a)

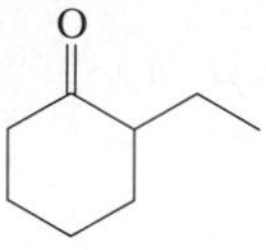

(b)

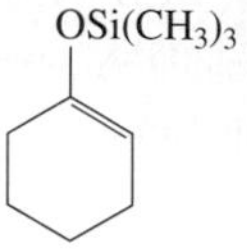

18-3

Base catalysis

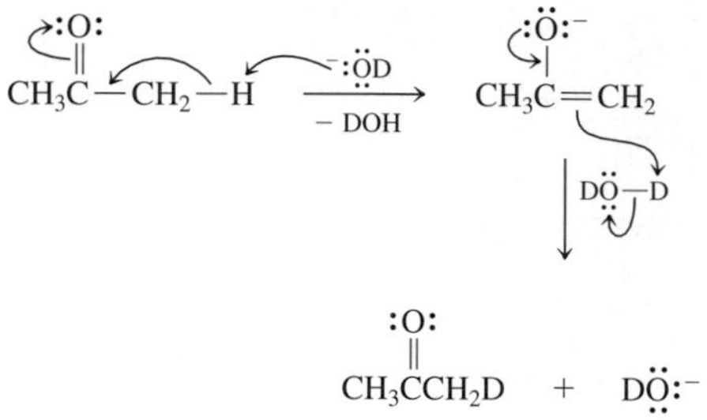

Acid catalysis

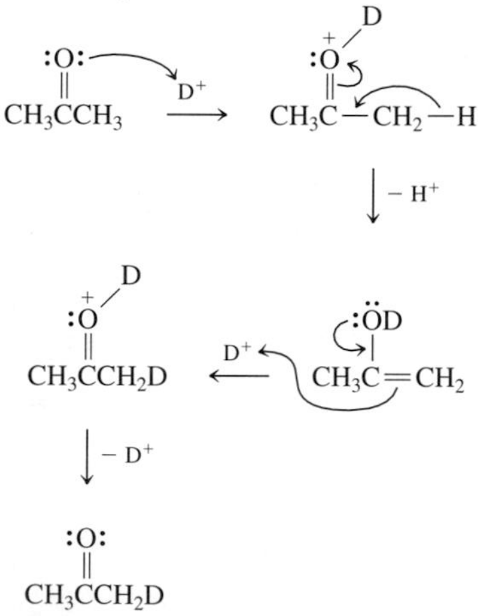

18-4

(a)

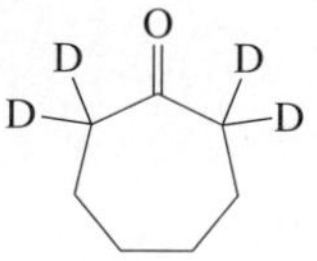

(b) $(CH_3)_3CCH$
No enolizable hydrogen

(c) $(CH_3)_3CCCD_3$

(d)

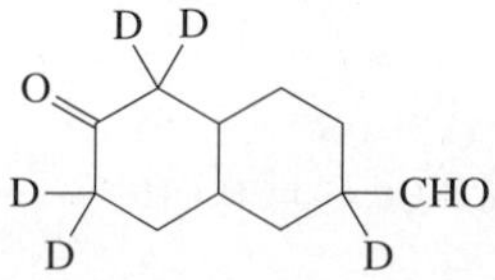

18-5

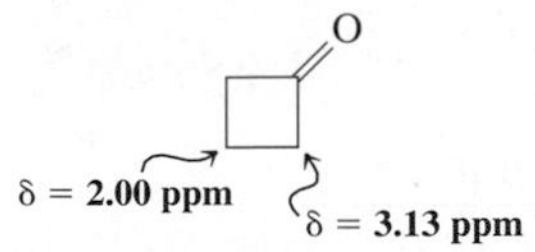
O
δ = 2.00 ppm
δ = 3.13 ppm

The peak at δ = 3.13 ppm would disappear on treatment with D_2O, NaOD.

18-6

Ketone A undergoes cis-to-trans isomerization by inversion (*R* to *S*) at the tertiary α-position. In ketone B, this position is not enolizable, because it is quaternary.

18-7

Acid-catalyzed:

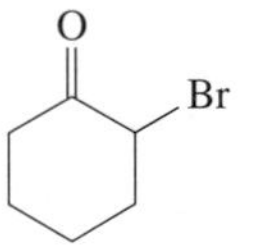
O
Br

Base-catalyzed:

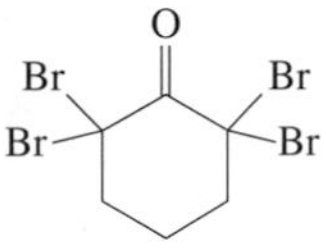
O
Br
Br
Br
Br

18-8

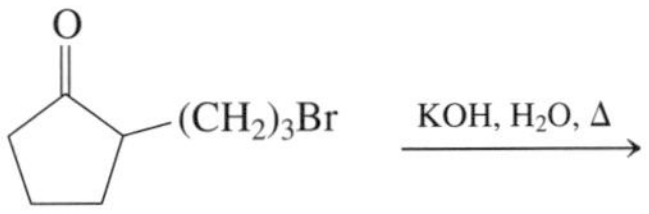
O
(CH2)3Br
KOH, H2O, Δ

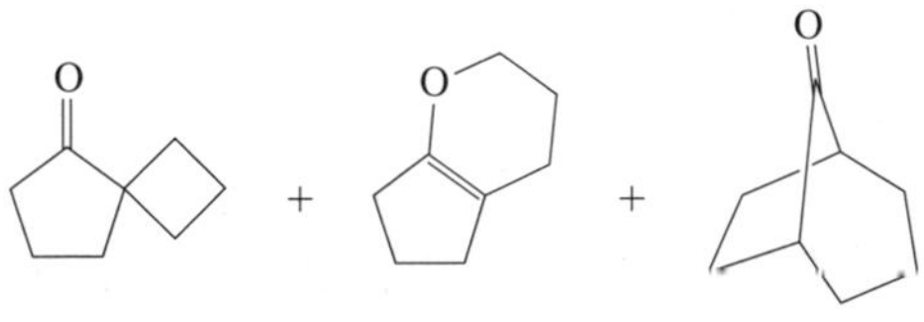
O
+
O
+
O

13% 15% 6%

18-9

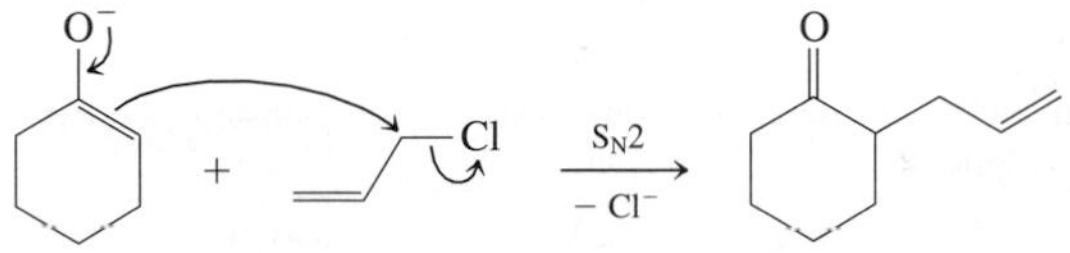
O−
+
Cl
SN2
− Cl−
O

The transition state is stabilized by overlap of the inverting C_{sp^3} orbital with the π system.

(a)

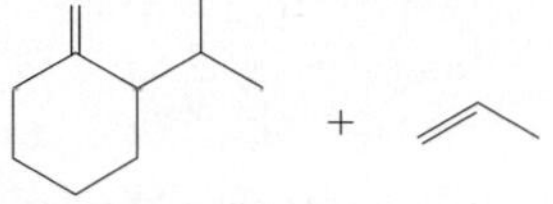
O
+

S_N2 E2

(b)

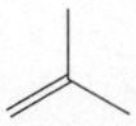

only

E2

18-10

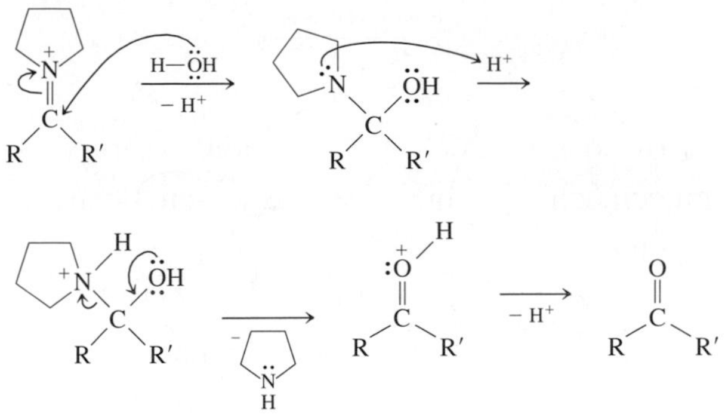
N
C
R
R′
H—ÖH
− H+
N
ÖH
C
R
R′
H+
H
N
ÖH
C
R
R′
N
H
H
:O
C
R
R′
− H+
O
C
R
R′

18-11

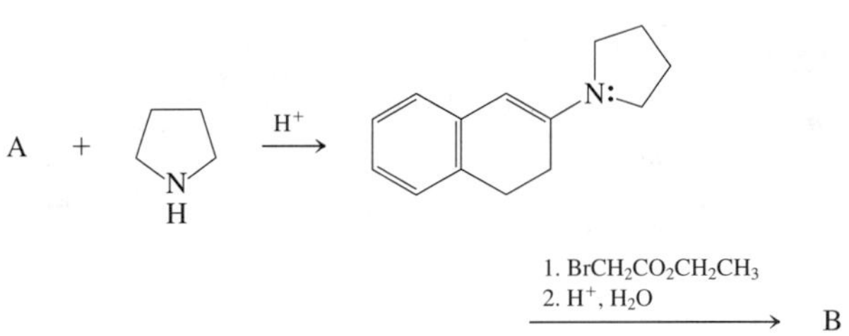
A
+
N
H
H+
N:
1. BrCH2CO2CH2CH3
2. H+, H2O
B

18-12

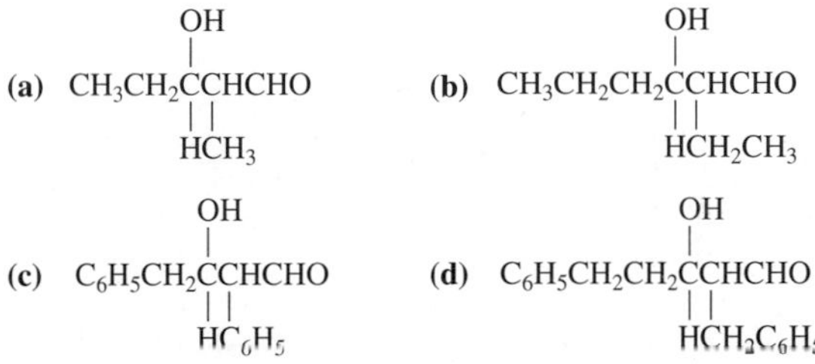
(a)
OH
CH3CH2CCHCHO
HCH3
(b)
OH
CH3CH2CH2CCHCHO
HCH2CH3
(c)
OH
C6H5CH2CCHCHO
HC6H5
(d)
OH
C6H5CH2CH2CCHCHO
HCH2C6H5

18-13

It cannot with itself, because it does not contain any enolizable hydrogens. It can, however, undergo crossed aldol condensations (Section 18-6) with enolizable carbonyl compounds.

18-14

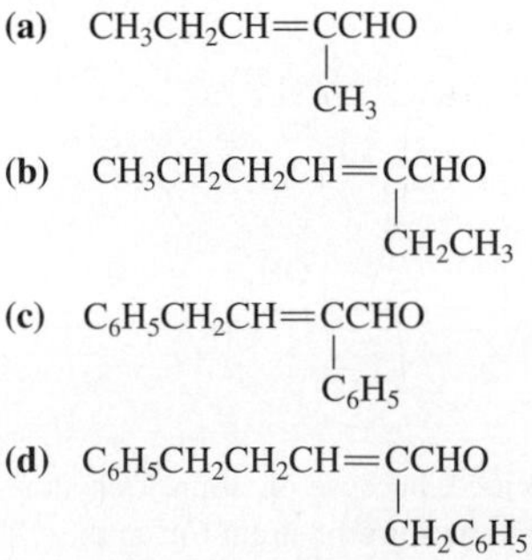
(a)
CH3CH2CH=CCHO
CH3
(b)
CH3CH2CH2CH=CCHO
CH2CH3
(c)
C6H5CH2CH=CCHO
C6H5
(d)
C6H5CH2CH2CH=CCHO
CH2C6H5

18-15

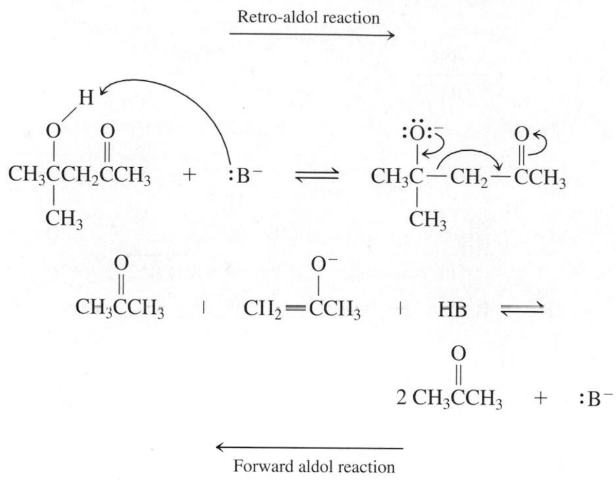

Retro-aldol reaction

$$CH_3CCH_2CCH_3 \quad + \quad :B^-$$
$$\overset{CH_3}{\underset{}{}}$$

$$CH_3C—CH_2—CCH_3$$

$$CH_3CCH_3 \quad + \quad CH_2=CCH_3 \quad + \quad HB \rightleftharpoons$$

$$2\ CH_3CCH_3 \quad + \quad :B^-$$

Forward aldol reaction

18-16

(a)

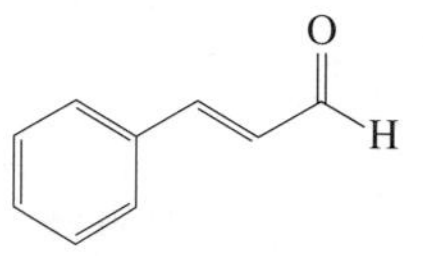

(b)

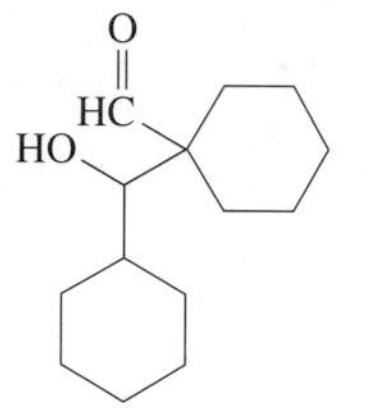

(c) $CH_2{=}CHCH{=}\overset{CH_3}{\underset{}{C}}CHO$

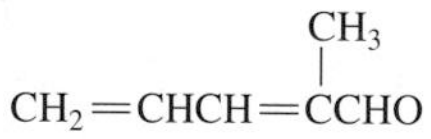

18-17

(a) $\xrightarrow{Na_2CO_3,\ 100°C}$

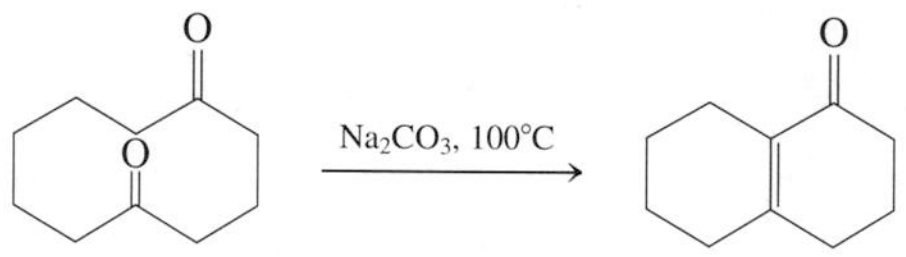

(b)

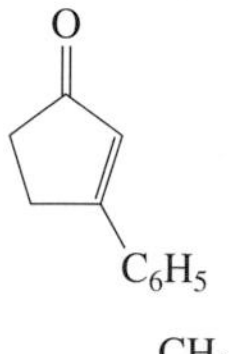

(c)

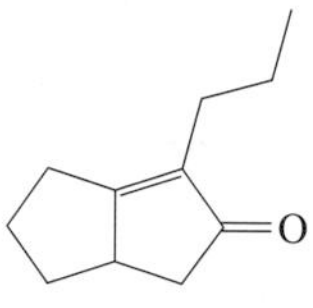

(d)

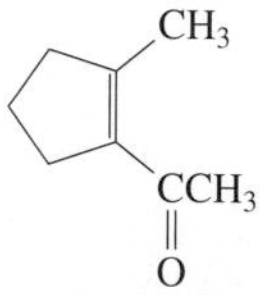

18-18

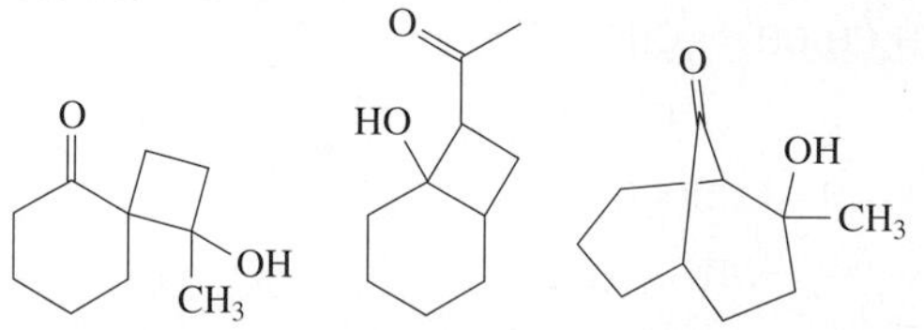

These three compounds are not formed, because of strain. Dehydration is, in addition, prohibited, again because of strain (or, in the

first structure, because there is no proton available). The fourth possibility is most facile.

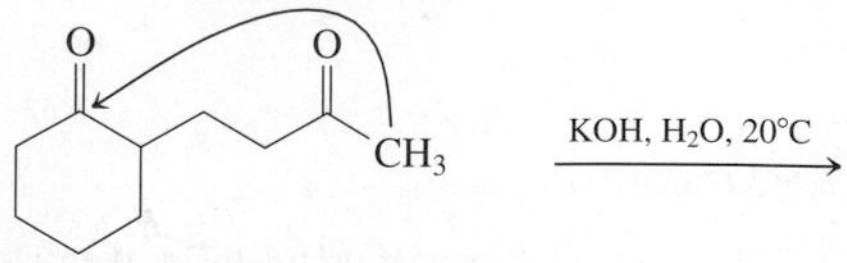

$\xrightarrow{KOH,\ H_2O,\ 20°C}$

2-(3-Oxobutyl)cyclohexanone

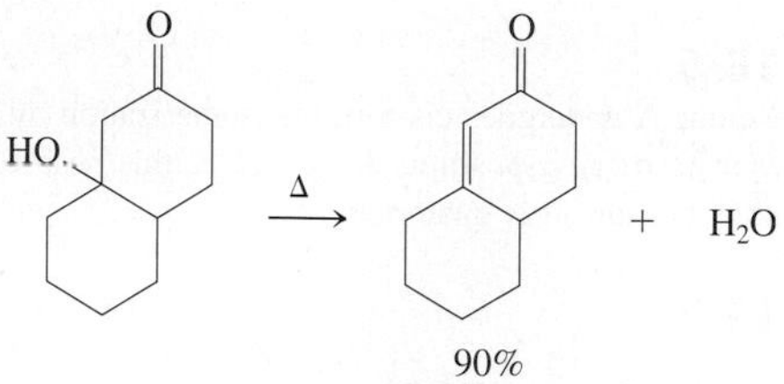

$\xrightarrow{\Delta}$ $+\ H_2O$

90%

18-19

(a)

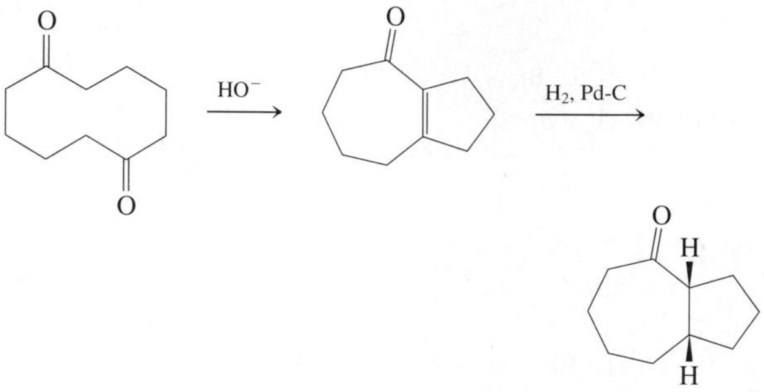

$\xrightarrow{HO^-}$ $\xrightarrow{H_2,\ Pd-C}$

(b) $+\ CH_3CCH_3$

(c)

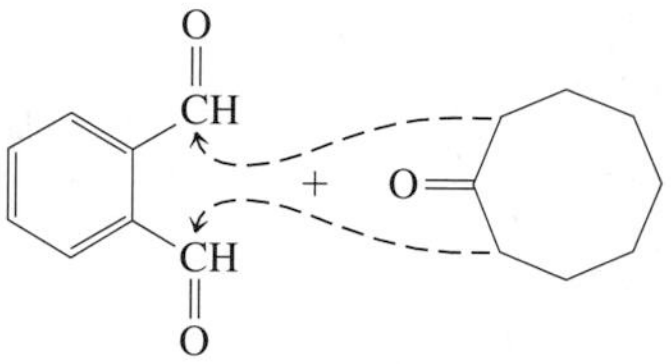

18-20

Mechanism of acid-mediated isomerization of β,γ-unsaturated carbonyl compounds:

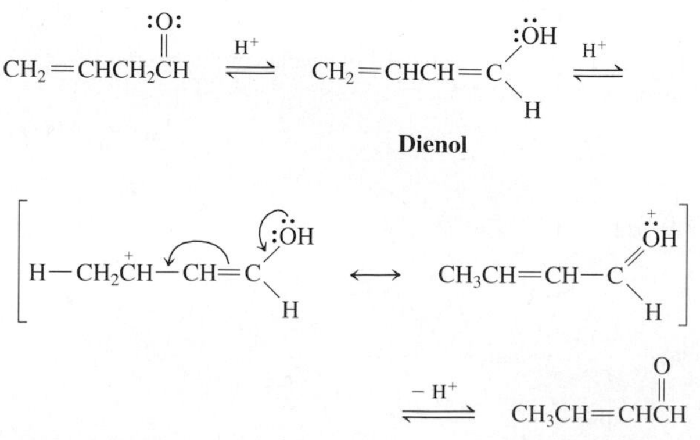

$$CH_2{=}CHCH_2\overset{:O:}{\underset{}{C}}H \overset{H^+}{\rightleftharpoons} CH_2{=}CHCH{=}\overset{:\overset{..}{O}H}{\underset{H}{C}} \overset{H^+}{\rightleftharpoons}$$

Dienol

$$\left[H{-}CH_2\overset{+}{CH}{-}CH{=}\overset{:\overset{..}{O}H}{\underset{H}{C}} \longleftrightarrow CH_3CH{=}CH{-}\overset{\overset{+}{\overset{..}{O}H}}{\underset{H}{C}} \right]$$

$$\overset{-H^+}{\rightleftharpoons} CH_3CH{=}CHC\overset{O}{\underset{}{\parallel}}H$$

18-21

$$CH_3CH_2CHO \xrightarrow{(C_6H_5)_3P=CHCHO}$$

$$CH_3CH_2CH{=}CHCHO \xrightarrow[2.\ LiAlH_4]{1.\ H_2,\ Pd\text{-}C} CH_3CH_2CH_2CH_2CH_2OH$$

18-22

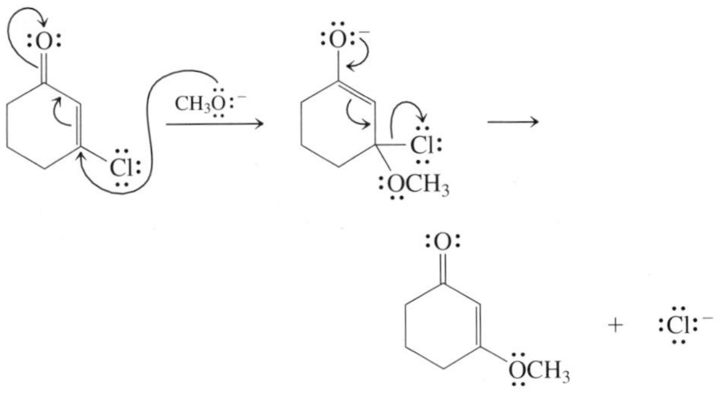

18-23

1. Protonation

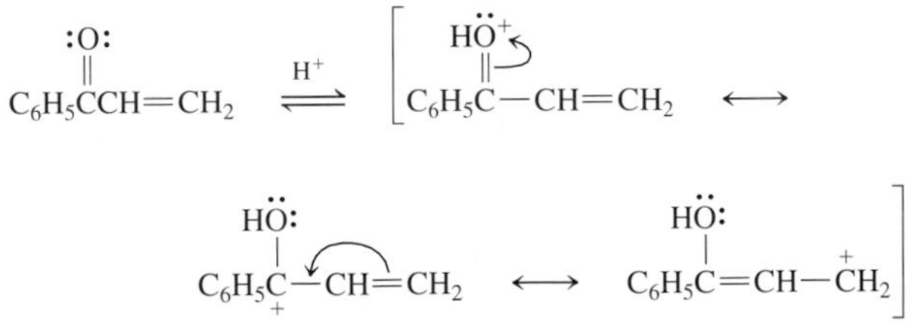

2. Cyanide attack

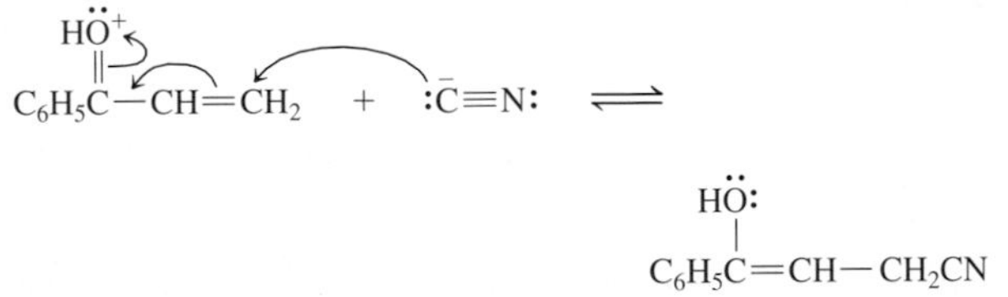

3. Enol–keto tautomerization

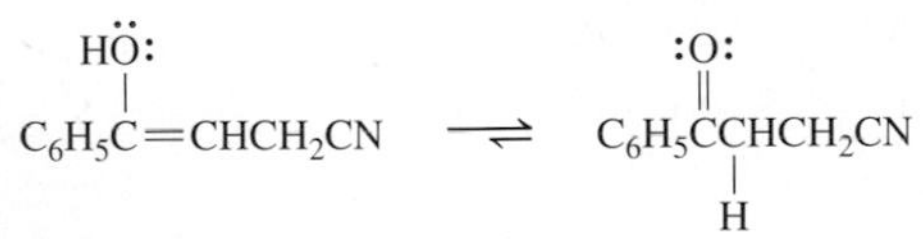

18-24

(a)

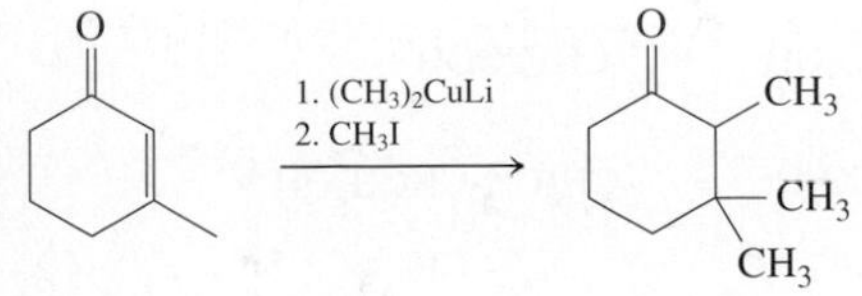

(b)

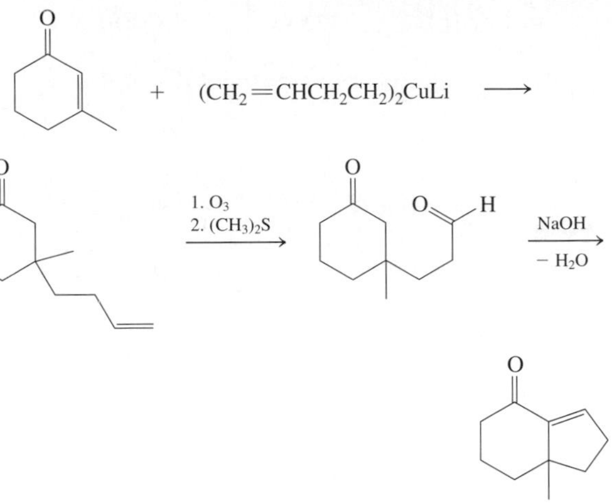

18-25

(a)

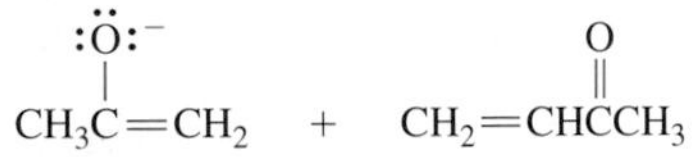

(b)

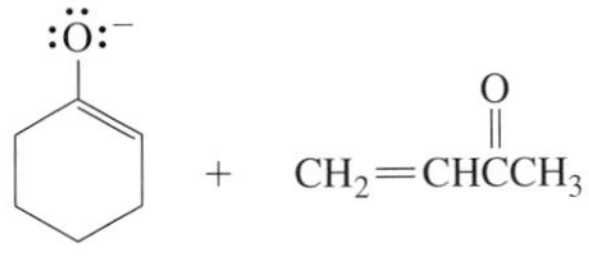

(c)

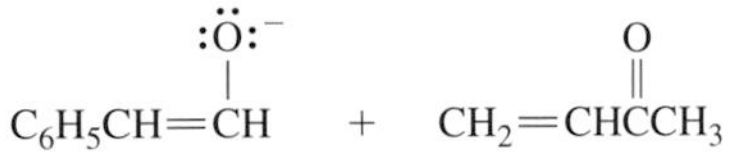

CHAPTER 19

19-1

(a) 5-Bromo-3-chloroheptanoic acid
(b) 4-Oxocyclohexanecarboxylic acid
(c) 3-Methoxy-4-nitrobenzoic acid

(d)

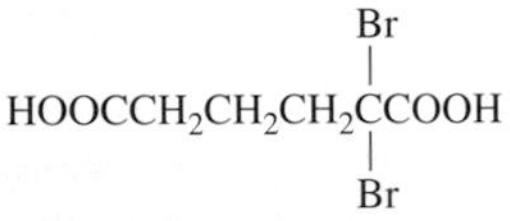

(e)

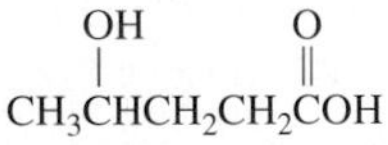

(f)

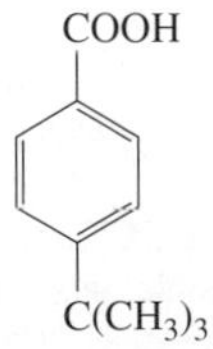

19-2

$CH_3CH_2CH_2COOH$, butanoic (butyric) acid

19-3

(a) $CH_3CBr_2COOH > CH_3CHBrCOOH > CH_3CH_2COOH$

(b)

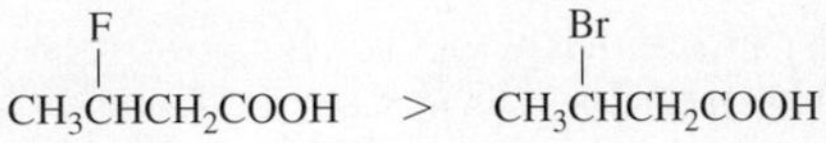

(c)

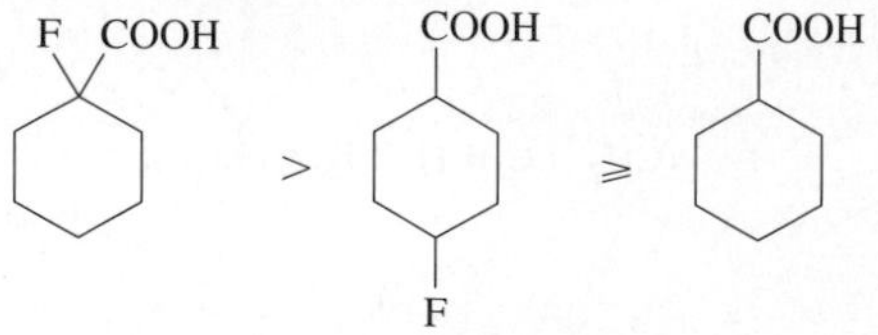

$$\text{F} \quad \text{COOH} \quad > \quad \text{COOH} \quad \geqq \quad \text{COOH}$$

(b)

$$\xrightarrow{\text{HBr}} \quad \text{H}_3\text{C} \quad \text{Br} \quad \xrightarrow[\substack{2.\ CO_2 \\ 3.\ H^+,\ H_2O}]{1.\ Mg} \quad \text{H}_3\text{C} \quad \text{COOH}$$

19-4

Protonated propanone (acetone) has fewer resonance forms.

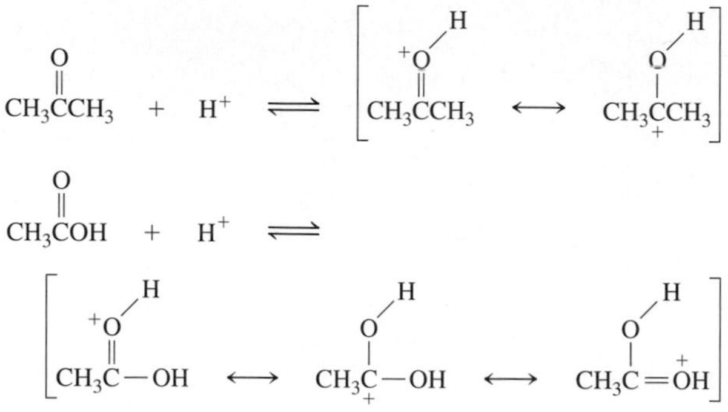

(c)

$$\xrightarrow[\substack{-\ Br^- \\ S_N2}]{^-CN} \quad \xrightarrow[\substack{1.\ HO^-,\ H_2O \\ 2.\ H^+,\ H_2O}]{}$$

19-8

(a) 1. $CH_3\overset{O}{\overset{\|}{C}}Cl \ + \ Na^{+\ -}O\overset{O}{\overset{\|}{C}}CH_2CH_3$

2. $CH_3\overset{O}{\overset{\|}{C}}O^-Na^+ \ + \ Cl\overset{O}{\overset{\|}{C}}CH_2CH_3$

(b) $CH_3\overset{H_3C}{\overset{|}{C}}H\overset{O}{\overset{\|}{C}}OH \ + \ SOCl_2$

19-5

(a) $CH_3(CH_2)_3COOH$ **(b)** $HOOC(CH_2)_4COOH$

(c)

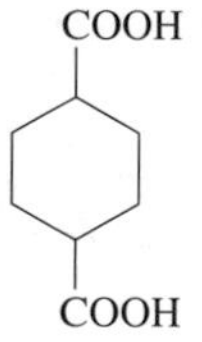

19-9

The reaction is self-catalyzed by acid.

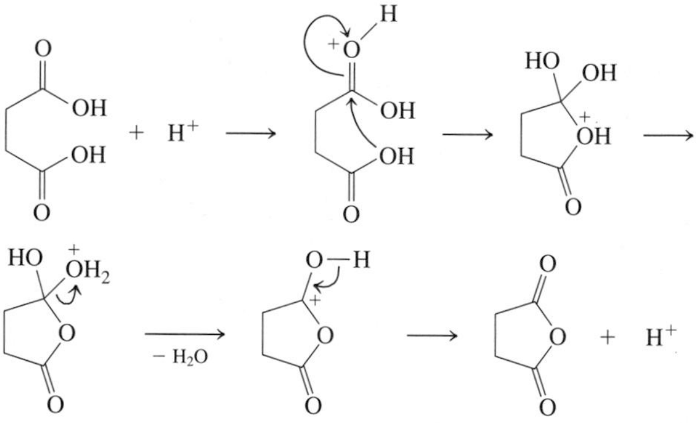

19-6

(a)

1. Mg, 2. CO_2, 3. H^+, H_2O
or 1. ^-CN, 2. H^+, H_2O

(b)

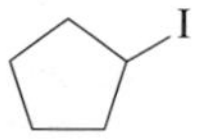

same conditions as in (a)

(c)

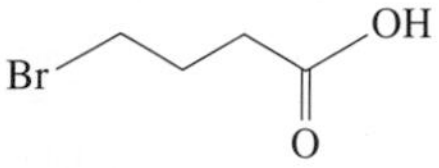

1. ^-CN, 2. H^+, H_2O

The alternative (1. Mg, 2. CO_2, 3. H^+, H_2O) fails because Grignard formation is not possible in the presence of a carboxy group.

(d)

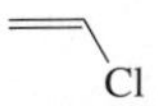

1. Mg, 2. CO_2, 3. H^+, H_2O

The alternative (1. ^-CN, 2. H^+, H_2O) fails because S_N2 reactions are not possible on sp^2-hybridized carbons.

(e)

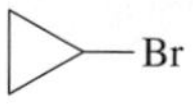

—Br 1. Mg, 2. CO_2, 3. H^+, H_2O

The alternative (1. ^-CN, 2. H^+, H_2O) fails because S_N2 reactions of strained halocycloalkanes are very slow.

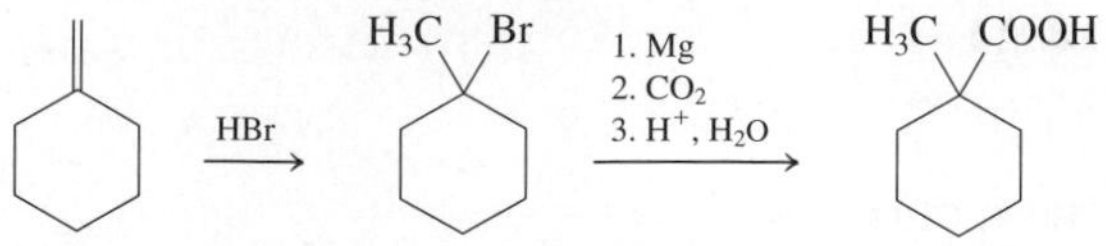

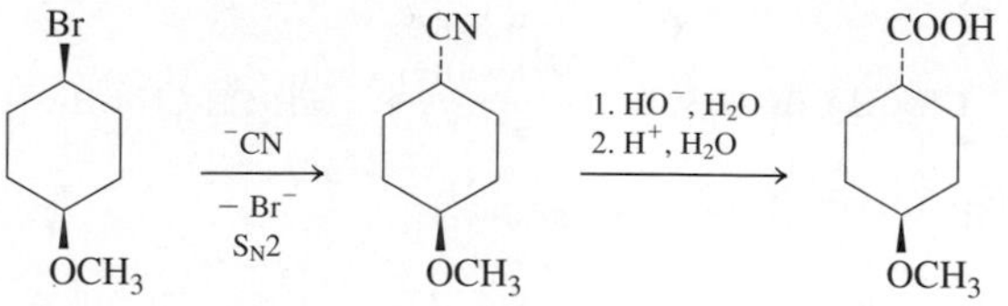

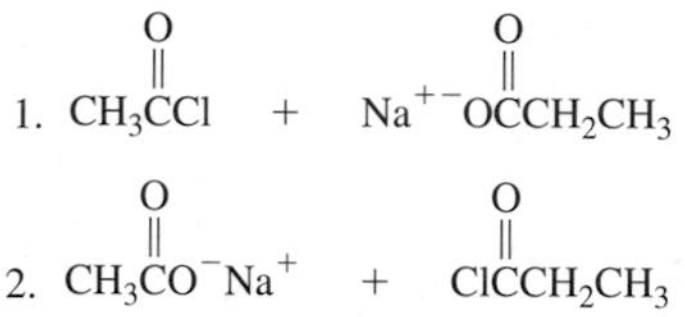

19-10

(a)

(b)

(c)

(d)

19-7

(a) 1. HCN, 2. H^+, H_2O

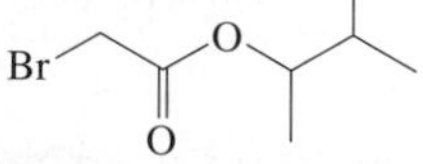

19-11

(a) $CH_3(CH_2)_3C\equiv CCH_2CH_2CO_2H + (CH_3)_2CHOH$
(or $CH_3CH=CH_2$)

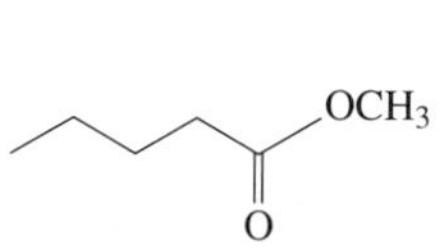

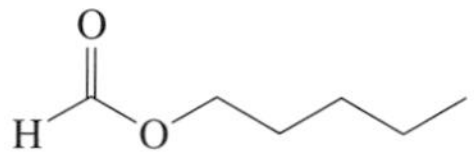

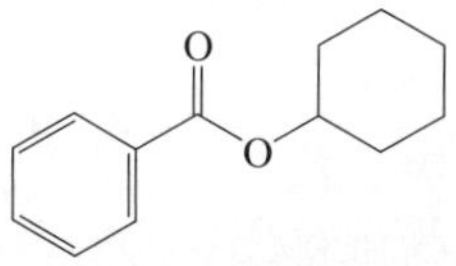

(b) 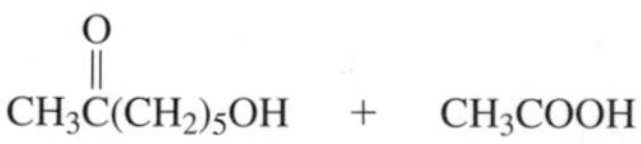 $CH_3\overset{O}{\overset{\|}{C}}(CH_2)_5OH \ + \ CH_3COOH$

(c)

$-CO_2H \ + \ CH_3CH_2\overset{}{C}HCH_2OH$
$\overset{|}{CH_3}$

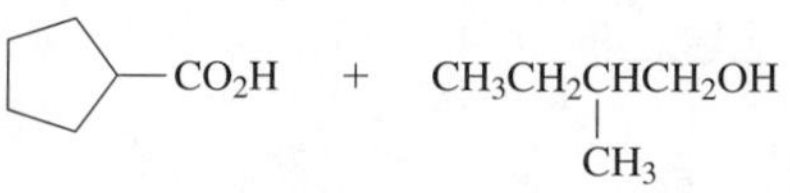

19-12

Label appears in the ester.

$$\underset{\substack{\|\\O}}{RCOH} \ + \ H^{18}OCH_3 \ \underset{\longleftarrow}{\overset{H^+}{\rightleftharpoons}} \ \underset{\substack{\|\\O}}{RC^{18}OCH_3} \ + \ H_2O$$

19-13

19-14

19-15

(a) 1. H^+, H_2O, 2. $LiAlH_4$, 3. H^+, H_2O
(b) 1. $LiAlD_4$, 2. H^+, H_2O

19-16

2. $RCH_2\overset{\substack{\|\\O}}{C}Br \ + \ H^+ \ \rightleftharpoons \ RCH_2\overset{\substack{\overset{+}{O}-H}}{C}Br$

3.

19-17

Self-explanatory.

CHAPTER 20

20-1

These compounds show decreasing reactivity in the order depicted in the problem. All of them are "double" carboxylic acid deriva-

tives of the unique "dioic" acid carbonic acid $\underset{\substack{\|\\O}}{HOCOH}$, and the arguments about relative reactivity of carboxylic acid derivatives apply here in the same way. The only difference is a somewhat attenuated influence of the second substituent on the carbonyl, because it has to compete with the first for resonance with the π bond.

20-2

At room temperature, rotation around the amide bond is slow on the NMR time scale and two distinct rotamers can be observed.

Heating makes the equilibration so fast that the NMR technique can no longer distinguish between the two species.

20-3

Not a strong contributor

The reaction scheme showing protonation of acetamide with the resonance structures, with "Strong contributor" labeled under the rightmost structure.

20-4

Section 19-12, step 2 of the Hell-Volhard-Zelinsky reaction.

20-5

$$CH_3COOH \xrightarrow{SOCl_2} CH_3\overset{O}{\overset{\|}{C}}Cl \xrightarrow{(CH_3)_3COH, (CH_3CH_2)_3N}$$

$$CH_3\overset{O}{\overset{\|}{C}}OC(CH_3)_3 + (CH_3CH_2)_3\overset{+}{N}HCl^-$$

20-6

Use *N,N*-diethylethanamine (triethylamine) to generate the alkanoyl triethylammonium salt; then add the expensive amine.

20-7

Use the following reagents.

(a) H_2O (b) cyclohexanol (OH on cyclohexane)   (c) $(CH_3)_2NH$

(d) $(CH_3CH_2)_2CuLi$ (e) $LiAl[OC(CH_3)_3]_3H$

20-8

Structure of succinimide (five-membered ring with two C=O groups and NH).

Butanimide (succinimide)

(See Section 19-10)

20-9

Mechanism scheme starting from $CH_3\overset{O}{\overset{\|}{C}}O\overset{O}{\overset{\|}{C}}CH_3$ with $+H^+$ / $-H^+$, then reaction with $HOCH_3$, $-CH_3OH$, etc., leading to $CH_3\overset{O}{\overset{\|}{C}}OCH_3 + CH_3\overset{O}{\overset{\|}{C}}OH$, and finally $CH_3\overset{O}{\overset{\|}{C}}OCH_3$.

20-10

(a) Propyl propanoate
(b) Dimethyl butanedioate
(c) Methyl 2-propenoate (methyl acrylate)

20-11

Mechanism scheme for acid-catalyzed reaction of a cyclic ketone (lactone ring) with H^+, H_2O, etc., yielding:

4-Hydroxybutanoic acid

20-12

Mechanism scheme for base-catalyzed reaction with HO^-, $-HO^-$, H_2O, $-{}^-OH$, yielding the carboxylate product.

20-13

Acid catalysis: As in Exercise 20-11, but use $BrCH_2CH_2CH_2OH$ instead of H_2O as the nucleophile in the second step.
Base catalysis: As in Exercise 20-12, but use $BrCH_2CH_2CH_2O^-$ instead of HO^- as the nucleophile in the first step.

20-14

$$C_6H_5Br \xrightarrow{Mg} C_6H_5MgBr$$

$$C_6H_5\overset{O}{\overset{\|}{C}}OCH_3 + 2\,C_6H_5MgBr \xrightarrow{H^+,\,H_2O} (C_6H_5)_3COH$$

20-15

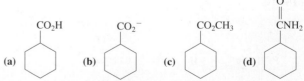

(a) CO_2H (b) CO_2^- (c) CO_2CH_3 (d) $\overset{O}{\overset{\|}{C}}NH_2$

(e) CH₃CH(OH)—cyclohexyl [CH₃CH OH with cyclohexane]

$$CH_3CH \!-\! OH$$

(f) CH₂OH—cyclohexyl

(g) H₃C—C(CO₂CH₂CH₃)—cyclohexyl

20-16

2,2-dimethylpyrrolidine (H₃C, H₃C on C2, N—H)

20-17

$$\xrightarrow{\text{LiAlH}_4}$$

20-18

In A, the negative charge can be delocalized over two carbonyl groups; not possible for B.

20-19

$$R\ddot{N}\!=\!C\!=\!\overset{..}{O} \;+\; H_2\ddot{O}\!: \longrightarrow$$

$$\left[R\ddot{N}\!=\!C\underset{\overset{+}{O}H_2}{\overset{\ddot{O}:^-}{|}} \longleftrightarrow R\ddot{N}\!-\!C\underset{\overset{+}{O}H_2}{\overset{\ddot{O}}{\|}} \right] \xrightarrow{\text{Double H}^+ \text{ shift}}$$

$$R\overset{H}{\underset{H}{N^+}}\!-\!C\underset{\overset{..}{O}:^-}{\overset{\ddot{O}:}{\|}} \longrightarrow R\ddot{N}H_2 \;+\; \ddot{O}\!=\!C\!=\!\ddot{O}$$

20-20

20-21

$$BrCH_2CH_2CH_2Br \xrightarrow[-\,Br^-]{^-CN}$$

$$BrCH_2CH_2CH_2C\!\equiv\!N \xrightarrow[-\,Br^-]{^-CN} NCCH_2CH_2CH_2CN$$

$$\delta = 117.6 \qquad\qquad 119.1 \;\; 22.6 \;\; 17.6 \text{ ppm}$$
and 3 other signals

20-22

The exact details of this reduction are not known. A possible mechanism is

$$R\!-\!C\!\equiv\!N \xrightarrow{\text{LiAlH}_4} \underset{R \quad H}{\overset{^-\text{AlH}_3\text{Li}^+}{\underset{\|}{N}}} \xrightarrow{\text{LiAlH}_4}$$

$$\underset{\underset{H}{\overset{|}{R\!-\!C\!-\!H}}}{\overset{Li^+H_3Al^- \quad ^-AlH_3Li^+}{N}} \xrightarrow{H^+, H_2O} RCH_2NH_2$$

20-23

(a) 1. H₂O, HO⁻, 2. H⁺, H₂O
(b) 1. CH₃CH₂CH₂CH₂MgBr, 2. H⁺, H₂O

$$\qquad\qquad\qquad CH_3$$
(c) 1. (CH₃CHCH₂)₂AlH 2. H⁺, H₂O
(d) D₂, Pt

20-24

(a) CH₃OCH₃, CH₃CH₂OH, O—O (oxirane), HCOH (with =O)

(b) H₂C=O **(c)** [oxetene], [methyl oxirane, CH₃], CH₂=CHCH (=O),

HC≡CCH₂OH, CH₃C≡COH, HC≡COCH₃,

[methylenoxirane], [dioxetane, O–O], [cyclopropanone], [cyclopropenol, OH]

20-25

CH₂Br₂: m/z = 176, 174, 172; intensity ratio 1:2:1

20-26

For most elements in organic compounds, such as C, H, O, S, P, and the halogens, the mass (of the most abundant isotopes) and valence are either both even or both odd, so even molecular weights always result. Nitrogen is a major exception: the atomic weight is 14, but the valence is 3. This phenomenon has led to the **nitrogen rule** in mass spectrometry, as expressed in this exercise.

20-27

Mass spectrum of 3-methyl-3-heptanol

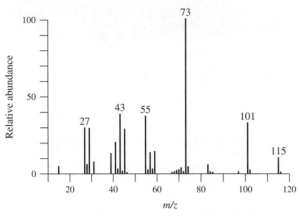

The major primary fragments are due to cleavage of the bonds α to the hydroxy group. Why? Consider their strength and the electronic structure of the resulting radical cations. (Draw resonance forms.) Do these cations fragment by loss of water?

20-28

(a) Both show the same α cleavage patterns but different McLafferty rearrangements.

$$\left[\begin{array}{c} \text{H} \\ \text{O} \quad \text{CH}_2 \\ \text{C} \quad \text{CH}_2 \\ \text{H}_3\text{C} \quad \text{CH} \\ \text{CH}_3 \end{array}\right]^{+\cdot} \longrightarrow \left[\begin{array}{c} \text{HO} \\ \quad \text{C}=\text{CHCH}_3 \\ \text{H}_3\text{C} \end{array}\right]^+$$

$$m/z = 100 \qquad\qquad m/z = 72$$

$$+$$

$$[\text{CH}_2=\text{CH}_2]^+$$
$$m/z = 28$$

$$\left[\begin{array}{c} \text{H} \\ \text{O} \quad \text{CH}_2 \\ \text{C} \quad \text{CHCH}_3 \\ \text{H}_3\text{C} \quad \text{CH}_2 \end{array}\right]^{+\cdot} \longrightarrow \left[\begin{array}{c} \text{HO} \\ \quad \text{C}=\text{CH}_2 \\ \text{H}_3\text{C} \end{array}\right]^+$$

$$m/z = 100 \qquad\qquad m/z = 58$$

$$+$$

$$[\text{CH}_3\text{CH}=\text{CH}_2]^+$$
$$m/z = 42$$

(b) Both show the same α cleavage patterns, but only 2-ethylcyclohexanone has an accessible γ-hydrogen for the McLafferty rearrangement.

$$\left[\text{(cyclohexanone with H)}\right]^{+\cdot} \longrightarrow \left[\text{(phenol OH)}\right]^+ + [\text{CH}_2=\text{CH}_2]^+$$

20-29

Pentanal: $m/z = 57$, $[\text{CH}_3\text{CH}_2\text{CH}_2\text{CH}_2]^+$
(**α cleavage**)

$$m/z = 44, \left[\begin{array}{c} \quad\quad \text{OH} \\ \text{H}_2\text{C}=\text{C} \\ \quad\quad \text{H} \end{array}\right]^+$$

(**McLafferty rearrangement**)

$m/z = 29$ $[\text{HCO}]^+$
(**α cleavage**)

Pentanoic acid: $m/z = 60$, $\left[\begin{array}{c} \quad\quad \text{OH} \\ \text{H}_2\text{C}=\text{C} \\ \quad\quad \text{OH} \end{array}\right]^+$

(**McLafferty rearrangement**)

Methyl pentanoate: $m/z = 85$, $[\text{CH}_3\text{CH}_2\text{CH}_2\text{CH}_2\text{C}=\text{O})]^+$
(**α cleavage**)

$$m/z = 74, \left[\begin{array}{c} \quad\quad \text{OH} \\ \text{H}_2\text{C}=\text{C} \\ \quad\quad \text{OCH}_3 \end{array}\right]^+$$

(**McLafferty rearrangement**)

$m/z = 57$, $[\text{CH}_3\text{CH}_2\text{CH}_2\text{CH}_2]^+$
(**α cleavage**)

20-30

(a) $C_7H_{12}O$ (b) C_6H_{14}

CHAPTER 21

21-1

(a) 2-Butanamine, *sec*-butylamine
(b) *N,N*-Dimethylbenzenamine, *N,N*-dimethylaniline
(c) (*R*)-6-Bromo-2-hexanamine, (*R*)-(5-bromo-1-methyl)pentylamine

21-2

(a) $\text{HC}\equiv\text{CCH}_2\text{NH}_2$

(b) $\text{CH}_2\text{NHCH}_2\text{CH}=\text{CH}_2$ (attached to benzene ring)

(c) $(\text{CH}_3)_3\text{CNHCH}_3$

21-3

The lesser electronegativity of nitrogen, compared with oxygen, allows for slightly more diffuse orbitals and hence longer bonds to other nuclei.

21-4

Less, because, again, nitrogen is less electronegative than oxygen. See Tables 10-2 and 10-3 for the effect of the electronegativity of substituent atoms on chemical shifts.

21-5

IR: Secondary amine, hence a weak band at $\sim$3400 cm^{-1}.

^{1}H NMR: s for the 1,1-dimethylethyl (*tert*-butyl) group at high field; s for the attached methylene group at $\delta \sim 2.7$; q for the second methylene unit close to the first; t for the unique methyl group at high field, closest to the 1,1-dimethylethyl (*tert*-butyl) signal.

^{13}C NMR: Five signals, two at low field, $\delta \sim 45$–50 ppm.

Mass: m/z = 115 (M^+), 100 $[(CH_3)_3CCH_2NH{=}CH_2]^+$, and 58 $(CH_2{=}NHCH_2CH_3)^+$. In this case, two different iminium ions can be formed by fragmentation.

21-6

As discussed in Section 21-4 (and Sections 16-1 and 16-3), the nitrogen lone electron pair in benzenamine is tied up by resonance with the benzene ring. Therefore, the nitrogen is less nucleophilic than the one in an alkanamine.

21-7

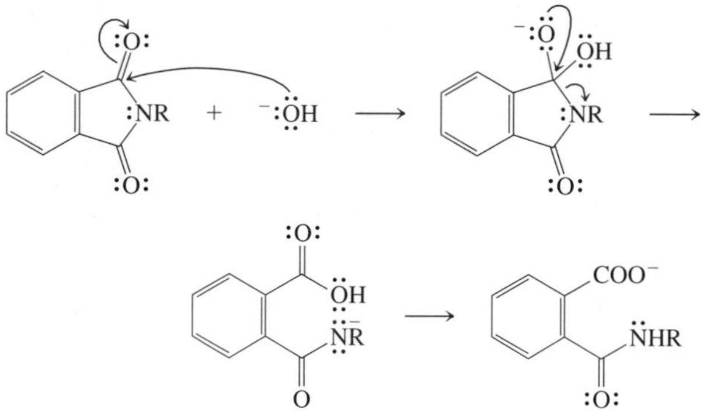

Continue as in the normal amide hydrolysis by base (Section 20-6).

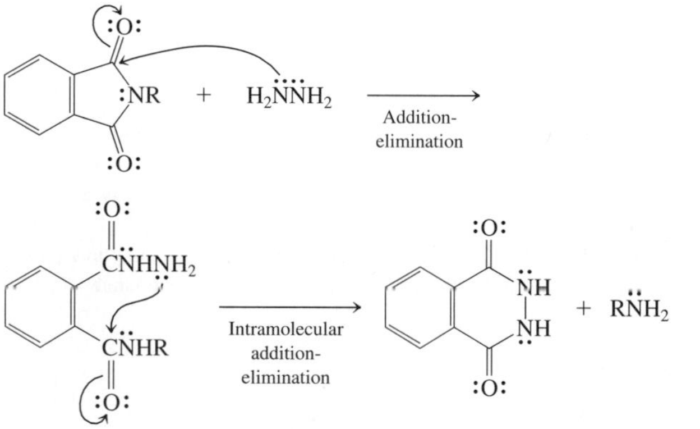

21-8

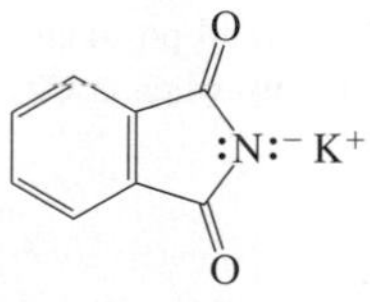

A

(a) 1. A + $CH_3(CH_2)_5Br$; 2. H^+, H_2O: 3. HO^-, H_2O

(b) 1. A + 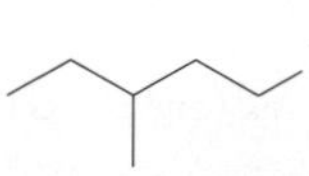 Br ; 2. H^+, H_2O 3. HO^-, H_2O

(c) 1. A + 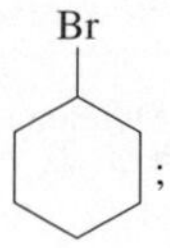 ; 2. H^+, H_2O; 3. HO^-, H_2O

(d) 1. A + $BrCH_2CO_2CH_2CH_3$; 2. H^+, H_2O

Protection of the carboxy group is necessary to prevent the acidic proton from reacting with A (see also Section 26-2). The azide method should work well for (a)–(c). For (d), the reduction step requires catalytic hydrogenation because $LiAlH_4$ would also attack the ester function.

21-9

In abbreviated form:

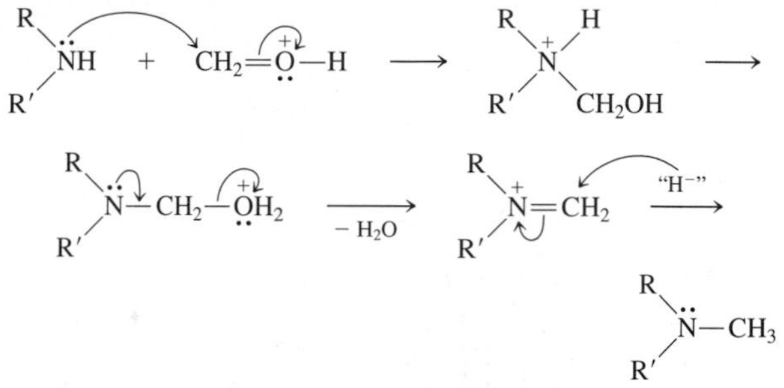

21-10

Not all intermediates are shown.

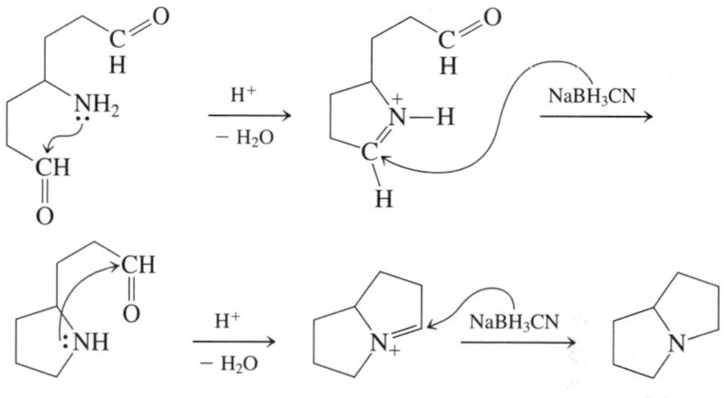

35%

21-11

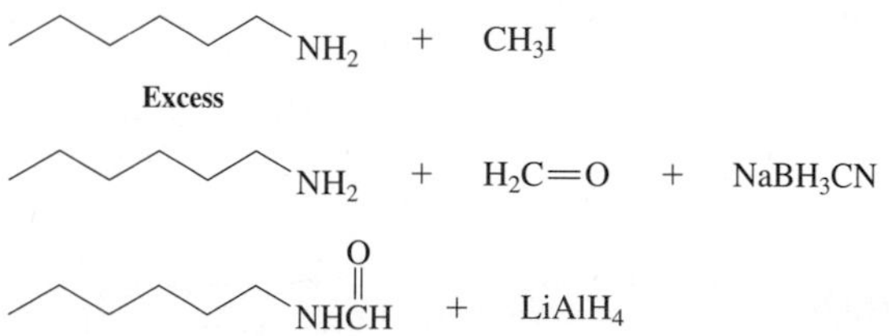

21-12

(a) $CH_3CH{=}CH_2$ and $CH_2{=}CH_2$

(b) $CH_3CH_2CH{=}CH_2$ and $CH_3CH{=}CHCH_3$ (cis and trans). The terminal alkene predominates. This reaction is kinetically controlled according to the Hofmann rule (Section 11-8). Thus, the base prefers attack at the less bulky end of the sterically encumbered quaternary ammonium ion.

21-13

The ethyl group can be extruded as ethene by Hofmann elimination [see Exercise 21-12(a)], thus giving product mixtures. Generally, any alkyl substituent with hydrogens located β to the quaternary nitrogen is capable of such elimination, an option absent for methyl.

21-14

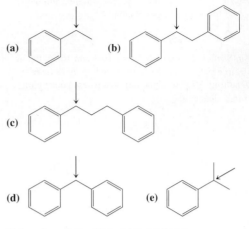

21-15

(a) [structure: cyclopentanone with CH₂NH₂ substituent]

(b) CH₃CHO with CH₃ and CH₂NH(CH₂)₅CH₃

(c) [structure: CH₃COCH₂CH₂N(CH₃)₂]

(d) [structure: cyclohexanone with CH₂–N(H)–cyclohexyl]

21-16

[structure: ketone] + CH₂=O + (CH₃)₂NH → 1. H⁺ 2. HO⁻

[structure: ketone with N(CH₃)₂] → 1. NaBH₄ 2. O₂N–C₆H₄–COCl

[structure: ester with (CH₃)₂N and NO₂] → Sn, HCl → tutocaine hydrochloride

21-17

Nucleophilic displacement of N₂ in RN₂⁺ by water occurs by an S_N2 mechanism with inversion.

21-18

[structure with H atoms]

Bicyclo[1.1.0]butane

CHAPTER 22

22-1

(a) [phenyl structure with arrow]

(b) [structure with arrow]

(c) [structure with arrow]

(d) [diphenylmethane with arrow]

(e) [tert-butylbenzene]

Order of reactivity: (d) > (e) > (a),(b),(c)

22-2

$(C_6H_5)_2CHCl$ solvolyzes faster because the additional phenyl group causes extra resonance stabilization of the intermediate carbocation.

22-3

Both react through an S_N2 mechanism that includes chloride attack on the protonated hydroxy group. The conversion of phenyl-methanol is accelerated relative to that of ethanol because of a de-localized transition state.

22-4

(a) $(C_6H_5)_2CH_2$, because the corresponding anion is better reso-nance stabilized.

(b) $4\text{-}CH_3OC_6H_4CH_2Br$, because it contains a better leaving group.

(c) $C_6H_5CH(CH_3)OH$, because the corresponding benzylic cation is not destabilized by the extra nitro group (draw resonance structures).

22-5

(a) 1. $KMnO_4$, 2. $LiAlH_4$, 3. H_2, Pd-C

(b) 1. $KMnO_4$, 2. H^+, H_2O, 3. Δ $(-2 H_2O)$

22-6

The nitro group is inductively electron withdrawing in all positions but can stabilize the negative charge by resonance only when at C2 or C4.

22-7

B, A, D, C

22-8

[structure: benzene ring with OCH₃, and two NO₂ groups]

22-9

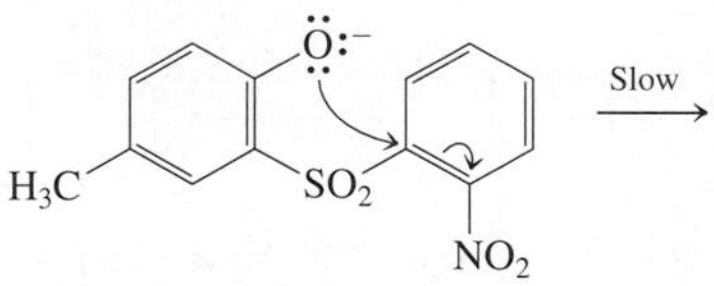

H₃C— (structure) —SO₂ ... NO₂ → Slow

A

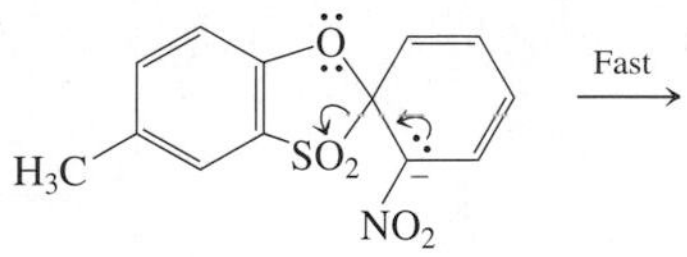

H₃C— (structure) —SO₂⁻ ... NO₂ → Fast

Intermediate B

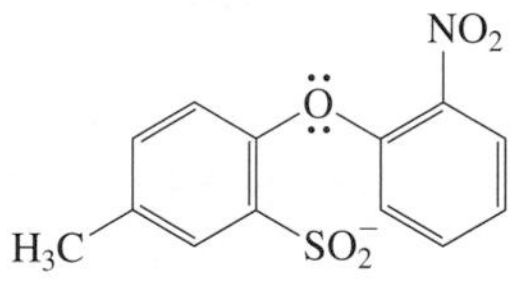

(structure with NO₂, O, SO₂⁻, H₃C)

C

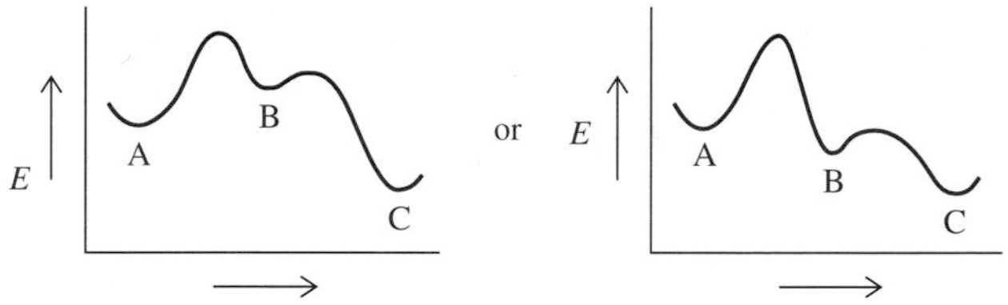

(Energy diagram with A, B, C)   or   (Energy diagram with A, B, C)

E ↑ E ↑

22-10

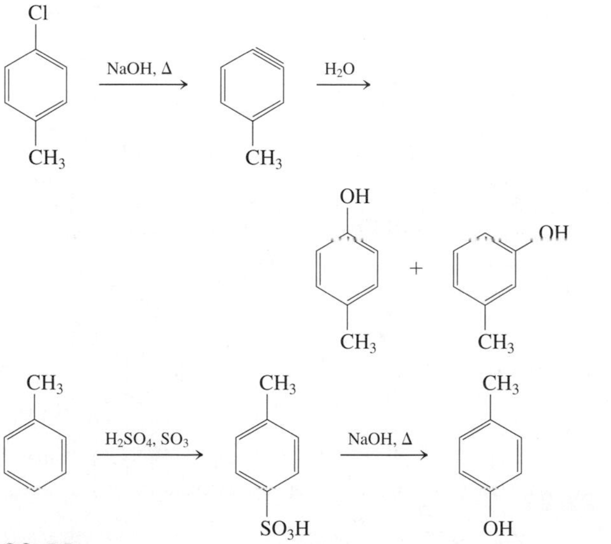

Cl (on ring, CH₃ para) — NaOH, Δ → (benzyne with CH₃) — H₂O →

OH (para CH₃) + OH (meta CH₃)

CH₃ (ring) — H₂SO₄, SO₃ → CH₃ (ring) SO₃H — NaOH, Δ → CH₃ (ring) OH

22-11

Br, ÖCH₃ (ring) — ⁻:ṄH₂ / − NH₃, − Br⁻ →

ÖCH₃ (ring) — ⁻:ṄH₂ → H₂N̈ ... ÖCH₃ (ring)

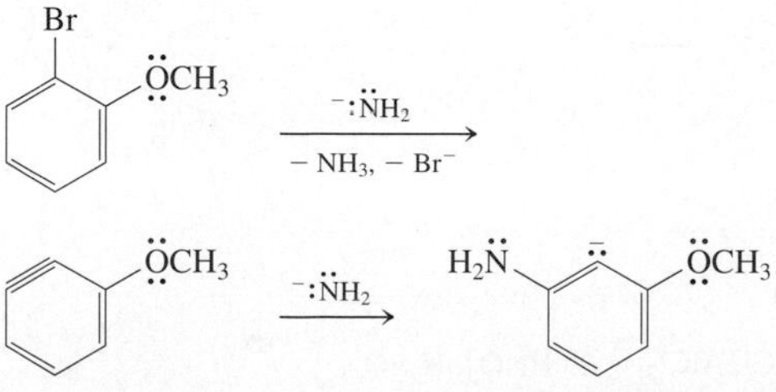

A

The product A of amide addition to the intermediate benzyne is stabilized by the inductively electron-withdrawing effect of the methoxy oxygen; therefore, it is formed regioselectively. Protonation gives the major final product. Note that there is no possibility for delocalization of the negative charge in A because the reactive electron pair is located in an sp^2 orbital that lies perpendicularly to the aromatic π system.

22-12

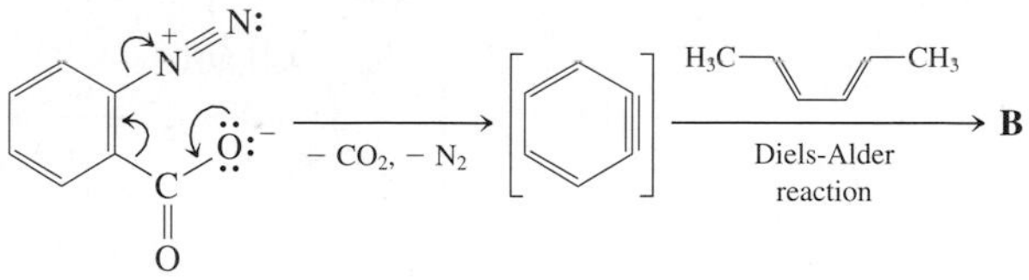

(structure) N⁺≡N: , Ö:⁻, C=O — CO₂, − N₂ → [Benzyne] — H₃C—CH=CH—CH₃ / Diels-Alder reaction → **B**

A   **Benzyne**

22-13

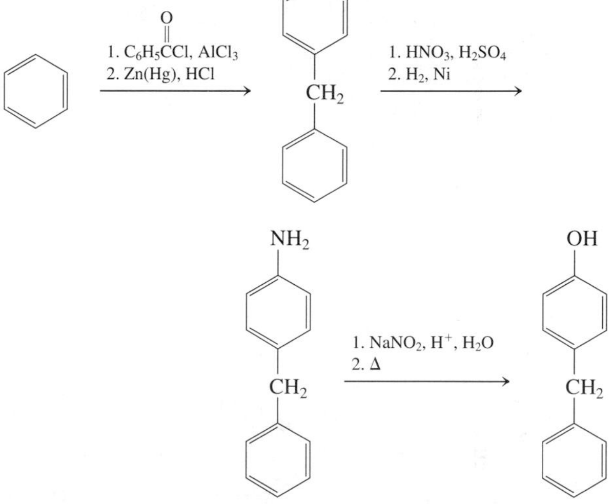

(benzene) — 1. C₆H₅CCl (with O), AlCl₃ 2. Zn(Hg), HCl → (diphenylmethane, CH₂) — 1. HNO₃, H₂SO₄ 2. H₂, Ni →

(NH₂ ring, CH₂, phenyl) — 1. NaNO₂, H⁺, H₂O 2. Δ → (OH ring, CH₂, phenyl)

22-14

Such a process would require nucleophilic attack by halide ion on a benzene ring, a transformation that is not competitive.

22-15

Amines are more nucleophilic than alcohols; this rule also holds for benzenamines (anilines) relative to phenols.

22-16

The 1,1-dimethylethyl (*tert*-butyl) group is considerably larger than methyl, attacking preferentially at C4.

22-17

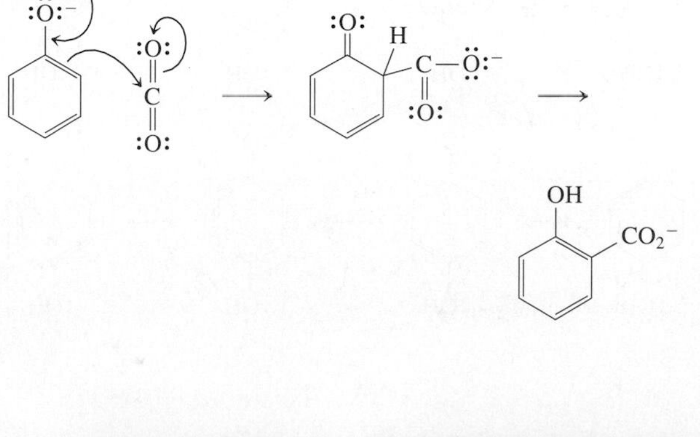

:Ö:⁻ (ring), :Ö: , C , :O: → :O: (ring), H, C—Ö:⁻, :O: →

OH (ring) CO₂⁻

22-18

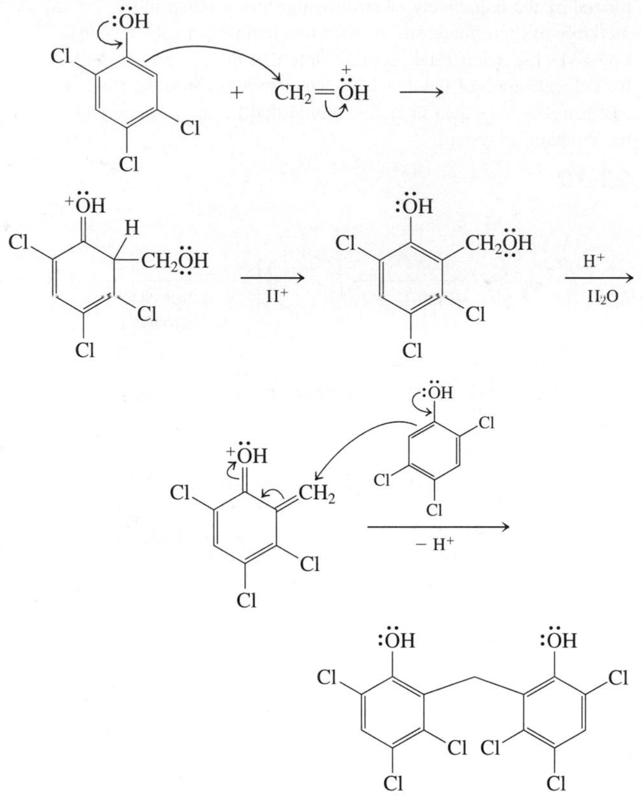

Hexachlorophene

22-19

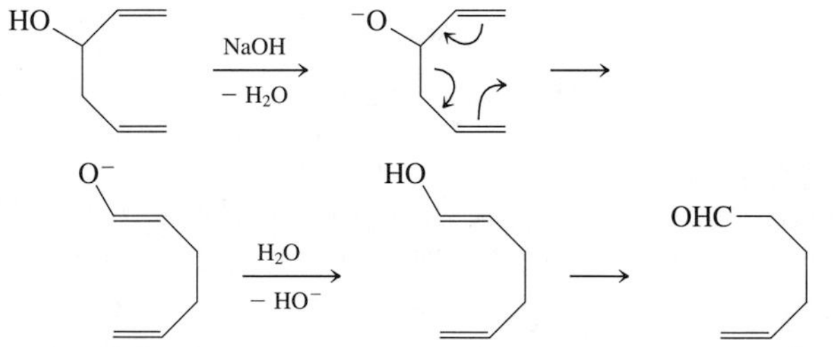

The Cope rearrangement is especially fast in this case because the negative charge in the initial enolate ion is delocalized.

22-20

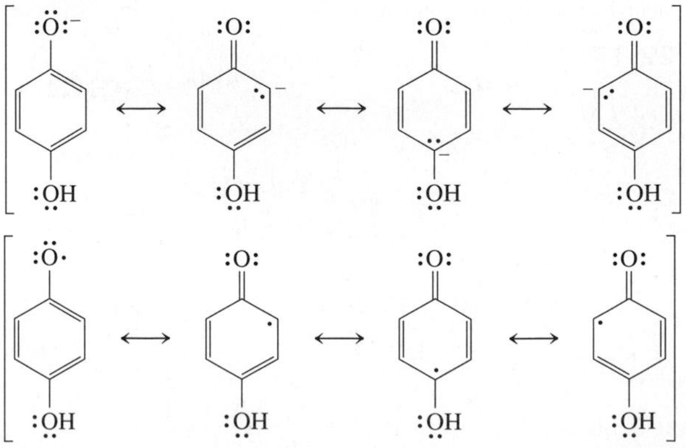

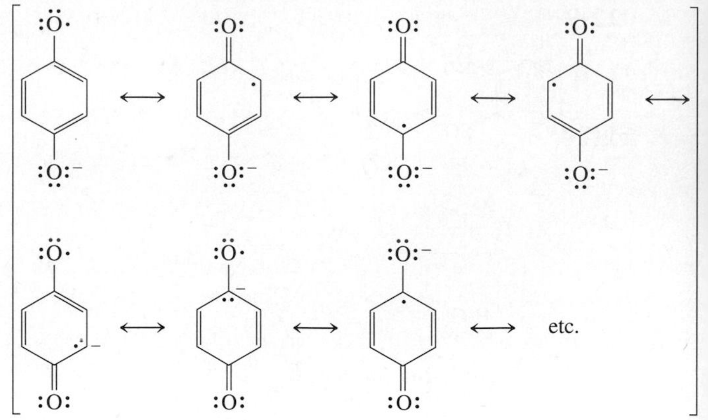

22-21

This exchange goes through two conjugate addition–elimination cycles.

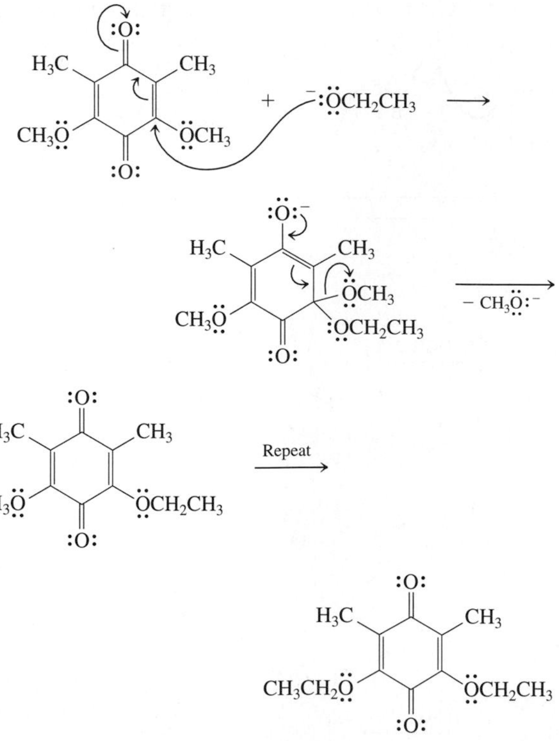

22-22

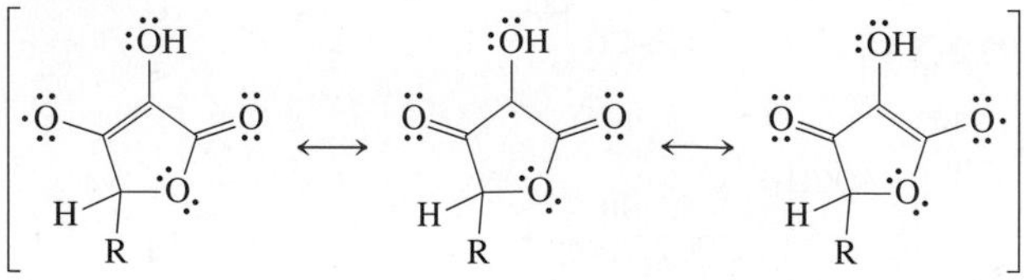

22-23

(a) 1. $CH_3\overset{O}{\overset{\|}{C}}Cl$, $AlCl_3$, 2. HNO_3, H_2SO_4, 3. H_2, Pd, 4. $NaNO_2$, HI

(b) 1. HNO₃ (2 equivalents), H₂SO₄, 2. H₂, Ni, 3. NaNO₂, HCl, 4. CuCN, KCN
(c) 1. HNO₃, H₂SO₄, 2. SO₃, H₂SO₄, 3. NaNO₂, HCl, 4. H₂O

22-24

$$\text{benzene} \xrightarrow[\text{2. H}_2\text{, Ni}]{\text{1. HNO}_3\text{, H}_2\text{SO}_4} \text{aniline (NH}_2\text{)} \xrightarrow{\text{3 Br}_2\text{, H}_2\text{O}}$$

2,4,6-tribromoaniline (NH₂ with Br at 2,4,6) $\xrightarrow[\text{2. H}_3\text{PO}_2]{\text{1. NaNO}_2\text{, H}^+}$ 1,3,5-tribromobenzene

22-25

(a) 4-OCH₃-C₆H₄–N=N–C₆H₅

(b) 2-Cl-4-OCH₃ azobenzene derivative + 4-Cl-2-OCH₃ azobenzene derivative

(c) (CH₃)₃C– / –N(CH₃)₂ substituted azobenzene

CHAPTER 23

23-1

(a) $\text{CH}_3\text{CH}_2\text{C}(\!=\!\text{O})\text{CHC}(\!=\!\text{O})\text{OCH}_2\text{CH}_3$
        with CH₃ substituent

(b) $\text{CH}_3\text{CHCH}_2\text{C}(\!=\!\text{O})\text{CHC}(\!=\!\text{O})\text{OCH}_2\text{CH}_3$
        with CH₃ and (CH₃)₂CH substituents

(c) $\text{CH}_3(\text{CH}_2)_3\text{C}(\!=\!\text{O})\text{CHC}(\!=\!\text{O})\text{OCH}_2\text{CH}_3$
        with (CH₂)₂CH₃ substituent

23-2

Retro-Claisen condensation

$$\text{CH}_3\text{C}(\!=\!\text{O})\text{--C}(\text{CH}_3)\text{--C}(\!=\!\text{O})\text{OCH}_3 \; + \; \text{CH}_3\ddot{\text{O}}\!:^- \longrightarrow$$

$$\text{CH}_3\text{C}(\text{O}^-)(\text{OCH}_3)\text{--C}(\text{CH}_3)\text{--C}(\!=\!\text{O})\text{OCH}_3 \longrightarrow$$

$$\text{CH}_3\text{C}(\!=\!\text{O})\text{OCH}_3 \; + \; (\text{CH}_3)_2\text{C}\!=\!\text{C}(\text{O}^-)\text{OCH}_3 \rightleftharpoons$$

$$\text{CH}_2\!=\!\text{C}(\text{O}^-)\text{OCH}_3 \; + \; (\text{CH}_3)_2\text{CHCOOCH}_3$$

Forward Claisen condensation

$$\text{CH}_3\text{C}(\!=\!\text{O})\text{OCH}_3 \; + \; \text{CH}_2\!=\!\text{C}(\text{O}^-)\text{OCH}_3 \longrightarrow$$

$$\xrightarrow[-\text{CH}_3\text{OH}]{} \text{CH}_3\text{C}(\!=\!\text{O})\text{CHCOOCH}_3$$

23-3

$$\text{CH}_3\text{CH}_2\text{COCH}_2\text{CH}_3 \; + \; \text{CH}_3\text{COCH}_2\text{CH}_3 \xrightarrow[\text{2. H}^+\text{, H}_2\text{O}]{\text{1. CH}_3\text{CH}_2\text{O}^-\text{ Na}^+\text{, CH}_3\text{CH}_2\text{OH}}$$

$$\text{CH}_3\text{CH}_2\text{C}(\!=\!\text{O})\text{CHC}(\!=\!\text{O})\text{OCH}_2\text{CH}_3 \text{ (CH}_3\text{)} \; + \; \text{CH}_3\text{CH}_2\text{CCH}_2\text{COCH}_2\text{CH}_3$$

$$+ \; \text{CH}_3\text{CCHCOCH}_2\text{CH}_3 \text{ (CH}_3\text{)} \; + \; \text{CH}_3\text{CCH}_2\text{COCH}_2\text{CH}_3$$

23-4

$$\text{HCOCH}_2\text{CH}_3 \; + \; \text{CH}_3\text{COCH}_2\text{CH}_3 \xrightarrow[\text{2. H}^+\text{, H}_2\text{O}]{\text{1. CH}_3\text{CH}_2\text{O}^-\text{Na}^+\text{, CH}_3\text{CH}_2\text{OH}}$$

Ethyl formate

$$\text{HCCH}_2\text{COCH}_2\text{CH}_3$$
80%
Ethyl 3-oxopropanoate

Ethyl formate is not enolizable and the carbonyl group is more electrophilic than that of the "methyl substituted" analog ethyl acetate.

23-5

This mechanism is abbreviated, showing only the most important steps.

$$\text{CH}_3\text{CH}_2\text{O}_2\text{C}(\text{CH}_2)_3\text{COCH}_2\text{CH}_3 \xrightarrow[-\text{CH}_3\text{CH}_2\text{OH}]{\text{CH}_3\text{CH}_2\text{O}^-}$$

$CH_3CH_2O_2CCH_2CH_2\overset{..}{\underset{}{C}}HCO_2CH_2CH_3$ $\xrightarrow[- CH_3CH_2O^-]{\overset{\overset{O\ O}{||\ ||}}{CH_3CH_2OCCOCH_2CH_3}}$

$\overset{\overset{O\ O}{||\ ||}}{\underset{\underset{CH_3CH_2O_2CCH_2CH_2\overset{|}{C}HCO_2CH_2CH_3}{}}{CCOCH_2CH_3}}$ $\xrightarrow[- CH_3CH_2OH]{CH_3CH_2O^-}$

$CH_3CH_2O_2C\overset{..}{\underset{}{C}}HCH_2CH_2\overset{|}{C}HCO_2CH_2CH_3$ ... $\overset{\overset{O\ O}{||\ ||}}{CH_3CH_2O\overset{..}{\underset{}{C}}C}$ $\xrightarrow{- CH_3CH_2O^-}$

$\xrightarrow{H^+, H_2O}$ $CH_3CH_2O_2C$ ⬡ (cyclopentane with two C=O and $CO_2CH_2CH_3$)

23-6

This mechanism is abbreviated.

(structure: phthalate diester with two OCH_2CH_3) $+$ $^-:CH_2CO_2CH_2CH_3$ $\xrightarrow[- H^+]{- CH_3CH_2O^-,}$

(structure) $\overset{..}{C}HCO_2CH_2CH_3$ ... OCH_2CH_3 $\xrightarrow{- CH_3CH_2O^-}$

$\xrightarrow{H^+, H_2O}$ (indane-1,3-dione with) $CO_2CH_2CH_3$

23-7

(structure: methyl 5-oxohexanoate) $\xrightarrow[2.\ H^+, H_2O]{1.\ CH_3O^-Na^+, CH_3OH}$ (1,3-cyclohexanedione)

Methyl 5-oxohexanoate   **1,3-Cyclohexanedione**

23-8

(a) (cyclohexanone) $+$ $CH_3CH_2O_2CCO_2CH_2CH_3$

1. $CH_3CH_2O^-$,   2. H^+, H_2O

(b) $CH_3\overset{O}{\overset{||}{C}}CH_3$ $+$ $HCO_2CH_2CH_3$

1. $CH_3CH_2O^-$,   2. H^+, H_2O

(c) (cyclooctanone) $+$ $CH_3CH_2O\overset{O}{\overset{||}{C}}CH_2CH_3$

1. NaH,   2. H^+, H_2O

(d) $H_3C\overset{O}{\overset{||}{C}}$—(chain)—$CO_2CH_2CH_3$

1. $CH_3CH_2O^-$,   2. H^+, H_2O

23-9

$CH_3\overset{O}{\overset{||}{C}}CH_2CH_2CH_2CO_2CH_3$ $\xrightarrow{\text{Exercise 23-7}}$

(1,3-cyclohexanedione) **100%** $\xrightarrow{2\ NaOCH_3, CH_3I, CH_3OH}$

(2,2-dimethyl-1,3-cyclohexanedione) **80%**

2,2-Dimethyl-1,3-cyclohexanedione

23-10

(mechanism structures)

$\longrightarrow$ (enol + CO_2 structures) $\longrightarrow$

$HO\overset{O}{\overset{||}{C}}CH_2$ ... CH_3 $+$ CO_2

23-11

(a) 1. $NaOCH_2CH_3$,   2. $CH_3CH_2CH_2Br$,   3. NaOH,
4. H^+, H_2O, Δ; **(b)** 1. $NaOCH_2CH_3$,   2. $CH_3(CH_2)_4Br$,
3. NaOH,   4. H^+, H_2O, Δ; **(c)** 1. 2 $NaOCH_2CH_3$,
2. 2 CH_3CH_2Br,   3. NaOH,   4. H^+, H_2O, Δ; **(d)** 1. $NaOCH_2CH_3$,
2. $C_6H_5CH_2Cl$,   3. NaOH,   4. H^+, H_2O, Δ

23-12

(a) 1. $CH_3CH_2OOC\overset{..}{\underset{\underset{CH_3}{|}}{C}}HCOOCH_2CH_3$,

2. $CH_3CH_2OOC\overset{\overset{(CH_2)_9CH_3}{|}}{\underset{\underset{CH_3}{|}}{C}}COOCH_2CH_3$,

3. $K^{+\ -}OOC\overset{\overset{(CH_2)_9CH_3}{|}}{\underset{\underset{CH_3}{|}}{C}}COO^-K^+$

(b) $CH_3CH_2OOCCH_2COOCH_2CH_3$ $\xrightarrow[\text{2. } CH_3Br]{\text{1. } NaOCH_2CH_3}$

$$CH_3CH_2OOCCHCOOCH_2CH_3$$
$$|$$
$$CH_3$$

cyclohexane with $CO_2CH_2CH_3$ / $CO_2CH_2CH_3$ $\xrightarrow[\text{2. } H^+, H_2O, \Delta]{\text{1. KOH, } CH_3CH_2OH}$

cyclohexane with H COOH

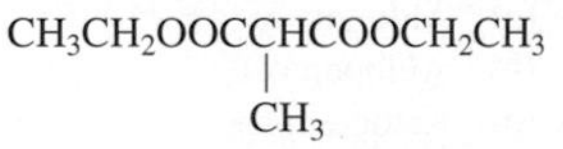

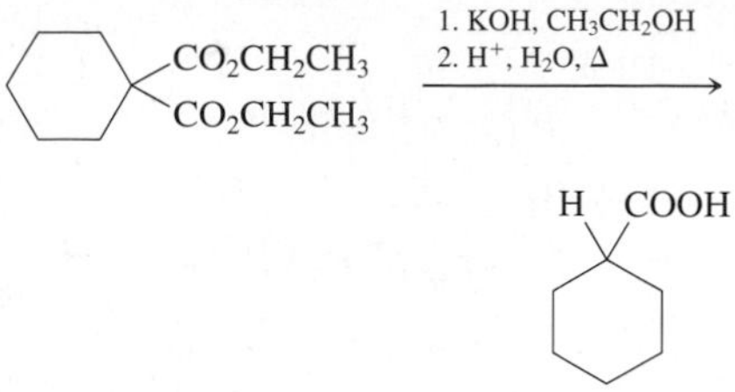

Cyclohexanecarboxylic acid

23-13

(a) 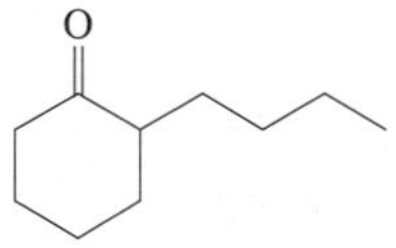 2-butylcyclohexanone structure

2-Butylcyclohexanone

(b) $CH_3CH_2CO_2H$
Propanoic acid

(c) 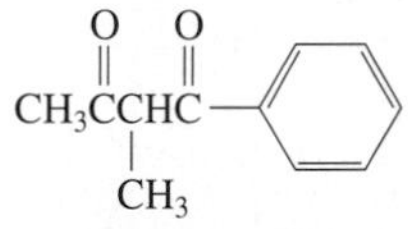
$$CH_3CCHC\!-\!\text{(phenyl)}$$
with O, O above and CH_3 below

2-Methyl-1-phenyl-1,3-butanedione

(d)   This sequence is general for 2-halo esters.

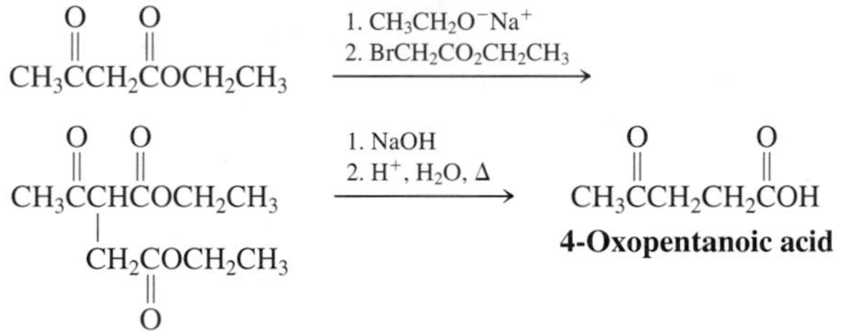

$$CH_3CCH_2COCH_2CH_3 \xrightarrow[\text{2. } BrCH_2CO_2CH_2CH_3]{\text{1. } CH_3CH_2O^-Na^+}$$
(with O, O above)

$$CH_3CCHCOCH_2CH_3$$
$$|$$
$$CH_2COCH_2CH_3$$
$$O$$
$\xrightarrow[\text{2. } H^+, H_2O, \Delta]{\text{1. NaOH}}$ $CH_3CCH_2CH_2COH$ (with O, O)

4-Oxopentanoic acid

Note that only the carboxy group attached to the α-carbonyl carbon can undergo decarboxylation.

(e) 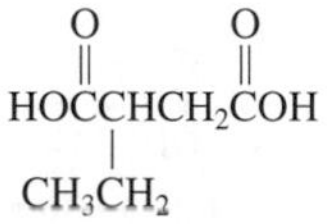
$$HOCCHCH_2COH$$
$$|$$
$$CH_3CH_2$$
(with O, O above)

2-Ethylbutanedioic acid

Excessive heating may dehydrate this product to the anhydride (Section 19-8).

(f) 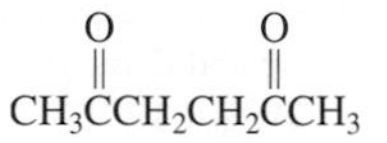 $CH_3CCH_2CH_2CCH_3$ (with O, O above)
2,5-Hexanedione

23-14

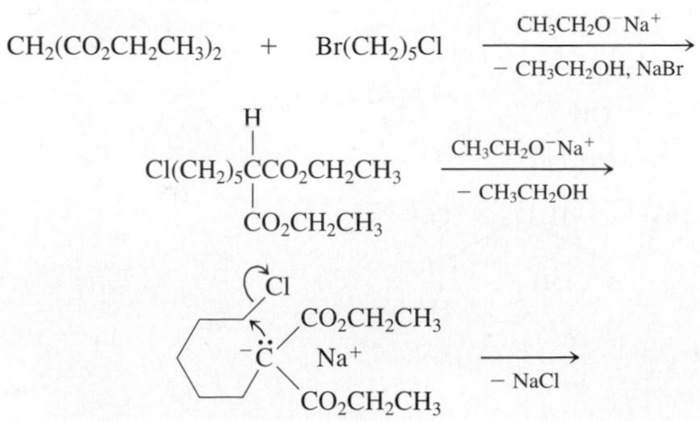

$$CH_2(CO_2CH_2CH_3)_2 + Br(CH_2)_5Cl \xrightarrow[-CH_3CH_2OH, NaBr]{CH_3CH_2O^-Na^+}$$

$$Cl(CH_2)_5CCO_2CH_2CH_3$$ with H above and $CO_2CH_2CH_3$ below $\xrightarrow[-CH_3CH_2OH]{CH_3CH_2O^-Na^+}$

cyclic mechanism structure with Cl, $CO_2CH_2CH_3$, C, Na$^+$, $CO_2CH_2CH_3$ $\xrightarrow{-NaCl}$

23-15

$$^-\!:\!CH(CO_2CH_2CH_3)_2 + CH_2\!=\!CH\!-\!CCH_3 \longrightarrow$$
(with O on acetyl)

$$(CH_3CH_2O_2C)_2CHCH_2CH\!=\!CCH_3$$ (with O$^-$) $\xrightarrow{CH_2(CO_2CH_2CH_3)_2}$

$$(CH_3CH_2O_2C)_2CHCH_2CH_2CCH_3 + {}^-\!:\!CH(CO_2CH_2CH_3)_2$$
(with O)

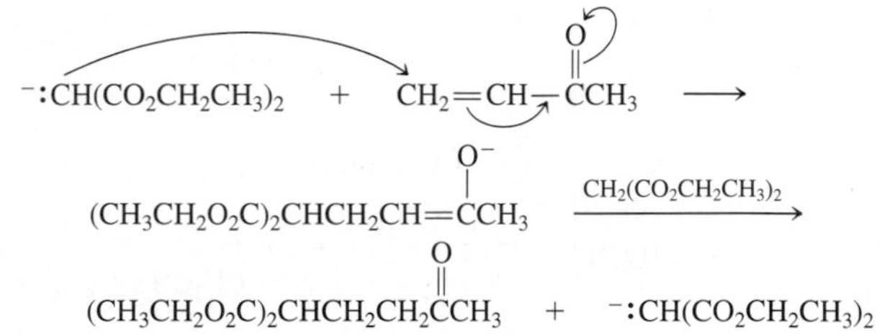

The enolate of the product regenerates the enolate of the starting malonic ester.

23-16

(a) 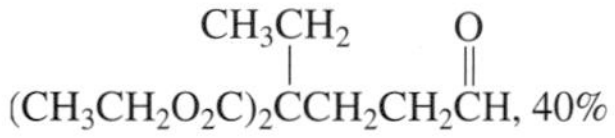 $(CH_3CH_2O_2C)_2CCH_2CH_2CH$, 40%
(with CH_3CH_2 above and O above the CH)

(b) 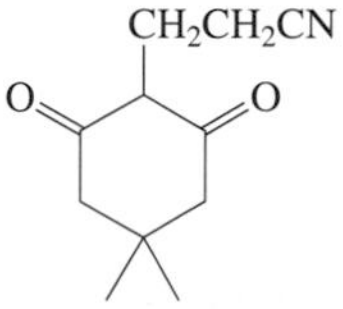 structure with CH_2CH_2CN , 56%

(c) 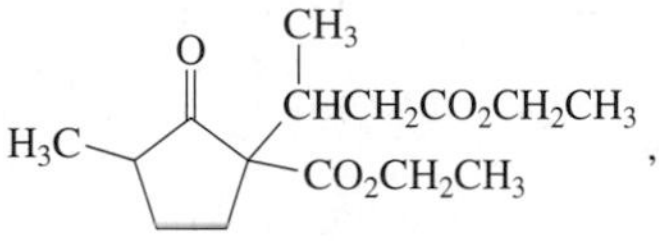 H_3C— cyclopentanone structure with CH_3, CHCH_2CO_2CH_2CH_3, $CO_2CH_2CH_3$, 66%

23-17

mechanism: dimedone enolate + $CH_2\!=\!CHCN$ $\longrightarrow$

intermediate with NC, H, Acidic $\longrightarrow$

structure with CN $\xrightarrow[\text{Second Michael addition}]{CH_2=CHCN}$ $\longrightarrow$ final structure with NC, CN

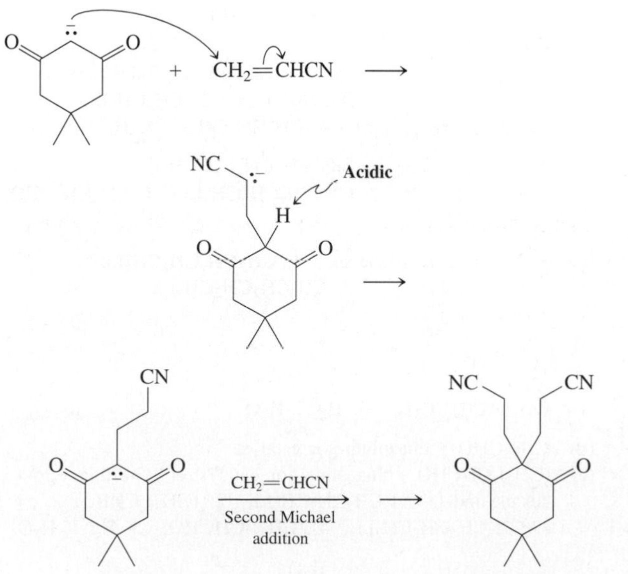

23-18

1. Michael addition

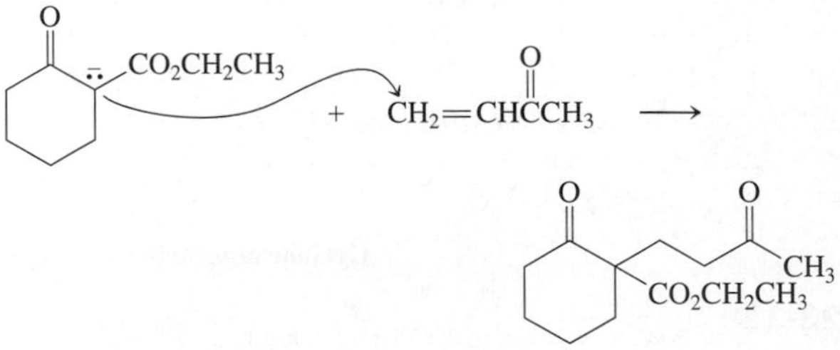

2. Aldol condensation

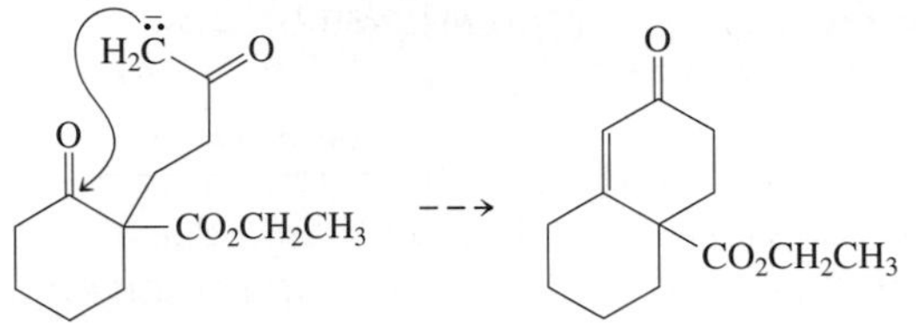

23-19

Nucleophiles add to the carbonyl group; bases deprotonate next to it.

23-20

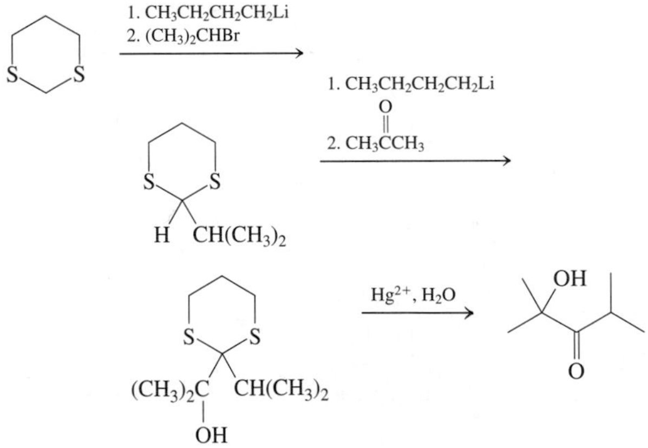

23-21

(a) CH_3CH_2CHO + thiazolium ion catalyst or 1,3-dithiacyclo-hexane and 1. $CH_3CH_2CH_2CH_2Li$, 2. CH_3CH_2Br, 3. $CH_3CH_2CH_2CH_2Li$, 4. CH_3CH_2CHO, 5. Hg^{2+}, H_2O

(b) 1,3-Dithiacyclohexane and 1. $CH_3CH_2CH_2CH_2Li$, 2. $CH_3(CH_2)_3Br$, 3. $CH_3CH_2CH_2CH_2Li$, 4. CH_3CH_2CHO, 5. Hg^{2+}, H_2O

(c) 1,3-Dithiacyclohexane and 1. $CH_3CH_2CH_2CH_2Li$, 2. $CH_3CH_2CHCH_3$, 3. $CH_3CH_2CH_2CH_2Li$,
$\qquad$ |
$\qquad$ Br

4. $CH_3CCH_2CH_3$ (ketone, O double bond), 5. Hg^{2+}, H_2O

(d) C_6H_5CHO + thiazolium ion catalyst

(e) $(CH_3)_2CHCHO$ + thiazolium ion catalyst or 1,3-dithiacyclo-hexane and 1. $CH_3CH_2CH_2CH_2Li$, 2. $(CH_3)_2CHBr$, 3. $CH_3CH_2CH_2CH_2Li$, 4. $(CH_3)_2CHCHO$, 5. Hg^{2+}, H_2O

CHAPTER 24

24-1

(a) Aldotetrose
(b) Aldopentose
(c) Ketopentose

24-2

(a) (2R,3R,4R)-2,3,4,5-Tetrahydroxypentanal
(b) (2R,3S,4R,5R)-2,3,4,5,6-Pentahydroxyhexanal

24-3

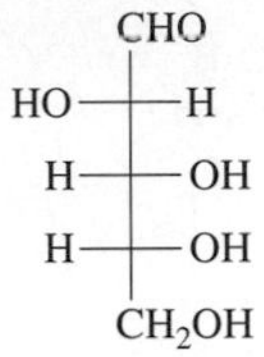

D-(−)-**Arabinose**

24-4

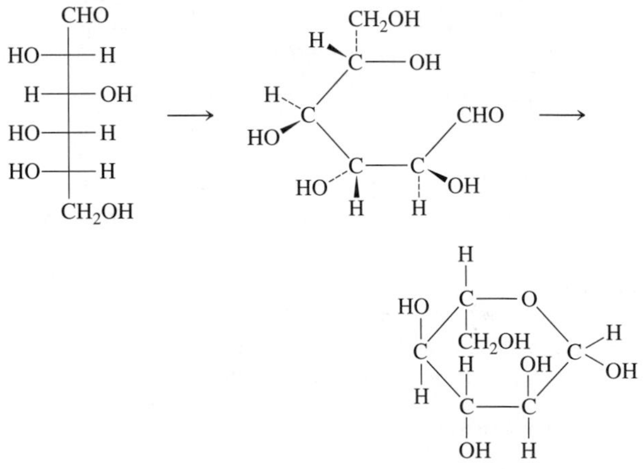

L-(−)-**Glucose**

24-5

False. They should form in unequal amounts because two diastere-omers are formed. In fact, the ratio of α to β is 36:64. Similarly, the relative amounts of α-D-, β-D-fructopyranose and α-D-, β-D-fructofuranose at equilibrium in aqueous solution are 3%, 57%, 9%, and 31%, respectively.

24-6

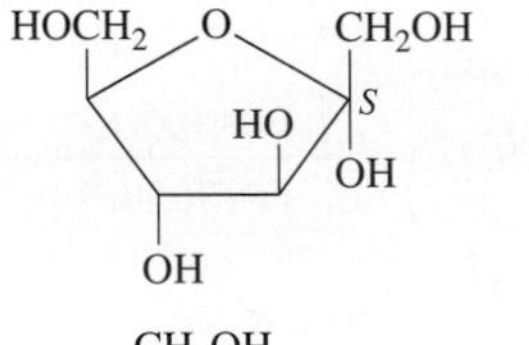

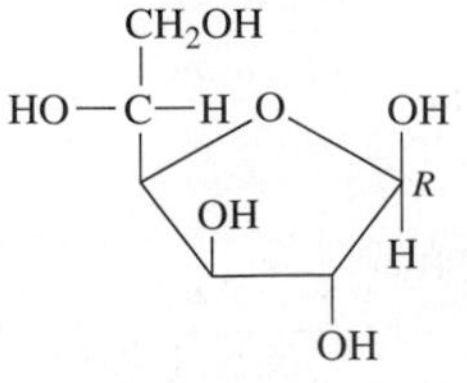

(c)

24-7

Four axial OH groups, $4 \times 0.94 = 3.76$ kcal mol^{-1}; one axial CH_2OH, 1.70 kcal mol^{-1}; $\Delta G = 5.46$ kcal mol^{-1}. The concentration of this conformer in solution is therefore negligible by this estimate.

24-8

Only the anomeric carbon and its vicinity are shown.

Planar

24-9

Pure α form, $+112°$; pure β form, $+18.7°$ ($\Delta\alpha = 93.3°$). After equilibration, $+52.7°$. Mole fraction of α: $(52.7 - 18.7)/93.3 = 0.364$. Hence, the mole fraction of $\beta = 0.636$; thus, the equilibrium mole ratio $\beta : \alpha = 0.636/0.364 = 1.75 : 1$.

24-10

$\Delta G°_{estimated} = 0.94$ kcal mol^{-1} (one axial OH); $\Delta G° = -RT$ ln $K = -1.36$ log $63.6/36.4 = -0.33$ kcal mol^{-1}. The difference between the two values is due to the fact that the six-membered ring is a cyclic ether (not a cyclohexane).

24-11

Oxidation of D-glucose should give an optically active aldaric acid, whereas that of D-allose leads to loss of optical activity. This result is a consequence of turning the two end groups along the sugar chain into the same substituent.

	COOH		COOH
H ——	OH	H ——	OH
HO ——	H	H ——	OH
H ——	OH	H ——	OH
H ——	OH	H ——	OH
	COOH		COOH

D-Glucaric acid   **D-Allaric acid**

Optically active   **Meso, not optically active**

This operation may cause important changes in the symmetry of the molecule. Thus, D-allaric acid has a mirror plane. It is therefore a meso compound and not optically active. (This also means that

D-allaric acid is identical with L-allaric acid.) On the other hand, D-glucaric acid is still optically active.

Other simple aldoses that turn into meso-aldaric acids are D-erythrose, D-ribose, D-xylose, and D-galactose (see Figure 24-1).

24-12

(a) $2 H_2C=O$ **(b)** $CH_3CH=O + H_2C=O$
(c) $2 H_2C=O + HCOOH$ **(d)** No reaction
(e) $OHCC(CH_3)_2CHO + CO_2$
(f) $3 HCOOH + H_2C=O$

24-13

(a) D-Arabinose $\longrightarrow$ $4 HCO_2H + CH_2O$
     D-Glucose $\longrightarrow$ $5 HCO_2H + CH_2O$
(b) D-Erythrose $\longrightarrow$ $3 HCO_2H + CH_2O$
     D-Erythrulose $\longrightarrow$ $HCO_2H + 2 CH_2O + CO_2$
(c) D-Glucose or D-mannose $\longrightarrow$ $5 HCO_2H + CH_2O$

24-14

(a) Ribitol is a meso compound.
(b) Two diastereomers, D-mannitol (major) and D-glucitol

24-15

All of them are the same.

24-16

The mechanism of acetal formation proceeds through the same intermediate cation in both cases.

24-17

24-18

Same structure as that in Exercise 24-17 or its diastereomers with respect to C2 and C3.

24-19

$\xrightarrow{H^+, CH_3COCH_3}$

Reactive

$\xrightarrow[\text{2. } H^+, H_2O]{\text{1. } PBr_3}$

24-20

(a) D-Ribose and D-arabinose

(b) D-Glucose and D-mannose

24-21

A, D-arabinose; B, D-lyxose; C, D-erythrose; D, D-threose.

24-22

^{13}C NMR would show only three lines for ribaric acid, but five for arabinaric acid.

24-23

(a)

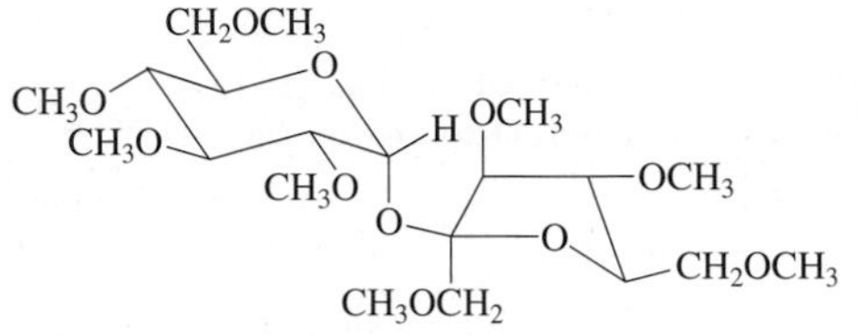

(b)

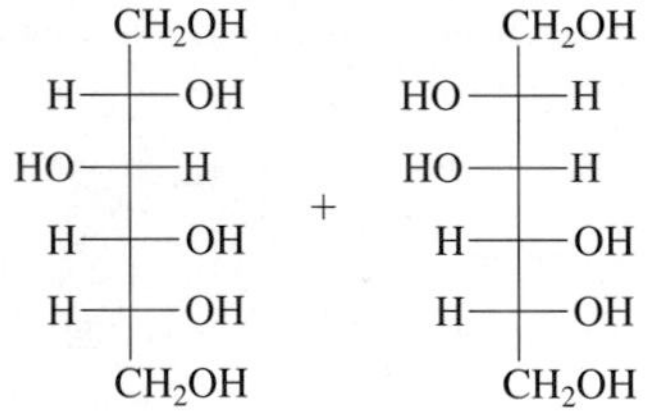

(c) No reaction.

24-24

(a)

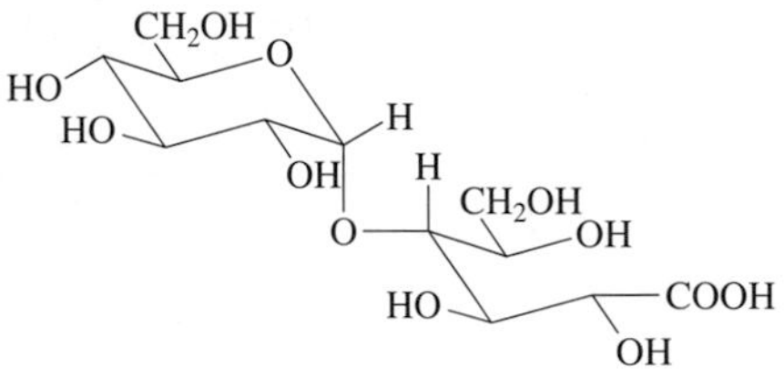

(b)

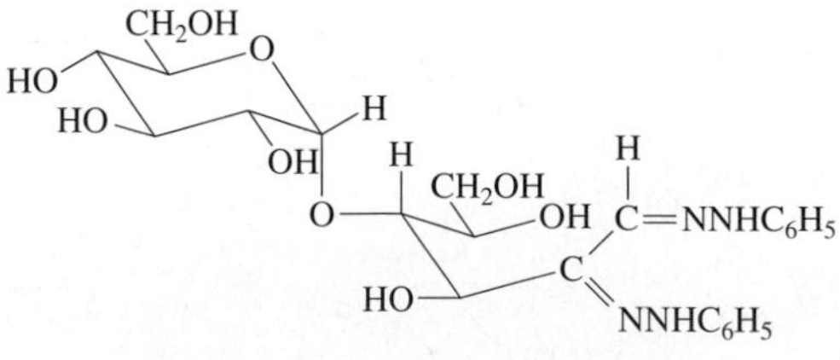

(c)

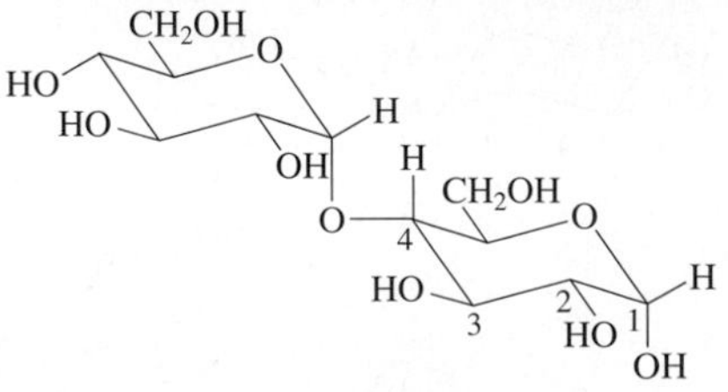

α-Maltose

CHAPTER 25

25-1

(a)

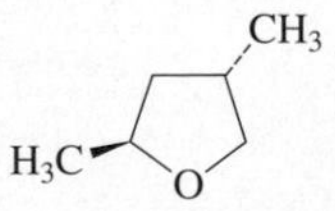

(b)

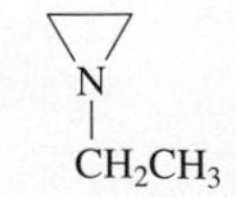

(c) 2,6-Dinitropyridine

(d) 4-Bromoindole

25-2

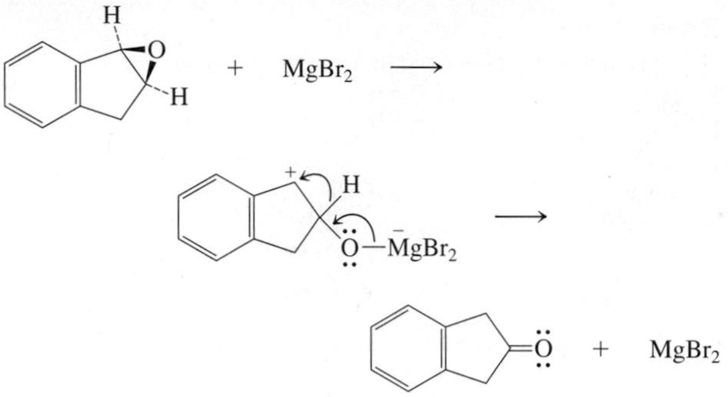

25-3

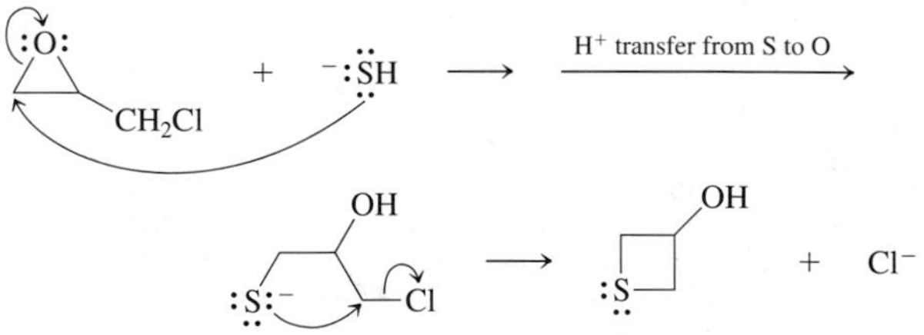

25-4

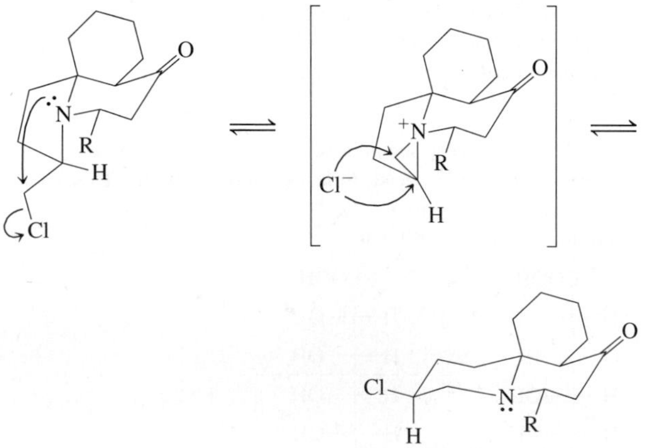

25-5

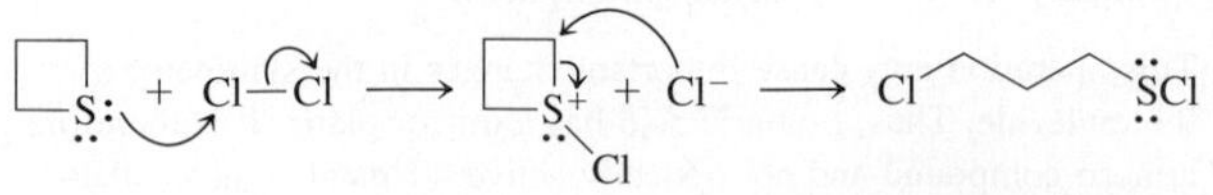

25-6

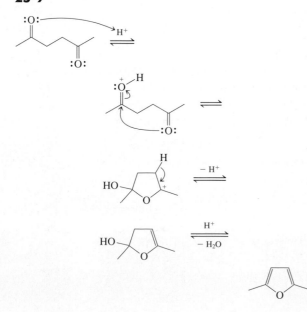

25-7

25-8

Nitrogen is more electro-negative than carbon

Because of resonance, the molecule is now polarized in the opposite direction

25-9

25-10

25-11

β-Keto amine   β-Keto ester

25-12

Attack at C5 avoids placing the positive charge on C3

69%

25-13

Protonation at the α-carbon generates a cation described by three resonance forms. Protonation of the nitrogen produces an ammonium ion devoid of resonance stabilization.

25-14

Only attack at C3 produces the iminium resonance form without disrupting the benzene ring

25-15

25-16

Because of the electronegativity of nitrogen, the dipole vector in both compounds points toward the heteroatom. The dipole moment of pyridine is larger than that in azacyclohexane (piperidine), because the nitrogen is sp^2 hybridized. (See Sections 11-3 and 13-2 for the effects of hybridization on electron-withdrawing power.)

25-17

(a) $CH_3\overset{O}{\underset{||}{C}}CH_2CO_2CH_2CH_3$, NH_3, , $CH_3\overset{O}{\underset{||}{C}}CH_2CN$

(b) $CH_3\overset{O}{\underset{||}{C}}CH_2CN$, NH_3, $(CH_3)_3CCHO$

(c) $CH_3CH_2\overset{O}{\underset{||}{C}}CH_2CO_2CH_2CH_3$, NH_3, CH_3CHO

25-18

C3 is the least deactivated position in the ring. Attack at C2 or C4 generates intermediate cations with resonance forms that place the positive charge on the electronegative nitrogen.

25-19

Attack at C2 and C4 produces the more highly resonance stabilized anions (only the most important resonance forms are shown).

25-20

Cl + $^-$:ÖCH$_3$ ⟶

Cl OCH$_3$ ⟶ OCH$_3$ + Cl$^-$

25-21

+ ⟶MgBr

1. (CH$_3$CH$_2$)$_2$O,
18 h, Δ
2. NH$_4$Cl

⟶

56%

2-(2-Propenyl)quinoline

+ ⟶MgBr

1. (CH$_3$CH$_2$)$_2$O,
18 h, Δ
2. NH$_4$Cl

⟶

57%

1-(2-Propenyl)isoquinoline

CHAPTER 26

26-1

(2S)-Aminopropanoic acid; (2S)-amino-3-methylbutanoic acid;
(2S)-amino-4-methylpentanoic acid; (2S)-amino-3-methylpentanoic
acid; (2S)-amino-3-phenylpropanoic acid; (2S)-amino-3-hydroxy-
propanoic acid; (2S)-amino-3-(4-hydroxyphenyl)propanoic acid;
(2S),6-diaminohexanoic acid; (2R)-amino-3-mercaptopropanoic
acid; (2S)-amino-4-(methylthio)butanoic acid; (2S)-aminobutane-
dioic acid; (2S)-aminopentanedioic acid

26-2

H$_2$N
H$_3$C—C—COOH
H

(S)-Alanine

H$_2$N
C$_6$H$_5$H$_2$C—C—COOH
H

(S)-Phenylalanine

H$_2$N
H—C—COOH
CH$_2$C$_6$H$_5$

(R)-Phenylalanine

H
N
C—COOH
H

(S)-Proline

26-3

In the amino acid carboxylates, the nitrogen of the amino acids exists as —NH$_2$, displaying *two* (symmetric and asymmetric stretching) $\tilde{v}_{NH_2}$ bands at ~3400 cm^{-1}, whereas proline carboxylate
shows only one.

26-4

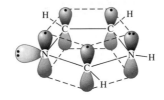

pK$_a$ ~ **13**

26-5

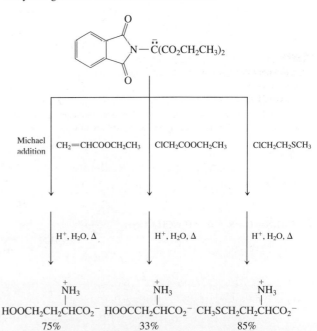

26-6

The yields given are those found in the literature.

Michael
addition

CH$_2$=CHCOOCH$_2$CH$_3$

ClCH$_2$COOCH$_2$CH$_3$

ClCH$_2$CH$_2$SCH$_3$

H$^+$, H$_2$O, Δ

H$^+$, H$_2$O, Δ

H$^+$, H$_2$O, Δ

$\overset{+}{N}H_3$
HOOCH$_2$CH$_2$CHCO$_2$$^-$
75%

Glutamic acid

$\overset{+}{N}H_3$
HOOCCH$_2$CHCO$_2$$^-$
33%

Aspartic acid

$\overset{+}{N}H_3$
CH$_3$SCH$_2$CH$_2$CHCO$_2$$^-$
85%

Methionine

26-7

These syntheses are found in the literature.

$$CH_2=O \xrightarrow{NH_4^{+-}CN, H_2SO_4} H_2NCH_2CN \xrightarrow{BaO, H_2O, \Delta} H_3^+NCH_2COO^-$$

2-Aminoethanenitrile

42%

Glycine

$$CH_3SH + CH_2=CHCH\!\!\begin{matrix}O\\ \|\end{matrix} \xrightarrow[\text{addition}]{\text{Michael}} CH_3SCH_2CH_2CH\!\!\begin{matrix}O\\ \|\end{matrix} \xrightarrow[\text{2. NaOH}]{\text{1. Na}^{+-}CN, (NH_4)_2CO_3} CH_3SCH_2CH_2CHCOO^-\!\!\begin{matrix}^+NH_3\\ |\end{matrix}$$

84%

3-(Methylthio)propanal

58%

Methionine

26-8

(structure of peptide)

26-9

Elimination of the amino end of the cleaved polypeptide:

Release of the carboxy end of the cleaved polypeptide:

26-10

Hydrolysis of the A chain in insulin produces one equivalent each of Gly, Ile, and Ala, two each of Val, Glu, Gln, Ser, Leu, Tyr, and Asn, and four of Cys.

26-11

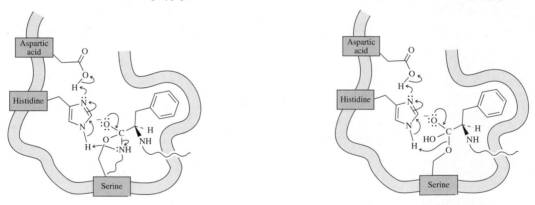

26-12

All peptide bonds are cleaved by "transamidation" to give all component amino acids as their corresponding hydrazides. Only the carboxy end retains its free carboxy function.

26-13

It is the A chain of insulin (see Figure 26-1).

26-14

26-15

1. Ala + $(CH_3)_3COCOCOC(CH_3)_3$ $\longrightarrow$ Boc-Ala + CO_2 + $(CH_3)_3COH$

2. Val + CH_3OH $\xrightarrow{H^+}$ Val-OCH$_3$ + H_2O

3. Boc-Ala + Val-OCH$_3$ $\xrightarrow{DCC}$ Boc-Ala-Val-OCH$_3$

4. Boc-Ala-Val-OCH$_3$ $\xrightarrow{H^+}$ Ala-Val-OCH$_3$ + CO_2 + $CH_2{=}C(CH_3)_2$

5. Leu + $(CH_3)_3COCOCOC(CH_3)_3$ $\longrightarrow$ Boc-Leu + CO_2 + $(CH_3)_3COH$

6. Boc-Leu + Ala-Val-OCH$_3$ $\xrightarrow{DCC}$ Boc-Leu-Ala-Val-OCH$_3$

7. Boc-Leu-Ala-Val-OCH$_3$ $\xrightarrow{\substack{1.\ H^+,\ H_2O \\ 2.\ HO^-,\ H_2O}}$ Leu-Ala-Val

26-16

26-17

The various dipolar amide resonance forms provide cyclic aromatic electron sextets.

26-18

5′-UAGCTGAGCAAT

26-19

(a) Lys-Tyr-Ala-Ser-Cys-Leu-Ser
(b) Mutation of C to G in UGC (Cys) to become UGG (Trp).

26-20

1. DMT–OR deprotection (R = sugar of nucleotide): hydrolysis by S_N1

$$\text{DMT—OR} \xrightarrow[-\text{ROH}]{H^+} \text{DMT}^+ \xrightarrow{^+NH_4\,^-OH} \text{DMT} \quad \text{OH}$$

2. Nucleic acid base deprotection (RNH_2 = base, $R'CO_2H$ = carboxylic acid): amide hydrolysis

3. Phosphate deprotection (R and R′ = sugars of nucleotides): E_2

4. Disconnection from the solid support (ROH = sugar of nucleotide): ester hydrolysis

26-21

The para nitro group is electron withdrawing by resonance, thus activating the carbonyl carbon with respect to nucleophilic attack, relative to a simple benzoate:

PHOTOGRAPH CREDITS

ORGANIC CHEMISTRY
Structure and Function

INDEX

Important terms are defined in the text on the pages appearing in **bold type.** Tables of useful properties can be found on pages whose reference is followed by a *t*.

Ester synthesis (*Continued*)
with diazomethane, 847
by esterification of carboxylic acids, 844–847
from haloalkanes, 294, 347–348, 350
from sugars, 1084–1085
by transesterification, 890–891
see also Ester reaction
Estradiol, 152
Estrogen, 152
Estrone, 910–911
Ethanal. *See* Acetaldehyde
Ethane, 2
heat of combustion, 120
conformational analysis of, 64–66
molecular-orbital picture of, 36
1,2-Ethanediamine, 552
Ethanedioic (oxalic) acid, 827, 833
1,2-Ethanediol (ethylene glycol), 292, 746
physical properties, 364
synthesis, 364
uses, 364
Ethanenitrile (acetonitrile), 903
Ethanethiol, 365
Ethanoate ion. *See* Acetate ion
Ethanoic acid. *See* Acetic acid
Ethanoic anhydride. *See* Acetic anhydride
Ethanol, 283, 825
condensed formula for, 38
fermentation, 363
industrial sources, 292, 363–364
physiological properties, 363
synthesis, 364
uses, 363
Ethanoyl group. *See* Acetyl group
Ethene (ethylene), 11, 52, 442, 483
addition reactions and, 519–520
bond angles in, 443
in Diels-Alder cycloaddition, 601
double bond in, 37, 443
industrial importance of, 519–520, 523t
molecular-orbital picture in, 443
molecular structure of, 443
oxidation of, 835
pi bond in, **443**–445
pK_a of, 446
sigma bond in, 443, 445
Ethenol (vinyl alcohol), 783
Ethenylbenzene (styrene), 640
Ethenyl ethanoate (vinyl acetate), synthesis, 519
Ethenyl (vinyl) group, 442
anion, 446
cation, 561
Ether, **52,** 283, 343–367
boiling points, 345t
cyclic, 344–350 (*See also* Acetal)
relative rates of formation, 349
molecular formula, 344
naming of, 343–344
physical properties, 344–345

physiological properties, 362
as protecting groups for alcohols, 354
as solvents, 344
symmetrical, **53**
unsymmetrical, **53**
uses, 362
water solubility, 345
Ether reactions
Claisen rearrangement, **1006**–1008
cleavage by hydrogenolysis, 988–989
nucleophilic opening of oxacyclopropanes, 354–355, 357–359
with oxygen, in peroxide formation, 352–353
with strong acids, 353
Ether synthesis
from alcohols, 350–352
by alcoholysis, 352
from alkenes, 500
from alkoxides, 347–348
cyclic, 347–350
from haloalkanes, 347–350
from phenols, 1001–1002
from sugars, 1084–1085
Ethidimuron, 1119
Ethoxide, 462–463
Ethoxyethane (diethyl ether), 344, 364
Ethyl, 58
Ethyl anion, 446
Ethylbenzene, 670
2-Ethyl-1-butene, 758
4-Ethyl-4-nonanol, 311
Ethylene. *See* Ethene
Ethylene glycol, *See* 1,2-Ethanediol
Ethylene oxide. *See* Oxacyclopropane
Ethylmethylamine. *See* N-Methylethanamine
Ethyl propenoate (ethyl acrylate), 602
Ethyl radical, 94, 98
Ethyne, 52, 545
acidity, 546
addition reactions to, 563
combustion, 550, 563
cyclooligomerization, 658
industrial chemistry with, 543
production of, 563
singly metallated, 553
structure, 37
17-Ethynylestradiol, 566–567
Ethynyl group, **544**
Eudesmane, 125
Exact mass, 918
Excitation, **386,** 617
Excited electronic state, **617**
Exhaustive methylation, **955**
Exo adduct, **605**
Exothermic reaction, **74**
Expanded octet. *See* Valence shell expansion
Extended pi system, 592, **598.** *See also* Conjugated pi system
External magnetic field, behavior of nuclei in, 388–389

Extinction coefficient, **618**
Extraction, as laboratory technique, 946–947

Faraday, Michael, 637
Farnesol, 615
Farnesyl pyrophosphate, 615–616
Fast proton exchange, **416**
Fats, **853, 894**
Fatty acid, **853**–857, **894**
biosynthesis, 854–855
unsaturated, 856–857
Fehling, Hermann C. von, 761
Fehling's test, **761,** 1079
Fermentation of sugars, 292
Ferritin, 1157
Fertility, control of, 153
Fexofenadine (Allegra), 1119
Fibrous protein, **1173**
Field strength, magnetic
and coupling constants, 406
and resolution, 412
and resonance frequency, 389
of spectrometers, 387–388
see also Nuclear magnetic resonance
Fingerprint region in IR spectra, **455**
Firefly, bioluminescence of, 349
First-order reaction, **76**
First point of difference principle, **60,** 177
Fischer, Emil, 179, 1089
Fischer projection, **179**–183
of amino acids, 1157
of 2-bromobutane, 180
of sugars, 1069–1072t, 1073, 1075–1076
Fischer proof, **1090**
Fleming, Alexander, 898
Fluorescence, 617
Fluorine, 8t, 13t, 107, 113
Fluoxetin (Prozac), 926, 1117
Fluracil (5-fluorouracil), 1188
Folic acid, 1138
Forces
coulombic, 63, 213
intermolecular, 63
London, **63,** 213–214
van der Waals, 6
Formal charge, formula for assigning, 16
Formaldehyde (methanal), 729, 957
industrial preparation, 736
synthesis of alcohols from organometallics and, 306
Formamide, 896
Formic (methanoic) acid, 380, 826, 851
industrial synthesis, 835
resonance forms, 22
structure, 828–829
Formula
bond-line, **39**
condensed, **39**
empirical, **38**
molecular, 38
Formyl (methanoyl) group, **730**